ISBN 978-0-265-86808-9
PIBN 10901874

POWER

DEVOTED TO THE GENERATION AND TRANSMISSION OF POWER

ISSUED WEEKLY

VOLUME LI

January 1 to June 30, 1920

McGRAW-HILL COMPANY, Inc.
10TH AVE. AT 36TH ST.
NEW YORK

POWER

GENERAL INDEX TO VOLUME LI

January 1 to June 30, 1920

EXPLANATORY NOTE

Illustrated articles are marked with an asterisk (*), book notices with a dagger (†), inquiries with a double dagger (‡). The cross-references condense the material and assist the reader, but are not to be regarded as complete or conclusive. So, if there were a reference from "Boiler" to "Trap," and if the searcher failed to find the required article under the latter word, he should look through the "Boiler" entries, or others that the topic might suggest, as he would have done had there been no cross-reference. Letters are indexed under title or subject, general articles under writer's name as well. Not all articles relating to a given topic necessarily appear under the same entries.

Following is a list of the pages included in the several numbers of the volume, by date:

Jan. 6	Pages	1-46
" 13	"	47-84
" 20	"	85-122
" 27	"	123-160
Feb. 3	"	161-206
" 10	"	207-246
" 17	"	247-284
" 24	"	285-318
Mar. 2	"	319-364
" 9	"	365-406
" 16	"	407-448
" 23	"	449-490
" 30	"	491-530
Apr. 6	"	531-574
" 13	"	575-614
" 20	"	615-652
" 27	"	653-692
May 4	"	693-750
" 11	"	751-790
" 18	"	791-830
" 25	"	831-866
June 1	"	867-906
" 8	"	907-948
" 15	"	949-988
" 22	"	989-1028
" 29	"	1029-1062

A

B

BOILER

M

N

O

AUTHOR INDEX

٢

POWER

Volume 51 New York, January 6, 1920 *Number 1*

ENGINEERS of course fully realize the value of lubrication. They would not attempt to run or allow their subordinates to keep in operation any piece of machinery that was not kept properly oiled. Yet in their daily lives and conduct, engineers-in-charge do not seem to realize the lubricating value of a smile. Brought up in the rough school of practice, by the time they have arrived at a position of authority a great many seem to have completely lost their sense of humor.

Looking backward after nearly a score of years of power-house work, I remember four men who seem to stand out, apart from the rest. The first and most successful of these had the gift or acquired the ability of mixing a smile with his daily contact with those over whom he had authority. From a technical viewpoint, I believe this man knew the least of the quartet, yet he was one of the most successful engineers that I have ever known. Perhaps he appreciated the human element a little more than the others and acted accordingly. His engine room and boiler room were neat and clean, and he was always able to get the hard, hot and laborious work done without friction with his subordinates. He was not a loud-voiced individual, neither was he a wooden Indian, but he had the happy faculty of smiling at the right time.

In the last fifteen years this man has been the engineer-in-charge in four different plants and he has had the same degree of success in each. It may be he possessed a certain amount of personal magnetism that made the pathway easy, but I am convinced that for the most part it was simply because he could smile with his men when occasion demanded it.

The second of these four men has been in charge of some large plants and is a thorough mechanic as well as a first-class engineer, that is, a first class engineer, judging him from his knowledge of power-plant equipment and its needs. A martinet for discipline, he has always erected a barrier between himself and his men. When conditions have arisen that should have caused him to open the door of his reserve, he always bolted it securely from the other side.

This man, after a lifetime in power-plant work, cannot be called a successful engineer. Like the old Bourbons of France, he has learned nothing and forgotten nothing.

The third is a man of quiet demeanor and good intentions and a capable engineer in many ways, yet he evidently hasn't smiled since he was a child. It would actually be easier for him to cry than to crack a smile, and for that reason he is no more capable of inspiring a man to a little extra effort than the boiler-feed pump would be.

The last of the four has more technical knowledge than the rest and is a bright, keen man in every way, yet he has been the least successful of them all. A man of nervous temperament, not only is he easily irritated, but he conveys that irritation to others with the result that there is constant friction between himself and his help.

A distinguished artist was once asked by a young enthusiast with what he mixed his paints. The great man, noted almost as much for his gruffness as famed for his greatness, replied, "With brains, sir! Brains!" Applying this test to power-house work I believe if the successful engineer were asked to what he attributed his achievements, the reply would probably be: "By mixing my work with smiles, sir."

This line of reasoning could well be made to fit any branch of work, but we who go down to the pit of whirring wheels and humming drums of concealed energy, are prone to think of things that concern only us and to judge men and conditions by what we have learned from daily contact with life as we find it.

Contributed by Frankland Cheney.

The Year's Progress

IN THE field of power the year 1919 is notable as it brought back to mind memories of James Watt. On August 19 it was just one hundred years since this genius passed away, leaving behind him the germ of the great industrial advancement of the present day.

Reviewing the past twelve months, it has been a period marked by "watchful waiting" for prices to lower, industrial disturbances caused by the radicals of labor and employers of the old school, and eventually an immense volume of business to make up demands accumulating during the period of the war. With labor less productive than before and sudden curtailment of effort due to the recent coal shortage, there has been little opportunity for development or even for special work of any kind. The big bulk of power equipment delivered has been along standard lines.

In engineering design it has been a period of crystallization. Operating results from high steam pressures and superheats are being weighed and digested, large turbines of different types are being compared, and designers are trying to settle in their own minds limits of speed and capacity. It is just a breathing spell, however, for the advancing cost of fuel makes action urgent.

As greater opportunities lie in higher pressures than in excessive temperatures, an early resumption of the advance in this direction may be expected. High pressures of yesterday are now standard, and a higher goal has been set. At no distant date the present maximum of 350 lb. gage coupled to a temperature approximating 650 deg. F. should be exceeded, with current practice following and slowly closing up the gap. Piping to carry pressures far above those now in common use is available, and welding will do considerable in eliminating leaks and similar troubles. Turbine designers do not anticipate insurmountable difficulties in building machines for any pressure that boilers will deliver. It is in the boiler itself, the economizer and the numerous auxiliaries of the plant that further investigation and development will be necessary. While boiler pressure might be increased to almost any desired point by the elimination of drums of large diameter, it is the steels now in commercial use that are the limiting factor. At about seven hundred degrees Fahrenheit the strength of the material seriously lessens, so that in operation it would be unsafe to exceed this temperature. More homogeneous material of higher heat-resisting qualities would solve the problem, but even the limit quoted would allow pressures up to six hundred pounds and low superheat, which in itself would give economies fifty per cent. greater than are obtained in present good practice, with due allowances, of course, for higher initial costs and probably higher upkeep resulting from the service. The ever-increasing price of coal has forced improvement in the average

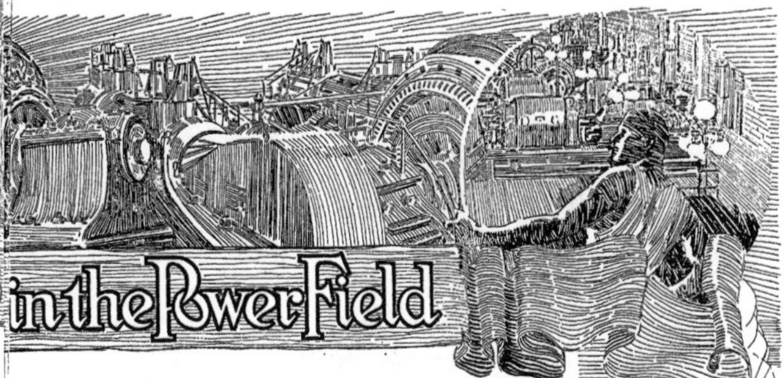

plant. The lessons taught by the Fuel Administration have not been forgotten. The possibilities of saving in the boiler room are more generally appreciated, and efficient combustion is being given more attention than ever before. Naturally, this has brought about a more general use of boiler-room instruments and by the manufacturer marked improvement in design, insuring greater accuracy and reliability.

Scarcity of coal as well as price has given impetus to the use of fuel oil, especially along the Atlantic coast; has brought about the use of inferior coals such as anthracite culm, coke breeze and lignites, and has stirred into renewed activity experimental work with powdered coal and colloidal fuel. Efforts toward conservation are turning more insistently to generation of electrical energy at the mines, to a more general use of water power and to the possibilities of low-temperature distillation of coal. Burning raw coal is extremely wasteful when it is possible to recover byproducts sufficient to pay for the fuel and still leave in the gas remaining, seventy per cent. of the original heat in the coal. Higher, but not excessive, initial cost of installation and more skilled attendance is the price of such recovery.

Scarcity of labor has enforced the installation of labor-saving devices in the power plant. The use of coal- and ash-handling equipment and of mechanical soot blowers has become more prevalent, and stokers cannot be made fast enough to meet the demand. The number of stoker manufacturers has increased at a surprising rate, and the capacity of each seems to be taxed to the limit. For the same reason and for the better economy that is possible, operators are resorting more frequently to automatic regulation of the damper, the supply of air to the furnace, the stoker engine and the water fed to the boilers.

High prices are also enforcing conservation along other lines. The profligate use of steel in the power-plant building, running as high as five to six pounds per pound of steam-generating capacity, is no more. The enormous heights above ground and depths below are giving way to more logical designs, inclosing only useful space, and economy in the use of construction material. A notable instance of this modern tendency is the new Moline boiler plant of the United Light and Railway Company. Reinforced concrete was used throughout with a considerable saving in first cost. A design studied to eliminate all but usable space and to combine bunker and roof support, was an additional factor in the economy.

A feature of the year was the remarkable activity among engineers. The characteristic reticence and retiring disposition of the average engineer gave way during the war to a realization of the value of coöperation and active

participation in public affairs. The result has been a desire for unification of effort among all engineering societies of the country. Plans of procedure have been remodeled and new services inaugurated to meet the new conditions. There has been a great movement toward coöperation between individual societies, and plans for amalgamation of engineering interests into one representative body, with the purpose of making more effective all work for the good of the public and the advancement of the profession.

Boiler Fuel

Briefly, the foregoing enumerates tendencies which require more illumination. Reverting to fuel, pulverized coal for the power plant has been in the spotlight again and notable improvement is evident. At Milwaukee the plant previously referred to in these columns has been extended and comprehensive tests have been conducted, but the results at this writing are not available for publication. Previous tests on the first unit were favorable, and encouraging word of equipment of this character comes from Seattle and Pittsburgh. At the spring meeting of the American Society of Mechanical Engineers the relative merits of powdered coal and modern stoker installations were given a thorough airing.

Urgent need for coal conservation has again brought to the front the great piles of culm in the anthracite regions. The coarser portion of this fuel under proper conditions can be burned directly in stokers. The finer culm passing through a 1/32-inch mesh is first pulverized and, as powdered coal, has been burned successfully under boilers. Another interesting experiment was the mixing of this pulverized culm with oil in the proportions of thirty and seventy per cent., respectively. Similar tests on a mixture of coal and oil known as colloidal fuel were made with excellent results by the Submarine Defense Association with a view to conserving the oil supply of the Navy. In numerous instances anthracite screenings have been mixed with bituminous coal in about the same proportions as with oil and have been burned satisfactorily in hand-fired furnaces depending upon natural draft.

FIG. 1. WESTINGHOUSE JET CONDENSER, WILL PASS 13,000,000 GAL. OF WATER PER HOUR

The rising price of coal and the shortage of labor has created a situation extremely favorable to oil. Where, formerly, regions remote from coal mines and contiguous to oil fields burned to advantage the latter fuel, quite recently there has been a steadily growing market for oil along the Atlantic coast, particularly in New England, which can be reached so conveniently by tank steamers. The movement is spreading to New York State, New Jersey and eastern Pennsylvania, and even in the bituminous coal regions of the Middle West there have been numerous transfers from coal to oil. The obvious reasons are the relative prices of the two fuels, the saving in labor effected by the use of oil and

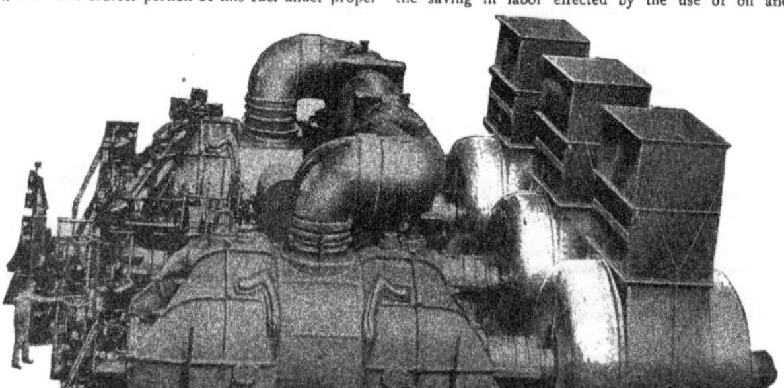

FIG. 2. WESTINGHOUSE 60,000-KV.-A. THREE-ELEMENT CROSS-COMPOUND TURBO ALTERNATOR

the more easily controlled combustion of the liquid fuel. Owing to the impetus given to the quest for petroleum during the war, the continued high prices for oil and the recent improved conditions in the market for drilling equipment, the domestic production for 1919 will reach probably 360 million barrels, as compared to 345 million barrels for 1918. The estimated consumption exceeds 400 million barrels, greater by a small percentage than the figure of 397 million barrels given for the previous year.

Mexico today is the great fuel-oil supply house. The 1919 production will likely be eighty million barrels; what the potential daily production is no one knows, but estimates by authorities place it at one million barrels. Oil interests are reported ready to spend one hundred million dollars in further developments there. The Mexican oil situation is now one of disputed ownership. In 1883 the Diaz government enacted the mining laws according to which the owner of the surface of oil-bearing land owned also the oil that the land might be made to yield. On these grounds the oil industry grew. At the constitutional convention of the Carranza government the famous "Article 27" of the constitution which became effective May 1, 1917, was written; it reads in part: ". . . the nation has direct dominion over . . . solid mineral combustibles, petroleum, and all solid, liquid or gaseous carbohydrates." Carranza, on Feb. 19 last, issued a statement defining the provision as meaning direct dominion. These decrees and others annulled the long standing previous legal status of all petroleum leases and titles.

FIG. 3. WORTHINGTON SURFACE CONDENSER, 56,000 SQ. FT. OF ACTIVE TUBE SURFACE

But, however the dispute is settled, the price of oil will surely chase the price of coal. This is indicated by the fact that in New England consumers signed long-term contracts for oil at ninety-two cents per barrel of forty-two gallons, while today such contracts cannot be made under about $2.50 per barrel, and then there is a sliding-price clause which may be to the advantage or disadvantage of the cansumer.

Notable Steam-Power Plants

Among the notable steam stations given publicity during the year were the United States Nitrate Plant No. 2 at Muscle Shoals, Alabama, and the Old Hickory plant near Memphis, Tennessee, both war projects conceived and built in record time. The former, which was completed shortly after the armistice was signed, has never been in operation. It was laid out for two turbine units to supply ninety thousand kilowatts for the manufacture of great quantities of ammonium nitrate. Only one unit of sixty thousand kilowatts capacity was installed, this being a three-cylinder reaction turbine with a water rate ranging between ten and eleven pounds of steam per kilowatt-hour. Boilers were of the Stirling type having 15,070 square feet of steam-making surface and fifteen-retort underfeed stokers. Initially it

FIG. 4. GENERAL ELECTRIC 45,000-KV.-A. SINGLE CYLINDER TURBO ALTERNATOR

was planned to use a total steam temperature of 590 degrees. Operating conditions were to be practically ideal. Supplying principally electric furnaces, the load factor would have been ninety-five per cent. Excellent coal, low in ash and moisture and high in volatile, was available. With a furnace volume of 11.6 cubic feet per square foot of active grate area, a boiler efficiency ranging from 76 to 78 per cent and a kilowatt-hour at the switchboard for seventeen to eighteen thousand British thermal units, was expected. Unfortunately, this plant has been shut down for nearly a year, waiting for the Government to decide what should be done with it.

At the Old Hickory plant are a large number of medium-sized boilers, sixty-eight of the Stirling type, rated at 823 horsepower each, having been installed. On the entire works over thirty-three thousand men were engaged in the construction. Everything was subordinated to speed and because of the probable short time of operation the cost was kept to the lowest possible point consistent with satisfactory operation. Boilers were installed on the basis of operation at 175 per cent. continuous rating and with the fifty-six thousand rated horsepower in boilers it was planned to carry day and night an average load of eighty-six thousand horsepower, this with Kentucky coal containing 14 per cent. ash, 1.8 per cent. sulphur and 12,600 B.t.u. per pound of dry coal. An estimated daily coal consumption of five thousand tons and about nine hundred tons of ash required an extensive system to store and handle, particularly as there was bunker capacity for only one day's run.

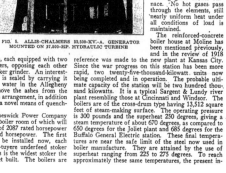

FIG. 5. ALLIS-CHALMERS 32,500-KV.-A. GENERATOR MOUNTED ON 37,500-HP. HYDRAULIC TURBINE

In its power plant at Springdale, Penn., the West Penn Power Company is installing four 1529-hp. boilers, each equipped with two thirteen-retort underfeed stokers, opposing each other and delivering the ash to a clinker grinder. An interesting feature is that the ashpit is sealed by carrying it down below the level of the water in the Allegheny River. It is intended to remove the ashes from the pits by bucket conveyors. This arrangement, in addition to sealing the ashpit, furnishes a novel means of quenching the ash.

At Cheswick, Penn., the Cheswick Power Company is building a power house, the boiler room of which will have twenty-eight boiler units of 2087 rated horsepower each, or a total of 58,436 rated horsepower. The first seven of these units are to be installed now, each equipped with a 17-retort, 21-tuyere underfeed stoker having a clinker grinder. This is the widest stoker the Westinghouse company has yet built. The boilers are 51 tubes wide and 18 tubes high. By the company's engineers it was found to be more desirable to use the wide boiler with single-end stoker equipment than the narrower boiler with a double fire.

It will be interesting to watch these two plants when they are both in operation, as they will use practically the same coal, carry similar loads and have the same make of stoker and boiler. The only difference will be that it is planned to operate the West Penn boilers at ratings of 250 per cent. up, while 175 per cent. will be the normal rate of operation in the Cheswick plant.

A relatively small but interesting plant is that of the American Smelting and Refining Company at Omaha. For simplicity, low initial cost, accessibility and minimum labor requirements, it is a model. At present it contains only one unit of two 406-horsepower boilers set over one eleven-retort underfeed stoker. Iowa coal containing on the average 14 per cent. moisture, 31 per cent. volatile matter, 35 per cent. fixed carbon and 20 per cent. ash, with a heat value of nine thousand British thermal units and a sulphur content of about six per cent., is burned at very high efficiency and with a minimum of operating troubles.

To utilize to the full the radiant energy from the fire, the boiler is made wider than usual and less tubes high. It is horizontally baffled at the fifth row of tubes. Because of its location back of the bridge-wall, the superheater also draws on the radiant energy of the fuel bed and the burning gases in the furnace. No hot gases pass through the elements, still nearly uniform heat under all conditions of load is maintained.

The reinforced-concrete boiler house at Moline has been mentioned previously, and in the review of 1918 reference was made to the new plant at Kansas City. Since the war progress on this station has been more rapid, two twenty-five-thousand-kilowatt units now being completed and in operation. The probable ultimate capacity of the station will be two hundred thousand kilowatts. It is a typical Sargent & Lundy river plant resembling those at Cincinnati and Windsor. The boilers are of the cross-drum type having 13,512 square feet of steam-making surface. The operating pressure is 300 pounds and the superheat 250 degrees, giving a steam temperature of about 670 degrees, as compared to 650 degrees for the Joliet plant and 685 degrees for the Buffalo General Electric station. These final temperatures are near the safe limit of the steel now used in boiler manufacture. They are attained by the use of superheat ranging from 225 to 275 degrees. To reach approximately these same temperatures, the present in-

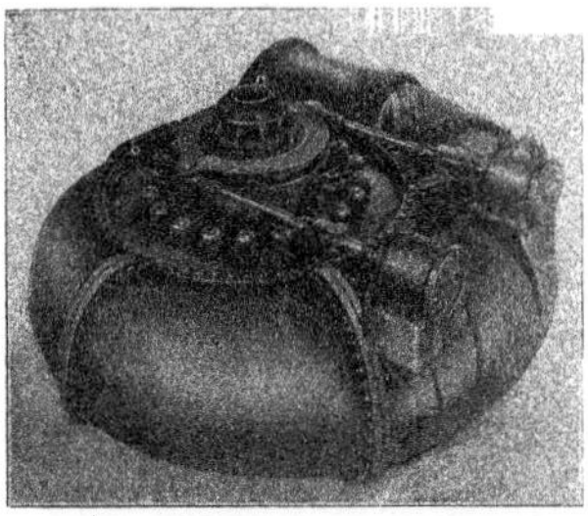

FIG. 6. I. P. MORRIS 37,500-HP. HYDRAULIC TURBINE

clination tends toward higher pressures and less superheat. As far as temperatures go, a pressure of six hundred pounds and superheat of one hundred and fifty degrees would not exceed the limits given in the cases cited.

The chain-grate stokers employed are unusually large, being 10 feet wide by 17.5 feet long. Large area and high stoker speed with relatively thin fires give the coal-burning capacity needed to operate the boilers at high rating.

For a high-pressure installation the economizer problem has been solved in an interesting manner. While cast-iron economizers have given safe and efficient service at pressures in the vicinity of two hundred pounds, conservative engineering opinion rather hesitates to subject the uncertainties of cast iron to pressures fifty per cent. greater. For economizers, steel is not so durable as cast iron on account of the greater corrosion, but it is safer. As a compromise, in the plant under discussion the economizer was made in two sections. A high-pressure element made of steel and resembling a section of a Stirling boiler, is placed above the boiler uptake and low-pressure cast-iron elements in the usual position above the boiler next to the induced-draft fan discharging to the smoke flue. As far as the water connections are concerned, the two sections are distinct. The feed water is pumped through the low-pressure section into a tank which supplies the suction of the feed pumps delivering through the high-pressure economizer to the boiler. In the low-pressure economizer the feed water is heated to a temperature somewhat less than two hundred degrees Fahrenheit to prevent the formation of steam and the resulting pressure, but still allowing it to pass to the high-pressure section at a temperature sufficiently high to prevent condensation on the tubes. The corrosion resulting from condensation is thus limited to the low-pressure elements, where it is better resisted by the cast iron and where leakage and deterioration are more easily controlled.

At Toledo the finishing touches are being given to the second large unit at the plant of the Acme Power Co. and at the Riverside station in Minneapolis the 25,000-kilowatt addition in one turbine unit is in operation. An interesting plant has been built by Armour & Company at South St. Paul. It consists of a refrigerating plant containing both compression and absorption systems and a mixed-pressure bleeder-turbine generating installation. The interesting feature is the balancing of the supply of exhaust steam with the requirements for process work in the packing plant. After careful selection of equipment in the first place, the final adjustment was left to the bleeder turbines, which make up a deficiency in the main exhaust or utilize an excess.

FIG. 7. FIELD COIL 32,500 KV.-A. ALTERNATOR

Mention of notable plants would not be complete without reference to the great expansion of power generation and distribution. Generating stations, hydraulic and steam, and even whole systems are being tied together, and water power, even of small capacity, is being utilized to the fullest extent, where before it was given only minor consideration.

To guide this movement and furnish to expanding industry and transportation an adequate, dependable and economical power supply, Secretary of the Interior Lane asked of Congress two appropriations, one of fifty thousand dollars to continue the power survey of the whole United States undertaken originally by the Geological Survey at the request of the Fuel Administration, and one of two hundred thousand dollars for a more intensive survey of the Atlantic coast section from Boston to Washington or Richmond, in which one-fourth of the entire generating capacity of the country is concentrated. The plan, which had the unqualified support of the engineering profession, was to enlist the coöperation of the electric industry in a careful economic and engineering investigation of the future power requirements of this region and of the best means of supplying its needs, taking into account the most efficient use of natural resources and linking all together in one great super-power system. With cheaper power and more general use of it, manufacturers could successfully compete in the world's markets and still main-

FIG. 8. RUNNER FOR 37,500-HP. TURBINE, FIG. 6

tain American standards of wages and living. Unfortunately, the proposal of Secretary Lane suffered the fate of the water-power bill and other well-merited legislation. The subject is too important to be set aside and should be given immediate consideration as soon as Con-

FIG. 9. RATEAU BATTU SMOOT 1700-HP. 3600-R.P.M. SYNCHRONOUS MOTOR-DRIVEN BLOWER

gress has disposed of its foreign-relations problems and the return of the railways.

Steam Turbines and Auxiliaries

A notable incident was the starting up of the great three-element cross-compound turbine unit of the Interborough Rapid Transit Company of New York. With a two-hour rating of 70,000 kilowatts and a normal output of 60,000 kilowatts, it is the most powerful prime mover in the world. In the turbine field much of the time seems to have been spent in digesting operating data obtained from the numerous large units recently installed, in determining and correcting the cause of accidents to a number of large machines, all of the same type, and in studying limitations as to speed and capacity of single-shaft units. The Committee on Prime Movers of the National Electric Light Association were of the opinion that with present constructional problems of prevailing frequencies and speeds and the recognized factors of safety, efficiency and cost, the size of the systems today will hardly warrant single units larger than 30,000 kilowatts capacity. In particular cases, and until increased reliability, as well as improvements in operating efficiencies, have been established, even extended systems will not justify the use of larger units. While on the whole this statement found support, there is general confidence that larger turbines having a greater proportionate size to the total

FIG. 10. RATEAU BATTU SMOOT 1000-HP. 3750-R.P.M. TURBO-BLOWER UNIT

station capacity will be developed gradually. Equally high efficiency could perhaps be secured in small machines, but the initial cost per kilowatt would be greater.

Later, the same question was discussed before the American Institute of Electrical Engineers. Turbine designers agreed that the limit of a single-shaft machine did not lie in the generator, but in the turbine—and in the last wheel of that. Concerning this limitation there were two sets of adherents, one favoring the division of the turbine into separate compounded elements, each driving its own generator and each capable of operating alone under high-pressure steam. Engineers of the opposing school preferred dividing the installation into a number of identical units of smaller capacity rather than into a number of dissimilar elements forming the successive stages of a compound unit. The accidents that had occurred to large single-shaft turbines were due to fluttering and vibration of the wheels, which had been made too thin in order to save weight and space. Stiffening the wheels and putting the openings in them nearer to the hub, where a suitable reinforcement could be provided, had overcome the trouble.

Electrical engineers were positive that generators could be designed conservatively and constructed for units of fifty thousand or even one hundred thousand kilowatts, but the higher the capacity the higher the voltage must necessarily become. For speeds of twelve hundred revolutions per minute and lower, the stator, and for higher speeds, the rotor, is the limiting member.

Relative to steam condensers, there is little to say, except that in size and refinements they have kept pace with the large turbine units. Notable among these is the surface condenser Fig. 3, which has fifty-six thousand square feet of active tube surface, and the jet condenser Fig. 1. The latter is capable of passing thirteen million gallons of water per hour. Certain large condensers have been placed on spring supports without expansion joints. Another method is to hang the condenser from the turbine foundation, which accomplishes practically the same results as the springs. Mention might be made of the prominent place in marine work given the steam air ejector and its increasing use in the stationary power plant. An over-all efficiency higher than that of the hydraulic air pump is claimed for this device. The disadvantage of a greater steam consumption is more than offset by the recovery of practically all the heat that it uses.

FIG. 11. WESTINGHOUSE 8333-KV.-A. SELF-COOLED TRANSFORMER

A new development is a four-compartment surface condenser in which tubes may be cleaned or repairs made while the turbine is in service. Each compartment is equipped with a set of valves to control the circulating water. Shutting off the valves allows a compartment to be cleaned. It is then placed in service and the same operation is repeated on the remaining sections in turn.

Rotary Blowers and Compressors

Although differing entirely from the steam turbine in design and use, but still a rotary machine, mention might be made here of a large synchronous motor-driven turbo-blower, Fig. 9, installed for the Tennessee Copper Company to discharge air into the main air line which directs the blast into several copper smelters.

Normally, 125,000 cubic feet of air per minute is discharged at a pressure of 50 ounces, this output probably being the largest ever delivered by a single unit. The blower Fig. 10 is driven by a one-thousand-horsepower steam turbine, runs at 8750 r.p.m. and compresses 15,000 cubic feet of free air per minute to a pressure of fifteen pounds. The dimensions of this unit are remarkably small for the capacity, the outside diameter of the blower casing being approximately thirty-six inches.

Word came recently of a high-pressure, rotary compressor of the eccentric rotor, telescopic blade type, per-

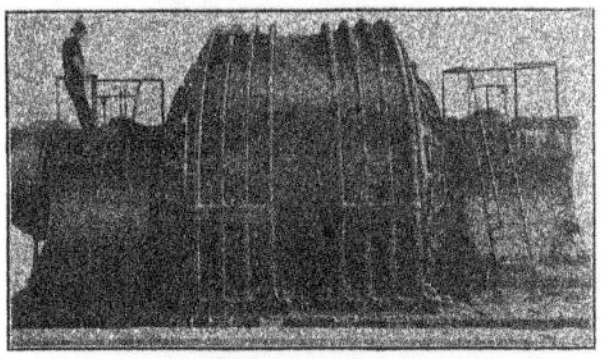

FIG. 12. GENERAL ELECTRIC 5000-HP. DOUBLE-UNIT DIRECT-CURRENT MOTOR

fected by W. G. E. Roloff, of Belleville, Illinois. According to report the leakage, friction and packing troubles inherent with this type have been eliminated to such an extent that a volumetric efficiency of 92 per cent. when compressing to one hundred pounds pressure is commercially obtainable. Other claims are that this machine without alteration will handle gas, air or liquid, or a gas and a liquid at the same time, and that it is possible to obtain a pressure of five hundred pounds in a single-stage machine and one thousand pounds in three stages with the elements on the same shaft—all this with a machine of about one-tenth the weight and occupying about one-twentieth the floor space required by a reciprocating compressor of the same capacity.

Steam Engines

A canvass among builders of reciprocating steam engines revealed that during the first half of this year, when rising costs of material and labor and the uncertainty as to whether these would soon reduce led to hesitation in new construction, there was not the usual demand for steam engines. But more recent conditions have indicated that the steam engine is by no means a back number, and the demand is active, especially in the small and medium sizes. In the large sizes, of course, there is still a large demand for water-works pumping.

The rising fuel and labor costs have created a market for the more economical types of engine, and have also resulted in a great impetus to the uniflow type. There are now nearly a dozen firms building these engines, many of whom have taken out licenses during the past year. This type, because of its flat load curve, makes an appeal where variable load conditions are encountered.

A giant Nordberg hoisting engine, even larger than the famous Tamarack hoist in the Calumet district, which had previously held the record, is being installed at the No. 2 shaft of the Quincy Mining Co. In general arrangement it is similar to the Tamarack hoist, the cylinders and frames being set at an angle of 45 deg., with the main bearing and drum shaft forming the apex of the triangle. The two high-pressure cylinders work on a common crankpin at one end of the crankshaft and the two low-pressure cylinders work on a common crankpin on the opposite end of the crankshaft. The high-pressure cylinders are 32 in. in diameter, the low-pressure cylinders 60 in. in diameter and all four cylinders have a common stroke of 6 6in. The hoisting drum is 30 ft. in diameter and approximately 30 ft. face. It is cylindro-conical in shape, having cones at each end and a center cylindrical section. The drum is built with a winding capacity of 13,000 ft. of 1⅝-in. rope in one layer. The inclination of the shaft is about 40 degrees.

The hoist will lift 20,000 lb. of copper ore per trip in skips weighing approximately 10,000 lb. each. The total rope pull is 52,000 lb., as compared to 42,000 lb. for the Tamarack hoist, and the torque when operating in balance is 275,000 ft.-lb. The hoisting speed will be 3200 ft. per min. The engine is equipped with a counter-current jet condenser and will be supplied with steam at 160 lb. boiler pressure. The auxiliaries, such as brakes, reversing gear and throttle, are operated by oil under pressure supplied from an accumulator.

Oil Engines

The past year has seen few radical changes in oil-engine design. The market for these engines is promising, and it is anticipated that the greater use of fuel oil for boiler purposes will carry with it a demand for oil engines in localities in which they have heretofore been used very little. This will be due to the modification of city ordinances regarding fuel-oil storage and to the fact that the oil companies will push the oil business in these localities to a greater extent.

Both the two- and the four-stroke cycle types are being built, although the latter is in the majority. Cer-

FIG. 13. GENERAL ELECTRIC 4000-HP. 6600-VOLT INDUCTION MOTOR

tain companies, by careful attention to design and workmanship, appear to have successfully overcome some of the previous troubles attributed to the two-stroke-cycle type.

The marine field is promising, and a considerable number of Diesel engines have been built and more are on order for cargo carriers. Twin engines of moderate power are apparently the more popular.

A considerable number of stationary units are being built for mining and oil-line pumping service, as well as scattered industrial applications. Those in the mining districts are mostly of the larger size, vertical type, and go to the Southwest. A number of these engines of the two-stroke-cycle type, in sizes up to twelve hundred and fifty horsepower, five cylinders, have been installed, and others are building in sizes up to twenty-five hundred horsepower, five cylinders. The largest four-stroke-cycle engine built for stationary practice in this country is sixteen hundred horsepower.

Fig. 14 shows a 1250-b.hp. two-stroke-cycle Diesel engine built by the Nordberg Manufacturing Company for driving a 6500 cubic foot variable-speed compressor at an elevation of 5000 feet. The engine has five working cylinders, each 20¾ by 26 inches, delivering 250 horsepower, and the sixth cylinder is the scavenging pump. The speed will vary from 180 to 80 r.p.m. The weight of the engine is 380,000 pounds, or slightly over 300 pounds per horsepower. A fuel consumption of one pound of oil per 1000 cubic feet of air compressed to 90 pounds is anticipated.

Boiler-Room Tendencies

Relative to the boiler room, previous reference has been made to the possibilities of higher pressures and to the more common use of instruments and labor-saving devices. In large plants the tendency seems to be toward boilers of moderate size rather than enormous capacities in one unit, the range running from one thousand to fifteen hundred horsepower. Boiler-setting design and materials used have received much attention of late. An innovation is the use of reinforced-concrete settings by the engineers of the Robert Gair Company, of Brooklyn. In the older boiler room of the company containing hand-fired return-tubular boilers, reinforced concrete was used for the settings, and cinder concrete containing no sand formed the lining for the furnaces. Eight years' use confirmed the reliability of these settings, so that recently, when adding additional boiler capacity, engineers of the Gair company had no hesitancy in prescribing concrete for the new settings and reinforced-concrete columns exposed to the heat of the furnace for support of the boilers, the latter being large stoker-fired return-tubular boilers. Results from the latest installations will be watched with interest, for if concrete can be made safe for boiler settings, it will lower the construction cost appreciably and no doubt come into wide use for the purpose.

Recent tendencies, particularly with high boiler ratings, is to build the walls of the furnace with as few different kinds of materials as possible and to make them solid, well-bonded and homogeneous. This development has been carried to the extent of building furnace walls of solid firebrick.

During the year a new marine boiler of the Heine cross-drum type was reported and publicity was given to a "double-service" boiler offering a new method of separating scale-forming solids from the feed water. The apparatus consists of a live-steam feed-water purifier located within a horizontal fire-tube boiler which may be of the locomotive, Scotch or return-tubular type. Carbonates, sulphates, magnesia, mud and oil are retained in the purifier and do not get into the boiler. As the fire strikes no portion of the purifier, the deposits do not bake on as they would in the boiler, and remaining in the form of a sludge, can be blown out through the blowoff at any time the boiler is under pressure.

Boiler explosions were about as numerous as usual, our old "enemy" the lap-seam boiler "hogging" the bulk of the accidents. The most disastrous accident of the year was the explosion occurring at Mobile, which resulted in four deaths and a property damage exceeding one hundred thousand dollars. Two water-tube boilers, one a two-hundred-horsepower Stirling and the other a Heine of nearly the same size, let go almost simultaneously.

Chain Grate and Underfeed Stokers

In the stoker field there has been great activity. Manufacturers of all types, and there are many new faces among them, have been busy trying to catch up with the demand. Orders held in abeyance during the war and the present labor situation within the power plant are the reasons. To meet the desire for an automatic stoker that can be forced to a point of capacity far in excess of the limits of natural draft, various chain-grate manufacturers have been attempting to apply forced draft to their stokers, with provisions to readily convert them to the natural-draft type during low-load periods. The stokers have been divided into distinctive compartments so that air pressure and volume may be suited to the requirements of the fuel passing over each compartment. The air is supplied by one large fan, with damper regulation at each compartment, or, as in one case, by a small turbo-blower for each compartment. Chain-grate stokers with forced draft have been operating satisfactorily for some time with various kinds of fine fuel such as anthracite screenings, culm, coke breeze, etc., and more recently successful installations have been made for low-grade fuels, such as bituminous coals high in ash, and lignites with excessive moisture.

Combustion rates up to seventy pounds of Illinois bituminous coal per square foot of grate area per hour have been maintained, and with the performance of the steam locomotive in view there are hopes of exceeding by a wide margin the figure given. The thickness of the fuel bed needed to prevent the air from blowing holes in the fire, the added clinker troubles with thick fires, the regulation of the various compartments, and furnace upkeep are factors holding back the advance. With a positive pressure in the furnace preventing infiltration of cooling air through the walls, deterioration of brickwork and refractory material is rapid.

For these very reasons a prominent manufacturer of chain grates is attacking the same problem from a different angle. Believing that it is final over-all results that count and seeing no advantage in a gain overcome by upkeep, the extra coal for high ratings is burned with natural draft by increasing the grate area and speeding up the stoker to the limit. Speeds of seven inches per minute, as compared to old standards of one to three inches per minute, have been attained. Thin fires are the rule, the fuel bed, depending on the character of the coal, being just thick enough for the coal to burn out completely during the active travel of the grate. With these thin fires there is no disturbance of the fuel bed and consequently no clinkers. The fire is bright and quick, and with a steady load and uniform coal, conditions are ideal. Fisk Street Station of the Commonwealth Edison Company and the new Kansas City plant furnish examples of the large stokers the above practice requires. In either station chain grates having the unusual length of 17.5 feet have been installed. Owing to the use of economizers, however, induced draft is employed.

In underfeed stokers the most important developments are the perfection of power dump grates and the clinker grinder, both of which are coming into more general use, the choice depending upon the character of the coal burned. Another praiseworthy improvement is the provision of sectional drive, wherein the total number of retorts is divided into sections of a few or several retorts, depending upon the size of the stoker.

Each section may be operated without interference with other sections, giving a flexibility highly desirable in maintaining a uniform fuel bed on the wide stokers now in use. Various methods of drive have been developed, and the acceleration or retardation of the feed of the coal is facilitated by means of two-speed power boxes. In stokers of extreme width this sectionalizing in a moderate degree has been carried to the chain grate.

SMOKE ABATEMENT

Combustion is a subject naturally suggested by stokers, and for the good of humanity as well as the pockets of plant owners, it is desirable that it be of the smokeless variety. Since the war period of dirty coal and fuel unfamiliar to the plant, there has been some recovery from the condition of smudge and smoke that was generally prevalent. Judging from records of the past, Chicago has taken a backward step in destroying the identity of its smoke department and making it an adjunct of the board of health. The old policy of education and coöperation that had proved so successful has been supplanted by arbitrary ruling based from the health standpoint. The violator is no longer shown how to mend his ways. Good combustion, which would solve the problem from all angles, is not considered.

A ray of hope comes from Salt Lake City, where the Bureau of Mines, operating under the provisions of the Sundry Appropriations Bill, in coöperation with the University of Utah, is making a thorough investigation of the smoke conditions. The most approved engineering methods are being used in securing data from which a working plan to reduce the smoke nuisance will be evolved. Authority is given the bureau to extend aid to other cities interested in smoke abatement.

FIG. 14. TWO-STROKE CYCLE, 1250-HP. DIESEL DRIVING AIR COMPRESSOR

All over the world there has been rapid expansion of the application of hydro-electric power. The shortage of coal during the war gave impetus to the movement, and today there is scarcely a country having hydraulic resources that is not considering the better use of its dormant water power or is already carrying out well-defined plans.

Norway has made exceptional progress in this line, and Canada is a close second with its 276 hydraulic horsepower per thousand of population, as compared to one hundred horsepower for the United States. In the former country the great Queenstown development of 300,000 horsepower on the Canadian side of Niagara is now under construction and orders have been placed by the Hydraulic Power Commission of Ontario for the first three units. Under a head of 320 feet each turbine will develop 52,500 horsepower at the point of maximum efficiency, and about 60,000 horsepower at full gate opening.

On the American side the Niagara Falls Power Co. is installing three units, each to develop 37,500 horsepower under a 214-foot head. The type of hydraulic turbine used to drive two of the generators is shown in Figs. 6 and 8. The first unit is nearly ready for operation and for the time being ranks as the most powerful turbine in the world. These machines are in line with the modern tendency of installing high-powered units, although Dr. C. P. Steinmetz has expressed the opinion that the country is approaching the limit of water-power development by present methods of concentrating in large stations.

Other tendencies in the field are the development of small water-power plants connected together and under remote control from the master station, the provision of piezometer equipment in each turbine intake so that operating efficiency may be checked and maintained, and general attention to details of design affecting the ease of operation and economy.

One of the latest developments is a new governor designed by M. Seewer, a Swiss engineer, for high-head hydraulic turbines of the Pelton type. The governing is effected by changing the shape of the jet by guide members located in the interior of the jet nozzle and so arranged that the jet may be practically cylindrical or may be partly or totally dispersed. A slight dispersion of the jet produces a material reduction in the hydraulic efficiency of the turbine and at complete dispersion no useful power is produced.

FIG. 15. RATEAU BATTU SMOOT 1000-KW. 2700-R.P.M. DIRECT-CURRENT TURBO GENERATOR

The Electrical Field

Although new types of electrical apparatus have been brought out during the past year, the greatest development has been along the line of large-scale operation with standard equipment. The year has seen the placing in operation of the largest turbo-generating unit ever built; the delivery of equipment for the world's largest furnace installation; the construction of the largest electric locomotives; the completion of the design for a transformer of higher rating than has yet been manufactured; and the construction of three of the largest hydro-electric generating units built up to the present time, although orders have been placed for machines exceeding these in size by about thirty per cent.; the export shipment of the electrical apparatus for what will be the first electrically driven steel blooming mill to be erected in the Far East; and the export shipment of the equipment for a large hydro-electric development in Europe.

The 60,000-kilowatt 25-cycle three-element cross-compound unit, Fig. 2, was placed in service at the Seventy-fourth Street Station of the Interborough Rapid Transit Company, New York City, in the early part of the past year, and the performance of this unit indicates that the construction and operation of such a machine and even one of larger capacity is within the premise of good engineering. These generators are so connected to the busbars that any combination of them can be operated in parallel. In practice, however, all three machines are brought up to speed together, and synchronized through a single oil switch connecting the generators to the main busses. An indication of the confidence in such machines is shown in the construction of the new plant at Cheswick, Penn., of the Duquesne Light and Power Company. In this plant it is planned to eventually install a total of five of these machines. The first and probably two of the units will be put into service during the coming year. A similar sixty-cycle unit was installed in the United States Government Nitrate Plant No. 2 at Muscle Shoals, Alabama. In this case the generator driven by the high-pressure turbine is operated at 1800 r.p.m. and the two driven by the low-pressure turbines run at 1200 r.p.m.; the generators, however, are all of the same capacity. The 45,000-kilowatt unit, Fig. 4, installed in the Detroit Edison Company's plant has also been in service for the past year.

Three of the most notable hydro-electric generators, as to capacity, ever built have been completed during the year, and the installation of these machines is well under way at this date. One of these generators is being built by the Westinghouse Electric and Manufacturing Company, one by the General Electric Company and the third by the Allis-Chalmers Manufacturing Company. The last-named company has built the complete unit both generator and hydraulic turbine as shown in Fig. 5. An idea of the dimensions of the generating unit may be gained from Fig. 7, which shows one of the field coils used in the Westinghouse machine. Each machine is rated at 32,500 kilovolt-amperes and will operate at 150 r.p.m. and generate 12,000-volt three-phase twenty-five-cycle current. The general outline of all three generators is similar to the unit shown in Fig. 5. As previously stated, these machines are being installed in the Hydraulic Power Company's plant of the Niagara Falls Power Company at Niagara Falls, New York. Among other important hydro-electric developments are the Caribou development of the Great Western Power Company, which is installing equipment and transmission lines for operation at 165,000 volts. Two 22,222-kilovolt-ampere 11,000-volt 60-cycle units are included in the initial installation. Orders were placed during the year for three of the six generators that will be installed at the Queenstown development of the Hydro-electric Commission of Ontario. These generators are rated at 45,000-kilovolt-ampere normal, and 50,000-kilovolt-ampere maximum continuous rating. The speed of these machines will be 187 r.p.m. and will generate 12,000-volt current.

FIG. 16. GENERAL ELECTRIC AUTOMATIC WELDING MACHINE BUILDING UP A SHAFT

The majority of installations during the year have followed along well-established lines in both capacity and design features. However, the unit that is now being built for the Southern California Edison Company may be considered a departure from standard practice. This unit is a 15,000-kilovolt-ampere 600-r.p.m. 50- to 60-cycle frequency changer set. Not only is this the largest in capacity thus far to be built, but its voltage is the highest that a revolving machine has been built for, motor and generator being designed for

15,000 and 18,000-volt normal operation respectively. An order has recently been placed by the Consumers Power Company, Jackson, Mich., for a 10,000-kw. frequency-changer set. This set is to operate so as to exchange power in either direction between a thirty-cycle and a sixty-cycle system.

Among the transformers constructed during the year are to be found a number of 8333-kilovolt-ampere 66,000-volt self-cooled type, one of which is shown in Fig. 11. These are the largest self-cooled units that have been built. The design has also been completed for a 23,600-kilovolt-ampere water-cooled transformer

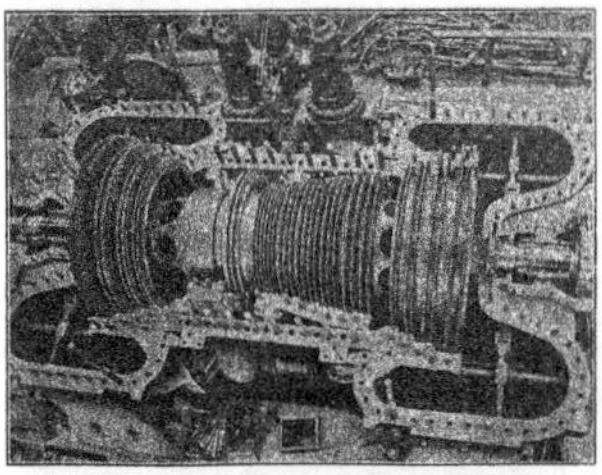

FIG. 17. INTERIOR OF WESTINGHOUSE TURBINE INSTALLED AT OSAKA, JAPAN

for 132,000-volt operation. This is the highest rated transformer yet attempted, and in physical size it will be considerably larger than its nearest rival. The building of eleven 11,500-kilovolt-ampere water-cooled transformers for 150,000-volt service is nearing completion. These transformers are designed for 72,000 volts on the low-tension side and for delta connection on the high-tension side.

A forty-inch reversing blooming mill at the Sparrows Point plant of the Bethlehem Steel Company was put in operation early in the year and has since been in successful operation. This equipment has a double-unit reversing motor, Fig. 12, with a normal continuous capacity of 5000 hp, at 50 r.p.m. and a momentary torque capacity of approximately 2,000,000 lb. at one-foot radius at any speed from zero to 50 r.p.m. During the early part of the year a 36-inch by 110-inch plate mill was started at the Fairfield Works of the Tennessee Coal, Iron and Railway Company. This mill is driven by a 4000-hp. 82-r.p.m. 6600-volt induction motor, Fig. 13. A forty-five-inch blooming mill for this installation is driven by a double-unit reversing motor with a normal continuous capacity of 5600-hp. at 55 r.p.m., having a momentary torque capacity of 2,300,000 pounds.

Considerable attention has been given to the possibility of using 220,000 volts for high-tension transmission of electrical power during the past year. The development of a 1100-mile 220,000-volt transmission line to be put in operation on the Pacific Coast, has been suggested, this system to have a capacity of 1,500,000 kilowatts eventually.

The tendency in the use of synchronous machines for power-factor correction and voltage regulation on high-voltage transmission systems is indicated in the 30,000-kilovolt-ampere synchronous condenser being built for the Southern California Edison Company. In the design of high-speed turbine-driven direct-current generators, the one shown in Fig. 15 is representative. This generator is of 1000-kilowatt capacity, operates at 2700 r.p.m. and generates 250-volt current. One of its radical features is the commutator, which of necessity must be of considerable length to provide the required brush contact surface to handle the 4000-ampere normal full-load current.

One of the most notable electrically driven mine-hoist equipments constructed during the year was that built for the Chicago, Burlington & Quincy Railroad Company and placed in operation at this company's coal mines operated by the Valier Coal Company. The hoist motor is 1350 horsepower direct-connected to a single-cylinder drum on which two ropes wind for balance operation with two dumping skips. The rope speed of this installation is 1500 feet per minute. A noteworthy feature of this installation is the semi-automatic operation, as the trip may be started either by an operator on the hoist platform in the usual manner or by the skip tender at the bottom of the shaft, and is automatically retarded and brought to rest, stopping accurately in the dump and at the loading chute.

Electric-arc welding during the year has been characterized by considerable effort to improve the art. The American Welding Society was formed and may be considered the successor to the Welding Committee of the United Emergency Fleet Corporation and the National Welding Council. It is to be hoped that the formation of this society will bring about the degree of coöperation between the various welding interests that is so

FIG. 18. ONE OF THE YORK VERTICAL COMPOUND COMPRESSORS, MERCHANTS REFRIGERATING CO., NEW YORK

necessary for the advancement of the art. The operator is still being considered one of the most important factors in the making of reliable welds, and it is being appreciated to a degree that some of the manufacturers of welding equipment have established schools for training operators sent by purchasers of their equipment. One of the most recent designs in electric welding with metal electrodes is that of an automatic machine shown in Fig. 16. Here the machine is shown building up a shaft that has been incorrectly machined.

Export of Electrical Machinery

Among the notable export shipments of electrical equipment during the year are: The equipment for the first electrically driven steel blooming mill to be erected in the Far East. This equipment is to replace a steam engine now operating a forty-inch reversing mill. The mill will form part of the plant of the Imperial Steel Works of Japan near Tokio, and will be operated by a 3500-horsepower single-unit motor, of the reversing blooming-mill type, taking direct current at six hundred volts. Two 25,000-kilowatt steam-turbine driven units, which, when installed, will complete the largest steam-driven electrical installation in the Far East, are now being erected at Osaka, Japan, for the Osaka Electric Light Company. The interior of one of these turbines is shown in Fig. 17. When these two units are installed, they will bring the capacity of this plant up to 75,000 kilowatts. Electrical equipment for twenty electric furnaces has been shipped to the Glomfjord Smeltverk Company, of Glomfjord, Norway. A 1300-kilovolt-ampere single-phase transformer is required to supply power to each of the equipments. Another notable shipment was the two 14,060-kilovolt-ampere 6600-volt hydro-electric units built for the Ebro Irrigation Company in Spain. An order has been placed in this country for the electrical equipment for a forty-inch reversing blooming mill, for the Tata Iron and Steel Company, India. This equipment is similar to the forty-five-inch blooming mill at the Fairfield Works of the Tennessee Coal and Iron Company, previously referred to.

Advance in Refrigeration

Refrigeration has witnessed a wider application and more scientific development of the high-speed compressor driven by the synchronous motor. Engineers and executives are anxiously awaiting reliable performance data of the high-speed plants which also embody refinements in design to improve heat transfer in condensers and coolers.

Two of the largest synchronous-motor-driven compound, compression refrigerating plants in the United States have been installed recently by the Merchants Refrigerating Company, of New York. The capacity of either plant is 550 tons of refrigeration. In one case three-cylinder compound vertical compressors are employed and in the other, machines of the horizontal double-acting type using feather valves. The speed exceeds 200 r.p.m.

Ammonia intercoolers are used between the low-pressure and high-pressure compressors in both installations. In one plant they are of the usual accumulator design, consisting of large cylindrical tanks with helical liquid cooling coils. In the intercoolers of the other plant the hot discharge vapors from the low-pressure compressor are cooled by means of a liquid rain over baffle plates.

The brine coolers are of the horizontal-shell straight-tube type in one plant and of the standard shell and helical coil design in the other. One of the horizontal coolers is supplied with helically wound iron strips in the tubes to maintain a nearly uniform brine temperature at any one point. Although this cooler has the same effective brine-cooling surface as the others, it has shown under the same conditions of operation, fifty per cent. greater capacity.

FIG. 19. THE DE LA VERGNE HIGH-SPEED HORIZONTAL COMPRESSORS, MERCHANTS REFRIGERATING CO., NEW YORK

Ammonia-condensing units in one of these plants are of the shell and straight-tube type erected vertically. The water-distributing device is so constructed that an annular ring of water inside of the tube flows down by gravity to the collecting point underneath the condenser. On account of the excellent distribution of the water the liquefaction temperature is never over five degrees above the outlet-water temperature. The advantages of this type of condenser are the small space occupied, the high efficiency and the ease with which they can be kept clean.

One of the plants was laid out to use not in excess of 1.8 brake horsepower per ton of refrigeration at 3 lb. back pressure and 165 lb. condenser pressure. The other was designed to produce a ton of refrigeration on not to exceed 1.65 brake horsepower at 8 pounds back pressure and 165 pounds condenser pressure. Present indications are that as soon as all necessary adjustments have been made, the power consumption will be less than the figures given, for the conditions stipulated.

The new synchronous-motor-driven high-speed refrigerating plant of the Quincy Market Cold Storage and Warehouse Company, of Boston, has received such recent mention in these columns that it will be unnecessary to enumerate details.

Engineering Societies Active

Never in their history have engineering societies been so active as in the past year. From the consideration of purely technical subjects they are now assuming greater activity in the industrial field, are following more closely engineering legislation, and directing engineering matters of local or national import. In other words, the plan is to render the maximum of national, social and political service. To carry on this work most effectively there is full realization of the value of coöperation and unity of action. A joint conference committee of the five great associations of the country has already outlined a plan to form a single comprehensive organization for securing united action of the engineering and allied technical professions in matters of common interest. It is to be made up of a federation of local and national engineering societies with the existing specialized professional associations still retaining their identity. Local societies and sections of the existing professional associations would thus be joined together for concerted action in their respective territories and the general engineering society, made up of representatives from the local groups, would care for matters of national scope. Action on this plan awaits the approval of the societies represented in the conference.

Another movement on foot is the formation of an American Engineering Standards Association, the objects being to unify and simplify the methods of arriving at engineering standards, to secure coöperation between various organizations, to prevent overlapping or duplication of work and to coöperate with similar organizations in other countries to promote international standardization. Realizing the value of coöperation from war work along these lines, the five national associations previously referred to invited the Government Departments of War, Navy and Commerce to appoint representatives to act with them to continue this work. The engineering standards committee so formed has drawn up plans of action that should not interfere with organizations already doing excellent work in this line and is awaiting ratification by the governing boards of all the societies and departments represented.

In the refrigerating industry the foremost societies of the field have united to form a parent body known as the American Association of Ice and Refrigeration, the objects being to avoid duplication of effort and by joint activity to do more effective work on the problems of the field.

The National Safety Council has approved the organization of an engineering section, so that civil, mechanical, electrical, mining and chemical engineers may contribute more effectively toward the solution of purely engineering problems encountered in safety work.

In passing, mention should be made of the effort to raise and standardize salaries of engineers. In this movement the American Association of Engineers took an active part.

The establishment of a Bureau of Research by the American Society of Heating and Ventilating Engineers and the arrangement by which a portion of the magnificent new laboratories of the Bureau of Mines at Pittsburgh was placed at its disposal, with Professor John R. Allen as director, should prove of inestimable value to a field lacking in constants and reliable data on heat losses.

Other work in which the aid of the Bureau of Mines is sought is the perpetuation of power economy by continuing the work of the Fuel Administration. An effort in this direction is being made by the International Power Economy Conference, but before the Bureau can assume this new rôle, funds are needed, and it is the primary intention to create sufficient public sentiment to influence favorable action by Congress.

Passage of the Jones-Reavis bill is another duty to be imposed on Congress as soon as the "back-home" campaign has had its effect. The bill proposes to reorganize the Department of the Interior into a Department of Public Works, consolidating under one head all building of the Government, amounting to twenty per cent of the appropriations made by Congress. Hitherto this work has been done by thirty different bureaus, twelve to fifteen of them of an engineering character tucked away in various departments foreign to their work. Concentration in one bureau will promote efficiency and enable Government building to be done at much less expense. Instigated by the Engineering Council, the movement is now being carried on by an organization perfected at a conference of seventy-four different engineering societies. State organizations with state directors are now being formed to conduct a strong publicity campaign on the merits of the bill.

Power Plant Legislation

Legislation affecting the power plant more directly, in addition to that already mentioned, was the creation in Massachusetts of a Department of Public Safety, organized into three divisions and incorporating the Board of Boiler Rules, boiler and building inspection and the examination and licensing of engineers and firemen into one of the three, known as the division of inspection. The repeal of the daylight saving law will have its effect on the general public as well as the power plant, unless general local action perpetuates in urban centers the advantages so evident during the war. There has been no indication of engineer's license legislation, but during the year bills for the legal adoption of the American Society of Mechanical Engineers' Boiler Code were introduced before state legislatures with the result that thirteen states have adopted the Code, namely: Massachusetts, Rhode Island, New York, New Jersey, Delaware, Pennsylvania, Ohio, Michigan, Indiana, Wisconsin, Minnesota, Missouri, Oklahoma and California.

The Honor Roll for 1919

During the year many received memorials for the excellent work they had done at home or abroad during the war or in the civil pursuits of life. The list is too extended for repetition and will be limited here to recipients closely connected with the power field. Early in the year Prof. Auguste C. E. Rateau, inventor of the Rateau turbine, was named a member of the Academie des Sciences. Major-General George Goethals was awarded the John Fritz medal in recognition of his distinction in the engineering profession and as the builder of the Panama Canal. At the recent annual meeting of the American Society of Mechanical Engineers a session was held in commemoration of the eightieth anniversary of the meeting of Cornelius De Lamater and Captain John Ericsson, their fifty years of service to this country and the thirtieth anniversary of their deaths. Fifteen technical societies and civic organizations were represented in the gathering. It is the intention to erect tablets in New York on the sites formerly occupied by the Phœnix Foundry and the De Lamater Iron Works, in which both men did so much for naval architecture and industrial engineering. During the annual meeting of the Mechanical Engineers a memorial tablet was unveiled and is to be placed in the society's rooms as a tribute to Frederick Remsen Hutton, who did such effective work for the society in its early days. Also a memorial brochure was presented in remembrance of William Kent, best known as the author of the handbook bearing his name.

By the Institute of Electrical Engineers the Edison Medal was awarded to Benjmain G. Lamme for invention and development of electrical machinery. At the annual meeting of the society a memorial tablet erected in acknowledgment of the services of William D. Weaver to the Institute was unveiled, following an appreciation of his efforts toward establishing the great technical library now quartered in the Engineering Societies Building. To the family of Andrew Carnegie a resolution expressing sincere appreciation of his great contributions to the advancement of engineering was passed by the leading engineering societies of the country and duly forwarded. The honorary degree of Doctor of Engineering was conferred upon Rear Admiral Robert S. Griffin, Engineer-in-Chief of the United States Navy, by the Stevens Institute of Technology. B. H. Peck was awarded the Chanute Medal for the best paper presented to the Western Society of Engineers during the year previous, and announcement was made by the same society of the award of the Alvord Medal to Herbert C. Hoover, who had been chosen by a committee of the society as the engineer who had done most for the comfort and well-being of humanity.

Engineering societies closely related to the field, honored the following men with the presidencies: Fred J. Miller, American Society of Mechanical Engineers; Calvert Townley, American Institute of Electrical Engineers; R. H. Ballard, National Electric Light Association; John J. Calahan, National Association of Stationary Engineers; F. E. Matthews, American Society of Refrigerating Engineers; Walter S. Timmis, American Society of Heating and Ventilating Engineers.

Necrology

Several men of prominence in the field passed away during the year. In chronological order the columns of *Power* has record of the following: Professor Rolla C. Carpenter, long associated with Sibley College of Cornell University; Abram T. Baldwin, president of the Precision Instrument Company; Burt O. Gage, a prominent inventor and pump expert and superintendent of the Warren Steam Pump Company; Frederick L. Hickock, president of the Reliance Gauge Column Company; Malcolm Gifford, president of the Gifford-Wood Company; Capt. Charles H. Manning, for many years prominent in engineering circles in New England as chief engineer of the Amoskeag Manufacturing Company, the largest cotton mill in the world; Professor Charles B. Richards, Emeritus Professor of Mechanical Engineering at Yale; George H. Phillips, president and treasurer of the Hewes & Phillips Iron Works; William T. Wheeler, president of the Diamond Carbonating Company; Charles Edwin Knox, one of New York's leading consulting electrical engineers; Samuel T. Wellman, past president of the American Society of Mechanical Engineers and an inventor best known by his patents on the Wellman hydraulic crane and the Wellman open-hearth charging machine; Oscar Otto, latterly general superintendent of the South Philadelphia Machine Works of the Westinghouse Electric and Manufacturing Company; Frederick Sargent, senior member of the firm of Sargent & Lundy and one of the most prominent consulting engineers in the United States; Andrew Carnegie, the great ironmaster and philanthropist; Charles E. Lord, general patent attorney of the International Harvester Company and at the time of his death chairman of the Chicago Section of the American Society of Mechanical Engineers; Richard Hammond, former president of the Lake Erie Boiler Works and the Lake Erie Engineering Works; H. L. Gantt, well-known industrial engineer and author of several works on industrial management.

Relationship Between Specific Gravity and Heating Value of Crude Oils

Speaking of tests conducted by Professor O'Neill, of the Department of Chemistry, University of California, C. R. Weymouth [Transactions A. S. M. E., Vol. 33, p. 68] says:

> It was found that there is no exact relationship between the specific gravity of crude oil and its hydrogen content, although there is a general tendency toward an increase in hydrogen content with a lighter oil. Examination of the ultimate analyses and calorimeter tests of a number of California oils indicates the rather startling fact that it is possible in oils having practically the same total quantity of inert constituents to have a variation in both hydrogen and carbon content with practically no variation in the calorific value of the oil.
>
> The inevitable conclusion that the calorific value of crude oil does not correspond to the heat of combustion of its elemental constituent and that formulas for calculating the total heat of fuels are not applicable to California crude oils is borne out by the fact that the calculated calorific value was, in one instance, 8.7 per cent. greater than that obtained from a calorimeter test.

While this relates to California oil, we have not seen anything to show that it does not equally apply to Mexican oil.

According to *Commerce Reports* it is announced that a great scheme of electricity supply for the north Midland counties is proposed, the chairman of the Nottingham corporation electricity committee having stated that government experts have recommended the erection of a superpower station on the banks of the Trent to meet the requirements of an area 50 miles north to south and 40 miles east to west, or 2000 square miles in all. Such area would include parts of the counties of Nottingham, Derby, Stafford and Leicester. The estimated cost of the first section is $25,549,125, while the whole scheme when complete is expected to require an outlay of $68,131,000. The ultimate annual saving to Nottingham tramways alone is put at $194,660. A further proposal, which is under inquiry by the Ministry of Health, relates to an application by the Manchester City Council for power to borrow $4,866,500 for extending its electricity works. In order to meet new calls for electricity for industrial purposes, it is stated that the extensions must be completed within three years.

In 1901 the first transmission line operating at a voltage above 70,000 was put into service. The use of high-voltage transmission has increased until today there are in this country and Canada about fifty companies operating systems at 70,000 volts and above. These systems represent an aggregate of approximately 14,000 miles of single-circuit line and have over 2,000,000 kw. of generating capacity connected to them.

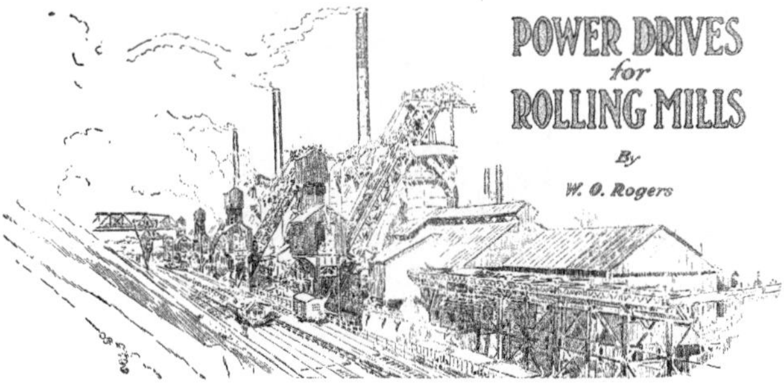

POWER DRIVES for ROLLING MILLS

By W. O. Rogers

"I UNDERSTAND that Youngstown, Ohio, is a great steel town," I remarked to a fellow traveler as the train on which we were riding rushed on its way toward that city.

"Steel man?" he asked, knocking the ashes from the end of his cigar.

I replied that I was not, but that I would be interested in going through some of the mills if permission could be obtained.

"That's a proposition that you can't tell anything about until you try. Now, you would find it next to an impossibility to obtain permission to go through some of the largest steel plants in Pittsburgh and through some of the plants in Youngstown and surrounding towns; that is, those operated by these same companies. Why, I don't know, and I doubt if they can tell unless they don't want to be bothered having an outsider around their works. It surely can't be because they expect to keep out men who might come from other works in order to obtain information regarding manufacturing processes, because it is assumed that such are employed in every mill.

"I would advise you to visit the Youngstown Steel and Tube Co.'s plants, the Republic Iron and Steel Co.'s mills, the Sharon Steel Hoop and several other that are situated in surrounding towns. You will be pretty sure to get through these mills because they are progressive and are not afraid of a little publicity."

Seeing that I was interested, my new-found friend went on to tell me something about steel mills. "Of course, you know that the iron ore that comes from the mines of the iron country looks to the layman a good deal like brown dirt. This ore is put in a blast furnace (Fig. 1) with a certain amount of coke and limestone; that is, with, say, 300 tons of ore about 200 tons of coke and about 65 tons of limestone, although the amount of fuel and limestone depends upon the nature of the ore and the grade of iron wanted. These furnaces are around 70 to 100 ft. high and are equipped with inclined elevators, each having two counterbalaned cages, the empty one descending as the loaded one ascends. At the top of the furnace the coke, limestone and ore are automatically dumped into a hopper and fed to the furnace in the order named.

"Upon melting, the nonmetallic matter separates from the reduced iron and unites with the lime, and the flux being lighter than the iron, floats on its surface and is drawn off as slag, through a suitable hole, after which the iron is drawn out through a lower tap-hole into pig-iron molds, or it is conveyed in ladle cars either to the bessemer converter or to the open-hearth furnace.

"All material charged into a blast furnace passes out either as a liquid or as a gas. The gas that comes off at the top passes down into ovens and is burned there. The liquid products pass off as iron and slag. Pig iron is often heated in a puddling furnace at a temperature somewhat above its melting point. The puddling furnace is a reverberatory furnace and the flame plays over the metal. The impurities are gradually burned out, and the pure metal is collected by means of long iron rods manipulated by the puddlers. The metal comes from the furnace in a pasty condition in the form of irregular balls (Fig. 3) and is then run through a squeezer to remove the main portion of the slag. It is finally passed through a rolling mill (Fig. 4). This metal, however, contains too much slag for ordinary purposes and it is sheared, and after reheating it is again passed through

FIG. 1. BLAST FURNACE AND ORE BRIDGE, MAIN PLANT YOUNGSTOWN SHEET AND TUBE CO.

a rolling mill, thus making it purer, and it is generally known as wrought iron."

"This is all mighty interesting," I said as my friend employed himself in wiping his glasses. "I suppose that steel is produced in much the same manner?"

He did not reply for a moment and then said: "I might say that the most common processes in making steel are known as the bessemer and the open-hearth processes. A pig-iron blast furnace produces a product known as pig iron, which is sold to such concerns as do not have a blast furnace or when the ore is obtained at a distance from the mill. For instance, some of the iron mined in the Michigan iron country is made into pig iron for shipment close to the mines, although most of it goes to the various Lake ports in its natural state and is put through the blast furnace at the mills. In the instance of the steel mill at Sparrows Point, Md., the pig iron comes from Cuba, and it is doubtless cheaper to handle it in that way than to bring in the raw iron ore.

FIG. 2. METAL LADLES READY FOR FILLING WITH LIQUID METAL

"Well, after the iron has been treated in the blast furnace at a steel mill, it is run into huge ladle metal cars (Fig. 2) and is carried either to the bessemer converter or to the openhearth furnace."

"What is the bessemer process? I have heard of it and seen mills where a sheet of flame and sparks was being blown apparently through the roof and they said it was a bessemer converter. What is the idea?"

"In the bessemer process the hot pig metal from the blast furnace is poured from the metal cars, of which I already spoke, into the converter, and air is blown through it. This is to remove the impurities by the combustion of the silicon, manganese and carbon. This process requires from seven to fifteen minutes. Both bottom-and-side-blow vessels are employed, but the side-blow is used mostly in this country. The air is introduced through tuyeres in one side of the vessel, and the pressure used is around three or four pounds per square inch.

"When charging, the vessel is tipped forward and the molten pig iron is charged in the proper amount. Then the converter is tilted back to an angle of five or six degrees and the air blast is turned on. You will first see a few sparks come from the mouth of the converter and after three or four minutes of blowing a slight flame will appear at the mouth. The vessel is in the meantime being slowly turned toward a vertical position and after from five to fifteen minutes the flame and sparks are shooting out at a greater rate. That is what you have seen, as you just mentioned."

"What is to prevent the metal from being blown out with the slag, sparks, etc.?" I asked.

"That is taken care of by reducing the air pressure as low as possible when the 'boil' takes place; but if the air pressure is reduced below about 1¼ lb., the metal or slag will run into the windbox."

"How does the operator know when the metal has been sufficiently heated?"

"An experienced blower can generally tell when the final flame has been had, but if he is not sure the vessel is turned down and the metal is inspected. If the last flame has occurred and the carbon has been eliminated, the metal and slag lie quiet and flat in the converter without bubbling, but if there is still carbon in the metal the slag boils somewhat."

"What happens after the metal has been blown sufficiently?" I next asked, as I handed out a cigar that cost two for a quarter, but probably formerly sold two for a nickel.

"It is formed into ingot molds," replied my friend as he suspiciously gazed at the cigar I had given him. "These ingots are heated and then run through rolling mills. You will see them if you can obtain permission to go through any of the works."

"You spoke of the openhearth process. How does that operate?"

"In the openhearth process scrap steel and some pig iron are melted in a bath type of furnace, and the carbon, silicon and manganese are oxidized out of the metal by means of certain additions of iron ore. Nearly pure iron is obtained in this type of furnace.

FIG. 3. PUDDLE FURNACE AND CAHALL WASTE-HEAT BOILER

FIG. 4. ROLLING PUDDLED IRON

"Coke fuel is not used as the heating medium, but instead producer gas or natural gas. Fuel oil can be used, and experiments have been made with powdered coal. However, whatever fuel is used the method of working and the conditions in an openhearth furnace are about the same."

"You mentioned that pig iron is put into the furnace with scrap steel. Doesn't that make a metal of less purity?"

"If the furnace metal is to be properly heated, it is necessary that it should bubble. The pig iron assists in this because it produces the essential carbon, but only enough pig iron is used to bring the carbon down to about what is required in the finished steel.

"With openhearth furnaces two processes are used, the acid and the basic. In either the steel is low in carbon and must be recarbonized by means of proper agents, as I have just stated. The process is carried on in either stationary or tilting furnaces, and some of the stationary furnaces have a capacity of 200 tons. It requires from six to twelve hours to run a heat."

"What is the difference in operation between a tilting and a stationary furnace?" I asked at this point of the conversation.

"The tilting furnace is tilted forward on rockers, similar to the motion of a rocking chair. The stationary furnace, as the name implies, does not move, but the metal is drawn from it through plugged tap holes into huge ladles, from which it is poured into ingot molds."

In Figs. 5 and 6 are shown the charging and tapping side of a 100-ton capacity openhearth furnace at the main plant of the Youngstown Sheet and Tube Co., which plant I visited later on, and many of the illustrations in this and following articles are reproduced through the courtesy of that company and of the Republic Iron and Steel Co.

"You spoke of acid and basic openhearth steel," I remarked as the train stopped at a station. "What is the difference?"

FIG. 6. GENERAL VIEW OF POURING FLOOR, OPENHEARTH FURNACE

"In acid openhearth steel the charge consists of pig iron and ore or pig iron and scrap having a low phosphorus content, and is melted in an openhearth furnace with an acid or siliceous lining. In the process the impurities in the pig iron are removed to a great extent by means of an oxidizing flame that is brought about by the mingling of producer gas and preheated air in a reverberatory regenerative furnace similar to the puddling-furnace process in making wrought iron; but the furnace temperature is carried much higher, and both the metal and the slag become molten.

"In making basic openhearth steel, the charge of either melted or pig iron or a mixture of pig iron and low carbon scrap is heated in a furnace similar to that used with the acid furnace, the lining, however, being of dolomite, lime magnesite or other basic material."

FIG. 5. A 100-TON OPENHEARTH FURNACE, SHOWING CHARGING CRANE AND CHARGING BOXES

"I don't quite get the idea regarding the application of the gas and air to the furnace."

"I'll try to make that plain to you," replied my companion. "The main parts of a regenerative furnace are the inclosed chamber or the hearth, the bottom of which is lined with refractory material, where the flame is produced and heats the charge of metal; the regenerative chambers, where the air is preheated, which are filled with brick checkerwork, and where the heat of the products of combustion leaving the hearth is absorbed, to be returned to the incoming gas and air when the direction of flow is reversed.

"For instance, gas from the gas producer is conducted through a suitable passage to a gas generator and is then conducted to the furnace at a point above the surface of the molten metal. Air also passes through an air regenerator, where it is heated, and is then delivered to the furnace with the gas. The gas burns, and the hot flame passes through a second set of gas and air generators, and the reverse of gas and air flow can be had by the use of a reversing valve. The gas and air are heated to about 1000 deg. before entering the combustion chamber.

"Of course I have hardly hit the high spots relating to blast furnaces, and the subject would require many books to cover, but I think I have given you an idea of what the process is and that was what you was after."

"You have been mighty interesting," I replied. "As I understand it, the bessemer process handles pig iron only and the openhearth handles pig and scrap. The products of both are poured into ingot molds and while hot are put through a blooming mill."

"That's about it," answered my friend, as he arose to get his hat, the train having slowed down for his station.

"There is one thing I don't understand and that is the charging of the openhearth furnace," I said as I went to the platform to bid him goodby.

"That is easily explained," was the answer. "A charging crane is used (Fig. 5). The crane runs on a very wide-gage track and is electrically operated. In front of the charging end of the furnace is another track on which cars run carrying the charging box. The crane is run along in front of these cars, an extension arm is run out and the head of the arm engaged with a suitably formed coupling head on the end of the charging box. The arm and box of metal are then elevated sufficiently to come in line with the furnace door, which is then opened and the arm with the charging box on the end is run into the furnace. Then it is turned half over and the contents of the box dropped into the furnace, after which the arm and box are turned back to their original position, the arm is withdrawn and the charging box is replaced on its car ready for hauling away for another load of scrap iron."

I returned to my seat as the train started on its way and in a short time reached my destination, Youngstown, where later, I saw most of the process of which my casual acquaintance had explained.

Contributory Negligence of Engineers

By A. L. H. Street

Plaintiff, as day engineer at the defendant's plant, mounted a ladder to make repairs on some steam pipes. The ladder slipped and he was injured. Suing for damages, he asserted negligence on the part of defendant employer in failing to provide a ladder so spiked as to avoid slipping. At the trial it was shown that the ladder was spiked at one end, and that the night before the accident the plaintiff had left the ladder in proper position. But during the night another employee used it in another place and then returned it wrong end up.

Denying the plaintiff's right to recover damages on the ground that plaintiff's own contributory negligence in failing to assure himself that the ladder was properly secured against slipping barred any valid claim on his part, the Pennsylvania Supreme Court says in the case of Finan vs. F. T. Mason Co., 107 Atlantic Reporter, 692:

An employer is bound to protect his employee from danger reasonably to be apprehended, but not against all possible danger, and least of all against danger occasioned by the employee's own negligence. Not only did the evidence come short of showing negligence chargeable to the defendant, but on the admitted facts the injury complained of was traceable directly to the plaintiff's negligence in failing to examine the position of the ladder before ascending it.

Measuring Diesel-Engine Crankpin Clearance

By L. H. Morrison

The average operator of Diesel engines "jumps" the connecting-rod big-end to determine the crankpin clearance. This is open to the objection that the result obtained is comparative, not exact.

To secure an accurate determination of pin clearance, either of two methods can be adopted. In Fig. 1, which outlines a big-end bearing and crankpin, a lead wire *B*, or soft soldering wire, about ⅛ in. in diameter, is inserted in the bearing between the pin and the babbitt. The bolts are next tightened until the halves of the bearing meet on the shims, or distance pieces, at the junctions *AA*. After loosening the bolts, the wire, which has been flattened in the tightening process, is

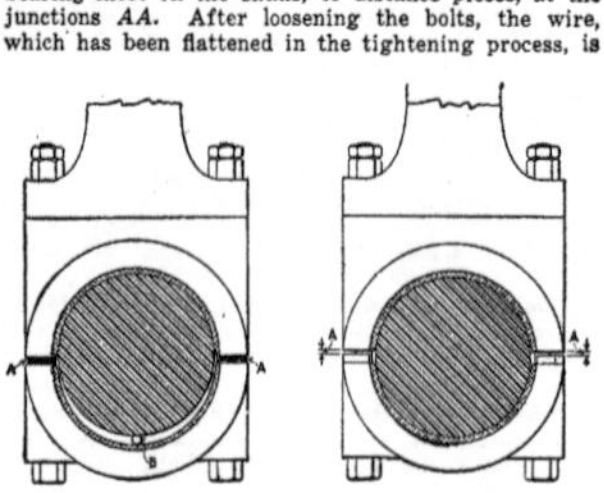

Fig. 1. By Lead Wire   Fig. 2. By Thickness Gage

METHOD OF MEASURING CRANK PIN CLEARANCE

measured by a micrometer. If the thickness is not between 0.007 in. and 0.010 in., shims can be removed or inserted at *A* to bring the clearance to this value. The objection to this method lies in the difficulty of removing the wire. Usually, the pressure exerted on the bolts with the consequent flattening of the wire causes the latter to adhere to the crankpin so that it is almost impossible to remove the wire without lowering the bottom half of the bearing.

A second method is outlined in Fig. 2. Here the bolts are drawn up as tight as possible and the opening at *A* is measured by a thickness gage. This width plus the desirable clearance of 0.007 in. is equal to the thickness of shims, or distance pieces, that must be inserted. If the halves of the bearing come together without shims, the bearing needs rebabbitting. In rebabbitting it is highly desirable to have ample shimming between the two halves. One-half inch on each side is not too much, since that will allow considerable wear before the bearing need be rebabbitted again.

Not only has increased efficiency been secured in the burning of coal in the furnaces under oil stills, as a result of Bureau of Mines experiments, but a considerable percentage of increased capacity has been attained. The work has been done in co-operation with the Sinclair Oil Company.

Steam-Turbine Governors—Small Shaft Governors

This, the first article, describes methods of governing and considers the construction and adjustment of a few turbine governors of the shaft type.

METHODS OF GOVERNING TURBINES—The manner of governing steam turbines is in some respects similar to that employed for reciprocating engines, with the chief difference that for large turbines the governing mechanism is usually more complex.

There are three types of turbine governors commonly used. The throttling type uses a balanced throttle valve operated through links and levers by weights or springs moved by centrifugal force, throttling the admission of steam to a chest from which the steam flows to the turbine nozzles. Nearly all small turbines are governed in this manner, the governor proper being attached directly to the end of the mainshaft and known as a direct, or shaft, governor.

The second, or intermittent admission, method is emplified in the Westinghouse and Parsons turbines. With this method of governing, the primary and secondary, or main, steam valves are held to their seats by springs and opened by the movement of pistons on the main-valve spindles, the pistons being actuated by steam or oil pressure admitted through relay valves.

The third method, wherein hydraulically moved relay valves or cams on a shaft operate a number of steam nozzle valves, as in the Curtis turbine, is a multiple-control method, or more correctly, control by a series of valves operated in rotation, for there are a number of poppet valves which admit steam to the nozzles. The chief advantage claimed for this manner of governing is that the working nozzles, or nozzles admitting initial pressure steam, do so continuously, thus they always work at highest efficiency.

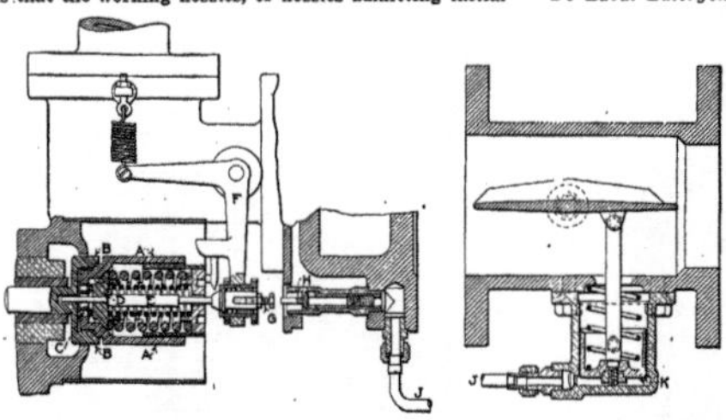

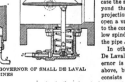

FIG. 1. GOVERNOR AND EMERGENCY GOVERNOR OF SMALL DE LAVAL STEAM TURBINES

De Laval Shaft Governor—In Fig. 1 is shown a sectional view of the shaft governor used on small De Laval turbines. This governor is mounted on the end of one of the gear shafts. (See Fig. 2.) Weights *AA*, Fig. 1, in the form of half-cylinders of bars, are mounted and hinged on knife-edges *BB*. The weights swing on their knife-edges and are held in place by two springs. As the speed increases, the weights fly out and cause the round-nosed pins *C* to move forward against the collar *D*, the action of centrifugal force in throwing out the weights being resisted by the springs. As the springs are compressed, the collar *D* pushes against the shoulder of the spindle *E*, which moves the bell crank *F* operating the balanced throttle valve, Fig. 3. There is a

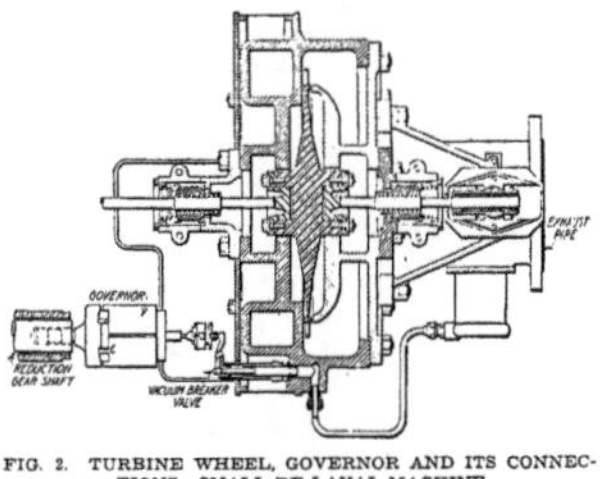

FIG. 2. TURBINE WHEEL, GOVERNOR AND ITS CONNECTIONS—SMALL DE LAVAL MACHINE

certain amount of "play" at *F* to reduce the amount of valve opening. In some sizes of turbines it requires but ⅛-in. movement of the spindle *E* to move the valve from full open to complete closure. The bell crank is balanced by a spring, as shown. The governor regulation is supplemented by hand-operated nozzle valves, shown in Fig. 4, which admit additional steam to the wheel during overloads.

De Laval Emergency Governors—Turbines designed to be run condensing cannot be operated against a relatively high back pressure. This fact is made use of in some of the De Laval machines. Referring to Fig. 1, it will be seen that there is an adjustable projection *G* that moves in and out with the spindle. In case the speed should increase beyond that predetermined, the projection *G* will move out, push open a valve *H* which admits air to the condenser through the hollow spindle of the valve, through the pipe *J* to the piston *K*.

In other types of small De Laval turbines the shaft governor is like the one described above, but the emergency stop consists chiefly of a butterfly valve in the exhaust pipe operated by the governor after the turbine has exceeded a predetermined speed.

The Kerr governor is indeed simple and in general like most shaft governors. The throw of the weights by centrifugal force is resisted by a heavy spring. The weights are in the form of half-cylinders, machined from a solid piece and split. They swing on hardened-steel knife-edges which are detachable. Those parts of the weights which push the spindle forward do so by making a rolling contact.

The spring on the governor-valve spindle tends to hold the valve open and resists the closing effort of the governor. To obtain slight speed variations, the compression of this spring may be increased or decreased. Normally, the governor is adjusted for faster or slower speeds by increasing or diminishing the compression of the main spring by screwing a nut down on or away from the spring. The spindle is well provided with oil channels for adequate lubrication. No emergency stop is provided, it being considered that the governor will always function to prevent excessive speeds.

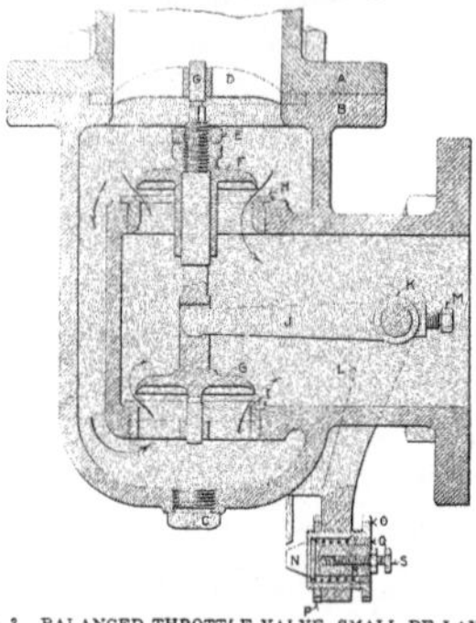

FIG. 3. BALANCED THROTTLE VALVE, SMALL DE LAVAL TURBINES

FIG. 4. GOVERNOR END OF SMALL DE LAVAL TURBINE

The small handwheels in front are used to open valves admitting more high pressure steam to the wheel to handle heavy loads

FIG. 5. LOOKING DOWN ON GOVERNOR AND GOVERNOR VALVE, STURTEVANT TURBINE

Sturtevant Turbine Governors—All Sturtevant turbines are provided with direct-type governors mounted on the end of the mainshafts. An emergency stop, independent of the main governor, prevents excessive speed.

Fig. 7 is a sectional elevation of a Sturtevant turbine, showing the relative position of the governor mechanism to the rest of the machine. As shown, the motion of the governor spindle is transmitted to the regulating valve through one bell crank, no other connecting levers being used. The illustration shows that the emergency stop valve *A* is of the butterfly type.

Fig. 5 is a view looking down on the governor mechanism, the top half of the casing having been removed. As the weights *A* and *B* are thrown out against the tension of a spring between them and on the spindle, the spindle is pushed forward, moving out the bell crank, lifting the regulating-valve stem and disks and decreasing the flow of steam to the turbine. Notice that the bell crank is pivoted on knife-edges, which should not be oiled. An oil cup drops oil to the end of the bell crank and the spindle. The centrifugal force

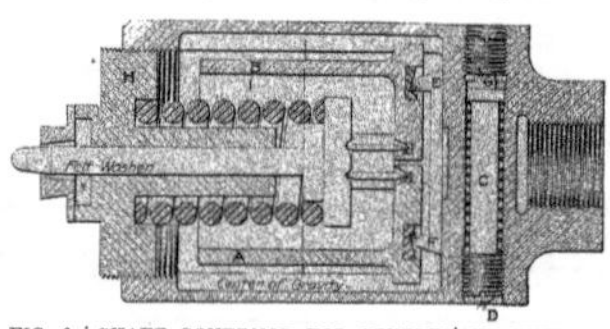

FIG. 6. SHAFT GOVERNOR FOR STURTEVANT TURBINE

causes the oil to flow along the spindle to a felt washer, shown in Fig. 6.

The Emergency Governor—As stated, the emergency governor valve is of the butterfly variety, placed between the regulating valve and casing. It is spring-loaded and held open by a notched bell crank, as shown in Fig. 5. Referring to Fig. 6, a bolt *C* is held in the casing (at a point *C*, Fig. 8) by a helical spring, and a threaded plug at one end of the hole. Should the regulating governor be deranged, the bolt will begin to move out of the casing at *D*, Fig. 6, when the speed has increased to about 8 per cent. above normal, and at 10 per cent overspeed will fly out and strike the inner end of the lever *C*, Fig. 5, which will release the resetting lever *D*, allowing the spring *E* to close the emergency valve and stop turbine. After the throttle has been closed, reset the emergency governor, a knob-handle being provided to assist in raising the resetting lever. Adjust the emergency governor when necessary by screwing in the plug *G*, Fig. 6, to make the governor more sluggish and screw it out to make it operate at less overspeed.

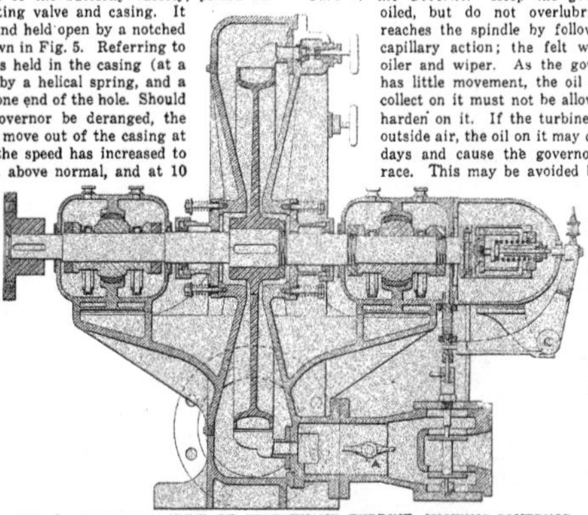

FIG. 7. SECTIONAL VIEW OF STURTEVANT TURBINE, SHOWING GOVERNOR

To prevent chattering of the emergency gear, a spring *F*, Fig. 5, rests against the bell-crank trip.

The weights are mounted on knife-edges at *E* and *F*, Fig. 6 (*D* and *E*, Fig. 8) and the spindle rests on knife-edges *J* and *K* (*F* and *G*, Fig. 8). Do not oil these knife-edges, for if oiled they may gather dirt and wear rapidly.

To increase the speed of the turbine, increase the compression of the spring by screwing in *H*, Figs. 5, 6 and 8; to decrease the speed, decrease the compression of the spring. In making adjustments to governors it is always well to go slowly.

Care of the Governor—Keep the governor spindle oiled, but do not overlubricate it. Oil reaches the spindle by following along by capillary action; the felt washer acts as oiler and wiper. As the governor spindle has little movement, the oil and dust that collect on it must not be allowed to gum or harden on it. If the turbine is exposed to outside air, the oil on it may congeal on cold days and cause the governor to jump or race. This may be avoided by moving the spindle in and out a few times by hand before starting the turbine.

At the end of runs stop the turbine by releasing the emergency valve. If this is not done, the stem may become so fast in the stuffing-box that the spring will not turn it when the resetting lever is released on overspeed.

No packing is used in the stuffing-box on the regulating valve. The stem is made a close fit in the box, which is quite long; leakage of steam around the stem is usually slight and an indication that there is sufficient clearance. Sometimes, when using superheated steam, the stem expands and sticks in the stuffing-box.

The regulating valve disks are not intended to fit tightly on their seats, but to reduce the area of valve opening enough to regulate the speed of the machine. The seats and ports are large in diameter and lift about $\frac{1}{16}$ in. from their seats. Of course when the governor weights fly all the way out, the valve should be closed.

FIG. 8. THE SHAFT GOVERNOR OF THE STURTEVANT TURBINE, ASSEMBLED AND DISMANTLED

The Cost of Low Power Factor

BY WILL BROWN
Engineer with the Electric Machinery Company

By using synchronous motors to correct power factor, large savings can be made in the cost of power, generator capacity and copper loss. Even in small systems it is now profitable to correct power factor, and the place to do it is at the motor end of the line. Pump and water-pipe diagrams make the article easy to understand.

THERE is a certain amount of lagging reactive kv.-a. (kilovolt-amperes) in every alternating-current power circuit where power factor is less than unity. In the analogies given in the previous article, "The Flow of Wattless Current Explained," in the Sept. 30 issue, the lagging reactive kv.-a. is the

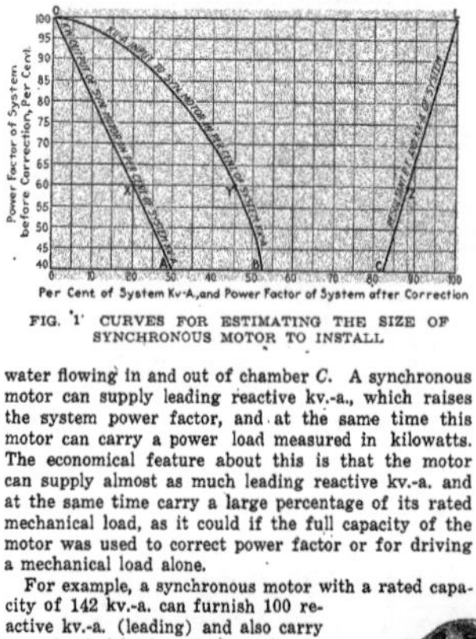

FIG. 1 CURVES FOR ESTIMATING THE SIZE OF SYNCHRONOUS MOTOR TO INSTALL

water flowing in and out of chamber *C*. A synchronous motor can supply leading reactive kv.-a., which raises the system power factor, and at the same time this motor can carry a power load measured in kilowatts. The economical feature about this is that the motor can supply almost as much leading reactive kv.-a. and at the same time carry a large percentage of its rated mechanical load, as it could if the full capacity of the motor was used to correct power factor or for driving a mechanical load alone.

For example, a synchronous motor with a rated capacity of 142 kv.-a. can furnish 100 reactive kv.-a. (leading) and also carry a mechanical load of 100 kw. On the other hand, if it is desirable that the motor carry the largest possible mechanical load, it can be operated at unity power factor, carrying 142 kilowatts, and no reactive kilovolt-amperes. The motor that is to operate at unity power factor can be built at less expense than one of the same capacity to operate at low leading power factor. There is less material used in the former case and the exciter is of smaller size.

Every synchronous motor requires direct current to energize its field windings—much the same as an alternator. An exciter, it is true, is an added complication but by no means a serious one. Any operator who has handled direct-current motors will see no great difficulty in a direct-current generator with a rheostat in series with its field windings. After an operator has seen these exciters operate for years with hardly any expense beyond the renewing of an occasional brush, exciter troubles cease to be something to worry over.

When a synchronous motor is supplying reactive kilovolt-amperes to the line, it is helping the whole system back to the generating station. That is the reason some power companies offer inducements to customers who will install synchronous motors and raise their power factor. As to how much low power factor costs consider the following case:

A 50-hp. induction motor consumed in one year a total of 83,000 kw.-hr., this being a load factor of 22 per cent. The power contract contained the following: "If the power factor of the energy taken is found to be less than 85 per cent., then the consumer shall pay for the electrical energy taken on the basis of 85 per cent. power factor. A further discount of 5 per cent. will be given to consumers installing synchronous motors."

The average power factor of the motor was 0.65, and the net power bill amounted to $1590. If this had been a synchronous motor, it would have met easily the requirements of 85 per cent. power factor and the bill would have been $1330. But in addition to this it would also have earned a bonus of $66.50, so that the net bill would have been $1263.50—a saving of $326.50 for one year's power bill on a 50-hp. motor that only averaged five hours and twenty minutes operation each day.

If the power factor and the kilovolt-amperes of a system are known, by the use of Fig. 1, the capacity of a synchronous motor to install can be found, which will give the maximum corrective effect to power factor while carrying mechanical load. On the vertical scale to the left locate the present power factor. From this point extend a horizontal line to the right and mark intersections of this line with curves *OA*, *OB* and *LC*. From these points of intersection drop vertical lines to the kv.-a. scale at the bottom. The point on curve *OA* will show the mechanical load in kilowatts that the synchronous motor can carry. The point on curve *OB* will show the kv.-a. capacity of the synchronous

FIG. 2. THREE-BEARING BELTED-TYPE SYNCHRONOUS DIRECT-CONNECTED EXCITER

motor. The point on curve *LC* will show the total kv.-a. of the system, also the new power factor of the system after such a synchronous motor has been added.

For example, suppose a line transmitting a load of 1000 kv.-a. is operating at 60 per cent. power factor; points *X Y* and *Z*, on the curves Fig. 1, determine answers to the following questions: What is the highest point we should try to raise the power factor of the line consistent with the economy of the apparatus? Answer, given by a vertical dropped from *Z* to the base line; namely, 88 per cent. power factor. What is the kv.-a. rating of the synchronous motor that will give this amount of correction? Answer, given by a vertical from *Y* to the base line; namely, 45 per cent. of the load on the system, in this problem $1000 \times 0.45 = 450$ kv.-a. What is the kilowatt load that the motor can carry, allowing the maximum correction for the system? Answer, given in per cent. of the kv.-a. load on the line

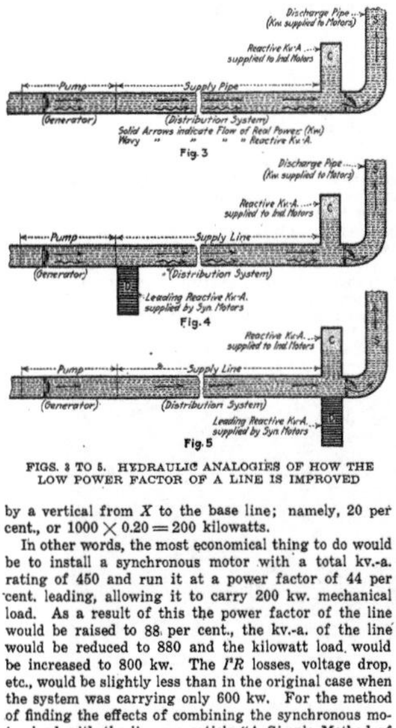

FIGS. 3 TO 5. HYDRAULIC ANALOGIES OF HOW THE LOW POWER FACTOR OF A LINE IS IMPROVED

by a vertical from *X* to the base line; namely, 20 per cent., or $1000 \times 0.20 = 200$ kilowatts.

In other words, the most economical thing to do would be to install a synchronous motor with a total kv.-a. rating of 450 and run it at a power factor of 44 per cent. leading, allowing it to carry 200 kw. mechanical load. As a result of this the power factor of the line would be raised to 88 per cent., the kv.-a. of the line would be reduced to 880 and the kilowatt load would be increased to 800 kw. The I^2R losses, voltage drop, etc., would be slightly less than in the original case when the system was carrying only 600 kw. For the method of finding the effects of combining the synchronous motor load with the line see article "A Simple Method of Determining Power-Factor Correction," published in *Power* June 15, 1919.

Low power factor of induction-motor loads has always been a source of trouble to power companies, but within the last year or two it has become such a serious proposition that most companies are preparing to enforce some kind of a penalty clause similar to that previously quoted. To a limited extent the power company can correct low power factor by installing large synchronous condensers in its substations, but the real place to correct power factor is right at the motors—the place where the low power factor is caused. In order to make this clear, consider again the pump and water-pipe diagrams. Fig. 3 indicates a pump and long supply pipe operating at lagging power factor due to the reactive chamber *C*. The power company is paid only for the water that passes through the discharge pipe *S*. But the company has to maintain a considerably larger pump and piping system to move all that excess water

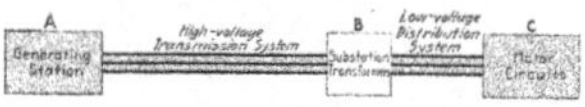

FIG. 6 DIAGRAM OF A GENERATION, TRANSMISSION DISTRIBUTION SYSTEM

when operating at low power factor. Also, there is an increased expense due to increased losses in the pump and pipes. In Fig. 4 the reactive chamber *D*, whose gas expands under pressure, as explained in the preceding article "The Flow of Wattless Current Explained," Sept. 30 issue, has been added to the line at a point near the pump. This corresponds to a synchronous motor operating at leading power factor placed in close proximity to the generators. Assuming that the action of this chamber exactly balances chamber *C*, the pump will be able to operate at 100 per cent. power factor, but the supply line or distribution system will still be at low power factor and burdened with the reactive current which flows back and forth between chamber *D* and chamber *C*.

Since there is a certain friction to every foot of pipe,

FIG. 7. SMALL SYNCHRONOUS MOTOR DRIVING CENTRIFUGAL PUMP

it is evident that there is considerable friction loss involved in moving this excess amount of water between *D* and *C*. This extra friction loss means a direct loss of power which must be made up by the prime mover driving pump.

In Fig. 5 the reactive chamber *D* is placed directly opposite chamber *C*. This corresponds to placing a synchronous motor close to the induction-motor load. The pump is now operating at 100 per cent. power factor and so is the whole supply line. The reactive current flows back and forth between chamber *C* and chamber *D* and relieves the whole line of the excess water and friction loss. The whole line, including the pump, now operates at practically unity power factor with the minimum of losses. It is evidently very desirable to correct low power factor as close as possible to its source.

FIG. 8. ENGINE-TYPE SYNCHRONOUS MOTOR DRIVING AMMONIA COMPRESSOR

Fig. 6 represents a power system from the generating station to the motors. Distances between *A* and *B* and *B* and *C* will vary in different systems. Assume this system to be operating at lagging power factor. If synchronous condensers are placed at *A* the power factor of the generating station will be improved, but all the balance of the system from *A* to *C* will remain at as bad power factor as before. If the synchronous condensers are placed at *B*, the power factor will be raised from *B* inclusive back to *A*, but from *B* to *C* the low power factor will still prevail as before. If synchronous motors are placed at *C* (on the motor system), the power factor will be raised on the whole system from *C* clear back to *A*, and these motors can also be driving mechanical loads without seriously interfering with their capacity for correcting power factor. There is great saving when the power factor is corrected at the motor system *C* instead of at the transformer substation *B*.

To illustrate the extent of the copper losses due to low power factor on a certain system, when tests were made, the following facts were discovered:

The motor load was practically 7500 kw. The power factor at the substation (because of synchronous condensers installed there) was near unity, but on the distribution circuit it was 70 per cent. The watts loss, or copper loss, on this circuit at this power factor was 1900 kw. Later, synchronous motors replaced part of the induction motors and the power factor of the total motor load was raised to unity. The motor load remained practically at the same figure, 7500 kw., but the whole circuit now operated near unity power factor and the copper losses were reduced to 900 kilowatts.

Thus a saving of 1000 kw. was secured by raising the power factor of this 7500-kw. motor system and feeder circuit from 70 per cent. to 100 per cent. And it was no longer necessary to run the synchronous condensers at the substation. Assuming the system to operate at this figure for 4000 hours per year, the total saving would be 4,000,000 kw.-hr. If we say the cost of generating and delivering this wasted power to the feeder circuit averages 0.5c. per kw.-hr., the total amount saved in dollars would be $20,000. This represents an actual saving at the coal pile, for copper losses take power from the prime mover just the same as if useful work was performed. This, however, is only a part of the loss. The increase in overhead necessary to take care of large amounts of reactive current on a system, also the greater percentage of trouble and upkeep expense, the lowered efficiency due to poor voltage regulation and other results of poor power factor make up a heavy total. Furthermore, poor voltage regulation at the motors causes a reduction in production, and in some cases an inferior quality of material produced, either one of which may amount to losses of large proportions in a year's operation.

Is it practical for a power consumer to correct the power factor of his own load? A few years ago the answer would have been no, it could not be done except in the cases of large users of power. But the time is now at hand when it will be possible for the great majority of plants to raise the power factor if they wish to. Synchronous motors of 50-hp. and up can be used for driving certain classes of loads that have formerly been driven by induction motors. The efficiencies of such motors even at leading power factors will compare favorably with induction motors. Even if the load to be driven is less than 50 hp., it may be practical to use a synchronous motor. A motor of larger kilovolt-ampere capacity could be installed and the excess capacity used for supplying leading reactive kilovolt-amperes to the induction motors. There is a slight loss in efficiency, and the first cost will be higher but not at all prohibitive. In fact, the savings on the power bill will in many cases pay this difference within a short period. After that it may be considered that the synchronous motor is a source of revenue to any plant that was formerly running at low power factor. The net result will be that power will cost less on such a basis than under any other method. The importance of a more economical use of power is becoming of greater concern each year, since the cost of producing it is on the increase and there is small prospect of a change in this condition.

FIG. 3. SYNCHRONOUS MOTORS WITH DIRECT-CONNECTED EXCITERS DRIVING CENTRIFUGAL PUMPS

Boiler Explosion at Rockport, Texas

DETAILS have just become available concerning a boiler explosion which, although occurring several months past, presents features that make an account of it interesting. The boiler, a 72-in. horizontal return-tubular, was in use at Heldenfeld Bros. Ship Yards at Rockport, Tex., and exploded at

FIG. 1. SAFETY VALVE AFTER ACCIDENT

7:20 on the morning of July 29, 1919, killing four persons, slightly injuring one, and causing a partial destruction of the boiler plant.

This boiler had been in continuous operation for two weeks and on the night preceding the explosion was steaming and seemingly in good condition until about 3 a. m., when the grates dropped. This resulted in cutting the boiler out, but not before it was filled with water until the gage-glass registered within two inches of the top and which was blown down to see if it was in good condition. There was also another boiler operating in battery with this boiler which was kept in operation until the end of the watch at 6 a. m. It was about 5:30 before the boiler was cool enough to work on, and the work of reinstalling the grates took until 6. When the day crew came on, there was still about forty pounds of steam, with the water in the glass at about the same position; so stated the night engineer.

At this time in the morning, after changing watches, there was no one but the day power-house crew and the watchman on duty, as the workmen were admitted at 7:25. The watchman stated that he was on his way to the main entrance gate to let the workmen into the yards and on passing the power house, he stopped to say a few words to the day crew. Their positions, he says, were as follows: One of the firemen was sitting directly in front of the boiler reading a paper, the other was a little to one side, and back of him, sitting on the ground, a laborer was cleaning some ashes from the entrance, and the engineer was between the boiler that had been cut off the line and the one that was kept steaming, a space of about twelve feet, working with a boiler-feed pump. A few words were passed and he proceeded to the front entrance gate and was just in the act of removing the lock to let the workmen into the yard when the explosion occurred.

When the bodies were found, the engineer was where he was last seen by the watchman, one of the firemen was in the same place, and the other was over on top of the mill house, about 125 ft. away. The laborer was found about 15 ft. in front and to one side of the boiler that was still in operation at 6 o'clock. The boiler was blown about 650 ft. north of the yards, wrecking part of the boiler house, leaving the other boiler intact, with the exception that all of the steam lines were wrecked. A blacksmith shop standing about 50 ft. to the north was also partly wrecked.

As all the day crew were instantly killed, it is impossible to determine with exactness the condition immediately preceding the explosion. The examination of the boiler, however, found all seams, dome, braces and shell intact. The tube sheets were bulged and all the tubes were out of the front sheet and some of them projecting through the rear sheet, bent over; others were still in the same position, with the beads in good condition. The tubes were beaded over at the rear, while only expanded at the front. There was no evidence of any part of the boiler being overheated. On examination of the bulged tube sheets it was found that the metal between the tube holes was not fractured. The circumferential seam holding the tube sheets showed no evidence of fracture. The shell at this point was slightly turned down and the rivet holes not pulled. Every rivet tested in the boiler was tight. The bottom of the shell was buckled in at the front and gradually down from 3 to 14 in. at the rear.

There are two theories as the possible, if not probable, cause of the explosion. It is stated by the night crew, that at the time the boiler was cut out for repairs it had 105 lb. steam and was not even known to simmer or partly pop. It is also stated that previously there had been a pressure of 115 lb. and the pop valve did not work. It is probable that the other boiler was kept steaming, as only one boiler was needed at this time (which, it is stated, has happened before). The day crew, realizing the need of the second boiler on short

FIG. 2. TUBES PROJECTING THROUGH REAR SHEET

notice, started a heavy fire in it, and according to one of the night firemen's statements it would take about twenty minutes to again get 100 lb. or more of steam in the boiler. The day crew, in the same position as when last seen by the watchman only about three minutes before the explosion, forgetting the boiler, permitted excessive steam pressure to build up and upon noticing this condition, one of the firemen went on top of the boiler to make the ball-and-lever safety valve work, and while trying to do so, the boiler blew up. This is the theory for the body of one of the firemen found on the mill-house roof. It is also probable that the day crew started a fire in the firebox when they came on duty. However, the owner, his superintendent and chief engineer stated that they saw the boiler pop the week previously, and one of the engineers said that he saw it pop the Saturday before the explosion. The superintendent asserts that he saw it pop on the morning of the day before the explosion. This is hardly possible, because on examination of the safety valve it was found that the valve stem, where it passes through the fulcrum or top of the valve body, was so corroded that it took a small sledge hammer to move it in either direction. The corrosion was found to be of old origin and about $\frac{1}{32}$ in. all around the stem of the valve where it passed through the cage, in which condition it could not have worked for some time, besides having a 100-lb. weight at least two-thirds the way out on the lever.

The other theory is that the pressure was not equalized, and upon trying to cut the two boilers together with unequal pressure, the explosion occurred. The owner states that at one time previously, with the pressures unequal, the boilers were cut together, resulting in blowing the fusible plug. This theory of the explosion is hardly probable, if the condition of the stop valve as found and demonstrated by the owner was the same as at the time of the explosion. The valve being closed, he took hold of the valve-stem wheel and moved it about one inch, stating that that was the position in which it was found only a few minutes after the explosion. This, however, would not open the valve, as it required a one-eighth turn of the wheel before the slack was taken up between the stem and valve. It then had to be moved about one-sixteenth of a turn more before the valve left the seat, which was good evidence that it had never been open far enough to admit steam. The prevailing opinion, which seems to be logical, is that an excessive pressure caused the front tube sheet to bulge and pull off the tubes at front, as they were not beaded.

This boiler was not insured and had never been inspected except at the time of the purchase, which was about a year previously. The boiler was second-hand and about twelve years old.

Equivalent Evaporation

By J. S. A. JOHNSON

Professor of Applied Mechanics and Experimental Engineering, Virginia Polytechnic Institute, Blacksburg, Va.

ONE of the principal objects of a boiler test is to determine accurately the total heat absorbed by the water in the boiler. This quantity of heat may be expressed in common thermal units, but owing to the large numerical values that would be involved, a larger unit of heat is used in expressing boiler capacities. This new heat unit in the British system is the amount of heat required to evaporate one pound of water from a temperature of 212 deg. F. to dry steam at the same temperature, being equal to 970.4 B.t.u. (recently found, it is claimed, to be 971.6 B.t.u), the B.t.u. being defined as $\frac{1}{180}$ of the total heat required to raise the temperature of one pound of water from 32 to 212 deg. F. The total heat absorbed by a boiler in terms of this larger unit is called the *equivalent evaporation*. It is the amount of water that would be evaporated to dry steam by the total heat actually absorbed by the boiler if the feed temperature, as well as that of the steam, were 212 deg. Therefore, to find the equivalent evaporation, the total heat in B.t.u. absorbed by the boiler is divided by 970.4.

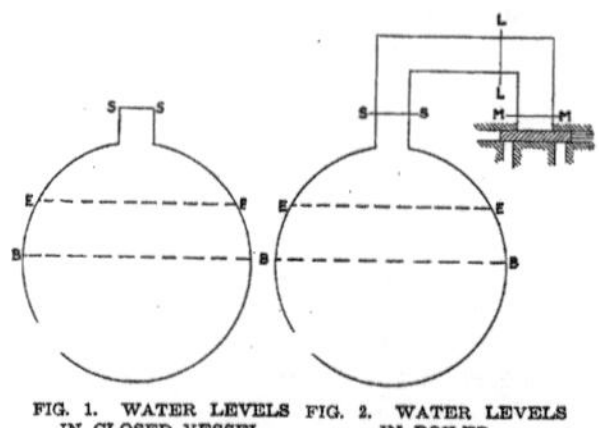

FIG. 1. WATER LEVELS IN CLOSED VESSEL FIG. 2. WATER LEVELS IN BOILER

As commonly stated, the equivalent evaporation is equal to the total feed water pumped into the boiler and apparently evaporated (corrected for difference in water levels at beginning and end of test), multiplied by a quantity called the "factor of evaporation."

In the new boiler code of the American Society of Mechanical Engineers equivalent evaporation "is obtained by multiplying the weight of water evaporated, corrected for moisture in the steam, by the 'factor of evaporation.' The latter equals $(H-h) \div 970.4$, in which H and h are respectively the total heat of saturated steam and of the feed water entering the boiler."

In order to correct for the quality of the steam, the following instructions are given in this code:

When the percentage of moisture is less than 2 per cent., it is sufficient merely to deduct the percentage from the weight of water fed, in which case the factor of correction for quality is

$$1 - \frac{\textit{Per cent. moisture}}{100} \qquad (1)$$

When the percentage of moisture is greater than 2 per cent., or if extreme accuracy is required, the factor of correction is

$$1 - P\frac{H-h_1}{H-h} \qquad (2)$$

in which P is the proportion of moisture, H the total heat of 1 lb. of dry saturated steam, h_1 the heat in the water at the temperature of saturated steam, and h the heat in water at feed-water temperature.

When, therefore, the weight of water pumped into the boiler and apparently evaporated is multiplied by the product of the "correction factor for quality" and the "factor of evaporation," in other words, the product of expressions (1) and (2), the result is the equivalent evaporation according to the code's definition. Since $P = 1 - x$, the product of (1) and (2) is

$$\left\{1 - (1-x)\frac{H-h_1}{H-h}\right\}\left\{\frac{H-h}{970.4}\right\}$$

Now since $H = L + h_1$ and $H_a = xL + h_1$, this expression may be simplified to read

$$\frac{H_a - h}{970.4} \qquad (3)$$

In these expressions x is the quality of the steam, L is the latent heat of evaporation and H_a is the total actual heat of the steam (above 32 deg.), whether wet, dry or superheated. The other symbols have been already defined.

It will therefore be seen that the total equivalent evaporation is obtained from the expression,

$$W_e = \frac{H_a - h}{970.4} W_a \qquad (4)$$

in which W_e is the equivalent evaporation, and W_a the apparent evaporation. In this expression H_a is equal (a) $xL + h_1$; (b) $L + h$; (c) $L + h + m\ (T - T_s)$; for wet, dry and superheated steam respectively, where m is the mean specific heat of superheated steam, T the actual steam temperature and T_s the temperature of saturated steam at the pressure observed.

It is evident, therefore, that it is a waste of time, being confusing, to use for wet steam the method recommended in the code. No distinction should be made between the *forms* of expression for the equivalent evaporation for wet, dry or superheated steam, the general expression applying to all cases.

"Factor of evaporation" should therefore be defined as the total actual heat of one pound of steam (wet, dry or superheated) above the feed temperature divided by 970.4. When, however, the water levels and steam pressures are not the same at the beginning and end of the test, none of the expressions referred to gives the true value for the heat absorbed by the boiler, for the reason that none takes into account accurately the heat of the liquid above feed temperature of the weight of water represented by this difference in level, as well as other quantities of heat.

When the water level at the end is higher than at the beginning of the test, it is evident that more water has been pumped into the boiler than has been evaporated, but the temperature of *all* this water has been raised to the boiling point. If, on the other hand, the level be lower at the end than at the beginning, more water will have been evaporated than pumped in, but the temperature of all this has not been raised from the feed-water temperature. The weight represented by the difference of level in this case will have been supplied with only the heat of vaporization.

It is a matter of some importance and considerable interest to consider that weight of water is represented by the difference in level at the beginning and the end of a test. When, for instance, the level is higher at the end, the excess of water supplied to the boiler over that evaporated will be the volume between the two levels multiplied by the specific weight of water at boiler temperature, minus the same volume multiplied by the specific weight of steam at boiler pressure. The corrections must include not only the heat of the liquid of the excess above feed temperature which is to be added, but also the latent heat of vaporization of the steam which originally filled this space, which amount of heat is to be subtracted since the steam was condensed. This conclusion follows from the fact that if water is pumped into a closed vessel containing water and steam, the pressure being constant, some of the steam will condense. The application of this fact to boiler conditions is as follows:

Fig. 1 represents a "closed vessel," the steam pressure being constant, at 140 lb. per sq.in., for example, the initial water level being at *B*. Should the water level be raised to *E*, the steam originally in the space *B-E* will be condensed.

Fig. 2 represents a boiler supplying steam to a reciprocating engine, the initial water level again being at *B*. Now at a given point of the cycle there will be a definite quantity of steam between the two sections *S-S* and *L-L* (or *M-M*). Should the water level be raised from *B* to *E*, the pressure being constant, the engine will continue to take the steam at the same rate and the conditions at any given section will be the same as at the beginning. Consequently, the same amount of steam has left the boiler as would have gone out had the water level not been raised. The total amount of water supplied to the boiler in a given time will therefore be equal to that supplied the engine plus the amount pumped into the equal "closed" vessel of Fig. 1, or equal to the steam supplied the engine (from constant water level *B*) plus the weight of water in volume *E-B*, minus the steam condensed in space *E-B*.

An example will help to make the discussion clear: Let it be assumed, for instance, that the total water pumped into the boiler in a 10-hour test is 40,000 lb.; that the water level is one inch higher at the end of the test than at the beginning; that the boiler pressure is 115 lb. per sq.in. at the beginning and 140 at the end; that the average pressure is 130 lb. absolute; that each inch of difference of level represents a volume of 4 cu.ft.; that the feed-water temperature is 180 deg. F.; and that the quality of steam is 99.5 per cent.: to find the total equivalent evaporation.

Using the method represented in equation (4), which amounts to the same thing as a combination of (1) and (2), we find:

Total heat in 1 lb. steam above feed temperature at pressure of 130 and quality of 0.995, $H_a = 318.6 + 0.995 \times 872.3 - 147.88 = 1040$.

Specific weight of water at 115 lb. = 56 lb. per cu.ft.

Specific weight of water at 140 lb. = 55.4 lb. per cu.ft.

Specific weight of water at 115 lb. = 0.258 lb. per cu.ft.

Specific weight of water at 140 lb. = 0.31 lb. per cu.ft.

Let

W_1 = Total weight of water in the boiler at beginning of test;

W_2 = Total weight of water in the boiler at end of test;

V_1 = Total volume in cubic feet occupied by W_1;

V_2 = Total volume in cubic feet occupied by W_2

w_1 and w_2 = Corresponding specific weights of water and w_3 and w_4 of saturated steam (dry).

Since w_1 and w_2 are usually approximately equal, and likewise w_3 *and* w_4, the following expression will give results sufficiently close for that portion of the water pumped in which is not evaporated and each pound of which received an amount of heat equal to the difference between the heats of the liquid at boiler and feed temperature:

$$(V_2 - V_1)\frac{w_2 + w_1}{2} - (V_2 - V_1)\frac{w_3 + w_4}{2}$$

In accordance with this method, therefore, the correction of feed water for inequality of water levels will be $4\frac{56 + 55.4}{2} - 4\frac{0.258 + 0.31}{2} = 222.8 - 1.1 =$ 221.7 lb.

According to a strict interpretation of the code the total equivalent evaporation is $\frac{(40{,}000 - 222.8)\ 1040}{970.4}$ = 42,630.

In this expression, however, no account has been taken of the heat absorbed by the water above the temperature of the feed, which amount must be added; and no account of the 1.1 lb. of steam (condensed) multiplied by the latent heat of vaporization at 140 lb. pressure, 867.6, which amount is to be subtracted.

The total heat absorbed by the boiler will be $(40{,}000 - 221.7) \times 1040 + 221.7\ (327.4 - 147.88) - 1.1 \times 867.6$ and the corresponding equivalent evaporation, that amount divided by 970.4, which gives a value of 42,670.

While there results only a small difference of equivalent evaporation with a correspondingly small percentage of error (about 0.1 of 1 per cent.), the amount seems well worth considering, especially in view of the fact that the calculations are simple. If the difference in water levels were 8 inches, the error would be nearly 1 per cent. by the old method. Every possible error should be eliminated in determinations of this kind, especially when useless refinement will not be entailed.

If the water level at the end were one inch lower than at the beginning, the total heat supplied would be $40{,}000 \times 1040 + 222.8 \times 867.6$; and not $(40{,}000 + 222.8) \times 1040$, the value obtained by the old methods, which gives a value about 0.1 of 1 per cent. too large for a difference of level of one inch.

It would therefore seem advisable to define equivalent evaporation as the total heat absorbed by the boiler in B.t.u. divided by 970.4, and to abolish the term factor of evaporation, because it is seldom that the water levels are the same at the end as at the beginning of a test.

Some Refrigerating Plant Experiences

By Roland L. Tullis

The average steam engineer seems to consider refrigeration as an art in itself and is disinclined to bother his head about the matter. This is a grave mistake, for no steam engineer knows but that he may some day be called to operate a refrigerating plant. Many a competent steam engineer has been compelled to turn down an offer of a better position simply because he was unacquainted with the operation of a refrigerating unit that was a part of the plant.

I once worked with the chief engineer of a large hotel that was buying ice and refrigeration from a central ice plant that was distributing refrigeration through a system of underground piping to a number of hotels in the vicinity. One day the agent of an ice-machine company succeeded in convincing the hotel manager that a considerable sum of money could be saved by installing one of his machines. The manager was impressed by the agent's talk and the data with which the latter backed up his claims, but when he called the chief engineer into consultation he almost had a fight on his hands. The chief didn't want any ice machine. He complained that it would only cause a lot of extra trouble and work for the operating force. He was skeptical. He didn't believe the agent's figures, although he frankly admitted that he knew nothing of the subject. He was perfectly satisfied with the present refrigeration system and its cost. He was a steam engineer of the old school, and he didn't want anything to do with an ice machine.

If the real reasons for the chief's opposition were known, he was simply afraid of an ice machine—afraid he couldn't operate it. However, the manager was quick to surmise about what was the cause of the chief's opposition on the subject. He went right ahead and ordered the machine installed and politely informed the chief that he would have to study up on the subject a little and learn to operate the machine, or he would be compelled to employ a man who was either willing to learn or already possessed the required knowledge. The chief received the manager's ultimatum as a man should. In a few months the machine was installed and running along smoothly. The machine lived up to the agent's promises and saved the hotel company quite a sum of money.

I once took charge of the engine room of a large hotel, the equipment of which included a 35-ton refrigerating machine. The compressor was of the horizontal, double-acting, steam-driven type. You could smell ammonia before you got inside the engine room. I asked the operating engineer about the cause of the leakage. He explained that there had always been a small leak around the piston rod on the compressor, regardless of the fact that they packed it about every ten days. He went on to explain that it gave off a strong odor, but it really did not amount to much as long as you could not see the ammonia coming out. He said that it had been that way as long as he had been on the job, about a year and a half. After I took charge of the plant, I discovered that the piston rod on the compressor was out of line about a sixteenth of an inch and was badly scored, consequently it was impossible to pack it so that it would hold for any length of time.

The ammonia condensers were situated on the main floor of the building near the laundry plant. The laundry foreman complained that the odor of the ammonia was so strong at times that it would drive the employees out of the laundry. The ammonia condensers were of the double-pipe coil type, fitted together with rubber gaskets. Most of the gaskets were in poor condition and leaking badly, especially when the high pressure on the machine rose a little above normal.

The operating engineer explained that they had been running the machine twenty-four hours a day, and that it would hardly supply the required amount of refrigeration. He complained that the capacity of the machine was insufficient for the work that was required of it. However, when I examined the ammonia receiver I could find no indications of ammonia in the glass. I diplomatically questioned him about this and pretended to be a gullible listener, while he explained that he never paid much attention to the ammonia receiver, but always figured that there was plenty of ammonia in the system as long as he could get frost on the pipes. He also complained that he would get too much high pressure on the machine when the ammonia receiver was filled to the proper level. Whenever the expansion valves refused to frost properly, he would close the king valve and pump down to about ten inches of vacuum, then open the valves and start all over again.

After being in charge of the plant a few weeks, I had the ammonia condensers overhauled and all the leaks stopped. The piston rod on the condenser was turned down and properly aligned, and after purging a few thousand cubic feet of air (more or less) out of the system, I had injected a charge of ammonia. The operating engineer was amazed to find that the high pressure stood around 185 lb. with the system full of ammonia. Here is the whole thing in a nutshell. He had been pumping down to ten inches of vacuum, sucking air into the system around the leaky piston rod on the compressor. The air in the system caused the exceedingly high back pressure. While he imagined that the high back pressure signified that there was plenty of ammonia in the system, he was simply using good steam to churn air and gas around through the system.

I found that, when this installation was properly operated, it could be shut down nearly twelve hours a day. The temperatures were lower than ever before, and the records of operation show an almost unbelievable saving in ammonia.

EDITORIALS

Progress in the Power Field

LOOKING back over the preceding year to review progress in the field has been the practice in *Power* for some time. As is usual at this period, our summary may be found on the leading pages of this issue. For the year no great advance can be recorded. It was more a period of crystallization in which the results from previous effort were considered as a guide for future development. This is particularly true of the turbine field, in which limits of speed and capacity of single-shaft units were given much attention. To improve efficiency the trend is toward higher pressures and less superheat. Disturbances in the industrial field from high prices, the expectation of a decided drop which did not materialize and the demands of radical labor followed by a heavy rush of orders for all kinds of standard power-plant equipment, excluded much of the pioneer work that might have been expected.

In all kinds of labor-saving apparatus for the plant and instruments to conserve costly fuel, there was great activity. In stokers of both the underfeed and chain-grate types there has been progress. To the latter the application of forced draft for burning bituminous coal received much attention, as did the alternate method of burning more coal by increasing the grate area and speeding up the stoker. Improved power dump grates and clinker grinders, sectionalizing of the stoker drive to make more feasible a uniform fuel bed on wide stokers and the development of various methods of drive to accelerate or retard the feed of the coal show the advance in the underfeed stoker.

High-priced coal and the recent shortage has brought about greater use of inferior fuels, more intensive investigation of the possibilities of powdered coal, low-temperature distillation and colloidal fuel, more general use of water power and a great turning to oil, particularly along the Atlantic Coast. To conserve fuel, the tendencies toward concentration and interconnection are more marked. An indication is the proposed power surveys of Secretary Lane.

Of the numerous power stations made ready for operation during the year the steam plant at Kansas City, with its unique economizer installation, is of exceptional interest. In the water-power field the great Queenstown development in Ontario and the new plant of the Niagara Falls Power Company that is to contain three of the most powerful water turbines in the world are in the limelight. The characteristic, size, has been a feature of the year. The past twelve months has seen the placing in operation of the largest turbo-generator ever built, the delivery of equipment for the world's largest furnace, the construction of the largest electric locomotives, the erection of a giant hoisting engine, the design of transformers larger than any yet made and so on.

During the latter part of the year the demand for steam engines was active, more especially for the economical types such as the uniflow. With the increasing use of oil, engines consuming this fuel are in more general use. Both the two- and four-cycle types are being built, and some of the units manufactured are of exceptional size.

In the refrigerating field the high-speed compressor, synchronous-motor driven, is the latest development. Several units of this type have been installed, and reliable data on economy, upkeep and operation is the question of the hour. Great activity marked the electrical profession, heavy export trade being a feature.

The year has been one of activity for the engineer. Lessons of the war had pointed the way. To raise the standing of the profession it was realized that the services rendered to the country should be continued during times of peace. In other words, the engineer has assumed his true rôle, and to make most effective his services to mankind he is realizing as never before the value of coöperation and unity of action. The movement is country-wide. Just what has been done and what is contemplated is told in the review.

Why Boilers Explode

IN a recent issue of *Power* C. E. Stromeyer, chief engineer of the Manchester, England, Steam Users' Association, commented upon the large number of boiler explosions in the United States as compared with those in other countries. The reason has long been known. Other countries exercise intelligent supervision over the design, construction and operation of boilers, and when one explodes a real investigation is made of the condition, and the responsibility fixed. In the majority of our states one may build a boiler any old way that he pleases, out of any kind of stock he can get, carry any pressure upon it that he has a mind to, and hire anyone who comes along to run it.

To those who are accustomed to expert supervision, and acquainted with its effect, this state of affairs is unbelievable. The inhumanity and economic ruthlessness of permitting a hundred or two of preventable boiler explosions a year, rather than to tolerate "Governmental interference with private property," and to maintain the "let us alone" attitude of "we will do as we please with what is our own. The risk is ours—we will pay the damage if there is any," is appalling. There is a gradual weakening in progress, and now fifteen out of our forty-eight states have provided more or less thorough means for the inspection and control of boilers, and four states and twenty-nine cities have laws providing for the certification of the men into whose charge they are given.

A pertinent example of the result of unsupervised operation is given in this issue. The boiler, the explosion of which is there described, was second-hand, a dozen years old, and had never been inspected, except at the time of its purchase a year ago. After the explosion the stem of the safety valve was found so crooked that it took a blow from a small sledge to move it through the cap in either direction. One would suppose that the men in charge of a boiler, whose lives are menaced by any risk incurred in its operation, would be on their guard against such a condition. But they do not know, and innumerable instances of this kind have no weight with the average legislator against the manufacturing constituent who wants to be let alone, or the lobbying dealer.

Lift the Bushel

A JUST complaint of professionals is that their services are not appreciated, and with equal justice it may be said that the general public is allowed to remain in ignorance of what services the professionals are capable of rendering. Their light has been kept under a bushel. Among men who attain the direction of affairs, ignorance of comprehensiveness of the different branches of engineering is astounding.

Probably not more than five men in every thousand are called upon in a lifetime to make arrangements for building a house, erecting a mill or ordering an important power-plant improvement. Accustomed to satisfy their material wants from stocks of shelf-goods, the first thoughts of a project are directed to finding the builder or dealer who will supply the want, although it may be known from experience that there are times and occasions when the drug clerk is to be superseded by the physician and that a man who is his own lawyer has a fool for a client. But without previous experience the prospective investor in engineering work has little or no conception of the advantages or necessity of guidance by services of trained specialists.

The World War has demonstrated that the material, physical and financial strength of nations depends mainly on technical development. The rapid and efficient preparations that we made for the conflict are commonly attributed to American ingenuity, versatility and initiative, but without the breadth of our technical training we could not have risen to the occasion. The opportunities were presented and the achievements of our technically trained men were no less a surprise to our own people than to the rest of the world. Are we now to return to the piping times of peace without better public knowledge or recognition of engineering services? The technical schools and colleges of the United States annually launch thousands of graduates upon the public, and in the present busiest era of our history in many quarters there is to be seen the same old order of affairs that was responsible for crudeness, obsoleteness, waste of time and extravagance in public and private works, due to the neglect of available technical services.

The public always is right, divinely right, because the public must be taken as we find it. But the apathy of which the scientific schools and societies often complain would be replaced by appreciative recognition of the professions if only the public were informed, not by grandiloquent insinuations that so rarely reach the rank and file of active Americans, but by educational advertising.

If our college professors wish to make their labors of actual material benefit and be in a position to demand instead of begging support like timid clerical mendicants, they would do well not only to prepare their students for the world but also to get to work with a vim to prepare the world for their students.

It is a pretty uphill job for a young graduate, after spending the most important years of his life in training for a useful profession, to be turned loose to create a public demand for his services. That is just what the majority of our young graduate engineers must do, and far too many are forced to seek occupations in which their years of special training are of little or no value. It is as criminal for college faculties to dazzle false prospects before parents and students as to salt a gold mine. While it is true—thanks to special opportunities—that hundreds of graduates occupy remunerative positions and are rendering valuable services, the country does not avail itself of the service of thousands of capable young men from a sheer lack of appreciation that results from abject ignorance of the places they are capable of filling in the industrial world.

Fancy how many men of a hundred, as met on the street, are informed of the province of civil, mechanical, electrical or mining engineers. These occupations should be as well known as those of doctors, dentists or clergymen. If professions are unknown to the public, how can it be expected to extend its patronage? If the colleges would have their work appreciated, let them spend a liberal proportion of their time and means spreading information of the advantages of engineering services at least through the public schools. Engaged in such a propaganda, college faculties will be doing more good for the public, for their students and for themselves than to be employed in original research or reading each others' books.

Good Things Coming

READERS will be glad to learn that *Power*, by combining issues in December, has caught up with its schedule and in the new year the paper will be published regularly once a week as usual. Printing difficulties have not been eliminated entirely, but no undue delay is anticipated. For the new year plans have been formulated to make the paper more useful and readable than ever before; and its scope in the industrial field will be broadened. One of the good things in store for the reader is a comprehensive study course on refrigeration. This will be elementary as the title implies and will cover the subject thoroughly. Principles of operation will be explained in a simple way. There will be articles on the compression and absorption systems and other articles dealing with the various elements of each system, such as the compressor, the condenser and the low-pressure end with reference to direct expansion and brine circulation. The series will alternate with an interesting group of articles on steel-mill power plants, discussing equipment and practice in the boiler and engine rooms and the features of the power drives in the mill. There will be other articles dealing with power requirements in textile mills, articles describing governors for steam turbines, small and large, and many more of equal interest to keep the reader abreast of the times in all branches of the profession.

The steel-mill articles start with this issue, and in the next number the study course on refrigeration will begin. The two series mentioned will supplant the electrical study course which was brought to a close in the last issue of the old year. No alarm need be felt, however, for there is a store of good practical electrical material that will make this department more valuable than ever to the operator in the plant.

That the articles mentioned, and in fact the entire contents of the paper from cover to cover, may be of maximum benefit to our readers and help them on their way toward the top is our one best wish for the new year.

It is announced that an engineer at Middlesborough, England, has succeeded in abstracting commercial alcohol and its derivatives from coke. It is asserted that if the process, which requires the use of gas, is applied to all the coal carbonized in Great Britain, a yield of fifty million gallons of motor spirits will be obtained annually. Prohibition papers please copy!

German miners have volunteered to introduce a seventh shift in the week's schedule in order to increase the output of coal. American miners, or rather miners of American coal, have —— Oh, well!

Superheated Steam—Why Not?

The statements of the editorial, "Superheated Steam —Why Not?" in a recent issue of *Power*, apparently were not entirely clear, inasmuch as there seems to have been a misinterpretation on the part of Robert W. Wyld in his letter on page 475 of the Sept. 16 issue. The editorial states that one superheater manufacturer has found that from 200 to 250 deg. superheat can safely and satisfactorily be employed with successful lubrication and without material change in the metal used in the cylinders, valves, fittings and piping.

This statement is most assuredly correct. There are many engines that can and do use highly superheated steam, and it is my belief that superheated steam at a temperature as high as practical conditions will permit should always be employed where steam engines are used. No one at the present time would criticize the use of highly superheated steam in turbines, and there are many instances where piston-valve and poppet-valve reciprocating engines are operating with superheated steam of 200 to 250 deg., and these engines are showing a very high economy.

The use of 75 to 100 deg. of superheat in connection with Corliss-valve engines, as pointed out by Mr. Wyld, is certainly conservative, as 100 to 125 deg. not only would be within the limits of safety, but would have the advantage of increased economy due to the high superheat.

As to the use of high degrees of superheat in high-pressure cylinders of marine engines having slide valves, I know of a number of ships that are operating with a superheat of 200 deg. These installations are highly satisfactory in every respect.

While, of course, I am not entirely sure of the company that is credited with having 18,000,000 hp. of superheater installations, I am rather of the opinion that it is possible Mr. Wyld may have had in mind a company that has 80,000,000 hp. of superheater installations in operation. The 80,000,000 hp. referred to is based on the policy of highly superheated steam and designed to produce a degree of superheat adapted to each individual installation. R. A. HANNING.

Yonkers, N. Y.

Using Compression Grease Cup as Wick-Feed Oiler

How an ordinary pressed-steel or brass screw grease cup may be made into a wick-feed oiler similar to the type used on small electric motors is shown in Fig. 1. This has the advantage of a larger oil container than is usually supplied. The base should be drilled out smoothly, if there are any burrs present, and a round felt wick cut to fit the hole snugly but so that it will move freely. A tapered spring is made of brass spring wire with a large-diameter base to fit the cap of the grease cup. This is attached to the wick as shown in Fig. 2. To use as an underfeed oiler, it is only necessary to adjust the wick and spring so that the former will project about a quarter of an inch from the nipple when the whole cup is removed from the bearing, filled with oil and screwed into place.

To use as an overfeed oiler, the cup should be packed with waste and the wick used without a spring, adjust-

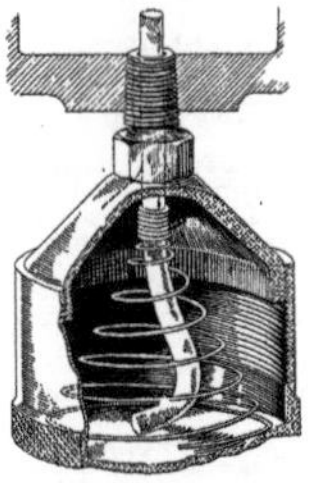

FIG. 1. WICK FEED OILER MADE FROM GREASE CUP

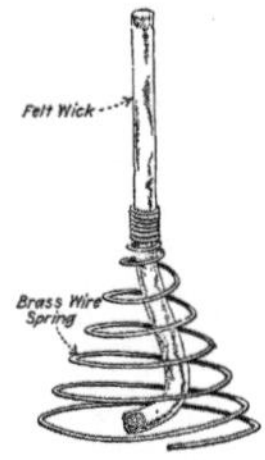

FIG. 2. METHOD OF JOINING SPRING AND WICK

ing its pressure on the bearing by screwing down on the cup. The waste will retain the oil in this case, provided the cup threads make a fairly good fit on the base; if too loose, the threads on both cup and base may be coated with shellac and this allowed to become hard before screwing together. H. H. PARKER.

Oakland, Calif.

Smokeless Chimneys Not Conclusive Proof of Furnace Economy

My attention has been called to a letter by W. E. Porter, Chief of Bureau of Smoke Regulations, Pittsburgh, relating to an article appearing under "Inquiries of General Interest;" entitled "Smokeless Chimney Not Conclusive Proof of Furnace Economy." Naturally, the smokelessness of a stack appeals to Mr. Porter in his professional capacity, but the fact remains that a smokeless stack is not conclusive proof of furnace economy. In fact, in a coal-burning plant a stack that is smokeless twenty-four hours a day is indicative of excess air beyond the requirements.

The actual loss of unburned combustible in a smoky stack is probably not more than 1 or 2 per cent. of the heat in the coal. The really objectionable features of smoke are, first, the sooting of the surfaces of the boiler tubes and, second, what is commonly known as the smoke nuisance in a community. The first objection is removed by the daily use of soot blowers, mechanical or hand. The second objection is a matter of the health and comfort of the community and is the only logical reason for a smokeless stack in a coal-burning plant. When a stack is smokeless continuously every hour in the day, the loss due to the excess air more than counterbalances the gain of a small percentage of the unburned combustible.

The foregoing remarks apply particularly to industrial plants. In central-station practice and in recent, well-designed industrial plants, sufficient regard has been taken for proper combustion space by high boiler settings and correct furnace design, so that it is possible to have a smokeless stack as well as an economical plant. In the average industrial plant, where the settings are usually low and no particular attention has been paid to furnace design, a smokeless stack usually means an uneconomical plant. Therefore, we can safely say that the combination of a smokeless stack and an economical plant is more a matter of proper design than of operation, although neither can be neglected at the cost of the other.

JAMES T. BEARD, JR., Chief Engineer,
Fuel Engineering Company of New York.

Remedy for Present Valve Setting

With reference to the indicator cards, by S. L. Gilliam, on page 518 of the Sept. 23 issue of *Power,* I believe the head-end steam valve is leaking badly as shown by the expansion curve and high terminal pressure on the card, Fig. 1, and also by the rapid building up of the compression curve to a point equal to the initial pressure before the piston reached the end of its stroke, although the steam valve has no lead.

The card in Fig. 2 shows a much improved expansion curve, due to lengthening the wristplate connection, giving the valve more seal when closed. The head-end steam valve evidently does not seat properly, and the high spots should be dressed off carefully with a fine file. Having seated this valve properly, adjust the radial rods to give the valves the same lap, with the wristplate central; adjust the hook and eccentric rods to give equal travel, each side of central position, to wristplates and rocker-arms; and, finally, set the eccentrics to give the proper lead. This lead may vary from $\frac{1}{32}$ in. to $\frac{1}{4}$ in., depending, principally, upon the speed of the engine, and should be determined by the appearance of the card. With a long range, Corliss gear, where the steam ports are, say, each $\frac{7}{8}$ in. wide, the valve should have approximately $\frac{9}{32}$ in. lap when closed, and be in the center of its travel, having about $\frac{13}{32}$-in. opening, when the wristplate is central. F. P. STODDART.

Regarding the double-eccentric long-range cutoff engine diagrams by S. L. Gilliam, in the Sept. 23 issue of *Power,* I understand that this type of engine requires a steam-port opening of $\frac{9}{16}$ in. with wristplate on center, instead of $\frac{9}{32}$ in. as he has it. Mr. Gilliam states that there is no lead on the head valve, "yet it is open with engine on the center more than it should be," and that, to decrease the port opening, the rod for that valve is lengthened to its limit.

Possibly the fault is too early admission, caused by the steam eccentric being too far ahead of crank. It should be about 30 deg. ahead, or just enough to open the admission valve $\frac{1}{32}$ in. when the crank is on center. Too early admission would cause the engine to pound. E. M. CALKINS.

Johnsonburg, Penn.

Operating a 220-Volt Plant from 500-Volt Service

Fire had destroyed the mill and power plant of a box factory, and I was called in to see what could be done to get things going again. The factory, which was separate from the mill, was unharmed. It was equipped with 220-volt direct-current motors, current being supplied from the power plant. There was no other 220-volt direct-current service to be had, and the only current available was from a 550-volt direct-current street-railway trolley line.

Arrangement was made with the trolley company to use one hundred horsepower from their line, and an 85-hp. 500-volt shunt-wound motor without a starting box was secured. This motor was installed and belted to a lineshaft in the factory. Leaving the belt on the 40-hp. 220-volt motor formerly driving the shaft, the latter was driven as a 220-volt generator to supply power to the other small motors and lights about the factory.

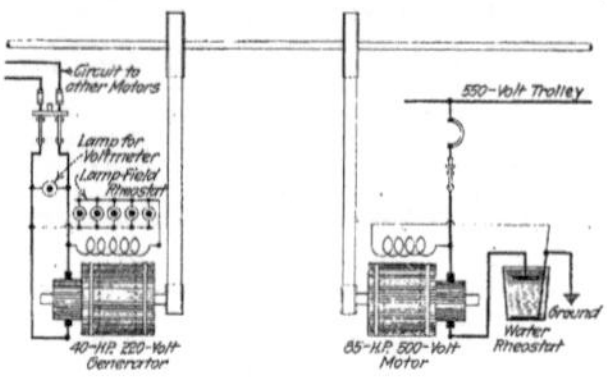

CONNECTION OF MOTOR AND GENERATOR

As shown in the figure, a set of lamp sockets grouped in parallel was connected in series with the generator's shunt field, for voltage regulation, and a water rheostat was used to start the 85-hp. 500-volt machine. Various-sized lamps were screwed in and out of the socket until the proper voltage was obtained. The switch and fuses used to protect the 40-hp. machine as a motor also served to protect it as a generator. There was no voltage release on the 85-hp. motor since it was started through the water rheostat; therefore it was necessary to keep an attendant near it all the time, as the trolley voltage was very unsteady. We installed a circuit-breaker for the protection of the 500-volt motor and used a pilot lamp temporarily for a voltmeter on the 220-volt generator.

This makeshift operated satisfactorily until the power plant was rebuilt, excepting that occasionally the trolley voltage would fail and shut the plant down. Once the power came on before the attendant got the water rheostat open, and caused the brushes to flash over on the 85-hp. motor, which set on fire an accumulation of sawdust inside the armature where it could not be reached.

Tampa, Fla. J. C. BUERKE.

Piston-Rod Ends Worked Loose

The piston rods in a large double-acting tandem gas engine kept working loose in the intermediate crosshead. No matter how tightly the cap was pulled down, the rear rod would work loose and turn far enough to one side, to let the tail-rod drain pipe rub the side of the slot in which it worked, to say nothing of spoiling the threads on the rod ends and in the crosshead. The construction of the crosshead was as shown in the cross section of the rod ends and the crosshead.

The usual practice in putting up this crosshead had been to raise the rod ends as high as possible, put the bottom half of the crosshead under, insert enough sheet-iron shims *A* between the rod ends to hold them in the proper position so that the threads in the bottom half of the crosshead would incline to pull the rod ends together when they were let down into place. The top half of the crosshead was then put on and pulled down as tight as possible with the nuts and studs *B*.

The engine would run quietly for some little time, but eventually the rods would work loose again and unless

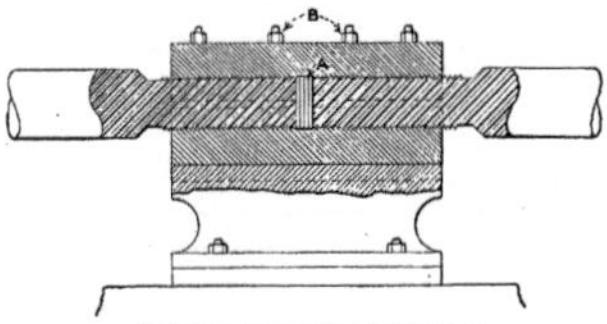

SECTION THROUGH CROSSHEAD

the engine was taken off the load, something would be ruined in a short time.

It was noticed while taking down the crosshead that the sheet-iron shims would be beaten out flat and although there had been as many put in as possible, they were loose when taken out. From this it was reasoned that there was too much give or spring to these shims and that as soon as the strain came on they flattened out and compressed together enough to allow the rods to come loose again.

The trouble was remedied by making cast-iron shims of the proper thickness to hold the rod ends just far enough apart that the back half of the thread on the rod would ride hard on the front half of the thread in the crosshead; the cap when put on was pulled down as tight as possible. EARL PAGETT.

Cherryvale, Kans.

Freezing of Brine a Puzzle

A recent puzzle in refrigeration proved interesting and might be helpful to engineers who do not as yet know everything about the subject. We have a fifteen-ton ice machine of the ammonia-compressor type. A small pump circulates the brine, a calcium-chloride solution, through the cooling or refrigerating coil and also through some large kitchen iceboxes.

Frequently, during the summer we were bothered with the brine pump coming to a standstill as the result of the brine freezing in the coil passing through the brine tank. Tests were made of the brine and showed a much lower freezing point than that indicated just before the freezing took place. In particular, after dissolving a drum of calcium for two days and testing with salimeter I found the brine should freeze at six above zero. It was a matter of great surprise the next day when it became known that the brine had frozen at twelve above zero.

Not knowing everything about physics, we wondered at this. I theorized that the thermometers might not be recording properly. One thermometer shows the temperature just before entering the refrigerating coil, and the other shows the brine temperature just after leaving the brine coil. Another possibility seemed to be that the calcium might have been adulterated to give the proper salimeter readings without being pure. This may have been fanciful, but it occurred to me nevertheless.

Before I had found out the truth or untruth of these ideas, a more reasonable solution presented itself. The brine tank had been leaking considerably, and it was impracticable to empty it during the hot weather. We allowed the depth of brine to decrease gradually for a few weeks and then filled up again with more water and additional calcium. This took place several times, and finally it occurred to me to connect it with the freezing of the brine at an apparently high temperature above the proper freezing point.

From one to two turns of the refrigerating coil would at times be exposed above the brine level in the tank. Here the freezing of the brine must have taken place. But how was it that the thermometer recorded a temperature six degrees too high for freezing at the departure end of the coil, where normally we find the very lowest temperature attainable in the coil?

The explanation seems as follows: In the refrigerating coil, which is double, consisting of an inside pipe containing the cold expanded ammonia gas and the space between it and the outside pipe containing the moving brine, the temperature of the gas is very low, much lower than the temperature that can be imparted to the moving brine, particularly when the brine pipe is immersed in and, so to speak, warmed by contact with the brine in the tank. But if one or two turns of the coil are not immersed in the brine, the temperature of the brine in these exposed parts of the coil falls very low, not being able to impart any of its low temperature to the surrounding medium—that is to say, not being able to absorb more than a trifle of heat from the air that surrounds the exposed parts of the coil, for air is not a good conductor of heat. Doubtless, before freezing the brine in this part of the coil had been cooled even below its freezing point, but the fact that it was kept in rapid motion by the pump still further reduced it freezing point.

Until the freeze-up became total, the brine of course continued passing through into the immersed part of the coil where it began to absorb heat from the surrounding medium and thereby became warmed so that at the point of departure from the coil its temperature was again practically normal and but little below the entering temperature, say two degrees. In this way the temperature as shown was deceiving concerning the extreme cold which the brine had experienced in one part of the coil. Of course when the general temperature of the brine had reached twelve degrees, the temperature in the exposed part had fallen much below the six degrees required, and brine crystals began to form on the surface of the ammonia, or inner, pipe faster than the moving brine could tear them off. These crystals gradually built up a solid body of frozen brine or brine-ice until the pump was no longer able to force a passageway. The slower the pumping became the faster the brine froze in that portion of the coil. Frost forming on the outside of the coil also acted as an insulator, further preventing the brine from absorbing heat from the air surrounding the exposed part of the coil.

Buffalo, N. Y. C. A. GOODWIN.

Reverse-Current Relays Installed on Emergency Service

The article entitled "Reverse-Current Relays—Principles of Operation," by Victor H. Todd, published in *Power*, July 15, 1919, is both interesting and instructive and should prove valuable to many engineers, whose plants would be benefited by the installation of this simple but effective protective device.

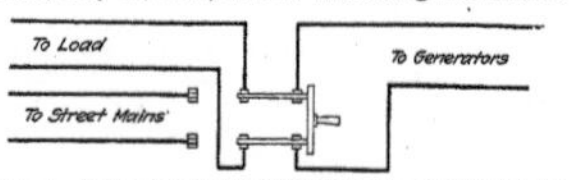

FIG. 1. EMERGENCY-SERVICE SWITCH CONNECTIONS

There is one application of this apparatus, however, not mentioned in the article, and that is its use in an isolated plant in connection with a service switch from the central station. The usual method of installation is to use a double-throw switch, connecting the electrical load to the blades, the plant busbars to one set of clips and the central-station supply to the other, as in Fig. 1. This is not altogether satisfactory, as it is necessary to disconnect the load in most cases before the double-throw switch can be operated.

About a year ago, when many plants, because of labor and coal conditions, were contracting for breakdown service with their local central stations, the plant I am connected with did likewise. However, it was deemed desirable to avoid the installation of the usual double-throw switch if possible. The usual practice when shifting the load from one generator to another is to parallel the machine so as to cause no interruption to the service. It therefore seemed practicable to do the same thing when changing to the outside current or back again; namely, to parallel the street service and the generators and thus avoid a service interruption. In the course of our investigation a representative of the Fire Underwriters was called in, who informed us that to comply with their regulations we would have to install a double-throw switch, as is usually done, or equip the circuit-breaker of our generators with reverse-current relays. In view of this ruling it was decided to install a reverse-current relay on the main circuit-breaker of each generator, as well as on the central-station service circuit-breaker. The outside current would then be connected to the plant busbars through a circuit-breaker, and shifting the load from the generators to the street supply or vice versa would be no different from changing from one generator to another.

The reverse-current relays would prevent the plant from supplying power to the central-station service supply mains and it would prevent the motorizing of our generators by the outside current supply, which I was given to understand was the underwriters' main reason for insisting on either reverse-current relays on the circuit-breakers or a double-throw switch. Fig. 2 gives a schematic diagram of connection, as we had intended for using the reverse-current relays in connection with the generators and service circuit-breaker. However, owing to the press of war work it was found impossible to get relays applied to the circuit-breaker by the manufacturers in a reasonable time, so the project was abandoned.

By equipping each generator's circuit-breaker, as well as the one on the outside-service mains, with its individual reverse-current relay, any opposition to connecting the street service directly to the busbars is eliminated. As the central station is mostly interested in preventing a reversal of current in the power leads they install in a building, by installing and maintaining a relay on their own service switch they would be amply protected from trouble originating in the private plant. The benefits to be derived from this type of installation would be favorable to both parties. For example, it would be possible to shift the load quickly to the service main when cleaning fires and eliminate the loss due to carrying an extra boiler on the line for this purpose, as is usually done. Then, in case of plant trouble there need be no hesitancy in shifting the load to the central station, where with the double-throw switch scheme many engineers take a chance because of the service interruptions while changing over. As can readily can be imagined, the more flexible

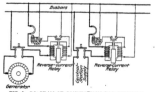

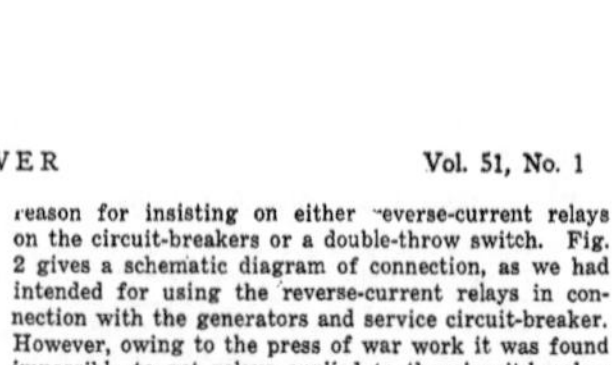

FIG. 2. DIAGRAM OF CONNECTIONS FOR REVERSE-CURRENT RELAYS ON CIRCUIT-BREAKERS

the service connection is the more often it will be used to the mutual advantage of the private plant and the central station.

I have never heard of a breakdown service being installed in this way and regret that it was not possible to try it out as contemplated in our plant. However, the idea seemed sound, and the possible benefits are such that I would like to know if such an installation has already been made. Comments on this subject would no doubt prove instructive to readers of *Power*.

New York City. WARREN D. LEWIS.

An employee of a railway company engaged in pumping water into a tank for use by engines pulling interstate trains is entitled to maintain an action under the Federal Employers' Liability Act—for injury received through an unguarded condition of gearing connected with an engine, negligently permitted by the company to exist—on a theory that he was engaged in interstate commerce. (New York Supreme Court, Appellate Division; Kelly vs. Erie Railroad Co.. 177 New York Supplement, 278.)

INQUIRIES OF GENERAL INTEREST

Lead for Direct Duplex Pump—Why is not lead given to the steam valve of a direct duplex pump? P. K. D.

The steam valves are without lead in order that the pump may be started at any point of the stroke; and also to enable the piston to be at slow speeds to complete enough of the stroke to obtain reversal of the valve gear for a return stroke.

Relative Pressure for Convex and Concave Heads—What is the relative strength of convex and concave head for a steam drum? B. J. T.

A concave head, or one dished inward so the drum pressure shall come on the convex side, should be used for a maximum working pressure of only 60 per cent. of that allowed for a head of the same dimensions with the pressure on the concave side.

Reduction of Thick Plates at Girth Joints—Why is it necessary to plane or mill down the thickness of plate at the top of the circumferential joints of return-tubular boiler sheets when the main plate thickness exceeds ½ in? J. S. F.

Where exposed to the fire or products of combustion, the plate is to be reduced to one-half thickness to obtain thin enough metal and short enough rivets to prevent burning of the material from not being kept cool by the water of the boiler.

Ammonia Coil for Cooling Water—What length of pipe coil in which there is circulation of zero ammonia should be employed for cooling 200 gal. of water per hour from a temperature of 80 deg. F. to 35 deg. F.? W. E. U.

The mean temperature of the water would be $(80 + 35) \div 2 = 57.5$ deg. F. From data of tests under similar conditions, it is estimated that there would be transfer of heat through an iron pipe coil at the rate of 105 B.t.u. per hour per square foot of pipe surface, for each degree difference of mean temperature. Hence one square foot of pipe surface would absorb $105 \times 57.5 = 6037.5$ B.t.u. per hour. In cooling the water from 80 to 35 deg. F., each pound of water would part with 45 B.t.u. and for cooling 200 gal., or about 1666 lb., of water per hour, there would need to be absorption of $1666 \times 45 = 74{,}970$ B.t.u. per hr., requiring $74{,}970 \div 6{,}037.5 = 12.4$ sq.ft. of coil surface, or $12.4 \times 2.3 = 28.5$ lin. ft. of 1¼-in. pipe. This is without any allowance for gain or loss of heat due to the temperature of surrounding atmosphere, extent of tank surface or character of insulating material employed.

Compression Valves on Steam End of Pump—How are the compression valves arranged on a pump that is equipped with them? F. C.

A compression valve at each end of the steam cylinder of a pump is so arranged as to vary the size of a small opening in the partition between the steam passage and the exhaust passage. The piston covers the exhaust passage before the stroke has been completed, thus compressing any exhaust steam then remaining in the cylinder and steam passage at the end of the cylinder, and if the piston cushions too heavily, the compression valve is adjusted to permit escape of excessively compressed exhaust in the end of the cylinder, to the exhaust passage, by way of the steam passage.

Click in Engine With a Particular Pressure—What may be the cause of a clicking noise that seems to be in the crankshaft of a large Corliss engine whenever the steam pressure drops to 85 lb.? This has taken place for many months, apparently without change, and after careful inspection of the shaft, flywheel, crank and crankpin there is no apparent cause for noise from movement of those parts of the engine. P. F. C.

The cause of clicking or knock in an engine is frequently difficult to locate on account of transmission of sound from one part of the engine to another. If the noise occurs only with change to a particular initial pressure, regardless of load, it is most likely due to play of a piston packing ring or of the bull ring. In any event, the cause of a click, knock or rattling that is coincidental with certain pressure conditions should be sought for in the cylinder of the engine.

Return Trap Connections—We are installing some machinery that requires a supply of live steam and will have return traps for returning the condensation to the boiler, carrying 85 lb. pressure. One of the machines is to be supplied with steam reduced to the pressure of 30 lb. gage. How should the returns and trap be arranged and connected? F. K.

All returns connected to a trap or receiver should be at the same pressure; otherwise those of higher pressure will prevent the trap from simultaneously receiving returns of lower pressure. Systems of returns that are at different pressures will require separate traps. The return pipes should be sloped so they will be drained perfectly clear of condensation and discharged by gravity to the trap with the trap placed four feet or more above the water line of the boiler. It is well to place an automatic air valve on the return line near the trap and also provide a pet-cock for testing whether the returns are being emptied. When the trap becomes filled with return water, the mechanism automatically admits live steam from the boiler for siphoning the water into the boiler, but cannot do so unless the receiver is elevated sufficiently for obtaining the necessary head of discharge water for overcoming friction of the trap discharge pipe and fittings. There should be a separate return connection to the boiler, with check valve or connection so arranged that the discharge will not be affected by pulsations of a boiler-feed pump. The inlets and outlets for admission and discharge of return water, and for admission of live steam and discharge of displacing steam, are usually described in directions issued by the manufacturer or are indicated on the trap. If the water inlet of the trap is not provided with a check valve, one should be placed in the return line near the trap. Most return traps are provided with a small outlet for discharge of the displacement steam. This should be provided with a stop valve and pipe connection for discharging the steam in the boiler ashpit or at some other point where a small quantity of air steam or water discharge from the trap would not be objectionable.

The Hvid Engine and Its Relation to the Fuel Problem

By E. C. BLAKELY

This paper describes in detail the arrangement and operation of an internal-combustion engine which, the author claims, has all the advantages and none of the disadvantages of the Diesel type. It can be made in small sizes and has no electric or other ignition devices to cause complications. Curves from actual tests are given to show its fuel economy as compared to gasoline and kerosene engine of the same size. Curves showing the results of heat balance tests are also given.

MUCH has been written and said during the last two or three years on the subject of using kerosene as fuel in conventional gasoline engines of both low- and high-speed types. While undoubtedly much has been learned concerning the characteristics of kerosene under such conditions, the burning of kerosene in gasoline engines has not been accomplished with complete success. By complete success is meant starting the engine on kerosene in atmospheric temperatures approximating zero degrees Fahrenheit and below, without preliminary heating of any sort, and burning kerosene so as to eliminate troublesome carbonization and complicated accessory apparatus, and obtaining high economy.

Kerosene and gasoline are widely different substances chemically, having nothing in common but the base from which they are derived. Their initial boiling points are wide apart, that of commercial gasoline being about 100 deg. F., while that of kerosene is about 330 deg. F. Under these conditions a carburetor designed for vaporizing gasoline cannot be expected to vaporize kerosene. The best it can do is to atomize it.

In order to vaporize, as well as to prevent precipitation or condensation of the atomized kerosene in the combustion chamber, it is necessary to heat the charge; and since the power output of the engine depends on the amount of charge taken in and burned during each cycle, it is clear that the more the charge is heated the less mixture we can get into the cylinder and the less power we can obtain. This forces us to a compromise between two conflicting conditions—the maintenance of the incoming charge at the lowest possible temperature that will vaporize the kerosene, and the prevention of precipitation in the combustion chamber. This compromise is especially difficult in the case of an engine running at varying speeds and loads.

In order to obtain maximum power from the engine, we must have maximum mean effective pressure; and since mean effective pressure depends largely upon compression pressure, we must use the highest compression pressure possible. In order to prevent so-called preignition when burning kerosene, we are forced to a relatively low compression pressure, which lowers the mean effective pressure and therefore the power output.

There have been built in the last ten years numerous engines capable of running consistently on the various crude and fuel oils, as for instance Diesel and the so-called semi-Diesel engine, hot-bulb and hot-surface ignition engines. These, however, have been used mainly in marine work and in relatively large units. The latter types have the disadvantage of requiring external preheating before they can be started, and the torches used for this purpose are a source of constant danger. Electric preheating has been tried, but with little success.

There is an engine, however, that has all the advantages of the aforementioned types and, we think, none of the disadvantages. This is the Hvid engine. It can be started cold on any liquid fuel that will flow through a pipe. It has no complicated air-compressure system for injecting the fuel, no electrical devices, no carburetor or mixer, no hot-bulb or torches, and runs with a fuel economy on a par with that of the Diesel engine. The Hvid engine can be, and is being, produced in units as small as 1½ hp., and so economically as to be able to compete with gasoline engines of the same size.

Briefly enumerated, the chief advantages of the Hvid engine are: Mechanical simplicity, low fuel consumption at all loads, ability to start and run on any oil that will flow, low water-jacket losses, no lubricating difficulties, constant compression, remarkable torque characteristics, absence of all electrical devices, hot-bulbs and torches for ignition purposes, absence of all carbureting mechanism, and finally, no carbon troubles. The Hvid engine is of the conventional four-stroke-

FIG. 1. SECTION THROUGH CYLINDER HEAD OF HVID ENGINE

cycle type embodying the usual inlet and exhaust valves, which open and close as in any four-stroke-cycle engine. The compression pressure is carried to between 425 and 475 lb. to the square inch, which heats the compressed air between 900 and 1000 deg. F. In the cylinder head there is a fuel-admission valve terminating in a small steel cup, by means of which a preliminary explosion is made to force the fuel into the combustion space. Referring to Fig. 1, the Hvid cycle is as follows: During the suction stroke pure air is admitted to the cylinder through the intake valve *A*. Fuel valve *B* is opened at the same time with intake valve *A* and some fuel flows into the cup *C* out of the hole *D*, which is uncovered by the opening of valve *B*. The fuel enters cup *C* partly by gravity

*Abstract of a paper presented to the Annual Meeting of the American Society of Mechanical Engineers, Dec. 2-5, 1919.

and partly by suction. The amount of fuel admitted is controlled by the metering pin *E*, which in turn is controlled by the governor. At the same time that the fuel is being taken into the cup, a small amount of fresh air has also gone through an auxiliary air-hole *F*, down past a fluted guide *G* into the cup *C*. At the end of the suction stroke, the fuel valve *B* and the air-intake valve *A* close, valve *B* covering the fuel-admission hole *D*. During the compression stroke all valves are closed and the air admitted to the cylinder on the suction stroke is compressed and its temperature raised. There is now a mass of highly heated air under high pressure in the combustion chamber, and this rushes into the cup *C* through the small holes *H*, near its bottom, until the pressure in the cup is practically equal to the pressure in the combustion chamber. The conditions in the cup are now most favorable to "cracking" the oil, and as it cracks, the lighter and more volatile components are ignited by the high temperature and the resulting high pressure within the cup forces the rest of the oil out into the air in the cylinder. The amount of fuel consumed in the cup per cycle is infinitesimal because there is only a very small amount of air present in the cup to support combustion. As the fuel, in an atomized and vaporous state, comes into contact with the heated air in the combustion space, very rapid combustion takes place and the pressure arising from it drives the piston. As in any four-stroke cycle, the exhaust valve opens and the products of combustion are forced out of the piston on the exhaust stroke.

It is interesting to note the waves in the expansion line of indicator cards taken from these engines. These are typical of all cards taken from Hvid engines and are caused by the introduction of fuel into the combustion chamber in waves. At the moment when the preliminary combustion takes place in the cup, we have in the cup a pressure approximately 800 lb., while in the combustion chamber there is a compression pressure of only about 425 lb. Some fuel is consequently sprayed out of the cup into the combustion chamber by the attempt at pressure equalization. When the fuel comes in contact with the highly heated air in the combustion chamber, the pressure rises in the combustion chamber and falls in the cup until equalized, then

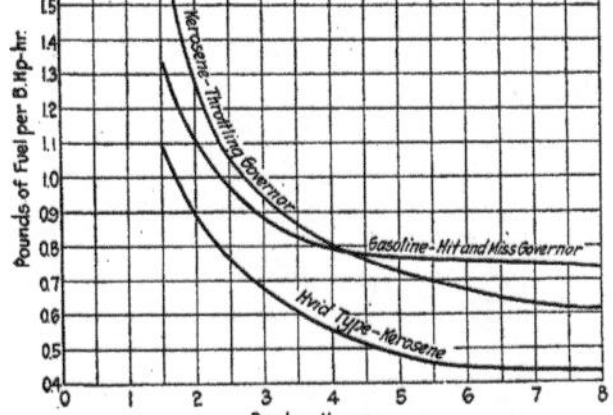

FIG. 2. COMPARATIVE FUEL-CONSUMPTION CURVES

no more fuel can get out of the cup until the piston moves forward and the pressure of the cylinder drops below that in the cup, when more fuel is ejected.

The fuel consumption of small Hvid-type engines is very good, being in general on a par with Diesel engines of large size. In the comparative fuel-consumption curves shown in Fig. 2, the fuel economy of the Hvid engine as compared with two other types of the same size stands out very plainly, particularly at the lower fractional loads. The "hit-and-miss" gasoline-engine test was made by Professor Dickinson, of the University of Illinois, and the throttling governing kerosene-engine test by Mr. MacGregor, of the Hercules Gas Engine Co., under the supervision of Government experts. The Hvid-engine test was made by the writer under the supervision of Professor Roesch, of the Armour Institute. The engine used in each case was a 5¼-in. by 9-in., running 450 r.p.m. and rated at 8 horsepower.

A small 3-in. by 4½-in. Hvid engine was designed to run at 1100 r.p.m., and at first the results to be expected from it were not looked upon with favor because of its high speed. This, however, showed that this engine could be run at a speed as high as 1500 r.p.m. without any appearance of interference with the perfect operation of the Hvid principles. Based on its performance, several engineers connected with the production of Hvid engines raised the normal speed of their engines with very beneficial results. The result of some heat-balance tests is shown in Fig. 3. These tests were made upon a 5¼-in. by 9-in. single-cylinder Hvid engine, which was flexibly

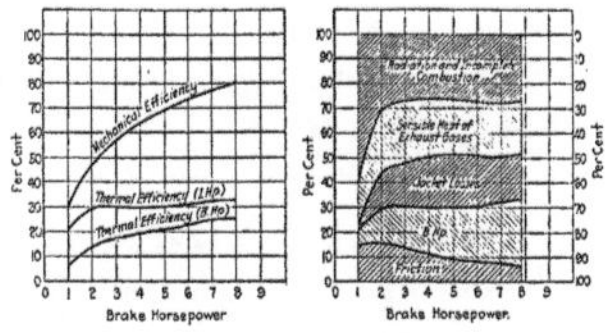

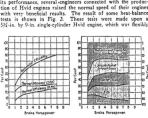

FIG. 3. EFFICIENCY AND HEAT BALANCE CURVES FROM 8-HP. HVID ENGINE

connected with a Sprague electric-cradle dynamometer by means of two universal joints. Arrangements were made for determining the jacket-water loss and the sensible heat in the exhaust gases, the latter by the calorimeter method. The developed and friction horsepower were determined by means of the dynamometer. Indicator cards were taken, but because of the probable errors due to the high pressure involved and the comparatively high speed of the engine, these cards were used merely to study the valve setting and general events of the cycle and not for indicated power measurements. The heat value of the fuel expressed in B.t.u. per pound of kerosene is calculated from the following accepted formula:

$$B.t.u. = 18{,}440 + 40\ (deg.\ B\acute{e} - 10)$$

and is 19,740 B.t.u. for the quality of fuel used in test runs.

The torque characteristics of the Hvid engine are of considerable interest. When a gasoline engine of conventional design is overloaded so that the speed drops below a certain point, its torque drops rapidly, because a certain velocity of air must be maintained through the carburetor to pick up and vaporize the fuel and carry it in the cylinder. In a Hvid engine, however, since the introduction of fuel in the cup and into the cylinder is not dependent upon the velocity of the air taken in, as the speed drops due to overload, more fuel is admitted than at normal speed, because the time element for the introduction of fuel is lengthened. The engine consequently shows remarkable "hanging-on" characteristics. Under these conditions it is very wasteful of fuel without a doubt, but there are certain conditions where this "bulldog" characteristic is desirable even though it be at the expense of fuel economy.

In conclusion the writer says that while the Hvid engine has by no means reached its ultimate state of development, it nevertheless possesses a number of wonderful characteristics which ought to attract many internal-combustion engineers by the possibility they hold out of helping to solve some of our fuel problems.

At hearing by Joint Congressional Commission on Reclassification of Salaries to committee representing Federal employees it was disclosed that a "turnover" in personnel among scientific and technical workers of the Government amounting in some offices to one hundred per cent. a year prevailed.

Steam Boiler and Flywheel Service Bureau Accepts Autogenous Welding

The committee referred to is made up of engineers of all the insurance companies writing steam-boiler insurance in the United States. For the purpose of obtaining uniform practice in the acceptance of autogenous welding the sub-committee reports:

By "autogenous welding" is meant any form of welding by fusion; that is, where the metal of the parts to be joined or added metal used for the purpose is melted and flowed together to form the weld. Such welding is accomplished by the oxyacetylene, hydrogen or other flame processes, or by the electric arc; no distinction is made between any of these processes. The general rule to govern the acceptance of such welds in insured vessels is that prescribed by the Boiler Code of the American Society of Mechanical Engineers, Par. 186, as follows:

> Autogenous welding may be used in boilers in cases where the strain is carried by other construction which conforms to the requirements of the Code, and where the safety of the structure is not dependent upon the strength of the welds.

The following illustrations will serve to point out where such work should be accepted or rejected.

1. Any autogenous weld of reasonable length will be permitted in a stay-bolted surface or one adequately stayed by other means so that should the weld fail, the parts would be held together by the stays. It is necessary for the inspector to use judgment in interpreting the meaning of "reasonable length" as given above, since it may vary in different cases. In the average case it should be not more than three feet. Autogenous welding will not be accepted in unsupported surface.

2. The edges of the inner and outer sheets of vertical firebox boilers or boilers of the locomotive type may be joined by autogenous welding to from the door openings, if the surrounding surfaces are thoroughly stayed. This would also apply to other openings of a similar character in such surfaces.

3. For low-pressure plate-steel boilers operated at a pressure not exceeding 15 lb. per sq.in., or for higher pressures in unfired vessels subjected to water pressure only, rectangular headers may be autogenously welded at the edges if the sheets are properly held together by stays. Autogenous welding of cracks and fractures in cast-iron boilers will not be permitted.

4. Fire cracks in girth seams extending from the edge of the plate to the rivet hole may be autogenously welded provided the cracks are properly prepared by cutting out the metal at the crack in the form of a letter V to permit fusion through the entire thickness of the plates. Similar cracks in girth seams located between the rivet holes may also be autogenously welded, provided the cracks do not extend beyond the edge of the lap of the inner plate. In the latter class of cracks it is advisable to drill a hole not exceeding ⅜ in. in diameter at the end of the crack before the weld is made. Cracks extending from rivet hole to rivet hole on girth seams cannot be welded. Calking edges of girth seams may be built up by autogenous welding where the original section of the metal between rivet holes and calking edge to be built up, averages equivalent to one-fourth the diameter of the rivet hole and the portion of calking edge to be replaced does not exceed 30 in. in length in a girthwise direction. In all repairs to girth seams by autogenous welding the rivets must be removed over the portions to be welded and for a distance of at least 6 in. at each end beyond such portions. After repairs are made the rivet holes should be seamed before the rivets are redriven.

5. Stayed sheets which have corroded to a depth of not more than 40 per cent. of their original thickness, may be reinforced or built-up by autogenous welding. In such cases the stays shall come completely through the reinforcing metal so as to be plainly visible to the inspector.

6. Where tubes enter flat surfaces and the tube sheets have been corroded or where cracks exist between the tube ligaments, autogenous welding may be used to reinforce or repair such defects. The ends of such tubes may be autogenously welded to the tube sheets. The above-mentioned repairs for tube sheets and the welding in of tubes in the sheets are not to be permitted where such sheets form the shell of a drum or boiler, such as in the case of Stirling type of boiler.

7. When external corrosion has reduced the thickness of plate around handholes to not more than 50 per cent. of the original thickness and for a distance not exceeding 2 in. from the edge of the hole, the plate may be built up by autogenous welding.

8. Pipe lines will be accepted where the flanges or other connections have been welded autogenously, provided the work has been performed by a reputable manufacturer and the parts properly annealed before being placed in position. Such welding, when made with the part in place and unannealed, will not be acceptable.

9. Autogenous welded-in patches in the shell of a boiler will not be acceptable regardless of the size of such patches. Autogenous welding of cracks in the shell of a boiler—except those specified in Par. 4—regardless of the direction in which they may lie, will not be permitted unless such welding is only for the purpose of securing tightness and the stresses on the parts is fully cared for by properly riveted-on patches or straps placed over the weld. The plates at the ends of joints may be welded together for tightness, provided the straps or other construction is ample to care for the stresses on the parts so welded.

10. Re-ending or piecing of tubes for either fire-tube or water-tube boilers by the autogenous process will not be permitted.

Electric Undertakings in Japan

The accompanying tables and charts are arranged from data given in the Statistical Report of Electric Undertakings in Japan, compiled by the Director General of Electric Exploitations, Department of Communications, Tokyo, Japan. This report is dated June, 1919, and is a summary of the chief statistics showing the condition of existing electric undertakings at the end of 1917.

Table I gives the number of undertakings by Kind of Primary Power for various classes of plants, while Table II gives

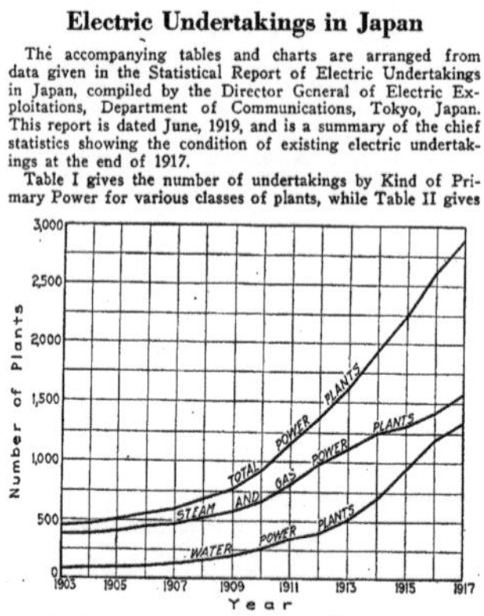

FIG. 1. GROWTH IN NUMBER OF ELECTRIC UNDERTAKTAKINGS IN JAPAN FROM 1903 TO 1917

the kilowatt capacities under the same classifications. It is interesting to note the relative number and capacity of the water-power plants as compared with the steam and gas plants.

The charts show the growth of the industry since 1903. Fig. 1 gives the number of plants and Fig. 2 their capacity at the end of each year from 1903 to 1917. The values in Fig. 2 include the reserve capacities, so they are somewhat higher at the end of 1917 than those given in Table II.

The report further classifies the plants as to voltage. There were 1865 plants having 51,170 kw. capacity carrying "low" voltage or less than 600 volts d.c. or 300 volts a.c.; 843 plants with 211,600 kw. carrying "high" voltage or higher than "low" voltage and lower than 3500 volts, and 183 plants with 490,000 ky. capacity carrying "extra high" voltage of more than 3500.

There were 37,120 miles of electric line using 132,794 miles of wire. The electric railways covered 434 miles and used 4539 cars.

Electric-light consumers numbered 4,243,000, or 41.2 per cent. of the number of households. They used over ten million lamps having a total equivalent electric power of 173,000 kw. Power consumers, including isolated plant users as well as public-service users, used 85,860 motors requiring practically 600,000 kw. Other electric appliances required 100,000 kw. additional.

The average price to small consumers for electric light was 9.9c per kw.-hr. from water-power plants and 11c from steam

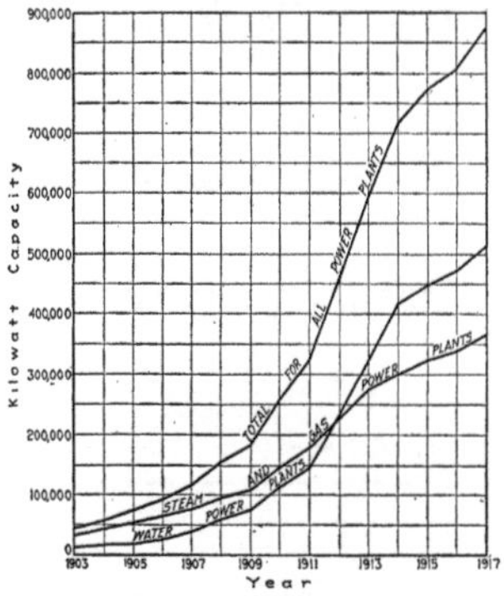

FIG. 2. GROWTH IN CAPACITY OF ELECTRIC UNDERINGS IN JAPAN FROM 1903 TO 1917

and gas plants. The corresponding average power rates were 3.15c and 4.46c respectively. Including traction plants there was a total paid-up capital of $180,650,000 in public-service plants on which amount a profit of 10.7 per cent. was earned.

TABLE I. NUMBER OF UNDERTAKINGS USING WATER, STEAM AND GAS POWER

Undertakings	Water	Steam	Gas	Total
Electric supply	360	44	93	497
Electric traction	23	5	..	28
Electric supply and traction	32	15	1	48
Isolated plants	869	786	533	2188
Official plants	47	73	10	130
Total	1331	923	637	2891

TABLE II. KILOWATT CAPACITY FROM WATER, STEAM AND GAS POWER

Undertakings	Water	Steam	Gas	Total
Electric supply	338,682	43,675	5,976	388,332
Electric traction	180	1,000	....	1,180
Electric supply and traction	71,624	68,809	655	141,088
Isolated plants	56,357	123,782	15,567	195,706
Official plants	400	20,630	5,665	26,695
Total	467,243	257,896	27,863	753,002

National Fuel-Inspection System

A fuel-inspection system that will give the consumer and the producer of coal accurate information concerning the commodity is proposed by the Secretary of the Interior. In an estimate recently transmitted to Congress, he asks for an appropriation of $725,000. The system is to be administered by the Bureau of Mines. The work is to be under the immediate direction of a chief coal inspector, who is to receive a salary of $5000. Among the other positions which are specified in the estimate are: Assistant chief coal inspector, $4000; 6 district coal inspectors, $4000; 100 field coal inspectors, $2400. In explaining the need for the appropriation, Secretary Lane said:

Inspection, sampling analysis, and certification of coal should be made by the Government. This practice is followed in other commodities and would be a benefit to the producers of clean coal, the consumers, and to the interest of foreign commerce. This will assure the consumer the grade of coal which he pays for, protect the interest of those preparing clean coal, without discouraging the mining of poorer grades of coal and should exert an influence toward good coal preparation.

This coal inspection system proposed by the Bureau of Mines contemplates as a salient feature advice as to the quality of coal shipped. It is proposed that each mining company set its own standard of quality, consistent with the particular vein preparation, and the market which the business affords, and that the Government shall publish such standard and certify as to whether such standard is being maintained by the mining companies. There is not contemplated a certification as to each and every shipment, but the system does provide for the accurate sampling of full carload lots and for the inspection and sampling at irregular intervals of a sufficient number of cars of coal to indicate whether the declared standard of the mining company is being maintained. The coal-mining companies would coöperate and enter the system and be privileged to advertise that their product was from a mine whose standard of preparation was certified by the Government. In case the coal shipment was below the standard, the mine owner would be promptly advised of the fact, and where possible, the consumer would receive a notice to the same effect. If the coal continues to fall below the standard, it would be removed from the certified list established by the Government and the mine would be required to declare a new and different standard which its product could meet.

The Ohio Board of Boiler Rules, at its last regular meeting held at Columbus, Ohio, Feb. 12-13, 1919, unanimously adopted the following resolutions: Whereas, Since the passage of the Ohio Boiler Inspection Law by an Act of Legislature, May 31, 1911, miniature boilers of less than two horsepower have appeared in large numbers; and Whereas, These miniature boilers are used in connection with tire vulcanizers, clothes presses, matrix tables, etc.; and Whereas, It is the opinion of this board that the purpose of the boiler inspection law is to apply only to steam boilers containing a volume of water sufficient to cause a serious explosion; Therefore, Be It Resolved, That Section 18 (1058-23 Ohio Code) be interpreted to exclude from shop inspection and stamping, miniature boilers referred to in Part 2, Sec. 4, p. 43, 1918 edition Ohio Boiler Rules.

SWAT THE WASTER

THE COAL MARKET

BOSTON—Current prices per gross ton f.o.b. New York loading ports:

Anthracite

	Company Coal
Egg	$8.20@$8.65
Stove	8.45@ 9.05
Chestnut	8.55@ 9.05
Pea	7.05@ 7.40
Buckwheat	3.60@ 3.90
Rice	2.30@ 2.50
Barley	1.25

Bituminous

	Clearfields	Cambrias and Somersets
F.o.b. mines, net tons	$2.85@$3.35	$3.15@$3.50
F.o.b. Philadelphia, gross tons	5.05@ 5.60	5.35@ 5.90
F.o.b. New York, gross tons	5.40@ 5.95	5.75@ 6.25
Alongside Boston (water coal), gross tons	7.00@ 7.75	7.60@ 8.00

Pocahontas and New River are practically off the market for coastwise shipment, but are quoted at $6.25@$7.00.

NEW YORK—Current quotations, White Ash, per gross tons, f.o.b. Tidewater, at the lower ports are as follows:

Anthracite

	Company Coal
Broken	$7.80@$8.25
Egg	8.20@ 8.65
Stove	8.45@ 9.05
Chestnut	8.55@ 9.05
Pea	7.05@ 7.40
Buckwheat	5.15
Rice	4.50
Barley	4.85
Boiler	4.85

Bituminous

	Spot
South Fork (best)	$3.25@$3.50
Cambria (best)	3.00@ 3.25
Cambria (ordinary)	2.60@ 2.90
Clearfield (best)	3.00@ 3.25
Clearfield (ordinary)	2.60@ 2.90
Reynoldsville	2.85@ 3.00
Quemahoning	3.25@ 3.50
Somerset (medium)	3.00@ 3.25
Somerset (poor)	2.50@ 2.75
Western Maryland	2.50@ 2.75
Fairmont	2.25@ 2.50
Latrobe	2.60@ 2.90
Greensburg	2.75@ 3.00
Westmoreland ¾ in.	3.40@ 3.50
Westmoreland run-of-mine	2.75@ 3.00

PHILADELPHIA—Anthracite prices are practically the same as those listed above for New York. Bituminous coal prices vary according to district from which they are mined. For ordinary slack the price is $2.45@$2.55; lump, $3.00@$3.35.

BUFFALO—

Anthracite

	On Cars, Gross Ton	At Curb, Net Ton
Grate	$8.55	$10.80
Egg	8.80	10.65
Stove	9.00	10.85
Chestnut	9.10	10.85
Pea	7.75	9.30
Buckwheat	5.70	7.75

Bituminous

Allegheny Valley	$4.80
Pittsburgh	4.65
No. 8 Lump	4.65
Mine Run	4.80
Slack	4.10
Smokeless	4.60
Pennsylvania Smithing	5.70

CLEVELAND—Prices of coal per net ton delivered in Cleveland are:

Anthracite

Egg	$11.75@$11.90
Chestnut	12.00@ 12.05
Grate	11.75@ 11.90
Stove	11.90@ 12.10

Pocahontas

Mine-run	$7.50

Domestic Bituminous

West Virginia splint	$8.50
No. 8 Pittsburgh	$6.60@ 6.90
Massillon lump	8.25@ 8.50
Coshocton lump	7.15

Steam Coal

No. 6 slack	$5.25@$5.50
No. 8 slack	5.10@ 5.50
Youghiogheny slack	5.25@ 5.50
No. 8 ¾ in.	5.70@ 6.00
No. 6 mine-run	5.25@ 5.50
No. 8 mine-run	5.75

Only coal available is mine-run Pocahontas.

MIDDLE WEST—Chicago quotations, F.o.b. cars at mine:

	Springfield Carterville Williamson Franklin Saline Harrisburg	Fulton Peoria	Grundy La Salle Bureau Will
Lump	$2.55@$2.70	$2.95@$3.10	$3.25@$3.40
Washed	2.75@ 2.90	....	3.45@ 3.60
Mine Run	2.35@ 2.50	2.75@ 2.90	3.00@ 3.15
Screenings	2.05@ 2.20	1.35@ 1.50	2.75@ 2.90

New Construction

PROPOSED WORK

Mass., Amherst—The Massachusetts Agricultural College will soon award the contract for the construction of a 2 story, 60 x 120 ft. memorial building on the Campus. A steam heating system will be installed in same. Total estimated cost, $150,000. James H. Ritchie, 8 Beacon St., Boston, Arch.

Mass., Brookline (Boston P. O.)—Louis Shapiro, 27 School St., Boston, will build a 2 story, 85 x 270 ft. auto sales building on Commonwealth Ave. A steam heating system will be installed in same. Total estimated cost, $250,000. Work will be done by day labor.

Mass., New Bedford—The City Hospital Commission received lowest bid for the installation of a steam heating system in the proposed hospital, which includes a 3 story, 46 x 175 ft ward, 2½ story, 46 x 180 ft. operating building and a 2 story, 50 x 160 ft. laundry, on Mt. Pleasant St., from the Power Heating & Ventilating Co., 49 Travers St., Boston, $121,750.

Mass., Wakefield—Lee & Hewett, Archs. and Engrs., 1123 Broadway, New York City, will soon award the contract for the construction of a 3 story factory addition for Heywood Bros. & Wakefield Co., Gardner. A steam heating system will be installed in same. Total estimated cost, $150,000.

Conn., Hartford—Bliss, Hill & Steiger are having plans prepared for the construction of a 6 story department store. A steam heating system will be installed in same. Buchman & Kahn, 56 West 45th St., New York City, Archs. and Engrs.

Conn., New Haven—Brown & Von Beren, Archs., Chamber of Commerce Bldg., will soon receive bids for the construction of an 8 story, 80 x 80 ft. store and storage building on Meadow St., for John Kilfeather, Meadow St. A steam heating system will be installed in same. Total estimated cost, $125,000.

Conn., Thomaston—The Seth Thomas Clock Co., c/o Lockwood, Green & Co., Archs. and Engrs., 101 Park Ave., New York City, will soon receive bids for the construction of a 2 story, 60 x 120 ft. and a 4 story, 60 x 120 ft. factory buildings. A steam heating system will be installed in same.

N. Y., Brooklyn—B. W. Dorfman, Arch., 26 Court St., is preparing plans for the construction of a 5 story factory on Grand Ave. and Dean St. A steam heating system will be installed in same. Total estimated cost, $200,000. Owner's name withheld.

N. Y., College Point—The American Hard Rubber Co. is having plans prepared for the construction of a 1 story, 120 x 130 ft. factory addition. A steam heating system will be installed in same. Walter Kidde, 140 Cedar St., New York City, Arch. and Engr.

N. Y., Long Island City—The B. J. Johnson Soap Co., c/o Lockwood, Greene & Co., Archs., 101 Park Ave., New York City, is having plans prepared for the construction of a 4 story warehouse. A steam heating system will be installed in same. Total estimated cost, $250,000.

N. Y., Milbrook—H. Ross, 47 West 34th St., New York City, is having plans prepared for the construction of a group of 1 and 2 story buildings for a sanitarium in Duchess County. A steam heating system will be installed in same. Total estimated cost, $300,000. Alfred Bodker, 62 West 45th St., New York City, Arch. and Engr.

N. Y., Mineola—The Knickerbocker Ice Co., 1480 Broadway, New York City, will soon award the contract for the construction of a 1 story ice plant. Estimated cost, $110,000. W. Mortensen, 209 West 76th St., New York City, Arch. and Engr.

N. Y., New York—The Board of Education, 500 Park Ave., is having plans prepared for the construction of a 5 story, 120 x 195 ft. school on St. Anns Ave. and Hagney Pl. A steam heating system will be installed in same. Total estimated cost, $500,000. C. B. J. Snyder, Municipal Bldg., Arch. and Engr.

N. Y., New York—The Consolidated Amusement Enterprises, c/o Arena Theatre, 8th Ave. and 40th St., is having plans prepared for the construction of a theatre on West 50th St. and 8th Ave. A steam heating system will be installed in same. Total estimated cost, $130,000. S. B. Eisendrath, 18 East 41st St., Arch. and Engr.

N. Y., Rochester—The E. W. Hamilton Manufacturing Co. plans to construct a 2 story, 55 x 195 ft. factory on Hallenbeck St. An electric power plant and machinery for the manufacture of paper boxes will be installed in same. Total estimated cost, $25,000.

N. Y., Syracuse—The Board of Contract and Supply will receive bids about January 1 for the construction of a 3 story school on South Alvord St. A steam heating system will be installed in same. Total estimated cost, $300,000.

N. Y., Syracuse—The Mills Oil Co., 263 Walton St., plans to build 4 or 5 buildings, including a boiler house, office building, stable, garage, storage tanks, etc. Total estimated cost, $200,000.

N. Y., New York—The Simrud Holding Corporation, 135 Broadway, will build several stores on a 74 x 94 ft. site on Fordham Rd. and Walton Ave. A steam heating system will be installed in same. Total estimated cost, $125,000. Work will be done by day labor.

N. Y., Syracuse—The Battery Service Co., Inc., 562 East Genesee St., will receive bids about March 1 for the construction of a 1 story battery service building on East Genesee St. Estimated cost, $50,000. M. E. Granger, Gurney Bldg., Arch. and Engr.

N. Y., Wards Island—The State Hospital Commission, Capitol, Albany, received bids December 17 for the installation of a heating system in the Manhattan State Hospital, here, from Leslie & Tracy, Inc., 5 Columbus Circle, $6300; Gillis & Geohegan, 537 West Broadway, $7180; Charles H. Darmstadt 352 West 43rd St., $7372. Contractors all of New York City.

N. J., Asbury Park—The Rydon Tire & Rubber Co., c/o E. A. Arend, Arch. and Engr., 105 West 40th St., New York City, will soon award the contract for the construction of a factory. A steam heating system will be installed in same. Total estimated cost, $100,000.

N. J., New Brunswick—The city plans to issue $10,000 bonds for the installation of a water tube boiler in the pumping station and $8000 bonds for a water wheel, rotary pump, etc.

Del., Wilmington—The Daily News Co., 511 Market St., is having plans prepared for the construction of a 6 story publishing plant. A steam heating system will be installed in same. Dennison & Hirons, 477 5th Ave., New York City, Archs. and Engrs.

Md., Baltimore—Louis Miller, 5 Lloyd St., will receive bids about January 15 for the construction of a 2 story, 45 x 60 ft. dairy and cold storage building at 5-7-9 Lloyd St. Separate bids will be received for installing steam heating system, hand power elevator and refrigerators, etc. Total estimated cost, $50 000. G. R. Gallis, Melvin Ave., Catonsville, Arch.

Md., Crisfield—The McGready Memorial Hospital, c/o Dr. R. R. Norris, is having plans prepared for the construction of a 1 and 2 story hospital on Main St. A steam heating system will be installed in same. Total estimated cost, $100,000. H. G. Jory, 1409 Munsey Bldg., Arch.

Tenn., Memphis—Jones & Furbringer, Archs., Porter Bldg., wil soon award the contract for the construction of a sanitarium for tuberculosis patients at Oakville, including a power and heating plant, etc., for Memphis and Shelby Counties. Total estimated cost, $500,000. W. B. Cleveland, Chn., Board of Trustees.

Ohio, Bedford—The Board of Education will receive bids about February 1 for the construction of a 2 story, 80 x 164 ft. high school. A steam heating system will be installed in same. Total estimated cost, $200,000. Hamilton & Watterson, Guardian Bldg., Cleveland, Archs.

Ohio, Cleveland—The Citizens Savings & Trust Co. and the Union Commerce Bank had plans prepared for the construction of a 20 story, 145 x 258 x 385 ft. bank and office building on East 9th St. and Euclid Ave. Four 300 h.p. boilers and stokers will be installed in same. Total estimated cost, $8,000,000. Graham-Anderson-Probst & White, Railway Exchange Bldg., Chicago, Engrs.

Ohio, Cleveland—The city received bids for the construction of a sewage disposal plant in the eastern section of the city, from the American Construction Co., Marion Bldg., $258,957; McHugh Brothers, West 37th St., $338,730; Masters-Mullen Construction Co., Electric Bldg., $400,486. One centrifugal pump with 1-2½ h.p. and 1-5 h.p. motors will be installed in same.

Ohio, Cleveland—The Hinkel Motor Truck Corporation, 6521 Euclid Ave., has purchased a site and plans to build a 2 story, 180 x 200 ft. auto salesroom and garage on East 71st St. and Carnegie Ave. A steam heating system will be installed in same. Total estimated cost, $100,000. Whitworth & Johnson, Engineers Bldg., Archs.

Ohio, Cleveland—J. W. Holcomb, 916 Citizens Bldg., plans to build a 4 story commercial building at 3812 Euclid Ave. A steam heating system will be installed in same. Total estimated cost, $100,000.

Ohio, Cleveland—The Bender Auto Body Co. will soon award the contract for the construction of a 1 story, 100 x 250 ft. factory on West 65th St. and Denison Ave. Two 250 h.p. boilers with stokers will be installed in same. Total estimated cost, $150,000. R. M. Hulett, Fire Bldg., Arch. Address H. Bender, c/o Northern Blower Co., Barberton Ave.

Ohio, Cleveland—A. B. Friedl, 3328 Vega Ave., has purchased a site on East 93rd St. and Kinsman Rd., and plans to construct a 3 story commercial building on same. A steam heating system will be installed in same. Total estimated cost, $100,000. W. S. Lougee, Marshall Bldg., Arch.

Ohio, Silver Lake—The village will soon award the contract for the construction of two 1 story, 17 x 17 ft. pump houses and is in the market for two deep well, 175 gal. per min. pumps, one pressure tank and seven 8-inch gate valves. Total estimated cost, $15,000. E. A. Tewksbury, Clk. R. Winthrop Pratt, Hippodrome Bldg., Cleveland, Engr.

Mich., Battle Creek—The Kellog Toasted Corn Flake Co. plans to construct a 3 story, 100 x 218 ft. dormitory for employes. A steam heating system will be installed in same. Total estimated cost, $100,000.

Mich., Detroit—The Board of Education, 50 Broadway, will soon award the contract for the construction of a 2 story, 84 x 134 ft. school building on Vancourt and Milford Aves., cost $250,000; also for the construction of a field school No. 1154, Agnes and Field Aves., $200,000, and Davidson School No. 1147, Roman Ave., near Davidson Ave., $200,000. Malcolmson, Higginbotham & Palmer, 405 Moffat Bldg., Archs.

Mich., Detroit—Louis Kamper, Arch., Book Bldg., will soon award the contract for the construction of a 1 story, 110 x 110 ft. pavilion on East Jefferson Ave. and Grand Blvd., for the Kling Products Co., 1424 East Jefferson Ave. A steam heating system will be installed in same. Total estimated cost, $70,000.

Mich., Detroit—The Associated Amusements Co., c/o Christian W. Brandt, Arch., Kresge Bldg., is having plans prepared for the construction of a 2 story, 120 x 135 ft. theatre on Hamilton Blvd. Steam heating equipment and a forced ventilation system will be installed in same. Total estimated cost, $250,000.

Mich., Detroit—The Board of Fire Commissioners received bids for the installation of heating, ventilating and plumbing systems in the proposed 2 story, 40 x 80 ft. fire engine house on 12th and La Belle Sts., from O'Connor Bros., 34 Roland Bldg., $9209; Irvine & Meier, 2289 Woodward Ave., $10,270; W. J. Rewoldt, Owen Bldg., $10,553.

Mich., Detroit—The Clayton & Lambert Manufacturing Co., Beaubien St., plans to construct a 1 and 2 story factory for the manufacture of automobile parts, along the Detroit Terminal Railway. Plans include a power plant. Total estimated cost, $500,000.

Mich., Detroit—The Columbia Motors Corporation, 1256 East Jefferson Ave., engaged Horace H. Lane, Arch. and Engr., Dime Bank Bldg., to prepare plans for the construction of a 2 and 3 story, 300 x 720 ft. automobile factory on Fort and Boyd Sts. Steam heating equipment will be installed in same. Total estimated cost, $600,000.

Mich., Detroit—Albert Kahn, Louis Kahn and Julius Kahn, Marquette Bldg., have purchased a site on Cass Ave. and Peterboro St., and will prepare plans for the construction of a 4 story, 130 x 185 ft. commercial building on same. Steam heating equipment will be installed in same. Total estimated cost, $400,000. Albert Kahn, Arch.

Mich., Eaton Rapids—The Board of Education engaged S. D. Butterworth, Arch., 432 Tussing Blk., Lansing, to prepare preliminary plans for the construction of a 2 story high school. Steam heating equipment will be installed in same. Total estimated cost, $200,000.

Mich., Grand Rapids—Willard M. Burleson, 148-150 Fulton St., E., engaged Osgood & Osgood, Archs., Herald Bldg., to prepare plans for the construction of an 11 story, 105 x 146 ft. sanitarium on Island St. and Jefferson Ave. Steam heating equipment will be installed in same.

Mich., Redford—The Michigan State Auto School, Woodward Ave., Detroit, is having plans prepared for the construction of eight buildings, including a 1 story, 80 x 150 ft. power house and pumping station. Total estimated cost, $2,000,000. Christian W. Brandt, Kresge Bldg., Detroit, Arch.

Ill., Chicago—W. G. Uffendel, Arch., 39 South St., will soon award the contract for the construction of a manufacturing plant, including a 2 story, 125 x 190 ft. main building, a 2 story, 47 x 95 ft. office building, and a 1 story, 47 x 66 ft. boiler house and garage, at 2701 Greenview Ave., for the Economy Fuse & Manufacturing Co., 328 West Kinzie St. A steam heating system will be installed in same. Total estimated cost, $150,000.

Ill., Chicago—Charles S. Frost, Arch., 105 South La Salle St., will soon award the contract for the construction of a 2 story, 150 x 175 ft. printing plant at 1101 North La Salle St., for the Uniform Printing & Supply Co., 351 West Chicago Ave. A steam heating system will be installed in same. Total estimated cost, $225,000.

Ill., Chicago—Holabird & Roche, Archs., 104 South Michigan Ave., will receive bids in January for the construction of a 3 story, 160 x 180 ft. hotel addition on Clark and Randolph Sts., for the Hotel Sherman Co., Hotel Sherman. Additions will be built to the present steam heating system in same. Total estimated cost, $500,000.

Ill., Jacksonville—The city plans to construct additions to the existing steam power plant. Estimated cost, $40,000. This work is in connection with the improvements to the waterworks system. C. M. Borchers, Mayor.

Wis., Rio—The city will soon award the contract for the construction of an 18 x 60 ft. power house and for furnishing one 15 h.p. engine, one 150 gal. deep well pump, one 10 in. well, 200 to 300 ft. deep, one 40,000 gal. elevator tank and one 80 ft. tower. Total estimated cost, $25,000. W. Colins, Clk. W. Kirchoffer, Madison, Engr.

Wis., Manitowoc—George Brothers, 913 South 8th St., have engaged W. J. Roueber, Arch., to prepare plans for the construction of a 3 story, 100 x 125 ft. theatre building on South 8th St. A steam heating system will be installed in same. Total estimated cost, $150,000.

Neb., Scribner—The Scribner Artificial Ice Co. plans to build an ice plant and an electric plant. Total estimated cost, $65,000.

Minn., Aitkin—The city received bids for furnishing electric light and waterworks equipment from J. G. Robertson, Wabash and Eustis Sts., St. Paul, $9489.

Minn., New Ulm—The city plans to construct a 90 x 90 ft. power plant. Estimated cost, $50,000. A. J. Mueller, Supt.

Minn., Duluth—F. G. Beamsley, Cashier, will receive bids until February 1 for the construction of a 2 story, 50 x 140 ft. bank and office building on Superior st. and 20th Ave., W., for the Duluth State Bank. A steam heating system will be installed in same. Total estimated cost, $75,000. Halstead & Sullivan, Palladio Bldg., Archs. and Engrs.

Minn., East Grand Forks—J. A. Noonan, Secy. of Independent School District No. 3, will receive bids until January 15 for the construction of a 3 story, 130 x 170 ft. high and grade school on Kiltson Ave. and 6th St. A steam heating and mechanical ventilating system will be installed in same. Total estimated cost, $375,000. Bert & Beck, Grand Forks, N. D., Archs. and Engrs.

Minn., Robbinsdale—The Board of Education will soon receive bids for the construction of a 2 story school. A steam heating system will be installed in same. Total estimated cost, $125,000. C. A. Garner, Clk. H. H. Livinston, Arch.

Minn., St. Paul—The Emporium Mercantile Co., Robert St., between 7th and 8th Sts., will soon award the contract for the construction of a 6 story, 145 x 160 ft. department store on 7th and Robert Sts. Separate bids for installing steam heating system will be received. Total estimated cost, $650,000. C. L. Pillsbury Co., 716 Capital Bank Bldg., Engrs. Buechner & Orth, 500 Shubert Bldg., Archs.

Tex., Houston—The Hulbert-Still Electric Co. received only bid for the construction of a 1 story, 75 x 126 ft. service station at McKinney and San Jacinta Sts., from Andrew Hess, 4630 Park Drive, $22,000.

Colo., Boulder—The University of Colorado will soon award the contract for the construction of a 4 story woman's dormitory with two wings, 1 story, 58 x 162 ft. and 1 story, 32 x 193 ft., on the campus. A Warren Webster steam heating system will be installed in same. Total estimated cost, $250,000. F. H. Wolcott, 526 Kittderge Bldg., Denver, Secy. of the Board of Regents. W. E. and A. A. Fisher, 337 Century Bldg., Denver, Supervising Archs.

Colo., Mineral Hot Springs—The Mineral Hot Springs Resort Co. plans to construct several buildings. An electric light plant will be installed in same. Total estimated cost, $100,000. W. N. Bowman, Central Bank Bldg., Denver, Arch.

Colo., Montrose—Montrose County plans to build a 3 story, 65 x 130 ft. court house. A vacuum steam heating system will be installed in same. Total estimated cost, $150,000. W. N. Bowman, Central Savings Bank Bldg., Denver, Arch.

Idaho, Challis—W. W. Adamson, hydro-electric engineer, Boise, has been granted authority by the Idaho State Public Utilities Commission to construct a hydro-electric plant and system here.

Ore., Bend—The Bend Water, Light & Power Co. has filed an application with the state engineer for permission to appropriate 50 second ft. of water from Tumalo Creek, for the development of 4225 h.p. The company also plans to construct dams and canal. Total estimated cost, $229,000.

Cal., Fresno—The Board of Education will soon award the contract for the installation of a heating system, etc., in the new John Muir, Arlington Heights, Webster, Longfellow and Kirk School buildings. J. R Fontaine, Secy.

Cal., San Jose—The Board of Supervisors will soon award the contract for digging a well and installing a pump and motor at the County Almshouse. Total estimated cost, $7000. H. A. Pfister, County Clk.

Cal., Ventura—The Sespe Light & Power Co., Merchants Trust Bldg., Los Angeles, plans to build hydro-electric plants and irrigation projects on the Sespe and Pivu Rivers; 1st unit to include 2 reservoirs of 51,000 acre ft., 46,000 ft. power conduit, 3000 ft. pressure pipe line, 1000 ft. head, 10,000 kw. hydro-electric plant, to be increased to 18,000 kw. when 2nd unit is built; 2nd unit, 100,000 acre ft. storage reservoirs, 90,000 ft. conduit, 4000 ft. pressure pipe line 2300 ft. head, and 8000 kw. plant; 3rd unit, additional storage to deliver 100,000 acre ft. per annum for irrigation. Cost for preliminary surveys about $100,000. F. Buren, Pres.

Ont., Hagersville—D. N. Almas is in the market for two 5 h.p. electric motors in his proposed seed cleaning building.

Ont., Almonte—The taxpayers will vote on by-law to raise $25,000 for the purchase of the Wylie Electric Power Plant. About $10,000 will be spent in development. W. Watchorn, Engr.

Ont., Alton—The Beaver Knitting Mills, Ltd., are in the market for one 5 h.p. and one 3 h.p. 60 cycle, 220 volt single phase motor.

Ont., London—The Bell Telephone Co., 118 Notre Dame St., W., Montreal, plans to construct a 1 story, 60 x 80 ft. addition to the telephone exchange building here. The present steam heating system in same will be enlarged. Total estimated cost, $100,000. W. J. Carmichael, 29 Hospital St., Montreal, Arch.

Ont., Mallorytown—The Consolidated School Board is having plans prepared for the construction of a 2 story school. A steam heating system and an electric operated ventilating fan will be installed in same. Total estimated cost, $160,000. B. Dillen, King St., Brockville, Arch.

Ont., Newmarket—The County of York, 57 Adelaide St., E., Toronto St., is having plans prepared for the construction of a 3 story, 80 x 150 ft. industrial home on Eagle St. A vacuum steam heating system will be installed in same. Total estimated cost, $28,500. E. A. James Co., Ltd., 36 Toronto St., Toronto, Engr.

Ont., Toronto—The city will soon receive bids for the installation of four 3 stage capacity, electrically driven pumps. Estimated cost, $75,000. R. G. Harris, City Hall, Engr.

Ont., Sudbury—The High School Board plans to construct a 3 story high school. A steam heating and ventilating system will be installed in same. Total estimated cost, $200,000.

Ont., Windsor—J. & J. J. Allen, Allen Theatre, Richmond St., Toronto, will soon award the contract for the construction of a moving picture theatre. A steam heating and blower ventilating system, etc., will be installed in same. Total estimated cost, $200,000. C. H. Crane, 2325 Dime Bldg., Detroit, Mich., Arch.

CONTRACTS AWARDED

Mass., Cambridge—Lever Brothers Co., manufacturers of soaps, have awarded the contract for the construction of a 2 story, 86 x 134 ft. boiler house at their plant to the Stone & Webster Engineering Corporation, 147 Milk St., Boston.

Conn., Greenwich—The town has awarded the contract for the construction of a 2½ story, 143 x 155 ft. school at Greenville, to the Rangeley Construction Co., 56 West 39th St., New York City. A steam heating system will be installed in same. Total estimated cost, $250,000.

Conn., Hartford—The South School District has awarded the contract for the construction of a 3 story, 87 x 87 ft. school addition on Wethersford Ave., to Wise & Upson, 36 Pearl St. A steam heating system will be installed in same. Total estimated cost, $210,000.

Conn., Hartford—C. H. Talcott, 19 Woodland St., has awarded the contract for the construction of a department store to the Republic Fireproofing Co., 26 Cortland St., New York City. A steam heating system will be installed in same.

Conn., New Haven—Kusterer & Smith, 375 Yale Ave., have awarded the contract for the construction of a 6 story, 80 x 116 ft. and 60 x 80 ft. business buildings on Court St., to DeBussey-Kusterer Co., 116 Church St. A steam heating system will be installed in same. Total estimated cost, $100,000.

Conn., Waterbury—The American Pin Co., Waterville, has awarded the contract for installing a steam heating system in the proposed 4 story, 50 x 204 ft. brass factory at Waterville, to M. J. Daly & Sons, 543-555 Bank St., Waterbury. Total estimated cost, $200,000.

N. Y., Long Island City—The Queens Electric Light & Power Co., 347 Central Ave., Far Rockaway, has awarded the contract for the construction of a 1 story, 40 x 200 ft. garage on Reade and Jane Sts., to R. Keeves & Son, 309 Broadway, New York City. A steam heating system will be installed in same. Total estimated cost, $150,000.

N. Y., Long Island City—The Transport Service Co., 125 Harris Ave., has awarded the contract for the construction of a 5 story, 75 x 100 ft. garage and service station, on Harris and Sherman Aves., to L. Gold, 44 Court St., Brooklyn. A steam heating system will be installed in same. Total estimated cost, $200,000.

N. Y., New York—The Lawyers Title and Trust Co., 160 Broadway, has awarded the contract for the construction of an 11 story, 50 x 106 ft. office building at 160 Broadway, to G. B. Beaumont, 286 5th Ave. A steam heating system will be installed in same.

N. Y., New York—George Backer, 20 East 32nd St., will build a 30 story, 100 x 162 ft. office, store and theatre building on 57th St. and 5th Ave. A steam heating system will be installed in same. Total estimated cost, $6,000,000. Work will be done by day labor

N. Y., New York—Fred T. Ley, 19 West 44th St., will build a 20 story office building on 42nd St. and Madison Ave. A steam heating system will be installed in same. Total estimated cost, $2,000,000. Work will be done by day labor.

N. Y., Syracuse—The Greenway Brewing Co. has awarded the contract for the construction of a 3, 5 and 7 story, 80 x 500 ft. garage on West Water St., to the Syracuse Bridge Co., University Bldg. A steam heating system, including two 80 h.p. boilers and fuel oil burners, will be installed in same. Total estimated cost, $250,000.

Pa., Philadelphia—Gomery & Schwartz, 128 North Broad St., have awarded the contract for the construction of a 6 story, 220 x 440 ft. sales and service building on 24th and Market Sts., to M. J. O'Meara, 6421 Overbrook St. A steam heating system will be installed in same. Total estimated cost, $700,000.

Md., Cumberland—The George Washington Hotel Co. has awarded the contract for the construction of an 8 story hotel to the Wise Granite & Construction Co., American National Bank Bldg., Richmond, Va. Contract for installing heating system in same will be sub-let. Total estimated cost, $350,000.

Ohio, Cincinnati—The Ohio Butterine Co., 50 Walnut St., has awarded the contract for the construction of a 1 story, 15 x 90 ft. boiler house, to F. W. Foltz & Co. and the Detroit Steel Products Co., 1460 Penobscot Bldg., Detroit.

Ohio, Cleveland—The Association Building Co., c/o Dr. W. E. Power, Osborn Bldg., has awarded the contract for the construction of a 4 story, 86 x 240 ft. office building, on Euclid Ave. and East 93rd St., to the Crowell-Lundoff-Little Co., 5716 Euclid Ave. A steam heating system will be installed in same. Total estimated cost, $300,000.

Ohio, Cleveland—The Parish-Pool Copper Co., c/o Paul E. Cleveland, Guardian Bldg., has awarded the contract for the construction of a 1 story, 150 x 500 ft. smelting and refining building on Clinton Rd., to A. W. Kilbourne Co., 6532 Euclid Ave. Two boilers will be installed in same. Total estimated cost, $300,000.

Ohio, Cleveland—J. W. Wilson, 450 Leader-News Bldg., has awarded the contract for the construction of a 4 story, 76 x 270 ft. apartment hotel, at East 96th St. and Euclid Ave., to Paul Bros., 528 Citizens' Bldg. A steam heating system will be installed in same. Total estimated cost, $400,000.

Ind., Indianapolis—The Cole Motor Car Co., 750 East Washington St., has awarded the contract for the installation of a heating system in the proposed 4 story, 179 x 225 ft. auto assembly factory, on Market and Davidson Sts., to Hayes Brothers Co., 236 West Vermont St. Total estimated cost, $300,000.

Mich., Detroit—The Board of Education, 50 Broadway, has awarded the contract for installing heating, ventilating and plumbing systems in the proposed 2 story, 150 x 211 ft. school on Maplewood and Spokane Aves., to Irvine & Meier, 2289 Woodward Ave., $81,682.

Mich., Detroit—The city has awarded the contract for the construction of seven mile sewer and pumping station and for furnishing equipment for same to C. A. Handyside, 833 Lathrop Ave., $26,600.

Mich., Detroit—H. C. Keywell, 501 Old Whitney Bldg., has awarded the contract for the construction of a 2-story, 38 x 115 x 190 ft. theatre on Grand River Ave., to Daniel Fidler, 242 Melbourne Ave. A steam heating boiler with forced ventilation equipment will be installed in same. Total estimated cost, $90,000.

Mich., Detroit—The La Salle Gardens Theatre Co., c/o C. W. Brandt, Arch., 1114 Kresge Bldg., has awarded the contract for the construction of a 2 story, 100 x 160 ft. theatre at La Salle Gardens, to Frank Farrington Co., Scherer Bldg. A steam heating system, consisting of boiler, motor-driven fan, etc., will be installed in same. Total estimated cost, $200,000.

Mich., Flint—St. Mathew's Parish has awarded the contract for the construction of a 2 story school on Beach St., to the Duboise Construction Co., 1519 North Saginaw St. A steam heating system will be installed in same. Total estimated cost, $150,000.

Mich., Lansing—The Atlas Drop Forge Co., 209 West Mt. Hope Ave., has awarded the contract for the construction of a 150 ft. extension to its steel storage yard and plant and for furnishing equipment, consisting of 3 heavy trip hammers, addition to furnace capacity, additional 300 h.p. boiler and traveling crane, to the Wisconsin Bridge & Iron Co., Penobscot Bldg., Detroit.

Ill., Chicago—The Channel Chemical Co., 58 East Washington St., has awardtd the contract for the construction of a 4 story, 150 x 300 ft. factory on Western Ave. and West 45th St., to Davidson & Weiss, 53 West Jackson Blvd. A steam heating system will be installed in same. Total estimated cost, $450,000.

Ill., Chicago—The Playerphone Talking Machine Co., 338 North Kedzie Ave., has awarded the contract for the construction of a 3 story 110 x 160 ft. phonograph factory on Lake St. and Kildare Ave., to J. C. Stellwagen, 1583 Clybourn Ave. A steam heating system will be installed in same. Total estimated cost, $110,000.

Ill., Chicago—The Fulton Market Cold Storage Co., c/o J. Byfield, Pres., Hotel Sherman, has awarded the contract for the construction of a 10 story, 250 x 325 ft. cold storage warehouse on Fulton and Morgan Sts., to Blome Sinick Co., 139 North Clark St. Estimated cost, $4,000,000. Noted Aug. 26.

Mo., St. Louis—The Bridge & Beach Manufacturing Co., 1st and Valentine Sts., has awarded the contract for the construction of seven buildings, including a 1 story, 25 x 60 ft. boiler house, on Union and Belt Sts., to the Fruin-Colnon Contracting Co., Merchants Laclede Bldg. Total estimated cost, $430,000.

Ont., Toronto—W. Davies Co., 29 Queen St., W., has awarded the contract for the construction of a 4 story, 36 x 103 ft. store and office building to the Corswell Construction Co., 58 Wellington St. A steam heating system will be installed in same. Total estimated cost, $100,000.

Ont., Toronto—F. R. Pember, Jones Ave., has awarded the contract for the construction of a 600 ton circular coal handling plant to J. B. Nicholson, Ltd., Excelsior Life Bldg. Estimated cost, $30,000.

Ont., Peterborough—The Board of Education has awarded the contract for the construction of a 2 story school building to Richard Sheeby, 751 George St., $167,963. A steam heating and ventilating system will be installed in same.

Ont., Weston—The Canada Cycle & Motor Co. has awarded the contract for the construction of a 2 story, 80 x 160 ft. bicycle factory addition on Weston Rd., to the John V. Gray Construction Co., Confederation Life Bldg., Toronto. A vacuum steam heating system will be installed in same. Total estimated cost, $150,000.

Ont., Kitchener—The Dominion Tire Co., 149 Strange St., has awarded the contract for the construction of a 4 story, 80 x 250 ft. rubber tire factory on King St., W., to the MacKinnon Steel Co., Sherbrooke, Que. A steam heating system will be installed in same. Total estimated cost, $300,000.

PRICES -- MATERIALS -- SUPPLIES

These are prices to the power plant by jobbers in the larger buying centers east of the Mississippi. Elsewhere the prices will be modified by increased freight charges and by local conditions.

POWER-PLANT SUPPLIES

HOSE—

	Fire	50-Ft. Lengths
Underwriters' 2½-in		75c. per ft.
Common, 2½-in		40%

Air

	First Grade	Second Grade	Third Grade
¾-in. per ft	$0 50	$0.33	$0.22

Steam—Discounts from List

First grade.........30% Second grade.......40% Third grade......45%

RUBBER BELTING—The following discounts from list apply to transmission rubber and duck belting:

Competition.........65% Best grade.........35%
Standard.........50%

Note—Above discounts apply on new list issued July 1.

LEATHER BELTING—Present discounts from list in the following cities are as follows:

	Medium Grade	Heavy Grade
New York	45%	35%
St. Louis	40%	30%
Chicago	20%	10%
Birmingham	15%	20%
Denver	30%	20%

RAWHIDE LACING—20% for cut; 45c. per sq. ft. for ordinary.

PACKING—Prices per pound:

Rubber and duck for low-pressure steam	$0.90
Asbestos for high-pressure steam	1.50
Duck and rubber for piston packing	1 00
Flax, regular	1.20
Flax, waterproofed	1.60
Compressed asbestos sheet	.90
Wire insertion asbestos sheet	1.10
Rubber sheet	.50
Rubber sheet, wire insertion	.70
Rubber sheet, duck insertion	.50
Rubber sheet, cloth insertion	.30
Asbestos packing, twisted or braided and graphited, for valve stems and stuffing boxes	1.20
Asbestos wick, ½- and 1-lb. balls	.85

PIPE AND BOILER COVERING—Below are discounts and part of standard lists:

PIPE COVERING Pipe Size	Standard List Per Lin. Ft.	BLOCKS AND SHEETS Thickness	Price Per Sq. Ft.
1-in.	$0.27	½-in.	$0.27
2-in.	.36	1 -in.	.30
6-in.	.80	1½-in.	.45
4-in.	.60	2 -in.	.60
3-in.	.45	2½-in.	.75
8-in.	1.10	3 -in.	.90
10-in.	1.30	3½-in.	1.05

For low-pressure heating and return lines: 4-ply .... 58% off; 3-ply .... 60% off; 2-ply .... 62% off

GREASES—Prices are as follows in the following cities in cents per pound for barrel lots:

	Cincinnati	Pittsburgh	Chicago	St. Louis	Birmingham	Denver
Cup	7	9	6.6	6.7	8½	13
Fiber or sponge	8	10	8.6	13	8½	18
Transmission	7	9	8.1	13	8½	15
Axle	4½	6	4.8	4.75	4½	5½
Gear	4½	9	6.1	7.5	8½	8
Car journal	22(gal.)	21(gal.)	4.7	4.7	8½	15½

COTTON WASTE—The following prices are in cents per pound:

	New York Current	New York One Year Ago	Cleveland	Chicago
White	13.00	11 00 to 13 00	14.00	11.00 to 14.00
Colored mixed	9.00 to 12.00	8.50 to 11.00	11.00	9.50 to 12.00

WIPING CLOTHS—Jobbers' price per 1000 is as follows:

	13½ x 13½	13½ x 20½
Cleveland	$52.00	$58.00
Chicago	43.00	45.00

LINSEED OIL—These prices are per gallon:

	New York Current	New York One Year Ago	Chicago Current	Chicago One Year Ago
Raw in barrels (5 bbl. lots)	$1.90	$1.50	$2.09	$1.90
5-gal. cans	2.07	1.64	2.34	2.00

WHITE AND RED LEAD—Base price per pound:

	Red Current Dry	Red Current In Oil	Red 1 Year Ago Dry	Red 1 Year Ago In Oil	White Current Dry and In Oil	White 1 Yr. Ago Dry and In Oil
100-lb. keg	14.00	15.50	14.00	14 50	14 00	14.00
25- and 50-lb. kegs	14.25	15 70	14.25	14 75	14.25	14.25
12½-lb. keg	14 50	16.00	14.50	15.00	14.50	14.50
1-lb. cans	16.00	17.50	.....	.....	16 00	16.00
5-lb. cans	17.00	18.50	.....	.....	17.00	17.00

500 lb. lots less 10% discount; 2000 lb. lots less 10-2½%.

RIVETS—The following quotations are allowed for fair-sized orders from warehouse:

	New York	Cleveland	Chicago
Steel 7/16 and smaller	40%	55%	50%
Tinned	40%	55%	50%

Boiler rivets, ¾, ⅞, 1 in. diameter by 2 in. to 5 in. sell as follows per 100 lb.

New York....$5.00 Cleveland....$4.00 Chicago......$4 87 Pittsburgh....$4.72

Structural rivets, same sizes:

New York....$5.10 Cleveland....$4.10 Chicago......$4.97 Pittsburgh....$4.82

REFRACTORIES—Following prices are f. o. b. works, Pittsburgh:

Chrome brick	net ton	$80-90	at Chester, Penn.
Chrome cement	net ton	45-50	at Chester, Penn.
Clay brick, 1st quality fireclay	net M	35-45	at Clearfield, Penn.
Clay brick, 2nd quality	net M	30-35	at Clearfield, Penn.
Magnesite, dead burned	net ton	32-50	at Chester, Penn.
Magnesite brick, 9 x 4½ x 2½ in	net ton	80-90	at Chester, Penn.
Silica brick	net M	41-45	at Mt. Union, Penn.

Standard size fire brick, 9 x 4½ x 2½ in. The second quality is $4 to $5 cheaper per 1000.

St. Louis—Fire Clay, $35 to $50.

Birmingham—Silica, $46.50-$51.50 per M; fire clay, $42.00-$45.00 per M; magnesite, $80 per ton; chrome, $80 per ton.

Chicago—Second quality, $25 per ton

Denver—Fire clay, $11 per ton.

BABBITT METAL—Warehouse prices in cents per pound:

	New York Current	New York One Year Ago	Cleveland Current	Cleveland One Year Ago	Chicago Current	Chicago One Year Ago
Best grade	90.00	95.00	70.00	93.00	60.00	96 00
Commercial	55.00	50.00	16.50	23.00	13.00	25.00

SWEDISH (NORWAY) IRON—The average price per 100 lb., in ton lots, is:

	Current	One Year Ago
New York	$21.26	$19.00
Cleveland	20.00	20.00
Chicago	16.50	19.00

In coils an advance of 50c. is usually charged.

SHEETS—Quotations are in cents per pound in various cities from warehouse; also the base quotations from mill:

	Mill Carloads Pittsburgh	New York Current	New York One Year Ago	Cleveland	Chicago
No. 28 black	$4.35-4.85	$7.00-8.00	$6.52	$5.77	$6.00
No. 26 black	4.25-4.75	6.90-7.90	6.42	5.67	5.90
Nos. 22-24 black	4.20-4.71	6 85-7.85	6.37	5.62	5 85
Nos. 18-20 black	4.15-4.65	6.80-7.80	6.32	5.57	5.80
No. 16 blue annealed	3.75-3.20	6.02	5.72	5.17	5.02
No. 14 blue annealed	3.65-3.10	5.92	5.62	5.07	4.92
No. 10 blue annealed	3.55-4.00	5.50-6.00	5.52	4.97	4.82
No. 28 galvanized	5.70-6.20	7.50-9.00	7.77	7.12	7.25
No. 26 galvanized	5.40-5.90	7.20	7.47	6.82	6.95
No. 24 galvanized	5.25-5.75	7.05	7.32	6.57	6.52

PIPE—The following discounts are for carload lots f. o. b. Pittsburgh; basing card of Jan. 1, 1919, for steel pipe and for iron pipe:

BUTT WELD

Inches	Steel Black	Galvanized	Inches	Iron Black	Galvanized
⅛, ¼, and ⅜	50½%	24%	¾ to 1½	30%	23½%
½	54½%	40%			
¾ to 3	57½%	44%			

LAP WELD

2	50½%	38%	2	23½%	18½%
2½ to 6	53½%	41%	2½ to 6	24½%	21½%

BUTT WELD, EXTRA STRONG PLAIN ENDS

⅛, ¼ and ⅜	46½%	29%	¾ to 1½	30½%	24½%
½	51½%	39%			
¾ to 1½	55½%	43%			

LAP WELD, EXTRA STRONG PLAIN ENDS

2	48½%	37%	2	33½%	20½%
2½ to 4	51½%	49%	2½ to 4	35½%	23½%
4½ to 6	50½%	39%	4½ to 6	34½%	23½%

Stock discounts in cities named are as follows:

	New York Black	New York Galvanized	Cleveland Black	Cleveland Galvanized	Chicago Black	Chicago Galvanized
¾ to 3 in. steel butt welded	47%	31%	43½%	34½%	57½%	44%
2½ to 3 in. steel lap welded	42%	27%	45½%	39½%	53½%	41%

Malleable fittings. Classes B and C, from New York stock, sell at list +12½%

Cast iron, standard sizes, 10-5% off.

BOILER TUBES—The following are the prices for carload lots, f. o. b. Pittsburgh:

Lap Welded Steel		Charcoal Iron	
3½ to 4½ in.	40½	3½ to 4½ in.	15½
2½ to 3¼ in.	30½	3 to 3¼ in.	1
2¼ in.	24	2½ to 2¾ in.	+ 1
1¾ to 2 in.	19½	2 to 2¼ in.	15
		1½ to 1¾ in.	+29

Standard Commercial Seamless—Cold Drawn or Hot Rolled

	Per Net Ton		Per Net Ton
1 in.	$327	1⅞ in.	$207
1¼ in.	267	2 to 2¼ in.	177
1½ in.	257	2½ to 3¼ in.	167
1¾ in.	207	4 in.	187
		4½ to 5 in.	207

These prices do not apply to special specifications for locomotive tubes nor to special specifications for tubes for the Navy Department, which will be subject to special negotiations.

ELECTRICAL SUPPLIES

RMORED CABLE—

B.& S. Size	Two Cond. M Ft.	Three Cond. M Ft.	Two Cond. Lead M Ft.	Three Cond. Lead M Ft.
No. 14 solid	$104.00	$138.00	$164.00	$210.00
No. 12 solid	135 00	170 00	225.00	265.00
No. 10 solid	185.00	235 00	275.00	325.00
No. 8 stranded	285 00	375.00	520 00	500.00
No. 6 stranded	400.00	500.00	560.00	......

From the above lists discounts are:

Less than coil lots	Net List
Coils to 1,000 ft.	10%
1,000 ft. and over	15%

BATTERIES, DRY—Regular No. 6 size, red seal, Columbia or Ever Ready:

	Each, Ne
Less than 12	$0.40
12 to 50	.39
50 to 125 (bbl.)	.36
125 (bbl.) or over	.34

CONDUITS, ELBOWS AND COUPLINGS—Following are warehouse net prices per 1000 ft. for conduit and per 100 for couplings and elbows:

	Conduit		Elbows		Couplings	
Size, In.	Black 1,000 Ft. and over	Galvanized 1,000 Ft. and over	Black 100 and Over	Galvanized 100 and Over	Black 100 and Over	Galvanized 1,000 and Over
⅜	$71.15	$76.25	$17.04	$18.18	$5.38	$5 74
½	71.15	76.25	17.04	18.18	6.25	6.70
¾	93.96	100.86	22.23	23.93	8.97	9.57
1	138.39	149.09	33.19	35.41	11.66	12.44
1¼	187.91	201.71	42.62	45.31	16.10	17.12
1½	224.68	241.18	56.82	60.62	19.89	21.15
2	302.29	324.49	104.17	110.77	26.52	28.20
2½	477.95	513.05	170.46	181.26	37.88	40.28
3	625.00	670.91	454.56	483.36	56.82	60.42
3½	779.24	834.44	1,003.82	1,067.42	75.76	80.56
4	945.03	1,010.43	1,160.08	1,223.57	94.70	100.70

5% cash 10 days.
From New York Warehouse—5% cash.
Standard lengths rigid, 10 ft. Standard lengths flexible, 100 ft. Standard lenghts flexible, ¾ to 2 in., 50 ft.

CONDUIT, NON-METALLIC, LOOM—

Size I. D., In.	Feet per Coil	List, Ft.	
¼	250	$0.05½	Coils.......55% off Less coils...45% off
5/16	250	.06	
⅜	250	.09	
7/16	200	.12	
½	200	.15	
⅝	150	.18	
¾	100	.25	
1	100	.33	
1¼	Odd lengths	.40	
1½	Odd lengths	55	
2			

CUT-OUTS—Following are the net prices each in standard-package quantities:

CUT-OUTS, PLUG

S. P. M. L.	$0.11	T. P. to D. P. S. P.	$0.24
D. P. M. L.	.11	T. P. to D. P. T. B.	.38
T. P. M. L.	.26	T. P. S. B.	.33
D. P. S. B.	.19	T. P. D. B.	.54
D. P. D. B.	.37		

CUT-OUTS, N. E. C. FUSE

	0–30 Amp.	31–60 Amp.	60–100 Amp.
D. P. M. L.	$0.33	$0.84	$1.68
T. P. M. L.	48	1.20	2.40
D. P. S. B.	.42	1.05	....
T. P. S. B.	.81	1.80	....
D. P. D. B.	.78	2.10	....
T. P. D. B.	1.35	3.60	....
T. P. to D. P. D. B.	.90	2.52	....

FLEXIBLE CORD—Price per 1,000 ft. in coils of 250 ft.:

No. 18 cotton twisted	$16.75
No. 16 cotton twisted	22.00
No. 18 cotton parallel	19.80
No. 16 cotton parallel	25.75
No. 18 cotton reinforced heavy	30.00
No. 16 cotton reinforced heavy	36.50
No. 18 cotton reinforced light	23.40
No. 16 cotton reinforced light	27.50
No. 18 cotton Canvasite cord	26.00
No. 16 cotton Canvasite cord	28.50

FUSES, ENCLOSED—

250-Volt	Std. Pkg.	List
3-amp. to 30-amp.	100	$0.25
35-amp. to 60-amp.	100	.35
65-amp to 100-amp.	50	90
110-amp to 200-amp.	25	2.00
225-amp to 400-amp.	25	3.60
425-amp to 600-amp.	10	5.50

600-Volt	Std. Pkg.	List
3-amp. to 30-amp.	100	$0.40
35-amp. to 60-amp	100	.60
65-amp. to 100-amp	50	1.50
110-amp. to 200-amp	25	2.50
225-amp. to 400-amp	25	5.50
450-amp. to 600-amp	10	8.00

Discount: Less than 1-5th standard package	30%
1-5th to standard package	40%
Standard package	52%

FUSE PLUGS, MICA CAP—

0–30 ampere, standard package	$4.75 C
0–30 ampere, less than standard package	6.00 C

LAMPS—Below are present quotations in less than standard package quantities:

Straight-Side Bulbs Mazda B— Watts	Plain	Frosted	No. in Package	Pear-Shape Bulbs Mazda C— Watts	Clear	Frosted	No. in Package
10	$0.35	$0.38	100	75	$0.70	$0.75	50
15	.35	.38	100	100	1.10	1.15	24
25	.35	.38	100	150	1.65	1.70	24
40	.35	.38	100	200	2.20	2.27	24
50	.35	.38	100	300	3.35	3.35	24
60	.40	.45	100	400	4.30	4.45	12
100	.85	.92	24	500	4.70	4.85	12
				750	6.50	6.75	8
				1000	7.50	7.75	8

Standard quantities are subject to discount of 10% from list. Annua contracts ranging from $150 to $300,000 net allow a discount of 17 to 40% from list.

PLUGS, ATTACHMENT—

	Each
Hubbell, porcelain No. 5406, standard package 250	$0.24
Hubbell composition No. 5467, standard package 50	.32
Benjamin swivel No. 903, standard package 250	.20
Hubbell current taps No. 5638, standard package 50	.40

RUBBER-COVERED COPPER WIRE—Per 1000 ft. in New York

No.	Solid Single Braid	Solid Double Braid	Stranded Double Braid	Duplex
14	$12.00	$14.00	$13.90	$25.00
12	13.25	15.70	18.05	30.70
10	18.30	21.00	23.85	41.50
8	25.54	28.60	32.70	56.77
6	......	......	51.40	......
4	......	......	70.00	......
2	......	......	101.80	......
1	......	......	131.86	......
0	......	......	160.00	......
00	......	......	193.50	......
000	......	......	235.20	......
0000	......	......	288.50	......

SOCKETS, BRASS SHELL—

¼ In. or Pendant Cap Key, Each	Keyless, Each	Pull, Each	¾ In. Cap Key, Each	Keyless, Each	Pull, Each
$0.33	$0.30	$0.60	$0.39	$0.36	$0.66

Less 1-5th standard package	+20%
1-5th to standard package	+10%
Standard package	−15%

WIRE, ANNUNCIATOR AND DAMPPROOF OFFICE—

No. 18 B.& S. regular spools (approx. 8 lb.)	36c. lb.
No. 18 B.& S. regular 1-lb. coils	37c. lb.

WIRING SUPPLIES—

Friction tape, ¾ in., less 100 lb., 50c. lb., 100 lb. lots	55c. lb.
Rubber tape, ¾ in., less 100 lb., 65c. lb., 100 lb. lots	60c. lb.
Wire solder, less 100 lb., 50c. lb., 100 lb. lots	46c. lb.
Soldering paste, 2 oz. cans Nokorode	$1.20 doz.

SWITCHES, KNIFE—

TYPE "C" NOT FUSIBLE

Size, Amp.	Single Pole, Each	Double Pole, Each	Three Pole Each	Fore Pole Each
30	$0.42	$0.68	$1.02	$1.36
60	.74	1.22	1.84	2.44
100	1.50	2.50	3.76	5.00
200	2.70	4.50	6.76	9.00

TYPE "C" FUSIBLE, TOP OR BOTTOM

Size, Amp.	Single Pole	Double Pole	Three Pole	Fore Pole
30	.70	1.06	1.60	2.12
60	1.18	1.80	2.70	3.60
100	2.38	3.66	5.50	7.30
200	4.40	6.76	10.14	13.50

Discounts:

Less than $10.00 list value	+ 5%
$10 to $25 list value	− 8%
$25 to $50 list value	−14%
$50 to $200 list value	−20%
$200 list value or over	−25%

POWER

Volume 51 New York, January 13, 1920 *Number 2*

The Chronic Grouch

By Rufus T. Strohm

HIS face, which bears a look of pain, is longer than a country lane, and both the corners of his mouth curve sharply downward to the south. He has a slow, dejected gait, as though he had to tote a freight of over-ripe and aged woes that had begun to decompose. His aura shows a shade of blue that's just as cheerful as the flu, and one could almost catch the pip by looking at his lower lip.

The passers-by step wide

His fellow-workmen never hear him speak a kindly word of cheer, nor do they ever see a grin between his forehead and his chin. For when it comes to cracking smiles, Job had him beat by forty miles. He's just as pleasant as the itch, the hydrophobia and sich; which means, of course, that none would weep if he should fail to wake from sleep—though those who know him best would feel that Satan got a rotten deal.

No matter what the job he gets, he stews and fumes, he rants and frets, and calls the boss the meanest man that's walked on legs since time began. But everybody's well aware that all his talk is heated air and that he simply must unload or else he'd probably explode. Still, no one pauses at his side; instead, the passers-by step wide. Throughout the plant, from end to end, there's not a man he calls his friend; and all he gets for being mean and turning loose his spite and spleen is all the lonesomeness there is, and sullen wrinkles on his phiz.

New Armour South St. Paul Plant

A plant of varied equipment supplying power, light, heat, hot water, compressed air and refrigeration to a large packing house. High economy is insured as exhaust steam requirements are practically balanced with exhaust steam produced.

A NEW packing plant to serve the Northwest has been built by Armour & Co. at South St. Paul, Minn. It will be somewhat larger than the existing plant at Omaha, and, like all institutions of the kind, there will be large requirements for water, steam, refrigeration, light and power and compressed air. Consequently, a power plant with diversified equipment was essential and much of it steam using, to meet the large demand for exhaust steam by the various processes in the packing house. To insure high economy it was necessary to so select the equipment as to make it feasible to effect a close balance between the amount of exhaust steam that might be produced and the requirements of the packing plant.

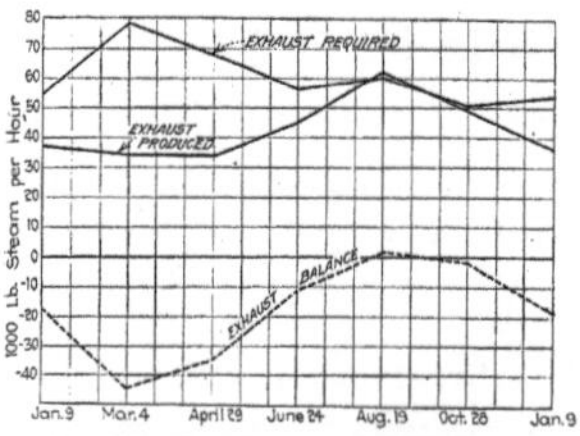

FIG. 1. EXHAUST STEAM BALANCE FOR DAY LOAD EXCLUSIVE OF GENERATING UNITS

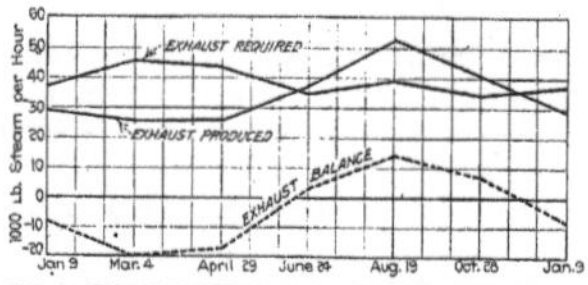

FIG. 2. EXHAUST STEAM BALANCE FOR NIGHT LOAD EXCLUSIVE OF GENERATING UNITS

To provide flexibility, the refrigerating plant was divided into compression and absorption systems, one producing exhaust steam and the other using it. The ammonia compressors are served by a barometric condenser that may be utilized if the demand for exhaust steam is low, the water from the tail pipe being used for domestic purposes or for boiler feed, so that the heat gained from condensing the steam is not lost. Normally, however, the exhaust goes direct to a 5-lb. exhaust-steam line that makes a circuit of the entire plant and supplies all requirements for low-pressure steam. The air compressors are operated noncondensing, as are

FIG. 3. MIXED PRESSURE EXTRACTION TURBINES

FIG. 4. AMMONIA COMPRESSORS WITH TURBINES AND AIR COMPRESSORS IN BACKGROUND

all auxiliaries in the plant for the exhaust steam that they will furnish. The prime movers of the generating units are turbines of the mixed-pressure extraction type from which steam may be bled into the exhaust line or

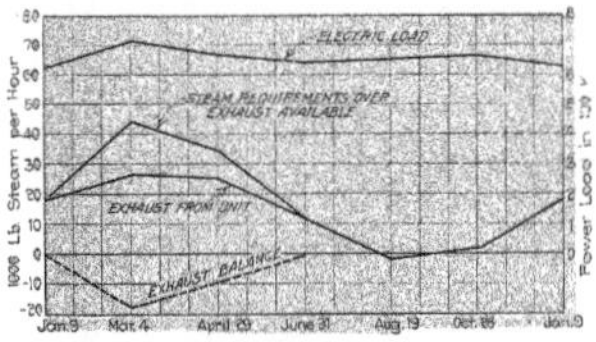

FIG. 5. EFFECT OF GENERATING UNITS EXHAUST DAY LOAD

the flow may be reversed automatically from the line to the turbine, depending upon the pressure in the exhaust line, which is influenced by the supply of and the demand for exhaust steam. There is thus sufficient

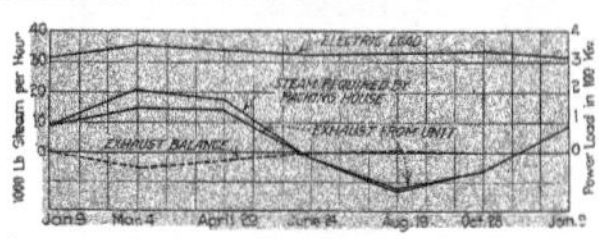

FIG. 6. EFFECT OF GENERATING UNITS EXHAUST NIGHT LOAD

flexibility so that an exhaust steam balance may be approximately maintained for the greater part of the year. During the winter months, with the heating load added, a deficiency in exhaust steam requires supplemental use of live steam reduced to the proper pressure. At no time is a surplus of exhaust expected, owing to the type of turbo-generators employed.

Such desirable operating conditions were the result of careful preliminary analyses, based on unit data ob-

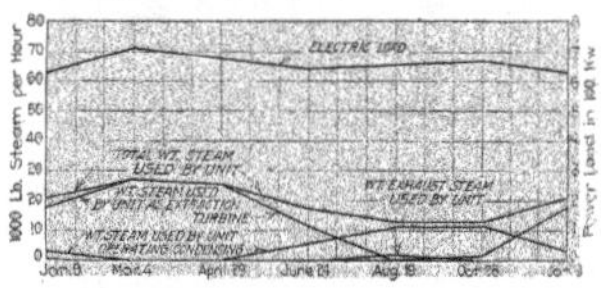

FIG. 7. STEAM USED BY GENERATING UNITS, DAY LOAD

tained from other plants of the company. With the types and sizes of the varied equipment of the plant tentatively selected, the hours of service and the requirements of the packing house determined, a careful estimate was made of the exhaust steam required and that

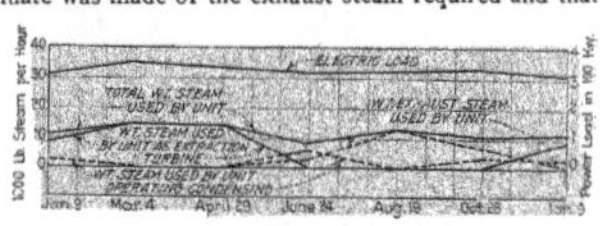

FIG. 8. STEAM USED BY GENERATING UNITS, NIGHT LOAD

produced. The figures were compiled covering each of the six periods into which the year had been divided and then referred to an hourly basis for both the day and the night loads. These data were then plotted, and the

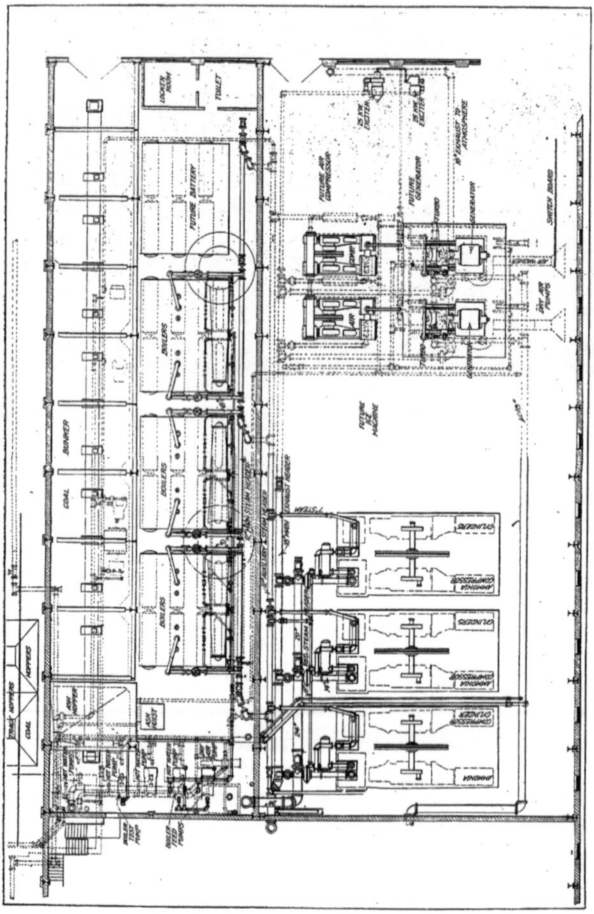

FIG. 9. GENERAL PLAN OF BOILER AND ENGINE ROOMS AND COMPRESSION REFRIGERATING PLANT

results of the analyses are shown in Figs. 1, 2, 5, 6, 7 and 8.

Figs. 1 and 2 show the estimated amounts of exhaust steam required, the exhaust produced and the balance between the two for the day and night loads, respectively, exclusive of the generating units. The day curve covers conditions between 7 a. m. and 5:30 p. m., and the night period continues from 5:30 p. m. to 7 a. m. These curves show a deficiency in exhaust steam during the greater part of the year. To relieve this situation, the mixed-pressure extraction turbines were installed. Figs. 5 and 6 show how the deficiency of exhaust steam was reduced by bleeding steam from the turbines when required. The electric power load had been estimated previously from load curves taken during three months' operation at Omaha. The steam consumption of the power units was figured on a low-pressure steam-consumption basis, when the deficiency of the exhaust steam could not be taken care of by the turbine operating partly as an extraction machine and partly as a condensing unit. The conditions were analyzed, and a division was made between live and exhaust steam. The different conditions under which the power unit would have to operate are indicated in Figs. 7 and 8. It will be evident that at times the machine will operate as an extraction turbine, sometimes as a condensing unit and at other times as a low-pressure turbine or as a combination unit using both high- and low-pressure steam.

In the plant two of these turbo-generators have been installed, each rated at 750 kw. and designed to deliver 440-volt three-phase 60-cycle currents. The speed is 3600 r.p.m., and separate excitation is supplied by either one of two turbine-driven 25-kw. exciters. Ordinarily, one of the main units is expected to carry the electric load.

The turbine is a three-stage machine with two rows of moving buckets in the first stage and one in each of the other two stages. Steam is extracted from the first stage after it has passed through the moving buckets or, if the operation is reversed, flows from the exhaust line into the turbine through the same passage and with the high-pressure steam enters the second stage through a diaphragm valve. The latter consists of a flat disk or nozzle plate with a number of different-sized openings. Through a ratchet and a series of levers leading to an isochronous governor the covering disk or valve proper is revolved to uncover or close the various openings in the nozzle plate in accordance with the pressure in the exhaust line. Simultaneously, the same governor regulates the flow through the high-pressure steam valve of the turbine. It is thus evident that when the pressure in the exhaust line exceeds 5 lb., the valve admitting high-pressure steam to the turbine tends to close and the surplus in the exhaust line flows into the turbine. When the pressure in the exhaust line drops below 5 lb., the operation is reversed and steam is bled from the turbine. All this takes place automatically so that within the limits of the turbines there is no difficulty in maintaining an exhaust balance. From the turbine the steam enters a jet condenser, and the heat in the mixture of condensation and cooling water is conserved by conducting the condenser discharge to the hotwell for boiler use or to a sump supplying domestic hot water for the packing house.

On the electrical side economy was effected by the judicious use of synchronous motors. The connected load consisted of 3000 hp. in three-phase 60-cycle 440-volt motors of the slip-ring and plain induction types varying in size from 3 to 50 hp. Depending upon how these various motors were loaded, an average power fac-

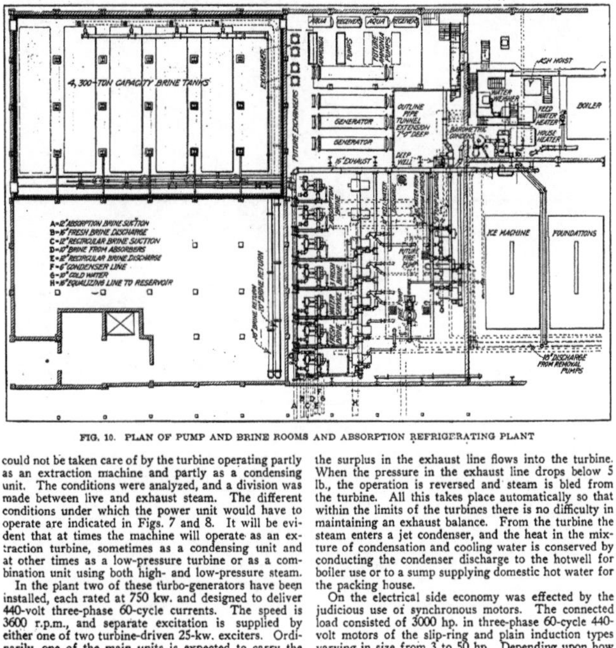

FIG. 10. PLAN OF PUMP AND BRINE ROOMS AND ABSORPTION REFRIGERATING PLANT

tor higher than 65 to 70 per cent. could not be expected. Four large motors were needed—three to drive deep-well pumps of 2000 gal. per min. capacity each and the other, machinery in the fertilizer plant. As the entire supply of water for the plant was to be obtained from wells and there would be need for continuous service, sturdy and efficient motors were desired and in addition the question of power factor correction had a bearing. To drive the pumps, which were single-stage operating against a total head of 95 ft., three 150-hp. 2300-volt three-phase vertical self-starting self-synchronizing synchronous motors with direct-connected exciters were selected and a 115-hp. motor of the same type for the fertilizer plant. The speeds are 1200 and 900 r.p.m., respectively. The pumps were over-motored, the actual horsepower required in each case being 75 to 80. The remainder of the synchronous-motor capacity was to be utilized to raise the power factor of the system to an estimated resultant of 91 per cent. In addition to power-factor correction a careful checking up of the efficiency of these large motors as compared with that of induction motors of the proper size to handle the mechanical load showed an advantage in their favor so decided as to warrant the extra expenditure. The motors are sturdy, have a comparatively large air gap, and owing to direct connection to the pumps and exciters, are self-contained units independent of any other apparatus.

A brief review of the other departments of the plant will give some idea of the size of equipment and the general layout, which is shown specifically in Figs. 9 and 10. To supply compressed air for general packing-house use and also for requirements in the power house, two duplex horizontal air compressors each having a capacity of 1531 cu.ft. per min. were installed. The air is compressed to 80 lb. pressure, and the steam from the compressor cylinders is discharged directly to the 5-lb. exhaust line for general use.

High-temperature cold storage in which the temperature of the rooms ranges from 30 to 50 deg. is taken care of by the compression system, consisting of three 400-ton duplex, opposed, double-acting compressors driven by cross-compound steam cylinders. These machines cool 6,040,000 cu.ft. of refrigerating space containing 250,000 lin. ft. of 2-in. pipe through which brine is circulated at temperatures of about 25 deg. Of this space 20,000 cu.ft. is cooled by return brine from the freezers. For normal service two of the compressors will be required and the third held as a reserve. The three machines are served by a barometric condenser having a capacity to condense 36,000 lb. of steam per hour. It is expected that the condenser will be in use only at infrequent intervals and that for the greater part of the year the compressors will exhaust directly into the 5-lb. exhaust line.

For the sharp freezers, in which temperatures of 0 to 15 deg. are maintained, cooling is also effected by brine circulation, the refrigeration being produced by two 150-ton absorption units. The auxiliaries are turbine-driven centrifugal pumps, exhausting into the 5-lb. line from which steam is taken directly to supply the needs of the absorption plant. These two units serve 698,000

FIG. 11. BOILER FRONTS SHOWING UNDERFEED STOKERS AND TRAVELING WEIGHING LARRY

cu.ft. of refrigerating space containing 150,000 lin.ft. of 2-in. pipe for brine circulation, also a chill-water cooling tank for oleo fat and crystallizing water for butterine. For the higher-temperature system there are four steel brine tanks 50 ft. long, 11 ft. wide and 15 ft. 6 in. deep, each containing 20 flat coils, 44 ft. long, of 1¼-in. pipe. Shell-type coolers are provided for the low-temperature work. Condensers for the absorption system are of the double-pipe type, and atmospheric condensers are provided for the compression system. Cooling water is obtained from the sump, which will receive mention later, and is returned to the hotwell for further use, the surplus, if any, being wasted to the sewer.

In the boiler rooms there are installed six 502-hp. boilers of the Stirling type arranged in batteries of two, with space for another battery of boilers and plenty of room for extension of the building in the future. The additional battery is to be installed shortly, making a total of eight boilers, which will be served by two radial brick stacks 9 ft. diameter by 185 ft. high. Thus four boilers are connected to each stack, and its location is midway between the two batteries served. The operating pressure is 150-lb. gage and the superheat 50 deg. F.

Anticipating variable loads and high peaks frequently throughout the day and sometimes all day, as compared to other days, it was decided to install underfeed stokers. Normally, the boilers are to be operated at 75 to 100 per cent. overload. High capacity and high efficiency were the influencing factors, and with the fuel that will be burned, either Illinois or Youghiogheny bituminous coal, no particular difficulty is anticipated, as inferior fuels have been burned in the Omaha plant in stokers of the same type. The stokers under consideration have an equivalent grate area of 99 sq.ft., giving a ratio to the steam-making surface of practically 1 to 51. They are installed in standard settings with no arch in the combustion chamber. Owing to the character of the coal hand-operated dump plates rather than clinker grinders are provided.

Forced draft is furnished to the stokers by two turbine-driven fans each having capacity to supply six boilers at 25 per cent. overload. The fans discharge into a duct running the full length of the boiler installation, which is provided with a damper at the middle, so that the service may be divided between the two fans, or with the damper open one fan may fill the requirements. Dampers are also provided in the branch ducts to each stoker. Pressure regulation controls the quantity of air to each boiler. The vertical engines, connected to the stokers through a lineshaft and chain drives, are under hand control.

Coal for the boiler room is dumped into a track hopper from railway cars. It is transferred to a crusher by an apron conveyor and is then conveyed to an overhead bunker by a continuous bucket carrier. The bunker, which is of the continuous steel type with parabolic bottom, is partitioned off for each battery of boilers. The capacity is about 400 tons. A traveling weigh larry, electrically driven, is interposed between the bunker and the stoker hoppers.

On account of the low elevation of the site and the presence of water it was impossible to excavate deep enough to provide for a suitable type of ash-handling equipment. An ordinary hand cart is used to transfer the ashes from the hoppers under the stokers to a skip-hoist which raises the ashes to an overhead concrete bunker, and through an under-cut gate and spout they are discharged into railway cars.

In all packing plants the water supply is of prime importance. As constant supply in large quantities is needed, the pumping equipment must be substantial to withstand the service and should be highly economical. Although the plant is located on the bank of the Mississippi River, the water is so contaminated by sewage

PRINCIPAL EQUIPMENT OF ARMOUR & COMPANY'S PLANT, SOUTH ST. PAUL, MINN.

No.	Equipment	Kind	Size	Use	Operating Conditions	Maker
6	Boilers	Stirling	502-h.p.	Generate steam	150 lb. pressure 50 deg. superheat	Babcock & Wilcox Co.
6	Stokers	Underfeed	99-sq.ft.	Serve boilers	Forced draft	American Engineering Co.
2	Fans	lorizontal	No. 7	Forced draft	Driven by turbines	Buffalo Forge Co.
2	Chimneys	Radial brick	9x185-ft.	Draft		Custodis
2	Turbo-Generators	A. C. 3-phase	50-kw.	Power generating units	3600 r. p. m. 2300 volts 60 cycles. 5 and 150 lb. Mixed-pressure	General Electric Co.
3	Ammonia Compressors	Duplex opposed. Double-acting	400-ton	Refrigeration	Steam cross-compound	Vilter Manufacturing Co.
2	Absorption ice machines	Absorption	150-ton	Refrigeration	Exhaust steam	Carbondale Machine Co.
2	Air compressors	Duplex horizontal	1531-cu. ft. per min.	General purposes	Steam, heavy-duty	Ingersoll-Rand
1	Pump	Hor. centrifugal	4200-g.p.m.	Sewage disposal	Motor-driven	Worthington
3	Pumps	Vert. centrifugal	2000-g.p.m.	Deep-well circulation	Motor-driven	American Well Works
1	Pump	Hor. centrifugal	1500-g.p.m.	Water circulation	Motor-driven	Union Steam Pump Co.
5	Pumps	Hor. centrifugal	1500-g.p.m.	Brine circulation	Turbine-driven	Hill Pump Co.
6	Pumps	Hor. centrifugal	1500-g.p.m.	Water circulation	Turbine-driven	Hill Pump Co.
2	Pumps	Hor. centrifugal	1500-g.p.m.	Water circulation	Turbine-driven	Hill Pump Co.
1	Pump	Hor. centrifugal	1500-g.p.m.	Circulation	Turbine-driven	Hill Pump Co.
4	Pumjs	Hor. centrifugal	550-g.p.m.	Hot-water circulation	Turbine-driven	Hill Pump Co.
2	Pumps	Centrifugal	500-g.p.m.	Boiler feed	Turbine-driven, three-stage	Cameron
1	Pump	Hor. centrifugal	400-g.p.m.	Water circulation	Turbine-driven	Hill Pump Co.
1	Pump	Duplex	20x12x16-in.	Fire	Steam, 1500 g.p.m.	Worthington
1	Pump	Hor. centrifugal	10-in.	Sewage disposal	Steam, Engine 9x9-in.	Morris Machine Works
2	Pumps	Hor. centrifugal	12-in.	Sewage disposal	Steam, Engine 12x10-in.	Morris Machine Works
1	Pump	Duplex	8x5x12-in.	Boiler test		Dean
1	Heater	Open	200,000 lb. per hr. to 220 deg. F.	Feed water	Exhaust steam	Hoppes
1	Heater	Open	210,000 lb. per hr. from 50 deg. F. to 220 deg. F.	House water	Exhaust steam	Vater
4	Brine tanks	Steel	50 x 11 x 15.5 feet deep with 20 flat coils 1¼-in. x 44 ft. long	Brine cooling		Armour Mechanical Co.
1	Condenser	Atmospheric	113x24x20ftx2 in.	Refrigeration		Vilter Manufacturing Co.
1	Condenser	Double-pipe	20x5x20ftx2 & 3 in	Refrigeration		Carbondale Machine Co.
2	Condensers	Jet	17,000-lb. per hr.	Steam		Alberger Pump & Condenser Co.
1	Condenser	Barometric	36,000-lb. per hr.	Steam		Alberger Pump & Condenser Co.
1	Conveyor	Apron and bucket	67-tons per hr.	Coal from track hopper to conveyor	Motor-driven	Jeffrey Manufacturing Co.
1	Coal crusher	30x30 in.	67-tons per hr.	Crush coal	Motor-driven	Jeffrey Manufacturing Co.
1	Weigh larry	Traveling	2000 lb.	Weigh coal	Motor-driven	Phillips Lang
1	Skip hoist	Bucket	18-ton	Ash-handling equipment	Motor-driven	R. H. Beaumont
250	Motors	Induction	3-to 50-hp.	General		General Electric Co.
4	Motors	Synchronous	115-and 150-hp.	Drive deep-well pumps and fertilizer plant		Electric Machinery Co.

that it is not fit for use in the packing plant or for boiler feed. As a consequence, the water supply is all drawn from the three wells previously mentioned, and is discharged into a large concrete cistern. From the cistern the water through pipe connections seeks its own level in a sump provided under the pumproom. This sump is divided into two sections, one being devoted entirely to fire service and the other to the supply of domestic water and boiler feed. There are three 24-in. lines between the cistern and the two sections of the sump. One leads to the section for fire service, another leads directly from the cistern to the other section, and the third leads from the cistern to the jet condensers serving the main turbo-generators. The condensers, however, may be bypassed and the water in the third line flow directly to the sump. A certain amount of the returns from various sources comes back to the division of the sump used for general purposes, so that the temperature ranges from 56 to 70 deg. Other returns go back directly to the feed-water heater and a certain amount to the hotwell. The latter also receives the cooling water from the barometric condenser serving the ammonia compressors.

From either of these sources the water is pumped under float control to either one of two large open feed-water heaters, one designed for house purposes and the other for boiler feed, although the two are cross-connected and can be used for either service. As water at 110 deg. F. is hot enough for most purposes throughout the packing house, with a maximum of 150 deg. for specific cases, each heater is provided with an attemperatur. Water from the heater which may be at a temperature of, say, 208 deg., enters the attemperatur, and by means of a thermostatic valve the proper amount of cold water is allowed to mix with the hotter water to give the desired temperature.

For boiler feed, water from the heater passes through a venturi meter into turbine-driven centrifugal pumps supplying the boilers. In case of necessity water can be drawn directly from the sump or the hotwell, and if the pumps should fail, high-pressure steam injectors are provided for emergency use. For testing purposes a weigh-meter and reciprocating pump are installed. Outside of the fire pump this is the only reciprocating pump in the plant.

The new packing plant is about ready for operation and should prove most economical in the generation and use of steam. All the engineering was done by the motive-power department of the company, of which A. McKenzie is superintendent.

The illustration at the bottom of this page, showing the concrete oil tank and oil-burning equipment of the Burnet Street School, Newark, N. J., will interest many *Power* readers who are considering oil fuel for small plants. As shown, this equipment supplies two 85-hp. horizontal-tubular boilers; the grate bars have not been removed. Air is used for atomization in this plant, though the more general practice is to use steam. The drawing so well shows the general layout that further text explanation is unnecessary.

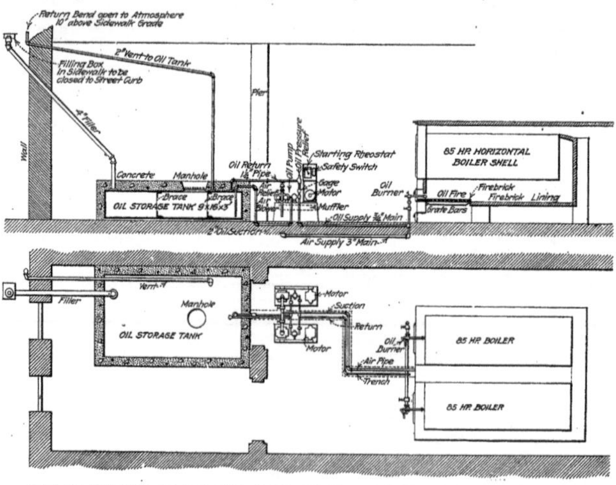

SECTIONAL ELEVATION AND PLAN VIEWS OF OIL STORAGE AND BURNING EQUIPMENT, BURNET STREET SCHOOL, NEWARK, N. J.

Drawing by courtesy S. T. Johnson Co., San Francisco, Calif.

Reconnecting Induction Motors—Reconnecting Problems

By A. M. DUDLEY
Designing Engineer, Westinghouse Electric and Manufacturing Company

Problems are worked out to show that any change in an induction motor's operating conditions may be considered as a voltage change.

CONSIDERATION of a simple case under each of the five characteristics of horsepower, poles, cycles, phase and voltage will bring out the manner of applying the "voltage method" to any and all changes in the motor-operating conditions.

1. Change in Voltage—A motor is connected series-star for three-phase 440 volts, as in Fig. 2. How should it be connected for 220 volts? [For convenience the table on page 725 of the issue of Nov. 27, 1917, is

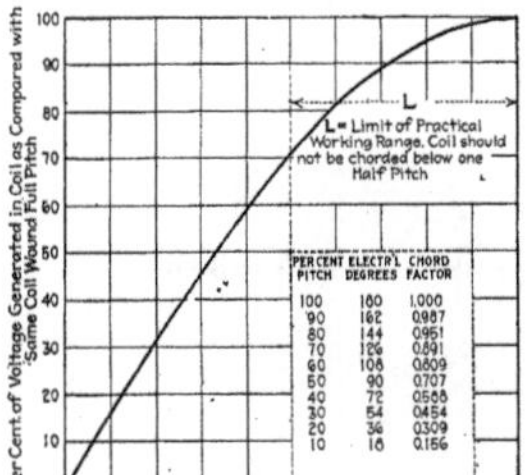

FIG. 1. CURVE SHOWING THE EFFECTS OF CHORDING A COIL UPON THE VOLTAGE GENERATED

here reproduced.] Looking at the table and following the horizontal line "Three-phase Series Star," there appears under vertical heading "Three-phase Series Star," also, the figures "100." That is to say, the motor as it stands on 440 volts is considered 100 per cent. The new voltage is to be 220, which is 50 per cent. of 440. Hence, the same horizontal line in the table, namely, "Three-phase Series Star," is followed along until the desired figure of 50 is found, which is under the vertical heading "Three-phase 2-Parallel Star." This is the correct answer: that is, if a motor is connected three-phase series-star for operation on 440 volts, it must be connected three-phase 2-parallel star, as in Fig. 3, to operate correctly on 220 volts.

2. Change in Phase—Refer again to the table and assume that a three-phase 440-volt motor is to be reconnected for two-phase 440 volts. Inspection shows that the winding as it stands on 440 volts is four-pole three-phase series-delta, as in Fig. 4. Select the horizontal column in the table marked "Three-phase Series Delta" and follow it across, looking for a vertical column showing the value "100," since the desired two-phase voltage is the same as the present three-phase voltage, or 100 per cent. Inspection shows that there is no "100" under any two-phase connection. This indicates at once that a three-phase series-delta connected motor which is normally operated on 440 volts cannot be changed and operated on two-phase 440 volts, without rewinding. The nearest value to "100" under a two-phase column is "70," shown under "Two-phase 2-Parallel." This means that if a three-phase 440-volt motor which is connected series delta, be reconnected for 2-parallel two-phase, as in Fig. 5, it should be operated on 70 per cent. of 440, or 308 volts.

3. Change in Frequency—It is desired to operate a three-phase 440-volt 60-cycle motor on 50 cycles at the same voltage. What change should be made in the connections? Inspection indicates that as the motor stands it is connected for three-phase 5-parallel star on 60 cycles. A change in frequency should be offset by a change in voltage in the same direction and by the same amount; hence, if a motor is operated on 100 per cent. voltage on 60 cycles, it should be connected for $\frac{5}{6}$ of 100, or $83\frac{1}{3}$ per cent., voltage on 50 cycles. However, the voltage is to remain the same on

COMPARISON OF MOTOR VOLTAGES WITH VARIOUS CONNECTIONS AND PHASES

If a motor connected originally, as shown in any horizontal column, had a normal voltage of 100, its voltage when reconnected, as indicated in any vertical column, is shown at the intersection of these two columns

Form of Connection	Three-Phase Series Star	Three-Phase 2-Parallel Star	Three-Phase 3-Parallel Star	Three-Phase 4-Parallel Star	Three-Phase 5-Parallel Star	Three-Phase 6-Parallel Star	Three-Phase Series Delta	Three-Phase 2-Parallel Delta	Three-Phase 3-Parallel Delta	Three-Phase 4-Parallel Delta	Three-Phase 5-Parallel Delta	Three-Phase 6-Parallel Delta	Two-Phase Series	Two-Phase 2 Parallels	Two-Phase 3 Parallels	Two-Phase 4 Parallels	Two-Phase 5-Parallels	Two-Phase 6-Parallels
Three-phase Series Star	100	50	33	25	20	17	58	29	19	15	12	10	81	41	27	20	16	14
Three-phase 2-Paralle Star	200	100	67	50	40	33	116	58	39	29	23	19	162	81	54	40	32	27
Three-phase 3-Parallel Star	300	150	100	75	60	50	173	87	58	43	35	29	243	122	81	60	48	41
Three-phase 4-Parallel Star	400	200	133	100	80	67	232	116	77	58	46	39	324	163	108	80	64	54
Three-phase 5-Parallel Star	500	250	167	125	100	83	289	144	96	72	58	48	405	203	135	100	80	68
Three-phase 6-Parallel Star	600	300	200	150	120	100	346	173	115	87	69	58	486	243	162	120	96	81
Three-phase Series Delta	173	86	58	43	35	29	100	50	33	25	20	17	140	70	47	35	28	23
Three-phase 2-Parallel Delta	346	173	115	87	69	58	200	100	67	50	40	33	280	140	94	70	56	47
Three-phase 3-Parallel Delta	519	259	173	130	104	87	300	150	100	75	60	50	420	210	141	105	84	70
Three-phase 4-Parallel Delta	692	346	231	173	138	115	400	200	133	100	80	67	560	280	188	140	112	93
Three-phase 5-Parallel Delta	865	433	288	216	173	144	500	250	167	125	100	83	700	350	233	175	140	117
Three-Phase 6-Parallel Delta	1,038	519	346	260	208	173	600	300	200	150	120	100	840	420	280	210	168	140
Two-phase Series	125	63	42	31	25	21	72	37	24	18	15	12	100	50	33	25	20	17
Two-phase 2-Parallels	250	125	84	63	50	42	144	73	49	37	29	24	200	100	67	50	40	33
Two-phase 3-Parallels	375	188	125	94	75	63	216	111	73	55	44	37	300	150	100	75	60	50
Two-phase 4-Parallels	500	250	167	125	100	84	288	148	97	73	58	49	400	200	133	100	80	67
Two-phase 5-Parallel	625	313	208	156	125	105	360	165	122	91	73	61	500	250	167	125	100	84
Two-phase 6-Parallels	750	375	250	188	150	125	433	217	144	108	87	72	600	300	200	150	120	100

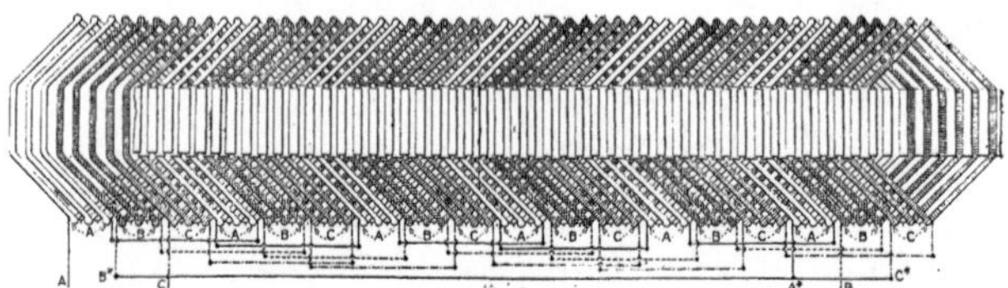

Fig. 2

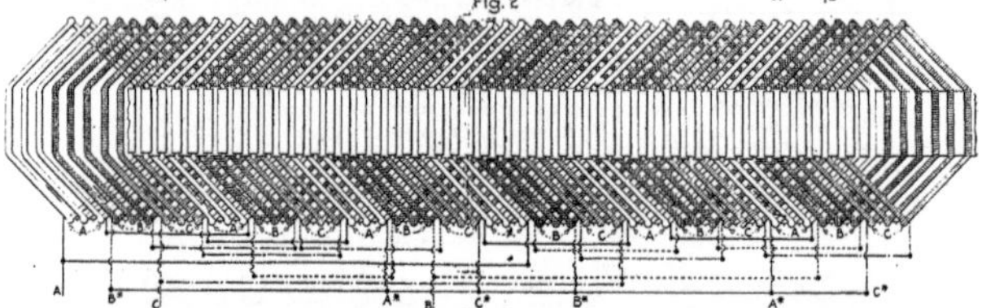

Fig. 3

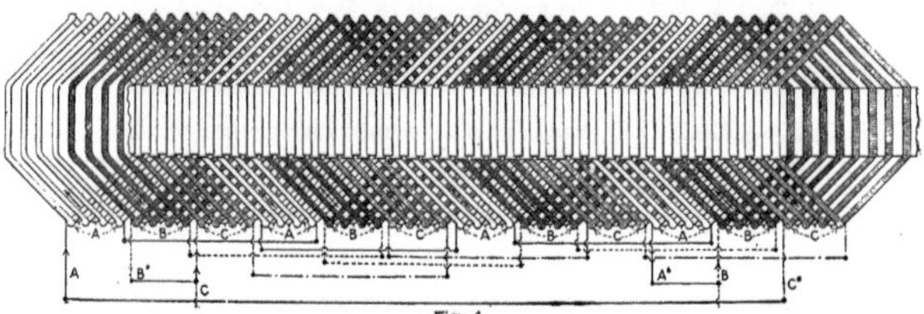

Fig. 4

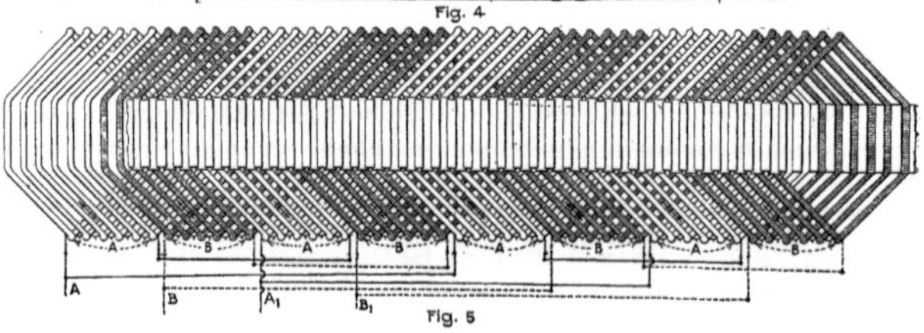

Fig. 5

FIGS. 2 TO 5. CONNECTION DIAGRAMS OF INDUCTION-MOTOR WINDINGS

50 cycles as on 60 cycles, so this result must be obtained in another way. If the voltage cannot be decreased, the number of turns in series can be increased. Another way of saying this is that we can reconnect the winding so that ordinarily it would be good for a higher voltage, and then if it is operated on the same voltage the effect will be the same as if a lower voltage had been applied to the original connection. In the case in hand the motor should, when connected on 50 cycles, be operated on $83\frac{1}{3}$ per cent. of the 60-cycle voltage. Only 100

per cent. is available, so the winding will have to be reconnected with $\frac{100}{83\frac{1}{3}} = 120$ per cent. of the original number of turns in series. This would ordinarily mean the winding was good for 120 per cent. of the original voltage. Hence, in looking up the change in the connection table the figure "120" is located instead of $83\frac{1}{3}$.

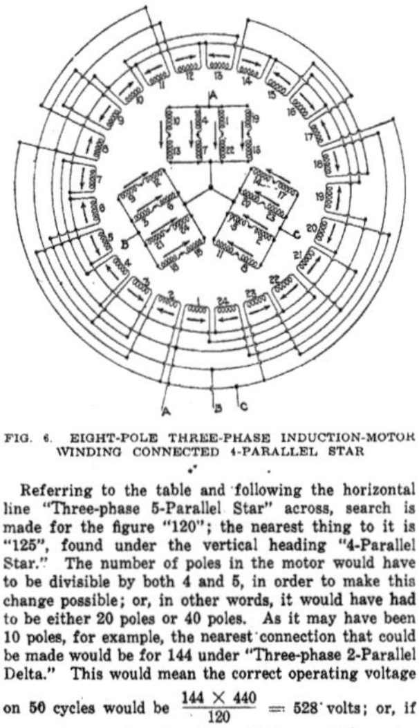

FIG. 6. EIGHT-POLE THREE-PHASE INDUCTION-MOTOR WINDING CONNECTED 4-PARALLEL STAR

Referring to the table and following the horizontal line "Three-phase 5-Parallel Star" across, search is made for the figure "120"; the nearest thing to it is "125", found under the vertical heading "4-Parallel Star." The number of poles in the motor would have to be divisible by both 4 and 5, in order to make this change possible; or, in other words, it would have had to be either 20 poles or 40 poles. As it may have been 10 poles, for example, the nearest connection that could be made would be for 144 under "Three-phase 2-Parallel Delta." This would mean the correct operating voltage on 50 cycles would be $\frac{144 \times 440}{120} = 528$ volts; or, if operated on 440 volts, it would be working under $\frac{440}{528} = 83\frac{1}{3}$ per cent. normal voltage, which would usually not be permissible on account of lowered torque and increased heating.

4. Change in Number of Poles or Speed—A 60-cycle three-phase motor is operating on 550 volts at 850 r.p.m.; it is desired to operate at 690 r.p.m. on the same voltage. What change in connections should be made, if any, in addition to changing the number of poles? Inspection shows the motor is connected 4-parallel star for 8 poles, as in Fig. 6. To get 690 r.p.m. would require to connect for 10 poles, since this would give a no-load speed of about 720 r.p.m. and a full-load speed of about 690 r.p.m. Since the motor is a generator also, it will generate only $\frac{690 \times 550}{850} = 446$ volts when connected for 10 poles and a slower speed. However, it is desired to continue at 550 volts, so that the connections will have to be changed to get the effect of $\frac{550}{446} = 123$ per cent. of the old voltage. In the table opposite the horizontal line "Three-phase 4-Parallel Star," the nearest figure to "123" is "116," which is found in the vertical column headed "Three-phase, 2-Parallel Delta." Hence, the conclusion is drawn that if an 8-pole motor, Fig. 6, is connected three-phase 4-parallel star and operated on 550 volts and it is reconnected for 10-poles 2-parallel delta, Fig. 7, it may be still operated on 550 volts, although, strictly speaking, its normal voltage would be $\frac{116 \times 550}{123} = 520$ volts. In this example no consideration was given to the fact that the throw of the coil in electrical degrees was changed in changing from 8 poles to 10 poles. This can be taken account of in the following way:

Suppose the motor as it stood had 120 stator slots and the coils lay in slots 1 and 13. Full pitch would be 1 and 16, since $\frac{120}{8} = 15$. Since the coils throw 12 slots and full pitch is 15 slots, the per cent. pitch $= \frac{12}{15} = 80$ per cent., and from Fig. 1 the chord factor for 80 per cent. pitch $= 0.95$. When reconnected for 10 poles, the throw of the coils is still 1 and 13, but this is now 100 per cent. pitch since $\frac{120}{10} = 12$ and 1 and 13 does span 12 slots. Therefore, when connected for 10 poles the coils are more effective in the ratio of $\frac{1.00}{0.95}$ since the chord factor for 100 per cent. pitch $= 1.00$ from Fig. 1. Therefore, when the change in chord factor is also taken account of, the new normal operating voltage is 520, as obtained in the foregoing, multiplied by $\frac{1.00}{0.95} = 548$ volts, or almost exactly right for operation on 550 volts.

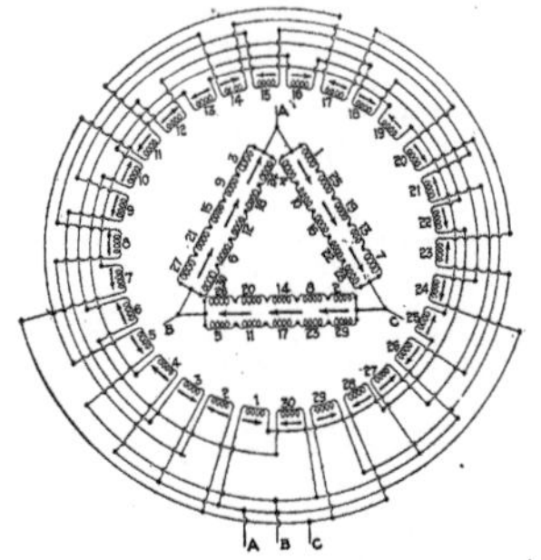

FIG. 7. TEN-POLE THREE-PHASE INDUCTION-MOTOR WINDING CONNECTED 2-PARALLEL DELTA

5. Change in Horsepower—A 10-hp. 220-volt motor

is operating above the allowable safe temperature, on its normal voltage, and it is found by experiment that when the voltage is raised to 250 its temperature is reduced to within safe limits. Can any change be made in the connections which will allow the motor to be operated still on 220 volts and duplicate the conditions when operating on 250 volts? An inspection of the winding shows the motor to be connected three-phase series delta, as in Fig. 4. The experiment which was made showed that the voltage should be increased to $\frac{250}{220} = 114$ per cent. of its original value. It has been pointed out that reducing the number of turns in series in a winding has the same effect as increasing the voltage on the same number of turns. In this case if the voltage was raised to 114 per cent. the same effect could be obtained by reducing the turns to $\frac{1.00}{1.14} = 87.5$ per cent. Consequently, in referring to the voltage change table, in this case, search is made for "87.5" and not "114."

Selecting, therefore, the horizontal line "Three-phase Series Delta" in the table and looking across the nearest figure to "87.5" is "86," which occurs under the vertical heading "Three-phase Parallel Star." Consequently, the conclusion is at once drawn that if a 220-volt motor has its connections changed from series-delta, Fig. 4, to parallel-star, Fig. 3, it will act in every way as though $\frac{220}{0.86} = 256$ volts had been applied to the series-delta connection. This is equivalent to increasing the horsepower of the motor, since on the original connection the motor was overloaded when carrying its rated load, but when the connections of the winding were changed the machine could drive its rated full load without distress. The reason for this is that, although the density of the magnetic flux was increased the cross-section of the copper in the winding was increased, consequently the copper losses were reduced. The latter being considerably greater than the former resulted in a reduced temperature. The capacity in horsepower has actually been increased to $\frac{256}{220} = 116$ per cent. of its original value.

From these five examples, which could be multiplied many times and from all sorts of combinations that could be made by changing the characteristics in pairs, it can be readily seen that any contemplated change can be reduced to an equivalent change in the applied voltage and the proper connection, if it is a feasible and rational change, selected from the table of phase and voltage given herewith.

(A later article will deal with a means of roughly figuring a new winding for an old core. This in turn will be followed by a description of methods for locating faults in windings.)

Treatment of Power-Plant Trouble

By M. A. SALLER

ONE OF the greatest bugaboos confronting the manufacturer and installer of power-plant equipment, as well as the purchaser or user thereof, is the "trouble case" where the apparatus does not do the work required of it, does not stand up, or in general proves unsatisfactory in operation. It is a well-known fact that very few large or small new plants are put in operation without a "trouble case" being developed in connection with some part or parts of the equipment.

Just where the difficulty will be encountered in any plant cannot usually be foreseen, for in a number of plants designed and erected by the same engineers, the operation of the equipment in the one plant will give extremely satisfactory results, but in another plant, under almost the same general conditions, a persistent trouble case will develop in connection with similar equipment.

Operator Should be Prepared to Act Promptly in Case of Trouble

The trouble case therefore cannot be anticipated either by the designer or by the operator, and consequently when difficulty does arise in connection with the operation or performance of any equipment or apparatus, steps should be taken promptly in order to study and rectify the trouble, and in practicably every case this can be done to the best advantage with the hearty coöperation of the designer, installer, operator and owner as the first essential in order to prevent a long-drawn-out controversy, because when the responsibility is shifted, and "the buck is passed," little actual progress in adjusting the difficulty can be accomplished.

A study of many power-plant troubles of various kinds points to the following as the average causes thereof: Misunderstanding of the functions of the apparatus; improper installation; improper operation or overloading of the equipment, and the mental attitude of the operators.

The first cause gives room for the development of all sorts of trouble cases. The designer or engineer may not properly or completely understand the detailed requirements of the user of the apparatus as it affects his own peculiar conditions of operation; or again, the purchaser or operator may assume, correctly or incorrectly, that the apparatus that is being installed by the engineer or the builder will do certain things that the maker does not expect or contemplate will be required of it. In other words, there is always present as a possible source of difficulty the human factor, which on one side underestimates or is ignorant of some special conditions which must be met, or on the other side expects too much or the "impossible" from the apparatus that is installed.

One method of guarding against such misunderstandings is the preparation, by the purchaser, of complete specifications as to the requirements, including all special features, and the submission by the designer or builder of equipment clear-cut statements as to the performance of the apparatus under certain conditions or under the special requirements, with guarantees and specifications as to just what he is to do; it being understood, of course, that the resort to such specifications and guarantees is not primarily for the purpose of allowing the builder or the purchaser to "get out from under," but more particularly for the purpose of developing the details as to what is wanted and what is to be furnished, so that both elements may finally be made to hamonize.

Frankness on the part of the purchaser as to what he really wants to accomplish, even though it does involve something out of the ordinary or difficult to attain, and frankness on the part of the engineer or equipment builder as to just what can be accomplished and done with the apparatus he proposes, and an acknowledgment that something special or more expensive is required in case conditions make necessary, are essential if the "trouble case" from this score is to be avoided.

Trouble Caused by Improper Installation of Equipment

Improper installation, of course, is responsible for a larger percentage of trouble with apparatus and equipment because of faulty or inefficient operation. Practically every manufacturer of standard power-plant equipment or material has available clear and complete instructions and directions as to the proper method of installation of the apparatus he supplies, but frequently, because of incompetence or ignorance on the part of the men doing the actual installation and erection work, many important points are overlooked which may later develop to be the causes of improper operation.

Again, it is possible that the individual piece of apparatus may be installed in accordance with the maker's standard directions, but in spite of this does not fit in satisfactorily with the general scheme of the plant, which may be along some specialized line. In other words, while standard directions for installation and operation may be complied with to satisfactorily cover the general run of cases, there is always arising in engineering problems the special case which demands treatment in other than the ordinary way.

Therefore, it seems that the most desirable course to pursue is for the purchaser or installer of equipment to submit to the manufacturer blueprint or details of the manner in which the apparatus is to be installed in his particular plant, with a statement of the general operating conditions, so that the benefit of the maker's wider experience in the arrangement of equipment can be obtained.

Along the same line, trouble can also be generated because of improper operation or overloading of equipment. Apparatus that is built for a certain work with provision for a certain overload cannot reasonably be expected to take care of a load considerably above that point, and manufacturers of equipment are often confronted with dissatisfied customers who condemn the equipment they have in service because it does not operate with complete satisfaction under a tremendous overload, when the obvious course to pursue is to install additional capacity. The same condition also applies to the operating factor. Changes in engineers or the employment of incompetent operators will often he reflected in the improper operation and adjustment of equipment, which lack of attention naturally means a falling off in efficiency or capacity. Here again the operator should find it profitable to resort to the builder or installer for guidance and information for securing the best results from the installation of his particular apparatus.

Mental Attitude Is Often a Source of Trouble

Another fruitful source of trouble cases is the mental attitude of the purchaser, the manufacturer or the operating force. Not infrequently the operating engineer has a prejudice against a certain class or type or make of equipment, with the result that often he will go out of his way to condemn the operation of the apparatus or magnify any difficulties encountered in its operation. Here again is the personal element, which can only be dealt with according to the circumstances of the case.

Again, the maker or the purchaser may get it into his head that the other fellow is trying to put one over on him in the way of shifting responsibility for some troubles or difficulties, and hence instead of the close coöperation and harmony between them that is essential and desirable to insure the best service and results, they continue at loggerheads, each condemning the other and his apparatus or plant, with the result that nothing definite is done to remedy the faults and dissatisfaction prevails until the apparatus is taken out of service and junked.

This almost inevitably leads to large expense and great inconvenience for both the purchaser and the maker, all of which could easily have been avoided by a spirit of closer coöperation and a better appreciation by each of the position of the other.

The engineer is wrong when he is not open to conviction that some other design of apparatus is as well fitted for the work as is the machine he has decided is best. The manufacturer is wrong when he attempts to sell a machine that is not adapted to the work or has features that are objectionable to the engineer. The purchaser is wrong when he is suspicious of everyone concerned.

Therefore, when a trouble case with equipment is encountered in any plant, it is well to bear in mind the four essential features as outlined, and to remember particularly that close coöperation between the maker and the user is essential in satisfactorily clearing up the difficulty.

Secretary Lane Reviews Fuel and Power Situation

Believes We Have Too Many Mines and Miners for Domestic Demands, and That We Should Cultivate a Greater Foreign Market

In his annual report to the President, Secretary Lane expresses the belief that the measure of a people's industrial capacity is fixed by the power possibilities and that America may well take stock of her own power possibilities and concern herself more actively with their development and wisest use. Excerpts from his report follow.

THE strike has brought concretely before us the disturbing fact that modern society is so involved that we live virtually by unanimous consent. Let less than one-half of one per cent. of our population quit their work of digging coal, and we are threatened with the combined horrors of pestilence and famine. There was an immediate demand for facts as to how much coal is normally mined in this country; by whom it is mined; its quality; to what uses it is put; what better methods have been developed for using coal than those of ancient custom; who is to blame that so small a supply is on the surface; why we should live from day to day in so vital a matter as a fuel supply; what substitutes can be found for coal, and how quickly these can be made available.

Many of these questions we were able to answer, but if coal operators had not carried over the statistical machinery developed during the war we would have been forced to the humiliating confession that we did not know facts which at the time were of the most vital importance.

In a time of stress it is not enough to be able to say that the United States contains more than one-half of the world's known supply of coal, and that we, while only 8 per cent. of the world's population, produce annually 46 per cent. of all the coal that is taken from the ground; that 35 per cent. of the railroad traffic is coal; that in less than one hundred years we have grown in production from 100,000 to 700,000,000 tons per annum.

The Government should have a more complete knowledge of the coal and other foundation industries than can be found elsewhere, and we should not fear national stock-taking as a continuing process. The war revealed how delinquent in this regard we have been in the past. To know what we have and what we can do with it is a policy of wisdom and of lasting progress. In furtherance of such a policy the first step is to know our resources; the second, to know their availability; the third, to guard them against waste, either through ignorance or wantonness, and the fourth, to prolong their life by invention and discovery.

Next to the fertility of our soil we have no physical asset as valuable as our coal deposits. Although sometimes alarmed because those deposits nearest to the industrial centers are rapidly declining and we can already see within this century the end of the anthracite field if it is made to yield as much continuously as the present, yet it is a safe generalization that we have sufficient coal in the United States to last our people for centuries to come, but the recent strike made it clear that it is not the coal in the mountain that is of value, but that which is in the yards.

We have been content to go without insurance as to a coal reserve, because each day has brought its daily supply. There was no thought of railroads stopping or mines closing down, so that large storage facilities had not been provided, and indeed we would rebel at paying for our coal the added cost of caring for it outside its native warehouse. We have not thought in terms of apprehension, but, as always, in the calm certainty that the stream of supply would flow without ceasing.

Average for Year Low

The record year 1918, with everything to stimulate production, had an average of only 249 working days for the bituminous mines of the country. This average of the country included a minimum among the principal coal-producing states of 204 days for Arkansas and a maximum of 301 days for New Mexico. In such a state as Ohio the average working year is under 200 days. In 1917 the miners of New Mexico reached an average of 321 days, and in the larger field, the Raton field, it was actually 336 days, probably the record for steady production.

This short year in coal-mine operation is due partly to seasonal fluctuation in demand. The mines averaged only twenty-four hours a week during the spring months. The weekly report of that date showed that 80 per cent. of the lost time was due to no market, and only 15 per cent. due to labor shortage, while car shortage was a negligible factor. In contrast with this was the last week before the strike, when the average hours operated were thirty-nine and no market was a negligible item in lost time, while car shortage was by far the largest item. It follows that the short year is a source of loss to the operator, the mine worker, and a tax on the consumer.

Plants and Labor Less Productive

With substantially the same number of mines and miners working this year as last, the accumulated production for the first ten months of this year is 100,000,000 tons less than that mined in the same period last year. This 25 per cent. loss in output means that both plant and labor have been less productive, and in terms of capital and labor coal has cost the nation more than this year than last.

The public must accept responsibility for the coal industry and pay for carrying it on the year round. Mine operators and mine workers of whatever mines that are necessary to meet the needs of the country must be paid for a year's work. The shorter the working year the less coal is mined per man and per dollar invested in plant, and eventually the higher priced must be the coal. It is obvious that the 236 tons of coal mined by the average British miner last year could not be as cheap per ton as the 943 tons mined by the average American mine worker, backed up as he was with more efficient plant equipment.

Have We Too Many Mines and Miners?

The problem of the miner and his industry may be stated in another way. We consume all the coal we produce. We produce it with labor that, upon social and economic grounds, works, as a rule, too few days in the year. We therefore must have a longer miners' year and fewer miners, or a longer miners' year and additional markets. One or the other is inevitable, unless we are to carry on the industry as a whole as an emergency industry, holding men ready for work when they are not needed in order that they may be ready for duty when the need arises. There are too many mines to keep all the miners employed all the time or to give them a reasonable year's work. This conclusion is based on the assumption that we now produce only enough coal from all the mines to meet the country's demands, which is the fact. More coal produced would not sell more coal, but more coal demanded would result in greater coal production. With the full demand met by men working two-thirds or less of the time in the year, there cannot be a longer year given to all of the miners without more demands for coal. Therefore, the miners must remain working but part time, as now, or fewer miners must work more days, or markets must be found for more coal, and thus all the miners given a longer year. If we worked all our miners in all of the mines a reasonable year, we would have a great overproduction, and to have all our mines worked a longer period means that we must find some place in which to sell more coal, either at home or abroad.

Present Methods Uneconomical

One naturally asks why the long haul over mountains and through tunnels and across bridges, and along streets and into houses by railroad, trucks and on the backs of men, when at the very pit mouth, or within the mine itself, this same coal might be transformed into electricity and by wire served into factories and homes one, two and three hundred miles from the mines? Why burden our congested railroads with this traffic? This may be a practical thing; it deserves study if coal is worth conserving.

It has been everyone's business to save coal. The railroads have experimented with some success. They get perhaps 10 per cent. of the heat energy from a ton shoveled beneath the locomotive boiler. They use one-quarter of all the coal mined. Next to labor this is the greatest expense which our railroads have. We can overlook the stoking of the domestic furnace as a national concern, for the amount of coal used in this way amounts to not more than 17 per cent. of the national coal bill, and this whole charge could be saved, it is estimated, by giving care to the bulk of our coal which is burned under boilers to make steam.

The Government should sample and certify coal. We do this as to wheat and meat, and it should be just as necessary to avoid injustice in the case of coal. The public should know the kind of coal it is buying. There need be no prohibition against the mining or selling of any coal, but coal should sell in terms of its capacity to deliver heat.

Boiler-Room Instruments Should Be Provided

It is not essential for the plant manager to be a fuel expert, but he should be familiar with the instruments that give a check on the daily operations. It is a mistake not to provide proper instruments, for they guide the firemen and show the management what has taken place daily. Instruments provided for the boiler room manifest the interest taken by the management toward conserving fuel. It indicates coöperation and encourages the firemen to work harder to increase efficiency.

It has been said that we have too many mines in operation, as we appear to have too many miners, if we are to maintain only our present output. Rapid expansion in the development of industry may justify the existence of such mines and so large a corps of workers. If, however, this should not be so, there is a foreign demand for the best of our bituminous coals, which at present we are altogether unable to meet through lack of credits on the part of those who wish the coal and lack of ships to carry it. England's annual production has fallen 100,000,000 tons, and European demand next year will be more than 150,000,000 tons above her production. As this country, prior to the war, sold abroad no more than 4,500,000 tons as against England's 77,000,000, it is quite manifest that here will be a new field for American enterprise.

Saving Coal by Saving Electricity

It is three years since Congress was urged that we should be empowered to make a study of the power possibilities of the congested industrial part of the Atlantic Seaboard, with a view to developing not only the fact that there could be effected a great saving in power and a much larger actual use secured out of that now produced, but also that new supplies could be obtained both from running water and from the conversion of coal at the mines instead of after a long rail haul. A stream of power paralleling the Atlantic coast from Richmond to Boston, a main channel into which run many minor feeding streams, and from which diverge an infinite number of small delivering lines—the whole an interlocking system that would take from the coal mine and the railroad a part of their present burden and insure the operation of street lights, street cars, elevators and essential industries in the face of railroad delinquencies—this is the dream of our engineers, and a very possible dream it seems to me. To tie together the separated power plants of ten states, so that one can give aid to the other, so that one can take the place of the other, so that all may join their power for good in any great drive that may be projected—this would be the prime purpose of the plan.

It is likely that the long-pending power bill which will make available the dam and reservoir sites on withdrawn public lands, and make feasible the financing of many projects on both navigable and unnavigable streams, will soon have become law. We shall then have an opportunity that never before has been given us to develop the hydro-electric possibilities of the country. However, as a matter of fact, the water-power resources of the country are by no means evenly distributed. Over 70 per cent. of the available water power is west of the Mississippi, whereas over 70 per cent. of the total horsepower now installed in prime movers is east of the river. Therefore, unless the East is to lose its industrial supremacy, it must press, and press hard, for the development of all water-power possibilities.

The Age of Petroleum

In 1908 the country's production of oil was 178,500,000 bbl., and there was a surplus above consumption of more than 20,000,000 bbl. available to go into storage. In 1918, ten years later, the oil wells of the United States yielded 356,000,000 bbl.—nearly twice the yield of 1908—but to meet the demands of the increased consumption more than 24,000,000 bbl. had to be drawn from storage. The annual fuel-oil consumption of the railroads alone has increased from 16⅔ to 36¾ million barrels; the annual gasoline production from 540,000,000 gal. in 1909, to 3,500,000,000 gal. in 1918. This reference to the record of the past may be taken not only as justifying the earlier appeal for Federal action, but as warranting deliberate attention to the oil problem of today. There are large bodies of public land now withdrawn, which, under the new leasing bill, which seems so near to final passage after seven years of struggle and baffled hope, will in all likelihood make a further rich contribution to the American supply.

And beyond these in point of time lie the vast deposits of oil shale which by a comparatively cheap refining process can be made to yield vastly more oil than has yet been found in pools or sands. The value of this shale oil will depend upon the cheapness of its reduction, and this must be greatly lessened by the value of byproducts before it can compete with coal or oil from wells.

Yet with all the optimism that can be justified there should be a policy of saving as to petroleum that should be rigid in the extreme. We must save the oil before it leaves the well, keep it from being lost and keep it from being flooded out and driven away by water.

There are possible substitutes for some petroleum products, but not for the whole barrel of oil; furthermore, petroleum is the cheapest material, speaking quantitatively, from which liquid fuels and lubricants can be made; therefore, substitutes obtained in quantity must cost more. Alcohol can be substituted for gasoline, but only in limited quantity and at increased cost. Benzol from byproduct coking ovens can also be used, but quantitatively is totally inadequate. For kerosene no quantitative substitute is known.

Wanted—A Foreign Supply

Already we are importers of petroleum. We are to be larger importers year by year. The situation calls for a policy prompt, determined, and looking many years ahead. For the American navy and the American merchant marine and American trade abroad must depend to some extent upon our being able to secure, not merely for today, but for tomorrow as well, an equal opportunity with other nations to gain a petroleum supply from the fields of the world.

We must look abroad for a supplemental supply, and this may be secured through American enterprise if we do the following things:

(1) Assure American capital that if it goes into a foreign country and secures the right to drill for oil on a legal and fair basis (all of which must be shown to the State Department) that it will be protected against confiscation or discrimination. This should be a known published policy.

(2) Require every American corporation producing oil in a foreign country to take out a Federal charter for such enterprise, under which whatever oil it produces should be subjected to a preferential right on the part of this Government to take all its supply or a percentage thereof at any time on payment of the market price.

(3) Sell no oil to a vessel carrying a charter from any foreign government, either at an American port or at any American bunker, when that government does not sell oil at a nondiscriminatory price to our vessels at its bunkers or ports.

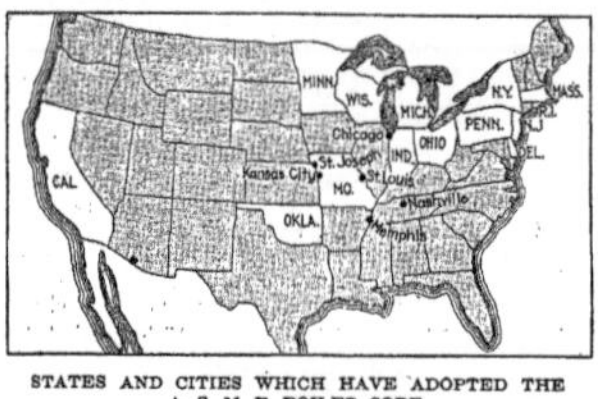

STATES AND CITIES WHICH HAVE ADOPTED THE A. S. M. E. BOILER CODE

Riva Pelton Turbine Governor

At the shop of the Costruzioni Meccaniche Riva, Milan, there is in process of construction an interesting and rather unusual type of turbine, says Norman Dagui, in *The Engineer*. This turbine is to be employed to produce power for the Italian Electric State Railways. It consists of two Pelton wheels of different diameter, on the same shaft, one operated by a head of 2000 ft. and the other of 670 ft. Because of the conditions under which the power is acquired, a large variation of load occurs throughout the twenty-four hours, and the plant is fitted with an automatic device by means of which the smaller wheel, which is only used when

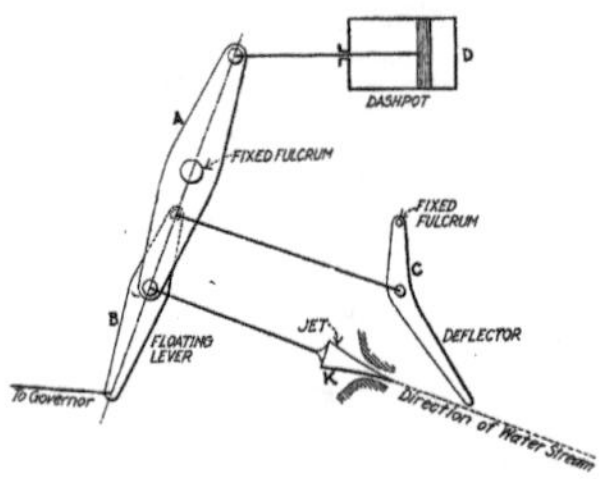

RIVA SPEED CONTROLLING ARRANGEMENT FOR PELTON-WHEEL TURBINES

the load rises to a certain point, may be put to work. The diameters of the two wheels, taken at the centers of the buckets, are 6.4 ft. and 3.67 ft., respectively, and the speed is 500 r.p.m. The larger-diameter wheel develops 3500 hp. with one jet and the smaller wheel 2500 hp. with two jets. The two Pelton wheels are keyed to the shaft between the two bearings.

The distinguishing feature of the Riva turbine of the Pelton type is the method of governing the speed. The governor or, rather, the actual control of the water jet, is effected in two ways—first, by deflection, and secondly, by throttling. It is necessary, of course, to produce deflection as an initial means of removing power from the rotor, for, were the necessary power removed to be made simply by the closing of the orifice, the water shock produced would become dangerous. To avoid this it is necessary that the closing of the jet be made slowly, but at the same time it is equally necessary that the power should be taken from the turbine with the least possible delay. This dual object is achieved by a system of mechanism shown diagrammatically herewith. In it A is a lever working on a fixed fulcrum. To one end of this lever is attached a dashpot D; on the other end is fulcrumed the floating lever B. On one end of this floating lever operates the deflector C; the other is connected with the governor. The fulcrum itself is connected with the jet K.

Supposing the load to be suddenly removed from the turbine and racing begins, the governor immediately reacts on the floating lever B, and the deflector comes into action on the water stream. Since resistance, however, is offered in its operation to the turning of the lever B about its fulcrum, the fulcrum itself becomes the moving point about the fulcrum of the lever A, and the jet K begins to close.

Refrigeration Study Course—I. What Is Mechanical Refrigeration?

By H. J. MACINTIRE

Professor of Mechanical Engineering, University of Idaho

REFRIGERATION is an old acquaintance, one that we meet face to face every day in one or another of its forms. The wind from the snow-capped mountain peaks, or from a flow of icebergs, or from a body of glacial water, gives us the sensation of cold; that is, the air becomes colder than our bodies and therefore it seems cold to us. In another manner the watering wagons, passing down the hot street and wetting the sizzling pavements, give us relief and a pleasant, cool sensation, which is again an application of refrigeration inasmuch as a reduced temperature has been produced which, for the time being at least, is colder than it was before. This is similar in action to that taking place in spray ponds and cooling towers, for a cooling action takes place which lowers the temperature of the water. Finally, we have the action of some volatile liquid like ether on one's hand. The ether evaporates, leaving a feeling of cold. Now, how do these different means of refrigeration take place, and what has it to do with mechanical refrigeration? An attempt in the following will be made to show analogies to the common means of cooling to be seen in nature.

Cooling With Ice

The most common form of refrigeration is that produced in iceboxes. Now, ice possesses a peculiar property of being able to store up "cold"; that is, it requires a large amount of heat units to change ice at

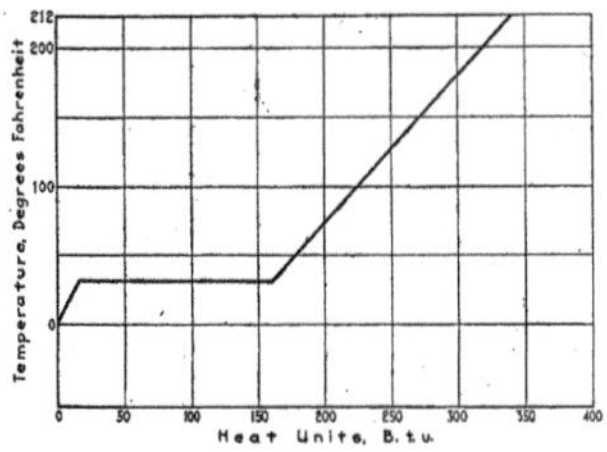

FIG. 1. SHOWING THAT IT TAKES PRACTICALLY AS MUCH HEAT TO MELT A POUND OF ICE AS TO INCREASE THE TEMPERATURE OF A POUND OF WATER FROM 32 DEG. TO 176 DEG.

32 deg. F into water at 32 deg. F. In fact, it will require as much heat to melt a pound of ice as would be required to raise the temperature of a pound of water from 32 deg. F. to 176 deg. F. This is shown diagrammatically in Fig. 1. But how does ice in an icebox or a bunker produce refrigeration?

First, we must understand that refrigeration is a means of lowering the temperature of something below the temperature of the surroundings or of keeping the temperature of commodities below the temperature of the surroundings. Melting ice would not be of any use in a fish freezer, where a temperature of —10 deg. F. is required, because it melts at 32 deg. F. Melting ice is only good for refrigeration requiring a temperature of 38 deg. or 40 deg. F. Now, why is this?

First, heat will not "run uphill" any more than water will. Heat will only go from a hotter to a colder substance, and the tendency is for a common temperature to be obtained between the hot and the cold substance.

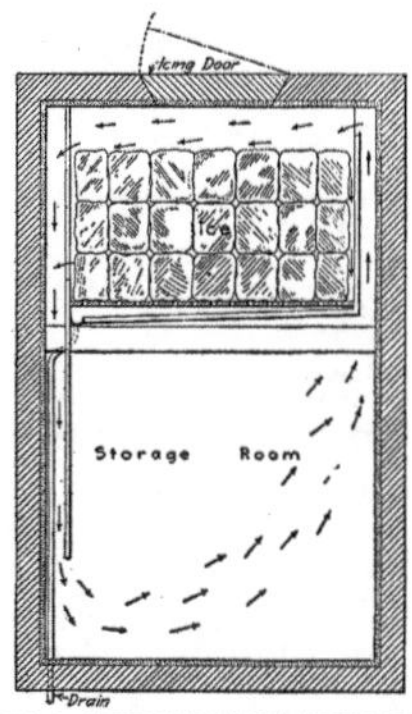

FIG. 2. SECTION OF ROOM WHERE ICE ALONE IS USED FOR REFRIGERATION

Now, ice will melt at 32 deg. F., and in so doing a pound of ice will absorb 144 heat units (144 B.t.u.). This heat must come from some other source—not goods or air at zero deg. F., because these are colder, but something hotter than 32 deg. In the ice bunker heat from the goods being stored or heat from the outside gives the air in the storage room some particular temperature. The air in the immediate vicinity of the ice in the bunker is very nearly 32 deg. F., say 34 deg. F. This cold air in the upper part of the storage room —the bunker part—is heavier than the air underneath, and in consequence (if proper vent openings are provided) a flow of air will take place, the cold air will descend and the hot air will ascend to give some of its heat to the ice which, in absorbing this heat, becomes melted. In other words, air in the room is a carrier of *heat*, but if the box is designed carefully for the conditions that prevail and sufficient ice is maintained, then the box will never get too warm and the meats and vegetables will be preserved from rapid decay.

On analyzing the foregoing, it will be seen that a temperature below that of the atmosphere may be obtained by securing a substance that can absorb heat *below* the temperature desired in the cold room (in the case cited it was ice), that there must be a fluid carrier of heat which will take the heat units from the goods to be cooled (the walls, ceiling, meats and vegetables) and carry this heat to the refrigerator (the ice), and finally, that the separate parts must be proportioned satisfactorily if success is to be attained. In heating by the hot-air method we have a similar process. The air is heated in a furnace or bunker coil, ascends by its own low density or is circulated by means of a suitable fan

FIG. 3. LIQUID AMMONIA IN A TEST TUBE BOILING AT ATMOSPHERIC PRESSURE

and comes into contact with the colder surroundings in the rooms to be heated. The warm air gives up a part of its heat to the room and may be sent again to the furnace or bunker for more heat units.

Cooling by a Volatile Liquid

However, the limitations to the use of ice may be seen readily for refrigeration of any except very small installations because of its expense, annoyance, dirt and its limited cooling temperature. For practical purposes it is necessary to obtain some system which will give any desired temperature and which is economical and reliable. As already suggested, there is another method of refrigeration, using a volatile liquid that produces an effect similar to that of an alcohol sponge bath for a fevered patient, ether on one's palm or water on a hot pavement, each in turn absorbing heat from the surroundings (boiling) and thereby becoming a vapor. This rather difficult phenomenon may be explained by considering the action of water and steam, remembering that *all vapors* act in a similar manner and have their own physical properties.

For example, we all know that water *boiling* at a pressure of 150 lb. gage has a temperature of 365.9 deg. F., but at atmospheric pressure (sea level) it boils at only 212 deg. F. Likewise, steam in a condenser at 26 in. vacuum *condenses* at a temperature of 125 deg. F., but at 29 in. vacuum it condenses at a temperature of 79 deg. F., assuming a standard barometric pressure of 29.92 in. mercury. Why is this? one might ask. Simply because this is the law governing the relation of pressure to temperature for saturated water vapor. It is absolutely constant for any particular liquid, and these relations are formed by experiment. If we have the temperature, we may find the corresponding pressure (provided the vapor is *saturated*) by referring to the proper tables where these values are given. However, water vapor is not suitable for low-temperature refrigeration and so some other vapor must be used, and as everyone knows, the most common refrigerant is *ammonia*.

The advantages of ammonia over those of water vapor or other refrigerants consist of the *low boiling* temperature of the substance and the reasonable pressures at which this takes place. For instance, if ammonia is allowed to boil in a glass, it will do so (at sea level) —27.2 deg. F. If it were possible to keep the pressure constant at 15.23 lb. gage and to have it boil, it would do so at zero degrees F., and at 20 lb. gage it would be 6.1 deg. F. These pressures are convenient to deal with, but, equally important, there is also the *heat capacity* of the refrigerant.

The heat capacity may be spoken of as the heat-carrying ability of a substance. Now, air has little heat-carrying capacity, for air at atmospheric pressure would require one cubic foot with a change of about 50 deg. F. to give one unit of heat (or refrigeration). On the other hand, one cubic foot of liquid ammonia would give approximately 20,000 heat units without a change of temperature. The comparison speaks for itself and shows one of the reasons why ammonia has been chosen as the refrigerating medium. But the problem is much more involved than has yet been developed. Ammonia is expensive (30c. per lb. now) and cannot be allowed to go to waste. We cannot, for economical reasons, allow this chemical to be thrown away after but a single round of duty. We must make it perform its work again and again. How can that be done?

The Closed Cycle

Evidently, we cannot allow the vapor to escape into the atmosphere. It must be kept confined all the time. But how can we accomplish this and yet keep the boiling pressure constant? And how can we change a gas back into a liquid again? The answer is, by means of a machine. Let us consider the matter and see if we haven't an old friend disguised in a new make-up.

Our old friend the steam engine takes steam which has been made by heating water in a boiler and then

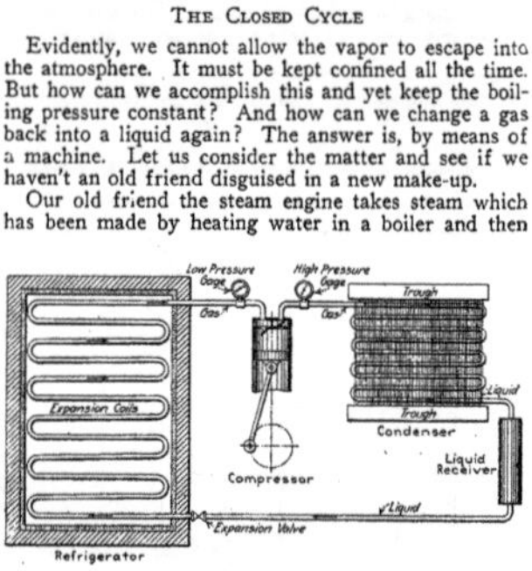

FIG. 4. DIAGRAMMATIC LAYOUT OF SIMPLE AMMONIA REFRIGERATING PLANT

evaporating it into steam. This steam was generated under, say, 100 lb. gage pressure and absorbed the necessary heat units from the heat of combustion of the fuel burning under the boiler. Then the steam was expanded in the steam engine and finally exhausted under greatly reduced pressure to a surface condenser where cooling water or some other *refrigerating* agent took away the heat of condensation and left water at a temperature corresponding to the pressure in the condenser. Let us repeat the cycle. Water is heated until a temperature is finally reached where evaporation begins, then evaporation continues at constant temperature (and constant pressure) until all the water is steam, then *work* is done on the steam engine, accompanied by a drop of pressure (and temperature) and finally the resulting low-pressure steam is condensed in a water-cooled condenser refrigerator. Note that condensation and evaporation were separated by *work done* on a steam engine. In refrigeration the opposite cycle must be used. For example, suppose the boiling ammonia absorbs heat and becomes vaporized (this corresponds to the steam boiler, only it is working the opposite way); then (to reverse the steam-engine cycle) the vapor must *have work done on it* and the pressure raised, and finally heat is taken away from the compressed gas until the latter becomes liquid.

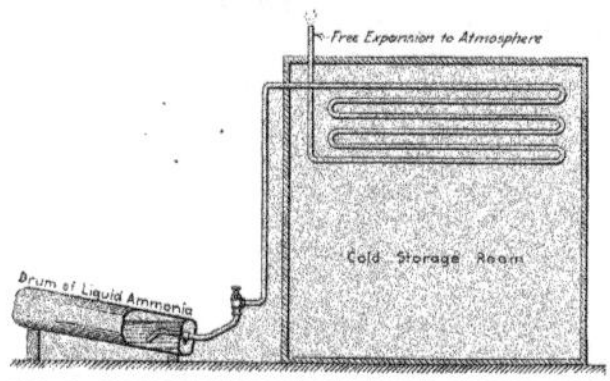

FIG. 5. ILLUSTRATING EXPANSION OF AMMONIA FROM SHIPPING DRUM

Let us see if this can be explained another way. Suppose we look at the problem face to face. We want to use the low-pressure ammonia vapor again, then how can we condense it?

Referring to tables of saturated ammonia, it will be seen that ammonia temperatures and pressures are as follows:

Pressure, Lb. Gage	Temperature, Deg. F.
75	43.2
100	63.4
125	74.4
150	84.0
175	92.5
200	100.3

Liquid ammonia at 100 lb. gage will boil at 63.4 deg. F., but also compressed ammonia gas at 100 lb. will *condense* at the same temperature. Likewise, at 150 lb. the compressed gas will condense at 84 deg. F. and at 200 lb. it will condense at 100.3 deg. F. So, then, if we have 80 deg. F. water, we can use enough of this so that the liquid ammonia after condensation is 92.5 deg., in which case the pressure will have to be 175 lb.

According to the report issued by the Department of Commerce, Bureau of Foreign and Domestic Commerce, Washington, D. C., the value of domestic exports of electrical machinery and appliances (excluding locomotives) from the United States during the month of October, 1919, to foreign countries was for the following items: Batteries, $507,704; carbons, $121,621; dynamos and generators, $452,946; number of fans, 4599, at a value of $79,406; heating and cooking apparatus, $164,114; insulated wire and cables, $487,444; interior wire and supplies, including fixtures, 168,902; number of arc lamps, 145, value $2239; number of carbon incandescent lamps, 26,791, value $6075; number of metal-filament incandescent lamps, 1,073,090, value $334,835; magnetos, spark plugs, etc., $222,972; meters and measuring instruments, $237,077; motors, $1,007,930; rheostats and controllers, $63,153; switches and accessories, $347,152; telegraph apparatus, including wireless, $44,566; telephones, $388,428; transformers, $218,789; all other, $2,425,310; total, $7,280,963.

Safety Motor-Starting Switch

A new type of motor-starting switch, which is low in first cost but provides complete protection to both the operator and the motor, has just been placed on the market by the Westinghouse Electric and Manufacturing Company. It is designed for connecting 250- and 550-volt single-phase or polyphase alternating-current motors, of from one to ten horsepower, directly to the line without the use of auto-transformers or resistance. As shown in the figure, all the mechanism with the exception of the operating handle is inclosed in a steel box. The protective devices are easily accessible on opening a door in the cover, as shown; but this door cannot be opened except when the switch is in the off position. Therefore it is impossible to receive injury from contact with live parts when either operating the switch or making ordinary adjustments.

No-voltage protection to the motor is provided by a magnetic release *M*, which opens the switch automatically on a failure of voltage. Overload protection is provided by means of relays *R* resembling cartridge fuses in appearance, each of which contains a contact connected in series with the release magnet. Harmless momentary overloads have no effect on the relays, but as soon as the load on the motor becomes dangerous, the relay contacts break the magnet circuit and the switch automatically opens. The relays then automatically reset themselves. The switch is so arranged that it cannot be kept closed on an overload or no-voltage, even if the handle is held in the on position.

SAFETY SWITCH WITH COVER OPEN

The October 15 quarterly number of the Bulletin of the National District Heating Association contains rates of heating companies in various cities of the United States brought down to date according to the best information available. The differences in rates charged in different places are evidently due to variations of conditions and go to emphasize the inadvisability of fixing rates in comparison with what others are charging. Charges should be based on a careful appraisal of the property used and useful in the business, together with consideration of operating expenses and the rate should obtain a fair return upon the capital necessarily invested. The method is equally applicable in arriving at fair charges for steam or heating service furnished to customers of isolated plants.

Practical Lubrication—Oils and Their Physical Characteristics

By W. F. OSBORNE

WHAT is the best way for the engineer to select the most efficient and economical lubricants for any particular purpose? There are so many lubricants, suitable and unsuitable, that if he depends solely upon trying first one and then another until he hits upon something satisfactory, he is quite likely to find it both expensive and dangerous. If he can eliminate the unsuitable ones first and then try out a few of the suitable ones until an efficient and economical lubricant is selected, he will save many a dollar and a great deal of effort.

No matter how carefully machines are built, there is always some little difference between them. These slight mechanical differences frequently affect the lubrication, making it hard to specify a single lubricant as being the correct one to use on the machine under all operating conditions. It is therefore necessary to investigate all the factors that may have any influence on the lubricants, and by knowing what will happen when lubricants of certain characteristics are used, it is generally possible to make a fairly accurate selection. This article will explain the characteristics of the principal types of lubricants and also some of the methods that can be used in testing for them. The subsequent articles will point out the principal factors that should be looked into in picking out lubricants for the most widely used types of machinery.

American lubricating oils of mineral origin can be divided into two general classes readily distinguished by their physical characteristics —paraffin-base oils and asphaltic-base oils. There is also a mid-continent or semiparaffin crude from which oils are made possessing some of the characteristics of both of the other types. Many grades of lubricants are made from each of these crudes, and they are also frequently blended in varying percentages to obtain certain results. Lubricating oils may be of animal, vegetable or mineral origin, or combinations of any or all of them for various purposes.

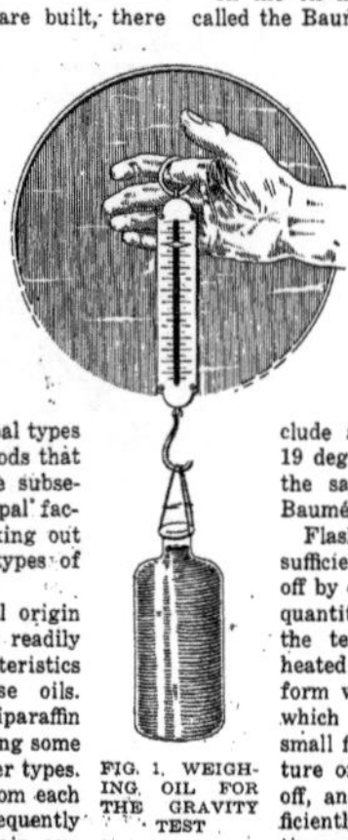

FIG. 1. WEIGHING OIL FOR THE GRAVITY TEST

Like many other materials it is practically impossible to judge from appearance alone whether an oil or grease will lubricate a machine as desired; accordingly, many attempts have been made to devise a series of chemical and physical tests to check the suitability of a lubricant before actually trying it out in the machine. The physical tests which are ordinarily made on oils are, gravity, flash, fire, pour, color, viscosity, carbon residue, emulsification, acidity.

Gravity—The specific gravity of an oil is the ratio of its weight to the weight of an equal volume of distilled water. As oils contract and expand with temperature variations, the standard of comparison is set at 60 deg. F. Specific gravities can be determined in a number of ways, but the easiest method for the engineer who has no laboratory equipment is for him to fill a clean bottle with distilled water at 60 deg. F. Weigh the full bottle and after emptying and drying it, weigh the bottle itself. Now fill the bottle to the same level with the oil to be tested heated to 60 deg. F. and weigh it (see Fig. 1). After deducting the weight of the bottle in each case, the specific gravity of the oil is found by dividing the weight of the oil by the weight of the water.

In the oil industry an arbitrary scale of gravities called the Baumé scale is used. On this scale the specific gravity of water corresponds to 10 B. and a specific gravity of 0.667 equals 80 B. To the engineer the gravity of an oil is of value only in so far as it indicates the kind of crude from which the oil was made. High-gravity oils ranging from 26 deg. B. to 32 B. are made from paraffin-base crudes, the low-gravity oils from 19 deg. to 22 deg. B. being made from asphaltic-base crudes. Excellent oils are made from both crudes for practically all purposes, and a gravity specification that would include all high-grade oils would have to read 19 deg. to 32 deg. B. Of two oils made from the same crude, the one having the higher Baumé gravity will be the thinner in body.

Flash and Fire—Any liquid on being heated sufficiently will give off vapor. The vapor given off by oil will burn when mixed with the proper quantity of air. The flash point of an oil is the temperature at which the oil on being heated slowly will give off enough vapor to form with the air over its surface a mixture which will ignite upon the introduction of a small flame and then go out. As the temperature of the oil is raised more vapor is given off, and when the production of vapor is sufficiently rapid the oil takes fire and burns continuously. This temperature of the oil is called its fire point.

A number of instruments have been made to determine flash and fire points, the most widely used being the Cleveland and Tagliabue open-cup testers. The Cleveland tester is shown in Fig. 2. The oil to be tested is placed in the cup *c* surrounded by an oil bath *b* to regulate the temperature. The heat is supplied by the gas jet *h*. The thermometer *th* shows the temperature of the oil when the flash and fire occur.

Although the flash point shows the temperature at which enough oil can be vaporized to cause a flash, it does not show how much oil will evaporate at that temperature nor does it show the temperature necessary to evaporate the remainder of the oil. It is quite possible for an oil to contain a small percentage that will evaporate at a low temperature, leaving in the cup a much greater portion requiring a high temperature for complete vaporization. Neither do the flash and fire tests show the actual lubricating qualities of the oils, although they may point out the possible unsuitability of an oil for certain purposes where a minimum evaporation is necessary. It should be remembered that an oil will not flash or burn spontaneously at temperatures anywhere near the flash point; an actual flame or "hot spot" must be introduced from some outside source, and then there must be a sufficient amount of oxygen present for combustion.

Paraffin oils have higher flash and fire points than asphaltic oils of equal body and quality. These oils were in use many years before asphaltic oils were developed, and numbers of engineers have become accustomed to specifying a high flash on this account. There seems to be no good reason for requiring a high flash, nor are there any special objections to it provided other characteristics are suitable, as practical experience has shown that both high- and low-flash test oils are equally satisfactory for many purposes. The lower-flash oils are usually more suitable for air-compressor or internal-combustion cylinders, as they evaporate more quickly without decomposition with its resulting carbon formation.

Pour Test—All oils contract when cooled. and if the temperature is reduced low enough they will cease to flow. The lowest temperature at which an oil will flow is called its pour test. Oils made from paraffin crudes contain wax in solution and as wax alone is solid at temperatures below 110 deg. F., these oils congeal at higher temperatures than the asphaltic oils containing no paraffin wax. Asphaltic oils have pour tests below 0 deg. F.; paraffin oils have pour tests ranging from 25 deg. to 80 deg. F., although they can be lowered by a rather expensive process of chilling and filter pressing to remove the wax.

The pour test should be given due consideration when selecting oils for refrigerating equipment, motor cars in winter, railway-signal mechanisms, ring- or chain-oiled bearings or any other place where the lubricant must remain liquid at low temperatures.

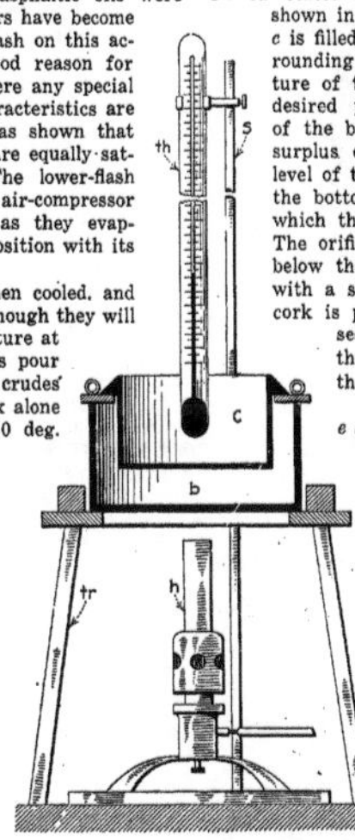

FIG. 2. DIAGRAM OF A CLEVELAND OPEN CUP TEST

Color—Outside of the laboratory the color of an oil is usually determined by looking at the light through a four-ounce sample bottle filled with the oil. Oils vary in color from water white, straw, yellow, amber and red, to dark green, brown or black. Light-colored oils are not always light-bodied, as very viscous asphaltic oils may have a pale amber color; some red or greenish oils are comparatively thin. Many pale oils are high-grade filtered oils while others are merely bleached; on the other hand, red oils are frequently filtered and when properly refined many of them are high-class lubricants. A greenish tinge to an oil usually indicates that steam-cylinder stock has been blended with a lighter oil to raise its viscosity.

The color of any oil becomes dark with use because of slight oxidization and by picking up minute particles of foreign matter. Red oils may change in characteristics just as much as the pale oils, but it is not so noticeable, owing to their darker color.

Viscosity—The viscosity of an oil is the resistance offered by its molecules in sliding past one another and indicates the body or fluidity of the oil. The relative body of two oils may be compared roughly in sample bottles by turning them upside down and watching the drop from the bottoms. The oil which drops first is the thinner.

Viscosities are determined accurately by instruments known as viscosimeters, of which there are several forms. The instrument in most general use in the United States is the Universal Saybolt viscosimeter shown in Fig. 3. To make the test the cylinder *c* is filled with the oil to be tested and the surrounding bath *d* with water or oil. The temperature of the oil to be tested is brought to the desired point by regulating the temperature of the bath. The gallery *g* is emptied of all surplus oil with a pipette, thus bringing the level of the oil to a positive starting point. At the bottom of the cylinder is a jet *j* through which the oil flows into the receiving flash *f*. The orifice is inclosed by the tube *t* extending below the bottom and is closed by the cork *l* with a string attached. To start the test the cork is pulled out quickly and the number of seconds required for the oil to rise up to the 60-c.c. mark on the flash *f* indicates the viscosity.

In the illustration, *u* is a steam heater, *e* an electric heater and *r* is a gas heater, and any of these three may be used; *s s* are stirrers used to evenly circulate the heating bath.

The body, or viscosity, of any oil changes with the temperature, and a viscosity reading should always be accompanied by the temperature at which it was taken. Spindle, engine, turbine, machine, and motor oil viscosities are taken at 100 deg. F. Steam-cylinder and gear oils, because they are so viscous, are heated to 210 deg. F. before taking the viscosity reading.

The two curves in Fig 4 show how the viscosity of oil varies with a change in temperature. These two oils are high-grade turbine oils and have about the same body at the working temperatures of the bearings, 125 deg. to 150 deg. F. As the curves show, an asphaltic oil having the same viscosity as a paraffin oil at 130 deg. F. will be quite a bit thicker at ordinary temperatures. If these two oils were cooled still further, the asphaltic oil would be much the heavier until about 35 deg. F. was reached, then the paraffin oil would thicken very rapidly and congeal at 30 deg. F. while the asphaltic oil would remain liquid at temperatures below zero. When deciding upon the oil to use, the operating temperature should be determined if possible and the right viscosity at that temperature specified.

The viscosity test is most important in selecting lubricating oils for almost all purposes. If the oil is too light in body at the working temperature to withstand the pressures, the metal surfaces will come into contact with friction and wear. On the other hand, if the oil is too heavy it may not flow to the working surfaces fast enough to maintain the oil film and wear

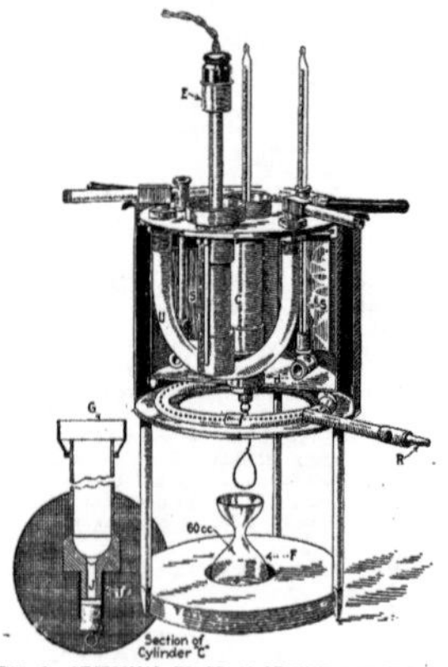

FIG. 3. SECTIONAL DIAGRAM OF THE UNIVERSAL SAYBOLT VISCOSIMETER

will result through lack of lubrication. Heavy-bodied oils also cause power losses through the internal friction of the oil itself. High-viscosity oils will sustain heavy loads and are used for slowly moving machinery; light-bodied oils are used for high-speed machines with light loads.

Carbon Residue—If oil is heated to a high temperature in the absence of a free supply of air, the greater part of the oil will distill, leaving a residue of carbon in the receptable. This carbon is formed by the partial decomposition of the oil, and the amount deposited by this test is an indication of the extent to which the oil will decompose in actual service. The oils which require the greatest heat to evaporate them usually form the largest carbon residue, as the high temperatures have a greater decomposing effect. Heavy-bodied oils leave a greater residue than lighter-bodied oils made from the same crudes with the same degree of refinement.

The carbon-residue test is helpful in selecting oils for any service where the oil is evaporated or burned, as carbon deposits, if formed, may seriously affect the operation of the machine. This applies particularly to air compressor and internal-combustion cylinders for which a low carbon residue is highly desirable. The carbon-residue test will also give some idea of the relative amount of sludge which might be formed in a steam turbine or engine oil-circulating system.

Emulsification—This test is made to show the rapidity with which oil will separate from water. Any engineer can test his steam-engine or steam-turbine oil by mixing equal parts of oil and water in a bottle and, after thoroughly shaking, observing how quickly the oil separates. The ability of any oil to separate depends principally upon its quality and the care taken in its refinement. Thin oils naturally separate more quickly than the more viscous ones, a characteristic that is taken advantage of by heating the oils in steam-engine circulating systems to improve their separation from water and dirt.

This test is valuable when selecting oils for the circulating systems of steam turbines and engines or for any purpose where there is a possibility of water getting into the oil. It will also point out the deterioration of oil with use, and if tests are made regularly it should be possible to know when a batch of oil in a system is approaching the end of its usefulness.

Acidity—A practical test to show the presence of acid can be made by coating a highly polished piece of brass with the oil and exposing it to the air for twenty-four hours. The quantity of acid in an oil can be determined accurately only in the laboratory.

All high-grade oils when new are free from acid, but they will all develop a certain amount of acidity in use. Oils containing animal or vegetable oils decompose rapidly when heated, and the acid developed is a large factor in the limited use of such lubricants. Mineral oils will also show an acid content after they have been in use for some time, due to the breaking down or decomposition of the oil. The acidity test of used oils is of value in determining when to throw away the oil in a turbine or engine oiling system and replace it with new oil. A good plan is to take samples about once a month and have the laboratory test for the amount of acid. When it is found that the acid begins to increase rather rapidly, the oil should be taken out, as there is a great possibility of its emulsifying.

All these tests taken together will usually eliminate unsuitable oils provided the service conditions under

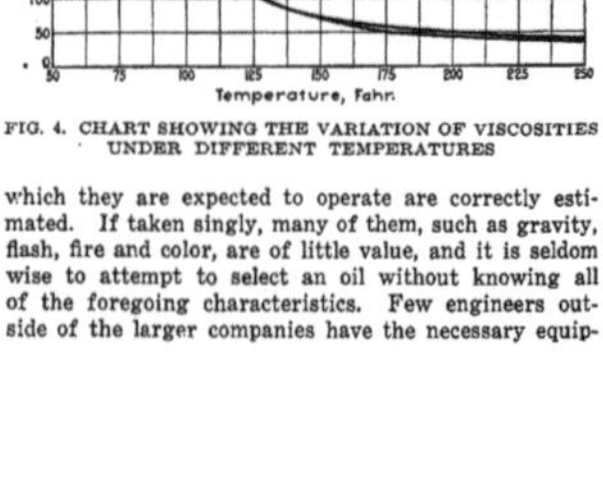

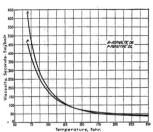

FIG. 4. CHART SHOWING THE VARIATION OF VISCOSITIES UNDER DIFFERENT TEMPERATURES

which they are expected to operate are correctly estimated. If taken singly, many of them, such as gravity, flash, fire and color, are of little value, and it is seldom wise to attempt to select an oil without knowing all of the foregoing characteristics. Few engineers outside of the larger companies have the necessary equipment to make scientific tests, but with a little care and experience many of the practical tests outlined will be of considerable assistance.

Bearings are sometimes located so that they are difficult to reach or are of such design that they waste oil through leakage. In such cases a lubricant thicker than oil is more economical and requires less labor in applying. Greases are usually made by thickening a mineral oil with a soap formed by saponifying animal or vegetable oils with an alkali. They are made of varying consistencies, ranging from the thin fluid grades to extremely hard greases with high melting points. Greases of the softer grades can also be made from pure petroleum and are used where temperatures are not excessive. The melting points vary with the materials used and the method of manufacture.

All good greases are homogeneous, free from excessive alkali, acid or moisture, and contain no fillers or stiffeners of a nonlubricating nature. Some greases also contain graphite which serves to fill up the pores of the metal, making the surfaces smoother and thus reducing friction. The method of manufacture of any grease has a far greater influence upon its quality than its chemical composition.

After the poor grades have been eliminated and a number of apparently suitable lubricants have been found, the most efficient one can be determined only by a practical test under actual operating conditions. The articles to follow cover the principal points that the engineer should investigate when picking out lubricants for various types of service as well as the effect they have on certain characteristics in lubricants.

Absorption Machine Merely Needed Better Steam Circulation

By W. Everett Parsons*

Several years ago I was retained by a manufacturer of absorption refrigeration machines in a lawsuit where the purchaser of one of his machines refused to pay for the installation on its failure to fulfill the guarantee as to capacity. The machine was of the low-pressure type and was supposed to be operated by exhaust steam at 5 lb. pressure.

The purchaser (a hospital) employed several well-known experts to make tests of the refrigerating system, and they reported that the machine failed to develop the capacity guaranteed. I did not see the machine. The matter was taken to several courts and finally was tried before a referee after about two years of litigation.

The manufacturer of the refrigerating machine claimed from the first that the generator heating coils were not properly connected to the exhaust-steam supply system; that when the machine was installed instructions had been given as to how this should be done and had been ignored. The heating system and connections from the source to the generator were installed under the plans and supervision of one of the best-known heating engineers in this country. The purchaser therefore naturally concluded that the heating expert probably knew more about the heating than the refrigerating-machine manufacturer did.

At the trial before the referee there were retained on the two sides at least eight of the foremost experts in this country.

*Consulting Engineer, Newark, N. J.

The experts for the purchaser were first to be called to testify. It soon developed that the coils of the generator of the refrigerating machine were connected on the induction system; that is, with one end of the coils connected to the exhaust main by a branch pipe and the other end to a header to which a trap and a small air-vent valve were connected. The several experts for the purchaser were asked whether the coils in the generator were likely to become airbound and how much of an opening there should be to get rid of the air. One expert for the purchaser said that a ½-in. globe valve just cracked open would be sufficient; another said a pinhole would be sufficient after the generator was in full operation. The latter expert, however, at the next sitting, asked to have the "pinhole" testimony stricken out.

At the time of the trial the writer was supervising the remodeling and enlarging of one of the largest combination power and refrigerating plants in New York, near where the trial was going on.

In the basement of this power plant there was a large horizontal Berryman feed-water heater. This heater had been originally connected so that practically all the exhaust steam from the engines and pumps entered it through a 16-in. exhaust main, and whatever was not condensed passed on to the steam condensers operated at atmospheric pressure to make distilled water for an ice-making plant. Some time before the trial began the steam condenser had been moved to another part of the basement and the steam outlet of heater had been closed with a flange for a 3-in. pipe, but bushed down to 1-in. Then a 1-in. nipple and globe valve were connected as an air vent.

The heater as connected was, of course, similar in principle and arrangement to the generator of the absorption machine. I decided to have a series of tests made with different sized air-vent openings. These tests were made while the examination of the experts was in progress. About the time this examination was completed, I handed to the lawyer for the refrigerating-machine manufacturer a copy of a table of the results of tests. From the series of tests the results of those in which all of the conditions were as nearly alike as possible excepting, of course, the temperature of the feed-water leaving the heater, were used. With the 1-in. air-vent valve wide open the results were 6.28 per cent. better than with the 1-in. valve open only one-half turn and the results obtained were 32.41 per cent. better with a 3-in. valve wide open.

With the 3-in. valve wide open some steam passed out with the air. This indicated a free circulation of steam over all the heat-absorbing surfaces, and this without a doubt helped to make the results so much better.

A free circulation of steam in the coils of the generator kept free from air is what the refrigerating-machinery manufacturers had provided for and insisted upon when the machine was installed.

The lawyer for the machine manufacturer decided to show the results of my tests to the opposing lawyer. When he did so, the case was promptly ended without taking further testimony and the machine manufacturer was promptly paid for his machine.

I might, however, add that I was informed that during the litigation the purchaser arranged with a competitor to remodel the refrigerating plant at a considerable expense, which, it would appear, was unnecessary.

Steam From the Earth

About a year ago mention was made in *Power* of a station in Italy in which use was made of steam extracted from the earth. Ing.-Prof. Luigi Luiggi, of Rome, who has recently been in this country, has kindly furnished *Power* with the particulars of the installation.

The regions south of Volterra are rich in volcanic manifestations, among them the "soffioni," or "whisperers," of Larderello, Fig. 1, and in the neighboring communities of Castelnuoeo, Sasso, Monterodondo, Lago, Lustidnano and Serrazzano. In all these regions, extending for many square kilometers, there break out of creases in the earth jets of steam at high temperatures, superheated vapors rich in boric acid, ammonia, carbonic

FIG. 1. NATURAL STEAM JETS OF LARDERELLO

acid, hydrogen sulphide and other gases. This vapor is condensed for the extraction of boric acid and other byproducts. By suitable borings from about 8 to 16 in. in diameter, and from 300 to 500, or even 750 ft. deep, lined with suitable steel tubes, there can be obtained powerful jets of superheated steam of from 10,000 to 40,000 lb. per hour capacity, and at a pressure of from two to five atmospheres, one of which is shown in Fig. 2. These wells remain unchanged in yield and temperature for years and are not affected by the driving of other wells if the latter are not too near to them.

The first experiments in the use of this steam for power-producing purposes were made by Prince Ginori-Conti, the leader of the borax industry. By directing a jet against a bucket wheel, an ordinary motor was made to drive a little dynamo capable of supplying a few electric lamps. Encouraged by these results, in 1905 he applied the natural vapor of the soffioni to a 40-hp. motor, utilizing in this way a small part of the jet from the Venella well, which issues at a pressure of about five atmospheres, a temperature of about 330 deg. and a quantity of about 11,000 lb. per hour.

The results obtained during several years were most satisfactory from the viewpoint of the power produced, but somewhat less so with respect to the action of the steam upon the motor, on account of the sulphuric and other corrosive products accompanying the vapor.

Meanwhile new wells were constructed, larger and deeper, and jets of various powers were obtained, among others one which produced 55,000 lb. of steam per hour at a pressure of about two atmospheres absolute, representing a theoretical capacity of 4000 hp. under conditions which made 40 per cent. of the theoretical realizable. Bringing together a number of these jets would have made it possible to supply turbines of several thousand horsepower capacity. Caution led, however, to making an experiment on a moderate scale sufficient for conclusive results by means of a turbo-alternator of 300 hp. set up in 1912 for supplying lighting and current to the Larderello establishment.

The results of this experiment were so satisfactory as to induce Prince Ginori-Conti to establish a plant of larger capacity, the demand for which was made more imperative by the fuel shortage due to the war. The plans were made and carried out by Ing. Bringhenti. Three Tosi turbo-alternators of 3000 kw. each were installed, with surface condensers supplied with circulating water from cooling towers.

To avoid the chemical action of the natural vapor mixed with corrosive gases on the metallic parts of the turbine and at the same time to avoid the possibility that the noncondensable gases issuing with the vapors should make it impossible to maintain a high vacuum in the condenser, the steam was not supplied directly to the turbine, but served to heat three multi-tubular boilers working at a pressure of 1¼ atmospheres absolute, which supplied the turbine with steam from the condensation water of the turbine itself. The steam from

FIG. 2. JET FROM A BORING CONFINED IN A PIPE

the soffioni condensing, while giving off its heat to the boilers, is then utilized, as previously, for the extraction of boric acid and other secondary products. The boilers are of the vertical-tubular type with aluminum tubes, this metal being the most resistant to corrosion from the gas of the soffioni.

The three turbines can develop about 4000 hp. each. They are direct-connected to three alternators of 3000 kw., working normally at 2759 kw. capacity. They are three-phase units of ordinary type with power factor 0.7 at 4500 volts and 50 cycles. The electrical energy is transformed to 36,000 and 16,000 volts, and distributed by five separate lines to various main centers of Tuscany.

EDITORIALS

Power Supply and Generation of the Future

IT HAS been estimated that with steam at three hundred pounds and five hundred and twenty degree superheat, used in a turbine operating on twenty-nine inches of vacuum and eighty per cent. Rankine-cycle efficiency, a kilowatt-hour may be produced on eight pounds of steam. Even this condition, when it will be possible of attainment, will utilize only about twenty-five per cent. of the energy in the coal. At present steam pressures are being used and values of superheat employed that make the development of fittings for boilers and steam lines a difficult problem. The stoker has been developed to where we are getting about all that can reasonably be expected from the boilers in the way of efficiency, and turbines and generators have been built in compound units up to sixty thousand kilowatts. Even before these sizes are reached, there is comparatively little to be gained in economy by increasing the size of the machine, and machines of this capacity are applicable only to the largest systems. Looking at these facts, it is evident that the development of power-plant equipment, so far as concerns the conversion of chemical energy in the coal into mechanical energy, has apparently almost reached the practical limit.

Dr. Charles P. Steinmetz suggests that the solution of the problem may be found in installing a simple steam-turbine-driven induction generator between the high-pressure steam boiler wherever steam is used for heating, and converting a percentage of the energy into electrical power and supplying this to the power-distribution system of the community. There are many industries that use large quantities of steam in their processes, where the foregoing has been given consideration for some time past. However, a large percentage of the heating plants are in operation not over six months per year, and the size of the equipment that could be operated from them makes the proposition economically doubtful. However, this objection might be satisfactorily overcome by supplying heat to all the buildings in one city block, or a number of blocks, from the same heating plant. This scheme would allow the use of economical equipment, but even such an arrangement has its limitations. When this is done under the most favorable conditions, it is at present possible to convert not more than sixty to sixty-five per cent. of the chemical energy in the coal into heat and mechanical energy. This is a decided improvement over the best efficiency of conversion now obtained in the most economical power plants, but conditions where this is possible for over six months of the year are very limited, consequently it does not offer any very promising solution to the problem.

With the present development in the heating and power equipment, it is apparent that some very radical change must be made in steam power-plant equipment before any great improvement can be made in the efficiency of converting the chemical energy in coal into mechanical or electric power.

There are two ways in which large conservation of coal can be made; namely, the more complete development of our water powers, and the electrification of our railways. It has been estimated that the steam railways of this country burn, each year, fuel equivalent to between one hundred and forty and one hundred and fifty million tons of coal. Over two-thirds of this can be saved by electrifying our roads, which is equivalent to saving about fifteen per cent. of our present coal production. Furthermore, if our railways were electrified, much of this power would be supplied from hydro-electric stations, resulting in even a greater saving of fuel. Probably not over fifteen per cent. of the fifty million kilowatts of our total potential water power is developed, leaving some forty million kilowatts undeveloped —more than enough to supply all our power requirements at the present time. However, a considerable percentage of this is situated where it is not available, and much of that which is available cannot be utilized by our present method of concentrating of large blocks of power in a small number of plants. Dr. Steinmetz's suggestion that the small water powers be developed by the installation of small water turbines and induction generators, and tied in through simple switching equipment to the transmission system, holds wonderful possibilities, especially as such plants are already in satisfactory operation. Although the prospects are not very favorable for any epoch-making development in the efficiency of our modern power plants in the near future, the electrification of our railways and the development of our water powers, if taken full advantage of during the next twenty-five years, will make this period equal in engineering achievement to that of the past twenty-five.

Secretary Lane's Recommendations

ENGINEERS and executives should read what Secretary Lane has to say about the fuel supply and power problems of the country. Admittedly, government statistics are very incomplete in this respect and a survey of power resources would repay the nation many times the expenditure involved. With singular far-sightedness Mr. Lane has grasped the necessity of altering our present wasteful methods of hauling and utilizing coal for power purposes if we are to conserve the supply and meet the demands of the future. Further development of our water powers, the location of great power plants at the mines and greater electrification of our railroads is a program that will soon become imminent.

"We have too many mines and too many miners to meet the domestic demand," says Mr. Lane, as a result of which the mine year, even for 1918 with everything to stimulate production, averaged only two hundred and forty-nine working days. Moreover, during 1919 the production decreased by twenty-five per cent. Undoubtedly this is a factor in the present high cost of coal, because both the labor and the overhead are proportionately greater on the smaller production.

But when we have effected the purposed conservation measures through giant power plants at the mines and electrification of the railroads, the domestic demand will have decreased still further and the average miner's year will have shrunk to fewer days. The remedy for this, we are told, is to cultivate a large foreign market for coal. This will undoubtedly make for increased business in the coal industry, but what about conserving our available coal supply for the future? In other words, Can we afford to live wholly in the present and let the future take care of itself, trusting to the development of ways to utilize the atomic energy which Sir Oliver Lodge has lately been discussing?

Again, the average purchaser of coal will be led to ask what assurance he is to have that the old law of supply and demand will not be invoked against him as an argument for further increases in price, if we are to cultivate a great foreign market.

But, putting aside the coal situation, which can be viewed from so many angles as to become positively confusing, it is most reassuring to read the recommendations, coming as they do from a Cabinet Officer, that the Government should adopt a definite policy of protection to all legitimate developments of oil properties by Americans in foreign countries. Oil fuel is a very live topic at present, and the problem of an adequate supply is pressing. Such action as urged by Secretary Lane should, therefore, receive immediate consideration.

Reasonable Administration of Boiler Laws

THE Massachusetts Board of Boiler Rules requires that "Each boiler shall be stamped by the builder in the presence of the inspector with a serial number and with the style of stamping shown in facsimile previously approved by this Board."

Occasionally, the representative of the builder picks up the wrong stamp or gets it upside down and the inspector does not notice it, with the result that when the boiler arrives in Massachusetts the number so stamped does not agree with that upon the certificate.

It would seem a simple matter for some member or agent of the Boiler Inspection Department in Massachusetts to correct the erroneous stamping when the chief was satisfied that an error had actually been made and that the boiler in question was really the one covered by the certificate with which the correction would make it agree as to number. The Massachusetts authorities have, however, insisted that the inspector under whose supervision the error occurred should go to Massachusetts and do the restamping himself. Under a former department head the employee of the boiler shop who did the stamping was obliged to go there, too. Inasmuch as the inspector's word and that of the employee who made the error is all that the department has to go by in case they are dragged to Boston, affidavits sent by mail ought to be equally convincing and, in connection with the bill of lading and description, ought to be sufficient to identify the boiler and warrant the official correction.

Insistence upon a personal correction by the man who made or passed the error as a punitive or deterrent measure is deplorable. To send one or two men from a distant state just to stamp a number is an economic crime. It wastes skilled labor, makes boilers cost more and serves no good purpose. Compliance by the Board of Boiler Rules with the request made at its last hearing to revise its practice in this respect would alleviate the apprehension with which business men regard the conferring upon governmental factors powers which permit them to interfere, even in a beneficent way, with private enterprise.

The Future of Refrigeration

SOME years ago we sat at a banquet of the American Society of Refrigerating Engineers in New York and heard a stirring address by Dr. C. E. Lucke, in which he expressed deep regret that there had been so little progressive development in the art of mechanical refrigeration. He took occasion to cite the forward strides in many other branches of the mechanical arts in comparison with refrigeration. With appealing emphasis he said the field for greater development was even then crying for attention. While the greater number of those present sympathized with the Doctor's views, there were some few who felt that the criticism was too sharp.

This year the society again had a banquet, and again Dr. Lucke was one of the speakers. He recalled his comments of the years before and expressed happiness that the situation was so favorably different. During the three days previous to this banquet the society had listened with profound gladness to several papers, some scientific, some descriptive of machines that marked a really great event in the art. The technical staff of the National Bureau of Standards had contributed data dealing with the properties of ammonia so sorely needed from a scientific point of view if less urgent from shop and operating viewpoints.

The Doctor saw in this meeting, as many others saw, that there was manifested a truly earnest effort for closer coöperation of the engineer and the scientist. Needless to say there must be an exercise of this spirit before the great needs of the industry are met.

Dr. Lucke sees that refrigeration is due for an epochal era. He thinks it may be here now in the form of the high-speed ammonia compressor electrically driven. Certainly, those who view the art and its potentialities in a broad yet enthusiastic way are in accord with the distinguished Doctor's views.

The American Order of Steam Engineers, through its organ, *The Popular Engineer,* voices editorially the sentiments of safe and sound unionism. Condemning the Boston police strike and the many other recent evidences of radicalism, it calls upon the engineering fraternity to come forward to assist capital and labor in understanding each other as partners in common endeavor, to the end that brains and technical training may receive just recognition in contradiction to the present tendencies towards subsidizing only those who work with their hands. It is gratifying to see such expressions from an association in a position to exert wholesome influence from within.

A bit of trouble-making legislation is proposed for the coming session of the New Jersey legislature in a bill which provides that water and electric meters shall be made so that they can be turned back to zero each time a reading is made by the corporation. The proposer evidently has no idea of the mere physical difficulties involved. If a man cannot take the reading of his meter and subtract the last month's reading from it, he will not be able to check his account if the meter starts from zero each time.

Representative Keller has introduced a resolution providing that, "as the foundation of our economic existence is endangered," the President should be instructed by Congress to take over and operate the bituminous-coal mines.

CORRESPONDENCE

Believes Purge Tank Won't Work

Having read H. J. Macintire's article on "The Refrigerated Purge Tank" in the Aug. 12 issue of *Power,* I cannot agree with his proposed method of purging foreign gases from a refrigerating system. I find his figures to be mostly correct, and if they could be put into practice a lot of ammonia that now goes out through the purge valve could be saved.

But the trouble is, if we cooled this mixture of foul gas and ammonia gas to 32 deg. F. it is likely that the same percentage of the permanent gas would condense as of the pure ammonia gas. This, of course, would depend on what this permanent gas was. This permanent gas or foul gas is often thought to be a noncondensable gas, while, in fact, all gas will liquefy if brought to the right pressure or temperature. I have found some permanent gas to be lighter and some heavier than ammonia gas.

The lighter will freely burn when emerging from the purge valve, while the heavier will not. Some of this gas (while it is not ammonia gas) will condense at about the same pressure and temperature as ammonia, but usually at a little higher pressure. This kind of gas is a source of trouble and is hard to deal with. It is undesirable because of its higher condensing temperature or pressure, which keeps the entire condensing pressure higher, and again, its latent heat of vaporization is usually not so much as that of ammonia, and its evaporating temperature is usually higher.

However, this sort of gas rarely accumulates in the ammonia system, but comes with the ammonia, consequently only the best ammonia should be used and it should be tested. The gas that does get in the system is air caused by pumping a vacuum on the system when there is a leak. The leaks are usually in the compressor-rod packing and stop-valve stem packings. Even if the system is tight and free from ammonia leaks, these packing leaks will start as the vacuum is formed, because the vacuum causes the temperature to drop 25 or 30 deg. F. on the "low" side and 75 to 100 on the "high" side, below usual working temperature. The lower temperature causes the packing to shrink and let in the air. Another "permanent" gas is generated from the compressor lubricating oil, caused by high compressor temperature, poor oil or both. With the plant temperatures I advise a discharge temperature from 150 to 220 deg. F., depending on the temperature of the brine and of the cooling water and the type of compressor.

But getting back to purging an ammonia condenser, the most practical method is to have a stop valve on both the gas inlet and liquid outlet to each stand of the ammonia condenser, likewise a purge valve for each stand. To purge, shut the gas-inlet valve to one stand of condensers, having the plant operate as usual, slow the compressor or compressors if necessary to keep the condensing pressure the same as the usual working pressure.

This stand with the gas shut off will gradually fill with liquid as the gas in it condenses. All the permanent gas will be forced to the top by the liquid coming in from the bottom. This gas can then be purged off. When the liquid reaches the purge valve, it is easily detected, as "it will smoke and stink." This stand can then be cut in and the other stands purged in a like manner. I would call Mr. Macintire's attention to the fact that cooling water at 70 deg. F. does not mean that the condensing pressure should be 114½ lb. (gage).

The pressure should correspond with the temperature of the water leaving the condenser and not with the temperature of the water going to it. But it seldom does so in practice. It usually runs 10 to 40 lb. higher unless the plant has lots of condensers or a small compressor or both. Supposing the 70-deg. water was raised in temperature to 80 deg. F. while going over the condenser. The condensing pressure would be 139.6 lb. gage instead of 114.5 lb. That makes a difference, and in Mr. Macintire's case the excess pressure would be 40.4 lb. instead of 65½ lb.

But, further, it is good practice with 160-lb. gage pressure and cooling water at 70 deg. F. And even though 180 lb. is carried, it is not necessarily permanent gas. It may be "short" water or "short" condensing surface or both. But the permanent gas is a good thing to look out for. With a good tight system charged with pure NH_3—lubricated with the proper kind of oil—and never exceeding 220 deg. F. discharge temperature and pumping to a vacuum only when necessary, and then only after all packing glands have been tightened, and if all pipe and fittings in the system are made for ammonia use, the system will require no purging. Of course, if we could have a rectifier in the system charged with something that would absorb everything but NH_3, then we would have no further trouble with foreign gas in the system if we kept the rectifier clean and adequately charged with the absorbent.

Winkelman, Ariz. GEORGE TROSPER.

Precaution Against Burns When Handling Lime

The increase in the use of boiler-water purification plants compels many engineers to handle the lime and caustic soda and other chemicals required for their operation. Ordinary care will safeguard against danger from injury, but it has been my experience that even with careful handling the operator is constantly open to the danger resulting from the chemicals spattering into the operator's eyes.

In our plant we have arranged to keep a pair of glass goggles at the chemical tank so that when the chemicals are being handled the operator can put them on. It is worth while to provide this safeguard in every plant handling material of this kind. M. A. SALLER.

Philadelphia, Penn.

Signal System for Use in Place of Telephone

Owing to the difficulties attending the construction and maintenance of a private telephone line paralleling a transmission line on the same poles as the latter, the writer designed a signaling system which has accomplished everything desired.

A single iron wire is run on double-petticoat deep-groove glass insulators and brackets five feet below the crossarm that carries the power wires. The end of this iron wire is grounded at the end of the transmission line

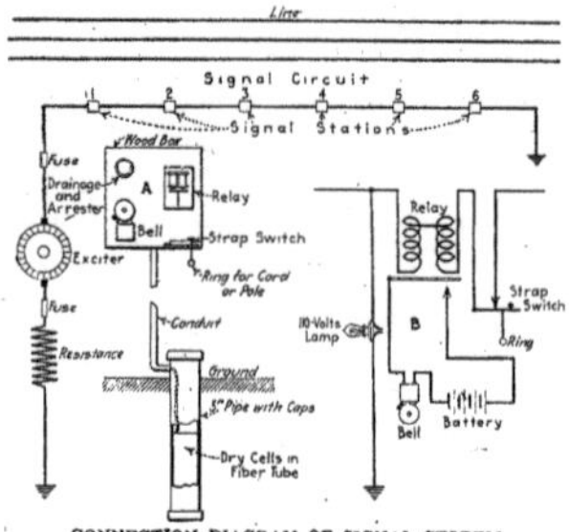

CONNECTION DIAGRAM OF SIGNAL SYSTEM

and is fed from the exciter bus at the power house, the opposite bus being ground through a fuse and fifty-ohm resistance tube, as shown in the figure.

At each substation and at each sectionalizing switch, a twenty-ohm telegraph relay is connected in series with this signal wire, the contact points being reversed so that the relay makes contact when there is no current flowing through it. An ordinary vibrating bell and three dry batteries on the "local" side of the relay complete the arrangement, and a strap switch, normally closed, is inserted in the signal wire at each relay. The general arrangement of a signal box is shown at *A* and the box circuits at *B* in the figure.

As a drainage for static a 50-watt 110-volt lamp is connected from line to ground at each signaling point. Instead of an ordinary lamp socket, a mogul series-lighting-socket is used, having a film cutout device, but the film cutout is in this case replaced by perforated sheet mica 0.02 in. thick, to serve as a lightning arrester. The lamp is held in the mogul socket by a reducer. A system of signals and calls has been adopted, which anyone can easily read. For example, the signal 3-1-4 means "Open disconnecting switches (at station called)"; the signal 6-4-1-pause-1-4-7 means "Tree ground, pole 147," etc. In case there is a contingency that cannot be adequately covered by the signal system, the signal 5-1-2, "Get calling station on nearest telephone," is used.

The system described is considerably cheaper than the simplest telephone system, even when signaling stations are spaced at frequent intervals. Since operators handle no live parts, the device is perfectly safe to handle, even during a storm or line trouble. As some of the signaling stations are outdoors, the company has constructed a standard outdoor signaling station which is used at all points, the battery being underground to prevent freezing. R. S. SEESE.

Carthage, Tenn.

Remedy for Present Valve Setting

Regarding the diagrams submitted by S. L. Gilliam on page 518 of the Sept. 23 issue of *Power*, I would say that the head-end steam valve is opening too early. The card shows that admission takes place two or three inches before the piston reaches the end of the stroke, which would cause a severe pound and would also increase the amount of steam used.

The following instructions will answer Mr. Gilliam's request for advice on double-eccentric Corliss engines: Set the steam wristplate and intermediate rocker in central position, and see that with a revolution of the eccentric they travel equally each side of the central positions. Proceed in a similar manner with the exhaust wristplate and the exhaust intermediate rocker. With the steam and exhaust wristplates in central positions, set the steam valve $\frac{5}{16}$ in. *open* and the exhaust valves ¼ in. *closed.* Unhook the reach rod, move the steam wristplate by hand and adjust the dashpot rods so that at each extreme position of the steam wristplate there will be $\frac{3}{32}$ in. clearance between the latch and tow plate. Put the engine on the crank-end center, hook up the reach rod and turn the steam eccentric on the shaft, in the direction of rotation of the engine, until as the crank-end steam-valve opening is increasing, it becomes $\frac{1}{16}$ in. Tighten the setscrews firmly in the eccentric and place the engine on the head-end center. With this position of the engine the head-end steam valve should be open the same amount, $\frac{1}{16}$ in., but if it is not, adjust the steam-valve rod so as to make it so, and at the same time adjust the dashpot rod an equal amount. Mark on the engine guides the extreme position of the crosshead. Set the crosshead 92 per cent. from the head end of the stroke, move the exhaust eccentric on the shaft in the direction of rotation of the engine until the crank-end exhaust valve is just closing. Tighten the setscrews in the eccentric *firmly,* and turn the engine over until the head-end exhaust valve is just closing. Measure the distance the crosshead is from the crank end of the stroke, and if the distance varies much from 92 per cent., adjust the valve rod to correct it. Put the governor on the stop, unhook the reach rod, move wristplate by hand and adjust the governor rods so that each latch plate will unhook when the wristplate is $\frac{3}{32}$ in. from the corresponding end of its travel. Raise and block up the governor to such a point that when the wristplate is moved its full travel by hand, each valve-gear latch will unhook just before the corresponding steam valve opens to steam. Fasten the governor collar in this position. Throw out the governor stop and let the governor down to the bottom, move the wristplate by hand and adjust the safety cams on the valve gear so that the latch plates will not hook up. Apply the indicator and correct for any slight defects.

Avoca, Penn. CHARLES J. HALLORAN.

Ideal Methods of Boiler-Room Operation

Volumes have been written on the subject of boiler-room economy, yet lack of system and the reign of chaos may be witnessed in many large and modern power plants. Although a few are giving details close attention, no uniform method of routine operation has been adopted, and one way of doing things is in vogue in one plant and an altogether different method in others.

Most up-to-date engineers agree that coal scales, draft gages, flue-gas pyrometers, CO_2 analysis, steam-flow meters, stoker-speed indicators or counters, automatic-draft control and daily sampling of coal and ash as well as correct tabulations of daily results are not frills and fads, but everyday necessities if ultimate economy is to be secured.

The instruments should be located at convenient and suitable points in the boiler rooms, on boards similar to switchboards, preferably at one central point under the control of an engineer who may be designated as the boiler-control engineer or dispatcher.

The dispatcher should also be in touch with the switchboard operators as to probable changes in the load, so as to prevent the loss of steam pressure or the blowing of safety valves. The chief error in the past, as I see it, is that instruments have been for the use of the testing departments and not for the men that are depended upon to actually operate and get results every day. Too many mere college boys have been playing around, testing a little while at this, then a little while at that, all of which is disconnected and without results. What is most needed, in my opinion, is not testing departments, but results departments, which are allied with the operating and maintenance departments and not separated as by a stone wall. Coöperation is the prime essential all the way through and I am convinced that there is but one way to get it, and that is, to call all of the various executives together first to formulate a plan of operation, then call in the actual operatives and explain to them what you are trying to do and assure them that a certain bonus will be paid to all employees if the results toe up to a certain mark.

Experience has proved to me that an individual bonus for high CO_2 or evaporation is not successful in securing the desired results or economy. The bonus should be paid to all the men concerned, thus eliminating the tendency to practice stunts and to withhold good ideas that would be of value to all. The individual bonus system also produces ill feelings among the men and leads to disorganization.

Other bad results of the individual bonus system are dishonesty, plotting and increased repair bills. The pool system covers the ground more thoroughly, is more flexible, and requires much less accounting, etc.

The motto of efficiency devices should be about as follows: "That which produces coöperation and results is fostered; and that which produces disorganization is tabooed."

The prime essential in producing coöperation, or the family spirit, is to get the men together occasionally for the purpose of discussing subjects pertaining to the work in the boiler room. There are other means of causing men to be interested and satisfied and loyal to the company besides the pay envelope. Some of them are as follows: Make them feel that they are more than mere hands and that they are responsible; teach them from the start that discipline is not humiliation or hardship, but necessary to prevent overwork; spring a variety of surprises in the way of entertainments allied with various forms of educational programs, as a reward for results. WALDO WEAVER.

Franklin, Ohio.

Synchronous-Motor Trouble

A certain synchronous motor began giving trouble at starting about a year ago, and continued to do so until recently, when the trouble was located and remedied. The motor is rated at 135 hp., 220 volts, 900 r.p.m., three-phase, 60 cycles. The stator winding is star connected and contains 24 coils, one coil per slot per pole, commonly known as chain winding.

The revolving field poles are laminated, and each pole is surrounded by a bronze collar that supports the field coils and retains the laminations. The poles do not contain an amortisseur or squirrel-cage winding such as is found on motors of present-day design. This motor is direct-coupled to a 100-kw. direct-current generator and an exciter.

Starting is accomplished by connecting the motor to low-voltage taps on a two-coil compensator. The motor started and ran satisfactorily for years, when all at once it became sluggish and refused to accelerate to a speed anywhere near synchronism when on the starting side of the compensator; nor would it come up to speed when supplied with full voltage. The starting voltage was found to be 1500 volts and all phases practically balanced. However, after several attempts at starting the motor finally came to speed. Intermittently it behaved this way for several months, and was subject to severe treatment that burned out one of the starter coils, and later the stator winding.

After these parts were renewed, the motor failed to accelerate beyond 300 to 400 r.p.m. It was tried with the field winding open and then short-circuited, but all attempts to bring it to speed were futile. As it was necessary to operate the set, provision was made to start from the direct-current end, using a water rheostat to limit the current. In this way it was accelerated to a speed of 500 to 600 r.p.m., when the direct-current was disconnected and the motor connected on the starter and brought into synchronism in the usual way. When on the line it carried a 100 per cent. momentary overload. The exciter voltage, field and line current were the same as before, when rheostats were placed on their old marks.

The problem was discussed with persons of broad experience. A motor having one slot per pole per phase has poor starting ability. Laminated poles are inferior to solid poles as regards starting. It should have an amortisseur winding. But, it started for years, why not now? All agreed that the motor had altered starting characteristics and were interested in how I should eventually come out. It was finally decided to install new field coils. These solved the problem. The insulation on the old coils was punctured between turns and between layers. The high induced voltage in the fields at starting had deteriorated the insulation and burned some of the wires nearly in two. Externally, the coils were apparently in good condition. I read a great deal on synchronous-motor troubles, but nothing that I could find dealt with a similar problem and pointed to field coils exclusively as the cause of trouble.

Rock Spring, Wyo. D. C. MCKEEHAN.

Springing a 15-Ton Crankshaft Into Alignment

The main bearings in a large three-cylinder oil engine gave the operator considerable trouble. The bearing shells, which were of the cylindrical steel, babbitt-lined type, could not be kept locked rigid in the frame, and two of them always operated at a dangerously high temperature. I was called in to make an inspection. The operating force was attempting to roll the bottom shell out of No. 3 bearing, Fig. 1, when I arrived. The shaft weighs about 15 tons, and I decided to set a jack under the nearest crank cheek.

Great was our surprise when, after turning the shaft 180 deg. for the purpose of placing the jack, we dis-

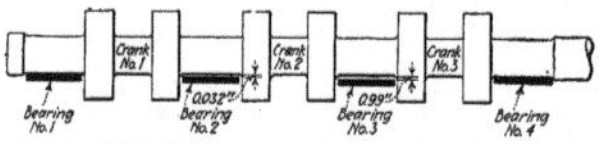

FIG. 1. SHOWING BEARING POINTS OF CRANKSHAFT

covered that the bearing shell, which had just been tight, was relieved and was easily rolled out by hand. Apparently, the bottom shell carried a heavy weight when No. 3 crank was in the inner dead-center and was relieved when this crank was placed on the outer dead-center. It was quite safe to assume that the crankshaft was sprung.

My next step was to secure a surface indicator with a range of 0.250 of an inch. The bearing shells were then removed from Nos. 2 and 3 main bearings so that the shaft rested in the two end bearings Nos. 1 and 4. The surface indicator was placed on top of No. 3 main-bearing journal and the shaft revolved one complete revolution. The readings showed an eccentricity in this journal of 0.162 in. In the same manner No. 2 journal was found to run 0.065 in. out of true. The maximum eccentricities of both journals were found to be in the same plane and in the highest vertical position when No. 3 crank was on the outer dead-center.

It seemed necessary to dismantle the engine, which involved taking off the rotor, and to ship the shaft to the factory to be trued up and have all main bearings rebabbitted to fit the size of the journals after they were turned down. However, the engine furnished power for a large industrial mill, and a long shutdown necessary to do this work would result in large losses to the firm and its employees. All the repair work so far had been done on Saturday afternoon and Sunday, and it caused no delays.

The crankshaft, as shown, is built up and there are no keys or pins in the shrunk joints. I decided, therefore, to put No. 3 crankpin under torsion and slide it 0.031 in. in the joint where it is shrunk into the cheek. This joint was 13 in. in diameter. The distance from the center of the crankpin to the center of the crankshaft was 17.25 in., therefore it would be seen that about 0.031 in. in the crankpin joint circumference would move the main journal 0.081 and would cause it to run true, as its total eccentricity was 0.162 in. It is true that by slipping No. 3 crankpin joint 0.031 in., not only No. 3 journal but Nos. 2 and 1 as well, would be shifted 0.081 in. off center relative to the remaining part, or the other end of the shaft, while the surface indicator showed only 0.065 eccentricity in No. 2 journal and No. 1 journal was turning true. However, it was reasoned that had it been practicable to support the shaft in No. 4 journal and the outbored bearing only with the shells removed from No. 1 journal also, the surface indicator would have shown 0.162 in. eccentricity in Nos. 1 and 2 journals as well. The weight of the crankshaft caused it to rest always

MICROMETER READINGS

At Start	Under Strain	Strain Off
302	455	396
396	465	405
405	492	432
432	500	448
448	507	450
450	510	451
451	520	460
Checked No. 4 journal showed 0.020 in. eccentricity.		
449	484	450
450	523	462
462	530	480
Checked No. 4 journal, 0,006 in. eccentricity.		
479	514	479
479	535	479
479	545	485
Checked No. 4 journal, 0.003 in. eccentricity.		

in No. 1 bearing and therefore run true at this point and at the same time decrease the eccentricity in No. 2 journal.

A large steel Z-shaped hook was made up and used in connecting with a 100-ton hydraulic jack as shown in Fig. 2. This rigging enabled us to apply about 1,020,000 lb. torsion in the 13-in diameter shrink joint which it required to start it. The pressure gage on the hydraulic jack made it easy to compute the pressures.

In order to be able to measure accurately the amount which the joint slipped, an inside micrometer was used between two points located 32.5 in. from the center of the crankpin. At this point 0.005 in. would correspond to 0.001 in. movement in the circumference of the 13-in. diameter shrink joint.

The table shows how the joint was moved gradually a few thousandths at a time. The strain on the jack

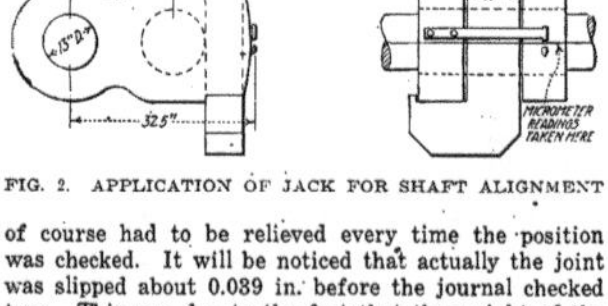

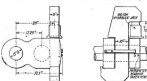

FIG. 2. APPLICATION OF JACK FOR SHAFT ALIGNMENT

of course had to be relieved every time the position was checked. It will be noticed that actually the joint was slipped about 0.039 in. before the journal checked true. This was due to the fact that the weight of the crankshaft decreased the eccentricity by causing it to run true in No. 1 journal.

When the bearings were again assembled they were given 0.006 in. clearance, and when the engine was started the shaft ran true and bearings remained cool.

Newark, N. J. RALPH MILLER.

It has always been considered dangerous to those handling a hose line to allow the stream to come in contact with an electric circuit. Prof. F. C. Caldwell has found by experiment that there is no danger of shocks when the distance between the nozzle and circuit is more than 25 ft. irrespective of the voltage.

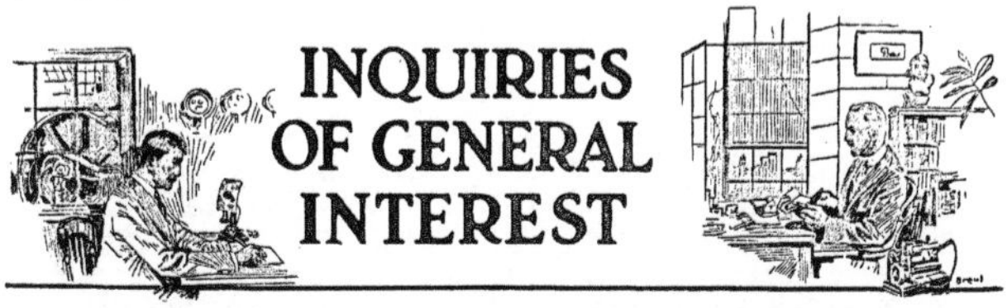

Holding Power of Flared vs. Beaded Tube Ends—What is the relative holding power of boiler tubes having the ends simply flared by the expanding tool, and when beaded over? F. C.

Beading over the tube ends, unless done with great care, has a tendency to weaken the hold of the tube and also to injure the texture of the material. With ordinary workmanship, the holding power is about 10 per cent. less than when the tube end is simply flared, but when beading is carefully performed upon metal of good ductility, the holding powers of beaded and flared tubes are sensibly equal. The durability of beaded tubes in resisting the action of fire is so much greater than when the tubes are simply flared that beaded ends are to be preferred.

Reason for Greater Lap on Head End of Valve—Why should there be more lap of a D-slide valve for the head end than for the crank end of an engine cylinder? W. R.

On account of the angularity of the connecting-rod there is less number of degrees rotation of the crankshaft and eccentric during the half-stroke in the head end than in the crank end of the cylinder. Therefore with equal laps, cutoff would occur later in the stroke from the head end and, to more nearly equalize cutoff, more lap must be given to the valve on the head end. But with a direct-acting eccentric rod, or valve gear provided with a straight rocker, it is not practicable to add sufficient lap to entirely equalize the points of cutoff, because the addition of lap on the head end reduces the lead on that end and lead is of more importance than equalization of cutoff. The unequal lap should not be more than will result in the greatest lead permissible for the crank end and the least necessary for the head end. Equalization of both lead and cutoff can be obtained by employment of a properly designed bent rocker.

Trouble from Long Pump Suction Pipe—Trouble is experienced in operating a 6 x 7 x 12-in. light service pump on account of leakage in the 4-in. suction pipe, which is 1300 ft. long. The suction pipe is buried 4 ft. below the surface of the ground. What would be the best way to locate the leak? F. B. S.

A tight suction pipe, of the length stated, is likely to become airbound in a short time from liberation of air out of the water during the great length of time the water is traversing the pipe while under less than atmospheric pressure and that trouble is commonly mistaken for a leak or solid stoppage of a long suction pipe. Before assuming that a leak has developed, it would be well to stop the entrance end of the pipe and use water from a tank under pressure or employ the pump for testing whether the pipe is tight against pressure. If it is found to be pressure-tight, the air "plugs," which usually consist of accumulations of air bubbles, will have to be pushed out of the pipe by forcing water through it. If the pipe is found to have a leak, its location can be determined by making pressure tests of different sections, beginning, for instance, at the middle of its length and testing each half, thus narrowing down the search to a small section of length that contains the leak. In most soils where the leak is considerable, its location will become revealed on the surface of the ground by keeping the pipe under water pressure for 12 to 24 hours.

Salvage of Blowoff Water—We have sixty-six 84-in. x 20-ft. horizontal return-tubular boilers and cut out three boilers daily for cleaning, discharging the boiler water to the public sewer. In addition to this waste of water, the boilers in use are blown down about three inches three times daily. Could not this water be discharged to a settling tank and be used over again as feed water instead of replacing it with fresh water from the city supply, which costs about one cent per 50 gal.? J. C. M.

Blowoff water usually contains a concentration of objectionable impurities and, without purification, would not be desirable for use, excepting in an emergency or where there is great scarcity of water. The waste of water described amounts to about 45,000 gal. per day, and at the excessive charge of one cent per 50 gal., or $200 per million, the expense of replacing the blowoff water is $9 per day, or only about 14 cents per boiler, an amount that does not appear to warrant salvage and purification of the blowoff water to render it suitable for boiler feed.

Testing Parallelism of Engine Guides—How can a test be made to determine whether the guides of a horizontal engine are parallel with the center of bore of the cylinder without removing the piston, piston rod or crosshead? D. A. S.

The center line of an engine cylinder usually is considered to be the line passing through the centers of the counterbores. However, if it is desired to make comparison of the guides with the center of the bore, it would be well to assume the center line of the bore to be a line through the centers of the bore at each end of the stroke of the piston. To establish such a center line without stripping the engine of the piston, piston rod or crosshead, remove the cylinder head and place the crank on dead-center away from the cylinder, to bring the piston at the crank end of the stroke. Wedge a wooden rod across the cylinder bore near the piston for holding one end of a light but strong line at the center of the bore, and secure the other end of the line to a stationary object at a distance of about six diameters of the bore beyond the head end. Draw the line as tight as possible and adjust the ends of the line so it will be central with the bore vertically and horizontally at both ends of the stroke of the piston. Then tightly stretch a second line of the same kind from the top of the cylinder, vertically over the first line and parallel to it, by careful measurements and plumbing from one line to the other. Stretch a third line parallel to the first and second lines, over or to one side of the cylinder the full length of the first lines, and beyond the guides, for a reference line. Carefully test the parallelism of the reference line with both the first and second lines, using a distance piece made of a light wooden rod with ordinary pins in each end. If in any case careful sighting along a line reveals sag, stretch another line a short distance above the sagging line by careful plumbing and, by replumbing, locate a supporting point for the line at about the middle of its span. Fasten to the crosshead a pointer terminating in a V opening that all but touches the reference line and after rechecking all adjustments remove the line and blocked rod from the cylinder. Then to test parallelism of the guides slowly turn the engine over, preferably backward, and observe whether the crosshead carries the pointer uniformly at the same distance from the reference line for the full length of a stroke.

Mexico, a Petroleum Storehouse*

MEXICO is the veritable land of promise and broken promises, of great natural wealth and abject poverty. From the days of Cortez the allurement of its natural resources has made an irresistible appeal to foreign capital and enterprise. Unfortunately, it is characteristic of a Mexican impulsively to stake all he owns on a cock fight, but to be unwilling to risk his savings in testing the mineral possibilities of his own farm. Consequently, nearly all the large public utilities, as well as the industrial developments of the country, have been promoted by foreigners.

The oil and asphalt seepages of the "Huasteca Region" in the states of Vera Cruz and Tamaulipas are frequently mentioned in the writing of the early Spanish explorers. The first attempt to drill for oil was made at Furbero by an Englishman named Autrey in 1869, but no important discovery was made until 1904, when a group of California capitalists brought in the first gusher at Cerrito de la Pez, near Ebano. From then on, however, oil field development has been rapid. About four years later the Pennsylvania Oil Co. drilled the famous Dos Bocas well, which burned for 57 days and could be seen at night by ships 30 miles out.

Big strides were made in 1910. Among the principal events of that year were the discovery of the Panuco field and the completion of the Juan Casiano and Potrero del Llano wells. It is safe to say that these two wells combined have already produced nearly 200,000,000 bbl. of oil, which is a record that has never been surpassed anywhere in the world.

In 1911 the first shipments of Mexican crude were made from Tampico to the United States, and the production jumped that year from four million to more than thirteen million barrels. At present Mexico is making an annual production of 75,000,000 bbl. from not more than 200 wells, or about one-fifth as much oil as is produced in this country from over 225,000 wells.

Today there are nearly 500 companies operating in the oil fields of Mexico, but less than a score of them can be rated as important producers. The ones which have now attained a solid footing expended unlimited sums of money for development when their future looked precarious. They have overcome innumerable difficulties and maintained an optimistic outlook all through the revolutionary disturbances.

Realizing years ago that Mexico promised to become the world's greatest producer of low-grade oil, the Standard Oil Company (N. J.) long since prepared to meet the new conditions as they materialized. At present the company is not by any means the largest producer in Mexico, but it is well established with a growing organization under the name of the Compania Transcontinental de Petroleo, S. A. The Standard Oil Company (N. J.) put into service the first large fleet of tank steamers to carry Mexican crude. At Tampico, now one of the important oil depots of the world, the Mexican company has an immense tank farm and a topping plant. It has tugboats on the river and is building a terminal loading station on the Gulf Coast, where the oil is piped out to sea and loaded in ships standing a mile or more off shore.

Potentialities of Mexico

The importance of Mexico in the oil business of the world today is but little appreciated. About 25 per cent. of its potential production is now being shipped.

Publications have estimated the possible daily production of the Mexican fields for the last few years as follows:

	Barrels		Barrels
1915	600,000	1917	850,000
1916	725,000	1918 and at present	1,000,000

At the same time the average daily shipments from Mexico during this period are shown below:

	Barrels		Barrels
1915	66,798	1917	126,175
1916	82,209	1918	152,363
First eight months, 1919			207,800

The extraordinary record of Mexican fields is more surprising when we appreciate that the real development is only fifteen years old. The producing business in Mexico can be said to have started on a practical basis in 1904 with the drilling operations of Messrs. Doheny, Canfield and others at Ebano. These operations were prompted by a desire to find oil for supplying railroads in the United States with fuel. The second important step in the development of the producing business was the opening of the so-called Southern field in 1910, when the Doheny interests drilled in the famous Juan Casiano well and the Pearson interests drilled in the equally famous Portrero del Llano. The latter was destined to produce 100,000,000 bbl. before it was drowned out by salt water in the past year, and the former, with a record of more than 80,000,000 bbl., is still producing 23,000 bbl. a day through a damaged gate valve which has not been moved for years.

The Mexican oil fields are divided into three distinct fields as regards location as well as quality of crude produced.

The Isthmus of Tehuantepec has not been a profitable field. The oil found is of good quality, about 36 Bé. In this field there have been a total of 221 wells drilled, which today have a total production of about 500 bbl. from 75 producing wells, all others having been abandoned. The Pearson interests initiated their efforts in the Mexican field in this region, and built a refinery at Minatitlan, which now depends principally on the other Mexican fields for crude, which it receives by tank steamer.

Another oil-producing region is generally known as the Northern field and produces a heavy crude varying from 8 to 15 gravity Bé. This field lies from twenty to fifty miles in a westerly direction from Tampico. It includes the fields of Ebano, Panuco and Topila. Panuco furnishes about 85 per cent. of the production of the field, and its crude has a gravity of 12 Bé. In the Northern field approximately 250 wells have been drilled, of which 150 are at Panuco.

The most important Mexican field is the Southern, or so-called "light oil" district, which starts at Tepetate, about 70 miles south of Tampico, and extends in a southerly direction to the Tuxpam River, a distance of about 35 miles. In this field the oil is about 21 gravity Bé. and the potential production is over 500,000 bbl. a day. This production is the result of drilling 185 wells, of which 68 are productive.

In the Southern field, the oil is transported by pipe line, but it is the usual practice to maintain a railroad from tidewater to the field. The railroads are about twenty miles long, fully equipped for handling freight direct from the wharves to the fields.

An extraordinary characteristic of the wells in both the Northern and Southern fields is the steadiness of the flow. The production is obtained, more especially in the Southern field, in the very porous limestone known as the Tamasopa. Where this formation outcrops, some 60 miles farther west, enormous caverns are found. The samples of the rock which are flowed from the wells when drilling them in look almost like honeycomb. The thickness of the formation is generally spoken of as from one to three miles, which is, in a measure, an explanation of the unusual staying qualities of the wells.

The wells, when shut in, show a pressure of from 300 to 1000 lb., depending on the locality. When the wells are being flowed, the gate valves on the big ones are generally opened from one to three turns, where full opening of the 8-in. valve used is about 27 turns. While the wells are being flowed under such conditions, they hold back pressure of from 200 to 700 lb. The oil as it comes from the well is 120 to 150 deg. F.

The water that flows from these wells following inundation is of even higher temperature, this situation being peculiar to limestone formations. The hot oil is much easier for the pumps to handle, an advantage which is fully appreciated.

The most extraordinary natural seepages of oil in the world are found in Mexico. Long before oil was produced from drilled wells, these seepages supplied asphalt, and the streets of Vera Cruz are paved with asphalt from seepages near the Tuxpam River. These seepages made a death trap for cattle and wild animals in the vicinity, and the numerous skeletons of all kinds bedded in the soft asphalt are mute witnesses of the effectiveness of these pitfalls afforded by a freak of nature.

*From *The Lamp*, published by the Standard Oil Co. of New Jersey for its employees.

In the Northern field the heavy oil from the Panuco and Topila districts is generally brought to Tampico with barges and stern-wheel steamers similar to those seen on the Mississippi and Ohio rivers. This is due in part to the opportunity afforded by the Panuco River, which at all seasons permits barges of seven-foot draft to operate 65 miles by water from Panuco. These barges hold from 6000 to 8000 bbl. each and are approximately 40 ft. wide and 200 ft. long. A large steamer generally has seven barges as her complement. She takes three empty barges from Tampico to Panuco, waits two or three hours for them to be filled, and returns to Tampico, making a round trip in 24 hours. These barges are left at the deep-water terminals to be pumped out into tanks ready for delivery to the tank steamers. The steamer leaves the three full barges, takes three empty barges and proceeds on another voyage. While she is away, the barges are pumped out in the course of six or eight hours, ready for the steamer on her return trip. One barge is generally under repairs, principally due to work of the teredo, which attacks any timber in the water, eating holes through it until it looks like honeycomb. An unusual characteristic of this pest is that it will not go from one plank to another, crossing a seam. The defective planks are removed from the bottom of the barge, frequently their weight being only 15 or 20 per cent. of what it was when new, the rest having been consumed by the teredo.

The loading of steamers in the Southern field is interesting and a departure from the usual practice. The steamers are loaded on the Gulf Coast by pipe lines which are run from one to two miles out from the coast to the steamer's moorings and the oil is delivered to the vessel while she is at sea.

The so-called sea lines are first laid ashore. The entire length of the line is then put on small cars on a railroad and pulled out to sea by tugs. When it is appreciated that these are 8-in. or 10-in. lines more than a mile long it is surprising to see the ease with which they are hauled out, provided no accident happens during the critical half-hour that the line is moving. If a considerable part of the line is in the water and an accident causes it to stop moving, it is frequently impossible to continue hauling it out, as it sticks in the sand and has to be recovered in sections by the aid of divers.

These lines are hauled out to about 40 ft. of water. On the end is an 8-in. hose about 100 ft. long attached to a small wooden buoy. Around the end of the pipe line, about 400 yd. apart, are placed four or five buoys to which the tank steamer secures before connecting up her hose to receive oil. After being moored she connects up the two hose, as it is the practice to lay two sea lines to each steamer berth, and signals to the pump station ashore to start pumping oil. This oil is then delivered at a rate of from 4000 to 6000 bbl. an hour. As the coast is little protected, the surf always makes communicating with the ship a difficult problem, and the men who handle the surf boats have many lively experiences, particularly in the winter when bad weather is the rule, rather than the exception.

When a Mexican Gusher Comes in

Ordinarily, the hole is drilled to within 300 or 400 ft. of the pay with a rotary. Pipe is then cemented, tested to 750 to 1000 lb. pressure, and gate valves are put on, which can be closed any time the well comes in. When it is expected that the pay may be reached, steam is kept up in a reserve boiler at a safe distance from the well and a man is stationed at the drilling boilers to turn out the fire at the first alarm. A man is stationed to close the gate valves and another one ready to pull out the tools.

The first indication is that water used in drilling starts to flow and generally within ten minutes the whole story is told if no accident occurs. The well throws the water and drilling tools out of the hole over the derrick. The man at the boiler shuts out the fire and turns steam into the firebox to prevent a conflagration. The man at the gate valves stands ready to close them the instant the tools have been blown out. The derrick is reinforced with heavy timber and tank steel so that the tools cannot fall back on the valves and break them off.

Frequently, in spite of all precautions, the wire drilling cable fouls the tools so that they stick in the pipe, resulting in a fishing job, while the well is flowing several thousand barrels. While the cable is still in the hole, of course, the gates cannot be closed. Where the preparations have not been sufficient or an accident occurs, it is frequently followed by disaster to the well as well as to the men.

A world-famous example is the Dos Bocas well, about ten miles north of Tepetate, which came in without being controlled, flowed thousands of barrels a day, caught fire and burned for some sixty days until it extinguished itself, the flow of oil having been replaced by a geyser of hot salt water. This is now a sort of salt-water volcano, the crater being half a mile in diameter and producing about 1,000,000 bbl. of boiled salt water a day.

In the famous Potrero del Llano well an accident to the casing made it impossible to shut the well in and resulted in seepages coming up through the ground for a radius of half a mile. While the well was producing 46,000 bbl. a day for years, half of the production came through the casing and the balance from these seepages, due to the enormous pressure on the well and the leaking through the defective casing.

The country is largely covered with jungle full of all manner of game, although the jungle is so dense that animals can only be seen in the narrow trails or along the rights-of-way which have been cleared by the operators for their pipe lines and railroads. The waters abound with fish, so that only the choicest varieties are even worth offering for sale.

The climate, although just within the tropics, is not as intensely hot as would be imagined, thanks to the trade wind which blows during the hot months from 10 o'clock in the morning until after midnight. From June until September is the rainy season, during which time there is a very heavy rainfall and passage through the country is extremely difficult on account of the mud. It is this difficulty which has compelled the operators to build railroads rather than roads for transportation between the waterways and the fields.

The drillers live in camps with Chinese servants. It would almost seem that without these Chinamen work in the remote districts would be next to impossible. The commissary supplies come largely from the United States, although fruit and vegetables are abundant in Tampico. The soil is extremely fertile, and vegetables of almost any description can be planted any day in the year.

Any article touching on the producing business in Mexico, however general, would be incomplete and unjust without paying tribute to those men who live in the jungle, exposed to the greatest hardship and dangers, removed from all bases of supplies, who by their perseverance, ingenuity and courage bring in and master such enormous wells, accompanied with a flow of deadly sulphur gas. Particularly was this true during the war, when the conditions were the worst from all standpoints, and the importance of the Mexican production, which is essentially fuel oil, for the use of the Allies, was of vital importance.

RUB-A-DUB. TWO MEN IN A TUB

Steamboat Inspection Service Amends Its Rules

Section 4433 of the revised statutes, as it at present reads, is unsuited to modern practice, and the Bureau recommends that it be amended to read as follows: "The working steam pressure allowable on all boilers inspected as required by the L. II. shall be determined by the rules of the Board of Supervising Inspectors with the approval of the Secretary of Commerce." The purpose of the proposed amendments in this section are, first, to do away with the obsolete rules contained in the present law which prescribes a working steam pressure for single-riveted joints without taking into consideration the percentage strength of the riveted joint, and allows 20 per cent. additional pressure for double-riveted joints, but which does not allow a greater working pressure for triple- and quadruple-riveted, etc., lap and butt joints, for which greater working pressure should be allowed on account of the greater strength of the triple-riveted and other joints of greater strength than the double-riveted joints. There is a provision of the section which reads: "No split-calking shall in any case be permitted." This has been omitted in the amendment, as the rules for calking would be prescribed by the Board of Supervising Inspectors in connection with the rules for riveted joints.

Section 4418 of the revised statutes with reference to hydrostatic pressure tests of boilers reads as follows: "All boilers used on steam vessels as constructed of iron or steel plates inspected under the provisions of the section 4430 shall be subjected to a hydrostatic test in the ratio of 150 lb. per sq.in. to 100 lb. per sq.in. of working steam pressure allowed." In the recommended amendment this part of the section is made to read: "All boilers used on steam vessels and inspected as required by L. II. shall be subjected to hydrostatic tests as shall be determined by the rules of the Board of Supervising Inspectors with the approval of the Secretary of Commerce." It is the desire of the Bureau that the amendments be enacted into law, so as to leave with the Board of Supervising Inspectors the authority for determining the hydrostatic test in connection with the rules for riveting which would be adopted by the Board, all of such procedure, of course, to be approved by the Secretary of Commerce.

The purpose of these and other amendments is to bring the Supervising Inspector's rules up to date and facilitate the rehabilitation of the American Merchant Marine.

New York Engineers Loyal

The largest meeting of engineers or firemen ever held in this country took place at Cooper Union, New York City, Dec. 11, 1919. The meeting was arranged by Locals Nos. 20, 56, 96, 319, 379, 608, 615 and 670, International Unilon of Steam and Operating Engineers, and Local No. 56, International Brotherhood of Stationary Firemen.

The purpose of the meeting, as expressed in the invitation extended to all engineers and firemen of New York, by the arrangement committee, was "the formulation of a wage scale for engineers, firemen, oilers, etc., in all steam plants of Greater New York and for the betterment of the crafts in general." The committee invited the general officers of both organizations and representatives of the American Federation of Labor.

At 8 p.m., when the general and local officers, members of the arrangements and invited guests appeared on the platform, the great hall, having a seating capacity of 2500 or more, was packed.

Joseph Monterfering, of Brooklyn, called the meeting to order and stated the first business was the selection of a chairman. Matt. Comerford, past general president of the I. U. S. & O. E., now with POWER, was selected as permanent chairman. P. J. Horan, of Local No. 20, I. U. S. & O. E., and Mr. Flannagan, of Local No. 56, I. B. S. F., were selected secretaries. After briefly reciting the history of the organizations and the principles on which they are based, the chairman introduced the following speakers:

James P. Holland, president New York State Federation of Labor; W. J. Collins, organizer, American Federation of Labor; Milton Snellings, general president, I. U. S. & O. E., Washington, D. C.; Timothy Healy, general president, International Brotherhood of Stationary Firemen, New York; Arthur M. Huddell, first vice president, I. U. S. & O. E., Boston, Mass.; H. M. Comerford, general secretary-treasurer, I. U. S. & O. E., Boston, Mass.; Edw. F. Moore, examiner of engineers and president of Local Union No. 569, of Chicago, Ill., and Joseph Monterfering, third vice president, I. U. S. & O. E., Brooklyn, N. Y.

The following resolutions were submitted to the meeting and unanimously adopted:

RESOLUTIONS

Whereas, There are today in the Greater City of New York about forty thousand steam-plant employees, comprising engineers, firemen, oilers, etc., employed in factories, public institutions, office buildings, loft buildings, hotels, apartment houses and various other plants; and

Whereas, These men, who must be American citizens, and pass examinations as to their fitness in order to receive licenses, all hold very responsible positions, being intrusted with the care and operation of plants that are essential to the preservation of the lives and property as well as the safety and comfort of millions of people every day; and

Whereas, There prevails today in the trade and calling of steam-plant operatives a scale of wages far below that of the average American mechanic and in some instances below that of the ordinary day laborer, the wages received being out of all proportion to the present high cost of living, while the working hours are much longer than those of the average mechanic; therefore,

Be It Resolved, That this mass meeting of engineers and firemen assembled in Cooper Union, Thursday evening, Dec. 11, 1919, authorize our various local unions in the City of New York to select a committee to be known as a Wage Committee, which shall be vested with full power to present to the Hotel Men's Association, Real Estate Men's Association, Merchants' Association and other groups of employers the demands of the men of our crafts for a uniform scale of wages, reasonable working hours and one day's rest in seven.

That we pledge our undivided support to any committee that may be selected, and promise to use all honorable and legitimate means within our power to induce employers to adjust matters in dispute so that engineers, firemen and other steam-plant operatives may be placed on an equal footing as regards wages and hours with all other mechanics in the City of New York.

Be It Further Resolved, That we reiterate our loyalty and devotion to this American Republic and denounce any attempts to subvert our established forms of government. We hereby pledge ourselves to coöperate with any agency patriotically striving for industrial peace and endeavoring to bring about harmonious relations between employers and employees.

Steam Versus Electric Main-Roll Drives for Steel Mills

On Saturday, Dec. 20, 1919, at the Hotel Chatham, Pittsburgh, Penn., the Association of Iron and Steel Electrical Engineers held its December meeting of the Pittsburgh Section. The meeting was preceded by a dinner and entertainment, which proved to be an enjoyable function. President D. M. Pettie presided at the Technical Session, which was devoted to a discussion of Steam Versus Electric-Driven Mills.. The discussion was opened by G. E. Stoltz, who explained by the use of moving pictures the operation of both steam and electrically driven reversing blooming mills. It was the opinion of Mr. Stoltz that steam was used to the least advantage when the mill was driven by steam engines, and that due to many advantages of the electric drive over the steam drive for rolling mills, the last ten years have seen a wide application of electric motors to the driving of these mills.

Mr. Seibert, of the Bethlehem Steel Co., felt that in almost all cases the electric drive could produce steel at a lower cost per ton than could be done with the steam-engine drive, although in some small isolated steel mills it might be found more economical to use steam drive. However, a careful study of all factors should be made in every case before deciding upon the type of drive. The speaker called attention to the lack of reliable data on the power requirements for steam drive, and presented by the use of lantern slides the results of investigations which he has made on steam and electric-driven reversing blooming mills. The results of these investigations showed a power cost per ton of $1.33 for the steam-engine drive against $0.76 for the electric drive. However, no fixed rule can be laid down as to the relative economy of steam and electric drive.

The uniflow engine was discussed as a possible competitor of the electric motor for driving steel mills. The non-reversing engine has been developed to where it can be used and is being used on main-roll drives, but the reversing type has not reached this stage of development.

K. A. Pauly pointed out that the question of steam engine versus electric drive for rolling mills resolved itself into a reversing blooming-mill problem, and if the electric drive could be developed to a point where it could produce the tonnage that is being obtained with steam-engine-driven reversing blooming mills, then there would be little question of the superiority of electric drive over the steam drive. Mr. Pauly no doubt created somewhat of a surprise for many when he said that all high-production reversing mills were steam-driven, and that in no case was there an electric-driven reversing mill that could compete with the steam-driven mill in production. However, the low-cost-production mills are electrically driven, but although cost of production is an important factor, it is frequently outweighed by high production in the minds of steel-mill men, since tonnage is the thing that they are after. The speaker called attention to the steam-driven mill rolling sixty or more ingots per hour, and said that in no case was there an electrically driven mill that was doing this. Mr. Pauly then told of some of the work that was being done at the Trumbull Steel Co.'s plant with an electrically driven reversing mill and said that they eventually expected to have this drive in such shape that it would be capable of a production equal to any steam-driven mill. This is also the case at the Sparrows Point Mill of the Bethlehem Steel Co., where they are rolling as high as 56 ingots per hour.

E. S. Jeffreys, electrical engineer for the Steel Company of Canada, gave the record of a 34-in. electrically driven reversing mill that had rolled 692,012 net tons without one minute's delay on account of the main-roll motor. In this mill the power requirement to drive the roll is about three times that required by the auxiliary drive. The cost of rolling a ton of steel in this mill averages 32.7 cents.

A great amount of detailed data was given in the discussion on production and cost of production with the two types of drives. Unfortunately, this material is not available for publication at this time, but no doubt will be in the near future, and a digest of it will appear in these columns.

A New Type of Hydraulic Turbine Runner*

By Forrest Nagler

The progress in waterwheel development has been more nearly connected with increase in speed than with improvement in efficiency. The development of high-speed electric generating machinery has continually called for increase in speed, especially for low-head sites.

The designers of waterwheels have had to meet a very wide variety of conditions due to nature's inconsiderate failure to standardize waterfalls. In practice waterwheel runners vary in dimensions, horsepower, speed and the head under which they operate.

To compare two wheels operating under different conditions, the first step would be to compute their power and speed when under the same head. This still leaves variations in power, speed and dimensions, so a second step is necessary. This second step may be:

A. Recompute and compare their powers on the basis of being so changed in dimensions as to have the same speed; or,

B. Recompute and compare their speed on the basis of the same power.

*Abstract of a paper presented at the Annual Meeting of the American Society of Mechanical Engineers, New York, Dec. 2-5, 1919.

Basis A would give a *characteristic power* and is the method used by Professor Jowski in the *Engineering Record*, Dec. 26, 1914. Basis B gives a *characteristic speed* and is the more general practice. This characteristic may be defined as follows:

The *characteristic speed* of a runner is the speed in r.p.m. which a model of that runner would have if operated under a head of 1 ft., this model to be reduced proportionally in all dimensions from the original until it will develop 1 hp. under 1-ft. head.

From the hydraulic laws governing the variation of power and speed of runners, it is shown that

$$N_e = N_1 \sqrt{hp_1} = \frac{N\sqrt{hp.}}{\sqrt[4]{H^5}}$$

hp = The horsepower of a runner;
N = The revolutions per minute;
H = The head in feet;
hp_1 = The horsepower under 1 ft. head;
N_1 = The r.p.m. under 1-ft. head;
N_e = The characteristic speed.

As an illustration of the universal application of the basis of comparisons, the accompanying table is given, which contains examples from actual installations.

Historically, the progress in hydraulic-turbine building may be illustrated by noting the increases in characteristic speed. The earliest types of waterwheels, antedating the Christian Era, were the current wheels, which developed into overshot, undershot and breast wheels. Their characteristic speed had

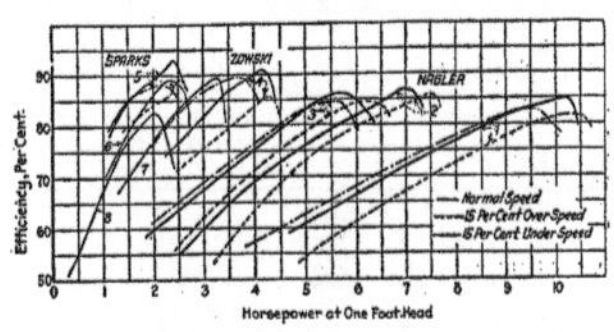

POWER-EFFICIENCY CURVES FOR WHEELS HAVING A NORMAL SPEED OF 50 R.P.M.

a maximum of possibly three, and as a consequence of their low heads their r.p.m. averaged probably under 20.

Between 1825 and 1840 Fourneyron developed the radial-outward-flow type and Jonval developed the axial-flow type, but demands for speed and capacity were such as to limit characteristic speeds to between 20 and 40 and capacities to considerably under 1000 hp. About 1870 Francis developed the radial-inward-flow type, which gradually changed until the water entered radially and was discharged axially. This type is especially an American product, and its characteristic speed has gradually increased from about 40 in 1870 to a little over 100 in 1914. The last increases, made by Zowski, were the basis of great advance in the design of low-head units. The new type of runner described by the author will give a characteristic speed of about 160.

In 1907 the author was engaged in field work on large-sized axial-flow pumps. Impressions gained in this work influenced his trend of thought in his later hydraulic-turbine work so that the Francis-type runner seemed unnecessarily complicated. From these ideas a definite application of the axial-flow principle was developed in 1913. The new design is based on a straight radial blade somewhat similar to a marine-propeller blade, which offers a minimum of wetted surface, and of bending moment at the root of the blade. Models were made, tests run, and such refinements added as seemed advisable. Tests on models were verified by tests on real runners at Holyoke in 1917.

Some characteristic curves are shown in the accompanying plot, where characteristic horsepowers are plotted against efficiencies. Curves Nos. 1, 2 and 3 are from the new-type runner, while No. 5 is from the runners of Professor Zowski.

From these curves the following may be deduced: For any given conditions of head and speed the runner No. 1 will give 140 per cent. and No. 2 will give 65 per cent. more power than types previously available as typified by No. 4.

Comparison on a speed basis, using equal power, shows that runner No. 1 will give over 50 per cent. and No. 2, 30 per cent. higher speed than No. 4.

The primary advantages shown by the new type-runner are as follows:

(a) Lower generator cost due to increased speed. This may be from 15 to 35 per cent.

(b) Lower turbine cost due to simpler runner. This averages around 10 per cent.

(c) Smaller generator diameter and therefore smaller power house.

(d) Higher generator efficiency due to better design possible with the higher speed.

(e) Greater turbine flexibility, which permits more power under flood conditions when the head is reduced.

This new type of runner is from the direction of flow a pure Jonval type, although Jonval's wheels seldom had characteristic speeds in excess of 20 or 30 and his blades were different. The primary essential of high characteristic speed is a reduction of hydraulic friction and centrifugal forces. Theoretically, high characteristic speed may be obtained by flattening the blade angles. Practically, this results in increasing the wetted surface and so reducing the efficiency due to friction. In the author's design these effects are counteracted by cutting out blades, which does not result in decrease of power, as might be expected.

In conclusion the development of this runner would offer such possibilities as the use of a 30,000-lb. casting, very simply made and easily transported instead of one weighing 130,000 lb. and ranking among the most complicated castings ever made, which was used at the Keokuk plant. With any given minimum generator speed the new runner will permit operation under heads one-half as high as those required in prior practice and for large powers units may be developed at lower costs.

COMPARISON OF OLD- AND MODERN-TYPE HYDRAULIC TURBINES

Item	—Old Types— Overshot Wheel	Fourneyron (Tremont) Turbine	——Modern Types—— Nagler High Speed Low Head	Usual Mixed-Flow or Francis Medium Speed	Usual Mixed-Flow or Francis Low Speed Dbl. Run	Impulse (Pelton) Twin
Head in feet, H	14	14	14	200	400	2,000
Horsepower, total	50	180	500	40,000	20,000	20,000
Runner horsepower, HP....	50	180	500	40,000	10,000	10,000
Speed, r.p.m., N..	10	53	200	150	360	375
Runner diam., in., d.........	144	40	72	130	72	96
Unit horsepower, HP_1...........	0.95	3.44	9.55	14.14	1.25	0.11
Unit speed, r.p.m. N_1	2.67	14.16	53.50	10.61	18.00	8.39
Characteristic speed, r.p.m., N_c	2.66	26.30	165.00	40.00	20.00	2.78

The Steam Turbine*

By E. H. Sniffen

Manager, Power Section, Westinghouse Electric and Manufacturing Company, East Pittsburgh, Penn.

Why was it that the steam turbine, being the earliest form of heat engine, remained in an undeveloped stage, while for a period of more than a hundred years the reciprocating engine became the only type of prime mover? The answer is that the development of the steam turbine as we know it today was delayed until the mechanical arts and the knowledge of materials had progressed sufficiently to make its construction possible. Comparatively speaking, the reciprocating engine was an easy thing to build. The steam turbine is in reality a metallurgical achievement. It could not well have come earlier.

We find in the British patent records many steam-turbine inventions occurring between the years 1800 and 1850. Prophetic inventions most of them, the kind which characterize most great developments. There is a class of men who dream dreams and another class who put their dreams to some use. I think perhaps our American inventors have excelled in the latter respect, for theirs has been a wonderful record, with the telegraph, telephone, airplane, submarine, talking machine, steam navigation, machine gun, cotton gin, harvesting machinery, etc. But we must not forget that Europe, too, has had her victories of invention. Here was the steam engine, the gas engine, the dynamo, both direct and alternating-current, the steam turbine, the locomotive, spinning machinery, the X-ray, radium and many others.

The steam turbine began its practical use on the other side. Parsons blazed the way, and at about the same time Delaval. Parsons built his first turbine in 1884 and for many years thereafter built a goodly number of what we would now call very small machines. In the year 1896 Mr. Westinghouse acquired the American rights for the Parsons turbine. After some years of development work we installed in 1889 three 400-kw. machines in the air-brake company's plant at Wilmerding. They are running yet.

In 1900 we put our first turbine out into the customer's hands. This was a 2000-kw. machine which we furnished to the Hartford Electric Light Co. It was double the capacity of any other turbine then in use, and European engineers looked upon it as a daring piece of work to put so large a capacity in one cylinder construction. Well, the business grew, and in truth the steam turbine made possible the tremendous strides that our electrical industry has made. This I shall try to make clearer in a moment. By 1905 our largest unit had grown to 7500-kw.; in 1909, 10,000; in 1913, 20,000; 1917, 35,000, and in 1918, 60,000 kw. Where will it end?

We cannot properly consider the steam turbine and the place it occupies without thinking about the electrical industry as a whole, for the turbo-generator is an electrical device; it generates electric current. I am going to show you in a few moments pictures of the largest power plants in the country some fifteen years ago. Beautiful plants they were, operating with reciprocating engines. One of them had eleven 3500-kw. units, the other eight 7500-kw. units. These plants were in New York City. The piston engine had reached its maximum size at 7500 kw. It was a ponderous thing. The engine and generator combined weighed nearly 2,000,000 lb. The diameter of the generator was 40 ft. We had to build a special shop in which to make it. And think of it! Only 7500 kw.! The engine had four cylinders, two horizontal and two vertical. Even then, divided that way, the low-pressure cylinders were 88 in. in diameter, which was felt to be about as large as a steam cylinder could be safely built. This plant of 60,000-kw. capacity was the last word in power-plant design. It had a fuel record of 2½ lb. of coal per kilowatt-hour. Those engine units cost originally about $175,000 each. After running about 15 years they were broken up and scrapped at a salvage of about $10,000 apiece and replaced with turbines of 30,000 kw. each, the plan being to install 240,000 kw., an increase in power of 4 to 1. Their original boiler equipment was not increased, though the furnaces were rearranged. The engines had required about 17.5 lb. of steam per kw-hr. The turbines took about 11 lb. The engine units weighed 280 lb. per kilowatt, the turbines about 45 lb. The comparative floor space was in favor of the turbine in the ratio of 1 to 10. The original station cost $128 per kilowatt installed capacity. On the basis of writing off the engine units entirely, the cost of the plant with the full complement of turbines was $42 per kilowatt.

The electrical industry has grown beyond our dreams until we have in this country today some ten million kilowatts of central-station capacity—about two-thirds of it steam power and one-third water power. It ministers to the necessities and the comfort of perhaps 65 per cent. of our population, which is about the proportion of those who live under electric wires. These people average to consume about 400 kw.-hr. per annum per capita. Here in Pittsburgh the consumption is around 600. The population of our country doubles about every twenty years. During the past two decades the per capita consumption of electricity has doubled about every five years. It takes now about 1/7 kw. installed capacity to serve the needs of each man, woman and child in American communities. This result would have been economically impossible with any other form of prime mover.

*From a talk before the Duquesne Light and Power Company's section of the National Electric Light Association.

Canadian Commission of Conservation on Pulverized Fuel

The Commission of Conservation of the Dominion of Canada has just issued a 57-page pamphlet dealing with the subject of powdered fuel and compiled by William J. Dick. The first 12 pages are devoted to data setting forth the shortage of coal in Canada and the increasing difficulties of obtaining fuel oil. It is to be noted that little is said by the commission itself about powdered fuel, the following being the most pertinent statement in the pamphlet and expressive of the commission's opinion:

"The economy secured by the use of pulverized fuel in stationary boilers instead of hand-fired coal is not so great in comparison as that derived from its use in locomotives. This is due largely to the fact that it is possible to equip stationary plants with the best mechanical stokers. There is an advantage of 2 or 3 per cent. in combustion efficiency in favor of pulverized coal, but this is offset by the additional cost of fuel preparation. While the above comparison is made from the standpoint of efficiency, and where almost similar coal is used, there are many localities, especially in northern Ontario, and portions of Manitoba, Saskatchewan and Alberta, where pulverized peat or pulverized coal could be used to decided economic advantage instead of higher-priced imported coal."

The remaining 44 pages of the pamphlet are devoted chiefly to reprints of descriptions of pulverized coal-burning equipment.

Industrial Safety Conference

In the absence from Washington of the Director of the Bureau of Standards, Dr. S. W. Stratton, the meeting of the Industrial Safety Conference on Dec. 8, 1919, was called to order by Dr. E. B. Rosa, who summarized at some length the events leading up to this conference and referred especially to the proceedings of the similar conference held on Jan. 15, 1919, of which this in a sense was an adjourned meeting.

The principal subjects which came up at the January conference were the reorganization of the American Engineering Standards Committee and the question of whether the safety work of the Bureau of Standards should be conducted under the scheme of procedure laid down by that committee. The result of a letter ballot was a decided majority in favor of procedure under the plan of the American Engineering Standards Committee. This committee has adopted, since the January conference, a revised constitution which opens its membership to other organizations in addition to the original five founder societies and three Government departments.

Prof. Comfort A. Adams, chairman of the American Engineering Standards Committee, spoke on the work of that committee and its recent reorganization. Membership in the committee is now open to such organizations or groups of organizations of national scope as may be approved; there shall be no more than three members from each such organization, and the annual dues are $500 for each representative. The speaker stated that he would be superseded as chairman of the committee by A. A. Stevenson, and that the permanent secretary will be Dr. P. G. Agnew, at present in the Bureau of Standards. Headquarters will be in New York City.

Chester C. Rausch, of the Safety Institute of America, introduced the following resolution which was adopted by the Conference:

Resolved: (1) That the American Engineering Standards Committee be asked to request the International Association of Industrial Accident Boards and Commissions, the Bureau of Standards and the National Safety Council to organize a joint committee on safety codes, this committee to include representatives of these bodies and such others as they may consider advisable; (2) that this joint committee report upon the safety codes required, priority of consideration of the codes, and sponsor bodies for their preparation; (3) that this report be put in writing and placed not later than Feb. 1, 1920, in the hands of the American Engineering Standards Committee.

Before taking the vote on this, however, another motion was passed confirming the result of the letter ballot taken last spring and expressing the decision of the conference that safety codes should be established under the procedure of the American Engineering Standards Committee.

In the discussion on this subject it was pointed out that the American Engineering Standards Committee was not primarily interested in safety matters and that the committee contemplated in the resolution of Mr. Rausch would be directly concerned in such matters and might well serve as a steering committee on safety-code work. The opinion was freely expressed that such a committee should be a permanent one, that it should contain representatives of all interests involved in safety codes and that it might well be called a National Safety Code Conference and hold annual meetings. Such a committee would be in a position to coördinate work on safety codes, to arrange for necessary interpretations, to initiate new codes as they become necessary, and to form a central agency to insure coöperation.

Hydro-Electric Development in Czecho Slovakia

A scheme for the utilization of the water powers in Czecho-Slovakia, which are estimated at 800,000 hp., has been approved by a bill adopted by the National Assembly. The installations necessary for laying out all the territory for the use of electricity will require an expenditure of 2,000,000,000 crowns ($400,000,000) for the hydraulic work, 500,000,000 crowns ($100,000,000) for the central stations, and 1,000,000,000 crowns ($200,000,000) for the distribution of electricity. It is calculated that the total revenue from the consumption of energy will reach 1,000,000,000 crowns ($200,000,000) per annum.

According to the law in question an annual credit of 75,000,000 crowns ($15,000,000) is to be entered in the budget estimates for the years 1919 to 1923, and two-thirds of this amount will be devoted to the construction of hydro-electric works, and the balance will represent the financial participation of the State in the working of undertakings for the production and distribution of energy. The State, departments, and the communes will alone or in coöperation with existing electricity companies take up 60 per cent. of the capital so as to secure influence in the undertakings, while private capital will be admitted only up to 40 per cent. Part of the capital to be raised will be covered by the issue of bonds and debentures. The sum of 8,000,000 crowns ($1,600,000) out of the total amount voted by the National Assembly will be placed at the disposal of the undertakings for the current financial year. The scheme aims at the establishment of nine large electrical undertakings, of which four would be in Bohemia, three in Moravia, one in Silesia, and one in Slovakia.—*The Electrical Review, London.*

Gadsden-Huntsville Extension

Capt. W. P. Lay, of the Alabama Power Co., states that the company's high-tension line from Gadsden to Huntsville, construction of which has just begun, is part of a plan to build a power loop, which will be one of the longest and most important in the world.

The company has a line from Lock 12 on the Coosa River to Birmingham, and from that point to Sheffield and Florence. From the same generating station another line branches out to the eastern side of the state, passing through Sylacauga, Talladega and Annison to Gadsden. With the completion of the extension to Huntsville it will be necessary to build between that city and Florence, a matter of only a few miles, to complete the loop which will belt the entire mineral and industrial section of Alabama.

This loop will be completed as early as possible, Captain Lay states, in order to connect a complete system of electric-power transmission, a system that will tap all towns and cities from the Tennessee River to Central Alabama and as far down as Selma.

Two emergency steam plants, together capable of taking care of any emergency from breaks or accidents, are stationed on this loop. They are located at strategic points, one being on the Warrior River below Birmingham and the other on the Coosa River at Gadsden.

Ten Hints for the Storage of Bituminous Coal

Tests conducted by the Bureau of Mines lead to the following conclusions regarding the storage of any bituminous coal:

1. Piles not to be over 12 ft. deep and no part of the interior to be over 10 ft. from the surface.
2. Store only screened lump coal—if possible.
3. Keep out dust as much as possible, and to do this avoid handling.
4. Have lump and fine evenly distributed. Do not let lumps roll to the bottom and form air passages.
5. Rehandle the screenings after two months, if possible.
6. Store away from any sources of even moderate heat and well away from the main buildings of the plant; never against a frame building.
7. Allow six weeks' seasoning after mining before putting into storage piles.
8. Avoid alternate wetting and drying.
9. Avoid admission of air to the interior of the pile through interstices around timbers, irregular brickwork or a porous bottom, such as coarse cinders.
10. If wet coal is received, dump in small piles around the edges, where air can get to it freely to carry away moisture, and where other coal will not be packed on top of it.

Oil Fire Efficiency Boosted by Orsat

Guy L. Bailey, manager, Municipal Light and Power Co., San Francisco, Calif., states that to get information as to the efficiency of the furnaces in the Municipal Light and Power Co.'s plant on Stevenson St., Orsat apparatuses were installed in the fireroom. The four boilers were of Stirling make, and the flue-gas samples were taken about a foot below the damper. The determinations were of great value at this time in educating the firemen as to the amount of air required.

"All of our men," says Mr. Bailey, "were experienced in the burning of fuel oil, but they were firing with from 80 to over 150 per cent. excess air. By shutting down on the draft and making frequent determinations with the Orsat until a high percentage of CO_2 was obtained, the firemen were taught how the fire should appear with the minimum amount of air for complete combustion. The men were unable to get much better than 12 per cent. CO_2 as above this point the fire was likely to smoke. The second month showed an increase of 19.3 per cent. in kilowatt-hours per barrel of oil obtained the month before installing the Orsat. This improvement was responsible, in small part, to the fact that the load factor on the turbine increased from 43 to 53½ per cent."—From *Transactions A.S.M.E.*

New Publications

A STUDY OF THE FORMS IN WHICH SULPHUR OCCURS IN COAL. By A. R. Powell with S. W. Parr. Bulletin No. 111, Engineering Equipment Station, University of Illinois, Urbana, Ill. 62 pages.

This bulletin is a report of investigations made to study the nature of the sulphur-containing compounds in coal, the quantity of each form present and the change which the form undergoes when the coal is allowed to stand or when it is coked. A knowledge of these points would have a practical bearing on the spontaneous combustion of coal and on the control of the sulphur content of coke. The bulletin gives a detailed account of the experimental chemical work performed in the development of a method for analyzing the different forms of sulphur in coal and a description of the changes in the forms of sulphur. The appendices contain historical matter on the constitution of coal and on the work of other investigators along the same lines.

ROBISON'S MANUAL OF RADIO TELEGRAPHY AND TELEPHONY. By Captain S. S. Robison, U. S. Navy. Published by the Lord Baltimore Press, Baltimore, Md., 1919. Cloth, 9x6 in.; 307 pages; 144 illustrations.

This radio manual was written for the use of naval electricians, student operators and others, and presents the elementary principles of the art of radio telegraphy and telephony, together with descriptions of apparatus commonly used. This is the fifth edition of this work, which was originally written by Lieut. (now Rear Admiral) S. S. Robison, U. S. Navy, in 1907. A comprehensive treatment is given on the construction and operation of radio equipment. The subject is presented from the practical viewpoint. Particularly noticeable is the absence of mathematical treatment. The major portion of the mathematics is given in one chapter of formulas and tables, so that the book may be read with interest by even the nontechnical reader. Those that are interested in radio subjects will find a wealth of practical information in this manual.

PULVERIZED FUEL, ITS USES AND POSSIBILITIES. By William J. Dick, M. Sc. Published by the Commission of Conservation, Ottawa, Canada. Paper, 57 pages.

This pamphlet begins with a review of the general fuel situation in Canada. In this connection the fact that although Canada has over 17 per cent. of the world's coal reserve, 55 per cent. of its consumption in 1916 was imported. A history of the pulverized-fuel industry, beginning with experiments in connection with cement burning, notes its gradual extension for railroad work, in the metallurgical industries, and in stationary boilers. The difficulties encountered in burning this fuel in steam boilers are summarized as follows: (1) The difficulty of maintaining continuous and steady ignition; (2) of finding a material to stand the furnace temperatures; (3) of producing a homogeneous mixture with varying grades of coal; (4) of maintaining a homogenous mixture of fuel and air during the period required for complete combustion; and (5) of handling the molten ash. A description of the Bettington boiler, a combined boiler and pulverizing plant, outlines some of the successfull methods of overcoming these difficulties. Reports on experience with pulverized c al from a large number of cement, metallurgical and power plants are included. These reports give considerable detail, and a number of them are accompanied by data and results from tests.

Personals

A. H. C. Dalley, formerly district sales manager of the Pulverized Fuel Equipment Corporation, has been promoted to the position of assistant vice-president of the same company. As before, Mr. Dalley's headquarters is in the Peoples Gas building, Chicago.

Engineering Affairs

The American Society of Mechanical Engineers will hold its next meeting at St. Louis, Mo., May 24-27, inclusive.

The Marine Engineers' Beneficial Association will hold its 45th annual convention at Washington, D. C., Jan. 19-24, with headquarters at the Hotel Raleigh.

Miscellaneous News

Consideration of the Water Power Bill in the open Senate began Jan. 5. The time was consumed principally by Senator Lenroot of Wisconsin, who is opposing the measure.

Edison Medal Awarded to W. L. R. Emmet.—At the meeting of the Board of Directors of the American Institute of Electrical Engineers held December 12, the Edison Medal Committee reported that the Edison Medal for the year 1919 had been awarded to Mr. W. L. R. Emmet, "for inventions and developments of electrical apparatus and prime movers." Arrangements will be made for the presentation of the Medal to Mr. Emmet at a convenient later date.

Business Items

Union Renewable Fuses, manufactured by the Chicago Fuse Manufacturing Co., Chicago and New York, have been approved by the Underwriters Laboratories of the National Board of Fire Underwriters, in capacity from 1 to 600 amperes. This action by the Underwriters will doubtless meet the approval of users of fuses.

The Reading Iron Co., Reading, Penn., has recently opened new district sales offices in Pittsburgh and Chicago. F. M. English has been appointed district sales manager in charge of the Pittsburgh office and R. A. Griffin, district sales manager in charge of the Chicago office. Craig Geddis has been appointed advertising manager and W. E. Dunham, production manager.

Trade Catalogs

The Schutte & Koerting Co. has lately issued a pamphlet entitled "Our Part in the War," and will take pleasure in mailing a copy to those applying for same.

Schutte & Koerting Co. desires to announce that it has issued a revised 8-B catalog on "Stop, Stop Check and Emergency Valves," and will be pleased to forward a copy to those interested in same.

The McAlear Manufacturing Co., Chicago, Ill., announces that its new catalog No. 25, entitled "25 Years of Know How," covering vacuum, vapor and air line heating specialties, is now ready and will be sent to anybody free upon request.

The Bailey Meter Co., Cleveland, Ohio, has recently issued its Bulletin No. 30, on "Fluid Meters for Low Pressure Gas and Air." A very complete description, with illustrations, of this type of meter. A copy of the bulletin may be had free upon request.

The second edition of "12 Reasons Why" is a bulletin describing the Hagan Steam Jet Ash Conveyor, made by the Hagan Corporation, Pittsburgh, Penn. Each of the "12 Reasons Why" is illustrated with line drawings and detailed description. A copy of this bulletin may be secured on request.

The Hagan System of Boiler Regulation is the title of a new 32-page pamphlet issued by the Hagan Corporation, Pittsburgh, Penn. The pamphlet goes minutely into the question of good and bad boiler regulations, showing how regulators improperly designed so often defeat the purpose for which they were installed. The Hagan idea of Regulating Boilers by Control from Steam Flow rather than from Steam Pressure is described in detail, using some interesting charts and curves to illustrate its advantages. Copies of this pamphlet are available on request.

POWER

Volume 51 New York, January 20, 1920 *Number 3*

Equipment Versus Men

By J. T. BEARD, Jr.

NOT long ago I was authorized by the president of a well-known manufacturing concern in New York City to obtain for him the services of the right kind of a man to superintend the operation of his boiler room. I do not mean to say that his boiler room was any worse than the average, but I do mean that it was average and only average. In other words, he was using and paying for sixty tons of coal every day, when his mill could have been run with forty tons.

It had taken a long time to get this idea across to him. He had spent money like water on new equipment and was at a loss to understand why he was no better off than before. He actually believed that when a stoker man guaranteed 74.5 per cent. efficiency at 150 per cent. rating, all he had to do was to put in that stoker and his troubles were over. Or when some prosperous salesman guaranteed him a 20 per cent. saving if he would install the Blank System of returning condensate to the boiler, he really believed he would automatically get that saving, once the equipment was installed.

In this respect he was not unique among his business contemporaries. In fact, the big complaint I have to make against business men in general is their failure to realize that equipment is no better and can be no better than the men who handle it. A common belief among them is that automatic stokers do away with intelligent firemen in the boiler room. No mistake can be more criminal than this. Automatic stokers mean a reduction in the quantity, but an increase in the quality of the men who handle the fires.

Recently, I had occasion to inquire of a plant superintendent how often he blew the soot from his boiler tubes. He promptly replied, "Twice a day." I asked him to come out to the boiler room with me. His boilers had been equipped for the last six months with one of the best types of soot blowers on the market. I pointed to the steam valve supplying the blowers on one of the units and asked him to open it. He couldn't budge it. The chief tried a 12-in. stillson on it, but still it wouldn't budge. It was frozen tight. And yet the superintendent really believed that the men had been blowing the tubes twice every day. A week later I had the opportunity to see this boiler when it was down for cleaning. The tube surfaces were well insulated from the furnace gases by an excellent coat of soot nearly an inch thick. And yet these tubes were blown "twice every day."

Not so very far from New York there is another plant where they have a poor quality of feed water. They had always had no end of tube trouble from scale. They installed an elaborate water-softening outfit, one of the best of its kind. Analysis of the water after it had passed through the softener showed that the scale-forming ingredients had been removed, and yet every two months a one-eighth-inch scale had to be removed from the tubes. This sounds like a mystery, but the solution was perfectly simple. Raw water was certainly entering the boilers. How? When I suggested to the manager that the softener was being bypassed, he laughed at me and said there was no bypass. His chief engineer and superintendent backed him up. And yet that was the only way that untreated water could have gotten into these boilers in such quantities. But murder will out, and it was finally discovered not only that there was a bypass, but that it was being used more than the softener. The bypass valve is now sealed so that it cannot be opened without the knowledge of the chief engineer.

These are not unusual cases. They are happening every day and in every plant in some form or other. If you do not believe they are happening in your plant, do a little "gumshoe" work next week. You may learn some real interesting things.

Equipment is inanimate; it is purely mechanical. It will do anything you make it do within its own physical limitations. It has no say in the matter; you are the boss. But with men it is different. They are often inclined to be human. If it is unhandy to blow the tubes in the last pass of No. 1 boiler, they very often remain unblown. If the air duct damper is in the rear of the boilers, it is not likely that it will be adjusted to suit the varying conditions of fuel bed and load. The feed-water heater is likely to stay dirty until someone remembers that it has not been cleaned for a couple of years. Your equipment is no better or more efficient than the men who handle it.

MECHANICAL STOKERS ABOARD SHIP

By Charles H. Bromley

Tells what is being done here and abroad to apply mechanical stokers aboard ship. Oil a factor in holding back their wide adoption at present.

POWER for several years has contended that mechanical stokers could with advantage be applied aboard ship. First, this means that the Scotch boiler is not anywhere nearly as well suited to stoker installation as the usual form of water-tube boiler. So, naturally, with the marine man's worship of tradition, he is not enthusiastic about severing connection with a type of boiler which all his forefathers in shipyards, in sea, at home and abroad have taught him to respect and admire.

Now this is not without reason. A ship plowing the sea is not a power plant on firm earth within walking distance of the equipment builders and the supply house. The Scotch boiler will stand more abuse than a water-tube boiler. The Scotch boiler made its reputation before the days of wireless telegraphy and telephony and before the days of high pressures and superheat. The mechanical equipment in a ship's power plant must be reliable. And who has the daring to question the safety and reliability of the venerable Scotch boiler?

Though the sea is always the same throughout the centuries, the ships that ride it are the product of the genius of the age in which they were built. And this is an age when things are successfully done by machine. Even the Atlantic has been crossed by machines of the air. But every news article published prior to the flights of the NC's and the Vickers-Vimy bomber related how Old Salt shook his head while gravely announcing that Read, Towers, Alcock, Brown and other brave fellows were going to their certain death. He would not have it. It had not been done before and again and again, therefore it could not be done now.

The water-tube boiler may have its faults, compared

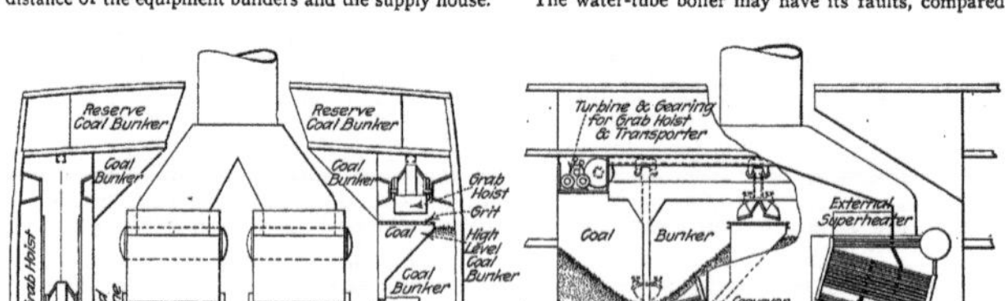

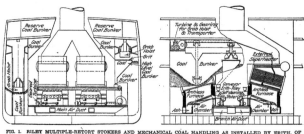

FIG. 1. RILEY MULTIPLE-RETORT STOKERS AND MECHANICAL COAL HANDLING AS INSTALLED BY ERITH, OF LONDON. INSTALLED ON NEW SHIPS ONLY

with the Scotch. It is "tenderer." But there have not been in the vessels of the Emergency Fleet Corporation the "awful" number of tube renewals whispered about here and there. We in this country do not use as heavy a tube as foreign countries do. Even if we did it would not matter, as the thickness common in American practice is heavy enough for the service imposed, provided always that the crew knows its business, which holds true for tubes whatever the thickness. The 2-in. tubes are of No. 8 gage and the 4-in. are of No. 6. A thin tube is wanted anyway.

For its war vessels the American Navy has used the water-tube boiler for about a quarter of a century and is still using it. Why, steam whalers in the Pacific, a service rough and tough enough to suit any Wolf Larson, uses the water-tube boiler.

It seems necessary to say something about the water-tube boiler here because it is essential to the adaptation of the stoker aboard ship. In no case that the writer has been able to learn about has failure of tubes or boiler or service been caused by the boiler alone. Of course, if a green crew gets the bilge full of oil and through a connection to the feed tank gets a boiler full of oil and the tubes melt down, something has happened. But the cause is the crew; not the boiler. If leaks in the oil-heating system cause oil to get into the

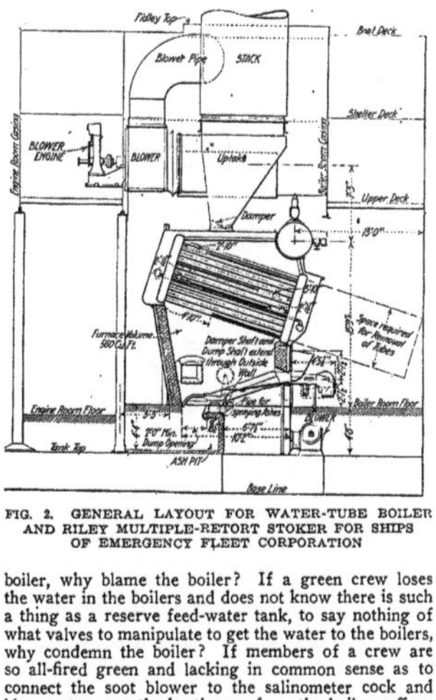

FIG. 2. GENERAL LAYOUT FOR WATER-TUBE BOILER AND RILEY MULTIPLE-RETORT STOKER FOR SHIPS OF EMERGENCY FLEET CORPORATION

boiler, why blame the boiler? If a green crew loses the water in the boilers and does not know there is such a thing as a reserve feed-water tank, to say nothing of what valves to manipulate to get the water to the boilers, why condemn the boiler? If members of a crew are so all-fired green and lacking in common sense as to connect the soot blower to the salinometer cock and blow water over the heating surface, the boiler suffers, but is blameless. These things have happened.

Of course the exigency imposed by the war was great, and it was not expected that all the crews put aboard the numerous ships under the Shipping Board would run things as well as seasoned crews.

But here is a record which, so far as the writer can learn, has never been equaled by a Scotch boiler ship: The "Great Northern" is an oil-fired water-tube boiler ship. Prior to our entry into the war she plied the Pacific; later she came to the Atlantic and was made a transport. For speed and number of trips she excelled all others in the service. On one of her runs she left

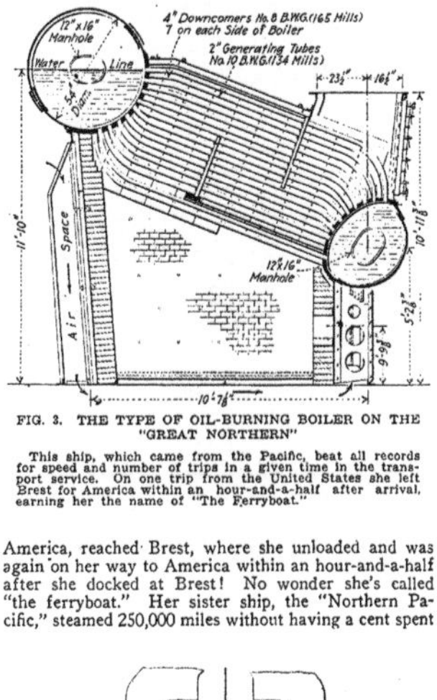

FIG. 3. THE TYPE OF OIL-BURNING BOILER ON THE "GREAT NORTHERN"

This ship, which came from the Pacific, beat all records for speed and number of trips in a given time in the transport service. On one trip from the United States she left Brest for America within an hour-and-a-half after arrival, earning her the name of "The Ferryboat."

America, reached Brest, where she unloaded and was again on her way to America within an hour-and-a-half after she docked at Brest! No wonder she's called "the ferryboat." Her sister ship, the "Northern Pacific," steamed 250,000 miles without having a cent spent

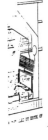

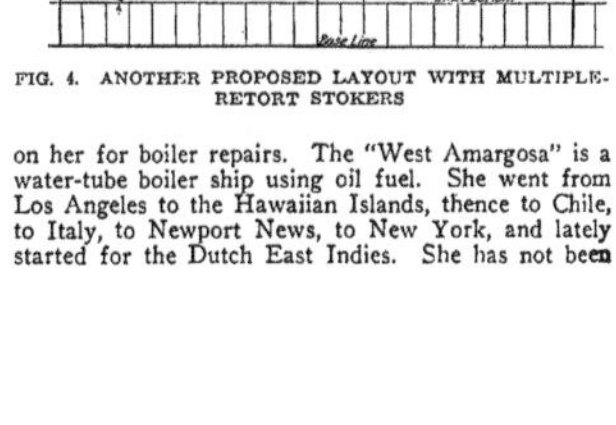

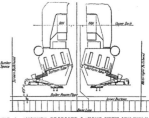

FIG. 4. ANOTHER PROPOSED LAYOUT WITH MULTIPLE-RETORT STOKERS

on her for boiler repairs. The "West Amargosa" is a water-tube boiler ship using oil fuel. She went from Los Angeles to the Hawaiian Islands, thence to Chile, to Italy, to Newport News, to New York, and lately started for the Dutch East Indies. She has not been down a minute for repairs to her boilers or other machinery. Her sister ship, the "West Arvada," has an equally noteworthy record.

During the great troop movement from Europe to America the water-tube boiler, oil-burning transports were run as veritable ferryboats, many leaving French

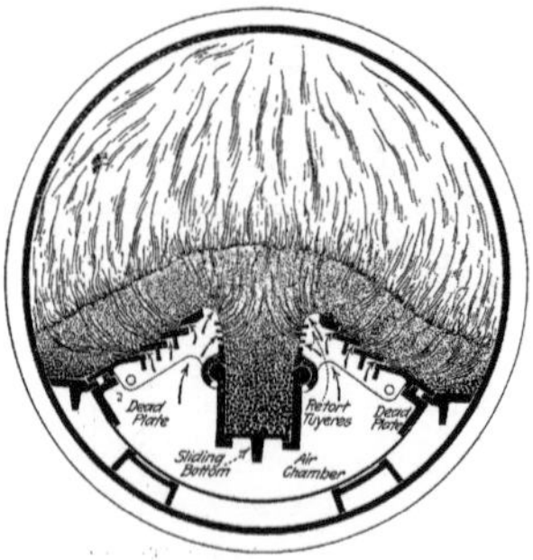

FIG. 5. FRONT SECTION SINGLE-RETORT STOKER IN SCOTCH BOILER

ports, loaded, within three hours after arriving from the United States.

Yes, the "1919 model office-building-basement boiler," as the old marine man calls the water-tube boiler, has an enviable record. The water-tube boiler will not take as large a dose of salt as the Scotch boiler. But if those in charge will keep the condensers tight and not push the evaporators too hard, salt will not give trouble; this is particularly true if the crew will run a cleaner through the three bottom rows of tubes whenever opportunity presents.

The status of the stoker aboard ship has progressed far. Take the practice of one English engineering company, the Erith Engineering Co., London. Erith handles the Riley multiple-retort underfeed stoker abroad. Without modification he has adapted it to the Babcock & Wilcox type cross-drum water-tube boiler, usually built there with tubes 14 ft. long. As seen in Fig. 1, the boilers are, in this case, installed athwartship. Owing to the moderate, continuous load very high combustion chambers are not required, the long-pattern stoker operating at comparatively low combustion rates. The stoker, of course, cleans itself of ash; the coal is mechanically handled, making for shipboard an installation closely similar to a modern stationary steam plant. As installed, the equipment does not occupy so much space as equivalent steaming capacity with Scotch boilers as usually installed. The weight is much less. For the present, at least, Erith will not make installa-

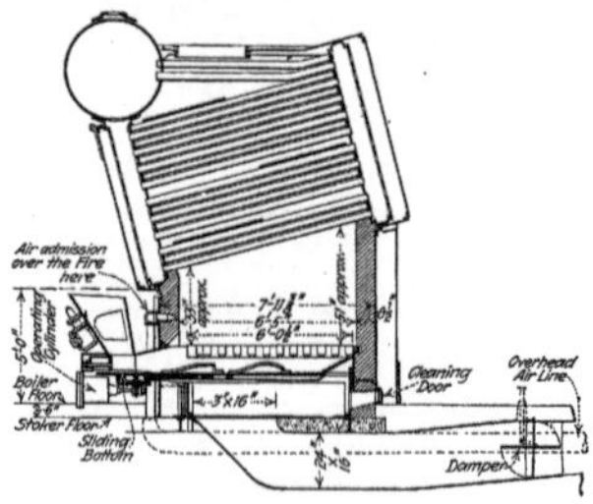

FIG. 7. SECTION OF TYPE E STOKER UNDER EMERGENCY FLEET BOILER

tions in other than new ships; several installations are now being made.

American practice has gone about as far as that just described. The types of boilers are shown in Figs. 2 and 3. The Emergency Fleet Corporation, with the assistance of the Bureau of Mines, has found by experiment the way to baffle this boiler to get the most heat out of the flue gases and into the boiler. These experiments are most interesting, but space here does not permit of tell-

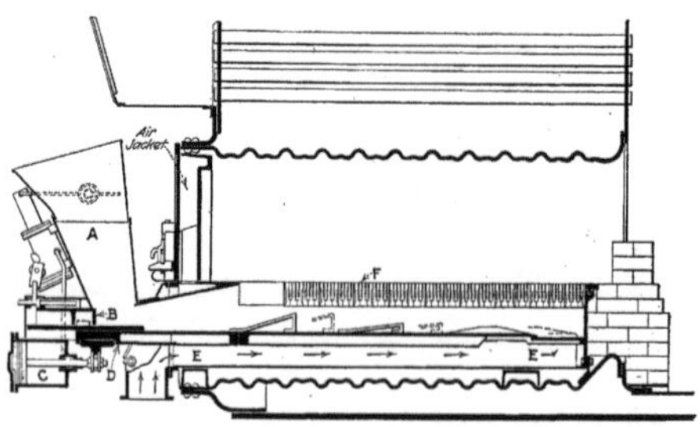

FIG. 6. SECTIONAL VIEW OF SINGLE-RETORT UNDERFEED STOKER IN SCOTCH BOILER

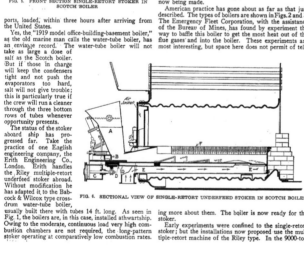

ing more about them. The boiler is now ready for the stoker.

Early experiments were confined to the single-retort stoker; but the installations now proposed use the multiple-retort machine of the Riley type. In the 9000-ton ships, which is the tonnage now being turned out at the rate of four ships a day—six later, perhaps—there are five retorts in the stoker for each boiler. Notice from Figs. 2 and 3 that plenty of room is provided for ash deposit and removal at the refuse end of the stoker. There is plenty of room for the whole installation including the coal-handling machinery, not shown in the drawings. Both forced and induced draft is used. Notice that the baffling in the Heine boiler is different

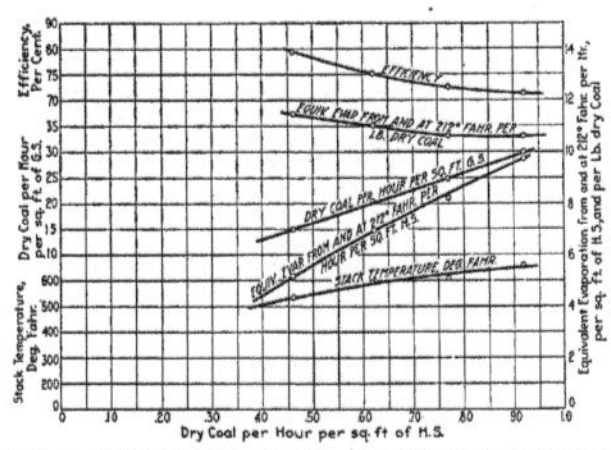

FIG. 8. PERFORMANCE RESULTS WITH TYPE E STOKER (TWO RETORTS) AND STANDARD MARINE WATER-TUBE BOILER OF THE EMERGENCY FLEET CORPORATION

The furnace was not suited to the stoker in that it was too wide and too low. Yet encouraging results were had.

from standard practice. Admiralty coal—that is, New River or Pocahontas, low in ash—is exclusively used. There is an open deck above the boilers, thus providing adequate combustion volume in the boiler furnaces.

These boilers are very much lighter for a given steaming capacity than the Scotch. The Heine boiler of 3100 sq.ft. heating surface, 225-lb. pressure, hand-fired, without grates but inclusive of water, weighs 60 tons. The Scotch of 3032 sq.ft., 220 lb. pressure, weighs dry 68.48 tons and requires 29.62 tons of water, making a total of 98 tons, as against 60 tons for the water-tube. The 8,000- to 10,000-ton ships of 11 knots speed need three such boilers. The saving in weight alone per ship is, therefore, appreciable.

At Erie, Penn., the Emergency Fleet Corporation tried out the Type E stoker under the Fleet Standard marine water-tube boiler. The furnace was not suited to the stoker, and no effort was made to make it so, yet encouraging results were had, as may be seen from the performance curves, Fig. 8.

The single-retort stoker has for some time been applied to Scotch boilers aboard ship. Some installations, where the right kind of care has been given to the selection of coal, operation and maintenance, have been successful. It is obvious with present knowledge of combustion volume or other means of bringing about intimate mixture of gas and air, that the Scotch boiler lacks the necessary space above the fuel bed to get thorough combustion. The higher the combustion rate the more marked becomes this disadvantage. The drawings in Figs. 5, 6 and 7 show the single-retort underfeed stoker as applied to ships by the Underfeed Stoker Co., Ltd., London. This company developed the original Type E stoker, well known in this country.

The stoker is a necessary development to meet the labor conditions, which continue to grow worse with time. These 8,000- and 10,000-ton ships, hand-fired, require 9 firemen, 6 coal passers and 3 water tenders. With stokers 3 firemen, 3 coal passers and 3 water tenders—half as many men—will easily handle the boilers.

For the present the stoker will not be applied to ships of the Emergency Fleet Corporation, owing to the low price of fuel oil to the corporation. During the war the Government had to pay 7c. to 9c. per gal. for fuel oil. At this figure coal could out-compete it. But the current price of oil to the corporation is 2⅛c, and coal cannot compete, so oil is used. However, opinion is that oil will not for long be had at this figure, and that we must, in a few years, use coal, stoker-fired.

The navies of the world will use oil regardless of price. So, too, will fast passenger ships. It is believed by many that there is not enough oil to warrant its wide use in the merchant marine for many years, although within recent weeks rich oil finds are reported from Colombia. Just now American coal is selling in Italy for $36 a ton. The freight rate is $22. So it is not bad business now to use oil-burning ships in the coal trade to Europe. Yes, the mechanical stoker for ships is here, and when oil again becomes high-priced or its supply alarmingly diminished, it will come into its own.

Removing Heads from Small Condensers

By M. A. Saller

We had in service at one plant several small surface condensers that had to be cleaned out frequently, owing to the dirty circulating water used. When we came to take off the condenser heads, we frequently found that the packing had cemented the two surfaces together so that it was necessary to drive in a cold chisel to pry

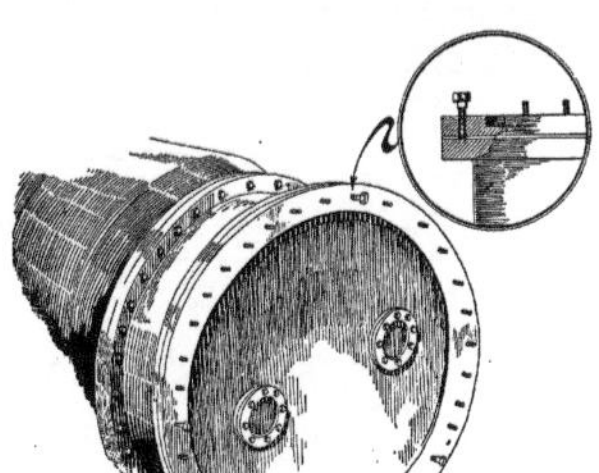

USING BOLTS FOR REMOVING CONDENSER HEAD

the two sections apart. This operation usually meant that a new gasket had to be cut.

In order to avoid this expense and inconvenience, we worked out the following plan: On opposite sides of the condenser-head plate we drilled holes, which were threaded for ¾-in. bolts. Into these holes we then screwed round-ended ¾-in. bolts, as shown in the sketch. Then when the head was to be drawn off we merely pulled up on the bolts, which pressed against the flange of the condenser and pushed the head plate away from the condenser body without difficulty.

Synchronous Motors for Driving Centrifugal Pumps

Vital Points of Synchronous-Motor Construction and Operation. Difference Between Self-Starting and Plain Synchronous Motor. Where It Should Be Used in Preference to Induction Motor. Direct-Current Ammeter Method of Determining When Motor Reaches Synchronism

By SOREN H. MORTENSEN*

SOREN H. MORTENSEN

THE development through which the self-starting synchronous motor has passed during recent years has made this type of machine suitable for constant-speed industrial drives, where frequent starts and stops are not required and where the starting torque does not exceed a certain percentage of the full-load motor torque.

To this class of drive belongs the centrifugal pump whose "shutoff horsepower," as explained in a preceding article,[1] can be reduced to the value of from 30 to 50 per cent. of the rated horsepower by starting it with a closed valve in the pump discharge.

If the characteristics of the centrifugal pumps are such that, during the starting period, they require 70 to 80 per cent. of full-load torque, it will be necessary to make the synchronous motor larger than would be the case if its capacity was limited through heating alone. For such cases overexcited synchronous motors for power-factor correction are particularly suitable.

The operation of the self-starting synchronous motor resolves itself into three periods: First, the starting; second, the pulling into step or synchronism; and third, the operation as a synchronous motor. In the following each of these periods will be treated separately.

The feature in which the self-starting synchronous motor differs from the synchronous motor that has to be brought up to speed by an external source of power and synchronized before it can carry a mechanical load, consists of an auxiliary winding embedded in its field poles. Such windings are shown in Figs. 1 and 2. Conductors or bars are located in the pole faces and interconnected by means of rings or links, thus forming a short-circuited winding similar to the squirrel-cage on an induction motor. The revolving field, created by the current in the stator of a polyphase motor, induces currents in the squirrel-cage winding in the rotor poles as well as in the pole iron itself and the field coils if the latter are short-circuited during the starting period, and the inter-action between the resulting fields produces the starting torque of the motor.

During the first part of the starting period the action of the self-starting synchronous motor is similar to that of a squirrel-cage induction motor. It is started in a similar manner by applying a reduced voltage to its armature terminals, which is sufficient to start it from rest and accelerate the motor to synchronous speed or within a certain percentage of synchronous speed, depending upon its load. If the load is light, the motor may lock into step on the application of the starting voltage, and after it has been excited full load may be applied. Such cases, however, are exceptional. For loads in which the torque increases with the speed of the motor, as is the case with centrifugal pumps, the motor will not reach synchronism, but will rotate at a lower speed. The difference between this running speed and synchronous speed is called the "motor's slip." After this point of the starting process is reached, the similarity between the operation of the synchronous motor and the squirrel-cage induction motor ceases, and

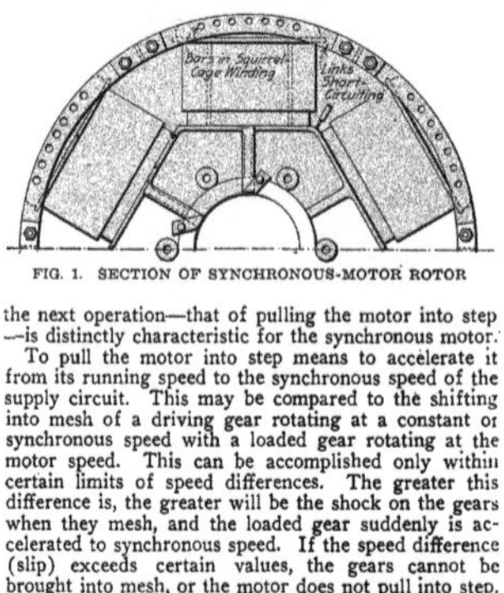

FIG. 1. SECTION OF SYNCHRONOUS-MOTOR ROTOR

the next operation—that of pulling the motor into step—is distinctly characteristic for the synchronous motor.

To pull the motor into step means to accelerate it from its running speed to the synchronous speed of the supply circuit. This may be compared to the shifting into mesh of a driving gear rotating at a constant or synchronous speed with a loaded gear rotating at the motor speed. This can be accomplished only within certain limits of speed differences. The greater this difference is, the greater will be the shock on the gears when they mesh, and the loaded gear suddenly is accelerated to synchronous speed. If the speed difference (slip) exceeds certain values, the gears cannot be brought into mesh, or the motor does not pull into step. In that case, the motor torque must be increased until it is sufficient to accelerate it to the point where it can lock into step. Within certain limits this can be accomplished by applying direct-current excitation to the motor's fields. However, if the motor does not lock

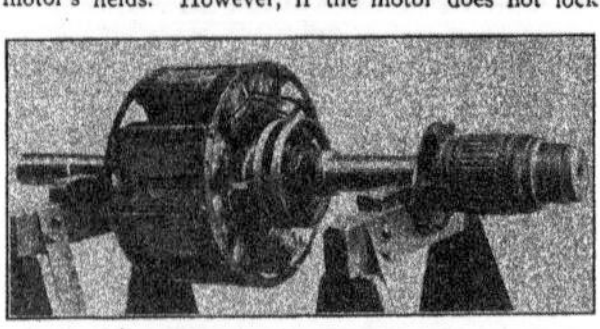

FIG. 2. SYNCHRONOUS-MOTOR ROTOR AND ARMATURE OF DIRECT-CONNECTED EXCITER

*Electrical Engineer, Allis-Chalmers Manufacturing Company.
[1] By T. M. Heermans in *Power*, May 20, 1919, p. 763.

into step after excitation is applied, it will be necessary to increase the applied alternating-current voltage until synchronism is obtained.

In this connection it might be of interest to point out the relation existing between the starting and pull-in torques and the resistance of the squirrel-cage winding of the motor. The high-resistance squirrel-cage wind-

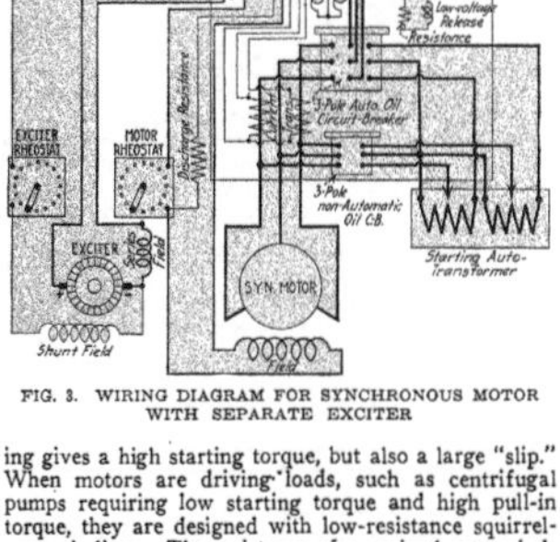

FIG. 3. WIRING DIAGRAM FOR SYNCHRONOUS MOTOR WITH SEPARATE EXCITER

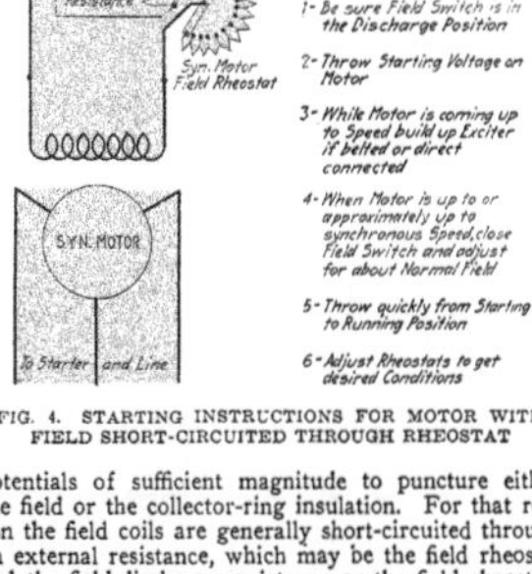

FIG. 4. STARTING INSTRUCTIONS FOR MOTOR WITH FIELD SHORT-CIRCUITED THROUGH RHEOSTAT

ing gives a high starting torque, but also a large "slip." When motors are driving loads, such as centrifugal pumps requiring low starting torque and high pull-in torque, they are designed with low-resistance squirrel-cage windings. The resistance of a squirrel-cage winding on a synchronous motor is, of course, without influence on the efficiency of the motor after it is in synchronism. For that reason this type of motor can be designed to meet efficiently more severe starting duty than a squirrel-cage induction motor, whose full-load efficiency decreases with the increase in resistance of its squirrel-cage winding.

During the starting period the motor's fields can either be left open or short-circuited upon themselves or through an external resistance. If the field circuit is left open, the motor will develop a higher starting torque than with the fields short-circuited, but this arrangement has the disadvantage that high potentials are induced in the field coils, due to the transformer action between the stator and rotor windings. This, in field coils with a large number of turns, may lead to potentials of sufficient magnitude to puncture either the field or the collector-ring insulation. For that reason the field coils are generally short-circuited through an external resistance, which may be the field rheostat and the field-discharge resistance or the field rheostats and the exciter armature, etc.

Fig. 3 is the connection diagram for a synchronous motor with a separate exciter. The motor is started on reduced voltage obtained from taps on an auto-transformer. In Fig. 4 is given the field connections and starting instructions for a motor with the field short-circuited through its field rheostat and field-discharge resistance.

It is important that the throwing from starting to

FIG. 5. TESTING SYNCHRONOUS MOTOR DRIVING TWO SINGLE-STAGE PUMPS

running voltage be accomplished in a short space of time (in three to five cycles), as during this period the motor is disconnected from the line and is consequently losing speed. If sufficient time elapses to permit it to slip a pole, line disturbance will result that may be sufficient to cause the tripping of overload relays and necessitate a repetition of the whole starting process.

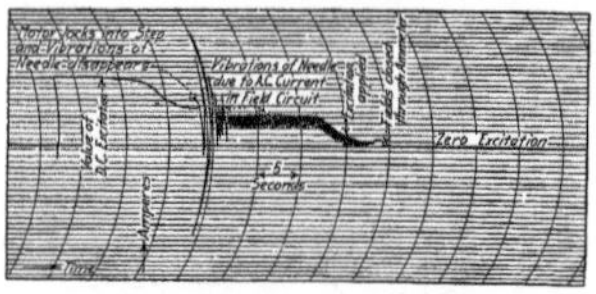

FIG. 6. RECORD OF CURRENT IN FIELD CIRCUIT DURING STARTING AND PULL-IN PERIOD

During the synchronizing period it is of importance to know when the motor approaches and attains synchronous speed. This will be indicated by the needle of an ammeter of the permanent-magnet type connected in the field circuit. While the motor speed is considerably below synchronism, the needle, unable to follow the rapid current pulsation in the field, will only vibrate, but as the slip decreases, the rapidity of the pulsation also decreases and the needle will swing slower and slower but with increasing deflections until a maximum is reached when the motor locks into step. After that the needle will become stationary and indicate the excitation current taken by the motor. This may be seen from the curve in Fig. 5, which was obtained with a recording direct-current ammeter connected in the field circuits of a synchronous motor during the starting period.

The maximum value of the current taken by a synchronous motor in starting depends upon the starting voltage. For motors driving centrifugal pumps the starting voltage might be kept low, as the starting torque required is small. The starting voltage must be sufficient to bring the unit into synchronism, or reasonably near synchronism, as otherwise the current taken when the motor locks into step by the application of full voltage will be excessive. For that reason the starting voltage is frequently chosen of such a value that the maximum currents during the starting and pull-in periods are approximately the same.

Fig. 7 is a graphic-ammeter record of the starting and pull-in- currents taken by the synchronous motor driving the centrifugal pumps in Fig. 5. These pumps require approximately 50 per cent. of the full-load torque, when running at synchronous speed with the discharge valves closed.

Referring to Fig. 7, the motor is started at *A* but does not lock into step after its fields are excited at *B*. Full voltage is applied at *C*, and at *D* the motor locks into step. After the discharge valve on the pump is opened and the motor excitation is adjusted to its proper value, the motor current drops to normal. For this particular set the starting current reaches for a short time the value of 1.8 times the rated current, and during the pull-in period peaks as high as 1.78 times rated current are recorded. These values would probably have been still higher if the recording instrument had been an oscillograph instead of a recording ammeter.

From these records it is apparent that the power system behind the self-starting synchronous motor must have sufficient capacity to maintain operating voltage, when the motor during starting draws a large, low-power-factor current. The power factor of the starting current with this type of machine will vary between 35 and 50 per cent.

After the synchronous motor is excited and in synchronism, its characteristics are well understood—it is a constant-speed motor rotating at synchronous speed independent of the supply voltage within the limits of its pull-out torque.

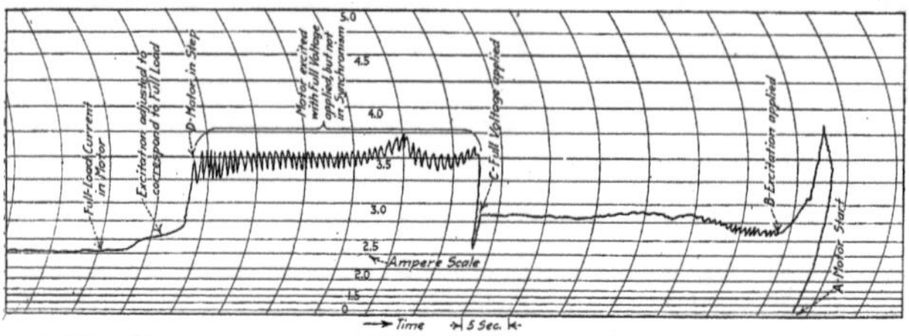

FIG. 7. RECORD OF CURRENT TAKEN BY SYNCHRONOUS MOTOR DRIVING CENTRIFUGAL PUMP DURING STARTING AND PULL-IN PERIODS

As most synchronous motors, with the proper excitation, will carry from 200 to 300 per cent. overload, they will maintain the rated speed and output of the centrifugal pump even after the voltage of the supply circuit drops to values that would cause considerable speed reduction if the pump were induction-motor driven.

In general, it can be said that the synchronous motor has a beneficial effect on the supply system. It generates a definite counter-electromotive force and tends to maintain constant voltage on the system during voltage fluctuations, and within its capacity it can likewise be adjusted to compensate lagging current and thereby improve the power factor of the system. This is done by adjusting the field excitation so as to make the motor draw either a lagging, in phase, or a leading armature current. By underexciting the fields, the balance of the motor excitation required is supplied by a lagging or magnetizing component of the armature current.

With the strengthening of the field of excitation, this magnetizing component decreases until it disappears when the motor current is in phase with the motor voltage, or its power factor is 100 per cent. If the field excitation is further increased, its excess ampere turns are compensated by a demagnetizing or leading component of the armature current. In short the motor-armature current is a minimum for 100 per cent. power factor and increases when the motor is either underexcited or overexcited. The field excitation is a minimum when the motor is underexcited. It has to be increased to obtain 100 per cent. power-factor operation and requires a further increase if the motor is to draw a leading current. The leading corrective effect of synchronous motors is thus limited not alone by the kilovolt-ampere capacity of the armature winding, but also by the heating and voltage margins of the field coils. Specifications for motors that are to be used for part mechanical loads and part overexcitation, should for that reason both specify the horsepower rating of

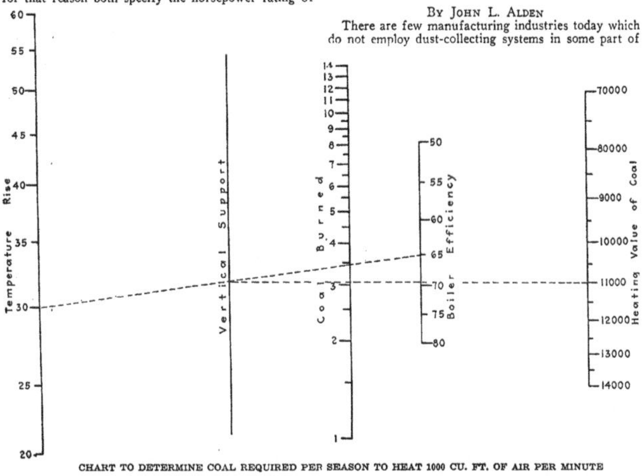

CHART TO DETERMINE COAL REQUIRED PER SEASON TO HEAT 1000 CU. FT. OF AIR PER MINUTE

the motor as well as the power factor at which this horsepower is developed.

Mechanically, the synchronous motor is of sturdy construction; it has a long air gap, thereby eliminating the danger of the rotor pulling over, if it is eccentric in the stator due either to wear of the bearings or to defective machine work.

Excitation of the synchronous motor may be obtained either from some outside source or from an exciter driven by the motor itself. This can be done either by mounting the exciter armature on the end of the motor shaft in the manner shown in Figs. 2 and 5 or by driving it from the motor shaft by means of a belt or chain drive. In conclusion it can be said that the self-starting synchronous motor is suitable for constant-speed drives that do not require phase-wound induction motors. In comparing the first cost of the synchronous motor with that of an induction motor of the same speed and horsepower, it will be found that with the slow-speed machines the synchronous motor is cheaper, but for medium- and high-speed machines the margin is in favor of the induction motor. This, combined with the simplicity of excitation, starting and the high efficiency obtainable on small high-speed induction motors of 125 hp. or less, gives this motor the advantage over the self-starting synchronous motor, but where larger sizes of motors are involved the efficiency of the synchronous motor is generally sufficiently better than that of the corresponding induction motor to warrant its installation even at the higher first cost.

Losses Due to Removing Hot Air from Factories

By John L. Alden

There are few manufacturing industries today which do not employ dust-collecting systems in some part of the manufacturing process. The literature of exhaust fans and blowers, together with their accompanying systems, is altogether inadequate.

One of the economic features which has been practically neglected is that of the heat loss for which such systems are responsible. This is the loss that takes place when warm air is driven out by the fans and is replaced by cold air from outside, which must be heated to room temperature. It is seldom that any attention is paid to the waste of this heated air except when it becomes so great as to make it difficult or impossible to heat the building. Situations of this sort are not uncommon. One of the most striking examples that has come under the writer's observation is that of the polishing room of a large arms factory; 180,000 cu.ft. of warm air per minute was thrown out of the building—enough to provide ample ventilation for 6000 men. The enormous volume completely upset the heating arrangements, not only of this department, but of all the rooms connected with it by doors and elevator wells. An extreme case of this kind takes care of itself. When it becomes impossible to heat a building the management will certainly inquire into the cause, and it is unnecessary in this article to point out the need for action. However, the loss of heat is not always so apparent, especially when the heating system is adequate to replace the heat abstracted from the building. If it is recognized that such wastes are present, and if their magnitude can be estimated, simple changes may often be made which will show a substantial profit, both in the first cost of the system and in the operation.

The accompanying chart has been prepared to show the effect on the coal pile of the loss of 1000 cu.ft. of warm air per minute. On the chart the "Temperature Rise" is the difference between the temperature of the inside and the outside air, in degrees Fahrenheit, and the "Coal Burned" is in short tons per heating season of seven months of 24 eight-hour days. The "Boiler Efficiency" is given in percentage and includes the efficiency of the entire transmission system, and the "Heating Value of Coal" is in B.t.u. per pound "as fired." The "Vertical Support" is simply a reference line for use in working the chart. To use the chart, it is necessary to know the difference in temperature between the outgoing air and that replacing it, the efficiency of the boiler, transmission system, etc., and the heating value of the coal as fired. With a straight-edge connect the temperature difference with the boiler efficiency. From the intersection of this line and the vertical support, draw a line to the heating value of the coal. The intersection of this line with that marked "Coal Burned" gives the tons of coal burned per heating season to heat 1000 cu.ft. of air per minute. Since the heat loss takes place only during cold weather, the heating season is taken as seven months of 24 eight-hour days each. This represents very fairly the conditions in most manufacturing plants in the northern part of the United States. A close estimate of the cost of wasted heat may be had by multiplying the chart figure of "Coal Burned" by the number of thousands of cubic feet of air per minute and this result by the current price of coal. To aid in determining the temperature difference, the average outdoor temperature from Oct. 1 to May 1 for the colder manufacturing areas is given in Table I. These data are based on Weather Bureau reports.

TABLE I. AVERAGE TEMPERATURE FROM OCT. 1 TO MAY 1.

State	Av. Temp., Deg. F.	State	Av. Temp., Deg. F.
Connecticut	36	New Hampshire	33
Delaware	42	New Jersey	41
Illinois	37	New York	37
Indiana	40	Ohio	38
Maine	32	Pennsylvania	41
Maryland	43	Rhode Island	38
Massachusetts	37	Vermont	28
Michigan	32	West Virginia	40
Minnesota	26	Wisconsin	32
Missouri	41		

TABLE II. DIRECT RADIATION REQUIRED PER THOUSAND CUBIC FEET OF AIR PER MINUTE.

Temp. Diff. Deg. F.	Radiation, Sq. Ft. Cast Iron	Pipe Coils
20	86	72
25	108	90
30	130	108
35	151	126
40	173	144
45	195	162
50	216	180
55	238	198
60	260	216

When a considerable quantity of heat is exhausted from a heated room, additional radiation must be provided. Table II shows the amount of direct radiation necessary to supply the heat carried off by 1000 cu.ft. of air per minute at various temperature differences. This table is based on a steam pressure of 2 lb. per sq.in. gage and a room temperature of 70 deg. F. For pipe coils the hourly heat transmission per square foot is 300 B.t.u. and for cast-iron radiation 250 B.t.u. The first cost per square foot of radiation is about 25c. for cast iron and 30c. for pipe coils. The labor and incidental expenses of erecting and connecting are from 40c. to 75c. per square foot for the average factory job. A fair average for radiation erected and in place is from 75c. to $1 per square foot. It is plain, then, that the necessary extra radiation adds materially to the cost of the blower system and should be included when making estimates. As an example of such extra first cost and waste, a 13,000 cu.ft. per min. system in New Jersey wastes annually about $125 worth of coal and requires $1200 extra investment in pipe coils.

Gasoline Substitute Makes Good Showing

Following apparently satisfactory test-block studies of a synthetic airplane engine fuel known commercially as "Alcogas" and composed of 38 parts alcohol, 19 parts benzol, 4 parts toluol, 30 parts gasoline and 7½ parts ether, the Post Office Department arranged for a test of the fuel under service conditions in the air mail. Mail plane No. 35, a Curtiss Model R4 machine equipped with a high-compression Liberty 12 motor, was assigned for the work, the check plane, flying the opposite trips during the same period with high-test aviation gasoline, being mail plane No. 34, also a Curtiss Model R4 plane, equipped with a low-compression Liberty 12 motor.

Thirty-one trips were made between New York and Washington—218-mile nonstop flights, on the regular Air Mail schedule between Aug. 4 and Sept. 19, 1919.

The tests indicated a saving of 3.3 gal. of fuel an hour in favor of the alcohol fuel. Noting the revolutions per minute, however, the saving is even greater, as alcohol fuel shows 1,514.3 r.p.m. as against 1,507.8 with gasoline. This means not only that there is a saving of 3.3 gal. of fuel per hour, but that 6.5 r.p.m. are gained by the use of the alcohol fuel.

Alcohol fuel also shows a saving in lubricating oil. The average for this fuel was 4.4 quarts per hour as against 4.98 quarts per hour for gasoline, a net saving of 0.58 quart per hour. This saving is thought to be due to greater thermal efficiency displayed by alcohol fuel as against gasoline, due to the fact that high-compression motors usually run considerably warmer than do low-compression motors.

The following is the report of the field manager on the condition of the water in plane 35 after the Alcogas tests:

> Carbon deposit was found to be from one-thirty-second to one-sixteenth inch thick, soft and flaky. Carbon was thickest on outside of piston crown, showing it to be caused from oil rather than incomplete combustion of fuel. Valves were all in good shape. Valve seats showed no signs of pitting or warping. No. 6 connecting-rod babbitt bearing cracked in both cap and rod. Two piston rings were broken and six stuck in grooves. Motor in very good shape considering number of hours run.

The high-compression motor used in plane No 35 during all its flights on alcohol fuel was torn down after approximately 125 hours and was found to be in excellent condition. The carbon deposited was less than that found in a motor using gasoline over a similar period of time.

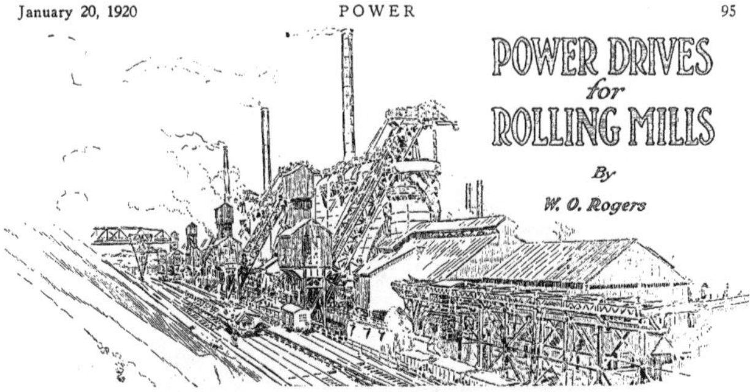

POWER DRIVES for ROLLING MILLS

By W. O. Rogers

A visit to the rolling mills discloses additional facts regarding the bessemer and openhearth furnaces. Their operation is briefly explained and also that of the blast furnace and heating ovens.

WHEN I arrived at Youngstown, it was well along in the afternoon and too late to make arrangements to visit any of the mills that day, but the next morning I readily obtained permission to go through the works of the Youngstown Sheet and Tube Co. and the Republic Iron and Steel Co.'s plant.

With the information that I had gained from my friend on the train, I determined to begin at the blast furnaces and follow the process as best I could. Fig. 1 shows a semisectional view of a blast furnace and the heating stoves which are used for heating the air blasts. The inclined skip hoist is shown at the left and the heating stoves at the right. If a furnace has a capacity of, say, 300 tons of iron per day, it will require several stoves, each about 20 ft. in diameter and around 100 ft. high. The temperature of the blast is usually kept around 1000 deg. F., otherwise there is danger of the contents of large furnaces hanging up and then slip-

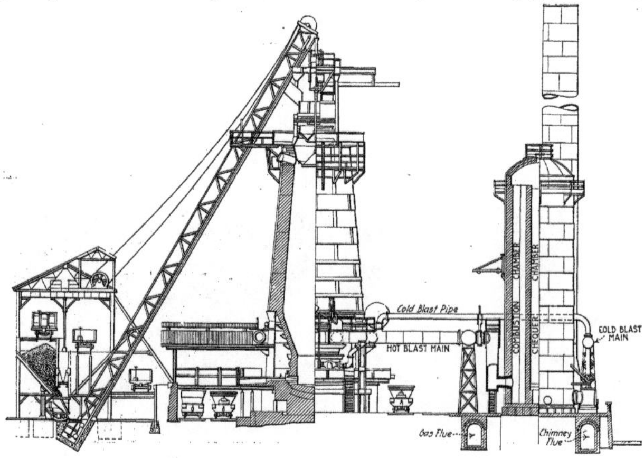

FIG. 1. SEMI-SECTIONAL VIEW OF BLAST FURNACE AND HEATING STOVES

ping, owing to the intense heat at too great a distance above the tuyere zone. The temperature is kept down by admitting a proper proportion of cold air with the blast. With each furnace there are generally four or five hot-blast stoves, which contain two or more firebrick chambers, one of which is open and the others are filled with a number of small flues. Gas and air are admitted through the bottom of the open chamber, in which they burn, and afterward pass from the top of the chamber and, dividing, go downward through the flues in the other chambers which surround the center one and escape at the bottom of the chimney as waste-heat gases. In passing through the stove the burning gases give out the greater part of their heat to the brickwork, and after the brickwork has become heated, air from the blowing engine enters the bottom of the stove and passes through them in a reverse direction to that taken by the heating gases. The reason four or five stoves are required for one blast furnace is because one stove is heating the blast air while the others are being heated by the gas and air simultaneously to get the brickwork hot for their turn at heating the blast. By changing the blast from one stove to another at intervals of about an hour, a satisfactory blast temperature is maintained.

FIG. 2. BLAST FURNACE STOVES AND GAS CLEANING APPARATUS

All waste gas used in heating the brickwork in the stoves is obtained from the blast furnaces. The gas has a temperature around 450 deg. F. and a heating value around 95 B.t.u. per cu.ft. About one-third of the gas coming from the furnace is used in the stoves, and the remainder is used in steam-boiler furnaces and in gas engines.

Gas from the blast furnace escapes at the top into a "downcomer" where considerable of the dirt carried with it is deposited in a dust catcher at the bottom of the pipe. The gas then passes into the gas-cleaning apparatus, Fig. 2, where it is cleaned and scrubbed and is then ready for use in the furnace stoves, boiler furnaces, or in gas engines. Generally, enough gas is available to operate the blowing engines and the apparatus operating with the furnace.

To one not familiar with the unloading of a blast furnace, the operation is interesting and especially so at night. The slag is tapped from the furnace first. As it has a lower specific gravity than iron, it floats on the liquid metal. When ready it is drawn into huge iron car ladles which run on a standard-gage track, and is then drawn away by a locomotive and dumped on the slag pile. Fig. 3 is a view of slag being tapped from a blast furnace.

After the last of the cinder has been removed from the iron, the furnace is tapped and its contents of 100, 150 or more tons of liquid pig iron made to flow into brick-lined ladles, which also run on a track. It is then drawn away to the bessemer blower or cast into pig iron. Fig. 4 shows the end of a cast at a blast furnace.

In the bessemer process about two ordinary-sized furnaces are required to supply metal for one converter. As the metal from any one furnace is a little different from that of another, the melts are poured into what is

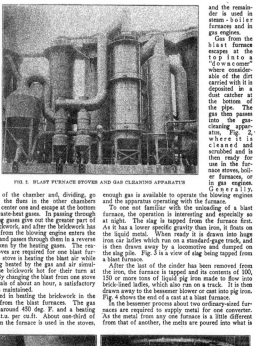

FIG. 3. SLAG FLOWING FROM A BLAST FURNACE

FIG. 4. END OF A CAST AT A BLAST FURNACE

known as a mixer, which is capable of handling up to 500 tons. The mixed metal is then charged into the converter. Fig. 6 is a good illustration of a mixer, which is of 250 tons capacity. Pig iron from the blast furnace is being poured into it from a ladle suspended by a crane. One would suppose that the metal would become cooled, but owing to the capacity of the mixer, in a case of delay at the furnace or at the converter the cooling is but slight.

From the mixer the metal is poured into the converter, which consists of a steel riveted shell and is hung on two trunnions upon which it rotates. One trunnion is hollow, and the air from the blowing engine enters through it to the windbox at the bottom of the converter. The other trunnion engages with a rack that is operated by a hydraulic piston by which means the converter can be made to rotate through an angle of 270 deg. or more. The vessel is lined, and the bottom lining contains about 270 half-inch holes through which the air blast from the windbox passes to the inside.

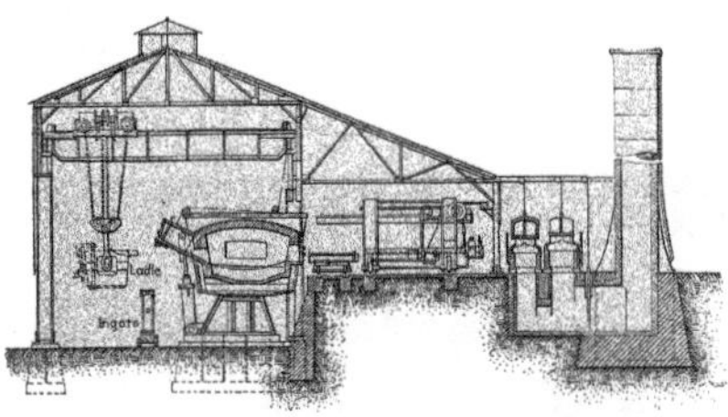

FIG. 5. OPEN HEARTH FURNACE LAYOUT

With the converter on its side the metal does not cover any of these holes, but when it is turned in a vertical position and the air blast is turned on, air is forced through the metal in a spray of very fine bubbles until the impurities are oxidized, after which the metal is run into a ladle from which it is poured into ingot molds.

These molds are about 7 ft. high and 2½ in. thick, about 16 in. square at the top and 19 in. square at the bottom. A mold will last about 100 heats, after which it becomes cracked on the inside, which makes it difficult to lift it from the metal after it has sufficiently cooled to solidify. The train of filled ingot molds is run to the stripping house, where the molds are removed, leaving the ingots standing on the car ready to run to the rolling mill.

Leaving the process at the point where the ingot left for the rolling mill, I next visited the openhearth furnaces. Fig. 5 shows a cross-section and Fig. 7 is a general view of such a plant. As my friend on the train told me, there are two types of openhearth furnaces, the stationary and the tilting, but I did not see any of the tilting type.

The melter's job is to keep the material in a very liquid state and at the same time to have an excess of air so that the atmosphere of the furnace will be slightly oxidizing to burn the impurities of the metal. The bath of, say, a 50-ton furnace has a length of about 30 to 35 ft. and a width of 12 to 15 ft. Eighty- and 100-ton furnaces are common. The metal from the openhearth furnace is run into ladles and then poured into ingot molds, and when sufficiently cooled is carried to the rolling mill. Bessemer ingots generally weigh two tons or more, and openhearth ingots weigh from three to ten tons.

I was told that various designs of openhearth furnaces were used at different mills. They are arranged in a single row with the level of the hearth several feet above the ground level of the plant. Gas producers are placed outside of the furnace building in a line parallel with it. By this arrangement the distance the gases have to pass between the producers and the regenerators is reduced to a minimum, the regenerators being placed below the working platform of the furnaces.

The four regenerator chambers are filled with brick checkerwork around which the gases and air pass. Before a furnace is started, these bricks are heated with a wood fire and then the gases from the producers are admitted. They enter the furnace through an inner chamber on one side, and air enters through the second chamber on the same side. They meet and, uniting, pass through the furnace and thence to the chimney through the two regenerative chambers on the opposite end. Thus, the brickwork of the outgoing chamber is heated by the waste heat from the furnace. About every twenty minutes the current of air, gas and waste-heat gases changes, and in this way the four chambers are kept hot. The direction of gas and air flow is controlled by means of suitable valves. With each reversal of gas flow the temperature of the furnace becomes higher. The generator furnace temperature is maintained at around 1800 deg. F. The temperature that can be obtained in the openhearth furnace by frequent reversals of gas and air is extremely high, but will average around 3000 deg.

I was fortunate in regard to seeing the drawing of

FIG. 6. A 250-TON HOT METAL MIXER

an openhearth furnace melt, but aside from being a different process from the bessemer, the method of handling the ladles and pouring into ingot molds seems to be similar.

While at the steel mill I did not see any electrical furnaces, but in talking with one man I learned that electric energy is used to bring the ore and reducing agents to the temperature at which reduction takes place and then melts the metal together with the flux. He was of the opinion that the success of such a furnace depended upon the price of electricity as compared with that of fuel, and that the application of the electric process was limited to districts where cheap electric energy could be obtained.

An electric reduction furnace is designed similar to the blast furnace. The stocking charge and hoisting apparatus are practically the same, but the height of a furnace is less, being about 30 ft. At a point corresponding to the tuyere zone in a blast furnace the shaft proper of the electric furnace ends, and in place of the narrow blast-furnace hearth the electric furnace is provided with a shallow hearth or crucible which forms a melting chamber.

Heat is generated by the resistance to the passage of current between the electrodes that project through the roof of the crucible and are embedded in the charge. In the electric furnace the volume of gas generated is about one-eighth of the blast furnace handling the same tonnage and using fuel, but its calorific value is about three times as high.

After leaving the furnaces, attention was devoted to the various types of boiler installations, which will be dealt with in the next installment.

In announcing that it would sell its sulphuric acid plant at Mount Union, Penn., the War Department states that Mount Union has the cheapest electric power in the State of Pennsylvania. The town of Mount Union draws its coal supply from a mine that is within the town limits.

Comparative Value of Coal and Oil as Fuel

In the best oil there are, roughly, 1.3 as many B.t.u. as in an equal weight of the best coal. A gallon of oil weighs about 7.7 lb., so there are about $7.7 \times 1.3 = 10$ (approximately) times as many B.t.u. in a gallon of oil as in a pound of coal.

From the point of view of heating value, then, one can afford to pay ten times as much for a gallon of oil as for a pound of coal; or $\frac{10}{2000} = \frac{1}{200}$ times as much for a gallon of oil as for a ton of coal; or $\frac{100}{200} = \frac{1}{2}$ as many *cents* per gallon as *dollars* per ton of coal; the best coal that can be procured being intended.

For the computation of the more specific case this reduces to

$$\frac{\textit{Cents per gal.}}{\textit{Dollars per ton}} = \frac{5 \times \textit{sp. gr. oil} \times \textit{B.t.u. per lb. oil}}{12 \times \textit{B.t.u. per lb. coal}}$$

which, with the values assumed—that is, specific gravity 0.93; B.t.u. per lb. oil, 20,000; B.t.u. per lb. coal, 15,000—works out as follows:

$$\frac{5 \times 0.93 \times 20{,}000}{12 \times 15{,}000} = 0.5 \textit{ cent}$$

William T. Donnelly, of New York City, is accredited with having developed a system of electric propulsion to do away with tugboats and make each barge or lighter capable of being propelled by power taken from a floating power plant. The system, as developed, consists of a boat on which an electric power plant is built, then power is supplied from this plant through cables to the boats that are to be towed. The latter are propelled by their own motors supplied with power from the floating power plant.

FIG. 7. OPEN HEARTH FURNACES AT AN 18 IN. HAND MILL, MAIN PLANT, YOUNGSTOWN SHEET AND TUBE CO.

POP'S WATER-POWER COURSE

By

John S. Carpenter

"POP, there are a few points I didn't quite get in the last talk on governors," said Jimmy, opening the conversation.

"I mighta thought as much! What's the tale o' woe now?" asked Pop.

"You better deliver your speech while you're feeling like it and let me break in where I want to," was Jimmy's reply, and he put his feet up on a chair and made himself comfortable. "You may answer all my questions as you go along, Popsie, but first have a cigar."

Pop took the cigar gingerly, smelled it and then bit into it before he had the courage to light up.

"Wall, son, I has told you that they is sich a thing as a hydro-electric plant a-gittin' outa whack. I has told yer about all the troubles that a water plant is heir to, an' when yer grows up into a chief's job see that they don't sell your plant out fer a song! Seein' as how I got that much off my chest, we'll now proceed with the real work. Glue your lamps to Fig. 1."

Jimmy did as directed, while Pop threw away Jimmy's gift cigar and lit up one of his own.

"I shows yer in Fig. 1 a modified steam-turbine governor. Now them flyball weights I shows is about the real thing if they is made drop-forged. The great trouble with castings, be they cast iron or cast steel, is that they is liable to have a sight o' blowholes in 'em that maybe could be prevented if yer cast them with both sides down; but apart from that bein' a impossibility, we is confronted with the facts. Sich a set o' flyballs, one heavy an' one light, is about as healthy to a good runnin' governor as a unbalanced flywheel would be to a engin'. When the weights is hung as Fig. 1 shows them, you have as near a frictionless flyball head as you can git. The principal source of friction is in the joints. On a line contact like a knife-edge, yer comes pretty near ideal conditions.

"Some Swiss builders, our lecturer told me, went so far as to use knife-edge bearing struts 'tween the spring collar an' the weights. If the lower edge of the collar is smooth-turned, an' the rollers I shows is hardened steel and runs in vaseline, they'll be little friction there. Other builders makes the weights with a spring to each weight and a groove cast on the weight to seat the spring. My objection to that is that the seat is not always at right angles to the center line of the spring on account of the oscillations of the weight, an' as them springs is purty heavy wire, it takes quite a bit o' force to bend the spring outa line. With the old-fashioned links—I means the ones which connects the weights to the crossbar collar—yer can't do very much else. One company, I'm told, uses small ball bearings on these joints, but to me, son, it looks like carryin' a chip on yer shoulder."

"Like the feller that can tell a joke when it's labeled, Popsie, I can tell that Fig. 2 is an oil brake. What's that bunk about 'constant bypass area?'" demanded Jimmy.

"S'posin' yer first tell me what the objict of a oil brake is," said Pop.

"You told me, or rather my impression is that it is to brake the tendency of the governor to act at every little speed change, in that way preventing constant speed oscillations. That's the same as making the governor more stable."

"Yep, that's it. I don't see why yer asked me then, yer almost got the idee right there. Well it's this—a small speed change means a small governor movement. Now if the flyballs move only a small distance hardly any braking effect will be got 'cause sich a small amount of oil is displaced. So it follows from that, that yer must screw that adjustin' screw way down before you'll git any brakin' effect at small speed changes. On the other hand, when the governor makes a big

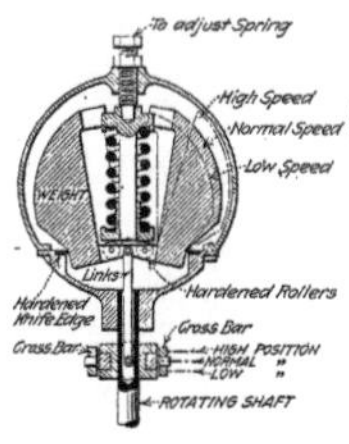

FIG. 1. GOVERNOR FLYBALL HEAD

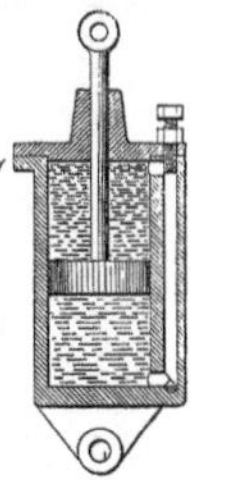

FIG. 2. OIL BRAKE WITH CONSTANT BYPASS AREA

move, it'll be dollars to doughnuts that the brakin' effect will be too much. So yer has to fiddle around until yer gits a compromise. Do yer git it now?"

"Almost. Keep on with your explanation."

"Now then, let's see what happens. We wants an oil passage—we wants it to be of no area at all for very small load changes—we wants it a wee bit for small changes, and gradually increasin' until at big changes in load they's hardly any brakin' a'tall. In other words, the amount of brakin' should vary nearly inversely as the load change."

"I'm beginning to see daylight," said Jimmy.

"So I shows the oil passage like that in Fig. 3; fer very small changes the pilot valve will not be moved a'tall, 'cause the force from the flyballs is not enough to move the brake piston past that small lap I shows. O' course they'll be some leakage around the piston rod an' on the outside of the piston, but even then I doubts whether the force from the flyballs would move the brake piston. You sees also that I makes the oil-passage groove taperin' so that more oil can be passed in the same time with larger changes in load. When the piston has moved a certain amount, the passage is suddenly enlarged so that they is hardly any brakin' or stoppin' the pilot valve from movin' a'tall."

"But what's them springs for on the piston?"

"Them springs is there for a good purpose. If you loosens them nuts on the spring collars, the governor head will not be held back as much as if they is tightened. Them springs plays the same part as the adjustin' screw on the constant-area oil brake. I can tell you another way to git the idee. Supposin' you has a weight on the end of a string an' you swings it like a pendulum; in the air that weight will swing pretty free, but if you puts the weight in a trough o' water an' then tries to swing it, you'll know what the oil brake does to the governor head."

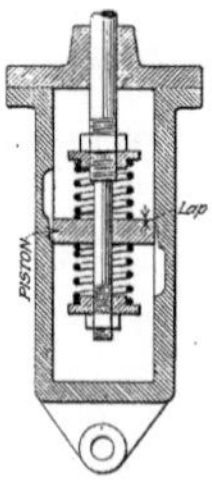

FIG. 3. OIL BRAKE WITH VARIABLE BYPASS AREA

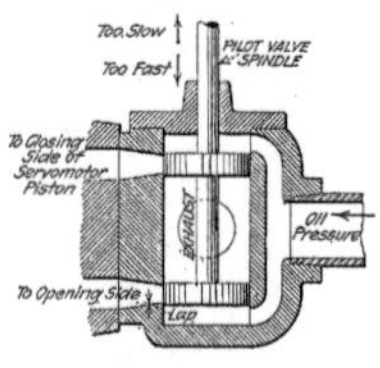

FIG. 4. RUDIMENTAL IDEA OF PILOT VALVE

"I suppose the kind of oil used would make a difference too, wouldn't it?" asked Jimmy.

"Sure. If I was to use somethin' about the consistency of asphaltum, I doubts if you could move the piston a'tall. That's about the same as screwin' the springs on the oil-brake piston down as far as they will go; in this way you makes the governor a hull lot more stable. In sich a case it wouldn't be worth while havin' a sensitive flyball head at all."

"What kind of oil is used, then? I should think a heavy oil would make the governor act more sluggish, too," Jimmy insisted.

"The manufacturers don't all recommend the same thing. They finds out the kind o' oil that works best with their spring adjustments, an' they sticks to that. One company insists on a very light grade of sperm oil, which tells me that he uses purty heavy springs; another uses a dynamo oil, still another uses quite a heavy grade of engine oil. Like you says, a heavy oil will make the governor work slower, 'cause it's so hard to push around through them ports, etc."

With that much finished, Jimmy turned to Fig. 4 with curiosity and spent several moments with it until Pop relighted his cigar butt and sent great clouds of smoke into the room.

"Well, what do yer make of it?" inquired Pop, bending over the table.

"Seems simple—nothing to it at all!" said Jimmy.

"Not much to that one, no. But did you see what I called it? I calls it a 'rudimentary idee,' like all those books on the gentle art of engineering that has the word 'easy' in the title.

"Now this is what yer might call a balanced pilot valve an' which prob'bly would be good enough fer a small governor. The reason that it ain't jes' the right thing for a big governor is that by the time you got it big enough so it could distribute oil enough for a big servomotor cylinder, it would be so heavy that it would make the hull governor unsensitive. As I said before, and from the looks of the critter, you'd think it was balanced, but in actual work they's quite some drop in pressure on the side of the piston that's feedin' oil to the servomotor cylinder, so they is a tendency for the pressure to push from the opposite side.

"The next time we dooly considers a real pilot valve among other things an' then winds up the subject of governors."

"But before we quit tonight, tell me something about the patent situation on these things we've been talking over," said Jimmy.

"Are you a-thinkin' of goin' into the governor business, or is it jes' curiosity? Well, anyhow, the governor head has two or three badly conflicting patents on it, so mebbe it's good not to say names. As for the constant-area oil brake, everybody can use that, but the variable-area oil brake is patented. They's no patents on the pilot valve, so everybody can use it.

"The lecturer told me that the waterwheel business has kept many a lawyer goin' an' that a big slice of the profits has been spent payin' up settlements on suits."

Stopping Leaks in Hydraulic Gates

By L. W. Wyss

Considerable water can be saved by stopping the leaks in a dam, and at many installations where all the water is used this leakage is equivalent to wasting many tons of coal during a year. Stopping the small leaks at the sides and bottoms of gates can be accomplished best with cinders. Sometimes special spouts must be made to guide the cinders to the leaks when obstructions prevent dropping the cinders directly above the leaks. Chips from water-logged wood will also sink slowly and stop leaks. When using cinders in a forebay near where small pipe intakes, such as for transformer-cooling water, etc., are located, it is well to close the valves temporarily or the pipes may be clogged.

It is frequently said that executives in municipal work find their hands tied by civil-service restrictions when it comes to the promotion of subordinates—that the seniority rule often works to the detriment of efficiency in such cases. Be that as it may, there appears to be ways to circumvent the effects without evading the rule itself. At least the city engineer of one of our largest municipalities seems to have gone a long way toward solving the problem by periodically transferring his engineers. Confronted by the fact that some of his oldest engineers, in point of services, were in the smaller and more obsolete pumping stations, he arranged to transfer them for a time to the most modern stations, so that should a vacancy occur in one of the latter the man designated by the seniority rule would be thoroughly familiar with the duties to be taken up. The plan is one that might well be emulated, not only by other municipalities, but by any system having a number of plants.

Care and Maintenance of Plunger Hydraulic Elevators

By A. B. Burgess

Every operating engineer having to do with plunger-type elevators has certain rules which he follows in their inspection and care. A few general suggestions which, if followed, will keep the maintenance cost of this type of elevator at a minimum, are given here.

THE most expensive operation in the construction of a plunger-type elevator is drilling of the cylinder hole and setting the cylinder. Hence it is of vital importance to see that this part of the machine is given a periodic inspection to prevent it from wearing out and thereby necessitating a replacement.

The plunger, passing through the stuffing-box at the top of the cylinder, is guided by several types of plunger bottoms. In the short-rise cars two types of bottoms are used, either the button plug, Fig. 1, or the wing plug, Fig. 2. These cars, traveling at a slow rate of speed, do not offer much chance for wearing through the bottom of the plunger or of grooving the cylinder.

Quite a different condition exists in the high-rise, high-speed passenger and freight elevators. Two general types of bottoms are here used to guide the plunger and to hold it central in the cylinder. The brush-type, Fig. 4, consists of three or four brushes of wire, held in a special form of casting. The other type of bottom, Fig. 3, has composition skates substituted for the brush. Springs are used to hold these parts away from the bottom, causing them to bear on the inside of the cylinder, thus holding the plunger in a central position in the cylinder, throughout the travel. The continual motion of the plunger causes the brush or skate to wear away gradually, and unless these parts are inspected at least once a year, preferably every six months, the chances are that the brush or skate will be found entirely worn out and the plunger-bottom cutting or

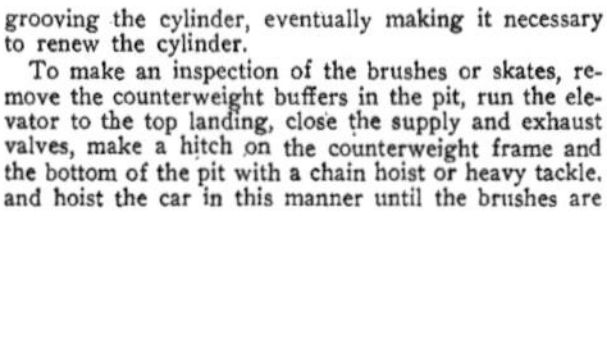

FIG. 1 FIG. 2

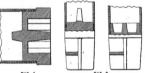

FIGS. 1 AND 2. PLUNGER BOTTOMS FOR SHORT-RISE PLUNGER ELEVATORS

grooving the cylinder, eventually making it necessary to renew the cylinder.

To make an inspection of the brushes or skates, remove the counterweight buffers in the pit, run the elevator to the top landing, close the supply and exhaust valves, make a hitch on the counterweight frame and the bottom of the pit with a chain hoist or heavy tackle, and hoist the car in this manner until the brushes are exposed. It will be necessary to remove the top automatic hitch.

The plunger itself should be lubricated once a week with graphite grease or an especially prepared compound. Oil is too light a substance to use when greasing the plunger. To lubricate the plunger, run the car slowly to the top, wiping off all the water. Then allow the car to descend slowly, applying the grease evenly with the hands. In some cases a lubricating ring

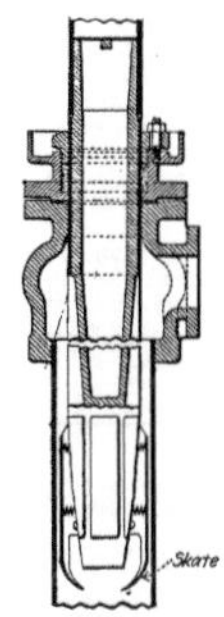

FIG. 3. SKATE-TYPE PLUNGER BOTTOM

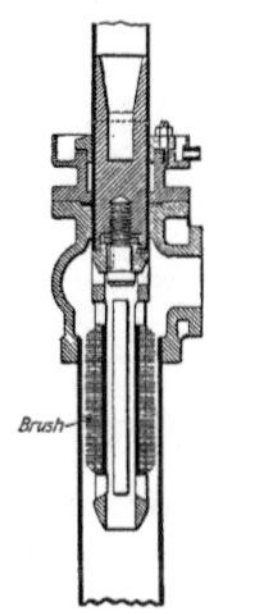

FIG. 4. BRUSH-TYPE PLUNGER BOTTOM

or cup has been attached to the stuffing-box, in which grease is put at regular intervals, every trip of the elevator causing it to be lubricated automatically.

A daily inspection of the stuffing-box, Fig. 5, through which the plunger travels, should be made, and in case of leakage the gland nut should be tightened evenly, say a quarter of a turn at a time, to insure the same relative bearing, until the packing needs renewing.

To repack the plunger, run the car up sufficiently high to work under it and block it. Shut off the valve in the supply line, then raise the stuffing-box gland to get at the packing space. Remove the old packing and cut the new in proper lengths to fit the plunger snugly. When the space is filled and the packing firmly in place, put the gland back and take up the nuts hand tight. Square flax or brake tubing is commonly used, although there are a number of special packings on the market.

Car and counterweight rails should be wiped down and greased with cylinder oil, or a very light grease, once a week. Fiber runners are furnished in both car and counterweight guide shoes, and should be subjected to weekly inspection. They should be renewed when they begin to wear, for they very decidedly improve the running of the car, and eliminate noise.

All sheaves, over which the automatic ropes, operating ropes and counterweight ropes pass, should be carefully examined daily and the grease candles replaced before being worn out. Any sheaves that are not provided with grease candles or oil cups should be oiled each day. Counterweight-sheave shaft boxes are heavily loaded, and the surface must be kept thoroughly smooth. A daily inspection here is well worth while.

The boxes which carry the operating shafts under the car platform should have their oil cups kept full. A daily inspection here is necessary. All bolts and bearings must be kept tight.

New ropes, when installed, are subject to considerable stretch and should be carefully watched. Screw connections for adjusting these ropes are provided, and when they have been taken up to the limit they are let out as far as possible and a fresh bite taken in the ropes by means of clamps. Care should be taken to have the automatic ropes adjusted so that the cars come to rest flush with the landings at both bottom and top terminals. Inspect the ropes weekly for broken wires, and rub them down with oil or some lubricant.

Daily inspect the valve, Fig. 6, by which the car is operated. All working parts must be thoroughly lubricated, keeping oil cups and grease cups full at all times. Repack the valve as soon as the elevator begins to creep. This may be accomplished in the following manner:

The car should be run to the lowest landing or, when there is no working space underneath the car, to the floor above the lowest landing and blocked, the supply and exhaust and to-and-from valves being shut. If the valve is of the pilot-controlled type, the handhole plates

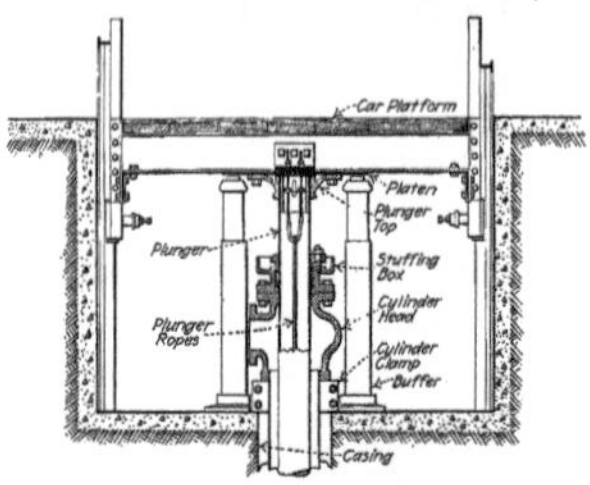

FIG. 5. SECTION THROUGH ELEVATOR-SHAFT PIT AND STUFFING-BOX

on the hood supporting the pilot valve and the end flange at the main valve should be removed. A stem puller is provided with all valves. Push the main valve toward the pilot valve as far as it will go, removing the pin connecting the main stem and rack, and pull the stem out of the valve steadily without wrenching or jarring, keeping the stem concentric with the main valve throughout the operation.

To renew the leathers, take off the nuts on the stems and remove disks and quills. See that they are laid aside in consecutive order. Be sure to note the relative positions of the leathers and see that the lips always point toward the water pressure. Inspect the valve linings and the ports. See that they are free from scratches and no foreign matter is hidden in them. See that the brass surfaces are not damaged after the packings are renewed. Insert the valve stem, push it far enough into the valve to get the pin which connects it with the rack into place, then pull the stem backward until the end piston comes flush with the end of the valve.

In replacing the main valve stem, see that the rack and pinion mesh according to the marks which indicate their location. There is a possibility of the valve

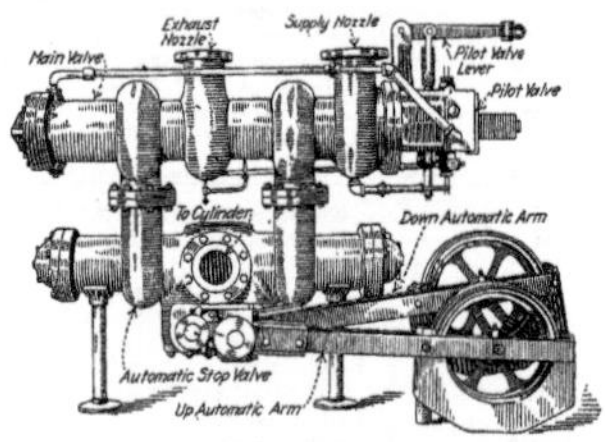

FIG. 6. CAR-OPERATING VALVE

becoming full of air during repacking. The air cocks on the stuffing-box should be opened, as well as those on the valve iself, to get the car running smoothly.

To pull the automatic stops, remove the large flange on the valve body and disconnect from the crank, taking them out from the ends of the valve after first removing the flanges. The lips of the leathers on the "down" automatic look from each other. The reverse is true on the "up" automatic. Before replacing the stem, be sure that the valve chambers are clean and free from all foreign substance. Replace all worn-out gaskets.

The pilot-valve stem is removed by taking off the bottom nuts and swinging the arm on the bottom of the pinion shaft clear. Then pull the stem, noting the position of the leathers, and renew same if necessary.

The automatic lever shafts are packed by loosening up the cap bolts on the lever head so that the driving and driven gears can be detached from the shafts easily. Before removing the levers and gears, block up the weight arm at the extreme outside end and sling weights and arms so that they can easily be slung into position after the shafts are packed. The inside collars and glands can then be removed and new hat packings put in, first seeing that the shafts are thoroughly cleansed and greased.

It is important that the valve leathers be inspected and renewed when necessary. If leathers are allowed to become worn out or if new leathers are faulty, these facts will be made apparent to the engineer by an unusual leakage of water through the drips, possibly by creeping of cars after the operating valve is closed or by unusual noise or knockings in the valve chambers.

According to the latest Labor Market Bulletin, issued by the New York State Industrial Commission, there are 10 per cent. more employees in the water, light and power industries than there were in June, 1914. The total wages paid these employees is 210 per cent. of the total wages in June, 1914. This represents an average increase of wages during that period of over 90 per cent.

CO_2 in Flue Gas When Burning Oil

An error in interpreting the CO_2 content in flue gas when burning oil is likely to arise among engineers who have been familiar with coal burning. This comes from the fact that perfect combustion of coal would give a higher CO_2 reading than perfect combustion of oil.

An example will help to show the reasons for this. Assume first a sample of coal having 73 per cent. carbon, 4 per cent. hydrogen, 8 per cent. oxygen and the remainder ash, etc. The carbon in each pound of coal will require for complete combustion $0.73 \times 2\frac{2}{3} = 1.95$ lb. of oxygen. The hydrogen will require $0.04 \times 8 = 0.32$ lb. of oxygen. The total oxygen required will be $1.95 + 0.32 = 2.27$, less the 0.08 lb. already in the coal, which leaves 2.19 lb. to be furnished by the air. This amount of oxygen is contained in $2.19 \div 0.23 = 9.53$ lb. of air, which will bring with it 7.34 lb. of nitrogen. The diagram, Fig. 1, shows these values graphically. The 0.73 lb. of carbon in 1 lb. of coal will produce $0.73 \times 3\frac{2}{3} = 2.68$ lb. of CO_2 in the flue gas. The hydrogen will form 0.36 lb. of water vapor, which will condense before reaching the flue-gas apparatus and so will not appear in the analysis. Therefore, for each pound of coal the flue gas will contain 2.68 lb. of CO_2 and 7.34 lb. of nitrogen, or a total of 10.02 lb. of gas. This gives $(2.68 \div 10.02) \times 100 = 26.7$ per cent. CO_2 and $(7.34 \div 10.02) \times 100 = 73.3$ per cent. of N *by weight*. Since the ratio of the weights of N to CO_2 is 14 to 22, the relative volumes will be $26.7 \div 22$, or 1.21, for CO_2 and $73.3 \div 14$, or 5.24, for N. The percentages by volume will then be $1.21 \div 6.45$, or 18.8 per cent. CO_2 and $5.24 \div 6.45$, or 81.2 per cent. N.

Assume, now, a typical sample of oil containing 85 per cent. carbon, 12 per cent. hydrogen and 3 per cent. oxygen. The carbon will require 2.27 lb. of oxygen and will produce 3.12 lb. of CO_2, while the hydrogen will require 0.96 lb. of oxygen and produce 1.08 lb. of water vapor, per pound of oil burned. The net oxygen required will then be $2.27 + 0.96 - 0.03 = 3.20$ lb. This means the introduction of 13.91 lb. of air containing 10.71 lb. of nitrogen. Remembering that the water vapor will be condensed, the flue gas per pound of oil will be $3.12 + 10.71 = 13.83$ lb. having a composition by weight of 22.5 per cent. CO_2 and 77.5 per cent. N. Reducing this to percentage by volume gives 15.6 per cent. CO_2 and 84.4 per cent. N. The diagram in Fig. 2 illustrates these computations for oil.

In comparing the two cases figured it must be remembered that they refer to the ideal condition where all the combustible is completely burned and no excess air is supplied. In other words, they represent the best

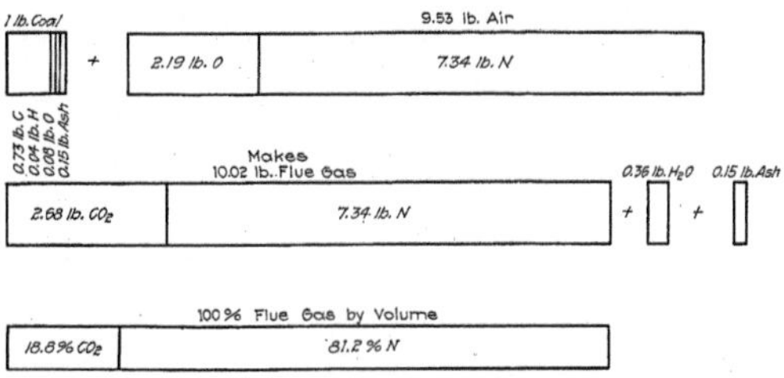

FIG. 1. DIAGRAM SHOWING FLUE GAS PRODUCED BY PERFECT COMBUSTION OF COAL

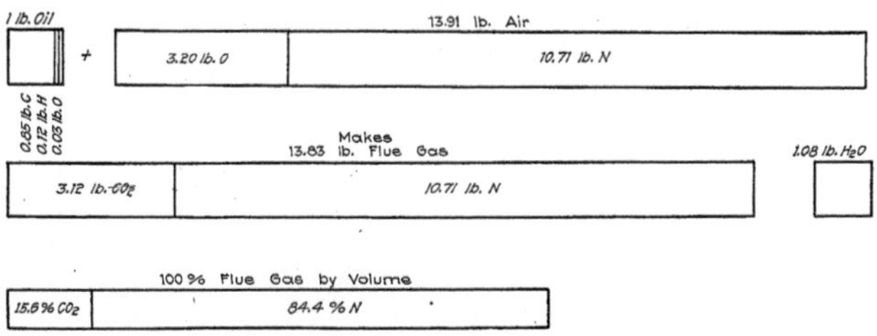

FIG. 2. DIAGRAM SHOWING FLUE GAS PRODUCED BY PERFECT COMBUSTION OF OIL

theoretical conditions. Comparing the particular samples taken, oil could not possibly give a higher CO_2 reading than 15.6 per cent., while the coal theoretically might give 18.8 per cent.

The explanation of this lies in the greater amount of hydrogen in the oil. The hydrogen requires oxygen for its combustion, and this in turn brings in nitrogen, which appears in the flue gas. The hydrogen does not produce CO_2, and the water vapor that it does produce does not appear in the flue-gas analysis. The result is that the higher the hydrogen content in the fuel the lower the theoretical CO_2 percentage in the flue gas.

Allowable Pressure in Thick Hollow Cylinders

A formula for computing the allowable pressure in thick hollow cylinders was given in London *Engineering* for Dec. 15, 1911. This formula expresses the experimental results of Gilbert Cook and Andrew Robertson and may be written,

$$\frac{P}{S_t} = 0.6\ \frac{K^2 - 1}{K^2}$$

where P = the allowable pressure in pounds per square inch, S_t = the allowable working stress of the mate-

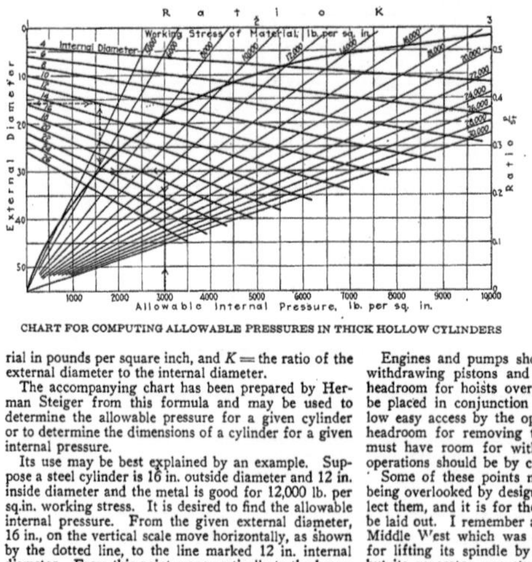

CHART FOR COMPUTING ALLOWABLE PRESSURES IN THICK HOLLOW CYLINDERS

rial in pounds per square inch, and K = the ratio of the external diameter to the internal diameter.

The accompanying chart has been prepared by Herman Steiger from this formula and may be used to determine the allowable pressure for a given cylinder or to determine the dimensions of a cylinder for a given internal pressure.

Its use may be best explained by an example. Suppose a steel cylinder is 16 in. outside diameter and 12 in. inside diameter and the metal is good for 12,000 lb. per sq.in. working stress. It is desired to find the allowable internal pressure. From the given external diameter, 16 in., on the vertical scale move horizontally, as shown by the dotted line, to the line marked 12 in. internal diameter. From this point move vertically to the heavy curved line and then horizontally to the line marked 12,000 lb. working stress. Directly below this point read off the allowable pressure on the horizontal scale, 3000 lb. per square inch.

Suppose it is desired to know the thickness of a cylinder to stand 3000 lb. per sq.in. pressure when the internal diameter is 12 in. and the allowable stress is 12,000 lb. per sq.in. From the 3000-lb. point on the horizontal scale move upward to the 12,000-lb. stress line and then horizontally to the curved line. From here move vertically to the 12-in. internal diameter line and read off the external diameter, 16 in., on the left-side scale.

Steam Electric Power-Plant Layouts

By B. W. Dennis

A few of the practical problems that are not always satisfactorily solved in existing power plants may be of interest and benefit to those who are engineering the location of new equipment or the alteration of present plants. The best results are obtained by making scale drawings of the building to be occupied by the equipment in question, and then fitting to this space in both plan and elevation the machines that are to be installed.

The lower limit in excavation is determined by drainage problems to sewers or sump pumps, and one of the usual mistakes is to limit the basement headroom for piping and allow too little head on drain traps and pumps. Many basements are designed without a thought for light, air or accessibility. If the operator must be lowered into a basement dugout, he surely ought to have room to work without landing on top of the pump or equipment to be repaired or inspected. Another fault with some modern plants is the careless placing and pointing of valves which must be quickly operated. Another fault is to unnecessarily crowd the side clearance around the power equipment. This is not only a fault from the operator's point of view, but from the manager's viewpoint it is costly as well. Boilers should not only have clearance for firing and withdrawing tubes, but room to inspect and repair the setting, piping and accessories, and room and ventilation for the ashman.

Engines and pumps should have not only room for withdrawing pistons and plungers, but the necessary headroom for hoists over the heavy parts and should be placed in conjunction with other equipment to allow easy access by the operator. Turbines must have headroom for removing their spindle, and generators must have room for withdrawing the rotors. These operations should be by crane whenever possible.

Some of these points might be considered as never being overlooked by designers, but operators will recollect them, and it is for the operation that plants should be laid out. I remember a large turbo-generator in the Middle West which was erected to allow access only for lifting its spindle by the electric crane overhead, but its generator was stuck back in a leanto building, evidently intending that it would never need repairs or handling of its rotor by the crane. The only object which could have caused placing it in this manner must have been the ventilation of the generator.

Designers cannot bury their mistakes as do physicians, and a little extra thought will be amply repaid.

EDITORIALS

Industrial Unrest

MUCH was said at the recent meeting of the American Society of Mechanical Engineers about the part which the engineer, with his analytical habit of thought, his judicial methods and his intermediate point of view, should play in allaying the prevailing social unrest and reconstructing industry upon the basis of a just and lasting peace.

If the engineer is to do this, if the engineering press is to help him, there should be some understanding as to the terms upon which most of the warring factions can agree, and, so agreeing, take up unitedly the common struggle of man against his environment, and of orderly procedure and security against lawlessness, confusion, self-interest and license.

Whatever the future may develop, we believe that at present the majority will agree:

That the capitalist is entitled to a fair amount of interest upon money actually invested.

That the promoter and conductor, the man with initiative and enterprise, the man with an idea and the ability to turn it into a service, is entitled to a fair profit upon the actual cost of the product sold or of the service rendered.

That the toiler with hand or brain is entitled to emolument commensurate with the service rendered; and that such emolument for such service as is within the capacity of any able-bodied, mentally sound person to render, should be sufficient to maintain him and those dependent upon him in healthful, comfortable and enjoyable circumstances and enable him to accumulate a competence for his nonproductive and declining years.

That the consumer is entitled to the necessities of such an existence at the cost of their production and distribution plus a fair profit to producer and distributor.

That anybody who is able and willing to work is entitled to a place in the industrial organization, with a secure tenure upon such opportunity; that hours of labor should be so adjusted that the offered man-power may be entirely utilized and that each may profit, in reduced hours or years of necessary service, by the full utilization of the available force and the improvement of its efficiency through greater diligence or better machinery, tools and methods.

It is urged that one cure for the high cost of living that is at the bottom of most of the present discontent is greater productiveness. If each is to have more, each should produce more. But when the individual workman is urged to greater diligence, he says: "What is the use? If I make twice as many pieces a day I will put some other fellow out of a job, restrict the demand for my own labor and get no more for working harder." If those who toil could be assured that the results of their greater activity and efficiency would come back to them in a diminished cost of the necessities of life, in an increase in the purchasing power of a day's work, in a diminution of their period of toil either in a shortened day or in their years of necessary service, and not go to swell the profits of the manufacturer and middleman, there would be a natural end to sabotage and slacking and strikes.

The man who buys a suit of clothes is paying, not for so many pounds of material, but for the labor of the men who reared and tended and fed the sheep while it was growing the wool; the men who sheared and collected and transported and sorted the wool; the men who wove it into cloth, the men who built the factory in which it was spun and woven and dyed and those who conducted those processes, even the men who mined the coal that operated the power plant of the factory; the men who marketed the cloth, and those who converted it into garments, and the men who operated the railroads by which it was transported and the buildings in which it was handled through all these processes. What the man who buys the suit is paying for is, labor plus profit; and assuming a given percentage of profit, the cost to him is fixed by the price paid to labor. If everybody along the line slacked and loafed and wasted and struck and demanded an exorbitant wage and short hours, his suit would cost him much more than as though everybody worked at his best efficiency for a reasonable number of hours at a fair price and made the most economical use of the material he handled.

The war has made us rich as a nation, but no nation can stay rich or get richer by laying off and spending its substance. There never was a better chance. The storehouses of the world are empty. The war-stricken nations are settling down to retrieve their broken fortunes. They need capital and raw materials and finished products. There is work enough for everybody in repairing the ravages of the years of war and getting the world upon its feet again. If labor will settle down to this task and work at its best efficiency, and if producers will be satisfied with the return which a fair rate of profit upon such a volume of business would yield, the workingman will be able to live respectably upon a wage which will allow those who neither profiteer nor bargain collectively to live also.

A High-Grade Man

ONE of our contemporaries recently published an editorial under the title "What Is a High-Grade Machine?" which brought forth considerable comment. In a discussion of this question another was brought up: "What Is a High-Grade Man?" We hear both these expressions, "a high-grade machine" and "a high-grade man," used rather generally, but just what do they mean? More especially, what do you mean when you speak of a high-grade man?

When considering machines, such qualities as design, materials, workmanship, etc., are used in an endeavor to classify or grade them as high or low. In all these ideas, however, suitability for the intended use is emphasized.

When considering men, we will find considerable difference of opinion as to what qualities should be used to measure a man's grade. The banker is likely to look on the ability to be a successful busines man as the correct measure of grade. The college president may consider the number of degrees he has as the important thing. The engineer may look into his ability to clean a fire or pack a pump. So we might go on indefinitely. One point seems to be common to all—the man must be able to do his job. The real reason for most of the differences is that unless you take a certain definite individual, each observer pictures a man of his own class or profession.

Another measure of a man's grade is the standing he has when compared with his immediate superiors. In the method of rating army officers during the war, each individual was compared, at intervals, with five officers of the next higher rank. This gave a measure of his fitness for promotion. In a power plant a man may be a high-grade watch engineer, but only comparison with a high-grade chief will show whether he can rise.

To grade the man in terms of machine qualities, he should have proper design, materials and workmanship. He should be designed or fitted for his work. A man with the necessary mental qualities to be a high-grade doctor would probably make a very low-grade engineer. He should be made of the proper stuff. He must have a clear mind and physical strength enough to do his work. And he should be the product of good workmanship. That is, he should have had the proper training and experience. A most carefully designed cylinder, cast from the best of iron, is not a high-grade piece of apparatus until it has been machined with correspondingly good workmanship. So the man may be suited for his work, may have the right stuff in him, but cannot be called a high-grade man until he has been fitted into his job by training and experience.

"Finish" is another quality that is often used as a measure of grade, both for machines and men. A high-grade machine and a high-grade man are likely to show a high-grade finish as a direct result of the presence of other high-grade qualities. On the other hand, many a machine or man shows a high-grade finish when that is the only high-grade quality it or he possesses. Appearance or personality in a man may cover up many a fault, but it wears off after a while just as the paint or polish wears off the machine. We don't criticize these qualities—they are good things to have, but alone they do not put a man in the high-grade class.

Our definition, then, of a high-grade man would be one who is designed for his job, has the mental and physical stuff in him to handle the job and who has had the proper experience and training for the job—in short, the man who can do his job well no matter what that job may be. If he has a high-grade appearance or personality, so much the better.

October Electrical Exports and Canadian Trade

ACCORDING to the report issued by the Department of Commerce, Bureau of Foreign and Domestic Commerce, Washington, D. C., the total value of exports of electrical machinery and appliances (excluding locomotives) from the United States during the month of October, 1919, to foreign countries amounted to approximately $7,280,000; of this, $1,205,000 was sent to Canada. These figures are significant for several reasons. First, here is a country in which a number of the largest United States electrical companies have established manufacturing plants, and yet it imported from this country electrical equipment that amounted to approximately one-sixth of our total electrical exports. If such an export trade can be developed with a country of approximately eight million population, how great must be the possibilities with the rest of the world if proper methods are adopted! Comparing our export figures to Canada with those to Mexico, both countries on our borders, the latter having about double the population of the former, we find the electrical exports to Mexico amount to only $215,358, or approximately one-sixth the value of our Canadian trade. This is all the more significant since Mexico is a republic and is responsible to no other country, whereas Canada is a British colony. Although both have vast natural resources, one has a stable and responsible government, and the other has only a semblance. In this we need not go to Europe to find what a chaotic government means to a nation; we have it at our southern border.

Canada, owing to her limited fuel supply and its location, must depend more and more upon her waterfalls for power to make possible her industrial development. The development of water power in most cases creates a demand for large quantities of auxiliary electric equipment. Another significant feature of the October electric export figures to Canada is that $236,467 of this was for electrical motors. A large proportion of these motors are used for industrial purposes. Do these figures not indicate that Canada is preparing for a big industrial boom, especially when so many other reports have come from over our northern borders of the approaching birth of a period of industrial development in that country? American manufacturers should read the handwriting on the wall and help to develop this next-door market.

When the Coal Is Gone

FARSIGHTED managers and engineers often speculate upon what will become of us when the fuel supply is exhausted. According to *The Engineer,* this condition has come to France already. "Supplies," it says, "may become more plentiful with a better organization of transport arrangement, but it will be a long time before they are sufficient for all purposes, and *the price of coal will never again make it economical as a source of motive power.*" The italics are ours.

This condition has compelled the government to open up various undertakings for utilizing the Rhine, the Rhone and the Pyrenees, the electrical energy so generated being distributed throughout the country by a network of high-tension mains. At the same time the government is empowered to oblige every owner of a waterfall to install an electrical generating apparatus, either for himself or for the use of his neighbors. Subsidies will be granted to groups of agriculturists and others for the laying out of hydro-electric generating plants.

And we are letting millions of horsepower go to waste while we merrily burn up the remains of our coal supply.

During the recent coal shortage, *Power* received several inquiries concerning the use of gasoline electric-generating sets for emergency service. As is usually the case, the demand in such cases comes very unexpectedly, and before it can be met the acute stage of the conditions creating it have usually passed. But this leads to the question whether there are not a great many plants operating under conditions where it would be economy to install such sets for the night load and occasionally to help over a peak. The stand-by losses are nil, and it is merely a question of fixed charges on the first cost balanced against stand-by fuel and labor costs of the steam plant during the period of very little load.

F. F. Uehling, writing to the editor of the New York *Evening Sun,* says that the burning of cinders by the aid of rock salt and oxalic acid is like going to the garbage pail for food. But—the garbage *has* some food value.

Neglected Parts of Water-Tube Boilers

In the Sept. 9 issue of *Power* I read an article entitled "Neglected Parts of Water-Tube Boilers." After over nine years' experience with twelve 500-hp. water-tube boilers, I cannot agree with some of the conclusions arrived at by Mr. Lewis. It is true that the cleaning of caps and headers is a hot, dirty job, but why the job should be prolonged by having to scrape off sticky gaskets from both caps and headers, I cannot understand.

Boilers equipped with outside caps, when new, have metal-to-metal joints and with ordinary care can be kept tight without the use of gaskets. In this plant with 2592 tubes, or 5184 caps, we have three caps with gaskets, the necessity for these being faulty metal in the headers.

Mr. Lewis' advice regarding tightening of the caps I consider dangerous to life, especially when dealing with outside caps. It is a wise procedure to watch the caps when the boiler is being filled with water and to rectify any leaky caps then, but when any man goes into the back chamber of a boiler that has full steam pressure on and attempts to tighten up a leaky cap he is surely courting trouble for himself.

While making any sort of a joint an engineer endeavors to make it tight, consequently he is likely to stress the bolts as much as is good for them. The outside cap of certain water-tube boilers is 4 in. inside diameter, and deducting 1⅛ in. for the hole, it means that at 175 lb. pressure the bolt has to take an added stress of 2025 lb., and to attempt to tighten up a leaky cap against that pressure is not good practice. Besides, if a cap leaks after having been tightened up, it clearly shows that some foreign substance has been left between the two faces, and the only remedy is to clean the joint.

My object in writing is to call attention to what I, at least, consider to be a very dangerous practice, and the slogan "Safety First" might well be emphasized here.

Vancouver, B. C. JOHN MELVILLE.

Faulty Soldering Caused Trouble on Direct-Current Armature

When soldering the armature connection of direct-current machines to the commutator risers, it is advisable to tilt the armature slightly as shown in the illustration, so that the solder will have a tendency to run toward the front of the risers. If this precaution is not taken, trouble is likely to occur due to loose pieces of solder lodging behind the commutator lugs. The armature may test all right and be put into service without the trouble showing up for a considerable time, but eventually it will do so, as the following example will illustrate:

This particular machine had given no trouble for two years or more. Then the commutator began to get excessively hot in places, due apparently to carbonized mica, as the bars were badly discolored by the heat. It was decided to re-mica the bad sections of the commutator. In carrying out this operation it was unnecessary to disturb the connections, as with the commutator loosened up it was a simple matter to slip the new mica between the segments. After tightening up, a bar-to-bar test was made with a millivoltmeter and everything appeared to be all right. The commutator was then trued up in the lathe, the surface shorts picked out and tested once more. The result this time was far from satisfactory. Four low readings were found between segments,

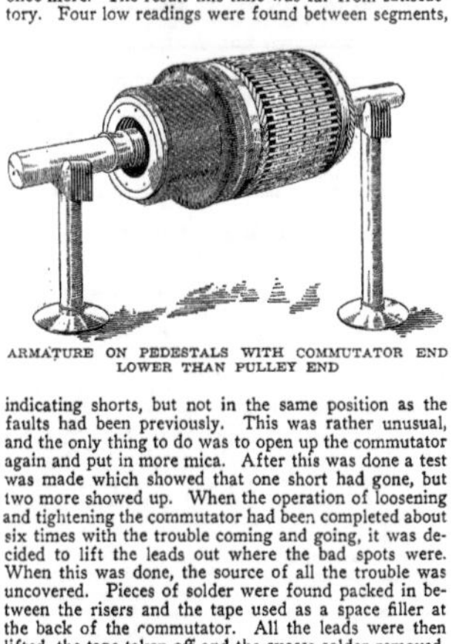

ARMATURE ON PEDESTALS WITH COMMUTATOR END LOWER THAN PULLEY END

indicating shorts, but not in the same position as the faults had been previously. This was rather unusual, and the only thing to do was to open up the commutator again and put in more mica. After this was done a test was made which showed that one short had gone, but two more showed up. When the operation of loosening and tightening the commutator had been completed about six times with the trouble coming and going, it was decided to lift the leads out where the bad spots were. When this was done, the source of all the trouble was uncovered. Pieces of solder were found packed in between the risers and the tape used as a space filler at the back of the commutator. All the leads were then lifted, the tape taken off and the excess solder removed.

After soldering up again, the armature tested all right, no further trouble occurring. It was apparent that in loosening the commutator to insert the new mica, the solder had been disturbed, allowing it to come in contact with the segments. The fact that the armature had been well varnished and baked probably prevented the trouble from taking place during the time the machine was in use. H. WILSON.

Toronto, Ont., Canada.

Boiler Bagged Due to Muddy Water

A small return horizontal-tubular boiler, 48 in. in diameter and 10 ft. long, made with $\frac{7}{16}$-in. thick plate and working with a pressure of 90 lb., was recently bagged due to muddy water that was obtained from a mine, although good water could have been obtained from the river, only 150 ft. distant from the plant.

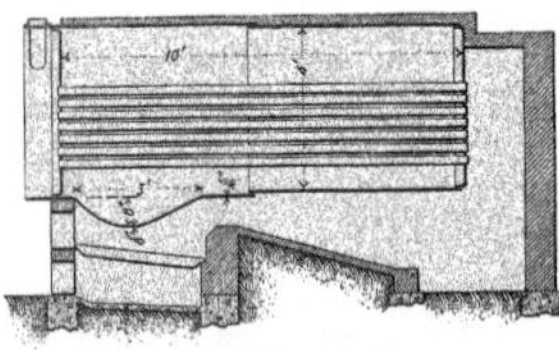

SHOWING WHERE BOILER BAGGED

The boiler has a bag 3 ft. in length and sagging at the lowest point 8 in. from its original position. The stretch of the plate was from $\frac{7}{16}$ in. to $\frac{3}{8}$ in. in thickness.

The question arises as to who is to blame for the trouble, the engineer in charge of the boiler or the manager in charge of the mine? There does not seem to be much sense in using muddy water from the mine when clean water could have been obtained without much trouble from the river which was so close by.

Wayne, Alberta, Canada. EDWARD JESSE.

Repairing a Commutator Under Difficulties

In many cases the motors in printing establishments are situated under the presses, where they collect oil and dirt, little attention being paid to them except to oil the bearings; and since the oil can is much in evidence around a press, the motors generally come in for an over supply. Often the oil-well overflow is choked with dirt and the excess oil is thrown out on the commutator, where it collects dust and dirt. This is a fruitful cause of grounds and short-circuits, since it produces sparking at the brushes, heats the commutator and carbonizes the mica in spots. The charred mica, if not picked out in time, will eventually work through to ground, if it does not cause one or more coils in the winding to burn out before this occurs.

On one occasion an armature was grounded and upon inspection the trouble was found to be at the back of the insulating ring on the front end of the commutator. It was very important that the machine be kept in service, therefore a hurry-up job in making the repairs was necessary. However, three things made it rather difficult, namely: The commutator tightening ring was on the inside, which meant lifting all the leads in order to loosen it up; the insulation on the winding was in very bad shape, consequently the coils would not bear removing; third, the commutator bars were worn so thin that it was out of the question to attempt opening the commutator and truing it up again in the lathe. The whole armature was in bad condition and needed to be rebuilt. Another motor to replace the machine in trouble could not be obtained on short notice, therefore it was decided to patch up the commutator somehow.

At first I endeavored to pick out the charred mica at the top of the commutator with a steel picker, but the trouble extended back to the corner of the insulating ring, which made it impossible to remedy the trouble by this method. As a last resort it was decided to cut the end of two bars off over the bad spot. This method worked all right and was carried out as follows:

A thin piece of steel was pushed under the bars back to the corner of the V-ring to get the depth, and a mark was made on the top of the commutator, as at *D*. Then with a drill slightly smaller than the width of a bar I drilled through at a point in front of the mark so as to leave a slight overlap to support the bar. The end was then broken off, exposing the bad spot, which was comparatively easy to clean out and reinsulate with new mica. With a small, sharp chisel the ends of the bars were shaped as at *E*, and two pieces of copper cut to replace the parts broken off. These had an extension *A* at the end, so that the cord band would help in holding them down. The parts are tinned and sweated together, care being taken to prevent solder working inside the commutator. Mica with a little shellac was then pushed in between the new pieces of bar, and calked in slightly to keep it in place. After smoothing down, the

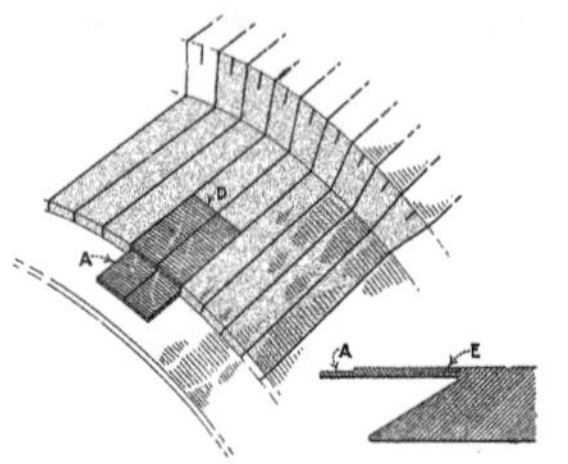

DETAILS OF COMMUTATOR REPAIR

points were hardly perceptible. The machine ran all right for some time until the armature was taken out and rewound, and at the same time a new commutator was installed. H. WILSON.

Toronto, Ont., Canada.

Wants a Continuous Indicator

A neglected feature in the power plant is what should be known as the self-recording cylinder indicator. Engineers indicate their engines once in 24 hours, three times a week, once a month, and in other cases not until an engine begins to pound or a bearing begins to heat. An indicator diagram shows the working condition of steam in the cylinder for that one revolution but nothing more.

Why does not someone invent an instrument that will record not only the exact condition of the valves at every revolution, but at the same time show the irregularities of the amount of horsepower being used at all times?

If the instrument has a clock dial, the engineer, upon noting a varied load when it should be normal, could go over the plant and thereby ascertain what machine was off and the cause for the delay of time or loss in output. As matters are now, nothing is known about it unless the fireman cuts his fires or the operating engineer notices the position of the governors of his engines or the reading of the switchboard instruments.

Brenham, Tex. H. W. Rose.

Repairing Cracked Pump Casing

The discharge casing of a large centrifugal pump developed a crack while the pump and its 30-in. suction line were being tested for leaks with a water pressure of 30 lb. This pump supplies the circulating water for

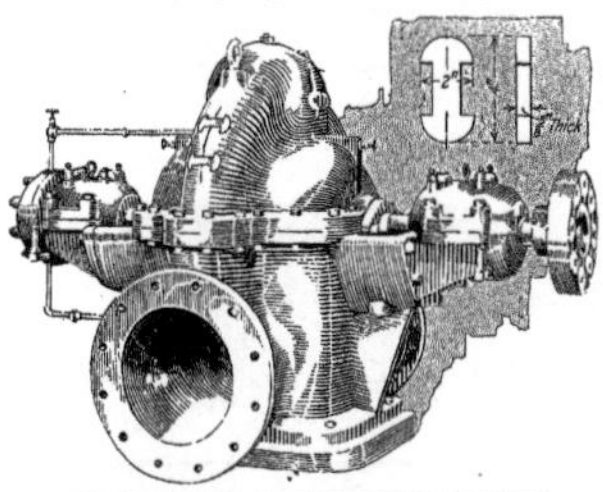

HOW THE CRACKED CASING WAS REPAIRED

a surface condenser and operates against an 8-ft. head, the suction lift being about 10 ft., so it is not plain why the 30-lb. test pressure was thought necessary.

The crack was on the top of the upper half of the casing, running lengthwise and practically in a straight line for 18 in. To repair this fracture, the casing was chipped at two equidistant points from the ends of the crack to receive the links as shown in the figure. These links were made of ⅝-in. boiler plate and constructed to give a snug fit when cold. The crack was then treated to an iron-cement preparation while a low vacuum was kept on the pump; the links, after being given a good red heat, were dropped into place and a piece of a hacksaw blade slipped between the lips of links and casing, giving a good fit.

The crack was drawn together firmly as the links cooled and has given no trouble, the pump having been in almost daily operation for the last five years.

Chicago, Ill. Arnold Erickson.

Why Not Express Vacuum Values in Per Cent.?

The method proposed by Mr. Weaver (*Power*, Sept. 2, 1919, p. 396) has been used in Germany for many years and seems to have been acceptable to the German engineers. It has not been adopted by American engineers, however, on account of certain objections.

In the first place "93 per cent. vacuum" does not mean anything to an engineer unless he also knows the barometer, for it is perfectly obvious that a "93 per cent. vacuum" and 30-in. barometer means a greater absolute pressure and temperature in a condenser than a "93 per cent. vacuum," say in Colorado, with only a 25-in. barometer. In the first case, the absolute pressure would be 2.1 in. of mercury with a corresponding steam temperature of 103 deg. F. and in the second case of 1.75 in. of mercury with a corresponding temperature of 97 deg. F. It is therefore obvious that vacuum can be expressed in "per cent." only in a section of country like Holland, where there is practically no change in altitude. Even in that case the daily variation of barometer would have to be taken into account, for a 27.8-in. vacuum with a 29.8-in. barometer is just as good a vacuum as a 28.3-in. vacuum with a 30.3-in. barometer, although the per cent. vacuum is different.

Furthermore, all our steam quantities, as given in the steam tables and on graphical diagrams, are based on absolute steam-pressure measurements. Therefore condenser pressures must finally be reduced to these absolute pressures if any computations are to be made.

Mr. Weaver is correct in stating that "vacuum in inches means nothing to me unless I know the actual barometer." If this information is at hand, it is unnecessary to calculate "per cent. vacuum," for reference can be made directly to the steam tables. It is therefore apparent that no appreciable gain would be made by the adoption of the proposed "per cent. vacuum" in this country.

A. G. Christie.

Baltimore, Md.

Fractured Oil-Engine Piston Head Repaired

On Diesel engines it is customary to "sew" cracks or fractures in the piston heads. This is a somewhat difficult proceeding and too expensive for hot-ball engine pistons. The illustration shows a method that has been followed with success in several instances. The fracture is covered with a drilled ¼-in. boiler-steel plate and the piston head has tapped holes for ⅜-in. capscrews to match the holes in the plate. A thin layer of

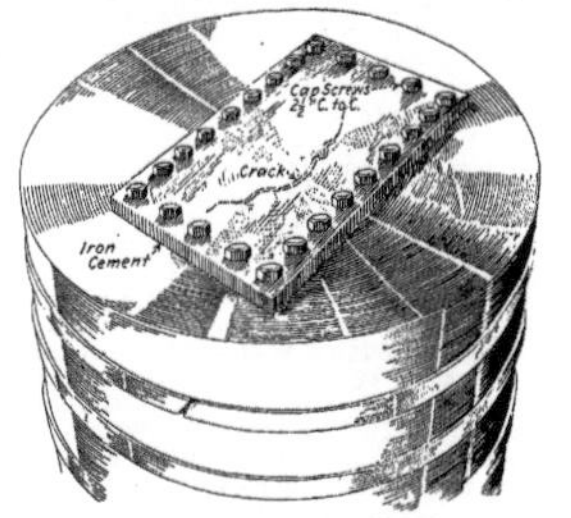

PATCHED OIL ENGINE PISTON

iron cement is placed between the piston head and the plate, after which the capscrews are tightened.

To eliminate danger of preignition, due to the capscrew heads becoming incandescent while the engine is running, the heads can be cut off and the shanks of the screws riveted over the plate. If this is done, the holes in the plate should be counterbored to permit a better riveting.

L. H. Morrison.

San Pedro, San Domingo.

Boiler Blowoff Lines and Valves

Having read with interest the various articles published in *Power*, relating to boiler blowoff valves and lines and having had experience with similar equipment, a few cases and their defects and the remedy as applied may be of interest.

In one plant containing ten vertical water-tube boilers set in batteries of two boilers each and nine horizontal return-tubular boilers set a solid battery, the blowoff system consisted of a 4-in. header run in an open trench back of the vertical boilers and about three feet below the mud-drum outlet, screw-end reducing tees being used in the header to take care of the blowoff from each boiler. The tees were 2½ x 4 x 4 in. with 2½-in. outlet looking up. Into the tee was screwed a short nipple with flange, on top of which was bolted the blowoff valve, of angle type; next to the inlet of the valve was placed an asbestos-packed plug-cock with screw ends, fitted with a nipple and flange connections between valve and boiler piping. From this cock piping was run direct to the mud-drum.

The blowoff line from the return-tubular boiler was run in the same manner with the exception that the 4-in. header was placed 5 ft. above the header for the vertical boilers. The two headers were connected by the use of two ells and a drop leg.

In blowing down the return-tubular boilers it was necessary for the men to walk along on top of the 4-in. header and stand over each valve and cock when opening them, as there was not space enough between the wall of the building and the boiler setting for a passageway. In blowing down the vertical boilers it was necessary to stand over the valves and on the pipe line or on the edges of the open trench, which, needless to say, was always filled with water and dirt.

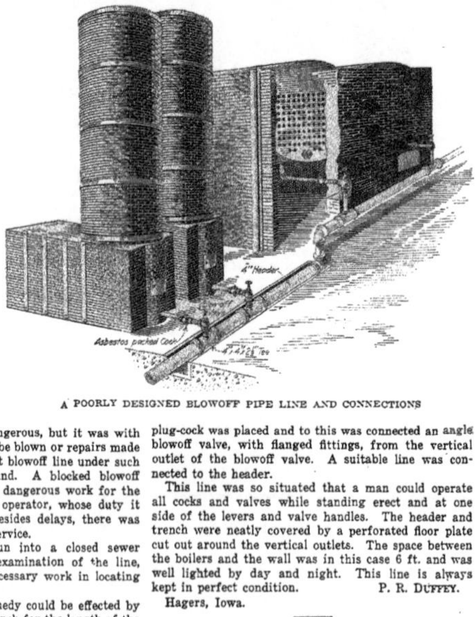

A POORLY DESIGNED BLOWOFF PIPE LINE AND CONNECTIONS

Not only was this layout dangerous, but it was with difficulty that the boilers could be blown or repairs made to the line, consequently a tight blowoff line under such conditions was not often found. A blocked blowoff was to be dreaded as it meant dangerous work for the mechanic and the boiler-room operator, whose duty it was to remedy the trouble; besides delays, there was the keeping the boiler out of service.

The blowoff header was run into a closed sewer which did not permit ready examination of the line, and it often necessitated unnecessary work in locating trouble in the line.

In this particular case a remedy could be effected by putting in a deeper concrete trench for the length of the boiler room and laying a new header supported on cast-iron chairs, high enough in the trench to remain dry; and by using flanged fittings instead of screw fittings, and placing a suitable gate valve between the end of the vertical boiler header and the header serving the return-tubular boiler; also providing at the outlet a suitable blowoff dump or tank, where an examination of the outlet line could be made at any time and a leaky valve or cock detected at once. A more compact and up-to-date type of valve and cock should also be used instead of one requiring a long-handled wrench for opening.

In another plant, having two batteries of horizontal water-tube boilers, the blowoff system consisted of a 4-in. header that ran from one end of the boilers to a surface-blowoff tank set on the outside of the building. This tank was provided with a vent pipe, through which any leakage was readily detected and with a manhole and surface drain for cleaning.

The blowoff lines consisted of two 2½-in. lines run from the back leg of each boiler downward and then connected horizontally by means of flanged ells. In the horizontal pipe a one-quarter-turn flanged-connection plug-cock was placed and to this was connected an angle blowoff valve, with flanged fittings, from the vertical outlet of the blowoff valve. A suitable line was connected to the header.

This line was so situated that a man could operate all cocks and valves while standing erect and at one side of the levers and valve handles. The header and trench were neatly covered by a perforated floor plate cut out around the vertical outlets. The space between the boilers and the wall was in this case 6 ft. and was well lighted by day and night. This line is always kept in perfect condition.

P. R. DUFFEY.

Hagers, Iowa.

The City of Tacoma has under consideration at the present time a scheme to develop 75,000 hp. from the Lake Cushman power site on the Snohomish River. This hydro-electric development will utilize a head of 580 feet.

INQUIRIES OF GENERAL INTEREST

Riveting of Boiler Dome for 125 lb. Working Pressure.—The 30-in. diameter dome of a boiler that is intended to carry 125-lb. pressure has a double-riveted lap joint and single riveting of the flange to the shell. Is this construction permissible under rules that follow the A. S. M. E. Boiler Code? J. S. F.

The construction would not be permissible for, according to paragraph 194 of the A. S. M. E. Boiler Code, the longitudinal joint of a dome 24 in. or over in diameter should be of butt and double-strap construction and its flange should be double-riveted to the shell when the maximum allowable working pressure exceeds 100 lb. per square inch.

Conversion of B.t.u. Into Horsepower—How many horsepower will be developed by an expenditure of 210 B.t.u. per minute? S. C.

The mean value of most trustworthy investigations for determination of the mechanical equivalent of heat, founded on experiments in which mechanical work was transformed directly into heat and electric energy was changed into heat, when reduced to mean B.t.u., gives the result,

$$1 \text{ mean B.t.u.} = 777.54 \text{ standard ft.-lb.}$$

For ordinary calculations the value 777.5 ft.-lb. is taken.

As a horsepower is development of energy at the rate of 33,000 ft.-lb. per min., an expenditure of 210 B.t.u. per min. would develop

$$(777.5 \times 210) \div 33{,}000 = 4.95 \ hp.$$

Advantages of Modern Over Old-Style Flywheel Pumps—In modern designs of direct steam pumps, why is it possible to dispense with a flywheel, and what is the disadvantage of a single pump with flywheel? R. B.

In the earlier designs of steam pumps a flywheel was necessary for obtaining momentum to reverse the direction of the stroke, and a single pump could not be started up with the crank on a dead-center or be run very slowly. In modern direct single steam pumps, the steam admission valve is steam-thrown. It is reversed at the end of the stroke by pressure of live steam under the control of an auxiliary valve that is mechanically operated by the main piston upon the completion of the stroke. In some forms of pumps the main piston and auxiliary valves are combined. In duplex pumps one engine operates the valve gear of the other; that is, reversal of one side is effected by the other side before reaching the end of its stroke. In modern pumps, besides dispensing with a flywheel for reversing, single as well as duplex pumps can be started at any point of the stroke and can be operated as slowly as desired.

Reduction of Plate Thickness at Circumferential Joints—Why is it necessary to plane down that part of boiler plates forming the laps of the circumferential joints of horizontal return-tubular boilers when the thickness of plates exceeds $\frac{9}{16}$ in.? Does this not reduce the strength of the joint? P. F.

The portion of the plates forming the laps of the circumferential joints where exposed to the fire or products of combustion, should be planed or milled down to ½ in. in thickness to render the plate thin enough for the plates and rivets at the joints to be kept sufficiently cooled by the water of the boiler to prevent their becoming burned. While this reduction of thickness lowers the strength of the joint, it may be reduced to only 35 per cent. of the strength of the longitudinal joints, provided 50 per cent. or more of the load that would act on an unstayed solid head of the same diameter is relieved by the effect of tubes or through stays, and in that way plates of unusual thickness and longitudinal joints of highest efficiency may be employed with circumferential joints planed down to ½ in. in thickness without reduction of working pressures below those appropriate to the strength of the longitudinal joints.

Removing Oil from New Boilers—How can new boilers be boiled out to get rid of considerable oil and grease in them? Will oil and grease left inside of the boilers by the boilermakers have any injurious effect in operation? L. E. G.

A simple method of removing grease and oil is to boil off the grease with about one-tenth pound of soda ash to a horsepower of boiler capacity. Pulverized soda ash can be introduced through any opening above the water level in the boiler when the gage pressure is zero. After the soda ash is added, fill the boiler nearly full of water and, with the safety valve raised so there cannot be accumulation of pressure above that of the atmosphere, hold the water to the boiling point with a slow fire for about 24 hours, and then draw off the boiler and thoroughly wash the interior with clean water and wash out the safety valve and other valves and connections, to remove all gumming or other traces of the soda ash. More than one such boiling off and washing may be necessary for removal of unusual quantities of oil and grease. Generally, the amount left by the makers is not enough to cause boilers to become burnt, but if not boiled off, the oil is likely to cause priming and trouble from leaks that may be started by the oil or prevented by its presence from becoming rusted tight.

Raising Water by Movable Bent Tube—What method is used in calculating the height to which water can be raised by an object passing through the water, such as a locomotive traveling at the rate of fifteen miles per hour, taking water from a track tank? D. J. J.

The theoretical height, or height to which standing water would rise in a bent tube with the open end presented in the direction of its motion, without frictional resistance, is given by the formula:

$$h = \frac{v^2}{2g}$$

where h = height in feet, v = velocity in feet per second, g = acceleration of gravity = 32.174 (ordinarily taken = 32.2).

For a velocity of fifteen miles per hour we would have $v = (15 \times 5280) \div 60 \times 60 = 22$ ft. per sec., and by substituting the values of v and g in the formula, the theoretical height is found to be

$$h = (22)^2 \div 2 \times 32.2 = 7.5 \text{ ft.}$$

But when a flow takes place, part of the pressure at the submerged end of the tube is lost in overcoming friction at the entrance of the tube and along its length, the amount depending on the form of the entrance and the form, length and roughness of the tube, the number and character of bends and rate of discharge, all tending to reduce the actual below the theoretical height. These sources of loss usually are so complex that the height to which water can thus be raised with a given rate of discharge can be determined only by trial under the actual conditions.

Condensed-Clipping Index of Equipment

Clip, paste on 3 x 5-in. cards and file as desired

Condenser, New Compartment
Wheeler Condenser and Engineering Co., Carteret, N. J.
"Power," Nov. 22, 1919.

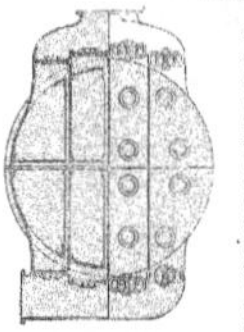

This condenser is divided into four compartments, each being equipped with a set of valves to control the circulating water. This permits any tube or tubes to be temporarily plugged and other repairs made without taking the condenser out of service. To clean while the turbine is delivering full power, the operator shuts the water off from one compartment, removes the cover, cleans the tube and replaces the cover, turns on the water again and then passes to the next compartment. The operation is repeated until all are clean. While one compartment is being cleaned the other compartments temporarily take over the turbine load.

Stoker, Illinois Forced-Draft Chain-Grate
Illinois Stoker Co., Alton, Ill.
"Power," Nov. 4, 1919.

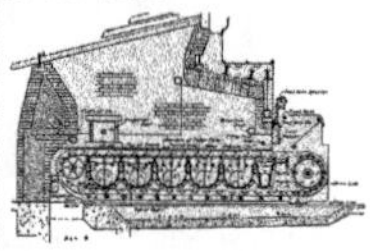

This stoker, of the chain-grate type, is fitted with U-shaped tuyere boxes that extend across the furnace. Air is admitted to the tuyere boxes from a windbox through a cast-iron sleeve, the air being controlled by hinged dampers or by cast-iron dampers operating through the side wall and located in the top of the tuyere boxes. They are operated from the outside of the boiler setting. Longitudinal girders on skids carry the grates and maintain the surface of the grate level and reduce the amount of drippage and eliminate the use of rollers. Stokers 10 ft. wide and over are provided with double drives. Siftings are removed from the tuyeres by means of a steam- or air-jet blower.

Setting, Uniflow Boiler
Uniflow Boiler Co., Inc., Philadelphia, Penn.
"Power," Nov. 4, 1919.

This furnace and setting is designed for each individual boiler, and drawings are furnished with each installation. Air enters through the bridge-wall in such a way that it is not preheated and smokeless combustion and secondary ignition from the fuel result. An inverted arch extends from the bridge-wall to the rear of the boiler, and the gases are made to impinge against the shell with a scouring action which keeps that portion of the shell clean. The construction of the setting prevents internal expansion from disturbing the entire brickwork. Red brick is retained as a structural element, and the firebrick functions as the heat-insulating element alone. With this setting the boiler is suspended from I-beams supported by the red brickwork.

Regulator, Hagan Combustion Control
Hagan Corporation, Peoples Bank Building, Pittsburgh, Penn.
"Power," Oct. 21, 1919.

This system consists of a regulator and a rotor reciprocating valve. The regulator controls the damper, the rotor valve, the steam to the stoker and fan engine. Both work in unison. The regulator is fitted with an auxiliary air cushion and a secondary diaphragm by means of which the mechanism is made to take an incremental movement for variations in rate of flow of steam pressure between two extremes. The sensitiveness of the apparatus is secured by varying the pressure in the air chamber and by changing the angularity of an adjustable compensator. These adjustments make it possible to secure a full stroke of the regulator with a differential pressure as small as 1 oz. or as great as 15 lb. By means of a bypass valve, the regulator can be thrown into its wide-open position and kept there independent of the flow of steam, or the bypass valve can be thrown in a reverse position and the regulator moved to a closed position and held there. The rotor valve has a combined reciprocating and rotating motion, the reciprocating motion being imparted to the valve by the regulator and the rotating motion adjusted from the outside of the valve.

Deoleizer, Oily Condensate
Oleite Corporation, 95 William St., New York City
"Power," Oct. 21, 1919.

This apparatus depends upon absorption by a property called "Oleite." It consists essentially of four parts, the base, the shell, the head, and the cover. It is divided into four spaces by perforated diaphragm. The oily returns of condensate enter at the top and the deoleized water leaves near the bottom through a similar nozzle. The oleite, with an intermediate cross-screen, rests on a diaphragm of noncorrosible metal. Any oleite that passes the perforated diaphragm sinks through the dead-water space at the base to the pocket below. Oleite has an affinity for oil, and as the water passes through the bed the oil is absorbed by it and the water passes to the boiler in a refined state. The oleite can be used over and over by burning the oil out of it, which makes it as good as new.

Cleaners, Atlas Boiler Tube
Atlas Manufacturing Co., San Francisco, Calif.
"Power," Nov. 22, 1919.

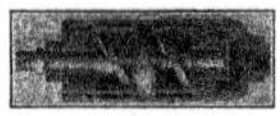

Cleaners are made for return-tubular and for water-tube boilers. The first type of cleaner operates with either air or steam; the other uses either water, air or steam. The plunger type of cleaner is made for 2-in. up to 6-in. tubes and has two plungers operating at right angles, delivering approximately 10,000 direct taps per minute. The vibrations loosen the scale. The cleaner for water-tube boilers consists of a rotor having no dead-centers. Pressure causes the rotor to revolve and the air, water or steam escapes through suitable ports to the front of the cleaner and around the cutter head.

Cement, "Hot Patch" Furnace
F. Obermayer Co., Pittsburgh, Penn.
"Power," Sept. 30, 1919.

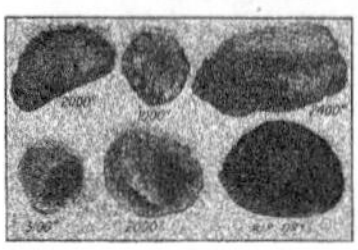

"Hot Patch" furnace cement is a new idea in high-temperature cement that will remain permanently adhesive at high temperatures. It expands and contracts closely with the brickwork, is applied thinly, and upon drying becomes hard and forms a highly refractory wall. The cement is made up with a new basis of green-sand binder; that is, a vegetable bond is used. This vegetable or inorganic material does double work in that, on burning out, it carbonizes in the center, taking care of expansion and contraction and leaving the outside of the cement a fused body capable of withstanding a temperature of 3100 deg. F. This vegetable compound is of plastic nature. When applied, the bricks are laid with a close joint and the work is then covered with a coat of cement applied with a stiff brush to the face of the brickwork.

Pump, The Rollway
Michigan Machine Co., Detroit, Mich.
"Power," Oct. 7, 1919.

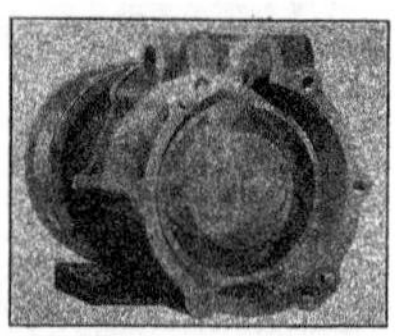

This pump consists of two rollers that rotate eccentrically in the pump chamber. The pump shaft is square. The openings in the roller fitting the shaft are rectangular and so placed in the shaft that it is slightly off center, bringing one side of the roller almost in contact with the side of the pump chamber. As the rollers revolve this contact continues, but at a different point. After the contacting point passes the intake of the pump, it creates a suction gradually filling up the chamber in back of the roller. When this point again reaches the top of the chamber the liquid passes out through the discharge.

Patented Aug. 20, 1918

Worcester Section A. S. M. E. Discusses Industrial Power

ON TUESDAY evening, Dec. 16, 1919, Reginald J. S. Pigott talked to the Worcester Section of the American Society of Mechanical Engineers on Industrial Power. There is a great and going movement, he said, in favor of centralization in power production, not only in the supply of numerous small plants from the local central stations, but in the tying in of the large çentral stations themselves.

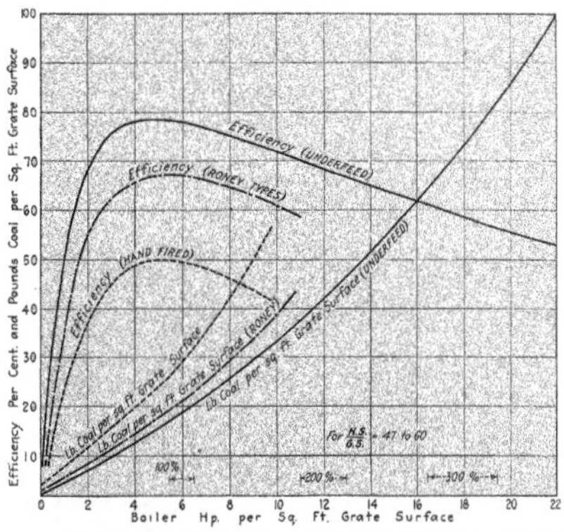

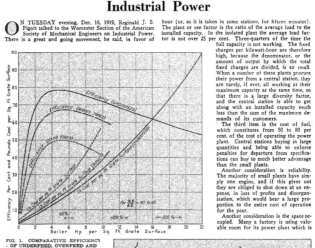

FIG. 1. COMPARATIVE EFFICIENCY OF UNDERFEED, OVERFEED AND HAND STOKING WITH DIFFERENT RATES OF FIRING

The lower curves show the amount of coal burned per square foot of grate surface at the corresponding capacities and efficiencies. The upper curves show the efficiencies when the number of pounds of coal indicated by the ordinates produce the capacity indicated by the abscissas.

In considering whether it is worth while to install or continue to run an isolated plant, one of the first factors to be considered is its size. This affects the thermal efficiency, the Rankine cycle efficiency, and the number of British thermal units per kilowatt-hour, which is becoming more and more the unit by which such stations are rated. A 500-kw. turbine is able to run with a Rankine efficiency of 50 or 51 per cent., a 5000-kw. turbine from 60 to 62 per cent. and a 30,000-kw. turbine as high as 75 per cent.

The second factor affecting the economy of such a plant is the load factor, or the average load of the plant divided by the maximum load for one hour (or, as it is taken in some stations, for fifteen minutes). The plant or use factor is the ratio of the average load to the installed capacity. In the isolated plant the average load factor is not over 25 per cent. Three-quarters of the time the full capacity is not working. The fixed charges per kilowatt-hour are therefore high, because the denominator, or the amount of output by which the total fixed charges are divided, is so small. When a number of these plants procure their power from a central station, they are rarely, if ever, all working at their maximum capacity at the same time, so that there is a large diversity factor, and the central station is able to get along with an installed capacity much less than the sum of the maximum demands of its customers.

The third item is the cost of fuel, which constitutes from 50 to 80 per cent. of the cost of operating the power plant. Central stations buying in large quantities and being able to enforce penalties for departure from specifications can buy to much better advantage than the small plants.

Another consideration is reliability. The majority of small plants have simply one engine, and if this gives out they are obliged to shut down at an expense, in loss of profits and disorganization, which would bear a large proportion to the entire cost of operation for the year.

Another consideration is the space occupied. Many a factory is using valuable room for its power plant which is

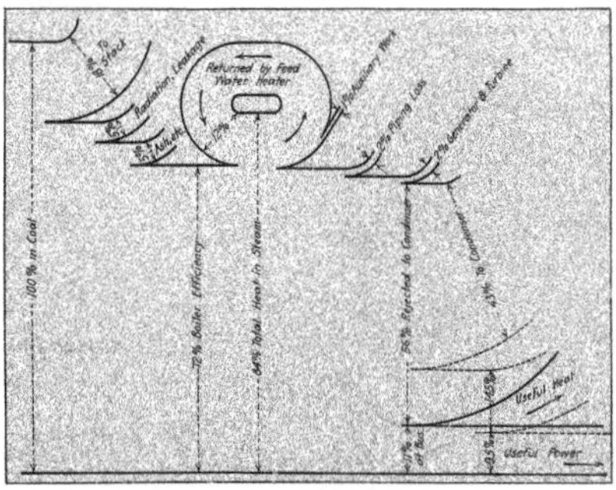

FIG. 2. HEAT-FLOW DIAGRAM

Starting with 100 per cent of the heat from the coal on the left-hand side, 18 per cent goes to the stack in the heat of the gases; 5 per cent is lost by radiation; 5 per cent in the ash pit; 13 per cent goes to the auxiliaries, of which 12 per cent is returned by the feed-water heater, leaving 1 per cent to go into the work of the auxiliaries; 2 per cent is lost in the piping; 2 per cent in the generator and turbine; 56 per cent is rejected to the condenser (of which 14.5 per cent may be recovered for useful purposes); 11 per cent goes to the busbars and 9.5 per cent comes off as useful power: this is for the industrial plant using low pressure steam for heat and processing.

making power for perhaps 10 per cent. less than it could be purchased for from the central station, which space might be employed in manufacturing the goods for which the factory is operated, and which would probably pay a profit of two or three times that made by the power plant.

Heat is the main reason why the isolated plant can justify itself on a basis of cost. If a large proportion of the exhaust steam can be used for heating or manufacturing processes, the cost of the power will be greatly reduced.

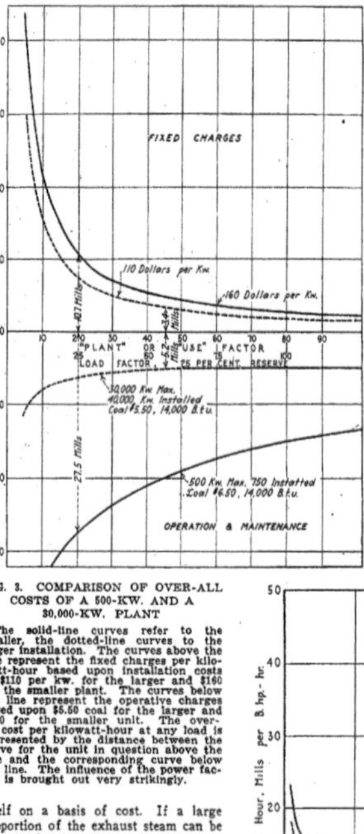

FIG. 3. COMPARISON OF OVER-ALL COSTS OF A 500-KW. AND A 30,000-KW. PLANT

The solid-line curves refer to the smaller, the dotted-line curves to the larger installation. The curves above the line represent the fixed charges per kilowatt-hour based upon installation costs of $110 per kw. for the larger and $160 for the smaller plant. The curves below the line represent the operative charges based upon $5.50 coal for the larger and $6.50 for the smaller unit. The over-all cost per kilowatt-hour at any load is represented by the distance between the curve for the unit in question above the line and the corresponding curve below the line. The influence of the power factor is brought out very strikingly.

The cost of installation of a small plant, per kilowatt installed, is much greater than that of the large plant. It costs, roughly, $160 per kilowatt to install a 500-kw. plant. (It could be done for $80 or $85 a few years ago.) A 30,000-kw. plant can be put in for about $110. The larger the plant the less installation cost and consequently less fixed charges. In fixing depreciation, obsolescence, which is simply accelerated depreciation, must be taken into account, so that the life of a plant cannot be estimated upon the number of years which the machinery may be physically expected to last. The fixed charges for gas and oil plants are taken at about 14 per cent., 6 per cent. interest, 1½ per cent. taxes and insurance, 6½ per cent. depreciation and obsolescence; those for turbine plants at 12 per cent., 6 per cent. interest, 1½ per cent. taxes, 4½ per cent. depreciation and obsolescence. Whether the values given for the last items are high enough is still an open question. Take as an example the Seventy-fourth Street Interborough Manhattan type engines installed in 1900. They were taken out and scrapped in 1913, to be replaced by 30,000-kw. turbines. They have been maintained for thirteen years in 100 per cent. condition, but went on the scrap heap because the superior economy and low labor cost of the new turbine rendered the operation of the engines an economic loss, and nobody would buy them as engines; so their value depreciated in thirteen years from $24 per kilowatt to $1.35. In the small plant the labor, as well as the fuel item, is greater.

Reference to Fig. 3 will show the large effect of use factor on the fixed charges entering into the total cost per kilowatt-hour.

Of the heat units in coal 18 per cent. went to the stack, 5 per cent. to radiation and leakage, 5 per cent. in the ash and unburned combustible, and 72 per cent. on an average into the steam.

Another consideration which entered into the decision as to whether one should put in an isolated plant is the local cost of power. The boiler room is the place to make money.

Mr. Pigott showed the diagram illustrated in Fig. 1, in which the boiler horsepower per square foot of grate surface is plotted against the efficiency and the coal burned per square foot of grate surface. The Interborough Station increased its efficiency 12 per cent. by changing from over- to underfeed stokers. The stoker pays even in small plants. If it does not save labor, it saves coal, and coal is more costly than labor. Eleven to twelve boilers per stoker operator and helper is the practice of the Interborough. The maintenance cost on the old Roney overfeeds was from 11 to 12c. per ton of coal fired, and 2 to 3c. for underfeed stokers. By this is meant furnace cost, the keeping up of the brickwork, etc.

Four-dollar coal costs $0.1265 per million B.t.u., and 4c. per gal. oil 29c. per million B.t.u. Oil has little advantage in efficiency; that is, a furnace fired by oil is not likely to be much, if any, more efficient than that fired by a well-run stoker.

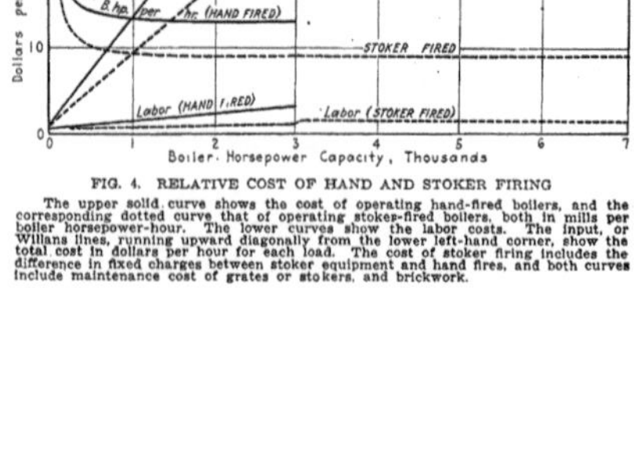

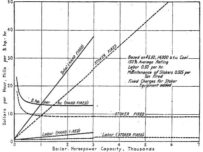

FIG. 4. RELATIVE COST OF HAND AND STOKER FIRING

The upper solid curve shows the cost of operating hand-fired boilers, and the corresponding dotted curve that of operating stokes-fired boilers, both in mills per boiler horsepower-hour. The lower curves show the labor costs. The input, or Willans lines, running upward diagonally from the lower left-hand corner, show the total cost in dollars per hour for each load. The cost of stoker firing includes the difference in fixed charges between stoker equipment and hand fires, and both curves include maintenance cost of grates or stokers, and brickwork.

For an equivalent heat value, oil ought to be bought at one-half as many cents per gallon (roughly) as coal costs dollars per ton; for example, if coal is $5 per ton, oil ought not to be more than 2½c. per gallon, to break even on cost per million B.t.u. The reasons for the adoption of oil in New England will be other than economy of fuel in general.

One of the principal troubles in the use of powdered coal is getting rid of the slag or melted ash. Another is the enormous size of furnace required. The old practice was 6 or 7 cu.ft. per sq.ft. of grate. With the modern stoker 11 or 12 cu.ft. are used, while 25 to 30 cu.ft. are at present required for powdered coal. Mr. Pigott then spoke of the colloidal fuel made by suspending powdered coal in fuel oil by means of a fixator, as was done on the steamer "Gem" of the Submarine Defense Association. He said that there might be good possibilities in this fuel, but as yet nobody has been educated to use it, and it is still in the experimental state. It will probably be somewhat cheaper than straight oil fuel.

The fuel situation today calls for the saving of every pound of steam, every pound of fuel, and every foot-pound of power possible.

Mr. Pigott then discussed some of the possibilities of compressed air as a subsidiary service from the industrial power stations.

One way of saving power is by the more intelligent installation of motor drives. The usual procedure is to take the builder's catalog as an authority for the power consumption of the various machines, add the maximum power required by all of the machines in a group, and then put in a motor large enough to drive them all at their maximum demand, taking no account of the diversity factor. It is seldom that all of the machines will be calling upon the motor for their maximum power at one time, and induction motors can be run at 25 per cent. over rating for a while. A little attention to this matter would result in a smaller investment for motors and an improvement in the power factor.

Of the other sources of power available the hydraulic plant is a special case. Few isolated plants have their own water power.

The higher local cost of oil fuel generally militates against the Diesel engine, and the fixed charges are very much higher than for a steam plant. Pre-war figures were $65 to $75 per kw. for turbine plants, as against $110 to $125 per kw. for the Diesel plants. The reliability of the Diesel engine is not ideal, and there is a dearth of trained attendants; higher skill is unquestionably required to operate it.

The Ford gas-steam plant was referred to as a freak. It occupies six or seven times as much space as would be required for an equivalent turbine plant, and while such a plant might be designed to be very economical in heat units, it could not be made to balance under varying loads, and the high fixed charges will more than overbalance operating savings.

The talk was illustrated with diagrams, some of which are here produced with explanatory notes.

Electrical Drives for Steel Mills

At the Engineers' Club in Philadelphia, Penn., the Philadelphia Section of the Iron and Steel Electrical Engineers held its regular monthly meeting on Jan. 3. The meeting was preceded by a dinner served at the club. At 8 p. m. F. H. Woodhull, chairman of the Philadelphia Section, called the meeting to order, and after announcing that R. B. Gerhardt, electrical superintendent of the Sparrows Point Plant, Bethlehem Steel Co., would present a paper at the next meeting, introduced J. D. Wright, who presented his paper on "Electrical Drives for Steel Mills." Mr. Wright, by the use of a large number of lantern slides, discussed the characteristics of the load produced by, and the capacity of motor required to drive continuous bar mills, merchant mills, sheet and tin mills, plate mills, and rail and structural-shape mills.

The question of 60-cycle versus 25-cycle motors was given considerable attention during the discussion. Mr. Wright said that an estimate made of the cost of 50 slow-speed 60-cycle motors and the same number of 25-cycle motors of the same speed and capacity for direct connection to main-roll drives, showed an average of 60 per cent. higher cost for the 60-cycle motors. For geared motors the cost averaged 4 per cent. higher for the 60-cycle equipment of the same size and speed. In the case of two identical equipments one 60-cycle and the other 25-cycle, the 25-cycle motor had a power factor of 82 per cent. against 65 per cent. for the 60-cycle equipment. To improve the low power factor that is inherently a characteristic of very slow-speed 60-cycle motors, it is necessary to gear the motor to the rolls. With motors of 1000-hp. and above, operating at 250 r.p.m. and less, the 25-cycle machine was to be preferred.

The use of synchronous motors was suggested as a possible drive for certain classes of mills as a means of eliminating the low power factor, the motor to be connected to the mill through a magnetic clutch so designed that the clutch would slip before the load reached the pull-out torque of the motor. The chief objections raised to this was the use of a magnetic clutch. It was the consensus of opinion that where clutches had been used in steel mills, they had given considerable trouble even on auxiliary drives. Furthermore, when a billet cobbles in a mill, it is frequent practice to reverse the motor and back it out of the mill, and the synchronous motor would not be applicable to this practice. There is one brass rolling mill in this country driven by a synchronous motor, but no steel rolling mill. It was brought out that in one case a synchronous motor was considered as a drive for a tin mill on account of the bad power factor of this type of load, but after careful consideration it was recommended against at that time, although a large rolling mill in England is driven by a synchronous motor.

For mills rolling rails and structural shapes, such as I-beams from 24 in. down to 10 in., variable speed is necessary. In this case the power required to drive the rail mill will determine the size of the motor.

Although the question of placing the flywheel on the pinion or gear shaft, in geared motor drives, came up for discussion, no definite opinion was expressed on the subject. It was thought by some that in certain cases of motor shafts breaking, it was traceable to the flywheel being placed on the motor shaft.

In answer to a question on the power consumption per ton of steel rolled, Mr. Wright said that this varied from as low as 6 kw.-hr. per ton for bar mill to as high as 200 kw.-hr. per ton for sheet mills. However, sheet mills are in operation, the power consumption of which is as low as 100 kw.-hr. per ton. The speaker pointed out that the friction load on a sheet mill was very high, amounting in some cases to 50 per cent. of the total power consumption; therefore for efficient operation with this type of mill it was necessary to watch the friction load and keep it as low as possible.

Shortage of Engineering Specialists

A veritable stampede of buying engineering services characterized the year 1919 as viewed by the service department of the American Association of Engineers. Employers are starting this year under pressure of unprecedented demand for technical men to increase production. The acute shortage of specialists in engineering looms larger in American industries just now than does the fate of the Peace Treaty. Information is received from many sources of professional engineers attaining the distinction of leadership, which they won under difficulties of production, transportation and labor, some of which at the time of their development seemed insurmountable.

Manufacturing has not attained the volume that would have been possible but for labor and fuel handicaps, but the total of production has been large and the demand for engineers is increasing steadily. The technical man has never been upon a higher plane of esteem or on a more satisfactory basis than at the present, and this is resulting in higher professional remuneration.

The anxiety over the question of employment for the large number of engineers returning from military and naval service has disappeared. Conditions have taken care of that promptly and comfortably and without overcrowding the market except for field engineers and superintendents of construction. The present demand is for engineering draftsmen, especially designers with experience in mechanical, structural or concrete work. Location is mostly in the Central States, particularly Ohio, Minnesota and Michigan.

High-Speed Turbine Gears

A paper on "High-Speed Turbine Gears" was read before the Manchester Association of Engineers on November 22, 1919, by Prof. Gerald Stoney, M.A.I., F.R.S. After a few remarks on the evolution of the steam turbine and its adaptations, the author remarked that in order to obtain silence and smooth running in a high-speed gear the utmost accuracy in the cutting of the teeth was requisite. He said there were three principal methods of generating such teeth: Cutting the teeth with an end miller, as in the Citroen gears; planing or shaping out the teeth, as in the Fellows and Sunderland methods; and cutting the teeth from a hob, which was really a cutter in the shape of a worm. The two latter processes had the advantage that the teeth were automatically generated from a straight-sided cutter. The last-named process was almost universally adopted, and it was used in the gear-cutting machines made by Muir & Co., of Manchester, who had supplied large numbers of such machines.

At first considerable trouble was caused by the fact that the master wormwheel, which rotated the gear wheel, was not quite accurate. This difficulty caused Sir Charles Parsons to devise what was known as the "creep," under which the gearwheel to be cut was caused to rotate at a slightly lower speed—about 10 per cent.—than that of the master wormwheel. By means of that device the errors were not only reduced to about one-eighth, but were distributed in spirals on the gearwheel instead of being parallel with the axis—a most important point. Even if the master wormwheel were true to begin with, it tended to wear out of truth, and the effect of such wear was minimized by the creep. As a result this addition to the ordinary gear-cutting machine had been very largely adopted, and of the gears at present in use in this country about 85 per cent. had been cut on machines fitted with the creep.

Helical gears for geared turbines, the author said, were invariably made with involute teeth, which had this advantage among others, that the centers could be slightly varied so as to give the desired clearance between the teeth, for it was found that this clearance had to be suitably adjusted so as to be neither too little nor too much. Another point was that such gears could be generated from a hob with straight sides; thus it was simple to manufacture them accurately. The angle of the hob varied somewhat in practice, but a very common tooth had the same angle for the hob as the Brown & Sharp tooth; namely, 14½ degrees. It was requisite that the bottom of the tooth be well rounded and have no sharp angles, and it was surprising that this precaution was so often neglected in gears of all classes. The angle of the first gears was about 23 degrees, but 45 degrees had been extensively used, for the larger angle had been found to conduce to silent running. There seemed now to be a general opinion, however, that 45 degrees was rather too large, and the angle now generally adopted was about 30 degrees.

With reference to pitch the author said that fine pitches in most cases gave more silent running at high speeds, and a normal pitch—that was to say, the pitch measured at right angles to the teeth of 0.583 in. (or 7/12 in.)—had been largely used in this country, with about 0.4 in. for small pinions or where extreme silence was of importance. In the United States larger pitches of 0.9 in. or even more were often used.

The minimum number of teeth in a pinion was 19, with normal addendum and dedendum, but 22 to 25 were preferable. A peculiarity of involute gear was, he said, that by varying the addendum and dedendum interference could be avoided, and thus smaller numbers of teeth than these could be used in certain cases.

The whole gear had to be supplied with an ample amount of oil, which was generally squirted from nozzles under pressure against the teeth where they came into contact with one another. This large supply of oil was necessary not only to lubricate the teeth, but also to carry off the heat that was generated. It was usual to have ¼-in. to $\frac{3}{16}$-in. nozzles of about 5-in. pitch, squirting oil under a pressure of 10 lb. to 20 lb. per square inch on to the line of contact between the gear wheels. The quantity of oil used was about one gallon per minute for each 100 to 150 hp. transmitted. The gear case had to be well drained, for if the wheels were allowed to dip into oil at the bottom, much heat would be generated by the churning of the oil. The power that could be transmitted by a pinion depended upon the allowable tooth speed, the allowable pressure per inch run, and the allowable distortion caused by the deflection and twist of the pinion, which had to be so small that the distribution of pressure on the teeth remained practically constant. Further, the stresses on the material had to be within safe limits. All these limits could be determined only by actual practice; and in this country a conservative policy was wisely decided on, and as experience was gained the allowable power transmitted by any given pinion was increased. One result of this policy had been that failures of gears were few and far between; in fact, the gear was quite as reliable as other parts of an installation, if not more so.

Tooth speeds up to 120 ft. per sec. were common and had in many cases been exceeded. What was exactly the limit of speed was uncertain, but so long as lubrication could be maintained there did not seem to be any reason why the figure given should not be considerably exceeded.

The pressure on the teeth was generally reckoned in pounds per inch run, and was of course the same whether per axial inch or per inch of tooth along the helix. It was determined by the safe limit of bending stress on the tooth and of the pressures at the points of contact, as well as by danger of the oil film failing. The bending stresses were in practice very low, and could in general be ignored. For similar crushing stress on the material of the tooth the pressure per inch run varied directly as the diameter; or $p = ad$, where p was the pressure per inch run in pounds, d the pitch diameter of the pinion in inches, and a a constant. Similarly, the stress on the oil film would obey some such law as $p = bd^n$. The whole theory of the lubrication of gear teeth was a most difficult one, and one that had been only partly worked out. Reference was made to a paper by H. M. Martin in which reason was given for the assumption often made that $p = b\sqrt{d}$. These two conditions meant that up to a certain diameter $p = ad$, and that above it $p = b\sqrt{d}$.

In this country Messrs. Parsons, who were the pioneers in these gears, very wisely resolved not to exceed $a = 80$ and $b = 175$; but as experience was gained b had been increased, first to 220 and now to 250, and other makers of gears seemed to have followed the same line closely. In America Mr. McAlpine had adopted a value for a of 105, and had ignored b. He claimed to be able to use these high pressures by the use of his floating frame gear.

The distortion of the pinion, the author said, was made up of two items—the twist of the pinion caused by the torque it transmitted and the bending caused by the pressure on the teeth. In estimating these factors, it might be assumed that the pressure on the teeth was uniform along the pinion—or, in other words, that the torque diminished uniformly from the driven end to the free end of the pinion. It might be further assumed that the effective diameter of the pinion was the pitch diameter.

There were two cases to consider—that in which there was no center bearing, and the pinion was supported from the ends, and the case in which there was a center bearing. The deflection of the pinion could be calculated in the usual way. It was found that in the case of a pinion without a center bearing the bending or deflection was the chief thing to be considered, while in the case of a pinion with a center bearing the twist was the chief thing. The allowable amount of distortion was very small, and in good design did not exceed one-thousandth of an inch. Of course the stresses on the material arising from bending and shearing had to be kept within safe limits, but these were generally quite low.

The alignment of the gears in the gear case was of the greatest importance, as also was the construction of the gear case. There were two principal types—the rigid gear case originally introduced by Sir Charles Parsons, which was the type generally adopted in this country; and the floating frame type introduced by John H. McAlpine in America. In actual practice the principal thing to be taken into consideration was the twist in the case of a pinion with a center bearing and the bending in one without such a center bearing. The exact angle that a floating frame of the McAlpine type would take up was dependent upon the distribution of pressure along the teeth. The general practice in this country was to have a rigid gear case, and it was found that the distribution of wear along the teeth was quite uniform, without the use of such an arrangement as the floating frame. In this respect it had to be remembered that the distortions in question were very small, never exceeding one-thousandth of an inch, and that the inevitable elasticity of a rigid gear box helped to diminish their effect. Besides, the oil film in the bearings was of considerable thickness, probably of the order of several mils. Little was known about the thickness of the oil film in a high-speed bearing; but there was evidence to show that it was very considerable, and that it must be fairly thick was further shown by the clearance that had to be given in a high-speed bearing. This clearance was from 2 to 3 mils per inch diameter, and thus was very large in comparison with the distortions to be dealt with. A very small variation in the thickness of the oil film would therefore compensate for the distortion of the pinion.

The rigid gear was the usual design adopted in this country, and the experience with some sixteen millions of horsepower of gears that had been constructed or were on order was that it worked excellently and was most reliable and efficient. On the other hand, the floating frame type of gear was used largely in America, and was said to give excellent results.—*The Engineer.*

Social Function of the Engineer

The New School for Social Research of New York has arranged for a series of weekly lectures on "The Social Function of the Engineer," to begin Feb. 2 and be given each Tuesday evening thereafter for twelve weeks at the school, 465 West 23d street, New York. The course will be conducted by Guido H. Marx, professor of mechanical engineering at Leland Stanford University. The theory upon which the course is based is that the production and distribution of goods is now fundamentally an engineering as well as a social problem. The engineer's function is to insure the full use and unimpeded operation of the material resources and industrial equipment of the nation, with a view to securing a larger volume of production and a more adequate distribution of the output. The course will be directed primarily to technical men, and its purpose will be to reach a common understanding of the existing industrial and social problems.

The New School of Social Research at which this course will be given was founded last year by a group of men who believed that there was educational need for an institution of the kind. Its purpose is to supply instruction of a post-graduate character, although an academic degree is not required for admission. Each student is assumed to be pursuing his particular studies for their own sake for his own aims, so there is no prescribed courses of study or fixed curriculum. There is also no grant of degrees.

Ontario A. S. M. E. Meeting

The American Society of Mechanical Engineers, Ontario, Canada, Section, held a meeting on Dec. 18, 1919, at the Engineers' Club. Before the paper of the evening was read, Prof. R. W. Angus, the chairman, gave a report of the convention in New York of the parent society. He was Ontario representative at the Sections conference held coincidently with the engineering convention and gave an interesting account of the progress made in section organization, showing that since the A. S. M. E. had begun to systematically develop in this way and capitalize the local interest of engineers in their home problems, it has stepped ahead till it now leads in membership all other national professional engineering societies. The keynote of the New York convention itself was organization for coöperation with labor, so as to cut the ground from under the feet of the agitating troublemaker by giving him no excuse for existence.

Mr. Quesnal read a paper prepared by himself and Mr. Lewis, both of the John Inglis Co., on "Thirty Years' Progress in Boiler Construction in Canada." Mr. Quesnal spoke of the shop practice of earlier days and traced the progress of improvements in methods and quality to the present day. He gave illustrations of the great speed of modern construction due not only to the large and powerful machine equipment but to the excellent small tools, mostly air-driven. These had revolutionized the practice of the old-time shops, even before the giant rolls and drilling machines became common. He spoke feelingly of the troubles that centered about the old-time flanging fire and of the various expedients that were used to get the head flat and the right size and contrasted this with modern hydraulic sectional and four-post machines. Another feature that amused the younger members was a description of old-time methods of dishing a tank head, by digging a concave hole in the dirt floor to act as a die under the plate, and the dropping on to the red hot plate, a large cast-iron ball which was manipulated by means of a rope wrapped about an overhead shaft. A variation of this, which involved the use of great wooden mallets swung by men standing around the circle of the plate, seemed even more antique.

No Shortage of Mexican Oil

From the Mexican Embassy comes the following official report, issued by the Department of Industry, Commerce and Labor:

The production of Mexican petroleum, far from having decreased, is progressively increasing. At the beginning of 1919 the potential daily capacity of all wells was 1,400,000 bbl. In the middle part of that year it was 1,800,000 and today it is over 2,000,000 bbl. Notwithstanding the fact that four petroleum wells have gone dry, the owners are using only 12 per cent. of the potential capacity, owing to the lack of vessels for transportation, a cause entirely apart from the Mexican petroleum policy, and the regulations of the government, which seeks only the compliance of its laws. The petroleum supply could be increased eight times if there were vessels, even without drilling new wells. The production of petroleum during 1918 was 64,000,000 bbl., and last year it was over 80,000,000 bbl.

In view of the foregoing information, the Mexican Embassy is in a position to deny emphatically the rumors which are being spread with the idea of making the American people believe that the Mexican petroleum industry is decreasing, whereas statistics prove the contrary.

The Mexican government is not preventing the production or the exportation of oil by any owners thereof, and while on four or six cases it has been deemed expedient to exact compliance from those who have disregarded the laws and regulations bearing on the boring of new wells, there is no cause to fear a shortage of Mexican oil supply attributable to any action of the Mexican government.

Light-Load Running Device for Semi-Diesel Engines

When a semi-Diesel crude-oil engine is run on light load, it consumes an extremely small quantity of fuel, and the heat produced is inadequate to vaporize the fuel oil for the succeeding injections and to give complete combustion. Messrs. Vickers-Petters, Ltd., have achieved a satisfactory solution, which, in addition, enables the engine to run at slow speed with perfect combustion and regularity of ignition. With the Petter patent light running device means are provided for giving a supplemental injection which is forced into the combustion chamber earlier than the normal injection. This is effected by giving an extra stroke to the fuel pump on single-cylinder engines, and on multi-cylinder engines a device is fitted whereby any fuel pump may be made to deliver fuel to two cylinders at each stroke, one portion of the charge being injected into the cylinder normally supplied and at the correct timing, and the other portion is injected in advance of the normal timing. In this way each cylinder is supplied with an advanced injection.

The earlier injection of the light charges gives more time for the vaporization of the fuel oil and maintains sufficient heat in the combustion chamber to produce perfect combustion, and the engine will continue to run at no load firing regularly without the aid of the starting lamp. The device also enables the engine to be run at low speed with the same result.—*The Electrical Review*, London.

Abatement of Corrosion in Central Heating Systems*

The main source of corrosion in hot-water or steam pipes has been traced to dissolved oxygen brought into the system either with the feed water or through leaks in return lines under a vacuum. Both wrought-iron and steel pipe are inherently subject to corrosion, and records seem to show practically no difference between reputable makes of either kind.

Referring to a paper by Speller and Knowland in the Proceedings American Society of Heating and Ventilating Engineers, Vol. 24, p. 217, it is pointed out that water in a liquid state may be considered as made up of positive hydrogen ions and negative ions. When this water comes in contact with iron, a number of minute electrolytic cells are established and a weak current flows from point to point. This action results in the removal of metallic iron from the positive points of the surface, or the "anodes," and the positively charged hydrogen ions are deposited on the negative points of the surface, or "cathodes." As soon as the positive hydrogen ions reach the cathode, they lose their charges and form a protecting coating which, when dense enough, stops further electrolytic action. If no oxygen is present this, so-called polarizing, coating remains on the cathode and no further rusting can take place. If free oxygen is present, however, the discharged hydrogen ions at once unite with it, leaving the cathode points clear for continued action. The presence of various salts in the water may help or hinder the action by their properties of increasing or decreasing the electric potential or voltage. The important point to remember is that corrosion is an electrolytic action and that it can be stopped by polarizing the cathode by any means whatever. The easiest general method seems to be by keeping oxygen out of the water.

Recirculation of the same water in the piping systems has been found to help unless a considerable proportion of makeup water, bringing with it a constant fresh supply of air, is introduced. Experiments in reducing the activity of the water by allowing it to come to rest several minutes at a temperature of 180 deg. F. in a vented tank result in the liberation of the bulk of the gases and in a material reduction in corrosion. The introduction of expanded steel lathing, and of a filter to clarify the water, eliminated practically all the oxygen at that point and after three years practically no corrosion was found in the piping.

Where exhaust steam from reciprocating engines is used, the coating of oil that usually forms on the inside of the return pipes gives considerable protection. In such a case a pure mineral oil with a paraffin base should be used. A case is cited where such an oil was introduced into a turbine to prevent corrosion of the blades. This resulted in considerable improvement, although it is admittedly a makeshift.

*Abstract of Technical Paper 236 of the Bureau of Mines, by F. M. Speller.

Artificial Daylight Achieved

The American Chamber of Commerce in London reports that a light has been perfected in Great Britain which far surpasses any existing arrangement of artificial light and is the closest approximation to actual daylight ever accomplished.

The apparatus consists of a high-power electric light bulb, fitted with a cup-shaped opaque reflector, the silvered inner side of which reflects the light against a parasol-shaped screen placed above the light. The screen is lined with small patches of different colors arranged according to a formula worked out empirically by Mr. Sheringham, the inventor, and carefully tested and perfected in the Optical Engineering Department of the Imperial College of Science and Technology.

The light thrown down from the screen is said to show colors almost as well as in full daylight. A test was made with such articles as colored wools, Chinese enamels, pastels and color prints, each being subjected successively to daylight, ordinary electric light and the new Sheringham light. Under the new light delicate yellows were quite distinct, indigo blues were blue, cobalts had their full value and violets lost the reddish shade which they display in electric light.

Outlook for Engineering Service

The demand for men of special experience is greater today than ever before since the service department has been in operation, so the American Association of Engineers reports. The greatest demand for men runs along mechanical lines, especially that of designers, but there is not an abundant supply of structural designers men familiar with reinforced concrete, power plants or electrical engineers. The salaries paid to new men are good, except with a few large companies. Most of the positions open at this time are east of Chicago, particularly in northern Indiana, Ohio and Pennsylvania.

Positions in sales engineering are to be obtained, but in most cases the firms desire a man with sales experience and engineering training. General business conditions viewed by the Service Department are improving. There is very little unemployment among technical men, the demand being far greater than the supply. Out of the 600 registered as open for positions less than 2 per cent. are out of employment.

Proposed $5,000,000 Central Station

It has been reported that the Hartford Electric Light Co., of Hartford, Conn., is planning to meet the power requirements for the greatly increased manufacturing in the city of Hartford within the next few years and has decided to erect and equip what is known as Meadows power plant, which, when completed, will have 100,000 kw. capacity.

The new plant will be erected and equipped in sections, each for the generation of 20,000 kw., the first section to be completed and in operation on or before the close of 1921, the others to be completed as rapidly as possible or as the new need of new manufacturing plants may warrant. It is understood that more than five million dollars will be spent by the company in carrying out its project.

Of the total 3,000,000 hp. of water power in Finland, about 1,050,000 hp. is available for industrial development. Of this only approximately 150,000 hp. is at present developed. There are 79 of these waterfalls with capacities ranging from 5,000 to 300,000 hp., 42 of which range from 5,000 to 20,000 hp.; 25 from 20,000 to 50,000 hp.; 6 from 50,000 to 100,000 hp.: and 6 from 100,000 to 300,000 horsepower.

Experimental Retort Test of Orient (Ill.) Coal

The following is from Technical Paper No. 134, of the Bureau of Standards. The Bureau, in connection with two special coke-oven tests, found it desirable to investigate the influence of the temperature of coking upon the quality of the coke produced and upon the quantity and quality of gas made from the coal. Although several very similar coals were used in the oven tests, the experimental retort work was limited to coal obtained from the Orient mine of the Chicago, Wilmington & Franklin Coal Co., Franklin County,

TABLE I. ANALYSIS OF THE COAL

	—Per Cent.—	
Volatile matter	37.8	38.0
Fixed carbon	52.3	53.4
Ash	9.9	8.6
Sulphur	.87	1.47
Phosphorus	.007	.002
Total carbon	73.0	76.3
Hydrogen	5.1	5.3
Oxygen	9.9	8.5
Ratio: hydrogen-oxygen	51.6	62.6
B.t.u. per pound (dry)	12,800	13,000
B.t.u. per pound (dry and ash-free)	14,200	14,200

TABLE II. SIZES OF CRUSHED COAL AS USED IN EXPERIMENTAL RETORT TEST

Mesh	Percentage Through Sieve	Mesh	Percentage Through Sieve
8	99.7	40	30.3
10	89.5	60	19.3
20	50.8	80	13.3
30	36.7	100	11.0

Ill. The coke was fairly good, and the charge was thoroughly coked, showing considerable contraction in the sample box. The pieces were of medium size with no evidence of sponge and only a small amount of fine pebbly material on top of the coke. The coke had very small and regular cell structure, but was soft, shattering rather easily, having a longitudinal fracture and showing a tendency to finger. It was light in weight and of a dark silver color.

TABLE III. EXPERIMENTAL RETORT TESTS OF ORIENT COAL

	Test No. 1	2	3	4	5
Temperature average T_1 (deg. C)	840	775	700	600	605
Weight of dry coal (pounds)	4.48	4.00	4.00	4.00	3.20
Gas generated (cubic feet at 30 inches 60 deg. F.)	26.3	18.5	13 6	8.5	7.8
Gas per pound of dry coal (cubic feet)	5 87	4.62	3.40	2.12	2.44
Gas per net ton of dry coal (cubic feet)	11,750	9250	6800	4250	4850
Candlepower of gas	5.4	10.0	12.2	15.7	15.9
Candle-feet of gas per pound of dry coal	31.7	46.2	41.5	33.3	38.8
Specific gravity of gas	.465	.510	.625	.660	.655
Analysis of gas:					
CO_2	4.7	5.1	8.2	7.2	8.2
O_2	1.0	1.3	1.0	.8	1.6
Illuminants	2.2	3.1	3.8	5.0	5.2
CO	17.3	13.6	11.3	8.1	8.3
CH_4	25.1	30.6	36.6	37.1	39.5
H_2	46.3	44.7	33.6	30.3	26.0
N_2	3.4	1.6	5.5	11.5	11.2
Heating value, calculated from analysis (B.t.u.)	510	570	600	610	625
B.t.u. in gas per pound of dry coal	3000	2630	2040	1300	1520
Coke formed:					
Large (pounds), dry	2.99	2.66	2.79	2.72	2.19
Loose material (pounds), dry	0.04	0.06	0.17	0.21	0.19
Total (pounds), dry	3.03	2.72	2.96	2.93	2.38
Coke yield:					
Large (per cent. of dry coal)	67.0	66.5	70.0	68.0	68.5
Loose material (per cent. of dry coal)	1.0	1.5	4.0	5.0	6.0
Total (per cent. of dry coal)	68.0	68.0	74.0	73.0	74.5
Coke analysis (dry basis):					
Volatile	5.3	7.1	7.1	11.5	14.0
Fixed carbon	80.2	79.1	....	....	74.2
Ash	14.5	13.8	....	....	11.8
Sulphur	.78	.78	....	....	1.23
Phosphorus	.014	0.14	....	....	.004

Test No. 2 was made at as nearly as possible constant vapor temperature, T_1, of 775 deg. C.

Obituary

Charles Keith Blackwood, vice-president, assistant treasurer and a director of the Sullivan Machinery Co., of Chicago, died on Dec. 14, 1919.

Personals

Arthur M. Crane, formerly general sales manager, has recently been appointed general manager of the New York Continental Jewell Filtration Co., Nutley, N. J.

Hugh L. Siegel, formerly general sales manager for the Ford Roofing Products Co., is now with Walter A. Zelnicker Supply Co., as assistant to the president.

J. W. Ledoux has resigned as chief engineer of the American Pipe and Construction Co., to open consulting engineering offices at 112 North Broad St., Philadelphia, Penn.

Burt A. Waltz, formerly assistant mechanical engineer for the Osborn Engineering Co., Cleveland, resigned to become chief engineer of the Portage Rubber Co., Barberton, Ohio.

W. P. Starkey, formerly engineer of the Harrisburg (Penn.) Pipe and Pipe Bending Co., and general superintendent for the last six years, has resigned his position, but will continue his interest in the company as a member of the board of directors.

Rol B. Woolley, who has been director of publicity of the Society for Electrical Development, Inc., for the past two years, is now with Thos. F. Logan, Inc., Advertising Agency, New York City. Mr. Woolley, who has been seventeen years in sales and advertising capacities, joined the society after service overseas; he will specialize on electrical and merchandising accounts.

Francis H. M. Riley, formerly associated with Vaughn & Meyer, consulting engineers, of Milwaukee, announces his appointment as Wisconsin representative for the Vulcan Soot Cleaner Co., Du Bois, Penn.; Vulcan Fuel Economy Co., Chicago; Green Engineering Co., East Chicago, Ind. His offices are at 614-615 Security Building, Milwaukee, Wis.

I. F. Baker, of the Wesinghouse Electric International Co., who has been located in the New York office of the company for the past two years, is now on his way to Tokio, Japan, where he will act as a special representative of the Westinghouse International Co. Mr. Baker entered the Graduate Student Course about 1908, later serving in various capacities in the resale and contract sections of the sales department. In 1913 he was transferred to the Foreign Department, which later became the Westinghouse Electric International Co.

George M. Brill has become established as a consulting engineer in New York City, with offices in the Singer Building. Mr. Brill's engineering experience has been wide and varied. He has been successively chief engineer of the Solvay Process Co., general engineer of Swift & Co., consulting industrial engineer in Chicago for a number of years, consulting mechanical engineer of the Guggenheim interests, commissioned as major in Ordnance Reserve Corps, and in the service was in charge of plant facilities; resigned commission to take charge of requirements for the Emergency Fleet Corp., of which he was later consulting engineer. He has also designed and supervised the construction of several high-tension transmission lines.

Edward D. Kilburn, who since March 15, 1917, has been New York district manager of the Westinghouse Electric & Manufacturing Co., was recently elected vice-president and general manager of the Westinghouse Electric International Co. Mr. Kilburn was graduated from Cornell University, and immediately after leaving college entered the employ of the Westinghouse Electric and Manufacturing Co., for a number of years being located at the Syracuse office of the company. Subsequently he was transferred to the Westinghouse Machine Co., with headquarters at New Haven, Conn. In 1915 he returned to the Electric Company as manager of the power division of the New York office. A year later he was also made manager of the railway and lighting divisions, subsequently becoming manager of the office.

Engineering Affairs

The American Society of Mechanical Engineers, St. Louis Section, will hold a meeting on Feb. 20, and at a dinner at the Midland Valley Country Club C. B. Lord, mechanical engineer of the Wagner Electric Co., will talk on "Industrial Fatigue."

The Association of Iron and Steel Electrical Engineers, Philadelphia Section, will hold a meeting on Feb. 7. R. B. Gerhardt, superintendent of the electrical department of the Bethlehem Steel Co., Sparrows Point, Md., will present a paper on the "Electrical Features of a Modern Mill."

The Manufacturers' Aircraft Association will hold its second annual aëronautical exposition in the Seventy-first Regiment Armory, 34th St. and Park Ave., New York City, Mar. 6-13, at which will be exhibited commercial airplaines, planes for private use, planes for transportation of freight and mail, etc.

The American Society of Mechanical Engineers, St. Louis Section, will hold a joint meeting with the Associated Engineering Societies of St. Louis on Jan 28, at which A. J. Brandt, engineer of the St. Louis Manufacturing Co., will deliver an illustrated paper on the "Construction of the Old Hickory Smokeless Powder Plant at Nashville, Tenn."

The American Society of Mechanical Engineers announces the following Section meetings: Jan. 21, New Haven, Conn. Joint meeting with the Winchester Engineering Club at the Mason Laboratory, Yale University: Subject: "Heat Treatment of Steel." Jan. 22, Boston, Mass., at the Boston Engineers' Club. Subject: "Research." Speakers: Dean John R. Allen and Prof. A. M. Greene. Jan 23, Schenectady, N. Y., at Edison Hall. C. E. Johansson will talk on "Gages." Jan. 27, Atlanta, Ga., in Carnegie Library reading room. Charles M. Rogers will talk on "Oil as a Fuel." Jan. 27, Philadelphia, Penn., at the Philadelphia Engineers' Club. Norman Reinicker will talk on "Modern Boiler Room Practice." Jan. 27, Cleveland, Ohio, in the rooms of the Cleveland Engineering Society, Roof Garden, Hotel Statler. Subject will be announced later.

The Engineers' Club of Philadelphia will hold a meeting on Jan. 22, at which the Society of Automotive Engineers, Philadelphia Section, will hold a joint meeting with the New York Section. The afternoon session will be devoted to making a special trip to the Philadelphia Navy Yard to inspect a Diesel engine taken from a German submarine interned at the navy yard. The evening session will be held at Kugler's. Hubert C. Verhey, head of the Diesel Engine Unit of the Emergency Fleet Corp., will talk on "Modern Practice in Heavy Oil Marine Engines and Possible Developments." There will also be discussions by well-known builders of Diesel engines.

Business Items

The Massachusetts Blower Co. announces that it has opened an office in the Kimball Building, 18 Tremont St., Boston, Mass., with A. C. Bartlett in charge.

The Nickle Engineering Works, Saginaw, Mich., is installing additional machine shop equipment to increase the scope of its operations.

Grinnell Co., Inc., Providence, R. I., has taken over all of the sales and contracting business carried on for years by the General Fire Extinguisher Co.

H. W. Johns-Manville Co. announces the removal of its Des Moines office to more modern quarters at 213 Ninth St., of which W. B. Roberts is in charge.

The American Steam Conveyor Corp., Chicago, announces that the Atlas Machinery and Supply Co. is now handling the sale of its conveyors in the St. Louis territory, with offices at 1416 Syndicate Trust Building. Wm. H. Patton is in charge, and has associated with him his brother, W. R. Patton.

The Westinghouse Air Brake Co., Wilmerding, Penn., in order to provide facilities adequate to handle the increasing export business and to develop its foreign trade to a greater extent, has organized an export department with headquarters in the Westinghouse Building, Pittsburgh, Penn., with E. A. Craig as export manager. Mr. Craig has been associated with the Westinghouse Air Brake Co. for thirty-two years, successively as auditor and assistant secretary and Southeastern manager of the company. This department will be represented in the New York office by W. G. Kaylor and in South America by R. M. Oates.

Pratt & Cady Co., Inc., Hartford, Conn., announces the following appointments: E. Coit Magens, director of sales; O. Lamson Beach, manager, Metropolitan Store, 259 Canal St., New York City; Quay T. Stewart, sales representative, Minneapolis, Minn.; Henry J. Bride, sales representative, Hartford, Conn.; also the opening of branch stores at 529-531 Arch St., Philadelphia, with H. H. Freund, Jr., manager; 505 Mission St., San Francisco, Calif., C. R. Mendelson, manager. In addition to its other lines, the company now also manufactures the Davis and Berryman line of feed water heaters, hot water generators and power pumps, having purchased this business from I. B. Davis & Son, the original manufacturers. The company has further extended its line by the more recent purchase of a welded steel tubing business and will continue manufacturing welded steel tubing in the various diameters and gages.

Trade Catalogs

The Illinois Stoker Co., Alton, Ill., has issued Catalog K, a publication of sixteen pages devoted to automatic stokers for oil refineries. As the shell type still is used in the majority of installations, the catalog deals principally with the general arrangement, construction, operating principles and advantages of the company's latest type of forced-draft stoker as applied to shell stills of all kinds. It is of interest to point out that the stoker passes entirely through the setting and discharges the ashes at the rear, so that no ashpit is required. The concluding pages contain a summary of the advantages of stoker vs. hand firing.

THE COAL MARKET

BOSTON—Current prices per gross ton f.o.b. New York loading ports:

Anthracite

	Company Coal
Egg	$8.20@$8.65
Stove	8.45@ 9.05
Chestnut	8.55@ 9.05
Pea	7.05@ 7.40
Buckwheat	3.00@ 3.80
Rice	2.30@ 2.50
Barley	1.25

Bituminous

	Clearfields	Cambrias and Somersets
F.o.b. mines, net tons	$2.85@$3.35	$3.15@$3.60
F.o.b. Philadelphia, gross tons	5.05@ 5.60	5.35@ 5.90
F.o.b. New York, gross tons	5.40@ 5.95	5.75@ 6.25
Alongside Boston (water coal), gross tons	7.00@ 7.75	7.60@ 8.00

Pocahontas and New River are practically off the market for coastwise shipment, but are quoted at $6.25@$7.00.

NEW YORK—Current quotations, White Ash, per gross tons, f.o.b. Tidewater, at the lower ports are as follows:

Anthracite

	Company Coal
Broken	$7.80@$8.25
Egg	8.20@ 8.65
Stove	8.45@ 9.05
Chestnut	8.55@ 9.05
Pea	7.05@ 7.40
Buckwheat	5.15
Rice	4.50
Barley	4.85
Boiler	4.85

Bituminous

Government prices at mines:	Spot
South Fork (best)	$3.25@$3.50
Cambria (best)	3.00@ 3.25
Cambria (ordinary)	2.60@ 2.90
Clearfield (best)	3.00@ 3.25
Clearfield (ordinary)	2.60@ 2.90
Reynoldsville	2.85@ 2.90
Quemahoning	3.25@ 3.50
Somerset (medium)	3.00@ 3.25
Somerset (poor)	2.50@ 2.75
Western Maryland	2.50@ 2.75
Fairmont	2.25@ 2.50
Latrobe	2.60@ 2.80
Greensburg	2.75@ 3.00
Westmoreland ¾ in.	3.40@ 3.50
Westmoreland run-of-mine	2.75@ 3.00

PHILADELPHIA—Anthracite prices are practically the same as those listed above for New York. Bituminous coal prices vary according to district from which they are mined. For ordinary slack the price is $2.45@$2.55; lump, $3.00@$3.35, at the mines.

BUFFALO—

Anthracite

	On Cars, Gross Ton	At Curb, Net Ton
Grate	$8.55	$10.80
Egg	8.80	10.65
Stove	9.00	10.85
Chestnut	9.10	10.85
Pea	7.75	9.30
Buckwheat	5.70	7.75

Bituminous

Allegheny Valley	$4.80
Pittsburgh	4.65
No. 8 Lump	4.65
Mine Run	4.80
Slack	4.10
Smokeless	4.60
Pennsylvania Smithing	5.70

CLEVELAND—Prices of coal per net ton delivered in Cleveland are:

Anthracite

Egg	$11.75@$11.90
Chestnut	12.00@ 12.00
Grate	11.75@ 11.90
Stove	11.90@ 12.10

Pocahontas

Mine-run	$7.50

Domestic Bituminous

West Virginia split	$8.50
No. 8 Pittsburgh	$6.60@ 6.90
Massillon lump	8.25@ 8.50
Coshocton lump	7.15

Steam Coal

No. 6 slack	$5.25@$5.50
No. 8 slack	5.10@ 5.50
Youghiogheny slack	5.25@ 5.50
No. 8 ¾ in.	5.70@ 6.00
No. 6 mine-run	5.25@ 5.50
No. 8 mine-run	5.75

Only coal available is mine-run Pocahontas.

MIDDLE WEST—Chicago quotations, F.o.b. cars at mine:

	Springfield, Carterville, Williamson, Franklin, Saline, Harrisburg	Fulton, Peoria	Grundy, La Salle, Bureau, Will
Lump	$2.55@$2.70	$2.95@$3.10	$3.25@$3.40
Washed	2.75@ 2.90	....	3.45@ 3.60
Mine run	2.35@ 2.50	2.75@ 2.90	3.00@ 3.15
Screenings	2.05@ 2.20	2.35@ 2.50	2.75@ 2.90

New Construction

PROPOSED WORK

Mass., Holyoke—The Century Machine Co. will soon receive bids for the construction of a 2 story, 50 x 90 ft. and a 1 story, 90 x 200 ft. factory building on Main St. A steam heating system will be installed in same.

Mass., Peabody—A. E. Bump, Engr., 60 North Market St., Boston, will soon award the contract for the construction of a 4 story, 70 x 180 ft. storehouse on Webster St. A steam heating system will be installed in same. Total estimated cost, $150,000.

Mass., Worcester—The Charles H. Tenney & Co., 201 Devonshire St., Boston, will build an 8 story building on Hygeia St. for the Worcester Cold Storage Co. Estimated cost, $300,000.

Conn., New London—The Sheffield Dentifrice Co. is having preliminary plans prepared for the construction of a power station. Daly Co., Waterbury, and Flagg & Co., 27 State St., Meriden, Engrs.

N. Y., Albany—The White Automobile Co., East 79th St. and St. Clair Ave., Cleveland, is having plans prepared for the construction of a 2 story, 75 x 200 ft. sales room and garage on North B'way. A steam heating system will be installed in same. Total estimated cost, $100,000. Watson Eng. Co., Hippodrome Bldg., Cleveland, Archt.

N. Y., Brooklyn—The William Fox Amusement Co., 126 West 46th St., New York City, plans to build a 100 x 200 ft. theatre on Flatbush and Tilden Aves. A steam heating system will be installed in same. Total estimated cost, $500,000. I. W. Lamb, 644 8th Ave., New York City, Archt.

N. Y., Farmingdale—The Raymond Eng. Corp., 309 Lafayette St., New York City, will build two 1 story, 60 x 120 ft. factories for the manufacture of motors, power tire pumps, etc. A steam heating system will be installed in same. Estimated cost between $85,000 and $90,000. Work will be done by day labor.

N. Y., College Point—J. B. Kleinert Rubber Co., 725 Broadway, New York City, is having preliminary plans prepared for the construction of a 4 story factory. A steam heating system will be installed in same. Total estimated cost, $150,000. R. G. Cory, 39 Cortland St., New York City, Arch. and Engr.

N. Y., Gabriels—Gabriels Hospital Commission is having plans prepared for the construction of a group of hospital buildings. A steam heating system will be installed in same. Total estimated cost, $3,000,000. J. R. Pope, 527 5th Ave., New York City, Arch. and Engr.

N. Y., Long Island City—J. Clark, c/o B. R. Swartburg, Arch. and Engr., 103 Park Ave., New York City, is having plans prepared for the construction of an 8 story, 100 x 150 ft. factory. A steam heating system will be installed in same. Total estimated cost, $380,000.

N. Y., Long Island City—L. Gold, 44 Court St., Brooklyn, will build a 3 story, 150 x 190 ft. garage and service station on Ely Ave. and South Jane St. A steam heating system will be installed in same. Total estimated cost, $300,000. Work will be done by day labor.

N. Y., New York—The Congregation B'nai Israel of Washington Heights, 535 West 148th St., is having plans prepared for the construction of a 1 story, 75 x 100 ft. synagogue at 617-621 West 149th St. A steam heating system will be installed in same. Total estimated cost, $200,000. E. Roth, 119 West 40th St., Archt. and Engr.

N. Y., New York—The Continental Athletic Club, c/o R. C. Lafferty, Archt. and Engr., 347 5th Ave., is having plans prepared for the construction of a 14 story, 100 x 200 ft. club house on Central Park West from 61st to 62nd Sts. A steam heating system will be installed in same. Total estimated cost, $1,500,000.

N. Y., New York—The Emerson Bldg. Corp., 103 Park Ave., will build a 12 story, 75 x 100 ft. office building at 35-37 West

45th St. A steam heating system will be installed in same. Total estimated cost, $375,000. Work will be done by day labor.

N. Y., New York—The International Mercantile Marine, 1 Bway., will rebuild its 11 story office building. A steam heating system will be installed in same. Total estimated cost, $1,000,000. Work will be done by day labor.

N. Y., New York—The 145 West 55th St., Inc., 1 West 67th St., is having plans prepared for the construction of a 15 story, 80 x 100 ft. hotel at 139-145 West 55th St. A steam heating system will be installed in same. Total estimated cost, $750,000. J. E. Eaton, 19 West 44th St., Archt. and Engr.

N. Y., New York—The William Fox Amusement Co., 126 West 46th St., is having plans prepared for the construction of a theatre on 14th St. and Irving Pl. A steam heating system will be installed in same. Total estimated cost, $300,000. T. W. Lamb, 644 8th Ave., Archt.

N. Y., New York—The Board of Education, 500 Park Ave., received bids for the installation of a boiler and condenser tubes, from the Wheeler Condenser & Engineering Co., 149 Broadway, $5685; E. F. Keating Co., 446 Water St., $7040; Parkesbury Iron Co., 30 Church, $9360. Noted Apr. 15.

N. Y., New York—The Harriman National Bank, 527 5th Ave., is having plans prepared for altering its present building. A steam heating system will be installed in same. Total estimated cost, $500,000. J. R. Pope, 527 5th Ave., Arch. and Engr.

N. Y., New York—Frank Koester, Engr., 50 Church St., is in the market for two 60 ton destructors, two 300 h.p. boilers, condenser pumps, one 200 kw. generator, one 300 kw. generator, one 50 kw. generator, transformers and switchboards for the new municipal lighting plant at Scranton, Pa.

N. Y., New York—The 195 Broadway Corporation, 195 Broadway, is having plans prepared for altering a 27 story office building. A steam heating system will be installed in same. Total estimated cost, $5,000,000. W. W. Bosworth, 527 5th Ave., Arch. and Engr.

N. Y., White Plains—W. S. Bessell, Archt., 56 West 45th St., New York City, is preparing plans for the construction of a theatre here. A steam heating system will be installed in same. Total estimated cost, $100,000. Owner's name withheld.

N. J., Bloomfield—The American La France Fire Engine Co., 250 West 45th St., New York City, is having plans prepared for the construction of a 1 story, 150 x 500 ft. factory. A steam heating system will be installed in same. Total estimated cost, $350,000. Starrett & Van Vleck, 8 West 40th St., New York City, Archts. and Engrs.

N. J., Newark—The Celuloid Co., 290 Ferry St., is having plans prepared for altering and building additions to its power house. Estimated cost, $100,000. Lockwood, Green & Co., 101 Park Ave., New York City, Archt. and Engr.

N. J., South River—The Town Council has authorized E. B. Hedden, Supt. Pub. Wks., Town Hall, to draw plans for the installation of two 500 kw. steam turbine engines, also building to house same. Total estimated cost, $140,000.

N. J., Westwood—The Bd. Educ. plans to build a 2 and 3 story school building. A steam heating system will be installed in same. Total estimated cost, $125,000. Rasmussen & Wayland, 1133 Bway., New York City, Archts. and Engrs.

Penn., Elkins Park—Herbert C. Wise, Archt., 925 Chestnut St., Philadelphia, will soon award the contract for the construction of a 3 story, 75 x 255 ft. high school here for Cheltenham Township. A steam heating system will be installed in same.

Penn., Kingston—The Wales Adder Machine Co., Walnut and Hoyt Sts., plans to construct a 3 story, 100 x 100 ft. factory on Hoyt St. A steam heating system will be installed in same. Total estimated cost, $100,000.

Penn., Lewisburg—The United Evangelical Churches plan to build a 2 story, 80 x 150 ft. orphanage. A steam heating system and electric power will be installed in same. Total estimated cost, $150,000.

Penn., Miners Mill (Wilkes-Barre P. O.)—The Blessed Sacrament Congregation plans to construct a 2 story, 40 x 100 ft. church on Main St. A steam heating system will be installed in same.

Penn., Philadelphia—McIlvain & Roberts. Archts., 112 South 16th St., will soon award the contract for the construction of an 8 story, 85 x 120 ft. sales room and office building on Broad and Vine Sts., for Bigelow Willeys Co., 304 North Broad St. A steam heating system will be installed in same. Total estimated cost, $100,000.

Penn., Scranton—The city has engaged Frank Koester, Electrical Expert, 50 Church St., New York City, to prepare plans and submit estimates for the construction of a 1 story, 76 x 76 ft. lighting plant. Alex T. Connell, Court House, Mayor.

O., Canton—W. I. Zink, Pub. Serv. Dir., will soon award the contract for the construction of a 1 story, 30 x 96 ft. pump house. Estimated cost, $20,000. R. Winthrop Pratt, Hippodrome Bldg., Cleveland, Engr.

O., Cleveland—The Acorn Refining Co., 8001 Franklin Ave., will receive bids about Feb. 10 for the construction of a 1 story, 30 x 50 ft. boiler house. Two 200 hp. boilers will be installed in same. Total estimated cost, $25,000. E. M. Katz, Pres.

O., Cleveland—The Anshe Emeth Beth Texilo Congregation, 10615 Grantwood Ave., plans to build a 1 story synagogue on East 105th St. and Drexel Ave. A steam heating system will be installed in same. Total estimated cost, $750,000. Louis Cahn, Treas.

O., Cleveland—The Bd. Educ., East 6th St. and Rockwell Ave., will receive bids until Jan. 26 for the construction of two 1 story school additions, each 55 x 150 ft., on St. Clair Ave. and Royal Rd. One heating boiler will be installed in same. Total estimated cost, $150,000. F. G. Hogan, Dir. W. R. McCormack, Archt.

O., Cleveland—The Bd. of Educ. will receive bids in the spring for the construction of a 2 story, 128 x 254 ft. high school on East 118th St. and Hopkins Ave. A steam heating system will be installed in same. Total estimated cost, $500,000. Frank G. Hohan, East 6th St. and Rockwell Ave., Dir. W. R. McCormack, East 6th St. and Rockwell Ave., Archt.

O., Cleveland—The Cleveland Twist Drill Co., East 49th St. and Lakeside Ave., plans to build a 1 story factory on Euclid Ave. and tracks of Nickel Plate R. R. A boiler plant will be installed in same. Total estimated cost, $100,000. J. D. Cox, Mgr.

O., Cleveland—The Eaton Axle Co., 10017 Euclid Ave., is having plans prepared for the construction of a 1 story factory at East 140th St. and the New York Central R. R. Two 300 hp. boilers will be installed in same. Total estimated cost, $100,000. J. O. Eaton, Pres. G. S. Rider & Co., Century Bldg., Archt. and Engr.

O., Columbus—D. Riebel & Sons, Archts., New First National Bank Bldg., is preparing plans for the construction of a 2 story school building for the Bd. Educ. in Linden, a suburb of Columbus. A steam heating system will be installed in same. Total estimated cost, $165,000.

O., Lakewood—C. A. Cotabish, c/o The National Carbon Co., West 117th St. and Madison Ave., Cleveland, is having plans prepared for the construction of a 3 story, 200 x 460 ft. commercial building and moving picture theatre at 13606 Detroit Ave. A steam heating system will be installed in same. Total estimated cost, $300,000. S. H. Weis, Schofield Bldg., Cleveland, Archt.

O., Youngstown—The Ohio Hotel Operating Co., Boardman and Hazel Sts., is having plans prepared for the construction of a hotel addition. A steam heating system will be installed in same. Total estimated cost, $350,000. Clinton & Russell, 32 Nassaw St., New York City, Archts. and Engrs.

Mich., Detroit.—The Detroit Library Commission, Public Library Bldg., plans to build a 2 story library on Mack and Montclair Aves., also a branch library of similar construction on West Fort St., near Waterman Ave., and one on Davidson Blvd. and Lumpkin Ave. Steam heating plants will be installed in same. Estimated cost, $200,000 each.

Mich., Detroit.—M. J. Schneider, 601 Dix Ave., will soon award the contract for the construction of a 1 story, 50 x 130 ft. battery service station on Dix Ave. A steam heating boiler, forced ventilation unit, fan and blower will be installed in same. Total estimated cost, $25,000. J. W. Townsend, 709 Empire Bldg., Engr.

Mich., Escanaba.—Philip L. Utley and John G. Sutherland have asked the Bd. Supervs. for permission to construct 2 reinforced concrete dams across the Escanaba River in Cornell and Baldwin Twps., one to be 50 ft. high, with concave face to serve as apron about 50 ft. wide, waste gate 12 ft. wide and 12 ft. deep, with log chute 5½ ft. wide and 4 ft. deep from the top of the dam. Other dam to be similar in all details except height, which is to be 60 ft. Project includes hydro-electric plants.

Mich., Port Huron—The City Comn. plans to install an additional steam driven pumping unit with a capacity of 6,000,000 gallons per day; also to replace 2 present boilers and adding 1 in reserve. E. R. Whitmore, City Hall, Engr.

Mich., Saginaw—The Saginaw Table & Cabinet Co., Wheeler St., is having plans prepared for the construction of a 2 story, 61 x 475 ft. furniture factory on Wheeler St. A steam heating system and electric power will be installed in same. Total estimated cost, $100,000. Cowles & Mutscheller, Saginaw, Archts.

Mich., St. Clair—The Bd. Educ. engaged Perkins, Fellows & Hamilton, Archts., 814 Tower Court, Chicago, to prepare plans for the construction of a 2 story high school on Main St. A steam heating system will be installed in same. Total estimated cost, $250,000.

Ill., Chicago—The Bd. Educ., 7 South Dearborn St., is having plans prepared for the construction of a 3 story, 150 x 250 ft. addition to the technical high school on Van Buren St. and Oakley Blvd. A steam heating system will be installed in same. Total estimated cost, $3,000,000. John Howatt, Engr., and A. F. Hussander, Archt., both of 7 South Dearborn St.

Ill., Chicago—Clarence Hatzfield, Archt., 7 South Dearborn St., is preparing plans for the construction of a 5 story, 104 x 117 ft. Masonic temple on 64th and Green Sts. A steam heating system will be installed in same. Total estimated cost, $250,000. Owner's name withheld.

Ill., Chicago—Clarence Hatzfield, Archt., 7 South Dearborn St., will soon award the contract for the construction of a 4 story, 75 x 125 ft. auto sales building on Milwaukee Ave. near Kedzie Ave. A vacuum steam heating system will be installed in same. Total estimated cost, $150,000.

Ill., Chicago—Z. E. Smith, Archt., 305 East 55th St., will soon award the contract for the construction of a 3 story, 95 x 180 ft. laundry and boiler house at 2135 West Madison St., for the Great Western Laundry, 2319 West Madison St. A 200 hp. steam power and heating plant will be installed in same. Total estimated cost, $225,000.

Ill., Oak Park—Holmes & Flinn, Archts., 8 South Dearborn St., Chicago, will soon award the contract for the construction of a 4 story, 144 x 157 ft. addition to the high school on Ontario St. and East Ave., for the Bd. Educ. A steam heating system will be installed in same. Total estimated cost, $200,000.

Wis., Green Bay—The John Hoberg Paper Co., Elm St., has selected a site and plans to build a 4 story, 60 x 250 ft. paper mill on 12th and Main Sts. A steam heating system and power equipment will be installed in same. Total estimated cost, $250,000.

Wis., Kiel—Stoetling Bros. Co. will soon award the contract for the construction of a 3 story, 60 x 120 ft. power plant on Main St. Estimated cost, $50,000.

Wis., Racine—Lambert Bassindale, Archt., Capitol Bank Bldg., St. Paul, Minn., will receive bids until March 1 for the construction of a 2 story, 60 x 110 ft. office building for the Journal News Publishing Co., 328 Main St. Separate bids will be received for installing heating, plumbing and lighting systems in same. Total estimated cost, $100,000.

Wis., Saukville—H. J. Cary is organizing a company which plans to build a 1 and 2 story, 36 x 376 ft. pea cannery. A steam power plant will probably be installed in same. Total estimated cost, $50,000.

Wis., Tomah—The Tomah Electric Light Co., c/o L. Barney, Supt., plans to construct a 1 story, 60 x 135 ft. light and power plant on Main St. Power machinery will be installed in same. Total estimated cost, $70,000.

Minn., Benson—The city is having plans prepared for the installation of a generator, switchboard and boilers in the electric light plant. Estimated cost, $40,000. L. M. Peterson, Clk. W. E. Shinner, 15 South 5th St., Minneapolis, Engr.

Minn., Grey Eagle—The Bd. Educ. is preparing plans for the construction of a 2 story, 68 x 120 ft. high school. Separate bids will be received for installing heating and plumbing systems in same. Rose & Harris, 471 Auditorium Bldg., Minneapolis, Engr. Stebbins & Haxby, 445 Auditorium Bldg., Minneapolis, Archt.

Minn., Lake Crystal—Independent School District No. 19 is having plans prepared for the construction of a 2 story, 90 x 140 ft. high school. A steam heating system will be installed in same. Total estimated cost, $275,000. Guy Lamoreau, Clk. Rose & Harris, 471 Auditorium Bldg., Minneapolis, Engrs. Stebbins & Haxby, 445 Auditorium Bldg., Minneapolis, Archts.

Minn., Luverne—Independent School Dist. No. 2 is having plans prepared for the construction of a 2 sotry, 68 x 120 ft. high school. Separate bids will be received for installing heating and plumbing systems. Total estimated cost, $300,000. H. W. Bertram, Clk. Tyrie & Chapman, 320 Auditorium Bldg., Minneapolis, Archts.

Minn., Northfield—M. Edward Wohn, Archt., 596 Endicott Bldg., St. Paul, will receive bids until Jan. 15 for the construction of a 3 story, 73 x 186 ft. science hall for St. Olaf's College. Separate bids will be received for installing heating and plumbing systems in same. Total estimated cost, $200,000.

Minn., St. Paul—Justina Sanford, 486 Otis St., is having plans prepared for the construction of a 12 story, 122 x 420 ft. hotel on Summit Ave. A steam heating system will be installed in same. Total estimated cost, $2,000,000. Mark Fitzpatrick, 17 West 9th St., Archt.

Minn., St. Paul—The Western Chemical Co., Hutchinson, is having plans prepared for the construction of a 3 story, 90 x 120 ft. chemical factory on Malcolm Ave. A steam heating system will be installed in same. Total estimated cost, $125,000. Downs & Eads, 803 Phoenix Bldg., Archts.

Colo., Seibert—The town will receive bids in the latter part of January for the construction of electric light and water plants. Estimated cost, $40,000. R. D. Salisbury, 1415 East Colfax Ave., Denver, Engr.

Cal., Independence—The Bd. of Supervs. of Inyo County will receive bids until Feb. 2 for the construction of a 2 story courthouse. A steam heating system will be installed in same. Total estimated cost, $150,000. W. H. Weeks, 75 Post St., San Francisco, Archt.

Cal., Pittsburgh—A. C. Cardinalle et al is having plans prepared for the construction of a 2 story, 70 x 150 ft. cold storage plant and ice factory on 4th St. I. Zanolini, 604 Montgomery St., San Francisco, Archt.

Cal., San Pedro—The Van Camp Sea Food Co., 14th St. and waterfront, will probably be granted a lease by the City Harbor Comn., Los Angeles, on 30 acres of tideland on which it proposes to erect an ice and cold storage plant, also a wharf 250 ft. long. Total estimated cost, $300,000.

CONTRACTS AWARDED

Conn., Danbury—The Danbury Felt Mills have awarded the contract for the construction of a 2 story, 60 x 60 ft. manufacturing building to Wescott & Mapes, Inc., 207 Orange St., New Haven. Plans include an 18 x 36 ft. boiler house. Total estimated cost, $80,000.

Conn., Manchester—The town has awarded the contract for the construction of a school on Spruce St. to W. A. Knosla, South Manchester. A steam heating system will be installed in same. Total estimated cost, $150,000.

Conn., West Haven—The town of Orange, Northern School Dist., has awarded the contract for installing a steam heating system in the proposed 2 story, 90 x 100 ft. school on 1st Ave. and Lanison St. to the Stone & Underhill Heating & Ventilating Co., 139 Pearl St., Boston. Total estimated cost, $100,000.

N. Y., Brooklyn—The Knickerbocker Ice Co., 1480 Bway., New York City, has awarded the contract for the construction of a 1 story, 70 x 100 ft. ice plant in Bay Ridge to F. G. Fearon, 280 Madison Ave., New York City. Estimated cost, $80,000.

N. Y., New York—The Chase National Bank, 57 Broadway, has awarded the contract for altering the present 4 story, office, store and bank building at 69-79 Greenwich St. to Marc Eidlitz, 30 East 42nd St. A steam heating system will be installed in same. Total estimated cost, $300,000.

Penn., Philadelphia—The Philadelphia Paper Mfg. Co., River Rd., has awarded the contract for the construction of a 1 story, 160 x 295 ft. administration building at Manayunk to Hughes, Foulkrod & Co., Commonwealth Trust Bldg. A steam heating system will be installed in same. Total estimated cost, $145,000.

Penn., Wilkes-Barre—The Kitsee Battery Co., 262 North Main St., has awarded the contract for the construction of a 3 story, 70 x 90 ft. storage battery manufacturing plant, to John A. Schmitt, Bennett Bldg. A steam heating system will be installed in same. Total estimated cost, $106,000.

O., Akron—The Goodyear Tire & Rubber Co., East Market St., has awarded the contract for the construction of a 3 story, 100 x 130 ft. dormitory for girls on East Market St., to the Hunkin-Conkey Construction Co., Century Bldg., Cleveland. A steam heating system will be installed in same. Total estimated cost, $200,000.

O., Cleveland—The Cleveland Motor Sales Co., 1628 Euclid Ave., has awarded the contract for the construction of a 2 story, 75 x 150 ft. sales building and show room on East 40th St. and Prospect Ave., to Ed. Paulsen, Erie Bldg. A steam heating system will be installed in same. Total estimated cost, $100,000. J. C. Simon, Mgr.

O., Cleveland—The Euclid Malta Co., c/o Colcher & Smith, Archts., Lennox Bldg., has awarded the contract for the construction of a 3 story, 50 x 175 ft. lodge and office building on Euclid Ave., near East 105th St., to the Du Peron Const. Co., East 22nd St. and Prospect Ave. A steam heating system will be installed in same. Total estimated cost, $100,000.

O., Cleveland—The Federal Packing Co., 3207 West 65th St., has awarded the contract for remodeling a 4 story, 40 x 160 ft. cold storage plant on East 4th St. and Bolivar Rd., to J. H. Dease, 520 Erie Bldg. Estimated cost, $50,000.

O., Silver Lake—The village has awarded the contract for the construction of two 1 story, 17 x 17 ft. pump houses to the Alger-Tilghman Co., Guardian Bldg., Cleveland. Deep well pumps with a 175 gal. per min. capacity will be installed in same. Total estimated cost, $15,000.

Mich., Lansing—The Detroit Beef Co., 1214 Washington St., has awarded the contract for the construction of a 2 story, 30 x 92 ft. cold storage plant and a 19 x 42 ft. garage, on Washington St. and the Grand Trunk Ry., to Pine & Municke, 103 Marquette Bldg., Detroit. A small steam heating boiler and electric motors for power will be installed in same. Total estimated cost, $50,000.

Ill., Chicago—The Continental Can Co., 111 West Washington St., has awarded the contract for the construction of a 1 story, 90 x 170 ft. factory on Ashland Ave. near 38th St., to E. W. Sproul Co., 2001 West 39th St. A steam heating system will be installed in same. Total estimated cost, $175,000.

Ill., Chicago—The Packard Auto Open Body Co., c/o S. Scott Joy, Archt., 2001 West 39th St., has awarded the contract for the construction of a 3 story, 120 x 400 ft. factory on East 104th St. and Erickson Ave., to E. W. Sproul Co., 2001 West 39th St. A steam heating system will be installed in same. Total estimated cost, $1,000,000.

Ill., Rock Island—Rock Island County has awarded the contract for the installation of steam heating and plumbing systems in the proposed 3 story, 44 x 99 ft. jail, on 15th St., to Channon & Dufra, Rock Island. Total estimated cost, $125,000.

Minn., Hastings—The State Asylum for the Insane has awarded the general contract for the construction of a 1 story, 45 x 140 ft. power plant to C. Ash & Son, 202 Maria Ave., St. Paul, $58,269; mechanical equipment to M. J. O'Neil, 60 East 6th St., St. Paul, $40,000.

POWER

Volume 51 New York, January 27, 1920 *Number 4*

Every Man an Efficiency Engineer

John Leitch in his excellent book, "Man-to-Man, the Story of Industrial Democracy," relates that in one plant at a meeting of the employees the suggestion was made by one that an efficiency engineer be called in to instruct the men in better methods for doing their work, and to increase production. Much discussion followed, and finally one of the men blurted out, "We have 268 efficiency engineers right here now!" (There were 268 employees in the plant, and they were all present at the meeting.)

There is a tremendous amount of sound reasoning and good judgment expressed in that short statement, and it should be true of every plant. When by instruction and coöperation each man is induced to take the interest that he should in the plant in which he is employed, then each man will become an efficiency engineer. It will be unnecessary to call in outside aid to find the weak spots, and to devise better methods of doing things. No man can be as familiar with the workings within a shop as he who spends day after day at his bench, at his lathe, or on whatever his job may be.

Every workman should constantly keep before his mind these three questions:

Am I doing my work in the most economical way for the interests of my employer? (Which, by the way, generally prove to be the employee's interests also.)

Am I eliminating all possible wastes of time, energy and material?

Am I doing everything possible to advance myself in my work, both to my own profit and for the interests of my employer?

If a man can answer "yes" to all these questions, then however humble may be his position, however rough and unskilled may be his labor, that man is an "efficient engineer," and as such is a most valuable asset to his employer. On the other hand, the reverse is true. Any man, no matter how high a position he may hold, who can not answer "yes" to the foregoing questions not only is depriving his employer of services that are his due, but is decreasing production, increasing costs and—what should be of more interest to him—retarding his own development and advancement. And it is such men that make it necessary to bring in efficiency experts from the outside.

The profession of efficiency engineering has been greatly abused, until in many plants an efficiency engineer is looked upon with suspicion by employer and employee alike. The mere use of the words arouses a feeling of hostility at once. This works a hardship upon the great number of men who are qualified by training and ability to give expert advice in matters of cutting costs, increasing production and reducing wastes, and who have made this work their profession.

But if the employees of every plant, large or small, would adopt for their motto the words, "Every Man an Efficiency Engineer," there would be no occasion for outside experts to be called in. Labor and Capital would alike benefit thereby, fuel administrators would be unnecessary, and power plants would operate at efficiencies heretofore unheard of.

EXTENSION to L STREET STATION, BOSTON

By CHARLES H. BROMLEY

The station represents the latest practice in large power-plant design. The boiler room is unequaled in provisions for convenient, economical performance. One 30,000-kw. single-cylinder impulse turbine is installed. A second 30,000 will be ready soon. Barometric condenser is used as heater. Main steam valves are remote-controlled electrically.

THE latest practice in power-station design is the new extension to the L Street Station of the Edison Electric Illuminating Co. of Boston, Mass. When more power was demanded of this station, the engineers had the advantage of making an extension along the most modern lines in building and equipment, instead of remodeling an old station, which in recent years has been the usual way of increasing a station's capacity. The headpiece is from the most recent photograph of the exterior of this station, and Fig. 13 shows a sectional view of the buildings and equipment of the recent extension. The station is exceptionally roomy, well-lighted and ventilated, and excellently equipped.

In the extension there are four Babcock & Wilcox cross-drum water-tube boilers (Fig. 2), each of 12,320 sq.ft. of heating surface. The turbine room now has one 30,000-kw. single-cylinder impulse turbine embodying the recent improvements made on this design by the General Electric Co. and another will soon be ready. The boiler room without question has no equal in point of design and equipment. Every provision that obviates or lightens heavy labor, that makes for convenience of operation and repair, that conduces to efficiency and checks and records it, is made. See Figs. 1 to 5, also Fig. 11. Each boiler is inclosed in a steel casing, its walls insulated; each has a Westinghouse 13-retort underfeed stoker with a clinker grinder. The retorts are sectionalized in three units of three retorts each and one unit of four retorts. The drive is by dustproof motor. Each stoker has four motors, one for each section of retorts. The forced-draft fans are on the boiler-room floor, and so, too, are the motor-driven centrifugal boiler-feed pumps.

With the type of settings under these cross-drum boilers the front of the boiler becomes the back as far as observation and control of fire goes. One man attends to the feeding of coal to all stoker hoppers and retorts by the use of 10-ton weighing lorry. Between the stoker side of the boilers and the back or control side there is a partition that keeps out the dust. The firemen are on the control side. The rheostats for the various motor controls are in the basement, chains connecting them with the control-room floor level at the boilers.

Each stoker has 13 retorts, the drive of which is sectionalized by four gear boxes with an individual motor for each section. This, of course, gives the crew positive mechanical control of the whole or any part of the fuel bed. The furnace has a combustion volume of 3000 cu.ft. and 930 lb. of coal per hour per retort. The ashpit volume is 1050 cubic feet.

The clinker grinder is of considerable interest, but as it was described in *Power* for Apr. 22, 1919, details here are unnecessary. Its performance is remarkable. New River coal, which averages 6.5 per cent. ash by analysis, which fuses at about 2500 to 2800 deg. F., is burned in this station. Analyses of the refuse passed through the grinder show it to contain from 6 to 7 per cent. carbon for the average over the usual operating day. The refuse may be emptied from the ash hoppers directly into motor trucks.

Each boiler has 50,000 cu.ft. per min. capacity at 7-in. static pressure in forced-draft fans, motor-driven, and 80,000 cu.ft. per min. in induced-draft fans, also motor-driven. The forced-draft fans are of straight-blade double-inlet type. The induced fans are of the multivane type.

Each boiler has an economizer of 7728 sq.ft. heating surface, built for 300 lb. pressure.

One of the features which impresses the visitor to the station is the adequate room, excellent ventilation and lighting everywhere provided. "It's a fine place to work in," is what one exclaims on seeing it. Unusual,

FIGS. 1 TO 5. BOILER-ROOM VIEWS IN THE L STREET STATION EXTENSION

Fig. 1—The control side of one of the boilers. Fig. 2—Showing the stoker drive side. Fig. 3—Looking down the new boiler room, control side. Fig. 4—Combustion chamber of one of the boilers. Fig. 5—Looking down upon the clinker grinder.

FIGS. 6 TO 10. DEAN ELECTRICALLY CONTROLLED VALVES IN L STREET STATION

Fig. 6—A large gate valve. Fig. 7—One of the high-pressure steam valves. Fig. 8—The high-pressure steam header. Fig. 9—A connection to the header, showing three controlled valves. Fig. 10—A "close up" of one of the large valves in the steam line.

indeed, are the number of iron grating platforms and runways about the boiler room. There are few flanges or fittings that a man cannot work upon while standing on these gratings. One with station experience does not question this investment.

Automatic feed-water regulators are used, as is now general practice in all modern power plants.

The boiler room of the extension is tied into the steam distribution system of the older boiler room. The pressure on the lines of the new extension is 300 lb., that in the older plant 200 lb. Fig. 12 shows the tie-in connections. Note that the General Electric pressure-reducing valve is the connecting point between the 300-lb. and 200-lb. systems. To avoid the possibility of 300 lb. on the 200-lb. system, a bank of twelve 4.5-in. pop safety valves are provided. These relieve into a 20-in. spiral-riveted pipe which connects with the main atmospheric relief pipe. This application of this pressure-reducing valve is unique. The valve is made for, and for a number of years has been used on, Curtis mixed-pressure turbines, but its use as at this station is new. The advantages of remote control of valves, steam and water, is now so generally recognized that it is being applied in all modern power stations. The remote control of valves at L Street Station is a feature. See Figs. 6 to 10. The following description and explanation of valve control at this station was written for this article by P. P. Dean, the engineer who has developed this system of control:

A noteworthy feature of this installation is the piping and valve equipment, which is especially designed for the high pressure and superheat used. The majority of the valves are of Schütte & Koerting manufacture and are built with cast-steel bodies and fitted with monel-metal seat rings. The stem is also of monel, and this metal is used on account of its ability to stand excessive heat and also on account of its low coefficient of expansion. The valves are of the double-disk taper type, having a toggle form of wedge mechanism for keeping the disks tight against the seats. The thickness of metal in the valve body is greater than the manufacturer's standard, and the flanges also are of extra thickness. A number of the valves, however, on the same line, are built by the Nelson Valve Co.

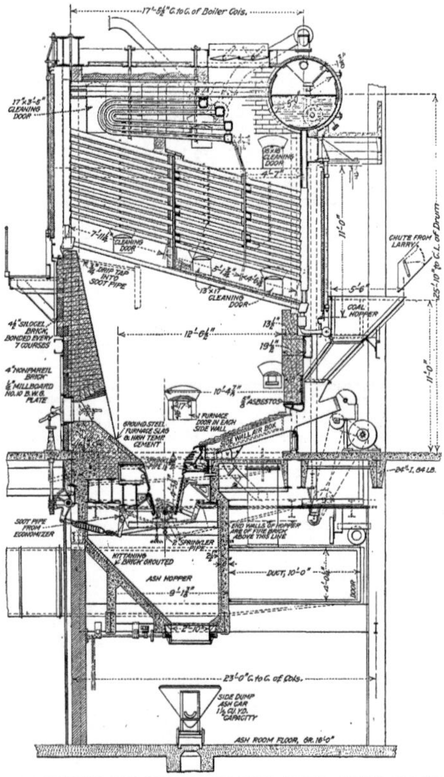

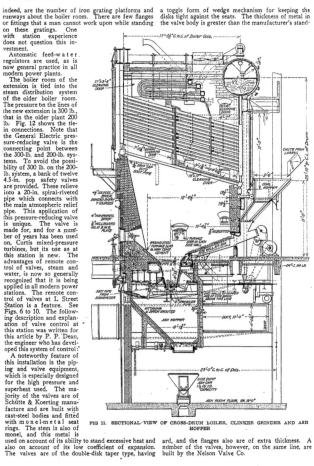

FIG 11. SECTIONAL VIEW OF CROSS-DRUM BOILER, CLINKER GRINDER AND ASH HOPPER

Eighteen of the important header and sectionalizing valves are equipped with electrical control containing a number of noteworthy features. This control is built by the Cutler-Hammer Manufacturing Co. and is known

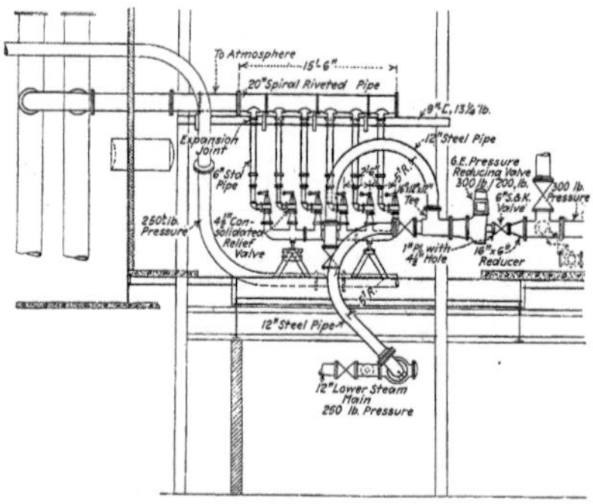

FIG. 12. ELEVATION OF NEW STEAM HEADER

as the Dean control system. It is, so far as known, the only electrically operated valve-closing device that is built especially for valve operation. The operation is by electric motor geared to a system of reduction gearing to reduce the speed of the valve stem to 12 in. per

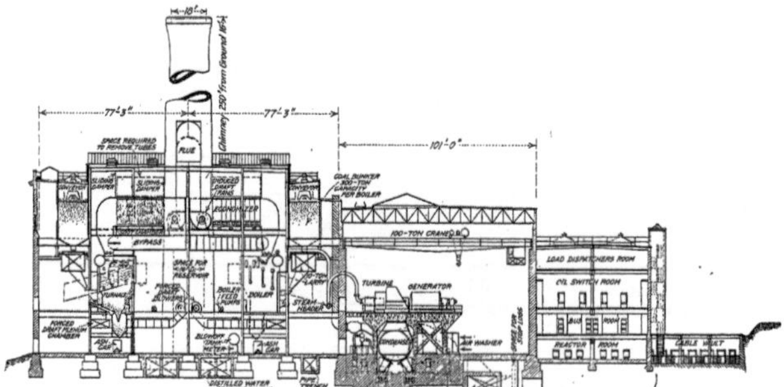

FIG. 13. SECTIONAL ELEVATION OF RECENT EXTENSION TO L STREET STATION

min. The motor is of the series-wound multi-polar, high-torque type, totally inclosed and weatherproof, and is connected to a system of combined planetary and worm gearing for the necessary reduction. The gear is continually immersed in oil and is fitted into an oil-tight cast-iron case forming the main housing of the unit. Contained within the unit is a limiting device which stops the valve the moment the predetermined limits have been reached and eliminates drift or overtravel, which is usually the cause of so much jamming and consequent breaking of stems in valves of this description.

A mechanical clutch automatically releases the high-speed motor and gears from the valve stem, and at the same moment an electrical limit opens the motor circuit.

The whole of the limit mechanism, gearing, etc., is totally inclosed, oil-tight and moisture-proof, while the cables from the motor to the switch housing are contained in cast-iron housing, making the unit waterproof and steamproof. The construction of the unit is such that the motor is a considerable distance from the valve yoke and free from the excessive heat consequent to the pressure and superheat used.

The Dean unit is fastened to a specially constructed valve yoke by four large studs, and the slow-speed shaft is geared to a valve stem by cast-steel cut gears which are completely inclosed by a sheet-iron casing. Handwheels are provided and keyed to an extended shaft so that manual operation may be performed in the event of failure of current. Some of the valves are controlled from one station and others from two remote points by the aid of standard Dean control stations which are fitted with red and green indicating lights. The electrical limit switch on the Dean unit is constructed to break the main circuit, which system, with single-point control, obviates the necessity of any relay panel and makes the complete control possible by the aid of five wires between the Dean unit and the control station.

The indicating lights on the control station show the full-open and full-closed position of the valve, and the connections are such that should a fuse blow or an electrical connection become loose, the light will disappear and immediately indicate trouble. Either the red or green light always glows according to the position of

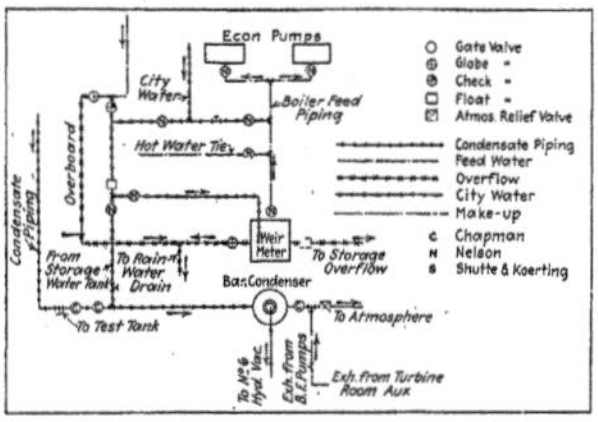

FIG. 14. DIAGRAM OF FEED-WATER PIPING

the valve, and forms a pilot showing the condition of the wiring, driving unit and all electrical connections.

The source of electric supply for the operation of all the valves is from a motor-generator set used for this purpose only, which is kept running continuously and has sufficient capacity to operate a majority of the valves at one time.

The wiring between the valves and source of supply is through steel conduit with water-type joints, as is the whole system and is, therefore, not affected by escaping steam or moisture.

An unusual feature of the station is the use of a barometer condenser as a heater for feed water. See the diagram, Fig. 14. The exhaust from the auxiliaries comes to this condenser, the condensing or injection water supply coming from the hotwell pumps. A 6-in. air connection from the top of the condenser connects with a hydraulic vacuum pump similar to those for the surface condensers for the main units; that is, the 30,000-kw. turbines. The tailpipe drops into a Bailey weir meter from which the water entering it through the tailpipe is picked up by the economizer pumps. Makeup water connections from a storage tank and from the city mains are of course provided.

The condenser-heater is located close to the 36-in. free exhaust pipe of the main turbines, a 24-in. free exhaust valve capable of being opened or closed from the operating floor being between the latter and the 30-in. auxiliary exhaust pipe.

The turbine room, in keeping with construction of the Edison Electric Illuminating Co., is roomy and beautiful. There will be three 30,000-kw. single-cylinder impulse machines. The generators are provided with water connections and nozzles which may be used to put out a fire should the generator get afire.

Boston has reason to be proud of this station. The writer is appreciative of the excellent coöperation given in the preparation of this article by Charles L. Edgar, president; I. E. Moultrop, superintendent of construction, and George E. Seabury, superintendent, station-engineering department, Edison Electric Illuminating Co., Boston

Calculations from a Flue-Gas Analysis of Loss Due to Incomplete Combustion

By H. M. Brayton, M.E.

Loss due to incomplete combustion which results in low boiler efficiency is a problem that has confronted combustion engineers for many years. It is still a vital problem, and many mechanical and chemical means have been employed in the effort to detect and remedy this loss.

The average engineer in the up-to-date power plant is accustomed to gage the efficiency of his boilers by the percentage of CO_2 which his recording apparatus shows. This record, however, is really not complete, and with this information alone it would be impossible to calculate the loss due to incomplete combustion no matter how much mathematics might be used. An analysis from an Orsat apparatus which shows the percentage of CO is also necessary.

The object of this article is to show the practical man in as simple a way as possible just how to calculate this loss in his boilers and then to go a step farther and present a chart from which he can instantly read off the loss when once he knows the CO_2 and CO content of the flue gases. The analysis of flue gas is always stated in percentage of volume.

In order to fully explain each step as we proceed, it will be necessary to give some theory, but we shall make this theory as simple as possible. If the practical man

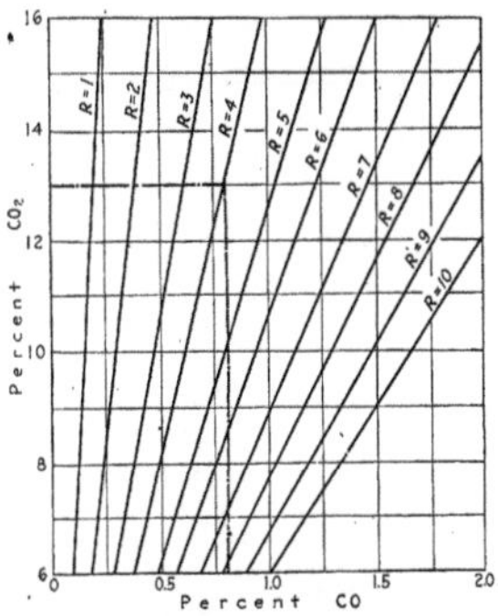

CHART TO DETERMINE PER CENT. OF HEAT LOSS FROM FLUE-GAS ANALYSIS

meets some new things, he must not think he cannot understand any of it, because we all have to meet and master new things every day. The vapor density of a gaseous compound is its weight referred to the weight of hydrogen as unity and is equal to one-half of its molecular weight. We all know that carbon dioxide (CO_2), as the formula indicates, is made up of one part of carbon which has an atomic weight of 12 and two parts of oxygen which has an atomic weight of 16. The molecular weight of a compound such as our CO_2 really represents is found by adding together all the atomic weights. In this case the molecular weight of CO_2 becomes $12 + 16 + 16 = 44$. It was stated above that the vapor density was just one-half the molecular weight, so that the vapor density of CO_2 would be 22.

Now let us turn our attention to the other compound with which we are to deal; namely, carbon monoxide (CO). The molecular weight of this substance is $12 + 16 = 28$, and from this we determine that the vapor density is 14.

Our flue-gas analysis tells us just what part of the gas is composed of CO_2 and just what part is composed of CO. Let us assume that out of 100 cu.ft. of the flue gas the analysis shows P per cent. of it to be CO_2 and P' per cent. of it to be CO. The remainder consists, of course, of air and other gases. Nitrogen comprises the major portion of what is left, but we are not now interested in that phase of the problem. The question is, How much do the CO_2 and the CO in our 100 cu.ft. of gas weigh when *compared to hydrogen as unity?* We may write

$$22 \times P = W \text{ } \textit{Weight of } CO_2 \quad (1)$$

$$14 \times P' = W' \text{ } \textit{Weight of CO} \quad (2)$$

From the atomic weights as stated we know that in each pound of CO_2 gas there is $\frac{12}{44}$ or $\frac{3}{11}$ of a pound of carbon. In each pound of CO there is $\frac{12}{28}$ or $\frac{3}{7}$ of a pound of carbon. So that in the W parts by weight of CO_2 as shown, we have $\frac{3}{11}$ W parts of carbon and of the W' parts by weight of CO we have $\frac{3}{7}$ W' parts of carbon. Then

$$\frac{3}{11} W + \frac{3}{7} W' = W'' \quad (3)$$

where W'' is the total weight of carbon in the 100 cu.ft. of flue gas.

The foregoing discussion leads up to the actual determination of the loss due to incomplete combustion of the carbon in the coal. If the practical man has been thinking that the percentage of CO_2 in his flue gases was the whole story, he must change his habits of thought. A very small amount of the CO gas will cause a very serious loss.

If we will substitute the values of W and W' as found in equations (1) and (2) into equation (3), we shall have an expression showing the total weight of carbon in 100 cu.ft. of flue gas in terms of the percentage of CO_2 and CO. This expression reads as follows:

$$\frac{3 \times 22 \times P}{11} + \frac{3 \times 14 \times P'}{7} = 6P + 6P' = W'' \quad (4)$$

Then $6P$ parts of carbon were burned to CO_2 and $6P'$ parts of carbon were burned to CO.

Heat of combustion is that amount of heat which will be given off when a fuel burns or unites with oxygen. When carbon burns to CO_2 in the furnace, it gives off exactly 14,650 B.t.u. per lb. When carbon burns to CO in the furnace, it gives off only 4,400 B.t.u. per lb. The enormous loss in the latter case is evident.

Now $6P$ parts of carbon burns to CO_2, hence the amount of heat generated by this carbon will be $6 \times 14{,}650 \times P$. (B. t. u.) and as $6P'$ parts of carbon burns to CO, the amount of heat generated by this carbon will be $6 \times 4{,}400 \times P'$(B.t.u.).

If all the carbon had burned to CO_2 the amount of heat generated would be $6\,(P + P')\,14{,}650$ (B.t.u.). From the foregoing equations we may obtain at once the amount of heat generated by the carbon which is burned to CO_2 and that generated by the carbon which is burned to CO. All we need to know is the values of P and P', which are given by the Orsat apparatus.

The percentage of loss due to incomplete combustion is given, of course, by the ratio between the actual amount of heat generated and the heat which could have been generated had all the carbon united with the oxygen of the air and formed CO_2. This ratio actually gives the efficiency, and to get the loss we must subtract from 100. The efficiency of the furnace may be written in general terms as follows:

$$\frac{100 \times 6\,(14{,}650\,P + 4400\,P')}{6\,(P + P')\,14{,}650} = \textit{per cent. efficiency} \quad (5)$$

or *per cent. efficiency* + *per cent. loss* = 100

$$\text{or } 100 - \textit{per cent. efficiency} = \textit{per cent. loss} \quad (6)$$

If now we expand equation (5) and boil it down, we secure the following workable equation which gives us the percentage of loss.

$$\frac{70\,P'}{(P + P')} = \textit{Per cent. loss due to incomplete combustion} \quad (7)$$

Equation (7) is of great value to the power-plant engineer. With it he can quickly determine how the efficiency of his plant is running.

For the engineer who may have occasion to solve equation (7) many times each day, it may be interesting and valuable to know how to make this formula into a chart from which the value of the percentage of loss may be instantly determined without calculation. Let R equal the percentage of loss. Equation (7) may then be written thus:

$$70 \times P' - P' \times R = P \times R$$

$$P' \times (70 - R) = P \times R \quad (8)$$

which is an equation of the first degree in three variables. We may therefore proceed to plot P against P' for different values of R. When plotted, we shall simply have a series of straight lines each marked with the respective value of R which was used in determining it.

The accompanying chart illustrates the foregoing statements. The percentages of CO_2 *(P)* are plotted as ordinates on the vertical scale and the percentages of CO (P') as abscissas on the horizontal scale. Each straight line is marked with its respective value of the per cent. loss R.

To locate these lines substitute some value of R in equation (8). For example when $R = 2$ equation (8) becomes

$$P' \times (70 - 2) = 2 \times P$$

therefore

$$P' = \frac{2}{68} P = 0.0294\,P$$

Now if $P = 6$, $P' = 0.0294 \times 6 = 0.176$; and if $P = 16$, $P' = 0.0294 \times 16 = 0.470$.

This will locate two points on the line $R = 2$ and the line may now be drawn. The other R lines are located in the same way.

The use of the chart is simply a matter of finding the percentage of CO_2 on the vertical scale, the percentage of CO on the horizontal scale, following each as indicated by the dot-dash line on the chart until they meet, and the R line nearest this point of intersection is the percentage of loss due to incomplete combustion with this analysis. The illustration on the chart was taken with the $CO_2 = 13$ per cent. and the percentage of CO at 0.8. These lines extended meet almost exactly on the $R = 4$ line. This is merely a coincidence. Note that if we have an analysis of $CO_2 = 14$ per cent. and $CO = 1$ per cent., the lines meet between $R = 4$ and $R = 5$ and it would be close enough to call it a loss of 4.5 per cent.

This chart will be found useful in the boiler room and also in the laboratory where analysis of the flue gases is studied. The modern demand for efficiency makes it imperative that every man in the organization shall know how to calculate loss due to incomplete combustion.

Overload Relays—Principles of Operation

By VICTOR H. TODD

The principles of operation of different types of simple overload relays are discussed, and then consideration is given to such characteristics as inverse-time-limit, definite-time-limit to control the time of operation of relays, and the use of relay switches to relieve the relay contacts of handling heavy currents.

THE rapid strides taken toward the perfection of apparatus for the generation, transmission and utilization of electric energy have been closely followed by improved instruments to protect the apparatus and preserve continuity of service under all conditions. Like a silent sentinel, the protective relay stands on guard on the lines, day and night, ready to detect trouble instantly and disconnect faulty apparatus or sectionalize defective lines with almost human intelligence and with more than human accuracy.

Protective relays are very sensitive but ruggedly constructed electrical instruments interposed between the lines and the operating mechanism of a circuit-breaker in such a way that any disturbance on the line or in the apparatus does not act directly on the breaker, but operates the relay, and the relay, in turn, causes the circuit-breaker to open.

The function of the simple overload relay is to trip the circuit-breaker when the current exceeds a definite value. Incidentally, it may be stated that the term "overload" is used synonymously with "excess current." This latter term is the correct one, but general practice has been to use the term "overload" or "overload relay," when in reality what is meant is "excess current" or "excess-current relay." When a relay operates on actual load—that is watts—it is called a "watt relay" or "power relay."

A relay operating on the effect of a solenoid to raise an iron plunger, thus closing or opening contacts, is shown in Fig. 1. Referring to the diagram of parts shown in Fig. 2, winding *A* is wound around the iron core *B*. Supported at the two poles *N* and *S* is an iron plunger *C* arranged so that it may slide up and down. When the current in *A* reaches a certain value, the iron core *C* is lifted, thus closing the contacts *D* and *E* with bridge *F*, which will immediately trip the breaker, as previously described.

When the current is greater than 1000 amperes, a winding is not necessary, as the magnetism from the straight bar or cable produces sufficient flux to operate the relay. The relay may then take the form shown in Fig. 3, the cable simply passing through the large hole *H*, which is surrounded by insulating material. In Fig. 4 is another modification which may be used on a busbar that runs vertically instead of horizontally. In this case the magnetic circuit is simply clamped around the busbar, which is of course insulated.

In order to adjust these relays to operate on various loads, the plunger *C*, Fig. 2, is arranged with a nut *G* by means of which the plunger may be raised or lowered on the stem *H*. Thus if the plunger is at the lowest point, it will take a maximum of current to raise it, but if it is set high, then it will rise on a minimum of current.

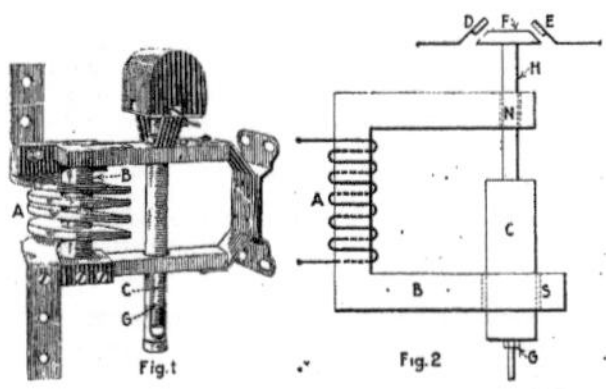

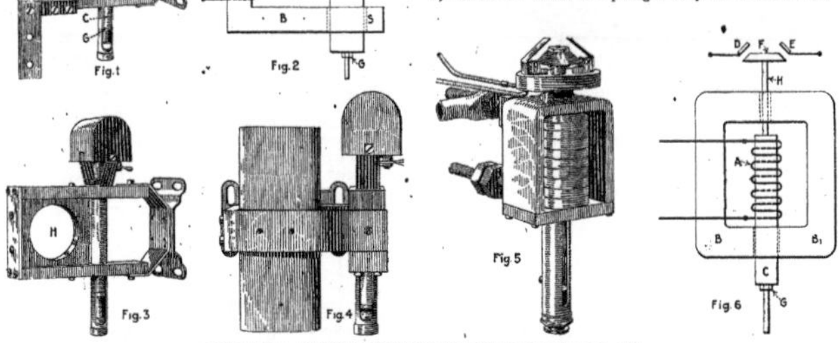

FIGS. 1 TO 6. TYPES OF INSTANTANEOUS OVERLOAD RELAYS

Another form of overload (excess-current) relay, utilizing the same principle of operation as those already described, is shown in Fig. 5, and the arrangement of parts in Fig. 6. Coil *A* is wound on the central part over the iron plunger *C*, and the magnetic circuit is completed by the two parts *B* and B_1. The action is identical with the previously described relay; namely, when the current reaches a certain value, the plunger *C* is lifted upward, thus causing the contact disk *F* to short-circuit the two contacts *D* and *E*, which complete the circuit that trips out the breaker. In another type of small capacity, adjustment is made by using taps on the winding; however, this cannot be done in capacities of several hundred amperes. The great advantage gained by the simple relay described further on in this article has discouraged the use of plunger-type relays on direct-current circuits. If a plunger-type relay is to be used with a shunt, as has been done in rare cases, the adjustment for load is made by varying the drop of the shunt.

In Fig. 7 is shown a type of simple overload relay which is connected by circuit to a shunt. In the diagrammatic scheme of parts, Fig. 9, the iron armature *A*, carrying the contact *B* and pivoted at *C*, is held in its normal position (contacts *B* and *D* open) by the tension

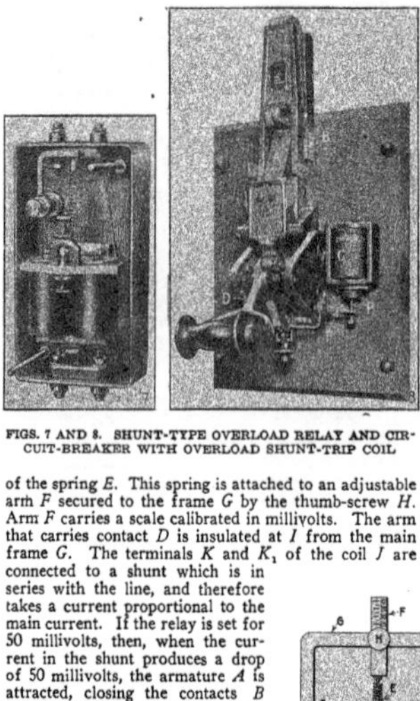

FIGS. 7 AND 8. SHUNT-TYPE OVERLOAD RELAY AND CIRCUIT-BREAKER WITH OVERLOAD SHUNT-TRIP COIL

of the spring *E*. This spring is attached to an adjustable arm *F* secured to the frame *G* by the thumb-screw *H*. Arm *F* carries a scale calibrated in millivolts. The arm that carries contact *D* is insulated at *I* from the main frame *G*. The terminals *K* and K_1 of the coil *J* are connected to a shunt which is in series with the line, and therefore takes a current proportional to the main current. If the relay is set for 50 millivolts, then, when the current in the shunt produces a drop of 50 millivolts, the armature *A* is attracted, closing the contacts *B* and *D*, which closes a circuit from *L* and L_1 to the circuit-breaker's trip coil.

In the circuit-breaker, Fig. 8, contacts *A* and *B* carry the main current and are arranged to open the circuit in the regular manner. But on the side of the breaker is an electromagnet *C* wound with a large number of turns of fine copper wire. When this coil is energized, it pulls up the plunger *P*, which strikes the trigger *F* and releases the operating arm at *D*, thus releasing the moving contact *A* and allowing the breaker to open. Fig. 10 gives the diagram of connections of the relay and circuit-breaker under normal load, the path of the currents being shown by arrowheads. An overload causes the relay contacts to close the trip-coil circuit to the breaker and the latter opens the circuit, thus relieving the overload on the system. In installations where the potential trip-coil circuit is connected to the circuit to be controlled, the overload trip attachment on the breaker should always be connected in the circuit, since dead short-circuit on the line may cause the voltage to drop so low that it will not operate the potential trip coil on the breaker. This allows the overload attachment on the breaker to be set high, for protection against short-circuits or other violent disturbances, but the relay is set so as to give protection against moderate overloads.

The relay has a more important duty, however, than simple excess-current protection. Suppose, for example, a motor is driving a machine that strikes a hard spot in the work and then is released after a very short period, or an accidental short-circuit occurs on the line and is burnt off in a flash. Either one of these causes produces an excess current which might trip the breaker, thus stopping all work until it could be reset and the machines started again. But the excess current came on and off in so short a period that no damage was done. It would have been better if the circuit-breaker had stayed in two or three seconds, and then if the current dropped to normal in this time there would be no interruption. On the other hand, if the overload continued there would still be ample time to open the circuit before any damage could be done. This is the function of the time-limit relay; it delays the opening of the circuit-breaker a definite period after the instant of excess current, thus allowing time for the current to drop to normal, if the fault is cleared within the time setting of the relay.

Fig. 11 shows a definite-time relay with the cover removed and Fig. 13 gives a schematic diagram of parts. The solenoid *A* has an iron plunger *B* which under normal condition rests on the moving arm *C*, pivoted at *F*, which carries a contact *D* and a counterweight *E*. When the solenoid *A* is energized, the core *B* is raised upward instantly; relieved of this weight, the counterweight *E* now causes the contact *D* to start upward to meet the upper contact *G*. However, attached to the arm *C* is a piston *H* working within a cylinder *I*, which retards the movement of arm *C*, making it move very slowly as the air escapes around the plunger. Then, after a definite time, from 1 to 5 seconds, depending in the initial distance between contacts *D* and *G*, the contacts *D* and *G* close, thus closing the circuit to the shunt-trip coil of the circuit-breaker, causing the latter to open.

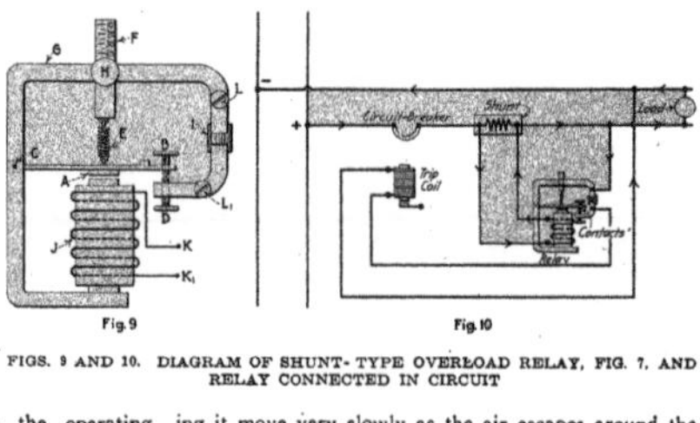

Fig. 9 Fig. 10

FIGS. 9 AND 10. DIAGRAM OF SHUNT-TYPE OVERLOAD RELAY, FIG. 7, AND RELAY CONNECTED IN CIRCUIT

If the current drops to normal before contacts *G* and *D*, Fig. 13, close, the solenoid allows the plunger *B* to drop, thus forcing the arm *C* downward into normal position. In order that the relay may reset quickly, a valve is provided in the dashpot plunger. This valve consists of a little steel ball *J*, which closes the air ports *K* when the piston moves upward and attempts to force air out of the port, but raises and allows the air to enter readily when the piston moves down, as in resetting. An outside view is given in Fig. 12 of a definite-time-limit relay similar to that in Fig. 11. This relay will close the circuit in the number of seconds that arm *A* points to on scale *S*.

The types of relays Figs. 1 to 5, if desired, may close the circuit to a definite-time-limit relay instead of tripping the breaker instantly, but then, while the action is selective the cost renders its use prohibitive. Selective action, except in very heavy short-circuits, may be obtained by lagging the time of the tripping, and making the relay an inverse-time-limit device. That is, the greater the overload the quicker the time. In fact, in actual practice, the instantaneous relay has a very limited use; an inverse-time-limit relay costs only slightly more, gives the same protection and will not interrupt service on transient short-circuits. The latter type is shown in Figs. 14 and 16. The plunger, in rising, compresses the air in the leather bellows *B*, which resists its upward movement. In the top of the casting to which the bellows is attached, is an air passage *C*, which may be partly or entirely closed by turning the screw *S*. If the air passage is entirely open, the action is almost instantaneous on moderate overloads, but if closed, the time may be anywhere from 1 or 2 sec. to 20 or 30 sec.

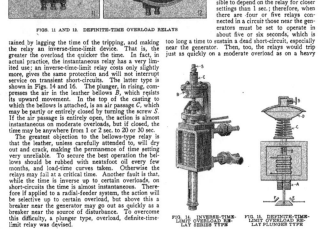

FIGS. 11 AND 12. DEFINITE-TIME OVERLOAD RELAYS

The greatest objection to the bellows-type relay is that the leather, unless carefully attended to, will dry out and crack, making the permanence of time setting very unreliable. To secure the best operation the bellows should be rubbed with neatsfoot oil every few months, and load-time curves taken. Otherwise the relays may fail at a critical time. Another fault is that, while the time is inverse up to certain overloads, on short-circuits the time is almost instantaneous. Therefore if applied to a radial-feeder system, the action will be selective up to certain overload, but above this a breaker near the generator may go out as quickly as a breaker near the source of disturbance. To overcome this difficulty, a plunger type, overload, definite-time-limit relay was devised.

A definite-time-limit relay is shown diagramatically in Fig. 15. The plunger *A* is not rigidly attached to the

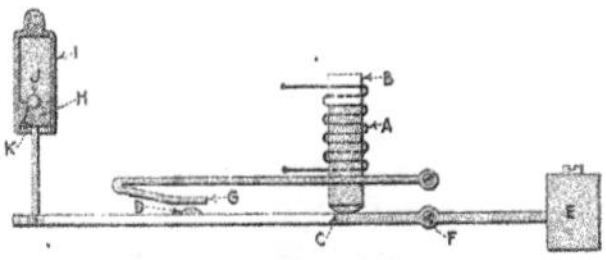

FIG. 13. SCHEMATIC DIAGRAM OF THE DEFINITE-TIME-LIMIT OVERLOAD RELAY, FIG. 11

stem *B*, as in the type previously described, but slides freely on it. If an overload occurs the plunger is raised and compresses the spring *C*, which in turn forces the stem *B* upward against the resistance of the bellows *D* and finally closes the contacts *E* and *F* with the disk *G*. It will be readily seen that no matter how severe the overload may be, it can only compress the spring *C*; consequently, the upward pressure on the bellows stem is constant regardless of overload, and the time is therefore constant. The duration of time is varied by opening or closing the air valve *S* as described for inverse-time-limit types. At the first glance this might appear the solution of radial protection, but it is impossible to depend on the relay for closer settings than 1 sec.; therefore, when there are four or five relays connected in a circuit those near the generators must be set to operate in about five or six seconds, which is too long a time to sustain a dead short-circuit, especially near the generator. Then, too, the relays would trip just as quickly on a moderate overload as on a heavy

FIG. 14. INVERSE-TIME-LIMIT OVERLOAD RELAY SERIES TYPE

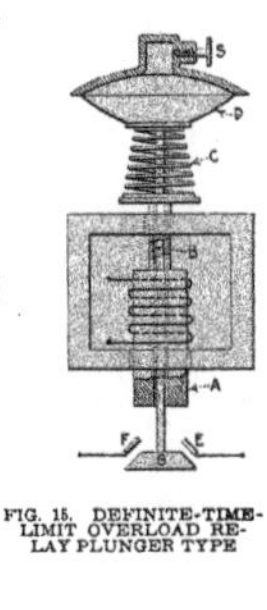

FIG. 15. DEFINITE-TIME-LIMIT OVERLOAD RELAY PLUNGER TYPE

overload, which is not at all desirable. Were the foregoing of great importance, it would be necessary to perfect a relay accurate within small percentage of sustained accuracy and one whose curve was inverse up to certain overloads, after which it would become a definite-time-limit device. However, owing to the unquestioned superiority of alternating current for high-tension long-distance transmissions and the comparatively small size of most direct-current radial systems of transmission, relay engineers have devoted most of their energies to the perfection of alternating-current relays which are to a high degree perfect in their protection. In large power plants or factories, however, where there are numerous machines that must be kept running unless actually damaged, a radial system of protection may be adopted with success.

FIG. 16. INVERSE-TIME-LIMIT OVERLOAD RELAY

This brings up an important use for definite-time-limit relays. Consider the distribution system shown in Fig. 17. Each time the line divides to supply a set of feeders, a definite-time-limit relay is supplied to operate a double-pole circuit-breaker. For instance, the feeder from the busbar is protected by breaker *A*, the next subdivisions are protected by breakers *B* and *C*, and the next by circuit-breakers *D*, *E*, *F* and *G*. Suppose a heavy overload occurs on the feeder protected by breaker *D*. The excess current extends all the way

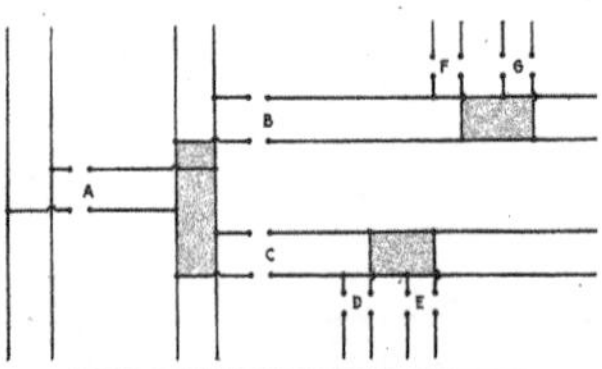

FIG. 17. DIAGRAM OF DISTRIBUTION SYSTEM

back to the main bus, and were definite-time relays not used, breaker *A* would go out as soon as breaker *D*, thus interrupting every circuit connected to the feeder protected by breaker *A*. But this is where the definite-time-limit relay enters in. The relay at *E* is set, say, for 1 sec., *B* and *C* for 2 sec., and *A* for 3 sec. Thus when the disturbance occurs, all the relays of breakers *A*, *B*, *C* and *E* start to operate, but at the end of 1 sec., breaker *E* opens, relieving the excess current, and all the other relays reset quickly, confining the disturbance to the one line on which it occurred. Had the disturbance occurred on feeder *C*, then the breaker at *C* would have gone out in 2 sec.; breaker *A* would not have had time to open and feeder *B* would not have been interrupted. In all the relays so far, we have assumed that the contacts themselves closed the circuit to the trip coil of the breaker, but when the breakers are large and require considerable current to trip them, the contacts of the delicately constructed relays are not heavy enough to safely close the heavy current required. To overcome this difficulty, a relay switch as shown in Fig. 8 is used. This switch consists of a solenoid *S* with an iron plunger *P*, to the bottom of which is attached a loosely held carbon disk *C*, insulated from the metallic plunger. When the overload or the definite-time-limit relay closes its contacts, it closes the circuit to the relay switch, and the energized solenoid instantly pulls its plunger upward, thereby pressing the carbon disk *C* against the two stationary carbon contacts *D* and *D*. Short-circuiting these contacts closes the circuit to the shunt-trip coil on the circuit-breaker. The contacts being of carbon, will carry a heavy current and will not stick. In another form of relay switch, Fig. 19, the plunger simply pushes up a pivoted arm, thus closing the two contacts *D* and *D*.

FIG. 18. RELAY SWITCH WITH CARBON CONTACTS

Sometimes it is desirable to trip two or more breakers at once with the same relay switch. In this case the disk is generally made of copper, and two or three sets of stationary contacts are used, thus closing two or three circuits simultaneously. It must be remembered that the arcing at the relay contacts will always be a great deal more severe when opening a circuit than when closing one. For this reason a relay should never open the trip circuit once established. If the trip circuit is fed from the load side of the breaker, it will be opened automatically when the breaker opens and the circuit will be dead when the relays reset. Should it be necessary to connect the shunt-trip circuit to the line side of the breaker, or if a separate circuit is used, then a switch must be arranged to open the trip circuit as soon as the breaker opens, thus relieving the relay contacts of this duty.

FIG. 19. RELAY SWITCH CONTACTS AT TOP

The *Compressed Air Magazine* has attained its 25th birthday and the position of the leading, if not the only, exponent of that important industry. May it live to see many happy and abundant returns.

Steam-Turbine Governors—Governors of Small Westinghouse Turbines

This article deals with direct-shaft governors and vertical-shaft governors for small Westinghouse steam turbines such as are used for pumps and electric generators.

THE Westinghouse shaft governor proper differs little from the usual shaft governor, several of which have been described in the first article of this series. Fig. 1 shows a small Westinghouse turbine, its governor and generator. The governor spindle is moved out with an increase in speed of the turbine and in doing so pushes forward the lever A, Fig. 2. The lever is pivoted at *B* and connects with the stem of the governor throttle valve *C*. On these small machines check nuts are provided on the valve stem to adjust the "play" of the lever and allow a greater or less movement of the governor lever for a given movement of the governor valve. A spring holds the lever against the governor spindle, and the tension of this spring may be adjusted by means of the handwheel.

As with all governors, the operator should see that the governor lever is always free to move easily and that the governor valve stem moves freely in its stuffing-box. Of course the governor throttle should be closed before the governor weights have reached their full travel.

Most small Westinghouse turbines use a vertical shaft, flyball type governor, driven from the turbine shaft through spur bevel gearing or worm and wheel. An exterior view of this governor is shown in Fig. 1, and another view is shown in Fig. 3. The spindle for the oil pump is continued from the governor spindle; the lower casing shows the handhole with cover.

Fig. 4 shows a section of this governor; this view shows only one governor weight (there are two, of course) so that the governor spring bolt may be shown. The weight *A* together with the weight arm *B* is mounted on knife-edges *C* and *D*. The action of centrifugal force is opposed by the helical spring, the tension on which is adjusted by the nut *E*. The governor weights are steel cylinders riveted between the weight arms. Notice that the governor arm block *F* is riveted to the lower part of the governor arm, that it has two knife-edges—*C*, resting on a steel block on the governor weight disk, and *D*, held between two blocks, the lower one in *F* and the upper one in the under side of the governor-spring sleeve *G*, which is loose on the governor spindle and revolves with the spindle. The purpose of the sleeve is to transmit the pressure from the knife-edges to the spring.

The sleeve *G*, in moving up and down under the action of the weights and spring, transmits motion to the governor clutch *H*, which carries the trunnion to which the governor levers which operate the governor throttle valve are attached.

A large nut *J* is screwed on the lower end of the sleeve and takes the upward thrust of the sleeve upon the clutch *H*; this nut has a pin and cotter nut to retain it in position.

There are quite a number of different arrangements of levers for transmitting motion from the governor sleeve to the governor throttle. Fig. 3 is a good view of a typical arrangement. Here the governor case is removed, showing the weights, the top part of the governor sleeve, the spindle, spring, adjusting nut and the governor weight disk.

Adjusting the Governor—The important thing to be sure of always is that the governor valve closes before the governor weights have reached their extreme throw or outer position. This will avoid a runaway.

Changes in speed are had by means of the nut *E* and the governor-spring adjusting-nut seat. The flange of

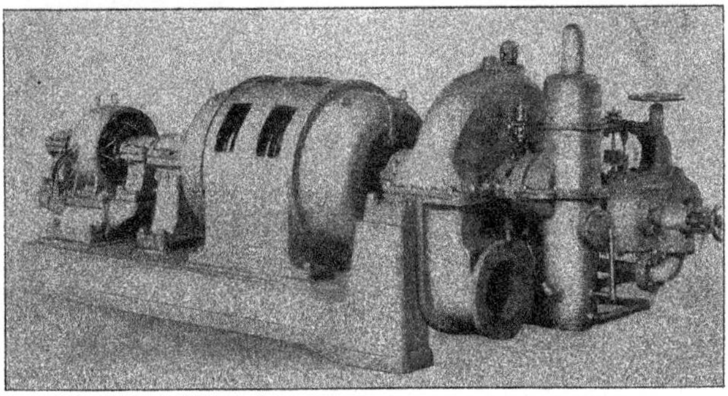

FIG. 1. SHOWING SMALL WESTINGHOUSE TURBINE AND GOVERNOR

the governor-spindle adjusting nut and the adjusting-nut seat have a number of holes into which a pin may be screwed and held in place by a cotter pin. This fixes the adjustment made to the governor. The adjusting-nut seat is held from turning on the spindle by a pin and keyway.

Usually, the first indication of trouble is that the governor hunts. Before disturbing the adjustment of

FIG. 1. SHAFT GOVERNOR, WESTINGHOUSE TURBINE

the nut *E* or the adjusting-nut seat, make certain that no parts of the governor, clutch, levers and governor valve stick because of gummed oil, dirt, lack of lubrication, or dry packing or too tight a gland nut on the valve spindle. This means a close examination of the governor.

If hunting is not because of any of the foregoing reasons, try to get the proper regulation by compressing the governor spring by means of the nut *E*. If this cannot be had by this means, cut out some of the spring coils by screwing the spring farther on to the governor-spring adjusting seat. This may be overdone, and if so, the governor will not be sufficiently sensitive.

More coils of the spring are then required, and the spring must be unscrewed from the adjusting-nut seat, a little at a time.

Close regulation may now be had by means of the nut *E provided there is not excessive lost motion in the levers between the governor valve stem and the governor clutch sleeve.*

The Governor Valve—Fig. 5 shows the governor valve and the automatic throttle valve. The former is of the double-disk poppet, balanced type. Notice that the valve is held to its seats by a spring.

The throttle valve is of the automatic type. The valve has a bypass which, when the throttle-valve wheel is turned to open the valve, lifts the bypass valve from its seat, thus putting steam line pressure on top of the throttle-valve seat, putting the valve in balance. Further turning of the handwheel will bring the throttle-valve stem collar *A* against the lower valve-stem nut *B*. This nut is loose and will move up and down.

In Fig. 5 the throttle is closed. When the valve is open it is held in that position by the latch *C*, which may be released by the automatic stop or emergency governor flying out, striking the trigger *D*. When the latch has been released, the spring on the throttle-valve nut forces the valve to its seat, closing it.

If the emergency governor has tripped the throttle or if it has been tripped by hand, the throttle cannot be opened until after the handwheel has been turned in the closed direction until the valve-stem nut is in its upper position and the valve-latch guide rod engaged the latch.

The Automatic Stop Governor—In the lower left-hand corner of Fig. 5 is shown the automatic or emergency governor, which is located in an extension of the governor driveshaft. It consists chiefly of a spring *A* and plunger *B*. Centrifugal force tends to make the plunger fly out; but the spring resists this tendency. When the speed increases the centrifugal force sufficiently to overcome the spring, the plunger flies out, striking the trigger *D*, which will release the automatic throttle valve.

To change the speed at which the emergency governor operates, insert or remove liners from the top of the spring at *C* in the lower left-hand corner of Fig. 5. There is a cotter pin that prevents the nut, which in-

FIG. 3. GOVERNOR WEIGHT CASE REMOVED; SHOWS SPRING, WEIGHTS AND LEVERS CONNECTING WITH GOVERNOR VALVE

closes the spring and plunger, from turning. Be sure this pin is in place after adjusting or taking the governor apart. Pulling up on the flyball-governor valve stem will overspeed the turbine, and if working properly the emergency governor will trip the throttle within 2 or 3 per cent. of the same speed every time it trips. Note the speed at which the emergency trips; it should be at 10 per cent. above the normal speed of the turbine. If it does not do this, make sure that the throttle does not stick. If it does not stick, then adjust the emergency to trip at 10 per cent. overspeed by the use of the liners already mentioned.

If during usual runs the emergency trips, it is reasonable to conclude that the turbine was overspeeding when it tripped. Take time to find out the cause before resetting the throttle and starting the turbine. This is important, for it may be that the turbine may soon overspeed again and the emergency fail to trip. This will likely cause a serious accident.

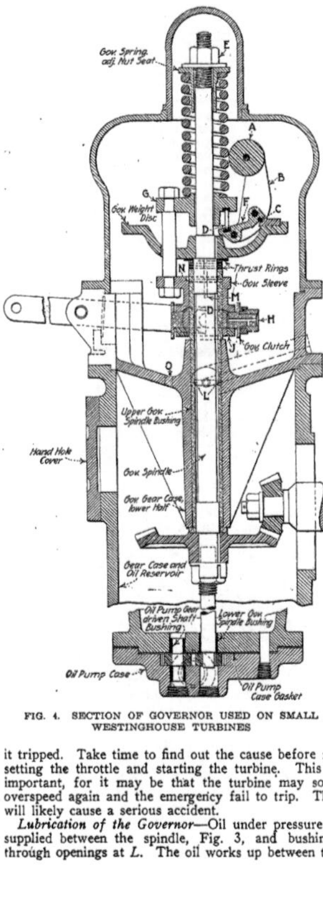

FIG. 4. SECTION OF GOVERNOR USED ON SMALL WESTINGHOUSE TURBINES

Lubrication of the Governor—Oil under pressure is supplied between the spindle, Fig. 3, and bushings through openings at *L*. The oil works up between the spindle and the bushings to a point *M* and then out *N* between the spindle and the sleeve. From here it finds its way through other parts to the trunnions. Some oil also reaches the thrust rings of brass and steel, immediately under the governor weight disk. The oil collects at *O* and from here is led to the gears through a small pipe.

The importance of having a clean and adequate supply of oil and of keeping all oil parts and channels free of dirt and gummy oil is obvious. The success of the whole turbine depends upon this, as with any similar mechanism.

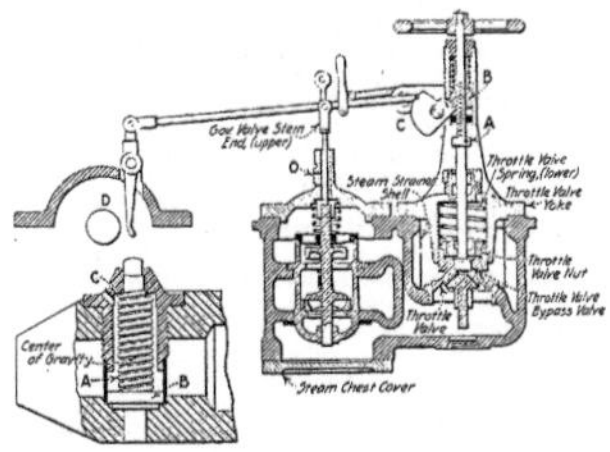

FIG. 5. GOVERNOR VALVE AND AUTOMATIC THROTTLE VALVE

Change in the Design of Rotary Converters and Generators

During the year 1919, the Westinghouse Electric and Manufacturing Co. completed service tests to determine the value of oscillators on rotary converters. The tests indicated that the oscillators should not be used. This feature is, therefore, being omitted from this company's standard synchronous converters, and improved results are being obtained.

The compensating winding has been incorporated in the design of all of the generators of the larger motor-generator sets built by this company. The use of this winding has considerably improved commutation over that of the machines having only commutating poles. The compensating winding, in addition to the commutating poles as incorporated in the recent designs, is particularly valuable from an operating standpoint on very heavy peak loads.

Flywheel Governor Bursts

Shortly before five o'clock on Jan. 5, the flywheel governor on a Harrisburg Standard engine at the Lock Haven Silk Mill, Lock Haven, Penn., burst. Fortunately, no one was injured and little damage was done except to the engine.

The governor was of the centrally balanced centrifugal inertia type and had been pounding slightly since a new piston valve with expansion rings was installed on Dec. 21. The engine was direct connected to a 40-kv.-a. generator which had been carrying about 15 per cent. over its rated capacity. The engine was in charge of C. M. Glossner, the operating engineer, at the time of the accident.

Refrigeration Study Course—II. The Ammonia-Compression System

By H. J. MACINTIRE
Professor of Mechanical Engineering, University of Idaho

IN THE first article of this study course an attempt was made to show what refrigeration really is and what means are resorted to to achieve this end. It was shown that refrigeration in nature is common, but in its practical application in appreciable capacity it was necessary to find a new method, one utilizing a vapor like ammonia, which, in changing from a liquid to a gaseous state (at the proper pressure), would boil at a relatively low temperature. Also that it was necessary to use the same refrigerant over and over again, which meant that a pump or compressor would have to be provided to increase the pressure sufficiently high that the gas might be condensed at atmospheric temperature. This process was shown to be similar to the steam-engine cycle, only with the events in the reverse order. The object of the gas pump, or compressor, is simply and solely to increase the pressure, thereby making it possible to condense the compressed ammonia by cooling with water at 60, 70 or 80 deg. F. temperature. Fig. 1 shows a typical refrigerating plant.

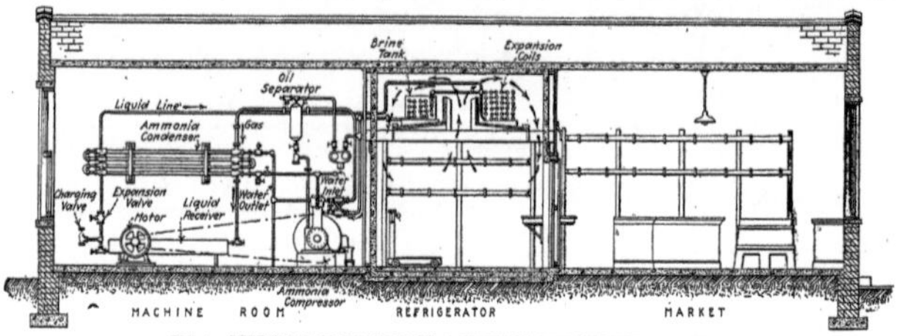

FIG. 1. SECTIONAL ELEVATION OF A TYPICAL REFRIGERATING PLANT

The High-Pressure Side

The reader should understand the reason for the use of the compressor—what it accomplishes and what is expected of it. There is really no unusual feature about it. The dry-air vacuum pump draws in the rarefied air at the pressure maintained in the surface condenser and pumps it up to an amount a little in excess of atmospheric pressure, when it is exhausted out of the condenser. The deep-well pump takes water at the level in the well and lifts it up to the cistern or overflow. In each case work is done on the substance, it is lifted in level (pressure level in one case and datum level in the other). In the case of ammonia we are able to lift it in still another sense—in the temperature level, because by increasing the pressure we can condense the vapor at a relatively high temperature.

As in compressing air, the ammonia gas becomes hot during compression and the discharge gas has a pressure of 125 to 185 lb. per sq.in. and a temperature of between 150 and 250 deg. F. This compressed hot gas is allowed to leave the compressor and pass into a kind of aftercooler, which is a coil of pipe suitably water-cooled so that finally the temperature of the gas is reduced down to within 15 to 20 deg. of the initial temperature of the water. From what has already been said it is clear that long before this has occurred condensation of ammonia has been completed and we have a liquid ammonia under a pressure approximately that of the discharge from the compressor. This liquid is collected into a drum, or header (called a liquid receiver), and is now ready for a second round through the refrigerating coils. The compressor, condenser and receiver are often called the high side.

The liquid ammonia stored in the receiver is under pressure and will flow through the pipe line to the cold-storage room similar in action to the manner of water in a pipe. At the cold-storage room, or brine tank, there is a valve called an expansion valve. It is not an expansion valve at all; it is only a pressure-reducing valve. It has been shown that to get a boiling temperature of zero degrees F., the ammonia must be kept constant at 15.23 lb. gage. So then, the expansion valve allows the pressure to drop from 150 lb. to 15 lb. and the liquid squirts into the coils like water under pressure through a sand hole in a pipe fitting. If the piping is properly arranged and proportioned, it will immediately begin to boil because of absorbing heat through the pipe coils from the air in the room, in consequence of which it becomes vaporized. As a vapor the ammonia has little refrigerating value, and it is allowed to pass out of the coils promptly and return to the compressor, which compresses it again.

The Refrigeration Cycle

And so it will be seen that the refrigeration cycle has three parts—the refrigerating coils, the compressor and the condenser. The refrigerating coils in action are much like the boiler in a steam-engine plant. The coils of pipe are so arranged as to be in contact with relatively hot air. This heats the pipes and boils the ammonia, vigorously or not, depending upon the temperature difference between the air temperature in the cold-storage room and the boiling temperature of the ammonia. Now, in a steam boiler, if boiling took place the steam would accumulate, and unless the engine or some other apparatus took the steam away as fast as it was generated, the pressure would rise until finally the safety valve would blow. On the other hand, if the steam were used too fast the steam pressure would drop and the temperature would be lowered in consequence. That the pressure may stay constant in a steam boiler we have to have means of generating just the same amount of steam as is used each minute.

Likewise the refrigerating plant must be so ordered that the vapor boiled from the refrigerating coils will be just enough for the compressor. If the boiling action is too rapid and more vapor is formed than can be pumped by the compressor, the boiling *pressure* will increase and the boiling *temperature* will likewise increase, perhaps to a point where satisfactory results will not be obtained. On the other hand, too high a speed of the compressor will lower the boiling pressure and temperature and poor results again will be encountered. The cubic feet of piston displacement of the compressor must be inter-related.

For example, suppose we have a room to be held at 36 deg. F. and 500 sq.ft. of cooling coils is supplied. The design allows a suction pressure of 37½ lb., which corresponds to 24 deg. F. boiling temperature of the ammonia. The temperature difference on the two sides of the pipe is 12 deg. F., and in consequence a certain amount of heat will flow through the pipe from the air to the ammonia. The amount will vary directly in proportion to the product of the area of the pipe and the temperature difference. The result is that ammonia is evaporated and it will be given off at the rate of 1.8 cu.ft. per minute.

It may not be clear to everyone that a gas or vapor has weight. Yet in a steam engine the steam condensed may be weighed, and we know that it requires 15, 20 or 30 lb. of steam to produce a horsepower for one hour. Likewise, air has weight, as we know from the successful accounts of the lighter-than-air or "Blimp" airship. So each pound of ammonia evaporated will occupy a certain particular volume, depending on the pressure. For instance, a pound of ammonia at—

200-lb.	gage	will	occupy	1.40	cu.ft.
150-lb.	gage	will	occupy	1.82	cu.ft.
100-lb.	gage	will	occupy	2.57	cu.ft.
30-lb.	gage	will	occupy	6.3	cu.ft.
20-lb.	gage	will	occupy	8.0	cu.ft.
10-lb.	gage	will	occupy	11.0	cu.ft.
0-lb.	gage	will	occupy	17.8	cu.ft.

Refrigeration a Matter of Heat Transfer

In the steam boiler it is usual to specify 10 sq.ft. of heating surface per boiler horsepower, and a boiler horsepower as the evaporation of 34½ lb. of water from 212 deg. to dry steam at 212 deg. F. In refrigeration we cannot proceed just like that. The refrigerating coils are really a boiler, but conditions are not always the same. For instance, we may design the coils to have 10 deg., 15 deg., or 20 or more degrees difference in temperature between the inside and the outside of the pipes. The greater the difference in temprature the more capacity may be obtained from the pipe refrigerating surface. Again, the pipe surface may be wet or frosted slightly, or heavily coated with ice. It is not reasonable to expect that the same amount of boiling action will occur in each of these cases, nor in the case where the inside of the pipe is filled with oil. In other words, there is a so-called coefficient of heat transmission for refrigerating piping as there is in all heat transmission piping.

As refrigeration is almost entirely a matter of heat transfer, it is desirable to be able to calculate how this takes place. The equation which we use is

B.t.u. = area of surface in square feet × coefficient of heat transfer × the temperature difference in degrees F. on the two sides of the surface.

The coefficient of heat transfer has to be found experimentally or by experience on similar kinds of work. The following problem will show the method of calculation. Suppose we have a direct expansion pipe composed of 1200 lin. ft. of 1¼-in. pipe. The ammonia boils at 25 lb. gage (11.7 deg. F.), and the room temperature is to be maintained at 34 deg. F. It is required to find the refrigeration to be supplied. The usual value of the coefficient of heat transfer under these conditions is 2.0 B.t.u. per hour, then

$$\text{Total B.t.u} = \frac{1200}{2.3} \times 2 \times (34 - 11.7) = 23{,}200 \text{ per hr.}$$

Steam boiler capacity is defined as the rate of *heating*, or at 33,479 B.t.u. per hr. In like manner we define the standard of refrigeration (the ton) as the rate of *cooling*, but in this case at the rate of 12,000 B.t.u. per hour (or 200 B.t.u. per minute). So, in the foregoing problem the tonnage will be $\frac{23{,}200}{12{,}000} = 1.94$ tons of refrigeration, and a machine would have to be selected which would compress enough vapor to supply this amount of refrigeration. How this may be determined for various conditions will be shown later.

Soldering a Handhole Seat

By Melford Kurtz

On one of the boilers at my plant, having a beveled ground joint seat at the handholes, the covers can be changed only when a tube is taken out, and it is some job to keep this boiler tight. I have used a number of lead gaskets which work very well where the pitting is not too bad in the seat, but a few days ago, after cleaning the boiler, one of the seats was found so badly pitted that the lead gasket could not be compressed in the bottom of the hole. I tried putting in an extra piece of lead, but it leaked and I was forced to drain that element.

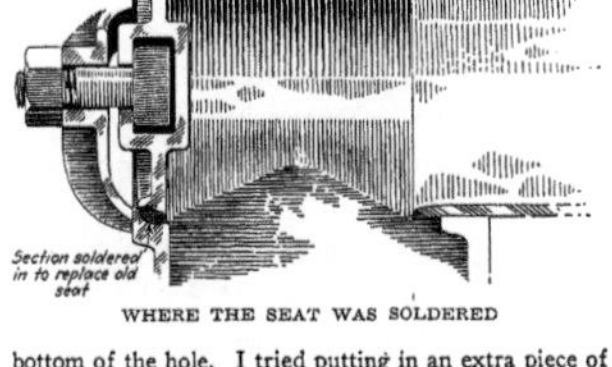

WHERE THE SEAT WAS SOLDERED

After scraping the bottom part of the seat, it was tinned and a shoulder of solder built up. It was then peened and scraped down to the proper shape. The necessary lead gasket was then put in place, the boiler closed up and, after getting up steam, cut in on the line. My troubles are over with that boiler, at least for some time to come.

Steam-Cylinder Lubrication

By W. F. OSBORNE

How to determine whether a steam-engine cylinder has either too much or too little oil. A discussion of the factors which affect the lubrication of the cylinder and of the best kinds of oil for the various conditions. The article concludes with a summary of the points brought out, arranged in numbered statements, and shows by examples how to use this summary to select the proper oil for any given condition.

IT IS not as easy to know when a steam cylinder is efficiently and economically lubricated as it is to check up bearings, because the inside surface of the cylinder cannot be seen or felt if the engine is running. Lack of lubrication in a bearing very quickly shows up by increased temperature. Wear in a cylinder is usually not detected except on an examination when the cylinder head is removed. Cylinder heads cannot be taken off every day, so the tendency is to use plenty of cylinder oil in order to be sure of getting enough.

A perfectly lubricated cylinder will be highly polished and have a glaze over its surface. It will have no rough or dull spots, nor should there be any sign of rust on examination immediately after stopping the engine. The color will vary from a bright iron-white to a light brown or steel blue. When the cylinder head is removed immediately after stopping the engine, there should be a film of oil over all the surfaces thick enough to saturate three or four thicknesses of cigarette papers of uniform thickness of 0.001 in. Even when the engine has stood for several hours, the oil should not be completely evaporated, although of course the film will not be so thick. The stain on the paper should be a clear brown; if it is black or has black particles or streaks, the cylinder or rings may be wearing or the oil is carbonizing. If there are pools of oil lying in the bottom of the cylinder or in the counterbore, too much oil is being fed and the quantity may be reduced. If the cylinder walls are dry in spots or show signs of wear, either too little oil is being fed or the wrong kind of oil is being used.

Examine Cylinder After Engine Stops

It is always important to examine the cylinder as soon as possible after the engine has been stopped, because the heat of the metal may evaporate the oil and the cylinder will appear dry, even though it may have been perfectly lubricated when the engine was running.

Lack of lubrication will sometimes be shown when the engine is running, by sticking valves or groaning sounds from the cylinder. The remedy is to use more oil until an opportunity comes for an examination of the inside of the cylinder to find out the real cause. Economy of lubrication can also be checked up by examining the exhaust steam or the piston-rod leakage. If the condensed steam shows considerable quantities of liquid oil, either too much is being used or the oil is not being properly atomized; if it shows minute drops of oil and is milky in color, we may feel confident that atomization is complete and that we are not feeding too much. When no oil is fed to the piston rod externally and it has a good film of oil on it, which can come only from within the cylinder, we are safe in assuming that we are getting satisfactory atomization of the cylinder oil.

Steam-cylinder oil is used to lubricate the admission and exhaust valves, the cylinder walls, the piston and piston rings, the throttle valve and on some engines the piston- or valve-rod packing. Two methods are in common use, the direct system and the atomization system. With the direct feed the cylinder oil is introduced directly into the steam chest, on top of the valves, or directly into the top of the cylinder. The oil fed to the top of a Corliss valve, for instance, is supposed to spread along the valve surface and lubricate it and the seat. Oil fed into the top of a cylinder is expected to flow slowly down the sides of the cylinder until it is caught by the sweep of the piston and is wiped over the entire surface.

The second method is based on the fact that the steam reaches every part requiring lubrication. The steam is therefore made the carrier for the oil to all the parts. For efficient and economical lubrication by this method the oil must be introduced into the steam at such a point that it will be thoroughly and completely atomized before it reaches the parts requiring lubrication. If any of the oil is carried into the cylinder in a liquid state, it is promptly carried out again by the exhaust steam and is wasted, but if it reaches the cylinder in a finely divided condition, part of it is deposited on every surface touched by steam. The remainder is, of course, carried out through the exhaust with the steam. Practical experience and many comparative tests have demonstrated beyond doubt that the atomization method will save from 25 to 50 per cent. in quantity of oil over the direct-feed method. With this fact firmly fixed in our minds let us consider the various operating conditions that affect the atomization of the oil and the characteristics of cylinder oils that can be used most economically under the different operating conditions.

Point of Introduction of Oil

The first point to investigate is always the point of introduction of the cylinder oil. The ideal place is in the steam line to the cylinder six or eight feet above the throttle valve. If the point of introduction is too close to the cylinder, complete atomization may not be secured; if it is too far away, there is a possibility that the oil will be deposited on the sides of the steam pipe and flow down to the cylinder in a liquid form. On some engines the oil-feed line is extended to the center of the pipe and flattened out in the shape of a spoon, so that the steam will strike it and blow the oil off, thoroughly mixing it with the steam. Another method is to extend the feed pipe across the steam line and perforate it with several small holes. Such an arrangement will provide ideal opportunities for proper atomization of the kind of oil that will best lubricate the valves and cylinders.

Sometimes the location of steam separators or sharp bends in the pipe close to the cylinder is such that the feed line must be introduced much closer to the engine than is desirable, and in many cases it is necessary to place it just above the throttle. This arrangement gives much less time for the oil to atomize before it reaches the cylinder, and larger quantities are required for good lubrication. The efficiency of atomization for such an installation can be improved by using oils that atomize more quickly, but as such oils also possess the characterisuc of evaporating more quickly from the cylinder walls, they are not always the most economical.

The characteristics of the steam are next in importance, and the pressure, temperature, percentage of moisture, degree of superheat, priming or foaming of boilers, amount of boiler compound carried over with the steam, etc., all have an effect on the kind of oil to be used and the efficiency of the lubrication.

Steam pressure in itself has no effect on the oil, but the temperatures corresponding to the various steam pressures do have a very marked effect. High steam temperatures will evaporate the oil film off the valve and cylinder surfaces faster than low temperatures, thus requiring the use of heavy-bodied oils or larger quantities of the light-bodied oils. The high temperature also thins the oil down and atomizes it more quickly; consequently, the heavy-bodied oils can be atomized satisfactorily and, as they stay on the cylinder walls longer than the light-bodied oils, are more economical. Low temperatures accompanying low pressures will not atomize the heavy oils properly, and low viscosity cylinder oils are necessary. The low temperatures do not evaporate them excessively.

Steam unless highly superheated always contains moisture by the time it reaches the cylinder. Condensation also occurs when the steam is cooled on entering the cylinder and during the expansion stroke. Moisture in steam will wash a pure mineral oil off the cylinder walls so that very large quantities would be necessary to get any kind of lubrication under bad water conditions. It is customary, therefore, to compound the oil with acidless tallow oil or degras when it is to be used for wet steam. The tallow oil or prepared degras saponifies with the moisture in the steam, forming a soap which lubricates the cylinder walls and assists the mineral oil in adhering. If a large quantity of water is drawn suddenly into the cylinder, practically all of the oil will be washed away temporarily and the cylinder will give off the peculiar groaning sound. When the sudden rush of water is over, the saponified oil will form a new film over the cylinder walls much more quickly than a pure mineral oil.

Amount of Compounding Required

The amount of compounding required in the cylinder oil depends upon the percentage of water or moisture in the steam. Compounding adds to the cost of the oil, and as excessively large percentages have no better lubricating value, the lesser compounded oils are used except for the most severe conditions.

It has been the general opinion that engines using a highly superheated steam are hard to lubricate, a situation created very probably by a failure to understand the actual operating conditions of a superheated steam engine. The general practice in the past has been to use a very heavy-bodied pure mineral oil, and uniformly good results have not always been obtained. Complaints have arisen that lubrication has not been good and that there have been carbon deposits.

With a steam pressure of 150 lb. and a cutoff at one-quarter stroke it requires about 200 deg. of superheat at admission to show any superheat in the exhaust. With any less degree of superheat the exhaust steam will be wet, and as we have to lubricate the cylinder on the exhaust stroke as well as on the expansion stroke, we should use the same kind of oil that would be used on any engine operating on saturated steam. When the heavy-bodied oils are used, they are not properly atomized, because most of these engines have the feeds very close to the cylinders.

Large quantities of pure mineral oils are necessary to lubricate the cylinders properly when wet, and this combination of the large amounts of heavy-bodied mineral oil results in the oil being fed to the cylinder faster than it can be atomized and carried away with the steam. Naturally, carbon deposits result from the collection of oil within the cylinder. In any event there is always the possibility that the superheaters may have to be discontinued temporarily and the engine operated on saturated steam, so it is always safer to use only a small percentage of compounding in the oil. In this way satisfactory lubrication can be obtained with a minimum consumption of oil.

Another thing that has a very bad effect on the oil film is alkali which may be carried over from the boiler with the steam. Excessive quantities of boiler compounds, if they get into the cylinder, destroy the oil film even with the most highly compounded oils, and if the condition exists for very long, the cylinders will begin to wear and score rapidly. Careful attention should always be given to the amount of boiler compound in the boiler water.

Sharp piston rings will also sometimes scrape the oil film off the cylinder surfaces and destroy the polish or glaze. Excessively tight piston rings will do the same thing.

Uniformly Loaded Engine Easy to Lubricate

A uniformly loaded engine is always easier to lubricate than one operating under a variable load. A rapidly changing load is more likely to cause boiler priming and a washing away of the oil from the cylinder walls. Lightly loaded engines have a greater cylinder condensation than fully loaded engines and may require a more highly compounded oil. The velocity of the steam through the steam pipe will be slow, and while it will give more time for atomization of the oil it will also give a greater opportunity for increased condensation.

A compound engine offers a very wide range of conditions which it is sometimes difficult to meet with one oil. It is very easy to pick out the right oil for the high-pressure cylinder, and if it is properly atomized the steam will carry the greater portion of it through the receiver into the low-pressure cylinder, lubricating it also. If the oil is not completely atomized, all the liquid portion will be separated with the condensed steam in the receiver and is lost. It is then necessary to feed additional oil to the low-pressure cylinder. It is hard to do this economically because there is seldom any opportunity to feed the oil into the steam line any distance from the cylinder, it being necessary usually to feed it directly into the steam chest, making atomization difficult. On this account some engineers prefer to use a light-bodied highly compounded oil for the low-pressure cylinder if they cannot get proper atomization of the oil before it enters the high-pressure cylinder. This is particularly true when the engine runs condensing.

Kind of Engine Should Have No Influence on the Oil Selected

If proper atomization of the oil is secured before the steam enters the cylinder and if the oil has the necessary compounding and other qualities for lubricating the valves and cylinders, the kind of engine to be lubricated should have no influence upon the oil to be selected. Simple slide-valve, piston-valve, Corliss-valve, compound, horizontal or vertical cylinders, uniflow or counterflow engines, can all be lubricated with equal efficiency if proper attention is given to the steam conditions, completeness of atomization and the operating conditions of the engine. There is a possible exception in the Corliss valve when the steam does not reach the ends of the valves, and it is then necessary to feed oil directly to them.

When the exhaust steam is used for manufacturing processes or for boiler-feed purposes where the presence of oil woull have a very bad effect, it is necessary to install an oil separator and use an oil that will separate from the condensed steam most efficiently. The compounded oils do not separate well because of the saponification of the animal oil content, and it is customary to use pure mineral oils on such engines and cut the feed down as low as possible.

The exhaust steam from the cylinders of marine engines is condensed and used for boiler feed. These cylinders are vertical and there is very little pressure against the cylinder walls. Usually, they receive no oil at all and depend on the moisture in the steam to lubricate the cylinders. When oil is used, mineral oils are supplied and fed to the cylinders very sparingly.

Having considered the principal operating conditions that affect lubrication, let us now investigate the various kinds of cylinder oils and see what sort of lubrication can be obtained from different grades.

Classification of Steam-Cylinder Oils

Steam-cylinder oils are classed as fire-stock oils, steam-refined oils and filtered oils. The fire-stock oils are the most adhesive, usually have the heaviest body, and are not likely to be as uniform in characteristics or qualities as the other grades. The steam-refined oils are better lubricants, more uniform in quality, are less likely to decompose in use or form objectionable deposits, and are more readily atomized. They are made in a wide range of viscosities, from 90 sec. to 225 sec. at 210 deg. F., and are dark green or brown in color.

The filtered cylinder oils are made by filtering the steam-cylinder oils through charcoal, which brightens the color, making them clear, and removes some of the more adhesive hydrocarbons. This process makes the oils less adhesive on the cylinder walls and renders them more quickly atomized. They also separate from condensed steam more readily than the steam-refined or fire-stock oils.

All three kinds of cylinder oils are used in the pure mineral state and also when compounded with tallow oil or prepared degras. The compounded oils are most satisfactory for wet steam condition as they are not so quickly washed off the cylinder walls. They atomize more quickly than the pure mineral oils, but do not separate from the exhaust steam as readily.

Considering viscosities alone, the lower viscosities atomize more quickly than the higher viscosities or heavier-bodied oils. Incidentally, they evaporate from the cylinder walls quicker. Heavy-bodied oils require higher steam temperatures for complete evaporation but are more adhesive. When properly atomized, they are very satisfactory; but if not thoroughly atomized, too heavy an oil will not lubricate a cylinder unless excessive quantities are used, with the possibility of carbon deposits.

How to Select Oils

With all of these factors to consider, it is impossible to make any fixed recommendations. Determine your operating conditions, study the following data and try to get the kind of oil which best fits your individual engine.

1. Economical and efficient lubrication depends upon the quality of the oil and its proper atomization.

2. Proper atomization depends upon the characteristics of the oil, the point of introduction, the condition of the steam and also on the operation of the engine.

3. Within certain limits the farther the point of introduction of the oil is from the cylinder the more thoroughly will be the atomization.

4. Oil fed directly to valves or cylinders has little opportunity for atomization.

5. On the other hand, if oil is introduced too far away, it condenses on the sides of the steam pipes and may be trapped out in pockets or drains.

6. Any device placed within the steam line which assists in atomizing the oil is of benefit.

7. High steam pressures because of their high temperatures atomize the oil quicker than low pressures and low temperatures.

8. High steam velocities atomize oil quicker than low velocities.

9. Low velocities give a longer time for atomization before the oil reaches the cylinder.

10. Low-viscosity cylinder oils atomize more easily than high-viscosity oils.

11. Filtered cylinder oils atomize more quickly than non-filtered cylinder oils.

12. Nonfiltered cylinder oils adhere to cylinder walls better than filtered cylinder oils.

13. Heavy-bodied oils stick to the cylinder walls longer than light-bodied oils.

14. Light-bodied oils evaporate from hot cylinder walls when engine is shut down quicker than heavier oils.

15. Water washes oil from valve and cylinder surfaces.

16. Moist steam therefore washes the oil off the valves and cylinder walls.

17. Light loads are accompanied by greater condensation in the cylinder and require more oil.

18. Heavy loads may cause boiler priming, and the water carried over with the steam washes the cylinder oil off the walls and valves.

19. Boiler water heavily loaded with boiler compounds carried over by the steam quickly destroys the lubricating oil film.

20. Any means of keeping excessive water out of the cylinders will improve lubrication.

21. Compounded cylinder oils containing tallow oil or degras saponify in the presence of moisture and do not wash off the cylinder walls as quickly as pure mineral oils.

22. Compounded oils therefore are more economical for saturated steam conditions.

23. Mineral oils may be used for dry or superheated steam when exhaust steam must be freed from oil.

24. Compound engines using superheated steam should use compounded oil on account of condensation in the low-pressure cylinders.

25. Compounded oils do not separate quickly from exhaust steam.

26. Mineral cylinder oils separate from exhaust steam more quickly than compounded cylinder oils.

27. Filtered cylinder oils separate from exhaust steam more quickly than nonfiltered oils.

28. Mineral oils are used where exhaust steam must be freed of oil.

29. Cylinder oils may be classed according to viscosity as follows:

Light bodied	95 sec.-115 sec. viscosity at 210 deg. F.
Medium bodied	115 sec.-135 sec. viscosity at 210 deg. F.
Heavy bodied	135 sec.-160 sec. viscosity at 210 deg. F.

As an example of the use of these items in selecting oils, let us assume the following conditions: Steam pressure, 150 lb.; moisture, considerable; point of introduction of oil, 6 ft. above throttle; boiler water, bad; load conditions, variable; type of engine, Corliss cross-compound condensing.

Analysis—Conditions are good for atomization as steam temperature due to high pressure is fairly high, and ample time is given with feed six feet from throttle.

Conditions are bad for maintenance of oil film on cylinder walls on account of variable loads, bad boiler water, wet steam and operating condensing.

Consequently we would use nonfiltered (12) highly compounded (21-22) heavy-bodied (13) oil and feel confident that it will be economical and efficient (3-7).

Another set of conditions might be as follows: Steam pressure, 100 lb.; moisture, very little; boiler water, good; point of introduction, directly on valve; load conditions, steady-rated capacity; type of engine, simple noncondensing; use of exhaust steam, boiler feed, open heater.

Analysis—Conditions are poor for atomization. Steam temperature due to low pressure is low, and little time is given with feed directly on valve.

Water conditions are good, and there is comparatively little moisture in the cylinder. Exhaust steam must rapidly be freed from oil. Accordingly, we would use a light-bodied (10) filtered (11-27) cylinder oil having low percentage of compounding.

These illustrations serve to show the impracticability of making a general positive recommendation of one or two brands of oil. The most satisfactory way is to examine the individual installation and select the kind of oil to meet that particular class of conditions and operation.

Emergency Automatic Controller

By B. A. Briggs

When an automatic controller on a pump or air-compressor motor must be taken out of service for repairs, a temporary controller can be made from an ordinary starting box having four binding posts. In this type of starter the no-voltage release coil is connected across the line and not in series with the shunt-field windings as in the three-terminal type. In Fig. 1 is shown a type of automatic starter in common use for automatically stopping and starting pump and air-compressor motors. The figure gives the connection for controlling a pump motor through a float switch operated from a float in the tank. When the float switch closes, the solenoid coil on the controller is energized, which in turn closes the contact and brings the motor up to speed automatically. When the tank is full the switch opens, de-energizes the coil, and allows the contactors to open and shut down the motor.

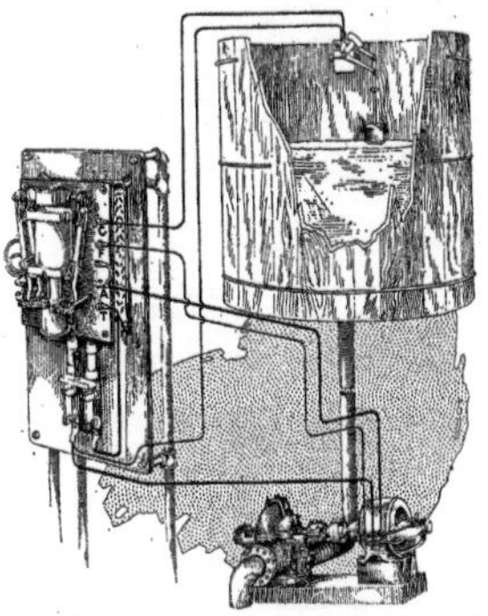

FIG. 1. CONNECTIONS FOR CONTROLLING PUMP MOTOR AUTOMATICALLY

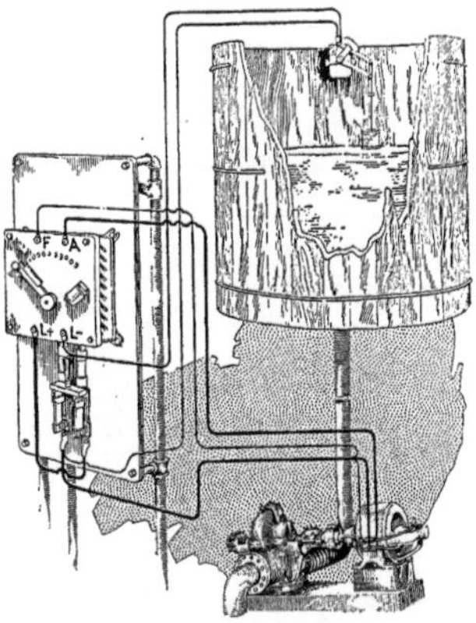

FIG. 2. CONNECTIONS FOR CONTROLLING PUMP MOTOR WITH STARTING BOX

In Fig. 2 the controller of Fig. 1 has been replaced by an ordinary starting box with four terminals. With this arrangement the motor will not start up automatically when the float switch closes, but will be shut down automatically when the float switch opens. If the water in the tank is at a level where the float switch is closed, when an attempt is made to start the motor, the starting arm will be held in the full-on position by the no-voltage release coil. When the tank has been filled and the float switch opens the no-voltage release-coil circuit, the starter arm will be released and the motor shut down automatically. However, if when the attendant started the motor, the water in the tank was at a level where the float switch was open, then the no-voltage release coil would not hold the starter arm, thus indicating to the attendant that the tank was still full. This scheme, although not starting the motor automatically, does stop it automatically and also gives the attendant an indication of the height of the water in the tank when he goes to start the motor. This is generally all that is required of a temporary controller on a pump or air compressor. Therefore, the foregoing scheme can usually be used to good advantage when the automatic controller is being repaired.

Developing the Chain-Grate Stoker

By H. F. GAUSS

Chief Engineer, Illinois Stoker Company

The application of forced draft has resulted in much higher rates of combustion, reduction of ignition troubles, better air control to the various sections of the grate and less leakage of air into the furnace. Securing of prompt and vigorous ignition and the reduction of air leakage have been interesting problems.

IN THE design of a successful stoker the most pertinent problem is to obtain prompt and vigorous ignition. It is a common assumption that this is entirely up to the ignition arch. This function of the ignition arch is by no means to be depreciated, but the entire responsibility for lack of good ignition is not to be placed upon the arch alone.

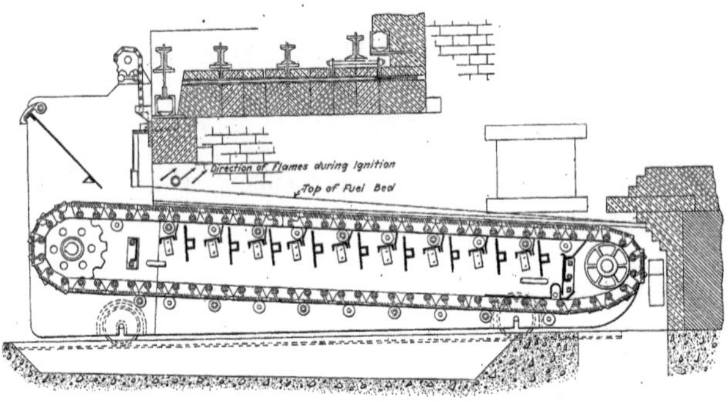

FIG. 1. CHAIN-GRATE STOKER WITH FLAT ARCH, SHOWIN G DIRECTION OF FLAMES

An inherent characteristic of a fuel bed ignited from the top is that the ignition must penetrate down in an opposite direction to the flow of gases liberated by the combustion of the upper layers. This condition imposes the real difficulty of obtaining successful ignition in the natural-draft chain-grate stoker, and it at once suggests that the draft intensity immediately under the front part of the ignition arch may have a great deal to do with good ignition. Fig. 1 is a typical stoker layout. It is evident that the products of combustion during ignition pass off diagonally as indicated by the arrows. They do not pass straight up and impinge against the arch. In fact, the corner between the feed gate and the front of the arch has no flames whatever.

If one lights a match and holds it in a vertical position with the lighted end up, it burns feebly for a short time and then goes out. Now light a match and support it in the same position as before, but place a piece of red-hot iron a few inches above it. The match now does not go out. However, a slight current of air from the side, giving the products of combustion a diagonal direction in leaving the match, will easily extinguish the flame. This latter condition is obtained in the ordinary natural-draft chain-grate stoker, and unless proper provision is made to overcome it, the result is invariably poor ignition.

There are a number of ways in which the necessary provision to maintain good ignition can be made. It is primarily essential that the fuel be properly prepared. If a furnace is designed and intended to burn 1½-in screenings, good results cannot be obtained by attempting to use 2-in. nut coal. Neither can good results be obtained by dumping coal for this stoker, after crushing it to 1½-in. size, into an overhead bunker, so arranged that all the fine coal accumulates in the center of the bunker directly over the outlet, while all the coarse coal rolls off the sides to be fed down last with no fine coal whatever. When this condition exists, as it frequently does, the first of the fuel fed down from a freshly filled bunker is so fine that it is almost impossible to ignite it except in a thin fuel bed, owing to its resistance to the passage of sufficient air to support combustion. The fuel then gradually approaches the proper condition, and it becomes necessary to thicken up the fire until the depth originally intended is reached. If conditions would stabilize here, all would be well, but the fuel continues to become coarser and the thickness of the fire must be increased correspondingly until the limit is reached, and then trouble with ignition begins again. It is understood, of course, that the trouble experienced in igniting coarse coal is caused by the fuels becoming so coarse that the small resistance of the bed to the passage of air permits too much air to rush through. Ignition is retarded instead of being maintained at a sufficiently rapid rate to permit the necessary grate speed for the capacity demanded. Adjustable dampers extending under the coal hopper in front and under a few feet of the ignition zone of the stoker will aid materially in controlling the air for ignition and permit burning coarse coal under the bunker conditions just described.

Someone will say, "Why not check the damper in the breeching?" This is a good suggestion, but the capacity of the boiler would also be cut down, and no stoker company could afford to design a machine that had to be operated in this way. The real solution of the difficulty is to provide a bunker and coal conveyor so designed that the coal will be uniform in size and condition when fed to the stoker. Unfortunately, the stoker designer seldom has much to say in regard to the arrangement of bunkers and conveyors.

The writer recently saw an expedient resorted to for the purpose of overcoming the difficulty just described, which consisted of a ¾-in. steam pipe introduced under bridge-wall and front boiler-room wall. This applies particularly to existing settings in which it is desired to put a stoker.

Generally speaking, the function of the arch is, first, to aid in maintaining ignition, and, second, to provide a sufficiently hot zone to consume completely the volatile constituents liberated during ignition. More than this an arch cannot do. Once the coal is thoroughly ignited, it burns when supplied with the proper amount of air, and part of the heat so evolved is utilized in keeping the arch hot. It is the writer's opinion that the shorter the arch in keeping with these two objects, the better. A long arch adds unnecessary restriction to the flow of the gases through the furnace.

Fuel ignited on top must burn from the top. Air passing through from the bottom makes the penetration of ignition slow. The top layer of fuel, first ignited, is first to burn up, and the ash forms on top first. This layer of ash renders the fuel bed more impervious to the passage of air, and frequently retards the combustion of

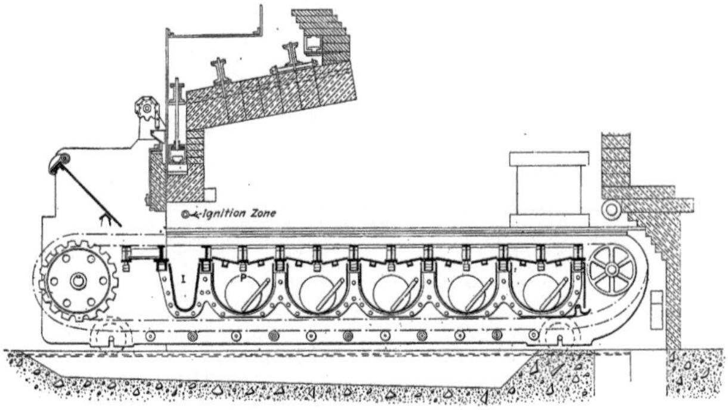

FIG. 2. CHAIN-GRATE STOKER, WITH FLAT TILTED ARCH

the ignition arch about one foot from the front end. Steam admitted at this point caused pressure inside of the furnace sufficient to keep out excess air, and in addition caused the air that was admitted to pass vertically up and impinge against the ignition arch. The device was successful in improving ignition of coarse coal, but its economy is doubtful, when the probable steam consumption of the blower is considered.

The design of an arch is not entirely a matter of mathematical analysis. It is a problem that must be tackled in the light of previous experiment and experience. When a stoker man speaks of designing an arch for burning a particular fuel, he means that he has taken out his notebook, looked up data on arches, which he knows have been used successfully under similar conditions and for similar fuel, and possibly has made a few changes suggested by good judgment. Once successfully designed, it is easy to explain scientifically just why the arch should give the good results sought.

A factor often controlling the length of an arch is the position of the boiler with reference to the ashpit, the lower layers of fuel to such an extent that unburned coal is carried over the rear of the stoker into the ash pit, resulting in a high carbon content in the ash. When this condition obtains, the remedy is a thinner fuel bed and a higher grate speed. The writer has seen stokers burning crushed mine-run bituminous coal of high ash content in which the appearance of the rear end of the grate was such as to lead one to believe that combustion was complete. Upon stirring up the refuse on the rear end of the grate with a rake, a layer of unburned coal was found underneath the ashes, and upon agitation it began to burn fiercely. There is no question but that the ash formed first on top. The writer had the same experience in burning lignite and anthracite culm.

Anyone who has the opportunity to observe the process of ignition through the small inspection door generally provided in the side wall for this purpose, will be able to see with the aid of a darkened glass exactly how ignition takes place.

When just inside the feed gate, the pieces of fuel on top give off tiny volumes of gas and vapor, become red, burst into small flames, and finally become incandescent. At the same time the particles in the layers of fuel underneath begin to undergo the same operation. The process necessarily requires time, and the fuel cannot be considered thoroughly ignited until it has traveled into the furnace at least 4 to 6 in., and more often as much as 12 in. is necessary. The process is not in the nature of an explosion. An explosion is intensely rapid combustion, which is completed in a fraction of a second. If ignition is complete within 4 in. of grate travel, at the rate of 4 in. per minute, it would require an interval of exactly one minute.

As the fuel advances into the furnace the fixed carbon is ignited, burns up and leaves on the grate a residue which often is thin in spots and likely to permit excess air to enter unless the rear sections of the stoker are provided with adjustable dampers to prevent it. Dampers of this kind are valuable for the purpose just mentioned, but when installed they must receive proper care from the attendant in order to be effective.

After the rear air baffles in the stoker and setting have been made as tight as possible, and even though the ashpits are well made and provided with tight doors, there is likely to be some air leakage between the bridge-wall and the rear end of the grate. To reduce this leakage to a minimum a waterback is recommended. As far as air leakage is concerned, either a closed or an open type of waterback is equally effective. From the standpoint of economy a closed waterback is to be preferred. In cases where there is a demand for hot water, an open waterback is probably the cheapest to maintain and is preferable. In addition to preventing air leakage the waterback is valuable in reducing bridge-wall maintenance by preventing the adherence of clinkers to the overhanging nose and the breaking of the nose tile when these clinkers are sliced off.

In locating the waterback it is essential that it be placed sufficiently high above the rear end of the grate to allow the ash to pass under without touching it. When the ash is allowed to roll up against the waterback, it piles up against the bridge-wall. The writer has seen cases in which just this condition has caused ashes to climb entirely over the bridge-wall of a horizontally baffled boiler and drop into the combustion chamber. A condition like this evidently cannot be permitted to exist, and there must be sufficient clearance under the waterback to let the ashes through. It is not possible to compress an ash bed by attempting to crowd it under a waterback.

A waterback so designed that its vertical height may be adjusted to suit the thickness of the ash bed is theoretically desirable, but not practically possible, as its location is such that the necessary guides, connections and appurtenances would be a source of endless annoyance and trouble. The alternative would be to arrange the stoker so that the rear end can be raised or lowered. This is more desirable construction than an adjustable waterback, but it, too, presents undesirable features and difficulties that render it impractical.

It is not possible to lay too much stress upon the necessity of using the utmost skill and care in preventing air leakage. All ledge plates, seals and joints must be made and kept tight with this end in view.

Attention was drawn at the outset to the recent development of the forced-draft chain-grate stoker in the statement that it had eliminated many of the difficulties formerly experienced in the design and operation of the chain-grate stoker. These may be enumerated as follows: Ignition troubles overcome; better air control to the various sections of the fuel bed; less leakage of air into the furnace.

Incidentally, it becomes possible to carry continuous loads of from 250 to 300 per cent. of normal rating and maintain 70 per cent. operating efficiencies.

At the present writing no data are available indicating the maximum rate of combustion per square foot of grate area per hour, that can be secured. Tests on a stoker containing 130 sq.ft. of active grate area under a 500-hp. water-tube boiler established a rate of 50.2 lb. of coal per square foot of grate area per hour, during a test embracing a period of 16.4 hours. The rate thus established was limited by the ability of this boiler to take the gases liberated, and not by the ability of the stoker to burn more coal. Any attempt to burn more fuel by speeding up the stoker and increasing the air supply resulted in building up a pressure in the furnace and causing the gases to be forced out of the setting wherever a crack or crevice would permit. Shorter tests have indicated a maximum rate of combustion of 70 lb. per square foot of grate surface per hour.

Returning now to analyze the three advantages of the forced-draft stoker, attention is called to the dead-plate directly under the hopper in Fig. 2. This has a two-fold purpose—it prevents leakage of air from the first tuyere to the atmosphere and incidentally prevents leakage from the atmosphere into the first tuyere.

When burning bituminous coal, the first small tuyere marked *I* is omitted and the dead-plate is extended up to the first large tuyere *P*. It is understood, of course, that the grate slides over the dead-plate. No air is supplied to the coal above the dead-plate. The space above the dead-plate and under the ignition arch is then the ignition zone of the furnace.

Sufficient air is supplied through the remaining tuyeres to burn the required amount of fuel, and the main damper in the breeching is closed until a balanced draft condition exists in the furnace. By this is meant a condition in which atmospheric pressure exists in the furnace. There will then be no tendency for excess air to be drawn into the furnace through any unavoidable crevices around the ignition zone. The products of combustion resulting from ignition pass vertically up and impinge against the ignition arch, which materially aids in maintaining vigorous ignition, as illustrated in the match experiment.

In the final analysis what is accomplished here is the gentle mixing with the fuel of sufficient air for ignition, with a time interval of sufficient duration to allow this ignition to penetrate down into the fuel bed without being interfered with by a rush of air in the opposite direction. Once thoroughly ignited, the fuel burns fiercely upon passing over the first pressure tuyere, *P*, Fig. 2.

When burning such fuels as anthracite culm and lignites, which are hard to ignite, tuyere *I* is retained, but it is connected with an exhaust fan which draws some of the hot gases from the furnace down through the ignition zone of the grate, and this facilitates ignition.

The fact that a balanced pressure is maintained in the furnace accounts for the reduced infiltration of air through cracks or crevices in any part of the setting. Balanced draft is a material aid to the waterback in performing part of its duty. It has been possible to obtain CO_2 as high as 16 per cent. with this type of chain-grate stoker. The flat dampers indicated in the tuyeres in Fig. 2, or dampers frequently located in the thimbles connecting the windboxes into the tuyeres, are the means of regulating the air supply to the various zones of the grate.

The advantages of the forced-draft chain-grate stoker are apparent, and from the rapid strides being made in its efficient development it would appear that there is no end to the possibilities.

EDITORIALS

Hunting for Fuel-Oil Profiteers

WHEN a great industrial nation like ours is confronted with a national coal-strike situation such as we recently partly passed through, everybody becomes more or less panicky. With the experience of the war so lately acquired, the Government, to head off unconscionable coal prices, revived the dormant Fuel Administration—and with good effect. But the fuel-oil division did not again go into action, and as a result our Uncle Sam has been made suspicious, and now he is hunting for fuel-oil profiteers through Attorney General Palmer, whose office is carrying a peak load with Reds and the familiar H. C. of L.

The vigorous fashion in which the attorney general is handling Reds and other law violators has evidently made fuel-oil interests remember the "Stop, Look and Listen" sign. The National Petroleum Association warns its members through its consul general, C. D. Chamberlin, not to charge "abnormally high prices" for fuel oil; the refiners of conscientious Kansas say the brokers and not themselves are to blame for high fuel-oil prices. Clifford Thorne, of whom we had not heard much since the Government took over the railroads, pleaded the case of the Western Petroleum Refiners before the Fuel Administration.

The Government recently began prosecuting the Trans-Continental Oil Company for selling fuel oil at "unjust and unreasonable prices." The prices, it is claimed, were three and one-half to four dollars per barrel.

It is easier to accuse than to convict. The inclination to exploit a crisis is human, though not necessarily just, fair or patriotic. The profiteer fever is in the air, and anybody is likely to get it. It is too early to say whether there has been profiteering in fuel oil. But the "man in the street," the "average man," the persons you refer to when you mean all with whom you talk and hear talk, feel that advantage is taken of high costs to boost prices out of the equitable ratio of costs and prices. Whether it is or not matters not; the public believes it is so, and as long as it finds food to nourish the belief, public officers will have to prosecute and "snoop into one's business."

When one knows that until not so long ago Oklahoma fuel oil sold for a dollar a barrel and that during the coal strike it sold for four dollars a barrel; when one knows that but a short time ago Mexican fuel-oil contracts of three to five years' terms were closed at ninety-two cents and a little over a dollar a barrel delivered to northern Atlantic Coast consumers, while now one cannot close a contract for even three years at less than two dollars and a half a barrel without a sliding price clause which may or may not favor the consumer—when these facts are known, it is a judicious mind indeed that does not conclude that somebody is exploiting a national crisis.

But such evidence is not of itself "worthy," as the lawyers say.

First, the law of supply and demand is up and doing despite parlor economists. Fuel-oil consumption for 1919 is officially put at four hundred million gallons. The estimated maximum production for the United States is reckoned to be under three hundred fifty million gallons. You will pay more for a product than you used to when you want more of that product than you can get.

The Mexican political mess, as it bears on fuel-oil production by foreign oil interests, grows worse instead of better. E. L. Doheny, president of the Mexican Petroleum Corporation, claims that the suspension of operation by Carranza of at least eighteen American-worked wells means that the Shipping Board will not get adequate supply for its ships. There is evidently considerable American capital going into investments in foreign-owned and foreign-worked oil fields, particularly so now when foreign exchange values are so low compared with the dollar.

The refiners at a recent meeting in Washington predicted the Government soon would be willing to pay at least three dollars a barrel for fuel oil. Statistics presented by the Bureau of Mines show the failure of many refiners. The coal market does not seem to warrant belief in any reasonably immediate betterment. The Cummins and Esch railroad bills now before Congress look as though higher rates, perhaps appreciably higher rates, will follow the passage of either.

So conditions tend toward higher and higher fuel-oil prices. If there has been profiteering, the Government may uncover and punish those responsible. It is too soon now to cast stones.

Electrical Maintenance Records

THE value of reliable cost records in manufacturing processes has long been appreciated; this is also true in power-plant operation. However, in the distribution and application of power in manufacturing plants the necessity of a detailed record of each feeder, motor and controller has not been given the consideration that it should have, excepting in isolated cases. In many plants where very carefully kept detail records are the practice in the power plant, the records of motor operation in the plant are practically lacking. Or methods of record keeping that, at first, may to a degree suffice, are now, owing to plant expansion, inadequate to give any intelligent record of what the actual cost of maintenance and repairs of the motors and controllers are or of the change in power requirements by the machines driven by the motors.

A record of motor operation should include the name-plate data of both motor and controller, details regarding the different parts of the machine so that they can be readily ordered when needed. A record should be kept of the amount of time spent in repairing the equipment, the cost of materials used, etc., the length of time the motor is out of service for repairs, also the name of the workman who made the repairs. In addition to this a record of current consumption taken at different periods, especially if taken with a curve-drawing instrument, will be of inestimable value where the equipment is giving trouble. This is all the more valuable if it includes a record of the voltage for the same period and also the power factor if the motor happens to be of the alternating-current type. From this information as comprehensive a report can be made as may be desired.

Well-kept records will be found of great value not only in isolating the maintenance cost of each piece of equipment, but in arriving at the cause of high maintenance costs. The latter may be due to improper applications, changed conditions after the motor has been installed, careless operators, negligent repairmen, etc., any of which are easily determined if proper records are kept. If the cost of maintenance is excessively high because of improper application, a reliable record of the conditions will be one of the best arguments to present to the management to obtain an appropriation to make the necessary changes. It will frequently be found that a change of operators on elevators, cranes, etc., will have a marked influence on the upkeep of the equipment. Just what the exact results are and the remedy can be most readily ascertained when a record of operation is kept. Then, again, the work done by one repairman may not be standing up the same as that done by others. This can be proved only by a record of those who do the work and when it is done. The quality of the same kind of material purchased from different firms must have carefully kept records of performance if an intelligent decision is to be made as to which material is best-suited to the conditions to which it is applied.

Properly kept records are as essential to the intelligent and economical operation of the electrical equipment in a manufacturing plant as they are in the power plant. And the former are much more easily obtained than the latter. The most difficult part is to get such a system started; after it is in operation, it will be found to be well worth any effort that may be required for its maintenance.

Scarcity of Men for an Essential Branch of Industry

A GREAT deal has been written recently about the troubles of the colleges and universities in finding or keeping competent men for the teaching forces at the salaries that such institutions can pay. A new phase of this situation has recently come to notice in connection with engineering-research work. The Engineering Experiment Station of the University of Illinois maintains a number of research graduate assistantships. The duties of these positions require not more than half time, the remainder being available for graduate study toward the degree of master of science. In return for his services to the station the research graduate receives five hundred dollars a year and freedom from tuition fees for any courses he may take at the university.

Recently, largely because of the small salary, the experiment station has had great difficulty in securing competent men for these positions. This situation is regrettable for two reasons. The research activities of the station will be seriously interfered with unless properly qualified assistants are available. Since during the two years' work in these positions the men receive valuable research training, present conditions are likely to curtail the supply of trained research workers.

Just at this time the need for skilled investigation along engineering lines is greater than ever before. Increased productive results from each individual's efforts and more extensive conservation of our natural resources are becoming more and more important. The research worker is the man to whom industry must look for the solution of these problems. Much has been done, but more must be done, and as has been previously pointed out in these columns, the more we have progressed along these lines the more difficult further progress becomes.

In the face of these facts the situation becomes really serious. Apparently, the financial conditions are not likely to improve greatly, as this is only a part of the general college situation. The training received, and the satisfaction that always accompanies investigations of engineering problems, must be relied upon to attract the right sort of men to this work. It is to be hoped that enough men will be willing, and able, to make the temporary financial sacrifice so that the important work of such experiment stations and the supply of trained research men may not be interrupted.

Reinforced Concrete For Stator Frames

WHEN concrete was first suggested as material for ship construction, the idea was looked upon as the dream of a romancer. Nevertheless, concrete ships are now in successful operation. Now the suggestion is made to construct the stator frame and thrust-bearing support of large slow-speed waterwheel generators of reinforced concrete. There are apparently no engineering problems involved which make such an application impossible. When made of cast steel or iron, these parts are not supposed to carry any of the magnetic flux, but to act only as a support for the stator core and in some cases the thrust bearing. Therefore, if the casing can be made of reinforced concrete, it will eliminate one of the difficult problems in manufacturing very large slow-speed waterwheel-driven generators; namely, the construction and transportation of the stator casing.

There is no doubt that reinforced-concrete construction is limited to machines of large dimensions, but these are the machines that present serious mechanical and transportation problems when they are made of cast iron or steel. However, since most of our water powers can be economically developed only by the use of comparatively small units, the use of reinforced concrete for stator frames is apparently very limited, even if its use should be adopted where it could be employed to advantage.

Long vs. Short Boiler Courses

A POINT that has not been brought out by any of our correspondents who are discussing the allowable length of longitudinal seams in boilers was mentioned, not at the hearing of the Boston Board of Boiler Rules, at which the question was also discussed, but in the discussion among the spectators following the hearing. This point was that when a bag occurs, it is likely to go a good deal farther in a large than in a small sheet; and that if the sheet has to be replaced it is much more easily and cheaply done in a boiler made up of small courses than in one having few courses made up of large plates.

The engineer who placed a lantern on top of an old-type bipolar generator and short-circuited the terminals found that it did not require the bursting of a flywheel or the blowing off of a cylinder head to have the machine start something, as with the steam engine, but like the tail end of a hornet, the electric generator, when it is alive, is always ready to sting whenever given the opportunity.

Somebody is determined that there shall be trouble between the United States and Mexico. If it starts, fill up your oil tanks!

CORRESPONDENCE

Regarding Diesel-Engine Trouble

Mr. Fagnan's article on Diesel engines in *Power* of Sept. 2, 1919, page 390, is interesting, portraying, as it does, the difficulties a "trouble man" encounters.

However, I believe that he is somewhat in error in some of his statements. The engines he refers to are evidently old American Diesel or "Type A" Busch-Sulzer engines, since no other manufacturer has built units of the dimensions given. In these engines the clearance between the piston and the cylinder head at the top dead-center is $\frac{1}{16}$ in., consequently it is impossible that the thumping he mentions as occurring in the middle cylinder was occasioned by the carbon deposit on the piston striking the cylinder head as he states.

Without doubt this cylinder did not receive its fuel charge at the proper time, and I believe that the pound was produced either by preignition due to early fuel admission or by delayed combustion resulting from late admission. Since he terms it a thump, probably the latter was the case; as the experienced engineer knows, the sound of preignition is quite sharp and vastly different from the dull thump arising from delayed combustion.

Mr. Fagnan's method of increasing the cylinder compression by shimming up below the piston-pin brasses is open to criticism and cannot be recommended to the Diesel operator. The connecting-rods of the "Type A" Busch-Sulzer were of the design appearing in Fig. 1, and the American Diesel used both Fig. 1 and Fig. 2 designs. Both rods have a marine big end which is separate from the connecting-rod proper. The correct way to alter the cylinder compression is by placing sheet-steel shims between the big end and the rod at the point *A*. Incidentally, this is the method followed on practically all Diesel engines. It avails but little to shim up under the piston-pin brass. The only adjustment to be made here is for the purpose of taking up the wear in the piston-pin brass.

A word of caution to the Diesel operator as to the use of a pencil to locate pounds in an engine may not be amiss. In engines with inclosed frames, such as discussed by Mr. Fagnan, it is almost impossible to determine the location of a pound by using a pencil or similar device. The frame acts as a sounding box, transmitting sounds from one cylinder to another. While it is questionable as to following the advice of the old railroad master mechanic who said that he always located a pound in the side rods of a locomotive by first deciding from which side the sound came and then taking up on the other side, still the pencil method is not much better.

The indicator will reveal whether the pound is the result of faulty combustion, and jumping the several pistons and rods will show if the sound is due to bearing play.

Mr. Fagnan states that on checking the camshaft gear train it was found that the gear had been advanced two teeth. He attributes the faulty fuel ignition in No. 2 cylinder to this. Undoubtedly the gears should have been meshed properly when the engines were installed. Nevertheless, since the fuel valves of No. 1 and No. 3 cylinders were operating properly, the fault was not in the mismeshed gears. By altering the gear timing Mr. Fagnan threw the No. 1 and No 3 cylinder valves out of correct timing. These had to be altered before the engine would function in a proper manner. With this type of engine the fuel cam nose, when only one valve is improperly timed, should be shifted to obtain correct functioning. If this alteration is not great enough, an offset key in the cam will shift it. The Diesel operator should hesitate a long time before changing the gear timing to obtain a correction for one fuel valve, since in so doing he is altering the timing of all the other valves—in the case under discussion this totals nine valves, including the air-starter valve.

San Pedro, San Domingo. L. H. MORRISON.

Determination of Volatile Matter in Coal

Of all the determinations made on a proximate or an ultimate analysis of coal, that least susceptible to close checks is the volatile determination. Variations of 1.5 per cent. to 2 per cent. in running checks on the same sample of coal at the same time are not at all uncommon, and checks of less than 0.3 to 0.4 per cent. are rarely obtained. Much work has been done on this, and Prof. S. W. Parr has shown that most of the variation can be traced to mechanical loss, or to oxidation under the cover. Suggestions have been made for minimizing these losses, but chiefly by means that involve the use of expensive apparatus or of considerable time, neither of which is usually available in the small commercial or power-plant laboratory.

I have, however, found that the suggestions of Professor Parr and of others may be combined to give close checks without sacrificing speed or simplicity, and the determination may be performed without the use of platinum crucibles.

Use a 10c.c. high-form crucible instead of a 25c.c. crucible and weigh into it a one-gram sample. Run three samples for each coal analyzed and add ten drops of kerosene to the sample. Heat slowly at first for two to three minutes with preferably a Meker burner instead of a bunsen burner, tapping the lid occasionally to insure that it sits tight. The crucible should be chosen with tightly fitting lids. Then raise the flame to full heat and heat the crucible for five minutes longer. Cool and weigh, and from the percentage loss in weight subtract the percentage of moisture. The result will be the percentage of volatile combustible.

I have almost invariably found that this method gives two results that check within 0.1 per cent. and often within 0.02 to 0.03 per cent., and I take the average of these two. The third result will probably be nearly 1 per cent. off, for reasons that I cannot explain.

Decatur, Ill. W. J. RISLEY, JR., Chemist, McLaughlin Foundries and Machine Shops.

Installing Disconnecting Switches

The article of Mr. Matthews in the July 25 issue of *Power*, in which he takes exception to statements in previous issues on the operation of disconnecting switches, is not entirely correct and is rather misleading. Disconnecting switches may be safely opened under load under certain conditions, but those conditions are not the ones granted by Mr. Matthews; namely, the breaking of the charging current of transformers and transmission lines. This operation more than any other should be held as absolutely beyond the scope of the ordinary disconnect switch and done only with the oil and air-break circuit-breakers designed to make and break such circuits. Even then, breakdowns of the latter type of switches, while interrupting a highly-inductive load, such as that of transformers, have occurred.

In one case the sectionalizing disconnecting switches of an 11,000-volt bus was opened, which had load on either side of the "disconnects," but a supply feeder on only one bus section. The load on the end of the bus was the charging current for a 500-kv.-a. transformer bank. The disconnects, without any barriers between phases, were located in the air chamber underneath the transformers. When the second of the three switches was opened, the arc spread to ground and across all

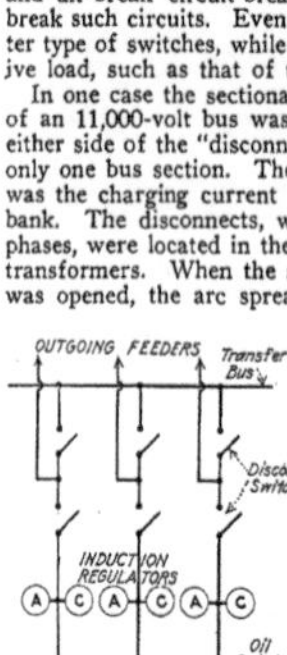

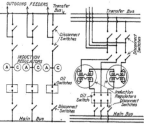

FIG. 1. ONE-WIRE DIAGRAM OF THREE CIRCUITS

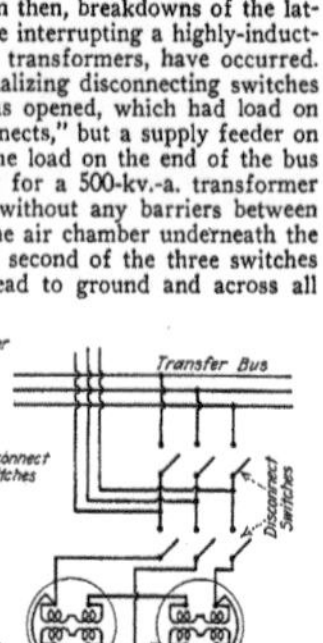

FIG. 2. DIAGRAM OF ONE CIRCUIT AND REGULATORS

three phases, burned the switches, shattered the insulators and tripped out the supply transmission line at the generating station more than ten miles away.

Even the charging current of a 25-kv.-a. induction regulator has proved too much for a disconnecting switch. In a substation, where each 2500-volt distribution feeder had selective switching connection through disconnects to a transfer bus, as shown in Figs. 1 and 2, after paralleling two circuits in order to cut out the station apparatus of one of them so that he might work on its oil switch, the operator broke the load current of the circuit at the oil switch instead of at the disconnecting switches, then attempted to break the charging current of the regulators with the latter. The first disconnect he opened broke down under the arc, burned off the cable connected to it, and only because there were barriers between phases, did no further damage except to trip out the oil switches and lose the load of both feeders.

Breaking the charging current of a transmission line has repeatedly proved disastrous to the disconnecting switch at which it was done. In two substations undergoing extensions, temporary 13,200-volt feeders were

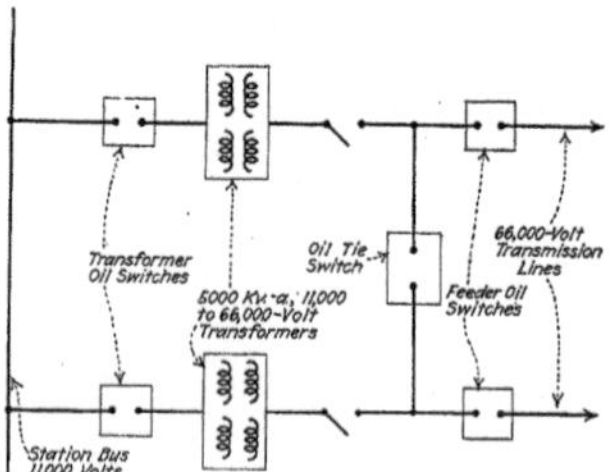

FIG. 3. DIAGRAM OF TRANSFORMER AND SWITCH CONNECTIONS FOR 11,000 TO 66,000 VOLTS

looped into and out of the stations and equipped only with disconnecting switches. Each feeder was made up of three-conductor 350,000-circ. mil. lead-covered cable, and no cable was over five miles long. On several occasions before the real cause was discovered the disconnects broke down while being opened to kill a cable which was carrying no load but its charging currents.

Disconnecting switches may be safely opened under load on a branch of a parallel circuit, where the load of that branch is immediately taken up by another branch and the main load-carrying circuit is not broken; such, for instance, as parallel transformer circuits. Hardly any high-tension alternating-current circuit is noninductive; where an inductive circuit is broken an intense arc is set up, and the disconnecting switch has no way of dissipating this arc.

Fig. 3 shows a transformer installation for stepping the voltage from 11,000 to 66,000 volts. Assume that it is desired to cut out one transformer and still keep both transmission lines energized. After the tie oil switch is closed, the "disconnects" of that transformer are opened, dropping its load over on the remaining unit, after which its magnetizing current is broken at its oil switch.

No accidents would happen if operators properly manipulated their disconnecting switches. The disconnecting switch should be opened very slowly during the early part of its stroke, and when the blade has been brought 0.25 in. to 0.5 in. away from the clips, the movement should be halted to observe the spark at the switch contacts to determine whether it is a load arc or a static discharge; if it is the former, the switch may be immediately closed. When an arc has been drawn in opening a disconnecting switch, it is always safer to close the switch than to attempt to break the arc by opening the switch wider, as that only enlarges the arc.

Barriers should certainly be installed between disconnecting switches, current transformers, potential transformer fuses, and all such high-tension apparatus, since all of it has been known to break down in service. The modern type of construction used in the big central-station systems is to have separate inclosed concrete or albarine cells for each phase on all apparatus, with both the doors of the cells and the cells themselves lettered with the phases and the designation of the apparatus. FRANK GILLOOLY.
Philadelphia, Penn.

Motor Trouble Was Caused by Brush-Contact Resistance

A single-phase repulsion induction motor gave trouble by not driving its load up to speed as it had done successfully for a considerable period. As the voltage and frequency were correct, the machine was taken apart and thoroughly cleaned and oiled, but this failed to result in an improvement.

An examination of the brushes showed that they were still of sufficient length, but on account of vibration the copper plating had worn off. After putting in a new set of copper-plated brushes, the machine worked satisfactorily, showing that poor contact between the brushes and holder can cause trouble, especially when the starting conditions are very severe. This at once suggests the use of pigtails on the brushes in order to positively remove trouble from contact resistance in the holders. HENRY MULFORD.
Patchogue, N. Y.

Reseating an Angle Valve

The exhaust valve that is connected to the top of a pulp-digester cover, through which the acid and steam from the digesters escape, rapidly deteriorates. No acid gets to the lower, or steam, valve and it wears a long

NIPPLE FORMS VALVE SEAT

time, but the life of the relief valve is only about four weeks. Sometimes the body of the valve gives out first, but generally it is the seat. The valves are then scrapped on account of the expense of putting in a new seat.

I have recently devised a cheap way of reseating such valves, which gives them from eight to ten days longer life. I simply tap the seat end of the valve all the way through and screw in a brass nipple until it protrudes about one-eighth inch. The end of the nipple being squared off, serves as the seat, as shown in the illustration. LOUIS J. FARLAND
Ausable Forks, N. Y.

Changing a Double-Connection to Single-Connection Lubricator

The illustration shows a standard double-connection lubricator changed to be used as a single-connection lubricator. It was used on the steam-extraction gear

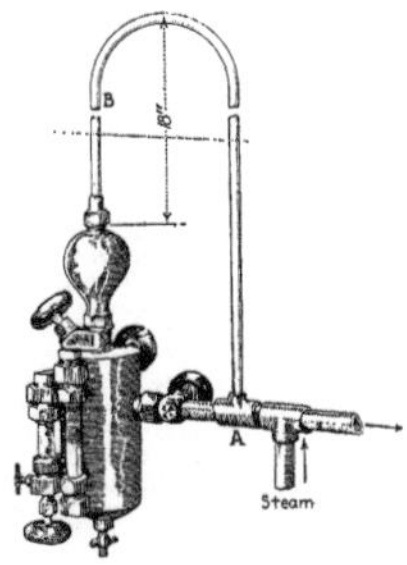

CONNECTION ON LUBRICATOR

of a steam turbine, and as the steam line came in from below, there was no overhead line to tap in on.

The ½-in. by ¼-in. tee *A* was used as shown. A ¼-in. pipe was put in from this tee to the water chamber, this pipe being 18 in. above the top of the water chamber of the lubricator. This gave a sufficient head of water in the leg *B* of the pipe.
Greenville, Penn. GEORGE B. LAMPKIN.

Electrically Welded Stay-Bolt Holes

On a certain railroad we experienced a good deal of trouble several years ago from leaks around stay-bolts due to bad water. If a locomotive was equipped with a new firebox, it frequently happened that leaking developed two weeks after the boiler had been returned to service. Inspections of removed stay-bolts showed that the threads were more or less corroded. If the water once has a chance to get around the stay-bolt threads, the corrosive action will be cumulative and will not only lead to a rapid destruction of the threads but also attack the firebox and wrapper sheets and grooving and small cracks will appear, as shown.

In such cases it became necessary to remove the stay-bolts and to drill and lap the holes to a larger diameter. If, after awhile, cracks and grooving developed again, there was nothing left but to remove and replace the whole firebox or wrapper sheet, as it is not feasible to use stay-bolt threads larger than 1⅛ or 1¼ in. in diameter, if the original thread was ⅞ inch.

The introduction of electric welding made it possible

to use a cheaper and more satisfactory method, which may be described as follows:

After removal of the old stay-bolts the holes are milled out in the usual manner; that is, to a diameter large enough to remove all corroded and grooved parts adjacent to the holes. These are then filled up by welding, as shown by the center illustration. It is not necessary to fill them solid, and it is usual to weld until the diameter has been reduced to, say, ¼ or ⅜-in. The opening thus left is then closed on one side by welding a bridge across the hole, so that a punch mark can be made and in order to have an easy start for the subsequent drilling of the new hole, which will be of the same diameter and thread as the original hole.

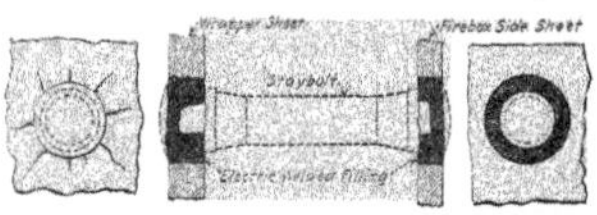

SHOWING HOW THE STAY-BOLT HOLES WERE WELDED

Compared with the old method, welding offers the advantage that it can be repeated several times without increasing the size of the stay-bolt holes and threads. It is therefore possible to use side sheets and wrapper sheets much longer than heretofore, provided there are no other reasons for replacing them. Welding also obviates the necessity of carrying in stock taps and stay-bolts of various thread diameters. The welding method is now used quite extensively in railroad shops and has given satisfaction from the mechanical and operating standpoint.
H. Bleibtreń.

Philadelphia, Penn.

Refrigerating Effect of Natural and Artificial Ice

Referring to the letter by H. S. MacDonald on the "Refrigerating Effect of Natural and Artificial Ice," in the Oct. 28-Nov. 4 number of *Power*, I heard a vender of lemonade on a hot summer's day in New York City confidentially declare to a friend of his that natural ice is colder than artificial ice.

There are a number of loopholes to watch out for in a discussion on this subject, one of which is that the meaning of refrigeration be clearly defined. Therefore when Mr. MacDonald asks, "Is there any more refrigeration in a pound of ice that is frozen by nature on a pond, lake or river, than there is in a pound of ice frozen by artificial methods in an ice plant?" the question arises, Just what is meant by refrigeration? Will a pound of natural ice cool an icebox to a temperature of 40 deg. for one hour, or will the pound of natural ice cool the same icebox to a temperature of 35 deg. for three-fourths of an hour? In one case we have longer duration, while in the other we have lower temperature. Which gives the most refrigeration? In the writer's mind the answer depends entirely upon the questioner's meaning.

A pound of snow placed in the icebox will reduce the temperature to a lower degree than a pound of ice, because the snow will melt more rapidly, and snow melts more rapidly because it is more porous.

When considered on a total B.t.u. basis, however, there is no difference. The pound of natural ice contains just as many heat units as the pound of artificial ice provided their temperatures are equal and provided the chemical composition is the same. This chemical composition factor, by the way, was omitted by the editor of *Power*. Perhaps there are other factors or specifications that must be borne in mind while discussing this slippery subject.

Anyway, of two pieces of ice, one natural and one artificial, of equal volume and equal weight and identical chemical composition and both at the same temperature, the same number of B.t.u. will be required to melt the one as will be required to melt the other.

New York City. W. F. Schaphorst.

Saving Oil at the Sump

In many power plants the oil filters are located in a section of the basement, the floor of which is sloped to drain into a sump, from which the water is pumped at intervals by a siphon or pump. The filter-water drains are connected to this sump instead of to the sewer direct, so that any oil leakage from the filter will be retained in the sump. Small leaks at the gage-glasses around the rods of the oil pump and spillage of oil while handling cause a continued seepage of oil to the sump. Owing to the large surface of the sump, a considerable amount of oil will be present in a thin film, and as skimming is a tedious job it will be pumped into the sewer.

To retain this oil in the sump until a sufficient quantity collects to make it worth while to dip it up, we installed a trap made from pieces of pipe, as illustrated. When the siphon is working, the water enters the 3-in. pipe *A* through the square openings in the lower end and flows to the siphon. When the level in the sump falls below the suction pipe *B* leading to the siphon, air is drawn in through the ¾-in. vent *C*, breaking the vacuum and preventing the oil between the square openings and the suction line from being drawn out of the sump. The storage capacity can be regulated by raising or lowering the position of the suction pipe. An added advantage is that, when using a hose to wash out filters or clean up the floor, the siphon may be left on, and when the cleaning is finished all oil still remains in the sump. This little device saves about two gallons of engine oil a day. At the price we pay, this amounts to about $250 a year.
Adolph Quadt.

Perth Amboy, N. J.

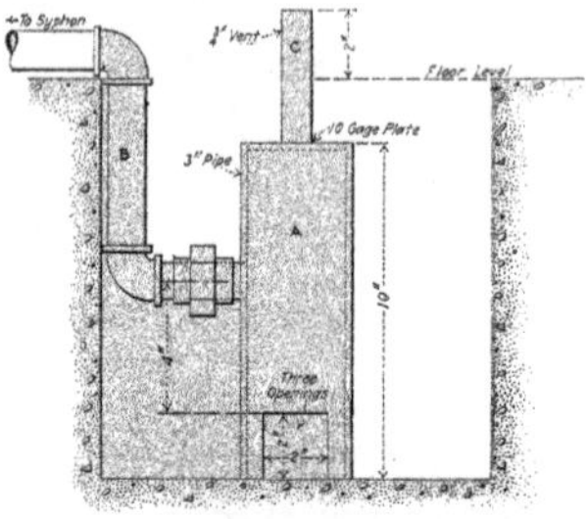

HOW THE OIL TRAP WAS MADE

INQUIRIES OF GENERAL INTEREST

Replacing Corliss Valve Stems—How would the key seats be laid off for new valve stems for a Corliss engine? J. E. E.

With the old stems in place, uncover the valve bonnets and, with the center mark on the wristplate hub brought opposite to the center mark on the wristplate stud, make a mark on the end of each valve to register with a mark made on the valve seat. Then replace the old valve stems with the new ones, and with the wristplate at the central position and the marks on the ends of the valve registering with the marks on the valve seats as before, and the valve arms in place, scribe off the positions of the keyways from the key seats in the hubs of the valve arms. If the work of marking and cutting the keyways is accurately done, the valve setting will be the same as it was with the old valve stems.

Equal Width Straps for Boiler Joints—What are the advantages and disadvantages of having the same width for the inside and outside straps of a longitudinal butt and double-strap boiler joint? E. L. L.

When both straps are of the same width, each held with the same number of rows of rivets, there is an advantage over unequal width straps in obviating eccentricity of shearing stress of the rivets due to unequal width of straps which has a tendency to bend the shell out of true circular form and thereby introduce longitudinal fracture of the main plate along the edges of the straps.

In American practice this disadvantage of unequal-width straps is assumed to be overcome by the stiffening effect of the broader inside strap. The principal disadvantages of employing straps of equal width are that the joints cannot be designed of as high percentage of efficiency, since for the best efficiency the outer rows of rivets must be spaced too far apart to form a good calking pitch. The latter objection can be overcome only by the expensive process of making the calking edges of saw-tooth form, with rounds and fillets that are more difficult to calk, and the calking edges must be of considerably greater length than straight calking edges.

Chord Factor—What is the chord factor of a coil in an induction-motor winding and how is it figured? C. A.

The chord factor of any coil in an electrical machine generating an electromotive force is a factor representing the effectiveness of the coil in generating voltage. The chord factor is equal to the sine of one-half of the angle in electrical degrees which the coil spans. In a stator core having 72 slots wound for 6 poles, for the coils to span full pitch they would have to span $72 \div 6 = 12$ slots, that is, the sides of a coil would have to lie in slots 1 and 13. This would represent 180 electrical degrees and the sine of one-half of 180 degrees is 1, hence, the chord factor of the coil is 1, and the coil is 100 per cent. effective in generating a counter-electromotive force. Each slot equals $180 \div 12 = 15$ electrical degrees. Then if the coil lay in slots 1 and 11 it would span only 10 slots or $15 \times 10 = 150$ electrical degrees, and the chord factor would be the sine of one-half of 150 deg., or 75 deg., equals 0.96. This means that the winding is only 96 per cent. as effective in generating a counter-electromotive force in the latter case as it was in the former. If with full pitch, under a given condition, the winding was good for operation on 500 volts, then with the coils in slots 1 and 11 the winding would be good for only $500 \times 0.96 = 480$ volts.

Flame in Smoke-Stack of Locomotive-Type Boiler—What is the cause of smoke burning in the stack of a locomotive-type boiler? C. E. S.

In burning solid fuels, the appearance of flame in the stack, or issuing from it, usually is due to particles of burning fuel or heated ash carried by the draft from the furnace and swept through the tubes to be discharged as cinders or sparks. The appearance of combustion is more pronounced when the fire surfaces of the boiler need cleaning of soot, for then the temperature of the gases may remain high enough for the heated particles to continue to glow and present the appearance of burning as discharged from the stack. Flame, with appearance of gases burning at the outlet of a stack, may result from combustion of volatile gases liberated from the fuel by the heat of the furnace, but which have not been mixed with enough air for their combustion until reaching the very point of discharge to the atmosphere.

Capacity of Cylindrical Tanks—What is a short rule for computing the number of gallons contained by a cylindrical tank 72 in. in diameter by 18 ft. long, with flat heads; and what allowance should be made when the heads are bumped? W. R. J.

The cubical content in American gallons of the tank with flat heads is given by the computation:

$$\frac{(72 \times 72 \times 0.7854) \times (18 \times 12)}{231} =$$

$$(72)^2 \times 18 \text{ ft.} \times 0.0408 = 3{,}807.13 \text{ gal.}$$

that is, where d = diameter of the tank in inches and L = length in feet, the general formula for capacity of a cylindrical tank with flat heads would be given by the formula,

$$\textit{Gallons capacity} = d^2 \times L \times 0.0408$$

Where h = the height of a bumped head in inches and d = its diameter in inches, the volume of a bumped head is given in gallons by the expression:

$$\left[\left(\frac{h}{2} \times 0.7854\, d^2\right) + (h^3 \times 0.5236)\right] \div 231 =$$

$$\frac{h \times 0.7854\, d^2}{2 \times 231} + (h^3 \times 0.00226).$$

Hence two bumped heads would contain the same number of gallons as a cylinder of diameter d and length h, $+ 2(h^3 \times 0.00226)$ gal., and the formula for complete capacity, including both bumped heads would be

$$\textit{Gallons capacity} =$$

$$\left(d^2 \times \textit{length of side in feet} + \frac{h}{12}\right) \times 0.0408 + 2(h^3 \times 0.00226)$$

Tanks for holding liquids under atmospheric pressure usually are provided with heads that are bumped so little that the value obtained for $2(h^3 \times 0.00226)$ is only a small percentage of the total tank capacity, and when there are two equal bumped heads the capacity is given close enough for most practical purposes by the formula

Number of gallons = *dia. in in.* × *dia. in in.* × *length of side plus height of one head in feet* × 0.0408.

[Correspondents should sign their communications giving full names and post office addresses.—Editor.]

Kerosene as a Fuel for High-Speed Engines*

By LAWRENCE SEATON

An account of tests run to obtain information regarding the correct design of a vaporizing system to successfully handle kerosene for high-speed engines. The description is given to the apparatus used, the method of making the tests, and the conclusions reached. Some consideration is also given to the question of the effect on the crank-case oil of unvaporized fuel leaking by the piston into the crank case.

AT THE present time there is a great deal of controversy regarding the correct design of a vaporizing system to successfully handle kerosene in high-speed engines. There has also been very little work done in determining the distribution of the heat in the fuel in such engines, and as to the effect on the crank-case oil due to unvaporized fuel passing by the piston and entering the crank case. It was therefore decided to fit up a high-speed engine so that the foregoing conditions could be carefully and accurately studied.

The apparatus used is shown in Fig. 1. The motor was a 4½ x 5½-in. high-speed heavy-duty type and was equipped with a Kingston carburetor connected to a specially designed vaporizer. The exhaust gases were piped directly to a calorimeter, but there was also provided a bypass pipe through the calorimeter, so that by means of dampers any desired amount of exhaust gas could be sent through the vaporizer and therefore any desired temperature maintained. Pyrometers were placed in the bypass and in the main-line exhaust pipe, and a thermometer was placed in the intake manifold. The air leading to the carburetor was maintained at a constant temperature by an electric heater. A manometer was connected to the intake opening into the carburetor.

Two Tabor engine indicators were used, one with a 240-lb. spring for measuring the combustion pressure and the other with an 80-lb. spring for measuring the compression pressure. A Dixie high-tension magneto was used for ignition. A water-spray nozzle was located in the intake manifold so that water might be introduced at high loads to prevent preignition.

FIG. 1. APPARATUS USED IN CONDUCTING TESTS

The cooling water was weighed every fifteen minutes. The fuel was weighed on a specially designed scale connected with a beam-bell contact, which made accurate reading possible. The water used in the manifold to prevent preignition was weighed on a sensitive scale.

The calorimeter for exhaust gases consisted of a Wheeler surface condenser, the water passing through the tubes and the exhaust gas entering the shell. Since the moisture in the products of combustion was condensed, it was necessary to provide for draining this moisture from the bottom of the condenser. The water used in the calorimeter was weighed on a set of scales having a range of 0 to 12,000 lb. The load carried by the engine was maintained by means of a Sprague electric dynamometer. The load maintained was so constant that no governor was necessary on the engine, and it was possible to study its action under a constant throttle opening.

After bringing the engine up to a speed of 1100 r.p.m., the speed recommended by the makers, the load was put on the generator of the dynamometer and the reaction of the pole pieces measured through a series of calibrated beams to the dial on the dynamometer.

A series of tests was first run with the air entering the carburetor maintained at a constant temperature of 115 deg. F., and the temperature around the vaporizer was increased from an initial temperature of 150 deg. F. for the first test, by 100 deg. increments, until a temperature of 650 deg. F. was reached. This corresponds to a temperature of 785 deg. inside

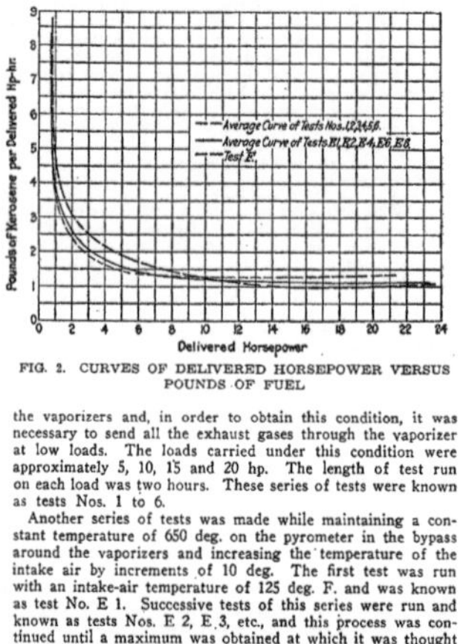

FIG. 2. CURVES OF DELIVERED HORSEPOWER VERSUS POUNDS OF FUEL

the vaporizers and, in order to obtain this condition, it was necessary to send all the exhaust gases through the vaporizer at low loads. The loads carried under this condition were approximately 5, 10, 15 and 20 hp. The length of test run on each load was two hours. These series of tests were known as tests Nos. 1 to 6.

Another series of tests was made while maintaining a constant temperature of 650 deg. on the pyrometer in the bypass around the vaporizers and increasing the temperature of the intake air by increments of 10 deg. The first test was run with an intake-air temperature of 125 deg. F. and was known as test No. E 1. Successive tests of this series were run and known as tests Nos. E 2, E 3, etc., and this process was continued until a maximum was obtained at which it was thought practicable to run the engine.

At the end of the series of "E" tests, which was thought to cover all conditions commonly met in practice, it was decided to run a test in which the limit of the apparatus was reached as far as furnishing heat to the mixture was concerned. This test was known as test F. All heat from the exhaust gases was passed through the vaporizer, and all possible current was passed through the heating coils in the air heaters. A so-called F test was run for each of the various loads, as was the case with the other series.

Indicator cards were taken at fifteen-minute intervals in order to ascertain the combustion pressure and to determine the evenness with which the fuel was burning in the cylinder, also to get some information as to the compression pressure. The indicator drum was not connected to the engine, but simply turned by hand while leaving the indicator cocks open.

The coil was cleaned out of the crank case at the end of each run, when its temperature, gravity and weight were determined. This was done to gain some definite information as to the effect of kerosene on the lubricating oil, especially

*Abstract of a paper presented at the annual meeting of the American Society of Mechanical Engineers, New York, Dec. 2 to 5, 1919.

when run without sufficient heat to more or less perfectly vaporize the fuel. From a summary of the results, the curve shown in Fig. 2 was plotted. This curve shows the weight of kerosene per horsepower per hour for the various horsepower outputs. It will be seen that the various heat changes made as described have practically no effect on the fuel consumption of the engine. It was observed, however, that the motor became much more flexible when the temperature of the ingoing air was raised. It was found that at low temperature the motor would not idle down below 800 or 900 r.p.m., while at the exceptionally high temperatures of the ingoing gases—about 375 deg. F.—it was found that the engine could be idled down to 150 r.p.m. and would pick up almost as well as when burning gasoline. It was also possible to start the engine on kerosene under these conditions.

The observation on the amount of kerosene deposited in the crank case was practically constant for all conditions, so that apparently the changes in heating the fuel had no effect in this matter. The only advantage apparent in heating the ingoing air and gas was that it made the motor more flexible.

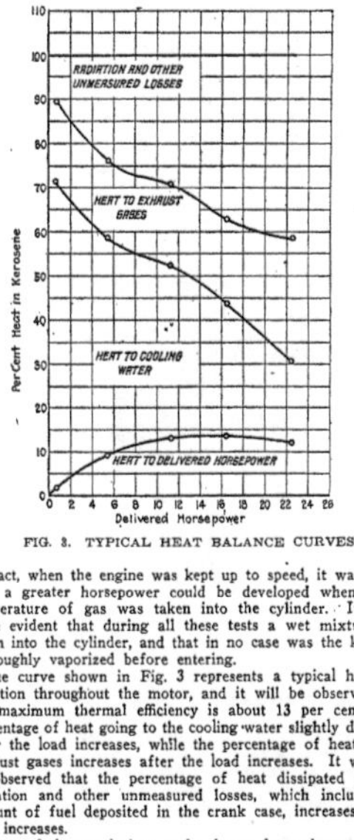

FIG. 3. TYPICAL HEAT BALANCE CURVES

In fact, when the engine was kept up to speed, it was found that a greater horsepower could be developed when a low temperature of gas was taken into the cylinder. It seems quite evident that during all these tests a wet mixture was taken into the cylinder, and that in no case was the kerosene thoroughly vaporized before entering.

The curve shown in Fig. 3 represents a typical heat distribution throughout the motor, and it will be observed that the maximum thermal efficiency is about 13 per cent. The percentage of heat going to the cooling water slightly decreases after the load increases, while the percentage of heat to the exhaust gases increases after the load increases. It will also be observed that the percentage of heat dissipated through radiation and other unmeasured losses, which includes the amount of fuel deposited in the crank case, increases as the load increases.

Some of the conclusions to be drawn from these tests are as follows: (1) It seems evident that to successfully burn kerosene in the type of internal-combustion engine now in general use, higher heat must be used so that the motor may be made sufficiently flexible to be practicable; (2) while the tests do not show the motor to be as efficient as when operating on gasoline, it is nevertheless believed that this defect can be overcome by designing the motor with high compression and large intake ports; (3) it was observed that the lubricating-oil temperature was higher when additional heat was added to the mixture, which means that a high flash-point oil would have to be used.

It is the belief of the writer that a motor designed as follows would handle kerosene at all loads successfully: The piston displacement should be greater per horsepower than that commonly used, a higher compression pressure should be obtained, the intake passages should be large and short, and the intake air should be heated to a temperature considerably above the boiling point of kerosene. This probably would be done with the exhaust gases, necessitating an automatic control at all loads.

New Engineering Publicity

One of the means that will be employed by the American Association of Engineers to promote engineering publicity throughout the United States is through a coöperative arrangement with the producer of several moving-picture weeklies. Arrangements have been made with the producers of six screen weeklies whereby the association will furnish lists of engineering projects and works, photographs of which will interest the public. After the film people have been notified of some engineering structure or event of sufficient interest, they will communicate with their correspondent living nearest the scene of the work or event, and he will either take still photographs or 100 or so feet of film, as the situation merits.

To make this project successful the association will require the active assistance of every professional engineer. Officials of corporations or construction companies who are doing or having done construction work or who are building or installing unusual or otherwise interesting machinery, engineers in charge of construction work or who are developing new methods or who know of spectacular or otherwise interesting features in which the public will be interested, can make this service a success by advising of such projects as will warrant photographing, the *Professional Engineer,* 63 E. Adams St., Chicago.

Do you know of any construction work, say of a large dam or bridge, or power plant or industrial works, or blast furnace, or cotton mill, which would be of sufficient interest to attract the attention of the general public? Do you know of a new process which has appealing scientific features and which can be photographed readily? Do you know of a survey which is to be attended by physical hardship or actual danger? Is there a building in your town which failed today because it was not designed or constructed under proper engineering supervision? Has your county or your state recently spent an enormous sum for highways? If it has, have the results been worthy of photographing? if you can answer any of these questions or any similar questions in the affirmative, send a postcard or brief letter or telegram advising the location of the work and give a brief description of its purpose with the approximate cost or other particular features which make it worth photographing. If the location is out of the usual course of travel, tell how the photographer can reach the work. The letter need not be long. Just outline the essentials as mentioned. Although the publicity committees of the chapters and clubs of the American Association of Engineers can be of great service in this matter, the foregoing appeal is not made wholly to the members of the American Association of Engineers. Engineers, public officials, corporation officials and others who are acquainted with engineering features of interest to the public are earnestly requested to render whatever assistance they are able in this development of engineering publicity.

Upset conditions in the coal industry in Great Britain are causing more and more interest in the substitution of oil fuel for coal, says the American Chamber of Commerce in London. It points out that the trend toward the use of oil fuel is well illustrated by the conversion of two of Britain's greatest liners to oil burners. Both the "Olympic," of the White Star Line, and the Cunarder "Aquitania" are being reconditioned and fitted with oil burners. Greater efficiency is expected and the engine-room personnel will be reduced perhaps from 350 to 40 or 50.

A Perfected High-Pressure Rotary Compressor*

By Chester B. Lord

The well-known disadvantages of a reciprocating machine have led to much research and many experiments in the hope of developing rotary machines both as prime movers and as pumps. In the compressor field three types have been developed—the centrifugal blower, the gear type and the eccentric rotor with a telescopic blade. The two former types are limited by practical considerations to low pressures, probably not over eight or ten pounds per single stage. The third type is capable of higher pressures, but up to the present has shown very low efficiencies.

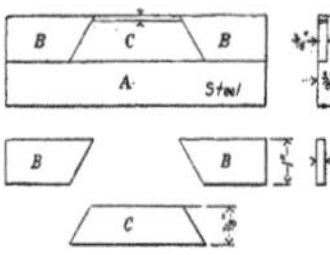

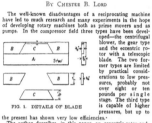

FIG. 1. DETAILS OF BLADE

The author describes, in this paper, an eccentric-rotor-and-telescopic-blade type compressor which has given 92 per cent. volumetric efficiency at 100 lb. pressure and which will pump a gas or liquid or a gas and liquid at the same time.

The main difficulties with this type of machine have been the necessity of maintaining a seal at one point between the rotor and the cylinder, the friction caused by the blade pressure against the end plates and the inner surface of the cylinder (the pressure being on one side of the blade, caused difficulty in moving it in and out of the slot in the rotor), and the difficulty of packing square corners.

The solution of the first difficulty has been accomplished by allowing the cylinder to rotate with the same surface speed as the rotor so that there is no sliding motion at the point of contact. In this machine the contact of the rotor with the cylinder carries the cylinder around, but since the circumference of the rotor is less than that of the cylinder, the angular velocity of the latter is less than that of the former. This principle is not original with the designer of this compressor, but since it solves only one of the difficulties mentioned, it was valueless by itself.

Excessive end pressure is overcome by floating end plates, held in contact with the cylinder by the pressure generated by the machine itself. These end plates, which are simply flat circular plates, are subject to the discharge pressure on the outside, and since the area of the outer surface is greater than the area of the inner surface, are held against the cylinder ends. These plates of course revolve with the cylinder or rotor or both, so there is little or no sliding to cause friction losses. The blade, moving in and out of its slot in the rotor, rubs against these floating cylinder ends, and this is the principal point of friction loss. From dimensions of a machine capable of compressing 100 cu.ft. of free air per minute, this total travel is calculated to be 93 ft. per minute, so the power lost is not great even for fairly high pressures.

The blade used is shown in Fig. 1. The blade proper is marked *A* and is cut away as shown. The pieces *B* and *C* are carried in the main blade, as shown. Centrifugal force throws the wedge-shaped piece *C* out, so pushing the packing pieces *B* against the ends of the cylinder and into the corners. The force with which these pieces press into the corners is regulated by the angle of the wedge sides. The main blade *A* has about $\frac{1}{32}$ in. clearance at each end.

For adjusting contact between the rotor and cylinder, a so-called wedge ring is used. This is simply an eccentric cradle carrying the bearings for the revolving cylinder and is turned by means of a threaded stud led through the outside casing.

Fig. 2 shows a section of a 20-ton refrigerating machine.

Among the advantages claimed for this machine is the possibility of direct-connecting it to the driver, as it can be run at any speed that is safe for the prime mover. Another advantage is its small space requirement as compared with reciprocating machines. As an example, a 100-ton rotary refrigerating machine requires only 60 sq.ft. floor space as against 1,200 sq.ft. for a horizontal reciprocating unit. The weights of the machines are given as eight pounds per ton capacity for the rotary machine and one hundred pounds per ton capacity for the reciprocating type.

The use of the floating cylinder ends, besides cutting down friction, enables the machine to handle liquids either alone or with gases. In a reciprocating compressor, if any liquid present cannot go out through the discharge valve fast enough, it simply breaks something. In this machine the end rings move away from the cylinder, allowing the liquid to escape, and as soon as the pressure is equalized, close up again.

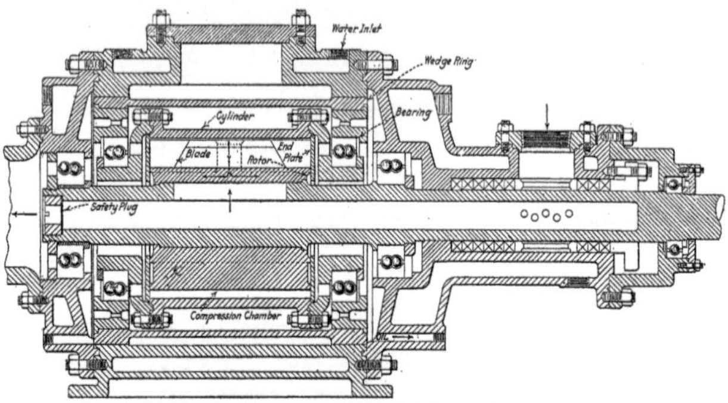

FIG. 2. CROSS-SECTION ASSEMBLY OF A 20-TON REFRIGERATING COMPRESSOR.

*Abstract of paper presented at the Annual Meeting American Society of Mechanical Engineers, Dec. 2-5, 1919.

The field for this type of machine embraces every use possible for the reciprocating type and some that are not. It is possible to get 500 lb. pressure single-stage and 1000 lb. in three stages. It is possible to pull a 23½-in. vacuum against 180 lb. pressure. These conditions are not commercial requirements, but are given to indicate the machine's possibilities. It has been used as a blast-furnace blowing engine, as a 0.5 cu.ft. per minute refrigerating machine for household use and for street car compressors.

The author says that all the statements in the article have been demonstrated commercially, and hence it would seem that the long-looked-for rotary compressor has arrived.

Moisture in Insulation

In a paper, "The Story of Insulation," presented at the 12th annual convention of the Pennsylvania Electric Association, C. E. Skinner tells the following story regarding the amount of moisture in the insulation:

At the time of the construction of the Ontario Power Co.'s transmission plant at 60,000 volts, we had arrived at the stage of the art when we considered it advisable to thoroughly vacuum dry complete transformers before immersing them in oil. Equipment for this purpose was not available at the factory for transformers of the size used by the Ontario Power Co., and it was therefore decided to make the transformer tank itself the vacuum tank and provide means for vacuum drying of transformers after being placed in the tanks at the point of installation. After drying for a week or ten days, following the course of the drying by measuring insulation resistance of the transformers until this was considered satisfactory, oil was admitted to the first transformer while still under vacuum. As a final precaution a sample of oil was drawn from the bottom of the tank only to find that it was about 98 per cent. water. Investigation of a second transformer into which the oil had not been admitted, discovered 6 or 7 gallons of water in the bottom of the transformer tank, although the transformer itself was undoubtedly satisfactorily dried by the vacuum process. The water doubtless came partly from the transformer coils, partly from the laminated iron and partly from the air which was inevitably drawn into the transformer tank on account of the impossibility of making the tank absolutely vacuum-tight. This moisture, condensing on the cooler top and sides of the tank, collected at the bottom. The cure was simple and consisted in a double valve arrangement with a small tank between, allowing the condensed water to be drawn off as the drying proceeded. This incident shows the very large quantities of water which could be collected under the circumstances described and indicates that apparatus not satisfactorily protected by a moisture-proof coating or by immersion in oil requires special precautions in installation and starting. Later studies have indicated that moisture in extremelv minute amounts may play a very important part in impairing insulation where the requirements are severe, as it may increase the dielectric losses and lower the insulation resistance and the dielectric value.

Canadian Engineers Will Meet in Montreal

The annual business and professional meeting of the Engineering Institute of Canada will be held at the Institute quarters, Mansfield street, on Jan. 27, 28 and 29.

At this annual convention questions of interest to the engineering profession of Canada will be discussed as well as topics of special interest to the people of the province of Quebec.

Papers dealing with the commercial, industrial and financial problems of the province will be delivered, also on its water-powers, both developed and undeveloped, its highways, its forest products and a general survey of its commercial potentialities.

It is expected that the convention will be the largest and most representative that has ever been held and it is hoped that it may assist in giving the public a more comprehensive view of the essential work achieved by the members of the engineering profession in building up the prosperity of Canada.

A bill has passed the Oregon Legislature adopting the A. S. M. E. Boiler Code, and now awaits the signature of the governor of the state. The Seattle boiler rules are about to be amended, making the code apply to that city.

Boiler Explosion Kills Three

At the East Chicago works of the Interstate Iron and Steel Co., a 250-hp. vertical water-tube boiler exploded at 4:30 a. m. on the morning of Jan. 19. Three men were killed outright, twelve were injured more or less seriously, but none fatally. The property damage was estimated at $10,000. The boiler was one of three in an auxiliary battery that was used to supply the demand for steam in excess of that furnished by the waste-heat boilers which ordinarily carried the bulk of the load.

The boilers were equipped with hand-fired dutch-oven furnaces. Two of them were being fired up on the morning of the accident, preparatory for the morning shift which started at 6 a. m. Before the working pressure of 120 lb. gage had been reached, one of the boilers let go. The tubes, accompanied by the bottom tube sheet, rose with the top drum and descended some 700 ft. away from the initial location. The settings of the other two boilers and the boiler house were demolished.

Closed valves proved that the boilers had not been cut in on the line. Apparently the cause of the accident was a hidden defect, which at the time of writing had not been detected. The boiler had been inspected and reported in good condition by an insurance inspector just twelve hours before the accident. More complete details will be given in a later issue.

California Requires More Hydro-Electric Power

"More hydro-electric energy must be developed in California to meet the present requirements, and vastly more is necessary to accommodate future demands."

There are eighty-four electric utilities in the state of California, operating seventy-five hydro-electric plants, with an installed capacity of 465,000 kw., and 50 steam plants with an installed capacity of 305,000 kw., making a total of 125 plants, aggregating 770,000 kilowatts, says a newspaper report.

During the year 1918 these plants generated a total of 2,892,000,000 kw.-hr., of which 2,163,000,000 kw.-hr., or 75 per cent. of the total, were produced from water power.

This power is transmitted through 7,300 miles of high-tension transmission lines to points of distribution, from which 84,000 miles of secondary distribution lines extend. Electric service is supplied to 545,000 consumers.

The installed capacity of consumers' lights, motors and other power-consuming devices exceeds 1,800,000 horsepower. Nearly 900,000 kw. of distribution transformers are installed on these systems.

A. A. E. Claims Membership of 18,000

The results of the membership drive held during December by the American Association of Engineers have been tabulated, and notwithstanding the difficulties imposed by the coal strike, indications are that the total number of members will approximate 18,000. At the close of business on Dec. 31, the number of fully accredited members exceeded 10,400. About 3,000 additional members have been accepted, but were taking the time alloted to pay dues, and most of the 7,000 applications received during the month of December had not been acted upon. Allowing for contingencies, the net membership should approach closely the total figure previously given.

CORRECTION: Through an error the review for the year, under Power Plant Legislation, in the issue of Jan. 6, included Massachusetts among those states which had adopted the A. S. M. E. Boiler Code. Also, the map on page 62 of the Jan. 13 issue likewise included this state. As is well known, the state of Massachusetts has its own boiler code, which very closely resembles the A. S. M. E. Code, but is not identical, and the latter code has not been officially adopted as yet by the state of Massachusetts.

Through Stone & Webster, of Boston, the Hartford Electric Light Co. has arranged for the purchase of the Connecticut Power Co., operating a large hydro-electric power plant on the Housatonic River and in western Connecticut.

THE COAL MARKET

BOSTON—Current prices per gross ton f.o.b. New York loading ports:

Anthracite

	Company Coal
Egg	$8.20@$8.65
Stove	8.45@ 9.05
Chestnut	8.55@ 9.05
Pea	7.05@ 7.40
Buckwheat	3.00@ 3.80
Rice	2.30@ 2.50
Barley	1.25

Bituminous

	Clearfields	Cambrias and Somersets
F.o.b. mines, net tons	$2.85@$3.35	$3.15@$3.60
F.o.b. Philadelphia, gross tons	5.05@ 5.60	5.35@ 5.90
F.o.b New York, gross tons	5.40@ 5.95	5.75@ 6.25
Alongside Boston (water coal), gross tons	7.00@ 7.75	7.60@ 8.00

Pocahontas and New River are practically off the market for coastwise shipment, but are quoted at $6.25@$7.00.

NEW YORK—Current quotations, White Ash, per gross tons, f.o.b. Tidewater, at the lower ports are as follows:

Anthracite

	Company Coal
Broken	$7.80@$8.25
Egg	8.20@ 8.65
Stove	8.45@ 9.05
Chestnut	8.55@ 9.05
Pea	7.05@ 7.40
Buckwheat	5.15
Rice	4.50
Barley	4.85
Boiler	4.85

Bituminous

Government prices at mines:	Spot
South Fork (best)	$3.25@$3.50
Cambria (best)	3.00@ 3.25
Cambria (ordinary)	2.60@ 2.90
Clearfield (best)	3.00@ 3.25
Clearfield (ordinary)	2.60@ 2.90
Reynoldsville	2.85@ 2.90
Quemahoning	3.25@ 3.50
Somerset (medium)	3.00@ 3.25
Somerset (poor)	2.50@ 2.75
Western Maryland	2.50@ 2.75
Fairmont	2.25@ 2.50
Latrobe	2.60@ 2.90
Greensburg	2.75@ 3.00
Westmoreland ¾ in.	3.40@ 3.50
Westmoreland run-of-mine	2.75@ 3.00

PHILADELPHIA—Anthracite prices are practically the same as those listed above for New York. Bituminous coal prices vary according to district from which they are mined. For ordinary slack the price is $2.45@$2.65; lump, $3.00@$3.35, at the mines.

BUFFALO—

Anthracite

	On Cars, Gross Ton	At Curb, Net Ton
Grate	$8.55	$10.80
Egg	8.80	10.65
Stove	9.00	10.85
Chestnut	9.10	10.85
Pea	7.75	9.30
Buckwheat	5.70	7.75

Bituminous

Allegheny Valley	$4.80
Pittsburgh	4.65
No. 8 Lump	4.65
Mine Run	4.30
Slack	4.10
Smokeless	4.60
Pennsylvania Smithing	5.70

CLEVELAND—Prices of coal per net ton delivered in Cleveland are:

Anthracite

Egg	$12.25@$12.40
Chestnut	12.50@ 12.70
Grate	12.25@ 12.40
Stove	12.40@ 12.60

Pocahontas

Mine-run	$7.50

Domestic Bituminous

West Virginia split	$8.50
No. 8 Pittsburgh	$6 60@ 6.90
Massillon lump	8.25@ 8.50
Coshocton lump	7.15

Steam Coal

No. 6 slack	$5.25@$5.50
No. 8 slack	5.10@ 5.50
Youghiogheny slack	5.25@ 5.50
No. 8 ¾ in.	5.70@ 6.00
No. 6 mine-run	5.25@ 5.50
No. 8 mine-run	5.75

Only coal available is mine-run Pocahontas.

MIDDLE WEST—Chicago quotations, F.o.b. cars at mine:

	Springfield, Carterville, Williamson, Franklin, Saline, Harrisburg	Fulton, Peoria	Grundy, La Salle, Bureau, Will
Lump	$2.55@$2.70	$2.95@$3.10	$3.25@$3.40
Washed	2.75@ 2.90	....	3.45@ 3.60
Mine run	2.35@ 2.50	2.75@ 2.90	3.00@ 3.15
Screenings	2.05@ 2.20	2.35@ 2.50	2.75@ 2.90

New Construction

PROPOSED WORK

Me., Augusta—Lockwood, Green & Co., Engrs., 60 Federal St., Boston, Mass., will receive bids for the construction of a 6 story, 65 x 135 ft. storehouse on Water St., for the Edwards Mfg. Co., Water St. A steam heating system will be installed in same. Total estimated cost, $100,000.

Vt., Lunenburg—The Fitzdale Paper Mills are having preliminary plans prepared for the construction of a 1 story power house. A steam heating system will be installed in same. Total estimated cost, $1,000,000. George F. Hardy, 309 Bway., New York City, Archt. and Engr.

Mass., Boston—John H. Lyons, 15 State St., is preparing plans for the construction of a 5 story, 35 x 110 ft. mercantile building on Stanhope St. A steam heating system will be installed in same. F. A. Norcross, 46 Cornhill St., Archt.

Mass., Wellesley—F. J. Untersee, Archt., 585 Boylston St., Boston, will soon award the contract for the construction of a 4 story, 110 x 190 ft. college building and tower on Worcester St., for the Academy of the Assumption, Wellesley Hills. A steam heating system will be installed in same.

Conn., Meriden—The Griswold, Richmond & Block Co., 2 West Main St., is having plans prepared for the construction of a 2 story, 120 x 150 ft. business building on West Main and South Grove Sts. A steam heating system will be installed in same. Total estimated cost, $100,000. David Bloomfield, 129 State St., Archt.

Conn., New Haven—Alexander & Links, 138 State St., plans to remodel the present 4 story building on State St., into an ice plant for its own use.

N. Y., Brooklyn—Lockwood, Green & Co., Archts. and Engrs., 101 Park Ave., New York City; will receive bids until Jan. 23 for the construction of a 3 story, 175 x 200 ft. factory on 51st and 52nd Sts. and 2nd Ave., for Bemis Bros. Bag Co. A steam heating system will be installed in same. Total estimated cost, $500,000.

N. Y., Brooklyn—Shampan & Shampan, Archts. and Engrs., 50 Court St., are having plans prepared for the construction of a 4 story, 40 x 100 ft. laundry at 835-849 Myrtle Ave. A steam heating system will be installed in same. Total estimated cost, $150,000. Owner's name withheld.

N. Y., Brooklyn—The Y. M. H. A., 63 Morton St., plans to build a 4 story, 100 x 150 ft. building on B'way and Rodney St. A steam heating system will be installed in same. Total estimated cost, $300,000. H. J. Rosenson, Pres.

N. Y., Brooklyn—The Bd. Educ., 500 Park Ave., New York City, plans to build a 5 story school on Nostrand Ave. and Sandford St., cost, $400,000; 3 story school on East 92nd St. and Ave. L., Canarsie, $200,000; 5 story school on 70th St. and 18th Ave., $400,000; 5 story school on 62nd St. and Ft. Hamilton Ave., $400,000; 4 story school on Newport St. and Stone Ave., $450,000; 5 story school on Saratoga and Livonia Aves., $450,000; 5 story school on New York and Tilden Aves., $500,000; 4 story addition to P. S. 16, Wilson St., near Bedford Ave., $400,000; 4 story addition to P. S. 136, 4th Ave. and 31st St., $400,000; 4 story addition to P. S. 67, $350,000; 4 story school on 86th St. and 2nd Ave., 350,000. Steam heating systems will be installed in same. C. B. J. Snyder, Municipal Bldg., New York City, Archt. and Engr.

N. Y., Far Rockaway—The Bd. Educ., 500 Park Ave., New York City, is having plans prepared for the construction of a school addition, here. A steam heating system will be installed in same. Total estimated cost, $200,000. C. B. J. Snyder, Municipal Bldg., New York City, Archt. and Engr.

N. Y., Jamaica—The Bd. Educ., 500 Park Ave., New York City, plans to build a 4 story school on Liberty, Bryant and Jerome Aves. A steam heating system will be installed in same. Total estimated cost, $400,000. C. B. J. Snyder, Municipal Bldg., New York City, Archt. and Engr.

N. Y., Long Island City—The Bd. Educ., 500 Park Ave., New York City, plans to build a 3 story school on Pierce Ave. and Briell St. A steam heating system will be installed in same. Total estimated cost, $300,000. C. B. J. Snyder, Municipal Bldg., New York City, Archt. and Engr.

N. Y., Maspeth—The Bd. Educ., 500 Park Ave., New York City, plans to build a 2 story addition to P. S. 72 on Maspeth Ave. A steam heating system will be installed in same. Total estimated cost, $150,000. C. B. J. Snyder, Municipal Bldg., New York City, Archt. and Engr.

N. Y., Rockaway Beach (Far Rockaway P. O.)—The Bd. Educ., 500 Park Ave., New York City, is having plans prepared for the construction of a school addition at Rockaway Beach. A steam heating system will be installed in same. Total estimated cost, $300,000. C. B. J. Snyder, Municipal Bldg., New York City, Archt. and Engr.

N. Y., New York—The Bd. Educ., 500 Park Ave., is having plans prepared for the construction of a 5 story, 60 x 193 ft. school on Fox St. and Leggett Ave. (Bronx Boro). A steam heating system will be installed in same. Total estimated cost, $250,000. C. B. J. Snyder, Municipal Bldg., Archt. and Engr.

N. Y., New York—S. Finkelstein, c/o W. Koppe, Archt. and Engr., 935 Intervale Ave. (Bronx Boro), will build a 6 story, 90 x 100 ft. clothing factory on Leggett and Whitlock Aves. A steam heating system will be installed in same. Total estimated cost, $170,000. Work will be done by day labor.

N. Y., New York—L. Gold, 44 Court St., plans to build a garage on 55th St. and 10th Ave. A steam heating system will be installed in same. Total estimated cost, $100,000. Work will be done by day labor.

N. Y., New York—The Gotham National Bank, 1819 Bway., will build a 22 story, 85 x 100 ft. bank and office building at 303 West 59th St. A steam heating system will be installed in same. Total estimated cost, $3,000,000. Work will be done by day labor.

N. Y., New York—The New York Curb Market Assn., Wall St., plans to build a court exchange building on Trinity Pl. and Greenwich St. A steam heating system will be installed in same. Total estimated cost, $700,000. Starrett & Van Vleck, 8 West 40th St., Archts. and Engrs.

N. Y., New York—The Allerton House Corp., 143 East 39th St., is having plans prepared for the construction of a 14 story hotel on Lexington Ave. and 57th St. A steam heating system will be installed in same. Total estimated cost, $1,500,000. A. L. Harmon, 3 West 29th St., Archt. and Engr.

N. Y., New York—The Corn Exch. Bank, 13 William St., is having plans prepared for the construction of a 3 story, 50 x 100 ft. bank, office and store building at 525-527 B'way. A steam heating system will be installed in same. Total estimated cost, $175,000.

N. Y., New York—Shampan & Shampan, Archts. and Engrs., 50 Court St., Brooklyn, are in the market for a 350 hp. steam power boiler, generators and motors for the proposed new laundry and manufacturing building on Myrtle and Marcy Aves., Brooklyn.

N. Y., White Plains—The Bd. Educ. is having preliminary plans prepared for the construction of a high school. A steam heating system will be installed in same. Total estimated cost, $125,000. Tooker & Marsh, 101 Park Ave., New York City, Archts. and Engrs.

N. J., Jersey City—The Standard Theatre Co., c/o E. C. Horn Sons, Archts. and Engrs., 1476 B'way, New York City, will soon award the contract for the construction of a theatre on Central Ave. and Sherman Pl. A steam heating system will be installed in same. Total estimated cost, $250,000.

N. J., Hoboken—The White Metal Mfg. Co., 1006 Clinton Ave., is having plans prepared for the construction of a 6 story, 135 x 125 ft. factory and a 2 story, 20 x 80 ft. foundry. A steam heating system will be installed in same. J. C. Schaffler & Co., 38 West 32nd St., New York City, Archt. and Engr.

N. J., Trenton—H. R. Mallinson Co., Inc., 136 Madison Ave., New York City, plans to construct a 1 story, 50x 70 ft. power house on Parker Ave. Estimated cost, $15,000.

Penn., Bethlehem—The Lehigh Alumni Association, Lehigh University, is having plans prepared for the construction of a 2 story, 80 x 160 ft. memorial building. A steam heating system will be installed in same. Total estimated cost, $350,000. T. Visscher & J. Burley, 363 Lexington Ave., New York City, Archts. and Engrs.

Penn., Chester—The Cochrane Estate plans to build an 18 story office building. A steam heating system will be installed in same. Total estimated cost, $450,000. E. W. Brazer, 1133 Bway., New York City, Archt. and Engr.

Penn., Easton—The McCormick Co., Archt. and Engr., 41 Park Row, New York City, is having plans prepared for the construction of a bakery plant. A steam heating system will be installed in same. Total estimated cost, $100,000. Owner's name withheld.

Penn., Lewistown—The Viscose Co., Marcus Hook, plans to build a plant, consisting of 5 buildings for factory purposes, also a number of houses for employes. A steam heating system will be installed in same. Total estimated cost, $2,000,000. Ballinger & Perrot, 17th and Arch Sts, Philadelphia, Archt. and Engr.

Penn., Philadelphia—The city is having plans prepared for the construction of a 2 story, 200 x 330 ft. library at 19th, 20th, Vine and Wood Sts. A steam heating system will be installed in same. Total estimated cost, $3,000,000. Horace Trumbaner, Land Title Bldg., Archt. and Engr.

Penn., Scranton—The city plans to build a 2 story, 80 x 80 ft. school on Van Buren St. A steam heating system will be installed in same. Total estimated cost, $115,400.

Penn., Scranton—The city plans to build a 3 story, 120 x 150 ft. school on Madison St. A steam heating system will be installed in same. Total estimated cost, $402,300.

Penn., Scranton—The city plans to build a 3 story, 100 x 115 ft. school on Providence Rd. A steam heating system will be installed in same. Total estimated cost, $277,350.

Penn., Scranton—Mark K. Edgar, Bd. of Trade Bldg., plans to build a 6 story, 140 x 320 ft. incubator plant on Lock Ave. A steam heating system will be installed in same. Total estimated cost, $1,000,000.

Del., Wilmington—The White Automobile Co., East 79th St. and St. Clair Ave., Cleveland, O., has had plans prepared for the construction of a 2 story sales room and garage on 31st St. and Market Ave. A steam heating system will be installed in same. Total estimated cost, $100,000. Watson Eng. Co., Hippodrome Bldg., Cleveland, O., Engr.

Va., Norfolk—The Third Christian Church is having plans prepared for the construction of a 1 story, 86 by 127 ft. church on Llewelly Ave. A steam heating system will be installed in same. Total estimated cost, $100,000. W. H. Nicklas, 1900 Euclid Bldg., Cleveland, O., Archt.

N. C., Lexington—The city will receive bids until Jan. 20 for the construction of a pumping station and a 500,000 gal. storage reservoir, also for the installation of motor driven centrifugal pumps, etc. G. C. White, Durham, Engr.

Miss., Jackson—The city will receive bids about Feb. 15 for the construction of three 2 story school buildings. A low pressure steam heating system will be installed in same. Total estimated cost, $200,000. E. L. Bailey, Secy. Emmett J. Hull, Jackson, Archt.

O., Canton—Edward L. Smith, c/o Charles Firestone, Archt., Renkert Bldg., will receive bids until Feb. 2 for the construction of an 8 story hotel. A steam heating system will be installed in same. Total estimated cost, $400,000.

O., West Barberton—The Portage Rubber Co. is having plans prepared for the construction of a plant, including an office building and power house. Total estimated cost, $200,000.

Mich., Allegan—The Bd. Educ. plans to build a 3 story, 100 x 158 ft. high school. A steam heating system will be installed in same. Total estimated cost, $150,000. John D. Chubb, 109 North Dearborn St., Chicago, Archt.

Mich., St. Ignace—Cunning & Mulcrone plan to build a 3 story, 70 x 136 ft. hotel here. A steam heating system will be installed in same. Total estimated cost, $130,000. F. E. Parmelee, Iron Mountain, Archt.

Mich., Strathmoor—School Dist. No. 4, Wayne Co., c/o Horace H. Hart, will soon award the contract for the construction of a 2 story, 90 x 110 ft. school. A steam heating system will be installed in same. Total estimated cost, $125,000. W. H. Adams, Vinton Bldg., Detroit, Engr. Butterfield & Butterfield, 1113 David Whitney Bldg., Detroit, Archt.

Ill., Chicago—A. S. Alschuler, Archt., 28 East Jackson Blvd., will soon award the contract for the construction of a 3 story, 200 x 400 ft. auto sales and service building on Michigan Ave., near 22nd St., for the Hudson Motor Co., 1615 South Michigan Ave. A steam heating system will be installed in same. Total estimated cost, $300,000.

Ill., Chicago—The Bd. Educ., 7 South Dearborn St., will soon award the contract for the construction of a 3 story, 132 x 175 ft. grade school on Augusta St., near Laramie Ave. A steam heating system will be installed in same. Total estimated cost, $350,000. John Howatt, 7 South Dearborn St., Archt. A. F. Hussander, 7 South Dearborn St., Engr.

Ill., Chicago—The University of Chicago is having plans prepared for the construction of a 100 x 100 ft. chapel. A steam heating system will be installed in same. Total estimated cost, $1,500,000. B. Goodhue, 2 West 47th St., New York City, Archt. and Engr.

Wis., Milwaukee—The City Sewerage Comm. has retained W. J. Sands, Consult. Eng., 1338 Wells Bldg., to prepare plans for furnishing of power plant equipment for the proposed power plant on Jones Island; not decided whether coal burning, steam boilers, oil burning or internal combustion engine type. Total estimated cost, $850,000 to $1,700,000.

Wis., Milwaukee—H. W. Voelz, Archt. and Engr., 86 Michigan St., is preparing plans for the construction of a 2 story, 70 x 115 ft. dairy. A refrigeration system will be installed in same. Total estimated cost, $75,000. Owner's name withheld.

Wis., Racine—Louis Lehle, Archt., 3810 B'way, Chicago, will soon award the contract for the construction of a 4 story, 100 x 225 ft. warehouse for the Horlicks Malted Milk Co., Horlicksville St. A steam heating system will be installed in same. Total estimated cost, $300,000.

Wis., Sheboygan—The Standard Oil Co., Wells St., Milwaukee, has had plans prepared for the construction of two 2 story, 60 x 95 ft. filling stations and office building on Calumet Rd. and 15th St. Electric pumps, etc., will be installed in same.

Wis., Stevens Point—The Citizens' Natl. Bank will receive bids for the construction of a 1 story, 25 x 100 ft. bank. A steam heating and refrigerating system will be installed in same. Total estimated cost, $125,000. A. Moorman & Co., Chamber of Commerce Building, Minneapolis, Minn., Archt.

Wis., Tomahawk—The city has had plans prepared for the installation of 2 centrifugal pumps, capacities of 150 and 500 gal. per min.

Wis., Two Rivers—The Bd. Educ. is having preliminary plans prepared for the construction of a high school. A steam heating system will be installed in same. Total estimated cost, $200,000. John D. Chubb, 109 North Dearborn St., Chicago, Ill., Archt.

Ia., Cherokee—The State Bd. of Control, Des Moines, is having plans prepared for the construction of a pipe line, pumphouse and reservoir, also for furnishing pumping equipment in connection with the proposed waterworks improvements here. Total estimated cost, $70,000. W. E. Buell, Davidson Bldg., Sioux City, Engr.

Ia., Washta—The city will receive bids until Feb. 15 for the construction of a well, pumphouse, reservoir and distribution system in connection with the proposed waterworks improvements. Total estimated cost, $22,000. G. S. Strivers, Clk. W. E. Buell, Sioux City, Engr.

Ia., Woodbine—The Bd. Educ. will soon award the contract for the construction of a 3 story, 59 x 81 ft. school. Estimated cost, $100,000. Steam heating contract will be sub-let. E. E. Higgins, Secy. W. E. Hulse & Co., 210 Masonic Temple, Des Moines, Archt.

Minn., Minneapolis—The Bd. Educ. will soon award the contract for the construction of a 3 story, 61 x 163 ft. addition to the Calhoun School on Girard Ave. A low pressure steam heating system will be installed in same. Total estimated cost, $200,000.

Kan., Hugoton—The city will receive bids until Jan. 30 for the construction of a water system and electric light plant. Estimated cost, $60,000. R. L. Smith, Chn. Ruckel Eng. Co., Hutchinson, Engr. Noted Sept. 30.

Kan., Neodesha.—The Standard Oil Co. of Kansas, 5301 East 9th St., Kansas City, plans to construct a 1 story refinery. Project includes 42 additional stills, a modern boilerhouse that will develop 5000 hp., tanks with a capacity of 300,000 bbls. of oil, etc. Total estimated cost, $1,500,000.

S. D., Ipswich—The Bd. Educ. is having plans prepared for the construction of a 2 story, 75 x 126 ft. grade and high school. Steam heating and fan ventilation systems will be installed in same. Total estimated cost, $100,000. Edwins & Edwins, 911 Northwestern Bank Bldg., Minneapolis, Engrs.

Okla., Mangum—Hawk & Parr, Archts., 501 Security Bldg., Oklahoma City, are preparing plans for the construction of a 3 story, 100 x 175 ft. high school on Main St., for the city. A steam heating system will be installed in same. Total estimated cost, $130,000.

Okla., Norman—The city voted $125,000 bonds to construct an electric light and power plant.

Okla., Oilton—The city voted $40,000 bonds to construct 2 wells and install air lift system in the pumping station.

Wash., Puget Sound—The Bureau of Yards & Docks, Navy Dept., Wash., D. C., will receive bids until Feb. 11 for an extension to the central power plant here.

Cal., Pasadena—The Water Dept. will soon lay main to the Annandale Dist., at a cost of about $125,000, install booster pump, about $10,000; also build concrete and earth dam to cost $15,000. Work will be done by day labor. S. B. Morris, City Hall, Engr.

Tex., Dallas—A. C. Bossom, Archt. and Eng., 366 5th Ave., New York City, will soon award the contract for the construction of a 10 story office building, for the Magnolia Petroleum Co., Great Southwestern Life Bldg. A steam heating system will be installed in same. Total estimated cost, $1,000,000.

Tex., Dallas—Lowe's Enterprises, 1492 B'way., New York City, are having plans prepared for the construction of a theatre. A steam heating system will be installed in same. Total estimated cost, $300,000. T. W. Lamb, 644 8th Ave., New York City, Archt.

Okla., Pryor Creek—V. V. Long & Co., Consult. Engr., 1300 Colcord Bldg., Oklahoma City, is preparing preliminary plans for the construction of a pumping station, transmission line, etc., for the city.

Col., Yuma—The Town Council will soon award the contract for the construction of a steam electric light power plant. Estimated cost, $50,000. R. D. Salisbury, 1415 East Colfax Ave., Denver, Engr.

Wash., Yakima—The Board of Directors of Yakima Co. Agricultural Union plans to build a 2 or 3 story addition to the cold storage plant on North 1st St. The Union has authorized $150,000 bond issue to cover the cost of same.

Cal., Newport Beach—The Newport Heights Irrigation Dist. is preparing plans for the construction of a pumping plant, mains, reservoirs and distributing system. Total estimated cost, $160,000. P. E. Kressly, Engr.

Cal., San Jose—The Bayside Canning Co., Alnso St., is having plans prepared for the construction of a 1 story cold storage plant. Estimated cost, $70,000. C. S. McKenzie, Bank of San Jose Bldg., Archt.

Ont., Newmarket—The city is having plans prepared for the construction of sewers on Darcy and Yonge Sts., including an outfall sewer, disposal and an activated sludge plant, pumphouse and electrically driven compressed air pumps. Total estimated cost, $60,000. E. A. James Co., Ltd., 36 Toronto St., Toronto, Engr.

Ont., Owen Sound—The Bd. Educ. plans to build a 3 story Collegiate Institute. A vacuum steam heating system will be installed in same. Ratepayers will vote on $180,000 by-law in January to raise money to cover cost of same.

Ont., Toronto—The Auto Service Co., 139 Pears Ave., is having plans prepared for the construction of a 6 story garage on Richmond St. A vacuum steam heating system will be installed in same. Total estimated cost, $600,000. Banigan & Thompson, 7 King St., E., Engrs.

Ont., Toronto—The Board of Governors of the University of Toronto, Queens Park, plans to build six 2 and 3 story buildings, including an addition to the central heating plant. Total estimated cost, $2,000,000. Dailing & Pierson, 2 Leader Lane, Archts.

Ont., Toronto—T. L. Church, Chn. of Bd. of Control, City Hall, will receive bids until Feb. 10 for the supply and installation of one or more 16 to 20,000,000 gal. centrifugal pumps at the waterworks mains pumping station. Estimated cost, $75,000. R. C. Harris, City Engr.

Ont., Windsor—The Borroughs Adding Machine Co., 10 Chatham St., E., has engaged Albert Kahn, Archt., Marquette Bldg., Detroit, Mich., to prepare plans for the construction of an adding machine factory. A steam heating system will be installed in same. Total estimated cost, $250,000.

Ont., Woodstock—The De Long Hook & Eye Co., 41 Finkle St., plans to alter an existing 3 story building and convert same into a factory. A steam heating system, electric motors and generators for light and power will be installed in same.

Alta., Calgary—The Dept. Pub. Wks., Capitol, Ottawa, Ont., will receive bids until Feb. 5 for the construction of an 11 story, 154 x 160 ft. public building on 1st St., E. Three tubular boilers and automatic stokers, sprinkler type, will be installed in same. Total estimated cost, $400,000.

Alta., Edmonton—The Marshall-Wells Co., Lake Ave., S., Duluth, and German & Jenssen, Archts. and Engrs., Exch. Bldg., Duluth, will receive bids until March 1 at the Lindsay Bldg., Winnipeg, Man., for the construction of a 3 story, 100 x 150 ft. warehouse and office building here. A steam heating system will be installed in same. Total estimated cost, $375,000.

B. C., Vancouver—T. W. Lamb, Archt., 644 8th Ave., New York City, is preparing plans for the construction of a theater. A steam heating system will be installed in same. Total estimated cost, $300,000. Owner's name withheld.

CONTRACTS AWARDED

N. Y., Long Island City—The Standard Steel Co., 1920 Bway., New York City, has awarded the contract for the construction of a 4 story service station on South James St., from William to Ely Aves., to L. Gold, 44 Court St., Brooklyn. A steam heating system will be installed in samo. Total estimated cost, $150,000.

N. Y., New York—S. Weil, 194 Franklin St., has awarded the contract for the construction of a 9 story, 60 x 100 ft. store and loft building at 502-508 Canal St., to J. H. Taylor Construction Co., 110 West 40th St. A steam heating system will be installed in same.

N. Y., Troy—The state has awarded the contract for the construction of a 2 story, 183 x 256 ft. armory on 15th St., to Peter Keeler, 425 Orange St., Albany, at $379,140. Heating contract will be sub-let.

N. J., Atlantic City—The Atlantic National Bank, Atlantic and Penn Aves., has awarded the contract for alterations and additions to the present bank building, to Hoggson Bros., 485 5th Ave., New York City. A steam heating system will be installed in same. Total estimated cost, $200,000.

N. J., Atlantic City—The Holy Spirit Roman Catholic Church has awarded the contract for the construction of a 2 story, 80 x 138 ft. school, to Michael Melody & Son, 1645 North Broad St., Philadelphia. A steam heating system will be installed in same. Total estimated cost, $175,000.

Penn., Philadelphia—The Misericodia Hospital, 54th St. and Cedar Ave., will build a 6 story, 44 x 112 ft. building. A steam heating system will be installed in same. Total estimated cost, $300,000. Work will be done by day labor.

Penn., Philadelphia—The Neel-Cadillac Co., c/o Ballinger & Perrot, Archts., 17th and Arch Sts., has awarded the contract for the construction of a 7 story, 150 x 330 ft. vice building at 25th St. and Penn, Olive and Fairmont Aves., to Ballinger & Perrot, 17th and Arch Sts. A steam heating system will be installed in same. Total estimated cost, $1,000,000.

Penn., Pottstown—The Hill School has awarded the contract for the construction of a 2 story, 48 x 120 ft. memorial hall and library on High and Madison Sts., to W. E. Hale, 1636 Ludlow St., Philadelphia. A steam heating system will be installed in same. Total estimated cost, $100,000.

Penn., Steelton—The Peoples Bank will build a 1 story, 36 x 80 ft. building on Front and Locust Sts. A steam heating system will be installed in same. Total estimated cost, $100,000. Work will be done by day labor.

Md., Baltimore—The Fidelity Storage Co., 812 Equitable Bldg., has awarded the contract for the construction of a 6 story, 50 x 125 ft. warehouse at 2104-08 Maryland Ave., to the West Construction Co., 907 American Bldg. Contract for installing heating system will be sub-let. Total estimated cost, about $200,000.

O., Canton—The Pythian Castle Building Co., has awarded the contract for the construction of a 7 story, 75 x 115 ft. hotel and lodge building, to the Drummond-Miller Co., 4500 Euclid Ave. A steam heating system and electric motors will be installed in same. Total estimated cost, $350,000.

O., Cleveland—Joseph & Feiss Co. has awarded the contract for the construction of a warehouse, boiler room and factory addition on Haight Ave., to Stone & Webster Co., 1516 Stambaugh Bldg., Youngstown. Two 300 hp. boilers with stokers will be installed in same. Total estimated cost, $750,000.

Ill., Peoria—The Advance Rumely Co., Laporte, Ind., has awarded the contract for the construction of a 2 story, 147 x 200 ft. warehouse, to W. M. Allen Son & Co., 929-32 Jefferson Bldg. A steam heating system will be installed in same. Total estimated cost, $100,000.

Minn., Austin—The Bd. Educ. has awarded the contract for installing heating, ventilating and plumbing systems in the proposed high school, to J. P. Adamson, 1920 University Ave., St. Paul, $187,900. Noted July 1.

Kan., Ottawa—The Ottawa University has awarded the contract for the construction of a 2 story, 60 x 80 ft. central heating building, to N. B. Beeler, Ottawa.

Kan., Pittsburg—The Bd. Educ. has awarded the contract for the installation of steam heating and plumbing systems in the proposed 2 story high school, to the Salina Plumbing Co., Salina, $37,583.

Mo., Kansas City—George W. Neff, 1505 Genesee St., has awarded the contract for the construction of a 3 story, 120 x 142 ft. warehouse on 5th St. and Grand Ave., to the K. C. Construction Co., 822 Commerce Bldg. Contract for installing heating system has been sub-let to the U. S. Engineering Co., Kansas City. Total estimated cost, $100,000.

Tex., Dallas—The Advance Rumely Co., Laporte, Ind., has awarded the contract for the construction of a 2 story, 100 x 200 ft. warehouse, to the Hedrick Construction Co. A steam heating system will be installed in same. Total estimated cost, $125,000.

Tex., Mineral—The Crazy Well Water Co. has awarded the contract for the construction of a 1 and 5 story, 130 x 160 ft. water bottling works, to the Industrial Bldg. Co., 38 South Dearborn St., Chicago. A steam heating system will be installed in same. Total estimated cost, $125,000.

Cal., Richgrove—The Southern California Edison Co., Edison Bldg., Los Angeles, has awarded the contract for the construction of a 3 story, 67 x 130 ft. sub-station, to Macdonald & Driver, 608 Hibernian Bldg., Los Angeles. Estimated cost, $111,000.

Cal., San Jose—The Security Warehouse & Distributing Co. has awarded the contract for the construction of a warehouse and cold storage buildings to F. L. Hoyt, 566 North 16th St. Total estimated cost, $500,000.

Ont., Toronto—The Willard Chocolates, Ltd., 443 Wellington St., W., has awarded the contract for the construction of a 5 story, 80 x 150 ft. chocolate factory on Dupont St. and Manning Ave., to Wells & Gray, Confederation Life Bldg. A vacuum steam heating system will be installed in same. Total estimated cost, $150,000.

POWER

Volume 51 | New York, February 3, 1920 | Number 5

New Power Plants for Saw Mills

By John B. Woods, F.E.

THE old-time sawmill was a simple mechanical contrivance designed to cut logs into boards by the process of placing the log upon a movable carriage and shoving it past a rotating circular saw. Logs were large and cheap, so slabs were taken off without much regard to the quantity of good wood that came away with them, and saws were thick, resulting in great production of sawdust along with the boards. And because he wasted so much wood in slabs and dust, the old-time operator was possessed of unlimited fuel for his boilers. His greatest problem when installing his mill was to anticipate his power need by buying boilers and engines of sufficient capacity. Steam was the universal power, supplanting water almost altogether because the demands were very heavy and the load was extremely irregular. The old-time sawmill man claimed that the most important single item in his equipment was a quick-acting governor on his steam engine. In fact, when our troops were cutting lumber in France we found that the mill with a powerful engine whose governor was designed by American builders could and did produce lumber much faster than those plants that we were compelled to equip with less flexible European engines.

Today, however, the lumber industry has developed in so many different directions that the needs of a single plant may call for the transmission of energy over an area of many acres, while several miles away, in the woods, there may be half a dozen active power plants working on railroad and skidding ground. I have in mind a modern sawmill, built five years ago, and capable of producing five carloads of lumber each working day.

There is a central boiler house, containing eight boilers with a combined capacity of 1600 hp. There are four principal demands upon this installation: A 300-hp. twin cylinder engine, which runs the log mill; a ten-chamber, steam drying kiln for curing the total product; a smaller steam engine of 100 hp., which turns the half-dozen planing machines in the finishing mill, and a 500-kw. horizontal turbo-generator. The finishing-mill engine is fed by a steam line two hundred feet long, so the steam demand is greater than the engine's capacity because of the condensation and loss. At another point within the boundaries of this manufacturing group there is a stave mill, where oak billets are fashioned into barrel staves. Here is another boiler and engine, fed with offal from the oak billets.

Scattered about the various sheds and tramways are conveyors, lifting devices, small saws and planers for reclaiming defective boards, a machine shop for general mechanical repairs, a car shop for railroad rolling stock, and sundry pumps, not to mention the hundreds of lamps required for night work and the winter afternoons. To visit each of the aforementioned machines a man would walk not less than a mile, and yet the motive power must be furnished from some central source, else the cost would be prohibitive. These are all supplied from the turbo-generator. And not only does this electric plant supply the sawmill and its dependent mill town with current for small machines, lights and the household conveniences of an enlightened community, but also it serves a thriving small city a mile away.

Of course there are plants, even in this modern time, where electricity is used only for lighting, and small mills where it is used not at all. On the other hand, the most modern mills are tending more and more toward all-electric machinery, with a generating plant, steam fed from waste that comes from the saws. The main reasons why current is supplanting the steam engine are three: Flexibility and ease of transmission, efficiency of operation, and economy of power. If a special need arises, it is but a short task to string a feed line, install a motor and deliver the power to a pump or small engine, and it does not matter much whether the distance is ten feet or a thousand feet from the power house. When a machine is needed, the operator has but to throw a switch and he can perform his work; the power is neither too small nor too great if machine and motor have been installed by a man who understood what was required. And it is available as long as needed. Where individual machines are turned by separate motors, there is no waste of power resulting from idling pulleys and transmission shafts. One of the efficiency side issues of this type of power is the fact that when each machine has its motor the operator can keep individual performance records of his machines and thereby get a much more valuable insight into his costs. There are other considerations, important to the mill man. Fire is a great hazard, especially when two or three wood-burning boilers are scattered about midst shavings and sawdust. Cleanliness is related to the fire question and has numerous good points of its own, such as raised morale and pride among the workmen.

The field for electrical energy is wonderfully broad, and at the present time it consists of two general divisions. First there is the field of the giant mill, with its acres of buildings, its communities, and its far-flung woods department where men work to get the logs into the mill to be manufactured. Even in this last-mentioned work there is a place for electricity, as has been proved by the powerful skidding motors of the Western coast country. The general characteristics of this big mill field are such that the greatest opportunities for the designer of lumber-manufacturing power plants appear to be here.

On the western coast, where the timber is gigantic and the machinery in proportion, electric power has made its greatest strides. In the states of the Southern producing area electricity plays its part in nearly every mill for isolated machines and lighting, and in the case of two or three new operations is employed as the sole power. With the advent of wide-awake salesmen in this latter region interest has been awakened. The lumberman is not afraid to spend money for something new, if he can see a chance to profit thereby. His plant is expensive, anyway, and he is after the best equipment he can get.

However, if one looks ahead twenty years or more, he will be likely to wonder what is to become of these great plants now operating. When they exhaust their present supplies of standing timber, how will the lumber industry be conducted? And the obvious answer is that in the far future it is quite possible that the big mill will be replaced by small plants, just as is the case in Europe today. Of course, twenty years is not a long enough period to bring about the change entirely, for there still will be many big mills running then. But two decades will show the trend, as many mills now running will have produced their last carloads of lumber. Such big plants as are producing twenty years from this date undoubtedly will be equipped with the most economical machinery obtainable, for their raw material will cost more than it does now and will be subjected to a more thorough utilization.

In the second general division are the small mills, and they will offer the electrical engineer opportunity to work along various lines. The city plant taking power from an outside generating station has its own problems of economical production, but they are not power-plant problems. The portable mills, moved about from small timber tract to small timber tract, have their own power plants. In the past they have depended upon steam engines, with occasional attempts to install electric generating plants. In general these efforts have met with only partial success, chiefly because the mill man has not been sufficiently educated in the application of electricity to know what he needed, and as a result he has spent more money for his power plant than his business warranted. By attempting to keep down the initial cost, he has bought generators and motors too small for his needs and later added to his troubles and expense by going to the other extreme.

There are regions in this country where steam power is more expensive than kerosene, notably in the East, where population is great enough to create a demand for mill refuse, so there the portable mill man sells his slabs and sawdust at a good price and buys oil delivered in barrels at the nearest crossroads. In the hills of western Massachusetts farmers drive several miles into the back valley where the small mills whine, to get the refuse, and the oil truck brings liquid fuel from the nearest distributing station. The fuel cost is practically the same to the mill operator, but by using a gas or oil engine he removes the necessity of hiring a licensed steam engineer and thereby cuts his labor cost at least 20 per cent. Probably the owner is sawyer, so if there is engine trouble he attends to its personally, and he could do the same if he employed an electric generating plant turned by an oil engine. Portable-mill operators are waiting for the engineer who can design a portable power plant suitable for their needs, and there are enough of them, working under different fuel conditions

so that both steam and oil outfits will be welcomed. As the writer has stated earlier, the makers of electric equipment have invaded the lumber field and have sold numerous sets of great power and value. The thing they have not done to a broad enough extent is to employ engineers who are really acquainted with the manufacture of lumber. And because of this condition of affairs certain lumbermen who have bought such equipment have failed to realize the full value from their investments. In illustration of this a specific case might be cited. A Southern mill began operations with steam drive for the mill proper, but with a generator of considerable power to furnish current for numerous small machines and for lights. Along came an electrical engineer and convinced the lumberman that he could get better results from an all-electrical power plant, which was fair enough and which would have worked out happily but for one thing. The engineer told the mill man that he only need buy another generator about twice as large as the first one and hook it up alongside the first one, driving it from the main sawmill engine. The proposition appeared reasonable enough to the buyer, so they went ahead. Finally, however, it became evident that the combination would not work, notwithstanding the engineer's assurances. So they were obliged to run two separate circuits, with the resulting multiplicity of wiring and switchboard equipment.

In another instance a salesman designed a power plant for lumber operations, and because he did not know the manufacturing process well enough, he failed to provide enough power for the main saws and lumber-handling machinery to take care of the enormous overload. Consequently, when the big saw hit a particularly knotty log at the same moment that a few teeth lost

their cutting points, there was a ruined fuse, and a few lost minutes resulted. The same thing occurred when heavy loads were thrown upon several conveyors driven from the same motor.

Then there is the question of operating electrical equipment after it has been installed and tested and the engineer has gone on about his other affairs. While a great many young men are skilled in electrical work of routine nature, they are not always capable of keeping up a big plant. And most of the old-time steam engineers are men of a single faith; namely, steam. One of the most successful sawmill experts was called to a Southern plant to help locate the trouble with their power. He found a splendidly designed and installed set of equipment. There had been twice as much power available as was needed at first. Two generators of equal size had been installed so that the mill would not lack power in case of shutting one down. The electrician in charge stated that for several months the generator drove the entire sawmill plant. Then he went over to the other one and used it alone while working upon the first. Finally, he was obliged to run both together, and even then there was a lack of power. The expert overhauled both generators and then remained at the plant long enough to give the electrician a practical course of instruction in the handling of that particular type of equipment. This indicates that often it would be well for the builders to sell education along with their machinery.

Certainly, with the field so large and promising there is work for engineers who will study the manufacture of lumber and byproducts. The man with the type of

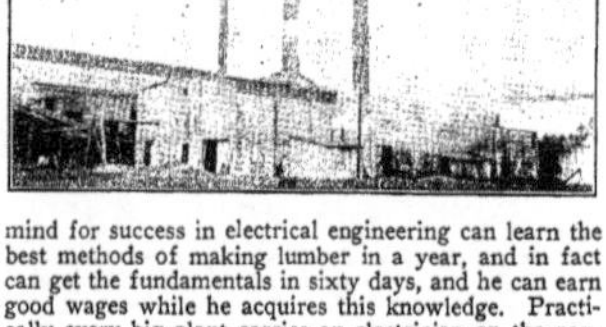

mind for success in electrical engineering can learn the best methods of making lumber in a year, and in fact can get the fundamentals in sixty days, and he can earn good wages while he acquires this knowledge. Practically every big plant carries an electrician on the payroll, and many of them carry two or more. And the knowledge thus gained would be of value to a man whether he became a salesman, a designer or an inventor.

Riveting Patches on Cylindrical Pressure Shells

By R. L. Hemingway* and F. A. Page

From time to time, in the different technical magazines, articles have appeared, with illustrations and tables, describing the method of patching cylindrical shells without decreasing their strength, by means of diagonal seams. The accompanying diagram, together with its explanation, appears to offer a more practical description of this method of patching than has hitherto been employed. Any practical boilermaker, with the aid of this diagram, should be able to patch almost any boiler and maintain the strength of the original construction, as a full description is included and the method of its application.

To Determine Roundabout Length of Patch

To find how far the patch must extend up the side of the boiler: First, determine length A. Then, multiply A by the constant in Table 3 corresponding to the angle obtained from Table 2. This gives the vertical height of the patch, as shown at V, V_1, V_2, etc. Mark this vertical on the boiler shell. It must be noted that this height is measured from a point level with the center line of the highest rivet in the short roundabout seam.

Example: Patch of the first seam on 54-in. boiler. A = 24 in.; shell, 5/16 in.; patch, 5/16 in., 55,000 lb. tensile strength. Longitudinal seam in boiler, double-riveted lap, ¾-in. rivet hole, 2⅞-in. pitch, 73.9 per cent. efficiency.

*Chief Boiler Inspector, Industrial Accident Commission of California.

Select pitch of rivets to be used in patch from Table 1. Say 1⅞-in. and 1 $\frac{1}{16}$-in. hole = 56 per cent. Then 73.9 ÷ 56 = 1.32 nearly. The next higher factor in Table 2 is 1.34, which corresponds to 40 deg. From Table 3 an angle of 40 deg. gives a constant, 1.192.

$$A \times 1.192 = 24 \times 1.192 = 28\frac{5}{8}\ in.$$

Therefore 28⅝ in. = height V and the per cent. of patch seam is 56 × 1.34 = 75.24, which is stronger than the original longitudinal seam. Now draw the diagonal line which gives the center line of rivets in the patch. Lay out all laps at 1½ times the diameter of the rivet hole used.

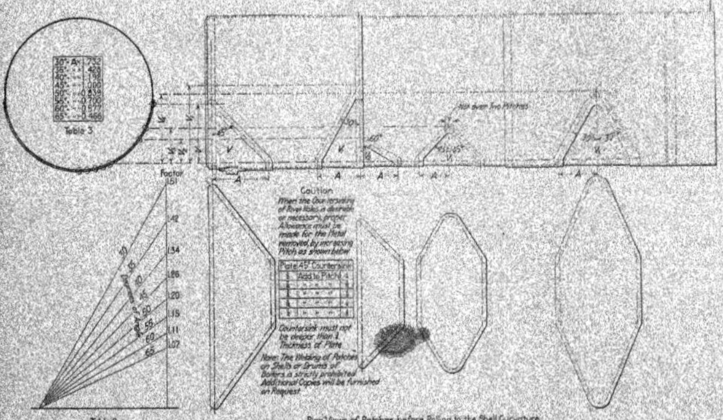

Plan Views of Patches before Rolling to the Shell Curvature

TABLE I. EFFICIENCIES OF SINGLE LAP JOINTS.

1½-In. PITCH

Rivet Dia., In.	Hole Dia., In.	Thickness of Plate, Inch ¼	5/16	⅜	7/16	½	Net Section Efficiency
[illegible]	[illegible]						46
[illegible]	[illegible]						53
[illegible]	[illegible]			45	54		60

1 11/16-In. PITCH

Rivet Dia., In.	Hole Dia., In.	Thickness of Plate, Inch ¼	5/16	⅜	7/16	Net Section Efficiency
[illegible]	[illegible]					48
[illegible]	[illegible]					55
[illegible]	[illegible]			43		62

1¾-In. PITCH

Rivet Dia., In.	Hole Dia., In.	Thickness of Plate, Inch ¼	5/16	⅜	7/16	½	Net Section Efficiency
[illegible]	[illegible]						50
[illegible]	[illegible]	44	50				56
[illegible]	[illegible]			42	50		63

2-In. PITCH

Rivet Dia., In.	Hole Dia., In.	Thickness of Plate, Inch ¼	5/16	⅜	7/16	½	Net Section Efficiency
[illegible]	[illegible]						53
[illegible]	[illegible]	41	47	55			59
[illegible]	[illegible]			39	47	59	65

2¼-In. PITCH

Rivet Dia., In.	Hole Dia., In.	Thickness of Plate, Inch ¼	5/16	⅜	7/16	½	Net Section Efficiency
[illegible]	[illegible]	52					55
[illegible]	[illegible]	39	44	52			61
[illegible]	[illegible]			37	44	55	67

2½-In. PITCH

Rivet Dia., In.	Hole Dia., In.	Thickness of Plate, Inch ¼	5/16	⅜	7/16	½	Net Section Efficiency
[illegible]	[illegible]	49	56				58
[illegible]	[illegible]		42	49	59		63
[illegible]	[illegible]			35	42	52	68

Tensile strength, 55,000 lb.
Shearing strength, 44,000 lb

Divide the efficiency of the longitudinal seam found on the certificate of inspection by the efficiency of the patch seam selected from Table 1. This gives the minimum factor required to maintain the original strength of the shell. Take from Table 2 the angle corresponding to this factor or next higher factor, which gives the required angle of the patch.

Note. Firebox steel, 55,000 T. S., must be used in repairing all Cal. Std. Boilers. Firebox or flange steel, 55,000 T. S. may be used in repairing noncode boilers. The use of tank steel in repairing boilers is strictly prohibited.

Rivet holes for patching shall be drilled full size from solid with patch in position, or they may be punched not to exceed ¼ less than full size for plates over 5/16 in. thick and ⅛ less than full-size for plates 5/16 or less in thickness and then drilled or reamed to full size with patch in place.

Properties and Products of the Combustion of Gasoline

Gasoline is a mixture of light hydrocarbons of the paraffin series. For it to ignite and explode in an engine cylinder, it must first be vaporized and mixed with a certain volume of air. At ordinary temperatures the mixture must contain not less than 1.5 per cent nor more than 5.2 per cent by volume of gasoline vapor in order to be explosive.

When the initial temperature is increased, the low limit is lowered somewhat until at 400 deg. C. it is 1.1 per cent. Hence a mixture containing less than 1.5 per cent is explosive in a gas-engine cylinder because of the high temperature therein caused by the initial compression and the incomplete cooling.

More than 174,000,000 tons of coal are available in the proved coal areas of Ireland, according to consular reports. Most of this coal is anthracite.

Synchronous-Motor Applications

By W. T. Berkshire*

The characteristics of the synchronous motor are compared with those of the induction motor. Applications for which the synchronous motor may be used are given and then the methods of starting are discussed.

FOR various classes of service the synchronous motor has several advantages over t[illegible]-tion motor, the recognition of which [illegible]-sulted in an ever-increasing demand for its application. These advantages include better efficiency and power factor and, particularly for slow-speed machines, less cost.

The efficiency of the synchronous motor is generally higher than that of the induction motor even when operating at leading power factors as low as 0.8. This is especially true of the more modern synchronous motors designed for unity power-factor operation, whose efficiency is practically the same from full load to half load, and is only slightly lower even at one-quarter load.

The synchronous motor can be designed to operate at either unity or any leading power factor, thus improving the power factor of the system. With normal field excitation, these machines will continue to improve the power factor when underloaded. In this respect the induction motor is always at a disadvantage; its power factor is always lagging, and although this power factor may be high at full load, it becomes rapidly lower at partial loads, consequently an underloaded induction motor further impairs the power factor of a system.

From the standpoint of dependability of operation the synchronous motor has a mechanical advantage over the induction motor by reason of the larger air gap of the former, which varie from five to eight times the length of the latter. he operating characteristics of an induction motor ma be seriously impaired by a slight change in air gap du to a little wear in the bearings. Owing to the large air gap of the synchronous motor the same change, o account of bearing wear, does not materially affect ts operating characteristics. In making a comparison o the relative starting ability of normally designed squirel-cage synchronous and induction motors, the folloung points must be understood:

First: If a motor has high initial starting torque, it must also have a low pul-in torque, and vice versa. This is true because the igh-resistance squirrel-cage winding which is require for high initial starting torque, produces low pull-i torque; whereas, the low-resistance squirrel-cage widing which is required for high pull-in torque, produes low initial starting torque.

Second: The inductio motor cannot [illegible] operate with a high-resistace squirrel-cage winding on account of high rotor loss under normal operation.

Third: The synchronos motor can use a high-resistance squirrel-cage winding because, when operating in synchronism, there is practically no loss in this winding. Therefore in cases where high initial starting with reasonably low pull-in torque is required, the synchronous motor has the distinct advantage that the high-resistance squirrel-cage winding, with its accompanying high starting torque and low kilovolt-ampere input, can be utilized a slight advantage.

There are a few cases of service requiring both high initial starting and

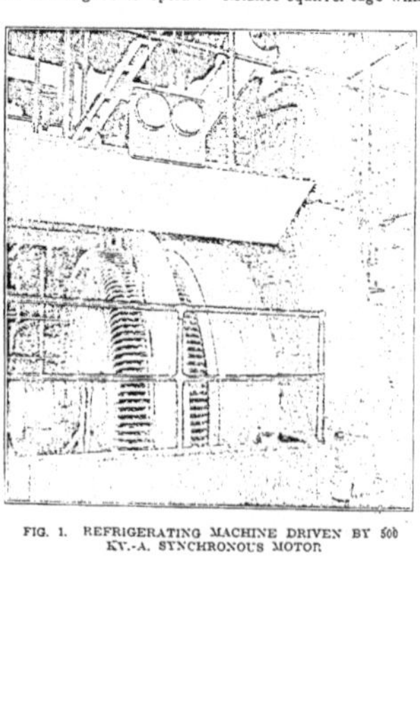

FIG. 1. REFRIGERATING MACHINE DRIVEN BY 500 KV.-A. SYNCHRONOUS MOTOR

*Alternating-current Engineering Department, General Electric Co.

high pull-in torque. In such cases double-squirrel-cage windings or other means are used in synchronous motors to obtain the required torque. Double squirrel-cage windings are also used on some of the larger induction motors. The starting of such loads by the synchronous motor, however, is usually attended by high current being drawn from the line. This is often objectionable both from the standpoint of the power company and that of the power consumer. This starting current can be more readily controlled by the use of the slip-ring induction motor, consequently this type is recommended for driving loads requiring both high starting and high pull-in torque.

Owing to certain starting-torque or speed requirements, there are four classes of service for which the normally designed synchronous motor is not recommended for direct drive. These are: Loads requiring the motor to start under full load, loads requiring variable speeds and loads requiring frequent reversal of the direction of rotation or requiring frequent starting and stopping.

FIG. 2. 750-KV.-A. SYNCHRONOUS MOTOR DRIVING RUBBER MILL THROUGH ROPE DRIVE

This first class includes flour mills, grain elevators or heavy line shafting where the torque required to overcome the static friction equals and often exceeds the full-load torque. In such cases the synchronous motor should be connected to the shaft through a clutch, thus permitting the starting and synchronizing of the motor before the load is applied. Where the service requires a variable speed, some mechanical means must be provided to obtain such speed variation.

Synchronous motors have been successfully applied for driving motor-generator sets, frequency converters, blowers, fans, air compressors, ammonia compressors, conveyors, tube mills, flour mills, pulp grinders, Jordans, rubber mills, stone crushers, cement mills, centrifugal pumps, plunger pumps, screw pumps, line shafting, steel and copper rolls and as synchronous condensers.

On account of the large current which would be drawn from the line at the moment of starting if full-line voltage were applied directly on the motor, a lower voltage is first applied and then after the motor has reached full speed, full-line potential is applied. This is accomplished by the use of a compensator, which is essentially an auto-transformer. The best method of starting synchronous motors is explained in the following paragraphs:

(a) Open the field switch completely if the excitation voltage is 125 volts. If the excitation voltage of the motor is higher than 125 volts, the field switch should not be opened completely, but left in the clips connected to the discharge resistance. This prevents any high induced voltage across the collector rings. However, if the motor is part of a motor-generator set, other than a frequency converter set, the field switch should be left in the clips connected to the discharge resistance, irrespective of what the field excitation is.

(b) Throw the compensator lever to "start" position. If oil switches are used, close the switch marked "start."

(c) After the motor has reached constant speed, close the field switch, the field rheostat having been previously adjusted to give a field current corresponding approximately to no load, normal voltage, with the machine running as a generator.

(d) Throw compensator lever quickly to the "run" position. If oil switches are used, open the switch marked "start," and after this, as quickly as possible, close the switch marked "run."

The attendant should be careful not to touch the collector rings or brushes when the motor is being started. An induced potential of about 2000 volts exists across the rings at the moment of starting. This voltage decreases as the motor speeds up, reaching zero at full speed. The motor should be started on the lowest tap

FIG. 3. SYNCHRONOUS MOTOR, 350-HP., DIRECT-CONNECTED TO TWO-STAGE AIR COMPRESSOR

FIG. 4. A 1200-HP. SYNCHRONOUS MOTOR DRIVING TWO FOUR-POCKET GRINDERS

Synchronous-Motor Applications

By
W. T. Berkshire*

The characteristics of the synchronous motor are compared with those of the induction motor. Applications for which the synchronous motor may be used are given and then the methods of starting are discussed.

FOR various classes of service the synchronous motor has several advantages over t[illegible]-tion motor, the recognition of which [illegible]-sulted in an ever-increasing demand for its application. These advantages include better efficiency and power factor and, particularly for slow-speed machines, less cost.

The efficiency of the synchronous motor is generally higher than that of the induction motor even when operating at leading power factors as low as 0.8. This is especially true of the more modern synchronous motors designed for unity power-factor operation, whose efficiency is practically the same from full load to half load, and is only slightly lower even at one-quarter load.

The synchronous motor can be designed to operate at either unity or any leading power factor, thus improving the power factor of the system. With normal field excitation, these machines will continue to improve the power factor when underloaded. In this respect the induction motor is always at a disadvantage; its power factor is always lagging, and although this power factor may be high at full load, it becomes rapidly lower at partial loads, consequently an underloaded induction motor further impairs the power factor of a system.

From the standpoint of dependability of operation the synchronous motor has a mechanical advantage over the induction motor by reason of the larger air gap of the former, which varies from five to eight times the length of the latter. The operating characteristics of an induction motor may be seriously impaired by a slight change in air gap due to a little wear in the bearings. Owing to the larger air gap of the synchronous motor the same change, on account of bearing wear, does not materially affect its operating characteristics.

In making a comparison of the relative starting ability of normally designed squirrel-cage synchronous and induction motors, the following points must be understood:

First: If a motor has a high initial starting torque, it must also have a low pull-in torque, and vice versa. This is true because the high-resistance squirrel-cage winding which is required for high initial starting torque, produces low pull-in torque; whereas, the low-resistance squirrel-cage winding which is required for high pull-in torque, produces low initial starting torque.

Second: The induction motor cannot efficiently operate with a high-resistance squirrel-cage winding on account of high rotor losses under normal operation.

Third: The synchronous motor can use a high-resistance squirrel-cage winding because, when operating in synchronism, there is practically no loss in this winding. Therefore in cases where high initial starting with reasonably low pull-in torque is required, the synchronous motor has the distinct advantage that the high-resistance squirrel-cage winding, with its accompanying high starting torque and low kilovolt-ampere input, can be utilized

In cases where the required starting and pull-in torques are about equal, but of a comparatively low value, the synchronous motor still has the advantage. If, however, a high pull-in torque, with a correspondingly low starting torque, is required, then the induction motor has a slight advantage.

There are a few cases of service requiring both high initial starting and

FIG. 1. REFRIGERATING MACHINE DRIVEN BY 500 KV.-A. SYNCHRONOUS MOTOR

*Alternating-current Engineering Department, General Electric Co.

high pull-in torque. In such cases double-squirrel-cage windings or other means are used in synchronous motors to obtain the required torque. Double squirrel-cage windings are also used on some of the larger induction motors. The starting of such loads by the synchronous motor, however, is usually attended by high current being drawn from the line. This is often objectionable both from the standpoint of the power company and that of the power consumer. This starting current can be more readily controlled by the use of the slip-ring induction motor, consequently this type is recommended for driving loads requiring both high starting and high pull-in torque.

Owing to certain starting-torque or speed requirements, there are four classes of service for which the normally designed synchronous motor is not recommended for direct drive. These are: Loads requiring the motor to start under full load, loads requiring variable speeds and loads requiring frequent reversal of the direction of rotation or requiring frequent starting and stopping.

FIG. 2. 750-KV.-A. SYNCHRONOUS MOTOR DRIVING RUBBER MILL THROUGH ROPE DRIVE

This first class includes flour mills, grain elevators or heavy line shafting where the torque required to overcome the static friction equals and often exceeds the full-load torque. In such cases the synchronous motor should be connected to the shaft through a clutch, thus permitting the starting and synchronizing of the motor before the load is applied. Where the service requires a variable speed, some mechanical means must be provided to obtain such speed variation.

Synchronous motors have been successfully applied for driving motor-generator sets, frequency converters, blowers, fans, air compressors, ammonia compressors, conveyors, tube mills, flour mills, pulp grinders, Jordans, rubber mills, stone crushers, cement mills, centrifugal pumps, plunger pumps, screw pumps, line shafting, steel and copper rolls and as synchronous condensers.

On account of the large current which would be drawn from the line at the moment of starting if full-line voltage were applied directly on the motor, a lower voltage is first applied and then after the motor has reached full speed, full-line potential is applied. This is accomplished by the use of a compensator, which is essentially an auto-transformer. The best method of starting synchronous motors is explained in the following paragraphs:

(a) Open the field switch completely if the excitation voltage is 125 volts. If the excitation voltage of the motor is higher than 125 volts, the field switch should not be opened completely, but left in the clips connected to the discharge resistance. This prevents any high induced voltage across the collector rings. However, if the motor is part of a motor-generator set, other than a frequency converter set, the field switch should be left in the clips connected to the discharge resistance irrespective of what the field excitation is.

(b) Throw the compensator lever to "start" position. If oil switches are used, close the switch marked "start."

(c) After the motor has reached constant speed, close the field switch, the field rheostat having been previously adjusted to give a field current corresponding approximately to no load, normal voltage, with the machine running as a generator.

(d) Throw compensator lever quickly to the "run" position. If oil switches are used, open the switch marked "start," and after this, as quickly as possible, close the switch marked "run."

The attendant should be careful not to touch the collector rings or brushes when the motor is being started. An induced potential of about 2000 volts exists across the rings at the moment of starting. This voltage decreases as the motor speeds up, reaching zero at full speed. The motor should be started on the lowest tap

FIG. 3. SYNCHRONOUS MOTOR, 350-HP., DIRECT-CONNECTED TO TWO-STAGE AIR COMPRESSOR

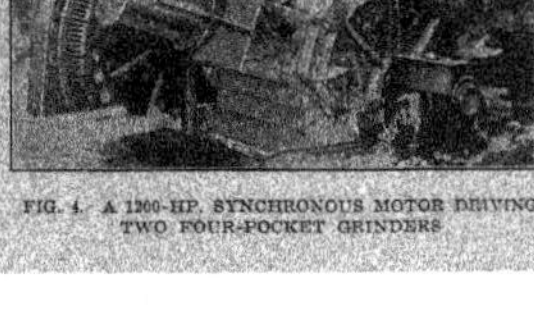

FIG. 4. A 1200-HP. SYNCHRONOUS MOTOR DRIVING TWO FOUR-POCKET GRINDERS

of the compensator that will start it promptly and bring it to full speed in about one minute. If two or three minutes are required in coming to full speed, there is danger of burning the squirrel-cage winding.

It has been pointed out that the field switch should be left completely opened at starting, as this insures the maximum initial starting torque with the minimum kilo-volt-ampere input. There are, however, two occasions for closing the field circuit through a resistance as part of the starting procedure. In one case the object is to increase the torque near full speed; in the other, to prevent high induced potential across the collector rings at starting. With the proper value of resistance across collector rings, the torque near full speed is increased. A change from this resistance in either direction will decrease the torque. At starting, however, any value of resistance will decrease the torque which the motor would develop with collector rings open. Hence, when a motor at the time it is purchased is required to pull into synchronism a large percentage of normal load, or when conditions arise in service where the "pull-in" torque requirements prove to be greater than were anticipated, the foregoing scheme is sometimes resorted to.

FIG. 5. STARTING COMPENSATOR

An accurate and convenient way of determining the proper resistance is to bring the motor to constant speed at full-line voltage with the load it has to pull into synchronism; then by means of a water-box connected across the collector rings, determine the resistance that will increase the speed to the highest value. This will be the proper resistance.

The field-discharge resistance in such case is increased to the proper value and capacity for this added service. Here, the switching procedure is only slightly modified. When the motor is running on the last compensator tap, or on the line, as the case may be, the next operation would be to close the field switch on the first point, thereby throwing the resistance across the field. A moment later, say 5 or 10 sec., close the field switch entirely.

FIG. 6. VERTICAL SYNCHRONOUS MOTOR, 400-KV-A. CAPACITY, DRIVING DEEP-WELL PUMP

On a given machine the higher the excitation voltage for which the field winding is designed the higher the induced voltage across the collector rings at starting. Motors that are designed for normal excitation voltages higher than 125 volts, or those which form part of motor-generator sets other than frequency converter sets should have the field winding short-circuited through the discharge resistance at starting. This will prevent the high induced voltage across the collector rings. It is standard practice to make all discharge resistances for synchronous motors of ample capacity for this service.

For a given impressed-voltage the torque of the synchronous motor has a definite value at the moment of starting, increases slightly as the speed increases, up to a certain point, and then decreases as synchronous speed is approached, reaching an extremely low value at synchronism. A higher impressed voltage would give a higher torque at all speeds, the torque increasing approximately as the square of the voltage. But since the kilovolts - amperes taken by the motor at starting also increase as the square of the impressed voltage, practical limits of kilovolts - amperes and therefore of torque, are soon reached.

FIG. 7. SYNCHRONOUS-MOTOR STARTING PANEL

Different kinds of service present different speed-torque requirements. Those requirements that conform approximately to the speed-torque characteristic of a synchronous motor, as outlined in the foregoing, can be met satisfactorily with standard compensators and standard starting operations. Motors of motor-generator sets may be mentioned as representative of this class. However, when the speed-torque requirements are different from what it is natural for the motor to develop—when, for instance, the torque required at the moment of starting is comparatively low, but high near synchronous speed (which is the requirement of a centrifugal pump) the condition is a much more difficult one. And it is made still more difficult by the additional requirement in some cases that the current drawn from the line during the starting operations be kept within low limits. These are therefore often treated as special cases.

Hence, there have been developed different methods

for starting motors that drive these two classes of apparatus. In the simple cases, such as motors of motor-generator sets or in those cases of motors driving air compressors or centrifugal pumps, where a low limit of line current at starting is not imposed, the starting operations described in the foregoing have become standard. For the difficult cases special methods have been developed, each applying to a particular duty.

Cleaning the Double-Pipe Ammonia Condenser

By A. G. SOLOMON

The author tells of the methods he uses to clean the coils of double-pipe ammonia condensers most quickly, conveniently and thoroughly.

An ammonia condenser is not efficient unless it is clean. It is easy to see and remove the dirt that collects on the pipes of an atmospheric condenser, so condensers of that type are, as a rule, clean and the water readily takes up the heat from the ammonia. Keeping a double-pipe ammonia condenser clean is not so easy, and for just that one reason the capacity of the whole refrigerating plant is often reduced.

As the circulating water passes through the inner pipes or tubes, the amount of dirt that collects cannot be seen. This being the case, the engineer must have some other means than sight to know the condition of the inner surface of the inner tubes of the double-pipe condenser.

To wait for increased condenser pressure to indicate the accumulation of dirt or the formation of scale is a mistake often made. When the water outlet of each separate stand is open for temperature taking, the decreasing difference between the temperature of the inlet and that of the outlet water will warn of the presence of dirt.

A thermometer in the liquid line is a good indicator to reveal the presence of dirt in the condenser tubes. When the condenser surfaces are clean, the exchange of heat from the ammonia to the circulating water is so complete that the liquid ammonia leaving the coils will usually have a temperature not over two degrees higher than that of the incoming water. This low temperature will not be attained if the charge of ammonia is insufficient to permit of slow travel.

Often, when the double-pipe condenser becomes dirty and the head pressure increases, a greater amount of water is circulated. This is an expensive way of operating, as the water and the additional pumping are costly, and it is only a temporary remedy at best.

The dirt must be removed; therefore regular cleaning must be resorted to. The frequency and method of cleaning the inner tubes will depend on the amount of sediment contained in the circulating water. There are few condensers that are not provided with a way for flushing out the inner tubes.

Fig. 1 shows the usual construction. By changing the position of the three-way cocks, the flow of water is reversed and the sediment washed out through the side opening of the bottom cock.

In some plants this manner of washing out is all that is required during the whole season. When the water outlets of all the coils are connected into one header and the header extends upward some distance, the pressure available for back-washing is sufficient to loosen and remove sediment that has not baked hard on the pipe surface. This back-washing should be done at least once a week even if the water shows only slight traces of sediment. The object is to remove all this sediment before hard scale is formed.

Where the circulating water contains a large amount of dirt, or in cases where the pressure is not great enough to give a good cleaning, a pump must be connected so as to give a sufficient force to the water to wash the tubes clean. Fig. 2 shows the connections. The water to the suction of this washout pump is taken from the outlet of the coils. The dirty water is allowed to pass to the sewer. A pressure of about 80 to 100 lb. is carried on the pump discharge and a reliable relief valve installed so that no damage is done when changing the water from one coil to another. There are many plants where this kind of tube washing is done daily.

Next to the back-washing and pressure washing of the inner tubes comes the brush or scraper cleaning. Some of the sediment contained in the water will stick

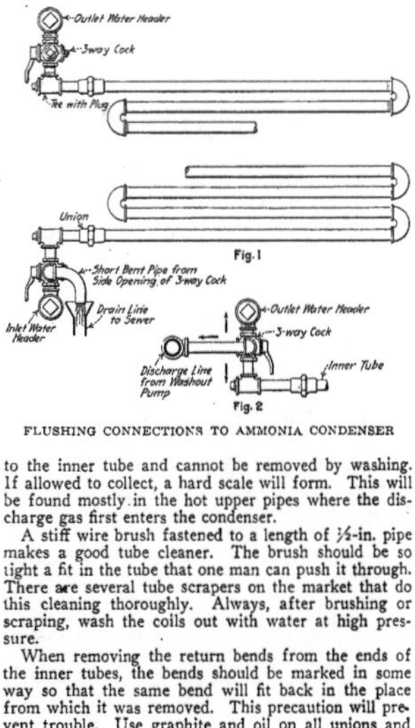

FLUSHING CONNECTIONS TO AMMONIA CONDENSER

to the inner tube and cannot be removed by washing. If allowed to collect, a hard scale will form. This will be found mostly in the hot upper pipes where the discharge gas first enters the condenser.

A stiff wire brush fastened to a length of ½-in. pipe makes a good tube cleaner. The brush should be so tight a fit in the tube that one man can push it through. There are several tube scrapers on the market that do this cleaning thoroughly. Always, after brushing or scraping, wash the coils out with water at high pressure.

When removing the return bends from the ends of the inner tubes, the bends should be marked in some way so that the same bend will fit back in the place from which it was removed. This precaution will prevent trouble. Use graphite and oil on all unions and bolts, so the bends can be easily removed the next time.

When replacing the return bends it is well to see that they are clear and not partly choked. If there is no strainer in the water line between the circulating pump and the condenser, there it always the liability of a broken pump valve, stud, a spring, or a piece of plunger packing becoming lodged in a tube or a return bend. Such obstructions must also be looked for in the valves and cocks.

Every winter the tubes of a double-pipe ammonia condenser should receive a real thorough cleaning. The winter's cleaning will be done with a twist drill welded on a short iron rod with threads provided so that a length of ½- or ¾-in. pipe can be used. The drill must be of the proper size so that it will fit snugly in the inner tube. This close-fitting drill will remove all scale that is in the tubes. The drill can be driven with an air or electric motor or can be turned by means of a pipe wrench.

The bottom tubes will be easy to clean, but those at the top of the coil will be harder. This is caused by the greater heat there baking the sediment to a hard scale.

There are many plants in which it does not pay to attempt to clean the top three or four inner tubes. The cost of removing hard, thick scale will be more than the cost of installing new tubes. In cases of this kind the inner tubes removed can be used for water lines or other pipework, as the pipe is as good as new except for the scale.

In putting in new tubes, care must be taken that defective pipes are not used. Test every new tube with 100 lb. air pressure while submerged under about six inches of water.

Forseille Liquid Level Gage

The following briefly describes a newly design liquid level gage that has recently been perfected b L. F. Forseille, of the Union Gas and Electric Co Cincinnati, Ohio, to accurately indicate the true lev of water, oil or other liquids.

It is especially adaptable to power stations and hydr electric plants situated along streams that are subje to variable stages or tidal changes. By inserting t operating float in a spiral pipe or other suitable tub the gage can be used in the swiftest water with exce lent results. It can be mounted directly over, or wit in any reasonable distance from the water, and by c erating as a direct measuring device (that is, with reducing mechanism) it is extremely simple and a curate.

By referring to the illustration, the dial is show graduated in feet and tenths, and the range is fro 0 to 72 in. The wheel is exactly 12 ft. in circumfe ence to the center of the operating cable, which is tached to a weighted float on the right-hand side; counterweight is suspended on the left-hand cable.

It is obvious that any rise in the liquid will cause t wheel to revolve to the left, or counterclockwise, a that a drop in the level will have the opposite effect.

The spiral lines on the front of the dial have t same pitch as the spiral on the back. A small roll supporting the radial moving indicator engages t latter spiral, keeping the point of the indicator direct over the proper figures, at all points within its range.

This device is being manufactured. The wheel, i cluding lines, figures and spiral, is of aluminum.

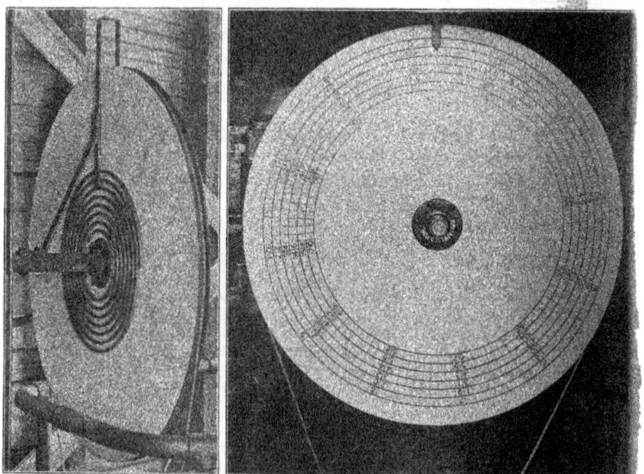

BACK AND FRONT VIEW OF THE LIQUID LEVEL GAGE

POWER DRIVES for ROLLING MILLS

By W. O. Rogers

It was discovered that many steel-mill boiler plants are up to date in design, using coal with blast-furnace gas as fuel, and in some instances the furnaces are arranged to burn either coal or gas. Waste-heat boilers are also used with the openhearth furnace.

AFTER putting in my first day at the blast furnaces, I returned to the hotel, and getting the best of a good meal, started out to look the town cer, with the hope that I might rub up against someoe who could give me a few pointers regarding the lcal mills, and perhaps prepare me for what I hoped to se during the next few days.

It was a pleasant evening, and I decided to take a sreet-car ride in the direction of Lowellville. I occupied a seat with another man in the smoking compartment of the car, and as he had the appearance of being a steel-mill man, I started a conversation with him; naturally, it drifted to the subject of rolling mills. He proved to be a man who had been connected with the steel industry for years, and evidently knew a good deal about local work and rolling-mill machinery in general. When he found out that I was interested in the subject, he said:

"You will find about everything in this town that you will find anywhere. Some of the power units are old, and others are new and up-to-date. Some of the boiler plants are of the long ago, and others are of the latest design. When you go through some of the steel mills you will see that great strides have been made, especially in the electrical operation of the machinery. As a matter of fact there are mills operated entirely by electricity."

"Which are the most economical to operate, steam-driven or electrically driven mills?" I asked.

"As on every other question, opinions differ. There are some who favor one, and some who are for the

FIG. 1. BATTERY OF WASTE-HEAT VERTICAL BOILERS

FIG. 2. BATTERY OF WASTE-HEAT WATER-TUBE BOILERS

When replacing the return bends it is well to see that they are clear and not partly choked. If there is no strainer in the water line between the circulating pump and the condenser, there it always the liability of a broken pump valve, stud, a spring, or a piece of plunger packing becoming lodged in a tube or a return bend. Such obstructions must also be looked for in the valves and cocks.

Every winter the tubes of a double-pipe ammonia condenser should receive a real thorough cleaning. The winter's cleaning will be done with a twist drill welded on a short iron rod with threads provided so that a length of ½- or ¾-in. pipe can be used. The drill must be of the proper size so that it will fit snugly in the inner tube. This close-fitting drill will remove all scale that is in the tubes. The drill can be driven with an air or electric motor or can be turned by means of a pipe wrench.

The bottom tubes will be easy to clean, but those at the top of the coil will be harder. This is caused by the greater heat there baking the sediment to a hard scale.

There are many plants in which it does not pay to attempt to clean the top three or four inner tubes. The cost of removing hard, thick scale will be more than the cost of installing new tubes. In cases of this kind the inner tubes removed can be used for water lines or other pipework, as the pipe is as good as new except for the scale.

In putting in new tubes, care must be taken that defective pipes are not used. Test every new tube with 100 lb. air pressure while submerged under about six inches of water.

Forseille Liquid Level Gage

The following briefly describes a newly designed liquid level gage that has recently been perfected by L. F. Forseille, of the Union Gas and Electric Co., Cincinnati, Ohio, to accurately indicate the true level of water, oil or other liquids.

It is especially adaptable to power stations and hydro-electric plants situated along streams that are subject to variable stages or tidal changes. By inserting the operating float in a spiral pipe or other suitable tube, the gage can be used in the swiftest water with excellent results. It can be mounted directly over, or within any reasonable distance from the water, and by operating as a direct measuring device (that is, with no reducing mechanism) it is extremely simple and accurate.

By referring to the illustration, the dial is shown graduated in feet and tenths, and the range is from 0 to 72 in. The wheel is exactly 12 ft. in circumference to the center of the operating cable, which is attached to a weighted float on the right-hand side; a counterweight is suspended on the left-hand cable.

It is obvious that any rise in the liquid will cause the wheel to revolve to the left, or counterclockwise, and that a drop in the level will have the opposite effect.

The spiral lines on the front of the dial have the same pitch as the spiral on the back. A small roller supporting the radial moving indicator engages the latter spiral, keeping the point of the indicator directly over the proper figures, at all points within its range.

This device is being manufactured. The wheel, including lines, figures and spiral, is of aluminum.

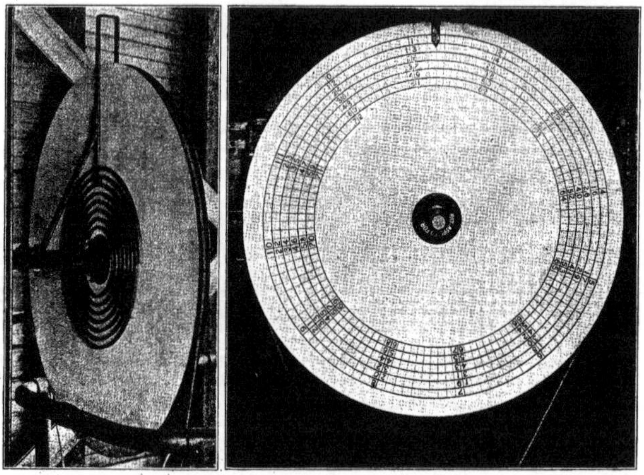

BACK AND FRONT VIEW OF THE LIQUID LEVEL GAGE

POWER DRIVES for ROLLING MILLS

By W. O. Rogers

It was discovered that many steel-mill boiler plants are up to date in design, using coal with blast-furnace gas as fuel, and in some instances the furnaces are arranged to burn either coal or gas. Waste-heat boilers are also used with the openhearth furnace.

AFTER putting in my first day at the blast furnaces, I returned to the hotel, and getting the best of a good meal, started out to look the town over, with the hope that I might rub up against someone who could give me a few pointers regarding the local mills, and perhaps prepare me for what I hoped to see during the next few days.

It was a pleasant evening, and I decided to take a street-car ride in the direction of Lowellville. I occupied a seat with another man in the smoking compartment of the car, and as he had the appearance of being a steel-mill man, I started a conversation with him; naturally, it drifted to the subject of rolling mills. He proved to be a man who had been connected with the steel industry for years, and evidently knew a good deal about local work and rolling-mill machinery in general. When he found out that I was interested in the subject, he said:

"You will find about everything in this town that you will find anywhere. Some of the power units are old, and others are new and up-to-date. Some of the boiler plants are of the long ago, and others are of the latest design. When you go through some of the steel mills you will see that great strides have been made, especially in the electrical operation of the machinery. As a matter of fact there are mills operated entirely by electricity."

"Which are the most economical to operate, steam-driven or electrically driven mills?" I asked.

"As on every other question, opinions differ. There are some who favor one, and some who are for the

FIG. 1. BATTERY OF WASTE-HEAT VERTICAL BOILERS

FIG. 2. BATTERY OF WASTE-HEAT WATER-TUBE BOILERS

other. Until recently there has been a considerable difference of opinion, especially regarding a steam and electrical drive for large reversing mills. This was because some held that electric drive was not a commercial proposition except when conditions were favorable, which would be governed by the cost of making electricity, using internal-combustion engines operating on blast-furnace gas. As a matter of fact, but few of the plants in this country use gas engines, the largest being the United States Steel Corporation at Gary, Ind.; the Lackawanna plant at Buffalo, N. Y., and the Bethlehem Steel plant at Sparrows Point, Md. I believe the Bethlehem company is particularly partial to gas-engine operation. There is this to consider, however: there is about twice the amount of energy recovered from the blast-furnace gas in using it in a gas engine as when burning it under a steam boiler.

FIG. 4. TYPICAL BLAST-FURNACE AND BOILER-PLANT INSTALLATION

"Some of the mills have regular central generating plants consisting of steam turbines or reciprocating engines, blowing engines and compressors. Other mills, as I have said, operate with gas engines. To my mind these various units all have a field of usefulness.

"There are at present some 600 motor-driven mills having a maximum of about 900,000 hp., and these are used with practically all types and sizes of steel mills; but at that, something like 75 per cent. of the total number of mills are not motor-operated. Some of the large plants are practically motor-driven from one end to the other. I think I am safe in saying that most of the mills built during the last few years are motor-driven. When you go through some of these mills you will readily see the difference between the steam and the electrically driven unit. The steam engine was the pioneer of rolling-mill drive. Many of them are still in use, and new engines are going in, but the electric drive has entered the field and is here to stay. The claim for the motor is economy and flexibility. Accompanying its use are cleanliness, absence of noise, reliability, and limited attention required. The main point, however, is lower cost and increased production."

FIG. 3. FOUR WATER-TUBE BOILERS USING BLAST-FURNACE GAS

"With a plant motorized, it would be necessary to have a regular central station to generate the energy," I remarked. "That would require the use of boilers, and I should think that better economy would be secured with modern installations than I understand they have as a rule."

"Now you are talking," replied my acquaintance. "Some of the boiler plants are frights, some are fairly good, and some are right up-to-date. You see, in the past but little attention has been given to the improvement of power plants, mainly, I suppose, on account of the cost, as improvements would mean a considerable change in the power plant. Recently, however, there seems to be a tendency to put aside the item of expense and go in for improvements. You can easily see that the mill capacity is limited by the main drive. I can show you many engines that have outlived their economical usefulness, because they are badly worn or not suitable for the demands now made upon them. With valves and piston leaking, an engine can easily consume 50 per cent. more steam to operate the mill than would be required to generate electrical energy and drive the mill by motors.

"You have read it, and have heard it stated, that the economy of the power plant must in the future come from the boiler room. Some of the boiler installations found in many steel mills are a crime against economy. If there were more gas from the blast furnaces than could be used, it would be different, but as there is not enough gas, coal is used in large quantities.

"The efficiency of a steam-electric plant ranges from 8 to around 16 per cent. of the heat value of the coal. The plant that is getting 16 per cent. is up in high C for economy, and such a figure can be obtained only when the plant is of the best design and scientifically operated. The average modern plant operates with an efficiency of around 10 to 12 per cent."

"How are these losses distributed?" was my next question.

"Well, we can say that the approximate loss in an up-to-date power plant, per pound of coal, is about 14 per cent. in the gas to the chimney and that, outside of that rejected to the condenser, which is around 61 per cent., is the highest of the remaining losses, which are about as follows: Requirement for all auxiliaries, 6.5 per cent.; radiation and leakage of boiler, 4 per cent.; ash, 1.5 per cent.; blowoff and leakage, 1.5 per cent.; radiation and leakage of piping, 1.5 per cent. This is with coal having an assumed heating value of 14,000 B.t.u. Taking a plant that has an efficiency of, say, 13 per cent., you can see that every increase in the percentage of economy, although it may seem very small, will improve the general results materially.

"Many of these losses can be easily reduced in the average plant by proper care and operation. The first item of importance is to have proper boilers properly set. That is something you will not find in some of the boiler plants around steel mills, any more than you will find it in other types of plants.

"It used to be, and is now for that matter, common practice to install boilers in the open. With their setting exposed to the elements, often with poorly insulated steam pipes, you cannot expect a brick setting to hang tightly together with a furnace temperature of around 2500 deg. and an outside atmospheric temperature below zero. The tendency nowadays is to put in better boiler plants and modern generating stations, as you will observe during your visit. Well, here's where I get off, so good night and good luck." So saying he left me, and at the next stop I got off the car and returned to the hotel.

The next morning I went to the mills again with the idea of going over the various boiler plants. The first thing of interest that I saw was a small battery of vertical boilers, Fig. 1; and close by were some water-tube boilers, Fig. 2. Both batteries were gas-fired, using gas from the blast furnace when in operation. These boilers were set in the open, with no protection from the weather, although there was a sort of leanto over the front, which protected the firemen from the rain or snow when it came straight down. This method of boiler installation would seem strange to most engineers who are used to a warm, dry boiler room. These boilers were dead and were evidently not used much except in emergency cases.

Going on a little farther, I came to a more modern boiler plant. Here were four 1265-hp. water-tube boilers, Fig. 3, fitted with gas burners. The gas came from blast furnaces and was unwashed, having a pressure of from 2 to 3 in. of water. It contained a heat value of about 95 B.t.u. per cu.ft. The boilers carried a pressure of 150 lb., and the steam was superheated from 50 to 100 deg. F. These boilers were used with what is known

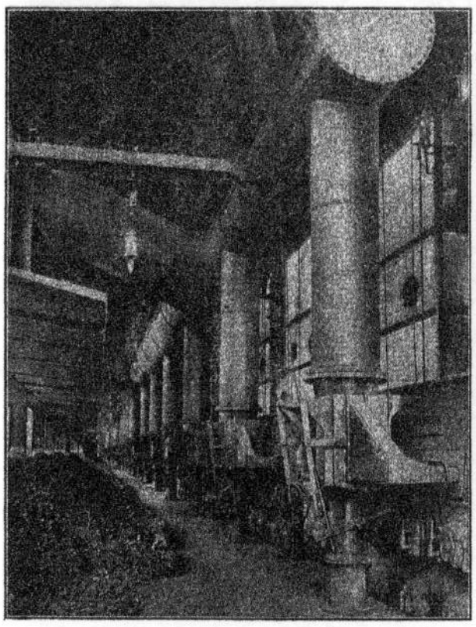

FIG. 5. BOILER FURNACES EQUIPPED FOR BURNING EITHER GAS OR COAL

FIG. 6. SHOWING BOILERS EQUIPPED WITH FRONT-FEED STOKERS

as No. 5 blast furnace of the Youngstown company's plant. Fig. 4 shows a typical blast-furnace and boiler-plant installation of the more modern practice.

It is frequently the case that a battery of boilers are arranged for burning either gas or coal. This was so at one of the large boiler plants of both the Republic and the Youngstown companies. At the latter plant there are 24 water-tube boilers, each of 464 hp. rating. They are equipped for burning either coal or gas.

When a blast furnace is down or when it is being emptied, there is not enough gas to supply steam for the steam-driven units, and when this occurs coal can be used to supplant the gas. A view of such a boiler plant at the Struthers mill is shown in Fig. 5.

It was found that practically four systems of boiler-firing were used—gas, gas and coal, coal alone hand-fired and coal stoker-fired. The boiler rooms of the stoker-fired and some of the hand-fired plants compare favorably with those found installed in large electric stations. In fact, many of them supply steam for large isolated generating plants. This is shown by Figs. 5 and 6. In one boiler house there are thirty-four 502-hp. boilers fired with blast-furnace gas. The gas is delivered from the blast furnace to gas-cleaning apparatus and is led into the boiler room through the main 72-in. delivery pipe, to the gas header, Fig. 5, from which it is fed to the boilers through the gas burners shown attached to the boiler front above the regular firing doors.

In the second boiler plant, Fig. 6, there are eight 763-hp. water-tube boilers, the furnaces of which are equipped with front-feed underfeed stokers and driven by either two of three stoker engines. These furnaces operate with forced draft supplied by four turbine-driven fans. Between the turbine and the fans a reduction gear is used. These fans supply the necessary air through a single air duct connected below the stokers

FIG. 7. BOILER ROOM AT MAIN STRUTHERS PLANT

of each boiler. The feed-water heater is a 10,000-hp. unit and is equipped with a V-notch meter. The plant is also equipped with steam-flow meters.

Fig. 7 shows five of eight 254-hp. water-tube boilers equipped with stokers and turbine-driven forced-draft fans. These boilers are at the main plant of the Youngstown company. The type of stoker engine used is shown at the fan, one engine running four stokers.

Later on I found another type of installation known as a waste-heat boiler, shown in Fig. 8. It was one of seven in service at the openhearth plant of the Trumbull Steel Co.'s works at Warren, Ohio, and was used in connection with the openhearth furnaces. It carried 150 lb. pressure. The steam was used by gas producers, openhearth pits, etc. Naturally, the power that can be obtained from waste gas depends on the temperature and weight. The gases from an openhearth furnace have a temperature of about 1000 and 1500 deg. F., which is available for steam generation. Of course these temperatures will vary at times.

The efficiency of a waste-heat boiler is governed largely by its design and setting. With high-tempera-

FIG. 8. WASTE-HEAT BOILER AT THE TRUMBULL CO.'S OPENHEARTH FURNACES

ture gas it is only necessary to have the amount of heating surface proportioned to the volume of gas to be burned, and in such cases a standard design of boiler is used with satisfactory results.

With low-temperature gases, however, a standard boiler will not serve so well, and in order to obtain a heat transfer that will approach ordinary boiler practice, it is found to be necessary to increase the velocity of the gases as they pass over the boiler-heating surfaces, which can be done by the use of mechanical draft sufficient to overcome the resistance of the gases due to friction in passing over the heating surface, in addition to that required to supply the draft necessary to the openhearth furnace. With a high temperature a chimney may be used having a height sufficient to produce the necessary draft, but with low-temperature gas a fan blower will be necessary.

It has been found that each waste-heat installation must be considered by itself, because there are so many factors that enter into the operation; that is, the character of the gases will vary in temperature and volume, and then there will be dust and a tar-like substance, all of which has a bearing on the figuring of the heating surface and of the gas passages.

Owing to the fact that openhearth gas is usually very dirty, the boilers using it must be provided with numerous cleaning-out doors, so that the heating surface can be thoroughly cleaned. There must also be a bypass for the gases, so that the boiler can be cut out of service for cleaning and repairs without interfering with the operation of the openhearth furnace.

It was brought home that steel mills occupy a large ground area, and that one could not be covered in a day if anything more than a casual glance was to be had. My day's experience was satisfactory, and there was no wakefulness that night on account of insomnia. The next day I found a new boiler plant that was modern in every respect, which will be described in the next article.

Frozen Safety Valve Causes Boiler Explosion

MORE information is now available on the boiler explosion at the East Chicago works of the Interstate Iron and Steel Co., reported briefly in last week's issue. Three men were killed by the accident, twelve were injured and the property damage was estimated at $10,000. As will be remembered from the previous notice, the boiler was in a battery of three used to supplement the supply of steam from the waste-heat boilers in the plant. It was of the Bass vertical water-tube type, rated at 250 hp. The boiler was equipped with 102 tubes 4 in. in diameter and 18 ft. long, with drums top and bottom 72 in. in diameter. Coal was the fuel used, and it was hand-fired into the usual dutch-oven furnace. The boilers were housed in a sheet-steel building just large enough for the battery. It kept out the weather, but the temperature approximated that outdoors. Two safety valves protected the boiler. They were set to blow at 110-lb. gage, although the boiler that exploded had been allowed a working pressure of 125 pounds.

From another plant the boiler had been moved to its present location in 1907. According to estimate it had seen eighteen years of service, but was reported to be in excellent condition. Just six weeks previous all the tubes had been renewed. The regular inspections by insurance-company inspectors and those of the steel-company inspector, which were made every two weeks just after the boiler was washed, showed the braces to be sound and taut and no evidence of cracks in the lower tube sheet or head. In fact, the boiler had been inspected on the day previous to the accident and no defect of any kind had been detected.

Reviewing the circumstances, on Sunday, Jan. 18, the boiler had been emptied, washed out and filled with cold water. At 9 p.m. a fire was started on the grates preparatory for the morning shift, which started at 6 a.m. At the usual intervals the boiler was fired, and at 4:30 a.m. it exploded. The bottom tube sheet fractured all the way around at the knuckle of the flange, so that the tubes, with the upper drum, shot up into the air several hundred feet and returned to earth, drum end down, about 700 ft. distant from the original location. The impact was so great that the upper tube sheet, with the weight of the tubes behind it, buckled the shell plate into the drum. The lower drum, minus its tube sheet, remained on the foundation. Naturally, the settings of the other two boilers and the building were demolished. The men working in the boiler room were killed, so that no information was available as to the action of the boiler previous to the accident, except that the watchman on his hourly trips through the plant had noticed that the steam gage remained at 40-lb. pressure.

FIG. 1. DEVASTATION CAUSED BY EXPLOSION OF 250-HP. VERTICAL WATER-TUBE BOILER AT EAST CHICAGO

As the temperature was below zero outside and nearly the same inside, it is quite likely that the 3/8-in. pipe connecting the pressure gage at the top of the furnace with the upper steam drum, had frozen up, so that the reading on the gage probably gave no proper indication of conditions existing within the boiler. It is certain that the boiler had not been cut in on the line, as the valve was found closed after the accident.

In the line drawing is shown the construction of the lower drum, the character of the bracing and the location of the fracture. The lower tube sheet was of 5/8-in. metal and the shell plate 1/2-in. metal. The longitudinal seam was quadruple-riveted and the circular seams had single rows of rivets. In either case the joint held, the fracture being in the metal at the turn of the flange, as indicated. Twelve stays held the lower tube sheet to the drum. Four of them were diagonal crowfoot shell braces, and the other eight were cluster sling stays

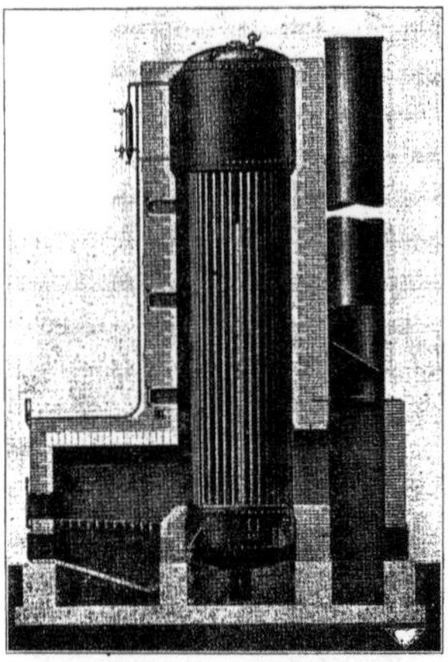

FIG. 2. TYPE OF BOILER THAT EXPLODED

attached to the bumped head. Each cluster consisted of two straps looped around the pin at each end, by which they were secured to double crowfoot supports. Consequently, the composite stay was made up of four elements 1 in. square. Four of these were spaced evenly on an outer circumference and four on a smaller concentric circle. Thus the lower tube sheet was stayed by 32 sq.in. of metal in the vertical stays and 6 sq.in. in the four diagonal braces.

A careful inspection of the boiler after the accident showed no defect that might have caused the initial rupture. The break all the way around was bright and clean, and before letting go the flange pulled up straight. The edge had feathered off, and the metal at the break had drawn out so that the reduction in area was approximately 30 per cent, indicating excellent material. Every rivet in the longitudinal and circular seams was intact, the joints showing no evidence of distress. The braces broke at various places; some at the knuckle of the crowfoot; others pulled the rivets securing the crowfoot to the bottom head, and in some cases the stays themselves pulled in two. Most of the diagonal braces broke at the first rivet in the shell. Reports of the various inspections gave no intimation of weakened stays. Apparently, the condition of the boiler was thoroughly good, and to complicate the diagnosis of the cause of the accident, the two safety valves had been blowing regularly in recent operation, so that there was every reason to expect them to function. What then caused the explosion?

It has been suggested that the safety valve froze up; that is, that ice accumulated above the disk in the body of the valve and in the vent pipe, which, with an elbow and a nipple turning upward, led out of the boiler house to discharge the steam outside. Owing to the frequent blowing off of safety valves on boilers of this type, the seats are generally wiredrawn and leaky, so that a small amount of steam is constantly escaping. When the boiler was being shut down, the cold weather existing at the time might have condensed this vapor and frozen the water, filling up the small drain at the elbow of the vent pipe. The day was stormy and additional water might have been collected from the sleet and snow that were falling. Apparently, this collection of ice blocked the discharge of the safety valve for the 7½ hours that fire was on the boiler and allowed the pressure to build up to such an extent that the boiler ruptured under the strain. The good condition of the boiler and the zero temperature existing lends credence to this suggestion, especially as the vent pipes of safety valves on other boilers that had been shut down for washing and inspection were found frozen solid. How this ice could remain in the presence of steam under high pressure for so long a time is a feature requiring careful analysis.

When the boiler was filled with cold water on Sunday, the cold air in the boiler would be pushed ahead of the water and compressed in the upper part of the top drum to perhaps the pressure of 40 lb. shown by the gage and retained after the pipe connecting it to the steam space had been frozen. As the pressure increased in the boiler, the air would become relatively lighter than the steam and stay in the top part of the drum, insulating the hot steam from the bumped head and keeping it away from the safety valve, air being a very poor conductor of heat. Would it do this or would the steam after it had risen above the 40 lb. pressure on the air, intermingle with the latter and come in contact with the top of the drum? By condensing here it would give up its latent heat and soon bring the metal to a high temperature, or by leaking through the valve, would eventually melt the ice in the valve body and the vent pipe. It will be remembered that the boiler had not been cut onto the line so that the steam space was dead ended. There was no air relief valve on the top drum or any other means for the air to escape. Conduction of heat along the metal of the drum would be very slow, as the space between the setting and the side of the drum forms a pocket in which the gases would be at a relatively low temperature, and the radiation from a bumped head exposed to zero weather would be excessive. The safety valves were connected to the top of the drum as indicated in the picture of the wreck, although the drum shown here belonged to one of the other boilers.

Another explanation of the accident, conforming more nearly to experience of the past, might be advanced. The breathing action commonly attributed to these boilers would naturally weaken the metal at the knuckle or turn of the flange. This would be the place subjected to the greatest strain. Continued working back and forth of the lower tube sheet would gradually weaken the metal at the turn so that it would eventually fail. The appearance of small cracks that would gradually develop is generally a forerunner of such a failure, but nothing of this character had been detected. On the other hand, the poor circulation in an airbound boiler and the cold weather would, no doubt, aggravate such a condition if it existed, and perhaps set up sufficient additional stress to cause the initial rupture in a boiler that had been subject to this action for years, particularly as the boiler had been used intermittently to fill in the peaks above the capacity of the waste-heat boilers. After the initial rupture the release of the potential energy in the 27,000 lb. of water contained in the boiler would readily inflict the damage pictured and blow the 18,000 lb. of boiler separated from the lower drum to a considerable distance.

In boilers of this character the tubes nearest the fire are hottest and those on the opposite side of the baffle comparatively cool. Consequently, the front tubes tend

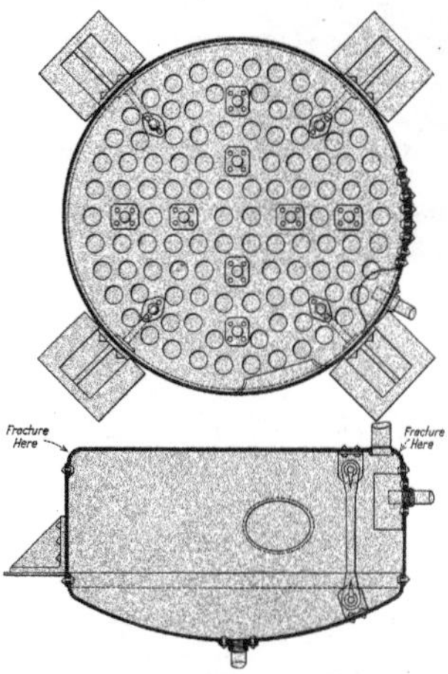

FIG. 3. SPACING OF BRACES AND LOCATION OF FRACTURE IN LOWER DRUM

to bow or buckle and pull up the tube sheet. There would also be the tendency to tilt the top drum in a direction opposite from the furnace and push down the lower tube sheet on that side. In other words, there would be a severe twisting movement with considerable leverage on the lower tube sheet, straining the metal particularly at the knuckle of the flange. Poor circulation would aggravate and prolong this action.

With stays of the character described, it is difficult to obtain uniform tension. One may be more taut than the others, and in assuming the greater part of the load the metal in the straps or under the pin in the crowfeet may stretch and weaken or the metal may fail in the crowfeet at the knuckle, which is commonly the weakest part of the type of stay used. The load would then pass to another stay and so on until all had been

FIG. 4. TOP DRUM AND TUBES 700 FEET DISTANT

weakened, allowing more or less play for the so-called breathing action of the tube sheet.

Owing to the fact that in the various inspections no defect had been noticed in these stays and no cracks had appeared at the turn of the tube sheet, the frozen vent pipe and valve was probably the direct cause of the accident, although the metal at the turn of the flange may have been weakened to some extent by the breathing action previously mentioned.

Possibilities of Alcohol As Fuel*

The fermentation industry, notably the branch having to do with the manufacture of industrial alcohol, has been strongly stimulated by war demands, and industrial machinery is now available for the production of considerable alcohol for fuel purposes.

Alcohol alone can be used to advantage only in engines especially adapted to this fuel, but various mixtures of alcohol, benzol, gasoline or other petroleum distillates and other materials have given promising results. It is of great significance from an economic standpoint that alcohol, benzol and the lighter petroleum distillates such as gasoline and kerosene can readily be rendered miscible. It is probable that alcohol, like benzol, will not come into widespread use as a single fuel, but has a broad significance, for the present at least, only as a blending agent in connection with liquid fuels obtainable in larger quantities.

The quantity of alcohol that will be produced in this country in the immediate future is much more difficult to forecast than in the case of benzol. The United States in 1916, 1917 and 1918, turned out about 50,000,000 gal. of denatured alcohol each year, having jumped from an output of 14,000,000 gal. in 1915 under the stimulus of a demand born of munitions requirements. Much of the industrial alcohol under manufacture today is made from sugar molasses and waste sulphite liquor; while garbage, fruit wastes and ethylene from coal-distillation plants have been suggested as supplementary resources. It is safe to estimate that for some time to come the available supply of alcohol will bear a close quantitative analogy to benzol, the two combined bulking small when compared with engine-fuel requirements which will approach 5,000,000,000 gal. in 1920.

*From a paper by J. E. Pogue before the Society of Automotive Engineers.

Novel Coal Handling at the McKeesport Tin Plate Company's Plant

THE McKeesport Tin Plate Co.'s plant, the largest single tin-plate plant in the United States, is located on the Youghiogheny River about 1½ miles above its junction with the Monongahela and opposite the City of McKeesport, Penn. This plant has always received its coal by river craft. The space available for the boiler house, Fig. 1, was limited and room had to be provided for the storage of coal on account of the transportation by water being interrupted during part of the winter months. Therefore, the undertaking involved not only the construction of the boiler house with its equipment, but also a coal-storage basin and wharfage.

FIG. 1. EXTERIOR OF THE NEW BOILER HOUSE

The provisions for wharfage consisted of a concrete wall built on the harbor line. This wall is about 34 ft. high, 19 in. wide at the base, and rests on wooden piles. Along the top of the wall, over that portion where coal is unloaded, a railroad track is provided for a locomotive crane. The coal is lifted from the barges with a clamshell bucket by the locomotive crane and deposited in a basin lying between the dock wall and another wall built parallel to it, about 85 ft. inshore. The space between these two walls is open, the bottom consisting of a concrete mat placed below normal river level.

A track supported on a concrete structure runs along the center of this basin, and a tunnel is built directly under the track and communicating with the storage space by means of a slot running the length of the tunnel and lying between the rails of the track which is supported on the tunnel walls. The track is provided in order to permit of the use of a traveling crusher equipment that discharges coal, crushed to the required degree of fineness, through the slot onto a belt conveyor that runs the length of the tunnel and discharges into a bucket elevator, by which the coal is elevated to the top of the boiler house and distributed into the overhead bunker by means of a suitable belt-conveyor arrangement and shuttle. The slot through which the coal is discharged as it leaves the crusher is provided with gates that are open only at the position where the crusher car stands, the open gates forming a hopper for guiding the coal into the slot. The crusher is of the four-roll type; the car is provided with standard railway-type motors, and the simplex contact system is used instead of trolley wires or third rail. The purpose of this arrangement is that the hopper into which the coal is to be discharged as it is lifted from the barges by the locomotive crane can always be in the right position to receive the coal without requiring the locomotive crane to make any extra movements to reach its point of discharge and also to avoid the loss of time and annoyance in the shifting of the coal barges. Prior to the coal strike the McKeesport Tin Plate Co. had between 50,000 and 60,000 tons of coal stored in this coal pocket and on the wharf. Under ordinary conditions, however, it it not expected to store over 15,000 tons in the coal pocket, wharfage space and overhead bunker combined.

The motor-driven coal handling machinery and arrangement consist of a conveyor in the tunnel, a bucket elevator, a stationary belt conveyor running half the length of the overhead bunker and a shuttle conveyor traveling on a track on top of the bunker for distributing the coal in the bunker. The system has a capacity of 120 tons per hour. The interlocking control is so arranged that in case any portion of the device is shut down, all other coal-handling apparatus, including the crusher, is automatically stopped. On the other hand, the crusher and its feeder cannot be operated until the balance of the coal-handling mechanism is set in motion. The bunker has a capacity of approximately 5,000 tons of coal. It is built of reinforced concrete.

There are twelve 606-hp. vertical type, high-arch Ladd water-tube boilers, each provided with an eight-retort front feed underfeed type of stoker, Fig. 3. The front draft air is led to the stokers through a plenum chamber communicating with each stoker through a suitable wind-gate arrangement set opposite the center of each boiler in the plenum chamber wall. The plenum chamber is accessible through an arrangement of lock doors. A similar provision is made for access to the wind chambers under the stokers. Both plenum chamber and the compartments under the stokers are provided with electric lighting so that any work that has to be done in this part of the plant can be done under as favorable conditions as possible. Ample space is provided in the wind-chamber compartment so that inspection or repair of the stokers as well as the disposal of siftings can be conveniently accomplished at any time.

The stokers are driven individually, each by an inclosed-type engine, and are arranged so that two stokers can be driven by the same engine in case of breakdown.

The stacks are supported on heavy structural steel bracing forming part of the boiler setting and are braced against wind by a series of struts combined with the building structural-steel framework. The stacks are 6 ft. in diameter and 150 ft. in height measured from the base up. The stack dampers are balanced with due provision for expansion and are operated by means of a control apparatus that is interlocked with a control for governing the speed of the stoker engines and controlling the wind gates, each boiler being provided with its own individual equipment. Fig. 4 shows a cross-section through the boiler house.

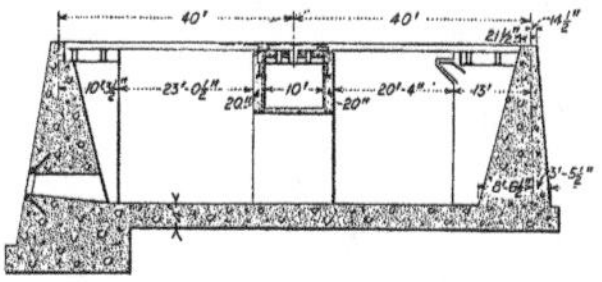

FIG. 2. COAL-STORAGE BASIN, CONVEYOR TUNNEL AT CENTER

Beneath the plenum chamber there is a subcellar for the accommodation of the side-dump ash cars. These cars are provided with storage batteries, are capable of making a turn inside of a 9-ft. radius and are used to convey the ashes from the hoppers under the boilers to an automatic ash skip hoist at the end of the building. The skip hoist discharges into a reinforced-concrete bin somewhat of the same construction as the coal bunker and located on the same level. This bin is provided with chutes and gates so arranged that the ashes can be loaded into either a railroad car or a motor truck.

The ash hoppers are built of cast-iron supporting plates throughout, lined with firebrick and provided with suitable spray pipes and curved ash gates operated by low-pressure hydraulic cylinders, the ash gates having sufficient extension beyond the necks so that, when filled with water, a reliable seal is formed against the passage of air into the ashbin.

FIG. 3. VIEW ALONG THE FIRING AISLE

Treated Youghiogheny River water is used and is delivered to the boilers by means of compound direct-acting pot-valve type feed pumps. The feed pumps and the water-treating system are located in a separate building, where provision is also made for the storage of chemicals.

The boilers are equipped with superheaters designed for delivering the steam at 125 deg. superheat. The boilers are designed for 200 lb. pressure. Each boiler is provided with an individual steam-flow meter, and there is a totalizing flow meter on the main line.

The coal-bunker arrangement with its relationship to the building-construction work and window openings is such that daylight is reflected to the boiler-house floor. The best of ventilation is provided both above the bunker and below. The space above is sealed off from the lower part of the building in order to prevent coal dust from sifting down into the boiler house. There are ventilating flues along the side of the building columns and extending above the roof of the building, so that the ash tunnel receives thorough ventilation. The buildings are brick clad up to the leanto roofs and above that point are covered with metal lath and cement. The roofs are of concrete and waterproofed with composition roofing.

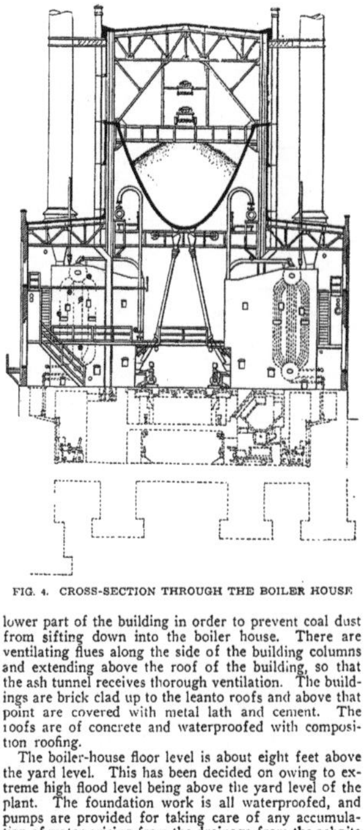

FIG. 4. CROSS-SECTION THROUGH THE BOILER HOUSE

The boiler-house floor level is about eight feet above the yard level. This has been decided on owing to extreme high flood level being above the yard level of the plant. The foundation work is all waterproofed, and pumps are provided for taking care of any accumulation of water arising from the drainage from the ashpits during floods.

How To Improve Boiler Efficiency

By H. F. Gauss

Chief Engineer, Illinois Stoker Company

An expert combustion engineer who can analyze and improve boiler-room conditions, capable firemen, and a simple equipment of instruments to give reliable operating data, are the first essentials. The author assumes a typical case, analyzes the losses and shows what a reduction of these losses would mean in dollars and cents.

"BURN coal efficiently" is a slogan that might well be adopted in every boiler room, large or small. Valuable advice and information have been given along this line, but often the material presented is not in a form readily applicable to the everyday operation of boiler-room equipment. On the other hand, boiler-room attendants are frequently inclined to discount the value of recommendations made by experts. Instead of lending their aid by endeavoring to make practical application of the information given, they are inclined to resent any implication that their old methods and ideas are not conducive to the best results. Such conditions are fatal to the realization of real boiler-room economy, and the boiler-room foreman who is awake to his opportunities will not, under any circumstances, tolerate such an attitude on the part of his assistants.

No man is worthy of his calling who is not willing to put forth every effort to advance it. The fireman who will not strive to burn coal more efficiently than his predecessors have done is not fit for his job. While the technical expert is indispensable in solving the boiler-room problems, the practical boiler-room man is no less important. In the final analysis it is up to the practical boiler-room attendant to carry out the ideas and suggestions of the combustion expert.

Importance of Results Must Be Realized

To bring about real, permanent results the start must be made by the man higher up. He must realize the importance of the results sought, and he must analyze his particular situation to determine how far he can go in investing money in new equipment or remodeling and fixing up present apparatus. A little figuring will disclose some amazing possibilities. To figure intelligently, however, the daily performance of the boilers in use must be determined. It is not a difficult matter to do this, but it must be done accurately to secure reliable data. The daily operating economy as the equipment stands is what is wanted, and not the results of an elaborate test run after special tuning up.

A test should embrace at least 24 hours, during which time an accurate log of the weight of coal burned and the total water evaporated should be kept and recorded. The coal burned must be sampled carefully and its heat value determined by proximate analysis. The temperature of the water fed into the boiler must be recorded at regular intervals and likewise the steam pressure, and steam temperature if superheaters are used. A pyrometer in the breeching will give the temperature of the escaping gases, and a single-chamber Orsat apparatus will suffice to determine the CO_2 in the products of combustion. It is necessary to keep a systematic log of all observations during the test, and the averages are the quantities to be used in the final computations. It is desirable to arrange apparatus for measuring the boiler-feed water by weight or by volume, but where this cannot be done, and meters are used for the purpose, they should be calibrated before and after the test. It is needless to say that before starting a test the blowoff valves and connections should be made tight.

Suppose, for example, that a certain plant has a boiler room containing four hand-fired 500-hp. boilers and it is desired to run a test on one of these. Apparatus is arranged for taking the observations previously mentioned, and the final averages for a period of 24 hours are as follows: Average steam pressure, 150 lb. gage; average steam temperature, 456 deg. F.; temperature of feed water, 180 deg.; flue-gas temperature, 650 deg.; total coal fired, 72,000 lb.; total water fed, 391,000 lb.; CO_2, 9 per cent. The analysis of the coal sample showed a heat value of 10,500 B.t.u.; and the following constituents per pound of coal as fired:

Moisure, per cent	9.2
Volatile, per cent	33.9
Fixed carbon, perc ent	39.4
Ash, per cent	17.5
Sulphur, per cent	2.0
Available hydrogen per pound combustible, per cent	2

The analysis of the ash sample showed 35 per cent. combustible in the ash.

In the final analysis a certain amount of available heat has been delivered to the furnace in the coal fired, which may be determined by multiplying the heat per pound of coal as fired by the total pounds of coal fired. Of the heat delivered to the furnace a certain percentage is absorbed by the boiler and used in converting water into steam. The heat so absorbed may be determined by multiplying the heat contents per pound of steam generated, by the total pounds of water fed to the boiler.

The heat contents per pound of steam is given in the steam tables, a convenient form of which is published by Marks & Davis. Turning to the table of properties of superheated steam, it will be found that at 150 lb. gage pressure, or 165 lb. absolute, and a temperature of 456 deg. the steam is superheated 90 deg. The total heat in the steam above 32 deg. F. is 1246.8 B.t.u. per lb. The heat per pound of feed water at the temperature of 180 deg. is $180 - 32 = 148$ B.t.u. It follows, then, that the heat absorbed by the boiler per pound of steam generated was $1246.8 - 148 = 1098.8$ B.t.u. Hence, the total heat absorbed by the boiler was $391,000 \times 1098.8 = 429,631,000$ B.t.u. On the other hand, the total heat in the coal fired was $72,000 \times 10,500 = 756,000,000$ B.t.u. Dividing the heat absorbed by the heat available gives an efficiency of $429,631,000 \div 756,000,000 = 56.8$ per cent.

With coal at, say, $3 per ton, the 72,000 lb. cost $108; of this only $61.34 worth was actually transferred into the steam. Such a condition as this exists in the majority of power plants.

Determining the Possible Improvement

The question naturally arising is, How much better efficiency can be expected? This can be answered by a careful analysis of the avoidable and unavoidable losses that take place in the operation of the boiler. For this purpose the charts, Figs. 1, 2, 3 and 4, are given. By analysis the coal contained 10,500 B.t.u. as fired. Using

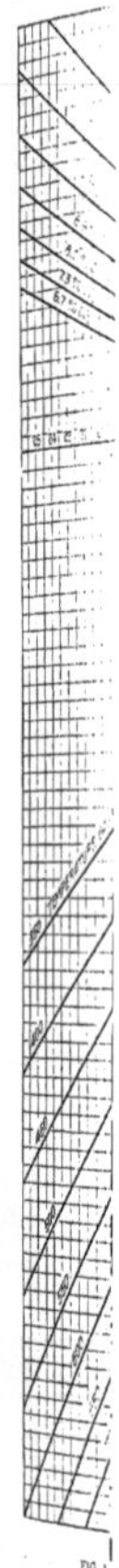

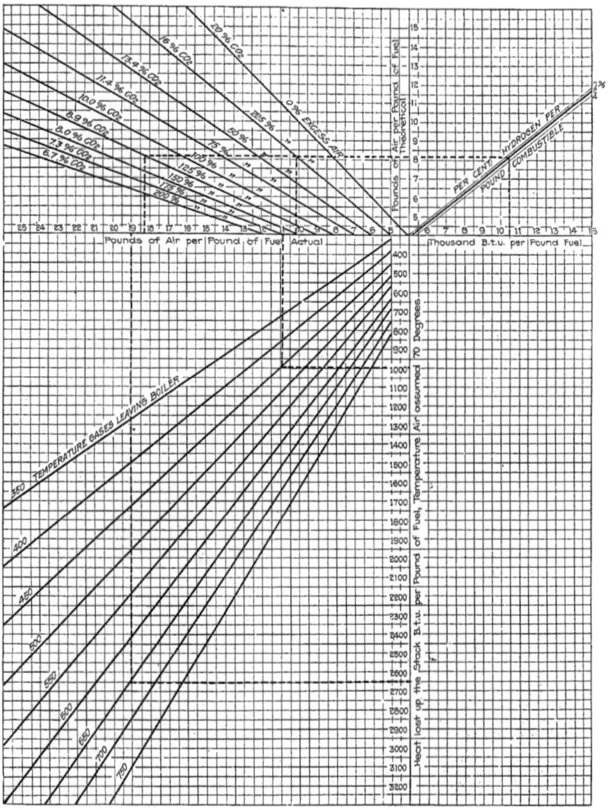

FIG. 1. CHART GIVING QUANTITY OF AIR PER POUND OF FUEL AND HEAT LOST UP STACK

Fig. 1, as indicated by the heavy line, it will be found that the gases leaving the boiler are carrying away 2650 B.t.u. per lb. of coal burned. After finding the actual air supplied per pound of coal fired in Fig. 1, it is necessary to add the combustible per pound of coal to determine the products of combustion passing up the stack. For the case in hand this combustible amounted to $1 - 0.175 = 0.825$ lb., hence the horizontal jog in the vertical line connecting the 8.9 per cent CO_2 line with the 650-deg. flue-gas line.

Turning now to Figs. 2, 3 and 4, it will be found that the following additional losses are taking place, those due to moisture and hydrogen being unavoidable.

Loss due to 9.2 per cent. moisture, B.t.u.	121
Loss due to 2 per cent. hydrogen, B.t.u.	235
Loss due to 35 per cent. carbon in ash, B.t.u.	1380

Adding these losses to that of the stack gives a total of 4386 B.t.u. The efficiency at which the coal was burned may be found by subtracting the heat lost per pound of coal from the heat available, and dividing the difference by the heat available, thus: $(10,500 - 4386) \div 10,500 = 58.2$ per cent.

This efficiency is slightly higher than was obtained by test, because there are certain unavoidable losses, such as radiation, which are difficult to account for with any

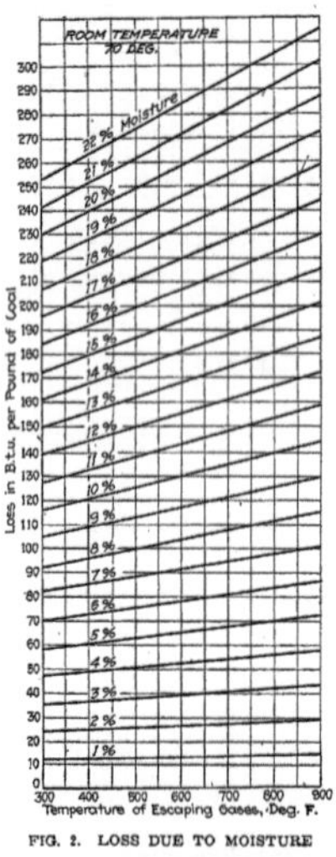

FIG. 2. LOSS DUE TO MOISTURE

degree of accuracy and, therefore, are either neglected or charged up as unaccountable losses. They vary from 1 to 5 per cent. in many cases.

As stated before, the losses due to hydrogen and moisture are fixed and cannot be reduced. The big losses, however, are those due to excess air and correspondingly low CO_2, high flue-gas temperature and carbon in the ash. These are the losses to fight against. Suppose, for example, that the operator succeeded in reducing the flue-gas temperature to 450 deg., raising the CO_2 to 16 per cent. and reducing the carbon in the ash to 20 per cent. Then the losses would be:

Loss due to 9.2 per cent. moisture, B.t.u.	112
Loss due to 2 per cent. hydrogen, B.t.u.	218
Loss due to 20 per cent. carbon in ash, B.t.u.	640
Loss due to stack gases, B.t.u.	990
Total loss, B.t.u.	1960

The efficiency now becomes $(10,500 - 1960) \div 10,500 = 81.3$ per cent. Deducting from this 1.3 per cent. unaccountable losses leaves a maximum efficiency of 80 per cent. to be striven for. In dollars and cents

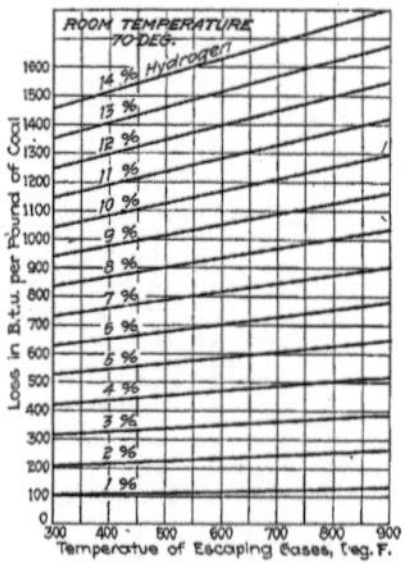

FIG. 3. LOSS DUE TO HYDROGEN

the accomplishment of this would mean a saving of \$24.95 per day for the one 500-hp. boiler, or per year of 300 working days, \$7485.

The owner or manager now has before him a figure that really means something. If necessary, he could spend enough money to buy automatic stokers to secure

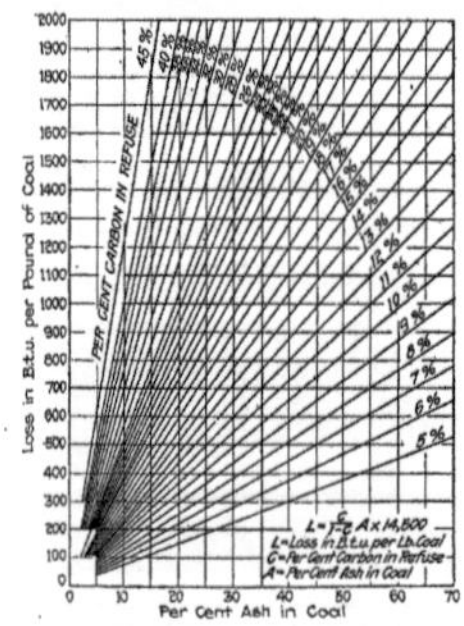

FIG. 4. LOSS DUE TO COMBUSTIBLE IN ASH

the desired results, and the saving secured would soon pay off the investment. But he also must realize from the possibilities that in no other department is there such an opportunity for saving as exists in the boiler room, therefore the necessity of competent men in charge. A high salary paid to a boiler-room attendant is money well invested, if by it a competent man, wide-awake to the responsibilities of his position, is secured.

In striving for high efficiency, the first step, then, is to put a competent man in charge. The next step is to see that he is equipped with the necessary instruments so that he can work intelligently. There should be draft gages, pyrometers, Orsat apparatus, and, of course, means of weighing the coal and measuring the water. A recording pyrometer in the breeching is desirable, but a hand Orsat for checking the CO_2 is sufficient.

After the maximum efficiency has been approached as nearly as possible with the existing equipment, it is time for the man higher up to consider stokers and coal and ash-handling machinery. In addition to reducing the coal there is the possible labor saving, all of which must be considered.

If the boiler-room problem is attacked logically, as outlined, and the solution suggested energetically followed up, good results are bound to follow.

Standard Meter Rates—Basis of Development

By LeROY W. ALLISON

Fundamental features of power-plant operation are discussed as regards rates. Such elements as station load, load factor, diversity factor, etc., are considered.

IN VARIOUS phases of power-station work, particularly in connection with the operating steam engineer, an intimacy with standard meter rates becomes distinctly essential. For instance, in the isolated plant, when the engineer is approached by the central-station power salesman, he can talk on an equal footing when rates or bases of rates are discussed. Likewise, he can help the management with its use of central-station energy and be of material assistance when monthly bills are checked up or consumption totaled. Again, such technical points arise as to why an "off-peak" consumer enjoys a lower rate than an "on-peak" customer; or the reason for a "demand charge," and so on. There is a constant call, often when least expected, for information on different points of rate development. And the engineer who is ready with the answer is the man who is prepared to move ahead.

In a consideration of electric-meter rates, there are defined elements of central-station operation that must be kept in mind, for these have a certain bearing on the subject. They include the station load, load factor and diversity factor from the actual operating end, while from the financial viewpoint there are the items of capital investment, fixed charges, operating expenses, depreciation and sinking funds.

The first group covers the fundamental features of operation, while the second deals more specifically with rate figures; that is, the actual charges to be made for service to the consumer to bring the desired return. This latter group can be neglected in the matter of exact significance or meaning of terms with regard to the particular branch of the subject at issue. A fair knowledge of the first noted, however, is important.

Station Load of the Power Plant

The station load means the character of service handled by the power plant. The maximum kilowatt demand that is made on a generating station under a yearly period determines its capacity. This is known generally as the yearly-peak load, and with the majority of central stations is evidenced in the month of December, when with short days, power and lighting requirements are exceptionally high. Moreover, this yearly peak is of short duration, from a half-hour to not more than an hour in extent in a 24-hour period, but this short, intensive demand indicates the plant capacity. If this rate of demand could be maintained at all times and the station loaded to capacity, a highly desirable condition would exist, making for utmost efficiency and lower rate schedules. But real situations and not ideal conditions must be faced.

The character of the station load of course varies considerably with different plants. In large communities the station carries lighting, power, possibly the local traction load, water-works, etc., while in a rural district the bulk of load will be found to deal with lighting service. A plant of this latter type, consequently, has little if any call during the day, and the few hours of darkness when service is at maximum demand must defray the expenses for the period of comparative idleness. Accordingly, higher rates for service are necessitated in a station of this nature, which frequently is closed down from midnight until well on into the next day, and sometimes until dusk.

Factors of station service—that is, the particular character of the load—are thus of considerable importance in the establishment of rates. It is the load curve from the daily log that tells the story.

The Load Factor in Power-Station Operation

There are, in reality, two load factors to be considered in power-station operation; namely, station load factor and consumers' load factor. The former is the ratio of average station load to the maximum station load over a certain period of time, as a day, a month or a year. It is determined by dividing the actual kilowatt-hours by the total of the maximum kilowatts multiplied by the hours. For example take a plant where the actual kilowatt-hours aggregate 1,200,000, with a maximum of 100,000 kw. in a 24-hour period. This would show a 50 per cent. load factor, or $\frac{1,200,000}{100,000 \times 24} = 0.50$.

A load factor of 50 per cent. would be considered fairly high, anything from 30 to 35 per cent. being good, while a load factor above 70 per cent. is deemed exceptionally good and is found only under conditions of a stable 24-hour load. As is generally understood, the nearer the load factor approaches to 100 per cent. the more effective is the operation and, consequently, the better the revenue derived from such plant operation. Under a 100 per cent. load factor the station would be operating at full load at all times. With a small plant operating under ordinary service, the load factor is usually low.

From this brief mention of station load factor, the exact significance or relation to rates will be noted readily, for even with as high a load factor as 50 per cent. a large amount of the plant investment is required for but a small percentage of the revenue. And this is why added service for off-peak periods is highly desirable; it improves the load factor and increases the percentage of return with the same capital investment.

As to consumers' load factor, this is the ratio of the average load to the maximum load in a 24-hour period. Very frequently, this term is used in connection with an "hours of work" basis, or usually 8 hours, but this is really a misnomer as applied to central-station operation. For example, assume that a consumer has a maximum demand of 300 kw. with an average load of 150 kw., the consumers' load factor on 8-hour basis would be 50 per cent., whereas, for a 24-hour period it would be $16\frac{2}{3}$ per cent., or $\frac{150 \times 8}{300 \times 24} = 16\frac{2}{3}$. In other words, on this basis, it is the percentage comparison between the kilowatt-hours used and the maximum use.

The Diversity Factor Defined

The diversity factor in central-station operation is the sum of all the maximum demands on a plant compared with the actual maximum demand. This is brought about through the variation in the time of various customers' maximum demand on the plant, as it is hardly necessary for the plant to furnish all customers' maximum demands at the same time. For instance, if a group of, say, 100 consumers have a maximum demand of 10 kw. each on a generator and the actual maximum load on the machine is 200 kw., the diversity factor on the unit is the total of the first two, divided by the latter, or $\frac{100 \times 10}{200} = 5$. In other words, the capacity of the generator would only have to be one-fifth of the total rating should all the consumers utilize their maximum power at the same time. As will be evident, the diversity factor is always more than 1, and the higher this factor in plant operation the lower the investment required to provide for a certain connected load. Different classes of consumers have quite a defined variation in diversity factors. For general power service, the average is usually about 1.5, while for ordinary lighting service it is from 3 to 6.

Classification of Rate Forms

There are certain rate forms that have become recognized as standard. These may be divided into two main classes covering so-called "meter rates" and "demand rates," while each of these is again subdivided into a number of different types of rate-computation schedules. Under the head of "meter rates" there are three classifications as follows: (1) Straight-line meter rate; (2) step meter rate; and (3) block meter rate. Under the subject of "demand rates" there are four different types or systems: (1) Flat demand rate; (2) Wright demand rate; (3) Hopkinson demand rate; and (4) Doherty, or three-charge, rate.

All these different rate forms are in use, some, of course, being more popular and serviceable than others. A study and understanding of the various fundamentals of computation give a comprehensive knowledge of this phase of rate-making and of sufficient extent for all practical purposes. Other methods of figuring electric charges have been developed for specific purposes, but speaking in the broad sense of general usefulness and utility, these are of minor attainments, taking their inception from recognized standard rate forms, and therefore hardly worthy of any extended study.

(*The two groups of standard-rate schedules mentioned—"meter rates" and "demand rates"—will be considered in detail in early issues of Power.*)

At the new six million dollar power plant of the West Penn Power Co., now under construction at Springdale, Penn., near Pittsburgh, a vertical shaft is being driven to a seven-foot vein of coal underlying the bed of the Allegheny River. The shaft will be 150 ft. deep, at which point entries will be run through the vein of coal which will be used for supplying this power plant with fuel. It is understood that there is approximately one million acres of coal land that will be mined before the supply becomes exhausted.

Power Growth in California

The normal growth of northern and central California will require the addition of power-generating facilities to the extent of 30,000 kw. per annum, according to estimates by Railroad Commission engineers of California. This will mean the expenditure of approximately $5,000,000 a year for power stations and transmission facilities alone. An equally large expenditure will be required for distribution and other facilities necessary to deliver this power.

It is estimated that the power demands for southern California—industrial, commercial, residential, and for irrigation—will reach a total approximating that necessary for the needs of the northern and central sections.

California's electrical needs, therefore—just enough to meet normal growth demands—will call for the expenditure for several years to come of approximately $20,000,000 annually. There are now under way four large hydro-electric projects, the completion of which will cost approximately $25,000,000. It is expected that these will be completed in the next year.

Completion of these four projects will render available 120,000 kw. of generated capacity. They are the Caribou plant of the Great Western Power Co. on the Feather River; Kerckhoff plant of the San Joaquin Light and Power Corp., on the San Joaquin River; Kern River plant No. 3, of the Southern California Edison Co., and the power plant of the City of Los Angeles, planned to utilize the waters of the aqueduct in San Francisquito Creek.

The Pacific Gas and Electric Co. recently enlarged its Lake Spaulding reservoir, purchased the power-production facilities and distributing system of the Northern California Power Co. and has under way a plan to lease the properties of the Sierra and San Francisco Power Co. A Pitt River power-development scheme made possible by the purchase of the Northern company is also planned by the Pacific Gas and Electric Co.

There are more than a half a million customers (545,000, to be exact) on the books of the existing electric utilities of the state, the commission points out in its annual report, just completed and forwarded to Governor W. D. Stephens.

At present the eighty-four companies engaged in the business of manufacturing and selling electric energy in California operate seventy-five hydro-electric plants. In addition there are fifty steam plants. The combined installation of the 125 plants totals 770,000 kw. During the year 1918 these plants generated a total of 2,892,000,000 kw.-hr., of which 2,163,000,000 kw.-hr., or 75 per cent of the entire production, was produced by water power. This production is still far below the energy required to meet the present demands.

According to George Otis Smith, Director United States Geological Survey, the United States produces about 95 per cent of the natural gas consumed in the world. Of this two-thirds is utilized in the industries, and one-third for domestic service. Pennsylvania ranks first as a consumer, while Ohio, Kansas, West Virginia and Oklahoma have contended for second and third honors. The increase in utilization of natural gas has been both rapid and steady. In 1908 there were 21,000 wells producing 402,000,000 cu.ft. Ten years later 39,000 wells had practically doubled that output. During the past 13 years the quantity used has approximated 8½ trillion cubic feet, and the present rate is nearly 9,000,000 a year. However, we are not warranted in expecting anything in the future but a general decline in the current supply of this fuel.

EDITORIALS

Power Costs To Increase and Not Decrease

ALMOST from the inception of the generation and transmission of power electrically it has been the dream of the engineers that some day the entire country would be tied together with a network of electrical transmission and distribution lines supplying power for every purpose at very low rates. There is little doubt that the first part of this conception is fast approaching a realization—transmission networks are spreading out over the country in every direction. As an indication of the trend in this development one has only to consider that there are over fourteen thousand miles of single-circuit transmission lines in service in this country operating at voltages of from seventy thousand to one hundred and fifty thousand, and super-power systems operating at two hundred and twenty thousand volts connecting up into one great network generating station having one million kilowatts or more capacity are the prospects of the near future.

However, these great networks for power transmission have not solved the problem of cheap power. There are localities in the immediate neighborhood of large hydro-electric developments where power is cheap, but when it is transmitted two hundred or three hundred miles the cost must be increased considerably to pay interest on the additional investment, maintain charges on the equipment, etc. The fact that power must be sold at three or four cents per kilowatt to make it a paying proposition, when developed in a hydro-electric plant and transmitted six hundred miles, by even the most highly developed means that the electrical art can command at the present time, shows that hydro-electric power to be cheap must be used close to the source of production.

Improvements in steam power-plant equipment have reduced the cost of electric energy generated from coal until in many cases it has become a competitor of the hydro-electric plant, especially where the hydro-electric power had to be transmitted a considerable distance. Previous to the war the large steam plant apparently had promising prospects of reaching a point in efficiency where it could compete with the hydro-electric plant on its own ground. But with everything that enters into the production of power rising in value, operating a steam plant has developed into what might be termed a struggle by the mechanical and the electrical engineers to improve the equipment and thus keep the expense down in the face of the rising cost of those elements required in power production, with the increasing cost of materials holding the winning hand, as evidenced by an advance in power rates allowed to many of the public utilities.

The comparatively low cost of power in steam plants in the past was made possible by high-grade coal at low cost, low labor cost, and comparatively low construction and maintenance cost. Coal at low prices was obtained by working the most accessible fields at low labor costs, with favorable freight rates and reasonable profits. Consideration of each one of these elements at the present holds forth very little encouragement for a decrease in the price of coal; in fact, if there is to be a change in the future, the indications are that it will be upward. Consequently, the price of power must go up. No doubt improvements in steam power-plant equipment will be made to improve the thermal efficiency, but unless something radical develops, which is not in sight at the present, the coal required per kilowatt-hour in our most efficient plants will not be greatly reduced. This being true, the prospect of power being supplied cheaply from large power systems, or any other systems for that matter, must remain a dream, except in localities near large hydro-electric sources that may be developed at low cost.

Progress in Water-Power Legislation

DESPITE the determined fight made on the Water Power Bill in the Senate by the conservation group, only eighteen votes were cast against it on final passage. Fifty-two Senators voted in the affirmative. Several important amendments to the bill that passed the House were approved by the Senate. Among these were the authorization of the twenty-five-million-dollar power development on the Potomac River at Great Falls and the conferring on the Federal Power Commission the duties already provided by statute for the Newlands Waterways Commission. This latter commission had been empowered to study and advise Congress on matters pertaining to reclamation, utilization of water power, flood control and other matters. It is regarded as likely that the Great Falls project and the rehabilitation of the Newlands Commission will go out of the bill in conference. It is also recognized that other very important changes will be made in the bill in conference.

Senator Lenroot of Wisconsin led the fight against the bill. His position and that of other conservationists is that the bill as amended by the Senate gives away the public's water-power resources for practically nothing by relieving licensees from paying what they regard a reasonable value for the power. They also claim that it limits the possible charge which may be made by the Government for the expenses of administering the act plus a small fee based on the value of government lands occupied, irrespective of their water-power possibilities. The Senate committee amendment, which is the basis of these objections, was voted into the bill because a majority of the Senators were convinced that no good could come from the legislation unless the development of the water power was made sufficiently remunerative to attract capital. Objection also was found to the Senate committee amendment which gives the public service commission of the state the power to decide alone that a contract may extend beyond the end of a lease.

A very determined fight was made against the wording of the bill which pertains to the terms of the license. As the Senate amended the bill, the effect will be, the conservationists declare, to make a license perpetual. The majority of the Senate, however, was unable to see any such danger as pointed out by Senator Lenroot and his associates and believe that the 50-years license and its renewal are amply safeguarded and provided for in a way that will induce capital to develop water power.

The bill provides for the creation of a Federal power commission which is empowered to issue licenses for the development of water-power projects at a reason-

able annual charge. State municipal power plants and industrial plants developing less than two hundred horsepower are not required to pay a license.

Among the beneficial results anticipated as certain to result from this legislation provided the Senate and House conferees are able to agree are a saving of coal and oil, more efficient transportation, development of new industries, the building up of new communities, the creation of new property values, added employment for labor and increased market for agricultural products. It also was pointed out that every year that the water powers of the country are undeveloped, the nation suffers a loss equal to the cost of their development.

Standard Meter Rates

THE subject of standard meter rates or schedules for electric service is one of great interest and importance to chief engineers, operating engineers and others engaged in power-plant work. How this particular rate charged is developed or that rate derived covers basic consideration of good central-station practice, and each such plant naturally employs a system of charge, utilizing a recognized standard meter schedule, that will make for entire uniformity and harmony in operation.

Rate making in itself has come to be a scientific study of broad proportions, and upon this phase of power-plant work is dependent, to a large degree, the successful attainment of the electric station. The prevailing charges for light and power service must be equitable and properly graded for different classes of construction, the whole showing a fair net return of profit. In these days exact service rates are governed to a large extent by utility-commission control in different states.

Beyond the direct application for specific character of service, or actual charge in force, a comprehensive knowledge of a proved system for rate computation as now in general practice is of great value and advantage to the engineer, not only in the light of broadened understanding in power-station activity, but for real service and utility in everyday operation. Whether engaged in central-station or municipal-plant work, or in the isolated-power plant, is not of great material moment—the engineer should know.

In this issue is published the first of three articles on "Standard Meter Rates." The purpose of these articles is to familiarize operating engineers with the various standard rates in common use by central-station companies in charging for power, so that the former will not be at a disadvantage when called upon to discuss these problems. The purchase of power from a central station does not necessarily mean the closing down of an isolated plant, since there are many conditions where power is purchased in conjunction where an isolated plant is in operation, such as to relieve the overload on the existing power plant, to eliminate some uneconomical mechanical drive, to drive a new mill when the capacity of the power-plant cannot be increased advantageously, to drive a new addition to a mill that cannot economically be reached by shafting, etc. There are so many of these conditions under which power may be purchased advantageously that the problem of purchased power versus isolated plant has become not so much a question of which shall exist, as how both may exist to the greatest advantage to industry as a whole.

It is not enough for the isolated-plant engineer in the industrial power plant to know that his plant is operated in the most economical way; he should be able to analyze the various conditions under which it might operate in combination with purchased power. The engineer who is willing to put forth the effort to make these analyses will find that his job is one that requires real engineering skill and ability, and the running of his plant will never develop into the following out of the regular routine. The effort is worth while, for it will relieve the monotony of the regular routine work that the operating engineer not infrequently complains of today, and the training will be of great assistance in obtaining the bigger job or the larger salary that is in the mind of every progressive power-plant engineer.

The Human Element

NOT with regard to the erratic and oftentimes irascible nature of the human element upon efficiency, but to the effect of the personality of the profession as a background for the popular conception and for the student's ideal of mechanical engineering as a profession do we raise the subject.

Professor Magruder, of the Ohio State University, writes: "Many of our students have very hazy ideas of what a mechanical engineer does. The most popular idea is that he is a high-grade machinist, hence I would be glad to dispel this ignorance and help the young men to form more accurate opinions of the profession and the men who have made it. As a means to this end I have had a number of photographs and halftone prints of some of our most prominent mechanical engineers framed and hung on the walls of our laboratory. These are referred to when the particular thing which they have done is the subject of the lesson. For example, a Corliss engine takes on personality when the picture of George H. Corliss is shown. The same is true of the Babcock & Wilcox boiler, and many other inventions and products of the brain of our great engineers. In my opinion this has an educational value which is worthy of the trouble. An expensive variant on the framed print is to have transparencies in the windows."

This is more than an ordinary suggestion. The student whose conception of his profession is illuminated by a knowledge of its history and peopled with concrete ideals of its personnel has more of the element of success than one the background of whose conception is the drawing board and whose brain is peopled only with principles and formulas.

Ice and a Boiler Explosion

ON other pages of this issue is given an account of a boiler explosion, and the cause is attributed to ice in the vent pipe of the safety valve and perhaps in the body of the valve itself. A careful review of the circumstances will make it evident that such an occurrence is entirely possible and discovery of frozen vent pipes on the other boilers in the same plant that had been shut down for washing and inspection, goes a long way toward confirming the foregoing supposition. With a complete set of new tubes and the boiler reported to be in excellent condition otherwise, and apparently so from the appearance of the metal at the breaks in the flange of the tube sheet and the braces, it would appear that two safety valves free to the atmosphere should have prevented a dangerous pressure. It is certainly an infrequent cause of boiler rupture, but one that should be taken into account in severe weather. Late types of safety valve would have relieved the boiler notwithstanding a choked vent pipe. From readers who may differ as to the cause of the accident or may have other suggestions to offer we will welcome a discussion. All the information possible on an accident of this character will help guard the future.

CORRESPONDENCE

Remedy for Present Valve Setting

In the issue of Sept. 23, 1919, page 518, S. L. Gilliam submitted indicator diagrams, a study of which indicates that most of the improvement in the second cards is due to higher boiler pressure and a decreased load. In my judgment a change was made in the setting of the exhaust valves, probably an adjustment of the reach rod from the eccentric to the wristplate, as the compression is more nearly equal in the second diagram than in the first.

The most important adjustment to be made on this engine is setting the exhaust valves. I believe that before a proper diagram can be obtained, the exhaust eccentric will have to be set back slightly.

The present pounding in the head end is caused by excessive compression which raises the head-end steam valve from its seat, thus causing it to slam. It would be advisable to draw this head-end steam valve and examine it and its chamber. The slamming may have chipped out pieces of the valve or port edges, which may be causing a leak through this valve. If the head-end steam valve and chamber are in good condition, replace the valve and note its position when closed. It may not travel far enough to give proper closing.

The position of the piston in the cylinder at both dead-centers should also be noted. The clearance at the head end may be considerably less than that at the crank end. From the looks of the second diagram I would expect to find either a leaky head-end steam valve or very little clearance at that end. These two points should be investigated before trying to set the valves.

To set the exhaust valves the first thing to do is to adjust the wristplate to rotate within its proper range of motion, as shown by the marks on the hub of the wristplate and the mark on the stud. If no marks can be found, then adjust the wristplate to move equally in both directions from the central position. The rods from the wristplate to the exhaust-valve arms should then be adjusted to give proper exhaust opening and closing. Judging from the diagrams shown it may not be possible to get the compression reduced sufficiently without causing too early exhaust opening with the present setting of the eccentric, unless the head-end steam valve is now leaking.

I believe that very little change in the setting of the exhaust valves will be found necessary if the head-end steam valve is tight and the clearance is equalized.

The second diagrams show the point of exhaust closure at the head end to be but a trifle earlier than that at the crank end, which is proper, as the head end should have more compression than the crank end because of the greater speed of the piston when traveling at the head-end half of the cylinder.

After getting the trouble with the head-end compression located and eliminated, the setting of the steam valves may be considered.

The steam wristplate should be adjusted for travel the same as the exhaust wristplate. From the statement that the threads run out of the link head, preventing further adjustment of the head-end valve, it would seem that the wristplate is now off center or else the valve stem was not properly keyseated or the stem has been twisted at some time.

Sufficient lead should be given to produce the desired straightness to the end of the indicator card. The amount of steam or exhaust opening when the wristplate is on the center has no effect on the steam rate of the engine, therefore it should not have very much weight in the setting of the valves.

The steam valves should be set to give proper lead with tight valves when closed and a proper working position of dashpot rods and hooks. If the hook rods have to be shortened too much in order to get sufficient lead, the valve will not close properly when released or the hook may pull back too nearly in a straight line with the rod for safe operation. In that case advance the steam eccentric.

It must be kept in mind when shifting the steam eccentric that advancing the eccentric will cut down the capacity of the engine if the engine is to run on the governor cutoff. That is, the limit of the cutoff is reduced by advancing the eccentric. The valve motion reaches its farthest point of travel earlier in the travel of the piston in the case of engines using the stationary knockoff blocks, or knockoff blocks fixed to rings shifted by the governor. In the case of the special gear sometimes used where the knockoff gear has an independent motion on centers controlled by the governor, then the point of cutoff is independent of the steam eccentric.

Peoria, Ill. C. E. MOULSON.

It is evident that there is too much lead on the head end, resulting in no work being done on that part of the stroke shown by the pointed tip of the diagram. Again, the crank end shows no terminal pressure, the expansion being carried too far for economy. The engine is evidently slightly underloaded.

I suggest that the valve might be changed on its spindle or that the eccentric be shifted. Adjustments of rod lengths do not change the general phenomenon of lead, since the angular advance influences all time elements during both strokes; so in this case it is best to delay the time of the motion rather than its amount. The compression line on the head-end shows that the exhaust valve closes too early. W. S. HOFFMAN.

Morgantown, W. Va.

Substitution of a Slip-Ring Rotor for a Squirrel Cage

A 100-hp. three-phase 60-cycle 220-volt squirrel-cage induction motor was used to drive a compound air compressor. Unless considerable care was used in starting, the inverse-time circuit-breaker would open. To improve the starting torque, a slip-ring type rotor and bearing housings from another motor were substituted for the squirrel-cage rotor. The original stator was left on its foundation.

For trial conditions, a water rheostat was connected in the rotor circuit, as shown in the figure. The starter

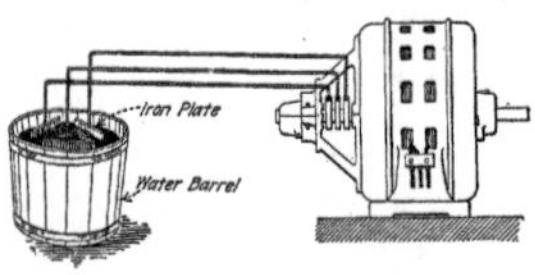

WATER RHEOSTAT CONNECTED TO ROTOR SLIP RINGS

used with the squirrel-cage motor was put in the running position and the main-line switch closed. A few trials showed that the substituted rotor would fulfill all requirements. The next step was to select a suitable permanent starter. A voltmeter was connected across the rings of the rotor while the brushes were lifted and the full line voltage applied to the stator. Voltage between rings obtained in this manner was 140. The secondary amperes were then calculated from the following formula:

$$\textit{Secondary amperes} = \frac{\textit{horsepower} \times 746}{\sqrt{3} \times \textit{voltage between rings}}$$

$$= \frac{100 \times 746}{1.732 \times 140} = 300$$

A starter designed to give full-load starting torque was selected. This starter had five points or steps in the secondary resistance. This number was used so that an even acceleration and maximum torque would be maintained during the starting period.

Washington, D. C. H. L. Hervey.

Electrical and Steam-Plant Operators

I wish to emphasize the importance of a distinction between the electrical and steam operation of power plants. It has long been the practice in the small and medium-sized plants to permit the engineer to have sole supervision over the electrical equipment. This being done chiefly from an economical point of view, the engineer makes all the minor repairs and acts as chief in general. His knowledge of the electrical line is sufficient to enable him to do practically all the work himself, calling for expert advice if the occasion arises.

This system without a doubt works quite satisfactorily in the small plant, but as the engineer assumes charge in a position with a greater responsibility, he will not be required to perform the duties of an electrician, but will have supervision over the electrical work. The fact that he holds a position with a title as steam engineer enables him to assume charge over a separate and distinct line. If this practice is to be adhered to, greater stress should be laid on the subject of electricity, in connection with the state examination.

It is clearly understood that the Underwriters' requirements for the safe operation and installation of electrical equipments are far in excess of those required for steam-plant operation, particularly in sections where a license is not required. The electrical worker qualified to perform the duties of an electrician must abide by these fixed rules. Nevertheless a complete knowledge of the Underwriters' requirements does not make an electrician, any more than a license makes a steam engineer.

I recall an instance where an engineer in charge of a plant directed an electrical worker to clean the exciter rings on a particular alternator. The machine happened to be of the stationary-field type, and the rings referred to consisted of four 2400-volt distribution rings. Had it not been for the experience of the workman, the result might be imagined.

On another occasion the circuit-breakers were blocked to prevent the automatic operation, thus preventing an interruption to the continuity of service. The fact was not realized that they were carefully set and tested for a maximum margin of safety. However, lightning struck the system, and resulted in a serious loss.

Parlin, N. J. J. W. Truesdell.

Operating 220-Volt Motors on 550-Volt Service

Fire destroyed a power plant from which two 220-volt-interpole motors were supplied, used to drive the machinery in a factory. Fortunately, the motors were exactly alike, and we were able to connect the load together mechanically, and the motors in series to operate from a 550-volt trolley line, as in the figure. The shunt fields were disconnected from the armatures and connected in series and through one starting box, no-voltage release coil only, as indicated. The armature resistance in the two starting boxes was connected in series with the armature circuits only. The return

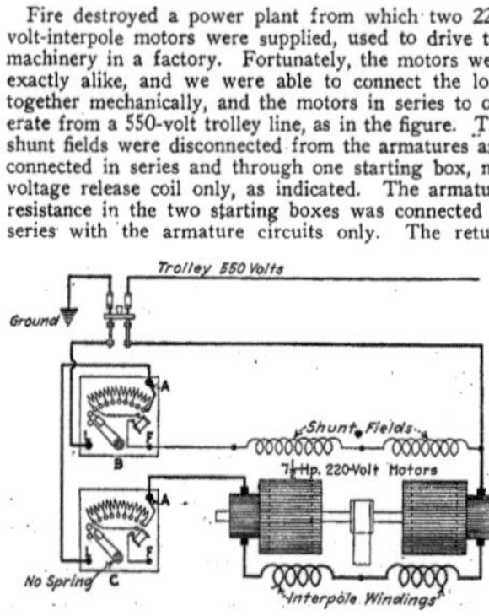

CONNECTION DIAGRAM FOR OPERATING TWO SHUNT INTERPOLE MOTORS IN SERIES

spring in the arm of box *C* was removed, and it was depended upon *B* alone to disconnect the motors from the trolley line if the voltage failed on the latter.

To prevent either motor from racing, their pulleys were removed and set on a timber frame, so that their ends were next to each other and lined up, butting the two shafts together. The armature lead on one motor was crossed to get the correct rotation. One pulley was used as a combination pulley and coupling, which clamped and held both shafts together. From this common pulley the motors were belted to their loads, which had been connected together mechanically. These machines operated for about a week when the wooden frame gave enough to allow the motors to get out of line and break open the pulley couplings, after which a couple of metal bars, run parallel to the shaft, were put under them, and we had no further trouble.

Tampa, Fla. J. C. BUERKE.

[It is apparent that Mr. Buerke went to considerable unnecessary work in coupling the motors together mechanically. Connecting the two loads together mechanically was equivalent to coupling the motors together. Therefore, all that was necessary was to connect the loads together mechanically and the motors in series. Of course, any difference in the size of the pulleys or thickness of the belts would have a tendency to make one motor take more load than the other. However, when both equipments are supposed to be identical, this unbalancing of the load between the two motors should be very small.—Editor.]

Repair Jobs in Central America

In Fig. 1 is shown an easily made lever arrangement to drain hoist-engine and pump cylinders when so placed as to be difficult to get at. To each drain-cock an evolvent-shaped lever is fastened by a rivet or cotter pin, and a convenient handle is attached to the rod which connects both levers. The handle is guided by a ring that is held under a stud nut on the cylinder. To this ring is fastened a hook for holding the lever in the open position; or when moved in the opposite direction, the cocks are held closed. On mine steam pumps which are occasionally submerged, I connected such a handle to a station ten feet higher, and many times the pumps could be started only by that arrangement.

One of the steam-port edges of a pump was damaged, as shown in Fig. 2, but little time could be lost, so I fitted a piece of iron by means of two screws and dowel pins to the slide valve. No trouble was experienced when the pump was started up.

An adjustable guide arrangement for the slide valve on a 10-hp. engine is shown in Fig. 3. This valve has such play that it was often forced from its seat when the governor, from any cause, closed too rapidly. The steam chest had to be opened each time this happened and the valve replaced. The two nuts, having been for some years in place, were so tight that to loosen them would overstrain the rod. A piece of ⅛ x 1-in. soft steel was shaped to give three points of contact only and was fitted to the valve and held to the rod by two small clamps, the bolts of which were locked. Then three holes were drilled in the steam-chest cover and tapped for ⅝-in. bolts, which screwed in with a close fit, and the threads were filed off at the end. The bolts and their nuts were screwed in place after applying graphite to prevent rusting. Three pieces, each drilled with two ¼-in. end holes, were made, and a center one in which the end of the bolt was fitted and the end riveted over, but leaving it free to turn. Then a piece of ¼ x 1-in. iron was made nearly the full length of the valve movement. To this piece the little clamp-like pieces were secured with copper rivets and the riveted ends smoothed with a file. When all was assembled, the

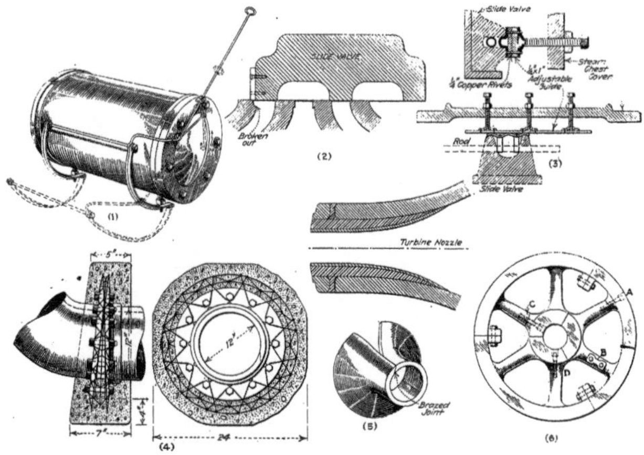

FIGS. 1 TO 6. ILLUSTRATING REPAIR JOBS THAT WERE DONE AT THE PLANT
Fig. 1. Cylinder Drain Arrangement. Fig. 2. A Valve-Seat Repair. Fig. 3. Engine Adjustable Guide Arrangement. Fig. 4. Pipe Joint Reinforced with Concrete. Fig. 5. Changed Water Turbine Nozzle. Fig. 6. Repaired Flywheel

engine was started slowly, feeding it fully with oil and a little graphite. Then the screws were so adjusted that the slide pieces just touched the guide. After having run for nearly two weeks, I opened the steam chest and found the surfaces in very good condition. We never experienced the least trouble, and the engine used about 30 per cent. less steam than before. After several weeks of service the screws were once more slightly readjusted.

Fig. 4 shows how a troublesome elbow joint was made tight with reinforced concrete made 4 in. thick and placed around the flange. The only reinforcing at hand was barbed wire. The water pressure carried in the pipe was about 100 lb. Some engineer had taken the angle at 30 deg., and upon assembling the 12-in. spiral riveted pipe line, it was found that the elbow did not fit. No amount of changing helped much, and a 1-in. opening resulted. A box was made to fit around the joint with about a 2-in. space between it and the metal and the wire secured under the bolts, making a netlike appearance of the reinforcing.

When the box was nicely fitted, the cement was mixed thoroughly with clean, sharp sand in proportion of 1 to 2, rather wet, and was then carefully tamped in place. A few pieces of broken stone were also used. The cement was kept wet for about two weeks before putting full pressure on this pipe. Remarkable as it may seem, this joint did not develop any cracks even when the 45-ft. bridge that supported the pipe line broke down and the joint had to withstand the weight of the pipe.

Considerable delay was had during the dry season with water turbines that were used to drive a mill. Any considerable expense seemed inadvisable, but more power was necessary. I ascertained the power required for each machine and the amount of water available and then calculated the nozzle opening for full load at 82 per cent. effective. Then, taking a piece of pipe one size larger than the required opening, it was heated and drawn out bell-shaped to fit the curve of the original wide nozzle, Fig 5. The opening formed in making the bell end was closed by brazing in a piece of metal. Holes were drilled in the old nozzle, and the new piece was set centrally in the nozzle and clay tamped all around. Then babbitt metal was cast to fill the space between the nozzles. This served so well that I made several sizes, so that we always have high-efficiency nozzles. No governor or regulating is required on the units, as the gate valve is opened wide and as calculated for the full load the speed is very near constant.

Fig. 6 shows a badly broken crusher flywheel, the result of a young engineer trying to force the wheel onto the shaft. Even when hot the bore was $\frac{3}{32}$-in. less than the shaft. However, the experiment failed when the wheel was driven nearly one-half on. Then it stuck, and instead of dismantling the wheel they tried to drive it off with an iron battering ram and after a few hard blows the wheel broke. To wait for a new one was out of the question; so I strengthened the arms with tight-fitting ½-in. bolts. The one at *A* was secured against loosening by a ⅜-in. screw, and the arm at *B* was secured by two pieces of ⅝-in. flat iron shaped and secured as shown. Two ⅞-in. holes, 4 in. center to center, were drilled in the arm, and corresponding holes were drilled in the two link pieces, but with $\frac{3}{16}$ in. less center to center. They were then heated and two ⅞-in. rivets were driven in hot. The spoke cracks at the hub were strengthened as shown at *C* and *D*. Then the bore of the wheel was made to fit the shaft when it was assembled on the shaft with two rings shrunk on the hub. This repaired wheel has been in service for seven years.

La Paz, Honduras, C. A. PAUL M. WOLF.

Repairing Cracked Tube Sheets

In a large manufacturing plant containing approximately one hundred boilers, of which a large percentage were of the vertical type, the steam demand had increased until it was necessary to run at high capacity both night and day to hold the steam; consequently each unit was kept on the line as long as possible and repairs were made only when absolutely necessary.

These boilers were equipped with chain-grate stokers, and when cleaning fires the firemen often knocked a few brick from the top of the bridge-wall, which allowed the flame to sweep down on the bottom tube sheet, causing in many cases cracks to develop between the tube hole and the edge of the plate. At first these were carefully chipped out and welded with fairly good success, but the strain of expanding the tubes often caused

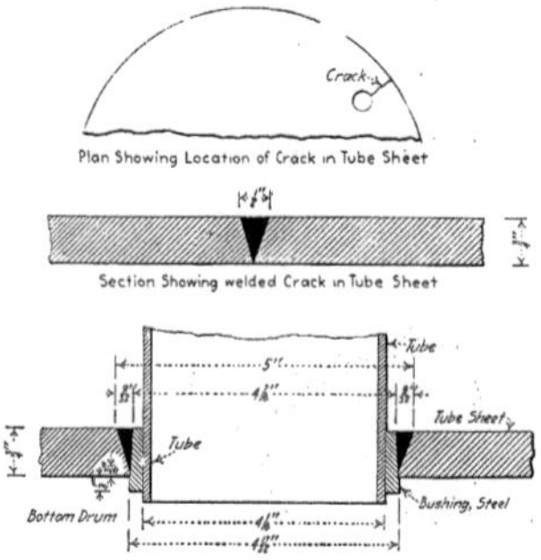

HOW THE REINFORCING RING WAS WELDED TO THE TUBE SHEET

the cracks to break out again, and they finally became so bad that they could no longer be welded satisfactorily.

About that time the inspector representing the insurance company recommended that new tube sheets be installed, but the expense involved was so great as to be almost prohibitive. There was considerable discussion as to whether these boilers should be replaced with new ones, inasmuch as they were over twenty years old, but after carefully considering the existing conditions it was decided that the cracked sheets could be repaired by a method which was submitted to the chief inspector for approval and adopted.

The tube hole was first reamed out on a taper from the top, then a ring was machined on the inside to slip over the tube easily, with the outside slightly tapered. This ring was driven into the tube hole from below and welded in place, the paper of the ring and hole leaving room between for filling, after which the crack was chipped and welded also. As the strain of expanding the tube then came on the ring instead of the sheet, no further trouble was experienced with the crack, and the sheet was practically as good as new, while the saving made by repairing in this manner the several boilers affected, over the cost of new sheets, was over $30,000. J. F. TUTEIN.

Pittsburgh, Penn.

Induction Generator for Small Water-Power Plants—What would be the advantages of an induction generator for a small water-power plant tied into a large electric system?
S. V. B.

The advantages would be simplicity and sturdiness of construction, small care for operation, requiring little more attention than necessary lubrication, no requirement of exciter, automatic adaptation of electric load to the variable water power available, and no requirement of a waterwheel governor.

Corrosion of Feed-Water Heater—What would cause rapid pitting of the shell of a closed exhaust-steam feed-water heater?
L. H.

Galvanic action is likely to take place with some corrosion of the shell when the heater is provided with tubes and shell of different metals, and corrosion of the kind is intensified by either an acid or an alkaline feed water when the feed water is in contact with the shell. In any form of heater corrosion may be caused by acid feed water or induced by electrolysis due to stray electrical currents.

Holes for Inspecting Thickness of Corrugated Furnaces—What means should be provided for inspecting the thickness of a corrugated furnace?
F. S. J.

For ascertaining the thickness of the material of a corrugated or ribbed furnace by actual measurement, the furnace should be drilled for a ¼-in. pipe tap and fitted with a removable screw plug. For Brown and Purves furnaces boiler codes usually provide that the holes should be in the center of the second flat, and for Morrison and other similar types, in the center of the top corrugation, at least as far in as the first corrugation.

Use of Flexible Coupling—Why is it necessary to employ a flexible coupling for direct connection of a motor and a centrifugal pump if both are mounted on the same baseplate?
B. F. K.

It is difficult to set the center of a centrifugal or other machine in true alignment with the center line of a motor, and still more difficult to maintain that alignment on account of difference in wear of the bearings. Hence, to obtain good running conditions when direct-connected, it is necessary to connect the motor to its load with a drag crank or some form of coupling that will adjust itself to the difference of alignment during operation.

More Uniform Pressure from Variable Boiler Pressure—We have a direct steam line from our 100-hp. return-tubular boiler to a small oven 60 ft. above the boiler, for which we wish to supply steam at a constant pressure of 5 lb. gage. The boiler pressure varies from 70 to 90 lb. The pressure-reducing valve placed near the oven does not operate as closely as we desire. Would it not be better to employ another valve in the boiler room for reducing the pressure to about 40 lb.?
W. C. H.

The trouble probably is due to the wide variation of the initial pressure presented to the reducing valve, and there should be delivery of more uniform pressure from the present reducing valve by placing another reducing valve in the line at the boiler, set to reduce the boiler pressure to an intermediate pressure as proposed.

Number of Cubic Yards in Block of Concrete—How many cubic yards of concrete are contained in an engine foundation 4 ft. x 12 ft. at the top, 6 ft. x 14 ft. at the base and 5 ft. 6 in. deep, with sides and ends battered? H. W. G.

Taking all dimensions in feet, the content in cubic feet is given by the prismoidal formula:

Number of cu.ft. = (area of top + area of base + 4 times sectional area at mid depth) ÷ 6 × total depth.

The area of the top would be 4 × 12 = 48 sq.ft., and that of the base would be 6 × 14 = 84 sq.ft. The section at mid-depth would be $\frac{4+6}{2} \times \frac{12+14}{2}$ or 5 ft. × 13 ft., and the area of the mid-depth section would be 65 sq.ft. Substituting these values and 5.5 ft. for total depth, the prismoidal formula gives,

$$\text{Number of cu.ft.} = \frac{48+84+(4\times65)}{6} \times 5.5 = 359.33 \text{ cu.ft., or}$$

$$359.33 \div 27 = 13.31 \text{ cu.yd.}$$

Determining Thickness of Bag in Boiler Shell—What formula or calculation would be used in computing the thickness of the metal at the center of a bag in a boiler shell?
J. W. M.

Without making actual measurements inside and outside of the bag, its form and the manner in which the original thickness is reduced are matters of speculation. But if it is assumed that there is a uniform reduction of thickness from the periphery to the center of the bag, the thickness at the center will be approximately twice the original thickness multiplied by the length of the bag taken in a straight line, divided by the curved length, less the original thickness of the shell. It may usually be taken for granted that the thinnest metal will be at the point where the greatest bulge has occurred. The proper method of determining the thickness is to make the measurement with a hook gage inserted through a hole drilled through the bag. Such a hole would be made for swaging the bag out of the sheet and, whether or not that is done, a hole thus made afterward may be filled with a rivet.

Compression Regulating Valves of Duplex Pump—What is the purpose of compression valves of a duplex steam pump and how are they connected?
F. C.

In ordinary duplex steam pumps, at each end of the steam cylinders, the opening of the exhaust passage into the cylinder wall is farther from the end of the cylinder than the opening of the steam passage. The latter usually is at the very end of the cylinder bore or wholly or partly in the cylinder head. Compression of exhaust steam is effected by the piston covering the exhaust passage before the exhaust stroke has been completed. Should the piston cushion too heavily, the compression can be reduced by opening a valve in a connection made between the end of the cylinder and the exhaust passage or, as commonly done, by regulation of a valve that controls the amount of opening of an aperture in the partition between the steam and exhaust passages. When the main valve shifts for reversing the pump, the exhaust port is covered at the same time the steam port is uncovered, so that during the return stroke of the piston the escape of live steam past the compression-regulating valves is as effectually prevented as from the cylinder itself.

Design of Railway Stationary Plants*

ANY plant having a capacity less than 1500 boiler horse-power rating cannot economically use water-tube boilers. The horizontal return-tubular boiler is best adapted for railroad stationary service in such plants. It is comparatively low in first cost, easy to operate and maintain, and in small plants is an economical steam producer when installed with a properly designed furnace.

These boilers should be suspended from lugs rather than set on them, the longitudinal seams should come well above the fire line, and transverse seams should be kept well back on the bridge wall. The units should be of such a size as to give flexibility and economy. The 72-in. x 18-ft. horizontal return-tubular boiler of 150 hp. rating meets all these requirements, but only if provided with a good setting. It should be set not less than 42 in. above the grates, and a combustion chamber built large enough to burn the gases. The bridge wall should be built with at least 14 in. clearance between it and the shell of the boiler, and built vertically downward to the combustion-chamber floor. The back wall should be at least 5 ft. from the end of the boiler, so as to allow plenty of room for the gases and for the workman when the boiler has to be retubed.

The flat suspended tile arch over the back end has given best satisfaction, as it relieves the side and back walls of outward thrust, which is common to the old arched type of construction. About 1½ in. clearance should be allowed between the arch and the rear end of the boiler when setting up, and this space filled with loose asbestos.

Build the side walls with plenty of clearance up to the closing-in point (2 in. above the top surface of the top rows of the flues), so that every square inch of water-heating surface will be effective. Put on a good cast-iron flush front and stay the setting with a few buck stays and through rods. Cover the brickwork with a plastic material that will retain its plastic properties under the action of heat.

MECHANICAL SOOT BLOWERS ARE FAVORED

Owing to the disagreeable task of keeping the boilers free from soot by means of the hand lances, and because on this account the work is not efficiently or frequently done, mechanical soot blowers are favored. Although the rear-end blower cleans the tubes by blowing with the draft, the temperature at the back end of the boiler is so high that it will burn out in time, and the boiler has to be shut down to repair it. On the other hand the front-end blower can be installed much nearer the flues, and the temperature at this point should be sufficiently low to avoid burning out the blower, and in case of trouble it is not necessary to shut down the boiler to make repairs. When installing mechanical blowers it is essential that the drain valve be placed after the main shutoff valve to be assured that no water is blown upon the flues.

Some success has been obtained by bricking up the combustion-chamber clean-out doors. The brick can be torn out easily at each cleaning of the combustion chamber and replaced with little expense, while infiltration of air is prevented, a uniform brick wall is presented to the impact of the gases, and they are not cooled by radiation through a cast-iron door. Boilers should never be set more than two in a battery, as division walls are always a point of trouble and expense to keep up.

Where the load requirements do not exceed 50 to 100 boiler horsepower, a locomotive-type boiler or a vertical boiler may be used. Plenty of ashpit space should be provided to give room for the admission of air and provide storage space for a few hours' run of ashes. More grates and mud rings have been burned out because the ashes piled up against them, particularly on vertical boilers, than for any other reason. All boilers of these types should be properly lagged with good insulating material. It is estimated that for every 100 sq.ft. of radiating surface 50 lb. of coal can be saved every 24 hours by lagging with 1-in. thickness of magnesia. A 48-in. vertical boiler has about 100 sq.ft. of radiating surface, and a 36-in. vertical boiler about 85 sq.ft. of radiating surface.

*Excerpts from report of Committee on Railroad Stationary Power Plants at eleventh annual convention of the Association of Railway Electrical Engineers.

Unfortunately, in hand-fired railroad stationary plants the design of the furnace cannot be arranged for a particular kind of coal, but installations must be such as to burn with average economy any kind of coal that is delivered. This can be accomplished by setting the boilers high, building a roomy combustion chamber and installing suitable grates.

There are on the market today many different kinds of grates, good, bad, and indifferent. A grate is merely a rack to hold the coal during combustion. Necessarily, when the coal is burned the ash and residue must be cleaned from the grate, and further, the grate must allow the proper amount of air to pass up through it to continue the process of combustion. The air spaces must be small enough to prevent the dropping through of fine coal and yet not become clogged up with clinker and ashes. The grate must be of such a design that the maintenance cost will be low, and repairs easy to make. Many experienced stationary firemen do not like a dumping grate, as they would rather clean the fires through the fire door. A rocking grate is to be recommended provided the grate bars do not lift far enough up into the fuel bed to tear it up and mix the ash and coal together, which is bound to give trouble from clinkering.

STEEL STACKS BEST FOR SMALL PLANTS

Steel stacks for the smaller sized plants are a better investment than either a brick or concrete stack. There is much in favor of separate steel stacks for each boiler or for each battery of two boilers. In general it might be said that for plants having more than six boilers, it would be good design to use a common stack. In any case the stacks should be of sufficient height, for it is easier to reduce the draft with a damper than it is to increase it with a steam blower. Each boiler should be equipped with a damper and a device for controlling the damper, which latter need not be complicated.

In considering boiler fittings and pipings, it would be well to make all installations conform to the A. S. M. E. code, as sooner or later it will be adopted generally. Since service is of first consideration, the first cost of the fitting should not be given too much weight in the selection. Valves that are easily repaired and give the best service for all conditions should be selected, but usually such equipment is not the cheapest in first cost.

Boiler nozzles should be connected to steam headers by steam loops of easy bends and should enter the header at the top. Steam lines taken from the headers at various points of distribution should also connect to the header at the top with a loop of easy curvature. Blowoff valves should consist of a good grade of extra heavy cast iron, or cast steel. Two valves in series should always be used for blowoff service.

OPEN TYPE FEED WATER HEATERS

Open-type feed-water heaters should be used, as this type is easy to operate and maintain. The steam and water are brought into intimate contact, and the steam is condensed and pumped back to the boilers, so that the amount of scale-forming compounds is reduced. Excess of pressure in the heater can be taken care of by an overflow loop or a U-pipe bend connected to the heater at the overflow level. It extends vertically downward for six or eight feet, turning upward again to the overflow level, and then downward to discharge into a sewer connection. A vent at the top of the second bend is necessary to prevent the loop siphoning all of the water out of the heater. To overcome difficulties with trapped air a ¾-in. or 1-in. pipe should be run directly out of the top of the heater and extended through the roof, with no valves or chances for water-pockets.

Besides the cold-water supply, provision should be made to take the discharge from steam traps, vacuum pumps and cooling water from the air compressors into the heater. Every lb. of condensed steam returned to the boilers reduces boiler-washing costs, and efforts should always be made to return all drips and drains to the heater.

Steam economy in boiler-feed pumps is a questionable saving in any railroad plant. The old duplex pump is always reliable and understood by everyone, and so far as plunger pumps for boiler feeding are concerned, the duplex is to be recommended

The ideal boiler feed is the centrifugal pump, which will some day entirely supersede the plunger pump for many classes of service.

The most economical coal-handling system for plants of this size is an overhead track, using dump-bottom cars discharging into a storage bin. The floor of the storage bin should be inclined so as to deliver the coal to the fireman in front of the boilers, without rehandling. In plants under 500-hp. rating, the wheelbarrow is about the most efficient ash-handling system. Steam-jet ash-handling systems may be considered in plants of greater capacity.

Air compressors for this size of plant should be of the two-stage engine-driven type. For compressors of 2000 cu.ft. capacity or over, the compound steam end is recommended.

Electrical generating equipment may be either reciprocating-engine-driven or steam-turbine-driven. For a railroad plant the low maintenance cost on turbines and the large amount of exhaust steam required for the winter heating season makes the high steam consumption of small turbines an advantage rather than otherwise. Where the generation is direct current, 250 volts should be used, and when alternating current, 440 volts three-phase is preferable. Alternating-current generation is recommended. The amount of direct current required is easily obtained through motor-generator sets.

An exhaust heating system is one of the most important and economical parts of those stationary plants that are located in the central and northern territories. It is the one reason for railroads generating their own power, even when a very attractive rate for power may be offered by central power stations. When included in the steam-using equipment, it raises the efficiency from 8 to 30 per cent., although when improperly installed or maintained it can become the source of more trouble and loss than anything else on the road. A well-designed exhaust-steam heating system will not require more than two to three pounds back pressure to operate, and will return condensed water to the boilers to the extent of over 50 per cent. of the total water required. It is well to bring the returns back to the power plant separately from each department so that trouble can be located quickly. All exhaust available should be discharged into the exhaust header. It has even been found that the vent pipe from the vacuum system can be connected to the exhaust header through reducing valves to good advantage. Vacuum pumps should be large enough to take care of the coldest season and should always be installed in duplicate.

Large-Sized Plants

A plant over 1000 boiler-horsepower rating can well afford the overhead cost of labor-saving devices, and some of the more refined power-plant equipment. The boilers should be of the improved safety water-tube type, of such size as to be flexible and yet sufficiently large to keep the idle investment charge down to a minimum. For the average heavy back-shop plant, the units should be of 200 to 350 rated boiler-horsepower capacity. Only the best material and workmen should be employed in the building of the boiler settings, as a poorly built system means serious losses in operation. If space is desired between the firebrick and common brick to allow for expansion, fill up the space with crushed cinders, fronts and clinkers, sand or other insulation. After the setting is built, it should be entirely covered with a plastic cement or other material that will remain plastic under temperatures to which the setting is subjected, and against time.

With this type of installation the question of soot blowers has not been satisfactorily answered. In plants of the size indicated usually, the force is large enough to keep the tubes well cleaned by the hand lance, and the supervision is sufficient to see that the tubes are kept cleaned.

Stokers are one of the labor-saving devices that a plant of 1000-hp. and over can well afford. The available coal supply will fix to a great extent the style of stoker. However, the chain or traveling grate seems to be well adapted to the Middle West territory, and has given good satisfaction in a large number of railroad plants. It is easy to operate and repair, and when the units are of such size as to give flexibility of operation, the inflexibility of the traveling grate is overbalanced by its advantages.

To supply draft, a well-built stack is always a good investment, as it can be depended upon at all times. Each boiler should be provided with a damper in the connection to the stack or the breeching, so that the boiler can be easily drafted for all conditions of load. These dampers should be provided with damper control that can be easily operated from the firing floor. Coal should flow to the stokers from overhead storage bunkers to which it is delivered from the railway car by either an overhead crane or a traveling bucket conveyor; with the latter the same conveyor can be used for disposing of the ashes. The ashpit should be located in the floor below the fireroom and should be of sufficient capacity to run at least twelve hours without cleaning. Other features of the plant have been covered in the discussion on the smaller station.

Fuel-Oil Shortage

Large refining companies are refusing to take any additional fuel-oil business. This is due to conditions in Mexico and shortage of transportation facilities, both ocean and land.

Oil production in Mexico, says the *Boston News Bureau*, which supplies most of the fuel oil, has been curtailed by the appearance of salt water in two of the large producing fields, necessitating the closing down of a number of wells. This situation could be corrected in time by bringing in new wells in other fields, were it not for the fact that the Carranza government has forced the cessation of drilling operations.

Delay in the delivery of steel for tank steamer and tank car construction is the reason attributed for the absence of adequate transportation facilities. One of the large independent companies has 150 tank cars on order on which deliveries should have been made two months ago.

This situation will undoubtedly operate against installation of oil-burning apparatus in the large buildings in New York City, which is now permissible under regulations approved by the Bureau of Standards and Appeals. Several of the large hotels and other buildings have deferred their intention of going ahead with the change of their apparatus.

All large companies say they will be able to make deliveries on their present contracts. This means there is not likely to be any interruption to fuel-oil supply for oil-burning ships and manufacturing plants in New England and other districts which now have contracts.

New Spanish Hydro-Electric Plant

The construction of a hydro-electric plant for the development of 7300 hp. has been started by the Compañía Andaluza [illegible] de Electricidad on the Guadalquivir River at [illegible], Province of Córdoba in Southern Spain.

The dam which is being started is expected to maintain a head of from 36 to 40 ft., from which can be obtained a minimum of 7300 hp. of electrical energy. Three turbines of 2500 hp. each will be installed. It is planned to connect these direct to three 5000-volt three-phase [illegible] generators. The voltage will be raised to [illegible] and [illegible] for transmission.

It is estimated that the construction will require over a year for completion. The Spanish engineer, Sr. Carlos Mendoza, is chief engineer of the project.

This is one of a number of such projects that are being planned along the Guadalquivir River. From the City of Córdoba, situated about [illegible] miles below the [illegible], to the City of Seville [illegible] are being [illegible] on a large scale and are being [illegible] by the Spanish government.

The Central Committee of France has set forth [illegible]

Standard Lubricating Oil Systems for Geared Turbines*

By J. EMILE SCHMELTZER and B. G. FERNALD

PRIOR to July 1, 1918, vessels propelled by geared turbines which had been designed and constructed for private owners had been delivered under requisition to the United States Shipping Board Emergency Fleet Corporation.

As operating troubles with geared propelling turbines were already being reported to an extent considered abnormal, and as many of these troubles were being attributed to faulty operation or failure of the lubricating oil system (in most instances the oiling system was not furnished by the builder of the turbine, but by the shipbuilder), it was decided to study all the oiling systems in use on American ships, and either to adopt or develop a standard oiling system for use on all geared turbine vessels then under construction or to be built by the U. S. Shipping Board Emergency Fleet Corporation.

A very small percentage of American marine engineers had had any previous operating experience with direct-connected marine turbines, and a negligible number had ever operated double reduction geared turbines, the type with which most of the vessels under construction were equipped.

It appeared that a standardized system would materially simplify the instruction of the new engineers. Lack of standardization, moreover, would have caused accidents and confusion, even with experienced engineers, as it was frequently necessary for engineering crews to go to sea on new vessels without first giving them time to familiarize themselves with the different piping systems on the vessels.

As a consequence of solicited criticisms, a revision of the system substantially according with the majority of the recommendations was made, and submitted again to a number of eminent marine engineers, and the final revised system adopted as a standard for vessels of the Emergency Fleet Corporation.

Coolers—Nearly all of the oil coolers previously used were inadequate to transfer the required number of British thermal units, which amounted in a number of instances to 10 per cent. of the propelling turbine horsepower. This was partly due to the fact that the reduction-gear manufacturers had overrated the efficiency of their reduction gears.

As a precautionary measure, it was considered essential that an auxiliary cooler be provided so that the oil could be cooled by one cooler in the event of the other being rendered inoperative by leaky tubes or other causes. The auxiliary cooler was also used in parallel with the main cooler when operating in tropical waters, where a cooling water temperature of 84 degrees or higher obtained. The technical order specified a 20-degree drop in oil temperature, and, while it was realized that the cooler would not be called upon to equal this temperature drop under normal running conditions, still this size of cooler provided for a considerable drop in the efficiency of the cooler, due to the fact that the tubes would in time become covered with a deposit which would materially decrease the efficiency as compared with a clean one.

Results obtained in operation have demonstrated that the size of cooler specified satisfies average operating conditions.

Gravity Tanks, Size and Location of—During the time the designs of the new system were in the course of preparation, one manufacturer claimed that it was necessary to have 12 lb. oil pressure on the turbine bearings and gear oiling devices on equipment manufactured by them, and still another used spray nozzles of such size as to make it almost impossible to maintain a higher pressure than 4 or 5 lb. The rest of the builders' requirements averaged about 10 lb. pressure at the oil manifold.

Due to the above facts, and based on the assumption that it was best to satisfy the higher pressure requirements, and also owing to the fact that all of the vessels on which installation of the system was contemplated would permit of the location, it was specified that the gravity tanks be located at a height of 30 ft. above center line of turbine in the engine-room casing. This location of the gravity tank has proved satisfactory in service, and no trouble has been experienced in securing the required pressure.

The size of the gravity tank, which is of 600 gal. capacity, was based on having three or four minutes reserve supply of oil available for the gears in the event of the oil supply to the gravity tanks being interrupted. This, it was thought, would provide sufficient time for action on the part of the operating personnel.

Kingsbury Thrust Bearing—Inasmuch as, on most of the various types of turbines purchased by the U. S. Shipping Board, the Kingsbury or similar type of thrust bearing is used, it is essential that there be a constant supply of cool, clean oil. Failure or interruption of the supply has caused the thrust bearing to burn out on several occasions, with consequent serious damage to the turbine.

This same type of thrust bearing was employed in the majority of cases for the main propelling shaft thrust, and therefore the same necessity for cool, clean oil obtained in this case.

Strainers—As a precautionary measure against sabotage, a twin suction strainer having a coarse mesh strainer basket was installed, but eventually omitted in the revised specifications at the end of the war.

A strainer having approximately 1/64-in. mesh was placed in the discharge line, and this is considered sufficient to remove all sediment incidental to ordinary operation after the system has been thoroughly cleaned before the trial trip as specified.

Filters and Separators—While all of the filters and separators specified are satisfactory for the purpose, it is considered that the results obtained from the centrifugal type of separator show greater efficiency, inasmuch as the oil, water and sediment are separated—the oil flowing from one nozzle, the water from another, the overflow mixture from another, and the sediment remaining in the bowl, from which it may be removed at will. Tests made at Annapolis prove that the rolling or pitching of a vessel will have no ill effects on its successful operation.

Alarm Systems—An electric alarm system actuated by a float switch, connected to the gravity tanks was provided (and since omitted), which would notify the chief engineer in his room and the engineer officer on watch, by means of an electric gong, that the oil level had dropped below normal operating level. This system, which was so designed as to not only give the necessary notification in case of oil level becoming low, but also to automatically increase the speed of the pumps, was later omitted, due to a desire to reduce the cost of the installation rather than because of any belief in a lack of necessity for it.

Drain Tank—The capacity of the oil drain tank was required to be from 800 to 1000 gal. for a 3000-hp. double-reduction gear turbine unit.

It was found that most of the drain tanks supplied at the time this system was originated were too small, and resulted in the pumps becoming vapor bound. In most cases of this kind, and also where the suction lift of the pump was high, the oil became dark in color. Oil experts who were consulted advised that the mixture of air with hot oil frequently resulted in the presence of sulphurous acid (H_2SO_3) in the oil. For this reason, the drain tanks were made sufficiently large and the pumps placed as low as possible to insure a short suction lift.

Pumps—While vertical pumps are more desirable for the purpose of securing low suction lift, it was not possible to procure them in the quantities required. Horizontal pumps were therefore used more extensively.

*Abstract of a paper read at the twenty-seventh general meeting of the Society of Naval Architects and Marine Engineers, held in New York, November, 1919.

Oils—Originally an oil having a viscosity of 300 sec. (Saybolt) at 100 deg. F. was specified, but later this was changed to 500 sec. viscosity. This latter viscosity was found to suit conditions better because the high-speed bearings demand oil having a low viscosity and the gears and slow-speed bearings a high viscosity oil. Inasmuch as two separate systems could not be considered on account of their cost, and due to the fact that oils ranging from 300 to 700 viscosity at 100 deg. have practically the same viscosity at temperatures above 140 deg., and also based on operating results, it was considered that the higher viscosity was more desirable.

The results obtained from the system in service have been very gratifying and show that the sizes of equipment specified were in keeping with the requirements, and it is believed that a system of lesser magnitude than that specified would endanger the successful operation of a double reduction gear turbine unit of 3000 hp.

There is no doubt that an adequate lubricating oil system is essential to the successful operation of the geared turbine, and while its maintenance cost is small, the consumption of oil through leaks and other sources amounting to about half a barrel per month, the first cost of installing such a system is high. It is believed, however, that an increase in efficiency will be obtained in the reduction gears of the future, which will permit of a considerable reduction in the size, and consequently the cost of the lubricating oil system.

Lubricating Oils—Extended experience with various oils has shown the necessity of using a heavy oil to protect the low-speed train of double reduction gears against pitting and other deterioration. Heavy oils, however, are not well suited for high-speed bearings, particularly where the clearance between bearings and journals is small, and will produce high bearing temperature and increase the friction loss. As it is not practicable to use either oils of two different viscosities in the same system, or to install separate systems for the gears and bearings, the only practical compromise is to increase the viscosity as much as possible without heating the pinion and turbine bearing. The viscosity of the oil at the point of entering the bearings can be controlled over a wide range by regulating the amount of water supplied to the oil coolers.

In the tropics, it may be necessary to use both coolers to keep the temperature down, and in northern waters, in the winter, it may be necessary to throttle the water supply to the coolers or cut it out entirely.

The general rule is to keep the oil as cool as possible without overheating the high-speed pinion bearings, paying also special attention to the bearing and thrust at the forward end of turbines.

The temperature may rise to and operate at 160 deg. F. without danger, with a good supply of oil and careful watching.

It cannot be too clearly emphasized how necessary it is to keep the oil pure and clear of water, by using the test cocks provided to denote its presence, by consistently heating and settling oil when the presence of water has been detected, and by keeping the oil pressure at coolers at all times above water pressure. Every opportunity should be taken to remove all sediment from the oil both by heating and settling, and strainer baskets should always be kept clean and filter cloths regularly renewed. If this is systematically carried out, the oil should last and remain perfectly lubricant indefinitely and make-up oil reduced to minimum.

By agreement with the Division of Operations, the following revised list of approved oils has been adopted for use on all installations:

Brand.	*Maker.*	*Viscosity of 100° F.*
D. T. E. extra heavy	Vacuum Oil Co.	525
Calol extra heavy	Standard Oil Co. of Cal.	700
Ursa	Texas Company	750
Algol	Texas Company	510

These brands have been selected with reference to distribution facilities of the manufacturers, as well as suitability of the oil.

Should other oils be approved from time to time after satisfactorily passing the required tests and after having proved by actual operations that they are suitable for the purpose intended, they will be added to the above list and all concerned will be duly notified.

Under ordinary conditions, the viscosity should be maintained at 450 sec. to 500 sec. Saybolt by keeping the temperature of the oil leaving the cooler at the figures shown below. and which are for 450 sec. and 500 sec. Saybolt viscosity respectively for the different brands.

Brand	Saybolt viscosity *Seconds*	Approximate degrees F.	Saybolt viscosity *Seconds*	Approximate degrees F.
D. T. E. extra heavy..	500	102	450	106
Calol extra heavy......	500	110	450	113
Ursa	500	111	450	114
Algol	500	101	450	104

Air Pumps for Condensing Equipment*

By Frank R. Wheeler

During the last few years changes in power-plant equipment have resulted in improved condensing apparatus which will maintain a vacuum of 29.5 in. referred to a 30-in. barometer, under proper operating conditions and with proper water temperature. This improvement is primarily due to the better design of air pumps and to the radical changes and innovations that have been introduced since 1914.

Condenser air pumps may be divided into two general classes, displacement pumps and impulse pumps. The former may further be divided between wet vacuum pumps, handling both the liquid and gas, and dry vacuum pumps, handling only gases. The impulse pumps may be classified as rotary or hurling-water pumps, and as ejectors, while the ejectors may be still further classified as water ejectors and steam ejectors.

Wet air pumps have proved their merits for more purposes and duties than is often realized and are today in many cases to be recommended and installed with a clear conscience. Improvements have been made in the better-known pumps of this type so that volumetric efficiencies have been materially improved and power consumption reduced. Succeeding the original bucket type with three valve decks, the Edwards air pump, the principle of which was described in 1845 by Bodmer, was introduced in 1900 by C. H. Wheeler and is in general the best-known pump of the wet vacuum type. The horizontal valveless pump has also been developed. On account of the distinct advantages of this type and particularly where the pump drive is combined with the circulating-water pump, the writer feels that the merits of these improved old-time designs are often overlooked, especially for units up to 1500 kilowatts.

Later the dry vacuum pump made with small clearances and handling air only, appeared, and this type has the same general advantages as the wet pump but with better volumetric efficiencies. Except for its complications, the number of moving parts, maintenance, size, etc., its use cannot be criticized. The increase in the size of turbines coming into general use, together with the continuous demand for higher vacuum, calls for dry air pumps of cumbersome size and excessive power, and the solution of this problem seems to point to an ejector air pump of some form.

The Hydraulic-Entrainment Air Pump

In 1862 Christian Schiels, of Oldham, England, invented a hydraulic-entrainment air pump, which expelled the air from a condenser by combining or entraining it with hurling water in its passage through a fan or centrifugal pump. The necessary water pressure was provided by means of a high-speed rotary device. This, together with the relatively small space required and the ability to maintain a high vacuum, are such desirable features that when the demand had exceeded the limitations of the dry air pump, development came very rapidly.

The other type of hydraulic-entrainment air pump is that in which the entrainment or compression of the air is performed in part of the apparatus entirely separate from that producing the pressure of the hurling or entraining water. These pumps follow the simple ejector principle using single or multiple jets of water discharging into a combining cone or diffuser and entraining the air due to the high velocity of the water

*Abstract from a paper presented at the Annual Meeting of the American Society of Mechanical Engineers, New York, December, 1919.

jet. The Koerting ejector is the best known of this type, the water being supplied from an exterior source and forced through a number of nozzles into a combining cone or diffuser.

In general, the power required by a hydraulic-entrainment air pump is 40 per cent. greater than that for a rotary dry air pump of the same relative capacity, and the air handling is subjected to strict limitations at moderate vacuums; consequently it cannot be counted on for overloads or unusual demands. The former is due to the power required to move the necessary hurling water against a discharge head, and the latter to the limiting weight of air that can be entrained by a quantity of water. The volumetric efficiency of the dry air pump falls off rapidly at high vacuum, owing to the rarefied air to be handled in a fixed cylinder volume, through fixed port areas, and to the detrimental effect of even the smallest clearances. The hydraulic pump is not so seriously affected; in fact, it improves in capacity at high vacuum.

The hydraulic-entrainment type of pump has given satisfaction in removing air and vapors from the condensing chamber and in maintaining as low a pressure as is consistent with circulating-water temperature. On account of the simplicity of the apparatus when compared with the rotary vacuum air pump, the admitted high power consumption has been accepted as a necessary evil, as has the difficulty of erosion.

The steam-ejector principle is an old one, but for high-vacuum service the ejector pump has not been in general commercial use in this country until the last three years, but it is now a definite engineering and commercial success. The reason for its adoption may be considered by reviewing the advantages claimed for it; namely, extreme simplicity, no maintenance or attention, reliability and flexibility, stability, minimum space and waste, low steam consumption with high efficiency. The first advantage enumerated, simplicity, while the most important, may be passed over as obvious. The second advantage results from the fact that there are no moving parts to wear out. The parts subjected to wear are steam nozzles and diffusers, and the medium used is the same as employed in the turbine; therefore the life of the pump, so far as its wearing qualities are concerned, should be as long as that of the nozzles of the turbine in service. Because there are no moving parts to oil or adjust, the only attention required is that of opening and closing valves controlling the pump.

Making Provisions for Leakages

As is well known, the air leakage varies greatly and provision must be made for leakages largely in excess of those expected with good operation. With other types of air pumps that are now being considered, it is usual to provide a pump large enough for maximum requirements and to operate it at reduced speed or displacement. With the ejector type, on account of its small size, weight and space requirement, the total estimated air-removal capacity can be provided by two or more units, one or more of which can be operated as required, thereby providing maximum efficiency and insurance against a shutdown.

As to stability, sufficient evidence in the form of test data is at hand to uphold this claim. The advantages in the line of space and weight are obvious.

The sixth claim requires the most consideration because it leads the list of disadvantages put forth by the few who have opposed the use of ejectors. Based on wasting all the steam used by the air pump, it is freely admitted that the ejector requires more steam than the rotary dry vacuum or wet vacuum pump. If there is a case where all the steam used by the air pump cannot be used in heating feed water, intercondensers can be used between the stages, resulting in a steam requirement less than that of the rotary dry vacuum or wet air pump. Excepting for the heat loss in radiation, the whole of the heat in the steam is available for heating boiler-feed water, and this arrangement will give a satisfactory heat balance. As a general rule plants for ordinary size ejectors without intercondensers require less than 1 per cent. of the total steam consumption for surface condensers and from 2½ to 3 per cent. for jet condensers. For intercondenser ejectors the percentages are approximately one-half this value.

The selection of air-pump capacity and size has always been shrouded with considerable mystery, and the methods to employ should be given more publicity. Air is admitted into the condensing chamber in three ways: First, from leaks into the system under vacuum; second, from air entrained in the boiler-feed water and carried over with the steam; and third, from entrained air in the ejection water, the third source of course being limited to jet or barometric condensers. The earliest methods of selecting air-pump capacity assumed that the displacement of the pump should be a function of the low-pressure engine cylinder displacement. Later, the rule making the displacement a function of the volume of condensed steam was generally used and proved to be satisfactory. The introduction of the ejector pumps required that some other method of selection be used.

For a number of years experiments and investigations have been made on the effect of air leakage on vacuum, heat transfer and so forth, all of which have indicated that the presence of only a small amount of air has a very detrimental effect. This caused large operators to introduce periodic tests of their

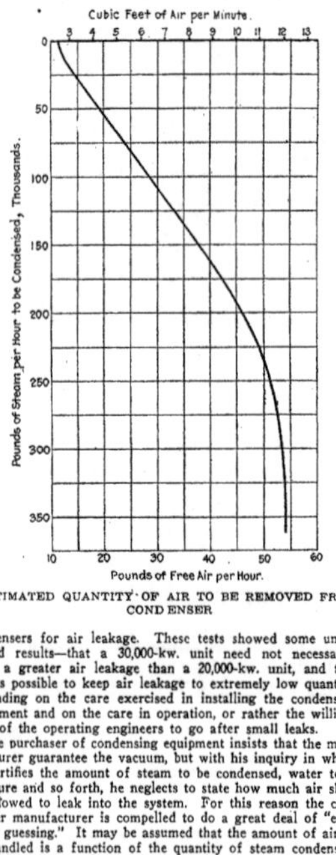

ESTIMATED QUANTITY OF AIR TO BE REMOVED FROM CONDENSER

condensers for air leakage. These tests showed some unexpected results—that a 30,000-kw. unit need not necessarily have a greater air leakage than a 20,000-kw. unit, and that it was possible to keep air leakage to extremely low quantity, depending on the care exercised in installing the condensing equipment and on the care in operation, or rather the willingness of the operating engineers to go after small leaks.

The purchaser of condensing equipment insists that the manufacturer guarantee the vacuum, but with his inquiry in which he certifies the amount of steam to be condensed, water temperature and so forth, he neglects to state how much air shall be allowed to leak into the system. For this reason the condenser manufacturer is compelled to do a great deal of "educated guessing." It may be assumed that the amount of air to be handled is a function of the quantity of steam condensed. In arriving at our figures, however, there are some very uncertain factors to be taken into consideration. Shall we assume the best operation possible or the operation of a shiftless attendant satisfied with a spring gage? Inasmuch as the manufacturer is interested in the prosperity of his customer, he assumes the former, and to be sure of his own guarantee, provides for the latter.

Based on a tight system and first-class operation, the amount of dry air leakage to be expected from the condensing equipment is shown in the accompanying curve. It should be noted that this is dry air and is less than the capacity to be provided for the air pump, because in passing through the condenser air becomes saturated and the air pumps must remove with the air an amount of water vapor dependent upon the pressure and the temperature of the mixture. It is to be understood that this chart is not from absolute data, but is simply an estimated quantity.

According to Dalton's Law, the total pressure in the condenser is the sum of the partial vapor pressure corresponding to the temperature of the mixture, taken from the steam and the partial air pressure due to the air or other gas present in the conductor. The volume of the air is directly proportional to its absolute temperature and inversely proportional to its pressure; therefore, if the mixture of air and water vapor in the condenser is cooled, the total pressure remaining the same, the vapor pressure is reduced and the partial air pressure increased, with the corresponding change in volume. The general problem may be considered as the determination of the total weight of saturated mixture which must be removed from the condenser to expel one pound of dry air. Now this capacity does not provide for emergencies, unusual operating conditions or excessive air leaks, and it is usual to select a larger pump and operate at reduced capacity, but the better way is to select two pumps of minimum size, as indicated from the capacity calculations. In this way a spare pump is provided as an insurance against shutdown, maximum economy of power is obtained, and the necessity of the use of both pumps is an automatic indication of excessive air leakage. The ejector, being so small and not requiring foundation, lends itself most admirably to this selection and can be used in two or as many more units as desirable, to provide for fractional loads and maximum economy.

In selecting an air pump for a jet or barometric condenser, capacity must also be provided for the entrained air and gases which enter with the injection water and immediately expand to the pressure being maintained in the condenser. Here we have a very uncertain item, for cold water will hold more air than hot water, and water flowing over leaves and vegetable matter will absorb CO_2. The air contents will be different in the water of a spray pond, a still lake, a slow moving river or a mountain stream. The quantity of air admitted into the condensing system with the injection water is greatly in excess of that entering with the steam and air leaks. A considerable quantity of air passes out of the system through the water-removal pump toward the tail pipe by entrainment. This latter quantity is sometimes as large as 40 per cent. of the total and seldom less than 20 per cent. It is, therefore, assumed that the air admitted with steam and by leaks in the system passes off with the injection water, and air-pump capacity is provided only for the air entering with the injection water.

Mr. Wheeler's paper also contains descriptions of a number of ejector air pumps, but as there is in course of preparation for *Power* an article that covers this ground, that part of the paper has been omitted in this abstract.

A. A. E. Organizes Ohio State Assembly

Representatives of all chapters of the American Association of Engineers in Ohio met in Columbus on Jan. 21 and formed the Ohio Assembly of the A. A. E. The objects of the state organization were defined "to initiate, coördinate and promote the interests and achievements of the A. A. E. in the State of Ohio."

It was tentatively decided that representation in the state assembly would be on the basis of one representative from each chapter having from 20 to 100 members, two from chapters having 101 to 500 members and three from those with from 501 to 1,000 members, and that voting power would be on the basis of units of 100 members represented. Thus the representative from a chapter having not over 100 members would have one vote, while the three representatives from a chapter having between 400 and 500 members would have five votes.

A committee was appointed to draft a constitution to submit to the chapters for discussion. Plans were discussed and an agreement reached that a state secretary should be employed in the near future and state headquarters opened in Columbus. Another development of the future is to bring together all members of the state in an annual get-together meeting, partly for social purposes and partly to discuss matters of common interest. Members and applicants of the A. A. E in Ohio now number about 1,500. Chapters are being formed at Canton, Youngstown, Newark and Marion. The existing chapters are Akron, Cleveland, Dayton, Cincinnati, Toledo, Ohio Valley, Steubenville, Ohio Northern, and Columbus.

Domestic Exports of Pumps and Pumping Machinery, to October, 1919

Countries	Dollars	Countries	Dollars
Belgium	21,904	Argentina	13,807
Denmark	2,342	Brazil	19,971
France	51,165	Chile	13,913
Greece	6,219	Colombia	1,796
Iceland and Faroe Islands	5	Ecuador	3,269
Italy	20,313	British Guiana	428
Netherlands	12,293	Peru	15,154
Norway	5,615	Uruguay	1,344
Portugal	946	Venezuela	7,584
Spain	8,550	China	13,985
Sweden	692	Japanese China	1,311
Switzerland	162	Chosen	92
England	34,808	British India	29,396
Scotland	330	Straits Settlements	2,839
British Honduras	12	Other British East Indies	7,233
Canada	102,320	Dutch East Indies	32,811
Costa Rica	2,792	Hongkong	2,248
Guatemala	1,313	Japan	5,675
Honduras	409	Russia in Asia	180,692
Nicaragua	424	Siam	1,529
Panama	767	Turkey in Asia	395
Salvador	310	Australia	9,605
Mexico	97,939	New Zealand	4,216
Miquelon, Langley, etc.	137	French Oceania	126
Newfoundland and Labrador	924	German Oceania	108
Barbados	152	Philippine Islands	26,654
Jamaica	696	Belgian Congo	3,506
Trinidad and Tobago	1,076	British West Africa	1,420
Other British West Indies	441	British South Africa	12,568
Cuba	55,193	Liberia	11
Danish West Indies	129	Portugese Africa	117
Dutch West Indies	1,855	Egypt	35
French West Indies	1,020		
Haiti	52	Total	852,394
Dominican Republic	5,251		

Exports, During October, 1919, of Air-Compressing and Refrigerating Machinery

Countries	Air-Compressing Machinery, Dollars	Refrigerating Machinery, Dollars
Belgium	404	......
France	119,313	......
Italy	2,667	......
Netherlands	59	......
Norway	4,126	......
Spain	12,510	26,564
Sweden	365	......
England	30,010	3,464
Scotland	3,842	......
Canada	36,592	27,355
Costa Rica	.....	217
Honduras	99	......
Nicaragua	25	......
Panama	.....	350
Mexico	30,977	65
Newfoundland and Labrador	58	2,012
Cuba	1,153	15,222
Dominican Republic	4,885	......
Argentina	1,718	8,148
Brazil	3,232	1,510
Uruguay	.....	64,077
Chile	18,246	......
Peru	3,629	2,370
Ecuador	.....	95
Venezuela	740	27
China	60	260
Japanese China	.....	17,052
British India	12,032	2,228
Dutch East Indies	4,035	......
Japan	38,140	1,261
Chosen	.....	15
Australia	3,978	......
Hongkong	.....	3,910
New Zealand	5,716	4,051
Philippine Islands	1,568	......
British South Africa	3,449	17,664
Total	343,228	197,917

Comparative Evaporative Values of Coal and Oil*

The power plant engineer in the East should be particularly interested in the calculations dealing with the comparative evaporative values of coal.

IN COMPARING the evaporative values of the two fuels, coal and oil, it is fair to select a coal that is considered standard and is largely used in evaporative tests. It would not be just to select a coal of the inferior value of Western coals, and Pocahontas coals and Kern (California) crude oil are therefore selected as a basis of comparison. These fuels have the following analyses:

	Pocahontas Coal, Per Cent.	Kern Crude Oil 14.5 Gravity, Per Cent.
Proximate Analysis		
Fixed carbon	73.30	.....
Volatile matter	17.61	.....
Moisture	0.49	0.15
Ash	8.60	.....
	100	
Ultimate Analysis		
Carbon	82.26	87.64
Hydrogen	3.89	10.48
Sulphur	0.49	1.02
Oxygen	4.12	0.08
Nitrogen	0.64	0.78
Ash	8.60	.....
	100	100
Heating value, B.t.u.	14,067	18,619

Assuming the complete combustion of the heat values in the two fuels as analyzed, and knowing that 1 lb. of carbon requires 11.6 lb. of air, 1 lb. of hydrogen requires 34.8 lb. of air, 1 lb. of sulphur requires 4.3, we have as the volume of air chemically required for the combustion of the coal and the oil the result shown in the following table:

VOLUME OF CHEMICALLY REQUIRED AIR FOR COMBUSTION OF COAL AND OIL†

	Weight of Air Required. Coal	Oil
Carbon		
Coal, 11.6 lb. air × 0.8226 lb. C	9.54	.....
Oil, 11.6 lb. air × 0.8764 lb. C	....	10.16
Hydrogen		
Coal, 34.8 lb. air × 0.0389 lb. H	1.35	.....
Oil, 34.8 lb. air × 0.104 lb. H	....	3.62
Sulphur		
Coal, 4.3 lb. air × 0.0049 lb. S	0.02	.....
Oil, 4.3 lb. air × 0.010 lb. S	....	0.04
Pounds of air required	10.91	13.82
Weight of combustibles (ash deducted from coal)	0.91	1.00
Total weight dry chimney gases, lb.	11.82	14.82

For comparison assume that the flue-gas temperature is the same in both cases, 500 deg. F.; moisture in the two fuels is the same; air-intake temperature (temperature of the room), 70 deg. F. The flue temperature is placed at 500 deg. F., which is probably the mean which obtains in practice.

Moisture as shown in the selected coal is low, and that in the selected oil also low. In most oils it is more nearly an average of 0.5, and for this reason we will consider that the moisture content is the same in this case.

The sensible heat carried away in the flue gases in the case of the coal is $0.24\ (500 - 70) \times 11.82 = 1219.8$ B.t.u.; in the oil, $0.24\ (500 - 70) \times 14.82 = 1529.4$ B.t.u.; 0.24 being the specific heat of the chimney gases.

In the combustion of hydrogen to water each pound of hydrogen results in 9 lb. of water, or in the coal the water formed will be $0.038 \times 9 = 0.34$ lb.

The heat lost, both latent and sensible, in the evaporation of this water is

$$0.34\ (142 + 966 + 288 \times 0.47) = 422.6 \text{ B.t.u.},$$

wherein 212 deg. — 70 deg. = 142 deg., assuming the temperature of the water in the coal to be that of the fireroom; 966 being the latent heat of vaporization of steam at 212 deg. F. 500 deg. — 212 deg. = 288 deg., being the difference between steam at 212 and the flue temperature as assumed; and 0.47 being the specific heat of superheated steam at atmospheric pressure.

In case of oil we have

$$0.104 \times 9 = 0.936 \text{ lb.}$$

The heat loss is

$$0.936\ (62 + 966 + 288 \times 0.47) = 1088.5 \text{ B.t.u.}$$

The loss in coal due to ash is 8.6 per cent. of the total, equaling 1209.7 B.t.u. A further loss in oil combustion is the heat absorbed by the steam used in atomizing, being in superheating the steam from 212 to 500 deg. F. This is

$$288 \times 0.47 = 135. \text{ B.t.u.}$$

Tabulating these losses, which for convenience have been called fixed losses, we have:

	Coal B.t.u.		Oil B.t.u.
Gases	1219.8		1529.4
Combustion of hydrogen	422.6		1088.5
Ash	1209.7	Burner	135.0
	2852.1		2752.9

While this calculation is not intended to convey any idea of the actual value of losses in the two fuels, it may afford a comparison of what we may term fixed losses within certain limits.

The amount of air required for the combustion of coal is much less than for oil, due principally to the much greater hydrogen content in the oil, the carbon content being, roughly speaking, the same. Furthermore, the combustion of the hydrogen requires a proportionately larger weight of air.

On the other hand, the greatest loss in the coal is in the ash, as in the comparison of the fuels we must consider that as received. In the oil the loss due to noncombustible is practically negligible. The loss due to superheating the burner (atomizing) steam is also small.

In examining the figures a remarkable equality of the total losses is immediately apparent. To what extent this relation would hold for coals of different composition would have to be determined. Subtracting these losses from the calorific value of each fuel we have:

$$18{,}619 - 2752.9 = 15{,}866.1 \text{ B.t.u.}$$
$$14{,}067 - 2852.1 = 11{,}214.9 \text{ B.t.u.}$$

These last qualities represent the available heat left in the fuel for evaporation of the water in the boiler. From this it would appear that the two fuels under examination had an evaporative value which is in the ratio of 11.2 for coal to 15.8 for oil. For coals equal in quality to that selected, the writer believes that this ratio will generally be found correct, lower-grade coals having correspondingly lower values. This ratio is also amply borne out in numerous actual tests.

It is shown in the foregoing that the theoretically required air supply is greater for oil than for coal. This means a greater volume of flue gases, therefore a greater loss in heat. If we consider, however, those coals having a large amount of volatile matter, it is safe to say that the air supply for their complete combustion will be much in excess of that for oil. Not knowing the composition of the volatile matter, we must necessarily figure with a certain indefinite quantity of excess air. In the combustion of oil we are able to supply more nearly that amount of air which is actually chemically required.

From the manner in which the oil is burned, it is evident that a more perfect mixture of the air with the gases is obtained. Not only is the air supply easily regulated by hand, but apparatus has been devised to effect the control mechanically and automatically to suit the momentary load demands. Although it does not necessarily follow that a smokeless chimney signifies complete or most economical combustion, it does mean much to neighbors, and one can safely say that such a chimney is more often seen in connection with an oil furnace than with one burning coal.

*Given by C. F. Wieland, of San Francisco, Transactions A. S. M. E., Vol. 33, 1911, p. 32, but considered to be particularly welcome to power-plant men at this time.

†In various textbooks the values given range from 16 to 18 lb. of air per pound of oil, but an average of 14 lb. is more nearly correct.—C. R. Weymouth.

The Diesel Engine

At a regular monthly meeting of the Engineers' Society of Milwaukee, held on Jan. 14 under the auspices of the Milwaukee section of the American Society of Mechanical Engineers, Max Rotter, consulting engineer of the Busch-Sulzer Bros. Diesel Engine Co., of St. Louis, delivered a most interesting lecture on the company's product. That the topic is of exceptional interest at the present time and the speaker a recognized authority on the subject was evidenced by an attendance exceeding two hundred enthusiastic engineers. Mr. Rotter was so full of the subject that there was no need for a written paper, his talk being well illustrated by numerous lantern slides.

For the benefit of those not particularly familiar with the internal-combustion engine, in his introductory the speaker reviewed briefly the principles of operation, giving the events of each stroke and explaining why an air compression of about 500 lb. had been found most economical. Theoretically, the higher the compression the greater the efficiency, but the mechanical losses increase and the lines of heat and mechanical losses cross at a point ranging between 450 and 500 lb. The temperature at the beginning of the power stroke was about 1,000 deg., the feed being regulated so that the gas burned uniformly instead of exploding as in the usual type of internal-combustion engine. Consequently the Diesel was a constant-pressure engine as distinguished from the constant-volume engine, such as the semi-Diesel or the gas engine.

In the semi-Diesel engine compression was carried to only one-half of that in the Diesel engine and some part of the cylinder in or near the head retained heat enough to ignite the fuel. This required a special form of ignition chamber, which was unnecessary in the Diesel engine.

As compression with the Diesel engine consumed from 8 to 10 per cent of the power developed, it was only logical to try to get rid of this loss. In this connection reference was made to the solid injection Vickers engine made in England, in which no compressed air was used, the fuel being forced directly into the cylinder. Unfortunately, this engine was not adaptable to all kinds of fuel and it required a special combustion chamber. It was the speaker's opinion that eventually this problem would be solved, but he did not think atomizing the oil mechanically without the air pressure would be a success. The compressed air had some other function and effect on the results than the mere mechanical action of forcing the oil into the cylinder.

By means of simple diagrammatic views the events of each cycle in two-stroke- and four-stroke-cycle Diesel engines were outlined. In the former particular reference was made to the valves in the head for scavenging or blowing the burned gases out of the cylinder.

Fuels for the Diesel Engine

As to fuels the best information on the subject was the Haas bulletin, published by the Bureau of Mines. Any liquid containing hydrocarbons could be burned in a Diesel engine, but all could not be burned with equal success. Crude petroleum and distillates were of course the fuels generally used, but there were instances where Diesel engines had been run on gashouse tar, fish oil, peanut oil, etc., but it was found that the byproducts were of more value than the use of such oils as fuel.

There was a great variation in petroleum. Some Mexican, Texas or California crudes were better than some Oklahoma distillates. For commercial purposes the viscosity of the fuel must be such that the oil will flow to the pump by gravity and from there be pumped to the cylinder. The serious limitations were the ash and sulphur. The former stays in the cylinder and grinds up the walls, while the sulphur forms sulphuric acid to corrode and pit the metal.

In the laboratory at the present time there is no reliable method for determining the suitability of an oil as fuel for the Diesel engine. Temperature and time are the two things they decide by. A sample of oil is heated in a test tube to 300 deg. and is maintained at this temperature for three hours. As a general rule this method will give good indications of the suitability of the fuel, but it is not infallible. The only sure way is to test the oil in the engine cylinder. Standard methods of determining asphaltum were numerous, but none was satisfactory. The chemical composition of the residue that enables it to be consumed in the engine is what counts. Physical tests, not analyses, are the thing. The name or brand of a fuel oil is no guarantee. Owing to variations every car must be tested. As previously explained it is the residue after distillation that counts, and this varies with each supply.

As to the fuels of the future, it will be remembered that the Diesel engine is not limited to petroleum. Because of the great increase in the use of automobiles, trucks, tractors, etc., the tendency will be to increase the supply of gasoline and lubricating oils. Lower down the scale there will be vast quantities of fuel oil that will be a drug on the market. This may be used with good satisfaction in the Diesel engine.

It had been the dream of Doctor Diesel to use tar oil recovered in the distillation of coal. This oil makes a good grade of fuel, but some of it is hard to ignite and it may be necessary to inject a small fraction of a lighter oil to start the ignition and raise the temperature sufficiently to insure prompt gasification and burning of the heavier fuel. This is easily done.

On the average, fuel oil contained about 18,800 B.t.u. per lb., 143,000 B.t.u. per gal. or 6,000,000 B.t.u. per bbl. On the bulk basis there was very little variation. The heat content in a gallon of oil was always so close to 143,000 B.t.u. that this figure may be accepted practically as standard.

High Efficiency Over Wide Range of Load

When it came to efficiency, no other method of generating power could approach the Diesel engine. The efficiency of the steam plant was lower and varied more with operating conditions, stand-by periods, variation of load, etc. Numerous curves were presented comparing the economy of small steam plants containing simple noncondensing units, compound condensing Corliss engines and steam turbines, with a Diesel-engine plant. No mention was made of the uniflow engine of recent design. The Diesel engine curve showing the steam rate at different loads was almost flat. It was approached by the steam turbine curve, but not by any other prime mover.

Including auxiliaries the average compound condensing Corliss engine plant required nearly 3 lb. of coal per kilowatt-hour or about 36,000 B.t.u. as compared to an average between 25 per cent and 100 per cent load of 10,300 B.t.u. for the Diesel engine, the fuel consumption of the former being 3½ times greater. In pumping oil by a steam plant burning oil or by a Diesel engine the ratio of fuel consumption approximated 3 to 1.

In marine work the flat curve of the Diesel engine was of great value. Curves were thrown on the screen showing the shaft horsepower versus the speed. The steam-engine curve rose more rapidly than the curve of the Diesel engine. In comparing fuel consumption the triple-expansion steam engine was credited with 1.2 lb. of coal or 0.93 to 1.23 lb. of oil per shaft horsepower as compared to 0.53 to 0.47 lb. of oil for the Diesel engine at three-quarter speed. In the steam plant the speaker went on to say that deficiencies did not always make themselves evident, so that there was always the prospect of operating considerably below maximum economy. In the Diesel engine any defect would be made known at once.

Heat from the exhaust gases was being used to some extent to heat the domestic water supply. As the temperature at the end of combustion was about 3,000 deg. F., water cooling was a necessity. The amount of heat carried off in the jacket water was almost equal to the heat converted into useful work. Owing to the severe service it was important to use the proper grade of cast iron. In certain kinds of iron considerable growth was known to take place. This could be avoided by careful selection of the metal.

As a climax to a most interesting talk the speaker described and thoroughly illustrated two of the latest designs of engine being developed by the company. One was a four-cylinder, two-stroke-cycle, slow-speed Diesel engine adaptable to land or marine work, weighing 300 lb. per horsepower, and the other a high-speed marine engine weighing 50 lb. per horsepower. In the unit first mentioned the scavenging pump and air compressor were integral with the engine. The tendency now in large units was to go over to the electric-driven blower for furnishing scavenging air, although the efficiency was somewhat less. The scavenging pump shown was of the tandem piston type using 8 to 10 per cent of the useful work of the engine. For this reason the two-stroke-cycle engine is less efficient than the four-stroke-cycle engine. If the amount of excess air that it is necessary to pump through the cylinder could be reduced and the pressure lowered, the efficiency of the two-stroke-cycle engine would be improved. In the later types of engine the scavenging pressure had been brought down to 3 lb., as compared to 6 or 7 lb. previously used, and its efficiency approached that of the four-stroke-cycle engine within 5 or 6 per cent. At underloads the disability of the two-stroke-cycle engine became less until at 20 per cent load the efficiencies of the two engines were equal. In the engine cylinder scavenging through valves in the head or through ports and a receiver at the side of the cylinder was pictured.

Other features brought out were an ordinary fuel valve and the starting valve located in one cage; also a timed starting valve supplemented by a compression relief valve to relieve the compression during the starting of the engine.

The question of atomizing was not so sacred as it was generally supposed to be. The speaker had tried numerous plans, but the results were all about the same, and he always came back to the usual method.

Curves giving the results of tests on a 1250-hp. two-stroke-cycle slow-speed heavy engine showed a fuel consumption of 0.418 lb. per brake horsepower-hour from full load to three-quarter load.

A brief description, with numerous slides, of the light high-speed marine engine that had been an important factor in subduing the submarine peril, followed. The engine had to be built so that everything could be removed from the bottom. The pistons were oil-cooled with the same kind of oil used for lubrication to avoid the necessity of using salt water for this purpose.

A luncheon that was much appreciated and generally enjoyed closed a most profitable evening.

Chicago Section A. S. M. E. Discusses Lubrication

On Jan. 23 at the Morrison Hotel, the Chicago section of the American Society of Mechanical Engineers held its second meeting of the season. An all-day blizzard reduced the attendance to about 100, but those who had the courage to venture out enjoyed a very profitable meeting. Preliminary to the subject of the evening, Arthur L. Rice, chairman of the section, reported the status of the Jones-Reavis bill which proposes to reorganize the Department of the Interior, change its name to the Department of Public Works, assemble in this department all engineering bureaus having anything to do with construction work of the Government and eliminating those bureaus that would be foreign to this work. The bill is still in committee, but is soon to be called out. At the present time two things are needed. One is as much publicity on the bill as possible so that the voters in the various states will appreciate the merits of the bill and favorably report it to their representatives at Washington. The second essential is money to carry on this publicity work. The state of Illinois still requires $2000 and it was intimated that a request for subscriptions would be made at an early date. It was the first time that the engineering profession as a whole had attempted to put over a great big thing of this character which would be of great benefit to the Government and to the country as a whole. Each engineer should take an active part in the movement and show what can be done by concentrated effort.

Lewis M. Ellison, chairman of the membership committee for the section, spoke of the drive that was being made for affiliate members. He was enthusiastic as to the possibilities, and in the matter of securing prospects he bespoke the aid of every member of the section.

William F. Parish, recently chief of the oil and lubrication branch, U. S. Air Service and now connected with the Sinclair Refining Company, gave a very interesting address on the proper balancing of fuel, lubricant and motor. According to the speaker, dilution of motor lubricating oil had always existed. It is caused by the gasoline mixture escaping past the piston rings and perhaps by drainage of the diluted oil from the cylinder walls. In the early years of highly volatile gasoline, the gas escaping past the piston rings did not enter extensively into the body of the lubricating oil and while mechanical losses of fuel took place in all four-cycle engines, this loss did not reflect itself in a problem of dilution. In the past few years, due to the heavy end components of the fuel, a larger amount after passing the rings, upon condensation, remained in the lubricating oil, thus creating a condition of dilution. Not until the period of high end point gasoline did the matter of fuel loss become apparent and then only in proportion to the amount held in solution in the lubricating oil. This condition, which has become especially accentuated in the past three or four years, has resulted in undue wear upon the engine parts, as well as affecting the use of fuel, lubricant and power. In the paper, Mr. Parish called attention to this underlying trend in the situation and suggested a method of application of motor oils to correct some of the present faults. It was his opinion that a true solution could be reached only by changes in mechanical equipment of the engine. He made a strong plea for better engine design and greater economy in the use of gasoline and lubricating oils. Numerous slides were presented which contained in curve form the results of tests made on airplane engines and commercial trucks. In an early issue the paper will be presented in more complete form.

Following the speaker of the evening, Dr. Pogue referred to the change in volatility of gasoline, due to the enormous demand for this product. The oil people were constantly turning more of the crude petroleum into the field of fuel and less into fuel oil and the heavy oils for lubrication. In this movement the increasing use of motor trucks and tractors was an important factor, for the fuel consumption of either was about six to one as compared with an automobile. Consequently the consumption was going up by leaps and bounds and it required the closest co-operation between engine construction and fuel and lubrication supply to allow the oil age to go on unmolested. Coördination and coöperation on the technical side of the question was particularly important. An extended discussion indicated the interest that engineers are now taking in the fuel supply for internal combustion engines and in proper lubrication.

A. I. E. E. Midwinter Convention

The Eighth Midwinter Convention of the American Institute of Electrical Engineers will be held in New York City, Feb. 18-20, 1920. While the details of the program have not yet been completed, the tentative arrangements may be summarized as follows:

On Wednesday, Feb. 18, the morning will be devoted to registration. In the afternoon members will be permitted a choice between inspection trips to points of engineering interest or attendance at the joint session with the American Institute of Mining Engineers. In the evening the president will deliver his address and three technical papers will be presented: "Daylight Saving," by Preston S. Millar, manager, Electrical Testing Laboratories, New York; "Essential Statistics for General Comparison of Steam-Power Plant Performances," by W. S. Gorsuch, engineer economics, Interborough Rapid Transit Co., New York, and "Standard Graphic Symbols," by E. J. Cheney, chief of Division of Heat, Light and Power, Public Service Commission, Albany, N. Y.

The Thursday morning session will be devoted to a discussion of an economical supply of electric power for the industries and railroads of the northeast Atlantic seaboard. A symposium by a number of authors will be presented on the subject. In the afternoon two parallel sessions will be held. The papers to be presented are: "Printing Telegraph Systems," by J. G. Bell, telegraph engineer, Western Electric Co., and "Maximum Output Networks for Telephone Substation and Repeater Circuits," by G. A. Campbell, research engineer, American Telephone and Telegraph Co., and Ronald M. Foster, assistant to research engineer, American Telephone and Telegraph Co.; "A Method of Separating No-Load Losses in Electrical Machinery," by C. J. Fechheimer, designing engineer, Westinghouse Electric and Manufacturing Co.; "Inherent Regulation of Direct-Current Circuits," by A. L. Ellis, Thomson Laboratory, General Electric Co.; "Measurements of Projectile Velocities," by P. E. Klopsteg, physicist, Leeds & Northrup Co., and A. L. Loomis, formerly major Ordnance Department.

Thursday evening is at the disposal of the Entertainment Committee.

At the Friday morning session three papers will be presented: "A New Form of Vibration Galvanometer," by P. G. Agnew, physicist, United States Bureau of Standards; "Precision Galvanometer for Measuring Thermo, Em.F.'s," by T. R. Harrison and P. D. Foote; and "Notes on Synchronous Commutators," by J. B. Whitehead, professor of electrical engineering, Johns Hopkins University, and T. Isshiki, engineer, Shibaura Engineering Works.

The last technical session will be held on Friday afternoon, at which two papers will be presented: "Oscillographs and Their Tests," by A. E. Kennelly, professor of electrical engineering, Harvard University; R. N. Hunter, Engineering Department, American Telephone and Telegraph Co., and A. A. Prior, instructor Case School of Applied Science, and "The Accuracy of Commercial Measurements," by H. B. Brooks, physicist, United States Bureau of Standards.

The convention will close on Friday evening, Feb. 20, with a dinner-dance at the Hotel Astor.

At the November meeting of the Diesel Engine Users' Association particulars were given of tests carried out on connecting-rod bolts of a Diesel engine after having been in use for 23,500 hours, corresponding to 214,000,000 revolutions of the engine. Tests of ultimate tensile strength, elastic limit, elongation, reduction in area, and hardness carried out before annealing and then after annealing at 900 deg. C. showed that the annealing process had practically no effect on the material. The bolts, which were $1\frac{1}{8}$ inch at the smallest diameter, were apparently of wrought iron and were the original bolts supplied with the engine in 1913.

Obituary

Albert Schmid, consulting engineer of the Westinghouse Electric and Manufacturing Co., died in New York City on Dec. 31, 1919. Mr. Schmid was born in Zurich, Switzerland, in 1857, and received his education in that city. He was an engineer of international reputation. He began his career with the French Westinghouse Air Brake Co., and in the early 80's was invited by Mr. Westinghouse to come to America. When Mr. Westinghouse became interested in the Union Switch and Signal Company he engaged Mr. Schmid as his chief designer and engineer in that field. Shortly after, he was transferred to the newly created Westinghouse Electric Co., becoming its first chief engineer in 1886 and in 1896 its general superintendent. He was director general of the French Westinghouse Co., director of the English Westinghouse Electric Co., Ltd., president of the Campagnie des Lampes a Filament Metallique, of France, and had general supervision of the Westinghouse Lamp Co.'s interests abroad. He had remarkable inventive genius, of which the Patent Office records much evidence.

Dr. Richard Cockburn MacLaurin, president of the Massachusetts Institute of Technology, died at his home in Boston, on Jan. 15, of pneumonia. Dr. MacLaurin was born in Edinburgh, Scotland, in 1870. His early boyhood was spent in New Zealand and his preliminary education completed in Great Britain. He entered Cambridge University in 1892 and there took the degrees of Bachelor and Master of Arts, besides winning the coveted Smith prize for excellence in mathematics. Upon graduation he was elected Fellow of St. John's College. In 1898 he was appointed professor of mathematics in the University of New Zealand. Later he became a trustee of that university and took an active part in the organization of technical education in the colony. For four years he served as dean of the faculty of law. In 1898 Cambridge University conferred upon him the degree of doctor of science and in 1904 the degree of doctor of laws. He came to this country in 1907 to take the chair of mathematical physics at Columbia University, and a year later was made head of the department of physics. In 1908 he was chosen as president of the Massachusetts Institute of Technology. When he went to the Institute there were great financial difficulties to meet, but he was remarkably successful in interesting men of wealth, and as a result he received a donation of $50,000 from T. Coleman du Pont for the purchase of a new site in Cambridge. As a result of an interview he had with George Eastman, of Rochester, N. Y., benefactions amounting to eleven million dollars were made to the Institute. In July, 1918, Dr. MacLaurin was appointed national director of college war training, and he was head of the Students' Army Training Corps under the Committee of Education of the War Department. He was a member of the American Mathematical Society, American Physical Society, American Academy of Sciences and Arts, and the American Philosophical Society. He was a member of many clubs in this city and in Boston. He is survived by his widow and two children.

Major Edward T. Walsh, of Plainfield, N. J., died at his home on Jan. 13, in his forty-eighth year. Major Walsh was born in South Orange. His technical education was obtained at Pratt Institute, Brooklyn. He was for some years connected with the maintenance department of the New York and Brooklyn Bridge. Later he went with the Atlas Portland Cement Co., at Catasauqua, Penn., where he was associated with the mechanical department in the design of power-plant and building equipment. After two years with this company he formed a connection with the power-plant department of the Interborough Rapid Transit Co., where he remained throughout the construction period. He next went to Dayton, Ohio, where he became a construction engineer with the National Cash Register Co. and superintended the construction of several large factory buildings. From there he went to Philadelphia where he continued his work in the design and construction of industrial buildings. In 1909 Major Walsh became associated with Frederick A. Waldron, Plainfield, N. J., where he had since made his home. He remained with Mr. Waldron for several years, and one of his notable undertakings at this time was the construction of the large service station for the Edison Illuminating Co. of Boston, Mass. Shortly after the opening of the world war Major Walsh became chief engineer of the Canadian Car and Foundry Co., of New York, which handled a number of large shell contracts for the Russian government. After this he spent nearly a year with the Maxim Munitions Corp. This experience proved of great value when the United States entered the war and he was given the commission of major in the Ordnance Department. He served during the war in some of the most active divisions of the work and following the armistice became connected with the salvage board of the Western New England district. At the time of his death he was chairman of this salvage board. He leaves a wife, a daughter and a son.

Personals

George A. Deibert, who has been assistant supervisor of safety for the Southern region with the United States Railroad Administration, has resigned to accept a position as director of public safety for the Georgia Railway and Power Co.

Arthur Elliot Allen has been appointed district manager at New York for the Westinghouse Electric and Manufacturing Co. to succeed Edward D. Kilburn, who has been elected vice president and general manager of the Westinghouse Electric International Co. Mr. Allen has been connected with the Westinghouse Electric and Manufacturing Co. since 1902, in various responsible positions.

John B. Miller, chairman of the Southern California Edison Co., of Los Angeles, has accepted the invitation of R. H. Ballard, president of the National Electric Light Association, to act as chairman of the convention committee of the national organization, which is to be held in Pasadena, Calif., May 18-22. The convention committee will have general supervision over all of the other committees.

F. L. Collins has been appointed power and electrical engineer of Freyn, Brassert & Co., Chicago, Ill. Mr. Collins was formerly connected with the Illinois Steel Co. as electrical engineer, later joining the Dominion Iron and Steel Co., Sydney, N. S., as chief electrical engineer. He was subsequently connected with the ordnance department of the United States Steel Corporation at Neville Island, and until recently was chief electrical engineer of the Pittsburgh Crucible Steel Co.

Engineering Affairs

The Association of Iron and Steel Electrical Engineers, Philadelphia Section, will hold a meeting at the Engineers' Club on Feb. 7. R. B. Gerhardt will talk on the "Electrical Features of a Modern Mill."

Business Items

The Fairbanks Co., New York City, has signed a contract with the Lincoln Electric Co., of Cleveland, which gives the former company the exclusive distribution of Lincoln electric motors for industrial applications. The Fairbanks Co. will also coöperate with the various Lincoln district offices in connection with the sale of the manufacturer's other products.

The Black & Decker Manufacturing Co. has further extended its organization by the establishment of a branch office at 6523 Euclid Ave., Cleveland, Ohio, with Garth A. Dodge, formerly connected with the Austin Co., as branch manager for the states of Ohio and Indiana, and of the Cleveland branch.

The Combustion Engineering Corp., New York City, announces that it has enlarged its facilities for handling business in the Philadelphia territory. The company's office is located in the Lincoln Building, with W. C. Stripe, formerly of the Pittsburgh office, in charge. Associated with him are C. L. Bachman, who was manager of the Chicago office before going into the United States service, and E. F. Kuehnle, formerly in the main office in New York.

The Westinghouse Air Brake Co. announces the following changes in its personnel: J. R. Ellicott, manager of the Eastern district, has retired after a long service, to devote himself henceforth to his Florida estate. He will, however, act with the officers of the company in a consulting capacity. He will be succeeded by C. R. Ellicott. C. H. Beck, heretofore special representative, Safety Car Devices Co., succeeds C. R. Ellicott as assistant eastern manager, with headquarters in New York City. With the promotion of E. A. Craig to the position of export manager, with headquarters in the Westinghouse Building, Pittsburgh, Robert Burgess, representative at Atlanta, becomes Southeastern manager, with headquarters in the Munsey Building, Washington, D. C. A. K. Hohmyer, heretofore representative attached to the Chicago office, is promoted to the position of assistant Western manager. J. B. Wright, assistant Southeastern manager, has been made assistant district manager at Pittsburgh. W. G. Kaylor, representative in the Eastern district, is appointed representative in the export department, with headquarters in New York City. F. H. Parke, resident engineer, Southeastern district, is now general engineer with headquarters in the Westinghouse Building, Pittsburgh. T. W. Newburn, assistant resident engineer, Southeastern district, becomes district engineer, Southeastern district, with headquarters in the Munsey Building, Washington, D. C. J. C. McCune, special engineer, Wilmerding, is appointed assistant to district engineer, Eastern district, with headquarters at 165 Broadway, New York City. J. H. Woods of the commercial engineering department, Wilmerding, is appointed engineer of the export department, with headquarters in the Westinghouse Building, Pittsburgh.

THE COAL MARKET

BOSTON—Current prices per gross ton f.o.b. New York loading ports:

Anthracite

	Company Coal
Egg	$8.20@$8.65
Stove	8.45@ 9.05
Chestnut	8.55@ 9.05
Pea	7.05@ 7.40
Buckwheat	3.60@ 3.80
Rice	2.30@ 2.50
Barley	1.25

Bituminous

	Clearfields	Cambrias and Somersets
F.o.b. mines, net tons	$2.85@$3.35	$3.15@$3.60
F.o.b. Philadelphia, gross tons	5.05@ 5.60	5.35@ 5.90
F.o.b. New York, gross tons	5.40@ 5.95	5.75@ 6.25
Alongside Boston (water coal), gross tons	7.00@ 7.75	7.60@ 8.00

Pocahontas and New River are practically off the market for coastwise shipment, but are quoted at $6.25@$7.00.

NEW YORK—Current quotations, White Ash, per gross tons, f.o.b. Tidewater, at the lower ports are as follows:

Anthracite

	Company Coal
Broken	$7.80@$8.25
Egg	8.20@ 8.65
Stove	8.45@ 9.05
Chestnut	8.55@ 9.05
Pea	7.05@ 7.40
Buckwheat	5.15
Rice	4.50
Barley	4.85
Boiler	4.85

Bituminous

Government prices at mines:	Spot
South Fork (best)	$3.25@$3.50
Cambria (best)	3.00@ 3.25
Cambria (ordinary)	2.60@ 2.80
Clearfield (best)	3.00@ 3.25
Clearfield (ordinary)	2.60@ 2.90
Reynoldsville	2.65@ 2.90
Quemahoning	3.25@ 3.50
Somerset (medium)	3.00@ 3.25
Somerset (poor)	2.50@ 2.75
Western Maryland	2.50@ 2.75
Fairmont	2.25@ 2.50
Latrobe	2.60@ 2.90
Greensburg	2.75@ 3.00
Westmoreland ¾ in.	3.40@ 3.50
Westmoreland run-of-mine	2.75@ 3.00

PHILADELPHIA—Anthracite prices are practically the same as those listed above for New York. Bituminous coal prices vary according to district from which they are mined. For ordinary slack the price is $2.45@$2.55; lump, $3.00@$3.35, at the mines.

BUFFALO—

Anthracite

	On Cars, Gross Ton	At Curb, Net Ton
Grate	$8.55	$10.80
Egg	8.80	10.65
Stove	9.00	10.85
Chestnut	9.10	10.85
Pea	7.15	9.30
Buckwheat	5.70	7.75

Bituminous

Allegheny Valley	$4.80
Pittsburgh	4.65
No. 8 Lump	4.65
Mine Run	4.80
Slack	4.10
Smokeless	4.60
Pennsylvania Smithing	5.70

CLEVELAND—Prices of coal per net ton delivered in Cleveland are:

Anthracite

Egg	$12.25@$12.40
Chestnut	12.50@ 12.70
Grate	12.25@ 12.40
Stove	12.40@ 12.60

Pocahontas

Mine-run	$7.50

Domestic Bituminous

West Virginia split	$8.50
No. 8 Pittsburgh	$6.60@ 6.80
Massillon lump	8.25@ 8.50
Coshocton lump	7.15

Steam Coal

No. 6 slack	$5.25@$5.50
No. 8 slack	5.10@ 5.50
Youghiogheny slack	5.25@ 5.50
No. 8 ¾ in.	5.70@ 6.00
No. 6 mine-run	5.25@ 5.50
No. 8 mine-run	5.75

Only coal available is mine-run Pocahontas.

MIDDLE WEST—Chicago quotations, F.o.b. cars at mine:

	Springfield, Carterville, Williamson, Franklin, Saline, Harrisburg	Fulton, Peoria	Grundy, La Salle, Bureau, Will
Lump	$2.55@$2.70	$2.95@$3.10	$3.25@$3.40
Washed	2.75@ 2.90	....	3.45@ 3.60
Mine run	2.35@ 2.50	2.75@ 2.90	3.00@ 3.15
Screenings	2.05@ 2.20	2.35@ 2.50	2.75@ 2.90

New Constructions

PROPOSED WORK

N. H., North Stratford—The Warner Sugar Refining Co., 79 Wall St., New York City, plans to build a sugar refinery. A steam heating system will be installed in same. Total estimated cost, $1,000,000.

Mass., Boston—The Federal Reserve Bank, 53 State St., has had plans prepared for the construction of a 5 story, 100 x 110 ft. bank on Pearl and Franklin Sts. A heating system will be installed in same. Total estimated cost, $5,000,000. C. R. Sturgis, 120 Boylston St., Arch.

Conn., Hartford—Berenson & Moses, Archs., 1086 Main St., will receive bids about Feb. 15 for the construction of a 4 story, 67 x 177 ft. warehouse on Winthrop St., for the Hartford Cooperage Co., 119 Portland St. A high pressure boiler and hydraulic or electric freight elevator will be installed in same. Total estimated cost, $200,000.

N. Y., Brooklyn—E. Riegelmann, Boro. Pres., received lowest 3 bids for installing a steam heating and ventilating system in the proposed court house, from Gillis & Geoghagan, 537 West B'way, $8179; Miller & Brady, 217 East 35th St., $8297; W. J. Olvany, 177 Christopher St., $8459; all of New York City.

N. Y., Brooklyn—A. A. Schwartz, 815 Flatbush Ave., will build a 1 story, 100 x 200 ft. theatre on Kings Highway and Coney Island Ave. A steam heating system will be installed in same. Total estimated cost, $350,000. Work will be done by day labor.

N. Y., Brooklyn—The Studebaker Corp., Main and Bronson Sts., South Bend, Ind., is having plans prepared for the construction of a 4 story office building and salesroom on Bedford Ave. and Sterling Pl. A steam heating system will be installed in same. Total estimated cost, $150,000. Tooker & Marsh, 101 Park Ave., New York City, Archs. and Engrs.

N. Y., Glens Falls—The Trustees of Warren Co., Lake George, are having plans prepared for the construction of a 2 story hospital near here. A steam heating system will be installed in same. Total estimated cost, $125,000. Tooker & Marsh, 101 Park Ave., New York City, Archs. and Engrs.

N. Y., New York—The Bd. Educ., 500 Park Ave., is having plans prepared for the construction of a 5 story school on Dawson St. and Rogers Pl (Bronx Boro.). A steam heating system will be installed in same. Total estimated cost, $500,000. C. B. J. Snyder, Municipal Bldg., Arch. and Engr.

N. Y., New York—The Mfrs. Cloak & Suit Trade, 136 Madison Ave., will build a 16 story office building at 36th and 38th Sts. and 7th and 8th Aves. A steam heating system will be installed in same. Total estimated cost, $8,000,000. Work will be done by day labor.

N. Y., New York—The Sidem Bldg. Co., c/o B. H. & C. N. Whinston, Archs. and Engrs., 2 Columbus Circle, will build a 5 story, 25 x 100 ft. office and store building at 117 West 33rd St. A steam heating system will be installed in same. Total estimated cost, $135,000. Work will be done by day labor.

N. Y., New York—The State Bank, 376 Grand St., is having plans prepared for altering the present 10 story office building at 280 5th Ave. A steam heating system will be installed in same. Total estimated cost, $500,000. H. R. Mainzer, 105 West 40th St., Arch. and Engr.

N. J., Athenia—The Mayor Car Co. will soon receive bids for the construction of a 1 story, 40 x 60 ft. power house. Two boilers, engine, compressor and pump will be installed in same. Total estimated cost, $35,000. W. Schoonover, 31 Clinton St., Newark, Engr. J .S. Pigott, 665 Broad St., Newark, Arch.

N. J., Deal Beach—Henry Ives Cobb, Arch. and Engr., 1465 B'way, New York City, is preparing plans for the construction of a 6 story hotel here. A steam heating system will be installed in same. Total estimated cost, $1,000,000. Owner's name withheld.

N. J., Lakehurst—The Bureau of Yards & Docks, Navy Dept., Wash., D. C., plans to construct a power plant and distribution system here. Total estimated cost, $550,000.

N. J., Newark—Loew's Enterprises, 1492 B'way, New York City, plan to build a theatre on Broad St. A steam heating system will be installed in same. Total estimated cost, $400,000.

N. J., Passaic—The Y. M. C. A., 2 West 45th St., New York City, is having plans prepared for the construction of a building. A steam heating system will be installed in same. Total estimated cost, $250,000. J. F. Jackson, 47 West 34th St., New York City, Arch. and Engr.

Penn., Chester—Ballinger & Perrot, Archs. and Engrs., 17th and Arch Sts., Philadelphia, will soon award the contract for the construction of a 3 story, 37 x 207 ft. hospital for the Surgical Wing Hospital. A steam heating system will be installed in same.

Penn., Dalton—The Scranton & Binghamton R. R. Co., 227 Wyoming Ave., is having plans prepared for the construction of a 35 story, 48 x 70 ft. addition to its power house. Three 500 hp. boilers and 15,000 volt electric units will be installed in same. Total estimated cost, $95,000.

Penn., Harrisburg—The state is having plans prepared for the construction of two 4 story office buildings at the State Capitol. A steam heating system will be installed in same. Total estimated cost, $3,000,000. Arnold W. Brunner, 101 Park Ave., New York City, Arch. and Engr.

Penn., Philadelphia—C. E. Oelschlager, Arch. and Engr., 1615 Walnut St., will soon award the contract for the construction of a 15 story, 40 x 131 ft. office building on 16th and Walnut Sts., for the Medical Arts Bldg., 16th and Walnut Sts. A steam heating system will be installed in same.

Penn., Philadelphia—LeRoy B. Rothchild, Arch., 1225 Sansom St., will soon award the contract for the construction of a 6 story, 58 x 125 ft. sales and service building on 16th and Vine Sts., for August Hartman, c/o architect. A steam heating system will be installed in same. Total estimated cost, $100,000.

Penn., Philadelphia—The Bd. Educ., Keystone Bldg., 19th and Chestnut Sts., will soon award the contract for the construction of a 3 story, 115 x 165 ft. grade school on 70th and Buist Sts. A steam heating system will be installed in same. Total estimated cost, $450,000.

Penn., Scranton—The city plans to construct a 3 story, 110 x 110 ft. technical high school on Wyoming Ave. A steam heating system will be installed in same. Total estimated cost, $277,700.

Penn., Scranton—The city plans to construct a 3 story, 260 x 300 ft. junior high school on Mulberry St. A steam heating system will be installed in same. Total estimated cost, $546,200.

Penn., Wilkes-Barre—The Diamond Drug Co. is having plans prepared for the construction of a 3 story, 22 x 180 ft. drug factory on Northampton St. A steam heating system will be installed in same. Total estimated cost, $170,000.

Md., Baltimore—The Diocese of Maryland, c/o B. Goodhue, Archt. and Engr., 2 West 47th St., New York City, is having plans prepared for the construction of a cathedral. A steam heating system will be installed in same. Total estimated cost, $5,000,000.

Ala., Mobile—J. H. and C. B. King, 264 Dauphin St., plans to build a 66 x 210 ft. theatre on Dauphin St. An auxiliary electric plant, steam heating and ventilating systems will be installed in same. Total estimated cost, $250,000. E. H. Slater, 101 Tuttle Ave., Archt.

La., Bastrop—The Municipal Electric Light & Water Wks. plan to enlarge their plant and are in the market for a 150 hp. natural gas engine to be direct connected to a 100 kw., 220 to 250 volts d.c., generator. Cary Robertson, Supt.

Tenn., Knoxville—The Signal Amusement Co. plans to construct a 75 x 150 ft. moving picture theatre on Clinch Ave. A typhoon heating and cooling system will be installed in same. Total estimated cost, $225,000. C. W. and G. L. Rapp, 190 North State St., Chicago, Archts.

Ky., Dawson Springs—J. A. Wetmore, Superv. Arch., Treasury Dept., Wash., D. C., received bids for the installation of mechanical and power equipment in the proposed 19 hospital buildings at the Sanatorium Grounds, also for the construction of a small power house, from S. W. Rittenhouse, Wash., D. C., $381,530; Thompson Bros., 125 West 5th St., Philadelphia, $416,045; F. A. Clegg & Co., 110-112 South 1st St., Louisville, Ky., $427,-378.

O., Canton—The Canton Ice Delivery Co., 601 4th St., will receive bids for the construction of a 1 and 2 story, 81 x 113 ft. ice storage plant on Navarre Rd. Estimated cost, $80,000. A. C. Bishop, 427 Guardian Bldg., Cleveland, Archt.

O., Canton—W. I. Zink, Pub. Serv. Dir., received bids for the construction of a 1 story, 30 x 96 ft. pump house from Keller Huff, $24,000, and Melbourne Bros., $38,000; both contractors of Canton.

O., Cleveland—The Estates & Investment Co. plans to construct a 4 story apartment hotel at 1906 East 105th St. A steam heating system will be installed in same. Total estimated cost, $250,000. G. A. Tenbusch, 1836 Euclid Ave.

O., Cleveland—The Cleveland Columbia Bldg. Co. will soon award the contract for the construction of a 3 story, 63 x 85 ft. club house at 3422 Euclid Ave. A steam heating and a vacuum cleaning system will be installed in same. Total estimated cost, $100,000. Address A. J. McGarrell, 507 Williamson Bldg.

O., Cleveland—The Forbes Chocolate Co. is having plans prepared for the construction of a 2 story, 100 x 200 ft. factory on Euclid Ave. and Ivanhoe Rd. Two boilers will be installed in same. Total estimated cost, $250,000. Address B. F. Forbes, 5100 Superior Ave. G. S. Rider & Co., Century Bldg., Archs. and Engrs.

O., Cleveland—The Realty Syndicate Co. is having plans prepared for the construction of a 3 story, 116 x 250 ft. public garage at East 79th St. near Hough Ave. A steam heating system will be installed in same. Total estimated cost, $400,000. Frank N. Riley, 309 Williamson Bldg., Secy. S. H. Weis, 1032 Schofield Bldg., Arch.

O., Lorain—The Bd. Educ. is having plans prepared for the construction of a 1 story grade school on 4th St. and Hamilton Ave. A steam heating system will be installed in same. Total estimated cost, $250,000. G. Bruell, 6th and Washington Ave., Clk. Franz C. Warner, Hippodrome Annex Bldg., Archt.

O., Youngstown—The Aultman Hospital Association is having plans prepared for the construction of a 3 story, 50 x 125 ft. hospital addition on Clarendon Ave. A steam heating system will be installed in same. Total estimated cost, $125,000. W. L. Alexander, Secy. W. W. Sabin, 1900 Euclid Ave., Cleveland, Archt.

Mich., Detroit—The Riviera Theatre Co., c/o C. Howard Crane and Elmer G. Kiehler, Archts., Huron Bldg., plans to build a 2 story, 50 x 185 ft. theatre and office building on Grand River and Linsdale Aves. A steam and forced air heating system, including fan, motor, etc., will be installed in same. Total estimated cost, $200,000.

Mich., Detroit—Albert Kahn, Arch., Marquette Bldg., will soon award the contract for furnishing 2 multi stage motor-driven centrifugal pumps, capacity 300 gal. per min., 375 ft. head, 60 hp. motors; 2 multi stage motor-driven centrifugal pumps, 200 gal. per min. against 200 ft. head, 20 hp. motors; 1 motor-driven centrifugal pump, 500 gal. per min. against 400 ft. head, 100 hp. motor, 240 volt.

Mich., Detroit—The E. D. Stickley Machine Co., Goebel Bldg., is in the market for three 6 in., 11 in. and 18 in. Gleason bevel gear generators.

Mich., Grand Rapids—The Bd. Educ. plans to construct a 3 story, 225 x 250 ft. school, to be known as the North High School, on Plainfield Ave. A steam heating system, including forced ventilating system, boiler and plenum chamber will be installed in same. Plans also include a power plant. H. H. Turner, 234 Division Ave., N., Arch.

Mich., Grosse Pointe—The Grosse Pointe Golf Club engaged Stratton & Snyder, Archs., Union Trust Bldg., Detroit, to prepare plans for the construction of a 2 or 3 story club house near Jefferson Ave. Steam heating equipment will be installed in same. Total estimated cost, $350,000.

Mich., Muskegon Heights (Muskegon P. O.)—The Shaw Electric Crane Works, McKinney St., plan to construct a 1 and 2 story foundry and machine shops on McKinney and Park Aves. The present power plant at the works will be enlarged. Total estimated cost, $2,000,000. Manning, Maxwell & Moore, McKinney St., Engrs.

Mich., Muskegon—The Occidental Hotel, c/o Edward R. Sweet, 3rd St., plans to build an 8 story, 132 x 140 ft. hotel on 3rd St. near Western Ave. A steam heating and ventilating system will be installed in same. Total estimated cost, $1,000,000.

Mich., Muskegon—The Superior Seating Co., Getty Ave., will soon award the contract for the construction of a 3 story, 100 x 320 ft. factory on Getty and Dales Aves. Project will include a power plant. Total estimated cost, $250,000.

Mich., Pontiac—The Great Lakes Hotel Syndicate engaged W. W. Alschleger, Archt., 111 West Washington St., Chicago, to prepare plans for the construction of an 8 story, 100 x 125 ft. hotel. Steam heating equipment will be installed in same. Total estimated cost, $750,000.

Mich., Sault Ste. Marie—The city has awarded the contract for furnishing and installing 4 electrically driven centrifugal pumping units in the present pumping station to Allis-Chalmers Mfg. Co., North Allis St., Milwaukee, Wis., at $11,707.

Ill., Chicago—The Chicago Reedware Mfg. Co., 1523 Carroll St., is having plans prepared for the construction of a 1 story, 200 x 350 ft. furniture factory on Grand Ave. and Franklin Blvd. A steam heating system will be installed in same. Total estimated cost, $165,000. Davidson & Weiss, 53 West Jackson Blvd., Archs.

Ill., Chicago—A. Emerman, c/o B. L. Steif, Arch., 30 North La Salle St., is having plans prepared for the construction of a 12 story, 115 x 130 ft. hotel at 918 Wilson Ave. A steam heating system will be installed in same. Total estimated cost, $1,500,000.

Ill., Chicago—Ludgin & Leviton, Archs., 53 West Jackson Blvd., will soon award the contract for the construction of a 2 story, 100 x 100 ft. dyeing and cleaning plant on Halsted and 76th Sts., for the Englewood Dyeing & Cleaning Co., 6343 South Halsted St.; 100 hp. high pressure boilers will be installed in same. Total estimated cost, $40,000.

Ill., Chicago—M. J. Morehouse, Arch., 343 South Dearborn St., will soon award the contract for the construction of a 3 story, 120 x 165 ft. factory on Ravenswood and Leland Aves., for the Johnson Fare Box Co., 236 South Rokey St. A steam heating system will be installed in same. Total estimated cost, $125,000.

Ill., Chicago—Robert T. Newberry, Arch., 108 South La Salle St., is preparing plans for the construction of a 2 story, 125 x 250 ft. corset factory on Crawford and Barry Aves., for the Lorette Corset Co., 510 Throop St. A steam heating system will be installed in same. Total estimated cost, $300,000.

Ill., Chicago—Henry L. Newhouse, Arch., 4630 Prairie Ave., will receive bids in March for the construction of a 1 story, 90 x 90 ft. theatre on State St. near Randolph St., for Ascher Bros., 220 South State St. A steam heating system will be installed in same. Total estimated cost, $1,000,000. W. C. Bennett, 327 South La Salle St., Engr.

Ill., Chicago—The O'Gara Coal Co., 332 South Michigan Ave., is in the market for two 8 x 10 in. horizontal center crank steam engines.

Ill., Chicago—The H. O. Reno Co., 123 West Madison St., is having plans prepared for the construction of a 4 story, 125 x 150 ft. publishing plant on Lincoln and Wrightwood Aves. A steam heating system will be installed in same. Total estimated cost, $225,000. Percy T. Johnson, 1254 Pratt Blvd., Arch.

Ill., Mattoon—Claude L. James, Supt. of Waterworks, is preparing plans for improving and extending the present waterworks. Boilers and pumping machinery will be installed in same.

Ia., Davenport—Clausen & Kruse, Archs. and Engrs., 316 Central Office Bldg., will receive bids for the construction of a 4 story, 75 x 135 ft. administration building on Brady St., for the Palmer School of Chiropractic. A complete ventilation system will be installed in same. Total estimated cost, $120,000.

Minn., Duluth—The Sister Superior of the Sisters of the Benedictine will receive bids about Feb. 15 for the construction of a 5 story, 40 x 200 ft. addition to St. Mary's Hospital, 3rd St., for the Catholic Diocese of Duluth, 3rd St. and 5th Ave. A steam heating and mechanical ventilating system will be installed in same. Total estimated cost, $200,000. F. H. Ellerbe, Endicott Bldg., St. Paul, Arch. and Engr.

Minn., McKinley—F. A. Walstrom, Clk. of Independent School Dist. No. 18, Gilbert, will receive bids until Feb. 3 for the construction of a 3 story, 75 x 150 ft. grade school here. A steam heating system will be installed in same. Total estimated cost, $150,000. W. T. Bray, Torrey Bldg., Duluth, Archt. and Engr.

Minn., Minneapolis—Magney & Tusler, Archts., Builders' Exch., will soon award the contract for the construction of a 4 story, 100 x 140 ft. wagon factory on 5th St., N., for the Schurmeier Mfg. Co., 320 East 9th St., St. Paul. A steam heating system will be installed in same. Total estimated cost, $150,000.

S. D., Watertown—The Bd. Educ. will receive bids until Jan. 30 for the construction of a 3 story, 151 x 178 ft. high school. A steam heating and ventilating system will be installed in same. Total estimated cost, $350,000. C. L. Pillsbury & Co., Metropolitan Life Insurance Bldg., Minneapolis, Engr. Tyrel & Chapman, 320 Auditorium Bldg., Minneapolis, Archts.

Ariz., Chandler—The Bd. Educ. will receive bids until Feb. 14 for the construction of a high school, including administration, domestic science, commercial and science buildings. Separate bids will be received for installing heating system. Total estimated cost, $150,000. Allison & Allison, 1405 Hibernian Bldg., Los Angeles, Archs.

Wash., Olympia—The State Capitol Comn. will soon award the contract for the construction of 3 buildings, including an office building and court house. A steam heating system will be installed in same. Total estimated cost, $500,000. Wilden & White, 50 Church St., New York City, Archs. and Engrs.

Cal., Del Monte—The Del Monte Bath House Co. is having plans prepared for the construction of a bath house, cottages, pier extension, etc., also for converting boat house into power house. Total estimated cost, $500,000. Work will be done by day labor under the supervision of C. L. Lyon, New York. P. V. Tuttle, 565 Lighthouse Ave., Pacific Grove, Arch.

Cal., Oakland—The Natl. Ice & Cold Storage Co., Postal Telegraph Bldg., San Francisco, is having plans prepared for the construction of a 1 story ice and cold storage building on North 1st St. Estimated cost, $250,000.

N. S., Halifax—The Comn. of Works & Mines of the Provincial Government will receive bids until Feb. 15 for the construction of a 3 story, 60 x 220 ft. hospital building, with two 30 x 100 ft. wings, on Tower Rd. Separate contract will be let for installing heating system. Total estimated cost, $220,000.

Ont., Merritton—The Canadian Steel & Forging Co. is having plans prepared for the construction of a steel plant addition including hydro-electric light and heating plants, also for installing equipment in same. Total estimated cost, $200,000. J. J. Wernette, 441 Houseman Bldg., Grand Rapids, Mich., Engr.

Ont., Scarboro—The town plans to construct 2 pump houses to be equipped with electrically driven pumps, distribution system, etc. By-law appropriating $300,000 for this project has been passed. R. E. James Co., Ltd., 36 Toronto St., Toronto, Engrs.

Ont., Windsor—The Bank of Hamilton plans to build a 10 story bank and office building on Quellette Ave. and Chatham St. Steam heating equipment will be installed in same. Total estimated cost, $750,000.

CONTRACTS AWARDED

Me., Lewiston—The Lewiston Buick Co. has awarded the contract for the construction of a 3 story, 95 x 130 ft. sales building on Sabattus and Main Sts., to F. T. Ley & Co., Inc., 185 Devonshire St., Boston. A steam heating system will be installed in same. Total estimated cost, $150,000.

R. I., Tiverton—The Sinclair Refining Co., Brockton, Mass., has awarded the contract for the construction of a 1 story, 32 x 90 ft. pump house, to Canning & Leary, Plant Bldg., New London, Conn. Estimated cost, $25,000.

Conn., Waterbury—The Waterbury Natl. Bank, Grand and Field Sts., has awarded the contract for the construction of a 4 story, 65.9 x 89.3 ft. building, to Irwin & Leighton, 126 North 12th St., Philadelphia. A steam heating system will be installed in same. Total estimated cost, $500,000.

N. Y., New York—Marcus Loew's Enterprises, 1492 B'way, has awarded the contract for the construction of a 16 story, 50 x 120 x 180 ft. theatre and office building on B'way and 45th St., to the Fleischman Constr. Co., 531 7th Ave. A steam heating system will be installed in same. Total estimated cost, $1,000,000.

N. Y., New York—The 145 West 55th St. Corp., Inc., 1 West 67th St., has awarded the contract for the construction of a 15 story, 80 x 100 ft. hotel at 139-145 West 55th St., to Fred T. Ley, 19 West 44th St. A steam heating system will be installed in same. Total estimated cost, $750,000.

N. Y., New York—The Sloane Estate, 316 East 65th St., has awarded the contract for the construction of a 10 story, 75 x 100 ft. factory on 39th St. near 8th Ave., to Fred T. Ley, 19 West 44th St. A steam heating system will be installed in same.

N. Y., Rochester—The city has awarded the contract for installing heating and power system in the proposed 3 story, 304 x 321 ft. Junior High School, at Wilson Park, to Barr & Greelman Co., 72 Exch. St., at $162,000.

N. J., Camden—The Ottawa Tribe of Red Men, c/o W. W. Sharpley, 1713 Sansom St., Philadelphia, has awarded the contract for the construction of a 3 story, 40 x 145 ft. clubhouse to Willis-Ludwich & Co., 1708 Sansom St., Philadelphia. A steam heating system will be installed in same. Total estimated cost, $140,000.

N. J., Elizabeth—The Simmons Co., Pearl St., Kenosha, Wis., has awarded the contract for the construction of a manufacturing plant, including a power house, etc., to Paul Riesen's Sons, 1018 Humboldt Ave., Milwaukee, Wis. Total estimated cost, $300,000.

N. J., Harrison—The Driver-Harris Co. has awarded the contract for the construction of a 6 story, 80 x 200 ft. factory on Middlesex St. to E. N. Waldron, Market St., Newark. A steam heating system will be installed in same.

Penn., Essington—The Westinghouse Electric Mfg. Co., Union Bank Bldg., Pittsburgh, has awarded the contract for the construction of a factory addition to Westinghouse, Church Kerr & Co., Inc., 37 Wall St., New York City. A steam heating system will be installed in same. Total estimated cost, $2,000,000.

O., Cleveland—Edward Shattuck, Comr. of Purchases, has awarded the contract for the installation of a 20,000,000 gal. first high service centrifugal pump to the Dravo-Doyle Co., Citizens Bldg., at $74,250.

O., Cleveland—The Sheriff St. Market & Storage Co. has awarded the contract for the construction of a 2 story, 200 x 885 ft. cold storage warehouse on Bolivar Rd. and East 4th St. to John Gill & Sons, Citizens Bldg. A steam heating system will be installed in same. Total estimated cost, $1,000,000.

Mich., Detroit—The Bd. Educ., 50 B'way, has awarded the contract for the installation of a heating system, etc., in the proposed 2 story, 85 x 134 ft. school on Vancourt and Milford Aves., to J. W. Partlan, 51 Park Pl., $53,800; in the proposed Field School, on Agnes and Field Aves., to W. J. Rewoldt, 506 Owen Bldg., $18,274; in the proposed Davidson School, near Davidson Ave., to O'Connor Bros., 34 Roland Bldg., $89,590. Noted Jan. 6.

Ill., Chicago—The Chicago Tribune, 1 South Dearborn St., has awarded the contract for the construction of a 15 story, 100 x 300 ft. loft building at 431 North Michigan Ave., to R. C. Wieboldt, 1534 West Van Burén St. A steam heating system will be installed in same. Total estimated cost, $2,500,000. Jarvis Hunt, 30 North Michigan Ave., Arch.

Ia., Bondurant—W. F. Dunkle, Secy. Bd. of Educ., has awarded the contract for installing a heating and plumbing system in the proposed 2 story, 63 x 105 ft. school to W. G. Madison, Ames, at $21,026.

Ia., Newton—The Automatic Washing Machine Co. has awarded the contract for the construction of a 4 story, 50 x 148 ft. factory and boiler house, to A. H. Neumann & Co., 7 Hubbell Bldg., Des Moines. Total estimated cost, $180,000.

Cal., Emeryville (Oakland P. O.)—The Amer. Rubber Mfg. Co., Park Ave. and Wait St., has awarded the contract for the construction of a 2 story factory and power house, to R. W. Littlefield, 565 16th St., Oakland. Total estimated cost, $100,000.

Cal., Los Angeles—The Pacific Mutual Life Insurance Co., 6th and Olive Sts., has awarded the contract for the construction of a 2 story, 170 x 218 ft. office and store building on 6th St. and Grand Ave. to the Pacific Marine & Constr. Co., 510 Wright & Callender Bldg. A steam heating system will be installed in same. Total estimated cost, $2,500,000.

Cal., San Francisco—St. Francis Hospital Association has awarded the contract for installing a heating system in the proposed 6 story hospital on Bush St., to A. W. Lawson, 180 Jessie St., at $8750.

Ont., Toronto—The Murray Printing Co., 9 Jordan St., has awarded the contract for the construction of a 4 story, 140 x 180 ft. printing shop, to L. E. Dowling, 167 Yonge St. Heating contract will be sub-let. Total estimated cost, $135,000.

Ont., Windsor—The Bd. Educ. has awarded the contract for the construction of a 3 story school, to Sam Dinsmore, 7 Royal Bank Bldg. A vacuum steam heating system with blower for ventilation will be installed in same. Total estimated cost, $400,000.

NEW INSTALLATIONS IN POWER

N. Y., Saranac Lake—Paul Smith's Electric Light & Power & R. R. Co. plans to construct a 14 mi. transmission line to the plant of the Chateagua Ore & Iron Co., Seynon Mountain, and install a 3 phase, 60 cycle 2500 to 4500 volts 1500 kva. generator. P. A. Gould, Mgr.

N. Y., Sanborn—The Sanborn Pekin Power Co. plans to extend its transmission line within the next three years; $35,000 will be expended for this project. R. E. Casselman, Secy.

Fla., Milton—The town plans to install 75 meters and a 100 kw. generator. I. I. Shepard, Supt.

O., Ada—The Ada Water & Light Co. plans to install new engines, generators and water pumps. P. O. Moore, Mgr.

Mich., Gladwin—The Gladwin Electric Light & Power Co. plans to build a 33,000 mi. transmission line from here to Beaverton. S. B. Wiggins, 115 Howard St., Saginaw, Secy.

Minn., Chisholm—The Minnesota Utilities Co., Eveleth, plans to install a 2500 kw. extension to its power plant. E. F. Kefgen, Treas.

Neb., Nebraska City—The Nebraska City Utilities Co. plans to construct a transmission line to Julian, Danbar and Syracuse. W. O. Dunn, Mgr.

S. D., Iroquois—The Iroquois Light Plant plans to install a 50 hp. Fairbanks Morse Type Y engine, belted to 42½ kva., 2300-3p. 60 cycle generator. W. H. Jordan, Pres.

Wyo., Gillette—The Municipal Electric Light Plant plans to install new engine and a 100 kva. generator. M. L. Thomas, Mgr.

Okla., Anadarko—The Anadarko Municipal Light & Water Plant plans to increase power plant and install 2 Diesel oil engines, 200 hp. each. N. A. Kunkel, Supt.

Colo., Eaton—The Farmers Electric Light & Power Co. plans to build 12 mi. of transmission line into irrigation territory. R. S. Farr, Mgr.

N. M., Carlsbad—The Carlsbad Light & Power Co., Chicago, Ill., and Carlsbad, N. M., plans to construct power house and install new 112 kva. generator and 2100 hp. oil engines for auxiliary plant. F. E. Hubert, Gen. Mgr.

Sask., Humboldt—The Municipal Electric Light & Power Plant plans to construct a new 250 kva. unit and install 300 hp. boilers. C. A. Cutting, Supt.

B. C., Cumberland—The Cumberland Electric Lighting Co., Ltd., plans a complete renewal of about 6 mi. of transmission lines. John Short, Secy. and Mgr.

PRICES -- MATERIALS -- SUPPLIES

These are prices to the power plant by jobbers in the larger buying centers east of the Mississippi. Elsewhere the prices will be modified by increased freight charges and by local conditions.

POWER-PLANT SUPPLIES

HOSE—

Fire	50-Ft. Lengths
Underwriters' 2½-in	75c. per ft.
Common, 2½-in	40%

Air

	First Grade	Second Grade	Third Grade
½-in. per ft.	$0.50	$0.33	$0.22

Steam—Discounts from List

First grade.......30% Second grade...40% Third grade....45%

RUBBER BELTING—The following discounts from list apply to transmission rubber and duck belting:

Competition.......65% Best grade.......35%
Standard.......50%

Note—Above discounts apply on new list issued July 1.

LEATHER BELTING—Present discounts from list in the following cities are as follows:

	Medium Grade	Heavy Grade
New York	45%	35%
St. Louis	40%	30%
Chicago	20%	10%
Birmingham	15%	15%
Denver	30%	20%

RAWHIDE LACING—20% for cut; 45c. per sq. ft. for ordinary.

PACKING—Prices per pound:

Rubber and duck for low-pressure steam	$1.00
Asbestos for high-pressure steam	1.70
Duck and rubber for piston packing	1.00
Flax, regular	1.20
Flax, waterproofed	1.70
Compressed asbestos sheet	.90
Wire insertion asbestos sheet	1.50
Rubber sheet	.50
Rubber sheet, wire insertion	.70
Rubber sheet, duck insertion	.50
Rubber sheet, cloth insertion	.30
Asbestos packing, twisted or braided and graphited, for valve stems and stuffing boxes	1.30
Asbestos wick, ½- and 1-lb. balls	.85

PIPE AND BOILER COVERING—Below are discounts and part of standard lists:

PIPE COVERING		BLOCKS AND SHEETS	
Pipe Size	Standard List Per Lin.Ft.	Thickness	Price per Sq.Ft.
1-in.	$0.27	½-in.	$0.27
2-in.	.36	1-in.	30.
6-in.	.80	1½-in.	.45
4-in.	.60	2-in.	.60
3-in.	.45	2½-in.	75
8-in	1.10	3-in.	.90
10-in.	1.30	3½-in.	1.05

85% magnesia high pressure		List
For low-pressure heating and return lines	4-ply	50% off
	3-ply	52% off
	2-ply	54% off

GREASES—Prices are as follows in the following cities in cents per pound for barrel lots:

	Cincinnati	Pittsburgh	Chicago	St. Louis	Birmingham	Denver
Cup	7	9	6.6	6.7	8½	14½
Fiber or sponge	8	10	8.6	13	8½	18
Transmission	7	9	8.1	13	8½	17
Axle	4½	6	4.8	4.75	4½	5½
Gear	4½	9	6.1	7.5	8½	8
Car journal	22 (gal.)	21 (gal.)	4.7	4.7	8½	8

COTTON WASTE—The following prices are in cents per pound:

	New York Current	New York One Year Ago	Cleveland	Chicago
White	13.00	11 00 to 13.00	14.00	11 00 to 14 00
Colored mixed	9.00 to 12.00	8 50 to 12 00	11 00	9 50 to 12 00

WIPING CLOTHS—Jobbers' price per 1000 is as follows:

	13½ x 13½	13½ x 20½
Cleveland	$52.00	$58.00
Chicago	41.00	43.50

LINSEED OIL—These prices are per gallon:

	New York Current	New York One Year Ago	Cleveland Current	Chicago Current	Chicago One Year Ago
Raw in barrels (5 bbl. lots)	$1.80	$1.49	$2 50	$1.98	$1.66
5-gal. cans	2.00	1.74	2.75	2.23	1.86

WHITE AND RED LEAD—Base price per pound:

	Red Current Dry	Red Current In Oil	Red 1 Year Ago Dry	Red 1 Year Ago In Oil	White Current Dry and In Oil	White 1 Yr. Ago Dry and In Oil
100-lb. keg	14.50	16.00	13.00	14.50	14.50	13.00
25- and 50-lb. kegs	14.75	16.25	13.25	14.75	14.75	13.25
12½-lb. keg	15.00	16.50	13.50	15.00	15.00	13.50
1-lb. cans	16.50	18.00	.....	.....	16.50	15.00
5-lb. cans	17.50	19.00	.....	.....	17.50	16.00

500 lb. lots less 10% discount; 2000 lb. lots less 10-2½%.

RIVETS—The following quotations are allowed for fair-sized orders from warehouse:

	New York	Cleveland	Chicago
Steel 7/16 and smaller	40%	55%	45%
Tinned	40%	55%	45%

Boiler rivets, ¾, ⅞, 1 in. diameter by 2 in. to 5 in. sell as follows per 100 lb.:

New York $5 [illegible] Cleveland $4 00 Chicago ....$4 97 Pittsburgh $4.72

Structural rivets, same sizes

New York.. $5 10 Cleveland $4 10 Chicago. ..$5 07 Pittsburgh $4.82

REFRACTORIES—Following prices are f. o. b. works, Pittsburgh

Chrome brick	net ton	$75–80	at Chester, Penn
Chrome cement	net ton	45–50	at Chester, Penn.
Clay brick, 1st quality fireclay	net M.	38–45	at Clearfield, Penn
Clay brick, 2nd quality	net M	33–35	at Clearfield, Penn.
Magnesite, dead burned	net ton	50–55	at Chester, Penn.
Magnesite brick, 9 x 4½ x 2½ in.	net ton	80–85	at Chester, Penn.
Silica brick	net M.	45–50	at Mt. Union, Penn.

Standard size fire brick, 9 x 4½ x 2½ in. The second quality is $4 to $5 cheaper per 1000.

St. Louis—Fire Clay, $35 to $50
Birmingham—Silica, $50; fire clay, $45; magnesite, $80, chrome, $80
Chicago—Second quality, $25 per ton.
Denver—Fire clay, $11 per ton

BABBITT METAL—Warehouse prices in cents per pound:

	New York Current	New York One Year Ago	Cleveland Current	Cleveland One Year Ago	Chicago Current	Chicago One Year Ago
Best grade	90 00	87 00	74.50	80 00	70 00	75 00
Commercial	50 00	42.00	20 50	21.50	15 00	15 00

SWEDISH (NORWAY) IRON—The average price per 100 lb., in ton lots, is:

	Current	One Year Ago
New York	$21–26	$19 00
Cleveland	20.00	20.00
Chicago	16.50	19.00

In coils an advance of 50c. usually is charged.
Domestic iron (Swedish analysis) is selling at 15c. per lb.

SHEETS—Quotations are in cents per pound in various cities from warehouse; also the base quotations from mill:

	Mill Carloads Pittsburgh	New York Current	New York One Year Ago	Cleveland	Chicago
No. 28 black	$4.35–4 85	$6.50–8 00	$6.52	$5.77	$6.
No. 26 black	4.25–4.75	6.40–7.90	6.42	5.67	5.
Nos. 22–24 black	4.20–4.71	6.35–7.35	6.37	5.62	5.
Nos. 18–20 black	4 15–4.65	6.30–7.30	6.32	5 57	5.
No. 16 blue annealed	3.75–3.20	6.02	5.72	5.17	5.
No. 14 blue annealed	3 65–3.10	5.92	5 62	5 07	4.
No. 10 blue annealed	3.55–4.00	5 50–6 00	5.52	4.97	4.
No. 28 galvanized	5.70–6.20	7 50–9 00	7.77	7.12	7.
No. 26 galvanized	5 40–5 90	7.20	7.47	6.82	6.00
No. 24 galvanized	5 25–5.75	7.05	7.32	6.57	6.[illegible]

PIPE—The following discounts are for carload lots, f. o. b. Pittsburgh; basing card of Jan. 1, 1919, for steel pipe and for iron pipe:

BUTT WELD

Inches	Steel Black	Galvanized	Inches	Iron Black	Galvanized
¼, ⅜ and ½	50½%	24%	¾ to 1½	30½%	23½%
½	54½%	40%			
¾ to 3	57½%	44%			

LAP WELD

Inches	Steel Black	Galvanized	Inches	Iron Black	Galvanized
2	50½%	35%	2	32½%	18½%
2½ to 6	53½%	41%	2½ to 6	34½%	21½%

BUTT WELD, EXTRA STRONG PLAIN ENDS

Inches	Steel Black	Galvanized	Inches	Iron Black	Galvanized
¼, ⅜ and ½	48½%	29%	¾ to 1½	39½%	24½%
½	51½%	39%			
¾ to 1½	50½%	43%			

LAP WELD, EXTRA STRONG PLAIN ENDS

Inches	Steel Black	Galvanized	Inches	Iron Black	Galvanized
2	48½%	37%	2	33½%	20½%
2½ to 4	51½%	49%	2½ to 4	35½%	23½%
4½ to 6	50½%	39%	4½ to 6	34½%	22½%

Stock discounts in cities named are as follows:

	New York Black	New York Galvanized	Cleveland Black	Cleveland Galvanized	Chicago Black	Chicago Galvanized
½ to 3 in. steel butt welded	47%	31%	43½%	34½%	57½%	44%
2½ to 3 in. steel lap welded	42%	27%	45½%	39½%	53½%	41%

Malleable fittings. Class B and C, from New York stock sell at list + 22½%. Cast iron, standard sizes, 10-5% off.

BOILER TUBES—The following are the prices for carload lots, f. o. b. Pittsburgh:

Lap Welded Steel		Charcoal Iron	
3½ to 4½ in.	40½	3½ to 4½ in.	18½
2½ to 3¼ in.	30½	3 to 3¼ in.	1
2¼ in.	24	2¼ to 2¾ in.	+1
1¾ to 2 in.	19½	2 to 2¼ in.	+18
		1¾ to 1¾ in.	+29

Standard Commercial Seamless—Cold Drawn or Hot Rolled

	Per Net Ton		Per Net Ton
1 in.	$327	1¾ in.	$207
1¼ in.	267	2 to 2¼ in.	177
1½ in.	257	2½ to 3¼ in.	167
1¾ in.	207	4 in.	187
		4½ to 5 in.	207

These prices do not apply to special specifications for locomotive tubes nor to special specifications for tubes for the Navy Department, which will be subject to special negotiations.

ELECTRICAL SUPPLIES

ARMORED CABLE—

B. & S. Size	Two Cond M Ft	Three Cond. M Ft.	Two Cond. Lead M Ft.	Three Cond. Lead M Ft.
No. 14 solid	$104 00	$138 00	$184.00	$210.00
No. 12 solid	135 00	170.00	225.00	265.00
No. 10 solid	185 00	235 00	275.00	325.00
No. 8 stranded	285 00	375 00	520.00	500.00
No. 6 stranded	400 00	500.00	560.00	

From the above lists discounts are:

Less than coil lots	Net List
Coils to 1,000 ft.	10%
1,000 ft. and over	15%

BATTERIES, DRY—Regular No. 6 size red seal, Columbia, or Ever Ready:

	Each, Net
Less than 12	$0 45
12 to 50	.38
50 to 125 (bbl.)	.35
125 (bbl.) or over	.32

CONDUITS, ELBOWS AND COUPLINGS—Following are warehouse net prices per 1000 ft. for conduit and per 100 for couplings and elbows:

Size. In.	Conduit Black 1,000 Ft. and Over	Conduit Galvanized 1,000 Ft. and over	Elbows Black 100 and Over	Elbows Galvanized 100 and Over	Couplings Black 100 and Over	Couplings Galvanized 1,100 and Over
⅜	$71.15	$76.25	$17.04	$18.18	$5.38	$5.74
½	71.15	76 25	17.04	18.18	6.25	6.70
¾	93.96	100 86	22.23	23 93	8.97	9 57
1	138.39	149 09	33.19	35 41	11.66	12.44
1¼	187 91	201.71	42.62	45 31	16.10	17.12
1½	224.68	241.18	56.82	60.62	19.89	21.15
2	302 29	324 49	104.17	110.77	26.52	28 20
2½	477.95	513 05	170 46	181.26	37.88	40.28
3	625 00	670.91	454.56	483 36	56.82	60 42
3½	779.24	834.44	1,003.82	1,067.42	75.76	80 56
4	945.03	1,010 43	1,160.08	1,233.57	94.70	100.70

5% cash 10 days.
From New York Warehouse—Less 5% cash.
Standard lengths rigid, 10 ft. Standard lengths flexible, ½ in., 100 ft. Standard lengths flexible, ¾ to 2 in., 50 ft.

CONDUIT NON-METALLIC, LOOM—

Size I. D., In.	Feet per Coil	List, Ft.
5/16	250	$0.05½
⅜	250	.06
½	250	.09
⅝	200	.12
¾	200	.15
⅞	150	.18
1	100	.25
1¼	100	.33
1½	Odd lengths	.40
2	Odd lengths	.55

Coils......45% off
Less coils, 35% off

CUT-OUTS—Following are net prices each in standard-package quantities:

CUT-OUTS, PLUG

S. P. M. L	$0.11	T. P. to D. P. S. B	$0.24
D. P. M. L	.11	T. P. to D. P. T. B	.38
T. P. M. L	.26	T. P. S. B	.33
D. P. S. B	.19	T. P. D. B	.54
D. P. D. B	.37		

CUT-OUTS, N. E. C. FUSE

	0-30 Amp.	31-60 Amp.	60-100 Amp.
D. P. M. L	$0.33	$0.84	$1.68
T. P. M. L	.48	1.20	2.40
D. P. S. B	.42	1.05	....
T. P. S. B	.81	1.80	....
D. P. D. B	.78	2.10	....
T. P. D. B	1.35	3.60	....
T. P. to D. P. D. B	.90	2.52	....

FLEXIBLE CORD—Price per 1,000 ft. in coils of 250 ft.:

No. 18 cotton twisted	$22.00
No. 16 cotton twisted	27.00
No. 18 cotton parallel	26.50
No. 16 cotton parallel	34.50
No. 18 cotton reinforced heavy	34.00
No. 16 cotton reinforced heavy	42.00
No. 18 cotton reinforced light	30.00
No. 16 cotton reinforced light	36.75
No. 18 cotton Canvasite cord	28.50
No. 16 cotton Canvasite cord	32 00

FUSES ENCLOSED—

250-Volt	Std. Pkg.	List
3-amp. to 30-amp.	100	$0.25
35-amp. to 60-amp.	100	.35
65-amp. to 100-amp.	50	.90
110-amp. to 200-amp.	25	2.00
225-amp. to 400-amp.	25	3.60
425-amp. to 600-amp.	10	5.50

600-Volt	Std. Pkg.	List
3-amp. to 30-amp.	100	$0.40
35-amp. to 60-amp.	100	.60
65-amp. to 100-amp.	50	1.50
110-amp. to 200-amp.	25	2.50
225-amp. to 400-amp.	25	5.50
450-amp. to 600-amp.	10	8.00

Discount:

Less 1-5th standard package	30%
1-5th to standard package	40%
Standard package	52%

FUSE PLUGS, MICA CAP—

0-30 ampere, standard package	$4.75C
0-30 ampere, less than standard package	6.00C

LAMPS—Below are present quotations in less than standard package quantities:

Straight-Side Bulbs Mazda B— Watts	Plain	Frosted	No. in Package	Pear-Shape Bulbs Mazda C— Watts	Clear	Frosted	No. in Package
10	$0.35	$0 38	100	75	$0.70	$0.75	50
15	.35	.38	100	100	1.10	1.15	24
25	.35	.38	100	150	1.65	1.70	24
40	.35	.38	100	200	2.20	2.27	24
50	.35	.38	100	300	3.25	3.35	24
60	.40	.45	100	400	4.30	4.45	12
100	.85	.92	24	500	4.70	4.85	12
				750	6.50	6.75	8
				1000	7.50	7.75	8

Standard quantities are subject to discount of 10% from list. Annual contracts ranging from $150 to $300,000 net allow a discount of 17 to 40% from list.

PLUGS, ATTACHMENT—

		Each
Hubbell porcelain No. 5406,	standard package 250	$0.24
Hubbell composition No. 5467,	standard package 50	.32
Benjamin swivel No. 903,	standard package 250	.20
Hubbell current taps No. 5638,	standard package 50	.40

RUBBER-COVERED COPPER WIRE—Per 1000 ft. in New York

No.	Solid Single Braid	Solid Double Braid	Stranded, Double Braid	Duplex
14	$12.00	$16 00	$13 90	$28.50
12	13 25	15.70	18 05	30.07
10	18.30	21.00	23.85	41.50
8	25.54	28.60	32.70	56.77
6	.....	.....	51 40	.....
4	.....	.....	70.00	.....
2	.....	.....	101.80	.....
1	.....	.....	131.86	.....
0	.....	.....	160.00	.....
00	.....	.....	193.50	.....
000	.....	.....	235.20	.....
0000	.....	.....	288.60	.....

Prices per 1000 ft. for Rubber-covered Wire in Following Cities:

	Denver Single Braid	Denver Double Braid	Denver Duplex	Birmingham Single Braid	Birmingham Double Braid	Birmingham Duplex
14	$11.25	$14.00	$31 00	$12.50	$15.39	$29 00
10	21.85	28.25	56.75	25.10	34.73	55.50
8	31.05	38 30	77.95	34 75	58 59	70.50
6	48.80	52.65	......	57.50	82.11	......
4	70.00	75.30	......	81.65	95.50	......
2	104 85	112.45	......	......	140 20	......
1	136.85	146.15	......	......	190.90	......
0	......	194.25	......	......	231.33	......
00	......	239.60	......	......	281.23	......
000	......	294.00	......	......	343.22	......
0000	......	358.25	......	......	416.80	......

Pittsburg—23c. base; discount 50% St. Louis—30c. base.

SOCKETS, BRASS SHELL—

¼ In. or Pendant Cap. Key, Each	Keyless, Each	Pull, Each	⅜ In. Cap. Key, Each	Keyless, Each	Pull, Each
$0.33	$0.30	$0.60	$0.39	$0.36	$0.66

Less 1-5th standard package	+20%
1-5th to standard package	+10%
Standard package	−15%

WIRE, ANNUNCIATOR AND DAMPPROOF OFFICE—

No. 18 B. & S. regular spools (approx. 8 lb.)	45c. lb.
No. 18 B. & S. regular 1-lb. coils	46c. lb.

WIRING SUPPLIES—

Friction tape, ¾ in., less 100 lb. 50c. lb., 100 lb. lots	55c. lb.
Rubber tape, ¾ in., less 100 lb. 65c. lb., 100 lb. lots	60c. lb.
Wire solder, less 100 lb. 50c. lb., 100 lb. lots	46c. lb.
Soldering paste, 2 oz. cans Nokorode	$1.20 doz.

SWITCHES, KNIFE—

TYPE "C" NOT FUSIBLE

Size, Amp.	Single Pole, Each	Double Pole, Each	Three Pole, Each	Four Pole, Each
30	$0.42	$0.68	$1.02	$1.36
60	.74	1.22	1.84	2.44
100	1.50	2.50	3.76	5.00
200	2.70	4.50	6.76	9.00

TYPE "C" FUSIBLE, TOP OR BOTTOM

Size, Amp.	Single Pole, Each	Double Pole, Each	Three Pole, Each	Four Pole, Each
30	.70	[illegible]	1.60	2.12
60	1.18	[illegible].60	2.70	3.60
100	2.38	[illegible] 66	5.80	7.80
200	4.40	6.76	10.14	13.50

Discounts:

Less than $10.00 list value	+15%
$10 to $25 list value	list
$25 to $50 list value	− 5%
$50 to $200 list value	−10%
$200 list value or over	−15%

POWER

Volume 51 New York, February 10, 1920

GONNA!

BY RUFUS T. STROHM

There is nothing that's so easy as a promise,
  As it only takes a little word or two,
But it's different when Harry, Dick or Thomas
  Is compelled to make the pesky thing come true;
For it's something of a job to shift the tenses
  'Twixt the simple phrase "it will be" and "it was,"
And it isn't very long before he senses
  That the man who's always "gonna" never "does."

When you find a fellow-mortal strong on talking
  Of the things he'll do tomorrow, it's a cinch
That you'll find him quoting alibis and balking
  When you come to test his courage in the pinch;
For the fellow who's extravagantly gabby
  Has a cranium that's filled with fluff and fuzz,
And the mainspring of his will is weak and flabby,
  Since the guy who's always "gonna" never "does."

Have you noticed how the bumblebee resembles
  Certain fellows you have happened to espy?
How the overarching welkin throbs and trembles
  When this fat self-advertiser lumbers by?
Yes, his disposition's lovable and sunny,
  But his days are largely spent in noisy buzz,
While his labor yields a minimum of honey,
  For the dub who's always "gonna" never "does."

When you've got a task to finish, go and do it!
  Don't procrastinate until the morrow's sun.
Save your breath to give you pep to struggle through it;
  There'll be time enough for talking when it's done.
Look at Job—he had a grand excuse for blowing
  As to how he'd raise a ruction there in Uz,
But he lived in patient silence, wisely knowing
  That the chap who's always "gonna" never "does."

Moline Station Adds to Steam and Generating Capacity

By THOMAS WILSON

New Boiler Installation with a Number of Interesting Features Housed in Building of Reinforced Concrete Throughout—Large Beam and Column Construction with Brickwork Filling in the Panels in the Side Walls

OWING to the rapid growth of the community embracing Rock Island, Moline and Davenport, an ideal location for various industries, the demand for power has grown in proportion. For the last five years the load on the big generating station of the United Light and Railway Co. has increased approximately 30 per cent. per year. In 1913 a 10,000-kv.-a. turbo-generator had been installed. Three years previously a 6000-kw. turbine had replaced an engine unit, and in addition there were some smaller turbines, giving an aggregate capacity of 21,500 kw. The turbines mentioned, having the old rating, were good for at least 50 per cent. overload. In 1916 it was found that this equipment would not be sufficient to meet the growing demand and that arrangements must be made at once to add both to the steam and the electrical capacity of the big Moline plant supplying the entire community.

Considering the present load, the plant was situated practically at the center of distribution, but the area of the site was limited, making difficult extensions requiring additional space. Rather than a new plant in another location, a way was found to make modern and economical additions to the present station. On this basis a 20,000-kv.-a. turbo-generator was ordered to replace some of the smaller and less efficient units, as had been the custom in the past, and space was found for a new boiler house containing at present three 1003-hp. Stirling boilers to serve the new turbine, with space for three more boilers of the same type for a future 25,000-kv.-a. unit. As outlined, these plans seemed to be the ultimate limit for this particular plant, and probably will take care of requirements for five years.

Reinforced Concrete Construction Adopted

Building a boiler house during the war period, when structural steel was at a premium and extremely difficult to obtain, necessitated plans other than standard. A decision was made to construct a building of reinforced concrete throughout. The materials required were more readily available, and according to estimates, the initial cost would be less. Laying out the boilers in two rows of three each, facing toward a central firing aisle, called for a building 82 ft. wide by 102.5 ft. long. A basement with a headroom of 18 ft. was provided, and it is 40 ft. from the boiler-room floor to the main roof. This gives just the desired clearance over the boiler tops, and there is no surplus building space. At the same time good light and ventilation are obtained by large window areas near the floor and roof lines in all elevations. By a patent operating device the windows are operated in sets, so that the openings are spaced at uniform intervals and the amount of air circulating in the room can be regulated closely.

The building and the roof proper are of simple column-and-beam construction, with the usual brick walls filling in the panels. Fortunately, the columns and foundation walls rest on solid rock, so that no expensive footings, such as might have been required by the immense concrete columns, were needed. In designing the structure, no attempt was made to attain architectural beauty at additional expense. The plant is situated where it is not readily observed, and the main objects were maximum capacity and efficient and convenient operating conditions at a minimum investment. Originally, the designer contemplated using a pilaster and elaborate brick cornice effect. As a final summary it was found that money could be saved by leaving the concrete columns and beams exposed, and using brick simply to fill in the panels. This construction presents a neat and substantial appearance and serves the purpose admirably. The roof proper, consisting of long-span hollow tile, covered with concrete, is supported by reinforced-concrete beams 14 in. wide by 38 in. deep.

Large Bunker Capacity Required

Because of irregularity in car service and limited storage facilities in the yards, an enormous bunker was designed to hold 10 tons per lineal foot, giving a total bunker capacity exceeding 1000 tons. It was erected over the firing aisle and, like the boiler room, has no waste space. The bottom of the bunker was located at a height that would permit drawing the tubes, and this came at an elevation that made it possible to support the main roof beams on the large concrete columns erected primarily for the bunker. Continuing this simplicity of construction, the bunker was made rectangular, the slopes required at the bottom to prevent dead pockets of coal being obtained afterward by means of cinder fills, as indicated in Fig. 4.

Concrete girders having a span of 25 ft. over the boiler fronts support the dead load of all the upper bunker construction as well as the coal in the bunker. The girders are 3 ft. wide by 9 ft. deep and are heavily reinforced. The main bunker columns, 18 x 50 in. in cross-section, have been placed at the front corners of the boiler settings and arranged to give maximum aisle space between boilers, the narrow dimension of 18 in. being crosswise with the boiler. To cut down the unsupported length of column, 40 ft. from floor to roof, a concrete strut is provided at mid-span between the columns on opposite sides of the aisle. In reality the two elements are a split column acting for a single purpose.

As the depth of coal in the bunker may approximate 30 ft., a considerable pressure on the side walls would be possible. To avoid using walls of extra thickness, vertical concrete columns were designed as beams to take this bulging action, so that the panels between, ranging from 14 to 16 ft. wide, could be filled with comparatively light brick construction. This design tended to keep down the dead load and, as it harmonized with the construction of the main part of the building, gave a more pleasing appearance than would walls entirely of concrete, which at one time were considered. At the bottom the bunker is made up of concrete slabs 8 in. thick and having a span of 6 ft. 3 in. These slabs are supported by concrete beams, 18 x 58 in., having a span of 19.5 ft. across the firing aisle and supported on the

FIGURE 1 FIGURE 2

FIG. 3. THE THREE 1,000-HP. BOILERS NOW INSTALLED

FIG. 1. NEW BOILER HOUSE UNDER CONSTRUCTION, SHOWING IMMENSE AMOUNT OF FORM WORK
FIG. 2. EXTERIOR VIEW OF COMPLETED BUILDING

main concrete bunker girders previously mentioned. To avoid the undesirability of having acid water from the coal permeate through the bottom of the concrete work, as had been the case in another plant of the company, it was decided to cover the lower sections of the bunker with a membrane waterproofing consisting of three independent layers of asbestos felt, secured together and to the initial concrete surface by layers of hot asphalt waterproofing cement. This treatment, applied to the top side of the slabs, and protected by brick, has given satisfaction.

At the operating-floor level of the bunker plenty of light and ventilation have been provided by rows of windows along either side. This arrangement gives ideal working conditions for the men having charge of the conveying equipment in this part of the plant. As is customary, the bunker is subdivided and each compartment has a gage at each corner, spaced in feet, so that the operator can determine the depth of coal in the bunker and, from a chart which is furnished, translate this depth into tons of coal. As the coal passes

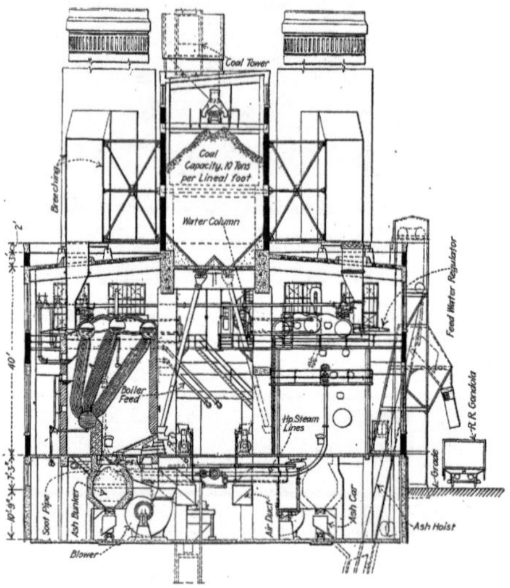

FIG. 4. TRANSVERSE SECTIONAL ELEVATION OF BOILER HOUSE

through a crusher before entering the bunker, it is of fairly uniform size, so that the initial calibration will remain reasonably accurate for the different grades of fuel that may be used.

At present the boiler room contains three units, each having 10,032 sq.ft. of steam-making surface and sufficient area in superheating coils to maintain a superheat of 125 deg., the operating pressure ranging from 200 to 225 lb. gage. These boilers will supply the new 20,000-kv.-a. turbine, which, at normal rating and 80 per cent. power factor, will deliver 16,000 kw., so that the ratio

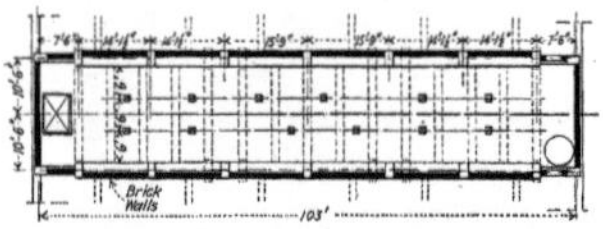

FIG. 5. PLAN OF BUNKER, SHOWING CONCRETE COLUMNS IN SIDE WALLS AND SUPPORTING GIRDERS

between boiler horsepower and kilowatt generating capacity is 1 to 5.3. This ratio is perhaps a little higher than the average, but the boilers may be operated up to 300 per cent. of rating. On the other side of the firing aisle every preparation has been made for the future units, so that their installation will be comparatively simple.

Each of the present boilers covers a floor space 23 ft. 6 in. wide by 21 ft. 1 in. deep, an area of 496 sq.ft., reducing to 0.5 sq.ft. per horsepower of boiler rating. From the floor to the top of the steam opening the distance is 26 ft. 6 in. The center of the mud drum is 7 ft. 6 in. above the floor line, an unusual height. Reference to the sectional view of the plant will show that there is no arch construction over the underfeed stoker serving the boiler, so that a straight front wall provides a furnace of large volume. High-grade firebrick laid in high-temperature cement line the front walls and the side walls back to the first pass. The balance of the brickwork in the setting is according to standard practice.

To guard against excessive radiation, the side walls back to the first pass are covered with insulation attached to ⅜-in. V-ribbed wire lath anchored to the walls by use of ⅜ x 2½-in. lagscrews, and expansion shields properly placed. Over the lath is applied a ½-in. thickness of No. 302 plastic asbestos cement, then 1 in. of No. 400 plastic asbestos cement, followed by poultry netting, over which is another ½-in. thickness of cement covered by a 6-oz. canvas jacket. The lower portions of the settings along the aisles between boilers are also protected.

Each boiler is equipped with a 22-tuyere underfeed

stoker, with extension grates and power-operated dumping mechanism. The stoker is driven by a vertical engine on the boiler-room floor, connected to the stoker through a jackshaft in such a way that it is possible to use two gear ratios. When operating at low rating, the smaller ratio is used, but at periods of overloads the gears of higher ratio are brought into action by means of a clutch. These provisions have been made in addition to the usual regulation of engine speed. Normally, the jackshafts are independent of each other, but in case of an engine failure may be linked together, so that the engines of the other two units may drive all three stokers. This entails operating all stokers at the same speed, which is not considered desirable, but is better than facing the possibility of shutting down an entire boiler unit of large rating, owing to failure of the drive.

Immediately under each stoker is a forced-draft turbine-driven fan, having capacity to supply sufficient air to operate the boiler at 300 per cent. of rating. Although each stoker has its individual unit, connections are made very gradual, so that little resistance is offered to the flow of gases.

Between the uptake connections to the breeching and the last connection and the stack, expansion joints are provided. To control the course of expansion, the breeching is anchored to the concrete roof at midpoints between the expansion joints, and to afford easy travel to or from the joints, the portions of the breeching near the joints are mounted on rollers.

With the breeching outside, insulation was considered desirable to avoid excessive temperature drops. In this particular case 2-in. vitrobestos was secured to the inside of the steel plate by ¼ x 3-in. bolts staggered on 18-in. centers. The joints between sheets were pointed up, and all exposed nuts or other metal thoroughly coated and protected from the gases of combustion. Placing the insulation on the inside did not affect its heat-resisting qualities and, besides, gave the additional advantage of protecting the metal. With the inside surface of the metal protected in this way and the exterior properly painted, deterioration should be less

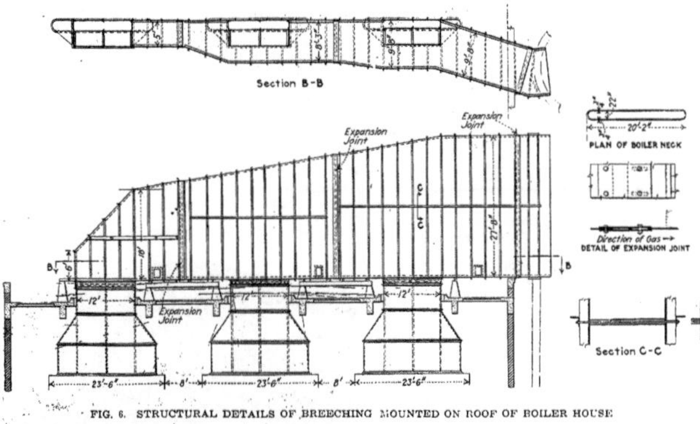

FIG. 6. STRUCTURAL DETAILS OF BREECHING MOUNTED ON ROOF OF BOILER HOUSE

between the air ducts of adjoining stokers, so that it is possible to operate three boilers on two fans during periods of shutdown of any particular unit.

One of the interesting features of the plant is the outside breeching mounted on pedestals supported by the roof beams, which were made heavy enough to take the additional load. The present breeching accommodates the three boilers now installed, and there will be a similar breeching on the other side, for which the foundations have been prepared. Either breeching will be served by a concrete stack 16 ft. 6 in. inside diameter at the top, and 238 ft. high from the base, which is 20 ft. below the boiler-room floor line. As will be apparent from Fig. 6, the breeching proper is rectangular in cross-section and tapers both in width and in height. Owing to the necessity of locating the chimney slightly off center with the uptakes, the breeching was increased in width toward the side of the offset, thus forming a natural path for the gases from the various boilers as progress is made toward the stack. The slopes are rapid than with the usual design using unprotected plates, and the life of the breeching greatly prolonged.

One of the noticeable features in the boiler room is the lack of complicated piping. Much of it has been carried to the basement, which, with headroom of 18 ft., offered excellent facilities for suspending the piping underneath the floor slabs, the construction having been designed amply strong for the purpose. This particular type of construction afforded an easy and economical method of support. At points of anchorage proper inserts were placed in the concrete, so that the line could be rigidly secured and forced to take the expansion as designed. All piping connections are direct, and with a minimum number of bends. Few fittings have been used and turns have been made by long-radius bends, which also provide for expansion. From each boiler branches feed into a short distribution header located midway under the central firing aisle. When the future boilers are installed, it will receive steam from either side. The header now feeds over to the old plant, but has been so valved that it may be sectionalized, and when supplying the future turbine it will be possible to operate on the unit plan.

It may be of interest to add that the high-pressure steam lines are insulated with double-asbestos-sponge-felt, 1½ in. thick, covered with canvas. The fittings are covered with ½ in. of fire-felt cement and 2½ in. of asbestos-sponge-felt cement covered with a canvas jacket. All high-pressure flanges have removable flange covers consisting of 2¼-in. asbestos-sponge-felt, lined with rolled firefelt, with covers to fit over the insulation. The exhaust piping is protected by 1-in. asbestos-cell insulation, with canvas jacket; the hot-water piping has 1 in. of asbestos-sponge-felt insulation, and all exhaust

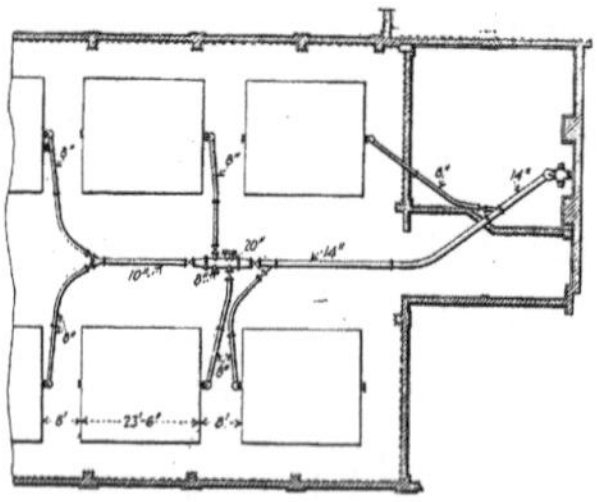

FIG. 7. HIGH-PRESSURE STEAM-PIPE LAYOUT

and hot-water fittings are insulated by a scratch coat of No. 352 cement finished with a No. 400 cement and canvas jacket. Boiler drum-heads have 2-in. magnesia blocks wired on and finished with ½-in. thickness of No. 400 cement and canvas cover.

Coal is brought to the plant in drop-bottom cars, which empty into two track hoppers 14 ft. wide by 28 ft. long. A motor-driven apron conveyor with a capacity of 70 tons per hour delivers the coal either to the crusher or direct to cross-conveyor belts, which in turn discharge onto another belt, carrying the coal to a 2-ton skip-hoist located at the end of the boiler house. The hoist is of standard passenger-elevator type, driven by a 35-hp. two-phase 440-volt motor and arranged so that the operation can be entirely automatic or controlled by hand. A skip-car travels from the loading point to the dump position, a vertical distance of about 108 ft., with full load, at a speed of 225 ft. per min. and an ultimate capacity of 80 tons per hour.

In the loading position the skip-bucket rests on a compression spring, adjustable for different loads. When the spring compresses to the desired point, a contact is made, actuating a relay which closes the cutoff valve controlling the coal feed from the hopper supplying the bucket. In closing, the valve actuates the starting relay for the hoist. The bucket is elevated to the dumping position, empties and returns to its original position, automatically opening the hopper valve to refill the bucket with coal and repeating the same cycle as before. The receiving hopper at the head of the belt conveyor is large enough to allow for small variations in the flow of coal. Another belt conveyor receives the coal from the skip-hoist and, by means of a traveling tripper, distributes it to any section of the bunker desired. All conveyors are driven from the discharge end by motors connected to spur-gear reducers which in turn are coupled to the conveyor shafts. These reducers consist of a series of gears entirely inclosed in a cast-iron housing and running in oil. With their motors the reducers are mounted on cast-iron bases with flexible couplings between motors, reducers and driven shafts. The unit is unusually small and compact, forming a pleasing contrast to the large and complicated gearing generally employed. A point worthy of mention is that all belt drives are identical, so that there is no possibility of one belt feeding to another at a rate faster than the receiving belt is able to handle.

Gates controlled from the floor admit the coal from the bunker into universal spouts, which deliver into chutes tricated at the discharge end. There are two spouts per stoker, and the six feeds give such excellent distribution of the coal that it is not necessary to do any shifting of the spouts or the coal. Owing to the difficulty of getting steel plate to make up these spouts, 10-in. pipe that had been used formerly as a heating main served the purpose. As the diameter of the pipe was considered a little small, an endless chain was passed through the chute so that it would be an easy matter to break up any clogging of coal that might take place. In case it should be desired later on to put in coal-weighing equipment, inserts have been provided for supporting the structure, and there is plenty of height to make the installation.

Under each boiler an ash hopper of sufficient capacity to care for a normal run of ten to twelve hours has been provided. To protect the concrete from excessive heat and abrasion from the ash and clinker, the interior is lined with one layer of paving brick laid in high-temperature cement. Underneath the hopper has a 24 x 36-in. discharge opening, closed by a low-body duplex cutoff valve, through which the ashes drop directly into a push-car. A loop track, taking in both rows of boilers, has been provided, and arrangements have been made so

FIG. 8. MOTOR AND SMALL REDUCER DRIVING BELT CONVEYOR OVER BUNKER

that all cars, if desired, may be passed over a weighing scale. The cars dump directly into a 1-ton ash skip-hoist, which raises the ashes to the outside of the building and dumps them directly into a railway car. A grating over the inlet to the skip-bucket limits the size of the clinkers so that lumps larger than 10 in. cannot be passed through without first being broken up by a sledge.

Boiler-feed water is taken from two 7500-hp. open feed-water heaters. Underneath the heaters is a gang of centrifugal pumps provided with flexible piping connections. A loop system has been provided for the suction lines from the heaters to the pumps, so that it is easily possible to sectionalize the feed water from any individual heater to a particular pump and thus make special arrangements when conducting boiler tests. The discharge lines also feed into a ring header giving similar flexibility. The feed-water system extends over into the old boiler plant and feeds into two loops, making in reality four feed-water lines to supply the older boilers. This was more duplication than was considered necessary for the new plant, as water conditions are fairly good and consequently a feed from either one of these loops was deemed sufficient protection against stoppage of the water supply. Under normal operation the feed to each boiler is controlled by a feed-water regulator. To insure a continuous feed to the boilers, an auxiliary loop with hand control has been provided.

Makeup water is obtained from the Mississippi River, but in case the house-service pumps should be down or the supply limited, connections have been made to draw on the high-pressure city service. Originally, the feed-water heaters were installed to give approximately a 6-ft. head on the pumps. Apparently this head was not sufficient, as the operating conditions resulting caused excessive maintenance charges. To relieve the situation, the heaters were raised to give a head of 10 ft., which was the maximum that could be obtained, owing to interference with other construction.

One of the interesting features of the boiler room is the series of runways provided to give ready access to all desired points. At the end of the boiler room steel stairways connect floor and roof. Over the center and running the full length of the firing aisle, a runway is suspended from the girders supporting the bunker. From this central location runways extend at right angles between the boilers and around to the rear, with offsets to give access to the water columns, which are mounted in front of the concrete columns rather than on the boilers, so that they will be in plain view from the firing aisle. At the rear access may be had to the feed-water regulators. Ladders lead to the tops of the boilers, where walks have been provided giving access to the safety valves and the feed-water lines. Lower down there is a second runway on a level with the check valves and the superheater clean-out doors. Another runway at the rear of the boiler is on a level with the upper clean-out doors. It will be evident that convenient and ready access were features given due consideration in the design. To insure safety to the operator, all runways are provided with railings.

FIG. 9. TURBO-GENERATOR UNITS; THE NEW 20,000-KV.-A. MACHINE IN THE FOREGROUND

One of the essentials of a modern plant is instruments, and with these the new boiler house is amply provided. Each boiler has a panel mounted on the front of a bunker column. It contains a steam-flow meter with a recording chart and also calibrated to read directly in thousands of pounds. There is a recording thermometer for the flue gases, a CO_2 recorder and a three-in-one draft gage giving readings from the windbox, over the fire and at the uptake. Low down on the boiler front are two steam-pressure gages connected to different points of the steam drum, the idea being that one shall serve as a check on the other, or as a duplicate in case one of the gages goes out of commission. For the same reasons there are two water columns per boiler brought out to the column fronts, as previously explained.

At the end of the firing aisle is automatic regulating equipment controlling the speed of the stoker engines, the forced-draft fan turbines, and a master damper in the breeching between the stack and the three boilers. Here also is a general instrument board containing recording steam-flow meters for each of the three principal generating units and a meter recording the steam going to the boiler-room auxiliaries. On the board there is a recording pressure gage, and an integrating meter on the steam line to the auxiliaries. With the data from these various instruments the boiler-room operator will be able to keep posted on the demands of the turbine room and separate the steam between that required for generation and that for the auxiliaries in the boiler plant. Duplicate flow instruments on individual panels in the turbine room give the operators there much the same advantage.

Changes in the Turbine Room

In the turbine room the only addition to the generating capacity is the 20,000-kv.-a. reaction turbine previously mentioned. It is supplied with steam at 200 lb. pressure and 125 deg. of superheat. The machine runs at a speed of 1800 r.p.m. and generates 4800-volt two-phase 60-cycle current. The condenser is of the surface type, containing 22,000 sq.ft. of surface, or approximately 1.4 sq.ft. per kilowatt of generator rating at 80 per cent. power factor. The auxiliaries are all of the rotary type and turbine-driven. The circulating pump has a capacity of 28,000 gal. per min., giving a ratio of approximately 70 lb. of water to 1 lb. of steam condensed. The cooling water is drawn from the Mississippi River just above the dam, and as the tailrace into which the water is discharged is ten feet lower, it is necessary to use the circulating pump only to start the flow. When it has been established, the difference in head will maintain the flow through a bypass around the pump.

A ventilating fan having a combination motor and steam-turbine drive supplies cooling air to the generator. Ducts are provided whereby the air can be supplied from the outside and returned through a separate discharge duct. In extremely cold weather, on account of the large temperature difference between the outside and inside air, undesirable operating conditions developed from condensation in the fan, forming into ice on the casing. To eliminate these difficulties, during the colder periods the cooling air is taken from the basement and discharged back into the generator room at a point some distance from the turbine. In this way the operator can control the temperature of the air that is being handled and also regulate the temperature in the turbine room, where a certain amount of heat during the winter months is desirable.

At the time of installing the air-inlet ducts, the great difficulty in securing plate steel made a change in the design to reinforced-concrete construction desirable. The walls of the duct were made of concrete 2½ in. thick and suspended from steel anchored in the concrete floor slab above. The bottom of the duct is reinforced and supported from the side walls. This type of construction made it possible to complete the work at the desired period for operation, and it has proved well adapted for the service.

Provision has been made for flooding with steam the generator ventilating system in order to quench such a fire as had developed previously in the 10,000-kw. turbo-generator. In this machine a short-circuit caused the armature to burn out, much of the damage being due to the inability of the operator to put out the fire. To guard against a similar emergency, steam connections were made into the air duct. Dampers in the inlet and the outlet were inserted to make a closed compartment possible, so that fresh air might be shut off from the machine during a fire. The main valve controlling the steam supply is sealed in a glass case to eliminate the possibility of an operator unintentionally opening the valve and admitting live steam into the generator during periods of normal operation. As a precaution against leakage of steam into the duct a second valve has been placed ahead of the main valve. In the pipe between the two valves a small hole has been bored, so that any leakage will be evident and the operator will be given the opportunity to make the necessary repairs.

On the electrical end there is no outstanding feature, as the construction is standard and made up of good, substantial equipment. The bus is of the ring type, to which the various generators of the plant are connected through circuit-breakers. From this ring power is supplied to the Rock Island Arsenal, the City of Rock Island and to circuits going to a new transformer building, where the current is stepped up from 4800 volts, generator voltage, to 13,200 volts for distribution.

All engineering design and construction work in the new additions to the station are credited to the engineering department of the company.

Home-Made Emergency Valve

By James E. Noble

The sketch shows a home-made emergency valve that will give fair service for a considerable period when used with water at a low pressure. The tee *A* is attached to an upright water pipe; *B* is a short nipple connecting the reducing tee *C* to *A*. A cone *D* is made from hard rubber, fiber, lead or other material. A hole

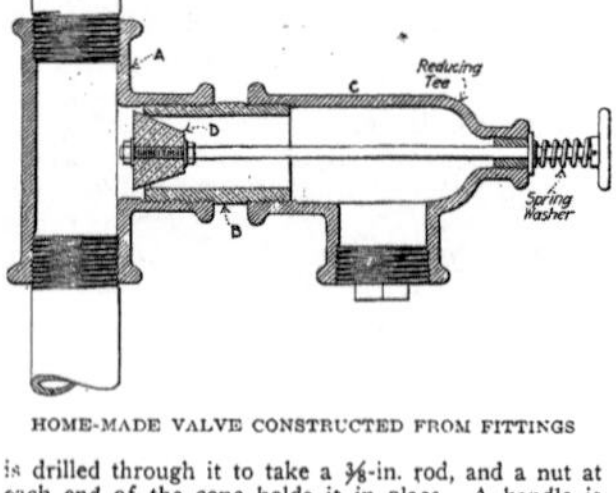

HOME-MADE VALVE CONSTRUCTED FROM FITTINGS

is drilled through it to take a ⅜-in. rod, and a nut at each end of the cone holds it in place. A handle is screwed on the outer end of the rod.

To open the valve, push in on the rod handle. If there is considerable pressure in the pipe, the valve will close as soon as the handle is released. Some water may escape past the handle, but this can be prevented to some extent by placing a leather washer around the rod, holding it in place with a light spring, as shown. A strong spring will help to hold the valve shut.

The Lubrication of Steam-Engine Bearings

By W. F. OSBORNE

Beginning with a discussion of the lubrication of bearings in general, this article continues with a description of the various systems of oiling steam-engine bearings, from the primitive "squirt-can" method to the modern force-feed system and the splash type. The factors affecting the operation of lubricants in each case are discussed and recommendations made as to the proper oils to specify. The use of grease is discussed briefly.

WE ALL know that bearings are lubricated to reduce friction and wear, thereby lowering the wasted power and increasing the life of the rubbing parts. Although any substance placed between

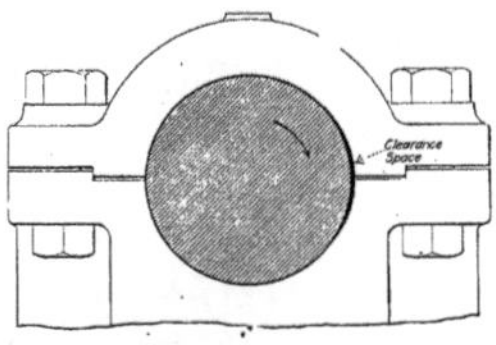

FIG. 1. WEDGE-SHAPED OPENING IN BEARING CLEARANCE

these rubbing surfaces and reducing the friction might properly be called a lubricant, we ordinarily think of lubricants as being oils or greases.

Perfect lubrication demands the complete separation of the rubbing parts, this separation depending upon the thickness of the lubrication film which must be maintained continuously. The friction of the bearing has now been changed to the sliding friction between the metal surfaces and the lubricant itself.

Efficient lubrication therefore depends upon the condition of the bearing surfaces and the body or viscosity of the lubricant. A properly designed bearing will have enough clearance between the shaft and the bearing metal to hold an oil film and will also have a wedge-shaped opening, as shown in Fig. 1, to allow the oil to feed into the clearance space. A reversing shaft should have this space on both sides. The oil is carried into the clearance by the friction between it and the shaft, and at high speeds it is carried in faster than at low speeds and the oil film can be maintained more successfully.

As has been pointed out before, the lubricant must keep the metal surfaces apart. The load on the shaft squeezes the lubricant out, the thinner-bodied oils being forced out more rapidly than the more viscous ones. It is quite practical to use thin oils when speeds are high and loads are light, but if the weight on the bearing is great enough to squeeze out the film and if the oil cannot be carried into the clearance space rapidly enough to maintain it, wear will result. Heavy oils do not squeeze out so rapidly and consequently stay in the bearing clearances longer than thin oils. On the other hand, if the shaft were running faster, sufficient thin oil might be drawn in to replace that squeezed out, thus maintaining the lubricating film.

This action might lead us to believe that heavy oils or greases were best for all bearings if it were not for other factors that must be considered. These are internal friction of the oil and the friction between the oil and the metal. Both vary with the viscosity of the oil, its quality and the character of the rubbing surfaces. Sometimes, if heavy oils are used on high-speed bearings, the friction is great enough to cause high temperatures, which can frequently be lowered by using a thinner oil. As the high temperature is a direct indication of power loss, we should make our specification read: "Use the thinnest oil that will prevent wear within a safe margin." For practical purposes oils of the same viscosity have very nearly the same coefficient of friction, although the better grades have higher lubricating values and a longer life.

Any friction existing in a bearing, however well lubricated, shows up in a rise of temperature. A hot bearing does not always mean that the metal is wearing, as the high temperature may be caused by the friction of the oil when the shaft is turning at high speeds. If the temperature rises slowly and remains below the softening point of the bearing metal, there is no need for worry; but if the rise is rapid, it is probably due to wear, and the cause should be investigated promptly and removed if possible. Bearing temperatures can be lowered by increasing the volume of the oil flowing over them.

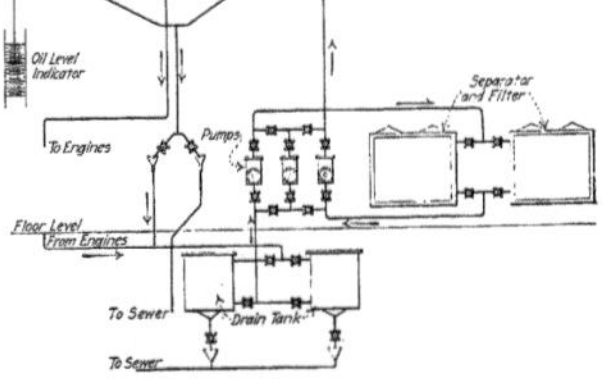

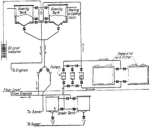

FIG. 2. GRAVITY OILING SYSTEM DESIGNED FOR STEEL-MILL ENGINES. NOTE DUPLICATE TANKS, FILTERS, ETC.

The oil absorbs the heat from the bearing and carries it away to cooler parts of the engine. It is quite possible to feed so much oil over the outside of the bearing as to carry away all the heat due to actual wear and the bearing may be wearing away without ever becoming hot to the touch. This might be caused by feeding large quantities of a thin oil to a slowly moving, heavily loaded bearing with large clearances. The large clearances would lead the oil away to the ends of the bearings, thus cooling the shaft without very much of the oil ever getting in between the wearing surfaces.

There are so many factors which affect the lubrication of a bearing—loads per square inch, rubbing speeds, radiated heat, clearances, character of surfaces, kind of bearing metal, variation of speed or load, quantity and frequency of oil supply, etc.—that it is rather difficult sometimes to select the most efficient lubricant without actual test. If we could always be sure of all the conditions, we could make very definite recommendations, but until we can determine them, we must necessarily allow a pretty large margin of safety.

The parts to be lubricated on steam engines are the main journals, crankpins, connecting-rod pins, crossheads and guides, eccentric straps, valve gear, governor, cylinders and valves. It is customary to use the same kind of lubricant on all of these parts with the exception of the cylinder and valves, the grade depending chiefly upon the operating conditions of the engine and the oiling methods and appliances.

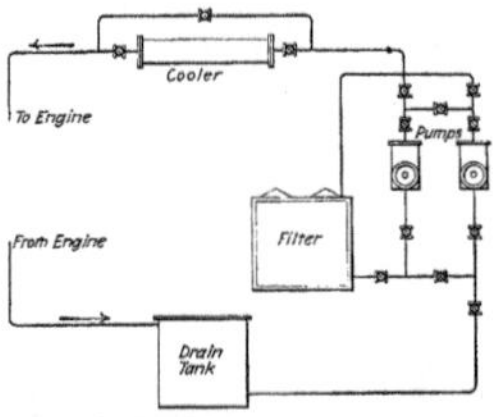

FIG. 3. FORCE FEED OILING SYSTEM

The most primitive way of lubricating an engine is to squirt a little oil from a squirt can into a hole in the bearing cup and let the oil run down to the surfaces. If the oil does not last until the oiler gets around again the bearings will run dry and wear. Lightly loaded engines sometimes use thin oil, but heavier oils are always more economical and are more likely to stay in the bearings until the parts are oiled again. Only small engines of the older types are now lubricated in this way, although minor bearings of small machinery are sometimes oiled by this method. Medium- or heavy-bodied oils of from 250 sec. to 500 sec. viscosity at 100 deg. F. are best suited for this service.

An improvement in lubrication was made when the oil cup with its drip feed was introduced. This method maintains a fairly regular supply of oil, requires less attention from the oiler and has been used for a great variety of operating conditions. Medium-bodied oils of between 175 sec. and 300 sec. viscosity at 100 deg. F. are usually selected for engine bearings lubricated in this manner. If the oil is used only once and is permitted to go to waste, the less expensive grades of oils are good enough, but if the oil is caught in pans, filtered and used over again, it pays to buy the better grades.

The manifold, with its reservoir and various leads to different bearings, is merely an extension of the drip-cup idea with the thought of cutting down labor and attention on the part of the oiler.

The ring and chain oilers were introduced to provide a continuous flood of oil to the bearing surfaces and to reduce the cost of lubrication by automatically returning the oil to the bearing. The flood of oil carried up by the ring or chain thoroughly lubricates the surfaces, washes out the dirt and minute bits of worn metal, and carries away part of the generated heat, thus cooling the bearing. The oil reservoir below the bearing should hold enough oil so that the dirt will have time to settle out to the bottom, and to permit the radiation of the absorbed heat. Light- or medium-bodied oils of from 150 sec. to 200 sec. viscosity at 100 deg. F., depending upon the loads, speeds, etc., will handle this sort of work all right. The better grades of oils are more economical, as they last longer, do not evaporate so quickly and maintain their original viscosity longer than the cheaper oils. This evaporation increases the viscosity of the remaining oil and causes the foreign matter to settle out more slowly.

An extension of the ring-oiler idea is the continuous circulating system with which most modern engines are now equipped. There are four types, commonly called gravity feed, forced feed, splash feed and automatic feed. The first one, shown in Fig. 2, consists of an overhead tank, commonly called a "gravity" tank, holding a supply of oil. Oil pipes lead the oil from this tank to the various parts requiring lubrication to which it is fed continuously in rapid drops or in a fine stream. The quantity needed for any bearing surface can be regulated very easily, and all parts may be flooded with oil if desired, washing out foreign matter and cooling the bearing by carrying away the generated heat. The oil flowing from the bearings is collected in troughs, pans or the engine bed and drained to a catch tank located below the engine. Sometimes a portion of the engine bed is arranged to serve for this purpose. From the catch tank the oil is pumped back to the gravity tank to be used again, having first been cleaned by passing through filters.

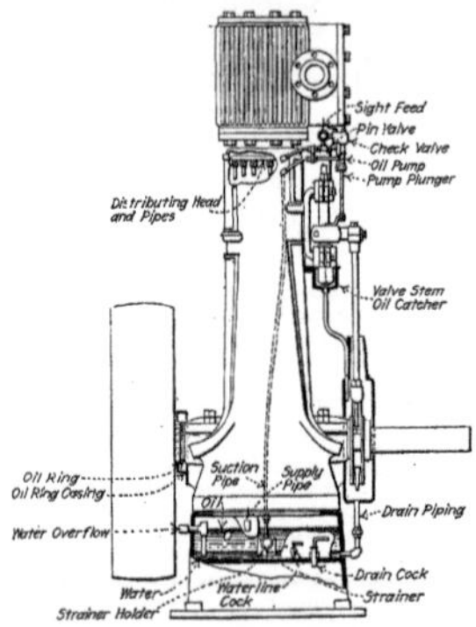

FIG. 4. AUTOMATICALLY LUBRICATED ENGINE

The force-feed system shown in Fig. 3 does not always have the gravity tank, and the oil is forced directly from the oil pump to the bearings under a pressure of from two to ten pounds. The gravity tank is always a good thing to have in starting the engine unless a separate steam pump is installed to be operated independently of the engine. The force-feed system has the advantage that the flow of oil through any bearing can be increased by raising the pressure. In the gravity system shown in Fig. 2 the flow of oil over the outside of the bearing can be increased greatly, but the amount of oil actually passing between the surfaces depends on the bearing clearances and the body of the oil, as the pressure forcing the oil through is fixed by the height of the gravity tank above the engine. It is customary also to install a sight-feed oiler just above the bearing, which absorbs the pressure due to the height of the gravity tank and reduces the pressure head on the bearings to a few inches.

In both of these systems the oil from the bearings carries small particles worn from the bearing metals, bits of dirt and foreign matter settling on the engine, condensation water leaking through the piston-rod and valve-rod stuffing-boxes and saponified cylinder oils carried in by the condensation water. If these foreign elements were allowed to accumulate, they would soon contaminate the oil to such an extent that it would no longer be of any value as a lubricant. Both systems should be equipped with settling tanks or separators for removing the water and heavy foreign matter, and filters to remove the lighter particles which will not settle out when the oil is at rest.

We now have something else to consider besides viscosity, and that is the ability of the oil to separate from water and foreign matter. As mentioned in the first article of this series, the quality of the oil, its viscosity and its specific gravity affect its separation from water. Of these the most important is quality, which can be regulated initially only by the refiner. The viscosity, which is second in importance, can be changed by heating the oil in the system until it becomes thin, thus permitting the water and dirt to settle out more readily. Heaters are placed in the separators, and the oil is heated to 110 to 125 deg. F., depending on its viscosity. The circulating system, by providing a flood of oil to all parts, permits the use of light-bodied lubricants which separate from water quickly and can be cleaned very satisfactorily by an efficient filtering system.

The quality of the oil that should be used depends to a certain extent on the efficiency of the system as a whole. If the system is tight and there is very little loss through leakage or by splashing, a high-grade oil can and should be used. In such a system very little new oil is added to make up for loss and the old oil remaining in the system should have all the qualities necessary for efficient cleaning. Again, the continuous evaporation of the oil when passing over the bearings drives off the lighter portions and the viscosity of the remaining oil gradually increases. On this account it is advisable to start out with a fairly light-bodied oil, even though the intitial evaporation may be greater, because the viscosity of the oil in the system can be maintained better by the addition of a light-bodied oil than by a medium oil.

If the system is constructed so that it loses considerable oil through leaky pipes, splashing from the engine as is the case with vertical engines, burning up of the oil splashed against hot cylinder heads, and other numerous causes, the quantity of the makeup oil is much greater, the oil does not remain in use so long and its viscosity does not increase as rapidly. Consequently, less expensive oils may be used if desired.

Splash Systems of Lubrication

Splash systems supply a flood of oil to all parts when the engine is running, but have the disadvantage that they must depend upon enough oil remaining in the bearings to lubricate them when first starting the engine. A splash-lubricated engine should be tightly inclosed—not only to prevent loss of oil, but to keep out moisture and dirt. In such systems the oil is constantly churned up and has very little opportunity to come to rest and permit the water or foreign matter to settle out. Consequently, a very high grade of oil should always be used, one that will separate from water quickly and one having a minimum evaporation during use. Excessive evaporation from an improperly refined oil will cause the viscosity to increase rapidly and the oil will not clean itself as it should. When the oil becomes loaded with foreign matter up to a certain point, it will emulsify or form a thick, heavy mass, which will clog up the oil pipes and grooves and prevent any oil getting to the working surfaces.

The automatically lubricated engine shown in Fig. 4 has a pump to take the oil out of the crank case and deliver it to the bearings. After lubricating them, the oil returns to the crank case for further use. The presence of water and the rapid agitation necessitate the use of a high-grade oil like that recommended for the splash method.

Some General Recommendations

In picking out oils for each of these systems we must consider all the individual operating conditions carefully and the effect they will have upon the oil to be selected It is possible, however, to make general recommendations as follows:

Splash and Automatic Systems—Use as light an oil as will lubricate the bearings. Viscosities 150-180 sec. at 100 deg. F. Highest quality obtainable and with best separating qualities as shown by emulsification tests.

Force Feed or Gravity Systems—Use light- or medium-bodied oils. Viscosities 150-200 sec at 100 deg. F. Light-bodied, high-quality oils are best for efficient systems, medium-bodied good-grade oils are suitable for less efficient systems.

Ring or Chain Oiled Bearings—Use medium-bodied oils of good quality. Viscosities 150-200 sec. at 100 deg. F., depending upon bearing loads and speeds.

Drip Cups and Manifolds—Use medium-bodied oils of 175-250 sec. at 100 deg. F. Unnecessary to use more expensive grades.

Squirt Can—Use medium- or heavy-bodied oil. Viscosities 250-500 sec. at 100 deg. F. Ordinary red engine oils are satisfactory.

A low cold test is desirable where the engine is exposed to low temperatures.

Some engineers prefer to have certain bearings lubricated with grease, which is not to be recommended as the best method of reducing friction to a minimum, although it does cut down the labor of caring for engines on which the expense of installing a circulating system is not warranted. The valve gear of many engines is equipped with compression grease cups which are easier to turn down when the engine is running than it is to try to lubricate the bearings with a squirt can without splashing oil all over the place. Whenever used, the grease should be of a good quality, homogeneous and free from any nonlubricating material which might separate and plug up the grooves.

Snow is the latest agency to be employed by officials in solving Salt Lake's smoke problem, according to H. W. Clark, Bureau of Mines, who is assisting in the smoke probe. Particles of dust in circulation are precipitated with snow, and consequently, the atmosphere is freed from impurities. Samples of snow have been taken by the Bureau of Mines and are now being analyzed. These results will be used to check the data acquired through the use of special apparatus for testing the atmosphere impurities, says Mr. Clark. The snow-testing method is based upon filtering melted snow from certain areas, the solid material containing acids, ammonia, sulphates and other impurities, being weighed and subjected to microscopic examination.

Standard Meter Rates—How Applied in Calculating Power Bills

By LeROY W. ALLISON

The previous article concluded with an explanation of the classification for "meter rates" and "demand rates" as now in common use. The first group, meter rates, is considered in the following, and includes (1) straight-line meter rate, (2) step meter rate and (3) block meter rate.

THERE are three general classes of central-station service—general service, regular power service, and large light and power service. The standard rate forms cover these different phases of operation in a manner that approximates equity to company and consumer. When it comes to a matter of actual contract for certain service, modifications may be made as required, say for temporary service in part, particular line extensions, and so on, but these changes or provisions have little, if anything, to do with the basic type of rate calculation.

Rate Definitions

A familiarity with the exact meaning of the different rate terms in current use is quite essential for an accurate understanding of the schedules. These, as noted in the text of this series of articles, are the definitions as approved by the Rate Research Committee of the National Electric Light Association, to the papers of which organization the writer is indebted for considerable information. The term "meter rate" is applicable to any method of charge for electric service which

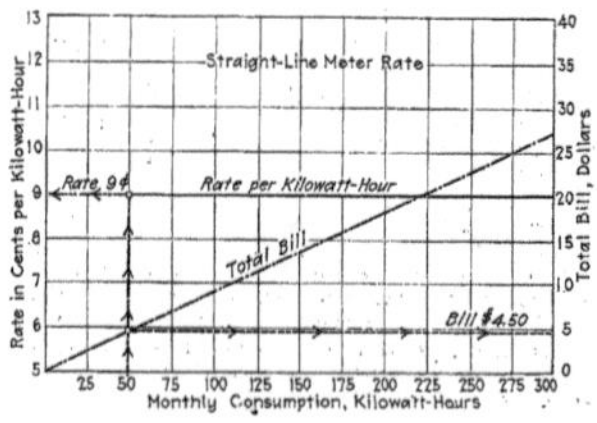

FIG. 1. GRAPHICAL REPRESENTATION OF A STRAIGHT-LINE METER RATE

is based on the amount used. This amount is expressed in units, as kilowatt-hours of electricity. Integrating meters or graphic meters are used.

Straight-Line Meter Rate

The term "straight-line," as used in connection with, and as applied to any method of charge, indicates that the price charged per unit is constant; that is, does not vary on account of any increase or decrease in the number of units used. The total sum to be charged is obtained by multiplying the total number of units by the price per unit.

The straight-line meter rate is extremely simple and is one of the standard forms in wide general use by central stations, particularly in the matter of commercial-lighting service. This type of rate schedule was adopted coincidently with metered method of measuring electric service, first under ampere-hour meters, and later, under watt-hour meters. The charge under

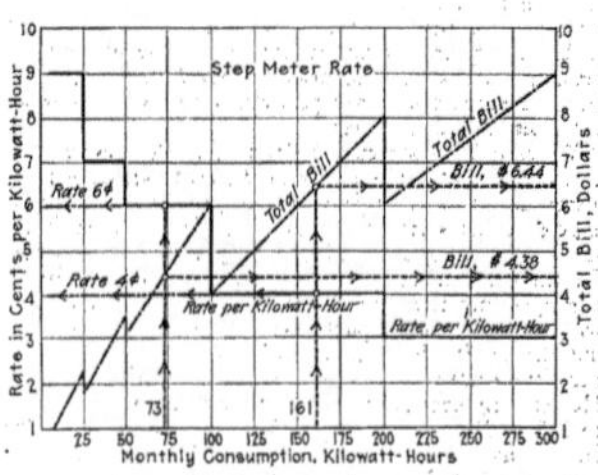

FIG. 2. GRAPHICAL REPRESENTATION OF A STEP-METER RATE

this type of metered service is made at a fixed rate per ampere-hour or watt-hour. Fig. 1 represents this meter rate graphically and shows at a glance the simplicity of this form of computation. As an example, assume a rate of 9 cents per kilowatt-hour and a total monthly service aggregating 50 kw.-hr., the charge would be $50 \times 9 = \$4.50$, as indicated on Fig. 1.

Step Meter Rate

The term "step," as used in connection with and as applied to any method of charge, indicates that a certain specified price per unit is charged for all or any part of a specified number of units, with reductions in the price per unit based upon increases in the number of units according to a given schedule. The total sum to be charged is obtained by multiplying the total number of units by the price applying for this number of units, or, in other words, by the primary price, and then deducting the discount applying for this number of units.

In further explanation of this standard definition the unit rate as charged depends on the particular "step" within which the total consumption comes. The term "step" typifies the character of the plotted rate, in its resemblance to a flight of steps, as will be noticed in Fig. 2. This diagram has been plotted for a step meter rate varying from 9c. to 3c. per kw-hr. on the following basis: For electric-energy consumption from 1 to 25 kw.-hr. per month, 9c.; from 26 to 50 kw.-hr., 7c.; from 51 to 100 kw.-hr., 6c.; from 101 to 200 kw.-hr., 4c.; and all over 200 kw.-hr. consumption in a month, 3c. Under this schedule the average rate per kilowatt-hour is 4½c., and as will be seen, the rate decreases as the consumption increases.

The figuring of the total charge under this rate system is a very simple matter. Assume that a customer has used a total of 73 kw.-hr. during the month; this would fall within the 6-cent step rate, and the total charge would be 73 × 6, or $4.38. Or, again, assume that a total of 161 kw.-hr. has been used during the month; the rate would be within the 4-cent step, with

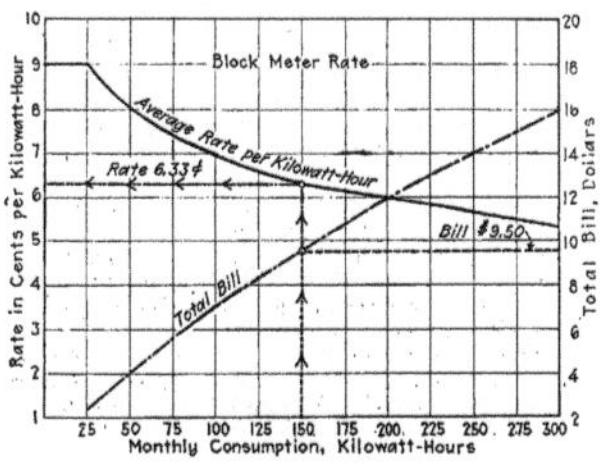

FIG. 3. GRAPHICAL REPRESENTATION OF A BLOCK-METER RATE

total charge, 161 × 4, or $6.44. The dotted lines, Fig. 2, show how the bills may be read from the curves. Owing to its simplicity, this method of rate computation is in wide general use both for wholesale and retail power service, as well as for general and commercial lighting work.

Block Meter Rate

The term "block," as used in connection with and as applied to any method of charge, indicates that a certain specified price per unit is charged for all or any part of a block of such units, and reduced prices per unit are charged for all or any part of succeeding blocks of the same or a different number of such units, each such reduced price per unit applying only to a particular block or portion thereof. The total sum to be charged is obtained by multiplying the number of units in the first block by the price per unit for that block and adding thereto the number of units in the second block times the price per unit for that block, and so on until the sum of the units falling within the different blocks equals the number of units to be charged for.

Under the block-meter rate system, certain blocks are arranged covering a specified number of kilowatt-hours in each block, as 25, 50 or 100 kw.-hr. or as the case may be. This type of rate computation is extremely simple, readily understood by consumers and allows for rapid calculation of bills in the clerical department of the central station. It has been adopted by many companies for general lighting, wholesale lighting and power, as well as for retail power and other character of service. This method of figuring charges corresponds in a general way to rates of the demand type.

This rate basis is shown diagrammatically in Fig. 3, the two curves covering the average rate per kilowatt-hour and the total bill being plotted for the specific conditions as apply. In this instance, the rate varies from 9 to 4 cents for the different blocks, as follows: For the first 25 kw.-hr. used per month, 9c.; for the next 25 kw-hr., 7c.; for the next 50 kw.-hr., 6c.; for the next 100 kw.-hr., 5c.; and for excess of all over 200 kw.-hr. used per month, 4 cents per kilowatt-hour. As an idea of the method of computing bills, say that a consumer uses a total of 150 kw.-hr. under this system of charges, in a monthly period. The first 25 kw.-hr. would be figured at 9c., the next 25 at 7c., the next 50 at 6c.; and the last 50 at 5 cents, or

25 x 9 = $2.25
25 x 7 = 1.75
50 x 6 = 3.00
50 x 5 = 2.50

Total bill $9.50 for 150 kw.-hr.

The average rate is equal to the total bill, in cents, divided by the total kilowatt-hours or, in this problem, equals 950 ÷ 150 = 6.33c. per kw.-hr. These values are indicated in Fig. 3 by the dotted lines.

The block-rate system is on a basis of metered service. The difference between this operating schedule and the step-meter rate plan, previously explained, will be readily noted from the examples given. Under the step-meter schedule, the rate at the particular step of consumption applies to the gross sum of the kilowatt-hours used, while with the block-meter rate, each block, or division of consumption rate charge, is computed on the unit basis as arranged for the block.

Power Service Under Block Meter Rate

An exact illustration of the use of the block meter rate system by a large Eastern central station will be interesting, this covering the application to the state utility commissioners for an increase in the existing uniform retail power rate, for which the block meter schedule is employed by the company. In the application and hearing, the data on this point as set forth were as follows:

Blocks	Uniform Retail Power Rate	Present Rate	Proposed Rate
First Block	10 cents to be paid per kilowatt-hour of consumption each month up to and including an amount of kilowatt-hours per horsepower of maximum demand of	20 kw.-hr.	30 kw.-hr.
Second Block	To be paid per kilowatt-hour for the next 50 kw.-hr. consumed in such month in excess of the first step of the rates	6 cents	8 cents
Third Block	To be paid per kilowatt-hour for the next 500 kw.-hr. consumed in such month in excess of the first and second steps of the rates	4 cents	5½ cents
Fourth Block	To be paid per kilowatt-hour for the consumption in such month in excess of the consumption mentioned in the first, second and third steps of the rates	2 cents	2.7 cents
	The customer to agree to a minimum bill each month equal to an amount per horsepower of the full rated capacity of the connected load amounting to	50 cents	70 cents

Power Service Under Step Meter Rate

This central station in connection with wholesale power distribution, uses the step-meter rate system for a phase of its uniform wholesale power service. In the schedule filed with the utility board for a proposed increase in such charges as in force, the different step classifications are made in the following manner:

Uniform Wholesale-Power Rate	Present Rate, Cents	Proposed Rate, Cents
A charge for electric power consumed to be paid each month as follows:		
For each kilowatt-hour of consumption in such month up to and including 3000 kw.-hr.	3 cents	4 cents
For each kilowatt-hour of consumption in such month over 3000 kw.-hr. up to and including 10,000 kw.-hr.	2 cents	2.7 cents
For each kilowatt-hour for excess consumption in such month over 10,000 kw.-hr.	1 cent	1.35 cents
The customer to guarantee a minimum monthly bill of	$300	$400

(The subject of "demand rates" will be discussed in detail in the next and last article on "Standard Meter Rates.")

Refrigeration Study Course—III. The Ammonia Compressor

By H. J. MACINTIRE

Professor Mechanical Engineering, University of Utah

IN THE preceding article it was shown that a refrigerating cycle was much like a steam-engine cycle, only that it was a cycle with the events in the opposite order. Also, it was shown that the refrigerating coils acted similarly to a steam boiler, but that the capacity of the coils depended on local conditions of operation. The compressor is the only part of the machine requiring an input of power. Whereas

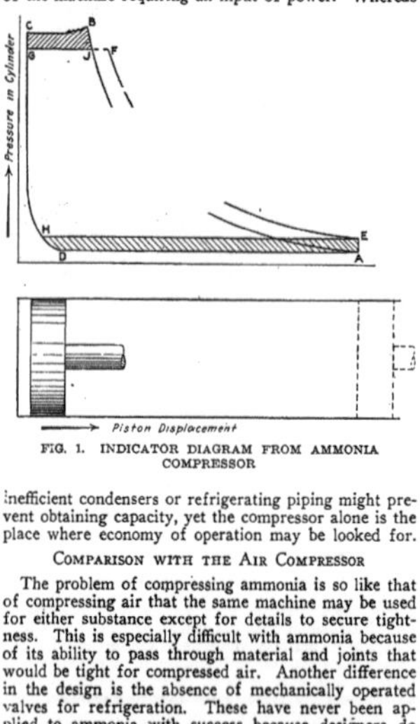

FIG. 1. INDICATOR DIAGRAM FROM AMMONIA COMPRESSOR

inefficient condensers or refrigerating piping might prevent obtaining capacity, yet the compressor alone is the place where economy of operation may be looked for.

Comparison with the Air Compressor

The problem of compressing ammonia is so like that of compressing air that the same machine may be used for either substance except for details to secure tightness. This is especially difficult with ammonia because of its ability to pass through material and joints that would be tight for compressed air. Another difference in the design is the absence of mechanically operated valves for refrigeration. These have never been applied to ammonia with success because designers do not care to increase the number of possible leaks, as would be necessary if mechanically operated valves were installed with the necessary stuffing-boxes into the cylinder. One other difference is to be found in that single-stage compression is used except in isolated cases—although one may venture to say that two-stage compression will be used more in the future. The use of the single stage is surprising when one remembers that in air compression multistage generally is employed if the discharge pressure is 70 lb. or more. To understand the problem of compressing ammonia a study of the valve action will be made, as shown in the indicator diagram, Fig. 1.

The reader will remember that the indicator diagram is simply an indication of the pressure in the cylinder for each and every portion of the piston. By means of special apparatus we are able to record to a special (known) scale the pressure inside the cylinder during the entire cycle of suction, compression and discharge of the gas. If the entrance into the cylinder is constricted, then the gas will have difficulty in getting in and the result is a partial vacuum formed during the suction stroke. If the discharge passages are constricted, there will be difficulty in getting the gas out of the cylinder promptly, and in forcing it out the piston has to increase the discharge pressure. Let us see how these points are represented on the diagram, remembering that every square inch of area on the indicator diagram represents a certain number of foot-pounds of work done on the gas by the piston, and so by the engine or electric motor driving the piston.

Effect of Valves and Ports

There are two separate indicator diagrams shown in Fig. 1. First we will discuss the one shown by the letters *ABCD*. This diagram represents the effect of suction and discharge valves and passages that are too small for the size and speed of the piston. The ammonia gas begins to enter the cylinder at the point *D* and continues to the end of the stroke represented by

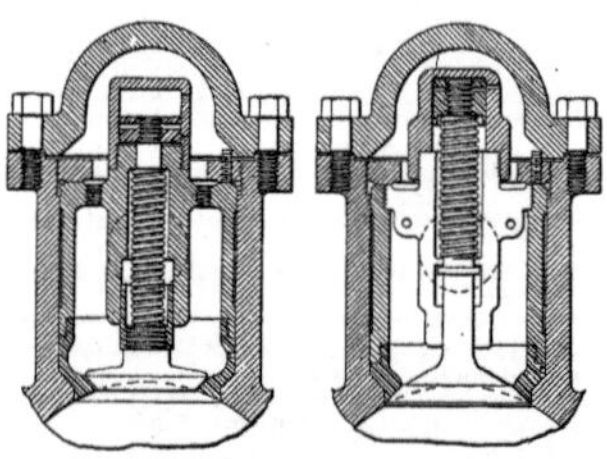

FIG. 2. THE TYPE OF POPPET VALVE SO SUCCESSFULLY USED ON SLOW-SPEED COMPRESSOR

the point *A*. On the return stroke the suction valve closes and the entrapped gas is compressed along the line *A B* until the discharge pressure *B* is reached, when the discharge valve opens and the compressed gas leaves the cylinder into the discharge pipe and the condenser. As the piston begins to move toward the right, the gas in the clearance volume expands until a pressure equal to the ammonia pressure in the suction bends is reached or some lower pressure, depending on conditions that prevail. This point is shown by *D* when the suction valve again opens for passage of a new volume of suction ammonia.

Now, if the suction gas could have entered freely into

the cylinder, the pressure would have been higher and the suction line would have been *HE*. Also, if the discharge out of the cylinder could have been free, the exit would have been shown by *FG*. In each case the cross-hatched area *DAEH* and *JBCG* is unnecessary work, but has to be supplied to get the gas into and out of the cylinder. In addition (as the suction pressure was less) the weight of ammonia pumped per revolution is much less, as shown partly by the length of the lines *GF, GJ*. It will be seen, if a study is made of ammonia-compressor design, that the main differences in each design (other than in safety of operation) are in an attempt to provide as free and unrestricted passage into and out of the cylinder as is possible for the type of construction that is used.

The Suction and Discharge Valve

By far the most successful type of valves used in refrigeration has been the so-called poppet valve shown in Fig. 2. Both suction and discharge valves have devices to prevent more than a predetermined travel. This point is especially important for the suction valve which might otherwise be caught in the cylinder, thereby causing a fracture of the cylinder and a release of the ammonia charge. Another feature is some dashpot arrangement generally used to prevent pounding. This is accomplished, usually, by means of a dashpot device—shown in the figure—which prevents seating of the valves with a shock. The poppet type of valve is reasonably safe, remains tight for a long period and is readily reground. There is always a helical spring provided to assist in quick closing. The suction valve is slightly different in the vertical single-acting compressor.

In the vertical single-acting compressor the return gas enters the crank case or the space between the top and bottom parts of the trunk piston as shown in Fig. 3. The valve is placed in the upper face of the piston and is usually so adjusted that its weight is just balanced by the spring and is then sensitive to any difference of pressure on the two sides of the piston. On the down stroke the valve opens and allows the return gas to pass through into the upper part of the cylinder. The valve itself is of the poppet type, and is popular, especially in the small-size compressors—say up to twenty tons of refrigeration.

Recently, the demand for high-speed machines, in order to secure greater capacity of direct connection to synchronous motors (in the larger sizes of refrigerating machines) or reciprocating engines of moderate speed, has promoted the development of a special form of valve of slight inertia and of large valve opening. These valves are known as plate valves or feather or ribbon valves. The feature of the high-speed valve is small inertia (usually of thin, special steel, ground to the seat) and they are flexibly held on their seats by light springs.

Safety Heads for Emergency Use

For some time it has been considered that clearance in ammonia compressors must be kept down to a minimum, and therefore special safety provisions must be made to prevent accident should liquid ammonia or a broken valve get into the cylinder. By clearance is understood both the striking clearance, which is frequently $\frac{1}{32}$ in., and volumetric clearance, which includes all the volume of entrapped vapor at dead-center. This latter is usually less than ½ per cent. of the total piston displacement. The dire results of moderate amounts of clearance as affecting economy of operation have been exploded for a long time, but the small-clearance machines still retain their popularity, particularly in the

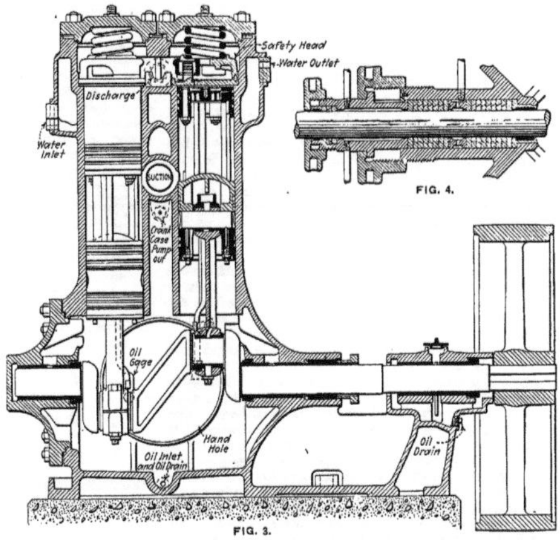

FIGS. 3 AND 4. SECTIONAL VIEW OF SMALL COMPRESSOR AND ITS LANTERN AND STUFFING-BOX

twin vertical single-acting compressors. The twin vertical machines, which are usually designed for moderate and sometimes for high speeds, lend themselves best to the construction of a safety head. This is simply a false head, shown in Fig. 3, held by stiff springs and designed for use only in cases of emergency. Should a slug of ammonia come over—like water in a steam engine—the entire head is expected to lift quickly enough to relieve the pressure, similar to the relief valves on a Corliss engine. At present no provision is made to apply the principle to the horizontal machine, which almost invariably is of a double-acting design.

For some time it has been generally conceded that operating engineers have been much too liberal with oil, the result being that this oil passed over to the condenser and finally to the refrigerating coils. Oil in the condenser pipe or evaporating coils is almost as bad as oil in boiler tubes—certainly as bad so far as transmission of heat is concerned. The result has been that many engineers have advocated little or no oil feed into the standard horizontal machine, believing that enough would carry over from the stuffing-box. The vertical machines of the inclosed type are "splash" lubricated, and the main difficulty with this form of design is to prevent the oil from passing out through the discharge and leaving the crank case with improper oil supply. The higher the speed the greater this difficulty.

Stuffing-Box Packing

For a machine properly erected, metallic packing is far the best. With soft packing the friction is much greater. The double-acting machine is usually constructed with a "lantern," as shown in the figure. On the bottom of the lantern is a connection to the oil supply, and the upper is connected to the suction heads. The result of this design is to prevent a greater pressure than the suction on the last part of the stuffing-box.

The Frame and the Bearings

In ammonia compression the forces developed are severe and alter so as to give repeated stress. The result is that the material entering into the construction must be of high quality and liberal in weight. The cylinders should be capable of several reborings. The bearings must not only be designed for low bearing pressures per unit of projected area, but should be capable of adjustment. We are now agreed on the policy of using a water jacket to prevent excessive temperatures in the discharge gas, but the tendency is to put them in the cylinder head only or the last portion of the cylinder barrel.

The manner of taking up the heat developed during compression was at one time a matter of bitter controversy, but engineers are now fairly well agreed on the subject. Some years ago it was the usual custom to bring back wet gas to the cylinder (ammonia with a certain percentage of liquid, similar to wet steam), and this liquid was expected to vaporize during compression and prevent undue rise of temperature. Provision was made to bleed liquid ammonia in the suction header so as to secure better adjustment of conditions. The other method was to provide a water jacket around the entire cylinder barrel and to hope that the jacket water would prevent excessive discharge temperatures, as well as achieve high efficiency of compression. We know now that water-jacketing on parts of the cylinder barrel does more harm than good, and that only in single-acting machines, where the jacket is placed near to the end of the cylinder (the third last), is it of appreciable value. In fact, the cylinder head is the only place where such advantage can be derived.

The reasons for the foregoing statements can be seen by studying what takes place. At the beginning of compression the gas has a temperature around zero degrees F., and it is not until after midstroke that a temperature above that of the cooling water is obtained. The end of the stroke is where the high temperature is obtained, and that is where the water jacket can make itself useful. In wet compression (where liquid injection is used for cooling) proper control cannot be maintained. The result is that we consider now that best efficiency is obtained by bringing back dry and saturated gas to the cylinder and using a water jacket where it will be of value.

One of the most serious considerations that refrigerating engineers have to remember is the volumetric efficiency of the compressor. Not only must one remember about the volume of a unit weight of ammonia and how this increases greatly as the suction pressure is decreased, but that the gas caught in the clearance space will expand and cause a piston to move 18 to 30 per cent. or even more before a new supply of ammonia is drawn into the cylinder. In addition, it is possible that the ammonia at the end of the suction stroke is superheated (like superheated steam) and therefore occupies a greater volume than it should. For instance, the volume of one pound of dry and saturated ammonia at 20 lb. gage pressure is 8.01 cu.ft.; but if the gas returns at the same suction pressure at 30 deg. F., then the volume will be 8.6 cu.ft., or an increase of 7½ per cent. Therefore, as useful refrigeration is proportioned to the weight of ammonia pumped, then the piston displacement will have to be increased a like amount or the capacity will be reduced. This shows the importance of bringing back the ammonia to the machine without superheat. Some engineers consider that the hot metal in the cylinder heats the gas during the suction stroke, but on account of the low temperatures encountered in refrigeration, the low coefficient of conductivity of the metal because it is dry, and the short time allowed to heat in, the amount of this superheating is very small. Leaky valves and piston rings are much more liable to cause trouble. The following table of volumetric efficiencies with no clearance and with 4 to 6 per cent. clearance, and the horsepower per ton of refrigeration, will be of interest:

HORSEPOWER PER TON OF REFRIGERATION

Condenser Pressure Lb. Gage	Suction Pressure Lb. Gage 5	10	15	20	25	30
145	2.02	1.69	1.40	1.26	1.11	[illegible]
155	2.11	1.76	1.48	1.30	1.19	1.05
165	2.20	1.85	1.57	1.40	1.25	1.10
175	2.34	1.94	1.65	1.48	1.34	1.18
185	2.48	2.08	1.75	1.57	1.40	1.27
195	2.55	2.14	1.84	1.67	1.48	1.32
205	2.69	2.24	1.91	1.75	1.55	1.40

VOLUMETRIC EFFICIENCY OF AMMONIA COMPRESSORS

Condenser Pressure, Lb. Gage		Suction Pressure Lb. Per Sq. In. Gage 0	10	20	30	40
120	A.	0.77	0.85	0.87	0.89	0.91
	B.	0.60	0.70	0.77		0.84
	C.	0.52	0.65	0.72		0.80
160	A.	0.74	0.80	0.83		0.88
	B.	0.54	0.65	0.72		0.80
	C.	0.44	0.58	0.66		0.75
200	A.	0.71	0.77	0.81		0.86
	B.	0.49	0.60	0.68		0.75
	C.	0.37	0.52	0.62		0.70

A, No clearance. B, 4 per cent clearance. C, [illegible] clearance.

Precautions To Observe When Operating Gates at a Dam

By L. W. Wyss

If for any reason a gate is thought to be frozen fast or some other obstruction prevents its free operation, it is well to try it by hand control, unless the operator is sure that he can handle the situation with the power without causing any damage. Hoisting cables or chains may become kinked or twisted. In a case where two chains or cables are used on a heavy gate, this may strain the gate if the twisted cable or chain carries the full weight, or the cable or chain may break, leaving the weight on the other side, which may also strain the gate. These kinks or twists are usually caused by unnecessary slack when the gate is resting at the bottom of its travel, so it is well to avoid this slack. It is sometimes advisable to attach spare chains or cables to gates in order to recover them, in the event of failure of the regular hoisting chains or cables. This is easily done and may save unnecessary wasting of water, or lowering the head-water elevation in order to recover the gate. When chain or cable-operated gates are raised to do such work as painting the gate or replacing gaskets, the workmen should not depend on the operating chains or cables, if their breaking would result in accident or loss of life. This is because the chains or cables may have been strained at some time and therefore cannot be considered dependable. Props, blocking or extra cables and chains will assure safety-first. When working on an electrically operated gate, where moving of the gate might result in accident, tag or lock the motor switch. This is especially important where gates are controlled from a distant point. Ice in exposed gear wheels may damage hoists if not chopped out. If no limit switches are provided on power-driven gates, the operator should be careful to prevent overtravel, which may result in damage, and if limit switches are provided it is well to inspect them occasionally.

Advantage must be taken of favorable load and water conditions, when work such as painting or replacing gaskets on gates is done. If a waste gate must be raised wide open, it should, of course, be done when water is plentiful, while if the head water must be lowered, it should be done at some period when the least inconvenience will result. The load dispatchers of the larger systems usually supervise such matters if the station capacity will be affected.

Dams built on soil foundations generally require more care than those built on rock. The correct supervision of wasting water may help considerably to prevent undermining or washing out. If the gates are all opened uniformly during flood periods, it will often be advantageous in preserving the dam. Then, again, at some dams there may be a certain section of gates located so that if the flood water is discharged through them it will result in a minimum of eddies or whirlpools; this will also help preserve the dam. Dams on soil foundations often have tumble bays for the waste gates to discharge into, to relieve the concrete of wear as well as to prevent eddies.

The reservoir capacity may often serve to relieve the dams from discharging the peak of a flood by wasting enough water in advance of the flood peak to lower the reservoir elevation and then let the peak of the flood bring the head back to normal. At some dams no advantage will be gained in doing this, but at others excessive floods wash away banks and endanger the dam structure, while at still others the spillway is not capable of discharging the peak of a flood, or the high tail water may flood the power house. The foregoing method of handling a flood may also be used to benefit a downstream dam, such coöperation being valuable regardless of whether the ownership is allied or not.

Paper Belting Made in Germany

During the latter part of the late war considerable was published in the newspapers regarding the scarcity of leather and rubber in Germany and of how the people were wearing paper clothing. Probably these newspaper reports were not exaggerated, as would seem to be indicated by the two samples of belting that have just been received.

In Fig. 1 is shown a sample of a 2½-in. "Sackolin" belting, ¼ in. thick. It came from Carl Scheip, Cologne, on the Rhine, who operates a supply house and carries, among other commodities, rubber articles, transmission belts, etc. This belting is made up of six plies of spun woven paper fiber. Two plies are stitched together and the thickness of the finished belt depends upon the number of these double plies. The belt shown in the sample is made up of three two-ply thicknesses as shown. One side is blackened and is the driving side. According to printed instructions, this kind of belt will transmit but 60 per cent of the load that a leather belt of the same thickness and width will, other conditions being the same. It cannot be tightened as

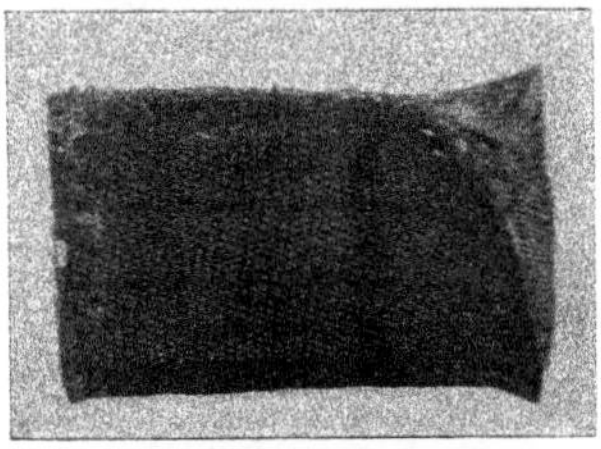

FIG. 1. A 2½-IN. BELT MADE OF WOVEN PAPER FIBER

much as a leather belt, and the instructions state that it is "absolutely" necessary to apply often the belt dressing called "Tush Durch" to the pulley side of the belt.

The belt shown in Fig. 2 is of wire-link construction surfaced with twisted fiber windings, as shown. Each metal link is made of one piece of wire and bent in a continuous 1¼-in. link, in this instance, forming a belt section 5¼ in. wide. Through the center of each link

FIG. 2. BELT CONSTRUCTED OF WIRE LINKS AND TWISTED PAPER FIBER

section is passed a paper strip about 1/16 in. thick and ⅜ in. wide, around which the twisted paper fiber is wound, each winding coming between two of the link windings. The various link sections are secured together by a stiff wire passing through the ends of the loop formed by the link winding. The outer end is bent parallel with the edge of the belt and then bent over the following tiepin, thus tying the belt together. The paper windings are for the purpose of providing a better nonslipping surface than would be afforded by the wire itself.

For some time it has been generally conceded that operating engineers have been much too liberal with oil, the result being that this oil passed over to the condenser and finally to the refrigerating coils. Oil in the condenser pipe or evaporating coils is almost as bad as oil in boiler tubes—certainly as bad so far as transmission of heat is concerned. The result has been that many engineers have advocated little or no oil feed into the standard horizontal machine, believing that enough would carry over from the stuffing-box. The vertical machines of the inclosed type are "splash" lubricated, and the main difficulty with this form of design is to prevent the oil from passing out through the discharge and leaving the crank case with improper oil supply. The higher the speed the greater this difficulty.

Stuffing-Box Packing

For a machine properly erected, metallic packing is far the best. With soft packing the friction is much greater. The double-acting machine is usually constructed with a "lantern," as shown in the figure. On the bottom of the lantern is a connection to the oil supply, and the upper is connected to the suction heads. The result of this design is to prevent a greater pressure than the suction on the last part of the stuffing-box.

The Frame and the Bearings

In ammonia compression the forces developed are severe and alter so as to give repeated stress. The result is that the material entering into the construction must be of high quality and liberal in weight. The cylinders should be capable of several reborings. The bearings must not only be designed for low bearing pressures per unit of projected area, but should be capable of adjustment. We are now agreed on the policy of using a water jacket to prevent excessive temperatures in the discharge gas, but the tendency is to put them in the cylinder head only or the last portion of the cylinder barrel.

The manner of taking up the heat developed during compression was at one time a matter of bitter controversy, but engineers are now fairly well agreed on the subject. Some years ago it was the usual custom to bring back wet gas to the cylinder (ammonia with a certain percentage of liquid, similar to wet steam), and this liquid was expected to vaporize during compression and prevent undue rise of temperature. Provision was made to bleed liquid ammonia in the suction header so as to secure better adjustment of conditions. The other method was to provide a water jacket around the entire cylinder barrel and to hope that the jacket water would prevent excessive discharge temperatures, as well as achieve high efficiency of compression. We know now that water-jacketing in parts of the cylinder barrel does more harm than good, and that only in single-acting machines, where the jacket is placed near to the end of the cylinder (the third last), is it of appreciable value. In fact, the cylinder head is the only place where much advantage can be derived.

The reasons for the foregoing statements can be seen by studying what takes place. At the beginning of compression the gas has a temperature around zero degrees F., and it is not until after midstroke that a temperature above that of the cooling water is obtained. The end of the stroke is where the high temperature is obtained, and that is where the water jacket can make itself useful. In wet compression (where liquid injection is used for cooling) proper control cannot be maintained. The result is that we consider now that best efficiency is obtained by bringing back dry and saturated gas to the cylinder and using a water jacket where it will be of value.

One of the most serious considerations that refrigerating engineers have to remember is the volumetric efficiency of the compressor. Not only must one remember about the volume of a unit weight of ammonia and how this increases greatly as the suction pressure is decreased, but that the gas caught in the clearance space will expand and cause a piston to move 18 to 50 per cent. or even more before a new supply of ammonia is drawn into the cylinder. In addition, it is possible that the ammonia at the end of the suction stroke is superheated (like superheated steam) and therefore occupies a greater volume than it should. For instance, the volume of one pound of dry and saturated ammonia at 20 lb. gage pressure is 8.01 cu.ft.; but if the gas returns at the same suction pressure at 30 deg. F., then the volume will be 8.6 cu.ft., or an increase of 7½ per cent. Therefore, as useful refrigeration is proportioned to the weight of ammonia pumped, then the piston displacement will have to be increased a like amount or the capacity will be reduced. This shows the importance of bringing back the ammonia to the machine without superheat. Some engineers consider that the hot metal in the cylinder heats the gas during the suction stroke, but on account of the low temperatures encountered in refrigeration, the low coefficient of conductivity of the metal because it is dry, and the short time allowed to heat in, the amount of this superheating is very small. Leaky valves and piston rings are much more liable to cause trouble. The following table of volumetric efficiencies with no clearance and with 4 to 6 per cent. clearance, and the horsepower per ton of refrigeration, will be of interest:

HORSEPOWER PER TON OF REFRIGERATION

Condenser Pressure Lb. Gage	Suction Pressure Lb. Gage 5	10	15	20	25	30
145	2.02	1.69	1.40	1.26	1.11	0.99
155	2.11	1.76	1.48	1.38	1.19	1.05
165	2.21	1.85	1.57	1.41	1.25	1.13
175	2.34	1.94	1.65	1.482	1.34	1.18
185	2.48	2.03	1.75	1.57	1.40	1.27
195	2.55	2.14	1.84	1.67	1.48	1.33
205	2.69	2 24	1.91	1.75	1.55	1.40

VOLUMETRIC EFFICIENCY OF AMMONIA COMPRESSORS

Condenser Pressure, Lb. Gage		Suction Pressure Lb. Per Sq. In. Gage 0	10	20	30	40
120	A.	0.77	0.83	0.87	0.89	0.91
	B.	0.60	0.70	0.77	0.81	0.84
	C.	0.52	0.65	0.72	0.77	0.80
160	A.	0.74	0.80	0.83	0.86	0.88
	B.	0.54	0.65	0.72	0.76	0.80
	C.	0.44	0.58	0.66	0.72	0.75
200	A.	0.71	0.77	0.81	0.84	0.86
	B.	0.49	0.61	0.68	0.72	0.75
	C.	0.37	0.52	0.62	0.67	0.71

A, No clearance. B, 4 per cent clearance. C, 6 per cent clearance.

Precautions To Observe When Operating Gates at a Dam

By L. W. Wyss

If for any reason a gate is thought to be frozen fast or some other obstruction prevents its free operation, it is well to try it by hand control, unless the operator is sure that he can handle the situation with the power without causing any damage. Hoisting cables or chains may become kinked or twisted. In a case where two chains or cables are used on a heavy gate, this may strain the gate if the twisted cable or chain carries the full weight, or the cable or chain may break, leaving the weight on the other side, which may also strain the gate. These kinks or twists are usually caused by unnecessary slack when the gate is resting at the bottom of its travel, so it is well to avoid this slack. It is sometimes advisable to attach spare chains or cables to gates in order to recover them, in the event of failure of the regular hoisting chains or cables. This is easily done and may save unnecessary wasting of water, or lowering the head-water elevation in order to recover the gate. When chain or cable-operated gates are raised to do such work as painting the gate or replacing gaskets, the workmen should not depend on the operating chains or cables, if their breaking would result in accident or loss of life. This is because the chains or cables may have been strained at some time and therefore cannot be considered dependable. Props, blocking or extra cables and chains will assure safety-first. When working on an electrically operated gate, where moving of the gate might result in accident, tag or lock the motor switch. This is especially important where gates are controlled from a distant point. Ice in exposed gear wheels may damage hoists if not chopped out. If no limit switches are provided on power-driven gates, the operator should be careful to prevent overtravel, which may result in damage, and if limit switches are provided it is well to inspect them occasionally.

Advantage must be taken of favorable load and water conditions, when work such as painting or replacing gaskets on gates is done. If a waste gate must be raised wide open, it should, of course, be done when water is plentiful, while if the head water must be lowered, it should be done at some period when the least inconvenience will result. The load dispatchers of the larger systems usually supervise such matters if the station capacity will be affected.

Dams built on soil foundations generally require more care than those built on rock. The correct supervision of wasting water may help considerably to prevent undermining or washing out. If the gates are all opened uniformly during flood periods, it will often be advantageous in preserving the dam. Then, again, at some dams there may be a certain section of gates located so that if the flood water is discharged through them it will result in a minimum of eddies or whirlpools; this will also help preserve the dam. Dams on soil foundations often have tumble bays for the waste gates to discharge into, to relieve the concrete of wear as well as to prevent eddies.

The reservoir capacity may often serve to relieve the dams from discharging the peak of a flood by wasting enough water in advance of the flood peak to lower the reservoir elevation and then let the peak of the flood bring the head back to normal. At some dams no advantage will be gained in doing this, but at others excessive floods wash away banks and endanger the dam structure, while at still others the spillway is not cabable of discharging the peak of a flood, or the high tail water may flood the power house. The foregoing method of handling a flood may also be used to benefit a downstream dam, such coöperation being valuable regardless of whether the ownership is allied or not.

Paper Belting Made in Germany

During the latter part of the late war considerable was published in the newspapers regarding the scarcity of leather and rubber in Germany and of how the people were wearing paper clothing. Probably these newspaper reports were not exaggerated, as would seem to be indicated by the two samples of belting that have just been received.

In Fig. 1 is shown a sample of a 2¼-in. "Sackolin" belting, ¼ in. thick. It came from Carl Scheip, Cologne, on the Rhine, who operates a supply house and carries, among other commodities, rubber articles, transmission belts, etc. This belting is made up of six plies of spun woven paper fiber. Two plies are stitched together and the thickness of the finished belt depends upon the number of these double plies. The belt shown in the sample is made up of three two-ply thicknesses as shown. One side is blackened and is the driving side. According to printed instructions, this kind of belt will transmit but 60 per cent of the load that a leather belt of the same thickness and width will, other conditions being the same. It cannot be tightened as

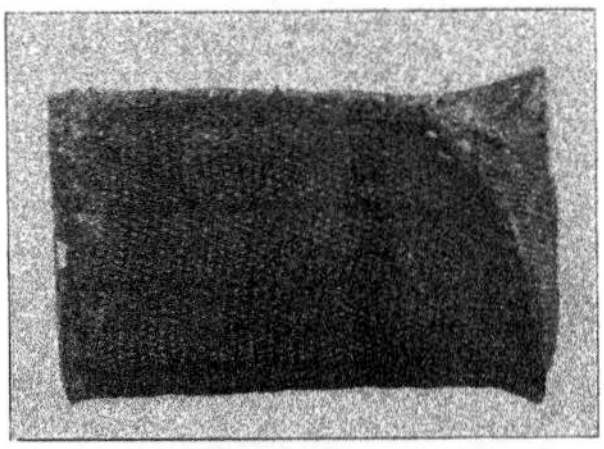

FIG. 1. A 2¼-IN. BELT MADE OF WOVEN PAPER FIBER

much as a leather belt, and the instructions state that it is "absolutely" necessary to apply often the belt dressing called "Tush Durch" to the pulley side of the belt.

The belt shown in Fig. 2 is of wire-link construction surfaced with twisted fiber windings, as shown. Each metal link is made of one piece of wire and bent in a continuous 1¼-in. link, in this instance, forming a belt section 5¼ in. wide. Through the center of each link

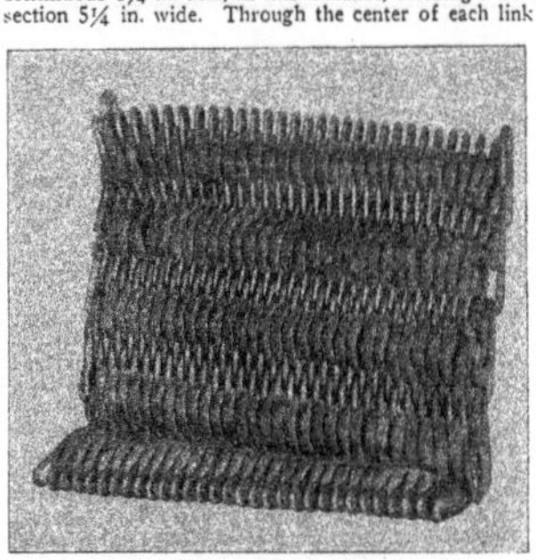

FIG. 2. BELT CONSTRUCTED OF WIRE LINKS AND TWISTED PAPER FIBER

section is passed a paper strip about $\frac{1}{16}$ in. thick and ⅜ in. wide, around which the twisted paper fiber is wound, each winding coming between two of the link windings. The various link sections are secured together by a stiff wire passing through the ends of the loop formed by the link winding. The outer end is bent parallel with the edge of the belt and then bent over the following tiepin, thus tying the belt together. The paper windings are for the purpose of providing a better nonslipping surface than would be afforded by the wire itself.

Explanation and Measurement of Power Factor

By VICTOR H. TODD

Power factor is defined. Why volts times amperes are not always the watts in an alternating-current circuit, the effect of low power factor, and the actual measurements of power factor in single-phase, two-phase and three-phase circuits are discussed.

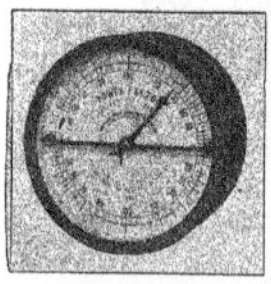

FIG. 1. SWITCHBOARD-TYPE POWER-FACTOR METER

In a direct-current circuit only two instruments are required to measure the volts, amperes and watts. A voltmeter will indicate the electromotive force in volts, an ammeter will give the current in amperes, and the product of the two readings will represent the watts. In an alternating - current circuit, however, this is not always true, as it is necessary to multiply the product of the volts and amperes by another number, called the "power factor," in order to determine the watts. The power factor may vary from 0 to 1, or, since it is often expressed in per cent, from 0 to 100 per cent. As will be shown later, the values between 0 and 1 may be known as lagging or leading power factor.

The simplest way to determine the power factor of a load is by the use of a power-factor meter such as is shown in Fig. 1. This gives a continuous indication, reading directly in per cent. power factor. It shows if the power is positive (from generator to load) by indicating in the upper half of the scale, and negative (load to generator) by indicating in the lower half, and if the current is lagging or leading the voltage. In other words, if the alternator is supplying power to the system, the instrument will read the power factor on the upper scale, but if the alternator is taking power from other generators on the system and operating as a synchronous motor, the meter will indicate the power factor on the lower scale. A power-factor meter of the portable type is shown in Fig. 2. This reads from 40 per cent. lag to 50 per cent. lead and includes all the values commonly met in commercial practice. Power-factor meters may be had to operate on single-, two- and three-phase circuits. They are generally wound for 5 amperes in the current coil and 110 volts on the potential coil, and therefore require current and potential transformers on circuits above these ratings.

To illustrate the use of power factor in the calculation of watts, assume that a single-phase induction motor is run from an alternating-current source, with instruments connected as in Fig. 3. Assume that the voltmeter reads 110, the ammeter 9 and the power-factor meter 0.80. The product of volts and amperes is $110 \times 9 = 990$, but this is not watts; it is called "apparent watts" or "volt-amperes." The power-factor meter indicates that the true watts are only 80 per cent. of the apparent watts; so in this case $990 \times 0.8 = 792$ true watts, or simply watts. This would be the amount indicated on a wattmeter. In dealing with quantities of sufficient magnitude, the units used are the kilowatt (1000 watts) and the kilovolt-ampere (1000 volt-amperes).

As is well known, the voltage and current values of alternating current are continually varying between 0 and maximum. If the current and voltage waves pass through the values of 0 and maximum at the same instants, they are said to be "in phase" and then the watts are equal to the volt-amperes. This condition is shown graphically in Fig. 4, where it will be seen that the current wave coincides with the voltage wave in both 0 and maximum values. This is the case when the power factor is 100 per cent. or unity. Under these conditions the watts at any instant are proportional to the product of the instantaneous values of current and voltage at that instant.

In an inductive circuit, the current change does not follow the change in voltage instantly, but "lags" behind the voltage. This condition is shown graphically in Fig. 5, where the full line represents the voltage, and the dotted line shows the current lagging 60 deg. behind the voltage. Now, the current and voltage are said to be "out of phase" and the angle of phase difference is 60 degrees. It will be noticed that there are now conditions when the voltage is in one direction and the current is in the opposite direction, and the product of these instantaneous values gives what is known as negative power. Negative power may be considered as power that has been delivered to the load, but is returned unused from the load to the generators or transformers, consequently does no useful work and only adds so much more load to the system. Obviously, this is the objection to low power factor; it is uneconomical and inefficient to transmit a certain amount of power with its

FIG. 2. PORTABLE-TYPE POWER-FACTOR METER

attendant losses in line and transformer, and then return a part unused. The excess of positive power over negative power appears as useful energy in the load; it is the true wattage as indicated or recorded by a wattmeter or watt-hour meter. A 60-deg. phase angle between the voltage and current is equivalent to a 50 per cent. power factor.

In a circuit that acts as a condenser, the current leads the voltage, but as far as the effect of power factor on the relation of watts and volt-amperes is concerned, the result is the same with either a lagging power factor or a leading power factor.

Induction motors, arc lamps, alternating-current welding outfits and furnaces are examples of inductive circuits; synchronous motors and converters operating with their fields overexcited, and unloaded high-tension transmission lines are examples of condensive circuits.

Fig. 6 shows graphically the condition of a circuit when the current lags 90 deg. behind the voltage. The positive power is now equal to the negative power; that is, the power is being returned from the receiving circuit as fast as it is delivered. This condition is called 0 power factor. In this case, no matter how much the voltmeter and the ammeter may read, their product when multiplied by 0 power factor will give 0 watts. For instance, in our first example, Fig. 3, if the power-factor meter had indicated 0, then $110 \times 9 \times 0 = 0$, and a wattmeter or watt-hour meter connected in circuit would not indicate or record irrespective of what the voltmeter and ammeter might indicate. This condition is seldom, if ever, encountered in commercial work. About the nearest approach is in the case of an induction motor running unloaded, when the power factor may drop to 15 or 20 per cent., but never to 0. There must always be enough excess of positive over negative power to supply the running losses in the machine, consequently the power factor can never drop to zero in an induction motor.

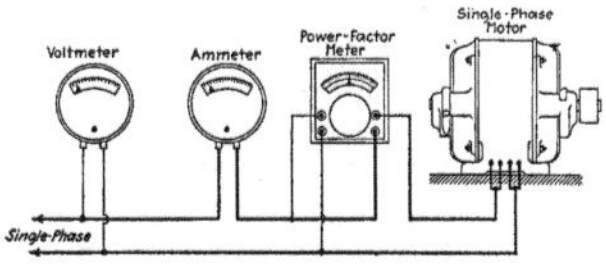

FIG. 3. VOLTMETER, AMMETER AND POWER-FACTOR METER CONNECTED IN A SINGLE-PHASE CIRCUIT

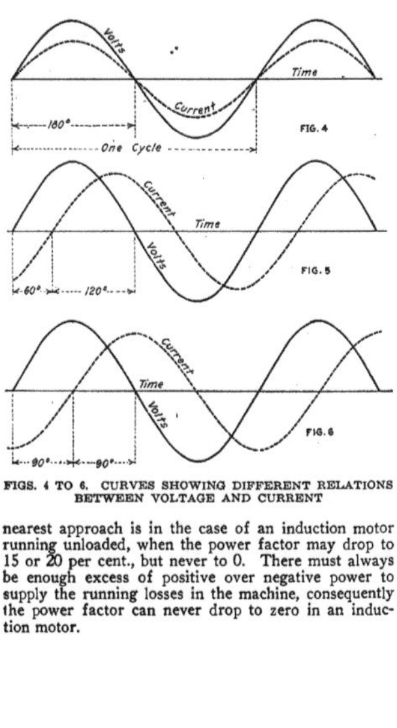

FIGS. 4 TO 6. CURVES SHOWING DIFFERENT RELATIONS BETWEEN VOLTAGE AND CURRENT

If it takes 100 amperes to transmit a given load at 100 per cent. power factor, then it will take 200 amperes to transmit the same power at 50 per cent. power factor, the voltage remaining the same. Consequently, the generator, transformers and lines must be larger to carry this increased current safely. The losses are greater in the generators, transformers and lines, since the heating loss is proportional to the square of the current, regardless of whether the current is "in phase" or "out of phase" with the main voltage. Still, the watt-hour meter will not record any more in the second case than in the first, although more expensive generating and transforming apparatus is required, and the loss is much greater in delivering the power in the second case than in the first. Evidently the customer who uses a certain load at a low power factor should pay more than a customer who takes the same load at a high power factor. In some cases power companies penalize the customer for low power factor by charging a higher rate per kilowatt-hour, the rate increasing as the power factor decreases. In some cases the customer is given a bonus for keeping a high power factor, the rate per kilowatt-hour decreasing as the power factor increases. Another method is to meter the active volt-amperes (watts) and the reactive volt-amperes (returned watts) and charge a separate rate for each, explaining to the customer that the reading of the reactive volt-ampere-hour meter is waste and can be eliminated by a proper manipulation of load.

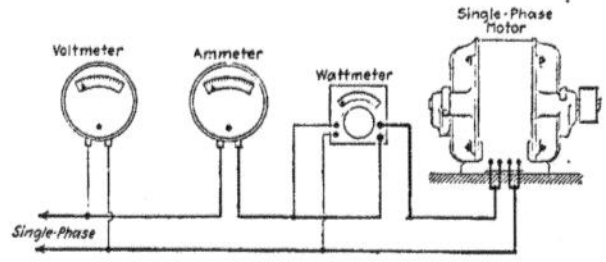

FIG. 7. VOLTMETER, AMMETER AND WATTMETER CONNECTED IN A SINGLE-PHASE CIRCUIT

The power factor is equal to the cosine of the angle of phase-difference between the voltage and current. In commercial work it is seldom necessary to know the actual angle of phase-difference, as the power factor can be calculated from the readings of a voltmeter, ammeter and wattmeter or read directly from a power-factor meter, such as shown in Fig. 1 or Fig. 2. Power-factor meters have the advantage of indicating if the current is lagging or leading, as this cannot be determined from the voltmeter, ammeter and wattmeter reading method. The following paragraphs give the methods of determining the power factor of various circuits:

Single-Phase, Two-Wire: A single-phase power-factor meter connected as in Fig. 3 will indicate the power factor of the circuit. If the voltage or the current is above the capacity of the meter, select a voltage transformer that will give about 110 volts on the secondary and a current transformer that will give between 1.5 and 5 amperes secondary on normal load. If, after connecting, the pointer indicates in the lower half of the scale, reverse the potential-coil connections. In Fig. 3, if we substitute a wattmeter for the power-factor meter, as in Fig 7, it will indicate the true watts. Then

$$\text{Power factor} = \frac{\text{true watts}}{\text{apparent watts}} = \frac{\text{wattmeter reading}}{\text{volts} \times \text{amperes}}$$

Assume that the meters in Fig. 7 read 110 volts, 9 amperes and 792 watts. Then 110×9 gives 990 apparent watts; and 792 divided by 990 gives 0.8, which is 80 per cent. power factor.

Single-Phase, Three-Wire: In this case, a three-wire current transformer should be used. This has two primaries and one secondary; a primary should be connected in each outside wire of the circuit, and the secondary to the wattmeter and ammeter or the power-factor meter. The voltage should preferably be taken across the outside wires. The connections are shown in Fig. 8.

Two-Phase, Four-Wire: The power factor may be taken on either phase with a single-phase power-factor meter, or by the voltmeter, ammeter and wattmeter method in exactly the same way as for a single-phase circuit. If the load is balanced, such as in the case of a two-phase motor load, the power factor may be measured on one phase and assumed to be the same on the other. If the load is unbalanced, the power factor must be taken separately on each phase. In reporting the power factor of an unbalanced two-phase load, it is preferable to state the power factor of each phase rather than attempt to strike an average. A two-phase power-factor meter will not give the average power factor, on an unbalanced circuit. They are designed for use on balanced loads only.

Two-Phase, Three-Wire: In this the power factor is determined in a similar manner to the single-phase. Care must be taken that the current coil is inserted in an outside wire, not the common, and that the voltage is taken between that wire and the common, not across the outside wires. The proper connections are shown in Fig. 9.

Three-Phase, Four-Wire: If the neutral of a three-phase circuit is available, the current coil may be connected in one line and the voltage taken from that line to the neutral. This applies to either a single-phase power-factor meter, or the voltmeter, ammeter and wattmeter method. The connections are as shown in Fig. 10. If the load is unbalanced, it is necessary to install three single-phase power-factor meters. A three-phase power-factor meter will not record the average power factor of an unbalanced three-phase load.

Three-Phase, Three-Wire (Balanced): A three-phase power-factor meter is preferable to find the power factor in this case. If a single-phase meter is used, it must be specially calibrated so that it reads 100 per cent. when the current and voltage are 30 deg. out of phase, because in a three-phase circuit, the current in each wire is 30 deg. out of phase with the voltage as measured between that wire and either of the others, at 100 per cent. power factor. Two single-phase wattmeters may be used to calculate the power factor, and two single-phase watt-hour meters may be used to calculate the average power factor for a long period of time; a month for instance. If two single-phase watt-hour meters are connected as in Fig. 11, they will record the total load. At 100 per cent. power factor both meters will record equal amounts, but as the power factor decreases, one meter will record more than one-half the load and the other less. At 50 per cent. power factor, one meter stops, while the other records the full load. Below 50 per cent. power factor, one runs backward, while the other records more than full load. The total kilowatt-hours consumption is the algebraic sum of the two readings in any case.

The power factor is determined by locating the ratio of the smaller to the larger reading on the curve in Fig. 12. The negative side of the curve is for power factors below 50 per cent. and the positive side for power factors above 50 per cent. Assume that the two watt-hour meters in Fig. 11 both start at 0000. At the end of a month's time assume the reading on one meter to be 4000 kilowatt-hours and the other 2400. Their sum would be 6400 kilowatt-hours, which represents the total power consumption. The ratio of the smaller to the larger reading is 0.6, that is, $2400 \div 4000 = 0.6$. As both readings are positive, it shows that the average power factor has been above 50 per cent. Locating where the 0.6 line in Fig. 12 is crossed by the curve, we find that the point is 93 per cent., which represents the average for the month.

At the end of the next month assume that the first meter reads 6000 and the other 1900. As the second meter reads less than before, we immediately know that it was running backward and that the average power factor was below 50 per cent. The first meter recorded

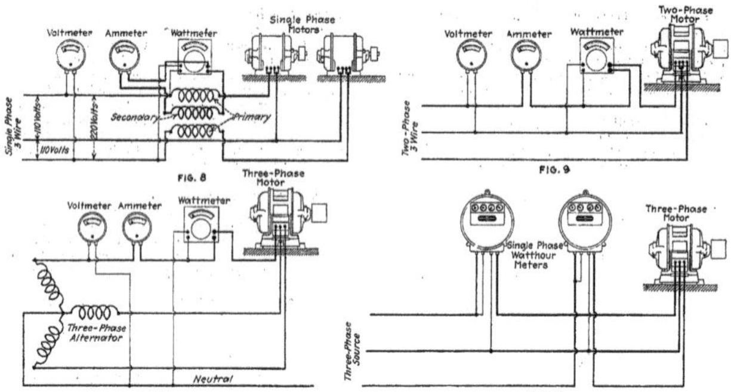

FIG. 10 FIG. 11

FIGS. 8 TO 11. INSTRUMENT CONNECTION ON DIFFERENT CLASSES OF CIRCUITS

6000 — 4000, or 2000 kilowatt-hours, and the other recorded 1900 — 2400, or —500 kilowatt-hours. The total consumption for the month would be 2000 — 500, or 1500 kilowatt-hours. The ratio of the smaller consumption to the larger is — 500 ÷ 2000 = — 0 25. Locating, on the negative side, where the curve crosses the 0.25 line, we find that the average power factor for the month has been 0.34, or 34 per cent. It is very important that this method be used only on balanced loads, as

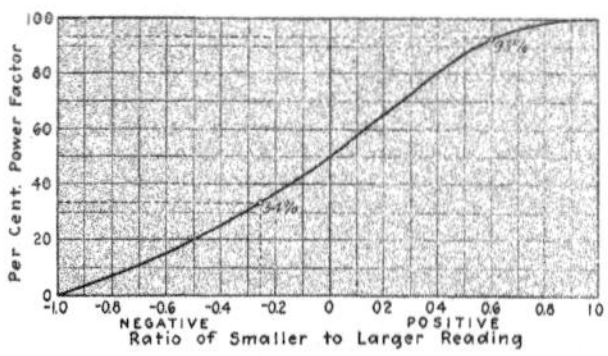

FIG. 12. CURVE SHOWING RELATION RATIO OF WATTMETER READING AND POWER FACTOR

an unbalancing of the load produces similar effects to a varying power factor.

Three-Phase, Unbalanced: The power factor of an unbalanced load may be defined as the ratio of the true watts to the apparent watts; but the apparent watts is a very elastic quantity. The most favored definition is that the apparent watts is the sum of the apparent watts of each phase, and the apparent watts of each phase is the product of the amperes in a wire and the voltage between that wire and an artificial neutral.

To measure this properly requires the use of three voltmeters, three ammeters and a polyphase wattmeter with connections as in Fig. 13. The formula is as follows:

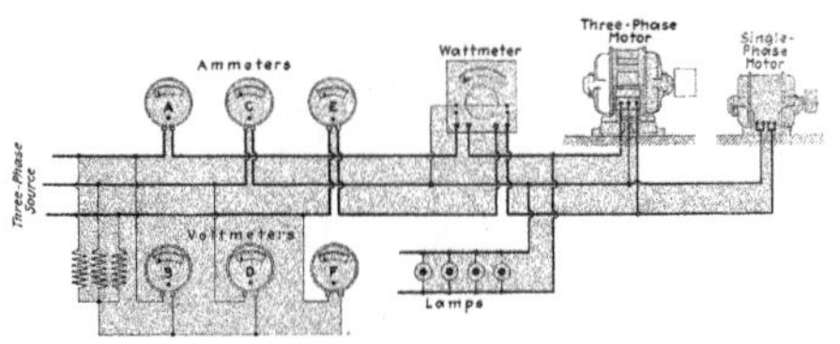

FIG. 13. LOAD AND METER CONNECTIONS ON AN UNBALANCED T...EE-PHASE CIRCUIT

$$\text{Average power factor} = \frac{W}{B \times A + D \times C + F \times E}$$

Assume $W = 1500$, $A = 5$, $C = 6$, $E = 7$, $B = 100$, $D = 110$, and $F = 120$; then substituting in the formula,

$$\text{Average power factor} = \frac{1500}{(100 \times 5) + (110 \times 6) + (120 \times 7)}$$
$$= 0.75 \text{ or } 75 \textit{ per cent.}$$

Owing to the importance of the measurement of power factor in a three-phase four-wire unbalanced circuit, which is the most commonly met case in commercial work, there is being developed at the present time an "apparent voltammeter" which will indicate this quantity on a three-phase circuit, thus requiring only two instruments to determine the power factor, and its calculation will then be as simple as for a single-phase circuit.

New Wheeler Auxiliary Tube-Plate Condenser

It was recently brought up at a meeting of marine engineers that many of the ships now in service are experiencing trouble with leaky condensers. These leaks have been almost invariably traced to the stuffing-box between the tube and the tube plate. It has been found that in spite of the kind of packing used, some

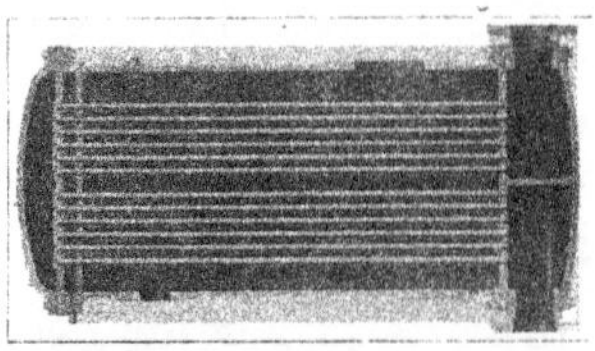

CONDENSER WITH AUXILIARY TUBE PLATE

leakage is practically unavoidable owing to the expansion and contraction of the tube with temperature change.

The illustration shows a new type of condenser developed by S. Morris Lillie, and manufactured by the Wheeler Condenser and Engineering Co., Cateret, N. J. This condenser is used chiefly in connection with salt circulating water on board ship or in regions where the water that is used for circulating purposes would be bad for boiler feed in case of leakage into the condenser and consequent mingling with the condensate.

The tubes are tightly expanded into the right-hand tube plate, which eliminates the possibility of leakage at this end. At the other end the tubes are provided with the standard ferrules and condenser packings. The packing permits the tubes to expand freely and yet maintains an almost leakless joint. It is therefore at the left-hand end that leakage occurs, if at all.

At the left end of this new condenser a thin auxiliary tube plate or baffle is provided, through which all the tubes are passed. The holes in this plate are of such size as to make a sliding fit with the tubes. It is, obviously, not necessary that these joints be water-tight.

In case of leakage of circulating water through the ferrules and packing, such leakage immediately drops to the bottom of the compartment between the two tube plates, and is then carried off by the drain shown at the bottom, to which is attached either a small direct-acting condensate pump or a vacuum trap.

Boiler-Room Instrument Boards

The necessity of having measuring devices for the purpose of controlling processes is gaining recognition, but much uncertainty still exists as to the benefits derived therefrom. Obviously, mere possession of measuring and indicating instruments is not sufficient to secure economic operation. Neither can reading and recording the indications possibly improve the operating efficiency unless the indications are understood, the effect on the results interpreted, and steps taken to adjust the operating conditions in accordance with established relations between conditions and effects. Unless the management is prepared to make proper use of the instrument equipment, the investment in such is of doubtful value.

The careful and thoughtful use made of instrument equipment invariably produces quick and gratifying results. A report of one of the power supervisors on the effect of the installation of some instruments permitting him to control more accurately the operating processes is interesting, as it is typical of all such cases.

BOILER-ROOM INSTRUMENT AND CONTROL BOARD

The board serves two separate boiler rooms. Two recording steam-flow meters are on the left, and each serves a main header in each boiler room; of the four recording thermometers, top row, there are two for each boiler room, where one serves the heater and one the economizer. Below are six indicating steam-flow meters. Under this row a three-in-one draft gage serves all boilers by means of selective valves operated by pressing any lever of the group under the gage. The pyrometer alongside the draft gage is used for the exit gases in the same way. The stoker and fan speed controls are shown.

. . . it is evident that in the four months preceding the installation of a boiler-control board, the savings on fuel due to various steps taken averaged \$435 per month, while in the four months following the installation of the instruments the savings, computed on the same basis, averaged \$1,135. In other words, the increased saving, due solely to the intelligent use of instruments on the boiler-control board, was \$710 per month, or \$8,620 on the annual basis, which means that the expenditure of \$2,080.46 for instruments of the board is an investment which in this case yields over 400 per cent. return.

The illustration shows the boiler-room instrument and control board as devised by Walter N. Polakov, the well-known consulting engineer. In passing it should be noted that the plant in question is a comparatively small mill boiler house consuming between 150 and 200 tons of coal per week.

The next question that comes up in connection with means for investigating existing operating efficiency and development of better methods is: What instrument equipment is necessary and sufficient for the purpose? Evidently, it is not possible to specify a set of instruments that meet the requirements of any plant. The character of generating equipment, its layout, service conditions, specific problems, etc., determine actual selection. In a paper before the 1918 spring meeting of the A. S. M. E., Mr. Polakov said:

> While the generation of power, and more specifically of steam, is the domain of the scientifically trained engineer, power-plant practice is conspicuous by the lack of accurate measurement of conditions and results.

Unless the results obtained are known, no opinion as to perfection of operation can be sound; furthermore, the practice is necessarily wasteful unless means are available to observe the conditions under which the process is performed. All instrument equipment of the boiler house can therefore be grouped into two classes: (1) Recording the results; (2) showing the conditions.

A plant of which it is not known how many pounds of coal are used per thousand pounds of steam, how the load is distributed among the units and throughout the day, etc., wastes fuel by necessity. The knowledge of these data does at least open the eyes of those responsible for its success, and further progress is thereby made possible.

The first group of boiler-room instruments thus comprises: *Quantity*—Recording coal scales, recording steam or water meters, coal calorimeter and moisture scales, feed-water thermometers, steam pressure gage and thermometer.

The second group of instruments is intended to direct the processes by controlling conditions: Flow indicators, draft gages, flue-gas thermometers, flue-gas analyzers.

Any investment in instruments is a pure waste of money and leads to demoralization unless means are provided for training men in their proper use, stimulation of men in their proper use, and complete, exact and continuous recording. The location and arrangement of instruments shall be such as to permit simultaneous readings and their comparison, permit plain view of units from instrument board and reverse, afford the opportunity to use one instrument for diverse units, eliminate unnecessary fatigue of observing scattered instruments, and assure ease and simplicity for testing.

These requirements are combined in the type of instrument boards shown herewith. A fallacy of economizing on instrument equipment or ignorant attempts to select only "the most important" ones has only one rival in absurdity—the tendency of installing instruments without giving the employees the opportunity to use them to advantage. Obviously, the operating men have no time, no ability for research work and little inducement to carry out investigations, standardization methods and set tasks. It should be the duty, therefore, of the management to render them necessary training and to assume responsibility for results.

The Bhakra Dam, which it is proposed to build across the Sutlej River, India, in the Bhakra Gorge, about forty miles above Rupar, the headworks of the existing Sirhind Canal, will be, says the *Madras Mail*, 395 ft. high from foundation level to roadway, and will be the highest dam in the world. The depth of water in front of the dam will be about 375 ft., good rock for the foundation having been found near the surface. The length of the top of the dam will be 1,015 ft. The water to be stored by the dam in the month of August annually will be 2½ million foot-acres.

EDITORIALS

Preventing vs. Correcting Low Power Factor

NUMEROUS surveys of industrial alternating-current power loads have found power factors as low as forty-five per cent and with sixty-five per cent as a common value. Under such conditions the voltage regulation is most seriously affected, owing to the excessive line drop and also to the demagnetizing effect of the lagging current upon the alternator field poles. Poor voltage regulation at the motors in all cases causes a reduction in production and not infrequently also lowers the quality of the product. Therefore, it is desirable that the voltage be maintained as constant as good engineering dictates, consequently the power factor of the load must be maintained at a fairly high value.

When the operating engineer considers improving the power factor of the system in his charge, he has two directions in which to turn his attention—the preventing of poor power factor and the correcting of it. Either one or a combination of both may be the solution of the problem. The prevention of low power factor is obtained by the proper application of motors, while power-factor correction is brought about by the use of corrective means such as synchronous machines used under various conditions, or by the installation of static condensers. One of the most serious causes of low power factor is over-motoring, and it has been well established by many power-load surveys that a large percentage of our industrial plants are over-motored. Over-motoring in the past has no doubt been due largely to the lack of reliable data on the power requirements for different classes of industrial machinery, and motors were installed that would be sure to do the work, which in general were anywhere from twenty-five to fifty per cent too large. However, reliable data on power requirements have now been obtained on most standard industrial machinery, and in these cases there is no engineering reason why such machinery, when driven individually, should not be properly motored. Where machines are driven in groups the problem takes on a different aspect. If the motor has capacity sufficient to drive all the machines in the group, which it must have in most cases, and if at times only a small portion of the machines are operated, the motor will be operating under-loaded. Then if the motor is of the induction type, it will be operating at a low power factor. Here is probably one of the best arguments for the use of synchronous motors, when they can be properly applied, on group drives. If only unity power-factor motors are installed, they will, at full load, operate at one hundred per cent power factor, and even under this condition help to improve the power factor of the industrial load. If the load is reduced on the motor, it will then begin to supply to the system a leading current that will help improve the power factor of the inductive load. Where the power factor of the average induction motor drops to about sixty-five per cent lagging at forty per cent load, that of a unity power-factor synchronous motor drops to fifty per cent leading, the latter a very desirable feature in most power systems.

Before power-factor corrective means are adopted, every effort should be made to prevent low power factor, by having the motors properly applied. Where motors of too large capacity have been installed, a power survey will frequently indicate how they can be exchanged so that they can be more nearly loaded to their full capacity; especially is this true where extensions are being made to the plant. Only after this has been done, or it is established that the low power factor cannot be prevented economically by a rearrangement of the motors, should corrective means be applied. Although the application of synchronous motors or static condensers will improve the power factor of the system as a whole, they do not prevent under-loaded induction motors from operating at low efficiency and low power factor, both of which increase power costs.

A National Society of Engineers

THERE is every likelihood of the formation in the early future of a broad national society which shall mobilize the engineers of the country in order that the engineering and allied technical professions may become a more active national force in economic, industrial and civic affairs.

The progressive element in the four great professional societies began to question, under the spur of criticism, whether organized action of engineers of all kinds for the public good and their own welfare could not be carried to a greater length than these specialized, exclusive and professional bodies were able or ready to go.

A Committee on Aims and Organization was appointed by the Mechanical, and Committees on Development by the Civil, Electrical and Mining engineers, out of which grew a Joint Conference Committee composed of representatives of the four societies named. This Joint Conference Committee recommended the formation of a single comprehensive organization, the purpose of which should be "to further the public welfare wherever technical knowledge and training are involved, and to consider all matters of common concern to these professions."

This great national organization of engineers is to be built from the bottom up, the plan being to organize a large number of local societies to which anybody who has legitimate interest in engineering may belong, and to tie these together in a national body, or federation, made up of representatives from the local bodies and from the professional and regional societies which are too large to be classed as locals. The relation of these societies, like the American Society of Civil Engineers, the American Institute of Electrical Engineers, the Institute of Mining and Metallurgical Engineers, and the American Society of Mechanical Engineers, to the larger organization should logically be the same as though that more heterogeneous organization had been formed first, and the specialized and exclusive institutes and societies crystallized out of it; but previous existence gives the present societies opportunities that will make their relations to the larger body more parental than filial. It should be through the activities of their local sections and affiliates that the constituent locals are incited into being and nurtured through their formative stages. They may need to finance the National Association in its early stages, although a moderate per capita tax upon the upward of one hundred thousand professional and semi-professional engineers available as members ought to provide ample funds for ordinary activities.

For the first time in history the governing boards of the Four Founder Societies, as they are called, met in joint session on Friday, January twenty-third, approved the recommendations of the Joint Conference Committee and authorized that committee to proceed to bring the national organization which it recommended into being. The next day the Council of the American Society of Mechanical Engineers approved and adopted the report of the Joint Conference Committee, authorized the society's representatives on the Joint Conference Committee to take the necessary steps to carry out the recommendations of the report and voted a generous appropriation for the use of the committee. The Civil and the Electrical engineering societies have already discussed the report in open meeting, and the Institute of Mining and Metallurgical Engineers is to discuss it in February.

All of the Four Founder Societies appear to be committed to the adoption and carrying out of the recommendation of the Joint Conference Committee for a National Organization, by the unanimous joint action of their governing boards on January twenty-third. This joint meeting of its four parent societies was called by Engineering Council, which was begotten by the civil, mechanical, electrical and mining societies out of United Engineering Society, which, being simply a holding company concerned with the administration of the buildings and property owned in common by the constituent societies, was not thought to be competent to undertake those activities, which appeared to belong to or to afford opportunities to engineers collectively rather than to any particular society. Council proved a precocious offspring, and before its parents realized what it was up to had committed them to a debt of twenty-five thousand dollars and laid out a program for the coming year which it would cost sixty thousand dollars to carry out. The family consultation was to increase its allowance or curtail its activities.

The ultimate effect of the action of the meeting upon the fate of Council cannot be foretold. As a medium for impressing the engineer upon current happenings, for obtaining recognition of the profession and organizing opportunities for its greater service, the larger, more democratic organization appears to offer the greater advantages. There may still be some functions that the more rigidly professional societies can better exercise in common through Council.

There is already in process of organization an engineering society upon the broad, democratic, inclusive plan under discussion—the American Association of Engineers. In a comparatively short time it has acquired a membership of over ten thousand. It deals with the human as well as the technical side of the profession. Its chapters are widely distributed geographically.

If it were so persuasive and convincing as to join the whole engineering profession and craft under its banner, it would be the big national society toward which we are all looking. What will be the effect of the new movement upon its growth and fate? Will it join the other societies in encouraging the members of its local chapters to join the community groups and through them the rival national organization, even though such a course imperils its own existence? It has already invited the coöperation of the older societies in a way that indicates that it is willing to do its part in the evolution of the large idea and the carrying out of the larger purpose.

Watch the Diesel

SELDOM has there been assembled on this side of the Atlantic so many recognized Diesel-engine experts as were seen at the recent meeting of the Society of Automotive Engineers in Philadelphia. That the discussion lasted over four hours is an indication of the thought being given to the subject, and it is quite evident that engineers are fully alive to the possibilities of the oil engine for certain classes of service and that much may be expected in the near future. Although the meeting brought out differences of opinion as to details, there was general agreement on the more important points. For instance, it was conceded that for some time at least the field for the marine Diesel will lie in its application to cargo vessels up to a maximum of twelve or thirteen thousand tons with speeds up to about fourteen knots. The popular size of units will be from one thousand to fifteen hundred horsepower, although where necessary they will be built up to twenty-five hundred or three thousand horsepower for the larger ships. The four-cycle engine now predominates, although it was prophesied that American designers, profiting by the mistakes and experiences of European builders, were on the way to overcoming certain features that caused trouble with the two-cycle type for marine service and that ultimately this type would come into its own.

It seems, also, the consensus of opinion that the auxiliaries should be independently driven.

War conditions gave a big impetus to the oil-fired marine boiler and geared turbine drive. Competition between this combination and the Diesel will be watched with interest. While the war was on ships were needed in a hurry; there were a number of firms in a position to turn out geared turbine sets in large quantity; there was no time for experiment or development. Besides, the limited capacity of our Diesel-engine builders was taxed to equip submarines. But now that condition no longer obtains and the Diesel builders are going after the mercantile business.

On a fuel-economy basis the Diesel will have an advantage at the start, and although it may now be at a slight disadvantage as regards first cost in some instances, this will disappear when standardization and quantity production have been effected. But it is not on fuel economy alone that the decision will be made; it is the greater radius of action and increased cargo-carrying capacity afforded by the Diesel installations that will be determining factors.

But not alone in the marine field are we witnessing growth of the oil engine. In the stationary field for mines and isolated stations in the South and Southwest, for oil-line pumping and irrigation work, and for many light and power plants of limited size in certain sections one notes increased use of this type of prime mover.

For land service consideration is being given also to the two-cycle engine and efforts are being made to increase the efficiency to that of the four-cycle engine. By reducing the excess quantity of air for scavenging and lowering the pressure on this air a small advance has been made, bringing the efficiency within five or six per cent of that possible in the four-cycle engine. In either type an outstanding feature is the sustained economy over wide ranges of load.

It is gratifying to note that the governor of Oregon has signed the bill which places that state in line with the thirteen other progressive states that have adopted the Boiler Code of the American Society of Mechanical Engineers.

CORRESPONDENCE

Field Winding in Opposition Caused Motor to Flash Over

It is always advisable to take nothing for granted when investigating trouble on motors or their control equipment, as the following example will show:

A 3-hp. direct-current 250-volt compound-wound motor operating a paper cutter began to flash over between the brushes on the commutator, at the same time blowing the fuse. It was decided that the motor's field was weak, but on making a drop test with a voltmeter across each coil, they all gave an equal reading. On examination the commutator showed signs of high mica, as it had the usual dark streaky marks found under this condition. Since this has frequently caused flashing, especially when a motor is overloaded, or when accelerated too quickly, the mica was undercut with a file, the commutator sandpapered, and the brush position checked, the latter being found to be correct. This reduced the sparking when the machine was running light, but on taking a cut with the machine, out went the fuses again.

The indications were that the motor was overloaded; so, in order to test it, the belt was taken off and an ammeter placed in the circuit. Holding a plank against

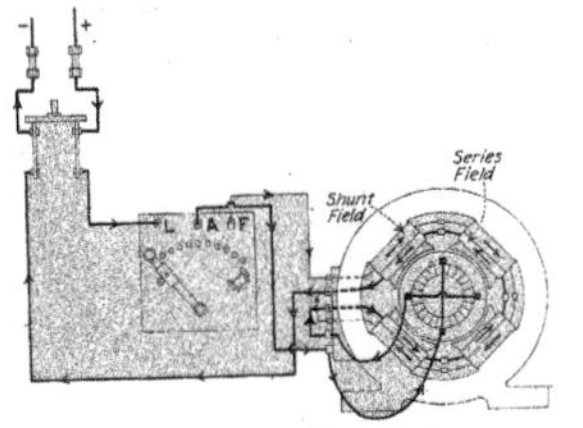

FIG. 1. COMPOUND-WOUND MOTOR SERIES- AND SHUNT-FIELD WINDING IN OPPOSITION

the pulley as a brake, the motor showed no signs of distress at 10 amperes, which was the full-load current. At 16 amperes the sparking was considerable but not excessive, and the flash over did not occur until about 28 amperes had been registered on the meter, representing an overload of about 180 per cent. With the cutter in operation the results were similar; running light the current taken by the motor was about 6 amperes, but on making a cut with the machine the current quickly increased to about 30 amperes, when the commutator flashed over and blew the fuses before the cut was finished.

After spending considerable time in trying to find some irregularity in the cutter to cause the trouble, it was decided that probably the series-field windings of

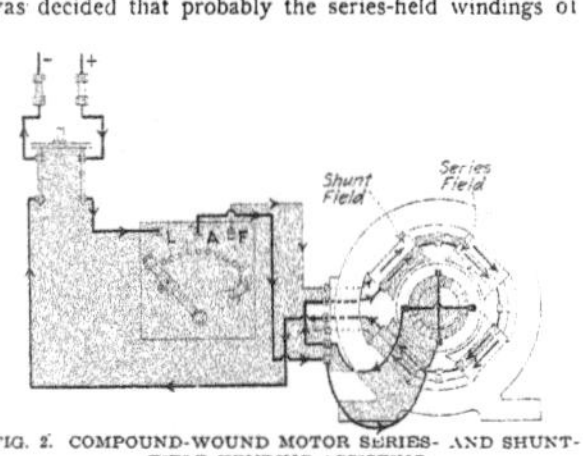

FIG. 2. COMPOUND-WOUND MOTOR SERIES- AND SHUNT-FIELD WINDING ASSISTING

the motor were connected in opposition to the shunt-field coils, as in Fig. 1. On making the change, as in Fig. 2, the results were satisfactory. The current was reduced to 24 amperes on taking the cut, and the motor stood up to this momentary overload without undue sparking at the brushes. H. WILSON.

Toronto, Ont., Canada.

[For a compound-wound motor, with the series- and shunt-field windings connected in opposition, not to flash over until subjected to a 180 per cent. overload, it must have a very small percentage of compounding.—Editor.]

Capacity of Steam Pipe with Superheated Steam

At a large pulp mill the temperature of the steam was originally 600 deg. F. at the point where it is used in the pulp digesters. The pressure in the boiler house averaged 120 lb. gage, and in the digester house 105 lb. By decreasing the heating surface of the superheater, the steam temperature was reduced to 400 deg. F. The digesters now take the same total quantity of heat, although the actual weight of steam consumed is, no doubt, greater.

I figured that the friction loss would be less using the less superheated steam, but actually this does not seem to be the case. One pound of the 600-deg. steam will occupy more space than the same weight at 400 deg., but of course more of the latter is required. As the result of my calculation of the friction losses in each case, I came to the conclusion that 20 per cent. less pressure drop should be expected after the superheater tubes had been shortened.

The size of the steam pipe is 10 in., extra-heavy, the total length being 600 ft., with two globe valves. There are no elbows or other fittings to restrict the flow. The weight of steam conveyed by this pipe was 70,000 lb. at the 600-deg. temperature. The pressures given were taken on the 10-in. pipe, and the drop in the superheaters is not considered.

Perhaps some readers of *Power* will throw some light on this subject. B. A. REMOL.

Hawkesbury, Canada.

Non-Upsetting Torch

The torch shown in the sketch is made from a piece of pipe, any size, threaded and bent as shown, capped at one end and an elbow and nipple screwed on the other end.

Before the cap and elbow are screwed on, cotton wicking is entered by means of a wire and pulled

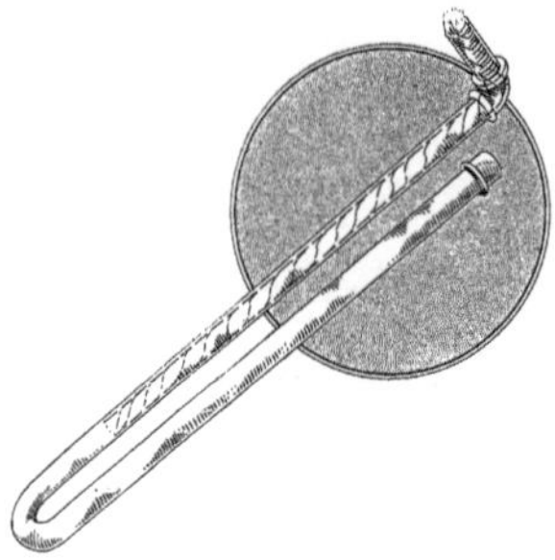

TORCH THAT WILL NOT UPSET

through the pipe until the tail end is about as shown. The projecting end is pulled through the elbow and nipple before it is screwed on. Oil is put in at the end *B* and the end capped.

It is impossible to upset the torch, it is easily carried, and it can be hung up by the hook.

Portsmouth, Ont., Canada. JAMES E. NOBLE.

That "Woman and Machine" Problem

I have read the letter of Jessie E. Minor, page 514, of the Sept. 23, 1919, issue, and also the article in the Aug. 26 issue, in which the statement was made that certain work could be done by a woman and a machine, etc., but I got an entirely different impression from the one expressed by the woman chemist. Having had charge of a manufacturing department employing about 75 men and boys and 50 women and girls for several years, this statement was to me merely an expression of the classification of the work—an expression of fitness and not at all an exclamation of the simplicity of the work or the inferiority of women workers.

In our regular routine work, if a new operation came up, we would classify it with a simple statement: "Give it to a 20-cent girl," meaning that, in our estimation, a girl who was capable of earning 20 cents an hour at other work was capable of doing the work in question. I would say just as quickly, "This work can be done by a 20-cent man" or "That job should be given to a 50-cent woman."

It is obvious that each person has work for which they are particularly adapted, and each class has its particular class of work. For instance, inspecting steel ball bearings is a girl's work. It is light, easily learned, and would be monotonous to most men and boys. Still, it is important. Most girls like a job which is easily learned and which repeats itself continually; its monotony never seems to pall, and they resent any change or transfer to new work, often taking it that the reason is unsatisfactory work and never thinking that the "boss" thinks they may be getting tired of its monotony.

Then, on the other hand, straightening cores or riveting bases of autometers is work much less important and is just as monotonous; but it is best done by a man or boy, because it involves the use of a hammer and this extra physical labor seems to offset the monotony. Punch-press work is usually adapted to men and boys, but if we had a punching job of light work requiring accuracy and cleanliness—for instance, fiber bushings—we would say, "This can be done by a woman," meaning that a woman could do it better than it could be done by a man or boy.

I recall one specific instance: On one operation we had a number of refractory girls whom the man in charge could not control. After we had warned him several times that his girls were not up to standard in behavior or in quality or quantity of work, I remember saying, "That job will take a woman," and consequently we made the change with surprising results. I could go on naming many operations, some requiring initiative, some skill, some accuracy, some reliability, some experience, all being held by women workers with wages on a par with those of the men, as we always worked on the basis, "equal wages for equal work," with a result that much of our work was done by women and girls receiving greater remuneration and doing more important work than many of our men.

Orange, N. J. VICTOR H. TODD.

Why Does the Water Lift When the Safety Valve Blows?

The letters about water lifting in the boiler remind me of an experience with a boiler using a water containing a high percentage of magnesia. Popping of the safety valve or a sudden increase of load would start the water to foaming violently and it would hump up under the safety valve and then level down and repeat until the water lowered four or five inches in as many minutes.

We know it did this, because we would open the surface blowoff on the back end of the boiler. At the same time the water went out of sight in the glass, the skimmer at the rear would blow steam; and when the water came back in the glass, the skimmer would blow water.

It required quick work to keep enough water to prevent pulling the fire. We kept a couple of wheelbarrow loads of wet ashes handy, and when the foaming started some of it would be dumped on the fire. We finally overcame the trouble. It was good water in one way, as it formed no scale. J. O. BENEFIEL.

Anderson, Ind.

Measuring Small Stack Velocities

I see from the article entitled "Experiments on Stack Performances," in *Power*, Sept. 16, page 464, that the problem of measuring small gas velocities in a stack is still open for solution. From experiments it was found that a vane placed perpendicular to the flow produces eddies and pulsations, thereby preventing good results.

I suggest to place a vane in the direction of the flow, the vane to be pivoted and provided with a counterweight to keep it stable in a horizontal position. Such a vane will offer scarcely any resistance to the flow of gas. If there is no flow of gas around the vane in the direction of the arrow, it will take a certain amount of energy to move the vane from its horizontal position to, say, about five degrees.

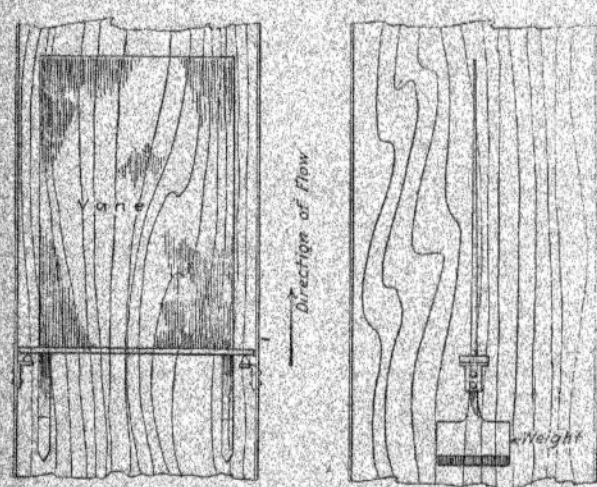

VANE FOR MEASURING STACK GAS VELOCITIES

If the vane is placed in a current of gas, it will take more energy to move it the same angle from the horizontal, and the energy required will increase with the flow of gas. Knowing the energy required will make a solution of the velocity possible. O. JANZEN.

Half Way Tree, Jamaica, B. W. I.

When Strikes Come

Mr. Speaker, Honored Legislators and all the Lobbyists within the reach of my voice, I rise to interrogate: Just where does labor leave off and capital begin? Who looks after the rights of the disorganized you-an'-me proletariats in between. I belong to the Majority Party of the Men-Who-Pay-the-Bills, and I am getting sore on languid labor and corpulent capital. I don't wear overalls and haven't a silk hat! I can't afford to own a yacht, and I can't afford to go on a strike, although I have worked hard for twenty years.

I have a family and I am more or less respectable. I pay my bills on the first of the month, likewise I pay town, state and national taxes. I have never been confined in a jail or in any institution supported by public funds. I am American born, with all the rights of citizenship, including protection—especially protection at home. I want to know what the Government of the U. S. A., apart from political "necessities," offers as a permanent and orderly cure for these periodic eruptions of strikes. If capital is to blame, what's the remedy? If labor is to blame, what's the remedy?

Or is the whole system rotten? Does the one squeeze the other when the opportunity comes, just because he can? I'm only the Man-Who-Pays-the-Bills. Yet I know my group of men is larger than the other two groups combined. I also know I'm the season-after-season goat and I want to ask, "Does this Government know I'm the goat and what are you going to do about it?"

A bunch of boomers get a hold on the natural water resources and overcapitalize; I pay twice for water. That's capital. When the weather turns cold, some coal miners in Pennsylvania or Colorado decide to strike—I don't know what for—and there isn't any coal and my baby gets the croup and nearly dies. That's labor.

I'm not jealous of what they get, but I think I ought to get something besides the honor of paying the bills and suffering the inconveniences.

I want to know when this Government, conceived in liberty, will give me a chance to plan my life just a little bit on a permanent basis. I can't strike for a raise in pay—I have to deserve it. I haven't accumulated capital enough to live on without work, and I want to know sure what the price of eggs will be tomorrow morning.

You know who I am. I'm the Average Man. I earn more than $1,300 a year, but not much more. I spend my salary for sane and sensible living. I have ambitions for my boy and girl, maybe a college education, and about every thirty days these two highwaymen, capital and labor, come along and hold me up! Whichever one wins, I lose, and I want a basis.

I say economic theories are all hunkydory, but a little common sense added ain't so bad at that! I say sympathy is all right until we develop professional sympathy grabbers. I say that the unions are all right until they become offensive instead of defensive.

THE "GOATS" WHILE LABOR AND CAPITAL FIGHT

I send my wife and kids on a vacation. Before the vacation is half over I have to wire them to rush back because there is going to be a railroad strike. Their good time is spoiled because they might be injured on the way back, maybe with a brick or a derailment. I have to spend $100 to go to meet them. One hundred dollars—the price of a winter suit and a pair of socks. Why? Because this country knows how to organize a strike, but not how to disorganize one.

And that's my kick, gentlemen. I want to know what you are going to do for us fellows who own the cottages and the Fords and who don't belong to the unions and don't wear overalls and don't own any banks. We buy what labor makes with capital's profit tacked on, an' I want to know where we get a square deal.

Rome, Ga. C. V. YOUNG

Loose Bolts Caused Induction Motor to Hum

It is not very often that peculiarities develop in a motor from mechanical derangements, which to all appearances seems to indicate that the electrical system is at fault. When such conditions develop, they are not always apparent at first, sometimes requiring an extensive investigation.

A 440-volt three-phase induction motor driving an air compressor gave trouble due to blowing its fuses. The line to the motor ran through conduit and was tested and found to be free from grounds. Next, the winding was inspected, the insulation found to be in poor condition, and it was decided to have the motor rewound. A potential of 600 volts was applied between the winding and frame, which punctured the insulation in a number of places. After the motor was wound and placed in service again, it ran for about a week, when it developed a loud hum. It was taken apart again, and it was found that the end shield bolts on both ends were not as tight as they should have been, but no attention was paid to this until the winding was inspected and found to be in first-class condition. Assembling the motor and tightening the end shields proved that the hum was due to the loose bolts. C. R. BEHRINGER.

Greenville, Penn.

National Engineers' License Law

I have read with interest the many discussions in engineers' magazines on the subject of engineers' license laws, and I wish to say a few words as one who holds an engineer's license in Massachusetts. I have become convinced that it would be advisable to have a law throughout the land making it necessary for all power-plant workers to pass an examination and obtain licenses graded according to their knowledge and ability. I include power-plant workers, because I do not believe the law should include only engineers and firemen, for there is practically as much reason why operators of gas engines and hydro-electric plants should also be licensed. Of course, laborers and helpers should be allowed to work without a license, or they would have no practical way to get enough elementary knowledge to secure one.

I started as hundreds of engineers have, firing in a steam sawmill, and worked my way up to larger plants. In some cases I had to wheel my own coal, fire, run the engine and do my own repair work and all that goes with the care and operation of a steam plant. In one case I fired four boilers and did all the work enumerated except wheeling the coal. I tell of these things to show conditions as I have found them in states where there are no standards of efficiency for engineers, and the same conditions exist in that plant today and in many others of which I know.

I held that job a couple of years and then worked along to a bigger plant, and so on until I became chief in a good-sized plant. I have held some fairly good positions in that capacity, but I was always more or less dissatisfied, owing to the fact that anyone who chose to call himself an engineer could be one, and if I tried to get the pay I deserved I was usually told that, while my work was satisfactory, there was someone who would take the job as it was. As a lot of plant owners never realize the difference between one engineer and another, there isn't much chance for an argument in such a case, and about all there is to do is move along.

I finally decided to go to Massachusetts and see if conditions were not better there. Upon finding a good location, I put in an application for a license, but I didn't get a first and found that I was lucky to get a second, for while I thought I had been studious and observing, there was really very little I knew well enough to be sure of, and no apologies are accepted by the board of examiners.

I am glad that I came to a license state, for I found how little I knew compared with what I had supposed I did. I also discovered that the time I had spent working in states where there were no license laws was about half wasted. The real trouble in my case before coming here was the same as in a lot of similar ones—there was no real incentive for me to get right down to the very bottom of things and find why they were thus and so, and I had therefore neglected to do so. When I went up before a board composed of men who knew their business I was simply stuck and was ashamed that I had claimed to know enough to apply for a license.

The only ones who are opposed to the license laws in this state are those who haven't been able to get a license as big as they think they ought to have, and also a certain class of plant owners who are afraid that the time is coming when they will have to pay power-plant operatives what they are worth.

It is the engineers' own fault to a great etxent, for had they stuck together as they should have done and helped one another to get a better education by belonging to an association founded upon the right principles, there would be a different state of affairs in the engineers' world today and every state would have a license law and the country would not be overrun with those who will work for what is offered them for the sake of being called engineers.

Unionism will bring about the desired conditions, but not the kind of unionism that most people think is going to do it, for there is not enough of the educational idea written into the constitutions of the labor organizations in this country, and there is not enough Americanism either, but there is a tendency for inefficient men to flock into the unions and lean on those who are efficient.

I am a member of the N. A. S. E and am a firm believer in the principles of the organization, and it is due to the efforts of that organization that we have an engineers' license law in this state. I believe that if every power-plant operative would join we would accomplish more for the benefit of our trade in a few years, by education and legislation, than has been accomplished by any means in the past. There is nothing in the constitution of the association that prevents a member from belonging to a labor union or that prevents a union man from joining the association, so there is no reason why they should not get together and work as a unit for the betterment of all concerned.

There are those who may be opposed to an engineers' license law because they doubt their ability to secure a license, but a provision could be made the same as that which was made in this state when the license law was passed, which allowed a man who had for a certain length of time successfully operated a plant that was rated as first-class to receive a first-class license, and so on down to the firemen. While this law did not have all the desired effect at once, it kept outsiders away and soon began to raise the plane of education among engineers and firemen, and at the present time the laws are much more rigid and a man has to pass a better examination to get a license now than at that time

Let us as engineers and firemen work toward the enactment of a National engineers' license law, which will be the greatest step we could take toward better educated men of our trade, which will raise our standing and lead to better wages and conditions.

Webster, Mass. LLEWELYN M. REED.

INQUIRIES OF GENERAL INTEREST

Height of Settings for Return-Tubular Boilers—How high above the grates should two 72-in. x 18-ft. return-tubular boilers be set to secure the best combustion? E. G.

The most perfect combustion would be attained with the boiler set so high above the grates that they would absorb no heat from the fires during the process of combustion. But with an ordinary return-tubular boiler setting, the greater the height the greater the loss of heat from the side walls and less benefit from absorption of heat by direct radiation from the fire to the under side of the shell. For average fuel and hand-firing, the best over-all fuel economy usually is obtained for 72-in. boilers when set 30 to 36 in. above the grates.

Setting Piston Clearance of Compressor—In which end of the cylinder of an ammonia compressor should the piston rod be set to give greater amount of piston clearance? M. J. M.

The end of the cylinder that should have the piston set with a greater amount of piston clearance depends on the design of the connecting-rod. If taking up the connecting-rod brasses tends to lengthen the rod, the piston should be set in the crosshead so there will be a greater piston clearance in the head end of the cylinder; and if taking up tends to shorten the connecting-rod, the piston should be set with greater piston clearance in the crank end of the cylinder.

Lame Running of Duplex Pump—What causes a duplex pump to run lame, and how can the fault be remedied? W. L. G.

If, after the valves have been set, the pump runs "lame," showing an inequality in the length or time of the strokes, it may be because the resistances of the water cylinders are different. They should be examined for leaky water valves, and tight packing of pistons and stuffing-boxes. If the pump continues to run lame after such faults have been corrected, the lost motion of the valves must be readjusted for obtaining the desired equality of strokes, bearing in mind that reducing the lost motion obtains more throw of the valve and provides wider steam and exhaust openings.

Velocity Acquired by Drop of Elevator Cage—If an elevator cage weighing 4,000 lb. should drop 150 ft., at what speed in feet per second would it be traveling and how long would it take to fall? J. A. C.

The velocity of a body freely falling in vacuo and the time required to fall are independent of the weight of the body. The velocity freely falling in vacuo is given by the formula $v = \sqrt{2gh}$, where v = the velocity acquired in feet per second; g = the acceleration of gravity, usually taken as 32.2, and in freely falling in vacuo from a height of 150 ft. the velocity would be

$$v = \sqrt{2 \times 32.2 \times 150} = 98.3 \textit{ ft. per sec.}$$

If t = the number of seconds in falling, then the time occupied falling freely in vacuo is given by the formula

$$t = \sqrt{\frac{h}{\frac{1}{2}g}},$$ or when $h = 150$ ft., the time would be

$$t = \sqrt{\frac{150}{16.1}} = 3 \textit{ sec.}$$

The velocity acquired by an elevator cage dropping would be less and the time required would be greater than falling in vacuo, depending on the frictional resistance of the guides and resistance of the air.

Testing Quality of Steam with Separating Calorimeter—For determining the quality of steam with a separating calorimeter, the graduated glass gage showed that the amount of water collected was 0.21 lb., and during the same time there was 2 lb. 3 oz. of condensate added to the condensing water. What was the quality of the steam? A. N. F.

Where W represents the weight of water that the calorimeter separates from the steam, and W_1 represents the weight of dry steam condensed after separation, then the total weight of steam tested would be $W + W_1$, and the quality q, or dryness, would be given by the formula

$$q = \frac{W_1}{W + W_1}$$

In the example, $W_1 = 2$ lb. 3 oz., or $2\frac{3}{16}$ lb. = 2.1875 lb., and $W = 0.21$ lb., and by substitution the quality is found to be

$$q = \frac{2.1875}{0.21 + 2.1875} = 0.912 \textit{ or about 91 per cent.}$$

Open Connection in Transformer Bank—Three single-phase transformers are connected delta as in the figure and are supplying a load of three-phase induction motors. If one

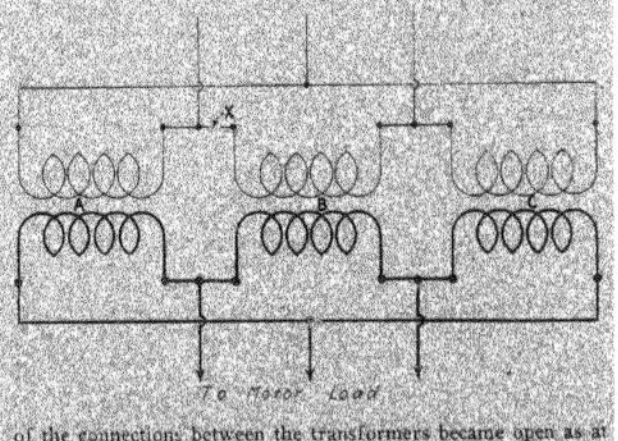

of the connections between the transformers became open as at X, what effect would it have on the operation of the motors? E. M. W.

With an open-circuit at X transformer B will be cut out of circuit and transformers A and C will be connected in open delta. Transformers A and C will then be capable of supplying only 58 per cent. of the normal capacity of the three transformers; that is, if the bank of three units has a capacity of 100 kv-a., two of them connected open delta will only have a capacity of 58 kv-a. As far as the operation of the motors is concerned, it will be about the same with the open-delta connection as with the closed-delta. However, if the three transformers were operating near full load before the open occurred, then the two transformers operating open delta will be overloaded and it will be necessary to shut some of the motors down to relieve the overload until the defect is repaired.

Excerpts From Elevator Safety Orders State of California

By EDWARD A. SMITH

Mechanical Engineer, Otis Elevator Co., New York, N. Y.

EXCERPTS from elevator laws of the State of California for power-driven passenger, freight and sidewalk elevators, effective Oct. 1, 1916; revised Apr. 1, 1918. For all elevators in the state except private houses and United States Government buildings. The excerpts are from the laws pertaining to both existing and future installations. Where the law is for future installations only, it is not quoted except where noted.

Load: Passenger—Passenger capacity limited to 2½ sq.ft. per person, provided this does not exceed load capacity.

Speed: Passenger—Not to exceed 600 ft. per min., except express elevators running 80 ft. or more without a stop, may run 800 ft. per min. (First stop 80 ft. and local above). Freight—Same as passenger, except without a regular operator limited to 100 ft. per min., but does not apply to operatorless, push-button or outside switch control.

Safeties and Governors: Passenger—Over 150 ft. per min. gradual-operating type safety with speed governor at top of hatchway to trip at 40 per cent above normal up to and including 550 ft. per min; above 550 ft. per min. trip 30 per cent above normal. Speed to be stamped on governor; 150 ft. per min. or under, instantaneous, or roll safeties with governor. Future governor to be equipped with device for cutting off current for electric machines 300 ft. per min. or over. No safety required on existing elevators rise 12 ft. or less. Instantaneous broken-rope safeties to be provided on counterweight located over passageway or vault. All future safeties to be under platform and between safety plank. Safety drums to have twice as much rope as required to operate. Governor ropes to have tension frames. Safe-lift elevators to have locking device. No safeties permitted on plunger elevators. Freight—Same as passenger except broken-rope safeties allowed for 15 ft. rise or less.

Car enclosures and platforms and slings: Passenger—Grille covered with ¾-in. wire mesh, No. 16 gage on outside. Enclosure to be bolted or screwed to car floors also bolted to sling. Existing grille tops to be made solid or covered with ½-in. screen No. 18 gage. Two openings in car maximum number allowed, with gate farthest from operator under operator's control (hydraulic elevator). For electric elevators gate contact or other equal device required; no gate is required for single opening car except push-button. No vertical compartments permitted, but two horizontal permitted with access to hatch through one only. Over-all dimension of vertical guide shoes to be not less than one-half largest car dimension except plunger elevators. No mirrors or glass permitted in car construction, except for covers for annunciators, lamps or over certificates and 36 sq.in. maximum mirror for operators. Cars to be lighted. Threshold light recommended but not required. Push-button elevator gates at all openings with contacts on future gates 3 in. maximum between vertical bars; alternate bars to act as guides. Future clearance between car and sill or gate, or door, minimum ¾ in., maximum 1½ in., minimum between any other part 1½ in. If impractical to maintain 1½ in. maximum, semiautomatic gates must be furnished on landing or gate on car with contact. Freight—Wood or metal 6 ft. high on unused sides, solid or open 2-in. mesh. None required on uncounterbalanced plunger cars rise one floor. Opening in side of car enclosure permitted for hand rope. No gates required. Tops of cars to be wire 1-in. mesh No. 12 gage or other construction of equal strength. Cut back not more than 6 in. from sill at entrance and hinged not less than 18 in. Car tops not required for following conditions: (1) Where trap doors are used: (2) where semiautomatic gates extend to floor, except at lowest floor; (3) where hatchway doors are provided opening from shaft side only and are kept closed; (4) uncounterbalanced plunger elevator; (5) elevators whose platforms exceed 96 sq.ft., except that hinged screens must be provided 4 ft. back and extend across entrance. No mirrors or glass permitted except for covers for annunciator, lights or certificates and a 36-sq.in. mirror for operator. No vertical compartment permitted, but two horizontal compartments permitted with access to hatch through one only. Car to be lighted. Over-all dimension of vertical guide shoes to be not less than one-half largest car dimension, except plunger elevator. Minimum clearance ¾ in. between sill and car, or gate, or door, 1½ in. maximum. Minimum between any other parts, 1½ in. If impractical to maintain 1½ in. maximum, semiautomatic gates must be furnished on landing or gate on platform with contact. Sidewalk—No enclosure required. No light required. Bow irons required or some automatic means of stopping car in case sidewalk doors refuse to open.

Control: Passenger and Freight—Centering device required where hand rope is on opposite sides of hatchway. Electric car switches to be self-centering. Push-button must have emergency stop, also switch control. Car switches or levers made to operate alike where two or more elevators are installed. Sidewalk—Centering ropes not required on hand ropes, otherwise same as passenger and freight.

Hatchway gates: Passenger—Not permitted. Freight and sidewalk—Gates or doors to be furnished at all hatch openings. Semiautomatic wood or metal gate to stand 250 lb. at any point without springing or binding, 5 ft. 6 in. high, 6 in. underclearance, 3-in. space between vertical members or wire mesh. Where future hand-rope operation an opening 3 ft. long and 5 in. wide to be provided in 5-ft. 6-in. gates only. Gates 3 ft. 6 in. high permitted, if above impractical due to bad conditions, at discretion of commissioner, provided telltale chains or ropes 4 ft. long and 5 in. apart on platform are furnished. Gates to begin to close before car moves one foot. Double-blade gates permitted only if solid panels are furnished or panels are covered with fine wire mesh. Full-automatic gates permitted at terminal landings only and 5 ft. 6 in. high. Manually operated gates to have contacts. Horizontal folding gate not permitted except at discretion of commissioner. Gate counterweight to be boxed in or run in guides to be closed at bottom to prevent counterweight from falling.

Ropes: Passenger—Iron or steel permitted. Steel ropes to be labeled, "Use steel cables only" on metal plate on crosshead and top counterweight. Metal tags on all ropes giving date of installation, at crosshead and counterweight. Factor of safety, 8 manufacturer's list. Future minimum size, ½ in. Future drum machines, 2 ropes to a car and each counterweight 1½ turns on drum. Future traction machines require 4 ropes; 2 ropes on progressive-groove traction machines. Future hydraulic machines, 4 ropes and 2 counterweights. No splicing of hoist or counterweight cables. Governor ropes must be iron or steel ½ in. minimum diameter over 150 ft. per min. car speed. Manilla or hemp, 150 ft. per min. or under, ⅝ in. minimum diameter. Ropes to drum must be clamped or have tampered sockets inside drum. Ropes in socket may be fastened by "turned in method," "turned out method" or "straight method." Hand ropes to be metal tiller ½ in. diameter, with electro-mechanical brake ⅜ in. permitted. Counterweight ropes passing through an upper counterweight to be protected from injury. Freight—Same as passenger except factor of safety of 6 permitted. Future traction freight elevators not limited to four ropes if speed is 150 ft. per min. or under. Future hydraulic freight elevators, 2 hoisting and 2 counterweight ropes. Sidewalk—Same as freight except not limited to minimum size of ½ in. Chains not permitted.

Equalizers: Passenger and Freight—Must be used on car and counterweight where overhead drum machines are used, when grooved right and left hand.

Grating: Passenger and Freight—Substantial grating 2-in. mesh or bars 2-in. centers with wire screen No. 20 gage, ½-in.

mesh, of metal, or a flooring to sustain a load of 300 lb. at any point, except where trapdoors are used. Cover entire hatch if 50 sq.ft. or less, otherwise 2 ft. outside of all sheaves and machinery with baseboard 6 in. high, and suitable handrail if space between grating and wall exceeds 12 inches.

Counterweight Screens: Passenger and Freight—Bottom screen 7 ft. high, fireproof material in fireproof shaft; if mesh ½ in. by No. 20 gage, not required with counterweight chains or cables. If counterweight is adjacent to door opening, screen to extend 7 ft. above lowest floor. No screen required for push-button elevator in single shaft or elevators with contacts on basement doors. Sidewalk—Same as passenger if counterweights are used.

Stop Motions and Shaft Limits: Passenger and Freight—Machine limits required on drum machines. Passenger electric machines (existing and future) to have electric shaft limits and future freights. Hand ropes to have stop balls to stop machines at top and bottom. Hydraulic elevators to have stop balls on hand ropes to stop machines at top and bottom. Sidewalk—Machine limits on drum machines and stop balls on hand rope or chain.

Interlocks: Passenger—Standard No. 3 lock or Wells-VanAlstyne or equally good for horizontal sliding doors to be used on all push-button operations. All future electric passenger elevators to have contacts on doors or other suitable interlocks with emergency releases in car under glass within easy reach of operator. Future safe lifts to have locking device. Freight and Sidewalk—Interlocks required on manually operated doors or electric contacts. Doors arranged to open outside of hatch by means of key only, where there is a regular operator, permitted provided doors have contacts or equivalent devices.

Emergency Switches: Passenger and Freight—Where interlocks are used (see interlocks), all future power-driven switch-controlled and push-button-controlled elevators to have emergency switch or button in car to cut off current independently with potential switches. Proper fuses or circuit-breakers to be furnished. Breaking or holding-open safety circuit not to depend on springs only. Existing and future passenger and power freight elevators (except with existing trapdoors and then gong to be furnished) to have signal system except push-button and serving two continuous floors only. Slack-cable device, covered and insulated, furnished on all drum machines.

Brakes: Passenger—Electric brakes required on all passenger elevators. Mechanical brakes not permitted. Freight and Sidewalk—Mechanical brakes permitted only on existing freight elevators speed 75 ft. per min. or under.

Rope Locks: Passenger and Freight—To be furnished on car for all hand-rope operations.

Hatchway Enclosure: Passenger—(See local laws for enclosures.) Enclosed on all sides at least 6 ft. high and full height on counterweight portion. Counterweight enclosure may be solid or open ¾-in. mesh. All unguarded sides inside of hatchway to be flush. All other sides flush or all projections beveled at angle of 60 deg. with horizontal. This applies to future and existing elevators, since Oct. 1, 1916, or other cases considered unsafe by accident commission. Safe access must be provided in penthouse and lighted. Freight—Same as passenger except enclosure only on unused sides 6 ft. high and over counterweight to run to ceiling solid or ¾-in. mesh. Sidewalk—Enclosure on unused sides 6 ft. high. Future openings in sidewalk not to exceed 5 ft. at right angles to, or 7 ft. parallel to, curb.

Hatchway Doors: Passenger—Doors must be furnished at all openings in accordance with fire ordinances in locality of installation. Minimum requirements wood or metal solid 3 ft. high with grille above 2-in. mesh or vertical bars 2-in. centers. Doors to be hatched on shaft side and lock to be inaccessible from outside except with a key, push-button elevators excepted. Latch of door, when open, to be free and accessible to operator. Floor space in front of door openings on future installations to have non-slip surfaces 20 in. wide. Doors should not be opened or closed unless car is stationary at landing. Freight—Semiautomatic doors or gates to be furnished at all entrances. Manually operated doors permitted if provided with interlocking device. Doors arranged to be opened from outside with a key must have contacts or equivalent and a regular operator. Future Peelle type doors must have contacts. Sidewalk—Same as freight. When sidewalk doors are open, must have screen or gate from top of door to sidewalk. Attendant must be posted at sidewalk before opening doors.

Motor-room enclosure: Passenger, freight and sidewalk—For penthouse, see paragraph on hatchway enclosures. Future machine-room enclosure of size to give free and safe access to all machinery. Existing and future must be lighted, machinery to be guarded according to general safety orders, electric machines to be grounded.

Operator: No regulation.

Hydraulic Tanks, Valves, Cylinders and Parts: Valve chambers to have means of removing air. Must have air-siphon relief valve on discharge pipe. Future vertical traveling-sheave straps to be made of two structural shapes, no "U" straps permitted. Existing and future pressure tanks designed so that air will not pass to cylinder. Reciprocating pump to be equipped with water safety valve only between pump and pressure tank to prevent increase of pressure more than 25 per cent when pump is at full speed. Valve to be piped to discharge tank above water line. Existing installations may be discharged into pump suction. Existing and future steam pumps connected to air-pressure tanks must have steam-pressure regulators. Electric pumps must have automatic pressure regulator controlling motor or with automatic bypasses. Pumps to accumulators other than regulating type not containing air to have automatic shutoff valves or bypasses. Installations prohibited: Belt or chain-driven machines prohibited on new installations. Power-driving mechanism applied to hand power. Friction-gear or clutch mechanism between drum or traction sheave and main driving gear. Worm gear cut into cast-iron rim not permitted, except dumbwaiter and for freight and sidewalk of 2,000-lb. capacity or under and 75 ft. per min. or under. Cast-iron guide rails prohibited. Two elevator cars, balancing each other and operated by the same machine, except for dumbwaiters having terminal landings only. Cast iron not permitted for safeties except drum. Gripping portions of safety device not to be used as guides to car. The use of control systems depending on electric batteries for interruption of operating circuit. Future mechanically controlled elevators (except dumbwaiters) opening of direction switches or operating valves not to depend solely on sprockets or chains. Existing and future machine and car counterweights not independent and separated prohibited.

Repairs and alterations: Changes other than ordinary repairs are considered new installations and must conform to above orders. An existing elevator moved to new location is a new installation.

Tests: Load test for every new elevator. Existing and future—Every car safety to have drop test before accepted by commissioner. New installations to have drop tests in presence of authorized representative of industrial accident commission. Safety devices to have actual running test at full load.

Fifteen Firms Unite in Plant Display at Machinery Center

Service to the prospective buyer of construction plants was made available in a new and progressive form when the Allied Machinery Center, comprising a permanent exhibit of the products of fifteen noncompeting manufacturers, was formally opened by the Allied Machinery Company of America in New York City. The new center, occupying 30,000 sq.ft. of exhibition and office space at Center and Walker streets, New York, is the first of its kind to be established as part of an educational and sales-promotion plan in the field of heavy construction plants. It is comparable, in a way, to the department-store idea of merchandising drygoods by assembling a variety of products in one building so that the buyer may supply his needs without the inconvenience of journeying to half a dozen stores.

But the idea behind the machinery-center project is a broader one than merely making it easy for the machinery buyer to do his shopping. A fundamental aim of the enterprise is the education of the user in sound construction economics—in using the right machine and particularly, where the size of the job warrants, the right combination of equipment for doing the work with the maximum of economy and speed. This feature of the plan, it is pointed out, has a special significance at present when the cost of common labor has soared to heights not dreamed of half a dozen years ago.

The machinery center is designed to serve the needs both of the export buyer and the domestic purchaser. With respect to the former the great need abroad is generally conceded to be an appreciation of the results obtainable by American construction methods, which means, of course, the substitution where practicable of machinery for hand labor. This lesson is still to be learned, a fact not fully appreciated by anyone who has not actually been on foreign soil and studied conditions at first hand.

The following firms are represented at the Machinery Center: Austin Manufacturing Co.; Austin-Western Road Machinery Co.; Barber-Greene Co.; Carbic Manufacturing Co.; C. H. & E. Manufacturing Co., Inc.; Clyde Iron Works; Cook Motor Co.; A. B. Farquhar Co., Ltd.; Hydraulic Pressed Steel Co.; Lakewood Engineering Co.; Parsons Co.; Sterling Wheelbarrow Co.; Thew Automatic Shovel Co.; Western Wheeler Scraper Co.; Wyoming Shovel Works.

At a luncheon given Jan. 27 to celebrate the opening of the Allied Machinery Center the speakers included Joshua W. Alexander, Secretary of Commerce; Senator Edge, W. L. Saunders, George Edward Smith and former Secretary of Commerce W. C. Redfield.

Measuring Flow of Water by Orifice*

By Raymond E. Davis and Harvey H. Jordan

A thin plate orifice inserted in a pipe line is looked upon as a temporary or field device for measuring the flow of water, being simple to construct and easy to install. In flanged pipe systems such an orifice may be inserted with little or no disturbance of existing piping, and in long pipe lines the loss in head caused by the orifice will be inconsiderable as compared with other losses.

An orifice causes an abrupt change in conditions of flow, and this change is accompanied by a drop in pressure head

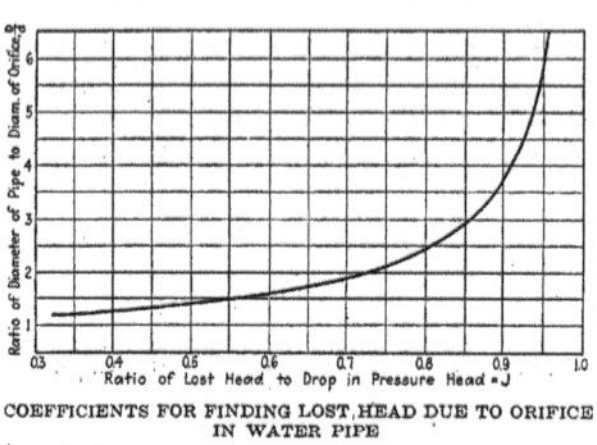

COEFFICIENTS FOR FINDING LOST HEAD DUE TO ORIFICE IN WATER PIPE

which may be easily measured and which varies with the velocity of flow in the pipe. The experiments herein described were made to determine (1) the positions of two cross-sections of the pipe where the drop in pressure head may be most favorably determined, (2) the relations between this drop in pressure head and the rate of discharge through the pipe, (3) the lost head caused by the orifice, (4) the effect of small deviations from what may be called standard conditions, and (5) the proper size of orifice for given conditions.

This method of measuring flow of water in pipes is by no means new, but the authors have aimed to use methods that are applicable to ordinary practice rather than to refined laboratory practice.

The principal tests were made on 4-in., 6-in. and 12-in. commercial steel pipe which had been in use for some years, so that, owing to roughness and other variations, the error of measuring diameters was perhaps 0.02 in. Eight orifices were used with each pipe ranging from one-eighth to five-sixths of the pipe diameter and the eight ratios of pipe diameter to orifice diameter for the 4-in. series were approximately the same as for the 6-in. and 12-in. series.

To determine the proper cross-sections of the pipe at which to take pressure readings, gage connections were made at a number of positions upstream and downstream from the orifice. A study of a large number of observations leads to the conclusion that for flow measurement, gage readings should be taken with connections at the section of the beginning of convergence of the stream and at the section of greatest contraction of the jet. These points were determined to be at eight-tenths of the pipe diameter upstream from the orifice and at four-tenths of the pipe diameter downstream from the orifice respectively. It was also shown that normal flow was again resumed at a distance of three or four pipe diameters downstream from the orifice.

Since the determination of the rate of flow is made from the drop in head between the points eight-tenths pipe diameters upstream, and four-tenths diameters downstream from the orifice, a wide range of tests was made to determine the relations between these values. The theoretical equation for the quantity of water discharged may be written as follows:

$$Q = A\sqrt{\frac{2gh}{\frac{\left[\frac{D}{d}\right]^4}{C^4} - 1}}$$

where Q = the discharge in cubic feet per second, A = the area of the pipe in square feet, $g = 32.16$, h = the drop in pressure head in feet of water, D = the diameter of the pipe in inches, d = the diameter of the orifice in inches, and C = the coefficient of discharge.

This equation may be written in the form

$$Q = AK\sqrt{h} \text{ where } K = \sqrt{\frac{2g}{\frac{\left[\frac{D}{d}\right]^4}{C^4} - 1}}$$

The value of K to which the authors give the name, "velocity modulus," is seen to depend on the ratio of pipe to orifice diameters and on the coefficient of discharge C.

From the results of these tests the values of K are given in Table I for a number of cases. These values of K are based on value of C for a pressure drop of one foot. Now

TABLE I. VELOCITY MODULUS—K

$\frac{D}{d}$	Velocity Modulus 4-In. Pipe	5-In. Pipe	6-In. Pipe	8-In Pipe	10-In. Pipe	12-In. Pipe
2.00	1.275	1.250	1.240	1.240	1.240	1.240
2.10	1.150	1.123	1.111	1.110	1.110	1.110
2.20	1.030	1.015	1.010	1.010	1.010	1.010
2.30	0.944	0.930	0.925	0.925	0.925	0.925
2.40	0.865	0.850	0.845	0.845	0.845	0.845
2.50	0.798	0.785	0.779	0.779	0.779	0.779
2.60	0.736	0.725	0.720	0.720	0.720	0.720
2.70	0.684	0.672	0.667	0.667	0.667	0.667
2.80	0.635	0.625	0.620	0.620	0.620	0.620
2.90	0.592	0.583	0.578	0.577	0.576	0.576
3.00	0.553	0.545	0.540	0.540	0.539	0.539
3.25	0.473	0.465	0.462	0.461	0.461	0.461
3.50	0.405	0.399	0.395	0.395	0.395	0.395
3.75	0.351	0.345	0.342	0.342	0.342	0.342
4.00	0.310	0.305	0.303	0.302	0.302	0.302
4.50	0.244	0.240	0.238	0.237	0.237	0.237
5.00	0.197	0.194	0.192	0.192	0.192	0.192
5.50	0.165	0.161	0.160	0.160	0.160	0.160
6.00	0.138	0.136	0.135	0.135	0.135	0.135

since C was found to decrease slightly as h increases, the values of K are too large for a large drop in head and slightly too small for a small drop in head. However, the practical limits for h seem to be between 0.1 ft. and 10 ft., and between these limits the errors introduced by the use of this table are negligible.

Another equation, based upon the observations made, is suggested, namely:

$$V = 4.85 \frac{h}{\left[\frac{D}{d}\right]^2 - \left[\frac{d}{D}\right]^2}$$

where V is the velocity in the pipe in feet per second and the other letters have the same meaning as in the former equation This equation has the advantage that the variable coefficient of discharge is eliminated and will give practically the same

*Abstract from Bulletin No. 109, Engineering Equipment Station, University of Illinois.

results as the other equation when values of $\frac{D}{d}$ are between 2 and 6.

The loss of head due to the orifice is the difference between the pressure head where the flow is normal upstream from the orifice and that where normal flow is again resumed below the orifice. The theoretical equation for the relation between this lost head and the drop in pressure at the orifice is too complicated for general use but the authors have suggested the simple form

$$h_a = J\,h$$

where h_a is the loss in head, J is a coefficient depending on the ratio of diameters, and h is the drop in head at the orifice. Values of J may be obtained from the accompanying curve.

The choice of the size of orifice to be used may depend upon four factors: The rate of flow in the pipe, the lost head that may be allowed, the desired precision of discharge measurement, and the maximum drop in pressure head the differential gage will register. In general it is best to use an orifice not larger than half the pipe diameter, or in other words, $\frac{D}{d}$ not less than 2.

Table II shows approximate values of drop in pressure head and lost head for several velocities and values of $\frac{D}{d}$.

As an example let it be required to measure the rate of discharge through an 8-in. pipe in which the velocity may vary

TABLE II. APPROXIMATE VALUES OF DROP IN PRESSURE AND LOST HEAD

h = drop in pressure head in feet; h_a = lost head in feet.

V	$\frac{D}{d} = 1.2$		$\frac{D}{d} = 1.5$		$\frac{D}{d} = 2.0$		$\frac{D}{d} = 2.5$		$\frac{D}{d} = 3.0$		$\frac{D}{d} = 4.0$	
	h	h_a	h	h_a	h	h_a	h	h_a	h	h_a	h	h_a
0.2	....	....	.....	.....	.....	.....	.....	.....	.....	.....	0.43	0.39
0.5	....	....	.....	.....	0.16	0.12	0.41	0.33	0.86	0.73	2.7	2.5
1	....	....	0.17	0.09	0.64	0.46	1.6	1.3	3.4	2.9	10.8	9.8
2	....	....	0.66	0.36	2.6	1.9	6.5	5.3	14	12	43	39
3	0.37	0.12	1.5	0.8	5.8	4.2	15	12	31	26	.....	.....
4	0.65	0.21	2.6	1.4	10	7	26	21	55	47	.....	.....
5	1.0	0.3	4.1	2.3	16	12	41	33	.....	.....	.....	.....
6	1.5	0.5	5.9	3.2	23	17	59	48	.....	.....	.....	.....
10	4.1	1.3	17	9	64	46	.....	.....	.....	.....	.....	.....

from one to six feet per second. What size orifice should be used? From Table II, with a diameter ratio $\frac{D}{d} = 2$, the lost head is 0.46 ft. and the drop in pressure head 0.64 ft. for a velocity of 1 ft. per sec., while they are 17 ft. and 23 ft. respectively for a velocity of 6 ft. per sec. An orifice with a diameter ratio $\frac{D}{d} = 1.5$ gives a range of lost head from 0.09 to 3.2 ft. and of drop in pressure head from 0.17 to 5.9 ft. The first orifice gives a rather high lost head and pressure drop at high velocities, while the second one gives rather low values at low velocities and will not be so accurate. Probably an orifice giving $\frac{D}{d} = 1.6$ or 1.75 would be advisable.

In using orifices for measuring flow of water in connection with the data given here, precautions should be taken to limit the avoidable errors. The edges of the orifices should be sharp and square. While these experiments were made with orifice plates $\frac{3}{16}$ in. thick, the authors think no error would be made in using thinner plates. The pipe should be straight and uniform for ten pipe diameters or more above the orifice. Gage connections should be not smaller than ¼-in. pipe, and care should be taken to remove all burrs from the pipe and that the nipples should not protrude beyond the inner surface of the pipe.

When $\frac{D}{d}$ is equal to or greater than two, a single nipple at each section will be sufficient and the exact location of the section may be varied slightly from the positions specified. When $\frac{D}{d}$ is less than two, however, two nipples should be used and no variation in their location should be allowed. Provision should be made for removing air from the pipe and gage connections. If proper precautions are taken, the authors claim that the rate of discharge as measured by a pipe orifice should be accurate within 2 per cent.

Throttling the gage connections was found to help only in the case of large orifices and had to be done very carefully. Small eccentricity in the location of the orifice had no effect on smaller orifices and could be taken care of by using two gage connections when using the larger orifices. The coefficient of discharge for bevel-edged orifices is much more variable than for square-edged orifices, and the use of bevel-edges is not recommended. In general, orifices larger than two-thirds of the pipe diameter should not be used except for very rough measurements.

Western Society Foresees Great Developments

On Jan. 28 the Western Society of Engineers held its annual dinner at Hotel Morrison, Chicago. About 350 were in attendance, and under the direction of Frederick P. Vose, a very capable toastmaster, the event was a great success. An orchestra, community singing, the Universal Glee Club and Quartette added zest to the meeting and kept the enthusiasm at the proper pitch to enjoy the many good things on the program.

Major General Leonard Wood spoke briefly, referring to the constructive engineer and the sanitarian as the principal aids of colonial administration. He reviewed the work they had done in the Philippines, at the Panama Canal and in Cuba. In the territory last named, an independent department of public works had been formed under his administration, and it did all work under a fixed budget. It was the speaker's opinion that a similar plan would be a good thing in this country. In the world war the engineer went hand in hand with the sanitarian and the doctor, each making an enviable record in his particular sphere. The general was especially interested in the building up and keeping alive of the Engineers Reserve Corps, so that such an important branch of the service would be ready when trouble came, and in times of peace the various units could prepare plans of defense for their respective cities and have all in readiness for construction when the occasion arose. A defensive alliance with the architects was desirable. There should be harmony between the two professions, and the combination should take an active part in public affairs, guiding and keeping constructive civic building and preventing the monstrosities that creep in when industrial interests are allowed free reign.

A. Stuart Baldwin, retiring president, delivered a most eloquent and inspiring address, dealing in particular with the broadening field of the engineer and his great work of the future in bringing about coöperation and coördination between the two extremes representd by capital and labor. It was his opinion that there would soon be a marked development of the power of the great middle class in exercising control over the relatively small groups of capital and labor. If either of the latter factions ruled implicitly, then civilization would fail. There must be liberty, equality and fraternity, with the Golden Rule as the guiding spirit. Lack of mutual consideration had been the chief cause of the growing antagonism and unrest. Good will was the most potent and practical force at the disposal of the engineer, and it is a force of which the world is in desperate need. Its recognition and acceptance would lead to development beyond any present conception.

During the year the society had trebled its membership and had obtained an excellent representation from the large body of engineers in the great industrial center about Chicago. Organization of the new members was the first duty, with an engineering building in the background. With so large and so diversified a membership including all kinds of engineers, in his opinion the work of the directors must be decentralized and assumed more by the chairmen of the various sections, each section becoming to a certain extent a self-contained society, but all coöperating as a part of the main body. If each section were the local branch and representative of the national society with which most of its members are affiliated, and if each national society would be the willing associate of the section, such a plan would strengthen both the sections of the Western Society of Engineers and the national associations, the latter having need of greater local activities, which they would gain through the support of the sections.

In his inaugural address F. K. Copeland, president-elect, touched upon future lines of activity for the society. In the past the society had stuck too closely to technical matters. The recent growth indicated that it had a place to fill. In civic affairs the society should take a part equally important as the Chamber of Commerce. There were questions of city planning and zoning, railway terminals, etc., in which the engineer would be of the greatest value. To promote social development the speaker urged the advantages of wider acquaintanceship through meetings, excursions, noonday luncheons and in particular urged the engineers to be more sociable among themselves at these gatherings. The younger men may not think that they can spare the time, but it was to the advantage of the employer to encourage them to make use of these opportunities of mixing with the older men of wider experience. Each man must think what he can do to improve and take his full quota of the advantages offered by the society. The many new members must be made to feel that their joining has been the most worthwhile thing they have done for years. The officers cannot do all. It is their plan to set each and every member to work. If every man will take part and carry on, the society has before it a most successful year.

Announcement was made of the following incoming officers: President, F. K. Copeland; first vice president, C. F. W. Felt; second vice president, J. L. Hecht; third vice president, Linn White; treasurer, F. F. Fowle; secretary, Edgar Nethercut; trustee for three years, J. H. Libberton; Washington Award Commission for three years, Albert Reichman and Major E. W. Allen. Chairmen of the various sections are: Hydraulic and municipal, William Artingstall; mechanical, Arthur L. Rice; bridge and structural, A. W. Dilling; electrical, A. F. Riggs; gas, C. C. Borden. Chairmen have yet to be elected for the three new sections; telegraph and telephone, industrial and railway.

Marine Engineers Hold Annual Meeting

The Marine Engineers Beneficial Association held its forty-fifth annual convention at Washington, D. C., during the week beginning Jan. 19, with headquarters at the New Ebbitt Hotel.

There were present 160 delegates, representing 231 votes. The opening session took place at ten o'clock on Monday morning, with every officer at his proper station. The large volume of business transacted made it necessary to hold night sessions, and during the week many important measures tending to the uplift of the organization were discussed and voted upon. The report of Albert L. Jones, the treasurer, showed that the association is on a sound financial basis. During the past year it was decided by a referendum vote to combine the offices of secretary and treasurer.

At the meeting on Monday morning the delegates were addressed by Darragh de Lancey, Director of Docks and Industrial Relations of the United States Shipping Board.

On Tuesday morning the convention listened to an instructive and entertaining address by George Uhler, Supervising Inspector General of the United States Steamboat Inspection Service.

Notwithstanding the inclement weather which prevailed through the week, there were theater parties, shopping trips and sight-seeing auto drives for the ladies.

On Wednesday evening the ladies of the convention were entertained by A. Warren France, of the France Packing Co., in his suite at the New Ebbitt Hotel. Refreshments were served in abundance. The entertainers were George E. Andrews, Bob Jones and Jack Armour.

The big feature of entertainment was the smoker on Friday night tendered the delegates by the Marine Engineers Supplymen's Association in the ballroom of the New Willard Hotel. There were upward of 500 in attendance, including army and navy officers, heads of the various departments, and engineers from the local associations. Briarwood pipes, tobacco and cigarettes were distributed, and an excellent vaudeville show was furnished.

At the afternoon session of the convention on Tuesday the following officers were reëlected for the term of three years: William S. Brown, president, Buffalo; John S. Purdie, first vice president, San Francisco; William H. Hyman, second vice president, Baltimore; Edward C. Killian, third vice president, New Orleans; George A. Grubb, secretary-treasurer, Chicago. The National Executive Committee comprises: Robert Goalett, Atlantic Coast District; Charles Tollett, Pacific Coast District; William J. Garrett, Great Lakes District; Charles M. Sheplâr, West Rivers District; Edward C. Killian, Gulf Coast District.

The Marine Engineers Supplymen's Association elected officers as follows: Andrew Lauterbach, president, the Lunkenheimer Co.; Charles A. Remsen, vice president, Dearborn Chemical Co.; Bob Jones, secretary-treasurer, France Packing Co.

At the final meeting of the supplymen, Joseph J. Cirek, the retiring president, was presented with a leather traveling bag.

Automotive Engineers Discuss the Diesel Engine

Diesel engines formed the topic for discussion at the joint meeting of the Pennsylvania and the Metropolitan (New York) Sections of the Society of Automotive Engineers at Philadelphia on Jan. 22. Leaving New York by special car in the morning, the Metropolitan members joined the Pennsylvania members at the Philadelphia Engineers' Club for luncheon, and were later conducted to the League Island Navy Yard to witness a run on one of a pair of Diesel engines recently removed from the German submarine U-117, which is there being tested out. After a supper at Kugler's in the

DELEGATES TO THE FORTY-FIFTH ANNUAL CONVENTION OF THE MARINE ENGINEERS' BENEFICIAL ASSOCIATION

evening, the members listened to an instructive paper on "Modern Practice in Heavy Oil Motor Installations, and Development Possibilities," by Hubert C. Verhey, head of the Diesel Engine Unit, Emergency Fleet Corporation. Many prominent Diesel-engine men were present, and a most profitable discussion followed.

The submarine U-117, it will be recalled, was the one which, early in 1918, operated off the Atlantic Coast. The engine tested is of the high-speed (450 r.p.m.) four-cycle Augsberg design, rated at 900 b.hp. in six cylinders, and presented some very interesting features. For instance, the housing and cylinder jackets are of cast steel with cast-iron cylinder liners. Owing to the scarcity of copper in Germany during the war, all the piping was made of iron and many of the connections welded. The compressor is a four-stage machine driven direct, and all auxiliaries likely to give trouble are in duplicate; there are also two fuel valves. Cross-connections were provided between the two engines, so that if certain auxiliaries on one failed, those on the other could be used while repairs were being made. The cylinder diameter is 17¾ in., and there was no taper on the skirt of the piston, as is very common practice. The stroke of 16¼ in. makes a low engine, which of course was necessary because of the limited headroom. Oil cooling is employed for the pistons through a link connection. It is not known how many miles this submarine had cruised in the hands of the Germans, but it has cruised over 10,000 miles since this country took it over; yet upon removing the pistons from one of the cylinders and calipering them, no perceptible wear was discovered. Moreover, the engine ran very quietly, with practically no vibration, as all of the nuts were found to have been put on with a driving fit. The workmanship throughout showed evidence of careful attention to every detail, which was characteristic of the Germans.

Diesel-Engine Discussion

Mr. Verhey, after following through, by the aid of slides and diagrams, the workings and complete installation of a marine Diesel engine outfit, proceeded to discuss the relative merits of different types and designs. He was of the opinion that although the four-cycle engine has largely displaced the two-cycle for marine purposes, ultimately the two-cycle design will predominate, because American engineers, profiting by the experience of European designers, are in a position to avoid the pitfalls of their predecessors and will overcome certain features that were responsible for troubles with the two-cycle design in the past. He pointed out the advantages of top scavenging for two-cycle engines, but showed that certain manufacturers were now bringing out designs embodying a double scavenging port, which would obviate the objections to the former method of port scavenging. He also referred to the advantages of the auxiliary exhaust.

Referring to fuel injection, he pointed out that it is possible to eliminate an air compressor entirely by injecting the fuel direct into the working cylinder under high pressure by the so-called solid injection system, which is now being employed by several builders; but with this system one is likely to get smoke and incomplete combustion, with resulting carbon when operating under light load. He favored very strongly independently driven auxiliaries, especially for marine purposes, for with this arrangement not only could there be employed greater duplication in auxiliaries, but the main engine would not be shut down if there should be trouble with any of the auxiliaries; and it permits continuing the circulating water for some time after shutting down the engine, a precaution that may save cracked cylinder heads.

Referring to the immediate field for the marine Diesel, he believed that it is limited at present to the propulsion of vessels up to 13,000 tons at from 10 to 12 knots, within which limit the motor installation will be able to hold its own under all circumstances. A vessel having about 1,500 hp., carrying 625 tons of fuel oil, can attain a cruising radius of about 20,000 miles with oil engines, as compared with 8,400 miles when equipped with steam-turbine installation, or 7,500 miles with a reciprocating steam installation. Passenger ships of greater tonnage and speed would not receive much competition from the Diesel engine, but would be equipped with geared-turbine installations for the time being.

Commenting upon the plans of the Emergency Fleet Corporation, Mr. Verhey stated that two 2,250-hp. four-cycle Burmeister & Wain engines have been ordered for a new 12,000 ton cargo boat that is now being built. In addition to this there has been installed a 750-hp. McIntosh & Seymour engine in one of the smaller boats, and another engine of 900 hp. is on order. The corporation is going slowly in the matter, but is making very definite progress.

In the discussion which followed it was brought out that the 750-hp. McIntosh & Seymour engine referred to had 22-in. cylinders without piston cooling, and was of the trunk piston type. This it was agreed was about the limit that one should go without piston cooling and without resorting to a crosshead

With reference to the merits of solid injection, it was pointed out that the Vickers Co., of England, are putting out a new engine using solid injection, and thereby saving 10 per cent of the compressor work. This engine is provided with an accumulator, and so far has shown very good results. It was agreed that forced-feed lubrication was desirable with a crosshead type of engine, although gravity feed was permissible with the trunk type. The crosshead type permitted the piston rod to pass through a stuffing-box and thus prevent the oil from becoming fouled with the gases leaking past the piston.

It was also mentioned that less trouble is experienced from a high-sulphur oil in a two-cycle engine than in a four-cycle, because of the absence of exhaust valves in the former.

The building of a large hydraulic electric power plant at the mouth of the Chelan River is now regarded as almost certain. Great Northern officials have been spending several days at Lake Chelan looking over the final details preparatory to the installation of the plant. However, no definite announcements have been made as to when this work will be commenced. The Great Northern was ready to build this plant at the time the United States went to war. It is proposed to build a 200-ft. dam 25 feet high at the point where the Chelan River flows out of the lake, thus raising the entire level of the lake. Between 40,000 and 50,000 hp. is to be developed at first and a total of 100,000 hp. is said to be available.

It is said that salt water infiltration in some oil wells in Louisiana amounts to as much as 40 per cent and affects a considerable portion of the Homer field, and that some wells have stopped flowing and others have been pinched in. Water thus far affects those wells in the deep sand and in a southeasterly direction from fault line which cuts diagonally through the Homer field from the northeast to the southwest. Wells to the northwest of the fault line in shallow sand, it is asserted, have not been affected and are expected to be of much longer life than those in deep sand.

SOMETHING TO THINK ABOUT

Personals

Joseph C. Reagan, who has been superintendent of the Trumbull Electric Manufacturing Co., Plainville, Conn., for the last three years, has severed his connections with that company.

Harry Streik, who was erecting engineer for Frick Co., the world's largest manufacturer of ice-making machinery, for twenty-seven years, is now superintendent of the new Russ Bros. plant at Harrisburg, Penn., which he designed and erected himself.

John E. Hubbell, counsellor at law and solicitor of patents, formerly of Chambers & Hubbell, Philadelphia, announces that he has opened offices at 469 Fifth Ave., New York City, where his practice will be devoted exclusively to patent and trademark matters.

Louis R. Chadwick, hitherto branch office manager of the Sullivan Machinery Co. at Spokane, Wash., has been appointed manager of the company's branch at 30 Church St., New York City. Robert T. Banks, who has been the company's sales engineer at its El Paso (Texas) office for several years, has been appointed manager at Spokane to succeed Mr. Chadwick.

J. W. Ledoux, who has been connected with the American Pipe and Construction Co., since July, 1891, has resigned to engage exclusively in consulting engineering on his own account, at 112 North Broad St., Philadelphia. His practice will embrace designs, supervision of construction, valuation, rate-making, arbitrations, etc., particularly on water works and water-power lines.

George L. Washington, of the negotiation division of the Westinghouse Electric International Co., has left East Pittsburgh for Cuba, where he will be located as salesman in the Havana office of the company. Mr. Washington entered the employ of the export department of the Westinghouse Co. in May, 1917, after completing the college graduate apprentice course. In September, 1917, he entered the service and was discharged Jan. 28, 1919, when upon his return he was placed on order work in the International company, and in May was transferred to the negotiation division.

R. H. Ballard, president of the National Electric Light Association, left Los Angeles on Jan. 11 for another tour of the East in the interests of the organization, and to perfect arrangements for the convention of 1920, which will be held in Pasadena, Calif., May 18-21. At San Francisco he was joined by John A. Britton, chairman of the Public Policy Committee, who accompanied him on his trip. They were given a complimentary dinner by the Electrical Development League of that city on Jan. 12, which was attended by the prominent electrical men of central California. Much enthusiasm was expressed concerning the convention and plans were outlined for the work preparatory to the convention which will be carried on during Mr. Ballard's absence. On his way East Mr. Ballard will attend meetings at various places, and will attend several company sections before his return home.

Engineering Affairs

The Material Handling Machinery Manufacturers' Association has found it necessary to change the date of its open convention in New York City at the Waldorf-Astoria hotel from Jan 29-30 to Feb. 26-27.

The Mohawk Valley Engineers' Club was organized on Jan. 7, at a dinner given at the Hotel Utica, Utica, N. Y., at which about 125 engineers, architects and chemists were gathered. The membership includes civil, chemical, electrical, mechanical, mining, textile, heating and ventilating, automotive and all technical engineers and architects. Regular meetings will be held on the first Tuesday of each month and special meetings at intermediate times. The officers for the first year are: President, Byron E. White, engineer of Utica Gas and Electric Co.; first vice president, Hubert E. Collins, consulting engineer; second vice president, Roy F. Hall, division engineer, State Highway Department; third vice president, Horace B. Sweet, consulting engineer; secretary, F. E. Beck, engineer of Consolidated Water Co.; treasurer, Clifford Lewis, Jr., civil engineer. The territory embraced by the activities of the club includes a radius of 40 miles from Utica, and it is expected the membership will reach above the 200 mark.

Business Items

The Industrial Engineering Co. is now located in its new quarters at 1536 Arcade Building, St. Louis, Mo.

The Busch-Sulzer Bros.-Diesel Engine Co., St. Louis, Mo., announces the opening of its Eastern sales office at 60 Broadway, New York City, with George D. Pogue as Eastern sales agent and Stanley Wright, assistant.

The Green Fuel Economizer Co., Beacon, N. Y., announces the opening of its own office at 1006 Finance Building, Philadelphia, in charge of W. F. Wurster, assisted by G. E. Kille. The company was formerly represented in this territory by the Baker-Dunbar-Allen Co.

The Ohio Body and Blower Co., a newly formed corporation, is an outgrowth of the old and well-known Ohio Blower Co., of Cleveland. The new company has taken over three plants from the older company, providing employment for over 600 workers in 150,000 sq. ft. of floor space. The officers of the older company will assume the same standing in the new company.

The General Chemical Co. has entered into a contract with the J. G. White Engineering Corp. covering the future designing, engineering and construction work incident to improvements and extensions to existing manufacturing plants, buildings and other property, and in connection with any new projects or developments which may be undertaken from time to time. Under this arrangement the J. G. White Engineering Corp. will handle their engineering and construction work.

The G. M. C. Engineering Co., Worcester, Mass., with offices in the Graphic Arts Building, has taken over the exclusive sales of Coppus turbo-blowers in the New England States. This company is composed of several men who were formerly in the employ of the Coppus Engineering and Equipment Co. John L. Gallivan, formerly sales manager of the company, is president and treasurer of the G. M. C. Engineering Co. Howard Mason, who is consulting engineer for McCleary, Wallin & Crouse, will also act in the same capacity for this company. Its service department will be at the disposal of all who call upon it for information.

Westinghouse Electric International Co. has recently opened a new office in the Royal Bank of Canada Building in Havana. This was found necessary because of the growing use in sugar mills and other enterprises in Cuba of Westinghouse electric equipment. J. W. White, who has been located in Cuba for some time past as Westinghouse representative, will be in charge of the new office. Mr. White's intimate knowledge of, and close relation with, the industries in Cuba peculiarly fit him to cope with all the problems that may be encountered. He will be assisted by other special engineers. A complete stock of catalogs and other descriptive matter will be maintained in the new office, where for the asking can be obtained complete information on everything electric.

Trade Catalogs

The Cutter Co., Philadelphia, Penn., has just issued a new handbook of the I-T-E circuit-breaker. This is a handsome cloth-bound book of 352 pages, 6 x 9 in. It illustrates and describes many of the largest or most significant circuit-breaker installations in the country and is of more than passing interest to the consulting engineer and to those responsible for the successful operation of important electrical plants. Applications, if approved, will be filled in the order of their receipt, if made upon the applicant's business letterhead.

"Westinghouse Opportunities for Technical Graduates" is the title of a 32-page illustrated pamphlet recently issued by the Westinghouse Electric and Manufacturing Co. This booklet describes in considerable detail the plan that has been developed by this company for the training of the graduates of technical schools at all of its various works. In it is included a list of prominent Westinghouse men who originally entered the company as graduate students, as well as a complete list of schools from which over 5,000 students have entered the employ of the company. Copies of the booklet will be sent to anyone interested, on application to the Educational Department of the company at East Pittsburgh.

THE COAL MARKET

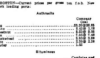

BOSTON—Current prices per gross ton f.o.b. New York loading ports:

Anthracite

	Company Coal
Egg	$8.20@$8.65
Stove	8.45@ 9.05
Chestnut	8.55@ 9.05
Pea	7.05@ 7.40
Buckwheat	3.60@ 3.80
Rice	2.30@ 2.50
Barley	1.25

Bituminous

	Clearfields	Cambrias and Somersets
F.o.b. mines, net tons	$2.85@$3.35	$3.15@$3.60
F.o.b. Philadelphia, gross tons	5.05@ 5.60	5.35@ 5.90
F.o.b. New York, gross tons	5.40@ 5.95	5.75@ 6.25
Alongside Boston (water coal), gross tons	7.00@ 7.75	7.60@ 8.00

Pocahontas and New River are practically off the market for coastwise shipment, but are quoted at $6.25@$7.00.

NEW YORK—Current quotations, White Ash, per gross tons, f.o.b. Tidewater, at the lower ports are as follows:

Anthracite

	Company Coal
Broken	$7.80@$8.25
Egg	8.20@ 8.65
Stove	8.45@ 9.05
Chestnut	8.55@ 9.05
Pea	7.05@ 7.40
Buckwheat	5.15
Rice	4.50
Barley	4.85
Boiler	4.85

Bituminous

Government prices at mines:	Spot
South Fork (best)	$3.25@$3.50
Cambria (best)	3.00@ 3.25
Cambria (ordinary)	2.60@ 2.90
Clearfield (best)	3.00@ 3.25
Clearfield (ordinary)	2.60@ 2.90
Reynoldsville	2.85@ 2.90
Quemahoning	3.25@ 3.50
Somerset (medium)	3.00@ 3.25
Somerset (poor)	2.50@ 2.75
Western Maryland	2.50@ 2.75
Fairmont	2.25@ 2.50
Latrobe	2.60@ 2.90
Greensburg	2.75@ 3.00
Westmoreland ¾ in.	3.40@ 3.50
Westmoreland run-of-mine	2.75@ 3.00

PHILADELPHIA—Bituminous coal prices vary according to district from which they are mined. For ordinary slack the price is $2.45@$2.55; lump, $3.00@$3.35, at the mines.

BUFFALO—

Anthracite

	On Cars, Gross Ton	At Curb, Net Ton
Grate	$8.55	$10.30
Egg	8.80	10.65
Stove	9.00	10.85
Chestnut	9.10	10.25
Pea	7.75	9.30
Buckwheat	5.70	7.75

Bituminous

Allegheny Valley	$4.80
Pittsburgh	4.65
No. 8 Lump	4.65
Mine Run	4.80
Slack	4.10
Smokeless	4.60
Pennsylvania Smithing	5.70

CLEVELAND—Prices of coal per net ton delivered in Cleveland are:

Anthracite

Egg	$12.25@$12.40
Chestnut	12.50@ 12.70
Grate	12.25@ 12.40
Stove	12.40@ 12.60

Pocahontas

Mine-run	$7.50

Domestic Bituminous

West Virginia split	$8.50
No. 8 Pittsburgh	$6.60@ 6.90
Massillon lump	8.25@ 8.50
Coshocton lump	7.15

Steam Coal

No. 6 slack	$5.25@$5.50
No. 8 slack	5.10@ 5.50
Youghiogheny slack	5.25@ 5.50
No. 8 ¾ in.	5.70@ 6.00
No. 6 mine-run	5.25@ 5.50
No. 8 mine-run	5.75

Only coal available is mine-run Pocahontas.

MIDDLE WEST—Chicago quotations. F.o.b. cars at mine:

	Springfield, Carterville, Williamson, Franklin, Saline, Harrisburg	Fulton, Peoria	Grundy, La Salle, Bureau, Will
Lump	$2.55@$2.70	$2.95@$3.10	$3.25@$3.40
Washed	2.75@ 2.90	....	3.45@ 3.60
Mine run	2.35@ 2.50	2.75@ 2.90	3.00@ 3.15
Screenings	2.05@ 2.20	2.35@ 2.50	2.75@ 2.90

New Construction

PROPOSED WORK

Me., Portland—J. C. and J. H. Stevens, Archts., 187 Middle St., will soon receive bids for the construction of a 4 story manufacturing building on Forest Ave. for A. S. Hinds, West St. A low pressure steam heating system will be installed in same. Total estimated cost, $500,000.

R. I., Providence—C. H. Lockwood, Archt., 171 Westminister St., will receive bids for the construction of a 1 story, 96 x 320 ft. service station on Reservoir Ave. for the Federal Truck Sales Co. of Rhode Island, 276 West Exch. St. A steam heating system will be installed in same. Total estimated cost, $100,000.

R. I., Westerly—F. P. Sheldon & Sons, Engrs., 1008 Hospital Trust Bldg., Providence, plan to construct plant additions, including a 4 story, 106 x 108 ft. mill, 1 story, 150 x 350 ft. weave shed and a 1 story, 120 x 120 ft. warehouse, on Mechanic St., for the Lorraine Mfg. Co., 547 Mineral Springs Ave., Pawtucket. Steam heating systems will be installed in same. Total estimated cost, $300,000.

N. Y, Albany—John D. Keeler, c/o Warren & Wetmore, Archs., 10 East 47th St., New York City, is having plans prepared for the construction of a 14 story hotel on B'way and Maiden Lane. A steam heating system will be installed in same.

N. Y., Brooklyn—W. Higginson, Archt., 18 East 40th St., New York City, will soon award the contract for the construction of two factory buildings, one 5 story, 75 x 140 ft., and other 1 story, 25 x 125 ft., a 2 story, 45 x 50 ft. garage and a 1 story, 25 x 50 ft. boiler house, for the Amer. Safety Razor Co., 303 Jay St. A steam heating system will be installed in same.

N. Y., Brooklyn—The Kings Co. Hospital, Clarkson St., is having plans prepared for altering existing hospital buildings and for the construction of an addition to same. A steam heating system will be installed in same. Total estimated cost, $3,000,000. Helmle & Corbett, 189 Montague St., Archts. and Engrs.

N. Y., Carthage—The West End Paper Co., West End Ave., Carthage Sulphide Pulp & Paper Co., Front St., Carthage Tissue Paper Co., Ryther & Prinyle Mfg., machinists, Carthage Machine Co., and others, owning power rights here, plan to construct a large central power station which will yield 18-20,000 hp.

N. Y., Dunkirk—The Niagara & Erie Power Co., Lamphere St., plans to construct a second 3 ph. 60,000 volt transmission line from here to Athol Springs substation.

N. Y., New York—Clinton & Russell, Archts. and Engrs., 32 Liberty St., are preparing plans for the construction of a 108 x 115 ft. office building from 38th-39th Sts., near 5th Ave. A steam heating system will be installed in same. Total estimated cost, $2,225,000. Owner's name withheld.

N. Y., New York—The New York Edison Co., Irving Pl., will soon award the contract for the construction of a 3 story, 45 x 75 ft. transformer station at 421-423 East 6th St. Estimated cost, $60,000. W. Whitehill, 32 Union Sq., Engr. Noted July 1.

N. Y., New York—D. Wortman, Arch. and Engr., 114 East 28th St., will receive bids about Feb. 15 for the construction of a 9 story, 50 x 100 ft. storage building at 108-110 West 107th St., for the Chelsea Fireproof Storage Warehouse, Inc., 426 West 26th St. A steam heating system will be installed in same. Total estimated cost, $125,000.

N. Y., Phelps—The Wilson Canning Co., Mexico, plans to construct a factory, warehouse and separate office building, here. Two 100 hp. boilers, transmission equipment, etc., will be installed in same. Total estimated cost, $15,000 to $18,000.

N. Y., Rochester—The city will soon award the contract for the construction of a 3 story, 304 x 321 ft. Junior High School at Wilson Park. An indirect radiation system, including boilers, engines, etc., will be installed in same. Total estimated cost, $1,500,000. Address J. M. Tracy, Supt. of Pub. School Bldgs. Edwin S. Gordon, c/o Gordon & Madden, Sibley Block, Arch.

N. Y., Rye—J. E. Bowman, Hotel Biltmore, New York City, has awarded the contract for the construction of a club and hotel building, here, to George A. Fuller, 175 5th Ave., New York City. A steam heating system will be installed in same. Total estimated cost, $2,000,000.

N. Y., Schenectady—The Genl. Electric Co., River Rd., has engaged Harris & Richards, Archs., Drexel Bldg., Philadelphia, to prepare plans for the construction of a 6 story, 50 x 200 ft. office building. A steam heating system will be installed in same. Total estimated cost, $300,000.

N. Y., Syracuse—The Syracuse Washing Machine Co., 507 East Water St., will soon award the contract for the construction of a 4 story, 200 x 250 ft. washing machine factory, including an 80 x 100 ft. moving picture auditorium for employes, on Main St. A steam heating system will be installed in same. Total estimated cost, $1,000,000.

N. Y., Watertown—The Brier Hill Electric Light & Power Co., Inc., St. Lawrence Co., plans to construct 4 mi. of high tension transmission line from here to Brier Hill. Estimated cost, $32,000.

N. J., Bayonne—The Knights of Columbus, Star of the Sea Council, c/o G. A. McCabe, Arch. and Engr., 96 5th Ave., New York City, is having plans prepared for the construction of a club house on Ave. C. A steam heating system will be installed in same. Total estimated cost, $250,000.

N. J., Verona (Montclair P. O.)—The Eagle Rock Mfg. Co. is having plans prepared for the construction of a 2 story, 40 x 350 ft. factory, for the manufacture of moving picture films. A steam heating system will be installed in same. Total estimated cost, $250,000. Lockwood, Green & Co., 101 Park Ave., New York City, Archs. and Engrs.

Pa., Allentown—The Immanuel Evangelical Congregation plans to build a 12 story, 50 x 100 ft. church on Main St. A steam heating system will be installed in same. Total estimated cost, $100,000.

Pa., Chester—C. W. Brazer, Arch. and Engr., 1133 B'way, New York City, will receive bids about May 1 for the construction of a 1 story church on 8th and Butler Sts., for the Trinity Methodist Episcopal Church. A steam heating system will be installed in same. Total estimated cost, $100,000.

Pa., Chester—The Frohn & Robb Association, Archs., will receive bids about May 1 for the construction of a church for the First Presbyterian Church. A steam heating system will be installed in same. Total estimated cost, $125,000.

Pa., Media—C. W. Brazer, Arch. and Engr., 1133 B'way, New York City, will receive bids about May 1 for the construction of an office and bank building for the Media Title & Trust Co., 11 South St. A steam heating system will be installed in same. Total estimated cost, $300,000.

Pa., Philadelphia—G. Edwin Braumbaugh, Arch., Real Estate Trust Bldg., will receive bids until Feb. 15 for the construction of a 1 story, 34 x 120 ft. market house at 1430 South St., for Robert Venturi & Bros., 1430 South St. A steam heating system and a refrigerating plant will be installed in same.

Pa., Philadelphia—The Chestnut Hill Hospital, 8815 Germantown Ave., engaged J. P. Simms, Arch., 1627 Sansom St., to prepare plans for the construction of a hospital in Chestnut Hill. A steam heating system will be installed in same. Total estimated cost, $100,000.

Pa., Philadelphia—H. B. Weldon, Arch., Fuller Bldg., is preparing plans for the construction of a 1 and 2 story, 40 x 160 and 25 x 160 ft. sales and service building for L. J. Kolb, 901 North Broad St. A steam heating system will be installed in same. Total estimated cost, $50,000.

Pa., Pittsburgh—The Mellon Natl. Bank, 514 Smithfield St., is having plans prepared for the construction of a 4 story, 120 x 130 ft. bank and office building on Smithfield St. between 5th Ave. and Olive St. A steam heating system will be installed in same. Trowbridge & Livingston, 527 5th Ave., New York City, Archs. and Engrs.

Pa., Plains (Wilkes-Barre P. O.)—The Plains School Bd. will receive bids until Feb. 23 for the construction of a 2 story, 70 x 130 ft. school on Moffitt St. A steam heating system will be installed in same. Total estimated cost, $150,000. Austin Rielly, Bennett Bldg., Wilkes-Barre, Arch.

Pa., Scranton—The Scranton Dress Co., South Webster Ave., will receive bids until Feb. 20 for the construction of a 3 story, 100 x 125 ft. dress factory on South Webster and Hickory Sts. A steam heating system and electric power will be installed in same. Total estimated cost, $150,000. F. W. Borgenicht, Pres.

Pa., Scranton—The Ahavath Sholem Congregation, Providence (Scranton P. O.), plans to build a 15 story 50 x 100 ft. synagogue on North Main Ave. A steam heating system will be installed in same. Total estimated cost, $100,000.

Pa., Scranton—O. D. Dewitt Auto Co., Wyoming Ave., is preparing plans for the construction of a 2 story, 90 x 170 ft. garage on Wyoming Ave. A steam heating system will be installed in same. Total estimated cost, $100,000.

Pa., Wilkes-Barre—The State and Luzerne Co. are having plans prepared by T. H. Atherton, Arch., Coal Exch. Bldg., for the construction of a 4 story, 90 x 100 ft. armory on Market St., along the Susquehanna River. A steam heating system will be installed in same. Total estimated cost, $350,000.

Pa., Wilkes-Barre.—Geo. Gildersleeve, Gildersleeve St., will soon award the contract for the construction of a 2 story, 45 x 225 ft. garage on South Main St. A steam heating system and electric power will be installed in same. Total estimated cost, $100,000. Thomas Atherton, Jr., Coal Exch. Bldg., Arch.

Pa., Wilkes-Barre—George S. Welch, Arch., Coal Exch. Bldg., will receive bids until Feb. 24 for the construction of a 3 story, 200 x 225 ft. silk mill for the Duplan Silk Corp. A steam heating system and electric power will be installed in same. Total estimated cost, $500,000.

Pa., Yeadon—S. L. Baraumbaugh, Hanison Bldg., Philadelphia, has engaged Jesse T. Hoekstro, Arch., 1713 Sansom St., to prepare plans for the construction of a 1 story ice plant at Presser Park. Estimated cost, $10,000.

Md., Baltimore—David Garrett & Sons, 1210 Bouldin St., are having plans prepared for the construction of a 1 story, 37 x 75 ft. boiler house and a 1 story, 65 x 65 ft. garage on East Ave. near Ellwood St. Total estimated cost, $35,000. F. Thomas, 135 North Kenwood St., Arch.

Md., Fairfield (Baltimore P. O.)—The Globe Shipbuilding & Dry Dock Co. of Maryland, Fidelity Bldg., Baltimore, is having plans prepared for the construction of a 1 and 2 story shipbuilding and ship repair plant along the water front. Contract for auxiliary power house equipment will be sub-let. Total estimated cost, $5,000,000. B. C. Cooke, Pres.

Va., Norfolk—The Beth El Congregation, c/o Herts & Robinson, Archs., 331 Madison Ave., New York City, is having plans prepared for the construction of a synagogue. A steam heating system will be installed in same. Total estimated cost, $250,000.

Va., Roanoke—H. C. Richards, Arch., 1713 Sansom St., Philadelphia, will soon award the contract for the construction of a 3 story school, here, for the Bd. Educ. A steam heating system will be installed in same. Total estimated cost, $175,000.

N. C., Aberdeen—The Bd. Comrs. will soon award the contract for the construction of a waterworks and sewerage system. Pumping equipment includes an oil engine, deep well pumps, etc. Total estimated cost, $40,000. J. B. McCrary Constr. Co., 3rd Natl. Bank Bldg., Atlanta, Ga., Engr.

Tenn., Knoxville—The Anderson Dulin & Varnell Co. plans to construct a 10 story, 90 x 100 ft. warehouse on Gay St. A steam heating system will be installed in same. Total estimated cost, $200,000.

O., Dayton—The Delco Light Co., East 1st St., plans to construct a 2 story power plant on Keowee and Pitt Sts. Estimated cost, $75,000. A Pressler, 329 East 1st St., Engr.

O., Elyria—The Elyria Savings & Trust Co. is having preliminary plans prepared for the construction of a 12 story bank and office building on Broad and Court Sts. Three boilers will be installed in same. Total estimated cost, $1,000,000. Walker & Weeks, 1900 Euclid Ave., Cleveland, Archs. and Engrs.

Mich., Detroit—The Bd. of Water Commissioners, 232 Jefferson Ave., will receive bids until Feb 17 for the construction of the sub-structure for a low lift pumping station, involving about 24,000 lin. ft. of concrete piles and 6,000 cu. yds. of reinforced concrete. Estimated cost, $300,000 to $350,000. T. A. Leisen, Engr.

Mich., Detroit—H. S. Koppin, Breitmeyer Bldg., is having plans prepared for the construction of a 2 story, 100 x 120 ft. theatre on Catherine and St. Antoine Sts. A steam heating plan, including boiler and ventilating equipment, will be installed in same. Total estimated cost, $250,000. C. H. Crane and E. G. Kiehler, Huron Bldg., Archs.

Mich., Detroit—C. H. Miles, Miles Theatre Bldg., Griswold St., is having plans prepared for the construction of a 2 story, 120 x 163 ft. theatre on Grand River and Roosevelt Aves. Steam heating system, with forced ventilation, will be installed in same. Total estimated cost, $1,000,000.

Mich., Detroit—B. F. Mortenson, 1119 Dime Bank Bldg., has purchased a site and plans to construct a 4 story, 130 x 200 ft. commercial building on Lafayette Ave. and 6th St. Steam heating equipment will be installed in same.

Mich., Flint—The Independent Cold Storage Co., 120 West Kearsley St., plans to construct a 3 story cold storage plant on Industrial Ave. Electric power and refrigeration equipment will be installed in same. Total estimated cost, $300,000.

Mich., Kalamazoo—The Clarage Fan Co., North and Porter Sts., plans to build a 1 story, 28 x 170 ft. and 66 x 70 ft. additions to its foundry and pattern shop on North St. Electric motors to supply power will be installed in same. Total estimated cost, $100,000.

Mich., Muskegon—The Bd. Educ. is having plans prepared for the construction of a 2 story school on Acorn St. A steam heating plant, including boiler, fan, pump and ventilating system, will be installed in same. Total estimated cost, $145,000. H. H. Turner, 234 Division Ave. N., Grand Rapids, Arch.

Mich., Saginaw—The Lufkin Rule Co., Hess Ave., plans to construct a 1 story, 115 x 219 ft. factory on Hess Ave. and Prescott St. Steam heating equipment will be installed in same. Total estimated cost, $100,000. Esselstyn, Murphy & Hanford, 810 Marquette Bldg., Detroit, Archs. and Engrs.

Mich., Three Rivers—E. H. Andrews, Secy. of School Dist. No. 1, will soon award the contract for the construction of a 2 story school. A steam heating system will be installed in same. Total estimated cost, $175,000. R. A. LeRoy, 122 Pratt Blk., Kalamazoo, Arch.

Mich., Ypsilanti—H. X. Wilder, Secy. of the Bd. Educ., will receive bids until Feb. 17 for the construction of a 2 story grade school. A steam heating system will be installed in same. Total estimated cost, $150,000. Robinson & Campau, 715 Michigan Trust Bldg., Grand Rapids, Archs.

Ill., Chicago—Paul Gerhardt, Arch., 64 West Randolph St., will soon award the contract for the construction of a 1 and 2 story, 50 x 200 ft. auto sales service building on Sheridan Rd. and Irving Park Blvd., for Hampton Winston, 749 North Michigan Ave. A steam heating system will be installed in same. Total estimated cost, $100,000. The Oakland Phillips Motor Co., 931 Wilson Ave., Lessee.

Ill., Chicago—The Jackson Express & Van Co., 3611 West 22nd St., is having plans prepared for the construction of a 6 story, 75 x 175 ft. storage warehouse on Madison St. and 60th Ave. A steam heating system will be installed in same. Total estimated cost, $150,000. G. S. Kingsley, 109 North Dearborn St., Arch.

Ill., Chicago—D. S. Klafter, Arch., 64 West Randolph St., will soon award the contract for the construction of a 4 story, 50 x 125 ft. auto sales building at 2449 Michigan Ave. for C. Schmalhausen, 2448 Michigan Ave. A steam heating system will be installed in same. Total estimated cost, $150,000.

Ill., Chicago—The Monighan Machine Co., 2036 West Carroll Ave., is having plans prepared for the construction of a 110 x 504 ft. manufacturing plant on Augusta, Kilpatrick and Belleplaine Aves. A steam heating system will be installed in same. Total estimated cost, $250,000. Alfred P. Weber, 111 West Washington St., Arch.

Ill., Chicago—Henry J. Schlacks, 721 North Michigan Ave., plans to construct an 8 story, 100 x 200 ft. apartment hotel on Chestnut St. and Delaware Pl. A steam heating system will be installed in same. Total estimated cost, $2,000,000.

Wis., Milwaukee—Cahill & Douglas, Engrs., 217 West Water St., are receiving bids for the installation of a 150 hp. high pressure tubular boiler and a 100 ft. steel stack, for the Barnett Woolen Mills Co., 200 Muskego Ave.

Wis., Tomah—The Bd. Educ. will soon award the contract for the construction of a 3 story, 50 x 80 ft. addition to the high school and a 30 x 55 ft. boiler house. Total estimated cost, $60,000. C. B. Drowatsky, Clk. H. C. Haeuser, Colby-Abbot Bldg., Milwaukee, Arch.

Minn., Hibbing—Victor L. Power, Mayor, will receive bids until March 15 for the construction of a 2 and 3 story, 200 x 300 ft. recreation building in the Central Division. A steam heating system will be installed in same. Total estimated cost, $325,000. Halstead & Sullivan, Palladio Bldg., Duluth, Archs. and Engrs.

Minn., St. Paul—The Hamline University, Snell and Capital Aves., is having plans prepared for the construction of a 3 story, 50 x 125 ft. dormitory. A steam heating system will be installed in same. Total estimated cost, $100,000. F. H. Ellerbe, Endicott Bldg., Arch.

Minn., Waseca—H. C. Gerlach, Arch., Mankato, will receive bids until Feb 17 for the construction of a 2 story, 60 x 100 ft. hospital building. A steam heating system will be installed in same. Total estimated cost, $80,000. Terry Schulte Eng. Co., Endicott Bldg., St. Paul, Engrs. Owner's name withheld.

Neb., Gering—Scotts Bluff Co. plans to construct a 3 story, 68 x 112 ft. court house. A vacuum steam heating system will be installed in same. Total estimated cost, $190,000. W. N. Bowman Co., Central Savings Bank Bldg., Denver, Colo., Archs.

Neb., Omaha—John Latenser & Sons, Archs., 632 Bee Bldg., will soon award the contract for the construction of a 3 story, 126 x 136 ft. warehouse on 8th and Jackson Sts., for Trimble Bros., 11th and Howard Sts. A steam heating system will be installed in same. Total estimated cost, $200,000.

Neb., Omaha—Clark G. Powell, 2051 Farnam St., is having plans prepared for the construction of a 4 story, 45 x 132 ft. auto supply building on 21st and Harney Sts. A steam heating system will be installed in same. Total estimated cost, $125,000 to $140,000. G. L. Fisher, 1604 City Natl. Bank Bldg., Arch.

Tex., Goose Creek—The Gulf Production Co. plans to construct 2 large electric power stations and transmission systems in the Goose Creek oil fields. Proposed plant will furnish electric power for operating the pumps of all of its wells in the field. Total estimated cost, $500,000.

Okla., Carnegie—The City Clerk will receive bids in two or three months for the construction of a 100,000 gal. surface reservoir in connection with the proposed development of a new water supply from deep wells. Air lift and motor-driven centrifugal high pressure service pumps will be installed in same. The city voted $24,000 bonds for the project. V. V. Long & Co., 1300 Colcord Bldg., Oklahoma City, Engrs.

Colo., Yuma—The city will soon award the contract furnishing one steel smokestack, one 250 hp. steam engine, one 125 hp. steam engine, two 200 hp. steam boilers, two alternators d.c. to above engines, switchboard and instruments, feedwater pumps, feedwater heater, etc. R. D. Salisbury, Denver, Engr.

Idaho, Harrison—The city plans an election to vote on $42,500 bonds for the construction of a waterworks system and installation of an electric lighting plant.

Wash., Vancouver—Rasmussen-Grace Co., Archs., Bates Bldg., Portland, Ore., will receive bids about Feb. 15 for the construction of a 2 story, 25 ton capacity, cold storage plant, here, for the Columbia Dairy Products Co. Estimated cost, $75,000.

Ore., Arlington—The city voted $15,000 bonds for the construction of a sewerage system and $5,000 for the extension of the electric light system.

Ore., Chemawa—The Supt. of Indian Schools, Box 156, received an appropriation of $15,000 from the U. S. Government, Wash., D. C., for the installation of a new heating plant at the Salem Indian Training School, here.

Ore., Fossil—The Butte Creek Land, Livestock & Lumber Co. plans to install a water system, including one 10 in. centrifugal pump, dam, etc. Total estimated cost, $5,000.

Cal., Davis—The City Trustees will soon receive bids for the construction of 2 drilled wells with 2 pumps of 500 gals. per min., operated by motors, 100,000 gal. tank on 100 ft. tower, auxiliary fire pressure pump of 1,000 gals. per min., 25 fire hydrants, and distributing system of 6 to 2 in. metal pipe. Total estimated cost, $75,000. Galloway & Markwart, First Natl. Bank Bldg., San Francisco, Engrs.

Cal., Salinas—Brown Bros. are having plans prepared for the construction of a theatre to be leased to Turner & Dahnken, San Francisco. A steam heating system will be installed in same. Total estimated cost, $125,000. A. W. Cornelius, Merchants Natl. Bank Bldg., San Francisco, Arch.

Cal., San Diego—The Bureau of Yards & Docks, Navy Dept., Wash., D. C., and Pub. Wks. Officers, 12th Naval Dist., Timken Bldg., San Diego, will receive bids for furnishing and installing complete power plant and electric generating equipment, including distribution systems for electricity and steam, at the Marine Corps Base, here.

Cal., Whittier—The city received bids for furnishing and erecting boilers and generator as follows: Terry & Carpenter, Sterling boilers, $18,070; Murray pump, $48,893; Kerr turbine and Allis-Chalmers generator, $16,911; switchboard, $2,400; total, $81,274. Thomas Haverty Co., 8th and Maple Sts., Los Angeles, Foster boilers, $15,250; Kerr turbine and Allis-Chalmers generator, $14,800; Ludlow engine, $57,900; switchboard, $2,321; total, $90,271. This work in connection with proposed waterworks system.

Cal., Whittier—The city received bids for furnishing and erecting pump and boiler from the Byron Jackson Iron Wks., 336 East 3rd St., Los Angeles, Thomas Haverty Co., 8th and Maple Ave., Los Angeles, $8,470; C. Terry, $8,588. This work in connection with the proposed waterworks system.

Ont., Galt—The city engaged H. H. Angus, Consult. Engr., 23 River St., Toronto, to prepare plans for the installation of a central heating plant to heat the 4 buildings at the hospital, here. A by-law appropriating $30,000 for this project has been passed.

Que., Montreal—The Congregation of St. Augustine plans to build a 50 x 100 ft. church in the north end of the city. Contract for heating will be sub-let. A sum of $200,000 has been raised for this project.

Que., Sherbrooke—The Howard Pulp & Paper Co., Montreal, has purchased 9½ acres on the bank of the San Francis River and is having plans prepared for the construction of a wood pulp mill on same. A 3,500 hp. power development will be added. Total estimated cost, $750,000.

CONTRACTS AWARDED

Me., Portland.—Hannaford Bros., Commercial St., have awarded the contract for the construction of a refrigerator plant to the Landers Eng. Co., 102 Exch. St. Estimated cost, $150,000.

Conn., West Haven—The West Haven School Bd. has awarded the contract for installing a steam heating system in the proposed 2 story, 45 x 137 ft. school on Noble St. to Charles W. Neumann, West Haven.

Mass., Boston—The New England Telephone & Telegraph Co. has awarded the contract for the construction of a 3 story, 80 x 100 ft. office building on Essex St., to I. F. Woodbury & Co., Summer St. A general heating system will be installed in same. Total estimated cost, $100,000.

Mass., Fall River—The Hotel Mohican, North Main St., will build a 4 story, 48 x 120 ft. addition. A steam heating system will be installed in same. Total estimated cost, $125,000. Work will be done by day labor.

N. Y., Brooklyn—The Congregation of Bikin Cholem, Fulton St. and Shepard Ave., will build a 2 story, 50 x 85 ft. synagogue on Arlington Ave. and Bradford St. A steam heating system will be installed in same. Total estimated cost, $100,000. Work will be done by day labor.

N. Y., Brooklyn—The Lincoln Hygeia Ice Co., 95 William St., New York City, will build a 2 story, 100 x 125 ft. ice plant addition on Pacific St. and Carlton Ave. Estimated cost, $100,000. Work will be done by day labor.

N. Y., New York—The New York Telephone Co., 15 Dey St., has awarded the contract for the construction of a 4 story, 100 x 145 ft. telephone exchange on Troutman Ave. (Bronx Boro.) to Gillies & Campbell, 103 Park Ave. A steam heating system will be installed in same. Total estimated cost, $200,000. McKenzie, Voorhies & Gmelin, 1123 B'way, Archts. and Engrs.

N. Y., New York—E. J. Nathan, Jr., 128 B'way, has awarded the contract for the construction of a 2 story, 77 x 106 ft. factory and store building on 7th Ave. and 27th St., to G. Richard Davis & Co., 30 East 42nd St. A steam heating system will be installed in same. Total estimated cost, $850,000.

N. Y., New York—The Round Hill Mercantile Co., 98 Bleeker St., has awarded the contract for the construction of a 10 story office building on 19th St. and 4th Ave., to W. Crawford, 7 East 42nd St. A steam heating system will be installed in same. Total estimated cost, $500,000.

N. Y., New York—L. K. Schwartz, 110 West 40th St., has awarded the contract for the construction of a 10 story, 41 x 98 ft. loft building at 41-43 West 28th St., to Barney-Ahlers Constr. Co., 110 West 40th St. A steam heating system will be installed in same. Total estimated cost, $250,000.

N. Y., New York—The Textile Bldg., Inc., 15 West 55th St., has awarded the contract for the construction of a 16 story, 164 x 198 ft. office building at 285-293 5th Ave., to the George Backer Constr. Co., 33 East 33rd St. A steam heating system will be installed in same. Total estimated cost, $3,000,000.

N. J., Passaic—The Passaic Cotton Mills, Brighton Ave., have awarded the contract for the construction of a 4 story, 100 x 700 ft. factory, to Fred T. Ley, 19 West 44th St., New York City. A steam heating system will be installed in same. Total estimated cost, $750,000.

N. J., Woodbridge—The Reliable Chemical Co., c/o Westinghouse, Church, Kerr & Co., Inc., Engrs., 37 Wall St., New York City, has awarded the contract for the construction of a ftcory and storage building between here and Rahway to Westinghouse, Church, Kerr & Co, Inc., 37 Wall St., New York City. A steam heating system will be installed in same. Total estimated cost, $150,000.

Penn., Chester—The Cochrane Estate has awarded the contract for the construction of an 18 story office building to Ward Bros. Constr. Co. A steam heating system will be installed in same. Total estimated cost, $450,000.

Penn., Lock Haven—The Lock Haven Trust & Safe Deposit Co., East Main St., has awarded the contract for the construction of a 1 story bank to Hoggson Bros., 485 5th Ave., New York City. A steam heating system will be installed in same. Total estimated cost, $300,000.

Pa., Philadelphia—W. S. Barnes, 20th and Erie Sts., will build a 4 story, 100 x 300 ft. assembly building. A steam heating system will be installed in same. Total estimated cost, $200,000. Work will be done by day labor.

Penn., Philadelphia—The Crescent Ice & Coal Co., 52nd St. and Woodland Ave., has awarded the contract for the construction of a 1 story, 100 x 145 ft. ice plant on Powers Lane and Island Rd. to W. F. Koelle & Co., 26th and Oxford Sts. A steam heating system will be installed in same. Total estimated cost, $35,000.

Penn., Philadelphia—The West Philadelphia Catholic High School, 49th and Chestnut Sts., has awarded the contract for the construction of a 3 story, 49 x 125 ft. school to Doyle & Co., 1519 Sansom St. A steam heating system will be installed in same. Total estimated cost, $155,000.

Penn., Pittsburgh—Alleghany Co. has awarded the contract for installing heating, ventilating and refrigerating systems in the court house, jail and morgue to L. G. Emery Co., New Castle, at $123,329.

Va., Danville—The Riverside & Dan River Cotton Mills Co. has awarded the contract for the construction of a 4 story, 133 x 800 ft. spinning mill, to the Aberthaw Constr. Co., 27 School St., Boston, Mass. A steam heating system will be installed in same. Total estimated cost, $2,000,000.

Va., Richmond—The Rennie Dairy Co., 7th and Leigh Sts., has awarded the contract for the construction of a 2 story, 45 x 70 ft. dairy, including cold storage rooms, on North 7th St., to J. C. Beasley, North 9th St.

Ala., Florence—The U. S. Engr.'s Office, War Dept., Wash., D. C., has awarded the contract for the installation of 2 air compressors to the Worthington Pump and Machinery Corp., 115 B'way, New York City, at $29,450.

O., Canton—The city has awarded the contract for the construction of a 1 story, 47 x 96 ft. pumping station in the North End, to Keller-Huff Co. Estimated cost, $24,900.

O., Cleveland—The Masonic Temple Bldg. Co., Rockefeller Bldg., has awarded the contract for the construction of a 6 story, 130 x 300 ft. masonic temple at 3515 Euclid Ave., to Masters & Mullen Constr. Co., Electric Bldg. Motors will be installed in same. Total estimated cost, $750,000.

O., Cleveland—The Eaton Axle Co. has awarded the contract for the construction of a 1 story, 300 x 450 ft. factory and a 40 x 80 ft. power house and heat treating building on East 140th St. and the New York Central R. R. to the Crowell-Lundoff-Little-Bicknell Co., 5716 Euclid Ave. A steam heating system, including H.R.T. boilers, will be installed in same. Total estimated cost, $600,000.

O., Cleveland—J. S. Kohn, Engineers' Bldg., has awarded the contract for the construction of a 3 story, 54 x 112 ft. show room and warehouse on East 14th St. to the Bolton Pratt Co., Columbia Bldg. A steam heating system will be installed in same. Total estimated cost, $100,000.

O., Parma—The Bd. Educ. has awarded the contract for the installation of a hot air heating system in the proposed three 1 story, 60 x 118 ft. school buildings to the Amer. Warming & Ventilating Co., Rose Bldg., Cleveland.

O., Salem—The Mullens Body Co. has awarded the contract for the construction of a factory to Stone & Webster, 120 B'way, New York City. A steam heating system will be installed in same. Total estimated cost, $500,000.

Mich., Detroit—The Stroh Products Co., 253 Elizabeth St., E., has awarded the contract for the construction of a 3 story, 22 x 85 ft. ice cream plant on Hastings and East Elizabeth Sts. to Pine & Munnicke, Marquette Bldg. Circulation pump, tanks, etc., will be installed in same.

Mich., Detroit—The Studebaker Corp., Brush and Piquette Sts., has awarded the contract for the construction of a 4 story, 200 x 250 ft. automobile factory addition to the A. J. Smith Const. Co., Campau Bldg. A steam heating system and electric motors for power will be installed in same. Total estimated cost, $300,000.

Ill., Chicago—The Boyle Ice Co., 136 West Lake St., has awarded the contract for the construction of a 1 story, 125 x 175 ft. ice plant at 855 Larrabee St. to the Menke-Thielberg Co., 139 North Clark St. A steam heating system will be installed in same. Total estimated cost, $65,000.

Ill., Chicago—The Exel Motor Truck Co., c/o Henry Raeder, Archt., 20 West Jackson Blvd., has awarded the contract for the construction of a 1 story, 350 x 500 ft. factory on Menard and Dickens Ave. to H. A. Peters Co., 19 South La Salle St. A steam heating system will be installed in same. Total estimated cost, $300,000.

Ill., Chicago—The Bnai Abraham Zion Congregation, c/o A. L. Levy, 11 West Washington St., has awarded the contract for the construction of a 1 story, 85 x 110 ft. synagogue and a 2 story, 25 x 125 ft. community house on Washington Blvd. and Karlov Ave., to Kadeshewitz & Buchel, 1342 Independence Pl. A steam heating system will be installed in same. Total estimated cost, $225,000.

Ill., Chicago—The Shore View Hotel Co., c/o Percy Johnston, Arch., 1254 Pratt Blvd., has awarded the contract for the construction of an 8 story, 60 x 145 ft. hotel at 4910 Sheridan Rd., to Olson-Carson Co., 7417 Rhodes Ave. A steam heating system will be installed in same. Total estimated cost, $800,000.

Ill., Springfield—The Springfield Marine Bank, 114 South 6th St., has awarded the contract for the construction of a bank, to Hoggson Bros., 485 5th Ave., New York City. A steam heating system will be installed in same. Total estimated cost, $500,000.

Minn., Hibbing—School Dist. 18 has awarded the contract for installing heating and plumbing systems in the proposed 3 story, 74 x 150 ft. addition to Ainsley School, to the Schirmer Plumbing Co., Hibbing, at $47,719.

Wis., Hartford—The city has awarded the contract for furnishing and installing stokers and soot blowers to the Combustion Engineering Co., 38 South Dearborn St., Chicago. Estimated cost, $11,725.

Wis., Rio—The city has awarded the contract for waterworks improvements, including the construction of a tank and tower, to the Wausaw Iron Works, Wausaw; well, to W. L. Thorne, Platteville; pump, to the Wisconsin Foundry Co., Madison. Total estimated cost, $30,000.

Kan., Haven—The Bd. Educ. has awarded the contract for installing heating and plumbing systems in the proposed 2 story high school to Stevens, Gill & Co., Hutchinson, at $16,375.

Kan., Galena—The city has awarded the general contract for the construction of a waterworks extension, to Ray & Son, $20,940; for installing pump equipment, to the Merkle Machinery Co., 508 Interstate Bldg., $5,958.

N. D., Milton—The city has awarded the contract for the construction of an electric light plant and equipment, to H. T. Lewis, Milton, at $6,399.

Mo., St. Louis—The Century Electric Co., 19th and Pine Sts., has awarded the contract for the construction of a factory, to Stone & Webster, 120 B'way, New York City. A steam heating system will be installed in same. Total estimated cost, $500,000.

Mo., St. Louis—The J. H. Forbes Tea & Coffee Co., 9th and Clark Sts., has awarded the contract for the construction of a 5 story, 90 x 120 ft. mercantile building at 922 Clark and 306 South 10th Sts., to S. S. Callin, Clayton. A steam heating system will be installed in same. Total estimated cost, $200,000.

Neb., Omaha—The Burgess-Nash Co., 16th and Harney Sts., has awarded the contract for the construction of an 8 story, 88 x 152 ft. department store building on 17th and Harney Sts. to James Stewart &

Co., 30 Church St., New York City. A steam heating system will be installed in same. Total estimated cost, $900,000.

Mo., St. Louis—The Halwe Investment Co., 3128 Locust St., has awarded the contract for the construction of a 4 story, 90 x 120 ft. mercantile building at 2805 Locust St. to A. D. Gates Constr. Co., Chemical Bldg. Two elevators and a steam heating system will be installed in same. Total estimated cost, $250,000.

Tex., San Antonio—The San Antonio Diocese has awarded the contract for installing a heating system in the proposed 3 story, 100 x 155 ft. theological seminary near Mission Conception to Chalkley Bros., 312 Main Ave., at $12,000.

Okla., Tahlequah—The city has awarded the contract for the construction of a lighting plant to A. M. Byrnes, Fayetteville, Ark., $13,149; for the installation of steam heating system and furnishing of electric equipment to Smith & Whitney, Southwestern Life Bldg., Dallas, Tex., $59,477; distribution system to the Empire Electric Co., 419 West Okmulgee St., Muskogee, $39,200.

Idaho, Plummer—The city has awarded the contract for the construction of a 14 mi. electric transmission line from here to a connection with the line of the Washington Water Power Co. at Teko, Wash., to R. H. Mercer, Plummer, at $11,975.

NEW INSTALLATIONS IN POWER

Conn., Hartford—Hotel Garde, Asylum and High Sts., is in the market for equipment for new kitchen apparatus and refrigerating plant. F. C. Walz, 348 Trumbull St., Engr.

N. Y., Long Island City—H. R. Mallison & Co., Woolsey Ave., is in the market for power house equipment, elevator, etc.

N. Y., New York—The Amer. Gas & Electric Co., 30 Church St., is in the market for second-hand 500, 600 and 750 hp., 250 lb. B. & W. boilers, also some 750 hp. 250 lb. Sterling boilers, preferably with superheat. Address N. M. Argabute.

Pa., St. Marys—The St. Marys Electric Light Co. plans to construct about 10 mi. of 22,000 volt transmission line. W. H. Brown, Genl. Supt.

S. C., Walterboro—The Walterboro Water & Light Plant plans to install one 75 kva. direct connected engine and generator Cozart Binns, Supt.

Ga., Albany—The Georgia-Alabama Power Co. plans to construct a 10,000 hp. plant along the Flint River and 60 mi. of 45,000 volt transmission line. L. H. Hardin, Chf. Engr.

Ga., Tallapoosa—The Municipal Electric Light & Water Plant plans to install one 150 kva., and one 100 kva. generators. W. A. Melvin, Supt.

Ga., Waynesboro—The city plans to install a 250 kva., direct conected generator and construct 3 mi. of transmission line. L. J. Porter, Chf. Engr. and Supt.

Tenn., Memphis—J. J. Ross, Consult. Engr. and Contr., 380 Randolph Bldg., is in the market for one small hoisting engine either 8 x 10 or 10 x 12 ft.

O., Greenville—The Franklin Tractor Co. is in the market for seven hundred and fifty 30 hp. 4 cylinder tractor motors, 750 magnetos, 750 complete sets transmission gears, 750 pressed, split steel pulleys, or power take-off systems complete, 750 motor fans and belts, 4,000 spark plugs, 750 carburetors and 750 governors. G. Nills, Pres. and Genl. Mgr.

Ill., Biggsville—Henderson Co. Pub. Serv. Co. plans to construct a 15 mi., 33,000 volt transmission line. D. W. Lee, Mgr.

Ill., Casey—The Municipal Electric Light Plant plans to construct 5 mi. of transmission line. J. C. Pfister, Supt.

Ill., Chicago—The Lincoln Park Comrs., Clark and Center Sts., plan to build a substation. Work involves a 12,000 volt, 3 ph., 60 cycle transmission line; two 3,000 gal. 64 lb. centrifugal pumps driven by 1,800 r.p.m., 4,000 volt, 3 ph. induction motors; two 300 kva., 12,000-4,000-2,300 volt, 3 ph. transformers; one 3 ph., 4 amp., 5,000 volt series ckts; storage battery; motor generator unit; local power and lighting panels and transformers, probably large floodlighting installations and concrete post drive lighting. C. H. Shepherd, Elec. Engr.

Ill., Divernon—The Municipal Light Plant plans to install a 50 kva. direct connection generator set. J. E. Rigg, Supt.

Ill., Farmer City—The Municipal Electric Light & Power Plant plans to install one 150 hp. boiler. O. B. Osborn, Supt.

Ill., Marengo—The Marengo Pub. Serv. Co. plans to construct several miles of transmission lines. R. T. Fry, Pres.

Ill., Marshall—The Marshall Ice & Power Co. plans to construct a transmission line from here to West Union. G. C. Hallaner, Mgr.

Ill., Noble—The Noble Electric Light & Power Co. plans to install a complete 25 hp. unit and 2,300 volt transformers. H. T. Flanders, Pres. and Mgr.

Ill., Peoria—The Central Illinois Light Co., 316-20 South Jefferson Ave., plans to install a one 12,500 kva. turbo-generator. L. Owen, Supt.

Ill., Princeton—The Municipal Water & Light Dept. plans to install a 300-400 kva., 2,300 volt, 3 ph., 60 cycle, 3 wire generator with direct connection to simple engine. W. N. Remsbury, Supt.

Ill., Rankin—The Rankin Electric Light Co. plans to construct about an 11 mi. 6,600 volt transmission line from here to Bocton. Charles J. Crump, Chf. Engr.

Ill., Savanna—The Peoples Gas & Electric Co., 214 Main St., plans to install boiler and generator and sub-station equipment, also construct a 10 mi. transmission line. F. P. Bowen, Engr.

Ill., Vermont—The Vermont Municipal Electric Light Plant plans to install an oil engine and generator. P. J. Tingley, Supt.

Ill., Xenia—The Village plans to install a 25 hp. oil engine and a 20 kva. generator. N. T. Pierce, Mgr.

Kan., Long Island—The Phillips Co. Light & Power Co. plans to construct a ½ mi., 2,300 volt transmission line into the farm service. Ira C. Young, Chf. Engr.

Neb., Lindsay—The Lindsay Electric Light & Power Co. plans to complete the installation of 50 hp. units and remodelling of present building. F. W. Edwards, Mgr.

Mont., Terry—The Terry Light & Power Co. plans to install a new generating unit. A. J. Enteness, Pres. and Mgr.

Okla., Kingfisher—The Kingfisher Electric & Water Plant, City Bldg., contemplates the installation of one 500 kw. turbine generator unit. E. Fisher, Supt.

Colo., Estes Park—The Stanley Power Dept. plans to install a new water wheel and enlarge generator to 450 kva., also change the 5 mi. transmission line extending from power house to village from single ph. to a 3 ph. W. C. Humphreys, Genl. Supt.

Colo., Golden—The Jefferson Co. Power & Light Co. plans to construct 8 mi., 6,600 volt transmission line to Montdale and Evergreen. E. A. Phinney, Pres. and Mgr.

Colo., Grand Junction—The Grand Junction Electric, Gas & Mfg. Co. plans to remodel its boiler house. A. E. Anderson, Genl. Supt.

Colo., Gyssum—The Casper Schumm Electric Light & Power Co. plans to construct transmission line extensions. W. C. Schumm, Pres. and Mgr.

Colo., Peetz—The Municipal Lighting Plant plans to install a 90 hp. engine and a 60 kv. generator. E. G. Ribble, Chf. Engr.

Idaho, Weiser—The Municipal Water & Light Dept. plans to rebuild approximately 2 mi. of 4,000 volt, 3 ph. transmission line. Lyle Wood, Supt.

Ore., Portland—The Dept. of Pub. Wks. plans to install one 75 hp. electric motor at the bunkers. P. McIntosh, Purch. Agt.

Ore., Shady (Roseburg P. O.)—The Perkins Sand & Gravel Co. plans to replace electric equipment of rock crusher plant with machinery adapted to burning oil. Estimated cost, $10,000.

Ore., Tillamook—The Coast Power Co. plans to install a 1,000 kw. turbine generator. C. J. Edwards, Pres. and Mgr.

Cal., Avalon—The city plans to install one 200 kw. generator. A. B. Waddingham, City Mgr.

Cal., Tehachapi—The Municipal Electric Light Plant plans to construct a 2 mi., 11,000 volt, 3 ph. transmission line for agricultural purposes. M. P. Evans, Genl. Supt.

Ont., Guelph—The Northern Rubber Co. is in the market for a new or used motor generator, 4 kw., d.c. voltage 110, a.c. voltage 550, 3 ph. or 110 single ph., 25 cycle. A. Kennedy, Engr.

The SEARCHLIGHT SECTION of this issue is on pages 60A to 78

POWER

Volume 51 New York, February 17, 1920 Number 7

Over the Top with the Boiler Room Crew

By Frankland Cheney

Jim Wright, the chief of the Porpoise Point Power Co., tamped the tobacco in his old corn cob, put his feet on the desk and tilted back his chair with a contented sigh. His visions of earthly joy, however, were soon interrupted, as the door slid open and in walked the boiler-room crew. A husky son of Erin seemed to be the acknowledged spokesman, and he was not long in making their mission known.

"Chief," said he, "we was thinking we would like to have a raise."

Jim looked them over for a second or two and decided the chances for a jolly were rather slim.

"Well, boys," he said, "I'll see what I can do and will let you know in a day or two."

The men thanked him and filed out of the engine room. Jim sat still for some minutes, wondering just what he was going to do. The Porpoise Point Power Co. was barely paying expenses and to increase the wages of the six men in the boiler room would bring a roar from the management that would put the turbines out of balance. But something had to be done.

As he straightened up in his chair, his eyes rested on the CO_2 machine in the corner of the engine room. This machine had been installed for more than a year, but somehow it had failed to come up to expectations. A bonus had been offered to the watch delivering the highest CO_2, but it had only created jealousy. No. 2 watch with a constant load had invariably won the money, and after the third week the others lost interest and refused to make an effort. Yet the chief realized that a raise would have to come through the medium of this machine.

The day was divided into three eight-hour shifts. No. 1 watch, from 12 to 8 p. m., had a light load. The furnaces were all thoroughly cleaned and clinkered, and it would be impossible to maintain any sort of CO_2 during this time. Watch No. 3 were on duty from 4 p. m. to midnight. The load on this run began to fall away around 8:30 and continued to drop until midnight, when it reached its minimum. This being the case, it was difficult to maintain a satisfactory CO_2 during the latter part of this run. The men of these two watches were efficient and capable and as worthy of consideration as those of watch No. 2, where a constant load made possible a very high CO_2 during the entire watch. A fireman and helper were on the boiler-room floor during each watch, and after considerable thought the chief tacked up the following notice:

Beginning the first of the week two bonuses will be regularly offered the boiler-room help. The first, a sum of nine dollars, to be equally divided among the firemen. The second, a sum of six dollars, to be equally divided among the helpers. The above amounts will be paid each week providing that the CO_2 for the entire week will average 10 per cent. The daily CO_2 charts will be measured by a planimeter for that purpose.

The boiler-room help read the notices, but did not appear to grow very enthusiastic. The chief waited for the idea to sink in and then summoned them. In a straight-from-the-shoulder talk he told them how the increasing cost of everything had affected the company; how the stockholders had received nothing on their investment for a year; how the light commissioners had refused to allow them to raise the rates. He also told them that he realized they were entitled to an increase in salary and if they were willing to do teamwork it could be squeezed out of the coal pile. He urged them to give his plan a trial, telling them if they would make an honest effort for two months and were still dissatisfied, he would make a fight for a straight raise all around.

The men were at first skeptical, but the earnestness of the chief's manner won them over and they finally agreed to give it a trial. The first two days the planimeter was put on the charts, it showed an average of less than 9; the third day it jumped to 10½; at the end of the week the average for the seven days was 10¾.

In going over his back records the chief found the pounds of coal per kilowatt had been from 3.2 to 3.5 and the coal consumed each day had ranged from 59,000 to 64,000 lb. For the week just ended the highest rate had been 3.3, which was at least encouraging; but the attitude of the men themselves gave the most satisfaction. At the end of the first month the charts were averaging better than 12 per cent. The average cost per kilowatt at the board for the preceding month had been 0.017; for the month just ended it was 0.015. Here was something real, and the chief determined that they should not lose interest in their work. As the second month was drawing to a close, the old notices were replaced by new ones which read:

The weekly bonus to be equally divided among the firemen will be fifteen dollars. The helpers' bonus will be ten dollars. To earn the above reward the management expects a weekly average of 12 per cent CO_2.

And the end is not yet.

Model Small Ice Plant in Cuba

By J. KILIAN

ALTHOUGH the climate of Cuba is tropical, ice-making and refrigerating plants have not been introduced in the smaller communities of the Island as extensively as might be supposed. It is true that the capital city, Havana, has a number of large ice plants, but in the interior towns they are few and far between. However, where plants have been installed, invariably they have proved successful.

During the war the Vuelta Abajo district enjoyed a period of great prosperity due to the increased demand for tobacco and the high price obtained, and with improved conditions there arose a demand for ice in the City of Pinar del Rio, which is the capital of the province of the same name and the trade center of the tobacco-growing district. The matter was taken up by the Hidro Electrica Pinarena, and after investigation a raw-water plant of 12 tons daily capacity was decided upon as being of sufficient size for the city, which has a population of about 12,000.

In choosing a site, it was desired to locate the plant as near a source of water supply as possible, so the building was placed about 600 ft. from the river, which was as close as local conditions permitted. The building is of tile construction, one story high, faced with a smooth cement finish. It is designed along simple lines and presents a rather pleasing appearance. The hip is supported by wooden trusses resting on pilasters in the walls, with a row of columns through the center of the building for additional support. The floors are concrete throughout. There are no interior partitions in the manufacturing room and large windows provide ample light throughout. The manufacturing room is 76 ft. 9 in. long and 47 ft. 7 in. wide, inside measurements, by 16 ft. clear height under the trusses. The ice-storage room is 34 ft. 7 in. and 18 ft. 6 in. by 10 ft. high, and the shipping room is 18 ft. 6 in. and 11 ft. 6 in. by 10 ft. high. The building is laid out with a view to doubling the capacity of the plant in the future, space being provided for duplication of all equipment.

The ice tank is insulated with pure sheet cork for the bottom, the sides and ends being insulated with granulated cork. The ice-storage room is lined with pure sheet cork plastered with portland cement. This room is maintained at a temperature of about 28 deg. F. by means of 500 ft. of 2-in. ammonia evaporating piping. This piping is connected to the ice-making compressor and also to a small vertical compressor provided to maintain the temperature when the ice plant is not in operation.

In the ice tank are 180 standard 300-lb. cans in which a remarkably fine quality of clear ice is obtained from the river water by means of a low-pressure air agitating system and a water softener. The latter is necessary to eliminate the objectionable mineral salts present in the river water used. The ice tank is 35 ft. long by 16 ft. 3 in. wide. A temperature of 16 deg. F. is maintained by ammonia evaporating piping immersed in brine. Brine circulation is effected by a 20-in. horizontal propeller and a suitable tank partition.

From the river, 600 ft. distant, the water is pumped into a 15,000-gal. storage tank by a 70-g.p.m. electrically operated pump, and from the tank it is distributed to the ammonia condenser, water softener, etc., as required. The tank is always kept filled, providing a four-hour reserve supply in case of stoppage of the pump.

All the machinery, with the exception of the small compressor, is motor-driven, polyphase low-tension current being used. The motors are of Swedish manufacture and were secured out of dealers' stocks in Havana, it being next to impossible to obtain shipment of American motors. The main motor driving the compressor is of the slip-ring induction type, 580 r.p.m., and drives the compressor at 100 r.p.m. through a double leather belt. The smaller motors driving the auxiliaries are of the squirrel-cage type. The small compressor is driven by a 5-hp. oil engine, an independent source of power for this service being considered advisable.

The water softener is of the intermittent type, the continuous type being impractical in this case, owing to the water supply varying in quality. This apparatus treats a definite quantity of water with a definite charge of chemicals. It consists of two treating tanks and a chemical tank with necessary connections. While one tank is treating the water to be softened, water which has already been treated is used from the other for ice-making purposes. Each tank is 7 ft. diameter by 6 ft. high, the whole equipment being mounted on a platform in the machine room. Hydrated lime and soda ash are the chemicals used. The lime precipitates the calcium and magnesium bicarbonates in the water by chemically combining with the free and half-bound carbon dioxide gas, and converts them to carbonates. The soda ash reacts with the magnesium sulphates and forms magnesium carbonate and sodium sulphate, the whole being precipitated as insoluble sludge to the bottom of the tank. The treated water is removed from the tanks by means of a floating pipe which receives the water from near the surface.

No expense was spared to make the plant as nearly perfect and permanent as possible, and the ice produced is the finest to be seen anywhere in Cuba. The plans were prepared for the Hidro Electrica Pinarena by Lombard & Co., Havana, and the entire installation was made under their supervision. While no speed records were made in the building and equipment of the plant, its complete erection was accomplished in a reasonable length of time, and in the face of conditions, due to the world war, which seemed almost insurmountable.

PRINCIPAL EQUIPMENT OF HYDRO-ELECTRIC PINARENA ICE PLANT

Equipment	Kind	Size	Use	Operating Conditions	Maker
1 Compressor	Simple, double-acting	8½ x 16 in.	Ice-making and cold storage	Motor-driven, 100 r.p.m.	Vilter Manufacturing Co.
1 Compressor	Vertical, single-acting	4 x 6 in.	Cold storage	Oil-engine driven, 200 r.p.m.	Vilter Manufacturing Co.
1 Condenser	Double-pipe	3 stands, 12 p.h.	Condense ammonia vapor		Vilter Manufacturing Co.
1 Ice tank	Steel	35 x 16.25 x 4 ft.	Ice making	Ammonia evaporating pipe in brine	Vilter Manufacturing Co.
1 Air-agitating System	Low-pressure	100 cu. ft. free air per min. at 3 lb.	Agitating water in ice cans	Blower motor-driven	Vilter Manufacturing Co.
1 Water softener	Intermittent	210 gal. per hr.	Water purification	Precipitation by lime and soda ash	Wm. Graver Tank Works
1 Motor	Slip-ring	50 hp.	Drive large compressor	Polyphase, low-tension, 580 r.p.m.	Allmänna Svenska Electriska Aktiebolaget Västeras, Sweden
1 Motor	Squirrel-cage	5 hp.	Drive propeller and core pump	Polyphase, low-tension, 940 r.p.m.	Allmänna Svenska Electriska Aktiebolaget Västeras, Sweden
1 Motor	Squirrel-cage	5 hp.	Drive blower	Polyphase, low-tension, 1420 r.p.m.	Allmänna Svenska Electriska Aktiebolaget Västeras, Sweden
1 Oil engine	Vertical	5 hp.	Drive small compressor	Intermittent, 300 r.p.m.	Bolinder, Sweden

FIG. 1 - SHOWING PLEASING ARCHITECTURE OF THE PLANT

FIG. 2 - THE EQUIPMENT IS ELECTRICALLY DRIVEN

FIG. 3 - VIEW OF THE FREEZING ROOM

Drying Out a Generator, Switchboard and Motors After a Flood

By KENNETH A. REED

DURING the flood through the Ohio and Indiana district, some years ago, a large number of power plants were flooded, and in many cases generators, motors and switchboards were under water for a week or longer. Before the water had subsided, practically all of the large manufacturers had gotten every available engineer lined up awaiting the call from their customers for assistance in getting these plants back into commission. Such assignments were not enviable ones by any means. Many animals—such as cats, dogs, rats, etc.—had been drowned in the buildings, and the sewers had filled the basements with refuse, so that the air was very foul and injurious.

The work to which I was assigned consisted of drying out a small direct-current engine-type generator, several motors, and the switchboard and cables of a plant. Upon looking the job over, I found that the apparatus, which was in a basement, had been entirely under water for several days, and the basement floor was still covered with about a foot of water and six inches of mud and slime. There was no outside source of power for drying out the generator electrically, owing to the fact that the city light plant had also been flooded. The customer's boiler room was on the same level as the power plant, therefore it was impossible to raise steam to run the engine and dry out the generator by the short-circuit method. It then became necessary to devise original means whereby the necessary heat could be produced and properly applied. After discarding several schemes, it was decided to build a corrugated sheet-iron house around the entire engine and generator, and place two wood-burning stoves inside, with openings made in the walls of the house for supplying fuel to the stoves. This was an entirely feasible method, inasmuch as the floor of the engine room was concrete; and it had the additional advantage of providing room for drying one motor, thereby greatly facilitating the total time required to dry out all the apparatus. Fig. 1 shows the general plan of the layout. The openings formed by the corrugations were sufficient to allow the moisture to escape from the box.

After completing the house and drilling holes in the top for the insertion of thermometers, fires were built and the temperature inside raised to about 90 deg. C. Arrangements were made for a night fireman, who was required to record the temperature readings every half hour, to avoid overheating and damage to the apparatus. Practically constant temperature was maintained in this housing for about five days and nights.

FIG. 1. IRON BOX AND ARRANGEMENT OF EQUIPMENT INSIDE FOR DRYING

At the end of this time the generator was fairly dry excepting for some excessive dampness inside the commutator. During this time, however, all the motors in the plant had been dried out by placing them one at a time in the oven with the generator, the drying of the motors of course proceeding much faster than that of the generator, because of the great difference in size.

Since the commutator moisture was inside the V-rings, external heat did not seem to have much effect on eliminating it, and heating the commutator itself was decided upon. This was done by running the generator with the shunt field disconnected and the armature short-circuited through the series winding, causing a heavy current to flow and generating a large amount of heat at the commutator. The voltage under these conditions was very low, consequently there was no danger of the machine breaking down to ground. However, in order to avoid generating sufficient current to damage the armature coils, a portable ammeter was placed in the circuit, as in Fig. 2, and the current kept at a safe value by throttling the engine. Blowtorches were applied to the commutator, and after short-circuit operation for about a day and a half, the machine was considered to be in serviceable condition.

While the generator and motors were being dried out, the switchboard was stripped of all equipment, including the switches, instruments, circuit-breakers, etc., after which the panels were washed and the iron framework given a coat of paint. The field rheostat was

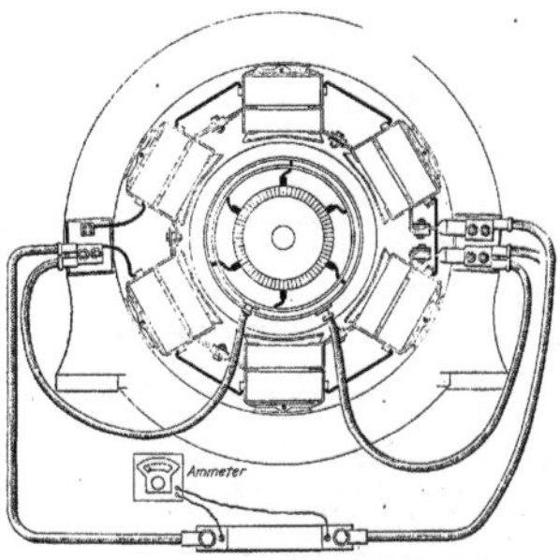

FIG. 2. ARMATURE SHORT-CIRCUITED THROUGH SERIES-FIELD WINDING

cleaned of mud and slime and placed on top of the oven over the generator to dry. Upon taking the covers off the meters, it was found that they were damaged beyond repair in the plant, and it was therefore necessary to return them to the manufacturer. The circuit-breakers were also badly damaged, but we were able to put them in shape for operation, and after cleaning and polishing the switches and busbar copper, the board was reassembled in practically new condition.

The cable ducts were, of course, full of water, and when the cable was pulled out the insulation was found to have deteriorated to such an extent that it would not stand being pulled back into the ducts after drying. New cable could not be secured locally, and in order to avoid delay in starting up the plant, the old cable was run across the floor on boards. The customer was advised to purchase lead-covered cable for the new leads, for no matter how much we might swab and dry out the duct lines, there would remain under the floor for a very long time a considerable amount of dampness, which would have proved very detrimental to braid-covered cable.

All the details having been completed, the machine was brought up to speed and the voltage gradually raised. Before reaching normal voltage, however, severe sparking developed in the form of glowing fire around the entire commutator. This, obviously, was due to short-circuits between bars, and when the machine was shut down such was found to be the case. About twenty places were discovered where the mica had been oil-soaked, but not bad enough to break down until after it had been under water. These places were carefully cleaned out until good clear mica was reached and then filled with dentists' cement, which sets very quickly. No further trouble was experienced in securing normal voltage, which was ascertained by means of a portable voltmeter borrowed from the city light plant. No-load and full-load voltage points were marked on the field rheostat, as there was no other means of determining the voltage until the instruments were received from the manufacturer, except by using judgment as to the brilliancy of the lights. Full load was put on the machine the following day, and all the apparatus operated satisfactorily. Nothing remained to be done except to mount the meters and install the cable when received.

Losses in Belt Transmission

By Charles H. Herter

Regarding the loss of power occasioned by having driving belts either slipping or tightened too much, tests enable the writer to qualify statements that the loss through slip, under ordinary conditions, ranges from a minimum of 5 to more than 15 per cent. If 5 per cent. and more was the loss suffered normally in a belt drive, leather belting would not be tolerated in an efficient power plant. Many examples can be cited where the loss in speed change from one shaft to the other does not exceed 1.1 per cent., creep and slip included. Whenever the combined loss exceeds 2 per cent., there is something wrong with the drive. So many misleading horsepower formulas exist for belt transmission that in many cases insufficient width of belting is provided, compelling the user to resort to excessive tightening to get the work done. To correct this evil wide-awake leather belt manufacturers now furnish expert advice gratis to intending purchasers.

Occasionally, one is called in to find a remedy where the speed loss is great enough to be disclosed by a corresponding loss in production. On a wire-cloth loom the slippage was found to be 4 per cent., causing the mesh to be irregular, the tension of the stiff canvas belt fluctuating with the hygrometric condition of the atmosphere. A good waterproof leather belt solved the problem. In a chocolate factory with an old three-ply leather belt 23 in. wide, the slippage reached 12 per cent. This was remedied by cleaning it and running a 16-in. heavy double belt independently on top of it. In an ice factory a 125-ton ammonia compressor with an old 30-in. double leather belt, saturated with oil, ran at 46 r.p.m. when it should have made 56 r.p.m. The loss was nearly 18 per cent., equivalent to $70 worth of ice per day. This was remedied by putting on a new belt. The belt efficiency of an identical unit alongside was found to be 98.8 per cent.

A simple way for checking the performance of a drive is to ascertain the loss in speed from the driving shaft to the driven shaft. Suppose we have an air compressor driven by belt from a constant-speed induction motor, diameter of compressor belt wheel, 50 in.; of motor pulley, 10 in.; speed of motor, according to name plate, 1,170 r.p.m., but with an accurate tachometer or speed counter we find it to be 1,230 r.p.m. at the time of test. The belt used is ¼ in. thick. Under favorable conditions and with sufficient width of leather belt, the speed of air compressor should be equal to $1230 \times \frac{10.25}{50.25} = 251$ r.p.m., which multiplied by 0.988 to allow for creep and slip equals 248 r.p.m. net. If, by actual count, the speed of the compressor is found to be only 241 r.p.m., the speed loss is $\frac{10}{251}$, or 3.98 per cent., showing that the belt is in poor condition, or probably not large enough for the power required. A flexible idler properly placed will rectify the trouble and avoid the need of excessive tension. A thicker belt is not advisable because it would not wrap well around the small pulley, the outer ply would be overstretched, its circumference being appreciably greater than that of the pulley. It would be hard on the cement, which is expected to hold the two plies together. For this reason it is good practice to have a belt no thicker than 2 or, at most, 3 per cent. of the diameter of the smallest pulley used with it.

In calculating the speed change from one shaft to another it is necessary to figure always from the driver to the driven, and to add one belt thickness to each diameter.

For example, a 4-in. motor pulley at 1000 r.p.m. will drive a 40-in. pulley at what speed? 100 r.p.m.? No. If the belt is 0.25 in. thick, we have

$$\frac{4.25 \text{ in.} \times 1000 \text{ r.p.m.}}{40.25 \text{ in.}} = 105.6 \text{ r.p.m.}$$

If the 40-in. diameter pulley is driving at 100 r.p.m., the speed of the 4-in. driven pulley be 1000 r.p.m.? No. It will be

$$\frac{40.25 \text{ in.} \times 100 \text{ r.p.m.}}{4.25 \text{ in.}} = 947.05 \text{ r.p.m.}$$

The pulleys are of the same size in both cases. In the first case, where the speed is being reduced, the slip may be 2 per cent., giving a true speed of $105.6 \times 0.98 =$ 103.5 r.p.m., while in the second case the slip may be 1 per cent., the elongation of the belt while on the 40-in. driving pulley being less than on the 4-in. pulley because of its greater leverage and grip. Therefore the true speed of the 4-in. driven pulley will be about $947 \times 0.99 =$ 937 r.p.m.

One writer (Mr. Engber) in *Power* of July 8, 1919, cites an unusual case with a 4-in. double leather belt driving a fan. He says it was at first so tight that the power input was equivalent to $\frac{22 \text{ amp.} \times 440 \text{ volts}}{746}$ $= 12.98$ hp. With a supposed 6-in. diameter motor pulley at 1200 r.p.m. the rim velocity is $6 \text{ in.} \times 0.2618 \times 1{,}200 = 1{,}885$ ft. per min. The pull per inch width of belt would therefore be $\frac{12.98 \times 33{,}000}{1{,}885 \text{ ft.} \times 4 \text{ in.}} = 56.8$ lb.

One is astonished to learn that, "with this high tension on the belt the loss through slip was from 7 to 12 per cent." The belt must have been oily and slippery. After it had been put in better condition, it was found to do its work with a power input equivalent to only $\frac{9.1 \text{ amp.} \times 440 \text{ volts}}{746} = 5.37$ hp., which is but 41.4 per cent. of the power previously expended. Thus the pull per inch width of double belt dropped to 56.8×0.414, or to 23.52 lb., and still "the slip under these conditions was from 2 to 5 per cent."

(The use of the factor $0.2618 = \frac{\pi}{12}$ saves multiplying by 3.1416 and then dividing by 12 to find velocity in feet per minute.)

These observations are rather at variance with those made by others who find that by keeping the useful tension down to 40 lb. with single belts, or 60 lb. with double leather belts, the loss by creep and slip does not exceed 1 per cent.

In transmitting 5.37 hp., the number of feet of double belt 1 in. wide per minute per horsepower figures out, for the motor pulley assumed,

$$\frac{1885 \text{ ft.} \times 4 \text{ in. double}}{5.37 \text{ hp.}} = 1{,}404 \text{ ft.}$$

which certainly is sufficient leather to secure an efficiency of 98 per cent. if the belt is in normal condition. In every plant there is a possibility of losses being suffered in the transmission of power by belting, but there is no excuse for permitting such losses to continue. By simply comparing the actual speed change of two shafts with the calculated speed change as indicated above, the efficiency of the drive can be easily determined and brought up to 98 or 99 per cent. Thus a substantial saving can be effected at a slight expense, because properly proportioned leather belt drives give satisfactory service for periods of ten to forty years.

It is remarkable how well the advocates of direct-connected motors and prime movers have succeeded in creating a wide impression that belt drives are only about 85 per cent. efficient. To cite but one convenient example, in a paper "The Oil Engineer for Ice Plant Service," read at the International Engineering Congress, San Francisco, September, 1915, and printed in the *Journal* of American Society of Refrigerating Engineers, September, 1915, page 38, it is recommended that the efficiency of the belt or rope drive be taken at 85 per cent. In reality, with rational design, using sufficient surface, speed and leather, the efficiency of a single drive can be made 97 to 99 per cent.

Power Plant Built in Germany During the War

The photographs on the opposite page are of a 128,000-kva. steam-power plant built in the village of Zschornewitz-Golpa, near Bitterfeld, in Prussian Saxony, 75 miles southwest from Berlin. This plant is in the center of the lignite district and was put in operation in the fall of 1918. At present it is supplying 30,000 kva. to Berlin, but as soon as the necessary transmission lines are completed the load will be increased to near the plant's capacity.

In the generator room, upper right-hand corner, are eight 16,000-kva. steam-turbine driven alternators, supplied with steam from 64 boilers. Draft for the boilers is obtained from eight stacks 325 feet high. Coal is obtained from the Golpa mine, near the plant, by an automatic system of cars, as shown in the figure across the center of the page. The arrangement of the turbines in the power plant differs from American practice in that the machines are placed end to end down the center of the turbine room, while in American plants they are usually placed crosswise. The frustum-shaped towers, at the right in the bottom figure, are evidently cooling towers for the circulating water. The German government in line with the plans for the socialization of the electric industry has taken over the stock of the company that built this plant.

Exports of American coal in much greater quantity than has been anticipated is expected to result from the determination of Denmark and Sweden to divert important portions of their tonnage to the American coal trade. This action on the part of two of the European countries is thought to be the forerunner of similar action by others. The Shipping Board recently called attention to the tendency on the part of some European countries to keep their shipping in profitable trade routes despite the dire need in Europe for coal supply.

During the entire winter, it is stated by Shipping Board officials, the vessels in the New England coal trade have been operated at a loss. In order to make this service self-sustaining an increase of 75c. per ton in the coal rates from Hampton Roads and Baltimore to Boston and other New England ports has been put into effect. The increased rates will permit of no profit on

Photos from Wide World Photos.

EXTERIOR AND INTERIOR VIEWS OF POWER PLANT NEAR BITTERFIELD, PRUSSIAN SAXONY need of excessive tension. A thicker belt is not advisable because it would not wrap well around the small pulley, the outer ply would be overstretched, its circumference being appreciably greater than that of the pulley. It would be hard on the cement, which is expected to hold the two plies together. For this reason it is good practice to have a belt no thicker than 2 or, at most, 3 per cent. of the diameter of the smallest pulley used with it.

In calculating the speed change from one shaft to another it is necessary to figure always from the driver to the driven, and to add one belt thickness to each diameter.

For example, a 4-in. motor pulley at 1000 r.p.m. will drive a 40-in. pulley at what speed? 100 r.p.m.? No. If the belt is 0.25 in. thick, we have

$$\frac{4.25 \textit{ in.} \times 1000 \textit{ r.p.m.}}{40.25 \textit{ in.}} = 105.6 \textit{ r.p.m.}$$

If the 40-in. diameter pulley is driving at 100 r.p.m., will the speed of the 4-in. driven pulley be 1000 r.p.m.? No. It will be

$$\frac{40.25 \textit{ in.} \times 100 \textit{ r.p.m.}}{4.25 \textit{ in.}} = 947.05 \textit{ r.p.m.}$$

The pulleys are of the same size in both cases. In the first case, where the speed is being reduced, the slip may be 2 per cent., giving a true speed of 105.6 × 0.98 = 103.5 r.p.m., while in the second case the slip may be 1 per cent., the elongation of the belt while on the 40-in. driving pulley being less than on the 4-in. pulley because of its greater leverage and grip. Therefore the true speed of the 4-in. driven pulley will be about 947 × 0.99 = 937 r.p.m.

One writer (Mr. Engler) in *Power* of July 8, 1919, cites an unusual case with a 4-in. double leather belt driving a fan. He says it was at first so tight that the power input was equivalent to $\frac{22 \text{ amp.} \times 440 \text{ volts}}{746}$ = 12.98 hp. With a supposed 6-in. diameter motor pulley at 1200 r.p.m. the rim velocity is 6 in. × 0.2618 × 1,200 = 1,885 ft. per min. The pull per inch width of belt would therefore be $\frac{12.98 \times 33{,}000}{1{,}885 \text{ ft.} \times 4 \text{ in.}} = 56.8$ lb.

One is astonished to learn that, "with this high tension on the belt the loss through slip was from 7 to 12 per cent." The belt must have been oily and slippery. After it had been put in better condition, it was found to do its work with a power input equivalent to only $\frac{9.1 \text{ amp.} \times 440 \text{ volts}}{746} = 5.37$ hp., which is but 41.4 per cent. of the power previously expended. Thus the pull per inch width of double belt dropped to 56.8 × 0.414, or to 23.52 lb., and still "the slip under these conditions was from 2 to 5 per cent."

(The use of the factor $0.2618 = \frac{\pi}{12}$ saves multiplying by 3.1416 and then dividing by 12 to find velocity in feet per minute.)

These observations are rather at variance with those made by others who find that by keeping the excess tension down to 40 lb. with single belts, or 60 lb. with double leather belts, the loss by creep and slip does not exceed 1 per cent.

In transmitting 5.37 hp., the number of feet of double belt 1 in. wide per minute per horsepower figures out, for the motor pulley assumed,

$$\frac{1885 \textit{ ft.} \times 4 \textit{ in. double}}{5.37 \textit{ hp.}} = 1{,}404 \textit{ ft.}$$

which certainly is sufficient leather to secure an efficiency of 99 per cent. if the belt is in normal condition. In every plant there is a possibility of losses being suffered in the transmission of power by belting, but there is no excuse for permitting such losses to continue. By simply comparing the actual speed change of two shafts with the calculated speed change as indicated above, the efficiency of the drive can be easily determined and brought up to 98 or 99 per cent. Thus a substantial saving can be effected at a slight expense, because properly proportioned leather belt drives give satisfactory service for periods of ten to forty years.

It is remarkable how well the advocates of direct-connected motors and prime movers have succeeded in creating a wide impression that belt drives are only about 85 per cent. efficient. To cite but one convenient example, in a paper "The Oil Engineer for Ice Plant Service," read at the International Engineering Congress, San Francisco, September, 1915, and printed in the *Journal* of American Society of Refrigerating Engineers, September, 1915, page 38, it is recommended that the efficiency of the belt or rope drive be taken at 85 per cent. In reality, with rational design, using sufficient surface, speed and leather, the efficiency of a single drive can be made 97 to 99 per cent.

Power Plant Built in Germany During the War

The photographs on the opposite page are of a 128,000-kva. steam-power plant built in the village of Zoschovnewitz-Golpa, near Bitterfield, in Prussian Saxony, 75 miles southwest from Berlin. This plant is in the center of the lignite district and was put in operation in the fall of 1918. At present it is supplying 30,000 kva. to Berlin, but as soon as the necessary transmission lines are completed the load will be increased to near the plant's capacity.

In the generator room, upper right-hand corner, are eight 16.000-kva. steam-turbine driven alternators, supplied with steam from 64 boilers. Draft for the boilers is obtained from eight stacks 325 feet high. Coal is obtained from the Golpa mine, near the plant, by an automatic system of cars, as shown in the figure across the center of the page. The arrangement of the turbines in the power plant differs from American practice in that the machines are placed end to end down the center of the turbine room, while in American plants they are usually placed crosswise. The frustum-shaped towers, to the right in the bottom figure, are evidently cooling towers for the circulating water. The German government in line with the plans for the socialization of the electric industry has taken over the stock of the company that built this plant.

Exports of American coal in much greater quantity than has been anticipated is expected to result from the determination of Denmark and Sweden to divert important portions of their tonnage to the American coal trade. This action on the part of two of the European countries is thought to be the forerunner of similar action by others. The Shipping Board recently called attention to the tendency on the part of some European countries to keep their shipping in profitable trade routes despite the dire need in Europe for coal supply.

During the entire winter, it is stated by Shipping Board officials, the vessels in the New England coal trade have been operated at a loss. In order to make this service self-sustaining an increase of 75c. per ton in the coal rates from Hampton Roads and Baltimore to Boston and other New England ports has been put into effect. The increased rates will permit of no profit on these vessels, it is said.

Photos from World Wide Photos.

EXTERIOR AND INTERIOR VIEWS OF POWER-PLANT NEAR BITTERFIELD, PRUSSIAN SAXONY

Steam-Turbine Governors—Shaft Governors of Terry Turbines

Describes the shaft governor, emergency governor and governor valves used on the Terry turbine. Directions are given for their adjustment and care.

ALTHOUGH the Terry Turbine is built in many sizes, the general design of the governing mechanism employed is the same. Greater powers required for the larger frames make it necessary to use larger parts throughout, but its design is not altered. The shaft governor is of the usual flyball type and is mounted on the end of the turbine shaft as shown in Fig. 1. The detail structure of the governor itself is shown diagrammatically in Fig. 2. Under the action of centrifugal force the governor weights *A* tend to separate. Rocking about the knife-edges *B* the weights push against the governor slide *J*, which is a movable member inserted in the end of the hollow shaft. Its motion is resisted by the main governor spring *C* acting against the governor spring yoke *H*, the yoke passing through milled slots in the shaft and acting against a hardened steel pin. The motion of the governor slide *J* is transmitted through a ball or ball race to a lever or series of levers which operate the governor valve. The governor slide *J* is protected against wear from the governor weights by a hardened steel shoe, which is secured by the governor yoke pin. An adjusting nut *G* is provided to vary the compression of the spring *C*.

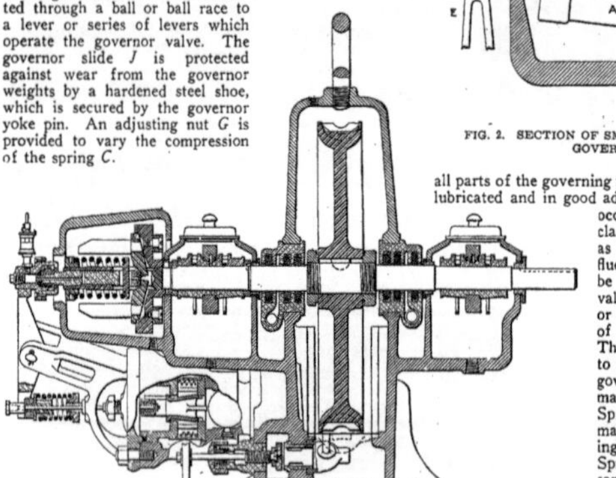

FIG. 1. SECTION OF TERRY TURBINE SHOWING STANDARD FORM OF GOVERNOR AND GOVERNOR VALVE

By altering the adjusting nut *G* a change of speed may be obtained—tightening the nut increasing the speed, slacking it off decreasing it. Adjustment from the nut *G* is only designed to alter the speed within the range of the governor capacity, as greater changes in speed would be liable to tax the governor weights and spring beyond their safe limit of strength. To obtain greater changes in speed, it is therefore necessary to procure heavier or lighter weights with the corresponding strength of spring. These may be obtained from the builder. After making adjustments, it is necessary to move the governor spring yoke *H* back to a central position so that it does not rub on the shaft.

Most governor troubles can be avoided by keeping all parts of the governing mechanism clean, properly lubricated and in good adjustment. When troubles occur, they can usually be classified as overspeeding or as hunting (periodic speed fluctuations). Hunting may be produced by improper valve setting or by sticking or lost motion in some part of the governor mechanism. There is a spring provided to take up lost motion in the governor mechanism, marked "Governor Valve Spring" on Fig. 5, which may be tightened by turning the "Lower Valve Spring Seat," shown in the same figure. Sticking in the valve or stem may easily be detected by pushing in on the valve stem and observing if it moves steadily. The valve-stem spring should readily pull the valve out again. A sticking valve or stem may be caused by a bruise or a bend in the stem, too tight valve-stem packing, the presence of rust or scale on the valve itself, or misalignment between the valve and the governor lever. The corrections for these difficulties are obvious. Sticking in the governor is usually produced by a gummy slide or by the rubbing of the governor-spring yoke against the shaft. A thorough cleaning with kerosene

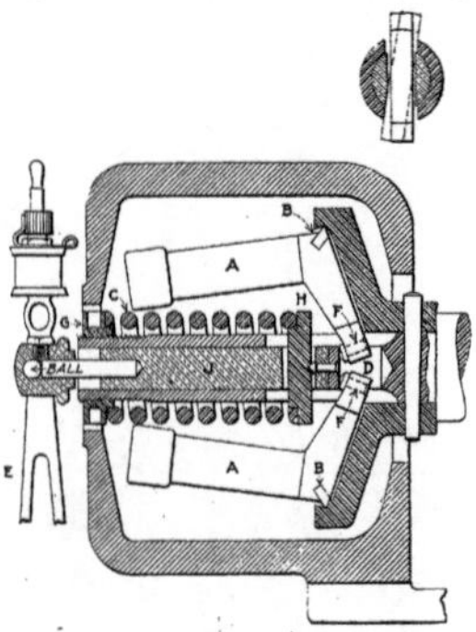

FIG. 2. SECTION OF SMALL TERRY SHAFT GOVERNOR

will usually eliminate the difficulty. If, when running at very light loads with all the jets open, the governor valve has a tendency to draw shut, the governor-valve spring shown in Fig. 5 should be tightened. Overspeeding at very light loads is caused by a poorly set

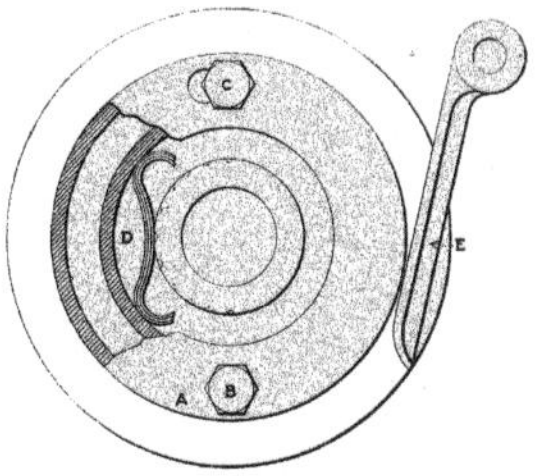

FIG. 3. TERRY EMERGENCY GOVERNOR

or a leaky valve. The setting of the valve will be explained later.

Terry emergency governors are usually set to shut off the steam to the turbine when the speed reaches about 10 per cent. above the operating rate. The form of emergency governor used on smaller Terry turbines is called the ring type, which is shown in Fig. 3. The ring *A* mounted on the back of the governor disk is pivoted at *B* and is limited in its movement by the stud

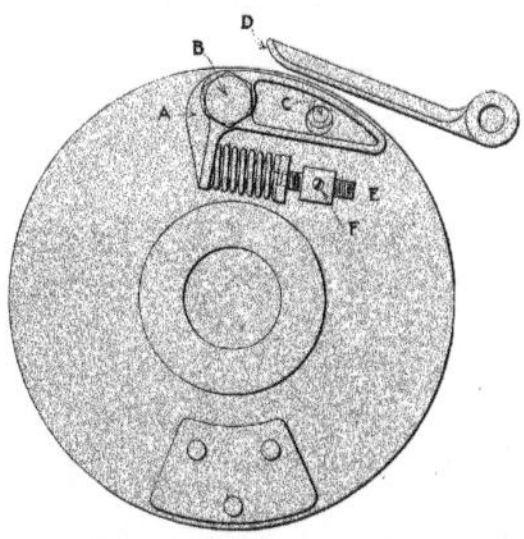

FIG. 4. ANOTHER TYPE OF TERRY EMERGENCY GOVERNOR

C. Normally, the ring is held concentric against the stud *C* by the spring *D*. As the ring is slightly unbalanced, centrifugal force tends to make it run eccentric. When the centrifugal force becomes great enough to overcome the compression of the spring *D*, the ring flies out to the limit of its travel and runs eccentric, hitting the trigger *E*, which closes the emergency valve.

The spring is held in position by a projection at the center which rests in one of the holes (shown dotted) on the hub of the governor disk. To lower the speed at which the emergency governor will operate, move the spring toward the pivot *B* with a screwdriver, and to increase the speed move it away from the pivot.

Another type of emergency governor is shown in Fig. 4. A lever *A* is pivoted on the stud *B* and is limited in its movement by a pin placed at the point *C*. This pin has about $\frac{3}{16}$ in. clearance. When the speed becomes high enough, the centrifugal force overcomes the pressure of the spring and the lever *A* flies out, striking the trigger *D*, which closes the emergency governor

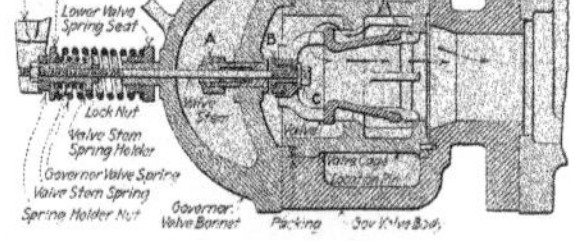

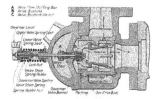

FIG. 5. STANDARD TYPE OF TERRY GOVERNOR VALVE

valve. This governor may be adjusted by altering the pressure of the spring with the adjusting screw *E*. A setscrew *F* is provided to lock the adjusting screw.

Emergency valves are of the butterfly type and are operated by a strong spring when released by the emergency governor. It is important that the packing in the stuffing-box be well lubricated and the stem always free to move.

Once a week, if the turbine is shut down that often, trip the emergency governor by hand. The turbine should stop or slow down considerably with the throttle

FIG. 6. THE TERRY SHAFT GOVERNOR

FIG. 7. ANOTHER FORM OF TERRY SHAFT GOVERNOR.

wide open. The trigger should clear the governor lever *A*, Fig. 4, or ring *A*, Fig. 3, by not more than $\frac{1}{16}$ in. when the turbine is turned over by hand.

The governor valve now used on all Terry turbines is known as the bushing type (see Fig. 5). The construction is such that it permits removal of the valve and seats for inspection or repairs without removing the steam connections. A strainer is always supplied to prevent solid matter from injuring the turbine nozzles and wheel. Its principle of operation can be readily seen from the figure and needs no further explanation.

All Terry governor valves should be so adjusted that their maximum travel is one-eighth of the valve diameter; that is, a 3-in. valve should have a $\frac{3}{8}$-in. maximum opening. To check this travel, push in on the valve stem until the valve seats when the turbine is not running. Make a mark on the valve stem opposite the edge of the stuffing-box nut, then pull out on the stem as far as the governor mechanism will permit and make another mark on the valve stem opposite the stuffing-box nut. The distance between these marks gives the valve opening and may be adjusted by turning the valve-stem in the valve-stem spring holder, being sure to readjust the locknuts at each end of the spring holder. Always be sure that the governor valve closes tight when the governor weights are all the way out. This is important with any governor.

The general appearance of a Terry shaft governor is shown in Fig. 6, and a small turbine for generator drive with motor connection to the governor for independent speed control is shown in Fig. 7.

Governing Turbine-Driven Pump

Turbine-driven centrifugal pumps, particularly for boiler feed, such as Fig. 8, are widely used. The governor valve is on the steam line to the turbine, is actuated by the pressure in the pump discharge line, and acts on the turbine as a throttling governor. In addition, the turbine is provided with the usual flyball governor, which is set for higher than maximum normal pressure. This protects the turbine against overspeed.

By a ruling of the Industrial Board, Department of Labor and Industry, Commonwealth of Pennsylvania, Harrisburg, Penn., adopted Jan. 13, 1920, and operative Feb. 12, 1920, no person shall be allowed to carry matches or other flame-producing device into any workroom or other portion of a building wherein people are employed and where ether is manufactured, rectified or handled daily in quantities exceeding one-half gallon. In such places where there is a possibility of ether vapor being present, electrical or other apparatus, such as switches, motors, etc., capable of giving off sparks, shall not be permitted. Electric lighting systems shall be placed in conduits with vapor-proof globes. Open flames of any kind shall also be prohibited.

It is estimated that the output of the central stations in the United States in 1919 was approximately 30,300,000 kw.-hr. The total revenue is estimated at approximately \$601,000,000. This is an increase of 2.1 per cent in output and 13 per cent in income over 1918.

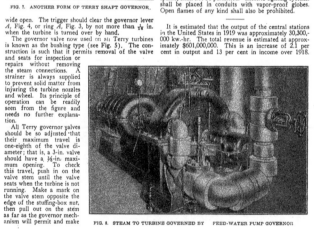

FIG. 8. STEAM TO TURBINE GOVERNED BY FEED-WATER PUMP GOVERNOR

Standard Meter Rates—Demand Rates

By LeROY W. ALLISON

This is the third and last of three articles on standard rate forms and deals with the second group of standard rate forms, known as demand rates. Four different standard demand-rate systems are discussed. Examples are given to show how these rates are applied in calculating bills.

UNDER the subject of "demand rates," there are four different standard rate systems in general use, these being (1) flat demand rate, (2) Wright demand rate, (3) Hopkinson demand rate, and (4) Doherty, or three-charge rate. These will be considered in the order noted. As mentioned in a previous article, the definitions given are those as approved by the Rate Research Committee of the National Electric Light Association. The term "demand rate" is applicable to any method of charge for electric service that is based on the maximum demand during a given period of time. The demand is expressed in such units as kilowatts or horsepower. Maximum-demand indicators or graphic meters are used. The demand rate has its basis in the customers' installations or maximum requirements in a certain period of time. It deals with the matter of connected load on the job.

Flat Demand Rate

The term "flat demand rate" is applicable to any method of charge for electric service that is based on the consumer's installation of energy-consuming devices, or on a fixed sum per customer. This rate is usually arranged on a basis of so much per watt or kilowatt, per month or for a longer term. Again, this type of rate is developed frequently at so much per consumer per month or per year, segregating the different classes of consumers into unit groups, each group covering a certain character and extent of service. The estimated demand, or the quantity of electric energy to be used, is an important consideration in the establishment of the flat rates. No meters are used.

The history of the flat-rate charge system dates back to the beginning of commercial electric service, for the first rates were arranged on this basis. For house and residence service a certain charge was made per 16-candlepower lamp per month, varying in different localities, as 40, 50 and 60 cents, and with minimum bill to be one dollar. As to be expected, many difficulties were encountered in the use of this type of rate in a general way. Careless people would allow their lights to burn continuously and beyond the point of needed utility; dishonest persons would burn more lamps than entitled to under their contract with the utility company; while in the matter of motor and other industrial service, like conditions prevailed. The fundamentals of operation dealt with the consumer being thoroughly fair and honest, but this wasn't always so.

Thus, when meters were introduced, the employment of flat rates decreased, and for a number of years they were almost in discard. The flat demand rate, however, is now regaining considerable popularity and is being used particularly for small consumers, such as residence and similar service. The type of rate makes it applicable to modification with the use of the "step" or "block" method of computation, as discussed in the issue of Feb. 10. A typical schedule from under the flat demand rate would be worked out along the following lines: $1 per month for each 50 watts of installation connected, applying primarily to small consumption; $30 per year for each horsepower of connected load; $45 per year per kilowatt for each of the first 25 kw. of maximum demand; $35 per year per kilowatt for each kilowatt in excess of 25.

Taking a typical example to show the method of calculating, assume that a consumer has a connected load of 400 watts. Under the flat rate of $1 per month for each 50 watts of installation connected, the monthly charge would be

$$\frac{400 \times \$1}{50} = \$8$$

Considering another instance, assume a maximum demand of 50 kw. per year for an installation, on the basis of $45 and $35, previously noted. This would work out

$$25 \times 45 = \$1,125.00$$
$$25 \times 35 = \quad 875.00$$

Yearly total $2,000.00

An example of a modification of the flat demand rate by the use of the "step" system is afforded in the present plan of lighting service of a Middle West central station. The basis of operation is as follows: For the first 30 hours' use of the customer's demand, a charge of 8c. net per kilowatt-hour; for the next 60 hours' use of the customer's demand, a charge of 6c. net per kilowatt-hour; for all further use of the customer's demand, the rate is 3c. per kilowatt-hour.

The customer's demand is based on the number of lamp outlets, the first 10 lamp outlets being rated at 30 watts each, the next 20 lamp outlets at 20 watts each, and all other outlets at 10 watts each. An an illustration, the demand for a house with 35 lamp outlets would be 750 watts, and if 100 hours' use was made of this demand, the total charge would be as follows:

Hours' Use		Watts Demand		Kw.		Cents Per Kw.-Hr.	Total
30	×	750	=	22.5	at	8	$1.70
60	×	750	=	45.0	at	6	2.70
10	×	750	=	7.5	at	3	0.225
Total 100				75			$4.625

Accordingly, on the foregoing basis a consumption of 75 kw.-hr. would average 6.16c. per kilowatt-hour.

Wright Demand Rate

The term "Wright demand rate" applies to that method of charge in which a maximum price per unit is charged for a certain amount of electric energy, and one or more reduced prices per unit are charged for the remainder, on the block principle, in accordance with a schedule based on the use of the maximum demand. This method of charge was developed by Arthur Wright and has come into extensive use for general lighting, commercial lighting, retail power and wholesale lighting and power. It is based on making a charge according to the number of hours' use of the maximum demand requirements. In other words, for the first hour's use per day, or aggregating 365 hours per year of maximum demand, the consumer pays a higher rate than for electric energy used in excess of the maximum. The Wright maximum-demand meter has become popular with central stations in connection with this now standard rate form. This method of rate computation in graphical form, plotted for a rate variation of from 9c. to 3c., is shown in Fig. 1. This curve is based on the following schedule:

For electric energy used equivalent to or less than the first 30 hours' use per month of maximum demand, 9c. per kw.-hr. For consumption equivalent to or less than the next 30 hours' use per month of maximum demand, 5c. per kw.-hr. And for all electric energy used per month in excess of the equivalent of 60 hours' use of the maximum demand, 3c. per kilowatt-hour.

As an example, assume that a consumer has a total connected load of 1000 watts, with a demand, estimated

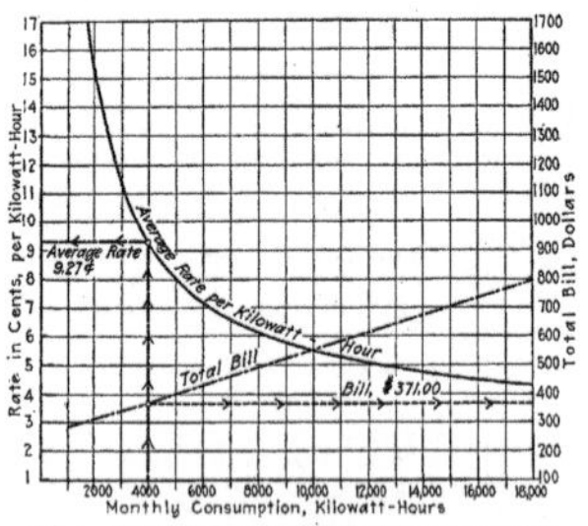

FIG. 1. GRAPHICAL REPRESENTATION OF WRIGHT DEMAND RATE

or measured, of 800 watts. On the basis of 30 hours' use of this demand (800 watts), the total is 800 × 30 = 24,000 watt-hours or, 24 kw.-hr.; this is the full amount to be charged for at the 9c.- and 5-c. maximum-demand rates. Say that the monthly consumption is 60 kw.-hr., the charge would be calculated as follows:

24 kw.-hr.	@	9c.	=	$2.16
24 kw.-hr.	@	5c.	=	1.20
12 kw.-hr.	@	3c.	=	0.36
60 kw.-hr.		Total....		$3.72

The average rate equals total cost in cents divided by total kilowatt-hours or, in this problem, equals 372 ÷ 60 = 6.2c. per kw.-hr. These values are indicated in Fig. 1 by the dotted lines. One of the prominent advantages of this method of rate calculation is found in the fact that it tends to encourage greater hours' use of the maximum demand, as the energy excess charges are low.

Hopkinson Demand Rate

The term "Hopkinson demand rate" applies to that method of charge which consists of a "demand charge," (a sum based on the demand either estimated or measured, or the connected load) plus an "energy charge" (a sum based on the amount of energy used). This comes under the term of a "two-charge rate." This rate, officially defined, is applicable to any method of charge for electric service in which the price per unit of metered electric energy for each bill period is based on both the actual or assumed quantity of electric energy consumed and the actual or assumed capacity or demand of the installation.

The Hopkinson rate takes its name from Dr. John Hopkinson, who presented its possibilities in 1892, after extensive central-station investigation and covering practically every phase of operation and expense. He held to the theory of the necessity for a fixed charge to be defrayed by the consumer, and according to his demand for power to cover the "readiness to serve" cost of production and distribution. The value of Dr. Hopkinson's work can be fully appreciated when it is stated that the majority of rate forms in use today are based, either in whole or in part, on his basis of assumption and computation.

In the Hopkinson demand rate the "demand charge" or the "energy charge," or both, may be of the block-rate type. The diagram, Fig. 2, sets forth this rate basis, taking 6, 4, 2 and 1c. per kilowatt-hour, for respective amounts of consumption.

Considering a practical example of the method of figuring rates under this system, assume a "demand charge" and an "energy charge" as follows:

Demand charge—$2.50 per month per kilowatt for the first 50 kw. of maximum demand in the month; $2 per month per kilowatt for the excess of maximum demand over 50 kilowatts..

Energy charge—6c. per kw.-hr. for the first 1000 kw.-hr. used per month; 4c. for the next 2000 kw.-hr.; 2c. for the next 2000 kw.-hr.; and 1c. per kilowatt-hour for excess over 5000 kw.-hr. used per month.

Now, assume that the consumer has a demand of 100 kw. and uses a total of 8000 kw.-hr. in a month, the bill would be calculated as follows:

50 kw.	×	$2.50	=	$125	
50 kw.	×	2.00	=	100	
100	Total demand charge				$225
1000 kw.-hr.	×	6c.	=	$ 60	
2000 kw.-hr.	×	4c.	=	80	
2000 kw.-hr.	×	2c.	=	40	
3000 kw.-hr.	×	1c.	=	30	
8000 kw.-hr.	Total energy charge				210
				Total bill......	$435

The average rate is 43,500 ÷ 8000 = 5.437c., and is indicated with the other items on the curves, Fig. 2, by dotted lines. As will be understood, the "demand charge" holds at the same figure from month to month, or as long as the maximum demand remains unchanged.

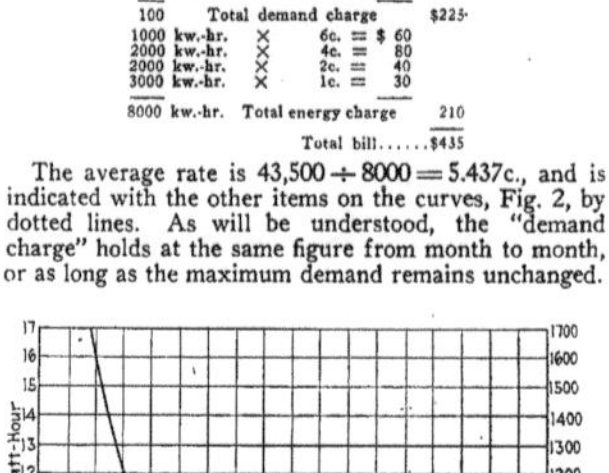

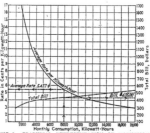

FIG. 2. GRAPHICAL REPRESENTATION OF HOPKINSON DEMAND RATE

The "energy charge" is the fluctuating element, varying from month to month in accordance with consumption.

The term "three-charge rate" is applicable to any method of charge for electric service in which the charge made to the consumer for each bill period consists of, (a) a sum based on the quantity of electric energy consumed, (b) a sum based on the actual or assumed capacity or demand of the installation, and (c) a charge per consumer.

Accordingly, a definition of the Doherty or "three-charge" rate would be: That method of charge which consists of a "customer charge," or a charge per customer per meter; plus a "demand charge," a sum based on the demand, either estimated or measured, or the connected load; plus an "energy charge," a sum based on the quantity of electric energy used. This rate may be expressed in either the "block" or the "step" form, terms that have been previously explained. As will be noted, this system adds a customer charge, or meter

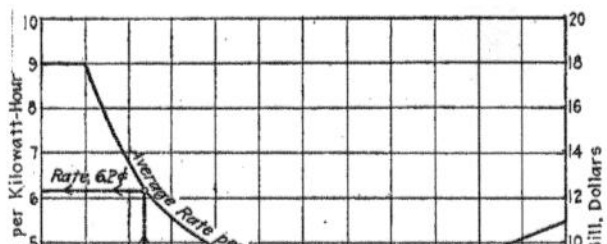

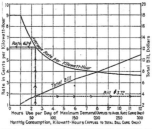

FIG. 3. GRAPHICAL REPRESENTATION OF DOHERTY OR THREE-CHARGE DEMAND RATE

charge, to the demand and energy charges, which are covered in the Hopkinson demand-rate schedule.

The Doherty charge was developed by Henry L. Doherty, New York, and presented before the National Electric Light Association in 1900, and has become useful for certain phases of power service. The first two charges enforced, "customer charge" and "demand charge," are designed to cover the cost of "readiness to serve," while the third, or "energy charge," provides for the actual consumption. Again, in further explanation, the customer and demand charges cover the average cost of service connections to the central station.

The typical curves in the diagram, Fig. 3, show the basis of the Doherty rate, at 5c. and 3c. per kilowatt-hour for amounts of consumption as expressed in the following example; the curve covering the average rate per kilowatt-hour indicates effectively the decrease in charge as the consumption increases.

The method of figuring a bill under this rate is very similar to that used for the Hopkinson rate, previously described, simply adding another established charge to the form. Taking a typical example:

Customer Charge—$1 per month per meter.

Demand Charge—$2.75 per month per kilowatt for the first 40 kw. of maximum demand in the month; $2 per kw. for excess of maximum over 40 kilowatts.

Energy Charge—5c. per kw.-hr. for the first 1000 kw.-hr. used per month; 3c. per kw.-hr. for excess over 1000 kw.-hr. used per month.

On this rate basis assume that the consumer has a demand of 100 kw. and uses a total of 4000 kw.-hr. in a month, the bill would be computed as follows:

		$ 1.00	Customer charge
40 kw.	× $2.75 = $110.00		
60 kw.	× 2.00 = 120.00		
100		230.00	Total demand charge
1000 kw.-hr.	× 5c. = 50.00		
3000 kw.-hr.	× 3c. = 90.00		
4000		140.00	Total energy charge
	Total bill.......	$371.00	

The first two charges noted are uniform from month to month unless a change is made in the installation and, consequently, in the maximum demand. The last charge varies from month to month as the total amount of electric energy fluctuates.

For additional explanation of the reason for the demand and energy charges brought out in the Hopkinson and Doherty rate schedules, it should be noted that these cover specific functions of central-station operation. The demand charge represents the return on the necessary investment to carry the maximum load that the consumer will have. It does not include any charge for the actual electric energy used. On the other hand, the energy charge is a charge proportional to the kilowatt-hours, horsepower-hours or lamp-hours, consumed; it does not contain any investment charge or overhead commercial charge, but such items as labor, fuel cost and supplies.

Season Fluctuations in Boiler Industry

It may surprise many to know that there are seasonal fluctuations in the boiler business, but such appears to be the case from statistics compiled by A. G. Pratt, vice president, Babcock & Wilcox Co., and recently presented before the meeting of the American Boiler Manufacturers' Association in New York. The data, as shown by the accompanying chart, cover water-tube boilers for the years 1907 to 1913, inclusive, this covering a period of normal or nearly normal business activity throughout the country. Any figures for 1914-15-16 would show the effects of the war in Europe, and 1917-18 would, of course, be extremely abnormal, as boilers for war purposes were required in excess of the maximum production of the manufacturers. Sales during 1919 were not included, as that year was considered abnormal; in the first five months general business conditions were unsettled, and practically no buying of any character was done; during the last seven months business was extraordinarily good, due to returning confi-

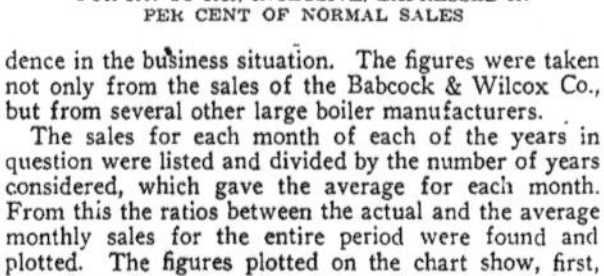

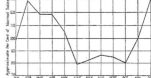

AVERAGE MONTHLY SALES OF WATER-TUBE BOILERS FOR 1907 TO 1913, INCLUSIVE, EXPRESSED IN PER CENT OF NORMAL SALES

dence in the business situation. The figures were taken not only from the sales of the Babcock & Wilcox Co., but from several other large boiler manufacturers.

The sales for each month of each of the years in question were listed and divided by the number of years considered, which gave the average for each month. From this the ratios between the actual and the average monthly sales for the entire period were found and plotted. The figures plotted on the chart show, first, that the sales were equal to or above normal from February through July and again in December; second, that sales were below normal in January and from August through November; third, that sales in different periods vary from a maximum of 120 per cent to a minimum of 70 per cent normal.

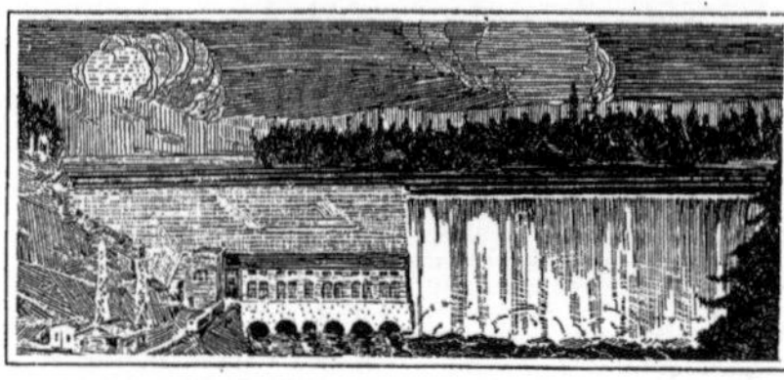

POP'S WATER-POWER COURSE

By John S. Carpenter

"WELL," announced Pop as he walked into Jimmy's room and threw his hat on the bed and himself in the softest rocker, "we're fired!" The old chief let his lower jaw hang to give, as the advertising men say, "atmosphere" to his scene.

"Fired? What do you mean?" blurted Jimmy.

"Well, we're fired outa the ol' steam plant, an' tomorrer we goes up the creek an' we takes the new hydro plant in hand. They's also the mere trifle of a fivespot a week raise in it for you." Pop put his thumbs under his suspenders while he watched the effect on Jimmy.

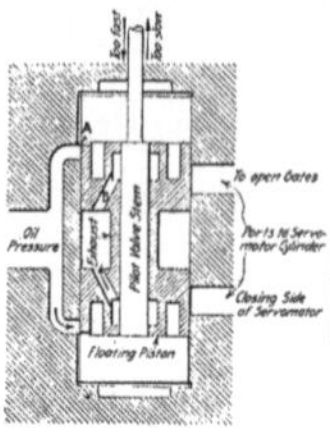

FIG. 1. PRINCIPLE OF MODERN PILOT VALVE

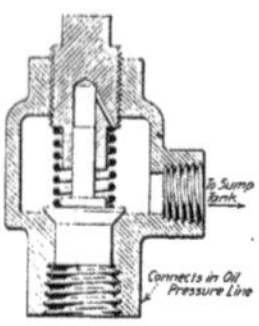

FIG. 2. VIEW OF BYPASS VALVE

"As yer might expect, I'm gonna be the boss an' you the feller wot does the work!"

"Which means, Popsie, that in about a year I'll be boss!"

Pop hurriedly changed the subject, as things were not taking the proper path.

"We'll git busy while th' gittin's good. We was a-talkin' the last time about pilot valves an' such like. I made a skitch of a pilot valve simplified so that you could git th' idee easy. Now, here (refer to Fig. 1) is a drawrin' that I made which shows the principle of the modern valve. Now git me when I says that this one ain't the only pebble on the beach, nor is it the only way or style that they can be made in. The fundymental idee of a pilot valve is that a dingus that must be delicate an' easily operated in order not to jim the sensitiveness of the governor flyballs, moves a little an' in so doin' calls into action a larger valve which has the necessary valve area to distribute the oil to the servomotor without excessive friction. There it is in a nutshell."

Jimmy fixed his gaze on Fig. 1, and a dozen possible cycles of operation passed through his mind.

"Now it works this way," resumed the old chief. "In the center of the drawrin' is located the exhaust port, which couldn't be shown very plain. On the left-hand side is the pressure-oil port. So, now, you sees that I has two litttle grooves cut in the body of the valve casing above and below the floating piston. Their purpose is to keep pressure oil always on both sides of the floatin' piston so that it is balanced an' therefore easily movable. The same ports or grooves helps keep the pilot-valve stem balanced although the balancin' idee ain't carried out in my design. To be right, the valve stem should be of the same size where it passes through the valve body at the top, an' the lower end ought go through the valve body, too. Then you has no end pressure on the valve stem an' the thing is balanced, seein' as how the side pressures neutralize themselves all around the valve stem.

"The floating piston has a few small grooves cut in the extreme faces in order to let the pressure oil git to the upper an' lower faces to carry out the balancin' idee, or otherwise the thing might settle down at the bottom of the valve body an' fergit they was a job to do. When the speed is normal, both valves assumes a central position as shown on the drawrin'. As soon as a change in speed occurs, let's say that some load has gone off the unit an' that the turbine is now runnin' too fast, then the pilot-valve stem will move downward, in the deerection of the arrow. When that happens, the hollow space at the top of the floating piston is uncovered an' the pressure oil that was in that space rushes through the small port B an' drains into the exhaust. This produces a severe case of what steam engineers calls wiredrawrin' at the little grooves marked A an' a unbalanced condition of pressure, the oil pressure bein' still workin' full on the bottom of the floating piston an' a hull lot less on the top. The result is that the floating piston is forced to the top of the valve body an' the trapped oil passes through the port B to the exhaust. In so movin' to the top of the valve body, the port marked 'to open gates' is connected with the exhaust, an' the other, marked 'closing side of servomotor,' is connected with the pressure oil port. The turbine gates is then moved an' the restoring mechanism returns the pilot valve stem to its proper place, which reverses the condition of affairs an' returns the floatin' piston to its normal place."

"I must admit that I understood that, only I guess that the actual work is done in about one-tenth of a second," said Jimmy.

"Yep, I reckon as how it don't take more then a dern small part of a second to do the hull thing."

"I suppose," commented Jimmy, "that those pilot valves have to be made just so to keep down leakage and that it takes very accurate work to produce a satisfactory valve. It seems to me that if the leakage was much more on one side of the floating piston than on the other, it would tend to make the governor act without the flyballs getting in on the deal at all."

"Yep, that could happen. Pilot valves an' floatin' pistons is usually ground to the exact size, an' the clearance ain't usually more than two or three thousandths of an inch. In most governors, the pilot valves has the ports for the pressure oil an' the exhaust ports chipped to a jig so that they comes in the right place, 'cause foundry work might make the ports come too much either one way or t'other. The valve stem is usually made of steel which has about 3½ per cent of nickel which gives it great toughness, an' the floating piston an' valve body is often made of Govamint bronze, the same bein' 88 parts copper, 10 tin an' 2 zinc, which is a good mix an' lasts long under the throttlin' an' wire-drawrin' that must an' does occur in sich a valve."

"You said something about there being a gear pump to these governors. Is there any particular reason that the builders prefer them to piston or plunger type pumps?"

"Yep, they's quite a few reasons. One is that a gear pump is a small, compact thing that can be driven at high speed, an' when they is made good they has good mechanical an' tightness efficiencies. Another, which kinder follows from the first, is that they are easy to get into the small spaces that people thinks governor pumps should occupy an' that they can be made quite cheap, although don't git the idee that them gears, which is usually made o' 3½ per cent nickel steel an' has cut teeth, comes free."

"What gets me is how they can get those gears tight enough so that they won't leak half the oil that they pump."

"Oh, that's toler'ble easy. Them gears is turned about three or four thousandths less than what the gear case is bored, an' the side clearance is usually made two thonsandths on each side. When the clearances is less than what I told yer, they is a strong tendency for the pump to heat up. If the heatin' continues long, the oil is liable to break down an' lumps of carbon be deposited in the pipes, jes' the same as carbon will be deposited in fuel-oil headers if the oil gits too hot."

"But governor oil is refined, isn't it?" insisted Jimmy.

"True enough! That don't prevent it from boiling if it gits too hot, does it?"

"S'pose not. I suppose, though, like water, that the fact of the oil being under pressure raises its boilin' point and that the decomposition point must be somewhere near there."

"I don't know about that, son. All I knows is what I told yer, that the oil does break down after long use at high temperatures an' high pressures. The use, itself, at high pressures ain't so bad; but it's the high temperatures that kills it purty quick."

"That brings back a thing I wanted to have thrashed out—how do they know how large to make these servo-motor cylinders?"

"It depends how you wants me to answer that question. If you wants to know just what force it takes to move them turbine gates at any point, I'm afraid that I can't answer it. The problem of the pressures on the front an' back sides of a gate is a deep one. Most builders uses an empirical formula which has enough leeway to it that they is safe. Their rule is about like this:

"Multiply the horsepower of the turbine by 40 if the turbine has inside gate mechanism an' divide the product by the square root of the head. This gives the foot-pounds of energy that the governor cylinder should have. If the governor handles a unit with outside gate mechanism, that meanin' one with the governor connections out of the water, use a constant of 50 instead of 40."

"The capacity of the governor in foot-pounds is the area of the cylinder in square inches times the oil pressure in pounds per square inch, times the stroke in feet, ain't it?"

"Certainly. You got it that time, Jimmy. They is a small loss due to friction, but it don't amount to much.

"Now I wants you to take a look at Fig. 2. This is a bypass valve. The idee is to lighten the work on the motor that drives the oil pump, or whatever source of power is used. When the pump has generated sufficient pressure to overcome the leakage in the different joints, if no bypass valve is used the pump pressure will build up an' increase, taking a lotta power for no good use. Just as soon as the right pressure exists in the pipe from the pump to the pilot valve, the bypass valve is opened an' the oil passes back into the sump tank, relievin' the pump of all the extra work that it would git from havin' the pressure build up on it. You see that the tension on the spring can be changed by screwin' up on the setscrew an' the pressure changed accordingly."

"Now tell me what that little port that is drilled through the setscrew is for."

"Nothin' more than a hole to let the trapped oil get through when the valve opens. They's always some little oil gits through the valve-stem clearance up into that space, an' that hole is jes' a drain."

"Now why couldn't the whole thing be done electrically? I mean the governing of the unit and all that sort of thing?" Jimmy was getting pretty deep into one of his inventive moods.

Pop leaned back in his chair and gazed steadily at something that was not where he gazed. "I heerd tell o' somethin' like that an' I'll tell yer about it. They once was an engineer feller who invented one o' them critters, an' I has the patent specifications down in my carpetbag some place. One of the engineers of a prominent turbine company said that to his best knowledge and belief the thing was divided into three parts. One part, the first, was quite simple to understand to the ordinary mechanical engineer with about ten or twelve degrees after his name an' who was born on the 31st of February. The second part, along with the first, was understandable to the inventor an' the Lord, but beyond any other human bein'. The last part the Lord has long since given up in despair, an' the inventor only had a shadowy idee of it himself. They never was any governor made after his specification, an' they most likely never will be."

"But, Popsie, can't it be done?" insisted Jimmy.

"It would seem that it can be done, son, but the time don't seem to be ripe now. It seems to me that it will be done that way some time in the future, and maybe it will be some solenoid-operated critter which will exert a pull or thrust proportional to the current flowing in the circuit. That might be connected to the gates in some way an' like a voltage regulatoı it may be the right trick."

The French Alpine regions of Dauphiny and Savoy have become the centers of electric-steel making; 1,045,000 hp. is supplied by the French Alps today, against 70,000 in 1890. This hydraulic power represents the equivalent of 6,270,000 tons of coal. Of this total, metallurgical operations absorb 303,000 hp., chemical operations 255,000, while the remainder is divided between other industries and light and traction service.—*Iron Trade Review.*

Dispatching Method of Operating Boiler Plant

When the quantity of steam required varies greatly from time to time, the rate of producing steam cannot be changed quickly enough in a plant with numerous large boilers to insure an adequate supply at highest boiler efficiency. Furthermore, steam required for any load curve can be produced by various combinations of boilers at different ratings, and the selection of the best combination is important.

To enable advantageous operation of boilers on a load which is subject to wide differences in daily requirements for energy, a plan has been developed for the Ashley Street plant of Union Electric Light and Power Company, of St. Louis, which is as follows:

The need was felt for a unit which was common to both prime movers and boilers and the name "kilber" was devised to stand for 1,000 lb. of steam per hour (450 kg. per hour). It is a combination of "kilo" (one thousand) and "lb." (pounds) with the euphonic abbreviation "er" representing "per hour." Since the adoption of this name the terms "kilowatts" and "boiler-horsepower-hours" are seldom heard in the boiler room. The steam-flow meters are now calibrated to read directly in "kilbers."

The system load dispatcher forecasts the load requirements and instructs the steam-plant operator what

FIG. 1. BOILER-ROOM DISPATCHING BOARD AT OPERATOR'S CONTROL DESK

The number of "kilbers" shown on the "kilberscope" here and on the one in the boiler room is regulated by the handwheel under the instrument. The illuminated figure near the top of the board gives the time when the "kilbers" will be required. The framed tabulations to the left give the number of "kilbers" required for any load with different turbo-generator combinations.

load, in kilowatts, will be required from that plant one-half hour later. The quantity of steam in "kilbers" which will be needed for that load with different combinations of turbo-generators is shown by a glance at the chart at the left in Fig. 1. The operator then orders from the boiler room the proper number of "kilbers" for delivery at a designated future time.

The apparatus for signaling orders and acknowledgments between the switchboard gallery and the boiler room is simple and effective. A circuit, containing a rheostat, runs from the operator's control desk to the boiler room and connects two illuminated dial voltmeters graduated in "kilbers." These "kilberscopes," as they are called, are mounted near time indicators which have twenty-four separately illuminated compartments numbered to correspond with the hours and half-hours.

The operator's control desk, the rheostat wheel below the kilberscope and the time indicator above this instrument are shown in Fig. 1. When this photograph was taken 340 kilbers had been ordered for nine o'clock. Having set the "kilberscopes" at the desired reading, the operator closes a circuit which illuminates a compartment of the distant time indicator and actuates a claxon alarm in the boiler room. The alarm continues

FIG. 2. SIGNALING BOARD IN BOILER ROOM

The "kilberscope" and time indicator are similar to those in Fig. 1. The chart here gives the output in "kilbers" from boilers in each of two groups when operated at different percentages of their rated capacity.

to sound until the boiler-room attendant closes a circuit which illuminates a corresponding compartment for the time indicator at the control desk, thus acknowledging the order. The apparatus shown in Fig. 2 is installed in the boiler room. The illustration shows that 265 kilbers will be required at 11:30. The chart shown in the illustration gives the "kilber" output of the boilers in each group when operating at different percentages of their rated capacity.

This system eliminates the use of an unnecessary number of boilers under bank or operating at reduced capacity. It reduces the probability of misunderstanding between boiler room and turbine operators, and is a convenient means of placing the load dispatcher in more effective control of results. The economies produced have been important and have stimulated the plant employees to greater efforts to add to their bonus earnings by effecting further economies with the use of this new tool.

A deposit of coal has been discovered at La Union in Chile. It contains 40,000,000 tons of good-quality fuel. Veins have also been struck at Mailef, near Valdivia, and in the Castro district, and these are being exploited.

The Uniflow Engine

THERE are certain large losses in the steam engine which cannot be overcome, but almost from the beginning two losses have been known which can at least be greatly reduced. These are the loss due to incomplete expansion of the steam in the cylinder and that occurring when the steam enters the cylinder which has been cooled below the temperature of the entering steam during the expansion and exhaust stroke. This loss is known as initial condensation.

The cooling of the cylinder is brought about largely by reëvaporation from the hot surfaces of moisture that

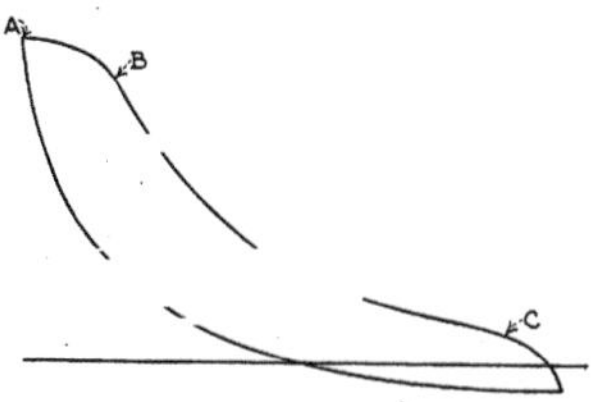

FIG. 1. UNAFLOW-ENGINE INDICATOR DIAGRAM

has been deposited upon them, and as most of this reevaporation takes place during the exhaust stroke, heat is directly transferred by it from the higher-temperature level to the condenser or atmosphere without doing useful work. The heat so abstracted must be restored by the incoming steam, which, if it is not superheated, can give up heat only at the expense of condensation.

A convincing demonstration of the cooling effect of evaporation may be had by standing before a hot-air register when you emerge from your bath. The current of air, which appeared so hot to your dry skin before, has now become a chilly draft. Its temperature is the same, but the rapid abstraction of heat from the body by the evaporation of the film of moisture upon it produces the sensation of cold. Artificial refrigeration is produced by the evaporation of ammonia and other volatile fluids. In tropical countries water is cooled in jars from the wetted felt covering of which moisture is evaporated by the heated atmosphere.

The loss due to incomplete expansion is reduced by cutting off the steam in the cylinder before the end of the stroke; the shorter the cutoff the less this loss, but the shorter the cutoff the greater the loss due to initial condensation. In the ordinary type of engine, on account of the greater cooling effect due to increased expansion, this latter loss equals the gain at about one-quarter cutoff, so that there is no economy in shortening the cutoff beyond this point.

Builders soon learned that by making two cylinders of different diameters and exhausting from the smaller into the larger, initial condensation was greatly reduced, as the difference between the initial and exhaust temperatures in each cylinder was cut in two, and as the total number of expansions is the product rather than the sum of the expansions in all the cylinders, the steam was exhausted at a low terminal pressure, so the loss due to incomplete expansion was also reduced—hence, the compound engine.

It might be reasoned that by putting a large number of cylinders in series, initial condensation could be reduced to that occurring in one of the cylinders, and this is theoretically true, but there is a loss in getting the steam from one cylinder into the next. The more cylinders in series the greater this loss, so that this would in time offset the gain by getting rid of initial condensation. This, combined with the extra cost, has precluded the use of more than four cylinders in series, and there were very few of these. In fact, excepting in marine and pumping engines, there are few compounds of more than two cylinders.

The compound engine was the only known way of getting high degrees of economy in a steam engine up to 1908, when Prof. Johannes Stumpf of Charlottenburg, conceived the "Una-Flow" engine with only the inlet end jacketed. This design practically eliminated initial condensation so that the steam might be cut off very short, permitting of a great deal of expansion in one cylinder, with consequent low terminal pressure. This reduced the incomplete expansion loss, saved the loss due to the transfer of steam from one cylinder to the other, and the greater radiation loss due to a multiplicity of cylinders inherent in the compound engine.

Professor Stumpf started with the idea of isolating so far as possible the parts of the engine heated by the incoming steam from the parts kept cool by the comparatively cool exhaust, and from the exhaust steam itself, in order that there might be a minimum of transfer of heat that did no work. Much of the superior economy of the four-valve over the single-valve engine is due to the use of a separate exhaust valve, instead of leading the incoming steam through a port and over surfaces that had just been bathed by the cool and moist exhaust. He adopted the idea which had been tried by Eaton in this country in 1857, and Todd in England in 1885, of taking the steam into the cylinder at the ends

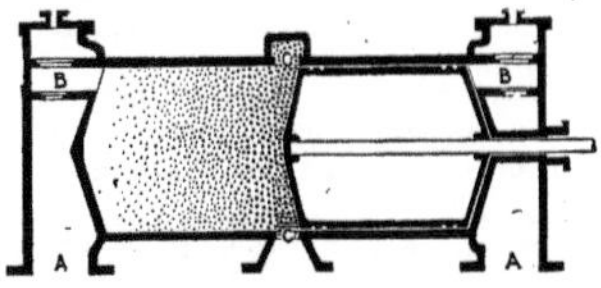

FIG. 2. SECTION THROUGH UNAFLOW-ENGINE CYLINDER

and exhausting it at the center, in order that there should be no back-flow of cold exhaust steam across the hot inlet end. It was this unidirectional flow which gave to this type of engine the name of Uniflow, or "Una-Flow," as the Stumpf people write it.

Neither Eaton nor Todd made a commercial success of his engine and it was evident that something was lacking. This missing link was supplied by Stumpf, who heated the inlet end of his engine by means of live-steam jackets.

The result of this heating was the forcing of the moisture caused by converting heat into work during expansion to the cold end of the cylinder, so that it would be wiped out of the cylinder when the piston uncovered the exhaust port. The most important effect of this is that the inlet end of the cylinder, being dry when the drop in temperature takes place, does not give off its heat by evaporation and so does not have to be reheated later. It will part with its heat but slowly by radiation and convection to a surrounding atmosphere of dry steam, while it would be rapidly chilled by evaporation if surrounded by a moist content. A good illustration of this action is seen in every blacksmith shop, where a piece of hot iron, if sprinkled with water, cools immediately, while it would retain its heat in a dry atmosphere for a long time.

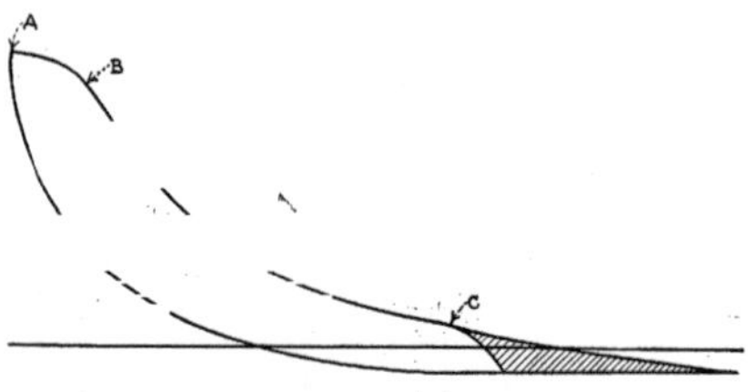

FIG. 3. UNAFLOW-ENGINE DIAGRAM AS IT WOULD APPEAR WITH CYLINDER LONG ENOUGH FOR COMPLETE EXPANSION

With the foregoing facts clearly in mind the Stumpf cycle can easily be followed. Referring to the indicator diagram, Fig. 1, steam enters at *A*, the valve remaining open until the piston reaches the point *B*, when it closes and cutoff occurs. From *B* to *C* the steam expands, dropping in pressure and acquiring moisture, due to the conversion of heat into work. The cylinder head being kept hot and dry by the steam jacket, the moisture is greatest nearest the piston, Fig. 2, so that when the piston uncovers the exhaust port at *C*, this moisture will be swept out by the flow of exhaust steam. At this time the inlet head, being dry, does not give off its heat, so that when the piston covers the exhaust port on its return stroke, there being no moisture nor cold cylinder head to absorb the heat of compression, and heat being added from the head jacket during compression, the temperature of the steam remaining in the clearance at the end of the compression stroke will be higher than the temperature of the boiler steam. When the piston moves away at the beginning of the next stroke, the high-pressure steam will enter a cylinder which is at a higher temperature than its own, hence there will be no initial condensation. Fig. 3 shows what the diagram in Fig. 1 would be like if the cylinder were long enough to carry the expansion down to the exhaust pressure. The shaded portion shows the loss due to incomplete expansion. Unfortunately, the diagram does not show the loss due to initial condensation, but from it can be computed the amount of steam that would be required to fill the cylinder up to the point of cutoff if there were no condensation; and if actual consumption is known, the difference is an approximate measure of condensation up to that point. This difference is rarely below 20 per cent.

Calculations From a Flue-Gas Analysis

BY H. M. BRAYTON

In this article the author shows in a simple way how to calculate combustion losses easily.

THERE are two principal losses in the furnace: First, the loss due to incomplete combustion, and second, that due to excess air. The first of these was taken up by the writer in a previous article in *Power*. The reason for this loss was pointed out. In addition an analytic and a graphical method were given for actually determining this loss. It was pointed out that a small percentage of CO would cause a serious loss. It was explained that the formation of this harmful CO could be largely prevented by admitting sufficient air through the fuel bed and by keeping the brick walls well sealed, thus preventing cold air from seeping through and cooling the hot gases.

Every engineer knows that he must admit an excess of air through the fuel bed to get good results. This is, unfortunately, necessary, because much heat would be saved could we get results by adding just the theoretical quantity of air. The reason why we cannot do this is that the air does not go through the fuel bed evenly. Some parts will get too much air and others not enough. Holes will develop in the bed through which air will flow and practically never come in contact with hot carbon. Many designs of boilers aim to thoroughly mix the gases as they leave the fuel bed.

The greater the excess of air given the furnace the lower the efficiency will be. The excess air merely carries off heat from which there is no return. The engineer is continually called upon to balance the draft between these two sets of losses. If he does not add enough air, the CO will run high; if too much air, the loss due to the excess will be prohibitive.

The previous article gave the methods of determining the loss due to incomplete combustion. In this article the writer wishes to show in as simple a way as possible how the loss due to excess air may be determined. This procedure is not difficult, and little theory is involved. Every practical engineer should know how to do this. He will find that it comes in handy, and a knowledge of all these things tends toward the bigger pay envelope.

To carry out the calculations, we need a certain amount of data. As in the case of incomplete combustion we shall have to know the analysis of the flue gases. This should consist of the percentage by volume of CO_2, CO and O_2. The amount of nitrogen can be found by subtracting the sum of these substances from 100. There will always be some water vapor or steam in the flue gases as well as some of the inert gases, but the volume is too small to receive practical consideration.

In addition to the foregoing information we must know the ultimate analysis of the coal used. By ultimate analysis is meant the actual percentage of each of the elements, such as carbon, oxygen, hydrogen, etc., which the coal contains. In contrast to this we have what is known as the proximate analysis, which gives the percentage of moisture, volatile matter, fixed carbon and ash which the coal contains. These values may be determined by a competent chemist, or for rough work, which would be good enough for practical purposes, they may be taken from a bulletin issued by the United States Geological Survey and which appears in many books for engineers. These tables also give the heat of combustion of the various coals, a value that is essential in all calculations pertaining to this work.

Several empirical formulas have been developed for the determination of the heat of combustion from the ultimate analysis of the coal. It will not be out of place to give them here because the engineer might have the analysis and not the heat of combustion. These formulas are not exact by any means, and yet they check well with experimental determinations. Dulong proposed the following:

$$\textit{Total heat} = 14{,}650\,C + 62{,}100\left(H - \frac{O}{8}\right)\ (\textit{B.t.u.})$$

in which C is the per cent. of carbon, H the per cent. of hydrogen and O the per cent. of oxygen in the coal by weight.

Mahler has proposed the following formula:

$$\textit{Total heat} = 14{,}650\,C + 62{,}100\,H - 5400\,(O + N)\ (\textit{B.t.u.})$$

in which N is the per cent. of nitrogen by weight and the other letters have the same significance as above.

Air Needed for Combustion

The moisture, carbon dioxide, argon, etc., in the air are too small in amount to be considered in a practical calculation. Neglecting these, air is made up of the following by weight: Oxygen, 23.2 per cent., and nitrogen, 76.8 per cent.; and by volume: Oxygen, 20.94 per cent., and nitrogen, 79.06 per cent. An easy way to remember the ratio between these two elements in the air is to think of air as a real compound (which it is not) and represented by the formula N_4O, which means four parts of nitrogen to one part of oxygen.

With this information it is easy to calculate the air required to burn a coal or fuel of known composition. To illustrate this point: When carbon burns to CO_2, it requires two parts of oxygen. The atomic weight of oxygen is 16 and that of carbon is 12. One pound of carbon will therefore require $32 \div 12 = 2.67$ lb. of oxygen. But air is 23.2 per cent. oxygen by weight, so that one pound of carbon requires $2.67 \div 0.232 =$ 11.5 lb. of air. This, of course, is for complete combustion to CO_2.

The same line of reasoning applies to the hydrogen content in the coal. The atomic weight of hydrogen is 1. One pound of hydrogen will require $16 \div 2 = 8$ lb. of oxygen, and $8 \div 0.232 = 34.5$ lb. of air. This hydrogen burns to H_2O, which is common water, and passes off in the form of superheated steam.

The foregoing may be stated in a single equation,

$$\textit{Lb. of air per lb. of fuel} = 11.5\,C + 34.5\left(H - \frac{O}{8}\right)$$

The calculation given does not consider the fact that most coals contain some oxygen. This action naturally reduces the amount of air needed for complete combustion. Most coals contain around 8 per cent. of oxygen. On this basis the air required per pound of coal will be reduced to about 10. The equation takes care of this in the last term.

Assume the analysis of a certain coal and go through the calculation on this basis. Suppose we have Pocahontas, a semibituminous coal with the following analysis: Carbon, 82.73 per cent.; hydrogen, 4.63 per cent.; nitrogen, 1.31 per cent.; oxygen, 4.42 per cent. The percentages of sulphur and ash are usually given, but are neglected here, as their effects are uncertain and small at the most.

From the atomic weights we determine that each pound of carbon will yield 3.67 lb. of CO_2 and each pound of hydrogen will yield 9 lb. of H_2O. We shall, therefore, have from the above analysis:

$$3.67 \times 0.8273 = 3.04\ \textit{lb. of}\ CO_2$$

$$9 \times 0.0463 = 0.42\ \textit{lb. of}\ H_2O;\ \textit{that is, steam}$$

To get the air required for combustion with this coal we shall have to apply the equation given above thus:

$$\textit{Lb. of cir per lb. of fuel} = 11.5\,C + 34.5\left(H - \frac{O}{8}\right)$$

$$\textit{Lb. of air per lb. of fuel} = 11.5 \times 0.83 + 34.5\left(0.046 - \frac{0.044}{8}\right) = 11.33$$

The air contains 76.8 per cent. of nitrogen by weight, so that the total weight of nitrogen in this case would be $0.768 \times 11.33 = 8.7$. The coal contains 0.013 lb. of nitrogen, so we have a total of $8.7 + 0.013 = 8.713$ lb. of nitrogen.

Specific Heat Defined

The specific heat of any substance is that amount of heat that must be added to one pound of it to raise its temperature one degree F. Any of the good handbooks will give a table showing the specific heats of all the common elements and substances. The value for CO_2 gas is 0.217, which means that this many B.t.u. must be added to one pound of CO_2 gas to increase its temperature 1 deg. F. Similarly, the specific heat of steam (H_2O) is 0.481, nitrogen 0.244, and air 0.238. If any CO is present its specific heat may be taken at 0.245.

Fifty Per Cen't. Excess Air

Suppose we assume that the flue gases show 50 per cent. excess air. Then calculate on this assumption. With this much excess air it is doubtful if there would be any CO. If the analyis showed any CO, then we would merely have to figure its weight and multiply this value by the specific heat of CO.

Heat Carried Off by Flue Gases

The following calculation is for the several components:

			B.t.u.
Carbon dioxide (CO_2)	3.04	× 0.217	= 0.660
Steam (H_2O)	0.42	× 0.481	= 0.202
Nitrogen	8.713	× 0.244	= 2.130
Air for dilution 50 per cent.	6.09	× 0.238	= 1.450
Total			4.442

The figures given consider that there is no CO in the flue gas. There should be none with this amount of excess air. If there is any, it shows poor firing. If CO is present, simply treat it the same way and add it into the total.

Now suppose the temperature of the external air is 50 deg. F. and that you have ascertained the temperature of the flue gases at the base of the stack to be 550 deg. F. The difference is, of course, 500 deg., and the heat in the chimney gases above the temperature of the air becomes $500 \times 4.442 = 2221$ B.t.u. The value 4.442 represents the amount of heat necessary to raise one pound of the flue gas 1 deg. F.

Dulong's formula gives the heat of combustion in B.t.u. of this coal to be:

$$\text{Total heat} = 14{,}650\,C + 62{,}100\left(H - \frac{O}{8}\right)$$

$$\text{Total heat} = 14{,}650 \times 0.8273 + 62{,}100\left(0.0463 - \frac{0.0442}{8}\right]$$

Total heat = 12,150 + 2330 = 14,480 B.t.u. per pound, which checks closely with the actual calorimeter test.

It is customary to assume in this work that 10 per cent. of the heat of coal is lost in radiation through the walls of the furnace. The 2221 B.t.u. which are allowed to pass up the chimney are also a loss, making the total loss of heat 2221 + 1448 = 3669 B.t.u. There will then, remain to be transmitted to the water in the boiler 14,480 — 3669 = 10,811 B.t.u. The efficiency of the furnace with the 50 per cent. excess of air would then be 10,811 ÷ 14,480 = 75 per cent.

One Hundred Per Cent. Excess Air

Suppose now we have 100 per cent. excess of air. The amount of CO_2, steam and nitrogen would remain the same. The air for dilution would be increased from 6.09 to 12.18, and the heat carried off into the chimney gases by this constituent would be doubled. The total heat carried off would then be 500 × (4.442 + 1.450) = 2946 B.t.u. per lb. of coal.

The heat of combustion of the coal, 14,480 B.t.u., and the 10 per cent. loss by radiation remain the same. The amount of heat available for the water then becomes 14,480 — (2946 + 1448) = 10,086 B.t.u. per lb. of coal. The efficiency of the furnace becomes 10.086 ÷ 14,480 = 70 per cent., which is a loss of 5 per cent. due to this excess of air.

It is highly necessary that the engineer keep the excess of air low. The right way to handle this problem is to balance the loss due to excess air against the loss due to CO. The two are directly opposed. If the excess of air is cut too low, CO will appear. The engineer should take the analysis of the flue gases and run through the foregoing calculation. If the loss seems high, cut down the excess air until CO makes its appearance. Then determine the loss due to this CO as outlined in a previous article. Try to find a point where this loss is a minimum.

The problems outlined in this and a previous article are vital ones in any power plant. The up-to-date engineer will know how to handle them. If he does not learn, the other fellow will, and he will soon find himself in a minor position. Keep the furnace efficiency high, as the whole manufacturing world now has its eye on the power plant.

Boiler-Feeding System for Paper Mills

Paper manufacturers have experienced considerable difficulty in removing water and air from their driers, which prevents the circulation of steam through the driers and causes uneven drying. To eliminate this trouble a new system has been developed by the Farnsworth Co., Conshohocken, Penn.

A high steam velocity through the driers is effected by a forced steam circulation system which eliminates all water and air and provides a constant, even temperature through the driers. By means of a closed loop boiler-feeding system the condensation from the paper machine and all other heating apparatus, and makeup water as well, is held under pressure and returned directly into the boilers at a high temperature.

The paper machine is divided into two sections, one of these sections to have 75 per cent of the driers and the other the remaining 25 per cent. The steam and return headers between the two sections are cut, and a steam separator *A* is placed on the end of the return header for the dry end section which separates the condensation from the steam which has blown out into this portion of the return header. This steam is passed over to supply the steam header for the remaining driers on the wet end. On the end of the return header for the wet end driers is placed at *B* a duplex condensation pump, condenser vacuum type. It has cold-water sprays in the top of the tank for condensing the vapors in the return line, producing a forced-steam circulation.

If insufficient steam should pass through the steam separator to maintain the required pressure in the wet-end section, steam is bypassed through a reducing valve *D* and the proper amount is supplied.

The condensation which is collected in the simplex condensation pump *C* and in the duplex condensation pump *B*, is pumped automatically to the receiving chamber of the duplex boiler feeder *E*.

The condensation from all high-pressure traps is discharged directly into the line leading to the receiving chamber of the boiler feeder *E*. The condensation from all low-pressure heating systems, fan coils, etc., is drained into the pumps *B* or *C* and is forced up to the receiving chamber of the boiler feeder *E* by the application of live steam on the surface of the condensate in the tank while it is in the receiving position.

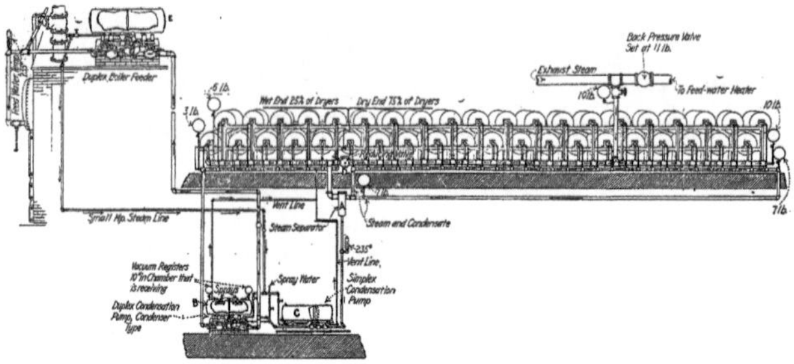

SHOWING ARRANGEMENT OF PIPING TO PAPER-MILL DRIERS AND BOILER FEEDERS

EDITORIALS

On a Paying Basis

IT will be generally conceded that many power plants are operated less efficiently than might be expected. As to the causes, the one fault that really covers all the others is careless operation, and more often than otherwise this originates with the owner or management. Too often the management does not realize the importance of the service rendered by the power plant and has not grasped the idea that it is not merely a necessary evil that must be endured, but is in a sense a manufactory taking coal and water as raw material and turning out a finished product or service that is necessary to the establishment served by the plant and must be purchased elsewhere, usually at a higher cost, if not made on the premises. Think what would happen if the power plant in a hotel failed to function. There would be no lights, no heat, no elevator service, no hot or cold water, and every department in the hotel would be paralyzed. Purchasing the service from an outside source does not help matters, for it only shifts the responsibility to others over whom there is no control.

As the plant manufactures various products, it should be considered as a revenue-producing department worthy of more attention than it usually receives, and should be run on the same general principles as other departments. The same attention to checking up results should be given to power-plant performances as to restaurant operation. It is perhaps more difficult to check results in the plants than in the kitchen or storeroom, for measurements must be made of intangible substances by the use of suitable instruments. Some of these are simple and inexpensive, while others are complex and cost considerable money. These instruments should be considered as observers more accurate than a human could be, and their wage is the interest and depreciation on the amount they cost. Looked at in this way, they are not really expensive compared with the wages of a man, who at best could only guess at what was going on. The wage of this man should pay the interest cost on several thousand dollars' worth of recording instruments.

Use of these instruments and the readings they give will show whether the plant equipment is operated, or is capable of being operated, at a fair efficiency; or the indications may be that different apparatus should be used or that different methods of operation should be adopted. Without a reliable meter there is no way of knowing how much steam is being generated, and if the coal is not weighed as it is fired, the evaporation of water per unit of coal cannot be checked. Analysis of the flue gases and their temperatures reveals the reasons for the high efficiency obtained or the lack of it. Temperature readings and water measurements show the amount of refrigeration done. Indicator diagrams show plainly whether steam engines and pumps are wasting steam. Electrical instruments are required to check up the current output and energy consumption of the various devices operated by electricity. Given the necessary means of determining the points mentioned, equipment adapted to the work and a demand from the management for the reports of operation, a credible efficiency must result.

It has been said that one of the requirements for economical operation is machinery adapted to the work. Add to this a place adapted to the machinery. Too often the architect and others allow inadequate space, and it becomes necessary to install apparatus that is adapted to the space rather than the service. The logical method is first to determine the service requirements, then select equipment best suited to the work and provide a proper space for it. Any mistake made in this respect is hard to remedy.

In one of the largest buildings in Chicago plans were completed and the building well under way before it was discovered that no space had been allowed for boilers. Space was found and the boilers were installed in such a location that the plant is permanently handicapped and the mistake is evident every time a coal bill is presented. It is important to arrange the equipment so that it may be cared for easily, because then it will be best cared for. This is important, for unless constant attention is given and repairs promptly made, a well-designed plant will degenerate and give poor results.

Summarizing the requirements, it is essential to give the engineer the proper equipment to work with, locate it in suitable space, and check up his records. The plant will then show a percentage of profit that will compare favorably with any other branch of the business.

Co-operation Between Educational Institution and Industry

REFERENCE was made in a recent issue of *Power* to the new plan, devised by the Massachusetts Institute of Technology, for coöperation with the industries. As further details of the plan are made public, its importance, not only to the Institute but to all other technical schools and to the manufacturers of the country, becomes more and more obvious.

One of the big handicaps of the engineering schools has been the difficulty of maintaining close contact between the instructing staff and the industrial world. Industry's problems have reached the schools slowly and often only after they had ceased to be problems. On the other hand, research work inside the colleges was often devoted to problems which, while of value, were far from helping the immediate needs of industry. This same lack of intimacy was, of course, reflected in the type of education provided and was responsible for many of the, often unjust, criticisms of the technical man.

Another great difficulty confronting the colleges is the financial problem. Much has been said of the difficulty of keeping the best men on the college faculties at the salaries the institutions could afford to pay. It was the serious financial situation at the Institute which led to the proposal of the "Technology Plan."

This plan provides for a contract signed by a manufacturing concern and the Institute. A definite sum, annually, for a certain number of years is agreed upon, for which the Institute agrees to place its staff, equipment and libraries at the disposal of the manufacturer for advice and aid in the solution of his technical problems. A new division of industrial service has been formed and placed under the direction of Prof. William

H. Walker, head of the Research Laboratory of Applied Chemistry.

The success of the plan seems assured. Already over one hundred and fifty contracts, aggregating more than a million dollars, have been signed. The list includes many of the leading manufacturing concerns in the country.

On the side of the Institute the selection of Doctor Walker as director of this part of the work assures a capable administration. He is probably best known to the public as the man who, as a colonel during the war, successfully commanded the Edgewood Arsenal and developed it into the largest poison-gas plant in existence.

Can this idea, brought forward to meet a crisis in a single institution, be broadened to apply to others? Undoubtedly it can.

The larger manufacturers have long since recognized the need for research and have met it with their expensive laboratories and high-salaried investigators. The smaller manufacturer sees the need, but cannot afford extensive investigations. Under this new plan he will be able to turn to the technical school and will have access to men, apparatus and records he could never hope to make wholly his own, but which can serve him as fully as the need requires. On the other hand, the large corporation will probably retain its private laboratory, but will supplement it by that of the university.

This plan, in addition to assuring the technical school of an adequate income, also affords it an excellent means of keeping in touch with the outside world and bridging the gulf between education and industry.

The Year 1919 Encouraging Regarding Power Costs

WITH coal, labor and other materials, except capital, that enter into the production of power soaring to undreamed of prices, it is believed by many that the cost of power must increase considerably above prewar rates. Already increases have been granted in some cases to help meet the higher operating expenses of central stations. Just what the future holds in regard to the rate per kilowatt-hour cannot be predicted at this time. However, the results of the last year hold forth some encouragement. An investigation made by the *Electrical World* (the findings of which were published in its January thirty-first issue) to obtain an accurate idea of the distribution of the gross income of central stations between manufacturing and distribution expenses and income available to pay interest and dividends, shows that after a gradual decrease in net income for nearly ten years, reaching its lowest in 1918, in the last year the net income per unit has increased twelve per cent. Increased output, higher rates and a more efficient use of coal have no doubt had a big influence in this increase. Although the total effect of increasing the wages of central-station company employees is not reflected in this figure, the outlook is nevertheless encouraging, and gives promise that the cost of power may still be maintained somewhere near prewar figures.

In this investigation steam plants were divided into two classes—those of twenty-five hundred kilowatts capacity and over, and those under twenty-five hundred kilowatts. Where, in 1919, thirty-five per cent of the gross income of stations of twenty-five hundred kilowatts and over was available for paying dividends and interest on indebtedness, the net income decreased to only thirteen per cent in stations of less capacity. Apparently, this is a strong argument against the small independent central station, and undoubtedly it favors the concentration of central-station power generation in large efficient units. This, however, should not be confused with the independent industrial plant, since the former, besides the cost of producing power at the switchboard, has the additional cost of doing business, which the latter does not have.

Keeping Out of the Rut

THE letter "Keeping Out of the Rut," by Thomas M. Gray, published in this issue, is worthy of serious consideration by every *Power* reader. What Mr. Gray has encountered has been the experience of many others; they have found themselves growing stale on the routine work of their jobs. This is one of the most critical periods of an engineer's career; his future depends very largely upon the way in which he meets this condition.

If he takes the course chosen by Mr. Gray and searches out the cause of his negligence in his work and then applies some remedy, his success is assured. On the other hand, if he allows himself to drift along with the current, he will soon find that his services around a power plant are not required.

In every power plant a considerable part of the work must of necessity be routine; nevertheless, in every plant, whatever the job may be, there is room for initiative and resourcefulness to a degree that will give opportunity to exercise the skill and ingenuity of anyone who will take full advantage of his work. In keeping power-plant records, it is one thing to record the readings of the various instruments and events about the plant and a far different thing to correctly analyze and interpret them; and unless those who are entrusted with the keeping of these records are capable of analyzing them, they are surely not making the best of their position.

In the industrial power plant there are so many conditions affecting economic operations that the engineer who is looking for an opportunity to develop his engineering talent can always find it. Not only does such study make the work more interesting, but also, as Mr. Gray discovered, it leads to talking the language of the chief and opens the road to advancement. Mr. Gray has apparently struck a keynote that should appeal to every operating engineer. No doubt the experiences of many other engineers, of how they have made their work more more interesting, would be instructive to others, and *Power* correspondence columns are open for a discussion of the subject.

Give the youngster a chance. Besides running errands, helping with the ashes and grooming the engine and boiler, let him go through all the motions of starting and stopping your engine. He may not do everything exactly right the first time, but don't forget that if he had your experience he would be holding your job, or perhaps a better one. There are numerous instances where men have grown up in power plants and for want of a little encouragement always remained too timid to ask for instruction. Looking back on your early experience, you must remember your own shortcomings, your ambitions and the reasons for your advancement. Undoubtedly, you cherish and respect, more than all else, the tips received from your superior. It will be pleasant to contemplate that at some future time you may look back with the satisfaction that you did as you would be done by.

CORRESPONDENCE

Why Does Water Lift When the Safety Valve Blows?

The lifting of water in a boiler when a safety valve blows is due to the expansive force of the steam that is released from the body of water in the boiler when the pressure is reduced by the opening of the safety valve. The dry steam in the steam space rushes to the valve opening ahead of the pent up steam bubble in the water, otherwise water would come out with the steam whenever the safety valve lifted.

When a boiler is operating under normal condition, there are several elements working harmoniously, but when the valve lifts there is a sudden interference and the water is driven into the current of steam flowing through the safety valve. The steam bubble pent up within the body of water can be compared to tiny compressed coiled springs submerged in a body of water, which, when released, would throw the water upward.

Water contained in a steam boiler has a tremendous force stored up ready to act when the right condition prevails, which is brought about when the safety valve suddenly releases the pressure. The operation of a coffee percolator gives a good illustration of the action of heat within a body of water. The expansive force of steam generated in the water space forces the water to the top of the small tube when the steam expands and throws water to the top of the percolator.

The remedy for lifting water is to provide as large a liberating surface and steam space as is practical. The place to separate water from the steam is in the boiler if it is possible. Among the devices employed are perforated or slotted pipes, and baffle plates, the latter being the most satisfactory and effective device. It prevents the straight lifting of the water under the steam opening and also catches the water that may creep up the sides of the shell. It is easily attached to the boiler, is readily inspected and kept in proper condition, and is not expensive to attach.

An important factor concerning this question is the size of the safety valve and the lift. If a boiler requires a large safety valve in order to relieve it of all the excess steam it is capable of making, two or more valves of small size should be used. Experience shows that a 4-in. valve is not too large.

Sioux City, Iowa. J. M. SHEA.

Locating Defective Field Coils

There are several ways of using a telephone receiver instead of a millivoltmeter to detect and locate faults other than those in an armature. As an example, some time ago, in an isolated plant operating a mine, a 150-kw. 550-volt, six-pole direct-current generator failed to maintain its voltage on full load. The speed, shunt excitation, etc., were the same as before the trouble. It was decided the difficulty was in the compound winding or probably part of a shunt-field short-circuited.

A set of batteries and a buzzer were taken from a Ford car, and a telephone receiver was taken off a near-by telephone. The buzzer and batteries were connected in series with the compound winding, and each series coil was bridged with the receiver, as in the figure. One failed to make the receiver vibrate, indicating that the coil was short-circuited. The shunt-field coils were also tested in the same manner, and it was found that a defective shunt coil had short-circuited part of the series coil that was on the same polepiece.

W. A. DARTER.
Brenham, Tex.

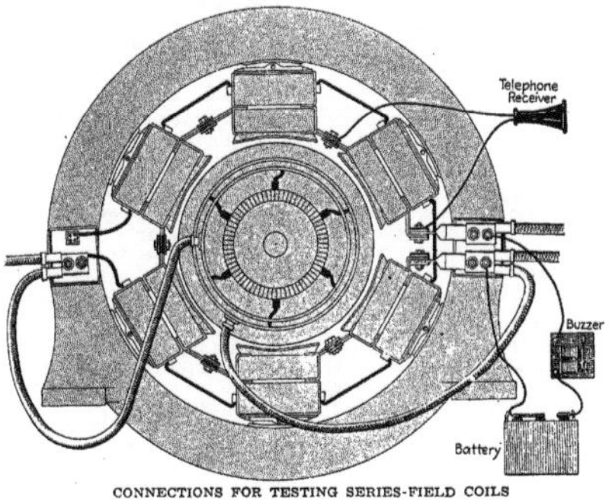

CONNECTIONS FOR TESTING SERIES-FIELD COILS

Scale Formed in Safety Valve

In the Sept. 16, 1919, issue, page 477, Thomas M. Gray relates his experience with scale forming in safety valves. This letter called to my mind trouble experienced along similar lines from the safety valves scaling on a high-pressure economizer installation.

In this case the water was forced through the economizer under a pressure of 160 lb. and from the economizer was fed to the boilers. No pressure regulators were used on the pumps feeding the economizer, and the feed-water regulation to the boilers was done by hand, the firemen looking after the water on their boilers. The ten boilers were of the vertical water-tube type, each of 300-hp. capacity. The load was variable and at times the plant was run at 225 per cent. rating. The stokers, forced-draft underfeed type, could maintain this rating continuously for a number of hours, and as the peak loads were frequent, dropping to less than 100 per cent. rating, water regulation by hand was far from satisfactory.

With poor regulation of the pumps and water to the boilers, the pressure on the economizer varied frequently, hence the entire feed-water lines depended for safety on the proper functioning of the four economizer safety valves, each 3-in. diameter connections, and the valves for a long time performed this work well.

The valves were placed on the discharge headers of the economizer and a 3-in. common drain header was run from the 3-in. discharge openings of each valve and a ½-in. drain line was run from the ½-in. drip of valve as shown by Fig. 1 into a 3-in. header at the tee. The waste water was run into the blowoff drain leading to the sewer.

The water used at this plant was drawn from a river seldom clear of mud. Even at its best it contained much incrusting materials, and as the economizer was really the first stage of heating the water and settling before it reached the boilers, it was not surprising that after six months' service one section of tubes and headers

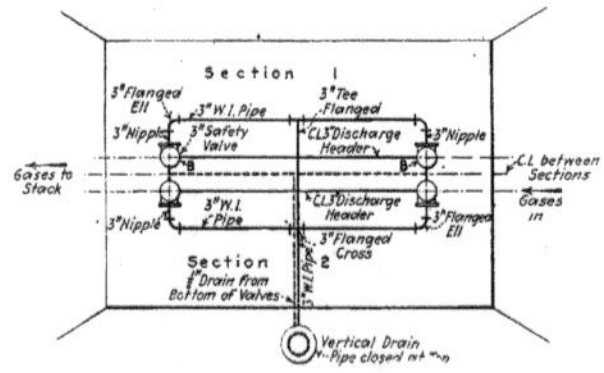

FIG. 1. ORIGINAL PIPING OF WATER CONNECTIONS

gave way, owing to a failure of the safety valve to release the excess pressure on that section of the economizer.

When the accident occurred, this economizer was cut out of service and inspected. It was found that hard scale had built up considerably in all tubes and that the two safety valves on this section (*A* and *B*, respectively) had become practically solid with scale around the seats and springs, and the discharge pipe was at least two-thirds full of scale.

After cleaning the section, making repairs to the headers and overhauling the safety valves, the section was filled and cut into service. Section 2 was then overhauled.

In order to remedy the trouble as far as the safety valves were concerned, I had the drain and discharge piping changed, as shown in Fig. 2, and run in such a manner that inspection could be readily made every day. Furthermore, a system of regularly cleaning the economizer was followed and safety valves were removed for cleaning and internal inspection not less than once in three months, as it was practically demonstrated that the safe limit of operation had been reached in that

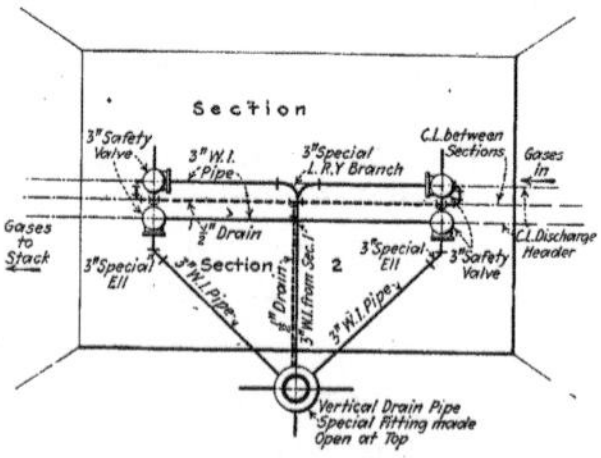

FIG. 2. NEW PIPING OF DRAIN PIPES

time under the existing operating condition. A further improvement was made by placing regulators on the feed pumps, so that excess pressure and undue lifting of the safety valves were stopped. After applying regulators to the pumps, the safety valves were lifted daily by hand to insure operation and to see that they were tight.

In this plant the twin safety valves on all boilers were allowed to blow through straight pipes run out through the side of the boiler house instead of running the safety valve discharge in a vertical line through the roof. This precaution was taken to prevent similar trouble, as explained by Mr. Gray. PAUL R. DUFFEY.

Hagerstown, Md.

Keeping Out of the Rut

Readers of *Power* have lately exhibited an interest in meter readings. Several good examples have been offered for solution. Would the following comments be of any interest on the subject?

Given a specific problem in meter reading, no matter how complicated, and in the majority of cases the answer will be made correctly by the experienced meter reader. That, however, is no test of the ability of the individual. The real test comes when one has to read meters daily. It would be surprising to note the number of mistakes that will crop up. One may ask why? There are two reasons: First, the human element, and secondly, the carelessness that is due to routine nature of the work.

Mistakes are bound to happen in a power plant. However conscientious one may be in doing his work, sooner or later something goes wrong. This cannot be regarded as intentional; it is simply the human equation and in the majority of cases cannot be avoided. Carelessness may be due to lack of interest or to the fact that the work has become routine. Lack of interest in the work is inexcusable and ought not to be allowed under any circumstance. The person is at fault and the remedy is self-evident. Now take the case where the work becomes routine. The monotony of reading meters day in and day out has driven more than one good man out of the business. There has to be an incentive, and no one can supply it except the man himself. Something new has to be injected into his work. The following tells my own experience, which may be of interest:

One of my duties when working in a central station was to get the midnight readings of the different meters in the plant. There were not many instruments, but after awhile errors cropped up and the matter would be called to my attention by the chief. For a while things would go smoothly, and then again errors would occur. That duty was getting to be routine. No matter how hard I tried, there were times when the meters would get barely a glance and the reading would be jotted on the daily log. It made things look bad for me, and it was not pleasing to the powers above. Who can blame them?

When it becomes necessary for the chief to check up the work of one his subordinates, the latter is going to fall very shortly. I faced that dilemma and planned how to remedy the fault. There was one way, namely, to keep one's own records in order that daily readings could be checked up. From simply checking up daily readings it was a slight step until these records included water rate, coal and other cost data. In a short time I had a new incentive. These records showed another side of the operation of the plant, which proved interesting and instructive. I was brought closer in touch with the chief, as I could talk the same language. The records were kept for a number of years, and although this meant a lot of extra work, I consider that it was worth while. THOMAS M. GRAY.

Bayonne, N. J.

Remedy for Present Valve Setting

The indicator cards submitted by S. L. Gilliam, in the Sept. 23 issue of *Power*, from an 18 x 30 double-eccentric engine, seem to show that the head end is taking steam too early. It is hard work to prescribe for an engine from a distance, but from the information at hand I would proceed as follows:

First, make sure that the eccentric rod and reach rod are properly adjusted to give equal wristplate travel on

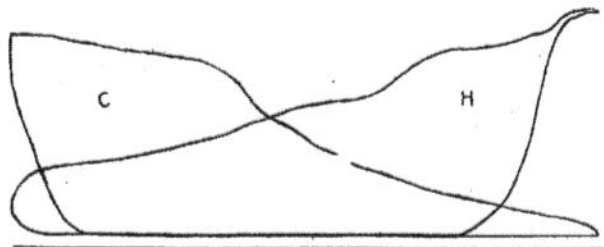

FIG. 1. DIAGRAM FROM ENGINE AS FOUND

both sides of center, then put the engine on head center and move the steam eccentric back till, with the head-end link-rod threads about halfway out, the valve is open about $\frac{1}{16}$ in. Lock the eccentric, then put the engine on crank-end center and check up the steam valve on this end, giving it the same lead. An indicator card should now show whether the load is equal on both ends of the cylinder, also whether admission is sufficiently early.

If the card shows improvement but still does not suit, change the adjustment slightly to give the desired effect. The fact that new valves have been fitted may be the cause of the trouble, so make sure that the marks on the ends of the valves correspond with the valve edges, also that the edges of the ports line up with the marks on the ends of the valve chambers.

I would not disturb the exhaust-valve setting unless necessary for quiet running, although the head-end ex-

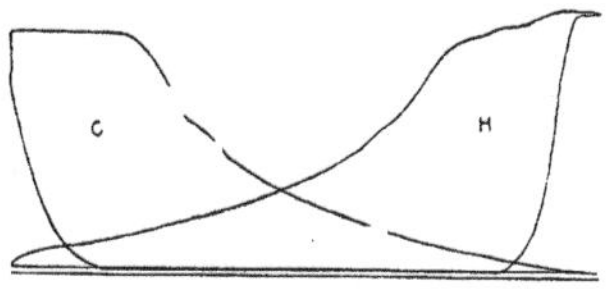

FIG. 2. DIAGRAM FROM ENGINE AS CHANGED

haust valve would stand a trifle earlier opening.

Syracuse, N. Y. A. H. BULLARD.

[The cards referred to in this letter are reproduced herewith as Fig. 1 and Fig. 2.—Editor.]

In any double-eccentric gear the exhaust valves operate the same as with a single-eccentric, with the exception that better results can be obtained by using the best setting for any particular case regarding release and compression percentages. The exhaust lines on Mr. Gilliam's diagrams are not entirely correct or equalized, but the valve setting, although deranged, is not enough so to cause serious trouble. The steam valves and the steam eccentric need adjustment, and there seems to be excessive leakage on the steam valves.

Double-eccentric steam gears differ considerably from single-eccentric gears. With the wristplate on the center the valves are open, but to what extent depends on the design of the valves, the valve travel and the range of cutoff to be obtained. The most important question is how much seal, or lap, the valves have when they are on the extreme closed position. In multiported valves for high-speed Corliss engines the seal is never very liberal and there is danger of the valve being partially open when in the extreme closing position.

Possibly when new valves were supplied for Mr. Gilliam's engine they were made wrong and do not match the ports. It is evident from the diagram that the head-end steam valve is open all the time, for when the exhaust valve closes the pressure rapidly rises, and on the steam line and expansion curve the evidence is that the steam continues to pass into the cylinder throughout the stroke. A sharp reversal of the pressure after exhaust closes causes the head end to pound. The crank end has excessive lead, and the expansion curve at the crank end indicates a leaky exhaust valve or piston.

Referring to the head-end diagram, why is it so much higher than the crank-end diagram and why does not the steam line rise above the admission line, showing the discharge of steam from the cylinder to the steam chest? It is possibly due to leaky exhaust on this end. It is possible that when the crank-end steam valve is open for admission and the head-end valve is open or leaking severely, the admission pressure on the crank end is reduced by a lowering pressure in the steam chest. F. JONES.

Cincinnati, Ohio.

Care of Pipe-Line Valves

When valves become hard to open and close, some engineers use a wrench or bar without any thought of locating and correcting the defect. In the first place, the packing should be kept lubricated and should be renewed often enough to prevent leakage. Many high-pressure gate valves can be packed when wide open and under pressure.

The stems of rising-stem valves should be given an application of graphite and cylinder oil at regular intervals. If this attention does not correct the trouble, the stem may be binding at the threads. To overcome this, the threads should be given an application of valve-grinding compound, and the valve should be opened and closed a few times. The grinding compound should then be cleaned off with kerosene.

Gate valves sometimes work hard on account of binding between the wheel hub or nut and the yoke. The nut should be removed from the yoke and cleaned and, if necessary, dressed off with a file. Before being replaced, the nut should be coated with graphite and cup grease. For positive assurance against the trouble occurring again, drill an oil hole in the yoke or put on a grease cup. The engineer who keeps the valves in his care in good condition will have less hard work and will have made life and property safer in his plant.

Providence, R. I. C. P. Lawton.

Maximum Length for Longitudinal Joints

It is with temerity that one would venture to criticize an opinion on boiler matters when given by such an eminent authority as the chief engineer of the Manchester Steam Users' Association. Steam engineers in America have felt greatly indebted to Mr. Stromeyer for the good work he has done in connection with boiler design and construction. However, some of the statements he makes in the article on page 626 of the Oct. 21, 1919, issue of *Power* would seem to need further explanation unless they are to be questioned by many practical engineers.

It may be true, as is stated, that the stress in the solid plate of an adjoining course near the end of a longitudinal joint is eight times that in the solid plate away from the joint, so long as the maximum stress concerned is within the elastic limit of the material. But it would seem to be evident from practice that this ratio between the stresses at the two points mentioned must undergo some marked change before the yield point or even the elastic limit of the material is reached.

It has been customary in this country to build nearly all shell boilers of any size in two or more courses. These boilers have generally operated at calculated factors of safety of from 4 to 5, with quite a few at a less factor than 4. Such safety factors, so far as the shell is concerned, are based on the calculated stresses assumed to exist in the longitudinal joints.

If the relation between the stresses at these joints and in the solid plate and opposite their ends were such as Mr. Stromeyer's test would indicate, the metal at the highest stressed points would fail. It would also appear from practical experience with boilers that the solid plate at the points specified cannot be stressed to or in any event beyond the elastic limit. The reason for making this statement is that wherever a portion of a boiler structure is excessively stressed, grooving or other form of corrosive action generally occurs. As far as the writer's own experience has gone, or from the experience of many other inspectors with whom he has been associated, he has never heard of any particular tendency to corrode at the location in question. From the behavior of boiler shells under hydrostatic test it would be surprising if a shell built in courses did not fail first at a longitudinal seam near the center of a course when placed under sufficient pressure to destroy it. Granted that this is so, it would appear that at some time between the first application of the pressure and the final rupture of the cylinder, the point of maximum stress would have to shift to the joint.

It is noted that reference is made to an unpublished destructive test experiment made by the Manchester Steam Users' Association on a full-sized boiler in the year 1874, the solid plate rupturing at the same time as the longitudinal seam failed, if the writer has properly interpreted the statement. Surely, Mr. Stromeyer does not intend to take the result of a single test of this kind to establish his theory! It is true that in boiler explosions the fractures—aside from the initial rupture—do not follow the lines of the greatest weakness, but generally occur in the solid plate. This result, however, has been assumed to be due to the fact that aside from the initial rupture the other breaks in the structure are supposed to be caused by water-hammer action produced by the rapid generation of steam throughout the mass of water, which is supposed to occur upon the sudden release of pressure.

It is stated in the article that no boiler with a factor of safety of 5 has ever exploded. This certainly cannot be true in the United States, and it would be interesting to learn what factors of safety existed in the few cases where boilers have exploded in England.

From Mr. Stromeyer's statement about the superiority of acid steel over that made by the basic process, it would seem high time that the laws and rules governing boiler construction in this country were being modified. If prohibiting the use of basic steel and the application of a safety factor of 5 in the construction of boilers will absolutely prevent boiler explosions, it is time we were getting on that basis. The safety factor of 5 is an accomplished fact now. It would, therefore, seem proper to use the right material and secure the same nearly 100 per cent immunity from boiler explosions that our English cousins enjoy.

As to the address which is referred to as read by an American engineer who qualifies as a boiler expert, owing to the fact that he has been present at a dozen boiler explosions and has experienced as many hair-breadth escapes, Mr. Stromeyer can rest assured that an engineer with such experience would be just as much of a curiosity in America as he would be in England. There have been a few explosions staged in this country for experimental purposes and some wrecks produced for movie thrills, but the audience in such cases has been amply protected. While it is possibly not polite to suggest it, it may be that this American was trying to put over a joke that his English audience would appreciate and made it a little too obvious.

It would seem to be futile to try to compare the number of boiler explosions in England and the United States without some idea as to the total number of boilers in operation in the two countries. However, it is likely true that there are fewer explosions per thousand boilers in England than in the United States. While better construction, which certainly must exist in the case of the average English boiler, may account in a measure for such differences, it would seem most likely that operating conditions were largely responsible for many of the explosions in this country. The lack of intelligent attention in the case of many boilers that are being operated in this country is appalling. No doubt the same may be said of many English plants.

Hartford, Conn. S. F. Jeter.

Averaging Pressure Charts

On page 669 of the Oct. 28-Nov. 4, 1919, issue of *Power*, a method of averaging circular charts by the polar planimeter is related by Mr. White. This method consists in finding the area of such a chart in square inches and then determining the radius of a circle of equal area. This radius is laid off on the chart from the center to locate the average chart reading.

The method is only approximate, by the following reasoning: Let

r_0 = Correct average radius of chart in inches.
r_1, r_2, r_3, etc. = Various radii at different times.
n = Number of radii taken to get an average.
r = Average radius as obtained by polar planimeter.

Now, obviously the correct average radius is

$$r_0 = \frac{r_1 + r_2 + r_3 + \ldots . r_n}{n}.$$

And

$$\text{Area of chart} = \pi \left[\frac{r_1^2}{n} + \frac{r_2^2}{n} + \frac{r_3^2}{n} + \ldots . \quad n \right]$$

By the method described by Mr. White,

$$\text{Area of chart} = \pi r^2$$

Comparing the last two equations,

$$r = \sqrt{\frac{(r_1^2 + r_2^2 + r_3^2 + \ldots . r_n^2)}{n}}$$

and is not equal to r_0 as given in the first equation. In other words, this method gives the square root of the mean of the squares of the radii instead of the arithmetical mean.

The method is good, however, as an approximation. It should be observed that the less the variation of the charted quantity the more nearly true is the result by the approximate method. If the variation is very wide, the method should not be used.

On page 260 of the Aug. 19, 1913, issue of *Power*, the writer gave some figures to show the percentage of error involved when getting mean values by the polar planimeter according to the system outlined by Mr. White.

JULIAN C. SMALLWOOD.
Baltimore, Md.

Boiler-Room Essentials

In the interesting article by H. F. Gauss, entitled "Boiler-Room Essentials," page 589, of the Nov. 22, 1919, issue of *Power*, the author made some statements about CO_2 equipment which lead me to suspect that he has not had experience with at least one of the efficient modern types of CO_2 recorders. Perhaps his experience was with early models, which admittedly do not compare with recent instruments in simplicity of operation and general satisfactory performance. Mr. Gauss appears to be laboring under the mistaken impression that CO_2 recorders involve extra labor, whereas my experience has been that, next to fuel saving, their principal advantage is in reducing the labor of the engineer and fireman, thus giving them more time to devote to other duties.

Mr. Gauss says: "In many cases the hand Orsat is more desirable than the CO_2 recorder, when the engineer on watch is required to make gas analyses at least once an hour and is required to see that these analyses show reasonable percentages of CO_2. That is, if he makes an analysis and finds the CO_2 low, he must at once determine the cause, have it corrected and secure the normal analysis. In large plants operating 10 to 20 boilers, this is not practicable and the recording apparatus must be resorted to with an expert devoting his entire time to the instruments in the boiler room."

Evidently, Mr. Gauss believes that the engineer should take CO_2 readings every hour when using an Orsat. At that rate there would be little time left to devote to other duties; in fact, taking readings so frequently would almost seem to require one man to devote his whole time to flue-gas analysis. Even then the time elapsing between the withdrawal of the gas sample and the completion of the analysis would prevent the engineer from correlating the information furnished by the analysis with the operating condition of the furnace. For instance, the CO_2 in the exhaust gases might be only 7 per cent at the instant that the gas is withdrawn and 12 per cent a few minutes later when the analysis is completed. A CO_2 recorder, on the other hand, furnishes the information continuously, thus enabling the fireman and the engineer to coöperate in keeping the percentage of CO_2 high all the time. The CO_2 recorder, working every second of the day and night, analyzes the gases better and cheaper than the engineer can do it, working in hourly periods.

The actual time required for attending to the CO_2 recording instruments with which I am familiar amounts to three or four minutes daily per unit, and less than one-half hour once a week for changing absorbent cartons and checking readings. The salary of a man required to attend an Orsat would pay for the installation and operation of an automatic recorder several times over in the course of a year.

Mr. Gauss states further that CO_2 recording equipment "must be resorted to" in large plants operating ten to twenty boilers, in which case an expert must be engaged to devote his entire time to the instruments. This statement is inaccurate so far as the successful types of instruments are concerned. Modern recorders are sturdy in construction and are no more difficult to use and keep in order than other types of boiler-room equipment. Ordinarily, they operate for years without other attention than occasional cleaning, say once or twice a year, and any mechanic can attend to this by following simple instructions. Far from requiring an expert, the successful modern types of CO_2 equipment are designed to avoid the necessity of expert attention. Any man who can understand and run a steam engine and its auxiliaries should have no difficulty with CO_2 equipment.

Mr. Gauss quite rightly approves of the use of the recording pyrometer. However, as the percentage of CO_2 fluctuates much more violently than the flue-gas temperature, the argument in favor of using recording equipment is even stronger in the case of CO_2 equipment than in the case of pyrometers.

C. J. SCHMID.
New York City.

Difference in Heating Systems

In answer to Mr. Molloy's letter on page 716 of the Nov. 11-18, 1919, issue of *Power* in regard to my heating-system letter in the June 3 issue, I will say that the radiators were controlled by a thermostat, which operated the radiator valves by compressed air. The vacuum was held steady after the building was once warmed up and the thermostats were operating.

The reason the pumps were set so high above the return main of the system was for show purposes only. This plant is used to furnish power, heat and light to a large manual-training school. Part of the course in training is operating engineering, and these pumps were put on the engine-room floor just for looks. This was a big mistake, which has been admitted since.

For the benefit of Mr. Downham I wish to say that I'll agree that the remedy would be to place the pumps in basement below the return main.

L. B. SHIELDS.
Chicago, Ill.

Size of Pipe for Boiler Feed

I have had considerable experience with boiler-feed lines in connection with the installation of a certain type of boiler-feed regulator, and it has been my experience that it is the usual custom to keep the velocity of water in the feed pipes below 200 ft. per min.

Answering Mr. Craft's question, page 781 of the Nov. 25, Dec. 2-9, 1919, issue, if a 4-in. pipe is used, a discharge of 60,000 lb. per hour will be equivalent to a velocity of 183 ft. per min., which is within the required allowance. If the pipe line is short and does not contain many fittings, it is possible that even a 3-in. pipe with a velocity of 325 ft. per min. might be used. This would be especially true if feed-water regulators having constant feed characteristics were used.

If an intermittent type of feed regulator or hand feeding is used, there are likely to be times when the feed is greatly in excess of the average, and the friction loss with the pipe, which will then be too small, will be considerable. Since this friction loss increases as the square of the velocity, it is evident that the discharge through the feed line should be kept as near as possible to the average. The best way is to install a regulator which feeds continuously in proportion to the load, and sometimes in inverse proportion to the load, depending upon conditions. This tends to reduce the friction losses in the pipe and admits of a smaller pipe being used for supplying the boiler.

Erie, Penn. V. V. Veenschoten.

What Happens When a Synchronous Motor Is Caused to Pull Out?

I have read with pleasure the interesting and instructive papers, by Will Brown, treating of the characteristics of the synchronous motor. The analogies are very clear and should enable the reader to readily comprehend the fundamental features both of construction and operation. There is one important point, however, which Mr. Brown, as well as others, touches upon very lightly, I may say even "gingerly." In his article in *Power*, issue of Sept. 2, 1919, he says:

> The force holding together the poles of the stator and the poles of the rotor suddenly snaps like a breaking rubber band, and the rotor stops, while the magnetic poles of the stator continue on at constant speed. This is what happens when a synchronous motor is overloaded and pulls out of step, and is called the motor's pull-out point. Of course, in actual practice the circuit-breakers will generally disconnect the line from the motor before it reaches this point, because of the enormous current that the motor would be pulling from the line.

He says further in the issue of Sept. 23:

> In Fig. 4 the load has literally pulled the rotor poles away from the stator poles. As soon as the point is reached where the north pole of the stator opposes the north pole of the rotor, the pull of the line snaps like a rubber band and the rotor is brought to a dead stop by the heavy pull of the load.

Now, it is known that something more than this really happens. It has been conceded by the electrical engineers of the two largest manufacturers of electrical equipment that there is a decided buck at this point, which, for the purpose of identification, we may call the "pull-out buck."

It is acknowledged by the electrical engineers before mentioned, that under certain conditions of line and load, which may and do occur in practical operation, this pull-out buck can take place at a point inside a reasonable setting of the circuit-breaker, say 150 to 175 per cent of full-load rating; that under these conditions the circuit-breaker cannot anticipate the pull-out, and because the buck is "instantaneous," the circuit-breaker does not relieve the reactive kick until too late; that the electrical "pile up" will be from eight to ten times the rated capacity of the motor; that the mechanical reactive stress impressed upon the cores on the periphery of the rotor will be from six to eight times the rated capacity of the motor.

Now this is of very great moment if it is true, and because no one seems to know just what does take place, it is of the utmost importance that the matter should be demonstrated to a conclusion that can be depended upon. Testing devices are being developed for this purpose and we may expect definite information on these points in the near future. In the meantime let us go over the "mechanics of the thing," which we can readily do, and see what may be the result. Let us take, for example, a 600-hp. synchronous motor, with a bad line swing, low voltage, a heavy load and a flywheel out of proportion for the conditions.

Now let us apply shock by introducing a liquid or a piece of metal into the compressor cylinder, and we have conditions under which, it is conceded, we may and do get an electrical angularity which will result in the motor "pulling out" inside the setting of the circuit-breaker. We will then have a pull-out buck.

Let us take the word of these electrical engineers and assume that the electric pile-up is ten times, and the electric reaction delivered across the air gap to the rotor is eight times the rated capacity of the motor. We then have at this instant a mechanical reactive stress at the "rim of the rotor," expressed as 4,800 hp., which is transmitted to the spokes of the rotor, thence through the hub and key to the shaft, then along the shaft, through the key, hub and spokes to the rim of the flywheel, where it is opposed by the inertia here provided, plus the compressor load.

It goes without saying, of course, that the initial stresses begin in the stator and are primarily impressed upon the "feet" and holding-down bolts. Now, then, if there is a weak link in this chain, and particularly if it is in the flywheel rim or spokes, something serious is bound to take place.

It is certainly not good engineering to let this matter drift along until we have an "accident" which may result in the loss of not only dollars and cents, but also human life. In the meantime it seems to me that we should anticipate by means of proper design and construction the stresses that may be logically expected. I want to say in this connection that I consider the flywheel an element of danger as well as expense, when used in conjunction with a direct-connected synchronous motor, and that, for the reason that it is entirely unnecessary, it should be eliminated from synchronous-motor drives. This can be accomplished with entire success if proper attention is given to rotor design.

F. L. Fairbanks, Chief Engineer,
Quincy Market Cold Storage and Warehouse Co.
Boston, Mass.

Reboring Corliss Valve Chambers

I have had quite an argument regarding the proper method of reboring Corliss valve chambers. When doing this work, should not the chambers first be lined up to show how much they are worn or out of line, and then have the boring bar set so as to rebore the chambers and bring them into line with the cylinder instead of just setting up the boring bar on the bonnet studs and calipering to bring the boring bar central?

How would readers of *Power* go about finding how much the valve chambers were out of line with the cylinder, and how would they set the boring bar?

New York City. Michael J. Murphy.

INQUIRIES OF GENERAL INTEREST

Left-Hand Adjusting Screw of Monkey Wrench—Why are monkey wrenches usually provided with left-hand adjusting screws? T. W.

A left-hand adjusting screw affords greater convenience in closing up the jaws of the wrench by operation of the milled collar with the thumb, while the wrench is held in the grasp of the fingers of the right hand.

Poor Circulation in Radiator of Vacuum System—In a vacuum heating system that supplies heat for two houses, there is heating up of only about one-third of the number of sections of a radiator in the building farther removed from the boiler. What may be the cause and remedy? R. A.

The cold sections probably remain airbound due to insufficient vacuum. If it is not possible to obtain a better vacuum at the return while the steam supply is open, the air might be dislodged, sufficiently for starting operation of all sections, by exhausting the radiator with the best vacuum obtainable for ten or fifteen minutes prior to admission of steam.

Percentage Improvement in Coal Economy per Kilowatt-Hour—Our fuel consumption for October was 1,960,010 lb. of coal costing $7.44 per ton of 2,240 lb. for an electrical output of 524,270 kw.-hr. For December the fuel consumption was 1,746,445 lb. of coal costing $7.44 per ton of 2,240 lb., and 690,260 lb. of screenings costing $1.50 per ton of 2,000 lb. for an electrical output of 574,040 kw.-hr. What was the percentage of improvement in coal economy? G. A. E.

For October the fuel cost was

$$\frac{1{,}960{,}010 \times \$7.44}{2{,}240} \div 524{,}270 = \$0.01242 \textit{ per kw.-hr.}$$

For December the fuel cost was

$$\left[\frac{1{,}746{,}445 \times \$7.44}{2{,}240} + \frac{690{,}260 \times \$1.50}{2{,}000}\right] \div 574{,}040 = \$0.01101 \textit{ per kw.-hr.}$$

Hence for December the improvement in coal economy over that for October was

$$\frac{(\$0.01242 - \$0.01101) \times 100}{0.01242} = 11.35 \textit{ per cent.}$$

Direction of Setting Globe Valve—What is the proper way to place a globe valve, to have the pressure under the disk or on top of the disk? H. P.

For most situations, especially for large globe valves or for high pressures, it is preferable to have the pressure under the disk, as that method permits packing the valve when closed and prevents the accident of the valve stem pulling out of the disk when the valve is opened under pressure. Another advantage of having the pressure under the disk is that the direction of flow results in less abrasion of the valve seat. The principal advantages of having the pressure on top of the disk are that the pressure holds the disk to its seat; also that when the line is allowed to cool down with the valve closed, the lost motion between the disk and the spindle permits the disk to follow up the contraction of the valve body and when steam is again turned into the line the valve is as tight as before, whereas with steam reintroduced on the seat side of the valve, there is leakage until the valve has become reheated and reëxpanded to its former dimensions, and to close the valve down tight before it is warmed up makes it harder to open after the expansion has taken place.

Reactive Factor—What is the reactive factor of an alternating-current system, how is it determined, and what relation has it to the power factor? M. O. W.

The reactive factor of an alternating-current circuit is the sine of the angle of lag or lead between the current and voltage. If the power factor is read from an instrument or calculated by dividing the watts by the volt-amperes, reference can be made to a table of natural sines and cosines and find the sine of an angle corresponding to the cosine that is the power factor. For example, if the power factor is 0.86, this is the cosine of a 30-deg. angle and the sine of 30 deg. is 0.5, therefore the reactive factor is 0.5 when the current is out of step with the volts by 30 degrees. When the current and volts are in phase, the power factor is 1.00 where the reactive factor is 0. Another way of determining the reactive factor is by the formula:

$$\textit{Reactive factor} = \sqrt{1 - \left[\frac{\textit{Watts}}{\textit{Volt-Amperes}}\right]^2}$$

The volt-amperes times the power factor equals true watts, where the volt-amperes times reactive factor equals the reactive component of the volt-amperes or, the volt-amperes that is being transmitted back and forth through the circuit doing no useful work.

Determining Size of Feed Pump—What size feed pump should be supplied for a 100-hp. boiler? E. A.

A standard boiler horsepower is the equivalent evaporation of 34.5 lb. of water from and at 212 deg. F. per hour, and as the actual conditions usually are such that the equivalent transfer of heat is accomplished by evaporation of a less quantity of water per hour, under normal conditions it will be on the side of safety to assume feed-pump capacity on the basis of 34.5 lb. But to cover operation of a boiler 50 per cent beyond its rated capacity, and extraordinary demands in case of low water or other emergencies, the rated capacity of the pump should be not less than 90 lb. of water per hour per rated horsepower of the boiler, and the pump capacity should be rated at no greater piston speed than about 50 ft. per minute.

At ordinary temperatures a cubic foot of water weighs 62.3 lb., so that a boiler-feed pump should have a displacement capacity of $\frac{1728}{62.3} \times \frac{90}{60}$ = about 42 cu.in. per min. per rated boiler horsepower capacity, and the pump piston speed of 50 ft., or 600 in. per min., would require $\frac{42}{600} = 0.07$ sq.in. of water-piston area per rated boiler horsepower. Hence, for a 100-hp. boiler the required area would be $100 \times 0.07 = 7$ sq.in., and neglecting the area occupied by the piston rod this corresponds to $\sqrt{\frac{7}{0.7854}} = 2.98$, or practically 3 in. diameter for a single pump, and $7 \div 2 = 3.5$ sq.in., which corresponds to $\sqrt{\frac{3.5}{0.7854}} = 2.11$, or practically 2 in. diameter for water cylinders of a duplex pump.

[Correspondents sending us inquiries should sign their communications with full names and post office addresses. This is necessary to guarantee the good faith of the communications.—Editor.]

Electrically Operated Pumping Plants in Indiana*

According to available records there are 206 public water-works systems in the State of Indiana. Seventy-six of these, more than one-third, are operated partly or wholly by electricity. During the past winter and spring twenty-seven of those plants were visited by members of the Engineering Experiment Station staff of Purdue University with a view to ascertaining the present status of electrical operation of such plants.

The plants visited supply water to communities ranging in population from 500 to 25,000 and were found to be operating under a variety of conditions. Fifteen plants used some form of elevated storage—tank, standpipe or reservoir; six were provided with compression tanks, the pressure being automatically controlled by pressure gage relays, while the remaining six had no provision for storage but maintained direct pressure by continuous operation of the pumps. This probably represents in a general way the distribution as to types of plant over the state as a whole.

About 20 per cent of the plants visited made use of deep-well pumps—air lift or displacement. The remainder used surface pumps, though in many cases the pumps were placed in pits from 4 to 15 ft. below ground level in order to reduce the suction lift.

The most commonly encountered source of water supply was the drilled well, 6 to 10 in. in diameter, and from 100 to 300 ft. deep. In a few cases the supply came from streams of spring-fed reservoirs, but these were exceptional. The water from wells appears to be, in general, clear and pure; requiring no outlay for purification equipment. In most localities the water stands from 10 to 20 ft. below ground level, but in a few places it flows to the pump under sufficient head to eliminate entirely the necessity for priming.

Centrifugal pumps and triplex displacement pumps were encountered in about equal numbers and the plant that contained neither of these types was exceptional. One rotary pump was observed, handling the second stage of a two-stage system; and the City of LaFayette has three direct-connected motor-driven propeller pumps forming the first stage of a two-stage system, the second being handled by steam pumps. These, however, are held for emergency use, the first-stage pumping being normally done by air-lift. Also in one case the first stage was handled by motor-driven deep-well pumps and the second stage by steam pumps. In 75 per cent of the plants inspected, however, the single-stage system was used.

Alternating-current energy was utilized in practically all cases, either because direct current was not available or because the alternating-current motor is cheaper, more rugged and requires less attention. Various types of alternating-current motors were found to be in use, however, the choice depending chiefly upon the type of pump to be driven. In most cases the available energy supply is 60-cycle three-phase current, and the choice lies between the squirrel-cage and the wound-rotor types of induction motor, the former being used for driving centrifugal pumps where the starting load is light, and the latter for driving displacement pumps where the starting load is relatively heavy. Exceptions were noted, however, to this general rule, and in one case synchronous motors were found giving very satisfactory service. Single-phase, 25-cycle motors were encountered in one plant, the energy being obtained from an interurban railway transmission line which passed through the town.

For domestic service the pressures in use were found to range from 25 to 90 lb. per sq.in., the average being about 40 lb., corresponding to a head of 93 ft. In many cases provision is made for raising this pressure for fire-fighting purposes to 90 or 100 lb. by cutting off the storage and bringing a special high-pressure unit into service. It is not uncommon to find steam pumps used for securing the extra pressure needed for fire service, though the motor-driven fire-service pump is popular in localities where auxiliary steam equipment would be expensive to provide and maintain. There appears to be a growing sentiment in favor of securing the extra pressure by using a portable pump at the location of the fire, rather than by boosting the pressure over the whole system, thereby making both mains and plumbing liable to damage or even complete failure at a time when such failure would be most serious. This method also eliminates the extra demand for water at service connections during the high-pressure periods when the greatest possible supply should be available for fighting fire.

For domestic service the average pump capacity provided was about 200 gal. per min., per 1,000 population, but this figure is not of great value since the capacity required depends largely upon local conditions and may vary widely for towns of the same population. Also, the system used has considerable influence upon the pump rating. A direct-pressure system must have sufficient capacity to supply the peak of the load, while in a storage system the pump needs to supply only the average demand, anything above this being drawn from the storage. The tendency is to provide too much rather than too little capacity, in order to be on the safe side and to provide for future requirements. The pumping units designed for special fire service average about 250 gal. per min. pump capacity per 1,000 population. These units also provide reserve capacity for abnormal demands other than for fire service.

Officials and operating men expressed themselves almost unanimously as well pleased with the results obtained through use of electrically operated equipment. Appreciation of the greater convenience and decreased operating expenses was especially marked in places where the pumps were previously steam driven. One plant was visited in which steam operation had been resumed after a trial of electric operation, but no one available at the time could give an adequate reason for the dissatisfaction with the electric drive.

Another case was encountered, where a direct-connected motor-driven centrifugal pump was giving unsatisfactory service, though no blame was placed upon the motor itself. The pump was obviously not designed for the conditions under which it was working, and the result was an energy consumption greatly out of proportion to the work done. Plans were being made to replace the centrifugal with a triplex pump, though it appears likely that a change in impellers would have accomplished the same result. This serves to emphasize the necessity for using great care in specifying a centrifugal pump for a given application, in order that its characteristics may be suited to the work it is expected to do.

Numerous cases of minor troubles with the equipment were reported. Of these directly chargeable to electric operation the only ones of serious import were those due to lack of adequate protective devices between motor and line. Some of them involved no serious consequences, while others involved considerable expense and inconvenience, as well as fire risk. It would appear that the necessity for adequate protection of the motor and control circuits is not generally appreciated. The cost of such protective apparatus is not great and will soon be repaid through decreased repair bills and greater reliability of the equipment. Troubles were reported due to lack of reliable low voltage and no-voltage cutouts and also to lack of proper protection against lightning. Good lightning-arrester equipment properly installed is essential where power is secured from a transmission line and is the only means of preventing damage from this source, which may vary all the way from a breakdown of insulation in one or more coils to destruction of the entire electrical equipment.

Next to having the apparatus itself it is of importance that attendants should understand the significance of the protective devices. One plant was visited where no voltage protection was provided, but a former attendant had decided that by wedging the starting handle in the full-on position he could start the pump from the power station and save himself a half-mile walk. Unfortunately, the motor windings did not survive the initial experiment, but had to be entirely replaced.

Other than electrical the most frequent source of trouble appeared to be due to freezing in severe winter weather. Bursting of pump cylinders was reported in numerous instances. In some cases these breaks had been repaired by

*Presented at the annual meeting of Indiana Engineering Society in Indianapolis, Jan. 23-24, by G. C. Blalock, instructor in electrical engineering, Purdue University, LaFayette, Ind.

welding; in others the cylinder had to be replaced. Troubles from this source can be charged only to neglect in providing some means of keeping the temperature sufficiently high to prevent freezing.

The installation of cheap equipment invariably results in troubles due to excessive wear and breakage; and poor judgment in the selection and use of equipment augments the difficulties along this line. In one plant the first stage was handled by an air lift and the second by a cheap rotary pump. The latter was driven at such high speed that it handled water faster than the air lift could supply it, so a portion was bypassed and pumped over again. In addition the wear on the impeller was so great as to necessitate its renewal two or three times each year.

Troubles due to flooding sometimes occur in plants having the pumps located in pits. Unless the motor windings have been thoroughly waterproofed, they are sure to be more or less permanently damaged, with consequent shortening of their useful life, although the damage may not be such as to require immediate rewinding.

Improper alignment of pump and motor causes undue wear and results in decreased efficiency and increased maintenance cost. The same cause sometimes results in broken rods in case of deep-well pumps. Corrosion and rapid wear due to sediment in the water being pumped were reported in one or two instances.

Only six of the towns visited had provision for metering the pump discharge, despite the obvious advantage of having this means of checking the performance of the pump and motor. With a watt-hour meter on the input side of the unit and a water meter and pressure gage on the output side, the duty and over-all efficiency may be determined readily at any time and any marked drop detected and accounted for. A centrifugal pump particularly needs to have its performance checked to avoid the marked drop in efficiency that occurs when the impeller is not properly designed for the conditions under which it is working. The substitution of an impeller of slightly different design may effect a saving that will pay for both meters and impeller within a comparatively short time.

Plant Operating Records and Costs

It was found rather difficult to obtain accurate records of plant operation and costs, particularly in the smaller towns. Combined power and pumping plants frequently do not provide separate meters for measuring the energy supplied to the pump motors nor divide the operating costs equitably between the power and pumping departments of the plant. Where energy is obtained from sources not owned or controlled by the interests operating the pumping plant, the meter provided often includes the energy used for street lighting, station lighting and heating or other purposes; and as previously mentioned, it is not often that provision is made for measuring and recording the amount of water pumped. The data obtained therefore involve estimates in many cases and should be considered as indicating plant performance in a general way only.

Data for calculating energy consumption per 1,000 gal. of water pumped were obtained for fourteen towns. In kilowatt-hours per 1,000 gal. this energy consumption ranged from a minimum of 0.20 to a maximum of 3.28, giving an average value for the fourteen towns of 1.31.

Since there is considerable variation in the total head against which these pumps are working, a better measure of performance may be obtained by reducing the figure given for each plant to a basis of 100 ft. total head. This gives values ranging from 0.125 to 1.96 with an average of 0.90 kw.-hr. per 1,000 gal. per 100-ft. head.

The operating cost per 1,000 gal. was found to range between 0.88 and 16.83c., with an average value of 4.97c. Reduced to a basis of 100 ft. head this average becomes 3.74c. per 1,000 gal.

Test Data from Specific Plants

Two plants equipped with meters for measuring the pump discharge were selected, and tests were run, with two objects in view: First, to determine the duty and over-all efficiency of the pumping unit, thus obtaining a basis for figuring energy costs per unit discharge; and second, to determine the nature of the power demand and thereby establish the degree of desirability of pumping loads from the central-station point of view. The test in each case extended over a period of 24 hours, and in addition to kilowatt-hour and water meters a graphic recording wattmeter was used for measuring and recording the power demand.

In the first plant the unit tested consisted of a 2½-in. single-stage, 175 gal. per min. centrifugal pump, driven by a 15-hp. squirrel-cage type induction motor, pumping continuously against the direct pressure of the mains. A total pumpage of 11,900 cu.ft. was observed—an average rate of 61.8 gal. per min. against a pressure of 40 lb. per sq.in. with an input of 9.38 hp., giving an over-all efficiency of 16.7 per cent. Estimates based upon pipe-friction tables give about 10 ft. loss of head in elbows, valves, meter, etc., between the pump intake and the point at which the pressure gage was attached to the main. Including this loss of head raises the efficiency about 2 per cent. This still gives a low over-all efficiency, but it should be noticed that the pump is discharging but little more than one-third of its rated capacity, this resulting in the motor running at considerably less than its normal rating. Neither pump nor motor can therefore be expected to show good efficiency. It is evident that this unit is not operating under the conditions for which it was designed, resulting in high energy consumption per unit output.

In the second plant the unit tested consisted of a 4-in. two-stage, 380 gal. per min., centrifugal pump, driven by a 25-hp. squirrel-cage type induction motor, pumping to mains, with standpipe storage. The frequency and duration of pumping periods were regulated by a pressure-gage relay and an automatic motor starter.

Three runs of approximately two hours each, at an average pressure of 55 lb. per sq.in., gave a pumping rate of 350 gal. per min. with an input of 32.25 hp., showing an over-all efficiency of 38 per cent. Estimated loss in head due to friction amounts in this case to about 35 ft., and inclusion of this loss as part of the total head raises the efficiency about 10 per cent. The over-all efficiency checks closely with manufacturers' estimates for this size and type of unit and the unit is evidently well adapted to its work.

The conclusions arrived at as a result of investigation and study of the present status of electricity in the pumping plants of the state may be summed up as follows:

1. Where local conditions have been carefully considered, the type and capacity of pump and motor which best suits those conditions intelligently selected, electrically driven pumps have proved highly satisfactory; they have been found to be easily handled, convenient, reliable and economical.

2. While the motor itself is unlikely to give trouble, and transmission lines have proved quite reliable, it appears advisable to provide some auxiliary source of power. The automobile-type gasoline engine seems particularly well suited to this purpose.

3. The energy consumption in properly designed plants may be expected to vary from 0.2 to 2 kw.-hr. per 1,000 gal. per 100 ft. head, with 1 as a fair average.

4. The operating expenses for an electrically operated plant are low—considerably below those for a steam plant and appreciably below those for an internal-combustion engine plant.

5. The pumping-plant load is a desirable one from the central power-station point of view.

Peter Cooper Hewitt has, says the Springfield, Ohio, *Sun*, developed a new machine which, he claims, will do even more than the required performances to win the $100,000 prize offered by the Michelin Co. for a machine that will rise straight into the air without a preliminary run and will land in a space 30 ft. square. According to Mr. Hewitt, with his new propeller it will be possible to reverse the plane while in flight without reversing the engine or turning about. It can also be made to 'eap or drop like a flying boat. These unusual movements are made possible by the odd propeller which may be pointed in any direction, and the entire force of the blades is applied to lifting. Once in the air, the shaft is pointed forward and the machine moves ahead with a speed that equals the best attained by other machines. At any time during the flight the position of the propeller may be changed.

Condensed-Clipping Index of Equipment

Clip, paste on 3 x 5-in. cards and file as desired

"No Kut" Pump Valve
Richardson-Phenix Co., Milwaukee, Wis.
"Power"

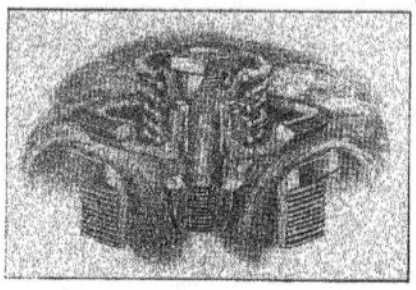

A new pump valve designed to eliminate leakage and slippage. The metal valve disk seats on a metal surface. The valve disk is fitted with a composition sealing ring around its outer edge and one around the valve stem. These sealing rings, being flexible, are forced against the valve seat by the pressure of the liquid as the valve closes. The greater the pressure of the liquid the tighter the valve is seated. Owing to the composition of which they are made, the sealing rings are durable, and no matter how badly the seating surface of the valve becomes worn, scarred or scored, the rings keep the valve tight.

Engine, Bannon Rotary Steam
Bannon Rotary Motor Co., 324 Linz Building, Dallas, Tex.
"Power"

The Bannon rotary steam engine has but three moving parts: An eccentric piston keyed to the driveshaft, and two sliding gates which operate with their ends forming a contact with the periphery of the piston. On the shaft of the two-cylinder engines there are two of these eccentric systems, set 180 deg. apart, which balance the shaft and eliminate the possibility of dust on the bearing surface. The principle of the lever is employed by having the steam expand against a revolving eccentric system so that the leverage of the system greatly multiplies the steam energy. Each impulse of steam drives the shaft three-quarters of a revolution, and as there are two impulses per revolution in each cylinder, an overlap of 90 deg. results. The engine can be coupled directly to the lineshaft.

Holes, Stine Screw
Stine Screw Holes Co., Waterbury, Conn.
"Power"

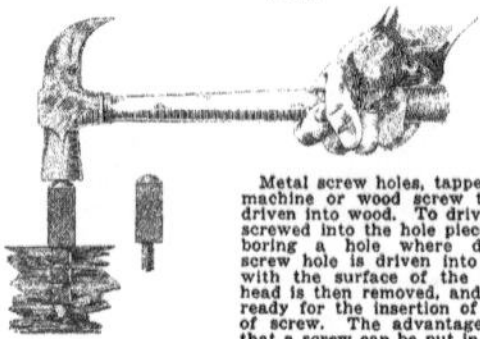

Metal screw holes, tapped for either machine or wood screw threads, are driven into wood. To drive, a head is screwed into the hole piece, and after boring a hole where desired, the screw hole is driven into place flush with the surface of the wood. The head is then removed, and the hole is ready for the insertion of either type of screw. The advantage of this is that a screw can be put in or removed as often as desired without enlarging the hole. These screw holes can also be molded in any material.

Cock, Double-Packed Stop
M. H. Treadwell Co., Inc., 140 Cedar St., New York City.
"Power"

This stop-cock is packed on top with asbestos, sprinkled with graphite, both ends of the packing rings fitting close together. The top gland is then tightened so that the plug will barely turn, after which the bottom is packed in the same manner, the packing being drawn tight enough to release part of the pressure at the top. This equalizing of the pressure will enable the stop-cock to turn freely. It is built for high pressure.

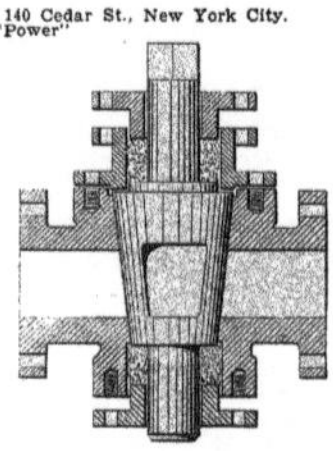

Valve, Dielmore Automatic Drain and Relief
Dielmore Valve Co., Box 674, Tacoma, Wash.
"Power"

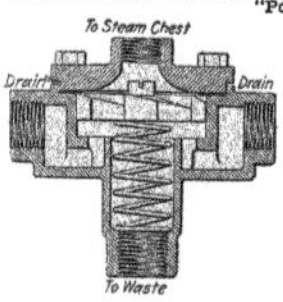

This valve is designed to automatically take care of the condensation in engine and pump cylinders after the steam has been shut off at the throttle valve. It is connected to the two ends of the cylinder and at the top to the steam chest. It contains a valve disk and spring. The pressure from the steam chest closes the valve against the pressure of the exhaust. When the throttle valve is closed, the pressure from the steam chest is removed and the spring opens the valve to allow the water to escape from the cylinder.

Union, Codd Double-Bronze Seated
The Codd Tank and Specialty Co., 406 W. Camden St., Baltimore, Md.
"Power"

This malleable-iron pipe union is made with two bronze seats. One seat is fitted in the main body of the union and the other is in the nut, thus giving a double security against leakage. The two bronze seats are compressed into recesses, as shown.

Regulator, Improved Copes
Northern Equipment Co., Erie, Penn.
"Power"

The new control valve used with the Copes regulator has abandoned the reciprocating stuffing-box and in its place is a horizontal shaft which rotates slightly, the frictional resistance having been reduced as much as 6 to 16 lb. The valve is of the balance type, and the weight of the valve lever exerts a constant closing force of 50 lb. on the valve piston. A weight is used instead of a spring. The thermostat tubing is hollow, the water level corresponding to that in the boiler acts on the temperature of the rod, thus lengthening and shortening the tube as the water rises and falls, the expansion actuating a control valve.

Valve, Nordstrom Lubricated Plug
Merrill Metallurgical Co., 121 Second St., San Francisco, Calif.
"Power"

The special feature of this valve is that it is lubricated, the lubricant being placed at the top of the plug. It is forced out through the grease ducts by the lubricating screw shown in the center of the plug head. As the grease is forced out of the grease ducts, it is also forced into the grease chamber at the bottom and the plug is thus lubricated.

Patented Aug. 20, 1918

Sierra and San Francisco Properties Leased

The California Railroad Commission has approved the lease by the Sierra & San Francisco Power Co. of its properties to the Pacific Gas and Electric Co. The lease is to run for a period of 15 years, dating from Jan. 1, 1920. In addition to paying all taxes and governmental charges, providing for special funds created by the lease and setting aside 2 per cent of the gross revenues received from the leased properties for depreciation, the gas company is to pay a rental of $50,000 a year for two years, $100,000 for the third and $150,000 for each of the remaining years of the lease.

Pointing out that for nearly two years the Sierra & San Francisco Power Co. has been unable to sell its bonds and thus handicapped has not been able to install very necessary power plants and transmission lines, Commissioner Irving Martin, who wrote the opinion accompanying the order of approval, calls attention to service improvements that will follow as a result of the lease. He says:

"Considerable preliminary work has been done toward the installation of the so-called Spring Gap plant on the Stanislaus River. Material has been ordered for a second transmission circuit from Port Marion to Salinas in order to improve the service conditions in Salinas and other territory served by the Coast Valleys Gas and Electric Co., which buys its energy from the Sierra company. Material has been ordered to strengthen the line from Manteca to Modesto, thence interconnecting from Modesto to Newman, with provisions for an additional line to Turlock. Work has already been started by the Sierra company on an additional pipe line to its Stanislaus plant. The estimated cost of these projects is in excess of $1,000,000."

The Sierra & San Francisco Power Co. sells and distributes water in Tuolumne County and electric energy in the "Mother Lode" district of Tuolumne and Calaveras counties; in Stanislaus, San Joaquin, Contra Costa, Alameda, Santa Clara, San Benito and Monterey counties and the City of San Francisco. It supplies power to the United Railroads of San Francisco; 62 per cent of its earnings form the sale of electricity coming from the street-railroad company.

Sampling Atmosphere by Airplane

In the investigation of the atmospheric conditions and the smoke problems of Salt Lake City progress is reported by Osborn Monnett, who has had direct charge of the work.

FIG. 1. FIELD EQUIPMENT FOR SO_2 DETERMINATION IN ATMOSPHERE

The fuel survey, begun in September, has been completed, together with Ringelmann chart readings of the various classes of plants giving out smoke. Laboratory experiments with household equipment have been carried on throughout the winter. The information obtained therefrom is to be used in personal instruction of householders in the operation of their furnaces. A house-to-house canvass of the city has been made by the Boy Scouts in coöperation with the Federated Women's Clubs, and the number of different types of heating equipment used in the city is being tabulated to facilitate the work of instruction.

In connection with the examination of the atmosphere with respect to the influence of the smelter gases, the airplane has been used. This is the first time on record where the heavier-than-air machine has been used for this purpose. Samples are collected at the 3,600-ft., 2,200-ft. and 1,000-ft. levels, and these, in connection with samples taken on the roofs of the buildings and at the street level, will give determinations at the various levels affecting the situation.

J. J. McKetteridge, an engineer from the United States Bureau of Mines Station at Minneapolis, has been temporarily

FIG. 2. CHEMIST WITH FIELD KIT READY TO TAKE SAMPLES AT HIGHER LEVELS

transferred to the Salt Lake Station. He is an expert domestic boiler and furnace man and will be used to organize the work of personal instruction to owners of domestic equipment. During the month of February engineers of the Salt Lake Station are conducting a campaign of demonstration of smoke-abatement possibilities in large heating plants.

Future Organization for "Welfare" Activities of Engineering Societies

By whom and how shall the "welfare" activities of American engineers be conducted henceforth? This question has been much debated. For several months the report of the Joint Conference Committee of the Founder Societies (Civil, Mining, Mechanical, Electrical) has been before the societies as the answer. Decision will vitally affect the Engineering Council.

The Engineering Council at its meeting Oct. 16, 1919, received the Joint Conference report, and voted unanimously that Council indorses the general plan for a national Engineering Council as therein outlined. Nevertheless, the Council as now constituted must "carry the torch" until taken by the next runner. The torch must be kept burning. The Council, in the desire to further the interests of engineers and to remove difficulties in its own operations, invited the governing bodies of its member societies to a conference. Members of the Joint Conference Committee and the Trustees of United Engineering Society also were invited.

This conference met in Engineering Societies Building, New York, Jan. 23, from ten in the morning until five in the afternoon. It was the first time that the members of the governing bodies of these national societies had been brought together. It was an educative and otherwise profitable meeting. Its discussions were made effective in the resolutions copied below, it being understood that these actions were in the nature of recommendations to the societies.

Resolved, That the amount contributed for Engineering Council by each founder society shall be $5,000 for the year 1920, in place of $3,000 as now provided, and by the American Society for Testing Materials, $1,000, instead of $600.

Whereas, The American Society of Mechanical Engineers and the American Institute of Electrical Engineers have adopted in principle the essential recommendations contained in the report of the Joint Conference Committee of the four founder societies; and

Whereas, It appears probable that the other founder societies will shortly also adopt the principles of this report; and

Whereas, It would be most unfortunate to permit certain

welfare measures of importance alike to engineers and to the public to be interrupted through lack of necessary immediate financial support; be it

Resolved, By this joint conference of the governing bodies of the American Society of Civil Engineers, the American Institute of Mining and Metallurgical Engineers, the American Society of Mechanical Engineers, the American Institute of Electrical Engineers, and the American Society for Testing Materials that appeals for contributions of $2 per person for the support of such welfare work during the interval before the recommendations of the Joint Conference Committee shall have been made effective, be sent at their discretion by the governing boards of the societies represented at this conference, to members of all grades, except juniors and student members; and be it further

Resolved, That the appeal make clear the fact that unless engineers lend financial support immediately, the public welfare work of engineers will be seriously discredited; and be it further

Resolved, That if the amount received by each society in response to the appeal be in excess of its appropriation voted for Engineering Council, such excess shall be appropriated by each society to Engineering Council or its successor.

Whereas, This Conference of national engineering societies has considered the recommendations of the Joint Conference Committee of the four Founder Societies; therefore, be it

Resolved, That the Conference adopts in principle that report and requests the Joint Conference Committee to call, without delay, a conference of representatives of national, local, state and regional engineering organizations to bring into existence the comprehensive organization proposed.

Voted: That if *this* Conference be called together again, it be called for a meeting in Chicago on Apr. 20.

Willful Waste Makes Woeful Want

There has been a fire burning on the waterfront of the Hastings Mills on Burrard Inlet for twenty-five years. Through rain and storm and winter and summer the flames burn brightly and the black smoke rises. Ever since the mills started there has been this bonfire, fed by the endless sawdust carriers. Hundreds of tons of wooden particles have been burned at this plant alone. At other mills throughout the province the sawdust is treated in the same way. As sawdust has a low commercial value and no one mill produces enough of the material to make a paying proposition of preparing it for various marketable uses, it has been allowed to go to waste for many years. It is thought now, however, that with many mills operating within a reasonably small radius there is enough being produced to make a sawdust-handling plant a good investment. It can be pressed into bricks or manufactured into dyes or other chemical properties extracted. The matter is being investigated by capitalists, and it is entirely probable that some steps will be taken this coming spring or summer to make use of sawdust waste from British Columbia mills.

World's Production of Petroleum

	Production, 1918			Total production, 1857-1918		
Country	Barrels of 42 gallons	Metric tons	Per cent of total	Barrels of 42 gallons	Metric tons	Per cent of total
United States..	355,927,716	47,457,029	69.15	4,608,571,719	614,476,230	61.42
Mexico ..	63,828,327	9,506,289	12.40	285,182,489	42,564,549	3.80
Russia ...	40,456,182	5,520,066	7.86	1,873,039,199	247,856,218	24.96
Dutch East Indies a.	13,284,936	1,836,914	2.58	188,388,513	25,465,114	2.51
Rumania .	8,730,235	1,214,219	1.70	151,408,411	21,058,193	2.02
India ...	b 8,000,000	1,066,667	1.55	106,162,365	14,154,982	1.41
Persia ...	7,200,000	b1,000,000	1.40	14,056,063	2,952,231	.19
Galicia ...	5,591,620	777,640	1.09	154,051,273	21,424,303	2.05
Peru	c 2,536,102	338,147	.49	24,414,387	3,255,251	.33
Japan and Formosa	2,449,069	326,543	.48	38,498,247	5,133,100	.51
Trinidad .	2,082,068	289,578	.40	7,432,391	1,033,712	.10
Egypt ...	2,079,750	277,300	.40	4,848,436	646,458	.07
Argentina.	1,321,315	192,612	.26	4,296,093	617,176	.06
Germany..	711,260	b 100,000	.14	16,664,121	2,254,974	.22
Canada ..	304,741	40,632	.06	24,425,770	3,256,769	.33
Venezuela.	190,080	26,400	.04 (Venezuela and Italy)	317,823	44,142	.02 (Venezuela, Italy, Cuba and Other countries)
Italy......	35,953	b 5,000		973,671	138,588	
Cuba	....	....	.....	19,167	2,662	
Other countries	....	....	.....	397,000	55,139	
	514,729,354	69,975,036	100.00	7,503,147,138	1,006,389,791	100.00

(a) Includes British Borneo. (b) Estimated. (c) Estimated in part.

Tests on Slag Wool for Steam-Pipe Covering*

In these tests dry steam is admitted to a 4-in. steam pipe of outside diameter 4.47 in. and length 13 ft. 8¾ in. between the insides of the flanges. The pipe is set nearly horizontal, a slight fall being given toward the outlet end to facilitate the draining off of the condensed steam. Steam is admitted at the upper end through a gravity separator to insure the supply of dry steam. The condensed steam is collected in a down tube at the lower end of the pipe and the level of water in this collecting pipe is shown by a gage-glass. During the test the water is maintained at constant level by adjusting the drain cock. The water is passed through a cooling coil before passing out into the open air in order to prevent reëvaporation on reduction of the pressure. To allow for the condensation due to the flanges, the test pipe is removed, and a pair of similar flanges, with a very short length of pipe, is put between the ends of the apparatus. The amount of condensation obtained under these conditions is subtracted from the amount obtained with the test pipe in place. After a preliminary heating of some hours, says *The Engineer*, the pressure of steam is kept as constant as possible and observations taken for one hour. Two such one-hour runs with the pipe bare and covered respectively constitute a test. The observations taken during each hour include (1) the temperature of the steam at the inlet to the test pipe; (2) the air temperature at two positions, about 4 ft. from the pipe; (3) the temperature of the outside of the cover; (4) the weight of steam condensed.

Tests were made with two different thicknesses of slag wool covering held to the pipe by metal sheets, having distance pieces to fix the thickness of the cover. In each case the length of cover and area of pipe covered was 13.69 ft. and 16.02 sq.ft. respectively. The details of the covers are as follows:

	Cover No. 1	Cover No. 2
Nominal thickness	2 in.	1 in.
Weight per foot run	7.99 lb.	4.04 lb.
Weight per square foot	6.83 lb.	3.46 lb.

The difference between the net loss from the covered pipe and the bare pipe gives the heat saved per hour and the efficiency of the cover is taken as the ratio of this to the net bare pipe loss. The heat loss per square foot per hour per degree difference of temperature between the steam and the outside air is nearly constant for medium and low pressures of steam, but generally rises for high pressure; it is, however, the most convenient number for calculation of the loss under different conditions, so long as the conditions are not greatly different from those of the test. It may be noted that there is no flow of steam through the test pipe, except a very small amount due to the condensation. Hence in practice, where there is flow, the quantity of heat lost for the same difference of temperature between steam and air would not be the same as in the test pipe, but this will not materially affect the efficiency figure, since this is the ratio of two tests under nearly identical conditions. In the results given below, the gage pressure is the pressure corresponding to the steam temperature less the atmospheric pressure. The results for the bare and covered pipe are indicated at A and B, and the difference between these results is shown at C.

Cover No. 1		Test No. 1	Test No. 2	Test No. 3
Gage pressure, lb. per sq. in.		54.1	107.8	258.9
Temperature of steam, deg. F.		301.5	342.5	408.9
Temperature of air, deg. F.		67.3	67.4	62.2
Temperature of cover, deg. F.		81.0	88.4	87.8
Difference between steam and air temperature		234.2	275.1	346.7
Loss per square foot per hour per deg. difference of temperature deg. F., B.t.u.	A	2.843	3.043	3.755
	B	0.235	0.257	0.281
	C	2.608	2.786	3.474
Saving of bare pipe loss, per cent		91.8	91.6	92.5

Cover No. 2		Test No. 1	Test No. 2	Test No. 3
Gage pressure, lb. per sq. in.		53.9	109.2	256.3
Temperature of steam, deg. F.		301.3	343.5	408.0
Temperature of air, deg. F.		64.4	66.2	72.8
Temperature of cover, deg. F.		98.9	99.8	112.8
Difference between steam and air temperature		236.9	277.3	335.2
Loss per square foot per hour per deg. difference of temperature deg. F., B.t.u.	A	2.854	3.056	3.601
	B	0.431	0.463	0.525
	C	2.423	2.593	3.076
Saving of bare pipe loss, per cent		84.9	84.9	85.5

*Abstract of National Physical Laboratory Report, Apr. 26, 1919.

Hydro-Electric Development Will Utilize 1,900-Foot Fall of Water

In a recent address before the Stockton Rotary Club, Samuel Kahn, vice president and general manager of the Western States Gas and Electric Co., referred to the additional 30,000-hp. hydro-electric development planned by his company. The new developments, which are located on the sduth fork of the American River, a few miles above the company's present hydro-electric development, include the construction of a reservoir for impounding waters in the Twin Lakes, Echo and Medley lakes at the headwaters of the south fork of the river.

As small dams were in existence at Silver Lake and Echo Lake, no further work was done at these points, but in the summer of 1917 construction was begun on an earth-filled dam with a concrete core wall at Twin Lakes, which was completed in 1918. This dam at the present time is only 8 ft. high, but eventually will be built to a height of 60 ft. In the summer and early fall of 1917 a rubble stone masonry dam was also completed at Medley Lakes. During the years 1918 and 1919 the surveys were still in progress and will probably be completed the early part of 1920.

A connecting ditch originally constructed for a capacity flow of 80 cu.ft. per sec. will be enlarged to carry 250 or 300 cu.ft. per sec. Furthermore, it may be lined with pre-cast concrete slabs to reduce friction losses and to minimize ice troubles, as during midwinter there are periods when the ditch freezes.

After the water flows down the ditch for a distance of 24½ miles, it will reach a proposed forebay and the waters from this forebay will be conducted through a high-pressure pipe line about 2½ miles in length, which will terminate in the south fork of the American River in the vicinity of Long Canyon. The difference in elevation between the proposed forebay and the proposed plant is 1,900 ft., which will make it one of the high-head developments of the country and, in fact, of the world. The pipe line will be constructed of wood and steel; that is to say, the upper end of the pipe line, where the pressure is relatively small, will be constructed of redwood staves, which have been found to be excellent for such purposes. The lower end of the pipe will be built of steel of varying thicknesses, the heaviest pipe, of course, being at the bottom of the canyon. As a matter of fact, the lower sections of the pipe will have to be constructed of the highest-grade steel and lap-welded to withstand a working pressure of 800 lb.

After the water passes through the proposed hydraulic plant, it will again flow down the American River several miles until it reaches the present dam and diversion, where it will be picked up again and used in the present plant. After the water has passed this plant, it will be used farther down stream again for irrigation and domestic purposes.

It is reported that there are thirteen sources of water power in Ceylon, which if developed would generate approximately 500,000 hp. Since the present demand for power is only 60,000 to 70,000 hp., it has been decided that only the Laxapana and Dickoya Rivers be exploited, as they would supply about 200,000 hp.—*London Electrical Review.*

New Publications

THE COAL CONSUMPTION OF POWER PLANTS AND BONUSES FOR COAL SAVING. By Robert H. Parsons. Published by the Electrical Review, London, England. Pamphlet; 8½ x 5½ in. 23 pages. Price 1 shilling.

The author sets forth some methods of establishing and utilizing standards of power-plant efficiency and describes bonuses for power-plant employees. The reader will perhaps get an idea of what the booklet contains by being advised that its contents is made up of text and charts like those in the paper by R. J. S. Pigott, "Graphic Analysis of Power Plant Performance," before the American Society of Mechanical Engineers a few years ago, though not as complete or detailed as Mr. Pigott's paper.

Obituary

Pierrepont Bigelow, who succeeded his father, Col. F. L. Bigelow, when the latter died a couple of years ago, as treasurer of the Bigelow Co., New Haven, Conn., died suddenly of pneumonia on Jan 28. Mr. Bigelow was 32 years old. He was a graduate of Sheffield Scientific School, a man of sterling qualities and high ideals, and had a good mechanical mind.

E. Fred Wood, formerly vice-president of the International Nickel Co., died suddenly at New York City on Jan. 5. Mr. Wood was born in Milwaukee on Aug. 28, 1858. He was educated in the public schools of his native city and later entered the University of Michigan, from which he was graduated. After leaving college he devoted himself to the study of metallurgy and in connection with his studies made extensive trips through the various mining camps of the West and lived for a year at Leadville and other mining towns, where he pursued his studies and obtained his practical experience. He later entered the employ of the Carnegie Steel Co. and rapidly rose to the position of assistant general superintendent of the Homestead plant, which position he filled for a number of years. During the period of the big strike at the plant, when Mr. Frick was shot, Mr. Wood was in entire charge of the plant. He was looked upon by Mr. Carnegie and his associates as one of the valuable men of the organization and was one of the so-called "Carnegie Veteran Associates." He was invited to join the International Nickel Co. upon its organization, becoming first vice-president of the company and a member of the board of directors and of its executive committee. He was an important factor in developing the mining, smelting and refining business of the company. When the United States entered the world war Mr. Wood resigned his official connection with the nickel company to devote himself to public work, and became a member of the Committee on Production of the War Industries Board. He served on this board during the entire period of the war without compensation and his work was fine and of inestimable value and won the admiration and commendation of those who were brought into contact with it. He was a member of the University of Michigan Club, the Automobile Club, the New York Athletic Club, the Society for Electro-Chemical Engineers, and the Railroad Club. He is survived by his widow and one daughter.

Personals

C. H. Gustafson, formerly with the St. Louis & San Francisco R. R. at Springfield, Mo., is now superintendent of motive power for the Wichita Falls, Ranger & Fort Worth Railroad.

A. S. Winter, formerly advertising and sales manager for the Wm. Powell Co., has joined the sales force of the Fairbanks Co., of Pittsburgh, Penn., and will represent the company in southern Ohio.

J. G. Miles has been appointed supply division manager of the Westinghouse Electric and Manufacturing Co.'s Seattle office, to succeed C. V. Aspinwall, now the company's representative at Spokane, Wash.

D. D. Tripp, vice president of the Pioneer Rubber Mills, San Francisco, has left for an extended trip through the Orient, to be gone six or eight months, to study the great possibilities for sales development in the East.

Hjalmar E. Skougor, who has been connected with Guggenheim Bros., Chile Exploration Co. and Braden Copper Co. for the last seven years as designing engineer, has severed his connection with these companies to return to private practice as consulting industrial engineer, with offices at 120 Broadway, New York City.

A. M. Quick, of Baltimore, Md., and Howard P. Quick, of New York City, consulting hydraulic power-plant engineers, are associated in the design and construction of a hydro-electric plant on the Shenandoah River, near Front Royal, Va., in connection with some industrial plants which have been under way for a year or more, having received special war grants for this work.

Lewis E. Ashbaugh, hydraulic engineer with the J. G. White companies and clients for the past twelve years, is now sales engineer at the New York office of John R. Proctor, Inc., of Bayonne, N. J., in charge of fuel-oil installations and other electrical equipment for power and lighting purposes.

Paul Wunderlich, consulting industrial engineer, will leave shortly for Japan as mechanical engineer for the George A. Fuller Co. He is endeavoring to get up a complete catalog and data file of mechanical, electrical and sanitary equipment, and will appreciate literature of this kind sent to him in duplicate, in care of the Geo. A. Fuller Co., Flatiron Building, New York City.

Aubrey G. Haven, Brooklyn, N. Y., has been appointed real estate engineer in charge of the valuation of right of way and real estate for the Gulf, Mobile & Northern R. R. Mr. Haven was for over five years with the Interstate Commerce Commission engaged on land valuations, and prior thereto was engaged on real estate appraisals for numerous other railroads.

Engineering Affairs

The American Society of Mechanical Engineers, St. Louis Section, will hold a meeting on Mar 19, at which P. E. Irvine will talk on the "New By-Product Coke Plant at Granite City."

The Engineers' Club of Philadelphia will hold a special meeting on Feb. 28, at which will be presented a "Report of Joint Conference of Committee on Development of the Four Founder Societies."

The American Society of Mechanical Engineers, Buffalo Section, will hold a meeting under the auspices of the Engineering Society of Buffalo on Feb. 24. A. L. De Leeuw will speak on the subject of "Wage Payments."

The Indiana Engineering Society, at the annual convention held in January elected the following officers: President, W. T. Titus, Indianapolis; vice president, O. C. Ross, Indianapolis; secretary, Charles Brossman, Indianapolis; trustees: D. D. Ewing, LaFayette; A. P. Melton, Gary; John W. Fulwider, Lebanon; Frank C. Wagner, Terre Haute; Edwin S. Pearse, Indianapolis.

Business Items

The Tacony Steel Co. is opening a Chicago sales office in the Marquette Building, with Frank B. Hillwick, until recently with the Crucible Steel Co., as district sales manager.

THE COAL MARKET

BOSTON—Current prices per gross ton f.o.b. New York loading ports:

Anthracite

	Company Coal
Egg	$8.20@$8.65
Stove	8.45@ 9.05
Chestnut	8.55@ 9.05
Pea	7.05@ 7.40
Buckwheat	3.60@ 3.50
Rice	2.30@ 2.50
Barley	1.25

Bituminous

	Clearfields	Cambrias and Somersets
F. o. b. mines, net tons	$2 83@$3.35	$3.15@$3.60
F.o.b. Philadelphia, gross tons	5.05@ 5.60	5.35@ 5.90
F.o.b. New York, gross tons	5.40@ 5.95	5.75@ 6.25
Alongside Boston (water coal), gross tons	7.00@ 7.75	7.60@ 8.00

Pocahontas and New River are practically off the market for coastwise shipment, but are quoted at $6.25@$7.00.

NEW YORK—Current quotations, White Ash, per gross tons, f.o.b. Tidewater, at the lower ports are as follows

Anthracite

	Company Coal
Broken	$7.80@$8 25
Egg	8 20@ 8 65
Stove	8 45@ 9.05
Chestnut	8 55@ 9.05
Pea	7.05@ 7.40
Buckwheat	5.15
Rice	4.50
Barley	4.85
Boiler	4.85

Bituminous

Government prices at mines:	Spot
South Fork (best)	$3.25@$3 50
Cambria (best)	3.00@ 3.25
Cambria (ordinary)	2.60@ 2.90
Clearfield (best)	3.00@ 3.25
Clearfield (ordinary)	2.60@ 2.90
Reynoldsville	2 85@ 2.90
Quemahoning	3.25@ 3.50
Somerset (medium)	3.00@ 3.25
Somerset (poor)	2.50@ 2.75
Western Maryland	2.50@ 2.75
Fairmont	2 25@ 2.50
Latrobe	2 60@ 2.90
Greensburg	2.75@ 3.00
Westmoreland ¾ in.	3.40@ 3.50
Westmoreland run-of-mine	7.75@ 3.00

PHILADELPHIA—Bituminous coal prices vary according to district from which they are mined. For ordinary slack the price is $2.45@$2.55: lump, $3.00@$3 35, at the mines.

BUFFALO—

Anthracite

	On Cars, Gross Ton	At Curb, Net Ton
Grate	$8.55	$10.80
Egg	8.80	10.65
Stove	9.00	10.85
Chestnut	9.10	10.85
Pea	7.75	9.30
Buckwheat	5.70	7.75

Bituminous

Allegheny Valley	$4.80
Pittsburgh	4.65
No. 8 Lump	4.65
Mine Run	4.80
Slack	4.10
Smokeless	4.60
Pennsylvania Smithing	5.70

CLEVELAND—Prices of coal per net ton delivered in Cleveland are:

Anthracite

Egg	$12.25@$12.40
Chestnut	12.50@ 12.70
Grate	12.25@ 12.40
Stove	12.40@ 12.60

Pocahontas

Mine-run	$7.50

Domestic Bituminous

West Virginia split	$8.50
No. 8 Pittsburgh	$6.00@ 6.90
Massillon lump	8.25@ 8.50
Coshocton lump	7.15

Steam Coal

No. 6 slack	$5.25@$5.50
No. 8 slack	5.10@ 5.50
Youghiogheny slack	5.25@ 5.50
No. 8 ¾ in.	5.70@ 6.00
No. 6 mine-run	5.25@ 5.50
No. 8 mine-run	5.75

Only coal available is mine-run Pocahontas

MIDDLE WEST—Chicago quotations, F.o.b. cars at mine:

	Springfield, Carterville, Williamson, Franklin, Saline, Harrisburg	Fulton, Peoria	Grundy, La Salle, Bureau, Will
Lump	$2.55@$2.70	$2 95@$3.10	$3 25@$3 40
Washed	2.75@ 2.90	....	3 45@ 3.60
Mine run	2.35@ 2 50	2.75@ 2.90	3 00@ 3.15
Screenings	2 05@ 2.20	2 35@ 2.50	2.75@ 2.90

New Construction

PROPOSED WORK

Mass., West Springfield (Springfield P. O.)—McClintock & Craig, Archts. and Engrs., 33 Lyman St., Springfield, will receive bids for the construction of a 1 story power house on Cold Springs Ave., for the Gilbert & Barker Mfg. Co.

Conn., Hartford—Bryant & Chapman, Woodland St., plan to construct a power plant. C. N. Flagg & Co., State St., Meriden, Engrs.

Conn., Hartford—A. Goldstein is having preliminary plans prepared for the construction of a 2 story, 101 x 119 ft. garage on Maple Ave. A steam heating system will be installed in same. Total estimated cost, $100,000. F. Walz, 348 Trumbull St., Archt.

Conn., Hartford—The Hartford Electric Light Co., Pearl St., plans to build a power plant at South Meadows. E. F. Lawton, Genl. Mgr.

Conn., Hartford—E. M. Stone, Engr., 327 Trumbull St., is having preliminary plans prepared for the construction of an office building at 55 Pratt St. A sprinkler and vacuum cleaning system will be installed in same. Owner's name withheld.

Conn., Meriden—The Mt. Carmel Italian Roman Catholic Church, Springfield Ave., c/o D. Ricca, Pastor, is having preliminary plans prepared for the construction of a church on Springfield Ave. A steam heating system will be installed in same. Total estimated cost, $100,000. Mardietti & D'Avina, 756 Main St., Hartford, Archts.

Conn., West Hartford—Ford, Buck & Sheldin, Engrs., 60 Prospect St., will receive bids about March 1 for furnishing and installing a steam heating plant including 125 lb. pressure H.R.T. boilers and 20,000 sq. ft. radiation for the Amer. School for the Deaf, 65 Garden St., Hartford. Total estimated cost, $60,000.

N. Y., Buffalo—The Bison Ice & Coal Co., Inc., plans to construct a 2 story branch ice plant at 200-216 Rapin Pl. Ice making equipment will be installed in same. Total estimated cost, $50,000.

N. Y., New York—The Bellevue & Allied Hospital, New York City, received bids for the installation of boiler pumps, etc., in the new boiler plant at the Fordham Hospital, from Chute, Thornton & Bailey Corp., 2 East 13th St., $36,500; Kings Co. Heating & Ventilating Co., 40 Boerum Pl., $38,750; Johnson Heating Co., $46,524, all of New York City.

N. Y., New York—M. E. Connelly, Pres. of Queens Boro, received bids for the construction of an automatic electric pumping station on Genesee St., from Fox & Renalls, 87 East 125th St., $177,281; J. L. Stegretto, 6103 18th Ave., $224,000; Ajax Drainage Contr. Corp., 60 Herkimer St., $227,000.

N. Y., New York—The Kelly Springfield Tire Co., 200 West 57th St., is having plans prepared for the construction of a 17 story, 100 x 104 ft. office building on 7th Ave. and 57th St. A steam heating system will be installed in same. Total estimated cost, $2,000,000. E. Necarsulmer, 507 5th Ave., Engr.

N. Y., New York—The Western Electric Co., 195 B'way, is having plans prepared for the construction of an 8 story, 200 x 325 ft. warehouse on Hudson, West Houston and Greenwich Aves. A steam heating system will be installed in same.

N. Y., New York—Zipkes, Wolff & Kudroff, Archts. and Engrs., 25 West 42nd St., is preparing plans for the construction of a factory. A steam heating system will be installed in same. Total estimated cost, $150,000. Owner's name withheld.

N. J., Pennsville—Charles R. Peddle, Archt., 136 South 4th St., Philadelphia, Pa., will receive bids in a month for the construction of a 2 story, 75 x 125 ft. school in Salem Co., for the Bd. Educ. A steam heating system will be installed in same. Total estimated cost, $100,000.

Pa., Johnstown—M. Lee Masterson, Supt. of Highways, will soon award the contract for the furnishing, setting up and erecting in place ready for service on ground owned by city, one 800 sq. yd., 3 unit asphalt plant, one 3,000 gal. portable steam melting kettle, 1 combination 40 hp. boiler with 25 hp. engine and one 5 ton tendem steam or motor roller. Bidders asked for prices with or without approved air lift and running trucks for any and all of said equipment.

Pa., Philadelphia—Meubauer & Supourtz, Archts., 929 Chestnut St., will soon award the contract for the construction of a 1 story, 92 x 100 and 36 x 115 ft. garage on Belfield Ave. and Broad St., for the Crosstown Realty Co. A steam heating system, pumps and tanks will be installed in same. Total estimated cost, $35,000.

Pa., Philadelphia—Stewartson & Page, Archts., 318 Walnut St., will soon receive bids for the construction of a 4 story addition to hospital for the Children's Hospital, 18th and Fitzwater Sts. A new power plant will be installed in same. Total estimated cost, $100,000. Noted Aug. 26.

Pa., Scranton—The Scranton Gas & Water Co., 115 Wyoming Ave., will soon award the contract for the construction of a 3 story, 80 x 100 ft. office building. A steam heating system will be installed in same. Total estimated cost, about $200,000. Edward H. Davis and George M. D. Lewis, Union Natl. Bank Bldg., Archts.

Pa., Shickshinny—The First Natl. Bank plans to construct a 3 story, 50 x 100 ft. bank on Main and Union Sts. A steam heating system will be installed in same. Total estimated cost, $100,000.

Va., Roanoke—The Bd. Educ. is having plans prepared for the construction of a 3 story high school on Park St. A steam heating system will be installed in same. Total estimated cost, $450,000. H. C. Richards, 1713 Sansom St., Philadelphia, Archt.

O., Cleveland—The city plans to construct a 2 story police station on East 37th St. and Orange Ave. A steam heating system will be installed in same. Total estimated cost, $100,000. F. H. Betz, 601 City Hall, Archt.

O., Cleveland—The City Club of Cleveland, c/o Francis Hayes, Hollenden Hotel, plans to purchase a site and build a club house on same. A steam heating system will be installed in same. Total estimated cost, $200,000.

O., Cleveland—The Cleveland Electric Illuminating Co., 75 Public Square, is having plans prepared for the construction of a 1 story boiler house at the foot of East 70th St. S. M. Gallagher, 408 Illuminating Bldg., Engr. Total estimated cost, $100,000.

O., Cleveland—W. S. Ferguson Co., Archts., 1900 Euclid Ave., will receive bids about Feb. 26 for the construction of a 6 story, 100 x 200 ft. commercial building on East 61st St. and Euclid Ave., for G. A. Tenbush, Union Bldg. A steam heating system will be installed in same. Total estimated cost, $400,000.

O., Cleveland—W. S. Ferguson Co., Archts., 1900 Euclid Ave., will soon award the contract for the construction of a 2 story, 75 x 100 ft. factory on Payne Ave. and East 37th St. for the Amer. Register Co., 403 Long Ave. A steam heating system will be installed in same. Total estimated cost, $100,000.

O., Cleveland—St. George's Church, c/o J. E. Potter, Archt., 947 Leader-News Bldg., is having plans prepared for the construction of a 2 story, 90 x 140 ft. school at East 67th and Superior Ave. A steam heating system will be installed in same. Total estimated cost, $100,000.

O., Cleveland—St. Paul's Episcopal Church, Rev. Walter R. Breed, pastor, East 40th St. and Euclid Ave., plans to construct a 1 story church. A steam heating system will be installed in same. Total estimated cost, $200,000.

O., Cleveland—The Second Presbyterian Church, c/o Rev. Paul F. Sutphen, pastor, 2525 Stratford Ave., plans to build a 1 story church. A steam heating system will be installed in same. Total estimated cost, $100,000.

O., Cleveland—The Cleveland Railway Co., Leader-News Bldg., plans to construct a 1 story power sub-station near East 105th St. and Euclid Ave. Estimated cost, $200,000. Mr. Radcliffe, Mgr. L. P. Crocelius, 700 Leader-News Bldg., Engr.

O., Shaker Heights (Cleveland P. O.)—The Bd. Educ., Carl A. Palmer, 10111 Euclid Ave., Cleveland, plans to construct a 2 story trade school. A steam heating system will be installed in same. A $300,000 bond issue has been passed for this project.

Ind., Gary—The Co. Comrs. will soon receive bids for the construction of a tuberculosis sanitarium including a boiler house, etc., for Lake Co. A high pressure steam heating system will be installed in same. Total estimated cost, $300,000.

Mich., Benton Harbor—The Congregation of Good Samaritan, c/o H. S. Gray, 8-10 Sonner and Gray Bldg., plans to build a 2 story community house. Steam heating equipment will be installed in same. Total estimated cost, $100,000.

Mich., Detroit—The Dept. of Pub. Wks. will soon award the contract for the construction of a 2 story, 58 x 121 ft. steam power building and a 190 ft. brick stack for heating city buildings, on Macomb and Clinton Sts. A 150 kw. motor generator set, two 300 hp. boilers and one 600 hp. water tube boiler and a battery set providing for 2 more will be installed in same. John Scott & Co., 2326 Dime Bank Bldg., Archt. Esselstyn, Murphy & Hanford, 810 Marquette Bldg., Engrs.

Mich., Detroit—The Detroit Lodge, Loyal Order of Moose, 40-43 Congress St., plans to construct a 3 story lodge temple on Cass Ave. and Elizabeth St. A steam heating system will be installed in same. Total estimated cost, $500,000.

Mich., Lansing—The Novo Engine Co., 702 Porter St., plans to construct a 1 and 2 story foundry on Porter St. Electric motor for power and probably a hot blast heating system will be installed in same. Total estimated cost, $250,000.

Mich., Pontiac—The Oakland Co. Bd. Supervs. plan to build a 3 story, 46 x 176 ft. jail addition on Wayne and Warren Sts. A steam heating system will be installed in same. Total estimated cost, $275,000. F. D. Madison, Royal Oak, Archt.

Ill., Chicago—The Commonwealth Edison Co., 72 West Adams St., plans to construct a power plant in the Calumet Dist. Plant will be largest in their system, having a capacity of over 185,000 kva. Total estimated cost, $10,000,000 to $12,000,000. Marshall & Fox, 721 North Michigan Ave., Archts. Sargent & Lundy, 72 West Adams St., Engrs.

Ill., Chicago—The Kewanee Boiler Co. is having preliminary plans prepared for the construction of a 10 story, 100 x 100 ft. salesroom and warehouse on Washington and Green Sts. A steam heating system will be installed in same. Total estimated cost, $500,000. S. N. Crowen, 30 North La Salle St., Archt.

Wis., Menomonee Falls—The Farley Candy Co., 720 North Franklin St., Chicago, Ill., plans to build a 1 story, 300 x 800 ft. chocolate factory. Steam heating and electric power will be installed in same. Total estimated cost, $300,000.

Wis., Sheboygan—The Mikadon Amusement Co., 1110 Washingon St., is having plans prepared for the construction of a 1 story, 70 x 200 ft. theatre. A steam heating system will be installed in same. Total estimated cost, $100,000. Stanley F. Kadow, 451 Mitchell St., Milwaukee, Archt.

Ia., Davenport—Roy France, Archt., Chicago, Ill., will soon receive bids for the construction of an 8 story, 128 x 152 ft. apartment hotel on 6th and Brady Sts., for the Hillcrest Apartment Hotel Co. A complete refrigeration system will be installed in same. Total estimated cost, $[illegible]00,000. Clausen & Kruse, 316 Central Office Bldg., Engrs.

Ia., Des Moines—R. E. Bales, Secy. Executive Council, State House, plans to build a 5 story, 65 x 175 ft. heating plant and warehouse with an 80 x 80 ft. wing, on the capitol grounds. Heating contract will be sub-let. Total estimated cost, $400,000. Canfield Eng. Co., 406 Flynn Bldg., Engrs.

Ia., Washta—J. K. McGonigae, Secy., will receive bids until March 16 for the construction of a 3 story, 58 x 102 ft. school. A steam heating system will be installed in same. Total estimated cost, $100,000. W. E. Hulse & Co., 210 Masonic Temple, Des Moines, Archt.

Kan., Arkansas City—The city is having plans prepared for the improvement of the waterworks system, including distributing reservoir and power unit for pumps and meters. Total estimated cost, $61,000. E. B. Wilkinson, Engr.

Kan., Coats—The city is having plans prepared for the construction of an electric transmission line from here to Pratt, a distance of 13 miles. Estimated cost, $24,000. G. L. Wright, Clk. W. B. Rollins & Co., 209 Ry. Exch. Bldg., Kansas City, Mo., Engrs.

Kan., Emporia—The State, c/o J. A. Kimball, will receive bids for the construction of a 1 story power plant. Estimated cost, $70,000. R. S. Gamble, 1415 Fillmore St., Topeka, Engr.

Mont., Dillon—Beaverhead Co. plans an election to be held April 23 to vote on $50,000 bonds for the construction of a central heating plant for county courthouse, high school, dormitory and other county buildings. W. E. Chapman, Co. Engr.

Mont., Helena—The Lewis & Clark Helena Valley Irrigation Project plans to irrigate 40,000 acres by means of 45 miles of canals, and the purchase of 2 pumping plants now in operation. Total estimated cost, $600,000. A. W. Verharen, 18 Kohrs Bldg., Engr.

Okla., Anadarko—The city voted $75,000 bonds to improve electric system and is in the market for Diesel engines to replace steam equipment.

Ore., Scio—The city plans to construct a cyclopean or rubble masonry dam flume and timber headgate, taking 47 second ft. of water from Thomas Creek, for pumping, light and power plant. Estimated cost, $38,427. A. G. Prill, mayor.

Ore., Wasco—H. S. Wall filed an application with P. A. Cupper, State Engr., Salem, for 500 second ft. of water to be taken from the John Day River near Early for the development of 600 hp. plant. Applicant will construct a canal, 20 ft wide, and install a turbine water wheel. Total estimated cost, $57,150.

Cal., San Diego—The Bureau of Yards & Docks, Navy Dept., Wash., D. C., will receive bids until March 1 for furnishing and installing complete power plant and electric generating equipment including distribution systems for electricity and steam at the Marine Corps Base, here. Noted Feb. 9.

Alaska, Petersbury—The city voted $75,000 bonds, $25,000 of which is to be used for the construction of a light and power plant and $50,000 for the construction of a 2 story school house. Noted Oct. 14.

Alaska, Port Chatam—The Pacific Sea Products Association, Seattle, Wash., plans to build a 2 story cold storage plant to have a 2,000,000 lb. capacity. Estimated cost, $100,000. O. O. Hytum, Pres.

Ont., Belleville—The Bd. Educ. plans to construct a 3 story collegiate building. A vacuum steam heating system will be installed in same. Total estimated cost, $350,000. A. McGil, Chmn.

Ont., Courtright—The town passed a $20,000 by-law for the construction of a waterworks system including deep wells, pumping machinery, pump house, mains, steel tank, etc. R. G. Stewart, Clk.

Ont., Dresden—The taxpayers authorizo an expenditure of $20,000 for the construction of waterworks system including deep wells, pumps, pump house and mains. Address J. T. Budgewater, Clk.

Ont., Sault Ste Marie—The School Bd. plans to build an addition to present high school and a 2 and 3 story technical school. A vacuum steam heating system will be installed in same. Total estimated cost, $412,000.

Ont., Toronto—The St. Clair Ave. Methodist Church, c/o F. L. Barker, 18 Oriole Gardens, plans to build a 1 story church. An electric operated organ and hot air heating system will be installed in same. Total estimated cost, $125,000.

Ont., Weston—The town plans to install a gasoline driven fire pump as a standkey, 750 imp. gal. pump against about 300 ft. head-3 stage driven by a 150 hp. marine engine. Estimated cost, between $8,000 and $10,000. E. A. James Co., 36 Toronto St., Toronto, Engr.

Ont., Windsor—The Border Cities Hotel Co. is having plans prepared for the construction of a 10 story hotel on Ouellette Ave. A vacuum steam heating system will be installed in same. Total estimated cost, $1,000,000. F. L. Howell, 17 Ouellette Ave., Secy. and Treas. Esenwein & Johnson, Ellicott Sq., Buffalo, N. Y., Archts.

CONTRACTS AWARDED

R. I., Providence—The Eastern Terminal Corp., 103 Federal St., New London, Conn., has awarded the contract for the construction of a 280 x 500 ft. cotton warehouse on Springfield Ave., to the Aberthaw Constr. Co., 27 School St., Boston, Mass. A steam heating system will be installed in same. Total estimated cost, $200,000.

N. Y., Brooklyn—The Standard Oil Co., 26 B'way, New York City, has awarded the contract for the construction of a 1 story, 50 x 90 ft. power house on Calyer and Kingsland Aves., to C. C. Woodruff, 213 10th St., Long Island City. A steam heating system will be installed in same. Total estimated cost, $26,000.

N. Y., Buffalo—The Philadelphia Rubber Co., 37th and Reed Sts., has awarded the contract for the construction of two 1 and 3 story factory, office, garage and buildings, including power house, to the Austin Co., 10112 Euclid Ave., Cleveland, O. Total estimated cost, $1,000,000.

N. Y., New York—The Bowman Realty Co., 225 West 49th St., will build a 4 story, 100 x 200 ft. garage on B'way between 131st and 132nd Sts. A steam heating system will be installed in same. Total estimated cost, $330,000. Work will be done by day labor.

N. Y., New York—The Fiske Rubber Co., Chicopee Falls, Mass., has awarded the contract for the construction of a 20 story, 64 x 150 ft. office building at 1765 B'way, to Fred T. Ley Co., 19 West 44th St. A steam heating system will be installed in same. Total estimated cost, $3,000,000.

N. Y., New York—The Natl. Park Bank, 432-434 Madison Ave., has awarded the contract for altering bank and office building, to Marc. Eidlitz, 30 East 42nd St. A steam heating system will be installed in same. Total estimated cost, $600,000.

N. J., Plainfield—S. Schwartz, 220 B'way, New York City, has awarded the contract for the construction of a 2 story, 100 x 150 ft. theatre on Front St. and Waterbury Ave., to M. Shapiro, 52 Vanderbilt Ave., New York City. A steam heating system will be installed in same. Total estimated cost, $250,000.

Pa., Philadelphia—The Evening Bulletin, Juniper and Arch Sts., has awarded the contract for the construction of a 7 story, 87 x 147 ft. office building addition, to Doyle & Co., 1519 Sansom St. Contract for installation of a steam heating system will be sub-let. Total estimated cost, $500,000. E. V. Seller, 101 Juniper St., Archt.

D. C., Washington—Frank C. Lewen, c/o Carrere & Hastings, Archts., 52 Vanderbilt Ave., New York City, will build a 12 story, 200 x 200 ft. hotel, here. A steam heating system will be installed in same. Total estimated cost, $1,000,000.

O., Cleveland—The Hydraulic Pressed Steel Co., Hydraulic Ave., has awarded the contract for remodeling the 1 story heat treating plant on Hydraulic Ave. to E. R. Russell, Sincere Bldg. Estimated cost, $10,000.

Ind., Indianapolis — The Indianapolis Power & Light Co. has awarded the contract for the construction of a 1 story power extension to H. K. Ferguson Co., 1532 Euclid Ave., Cleveland, O. Estimated cost, $75,000.

Mich., Detroit—The Amer. Electrical Heater Co., Woodward and Burroughs Sts., has awarded the contract for the construction of a 1 story, 60 x 417 ft. factory and boiler house on Commer and Dequindre Sts. to Conrrl Keller, 210 Moran St. One boiler will be installed in same. Total estimated cost, $35,000.

Ill., Chicago—Linstrom Smith & Co., 1104 South Wabash Ave., has awarded the contract for the construction of a 1 story, 130 x 250 ft. factory on Lake St. and Kedzie Ave., to Jacob Rodatz, 209 South La Salle St. A steam heating system will be installed in same. Total estimated cost, $125,000.

Wis., Milwaukee—The Westinghouse Lamp Co., 165 B'way, New York City, has awarded the contract for the construction of a 4 story factory addition, to Stone & Webster, 120 B'way, New York City. A steam heating system will be installed in same. Total estimated cost, $400,000.

Wis., Sheboygan—The Vollrath Co., North Michigan Ave., has awarded the contract for the construction of a 1 story power plant, to the Westinghouse, Church Kerr Co., Inc., 37 Wall St., New York City. A steam heating system will be installed in same. Total estimated cost, $200,000.

Minn., New Ulm—The Eagle Rolling Mill Co. has awarded the contract for furnishing rolling equipment, including boilers, crane and coal handler, to J. G. Robertson, 2542 University Ave., St. Paul. Total estimated cost, $35,000.

Minn., Minneapolis—The Bd. Educ. has awarded the contract for installing a heating system in the proposed 1 story, 60 x 194 ft. addition to the Whitney School, on 19th Ave., N. E., to Bjorkman Bros., 712 10th St. South, at $20,004.

Minn., Minneapolis — The Bureau of Buildings, Bd. Educ., has awarded the contract for installing a steam heating and ventilating system in the proposed 3 story, 61 x 163 ft. addition to Calhoun School, on Girard Ave., to H. Kelly & Co., Plymouth Bldg., at $58,597.

Kan., Cheney—The city has awarded the contract for installing pumps in connection with the waterworks system, to Lutweiler Pump Co., Rochester, N. Y., at $4,000.

Kan., Independence—The Commercial Natl. Bank has awarded the contract for the installation of a heating system in the proposed 6 story, 70 x 140 ft. bank, to the Sell Orr Co., Independence. Total estimated cost, $150,000.

Wyo., Casper—The city has awarded the contract for the construction of a water-gallery in the bed of the Platte River at pumping station intake and for the revision of pump station buildings, hauling and setting up machinery, to W. F. Henning, $47,650. A semi Diesel internal combustion engine will be installed in same.

Tex., Abilene—The Abilene Gas & Electric Co. has awarded the contract for the construction of a new central station to the W. C. Hedrick Constr. Co., Southwestern Life Bldg., Dallas, at $110,000.

Tex., Longview—The Longview Ice & Light Co. has awarded the contract for the construction of a new ice plant, to the W. C. Hedrick Constr. Co., Southwestern Life Bldg., Dallas. Estimated cost, $50,000.

NEW INSTALLATIONS IN POWER

Conn., Hartford—Wise & Npson, 36 Pearl St., is in the market for ½ yd. electric driven concrete mixer and a 50 hp., 3 ph., 220 volt, 60 cycle double drum electric hoist.

N. Y., Long Island City—The Metal Stamping Co., 13th St. and East Ave., is in the market for a steam driven air compressor, to have a 1,200 cu. ft. capacity and inclinable presses.

N. Y., Norwich—The New York State Gas & Electric Corp. plans to construct a 10 unit, 3 ph. transmission line from here to Oxford and install a new Skinner generating unit. O. E. Wessel, Mgr.

Pa., Gouldsboro—H. C. Reiling is in the market for two 60 hp. steam boilers, one 35 hp. steam engine and 1 dynamo.

Pa., Philadelphia—The Philadelphia & Reading R. R., 11th and Market Sts., is in the market for complete electrical power house equipment to replace present equipment in the 1119 Arch St. power house, also for air compressors, ditchers, wrecking crane and radiator valves.

Pa., Pittston—Thomas W. Bentley is in the market for 100 5hp. electric motors.

N. C., Moravian Falls—The Moravian Falls Milling & Power Co. plans to install a new turbine generator, etc. J. T. Humphries, Mgr.

Mich., Milford—The Milford Electric Co. plans to construct a transmission line through the country reaching 4 villages and to install motors and transformers and have further development to water power. F. S. Hubbell, owner.

Wis., Milwaukee—Robert L. Deisinger & Co., 913 1st Natl. Bank Bldg., is in the market for a 16 hp. hoisting engine.

Wis., Rome—The Bark River Electric Service Co. plans to construct a 20 mi. transmission line. A. N. Becker, Supt.

Minn., Alberta—The Southern Minnesota Gas & Electric Co. plans to install a 300 hp. boiler and construct a 40 mi., 13,200 and 6,600 volt transmission line. H. L. Nickels, Pres. and Mgr.

Minn., Sunrise—The Sunrise Milling & Electric Co. plans to construct a 25 mi. new 6,600 volt transmission line. F. A. Spivak, Engr.

Kan., Hanover—The Hanover Municipal Electric Light, Power & Water Plant plans to install a 325 kva. hydro-electric plant on the bank of the Little Blue River. Estimated cost, $75,000. Ray C. Walker, Chief Engr.

Neb., McCook—The McCook Electric Co. plans to install one 500 kva. turbo condensing generator. Alex. Speer, Secy. and Mgr.

Mont., Mansfield—The Mansfield Light & Power Co. plans to install a generating engine. Walter C. Coday, Mgr.

Okla., Laverne—Laverne City Light Dept. plans to install 2,400 volt alternators and transformers. Claude H. Arbuthnot, Supt.

Ore., Keno—The Keno Power Co. plans to construct a 6 mi. transmission line and install a 800 hp. turbo unit. J. W. Kerns, Pres. and Mgr.

R. R., Arecibo—The Municipal Electric Light Plant plans to construct an extension to transmission lines and install a 75 kw. generator and a 100 hp. Mietz & Weiss engine. J. M. Ginzaley, Supt.

Ont., Campbellford—The Municipal Electric Light Plant plans to increase transmission voltage to 6,600 in order to supply additional hp. to one of its present power users. J. F. McGregor, Acting Secy.

Ont., Ottawa—The Mortimer Co., 259 Sparks St., is in the market for one 15 hp. d.c. 500 volt motor compound wound, speed not more than 1,200 r.p.m.

POWER

Volume 51 New York, February 24, 1920 *Number 8*

Buckle Down

It was the noon hour, and the machinery in the great power house was still. The men, after finishing their midday meal, had formed into little groups. Some were gossiping, some disposing of the problems of the day according to their own particular ideas. The superintendent was leaning against a pillar, idly watching the smoke rise from his pipe. Near-by two men were talking earnestly. As their conversation waxed warmer, their raised voices brought their words plainly to the superintendent's ears.

"It's no use to try to get ahead," said one, moodily. "There's no chance for a man these days. I've worked here four years, and the other day when they wanted a foreman, they gave the job to a fellow who hadn't been here two. Ain't I as good as he is? Sure I am. I was here the longest and I ought to have had the job."

The superintendent walked slowly toward the men and stopped in front of them.

"Boys," he said, "I don't want to butt in, but I couldn't help hearing what you said just now, and I want to tell you something that happened to me.

"I came to these works ten years ago as a machine hand. I learned the ropes a little, and for three years I worked steadily. Then the foreman left and I applied for the job. I didn't get it, though. Another man went over my head. It gave me a shock, for I was pretty sure I'd get the job. I had been here the longest of any man in the shop, and I thought I was entitled to it for that reason alone.

"Then I began to ask myself why I had been turned down. The answer was plain. I had been on the job every day for three years, but I hadn't done a bit more than was asked me to do, and I hadn't learned to do anything but my special work. One night I took stock of myself; the result wasn't pleasing. I made up my mind if I was going to advance I had got to do something to make myself worthy of promotion.

"Then I buckled down to work. Five nights a week I went to a school, where I learned about electrical and steam engineering, shop practice and drafting, and things that I needed to know in my line.

"You can bet it was hard work after being in the shop all day, and many a time I was tempted to give it up. But I was ambitious and had my eye on the foreman's job, so I kept on. Often the boys would call me a fool for devoting my evenings to study instead of enjoying myself as most of them did.

"Well, I kept it up for two years and had just started on my third year when the foreman left. Boys, I didn't have to apply for the job that time. They offered it to me. I don't want to boast, but I'm superintendent of the whole plant because I was 'fool' enough to fit myself for the job.

"Beg your pardon, boys, for butting in, but when I heard you complaining about not getting ahead, I thought I'd tell you how I did it. There's only one way and that's to buckle down, get busy, and—"

Just then the whistle blew, the men went to their places, and the great wheels began to revolve once more. And all the afternoon you could hear them say, amidst the whir and noise:

"Buckle down! Buckle down! Buckle down!"

Contributed by Frank Dorrance Hopley.

ELECTRICALLY DRIVEN HOISTS

By
Fraser Jeffery*

Unbalanced single-drum hoisting and the more efficient balanced hoist employing a second cage or a counterweight. The character of load imposed by different types of hoist and different hoisting depths. Prevalent use of slip-ring induction motor. Details of electric braking and a comparison of full magnetic and liquid control.

ELECTRIC motors are fast superseding the steam engine for the operation of hoists. This is due to a great extent to the comparative ease with which electric energy can now be transmitted to the districts where the mines, quarries, etc., are located and to the greater convenience and economy of the electric motor over the steam engine.

The usual function of any hoist is to elevate material. The material may be moved either in a vertical shaft or on an incline, and the work may be accomplished by means of a "balanced" or an "unbalanced" hoist.

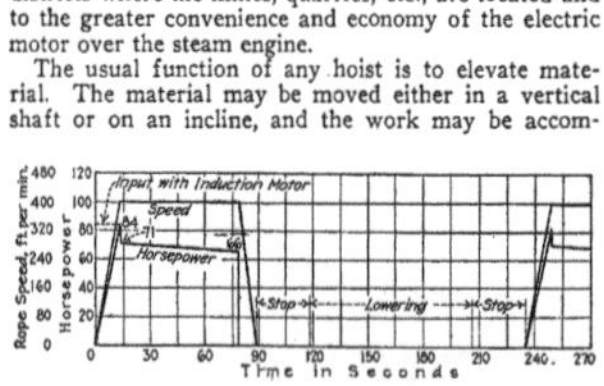

FIG. 1. PERFORMANCE OF SINGLE-DRUM HOIST OPERATING IN SHAFT 500 FEET DEEP

For the balanced hoisting the arrangement usually consists of two drums on a common shaft (see headpiece) or two drums on parallel shafts driven by a common pinion. Each drum is preferably driven through a clutch so that the relative position of the drums, hence the skips or cages, can be changed.

For unbalanced hoisting the arrangement generally consists of a single drum on which the hoisting rope is wound in one or more layers, as in Fig. 4. The drum may be keyed to the shaft supporting it or connected by means of a clutch. With the former arrangement the driving motor and gearing naturally revolve in lowering, while with the latter only the drum turns. When the hoist is operated unbalanced, the energy expended in hoisting the skip or cage, together with the car and rope, is wasted. Since the weight of these parts frequently amounts to more than the useful load hoisted, the efficiency with an unbalanced load is necessarily low.

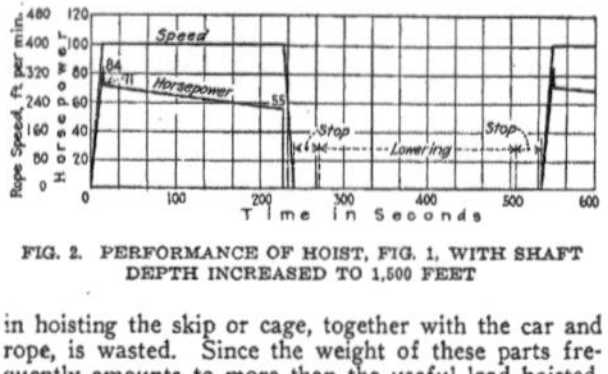

FIG. 2. PERFORMANCE OF HOIST, FIG. 1, WITH SHAFT DEPTH INCREASED TO 1,500 FEET

To increase the efficiency of hoists by the elimination of these losses, two skips or cages are arranged in balance, one to ascend while the other descends, the ascending skip or cage being loaded and the descending one being empty. This is known as the "balanced" hoist. Two skips or cages operating in this manner require two-compartment shafts. With a single-compartment shaft similar results may be obtained by using a counterweight, and in this manner the ascending skip or cage can be properly balanced. The counterweight is taken generally to equal in weight, one-half of the live load plus the dead load. This method of counterbalancing gives the same load on the driving motor when hoisting the material as when hoisting the counterweight.

There are numerous modifications to either of the foregoing types of hoists, each in its way affecting the power required to do the work. This necessitates a close investigation of each condition to be met. A study of Figs. 1, 2 and 3 will give some idea as to the kind of load imposed upon a motor by different kinds of hoists

*Electrical Engineer, Allis-Chalmers Manufacturing Co., Milwaukee.

working under the same conditions and for different hoisting depths.

In Fig. 1 a single-drum "unbalanced" hoist is assumed, having a maximum rope pull of 5000 lb. with a rope speed of 400 ft. per min. and a shaft depth of 500 feet. If the rope diameter is ¾ in., making the weight of the rope 440 lb., a total weight of 4560 lb. of skip or cage and material to be hoisted is permitted. If full speed is reached in 13 seconds, the curve of torque over time will be as shown. It will be noticed that the peak of the load is extremely high compared with

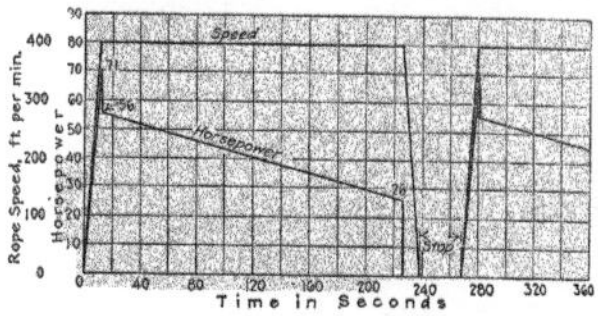

FIG. 3. PERFORMANCE OF BALANCED HOIST OPERATING IN SAME SHAFT AS FIG. 2

the average load and that the motor is in service only a short part of the total time, the rest period of the motor being formed by two stops and the lowering period, in addition to the small amount of time required for braking.

Fig. 2 shows the performance of this same hoist when used for hoisting from a 1500 ft. depth instead of 500 ft. Since the rope now weighs 1320 lb., the combined weight of skip or cage and material to be moved cannot be more than 3680 lb. The beginning of the load curve will be the same as for Fig. 1, until the 1000-ft. level is reached, when the combined weight of the cage, material and rope is equal to the weight of the cage and material in Fig. 2. After this the load will continue to decrease at the same rate as before until the top is reached, at which time the rope pull will be equal to the weight of the cage and material alone, or 3680 lb. The rest period of the motor now covers a smaller percentage of the total time than before, because the time of stops remains constant.

Consider this same hoist operating in "balance," which would be the case if a second drum and a second skip or cage were added to the equipment. The load due to the descending skip or cage and rope must be deducted from that due to the ascending skip or cage, material and rope, and a certain amount must be added during the acceleration period, which will be higher than the corresponding amount with the single-drum hoist, in proportion to the larger masses to be accelerated. Fig. 3 shows the performance under these conditions when hoisting from a 1500-ft. depth. It is obvious that the rest period of the motor with the "balanced" hoist is much shorter than with the "unbalanced" hoist, since no lowering is done by gravity, a load being hoisted every trip.

It is seen from the foregoing that the operating conditions of a hoist are not only intermittent, but the rope pull varies continually during the hoisting period. Therefore, the capacity of the driving motor must be selected according to the peak during acceleration and according to the heating of the motor when performing its cycle of operation at the regular intervals. This latter is determined by calculating the "root mean square" values of the load over the operating cycle.

With very large hoists the problem of furnishing power for a service of comparatively small average power demand with intermittent peaks sometimes becomes serious. Where the prime-mover equipment and the line conditions are such that intermittent fluctuations in load are permissible, the alternating-current slip-ring type of induction motor is invariably used. Direct-current motors are not so often used, mainly for the reason that the distribution of power is almost always by means of alternating current. For the hoisting conditions prevalent in this country the induction motor is used most extensively, while the number of direct-current installations is limited more or less to special applications or to conditions where the size of the unit is such as to demand a smoothing out of the load peaks.

If the reduction of line peaks becomes necessary, a system involving the use of flywheels for storing energy has been resorted to. Briefly, it comprises a direct-current hoist motor and a motor-generator set consisting of a

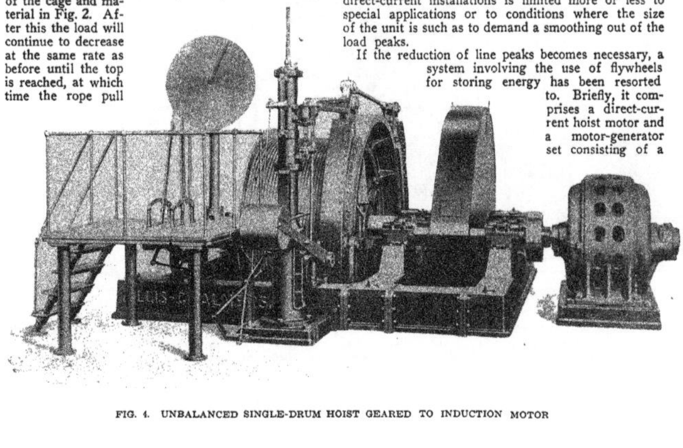

FIG. 4. UNBALANCED SINGLE-DRUM HOIST GEARED TO INDUCTION MOTOR

direct-current generator driven by a motor which is usually of the alternating type, and equipped with a flywheel. To avoid the starting losses inherent with any type of resistance control, voltage regulation between the direct-current generator and the hoist motor is used. By proper field control, there are times when the hoist motor may be advantageously operated as a generator and the generator of the flywheel set as a motor, thus returning energy to the flywheel and reducing the power demand from the line. This is commonly known as "dynamic braking" and is resorted to at the time when ordinary braking would occur. This is usually taken care of by the motor-control lever actuating the automatic control, and in this way a simplified means is provided the operator for performing an otherwise complicated function.

With the alternating-current induction motor a means of braking, commonly called "electric braking," is frequently resorted to. It is used mostly in places where hoisting is done on long inclines and more with the "unbalanced" type of hoist than with the "balanced" hoist. In such an arrangement the motor revolves during the lowering period. Under these conditions and at this particular time, if the motor, with all external secondary resistance cut out and secondary winding short-circuited, is thrown across the line, the weight of the descending parts tends to operate the motor above synchronous speed. This action is such as to return energy to the line and to develop a counter-torque sufficient to act as an effective brake, thus not only reducing the power demand, but increasing the longevity of the friction brakes. This functioning is also taken care of in a simplified manner to facilitate ease of handling for the operator.

In addition to a careful analysis of load cycles and selection of the proper size of driving motor, the question of control is of considerable importance. As

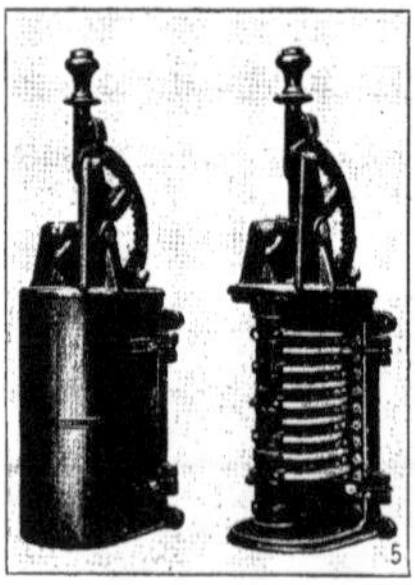

FIG. 5. MASTER SWITCH FOR CONTROLLER PANEL

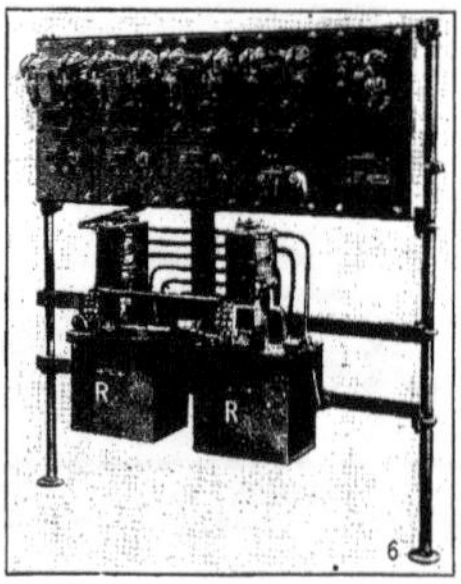

FIG. 6. ELECTRICALLY DRIVEN HOIST CONTROL PANEL

brought out before, the direct-current installations are limited to more or less special applications and to the large hoists where the question of line peaks is likely to become serious. In general, the slip-ring type of induction motor predominates for the average hoist installation, therefore it would seem logical to give some data on the control for this type of machine rather than to describe some one special direct-current application.

Outside of the usual drum-type controller in use with the smaller hoist motors up to about 100 hp., there is a choice between the full-magnetic type of control and the liquid type of control. The magnetic control con-

FIG. 7. LIQUID RHEOSTAT AND MAGNETICALLY OPERATED REVERSING SWITCHES

sists of a master switch or lever, Fig. 5, in the hands of the operator. This switch, for different positions, actuates relays mounted on a separate panel, as in Fig. 6, and in this manner controls the primary and secondary circuits, the speed, direction of rotation and the torque of the induction motor by opening or closing the primary switches and by cutting in or cutting out resistance in the secondary circuit.

For the liquid control, Fig. 7, the equipment consists essentially of two parts—a primary reversing-switch mechanism and a tank for the control of the secondary circuit. The primary of the motor is controlled by two magnetically operated reversing switches *R*, as shown on the small panel, interlocked so that only one can close at a time. This type of primary reversing switch is used for motors up to 550 volts. For higher voltages high-tension reversing oil switches, such as are shown at *R* and *R* in Fig. 6, are substituted.

The tank for the secondary control is divided into an upper and a lower compartment. The lower compartment is the storage and cooling tank for the electrolyte, which is pure water in which a certain amount of sodium carbonate has been dissolved. In the upper compartment the electrodes, consisting of a set of plates connected in each phase of the induction-motor's secondary circuit, are suspended from an insulated support. The amount of resistance in the secondary circuit, and consequently the speed and the torque of the motor, are governed by the height of the electrolyte in the upper compartment. This in turn is regulated by means of a multiple-shutter type weir located at one end of the upper compartment and controlled directly by the hand lever in the operator's pulpit. The opening and closing of the weirs regulates the height of the electrolyte.

Only one hand lever is used to operate the controller for both forward and reverse directions of rotation of the motor. The direction in which the hand lever is moved from the central, or "off," position determines the direction of rotation of the motor. The first movement of the operating lever from the central forward position actuates a small master switch *S*, which is mounted on the base of the tank, as shown in Fig. 7. This master switch closes the primary switch for forward direction of the hoist motor. The remaining travel of the hand lever gradually closes the weir, causing the electrolyte to rise, thereby cutting out resistance and bringing the motor up to speed. Returning the lever to the central position opens the weir, which in turn lowers the level of the electrolyte, inserting resistance in the rotor circuit and opening the primary switch, thus cutting off current entirely from the motor. By moving the lever in the opposite direction, the same cycle is obtained except for the reverse direction of the hoist motor. The small motor *M* on the base of the tank, Fig. 7, operates a pump *P* and constantly circulates the electrolyte from the lower to the upper compartment. It is thus seen that with the weirs closed the electrolyte level will rise, but with the weirs open it will remain at low level such as to insert full resistance in the secondary circuit of the motor.

For the smaller sizes of motors of 100 hp. and up to 300 hp. the full-magnetic type of control, such as is shown in Fig. 6, has been used extensively. For motors of this size the first cost of the magnetic control, not considering installation cost, is lower than that of the liquid control. The cost of installing the magnetic control is, however, greater than that of the liquid control. It is generally conceded that for smoothness of acceleration and minimum line disturbance, especially for the larger sizes of hoist motors, the liquid control is preferable. The following citation of an actual installation first using magnetic control, then changing to liquid control, will bring out some of these points.

A large coal and coke company had in operation a double-drum hoist equipped with a 350-hp. 550-volt 60-cycle slip-ring induction motor governed by a contractor-type magnetic control. The hoist operated two cages in balance in a vertical shaft, the approximate hoisting distance being 500 ft., and a trip was made every 30 seconds. The power for the hoist was supplied from the coal company's central station, where a 2500-kw. turbine carried the load of all the mines. In addition to the various hoisting loads there were some fairly large pump motors and small rotary converters.

On account of the rapidity of the trips, the accelerating peaks on the hoist motor were naturally high and amounted to approximately 720-hp. torque. This caused considerable line disturbance, which was further augmented by the action of the magnetic controller, being amplified as each successive switch closed. The line fluctuations when the hoist was operating were so great that the flickering of the incandescent lamps was pronounced.

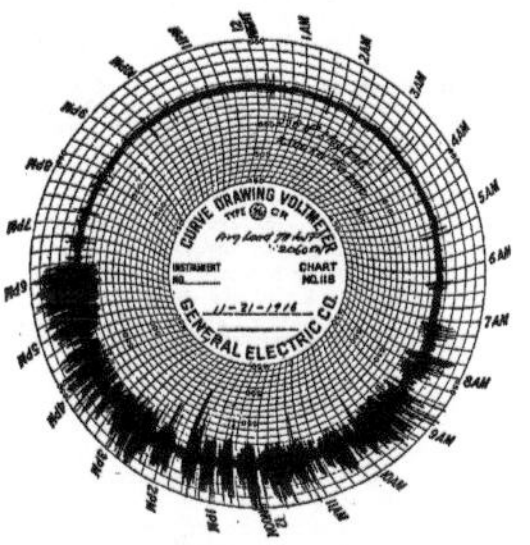

FIG. 8. VOLTAGE CHART WITH 350-HP. MOTOR OPERATED WITH MAGNETIC CONTROL

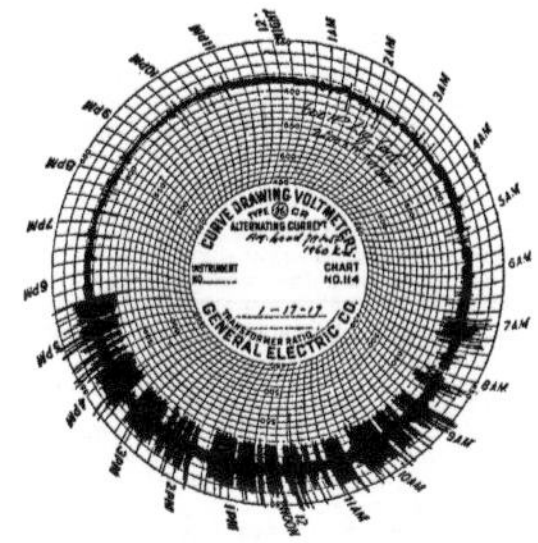

FIG. 9. VOLTAGE CHART WITH 600-HP. MOTOR OPERATED WITH LIQUID CONTROL

It was decided to increase the output of the mine by speeding up this particular hoist so as to make a trip in 22 seconds. Calculations show that the new conditions would require a 600-hp. hoist motor and that the peaks at acceleration would amount to approximately 1000 hp. torque. On account of the increased peak load, serious consideration was given to the advisability of using an equalizer flywheel set instead of an induction motor. This would have involved large first cost, considerably more floor space and a longer time of shutdown to make the change. After a careful analysis of all the operating conditions, it was finally decided to install a 600-hp. slip-ring type induction hoist motor with a liquid controller.

The two voltage charts, Figs. 8 and 9, were taken from the recording meter on the switchboard of the coal company's power plant. Chart, Fig. 8, was taken when the 350-hp. motor with magnetic control was operating the hoist just before the change was made. Chart, Fig. 9, was taken after the hoist had been speeded up and was being operated by the 600-hp. motor with liquid control. The marked regularity of operation obtained with the liquid controller is noticeable. This is due to the gradual and uniform acceleration with the liquid type. The actual peaks at acceleration under the new conditions amounted to nearly 1200-hp. torque, but even with the greater peak the sudden voltage fluctuation was less than obtained with the 720-hp. torque peak and the smaller motor. The acceleration with the liquid type of control is so gradual that the governor on the steam turbine has time to pick up and follow the load variations and respond to the line demand.

The change resulted in the elimination of the flickering of the incandescent lights. A graphic meter with minute feed would have shown more clearly that the voltage fluctuations in the case of chart, Fig. 9, were gradual and uniform rather than sudden and instantaneous, as was the case of chart, Fig. 8, but such a meter was not available at the time the tests were made.

The Origin of Petroleum

Many different authorities have given different reasons for the origin of petroleum. The question, however, has now come down to a controversy between the advocates of the theory which supposes petroleum to be of inorganic origin, and those who claim it to be of organic origin. The latter by far predominate.

In the year 1794 Haquet suspected that the oils of Galicia came from marine mussels "dissolved" in salt water. Many eminent geologists have accepted this theory. The assumption is that decomposition of these organic bodies took place at comparatively low temperature and high pressure. It is interesting to note that in a paper before the Institute of Mining Engineers, 1914, Dr. Hans von Höfer, of Vienna, states that C. Engler tested his (Höfer's) thesis experimentally by heating in retorts under a pressure of ten atmospheres and to a temperature of from 300 to 400 deg. C., first fish oil, and afterward fishes and mussels. He obtained in this way a product very like petroleum.

Some geologists assign an animal origin to the nitrogeneous California petroleum and a vegetable origin to the nonnitrogeneous oils of Pennsylvania. The greatest defender of the organic-origin theory in America is Eugene Coste, of Toronto, who sees in petroleum the product of "volcanic solfataric emanation."

The following is from a paper by Coste before the New York meeting of the Institute of Mining Engineers, February, 1914:

Sedimentary strata are composed of alternate beds of shales, sandstones and limestones, most of which are sufficiently close-grained to have retained from the time of their deposition enough water held in their minute pores by capillary action to fill all the spaces between their grains and render them entirely impervious to other fluids. Shales or consolidated clays form the bulk of the sediment, and are especially fine-grained saturated impervious rock, but many beds of sandstone or limestone are also quite impervious. A sufficient proof is the very strong natural gas pressures always recorded in the gas and oil deposits and the great difference in the pressure of the gas of different sands tapped at different depths of the same well or in near-by wells in the same pool or field.

To explain the retention of this gas under strong pressures (up to 1500 lb. per sq.in. in some cases) in so many thousands of separate spots and at depths sometimes as shallow as 100 ft. and varying from that to 4000 ft., we must certainly conclude that the sedimentaries forming the substrata of this vast region and ranging through many geological ages, are remarkably impervious. It is for this reason that the petroleum deposits have not escaped to the surface and have been preserved in a multitude of separate pockets, pools or fields with different pressures, though often quite close to one another and always comparatively near the surface.

Prof. Edward Orton, in his report on the occurrence of petroleum, natural gas and asphalt rock in western Kentucky, says in the "Kentucky Geological Survey," 1891:

The statements now presented .... bring before us two main views as to the origin of petroleum: First, petroleum is produced by the primary decomposition of organic matter, and mainly in the rocks that contain the organic matter. Second, petroleum results from the distillation of organic hydrocarbons contained in the rocks, and has generally been transferred to strata higher than those in which it was formed.

Discussing the age of petroleum deposits, Dr. Orton says: "Hunt believes petroleum to have been produced at the time that the rocks that contained it were formed, once for all." Newberry believes it to have been in process of formation solely and constantly since the strata were deposited. Peckham refers it to a definite but distant time in the past, but long subsequent to the formation of the petroliferous strata. He supposes it to have been stored in subterranean reservoirs from that time to the present.

Dr. David White, speaking of the rapid disintegration of animal matter, says:

This forcibly suggests that the algæ which almost certainly have exerted an attraction for certain bitumens of extraneous origin, may have played the important rôle, first of storing up some of the hydrocarbon products of the decaying animal matter, and later, under dynamic influences, of releasing a part at least of the stored as well as, presumably, of their own original content in the form of petrolic or gaseous hydrocarbon.

Dr. F. W. Clarke believes that this transformation took place after burial: "Whatever sediments are laid down, inclosing either animal or vegetable matter, where bitumen may be produced, the presence of water, preferably salt, the exclusion of air, and the existence of an impervious protecting stratum of clay seem to be essential conditions toward rendering the transformation possible.[1]"

Cunningham Craig says: "It seems probable—but here we enter into speculation—that it is the pressure that is the determining factor, as it is in so many chemical reactions. Given the vegetable matter from which petroleum can be formed, and inclosed in a well-sealed deposit, given the presence of a limited quantity of water and the necessary but by no means high temperature, as soon as the pressure reaches a certain point the action will begin.[2]"

Concluding a discussion of the many theories to account for the origin of petroleum, E. T. Dumple, in a paper before the Institute of Mining Engineers, February, 1914, states:

While many, if not all, of the theories advanced as to the production of petroleum have in them some element of truth, and while we believe that under diverse conditions petroleum may be and is formed in many different ways, the theory outlined, which recognizes the formation of petroleum under suitable paludal, bog or marine conditions at ordinary temperatures, by natural processes, the transformation being accomplished by the organic matter still retained essentially in its original character, and its deposition as petroleum with other sediments of its day seems well supported by observational facts, and its very simplicity and general applicability appear to offer strong grounds for the belief that it will sooner or later come to be recognized as one of the principal methods of the formation and accumulation of petroleum.

In such a case all indigenous or primary deposits of petroleum will be of the same age as the sediments in which they occur and those of secondary or migrating oil would be of a different age than the inclosing beds.

Crude oil is petroleum that is in its crude state.

Fuel oil, as now delivered, is petroleum from which the gasoline and other highly volatile constituents have been driven off by distillation.

[1]Data of Geochemistry, p. 703, 1911.

[2]Oil Finding, p. 28.

Power Drives for Rolling Mills

By W. O. Rogers

A new boiler plant equipped with the latest apparatus and planned to be the last word in steel-mill boiler installation. The ash- and coal-handling equipment is of especial interest. Stokers are motor-driven and forced- and induced-draft fans are turbine-driven through a reduction gear.

AN EXAMPLE of the modern boiler plant for steel mills was found during my stay around Youngstown. This was at the Republic Iron and Steel Co.'s works, where they were building a new boiler plant in which was incorporated the best ideas and equipment that the engineers of the company could devise and procure. The reason for building this new boiler plant was because a greater steam volume was required than the existing plants were capable of producing and because it was necessary to change from gas to coal at intervals, the gas supply from the blast furnaces being insufficient to supply the fuel necessary to generate sufficient steam to operate the steam unit. As a large number of small boilers cannot be operated as economically as a battery of large units, it was decided to install modern high-pressure boilers equipped with front-feed underfeed stokers, superheaters and economizers, the boilers to be operated at from 100 to 200 per cent above rating.

The new boiler house is situated at a distance of 250 ft. from the turbine house, to which the steam will be piped. The building is of brick, with one end shut in with a false wall of corrugated iron, thus providing for future extension.

At present there are four 1,130-hp. water-tube boilers having four steam drums each 42 in. in diameter and 22 ft. 4 in. long. The drums are made with three $\frac{11}{16}$-in. sheets with triple-riveted butt-strap joints. The circular seams are lap-joint, double-riveted. The 4-in. tubes are arranged 36 wide and 14 high, and are 20 ft. long, giving a total of 11,296 sq.ft. of heating surface. The tubes are arranged in two banks. The lower bank is two tubes high with a space between the top row and the bottom row of the 12-high top section equal to that occupied by two rows of tubes. On top of the lower bank of tubes a horizontal baffle is used, running from a short distance in front of the vertical baffle of the first pass to the rear header. This gives both vertical and horizontal baffling to the boiler. The arrangement described is shown in the sectional view, Fig. 1. All pressure parts are made of openhearth steel at from 52,000 to 58,000 lb. tensile strength. The main steam outlet is 10 in. in diameter, and the boilers are designed for 225 lb. pressure with 125 deg. of superheat. Each weighs 785,000 lb., or 392.5 tons, and has a width of 24 ft. 10 in. and a height of 24 ft. 8¾ in. from the foundation to the steam outlet. Each boiler is equipped with a double U-bend superheater made up of four headers. The tubes are 71 wide, and arranged four deep with two high U-shaped tubes connecting with each header. Steam from the boiler steam drums enters the top superheater header and, passing through its respective tubes, enters the boiler header which is connected by nipples with the lower center header. From this header the steam passes through its respective tubes to the upper center header and then goes to the steam main.

In designing the furnace, attention has been given to the problem of obtaining proper combustion of the fuel burned. The furnace is built 12 ft. high and the width of the boiler. There are four 12-retort underfeed stokers, each 21 ft. 3 in. wide, with a length of 10 ft. 10 in., and 22 tuyeres high. The windbox under each stoker is sectionalized, so that by the operation of three dampers if any one of four plungers gets out of order the air can be cut out from that zone and the remainder of the stokers used until repairs can be made, thus preventing the boiler as a whole from being put out of service. Each stoker is equipped with four power boxes, thus sectionalizing the stoker into four sections of three retorts each.

Bituminous run-of-mine coal is to be used, and the stokers are designed to burn sufficient coal of not less than 12,000 B.t.u. to deliver 215 per cent of normal boiler rating continuously and to deliver, with coal of the same B.t.u., 300 per cent of normal rating for a duration of four hours.

The following are data of the stoker performance:

Percentage of Rating	Boiler and Furnace Efficiency	Flue Gas CO_2 Not More Than Per Cent	Air Required per Stoker at the Stoke	
			Cu. Ft. per Min.	Second Pass
125	75	13.5	17,400	2.0
150	75	13.5	21,000	2.5
175	74	13.5	25,000	3.0
200	71	13.5	29,200	3.5
225	69	13.5	34,000	4.0
250	67.5	13.5	38,600	4.6
300	....	....	50,000	6.0

The stokers are driven by means of four 20-hp. motors, one for each unit, and having a variable speed of

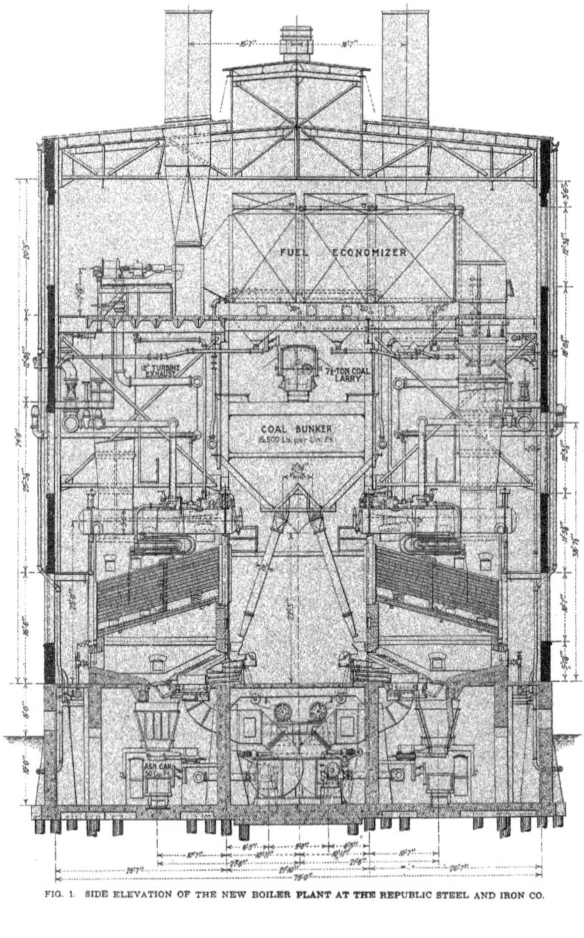

FIG. 1. SIDE ELEVATION OF THE NEW BOILER PLANT AT THE REPUBLIC STEEL AND IRON CO.

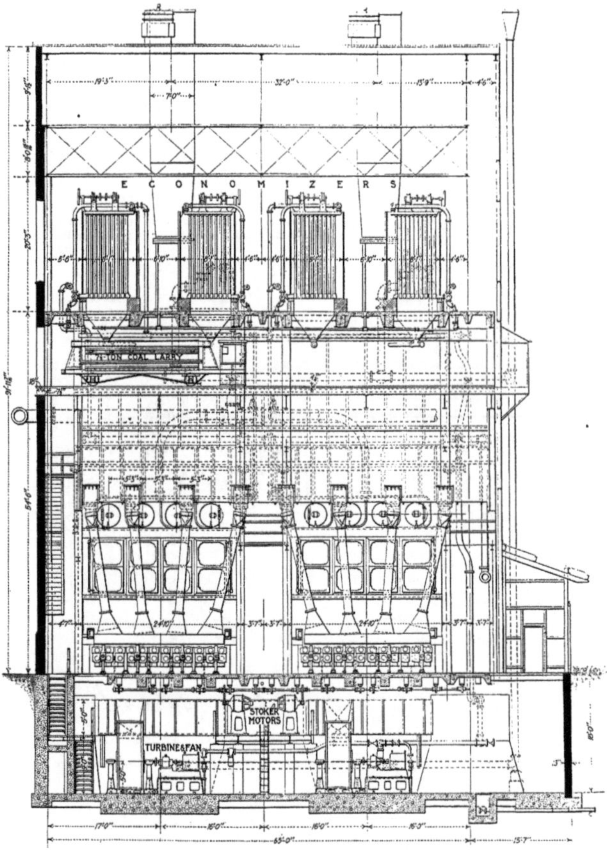

FIG. 2. FRONT ELEVATION OF THE NEW BOILER PLANT AT THE REPUBLIC STEEL AND IRON CO.

from 300 to 1,200 r.p.m., thus giving a stoker speed for any capacity from minimum to maximum. A single silent-chain drive is used between the motors and the driving shaft, but four chain drives are used with each stoker, there being a clutch for each four sections into which the stoker drive is divided.

All four motors are mounted in the basement on a platform built on I-beams inserted at the ends of the

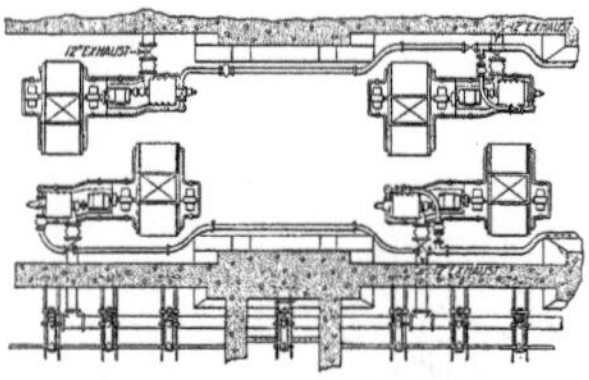

FIG. 3. PLAN OF FORCED-DRAFT LAYOUT

concrete tiers that support the front portions of the boiler (see Fig. 2). This gives a headroom below the motor of 8 ft. 10 in. The stoker driving shaft is also placed in the basement, being suspended in suitable bearings hung from the boiler-room floor.

As the stoker motors are located below the operating aisle between the boilers, two of which are on each side and mid-distance of the length of the boiler room, ample space is provided for the four turbine-driven forced-draft fans, which are also placed in the basement. Each turbine is of 130 b.hp. capacity and drives its connected fan through the medium of a reducing gear, the pinion having a speed of 3,510 r.p.m., and the fan 1,350 r.p.m. Each turbine is designed with eight stages to operate either condensing or noncondensing with 225 lb. steam pressure and 200 deg. superheat. A direct-driven governor controls the speed, and an emergency governor which will automatically close its governor valve when a predetermined speed limit has been passed.

The turbines are designed to operate either condensing or noncondensing. When operating condensing extra nozzle holes are plugged, but if the turbine is to operate continuously noncondensing, the plugs are removed. At least 10 per cent better economy will be obtained running condensing than operating noncondensing with the bypass valve open. Each turbine will develop full or partial load without the use of the bypass or overload valve and will also develop one-half greater overload without the use of the bypass, but with the overload valve open.

These turbines are designed to carry their load at their respective speeds with any steam pressure as low as 70 lb. gage, running noncondensing. With both bypass and overload valves closed, the turbines will carry their normal rated load and speed when operating condensing. With the overload valve open and the bypass valve closed, they will carry 25 per cent overload condensing, and with the bypass valve open and the overload valve closed each machine will carry its normal rated load and speed when operating noncondensing. With both bypass and overload valves open the turbines will carry 25 per cent overload operating noncondensing.

The following are the guarantees for condensing operation load or normal full load of 130 b.hp., and with a turbine speed of 3,510 revolutions per minute:

Load Per Cent	Horsepower	Steam Consumption per Horsepower-Hour Condensing	Noncondensing
125	162.5	21.3	45.6
100	130.0	20.5	43.4
75	97.5	21.6	47.5
50	65.0	24.3	51.5

The following horsepower guarantees are made with the speeds given:

Hp.	Turbine r.p.m.	Gear r.p.m.	Pounds Steam Condensing	Noncondensing
130.0	3,510	1,330	20.5	43.7
70.2	2,940	1,130	24.5	50.8
36.1	2,275	875	35.7	77.0
21.7	1,935	740	47.6	104.5

Each turbine is connected to its individual fan by a double-helical reduction gear and a flexible coupling, the latter being placed between the reduction gear and the fan. All four forced-draft fans are of the double side-inlet type. The full housing is horizontally split along the center line of the shaft, which permits of removing the rotating element without breaking the flue connection. The discharge outlet is at an angle of 45 deg. and connects to the air duct with easy bends, which reduce the resistance of the air in passing through

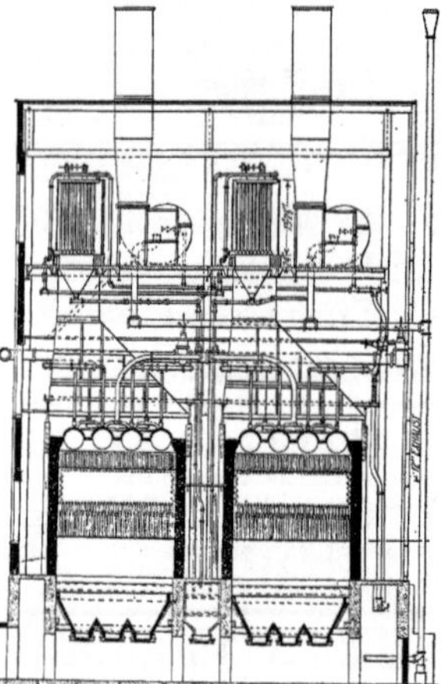

FIG. 4. SHOWING ECONOMIZER CONNECTION TO SMOKE FLUE

the connection. Two fans are connected to each of the two ducts, which extend below the boiler-room floor for the length of the boiler setting, one on each side of the operating aisle. Each air duct is divided in the center by a damper, so that either boiler can use either fan, or in case of emergency one fan can take care of both boiler furnaces under certain rating. Each air duct is 5 ft. high and 3 ft. 3 in. wide, inside dimensions. The four outlets to the windboxes of each stoker connect the main air duct with easy curves. In the discharge of each fan is mounted a louvre type of damper connected through the fan housing with a common lever for hand operation. The approximate rate of each forced-draft fan is 5,200 lb.; that of the induced-draft fan is 8,140 lb. Fig. 3 is a plan view of the forced-draft fan layout.

The capacity of each forced-draft fan at 1,350 r.p.m. is 60,000 cu.ft with air at 70 deg. F., and the capacity of the induction-draft fan at 690 r.p.m. is 97,000 cu.ft. per min., the static pressure being 6 in. of water. The draft loss is figured as follows: Draft loss through the boilers, 2 in.; loss through the economizers, 2.65 in.;

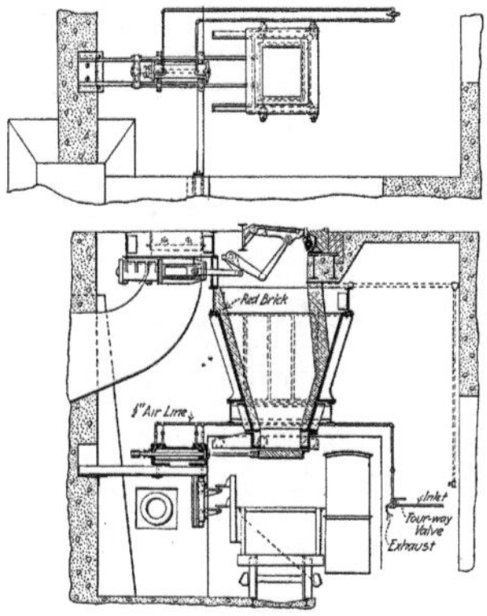

FIG. 5. DETAILS OF ASH HOPPER AND ASH GATES

through flues, 0.25 in.; total draft loss through boiler, 4.9 in.

Above the boilers is a floor on which the four economizers and the four turbine-driven induced-draft fans are placed, Fig. 4. Each fan discharges into a venturi stack, 7 ft. in diameter at the top and tapering to 5 ft. 4 in. at the bottom, the taper extending for a distance of 11 ft. The top of each stack is 34 ft. above the outlet connection of the fans.

Nothing could be more directly connected and compact than the arrangement of the economizers and the induced-draft fan. The flue connection from each boiler to its economizer is vertical and is of the venturi design, being 7 ft. in diameter at the top and 6 ft. 4 in. at the boiler end. Each economizer contains 520 pipes arranged in 52 sections, 10 pipes wide, approximately 12 ft. long, and having a total heating surface of 18,500 square feet.

Cross-circulation is used. The pipes are arranged so that each economizer is divided into three groups. Water enters the bottom cross-pipe in the first group, and the hot gases enter at the opposite end of the economizer, thus adopting the counterflow principle. From the bottom header the water flows upward through the tube of that section, and into the top headers, and then into the top cross-pipe which connects with a bottom cross-pipe of the second group; then, as before, to the top cross-pipe of the second group, and by a cross overpipe to the bottom cross-pipe of the third group. The water is delivered from the top cross-pipe to the third group of pipes at its maximum temperature, and is then fed to the boilers. With a gas-inlet temperature of 615 deg. F. and a gas-outlet temperature of 397 deg., the water, with an inlet temperature to the economizer of 190 deg., will have an approximate outlet temperature of 300 deg. The approximate weight of each economizer is 240,000 lb., or 120 tons.

Each economizer is fitted with three 2½-in. combination blowoff valves and three release valves, one of each for each section. They are inclosed with sectional russia iron, lined with 2-in. asbestos, each section being easily removable for inspection. The scraper drive for each unit is a 13-hp. motor.

Ashes from the stokers of the boilers are deposited in an individual three-outlet iron ash hopper, lined with red brick. Each is constructed so that the sections which are bolted together are 21 ft. 3 in. long and 6 ft. 8¾ in. wide at the top and 2 ft. wide at the bottom, the outlets being 2 x 3 ft. Pneumatic operated gates, details of which are given in Fig. 5, are fitted to the outlet of the ash hopper, the gate being operated on rollers and actuated by a plunger controlled by a four-way valve. The ashes are dumped into motor-operated cars having a capacity of 50 cu.ft. and run on a 2-ft. 6-in. gage track. The loaded car is run to a hopper, from which the ashes are elevated to the top of the hoist and automatically dumped into a chute leading to a bin, from which they are loaded into railroad cars for removal.

The operation of the ash hoists is of interest. When the ash handler desires to elevate a load of ashes that has been dumped into the hopper, he presses a starting switch to start the 50-in. d.c. motor. When the 70-cu.ft. capacity skip nears the top, a resistance coil is cut in on the circuit, which causes the motor to slow down. When the ash bucket reaches the dumping position, the circuit is automatically cut and the bucket stops long enough to discharge its contents. By means of a relay switch on the control board, after the bucket has remained at the dumping position a sufficient length of time to dump, the current is automatically applied to the motor, which has been reversed by means of a limit switch, and the bucket is sent to its original loading position ready to receive another load. The cycle comprises a period of 2½ minutes.

Coal is delivered from bottom-hopper railroad cars into a rack hopper having a top measurement of 13 x 36 ft., or 468 sq.ft. area. From this hopper the coal is conveyed to the crusher by means of a 65-ft. long center to center, apron type conveyor, with a width of 36 in. It has a capacity of 180 tons per hour at a speed of 40 feet per minute. Power to operate this conveyor is taken from the crusher-shaft transmission by means of a chain drive to the hoist countershaft of the conveyor drive. The crusher is of the four-roll type, each roll being 36 in. in diameter and 36 in. face. One pair of rolls is held rigidly in position, but the other pair is provided with heavy coil springs which exert enough tension on the rolls to press the coal, and yet permit of giving in case a foreign substance is encountered. The rolls and apron conveyor are operated by a 100-hp. compound-wound motor at 850 r.p.m. The capacity of the crusher with run-of-mine bituminous coal crushing to 90 per cent ¾-in. size, is 180 tons per hour.

Both the crusher and apron conveyor are located outside of the boiler house, as is also the double skip hoist that elevates the coal from a bifurcated hopper after it has passed through the crusher. Each skip bucket has a capacity of 120 cu.ft. The 42-in. double-drum hoist is operated by a 60-hp. compound-wound motor. The grooves in the hoist drum are for a ⅞-in. cable. When the skip descends, a lever on the front of the bucket engages the gate of the chute leading from the crusher hopper, thus opening the gate, which is held open by means of a latch. The skip bucket then comes to rest on a counterweight lever of the skip. When sufficient coal has been discharged into the skip, the counterweight lever is depressed, the latch holding the gate open is released, and the gate closes to shut off the flow of coal. When the lever is thus operated, a switch is thrown which starts the motor operating the hoist, and the bucket is hoisted until it reaches a predetermined point, when resistance is cut in the circuit. When the bucket reaches the dumping position, current is cut off from the motor, and the bucket stops.

A 7½-ton motor-operated distributing larry car carries coal from the skip-hoist hopper, located at the head of the hoist, and discharges from both sides simultaneously into a coal bunker placed central over the boiler-room operating aisle. This bunker has a capacity of 15,500 lb. per lin. ft., and as it is 64 ft. in length, the capacity is approximately 500 tons. From the overhead bunker the coal is fed through four 15-in. spouts to the stoker hopper.

Dalton's Law Applied to Air-Pump Capacities

By Frank R. Wheeler

The presence of air in a condenser has two very detrimental effects. It increases the total absolute pressure or reduces the vacuum dependent upon the amount present and, in a surface condenser, reduces the heat transfer. Josse[1] reported a relative coefficient of heat transfer, as referred to steam of one to seven hundred.

Although the air that enters the condenser with the boiler-feed water and through leaks in the system has a moisture content dependent upon its humidity and temperature, for practical reasons it is usual to consider the leakage as dry air. This air admitted to the condenser is heated to the temperature corresponding to the vacuum, and through contact with the steam becomes saturated. The air pump must, therefore, remove with the noncondensable gases, or air, an amount of water vapor, or steam, dependent upon the temperature and pressure of the mixture at the air-pump suction.

By Dalton's Law the total pressure of a mixture of gases is the sum of the partial pressures of the gases present. For example: A condenser maintaining a vacuum of 2 in. absolute (28 in. referred to a 30-in. barometer) with condensate at 90 deg. F. indicates that the steam present has, from the steam tables, a partial pressure of 1.417 in. absolute and the partial air pressure then is 2 — 1.417, or 0.583 in. absolute.

A mixture of steam and air, unlike a pure saturated mixture, may change in temperature without effecting a change in the total pressure of the mixture, the change in temperature depending upon the relative partial pressures of the steam and air.

Assume a mixture of steam and air at a temperature T deg. F. and a total pressure P lb. per sq.in. absolute made up of 1 lb. of steam at temperature T deg. F. and corresponding pressure P_s and occupying V cu.ft., and a quantity of air also at temperature T deg. F. and occupying V cu.ft., but at a partial pressure P_a.

One pound of dry air at 32 deg. F. and a barometric pressure of 29.92 in. of mercury, or 14.6963 lb. per sq.in. absolute, occupies 12.387 cu.ft. and at temperature T deg. F. and pressure P_a in. of mercury occupies:

$$\frac{29.92}{P_a}\times\frac{T+459.6}{32+459.6}\times 12.387 \text{ or } \frac{(T+459.6)\times 0.75452}{P_a} \textit{ cu.ft.}$$

V cu.ft. of the air weighs $\frac{P_a\times V}{(T+459.6)\times 0.75452}$ lb., and since this is the weight of air mixed with one pound of steam, the weight of steam per pound of air is

$$\frac{1}{\frac{P_a\times V}{(T+459.6)\times 0.7542}} \quad\text{or}\quad \frac{(T+459.6)\times 0.75452}{V\times P_a}$$

where V = specific volume per pound of steam at temperature T deg. F.; P = total pressure in condenser, inches Hg. absolute; T = the temperature of the mixture in degrees F.; P_s = partial steam pressure corresponding to the temperature of the water present, in inches of mercury absolute; and P_a = partial air pressure in inches of mercury absolute $(P\text{-}P_s)$.

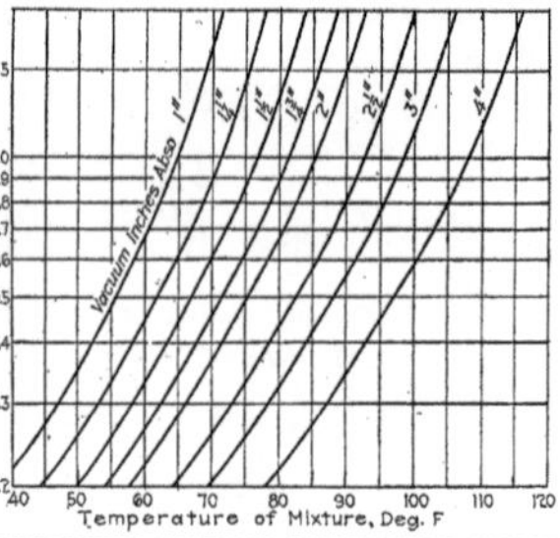

CHART SHOWING ADDITIONAL CAPACITY THAT MUST BE PROVIDED IN AN AIR PUMP TO REMOVE ASSUMED LEAKAGE OF DRY AIR INTO THE CONDENSING SYSTEM

From this formula the accompanying chart has been plotted to show the additional capacity that must be provided in an air pump to remove the assumed leakage of dry air into the condensing system.

For example: Assume a turbine installation where provision is to be made for an air leakage of 20 lb. of dry air per hour (approximately 4.45 cu.ft. per min.) through joints and from air entrained in the boiler-feed water. This condenser is to maintain a vacuum of 28 in. referred to a 30-in. barometer, or 2 in. absolute, with circulating water entering at 70 deg. F. and leaving at 88 deg. F. With good design the mixture at the air-pump suction can be assumed as a mean of these, or 79 deg. F.

From the chart it will be found that to remove one pound of air under these conditions it is necessary also to remove 0.6156 lb. vapor, so that the total air-pump capacity required is 1.6156 × 20, or 32.3 lb. of mixture per hour (approximately 7.18 cu.ft. per min.) in order to remove 20 lb. of dry air per hour.

[1] *Engineering*, London, Vol. 85, 1908.

Sampling Tanks Versus Continuous Recorders for Determining CO_2

By CHARLES C. PHELPS

A discussion of the correctness of results of CO_2 determinations by different methods. Explains some of the common errors and suggests proper method of procedure.

ENGINEERS and firemen who are striving to operate their power plants efficiently, try to maintain a high percentage of carbon dioxide (CO_2) in the flue gases from their boiler furnaces. Hence, frequent measurement of CO_2 becomes necessary. Three methods of taking these measurements are in common use, namely: By withdrawing samples of gas directly from the flue at regular intervals and making analyses with an Orsat or other hand analyzing apparatus; by collecting the gas in a sampling tank during a period of definite duration, for instance, a shift or a 24-hour run, and later analyzing the contents of the tank for the average percentage of CO_2 during the period covered; and by automatically recording the percentage of CO_2 continuously on the chart of a CO_2 recording instrument which is permanently connected with the flue.

The last-named method has many important advantages over the other two. It shows stack conditions at once, instead of some minutes or hours later, hence it is easier to correlate the CO_2 with such factors as change of draft, addition of coal and varying demand for steam, and to arrive at logical conclusions as to causes of inefficiency, while such knowledge can be used to improve the efficiency. Furthermore, it gives a continuous permanent record from which the heat losses up the chimney may be determined for any instant or for a total elapsed time. So-called snap tests made by hand analyzing apparatus give the data for determining chimney losses at only a given instant, but it is unlikely that such data would represent true average conditions. The CO_2 sampling tank, on the other hand, does not supply a means of determining CO_2 percentage at any given instant, but furnishes a sample from which it is assumed that the average heat lost up the chimney may be calculated. This last assumption, however, is not warranted by the facts.

Sampling Tanks Take No Account of Rate of Driving

Gas flows into the sampling tank at a uniform rate; that is, a constant volume is collected during each unit of time. One thousand cubic inches, say, of gas

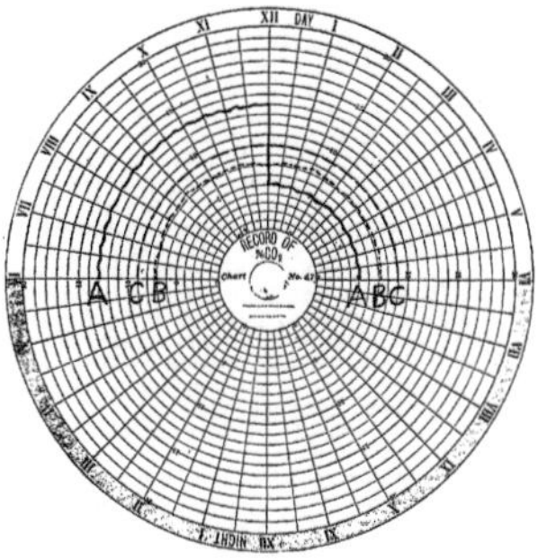

FIG. 1. CHART SHOWING AVERAGE CO_2 IS NOT GOOD INDICATION OF HEAT LOSS

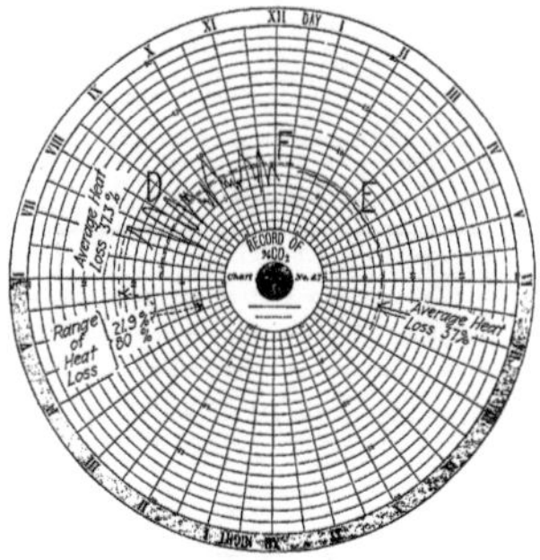

FIG. 2. CO_2 CHART SHOWING AVERAGE HEAT LOSS FOR VARYING CO_2

are withdrawn from the flue every hour irrespective of the total amount of gases passing up the chimney during each hour, the rate of flow up the chimney, of course, varying with the rate of driving the fire. Supposing the coal records show that five times as much coal is consumed during the second hour as during the first and a record from a continuous CO_2 chart shows a uniform CO_2 percentage for the two hours; we would then know from the coal and CO_2 records that the total loss was five times as serious during the second hour, although the percentage of CO_2 did not change. Analysis of the tank sample, on the other hand, would not give much enlightenment, for we would not know whether CO_2 was high during the first hour and low during the second (which would make the losses worse than they might appear to be) or whether low CO_2 occurred during the first hour and high during the second (which would mean better economy than the analysis would appear to indicate). It is apparent that the use of the sampling tank is likely to lead to serious errors in both directions, for the simple reason that it withdraws its sample at a uniform rate, whereas the flow of gases up the flue is constantly fluctuating. Another point is that the greatest opportunities for fuel saving occur while the furnace is being forced, and that a special effort should be made to keep CO_2 as high as possible at such times.

Why Average CO_2 Does Not Determine Average Heat Losses

In addition to the objection already pointed out, there is inherent in the sampling-tank method another fallacy which is perhaps even more serious. A definite relation exists between the percentage of CO_2 in the flue gases and the percentage of the total heat of the fuel that is wasted up the chimney. At a stack temperature 600 deg. F. above the temperature of the atmosphere this relation is approximately in accordance with the accompanying table. This table is based on the complete combustion of pure carbon as fuel. For fuel containing other constituents besides carbon, and for incomplete combustion, the ratios would of course vary somewhat. For stack temperatures other than that assumed in this case, the heat losses would be in proportion to the temperature; for instance, with a stack temperature 300 deg. above atmospheric the losses would be one-half as great as those noted in the table.

To illustrate the second fallacy involved in the sampling tank method, let us assume such a case as that represented by the CO_2 recorder chart shown in Fig. 1. Here the chart record shows by *A* that the percentage of CO_2 was kept steadily at 13 per cent. from 6 a. m. to noon and at 5 per cent. from then until 6 p. m. (an average of 9 per cent. CO_2 for the twelve hours). From the table the loss up the chimney corresponding to 13 per cent. CO_2 is found to be 19.3 per cent. and the loss corresponding to 5 per cent. CO_2 is 49 per cent. Hence, the true average loss up the chimney for the two six-hour periods is $\frac{19.3 + 49}{2} = 34.2$ per cent. if the flue temperature and rate of driving remain constant. The same loss would have resulted if the CO_2 had been maintained steadily at 7¼ per cent. (*B*, Fig. 1) for the full twelve-hour period, inasmuch as 34.2 per cent. loss corresponds to 7¼ per cent. CO_2 (see table).

The sampling tank would have given a result entirely different from the foregoing. One-half of the flue-gas sample would have contained 13 per cent. CO_2 and the other half 5 per cent., making the average percentage of CO_2 as determined by analysis $\frac{13 + 5}{2} = 9$ per cent. (*C*, Fig. 1). The loss corresponding to 9 per cent. CO_2 is 27.6 per cent. This false result would be certain to mislead an observer into believing that the chimney losses were $34.2 - 27.6 = 6.6$ per cent. lower than they really were. In other words, the error resulting from the use of the sampling tank in this case amounts to $\frac{6.6}{34.2} = 19$ per cent. In actual practice the error might be much greater or less than this, depending entirely upon the conditions of operation.

From the foregoing it is quite apparent that it is more important to know the extent of the fluctuations in percentage of CO_2 than the average values. One of the advantages of snap tests made by hand analyzing apparatus over averaging tests based on tank samples is that if the snap tests are made with sufficient frequency they do give some idea of the extent of fluctuations and at least make it possible to determine the exact loss at the instant the sample was taken. The continuous autographic record is, on the other hand, preferable to either of the other methods inasmuch as it really is equivalent to an infinite number of readings and, hence, all fluctuations of CO_2.

From the "Loss Increment" column in the table it will be noticed that the heat lost up the chimney increases rapidly as the percentage of CO_2 in the flue gases becomes less. Hence, on such a record as that shown at *D*, Fig. 2, the losses represented by the parts of the record near the center of the chart are out of all proportion to the losses on parts of the record farther out radially. This illustrates clearly why a flat record is more desirable than a fluctuating one and shows why, when CO_2 is held constantly at one point, the chimney loss is bound to be less than when CO_2 fluctuates in amount but is of an equivalent *average* percentage. Conversely, a flat curve, as *E*, might actually mean better efficiency than a fluctuating curve *D*, which represents a greater average percentage of CO_2. In this case the flat record *E*, showing 6.7 per cent. CO_2, indicates 37 per cent. chimney loss, whereas the fluctuating curve *D* shows an average chimney-loss percentage of 37.3 per cent. (the latter calculated from readings at 7½-minute intervals on the chart). Curve *D* fluctuates between losses of 21.9 and 80 per cent. heat losses. Here the record *D*, which shows the greater average percentage of CO_2, namely, 7.2 per cent. (as actually determined by radial planimeter and represented on Fig. 2 by the dotted line *F*) paradoxically represents a greater loss than the record *E* with a smaller percentage of CO_2.

HOW CHIMNEY LOSSES VARY WITH CO_2 AT STACK TEMPERATURE 600 DEGREES F. ABOVE TEMPERATURE OF ATMOSPHERE. BASED ON CARBON AS FUEL

Per Cent. CO_2 in Flue Gas	Per Cent. Loss Up Chimney	Per Cent. Loss Increment	Per Cent. CO_2 in Flue Gas	Per Cent. Loss Up Chimney	Per Cent. Loss Increment
21	12.2		11	22.8	1.9
20	12.8	0.6	10	25.0	2.2
19	13.5	0.7	9	27.6	2.6
18	14.2	0.7	8	31.0	3.4
17	15.0	0.8	7	35.3	4.3
16	15.9	0.9	6	41.0	5.7
15	16.9	1.0	5	49.0	8.0
14	18.0	1.1	4	61.0	12.0
13	19.3	1.3	3	80.0	19.0
12	20.9	1.6			

The cases cited in this article point to conclusions that should be followed in practice.

1. Strive to obtain the highest practicable percentage of CO_2 in the flue gas.
2. Keep the percentage of CO_2 as nearly uniform as possible. Uniformity is second in importance only to the *largeness* of percentage of CO_2.
3. The more nearly continuous the record of CO_2 the better, inasmuch as this enables the operator to determine the scale of fluctuation more accurately.
4. The average or total heat losses up the chimney cannot be calculated with accuracy from the average percentages of CO_2. The average percentage of heat loss is determined most accurately by finding the corresponding heat losses for as many points on the CO_2 record as possible and then averaging these figures.
5. In determining chimney heat losses for a "heat balance" or as a check on regular operation, keep in mind that the varying rate of combustion is of as great importance as the percentage of CO_2 in the flue gas.
6. The cost of installing and operating a continuous autographic CO_2 recorder is small in comparison with the possible savings in fuel in the average power plant. The United States Fuel Administration recommended the use of CO_2 equipment for any plant burning five tons of coal or more per day; in many plants where the fuel consumption is much less, the operation of such equipment pays handsome returns on the investment. In very small plants it will surely pay to use a hand analyzer, which, though possessing many shortcomings, is nevertheless a very valuable instrument.

Locating Faults in Induction Motors—

Noise *and* Vibration

By
A. M. Dudley *

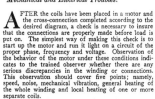

The First of Three Articles on Locating Faults in Induction-Motors. The Observations of the Behavior of the Motor To Be Made When Started for the First Time, and How To Distinguish Between Mechanical and Electrical Troubles.

AFTER the coils have been placed in a motor and the cross-connection completed according to the desired diagram, a check is necessary to insure that the connections are properly made before load is put on. The simplest way of making this check is to start up the motor and run it light on a circuit of the proper phase, frequency and voltage. Observation of the behavior of the motor under these conditions indicates to the trained observer whether there are any serious discrepancies in the winding or connections. This observation should cover five points; namely, speed, noise, mechanical vibration, general heating of the whole winding and local heating of one or more separate coils.

The speed, if correct, should be of nearly synchronous value when the motor is running without load; that is, equal to cycles times 120 divided by the number of poles.

The motor should give a low humming noise similar to that made by transformers, but there should be no irregular or "growling" noise. There may also be a considerable volume of air noise or whistle caused by the ventilating air passing through the air ducts in the rotor and stator. The magnetic noise may be distinguished from the air noise by the expedient of opening the switch for a second or two while the motor is running full speed without load. Opening the switch breaks the current and removes the magnetic field, and consequently the magnetic noise ceases, but leaves the rotor running at practically the same speed owing to its inertia or stored energy, and hence the windage, or air noise, is practically unaffected. In this way, by opening and closing the switch two or three times, it becomes readily apparent what part of the total sound made by the motor is magnetic and what part is windage. It also indicates whether either or both of these sounds are abnormal. If the speed is correct and the motor makes no more than a reasonable singing or humming noise, the hand should be placed on the frame to note the mechanical vibration.

If there is noticeable mechanical vibration, it may be due to purely mechanical causes or to magnetic causes or possibly to both. By opening and closing the switch, as described in the foregoing, the mechanical vibration due to the magnetic field can be easily separated from that due to strictly mechanical causes, because when the switch is open there is no magnetic field present. Suppose, for example, that when the motor is running at full speed there is a marked vibration or trembling that can be felt when the hand is laid on the frame of the motor. Suppose, then, that when the switch is opened for a second or two the vibration disappears and the motor rotates smoothly at nearly the same rate of speed. This, then, is evidence that the vibration was caused by the action of the magnetic field on the stator and rotor. However, if the motor vibrates whether the switch is open or closed, it is evidence that the action is purely mechanical and is affected little or not at all by the presence of the magnetic field.

When the trouble is traceable to the magnetic field, it may indicate improper connection of the winding or it may indicate that the mechanical clearance between stator and rotor is not symmetrical or that there is some similar combination of mechanical and magnetic features that is responsible for the vibration noticeable. The commonest mechanical causes for vibration are rotor out of balance, either standing or running; bent shaft; too great clearance between shaft and bearings; unbalanced or eccentric coupling or pulley or a combination of two or more of these faults. These mechanical conditions are easily determined and can be corrected. The commonest causes of mechanical vibration due to a combination of mechanical and magnetic conditions are rotor out of round, stator out of round, too great clearance in the bearings, or rarely, uneven or eccentric air gap or clearance between stator and rotor. The latter point seldom gives trouble and a polyphase motor will practically always run without giving any trouble until the bearing wear allows the rotor to strike on the stator. Single-phase motors are more sensitive to eccentricities in the air gap or clearance between stator and rotor and sometimes show a considerable variation in torque in motors otherwise duplicate due to such irregularities.

There are a number of elements that may cause the rotor or stator to be out of round. In the first place there is a slight variation due to the punch-and-die work, which may amount to 0.005 in. between individual punchings. In the second place some allowance around the outside of the punching must be made in the fixture or frame in which they are built up so that they will assemble readily, and this allows the punchings to stagger more or less. In the third place when the punchings are actually assembled in the frame, the frame may spring out of shape slightly after machining, owing to the release of casting strains when removing the material in the cut. Of course none of these variations is in itself large, but when they all accumulate in the same direction, perceptible eccentricity may result amounting to a good many thousandths of an inch. This is not serious, since it is present to some extent in all motors, but under extreme or extraordinary conditions it may cause mechanical vibration.

*Designing engineer, Westinghouse Electric and Manufacturing Co.

Mechanical vibration caused by the windings may be due to either the rotor or the stator. For example, in a squirrel-cage rotor there may be bad contacts between certain bars and the short-circuiting rings, resulting in more resistance in some parts of the winding than

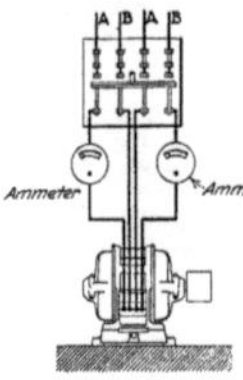

FIG. 1. AMMETER CONNECTIONS FOR FOUR-WIRE TWO-PHASE MOTOR

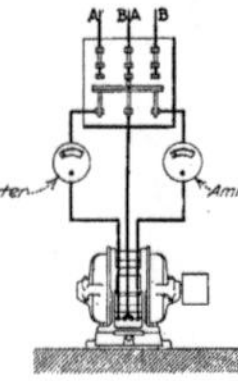

FIG. 2. AMMETER CONNECTIONS FOR THREE-WIRE TWO-PHASE MOTOR

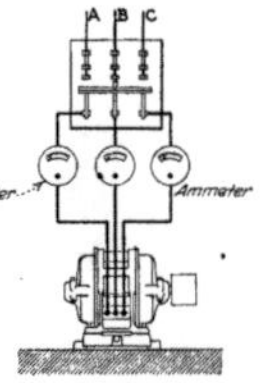

FIG. 3. AMMETER CONNECTIONS FOR THREE-PHASE MOTOR

in others. This in turn affects the distribution of current in the different bars and hence affects the magnetic field and varies the mechanical pull from point to point. Or if the winding on the rotor of a wound-rotor type motor is ground in a number of places, it will also cause unequal distribution of the current in the windings, which in turn causes severe vibration during the starting period. However, this generally disappears to a large extent after the motor comes up to full speed. From this it may be seen that where mechanical vibration is absent the conclusion may be drawn that the windings are symmetrical and are functioning properly, but where vibration is present it may be caused by a number of things, some of them obscure, and must not immediately be attributed to improper winding connections until a further examination is made.

The next point to be observed is the general temperature of the entire winding as determined by passing the hand around the ends of the windings. It is best practice in making this examination to shut down the motor after it has run three to five minutes. If the examination is made while the motor is running, care should be taken to avoid injury by coming in contact with moving parts and also to avoid injury from electric shock, if the circuit is 550 volts or over. If the winding as a whole is cool, inspection should be made for individual coils that are much hotter than the rest of the winding, as these may indicate short-circuits or improper connections in that particular coil.

If a motor is operating freely and easily at the proper speed without undue noise or mechanical vibration and if there is no general or local heating of the winding, the next step is to measure the current in each phase. This may be done as indicated in Figs. 1, 2 and 3. If possible an ammeter should be connected in each phase so that the readings of all phases may be taken simultaneously. For a two-phase motor two ammeters are required, as in Figs. 1 and 2, and for a three-phase motor three ammeters are required, as in Fig. 3. The no-load, or magnetizing current as it is called, will usually be somewhere between 15 and 35 per cent of the full-load current with an average value of perhaps 25 per cent. If the no-load current in all phases is equal and approximately 25 per cent of the full load, it is safe to assume that the winding connections are properly made. If a wattmeter is available, a further check might be made on the total watts taken by the motor running light, but that does not add greatly to the ammeter check. The connections for connecting two wattmeters in a two-phase four-wire circuit are given in Fig. 4 and for a three-phase circuit in Fig. 5. The connections for a three-wire two-phase would also be the same as those in Fig. 5; where only one wattmeter is available, it may be connected into a three-phase circuit with a single-pole switch, as in Fig. 6, so that the two readings may be taken by simply throwing the switch. In a two-phase circuit the total watts will always be the sum of the two readings, but in a three-phase circuit this is true only when the power factor is greater than 0.50. Where two wattmeters are used to measure the no-load watts of a three-phase motor, the difference of the two readings gives the correct value of the watts, since the power factor of an induction motor at no load is always less than 0.50. The watts taken at no load and full speed and voltage cover the iron loss, bearing friction, windage and a small amount of copper loss. The total no-load watts will in general be in the order of 5 per cent to 8 per cent of the rating of the motor in watts varying with the capacity and speed of the motor. The motor rating in watts would be the horsepower from the nameplate multiplied by 746.

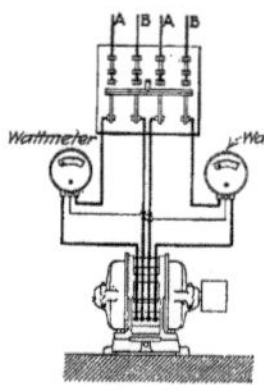

FIG. 4. WATTMETER CONNECTIONS FOR FOUR-WIRE TWO-PHASE MOTOR

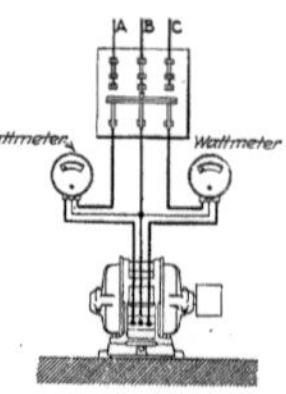

FIG. 5. WATTMETER CONNECTIONS FOR THREE-PHASE MOTOR

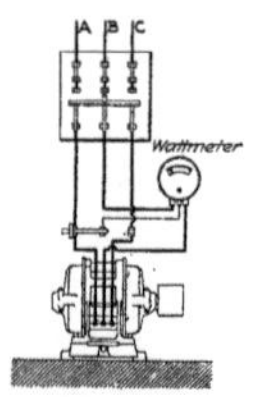

FIG. 6. ONE WATTMETER CONNECTED TO A THREE-PHASE MOTOR

If the foregoing checks indicate that the motor is not acting normally, they should also give some evidence that there is a fault in the coils of the winding or in the manner in which these coils are connected, and further search is made to analyze the nature of this fault so that it may be located and corrected.

Refrigeration Study Course—IV. The Ammonia Condenser

By H. J. MACINTIRE

Professor of Mechanical Engineering, University of Idaho

IN THE previous part of this course it was brought out that in the simple refrigerating plant there was but one moving piece of apparatus; that is, the compressor. The compressor is the pump which, by increasing the pressure of the gas from the refrigerating coils to one some six to twelve times as great, makes it possible to use the same ammonia continuously. The compressor pumps the compressed gas into a condenser, which is water-cooled, and which makes it possible to cool the gas to a point where it condenses (still under pressure) and becomes all liquid again.

Physical State of Ammonia

For some reason or other there seems to be some difficulty in understanding the condenser, and yet it is not difficult if the reader comprehends the principles involved. First, one should understand clearly the three physical conditions that are involved in the condenser and see if they, too, are not conditions met with in steam-engineering practice.

In steam-boiler practice we have to feed water to the boiler at, say, 210 deg. F., and it is heated as water to 370 deg. F., and then at the corresponding pressure it begins to boil. No boiling occurs until the temperature corresponding to the boiler gage pressure is obtained. As the boiler pressure is constant, boiling continues at 370 deg. F. and the steam given off is said to be saturated. It is not *dry* and saturated unless some means is provided to separate the particles of liquid which are thrown into the steam space by the violence of the boiling. If a superheater is attached to the boiler, or if the boiler design is such that the *steam* comes into contact with more heating surface, like the upper part of certain vertical boilers, the steam is heated some more and the temperature increases much beyond 370 deg. and it is said to be superheated.

In the ammonia refrigerating system the gas from the refrigerating coils (the steam boiler of this system) is supposed to be just dry and saturated. If some of the coils convey dry ammonia gas, or if dry gas passes through uninsulated return lines, then heat will pass through the pipes and will increase the gas temperature. (Really we try to perform refrigeration with a dry gas.) The ammonia then becomes *superheated,* which will be shown later to be detrimental to the compressor capacity. In any case, however, the discharge gas from the compressor is superheated, varying in amount from 60 to 200 deg. F. superheat—figured from the temperature of saturation at the condenser pressure. So in the refrigerating coils it is quite possible to have the conditions of liquid ammonia, wet gas, dry and saturated gas and even superheated ammonia. These same conditions occur in the condenser, but in the reverse order.

The Ammonia Condenser

The compressor, as mentioned, takes the gas either dry and saturated, or slightly superheated, and compresses it, thereby discharging into the condenser a superheated vapor. This condenser is an arrangement of piping, with suitable fittings, so that the gas may pass through along a carefully prescribed path and be in *metal* contact all the time with the condensing water. As the gas is made to pass through a long path kept cool by the water, the superheat is first removed, then the heat necessary to condense it, and finally the liquid may be cooled below the temperature of saturation.

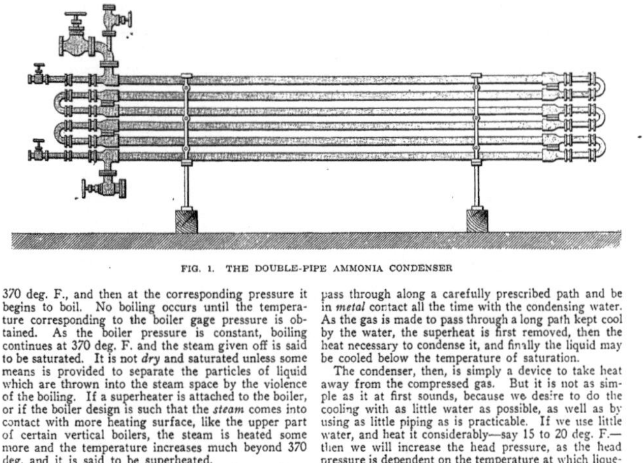

FIG. 1. THE DOUBLE-PIPE AMMONIA CONDENSER

The condenser, then, is simply a device to take heat away from the compressed gas. But it is not as simple as it at first sounds, because we desire to do the cooling with as little water as possible, as well as by using as little piping as is practicable. If we use little water, and heat it considerably—say 15 to 20 deg. F.—then we will increase the head pressure, as the head pressure is dependent on the temperature at which liquefaction takes place. If we use quite a little water, with a water temperature rise of 5 to 8 deg. F., then the cost of pumping water assumes larger proportions. But more important than all is the question of heat transfer. In the steam surface condenser the water usually passes through the tubes and the steam impinges perpendicular to the tubes. In this steam is a little air, which seems to stick to the tube and forms an insulation air film around the tube, as shown in the figure. The result is a much-reduced heat-transmission rate, and it takes two or three square feet to do the same amount of work that would be required of one square foot without the insulating film. In a steam condenser the air is removed constantly by means of an air pump, this being necessary on account of the high vacuum carried, and the continual air leaks which cannot be prevented.

The Gas Film in the Condenser

In the ammonia condenser there is always some air as well as decomposed ammonia and oil, all of these tending to the formation of a gas film. However, here there is a difference in that, as the flow of the ammonia is along the tubes, there is a tendency to scrape off this film if there is some place for the accumulated gas to go. For instance, in the bleeder type of condenser the gas enters at the bottom and passes upward, suitable provision being made so that the liquid ammonia may be drained or bled off as it is condensed. The idea here is, as it is in the steam condenser, that the best efficiency may be obtained by operating the condenser with as little liquid in the pipes as is possible. In the flooded condenser, the double-pipe, and the atmospheric types, the gas from the compressor enters at the top and works downward so that the lower pipes are partly or, possibly, completely filled with liquid ammonia. In the figure it is seen how this affects the action of the condenser, inasmuch as the effective area of the pipes is reduced in proportion to the area exposed to the liquid.

In addition to the effect of a gas film, or accumulated liquid, on the pipe surface, there is the effect of oil and scale. This is not a question of design, except that the right kind of condenser should be chosen for the kind of condensing water available at the plant. Boiler tubes require cleaning for economy as well as safety, and ammonia condensers have a similar need. The condenser must be easily cleaned, and cleaning should be performed frequently enough to prevent the solid matter forming a hard scale.

In condensers, as well as other heat transferring apparatus, there are three factors that affect the quantity of heat conducted through the walls of the pipes: The area, the coefficient of heat transfer, and the temperature difference in the two sides of the pipe. The product of these three quantities will give the number of heat units transferred per hour or per 24 hours. The area is taken in square feet, and the coefficient of heat transfer is found by experiment and is affected by the kind of metal used, its thickness, the condition of the surface, whether oily or scaly, or whether a gas film is around it. The temperature difference requires a little more explanation.

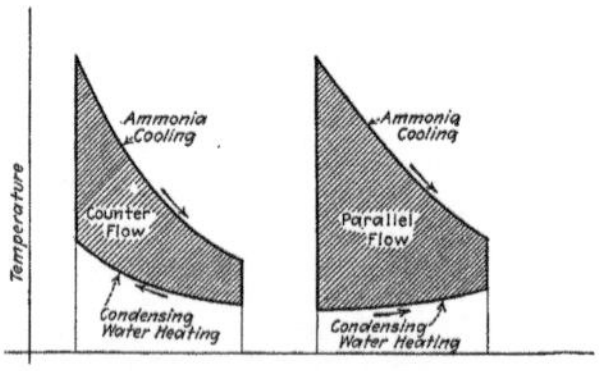

FIG. 3. EFFECT OF COUNTER FLOW AND PARALLEL FLOW THROUGH AMMONIA CONDENSERS

In Fig. 3 it is attempted to show the action taking place in the condenser. Two cases are shown, one representing parallel flow of the gas and water, as in the atmospheric type of condenser where the water flows out of a trough and then falls first on one pipe and then on another; the other is a counterflow of the gas and water as in the bleeder type of condenser or the double-pipe condenser. From the diagram it will be seen that parallel flow and counterflow do not give the same results. What is desired most particularly is to get as cold a liquid ammonia as is possible, because it has been found that more refrigerating capacity is obtained thereby, and the condenser pressure is lowered. Therefore it is most desirable to bring the coldest water in contact with the coldest ammonia, and the warmest water where the compressed gas enters. To show how this works out in practice it may be said that the usual specification for the atmospheric type of condenser is 60 lin. ft. of 2-in. pipe, but only 30 lin. ft. for the bleeder type, using the same liberal factor for safety.

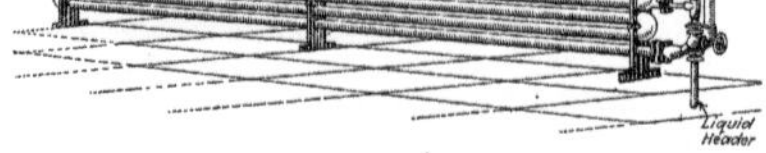

FIG. 2. SHOWING THE ATMOSPHERIC TYPE AMMONIA CONDENSER

A type of condenser that was regarded with considerable favor some time ago was the flooded condenser, a design which is now being discarded. The idea in this arrangement was that the coefficient of heat conductivity in the condenser from the gas to the water was very low in value, but if the surface could be wetted then results would be better. The great difficulty here, however, is in preventing what the name implies, because flooding the condenser means choking it up with liquid ammonia so that the conduction of heat is retarded instead of being accelerated. The action of the condenser is to install an injector so arranged that the gas from the compressor passes through and drags some liquid ammonia from the liquid receiver and the gas and liquid mixture passes through the condenser, as in the ordinary design.

Condenser Capacity

As regards the capacity of an ammonia condenser, the condition is similar to that of a steam boiler of the water-tube type. We usually rate such a boiler at 10 sq.ft. of heating surface per boiler horsepower; yet there are cases where such boilers have developed 2, 2½, 3 or even more boiler horsepower for every 10 sq.ft. of heating surface. It depends on the condition of the water, the tubes and the ability to burn fuel under the tubes. In like manner the condenser may be able to do cooling at the rate of 12, or 10, or even 6 sq.ft. per ton of refrigeration, and yet we would not be justified in using such small surfaces. The efficiency of the condenser may drop off, some stands may be laid down for repair, or perhaps storage volume may be required for the liquid ammonia during times when repair work on the low-pressure side is required.

GALLONS CONDENSER WATER PER TON REFRIGERATION

Condenser Pressure, Lb. Per Sq. Inch, Gage.	Corresponding Temp. Deg. F.	60 Degrees F. Water				60 Degrees F. Water				80 Degrees F. Water				90 Degrees F. Water			
		Range, Deg. F.	Water Per Ton of Refg'n. Gal. Per Min. Suction Pressure, Lb. Gage			Range, Deg. F.	Water Per Ton of Refg'n. Gal. Per Min. Suction Pressure, Lb. Gage			Range, Deg. F.	Water Per Ton of Refg'n. Gal. Per Min. Suction Pressure, Lb. Gage			Range, Deg. F.	Water Per Ton of Refg'n. Gal. Per Min. Suction Pressure, Lb. Gage		
			15	20	25		15	20	25		15	20	25		15	20	25
126.4	75	10	2.90	2.85	2.80												
131.4	77	12	2.40	2.35	2.30												
136.6	79	14	2.05	2.00	1.95												
141.8	81	16	1.85	1.80	1.75	6	4.87	4.80	4.70								
147.2	83	18	1.65	1.60	1.55	8	3.67	3.60	3.55								
152.7	85	20	1.47	1.45	1.41	10	2.95	2.90	2.85								
158.3	87	22	1.35	1.32	1.30	12	2.45	2.40	2.35								
164.1	89	24	1.25	1.22	1.20	14	2.10	2.05	2.00								
170.1	91	26	1.15	1.13	1.10	16	1.90	1.85	1.80	6	5.00	4.90	4.80				
176.2	93	28	1.08	1.05	1.03	18	1.68	1.63	1.60	8	3.75	3.70	3.65				
182.6	95	30	1.00	0.99	0.97	20	1.50	1.48	1.45	10	3.00	2.95	2.90				
189.1	97	32	0.95	0.93	0.90	22	1.38	1.35	1.32	12	2.52	2.48	2.43				
195.7	99	......	......	......	......	24	1.28	1.25	1.22	14	2.13	2.10	2.06				
202.5	101	......	......	......	......	26	1.17	1.15	1.13	16	1.92	1.85	1.83	6	5.10	5.00	4.90
209.5	103	......	......	......	......	28	1.10	1.07	1.05	18	1.70	1.67	1.63	8	3.83	3.75	3.70
216.5	105	......	......	......	......	30	1.02	1.00	0.98	20	1.53	1.50	1.48	10	3.07	3.00	2.95
223.7	107	......	......	......	......	32	0.95	0.93	0.02	22	1.40	1.37	1.35	12	2.57	2.51	2.47
231.1	109	......	......	......	......	......	......	......	......	24	1.30	1.28	1.25	14	2.19	2.13	2.10
238.7	111	......	......	......	......	......	......	......	......	26	1.20	1.16	1.15	16	1.95	1.90	1.88
246.5	113	......	......	......	......	......	......	......	......	28	1.12	1.08	1.06	18	1.74	1.69	1.65
254.5	115	......	......	......	......	......	......	......	......	30	1.05	1.02	1.00	20	1.56	1.53	1.50
262.7	117	......	......	......	......	......	......	......	......	32	1.00	0.98	0.95	22	1.43	1.40	1.37

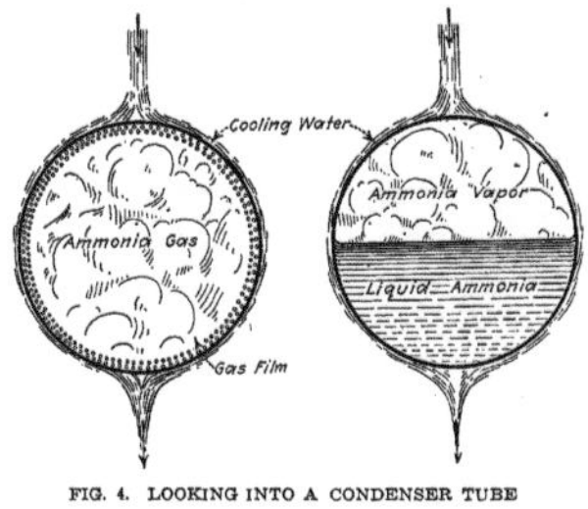

FIG. 4. LOOKING INTO A CONDENSER TUBE

The most important function of the condenser is the determination of the condenser pressure, which determines directly the work done by the steam engine in the electric motors. If no air were present, or decomposed oil or ammonia, then the condenser or head pressure would be determined by the temperature at which liquefaction took place. And the temperature of liquefaction is determined directly by the amount of cooling showered over the condenser and its temperature. As an example, we may say that in acquiring a ton of refrigeration, the liquid ammonia in the expansion coils will absorb heat at the rate of 200 heat units (200 B.t.u.) per minute. Additional heat, amounting to about 60 heat units, is added during compression, so that the total heat to be removed by the cooling water in the condenser is about 260 heat units per ton of refrigeration. This has to be taken away by the cooling water, which is heated in consequence. This could be done by $\frac{260}{10} = 26$ lb. (3.1 gal.) of water per minute heated 10 deg. in temperature, 19½ lb. (2.34 gal.) when heated 15 deg., or 13 lb. (1.36 gal.) when heated 20 deg. The table shows what can be obtained in practice.

How 25-Cycle Alternating Current Came Into Use

In a paper, "The Technical Story of the Synchronous Converter," presented before the St. Louis Section of the American Institute of Electrical Engineers, Jan. 28, 1920, B. G. Lamme said in part:

It was not until 1892 that deliberate efforts were made to transform, on a commercial scale, from alternating to direct current. It was becoming recognized that the polyphase system offered great future possibilities, and the synchronous converter began to be looked upon as one of the accessories of such a system. In consequence, arrangements were made to obtain tests on a relatively large scale to determine the possibilities of such a device. For this purpose, a standard 150-hp. 500-volt 850-r.p.m., four-pole belted-type railway generator was equipped with four collector rings placed over one end of its commutator. Suitable brush-holders were arranged for operating on these collector rings, in addition to the usual direct-current brush-holder equipment.

From the first this machine operated in a very satisfactory manner as a synchronous converter, and a long series of tests were carried on by the writer, assisted by N. W. Storer, at that time one of his associates in the testing room. It was determined that this machine would commutate in just as satisfactory a manner as when acting as a direct-current generator. As far as could be observed, the machine in every way was a practical one.

Meanwhile the whole synchronous-converter situation was being worked over, and it was decided that 3,600 alternations (30 cycles) was about the best frequency for such machines. This was slightly higher than the frequency at which the tests were being made, but it was felt that this increase was permissible, in view of the excellent results already obtained at 3,400 alternations (850 r.p.m., four poles). In those days everything was measured in alternations per minute instead of cycles per second as in present practice.

About this time a commission, which had taken up the development of the Niagara Falls power, had begun negotiations with the Westinghouse company, with a view to getting certain electrical machinery constructed according to the designs of the commission. The principal apparatus involved was an electric generator of 5,000-hp. capacity at 250 r.p.m. The design contemplated was the "umbrella type" with external rotating field and vertical shaft. This construction as a whole was considered practicable by the Westinghouse engineers, but the electrical proportions of these generators were not acceptable. The commission had decided upon a frequency of 2,000 alternations per minute (eight poles and 250 r.p.m., now 16⅔ cycles per second). One of the objects in this low frequency was to be able to operate commutator-type motors by means of alternating current, as certain members of the commission believed that this was the solution of the small alternating-current motor problem.

The Westinghouse company objected to this low frequency and proposed, as an alternative, 16 poles at 250 r.p.m., giving 4,000 alternations per minute, or 33⅓ cycles, this being the nearest that could be obtained to its own proposed standard of 30 cycles. There was considerable discussion on the merits and demerits of these two frequencies. One of the advantages claimed for the higher frequency was that it would be much more suitable for synchronous converters, as the possible combinations of poles and speed would be very much greater than for 16⅔ cycles. In fact, the lower frequency was considered to be more or less prohibitive as regards small or moderate capacity synchronous converters and induction motors, as their speeds would be too low. It is interesting to find how thoroughly this problem was appreciated even at that early date, when only one experimental converter had been placed on test.

A compromise was finally made on the question of frequency by selecting a 12-pole machine for the Niagara generator, thus giving 3,000 alternations, or 25 cycles. This was the origin of 25 cycles as a standard. The decision was made largely on the basis of future possibilities in the way of synchronous converters and induction motors.

Smith Door Protectors

One of the large items of upkeep expense with hand-fired boilers is that of maintaining in condition the front masonry, door linings and boiler fronts. If the boiler furnace is forced hard, new door linings, arches and side walls are necessary at least once a year.

A device designed to eliminate the necessity of these frequent repairs is known as the Smith door protector, manufactured by the International Engineering Works, Inc., Framingham, Mass. The protector is made of firebox steel of the same quality used in the boiler itself. The outside is circular in form and made with openings for pipe connections. They are set back at the front of the boiler and are piped so that the water is taken from the lowest part of the boiler, flows through the protectors and then back to the upper part of the boiler. This affords direct and constant circulation from the boiler to the protector and back again. The lower section of each protector is below the grate level, and a dead plate is set in the protector at the grate line. The bottom of each protector is connected to a blowout pipe. As the lower section of the boiler is below the grate line, a sediment chamber is formed so that any scale or other foreign matter in the water will be precipitated and easily blown out. Each protector is made of the proper size to fit the door opening for which it is to be used.

The inside section is rolled up and flanged out against the outside shell, and the two are welded together, making a practically seamless construction that is not affected by contraction or expansion and thus eliminates the use of bolts or rivets. The only alteration to the boiler in making this installation is the tapping of two holes in the tube sheet for the water connection and the piping from the blowout of the protector to either the ashpit or the boiler blowoff pipe.

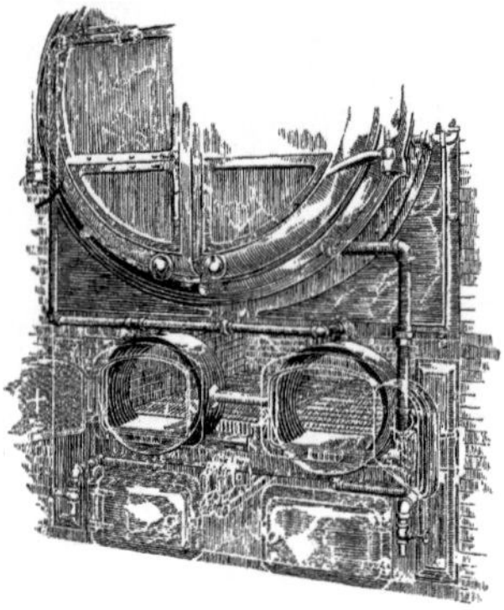

DOOR PROTECTORS APPLIED TO A BOILER FURNACE

For several years the United States has produced approximately 65 per cent of the world's output of crude oil and has contributed about 60 per cent of the total amount produced in the world to date. The United States Geological Survey estimates that our reserves are approximately 40 per cent exhausted. Development and exploration work is not affording any positive evidence of relatively large new discoveries, while demands for petroleum are increasing. On the other hand, large potential reservoirs are lying dormant in other parts of the world which, if they can be developed, will supply the world's demands for some time. When these sources fail or become too high-priced, the country can fall back on our tremendous reserves of oil shales.

EDITORIALS

Where Electricity Is a Luxury

THE recent decision of the Idaho Public Utilities Commission has apparently exploded the idea that the time would come when not only would the heavy parts of all our tasks be performed by energy transmitted electrically, but also our homes and buildings would be heated—in fact, everything that required light, heat or power would be done electrically. In the early developments of the large irrigation enterprises in the State of Idaho, the Great Shoshone and Twin Falls Power Company was formed, and later the United States Government constructed a hydro-electric plant in connection with its irrigation project on Snake River. In both of these developments there were large quantities of excess energy, and to provide a sufficient demand for the current during the winter months a very cheap rate was put into effect for current used for heating purposes. These rates took into account only the additional cost of equipment necessary to deliver the current to the customer, consequently, the charge was considerably below the actual cost, had the total investment been taken into consideration.

When, during the war, there was an increased demand for power and an increased cost of coal for fuel purposes, there was considerable discussion as to the advisability of compelling the hydro-electric power companies of the state to install sufficient capacity to take care of the demand for current for general heating purposes. To determine the cost of this service, the Public Utilities Commission started an investigation, the findings of which have been very clearly set forth in the January number of the N. E. L. A. Bulletin, by Commissioner George E. Erb.

The facts brought out during the hearings showed that it is impossible to furnish electricity for house heating in competition with coal or wood. For open-air heating it would cost from four to six times as much when electricity is used and charged for at a rate that would pay a reasonable return on the investment necessary to render the service. Only where there is an excess of power that may be supplied for heating can this be done, and even then it is questionable. In the case of the Idaho Power Company the taxes paid per kilowatt-hour sold were more than the gross return on a kilowatt of energy for house heating.

In many of the irrigation projects the water has to be supplied by pumps driven by electric motors, and in these cases it was found that the heating load of the winter season overlapped the summer pumping load from four to six weeks both fall and spring, making it necessary to install considerable additional capacity if both interests were to be adequately served, with an attending increase in cost of the service. The period of greatest demand for current for heating in the intermountain regions is at a time when there is a minimum flow of water. This is due not only to low water, but also to freezing. The increased demand that has developed during recent years for water for irrigation purposes has made it necessary to store the water in reservoirs during the winter months for use during the summer. This practice has materially limited the power possibilities of some of the streams of Idaho. When it is considered that the power plants in the State of Idaho are favorably located for supplying power to their customers, the foregoing facts apparently give very little hope of using electricity for general heating purposes even when supplied from favorably located hydro-electric plants, except as a luxury, and this, in substance, was the finding of the commission.

There is little doubt that electricity is one of the greatest servants at the command of man, but, like all other great agents, it is not now, nor will it ever be, a catholicon for all our industrial and economic problems; but if intelligently used it can be of great economic value. This stands out plainly in the State of Idaho as brought out in the testimony before the commission.

An enormous economic waste is constantly going on in the use of coal for power, especially for firing railway locomotives. This waste is not confined to the locomotive, but includes the loss of power necessary to transport the coal. Would it not be more in keeping with the economic utilization of our natural resources to use hydro-electric power where available for operating our railways, or where this is not available, to burn the coal in economic power plants and supply electric energy to drive the locomotive—when one pound of coal burned in the former will do the work of three burned in the latter—and until such times as there will have been discovered a cheaper method of generating and transmitting power than at command at present?

Power Shortage on the Pacific Coast

IT MAY be that some day the East in general and Congress in particular will know and grasp the power situation on the Pacific Coast. They do not know it now. It is not unfair to say that the East has not the faintest conception, not to say adequate knowledge, of what the West has in the way of potential and developed power, nor does it realize how vital is power to the very life of this great and rapidly growing part of the country.

Agriculture and horticulture are the two great pursuits of the Coast from Seattle on the north to San Diego on the south. The average American, who knows more about Russia than he does about the San Joaquin and the Yakima valleys, cannot for the life of him see why a farm needs power other than that for the tractor and the Ford. He does not know that at this very writing every power company on the Coast is wofully overloaded furnishing electricity to farms and has booked so much farm business that the matter of delivery is a veritable nightmare to those responsible. This is particularly true of California, though most of the plants in the Northwest also are overloaded with farm and lumber industry consumers.

At present the "rainy season" in California is about half over and there is a shortage of more than eight inches of rain. From a power standpoint this has created a desperate situation. Every kilowatt of steam-generating capacity that can be unearthed and set to work is eagerly sought. While it may be true that the question of reserve steam stations has not taken a new turn in the minds of engineers here, it is certain they are thinking about the question with an intensity which is more marked than ever. Oil is becoming so high in price that every day some additional engineer joins the ranks of those who are ready to look upon it as a prohibitive fuel. This oil is, of course, a California product, little or no Mexican oil having yet found use here. The natural gas available is of low heat value and heretofore so expensive that it found but small market in power plants. However, one large company in the southern part of the state is completing a large steam station using natural gas.

The serious power shortage may, nevertheless, have a happy ending in that it will direct such attention to the matter of fuels for steam stations that not only will there be a greater percentage of steam power, but the stations may become of such an efficient character both in design and operation that it will be worth while finding service for them that will be of regular instead of an emergency character. This much is certain—a given sum of money will go farther building steam stations to meet certain load conditions, the time element considered, than is true for hydro-electric developments.

The engineers of the Coast know better than outsiders can possibly know what is the solution of their power difficulties, but the problem is a Coast one in particular and a national one in general. It is this that the country as a whole should, but does not, realize.

"Better Safe Than Sorry"

FROM an advanced summary of the 1917 mortality statistics received by the National Safety Council from the United States Census Bureau, the total number of accidental deaths in the area of continental United States under registration amounted to 53,544, as against 60,072 during the previous year. The reduction of 6,500 in the number of deaths is attributed to the accident-prevention movement of the last few years. It is gratifying to learn that America is gradually becoming less careless. Standardizing and making uniform the large number and variety of safety codes now in existence, a movement that is now under way, should reduce the cost of accident prevention and make the work more effective.

Perfecting the safety codes and making them uniform, however, is only part of the battle, as statistics show that over one-half of the accidents in the commonly accepted hazardous class are controlled by the personal equation and are due largely to carelessness arising partly from a lack of discipline and a natural disinclination to follow orders.

In hazardous work, such as that of the power plant, it is interesting to note the relative percentages of accidents. Electric current, moving machinery and steam explosions contribute but twelve per cent. This figure may be compared to 16.2 per cent credited to slipping and falling, 18.6 per cent to handling tools and 19.5 per cent to handling material. Nearly a quarter of the injuries due to falls were not from scaffolds, poles or other elevations, but occurred on the level where least justified.

Personal carelessness, then, and contributory negligence on the part of associates, are the leading causes of accident. Eliminate these factors, and immediately the number of accidents is reduced by one-half. No elaborate or costly equipment is required. The attitude of the executives will have a bearing, but the real solution lies with the worker himself. Be careful! It is better to be safe than sorry.

Needed, a New Association

THERE has been in existence in England for several years past an association of Diesel-engine users, made up of men who build, own and operate Diesel engines. Its meetings not only afford an excellent opportunity for an exchange of views and mutual help among operators, but also an opportunity for the designer to profit by the experience of the men who actually own the engines. The results have more than justified the founders of the association.

The use of oil engines is growing so rapidly in the United States that such an association is needed badly. At present papers on oil engines are occasionally given before various engineering societies, and their usual reception is concrete evidence that engineers are keenly interested in the subject. But the men who take part in these discussions are usually concerned with the building of the engines, and seldom does the operating man enter into them. On the other hand, there is an occasional talk on the subject at a meeting of one of the operating engineers' meetings, but here they are necessarily restricted through not gaining the designers' viewpoint.

To get the best out of any piece of power-plant apparatus, and especially an oil engine, there should be the closest coöperation and confidence between builder and operator. Often, equipment is discredited because of misunderstanding upon the part of the operating man, whereas a frank exchange of ideas with the designer would have resulted in eliminating the cause of the trouble.

While it is true that geographical limitations in England make it easier for Diesel-engine men to get together than in this country, where they are more scattered, still there are certain well-defined regions where the use of the oil engine is most general, and it is within these regions that such a movement could be started.

Mr. Lane's Constructive Work

ON MARCH first Franklin K. Lane will relinquish his post as Secretary of the Interior and retire to private life. His going will be regretted by all, regardless of party affiliations.

Mr. Lane leaves a record, covering seven years of broad and constructive administration, that has entitled him to front rank in the Cabinet. In fact, he has generally been regarded as the most popular of the President's official advisers.

The Department of the Interior houses numerous bureaus dealing with many diversified subjects, and its head must possess versatility second to none. With the combined training of a lawyer and a journalist, Mr. Lane has been enabled to see the broad economic possibilities of our natural resources and the necessity for their further development in a way that will bring greatest prosperity to the country. Moreover, he has shown a keen grasp of the more or less technical and engineering details as pertain to our fuel supply, water-power development and power problems in general. He has been an ardent champion of the "super-power plan" and last year urged Congress for an appropriation to cover a survey of the power situation in the Atlantic Coast section. Had Congress acceded to this request, we would have been in a position to meet more intelligently the fuel conditions confronting that section today.

Mr. Lane was also solidly behind the water-power bills that recently passed both houses of Congress and are now in conference. If they finally become law, as indications point they will, much credit will be due him.

Averaging Pressure Charts

The article by W. V. White on "Averaging Pressure Charts," on page 669 of the Oct. 28-Nov. 4, 1919, issue of *Power*, should not pass unchallenged. It is a source of some surprise that the fallacy of endeavoring to average circular charts by means of the ordinary planimeter is not commented upon, since this subject was thoroughly threshed out years ago in *Power*. Reference is made particularly to the issues of Mar. 3 and Mar. 31, 1908; Oct. 2 and Nov. 24, 1913, which lead to the correct conclusion that the mean value of the squares is not the same as the average diameter.

Several special planimeters have been devised for solving the problem, including the Durand, described in *Power*, Mar. 23, 1911, and instruments manufactured by Builders Iron Foundry, Bristol Co. and General Electric Co.

Polar planimeters can be used on circular charts with practically accurate results only when the charted line is confined to a narrow zone; the error may be very large otherwise. CHARLES G. RICHARDSON.

Providence, R. I.

Why Does the Water in a Boiler Lift When the Safety Valve Blows?

On page 715 of the Nov. 11 issue I read Charles L. Anderson's version of "Why Does the Water in a Boiler Lift When the Safety Valve Blows?" Mr. Anderson believes that the A. S. M. E.'s removal of the limit as to the size of safety valves is not progress; I agree with him.

I have had boilers with three 5-in. safety valves, and I believe 5-in. valves to be too large. One day, by accident, two of these valves blew at the same instant on one boiler, with the result that a jet of water came from each escape pipe with tremendous force. As soon as the blowing was over, one valve spring was eased to allow that valve to blow a couple of pounds earlier. One of these valves blowing at a time did not carry water with the steam.

Some years ago I was engineer on a small steamboat that carried a water-tube boiler. On one bell the water would stay at two gages, but on the jingle bell, if caution was not used the water would lift out of sight in the glass. If the throttle was pulled open very gradually, the water would remain in the glass at perhaps 2½ gages and then settle to where it belonged.

I believe the large opening relieves the pressure too quickly, and then the water flashes into steam so rapidly it spills over, as it were. I know of a case where water was being taken over by the turbines to the extent that the blading had to be scraped perhaps twice a year. After the holes in the dry pipes in the boilers were reamed out a few sizes larger, no more water was taken over. This was a station of several thousand kilowatts, and it took some time to remedy the trouble, but the enlarged holes did the trick.

I know of a case where the water was being carried at one gage in a battery of horizontal-tubular boilers and the engine was getting large quantities. It was suggested that the water was so low that the top row of tubes was throwing it up with the steam. The water was raised to two gages and the engine took very little water after that. C. W. PETERS.

New York City.

Extending the Range of Voltmeter

Referring to the method of "Extending the Range of Voltmeter" in the Nov. 11-18, 1919, issue of *Power*, it is necessary to take into consideration the wattage of the lamps employed, as the method described is not theo-

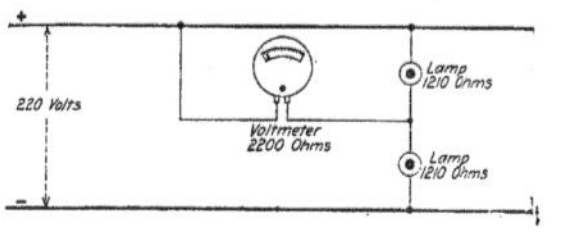

VOLTMETER CONNECTION FOR MEASURING DROP ACROSS ONE LAMP

retically correct and is only approximately correct when 200-watt lamps are used, the error decreasing as the size of lamps is increased.

In order to calculate the error, assume that two 10-watt lamps, each having a hot resistance of 1,210 ohms, are used on a 220-volt circuit. Assume the voltmeter resistance is 2,200 ohms, as indicated in the figure, a fair value for a dynamometer type. The joint resistance of the voltmeter and one lamp in parallel equals $\frac{1}{\frac{1}{2,200}+\frac{1}{1,210}} = 780$ ohms. This is in series with the other lamp of 1,210 ohms, making the total resistance of the circuit equal to $780 + 1,210 = 1,990$. By Ohm's law the current in the circuit will be $I = \frac{E}{R} = \frac{220}{1,990} =$ 0.11 ampere, and the volts drop across the voltmeter is equal to the resistance of this section of the circuit times the current or $780 \times 0.11 = 86.8$ volts. Since connecting the voltmeter across the other lamp will give identically the same results, then the sum of the two readings would be 86.8 + 86.8 = 173.6 volts, which Mr. Nichols says should equal the line voltage. These figures show that the meter in this case would read 21 per cent. low when using 10-watt lamps.

But if we substitute 200-watt lamps each having a resistance of 60 ohms, then the joint resistance of one lamp and voltmeter is 58.4 ohms, and the drop is 108.3, which is only 1.3 per cent. low. Using larger lamps still further reduces the error, but it can never be reduced to zero. PHILLIP A. LITTEL.

Orange, N. J.

[If the foregoing measurements had been made on a direct-current circuit with a voltmeter of the permanent-magnet type, having a range of about 150 volts, and using ordinary 25-watt tungsten lamps, which is a very common type, the percentage of error would be practically negligible. The hot resistance of the lamps would be about 302.5 ohms and the resistance of the voltmeter 15,000 ohms. Working this out, the voltmeter reading will be 109.4 volts, or 0.55 per cent. error. This would be in general the conditions we would expect to find on a direct-current circuit. However, on an alternating-current circuit the error will be greater on account of the lower resistance of the voltmeter; the error being somewhere between the extremes given by Mr. Littel, probably in general around 4 or 5 per cent. Of course if care is used to obtain high wattage lamps, the error would be small. In emergency makeshifts whatever comes handiest is most generally used, and in this case lamps of 40- or 60-watt capacity are the ones that would in general be most likely to be available. —Editor.]

Demagnetizing a Watch

An old-type Edison bipolar generator has an extremely large magnetic leakage. In working around one of these machines my stop-watch became magnetized so that when I snapped the watch to start, the resetting lever dragged the cam to its point before releasing, thus moving the hand from 0 to 30 sec. in a small fraction of a second, before the wheel engaged with the watch mechanism.

To demagnetize it, I took a coil about 5 in. in diameter, consisting of 600 turns of No. 30 B. & S. gage magnet wire, and connected it to 110-volt 60-cycle circuit. Then tying the watch to a piece of string and twisting it so that it would spin around, I passed it through the coil with the watch spinning rapidly. There was not a trace of permanent magnetism left in the watch after performing this operation once. V. H. TODD.

Orange, N. J.

Fuel Savings by the Economizer

Statements published in books that by the use of an economizer the fuel consumption can be reduced from 10 to 20 per cent. are of no great use to an engineer except as a matter for comparison. What an engineer wants to know is how much fuel can be saved by its use in his own plant.

Some years have passed since I came to my present plant, but it was only this year that I got a chance to fully realize what a clean economizer meant, when one was cut out to have the tubes bored.

We have a 28 series (8 pipes in each series) economizer attached to three Lancashire boilers operating with 160 lb. pressure. The average injection water temperature is 100 deg.; hotwell temperature, 130 deg.; economizer outlet temperature, 275 deg. F.

From the accompanying chart the average coal consumption per indicated horsepower per hour will be seen to be 2.26 lb. per i.hp. before the economizer was stopped; 2.40 lb.i.hp. per hour when the economizer was stopped; and 2.11 lb. i.hp. per hour when the economizer was started again.

Then, $2.40 - 2.11 = 0.29 \times 100 \div 2.11 = 13.7$ per cent. the amount of the fuel that can be saved by the use of the economizer. But $2.40 - 2.26 = 0.14$ and $0.14 \times 100 \div 2.26 = 6.15$ per cent. Therefore but 6.15 per cent. of the fuel was being saved before economizer tubes were bored. This was due to scale, which was about one inch thick in the tubes. Therefore, only half the percentage of the fuel can be saved by permitting the tubes to work with a thick layer of scale.

Further, $2.26 - 2.11 = 0.15$ and $0.15 \times 100 \div 2.11 =$ 7.1 per cent. Therefore, 7.1 per cent. greater saving in fuel is being made by the economizer after it was cleaned of scale than before being cleaned.

On the whole, I found that about 13 per cent. of the fuel can be saved by the use of the economizer in my own plant.

Moreover, the same was the case when our economizer was stopped for boring the tubes seven years ago,

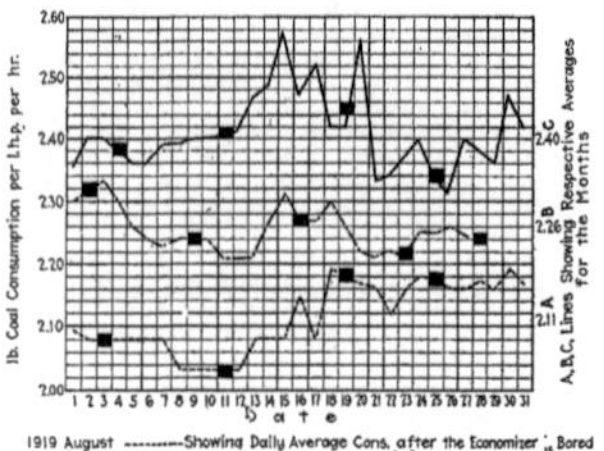

SHOWING AVERAGE COAL CONSUMPTION WITH THREE ECONOMIZER CONDITIONS

as will be seen from the accompanying monthly consumption averages per indicated horsepower per hour.

Before the Economizer Was Stopped for Boring

		Pounds
1912	Jan.	2.02
	Feb.	2.13
	Mar.	1.99
	Apr.	1.81
	May	1.89
	June	1.87
	July	1.83
	Aug.	2.04
	Sept.	1.70
	Oct.	1.72
	Nov.	1.75
	Dec.	1.90
	Total	22.65
	Average	1.88

When Economizer Was Stopped

		Pounds
1913	Jan.	2.07
	Feb	2.13
	Total	4.20
	Average	2.10

After Economizer Was Started Again

		Pounds
1913	Mar.	1.97
	Apr.	1.81
	May	1.89
	June	1.87
	July	1.96
	Aug.	1.85
	Sept.	1.74
	Oct.	1.71
	Nov.	1.69
	Dec.	1.74
	Total	18.23
	Average	1.82

The averages when the economizer was worked before and after being bored are nearly the same, $1.22 + 1.82 = 3.70 \div 2 = 1.85$ average. Average was 2.10 when the economizer was stopped, therefore $2.10 - 1.85 =$ 0.25 and $0.25 \times 100 \div 1.85 = 13.5$ per cent.

This percentage of savings corresponds to that shown by the chart, which was platted from data obtained this year.

Ahmedabad, India. S. B. R. PATEL.

Automatic Control for Siphon

A sump-hole in a mill building drained by a siphon, located in the engine-room basement 200 ft. distant, was a continual nuisance. The engineer would either neglect to start it or forget to stop it, and as a consequence, the mill basement was flooded or steam blew needlessly into the sewer. To remedy this condition, an automatic control made from parts picked up around

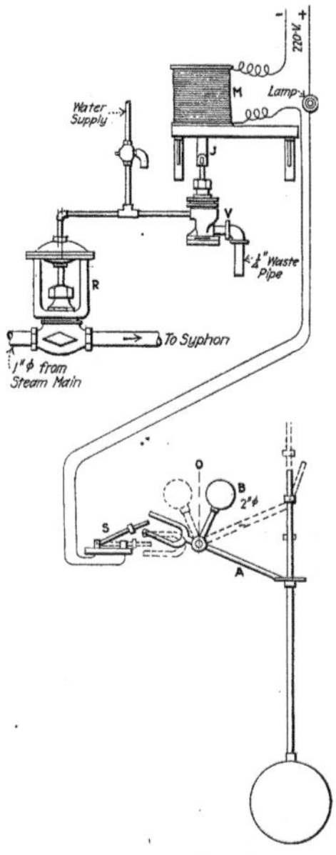

ARRANGEMENT OF SIPHON CONTROL

the plant was devised and installed, as shown by the illustration.

As the sump fills, the float raises the forked lever *A*, causing the lead ball *B* to move toward the center line *O*. When the ball *B* crosses the center line, it falls to the dotted position and closes the switch *S*. Current then flows from the power lines through a lamp in series with the magnet *M*, which causes the valve stem of the valve *V* to rise and allows water to flow through the waste pipe. The pressure on the diaphragm of the steam valve *R* thus being removed, allows the valve to open and steam to flow to the siphon.

The siphon continues to operate until the water level falls and the float pulls the weight *B* past the center line *O*, when it falls and opens the switch *S*. As the current is then cut off, the magnet core drops, closes the valve *V* and the water pressure builds up on the diaphragm of the valve *R*, closing the valve and shutting off the steam.

It is well to allow ⅛ in. lost motion in the joint *J*, so as to give a hammer-blow effect to the movement and overcome any tendency to stick in the packing on the valve stem. A globe valve could be used instead of the trap valve *V*, but it would not open so easily, owing to the pressure above the disk holding it to the seat. This would be no drawback if the magnet was capable of opening it.

Perth Amboy, N. J. A. QUADT.

Attention to Little Things Saved Coal

It may be of interest to readers of *Power* to know how we have effected a saving in coal slightly better than 20 per cent. Starting at the boilers, we keep the tubes clean, inside and out. The baffle bricks are kept in repair and tight, so that the gases cannot short-circuit and pass out without giving the boiler a fair chance to absorb the heat. We also adjust the damper according to the load and the conditions of the fire, so as to maintain a high percentage of CO_2 and keep the flue-gas temperature down to from 70 to 100 deg. above the temperature of the steam. Of course adjusting the damper alone will not produce high CO_2; the fires have to be closely watched to do that.

We have placed a steam-blow telltale on every steam trap. It consists of a valve and discharge pipe. As soon as a trap is found to be blowing steam, and they are tested often, it is repaired. All pipe coverings on steam, exhaust and return lines and also on all flanges, valves and traps are kept in proper repair, thus reducing condensation to a minimum.

I believe that in most plants there is a chance to make a fair showing by saving the steam after it is generated, as well as saving coal during the process of making steam. It is equally as important to keep all valve stems packed and to look after the little details, as it is to look after the economy of steam in the prime movers. We are only doing what we should do and what should be done in every power plant. To do this requires the coöperation of the chief with the men and the men with the chief.

In the larger plants the chief cannot keep in touch with all the details and know the condition of all the equipment, etc. There may be a trap blowing, due to a bypass valve leaking, or some other waste which he cannot observe on his trips through the plant. But if he has the good will and coöperation of his men, those things will be taken care of.

Binghamton, N. Y. EDWARD J. DOWD.

Distance of Bridge-Wall from Boiler Shell

I would like to obtain the opinion of engineers regarding the following:

A few weeks ago our superintendent had the bridge-walls of the boilers cut down. I told him that I thought he had made a mistake in doing so, but he didn't think so. At any rate, the firemen have had harder work to keep up steam, and they burn more coal since it was done. The boilers are hand-fired. One is 19 ft. long, 6 ft. 6 in. diameter and has 114 three-inch tubes; the other is 18 ft. long, 6 ft. diameter and has 138 two-

and-one-half-inch tubes. The rocking grates are 6 x 6 ft., and the chimney is 100 ft. high with a good draft.

The bridge-walls, as they are now, are 18 in. from the shell of the boilers and the coal burns out very fast on the back area, so much so that the firemen cannot keep any coal on the last half of the grates next to bridge-walls, no matter whether the fires are thick or thin. We carry from 75 to 100 lb. of steam.

I argue that if the bridge-wall was put back to within ten or twelve inches of the shell of the boilers, we would get better combustion, better results and burn less coal. I am not sure as to the distance, but as the superintendent is one who thinks he knows about everything and will not listen to me, I would like to have the opinions of two or three experts in regard to the distance the bridge-walls should be from the boilers in the average setting.

Ashland, N. H. ALBERT L. SMYTHE.

Right- or Left- Hand Shoveler—Which?

I wish to take exception to the explanation of a "Right-Hand and Left-Hand Shoveler," in "Inquiries of General Interest," in the Oct. 7, 1919, issue of *Power.* In this locality we have always understood that a right-handed shoveler is one who supports the load with his right hand, or has his right hand nearest the shovel, and a left-handed shoveler is one who supports the load with his left hand. When using a sledge for riveting or striking, the man who grasps the handle with his right hand nearest the sledge, is a right-handed man, and vice versa, and has nothing to do with whether a person is naturally right-handed or left-handed, and the same thing applies to shoveling.

Lima, Ohio. S. M. WILLIAMS.

WHICH IS THE RIGHT- AND WHICH THE LEFT-HAND SHOVELER?

Drain Trenches in Pump and Condenser Pits

Progressive power-station designers and erectors now realize that pump and condenser pits should be made fit places for human beings to work in and not merely spaces for machinery. Good lighting and drainage have long been a rarity in such places. It is human nature for a man to give better attention to a machine in a clean and dry place than in a sloppy, greasy pit. Many pits are provided with one or two floor drains, but the assistance of a broom is required to complete the drainage, and drain pipes have a way of getting stopped up.

Drain trenches should be run the full length of the pits and should be covered with cast-iron grids instead of plates. The trenches should be so located that no part of the floor is more than 6 or 8 ft. from a trench, and the trenches should be about one foot wide and deep. The floor should slope toward the trenches at least one inch for every three or four feet. Drains from bedplates, glands and water seals should be piped to the trenches and not allowed to run across the floor. If proper attention is given to lighting and drainage, there will be no excuse for the operators not keeping the pit and machinery within it clean and in good order.

Providence, R. I. C. P. LAWTON.

Preventing Ice Troubles

In severe winter climates heated gatehouses are often provided at head works or at the gates of spillways of hydro-electric plants. At waste gates these houses may not cover all the gates, as it may not be necessary to keep all the gates free. Steam-heated gate chambers are also used which cover individual gates.

During the low-water period of winter many plants use all the water, and if the gates are not kept free of ice they may freeze so solid that it may take considerable time to chop them free. Under these conditions the head elevation should be maintained a sufficient distance below the dam crest to allow ample time to free enough gates to waste the water in place of it going through the turbines, in case the latter lose their load in the event of transmission-line failure.

At a new installation it may be advisable to find out how long it takes to free certain gates and what tools are best adapted for the gates in question before the actual need arises. Saws, picks and axes are the tools generally used. Ice picks with three sharp prongs are excellent for such work and linemen's safety belts are handy when working on ice-covered gates.

Head-gate houses may be heated electrically. At one large plant a 50-hp. motor drives a fan which delivers air from the generator room through a chamber containing a 300-kw. electric heater to the head-gate house. Here some of the air discharges so that it blows against the racks and steelwork of the head gates at a temperature of 60 deg. C., which keeps the ice from forming.

Experience indicates that if the tops of the racks are exposed to cold air above water, the steel conveys the heat from the water to the air and results in the racks having a lower temperature than the water, causing ice formation on the racks. For this reason some rack installations are designed so that the racks do not project above the water. Others are removable to allow the water to pass freely in case of anchor or frazil ice.

Charles H. Bromley's article in *Power,* June 3, 1919, covered the ice situation at Holtwood so thoroughly that any further space given to this matter would necessitate duplication, as the conditions encountered at Holtwood are typical.

Iron Mountain, Mich. L. W. WYSS.

INQUIRIES OF GENERAL INTEREST

Inside and Outside Calking—Is it customary to calk boiler seams inside as well as outside of the boiler? L. E. F.

Calking usually is done only on the outside. In marine boilers and very carefully constructed boilers a preliminary calking is given to the inside laps, and absolute tightness is afterward secured by a calking on the outside.

Merits of Chain or Ring Methods of Lubrication—What are the merits of the chain or the ring method of supplying oil to a bearing? R. M.

These methods have the serious defect that the spent and dirty oil is returned again and again to the bearing and, unless replenished, the oil finally becomes so loaded with foreign matter that it is a less efficient method of lubrication than occasionally oiling with a squirt-can. The chain or the ring method is serviceable only in situations where the oil well is thoroughly cleaned at frequent intervals and refilled with oil free from dirt.

Blowing Off a Cut-Out Boiler—When a boiler is cut out from the main line for cleaning, should it be blown out when steam has become reduced to about 15 lb. pressure, or left standing for several days filled with cold water? A. F. B.

A boiler should not be blown off under pressure or while the brickwork is hot. Sudden reduction of pressure is injurious to the joints, and if the boiler and brickwork are not cool, the heat will bake the mud and scale in the boiler, making it difficult to clean the shell and tubes. After cutting out, there will be no advantage in allowing the boiler to stand longer than necessary for allowing it and the brickwork to become cooled.

Reporting Average Cost per Pound of Steam—The monthly report blanks of our vacuum-heating plant provide for "average cost per pound of steam in system" and "average cost per pound of steam in boiler." How should these items be computed? F. D. C.

For obtaining the "average cost per pound of steam in the system," divide the total cost of fuel, labor and other expenses for the month, by the total number of pounds of steam supplied to the coils or radiators, as ascertained by weighing or metering the water that has been discharged by the vacuum pump, less the weight of injected water. For ascertaining the "average cost per pound of steam in boiler," divide the total cost by the number of pounds of water fed to the boiler.

Trying Glass-Gage Connections—How often should the glass water gage of a boiler be blown out? M. D. T.

Glass gages should be blown out at least once a day, and as much oftener as may be found necessary to make sure they are in proper working order, and when this is done the operator should be careful to see that the cocks are open or the gage will not indicate correctly. If the lower connection is open and the top one closed, the boiler pressure will force the water too high in the glass, and if the bottom connection is closed and the top one open, the steam entering at the top will be condensed and gradually show a higher water line than actually in the boiler. Of course if both connections are closed, the gage will not indicate changes of the water level and high or low water may result without the knowledge of the fireman.

Preventing Siphonage of Oil—We are having trouble with leads on a number of starting compensators siphoning oil from the tanks. What method can be recommended to remedy this difficulty? A. A. F.

The siphoning of oil along the leads of transformers or compensators may be prevented by first removing the insulation from each lead above the highest oil level for about two inches. Then sweat the strands of the leads solidly together so that there will be no openings between them for the oil to pass through. After this is done, insulate the leads with treated linen tape, wrapping the tape very tightly around the conductor, and paint each layer with insulating varnish while wrapping. Allow the tape to extend about one inch down over the insulation on the conductor.

M. E. P. of Engine for Generation of Stated Kilowatt Load—What m.e.p. would be necessary for a 10x10-in. engine running 300 r.p.m. for operation of a 35-kw. direct-current generator when developing full capacity? G. B.

The efficiency of a generator of the given capacity under full load would be about 90 per cent, and the mechanical efficiency of an engine of the stated size and speed when driving the load given would be about 88 per cent, depending on the design, construction and adjustment. As 1 kw. = 44,240 ft.-lb. per min., the assumed efficiencies would require the steam acting on the piston to develop $\frac{44{,}240 \times 35}{0.90 \times 0.88} = 1{,}955{,}050$ ft.-lb. per min. Neglecting the reduction of piston area by the piston rod, 1 lb. m.e.p. would develop

$$(10 \times 10 \times 0.7854) \times \frac{10}{12} \times 2 \times 300 = 39{,}270 \textit{ ft.-lb. per min.}$$

so that development of 1,955,050 ft.-lb. would require

$$1{,}955{,}050 \div 39{,}270 = 49.78 \textit{ lb. m.e.p.}$$

Running Over or Under—Is more power derivable from an engine running over than running under, and if so, why? J. B.

When running over, pressure from the crosshead is on the lower guide and is due to reaction of the connecting-rod and weight of the crosshead, and as much weight of the piston rod and connecting rod as is supported by the crosshead. When running under, the pressure from reaction of the connecting-rod comes on the upper guide, but it is reduced by the weight of the crosshead and parts supported by the crosshead. Therefore in running under the guide friction is less and if anything the actual or brake power developed for the same mean effective pressure the piston would be greater. But under ordinary conditions the friction of guides is such a small percentage of engine friction that the difference of friction running over or under would be imperceptible. The principal advantage of running over is that with the pressure from the crosshead downward there is better opportunity for lubrication of the guide that receives the principal wear, better opportunity for adjusting and maintaining the alignment of the crosshead and less tendency of the crosshead to pound at the ends of the stroke.

[Correspondents sending us inquiries should sign their communications with full names and post office addresses. This is necessary to guarantee the good faith of the communications.—Editor.]

Production of Electric Power and Consumption of Fuel by Public Utility Power Plants in the United States for the Months of February, March, April and July, 1919

Compiled by the United States Geological Survey

	Thousands of Kilowatt-Hours Produced								Coal, Short Tons				Petroleum and Derivatives Barrels				Natural Gas Thousands of Cubic Feet			
	By Water Power				By Fuels															
State	February	March	April	July	February	March	April	July	February	March	April	July	February	March	April	July	February	March	April	July
Alabama	39,341	30,275	28,488	28,442	11,059	4,907	4,477	3,689	27,026	16,467	14,836	15,458	8	8	8	8	....	....	....	....
Arizona	3,967	6,689	7,577	7,478	17,448	21,807	16,488	18,877	6,316	6,160	5,360	7,315	124,520	88,011	65,824	61,030	....	....	....	....
Arkansas	79	72	69	88	6,252	6,625	6,030	7,692	13,386	11,177	9,902	8,364	447	489	456	450	63,430	85,398	97,726	182,848
California	187,486	215,914	228,775	200,528	38,268	33,939	33,645	116,804	....	....	....	....	199,388	185,127	161,651	550,799	21,611	181,785	273,642	242,228
Colorado	13,141	14,254	15,019	14,947	17,279	17,144	14,365	12,177	41,507	44,079	36,135	27,888	100	95	98	174	....	....	....	....
Connecticut	7,043	14,884	15,804	5,943	42,589	37,469	35,525	28,182	63,942	62,234	58,094	67,513	333	323	279	213	20,766a	12,615a	11,309a	....
Delaware	....	....	....	....	5,551	4,784	4,655	5,225	8,610	7,713	7,000	6,604	....	17	....	....	....	....	....	....
Dist. of Columbia	....	....	....	....	18,094	19,331	18,457	19,221	21,418	21,440	20,983	21,412	....	....	....	....	....	....	....	....
Florida	678	892	777	942	8,582	8,789	7,776	7,731	3,424	5,594	3,056	3,193	32,714	32,025	28,785	30,745	98	151	3,780	....
Georgia	35,909	37,264	31,404	34,224	6,054	6,082	6,206	6,595	16,529	11,934	12,009	13,560	120	120	167	2,386	....	....	....	....
Idaho	41,430	39,913	36,019	50,209	168	257	155	1,219	120	111	121	88	10	10	10	9	....	....	....	....
Illinois	15,611	15,282	16,134	15,323	198,899	207,673	194,109	183,518	335,309	350,824	318,378	304,160	3,918	2,198	2,142	984	....	....	....	....
Indiana	3,682	3,654	3,511	2,736	52,342	54,617	51,980	54,354	141,080	157,931	143,829	142,587	448	167	420	402	2,159	2,108	1,909	1,878
Iowa	44,182	48,696	48,737	51,802	23,993	24,964	23,646	24,531	85,745	79,694	72,211	69,703	787	730	743	720	....	....	....	....
Kansas	1,267	1,223	913	981	29,619	30,458	30,870	35,068	47,588	51,022	48,636	50,817	47,747	60,803	58,432	60,742	83,405	86,901	65,374	80,846
Kentucky	4	4	4	4	18,794	19,127	18,555	19,205	38,252	39,386	37,380	39,390	358	351	353	341	....	....	....	....
Louisiana	....	....	....	....	14,130	14,876	14,582	15,518	14,305	13,885	13,878	11,324	30,811	30,094	28,192	30,508	55,092	47,458	49,599	63,624
Maine	18,831	20,157	18,708	17,678	73	73	93	220	333	348	357	534	7	7	17	....	....	....	....	....
Maryland	284	371	358	329	18,435	13,898	11,442	16,378	28,452	23,657	20,755	25,951	19	18	30	30	1,500	1,500	1,500	1,350
Massachusetts	18,698	31,365	32,322	15,163	101,799	100,030	90,601	110,467	129,171	143,525	129,877	130,506	24	14	17	21	....	....	....	....
Michigan	52,236	59,649	65,211	50,043	97,285	96,125	88,363	108,402	129,417	124,941	113,623	129,612	191	104	139	204	....	....	....	....
Minnesota	18,303	32,740	43,204	37,436	26,826	15,220	7,261	10,929	57,282	39,810	25,038	25,866	1,138	1,016	854	832	....	....	....	....
Mississippi	....	....	....	....	5,115	5,393	4,927	5,206	16,381	16,542	16,008	13,272	370	320	311	443	....	....	....	....
Missouri	4,297	5,920	4,418	2,237	37,207	40,551	36,245	39,132	86,645	89,531	82,143	82,045	33,571	21,053	19,519	18,847	....	....	....	....
Montana	68,408	77,869	77,853	83,220	883	888	821	848	5,537	5,591	5,455	4,894	530	492	488	502	1,008	960	907	912
Nebraska	748	959	1,118	1,302	15,493	16,413	16,004	16,602	29,946	30,956	29,813	31,904	3,512	3,509	3,433	3,387	....	....	....	....
Nevada	3,198	2,812	2,470	2,559	137	121	727	720	198	180	140	195	900	1,044	997	1,780	....	....	....	....
New Hampshire	4,946	5,763	5,683	4,499	2,309	2,114	1,898	3,814	5,395	3,911	3,241	5,808	20	20	3	3	....	....	....	....
New Jersey	162	197	187	151	75,665	76,412	76,642	79,546	112,366	115,711	114,137	116,944	92	103	80	79	....	....	....	....
New Mexico	57	56	80	144	1,416	1,556	1,455	1,603	4,045	4,438	3,627	4,133	1,060	1,360	1,373	1,610	....	....	....	....
New York	195,219	223,078	220,677	212,033	288,271	288,537	270,644	276,828	368,569	363,338	325,474	336,004	544	546	501	1,224	149,831	164,408	147,432	56,949
North Carolina	42,645	46,318	45,755	45,317	7,149	7,426	7,082	7,243	15,648	16,234	15,612	14,915	20	20	20	20	....	....	....	....
North Dakota	....	....	....	....	2,312	2,400	2,103	2,449	16,638	17,153	14,473	14,808	539	514	530	553	....	....	....	....
Ohio	3,053	3,932	3,565	2,378	189,041	194,937	183,702	208,240	314,873	321,643	289,610	288,658	739	842	764	689	250,488	323,320	333,684	b512,586
Oklahoma	167	183	149	161	13,080	13,796	12,989	13,826	11,136	12,578	10,075	8,717	8,109	6,300	9,921	20,057	423,813	477,050	458,496	474,930
Oregon	27,876	29,113	27,248	29,130	4,754	3,759	3,539	4,323	223	433	223	222	27,684	19,459	15,456	5,664	....	....	....	....
Pennsylvania	55,098	62,509	58,466	52,648	238,970	246,930	233,597	238,021	401,874	412,561	396,096	388,957	22	14	16	297	42,301	54,450	53,533	40,531
Rhode Island	562	1,042	726	240	28,767	18,405	16,839	19,821	29,692	22,024	20,819	22,648	....	....	....	....	....	....	....	....
South Carolina	43,899	43,394	43,287	42,649	3,876	4,010	3,676	3,480	9,844	9,540	8,811	8,281	78	82	93	102	....	....	....	....
South Dakota	2,683	3,579	4,104	3,468	2,949	3,208	1,920	2,920	7,082	8,395	4,371	6,841	3,164	3,379	3,091	3,192	....	....	....	....
Tennessee	43,640	45,222	37,999	32,316	9,774	10,680	10,405	9,164	23,726	28,724	24,710	22,156	86	84	81	82	....	....	....	....
Texas	266	284	305	191	43,070	47,616	44,816	45,537	25,330	33,963	26,618	29,269	170,905	181,397	178,033	188,691	173,695	179,760	177,778	114,148
Utah	13,135	15,463	17,434	12,405	....	....	....	....	65	61	31	....	....	....	....	....	....	....	....	....
Vermont	13,428	18,574	19,482	11,748	173	179	187	976	542	412	246	3,909	2	2	2	2	....	....	....	....
Virginia	17,883	20,122	19,298	17,666	17,254	18,709	16,539	19,336	29,127	31,444	28,453	30,674	122	153	52	122	....	....	....	....
Washington	70,779	79,346	74,575	79,881	4,171	4,521	4,164	5,434	3,611	3,473	2,411	2,060	17,810	18,804	17,879	22,071	....	....	....	....
West Virginia	1,378	1,593	1,642	1,399	52,620	58,108	59,396	66,523	62,459	70,717	69,426	83,556	52	58	52	45	118,029	142,688	180,069	264,054
Wisconsin	32,534	47,531	48,880	41,512	32,916	33,075	24,682	27,637	69,225	74,987	53,037	46,221	644	662	565	574	....	....	....	....
Wyoming	248	172	138	179	3,315	3,603	3,249	3,386	12,450	13,384	12,242	12,752	8,470	9,062	8,840	4,008	5,278	6,208	3,129	3,481
Total	1,148,634	1,308,329	1,308,573	1,213,729	1,834,222	1,841,542	1,717,523	1,928,617	2,873,265	2,919,406	2,648,760	2,676,688	722,133	671,006	610,736	1,014,652	1,412,504	1,766,850	1,839,867	2,040,365
Total, by water power and fuels	....	....	....	....	2,982,856	3,149,871	3,026,096	3,142,346	....	....	....	....	....	....	....	....	....	....	....	....

(a) Artificial gas. (b) Includes 182,415.

These reports are based on returns received from about 3,000 electric-power plants engaged in public service, including central stations, electric railways, and certain other plants which contribute to the public supply. Returns were received from plants whose aggregate capacity of generators is about 90 per cent of the total installed capacity of generators of public-utility plants. Estimates of the output of plants which did not submit returns were made from available information. The figures given are subject to revision in subsequent statistical reports of the United States Geological Survey relating to power production.

Operation and Maintenance of Railroad Stationary Plants*

KEEPING soot from accumulating on tubes and flues is one of the most important steps toward economical operation. Flues should be blown at least once each eight hours, and when boilers are down for a washout they should be thoroughly cleaned with a brush or scraper. In water-tube boilers the tubes should be thoroughly blown when the combustion space is cleaned. The boilers should be blown down at least once each eight hours, and the men coming on shift should test the water columns, try-cocks and pop valves. Feed the boilers continuously, maintaining a constant water level. Regulate the draft with the stack damper rather than the ashpit doors.

When the boilers are down for a washout, be sure that they are cleaned. Rattle or turbine the tubes if necessary and scale the shell. Clean out the furnace and combustion chamber, the bridge-wall and the smokebox. See that all repairs to the furnace and combustion-chamber walls and arches are taken care of. Repair the blowoff valves and feed valves if necessary, and see that all doors are air-tight. Minor furnace and combustion-chamber repairs can be made with plastic cement, and air leaks can be stopped up with a plastic boiler covering. Stacks should be painted with an approved stack paint at least once a year.

See that nonreturn valves are kept in working order and that all valves are kept packed; paint valve stems with graphite and oil about once a week to save valve-stem packing. If the insulation is removed from the piping to make repairs, see that it is replaced promptly.

The time to clean feed-water heaters can be determined by watching the temperature of the feed water, which should never be allowed to drop below 200 deg. Carry the water level in the heater at such a point as to keep a head of water on the pump at all times, but not so high as to cause the heater to overflow continuously.

Feed pumps should be repacked once every 30 days. This insures one good pump on the job all the time.

Air compressors and power equipment should have the rods and bearings keyed up at all times. Indicate all engines and compressors frequently and take care of leaky valves and faulty settings as quickly as possible. Such equipment needs a general overhauling every three to five years. Inspect the water jackets on the compressors frequently, as they may fill up quickly with mud and scale.

The exhaust-steam heating system is the most economical device in the plant if properly operated. Almost any railroad exhaust-steam heating system should operate with satisfaction on two or three pounds pressure. A well-designed and well-operated exhaust-steam heating system will reclaim about 90 per cent of the heat in the steam, and the greater part of the other 10 per cent goes into the feed water. A vacuum of 5 to 10 in. of mercury should be sufficient, provided the return lines are tight and the traps are not leaking.

During the summer of each year all heating-system traps and sediment chambers should be removed and cleaned. See that the return lines are all in good shape and air-tight, and overhaul the vacuum pumps. If it becomes necessary during the winter to shut the heat off for a short time, be sure that all drains from the system are opened to the sewer to prevent freezing, and keep them open after the heat is again turned on until the system has been warmed up, then cut in the vacuum pump.

If loss of vacuum occurs, find out in what section the trouble lies, by isolating each section in turn; then test the pumps by shutting off the returns. Having located the section, go over the return lines in this section with a torch, which is about the simplest method for locating leaks in vacuum lines. If the leak is not located in the return lines, look for a leaky trap, by shutting the steam valve on each coil in turn.

See that all steam, air and water lines and valves are kept in repair. Do not allow pipe covering to become ragged and fall off. Do not allow pipe-line supports and hangers to become loose. Keep boiler and engine rooms clean and painted. Do not allow temporary shacks, piping and scaffolds to stand after they have served their purpose. Put in the broken window lights, rehang sagging doors, and paint the walls, piping and foundations. Make the stationary plant a clean and desirable place to work in.

Fuels and Combustion

When oil and gas are in active competition with coal, they can be burned with great economy. There are only a few places where gas is used, and the supply of natural gas is too small to make this fuel very active for commercial purposes. Burning of oil in boiler furnaces has assumed great importance during the last few years, and the continual rise in price of coal is making oil an active competitor of coal in the Middle Western territory.

Steam seems to be the favorite medium of atomization, although low-pressure air has been used with success. Any burner that will atomize the oil efficiently, with a minimum of steam consumption, will give economical results, although the design of the furnace has considerable to do with the economy. The use of preheated air (air usually heated by passing through ducts in the furnace and floor, side walls or arches) does much to increase the efficiency of oil-burning furnaces. When using oil in return-tubular settings, care should be taken to direct the flame so that it will not impinge on the boiler shell.

The use of powdered coal at this time is not recommended. It is still in the experimental stage and there is considerable danger in handling and transporting it. It is understood that the plant at Parsons, Kan., which was the pioneer in powdered-coal burning at stationary plants, has abandoned that fuel for oil.

The United States Shipping Board has been experimenting with "Colloidal" fuel for over a year, and according to all reports, has obtained good results. It consists of a fuel oil having powdered coal held in suspension by a secret "colloid" added to the mixture. "Colloidal" fuel burning presents about the same conditions as oil burning, and it can be burned in the same burners. It has all the advantages of powdered coal and none of the dangers in handling and transporting. "Colloidal" fuel is heavier than water, and when dumped upon the surface of water, it will sink, so that the fire hazard is reduced even over oil.

Smoke is usually an indication of inefficient operation, and at best is a nuisance. Smokeless combustion involves the proper and right relation of air and unburned gases, and the maintaining of a sufficiently high temperature during combustion of the gases. The proper mixing of air and gases usually involves a secondary admission of air, and the temperature is maintained usually by brickwork in the form of secondary arches, bridge-walls, etc. If secondary admission of air is used, great care should be taken to see that this admission is properly controlled, for great damage and loss of efficiency may result from its improper use. Also, remember that introduction of smoke and secondary arches results in a loss of draft, and that due allowance should be made. While "dutch oven" furnaces are recommended in some cases, there are many coals for which it is not suitable. Furnace temperatures with "dutch ovens" are always high. With a coal having ash of low fusing point, extremely high furnace temperatures are likely to result in the formation of clinkers. Keep the furnace temperatures as high as is consistent with the fusing point of the ash, but keep the combustion-chamber temperatures as high as can be obtained. There are many places where, if the "dutch oven" arch was moved out of the furnace and into the combustion chamber, a saving of fuel and higher boiler capacity would be effected.

There is no doubt that forced draft will increase the capacity of any furnace, for it increases the rate of combustion.

---

*Excerpts from report of Committee on Railroad Stationary Power Plants presented to eleventh annual convention of the Association of Railway Electrical Engineers.

But forced draft may become the source of great loss of economy when used inadvisedly. The maintenance cost of forced-draft equipment is also an item deserving consideration, as is furnace maintenance cost, which will be increased.

To overcome the bad features of forced draft, go a step farther and install balanced draft. There are several such systems on the market, and in choosing any one of them be sure to analyze the system completely. There are two points to consider as the basis of regulation—the load conditions, of which the steam pressure is an indication, and the furnace conditions indicated by the furnace draft. All the controlled forced-draft systems admit of regulation for load conditions by a pressure regulator which speeds up or slows down the fan, accordingly as the pressure drops or rises. As for the second point of regulation, the furnace condition is primary. So, in choosing any of the controlled- or balanced-draft systems, be sure that the furnace is used primarily as a basis of regulation.

Urges Greater Use of Niagara Power

Addressing the American Electrochemical Society at Buffalo on Feb. 3, John L. Harper, chief engineer of the Niagara Falls Power Co., gave a convincing argument for further development of power at the Falls, and reviewed briefly, by means of slides, the present development, especially as pertains to the Hydraulic Plant, the recent extension of which contains the three largest water-power units in existence; namely, 40,000 hp. apiece.

Water-power propositions, said Mr. Harper, may be divided into two general classes—those in which the rate of developing the water power may be controlled by full or partial water storage, and those in which no storage is possible. To the latter class belongs Niagara power. In the first class the use of the water is controllable over certain hours of the day, or even certain seasons of the year, whereas in the second class any diminution of use represents a total loss. The ideal load for the second class of power is represented by that of the electrochemical and metallurgical processes, where the use factor is large and where the cost of power entering into the process is large compared with the cost of labor. In fact, 95 per cent of the ferro alloys of domestic manufacture are made at Niagara. Other types of load, such as light manufacturing, where the cost of power is relatively a small factor, and lighting or street-railway service, where a variable load is encountered, should, in the speaker's opinion, be left for either steam power or water power having storage facilities; or, as an alternative, the base load might be cared for by the hydro-electric power and the peak by steam power.

New York State has two major sources of hydro-electric power—the Adirondack and St. Lawrence section, and the Niagara River section. The former is somewhat removed from good transportation facilities, and in this case it would be permissible to transmit power a considerable distance to manufacturing centers. The situation at Niagara, however, is different. It is near a shipping center, both rail and water, and it would be poor engineering to transmit this power over a long distance, with the accompanying transmission costs and losses, when the industries can be located right at the source of the power.

Mr. Harper was strongly opposed to any system that would tax the people of the state for the development of hydro-electric power, from which they would derive no direct benefit. He believes that the cost of development should be borne only by the people who benefit by it through rates which they individually pay for this service, or indirectly in the price of the commodities manufactured or distributed by means of it.

The speaker threw on the screen a map showing the first survey of Niagara Falls in 1764. This showed a distinctly different contour of the falls from that of the present map and illustrated how the falls are receding year by year; on the Canadian side this recession is estimated at six feet per year. It has served to increase the length of crest by about 50 per cent during this period, and this greater length, together with the fact that the recession is wearing a channel at the middle of the falls, accounts for considerable rock showing at

TWO OF THE NEW 40,000-HP. UNITS AT THE HYDRAULIC PLANT, NIAGARA

the edges. This, it was pointed out, is what is sometimes mistaken by conservationists as an indication that less water is going over the falls. The total flow of the river is 210,000,000 ft.-sec., and Mr. Harper believes that 60 per cent could be diverted for hydro-electric purposes without marring the beauty of the falls. It would be necessary to leave 40 per cent in order to maintain the river and discharge ice in the winter time.

With a diversion of 60 per cent of the water, and the placing of concrete booms in the upper river so as to divert the water from the channel and spread it over the falls, the rocks would be completely covered and the level of the water in the river below the falls would be lowered considerably, thus having the effect of increasing the height of the falls. At present it is estimated that the depth of the river below the falls is about 200 ft. A diversion of 60 per cent of the water would make available about 3,600,000 hp.

The extension to the Hydraulic Plant, which is now nearing completion, was undertaken during the war at the request of the War Department, for emergency power, it being desired to add approximately 100,000 hp., bringing the diversion of water up to the maximum allowable by the Burton Act. The time from starting construction to the putting the first unit in service, December, 1919, was 19 months, a really remarkable record, considering the difficulties that had to be overcome in constructing the penstocks through solid rock, constructing the plant at the edge of the gorge and lowering all material and equipment over the cliff a distance of 215 feet.[1]

The three new units, two of which are now in service, are each rated at 37,500 hp., although they have carried 40,000 hp. They are of the vertical type running at 150 r.p.m. under an effective head of 215 ft. and direct-connected to 32,500-kva. three-phase 25-cycle 12,000-volt generators. A view of two of these units is shown in the illustration.

[1]*Power* expects to publish a detailed description of this plant when construction has been completed and the information released. This, it is believed, will be within two or three months.

Personals

Thomas H. Macauley, who has been manager of construction and operation of the Calgary (Canada) Municipal Railway since 1909, is now general manager of the New Brunswick Power Co.

John A. Stevens, consulting engineer in power-plant design and construction at Lowell, Mass., has opened a branch office in Fall River, to handle business in the southern New England district.

Edward C. Jones, who has been chief engineer of the gas department of the Pacific Gas and Electric Co. since 1891, has resigned to go into business for himself in San Francisco, where he has opened offices.

John Clyde Oswald, president of the Oswald Publishing Co. and publisher of the "American Printer," has been appointed vice president of the Preston Trading Co., New York City. Mr. Oswald has also been elected president of the National Paper Trades Exchange.

Robert N. Hodgson has resigned as general manager of the Binghamton Light, Heat and Power Co., and as general sales manager of the controlling company, the W. S. Barstow Co., to give his time to his own business interests. He is succeeded by R. L. Peterman, who was formerly superintendent of the utilities company at Easton, Penn.

Frank R. Towles, formerly general superintendent of the York County Light and Power Co., has resigned to become superintendent of the Cumberland County Power and Light Co., at Portland, Me. William E. Shaw, local manager of the former plant, succeeds Mr. Towles as superintendent of the electrical division, and Ralph W. Hawks succeeds Mr. Shaw as local manager.

Harry Z. Bixler, formerly chief engineer for the Youngstown (Ohio) Steel Co., and in the same capacity with the Brier Hill Steel Co. since the latter's incorporation, has resigned to identify himself with the development of what promises to be a new era in the manufacture of iron and steel in this country and throughout the world. He is succeeded by W. H. Ramage, formerly assistant chief engineer. T. M. Phillips, who was superintendent of the washed metal plant during the joint régime of the two companies, has associated himself with the same development in which Mr. Bixler is interested.

Engineering Affairs

The American Association of Engineers will hold its annual convention at St. Louis, Mo., May 10-11, with headquarters in the Planters Hotel.

The Electric Club of Chicago has opened permanent headquarters on the mezzanine floor in the Hotel Morrison. A part of the quarters is devoted to a library, which will be equipped with all the popular magazines as well as those applying to the club's own industry.

The American Society of Mechanical Engineers announces the following section meetings: Atlanta, Ga., at the Georgia School of Technology, Feb. 24; subject, illustrated lecture on "Building a 30,000-kw. Turbo-Generator Set," by H. E. Bussey. St. Louis, Mo., April 7, joint meeting of the Associated Engineering Societies of St. Louis, at which F. O. Pahlmeyer, engineer of the Heine Safety Boiler Co., will deliver an illustrated paper on the "Different Processes in the Manufacture of a Heine Boiler." Baltimore, Md., Feb. 27; subject for discussion, "Vocational Education," and many prominent speakers are expected to attend.

The Association of Iron and Steel Electrical Engineers announces the following meetings: Cleveland, Ohio, Feb. 28, at Union Electric League Rooms, Hotel Statler; E. S. Lammers will talk on "Automatic Control of Electrically Operated Blooming Mills." Pittsburgh, Penn., section: March 20; T. D. Lynch, research engineer of the Westinghouse Electric and Manufacturing Co., East Pittsburgh, will talk on "Babbitt and Babbitting." April 17, Robert B. Treat, electrical engineer of the General Electric Co., will talk on "Grounded Neutral." On May 22 there will be an inspection of one of the largest steel plants in the state of Ohio. June 12, R. H. Keil, power engineer of Jones & Laughlin Steel Co., will talk on "Current Limit Reactance."

Miscellaneous News

The National Exposition of American Manufacturers, which was to have been held in April, 1920, at Buenos Aires, Argentine Republic, has been postponed until November, 1920, to allow more time for adequate preparations and shipments from other parts of North and South America.

The International Conference of Refrigeration, held at Paris last month, was convened in view of examining the creation of an international organism to insure common action as regards the economic affairs of each nation and efforts of scientists, economists and industrials engaged in the study of refrigerating questions. After January, 1920, the Institute will issue a monthly bulletin of information on refrigeration, edited in two languages, English and French, in which all scientific, technical and economical data sent to the Institute and appearing in all parts of the world will be classified.

The National Marine League has begun the promotion of a national publicity campaign to start the American people thinking seriously upon the problem and possibilities of a merchant marine. The campaign will culminate in National Marine Week, Apr. 12-17, and while nationwide in scope will center in New York City, where a series of celebrations and demonstrations will be given. The ceremonies for the week will be opened aboard an American built, owned, operated and manned ship in New York harbor, by Chairman Payne, of the Shipping Board, and Secretary of Commerce Alexander. The formal opening of the National Marine Exposition will be the same evening at Grand Central Palace. On Tuesday afternoon the free Marine Art Exhibit will be opened under the direction of the leading artists of the country. A feature of this exhibit will be the greatest gathering of ship models ever assembled in America. The annual dinner of the National Marine League will be held that night. Wednesday will be Shipbuilding day; Thursday, Merchant Mariners' day; Friday, Fuel and Engineering day; Saturday, Inland Waterways and Harbors day, when demonstration parades will be reviewed by Government officials.

Business Items

Yarnall-Waring Co., Philadelphia, Penn., has opened a branch office in Detroit, in the Builders' and Traders' Exchange, Penobscot Building, with Walter G. Heacock in charge.

C. F. Braun & Co., manufacturing mechanical engineers, of San Francisco, Cal., have removed their executive offices to the tenth floor of the Atlas Building, 604 Mission St., San Francisco.

The Automatic Fuel Saving Co., Philadelphia, has purchased from the Vapor Vacuum Heating Co. and George C. G. Gray the patent and goodwill of the Gray system of combustion control for power plants. The new company has adopted the name of automatic fuel saver for the system.

The Chicago Fuse Manufacturing Co. has purchased from the Otis Elevator Co. the property at the southwest corner of Fifteenth and Laflin Sts., Chicago. It is expected the company's two plants at West Congress St. and at South Throop St. will be transferred shortly to the new location.

The Maher Engineering Co., 30 North Michigan Ave., Chicago, announces the opening of a branch office at 708 Schofield Building, Cleveland Ohio, handling the distribution of Erie Engine Works engines, Dayton-Dowd Co. centrifugal pumps and Galland Henning hydraulic presses. The office will be under the management of Lincoln E. Maher.

The Collins-Lotz Co., recently formed, has opened an office at 26 Hartford Trust Building, Hartford, Conn., for the sale and application of heat and cold insulation materials of all kinds. A warehouse will also be maintined at Hartford. Mr. Collins was formerly the treasurer and manager of sales and contracts of Robt. A. Keasbey Co., New York City. Mr. Lotz was for many years manager of the Hartford office of the same company.

The Chicago Pneumatic Tool Co. announces the following appointments: Edward A. Woodworth as special railroad representative attached to the staff of manager of western railroad sales, with headquarters at Fisher Building, Chicago, Ill. Mr. Woodworth has been for several years secretary of the Committee on Standards, U. S. Railroad Administration, Washington, D. C. C. E. Laverenz as special railroad representative attached to the staff of manager of western railroad sales. Mr. Laverenz for several years was an inspector in the Ordnance Department of the U. S. Navy, and previously held positions as boilermaker and foreman of the Chicago & Northwestern and Illinois Central Railroads.

'rade Catalogs

and Old Fallacies" is the ...ge booklet just issued by ...tal Co., New York City. It contains intimate and instructive information concerning babbitt metals, not generally known, and users and buyers will find it a useful guide in selecting such products. A copy may be had free upon request.

Laclede-Christy, St. Louis, Mo., has issued a 42-page bulletin entitled "Boiler Capacities with Chain Grates," showing the importance of fuel-bed temperatures in increasing the maximum capacity attainable, in a way that is easily understood. The booklet was compiled and edited by Loyd R. Stowe, manager of the company's stoker department, and contains much useful information on this subject.

"Fuel Oil Applied to Power Boilers" is the title of an interesting 16-page booklet recently issued by the Engineering Dept. of the Coatesville Boiler Works, Coatesville, Penn. It contains some valuable information as to the relative merits of different fuels for use under boilers of all types, as well as on the dimensions and capacity in gallons of horizontal and vertical storage tanks. A copy may be had free upon request.

"Securing Better Combustion of Midwest, Western and Southwestern Fuels" is the title of a 31-page booklet recently issued by the Green Engineering Co., of East Chicago, Ind. It is a compilation of articles that have appeared in Power and Electrical World during the last year, discussing the many problems which arise in securing efficient combustion with above mentioned fuels, and pointing out their solutions. A copy of the bulletin may be had free upon request.

The De Pere-Burton Company, Milwaukee, Wis., is now sending out upon request a 32-page loose-leaf catalog dealing with the Burton semi-sectional and De Pere improved cross-drum water tube boilers. The publication enters into the construction and the many good points of design attributed to these types of boiler and pictures their application to stationary and marine service. The adaptability of the boiler to waste-heat service is featured, and mention is made of excellent facilities in a shop that has had years of boiler experience.

The Carrick Engineering Co., Chicago, has issued a 12-page catalog, letter size, describing the Carrick Combustion Control which automatically synchronizes fuel feed and damper position to correspond with the load, holding the proper relation for efficient results. Due to a graduated movement, it takes and holds any intermediate position that is required for the load carried and instantly moves to a new position if there is a slight change in the load or steam flow, in feed water, or in furnace conditions. It automatically correlates fuel and air supply by individual control of stoker engines, stoker motors, dampers, forced draft and induced draft fans. A copy of the catalog will be sent upon written request.

"Westinghouse Underfeed Stoker" is the title of a 36-page, 8½ x 11 in. booklet, with an attractive three-color cover, just issued by the Westinghouse Electric and Manufacturing Co., East Pittsburgh, Penn. Under the head of Details of Construction, it gives a comprehensive description of the leading features of design contributing to the efficiency and reliability of operation. The interconnection of grates and tuyeres and the arrangement of brackets, gears and fuel deflecting plates, are clearly described. Taking up the principles of operation, the circular explains, by use of sectional cuts and charts, the distinctive features of the Westinghouse zone system of air distribution. The subjects of efficiency and capacity are taken up together with flexibility. Curves setting forth extensive performance data relative to these subjects and others are used to amplify the text. Under the subjects of the coal-saving problem and fuel burning equipment of modern power stations, an interesting and instructive array of information is given for the benefit of those who have the operation of stokers in charge. A large list of representative installations making use of these stokers is given. The booklet is illustrated throughout.

THE COAL MARKET

BOSTON—Current prices per gross ton f.o.b. New York loading ports:

Anthracite

	Company Coal
Egg	$8.20@$8.65
Stove	8.45@ 9.05
Chestnut	8.55@ 9.05
Pea	7.05@ 7.40
Buckwheat	3.60@ 3.80
Rice	2.30@ 2.50
Barley	1.25

Bituminous

	Clearfields	Cambrias and Somersets
F.o.b. mines, net tons	$2.85@$3.35	$3.15@$3.60
F.o.b. Philadelphia, gross tons	5.05@ 5.60	5.35@ 5.90
F.o.b. New York, gross tons	5.40@ 5.95	5.75@ 6.25
Alongside Boston (water coal), gross tons	7.00@ 7.75	7.60@ 8.00

Pocahontas and New River are practically off the market for coastwise shipment, but are quoted at $6.25@$7.00.

NEW YORK—Current quotations, White Ash, per gross tons, f.o.b. Tidewater, at the lower ports are as follows:

Anthracite

	Company Coal
Broken	$7.80@$8.25
Egg	8.20@ 8.65
Stove	8.45@ 9.05
Chestnut	8.55@ 9.05
Pea	7.05@ 7.40
Buckwheat	5.15
Rice	4.50
Barley	4.85
Boiler	4.85

Bituminous

Government prices at mines:	Spot
South Fork (best)	$3.25@$3.50
Cambria (best)	3.00@ 3.25
Cambria (ordinary)	2.60@ 2.90
Clearfield (best)	3.00@ 3.25
Clearfield (ordinary)	2.60@ 2.90
Reynoldsville	2.85@ 2.90
Quemahoning	3.25@ 3.50
Somerset (medium)	3.00@ 3.25
Somerset (poor)	2.50@ 2.75
Western Maryland	2.50@ 2.75
Fairmont	2.25@ 2.50
Latrobe	2.60@ 2.90
Greensburg	2.75@ 3.00
Westmoreland ¾ in.	3.40@ 3.50
Westmoreland run-of-mine	2.75@ 3.00

PHILADELPHIA—Bituminous coal prices vary according to district from which they are mined. For ordinary slack the price is $2.45@$2.55; lump, $3.00@$3.35, at the mines.

BUFFALO—

Anthracite

	On Cars, Gross Ton	At Curb, Net Ton
Grate	$8.55	$10.80
Egg	8.80	10.65
Stove	9.00	10.85
Chestnut	9.10	10.85
Pea	7.75	9.30
Buckwheat	5.70	7.75

Bituminous

Allegheny Valley	$4.80
Pittsburgh	4.65
No. 8 Lump	4.65
Mine Run	4.80
Slack	4.10
Smokeless	4.60
Pennsylvania Smithing	5.70

CLEVELAND—Prices of coal per net ton delivered in Cleveland are:

Anthracite

Egg	$12.25@$12.40
Chestnut	12.50@ 12.70
Grate	12.25@ 12.40
Stove	12.40@ 12.60

Pocahontas

Mine-run	$7.50

Domestic Bituminous

West Virginia split	$8.50
No. 8 Pittsburgh	$6.60@ 6.90
Massillon lump	8.25@ 8.50
Coshocton lump	7.25

Steam Coal

No. 6 slack	$5.25@$5.50
No. 8 slack	5.10@ 5.50
Youghiogheny slack	5.25@ 5.50
No. 8 ¾ in.	5.70@ 6.00
No. 6 mine-run	5.25@ 5.50
No. 8 mine-run	5.75

Only coal available is mine-run Pocahontas.

MIDDLE WEST—Chicago quotations, F.o.b. cars at mine:

	Springfield, Carterville, Williamson, Franklin, Saline, Harrisburg	Fulton, Peoria	Grundy, La Salle, Bureau, Will
Lump	$2.55@$2.70	$2.95@$3.10	$3.25@$3.40
Washed	2.75@ 2.90	....	3.45@ 3.60
Mine run	2.35@ 2.60	2.75@ 2.90	3.00@ 3.15
Screenings	2.05@ 2.20	2.35@ 2.50	2.75@ 2.90

New Construction

PROPOSED WORK

Me., Mt. Desert Island—The Bureau of Yards & Docks, Navy Dept., Wash., D. C., will soon receive bids for the construction of a frame building and for the installation of heating, plumbing and electric light systems at the Naval Transmitting Station, here.

Conn., New Haven—The State Bd. of Health, 308 Church St., Hartford and Bilderbeck & Langdin, Archts., Manwaring Bldg., New London, will receive bids about March 1 for the construction of a 3 story 51 x 111 ft. laboratory and experimental station on Highland St. A vacuum steam heating system, including new L.P. boilers, hot water supply and refrigerating equipment, will be installed in same. Total estimated cost, $125,000.

N. Y., Brooklyn—The Amer. Mfg. Co., Noble and West Sts., is having plans prepared for the construction of a 2 story, 150 x 440 ft. warehouse at 24 Greenpoint Ave. A steam heating system will be installed in same. Total estimated cost, $155,000. W. Higginson, 18 East 41st St., New York City, Archt. and Engr.

N. Y., Brooklyn—The Brooklyn Retail Butchers Corp., 3285 Fulton St., plans to build a 3 story, 130 x 135 ft. ice plant and storage house on Atlantic Ave. and Ft. Greene Pl. A steam heating system will be installed in same. Total estimated cost $400,000. A. Rosen, Pres.

N. Y., Syracuse—The Onondaga Milk Producer's Coöperative Association, S. A. and K. Bldg., has awarded the contract for the installation of electrical work in its plant on Burnet Ave., to the Hughes Electrical Corp., 413 South Clinton St. One 75 hp., one 25 hp., two 15 hp., two 5 hp., seven 1 hp., one 7½ hp., four 10 hp., eight 2 hp., and four 2 hp. motors will be installed in same. Cost to exceed $10,000.

Pa., Scranton—Edward H. Davis and George M. D. Lewis, Associate Archts., Union Natl. Bank, are preparing plans for the construction of a 5 story, 25 x 100 ft. store building. A steam heating system will be installed in same. Total estimated cost, $125,000. Owner's name withheld.

Pa., Scranton—The Welfare League is having plans prepared for the construction of a 1 story, 100 x 100 ft. private school on Wabash St. A steam heating system will be installed in same. Total estimated cost, $85,000. Edward H. Davis and George M. D. Lewis, Union Natl. Bank Bldg., Archts.

Md., Highlandtown (Baltimore P. O.)—The Crown Cork and Seal Co., 1511 Guilford Ave., has awarded the contract for the construction of a 1 story, 30 x 40 ft. transformer house on Eastern Ave., to H. C. Barnes, 2631 Wilkens Ave. Estimated cost, $6,000.

S. C., Columbia—The State Legislature will appropriate $50,000 for repairs and installation of heating plants in the several buildings on the campus of the University of South Carolina.

Tenn., Chattanooga—The Thatcher Spinning Co., 2714 East 18th St., has awarded the contract for the construction of an addition to its spinning plant, to the Turner Constr. Co., 244 Madison Ave., New York City. A steam heating system will be installed in same. Total estimated cost, $1,200,000.

O., Bloomville—The Bd. Educ. plans to construct a 2 story, 80 x 126 ft. school. A steam heating system will be installed in same. Total estimated cost, $110,000. Glass & Austin, Grand Theatre Bldg., Columbus, Archts.

O., Cleveland—The Cleveland Railway Co., Leader-News Bldg., plans to construct a 1 story power sub-station on Denison St., Lorain Dist. Estimated cost, $200,000. Mr. Radcliffe, Mgr. L. P. Crocelius, 700 Leader-News Bldg., Engr.

O., Cleveland—The Cleveland Telephone Co., c/o R. Pate, Engineers Bldg., will receive bids in the spring for the construction of a 2 story, 97 x 145 ft. telephone exch. building on Dille Rd. A steam heating system will be installed in same. Total estimated cost, $250,000. A. N. Symes, Engineers Bldg., Engr.

O., Cleveland—S. Cowan, Society for Savings Bldg., is having plans prepared for the construction of a 4 story, 140 x 240 ft. family hotel on East 105th St. and Ansel Rd. A steam heating system will be installed in same. Total estimated cost, $400,000. S. H. Weiss, 1032 Schofield Bldg., Archt.

O., Cleveland—The city received only bid for the installation of a heating system in the bath house, here, from the Ohio Heating & Pipe Bending Co., 1063 West 11th St., $33,153.

O., Cleveland—The East End Chamber of Commerce had plans prepared for the construction of an 8 story, 80 x 120 ft. commercial building at East 105th St. and Superior Ave. A steam heating system will be installed in same. Total estimated cost, $400,000. Address S. H. Kennard, 10708 Superior Ave.

O., Cleveland—G. A. Tenbusch, Union Bldg., is having plans prepared for the construction of a 4 story hotel at 1906 East 105th St. A steam heating system will be installed in same. Total estimated cost, $250,000. W. S. Ferguson Co., 1900 Euclid Ave., Archts.

O., Columbus—G. R. Bassett, Archt., Central Natl. Bank Bldg., will soon award the contract for the construction of a 3 story, 60 x 188 ft. salesroom on 9th and Broad Sts., for the Ohio Sales Co., 771 North High St. A steam heating system will be installed in same. Total estimated cost, $150,000.

O., Grafton—The village will receive bids about April 15 for the construction of a waterworks system, including an 18 x 20 ft. pump house, 2 deep well pumps, etc. Total estimated cost, $35,000. R. Winthrop Pratt, Hippodrome Bldg., Cleveland, Engr.

O., Kent—The Bd. Educ. will receive bids about April 1 for the construction of a 3 story, 100 x 200 ft. high school. A steam heating system will be installed in same. Total estimated cost, $250,000. Mills & Millspaugh, 67 East Long St., Columbus, Archts.

Mich., Atlas—The city engaged Van Leyen, Schilling & Keough, Archts. and Engrs., 1115 Union Trust Bldg., Detroit, to prepare preliminary plans for the construction of a 2 story hospital. A steam heating system will be installed in same. Total estimated cost, $150,000.

Mich., Battle Creek—The city plans to install a centrifugal pump, air lift and electric motor at the pumping station.

Mich., Detroit—A syndicate headed by C. F. Clippert, 2051 West Grand Blvd., is having plans prepared for the construction of a 2 story, 130 x 200 ft. theatre and office building along the Grand River and West Grand Blvd. Steam heating apparatus, with forced ventilating system, etc., will be installed in same. Total estimated cost, $2,000,000. C. W. and G. L. Rapp, 190 North State St., Chicago, Archts.

Mich., Ecorse—G. W. Graves, Archt., 43 John R. St., Detroit, will receive bids until Feb. 20 for the construction of a 2 story, 96 x 200 ft. school on Webster and State Sts., for the St. Francis Xavier Parish. A steam heating system will be installed in same. Ammerman & McColl, 2348 Penobscot Bldg., Detroit, Engrs.

Mich., Lansing—The City Light & Water Bd. plans to build the first unit of an electric power station to have a capacity of 10,000 kw., ultimate capacity, 40,000 kw., on Division St. Modern power plant equipment, including conveyors for coal and ashes, automatic stokers, boilers with turbine engine to run generator, will be installed in same. Total estimated cost, $500,000. G. G. Crane, City Hall, Mgr.

Mich., Port Huron—A. B. Parfet, Military St., plans to construct a 2 story 100 x 250 ft. service station. A steam heating system will be installed in same. Total estimated cost, $100,000.

Mich., Redford—W. E. N. Hunter and Neil C. Sorenson, Archts., Chamber of Commerce, Detroit, will receive bids until March 20 for the construction of a 1 story, 60 x 150 ft. church on Grand River Ave., for the Redford Methodist Episcopal Congregation. A steam heating system will be installed in same. Total estimated cost, $150,000.

Mich., Wyandotte—The Bd. Educ. will soon award the contract for the construction of a 2 story, 150 x 350 ft. high school on Superior St. Steam heating and ventilating systems will be installed in same. Total estimated cost, $400,000. B. C. Wetzel & Co., 2317 Dime Bank Bldg., Detroit, Archts.

Ill., Vandalia—The city will soon award the contract for the construction of an electric light and power plant. Work involves the installation of one 300 hp. water tube boiler, breeching grates, fittings, one steam engine to operate 300 kva. generator, one 300 kva. directed connected generator and one 15 kw. exciter, etc. Fuller & Beard, Chemical Bldg., St. Louis, Mo., Engrs.

Ill., Wauconda—The city plans to build a system of 10 to 24 in. storm and sanitary sewers, also a water plant including a pumping plant, etc. J. L. Pickens, Kankakee, Engr.

Wis., Milwaukee—H. A. Kulas, Archt., 883 Windlake Ave., is preparing plans for the construction of a 2 story, 50 x 80 ft. factory and a 20 x 50 ft. boiler house. A high pressure boiler will be installed in same. Owner's name withheld.

Wis., St. Francis (Milwaukee P. O.)—The Milwaukee Electric Ry. & Light Co., Pub. Serv. Bldg., is having preliminary plans prepared for the construction of a 185 x 400 ft. power house. Generators, engines, boilers, etc., will be installed in same. Estimated cost, between $500,000 and $700,000. R. H. Pinkley, Engr.

Minn., Chisholm—F. Drew, Secy. of Bd. Educ., will soon award the contract for the construction of a 3 story, 72 x 147 ft. grade school on 1st Ave. A steam heating system will be installed in same. Total estimated cost, $200,000. W. T. Bray, Exch. Bldg., Duluth, Archt. and Engr.

S. D., Brookings—P. J. Erie, Clk. Bd. Educ., will soon award the contract for the construction of a 2 story, 152 x 176 ft. high school. A steam heating system will be installed in same. Total estimated cost, $200,000. Tyrie & Chapman, Auditorium Bldg., Minneapolis, Minn., Archts. C. L. Pillsbury Co., Metropolitan Life Bldg., Minneapolis, Minn., Engrs.

Cal., Benicia—The Eng. Dept. of Benicia Arsenal has prepared plans for motorizing the mechanical shop, requiring about forty 5 to 35 hp. motors. E. P. O'Hern, Commandant.

Cal., Los Angeles—The California Walnut Growers Assn., 1326 East 7th St., is having plans prepared for the construction of a 7 story, 100 x 150 ft. warehouse and packing building on 7th and Mill Sts. Conveyors, elevators, etc., will be installed in same. Total estimated cost, between $150,000 and $200,000. A. C. Martin, 430 Higgins Bldg., Archt.

Cal., Newport Beach—The Southern California Edison Co., 120 East 4th St., Los Angeles, will build a sub-station, cost, $20,000, and a 20 mi. auxiliary transmission line from here to San Jose Capistrano, cost, $30,000.

Ont., Havelock—The Hydro-Electric Commission of Ontario, 190 University Ave., Toronto, has submitted a by-law to the taxpayers for the construction of the hydro-electric lighting and power system. About 300 house service meters will be installed. Estimated cost, $28,000. F. A. Gaky, Engr.

Ont., London—The Greene Swift Co. will receive bids until March 15 for the construction of a 3 story, 100 x 148 ft. clothing factory on Talbot St. A steam heating system will be installed in same. Total estimated cost, $125,000. W. G. Murray, Dominion Savings Bldg., Archt.

Ont., Toronto—The Bd. Educ. plans to construct 2 and 3 story additions to the various schools in the city. Vacuum steam heating systems will be installed in same. Total estimated cost, $1,500,000. W. W. Pearse, 155 College St., Mgr.

Ont., Toronto—The Westminster Presbyterian Church, 47 Bloor St., E., plans to rebuild the 1 and 2 story church and Sunday school recently destroyed by fire. A steam heating system will be installed in same. Total estimated cost, $165,000.

Ont., Windsor—M. A. Dickinson, City Clk., will soon award the contract for furnishing a motor driven street flusher, 1,000 gal. capacity, with 4 flushing nozzles and 2 sprinkling heads, centrifugal or rotary gear pump to be operated by separate motor or engine. Estimated cost, $15,000. M. E. Brian, City Hall, Engr.

CONTRACTS AWARDED.

Mass., Boston—A. Rosenthal, 1035 O. S. Bldg., has awarded the contract for the construction of an 11 story, 100 x 120 ft. office building on Atlantic Ave., to Gascoigne & Lindenthal, 43 Tremont St. A steam heating system will be installed in same. Total estimated cost, $1,000,000.

N. Y., Arverne—A. Messer, 66 East 104th St., New York City, has awarded the contract for the construction of a 3 story, 80 x 100 ft. hotel on Cedar Ave., to Carucci & Wolpert, 186 Remsen St., Brooklyn. A steam heating system will be installed in same.

N. Y., Bronxville—The Village Investing Co., Arcade Studios, will build a 1 story, 25 x 30 ft. power house on Fairfield Rd. A steam heating system will be installed in same. Total estimated cost, $25,000. Work will be done by day labor.

N. Y., Brooklyn—Humbert & Andrews, 646 Dean St., will build a factory for manufacturing preserves, on Vanderbilt Ave. and Bergen St. A steam heating system will be installed in same. Total estimated cost, $120,000. Work will be done by day labor.

N. Y., Long Island City—The Astoria Light & Power Co., Shore Rd., Astoria, has awarded the contract for the construction of an office and loft building on Van Alst Ave., to Bartlett-Hayward Co., 100 B'way, New York City. A steam heating system will be installed in same. Total estimated cost, $1,000,000.

N. Y., New York—The Amer. Surety Co., 100 B'way, has awarded the contract for the construction of a 21 story office building at 96 B'way, to Cauldwell-Wingate Co., 331 4th Ave. A steam heating system will be installed in same. Total estimated cost, $1,500,000.

N. Y., New York—Clarence Davies, 51 East 42nd St., has awarded the contract for the construction of a 100 x 125 ft. theatre on Burnside Ave. and Walton St., to Patrick Murphy, 371 East 144th St. A steam heating system will be installed in same. Total estimated cost, $200,000.

N. Y., New York—The Exchange Buffet Corp., 52 William St., has awarded the contract for the construction of a 2 story restaurant and store building at 17-23 John St., to Cauldwell-Wingate Co., 331 4th Ave. A steam heating system will be installed in same. Total estimated cost, $250,000.

N. Y., New York—The Forty-eighth St. Co., 241 West 43rd St., will build an 8 story, 100 x 100 ft. printing building at 331 West 44th St. A steam heating system will be installed in same. Total estimated cost, $400,000. Work will be done by day labor.

N. Y., New York—The Interborough Rapid Transit Co., 165 B'way, will build a 1 story, 50 x 90 ft. power house on Westchester Ave (Bronx Boro.). A steam heating system will be installed in same. Total estimated cost, $50,000. Work will be done by day labor.

N. Y., New York—The T. Klein Constr. Co., Inc., 652 West 185th St., will build a 1 story, 199 x 385 ft. garage on 5th Ave., between 136th and 137th Sts. A steam heating system will be installed in same Total estimated cost, $150,000. Work will be done by day labor.

N. Y., New York—The Realty Managers, Inc., 200 B'way, will build a 1 story, 20 x 300 ft. store building on Jerome and Davidson Aves. (Bronx Boro.). A steam heating system will be installed in same. Total estimated cost, $200,000. Work will be done by day labor.

N. Y., Syracuse—The Onondaga Milk Producers' Coöperative Assn., S. A. and K. Bldg., has awarded the contract for the installation of heating and plumbing systems in its plant, to W. B. Goldie, 274 James St. Two 200 hp. boilers with a low and high pressure steam line for radiation will be installed in same. Total estimated cost, $40,000.

Pa., Altoona—The Altoona School Dist. plans an election to vote on a $1,000,000 bond issue, $600,000 of which will be used for the construction of a 3 story, 200 x 250 ft. junior high school on 14th St and 5th and 6th Aves., including a power plant capable of heating this building and present building, together with 4 steam systems for grade schools. C. M. Piper, Secy.

Pa., Johnstown—The Johnstown Terminal Warehouse, Johnstown Trust Bldg., has awarded the contract for the construction of a 5 story, 250 x 132 ft. warehouse and storage house on Maple St., to William Steele Sons Co., 16th and Arch Sts., Phila. Total estimated cost, $300,000.

Pa., Leola—N. E. Martin is in the market for 100 and 150 hp. return tube boilers with lap seam good for 50 lbs. working pressure.

Pa., Philadelphia—Sears Roebuck & Co., Arthington and Homan Aves., Chicago, has awarded the contract for installing six 500 hp. steam boilers, one 300 kw., engine drive, 220 volt d.c. generator, one 750 kw., engine drive, 220 volt d.c. generator and one 1,500 kw., geared turbine, 220 volt d.c. generator in the proposed new plant, here, to Comstock & Co., Inc., 30 North Michigan Ave., Chicago.

Md., Baltimore—A. H. Kuhlemaun Co., 2961 Frederick Ave., has awarded the contract for the construction of a 3 story, 75 x 200 ft. oleomargarine factory, to W. L. and G. H. O'Shea, 29 B'way, New York City. A steam heating system will be installed in same. Total estimated cost, $100,000.

W. Va., Elkins—The city plans to construct a pumping station and filtration plant. Estimated cost, $130,000.

Ga., Atlanta—R. S. Armstrong & Bro., 676 Marietta St., is in the market for hoisting engines, 7 x 10 and 1¼ x 10 cylinders.

O., Cleveland—The Bd. Educ., East 6th St. and Rockwell Ave., has awarded the contract for installing a steam heating system in the proposed school addition on St. Clair Ave. and Ruple Rd., to the J. Roemer Heating Co., East 22nd St., at $16,960. Noted Jan. 20.

O., Cleveland—The city has purchased a site on Starkweather Ave., opposite Lincoln Park, and plans to build a 2 story, 76 x 76 ft. bath house on same. A steam heating system will be installed in same. Total estimated cost, $100,000. Ed. Shattuck, Purch. Agt. F. H. Betz, 604 City Hall, Archt.

O., Columbus—H. C. Godman Co., 31 North 4th St., has awarded the contract for the construction of a 3 story, 45 x 187 ft. shoe factory on Thurinan Ave., to H. H. Carr, 2582 West Broad St. A steam heatink system will be installed in same. Total estimated cost, $100,000.

O., Dayton—The Peckham Coal & Ice Co., Daller St., has awarded the contract for the construction of a 1 story, 100 x 110 ft. ice plant on Linden Ave., to the A. C. Bishop Contg. Co., Guardian Bldg., Cleveland. Estimated cost, $50,000.

Ind., Vincennes—The City Clerk is in the market for additional machinery for the municipal light plant, including 2 high pressure boilers, one 500 kw. turbine driven generator, automatic coal stokers and conveyors. A bond issue for $70,000 has been authorized for the above. Ezra Mattingly, City Attorney.

Mich., Detroit—The First and Old Detroit Natl. Bank, Ford Bldg., has awarded the contract for the construction of a 24 story, 169 x 186 ft. office building on Woodward Ave. and Cadillac Sq., to the Foundation Co., 233 B'way, New York City. Steam heating equipment will be installed in same. Total estimated cost, $4,000,000.

Mich., Detroit—Ben Robinson, Penobscot Bldg., has awarded the contract for the construction of a 5 story, 80 x 100 ft. hotel on Abbott St., to Shank & Co., 38 South Dearborn St., Chicago, Ill. A steam heating system will be installed in same. Total estimated cost, $350,000.

Mich., Detroit—Ben Robinson, Penobscot Bldg., has awarded the contract for the construction of an 8 story, 144 x 200 ft. hotel on Woodward Ave., to Shank & Co., 38 South Dearborn St., Chicago, Ill. Steam heating, ventilating and refrigerating systems will be installed in same. Total estimated cost, $900,000.

Mich., Grand Rapids—The Knott Realty Co., Commerce Ave., has awarded the contract for the construction of a 5 story, 50 x 100 ft. mercantile building on Commerce Ave. and Island St., to Owen Ames, Kimball Co., 42 Pearl St. A steam heating system will be installed in same. Total estimated cost, $75,000.

Ill., Chicago—The Flexible Steel Lacing Co., 522 South Clinton St., is having plans prepared for the construction of a 1 story, 200 x 200 ft. factory on Polk St. and the Belt Line R. R. A steam heating system will be installed in same. Total estimated cost, $200,000. W. G. Carnegie, 189 West Washington St., Archt.

Ill., Marion—The Bd. Supervs. of William Co. is having plans prepared for the construction of a 3 story, 80 x 100 ft. court house. A steam heating system will be installed in same. Total estimated cost, $250,000. N. S. Spencer & Sons, 37 West Van Buren St., Chicago, Archts.

Ill., Moline—Charles D. Wyman, 11th Ave. and 8th St., has awarded the contract for the installation of a steam heating system in the proposed 4 story, 56 x 125 ft. store building on 5th Ave. and 11th St., to the Moline Heating Co. Total estimated cost, $125,000.

Wis., Milwaukee—Philip Schwab Co., 16th and Canal Sts., has awarded the contract for the construction of a 1 story, 60 x 200 ft. boiler shop on Canal St., to Riesen Bros. Co., 86 Michigan St. Estimated cost, $50,000.

Wis., Sheboygan Falls—The city, c/o S. DeLong, Mayor, has awarded the contract for furnishing 1 air pumping unit, piping, compressor motor, air receiver with automatic control and 2 ore centrifugal pumps, 450 gal. per min. direct connected to motor of proper design, to the Indiana Air Pump Co., 812 Indiana Pythian, Indianapolis, at $10,824.

Minn., Minneapolis—The Orpheum Theatre Co., 21 South 7th St., is having plans prepared for the construction of a 4 story theatre on Hennepin Ave. and 9th St. A steam heating system will be installed in same. Total estimated cost, $800,000.

Minn., Renville—The Bd. Educ. is having preliminary plans prepared for the construction of a 2 story school. A steam heating system will be installed in same. Total estimated cost, $250,000. A. R. Holmberg, Clk. Croft & Boerner, 833 Palace Bldg., Minneapolis, Archts.

Neb., Ord—Valley Co. has awarded the contract for the installation of a steam heating system in the proposed 3 story, 78 x 101 ft. court house, to John A. Anderson, 2404 Leavenworth St., Omaha, at $33,729.

Neb., West Point—The School Bd. will soon award the contract for the construction of a 2 story, 80 x 175 ft. high school. A separate contract for installing a heating system will be let. Ellery Davis, 515 Security Mutual Bldg., Lincoln, Archt.

Mo., Leadwood—The St. Joseph Lead Co., 61 B'way, New York City, has awarded the contract for the construction of a 1 sory power house, to Stone & Webster, 120 B'way, New York City. A steam heating system will be installed in same. Total estimated cost, $700,000.

Tex., Dallas—Reilly & Hall., Archts. and Engrs., 749 5th Ave., New York City, are preparing plans for the construction of a 20 story office building, here. A steam heating system will be installed in same. Total estimated cost, $2,500,000. Owner's name withheld.

Cal., Berkeley—The Bd. Educ. will soon award the contract for the construction of a 3 story, 64 x 301 ft. and an 80 x 301 ft. high school on Grove and Milvia Sts. A steam heating system will be installed in same. Total estimated cost, $400,000. William C. Hays, First Natl. Bank Bldg., San Francisco, Archt.

Cal., Whittier—The city has awarded the contract for furnishing 2 electrically driven centrifugal pumps for the proposed waterworks system, to the Byron-Jackson Iron Wks., 336 East 2nd St., Los Angeles, at $6,855. Noted Feb. 9.

Ont., Ottawa—The Bd. Educ. is having plans prepared for the construction of a 2 story school. A vacuum steam heating system with coils for heating ventilating air will be installed in same. Total estimated cost, $250,000. W. C. Beattie, Elgin St., Archt.

NEW INSTALLATIONS IN POWER

Ont., Ottawa—The city is in the market for one 75 hp. electric motor and an oil pump for road oil work.

N. F., Kingwell, P. B.—Rodway Bros are in the market for marine engines and motor engines.

Sask., Saskatoon—The Electric Light & Power Dept., City Hall, plans to install a 6,000 kva. west turbo generator. J. R. Cowley, Electrical Engr.

Alta., Calgary—The Municipal Electric Light & Power Dept. plans to construct an addition to boiler plant and ash handling. J. Curliss, City Hall, Mgr.

POWER

Volume 51 New York, March 2, 1920 *Number 9*

WHEN "DAD" GOES AFTER THE CHIEF'S JOB

Practical Refrigeration for the Operating Engineer —Getting Acquainted with the Plant

By B. E. HILL
Formerly Refrigerating Engineer, Armour & Company

WHEN once the engineer has the operation of the compression system of refrigeration as securely fixed in his mind as he has the steam plant, he will make a long step toward a better understanding of both systems, together with better results.

The actual operation of the compression system is so nearly like that of the steam-boiler plant, where the latter is in connection with a compound condensing engine, that when one system is thoroughly understood it is easy to understand the other, as the two are based on the economical evaporation and use of two liquids, one of which is evaporated to produce power and the other to produce refrigeration. Fig. 1 shows a small ice plant.

There are many engineers who appreciate this similarity of operation and are thoroughly familiar with the two systems. But as I have found many more, especially among the younger engineers and some of the older ones, who do not understand, this and other articles to follow will make comparisons and give experiences intended for the engineer who has not had the advantage or the years of experience in connection with the operation and correction of refrigerating-plant troubles.

With a steam boiler or steam-generating plant in connection with a compound condensing engine the water is pumped into the boiler and evaporated into steam by means of heat from the fire in the furnace. The steam is carried at a pressure ranging from 5 lb. to more than 250 lb. per sq.in. The steam is delivered to the engine in a saturated or superheated condition through the high-and low-pressure cylinders and from there to a surface steam condenser, where it is condensed to water and pumped back into the boiler. If there were no losses of water due to leakage, the same water could be used indefinitely. Now follow the compression system of refrigeration through a similar circuit, starting at the boiler for the liquid anhydrous ammonia. Suppose we place the liquid ammonia in the evaporating, or so called expansion, coils. The heat from the brine or surrounding temperature of the room, as the case may be, on the outside of the coils will evaporate the liquid ammonia to a saturated gas which passes from the evaporating coils to the compressor, through the compressor cylinder to the ammonia condenser, where the gas is condensed back to liquid, and from the condenser the liquid is carried back to the evaporating coils ready to be fed into them and start another complete circuit. If there were no losses due to leakage, etc., the same liquid ammonia could be used indefinitely.

So far as a point of operation is concerned, there is not much difference. One is a transfer of heat to produce power, and the other is a transfer of heat to produce cold.

When speaking of boiling points or heat, we unconsciously compare the term with something that has a higher temperature than the body, or to the sense of touch, but all temperatures *above* 450 deg. F. below zero, which is absolute zero, are designated by science as degrees of heat. It is easy to understand that the boiling points of liquid refrigerants are cold. The boiling point of liquid air is 312 deg. F. below zero, and science has not as yet discovered anything with a boiling temperature low enough to freeze it. We might consider that the boiling points of various liquid refrigerants range in temperature anywhere from 312 deg. F. below zero to 40 or 50 above; but as we are considering ammonia only, we will confine ourselves to this refrigerant.

With any liquid that boils under all ordinary temperatures and pressures where refrigeration is required, it is necessary to confine the gas (after being evaporated) under pressure and apply cold water to convert it back to liquid so that it may be again evaporated by taking up heat from surrounding brine or air. Ammonia being a liquid that boils at 28½ deg. below zero under atmospheric pressure, it is

FIG. 1. A TEN-TON RAW-WATER, FLOODED-ICE PLANT, YORK TYPE, AT WILSON, ARK.

necessary to use a condenser. These condensers are of many types and designs the same as steam condensers, and are following the same lines of development as the latter. Fig. 2 shows an atmospheric flooded-type condenser. To further consider from an operating standpoint, let us assume that the ammonia gas has gone through the process of liquefaction and is ready for use as a refrigerant. Suppose we now feed the liquid ammonia into a set of coils placed in an insulated tank and submerged in salt brine. The boiling point of the liquid ammonia entering these coils will vary in temperature with the rise and fall of pressure, the same as the boiling point of water under different pressures. It is necessary only to consult the tables in either case for the properties of saturated ammonia or steam. Assume that the brine surrounding the ammonia coils is at a temperature of 20 deg. F. and that the liquid ammonia and gas inside the coils are under a pressure of 15 lb. Then, by consulting the tables it will be found that the ammonia will boil at about 0 deg. F., which is 20 deg. F. *below* the temperature of the brine; or in other words, if the brine is 20 deg. F. *above* the boiling point of the liquid ammonia, the ammonia will readily be evaporated into a gas the same as water when fed into the boiler to be evaporated into steam.

FIG. 2. FLOODED AMMONIA CONDENSER, YORK ATMOSPHERIC TYPE, AT PHILADELPHIA

The various types of evaporating coils or brine coolers are as numerous as the types of steam boilers, but the only kind we will consider at present is the old type with the submerged coils in an insulated tank. The brine tank is insulated to save heat through radiation the same as the boiler walls, steam piping, etc., which are insulated for the same purpose. The brine on the outside of the coils, being at a temperature of 20 deg. F., has the same function to perform as the fire in the furnace in evaporating the water, and as the brine is 20 deg. F. above the boiling point of the liquid ammonia, the ammonia is evaporated into a saturated gas which is taken by the compressor and delivered to a set of coils (the condenser) under a pressure of 150 to 185 lb. These coils are supplied with condensing water ranging in temperature from 50 to 70 deg. F. ordinarily, which condenses the gas back to a liquid which drains from the coils to a tank below the condenser coils, called a liquid receiver; from the receiver it is ready to be delivered to the evaporating or brine cooler and is now ready to be reëvaporated. In the operation of the feed valves, or so-called "expansion valves," which regulate or feed the liquid ammonia into the evaporating coils, it is important that the engineer should not try to force the temperature of the brine down by feeding too much liquid into the coils, as the effect would be the same as trying to raise steam in a boiler by opening the feed valve till the water carried over to the engine cylinder. Many a compressor head has been knocked out by this process of operation.

The liquid in the coils at 15 lb. back pressure is boiling at about zero degrees F., and the amount of liquid that can be evaporated will depend on the evaporating surface of the coils (heating surface), difference in temperature between the boiling liquid inside the coils, the temperature of the brine outside, and the rate of circulation of the brine.

To prevent the liquid ammonia being carried over to the compressor, the engineer or "temperature man" should watch the outlet or return end of the coil or coils, if possible, and not depend on guesswork or wait until the compressor shows signs of distress by the valves pounding, discharge getting too cold, or until the compressor gets a "slug" and raises the engineer up on his toes and his hair on end while

FIG. 3. EXTERIOR OF A 75-TON ICE PLANT, PHILADELPHIA

he makes a mad rush for the throttle. By the time he reaches the engine and is lucky enough to get it slowed down without having the compressor head punched out, he generally finds himself alone. Occasionally, he will find one of the old-time oilers that will run to the compressor suction valves and get them closed.

Cases of this kind, especially in large plants, are generally due to the fact that the temperature or "expansion" man has not had enough experience to properly operate the feed valves, and the same applies to many engineers who have not had the proper training and experience; and although the operation of the feed valves is simple, I have found many young engineers and some of the older ones who do not know how or where to look for a valve that is feeding too much liquid and is "freezing back" too strongly, and will cut back on all the valves (where there are a number leading into one suction line) to get the right one, and in this way it is easy, and it generally is the case that several valves are cut off that are not freezing through the coils. Fig. 3 is an exterior view of a 75-ton plant.

If the engineer of a steam plant is getting a dose of water, he will hurry to the boiler room to see if the watertender is carrying the water too high. He has only to look at the gage-glass, which shows him the water level, but it is different in feeding the evaporating coils with liquid ammonia, as there is no glass to show where the liquid stands in the coils; but if the engineer will clean off a small place on the outlet or return end of each coil, after he has scraped the frost and ice off down to the clean iron surface, he can apply the "stick test," which is simple and effective. He can regulate the feed to the limit on any one or any number of coils without endangering the compressor, and besides, he can run the plant up to the limit of capacity so far as the evaporating coils are concerned.

The "stick test" is applied by cleaning a small place on the return end of the coil, as already explained, wetting the thumb and applying it to the clean surface of the pipe. If the moisture freezes instantly and sticks hard enough to almost take the skin off when removed from the pipe, the coil is freezing back to the machine, but if it barely sticks after holding the thumb on the pipe for a second or two, it is not doing any harm in the engine room. In case there is no stick at all, the feed valve should be opened a little more.

The "stick test" is to the evaporating coils what the water glass is to the steam boiler, and with a little practice, it may be made almost as effective in the proper regulation of "slugs" or overfeeding of the evaporating coils.

Do not try the stick test twice in the same place without first cleaning the pipe surface, as there will be a thin skin of ice left on the pipe from the freezing of moisture on the thumb, which acts as an insulation and will not give a good test.

The coils that generally give the most trouble are those located in a room where the lowest temperatures are held, such as sharp freezers (where direct expansion is generally used), which are carried at a temperature of zero to 20 deg. below, or storage coolers, which are held around 5 to 10 deg. above zero.

There is more danger from "slugs" in direct-expansion quick-freezers than from any other source, due to the fact that the liquid ammonia evaporates very slowly, and for this reason the coils often fill up; then the ammonia is referred to as being "sluggish" or of an inferior quality, or is "hiding out," "lost," etc., when in fact, the sharp or low-temperature freezer coils evaporate the ammonia so slowly that they fill up with liquid ammonia and in many cases will rob the rest of the coolers of the liquid they should get.

Steam-Turbine Lubrication

By W. F. OSBORNE

Discusses the lubrication problems presented by the high speeds and high bearing temperatures met with in steam turbines. As most of the larger turbines are equipped with oil-circulating systems, the layout and operation requirements of such systems are considered. The extra problems involved when reduction gearing is used are discussed. The description and viscosities of oils for the various conditions of turbine work are given.

THE high speeds at which steam turbines normally operate are responsible for many of the advantages secured through their use. At the same time they make necessary a greater care in the design of the bearings and a closer attention to their lubrication. When properly lubricated, high-speed bearings will cause no more trouble than low-speed bearings, but if anything should occur to prevent the flow of sufficient oil of the right body they may burn up very quickly. Consequently, the greatest care should be exercised in selecting turbine oils and in taking the proper care of them while in use.

Ring-oiled turbine bearings are no harder to lubricate than any other high-speed bearing operating at the same temperature. All turbine bearings naturally run warm on account of the speed of the shaft and the radiated and conducted heat from the turbine casing. On some of the smaller turbines which have the bearings located very close to the casing, they may get as hot as 250 deg. F., which, of course, thins down any oil that is used.

To reduce these high temperatures, the bearings are usually water-jacketed, and when so located as to prevent excessive heating from the casing they run from 125 deg. F. to 150 deg. F. For such turbines a high-grade light-bodied oil of 150-200 sec. viscosity at 100 deg. F. is about right. When the very high temperatures previously mentioned exist, it is sometimes necessary to use a heavier oil of perhaps 300-500 sec. viscosity, as the quantity of oil in the bearing reservoir is not usually enough to give time for very much cooling of the oil. It is always best to use the thinnest-bodied oil that will lubricate the bearings, because of the lower friction of such oils, particularly at speeds of 3,000 to 10,000 r.p.m.

Practically all steam turbines except those of small size are equipped with continuous circulating oiling systems. A pump geared to the turbine shaft supplies all the bearings with oil under a pressure varying from 2 lb. to 15 lb. per sq.in., depending upon the type of the turbine, although vertical turbines have much higher pressures. This pressure insures an ample supply of oil to all rubbing surfaces, washing them clean of any bits of worn metal and cooling them by carrying away the heat. The oil pump is always provided with a bypass with which the pressure and the quantity of oil going to the bearings may be controlled—a very valuable feature if it should be necessary to increase the flow of oil to a hot bearing.

These circulating systems are carefully inclosed to prevent splashing of oil from the turbine and to keep out all foreign matter. Any small bits of grit carried by the oil between the rubbing parts at high speeds could cause a lot of trouble very quickly. The oil drained from the bearings flows to a catch tank, thence through a cooler consisting of a number of coils of pipe in which cool water is circulated to absorb the heat taken up from the bearings.

The temperatures to which the oil is heated in the bearings frequently range from 150 to 175 deg. F. Heating to these temperatures causes the lighter portions of the oil to vaporize and in addition slowly decomposes the oil. Although the system may be ever so carefully inclosed, part of the vapor gets away and the viscosity of the oil remaining in the system is increased. The rapidity of this thickening depends a good deal upon the quality of the oil, the temperature, the amount of agitation and the length of service. Properly refined oils having a minimum evaporation will last a long time without a serious increase in viscosity, and there are records of turbines operating for years on the same oil, only a small amount being added from time to time to make up for evaporation and leakage.

Turbine Oils Grow Dark with Use

All turbine oils change in color, growing darker with use because of slow oxidation or decomposition at the temperatures to which they are exposed. This change in the oil is more noticeable in the pale oils than in the red oils, although the physical change in the latter is equally great.

The engineer may now ask, and quite properly: "How long should the oil remain in the system? How can I tell when to change it and what will happen if I don't?"

It is, of course, manifestly impossible to lay down any fixed rule about the length of time a turbine oil should be used because there are so many sets of operating conditions, all affecting the oil. If samples are taken from the system at regular intervals of two or three months and analyzed for oleic acidity, it is sometimes possible to foretell when the oil is reaching the limit of its usefulness. Oleic acid is caused by decomposition of the oil, the decomposition being accompanied by a very marked tendency to emulsify. As long as the samples show a slow uniform increase in acidity, the oil may be used with safety, but as soon as the increase becomes more rapid, the wise engineer will take out at least a portion of the oil and replace it with new. By this method the consumption may be reduced to a minimum without any danger of the oil being held in service so long as to cause trouble through emulsification. The amount of oleic acid can be determined only by laboratory tests. A description of the method of making these tests is given in Archbutt and Deeley's "Lubrication and Lubricants," on pages 241-2.

Emulsification Troubles

As previously stated, oil that has been used too long may emulsify if a small quantity of water should get into it from leaking cooling coils or from steam blowing through the turbine-shaft stuffing-boxes. High-grade new oils and pure water, however well mixed, will separate cleanly if given an opportunity. If some impurity, such as a small amount of alkali or boiler compound, is added to the water, a thorough agitation such as the oil receives in a turbine circulating system will form a mixture which does not separate as readily. If in addition, the oil is impure, through not being properly refined or from being used too long, an emulsion will be formed which is very difficult to break up. On being circulated through the system, this mixture gradually picks up foreign matter and more water, increasing in volume and becoming very thick—so thick and heavy sometimes that it clogs up the oil pipes and bearing grooves. Boiler compounds carried into the oiling system by steam leaks through stuffing-boxes are a common cause of this sort of trouble.

These emulsions are difficult to break up, although it can be done sometimes by drawing off a portion of the oil and heating it before settling. It is therefore of the greatest importance to secure the very best quality oil and to exercise the greatest care to prevent foreign matter of any kind getting into the system.

A very common source of trouble with turbine oiling systems is the small capacity of many installations. The force-feed pump geared to the turbine drives the oil through the system very rapidly and thoroughly mixes with it any impurities that may not have settled out when the oil was in the reservoir or coolers. The quality and life of the oil many times can be improved by increasing the volume of oil in the system through the addition of storage tanks or filters in which part of the oil may be brought to rest and the foreign matter allowed to settle out. Some oils have the necessary characteristics which permit them to be cleaned even when circulated at high speeds; other oils can be used only when the system contains enough oil so that each part of it reaches the bearings only a few times per day. This increased volume also allows the oil to cool off, thereby lowering the temperature of the bearings and decreasing the rate of evaporation. All these things naturally improve the quality of the oil and increase its life of good service.

Sometimes the oil in a circulating system will begin to foam for no apparent reason and will overflow the reservoir, spilling onto the engine-room floor. This foaming may be caused by water getting into the oil and being churned by the rapid circulation through the pump. If the water does not have a chance to separate so that it may be drained off, the total amount of the mixture of water and oil will soon become larger than the capacity of the system, and it overflows. The remedy is to keep the water out or to increase the capacity of the system by adding a separate tank into which a part of the oil can be removed from the system periodically and the water separated.

Leaking Oil Causes Foaming

Air getting into the system through leaks in the pump suction or forced into the bearings by the ventilating system of an electric generator will also cause foaming. When this air is churned up with the oil, minute bubbles are formed which do not separate quickly and may increase the volume of the oil in the system sufficiently to cause it to overflow. Sometimes the drain lines are too high above the oil level in the reservoir so that the fall of the oil into the reservoir is sufficient to cause a foam that rapidly accumulates. In either case, when the air is released it carries with it some oil vapor, thus increasing the evaporation and deterioration of the oil.

Deposits in coolers, on bearings and in oil lines are caused principally by the use of oils unsuited to the high temperatures and the agitation to which they are subjected. Oils that have not been highly refined and carefully filtered to remove any foreign matter or heavy hydrocarbons, that might decompose, will sludge or throw down a heavy slimy substance. If oils are used too long, the same sort of sludging is sometimes observed. When these deposits form on the coils of the cooler, they act as an insulation, greatly reducing the efficiency of the cooler and raising the temperature of the oil in the system. This necessitates a thorough cleaning of all the pipes and tanks.

Hot turbine bearings may be caused by lack of lubrication through stoppage of the oil lines or oil grooves or a failure of the pump. Sometimes they run hot because the clearance along the side of the bearing is not large enough to allow the oil to be carried between the working portions of the surfaces. In other instances too heavy an oil or an oil thickened through long use or partly emulsified will cause this trouble. Permanent temperature reductions may be made only by locating the true cause and applying the proper remedy.

When reduction gear drives were added to steam turbines, an entirely new demand was placed upon the oil.

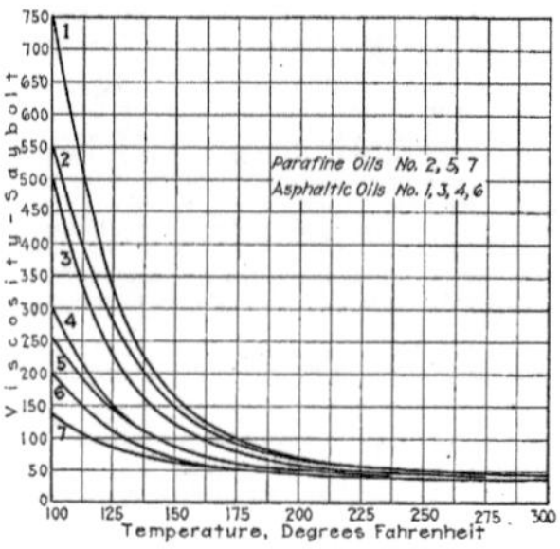

VISCOSITIES OF SOME WELL-KNOWN TURBINE OILS

The pressure per square inch on the teeth of the gears is many times greater than the load on any bearing, so that heavier-bodied oils are necessary to prevent a breakdown of the film and wear of the teeth. Furthermore, unless the gears are carefully made and lined up, there is a great likelihood of there being some noise as they mesh. Steam-turbine reduction gears are of the herringbone type which reduces the noise to a minimum, and when lubricated with a sufficiently viscous oil they should be comparatively quiet.

The reduction gears are usually inclosed within an oil-tight casing, and the gear bearings are so arranged that the oil draining from them goes into the gear case. On this account it is necessary to use to lubricate the gears something that can also be used on the bearings without causing trouble. When the gears and their bearings are isolated from the steam-turbine oiling system, it is possible to use a heavy-bodied oil of the proper viscosity to take care of the gears. On such installations a pump is provided to take the oil from the gear case and deliver it under pressure to the bearings and spray it onto the gears at the point of meshing. In this way there is nothing to affect the oil except the temperatures within the gear case, and a heavy-bodied oil can be used for a long time without any trouble.

A very common practice, however, is to include the lubrication of the gears and gear bearings in the circulating oiling system for the turbine bearings. These bearings normally require a light-bodied oil, and an exceptionally viscous oil will cause them to heat excessively. Moreover, the heavy-bodied oils do not separate from water and foreign matter as readily as the lighter oils; consequently, a compromise must be effected and an oil selected which will lubricate the gears as well as possible and at the same time be thin enough to operate in the circulating system without trouble.

As mentioned before, the heavier-bodied oils made necessary by the addition of gears do not separate from moisture as quickly as lighter oil, and therefore have a greater tendency toward sludging, emulsification and decomposition. This means that greater opportunities must be provided for rest and cleaning. This can be done by using a larger oiling system or by adding settling tanks to increase the capacity, thus reducing the frequency with which the oil passes over the bearings or gears. Efficient coolers are essential, because the high temperatures of the turbine bearings will thin down the oil so that if the temperature of that going to the gears is not sufficiently low, its viscosity will be reduced to such a point that it will be of little value.

The viscosities of some of the better known turbine oils are shown in the curves. Some of these oils are very well suited to the lubrication of turbines without reduction gears, while others are more suitable for the geared types. For instance, the oil shown by curve No. 3, when cooled down to about 100 deg. F., is quite heavy and if supplied to the reduction gears at this temperature, is very satisfactory. It also has the advantage of thinning down with increased temperatures more rapidly than some of the others, and when at the temperature of the turbine bearings, between 150 and 175 deg. F., it is thin enough to prevent any extra friction in the bearings. This quality makes it easily cleaned when heated, very desirable when heavy-bodied oils are used for circulating systems.

One of the principal troubles with reduction-gear lubrication has been a tendency of the gears to pit along the pitch line. This shows up at first in the form of very small pits, which grow larger as the wear increases until a serious groove or ridge is formed. There is still some question as to just what causes this pitting and the best method of overcoming the trouble. It may be caused by surface crystallization due to vibration of the gears through an insufficiently stiff framework or slightly worn bearings, a situation that could evidently be reduced by the use of higher-viscosity oils. In a number of installations the heavier-bodied oils have stopped the trouble completely, but whether they will do so in all cases yet remains to be demonstrated.

When we attempt to write a specification for a high-grade turbine oil, we must carefully consider all these various things that influence the life of the oil and the possibility of its emulsifying or sludging in the system.

A suitable turbine oil should be a highly filtered, properly refined, pure mineral oil, free from acid or alkali, having the best possible ability to separate quickly from water without the formation of emulsions or sludging, as shown by the emulsification test. The correct viscosity is determined by the oiling system of the turbine and the temperature conditions. Standard practice has found the following viscosities, all at 100 deg. F., to give satisfaction:

Ring oiled bearings with or without water jackets	150-200 sec.
Ring oiled bearings subjected to extreme radiated heat	300-500 sec.
Circulating systems, ordinary conditions	150-200 sec.
Circulating systems, with reduction gears	250-350 sec.
Circulating systems, with reduction gears and vibration	350-750 sec.
Circulating systems, marine reduction gears	450-750 sec.

Locating Faults in Induction Motors—

Most Common Defects

By
A. M. Dudley*

The Ten Most Common Faults in Induction-Motor Windings Are Enumerated, and the Methods Usually Followed in Locating and Correcting These Faults Are Discussed.

THE winding of an induction motor is made up of a number of similar coils connected into groups. These groups in turn are connected in such a manner that when an alternating current of the proper characteristics flows through them, a magnetic field having alternate north and south poles is set up and caused to rotate in the motor. The coil itself is usually made up of two or more turns of wire or strap so that there are at least ten chances for defects in the winding after the coils are all in place and connected. Some of these faults are simple and readily rectified, while others are more obscure and difficult to handle.

These ten most common defects in the order of their likelihood are:

1. The winding grounded on the core.

2. One or more turns in one or more coils short-circuited.

3. One or more complete coils short-circuited at the coil ends or at the "stubs."

4. A complete coil reversed or connected so that the current flows through it in the wrong direction.

5. A complete group of coils or pole-phase group is reversed; that is, connected so that the current flows through the group in the wrong direction, making a north pole where a south should be or vice versa.

6. Owing to lack of care in counting, two or more pole-phase groups may include the wrong number of coils.

7. A complete phase in a three-phase star or delta winding is reversed.

8. The winding connections may be properly made in themselves, but not right for the voltage upon which the motor is to be operated. That is, the motor may be connected properly for 110 or 440 volts, but the motor is to operate on 220 volts.

9. The winding connections are properly made, but

*Designing engineer, Westinghouse Electric and Manufacturing Co.

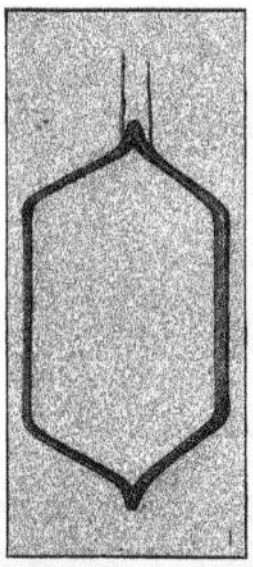

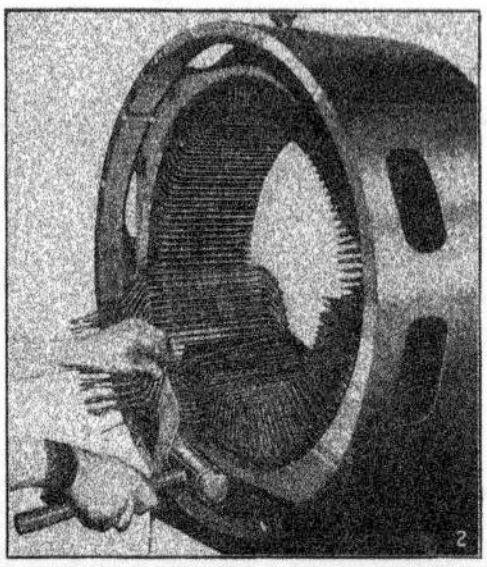

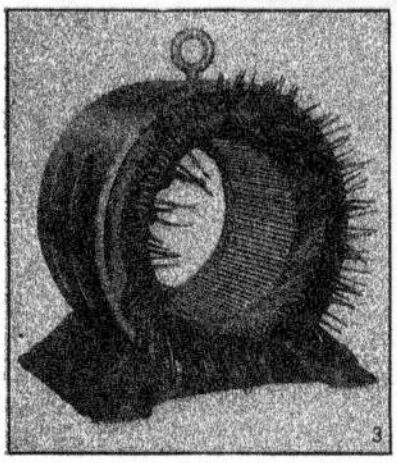

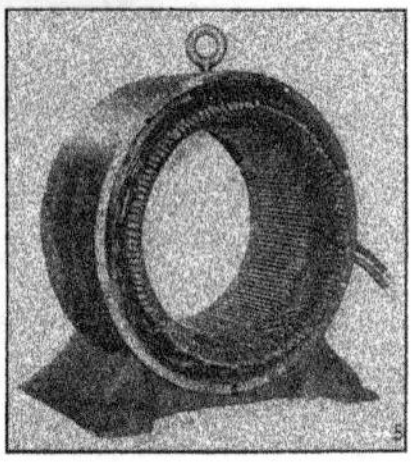

FIGS. 1 TO 5 SHOW DIFFERENT STEPS IN THE WINDING OF AN INDUCTION MOTOR

they are for the wrong number of poles, and hence the motor runs at a different speed from that which was intended.

10. An open circuit somewhere in the winding, or one or more coils are omitted and left out of the winding, known as "dead" coils.

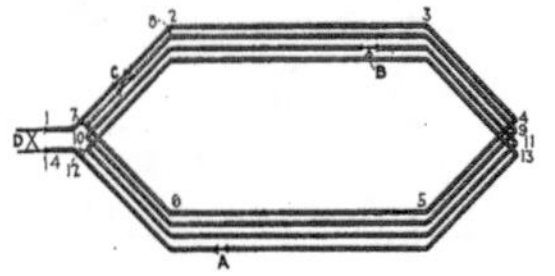

FIG. 6. COIL WITH TURNS SEPARATED

The manner in which these various faults occur can be best understood by referring to what takes place, first, in winding and insulating the coils, and, second, in placing them in the core and connecting them.

Fig. 1 shows a coil of the usual form wound up from several turns of wire and insulated ready to be used in the slot; Fig. 2, the operation of winding these coils in place in the core; and Fig. 3, the coils all in place ready for connecting. The coils connected into pole-phase groups, with the coil ends at the beginning of each group bent into the bore and the coil ends at the end

The first fault listed—grounding of the winding on the core—occurs when in some manner the insulation becomes stripped from the coil and also the cotton covering from the wire so that at some point, as at *A*,

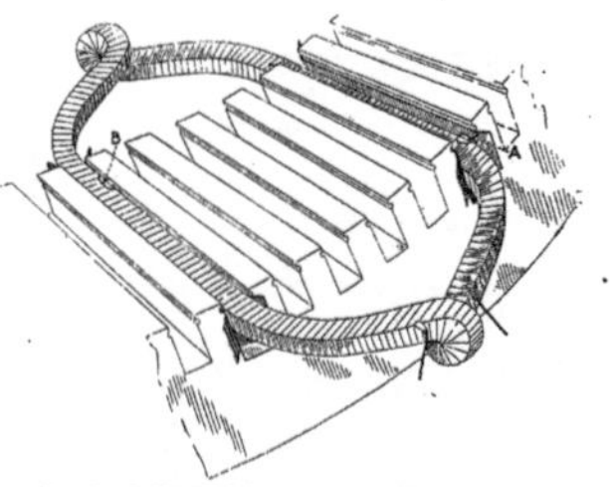

FIG. 7. SINGLE COIL PLACED IN STATOR SLOTS

Fig. 7, the bare-copper conductor touches the laminated-iron core and by so doing "grounds" the winding. This means that a live-current carrying part is touching the metal structure of the motor, and when this condition exists anyone who touches the frame of

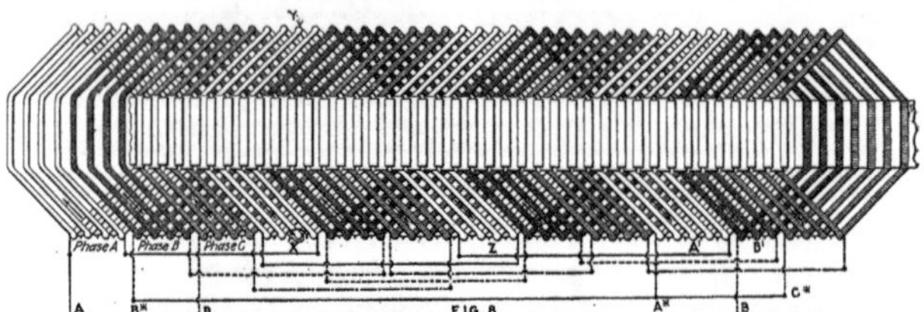

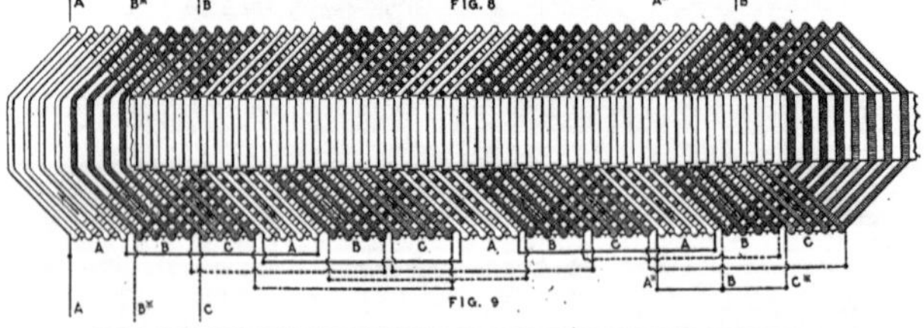

FIGS. 8 AND 9. STAR-CONNECTED INDUCTION-MOTOR WINDINGS SHOWING DIFFERENT FAULTS

of each group bent out toward the frame are shown in Fig. 4. The cross-connections are made in Fig. 5, thus completing the winding connections. Fig. 6 shows the coil in Fig. 1 as it would appear if the insulation were stripped off and individual turns of wire separated.

the motor actually touches a live conductor. This may not be detected if the entire winding and the supply circuit otherwise is free from grounds, but it often happens that other grounds are present somewhere in the system so that in standing on the ground and touching the frame of the machine the chances are very good of getting a shock at a voltage that may equal that of the supply circuit.

Referring to Figs. 6 and 7, should two grounds occur simultaneously, as for example, at *A* and *B*, a short-circuit would be formed in the loop, Fig. 6, from *B* through 11, 12 and 13 to *A*; and if the normal voltage remained on the motor, this short-circuited turn would immediately become hot enough to destroy the insulation on the complete coil. This is the second fault listed and may occur without grounding by the touching of the bare conductor of adjacent turns as at *C*, where the

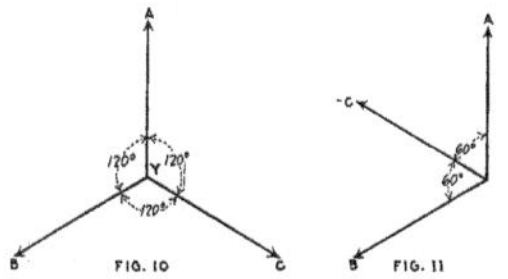

FIG. 10 FIG. 11

FIGS. 10 AND 11. VOLTAGE VECTOR DIAGRAMS

complete short-circuit follows the path of *C* 2, 3, 4, 5, 6, 7 and *C*.

The third fault—short-circuiting a complete coil—can also be seen from Fig. 6 and exists when the insulation of the ends of the coil 1 and 14 become damaged and allow these two wires to touch, as at *D*. A current then flows in the entire coil, in addition to and aside from the line current, equal to the voltage of the coil divided by its impedance. In other words, what happens is equivalent to removing that particular coil from the main winding where it is generating its share of the useful counter-electromotive force and using up this same generated or induced counter-voltage, simply, to force current through the coil itself. This coil would heat up practically as fast as would any induction motor winding if the rotor was held from rotating and full-line voltage applied to the stator winding.

The fourth fault occurs when the two leads of a coil are interchanged, as at *X*, Fig. 8. This has the effect of causing the one coil, or in this case coil *Y*, to "buck" all the other coils in the same pole-phase group. Expressing this in another way, the cross-connected coil is trying to produce a magnetic north pole when all the other coils in its group are producing a south pole. The effect of this is magnetic dissymmetry and manifests itself, as do most irregularities in winding, in noise and heating.

The fifth fault, and one that can occur readily in connecting, is when an entire pole-phase group is reversed, as at *Z*, Fig. 8. This can be understood from Fig. 4. The beginnings of all pole-phase groups are bent in toward the center of the bore, and the endings are all bent out. Should one of the ends bent out be used as a beginning and the other end as an ending, the entire group would be reversed with consequent magnetic distortion and trouble due to noise and heating.

The sixth fault is one due wholly to wrong counting in grouping the coils. In a three-phase four-pole motor with 48 coils there should be in each group $48 \div (3 \times 4) = 4$ coils, and the presence of 3 coils or 5 coils in any group constitutes the sixth fault as they are here listed. This is also shown in Fig. 8, where all the groups have 4 coils except A^1 and B^1, which have 5 and 3 coils respectively.

The seventh fault is present only in the case of three-phase motors and consists in reversing the ends of one-third of the winding so that one leg of the star or one side of the delta is connected in such a way that the voltages generated in the three phases are only 60 electrical degrees apart, whereas the currents supplied from any normal three-phase generator are 120 electrical degrees apart, and hence these three voltages and currents cannot combine to produce power as they properly should. This can be understood by referring to Figs. 10 and 11. Fig. 10 shows the three voltages generated in a three-phase winding as represented by three arrows or vectors arranged 120 deg. apart. If, however, one phase of the winding was reversed and the lead connected to the star point and vice versa, the back, or counter-electromotive, force generated in that winding would be reversed and would no longer be 120 deg. from the voltages in the other two phases, but would be 60 deg. from them, as in Fig. 11. This would mean that the magnetic field in the stator, instead of being a balanced succession of north and south poles rotating and pulling the rotor around, would become unbalanced and would no longer rotate properly. According to another method of looking at the matter, there would be one field rotating clockwise and another different kind of a field rotating counterclockwise, and the natural result would be that these two fields would interfere, and instead of rotating, the motor would remain at a standstill, emitting an unusual amount of noise and reaching a dangerous temperature in a very short time.

A four-pole three-phase winding with the *B* phase reversed is shown in Fig. 9. It will be observed that instead of the arrows on the pole-phase groups pointing in alternate opposite directions, as they should for a correct connection, they point in opposite directions in groups of three. Further consideration will be given this feature in the next article.

In the eighth fault the winding connections are all made properly to form the magnetic poles in their proper sequence, but there are only half as many turns in series or perhaps twice as many as there should be. When this is discovered, the winding may be such as to permit connecting in series instead of parallel or vice versa, but under the worst conditions it may be necessary to remove the entire set of coils and replace with a new set having the proper number of turns for the required voltage.

The ninth fault is sometimes overlooked unless the speed is taken with a tachometer or speed counter, in which case it is readily detected. Its correction is not always either evident or simple, but can often be accomplished without change in coils by following some one of the various methods described in the articles on "Reconnecting for Changes in Number of Poles," in *Power* for Apr. 9, 1918, Aug. 20, 1918, and Jan. 7, 1919.

The tenth fault, "open-circuits," may be due to failure to solder a joint properly or to a joint being broken mechanically after having been once made. "Dead" coils are usually purely inadvertent and are sometimes present without being discovered at all. Such an occurrence could hardly happen unless there were a large number of small coils crowded together.

(The next and last article will explain how to make the tests necessary to locate the ten faults outlined in this article.)

Henry Ford's new gasoline driven street car is nearly ready. If it is successful and Mayor Couzen's municipal ownership plan is adopted Apr. 5, the new car may become a fixture on Detroit streets. The motors are 75 and 150 hp. and will be fitted to burn alcohol in addition to gasoline or kerosene.

Steam-Turbine Governors

Shaft Governors of Small Curtis Turbines

Illustrates and describes the Curtis direct-type governor, its care and adjustments. Directions are given for putting in a new spindle and for testing the emergency stop.

CURTIS turbines from 25 and 35 kw. capacity, 3,600 r.p.m., are provided with direct, spring-tension type governors. Fig. 1 shows one of these machines connected to a generator. The governor is mounted on the end of the mainshaft and operates a double-seated balanced throttle valve through a bell crank directly connected to the valve spindle, as shown in Fig. 3, which shows the governor all the way out and the regulating or governor valve open. Weighted arms *AA* supported on knife-edges *DD* and *BB* pull outward against the spring due to centrifugal force. The governor transmits motion to the bell crank through a spindle *C*, which moves forward and backward with the spring and struts. The whole governor frame revolves, but only the spring, spindle and struts move forward and backward and these do so in a straight line. The spindle and struts are held in place by the spring and weights only. The spindle does not move in a close-fitting guide and is not lubricated. The knife-edges of the struts and weights have little movement *and must not be oiled.*

Formerly a so-called "lapseat" thrust was furnished on the outer end of the governor spindle, but this type has been discontinued and a ball thrust *E*, being more sensitive, is now used. As shown, the thrust is in the hollow end of the bell crank and has one large ball screwed on the end of the spindle. This ball is grooved for a guide, but the guide makes little or no contact in the groove as the large ball rolls on the small ones. The frictional resistance of this thrust is small, but the thrust depends on adequate lubrication to enable it to work well. The lower half of the bell crank is hollow and provided with braided yarn, oil being supplied to the yarn from an oil cup connected to the trunnion, which is provided with an oil passage. It is important that this cup be kept filled with a medium cylinder oil. If the turbine is exposed to very low temperatures, a lighter oil, as engine oil, should be used. The oil passages must, of course, be kept clear.

The Governor Valve—The valve stem is a loose fit in the governor valve to promote self-alignment. When full open, the valve is ⅛ in. off the seat and will seat tightly before the governor completes its travel. Adjustment of this difference may be made by raising or lowering the valve stem at *F*, Fig. 3. The travel of the governor spindle should be ⅛ in. and may be kept at that by adjusting shims held in the end of the mainshaft by a long screw *G*, Fig. 3.

Putting in a New Spindle—Refer to Fig. 3. To put in a new governor spindle, remove the governor casing, take out the plug *H* and unscrew the ball *E* from the spindle (right-hand thread) by using a screwdriver, the ball being slotted to receive one. Remove the bell crank, but do not remove the ball or its seats. The governor is next removed from the shaft (right-hand thread) by a special spanner furnished with the turbine. The governor is screwed on tightly, and it may be necessary to strike the spanner a rather sharp blow with a hammer to start the nut. Remove the nut *J* with the spanner; this will relieve the knife-edge seats, after which the pin *K* may be removed, the spindle taken out and a new one put in. Do not change the position of other nuts. Care must, of course, be taken not to han-

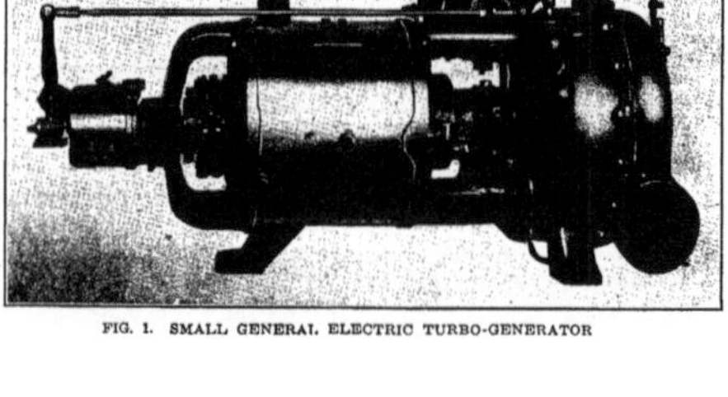

FIG. 1. SMALL GENERAL ELECTRIC TURBO-GENERATOR

dle the governor so roughly as to bend the spindle or dull the knife-edges.

The Emergency Governor—The emergency governor valve *M*, Fig. 1, is of the plain flap type, held open by a trigger. The valve and its mechanism are mounted on a separate chest, the valve fitting loosely on the resetting shaft *A*, Fig. 2. The latch holding the valve open is mounted on a tripping shaft *B*, having a lever keyed to the outer end. The square rod *C* riveted to the other end of the lever holds the tripping lever and the latch in position by being held up by a trigger *D*. A clock tripping the emergency stop, as the packed shaft or spindles will stick from non-use and may fail to work when most needed.

When these small Curtis sets are first started, "feel" the governor by pushing in on the adjusting nut *O*, Fig. 3, the governor casings having been removed. Test the emergency stop by holding the bell crank out with a piece of cord or wire. At about 8 per cent overspeed the spring *E*, Fig. 2, should begin to fly out, and at 10 per cent should trip the trigger. If the emergency governor needs adjusting, loosen the nut holding the spring.

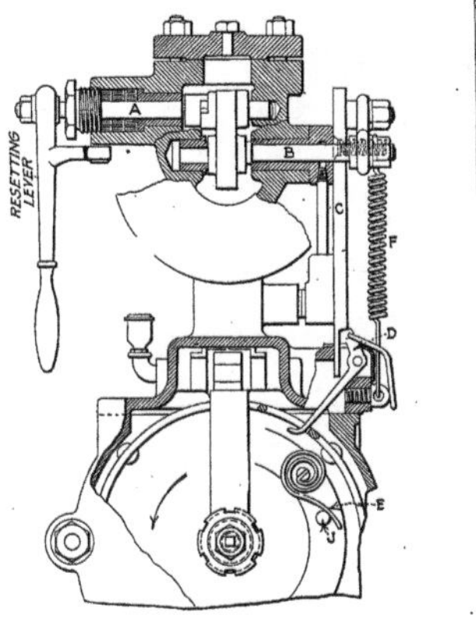

FIG. 2. EMERGENCY GOVERNOR, SMALL CURTIS TURBINES

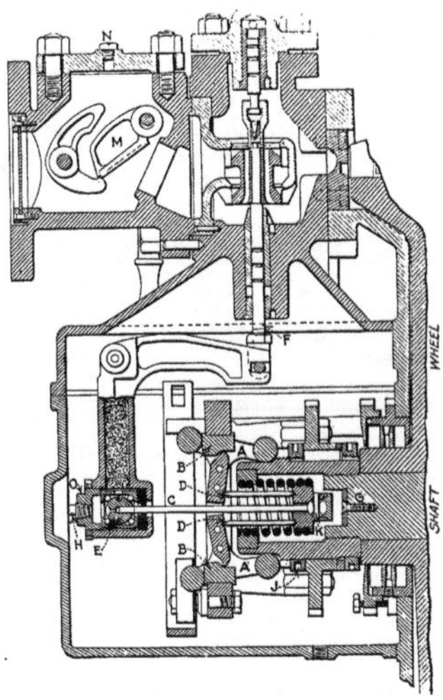

FIG. 3. SECTION OF SHAFT GOVERNOR AND GOVERNOR VALVE, SMALL CURTIS TURBINES

spring *E* revolves with the shaft, and at about 10 per cent overspeed will fly out, striking the trigger *B*, allowing the spring *F* to trip the mechanism which allows the valve to close.

Resetting the Emergency Stop—If the emergency stop trips and the turbine is to be started, close the throttle, bleed the steam from the chamber through the plugged hole *N*, Fig. 3, and raise the resetting lever as far as it will go. Always return the lever to its stop so that the emergency valve will not have to move this lever the next time it trips. If the lever is not returned to the stop, the next time the valve trips it must do so against the frictional resistance offered by the packing in the stuffing-box of the tripping shaft. The tripping-shaft bushing is not packed, and the little steam escaping is piped to the turbine casing.

It is good practice to stop frequently any turbine by

Turn the stud so that the spring bears harder against the stop *J* if the emergency trips at too slow a speed and not so hard if it trips at too fast a speed.

When tightening the nut, be careful that the adjustment of the spring is not disturbed.

Finally, as a word of caution, do not oil the knife-edges of the speed governor, as they will collect dirt, cause poor regulation and quickly wear.

The application of pulverized coal to copper blast furnaces has been made successfully, according to the *Compressed Air Magazine*. This is a comparatively new application of pulverized coal to industry. The pulverized fuel is introduced into the blast pipes above the tuyeres without disturbing seriously the existing conditions at any modern blast furnace plant.

Turbine-Driven *versus* Engine-Driven Alternators-I

*By S. H. Mortensen**

A comparison of the construction and operation of the rotor of these two distinctly different types of alternator. One slow speed with a short rotor of large diameter and many poles as compared to a long rotor of small diameter and few poles built for high peripheral speeds.

OPERATING speed in revolutions per minute of an engine-type alternator is necessarily much lower than the corresponding speed of a generator driven by a steam turbine, and its number of poles for the same frequency must then be correspondingly larger. The relationship between frequency F of an alternator, its rotating speed in number of revolutions per second N and its number of poles P is,

$$F = N \times \frac{P}{2}$$

For example, an alternator rotating at 120 revolutions per minute, or 2 revolutions per second, must have 60 poles to give a 60-cycle current, whereas an alternator driven by a steam turbine at 3,600 revolutions per minute, or 60 revolutions per second, requires only two poles to develop this frequency.

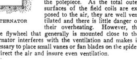

This difference in the number of poles and revolutions leads to distinctly different types of machines. The extreme operating speed of the turbo-alternator imposes problems in design and operation that are unknown in engine-type machines. Comparative features of engine and steam-turbine driven alternators are indicated in the accompanying table and are shown in the sketches,

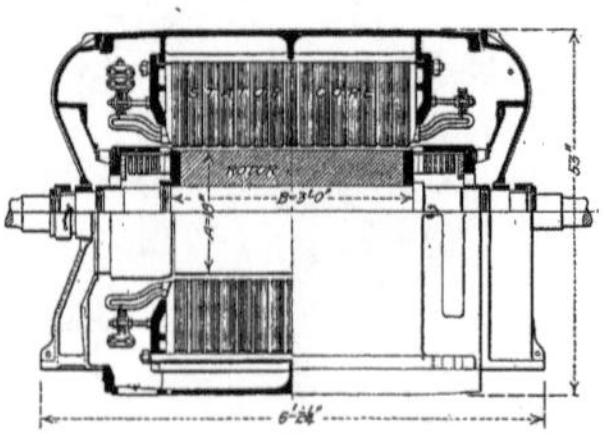

FIG. 2. SECTION THROUGH 1,000-KW. TURBINE-DRIVEN ALTERNATOR

Figs. 1 and 2, which depict the corresponding machines. Fig. 3 shows the rotor construction of a slow-speed alternator. In this type of machine the diameter has to be large enough to permit mounting the required number of poles on the rotor spider and allow the proper clearance between the adjacent coils. The axial width of the pole is determined by the cross-section required to carry the magnetic flux of the machine.

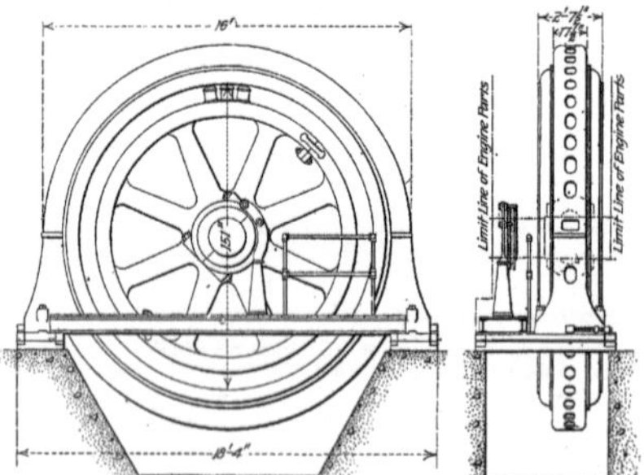

FIG. 1. CHIEF DIMENSIONS OF 1,000-KW. ENGINE-DRIVEN ALTERNATOR

As the operating speed of such machines is low, the mechanical features present no difficult problems. The pole may be held to the spider by means of bolts or dovetails, and the rotor field coils are supported between the spider and the projecting pole tips. The field coils are wound with copper strip bent on edge and insulated from each other by treated paper. Fig. 4 shows such a strap-wound coil before and after it is mounted on the polepiece. As the total outer surfaces of the field coils are exposed to the air, they are well ventilated and there is little danger of their overheating. However, the large flywheel that generally is mounted close to the alternator interferes with the ventilation and makes it necessary to place small vanes or fan blades on the spider to direct the air and insure even ventilation.

*Electrical engineer, Allis-Chalmers Manufacturing Company.

Flywheels may either be constructed as a part of the alternator spider or be mounted as a separate element on the rotor shaft. They are required to compensate for the variation in the turning effort during the revolution, inherent to reciprocating engines, and caused by the varying pressure on their pistons. The flywheel effect of the rotating parts of the alternator must then be sufficient to hold the variation in the engine speed during the revolution within certain limits fixed by the requirements of the connected machines. For alternators operating in parallel, these limits are necessarily narrow. If one rotor for any reason drops behind, power will flow from the faster to the slower machine, tending to accelerate it. If the displacement between the rotors exceeds certain values, parallel operation becomes impossible and the alternators fall out of step.

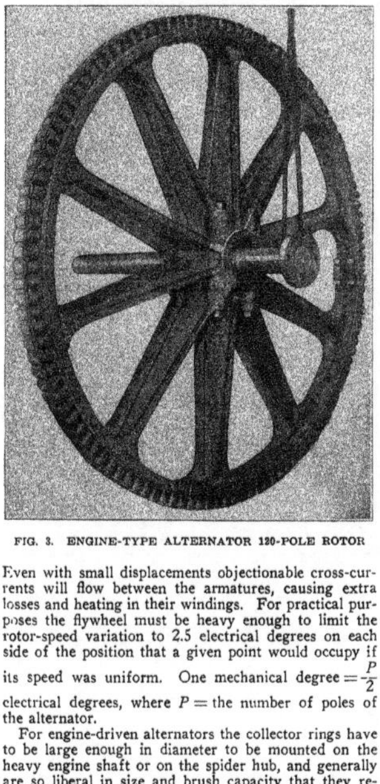

FIG. 3. ENGINE-TYPE ALTERNATOR 120-POLE ROTOR

Even with small displacements objectionable cross-currents will flow between the armatures, causing extra losses and heating in their windings. For practical purposes the flywheel must be heavy enough to limit the rotor-speed variation to 2.5 electrical degrees on each side of the position that a given point would occupy if its speed was uniform. One mechanical degree $= \frac{P}{2}$ electrical degrees, where $P =$ the number of poles of the alternator.

For engine-driven alternators the collector rings have to be large enough in diameter to be mounted on the heavy engine shaft or on the spider hub, and generally are so liberal in size and brush capacity that they require practically no care after once they are adjusted. They are made from cast iron or bronze alloy and work with carbon brushes. The rings for the machine of Fig. 2 are shown in Fig. 8.

Turbo-alternator rotors have to meet the same requirements as the fields on engine-type machines, but in addition the mechanical stresses inherent to high-speed operation limit their diameter and lead to rotors of great axial length. Figs. 5, 6, 7 and 8 show four different stages of completion of a rotor made from a solid steel forging. Fig. 8 shows the finished rotor. To take care of mechanical stresses due to the centrifugal force, the field coils are subdivided and embedded in radial slots in the rotor. Heavy metal wedges are fitted into grooves in the rotor slots for holding the rotor coils in position. The individual turn in such a field winding is placed in the rotor slots with great care and the insulation between the turns as well as between the completed coils is generally made from micanite or asbestos, rendering the field fireproof. This class of insulation will operate continuously at temperatures reaching 150 deg. C. (302 deg. F.).

FIG. 4. STRAP-WOUND FIELD COILS FOR SLOW-SPEED ALTERNATOR

COMPARATIVE DATA ON ENGINE AND STEAM-TURBINE DRIVEN ALTERNATORS

	Turbo-Generator	Engine-Type Generator
Maximum capacity, kw. at 80% p.f.	1,000	1,000
Frequency	60	60
Revolutions per min.	3,600	120
Number of poles	2	60
Diameter of rotor over poles (A) in	18	151
Length of rotor over poles (B) in	36	10
Peripheral velocity, ft. per min	16,900	4.750
Ratio A/B	0.5	15.1
Weight rotor, net, lb	3,750	22,000
Weight stator, net, lb	15,000	27,850
Losses due to windage and friction, kw	20	6
Copper loss and stray loss, kw	12	30
Excitation, kw	11	20
Air for forced vent., cu. ft. per min	5,500	none
Full-load efficiency	95	93.3

In the design of turbo-rotors the main problem is in the choosing of the rotor diameter together with the number and size of slots for the field coils. The diameter is to be kept as small as practicable to limit the centrifugal force and also the windage losses of the machine, but it has to be sufficient to permit the required amount of space for the field winding. For large machines the rotor velocity is limited by the strength of the material available for their construction. At the present time the peripheral velocity of

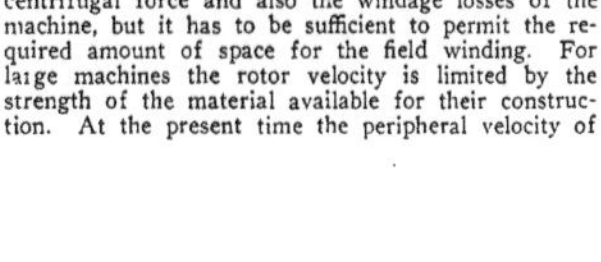

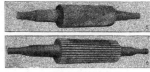

FIGS. 5 AND 6. FORGED ROTOR CORE BEFORE AND AFTER MACHINING

FIG. 7. TURBINE-DRIVEN ALTERNATOR'S ROTOR IN PROCESS OF WINDING

rotors is limited to approximately 25,000 ft. per minute.

The highest-stressed material is not found in the rotor body proper, but in the metal caps or rings that hold the projecting ends of the field coils in place (see *R* in Fig. 8). For large-sized machines these rings are manufactured from the strongest and toughest material available, which is alloy steel having an ultimate strength of 120,000 lb. per sq.in., or more.

Instead of being made from one forging, as shown in Fig. 5, smaller rotors may be built from punchings mounted on a shaft and clamped securely between heavy metal end plates, or from a number of steel disks either mounted upon the shaft or clamped between stub shafts. In either case the general appearance of the rotor will be as shown in Fig. 8. The choice between the solid and laminated rotors is generally decided by their tendency to vibrate, which for large high-speed machines necessitates the adoption of the most rigid construction available—the solid forged rotor. The fans shown on the ends of the rotor in Fig. 8 supply the air for ventilating the stator and rotor, and great care should be exercised in the shaping of the fan blades. Turbo-driven alternators do not need flywheels, as the turning effort of the steam turbine is uniform throughout the revolution.

On the turbo-generator the collector rings are necessarily run at much higher peripheral velocity than is the case with engine-type machines. Either alloy or steel rings are used with copper gauze or carbon brushes. Care must be used to insure the correct settings of the brushes and likewise the proper brush pressure. The steel rings and carbon brushes are used on most modern machines. With the proper setting of the brushes, the rings will acquire a polished surface and work with practically no wear on either rings or brushes. However, they should be examined at regu-

FIG. 8. FINISHED ROTOR FOR TURBINE-DRIVEN ALTERNATOR

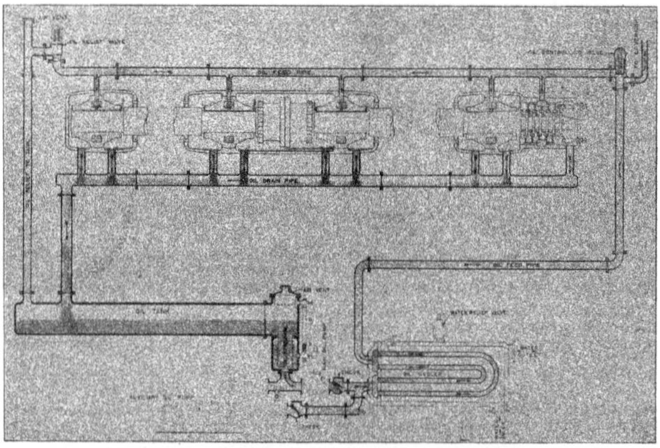

FIG. 9. LOW-PRESSURE SYSTEM OF LUBRICATION FOR TURBO-ALTERNATORS

lar intervals, and if the surface of the rings is found to be cut or rough, it should be turned and polished at the earliest opportunity, as otherwise sparking and heating will develop and cause excessive wear on the brushes.

On engine-type alternators the bearing and lubricating system is of the well-known oil-ring type. The amount of oil in the bearings, combined with their radiating surface, makes it unnecessary to provide for any special arrangement for carrying off the heat generated by the bearing friction. With the proper inspection of this class of bearing, accidents such as wiping of the bearing lining can be eliminated, as the increase in the temperatures of the bearings inherent to abnormal operation gives ample warning for the shutting down of the machine before any damage is done.

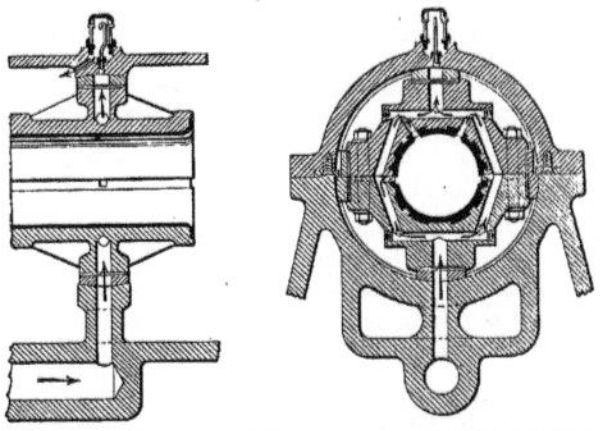

FIG. 10. SELF-ADJUSTING SPLIT BEARING OF BALL AND SOCKET PATTERN

Bearings and lubrication of the turbo-generators require careful attention in design and construction. In the design of the bearings on the early turbo-generators great ingenuity was displayed, but with the improvements in the accuracy of tools and workmanship attainable in modern shops, the self-adjusting split bearing of the ball-and-socket pattern lined with babbitt has proved itself satisfactory for this class of operation. Fig. 10 shows a bearing of this type with arrows indicating the oil flow. The lubrication of bearings is always of great importance, but especially in high-speed work, where even a momentary interruption in the oiling system invariably leads to grave results. Effective lubrication is obtained by admitting the oil to the bearings at a point of low pressure which, for horizontal turbo bearings, is on the top and sides as all the pressure on a properly erected machine is downward.

Several systems of forced oiling are in use, and in Fig. 9 one such system is shown diagrammatically. The oil is pumped from an oil tank through a strainer and an oil cooler and from there to the top and sides of the bearings. The rotation of the shaft establishes the oil film, and the heat generated by the rotation of the journals is absorbed by the continuously circulating oil carried to the cooling system and dissipated. This arrangement may be termed a low-pressure system, inasmuch as the oil pump develops only sufficient pressure to maintain an adequate flow of oil to the bearings. The system shown in Fig. 9 has, in addition to the main oil pump, a separately driven auxiliary pump for establishing oil circulation during the starting period.

Elements of Combustion of Fuel Oil

Perhaps nothing better dealing with the mechanics of combustion of fuel oil has been written than that by the United States Naval Liquid Fuel Board, whose report appeared a number of years ago. In view of the interest now taken in fuel oil in stationary plants, particularly along the Atlantic Coast, what the Board has to say about this subject is of interest.

WHILE the theory of combustion is well understood by all having a knowledge of the elementary principles of chemistry, there are particular considerations that enter into the matter of burning both oil and coal as a fuel, and therefore the economical consumption of fuel in large quantities can be effected only in boiler or furnace installations which have been designed by technically trained experts possessing a knowledge of what well may be termed the practical mechanics of combustion.

Hundreds of oil burners have been designed, which, viewed from a mechanical or theoretical standpoint, should have operated efficiently; and yet when such appliances were subjected to actual test the devices proved unsatisfactory. There are, therefore, practical conditions as well as chemical principles that must be considered in the solution of the liquid-fuel problem, and the Board thus regards the mechanical feature of the oil-fuel combustion question as a subject deserving special study and investigation.

Everyone is aware that with a charcoal or coke fire it is possible to maintain intense combustion within a comparatively small space, and with little smoke. This sort of fire was known to the smelters of the Bronze Age, and it is still used in blast furnaces and other operations where great concentration is required.

The explanation lies in the fact that the fuel is solid, even at the highest temperature. The solid particles in the smoke are probably particles of ash, but whether they are ash or unconsumed carbon, they are exceedingly small, as shown by the bright-blue color of the smoke. As a result of this solidity no carbon can leave the bed of hot coals except as a constituent of CO or CO_2. In either case the combustion will be free of soot or smoke, since both gases are colorless and transparent. If the carbon goes up the stack as a constituent of CO, it carries with it two-thirds of the heat that it is primarily capable of yielding up, and it is only with respect to the possible formation of CO instead of CO_2 that a charcoal fire fails to give perfect results. Were it not for this possibility there would be no reason why a charcoal fire should need more space than is sufficient to contain the fuel itself.

The obvious way to prevent the formation of CO_2 is to force a larger quantity of air through the bed of coal, but this alone is insufficient. Experience shows that even when the amount of oxygen passing through the bed of coal is twice that requisite for the complete combustion of carbon, there is still some of the carbon that leaves the furnace in a partly burned condition. The only way in which this carbon can be completely burned is by subsequent diffusion of the gases, whereby each molecule of CO sooner or later meets with an atom of free oxygen and so becomes CO_2. But this diffusion requires time, and as the gases are being constantly cooled as they are carried along it may happen that they will be cooled below the ignition temperature before union takes place. So that even in a charcoal fire the need for appreciable combustion space is obvious.

Thorough Mixture Essential to Good Combustion

Complete combustion requires that for every atom of carbon, and for every two atoms of hydrogen, there shall be at least one atom of oxygen brought in close proximity, and then and there subjected to a temperature sufficient for ignition. In other words, there must be a thorough mixture, and then ignition. It is doubtful if a mere mechanical mixture, however complete, could ever be perfect enough to bring about the desired result. This is well illustrated by contrasting the smoky combustion of black gunpowder, where there is a mechanical mixture, with the combustion of the so-called smokeless powders in which the mixture is so thorough and minute that similar proportions of oxygen, carbon and hydrogen occur in each separate molecule.

In all ordinary cases of combustion, however, where we draw our supply of oxygen from the atmosphere, it is only by virtue of the property of diffusion that a sufficiently intimate mixture is attained. As to the real nature of diffusion, it is known that at ordinary temperatures the particles of oxygen in the air are moving in every conceivable direction with velocities averaging over 1600 ft. per sec. Any one atom, however, moves only an inappreciable distance before being arrested by collision with another atom, so that although the average velocity of the atoms is probably equal to that of a rifle ball, it still takes an appreciable time for a particle to travel even a moderate distance. It is this time element that constitutes the great stumbling block when the attempt is made to burn a large amount of combustible in a small space.

Combustion of Hydro-Carbons

As before noted, the reason why intense combustion is easily attained with a charcoal fire is that the fuel is solid at the temperature of ignition. Being solid it can present a large surface for the oxygen to act upon, and an atom cannot break away and go up the chimney first without being united with at least one atom of oxygen. In the combustion of hydro-carbons, on the other hand, we have the following condition: The fuel is already on its way to the chimney before it is even partly burned. The first effect of the heat is to dissociate the carbon from the hydrogen. Whether or not the latter unites with the oxygen does not affect the soot or smoke question, since the constituents, and also the products of combustion of hydrogen, are like transparent colorless gases. But in any case the carbon left alone in the form of an impalpable dust is much less favorably circumstanced than that in a charcoal fire. If it were attached to a hot coal, as in the charcoal fire, so as to be capable of receiving a blast of air, its combustion would be easily accomplished. But instead of this it is carried along by the current of gases, and unless it is given plenty of time before being cooled it will be left alone as a particle of soot.

An examination of the nature of flaming leads to similar conclusions. The luminous part of a flame is caused by the white-hot particles of carbon. These particles have been robbed of the hydrogen with which they were formerly associated, and they have not yet met the oxygen necessary for complete combustion. This process of finding, or of being found by, the oxygen requires time, and if perchance the temperature falls below that of ignition before the process is completed, the carbon will be deposited as soot or else go on up the stack as smoke along with the excess oxygen, with which it should have been united. Thus an unmistakable symbol of the conditions that are necessary to burn a large amount of combustible in a small space is a short flame. The circumstances which conduce to shortness of flame are: First, pure carbon fuel, because the fuel cannot leave the grate or furnace until it is burned to CO at least. In any case it cannot deposit soot, since CO, when cooled, is a transparent gas. Second, intimate initial mixture of oxygen with the fuel, since the more intimate the mechanical mixture the less time will it take the gases, by the process of diffusion, to become perfectly mixed. Third, initial heating of the air, since the rate of diffusion increases with temperature. Fourth, large surface of fuel presented for impact of the oxygen.

Proper Size of Combustion Space

The desirability of supplying a combustion chamber whose volume is at least equal to the volume of the flame seems obvious. In this connection the fact should not be overlooked that a slight increase in the volume of the combustion space acts two ways to improve the solidity of combustion. One way—that having to do with the greater time permitted for diffusion—has already been touched upon; but apart from that there are influences that work, in consequence of which an increase in the volume of the combustion space actually diminishes the volume of the flame. This is because the temperature of the larger space is higher, and the higher temperature hastens the process of diffusion.

During the process of diffusion heat is being liberated at all points throughout the combustion space. Therefore, all parts of the space are being traversed by heat rays emanating from every other part of the combustion chamber. It is readily seen that the temperature within the space must under these conditions increase with the volume to an extent limited only by the transparency to radiant heat, and by the temperature of dissociation at which necessarily heat ceases to be liberated. Since the transparency of the combustion space is diminished by the presence of solid carbon (for whether black or incandescent, it is in any case opaque), it follows that the increase of temperature with a given increase of volume will be less in a space filled with luminous flame than in one filled with burning hydrogen or CO.

Incandescent Walls Hasten Diffusion

The question of the proper size of the combustion space is further complicated by the presence and condition of the solid walls of the furnace; whether, for instance, they are themselves incandescent, or merely black absorbers of heat. There seems no reasonable doubt, however, that incandescent walls will hasten diffusion and hence shorten flame.

Where it is possible for the diffusion to be completed before combustion begins as in the bunsen gas burner, the difficulties naturally disappear, and there is readily attained a very short flame, which, moreover, is incapable of depositing soot, even on a cold body.

In the case of liquid fuel, which is incapable of vaporization, the diffusion and ignition must occur simultaneously. With such a fuel there is bound to be considerable flaming. Another difficulty, and one from which all solid fuels are free, arises from this sort of fuel from the action of capillary or surface tension. Thus no matter how finely the liquid is pulverized, each tiny drop assumes a spherical shape and so presents the least possible surface for the impact of oxygen atoms.

From what has been said it is clear that a liquid fuel, such as crude petroleum, requires an ample combustion space—more, indeed, than does almost any other sort of combustible material.

The relative dimensions, like breadth and depth, of the combustion space are of minor importance. A primary requisite is volume, and that alone provided all parts of it are traversed by the same quantity of gas in a given time—in other words, provided the gases are not short-circuited through or across some parts of the space to the neglect of others. The advantages are in favor of the combustion space of large cross-section and short in the direction of the flow of the gases.

A primary requisite for the successful burning of a nonvolatile liquid fuel is the exposure of the fuel to the heat of the furnace in such a form that it presents the largest possible surface for the impact of the atoms of oxygen. From the principles of capillary action it is possible to establish a standard by which to judge of the efficiency of the various methods of increasing the free surfaces of the oil.

Oil in bulk has practically no surface. When broken up into fine drops, the surface is the aggregate surface of all of the drops. The smaller the drops the more perfect the spheres. Hence, drops of oil one-thousandth of an inch in diameter are known to assume the spherical form with a rigidity comparable to that of a steel ball one inch in diameter. The work necessarily performed by the atomizing agent is simply the work of stretching the surface.

Automatic Compensator for Starting Induction Motors

A compensator for automatically starting and stopping squirrel-cage type induction motors or for controlling them from remote points has recently been developed and placed on the market by the Industrial Controller Co., Milwaukee, Wis. The entire starting and stopping is accomplished by means of remote-control stations, one or more of which can be situated at convenient locations, while the compensator can be placed near the motor. A float switch or pressure regulator may also be used to operate the compensator. The compensator, shown with the cover removed in Fig. 1, consists of two oil-break magnetic contactors, two auto-transformers and two time-limit overload relays. As indicated in the figure, the switch mechanism which is immersed in oil occupies the upper portion of the casing. Each of the contacts is closed by the electro-magnets M and M, one connecting the motor to the line through the auto-transformers, and after a predetermined period the other magnet connects the motor directly across the line.

FIG. 1. AUTOMATIC COMPENSATOR WITH COVER REMOVED

The control stations, Fig. 2, are provided with a starting handle H, which is depressed for starting, and a stop button S, provided with a lock-out mechanism so that the control station may be locked open, and the motor cannot be started until the stop button is released. If desired, these stations can be equipped with an inching button, which is used only for this purpose. A simple adjustment is provided within the control stations so that the starting-time interval can be changed to suit each individual application, without going near the compensator. At starting, the operator presses down the starting levers on the control station, which automatically throws in the contactor connecting the auto-transformers into circuit. After the proper time interval a positive-geared clock mechanism in the control station, similar to that in an automatic messenger-call box, energizes the coil that closes the contactor connecting the motor directly to the line. It is impossible to have both contactors closed at the same time, since they are mechanically interlocked.

FIG. 2. CONTROL STATION

Overload protection is provided by two time-limit overload relays R and R. Current adjustment of the relays is accomplished by turning a thumbnut at the top of the relays, and time adjustment is made by changing the size of the valve opening in the dashpot plunger. Proper starting voltage is obtained at the motor's terminals by taps brought out to terminals T and T on the panel board. To make the voltage adjustment, all that is necessary is to change the transformer taps from one terminal to another. Since the compensator can be locked closed and the controller stations locked in the off-position, this piece of equipment is strictly a safety-first device.

England Adopts American Industrial Policies

The average generating capacity of the 600 establishments in England that sell electric current is only 5,000 hp. each. This is about one-fourth the capacity of a generating plant of economical size. So unorganized is the whole electrical industry that "one cannot purchase a simple electric bulb without specifying the particular type of socket" in which it is to be used. There is a strong movement in England today to divide the entire country into sixteen power zones, under the control of the government, and gradually to replace existing plants by sixteen superpower stations, capable of supplying industrial power to the whole of England. This movement is the outgrowth of the conviction on the part of many officials that the electrical industry is the "key industry" upon which depends Great Britain's future position in the trade of the world.

Pulverized Coal Under Central-Station Boilers

By JOHN ANDERSON*

Equipment Required, Method of Applying and Facts Concerning Operation. The Chief Merits of Powdered Coal Are High Efficiency and Flexibility To Meet Varying Demands

On Feb. 19 and 20 the Oneida Street plant of the Milwaukee Electric Railway and Light Co., with five of its boilers equipped for burning pulverized coal, was open for inspection for the first time to visitors who had previously registered for the occasion. Invitations had been sent out by the Technical League of the Employees' Mutual Benefit Association. Under the auspices of this company organization a most interesting paper was delivered Thursday evening by John Anderson. There was an introductory on the use of pulverized coal (reproduced in the article following), a complete report and analysis of a four-day test on the five boilers in the plant under average operating conditions, in which comparisons were made with underfeed stoker operation, followed by comments on the feasibility of burning pulverized fuel in the power plant by Paul W. Thompson, Technical Engineer of Power Plants, Detroit Edison Co., who had been an observer of the test. His opinion and the report of the test are given on other pages of this issue.—Editor.

UP TO within a few years ago, the use of powdered coal was most advantageous in those furnaces that operate under a slight pressure for the purpose of excluding air when the doors are opened and wherein a low velocity of gas travel is desired—typical requirements of furnaces in the cement, copper and iron industries. For these special and for similar reasons, pulverized fuel found its greatest use in annealing and forging furnaces; in puddling open-hearth steel and smelting furnaces; in cement and lime kilns; in the manufacture of refractories and substances that require drying or burning; and in surface treatment of many metals.

However, with the recent heavy increases in price of coal any process which promised greater efficiency found a broad field opened up before it for application to stationary boilers. Previous to this time some work had been done in developing its use in locomotives, but the reasons for its application to stationary-boiler furnaces, and the small saving that might be effected with coal at a low price, did not promise to offset the cost of pulverization. Such conditions did not encourage the use of fuel in powdered form, therfore, until the higher prices of the last three years forced economy in every direction.

In any industry the use of pulverized fuel necessitates a special plant for preparing the coal. It is the cost of installing and operating this plant that will always determine the advisability of adopting powdered coal for steam generation in any one location. In such a plant there must be equipment to separate trampiron from the coal, to crush it to a size suitable for drying and feeding it to the pulverizers, to remove the moisture, and then to grind it to the desired fineness.

As the coal is received at the plant in the form it left the mines, it is necessary to pass it over a magnetic separator which removes all trampiron, such as bolts, nuts, pick-points, etc., this serving as a safeguard against damage to the pulverizing and conveying machinery. A crusher must be used to reduce the green coal to a size suitable for proper drying and pulverizing, unless coal can be procured in sizes less than ½-in. screening without sacrificing heat value which is usually at its lowest in smallest size screenings.

The fuel is next dried, the object being to bring it to a condition in which it can most easily be pulverized and most uniformly controlled by feeders into furnaces. In a pulverized-coal fired furnace uniform temperature conditions are obtained with dry coal. Moisture in varying or large quantities in pulverized fuel definitely prevents a regular controllable flow of mixed fuel and air to the burners. For the removal of moisture a mechanical drier, which reduces the water content of the fuel to about one or two per cent, is installed. This drier usually consists of a double rotating cylinder, so arranged that hot gases from a small furnace are passed through the coal as it travels by gravity through the shell of the drier and is delivered to the pulverizers.

The dried coal is next fed into the pulverizers. The fuel is ground to an impalpable powder, varying in fineness from 80 to 85 per cent, through a 200-mesh screen. This process completes the preparation cycle, and the coal in this form, after being conveyed to storage bunkers adjacent to furnaces, is ready for firing.

Dust collecting and reclaiming equipment, although only auxiliary to the drying and pulverizing machinery, is of great importance in the economical and clean operation of a preparation plant.

In preparing coal the cost per ton depends upon the size of the plant and the quantity of fuel handled, but ordinarily will vary between the limits of twenty-five and fifty cents. It seems that a great deal of work toward reducing this cost remains to be done, especially along lines of utilizing waste gases for drying, in plants not provided with economizers. Development of a furnace design that will allow of burning a large percentage of finely divided coal, but capable of taking care of a smaller amount of larger-sized particles, presents an opportunity for further study along this line. This is possible in view of the fact that the pulverizers consume about 86 per cent of the total electric energy required for all moving parts of the coal preparation equipment. Both of the items mentioned above, if capable of being reduced as suggested, will also cause a re uction in the cost of the pulverizing-department labor.

Continuous and uniform operation of the pulverizing division of a plant is dependent largely upon the ability of the operators to recognize and meet the varying properties of the coal as it arrives. Drier operation especially must be changed frequently so as to handle varying sizes of coal (from high percentages of dust to high percentages of small nut) and varying amounts of moisture, both inherent and surface, and yet supply continuously a coal of uniformly low moisture. Irregularities in the drying process usually manifest themselves throughout the entire coal cycle, and very often with the result that the fuel-feeding system becomes plugged at points most readily affected by wet coal.

*Chief Engineer of Power Plants, The Milwaukee Electric Railway and Light Company.

Little need be said concerning the equipment which feeds the fuel to the furnaces. Feeder screws driven by variable-speed motors carry the fuel from the storage bins to the air-mixing chambers from which it is blown as a mixture through pipes to the burners and then into the furnaces. The feeder-motor controls at the Oneida Street plant are all provided with cutouts, and if at any time the air blast to the mixing chambers and burners is lost due to a blower shutdown, all coal feeders are automatically stopped, thereby preventing serious plugging with attendant interruption to the operation of the boilers for what could easily be a considerable period. Air additional to that used for carrying the fuel into the furnace is taken through auxiliary inlets located in the furnace front and can be varied in amount by manipulation of the stack damper, which serves to change the gas velocities through the boiler and thus furnace efficiency therefore resolves itself mainly into one of air supply, not to the furnace alone, but to each particle of coal. With a fuel finely pulverized, air can be made at once available to each particle, since the surface exposed to the air is increased many times that of a lump and combustion is rapid, complete and efficient. With fuel in lump form air must be supplied through a bed, the outer surface of the lumps must be burned off, the process of combustion eating into the lumps slowly, and as a result combustion is often gradual and unequal. Where the fire is thin, too much air is passed; where it is heavy, there is too little air. To insure complete combustion in any furnace some excess air must therefore be provided. The use of pulverized fuel reduces this excess to a minimum, gives a higher and more uniform percentage of CO_2 and therefore also a higher and a more uniform efficiency.

FIG. 1. PULVERIZED FUEL FURNACES AT ONEIDA STREET STATION

the volume taken into the furnace. With coal and air supplies easily adjustable, perfect fire control is assured and it becomes at once obvious why coal is burned so efficiently in pulverized form.

In further explanation of this it is well to consider briefly the indications of efficient combustion applied to the steam boiler. Chief among these is the percentage of carbon dioxide, or CO_2—the product of complete combustion—which is found in flue gases. Next in importance is the percentage of carbon monoxide, CO—a product of incomplete combustion—which may be found at the same time.

The condition desirable is that with the percentage of CO_2 as high as practicable there should be no CO—a condition obtainable to a greater degree in a pulverized-fuel furnace than in any other type. Complete combustion with the least amount of excess air requires the best distribution of that air and the problem of

In pulverized-fuel practice the percentage of CO_2 to be maintained is determined to a great degree by furnace limitations rather than combustion consideration. From 16 to 17 per cent CO_2 in the flue gases is easily obtainable, but it cannot be maintained in actual operation due to exceedingly high flame temperatures that result in the consequent destruction of the brickwork. The temperature of the furnace therefore must be regulated by varying the volume of excess air. The temperature of the flame should not be above 3,000 deg. F. and that of the brickwork in the flame zone not more than 2,400 or 2,500 deg. F. Higher temperatures than these usually result in fusion of the ash particles and formation of a molten slag that is very destructive to the brickwork with which it comes in contact.

In attempting to carry combustion conditions of a furnace so as to obtain exceedingly high percentages of CO_2, usually too little attention is given to the losses resulting from the formation of carbon monoxide or CO. Carbon monoxide, a highly combustible gas, is formed when the excess air admitted to the furnace is insufficient to properly surround and completely oxidize all the coal particles. Heat losses occurring when carbon monoxide is present in the gases of combustion outweigh any gains that may be effected by an exceedingly high CO_2 and are always to be avoided.

In the discussion of furnace efficiencies consideration should be given to the adaptability of a pulverized-fuel fired furnace to the use of widely varying grades of coal without resultant losses in economy. Further, boiler capacity, even under heavy overloads, is in no way affected when an inferior quality of coal is burned. It is known that the combined efficiency of a boiler and furnace does not decrease when the fuel is poor, which condition does not hold true for the stoker. In the case of the stoker, the dropping off in efficiency is at a more rapid rate than the B.t.u. value of the fuel would indicate as normal, and so much so that the point is rapidly reached when proper combustion cannot be maintained.

Operation of a pulverized-fuel fired boiler equipped with proper instruments can be varied to take big fluctuations in load over very brief periods of time. A heavy overload can be quickly taken on or dropped off by adjustment of the coal and air feeds, and without any waste of fuel as always occurs under like conditions in stoker practice. No losses occur due to clinkering of coal or cleaning of fires, this condition of operation being entirely eliminated. Irregularities caused by change in quality and variation in size of coal, such as the fireman cannot successfully cope with on stokers, are also eliminated. Furnace conditions necessary to most economical combustion are more perfectly obtained and hence a horizontal combined efficiency curve is possible of approximate attainment.

FIG. 2. INSTRUMENT BOARD FOR TWO BOILERS

Due to its easily regulated coal and air supply and its perfectly controlled rate of combustion, the pulverized-fuel furnace practically eliminates losses of combustible in ash. Ordinarily, this loss is relatively large and varies according to the nature of the coal, type of stoker and the boiler load carried. In pulverized-fuel practice the loss is very small and these variations do not occur.

The ease with which the fuel feed and draft is controlled, the ability to take on and drop off heavy overloads in a brief time, the thorough combustion of the coal and the uniformly high efficiency obtainable under normal operation constitute the chief advantages of pulverized fuel over other methods of coal burning.

An additional economy is effected during banked boiler hours. Banking conditions when operating with pulverized fuel are somewhat different from those obtained in stoker practice. By stopping the fuel supply and closing up all dampers and auxiliary air inlets, a boiler fitted for use of pulverized fuel can be held up to pressure for several hours. The furnace brickwork, having been heated to incandescence during operation, gives off a radiant heat which is almost all absorbed by the boiler rather than escaping up the stack intermixed with an excess of cooling air. Only radiation losses occur, as against radiation plus stack and grate losses in the case of the stoker.

FIG. 3. BLOWER UNIT AND FEEDER CONTROLS

Commenting for a moment on the maintenance features of such a plant as has been described, it is the writer's belief, based on two years' operating experience, that the furnace brickwork in a pulverized-fuel furnace will stand up equally as well as a stoker installation, with a very great advantage in favor of the former due to the elimination of all ironwork in the furnace or anywhere near the high-temperature zones of the boiler furnace. Regarding the maintenance of the pulverizing-plant equipment it has not been found that any great amount of maintenance is likely to be necessary, as all the equipment is of the slow-moving type and many opportunities are afforded of applying the same concentration of effort that has been typical of the stationary engineer's work in improving equipment when defects or fast-wearing parts are uncovered. The pulverized-machinery manufacturers have done a great deal along this line, but there are still matters that can be improved upon by the engineer looking for the least troublesome as well as the most economical plant from a maintenance standpoint.

Powdered-fuel installations are not feasible in every l cation. There is one limitation and that is the size of boiler plant to be served. A plant of less than 2,500 developed boiler horsepower on a 24-hour operating basis should not consider using powdered fuel. The amount of coal pulverized per day, the cost of installation, and the labor for operating the preparation plant, when properly studied, will bring before those interested the reasons therefor.

In general, we have found the powdered-fuel method applied to our furnaces a distinctly advantageous one. Our firemen prefer to operate such equipment rather than the stokers when it becomes a matter of choice.

It has proved more economical, as evidenced by the monthly coal bill. It seems to make the formation of scale in the boiler less than in stoker-fired boilers. There is absolutely no trouble from smoke, consequently no reduction in ability of the boiler to absorb heat due to soot on the tubes.

The use of high-sulphur coal, which is so destructive to boiler tubes, breechings, smokestack and all other steel equipment found in a boiler plant, is much more satisfactory as the low moisture content of the coal as fired reduces the opportunity for attack from sulphuric acid.

Although frequently cautioned against explosions, we have had no evidence that such caution is necessary. The reason for our freedom from such unpleasant occurrences is due almost entirely to a proper care in preventing coal from being dried too much and pulverized to a fineness beyond what is necessary. Matters of a kind similar to the foregoing are in the hands of the operating engineers and do not benefit by the high-brow application of theories. The engineer who is careful of his everyday equipment and keeps his plant free from accidents of every nature, from flywheel explosions to burned-out motors, can operate a pulverized-fuel plant successfully without other assistance than his own experience, as with the exception of the drier, the manipulation of the plant is simple.

An actual comparison of results obtained with pulverized fuel and with a stoker-fired furnace has been derived from tests conducted on both at the Oneida Street plant. [These tests are presented on other pages of this issue.]

Pulverized Fuel at Oneida Street Plant*

By PAUL W. THOMPSON

Technical Engineer of Power Plants, Detroit Edison Company

THERE is at the present time a considerable difference in opinion among engineers as to the feasibility of burning pulverized coal in central generating stations, and as to whether, taking all facts into consideration, the use of pulverized fuel will result in a net saving over the results obtained with modern stoker installations. It is to be presumed that a general comparison of any two plants, one equipped with stokers and one equipped for the preparation and burning of pulverized fuel, would probably not give results which could be readily comparable in determining the ultimate relative value of one method of burning fuel over the other because of the probable different conditions of operation and the difference in design, kind and cost of fuel, cost of labor, etc.

In order to determine the feasibility of burning pulverized fuel, the Milwaukee Electric Railway and Light Co., early in 1918, decided upon a trial installation on one of the 468-hp. boilers in the Oneida Street power plant.[1] The necessary equipment was installed and after preliminary operation and making certain changes which were found to be necessary, they found that the installation had proved to be entirely successful, so the remaining four boilers in the south end of the plant were equipped for burning pulverized fuel. By so doing they now have five 468-hp. boilers under which the pulverized-fuel system has been installed. This is practically one-half the boiler capacity of the Oneida Street plant and is a sufficiently large installation to compare directly with the other boilers in this station having Riley underfeed stokers, or similar boilers in the Commerce Street station also equipped with Riley stokers, which are subjected to practically the same operating conditions as are the boilers under which the pulverized fuel is burned.

To obtain data on the pulverized-fuel equipment for the purpose of comparing it with their stoker installation and determining the relative advantages, the Milwaukee Electric Railway and Light Co. conducted tests, presented elsewhere in this issue, at which the writer was present as an observer. The five boilers were operated at as nearly a constant rating as possible of 120 per cent for a period of four days from Nov. 11 to 15. Throughout the duration of the test data were taken from which the over-all efficiency of boilers and furnaces was computed and also attention was paid to the operating labor cost. It is unnecessary to go into the details of the test or methods employed, as these are given in the report of the test. It is sufficient to say that the test was properly conducted and particular care exercised in obtaining an accurate record of all quantities involved. The water and coal were weighed and the scales checked for accuracy at frequent intervals. All blowoff piping was disconnected from the boilers to preclude any possibility of error due to leakage. The boilers were subjected to regular normal operation except the rating was held at practically a constant figure, and no special effort was made to obtain an efficiency which would not be obtained under normal plant operation.

During the test the writer availed himself of the opportunity of watching carefully the boiler-room operation and also the operation of the pulverizing, drying and coal-handling equipment and was favorably impressed with the installation from an operating viewpoint.

The boiler-room operation was much simpler than is obtained with a stoker installation. The rate of steaming of the boiler is controlled by varying the speed of the feeder motor and adjusting the damper to take care of the different quantity of flue gas. It is unnecessary to look into the furnace at frequent intervals as is the case when firing with stokers and where holes in the fire or heavy spots must be corrected. In fact, when once the feeder speed is set to give a certain rate of steaming of the boiler, there seems to be no reason why this rating could not be maintained continuously so far as the furnace is concerned without it being necessary to make any changes whatever. Variations in the kind and quality of fuel burned seemed to have no effect on the operation except that when feeding at a constant quantity the rating of the boiler varied with the heat value of the coal. At one time during the test Youghiogheny was used, this coal having a higher heat value, higher volatile content, less ash and less sulphur than the mixture of Youghiogheny and Kentucky coal used throughout the remainder of the test.

Losses that are inherent in stoker practice, such as breakdowns in the stoker itself, breaking up clinkers, loosening clinkers, continually watching the fire to maintain correct and uniform thickness, watching the gas passes of the boiler for large sparks indicating the carrying away of combustible, dumping, and the many

---

*Presented before Technical League of Employes Mutual Benefit Association, T. M. E. R. & L. Co.

[1]See *Power*, Oct. 15, 1918, pp. 556-9.

other operations that are necessary in stoker operation, are eliminated. In other words, efficient combustion is obtained at all times without continual supervision by an experienced operator and from the standpoint of reliability of operation the odds are in favor of the pulverized fuel. This is an item for serious consideration in plants designed with 4.5 kw. capacity or more per installed boiler horsepower, where the losing of a boiler due to stoker trouble at the time of maximum load on the station, might seriously overload the remaining boilers or make it necessary to drop a portion of the load.

The handling of the ash resulting from combustion of the pulverized fuel is a simple matter due to the very small quantity that is deposited in the furnace. It is in the form of a fine impalpable powder which during the test was removed twice each 24 hours. On several occasions during the first and second days of the test, slagging occurred in a furnace due to not admitting a sufficient quantity of air. The direct cause of this was over-anxiety on the part of the men conducting the test to obtain a high percentage of CO_2 in the flue gas. The reduction in the excess air permitted the furnace temperature to rise to a point where slagging occurred. The removing of this slag from the bottom or floor of the furnace presented more difficulty than is usually experienced in removing the refuse from the ash hopper of a stoker-fired boiler. This slag had to be broken up and pulled out before it fused to the brick lining of the furnace. It appeared to the writer that this formation of slag could have been eliminated almost entirely by a more frequent inspection of the floor of the furnace and the admitting of more air through the openings in the front of the furnace if it was found that slag was beginning to form. Even at the time of removing this slag it was possible to maintain the rating on the boiler by increasing the coal feed. This, however, resulted in a decreased efficiency during the time required, amounting to about forty-five minutes in 24 hours for each boiler.

A large portion of the ash resulting from combustion is carried on through the passes of the boiler and out of the stack. Owing to the fineness of this ash it apparently is carried a considerable distance even in a moderate wind before being precipitated. Throughout the test there was a moderate wind blowing, probably between 4 and 8 miles per hour, and the writer was unable to find any noticeable deposit in the streets. Even at the time of blowing the deposit of ash from the boiler tubes, which was done three times a day, requiring about 20 minutes per boiler for each blow, no noticeable precipitation of ash could be found in the streets. At no time during the test was there any tendency for slag to form on the tubes of the boiler.

Strictly speaking, there is no such thing as a banked boiler when using pulverized fuel, as all that is necessary when it is desired to cut out a boiler is to shut off the coal feed and close all the dampers and auxiliary air inlets to the furnace. In this way the company has found by test that it is possible to hold the boiler up to pressure for about ten hours by the radiant heat stored up in the furnace and boiler setting which is gradually absorbed by the boiler. The loss which occurs and which can be compared to the banking loss in a stoker-fired boiler is the heat radiated from the boiler and setting, an equivalent in amount to that required to heat up the boiler and setting again to the temperature attained when steaming. In a plant where the ratio of boiler hours to boiler steaming hours averages 43 per cent or greater, which corresponds to an average daily plant load factor of 67 per cent, the saving resulting from the use of pulverized coal is worth considering. Assuming 0.2 lb. of coal consumed per boiler-horsepower-banking hour in a plant equipped with underfeed stokers, this loss amounts to about 1.5 per cent. In a pulverized-fuel-burning plant the loss should be reduced to one-half this figure, resulting in a net saving of 0.7 per cent on this one item alone.

Conveying and preparation of pulverized fuel presents a somewhat more complex problem, although the present equipment of the Oneida Street station is operating satisfactorily, and, during the test, operated without any serious interruptions. Moisture in the pulverized fuel, caused by sweating on the inside of the pulverized-fuel bins, resulted in some feeder troubles, but this was only a temporary condition and was overcome by closing the windows just above the bins, stopping the cold air from blowing directly on them. One of the feeder pipes between the bin and the furnace became partly plugged up due to a paper composition gasket becoming lodged in the pipe just above the burner. The boiler on which this occurred was operated for at least 24 hours at the desired rating by increasing the feeder speed on the other burner until the trouble was located and removed. During this period the efficiency of combustion was undoubtely below the average, as the coal that did come through the plugged feeder was not fed in with the correct quantity of air. No trouble was experienced with the drier or pulverizing mills at any time during the test.

The writer does not believe that under test conditions over a period of constant boiler rating the efficiency obtained with the use of pulverized fuel will exceed that which has been obtained from the best stoker practice under similar operating condition. However, under normal operation it is believed that the elimination of the many variable conditions entering into stoker operation will result in higher efficiency for the pulverized-fuel installation. Over-all efficiencies of boiler, furnace and grate as high as 82 or 83 per cent have been obtained on test with stoker-fired boilers, but normal operation day in and day out seldom exceeds 76 per cent in the very best practice where highly skilled help is employed in supervising the boiler-room operation. The gross over-all efficiency of boiler and furnace of 79.6 per cent, as obtained from the results of this test, would unquestionably have been higher had the boiler been cleaned prior to the test. As a matter of fact each boiler had been in operation prior to the test approximately 600 hours since being entirely cleaned, including approximately 300 hours on three of the boilers since cleaning the first four rows.

Certainly, the results obtained in the pulverized-fuel burning plant as a whole, where the equipment was installed and made to fit an old plant originally equipped with Jones stokers, are encouraging enough to warrant serious consideration of the use of this kind of fuel in stations to be built in the future. There are many improvements which can be made in a new plant, especially in the design of the furnace, location of the drying and coal-pulverizing equipment, method of coal handling and pulverizing, method of ash handling, slag prevention, possibility of using waste gases for the drying of fuel, all of which will have an effect on the efficiency which may be obtained. The application of pulverized fuel to central generating stations has been in use to a very limited extent for several years, but there still remains much experimental work to be done before we can hope to exhaust all the possibilities for increased efficiency, and bring it to as high a state of development as is the stoker at the present time.

EDITORIALS

Fatal Accidents on Low-Voltage Circuits

WHEN alternating-current systems first began to come into use, and the voltages of these systems were increased to eleven hundred or twenty-two hundred volts, there was considerable discussion of the question of limiting, by law, all electric circuits to one thousand volts and under, since an electric pressure above this value was considered too hazardous to work around. However, such legislation never was passed, and the voltages of electric circuits have increased until at the present time circuits operating at one hundred and fifty thousand volts are in service and the two hundred and twenty thousand-volt system is a possibility of the near future. Fatal accidents have occurred on circuits of less than one hundred volts, yet the average electrical worker will handle one hundred and twenty and two hundred and forty volt circuits with a sense of safety. Some industrial authorities recommend that not over two hundred and forty volts should be used on direct-current motor circuits.

A report made by G. S. Ram, electrical inspector of factories in England, contains some very interesting data on this subject. Out of ninety-nine fatal electrical accidents occurring in England during the period from 1915 to 1918, inclusive, twenty were on circuits of one thousand volts and over, while the other seventy-nine were on circuits of five hundred and fifty volts and less. Fifty-nine of these were on circuits having voltages between five hundred fifty and three hundred forty-six, and the remaining twenty happened on circuits of two hundred and forty down to as low as one hundred and five volts. Of the sixty-nine accidents occurring on circuits of four hundred and forty volts and less, it is doubtful if any of the fatalities were due to pressures greater than two hundred and fifty volts, since the voltage to earth on any of these circuits was not above the latter value. Ninety-five of the total number occurred on alternating-current circuits. This is probably due to the wide use of alternating current for industrial purposes. Since a direct-current machine or starting device is not in general as well insulated to prevent getting in contact with bare parts as alternating-current equipment, there is always greater danger of receiving a shock from the former than from the latter. The direct-current machine always requires more or less attention while in operation, such as sandpapering the commutator and adjusting the brushes, which is not required on polyphase alternating-current machines. However, in all low-voltage alternating-current circuits there is always danger of the insulation between the primary and secondary of transformers breaking down and subjecting the secondary to primary volts to ground, a dangerous condition unless the secondary winding is grounded.

Since the probable voltage of the shock in sixty-six of the foregoing cases was between two hundred fifty and two hundred volts, it would seem that these voltages are as dangerous as five hundred and fifty volts. One of the greatest dangers of accidents from electric circuits is the disregard that a large majority of workmen have for low-voltage circuits, which develops from knowing that under ordinary conditions a serious shock will not be received from these circuits. However, if a low-resistance circuit to ground is made and the individual cannot readily get clear of the circuit, there is danger of as low as one hundred and ten volts or even less causing fatal accidents, therefore such circuits should always be treated as hazardous.

Pulverized Coal at Milwaukee

THROUGH many vicissitudes pulverized coal in the power plant has at least reached the stage where it may be pronounced a commercial success, although there still remains much experimental work to be done. The numerous difficulties that have blocked the way have been overcome gradually, and it has been proved beyond peradventure that coal in the powdered form can be burned to advantage over the most modern and recent methods of combustion. The general increase in the price of fuel has been the immediate incentive. Under test conditions the efficiency of burning is only a trifle higher than in the modern type of stoker, but in daily operation the elimination of the many variable conditions entering into stoker operation, the flexibility and the ease of control and other inherent advantages are reported to give to powdered coal a margin that will more than offset the higher cost of fuel preparation and the additional investment in equipment.

The glory goes to Milwaukee and in particular to the men who have struggled for two years to make the previous claims possible. It has not been an easy task, and there were times when the difficulties seemed almost insurmountable. Five times the furnaces were changed in shape and volume and additional provision made for the admission of air to improve combustion and protect the brickwork in the furnace before the present satisfactory results were obtainable. The furnace volume selected, 0.359 cubic foot per square foot of water-heating surface, is not excessive when it is considered that it is equivalent to the furnace and ash-pit of a stoker.

Re-equipping five boilers in an old plant made the work more difficult than it would have been with an original layout. The story is told on other pages of this issue. A résumé is given of the requirements and the advantages. There is a summary of a four-day test on all five boilers under average operating conditions, and to confirm the reliability of methods and figures, there is comment from a competent observer of the test attached to one of the largest and best-equipped stoker plants in the country. The proof of the pudding is in the eating, and in this case the verdict has been so favorable that the new Lakeside plant of the Milwaukee company will be equipped to burn pulverized fuel. As an initial installation it is to have eight 1,306-horsepower boilers, two 20,000-kilowatt generating units and a separate building for the preparation of fuel. The coal consumption will approximate seven hundred tons per day. The added investment in building and equipment for preparing the fuel over a stoker plant has been estimated to exceed twenty dollars per boiler horsepower, but the saving possible, based on improvements suggested by operation at Oneida Street station, indicates that this additional expenditure will be well worth while. A more exact statement of the possibilities will be available at a later date.

Referring again to the opinions of the men responsible for the progress made and of those who had an opportunity to inspect the plant at its present stage of development, it would appear that boiler-room operation when burning powdered coal is much simpler than with a stoker installation. The rate of steaming is controlled by varying the speed of the feeder motor and adjusting the damper for the variation in the quantity of flue gas. It is unnecessary to correct the holes and heavy spots that develop in the fuel bed of a stoker; and variations in kind and quality of fuel burned seem to have no effect on operation except that with a constant feed the rate of steaming will vary with the heat value of the coal. Losses inherent in stoker practice, such as breakdowns in the stoker itself, breaking up and loosening clinkers, watching the fire to maintain uniform thickness, dumping and many other operations, are eliminated. Efficient combustion is maintained without continual supervision by an experienced operator, and from the standpoint of reliability the odds have been placed in favor of pulverized fuel.

With coal and air supplies easily adjustable, almost perfect fire control and proper air distribution are obtained. This results in thorough combustion, uniformly high efficiency under normal operation and the ability to change from no load to two-hundred per cent overload in a few seconds. Regularity of action, making it possible to anticipate results, is a factor that appeals to the operator.

With powdered coal, banking a boiler is a simple process, requiring no additional fuel to hold the fire. It is necessary only to stop the fuel supply and close the dampers and auxiliary air inlets. The radiant energy in the brickwork is retained and will hold pressure on the boiler for hours.

Observation for a period of two years in the installation under discussion has demonstrated that the brickwork in a pulverized-fuel furnace will stand up equally as well as in a stoker installation, with the added advantage that ironwork is eliminated from the high-temperature zone of the furnace. Maintenance exists principally in the pulverizing plant. With machinery of the slow-moving type it is not excessive, although there is room for improvement.

Owing to the soft heat and no impingement of the heat on the tubes caused by a strong draft, the formation of scale in the boiler seems to be less. There is absolutely no trouble from smoke and no soot to reduce the heat-absorbing ability of the boiler, although there is some collection of very light ash. Much of the ash apparently floats out of the stack, leaving a comparatively small amount to be removed from the combustion chamber. It is interesting to note that the low moisture content of the coal as fired reduces the opportunity for the formation of sulphuric acid, so destructive to all ironwork with which the flue gases come in contact.

That there has been no serious trouble from coal-dust explosions is attributed to proper care in preventing the coal from being dried too much or pulverized to a fineness beyond the requirements. The drying process requires careful supervision and ability in the operator to recognize and meet the varying properties of the coal as it arrives. Drier operation must be changed to meet variations in size and moisture if a uniform product is to be delivered. Coal that is too dry increases the danger from explosion, and too much moisture induces a tendency toward clogging the system at points most readily affected. At Milwaukee moisture has been one of the bugbears encountered.

No claim is made that perfection has been reached in the present installation or that other designs of equipment might not work equally well or possibly better. Valuable experience has been gained, however, and the next installation will contain a number of improvements. It is felt that something might be done toward reducing the cost of preparing the fuel, especially along the lines of utilizing waste gases for drying in plants not provided with an economizer. A furnace that would burn a large percentage of finely divided coal and still care for a smaller amount of larger-sized particles would be a decided advantage, as it would reduce the work of the pulverizer, which requires about seven-eighths of the total electrical energy consumed by all moving parts of the coal-preparation equipment, and also curtail the labor in this department.

Admission of excess air to prevent slag formation and the destruction of the brickwork tends to lower the CO_2 and reduce to some extent the efficiency of combustion, or at least the absorption by the boiler due to the lower heat head. There is an opportunity here for improvement. A more thorough mixing of the air and fuel in proper proportions before entrance into the furnace may be the solution.

The proper furnace volume per square foot of steam-making surface or per unit of coal burned, the position of the furnace relative to the boiler, the location of the burner, whether vertical or horizontal, the best methods of mixing coal and air, location of equipment, etc., are all factors requiring more study. These are the problems of the future. Suffice it to say that the plant which has been made a success at Milwaukee, has paved the way. The pioneer work has been done. There remain perfection of detail and experimental work to exhaust the possibilities for increased efficiency.

An interesting point is the conclusion that a plant developing less than twenty-five hundred boiler horsepower on a twenty-four hour operating basis should not consider using powdered coal. The reasons advanced for this limitation are the cost of installation, the relatively small amount of coal pulverized per day and the proportionately large labor charge for operating the preparation plant.

A few weeks ago in Milwaukee a federation of technical engineers, architects and draftsmen pledged themselves to uphold and sustain the principles adopted by the American Federation of Labor. They were determined to have a voice in the adjustment of disputes in all industries of which they were a part and use, if necessary, the powerful backing of labor. The reward of individual merit, coöperation or the policy of "suggestions" for mutual benefit, had no place in their sympathy. In brief, force is the guiding spirit. On the economic position of the engineer the American Association of Engineers takes the attitude expressed on other pages of this issue. Being a constructive agent, it is their belief that the engineer should be opposed to strikes, which in their very nature are destructive. For just reward of individual effort he should depend upon the justice of the facts, presented collectively if necessary, upon enlightenment of public opinion, upon loyalty between employer and employee and upon the fundamental desire of the great majority to do what is fair and right when the merits of the case are clearly presented. For a dignified profession there is only one choice between the tenets of the two organizations.

Engine Running Twenty Years

On page 668 of the Oct. 28-Nov. 4, 1919, issue of *Power,* Mr. Molloy states how satisfactorily his engine runs and mentions the fact that he does not intend to overhaul it if it does not need it.

I presume Mr. Molloy has lately run a test on this engine or keeps a close record of operating costs, and that his statement is based on the unit's performance. No doubt these data would prove interesting reading in comparison with the same relative values obtained with more up-to-date equipment, and I am sure that many readers besides myself would be glad to see these figures in *Power* if Mr. Molloy could be induced to submit them for publication. ROBERT W. RIORDAN.

Brooklyn, N. Y.

When the Ground Indicators Deceived

In a large generating station each alternator had two sets of ground lamps connected across its exciter; one set on the main switchboard, the other on a power-supply switchboard on the turbine-room floor. Each set consisted of two 250-volt lamps in series, with the wire common to both lamps grounded, as in Fig. 1. Shortly before peak-load time one afternoon, the switchboard

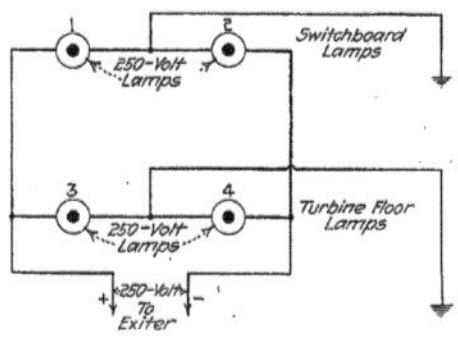

DIAGRAM OF GROUND-DETECTOR LAMPS

ground lamps of one of the units began to flicker, and a warning was given that the unit had a swinging ground on its field.

The ground seemed to clear itself in a short time, only to immediately return and become solid. The filament of lamp 1 was barely red, while lamp 2 was at full brilliance, indicating a positive ground. Both lamps normally burned at half brilliance since, being connected in series across a 250-volt circuit, they operated at only half voltage. At the time of the trouble no one gave thought to the scheme of connections, or remembered to check the switchboard lamps against the set in parallel with them on the turbine floor.

While arrangements were being made for a reduction of load on the station, in anticipation of trouble with this unit, an electrician passing the turbine-room lamps noticed lamp 4 dead and lamp 3 dim. He twisted the dead lamp in its socket and both sets of lamps came back to normal. Vibration had loosened lamp 4 in its socket, causing the other three lamps to flicker; and when the circuit was broken at lamp 4, lamps 1 and 3 became dim and lamp 2 bright. FRANK GILLOOLY.

Philadelphia, Penn.

Leaky Safety-Valve Seats

If a safety valve leaks after having been reseated or ground in, an examination may show that the leakage occurs between the seat and the valve body. When the valve is cold, the seat may be found to be quite loose. This may be accounted for by the fact that the unequal expansion of the bronze seat and iron body compresses the seat. If the seat seems to be tight, it is a good plan to invert the valve and fill it with water in order to locate any leakage.

The remedy for a loose seat is to expand it with an ordinary tube expander. If the seat is of sufficient length, the expanding should be done below the top of it. Afterward the valve should be reseated and ground in in the usual manner. Some types of globe and angle valves can be repaired in the same manner.

Providence, R. I. C. P. LAWTON.

Why Does the Water in a Boiler Lift When the Safety Valve Blows?

Referring to Mr. Fryant's inquiry in the Nov. 11-18, 1919, issue, would say that I have given the subject of water lifting in a boiler much thought and study, and it is my opinion that high or low velocity of the steam leaving a boiler has no effect on the moisture carried over; but it is the quantity of steam delivered in a unit of time, together with the liberating surface of water. The smaller the drum, or shell in which the steam is liberated from the water in proportion to the quantity of steam delivered, the greater the moisture. Then the position of the nozzle or outlet flange on the drum or drums of water-tube boilers has quite an effect on the moisture, as the farther the steam travels before it enters the outlet the drier it is. The kind of dry pipe also has quite an effect. Some dry pipes are collectors of moisture instead of eliminators. I have taken some dry pipes out over nine feet long, which delivered very wet steam.

The height of the water line in water-tube boilers in which the drums are set on an incline affects the moisture, and the higher the water is carried (within reason) the drier the steam. This is caused by the steam and water spouting up through the water in the drum when the water is low, which it cannot do when the water is high. The steam outlet is generally right over where the water and steam is liberated from the tubes in this type of boiler, therefore the steam has a very short travel to the outlet and does not have time to drop its moisture.

There is no doubt about the steam being drier where the velocity in the pipes is high and the drop in pressure is high, because this drop causes the entrained water to be evaporated by the excess heat in the steam over what it was at a higher pressure, just as is done in a throttling calorimeter. J. T. FENNELL.

Philadelphia, Penn.

Power-Plant Log Sheet

In the Sept. 16, 1919, issue of *Power* there was published an article entitled "Boiler- and Engine-Room Record Sheets." This article described and illustrated four record sheets designed by the author for use in the boiler and engine rooms of power plants under his supervision.

Later, it has developed that with a large number of plants reporting to one supervising officer, the amount of clerical work, checking and figuring on four reports could be dispensed with in favor of the revised form shown in Fig. 1, which has been put in use to replace the four forms previously mentioned.

Fig. 2 shows a running record form by months and years of the fuel consumed, value and per cent of increase or decrease in fuel per year, for comparison of four years. The figures are transmitted from the log

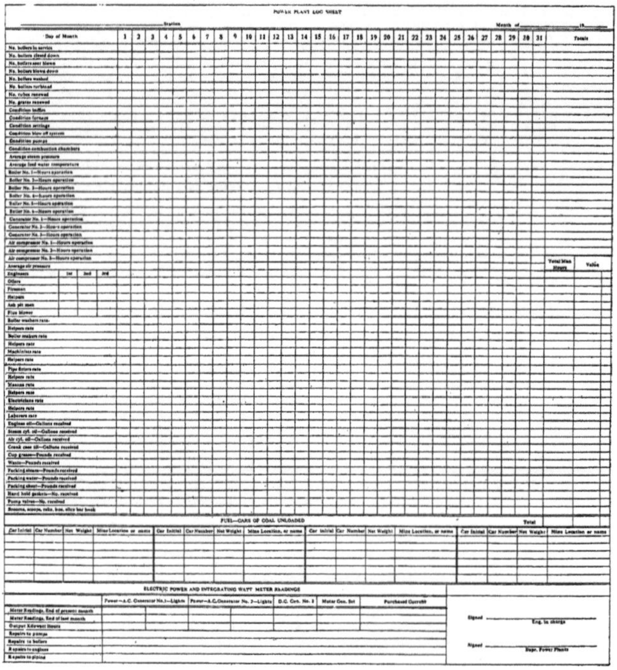

POWER PLANT LOG SHEET

______ Station Month of ______ 19__

Day of Month	1	2	3	4	5	6	7	8	9	10	11	12	13	14	15	16	17	18	19	20	21	22	23	24	25	26	27	28	29	30	31	Totals
No. boilers in service																																
No. boilers closed down																																
No. boilers soot blown																																
No. boilers blown down																																
No. boilers washed																																
No. boilers turbined																																
No. tubes renewed																																
No. grates renewed																																
Condition baffles																																
Condition furnace																																
Condition settings																																
Condition blow off system																																
Condition pumps																																
Condition combustion chambers																																
Average steam pressure																																
Average feed water temperature																																
Boiler No. 1—Hours operation																																
Boiler No. 2—Hours operation																																
Boiler No. 3—Hours operation																																
Boiler No. 4—Hours operation																																
Boiler No. 5—Hours operation																																
Boiler No. 6—Hours operation																																
Generator No. 1—Hours operation																																
Generator No. 2—Hours operation																																
Generator No. 3—Hours operation																																
Air compressor No. 1—Hours operation																																
Air compressor No. 2—Hours operation																																
Air compressor No. 3—Hours operation																																
Average air pressure																																Total Man Hours / Value
Engineers (1st, 2nd, 3rd)																																
Oilers																																
Firemen																																
Helpers																																
Ash pit men																																
Flue blower																																
Boiler washers rate																																
Helpers rate																																
Boiler makers rate																																
Helpers rate																																
Machinists rate																																
Helpers rate																																
Pipe fitters rate																																
Helpers rate																																
Masons rate																																
Helpers rate																																
Electricians rate																																
Helpers rate																																
Laborers rate																																
Engine oil—Gallons received																																
Steam cyl. oil—Gallons received																																
Air cyl. oil—Gallons received																																
Crank case oil—Gallons received																																
Cup grease—Pounds received																																
Waste—Pounds received																																
Packing steam—Pounds received																																
Packing water—Pounds received																																
Packing sheet—Pounds received																																
Hand hold gaskets—No. received																																
Pump valves—No. received																																
Brooms, scoops, rake, hoe, slice bar hook																																

FUEL—CARS OF COAL UNLOADED Total

Car Initial	Car Number	Net Weight	Mine Location or name	Car Initial	Car Number	Net Weight	Mine Location, or name	Car Initial	Car Number	Net Weight	Mine Location, or name	Car Initial	Car Number	Net Weight	Mine Location or name

ELECTRIC POWER AND INTEGRATING WATT METER READINGS

	Power—A.C. Generator No. 1—Lights	Power—A.C. Generator No. 2—Lights	D.C. Gen. No. 1	Motor Gen. Set	Purchased Current
Meter Readings, End of present month					
Meter Readings, End of last month					
Output Kilowatt Hours					
Repairs to pumps					
Repairs to boilers					
Repairs to engines					
Repairs to piping					

Signed ______ Eng. in charge

Signed ______ Supt. Power Plants

FIG. 1. A REPORT SHEET THAT DISPENSED WITH FOUR REPORT SHEETS sheet and are inserted monthly for ready reference of the management, to show at a glance what is being done each month in the matter of fuel economy. The load at all plants is not of such variable nature as to have much effect on fuels consumed.

Fig. 3 shows a statement of money expended for power-plant labor, together with the number of man-

POWER PLANT FUEL STATEMENT												
Station	Year	Gross Tons Jan.	Value	...e	Gross Tons Dec.	Value	Gross Tons Year	Value for Year	Gross Tons Increase or Decrease	Value of Increase or Decrease	Percent Increase	Percent Decrease

FIG. 2. RUNNING RECORD FORM BY MONTHS AND YEARS

POWER PLANT OPERATING STATEMENT												
Station	Year	Man Hours Jan.	Value	...ue	Man Hours Dec.	Value	Total Man Hours	Total Value	Increase or Decrease Man Hours	Increase or Decrease Value	Percent Increase	Percent Decrease
NOTE—Supplies, Material and Fuel Not Included												

FIG. 3. MONTHLY LABOR EXPENSE SHEET

hours required. This information is also obtained from the log sheet.

Checking statements such as this, together with similar statements, are helpful in indicating just what each plant is doing, and with recording flow meters, pressure and temperature recorders, etc., an efficient operation can be maintained.

Hagerstown, Md. PAUL R. DUFFEY.

Blowing Down Boilers

I read L. A. Cole's letter in the Oct. 21, 1919, issue of *Power*, page 265, and from my point of view and experience Mr. Cole displays very good boiler logic. The blowing down of a boiler should be governed according to the conditions of the water. Water that is not treated before it is fed to the boiler will naturally deposit more dirt, scale and sediment than water that does not have to be treated. Such loose scale and dirt as is liberated in the boiler should be removed, and the way to do this, when a boiler is being worked up to its capacity, is to blow down at least every three hours.

If a suitable boiler-feed water is used, blowing down once a day is sufficient; otherwise the boilers should be blown down at such intervals as will remove such mud, scale and sediment as will flow to the blowoff valves under an ordinary blowdown. There can be no fixed rule as to the amount of water that should be blown from a boiler at one time.

The noise mentioned by Mr. Cole in the blowoff pipe when he first begins to blow down is caused by an accumulation of hard pieces of scale around the blowoff pipe cutlet, and when the valve is first open this accumulation of scale is blown out through the pipe.

Brenham, Tex. H. W. ROSE.

Home-Made Baffle-Wall Mixture

I have read from time to time of trouble with boiler baffles, and I have had difficulty in keeping them from burning away at the first pass up to the fourth and fifth rows of tubes. We bought about everything on the market that was supposed to be able to stand up, but could not get the desired results. I therefore took it upon myself to create a mixture that has stood a very severe test, as our boilers are forced much above their rating for about six hours each night.

I used a mixture made up as follows: About 200 lb. of clinkers that were thoroughly burned and showed symptoms of having been heated to the melting point were crushed to a size not larger than rice coal. Then about 100 lb. of dry sand was thoroughly mixed with the ground clinker. Next I obtained 50 lb. of glass and put it into a box and powdered it as fine as possible. This was then mixed with the sand and cinders, after which 50 lb. of good-quality portland cement was added, and after being thoroughly mixed about 25 lb. of fine dry asbestos was added. The batch was turned over and over until the materials were thoroughly mixed in a dry state.

When ready to apply the mixture, a stick of ¾-in. stock as shown at *A* was obtained, in the head of which No. 10 nails were driven about two inches apart and

APPLYING THE BAFFLE MIXTURE

all the way through. The mixture was then wet just enough to make a stiff paste and a wad placed on the nails, run up between the tube and slapped into place, and the operation repeated until the job was finished. After a portion of the baffle was covered I used a tool made of a block of wood 2 in. wide, 5 in. long and 2 in. thick, having a handle about 3 ft. long, as shown at *B*. A bamboo fishpole makes a good handle. This was passed up between the tubes and used to hammer the mixture in tight and of a smooth and uniform thickness.

New Rochelle, N. Y. B. GADDIS.

Weld-Testing Device

In order to test welds in boiler tubes, the device illustrated was made. It consists of two pistons, made as shown, one to come on each side of the weld. A rubber hose is attached to the rod projection, and com-

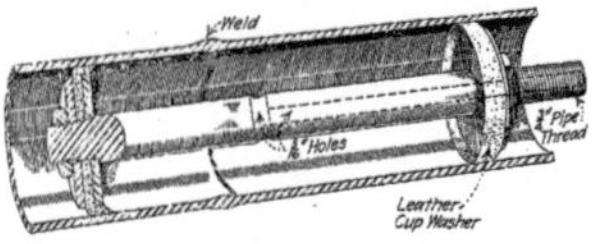

DESIGN AND APPLICATION OF TUBE TESTER

pressed air is admitted through the hollow rod and outlets. Thick soapsuds are applied to the tube over the weld, and if a leak is present bubbles will form. No defective welds have got past since the tester has been used, whereas before its use several tubes had to be removed after they had been put in the boiler and steam raised. R. McLaren.

Toronto, Canada.

Commutating-Pole Motor Trouble

A new commutating-pole shunt motor was installed to drive a centrifugal water pump. The motor was rated at 21.5 amperes, but frequently, when the machine was started, the 30-ampere fuses protecting the motor were blown. Investigation showed that the motor

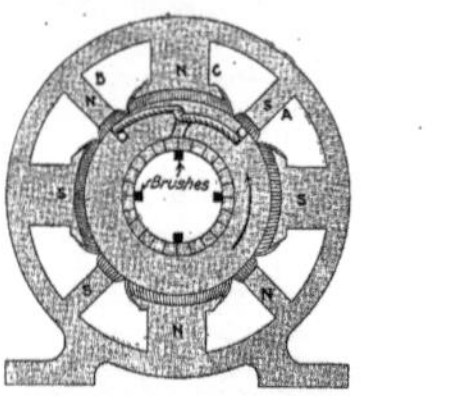

FIG. 1. INTERPOLE MOTOR WITH BRUSHES LOCATED IN CORRECT POSITION

speed was too high. The trouble was overcome by shifting the brushes forward, thus reducing the speed and current to their proper values. It is particularly important that the speed of a centrifugal pump be correct, as the power required to drive it varies with the cube of the speed. It is also important that the brushes of a commutating-pole motor be properly set. A greater effect is produced in a commutating-pole motor than in a noncommutating pole by incorrect brush setting. The theory of this change of speed in an interpole motor is as follows:

When the brushes are in the correct position, as shown in Fig. 1, the coils on the armature between commutating poles *A* and *B* are cutting flux from the main pole *C*. If they cut any of the flux from the commutating pole *N*, they also cut an equal number of lines of force from the *S* commutating pole, hence the voltage in these coils is not influenced by the commutating flux. But if the brushes are shifted backward to a position as in Fig. 2, the coils between the sides of the coil under commutation will be cutting the flux of the main *N* pole *C* and also from the flux from the *S* commutating pole *A*, thus decreasing the effective volts gen-

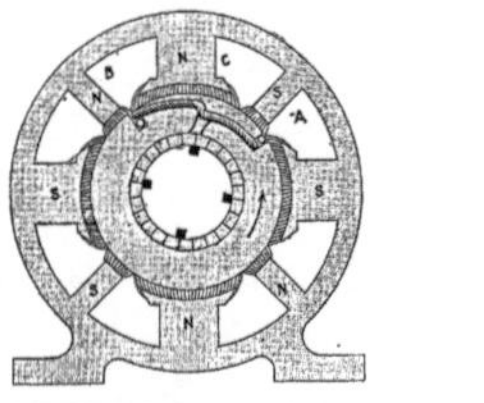

FIG. 2. INTERPOLE MOTOR WITH BRUSHES LOCATED OFF CORRECT POSITION

erated in these coils and causing the speed to increase. If the brushes had been shifted forward, the speed would have been decreased.

Washington, D. C. H. L. Hervey.

Removing Retaining Wedges and Coils from Induction Motors

The following method for removing slot wedges and coils from stators has been used with considerable success. Nearly all insulating varnishes and paints are softer when warm than when cold, hence the coils and wedges can be removed more easily if the frame is heated. To loosen the wedge a drift of iron or fiber about $\frac{1}{32}$ in. narrower than the slot and five or six

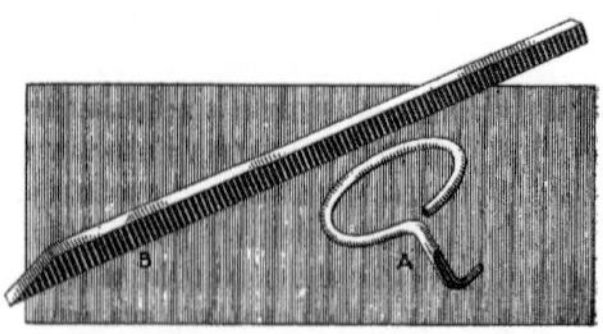

TOOLS FOR REMOVING COILS

inches long is placed on top of the wedge and tapped with a hammer. After the wedge has been loosened over its full length, it can be driven out by driving on its end with a metal drift.

In removing the coils the greatest difficulty is with the first one. In order to start this coil the hook tool, as shown at *A*, is used to lift the coil ends as high as they will go. The pointed tool *B* is then inserted under the coil at the end of the slot, and the coil is raised.

Washington, D. C. H. L. Hervey.

INQUIRIES OF GENERAL INTEREST

Enlargement of Crankshaft Diameter—Why are crankshafts of engines made larger at the center than at the ends? J. W. H.

Crankshafts have enlarged diameters at the middle of their length to afford greater strength and stiffness in resistance of bending stresses due to the weight of the flywheel, and tension of belt or thrust of driving gears, and also to compensate for the reduction of strength from removal of material for a keyway.

Handholes for Vertical Fire-Tube Boilers—How many handholes should there be in a vertical fire-tube boiler? G. L. S.

Boilers more than 24 in. in diameter should have not less than seven handholes; namely, three in the shell at or about the line of the furnace crown sheet, one in the shell at or about the water line or opposite the fusible plug when used, and three in the shell at the lower part of the water leg.

Additional Steam for Increased Back Pressure—How does increase of back pressure affect the quantity of steam required by an engine? J. R. A.

For the same load on the engine each pound of additional average back pressure will require an additional pound of average forward pressure, so that to meet an increase of back pressure, the supply of steam will have to be increased as much as though the load had been increased to require the same additional mean effective pressure.

Forward Pressure, Back Pressure, and M.E.P.—What is meant by the terms average pressure, mean forward pressure, mean back pressure and mean effective pressure of an engine? W. P. S.

The word "mean" is to be understood to signify simple average. The pressure that urges a piston forward in the direction that it travels in performing useful work is called forward pressure, and the average forward pressure is frequently designated as the mean forward pressure. Pressure that opposes the motion of the piston in performance of useful work is called back pressure, and the average back pressure is also called the mean back pressure. The difference between the mean forward pressure and the mean back pressure is called the mean effective pressure.

Carbon Brush Wear—In one of our direct-current motors that has four brushes on the commutator, two of the brushes wear away faster than the other two. What would cause this uneven wear when all four brushes are of the same grade? T. A.

The brushes from which the current flows into the commutator always wear away faster than those through which the flow is from the commutator. The brushes through which the current flows into the commutator would be positive on a motor and negative on a generator. The additional wear on these brushes is due to small particles of carbon carried by the current from the brush contact surface and deposited on the commutator. If two of four brushes, of the same polarity, were to wear away faster than the other two, a difference in resistance in the shunts from the brushes to the brush-holder, or a difference in the tension on the brushes, would cause an unequal distribution of the current between the brushes and unequal wear.

Starting-Compensator Oil—What is the best grade of oil to use in starting compensators employed on 220-volt induction motors ranging in size from 10 to 75 hp. Would a poor grade of oil cause the contacts to burn? M. C. R

In compensators employed for starting squirrel-cage motors a mineral oil having a high flash point should be used; it should be free from any trace of alkali or acid and have a low factor of evaporation. Under operating conditions the oil should not form deposits in the tank. One of the surest ways of securing the proper grade of oil is to obtain it from the manufacturers of the compensators. The design of the contacts, operating conditions and application of the compensator in general have a much greater influence on the burning of the contacts than the grade of oil. If the contacts are designed to give a rubbing make-and-break and the compensator is not abused either through improper handling or misapplication, little trouble should be experienced with burning and pitting, although a certain amount of this will occur under the best of conditions. In general the oil should be changed when it turns dark in color and sediment forms in the bottom of the tank.

Flashing Point of Fuel Oil—What is meant by the flash point of a fuel oil, and what do the "open" and "closed" tests consist of? W. T. D.

The flashing point of an oil is that temperature to which the oil must be heated, at a specified rate, to cause it to give off a vapor which will ignite when mixed with air and exposed to a flame near the surface of the oil. The "open test" consists in heating some of the oil, in which the bulb of a thermometer is placed, in a small open metallic cup, a porcelain crucible embedded in sand or some equivalent contrivance, heated by application of a gas flame or other source of heat that is under the control of the experimenter so that the temperature of the oil increases at the rate of 10 deg. F. per minute. A small flame from a taper or gas tube is passed across the surface of the oil for every five degrees rise in temperature until the vapor ignites or flashes. The temperature at which this flash is first observed is taken as the flashing point of the oil. The open-cup method is beset by uncertainties. The temperature at which the first flash is obtained depends on the presence or absence of air currents, the actual rate of heating, the size and shape of the vessel containing the sample, the distance of the flame from the oil surface and the skill of the experimenter.

The closed test, as the name implies, is performed with apparatus from which air currents are excluded and in which the oil and vapor are stirred and the heat applied may be so regulated that when the temperature nears the flash point, the rise in temperature is so reduced that the testing flame may be quickly applied to the surface of the oil every two degrees rise of its temperature until the flashing point is reached.

Open-cup testing usually gives 5 to 25 degrees higher flash point than closed-cup testing, and the presence of much moisture causes irregular and incorrect results by either method of testing.

[Correspondents sending us inquiries should sign their communications with full names and post office addresses. This is necessary to guarantee the good faith of the communications and for the inquiries to receive attention.—Editor.]

Emergency Fleet Corporation Water-Tube Boilers for Wood Ships*

By F. W. DEAN and HENRY KREISINGER

This paper describes and reports tests upon the Standard water-tube marine boilers planned by the United States Shipping Board, Emergency Fleet Corporation. It also reports the results of special investigations of the mixture of air and combustible gases in the furnace and of the temperature of the gases.

IN 1917 the Emergency Fleet Corporation intended to build one thousand wood ships of 3,500 tons dead-weight capacity. Each ship would require two boilers containing about 2,500 sq.ft. of heating surface each. The problem of securing two thousand good-sized boilers was a serious one, and naturally caused careful consideration of the relative merits of the water-tube and of the Scotch boiler. It was finally decided to use water-tube boilers, and to show the saving in steel caused by this decision, the weight of that metal in one of these boilers, with casing, is 41,200 lb., while for the equivalent Scotch boiler it is 110,000 lb. Although finally only 1,352 boilers were ordered, the saving of steel by the use of the water-tube type was more than nine million pounds. Furthermore, the adoption of the water-tube type rendered most of the inland boiler shops of the country available for this work, since this type could be transported easily. Another great advantage of this design came from the fact that all the boilers in the first lot of 706 units were alike, and differed only slightly from the later orders of 646 units, which were also alike. The differences in the design came from changing the number of baffles from three to four and in using Key handhole caps instead of plugs with copper ferrules.

The boiler as first designed is shown in Fig. 1. It consists of two headers each composed of two plates connected at the edges by channel-shaped pieces, the two headers being connected by tubes. The front header is surmounted by a saddle, which is riveted to both the steam drum and the header, and holes in the bottom of the drum, within the limits of the saddle, furnish the means of connecting the interior of the drum and the header, and two rows of so-called circulating tubes connect the upper part of the back header with the drum and serve to conduct the steam to it.

The header plates were stayed by means of hollow iron stay-bolts, the holes being ¾ in. in diameter. The handholes of the first 706 boilers were closed by means of taper plugs surrounded by thin copper ferrules. The plugs had a threaded shank secured by a yoke and nuts, as usual for handhole plates. While these plugs have in general given satisfactory service, it was recognized that the Key cap is better, and all boilers afterward ordered were provided with these caps.

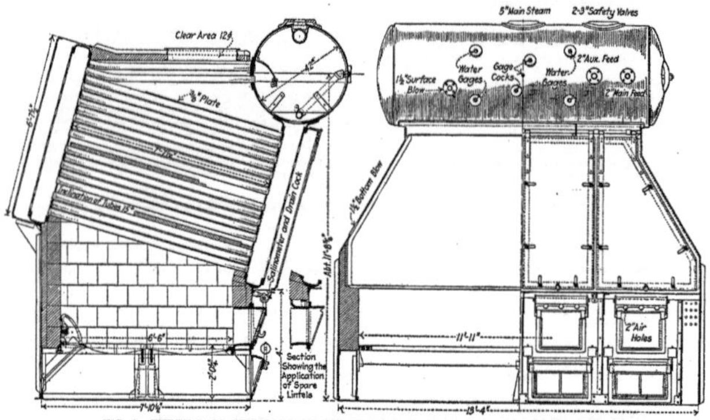

FIG. 1. THREE-PASS STANDARD WATER-TUBE MARINE BOILER FOR WOOD SHIPS

The drum was made with the longitudinal joint where the circulating tubes entered, there being here an inside and an outside strap, and the expanding occurred only in the straps, since the holes in the shell are larger than the tubes. The tubes are seamless hot-rolled steel. The baffles are of the longitudinal type, of which the first boiler mentioned has three, forming three passes, as shown in Fig. 1. After some testing, it was decided to place the lowest baffle on the lowest row of tubes in order to render the second and third rows of tubes more active as heating surface. When this change was being made, it was seen that there was room for four passes, and accordingly four baffles were inserted. This rendered the boiler somewhat more efficient, especially at the higher powers.

In the three-pass boiler the upper baffle was of steel plate and the others of tiles, but in the four-pass boiler the two upper baffles were of steel, and the indications are that the others might be of steel also if the tubes were sufficiently near together to touch the baffles on both sides. By using the steel

*Abstract from a paper read before the American Society of Mechanical Engineers at the Annual Meeting, December, 1919.

baffles in the four-pass boiler, 34 more tubes could be put in. In the drum of the four-pass boiler there is the usual deflector-plate, which compels the steam as it comes through the circulating tubes to pass to the end of the drum. This deflector was removed for a time, and no difference in the behavior of the boiler could be observed except that the water as it appeared in the glass was livelier. In a seaway other advantages might appear. In the top of the drum was a per-

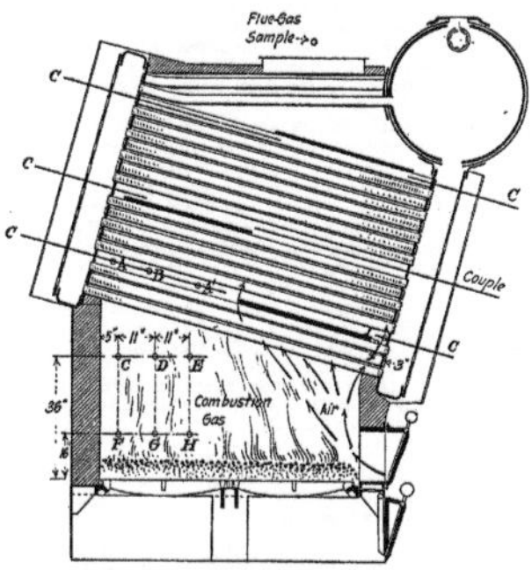

FIG. 2. DIAGRAM SHOWING METHOD OF TESTING THREE-PASS BOILER

forated dry pipe having eighty-eight ⅜-in. holes, and tests with and without this dry pipe proved that it had an important effect in separating moisture from the steam. At ordinary rates of working the loss in pressure amounted to little, but when forcing, it occasionally reached to six or eight inches of mercury for a short period.

The casing of the boiler is made up of steel plates joined together by external and internal shapes secured to a channel frame at the bottom, which rests on the keelson of the ship. At the front there are two vertical channels acting as columns, which are secured to the channel frame and bolted to tees riveted to the bottom of the header. The rear header rests on a shelf formed in the casing. The front column forms the only rigid connection to the channel base, and the boiler is free to slide in the rear part of the casing. In the front and back of the casing are sheet-iron doors covering the headers, and soot blowers are installed between the headers and the doors.

In the three-pass boiler there are four fire-doors and four ash-pit doors, but in the four-pass boiler there are three of each. These doors are of the inswinging type. Above each is a hollow perforated lintel with air entering through holes in the casing, and on the sides of the door there are perforated jambs with air supplies from the outside. These air openings are all useful in protecting the parts from burning and they also promote economy, for none too much air enters the furnace through them.

The grates are of the fixed type of double bars in two lengths and have ½-in. air and ½-in. iron spacing. Across the back end of the furnace there is a cast-iron bridge-wall with narrow air spaces, which saved nearly 2 per cent of coal, perceptibly diminished smoke, improved gas analysis, saved the brickwork and reduced the time and labor of cleaning fires nearly 50 per cent.

The leading dimensions of the boiler are given in Table I.

The tests described here were made for the purpose of studying the behavior of the boiler and, if possible, improving it. In connection with this work the Bureau of Mines has named Henry Kreisinger to take charge of the high-temperature measurements and the various investigations in connection with the paths and velocity of the gases, the draft magnitudes, losses through the passes, the advisable quantity of air and position of its admission.

Georges Creek, Cumberland coal from the Big View vein, was used on all tests in order to have a standard coal of uniform quality. Tests were made with fixed grates, shaking grates, without a bridge-wall, with the iron bridge-wall and with the brick bridge-wall covering the same area. Tests were also made with several kinds of oil furnaces using Mexican oil.

By the addition of the iron bridge-wall, admitting air around the fire-doors, lengthening and otherwise changing the baffles and studying the method of firing, the efficiency of the three-pass boiler was raised from about 60 per cent to about 71 per cent based upon dry coal, and the efficiency of the four-pass boiler was about 72½ per cent based upon dry coal.. These tests were made when firing by hand.

Results of the Tests

Tables II, III, IV and V show the general results of the tests of the three-pass and four-pass boilers under various conditions, both with hand and stoker firing.

Tests were run at various rates to determine the effect of forcing on the economy. In one the four-pass boiler was worked at 131 per cent of marine rating and 229 per cent of land rating. Even at this high rate the efficiency was about 71 per cent and but little below that of lower rates. With stokers the boiler was operated at 162 per cent of marine rating and 282 per cent of land rating, and at this rate the efficiency was 73.3 per cent based upon combustible. The test

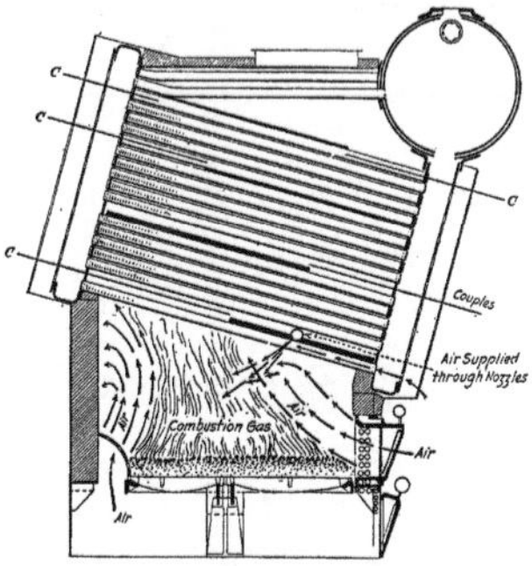

FIG. 3. DIAGRAM OF FOUR-PASS BOILER SHOWING METHODS OF INTRODUCING AIR OVER FUEL BED

of Apr. 8 shows a very high efficiency, but this is open to suspicion because the heat balance did not work out quite properly, due, it is thought, to an error in the coal weight of one hour.

The conclusion arrived at is that the boiler is of excellent efficiency and that the four-pass boiler is well adapted to overload. The only defect of the latter is, as might be expected, that there is considerable absorption of draft in the passes, and the greatest loss was in the third pass from the bottom.

As an experiment half of the circulating tubes and half the number of holes in the bottom of the drum were plugged, together and separately, to determine the effect on the circulation, but no effect could be observed. The conclusion was that an unnecessary number of such tubes and openings are used in this type of boiler and by reducing them the drum will be made a safer structure.

The quality of the steam was most satisfactory when water was carried at a proper level, and it was necessary to carry it near the top of the glass before the limit of a throttling calorimeter was reached.

TABLE I. DIMENSIONS OF BOILER

Width of casing at floor level	13 ft. 4 in.
Length of casing at floor level	7 ft. 10½ in.
Height of center of drum above floor	11 ft. 8 5/16 in.
Thickness of header plates	11/16 in.
Width of water spaces of headers	8 in.
Outside diameter of tubes	3 in.
Exposed length of tubes between headers	7 ft. 7½ in.
Number of tubes between headers	388
Number of tubes between rear header and drum	21
Inside diameter of drum	42 in.
Thickness of drum plates	9/16 in.
Width of furnace	12 ft.
Depth of furnace	6½ ft.
Height of furnace at center	3 ft. 8 in.
Firing doors	15 in. by 18 in.
Width of grate	11 ft. 11 in.
Depth of grate without bridge-wall	6 ft. 6 in.
Depth of grate with bridge-wall	5 ft. 8 in.
Grate area without bridge-wall	77½ sq. ft.
Grate area with bridge-wall	67½ sq. ft.
Heating surface fire sides	2518 sq. ft.
Thickness of brick lining	8½ in.

The Preliminary Tests

The preliminary tests to determine the best arrangement of baffles and distribution of air inlets consisted principally of a study of combustion by means of gas analyses and temperature readings. The gases were sampled with water-cooled sampling tubes and collected in glass holders over mercury. Each group of samples was collected simultaneously over a period varying from 15 to 40 minutes. The samples collected at the base of the stack were analyzed with the Orsat apparatus for CO_2, O_2 and CO. The gas samples taken in the furnace were analyzed by Hempel's method, for CO_2, O_2, CO, H_2, CH_4, and unsaturated hydrocarbons.

Fig. 2 is a diagram of the three-pass boiler as originally designed. After the first few tests the baffles were extended, making smaller gas passages between their ends and the water-legs. Other changes were also made as the investigation indicated the need for them. The points at which the gas samples were collected are indicated by the small circles designated by the capital letters *A* to *I*.

During the first six tests a large amount of combustible gas passed out of the furnace and either burned at the base of the stack or escaped unburned. It was apparent that not enough air was admitted over the fuel bed and that the air which was admitted was not sufficiently mixed with the gases to cause them to burn completely in the furnace. Some of the air admitted through the firing door found its way out of the furnace through the opening at *I*, between the first baffle and the front water-leg, as indicated by the arrow in Fig 2. A large part of the air also passed around the ends of the first baffle without mixing thoroughly with the combustible gases rising from the fuel bed. Evidently, the combustion

TABLE II. RESULTS OF EVAPORATIVE TESTS OF THREE-PASS BOILER EQUIPPED WITH WAGER BRIDGE-WALL AND HAND-FIRED

1	Date, 1918	June 12	June 13	June 14
2	Duration, hours	10	10	10
	Dimensions and Proportions			
3	Grate area, sq.ft.	65.54	65.54	65.54
4	Heating surface, sq.ft.	2,500	2,500	2,500
5	Ratio heating surface to grate area	38.20	38.20	38.20
	Average Pressures			
6	Gage pressure, lb.	179	178	179
7	Atmospheric pressure, lb.	14.28	14.41	14.41
8	Absolute pressure, lb.	193.28	192.41	193.41
9	Draft between damper and boiler, in.	0.53	0.58	0.63
	Average Temperature			
10	External air, deg. F.	59	64	64
11	Fire room, deg. F.	74	77	70
12	Feed water, deg. F.	204	209	207
13	Escaping gases, deg. F.	538	543	551
	Fuel			
14	Moist coal consumed per hour, lb.	1,455	1,411	1,521
15	Moisture in coal, per cent	2.19	2.39	2.02
16	Dry coal per hour, lb.	1,423	1,380	1,490
17	Dry refuse per hour weight, lb.	151	143	159
18	Dry refuse per cent	10.60	10.35	10.65
19	Combustible per hour, lb.	1,272	1,237	1,331
	Quality of Steam			
20	Moisture in steam, per cent	0.51	0.62	0.41
	Heat Value of Coal and Efficiency			
21	Heat value of a pound of dry coal, B.t.u.	14,380	14,396	14,350
22	Efficiency of boiler based on dry coal, per cent	70.30	71.50	70.80
23	Efficiency of boiler based on combustible, per cent	72.00	73.20	73.00
	Water			
24	Water supplied to boiler per hour, lb.	14,032	13,922	14,780
25	Dry steam generated, lb.	13,962	13,836	14,719
26	Factor of evaporation	1.057	1.052	1.054
27	Equivalent evaporation from and at 212 deg., lb.	14,758	14,555	15,514
	Evaporative Performance			
28	Water evaporated per pound of dry coal, lb.	9.81	10.05	9.90
29	Equivalent from and at 212 deg., lb.	10.36	10.56	10.42
30	Water evaporated per pound of combustible, lb.	10.95	11.22	11.05
31	Equivalent from and at 212 deg., lb.	11.56	11.78	11.68
	Rate of Combustion			
32	Dry coal burned per sq.ft. of grate per hour, lb.	21.80	21.10	22.80
33	Dry coal burned per sq.ft. heating surface per hour, lb.	0.57	0.55	0.60
	Rate of Evaporation			
34	Water evap. from and at 212 per sq.ft. h. s. per hour, lb.	5.90	5.80	6 20
35	Water evap. from and at 212 deg. per sq.ft. grate per hour, lb.	226	222	257
	Power of Boiler			
36	Commercial horsepower for land use	428	422	450
37	Excess above commercial rating of 250 hp., per cent	71	69	80
38	Marine rating, water evap. from and at 212 deg. per hour, lb.	15,000	15,000	15,000
39	Equivalent commercial horsepower of marine rating	435	435	435
40	Excess above or below marine rating, per cent	1.80 below	3.20 below	3.20 above

TABLE III. RESULTS OF EVAPORATIVE TESTS OF THREE-PASS BOILER EQUIPPED WITH WAGER BRIDGE-WALL AND HAND-FIRED

1	Date, 1918	June 17	June 18	June 19	June 20
2	Duration, hours	10	10	10	10
	Dimensions and Proportions				
3	Grate area, sq.ft.	65.54	65.54	65.54	65.54
4	Heating surface, sq.ft.	2,500	2,500	2,500	2,500
5	Ratio grate area to heating surface	....	....	....	....
	Average Pressures				
6	Gage pressure, lb.	175.00	178.00	182.10	180.30
7	Atmospheric pressure, lb.	14.49	14.54	14.56	14.53
8	Absolute pressure, lb.	189.49	192.54	196.66	194.83
9	Draft between damper and boiler, in.	0.64	0.64	0.61	0.39
	Average Temperatures				
10	External air, deg. F.	74	68	68	61
11	Fire room, deg. F.	83	80	80	69
12	Feed water, deg. F.	211	212	211	216
13	Escaping gases, deg. F.	538	531	556	510
	Fuel				
14	Moist coal consumed per hour, lb.	1,478	1,436	1,530	1,274
15	Moisture in coal, per cent	1.76	2.02	2.09	2.04
16	Dry coal consumed, per hour, lb.	1,450	1,406	1,498	1,248
17	Dry refuse per hour, lb.	152	138	132	109
18	Dry refuse in per cent	10.50	9.80	8.80	8.75
19	Combustible consumed per hour, lb.	1,298	1,268	1,366	1,139
	Quality of Steam				
20	Moisture in steam, per cent	0.40	0.40	0.30	0.30
	Heat Value of Coal and Efficiency				
21	Heat value of one pound of dry coal, B.t.u.	14,409	14,237	14,441	14,251
22	Efficiency of boiler based on dry coal, per cent	69.80	70.80	67.30	68.60
23	Efficiency of boiler based on combustible, per cent	72.00	71.50	68.30	69.00
	Water				
24	Water supplied to boiler per hr., lb.	14,345	13,915	14,242	12,094
25	Dry steam generated per hour, lb.	14,284	13,859	14,196	12,056
26	Factor of evaporation	1.049	1.048	1.050	1.045
27	Equivalent evaporation from and at 212 deg. per hour, lb.	14,984	14,524	14,906	12,599
	Evaporative Performance				
28	Water evaporated per pound of dry coal, lb.	9.87	9.84	9.50	9.65
29	Equivalent from and at 212 deg., lb.	10.32	10.33	9.96	10.02
30	Water evaporated per pound of combustible, lb.	11.00	10.90	10.40	10.55
31	Equivalent from and at 212 deg., lb.	11.55	11.45	10.92	11.05
	Rate of Combustion				
32	Dry coal burned per sq.ft. of grate per hour, lb.	21.10	21.50	22.90	19.00
33	Dry coal burned per sq.ft. of heating surface per hour, lb.	0.58	0.56	0.60	0.50
	Rate of Evaporation				
34	Water evap. from and at 212 deg. per sq.ft. h.s. per hour, lb.	5.94	5.81	5.96	5.04
35	Water evap. from and at 212 deg. per sq.ft. grate per hour, lb.	229	222	228	192
	Power of Boiler				
36	Commercial horsepower for land use	434	421	432	365
37	Excess above commercial rating of 250 hp., per cent	74	68	73	46
38	Marine rating in water evap. per hour from and at 212 deg., lb.	15,000	15,000	15,000	15,000
39	Equivalent commercial horsepower of marine rating	435	435	435	435
40	Excess above or below marine rating, per cent	0.20 below	3.20 below	0.70 below	15 below

space was too small and the path of the gases and air through the furnace was too short for the two to form an intimate mixture.

It was apparent that more air had to be admitted over the fuel bed and better means provided for mixing it with the combustible. It is a proven fact that in a hand-fired furnace, when the fuel bed is level and five inches or more deep, no free oxygen can be forced through it, no matter what air pressure is used in the ashpit. It is true that if the fuel bed contains holes or thin spots, air passes through them. However, admission of air through holes in the fire is undesirable because the size of holes cannot be controlled. It is therefore much better to admit air over the fuel bed through special openings in the firing door or in the bridge-wall, as the quantity of air admitted through such openings can be very easily controlled. The air should be supplied in a large number of small jets and as close to the fuel bed as practicable, so that as much as possible of the combustion space above the fuel bed is utilized for mixing and combustion.

Following these principles a number of ½-in. holes were made in each firing door; the opening in the first baffle at *I* was closed and the baffle extended, making the gas passage between the end of the baffle and the rear water-leg 36 in. wide. This had the effect of causing the air admitted through the firing door to flow farther to the rear of the furnace and to facilitate better mixing with the combustible gases. In addition to these changes Wager's bridge-wall was installed, as shown in Fig. 3, to supply additional air to the rear part of the furnace. This bridge-wall consists of a large number of cast-iron bars placed in the rear wall of the furnace and forms a structure similar to a plain grate. The air passes into the furnace in a large number of thin streams through narrow slots between the bars.

With the air admitted through the firing door and through the bridge-wall, the combustible gases rising from the fuel bed are squeezed between two streams of air coming in from two different directions, and the mixing is greatly aided. The total area for admission of air over the fuel bed, including the openings in the fire-doors, the space between the door frames and the fire-doors and the space in the bridge-wall, was about 160 sq in. This opening is approximately 2 per cent of the grate area.

The analyses of the gas samples taken from various parts of the furnace were studied to determine the effect of the different methods of admitting air. It was found that the use of the Wager bridge-wall provided the best condition for combustion, if too thick a fuel bed was not maintained next to the bridge-wall. If the fuel bed was too thick, it cut down the available area for admission of air over the fuel. In parts of these tests air was admitted through nozzles in the first baffle, as shown in Fig. 3. It was found in general that the admission of air through the bridge-wall under pressure proved to be undesirable in this type of furnace, and while the test using the nozzles proved satisfactory, the installation of these nozzles was omitted as entailing undesirable complications in boiler plants for ships.

TABLE IV. RESULTS OF EVAPORATIVE TESTS OF FOUR-PASS BOILER EQUIPPED WITH WAGER BRIDGE-WALL AND HAND-FIRED.

1	Date, 1918	Dec. 19	Dec. 16	Dec. 17	Dec. 18
2	Duration, hours	12.01	7.97	10.07	8.27
	Dimensions and Proportions				
3	Grate area, sq.ft.	66.53	66.53	66.53	66.53
4	Heating surface, sq.ft.	2,500	2,500	2,500	2,500
5	Ratio grate area to heating surface	37.58	37.58	37.58	37.58
	Average Pressures				
6	Gage pressure, lb.	200.50	200.00	197.30	198.60
7	Atmospheric pressure, lb.	14.67	14.67	14.68	14.70
8	Absolute pressure, lb.	215.17	214.67	211.98	213.30
9	Draft between boiler and damper, in.	0.34	0.44	0.78	1.56
	Average Temperatures				
10	External air, deg. F.	49	37	38	41
11	Fire room, deg. F.	63	59	57	59
12	Feed water, deg. F.	207.80	184.50	206.00	198.90
13	Escaping gases, deg. F.	506	513	557	648
	Fuel				
14	Moist coal consumed per hour, lb.	1,032	1,149.70	1,443.70	1,934
15	Moisture in coal, per cent	1.49	2.15	1.54	1.91
16	Dry coal per hour, lb.	1,017	1,125	1,411.50	1,897
17	Dry refuse per hour weight, lb.	111	168	143.40	166.50
18	Dry refuse, per cent	10.95	13.45	10.15	8.80
19	Combustible per hour, lb.	906	957	1,268.10	1,730.50
	Quality of Steam				
20	Moisture in steam, per cent	0.61	1.07	0.62	0.54
	Heat Value of Coal and Efficiency				
21	Heat value of one pound of dry coal, B.t.u.	14,292	13,990	14,213	14,216
22	Efficiency of boiler based on dry coal, per cent	72.60	72.80	71.00	70.90
23	Efficiency of boiler based on combustible, per cent	74.30	76.60	72.30	71.00
	Water				
24	Water supplied to boiler per hr., lb.	10,349	11,074	14,042	18,509
25	Dry steam generated per hour, lb.	10,286	10,956	13,955	18,409
26	Factor of evaporation	1.053	1.077	1.055	1.073
27	Equivalent evaporation from and at 212 deg. per hour, lb.	10,831	11,800	14,723	19,753
	Evaporative Performance				
28	Water evaporated per pound of dry coal, lb.	10.10	9.73	9.90	9.75
29	Equivalent from and at 212 deg. lb.	10.70	10.50	10.40	10.40
30	Water evap. per lb. combustible, lb.	11.30	11.45	11.00	10.62
31	Equivalent from and at 212 deg. lb.	11.95	12.30	11.61	11.40
	Rate of Combustion				
32	Dry coal burned per sq.ft. grate per hour, lb.	15.30	16.90	21.30	28.60
33	Dry coal burned per sq.ft. heating surface per hour, lb.	0.41	0.45	0.57	0.76
	Rate of Evaporation				
34	Water evap. from and at 212 deg. per sq.ft. h. s. per hour, lb.	4.30	4.72	5.90	7.50
35	Water evap. from and at 212 deg. per sq.ft. grate per hour, lb.	163	177.50	222	297
	Power of Boiler				
36	Commercial horsepower for land use	314	342	427	573
37	Excess above commercial rating of 250 hp., per cent	26	37	71	129
38	Marine rating in water evap. from and at 212 deg. per hour, lb.	15,000	15,000	15,000	15,000
39	Equivalent commercial horsepower marine rating	435	435	435	435
40	Excess above or below marine rating, per cent	28 below	22 below	2 below	31 above

TABLE V. RESULTS OF EVAPORATIVE TESTS OF FOUR-PASS BOILER EQUIPPED WITH TWO TYPE "E" UNDERFEED STOKERS

1	Date, 1919	April 8	April 9	April 13	April 12
2	Duration, hours	24	24	24	24
	Dimensions and Proportions				
3	Grate area, sq.ft.	77.50	77.50	77.50	77.50
4	Heating surface, sq.ft.	2,500	2,500	2,500	2,500
5	Ratio grate area to heating surface	32.26	32.26	32.26	32.26
	Average Pressures				
6	Gage pressure, lb.	199.00	199.00	197.00	198.40
7	Atmospheric pressure, lb.	14.51	14.51	14.50	14.44
8	Absolute pressure, lb.	213.51	213.51	211.50	212.84
9	Draft between damper and boiler, in.	0.35	0.85	1.40	1.43
	Average Temperatures				
10	External air, deg. F.	47	53	50	44
11	Fire room, deg. F.	61	68	63	62
12	Feed water, deg. F.	66	67	66	63.40
13	Escaping gases, deg. F.	535	591	618	659
	Fuel				
14	Moist coal consumed per hour, lb.	1,189	1,614	2,000	2,392
15	Moisture in coal, per cent	2.92	2.82	3.35	4.14
16	Dry coal consumed per hour, lb.	1,155	1,568	1,933	2,293
17	Dry refuse per hour, lb.	126	160	184	214
18	Dry refuse in per cent	10.90	10.20	9.50	9.30
19	Combustible consumed per hour, lb.	1,029	1,408	1,749	2.079
	Quality of Steam				
20	Moisture in steam, per cent	0.75	0.75	0.85	1.02
	Heat Value of Coal and Efficiency				
21	Heat value of one pound of dry coal, B.t.u.	14,040	14,342	14,234	14,404
22	Efficiency of boiler based on dry coal, per cent	79.20	75.00	72.50	71.40
23	Efficiency of boiler based on combustible, per cent	80.20	76.50	72.90	72.50
	Water				
24	Water supplied to boiler per hour, lb.	11,124	14,588	17,253	20,390
25	Dry steam generated, per hour, lb.	11,041	14,479	17,106	20,182
26	Factor of evaporation	1.20	1.20	1.20	1.204
27	Equivalent evaporation from and at 212 deg. per hour, lb.	13,249	17,375	20,527	24.299
	Evaporative Performance				
28	Water evaporated per pound of dry coal, lb.	9.55	9.23	8.87	8.80
29	Equivalent from and at 212 deg., lb.	11.47	11.08	10.62	10.60
30	Water evaporated per pound of combustible, lb.	10.85	10.21	9.80	9.66
31	Equivalent from and at 212 deg., lb.	12.90	12.34	11.70	11.68
	Rate of Combustion				
32	Dry coal burned per sq ft. grate per hour, lb.	14.90	20.23	23.80	29.60
33	Dry coal burned per sq.ft. heating surface per hour, lb.	0.46	0.62	0.77	0.92
	Rate of Evaporation				
34	Water evap. from and at 212 deg. per sq.ft. h.s. per hour, lb.	5.30	6.95	8.21	9.72
35	Water evap. from and at 212 deg. per sq.ft. grate per hour, lb.	170.90	234.20	264.00	313.60
	Power of Boiler				
36	Commercial horsepower for land use	384	504	595	704
37	Excess above commercial rating of 250 hp., per cent	54	102	138	182
38	Marine rating in water evap. per hour from and at 212 deg., lb.	15,000	15,000	15,000	15,000
39	Equiv. commercial hp. of marine rating	435	435	435	435
40	Excess above or below marine rating, per cent	12 below	16 above	37 above	63 above

The temperatures were measured with thermocouples inserted in the setting through the hollow stay-bolts and moved across the gas passages, as indicated by the straight lines labeled couples, or *C*, in Figs. 2 and 3. These temperatures were measured across the space between the end of the baffles and the water-leg.

An analysis of the temperature readings at various points in the passage gave considerable information as to the flow of the gases. In the three-pass boiler there appeared no regular variation in temperature in the first pass. This indicates that the distribution of the hot gases through this pass is nearly uniform. The second pass, however, shows the highest temperature one foot from the end of the baffle and the temperature drops down rapidly on each side of this point. This seems to indicate that a large part of the hot gases had passed through the space between 6 in. and 24 in. from the end of the baffle. The remaining space of about 24 in. next to the front water-leg contains only slowly moving and comparatively cool gases.

The same condition exists in the third pass, most of the hot gases passing through the 18-in. space next to the end of the third baffle. These temperature measurements show the desirability of lengthening the second and third baffles and thus causing the hot gases to pass over the greater part of the heating surface of the boiler. Although the temperature measurements did not show the desirability of lengthening the first baffle, this baffle was made 10 in. longer after the first six tests, for the purpose of improving the mixing of the air and combustible gases rising from the fuel bed. The variation in temperature measurements showed the difficulty of measuring temperature with a single couple. Such measurements might easily be several hundred degrees higher or lower than the average temperature in the pass. In general the temperature measurements in the four-pass boilers indicated the same characteristics as were found in the three-pass boilers. These measurements showed the desirability of decreasing the area of the passages so that more heating surface might be available even though it caused a greater frictional resistance to the passage of the gases.

Oil Fuel Versus Coal

By David Moffat Myers

Oil is almost an ideal fuel. It is the most beautiful fuel from an engineering standpoint with the exception of natural gas. Like natural gas, its supply is distinctly limited. One estimate gives 20 to 25 years as the probable period that oil fuel will last. It should, therefore, be applied to very special uses where its natural characteristics are distinctly and specially in demand, as in naval and certain classes of marine practice and for special processes, etc. It should not be burned indiscriminately and wasted, as our present coal supply is, and has been for many years.

Under equally favorable conditions, it is true that a somewhat higher thermal efficiency can be obtained with oil than with stoker-fired soft coal. On the other hand, an oil fire may be flooded with a tremendous excess of air without indicating the fact to the operator, as a coal fire would under the same conditions. It is really easier to fail in the efficient operation of oil than with coal, so that in the long run, especially in small plants, it is quite possible that the efficiency with oil might run below that which would probably obtain with coal. In a recent examination of a plant there was found an extravagant waste of oil on account of the reasons just stated.

The labor of ash handling with oil burning is, of course, entirely eliminated. In the small plant, however, it would frequently be impossible actually to save any money, as the time of one man could not be eliminated even if ash handling had become unnecessary. In large plants, on the other hand, it is possible that a considerable labor saving can be made by the elimination of the handling of ash. For the sake of safety in any boiler plant at least two men should be on the job so that the apparent saving of labor does not always reflect itself in money saving.

*From paper presented at annual meeting of the American Society of Heating and Ventilating Engineers, New York, January, 1920.

Owing to the extremely small supply of oil as compared to coal, the former fuel should be used, as has been previously indicated, for special purposes which require its advantages, and it is evident that it should not be consumed wastefully. In the average office building not over 4 or 5 per cent of the heat energy in the fuel as fired is converted into electrical energy at the busbars. It seems a crime to consume so rare and valuable a fuel as oil with a heat loss of 25 per cent. It is bad enough to do this with cheaper coal of which there is an infinitely greater supply. Of course during the heating season, when possibly as high as 60 per cent of the original heat value of the fuel is rendered useful by the application of exhaust steam to heating and process, the case takes on a different aspect.

It is probable, however, that not much "discouraging" will be necessary. I recently endeavored to secure a supply of fuel oil for one of the legitimate cases, for a plant just outside New York City limits, and found it impossible to obtain a contract from any of the oil companies. These companies were unable to say how soon they would be able to make any further contracts and in any event did not wish to consider guaranteeing a supply for more than a three months' period. One stated they might, at some distant time, give a one-year contract.

We have recently been called to report on the oil and coal situation in a very large office-building plant in Philadelphia, which is to be constructed. Our findings may be briefly summarized as follows:

A. For the safe storage of oil an additional $60,000 would have to be spent, of which excavation alone would cost $25,000.

B. Owing to the uncertainty in supply, it would not be safe to equip such a building to depend on oil exclusively. Consequently, a coal-burning equipment would have to be installed in any event, and if oil were used it would be used only as an auxiliary fuel at such times when prices might favor its use.

C. No saving in operation at existing prices of oil and coal could be shown in favor of oil. In fact, the difference in operating cost favors coal.

This does not mean that there may not be a great many cases, as there undoubtedly are, where oil would be the more economical fuel to employ. In fact, we have recently reported in favor of oil for a certain industrial plant where a long contract for oil supply could be secured and where the labor situation was the deciding factor in the problem. In this plant, however, the exhaust steam was entirely utilized so that the over-all efficiency of the plant would be high and the oil would be economically applied.

Ball Bearings in Steamship Engine Construction

The Swedish ball-bearing manufacturers, after introducing successfully ball bearings in Swedish rolling stock, especially the Swedish State railways, are now entering new fields of experimentation, which have for their object the introduction of the ball-bearing principle in ship construction, especially in connection with propeller shafting. The problem comprises the construction of a bearing able to withstand all axial pressure from the propeller and excluding all possibilities of hot-box and kindred troubles. The successful ball bearing, it is claimed, must under the most trying conditions be able to withstand the axial pressure automatically and require little or no attention and must be of such simple and reliable construction that friction within the same is minimal and that the ship's movements may not influence disadvantageously the ball-bearing mechanism.

The bearings hitherto used have proved unsatisfactory in these respects, it is said.

While ball bearings are used to a considerable extent in marine engineering, their application and their mounting have seen their highest development in Swedish shipyards. This is the contention of Swedish shipbuilders generally, but will doubtless meet challenge in the United States and Great Britain, where shipbuilding yards have not been slow to see the advantages offered by the ball bearings in modern ship construction.—*Commerce Reports.*

Condensed-Clipping Index of Equipment

Clip, paste on 3 x 5-in. cards and file as desired

Pump, Squeegee Rotary
Avamore Engineering Co., Ltd., Queen Victoria Street, London, England.
"Power," Dec. 16-23-30, 1919

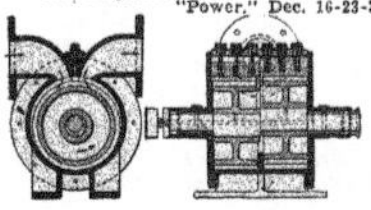

This pump consists of six main parts — the casing, two covers, shaft and two rotors, which in the standard design are covered with India rubber. The eccentrics, which are solid with the shaft and have only a small throw, impart movement to the rotors, which are a loose fit on the eccentrics. The displacement is caused by the rolling of the rotor when driven around by the eccentric on the main shaft. With the shaft rotating in an anti-clockwise direction, the rotor rolls around in the same direction, but on account of its being smaller in circumference than the track in the casing, it rotates in a clockwise direction in proportion to the difference between the circumference of the casing and that of the rotor. Small deflecting valves, one for each rotor, are placed in the upper portion of the casing, and are held in position against the rotor by small springs. The pumps are made in sizes from 1 to 4 in. diameter pipe connection, with a capacity of 700 gal. to 1,300 gal. per hour when run at 1,250 and 720 r.p.m. respectively.

Valve, Universal All-Service
Universal All-Service Valve Co., 12419 Lancelot Ave., Cleveland, Ohio.
"Power," Nov. 11-18, 1919

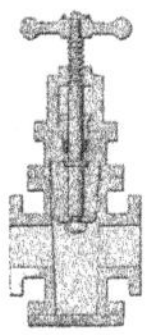

This valve is made so that it can be used as a straightway, three-way or angle valve by merely changing the position of a removable flange. The disk is round and taper-shaped and is removed from the action of the flow of liquid to the valve when in the open position. A pocket is found below the disk in which sediment is deposited and can be removed through a blowoff pipe or by removing the bottom flange. In large valves a by-pass is arranged, connecting with tha valve body and the bottom flange, so that when the pressure is equalized, the pressure also exerts itself against the bottom of the flange of the disk and assists in opening it. The vaive is adaptable for steam lines, blowoffs, etc.

Valve, Defiance Packless
Defiance Packless Valve Co., St. Louis, Mo.
"Power," Nov. 25, Dec. 2-9, 1919

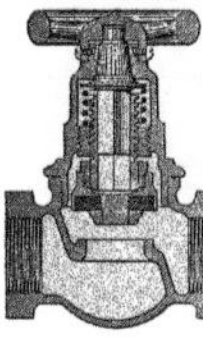

This valve is made with a metal-to-metal seat at the top of a nonrising stem. The one-piece bonnet is interchangeable with practically all valve bodies. No packing is employed. The stem acts as a key to raise and lower the valve disk. It is double-seated in the bonnet at angles of 10 and 30 deg., and is held against the seats by a spring. The valve stem is made square in cross-section, and of metal harder than the bonnet, to take care of expansion and contraction and to guard against cutting or grinding. Turning the handle will raise or lower the valve disk without imparting a similar movement to the stem.

Filter, Bowser Seven F
S. F. Bowser & Co., Inc., Fort Wayne, Ind.
"Power," Nov. 11-18, 1919

In this filter four operations are performed, in the order following: Screening, precipitation of water, precipitation of solids, and filtering and storing of the clarified oil. Foreign material in the oil is separated in a strainer, and the oil then passes down around the steam coils and up and around a precipitating tray, and then into a separate compartment into filtering bags, from which the oil runs to a storage base below.

Cleaner, Adriance-Bates Mechanical Flue
Adriance-Bates Steam Fuel Economy Corp., 80 Richards St., Brooklyn, N. Y.
"Power," Nov. 11-18, 1919

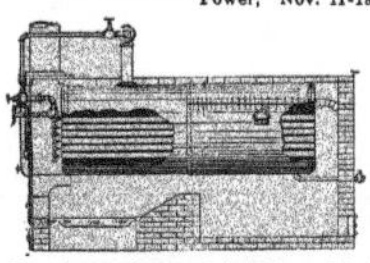

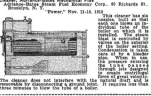

This cleaner has six nozzles, built so that each one blows an individual tube of the boiler on which it is installed. The steam blast is controlled by valves on the exterior of the boiler setting. Condensation is taken care of by a bleeder pipe. When in use, the pressure entering the tube passes through jets designed to create centrifugal force of great velocity. The cleaner does not interfere with the natural draft and is removable by disconnecting a ground joint. It requires less than three minutes to blow the tube of a boiler.

Meter, Boiler Coal
Lee Recorder Co., Ltd., 28 Deans Gate, Manchester, England
"Power," Sept. 16, 1919

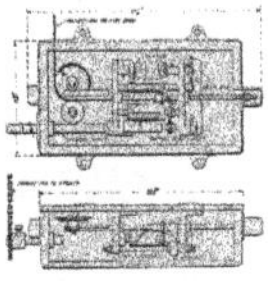

This device is especially intended for boilers operating with chain grates and works on same lines as the V-notch recorder and integrator for water measurements. The amount of coal passing under the fire-door is regarded as the flow, but the depth and velocity of the fuel are subject to variation. The reading is easily made, no skilled attention is required, and the makers give a guarantee of accuracy to within 5 per cent.

Bearings for High Speed Gears
British Westinghouse Electric & Mfg. Co., London, England.
"Power," Dec. 16-23-30, 1919

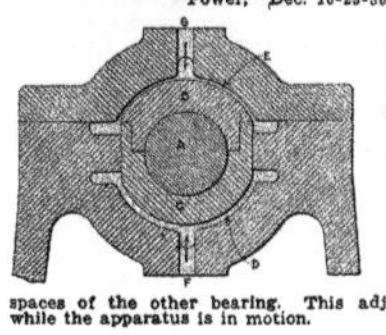

This bearing is supported in position by films of oil which are maintained by forcing the oil through one or more relatively small clearance spaces provided between the bearing and its support. The position of one bearing with respect to another may be adjusted by forcing a greater quantity of oil through the clearance space or spaces between the bearings and its support than through the clearance space or spaces of the other bearing. This adjustment may be made while the apparatus is in motion.

Latch, De Waters Safety
De Waters Safety Latch Co., Central Ave., Far Rockaway, N. Y.
"Power," Dec. 16-23-30, 1919

This latch is designed to prevent furnace doors from being forced open in case of ruptured boiler tubes. The latch cannot be lifted, except by external force applied to a weight, which always tends to force the latch nose down on the keeper. The latch is always in a position ready to lock. Its movement is limited by a double stop space to the amount required to clear the keeper. When released, it rests on a beveled stop face in a position to ride over the keeper when the door is closed. This latch is also applicable to elevators, fire-doors and other openings.

Patented Aug. 20, 1918

THE isometric sketch, Fig. 1, and the sectional view, Fig. 2, set forth the equipment used in the operation of the pulverized-fuel burning plant at the Oneida Street station. Screenings coal is delivered to the station by barge and run over an automatic scale, discharged onto a short belt conveyor equipped with a magnetic separator pulley for removal of trampiron and then into a bucket conveyor which carries it to the green-coal storage bunkers, the starting point as shown on the sketch. From there it is conveyed and elevated to a disintegrator, where it is prepared for drying by being crushed, discharged into a small storage bin and fed to the drier. The drier is of the double-shell type and is so arranged that the gases exhausted from it are discharged through a cyclone separator where coal dust, carried from the shell, is reclaimed and conveyed to the pulverizers. The gases discharged from the separator are vented into the smokestack. The dried coal is carried from the drier discharge by a bucket-elevator to the dry-coal storage bins, to which are connected the pulverizer feeders. The pulverizers are of the five-roll type. Pulverized coal from the mill outlets is conveyed by screw conveyors to the pulverized-fuel storage bins. Screw feeders take the fuel to the air-mixing chamber from where it is blown to the furnaces.

To obtain complete data on the pulverized-fuel installation for the purpose of making comparison with the stoker installation, no attempt was made to establish boiler-room conditions other than those maintained during regular operation.

To insure against any breakdowns during the test, the physical condition of the plant was carefully checked over, repairs being made wherever considered necessary. Boiler settings were examined and made airtight and all heat-radiating surfaces were properly insulated.

All necessary precautions against water leakage, from or to the boiler, were taken. Blowoff lines were disconnected and blanked. Water-column connections were broken and all regular feed lines to the boilers were disconnected and blanked.

All instruments were carefully calibrated and set, and arrangements for frequent checks were made wherever possible. Preliminary tests were made on the various parts of the system so as to insure proper operating conditions during the final test. A 24-hour preliminary test was conducted on the entire plant for the purpose of training the personnel as well as that of trying out the test equipment.

Apparatus and Instruments

The coal was weighed on an Avery scale, the automatic devices of which had been disconnected, making a hand scale, with dead weight on one side of the fulcrum and coal on the other. The capacity was 3,000 lb. per dump. The scale was checked at regular intervals, not more than 120 tons being weighed between calibrations. The accuracy was tested with 3,000 lb. in the form of 50-lb. standard scale blocks.

Four 2,000-lb. capacity tanks placed on platform scales were used for weighing water, these being mounted so as to empty into a receiving reservoir below, which was connected to the feed pump. The scales were balanced and checked at regular intervals.

Slag, ash and refuse were weighed on an ordinary platform scale. An aneroid barometer was used for the determination of atmospheric pressures. For obtaining flame temperatures, an optical pyrometer was used. Stack and steam temperatures were recorded by instruments. For all other temperature determinations mercury indicating thermometers were used. Drafts were read from liquid gages of the U-tube and multiplying types. For determination of the composition of flue gases, Orsat instruments were used. Solutions in these were renewed daily. A recording machine was also connected and used for comparison. Electrical data were obtained from integrating watt-hour meters. All pressure gages were set from a calibrated standard test gage, with proper allowances for water columns.

All data were recorded on prepared forms. Instrument readings were taken at regular intervals, varying from fifteen-minute periods on some instruments to one-hour periods on

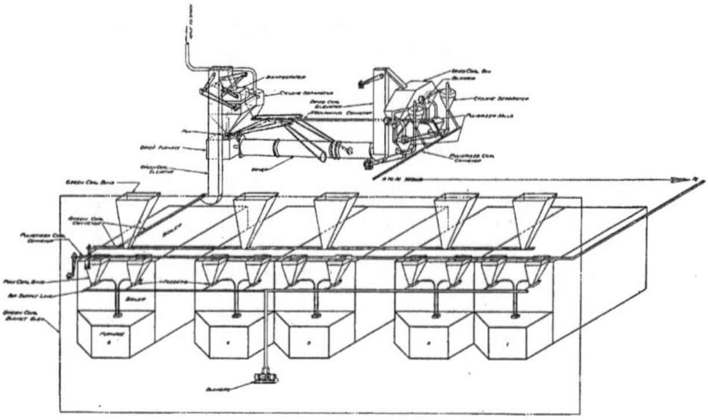

FIG. 1. LAYOUT OF PULVERIZED-FUEL PLANT AT ONEIDA STREET STATION

others. Wherever possible automatic counters were used to check tallies made by operators. In other instances a double tally was made, each independent of the other. Data were so arranged that approximate checks could be made after the burning of each 40.5 tons of coal. Hourly checks were made on the water in order to maintain a uniform rating.

Kind of Coal Used in the Test

With the exception of the first day, when 100 per cent Youghiogheny was used, the coal for the test was a mixture 50 per cent each Eastern Kentucky and Youghiogheny screenings, running approximately 25 per cent nut, 45 per cent pea and 30 per cent slack. This coal is the same as is used in daily operation. The coal as supplied to the drier after passing through the disintegrator was approximately 50 per cent slack and 50 per cent small pea and nut, not any of the pieces being larger than ½ inch.

During the progress of the test the coal was regularly sampled at five points along the line of fuel travel. Samples taken at the coal scale were for moisture, proximate and ultimate analyses, while those taken at the drier inlet and outlet, and burner outlets were for moisture determinations only. All pulverizer outlets, and burner outlets were for moisture determinations only. All samples were taken and made up according to the A. S. M. E. standards. All determinations and analyses were made at the laboratories of the company, according to approved practices of the Bureau of Mines. An Emerson Bomb was used for obtaining caloric values of the coal.

In pulverized-fuel test practice standard methods cannot be followed. The levels of pulverized fuel in the storage bins is the determining factor in starting, checking and stopping.

Previous to the starting signal for the final test the green-coal system was run clean of fuel and the pulverized-fuel system filled to capacity. The conditions of the respective systems were identical for each check (after the burning of 40.5 tons of coal) and at the close of the test. The amount of coal to be burned between checks was determined by the capacity of the green-coal storage bunkers. Three bunkers of 40.5 tons capacity each were available during the test, all of which were filled once in twenty-four hours.

Operating Conditions

Operation in the pulverizing room was changed somewhat during the test in order to fulfill conditions required in making the periodic checks of boiler operation. It was essential that the levels of the fuel in the pulverized bins should be controllable at certain hours of the day (at the time of check) and therefore operation of pulverizer equipment was extended over twenty-four hours, although, without these considerations, sufficient coal could have been pulverized during an eighteen-hour run. As a result, irregular operation of the equipment—frequent starts and short runs—increased power consumption and decreased hourly capacities.

No interruptions due to failure of equipment occurred during the test. The pulverized-fuel conveyor choked up on two occasions when the storage bins were allowed to overfill at a check hour. Uniform and satisfactory removal of moisture was affected by the drier without any unusual regulation. The firing of the furnace was varied, as it is ordinarily, depending upon the moisture content of the green coal. The pulverizers operated uninterruptedly and provided fuel of the desired fineness with little variation.

During the first twenty-four hours of the test, it appeared that moisture, with its attendant difficulties, was collecting in the storage binds. Cold-air drafts through windows along the side of the bins caused this condition by rapidly condensing the vapor in the entrained air. When the windows were tightly closed, it was eliminated.

Choking and plugging of the screw feeders and feeder pipes were the chief causes of interruption in the fire room. It was on the second day that the tendency of the feeder lines to choke was most noticeable, and this must be attributed to the moisture conditions encountered the night previous. In one instance, however, one of the lines to a burner stopped feeding when a piece of tarred paper lodged in it above the burner. No doubt this had been dropped into the pulverized-fuel system accidentally. Operation of the furnace on which this occurred had been noticeably affected during the twenty-four hours previous to the removal of the pipe and the discovery of the source of trouble. A total of four feeder hours were lost during the test.

A high percentage of CO_2 was easily obtainable, but could not be maintained for longer than an hour at a time, owing to excessive slagging on the hearth. This slagging on the hearth and furnace bottoms may be attributed to flame characteristics resulting from certain draft conditions and can be avoided only by air regulation. On the newer type of Lopulco furnaces, such as are in use at the Oneida Street plant, the method of air regulation is such that while admitting air for slag prevention, a large volume not needed for combustion, enters, bypasses the flame zone and is carried with the products of combustion in the form of excess air. The high percentage of excess air, together with a correspondingly low percentage of CO_2, as indicated by the flue-gas analysis, was not determined by combustion considerations but rather by furnace limitations.

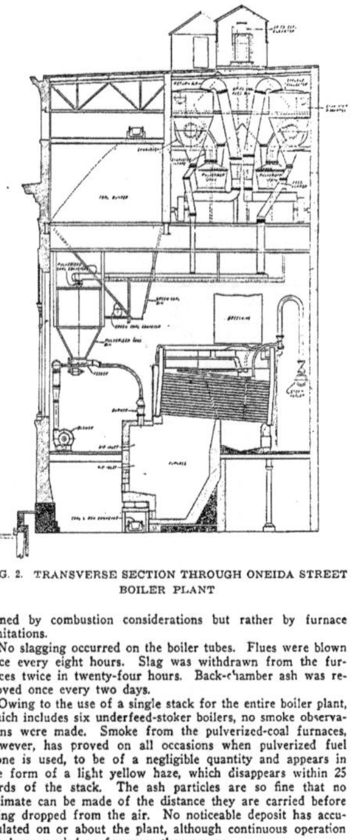

FIG. 2. TRANSVERSE SECTION THROUGH ONEIDA STREET BOILER PLANT

No slagging occurred on the boiler tubes. Flues were blown once every eight hours. Slag was withdrawn from the furnaces twice in twenty-four hours. Back-chamber ash was removed once every two days.

Owing to the use of a single stack for the entire boiler plant, which includes six underfeed-stoker boilers, no smoke observations were made. Smoke from the pulverized-coal furnaces, however, has proved on all occasions when pulverized fuel alone is used, to be of a negligible quantity and appears in the form of a light yellow haze, which disappears within 25 yards of the stack. The ash particles are so fine that no estimate can be made of the distance they are carried before being dropped from the air. No noticeable deposit has accumulated on or about the plant, although continuous operation has been carried on for more than a year.

TABLE I. TEST OF FIVE 468-HP. BOILERS
November 11-15, 1919

	Dimensions	
1.	Number and kind of boilers	Five Edge Moor water-tube boilers
2.	Kind of furnaces	Pulverized Fuel Equipment Corporation
3.	Volume of combustion space, per boiler, cu. ft	1678
4.	Water heating surface, sq.ft., per boiler	4680
5.	Superheating surface, sq.ft. per boiler (approximate)	594
	(a) Type of superheater	Foster
6.	Total heating surface, sq.ft., per boiler	5274
	(a) Ratio of water heating surface to volume of combustion space	1 to 0.359
	(b) Ratio of total heating surface to volume of combustion space	1 to 0.318
	Date, Duration, Etc.	
7.	Date	November 11-15, 1919
8.	Duration, hours	99
	(a) Boiler, hours	495
9.	Kind and size of coal——Mixture—50 per cent Yough. Scrgs. 50 per cent Eastern Kentucky Screenings.	
	Average Pressures, Temperatures, Etc.	
10.	Steam pressure by gage, lb. per sq. in.	167.8
	(a) Barometric pressure, in. of mercury	29.49
11.	Steam pressure, absolute, lb. per sq.in	182.3
12.	Temperature of steam leaving superheaters, deg. F.	441.9
	(a) Normal temp. sat. steam at above pres., deg. F.	374.2
13.	Temperature of feed water entering boiler, deg. F	156.3
14.	Temperature of escaping gases, deg. F.	496.6
	(a) Temperature of flame above hearth, deg. F.	2,767
	(b) Temperature of furnace bottoms, deg. F.	2,180
15.	Draft under damper, inches of water	0.173
16.	Draft in furnaces, inches of water	0.031
17.	Air pressure at blower, inches of water	6.36
	(a) Pressure of air mixing with coal at screw feeder, inches of water	6.00
	(b) Pressure of air and coal mixture above burner outlet, inches of water	1.00
18.	State of weather.	
	(a) Temperature outside, deg. F	28.
	(b) Relative humidity, per cent	72.
	(c) Room temperature, deg. F.	75.7
19.	Number of degrees of superheat	67.7
	Total Quantities	
20.	Total weight of coal, as received, pounds	958,074
21.	Percentage of moisture	7.23
22.	Total weight of coal, as fired, pounds	894,800
23.	Percentage of moisture	0.67
24.	Total weight of dry coal, pounds	888,805
25.	Slag, ash and refuse (dry, laboratory basis) per cent	11.90
	(A) Withdrawn from furnace bottom, pounds total	9,770
	(a) Withdrawn from furnace bottom, pounds per hour per boiler	19.8
	(B) Withdrawn from tubes, flues and combustion chamber, pounds total	9,862
	(b) Withdrawn from tubes, flues and combustion chamber, pounds per hour, per boiler	20.0
	(C) Blown away with gases, pounds (difference between laboratory and actual weighed)	87,549
	(c) Blown away with gases, pounds per hour per boiler	176.8
	(D) Percentage of total lost with gases	82.8
	(E) Percentage of combustible in slag and ash recovered, per cent (combined analysis)	6.9
26.	Total combustible burned, pounds	781,622
27.	Total weight of water fed to boilers	8,249,536
28.	Factor of evaporation	1.1473
29.	Total equiv. evap. f. and a. 212 deg. F., lb.	9,464,693
	Hourly Quantities and Rates	
30.	Dry coal per hour, pounds	8,978
	(a) Dry coal per hour, per boiler, lb.	1,796
31.	Water evap. per hour, lb. actual	83,328
	(a) Water evap. per hour, per boiler, lb. actual	16,666
32.	Equiv. evap. per hour f. and a. 212 deg. F., lb.	95,603
	(a) Equiv. evap. per hour, per boiler, f. and a. 212 deg. F., lb.	19,121
33.	Equiv. evap. per hour, f. and a. 212 deg. F., per sq.ft. of water heating surface, lb.	4.09
	Capacity	
34.	Evap. per hour, f. and a. 212 deg. F., per boiler, lb.	19,121
	(a) Boiler horsepower developed	554
35.	Rated cap. per hr. f. and a. 212 deg. F., per boiler, lb.	16,146
	(a) Rated boiler horsepower	468
36.	Percentage of rated capacity developed	118.4
	Economy	
37.	Water fed per lb. of coal, as received, lb.	8.611
38.	Water fed per lb. of coal, as fired, lb.	9.219
39.	Water evap. per lb. of coal dry, lb.	9.282
40.	Water evap. per lb. of combustible, lb.	10.554
41.	Equiv. evap. f. and a. 212 deg. F., per lb. of coal, as received, lb.	9.879
42.	Equiv. evap. f. and a. 212 deg. F., per lb. of coal, as fired, lb.	10.577
43.	Equiv. evap. f. and a. 212 deg. F., per lb. of coal dry, lb.	10.649
44.	Equiv. evap. f. and a. 212 deg. F., per lb. of combustible, lb.	12.109
	Gross Efficiencies	
45.	Calorific value of 1 lb. of dry coal by calorimeter, B.t.u.	12,810
46.	Gross Efficiency of boiler and furnace, per cent	80.67
47.	Efficiency of furnace, per cent	99.79
48.	Smoke data	See notes
	Analyses of Flue Gases	
49.	Carbon dioxide, per cent	12.26
50.	Oxygen, per cent	6.82
51.	Carbon monoxide, per cent	0.00

Analyses of Coal

TABLE I—CONTINUED

52. Proximate.	As Received	As Fired	Dry
(a) Moisture	7.23	0.67	.....
(b) Volatile	32.13	34.40	34.63
(c) Fixed Carbon	49.60	53.11	53.47
(d) Ash	11.04	11.82	11.90
	100.00	100.00	100.00
(e) Sulphur separately determined referred to dry coal			1.62

53. Ultimate Analyses.	
(a) Carbon	73.57
(b) Hydrogen	4.35
(c) Oxygen	7.22
(d) Nitrogen	1.34
(e) Sulphur	1.62
(f) Ash	11.90
	100.00

54. Analyses of—	Slag	Combustion Ash Retained	Combustion Ash Lost
(a) Moisture	0.00	13.00	
(b) Combustible	0.59	13.76	Unknown
(c) Earthy matter	99.41	86.24	

Heat Balance

	Method: A.S.M.E. B.t.u.	A.S.M.E. Per Cent.	Uehling B.t.u.	Uehling Per Cent.
(a) Heat abs. by boiler	10334	80.67	10334	80.67
(b) Loss due to evap. of moisture in coal	8	0.06	8	0.06
(c) Loss due to heat carried away by steam formed by the burning of hydrogen	486	3.79	463	3.61
(d) Loss due to heat carried away in dry flue gases	1527	11.93	1551	12.11
(e) Loss due to carbon monoxide	0	0.00	0	0.00
(f) Loss due to combustible in ash	23	0.18	23	0.18
(g) Loss due to heating moisture in air	39	0.30	39	0.30
(h) Loss due to combustible carried away with flue gases, unconsumed hydrogen, hydrocarbons, radiation and unaccounted for	393	3.07	393	3.07
(i) Total calorific value of 1 lb. of dry coal	12810	.....	12810	.....
(j) Total per cent	.....	100.00	.....	100.00

Analyses of Losses

Losses due to the evaporation of moisture in coal, when expressed in a heat balance, are dependent upon the quantity of moisture in the coal as fired, and ordinarily are independent of installation. Since in this case, the coal was dried to 0.67 of 1 per cent of moisture, only a small loss, 0.06 of 1 per cent, due to the presence of moisture, occurred in combustion. The actual loss resulting from this factor is set forth in the table showing net boiler efficiency. It exceeds the minimum theoretical loss, dependent upon quantity of moisture that is always assumed in a heat balance and which in this case would be (if coal had not been dried) equal to 0.6 per cent. The efficiency of the drier referred to this basis is but 40 per cent.

Losses due to hydrogen in the coal are independent of the installation. In view of the fact that the boilers were operated at the most economical rating, losses due to heat carried away in the dry flue gases were higher than expected. To account for them, it must be considered that up to the time of the test three of the five boilers had been operated about 300 hours since receiving a partial wash, and that each of the five had been in service 600 hours since the last full cleaning. This condition, together with the fact that 90 per cent of the feed water was untreated reduced the heat absorption of the boilers appreciably. One of the boilers examined subsequent to the test showed scale deposits sufficient to affect efficiency.

Further losses under this heading may be attributed to an excess of air, not needed for proper combustion, but essential to the control of furnace temperatures and the prevention of excessive slagging on the hearth and furnace bottoms.

Losses due to carbon monoxide were not measurable, since but few analyses showed CO present, and then only in traces.

The loss due to combustible in the ash is very small and denotes the completeness of combustion. Since but 17.2 per cent of the total ash (laboratory basis) was recovered and half of that—the slag—contained no combustible, it might be assumed that, with the 82.8 per cent of ash that escaped with the flue gases, a large amount of combustible was carried away. On the other hand, it is more likely that since the unaccounted for losses are less than in the average boiler test, most of the unburned combustible lodged in the combustion chamber, and was without doubt the heavier particles.

Losses due to evaporation of moisture in the air are independent of the installation. The losses due to combustible carried away with the flue gases, unconsumed hydrogen, hydrocarbons, radiation and unaccounted for losses are not great comparatively, and are not wholly preventable. Radiation was reduced to a minimum by properly covering and lagging the boilers.

Analyses of Coal During Test

In twelve proximate analyses of the coal made during the test, the moisture averaged 7.23 per cent in the coal as received at the plant. In the dry coal the following averages were obtained: Fixed carbon, 53.47; volatile matter, 34.63;

TABLE II. TEST ON FUEL PULVERIZING EQUIPMENT

November 11-15, 1919

	Average Temperatures	
1	Temp. of air entering drier furnace, deg. F	93.8
2.	Temp. of gases leaving drier, deg. F	181.8
3.	Humidity of outside air, per cent	72.0
4.	Draft through drier, in. of water	0.77
5.	Vacuum in pulverizers, in. of water — No. 1: 5.0; No. 2: 5.16; Av.	5.08
	Coal Temperatures, Moistures and Fineness	
6.	Temp. of coal entering drier, deg. F	88.2
7.	Temp. of coal leaving drier, deg. F	237.9
8.	Temp. of coal leaving pulverizers, deg. F	169.7
9.	Moisture of coal entering drier, per cent	5.59
10.	Moisture of coal leaving drier, per cent	1.61
11.	Moisture of coal leaving pulverizers, per cent	1.03
12.	Fineness of pulverized coal, 200 mesh, per cent	81.30
13.	Fineness of pulverized coal, 100 mesh, per cent	97.40
14.	Fineness of pulverized coal, 80 mesh, per cent	99.30
15.	Fineness of pulverized coal, 60 mesh, per cent	100.00
	Total and Hourly Quantities	
16.	Total coal crushed, as received at crusher, tons	479.0
17.	Coal crushed per hour, as received, tons	17.5
18.	Total coal dried, as received at drier, tons	471.2
19.	Total coal dried per hour of drier operation, as received, tons	6.7
20.	Total coal pulverized, coal from drier, tons	447.4
21.	Capacity of pulverizer per hour, tons	5.0
22.	Coal pulverized per hour, dry, tons total	7.90
23.	Coal pulverized per hour, dry, tons per mill	3.95
24.	Coal pulverized per hour, per mill, as received at plant, tons	4.23
	Consumption of Lubricants	
24.	Total grease consumed by elevators and conveyors, lb	6.0
25.	Grease per ton of coal, as received, lb	0.012
26.	Total grease consumed by pulverizers, lb	13.0
27.	Grease consumed per pulverizer per hour of operation, lb	0.112
28.	Grease consumed per pulverizer per ton of coal pulverized, lb.	0.028
29.	Grease consumed on all equipment per ton of coal, as received, lb.	0.040
30.	Total oil consumed on all equipment, quarts	17.0
31.	Oil consumed per ton of coal as received, quarts	0.036
	Electric Energy and Coal Consumption	
32	Total energy consumed by crusher and green coal elevator, kw.-hr.	220.0
33.	Energy per ton of coal, as received, kw.-hr.	0.47
34.	Total energy consumed by drier, kw.-hr	735
35.	Energy per ton of coal, as received, consumed by drier	1.53
36.	Total energy consumed by pulverizers, kw.-hr. (fan and drive motor)	8010
37.	Motor input per hour, hp — Mill No. 1: 93.8; Mill No. 2:	90.2
38.	Energy consumed by pulverizer per ton of coal, as received kw.-hr.	16.72
39.	Energy consumed by pulverizer per ton of coal, as pulverized, kw.-hr.	17.90
40.	Total energy consumed by pulverized coal conveyors, feeder blowers and feeders	1789
41.	Total energy consumed by pulverized coal conveyors, feeder blowers and feeders per ton of coal as received	3.73
42.	Total energy consumed by pulverized coal conveyors, feeder blowers and feeders per ton of coal as fired	4.00
43.	Total energy consumed by all equipment on preparation and firing of pulverized fuel, kw.-hr	10754
44.	Energy per ton of coal, as received, kw.-hr.—grand total	22.45
45.	Coal equivalent for this energy at 1.5 lb. coal per kw.-hr., lb.	33.68
46.	Total coal used in drier furnace	12291
47.	Coal per ton of fuel dried, lb. (based on coal as received)	25.66
48.	Total coal and equivalent consumed in preparation and firing of one ton of pulverized fuel, lb	59.34
	Cost of Preparation—Operation and Maintenance	
49.	Cost of labor per ton of coal—operation	$ 0.143
50.	Cost of fuel for drying plus fuel for electric energy—coal at $4 per ton	0.119
51.	Cost of lubricants per ton of coal—grease at 9c per lb	0.007
52.	Cost of labor per ton of coal—maintenance	0.036
53.	Cost of material—maintenance	0.020
54.	Total cost per ton of coal	0.325

Note: Item 49 is based on the labor required to pulverize coal sufficient for five boilers through a twenty-four hour run per day.

ash, 11.90. In the coal as received the B.t.u. per pound averaged 11,884 and in the dry coal, 12,810. The average of four ultimate analyses of dry coal showed 73.57 per cent carbon, 4.35 per cent hydrogen, 7.22 per cent oxygen, 1.34 per cent nitrogen, 1.62 per cent sulphur and 11.90 per cent ash.

From an average of twelve tests the per cent of moisture in the coal at various points along the line of fuel travel was as follows: As received at the plant, 7.23; at the drier inlet, 5.59; at the drier outlet, 1.61; at the pulverizer outlet, 1.03; in the coal at the burners, 0.67. The fineness of the coal averaged through a 200-mesh screen, 81.3 per cent; through a 100-mesh screen, 97.4 per cent; through an 80-mesh screen, 99.3 per cent, and through a 60-mesh screen, 100 per cent.

TABLE III. SUMMARY OF RESULTS

	Electric Energy and Fuel Consumption Per Ton of Coal Pulverized	
1.	Energy consumed by conveyors, crushers, elevators, drier, blowers and feeders, kw.-hr	5.73
2.	Energy consumed by pulverizer, kw.-hr	16.72
3.	Total energy, kw.-hr	22.45
4.	Coal equivalent at 1.5 lb. per kw.-hr., lb	33.68
5.	Coal consumed in drier furnace, lb. per ton of fuel dried	25.66
6.	Total coal and equivalent, lb	59.34
7.	Gross efficiency less deductions for total coal and equivalent—Item 6	78.36
	Cost of Fuel Preparation, Firing and Ash Disposal	
8.	Labor—Coal preparation	$ 0.143
9.	Labor—Firing	0.112
10.	Labor—Ash removal	0.025
11.	Drier fuel—Coal at $4 per ton	0.051
12.	Electric energy—Coal per kw.-hr. at 1.5 lb	0.068
13.	Maintenance (Labor at 3.6c, material at 2c, manufacturers' estimate, lubricants at 0.7c)	0.063
14.	Total cost of fuel preparation, firing, ash disposal and maintenance	0.462
15.	Price of coal as purchased, per ton	4.000
16.	Total cost	4.462
	Efficiencies	
17.	Actual gross efficiency, per cent	80.67
18.	Net efficiency after all incidental costs have been accounted for, per cent	72.32

Conclusions can be best drawn from a comparison between pulverized-fuel burning equipment and mechanical stokers such as is given in Table IV.

TABLE IV. COMPARATIVE RESULTS WITH PULVERIZED FUEL AND UNDERFEED STOKER PER TON OF COAL BURNED

	Pulv. Fuel System	Modern Stoker
Energy consumed by conveyors, crusher, elevators, driers, fans and feeders, kw.-hr	5.73	10.94
Energy consumed by pulverizer, kw.-hr	16.72	.....
Total energy, kw.-hr	22.45	10.94
Coal equiv. at 1.5 lb. per kw.-hr., lb	33.68	16.41
Coal consumed in drier furnace, lb	25.66	.....
Total coal and equivalent, lb	59.34	16.41
Cost of Fuel Preparation, Firing and Ash Disposal		
Labor—coal preparation	$0.143	$0.000
Labor—firing	0.112	0.140
Labor—ash removal (in plant)	0.025	0.064
Drier fuel—coal at $4 per ton	0.051	0.000
Electric energy—coal per kw.-hr. at 1.5 lb	0.068	0.033
Maintenance:	Labor at $0.036—material at $0.020, manufacturer's estimate—lubricants at $0.007 $0.063	Labor at $0.046, material at $0.049, lubricants at $0.002 $0.097
Total cost of fuel preparation, firing, ash disposal and maintenance	$0.462	$0.334
Price of coal as purchased per ton	4.000	4.000
Total cost per ton	4.462	4.334
Cost per ton of coal in P.F. system over modern stoker	0.128	.....
Efficiency		
Actual gross efficiency, per cent	80.67	76.80
Net efficiency after all incidental costs have been accounted for, per cent	72.32	70.88
Difference in favor of pulverized fuel system, per cent	1.44	.....

The tests were conducted by the engineers of the Milwaukee Electric Railway and Light Co., being directly supervised by Fred Dornbrook, chief engineer of the Commerce Street station, assisted by Messrs. Schubert and Mistele, test engineers. Paul W. Thompson, technical engineer of power plants, Detroit Edison Co., was present throughout the four days as an observer. The test was approved by John Anderson, chief engineer of power plants, for the company.

The expenditure which it is proposed to make on the development of the water power of the St. Lawrence is $26,000,000, while the output, it is said, will be 1,000,000 hp. The 80,000 hp. made available by the closing of the munition factories has already been used up, and there are applications for 110,000 hp., which cannot be supplied.

Trade Unionism in the Engineering Profession

On Feb. 24 the Milwaukee Chapter of the American Association of Engineers held an interesting meeting devoted to trade unionism in the profession. Dr. Isham Randolph and C. E. Dreyer, national secretary of the association, and A. E. Holcomb, president of the Milwaukee Chapter, addressed the gathering, consisting of some five hundred local engineers and architects. In reviewing the past and picturing the future of the engineer, Dr. Randolph expressed the firm belief that the engineer would win by hard work and by abiding by principles—not force. He urged the adoption of the Jones-Reavis bill. According to Secretary Dreyer, the unions and the American Association of Engineers have only one thing in common, and that is to improve economic conditions of their members. Violence and waste of strikes was a thing of horror to the engineers. In contradistinction from the unions the association is an incorporated organization responsible for its acts. On this question the attitude of the association is more fully expressed in the following printed statement used in circularizing the meeting.

"The engineer is the medium through which both capital and labor are used in production and in industrial development. The aim of the profession is to advance civilization and render the highest service to society. Except when their acts further this aim, it is an advocate of neither capital nor labor.

"Production should be increased—not limited. The profession cannot support strikes or lockouts or any other methods that may benefit any class at the expense of the nation as a whole. They are unsound and must inevitably lead to economic disaster. The law of supply and demand for men or material must ultimately prevail. Attempts may be made to limit the supply of either, but looking toward the upbuilding of civilization we believe rather in increasing the demand through the promotion of legitimate enterprises.

"Rewards should be according to ability, initiative and constructive effort. Men are not equal in these respects. Each man should be encouraged to do his utmost and be given compensation according to ability and will to increase production and to achieve large results.

"The engineer, as an educated professional man, believes in basing his claims for proper and just reward for his services upon the justice of the facts presented, upon enlightenment of public opinion, upon loyalty between employer and employee, and upon the underlying fundamental desire of the great majority to do what is fair and right when the merits of the case in question are clearly presented and demonstrated. We believe in organized representation for the correction of wrong, the advancement of the profession and service to the public but are opposed to methods inconsistent with the dignity of the profession which would lessen public confidence."

In his concluding remarks Mr. Dreyer told of the work of the association in the railroad and civic fields, of its service department and of its efforts to promote ethics. A. E. Holcomb reviewed local work along similar lines.

President Wilson has signed the oil-land leasing bill, which opens up for development millions of acres of land in the west. The total area thrown open for lease under the bill is estimated by the Geological Survey at more than 6,700,000 acres, while proven coal lands under Government withdrawal total approximately 30,000,000 acres, with 39,000,000 acres still to be classified.

IF DREAMS CAME TRUE

Obituary

Harold M. Davis, advertising manager of the Sprague Electric Works, died at his home in Brooklyn on Feb. 9, after a lingering illness. Mr. Davis was born in Jerseyville, Ill., Aug. 26, 1860, and was the son of Samuel W. and Mary J. McGill Davis. His early boyhood was spent in Paola, Kansas, and after finishing high school he secured a position as office boy in a lead and oil factory, later receiving an appointment as chief clerk of the St. Louis United States Assay Office. A government position was too slow and uncertain for him, however, so he came to New York to study architecture. The financial panic of 1893 offered a chance to get into the advertising line, in which he had had some experience while in St. Louis, and for four years he was advertising manager of a trade paper in New York, later becoming connected with an advertising agency and finally, in 1899, he was made manager of the advertising department of Sprague Electric Works of the General Electric Company.

Personals

F. M. Nourse joined the Chas. L. Benjamin organization at Chicago on January 1st. Mr. Nourse has been engineer in the advertising department of the Cutler-Hammer Manufacturing Co. at Milwaukee for the past two years and prior to that was associated with the Wisconsin Power, Light and Heat Co. at Portage, Wisconsin. He is a University of Illinois graduate in electrical engineering and worked for the Public Service Company of Northern Illinois both before and after his graduation.

W. K. Eicher, formerly with the General Electric Co. of Detroit, is now associated with the C. H. Wheeler Manufacturing Co., with headquarters at the Marquette Building, Chicago.

Engineering Affairs

The American Association of Engineers has adopted a policy of appropriating for the furtherance of license legislation a reasonable amount of money from the general fund equal to that raised by any state society, either alone or in conjunction with any other chapter or society. Under such an arrangement the proposed bill and the general plan of campaign must be submitted to the board of directors of the association for approval.

A. A. E. OFFERS NATIONAL UNITY PLAN

A plan for a unified engineering organization for the United States is contained in an announcement made by the American Association of Engineers that one-half the entrance fee of $10 will be waived to applicants who are members of recognized national and local engineering societies and clubs of the United States. This arrangement will be in effect only until the sixth annual convention to be held May 10, 11 and 12 in St. Louis, unless it is extended by the convention. The resolution is as follows:

"Inasmuch as the support in membership of every professional engineer in the United States is needed by the American Association of Engineers to accomplish professional unity, and inasmuch as many engineers have helped the profession by giving time and money for a number of years to that end through recognized societies and clubs:

"Be it resolved, that until the 1920 convention (May 10-11) the membership committee be authorized to waive one-half of the entrance fee of $10 to applicants who are in good standing in such recognized clubs and societies and to accept such applications when accompanied by an entrance fee of $5."

Business Items

E. R. Ladew, Inc., the large leather belting concern at Glen Cove, L. I., has been sold to Graton & Knight, leather belting manufacturers at Worcester, Mass. Millions of dollars are involved in the deal.

THE COAL MARKET

BOSTON—Current prices per gross ton f.o.b. New York loading ports:

Anthracite

	Company Coal
Egg	$8.20@$8.65
Stove	8.45@ 9.05
Chestnut	8.55@ 9.05
Pea	7.05@ 7.40
Buckwheat	3.60@ 3.80
Rice	2.30@ 2.50
Barley	1.25

Bituminous

	Clearfields	Cambrias and Somersets
F.o.b. mines, net tons	$2.85@$3.35	$3.15@$3.60
F.o.b. Philadelphia, gross tons	5.05@ 5.60	5.35@ 5.90
F.o.b. New York, gross tons	5.40@ 5.95	5.75@ 6.25
Alongside Boston (water coal), gross tons	7.00@ 7.75	7.60@ 8.00

Pocahontas and New River are practically off the market for coastwise shipment, but are quoted at $6.25@$7.00.

NEW YORK—Current quotations, White Ash, per gross tons, f.o.b. Tidewater, at the lower ports are as follows:

Anthracite

	Company Coal
Broken	$7.80@$8.25
Egg	8.20@ 8.65
Stove	8.45@ 9.05
Chestnut	8.55@ 9.05
Pea	7.05@ 7.40
Buckwheat	5.15
Rice	4.50
Barley	4.85
Boiler	4.85

Bituminous

Government prices at mines:	Spot
South Fork (best)	$3.25@$3.50
Cambria (best)	3.00@ 3.25
Cambria (ordinary)	2.60@ 2.90
Clearfield (best)	3.00@ 3.25
Clearfield (ordinary)	2.60@ 2.90
Reynoldsville	2.85@ 2.90
Quemahoning	3.25@ 3.50
Somerset (medium)	3.00@ 3.25
Somerset (poor)	2.50@ 2.75
Western Maryland	2.50@ 2.75
Fairmont	2.25@ 2.50
Latrobe	2.60@ 2.90
Greensburg	2.75@ 3.00
Westmoreland ¾ in.	3.40@ 3.50
Westmoreland run-of-mine	2.75@ 3.00

PHILADELPHIA—Bituminous coal prices vary according to district from which they are mined. For ordinary slack the price is $2.45@$2.55; lump, $3.00@$3.35, at the mines.

BUFFALO—

Anthracite

	On Cars, Gross Ton	At Curb, Net Ton
Grate	$8.55	$10.80
Egg	8.80	10.65
Stove	9.00	10.85
Chestnut	9.10	10.85
Pea	7.75	9.30
Buckwheat	5.70	7.75

Bituminous

Allegheny Valley	$4.80
Pittsburgh	4.65
No. 8 Lump	4.65
Mine Run	4.80
Slack	4.10
Smokeless	4.60
Pennsylvania Smithing	5.70

CLEVELAND—Prices of coal per net ton delivered in Cleveland are:

Anthracite

Egg	$12.25@$12.40
Chestnut	12.50@ 12.70
Grate	12.25@ 12.40
Stove	12.40@ 12.60

Pocahontas

Mine-run	$7.50

Domestic Bituminous

West Virginia split	$8.50
No. 8 Pittsburgh	$6.60@ 6.90
Massillon lump	8.25@ 8.50
Coshocton lump	7.15

Steam Coal

No. 6 slack	$5.25@$5.50
No. 8 slack	5.10@ 5.50
Youghiogheny slack	5.25@ 5.50
No. 8 ¾ in.	5.70@ 6.00
No. 6 mine-run	5.25@ 5.50
No. 8 mine-run	5.75

Only coal available is mine-run Pocahontas.

MIDDLE WEST—Chicago quotations, F.o.b. cars at mine:

	Springfield, Carterville, Williamson, Franklin, Saline, Harrisburg	Fulton, Peoria	Grundy, La Salle, Bureau, Will
Lump	$2.55@$2.70	$2.95@$3.10	$3.25@$3.40
Washed	2.75@ 2.90	....	3.45@ 3.60
Mine run	2.35@ 2.50	2.75@ 2.90	3.00@ 3.15
Screenings	2.05@ 2.20	2.35@ 2.50	2.75@ 2.90

New Construction

PROPOSED WORK

Mass., Boston—The Olympia Theatres, Inc., 35 Piedmont St., plan to construct an 11 story, 88 x 300 ft. theatre and office building on Washington and Milk Sts. A steam heating system will be installed in same. Total estimated cost, $1,000,000. G. U. Crocker, Pres.

Mass., East Springfield—The Westinghouse Electric Co., 6905 Susquehanna St., will soon award the contract for the construction of a 1 story machine shop at its plant. A steam heating system will be installed in same. Total estimated cost, $360,000.

Mass., Lowell—O'Connell & shaw, Archts., 18 Boylston St., Boston, will soon award the contract for the construction of a 7 story, 80 x 100 ft. hotel on Middle St., for F. E. Harris, c/o Hotel Harrisonia. A steam heating system will be installed in same. Total estimated cost, $175,000.

Mass., Newton Upper Falls—Lockwood, Green & Co., Engrs., 60 Federal St., Boston, will receive bids about March 1 for the construction of a 2 story, 200 x 350 ft. foundry, for the Saco-Lowell Shops, 77 Franklin St., Boston. Electric motors for power will be installed in same. H. M. Lane Co., 701 Owen Bldg., Detroit, Mich., Engr.

Mass., Northampton—Charles F. Atkinson, c/o Funk & Wilcox, Archts., 1148 Old South Bldg., Boston, plans to construct a 1 and 2 story, 85 x 190 ft. theatre and business building on King St. A steam heating system will be installed in same. Total estimated cost, $150,000.

N. Y., Albany—Charles F. Rattigan, Supt. of Prisons, Capitol, Albany, will receive bids until March 9 for the construction of underground tunnels and installation of underground heating mains, sewer and water connections, etc. Separate bids will be received for the construction, heating, sanitary and electric work.

N. Y., Hornell—The Wayne Power Co., Sodus, plans to construct a 30 mi. high tension transmission line from here to Wallace. Estimated cost, between $35,000 and $40,000.

N. Y., New York (Borough of Bronx)—The Bd. of Educ., 500 Park Ave., Manhattan, will receive bids about March 4 for the construction of a 5 story school at Crotona Park and Charlotte Ave. A steam heating system will be installed in same. Total estimated cost, $500,000. C. B. J. Snyder, Municipal Bldg., Manhattan, Archt. and Engr.

N. Y., New York (Borough of Bronx)—The Bd. of Educ., 500 Park Ave., Manhattan, will receive bids about March 4 for the construction of a 5 story, 230 x 280 ft. school on Southern Blvd. and Leggett Ave. A steam heating system will be installed in same. Total estimated cost, $600,000. C. B. J. Snyder, Municipal Bldg., Manhattan, Archt. and Engr.

N. Y., Brooklyn—Dodge & Morrison, Archts. and Engrs., 135 Front St., New York City, will soon award the contract for the construction of a 4 story, 50 x 100 ft. factory at 28 Waverly Ave., for the Empire Biscuit Co., 30 Waverly Ave. A steam heating system will be installed in same. Total estimated cost, $100,000.

N. Y., Long Island City—The Title Guarantee & Trust Co., 176 B'way, New York City, is having plans prepared for the construction of an office building on Hunter Ave. and the Queenboro Bridge Plaza. A steam heating system will be installed in same. Total estimated cost, $100,000. Severence & Van Allen, 111 East 40th St., New York City, Archts. and Engrs.

N. Y., Marcy—The State Hospital Comn., Capitol, Albany, N. Y., will receive bids until March 3 for the installation of a heating system in the tuberculosis pavilion and mortuary, at the Utica State Hosptal, here.

N. Y., New York—R. A. Gershee, c/o F. P. Kelly, Archt. and Engr., 477 5th Ave., is having plans prepared for the construction of a 25 story, 25 x 100 ft. office, show room and cafeteria at 138 West 43rd St. A steam heating system will be installed in same. Total estimated cost, $2,000,000.

N. Y., New York—The Canadian Bank of Commerce, 16 Exch. Pl., is having plans prepared for altering the present 8 story bank and office building, and building a 2 story addition. A steam heating system will be installed in same. Total estimated cost, $500,000. A. F. Gilbert, 80 Maiden Lane, Archt. and Engr.

N. Y., New York—R. Dunther and R. Law, Inc., 25 Broad St., are having plans prepared for the construction of a 7 story, 40 x 50 ft. office building at 131 Cedar St. A steam heating system will be installed in same. Total estimated cost, $160,000. A. F. Gilbert, 80 Maiden Lane, Archt. and Engr.

N. Y., New York—H. A. Jacobs, Archt. and Engr., 320 5th Ave., will soon award the contract for the construction of a 6 story, 25 x 100 ft. store and office building at 675 5th Ave., for H. D. Downs, 552 3rd Ave. A steam heating system will be installed in same. Total estimated cost, $85,000.

N. Y., New York—B. H. and C. N. Whinston, Archts. and Engrs., 2 Columbus Circle, will soon award the contract for the construction of three 5 story loft buildings at 117-119-121 West 33rd St., for the 121st West 33rd St. Corp. A steam heating system will be installed in same. Total estimated cost, $150,000.

N. Y., New York—The Munson Steamship Line, 82 Beaver St., is having plans prepared for the construction of a 25 story, 22 x 144 x 160 ft. office building on Pearl St., between Beaver and Wall Sts. A steam heating system will be installed in same. Total estimated cost, $2,000,000. K. M. Murchison, 101 Park Ave., Archt. and Engr.

N. Y., New York—The New York Telephone Co., 15 Dey St., is having plans prepared for the construction of a 5 story telephone exchange on 30th St. and 2nd Ave. A steam heating system will be installed in same. Total estimated cost, $1,000,000. McKenzie, Voorhies & Gmelin, 1123 B'way, Archts. and Engrs.

N. Y., New York—F. E. Vibolo, Archt. and Engr., 56 West 45th St., is preparing plans for the construction of a 10 story office building on Franklin St. and West B'way. A steam heating system will be installed in same. Total estimated cost, $200,000. Owner's name withheld.

N. Y., Oneida—The Adirondack Electric Power Corp., 16½ Lenox Ave., plans to extend its transmission line from here to Munnsville and Stockbridge. Estimated cost, between $12,000 and $15,000.

N. Y., Peekskill—St. Mary's Community is having plans prepared for the construction of a school and gymnasium. A steam heating system will be installed in same. Total estimated cost, $325,000. C. P. H. Gilbert, 1123 B'way, New York City, Archt. and Engr.

N. Y., Watertown—H. H. Babcock, Factory Sq., plans to increase the hydro-electric power by modern development of present rights. Plans include construction of power house and installation of generator, etc. Total estimated cost, $100,000.

N. Y., Wʌodhaven—The Merritt Hosiery Co., c/o Bloch & Hesse, Archts., 18 East 41st St., will soon award the contract for the construction of a 4 story, 60 x 180 ft. factory on 104th St. A steam heating system will be installed in same.

N. J., Glassboro—The state plans to build a 2 story, 42 x 79 ft. power house, to be used for the proposed normal school, including a 26 x 79 ft. coal bunker, 16 x 32 ft. water tank, fire pump, heater, three 360 hp. boilers and 2 generators. Total estimated cost, $100,000. F. H. Bent, 142 West State St., Trenton, Archt.

N. J., Jersey City—The Continental Can Co. is having plans prepared for the construction of a 4 story factory at 15th, 16th, Cole and Monmouth Sts. A steam heating system will be installed in same. Total estimated cost, $750,000. Francisco & Jacobus, 511 5th Ave., New York City, Archts. and Engrs.

N. J., Rahway—The state plans to build a 2 story, 45 x 120 ft. refrigerator plant, etc., at the Rahway Reformatory. Estimated cost, $20,000. F. H. Bent, 142 West State St., Trenton, Archt.

N. J., South River—The Borough Pres. will receive bids about March 1 for the construction of a 1 story, 55 x 100 ft. power house, for the Bd. Pub. Wks. Estimated cost, $125,000. Goss, Bryce & Johnson, 55 Liberty St., New York City, Archts. and Engrs.

Pa., Bethlehem—The Bethlehem Natl. Bank plans to build a 5 story, 40 x 150 ft. bank. A steam heating system will be installed in same. Total estimated cost, $150,000.

Pa., Philadelphia—The Arcadia Cafe, Widener Bldg., will have plans prepared for the construction of a 1 story chateau at Fairmount Park. A steam heating system will be installed in same. Total estimated cost, $300,000. Ralph E. White, Penna Bldg., Archt. and Engr.

Pa., Philadelphia—The Apperson Motor Car Co., Broad and Race Sts., is having plans prepared for the construction of a 6 story, 75 x 100 ft. service building on Girard and Broad Sts. A steam heating system will be installed in same. E. H. Yardley, 1713 Sansom St., Archt.

Pa., Philadelphia—Heacock & Hohansen, Archts. and Engrs., Bailey Bldg., will soon award the contract for the construction of a 4 story, 40 x 90 ft. warehouse at 810 Queen St., for the Hagy Waste Wks., 836 South Swanson St. A steam heating system will be installed in same.

Pa., Philadelphia—The Keller Eng. Serv. Co., Otis Bldg., is having plans prepared for the construction of a 4 story, 70 x 125 ft. store and office building on 16th and Sansom Sts. A steam heating system will be installed in same. Total estimated cost, $150,000. Simon & Simon, 249 South Juniper St., Archts.

Pa., Phillipsburg—George S. Idell, Archt., 1705 Chestnut St., Philadelphia, will soon award the contract for the construction of a 6 story, 80 x 100 ft. hotel building, here, for the Phillipsburg Hotel Corp. A steam heating system will be installed in same. Total estimated cost, $100,000.

Pa., Pittsburgh—William Allen Balch, Archt., 756 Woodward Ave., Detroit, is preparing plans for the construction of a 5 story, 57 x 224 garage and service station. Steam heating equipment will be installed in same. Total estimated cost, $200,000. Owner's name withheld.

Pa., Reading—S. S. Kresge Co., Kresge Bldg., Detroit, engaged J. E. Mills & Son, Archts., 1305 Kresge Bldg., Detroit, to prepare plans for altering and constructing a 4 story, 50 x 270 ft. addition to mercantile building. Steam heating equipment will be installed in same. Total estimated cost, $100,000.

Pa., Scranton—Edward H. Davis and George M. D. Lewis, Associate Archts., Union Natl. Bank, are preparing plans for the construction of a 3 story, 50 x 145 ft. warehouse on Locka Ave. A steam heating system will be installed in same. Total estimated cost, $110,000. Owner's name withheld.

Pa., Scranton—Edward H. Davis and George M. D. Lewis, Archts., Union Natl. Bank Bldg., are preparing plans for the construction of a 3 story, 55 x 150 ft. factory on Washington Ave. A steam heating system will be installed in same. Total estimated cost, $175,000. Owner's name withheld.

Pa., Windber—The Windber Hospital is having plans prepared for the construction of 2 story, 35 x 90 ft. and 30 x 50 ft. hospital buildings. A steam heating system will be installed in same. Henry L. Reinhold Jr., 1513 Walnut St., Philadelphia, Archt.

O., Cincinnati—The Second Church of Christ Scientist will soon receive bids for the construction of a 2 story, 90 x 120 ft. church on Clifton and Probasco Aves. An indirect heating and ventilating system will be installed in same. Total estimated cost, $175,000. Arthur Neal Robinson, Candler Bldg., Atlanta, Ga., Archt.

O., Cleveland—The Babies Dispensary & Hospital, 2500 East 35th St., is having plans prepared for the construction of a 7 story, 60 x 120 ft. hospital on Adelbert and Euclid Aves. A steam heating system will be installed in same. Total estimated cost, $600,000. E. B. Green, Treas. Abram Garfield, 915 Natl. City Bldg., Archt.

O., Cleveland—Christian, Schwarzenburg & Gaede Co., Archts. and Engrs., 1900 Euclid Ave., will soon award the contract for the construction of a 3 story, 96 x 221 ft. textile factory on Buckeye Rd. near East 104th St., for Schaffner Bros., 1013 Oregon Ave. Two boilers will be installed in same. Total estimated cost, $150,000.

O., Cleveland—The city will soon award the contract for installing two 850 gal. per min. centrifugal boiler feed pumps. Estimated cost, $10,000. Edward Shattuck, Purch. Agt. R. Hoffman, City Hall, Engr.

O, Cleveland—The Creswell Realty Co., Union Bldg., is having plans revised for the construction of an 8 story, 100 x 125 ft. garage on Huron Rd. and East 12th St. A steam heating system will be installed in same. Total estimated cost, $500,000. W. S. Ferguson Co., 1900 Euclid Ave., Archts. and Engrs.

O., Cleveland—The Detroit Lake Ave. Co., Sloan Bldg., will receive bids after March 1 for the construction of a 6 story, 190 x 232 ft. theatre and office building on Detroit and Lake Aves. A steam heating system will be installed in same. Total estimated cost, $600,000. G. W. Armstrong, Pres.

O., Cleveland—Emmett P. and John L. Dowling, Williamson Bldg., plan to build a 6 story commercial building on East 14th St. and Caton Court. A steam heating system will be installed in same. Total estimated cost, $250,000.

O., Cleveland—The Labor Lyceum Co., 2460 East 9th St., plans to build an 8 story Labor Temple on East 18th St. near Prospect Ave. A steam heating system will be installed in same. Total estimated cost, $300,000. Charles Smith, Secy. W. S. Longee, Marshall Bldg., Archt.

O., Cleveland—Charles Neuburger, Statler Hotel, Euclid Ave. and East 12th St., plans to build a 2 or 8 story, 75 x 200 ft. commercial building on Euclid Ave., near East 32nd St. A steam heating system will be installed in same. Total estimated cost, $200,000.

O., Cleveland—The Realty & Investment Co., c/o G. A. Tenbusch, Union Bldg., will soon award the contract for the construction of a 6 story, 100 x 200 ft. commercial building on East 65th St. and Euclid Ave. A steam heating system will be installed in same. Total estimated cost, $400,000. W. S. Ferguson Co., 1900 Euclid Ave., Archt. and Engr.

O., Cleveland—Frank N. Riley, 309 Williamson Bldg., is having plans prepared for the construction of a 4 story, 116 x 240 ft. commercial building on East 79th St. and Hough Ave. A steam heating system will be installed in same. Total estimated cost, $250,000. J. Milton Dyer, Amer. Trust Bldg., Archt.

O., Cleveland—St. Luke's Hospital, 6606 Carnegie Ave., is having preliminary plans prepared for the construction of a 6 story hospital on Fairmount Blvd. Three boilers with necessary equipment will be installed in same. Total estimated cost, $750,000. Hubbell & Benes, 4500 Euclid Ave., Archts.

O., Columbus—The Bd. Comrs. Franklin Co., Court House, will receive bids until March 4 for the construction of a 2 story tuberculosis hospital on the Infirmary grounds. A steam heating system will be installed in same. Total estimated cost, $125,000. W. H. Tremaire, 602 Chamber of Commerce Bldg., Archt.

O., Lakewood (Cleveland P. O.)—The Bd. Educ. plans to build a 2 story school to be known as the Lincoln School. A steam heating system will be installed in same. Total estimated cost, $300,000. G. W. Grill, Warren Rd., Clk. C. W. Hopkinson, 900 Rose Bldg., Cleveland, Archt.

O., Massillon—William Tilberth, Clk. Bd. Educ., will soon award the contract for the construction of a 2 story, 70 x 95 ft. school. A steam heating system will be installed in same. Total estimated cost, $125,000. O. D. Howard, 8 East Broad St., Columbus, Archt.

O., Norwood—The city plans to restore the municipal electric light plant which was destroyed by fire. Sidney Crew, City Hall, Engr.

O., Parma—The Bd. Educ. plans to build three 1 story, 60 x 120 ft. school buildings. Steam heating systems will be installed in same. Total estimated cost, $400,000. Charles W. Bates, Natl. Bank Bldg., Wheeling, W. Va., Archt.

O., Springfield—S. S. Kresge Co., High St., is having plans prepared for the construction of a 4 story, 50 x 130 ft. business building on Main St. A steam heating system will be installed in same. Total estimated cost, $150,000. Hall & Lethly, 807 Fairbanks Bldg., Archts.

O., Troy—The city plans to build a 1 story, 40 x 60 ft. electric light plant on East Water St. Estimated cost, $20,000. Address V. S. Deaton, Mayor.

Mich., Bay City—A. E. Munger, Archt., 420 Shearer Office Bldg., will soon award the contract for the construction of a 3 story, 250 x 350 ft. automobile body factory addition, for the Wilson Body Co. A steam heating system will be installed in same. Total estimated cost, $500,000.

Mich., Detroit—The Bd. Educ., 50 B'way, engaged Malcolmson, Higginbotham & Palmer, Archts., 405 Moffat Bldg., to prepare plans for the construction of a 2 story, 90 x 150 ft. school on the north side of Charles Ave. A Plenum steam heating system and ventilating system will be installed in same. Total estimated cost, $200,000.

Mich., Detroit—The Detroit Baptist Union, 1105 Vinton Bldg., plans to build a 2 story, 64 x 120 ft. church on Jefferson Ave. and Lakewood Blvd. A steam heating system will be installed in same. Total estimated cost, $125,000. Dalton R. Wells, 435 Woodward Ave., Archt. and Engr.

Mich., Detroit—The Standard Accident Insurance Co., 34 Fort St., W., engaged Albert Kahn, Archt., Marquette Bldg., to prepare plans for the construction of an 8 story office building on Bagg St. and Cass Park. A steam heating system will be installed in same. Total estimated cost, $500,000.

Mich., Flint—St. Michael's Parish, 6th St., is having plans prepared for the construction of a 2 story, 62 x 245 ft. school on 6th St. A steam heating system will be installed in same. Total estimated cost, $250,000. Van Leyen, Schilling & Keough, 1115 Union Trust Bldg., Detroit, Archts. and Engrs.

Mich., Northville—C. Frieburger, Secy. Bd. of Health, 233 St. Antoine St., Detroit, will receive bids until March 1 for the construction of a 1 story, 40 x 120 ft. power plant, in connection with the group of buildings to be constructed at the proposed Detroit Municipal Tuberculosis Sanitarium. A steam heating system, including feed water heater, pump, etc., will be installed in same. Stratton & Snyder, 1103 Union Trust Bldg., Detroit, Archts.

Mich., Northville—L. A. Young, 93 Rhode Island Ave., Detroit, is having plans prepared for the construction of a country estate, including a 150 x 350 ft. barn and a 2 story, 30 x 50 ft. garage. A steam heating system will be installed in same. Total estimated cost, $200,000. C. W. Brandt, Kresge Bldg., Detroit, Archt.

Ill., Chicago—H. C. Miller, Archt., 112 West Adams St., will soon award the contract for the construction of a 3 story, 120 x 140 ft. paper box factory at 1701 West Superior St., for the Kroeck Paper Box Co., 220 Institute Pl. A steam heating system will be installed in same. Total estimated cost, $175,000.

Wis., Manitowoc—The city plans to raise funds for the construction of a 3 story, 60 x 180 ft. armory. A steam heating system will be installed in same. Total estimated cost, $125,000. A. Zarder, Clk.

Wis., Milwaukee—Geuder, Paeschke & Frey, St. Paul Ave. and 15th St., are having plans prepared for the construction of a 1 story, 75 x 100 ft. power plant on St. Paul Ave. Three 300 hp. W.T. boilers, superheater, switchboard and handling equipment will be installed in same. Total estimated cost, $150,000. Cahill & Douglas, 217 West Water St., Engrs.

Wis., Saukville—H. J. Cary, Pres. of the Ozaukee Heating Co., has purchased a site and plans to build a 2 story, 36 x 376 ft. pea canning factory on Main St., and is in the market for several boilers, etc. Total estimated cost, $40,000.

Wis., Sheboygan—The Security Natl. Bank, 529 North 8th St., will soon receive bids for the construction of a 4 story, 60 x 110 ft. bank and office building on Center and 8th Sts. A steam heating system will be installed in same. Total estimated cost, $250,000.

Wis., Waupaca—J. G. D. Mack, State Chief Engr., Capitol Bldg., Madison, will receive bids until March 2 for the entire reconstruction of the power house at the Wisconsin Veterans' Home, here.

Iowa, Fairfield—The Bd. Educ. will soon receive bids for the construction of a 2 story, 68 x 118 ft. junior high school. A steam heating system will be installed in same. Total estimated cost, $150,000. W. Gordon, 317 Hubbell Bldg., Des Moines, Archt.

Iowa, Grinnell—The Bd. Educ. plans to build a 2 story high school. A steam heating system will be installed in same. Total estimated cost, $300,000. W. Ray, Secy.

Minn., Arlington—The village Clerk will soon award the contract for the installation of one power pump head, double or single stroke, 24 in. stroke, to pump water from well 317 ft. deep, 2 in. cylinder, being 160 ft. in well, into elevated tank 110 ft. from ground. Driven by 18 in. pulley running 327 r.p.m., pulley 1 pump head to conform to give 30 strokes per min.

Minn., Duluth—The Duluth Builders' Exch., 201 Glencoe Bldg., 1st St., plans to construct a 10 story, 100 x 140 ft. office building on 1st St., W. A steam heating system will be installed in same. Total estimated cost, $750,000.

Minn., Minneapolis—The Bureau of Engraving, 17 South 6th St., is having plans prepared for the construction of a 4 story, 100 x 130 ft. business building on Mary Pl. at 12th St. A steam heating system will be installed in same. Total estimated cost, $200,000.

Kan., Abilene—The city plans to extend the water mains and build a 100,000 gal. tank on an 80 ft. tower. Total estimated cost, $40,000.

Neb., Omaha—The Ames Realty Co., 2406 Ames Ave., will soon award the contract for the construction of a 3 story, 125 x 133 ft. theatre, store and apartment building on 25th St. and Ames Ave. A steam heating system will be installed in same. Total estimated cost, $200,000. Everett S. Dodds, 715 Brandeis Theatre Bldg., Archt.

Mo., Carthage—Shepard & Wiser, Archts., R. A. Long Bldg., Kansas City, will soon award the contract for the construction of a 5 story, 75 x 90 ft. building on 4th and Howard Sts., for the Chamber of Commerce. A hot water heating system and electric elevator will be installed in same. Total estimated cost, $140,000.

Mo., Kansas City—Keene & Simpson, Archts., 400 Reliance Bldg., are receiving bids for installing a steam heating system in the proposed 10 story, 110 x 115 ft. office building on 19th and Walnut Sts., for T. C. Bourke, 3520 B'way.

Okla., Pryor Creek—The city plans to hold an election to vote on $75,000 bonds for the construction of a filter plant, and 8 mi. of 200 kw. 6,600 volt transmission line transformers at both ends of line, also the installation of two 2,200 volt, 75 kw. high pressure motor driven centrifugal pumps and two 15 kw. and one 20 kw. low pressure motor driven centrifugal pumps. Total estimated cost, $75,000. V. V. Long & Co., 1200 Colcord Bldg., Oklahoma City, Engr.

Colo., Boulder—Walter S. and L. D. Cambell head company which has been formed to build a pipe line and power house, here. Total estimated cost, $150,000.

Wash., Spokane—The Bd. of Trustees of Whitworth Presbyterian College, Spokane Estates, plan to build a 1 and 2 story building to cost $100,000, women's dormitory, $75,000, and a central heating plant, $75,000, on the college campus. A. Beatie, Pres.

Ore., Canby—The city plans to reconstruct the pump house and install new pumping machinery for the waterworks system. Estimated cost, $7,500. A special election will be held April 19 to vote on bonds for this project.

Ore., Klamath Falls—The Ewauna Box Co. plans to install 1 gang outfit, with a 100,000 ft. capacity every 8 hours, one 9 ft. band saw, also 1 steam boiler. Cost to exceed $10,000. F. H. Hunter, Archt.

Ore., Lebanon—The Lebanon Electric Light & Water Co. plans to double the size of its power plant by constructing a new plant on the west side of Main St. and north side of Canal St., including 3 mi. flume, 2 water wheels, concrete dam, etc., also the installation of new pumping equipment and electric motors. Total estimated cost, $75,000.

Ore., Portland—The Comrs. Multnomah Co., Court House, plan to install one 40 hp. gas or oil burning engine at Hoyt Quarry. Estimated cost, $1,000. R. W. Hoyt, Chmn.

Cal., Bakersfield—The San Joaquin Light & Power Co., 122 Tulare St., plans to build a power plant 35 mi. west of here, where natural gas now going to waste will be used to generate electric power. First unit will cost $1,000,000; ultimate cost, $6,000,000.

Cal., San Diego—The Pub. Wks. Office, 12th Naval Dist., Timken Bldg., San Diego, will receive bids until March 3 for furnishing and installing complete power plant and electric generating equipment, including piping, distribution systems for electric and steam power, 2 oil burning boilers, oil pumps, feed water heater, 2 engine driven alternators, 1 engine driven and 1 electric driven excitor, transformers, vacuum pump, separating tank, etc., for the Marine Corps Base, here. Total estimated cost, $150,000.

Cal., Lancaster—The Bd. Supervs. of Los Angeles Co. will soon award the contract for the installation of a distributing system, well and pit, 75,000 gal. steel tank on a 90 ft. steel tower, vertical plunger pump with motor to lift 170 gal. per min., and construction of pump house, in Los Angeles Co. Waterworks Dist 4. Estimated cost, $30,000. W. Davidson, Hall of Records, Los Angeles, Engr.

Cal., Long Beach—The city voted $500,000 bonds to improve the water system, including reservoir, pumping equipment and mains. H. C. Waughop, City Clk. A. DeRuez, City Engr.

Cal., Orange—The City Trustees plan to hold an election to vote on bonds to install a well and pump. D. G. Wettlin, City Clk. W. J. Richardson, Water Supt.

Cal., Tujunga—The Haines Canyon Water Co. is having plans prepared for the construction of 2 earth reservoirs with 10 in. connecting pipe line and motor driven pumps. Total estimated cost, $17,000. H. B. Lynch, 314 South Brand Bldg., Glendale, Engr.

Ont., London—L. V. Carrothers, Archt., Hydro Offices, will receive bids until April 1 for the construction of two 3 story, 100 x 185 ft. schools in the southern part of the city, for the Bd. Educ., City Hall. A steam heating system will be installed in same. Total estimated cost, $275,000.

CONTRACTS AWARDED

Me., Deer Rips (Lewiston P. O.)—The Androscoggin Light & Power Co. has awarded the contract for the construction of a power house addition, here, to the Landeis Engr. Co., 102 Exch. St., Portland. Boilers, electrical equipment and heating and power machinery will be installed in same. Total estimated cost, $50,000.

Mass., Boston—The Federal Reserve Bank, 58 State St., has awarded the contract for the construction of a 5 story bank on Pearl and Milk Sts., to L. P. Soule & Co., 80 Boylston St. A steam heating system will be installed in same. Total estimated cost, $4,500,000.

Mass., Boston—S. S. Kresge Co., Kresge Bldg., Detroit, Mich., has awarded the contract for the construction of a 6 story, 90 x 90 ft. store building on Washington and Temple Sts., to George A. Fuller Co., 710 Bd. of Trade Bldg. A steam heating system will be installed in same. Total estimated cost, $400,000.

Mass., Lowell—The city has awarded the contract for the construction of a 2 story, 150 x 150 ft. memorial hall on River St., to George Drapeau. A steam heating system will be installed in same. Total estimated cost, $750,000.

Mass., Robert's Crossing (Boston P. O.)—W. C. Welch Co. has awarded the contract for the construction of a 2 story, 40 x 100 ft. factory, including boiling house, fire room and garage, to R. F. Morrison Co., Charlestown (Boston P. O.).

R. I., Providence—L. E. Danforth, 76 Dorrance St., will build a 3 story, 90 x 100 ft. factory and office building on Willard Ave. A steam heating system will be installed in same. Total estimated cost, $125,000. Work will be done by day labor.

Conn., Derby—The Derby & Ansonia Brewing Co., Derby Ave., has awarded the contract for the construction of a 2 story, 33 x 100 ft. ice plant on Derby Ave., to M. A. Durrschmidt, Main St. Estimated cost, $25,000.

Conn., Derby—The Derby Gas Co., 22 Elizabeth St., has awarded the contract for the construction of a 2 story, 42 x 100 ft. boiler house on Housatonic Ave., to M. A. Durrschmidt, Main St. Estimated cost, $35,000.

N. Y., Buffalo—The Dunlap Rubber Co., 1808 B'way, New York City, has awarded the contract for the construction of a factory consisting of twelve 1 and 2 story buildings, to the Foundation Co., 233 B'way, New York City. A steam heating system will be installed in same. Total estimated cost, $20,000,000.

N. Y., Long Island City—The Metropolitan Life Insurance Co., 1 Madison Ave., New York City, has awarded the contract for the construction of a 5 story, 50 x 400 ft. printing building on Court St. and Thompson Ave., to the Turner Constr. Co., 244 Madison Ave., New York City. A steam heating system will be installed in same. Total estimated cost, $1,000,000.

N. Y., New York (Borough of Bronx)—The Jackson Film Corp., 143 B'way, Manhattan, will build a 180 x 224 ft. film studio on Webster and Jackson Aves. A steam heating system will be installed in same. Work will be done by day labor.

N. Y., Brooklyn—The Kraslow Bldg. Co., Inc., 190 Montague St., will build a 12 story office building at 180-186 Montague St. A steam heating system will be installed in same. Total estimated cost, $1,000,000. Work will be done by day labor.

N. Y., Brooklyn—L. Mundet & Sons, 59 Pearl St., New York City, will build a 4 story, 200 x 200 ft. factory on King and Ferris Sts., Ridgewood. A steam heating system will be installed in same. Total estimated cost, $400,000. Work will be done by day labor.

N. Y., Brooklyn—J. Rubin & Son, 142 Greene Ave., will build a 3 story, 100 x 100 ft. factory on Grand and Park Aves. A steam heating system will be installed in same. Total estimated cost, $150,000.

N. Y., Brooklyn—The Union Elton Co., 1784 Pitkin Ave., will build four 4 story, 40 x 90 ft. factory buildings at 2402-2430 Atlantic Ave. A steam heating system will be installed in same. Total estimated cost, $200,000. Work will be done by day labor.

N. Y., New York—B. Altman, 5th Ave., 34th St., has awarded the contract for the construction of an 8 story, 60 x 100 ft. department store at East 35th St., to Marc Eidlitz, 30 East 42nd St. A steam heating system will be installed in same.

N. Y., New York—H. L. Schraffts Co., 157 West 54th St., has awarded the contract for the construction of a 4 story factory at 161 West 64th St., to the Bethlehem Eng. Co., 527 5th Ave. A steam heating system will be installed in same. Total estimated cost, $140,000.

N. J., Athenia—The Magor Car Co. has awarded the contract for the construction of a 1 story, 40 x 60 ft. power house, to W. Hassen, 625 Main St., Passaic. A steam heating system will be installed in same. Total estimated cost, $35,000.

N. J., Clifton—The Clifton Textile Co., 128 Hackensack Plank Rd., will build a 1, 2 and 3 story, 250 x 350 ft. factory, consisting of 6 buildings. A steam heating system will be installed in same. Total estimated cost, $600,000. Work will be done by day labor.

N. J., Jersey City—The W. M. Crane Co., Garfield Ave., has awarded the contract for the construction of two or three 1 story foundry buildings, to Ballinger & Perrot, 17th and Arch Sts., Philadelphia. A steam heating system will be installed in same. Total estimated cost between $200,000 and $300,000.

N. J., Trenton—The Agasote Millboard Co., Fernwood, has awarded the contract for the construction of a 1 story, 40 x 40 ft. boiler house and stack, 5 ft. base and 150 ft. high, to N. A. K. Bugbee, 206 East Hanover St. Total estimated cost, $15,000.

Pa., Johnstown—The Johnstown School Dist., 601 Swank Bldg., has awarded the contract for installing a steam heating system in the proposed 2 story junior high school, to the Swank Hardware Co., Johnstown, at $30,224.

Pa., Philadelphia—Friedberger & Arson, 18th and Courtland Sts., have awarded the contract for the construction of a 3 story, 74 x 100 ft. factory addition on Courtland and Gratz Sts., to W. Steele & Sons Co., 16th and Arch Sts. A steam heating system will be installed in same. Total estimated cost, $100,000.

N. C., Winston-Salem—P. H. Hanes Knitting Co. has awarded the contract for the construction of a 6 story, 80 x 200 ft. knitting mill and dye house to the Southern Ferro Concrete Co., Trust Co. of Ga. Bldg., Atlanta. A steam heating system and electric power will be installed in same. Total estimated cost, $300,000.

O., Bellevue—The Bd. Educ. has awarded the contract for the construction of a 3 story, 80 x 90 ft. school, to Willing Bros., Bellevue. A steam heating system will be installed in same. Total estimated cost, $120,000.

O., Toledo—The Standard Oil Co. of Ohio, East Ohio Gas Bldg., Cleveland, has awarded the contract for the construction of a 1 story 63 x 204 ft. boiler house for its oil refinery, to James Stewart & Co., 30 Church St., New York City. Estimated cost, $100,000.

O., Troy—The city has awarded the contract for furnishing and installing a 1,000 kw. electric turbine generator for the electric light plant, to the Westinghouse Electric & Mfg. Co., Schenectady, N. Y. Address V. C. Deaton.

Ind., Indianapolis—The U. S. Corrugated Fiber Box Co. has awarded the contract for the construction of a 3 story, 100 x 160 ft. factory on Roosevelt and Martindale Aves., to C. J. Wacker, 628 Law Bldg. A steam heating system and power will be installed in same. Total estimated cost, $125,000.

Mich., Lansing—The Federal Drop Forge Co. has awarded the contract for the construction of a 1 story, 60 x 120 ft. drop forge plant, to the Christmen Constr. Co., South Bend, Ind. A power plant will be included in the project.

Ill., Chicago—The Agar Provision Co., 310 North Green St., has awarded the contract for the construction of a 4 story, 125 x 190 ft. factory on Fulton and Green Sts., to Wells Bros. Constr. Co., 53 West Jackson Blvd. A steam heating system will be installed in same. Total estimated cost, $400,000.

Ill., Chicago—The William Davies Co., Ltd., 4101 South Union Ave., has awarded the contract for the construction of a 6 story, 100 x 130 ft. meat canning plant on Uion Ave. and 41st St., to Wells Bros. Constr. Co., 53 West Jackson Blvd. A steam heating system will be installed in same. Total estimated cost, $300,000.

Ill., Chicago—Charles Lange & Bros. Co., 2740 Armita-- Ave., has awarded the contract for the construction of a 4 story, 75 x 125 ft. auto sales and service building on Milwaukee and Kedzie Aves., to G. Kehl & Son Co., 1225 North Maplewood St. A steam heating system will be installed in same. Total estimated cost, $135,000.

Ill., Chicago—A. Plamondon Mfg. Co., 24 North Clinton St., has awarded the contract for the construction of a 1 story, 100 x 500 ft. machine shop, foundry and factory on 52nd St. and Western Ave., to the Overland Constr. Co., 29 South La Salle St. A steam heating system will be installed in same. Total estimated cost, $250,000.

Ill., Chicago—The Tuthill Spring Co., 760 West Polk St., has awarded the general contract for the construction of a 1 and 2 story, 210 x 420 ft. auto spring factory on 31st St. and Kilbourne Ave., to C. B. Johnson & Son, 111 West Washington St. A steam heating system will be installed in same. Total estimated cost, $225,000.

Iowa, Whiting—The Bd. of Educ. of the Consolidated Dist. has awarded the contract for installing a heating system in the proposed 2 story, 60 x 150 ft. school, to Thomas Reinhardt, Le Mars, Iowa, at $34,000.

Minn., Duluth—The United Holding Co. has awarded the contract for the construction of a 2 story, 74 x 140 ft. medical building and clinic for surgery on 1st St. and 10th Ave., to the Bowman Bldg. Co., 303 Exch. Bldg. A steam heating system will be installed in same. Total estimated cost, $125,000.

Neb., Omaha—Trimble Bros., 11th and Howard Sts., has awarded the contract for the construction of a 3 story, 126 x 136 ft. warehouse and office building on 8th and Jackson Sts., to Grant Parsons, 616 Keeline Bldg. A steam heating system will be installed in same. Total estimated cost, $200,000.

Mo., Kansas City—The Mutual Drug Co., 1200 West 9th St., Cleveland, O., has awarded the contract for installing steam heating and plumbing systems in the proposed 4 story, 50 x 115 ft. business building on 20th and Main Sts., to the Magill Plumbing & Heating Co., 1320 McGee St.

Ariz., Yuma—The city has awarded the contract for the construction of 13 mi. of pipe line in Dist. 3, including a sewage pumping unit and concrete pump house, to G. S. Benson & Sons, Stimson Bldg., Los Angeles, Cal., at $91,324.

Wash., Kennewick—The Columbia Yakima Storage Co. has awarded the contract for the construction of a 2 story, 100 x 250 ft. cold storage building, to Julian S. Geneva Co., Jefferson, Iowa. A steam heating system and a 10 ton ice plant and refrigeration machinery will be installed in same. Total estimated cost, $80,000.

Ore., Portland—Wells Bros. Contg. Co., 53 West Jackson Blvd., Chicago, contractors for the proposed new building for Montgomery Ward & Co., 581 Upshur St., has awarded the contract for furnishing and installing plumbing, heating and mechanical equipment in same, to the Dausch-Weber Heating & Eng. Co., Oregon Bldg., at $150,000.

NEW INSTALLATIONS IN POWER

Pa., Wilkes-Barre—The School Bd. is in the market for 6 low pressure boilers for the heating plant. Robert Ireland, Supt. and Archt.

Md., Baltimore—The Machinery Clearing House, Wolf Bldg., 110 East Lexington St., is in the market for ¼ hp., single phase, 110 volt, 60 cycle motors. Address A. Kenny, Mgr.

Ga., Cogdell—J. M. Morse is in the market for shingle mill and lath mill machinery and for a 75 to 100 hp. boiler.

Ill., Downs Grove—The Municipal Electric Light & Waterworks plans to rebuild its transmission line. F. W. Allen, Supt.

Iowa, Britt—The Britt Light & Power Co. plans to install a 150 hp. engine and a 100 kw. generator. L. M. Goodman, Pres. and Mgr.

Iowa, Columbus—The Louisa Co. Power Co., Columbus Junction, plans to build about 40 mi. transmission line. R. L. Van Meter, Pres. and Mgr.

Iowa, Earlham—The Municipal Electric Light & Water Plant plans to extend its rural transmission lines. M. A. Thorp, Supt.

Iowa, Farragut—The Farragut Pub. Serv. Co. plans to build 2 transmission farm lines from their station. J. J. Whisler, Mgr.

Iowa, Hawarden—The Electric Light, Power & Water Dept. plans to install a 200 kva. engine generator, set boilers and build rural transmission lines. Otto Pyles, Supt.

Iowa, Jewel—The Henderson Light & Power Co. plans to build 2 transmission farm lines with 50 consumers. R. D. Henderson, Treas.

Iowa, Lorimor—The Lorimor Light & Power Co. plans to build 8 mi. 2,300 volt transmission farm lines. J. F. Smith, Mgr.

Iowa, Manson Lake—The Manson Light, Heating & Power Co. plans to change its d.c. generating system to a.c. in the spring. F. W. Mack, Mgr.

Iowa, Odebolt—The Odebolt Elec. Serv. Co. plans to build 30 mi. transmission farm lines and 12 mi. 6,600 volt transmission line to serve another town. H. E. Russell, Mgr.

Iowa, Sibley—The town plans to build a transmission line from here to Melvin. J. J. Shoemaker, City Hall, Supt.

Iowa, Sigourney—The Sigourney Electric Co. plans to double the capacity of its plant. H. L. Mann, Supt.

Iowa, Traer—The Traer Electric Plant plans to purchase a 400 hp. boiler in March and also construct a 40 mi., 2,300 volt transmission farm line extension. Edward Noyer, Mgr.

Iowa, Winterset—The Winterset Electric Light & Waterworks plans to change its entire d.c. generating system to a.c. Charles H. Smith, Supt.

Kan., Hiawatha—The Hiawatha Light, Power & Ice Co. is in the market for ice plant equipment and Diesel engine. Total estimated cost, $60,000.

Mont., Bridger—The Bridger Water & Light Co. plans to build 10 mi., 50,000 volt transmission line. J. S. Emmett, Secy.

PRICES -- MATERIALS -- SUPPLIES

These are prices to the power plant by jobbers in the larger buying centers east of the Mississippi. Elsewhere the prices will be modified by increased freight charges and by local conditions.

POWER-PLANT SUPPLIES

HOSE—

Fire	50-Ft. Lengths
Underwriters' 2½-in.	75c. per ft.
Common, 2½-in.	40%

Air	First Grade	Second Grade	Third Grade
¾-in. per ft.	$0.50	$0.33	$0.22

Steam—Discounts from List

First grade.....30% Second grade...40% Third grade....45%

RUBBER BELTING—The following discounts from list apply to transmission rubber and duck belting:

Competition.....65% Best grade.....35%
Standard.....50%

Note—Above discounts apply on new list issued July 1.

LEATHER BELTING—Present discounts from list in the following cities are as follows:

	Medium Grade	Heavy Grade
New York	45%	35%
St. Louis	40%	30%
Chicago	20%	10%
Birmingham	15%	15%
Denver	30%	20%

RAWHIDE LACING—20% for cut; 45c. per sq. ft. for ordinary.

PACKING—Prices per pound:

Rubber and duck for low-pressure steam	$1.00
Asbestos for high-pressure steam	1.70
Duck and rubber for piston packing	1.00
Flax, regular	1.20
Flax, waterproofed	1.70
Compressed asbestos sheet	.90
Wire insertion asbestos sheet	1.50
Rubber sheet	.50
Rubber sheet, wire insertion	.70
Rubber sheet, duck insertion	.50
Rubber sheet, cloth insertion	.30
Asbestos packing, twisted or braided and graphited, for valve stems and stuffing boxes	1.30
Asbestos wick, ½- and 1-lb. balls	.85

PIPE AND BOILER COVERING—Below are discounts and part of standard lists:

PIPE COVERING		BLOCKS AND SHEETS	
Pipe Size	Standard List Per Lin.Ft.	Thickness	Price per Sq.Ft.
1-in.	$0.27	½-in.	$0.27
2-in.	.36	1 -in.	.30
6-in.	.80	1½-in.	.45
4-in.	.60	2 -in.	.60
3-in.	.45	2½-in.	.75
8-in.	1.10	3 -in.	.90
10-in.	1.30	3½-in.	1.05

85% magnesia high pressure		List
For low-pressure heating and return lines	4-ply	50% off
	3-ply	52% off
	2-ply	54% off

GREASES—Prices are as follows in the following cities in cents per pound for barrel lots:

	Cincinnati	Pittsburgh	Chicago	St. Louis	Birmingham	Denver
Cup	7	9	6.6	6.7	8½	14½
Fiber or sponge	8	10	8.6	13	8½	18
Transmission	7	9	8.1	13	8½	17
Axle	4½	6	4.8	4.75	4½	5½
Gear	4½	9	6.1	7.5	8½	8
Car journal	22 (gal.)	21 (gal.)	4.7	4.7	8½	8

COTTON WASTE—The following prices are in cents per pound:

	New York Current	New York One Year Ago	Cleveland	Chicago
White	13 00	11 00 to 13 00	14.00	11 00 to 14 00
Colored mixed	9.00 to 12.00	8.50 to 12 00	11.00	9.50 to 12 00

WIPING CLOTHS—Jobbers' price per 1000 is as follows:

	13½ x 13½	13½ x 20½
Cleveland	$52.00	$58.00
Chicago	41.00	43.50

LINSEED OIL—These prices are per gallon:

	New York Current	New York One Year Ago	Cleveland Current	Chicago Current	Chicago One Year Ago
Raw in barrels (5 bbl. lots)	$1.80	$1.49	$2.50	$1.98	$1.66
5-gal. cans	2.00	1.74	2.75	2.23	1.86

WHITE AND RED LEAD—Base price per pound:

	Red Current Dry	Red Current In Oil	Red 1 Year Ago Dry	Red 1 Year Ago In Oil	White Current Dry and In Oil	White 1 Yr. Ago Dry and In Oil
100-lb. keg	14.50	16.00	13.00	14.50	14.50	13.00
25- and 50-lb. kegs	14.75	16.25	13.25	14.75	14.75	13.25
12½-lb. keg	15.00	16.50	13.50	15.00	15.00	13.50
1-lb. cans	16 50	18.00	.....	.....	16.50	15.00
5-lb. cans	17.50	19.00	.....	.....	17.50	16.00

500 lb. lots less 10% discount; 2000 lb. lots less 10-2½%.

RIVETS—The following quotations are allowed for fair-sized orders from warehouse:

	New York	Cleveland	Chicago
Steel 7/16 and smaller	40%	55%	45%
Tinned	40%	55%	45%

Boiler rivets, ½, ¾, 1 in. diameter by 2 in. to 5 in. sell as follows per 100 lb.:
New York.. $5.00 Cleveland...$4.00 Chicago ... $4.97 Pittsburgh $4.72

Structural rivets, same sizes:
New York...$5.10 Cleveland...$4.10 Chicago.....$5.07 Pittsburgh $4.82

REFRACTORIES—Following prices are f. o. b. works, Pittsburgh:

Chrome brick	net ton	$75–80	at Chester, Penn.
Chrome cement	net ton	45–50	at Chester, Penn
Clay brick, 1st quality fireclay	net M.	38–45	at Clearfield, Penn
Clay brick, 2nd quality	net M.	33–35	at Clearfield, Penn
Magnesite, dead burned	net ton	50–55	at Chester, Penn.
Magnesite brick, 9 x 4½ x 2½ in.	net ton	80–85	at Chester, Penn.
Silica brick	net M.	45–50	at Mt. Union, Penn

Standard size fire brick, 9 x 4½ x 2½ in. The second quality is $4 to $5 cheaper per 1000.

St. Louis—Fire Clay, $35 to $50.
Birmingham—Silica, $50; fire clay, $45; magnesite, $80; chrome, $80.
Chicago—Second quality, $25 per ton.
Denver—Fire clay, $11 per ton.

BABBITT METAL—Warehouse prices in cents per pound:

	New York Current	New York One Year Ago	Cleveland Current	Cleveland One Year Ago	Chicago Current	Chicago One Year Ago
Best grade	90 00	87 00	74.50	80.00	70.00	75.00
Commercial	50 00	42 00	20.50	21.50	15.00	15.00

SWEDISH (NORWAY) IRON—The average price per 100 lb., in ton lots, is:

	Current	One Year Ago
New York	$21–26	$19 00
Cleveland	20.00	20.00
Chicago	16.50	19.00

In coils an advance of 50c. usually is charged.
Domestic iron (Swedish analysis) is selling at 15c. per lb.

SHEETS—Quotations are in cents per pound in various cities from warehouse; also the base quotations from mill:

	Mill Carloads Pittsburgh	New York Current	New York One Year Ago	Cleveland	Chicago
No. 28 black	$4.35–4.85	$6.50–8.00	$6.52	$5.77	$6.00
No. 26 black	4.25–4.75	6.40–7.90	6.42	5.67	5 90
Nos. 22–24 black	4.20–4.71	6.35–7.35	6 37	5.62	5.85
Nos. 18–20 black	4.15–4.65	6.30–7.30	6 32	5.57	5 80
No. 16 blue annealed	3.75–3.20	6.02	5.72	5.17	5.02
No. 14 blue annealed	3.65–3.10	5.92	5.62	5.07	4.92
No. 10 blue annealed	3.55–4.00	5.50–6 00	5.52	4.97	4.82
No. 28 galvanized	5.70–6 20	7.50–9.00	7.77	7.12	7.25
No. 26 galvanized	5.40–5.90	7.20	7.47	6.82	6.95
No. 24 galvanized	5.25–5.75	7.05	7.32	6.57	6.52

PIPE—The following discounts are for carload lots, f. o. b. Pittsburgh; basing card of Jan. 1, 1919, for steel pipe and for iron pipe:

BUTT WELD

Inches	Steel Black	Galvanized	Inches	Iron Black	Galvanized
⅛, ¼ and ⅜	50½%	24%	¾ to 1½	30½%	23½%
½	54½%	40%			
¾ to 3	57½%	44%			

LAP WELD

2	50½%	35%	2	32½%	18½%
2½ to 6	53½%	41%	2½ to 6	34½%	21½%

BUTT WELD, EXTRA STRONG PLAIN ENDS

⅛, ¼ and ⅜	48½%	29%	¾ to 1½	39½%	24½%
½	51½%	39%			
¾ to 1½	50½%	43%			

LAP WELD, EXTRA STRONG PLAIN ENDS

2	48½%	37%	2	33½%	20½%
2½ to 4	51½%	40%	2½ to 4	35½%	23½%
4½ to 6	50½%	39%	4½ to 6	34½%	22½%

Stock discounts in cities named are as follows:

	New York Black	New York Galvanized	Cleveland Black	Cleveland Galvanized	Chicago Black	Chicago Galvanized
¾ to 3 in. steel butt welded	47%	31%	43½%	34½%	57½%	44%
2½ to 3 in. steel lap welded	42%	27%	45½%	39½%	53½%	41%

Malleable fittings. Class B and C, from New York stock sell at list + 22½%. Cast iron, standard sizes, 10-5% off.

BOILER TUBES—The following are the prices for carload lots, f. o. b. Pittsburgh:

Lap Welded Steel		Charcoal Iron	
3½ to 4½ in.	40½	3½ to 4½ in.	15½
2½ to 3¼ in.	30½	3 to 3¼ in.	1
2¼ in.	24	2½ to 2¾ in.	+1
1¾ to 2 in.	19½	2 to 2¼ in.	+16
		1¾ to 1¾ in.	+29

Standard Commercial Seamless—Cold Drawn or Hot Rolled

	Per Net Ton		Per Net Ton
1 in.	$327	1¾ in.	$207
1¼ in.	267	2 to 2¼ in.	177
1⅜ in.	257	2½ to 3¼ in.	167
1½ in.	207	4 in.	187
		4½ to 5 in.	207

These prices do not apply to special specifications for locomotive tubes nor to special specifications for tubes for the Navy Department, which will be subject to special negotiations.

ELECTRICAL SUPPLIES

ARMORED CABLE—

B. & S. Size	Two Cond M Ft	Three Cond. M Ft.	Two Cond. Lead M Ft.	Three Cond. Lead M Ft.
No. 14 solid	$104 00	$138.00	$164 00	$210.00
No. 12 solid	135.00	170 00	225.00	265.00
No. 10 solid	185 00	235.00	275.00	325.00
No. 8 stranded	285.00	375 00	520.00	500.00
No. 6 stranded	400 00	500.00	560.00	

From the above lists discounts are:

Less than coil lots	Net List
Coils to 1,000 ft.	10%
1,000 ft. and over	15%

BATTERIES, DRY—Regular No. 6 size red seal, Columbia, or Ever Ready:

	Each, Net
Less than 12	$0.45
12 to 50	.38
50 to 125 (bbl.)	.35
125 (bbl.) or over	.32

CONDUITS, ELBOWS AND COUPLINGS—Following are warehouse net prices per 1000 ft. for conduit and per 100 for couplings and elbows:

	Conduit		Elbows		Couplings	
Size, In.	Black 1,000 Ft. and Over	Galvanized 1,000 Ft. and over	Black 100 and Over	Galvanized 100 and Over	Black 100 and Over	Galvanized 1,100 and Over
⅜	$71.15	$76.25	$17 04	$18.18	$5.38	$5.74
½	71.15	76 25	17.04	18.18	6.25	6.70
¾	93.96	100.86	22.23	23.93	8.97	9 57
1	138 39	149.09	33.19	35.41	11.66	12.44
1¼	187 91	201.71	42.62	45.31	16.10	17.12
1½	224.69	241.18	56.82	60.62	19.89	21.15
2	302 29	324.49	104.17	110.77	26.52	28.20
2½	477 95	513 05	170.46	181.26	37.88	40.28
3	625 00	670.91	454 56	483 36	56.82	60.42
3½	779.24	834 44	1,003.82	1,067.42	75.76	80.66
4	945 03	1,010.43	1,160.08	1,233.57	94.70	100.70

5% cash 10 days.

From New York Warehouse—Less 5% cash.

Standard lengths rigid, 10 ft. Standard lengths flexible, ½ in., 100 ft. Standard lengths flexible, ⅝ to 2 in., 50 ft.

CONDUIT NON-METALLIC, LOOM—

Size I. D., In.	Feet per Coil	List, Ft.
[illegible]	250	$0.05½
[illegible]	250	.06
[illegible]	250	.09
[illegible]	200	.12
[illegible]	200	.15
[illegible]	150	.18
1	100	.25
1¼	100	.33
1½	Odd lengths	.40
2	Odd lengths	.55

Coils.....45% off
Less coils, 35% off

CUT-OUTS—Following are net prices each in standard-package quantities:

CUT-OUTS, PLUG

S. P. M. L	$0.11	T. P. to D. P. S. B	$0.24
D. P. M. L	.11	T. P. to D. P. T. B	.38
T. P. M. L	.26	T. P. S. B	.38
D. P. S. B	.19	T. P. D. B	.54
D. P. D. B	.37		

CUT-OUTS, N. E. C. FUSE

	0-30 Amp.	31-60 Amp.	60-100 Amp.
D. P. M. L	$0.33	$0.84	$1 68
T. P. M. L	.48	1.20	2.40
D. P. S. B	.42	1.05	....
T. P. S. B	.81	1.80	....
D. P. D. B	.78	2.10	....
T. P. D. B	1.35	3.60	....
T. P. to D. P. D. B	.90	2.52	....

FLEXIBLE CORD—Price per 1,000 ft. in coils of 250 ft.:

No. 18 cotton twisted	$22.00
No. 16 cotton twisted	27.00
No. 18 cotton parallel	26 50
No. 16 cotton parallel	34 50
No. 18 cotton reinforced heavy	34.00
No. 16 cotton reinforced heavy	42 00
No. 18 cotton reinforced light	30.00
No. 16 cotton reinforced light	36 75
No. 18 cotton Canvasite cord	28.50
No. 16 cotton Canvasite cord	32.00

FUSES, ENCLOSED—

250-Volt	Std. Pkg.	List
3-amp. to 30-amp.	100	$0.25
35-amp. to 60-amp.	100	.35
65-amp. to 100-amp.	50	.90
110-amp. to 200-amp.	25	2.00
225-amp. to 400-amp.	25	3.60
425-amp. to 600-amp.	10	5.50

600-Volt	Std. Pkg.	List
3-amp. to 30-amp.	100	$0.40
35-amp. to 60-amp.	100	.60
65-amp. to 100-amp.	50	1.50
110-amp. to 200-amp.	25	2.50
225-amp. to 400-amp.	25	5.50
450-amp. to 600-amp.	10	8.00

Discount: Less 1-5th standard package.....30%
1-5th to standard package.....40%
Standard package.....52%

FUSE PLUGS, MICA CAP—

0-30 ampere, standard package	$4.75C
0-30 ampere, less than standard package	6.00C

LAMPS—Below are present quotations in less than standard package quantities:

Straight-Side Bulbs Mazda B— Watts	Plain	Frosted	No. in Package	Pear-Shape Bulbs Mazda C— Watts	Clear	Frosted	No. in Package
10	$0.35	$0.38	100	75	$0.70	$0.75	50
15	.35	.38	100	100	1.10	1.15	24
25	.35	.38	100	150	1.65	1.70	24
40	.35	.38	100	200	2.20	2.27	24
50	.35	.38	100	300	3.25	3.35	24
60	.40	.45	100	400	4.30	4.45	12
100	.85	.92	24	500	4.70	4.85	12
				750	6.50	6.75	8
				1000	7.50	7.75	8

Standard quantities are subject to discount of 10% from list. Annual contracts ranging from $150 to $300,000 net allow a discount of 17 to 40% from list.

PLUGS, ATTACHMENT—

		Each
Hubbell porcelain No. 5406,	standard package 250	$0.24
Hubbell composition No. 5467,	standard package 50	.32
Benjamin swivel No. 903,	standard package 250	.20
Hubbell current taps No. 5638,	standard package 50	.40

RUBBER-COVERED COPPER WIRE—Per 1000 ft. in New York.

No.	Solid Single Braid	Solid Double Braid	Stranded, Double Braid	Duplex
14	$12.00	$16.00	$13.90	$28.50
12	13.25	15.70	18.05	30.07
10	18.30	21.00	23.85	41.50
8	25.54	28.60	32.70	56.77
6	.....	.....	51.40	.....
4	.....	.....	70.00	.....
2	.....	.....	101.80	.....
1	.....	.....	131.86	.....
0	.....	.....	160.00	.....
00	.....	.....	193.50	.....
000	.....	.....	235.20	.....
0000	.....	.....	288.60	.....

Prices per 1000 ft. for Rubber-covered Wire in Following Cities:

	Denver			Birmingham		
	Single Braid	Double Braid	Duplex	Single Braid	Double Braid	Duplex
14	$11.25	$14.00	$31.00	$12.50	$15.39	$29.00
10	21.85	28.25	56.75	25.10	34.73	55.50
8	31 05	38.30	77.95	34.75	58.59	70.50
6	48 80	52.65	......	57.50	82.11	......
4	70 00	75.30	......	81.65	95.50	......
2	104.85	112.45	......	......	140.20	......
1	136.85	146.15	......	......	190.90	......
0	......	194.25	......	......	231.33	......
00	......	239.60	......	......	281.23	......
000	......	294.00	......	......	343.22	......
0000	......	358.25	......	......	416.80	......

Pittsburg—23c. base; discount 50% St. Louis—30c. base.

SOCKETS, BRASS SHELL—

¼ In. or Pendant Cap. Key, Each	Keyless, Each	Pull, Each	¼ In. Cap. Key, Each	Keyless, Each	Pull, Each
$0.33	$0 30	$0.60	$0.39	$0.36	$0.66

Less 1-5th standard package.....+20%
1-5th to standard package.....+10%
Standard package.....−15%

WIRE, ANNUNCIATOR AND DAMPPROOF OFFICE—

No. 18 B. & S. regular spools (approx. 8 lb.)	45c. lb.
No. 18 B. & S. regular 1-lb. coils	46c. lb.

WIRING SUPPLIES—

Friction tape, ¾ in., less 100 lb. 50c. lb., 100 lb. lots	55c. lb.
Rubber tape, ¾ in., less 100 lb. 65c. lb., 100 lb. lots	60c. lb.
Wire solder, less 100 lb. 50c. lb., 100 lb. lots	45c. lb.
Soldering paste, 2 oz. cans Nokorode	$1.20 doz.

SWITCHES, KNIFE—

TYPE "C" NOT FUSIBLE

Size, Amp.	Single Pole, Each	Double Pole, Each	Three Pole, Each	Four Pole, Each
30	$0.42	$0.68	$1.02	$1.36
60	.74	1.22	1.84	2.44
100	1.50	2.50	3.76	5.00
200	2.70	4.50	6.76	9.00

TYPE "C" FUSIBLE, TOP OR BOTTOM

Size, Amp.	Single Pole	Double Pole	Three Pole	Four Pole
30	.70	1.06	1.60	2.12
60	1.18	1.80	2.70	3.60
100	2.38	3.66	5.50	7.30
200	4.40	6.76	10.14	13.50

Discounts:

Less than $10.00 list value	+15%
$10 to $25 list value	list
$25 to $50 list value	− 5%
$50 to $200 list value	−10%
$200 list value or over	−15%

POWER

Volume 51 | New York, March 9, 1920 | Number 10

Ford Uses Large Refrigeration Plant in Making Automobiles

BY ERIC H. PETERSON

Four-unit motor-driven compression plant of 1,000 tons refrigerating capacity used to cool quenching oil and drinking water. Long sections of atmospheric condensers used for ammonia and stands of similar design, with the principle of action reversed, cool the oil, although a portion of the oil is precooled by double-pipe water coolers.

THE refrigerating plant here described is probably the largest of its kind in the country. It is used for keeping a constant temperature on the quenching oil in the various heat-treat departments in the factory and to cool the drinking water. It is made up of four ammonia compressor units, 110 sections of atmospheric ammonia condensers and three liquid receivers on the high-pressure side, with 36 sections of atmospheric oil coolers and five drinking water coolers on the low-pressure side. Besides, there are 32 sections of double-pipe oil coolers in the system and a large atmospheric cooling tower. The compressor room is 127 ft. long by 50 ft. wide, and the cooling tower with the condenser room underneath is 218 ft. long, 29 ft. wide and 66 ft. high.

Operated at 35 lb. suction pressure the ammonia compressors are rated at 1,000 tons of refrigeration per 24 hr. They consist of four 18 x 30-in. and two 13½ x 24-in. horizontal double-acting compressors. The four large machines, as shown in Fig. 1, are arranged in two units, twin compressors being driven by a 600-hp. three-bearing direct-current motor running at 243 r.p.m. The small compressors, shown in Fig. 3, are each driven by a 150-hp. three-bearing motor running at 450 r.p.m. Belts with idler pulleys are used for the drive in both cases. The belts for the large units are of unusual size, measuring 60 in. wide by ⅝-in. thick. Running smoothly over an 18-ft. flywheel and a 4-ft. pulley on the motor, their operation is entirely satisfactory.

Variation in the size of the units gives flexibility to meet the demand for cooling. In winter one of the smaller machines is large enough to carry the load. During the hotter months the machines are used as requirements dictate. One large and one small machine may carry the load and at intervals the two large machines may be required. The motor drive was selected to benefit from the economy of the main generating units. The boiler plant is located at some little distance and with a surplus of exhaust steam there was no good reason for installing steam-driven compressors.

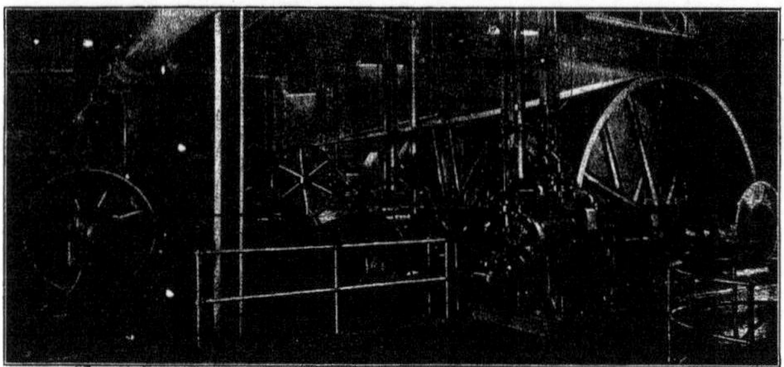

FIG. 1. TWIN AMMONIA COMPRESSORS, 18 x 30 IN., BELTED TO 600-HP. D.C. MOTOR

All these compressors have gravity oiling and telescopic oilers for crossheads and cranks. In the trusses above the compressors is a platform carrying the starting boxes for the motors and a 100-gal. oil supply tank. In the basement under the compressor room there is a 100-gal. oil filter with two pumps attached. The oil is drained from the compressors through a 2½-in. header to the filter. The pumps force the oil up to the overhead supply tank, from which it flows by gravity to the compressors.

Fig. 4 shows a cross-section of the refrigerating plant, including the cooling tower which is arranged with 11 decks, each 205 ft. long by 19 ft. wide. The isometric diagram at the left shows the piping arrangement from the circulating pumps up to the top of the cooling tower. Here the water is distributed, as shown in Fig. 2, by two hundred and eight 1¼-in. nozzles into one hundred and four 45-deg. troughs from which it overflows and spreads over the entire deck, trickling down by gravity from deck to deck into six sumps under the bottom deck. From these sumps the water is led by gravity through a 10-in. header to the troughs above the ammonia condensers, located underneath the cooling tower. It flows over the condensers into six sumps in the floor, from where the water runs back by gravity to the circulating pumps shown in Fig. 5. These pumps are five in number and have a capacity of 1,000 g.p.m. each against a 60-ft. head. They are driven by 30-hp. motors running at 1,750 r.p.m. In a 2-hr. test of the cooling tower conducted in August, 1918, with only three pumps running and circulating 3,250 g.p.m., the following results were obtained:

FIG. 2. TOP VIEW OF COOLING TOWER, SHOWING WATER DISTRIBUTING SYSTEM

Direction of wind	S 15 deg. W
Velocity of wind, ft. per min.	675
Temp. of dry bulb, entering, deg. F	79.4
Temp. of wet bulb, entering, deg. F	68.55
Temp. difference entering, deg. F	10.85
Relative humidity, entering, per cent	57.00
Temp. of dry bulb, leaving, deg. F	75.56
Temp. of wet bulb, leaving, deg. F	67.7
Temp. of difference, leaving, deg. F	7.86
Relative humidity, leaving, per cent	67.00
Temp of water leaving nozzles, deg. F	70.1
Temp. of water at bottom of tower, deg. F	68.37
Temp. difference of water or drop over tower, deg. F	1.73
G.p.m. of water circulated	3,250
B.t.u. extracted from water per min.	46,854

In addition to the circulating system for the cooling tower, just described, there are five distinct piping systems in the plant, namely: The ammonia suction and discharge; the liquid connections; the pump-out line, the quenching oil system and the drinking water system.

By the compressors the ammonia vapor is compressed from the suction pressure, generally about 15 lb., to a condenser pressure of about 150 lb., and is forced through a 70-in. header to three 24-in. oil traps, where the lubericating oil from the compressor cylinders is trapped, and from which the gas is distributed to 110 sections of atmospheric ammonia condensers, a portion of which are shown in Fig. 6. Here the vapor is converted into liquid runs by gravity through two 3-in. lines to receivers. The condensers are grouped in 12 batteries. Each section is made up of 2-in. pipe, 20 pipes high and 20 ft. long.

A general outline of the liquid connections is shown in Fig. 10. The oblique diagram shows how the liquid travels through a 3½-in. line from the receivers and is distributed to 36 expansion valves on the so-called atmospheric oil coolers, and also through a 2-in. line to the five drinking water cooling tanks, where the liquid is distributed to the expansion valves on their 20 coils.

Every ammonia coil and vessel needs a drain or pump-out connection to facilitate pumping out all ammonia in case anything needs to be done to the coil. There is also oil coming along with the ammonia from the compressor. Although the greatest part of the oil is removed by the oil traps before entering the condensers, some, however, passes along with the ammonia through the entire cycle. It is important to minimize the amount of oil in the ammonia system, because its presence reduces the efficiency of the refrigerator to a marked extent. The oil being heavier than ammonia, collects at the bottom of the oil traps, the receivers and the coils, and if it is drained or blown out into a pail, as is often done, a great deal of ammonia is lost. To eliminate this loss and to keep all the coils at the

FIG. 3. COMPRESSORS, 13¾ x 24-IN., BELTED TO 150-HP. D.C. MOTORS

topmost efficiency, an ammonia purifier has been made a part of the pump-out system.

Made up of a 16-in. pipe the purifier is 10 ft. long, as shown in Fig. 11. It has a 1-in. inlet in one end leading from the drain or pump-out connections from all of the coils, oil traps and ammonia receivers, and through this line the oil is blown into the purifier by the pressure in the ammonia system. From the other end of the purifier a 2-in. pipe extends into the drum for the full length. A plate welded onto the end closes the inner opening of the pipe. Inside the 2-in. pipe there is a $\frac{3}{4}$-in. open steam pipe. On the top of the purifier is a $1\frac{1}{4}$-in. suction connection to the compressors, a relief valve and a high-pressure gage. After the ammonia is blown into the purifier the steam is turned on, heating the contents of the drum until all

rotary pumps that raise oil from the receiving tanks to the top of the coolers.

The oil-cooling system, which carries about 60,000 gal. of oil, is provided with both ammonia coolers and double-pipe water coolers. It is designed to cool 1,600 gal. of oil per minute from 120 to 60 deg. F. Part of the oil is first pumped through the double-pipe coolers and cooled by water to 86 deg. F. and then led through a header to the troughs over the ammonia coolers, see Fig. 8. It flows over the refrigerating coils and is cooled to a final temperature of 60 deg. F. The balance of the oil is passed directly over the ammonia coolers.

Consisting of 32 sections of 2-in. and 3-in. pipe, 10 pipes high and 20 ft. long, the double-pipe coolers are grouped into four batteries. The oil travels in the

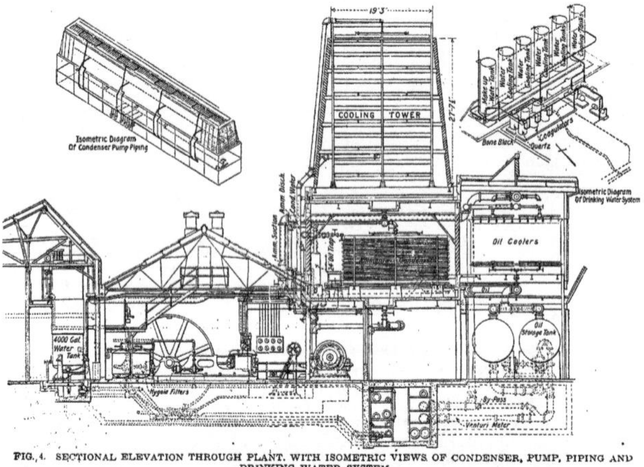

FIG. 4. SECTIONAL ELEVATION THROUGH PLANT, WITH ISOMETRIC VIEWS OF CONDENSER, PUMP, PIPING AND DRINKING WATER SYSTEM

of the ammonia evaporates and in the form of vapor is drawn to the compressors. The steam condenses and is drained from the 2-in. pipe through a steam trap to the sewer, while the free oil is drained off through a 1-in. connection at the bottom of the purifier. During the operation the pressure in the purifier is maintained at 75 to 80 pounds.

From the coolers to the tanks in the factory the quenching oil is circulated by one 1,000 g.p.m. and two 500 g.p.m. rotary pumps, shown in Fig. 7. It is returned by pumps located at the quenching tanks in the factory, taking the oil as it overflows from these tanks and forcing it back through a system of piping in tunnels to two large receiving tanks in the oil pit under the coolers. These tanks are vented to the atmosphere so as to relieve the oil of air before it is pumped through the cooling coils. In the oil pit there are two 1,000 g.p.m.

annular space between the two pipes, is fed in at the bottom and let out at the top. The cooling water runs by gravity from the cooling tower through the inner 2-in. pipe, entering at the top and leaving at the bottom of the cooler, so that the flow of the oil and the water is counter-current, this being characteristic for all double-pipe coolers.

The ammonia oil cooler, or atmospheric oil cooler as it is frequently called, consists of three open tanks, each holding 12 sections of oil coolers made up of 2-in. pipe, 20 pipes high, 20 ft. long, and arranged in two batteries. Under each pipe is a perforated strip for the purpose of maintaining an even film of oil over each pipe. To get a counter-flow between the ammonia and the oil the ammonia liquid is fed in at the bottom, evaporates inside the coils by absorbing the heat taken from the oil on the outside, and the gas is drawn off from the top of the cooler by the ammonia compressors.

In Fig. 4 the drinking water system is shown diagrammatically, and Fig. 9 is a view of the pumps and filters in the compressor room. The maximum consump-

From the city water mains the make-up pump takes the supply, pumps it through the filters to the make-up water tank, from where it joins the water returning from the factory and flows by gravity into the circulat-

FIG. 5. FIVE 1,000-G.P.M. CIRCULATING PUMPS FOR COOLING TOWER

FIG. 6. ATMOSPHERIC AMMONIA CONDENSERS, OIL TRAP AT RIGHT

FIG. 7. OIL PUMPS AND RECEIVING TANKS

FIG. 8. TOP VIEW OF AMMONIA ATMOSPHERIC OIL COOLERS

FIG. 9. PUMPS AND FILTERS OF DRINKING-WATER SYSTEM

tion of drinking water in the plant is 95 g.p.m., which is equivalent to an average of 3 gal. per man per 8 hr. The system consists of two 500 g.p.m. circulating pumps (one of them being a spare), one 100 g.p.m. make-up pump, six 42-in. Hygeia filters, one make-up water tank, five cooling tanks and about 550 fountains scattered throughout the plant.

ing pump. The latter forces the water through the five cooling tanks and out to the fountains. The temperature of the water leaving the coolers is 45 deg. F. The tanks are all 6 ft. in diameter and 18 ft. high, having a capacity of 4,000 gal. each. In each tank there are four ammonia expansion coils having a total length of 3,600 lin.ft. of $1\frac{1}{4}$-in. galvanized pipe.

The circulating pump operates against an 85-lb. head and is driven at 1,700 r.p.m. by a 40-hp. motor. The make-up water pump, operating at the same speed and against a 60-lb. head, is driven by a 7½-hp. motor. The pressure in the return line from the factory is 22 lb., but before the water enters the circulating pump this pressure is reduced by a pressure regulating valve to 8 lb., which is the static head in the make-up water tank.

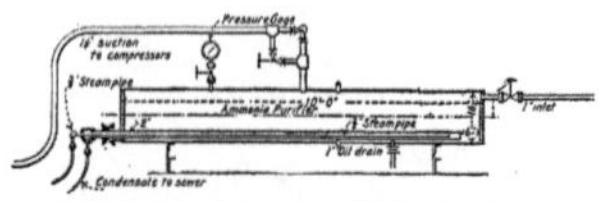

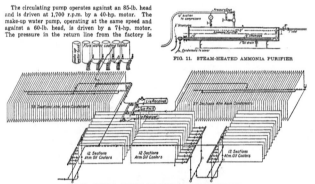

FIG. 11. STEAM-HEATED AMMONIA PURIFIER

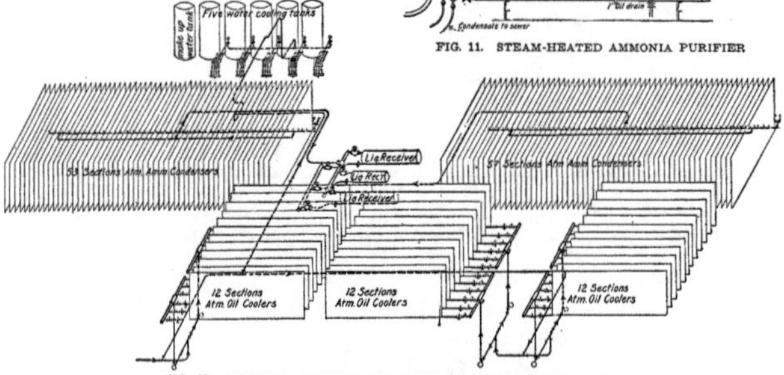

FIG. 10. GENERAL OUTLINE OF AMMONIA LIQUID CONNECTIONS

There is about 13,000 lb. of ammonia in the refrigerating system. To refrigerate the drinking water takes 185 tons of refrigeration per 24 hours, and to cool 1,000 gal. of oil per min. from 120 to 60 deg. F. takes $(1{,}000 \times 7.5 \times 0.4592 \times 60) \div 200 = 1{,}033$ tons of refrigeration per 24 hr., of which 460 tons is carried by the double-pipe coolers. Thus the load on the ammonia compressors at this rate is $185 + (1{,}033 - 460) = 758$ tons of refrigeration per 24 hr.

Lining Turbine and Generator Shafts

By O. G. A. Patterson

The various articles that have appeared in *Power* and other engineering papers on installing running and lining up turbine machinery clearly show the need for better methods of dealing with these problems. In my younger days lining up long marine propeller shafting with feelers and a rule was quite the thing and is to this day. A shaft a little out of line did not count for much as long as its bearings ran cool, and even if it was in line the ship would probably settle out of true, half an inch or more when laden, so at the best the results could only be called approximate. As most steamer propelling machinery runs at 200 r.p.m. and less the results can be classed as satisfactory.

Higher-speed machinery, such as forced lubrication enclosed engines, require still greater care to give satisfactory service and with turbine machinery running at anything up to 3,500 r.p.m. as a standard, with revolving parts weighing several tons, especial pains must be taken to get the turbine and general shafts in line.

The various accidents that keep occuring from vibration and other causes should not all be blamed on defective governors and material, as is generally done, as in some cases there are indications that the machines were more or less out of line or balance, except when first erected. Even if the turbine and generator shafts are lined up true when first installed, there is no certainty that they will remain so when warmed up and running. At one end of the unit there is the hot turbine that naturally warms up, and to a certain extent its foundation, and at the other end there is the comparatively cold generator and foundation, thus causing unequal expansion which tends to throw the shafts a little out of line. There appears to be no remedy for this trouble except to have the far end of the steam turbine a shade low

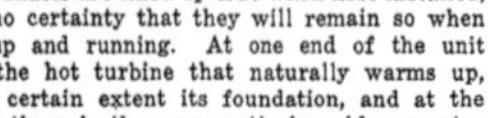

FIG. I

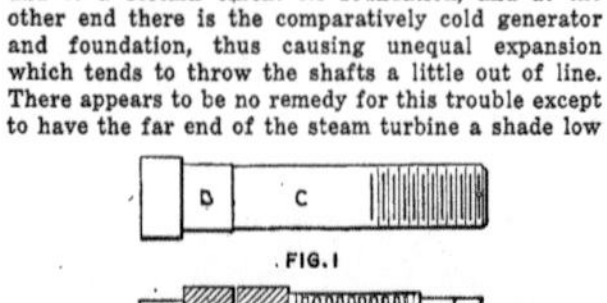

FIG. 2

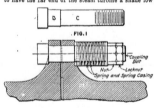

FIG. 1. TURNED COUPLING BOLT. FIG. 2. COUPLING BOLT AND SPRING

and check over the lineability of the machines after a run and while still hot.

One way of lining up the shafts is as follows: First the shafts are lined up with feelers and a rule in the usual way and then a set of turned bolts are put in the coupling flanges of the shape as shown in Fig. 1, the part *B* being a fair fit in one of the flanges and the part *C* about $\frac{1}{16}$ in. smaller than the holes in the other coupling. These bolts are threaded through two sets of spherical washers and springs as in Fig. 2, the nuts being screwed up equally so that the shafts are drawn firmly together into their running positions, but not sufficiently tight to spring the shafts or flanges. Both washers and bolts should be well oiled before inserting.

On the edge of one shaft coupling a clamp is bolted, as in Fig. 3. This clamp has an arm welded to it as shown, the overall length *A* depending on the amount of clearance between the shafts and the sides of the machine, bedplate, etc. The longer the length of the arm *A*, the greater will be the accuracy obtained, remembering always that there must be no shake, sag or spring of the arm in its different positions.

On the end of the arm a plate, bent at right angles, is bolted and having two tight fitting 1-in. set screws as shown. The ends of the set screws must be faced off square and should preferably be hardened and polished.

On the other shaft flange a similar clamp is bolted, Fig. 4, having an arm a little shorter than the one just described, and also having a bent plate bolted to it, as shown, and carrying two screws, preferably micrometer screws *D* and *E* for accurate measurements.

These clamps are secured to the turbines and generator shaft flanges as shown in Fig. 5 with arms

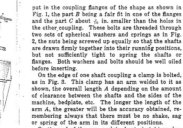

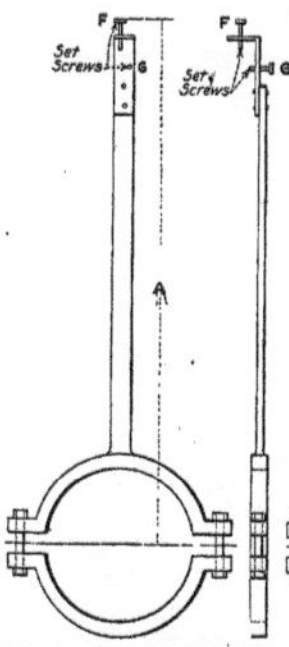

FIG. 3. CLAMP WITH SET SCREWS IN ARM

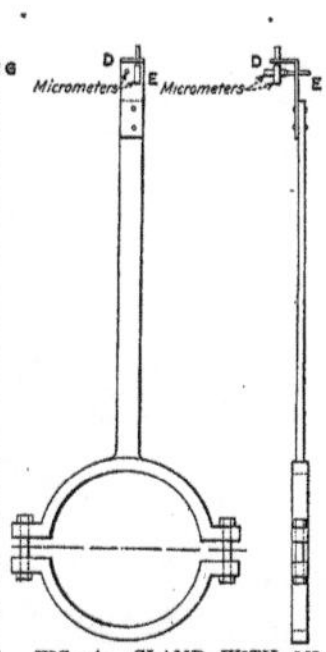

FIG. 4. CLAMP WITH MICROMETER SCREWS IN ARM

vertical, and the screws *F* and *G* and micromter screws *D* and *E* adjusted and measurements taken.

These readings being booked the shafts can be turned, taking readings every 90 deg. when the shafts have turned 180 deg. the difference between that reading and the first reading will show the amount the shafts are out of line with each other. When the shafts have been turned to 90 and 270 deg. the readings will show any error sideways.

By this method it should be possible to accurately align the shafts especially if cast-iron adjustable wedge blocks are used under the foundations. After the shafts have been leveled the turbine could be given a trial run, the measuring clamps being removed, but the bolts with the springs left in the couplings. After the machinery has got thoroughly warmed up, it should be checked over again, before it has time to cool, and any alteration made that is found necessary.

The micrometer screws can be made to suit the erecting engineer's taste, either flat, pointed, round or chisel shape. The micromter screws as usually sold are rather too delicate unless carefully used. These

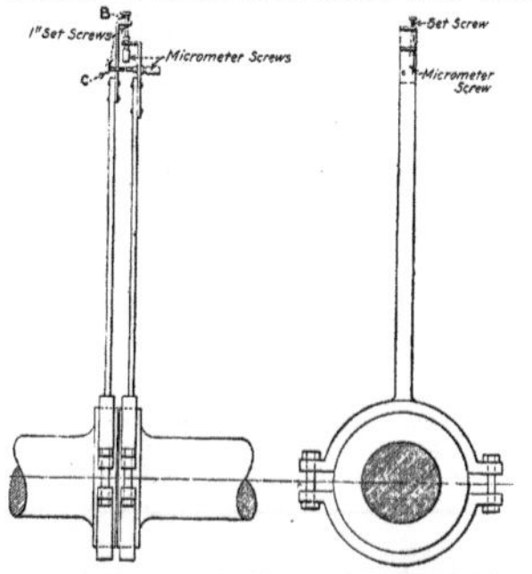

FIG. 5. CLAMPS ADJUSTED TO COUPLING FLANGES

screws can be turned up in a lathe having a screw 1 in. diameter bevelled down at the measuring end, and having an overall length of 5 or 6 inches.

Another way to take these readings would be to use an electric buzzer, bell or phone contact arrangement, so that when contact is made between the wire on the one side on one clamp and adjustable screw on the other bell would ring. This method is probably as accurate as any when well made and installed, as it does away with any mistakes in reading the micrometer, etc.

A Brick chimney 300-ft. high by 27-ft. 6-in. diameter has recently been built near Sudbury, Ontario, for a nickel smelting company. The chimney is chiefly noteworthy on account of the precautions necessary to prevent the sulphurous acid in the gases, amounting to from 3 to 4 per cent, from damaging the structure. It has been lined throughout with 4-in. perforated brick separated from the outer wall by a 2-in. space, says *The Engineer.* The lining is supported on corbels built out from the inside of the wall, and is divided into twelve sections ranging from 15-ft. to 30-ft. in height, so that in case any part of it becomes damaged that part may be removed and replaced without much harm to the remainder. In order partially to seal the air space between the top of each section of the lining and the bottom of the corbels directly overhead a sheet of lead was embedded in the corbel brickwork and bent downwards over the opening and extended an inch below the top of the lining.

Turbine Installed in Record Time

IN THE "*Synchroscope,*" the monthly publication of The Detroit Edison Co., G. K. Saurwein tells how a 10,000-kw. turbo-generator was installed at the Connors Creek plant in the record-breaking time of 36 field working days.

Since Feb. 2, 1917, two 30,000-kw. units had been on order. In ordinary times these machines would have been installed complete and carrying load by the early part of the past summer, but, due to the exigencies of the war the construction of these units was greatly delayed and the first machine was not shipped in time for the December peak. At first it was believed that this would be early enough, for industry was only slowly recovering from the first post-war slump in Detroit as elsewhere. Factories were running at half capacity and electric power consumption had dropped off greatly. Detroit's recovery was unexpectedly rapid, however, and when in August, last year's December peak was topped by several thousand kilowatts, it was decided that an additional 10,000-kw. of turbine capacity might just enable the plants to carry this year's maximum load, due then within several weeks.

The war had held up the construction of the up-river plant and this left available the two 10,000-kw. turbo-generators that had been bought for that plant and which were on hand. It was, therefore, decided to install one of these units at Connors Creek on the foundation for a second 45,000-kw. machine, now on order.

When on Sept. 12, the order came from headquarters to have this unit ready to take load in sixty days, it was believed well nigh impossible of accomplishment by most of those concerned, for there was all the design work for the steel foundation, piping and electrical layout to be done; material to be purchased, fabricated and erected; nearly all of the machine work on the condenser remained to be done—all this in addition to the actual erection of the unit and its auxiliaries.

But the thing was done with comparatively little overtime, and it proved to be just the thing to meet the winter's siege.

Drawings Ready in Ten Days

The story of this particular unit dates from Sept. 13, 1919. On that day, the electrical and mechanical drafting rooms started work on the drawings for foundations, piping, transformers, switch-gear and all the miscellaneous equipment involved in the operation of such a unit. The drawings were ready in ten days, and on Sept. 25 work actually began in the field.

As previously intimated the unit is installed temporarily on the foundation for a 45,000-kw. turbo-generator. The structural steel frame on which it is carried is so designed that it may be moved with the

FIG. 1. TEMPORARY 10,000-KW. TURBINE READY TO START 36 FIELD WORKING DAYS AFTER STARTING INSTALLATION

FIG. 2. STEEL FOUNDATION FOR THE TEMPORARY 10,000-KW. TURBINE

unit and used on the ultimate installation (probably at the Marysville plant). Due to the fact that this machine as installed will be operated for a short time only—less than a year—it was not deemed necessary to add extra height to the foundation steel to bring the machine up to the level of the turbine room floor. It was thought unnecessary to install a dry vacuum pump for a unit intended for temporary service. A cross-connection to the dry vacuum pumps of the other units was accordingly made and has proved entirely adequate. Other auxiliaries are a circulating pump and two hot-well pumps, all motor-driven. The machine has its own exciter mounted on the shaft. It generates at 4,600 volts, stepping up to the system voltage, 23,000 volts, by three single-phase oil-insulated, water-cooled transformers.

Company Fabricates Foundation

In the company's own shops the steel foundation, Fig. 2, was fabricated. It was erected and riveted up in three days. After the steel was up, the forms for the concrete top slab were built and the slab poured. On Oct. 12, the condenser was finally delivered from the contractor's shop, and its erection was started at once. Incidentally, this condenser was the first built after the company's own design.

On Oct. 14, erection of the turbo-generator was begun under the direction of the General Electric Company's foreman of construction and the unit was up in two weeks' time.

It can be readily understood that with the condenser and turbine gangs crowded around the machine and scaffolding and cribbing almost hiding it—the pipefitters and electricians did not have much of a chance to work. Most of their work had to wait until the unit and the auxiliaries were in place. As a matter of fact, the big share of the pipe lines, including all of the multifarious small lines for drips, oil, sealing water, gages, etc., that go with an installation of this kind, were put in during the last three days before the machine was put in service.

On Saturday, Nov. 1, the unit, Fig. 1, was practically ready for business. Except for some leaks in the 36-in. steel circulating water piping, the installation was complete. The drying out run was completed on the following Wednesday and on Friday, Nov. 7, the machine was on the line carrying load.

One of the stunts in connection with this installation, which should be mentioned, was the delivery of the large circulating pump with its motor. This was the last link in the chain. Even by crowding the manufacturer as much as possible, the pump could not be finished before Oct. 18. On that Saturday night, one of the Edison men got that pump into a train at Harrison, N. J., started it through the maze of railroad trackage at New York City and on Tuesday he landed at Windsor. A lost manifest caused delay in the customs office, but on Wednesday, Oct. 22, the carload of pump came into the Connors Creek yard and the equipment was soon ready for service.

Standardization of Boiler Plants

Involving the Use of a Unit Design That Covers a Wide Range of Capacities

By Louis R. Lee
Chief Engineer, E. W. Clark & Co., Management Corporation.

FOR some time it has been apparent that a standardization of the assembly of boiler-plant equipment would be required to effect further improvements in power-plant design. Much has been accomplished in the last few years in the improvement of power-plant equipment, and it has now reached a stage of development whereby the standardization of the assembly of such equipment could be safely and profitably made. The war prices of material and equipment are still in effect and bring increased emphasis on this question. To this end a unit design has been perfected, defined as "a design that may be built in units or sections, each unit being complete in itself and able to give efficient, reliable, and low-cost service with minimum investment, the first cost varying approximately directly with the installed capacity."

The most direct benefits that will be derived from the unit idea are flexibility of design, allowing for comparatively easy extensions to care for future growth, also a low first cost and a high plant efficiency for any desired capacity. It has not in most cases been difficult to lay out a plant that would secure high efficiency under specified conditions. However, variations of the conditions laid down often upset the schedule, so that the plant efficiency would be lower than expected. For example, many plants are so arranged that extensions are expensive and unsatisfactory. On the other hand, elaborate preparations for future extensions increase the cost of the initial installation and cause a low efficiency until such time as the load warrants the completion of the plant. The unit design meets these conditions as they arise without tying up investment.

Plant Efficiency to be Considered

It is important to note that in speaking of efficiency the only efficiency worth considering is plant efficiency, which should be defined as the ratio of the gross input to the gross output. The gross output may be expressed in pounds of steam, or in kilowatt-hours, where the boiler plant supplies steam to turbo-generators. In the first case the efficiency ratio should show the cost per thousand pounds of steam and in the latter case per kilowatt-hour, in each case the cost or input being actually all that is put in the plant, including cost of fuel, labor for operating, labor for maintenance and fixed charges on the total investment.

It is obviously impracticable for the initial installation of any plant to take care of all future requirements. Such a plant, if built, would have unnecessarily high fixed charges during its early period and might not be at all satisfactory when the future requirements materialized.

In considering power plants, it is recognized that the really important and necessary part of all power plants is the boiler plant. In this unit design, which applies to the boiler plant, each unit is practically dependent only upon itself. The unit must obviously be built around the boiler. The boiler complete, with furnace-draft equipment and feed-water supply, is essentially a complete plant, and it is only necessary to multiply the output of such a unit to secure any desired output. By making the unit a complete fractional part of a plant, extensions are taken care of with minimum trouble and cost.

Selection of Equipment Under Standardization

By devising the unit design with a limited selection of characteristic equipment and by careful assembly of the different parts of the unit, a finished plan is secured that may be extended for any given installation and a reliable estimate easily obtained of the cost of the total installation. This plan not only effects savings, due to the elimination of unnecessary parts, but reduces the cost of many of the parts themselves, due to the careful design of the whole unit so that each part receives its particular amount of thought in the design and is worked in to secure the best effect in the total assembly. In arranging the various pieces of equipment in the unit, consideration is given to the effect that the price of fuel will have on any installation. Fuel costs have increased throughout the country, and it is doubtful whether they will ever be returned to their old level. They certainly cannot until storage and freight rates have been reduced, and at the present time there is no promise of reduction along these lines. Fuel will continue to cost more in one locality than it does in some other; therefore, there must be variations in the unit to meet these different fuel costs. The variations should deal principally with economizer surface, draft equipment and furnace design.

That minimum first cost will be maintained while using the best standard equipment, it is recognized that the plant foundations and supporting steel for the equipment must be as simple as possible, and to this end the design is made with the idea of using more ground area rather than high elevation. High structures, where equipment is located on many different levels, can be justified only in congested city districts where ground area is not available or where land values are high. The many-level plant involves high operating costs, due to the increased number of operators required and the difficulty of making repairs. To secure low operating costs all equipment must be so located that all parts are quickly accessible and the disassembly is easily effected for making repairs and maintenance. The arrangement must be such that the repairs of one unit will not interfere with the operation of the others. Savings in first cost of construction must be effected only when they will not reduce the possible saving in operation and maintenance and also limit the arrangement to the increased use of certain equipment.

While the equipment manufactured by one company may have superior advantages under certain conditions, it must be borne in mind that the equipments manufactured by other concerns have their advantages and that a broad selection which may use any characteristic equipment will be a better design than one limiting itself to a small range of selection. As a rule, the general

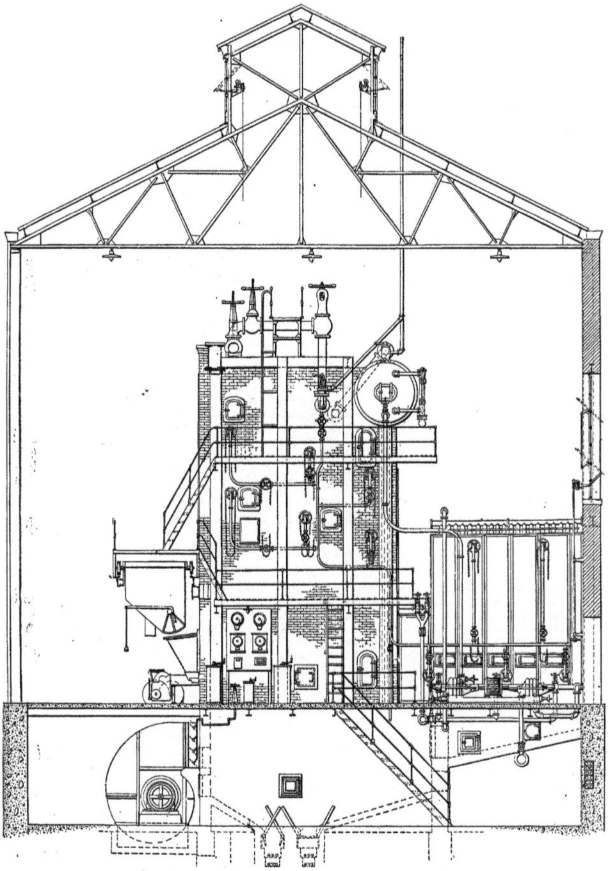

BOILER HOUSE AND ITS EQUIPMENT AS STANDARDIZED (ARRANGEMENT A).

arrangement of a plant is more often fixed by the ideas of the designer than by the peculiarities of the equipment. In other words, a successful standardization of well-known equipment can be accomplished without much difficulty.

The accompanying illustrations show the general assembly of the complete unit and the different possible building arrangements. Arrangement *A* and *B* shows a plant where the boilers, economizers and auxiliaries are housed by the main building, while the induced-draft equipment, coal bunkers and ash discharge are outdoors. With arrangement *C* and *D* the main building may be considerably reduced in size. This reduction is accomplished by providing each economizer unit with a self-containing weatherproof covering, this covering also extending over the aisles between economizers. It is economizer unit, its possibility of inspection and the ease with which it may be repaired. Economizers as well as boilers are provided with mechanically operated soot blowers using superheated steam.

Principal Details of the Unit

This unit consists essentially of a cross-drum boiler set singly with suitable stoker furnace and provided with a direct-connected economizer. In many cases it may be that the underfeed type of stoker, which is the one illustrated, will be most satisfactory. The draft equipment consists of a forced-draft fan and an induced-draft fan, both being motor-driven. The forced-draft fan motor is a constant-speed induction motor direct-connected to fan, while the induced-draft motor is a multi-speed induction motor silent-chain connected to fan. The

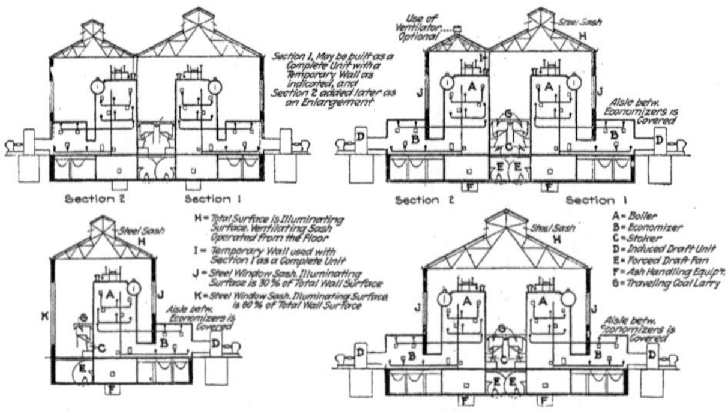

SECTIONAL VIEWS OF SOME STANDARDIZED BOILER PLANTS. UPPER LEFT: ARRANGEMENT B, SECTION 2 ADDED AFTER SECTION 1; UPPER RIGHT: ARRANGEMENT D; LOWER LEFT: ARRANGEMENT C; LOWER RIGHT: ARRANGEMENT E

believed arrangement *C* and *D* would be satisfactory in any climate and would make an appreciable reduction in the total cost. This is a step in the direction of a boiler plant with minimum building. Probably the actual minimum will be reached when only the firing aisles are inclosed, all the equipment being provided with weatherproof covering. In these designs the building itself is simple and the cost low.

A variation of any of these arrangements can be made where it is necessary to locate the plant in a congested district, so that the total ground area required will be that of the building required under arrangement *A* and *B*, the induced-draft equipment being supported over the economizer sections.

Comparison of any of these building arrangements with the best present-day practices will show much saving in cubical contents of building when compared on a rated horsepower basis. Also, it will be noted that great savings in material for supporting equipment, particularly the economizers and draft equipment, are effected. Note the entire absence of flue between boiler and economizer and consider the saving in first cost and heat loss which this feature will secure. Note assembly of the induced draft is regulated by an automatically operated damper located at the fan.

The induced-draft equipment, together with stack, is located outdoors where space permits, and inspection of the fan is readily obtained through a doorway opposite each unit aisleway. The width of these aisleways may be varied to suit local conditions, but ten feet is recommended as standard.

The stoker is driven by a multi-speed back-geared motor through silent chain.

Without changing the essentials of this unit, the rated capacity (see rating table) of the unit may vary from 700 boiler horsepower to 1,400 boiler horsepower, with a possibility of operating any boiler selected at a maximum of 350 per cent of its rating with an over-all thermal efficiency for the unit of 80 per cent, with possibly a maximum thermal efficiency somewhat higher, at about 275 per cent rating. Owing to the fact that there are practically no constant loads to be applied to boiler plants, the average rating obtained from boiler plants must be less than the maximum, hence the maximum efficiency point can be placed to advantage somewhat below the maximum capacity.

A central observing and control station is provided at the side of each boiler unit, these stations being arranged, where desired, right- and left-hand, so that one operator may observe and control two units. The control of these stations will cover the operation of the stoker, forced- and induced-draft fans, draft regulation, feed-water supply and ash dump. The operator will also be able to observe the steam pressure, temperature and flow, the draft loss through furnace, boiler and economizer and also the temperature of the flue gases at the boiler and economizer exits; the temperature of the feed water entering the economizer and boiler.

An indicating wattmeter shows the operator at all times the total station output. A recording wattmeter gives the energy used by the auxiliaries—that is, fans, stoker and feed pump—and record will also be made of the coal used. Therefore, a complete record may b3 made which will be an exact check and record of the performance of each unit. Practically all this information may be recorded on charts for future reference, the records being automatically kept so that the operator may give his best attention to the work.

All Equipment Accessible

Walkways are provided at the side and top of the boiler and economizer so that repairs, cleaning, inspection and such operation as is necessary on these higher elevations may be quickly and safely taken care of. It is to be noted that the boiler may be cleaned without interfering with the operation of the other units, and special arrangements are provided at the back of the boiler for carrying off the waste water in washing out the boiler.

All soot and ash deposited in the economizer is taken care of by the boiler ash disposal system. This system consists essentially of a duplicate drag-chain conveyor, the drag chain being made of extra-heavy malleable iron links traveling on flat cast-iron slabs laid in concrete pits. In this way the ash comes in contact only with malleable or cast-iron parts, which have a low cost and a low depreciation, even with the worst kind of ash. Similar equipment conveys the ash to the proper elevation for depositing in cars, there being no elevators, cars or skips required for elevating and depositing the ash in railroad cars.

Piping is of the best material, using steel flanges, vanstone joints, monel-metal gaskets, with welded nozzles for all branch lines, the aim being to make the arrangement of piping as simple as possible, locating main valves in the most convenient location so that little experience will be required of the operator to assume charge of a unit. Practically all of the piping will be located above the main floor and in plain view, where leaks will not escape attention and where repairs may be easily made.

The feed-water supply system consists principally of meters for measuring the return from turbine condensers and added makeup water; a surge tank and an open feed-water heater which receives a regulated supply of steam from the main turbines to raise the feed water to a predetermined temperature; for standard conditions 140 deg. F. is recommended.

For a plant supplying steam for other than steam-turbine purposes, a number of auxiliaries, preferably feed pumps, may be steam-driven, the exhaust steam being used in the feed-water heater.

Each unit has duplicate feed pumps which the economizer and boiler operate through automatic regulators. A small auxiliary feed-water heater allows water to be pumped direct to boilers in an emergency under hand regulation.

By reference to the diagram drawings, it is to be noted that boiler units are to be set in rows facing each other, two rows constituting a group, and it is recommended that five units be considered a full row; in other words, a station having two rows would have ten units, or a complete group. By referring to the rating table, the capacity of the station can then be determined according to the number of units and the size selected.

UNIT BOILER-PLANT RATING TABLE

Kilowatt capacities at 12.5 lb. steam per kilowatt-hour Steam pressure, 250 lb. Superheat, 200 deg. F.
Light figures, 275 per cent boiler rating. Heavy figures, 350 per cent.

No. Units.	Unit Sizes 7	8	9	10	11	12	13	14
1	4100	4700	5300	5900	6500	7100	7700	8300
	5400	**6100**	**6800**	**7500**	**8300**	**9000**	**9700**	**10400**
2	8200	9400	10600	11800	13000	14200	15400	16600
	10800	**12200**	**13600**	**15000**	**16600**	**18000**	**19400**	**20800**
3	12300	14100	15900	17700	19500	21300	23100	24900
	16200	**18[illegible]00**	**20400**	**22500**	**24900**	**27000**	**29100**	**31200**
4	16400	18800	21200	23600	26000	28400	30800	33200
	21600	**24400**	**27200**	**30000**	**33200**	**36000**	**38800**	**41600**
5	20500	23500	26500	29500	32500	35500	38500	41500
	27000	**30500**	**34000**	**37500**	**41500**	**45000**	**48500**	**52000**

There are eight standard sizes. For example, unit size No. 7 has a maximum capacity of 5,400 kw., while five of these units would have a capacity of 27,000 kw., and in same way unit No. 14 has a capacity of 10,400 kw. Therefore we have a range using two size No. 7 units of 10,800 kw., or for a group of ten units of the size No. 14, of 104,000 kw., it being practicable to obtain various combinations between these ranges. For example: For a station of 25,000 kw. capacity, with a possible future growth to 50,000 kw., four size No. 11 units, arranged in single-row building arrangement *A*, would give an arrangement that would provide 26,000 kw. average initial capacity. Later, a second arrangement of four units, building arrangement *B*, would give 52,000 kw. In the first case any three units in service would have a maximum capacity of 24,900 kw., thus allowing one unit for cleaning and repairs. The load characteristics and seasonal variations would govern to a large extent the selection of units. Generally it would be considered good practice to lay out all boiler plants with at least four units, this arrangement giving a reasonable insurance against power shortage due to a unit being out for cleaning or repairs.

Usually the units should be as large as the load conditions will permit. While there are several little troubles that may take a boiler off the line, most troubles can be anticipated, and the boiler is a reliable piece of equipment.

The building is a simple standard steel structure with basement and main floor, without galleries or elevated floors, the walls usually being made of plain, hard-burned red brick. The roof, which is provided with a monitor for ventilation and light, is constructed with a concrete slab water-proofed with asbestos fabric and asphalt. In warm climates it has been found satisfactory to cover the walls and roof with asbestos-covered corrugated iron, thus securing considerable saving in first cost. The side walls, being unobstructed by equipment, admit of the installation of large steel-frame windows which provide maximum light and ventilation, which assists materially in obtaining a satisfied and efficient operating force.

The cost of the building can be varied by selecting different arrangements. For example, if the economizer is housed the building will have a maximum size for a given number of units. Considerable saving can be made by allowing the economizer to be outdoors, but even in the outdoor location the operating aisle between the economizer units is housed so that the piping for soot blowers, drains, feed-water supply, etc., is housed and will not cause trouble by freezing. The housing in this case consists of a covering even with the top and outward side of the economizer unit. For standard construction the induced-draft fan units are outdoors.

When building arrangement *A* is selected with the expectation of later doubling the capacity of the plant, a temporary side wall can be constructed of corrugated asbestos-covered iron, this material later being used in roof and floor construction of the extension. Various combinations may be worked out to provide for future extensions, the details of the arrangements being controlled largely by local conditions.

Lubrication of Electric Motors

By W. F. OSBORNE

Gives descriptions and viscosities for electric-motor oils covering a very wide range of conditions, from very small high-speed motors to heavy-duty machines such as are used on rolling-mill drives, etc. The temperature conditions cover a range from outdoor operation in cold weather to use where high temperatures are caused by surrounding apparatus. The use of grease and oil-soaked waste is considered.

As IN the lubrication of any other machine the factors that influence the selection of lubricants for electric motors may be divided into two classes—those embodied in the design and those due to operating conditions. Bearing loads, clearances, and metal, method of applying lubricant, speeds, windage, etc., are usually more or less fixed when the motor is built. Operating conditions, such as heat generated by the motor, radiated heat, exposure to dust, grit, mud, water, care and attention given to the motor, as well as climatic conditions, vary with each installation, and they all affect the character of a suitable lubricant.

When oil is the lubricant used, the principal characteristic affected by the motor design is its viscosity, or body. The oil ought to be sufficiently fluid to allow it to flow into the clearances between the shaft and bearing rapidly enough to maintain the oil film that is being carried away constantly by the rotation of the shaft. On the other hand, it must be heavy enough to prevent actual contact between the bearing surface and the shaft.

Thin oils waste from loosely fitted bearings by creeping along the shaft and unless proper oil throwers are provided, the oil will work out onto the commutator and armature. Improperly refined light oils also tend to vaporize excessively, and the natural circulation of air through the motor draws this vapor into the field coils and armature. In addition to collecting dust and creating a fire hazard, this oil has a solvent action on certain insulating varnishes and compounds used in impregnating the field coils or armatures, thus destroying the insulation and causing short-circuits. Some kinds of varnishes are attacked quicker by certain oils than others, and there are some which seem to be practically impervious to any kind of oil.

Another natural characteristic of all oils that must be reckoned with is the internal, or fluid, friction of the oil itself. This friction increases with the viscosity of the oil, so that if heavy oils are used on high-speed bearings, they may possibly heat up unnecessarily. Accordingly, light-bodied oils are best for high-speed bearings where loads are not so great as to destroy the oil film.

Bearings equipped with drip cups or caps filled with waste, both of which feed small quantities of oil, should use a comparatively heavy-bodied lubricant to maintain the oil film and prevent wear. Such oils do not drain away as rapidly as lighter oils and are more economical for this service. Circulating systems and ring or chain oilers supply larger quantities of oil to the bearing surfaces continuously and therefore can use lighter oils satisfactorily. The oil returns to the reservoir with very little waste.

Temperatures, both internal and external, play a very important part. For instance, heavily overloaded motors frequently generate very high armature temperatures, and so cause high temperatures in the bearings, through conduction along the shaft and by radiation. Other motors are exposed to high radiated heat from heating furnaces or hot materials used in manufacturing processes. These high temperatures thin down the oil and necessitate the use of heavy-bodied lubricants to prevent wear and excessive loss through leakage. Motors driving steel rolling-mill table rollers, for instance, frequently use an oil of 500 sec. viscosity economically.

Effect of Low Temperatures

On going to the other extreme, when motors such as those on outdoor electric cranes or power pumps are exposed to very low temperatures, the congealing point and cold test of the oil must be considered. If the oil congeals, the oil rings do not operate, no oil is carried to the surfaces and they get dry and wear. Some motor oils congeal at temperatures as high as 35 to 40 deg. F., while others remain fluid below zero. Naturally, the low cold-test oils will work more satisfactorily under such conditions.

Other motors are frequently located in dusty places, such as coal handling and crushing plants, cement mills and flour mills. The dust works into the bearings and is carried into the oil reservoir of the ring-oiled types. Light-bodied oils permit this foreign matter to settle out more readily than the heavier oils, with the result that the oil reaching the bearing surfaces is much cleaner.

Motors used for driving centrifugal pumps, mixers, agitators, etc., are frequently splashed with water, which gets into the bearings. The oil used should separate quickly from such water without the formation of emulsions, which may prevent rings and chains from operating or hold foreign matter in suspension. Filtered oils, owing to their method of manufacture, separate more readily and are most desirable for such service.

When motors are located in places not readily accessible or where they are not likely to be oiled as often as they should bé, it is customary to lubricate them with grease. Other motors are sometimes attached to apparatus which would be damaged if any oil were to leak out onto it. Greases are found better than oils for these places also.

If greases are used, they should be smooth, clean, free from grit, lumps, excessive moisture, acid or alkali and should not contain any nonlubricating fillers or stiffeners. The softer greases cause less friction in the bearings than the stiff greases and are almost always used for ball bearings.

Summarizing, practical experience has led to the adoption for electric motors of filtered oils, properly refined to permit of rapid separation from water and of the correct viscosity to meet the various operating conditions.

The following viscosities, all at 100 deg. F., give satisfactory results:

Very small high-speed motors, for electric fans, sewing machines and jewelers machinery, 70-100 sec.

Light and medium all around service such as machine-tool drives, manufacturing machinery, lineshaft drives, when ring or chain oilers are used, 150-200 sec.

Heavy-duty motor drives and motors exposed to high temperatures, 300-500 sec.

Motors equipped with drip cups or oil-soaked waste, 250-500 sec.

When bearings are worn and loose or when motors are heavily loaded, the higher viscosities mentioned should be used.

How Power Factor Affects the Voltage Regulation of an Alternator

By QUINTEN GRAHAM
Engineering Dept., Westinghouse Electric and Manufacturing Co.

A simple explanation is given of how a load having a lagging power factor tends to decrease the voltage of an alternating-current generator, where a leading power factor load tends to increase the voltage.

IT HAS been shown in a previous article, "The Rotating Magnetic Field of Alternating-Current Motors," *Power*, Sept. 2, 1919, that the effect of the currents in a polyphase winding is to set up a magnetic field that revolves at exactly synchronous speed. With this as a starting point it will be shown that the excitation of a generator must be varied for different loads and power factors on account of the demagnetizing action of its armature magnetomotive force. Since the magnetomotive-force (m.m.f.) wave of the armature revolves at the same speed and in the same direction as the field poles and is, therefore, stationary with respect to them, both the armature magnetic field and the poles may be considered to be at a standstill and can be represented by a diagram. The first point that must be decided upon is the relative position of the center of the pole and the center or maximum point of the armatures m.m.f. wave. The maximum point of the m.m.f. wave occurs midway between the centers of any two successive groups of conductors of the same phase at the instant that the current in that phase is a maximum. This can be illustrated by referring to the diagram, Fig. 1. At the instant that the current in phase *B*, for example, is at its maximum value the maximum point of the moving m.m.f. wave is at *b*, midway between the centers of the two *B* phase groups. When the current in phase *C* is a maximum, the m.m.f. wave will have moved over so that its maximum point is at *c*, midway between the centers of the *C* phase groups. It is necessary then to know the instant at which the current in any phase reaches its maximum in order to determine the position of the m.m.f. wave at that instant.

The time at which the current in any phase group reaches a maximum is in turn dependent upon the instant at which the generated voltage in that group is a maximum and on the power factor of the load circuit. For instance, if the load on the generator has a power factor of 100 per cent, the current in any given group of conductors reaches its maximum at the same instant that the voltage of that group of conductors is a maximum; and if the power factor of the load is less than 100 per cent lagging, the current reaches its maximum at a certain interval of time after the voltage has reached its maximum. The next step then is to find the position of any given conductor group with respect to the pole when its voltage is a maximum. Knowing this and the power factor of the load, it will be possible to locate the position of the m.m.f. wave with respect to the pole.

During the time any group of conductors is being cut by the flux of a pole, the voltage generated in those conductors rises from zero to a maximum and

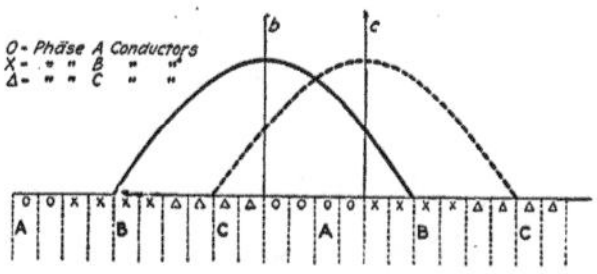

FIG. 1. SHOWS POSITION OF MAGNETOMOTIVE-FORCE WAVE RELATIVE TO ARMATURE CONDUCTORS

then falls to zero again. The maximum voltage occurs when the center of the pole is in line with the center of the group, for at that instant the conductors are being cut by more flux lines than at any other instant—the flux lines being more numerous under the center of the pole than toward the edges.

Assume, then, that the generator is carrying a load having 100 per cent power factor. Since the maximum voltage occurs in group *A*, Fig. 2, when it is in the position shown, and since the maximum current occurs at the same instant, the m.m.f. wave must have its center at *a*. Fig. 3 shows the position of the wave when the generator is carrying a load of about 85 per cent power factor, lagging. In this case the maximum current does not occur in phase *A* until a certain interval of time after the maximum voltage. During this interval the pole has had time to move over to the position shown. Fig. 4 illustrates the condition that exists when the generator is carrying current at a very low power factor, practically zero per cent. The maximum current at zero per cent power factor occurs practically 90 electrical degrees after the maximum voltage, or in other words, the interval between

By combining the force of the field with that of the armature at each point, the resultant magnetomotive force which sends flux across the air gap can be determined. It can be seen that with 100 per cent power factor the effect of the armature m.m.f. is to add to the field m.m.f. at one side of the pole and to decrease it at the other side. The result is that the distribution of flux under the pole is no longer symmetrical, as it was at no-load, although the total amount of flux under the pole is not necessarily reduced appreciably. With the same current but a lower power factor (lagging) the part of the armature

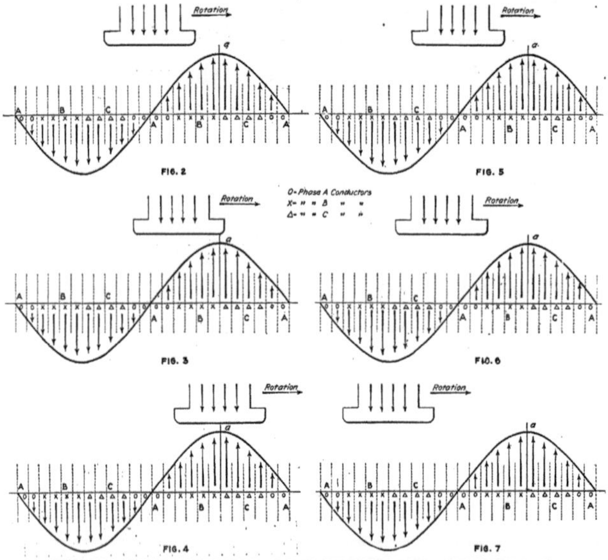

FIGS. 2 TO 7. SHOW HOW THE POWER FACTOR CHANGES POSITION OF MAGNETOMOTIVE-FORCE WAVE RELATIVE TO THE POLEPIECES

maximum voltage and maximum current in the group of conductors is sufficient time for the pole to move one-half of a pole pitch, traveling as it does at synchronous speed. For each power factor, then, there is a certain position for the m.m.-f. wave of the armature relative to the field pole.

If the direction of the magnetomotive force of the field pole be represented by the arrows, Fig. 2, and the rotation of the field is in the direction indicated, it will be found that the direction of the armature m.m.f. is represented by the arrows under the curve.

m.m.-f. wave which is opposed to the field m.m.-f. is more nearly under the center of the pole, as in Fig. 3, so that the total flux from the pole is reduced. It is evident that as the power factor is lowered this demagnetizing action is increased and is greatest when the power factor is zero lagging, as in Fig. 4.

The effect of a load having leading power factor is represented in Figs. 5, 6 and 7, and it is shown to have just the opposite effect of a lagging power-factor load. Since in this case the current in any phase group reaches its maximum before the voltage does, the armature m.m.f. wave must shift in the opposite direction to that due to a lagging power-factor load. This means that the part of the wave that tends to assist the field m.m.f. is more nearly under the center of the pole and therefore tends to increase the total

Fig. 5 is the same as Fig. 2, and what has already been said regarding the former also applies to the latter. In Fig. 6 the relation between the stator and field pole m.m.f. is shown at a power factor of about 85 per cent leading. In this case the maximum current occurs in phase *A* at a certain interval of time earlier than the maximum voltage. In Fig. 7 the power factor is practically zero and the maximum current occurs practically 90 electrical degrees before the maximum voltage. The result of this is that the part of the stator m.m.f. wave that assists the field m.m.f. is more nearly under the center of the pole so that the total flux is increased, the maximum effect being obtained when the power factor is zero leading, as in Fig. 7.

In order for the voltage of the generator to remain constant, the total flux must be maintained at a constant value. It is evident that in order to maintain constant flux with varying load and power factor, the field current must be varied, depending upon the value and position of the armature m.m.f. wave.

The preceding paragraphs have explained in a general way the effect of loads of different power factors and have shown why it is necessary to increase the field current in some cases and to decrease it in others. There are, however, a number of refinements of this theory which have not been dealt with and which are not essential to a general understanding of the operation of the machine. There is one fact, however, which will become evident after a consideration of the diagrams and which has not been taken into account so far. This is, that with an unsymmetrical distribution of flux under the pole the assumption that the maximum voltage is generated when the conductor group is under the center of the pole will not hold. For instance, with a load of unity power factor it was shown that the combined action of the armature and poles produces a distorting of the air-gap flux so that the maximum density occurs nearer one edge of the pole than before. The result is that the maximum voltage generated in a conductor group occurs when the center of that group is approximately at the point of maximum flux density and not at the center of the field poles.

Another fact that has not been taken into account is the self-induction of armature winding, the effect of which is to cause a slightly greater time interval between maximum current and maximum voltage than is indicated by the power factor of the external load. If the term "power factor" is taken to mean the cosine of the phase angle between current and generated voltage in the armature winding rather than the voltage at the terminals of the machine, the discrepancy due to self-induction of the armature winding will not exist.

Magnet-Type Pocket Lamp

IN GENERAL the source of power for pocket electric lamps has been a dry cell. However, there has recently been placed on the market by the Combined Engineering and Supply Co., 30 Church St., New York City, a pocket electric lamp, Figs. 2 and 3, that is supplied with power from a small electric generator built in as part of the lamp, and is known as the "Magnet Lamp." The generator, a back view of which is shown in Fig. 2, is a small revolving-field type alternator. A concentrated winding *W* is placed on a 6-pole laminated-iron frame and constitutes the stator or armature. Three permanent bar magnets *M* are mounted on a shaft so as to form a six-pole revolving field. By pulling on the ring *R* the field *M* is made to revolve past the stator coils and generate a voltage across the lamp terminals. Attached to the ring *R* is about 18 in. of small chain which is wound upon a wheel by a flat spiral spring. Attached to the chain wheel is a gear wheel of 112 teeth meshing into a pinion on the generator shaft having 14 teeth. The pinion is free on the rotor shaft and is connected to the field poles by a pawl, so that the rotor always turns in one direction. The lamp is intended to be carried as in Fig. 1, by a cord around the operator's neck, and is easily operated by one hand pulling out the chain and allowing the spring to wind it up again ready for another pull. In operation one sharp pull on the chain about every 3 sec. is sufficient to keep the lamp burning.

FIG. 1. SHOWS HOW LAMP IS USED

FIG. 2. LAMP GENERATOR WITH COVER REMOVED. FIG. 3. MAGNET-TYPE LAMP AND GENERATOR ASSEMBLED

Refrigeration for the Practical Engineer

Inflammability and Explosion of Ammonia Caused by Overheated Compressor

By B. E. HILL

FROM time to time one hears of explosions of ammonia compressors where the compressor heads have been blown out, and in some cases fire has resulted. These accidents have been referred to as "mysterious," as pure anhydrous ammonia will not burn or explode.

A refrigerating plant does not have to be run many years for the system to become impregnated with gases other than pure anhydrous ammonia, especially if from lack of knowledge and experience the plant is not properly handled. Regardless of knowledge or experience it is impossible to keep the system at all times clear of gases other than pure ammonia.

A new installation of a refrigerating or ice-making system that has never been in operation for a minute is never started with strictly pure ammonia, for in the first place the ammonia itself is not always absolutely pure, and in the second place it is impossible to pump a perfect vacuum in the coils before the ammonia is charged into them. After the plant is in operation and it becomes necessary to pump out a coil or set of coils for repairs, there is always a small amount of air left behind, as there never was a compressor made perfect enough to entirely expel the air or pump a perfect vacuum.

Accumulation of Impurities Dangerous

The air left behind in this way is of no practical consequence in the operation of the plant, but if an accumulation of air and other impurities such as oil vapor, decomposed ammonia (nitrogen and hydrogen) and many other gaseous impurities is allowed, there is no knowing what composition these mixtures will form, or at what temperature they are likely to explode; and as such explosions have occurred at various places throughout the country, causing serious damage to life and property, it behooves the operating engineer to think before allowing the plant under his supervision to accumulate too great a charge of impurities or to allow the discharge line and compressor to run at too high a temperature.

New York City has passed an ordinance allowing the fire department to open up connections from a refrigerating system and blow the entire charge of ammonia into the sewer in case of fire, the valves being located in a locker outside the building, the ammonia pipe lines discharging into water and then into the sewer.

The question of inflammability and explosion of ammonia has been definitely determined by Arthur Lowenstein and his associates, R. J. Quinn and S. Drucker, who conducted research and experimental work along these lines extending over a period of several years, the chief results of which are quoted herewith from a paper written by Mr. Lowenstein and read before the Chicago branch of the American Society of the Brewing Technology, March 29, 1916:

Briefly, however, it may be stated from these experiments that ammonia gas, mixed with air in certain proportions, when ignited, will propagate a flame. Special apparatus has been designed which enables us to determine the conditions under which ammonia will burn, when all known mixtures with air are brought together under suitable conditions; also known mixtures of oxygen and ammonia.

These tests were made in steps of 0.5 per cent, using mixtures varying from 1 to 100 per cent of ammonia in air, both dry and saturated with water vapor. The results under these conditions were that no visible form of burning was evident until the region of 11 to 13 per cent of ammonia was reached. It was found that a small yellow flame was produced at 11 per cent ammonia in air, which increased in size with increase of ammonia content until at the proportion of 13.25 per cent of NH_3 the burning was complete. At this concentration a yellow flame completely enveloped the glass containing vessel and the combustion was sufficiently violent to shatter the vessel.

Another set of experiments were made by using an electrically heated incandescent platinum wire in place of the spark. "The percentages of ammonia in air were increased up to 19.58 per cent, at which point the mixture violently exploded upon ignition. . . , Under these conditions mixtures of between 19.58 per cent and 26 per cent of ammonia in air were exploded. The explosions were very violent, sufficient pressure being generated to shatter the glass container."

There is no danger of an explosion unless heat is applied at some part in the system to cause one. The place where this can occur is at the compressor. If the compressor is kept in good condition there is little danger, but if there are a set of leaky rings, which may be broken or otherwise, causing the piston to churn the gas back and forth until the temperature reaches a point where the solder will melt and run out of the discharge connections, there is a possibility of one more "mysterious explosion" and if the engineer is lucky, he will be called upon to explain why.

There are many other reasons for the compressors running hot, and the engineer will find that there will be less explaining if he will see to it that the compressor is run cool at all times; there will be a better lubricating effect, less wear to the compressor cylinder, less expense of upkeep, longer life of the machine and a longer life to the engineer.

There was one experience that the writer is always rather timid and backward about mentioning. It happened at the Cudady Packing Company's plant at Kansas City. On passing through the engine room, he noticed a cherry-red spot on the side of a compressor piston rod about the size of a silver dollar. This spot was passing through the packing and into the cylinder, and when the machine was shut down and the rod allowed to cool, the metal was pulled away and roughed up so that the rod had to be filed off before starting the machine.

This condition was caused by the packing being allowed to remain in service until too hard, and the red spot was probably caused by a soft spot in the metal; anyway, it was rather a freak condition and one that would not have been "healthy" if an explosive mixture of impurities had been in the system.

As a final precaution against fire in case of a possible explosion, the engineer should insist on the elimination of arc lights, open gas jets or any other form of inflammable light in the room where the refrigerating or ice machines are located.

George H. Diman—Power Plant Pioneer

> *"Study the interests of your employer, and strive to accomplish the best results with whatever machinery or work is put in your charge; cultivate the habit of knowing how to handle men, in order to gain their confidence and respect."*—Mr. Diman's advice to young engineers.

THE story of the career of a man who has had a share in laying the foundation, and in the early development of an industry, is an inspiration to the men who carry on that industry. Mr. George H. Diman, of Lawrence, Mass., is a man who grew up with the power plant industry, who played an important part in its development and who until a short time ago continued to occupy a prominent position among power plant engineers.

Mr. Diman was born in Fall River, Mass., on Sept. 4, 1845. After receiving a public school education, he went to work on the railroad, where he labored alongside the late Henry Rogers, of the Standard Oil Co., when both were earning $1.25 per day. At this time, Mr. Diman's chief ambition was to become a locomotive engineer. He tells us that his mother very strongly objected to this ambition, so it was probably due to her urging that he finally turned his attention to mill engineering. His first work in this line was at the Union Mill in Fall River.

At that time the modern steam power plant was not even the dream of the most visionary. The maximum steam pressure was from 80 to 90 lbs., boilers were equipped with ball-and-lever safety valves and cast-iron piping with putty joints. No gaskets or glass water gages were used, and economizers and stokers were not even thought of. Most of the engines were of the Corliss type and the largest engine in New England was of less than 1000 hp. There was not a single indicator in the City of Fall River, and very few in New England. The question of coal saving was almost never given any consideration.

Mr. Diman says that his attention was first drawn to fuel economy by his employer, Mr. Foster Stafford. Although they were good friends, Mr. Stafford never visited Mr. Diman's department without objecting to the amount of coal used. This friendly nagging led him to a thorough study of combustion and of all possible means of fuel economy. Throughout his connection with the power plant industry, Mr. Diman has made many valuable contributions which are the result of his continued interest in this problem. He relates an incident in connection with Mr. Stafford's supposed desire to get all the work possible out of his employees.

"It was against the rules to read during working hours and one day Mr. Stafford came into the engine room and found me reading an English book on combustion. There were not many American books on this subject at that time. He asked what I was doing, and I replied that I was trying to better myself for my own benefit as well as that of my employer. He took up the book in a gruff manner and looked it over, then went out. Next pay day I found a half dollar per day raise."

While at Fall River Mr. Diman became acquainted with Mr. George H. Corliss. This acquaintance ripened into a close friendship which lasted up to the time of Mr. Corliss' death. In 1875 a pair of Corliss engines, totaling 1200 hp., were installed in the Social Mill, Woonsocket, Rhode Island, and Mr. Diman took charge of them for Mr. Corliss, whose contract required him to guarantee against all breakage for one year. It was about this time that he began to be interested in compound engines, as he noted that compounding was very common in England. The English engines were mostly equipped with slide valves and throttle governors, so they did not show exceptional economy, but Mr. Diman felt sure that Corliss engines could be made to show very good results if compounded. This idea was not very favorably received, but when additions were to be made to the mill where he was employed, he induced

Mr. Corliss to install the first compound engine to be put into a New England mill. It is interesting to note that while at this plant Mr. Diman came in touch with Deacon Kendall, who, with Mr. Leavitt, was the original designer of the butt strap boiler joint. At this time, also, he did his first consulting work, going to another state to make investigations and tests for Mr. Corliss.

After seven years at the Rhode Island plant he took charge of ten obsolete mills in Massachusetts, with the understanding that he was to put them in good condition for economical operation without buying any new appliances. That there was room for improvement is shown by the following description given by Mr. Diman of the plant. "There were five different boiler rooms without heaters or pumps, all boilers being fed by injectors. Anthracite, grate size coal was used, and new fires built every morning. Numbers of families received their winter coal supply from the ashes. Boilers were blown down every Saturday night and immediately filled with cold water. Engines were given a yearly overhaul which usually left them in worse condition than before." By intelligent reforms in operation and arrangement, Mr. Diman made appreciable reductions in operating cost and increased the capacity of the plant. One interesting experiment was to connect an engine, which had been exhausting to the atmosphere, to a condenser located about 1,000 ft. away. In spite of the long exhaust pipe, this arrangement resulted in considerable saving in steam.

About thirty-three years ago Mr. Diman became connected with the Washington Mill in Lawrence, Mass. In taking up his new duties he found himself in a rather delicate situation. A new power plant, designed by E. D. Leavitt and installed by the Dickson Co., of Scranton, had just been tested and had fallen down on the economy guarantee. Mr. Diman says, "I felt that the contract could be fulfilled but had just arrived at the plant and was to help force my employers to pay for apparatus they had refused to accept." Determined to do the right thing, however, he made sure that the new tests should be fair to everybody, with the result that the plant was paid for. Some of these boilers and engines are still in operation after thirty-three years continuous service. In this work Mr. Diman was associated with John T. Henthorn and A. M. Mattice, both well known to the older power-plant engineers.

The Washington Mill later became a part of the American Woolen Co., and Mr. Diman remained with this organization until his retirement from active duties a short time ago. During this time he has had a decided influence on the engineering policy of the company, and has made an enviable reputation for himself, not only among his immediate associates but among all engineers having a knowledge of his work. In referring to his work with this organization he took occasion to express his appreciation for the many courtesies he has received from William M. Wood, president of the company.

He is a member of the American Society of Mechanical Engineers and of the National Association of Stationary Engineers and is the only honorary member of the Engineers Blue Room Club of Boston.

Mr. Diman expresses the opinion that, great as has been the advance in power plant engineering in the half century during which he has been an engineer, the next half century will see as great or greater progress. In spite of all that has been accomplished, he still emphasizes the importance of economy in fuel consumption, especially in the boiler room, and the necessity for further improvement along this line.

Design and Construction of Power-Plant Piping

BY JOHN D. MORGAN

This article deals with the necessity of protecting pipe lines with suitable coverings and also with the construction of high- and low-pressure lines, and oil lines.

IN THE transmitting of steam through pipes there are two sources of energy losses, the radiation loss and the friction loss against the pipe surface which causes a pressure drop. The loss due to the drop in pressure is not nearly so great as that due to radiation. When a drop in pressure occurs in a pipe line there is no loss in heat and the drop in pressure tends to slightly superheat the steam and in general it can be said that the pressure drop varies with the square of the velocity.

The loss due to radiation is the major loss in the transmission of steam in pipe lines. This loss is dependent on the temperature of the external surface of the pipe rather than on the temperature of the steam flowing through the line, therefore in all problems pertaining to radiation from steam mains we have to consider the drop in temperature of the steam to the inside of the pipe and from the inside of the pipe to the outside of the pipe.

The drop in temperature through the walls of the pipe is so small that it is difficult to measure and so in most cases it is calculated and the method of calculating can be based with fair accuracy by using the figure of 417 B.t.u. per square foot per degree temperature difference F. per inch thickness per hour.

The drop in temperature from the steam to the inside of the pipe depends upon the temperature and the nature of the steam. It has been shown in several well-known tests that when superheated steam is used the drop in temperature will be about ten times greater than when saturated steam is used. This accounts somewhat for the fact that higher economy can be secured by the use of superheated steam in pipe lines for it is certain that the radiation loss depends upon the temperature of the outside of the pipe rather than the temperature of the steam and when superheated steam is used of the same final temperature as that of saturated steam the superheated steam line will have a lower outside pipe temperature than that of the saturated steam main and will therefore have a lower radiation loss.

There is no question as to the advisability of covering steam pipes with a good grade of pipe covering for this will reduce the condensation from 80 to 90 per cent. In determining the amount of heat that will be transmitted by a given insulating covering there are several factors to be considered some of the more important being, character of material, temperature difference between its two boundaries, thickness of the material, character of the surface and the velocity of the air. Another factor that must be considered in the losses due to radiation is that of the loss in pressure due to the presence of moisture. If it is a fair assumption to consider that each one per cent of moisture causes a drop in pressure of one per cent.

In the design of the exhaust steam lines the main consideration is to provide lines of sufficient size to properly do the work intended and to provide sufficient lines to insure the continuity of operation. The piping used is of standard material for the pressures or temperatures and does not warrant the use of special materials and in many cases cast-iron pipes are used for the larger size mains. If the exhaust steam is to be used for feed water or special process work it is necessary that oil separators be placed at suitable point in the system to remove the oil from the exhaust steam.

Auxiliary Piping Material

For the atmospheric exhaust lines it is advisable to have the piping made of riveted steel pipe owing to its lighter weight and provided with a suitable exhaust head. The valves owing to their great size should be hydraulically or motor operated.

In the design of the feed water line it is advisable to use extra heavy brass pipe or a good grade of wrought-iron pipe, to guard against corrosion. It is generally a good plan where continuous operation is required that the system be constructed on the ring plan.

The piping for the b'ow-off lines should be made of extra-heavy pipe and fittings and with the minimum number of fittings and bends. And where bends are used the curvature should be as free and easy as possible.

The drips and drains are of vital importance to good piping design and operation and it is advisable to have a drip and drain header and connect the various drips and drains to this header. This header should be amply supplied with traps. Wherever possible the main lines should have welded on them a drip outlet to avoid the necessity of drilling and tapping the main pipes, for a drilled and tapped drain outlet on a main steam line is the source of continuous trouble and will often necessitate the shutting down of a main unit for repairs.

Design of Oil Piping

It has been customary even in large power plants to pay little or no attention to the design of the oil piping. It is now well-known that when turbine oils or their equivalents have been exposed to heat, air and moisture as they are in the modern circulating oil systems, and when this system is composed of two or more metals of decidedly different potentiality such as brass pipe, copper pipe, brass netting or cloth, galvanized iron, etc., the oils are broken up rapidly due to electro-chemical or catalytic actions, as soon as an electrolyte is formed through the oxidation of oils. When this stage has been reached the oils begin to break up, and insoluables are thrown down in the form of muck so frequently found in oiling systems, although part of them remain in solution. Besides the chemical and physical changes that take place in the oils more or less corrosive action takes place in the metals. Numerous experiments have demonstrated the fact that the piping should be of one material and the one metal that is best suited to the purpose is iron.

In plants where considerable oil is thrown into the cooling water or sewerage it will pay to install concrete tanks and traps to catch the oil. These traps should be divided into compartments by means of baffles and also used for checking the flow of water and allowing the oil to raise to the surface; in the final compartment the oil can be scooped from the surface of the waters.

The estimated cost of the pipe and fittings can be taken from any pipe and fitting catalog, applying the necessary discount to give the net cost on the job for all material. In order to simplify the application of gasket and bolt prices unit costs should be made up practically as follows:

Flanges—unit price of single flange to include 1 single flange, faced and drilled; 1 gasket; 1 set of bolts with hex. nuts; 1 thread and make up. Unit price of pair of flanges to include: 2 flanges faced and drilled; 1 gasket; 1 set of bolts with hex. nuts; 2 threads and make up.

All screwed standard fittings up to and including 12 in. have one thread added to the price. All screwed extra heavy fittings have a thread added for each outlet, since this price is delivered with plain ends, likewise brass fittings have a thread added for each outlet. All flanged fittings are figured net as the gaskets, bolts and nuts are included with the flanges.

All standard pipe up to and including 12 in. is taken at random lengths threaded. All cutting and re-threading is covered by including this with the flanges and fittings. All standard pipe above 12-in. and extra heavy and brass pipe is delivered with plain ends and the threading is included with the flanges and fittings. Vanstone joint—include flange, lap of pipe, gasket and bolts. Welded necks—include pipe and flanges made up.

Calculating Prices

Valves are to be priced by the catalog method, the cost of one thread being added for screwed valves the same as with standard fittings. The gaskets and bolts for flanged valves are taken care of on the flanges.

The price of the hangers should be calculated on the per pound basis using a standard for each size of pipe and adding the turn buckles. In assuming hanger prices it is a fair assumption to assume all supports of all kinds as hangers and apply the standard pipe. Anchor costs can be taken as three times the hanger cost.

A price can be obtained for the welded drip outlet and for the plain tapped hole in pipe or fitting.

Prices can be obtained from the pipe covering manufacturers that will include the covering material, painting and labor or in brief the cost of the covering in place; covering costs should not be included in material for estimating erection labor of piping.

The estimation of the erection of piping is a most difficult item but from a large number of installations it is a fair assumption to assume the erection labor of the high-pressure lines to be about 20 per cent of the material cost and that of the low-pressure lines to be 25 per cent of the material cost. Material being understood to be pipe, fittings, hangers, gaskets, bolts, valves, etc.

It will be found that if proper attention is paid to the piping systems at large that it will greatly assist in the economical operation of the plants as well as improve the operation.

Fuel Efficiency in Power Production in Different States

By R. S. McBride

The electric power and fuel consumption data recently issued by the Division of Power Resources of the United States Geological Survey for four months in 1919* (all the data thus far compiled) furnished some very interesting material that has probably not been much appreciated. It is particularly striking to notice in these figures the very great difference in the efficiency of fuel use for public utility power production in different parts of the country. Unfortunately the original data do not show this without careful analysis and computation and thus some of the most valuable conclusions which one can draw from these statistics are likely to be overlooked.

For sake of comparison, it has been assumed that four barrels of oil are equal to a ton of coal for power production purposes and similarly that 20,000 cubic feet of natural gas equals a ton of coal. Of course, these relations are only approximately correct but they represent what is perhaps average practice. On this basis the figures in Table I have been prepared for total production of current and total fuel use. The ten States for which data are given in this table are the ten largest producers of power. They include all those where more than fifty million kilowatt hours were produced in any month of the four recorded.

TABLE I. EFFICIENCY OF POWER PRODUCTION IN TEN STATES

Results for four months, February, March, April and July, 1919, as compiled by U. S. Geological Survey, Division of Power Resources. Computed on the basis of four barrels of oil equals one ton of coal and 20,000 cubic feet of gas equals one ton of coal.

State	Current Produced Millions of Kw.-Hrs.	Fuel Used Thousands of Tons of Coal or Equivalent	Efficiency Kw.-Hrs. per Ton
New York	1,124	1,420	792
Pennsylvania	958	1,610	595
Illinois	784	1,311	597
Ohio	776	1,282	605
Massachusetts	403	533	756
Michigan	390	497	787
New Jersey	308	459	671
West Virginia	237	321	738
California	222	310	716
Indiana	213	585	364
Average	....	....	662

This first comparison shows that some States, such as New York, Michigan and Massachusetts, secure twice the current production per ton of coal as is obtained, for example, in Indiana. This at once raises the question as to why there are such great differences. Kind of coal might appear to be a factor were it not for the fact that Illinois and Ohio, Indiana's nearest neighbors and presumably users of about the same grade of coal, produce current as efficiently, judged on the basis of kilowatt hours per ton, as does Pennsylvania, where one would certainly expect the use of good coal.

The question of oil as a fuel is also interesting. In California it is noted that about 716 kilowatt hours per ton of coal equivalent is produced from oil and in Oregon the figures are over 890 kilowatt hours per ton on the average. On the other hand, Texas, which is the second largest user of oil for power production, and also one of the largest users of natural gas for this work, secures only about 550 kilowatt hours per equivalent of a ton of coal.

All of the ten States which are the largest users of power lie in the north-eastern quarter of the country except California. It is of interest, therefore, to notice particularly the results in some of the Southern and Western States. South Carolina, Georgia, Alabama and Kentucky similarly computed in comparison show, respectively, 417, 481, 324 and 493 kilowatt hours produced per ton of coal used. Thus the average for these four seems to be less than for the States which are larger power producers, but in general the average is not as low as Indiana, already mentioned. Some of the Western States seem to be the worst offenders in the way of low efficiency, for in North Dakota and Montana the average current production is only approximately 150 kilowatt hours per ton, or less than one-quarter of the average in the States which are the largest users.

TABLE II

Power production from fuels.
Results for entire United States for four months summarized from U. S. Gelogical Survey Data.

Month	Current Produced Millions of Kw.-Hrs.	Fuel Used Thousands of Tons of Coal or Equivalent	Efficiency Kw.-Hrs. per Ton
February	1,834	3,124	586
March	1,842	3,175	580
April	1,718	2,894	594
July	1,929	3,032	636
Total or average	7,323	12,225	599

The character of coal used here may be a considerable factor though the results for Colorado, which also uses Western coal with a production of over 400 kilowatt hours per ton, indicate that this fuel quality cannot altogether explain the difference.

Changes in load factor for the different seasons might, one would think, cause very large differences in efficiency of fuel use. However, the results shown in Table II, where the results for the four months reported are separately computed, indicate that this, taking the country as a whole, is not as large a factor as one might at first think.

With variations in efficiency from 790 for New York to 150 or even lower for some of the States, it is reasonable to ask what is the matter with our public-utility power plants at the present time. No one can expect every State to reach the figures now attained by the best, but it is not readily apparent why there should be a ratio of 5 to 1. And certainly if this large item of public-utility-company expenditure can be analyzed more critically, the result will be that the poorer will be brought up to standards demonstrated to be obtainable elsewhere and we will thus have one item of financial assistance for these hard-pressed public service corporations.

*See *Power*, Feb. 24, 1920.

Sweden's available water-power is estimated at 6,750,000 horsepower. In 1916 there had been developed and put into operation 1,105,000 horsepower. A good beginning has been made in the development of this power. During the war great strides were made in building new works, making the nation less dependent upon outside sources for power. The electrification of the railways has begun, and power is being transferred by cable to Denmark. Denmark's power for her industries is steam, generated from coal, imported from England. Coal has been delivered from the north of England to Copenhagen as cheaply as to London. The use of Sweden's waterpower will help to conserve England's coal resources.—*Bulletin of Pacific Power & Light Co.*

EDITORIALS

An Operating Code for Large Central Stations

THERE should be a code for the proper operation of large central stations similar, in general arrangement and scope, to the Power Test Codes of the A. S. M. E., which is now in the process of revision.

The matter of an operating code has been placed before the members of the Power Test Codes Committee of the A. S. M. E., and they have decided that this is outside the scope of their work, and furthermore, they do not feel that it is a function they should undertake. They purpose to limit their activities to a code which provides for the acceptance tests of equipment, and this leaves no one to father a code for the operation and maintenance of this equipment after it has once been put into service.

The large systems of the country are almost entirely limited to central stations; and if an industrial system has a large power station, it is usually in a complicated installation. This is evidenced by the fact that there are only one or two 10,000-kw. turbines throughout the entire industrial power stations of the country in actual operation, although one large industrial concern has a 15,000-kw. turbine on order. This means that without large turbines there is an absence of large condensers, boilers and other allied apparatus.

It is therefore suitable and proper that the men operating the large central-station systems should be the ones to write an operating code, as they have, for the most part, not only the problems involved in large units, but they have small stations on their property, so that they are in a position to cover this part of the code, which would be valuable for a large body of industrial power station men.

Following this matter to an ultimate conclusion, an operating code should be written by the Prime Movers' Committee of the National Electric Light Association. Such a code is under contemplation by central station men at the present time, and at the next meeting of the Executive committee of the Technical and Hydro-Electric Section of the N. E. L. A. authority will be asked for an assignment of this work to the Prime Movers' Committee for the next year. It is expected that it will take about two years to write the code, but the plan is to finish certain parts of it as rapidly as possible and submit it to the membership, through the technical publications, for constructive criticism.

There is also a desire on the part of operating men to define the limiting features in the design of apparatus to the manufacturer in order to prevent some of the very rapid steps that are made in development from time to time for commercial reasons instead of based on conservative engineering judgment. These are usually the result of strong competition, and it takes the form of producing something different in apparatus in order to use it as a selling point, rather than working with standardized apparatus and making quantity production and economical methods the basis of the competition. As it is impossible to do more than write an operating code during the next two years, it was the general feeling that no attempt at standardization of specifications at this time should be undertaken, as it would be something which could not be completed, but publicity of this thought might result in an expression sufficiently large to induce an extra effort to bring about an earlier start along these lines.

Stabilization of the Bituminous Coal Industry

AN OPEN forum on the problem of stabilizing the bituminous coal industry in this cocountry was one of the features of the recent meeting of the American Institute of Mining and Metallurgical Engineers in New York. In opening the discussion Herbert Hoover, who has just been elected president of the Institute, stated that the peak load in the mining and distribution of bituminous coal was at least twenty per cent above the normal load.

The American people have fallen into the habit of stopping to listen when Mr. Hoover talks, and here is a case where the power-plant industry may well listen and heed. The public, and especially the power consumer, has to pay the bills caused by this peak load. It must pay the cost of keeping mines in working condition when they are closed much of the time. It must pay the miner enough to keep him alive the year around, although statistics show he must remain idle on an average of ninety-three days out of a possible three hundred and eight working days per year. It must pay the cost of maintaining a large excess of transportation facilities which are only in the road when not required. And, finally, it must pay the cost—and this is the big item—of inability to get coal when needed, because everybody needs it at the same time.

Practically every coal user requires more coal in winter than in summer. The average user buys his coal as he needs it. The natural result is a rush of orders to the mine, a call from the mine for coal cars, a shortage of cars, and often about this time a big storm comes along and helps mess things up by delaying, if not entirely stopping, railroad traffic.

This problem is by no means new, but with the expansion of our industries it is becoming more and more serious. The mining engineers have tackled it in earnest and hope to help in the solution.

The obvious remedy is to run the mines steadily and during the summer months store up excess coal to meet the winter demand. That sounds easy, but where and by whom the coal is to be stored is the next question. Storage at the mines would solve only half the problem. Under present conditions the mines are often shut down in the busiest season because of lack of cars. Trying to do all the hauling in winter simply invites delays and extra cost, due to weather conditions. The storing must be done as near the point of consumption as possible.

It has been suggested that a lower price be made and a lower freight rate be allowed during the low-demand months. This should induce the consumer to buy in summer and store for winter demands, and would undoubtedly appeal to many buyers. Another feature would be the insurance against shutdown, which is represented by a coal pile handy to the boiler room.

There are many details of the problem yet to be considered. Local facilities for storage vary greatly. Can the individual user carry his own supply or are central storage yards better? What are the precautions necessary to prevent fires in storage? All these must be carefully thought out.

Meanwhile, any consumer of bituminous coal who can buy his winter supply in the spring and summer months will be helping to solve a great national problem and assuring himself of an adequate supply of fuel when the peak load comes.

What Frequency Shall be Adopted as Standard?

WHEN two or more polyphase alternating-current systems are to be inter-connected they must be alike in voltage phase and frequency, or means must be provided to adjust the dissimilarity between the systems. In the case of a difference in voltage, phase or in both, it is easily taken care of by the use of transformers. However, where a difference of frequency exists the dissimilarity can only be overcome by the use of motor-generator sets usually referred to as a frequency-changer set. Where the efficiency of a large transformer is ninety-eight or ninety-nine per cent. that of a frequency-changer set is only about ninety or ninety-two; besides, it requires expert attention in operation and maintenance. It is, therefore, evident that difference of frequency is to be avoided wherever possible when power systems are to be tied together.

In this country frequencies are now in use on important installations ranging from sixty-two and one-half to twenty-five cycles, although sixty and twenty-five cycles have been adopted in general as standard, and today form a large percentage of the alternating-current capacity installed in this country. To tie a twenty-five-cycle system into one of sixty-cycles the only commercial combination of machines available is a set having ten poles for the twenty-five-cycle unit and a twenty-four pole machine for the sixty-cycle element, which will give a speed of only three hundred revolutions per minute. This speed is only about one-fifth of what it should be for the most economical design, consequently the set must be much larger and more expensive than it would be if advantage could be taken of the most economical design. Therefore if alternating-current systems are tied together on a large scale by frequency-changer sets considerable of the advantage gained by interconnection will be offset by the loss in efficiency.

Within the zone to be supplied from the proposed super-power system for the northeast Atlantic seaboard frequencies of sixty-two and one-half, sixty, forty, and twenty-five cycles are in use, hence the question of what frequency shall be adopted for this system when it is put into operation is one that is of no small importance. Probably ninety-nine per cent of the central-station capacity now installed within the proposed super-power system zone at this time is either sixty or twenty-five-cycle equipment. This high percentage was brought about recently by the New York Edison Co. changing from sixty-two and one-half to sixty cycles on what is known as its sixty-cycle system—a change that is not only commendable but no doubt will be far reaching in its effects in the future. There is little doubt that the percentage of alternating-current motors served from central stations in the district approaches a high value of the total installed capacity since a very large portion of the twenty-five-cycle current is used for direct-current purposes through rotary converters. For this reason if for no other, sixty cycles is the frequency that would cause the least inconvenience and expense in its adoption as standard. The change from twenty-five to sixty cycles in twenty-five-cycle rotary-converter substations can be made gradually since most of this equipment is supplied from large and efficient generating points that will not be abandoned until at least they can be substituted with more economical machines and it would cause but small inconvenience to operate with twenty-five and sixty-cycles converters in the same substation.

In his contribution to the symposium on "Economical Supply of Electric Power for the Industries and the Railroads of the Northeast Atlantic Seaboard" presented recently before the midwinter convention of the American Institute of Electrical Engineers, H. G. Reist suggests that the use of fifty cycles be considered as the frequency for the proposed super-power system. The author's reason for suggesting the use of this frequency is that "this frequency is entirely satisfactory for lighting, for commercial motors of all sizes, and for most all other work and it would probably be found that ultimately its adoption would be of great advantage to this country commercially, since we would then have what is going to be world frequency outside the United States." Today all that can be claimed for a fifty-cycle system is equally true of a sixty-cycle, except, that it is different from what is expected will be standard outside this country. When we compare the commercial advantages of a fifty-cycle system with the difficulties it would put in the way of putting the super-power system into operation, the latter outweighs the former to such an extent as to leave little in favor of adopting fifty cycles, especially when such a large percentage of the load within the proposed super-power zone is already supplied at sixty-cycles. One of the largest industrial loads, of which a large percentage is operated on twenty-five cycles, is the steel mills. These mills to a large extent generate their own power and on account of their peculiar advantages in this respect will no doubt continue to do so. Consequently the frequency that apparently offers the greatest advantages to its general adoption and is also in accord with recent tendencies is sixty cycles, and at the same time involves the least expense and inconvenience in bringing about an adequate power supply for our railways and industries.

The popular trend toward fuel oil appears to have received a jolt through inability in many sections to secure any assurance from the oil companies that an adequate supply will be forthcoming.

The remarks of Mr. Hill in this issue on inflammability and explosion of ammonia will bear careful reading. Strict attention to the points noted may prevent disastrous results.

CORRESPONDENCE

Handy Fuse Puller

The figure shows a simple yet handy way to make a small cartridge-fuse puller from a couple of legs of an old boxwood rule. The section of the rule is closed

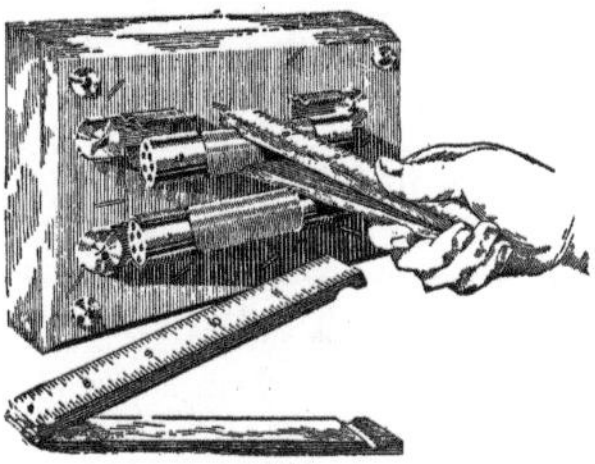

BOXWOOD RULE FUSE PULLER

together and a hole drilled through the ends to take the fuse. The figure readily conveys the idea.

Concord, N. H. CHARLES H. WILLEY.

Better Cylinder-Oil Testing Methods

I heartily agree with Mr. Weaver in what he says on page 625 of the Oct. 21, 1919, issue of *Power*; regarding Pennsylvania stock with 5 to 10 per cent acidless tallow making a mighty good lubricant, but I cannot agree with the statement: "However, it is not essential that the oil be tested on the engine in which it is used provided the steam conditions (moisture, superheat and priming) are approximately the same."

It is the condition of the unit itself that decides what oil to use and to what percentage the compounding should be carried. Two units on the same line—one within 100 ft. of the boiler and the other 400 ft. away—will require a differently compounded oil. Two engines of different makes and types of valve gears will require different oils. Of course, we can't keep a dozen different grades of oil for the various engines, but it is wise to keep two or three kinds—say a high-pressure, a low-pressure and a pump oil.

I have tried cylinder oil that worked well on a flat valve, on a Corliss gear, but the valves would not close properly; likewise, I have tried cylinder oil on a flat valve gear that was apparently working nicely on the Corliss gear, and the flat valve gear tried to pull itself to pieces.

It is true that the best cylinder oil is the cheapest, but I wish to cite a case that came up more than twenty years ago. The plant consisted of a small horizontal boiler and a 75-hp. four-valve piston-valve engine. The lubricating device had been made by the engineer, and the owners of the establishment corroborated the engineer's statement that more coal was being saved by this device than could be believed. The apparatus that saved the fuel is shown in the illustration.

I will not try to tell how the pipes were made fast in the cover of the Mason preserve jar or how many pounds pressure a Mason jar will withstand, but I saw the apparatus in operation and an immense quantity of the cheapest cylinder oil possible to buy was being used.

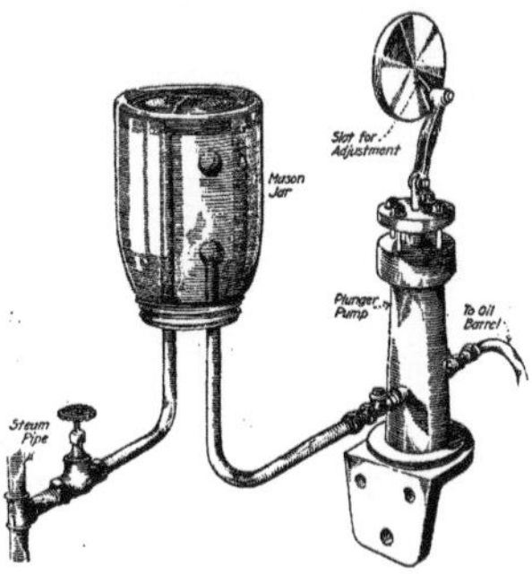

A HOME-MADE FORCE-FEED CYLINDER LUBRICATOR

Both the owner and engineer claimed that the fuel consumption had been reduced to a remarkable degree by this performance.

It is a safe guess that the exhaust steam was not used for heating or any other purpose.

This would perhaps convince some people that large quantities of very cheap oil are advisable, but I agree with Mr. Weaver that the best is about good enough.

New York City. C. W. PETERS.

An Interesting Metering Problem

I was very much interested in studying the discussion of "An Interesting Meter Problem," published on page 151 of the July 22, 1919, issue of *Power*, but on going over the connection diagram several points were not clear. As these will doubtless occur to other readers, an explanation will be greatly appreciated:

1. The demand meter is shown with potential connection only. Should not current connection be given?
2. From the diagram a single-phase demand meter is evidently used. Unless the demand meter is intentionally connected to show demand on an overloaded phase, on account of probable unbalance of load and greater load on one current transformer (which would probably impair to some extent the accuracy of the readings), would it not be better practice to use a polyphase demand meter connected to both current transformers, with proper potential connections either direct to line or through potential transformers, thus indicating true demand and not demand on one phase only?
3. The polyphase meter is evidently designed for direct connection to the 440-volt line, without the use of potential transformers. Of what use is the 110-volt connection brought to the middle terminals on both sides of this meter, with one leg from the potential transformer through demand meter, and the other leg direct from the other side of potential transformer?

Newark, N. J. W. E. Clemson.

[This letter was submitted to Victor H. Todd, and the following are the explanations given.—Editor.]

The three points brought up by Mr. Clemson can best be answered by explaining the principle of operation of this demand meter, which is the General Electric Co.'s Type G2. This demand meter cannot be used alone, but must be used in conjunction with a contact-making single-phase or polyphase or watt-hour meter. The demand meter itself is merely an electrical revolution counter, automatically counting the number of revolutions made by the watt-hour meter in a given time interval and recording the number as well as the time of occurrence on a chart. This is done as follows:

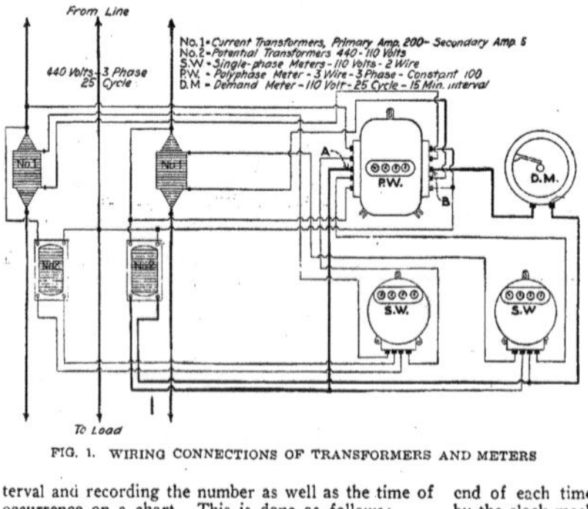

FIG. 1. WIRING CONNECTIONS OF TRANSFORMERS AND METERS

On the back of the watt-hour meter dial are two small contacts which may be closed and opened by a small toothed wheel operated by the revolving of the disk. These two contacts are connected to terminals A and B in the diagram, Fig. 1. Every time the disk makes a certain number of revolutions, these contacts are closed, and as shown by the heavy lines in Fig. 1, this closes the circuit from the transformer secondary (110 volts) to the demand meter.

The demand meter consists of a high-grade eight-day clock movement (hand-wound) which drives the circular paper chart and resets the stylus on zero at the end of each time interval, and an electromagnet and armature with a pawl and ratchet wheel arranged to drive the stylus across the paper, one division for each closing and opening of the solenoid circuit. This is shown more clearly by the schematic diagram of parts in Fig. 2.

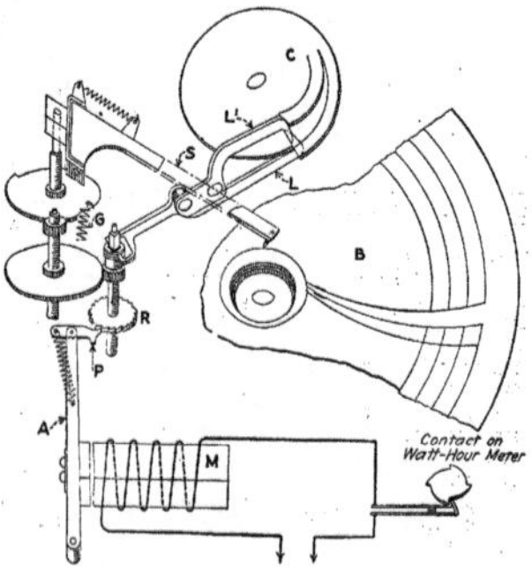

FIG. 2. SCHEMATIC DIAGRAM OF DEMAND-METER PARTS AND CONNECTIONS

When the contact device in the watt-hour meter closes the circuit through the electromagnet M of the demand meter, the armature A is moved forward and the pawl P at the end turns the ratchet wheel R. When the contact device opens the circuit, then a spring causes the armature to return to the normal position. The motion of the ratchet wheel is transmitted through proper gearing to the stylus S, causing it to move over the paper, one division for each closing of the circuit.

The stylus continues to move upward over the chart B as the circuit is alternately closed and opened by the contact device until the end of the period is reached. At the end of each time interval, a cam C which is driven by the clock mechanism, has rotated to such a position that the first of two trip levers L, resting on it, falls, thus disengaging a sliding pinion from the gear with which it meshes. This opening of the gear train allows a spring G to return the stylus mechanism to the zero position. Further rotation of the cam then allows the second trip lever L' to operate, thus returning the sliding pinion to its former position and reëstablishing the gear train between the armature and the stylus. The mechanism is now in position to again drive the stylus upward over the chart as the watt-hour meter registers the energy consumption.

From this it will readily be seen that this type of demand meter does not require any current connections and measures the actual demand of the three-phase system regardless of the unbalancing power factor.

In this particular installation the polyphase meter, a 440-volt 200-ampere three-phase watt-hour instrument, closes the circuit every five revolutions of the disk. Since in this case each revolution represents 100 watt-hours and one closing of the contacts represents one division on the chart, then each chart division represents 500 watt-hours, or 0.5 kilowatt-hour. This is a 15-min. interval meter; therefore, the rate of consumption, or the kilowatt demand per hour, will be four times this, or $0.5 \times 4 = 2$ kw. demand per division. Then if the meter showed 60 divisions as the maximum, the maximum demand would be 60×2, or 120 kw. per hour. If it showed 80 divisions, the demand would be 160 kilowatts.

Another way of considering this is to assume a certain load and from this determine the indication. For instance, with 200 amperes per phase at 400 volts, the load will be $400 \times 200 \times 1.73$, or 138.4 kw. at 100 per cent power factor. At 138.4 kw. the time in seconds that is required by the meter's disk to make one revolution is:

$$\textit{Seconds per revolution} = \frac{3{,}600 \times \textit{disk constant}}{\textit{watts}}$$

Therefore, the seconds required for one closing of the contacts (five revolutions) will be

$$\textit{Seconds} = \frac{3{,}600 \times 100}{138{,}400} \times 5 = 13$$

In a period of 15 minutes (900 seconds), the meter will make $900 \div 13 = 69.23$ closings. As we now know that 138.4-kw. demand makes 69.23 divisions, then it is evident that one division is equal to 2 kw. demand; and the demand on this installation is found by multiplying the divisions recorded by the constant of 2. The dials are divided into 100 divisions, and when the watt-hour meter and demand meters are furnished as a unit, endeavors are made to furnish a dial on the demand meter marked to read the demand direct without the use of a constant wherever possible.

VICTOR H. TODD.
Orange, N. J.

Simple Oil Filter for Reclaiming Lubricating Oil

Following is a description of an oil filter that we use to reclaim lubricating oil that is black with suspended carbon from the drip pans of two 50-hp. oil engines. It is first put into a barrel to settle. After a week's time the oil is skimmed and passed into a sand filter consisting of two oil barrels cut off 6 and 18 in. from the top respectively. Inserted into the lower barrel is a funnel-shaped galvanized iron cone. The top barrel fits into this cone and is filled with sand. Holes are drilled through the bottom, over which is put a layer of waste to prevent the sand from sifting through with the oil. The sand must be dry, otherwise the oil will not pass through. Embedded in the sand is a round close-mesh strainer into which the used oil is poured. As may be seen from the sketch, the oil passes through the funnel to the bottom of the lower barrel, then rises until it reaches the overflow pipe. This gives some of the suspended matter a chance to settle before the oil is drained off.

From the overflow pipe the oil is passed to the strainer of the bagasse filter, which is made of two old gasoline cans with their tops cut off. The upper one has holes punched in the bottom through which the oil passes. A strip of tin may be soldered to the under side of the upper can to prevent it from sliding off. The upper can also has a layer of waste put over the

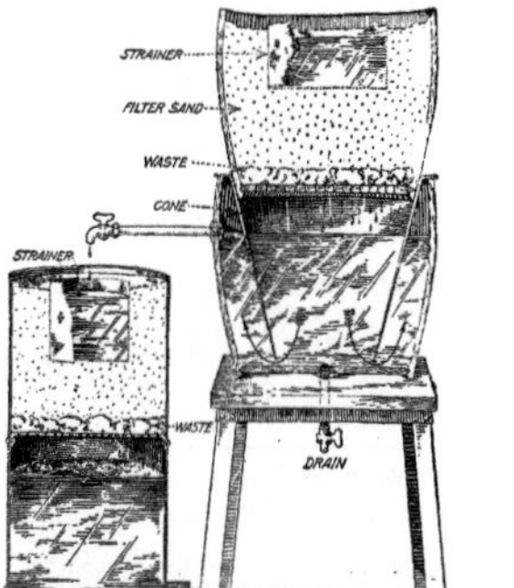

SIMPLE HOMEMADE OIL FILTER

bottom for the same reason as in the top barrel and is filled with bagasse ash. A fine sand probably will give the same results as bagasse ash where the latter is not obtainable.

A clean and exceedingly clear oil can be drained from the lower can and it well takes the place of dynamo oil previously used on the bearings of the machinery of the plant.

The sand around the strainer of the top filter must be renewed about once a month, as it will be saturated with carbon particles, which will prevent the passage of the oil. In doing this, all the sand need not be removed; but only a layer of about two or three inches nearest the strainer.

In first filling the filter with sand or bagasse ash, care should be taken that the sand nearest to the outside walls is well packed, as otherwise the oil may find its way down along the sides, in which case, of course, the effectiveness of the filter would be decreased.

The following are the results of a laboratory test of the filtered oil:

Kind of Oil	Where Used	Sp. Gr.	Flash Point Deg. F.
Ursa	Oil engines	0.926	428
Fuel oil	Oil engines	0.880	240
Filtered oil	Bearings	0.890	388

The filtered oil represents a mixture of Ursa oil used for cylinder lubrication, with admixtures of oil from bearings and refuse of unburned fuel oil from the cylinders.

A consideration of specific gravities of fuel and Ursa oil would indicate that 78 per cent of fuel oil with 22 per cent of Ursa oil made up the filtered oil. The flash-point consideration does not indicate these proportions, but shows that the more volatile oils have been burned out of the fuel oil, leaving an oil of about the same specific gravity but of a higher flash point.

A pure consideration of the flash points would indicate 78.7 per cent Ursa oil and 21.3 per cent fuel oil as making up the filtered oil. This is at direct variance with the specific-gravity indications. It is probable that neither consideration shows what actually takes place, but that the filtered oil represents a combination of the two effects.

The filtered oil is neutral, having taken up nothing in passing through the bagasse ash filter.

From the foregoing it may be seen that with little expense a very good oil filter can be constructed, as oil barrels and gasoline cans, as well as a couple of pipe fittings, are within most engineers' reach.

Fajardo, Porto Rico. E. GROSSENBACHER.

Allowable Length of Longitudinal Seams in Boilers

In connection with the discussion that has been going on in your columns regarding the allowable length of longitudinal seams in boilers, the question arises, Which of the two constructions shown in the accompanying

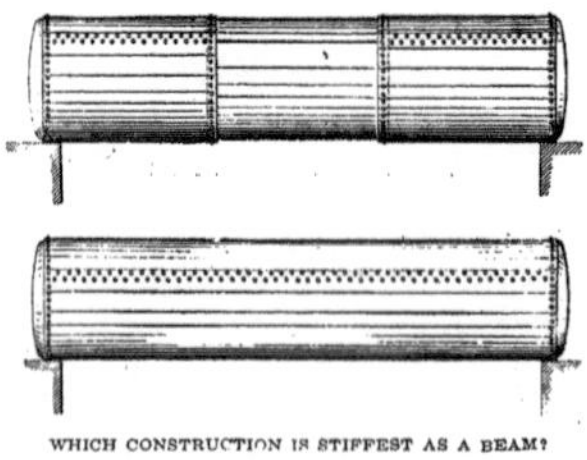

WHICH CONSTRUCTION IS STIFFEST AS A BEAM?

sketch is stiffest as a beam? The structure with the three courses has the stiffening effect of the circumferential seam, but the bottom of the beam, which is a continuous piece in the one-sheet boiler, is broken by two seams which have but little more than half of the value of the plate. I should be interested to have the opinions of your contributors upon the subject.

New York City. E. A. DALY

Off-Peak Schedule Motor Cutout

The device shown in the figure was developed to prevent motor users from using power during certain hours of the afternoon and evening, when such users are operating under an off-peak schedule. Heretofore, it was difficult to enforce the ruling without having arguments, and the motor business in the particular locality was so widely scattered that the cost of a separate motor circuit would have been prohibitive. The cost of time switches, to say nothing of the maintenance, would also have been considerable.

In the first place, all main switch-fuse combinations were replaced with carbon contact air-break circuit-breakers equipped with shunt strips. The expense of this change was borne partly by the customer, inasmuch as he would have no fuse expense after installing

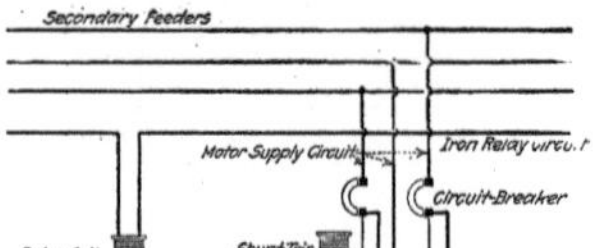

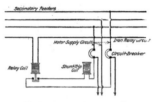

WIRING DIAGRAM OF OFF-PEAK CUTOUT

breakers. A series iron-wire circuit of No. 10 B. & S. weatherproof encircles the locality where it is required to control the service. A relay is inserted in the circuit on the premises of each consumer, and wired as shown in the diagram, so that when current is flowing through the iron-wire series circuit, the shunt trip coils of any installations where the circuit-breakers are closed immediately open such breakers and will not permit them to remain closed until the series circuit has been "killed." It will be noticed that since the shunt trips are wired up on the load side of the breakers, current flowing in the trip coils is of momentary duration only.

The relays used on this circuit were made up locally, although, at a greater cost, circuit closing relays that will answer the purpose can be purchased. The relays used on this job consist of a spool having the following dimensions: Hollow fiber core 3 in. long, 1/4 in. diameter inside, 3/8 in. outside; end washers 1/8 in. thick, 1 3/4 in. diameter; spools wound full of No. 24 enamel-insulated wire.

The arrangement sounds expensive, but where there are several motor users rather closely grouped, it is infinitely more satisfactory than the time-switch system and less expensive, the installation in question having cost less than $9 per consumer exclusive of circuit-breakers, which of course would be necessary in either case.

As a matter of interest the restricted period varies from 4:30 to 9 p. m. in December, to 7 p. m. to 9 p. m. in June. A liberal additional discount is allowed for motor operation on the restricted schedule, in localities where there is no separate motor circuit. The knife-switch controlling the series circuit is operated at the company's local office.

Carthage, Tenn. R. S. SEESE.

Refacing Caps, Nuts and Headers of Water-Tube Boilers

At the plant where I am employed we had trouble making caps tight on the Babcock & Wilcox type of boiler, and after trying various methods of refacing and cleaning, we arrived at what I think is the best way to do it.

Pumping and testing boilers after capping them up, especially where water has to be bought, is no small item in running up the costs. It makes no difference whether you have had one cap off or 250 of them—you do not know how many are going to leak until they have been tested. If the leak is on the bottom row, which is generally the case, it will be necessary to empty the boiler. The method we use is as follows:

To a one-horsepower single-phase motor is connected a five-foot length of flexible shaft, and to the end of the shaft is attached a wooden holder, Fig. 1, to which is secured emery cloth which is used to face the headers. This holder is made by turning a 1-in. thick piece of maple or oak down to 6 in. diameter and facing down to

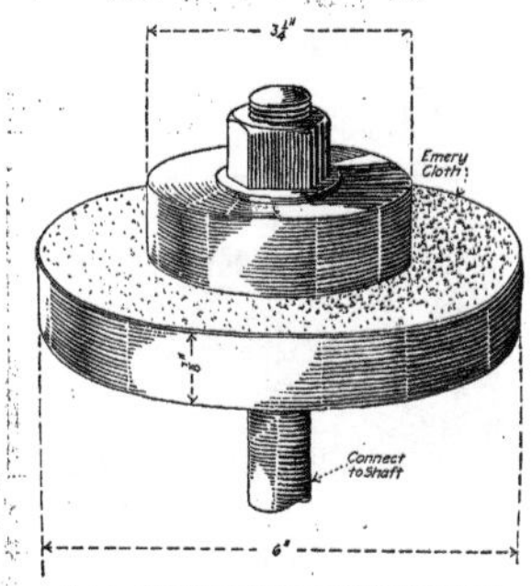

FIG. 1. WOODEN DISK FOR FACING HEADERS

⅞ in. thick. Another piece of the same wood is turned to 3¼ in. and ¾ in. thick, and a ½-in. bolt hole is bored through the center of both disks. Circular sheets of No. 1½ emery cloth are cut with holes in the center to fit over the bolt and between the larger and smaller wooden disks, the whole being tightened by means of the nut. This will hold the emery cloth on the facing disk and the rig is ready to face up the headers.

For facing the caps and nuts we have a different rig, which is made as follows: Take a blank flange 18 or 20 in. in diameter, drill a 1¼-in. hole in the center and fit in a piece of shafting about 3½ ft. long. The best way is to heat the flange and shrink in the shaft. Then put it in a lathe, face up the flange smooth and take a cut off the full length of the shaft at the time of setting up, so that the face of the disk will be true with the shaft. A strong table about 3½ ft. wide and 4½ ft. long with heavy 6 x 6-in. legs and 2-in. planking was made, bracing the sides well. Bearings were placed as shown in Fig. 2, for the shaft. A 5-in. diameter pulley was placed on the bottom end of the shaft for a driving belt. The disk runs at a speed of 300 to 350 r.p.m. A water pipe was run up to a point above the disk with a ½-in. pet-cock on the end to control the supply of water which is fed into a tin trough and which will drop on the center of the disk. Sand is thrown in the trough and the mixture of sand and water is run very slowly

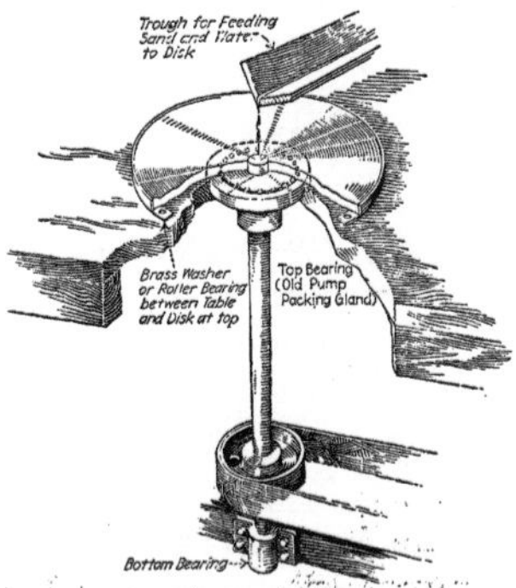

FIG. 2. TABLE AND CAP SURFACING MACHINE

on the disk. Caps and nuts are held on the disk, and they are quickly faced up. Fig. 2 shows the apparatus. We have used this method for some time, and in several cases where we have had all the caps off a 650-hp. boiler, we have had only slight leaks on one or two of

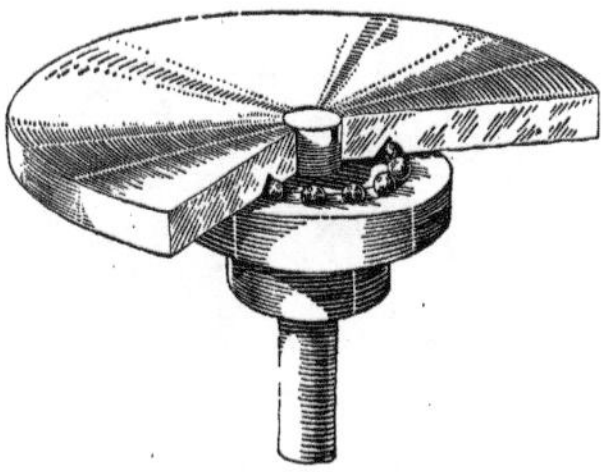

FIG. 3. BALL BEARING FOR SURFACING MACHINE

the caps. When we take off four or six caps to pull a tube or make a roll, we never have a leak.

An old pump gland makes a good top bearing by countersinking it in the table and grooving it out and placing balls in, as shown in Fig. 3. It should be babbitted to fit the size of the shaft if it is too large. Caps should be wiped and the finished face oiled immediately.

New Rochelle, N. Y. B. GADDIS.

Resistances in Series and Parallel

With reference to C. R. Behringer's scheme for making a quick change in the connection of resistances from series to parallel and vice versa, given on page 595, Nov. 22, 1919, issue of *Power*, I wish to call attention to another arrangement of connections, shown in the

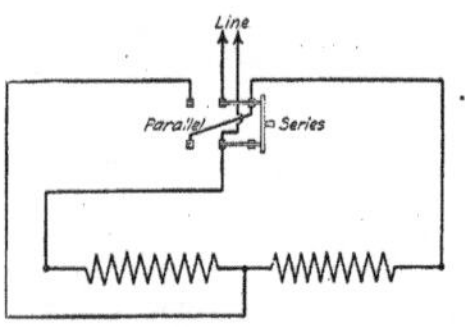

DIAGRAM FOR CONNECTING RESISTANCE IN SERIES OR PARALLEL

figure, for accomplishing the same thing. I used this scheme in an electric furnace where it was desirable to bring the temperature quickly up to a high degree and then maintain that temperature for a considerable length of time. The method has worked out very satisfactorily.

Elizabeth, N. J. E. OGUR.

Loose Piston Caused Knock

Our drop-valve type engine developed a knock that became audible only when a certain speed was reached; when it was run single-acting with steam shut off at the head end, the knock became inaudible. Upon taking the engine down for examination, we found that the piston was constructed as shown in Fig. 1. It was held

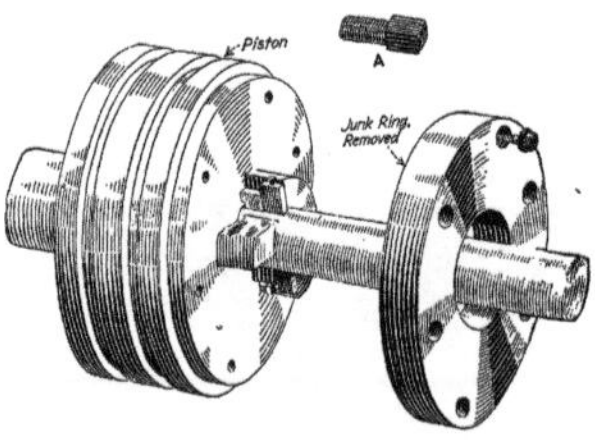

FIG. 1. DESIGN OF ENGINE PISTON

in position on the piston rod by two taper wedges, which had worked a little loose, hence the knock.

We overcame the trouble by drilling and tapping the large ends of the wedges, afterward screwing into each one a brass stud, as shown at *A*, the plugs being made a little longer than required. In reassembling, the wedges were first driven in tight, then the brass studs were screwed in down to the head and filed off to the same diameter as the bore of the junk ring, the fastening on of the junk ring completing the job. The studs, being of brass, will expand under heat a little more than the cast-iron ring and the steel wedges, thus tending to keep everything tight.

On the same engine the drop-valve piston was held on the rod, as shown in Fig. 2, by two nuts and a cotter pin on the under side. The split pin sheared off on one and the nuts becoming loose, ultimately came off, though,

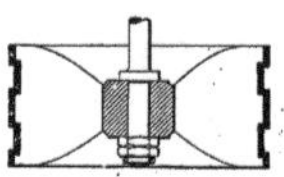

FIG. 2. FORMER DESIGN

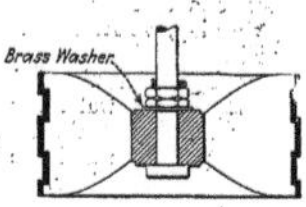

FIG. 3. AS CHANGED

fortunately, not causing any damage. This trouble was overcome by reboring the piston a little larger and changing the rod to the form shown in Fig. 3.

I would recommend the use of brass washers where heat is used and slight tension is required, brass having a larger coefficient of expansion than cast iron or steel.

Nagoya, Japan. F. H. COMBS.

They Ought To Steam Better

The illustration shows a pile of scale that was removed from five horizontal-tubular boilers by means of a tube cleaner. Naturally, after removing this amount

NOT COAL, BUT SCALE

of scale from the boilers the tubes leaked, but these were rolled, and then the trouble stopped. The pile of scale speaks for itself.

Bicknell, Ind. ROY TUCKER.

Licensing Steamfitters

I would like to say a few words regarding compelling steamfitters to pass an examination and secure licenses before they can install or repair heating or power-plant piping. I agree with C. S. Regan in a recent issue that a better plan would be to allow no new or repair work to be made until after an examination by a qualified state, municipal or insurance inspector.

In my twenty-five years' experience I have noticed that some of the most expert steamfitters were men who could hardly write their own names. They had no technical education, but had started in as apprentice boys and learned every step of their trade from hard practical experience in the steamfitting shop. According to my knowledge one of these mechanics is worth about two dozen of the technical-school men who have had but little steamfitting experience.

Irvington, N. J. FREDERICK FRANCIS.

INQUIRIES OF GENERAL INTEREST

Direction of Laps of Horizontal R. T. Boilers—Why do the inside laps on longitudinal seams of horizontal return tubular boilers point downward? H. S.

The laps are made so the outside lap will point upwards and may be calked with the greater convenience of "down hand" calking.

Advantages of Four-Valve Engines—Why may four-valve engine be more economical than single-valve engines? A. B. T.

Four-valve engines admit of design and construction with shorter steam passages and less percentage of cylinder clearance; and the valves can be adjusted independently of each other for obtaining more economical distribution of steam and control of the exhaust.

Regulating Pump with Variable Steam Pressure—Our boiler pressure varies from 80 to 100 pounds and we are desirous of maintaining a regular speed of a tank pump. Can it be accomplished with a pressure reducing valve or a pump governor? D. A. H.

If there must be a uniform number of strokes per minute it will be necessary to control the pump with some type of centrifugal throttling governor driven from a connecting rod and crank-shaft motion. But for controlling the supply of steam so the speed of the pump will be adapted to maintaining a constant discharge pressure, use a pump governor that is operated by the discharge or tank pressure.

Use of Gear Drives for Throttling Governors—Why are not gear drives used on throttling governors instead of belts? P. F.

Many of the early designs of throttle-governed engines were provided with governors driven by spur wheels and connected to the throttle valves by reach rods, or when placed near the engine cylinder the governor was driven from the engine shaft by bevel gears and a lay shaft. But gear drives are expensive to construct, more or less noisy, dangerous and easily disabled by the gear teeth becoming stripped and for most situations belt drives are preferable on account of their smooth and noiseless running, ease with which pulley diameters can be adapted to the desired speeds, and safety against stripping of necessary governor gearing by slippage of the belt.

Vacuum Gage Reading Referred to Barometer—What is the actual pressure in a condenser with a vacuum gage reading of 25 in. and barometer 29 in? G. W. D.

Inches of vacuum is the height that a mercurial barometer would indicate if the upper end of the barometer tube were connected with the condenser. Therefore the inches of vacuum plus the condenser pressure equals the pressure of the atmosphere, in inches of mercury column pressure; or, inches of barometer minus inches of vacuum is equal to inches of mercury column pressure in the condenser. Hence with the barometer reading 29 in., and 25 in. vacuum the absolute pressure in the condenser is $29 - 25 = 4$ in. mercury column pressure. As each inch of mercury column pressure equals 0.491 lb. per sq.in., the 4 in. mercury column pressure is equivalent to $4 \times 0.491 = 1.964$ or practically 2 lb. per sq.in. above a perfect vacuum.

Increasing Size of Water Motor Supply Pipe—A 24 in. dia. water motor that is supplied with water at a pressure 28 lb. per sq.in. is provided with a 4½-in. supply inlet, but was connected up with a 3 in. supply pipe. With an ample supply of water how much more power would be developed using a 4½ in. supply pipe? T. W. L.

With a larger supply pipe there would be less loss of pressure for overcoming pipe friction, depending on the length of the pipe, and development of more power for consumption of the same quantity of water. A larger pipe would also be capable of delivering a higher rate of flow, but the increase of power for a larger supply would depend on the capacity and efficiency of the motor.

Overstroking of Duplex Fire Pump—We have a 12 x 10 x 18-in. duplex fire pump with an 8-in. suction line about 60-ft. long. It seems impossible to run the pump faster than 15 strokes per minute without striking one head or the other. How can the stroke be shortened? L. G.

A good sized air chamber on the suction pipe near the pump would remove much of the trouble from over stroking of the pump. Duplex fire pumps usually are provided with cushion valves on the steam end for cushioning on the exhaust steam by retarding the escape of the exhaust when the piston is near the end of the stroke. If neither of these remedies for striking is available, the stroke of either side can be shortened, but with reduction of capacity per stroke, by reducing the amount of lost motion or clearance between its steam valve and valve rod.

Changing Voltage of Alernators—We have a 6,600-volt 3-phase 60-cycle alternating current generator operating at 3,600 r.p.m., with a 36-coil winding, connected series star. What voltage can be obtained by connecting the winding two-paralled star, series delta and two-parallel delta? We have also a second machine that runs 3,600 r.p.m. and generates 3-phase, 60-cycle 600-volt current. The winding in this machine has 48 coils connected two-parallel delta. What voltage will be generated when the winding is connected series star, two-parallel star and series delta? W. W.

Since the first machine generates 6,600 volts when connected series star it will develop one-half of 6,600 when connected two-parallel star or $6{,}600 \div 2 = 3{,}300$ volts. When connected series delta the voltage will be that of the series star connection divided by 1.732, in this case $6{,}600 \div 1.732 = 3{,}810$ volts. With a two-parallel delta connection the volts will be one-half that obtained with a series-delta connection or, $3{,}810 \div 2 = 1{,}905$ volts.

If the second machine generates 600 volts when connected two-parallel delta, then it will generate $600 \times 2 = 1{,}200$ volts when connected series delta. With a series-star connection it will generate a voltage equal to that obtained from a series-delta connection times 1.732, in this problem $1{,}200 \times 1.732 = 2{,}078$ volts. The two-parallel star connection will give one-half that obtained from the series-star grouping or, $2{,}078 \div 2 = 1{,}039$ volts. It may be possible to so change the speed, excitation and the winding pitch of these machines as to obtain a standard 2,200 or 2,300 volts, but before trying this it would be advisible to consult the manufacturers.

Valve Failure and Valve Steels in Internal-Combustion Engines*

By Leslie Aitchison

THE purpose of this paper is to state the principal troubles which arise in connection with internal-combustion engine valves and to indicate the best way to overcome them, paying particular attention to the selection of proper steel for use in these valves. The author calls attention to the fact that the valves work under very high temperature, up to a maximum in some cases of 860 deg. C., and then classifies under eight different heads the ways in which they may fail. This discussion of the methods of valve failure leads to the following list of properties which an ideal valve steel would possess: 1. High tensile strength at high temperature. 2. High "notched bar value." 3. Ease of forging. 4. Ease of manufacture without cracks. 5. Ease of heat-treatment. 6. Little tendency to scale. 7. Retain original properties after frequent heating and cooling. 8. Will not air-harden after heating to engine temperatures. 9. May be heat-treated after forging to relieve strain. 10. Sufficient hardness to withstand wear in the stem. 11. May be easily hardened at the foot of the stem. 12. May be easily machined.

It is evident that no particular steel will possess all these properties to a maximum degree. The steel must be selected, then, either as possessing all these properties to a reasonable extent or as possessing some of them to a remarkable extent. After considering some of the properties of various steels, the author discusses the twelve ideal properties outlined, in connection with five typical steels, namely: (a) Tungsten steel containing not less than 14 per cent tungsten with about 0.6 per cent carbon; (b) steel with about 13 per cent chromium and about 0.35 per cent carbon; (c) steel with 7 to 10 per cent chromium, 0.6 per cent carbon; (d) steel with about 3 per cent nickel, and (e) ordinary nickel-chromium steel.

With respect to tensile strength at high temperatures the steels may be placed in the following order: High-tungsten (*a* above), high-chromium, high-carbon (*c* above), high-chromium, low-carbon (*b* above) with 3 per cent nickel steel and the nickel-chromium steel about alike

With respect to notched bar values, these appear to be largely a reflection of the tensile strength, except that as temperature increases the notched bar value rises much more rapidly than the tensile strength falls.

As to ease of forging and also ease of manufacture without cracks, the order is almost exactly the reverse of that based on tensile strength. The high-chromium and high-tungsten steels are all rather poor. The high-chromium steels probably rank first as to ease of heat treatment. They can be air-hardened from about 900 deg. C., after which they can be tempered up to about 750 deg. C. The tungsten steels may be air-cooled from a temperature of 950 deg. C. and then may be tempered up to about 800 deg. C. The nickel and some nickel-chromium steels must be hardened in oil from about 830 deg. C. and tempered to not over 600 deg. C. Some of the latter may be air-hardened from 800 deg. C. and tempered at about 600 deg. C., but this is not to be recommended

The chromium steels scale very little, and this scale is in general not objectionable as it has little tendency to drop off. The other steels are much less satisfactory in this respect.

Almost all of the steels will retain their original physical properties under service, provided the temperatures to which they are exposed do not exceed those at which they were tempered durihg the heat-treatment.

The requirement that the steel should not harden when cooled in air from its running temperature strictly limits the application of the various steels. Tungsten steel hardens slightly when cooled from 850 deg. C. The chromium steels are completely hardened when air-cooled from 900 deg. C., the hardening feature first appearing when cooled from 820 deg. C. The nickel steels probably do not air-harden at all, but the nickel-chromium steels are distinctly air-hardening at temperatures that are not very high.

Freedom from distortion is easier to obtain in the air-hardened steels, although there is little to choose in this respect.

Wear in the stem seldom takes place with tungsten steel, but gives considerable trouble with the chromium steels.

With respect to case-hardening of the foot of the stem, tungsten steel has been most often treated in this way, although the chromium steel may be manipulated much more easily. The nickel steels require quenching in a liquid, which is not as simple as the air-hardening methods.

The high-chromium low-carbon steels are easiest to machine with the straight nickel steel next. The tungsten, nickel-chromium and high-chromium high-carbon steels may be satisfactorily machined, although the last named will give trouble if the carbon is higher than the 0.7 per cent.

The accompanying table gives a synopsis of the foregoing discussion of properties for ready reference. The steel having the desired properties in the greatest degree is marked 1, the worst 5, etc. Where there is little or no choice between two or more steels, the same number is given.

RELATIVE PROPERTIES OF STEELS FOR VALVES

Property	High-Tungsten High-Carbon Steel	High-Chromium Low-Carbon Steel	High-Chromium High-Carbon Steel	Three Per Cent Nickel Steel	Nickel-Chrome Steel
Tensile strength at high temperature	1	3	2	4	4
Ease of forging	4	3	4	1	2
Ease of sound manufacturing	4	3	4	1	2
Ease of heat treatment	2	1	2	4	3
Scaling	3	1	2	4	4
Retention of physical properties	1	2	2	3	4
Self-hardening in running	2	3	4	1	5
Freedom from distortion	1	1	1	2	2
Wear in stem	1	3	1	2	2
Hardening in foot	2	1	1	3	3
Ease of machining	3	1	5	2	4

Summarizing the results set out herein, the author considers that a reasonable selection of steel may be made from three types—a tungsten steel, a high-chromium steel and a nickel steel. The nickel steel is put into the list because it is desirable in many cases to use a good-quality steel which can be obtained fairly cheaply. Its tempering temperature in the heat treatment is never likely to be more than 650 deg. C., and the working temperature therefore must not exceed this. This limits the nickel steels to use in fairly low-temperature engines.

The high-chromium steel is about four times as expensive as nickel steel and has many advantages. It is usually tempered at approximately 750 deg. C. and has a considerably higher ultimate tensile strength than the nickel steel. It is too good for use in the inlet valves, and, generally speaking, is unnecessarily expensive for the exhaust valves of the cooler engines.

The great value of tungsten steel is that it retains, at high temperature, a greater proportion of its low-temperature strength than does any other steel. It is quite the strongest steels at high temperature and at the same time has such properties at ordinary temperatures as to make it easily handled in the machine shop. As it is a very expensive steel, it should be used only in the hotter engines for exhaust valves.

The author closes the paper with an appendix in which is given a close examination of some of the experimental results. The object of the tests carried out on the series of tungsten steel was to discover the comparative importance of carbon and of tungsten in regard to the properties of the steels at high temperature, and also to ascertain

*Abstract of a paper By Leslie Aitchison, read before the Institution of Automobile Engineers, England, November, 1919.

whether vanadium has any influence upon the properties of the steel. Making allowance for the difficulties of determining tensile strength at high temperatures, the conclusions were in general that the higher the percentage of carbon the higher the tensile strength, and also that the higher the percentage of tungsten the higher the tensile strength at high temperature. It was likewise concluded that it does not matter whether the steels contain vanadium or not.

In connection with the chromium steels a general comparison between steels containing 6.5 per cent of chromium and those containing 13 per cent of chromium shows that at a temperature of 700 deg. C. the steels containing the lowest percentage of chromium are stronger. At about 900 deg. C., however, both classes of steel have approximately the same tensile strength. Tests to show the effect of varying percentages of carbon upon steel containing about 6.5 per cent of chromium indicate that the higher percentage of carbon results in a distinctly greater tensile strength at high temperature. Tests to show the effect of silicon were entirely inconclusive. Other tests show that the presence of as much as 3 per cent of nickel has no appreciable effect on the tensile strength at high temperatures. A number of tests were made to show the effect of different tempering temperatures upon the properties of the steel. It was found that the addition of 3 per cent of nickel lowers considerably the temperature at which the steel begins to harden. The samples containing nickel begin to harden in air at a temperature 700 deg. C., but none of the others hardens in air until the quenching temperature has been raised to more than 750 deg. C. The variation in carbon content seems to make no difference in the effect of the nickel. It is well known that the addition of chromium to steel raises the temperature at which the steel hardens in air, and these results show that the addition of nickel has a powerful influence toward neutralizing this effect.

Another point noticed is that the raising of the carbon content in the steel has the effect of producing a higher ultimate strength, whatever tempering treatment has been given to it.

St. Lawrence Power*

By Alexander T. Vogelsang
First Assistant Secretary of the Interior

The Census of 1912 showed the developed water power of the United States to be 4,870,000 horsepower. According to that census, the installed capacity of stationary prime movers was 30,000,000 horsepower, furnished by water, gas, and steam. This does not include power generated in locomotives, marine engines, automobiles, and such apparatus. The average power furnished by these prime movers was probably not more than 20 per cent of installed capacity, so that power produced in 1912 was probably not more than 6,000,000 continuous horsepower.

Seventy per cent of the waterpower is west of the Mississippi, and over 70 per cent of stationary prime-movers horsepower is east of that river.

St. Lawrence River from Tibbets Point, at the foot of Lake Ontario, to deep water at the level of Lake St. Francis at St. Regis, N. Y., flows about 113 miles along the international boundary and has a fall at low water of about 92 ft., of which about 91 ft. is in the lower 48 miles, from Galop Rapids to St. Regis, there being a fall of but one foot in the 65 miles from Lake Ontario to Galop Rapids. The fall in the lower section of the river is concentrated in a number of rapids which are passed by means of canals and locks on the Canadian side. The Long Sault Rapids, with a fall of 48 ft., extend for about 12 miles along the lower end of this section.

From St. Regis to ocean navigation at Montreal there is a fall of about 130 ft. in a distance of about 70 miles, of which 129 ft. is concentrated in two stretches of 14 and 8½ miles, the former including Coteau, Cedar and Cascade rapids, in which the fall is 84 ft., and the latter, including Lachine Rapids, in which the fall is 45 ft.

*From an address before the Rivers and Harbors Congress, Washington, D. C.

The amount of theoretical power per foot of fall in the St. Lawrence River, as shown by the mean annual flow, is 27,360 hp., and as a total fall of 92 ft. is available in the international section of the river, the theoretical power in that section is about 2,520,000 hp. Assuming that only 70 per cent of this energy can be made available on account of losses in head and inefficiencies of waterwheels and generating equipment, this stretch of the river, owned equally by the United States and Canada, would furnish 1,764,000 hp., or 882,000 hp. to each. About 95,000 hp. of the portion belonging to the United States has already been developed by the St. Lawrence River Power Co. by means of a canal which diverts a maximum of 30,000 sec.-ft. from the river near the head of the Long Sault Rapids through the power house into Grass River, which joins the St. Lawrence below the Long Sault Rapids. Here a head of about 40 ft. is developed. After the power utilized at this development is deducted from the power of the portion belonging to the United States, the power not yet developed in that portion of the river amounts to about 790,000 horsepower.

The utilization of this power in public service to replace coal now consumed in the large manufacturing centers of New York and New England has an important bearing on the extension of the life of our coal deposits. On the basis of three pounds of coal per kilowatt-hour, this St. Lawrence River power is the equivalent of about 7,750,000 tons of coal annually. This is about 1.3 per cent of the total production of bituminous coal in the United States in 1918. The amount of coal consumed per year by public-utility plants in New England and New York is about 7,000,000 tons.

Will Foster Electrochemical Industries

If, on the other hand, as the demand for electrochemical and other products requiring large quantities of power increases, these water powers will give opportunity for greatly enlarged industrial development. Doubtless most of the power developed on the Canadian side, not only in the boundary portion of the river, but in the purely Canadian portion below, would be utilized for such purposes. This portion is within easy reach of the largest commercial centers, and after the St. Lawrence has been made more fully available for navigation, it would be possible for ocean shipping to reach the factory sites. Assuming, as seems entirely likely, that these developments are made at low cost and can deliver power at attractive rates, it is entirely unlikely that this power can be marketed as rapidly as it can be developed.

The saving in man power by producing energy by water power, as compared with steam power, is large, especially if the number of miners required to provide coal is considered. A large southern power company shows, by comparing the number of men needed to operate one of its hydro-electric plants on the Coosa River with the number, including the miners, needed to operate one of its steam plants, that in a water power plant one man is required per 5,250 kw. of capacity, whereas in a steam plant one man is required per 74 kw. of capacity.

Applying these figures to the water power available on the St. Lawrence and assuming that the 590,000 kw. of water power would replace the same amount of steam power, about 7,900 men would be released for work in other fields.

The power possibilities of the St. Lawrence have therefore very large economic importance, and any plans for developing the river for navigation should make provision for the maximum development of the energy that it will afford. This provision can be made only by working out a comprehensive plan in which both Canada and the United States seek the best engineering and economic solution of the problem. No doubt private capital is available to make all this improvement in consideration of a grant of the power privilege, but there is equally no doubt that this will not be done. Canada, as well as the United States, would not permit it.

It is now three years since Congress was urged to authorize a study of the power possibilities of the congested industrial part of the Atlantic seaboard with a view to disclosing not only the fact that a great saving in power could be effected and a much larger actual use of the power now produced could be gained but that new supplies could be obtained both from running water and from the conversion of coal into power at the mines instead of after a long haul. The feasibility of constructing a super-transmission line of electric power paralleling the Atlantic from Richmond to Boston, affording connections at New York and Boston with the vast water-power resources of the St. Lawrence on the north and Niagara on the west, a main transmission line into which would run many minor feeding streams of power and from which would diverge an infinite number of small delivering lines tying together the separated power plants of ten states, so that one could give aid to the other, so that one could take the place of the other, so that all may join their power for good in any great drive that may be projected, may well be investigated by the Government, not with the thought that the Government would construct or even operate such a trunk line, but that the project might so attract the attention of the engineering and financial world as to make it a reality.

The power production will pay the entire cost of all the work. As I understand it, our neighbor proposes to go into the enterprise on a fifty-fifty basis in spite of the apparent fact that the present advantage is enormously in our favor. Shall Canada's seven million people say "I will," and Uncle Sam's one hundred and tén million say "I won't"? Or, shall we rather coöperate and develop this waterway, obtaining thereby, practically as a by-product, this tremendous resource of power?

Additions to Pacific Gas and Electric Co.'s System

Revised official figures given out by the Pacific Gas & Electric Co. put the additional power which it will possess by reason of the lease of the Sierra & San Francisco Power Co. and purchase of the Northern California Power Co. at 138,000 hp., installed capacity. The Pitt River project will give a possible added 200,000 hp. installed capacity. The proposed expenditure of more than $3,000,000 on the Pitt river project, is to cover only the needs of 1920, and further expenditures are to be made. The company is now at work along the Pitt River and is pushing preparations for the erection of the first station with all speed. Ultimately a number of stations will be installed.

Officials regard the problem of long distance transmission as solved. The Southern California Edison Co. is transmitting power 250 miles into Los Angeles at 150,000 volts, and that is approximately the distance of the Pacific Gas & Electric Company's power site from San Francisco. The Northern California Power Co. has an installed capacity of 48,000 hp. The Sierra & San Francisco Power Co. has an additional 90,000 hp. of which 57,000 hp. is hydro-electric and 33,000 hp. is steam. The plants and their capacities the Pacific Company gets control of by the lease of the latter, it is announced, are: The main hydro-electric plant on the middle fork of the Stanislaus river about 14 miles from Angels Camp, Calaveras county; and the hydro-electric plants at Phoenix, about four miles from Sonora, Tuolumne county; a 100,000-volt transmission line from Stanislaus power house through the counties of Tuolumne, Calaveras, Stanislaus, San Joaquin, Alameda, Santa Clara and San Mateo to the bay shores; a step-down substation in San Mateo county; 100,000-volt substation at Copperopolis; 60,000-volt transmission line from Port Marion south through Santa Clara, San Bonita and Monterey counties to the city of Salinas; step-down substation near San Juan Bautista; 30,000-volt transmission line from the power house at La Grange to the cities of Modesto and Turlock and other cities and towns in Stanislaus county. Also a steam plant in San Francisco with 11,000-volt distribution lines connecting North Beach steam plant with the bay shore substation and with substations at Turk and Fillmore Sts., Eleventh and Bryant Sts., Geneva and San Jose avenue and the city of Burlingame. Secondary 17,000-volt transmission system in the counties of Tuolumne, Calaveras, Stanislaus, San Joaquin, Contra Costa and Alameda with electrical distributing systems in town and rural territory in these counties.

A. I. E. E. Holds Midwinter Convention

During the three days of Feb. 18 to 20 inclusive, the American Institute of Electrical Engineers held its eighth midwinter convention in the Engineering Societies Building, New York City, with a total of about one thousand members and guests in attendance. All during the convention period the members and their guests were given an opportunity to visit the research laboratories of the Bell Telephone systems at the Western Electric Co.'s building; Electrical Testing Laboratories; and the power plant of the New York Edison Co., Brooklyn Edison Co.; Brooklyn Rapid Transit Co., United Electric Light & Power Co., New York Central R.R., Interborough Rapid Transit Co., and the Public Service Co.'s Essex Power Station at Newark, N. J.

The dinner dance was held at the Hotel Astor on Thursday evening and proved to be a very enjoyable social function for the three hundred and sixty members and guests who attended.

The convention was formally opened on Wednesday evening, Feb. 18, by President Calvert Townley, who delivered an address in which he directed the attention of the engineer to a greater participation in public affairs. An abstract of president Townley's address appears in this issue. Three technical papers were also presented at this meeting. The first paper, "Daylight Saving," by Preston S. Millar was presented by the author, and not only considers the question of daylight saving from an ecomonic standpoint but also from the sociological aspect. An abstract of this paper will appear in a later issue.

"Essential Statistics for the General Comparison of Steam-Power Plant Performances" by W. S. Gorsuch was presented by the author. This paper briefly outlined a method for preparing statistical reports relating to power generation in steam-power plants, whereby a fairly close comparison can be made of the efficiency between different plants, without going into detailed study of the thermal characteristics of the plant, or the intricate subject of power cost. The third paper, "Standard Graphic Symbols" by E. J. Cheney was presented by the author and discusses a number of suggested symbols for representing electrical apparatus. The discussion on this paper indicated that although such standard symbols were something to be desired there is a wide diversion of opinion as to just what these symbols should be. It was clearly indicated in the discussion that daylight saving has become a class question, that where it is beneficial to the population living and working in towns and cities it was in general a hardship on the farmer. It was also pointed out that although different local standard times would be very objectionable there was no reason why each class of workers should not adjust working hours to best suit its own needs.

At the Thursday morning meeting a symposium by W. S. Murray, W. L. R. Emmet, J. F. Johnson, H. G. Reist, F. D. Newbury, W. D. Potter, Philip Torchio, Percy H. Thomas, W. D. A. Peaslee and A. O. Austin was presented on "Economical Supply of Electric Power for the Industries and Railroads of the Northeastern Atlantic Seaboard." A comprehensive abstract of this symposium and the discussion thereon will appear in an early issue of *Power*. At the close of the discussion a motion was passed that the Board of Directors be requested to appoint a committee with W. S. Murray as chairman to put the Institute squarely behind the movement to bring about the realization of this super-power system.

On Thursday afternoon parallel sessions were held at which five papers were presented: "Printing-Telegraph System" by J. H. Bell, described the printing-telegraph systems in use today which are designed for handling traffic at more than one hundred words per minute and discusses the operating features such as accuracy, speed of service, operator output, maintenance, line economies, flexibility. "Maximum-Output Networks for Telephone Substations and Repeater Circuits" by G. A. Campbell and R. M. Foster shows that ideal telephone substation and repeater circuits to present output and input requirements which can be met by a type of circuit containing four resistances, each of which has maximum output. "A Method of Separating

No Load Losses in Electrical Machinery" by C. J. Fechheimer proposes a method which makes use of idle operation of the machine as a motor, the voltage being varied and the speed kept constant, and after deducting the armature I^2R losses from the watts input the remaining watts are plotted against the voltage. "Inherent Regulation of Direct Current Circuit" by A. L. Ellis and D. W. St. Clair discusses the voltage changes in direct-current circuit upon change of load, and gives the results of an oscillograph and mathematical investigation into this subject. "The Measurement of Projectile Velocities" by Paul A. Klopsteg and Alfred L. Loomis, discusses the requirements imposed by proving-ground practice upon a chronograph, which is intended for general ammunition testing and describes an instrument which was developed and adopted as a standard ordnance chronograph.

Two technical sessions were held on Friday. Three papers were presented at the morning session and two at the afternoon session. "A New Form of Vibrating Galvanometer" by P. G. Agnew describes a type of instrument that consists essentially of a fine steel wire mounted on one pole of a permanent magnet and so arranged that the free end of the wire may vibrate between the poles of an electromagnet through which the current to be detected passes. "Precision Galvanometric Instrument for Measuring Thermoelectric E. M. F's" by T. R. Harrison and P. D. Foote describes an instrument in which provision is made whereby the total resistance of the circuit may be easily adjusted to the preassigned value for which the scale of instrument is graduated without requiring the use of any auxiliary instrument or a source of voltage other than that of the thermocouple being measured. "Notes on Synchronous Commutators" by J. B. Whitehead and T. Isshiki discusses the magnitude of the error that may arise in measuring crest values due to the relatively small capacity in the commutator and galvanometer circuit, when a synchronous commutator is used as a suppressor.

The two papers presented at the closing session Friday afternoon were "Oscillographs and Their Test" by A. E. Kennelly, R. N. Hunter and A. A. Prior; and "The Accuracy of Commercial Measurements" by H. B. Brooks. In the first paper a method and technique for the testing and calibration of oscillographs is described, using an auxiliary vibrator or oscillographmeter. The second paper discusses the accuracy required in commercial electrical measurements, and the means of obtaining it; namely, proper selection, installation, use, and maintenance of the instrument.

Public Utility Companies Side of the Coal Situation

That storage of coal by public utilities and by other large consumers is regarded by the President's coal commission as an important factor in stabilizing the coal industry, became very evident during the hearing of February 17 when Chairman Robinson asked M. H. Aylesworth, representing the National Electric Light Association, if he did not regard it as the duty of large consumers to store coal during the spring and summer months, even if it increased the price of their fuel. Mr. Aylesworth expressed the firm conviction that large consumers should move as much of their winter's fuel as practicable during the summer. He urged, however, that the Commission use its influence to have the railroads, the operators, and others who would be benefited as well, make a seasonable price, which would tend to offset the rehandling and other costs to which that practice would subject the consumer.

In that connection, Chairman Robinson asked George W. Elliott, of the National Committee on Gas and Electric Service, if public utility companies take advantage of the fact that they are second only to the railroads in the matter of priority and defer buying when there is a chance that a better price may be obtained. With the certainty that coal will be found for them it has been claimed that some of these companies do not hesitate to allow their surplus to dwindle past the danger line. Mr. Elliott expressed the opinion that such instances would be found to be so few as to be insignificant in considering the situation as a whole. His estimate of the annual coal consumption of the country's public utilities, was 50,000,000 tons.

All of the representatives of the public utility interests testified to the serious annoyances and losses which have come to them as the result of railroad confiscations and diversions. Great emphasis was also laid by each witness on the difficulties which they have had to meet because they have been forced to pay the 14 per cent advance in wages granted the mine workers by Dr. Garfield. They contended that the spirit of the Garfield order should prevail and the 14 per cent be absorbed by the operators. They pointed out that their income is not readily adjustable thereby working a particular hardship on that class of consumers.

Ralph Crews, representing the bituminous operators in the central competitive field, explained that the fuel administration had issued an order excluding from the general proposal that the increase be absorbed by the operators, any coal which had been contracted for prior to the date of the announcement, where the contract contained a provision that any increase in labor cost was to be added to the price. The idea is, Mr. Crews stated, that these contracts were entered into in good faith without any knowledge of what the future had in store, and that it would be unfair to interfere with such contracts. Since 98 per cent of the coal used by public utilities is purchased under contract, such consumers were particularly hard hit by the 14 per cent increase.

George R. Starrs

GEORGE R. STARRS

Following an illness of less than thirty-six hours, George R. Starrs, past president of Paterson Association No. 2, of the National Association of Stationary Engineers, died Tuesday, Feb. 24, at his home, Paterson, N. J. The deceased was born in Kingston, Ontario, on December 19, 1858, and at an early age went to Sheboygan, Wis., where he was engaged for fifteen years in operating engineering. He then located in Detroit for several years and later in Chicago. From 1893 to the time of his death he lived in Paterson, where he was employed as an operating engineer. About ten years ago he embarked in the engineering supply business, principally representing Green, Tweed & Co., and continuing to take an active interest in N. A. S. E. affairs, represented his local and state association at several state and national conventions, and will be remembered for his efforts in behalf of engineers' license laws.

At the time of his demise Mr. Starrs was chairman of the license board of the state association and a member of the National License Law Committee of the N. A. S. E. He had been a member of the Paterson Board of Education for over four years, first having been appointed to fill an unexpired term and subsequently reappointed to a full term of three years, of which he had served but one year.

The deceased was a member of Paterson Lodge, B. P. O. Elks, No. 60, and is survived by his wife, Mrs. Minnie Starrs, their two sons and two daughters. Funeral services were held on Thursday evening, Feb. 26, from his late residence.

A. I. E. E. to Meet in Pittsburgh

On Friday, March 12, 1920, the American Institute of Electrical Engineers will hold its 358th meeting, under the auspices, jointly, of the Pittsburgh Section and the Traction and Transportation Committee. The headquarters for the meeting will be the William Penn Hotel. During the morning there will be an inspection trip through the Westinghouse Works, concluding with a luncheon to be served to all as guests of the company.

PRICES - MATERIALS - SUPPLIES

These are prices to the power plant by jobbers in the larger buying centers east of the Mississippi. Elsewhere the prices will be modified by increased freight charges and by local conditions.

POWER-PLANT SUPPLIES

HOSE—

Fire	50-Ft. Lengths
Underwriters' 2½-in.	75c. per ft.
Common, 2½-in.	40%

Air	First Grade	Second Grade	Third Grade
¾-in. per ft.	$0.50	$0.33	$0.22

Steam—Discounts from List

First grade...... 30% Second grade.. . 40% Third grade.... 45%

RUBBER BELTING—The following discounts from list apply to transmission rubber and duck belting:

Competition	60%	Best grade	30%
Standard	45%		

LEATHER BELTING—Present discounts from list in the following cities are as follows:

	Medium Grade	Heavy Grade
New York	20%	25%
St. Louis	40%	35%
Chicago	45%	40%
Birmingham	35%	30%
Denver	10%	5%

RAWHIDE LACING—25% for cut; 86c. per sq. ft. for ordinary.

PACKING—Prices per pound:

Rubber and duck for low-pressure steam	$1.00
Asbestos for high-pressure steam	1.70
Duck and rubber for piston packing	1.00
Flax, regular	1.20
Flax, waterproofed	1.70
Compressed asbestos sheet	.90
Wire insertion asbestos sheet	1.50
Rubber sheet	.50
Rubber sheet, wire insertion	.70
Rubber sheet, duck insertion	.50
Rubber sheet, cloth insertion	.30
Asbestos packing, twisted or braided and graphited, for valve stems and stuffing boxes	1.30
Asbestos wick, ½- and 1-lb. balls	.85

PIPE AND BOILER COVERING—Below are discounts and part of standard lists:

PIPE COVERING		BLOCKS AND SHEETS	
Pipe Size	Standard List Per Lin. Ft.	Thickness	Price per Sq. Ft.
1-in.	$0.27	½-in.	$0.27
2-in.	.36	1-in.	.30
6-in.	.80	1½-in.	.45
4-in.	.60	2-in.	.60
3-in.	.45	2½-in.	.75
8-in.	1.10	3-in.	.90
10-in.	1.30	3½-in.	1.05

85% magnesia high pressure		List
For low-pressure heating and return lines	4-ply	50% off
	3-ply	52% off
	2-ply	54% off

GREASES—Prices are as follows in the following cities in cents per pound for barrel lots:

	Cincinnati	Pittsburgh	Chicago	St. Louis	Birmingham	Denver
Cup	7	9	6.6	6.7	8½	10.5
Fiber or sponge	8	10	8.6	13	8½	13.75
Transmission	7	9	8.1	13	8½	12.5
Axle	4½	6	4.8	4.75	4½	7.0
Gear	4½	9	6.1	7.5	8½	8.5
Car journal	22 (gal.)	21 (gal.)	4.7	4.7	8½	9.0

COTTON WASTE—The following prices are in cents per pound:

	New York Current	New York One Year Ago	Cleveland	Chicago
White	13.00	11.00 to 13.00	14.00	11.00 to 14.00
Colored mixed	9.00 to 12.00	8.50 to 12.00	11.00	9.50 to 12.00

WIPING CLOTHS—Jobbers' price per 1000 is as follows:

	13½ x 13½	13½ x 20½
Cleveland	$52.00	$58.00
Chicago	41.00	43.50

LINSEED OIL—These prices are per gallon:

	New York Current	New York One Year Ago	Chicago Current	Chicago One Year Ago
Raw in barrels (5 bbl. lots)	$1.80	$1.49	$1.98	$1.66
5-gal. cans	2.00	1.74	2.23	1.86

WHITE AND RED LEAD—Base price per pound:

	Red Current Dry	Red Current In Oil	Red 1 Year Ago Dry	Red 1 Year Ago In Oil	White Current Dry and In Oil	White 1 Yr. Ago Dry and In Oil
100-lb. keg	15.50	17.00	13.00	14.50	15.50	13.00
25- and 50-lb. kegs	15.75	17.25	13.25	14.75	15.75	13.25
12½-lb. keg	16.00	17.50	13.50	15.00	16.00	13.50
1-lb. cans	18.50	20.00	15.00	16.50	15.00	15.00
5-lb. cans	20.50	22.00	16.00	17.50	16.00	16.00

500 lb. lots less 10% discount; 2000 lb. lots less 10-2½%.

RIVETS—The following quotations are allowed for fair-sized orders from warehouse:

	New York	Cleveland	Chicago
Steel $\frac{7}{16}$ and smaller	30%	55%	45%
Tinned	30%	55%	45%

Boiler rivets, ¾, ⅞, 1 in. diameter by 2 in. to 5 in. sell as follows per 100 lb.:

New York. .$6.00 Cleveland...$4.00 Chicago $4.97 Pittsburgh...$4.72

Structual rivets, same sizes:

New York....$6.10 Cleveland.. $4.10 Chicago $5.07 Pittsburgh...$4.82

REFRACTORIES—Following prices are f. o. b. works, Pittsburgh:

Chrome brick	net ton	$75-80	at Chester, Penn.
Chrome cement	net ton	45-50	at Chester, Penn.
Clay brick, 1st quality fireclay	net M.	38-45	at Clearfield, Penn.
Clay brick, 2nd quality	net M.	33-35	at Clearfield, Penn.
Magnesite, dead burned	net ton	50-55	at Chester, Penn.
Magnesite brick, 9 x 4½ x 2½ in.	net ton	80-85	at Chester, Penn.
Silica brick	net M.	45-50	at Mt. Union, Penn.

Standard size fire brick, 9 x 4½ x 2½ in. The second quality is $4 to $5 cheaper per 1000.

St. Louis—Fire Clay, $35 to $50.
Birmingham—Silica, $50; fire clay, $45; magnesite, $85; chrome, $85.
Chicago—Second quality, $25 per ton.
Denver—Silica, $18; fire clay, $12; magnesite, $57.50.

BABBITT METAL—Warehouse prices in cents per pound:

	New York Current	New York One Year Ago	Cleveland Current	Cleveland One Year Ago	Chicago Current	Chicago One Year Ago
Best grade	90.00	87.00	70.00	80.00	70.00	75.00
Commercial	50.00	42.00	20.00	21.50	15.00	15.00

SWEDISH (NORWAY) IRON—The average price per 100 lb., in ton lots, is:

	Current	One Year Ago
New York	$21-26	25.50-30.00
Cleveland	20.00	20.00
Chicago	16.50	16.50

In coils an advance of 50c. usually is charged.
Domestic iron (Swedish analysis) is selling at 15c. per lb.

SHEETS—Quotations are in cents per pound in various cities from warehouse also the base quotations from mill:

Blue Annealed	Large Mill Lots Pittsburgh	New York Current	New York One Year Ago	Cleveland	Chicago
No. 10	3.55-4.00	5.57-6.80	5.17	5.35	5.27
No. 12	3.60-4.05	5.61-6.85	5.22	5.40	5.32
No. 14	3.65-4.10	5.67-6.90	5.27	5.45	5.37
No. 16	3.75-4.20	5.77-7.00	5.37	5.55	5.47
Black					
Nos. 18 and 20	4.15-4.80	6.80-7.30	6.02	5.95	6.30
Nos. 22 and 24	4.20-4.85	6.85-7.35	6.07	6.00	6.35
No. 26	4.25-4.90	6.90-7.90	6.12	6.05	6.40
No. 28	4.35-5.00	7.00-8.00	6.22	6.15	6.50
Galvanized					
No. 10	4.70-6.00	7.50-9.00	8.22	5.05	6.65
No. 12	4.80-6.10	7.60-9.10	8.27	5.10	6.70
No. 14	4.80-6.10	7.60-9.10	8.42	5.25	6.85
Nos. 18 and 20	5.10-6.40	7.90-9.40	8.72	5.55	7.15
Nos. 22 and 24	5.25-6.55	8.05-9.55	7.12	6.95	7.55
No. 26	5.40-6.70	8.20-9.70	7.27	7.40	7.70
No. 28	5.70-7.00	8.50-10.00	7.57	7.50	8.00

PIPE—The following discounts are for carload lots f. o. b. Pittsburgh; basing card of Jan. 1, 1919, for steel pipe and for iron pipe:

Inches	Steel Black	Steel Galvanized	Inches	Iron Black	Iron Galvanized
BUTT WELD					
¼, ⅜, and ½	50½%	24%	¼ to 1½	30½%	23½%
¾	54½%	40%			
1 to 3	57½%	44%			
LAP WELD					
2	50½%	35%	2	32½%	18½%
2½ to 6	53½%	41%	2½ to 6	34½%	21½%
BUTT WELD, EXTRA STRONG PLAIN ENDS					
¼, ⅜ and ½	48½%	25%	¼ to 1½	39½%	24½%
¾	51½%	39%			
1 to 1½	50½%	43%			
LAP WELD, EXTRA STRONG PLAIN ENDS					
2	48%	37%	2	33½%	20½%
2½ to 4	51%	49%	2½ to 4	35½%	23½%
4½ to 6	50%	39%	4½ to 6	34½%	22½%

Stock discounts in cities named are as follows:

	New York Black	New York Galvanized	Cleveland Black	Cleveland Galvanized	Chicago Black	Chicago Galvanized
¾ to 3 in. steel butt welded	40%	24%	40%	31%	54%	40½%
2½ to 3 in. steel lap welded	35%	20%	42%	27%	50%	37½%

Malleable fittings. Class B and C, from New York stock sell at list + 23%.
Cast iron, standard sizes, net

BOILER TUBES—The following are the prices for carload lots, f. o. b. Pittsburgh:

Lap Welded Steel		Charcoal Iron	
3½ to 4½ in.	40½	3½ to 4½ in.	15½
2½ to 3¼ in.	30½	3 to 3¼ in.	1
2¼ in.	24	2½ to 2¾ in.	+ 1
1¾ to 2 in.	19½	2 to 2¼ in.	+16
		1¾ to 1¾ in.	+29

Standard Commercial Seamless—Cold Drawn or Hot Rolled

	Per Net Ton		Per Net Ton
1 in.	$327	1¾ in.	$207
1¼ in.	267	2 to 2¼ in.	177
1½ in.	257	2½ to 3¼ in.	167
1¾ in.	207	4 in.	187
		4½ to 5 in.	207

These prices do not apply to special specifications for locomotive tubes nor to special specifications for tubes for the Navy Department, which will be subject to special negotiations.

ELECTRICAL SUPPLIES

ARMORED CABLE—

B.&S. Size	Two Cond. M Ft.	Three Cond. M Ft.	Two Cond. Lead M Ft.	Three Cond. Lead M Ft.
No. 14 solid	$104.00	$138.00	$164.00	$210.00
No. 12 solid	135.00	170.00	225.00	265.00
No. 10 solid	185.00	235.00	275.00	325.00
No. 8 stranded	285.00	375.00	520.00	500.00
No. 6 stranded	400.00	500.00	560.00	....

From the above lists discounts are:

Less than coil lots	Net List
Coils to 1,000 ft.	10%
1,000 ft. and over	15%

BATTERIES, DRY—Regular No. 6 size red seal, Columbia, or Ever Ready:

	Each, Net
Less than 12	$0.45
12 to 50	.38
50 to 125 (bbl.)	.35
125 (bbl.) or over	.32

CONDUITS, ELBOWS AND COUPLINGS—These prices are f. o. b. New York with 10-day discount of 5 per cent.

	Conduit		Elbows		Coup'ings	
Size, In.	Black 2,500 to 5,000 Lbs.	Galvanized 2,500 to 5,000 Lbs.	Black 2,500 to 5,000 Lbs.	Galvanized 2,500 to 5,000 Lbs.	Black 2,500 to 5,000 Lbs.	Galvanized 2,500 to 5,000 Lbs.
½	$80.50	$85 59	$19.13	$20 27	$7.04	$ 7.46
¾	105.51	113.51	25.17	26 67	10.07	10.67
1	157.59	167.79	37.25	39.47	13.09	13.37
1¼	213.11	227.01	48.46	50 26	18.30	18.98
1½	254.93	271.43	64.62	67.02	22.61	24.45
2	342.99	365 19	118.47	122.87	30.15	31.27
2½	542.30	377.40	193 86	200.00	43.08	44.68
3	709.16	755 06	511.96	536 16	64.62	67.02
3½	871.24	935 64	1,141 62	1,184 02	86.16	89.36
4	1,064.93	1,130.33	1,319.32	1,368 32	107.70	111.70
4½	1,240.79	1,318.99	1,997.83	2,072.03	161.55	167.55
5	1,445.96	1,534.76	2,773.27	2,866.27	177.70	184.30
6	1,875.84	1,991.04	3,446.40	3,574.40	258.48	268.08

Standard lengths rigid, 10 ft. Standard lengths flexible, ½ in., 100 ft. Standard lengths flexible, ¾ to 2 in., 50 ft.

CONDUIT NON-METALLIC, LOOM

Size I. D., In.	Feet per Coil	List, Ft.	
3/16	250	$0.05½	Coils..... 40% off; Less coils, 25% off
¼	250	.06	
5/16	250	.09	
⅜	200	.12	
½	200	.15	
¾	150	.18	
1	100	.25	
1¼	100	.33	
1½	Odd lengths	.40	
2	Odd lengths	.55	

CUT-OUTS—Following are net prices each in standard-package quantities:

CUT-OUTS, PLUG

S. P. M. L.	$0 11	T. P. to D. P. S. B.	$0.24
D. P. M. L.	.11	T. P. to D. P. T. B.	.38
T. P. M. L.	.26	T. P. S. B.	.33
D. P. S. B.	.19	T. P. D. B.	.54
D. P. D. B.	.37		

CUT-OUTS, N. E. C. FUSE

	0-30 Amp.	31-60 Amp.	60-100 Amp.
D. P. M. L.	$0.33	$0.84	$1.68
T. P. M. L.	.48	1.20	2.40
D. P. S. B.	.42	1.05	....
T. P. S. B.	.81	1.80	....
D. P. D. B.	.78	2.10	....
T. P. D. B.	1.35	3.60	....
T. P. to D. P. D. B.	90	2.52	....

FLEXIBLE CORD—Price per 1000 ft. in coils of 250 ft.:

No. 18 cotton twisted	$22.00
No. 16 cotton twisted	27.00
No. 18 cotton parallel	26.50
No. 16 cotton parallel	34.50
No. 18 cotton reinforced heavy	34.00
No. 16 cotton reinforced heavy	42.00
No. 18 cotton reinforced light	30.00
No. 16 cotton reinforced light	36.75
No. 18 cotton Canvasite cord	28.50
No. 16 cotton Canvasite cord	32.00

FUSES, ENCLOSED—

250-Volt	Std. Pkg.	List
3-amp. to 30-amp.	100	$0.25
35-amp. to 60-amp.	100	.35
65-amp. to 100-amp.	50	.90
110-amp. to 200-amp.	25	2.00
225-amp. to 400-amp.	25	3.60
425-amp. to 600-amp.	10	5.50

600-Volt	Std. Pkg.	List
3-amp. to 30-amp.	100	$0.40
35-amp. to 60-amp.	100	.60
65-amp. to 100-amp.	50	1.50
110-amp. to 200-amp.	25	2.50
225-amp. to 400-amp.	25	3.50
450-amp. to 600-amp.	10	8.00

Discount: Less 1-5th standard package 30%
1-5th to standard package 40%
Standard package 52%

FUSE PLUGS, MICA CAP—

0-30 ampere, standard package	$4.75C
0-30 ampere, less than standard package	6.00C

LAMPS—Below are present quotations in less than standard package quantities:

Straight-Side Bulbs Mazda B— Watts	Plain	Frosted	No. in Package	Pear-Shape Bulbs Mazda C— Watts	Clear	Frosted	No. in Package
10	$0.35	$0.38	100	75	$0.70	$0.75	50
15	.35	.38	100	100	1 10	1.15	24
25	.35	.38	100	150	1.65	1.70	24
40	.35	.38	100	200	2.20	2.27	24
50	.35	.38	100	300	3.25	3.35	24
60	.40	.45	100	400	4.30	4.45	12
100	.65	.92	24	500	4.70	4.85	12
				750	6.50	6 75	8
				1000	7.50	7.75	8

Standard quantities are subject to discount of 10% from list. Annual contracts ranging from $150 to $300,000 net allow a discount of 17 to 40% from list

PLUGS, ATTACHMENT—

		Each
Hubbell, porcelain No. 5406,	standard package 250	$0.24
Hubbell composition No. 5467,	standard package 50	.32
Benjamin swivel No. 903,	standard package 250	.20
Hubbell current taps No. 5638,	standard package 50	.40

RUBBER-COVERED COPPER WIRE—Per 1000 ft. in New York

No.	Solid Single Braid	Solid Double Braid	Stranded, Double Braid	Duplex
14	$12.00	$16 00	$13.90	$28.50
12	13 25	15.70	18.05	30.70
10	18.30	21.00	23.85	41.50
8	25.54	28.60	32.70	56.77
6	......	......	51.40	.....
4	......	......	70.00	.....
2	......	......	101.80	.....
1	......	......	131.86	.....
0	......	......	160.00	.....
00	......	......	193.50	.....
000	......	......	235.20	.....
0000	......	......	288.60	.....

Prices per 1000 ft. for Rubber-covered Wire in Following Cities:

	Denver			Birmingham		
	Single Braid	Double Braid	Duplex	Single Braid	Double Braid	Duplex
14	$19 50	$24.00	$39.50	$12.50	$19.50	$29.00
10	28.92	33.12	65.52	25.10	35.50	55 50
8	40.32	45.12	89.64	34 75	49.50	70 50
6	63.84	69.24	.....	57 50	68 60	.....
4	......	97.86	.....	81.65	95 50	.....
2	......	145.20	.....	......	140.20	.....
1	......	189.00	.....	......	190 90	.....
0	......	251.40	.....	......	231.33	.....
00	......	305.40	.....	......	281.23	.....
000	......	317.40	.....	......	343.22	.....
0000	......	451.20	.....	......	416.80	.....

Pittsburg—23c. base; discount 50% St. Louis—30c. base.

SOCKETS, BRASS SHELL—

¼ In. or Pendant Cap, Key, Each	Keyless, Each	Pull, Each	⅜ In. Cap Key, Each	Keyless, Each	Pull, Each
$0.33	$0.30	$0.60	$0.39	$0.36	$0.66

Less 1-5th standard package ..... +50%
1-5th to standard package ..... +20%
Standard package ..... −10%

WIRE, ANNUNCIATOR AND DAMPPROOF OFFICE—

No. 18 B.&S. regular spools (approx. 8 lb.)	55c. lb.
No. 18 B.&S. regular 1-lb. coils	56c. lb.

WIRING SUPPLIES—

Friction tape, ¾ in., less 100 lb. 50c. lb., 100 lb. lots	55c. lb.
Rubber tape, ¾ in., less 100 lb., 65c. lb., 100 lb. lots	60c. lb.
Wire solder, less 100 lb. 50c. lb., 100 lb. lots	46c. lb.
Soldering paste, 2 oz. cans Nokorode	$1.20 doz.

SWITCHES, KNIFE—

TYPE "C" NOT FUSIBLE

Size, Amp.	Single Pole, Each	Double Pole, Each	Three Pole, Each	Four Pole Each
30	$0.42	$0.68	$1 02	$1.36
60	.74	1.22	1.84	2.44
100	1.50	2.50	3.76	5.00
200	2.70	4.50	6.76	9.00

TYPE "C" FUSIBLE, TOP OR BOTTOM

Size, Amp.	Single Pole, Each	Double Pole, Each	Three Pole, Each	Four Pole Each
30	.70	1.06	1.60	2.12
60	1.18	1.80	2.70	3.60
100	2.38	3.66	5.50	7.30
200	4.40	6.76	10 14	13.50

Discounts:

Less than $10.00 list value	+25%
$10 to $25 list value	+10%
$25 to $50 list value	+ 5%
$50 to $200 list value	−10%
$200 list value or over	−15%

THE COAL MARKET

BOSTON—Current prices per gross ton f.o.b. New York loading ports:

Anthracite

	Company Coal
Egg	$8.20@$8.65
Stove	8.45@ 9.05
Chestnut	8.55@ 9.05
Pea	7.05@ 7.40
Buckwheat	3.60@ 3.80
Rice	2.30@ 2.50
Barley	1.25

Bituminous

	Clearfields	Cambrias and Somersets
F.o b. mines, net tons	$2.85@$3.35	$3 15@$3 60
F.o b. Philadelphia, gross tons	5.05@ 5.60	5 35@ 5 90
F.o.b. New York, gross tons	5 40@ 5.95	5 75@ 6 25
Alongside Boston (water coal), gross tons	7 00@ 7.75	7 60@ 8·00

Pocahontas and New River are practically off the market for coastwise shipment, but are quoted at $6.25@$7.00.

NEW YORK—Current quotations, White Ash, per gross tons, f.o.b. Tidewater, at the lower ports are as follows:

Anthracite	Company Coal
Broken	$7.80@$8.25
Egg	8.20@ 8.65
Stove	8.45@ 9.05
Chestnut	8.55@ 9.05
Pea	7.05@ 7.40
Buckwheat	5.15
Rice	4.50
Barley	4.85
Boiler	4.85

Bituminous

Government prices at mines:	Spot
South Fork (best)	$3 25@$3.50
Cambria (best)	3 00@ 3 25
Cambria (ordinary)	2 60@ 2.90
Clearfield (best)	3.00@ 3.25
Clearfield (ordinary)	2.60@ 2.90
Reynoldsville	2 85@ 2.90
Quemahoning	3 25@ 3.50
Somerset (medium)	3 00@ 3.25
Somerset (poor)	2.50@ 2.75
Western Maryland	2 50@ 2.75
Fairmont	2 25@ 2.50
Latrobe	2.60@ 2 90
Greensburg	2.75@ 3 00
Westmoreland ¾ in	3 40@ 3 50
Westmoreland run-of-mine	2.75@ 3 00

PHILADELPHIA—Bituminous coal prices vary according to district from which they are mined. For ordinary slack the price is $2.45@$2.55; lump, $3.35, at the mines.

BUFFALO—

Anthracite

	On Cars, Gross Ton	At Curb, Net Ton
Grate	$8.55	$10.80
Egg	8.80	10.65
Stove	9.00	10.85
Chestnut	9.10	10.85
Pea	7.75	9.30
Buckwheat	5.70	7.75

Bituminous

Allegheny Valley	$4.80
Pittsburgh	4.65
No. 8 Lump	4.65
Mine-run	4.80
Slack	4.10
Smokeless	4.60
Pennsylvania Smithing	5.70

CLEVELAND—Prices of coal per net ton delivered in Cleveland are:

Anthracite

Egg	$12.25@$12.40
Chestnut	12.50@ 12.70
Grate	12.25@ 12.40
Stove	12.40@ 12.60

Pocahontas

Nine-run	$7.50

Domestic Bituminous

West Virginia split	$8.50
No. 8 Pittsburgh	$6.00@ 6.90
Massillon lump	8.25@ 8.50
Coshocton lump	7.15

Steam Coal

No. 6 slack	$5.25@$5.50
No. 8 slack	5.10@ 5.50
Youghiogheny slack	5.25@ 5.50
No. 8 ¾ in	5.70@ 6.00
No. 6 mine-run	5.25@ 5.50
No. 8 mine-run	5.75

Only coal available is mine-run Pocahontas.

MIDDLE WEST—Chicago quotations, f. o. b. cars at mine:

	Springfield, Carterville, Williamson, Franklin, Saline, Harrisburg	Fulton, Peoria	Grundy, La Salle, Bureau, Will
Lump	$2 55@$2.70	$2.95@$3 10	$3.25@$3 40
Washed	2.75@ 2.90		3.45@ 3 60
Mine-run	2 35@ 2 50	2 75@ 2.90	3.00@ 3.15
Screen'gs	2.05@ 2.20	2.35@ 2.50	2.75@ 2.90

New Construction

PROPOSED WORK

Me., Augusta—The State Hospital Comn. has had plans prepared for the construction of a 3 story, nurses' home at the State Hospital, here. A steam heating system will be installed in same. Total estimated cost, $100,000. H. S. Coombs, 11 Lisbon St., Lewistown, Archt.

Me., Damariscotta—The Lincoln Co. Power Co., Inc., plans to install a 750 hp. wheel and a 560 kva. generator at Damariscotta Mills to replace present 200 kw. generator. E. S. Lincoln, Mgr.

Me., Skowhegan—The Central Maine Power Co. will receive bids until March 15 for the construction of a power house and the installation of 5 turbine water wheels, each having a capacity of 4,300 hp. Total estimated cost, $100,000. J. A. Lenard, Engr.

Me., Vinal Haven—Vinal Haven Light & Power Co. plans to construct a transmission line from here to North Haven. A. A. Peterson, Mgr.

N. H., Bradford—The Bradford Light & Power Co. plans to install new meters in the generating system. C. E. Hadley, Genl. Mgr.

Mass., Boston—A. H. Bowditch, Archt., 44 Broomfield St., will soon receive bids for the construction of a 5 story, 60 x 150 ft. storehouse on Brookline Ave., for the Goodyear Tire & Rubber Co., 61 Brookline Ave. A steam heating system will be installed in same. Total estimated cost, $100,000.

Mass., Boston—The Eastern Massachusetts Electric Co. plans to construct a transmission line extension from West Peabody to Danvers. R. S. Tenney, Asst. Treas.

Mass., Boston—L. C. Greenleaf, Archt., 101 Tremont St., will soon award the contract for the construction of a 1 story, garage and machine shop on Tremont St., for the W. B. Rice Estate. A steam heating system will be installed in same. Total estimated cost, $150,000.

Mass., Brockton—The Edison Electric Illuminating Co. plans to install two 600 hp. boilers and construct 10 mi., 13,000 volt transmission line. A. F. Nelson, Mgr.

Mass., Brockton—Jenks & Ballou, Engrs., 1035 Grosvenor Bldg., will soon award the contract for the construction of a 4 story, 60 x 400 ft. manufacturing plant, for George E. Keith Co., 23 Station Rd. A steam heating system and electric power will be installed in same. Total estimated cost, $350,000.

Mass., Concord—The Concord Municipal Light Plant plans complete renewal of boiler plant, piping, etc. A. W. Lee, Mgr.

Mass., Groton—The Municipal Electric Light System plans to construct new copper trunk transmission line from here to Aver. W. H. Dodge, Supt.

Mass., Lowell—The Lowell Electric Light Corp., 29 Market St., plans to install a 12,500 kw. turbine generator with boilers and auxiliaries and construct a 2 mi., 13,200 volt, 5,000 kw. capacity, transmission line. J. A. Hunnewell, Mgr.

Mass., New Bedford—The New Bedford Gas & Edison Light Co. plans to install eight 755 hp. boilers, one 20,000 kw. turbine generator and construct extension to station. C. R. Price, Treas.

Mass., Norwood—The Municipal Electric Lighting System plans to install two 2,000 kw. transformers to replace the two 1,000 kw. transformers. F. S. Barton, Supt.

Mass., Springfield—The Bd. Educ. will soon award the contract for the construction of a 2 story, 116 x 186 ft. high school on Spring St. A steam heating system will be installed in same. Total estimated cost, $200,000. R. B. Warner, 168 Bridge St., Archt.

Mass., Springfield—The United Electric Light Co., 73 State St., plans to construct a 3 story, 44 x 80 ft. generator sub-station. W. L. Mulligan, Mgr.

Mass., Taunton—The Municipal Lighting Plant plans to install one new 4,000 turbo generator. W. B. Lewis, Mgr.

Conn., Bridgeport—Bollinger & Perrott, Archts. and Engrs., 47 West 34th St., New York City, will soon award the contract for the construction of a 1 story, 100 x 340 ft. factory, for the Salts Textile Mfg. Co. A steam heating system will be installed in same. Total estimated cost, $250,000.

Conn., Bridgeport—Westcott & Mapes, Inc., Engrs., 207 Orange St., will soon award the contract for the construction of a power plant on East Main St., for the United Illuminating Co., 84 Temple St., New Haven. Total estimated cost, $3,000,000.

Conn., Seymour—Johnson & Burns, Archts., c/o A. Johnson, 174 Bond St., Hartford, will soon award the contract for the construction of a 2 story, 62 x 124 ft. high school, including a boiler room, etc., for the Bd. Educ. Steam heating and ventilating systems will be installed in same. Total estimated cost, $150,000. R. C. Clark, Seymour St., Supt.

Conn., Westbrook—Knothe Bros. Co., 124 5th Ave., New York City, plans to build a 2 story, 54 x 150 ft. factory. A sprinkler and steam heating system will be installed in same. Total estimated cost, $120,000. Fletcher-Thompson, Inc., 1089 Broad St., Bridgeport, Engr.

N. Y., Addison—The Addison Electric Light & Power Co. plans to install a new steam unit in the generating system. M. G. Hubbs, Secy.

N. Y., Arcade—The Municipal Lighting Plant plans to install a new 250 hp. W.T. boiler, pumps and heater, also extend its transmission line to Sandusky, and some rural lines. L. A. Mason, Clk.

N. Y., Bath—The Municipal Light & Power System plans to enlarge its plant and install a 350 hp. boiler with direct connected unit. C. Cotton, Mgr.

N. Y., Broadalbin—The Broadalbin Electric Light & Power Co. plans to install regulators in its transmission system. T. W. Drought, Mgr.

N. Y., Buffalo—The Buffalo General Electric Co. plans to increase its transmission line capacity. R. McMillan, Statistician.

N. Y., Cooperstown—The Southern New York Power Co., (formerly Colliers Light, Heat & Power Co.), plans to install a 500 kw. 2,300 volt, 60 cycle turbo generator at the Walton plant. C. L. Stone, Genl. Mgr.

N. Y., Dansville—The Dansville Gas & Electric Co. plans to change from 133 cycles to 60 cycles. C. H. Mason, Supt.

N. Y., Elmira—The Elmira Water, Light & Power Co, Hulett Bldg., Lake and Water Sts., plans to install a new 6,000 kw. turbine, 786 hp. boiler and also construct a new transmission line to Elmira Heights. F. H. Hill, Mgr.

N. Y. Franklin—The Delaware & Otsego Light & Power Co. plans to construct a 6 to 10 mi. transmission line and rebuild all old transmission lines. E. A. Mackey, Mgr.

N. Y., Jamestown—The city voted $250,000 bonds to enlarge the electric light plant. Address Bd. of Water & Lighting Comrs.

N. Y., Long Island City—Lockwood, Green & Co., Archts. and Engrs., 101 Park Ave., New York City, will soon award the contract for the construction of a 2 story, 125 x 207 ft. silk mill, for H. R. Mallison, c/o Archt. A steam heating system will be installed in same. Total estimated cost, $225,000.

N. Y., Newark—James Goosen, 28 East Ave., is in the market for a 12 hp. steam tractor, complete with tank.

N. Y., New York—The Cammeyer Shoe Co., 47 West 34th St., is having plans prepared for the construction of a 6 story, 50 x 100 ft. store and office building on 5th Ave. A steam heating system will be installed in same. Rouse & Goldstone, 512 5th Ave., Archts. and Engrs.

N. Y., New York—E. Friedman, 3rd Ave. and 119th St., is having plans prepared for the construction of a 5 story, store and loft building on 120th St. and 3rd Ave. A steam heating system will be installed in same. Buchman & Kahn, 56 West 45th St., Archts. and Engrs.

N. Y., Norwich—The New York State Gas & Electric Corp. (formerly the Norwich Gas & Electric Co. controlled by the Associated Gas & Electric Co., 43 Exch. Pl., New York City), plans to install a new Skinner generator unit and construct 10 mi. of 3 ph. transmission line to Oxford. B. Diamond, Supt.

N. Y., Norwood—The Norwood Electric Light & Power Co. plans to install a generator and wheel, etc. C. F. Vance, Mgr.

N. Y., Nunda—The Nunda Electric Light Co., Inc., plans to increase the standby from 125 kva. to 250 kva. A. J. Woodworth, Pres.

N. Y., Olean—Mowbray & Effinger, Archts. and Engrs., 56 Liberty St., New York City, will receive bids until March 15 for the construction of a 5 story bank addition, for the Exchange Natl. Bank. A steam heating system will be installed in same.

N. Y., Port Leydon—The Port Leydon Electric Light & Power Co. plans to install new generator. Address Louise E. Wilson, Owner.

N. Y., Rouses Point—The Municipal Electric Light Plant plans to install new generator system. H. A. Sabourin, Pres.

N. Y., Wellsville—The Wellsville Water & Light Dept., plans to install a 300 kva. turbo generator. C. F. Schermerhorn, Genl. Supt.

N. J., Belvidere—E. R. Collins & Son, Belvidere and Westfield, plan to install a new hydro generator. P. E. Widener, Supt.

N. J., Chatham—The Municipal Electric Light Plant plans to install one 180 kva. direct connected unit. T. T. Haldaman, Supt.

N. J., Lakewood—The Lakewood & Coast Electric Co., Point Pleasant, plans to install a 1,500 kva. G.E. generator and 3 ph., 2,200 volt turbine and increase the capacity of its transformers. A. S. Phelps, Genl. Supt.

N. J., Sussex—The Woodbourne Electric Light, Heat & Power Co., plans to extend its transmission line in rural section 1 mi. I. D. Shorter, Secy.

Pa., Dorranceton (Kingston P. O.)—The Duplan Silk Corp., Wilkes-Barre, is having plans prepared for the construction of a 3 story, 100 x 300 ft. addition to its silk mill. A steam heating system will be installed in same. Total estimated cost, $400,000. G. Welch, Coal Exch. Bldg., Wilkes-Barre, Archt.

Pa., Harrisburg—The Lemoyne Quarries Co., Hershey Bldg., plans to construct a complete addition to the present crushing plant at its quarry, here, and is in the market for six 15 to 60 hp. electric motors, 1 electric hoist, 4-ton friction drum 350 ft. air compressor, 2 jackhammers, one 2½ in. steam drill, etc. Total estimated cost, $20,000.

Pa., Philadelphia—F. F. Durang, Archt., 1220 Locust St., will soon award the contract for the construction of a 3 story, 70 x 125 ft. parochial school on Allegheny Ave. and 28th St., for the Corpus Christi Parochial School. A steam heating system will be installed in same. Total estimated cost, $150,000.

Pa., Philadelphia—The O'Brien Machinery Co., 119 North 3rd St., is in the market for two 500 hp. water tube boilers to have a capacity of 160 lb. Ohio inspection, and 400 kva. 220 volt, 3 ph., 60 cycle belted generators, etc.

Pa., Philadelphia — Seamens Institute, Front and Queen Sts., engaged Ballinger & Perrott, Archts., 329 South Broad St., to prepare plans for the construction of a home on 2nd and Walnut Sts. A steam heating system will be installed in same. Total estimated cost, $500,000.

Pa., Scranton—The Scranton Business College, Jefferson Ave., is having plans prepared for the construction of a 4 story, 40 x 82 ft. college at 305-307 Jefferson Ave. A steam heating system and electric power will be installed in same. Total estimated cost, $150,000. John H. Seeley, Pres.

Pa., Wilkes-Barre—The Epstine Produce Co., Northampton St., is having plans prepared for the construction of a 2 story, 40 x 210 ft. warehouse. Cold storage plant equipment will be installed in same. Total estimated cost, $100,000.

Pa., Wilkes-Barre—The Vulcan Iron Wks., 730 South Main St., is having plans prepared for the construction of a 40 x 90 x 200 ft. locomotive shop on Hazle Ave. A steam heating system will be installed in same. Total estimated cost, $130,000. E. D. Rhinehimer, Carlisle St., Engr.

Md., Hagertown—The Municipal Electric Light Plant, 21 East Franklin St., plans to change its plant from 2 ph. to 3 ph., 4,000 volt. A. B. Grubmeyer, Mgr.

Md., Ocean City—The Delmarvia Utilities Co. plans to install a new 125 kw. generator and construct a 1 mi., 2,400 volt, 3 ph., transmission line. H. T. Moore, Treas.

D. C., Washington—The General Purchasing Officer of the Panama Canal, Wash., D. C., will receive bids until March 16 for furnishing three 110 and 220 volt, 25 cycle, single ph., repulsion inducton motors, 1 hp., 1,500 r.p.m. each.

N. C., Burlington—The Mayor will soon award the contract for the construction of a pumping plant, rain water reservoir of 1,700,000 gal. capacity, coagulating basin, clear water reservoir of 500,000 gal. capacity, pumps, etc. The Ludlow Engineers, Winston-Salem, Engrs.

N. C., Burlington—The Piedmont Power & Light Co. plans to install a 3,000 kva. generator and construct a 4 mi. transmission line to Simpsonville. J. H. Harden, Mgr.

N. C., Greenville—The Water & Light Comn. plans to install a new turbine and condenser. F. A. Haskins, Acting Supt.

N. C., Madison—The town of Madison plans to install either a new water plant or a Larfe gas engine. R. P. Patterson, Supt.

N. C., Newbern—The city plans to install a new 1,250 kw. turbine generator. F. G. Godfroy, Supt.

N. C., Pinehurst—The Pinehurst Power Plant plans to construct new underground transmission wires. L. E. Pender, Purch. Agt.

N. C., Scotland Neck—The Municipal Electric Light & Power Plant plans to change its transmission systems. L. R. Mills, Jr., Supt.

N. C., Wilson—The City Comrs. will soon award the contract for the construction of a power house, transformer station and 4 mi. of 13,000 volt transmission line. L. J. Hessing, Chn. T. A. MacEwan, Charlotte, Engr.

N. C., Zebulon—The Municipal Light & Power Plant plans to install one complete generator unit and construct a transmission line to Wakefield. R. E. Ward, Mgr.

Ga., Alma—The City Council will soon award the contract for the construction of a power plant and an oil engine plant, etc. Total estimated cost, $15,000.

Miss., Clarksdale—The Clarksdale Water & Light Plant plans to install a 1,250 kw. turbine generator. W. S. Bobs, Supt.

Miss., Cleveland—The Home Light & Ice Co. plans to construct a new 25 ton ice plant in addition to its present 10 ton plant, a boiler room, etc. F. G. Proutt, Treas.

Miss., Durant—The Durant Municipal Electric Light Plant plans to install a new 100 kva. generator unit. J. D. Alsbury, Supt.

Miss., Meridian—The Meridian Light & Ry. Co. plans to extend its transmission lines. L. H. Archer, Secy.-Treas.

Miss., Oxford—The Oxford Light & Water Plant plans to install a new generator set, etc. D. H. Hull, Mgr.

Miss., Shelby—The Shelby Light & Power Co. plans to install one 170 kva. alternator directly connected to a 200 hp. oil engine. D. Bradera, Mgr.

Miss., Woodville—The Woodville Light & Water Plant plans to construct a new ice plant. J. A. Massey, Supt.

La., Alexandria—The City Electric Light & Water Wks. plans to install one 200 hp. motor driven air machine. J. R. Laney, Supt.

La., Bunkie—The Bunkie Ice Co., Ltd., plans to install a 150 hp. crude oil generating unit. C. J. Pope, Vice-Pres.

La., De Ridder—The De Ridder Light & Power Co. plans to install a 250 hp. S. Kenner engine to 200 kwa. G.E. generator. M. B. Morgan, Treas.

La., Jennings—The Pub. Servc. Co. of Southwest Louisiana plans to change from d.c. generating system to a.c. C. L. Pardee, Asst. Mgr.

La., Natchitoches—The Municipal Electric Light & Water Wks. plans to install two 150 hp. crude oil engines and a 125 kva. generator. G. J. Shehane, Supt.

Tenn., Knoxville—The Knoxville Ice & Cold Storage Co., 613 McGhee St., plans to construct an addition to the present cold storage plant to increase the daily capacity of same. Total estimated cost, $30,000.

Ky., Berea—The Berea College Heating & Power Plant plans to construct a 2,500 ft. transmission system. G. G. Dick, Supt.

Ky., Jackson—The Jackson Utilities Co. plans to install one 25 hp. oil engine direct connected to a 20 kva., 2,300 volt, 3 ph. generator. O. G. Gillum, Mgr.

Ky., Mayfield—The Municipal Electric Light & Water Plant plans to install a new 300 kw. generator. C. A. Orr, Mgr.

Ky., Murray—The Municipal Electric Light & Waterworks Plant plans to install a 200 kw. steam unit. J. W. Stett, Supt.

Ky., Owingsville—The Electric Light Co. plans to install a new 50 kw. d.c. generator. E. V. Brother, Mgr.

Ky., Paris—The Paris Gas & Electric Co. plans to construct a 9 mi., 6,600 volt transmission line. C. L. Steinberger, Mgr.

Ky., Stanford—The Stanford Water, Light & Ice Co. plans to construct a 20 mi. transmission line to serve 2 towns. B. T Penny, Secy.-Treas.

Ky., Walton—The Walton Electric Light Co. plans to install 25 meters for its transmission system. E. L. Kelly, Owner

O., Cleveland—The B'nai B'rith Club, 7103 Euclid Ave., plans to build a 2 story club house. A steam heating system will be installed in same. Total estimated cost, $100,000. Max Ozersky, Treas.

O., Cleveland—The Guarantee Bldg. Co., Union Bldg., is having plans prepared for the construction of a 4 story, 60 x 150 ft. auto sales building and show rooms on East 55th St. and Prospect Ave. A steam heating system will be installed in same. Total estimated cost, $15,000. Max Weis, 618 Union Bldg., Archt.

O., Columbus—The Bd. Educ. will soon award the contract for the construction of a 2 story school. A steam heating system will be installed in same. Total estimated cost, $165,000. D. Riebel & Sons, New Trust Natl. Bank Bldg., Archts.

O., Hamilton—The City Council is having plans prepared for the construction of an electric light plant. The city voted $650,000 bonds for this project.

O., Ravenna—A. S. Helbig, Dir. of Pub. Serv., will receive bids until March 12 for furnishing one 2,000 G.P.M. electrical pumping unit with pipe and electrical connections.

Ind., Seymour—The Seymour Water Co. plans to improve the power station. Work involves a new sedimentation basin of 2,000,000 gal. capacity and rebuilding of the steam plant. The company recently installed hydraulic power equipment and operates its steam plant only when the condition of the White River will not permit the operation of the water system.

Ind., Valparaiso—Henry L. Newhouse, Archt., 4630 Prairie Ave., Chicago, will soon award the contract for the construction of a 2 story, 75 x 80 ft. theatre and office building, here, for G. G. Shaner & Sons Co. A steam heating system will be installed in same. Total estimated cost, $100,000.

Mich., Cass City—The Municipal Electric Light & Water Plant plans to install a steam driven, 2,300 volt generator and possibly a waterworks pump. W. N. Straube, Supt.

Mich., Detroit—The Bd. Educ., 50 B'way., will soon award the contract for the construction of a 2 story school on St. Antoine St. between Palmer and Ferry Aves. A plenum steam heating system, including ventilation device, etc., will be installed in same. Total estimated cost, $200,000. Charles A. Gadd, Secy. Malcolmson, Higginbotham & Palmer, 405 Moffat Bldg., Archts.

Mich., Detroit—Brown & Preston, Archts., 406 Empire Bldg., will soon award the contract for the construction of a 1 story, 60 x 150 ft. factory, 2 story, 36 x 60 ft. office building and an 18 x 22 ft. boiler house, for the Pharo Mfg. Co., 1336 Woodward Ave. A steam heating boiler, etc., will be installed in a separate boiler house. Total estimated cost, $30,000.

Mich., Detroit—Albert Kahn, Archt., Marquette Bldg., will soon award the contract for the construction of a 95 x 209 ft. sales building on Jefferson Ave. and St. Antoine St., for the Packard Motor Car Co., East Grand Blvd. Steam heating boiler, etc., will be installed in same. Total estimated cost, $175,000.

Mich., Detroit—Murray W. Sales & Co., East Jefferson Ave., is having plans prepared for the construction of a 3 story, 80 x 190 ft. storage warehouse on 3rd Ave. Electric power to operate conveyors will be installed in same. Esselstyn, Murphy & Hanford, 810 Marquette Bldg., Archts. and Engrs.

Mich., Detroit—The Parker-Johnson Co., 715 Dix Ave., is in the market for 1 medium capacity air compressor.

Mich., Detroit—Spier & Gehrke, Archts., 1317 Chamber of Commerce, are preparing plans for the construction of a 2 story, 90 x 100 ft. church in the eastern section of city. Steam heating equipment will be installed in same. Total estimated cost, $100,000. Owner's name withheld.

Mich., Detroit—The Water Bd., 232 Jefferson Ave., received lowest 3 bids for furnishing labor and material for constructing a sub-structure for the proposed low lift pumping station, from the Walsh Constr. Co., 114½ West 3rd St., Davenport, Ia., $268,850; Raymond Concrete Pile Co., Empire Bldg., $140,000; Foundation Co., 209 South La Salle St., Chicago, $369,000.

Mich., Durand—The Bd. Educ. is having plans prepared for the construction of a 2 story school. A steam heating boiler, etc., will be installed in same. Total estimated cost, $150,000. Van Leyen, Schilling & Keough, Union Trust Bldg., Detroit. Archts. and Engrs.

Mich., Frontier—The Economy Electric Co., Camden, plans to construct a transmission line from here to Ray, Ind. S. P. Hull, Secy.

Mich., Grand Rapids—The Municipal Electric Light Plant plans to install a 1,000 kw. Allis-Chalmers Co., turbo to generator system. B. E. Knappen, Electrician.

Mich., Greenville—R. J. Tower Electric Co., 210 South Lafayette St., plans to install new and larger generators. S. L. Tower, Mgr.

Mich., Hamtramck Village (Detroit P. O.)—The Village Council retained Alvan E. Harley, Archt., 435 Woodward Ave., Detroit, to prepare plans for the construction of a 2 story hospital. A steam heating system will be installed in same. Total estimated cost, $150,000.

Mich., Highland (Detroit P. O.)—The Ford Motor Co., Woodward Ave., is in the market for 1 turbine driven exciter unit.

Mich., Holland—The Bd. Pub. Wks. plans to install one 1,500 kva. Allis-Chalmers turbine. W. Winstrom, Secy.

Mich., L'Anse—The L'Anse Hydro Electric Light Plant plans to reinforce the flume pipe and raise the dam 3 ft. J. E. Everton, Supt.

Mich., Lansing—The State is having plans prepared for the construction of an Industrial School for Boys, including 1, 2, 3, 4 and 5 story buildings. A steam heating system will be installed in same. Total estimated cost, $3,000,000. York & Sawyer, 50 East 41st St., New York City, Archts. and Engrs.

Mich., Marcellus—The village plans to install 2 new units for power and a 24 hr. service. F. L. Knight, Supt.

Mich., Montgomery—The Economy Electric Co. plans to construct a 4 mi. transmission line from here to Roy. S. P. Hull, Mgr.

Mich., Otter Lake—The Otter Lake Light Plant plans to construct a 7 mi., 13,200 volt transmission line from here to Mayville. C. C. Tinke, Mgr.

Mich., Oxford—The Municipal Electric Light & Water Wks. plans to rebuild its transmission distributing system. A. E. Detwiler, Clk.

Mich., Plainwell—The Eesley Light & Power Co. plans to install 1 Allis-Chalmers vertical wheel and a 700 kva. generator. J. F. Fesley, Pres. and Mgr.

Mich., River Rouge—The Bd. Educ., c/o A. R. Heuer, Pres., engaged Van Leyen, Schilling & Keough, Archts., Union Trust Bldg., Detroit, to prepare plans for the construction of a 3 story school on Dearborn Rd. Modern heating and ventilating equipment will be installed in same. Total estimated cost, $750,000.

Mich., Saginaw—The Saginaw Creamery Co., 209 North Water St., is in the market for one 40 to 60 hp. horizontal tubular or internal fire-box boiler.

Mich., Sturgis—The Municipal Electric Light Plant plans to install either auxiliary unit. Diesel oil or 500 or 700 kw. steam engine and construct a 16 mi. transmission line service connection with the Indiana Electric Co., Constantine. J. S. Flanders, Genl. Mgr.

Mich., Three Rivers—The Three Rivers Municipal Plant plans to construct 1 additional transmission circuit. O. O. Johnson, City Mgr.

Mich., Yale—The Municipal Electric Light, Power & Water Plant plans to install an additional 75 kw. generator. A. T. Kent, Supt.

Ill., Elgin—The Elgin Natl. Watch Co., 10 South Wabash Ave., Chicago, is having plans prepared for the construction of a watch factory, adjoining its present plant, including several 4 story buildings. A steam heating system will be installed in same. Total estimated cost, $1,000,000. Postle & Fisher, 140 South Dearborn St., Chicago, Archts.

Wis., Fond du Lac—The Commercial Natl. Bank is having plans prepared for the construction of a 6 story, 66 x 105 ft. bank and office building. A steam heating system will be installed in same. Total estimated cost, $250,000. Sidney Lovell, 30 North Michigan Ave., Chicago, Archt.

Wis., St. Francis—The Milwaukee Electric Ry. & Light Co., Pub. Serv. Bldg., Milwaukee, will soon award the contract for the construction of five 1 story power plant buildings, here. Total estimated cost, $500,000. R. H. Pinkley, Engr.

Iowa, Mt. Pleasant—The Henry Co. Hospital Association is having plans prepared for the construction of a 2 story, 50 x 100 ft. hospital building. A steam heating system will be installed in same. Total estimated cost, $100,000. G. L. Lockhart, Endicott Bldg., St. Paul, Minn., Archt.

Iowa, Webster City—The Commercial Club will soon receive bids for the construction of a 3 story, 70 x 118 ft. hotel. A steam heating system will be installed in same. Total estimated cost, $100,000.

Minn., Aurora—The Water & Light Comn. plans to install 2 boilers and one 312 kva. generator unit. C. Juhlin, Eng. Power Sta.

Minn., Benson—The City Water, Light, Heat & Power Plant plans to install a complete new Allis-Chalmers plant. T. K. Lee, Supt.

Minn., Buhl—The Bd. Educ. is having plans prepared for the construction of a 1 story, 24 x 93 ft. grade school with a 23 x 52 ft. wing. A steam heating system will be installed in same. Total estimated cost, $100,000. M. D. Aygard, Supt. A. W. Kerr & Co., Virginia, Archt.

Minn., Cass Lake—The Bd. Educ. is having plans prepared for the construction of a 2 story, 86 x 130 ft. high school. A steam heating system will be installed in same. Total estimated cost, $100,000. E. F. Broomhall, 710 Alworth Bldg., Duluth, Archt.

Minn., Ely—The Municipal Electric Light Plant plans to install three 213 hp. Heine boilers and a new condenser. W. Mitchell, Supt.

Minn., Forest Lake—The Bd. Educ. is having plans prepared for the construction of a 3 story, 80 x 120 ft. high school. A steam heating system will be installed in same. Total estimated cost, $100,000. Foss & Foss, St. Cloud, Minn., Archts.

Minn., Good Thunder—The Good Thunder Electric Co. plans to construct a 2½ mi. transmission line extension from the hydro-electric plant to the sub-station, $4,136. F. H. Griffin, Secy.

Minn., Granite Falls—The Granite Falls Electric Light & Water Plant plans to install more generating power. A. W. Thompson, Mgr.

Minn., Hopkins—A. Close, Village Recdr., will soon award the contract for the construction of a 750 ft. deep well and for the installation of pumping machinery. Total estimated cost, $20,000. John W. Schaffer, 501 New York Life Bldg., Minneapolis, Engr.

Minn., Kinney—Lee Ranchstadt will soon award the contract for the construction of a 2 story, 152 x 216 ft. school, for Independent School Dist. 35 of Buhl. A steam heating system will be installed in same. Total estimated cost, $250,000. A. W. Kerr & Co., Matheson Bldg., Virginia, Archts. and Engrs.

Minn., Lake City—The Municipal Electric Light & Waterworks Plant plans to construct further extensions to its transmission lines. W. J. Howe, Supt.

Minn., Madison—The Municipal Electric Light Plant plans to install one 18 x 72 ft. horizontal brass boiler. T. B. Leasman, Mgr.

Minn., Mankato—The State Bd. of Control, Capitol, St. Paul, will soon award the contract for the construction of a 3 story, 50 x 100 ft. dormitory at the Normal School, here. A steam heating system will be installed in same. Total estimated cost, $100,000. C. H. Johnston, Capitol Life Bldg., St. Paul, Archt.

Minn., Middle River—The Middle River Light & Power Co. plans to install one new unit in the generating system. E. Peterson, Mgr.

Minn., Milroy—The Milroy Electric Light & Power Co. plans to construct transmission lines into the country. C. F. Boyer, Mgr.

Minn., Minneapolis—The Bd. Educ. is having plans prepared for the construction of a 2 story, 225 x 225 ft. senior high school on Monroe St., N. A steam heating system, including a ventilating fan, etc., will be installed in same. Total estimated cost, $500,000.

Minn., Minneapolis—The Hamm-Hamm Realty Co., 7th and St. Peter Sts., St. Paul, is having plans prepared for the construction of a 2 story, 110 x 110 ft. business building, on Nicollet Ave. and 10th St. A steam heating system will be installed in same. Total estimated cost, $180,000. Toltz, King & Day, 1410 Pioneer Bldg., St. Paul, Archts.

Minn., New Prague—The Electric Light & Waterworks Dept. plans to construct 250 volt, d.c. transmission lines. J. T. Rynda, Supt.

Minn., Olivia—The Bd. Educ. is having plans prepared for the construction of a 3 story, 63 x 150 ft. high school. Separate bids will be received for installing heating and plumbing systems. Total estimated cost, $200,000. George Peterson, Clk. W. L. Alban, 347 Endicott Bldg., St. Paul, Archt.

Minn., Rochester—John E. Crewe, 12 West 4th St., is having plans prepared for the construction of a 3 story sanatorium. A steam heating system will be installed in same. Total estimated cost, $150,000. F. H. Ellerbe, Endicott Bldg., St. Paul, Archt.

Minn., Sacred Heart—The Bd. Educ. is having plans prepared for the construction of a 3 story, 60 x 164 ft. high school. A steam heating system will be installed in same. Total estimated cost, $160,000. H. Evenson, Clk. W. L. Alban, 347 Endicott Bldg., St. Paul, Archt.

Minn., St. Cloud—Nicholas Thomey, Aud. Sterns Co., will receive bids until March 25, for the construction of a 4 story, 100 x 140 ft. courthouse. A steam heating system will be installed in same. Total estimated cost, $600,000. Toltz, King & Day, 1410 Pioneer Bldg., St. Paul, Archts.

Minn., St. Paul—St. Mary's Hospital is having plans prepared for the construction of a 6 story, 45 x 140 ft. addition on 5th Ave., E. A steam heating system will be installed in same. Total estimated cost, $200,000. F. H. Ellerbe, Endicott Bldg., St. Paul, Archt.

Minn., Waconia—The Waconia Light & Power Co. plans to construct 6 mi., 13,000 volt transmission line. E. F. Strong, Pres.

Minn., Waubun—The Waubun Electric Light Co. plans to install an a.c. in place of the d.c. generator. L. A. Fancher, Pres.

Minn., Willmar—The Municipal Electric Light Plant & Waterworks plans to install a new boiler. N. W. Larson, Supt.

Kan., Attica—The Attica Municipal Water & Light Plant plans to install a 100 hp. oil engine. J. C. McCaddan, City Clk.

Kan., Burlingame — The Burlingame Municipal Electric Light Plant plans to install a 3 ph., 60 cycle, 220 volt motor circuit in the city. W. R. Howe, Supt.

Kan., Chanute—The Municipal Electric Light Plant plans to install one 500 kw. turbo generator and one 250 hp. Heine boiler, also construct new engine room and spray pond for cooling water. H. R. Williams, Supt.

Kan., Clay Centre—The Municipal Light & Water Plant plans to construct a 12 mi. transmission line. C. S. Allender, Supt.

Kan., Coffeyville—The Municipal Water & Light Plant plans to install two additional 375-500 hp. boilers. E. J. Walcott, Supt.

Kan., Elk City—The C. R. Long Light & Power Co., Box 117, plans to install a 100 hp. oil engine and a 75 kva. generator, also construct an 8 mi. transmission line. J. A. Logan, Supt.

Kan., Ellsworth—The Weber Electric Power Co. plans to extend its transmission lines to 5 or more small towns. B. A. Bearnes, Supt.

Kan., Emporia—The State, c/o J. A. Kimball, will receive bids until March 11 for the construction of power plant. New boilers will be installed in same. Total estimated cost, $70,000. Ray L. Gamble, 1415 Fillmore St., Topeka, Archt.

Kan., Grenola—The City Light Plant plans to install a new 75 kw. generator and a 100 hp. engine. Edgar Robertson, Supt.

Kan., Hutchinson—The United Water, Gas & Electric Co. plans to install two 1,000 hp. boilers, feed water pumps and construct a new boiler room and transmission lines to supply Arlington, Patridge, Langdon and Tierom, one to supply Alden, and one to supply Little River and Benesio. W. E. Quillan, Mgr.

Kan., Kanapolis—The Kanapolis Water & Light Plant plans to install 1 generator unit and one 10 ton ice machine. J. S. Holmquist, Supt.

Kan., LaCygne—The Municipal Electric Light Plant plans to install one 75 kw., 2,300 volt, 60 cycle generator and construct a pole line and transformer. W. W. Petifish, Supt.

Kan., Lebanon—The Municipal Light, Water & Power Plant, P. O. Box 145, plans to install one 200 hp. engine. I. V. Arnold, Supt.

Kan., Lyndon—The Lyndon Light & Water Plant plans to install a new 250 hp. unit and construct a transmission line. C. W. Rockey, Supt.

Kan., Meriden—The city voted $22,000 bonds to construct an electric transmission line from here to Topeka. W. B. Rollins & Co., Ry. Exch. Bldg., Kansas City, Mo., Engrs.

Kan., Moran—The Municipal Electric Light & Water Plant plans to rebuild its transmission line. M. M. Humphrey, Supt.

Kan., Oakley—The Municipal Water & Light Plant plans to install a complete 150 hp. steam or oil plant. W. B. Day, Supt.

Kan., Phillipsburg — The Phillips Co. Light & Power Co. plans to install a 200 hp. boiler, 250 hp. steam engine and generating unit, also construct a 10 mi., 13,200 volt transmission line to Almena. F. J. Spuhler, Mgr.

Kan., Pretty Prairie—The Pretty Prairie Light & Power Plant plans to install one 70 kva., 2,300 volt, 60 cycle, 3 ph. generator. J. J. Siebert, Mgr.

Kan., Rossville—The Rossville Electric Light, Ice and Elevator Co. plans to install a new 60 kw. generator, also construct a transmission line to Delia and Willard. A. H. Daw, Mgr.

Kan., St. Francis—The Municipal Water & Light Plant plans to install a new generator plant. W. Roelfs, Supt.

Kan., Wellington—The Wellington Municipal Light & Water Plant plans to install one 750 kva. turbine unit and a switchboard. R. Perry, Supt. Pub. Utilities.

Neb., Arcadia—The Arcadia Electric Light & Telephone Co. plans to install a new 75 hp. oil engine. J. L. Wait, Prop.

Neb., Auburn—The Intermountain Ry. Light & Power Co. plans to install one 310 hp. Nordberg Corliss engine with a 200 kw., 60 cycle, 3 ph., G.E. generator in place of the 125 hp. Hamilton Corliss generator. R. Y. Pool, Genl. Supt.

Neb., Avoca—The Avoca Light Plant plans to install an extra engine and generator. L. W. Fahnestock, Mgr.

Neb., Bartley—The Bartley Municipal Lighting System plans to construct a 24 mi. transmission line. J. A. Finnegan, Mgr.

Neb., Chappell—The Municipal Water & Light Plant plans to install a complete new 200 or 250 hp. oil engine in the spring. C. P. Tschabrun, Mgr.

Neb., Fairbury—The Municipal Electric Light & Water Plant plans to install one 400 kw, direct connected unit and construct a 6 mi., 3 ph., 6,600 volt transmission line. G. D. Myers, Mgr.

Neb., Harvard—The Harvard Electric Co. plans to construct a power line. J. E. Person, Mgr.

Neb., Kimball—The city plans to install a 200 hp., type "Y" engine and direct connected generator. F. C. Overton, Clk.

Neb., Lincoln—The Municipal Electric Light & Water Plant plans to build a new plant over the old and install new turbine, boilers, coal conveyor and all necessary equipment. P. W. Doerr, Supt.

Neb., Millard—The Millard Light & Power Co. plans to construct transmission lines to Chalco. C. W. Peters, Owner.

Neb., Morrill—The Municipal Electric Light & Power Plant plans to install a direct connected, 100 hp. unit. O. T. Moline, Clk.

Neb., Randolph—The Randolph Electric Light & Water Plant plans to install an a.c. generator this spring in place of the old one. R. J. Kirk, Supt. Dist.

Neb., Ravenna—The Ravenna Electric Light, Heat & Power Co. plans to extend its generator system. A. T. Shellenbarger, Supt.

Neb., Rushville—The Sheridan Electric Service Co. plans to increase the capacity of its generator system by 25 kw. T. C. Shipley, Mgr.

Neb., Schuyler—The City Light & Waterworks plans to construct a 2 unit generator room and install one 125 kva. and one 210 kva. generator. E. A. Schnid, Light & Water Comr.

Neb., Scotia—The Municipal Electric Light Plant plans to install a new engine, generator and 11,000 gal. distillation tank. Address B. W. Ammermann.

Neb., Scottsbluff—The Intermountain Ry. Light & Power Co., 1219 B'way., plans to install a 750 hp. W. T. boiler, double its ice plant equipment and increase the size of the copper gearing transformers. J. R. Murphy, Supt.

Neb., Seward—The Blue River Power Co. plans to build hydro plants and transmission lines. B. Boyes, Secy.-Treas.

Neb., Sidney—The Sidney Municipal Service plans to install a new boiler and generator unit. J. E. Poquette, Mgr.

Neb., Superior—The Southern Nebraska Power Co. plans to install one new 650 hp. water wheel and construct 45 mi. of transmission lines. D. Guthrie, Pres.

Neb., Sutton—The Municipal Electric Light & Power Plant plans to build minor extensions to its transmission system. H. J. Bauer, Light & Water Comr.

Neb., Syracuse—The Municipal Electric Light Plant plans to purchase curren from Nebraska City or enlarge power unit and sell 50 hp. type "Y" F. M. engine. R. E. Wood, Water & Light Comr.

S. D., Erwin—The Bd. Educ. is having plans revised for the construction of a 2 story, 100 x 132 ft. school. A steam heating and ventilating system will be installed in same. Total estimated cost, $150,000. G. A. Jenson, Clk. K. T. Snyder, 933 Plymouth Bldg., Minneapolis, Archt.

S. D., Sioux Falls—The Diocesan College is having plans prepared for the construction of a group of 4 buildings, including a 3 story class-room building, 2 story dormitory, 1 story gymnasium and auditorium and a power plant. A steam heating system will be installed in same. Total estimated cost, $500,000. Edwain S. Lundie, 530 Endicott Bldg., St. Paul, Minn., Archt.

N. D., Cooperstown—The City Council will soon award the contract for the construction of a combined well and 100,000 gal. reservoir, 1 pump house, 100 gal. 1 min. pump, 500 gal., 1 min. pump, tower and tank, etc. Total estimated cost, $79,000. Dakota Eng. & Constr. Co., Valley City, Engr.

N. D., Fessenden—The Fessenden Light & Power Co. plans to install one 200 hp. generator producer and also extend its transmission lines. H. C. Whitcomb, Mgr.

N. D., Fordville—The Fordville Garage & Light Plant plans to install one 25 hp. oil engine and generator. C. H. Thol, Mgr.

N. D., Garrison—The Garrison Coal, Light & Power Co. plans to install a new 150 hp. boiler and a 100 kw. direct connected generator. J. A. Kunkel, Secy.

N. D., Hettinger—The Hettinger Electric Light & Power Co. plans to install two 18 x 72 ft. return flue boilers, one 250 hp. engine, six 75 kw., 13,200 volt power transformers. G. Smith, Owner and Mgr.

Mo., Bethany—The Municipal Electric Light & Waterworks plan to install a new engine and boiler. J. F. Slinger, Supt.

Mo., Bolivar—The Municipal Electric Light & Waterworks Plant plans to install a new and larger generating system. A. F. Leavitt, Light & Water Comr.

Mo., Boonville—The Boonville Light, Heat & Power Co. plans to construct small transmission line extensions. T. W. Long, Mgr.

Mo., Bunceton—The Bunceton Ice, Light & Fuel Co. plans to construct additional transmission lines. W. K. Simmons, Mgr.

Mo., Galt—The Galt Light & Power Co. plans to construct a transmission line either from here to Trenton, or from here to Brookfield. D. H. Clark, Pres.

Mo., Lathrop—The Lathrop Light, Heat & Power Co. plans to rebuild part of its transmission system. E. S. Grant, Mgr.

Mo., Lincoln—The Lincoln Light Co. plans to install a 10 hp. engine and a 5 kw. generator. I. T. Sterrett, Supt.

Mo., Novinger—The Merchants' Light & Power Co. plans to install a new 150 kva. unit generator. C. H. Charlton, Secy.-Treas.

Mo., Odessa—The Municipal Light Plant plans to construct a 35 mi., 23,000 volt transmission line. Voltage now 220 d.c. will be 60 cycle, 110 a.c. J. G. Lightner, Mgr.

Mo., Pattonsburg—The Pattonsburg Light & Ice Co. plans to construct transmission lines to Kidder, Jameson, and McFall. O. L. Wright, Pres.

Mo., Parma—The McMullin Light & Ice Plant plans to install a 50 hp. engine and a 35 kw. generator. J. McMullin, Pres.

Mo., Richland—The Richland Light & Power Co., Inc., plans to install a 50 ph., 60 kw., 2,300 volt generator connected with a 60 hp. fuel engine, also a 60 kw., 3 ph. generator and a 60 to 75 hp. fuel oil engine. E. K. Woodward, Pres.

Mo., Unionville—The Municipal Electric Light & Waterworks Plant plans to install new boilers, etc., also rebuild all transmission lines. M. Flynn, Supt.

Mo., Windsor—The Windsor Light & Power Co. plans to install an a.c. generating system in place of the present one. R. F. Williams, Secy. and Mgr.

Tex., Houston—The Baylor University, 5th and Speight Sts., Waco, is having plans prepared for the construction of a 1 story power house. Estimated cost, $100,000. Birch D. Easterwood, 405 Praetorian Bldg., Waco, Archt.

Okla., Oklahoma City—Charles P. Nieder, Archt., Oklahoma City, will receive bids about March 15 for the construction of a 3 story, 80 x 190 ft. factory, for the New State Shirt & Overall Co. A steam heating system will be installed in same. Total estimated cost, $140,000.

Utah, Paragonah—The city plans to build a 500 kw. electric light plant. S. M. Winsor, Logan, Engr.

Ariz., Phoenix—David Goldberg and Associates plan to build a 6 story, 105 x 200 ft. theatre and office building on 1st and Washington Sts. A complete steam heating system will be installed in same. Total estimated cost, $1,000,000. Lyman & Place, Tucson, Archts.

Wash., Mansfield—The city will soon award the contract for the construction of a pumping plant to have a capacity of 200 gal. per min. C. O. Allen, City Clk.

Ore., Portland—E. G. Hopson, 205 Central Bldg., filed an application with P. A. Cupper, State Engr., Salem, to take 1,200 second ft. water from Metolius Creek, Crook Co., for power purposes; will also construct dam 20 ft. high and 650 ft. long and install 43,000 hp. electric generators. Total estimated cost, $1,500,000.

Ore., Vale—The City Council adopted a resolution to purchase a 2 acre site along Bully Creek south of the city and plans to build a permanent well and pump house on same. Estimated cost between $5,000 and $8,000.

Cal., Burbank—The Allen Burbank Motor Co., Marsh-Strong Bldg., Los Angeles, retained F. R. Dorn, Archt., Marsh-Strong Bldg., Los Angeles, to prepare plans for the construction of the first unit of its plant, for the manufacture of tractors and automobile. Electric power will be installed in same. Total estimated cost, $125,000.

Cal., Davis—The University of California, Berkeley, is having plans prepared for the construction of a 1 story creamery at the University farm, here. A boiler room, etc., will probably be installed in same. Total estimated cost $100,000. W. C. Hays, 1st Natl. Bank Bldg., San Francisco, Archt.

Cal., Pasadena—The Throop College of Technology will soon award the contract for the construction of a laboratory. A steam heating system will be installed in same. Total estimated cost, $150,000. B. Goodhue, 2 West 47th St., New York City, Archt. and Engr.

Cal., Westwood—The Red River Lumber Co. of Lassen Co. plans to construct a dam across the Fecther River at Clear Creek, for the purpose of impounding water for electric power, thereby adding 5,000 hp.

Que., Montreal — The Administrative Comn., City Hall, will receive bids until March 15 for the manufacture, delivery and construction of two 30,000,000 imperial gal. motor driven pumps. Total estimated cost, $100,000. A. E. Doucet, Dir. of Pub. Wks.

Ont., Acton—The town is having plans prepared for the construction of a pump-house and a 60,000 gal. steel tank. Electrically driven turbine pumps with possibly a gasoline fire pump will be installed in same. Total estimated cost, $60,000. E. A. James Co., 1004 Excelsior Life Bldg., Toronto, Engr.

Ont., Belmont—W. F. Hamlyn is in the market for saw-mill equipment including one 14 x 16 or 16 x 18 in. modern slide valve engine, one 12 x 12 or 12 x 14 in. valve engine, one locomotive type, 50-60 hp. boiler, one log jack loader, 1 heavy steel carriage with or without gunshot feed and 1 line of live gear driven rolls.

Ont., Dresden—F. W. Farncombe, Engr., Dominion Savings Bldg., London, will soon award the contract for remodelling the present steam operated plant into one operated by hydro power and is in the market for electrical pumping equipment. Total estimated cost, $30,000.

Ont., Hamilton—The Dept. of Pub. Wks., Ottawa, will receive bids about March 15 for the construction of a 3 story post office, here. A steam heating system will be installed in same. Total estimated cost, $800,000. A. Wright, Archt. and R. DeB. Corrivean, Engr., both of Parliament Bldgs., Ottawa.

Ont., Niagara Falls—W. J. Seymour, City Clk., will receive bids until March 15 for furnishing a Standard built motor pumper, 100 to 110 hp., rotary pump of 600 gal. capacity, etc. D. T. Black, Erie Ave., Engr.

Ont., Stratford—The Kindel Bed Co., Ontario St., is having plans prepared for the construction of a 1 story, 24 x 30 ft. boiler house. One 250 hp. horizontal return tubular boiler with boiler feed pump, water heater, etc., will be installed in same. Total estimated cost, $18,000. James Russell, Downie St., Archt.

CONTRACTS AWARDED

Mass., Springfield—The Cheney-Bigelow Wire Wks., Warwick Ave., has awarded the contract for the construction of a 2 story, 171 x 195 ft. and 48 x 271 ft. manufacturing plant, to L. S. Wood, 14 Stockbridge St. A steam heating system will be installed in same. Total estimated cost, $100,000.

Conn., Bridgeport—The Bd. of Contract & Supply has awarded the contract for installing heating and pumbing system in the proposed 1 story, 146 x 200 ft. garage on Madison Ave., to H. D. Fitzgerald Co., Bassick Ave. Total estimated cost, $150,000.

Conn., Bridgeport—The Bridgeport Screw Co., Union Ave., has awarded the contract for the installation of a steam heating system in the new 2 story, 60 x 150 ft. factory and the 2 story, 49 x 101 ft. factory addition, to the Files Eng. Co., 75 Westminster St., Providence, R. I. Total estimated cost, $100,000.

N. Y., Brooklyn — Loew's Enterprises, 1492 B'way, New York City, have awarded the contract for the construction of a theatre on B'way and Gates Ave., to John Auer, 648 Lexington Ave., New York City. A steam heating system will be installed in same. Total estimated cost, $300,000.

N. Y., Long Island City — Maurice F. Connolly, Boro Pres., has awarded the contract for the construction of an automatic electric pumping station at Genesee St., Sect. "A" and "B" to Fox & Renalls, 87 East 125th St., New York City, at $177,281.

N. Y., New York—The Carl Platon Realty Corp., c/o Shape, Bready & Peterson, Archts. and Engrs., 220 West 42nd St., has awarded the contract for the contsruction of a 12 story, 70 x 75 ft. office building on Broad and Front Sts., to C. L. Frazer, 103 Park Ave. A steam heating system will be installed in same.

N. Y., New York—The Circle Concrete Co., 570 5th Ave., will build an 18 story, 60 x 100 ft. office and loft building, at 49-53 West 45th St. A steam heating system will be installed in same. Total estimated cost, $500,000. Work will be done by day labor.

N. Y., Watertown—The Columbia Realty Bldg. Co., Poughkeepsie, has awarded the contract for the construction of a theatre, to Edgar V. Anderson, 39 Market St., Poughkeepsie. A steam heating system will be installed in same. Total estimated cost, $200,000.

N. J., East Orange—The Savings & Investment Trust Co., 15 South Orange Ave., has awarded the contract for the construction of a 2 story, 65 x 135 ft. bank building, to R. A. Howie, Newark. A steam heating system will be installed in same. Total estimated cost, $150,000.

N. J., Newark—C. W. Oathout, 2 Stratford Pl., has awarded the contract for the construction of a 3 story, 100 x 185 ft. factory on Highland St., to Tucker & Lewis, 101 Park Ave., New York City. A steam heating system will be installed in same. Total estimated cost, $135,000.

Md., Baltimore—Hotel Caswell, 1012 Keyser Bldg., has awarded the contract for altering and constructing a 6 story, 35 x 250 ft. addition, to J. H. Miller, Inc., Eutaw and Franklin Sts. Heating contract will be sub-let. Total estimated cost, $750,000.

Md., Baltimore—The Trustees of Johns Hopkins Hospital, Fidelity Bldg., have awarded the contract for the construction of a 6 story, 60 x 150 ft. warehouse addition at 536-542 West Pratt St., to the Price Constr. Co., 210 Maryland Trust Bldg. Steam heating contract will be sub-let. Total estimated cost, $150,000.

S. C., Manning — Lafaye & Lafaye, Archts., Columbia have awarded the contract for the construction of a 2 story hotel, to the Jackson Constr. Co., 1730½ Main St., Columbia. A steam heating system will be installed in same. Owner's name withheld.

S. C., Sumter—The town has awarded the contract for the construction of lighting and ice plants, to Tucker & Laxton, Realty Bldg., Charlotte, N. C. Total estimated cost, $100,000. G. C. White, Durham, N. C., Engr.

Ga., Atlanta—The Crane Co., 31 Garnett St., has awarded the contract for the installation of a steam heating sys..... in the proposed 4 story, 100 x 100 ft. warehouse on Washington St., to Fred Cantrell Co. Total estimated cost, $120,000.

La., New Orleans—The United Fruit Co., c/o Deball & Owen, Ltd., Archts. and Engrs., Interstate Bank Bldg., has awarded the contract for the construction of a 10 story office building, to G. A. Fuller, 175 5th Ave., New York City. A steam heating system will be installed in same. Total estimated cost, $550,000.

Mich., Detroit—Morgan & Wright (U. S. Rubber Co.), Jefferson Ave. and Bellevue St., have awarded the contract for the construction of a 2 story, 81 x 90 x 275 ft. power house on Jefferson Ave., to ..... & Webster, Book Bldg. Boilers, feedwater heaters, pumps, motor generator sets, etc., will be installed in same. Total estimated cost, $1,000,000.

Mich., Detroit—The United Brotherhood of Maintenance of Way Employees and Railway Shop Laborers, 27 Putnam Ave., has awarded the contract for the construction of a 12 story 93 x 160 ft. office building on Clifford and Columbia Aves., to W. H. Mueller, 1711 Ford Bldg. A central steam heating system will be installed in same. Total estimated cost, $500,000.

Ill., Chicago—The Splitdorff Electric Co., 1466 South Michigan Ave., has awarded the contract for the construction of a 5 story, 60 x 200 ft. sales and office building on 29th St. and Michigan Ave., to George A. Fuller Co., 140 South Dearborn St. A steam heating system will be installed in same. Total estimated cost, $300,000.

Mont., Helena — The Algeria Masonic Temple has awarded the contract for installing heating and plumbing systems in the proposed lodge building on Neill and Benton Aves., to the John Smyth Heating & Plumbing Co., Granite Blk., Spokane, Wash., at $42,600.

Mo., St. Louis—The Polar Wave Ice & Fuel Co., 3628 Olive St., has awarded the contract for the construction of a 1 story, 115 x 115 ft. ice factory at 450 Eichelberger St., to Hy. Dilschneider, 5408 Easton Ave. Estimated cost, $15,000.

Okla., Ada—The city has awarded the contract for the construction of a reservoir and a 2,000,000 gal. capacity pumping station, to J. E. Nelson & Son, 118 North LaSalle St., Chicago, Ill., at $49,939.

Ariz., Flagstaff—The State has awarded the contract for the construction of a 2 story, 45 x 115 ft. addition to the State Normal School, to Edwards, Wildey & Dixon Co., 515 Black Bldg., Los Angeles, Cal., at $97,983. A steam heating system will be installed in same.

Wash., Puget Sound — The Bureau of Yards & Docks, Navy Dept., Wash, D. C., has awarded the contract for the construction of "A" engine room at the central power plant, to Swensson & Co., 4099 Arcade Bldg., Seattle, at $48,800.

Cal., Healdsburg—The Healdsburg Natl. & Savings Bank has awarded the contract for installing a steam heating system in the proposed 1 story, 60 x 100 ft. bank on West and Powell Sts., to A. W. Garrett, Healdsburg. Total estimated cost, $100,000.

Cal., Los Angeles—The Charles Ray Production Co., 1425 Fleming St., has awarded the contract for the construction ... motion picture producing plant, to the Nance Constr. Co., 846 Linden St., Los Angeles. Generating equipment will be installed in same. Total estimated cost, $50,000.

Cal., Los Angeles—The Los Angeles Soap Co., 633 East 1st St., has awarded the contract for the construction of a 1 story 45 x 125 ft factory, including boiler room, distillation room and machine shop, on 1st and Banning Sts., to the Davidson Constr. Co., 1445 East 16th St., at $23,800.

Cal., Oakland—The Natl. Ice & Cold Storage Co., 2nd and Market Sts., has awarded the contract for the construction of an ice and storage plant on North 1st St., to Nelson & Forsyth, Union Savings Bank Bldg. Estimated cost, $200,000.

Ont., Peterboro—The Canadian Aladdin Co., Canadian Pacific Ry. Bldg., Toronto, has awarded the contract for the construction of a 3 story, 60 x 80 ft. factory, to the Dickie Constr. Co., Ryrie Bldg., Toronto. A vacuum steam heating system will be installed in same. Total estimated cost, $125,000.

Cal., Pomona—The Y. M. C. A. has awarded the contract for the construction of a 3 story, 122 x 162 ft. building, to Fred L. Somers, 2650 Santa Fe Ave., Los Angeles. Boilers, etc., will be installed in same. Total estimated cost, $120,000.

POWER

Volume 51 | New York, March 16, 1920 | *Number* 11

Alibi Al

BY RUFUS T. STROHM

In almost every plant you'll find this specimen of human-kind. He simply will not hear of taking any sort of blame that might besmirch his snowy fame as running engineer. It doesn't matter what goes wrong, as soon as trouble sprints along, he starts to dodge and shy. In spite of damning evidence as glaring as a white-washed fence, he pulls an alibi.

Just for example, let's suppose the pressure in a steam pipe blows a chunk of gasket out, and in a moment more enshrouds the boiler room in whitish clouds that roll and swirl about. Does Al speak up and frankly tell the flanges weren't parallel? He doesn't, you can bet! He vows to heaven he'll be cursed "if that there packin' ain't the worst" that he's been handed yet.

Perhaps a bearing gets so hot it scores the pin; or, like as not, turns babbitt into soup. There surely is the deuce to pay; the engine's shut down right away; the mill hands homeward troop. Yet Al disputes the honest chaps who venture to suggest, perhaps no oil was in the cup. Good heavens, no! That ain't his way! It wasn't only yesterday he filled the blamed thing up!

His boiler, in the heart of town, explodes and knocks six hen-coops down and wrecks the calaboose; but long before the echoes die, friend Al is working, swift and spry, to find a fair excuse. Suspicions of incompetence and carelessness he soon resents, or treats with loud guffaws. A boiler plate corroded thin or deeply grooved could not have been by any means the cause.

If you've a noodle packed with punk, and you believe his boastful bunk and his deceitful dope, you'll swear he's just as apt to make a conscious error or mistake as kaiser, king or pope. And so he strives with all his might to keep his record fair and white as Desdemona's cheek; but every action goes to show what all his fellow-workers know: He has a yellow streak!

ADJUSTABLE-SPEED MOTOR EQUIPMENT FOR ROLLING MILLS

By Fraser Jeffery*

Use of a combination of induction motor rotary converter and direct-current machine with control equipment known as the Kraemer system, makes it possible to obtain speeds independent of the load and economical operation at these speeds whether near synchronism or much lower.

FOR conditions necessitating a wide speed range for a given product and on a given steel mill, the ordinary slip-ring type induction motor used as the driving medium of the rolls is not suitable. The speed-torque characteristics of such a motor are such that speed control without load is not possible; in other words, to obtain a constant speed there must be a constant load. It is obvious that, in general, the rolling conditions in a steel mill are such that there is hardly ever a constant load, and at times it is highly fluctuating, varying instantly from zero to large overloads and vice versa. Assuming that the load conditions were constant and that a rolling speed such as to give a motor speed one-half of the synchronous speed was desirable, a slip-ring induction motor would prove costly from an economical operating point of view because the low speed could be obtained only by dissipating a large amount of energy in a resistance unit.

To obtain set-rolling speeds independent of the load and economical and efficient operation at these speeds, whether they are near synchronism or much lower, a combination of machines and control has been used that is well suited to the conditions to be met. A large number of these units are operating in various steel mills throughout the country today, and their application, while needing a careful analysis of all the rolling conditions to be encountered, has shown results highly satisfactory to the mill owner and operator.

FIG. 1. FLYWHEEL ON MAIN-MOTOR SHAFT

A slip-ring induction motor having its primary winding connected to the source of supply and its secondary winding connected to the alternating-current side of a rotary converter and mounted on a shaft coupled to a direct-current machine, the commutator of which is permanently connected to the direct-current side of the rotary converter, forms the combination of machines as used to meet the varied speed conditions demanded. Fig. 2 shows the simplified connections of such a combination while the headpiece shows the combination of actual machines as used. In the headpiece the machine in the center background is the slip-ring induction motor direct coupled to the direct-current motor.

The rotary converter is shown in the right-hand foreground. The drive to the steel mill is by means of a coupling on the induction-motor end of the set, which is generally known as the main-mill motor, or Kraemer unit. The fields of both the direct-current machine and the rotary converter may be excited from the direct-current mill-supply line, but a separate and independent motor-generator exciter set is much to be preferred.

When voltage is applied to the primary of an induction motor, the currents in the various coils produce a magnetic field which rotates at a speed depending directly upon the frequency and inversely upon the number of poles. If the rotor is at a standstill, the secondary voltage is at maximum value and the frequency is the same as the primary frequency. The currents which flow in the secondary windings exert a torque such as

*Electrical Engineer, Allis-Chalmers Manufacturing Company.

to cause the rotor to turn in the direction of the rotating field. As the rotor speed increases, the secondary voltage and frequency decrease until the voltage is just sufficient to send a current through the impedance of the windings to develop the torque required. At synchronism the secondary voltage and frequency are zero. If the speed is reduced (the load remaining constant) by inserting resistance in the secondary circuit, the secondary voltage and frequency are practically proportionate to the variation from the synchronous speed. If, in place of the external resistance, a means of providing a constant voltage to oppose the secondary voltage at slip frequency, is used, the speed of the rotor will adjust itself to a point where the secondary voltage will exceed the opposing voltage by an amount necessary to send a current through the impedance of the circuits, sufficient to develop the necessary torque. It is then obvious that if the opposing voltage is adjustable, the speed of the main motor readily may be changed. This is the principle of the Kraemer system for adjustable speed. The induction motor of the Kraemer unit receives its adjustable and opposing secondary voltage at slip frequency through the medium of the rotary converter.

To provide for emergencies such as a breakdown in any of the auxiliary machines, the induction motor and direct-current machine are mounted on separate shafts connected by means of a forged flange coupling. The complete unit is provided with a heavy base and three pedestal-type bearings, the center one of which, and the coupling, being so arranged that the direct-current machine may be shut down by disconnecting the coupling, thus allowing the alternating-current motor of the set to operate the mill.

A full-automatic type of magnetic control, Fig. 3, is most generally used. It is sometimes modified to include a liquid starter instead of the magnetic, and if the nature of the load is such as to impose objectionably high peak-load conditions on the line, then the use of a flywheel on the set, Fig. 1, is frequently desirable. The flywheel can be made to give up part of its energy by the proper adjustment of the field of the direct-current machine, the effect of a variation of this field being such as to slow down the direct-current machine of the Kraemer unit, thus allowing the flywheel to act and in this manner relieving the line peaks. The control may be governed through a drum-type master switch, Fig. 4, the manipulation of which starts the

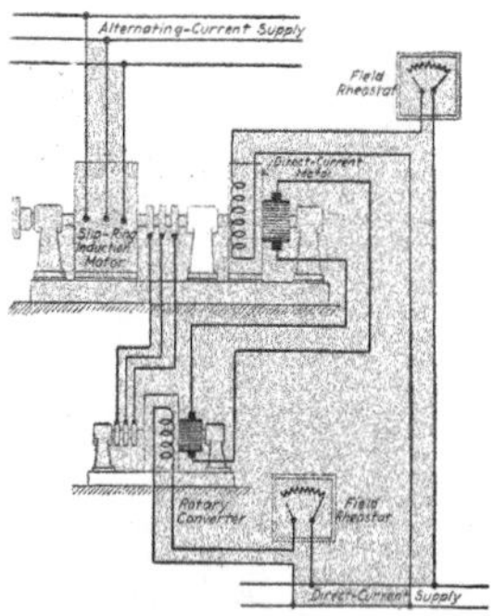

FIG. 2. SIMPLIFIED CONNECTIONS FOR ADJUSTABLE-SPEED CONSTANT-HORSEPOWER KRAEMER SET

main-mill motor and brings it up to normal speed under current-limit acceleration; connects the rotary converter into the circuit; disconnects the starting resistance and closes the field switches of both the rotary and the direct-current machine at the proper time. For

FIG. 3. ROTARY AND SECONDARY PANELS FOR REVERSIBLE KRAEMER SYSTEM CONTROL

emergency stops the master switch may be moved in the opposite direction to the starting position. The auxiliary equipment is then automatically disconnected from the circuit; the "plugging" resistance connected into the secondary circuit of the induction motor; the primary current reversed through the magnetically operated primary contactors and magnetically operated forward and oil reverse switches, Fig. 5, and the main-mill motor quickly stopped.

When the master switch is placed in the starting position, the synchronous converter is disconnected from the slip rings of the induction motor and is replaced by the starting resistance. The primary oil switch is then closed and the unit automatically comes up to the full speed of the induction motor. When rolling at this speed, the unit is operated as any ordinary induction motor. To operate at a lower speed, the field circuit of the synchronous converter is closed, and, with the field of the direct-current motor open, the slip rings of the synchronous converter are connected to the slip rings of the induction motor. The starting resistance is then disconnected in a weak field applied to the direct-current machine. The direct current generated by the direct-current machine is converted to alternating current and applied to the rotor of the induction motor at slip frequency. The speed of the rotary will increase and that of the Kraemer unit decrease until a point of stable operation is reached. By increasing the field on the direct-current machine, the speed of the unit may be reduced to the desired speed. For any particular setting the speed of the main-mill motor remains fairly constant under varying load conditions, a variation of not over 5 per cent. from no load being a fair average. The master switch may be used entirely for starting and the various functions mentioned in the foregoing. A hand-operated rheostat in the shunt-field of the direct-current machine is used for obtaining the different speeds that may be required. Unlike the master switch, as used in the Ilgner system of control for reversing mills and constantly under the guidance of the operator's hand, the master switch for the Kraemer control is seldom touched except in case of emergency after the desired speed setting has once been obtained.

FIG. 4. DRUM-TYPE MASTER SWITCH

FIG. 5. OIL REVERSE SWITCH

When rolling steel, the speed is governed to a certain extent by the quality and size of the material, the reduction in area and the power requirements. Small-sized material is usually rolled at higher speed than large-sized material, and in consequence it is desirable that the driving motor should deliver a constant horsepower within its speed range. This is a natural condition as the larger product rolled at lower speed requires a greater torque. It follows that the Kraemer combination, as described, is well suited to meet this particular requirement, as constant horsepower and increasing torque with decreasing speed can be obtained throughout the speed range. This is due to the fact that the slip energy of the induction motor is returned to the direct-current machine of the main-mill motor through the medium of a rotary converter. As the induction motor is in itself a constant-torque machine, the combined turning effort of the main-mill motor or Kraemer unit naturally increases as the speed decreases due to the added torque supplied by the direct-current machine. For this reason the Kraemer unit is essentially a constant horsepower outfit and as such is well suited to rolling-mill conditions as they are actually found.

Kraemer units are in general made for a speed variation as low as 50 per cent of normal and are as adaptable to 60-cycle as to 25-cycle circuits. The combination of induction motor, direct-current machine and rotary converter requires no special machines. Standard apparatus can be used throughout. Another desirable feature of such an installation is its capability of overcoming the inherently bad feature of the induction motor; namely, power factor lower than unity. In the same manner that speed control is obtained by adjusting the fields of the direct-current machine, the power factor of the system can be brought to unity or leading currents made to flow by the proper adjustment of the field of the rotary converter. When energy is bought, this may not only effect a considerable saving in the power bill, but if energy is generated locally it adds that much more ultimate capacity to the electrical end of the prime movers. This is especially advantageous if the generators are already being worked fairly well up to capacity, and in many cases might save the expense of adding more prime-mover equipment.

Fig. 6 shows the characteristics of a 1500-hp. 6600-volt three-phase 60-cycle 508-207 r.p.m. Kraemer unit when operating at half speed. It is interesting to notice the extremely high efficiency of the unit when operating at this low speed and the unity-power-factor feature of the induction motor which takes energy from the source of supply or the line. In comparison to this a study of Fig. 7 shows the characteristics of this same induction motor disconnected at the coupling from the direct-current machine and operating at a normal speed

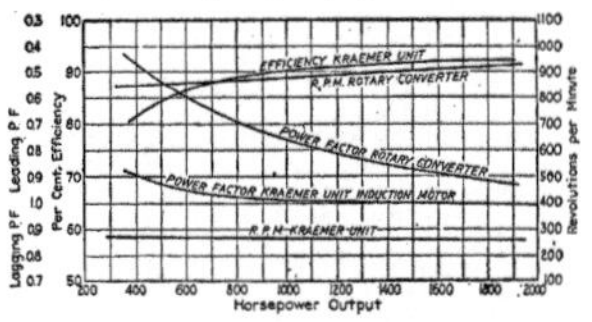

FIG. 6. CHARACTERISTICS OF 1500-HP. KRAEMER UNIT WHEN OPERATING AT HIGH SPEED

practically twice that of Fig. 6. When operating at a speed of one-half normal, the Kraemer unit has an efficiency slightly lower than that of the straight induction motor when operating at full speed. This is due to the added losses of the direct-current machine and rotary converter which have to be considered when operating at speeds below synchronism. Ordinarily, and where load conditions permit, a speed reduction of 5 per cent is obtained by inserting resistance in the secondary circuit of the induction motor, but for lower speeds the

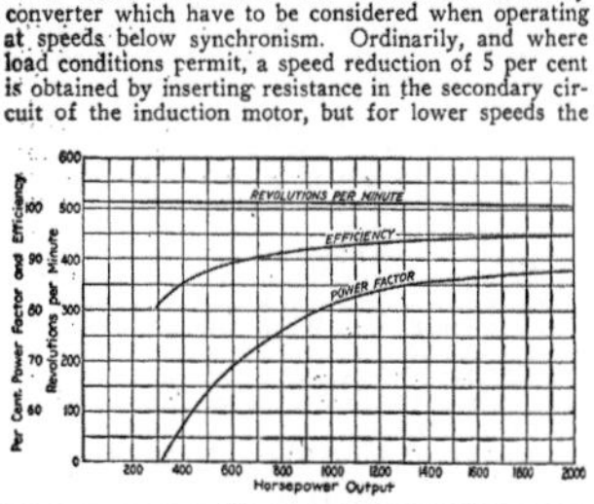

FIG. 7. CHARACTERISTICS OF INDUCTION MOTOR OF FIG. 6 OPERATING INDIVIDUALLY AT NORMAL SPEED

direct-current machine and the rotary converter are brought into action and the unit then operates as a strictly Kraemer combination.

Fig. 8 and the following formula show that the capacity of the direct-current machine and the rotary converter is dependent entirely on the nature of the load and the amount of speed control desired:

$$p = \frac{SP}{100}$$

where p equals the capacity in horsepower of auxiliary machines, S equals the per cent slip, and P equals the shaft horsepower of the Kraemer unit.

As shown in Fig. 8, for 50 per cent speed reduction and constant horsepower the direct-current machine and the rotary converter should be rated at one-half that of the induction motor. For 30 per cent speed reduction the auxiliaries should be rated at one-third that of the induction motor, etc., so that the less speed control required the smaller the auxiliaries become, consequently the lower the cost of the equipment.

A variation in speed of the main induction motor causes a corresponding variation in the voltage across its slip rings. For exact synchronous speed this voltage is a minimum, or zero, while at standstill it is a maximum in value depending on the design of the machine. Between the two points of minimum and maximum the voltage varies inversely as the speed, so it is seen that for changes in speed of the Kraemer unit a varying voltage is applied to the slip rings of the rotary converter. As there is a definite ratio of voltages between the alternating-current and the direct-current sides of a rotary converter, any change in the voltage of the former is immediately reflected by the latter. If a certain power throughout a definite speed range is to be delivered by the Kraemer unit, the commutators of both the

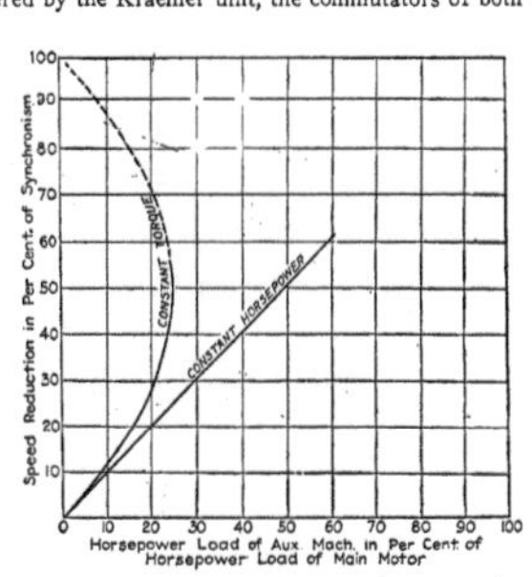

FIG. 8. SIZE OF AUXILIARIES FOR DIFFERENT SPEED REDUCTIONS

rotary and the direct-current machine have to be made capable of handling a certain amount of current with a varying voltage. When the Kraemer units operate at one-half speed and constant horsepower, the voltage across the slip rings of the rotary is one-half of the standstill voltage of the induction motor and there is a corresponding voltage and a definite amount of current on the direct-current commutator of this same machine. At a Kraemer speed three-fourths of normal the voltage across the rings is one-fourth of the "standstill" voltage (one-half as much as before), but inasmuch as the rotary under these conditions has to handle only one-half as much power as before, the current handled by the commutator remains the same, and therefore it has a constant value throughout the speed range. This is, then, the inherent feature of the Kraemer system. It is essentially a constant horsepower unit handling constant currents with varying voltages on the commutators of both the rotary and the direct-current machines.

The question of constant torque presents a different problem. For 50 per cent speed control and constant torque, Fig. 8, the theoretical size or physical dimensions of the auxiliary machines will be one-half the size they would have to be made if constant horsepower were required. This ratio, however, does not hold throughout the speed range as the larger the speed reduction the more the capacity of the auxiliaries gradually decrease, as shown by the dotted line, Fig. 8, until zero capacity at zero speed is required. It is also apparent that commercial units should never be made with the auxiliary equipment capacity for constant torque less than that required for 50 per cent speed reduction.

When the Kraemer unit is operating at one-half speed, constant torque, the voltage across the slip rings of the rotary is one-half of the standstill voltage of the induction motor and there is a corresponding direct-current voltage and a definite amount of direct current on its commutator. At a Kraemer speed of three-fourths normal, the voltage across the rings is one-fourth of the standstill voltage (one-half as much as before), but inasmuch as the rotary has to handle 0.75 (see Fig. 8) as much power as before, it is seen that the ratio of

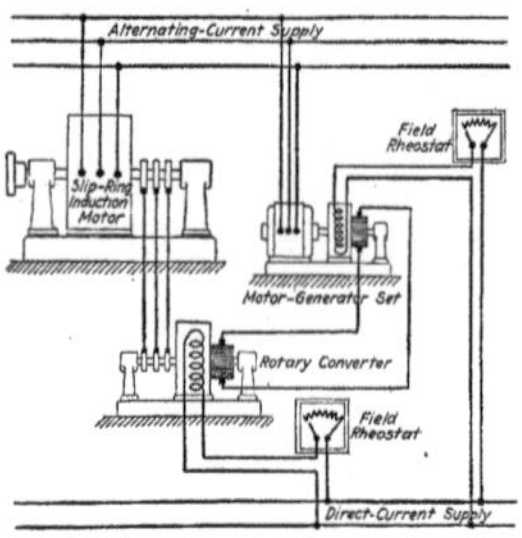

FIG. 9. SIMPLIFIED CONNECTION FOR ADJUSTABLE-SPEED CONSTANT-TORQUE KRAEMER SET

capacity to voltage decrease is not constant as in the case of the constant horsepower condition, but keeps on increasing the less the speed of the Kraemer unit is reduced. This means that the less speed reduction required, such as close to synchronism, imposes a condition where the auxiliary equipment delivers a relatively small amount of power with a relatively large amount of current to be carried by the commutators.

To meet such a condition if it were actually to prevail, and in view of the facts as already presented in the foregoing, the combination of machines shown in Fig. 9 could be used. It consists of a standard induction motor and a rotary converter, the same as in Fig. 2, but in place of the direct-current machine on the Kraemer set, a separate high-speed motor-generator set is used. This latter set comprises a direct-current machine and an induction generator returning energy to the line.

In reality, a constant-torque Kraemer unit combination of machines, such as the constant-horsepower combination shown in Fig. 2, would require auxiliary apparatus just as large for a constant-horsepower outfit on account of the currents to be handled by the commutators of the rotary and the direct-current machine. In the constant-torque arrangement shown in Fig. 9, the rotary converter is the same physical size as would be used in the constant-horsepower arrangement of Fig. 2. The reason for using a separate high-speed motor-generator set in place of the direct-current machine becomes apparent when it is considered that the Kraemer unit may run at a speed of, say, 200 r.p.m., while the motor-generator set may run at 1200 r.p.m. While the direct-current machine of the high-speed set has to handle just as much current under any load condition as the apparatus in Fig. 2, the high-speed set cuts down the initial cost of the whole combination over what it would be if the combination of Fig. 2 were used.

In Fig. 8 the losses of the main motor have been neglected, but this affects the capacity of the auxiliary machines very little. For a 1500-hp., 508- to 257-r.p.m. Kraemer unit of the constant-horsepower type, the direct-current machine should be rated at 750 hp., 1,200 amp., 508 to 257 r.p.m. and the rotary at 600 kw., 500 volts direct-current, 370 volts alternating current, three-phase, 30 cycles, 900 r.p.m.

It is often advisable in the larger-sized units to have the rotor of the induction motor wound six-phase with six slip rings, for this gives a more economical design of rotary converter than could be obtained with a three-phase machine.

From the operating point of view the Kraemer system for adjustable speeds involving the combination of standard simplified machines and simplified control apparatus, especially for steel-mill service, tends toward low maintenance cost, long life, maximum production, high factor of safety and, still more important, continuity of service.

Miners Earn $10 to $12 Per Day

Many of the miners, says *Coal Age*, have plenty of opportunity to work, but prefer to be idle a large part of the time, is a statement made by the bituminous operators of the Central Competitive field. A statement by the operators says that they purpose to make an effort to convince the commission that the miners have plenty of opportunity to work, but that many of them prefer to be idle a large part of the time; that their wages are high and that the cost of living to them has not gone up to the high figures which their officers have given.

"The operators have maintained," says the statement, "that the differential and similar issues raised by the miners have been dragged in in an attempt to befog the real issue, which, they say, is purely one of the straight wage advance. These other issues interjected by the miners have been characterized by the operators as attempts to obtain what amounts to an additional wage increase by a subterfuge.

"Payroll evidence will be offered by the central competitive operators to show that the miners, in most fields, can make from $10 to $12 a day, and that $15 a day can be earned readily in the richer mining districts, which operate from 250 to 275 days a year. It will be shown that comparatively few of the men in the field work steadily, and that if they did their yearly earnings would be increased by from 15 to 20 per cent. This, they will show, would not add to the present cost of producing coal.

"The cost of living in mining communities will be presented to refute the assertion of miners' representatives that the cost of living in mining centers increased as much as 125 per cent during the war. The operators will show that in no mining communities have rent and fuel increased at all, while the cost to miners for food and clothes has not been such as to create any such advance in living cost as the miners have maintained before the commission."

Propelling Machinery for U.S.S. Tennessee*

By: Wilfred Sykes

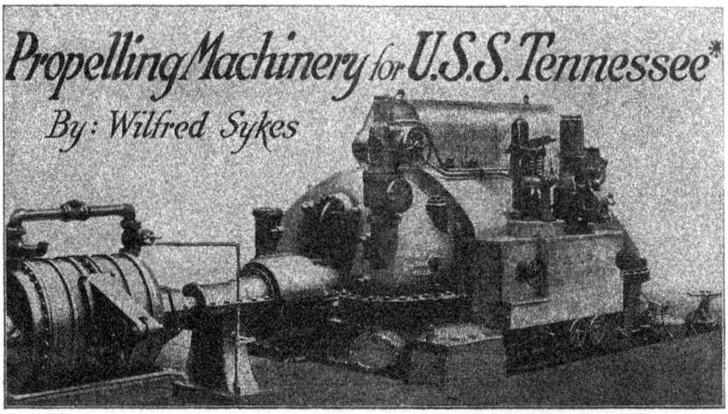

THE propelling machinery for the U. S. S. "Tennessee"† consists of two turbo-generators and four propelling motors. The switching apparatus is so arranged that either generator can drive all four motors or with two generators in operation each will drive two motors. The turbo-generators have a combined capacity of about 26,500 kw., which will be required to drive the motors when operating under the maximum conditions, each then delivering 8,375 hp. The speed of the motors is varied by varying the revolutions of the turbine within certain limits, and the motors also have two sets of poles, so that with the turbine running at full speed there is a speed reduction of about 12 to 1 and 18 to 1 to the propellers. The turbines, one of which is shown in the headpiece coupled to a water brake, are designed to operate with a steam pressure of 250 lb. per sq.-in., 50 deg. superheat and with a vacuum of about 28½ in. The turbine speed under normal operating conditions is about 2,130 r.p.m. maximum, and on test the machines have been run to 2,480 r.p.m. The speed of the turbines is varied by means of a specially designed governor which, instead of being loaded by a spring as is usual, is arranged so that the centrifugal weights act against an oil pressure on one side of a piston. With such an arrangement the speed at which the governor will affect the steam supply will vary with the oil pressure; the turbines can be set to run at any speed desired. A specially designed variable-pressure valve is provided in the control room, for the purpose of operating the governors, and the only connection between the control room, where the speed is varied, and the turbine room is an oil-pipe connecting to the governor piston.

MAIN PROPELLING MOTOR WITH BLOWERS ASSEMBLED

The generators are three-phase 3,400-volt machines, operating at about 35 cycles at full speed. They are standard construction, except that special provisions have been made to avoid salt deposits in the windings. This is a trouble that is likely to occur where machines are operated in salt-laden air.

The guarantees for this ship based on 250-lb. steam pressure, 60-deg. circulating water with no superheat, are at 21 knots, 11.9 lb.; 19 knots, 11.65 lb.; 15 knots, 12.1 lb.; 10 knots, 15.45 lb. These figures include all the power required for driving the circulating pumps, condensate pumps, ventilating blowers for machines and excitation; that is, the auxiliaries for the main units.

*A. I. E. E. Journal, Jan., 1920.
†Power, Aug. 7, 1917.

Steam-Turbine Governors—Governors of 75- to 3,000-Kw. Curtis Turbines

By CHARLES H. BROMLEY

Treats of the operation, care and adjustment of the single and double-deck inertia governors used on Curtis horizontal turbines of moderate capacities. This article applies to Curtis turbines of 75 to 3,000 kw., though other forms of valve mechanism have been used on units of these ratings.

SINGLE-DECK INERTIA GOVERNORS—It is important that the operating engineer acquaint himself with the Curtis inertia governor and combined throttle and trip valve. The automatic throttle shown in Fig. 10 is in wide use, though on machines now being furnished this particular type of valve is used only in special cases. It should be understood that the only difference between the single- and double-deck governors is that the latter is simply two single decks, one over the other. The operation of both is the same, the double-deck governor being used for larger units than those having the single-deck type; they are used also for low-pressure turbines. Both are driven by a secondary shaft geared to the mainshaft.

The governor is of the inertia type and functions because of the unbalanced force due to inertia and centrifugal effect of two weights *B* and *C*, Fig. 2, pivoted on ball bearings *D* and *E*. The centrifugal effect and inertia are opposed by the helical spring. The weights are mounted on knife-edges. The center of gravity of the weights is inside the radius of the point of support and in advance of it with respect to rotation. This decreases the tendency to "jump" on quick changes of load.

The effect of revolving the weights is to bring the points *J* and *H* together. At each of these points is a hollow link *K*, Fig. 3, and these links rest in the ball joints *H* and *J* and support a thrust block *A* at their upper ends. Therefore when the ball joints *H* and *J* move toward each other the thrust block is lifted, and is lowered when *H* and *J* move apart on the reduction of speed. The thrust block transmits motion to the levers that operate the governor throttle valve. The links *F* and *G*, Fig. 2, give the weights a parallel motion. Fig. 1 shows a view of the double-deck governor.

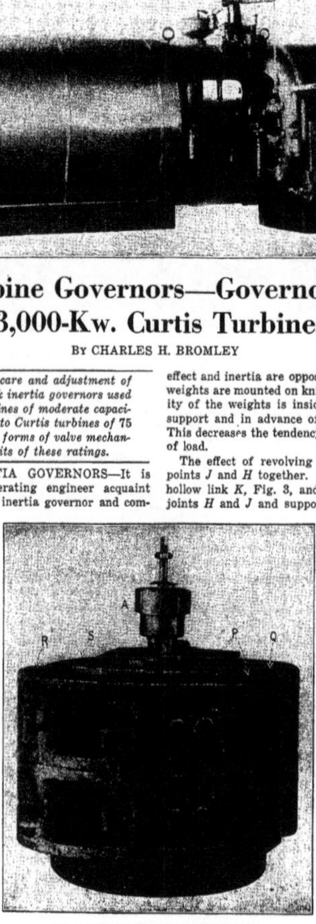

FIG. 1. THE DOUBLE-DECK GOVERNOR

Adjusting the Governors—Assume that the speed must be raised or lowered without materially changing the "range." To do this decrease or increase the weight by taking out, screwing in, or substituting lead or steel weights, which screw into the tapped holes, *P*, *Q*, *R* and *S*, Figs. 2 and 3. *Remember always to have plugs of the same weight in the diagonally opposite holes in both*

weights. For example, do not have two plugs of different weights in the holes *R* and *Q* or holes *P* and *S*. This upsets the balance and may cause trouble.

Any change in spring tension will affect the width of regulation; decreasing the tension broadens it and

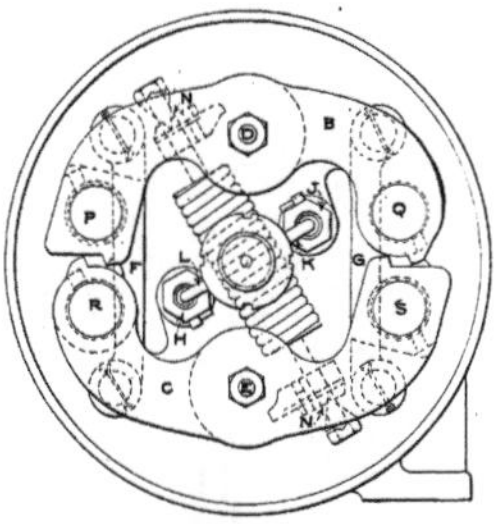

FIG. 2. LOOKING DOWN UPON "DECK" OF THE CURTIS GOVERNOR

increasing the tension narrows it. When the regulation is so affected the change in speed is about twice as rapid at the lower end of travel as at the upper. The regulation can be further broadened by screwing the plugs farther into the springs, but this practice is not

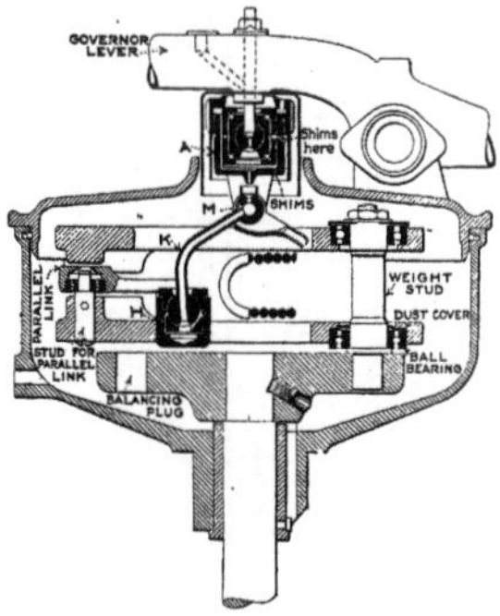

FIG. 3. SECTION OF THE GOVERNOR

advised except when the other two means are not sufficient.

By applying the two methods of adjustment in proper proportions all desired speeds possible with the design of the governor may be obtained. Suppose the regulation is narrow and high and it needs to be broadened, add weight and reduce the tension of the spring at the same time. Assume that the regulation is high and broad and it is desired to narrow it, add weight until the speed is reduced below normal (not too much) and then increase the speed by increasing the tension of the spring.

Making adjustments frequently, with a trial after each adjustment, is better than going to extremes. As with every governor, adjustments will be made quicker and easier as one becomes familiar with it. The tension broadens the width of regulation. To raise or lower the speed without affecting the width of regulation add to or take from the weights in the parallel links.

Lubrication of the Governor—The oil pump is driven from the turbine governor shaft, as usual, and consists essentially of two gears running in close mesh and hav-

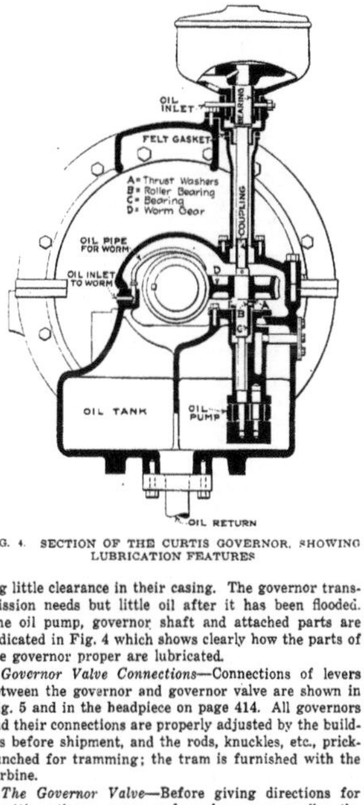

FIG. 4. SECTION OF THE CURTIS GOVERNOR, SHOWING LUBRICATION FEATURES

ing little clearance in their casing. The governor transmission needs but little oil after it has been flooded. The oil pump, governor shaft and attached parts are indicated in Fig. 4 which shows clearly how the parts of the governor proper are lubricated.

Governor Valve Connections—Connections of levers between the governor and governor valve are shown in Fig. 5 and in the headpiece on page 414. All governors and their connections are properly adjusted by the builders before shipment, and the rods, knuckles, etc., prick-punched for tramming; the tram is furnished with the turbine.

The Governor Valve—Before giving directions for resetting the governor rods and gear generally, the governor valve must be understood. Fig. 6 looks more complicated than the valve appears when dismantled.

Briefly, the valve consists chiefly of six individual valves and seats *F*, *B*, *A*, *C*, *D* and *E* all on one rod and all in one long cylinder, the valves becoming progressively larger from the bottom valve to the top valve. Each valve has its own separate passage for steam to the turbine. The valves are moved by steam pressure applied to the operating piston *N*, the steam admission to the operating cylinder being controlled by the pilot valve *G*. Notice that the steam passages from the pilot valve to the operating cylinder are criss-crossed. The valves open one at a time in the alphabetical order shown—that is, *A*, *B*, *C*, *D*, *E*, *F*—and the steam goes to the first-stage nozzle and wheel, as shown in Fig. 5, although the valve here is slightly different. As the lugs on the valve spindle engage the valves, each one is lifted from its seat. Fig. 7 shows the valve dismantled.

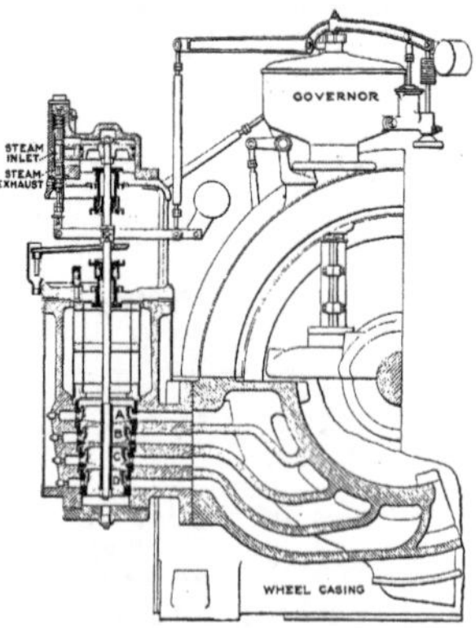

FIG. 5. SECTION OF THE GOVERNOR VALVE, SHOWING THE STEAM PASSAGE TO THE NOZZLES

When the turbine is at rest, the pilot valve is in a position to admit steam under the operating piston. When the throttle is opened and the turbine picks up speed, the governor, through its levers, moves the pilot valve so that it now admits steam above the operating piston. The governor valve now begins to close until the governor takes a definite position, that is, the floating lever *K*, Fig. 6, fulcrums on the pin at the extreme right, and the pilot valve is then in a neutral position. When the load is put on, the governor responds by shifting the pilot to admit steam *under* the operating piston, and this operation lifts the valves until the speed is restored.

The reader is now likely aware that this multiple valve has one or more of its six valves *full open or closed all the time*, avoiding throttling. Leakage from one valve to another is avoided by the use of asbestos packing (gaskets) under each valve-seat ring frame, as shown at *J*.

The steam piping to the pilot valve is connected to the entrance side of the throttle valve, and no valve should be put in this pipe. If a valve should be put in and by ignorance or accident closed, it would result in the turbine running away if the load were suddenly thrown off.

Resetting the Governor Gear—By gear is meant the rods, levers, knuckles, etc., attached to the governor and connecting it with the governor valve. These parts are properly set before the turbine is turned over to the purchaser, and the parts prick-punched so that they may be trammed; the tram is furnished by the builder.

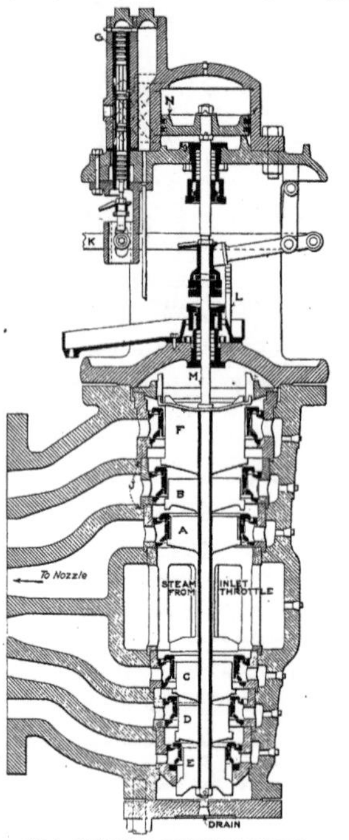

FIG. 6. SECTION OF THE CURTIS TURBINE GOVERNOR VALVE

As the governor lifts the valve rod the valves open full, one at a time, in the order A, B, C, D, E and F, thus there is no throttling by any one valve.

Suppose the tram marks have been obliterated or new parts furnished and the gear is to be reset. Here are the directions; See that all joints are free to move and that there is no serious backlash. Disconnect the synchronizing spring from the governor lever and adjust the dashpot piston, Fig. 8, within about ¼ in. from the dashpot cover. Remove the lagging from the steam chest and insert cocks in the ¼-in. tapped holes (shown at the right and plugged) leading to the upper valve *F*, Fig. 6, and to the third valve *A*. An arrow is stamped on the lower half of the coupling shown just to the left of *L*, Fig. 6, and a valve indicator *L* indicates the position and travel of the various valves. Open the throttle and bring the turbine to speed, mark the position of the arrow on the indicator when all valves are closed; that is, the instant when practically no steam is blowing out of the cock at the valve *A*. Throttle the steam until all the valves are open and again mark the position of the arrow on the indicator. The distance between these two marks should be about two inches.

It may be necessary to shorten the connection rod to obtain the travel needed for the valve stem to open all the valves or to lengthen it so it will close all the valves. It is desirable to have the pilot lever approximately horizontal for the mid-position or neutral position of the pilot valve. This position may be had by readjusting the length of both the governor connection and the pilot valve. Lengthening the governor connection will require lengthening the pilot valve and vice versa. Check the valve travel after any readjustment in active length of rods or connections. This is important.

Now attach the synchronizing spring and with it set in mid-position, check and, if necessary, make slight readjustments. The maximum and minimum travel of the synchronizing spring should be limited by pins to permit not over 2 per cent increase or decrease in speed. More than 2 per cent will throw too much effort on the main governor and consequently cause rapid wear of that governor.

Finally, check the no-load speed and if possible the full-load speed. If the speed is high or low, adjust the main governor as described in the early paragraphs of this article. Now tram the various parts and prick-punch them so as to guide you or other operators in later resetting the connections.

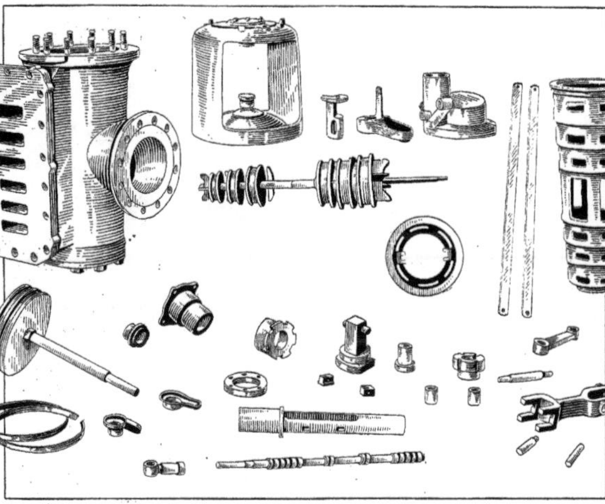

FIG. 7. THE MEDIUM-SIZED CURTIS TURBINE GOVERNOR VALVE DISMANTLED

As a guide in making adjustments it is to be noted that the maximum travel of the governor is about $\frac{7}{16}$ in. The working travel is limited to about $\frac{5}{16}$-in. The travel of the main valve stem *M*, Fig. 6, is 2 in. from the position of the first valve closed to the valve wide open. The travel of the pilot valve for full opening and closing of the ports is ½ in. To close the main valves the governor transmission travels upward, while the connection rod, pilot valve and main valve stem travel downward.

The Emergency Governor—Fig. 9 shows the emergency governor. This particular type is used on all Curtis turbines from 500 kw. to 2,500 kw. It consists essentially of two clock springs under tension against stops on the emergency disk *A*. The tension of the springs is such that at about 7 per cent above normal speed the centrifugal effect overcomes the springs, and at 10 per cent the ends of the spring move about $\frac{3}{8}$ in. from the nuts and trip the trigger *A*, Fig. 9. This allows the throttle, Fig. 10, to drop to its seat.

To reset the emergency, turn the handwheel on the throttle valve to its closed position, raise the resetting handle *B* and open the throttle.

The Synchronizing Mechanism—This is shown in Fig. 8. Increasing the tension of the spring by adjusting the screw into the spring holder raises the speed of the operating governor, slightly narrows the effective range of the synchronizing spring and lowers the speed of the operating governor. In case readjustments are made, note that the spring should not be shortened enough to come within one-eighth inch of closing when the upper limit of travel of the lower plug is reached. Keep the wormshaft and gear lubricated. The motor

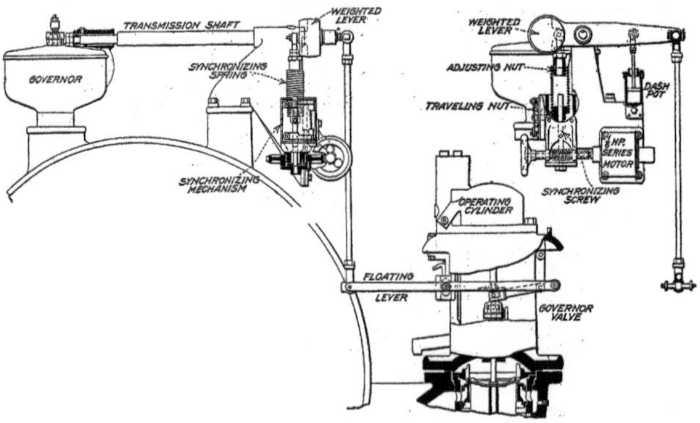

FIG. 8. SYNCHRONIZING MECHANISM AND CONNECTIONS WITH GOVERNOR AND GOVERNOR VALVE

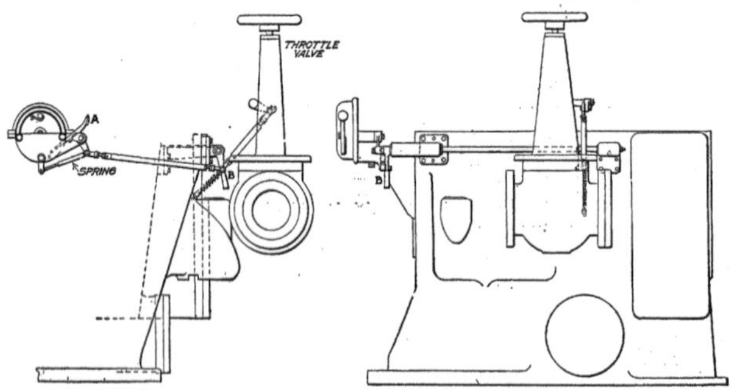

FIG. 9. SHOWING THE CONNECTIONS TO THE OVERSPEED STOP GOVERNOR OF THE CURTIS TURBINE

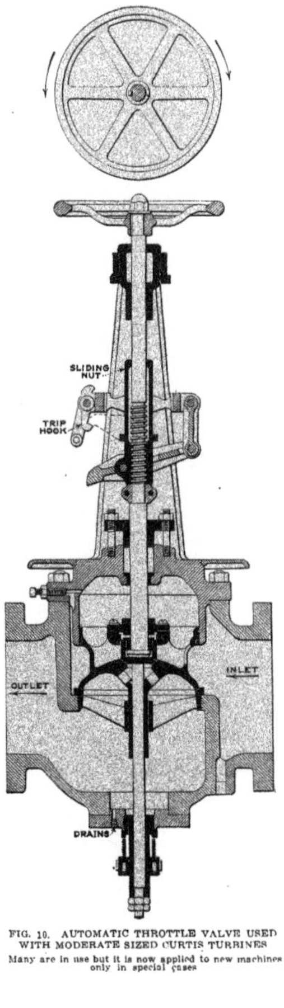

FIG. 10. AUTOMATIC THROTTLE VALVE USED WITH MODERATE SIZED CURTIS TURBINES

Many are in use but it is now applied to new machines only in special cases

journals are ring lubricated, therefore keep the oil reservoirs full.

Have in mind that this article applies particularly to Curtis turbines of capacities ranging anywhere from 75 to 3,000 kw., although on some machines of this type and these ratings other valve mechanism will be used.

Russell Storage Heater

Storage-type heaters which insure a reservoir of hot water have been used in their general form for years. In these heaters exhaust or live steam is supplied to a heating element, where it condenses and heats the water in the tank.

The Griscom-Russell Co., 90 West St., New York City, has recently placed on the market the Russell storage heater, intended to supply hot water in hotels, apartment houses, factories, etc., and this heater departs from the

SEMI-SECTIONAL VIEW OF THE RUSSELL STORAGE HEATER

general practice as regards the construction of the heating surface.

The Russell heater is designed with a number of straight tubes in the steam element, and of sufficient length to insure that the required heating will be taken care of and that all the steam will also be condensed when it reaches the rear end of these heating tubes. It is therefore only necessary to provide an area of drain-tube section sufficient to carry the condensation back to the steam head.

The heating surface consists of seamless drawn brass tubes expanded into a fixed tube plate at one end and a floating tube plate at the other end, thus permitting expansion and contraction of tubes without strain on the tube joints.

The shell is regularly furnished of welded construction, but can be riveted if preferred. A manhole is provided to permit easy access to the shell for inspection and cleaning.

Too much stress cannot be laid to the fact that the failure of high-head hydraulic pressure-regulating apparatus to function properly may cause pipe-line breaks which would often cause loss of life or considerable damage to property. The governor and governor connections of hydraulic pressure regulators should be well lubricated and examined regularly. Hydraulic pressure regulators that do not function during the normal operation of the plant should be operated occasionally to be sure that they do not stick. A good grade of packing should be used in the packing glands of hydraulic pressure regulators and the adjustment of the glands should not be trusted to anyone who is not practiced.

On the Road with the Refrigerating Troubleman

By J. C. MORAN

Called to a distant job, the present one is left in charge of an assistant and the purchaser's chief engineer. Hurriedly returning, he finds the compressor unable to be operated because of violent thumping, with suction pressure of 70 lb. He finds and corrects the trouble and breezily tells how.

I ONCE had charge of installing a large air-conditioning plant in a factory in the South. Before the job was finished, I was sent out on another case where they were having trouble and the work was turned over to an assistant. Before I left, we talked over testing and charging the system and the precautions that should be taken before starting the plant. I went away confident that there would be no trouble, as my assistant had worked with me for over a year and the chief of the plant assured me that he had had a long experience with refrigerating machines.

The work in the plant several hundred miles away finished, I wrote the home office for further instructions and received a somewhat sarcastic answer. I was informed that they had been unable to start the plant left in charge of the assistant and that I must get back there quickly because the purchaser was threatening to cancel the contract.

The Assistant and Chief Have Trouble in Starting the Plant

Arriving there, I found the assistant and the chief calling each other names, and I got my share. They had been trying for nearly a week to get the plant going. Sometimes the machine would start to pound so hard that it would nearly jump off the foundation, about the same as if there were a lot of water in the cylinder; at other times it would run hot. They had been unable to get the frost to come back on the suction line or even at the suction outlet of the expansion coils, except for a short time on two occasions. They had also been unable to do any cooling worth while in the water coolers through which the air passed. The original water temperature at this time of the year was about 80 deg., and it was required that the temperature be carried at about 40 deg. as it came off the cooling coils. So far they had been able to reduce the temperature only 10 to 15 deg. and in most cases only 4 or 5 deg.

Examination of the whole plant from top to bottom showed nothing wrong. We had about 600 lb. of ammonia on hand, and thinking that the trouble might be due to the charge being too light, we put in all of it.

There were four coolers, and to avoid complications I decided to break them in one at a time. Beginning on the one nearest the engine room, so as to be as close to the compressor as possible in case of trouble, we started the compressor slowly and then opened the expansion valve on this coil a little at a time every ten or fifteen minutes until the discharge line began to get cold and a few slight thumps in the cylinder indicated that there was liquid coming over with the suction gas. The suction pressure kept going up, and when the thump was heard in the cylinder, it was 70 lb. The assistant and the chief gave me the laugh when I continued to operate at this pressure, saying that no self-respecting machine had ever been known to do any work at such a suction pressure and no frost on the suction line. The suction line began to sweat for its entire length, and I stuck close to the compressor with hand on the discharge line. Soon the line began to get cool, and I promptly went out to the cooler and pinched back on the expansion valve, when the discharge again warmed. This was repeated regularly and in the meantime both the suction pressure and the water temperature were rapidly coming down.

The suction line began to show frost, and the suction pressure had dropped to 40 lb. The temperature of the water leaving the cooler was 38 deg., and they were able to start the fan for the first time and try the air conditioning and circulating system on one room.

We then cut off the feed on this coil to about half of what it had been up to this time and cut in one of the other coils a little at a time. It was not long before the suction pressure took a jump and the frost came off the suction line. In a few moments the discharge became cold again and the cylinder began to thump. The suction pressure by this time had climbed to 60 lb. There was no secret as to where the liquid came from. It was realized that for quick results it would be necessary to again cut out the cooler in which the temperature had come down. It was decided to bring each one down in succession, as the water temperature, once it was down, would not rapidly rise before all coolers were of uniform temperature.

The procedure with the other three coolers was the same as with the first one and with the same result. When the last one was down to 45 deg., the temperature of the water in the first one had risen to only 48 deg., while in the others it was lower. We then began to cut them in all together.

Each expansion valve was gradually opened until the discharge began to get cold, when they were pinched back slightly until the suction line frosted. It was then a simple matter to detect which one of the coils was getting the most liquid and adjust the expansion valves accordingly. The line that frosted back the most was, of course, to the cooler that was getting the most liquid, and all that was required was to pinch back on this valve and possibly open up a little more on some of the coils on which the frost was not so heavy until they were all working at full capacity. I then told the master mechanic to start the fan system on all the rooms.

The Cause of the Trouble

The owner and everyone else were pleased. The trouble arose because the chief and the assistant were not posted on fundamentals. In starting with the water temperature up around 80 deg., the temperature of the ammonia boiling on the inside of the coils would be somewhere between 65 to 75 deg., and of course the gas evaporated would be at this or higher temperature as it passed along through the suction line. If there was any liquid carried along, the temperature would be about the same as it left the cooler, and as a consequence it would be impossible to frost the suction line. The temperature of the gas and that of the suction line were between 30 and 40 deg. above the freezing point. This caused the high suction pressure at the start, which, by the way, was just what was wanted to get the most work out of the machine in the shortest time. The pressure corresponding to 70 deg. of the boiling ammonia in the coils is 114 lb. So that is the pressure that could be expected, or something near it if we had been able to adjust the feed to full capacity at once. Then, again, there was considerable friction loss in the line when taking all the gas out of one coil, as I did at first, and this will also in a large measure account for the fact that the pressure never went much above 70 lb. at the start. In addition the temperature of the water had dropped several degrees before the coils were working to advantage.

Here is where the assistant and the chief got into trouble. First they tried to hold the suction pressure down to about 25 lb. for the reason that they had never seen a machine operating any length of time at a higher pressure, and they concluded, therefore, that it was necessary to keep the pressure below this point; the lower the pressure the more work would be done, they believed. In doing this they could only open the expansion valves a little, and therefore but a small amount of ammonia would be evaporated and the gas would come back to the compressor highly superheated. So there was little work done and the suction line would not frost.

A few times they had tried to operate the machine at the higher suction pressure, but they began with all coils in on the line. With it being impossible to frost the suction line at this high suction pressure and temperature, there was no indication when one coil was getting too much liquid. And the only place where it could be checked—until it started a pound in the cylinder—was to feel of the discharge line, which would get cold. This neither of them had thought of, and even so, it would give no clue to which one of the coils was doing the mischief and throwing the liquid in the line. However, they could have pinched back on all of them until the pressure came down enough to frost the line, when they could have made the proper adjustment. As a consequence they nearly wrecked the machine a couple of times when the liquid came back in heavy slugs, and they decided that something was wrong with the plant instead of themselves.

The Troubleman Relates Another Similar Experience

I had a similar experience with a small direct-expansion system installed in a sanatorium for cooling the drinking water, of which a large quantity was used and the temperature before cooling was often near 85 deg., from which it was reduced to 45 deg. That is, they desired it to be at a temperature of 45 deg., but so far they had been unable to come anywhere near this figure. In addition to cooling the drinking water, the machine served a small refrigerator in the kitchen.

The man in charge was of the old school and could never think of anything being done any other way than the way he had seen it done. He had worked in a fish-freezing plant where they had been unable to do business at temperatures in the coolers much above zero. As a consequence he had formed the rockbound opinion that no machine could do work unless it was operated at or near zero suction pressure. "The lower the pressure the more the work," was his argument.

Well, he applied this reasoning to the cooling of the drinking water and it did not work. Neither did he get results in the kitchen refrigerator. And as usual he blamed the machine.

The chief stated that he wanted a temperature of about 40 deg. in the kitchen refrigerator, and I found it had an abundance of pipe for the work required. By keeping the pipes clean they ought to do the work with a temperature difference between the air on the outside of the coil and the ammonia liquid on the inside of about 5 to 8 deg. This would mean a temperature of the boiling ammonia in the coil of about 32 deg. at the lowest, which corresponds to a suction pressure of 45 pounds.

The coil in the water cooler was about double the size necessary for the work it had to do. If it was turned on full, it would do the required work all right and bring the water temperature down to the freezing point, but the suction pressure was too high for the kitchen refrigerator to do its work properly.

To balance the system and make it simpler to operate, I connected a bypass into the water-cooling coil so that by opening it one-third of the coil was eliminated, the gas going direct from two-thirds of the coil to the suction line, leaving the last third of the coil out of service. In other words, the two-thirds of the coil had to do the same work that the whole coil had done before, and as a consequence it had to be worked at a lower suction pressure. This corresponded to the suction pressure required to get the desired results from the refrigerator in the kitchen, and it steadied the operation of the system.

The Engineer Predicted Failure

When started up after making the changes, the engineer predicted all kinds of dire failures when the suction pressure was worked up to about 40 lb. I first started with only the water cooler connected and the whole coil working. When the water temperature came down to 45 deg., I cut in the pantry refrigerator and opened the bypass to the suction on the water-cooling coil. After playing with the outfit for about a day, I found that the water temperature could be held between 35 and 40 deg. most of the time and the refrigerator temperature about the same with the machine running moderately fast and carrying a suction pressure of 40 to 45 pounds.

It took some pretty hard explaining before I was able to show the engineer the relation between compressor capacity, coil capacity and suction pressure. The lower the suction pressure the lower will be the compressor capacity, the capacity varying practically as the absolute pressure. In other words, a compressor working at zero pressure has about half the capacity of one working at 15 lb. pressure. On the other hand, the coil capacity falls off as the suction pressure increases, because with an increase of pressure the temperature of the boiling ammonia increases and the temperature difference between the ammonia on the inside and the medium to be cooled on the outside becomes less and the heat-transfer rate will be slower. If the suction pressure is increased to the point where the boiling temperature of the ammonia is the same as the air or other substance on the outside of the coil, there will be no heat transfer and no work will be done by the coil. This happens when a heavy high-temperature load is suddenly thrown on a system where there are a few low-temperature coils connected. The suction pressure takes a jump and often high enough to stop the heat transfer in the low-temperature coils, the liquid working back to the compressor, starting trouble.

There are to be installed in Glasgow, Scotland, by the Glasgow Corporation, mercury arc rectifiers that have ratings of 170, 330 and 660 hp. The rectifiers are to be used for supplying direct current from an alternating-current supply for light and power purposes. Instead of the glass bulb commonly used in this type of equipment, a steel cylinder insulated on the inside is employed.

Fluctuations in Coal Production—Their Extent and Causes*

By GEORGE OTIS SMITH† AND F. G. TRYON‡

AN ELECTRICAL engineer has supplied us with the phrase that best expresses what's wrong with our coal industry—it is the "bad load factor." Whether we refer to full rated capacity or to average output, the operation of the soft-coal mines of the country from year to year, from month to month, and from day to day presents a load factor that has been too wasteful of plant and labor and too productive of high costs and uncertain supply.

The fluctuations in coal production can be regarded as annual, which in a large way reflect nation-wide business conditions; as seasonal, which express conditions of market and distribution; and as daily, which express conditions of labor and car supply. In Fig. 1, the facts of coal production, labor supply, mine capacity and average return to the industry per ton produced, are set forth in curves for the 30-year period 1890-1919. At the base of the diagram is a graphic statement of lost time in mine operation, which is the measure of wasted opportunity for the economic use of both plant and labor.

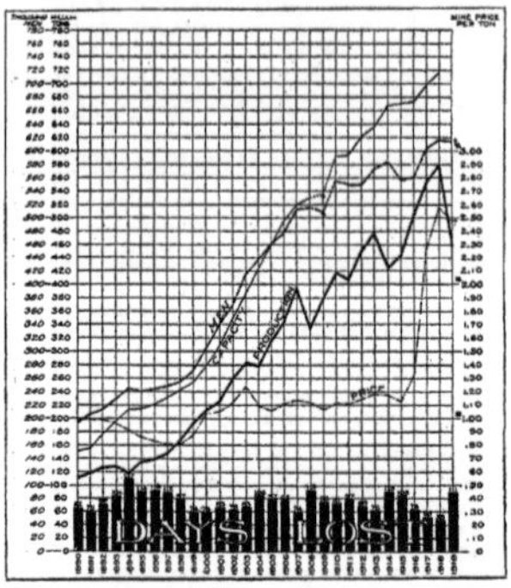

FIG. 1. PRODUCTION, CAPACITY, MEN EMPLOYED, MINE PRICE PER TON AND DAYS LOST AT BITUMINOUS COAL MINES, 1890-1919

The statistics thus presented are weighted averages for the industry spread over the country and carried on under conditions that are widely divergent from place to place.

Mine capacity is seen to have kept well in advance of output, and this capacity curve, so far above even the emergency output of the war period, suggests strongly that the industry has become overequipped and overmanned.

The blocks at the base of the diagram, representing lost time in the soft-coal industry, show that in only seven of these 30 years was such lost time less than 25 per cent of the working year. During this period, out of 308 possible working days a year, mines were idle on the average 93 days. Ten times during that period the time lost exceeded 100 working days. The smallest

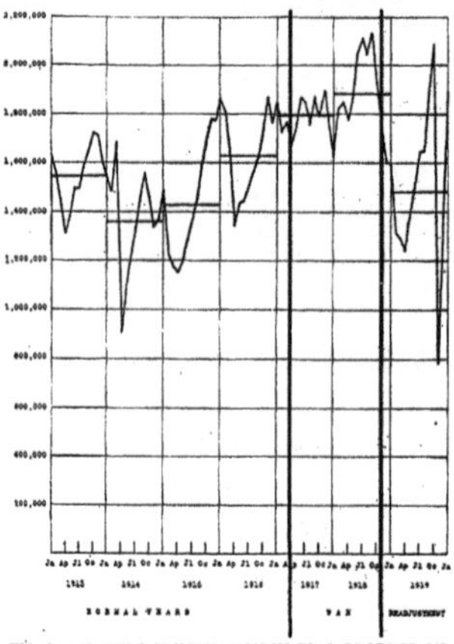

FIG. 2. MONTHLY FLUCTUATION IN COAL PRODUCTION

loss occurred in 1918, the year of record production, yet even during that year the mines were closed down for nearly one-fifth of the time. These figures for lost time, remember, show only the days that the mines were not operated, but absenteeism of a part of the force when the mines were running still further reduced the output. That is, they show only the average opportunity offered to labor by the mines, not the extent to which the individual miner took advantage of his opportunity.

From a study of the diagram the conclusion seems justified that losses above 78 or 80 days are the measure of the effects of annual fluctuations, and that losses below this figure are attributal to season and daily losses.

The annual as distinct from seasonal fluctuation is not confined to the coal industry, and no help can be found for it short of doing away with business cycles of good and bad years.

*Abstract of paper presented at the February, 1920, meeting of the American Institute of Mining and Metallurgical Engineers.
†Director, United States Geological Survey.
‡With the United States Geological Survey.

There remains, however, that residue of lost time from which no normal year is free, and light on its causes must be sought in the study of seasonal fluctuations. To help in this study Fig. 2 has been prepared, on which the annual averages are shown by the heavy horizontal lines and the monthly fluctuations by the broken line. The two pre-war years 1913 and 1914 are typical. The greatest extremes shown in the diagram for normal years occurred in 1914, when the rate of production rose in March to 123 per cent and fell precipitately in April to 66 per cent of the average for the year. In that year the normal seasonal fluctuation was intensified and distorted by the biennial wage negotiations. In one respect, however, 1914 was not typical. The autumn peak came in September and was followed, in the last quarter of the year, by a depression which marks the effect of the outbreak of the European war. In other years the peak was reached in November.

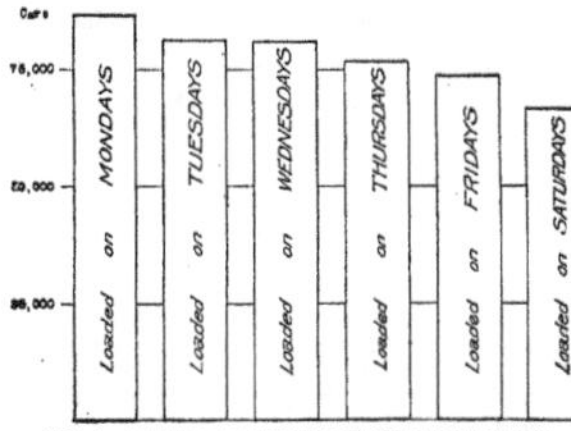

FIG. 3. DAILY FLUCTUATIONS IN PRODUCTION

The year 1913 may be accepted as a fair type of the odd year, when monthly fluctuations represent seasonal fluctuations in demand only, uninfluenced by labor disturbances. In such a typical year the capacity required during the month of maximum demand will be from 35 to 40 per cent greater than in the month of minimum demand.

To put it another way, even in years of active demand the present inequalities in the summer and winter buying of coal render inevitable a long period in which the labor and capital engaged in the industry cannot work more than 27 to 30 hours out of a 48-hour week. Let no one regard this as a condition to be accepted as the measure of working time necessary to meet demand. As a matter of fact the 30-hour week is the spring ailment of the bituminous-coal industry, not its cure.

In addition to the annual and seasonal fluctuations in production, a third set of fluctuations is exemplified in Fig. 3. The railroad works seven days a week; the mines work six days. Over Sunday the carrier catches up in its work of placing cars, and in consequence the car supply on Monday is by far the best of the week. As a result the miners work longest on Monday, but later in the week their hours of labor show a gradual decline, which is accentuated on Saturday by holiday absenteeism. Even if the mines should attain full time on Monday, they could not under the circumstances expect to work more than 86 per cent of the time on Friday and 79 per cent on Saturday.

The data on seasonal fluctuations so far presented apply to the country as a whole. The typical curve of production for the United States is in fact a composite of a large number of other curves, which differ widely from field to field.

A further contrast distinguishes the union from the nonunion fields. Illinois and much of the northern Appalachian region shows, in the even years, a profound drop in April, which marks the biennial wage negotiations. The slump is regularly preceded by a period of active buying, which often makes the March production the highest of the year. This effect is largely absent from the curves of the middle Appalachian region, which includes for the most part nonunion mines.

Another notable fact is the manner in which the fluctuations for different regions tend to neutralize one another. The demand in the regions served by the Chesapeake & Ohio and Norfolk & Western Railroads, for example, starts upward in February and March, at the very season when the Illinois demand has begun its dive downward. Combining the two gives a much flatter curve than either exhibits alone. The average for the country thus does not at all reveal the full extent of the disease from which the industry suffers. We have merely charted the average temperature of a number of patients in which the severe chill of one patient is offset by the high fever of another. This all-country curve seems discouraging enough, but it is a fact which conceals much local trouble, and any remedy must be applied in terms of local, not national, fluctuations.

The shape of the middle Appalachian curve is determined largely by the movement of coal to the Lakes for consumption in the Northwest. The Lake movement has as its limits April 15 and Dec. 1, and this necessity

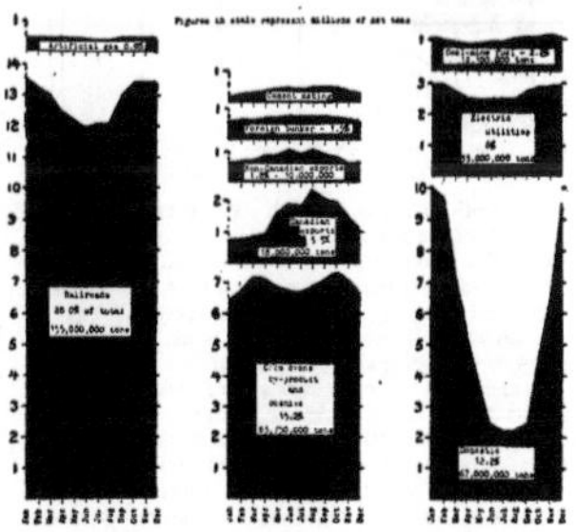

FIG. 4. TYPICAL CURVES OF MONTHLY BITUMINOUS CONSUMPTION

exercises a wholesome influence on working time on Ohio, western Pennsylvania, West Virginia and, to some extent, eastern Kentucky.

The demand for Illinois coal illustrates seasonal fluctuations at its worst. During the "odd" years the demand in the slack month sinks to half what it reaches at the peak, and during the "even" years the production in April approaches zero. The causes of the summer slump are twofold. First the natural market for Illinois coals, as limited by transportation costs, is one in which domestic consumers rely largely on bituminous coal, and of all classes of demand that of domestic consumers fluctuates most with the seasons. In that market, unfortunately, the steadying effects of the Lake and New England movements and of overseas exports do not enter. In the second place, Illinois coals do not store easily, and up to the present time they have not found favor with coke makers and therefore do not feel the steadying influence of the demand for coke.

It is perhaps significant that there is a rough relation between the loss of working time and the degree of unionization. Those bituminous regions in which interruptions to operation are most pronounced show a tendency to become union territory. The presence of the

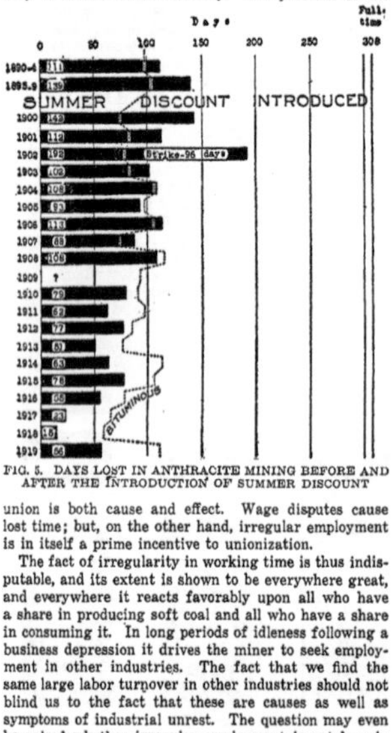

FIG. 5. DAYS LOST IN ANTHRACITE MINING BEFORE AND AFTER THE INTRODUCTION OF SUMMER DISCOUNT

union is both cause and effect. Wage disputes cause lost time; but, on the other hand, irregular employment is in itself a prime incentive to unionization.

The fact of irregularity in working time is thus indisputable, and its extent is shown to be everywhere great, and everywhere it reacts favorably upon all who have a share in producing soft coal and all who have a share in consuming it. In long periods of idleness following a business depression it drives the miner to seek employment in other industries. The fact that we find the same large labor turnover in other industries should not blind us to the fact that these are causes as well as symptoms of industrial unrest. The question may even be raised whether irregular employment is not largely responsible for the failure of coal miners to take full advantage of the opportunity to work when the mines are open.

The case of the miner against irregular operation has already been forcibly set before the public. What is not so generally realized is that the case of the operator is just as damaging to him. His capital is idle, and his mine is raidly depreciating. Although the mine shuts down, his fixed charges run on—not only interest charges and salaries, but a host of maintenance charges as well. And in the end the coal consumer pays the bill for idleness of miner and mine.

The effects of fluctuation in coal production on our transportation system can readily be appreciated. The coal mine is the railroad's largest shipper, and the railroad in turn is the largest consumer of coal; in fact it has been remarked that coal is the nucleus around which our railroad system is built. The railroad suffers from the seasonal fluctuations in coal production as well as the coal-mining industry, for it meets the same seasonal demand. Equipment sufficient to transport all the coal that the mines can produce in November would in large part lie idle during the slack season of summer. The capital investment in coal-carrying equipment alone is of the same order of magnitude as the capital investment in coal mining, and it is no less desirable to provide constant employment for the railroad capital than for the mine capital.

A complete analysis of the effects of the irregular operation of the coal mines would include references to the coal dealers, both wholesale and retail, but the consumer is, of course, the one whose purse suffers most, for in the long run he pays for all the wasteful practices. He must support the miner for not only the 215 days that coal is coming from the mines, but also for the 93 possible working days when the mines are closed.

An attempt to study the degree to which different consumers affect the seasonal market yielded the group of consumption curves given in Fig. 4; the values in this diagram are only tentative and are subject to sweeping revisions. The seasonal fluctuation in locomotive consumption is seen to be a factor, for the railroads consume 28 per cent of all soft coal mined. An even larger offender in degree is the domestic consumer, although the amount involved is much smaller. The general industrial user doubtless ranks next as a contributor to the seasonal fluctuations, but the public utilities present a much more even curve, and the curve for the iron industry is somewhat the same. Fortunately, certain other industries, export and bunker trade and the Lake shipments tend to smooth out the seasonal curve.

The problem of improving the load factor for the coal mines thus becomes a problem of encouraging the summer buying of coal. Two methods of solving this problem have been suggested—seasonal discounts of the coal price and seasonal freight rates. Fig. 5 presents a comparison of lost time in the anthracite and bituminous mines since the summer discount was introduced for anthracite. It will be noticed that the present advantage of the anthracite mines did not immediately follow the new system—the consumer has to be educated even to serve his own interests. A seasonal discount in freight rate seems wholly justified by the railroad's interest in its own load factor; it, too, can afford to bid for summer traffic in preference to the more expensive winter haul.

The general interest of our whole nation in bettering the load factor of the soft-coal industry is large. Society cannot view with equantity the spectacle of an excess mine capacity of 150,000,000 to 200,000,000 tons, and an excess labor force of perhaps 150,000 men.

High wages, due to the war, and steadily increasing demands for labor which cannot be supplied are the two outstanding reasons for soaring prices according to Calvert Townly, President of the A. I. E. E. at a recent address before a meeting of the Schenectady section of the Institute.

As soon as European indusries begin to function properly and the demands of commerce again become normal, American will cease to supply a large share of the world with necessities, the labor situation will adjust itself and prices will drop.

Turbine-Driven *versus* Engine-Driven Alternators—II

By S. H. Mortensen*

Features of stator construction and generator operation are compared, covering insulation, operating temperatures, voltage regulation, synchronizing, parallel operation and other operating features.

THE stator of an engine-type alternator consists of the laminated iron core, the stator coils and the supporting structure, which is called the yoke. The core is built from a number of thin iron stampings or punchings located inside the yoke and carrying the stator coils in slots in their inner cylindrical surface. Fig. 1 shows the longitudinal section of an engine-type alternator. The stator-core punchings *R* are dovetailed into grooves in the yoke *Y* and are securely held between clamping plates *C*. The core is subdivided by means of spacers or ventilating segments, which allow the air currents set up by the rotation of the rotor to pass through the core iron and escape through holes *H* in the yoke. In addition to carrying the weight of the stator punchings and coils, the yoke has to have sufficient stiffness to resist the magnetic pull to which it becomes subjected when the rotor gets out of the bore center. The stator coils are held in the slots by means of wedges fitted into grooves in the core iron. The projecting parts of the coils are protected against mechanical damage by means of shields bolted to the yoke casting. Openings in the shields permit the ventilating air to escape after sweeping over the coil ends. The free ends *E* of the stator coils seldom extend far enough beyond the core iron on machines of this type to require a special support, and in extreme cases it is generally sufficient to interlace the coil ends to enable them to withstand the shocks due to short-circuits on the armature or chafing due to vibration.

In engine-type alternators the coils are protected generally by A. I. E. E. class "A" insulation made from cellulose materials, such as treated cotton tape, paper and similar materials. This insulation will withstand continuously a temperature of 105 deg. C. (221 deg. F.) without losing its insulating qualities or becoming brittle. It is to be distinguished from class "B" insulation made from materials such as micanite, asbestos, etc., which can be operated continuously at temperatures from 125 to 150 deg. C. (257 to 302 deg. F.) without deteriorating.

Experience has shown that the durability of the insulation is not increased by operating it at temperatures below the maximum permissible values already stated, but if the temperature in any part of the coil exceeds 105 deg. C. (221 deg. F.) for class "A" insulation, it will carbonize rapidly, become brittle and suffer permanent damage, the extent of the deterioration depending upon the length of time and the excess of temperature to which the insulation is subjected.

As the life of a machine depends upon its maximum operating temperature, it is important to determine this temperature with as great accuracy as possible. A number of methods are in use, but for a machine whose diameter is large as compared with its length and whose parts are accessible, the thermometer method is generally applied. Such measurements, however, give only approximate results, as the maximum coil temperature is found inside the insulation, which, of course, is inaccessible. To offset this discrepancy, it is customary to add a corrective factor to the observable temperature, which approximates the difference between the temperature measured on the outside of the coil and the temperature on the inside of the coil. According to the rules of the Standardization Committee of the A. I. E. E., this factor is 15 deg. C., or 27 deg. F., if the temperature is determined by thermometer. With 105 deg. C. (221 deg. F.) as a maximum safe operating temperature and a corrective factor of 15 deg. C., the maximum observable temperature of the stator coil is 105 — 15 = 90 deg. C., or 194 deg. F.

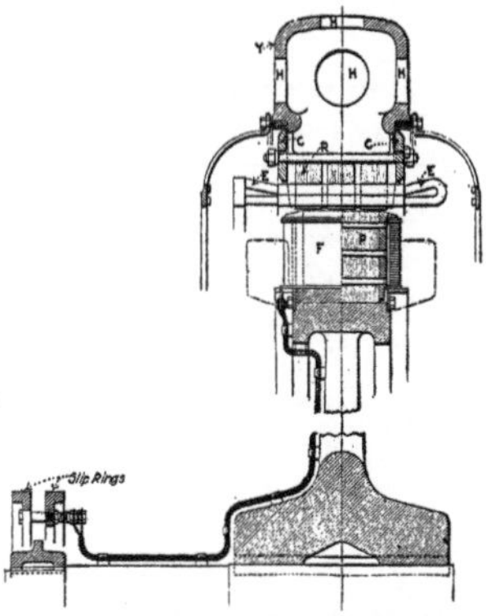

FIG. 1. SECTION THROUGH ENGINE-TYPE ALTERNATOR

*Electrical Engineer, Allis-Chalmers Manufacturing Company.

As a standard for the air temperature upon which the rating of electrical machines is based, the American Institute of Electrical Engineers prescribes 40 deg. C. (105 deg. F.), and this then limits the maximum temperature rise for a class "A" insulated coil to 50 deg. C., or 90 deg. F. This rule does not apply to field coils wound of copper strip, such as were shown in Fig. 4 of

FIG. 2. STATOR YOKE OF TURBO-ALTERNATOR

the preceding article dealing with the rotor, Mar. 2 issue. For this kind of coil, where the thermometer can be placed directly on the copper, the corrective factor, according to the rules of American Institute of Electrical Engineers, is 5 deg. C. This permits a maximum observable temperature of 100 deg. C. (212 deg. F.) or a temperature rise of 60 deg. C., or 108 deg. F., based on 40-deg. C. (104 deg. F.) air, for a class "A" insulated field coil. It is frequently more convenient to determine the temperature of the field winding by its rise in resistance, and to the value thus obtained a corrective factor of 10 deg. C., or 18 deg. F., is to be added to attain the maximum coil temperature. Class "B" insulation is seldom applied on engine-type alternators.

On a turbo-alternator the stator has to meet the same requirements as the stator of an engine-driven machine, but on account of its crowded construction and great width, additional problems have to be met in its design. Figs. 2, 3 and 4 show a turbo-stator in three different stages of completion, and Fig. 5, the finished stator with its end covers. To insure a tight stator core, it is necessary to clamp the core punchings together under great pressure, and the clamping plates and clamping bolts are correspondingly heavy, as may be seen from Fig. 3. Special provisions have to be made to rigidly support the projecting ends of the stator coils so that they will withstand the heavy shocks to which they are subjected should the generator become short-circuited or be thrown in parallel with other machines when out of phase with them. In either case the value of the short-circuit current during the first few cycles may reach from 10 to 40 times the rated generator current, depending on the characteristics of the machine. The duration of this heavy current is instantaneous, and it recedes, within two to three seconds after the short-circuit, to a value that is fixed by the excitation of the machine when the short-circuit occurs. This latter value is generally from 1.2 to 3 times the rated current of the machine.

The peak value of the current is of too short duration to affect the coils by heating, but the magnetic flux that accompanies the current develops attraction or repulsion between the stator coils in the different phases sufficient to twist and bend the parts of the coils not rigidly supported, which means the parts of the coils that are not embedded in the stator slot. For that reason, special provision must be made for bracing and supporting thoroughly the coil ends. In Fig. 4 one method of coil support is shown. Instead of using the bracket support shown, the coil ends may be laced to one or more supporting rings. Relays are not practicable for protecting a machine against the shocks due to short-circuit because they do not act quickly enough. The problem of coil support is particularly difficult in the construction of large 25-cycle machines, and has frequently led to the adoption of reactance coils connected in series with the generator terminals. The function of the reactance coils is to choke down the instantaneous short-circuit current and thereby reduce the shock to which the coils and the machine are subjected.

The electrical operation of engine-driven alternators and turbo-alternators is quite similar. The voltage regulation is accomplished by varying the field strength, either by changing the setting of the main alternator's field rheostat or by altering the exciter voltage. As the energy dissipated in the field rheostat is a total loss, the voltage variation should, wherever possible, be accomplished by changing the exciter voltage.

Of engine-type alternators the voltage regulation, as a rule, is better than with the turbo-alternator. However, the introduction of automatic voltage regulators

FIG. 3. STATOR YOKE OF TURBO-ALTERNATOR WITH CORE PUNCHINGS CLAMPED IN PLACE

has made this feature one of minor importance. At the present time machines whose voltage regulation from full load to no load is from 40 to 60 per cent are in successful operation.

In synchronizing, the same requirements regarding voltage, frequency and phase relationship are to be met by either type of machine. It is, however, much easier to synchronize turbo-generators with each other or with engine-driven alternators than it is to synchronize engine-type machines with each other. This is due to the uniform speed of the turbo-generator, which permits periods ranging from 1 to 15 seconds for synchronizing. After the machines are operating in parallel, the most noticeable difference is due to the speed regulation of the respective prime movers. As the steam-turbine regulator can be adjusted to give much closer regulation, the alternator endeavors to take all the load fluctuations on the system. The engine-driven machines will carry a steady load as long as the fluctuations do not exceed the capacity of the turbo-unit; in other words, the turbo-generator functions as a flywheel or a storage battery.

The uniform speed of the turbo-alternator makes it particularly suitable for supplying power to rotary converters or synchronous motors, and the trouble of hunting frequently experienced when machines of these types are supplied with power from engine-driven alternators is practically unknown where turbo-generators are used. The close speed regulation, together with the stabilizing effect of the turbo-unit, makes it a desirable addition to power systems with heavily fluctuating loads where the power is developed by reciprocating engine-driven alternators. On such systems, during the periods of light load the turbo-unit can be adjusted to operate lightly, and during the peaks, with the steam valves fully opened, it will be able to meet the additional power requirements which otherwise would cause appreciable drop in frequency and voltage and interfere with the operation of the synchronous machines on the system. If close speed regulation is not desirable, the regulator on the steam turbine of course can be adjusted for coarser governing.

Between alternators in parallel the load division in kilowatts depends on the power input of the prime movers, but the distribution of the wattless current required by the system between the different machines

FIG. 4. BRACKET-TYPE COIL BRACING IN TURBO-ALTERNATOR

is controlled by the field excitation, which is generally adjusted to give each machine its proper share of the wattless current in addition to its kilowatt load. Under special circumstances it may be desirable to let certain of the units carry the kilowatt load and then supply the wattless current from one or more units by overexciting their alternators. The turbo-driven alternator is particularly suitable for this class of service. It is started and synchronized in the usual manner, but after it is in parallel its excitation is increased until its current is limited either through the rotor or stator heating. At the same time the steam admitted to the turbine can be reduced until the amount is just sufficient to carry off the heat that is generated by the friction between the turbine blading and the surrounding steam. If no provisions are made for this purpose, the blading will overheat and suffer permanent damage and finally break off. If the amount of steam admitted is not sufficient to supply the losses in the unit, the generator will draw the balance of the power required from the system, and operate as an overexcited synchronous motor (synchronous condenser). The power required to drive a

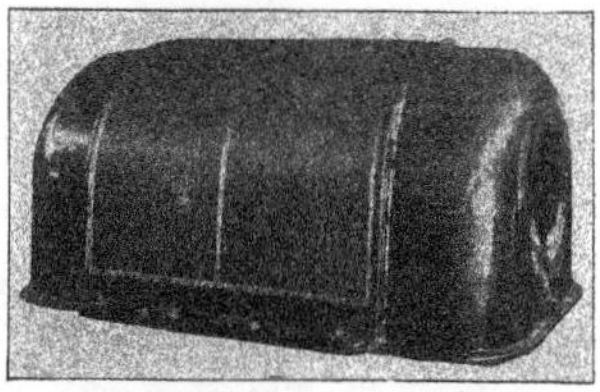

FIG. 5. STATOR OF TURBO-ALTERNATOR COMPLETE WITH END COVERS IN PLACE

turbo-alternator under this condition is small, and as its characteristics are such as to exclude the danger of hunting, this method of operation can be counted upon to give satisfactory results. Engine-driven alternators are less suitable for power-factor correction, as they require a considerable amount of power and are likely to hunt and cause disturbances.

In the foregoing a few of the salient points in construction and operation of engine-type and turbo-alternators have been covered. In conclusion it can be said that the operation, higher efficiency, small space required, etc., of the turbo-unit gives this type of machine an advantage so decided as to lead to its replacing the reciprocating unit except in the special cases where cheap gas or oil warrants the installation of the engine-type alternator.

In the course of the excavations for the new Welland Ship Canal, near Thorold, in Southern Ontario, the site of an immense prehistoric waterfall has been uncovered. The Chicago *Evening Post* says that the edge of the fall has been disclosed for a distance of 400 ft., but the depth has not been ascertained. Above the main fall there are indications of a series of lesser falls, with a total drop of 25 ft., which are plainly water-worn. The site is about half a mile from the present Falls of Niagara, and it is believed that the deep canyon-like valley through which the ship canal passes, where Eight Mile Creek once meandered on its way to Lake Ontario, must have been the bed of the prehistoric river which furnished the waters of the giant falls with their outlet to the sea.

There is a gold mine in Ouray County, Colorado, where the whole of its plant, including everything from the crushing mill to the settling tanks, is built in underground chambers. This is because of the severity of the snow slides, which wreck all buildings above the ground level.

An Advocate of the Junior Technical School

Arthur L. Williston, Who Has Devoted Himself to the Training of The Non-commissioned Officers of Industry

DURING the past half century, a great deal has been said and written about technical engineering education. Beginning with a few pioneers, the idea of technical training has swept over the whole country, until, now, a college or university seems incomplete if it does not have one, or more, engineering courses. No one now questions the fact that the technical men, trained in the engineering schools and universities, have played an important and essential part in the world's industrial progress. There is another field, however, which has attracted comparatively little attention, but which at last seems to be coming into its own. This is the field of technical education as represented by the trade schools, or those schools which do not go into the highly specialized technical work as is done in the colleges and higher schools of technology.

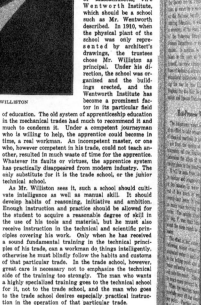

ARTHUR L. WILLISTON

One of the men most prominently identified with this work is Arthur L. Williston, principal of the Wentworth Institute in Boston. Mr. Williston is a real enthusiast when it comes to this type of industrial education. He was born October 18, 1868, in Cambridge, Mass., and comes from a family of educators, both his father and grandfather being engaged in this work. The latter, Mr. Samuel Williston, was the Founder of the Williston Seminary at Easthampton, Massachusetts.

After graduating from the Massachusetts Institute of Technology in 1889, Mr. Williston spent a year at graduate work in the Institute, and then worked on railroad construction, in Oregon and Washington, for the Union Pacific Railway. In 1891, he had his first experience in teaching as an instructor in steam engineering at Technology, but the next year we find him back in railroad work, in charge of bridges over about 6,000 miles of road for the Big Four Railway.

In 1893, he came back to Boston as a mechanical engineer with Lockwood, Green Company, but that fall went to Columbus, Ohio, and became Professor of Mechanical Engineering at the Ohio State University. While here, Professor Williston was also secretary of the Engineering Faculty of the University. After five years in Ohio, he was made director of the School of Science and Industry at Pratt Institute, Brooklyn, New York. Here began, what Mr. Williston describes as his life work, "in the pioneer field of creating a new type of technical education."

Arioch Wentworth, a citizen of Boston, expressed in his will an aim to establish a school "for the purpose of furnishing education in the mechanical arts." As a result, there was incorporated, under the laws of Massachusetts, the Wentworth Institute, which should be a school such as Mr. Wentworth described. In 1910, when the physical plant of the school was only represented by architect's drawings, the trustees chose Mr. Williston as principal. Under his direction, the school was organized and the buildings erected, and the Wentworth Institute has become a prominent factor in its particular field of education. The old system of apprenticeship education in the mechanical trades had much to recommend it and much to condemn it. Under a competent journeyman who is willing to help, the apprentice could become in time, a real workman. An incompetent master, or one who, however competent in his trade, could not teach another, resulted in much waste of time for the apprentice. Whatever its faults or virtues, the apprentice system has practically disappeared from modern industry. The only substitute for it is the trade school, or the junior technical school.

As Mr. Williston sees it, such a school should cultivate intelligence as well as manual skill. It should develop habits of reasoning, initiative and ambition. Enough instruction and practice should be allowed for the student to acquire a reasonable degree of skill in the use of his tools and material, but he must also receive instruction in the technical and scientific principles covering his work. Only when he has received a sound fundamental training in the technical principles of his trade, can a workman do things intelligently, otherwise he must blindly follow the habits and customs of that particular trade. In the trade school, however, great care is necessary not to emphasize the technical side of the training too strongly. The man who wants a highly specialized training goes to the technical school for it, not to the trade school, and the man who goes to the trade school desires especially practical instruction in the operation of that particular trade.

A system of education based on these principles has been Mr. Williston's dream for a number of years. His nine years of varied engineering experience and his twelve years in the Pratt Institute, in addition to his thorough technical training, gave him an excellent foundation for the practical development of this dream when the opportunity arose. As the active head of the Wentworth Institute from its beginning, Mr. Williston has had an excellent opportunity to convert his ideas into practical results. Whatever measure of success the Institute has achieved, must be credited, to a very large extent, to him. That the school has been a success is testified by the fact that the War Department established one of its schools for engineering troops there during the war. This was not a unit of the students' army training corps, but a school for the special training of a number of the engineering corps for their particular duty.

Enthusiastic as Mr. Williston is over the Wentworth Institute, he finds time for considerable work on the outside in the same direction. He is a member of the American Society of Mechanical Engineers, and the Society for the Promotion of Engineering Education, of which he is secretary and vice-president, and of the National Society for the Promotion of Engineering Education, of which he is a director, and also a secretary of the State branch. He is also president of the Industrial Educators Department and of the Science Department of the National Educational Association. In addition to the active part that he has taken in the work of these societies Mr. Williston, himself, has made surveys and investigations along the lines of industrial education for a number of important educational institutions throughout the country. During the war he was Educational Director for New England for the War Department Committee on Education and Special Training.

Hot-Process Water Softening

The hot-process water softening system is the result of an effort to take advantage of the increased rapidity of chemical reactions in hot water. The curves in Fig. 1 are from the results of tests made in the chemical laboratory of the Harrison Safety Boiler Works to determine the effect of temperature on the reactions taking place in the treatment of boiler feed water.

It will be seen from these curves that the reduction in calcium and magnesium salts is greater at the end of ten minutes in water at 210 deg. F. than at the end of five hours in water at 50 deg. F. In fact, twenty-four hours in cold water did not bring down as much scale forming matter as ten minutes in hot water. The sample which had stood for five hours at 50 deg. F. was then heated to the boiling point with about the same result as ten minutes' treatment in hot water.

The effect of hot water may be shown roughly by the following simple experiment. Place one pint of cold untreated water in each of two beakers, heat one sample to the boiling point and add the theoretical amount of water softening chemical to each sample. Stir thoroughly and note the result. The reaction and sedimentation will be much more prompt in the hot water.

When the reaction seems complete in the cold water, carefully pour off the clear liquid and heat it to the boiling point. Further precipitation will result, showing that the reaction was incomplete in cold water.

Another advantage of the hot process is in the increased velocity with which the particles of solid matter settle to the bottom of the liquid. From Sokes' law, the velocity with which a spherical particle will sink in a liquid is given by the following:

$$V = \frac{2}{9} \times \frac{D - d}{N} \times gr^2$$

where V is the velocity, D the density of the particle, d the density of the liquid, g the acceleration due to gravity, r the radius of the particle, and N the coefficient of viscosity of the liquid.

It is obvious from this formula that the smaller the coefficient of viscosity, the faster the particles will sink to the bottom. The accompanying table gives some

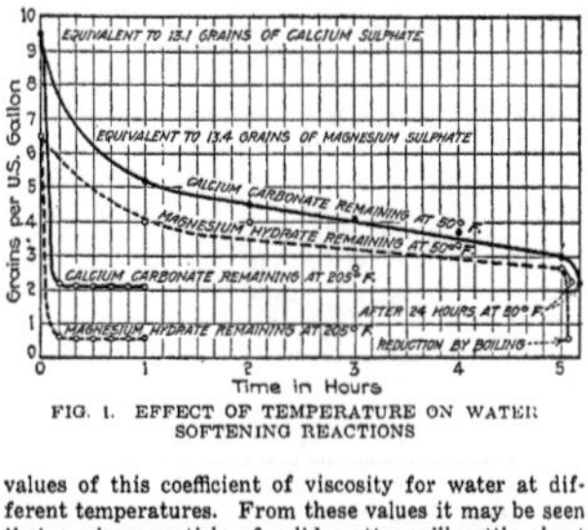

FIG. 1. EFFECT OF TEMPERATURE ON WATER SOFTENING REACTIONS

values of this coefficient of viscosity for water at different temperatures. From these values it may be seen that a given particle of solid matter will settle about four times as fast in water at 212 deg. F as in water at 68 deg. F.

Consider the effect of this fact in a tank of quiet water. A particle of a given size will fall four times as far in hot water during a given time as in cold water. In the case of a continuous water softener the velocity of flow through the sedimentation tank may be four times as much with hot water as with cold water, without carrying particles of a given size to the pump supply. It will be noticed further from the formula that the sinking velocity of a particle varies as the square of its radius. Therefore it is obvious that much smaller particles will have time to settle out in hot than in cold water if we keep the velocity of the water through the settling tank constant.

In any water softening apparatus it is necessary that the chemical reagents be supplied at a rate which is proportional to that at which the raw water enters the system. Many mechanical devices have been used to control the reagent supply. In some of these the raw water flowing to the softener turns a waterwheel or operates a tilting bucket which in turn operates dippers by which the reagents are ladled out. In other softeners a certain portion of the water, divided from the main supply by weirs or orifices, is led through chambers containing the reagents. In another form the water displaces the reagent from a tank, at the same time diluting the remaining reagent, this dilution being compensated for by adjusting the orifices.

The hot-process softener should carry a few pounds pressure so as to insure a temperature of at least 205 deg. F at all times. Under these conditions the methods of proportioning referred to are not entirely satisfactory. One method applicable to the hot-process is to drive a pump, supplying the reagent, by means of the raw water. Of course in this case both the water motor and the pump must be kept in first class condition to avoid poor proportions due to leakage.

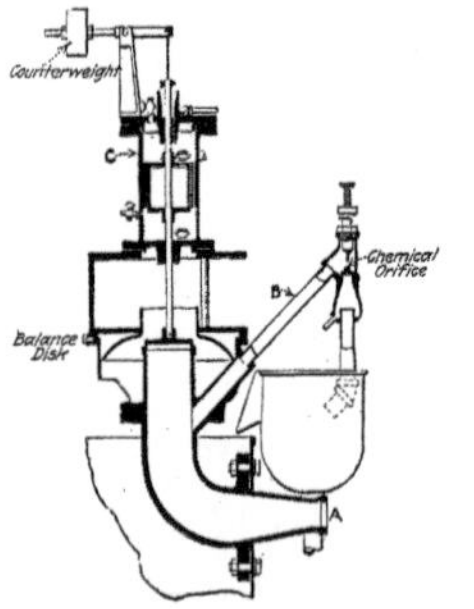

FIG. 2. REAGENT PROPORTIONING DEVICE

Another reagent proportioning device depends upon the difference in pressure on the two sides of an orifice through which the raw water flows. A diagram of this apparatus is shown in Fig. 2. The raw water flows through an orifice (not shown) on its way to the softener. Pipes are led from above and below this orifice to the top and bottom respectively of the cylinder *C*. The differential pressure on the orifice then tends to lower the piston. The chemical reagent enters at *A* under pressure and a part goes up through the pipe *B* to the *chemical orifice*. The remainder of the reagent flows out under the valve or *balance disk* which is held down by the piston referred to above. Thus the pressure on the reagent orifice is always equal to the differential on the raw water orifice and the amount of reagent fed is always proportional to the amount of raw water. The reagent which flows out at the balance disk is returned to the reagent storage tank and pumped into the apparatus again. The chemical passing the orifice is caught in the funnel shown and drains from there to the chemical feed pump which forces it into the softener against whatever pressure is maintained. The counter weight balances the weight of the piston and valve disks.

Feed Water in Contact With Steam

The hot-process apparatus should consist of an open type feed water heater, where the feed water is heated by direct contact with the steam. As the heating takes place the dissolved gases will be given off and then the chemical reagent should be mixed with the water. The water then should move slowly downward to allow settling of the precipitate and be drawn off near the bottom of the tank. By admitting the water at high temperature at the top of the tank and causing a slow downward circulation, no convection currents are present to stir up the mass of precipitate.

Some conditions require the use of sand filters after the chemical treatment. These are simply beds of sand through which the water flows by gravity. These filters are cleaned by reversing the direction of the water flow, "back washing," and at the same time blowing air through orifices embedded in the sand bed.

TABLE BY THORPE AND RODGER SHOWING VARIATION IN VISCOSITY OF WATER WITH TEMPERATURE

Temperature Degrees F.	Co-efficient of Viscosity
32	.01776
50	.01303
6[illegible]	.01002
86	.00798
104	.00654
122	.00548
140	.00468
158	.00406
186	.00356
194	.00316
212	.00283

Gallagher Flexible Staybolt

One of the chief troubles incidental to the use of staybolts between flat surfaces arises from the unequal expansion of the two sheets. One is usually in contact with the fire, and the other is heated to the temperature of the water. This difference in temperature causes considerable relative motion between the two sheets, which frequently results in broken staybolts in a section near the inner surface of the sheet.

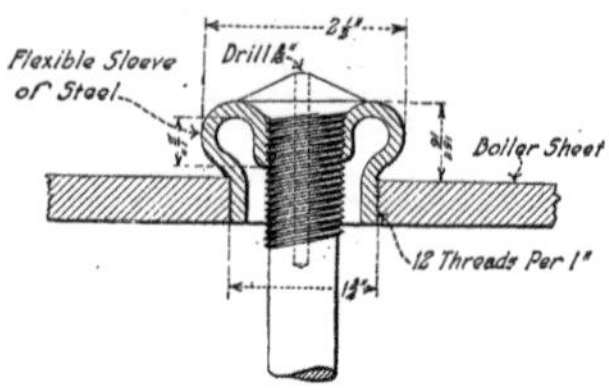

STAYBOLT AND FLEXIBLE SLEEVE APPLIED TO BOILER SHEET

The old non-flexible staybolt is especially susceptible to rupture, and, as a result, several forms of flexible staybolts have been used with success. A recent form of flexible staybolt is the invention of P. F. Gallagher, Clifton, S. I. It consists of a flexible steel sleeve 2½ in. at the largest outside diameter. This sleeve is threaded to screw into a 1¾ in. tapped hole in the boiler sheet, and is also threaded 12 threads per inch to take a 1 in. dia. drilled staybolt at the threaded portion. The end of the staybolt is riveted over as shown in the illustration. The flexible sleeve extends $\frac{15}{16}$ in. beyond the boiler sheet on the outside, and is $\frac{3}{16}$ in. thick. This allows of a staybolt 1 in. longer than could be used if the staybolt were screwed directly into the boiler sheet. The idea of this design of staybolt is to obtain flexibility in the steel sleeve, instead of having the strain, due to unequal expansion of the plates, come directly on the staybolt next to the boiler sheet.

EDITORIALS

Fluctuations in Bituminous Coal Production

IN THIS issue appears an abstract of the paper presented by George Otis Smith, Director of the United States Geological Survey, before the American Institute of Mining and Metallurgical Engineers. It gives statistics from the Survey's data regarding the irregularity of coal production and shows how serious is the situation. When an industry as important to the welfare of the country as this one is shut down an average of 93 out of 308 working days per year, the public ought to know the reasons. If the causes contributing to this condition can be removed, no time should be lost in doing so.

The immediate causes of seasonal fluctuations are car shortage, labor shortage, mine disability and "no market." A study of such figures as are available shows that by far the major portion is credited to car shortage in the winter months and "no market" in the spring months. For example, in April, 1919, the time lost due to the latter cause averaged over forty per cent of full-time operation. This of course results in a condition of car shortage when the demand rises in the fall and winter months. It is interesting to note that during the period from 1910 to 1918 only 10.6 per cent of the time lost in the bituminous mines was due to strikes.

The measure of responsibility of the domestic consumers for the fluctuation is shown very forcibly in the curves of monthly consumption. Although householders take only twelve per cent of the total production, the fluctuation from about two million tons in July to ten million tons in January is more serious than that due to any other class.

The picture Mr. Smith's paper paints is black, but the reality is worse than may appear from a hurried inspection of the data. The number of "lost days" as given represents only days when the mines are closed, days when no coal can be mined whether the miner wants to work or not. No statistics are given to show how completely the men take advantage of the days when the mines are open. That there must be considerable loss from individual absenteeism is evident. Moreover, the seasonal "slump" in production occurs at different times in different fields. Therefore, in many localities the conditions are much worse than the average curves for the whole country show.

It is pointed out that this condition adversely affects the miner, the operator, the railroad, and the consumer. They all suffer, but as always, the consumer has to pay the bill. The miner has to eat 365 days a year, the operator and railroads have to earn a reasonable return on their capital. If miner, operator and railroad could work continuously, they would receive a bigger return and at the same time the consumer would have cheaper coal.

As was pointed out in these columns last week, the solution of the problem must come, for the most part, from the storage of coal during the summer to help meet the extra winter demand. Furthermore, this storage must be at or near the point of consumption.

There have been several objections raised to the plan of summer storage. One is that an enormous amount of capital would be tied up in coal-storage equipment and in the stored coal. It does cost a lot of money to build a coal-storage plant. It is also true that storage of coal means an extra handling. On the other hand, a few shutdowns of any plant due to inability to get coal would more than make up for the cost. That such shutdowns are likely to occur the experiences of the last few years furnish sufficient evidence.

Another objection to the storage of bituminous coal in large quantities is its tendency to spontaneous combustion. It cannot be denied that this danger is a real one with improperly stored coal. It has been proved, however, that bituminous coal can be stored safely and that it will suffer little or no depreciation if reasonable care is used.

A reduced summer price at the mines and a reduced summer freight rate would undoubtedly lead many consumers to buy coal at the slack season and store it. This would divide the burden more evenly among those who would benefit. There has been a feeling that the coal interests welcomed the annual shortage in winter in order to raise prices, and such a differential rate would help materially to remove this friction. One of the diagrams in Mr. Smith's paper shows how successful this plan has been in the anthracite mines.

Whatever the final result of the various plans may be, it is obvious that the storage of coal in summer by the individual consumer will result in benefit to himself. At the same time he will be doing his part toward the solution of a serious national problem.

Economical Supply of Electric Power

SINCE the introduction of the steam turbine as a successful prime mover, advancement in large power-plant design has been so rapid that one scarcely has had time to recover from amazement at some great achievement in power-plant engineering before another of greater magnitude has been accomplished. In the opinion of many the limit in size of steam-turbo-generator units has been nearly reached; if not from structural limitations, from economical considerations. However, now comes the superpower plan, to provide an economical supply of electric power for the industries and the railroads of the northeast Atlantic seaboard.

This plan, the most stupendous conception in power generation and transmission ever conceived, and some of the engineering problems it involves, were presented at the midwinter convention of the American Institute of Electrical Engineers, in the form of a symposium, by ten of the leading consulting, designing and power-plant engineers and discussed by a number of others. An abstract of this symposium and discussion appears in this issue. The plan as proposed suggests an electric generation and transmission system running from Boston to Washington, D. C., from which power will be supplied to factories and railroads wherever economically possible in a zone extending from one hundred to one hundred and fifty miles inland from the coast. It is estimated that such a scheme would supply the necessary power from a plant capacity of five and a half million horsepower which at the present time requires seventeen million—ten million for the industries and seven million for the railroads. It is predicted that the load factor, at present not exceeding fifteen per cent, would be raised to more than fifty per cent. As a result there would be an estimated yearly saving of three hundred million dollars and thirty million tons of coal, besides relieving the intolerable congestion of the railroads in this zone. It will also make possible the economical development of many of the water powers within this zone that cannot be utilized economically under the present system. Whether this plan or some other is the solution of the power and transportation problem along our northeastern seaboard, it is apparent that a very serious condition exists in that zone as was emphasized by harassing experiences during the war.

The first logical step in such an undertaking is a complete survey by a representative commission of engineers. An appropriation by Congress to make the necessary investigation under the direction of the Department of the Interior, as has been requested, would be a step far reaching in the solution. There is little doubt that this system will consist of a tieing together of the large generating systems in the large industrial centers reinforced by water-power plants on the rivers and steam-power plants at the mines. Consequently, before anything constructive can be done it must be definitely known just what the essential features of this development will be and if the scheme suggested is economically correct. If this system is going to operate at sixty cycles and it becomes definitely known that this is to be the standard frequency, then it will be a matter for the various power companies in this zone standardizing as far as economically possible on this frequency, so that when the time comes for tieing all the large systems within this zone together, one of the most important features in regard to interconnection will be in a fair way to being solved. Prof. Malcom Maclaren, in discussing this problem, said:

"The picture, which might be formed of a general scheme to be followed during the next ten years to meet the growing need for power within this territory, is that the power companies in the large cities should in general complete their present construction programs but should not greatly increase their generating facilities beyond this point, additional capacity being obtained by the construction of steam plants at the mines, reinforced on the east by hydro-electric plants on the Delaware River and on the west by hydro-electric plants on the Susquehanna, the whole territory being interconnected with transmission circuits. Such a system should procure a cheap and reliable source of power under normal conditions and would have the greatest flexibility for meeting emergencies, for the generating facilities would be divided in approximately equal amounts among steam plants at the large cities burning bituminous coal, steam plants at the mines burning anthracite and hydroelectric plants on the Delaware and Susquehanna Rivers; the unified system of transmission circuits making it possible to transfer larger blocks of power to any point in distress without the use of rail transportation."

The foregoing apparently is one of the most logical schemes, since on account of certain physical problems there is little doubt that a large percentage of the power generated from coal in this zone must be produced locally near the large industrial centers, as at present.

Oil Land Leasing Bill a Law

PRESIDENT WILSON has signed the bill releasing for development millions of acres of Government oil, coal and phosphate lands in the West. California and Wyoming lead in the extent and richness of their deposits, although Louisiana, Arizona, Utah, Wyoming, Colorado and Alaska are represented in the oil field. In coal lands North Dakota leads the other twelve States with more than eleven million acres.

The bill provides for a royalty rental on the fuel produced, and in the case of oil this may be demanded in oil or in cash as the Government elects. This year, at least, the Government is expected to collect its royalties in oil so as to assure a supply for the Shipping Board at fair prices. The naval reserve oil lands are not to be opened unless the President shall so prescribe. That these lands will be promptly developed is evident from further reports that there were "rushes" to file location claims in a number of the fields. As most of these oil lands are in the fields producing fuel oils, this development will be of interest to all fuel-oil consumers. There has been a feeling that the largely increased use of oil instead of coal would result in a prohibitive increase in the price of the former. The signing of this bill may at least help to postpone the evil day.

In this connection, however, it is well to note that the country's supply of petroleum is not inexhaustible. Fuel oil offers many advantages for the power plant, but it is little short of criminal to waste it. Conservation of our natural fuel supply is a vital necessity, and conservation of our limited oil supply is especially important.

Referring to the proposed cut in appropriations for promoting foreign trade, from $1,658,000 to $490,000, Secretary Alexander of the Department of Commerce says, "This is the most serious blow ever aimed from within at our foreign trade." The sum allowed is only one-half of the appropriation now used. This cut in funds will abolish entirely the commercial attachés of embassies in a dozen different countries, will greatly reduce the number of trade commissioners who are reporting on rapidly changing commercial conditions in Europe and will close up almost all of the district offices in this country. The point considered most vital is the loss of the commercial attaché system, which has been so highly commended by many business men with a knowledge of foreign trade. In short, at a time when other countries are bending every effort to establish their foreign trade, this bill would cripple the machinery by which the United States keeps in touch with the commerce of the world. In the name of economy we are to smash our carefully built up, smooth-running, trade-promotion machine and then spend many times as much money to put it in running order again when the error of shutting it down has been demonstrated. And it will be demonstrated in a very few months.

CORRESPONDENCE

Frozen Boiler Connections

In reference to the boiler explosions published on page 175 of the Feb. 3rd issue, I would like to say that ice forming conditions are altogether too prevalent in some boiler rooms. In spite of all the safety precautions demanded by law and although there are careful well-trained boiler-room employees, there are thousands of plants that are exposed to the cold, due doubtless to the close-fisted attitude maintained by the owners.

I recently worked at a plant where the boiler blow-off pipe would freeze up during the day's run, between the valve and the cock, if both were closed and no blowing out was done oftener than once in 10 hr. in real cold weather. It is not uncommon for a steam gage connection to freeze and the gage register a false pressure. J. FOREST.

Norwood, Ohio.

Compressed Safety-Valve Springs

After reading the account of the Interstate Iron and Steel Company's explosion, page 175 of the Feb. 3 issue, and also the editorial in the same issue, it seems quite possible that the accident was due to the defective condition of the safety valves.

As suggestions are in order to prevent future accidents, the following may be of value to operators and owners of plants. I have had several years' experience in the inspection of boilers and have found several cases where safety valve springs, especially old ones, were so compressed that the normal lift was so reduced that even though a valve would open at the pressure at which it was set it would not discharge anything near the amount of steam generated in case of a shutdown of the engine. I have known of cases where the pressure increased from 75 to 100 lb. while the valves were blowing, and if in such a case the steam gage happened to be defective trouble would very likely result, possibly a boiler explosion.

This condition of safety valves has received very little attention from boiler operators or others, but I feel that it is worthy of due consideration. If boiler attendants would occasionally take the time to shut the stop valve while the boiler is run to full capacity and beyond, they could readily determine the capacity of the safety valves and if deficient a new spring or a larger valve could be provided.

Frozen steam gages and their piping are often met with during the winter weather. This is a source of danger particularly to the class of men who are ignorant or careless in their work. There is a remedy for this condition, which is inexpensive and easily carried out. If the piping and gage are thoroughly drained and then filled with a reliable freeze proof mixture such as can be purchased, there will be no occasion to drain the piping for the season and freezing will not occur. There may be cases where an alcohol mixture or kerosene will serve but the first mentioned seems to be better suited for such a purpose. JAMES COYNE.

Pittsburgh, Penn.

Air in Steam Boilers

The suggestion is made in the article relating to the boiler explosion at East Chicago that it was due to overpressure because the safety valves failed to function, due to ice that clogged the vent pipe, in spite of the fact that there had been a fire under the boiler for seven hours. The assumption is also made that air was trapped in the boiler when it was filled and that it acted as an insulator, keeping the steam in the boiler from reaching the top of the drum and thus thawing out the safety valve.

My experience tends to disprove this assumption. When getting steam up on one of eight 400-hp. boilers a $\frac{3}{4}$-in. bleeder valve, located under the stop valve was left wide open to allow the air to escape. On several occasions, when some one of the operating crew had shut this bleeder when the pressure reached 50 lb., because of the roar of the escaping steam, we would lose vacuum on both condensers, when the boiler was cut in on the line. Even though the bleeder was left cracked until the boiler went in, a noticeable drop in vacuum always occurred so that it became customary to speed up the vacuum pumps whenever a boiler was to be cut in.

If stratification takes place in the drum the air would collect in the nozzle under the stop valve and escape through the bleeder. The fact that air is present in such quantities in the boiler as to interfere with the condensers, in spite of the bleeding, proves that the air and steam intermingle, doubtless due to a counter-current being set up in the steam and gas by the circulation of the water in the drum.

In this particular case the explosion may have been caused by a frozen safety-valve vent pipe, the ice remained despite the heating from the steam, and not because air was present which prevented the steam from reaching and thawing the valves.

Perth Amboy, N. J. ADOLPH QUADT.

Frozen Safety Valve Discharge Pipe

The account of the boiler explosion at the Interstate Co.'s plant in the Feb. 3 issue brings to mind an experience I had with the same make of boiler, but used as a waste-heat boiler over a furnace and for heating material for steel forgings. The furnaces were started early one Monday morning so as to have a heat ready for the forgers when they came at 7 o'clock.

The boiler before that time furnished steam for a boiler-feed pump and for a small engine running a forced-draft fan for the furnaces. Shortly after 7 o'clock I went to the forge shop and the engineer called my attention to the steam gage which registered at 135 lb. The spring pop safety valve set at 125 lb. was not blowing. The hammers were at that time running, taking some steam.

I went on top of the boiler and found that the valve and the pipe leading from the valve to the outside through the roof were frozen. I worked the lever and hammered the pipe until the steam began to leak through the valve a little and finally blew. The engineer who was watching the end of the pipe from the outside said that a piece of ice shot out of the pipe for a distance of about 100 yd, out on the marsh. I think the pipe became full of snow and sleet on Sunday, while the furnace fires were banked, due to a sleet storm that blew right into the mouth of the pipe and froze solid.

I believe that every safety valve that blows outside of a building through a pipe, should have an elbow opened downwards on the end of pipe. This will prevent snow and sleet from getting into the end of the pipe and freezing. Be sure that there is a drip hole in the body of safety valve, above the seat that will drain any leakage from the valve and that the valve above the seat will not then be covered with water and possibly freeze.

Chester, Penn. FRANK MCLEAN.

National Engineer's License Law

I have read with much interest the letter by Mr. Reed—"National Engineer's License Law"—on page 234 of the Feb. 10 issue. Having been an engineer in Massachusetts when the license law was enacted, I feel that I should know something about it. It was an up-hill undertaking to get the wheels in motion that finally drafted a law, and after it became effective the engineers in general were more or less dissatisfied. It would be the same thing if a National engineer's license law were enacted.

There were engineers in Massachusetts who had been operating plants of perhaps 160 hp. for several years. A first-class engineer's license permitted the holder to have charge of an unlimited amount of horsepower; a second-class engineer's license permitted the holder to have charge only up to 150 hp. A clause was inserted in the law that provided for the engineer who was unable to pass an examination.

Operating a plant as chief for five years entitled the said operator to a license (without examination) sufficient to cover his plant. Therefore many an engineer operating a plant of slightly over 150 hp. for five years was given a first-class certificate. The same thing applied to the lower grade licenses and many a man unable to read or write was granted a license. The licensing of engineers and firemen is simply a "safety first" affair and is not (at least legally) an issue to raise power workers' wages.

At first the point was raised that it was the boiler that was dangerous, and only this piece of apparatus need be covered by the law, but eventually the majority ruled and engines were covered also.

If a National license law is enacted the illiterate will have to be taken care of in other states as they were in Massachusetts, so for a period of years no apparent results will obtain, but eventually great good will probably come from it. C. W. PETERS.

New York City.

Regulating the Oil Pressure

The step bearings of two vertical turbo-generators were supplied with oil at a pressure of 500 lb., the supply coming from a line under 750 lb. pressure through a reducing valve. An 18-ton dead-weight accumulator connected in with the oiling system supplied

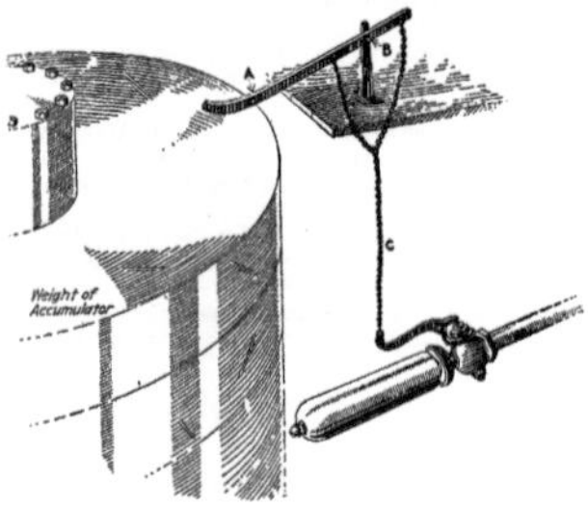

ACCUMULATOR TRAVEL ALARM WHISTLE

oil at the required pressure for a few minutes in case the pump in service became airbound, etc., to allow time to put a spare pump into operation.

The oil pressure was regulated by the height of the accumulator, a ¼-in. cable passing over two small pulleys from the accumulator to a weighted lever valve in the steam line to the pump. The arrangement was not satisfactory as the pumps were kept running with a jerky motion due to the action of the valve lever, which caused the accumulator to bob up and down incessantly with the consequent packing wear and leakage around the accumulator shaft.

To do away with this trouble, a globe valve was placed in the steam line with a 5-in. pulley fastened to the stem, the face of the pulley being made wide enough to allow several turns of the cable. The free end of the cable was weighted just enough to close the valve. The pump action was steadied, but the valve stem threads wore quickly. Upon substituting one of steel the arrangement proved satisfactory.

In case the accumulator rose or fell too far beyond a certain position, an alarm, as shown in the figure, would be put into operation. The curved lever *A* was pivoted at *B*, the curved end resting on the top weight of the accumulator, the alarm whistle being operated by the chain *C*. ARNOLD ERICKSON.

Chicago, Ill.

Emergency Repair to Lubricator Plug

The threads on the filling plug of a lubricator became so badly worn that the plug would not hold owing to countless removals and insertions which had blunted the threads. We were some distance from town, and with no repair shop to speak of, it was a problem to get oil into the engine cylinder.

A hole was drilled in the end of the plug to a distance a little beyond the threads, leaving a fair amount of stock. A taper pin was filed and driven into the hole to expand the lower end of the plug, thus enlarging the threaded portion. After being expanded until it was a snug fit in the threads in the body, the pin was driven in and the end of the plug peened slightly to hold it in place. Although the threads were worn, the enlarged plug worked very well for some months until the engine was no longer needed. Then the plug was sent to a machine shop for proper treatment. ARNOLD W. BENTLEY, JR.

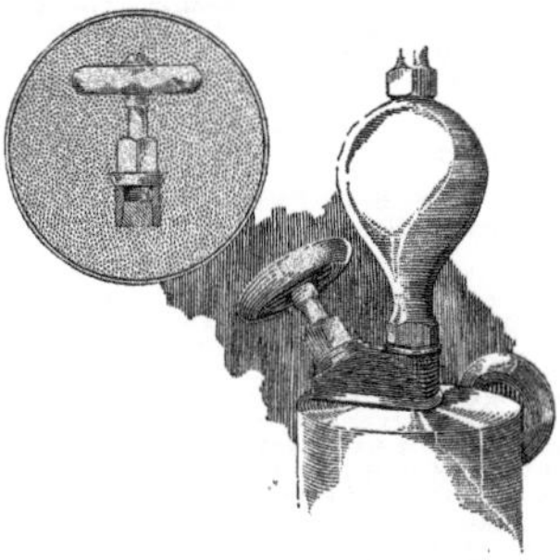

PLUGGED THE LUBRICATOR FILLING PLUG

Missouri Valley, Iowa.

How the Boiler Blowoff Water Was Utilized

In our plant water does not cost much, but as we have to blow down each boiler twice each 8-hour watch, it is evident that a considerable amount of water is wasted, but what is more serious, it is hot water and it costs money to heat it.

We have a hot pond that has to be kept full of hot water for softening the bark on pulp wood so that the bark can be more readily removed and also to thaw the frozen wood which makes it easier for the chippers. It takes about all a 2-in. steam line can deliver to keep this pond hot.

We piped the blow-off from the boilers into this pond and find that very little live steam is required to keep it hot in the winter and in the summer it does not require any live steam at all.

Van Buren, Me. WILLIAM B. SMITH.

Oil Indicator Gage

Recently the firm by which I am employed installed oil-burning equipment for their boilers. We desired a gage to show the quantity of oil in the tank which is below the ground, and for this purpose adapted a water-level indicating gage similar to those used for showing the elevation of water in reservoirs, etc. The illustration shows the resulting installation.

The gage has a 5-in. dial furnished in black with white enamelled figures and letters and sets on a 5-in. wrought-iron pipe which contains a float resting on the surface of the oil in the tank below. The small pipe at the right guides the counterweight. The dial may be graduated directly in barrels, gallons, or pounds; in the instrument illustrated each division represents 50 gallons. The amount of oil put into or taken out of the tank is readily determined by observing the amount of movement of the dial hand which also shows the amount of oil in the tank.

The measurement of fuel oil in power plants with the usual types of meters presents certain difficulties which often make its measurement impracticable. The method described herein obviously has great advantages over the old method of measuring black, sticky oil with a measuring rod.

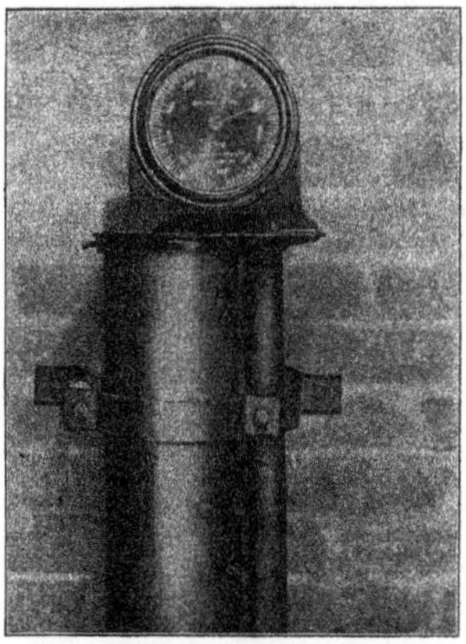

OIL GAGE, EACH GRADUATION REPRESENTING 50 GALLONS

If the float pipe cannot be placed directly over the tank it may be possible to connect the two at the bottom by a small pipe and thus the gage can be located at considerable distance away. If it is not possible to change the location of the float pipe the gage may be located at some distance from it by leading the float cord over special idler pulleys.

Providence, R. I. A. E. WATJEN.

Reinsulating Series Winding of Compound-Wound Motors

Where it is necessary to rewind the series coils of medium- or large-sized compound-wound machines, it may be found difficult to obtain the proper size of wire. In such cases the coils can generally be satisfactorily reinsulated.

First the old insulation must be entirely removed from the copper, which is easily done by subjecting the coil to fire and burning it off. This also anneals the copper and makes it easy to handle. Next the coil is placed on a long piece of pipe that is stiff enough to support it. Then the turns of the coil are very carefully separated, as shown in the figure, after which they can be conveniently insulated with linen tape and the turns may be reassembled as in the original coil. A. A. FREDERICKS.

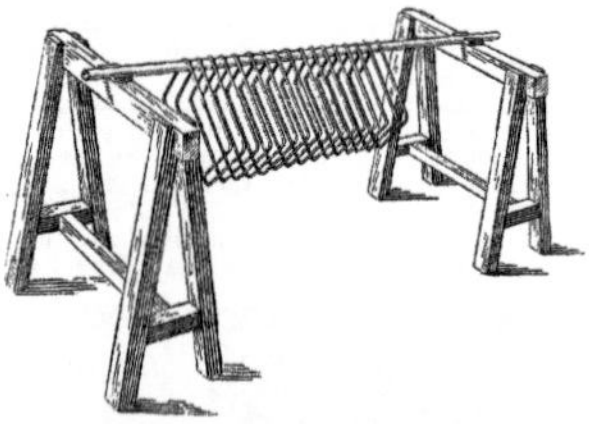

SERIES FIELD COIL SEPARATED READY FOR REINSULATING

New York City.

Operating Troubles With Steam Turbines

I have followed with interest the many articles in *Power* in regard to operating troubles with the larger sizes of steam turbines, and in particular the recent discussion which seems to indicate that there is more or less of a tendency for the large low-stage wheels to flutter on the rim. While I have never been in personal contact with these machines, this particular point in their construction has interested me.

When a young man I worked for several years around large circular saws in lumber mills, and while I have seen saws that ran at a rim velocity of 300 ft. per sec., anyone who suggested that they would stand speeds much in excess of that would have been considered crazy.

The important point about these high-speed saws was the amount of strain or tension, as the filer would name it, that was required to keep them from fluttering on the rim. The saw plate was stretched with a hand hammer, on a small anvil, for about one-half the distance between the eye and the rim until the saw was dished like a saucer when standing still. I have seen this dish amount to as much as 4 in. in a saw 5 ft. in diameter, yet when it came up to speed it was straight, the centrifugal force stretching the rim enough to equalize the initial looseness of the center.

While I realize that the present type of construction in regard to clearance, etc., would make it impossible to use this system to the extent of entirely equalizing the strain in the modern turbine wheel, it seems that a certain amount of this initial tension would be beneficial, inasmuch as present design almost fills the bill.

The influence of temperature changes was especially noticeable in these large saws. If the arbor bearing next to the saw collars warmed up even a small amount, the saw would fall over toward the log. Sometimes, when dull, it would crowd the guide pins at the rim, which would cause an increase in temperature that was scarcely noticeable to the bare hand, yet was sufficient to cause a pronounced flutter.

Perhaps if those big turbine wheels ran out in the open where their performances could be observed, some of the experts might discover something new.

Whitehall, N. Y. R. G. YAXLEY.

Faulty Heating System Piping

A certain five-story building 90 x 110 ft. had about twenty ceiling radiators of 1¼-in. pipe. The manner of connecting the drains to the trap is shown in the sketch. It will be noticed that the check valve and air vent on each raidator are at the same level, and as it requires a small head of water to lift the check-valve disk, it will be apparent that the air vents would have to operate on water. That they did this was evident from the fact that the outlets were plugged with bolts, screws and other things that would serve the purpose.

Live steam through a reducing valve was used for heating the building, although an ample supply of exhaust steam passed through the basement of the building to serve a dry kiln adjacent to it. In the illustration *A* is a drain from a special radiator that was served with live steam, but not through the reducing valve, regulation being effected by "cracking" the valve. That often resulted in having the valve wide open, with the inevitable result that a pressure was built up on the discharge side of the check valve, holding it shut and giving the air vent a supply of water on which to operate.

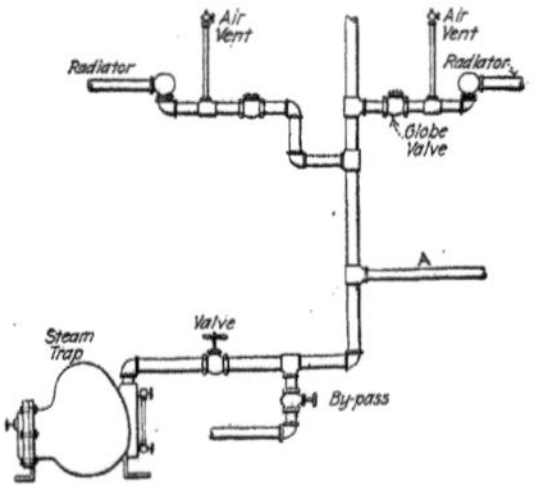

HOW THE PIPING WAS DESIGNED

The installation was made by a reputable firm of heating engineers who would be greatly peeved if their ability were questioned. R. McLAREN.

Toronto, Ont., Can.

Boiler-Tube Failures

In a recent issue of *Power*, a reader, Mr. Tousaint says that he does not agree that graphite is a scale remover. My experience with graphite is that it will remove scale, and very quickly at that. I know of a case where chemical treatment was unsuccessful. Boiler graphite was used, and the scale came off in large flakes and new scale was not formed. The graphite and scale-forming material was in the form of a soft, black mud and was easily washed out of the boilers. Before graphite treatment was started, no less than sixty tubes were lost in three months. After the treatment was well under way, but twenty tubes were lost in the same length of time.

However, to get good results boilers must be washed out frequently, to the fourth row of tubes at least, as the mud and graphite is sure to cause blisters. I have coated the inside of steam drums with graphite without any beneficial result. One boiler inspector recommends graphite and boiled linseed oil for coating tubes and steam drums. HERBERT B. BRAND.

Homestead, Penn.

The Fusibility of Ash from Eastern Coals

In *Power* for Nov. 4, 1919, page 655, there were published the results of over 120 determinations of the softening temperature of coal ash, together with the sulphur and ash content of the dry coal.

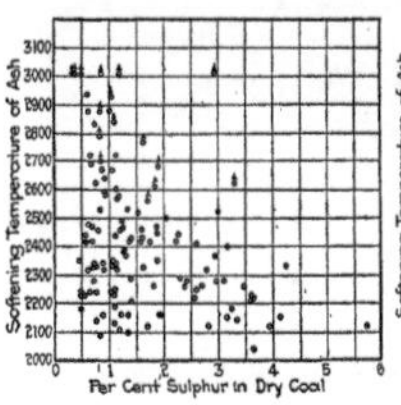

FIG. 1. SOFTENING TEMPERATURE OF ASH PLOTTED AGAINST PER CENT SULPHUR IN COAL

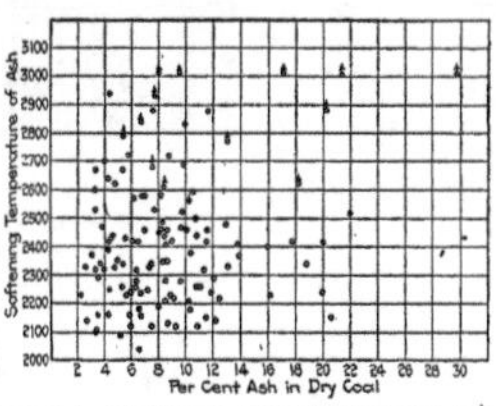

FIG. 2. SOFTENING TEMPERATURE OF ASH PLOTTED AGAINST PER CENT ASH IN COAL

In order to determine whether a relation exists between the softening temperature of the ash and the sulphur or ash content of the coal, I plotted these values as shown in Figs. 1 and 2. The points having arrows attached indicate that the softening temperature is higher than the temperature indicated by the point itself.

In Fig. 1, the softening temperature of the ash is plotted against the percentage of sulphur in the dry coal. From the manner in which the points are scattered, it is clear that there is no definite relation between the percentage of sulphur in coal and the softening temperature of its ash. It does seem to be true that the coals having the ash of highest softening temperature have low sulphur content, and that a high-sulphur coal is quite likely to have a low melting ash; but a low-sulphur coal is quite as likely to have an ash of low softening temperature. Thus the sulphur content of a coal gives us no reliable index of the probability of clinker trouble.

In Fig. 2 the softening temperature of the ash is plotted against the percentage of ash in the dry coal. From the very complete scattering of the arrow points it would seem that there is no relation whatever between the amount of ash and its softening temperature. (One sample included in the data was not plotted, because of its very high percentage of ash, 42.98 per cent, with a softening temperature above 3,010 degrees.)

The results are in no way surprising, but it is interesting to have so good an opportunity to check the facts from reliable data covering a wide range of coals.

Detroit, Mich. G. HAROLD BERRY.

Treating Boiler-Feed Water

I read with interest the communications on the subject of "Boiler-Tube Failures" that have appeared from time to time. It is not my purpose to comment on the theories advanced or objections made to the different ideas or suggestions presented in the articles in the Dec. 16-23-30 issue, page 838, but to call attention to one paragraph in which some unusual mathematical results are given in regard to the quantity of treatment prescribed, or recommended.

The author takes one gallon of treatment to a 100-hp. boiler operating a ten-hour day as the basis of his calculations. The assumption is made then that one gallon of compound weighs 10 lb. A mathematical computation then results in the statement that this would mean one ounce of compound to 6,000 lb. of water. This is a figure I can not understand. An evaporation of 3,750 lb. of water per hour is suggested, and ten times that quantity in ten hours. If 10 lb. of compound are recommended per day this would mean one pound of compound per hour, or one pound of compound to 3,750 lb. of water evaporated.

Assuming that a gallon of water weighs approximately 8⅓ lb., this would mean somewhere between 400 and 500 gal. per hour. According to conversion tables one pound per 1000 gal. is virtually the same as seven grains per gallon; one pound per 500 gal. would then be 14 grains per gallon, which would be the quantity of compound introduced into a boiler water on the basis expounded. Fourteen grains per gallon is a great deal; it is more than the total mineral residue in many waters. If properly put together, the reactions should take care of much more than this amount of material.

As I figure it, the dosage referred to is rather excessive, and I simply make the calculations this way, hoping that if the computations are incorrect they will be corrected, and if not that the author will explain the course of arriving at the dosage of one ounce per 6,000 pounds of water. Perhaps some reader of *Power* may have something of interest to say regarding the subject

Chicago, Ill. D. K. FRENCH.

Every Man Not an Efficiency Engineer

With reference to the foreword "Every Man an Efficiency Engineer" on page 123 of Jan. 27 issue of "*Power*," I wish to take exception to several statements made therein.

It is true that many individuals of meager experience hide behind the much abused name of "Efficiency Engineer," and it seems obvious that since the term engineer incorporates the meaning efficiency—for it is the engineer's function to introduce efficiency in all undertakings—the modifier, "efficiency," should be eliminated. All reputable engineers must first be efficient to call themselves engineers at all.

While it is also true that the importance of the coöperation of all concerned in the operation of any plant cannot be overestimated, I think that, generally speaking, such statements as—"It will be necessary to call in outside aid to find the weak spots, and to devise better methods of doing things," and "No man can be as familiar with the workings within a shop as he who spends day after day at his bench, at his lathe, or on whatever his job may be," leave a wide field for argument as to the soundness of this reasoning.

In the operation of a plant, the heating load, steam requirements for processes and energy generation must be treated as a closely interrelated unit and must be carefully balanced against each other. The individual interest of efficiency engineers can only result in individual economies, but cannot yield the bigger economies which usually result in working combinations. To make the latter more clear, I will mention a specific case, which must not be considered highly unusual.

In a certain wood-working establishment, a dry kiln, consuming as much steam as the Corliss engine unit that was operating the plant, was operated with live steam, whereas the Corliss engine was running condensing. No matter how much dry kiln or engine operation were improved, the broader economy to be secured by running the engine noncondensing and exhaust ing into the dry kiln had never occurred to anyone, since no knowledge had been obtained of actual steam quantities involved.

It has been my experience that the bigger and more intricate a plant becomes, the more likely will be the man in charge to fail to recognize combinations. It is not frequently the case, that a man with thorough general engineering knowledge, embracing heating, ventilating, steam, hydraulics and electricity generation, is in charge of such a plant for obvious reasons, and a man would have to "work at his lathe" for many a year before he could analyze such problems.

Combined Experiences Better Than Individual

The experience of a corps of organized engineers should always be able to treat such a combination of problems to better advantage than any one man who will not usually be master of all.

It does not seem true that he who is at the same bench or lathe knows better the workings of his routine than an outsider. We are all more or less slaves to habit, and the more we are confined to routine, the more we lose sight of possibilities. It is usually he who performs a routine task who is more blind to betterments, for in his routine he has developed certain habits and has formed partialities and prejudices. The outsider who can view matters in the perspective, and who has had wider experience in having been in intimate contact with the experience of others can serve to good advantage as a clearing house.

An engineer who can solve such problems must have trained his faculties of observation and developed an analytical mentality; and "working at a lathe," or at any routine task, tends in exactly the opposite direction.

This much in defense of the reputable consulting engineer, who is not a child of charity, but one who is as necessary in the economical operation of power plants as the physician is necessary to human life.

Milwaukee, Wis. CORNELIUS G. WEBER.

Peculiar Engine Design

In going through my Nov. 11-18, 1919, issue of *Power*, I note an engine of peculiar design. I would say that it is an "Estey" engine and was built about 1880 or 1890 by the Novelty Iron Works, of Dubuque, Penn. I have among my engineering files a copy of the catalog

AN ENGINE OF PECULIAR DESIGN

then issued regarding it. Such items are interesting in that they serve to recall to us old-timers the hysterics and convulsions the steam engine has gone through in the past forty years. W. V. WHITE.

Hamilton, Ohio.

Boiler Bagged Due to Muddy Water and Neglect

After reading the article on page 108 of the Jan. 20 issue of *Power*, by Edward Jesse, in regard to the boiler that was bagged owing to muddy water, I am of the opinion that the engineer in charge of the plant was to blame, because I considered that he had full charge and could have made the necessary changes required to get the water from the river, and also that, as he was using water from the mine, he should have taken care of the boiler by washing it out at least every fourteen days and used the boiler blowdown not less than once every eight hours.

Nearly every case of bagged boiler is caused by neglect. This case was no doubt neglect on either the engineer's or the manager's part. If the engineer really had charge of the plant, it was his neglect because he did not take proper care of the boiler and because he did not get better feed water when it was so handy.

Chicago, Ill. L. B. SHIELDS.

INQUIRIES OF GENERAL INTEREST

Girth Seams Lapped Toward Fire—What is the disadvantage of having the ring seams of a horizontal return-tubular boiler lap toward the fire?

The outer plate of the lap is more or less overheated and has a tendency to expand away from the inner plate and cause leakage, and more especially when the lap is toward the fire, as the calking is exposed to the direct impingement of the heated gases.

Banking Fire with Coarse Coal—We have been accustomed to use run-of-mine bituminous coal, employing the smallest sizes for banking fires. The last coal shipment received was nearly uniform stove size, and we are troubled to hold a banked fire. How should the banking be done?

Prepare a heavier bank of clean live coals sloped against the bridge-wall and covered with a layer of fresh coal. To hold the fire, blanket the bank with a thick enough layer of fine ashes to prevent a draft through the coals. For restarting the fire, drag off the ashes and only enough of the top layers of coal to leave a bank of clean kindling coal.

Increasing Length of Meter Connections—In connecting up alternating-current meters on the switchboard, does it affect the accuracy of the meter to splice four or five feet of wire to the potential- and current-transformer secondary leads so as to make them reach the meters? V. K. S.

The only case where the accuracy of electrical measuring instruments is affected by a change in the length of the leads is when the instrument is connected to a shunt, such as direct-current ammeters and in some cases the current coils of watt-meters. With alternating-current instruments connected to the secondaries of transformers the length of the leads may be changed without affecting the accuracy of the instrument.

Choke-Coil Voltage—What determines the voltage rating of choke coils such as are used with lightning arresters? W. A. M.

The voltage rating of a choke coil used with lightning arrester has in general no reference to the coil, but to the insulators on which the coils are mounted. A coil designed for 100 amperes would be practically the same whether used on a 11,000-volt circuit or 33,000 volts, but the insulators for the 33,000-volt coil would have to be designed to withstand three times the voltage of the 11,000-volt coil. From a protective standpoint the 33,000-volt coil would be perfectly satisfactory on a 6,600-volt circuit, but from the commercial side the cost of insulators would be necessarily high.

Conversion of Inches of Vacuum to Pounds per Square Inch—What pressure per square inch would be equivalent to 26 in. vacuum? E. C. H.

A column of mercury 1 in. high at ordinary temperature exerts a pressure of 0.491 lb. per sq.in., and "inches of vacuum" signifies the number of inches of mercury-column pressure less than the pressure of the atmosphere at the place of observation. When a vacuum gage indicates 26 in. vacuum, the pressure is $26 \times 0.491 = 12.766$ lb. per sq.in. less than the pressure of the atmosphere. The absolute pressure of the atmosphere varies with the elevation of the place. At sea level it usually is assumed to be 14.7 lb. per sq.in., so that at sea level the absolute pressure for 26 in. vacuum would be, $14.7 - 12.766 = 1.934$, or practically 2 lb. per sq.in. absolute.

Live Steam Wasted Through Back-Pressure Valve—Why is our boiler called upon to generate more steam when heating the building by adding live steam to the exhaust of the engine operated with a light load than when there is a heavy load on the engine and plenty of exhaust for the heating without addition of live steam? C. T. B.

The greater amount of steam used when the exhaust is supplemented by live steam undoubtedly arises from the escape of more steam to the atmosphere through the exhaust back-pressure-relief valve. When live steam is used to supplement exhaust for heating, it should be supplied through a pressure-reducing valve that will regulate the supply at a pressure several pounds below the relief pressure of the back-pressure valve. Otherwise there is likely to be great waste of live steam in attempting to satisfy the requirements for heating.

Galvanizing Process—How are water fittings and other articles galvanized for the purpose of resisting corrosion? B. N. M.

The articles to be galvanized must first be pickled in a dilute acid to remove or loosen all surface dirt and scale, and when removed from this bath, the articles are well brushed with brooms or steel brushes and washed by dipping in fresh water or with clean water from a hose. They are then placed in a weak bath of muriatic acid to insure a clean metallic surface and taken directly from this bath to a flux bath that consists of a hot solution of sal ammoniac covered with beef tallow. From the fluxing bath the articles are lifted dripping and immediately lowered into the galvanizing bath. This consists of molten zinc, contained in a pot or iron tank surrounded by brick walls, with space enough between the walls and the tank to maintain a coke fire or heat of gas flame for keeping the zinc melted. The articles to be galvanized, suspended from wires or supported in a wire basket, are lowered in the zinc bath for a few minutes and then lifted out and allowed to cool.

Equalization of Draft of Boilers Set in Battery—We have three horizontal return-tubular boilers set in battery, each provided with a 48 by 16-in. uptake, and all discharging into a 48-in. diameter stack connection. There is difficulty in obtaining equalization of the draft, especially when one of the boilers is operated with a forced draft. How can the trouble be remedied? W. G. J.

For obtaining the most benefit from smoke connections of a given size, all junctions of the uptakes with the main stack connection, and junction of the latter with the stack flue, should be made with easy bends so that flow of the gases may be smooth and continuous. To obtain equalization of the draft, each boiler uptake except the one farthest from the stack should be provided with a baffle for so directing the course of the gases that, when discharged into the main connection, their flow will be toward the stack before coming in contact with gases discharged by other boilers. By this arrangement the gases discharged from a boiler with stronger draft exert a jet-pump action rather than an obstruction to the discharge of the furnace gases from the other boilers.

[Correspondents sending us inquiries should sign their communications with full names and post office addresses. This is necessary to guarantee the good faith of the communications and for the inquiries to receive attention.—Editor.]

Economical Supply of Electric Power

IN A symposium entitled "Economical Supply of Electric Power for the Industries and the Railroads of the Northeastern Atlantic Seaboard," by W. S. Murray, W. L. R. Emmet, J. F. Johnson, H. G. Reist, F. W. Newbury, W. B. Potter, Philip Torchio, Percy H. Thomas, W. D. A. Peaslee and A. O. Austin, presented at the midwinter convention of the American Institute of Electrical Engineers, held in New York City, Feb. 18-20, are discussed the problems involved in the superpower system proposed for the industrial district along the North Atlantic Seaboard. Abstracts from this symposium and discussion thereon follow and show that such a system as proposed can be installed and operated at pressures as high as 250,000 volts.

In outlining what the proposed superpower plan is, W. S. Murray said in part: "The superpower plan, briefly summarized, provides a means by which a present estimated machinery capacity of 17,000,000 hp., divided 10,000,000 for industrial purposes and 7,000,000 for the railroads, in a region between Boston and Washington and extending inland from the coast 100 to 150 miles, now operated with a load factor not exceeding 15 per cent, can be lifted to a load factor greater than 50 per cent and possibly to 60 per cent, and a means by which, conservatively speaking, one ton of coal will do the work of two, and the railroads

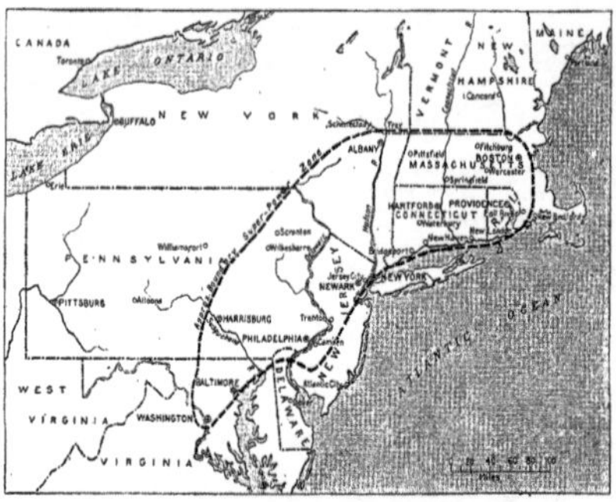

FIG. 1. MAP SHOWING APPROXIMATE SUPERPOWER ZONE

within the above zone and those carrying coal into that zone will be relieved of transporting one-half of the amount required for power and lighting purposes. In short the value of machine capacity from a utilization standpoint will be increased from threefold to fourfold, and coal resources for the purpose named being conserved twofold. This means that a present plant capacity of 17,000,000 hp. can be replaced by one not greater than 5,500,000 hp., and that not less than 30,000,000 tons of coal per annum can be saved, which at $5 per ton will represent $150,000,000 per year.

"Besides the foregoing savings, two great departments of economy will be created; one applying both to the railroads and industries, in the reduced cost of maintainence of machinery, and the other applying to the railroads alone in the reduction of train miles. It is estimated that these latter economies will effect a saving of another $150,000,000 annually, thus making a saving of $300,000,000. The foregoing are the direct savings as estimated from data collected from actual past operation of the specific order contemplated to be put into force in the zone under consideration.

"This plan offers immediate relief from the present intolerable congestion of our railroads by automatically increasing rail capacity without increasing track mileage, and reducing power equipment to a minimum. By the creation of an overhead common carrier system of power, the present cargo space now required for industrial coal will be cut in half; train equipment in all classes will have its service practically doubled, and the present steam-power equipment, replaced by electrical equipment, can be transferred to other divisions where it is so vitally needed. The superpower plan contemplates the general application of electricity wherever economically possible in the factories or on the railroads in the zone described."

W. L. R. Emmet, in discussing large steam turbines, pointed out that "the best steam-turbine station equipment, operating under favorable conditions can deliver a horsepower-hour in the form of electricity with an expenditure of one pound of coal, where four pounds are required to deliver a horsepower-hour to the draw bar of a good locomotive. The locomotive is subject to many disadvantages; its fuel supply must be delivered and stored on a relatively small scale, in many inconvenient places, and its efficiency is greatly affected by conditions of temperature. The water supply is also a matter of much trouble and expense. While the comparison of efficiency between the large power station and the smaller engine or turbine equipment used in small stations and isolated plants is less striking than that with the locomotives, it is nevertheless highly unfavorable to the small plant."

J. F. Johnson expressed the opinion that "The design of the steam-generating stations such as would be required, would probably not differ materially from those being employed for the larger and most modern stations now building, and their efficiency would not differ materially from those now obtainable except as affected by the higher load factor under which they would operate.

"Reliability would, of course, have to be the first aim of the designer, and this requisite would preclude the adoption of experimental apparatus or conditions widely removed from those now used in order to produce satisfactory results. Generating stations of some 200,000- to 300,000-kw. capacity, employing generating units of some 60,000- to 75,000-kw. capacity, each operated on steam conditions of 300-lb. pressure, 200-deg. F. superheat and 29-in. vacuum referred to a barometer of 30 in., would involve no difficulties of design, construction or operation, and from such stations a steam-consumption rate of 10 lb. or less per kilowatt-hour and a total station rate of less than 1.5 lb. of good-quality coal per kilowatt-hour output should be obtained."

"The size of the generating unit for such a project," said H. G. Reist, "will undoubtedly be determined largely by the output required for individual power houses. It seems probable that whatever frequency is selected for this system of distribution, turbines and generators can be supplied to meet the most desirable number of units to be placed in each installation. So many generators of from 30,000 to 50,000 kva. are at present in operation, giving satisfactory service, that there need be no hesitancy in considering generators as large as those now in use or larger. The potential of these generators should be within the limits of our experience; that is, not above 13,200 volts, and preferably lower, if this does not cause inconvenience in the lines leading from the generators to the step-up transformers. The efficiency of large modern generating units is very high. The losses may be expected to be less than 2 per cent of the output of the machine. On account of the great capacity of the lines in such a large system, there may be some advantages in the use of induction generators. Roughly, the size of induction generators would probably be from 25 per cent to 30 per cent larger than a synchronous generator and the cost of the machine in proportion.

F. D. Newbury gave his point of view, in regard to the generating end of this subject, by saying that "the generating element in the proposed Boston-Washington power-supply system does not involve anything new or untried. The individual power station need not, and probably would not, be any larger than stations now in operation or under construction. Certainly, any probable station involved in carrying out the superpower development could be designed with steam or electric generating units of a size now available. Suppose we assume a station of 300,000 to 500,000 kw. total capacity. We have available single-shaft generating units up to 40,000 kw. with generators of 50,000-kva. capacity. The preferred speed for such units would be 1,200 r.p.m., 60-cycles. Eight to twelve such units would give the assumed total station capacity. There have been developed and built triple-shaft compound units of 60,000 kw. Five to eight such units would constitute as large a station as has been suggested. All the generating units referred to are actually conservative in size and by no means represent the largest unit of the speed in question that could be designed."

From a consideration of data available on railway operation W. B. Potter draws the conclusions that "of the whole mileage included in the zone, a not very large proportion has been electrified, but main-line electrifications now in operation are of sufficient extent to carry tonnage of a character to present data that can be fairly applied to the traffic of the whole district. The traffic within the zone now handled by steam locomotives, if handled electrically, would require an average output of less than 750,000 kw. and if produced entirely by coal-burning electric-power station, would reduce the coal requirements for transportation purposes from 21,000,000 tons to 7,000,000 tons annually. As a certain proportion of electric power would be produced from hydraulic-power stations, this coal requirement would be reduced in proportion as advantage is taken of hydraulic operations. The reduction in cost of maintaining the motive-power units would be a large amount, which, estimated from the locomotive mileage, would be in the order of $15,000,000 or more annually."

Philip Torchio, in his contribution, discussed the relative values of water- versus steam-power development. From a review of the resources and consumption of coal and the availability of potential water power in the United States, Mr. Torchio said: "The Western States, Mountain and Pacific, have resources in both coal and water to meet indefinitely all the heat and power required from either source of supply, the potential water powers alone being large enough to supply over six times all the heat and power requirements in 1915. The other states, with corresponding heat and power requirements forty times greater than the Western States, have actually smaller resources in coal and potential water power, the latter capable of supplying only 8 per cent of their total heat and power requirements. It follows, then, that these states must indefinitely, as far as present human knowledge can forsee, depend upon the use of coal to supply the great bulk of these needs. These states, comparatively so deficient in water powers, consumed in 1915, 528,000,000 tons of coal, about one-half for generating power and one-half for generating heat. They obtained an average efficiency from coal of 5 per cent for the power and 50 per cent for the heat."

In a comparison of the cost of steam-power plants and hydro-electric developments Mr. Torchio pointed out that the former cost about one-third to one-quarter that of the latter, consequently one dollar invested in steam plants will make available several times more power for industries than a dollar invested in hydro-electric plants. The author urges development of our water powers especially in connection with electrochemical processes and metal refining using immense amounts of power continuously, where the net cost of hydroelectric power may be considerably less than steam power. "There is a great advantage to the whole nation to devoting as much as possible of the economical water powers, particularly those which are continuous or nearly so and those without regulatory daily storage, to intensify the development of these industries which can prosper and render their full benefit to the Nation only under the conditions of mass production and most favorable cost of power. The harassing experiences of coal famines have brought forth repeatedly pressure that immediate efforts be directed to the development of the potential water powers of the country. While from the general standpoint of conserving the national coal resources the proposition deserves the most favorable consideration, it is questionable whether quicker and greater results could not be obtained by concentrating the effort upon improving and changing wasteful methods of utilizing coal."

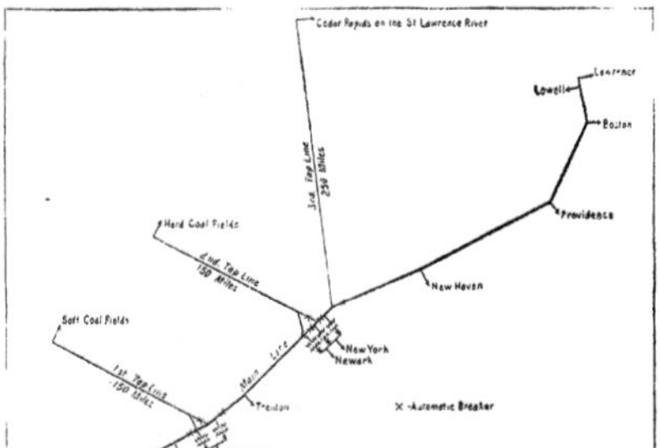

FIG. 2. SUGGESTED POWER-TRANSMISSION SCHEME FOR SUPERPOWER ZONE

Tables are given comparing the relative values of hydro-electric and steam power for a number of specific conditions of service which, cover all possible ranges of conditions that may exist for different situations of power production from very large plants, equipped with modern apparatus. In another table "Review of the Supply and Consumption of Coal by Classes and Territories and the Relation Which the Potential Water Power Bears to the Amount of Coal Used in the United States," very interesting information and data on potential water power and coal saving is given.

Percy H. Thomas, in discussing this subject, assumed a transmission line extending from Lawrence, Mass., to Washington, D. C., as in Fig. 2, operating at 250,000 volts, with tap lines running to the various sources of power and large centers of power consumption, and shows that it is within the premise of present engineering practice to construct and operate such a system.

In discussing the capital cost of such a system, the author points out that it may be assumed that a 400,000-kw. power station may be built under reasonably favorable price conditions for $100 per kw., including overhead necessary for design and construction; that the transmission line as proposed (namely, two 150-mile 250,000-volt 400,000-kw. tap lines, and 100 miles of similar capacity line for the central part of the main connecting line, and 250 miles of 200,000-kw. line) can be built at an average cost of $47.50 per kw. generated, including overhead; also, that the necessary 250,000-volt step-up transformers with 250,000-volt switches can be installed for $15 per kw., allowing $1.50 per kw. for miscellaneous cost, gives the total cost for 900,000 kw., allowing 12½ per cent spare as $150,000,000 or $164 per kilowatt. A fair estimate of the cost of installing plants to generate the same amount of power at the several large centers, supposed to be supplied, would be about $125 per kw., leaving an excess capital cost on account of the transmission scheme of $40 per kw., a very small amount for the benefits offered.

Insulators for Proposed System

W. D. Peaslee said that, "The utilization of 220,000 volts or higher for the transmission of power is not a problem that holds any terror for insulator manufacturers." He also called attention to a disk insulator that was being developed and said that "this disk insulator will be very well suited for this particular work, and our research work has progressed far enough so that we feel confident that by the time such a line is an active proposition, we will be able to undertake the installation problem."

A. O. Austin, in speaking of the problem of insulating such a system as proposed, said: "The large conductors necessary for a system such as this, would require heavy working stresses for the insulators. The insulator art, however, is such that, high ultimate insulators with long life can be produced at a much lower price than they could formerly. It is now possible to produce insulators such that a single string will carry the line at a cost probably not more than 50 to 60 per cent of that required for a 110,000-volt line. From this it will be seen that the cost of insulators for a 220,000-volt project will probably be less than the cost of 150,000-volt line previously installed. The transmission art has advanced rapidly and many an economy could be effected by carrying out a comprehensive scheme, which is not possible in many of existing systems."

Discussion on Super Power System

In discussing this project Prof. Malcolm Maclaren suggested that some independent body make a careful study of this situation, and pointed out that the institute could render no greater service to this community at the present time than by urging the creation of a commission which should direct the course that these developments should take.

The speaker called attention to the power requirements of a portion of this district, namely, eastern Pennsylvania, New Jersey and New York City. This portion he divided into four industrial districts, one including Philadelphia and extending along both banks of the Delaware River from Chester to Trenton; another centered around Newark, reaching from Paterson in the north to Perth Amboy in the south; the third is confined within the limits of greater New York; and the fourth includes the anthracite-coal fields extending south to Allentown, Bethlehem and Easton, which may be designated as the Lehigh district. From a recent analysis made by Prof. Maclaren the results show that for the Philadelphia district the maximum demand would be approximately 700,000 kw., for the Newark district 550,000 kw., for the Lehigh district 650,000 kw., and for the New York district approximately 700,000 kw., making a total of 2,600,000 kw. The most promising methods for obtaining additional capacity in this district, says Prof. Maclaren, are "by hydroelectric developments on the Susquehanna River, hydroelectric developments on the Delaware River and by construction of steam plants at the anthracite mines. The largest single development projected for the Susquehanna River, at Conowing, near the mouth of the river, will have an initial capacity of 125,000 kw. and will be capable of producing approximately 700,000,000 kw.-hr. in a dry year. This 700,000,000 kw.-hr. represents only about one-third of the energy of the river at Conowing, and with a relatively small amount of river control it should be possible to increase the output at this point and develop other sites that would yield between 200,000 and 300,000 kw. and 1,500,000,000 kw.hr. annually. By plans that are well advanced for the development of a series of hydroelectric plants on the Delaware River, this river would be capable of producing 250,000 kw. of primary power and 1,000,000,000 kw.hr. annually.

"Considering next the construction of steam plants at the mines, it should be noted that it may not prove feasible in the present state of the art to install in the anthracite region superpower plants of 300,000 kw., or more, such as has been proposed for other districts, as there are few if any sites in this district where there is combined an adequate supply of the smaller sizes of coal with water conditions suitable for operating a plant of such magnitude.

"The report of the Bureau of Mines for Pennsylvania shows that for a total annual production of 80,000,000 tons of anthracite for this district, over 8,000,000 tons was used in mine operation. Data obtained from the electrified mines of the district indicate that if all the mines were electrified from a central-supply system, this 8,000,000 tons now being used for mining coal would be capable of supplying all the needs of the coal companies and leave a balance sufficient to operate a generating system of over 600,000 kw. capacity at 50 per cent load factor.

A Picture of a General Power Scheme

"The picture that might be formed of a general scheme to be followed in the next ten years to meet the growing needs for power within this territory is that the power companies in the large cities should in general complete their present construction programs, but should not greatly increase their generating facilities beyond this point, additional capacity being obtained by the construction of steam plants at the mines, reinforced on the east by hydro-electric plants on the Delaware River and on the west by hydro-electric plants on the Susquenhanna, the whole of the territory being interconnected with transmission circuit."

Prof. Charles S. Scott reviewed the growth of electrical power devices, during the past 13 years and said that with the load doubling every five years it is absolutely necessary to consider some economical means of providing an adequate power supply to meet the requirements of the future. The superpower scheme offers one of the solutions to the problem.

N. W. Storer referred to the shortage of labor and fuel being experienced by the railway companies at the present time and said that they should take a lesson from the housewife and do it electrically. The speaker said that the electric locomotives can do more work than the steam locomotive and do it better, and do it on one-third of the coal burned in a steam locomotive.

Electrifying of All Industries

D. B. Rushmore reviewed the changes that led up to the present industrial development and said that industry in this country is bound to increase, and with labor becoming more expensive, the situation can be met only by an increased use of power, which means the electrifying of all industries. On the suggestion of Professor Maclaren and Mr. Rushmore a motion was put by H. W. Buck that the board of directors be requested to appoint a committee with W. S. Murray as chairman to give the support of the American Institute of Electrical Engineers to the superpower plan. The motion was carried.

W. D. Peaslee announced that from the result of an investigation his company was now ready to provide insulators up to 350,000 volts and ready to guarantee them.

E. E. F. Creighton said that in the very near future means would be available whereby any disturbance from lightning would be removed from the line in a few hundredths of a second.

J. W. Lieb called attention to the fact that an adequate power supply is one of the most important economical problems before the nation at the present time, therefore it should have the support of the American Institute of Electrical Engineers. The engineer should view the problem with a broad national outlook and not make the mistake of the railroads by a battle over the different systems. Every day's delay makes the consummation of the superpower plan more difficult. The speaker referred to the question of both voltage and frequency making the problem involved and illustrated this point by our experience during the war, when New Jersey was short of power, and in New York City 100,000 hp. was available, but could not be connected with New Jersey's power system on account of a difference of 2½ cycles between the two systems. Mr. Lieb announced that this difference, however, no longer existed, since the New York Edison Co. has recently changed the frequency of their system from 62½ to 60 cycles. Mr. Lieb laid great stress on the necessity of continuity of service and said that in a community like New York City, the seriousness of interrupted service, even for a short time, would be impossible to exaggerate, therefore it should be assured by everything that money can buy, that human foresight can propose and that engineering ability can provide.

Chicago N. A. S. E. Welcome National Officers

On Saturday night, Feb. 28, the combined N. A. S. E. associations of Chicago assembled 350 strong to greet the national officers at a banquet held in their honor at the central Y. M. C. A. Auditorium. The officials were on their way to Milwaukee to make arrangements for the next annual convention. With representatives of the Chicago Association of Commerce, the Manufacturers' Association and the Building Managers, present upon invitation, the banquet was made a booster meeting. For forty years the association has been working along quietly, hiding its light under a bushel, and it was felt that it was high time for publicity on the aims and objects of the association, what it stands for and what it has been doing for the last forty years. The various officers taking part in the program brought out from various angles the features in the preamble and one and all eloquently expressed the ideals of the association. Education, economy, the enactment of license laws and harmony between employer and employee were the features brought to the fore. "To earn more, learn more" is the guiding spirit of the association, which sets its face against the furtherance of strikes or interference of any kind between its members and their employers. In these days of social unrest, bringing sudden changes and reversals of pre-war conditions, an association founded on the principles previously outlined is an exception worthy of the highest commendation. The men who pay the wages of its members should know all about it, and the various speakers took pride in laying their claims before them.

Tom McNeill, chairman of the entertainment committee, introduced Alfred Johnson as toastmaster for the evening, and in turn he called upon President Callahan, who was followed by W. M. Ellis, of the Building Managers' Association, Homer J. Buckley, of the Chicago Association of Commerce, the following officers of the association: Joseph F. Carney, William Reynolds, John A. Wickert, C. W. Naylor, Joe O'Connell, Fred Raven, "Dad" Beckerleg; and Homer Smith, of the Exhibitors' Association.

During the evening it was brought out that the association is in a better position than ever before. The paper is prospering and the drive for a 25 per cent increase in membership is progressing favorably. New Jersey has high hopes of pulling down the prize, but New York and other states expect to offer close competition, and Wisconsin, with the convention coming to Milwaukee, has already shown an increase throughout the state of 25 per cent, with several months to go.

Passing on to Milwaukee, the officers were given a rousing reception at a beefsteak dinner Monday night and participated in the formalities attending the initiation of 75 new members. On the following evening they had the pleasure of initiating almost as many in the rooms of No. 1 association, Chicago.

Military Engineers' Society Is Being Organized

A national association of present and former officers of engineers and civilian engineers who have served in any arm or branch of the U. S. Army—to be called the Society of American Military Engineers, is being organized by a committee appointed by the Chief of Engineers. The society's objects are to promote the science of military engineering and to foster the co-operation of all arms and branches of the service, and of civilian engineers, in that science. The objects and a provisional constitution of the society have been approved by the Chief of Staff.

A letter ballot from all engineer officers in the service and a canvass of representative opinions among officers who had been in the service during the war showed a strong majority in favor of such a society. The committee originally appointed to consider the feasibility of a technical organization, therefore, drafted a provisional constitution and created a temporary board of directors from its membership.

The annual meeting is to be held in Washington, and its date is fixed with reference to that of the American Society of Civil Engineers in order to make possible attendance of members at both meetings. The dues are fixed not to exceed $5 per year, those for the present year being $4.50.

The society is to publish, bi-monthly, a journal to be called "The Military Engineer," which will supplant "Professional Memoirs" heretofore published by the Corps of Engineers.

Further information regarding the society may be secured from Col. G. A. Youngberg, Office, Chief of Engineers, U. S. Army, Washington, D. C.

A. I. E. E. Boston Meeting

The Boston meeting of the American Institute of Electrical Engineers will be held, April 9, 1920, jointly with the American Electrochemical Society and will be under the auspices of the Boston Section and the Committee on Electrochemistry and Electrometallurgy. Tentative plans have been arranged as follows:

On Thursday, April 8, the American Electrochemical Society will hold meetings, morning and afternoon, at the Massachusetts Institute of Technology, and members of the A. I. E. E. are cordially invited to attend. On Thursday evening at 8:30 a get-together smoker is planned by the American Electrochemical Society at the Copley Plaza Hotel, to which all Institute members are invited.

On Friday, April 9, a joint session will be held at 9:30 a.m. for the presentation and discussion of a symposium, "Electrically Produced Alloys." In the afternoon an inspection trip will be made to the laboratories of the General Electric Co. at Lynn, where talks on research work at the laboratories will be delivered by Prof. Elihu Thomson and others. At 6:30 p.m. a subscription dinner will be served at the Copley Plaza, at $4 per plate. The evening session will be held in the Copley Plaza, at 8:30 p.m., where a symposium will be presented on "Power for Electrochemical Purposes." The American Electrochemical Society will continue its meeting on Saturday.

In order to provide a dependable water supply against the threatened dry season this year, the land owners under the Government project in the Okanogan Valley, Washington are preparing to install one of the most powerful steam pumping plants ever used for irrigation purposes. It is proposed to lift sufficient water out of the Okanogan River to irrigate 5,300 acres of land now in bearing orchards in the vicinity of Omak. This water will have to be lifted to the total height of 470 ft. and pumps having a capacity of 25,000 gal. per min. will be required for the work.

Fatal Boiler Explosion at Seattle

Five men are dead and three others are injured, two possibly fatally, following an explosion, at noon, Saturday, Feb. 14, of a mud drum of a water-tube boiler at the steam plant of the Puget Sound Traction, Light and Power Co.,

FIG. 1. FRONT AND SIDE VIEW OF THE EXPLODED BOILER

Seattle, Wash. Windows were blown out, brick walls were shattered and the employees of the plant were showered with steam, boiling water and bricks.

The exploded boiler, which had been inspected on Feb. 6, was located in a one-story building adjacent to a five-story building housing five large main boilers. The small building contained two auxiliary boilers, used only in cold weather, or as relief for one of the main boilers when necessary.

Steam had been cut off from one of the main boilers, and the auxiliary boiler that exploded was fired up to take its place. It had been in service about an hour when the explosion occurred.

The boiler was of oil-burning type, encased in a brick setting. The initial fracture is believed to have been in the cast-iron mud drum, pieces of which were found more than thirty-five feet in the rear of the boiler.

Asked for a statement regarding the explosion, Superintendent Santmyer said:

"The explosion, as we now understand it, occurred when the mud drum let loose. The boiler had not been in use since Feb. 6, when it was inspected by the insurance company. The inspector pronounced it in first-class order.

"The boiler was installed in 1906, but was considered still as good as new, as it has not been used except for relief of one of the regular boilers. It was rated at 500 horsepower."

C. D. Paddock, chief boiler inspector for the Hartford company, also stated that his company had made an inspection of the boiler on Feb. 6 and had found it in good order, He said:

"A thorough investigation will be made by City Boiler Inspector William E. Murray and myself to determine just what was responsible for the explosion. I cannot say from the preliminary investigation that I have made that the bursting drum was responsible. It may be that through some sudden change of the pressure the drum was broken, or it may be that the explosion was not in or near the mud drum, but that it was a result of some other condition."

The dead are: William Santmyer, machinist, son of Walter J. Santmyer, superintendent of the Puget Sound Traction, Light and Power Co.; George O. Bunnell, storekeeper; David T. Spurgeon, engineer; T. A. Nolan, steamfitter, and P. H. Jacobs, carpenter. The injured are: T. D. Parr, foreman; Edward H. Huntley, machinist, and Fred Offield, steamfitter.

Spain's available water-power is estimated at 5,000,000 hp., only a tenth of which has been developed. There have been recent developments in hydro-electric power transmission. In Catalonia, the principal industrial district of Spain, works already under construction will add 200,000 hp. to that now in use, making a total for the country of about 700,000 hp. The country has large resources in coal. However, inadequate transportation in the interior has prevented a development of these resources, and while coal might have been exported, it has been imported.

FIG. 2. SIDE VIEW OF THE BOILER AFTER EXPLODED

FIG. 3. BACK-END VIEW OF BOILER

One of the World's Largest Steel-Mill Electrical Equipments

AT THE regular monthly meeting of the Philadelphia section of the Association of Iron and Steel Electrical Engineers, R. B. Gerhardt, superintendent of the electrical department of the Sparrows Point plant of the Bethlehem Steel Co., presented a paper on "Electrical Features of a Modern Steel Plant." The author gave a very comprehensive description of the electrical equipment of the Bethlehem Steel Co.'s plant at Sparrows Point, Md., which is now nearing completion after four years of work and involving an expenditure of $50,000,000. In this plant there is installed a total of 2,169 motors having an aggregate capacity of 117,850 horsepower.

The plant consists of an ore-handling plant; coal-handling plant; coke ovens; blast furnace plant; open-hearth plant; bloom and rail mills; bloom, billet and bar mills; 36 x 110 in. sheared-plate mill; 60-in. universal plate mill, and sheet and tin-plate mill. The ore-handling plant is equipped with four motor-operated unloaders, which with three of them working are capable of unloading 10,000 tons of ore from a boat into the storage yard in 16 hours. The electrical equipment of this part of the plant consists of 60 motors having a total capacity of 3,618 hp. A ton of ore can be transferred from a boat into the storage yard and then from here into the furnace stock bins, on an average power consumption of 1.5 to 2 kilowatt-hours.

The coal-handling plant is equipped for unloading either from boats or railway cars. There are two breakers and four crushers in this plant all electrically driven. The electrical equipment in this plant comprises 56 motors totalling 4,100 horsepower.

Coal after passing through the crushers is delivered to the storage bin over six batteries of by-products coke ovens, each comprising 60 ovens. Each battery has a capacity of 750 tons of coke per 24 hours. For the total coke-oven plant including the coal unloading, crushing and by-products plant, there are required a total of 174 motors representing 8,035 hp. The average power consumption per ton of coke is 1.3 kilowatt-hours. The foregoing does not include the power required to drive the gas exhausters, which are driven by steam turbines.

Blast-Furnace Plant Equipment

In the blast-furnace plant there are four 450-ton furnaces in operation and two 500-ton units nearing completion. For blowing the furnaces there are two gas-blowing engine plants arranged for a total of fourteen Bethlehem 47 x 60 in. gas end by 84 x 60 in. air end, including all necessary auxiliaries. There is also a steam blowing plant which contains four Corliss engines, 42 x 88 x 60 in. steam end by 84 x 84 in. air end. Also two blowing engines 38 x 38 x 60 in. steam end by 84 x 84 in air end. There is also a 10,000 hp. gas-fired boiler plant, containing forty 250 hp. boilers. Most of the boilers are not operated unless there is a surplus of gas from the gas engines. A total of 10,887 hp. of motors is required in this plant. Excluding the casting plant and sintering plant, the average power consumption of 6.9 kilowatt-hours is required per ton of pig iron. In the sintering plant an average of 15.5 kilowatt hours per ton of sinter is required.

There are two open-hearth plants in which are installed a total of 237 motors having an aggregate capacity of 12,658 horsepower.

The rolls of both the bloom and rail mills are driven by steam engines operating condensing and supplied with steam from twenty 250 hp. oil-fired boilers. For auxiliary drives in this department there is a total of 136 motors representing 5,190 horsepower.

The bloom, billet and bar mills are located in one building and aranged to roll billets, slabs or bars direct from ingots. A 5,000-hp., 600 volt direct-current reversing motor drives the bloom mill. This motor is supplied with power from two 2,000 kw. 600-volt direct-current generators driven by a 3,000-hp. 6,600-volt induction motor. This mill rolls ingots weighing 16,500 lb., into slabs 9 x 38 in. in 16 passes. Power consumption averages 10.4 kilowatt-hours per ton of finished slabs. Ingots weighing 12,000 lb. are rolled down to 8 x 8 in. blooms in 20 passes in this mill on an average power consumption of 20.8 kilowatt-hours per ton of finished blooms.

A six-stand 24-in. continuous mill driven by a 4,000-hp., 6,600-volt three-phase motor rolls 4 x 4 in. billets or 8 x 2 in. slabs on a power consumption of 9.55 kilowatt-hours per ton of finished product. The 8 x 2 in. slabs go to a six-stand 18-in. continuous mill driven by a 3,250-hp. motor where they are rolled into sheet bars, on an average power consumption of 26.8 kilowatt-hours per ton of finished product. In the bloom, billet and bar mills a total of 18,774 hp. is required in 181 motors.

A 4,000-hp., 6,600-volt slip-ring induction motor drives the 36 x 110-in. sheared-plate mill. The average slabs rolled in this mill are 6 x 38 in. weighing 4,000 lb. and are rolled down to $\frac{1}{2}$ x 72-in. plates with an average power consumption to the main motor of 21.0 kilowatt-hours per ton of slabs, and on 44.0 kilowatt-hours for the total mill. The total horsepower required in this mill is 8,653 in 122 motors.

Two-High Reversing Plate Mill

A 60-in. two-high reversing universal plate mill is under construction and will be in operation in the near future, and is designed to roll 13 x 62 in. slabs weighing 10,000 lb. down to a $\frac{3}{8}$ x 60 in. plate in 21 passes. This mill will be driven by a 700-volt direct-current reversing motor having a maximum peak rating of 17,500 hp. and will operate at a speed of from 0 to 100 r.p.m. in either direction. Power is supplied to this motor from a flywheel motor-generator set consisting of two 3,000-k.w. 700-volt direct-current generators directly connected to a 5,000-hp. 6,600-volt motor and a 75-ton flywheel all mounted on a common bed plate.

In the sheet mill sheet bar is rolled in hot mills arranged in four groups of six-roll stands each. Two of the groups are driven by 1,200-hp. and two 1,000-hp. wound-rotor induction motors equipped with double flywheels and liquid slip regulator. The total connected load in this mill is 6,070 hp. made up of 96 motors. In the tin-plate mill 116 motors are used totalling 5,441 horsepower.

Power for the plant is supplied from two sources, namely, the Consolidated Electric Light and Power Co. and from the Bethlehem Steel Company's own gas-engine driven plant at the mill. Seventeen thousand kilowatts are supplied from the former source at 26,000 volts, over a duplicate transmission line from Baltimore, Md. All the rest of the power required in the mills is supplied from the company's own gas-engine driven plant, in which at the present time five 5,000 kva. generators are being installed. Each generator is driven by a Bethlehem 47 x 60 in. twin-tandem gas engine. The power plant is designed for an ultimate capacity of 100,000 kva. Power is distributed about the plant at 6,600 volts and this distribution system takes the form of a double-ring bus, which originate and ends at the gas-engine plant. This bus consists of 1,000,000-circ.mil. weatherproof cable carried overhead on special cable-top 33,000-volt insulators, mounted on 85-ft. steel towers. On this bus there are located four switching towers with taps to the four principal substations located at the 40-in. blooming mill, 60-in. universal plate mill, 110-in. plate mill, rail mill and coke ovens.

In closing, Mr. Gerhardt paid high tribute to the members of his electrical organization at Sparrows Point, and said it was only through the hearty co-operation of every one in this organization that made possible the result obtained at Sparrows Point during the past four years.

Temporary Fuel Control

By an executive order the President has extended the powers of the Fuel Administration for a period of sixty days from March 1. The control, however, is divided between Director General Hines and a commission of four, consisting of A. W. Howe, of the Tidewater Coal Exchange, Rembrandt Peale, representing the coal operators, and F. M. Whittaker and J. F. Fisher, representing the railroads.

Mr. Hines will retain the jurisdiction over domestic distribution, and the commission, which will function through the Tidewater Coal Exchange, will look after bunker and export coal. Although the office of Director General of Railroads expired March 1, Mr. Hines' authority as concerns coal distribution has been extended for the period stated.

It is specifically directed that the order issued by the United States Fuel Administration Nov. 6, 1917, relative to tidewater transshipment of coal at Hampton Roads, Baltimore, Philadelphia and New York, and for the employment of and co-operation with the Tidewater Coal Exchange, as a common agency to facilitate transshipment and to reduce delays in the use of coal cars and coal-carrying vessels, suspended by Dr. Garfield on Feb. 20, 1919, be reinstated.

New Publications

HENDRICKS' COMMERCIAL REGISTER of the United States for Buyers and Sellers, 1919-1920. S. E. Hendricks Co., Inc., New York City. Cloth, 2,703 pages. Price $12.50.

The 28th Annual Edition of "Hendricks" has just been published, after having been delayed for two months by the strike of the printers. The new edition contains several improvements, the most noticeable being the new method of exterior indexing by coloring the front edge red, white and blue to indicate the different main sections of the book. First is blue, on which is stamped the words "Trades Index." This is a section of 162 pages in which every product listed in the book is indexed and cross-indexed for ready reference. The red section is the main classified trades list. It contains 1,813 pages, listing over 18,000 different products. In the present edition there are more than 1,200 new headings, including many covering the chemical industry. The third section, as indicated by the white edges, contains 216 pages listing the trade names under which products are manufactured, with the names and addresses of the manufacturers. The second blue section is the alphabetical section of 487 pages containing all the names in the book in one alphabetical list, with addresses and their main line of business. This is followed by a 20-page index to advertisers, containing a full list of branch and foreign offices following each name. The volume contains 2,703 pages.

Obituary

Cyrus Thurston Johnston, mechanical and electrical engineer, died at his home in St. Paul, Minn., Wednesday, Feb. 25, after a brief illness. He was a graduate of the Massachusetts Institute of Technology, class of 1909, and had been associated with his father, Clarence H. Johnston, architect, of St. Paul. At the time of his death he had entire charge of the heating, plumbing and ventilating work of the office. A host of friends will lament the untimely ending of his promising career.

John A. Randolf, of the Westinghouse Electric and Manufacturing Co., died at the Columbia Hospital, Wilkinsburg, Penn., on Friday, Jan. 30, 1920. Death was caused by pneumonia. Mr. Randolf was a graduate of Syracuse University with the class of 1903. Following his graduation he engaged in engineering and writing technical literature for trade magazines; old readers of *Power* will no doubt recall some of his articles dealing with electrical subjects. He was also an active worker for a time with the Society for Electrical Development, located in New York.

Personals

David J. Armour, formerly associated with the United States Rubber Co., has joined the New York City sales force of the McLeod & Henry Co.

Miles B. Lambert was appointed manager of the railway department of the Westinghouse Electric and Manufacturing Co. He has been connected with the Westinghouse forces since 1901.

W. S. Rugg, manager of the railway and marine departments, and Charles Robbins, assistant sales manager, of the Westinghouse Electric and Manufacturing Co., have been appointed assistants to the vice president.

George M. Ogle, formerly chief electrical engineer of the United States Shipping Board, Emergency Fleet Corporation, is now a member of the Engineering Organization of the Vulcan Iron Works, Inc., Jersey City, N. J., in charge of electrical contracting and the consulting engineering department.

A. T. Shurick, of the business staff of the *Coal Trade Journal* has been elected vice president of F. C. Thornley & Co., Inc., at 31 West 43rd St., New York City. Mr. Shurick is a graduate of the Virginia Polytechnic Institute. He was actively engaged in his profession as engineer, for ten years with the Rock Island Coal Co., the Mexican Coal & Coke Co. and the Anaconda Copper Mining Co. In 1911 he joined the editorial staff of *Coal Age*.

Fred W. Herbert has been appointed superintendent of the service department of the National Electric Light Association Headquarters, 29 West 39th St., and George F. Oxley has been selected as director of the publicity department. Mr. Herbert will take charge of interpreting the uniform classification of accounts and rate research and general statistical service for member companies of the association. Mr. Oxley will have charge of the National Association Magazine and all publicity in the interests of the electrical industry.

James T. Hutchings has resigned as president of the Rochester Gas and Electric Corp. to accept the position of assistant general manager with the United Gas Improvement Co., of Philadelphia. Mr. Hutchings is a graduate of the Amherst Agriculture College. He was connected at various times with the Thomson-Houston Electric Co.; the Germantown Electric Co., of Philadelphia as assistant superintendent; the West End Electric Co., of Philadelphia, as superintendent, and with the Rochester Railway and Light Co., of which he became president.

Engineering Affairs

Buffalo Chapter American Association of Engineers, 400 members, has voted to join the Buffalo Chamber of Commerce.

The Pennsylvania Congress for 1920 will be held at Harrisburg March 21-25 under the auspices of the Department of Labor and Industry of the Commonwealth of Pennsylvania. With the exception of Sunday sessions all meetings will be held in the Hall of the House of Representatives.

The American Institute of Electrical Engineers, at the meeting of its board of directors held Feb. 18, decided, in accordance with the recommendation of Vice President Fisken and the Portland section, to hold the 1920 Pacific Coast convention at Portland, Ore., July 21-23.

The Eastern Pennsylvania Chapter of the American Society of Heating and Ventilating Engineers will hold a meeting at the Engineers Club of Philadelphia, 1317 Spruce St., Thursday evening, April 8. An illustrated lecture on "Economizers" will be given by M. C. Sherman, mechanical engineer, of the Green Fuel Economizer Company.

The American Institute of Electrical Engineers, in accordance with the action of its board of directors at a meeting held Feb. 18, and negotiations carried on since, has decided to hold its annual convention at The Greenbrier, White Sulphur Springs, W. Va., June 29-July 2. This is a week later than it was originally planned to hold the convention. Details will be announced later.

The Association of Iron & Steel Electrical Engineers, Philadelphia Section, will hold the following meetings: On April 3, James E. Wilson, electrical engineer of the Treadwell Engineering Co., will present a paper on "Electric versus Gas Welding." On May 1, every active member is invited to prepare a paper on the subject pertaining to the application of electricity to the iron and steel industry.

Business Items

The Dayton-Dowd Co., of Quincy, Ill., has opened a branch office in Cleveland, Ohio, under the management of L. E. Maker, of the Maker Engineering Co., of Chicago.

The Jeffrey Manufacturing Co., of Columbus, Ohio, has opened a new branch office in Los Angeles, Cal. The office will be in charge of F. R. Field, formerly manager of the Denver office.

The Black & Decker Manufacturing Co. has been sold to the Dieffenbach Westendorf Manufacturing Co. The new company is composed entirely of former employees of the Black & Decker Manufacturing Co.

The Hays Engineering Co., of Columbus, Ohio, has a district office at 1008 Scofield Building, Cleveland. The district office manager is Henry Kelsey. The company handles the Joseph W. Hays Corp. line of combustion apparatus.

The Buffalo Forge Company, Buffalo, N. Y., at a recent stockholders' meeting appointed the following new officers: Henry W. Wendt, president; Edgar F. Wendt, Jr., vice president and secretary, and C. A. Booth, vice president and sales manager. The new directors include the above-named officers and, in addition, H. S. Whiting.

Trade Catalogs

G-R Instantaneous Heater. Bulletin No. 221. The Griscom-Russell Co., New York City. 11 pages, 6 x 9 in. illustrating and describing the construction and uses of this heater.

The Sterling Engine Co., Buffalo, N. Y., has recently published a 23-page, illustrated catalog on Sterling engines of the commercial type for boats, pump and generator drives. A page is devoted to gasoline motors for stationary work for driving centrifugal pumps and electric generators; several pages to various Sterling engine models, specifications and installations, and charts showing the guaranteed fuel and horsepower curves of two models. A copy may be had upon application.

Worthington Pump and Machinery Corp., New York City, has issued a new and complete 115-page, illustrated catalog on "Worthington Condensing Apparatus." All types of condensers are described—vacuum and circulating pumps, steam ejectors, relief valves, flow of water in pipes, etc. It has also several pages of low pressure steam tables, a page of conversion tables, with other notes and information.

THE COAL MARKET

BOSTON—Current prices per gross ton f.o.b. New York loading ports:

Anthracite

	Company Coal
Egg	$8.20@$8.65
Stove	8.45@ 9.05
Chestnut	8.55@ 9.05
Pea	7.05@ 7.40
Buckwheat	3.60@ 3.80
Rice	2.30@ 2.50
Barley	1.25

Bituminous

	Clearfields	Cambrias and Somersets
F.o.b. mines, net tons	$2 85@$3 35	$3 15@$3 60
F.o.b. Philadelphia, gross tons	5.05@ 5 60	5.35@ 5 90
F.o.b. New York, gross tons	5 40@ 5.95	5 75@ 6 25
Alongside Boston (water coal), gross tons	7 00@ 7 75	7.60@ 8 00

Pocahontas and New River are practically off the market for coastwise shipment, but are quoted at $6.25@$7.00.

NEW YORK—Current quotations. White Ash, per gross tons, f.o.b. Tidewater, at the lower ports are as follows:

Anthracite

	Company Coal
Broken	$7.80@$8.25
Egg	8.20@ 8.65
Stove	8.45@ 9.05
Chestnut	8.55@ 9.05
Pea	7.05@ 7.40
Buckwheat	5.15
Rice	4.50
Barley	4.85
Boiler	4.85

Bituminous

Government prices at mines:	Spot
South Fork (best)	$3.25@$3.50
Cambria (best)	3.00@ 3.25
Cambria (ordinary)	2 60@ 2.90
Clearfield (best)	3.00@ 3.25
Clearfield (ordinary)	2.60@ 2.90
Reynoldsville	2.85@ 2.90
Quemahoning	3 25@ 3.50
Somerset (medium)	3 00@ 3.25
Somerset (poor)	2.50@ 2.75
Western Maryland	2.50@ 2.75
Fairmont	2 25@ 2.50
Latrobe	2.60@ 2.90
Greensburg	2 75@ 3.00
Westmoreland ¾ in.	3.40@ 3.50
Westmoreland run-of-mine	2.75@ 3 00

PHILADELPHIA—Bituminous coal prices vary according to district from which they are mined. For ordinary slack the price is $2.45@$2 55; lump, $3.35, at the mines.

BUFFALO—

Anthracite

	On Cars, Gross Ton	At Curb, Net Ton
Grate	$8.55	$10.80
Egg	8.80	10.65
Stove	9.00	10.85
Chestnut	9.10	10.85
Pea	7.75	9.30
Buckwheat	5 70	7.75

Bituminous

Allegheny Valley	$4.80
Pittsburgh	4 65
No. 8 Lump	4.65
Mine-run	4.80
Slack	4.10
Smokeless	4 60
Pennsylvania Smithing	5 70

CLEVELAND—Prices of coal per net ton delivered in Cleveland are:

Anthracite

Egg	$12.25@$12.40
Chestnut	12.50@ 12.70
Grate	12.25@ 12.40
Stove	12.40@ 12.60

Pocahontas

Nine-run	$7.50

Domestic Bituminous

West Virginia split	$8.50
No. 8 Pittsburgh	$6.60@ 6.90
Massillon lump	8.25@ 8.50
Coshocton lump	7.15

Steam Coal

No. 6 slack	$5.25@$5.50
No. 8 slack	5.10@ 5.50
Youghiogheny slack	5.25@ 5.50
No. 8 ½ in.	5.70@ 6.60
No. 6 mine-run	5.25@ 5.50
No. 8 mine-run	5.75

Only coal available is mine-run Pocahontas.

MIDDLE WEST—Chicago quotations, f. o. b. cars at mine:

	Springfield, Carterville, Williamson, Franklin, Saline, Harrisburg	Fulton, Peoria	Grundy, La Salle, Bureau, Will
Lump	$2.55@$2.70	$2.95@$3.10	$3.25@$3.40
Washed	2.75@ 2 90	....	3.45@ 3.60
Mine-run	2.35@ 2.50	2.75@ 2.90	3.00@ 3.15
Screen'gs	2.05@ 2.20	2.35@ 2.50	2.75@ 2.90

New Construction

PROPOSED WORK

Conn., Bridgeport—C. J. Lewis, 166 Fairfield Ave., will receive bids for the construction of a 3 story, 67 x 87 ft. bowling alley addition and altering of present building on Elm St. A steam heating system will be installed in same. Total estimated cost, $100,000. Davis & Done, 1024 Main St., Archts. and Engrs.

Conn., Torrington—J. A. Jackson, Archt., 1133 B'way, New York City, will soon award the contract for the construction of a 1 story, 125 x 120 ft. church on Union St., for the St. Marys Polish Roman Catholic Society, Rev. J. Kowalski, Pastor A steam heating system will be installed in same. Total estimated cost, $125,000

N. Y., Arcade—The Bd. Educ. is having plans prepared for the construction of a school building. A steam heating system will be installed in same. Total estimated cost, $125,000. Tooker & Marsh, 101 Park Ave., New York City, Archts. and Engrs

N. Y., Little Falls—The Bd. Educ. is having plans prepared for the construction of a school building A steam heating system will be installed in same Total estimated cost, $125,000. Tooker & Marsh, 101 Park Ave., New York City, Archts. and Engrs.

N. Y., New York—The Bd. Educ., 500 Park Ave., is having plans prepared for the construction of a 5 story, 70 x 145 x 145 ft. school on 87th St. A steam heating system will be installed in same. Total estimated cost, $470,000. C. B. J. Snyder, Municipal Bldg., Archt. and Engr.

N. Y., New York—The Merchants & Shippers Insurance Co., 49 Wall St., is having plans prepared for the construction of a 12 story office building at 14 South William St. A steam heating system will be installed in same. Charles H. Higgins, 19 West 44th St., Archt. and Engr.

N. Y., New York—The Post Office Dept., U. S. Government, Wash., D. C., is having plans prepared for the construction of a 4 story, 125 x 200 ft. post office building, here. A steam heating system will be installed in same. Total estimated cost, $400,000. John H. Sheier, 25 West 42nd St., Archt. and Consult. Engr.

N. Y., Ogdensburg—Williams & Johnston, Archts., will receive bids for the construction of a 1 story, 73 x 289 ft. factory building, a 70 x 130 ft. foundry and a 25 x 45 ft. garage on Green St. Power boilers and a complete steam heating system will be installed in same. Total estimated cost, $100,000.

N. Y., Ossining — Charles F. Rattigan, Supt. of the State Prisons, Hall of Records Bldg., Capitol, Albany, will receive bids until March 23 for the construction of a pump house and installation of heating systems, etc., at the Sing Sing Prison, here.

N. Y., Randalls Island—The Dept. of Pub. Charities, Municipal Bldg., New York City, received bids for delivering and installing 4 new boilers in the power house, here, from Chute, Thornton & Bailey, 2 East 13th St., New York City, $53,800; Boston Eng. Co., 308 Atlantic St., Boston Mass., $69,875; W. J. Olvany, 177 Christopher St., New York City, $74,380.

N. Y., Schenectady—Dubois B. Bennett & Son, Contractors, Engrs., and Builders, 413 McClellan St., are in the market for second hand equipment including one 18 or 20 C. F. Mixer, Hopper Loading with engine, with or without boiler, 1 hoisting engine D. D. d. c. skeleton, 1 stiff-leg derrick 80 ft. boom 10 ton capacity, fittings to be for 3 lines, 1 C. Y. clamshell bucket, independent swinging engine, 1 40 to 60 hp. locomotive type boiler and 3 line derrick fittings.

N. Y., Seneca Falls — The Seneca Co. Products Creamery Co. Inc., will install an ammonia cold storage plant and ice cream making machinery. Total estimated cost, $5,000. Address William L. Clark.

N. J., Vineland—The Catholic Church of the Sacred Heart will soon award the contract for the construction of a 2 story, 60 x 120 ft. school. A steam heating system will be installed in same. Total estimated cost, $100,000. Berry & Cummiskey, Real Estate Trust Bldg., Philadelphia, Archts.

Pa., Boiling Springs—The Boiling Springs Electric Light & Water Co. plans to install one 100 kva., 2,400 volt, 3 hp., 60 cycle, 1,200 r.p.m. Westinghouse a.c. generator for the lighting system. J. C. Bucher, Mgr.

Pa., Dushore—The Sullivan Co Electric Co. plans to install a new generator and boiler and construct a 3 mi. transmission line. H. L. Troxell, Mgr.

Pa., Ford City—The Municipal Electric Light Plant plans to install one more unit. P. J. Winigens, Supt.

Pa., Grove City—The Municipal Electric Plant plans to install one 125 kw. unit directly connected to oil engine and construct 2 mi., 3 ph., 3 wire, 2,200 volt transmission line. Frank H. Poehlman, Supt.

Pa., Hazleton—The Hazleton School Bd. plans to build a 2 story, 100 x 100 ft. junior high school at Hazleton Heights. A steam heating system will be installed in same. Total estimated cost, $200,000.

Pa., Larksville—(Wilkes-Barre P. O.)—St. John's Church Congregation plans to build a 3 story, 50 x 100 ft. school building. A steam heating system will be installed in same. Total estimated cost, $100,000.

Pa., Monessen—The Borough of Monessen will soon receive bids for the construction of a deep well pump Chester & Fleming, Union Bank Bldg., Pittsburgh, Engrs.

Pa., Orrtanna — The Orrtanna Electric Light & Power Co. plans to construct a transmission line from here to Biglerville. S. Z. Musselman, Mgr

Pa., Oxford — The Oxford Electric Co. plans to construct an extension to transmission system. Estimated cost $100,000. J. H. Ware, Mgr.

Pa., Philadelphia—Proctor & Schwartz, Tabor Rd. and 6th St., will build a 4 story, 77 x 150 ft. factory addition. A steam heating system will be installed in same. Total estimated cost, $150,000. Day & Zimmerman, 611 Chestnut St., Archts. and Engrs.

Pa., Pitcairn — The Municipal Electric Light & Power Plant plans to install a new 200 kva. gas unit. M. H. Stout, Secy.

Pa., Ridgway—The Ridgway Electric Light Co. plans to install one 5,000 kva. turbo-generator, boiler and condenser equipment and construct about 2 mi., 22,000 volt transmission line. W. H. Brown, Supt.

Pa., Scranton — Worthington Scranton, Monroe and Linden Sts., will soon award the contract for the construction of a 1 story, 85 x 195 ft. school on Clay Ave. and Poplar St. A steam heating and ventilating system will be installed in same. Total estimated cost, $100,000. Edward H. Davis and William G. D. Lewis, Union Bank Bldg., Archts.

Pa., Williamstown — The Lykens Valley Light & Power Co. plans to construct a 5 mi., 3 ph. transmission line. J. B. Whitworth, Mgr.

Md., Baltimore—The Hess Steel Corp., Biddle St. and Loney's Lane, is in the market for a 75 ft. stack measuring 40 in. in diameter.

Md., Baltimore — Samual T. Williams, 223 North Calvert St., is in the market for a 100 hp. steam engine, either throttling or automatic (used).

N. C., Wake Forest—The Mayor will receive bids until March 25th for the construction of a pumping station, tower tank, etc. W. M. Piott, Durham, Engr.

S. C., Bamburg—The Comrs. of Pub. Wks. plan to install one 125 kw. generator. J. F. Black, Supt.

S. C., Chesterfield—The Teal Light & Power Co. plans to install a duplicate unit. D. T. Teal, Mgr.

O., Cleveland—The city will soon award the contract for installing two 125 hp. boilers with stokers. Total estimated cost, $10,000. C. C. Smith, Engr.

O., Cleveland—The city received bids for the installation of two 850 gal. per min. centrifugal feed water pumps, from Dravo-Doyle Co., Citizens Bldg., $9,850 and $10,598; J. Sheldon Engine Co., Toledo, $10,820 and $11,024.

O., Cleveland—St. Joseph's Church, Rev. J. W. Bell, Pastor, 14422 Aspinwall Ave., is having plans prepared for the construction of a 3 story, 80 x 145 ft. school on St. Clair Ave. between East 144th and 146th Sts. A steam heating system will be installed in same. Total estimated cost, $300,000. Charles Greco, Guardian Bldg., Archt.

O., Cleveland—St. Stanislaw's Church, Rev. W. Krzycki, Pastor, 3649 East 65th St., plans to build a 3 story school building. An extension to steam heating system in present plant will be installed in same. Total estimated cost, $150,000. W. C. Jansen, Electric Bldg., Archt.

O., Cleveland—M. Silverberg, 9305 Hough Ave., is having preliminary plans prepared for the construction of a 2 story theatre and commercial building on East 152nd St. and St. Clair Ave. A steam heating system will be installed in same. Total estimated cost, $100,000. F. J. Prochaska, 11111 Buckeye Rd., Archt.

O., Cleveland—W. Taylor Son & Co., 630 Euclid Ave., plans to construct a 130 x 160 ft. warehouse on East 17th and Superior Ave. A steam heating system will be installed in same. Total estimated cost, $100,000.

O., Columbus—R. L. Wirtz, 2080 Luka Ave., is having plans prepared for the construction of a 4 story, 62½ x 131½ ft. office building on East State St. A steam heating system will be installed in same. Total estimated cost, $250,000. E. E. Pruitt, 209 South High St., Archt.

O., Jewett—The village will receive bids until March 25 for the construction of a municipal waterworks plant including deep well pumps, steel reservoir, etc. George H. Collins, Pres. of the Bd. of Trustees of Public Affairs.

O., Marion—The Bd. Educ. will receive bids until March 29 for the construction of a 2 story, 70 x 90 ft. school on Grand Ave. A steam heating system will be installed in same. Total estimated cost, $100,000. H. J. Merchant, Clk. O. D. Howard, 8 East Broad St., Columbus, Archt.

Ia., Ames—The Dept. of Eng., will soon receive bids for the construction of a sewage disposal plant consisting of a pumping plant, etc., for the city. Total estimated cost, $100,000. P. E. Hopkins, Mgr.

Ia., Aurelia—E. S. Kiernan, Secy. of the Bd. Educ. received bids for the installation of a heating and plumbing system in the proposed 3 story, 75 x 166 ft. school building from the C. F. Rock Plumbing & Heating Co., St. Joseph, Mo., at $41,895.

Ia., Des Moines — Frank Jeffries, City Clk., will soon award the contract for the installation of 1 single unit electric sump pump in connection with the Northeast Sanitary Sewer system project. Total estimated cost, $35,000. K. C. Kastberg, City Engr.

Ia., Sutherland—F. E. Tellier, Pres. of the Bd. Educ. is having plans prepared for the construction of a 3 story, 60 x 140 ft. consolidated school building. A steam heating and ventilating system will be installed in same. Total estimated cost, $125,000. Keffer & Jones, 204 Masonic Temple, Des Moines, Archts.

Minn., Nashwauk—R. M. Blockburn, Clk., will receive bids until March 20 for the construction of a 3 story, 160 x 174 ft. high school for the School Dist. 9, Itasca Co. A steam heating and ventilating system will be installed in same. Total estimated cost, $400,000. W. T. Bray, Torrey Bldg., Duluth, Archt. and Engr.

Kan., Ft. Scott—Thogmartin & Gardner, Contractors, are in the market for one 75 hp. traction engine.

Neb., Kearney—The State Bd. Educ. will receive bids until April 11 for the construction of a 3 story, 64 x 160 ft. dormitory. A steam heating system will be installed in same. Total estimated cost, $100,000. J. H. Craddock & Co., Elks Bldg., Omaha, Archt.

S. D., Armour—The South Dakota Light & Power Co., plans to install one 500 hp. oil engine Diesel or semi-Diesel type and one 300 hp. producer. G. W. Welch, Supt.

N. D., Fargo—The Public Service Garage Co. will soon receive bids for the construction of a 2 story, 140 x 200 ft. garage. A steam heating system will be installed in same. Total estimated cost, $125,000. Braseth & Rosatti, 105½ B'way, Archts.

Wyo., Riverton — Dubois & Goodrich, Archts., Townsend Bldg., Casper, will receive bids until April 15 for the construction of a 2 story public school for the city. A steam heating and ventilating system will be installed in same. Total estimated cost, $100,000.

Mo., Fulton—Holmes & Flynn, Archts., 8 South Dearborn St., Chicago, will receive bids in March for the construction of a 3 story 65 x 140 ft. academy, for the William Woods College. A steam heating system will be installed in same. Total estimated cost, $125,000.

Mo., St. Louis—The Lehmann Machine Co., 606 South B'way, plans to construct a 2 story, 110 x 195 ft. factory on La Salle and Grand Sts. A steam heating system will be installed in same. Total estimated cost, $450,000.

Mo., St. Louis—The Wiedmer Eng. Co., Archts., Syndicate Trust Bldg., will soon receive bids for the construction of a 3 story, 180 x 200 ft. factory for the manufacture of automobiles, on Forest Park Ave. and Sarah St., for The Dorris Motors Corp., Sarah St. and Laclede Ave. A steam heating system will be installed in same. Total estimated cost, $350,000.

Tex., Abilene—The Abilene Gas & Electric Co., plans to install a new generating plant of a 25 kw. capacity. J. W. Dawley, Mgr.

Tex., Archer—The Municipal Electric Plant plans to construct an entire new plant for the generating system. A. Minnich, Mgr.

Tex., Bastrop—The Bastrop Water, Light & Ice Co. plans to install one 150 kva. generator. J. E. Haines, Supt.

Tex., Cisco—The Cisco Gas & Electric Co. plans to construct a transmission line from here to Abilene. P. W. Campbell, Mgr.

Tex., Gordon—The Gordon Mingus Light Co. plans to install a 50 hp. unit in the generating system. S. W. Daws, Mgr.

Tex., Henderson—The Henderson Cotton Oil & Gin Co. plans to install a complete new generating plant. O. E. Morris, Mgr.

Tex., Lometa—The Lometa Light & Power Co. plans to install an additional oil engine in the generating system. M. D. Melear, Mgr.

Tex., Marshall—The Marshall Electric Co. plans to install one 1500 kw. turbo generator. W. Campbell, Mgr.

Tex., Nocoma—The Nocoma Ice & Light Co. plans to install a new 75 kw. 250 volt d.c. generator, and construct a 3 wire, 250 volt transmission system. J. W. Lehman, Mgr.

Tex., Palacios—The City of Palacios plans to install one 75 kva. generator. M. Lipscomb, Supt.

Tex., San Benito — The Commonwealth Water & Electric Co. plans to install one complete 200 hp. unit in generating system. A. L. Harris, Mgr.

Tex., Troup—The Troup Power & Light Co. plans to replace the power plant with a transmission line. H. E. DeLawe, Vice Pres.

N. M., Silver City—The Silver City Natl. Bank is having plans prepared for the construction of a 6 story, 102 x 146 ft. hotel. A vacuum steam heating system will be installed in same. J. J. Frauenfelder, 1116 Story Bldg., Los Angeles, Archt.

Utah, Myton—The Uintah Power & Light Co. plans to construct a new power plant and a transmission line to connect with 6 small towns. J. H. Colthorp, Mgr.

Ariz., Phoenix—Charles Korrick, c/o L. M. Fitzhugh, Archt., 210 Noll Bldg., is having plans prepared for the construction of a 4 story, 123 x 130 ft. office building on 3rd Ave. and Washington St. A steam heating system will be installed in same. Total estimated cost, $250,000.

Ariz., Phoenix—The Salt River Valley Water Users Association plans an election in April to vote on proposed installation of pumps, etc. Total estimated cost, $400,000. D. W. Murphy, 606 South Hill St., Los Angeles, Cal., Consult. Engr.

Wash., Chelan—The Chelan Electric Co. plans to install a 75 kw., 3 ph. generator and construct a 3 mi. transmission line. F. V. Harper, Mgr.

Wash., Clarkston—The Washington, Idaho Light & Power Co. plans to extend a 9 mi. transmission line from here to Cul de Sac, Idaho. F. J. Keyes, Mgr.

Wash., Connell—The Town Council plans to construct a waterworks system for domestic purposes including the installation of a pumping plant, etc. Total estimated cost, $8,000. C. K. Bell, Clk.

Wash., Ephrata—The Ephrata Light & Power Co. plans to change generating system from a 110 to 250 hp. L. R. Nelsen, Mgr.

Wash., Republic—The Republic Light & Power Co. plans to construct a new pipeline at power house in generating system. A. L. Capper, Mgr.

Wash., Seattle—The Westminster Presbyterian Church, B'way and East Columbia St., has retained R. H. Orr, Archt., 1301 Van Nuys Bldg., Los Angeles, to prepare plans for the construction of a 126 x 140 ft. church. Steam heating and ventilating system will be installed in same. Total estimated cost, $125,000.

Wash., Walla Walla—The city plans to construct a hydro electric plant based on 50 second ft. water to generate 1,000 kw. of power, capable of furnishing 900 kw., 1,200 hp. at city line. Total estimated cost, $381,000. E. B. Hussey, Alaska Bldg., Engr.

Wash., Yacolt—The Northwestern Electric Co., Portland, Ore., is making surveys for an electric power plant site near here. Estimated cost, $5,000,000.

Ore., Portland—The Multnomah Hospital Comn. plans to build a 5 story, 150 x 210 ft. building. The boiler house and laundry departments are to be in a separate building. Total estimated cost, $800,000.

Ore., Stayton—The Stayton Light & Power Co. plans to install a 125 hp. engine. H. J. Rowe, Mgr.

Cal., Burlingame—The City Trustees received bids for furnishing and installing two pumps for the city waterworks system, from the Simonds Mchy. Co., 117 New Montgomery St., San Francisco, $11,107; L. E. Penniman, San Mateo, $12,581.

Cal., Huntington Park (Los Angeles P. O.)—The City Trustees are having plans prepared for the installation of pumps, construction of a 1,000,000 gal. reservoir, etc. The city voted $160,000 bonds for this project. F. E. Trask, 616 Union Oil Bldg., Las Angeles, Engr.

Cal., Redlands—The Redlands University retained N. F. Marsh, Archt., 211 B'way Central Bldg., Los Angeles, to prepare plans for the construction of 3 buildings including a 2 story dormitory and 2 smaller buildings for the Art Dept. An extension to the steam heating system will be installed in same. Total estimated cost, $120,000.

Ont., Ingersoll—W. G. Murray, Archt., Dominion Savings Bldg., London, will soon award the contract for the construction of a 2 story school for the School Bd. A steam heating system will be installed in same. Total estimated cost, $125,000. R. B. Hutt, Chn.

Ont., Kingston—The Paramount Theatres, Ltd., Temple Bldg., Toronto, plan to build a 1 and 2 story theatre. A steam heating system will be installed in same. Total estimated cost, $350,000. N. L. Nathason, Temple Bldg., Toronto, Mgr.

POWER

Volume 51 New York, March 23, 1920 *Number* 12

Moderation

HAVE you ever stopped to consider the importance of the idea conveyed by the word "Moderation"? That one word would make a good sign to hang up beside the "Safety First" card, where everyone would see it.

There are two men likely to be found around any power plant who have to be watched. One is the man who thinks that things "that are" must be the best because they "are." The other thinks that anything new must be best because it is new. The man who picks his way between these two is the one who gets results.

The "Let's-try-something-new" man thinks he stands for progress, and often he does, but unless he is very careful his progress may lead to chaos. The only kind of progress that pays is progress in the right direction.

The "Old Fogy" thinks he stands for what is sure to be right because it has seemed to be right in the past. If his advice was followed entirely, no progress would ever be made. In fact, we wouldn't have any power plants to make progress in.

The man whose motto is "Moderation" listens to both stories. He knows that only by using the new things and the new ideas can improvements be made. He also knows that a modern plant cannot live if it stands still. It must go ahead. On the other hand, he realizes that a new idea is not bound to be good just because it is new. If an idea has been used, and proved its worth, it should not be discarded simply to make a change. As the lawyers say, "The burden of proof is on the advocate of a change."

Enthusiasm is a great thing, but enthusiasm tempered by moderation is much greater.

Contributed by Waldo Weaver

High-Speed Turbine-Driven Blowers

By C. H. Smoot*

DEVELOPMENT of the centrifugal blower into its modern form permits high speeds of rotation and economical machines handling small volumes to high pressures. When the speed of rotation is limited, a given pressure generally determines a given minimum horsepower volume which can be handled economically by a turbo-blower. By using high rotative speeds, this limit can be removed, and turbo-blowers built in a range of capacities and pressures otherwise impossible.

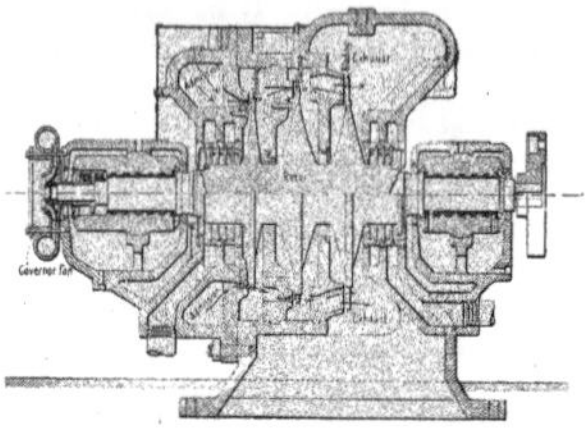

FIG. 1. SECTION THROUGH TURBINE.

On raising the speed of rotation for a given horsepower output, the first limit encountered is at the turbine. It is more difficult to find means of constructing a high-speed turbine than to find means of constructing the turbo-blower which the former will drive, and in almost every instance the limit of rotative speed is found in the turbine end of the unit and not in the blower. The type of high-speed turbine and blower which the Rateau Battu Smoot Co., of New York City, has developed is shown in the headpiece and also in Figs. 1 and 5.

The most significant feature in the turbine construction is the use of a single steel forging for the entire rotating element, as indicated in Fig. 1. The shaft and disk wheels are made integral, with turbine buckets mounted in the periphery of the wheels; the latter are virtually collars on the shaft. As a result the shaft diameter can be made as large as desired to accomplish the almost necessary condition of having its critical speed above the speed of rotation. Since the disk wheel is integral with the shaft and is not a perforated ring of metal, its stresses can be maintained low, as the shaft where it passes through the wheel contributes its full share in holding the wheel against centrifugal strains. A wheel that is made separate from the shaft, on the contrary, requires either material that will stand much higher stresses, or, on the other hand, the shaft must be made so small that the machine passes through its critical speed, both conditions being objectionable and both conditions being eliminated by the use of the integral construction.

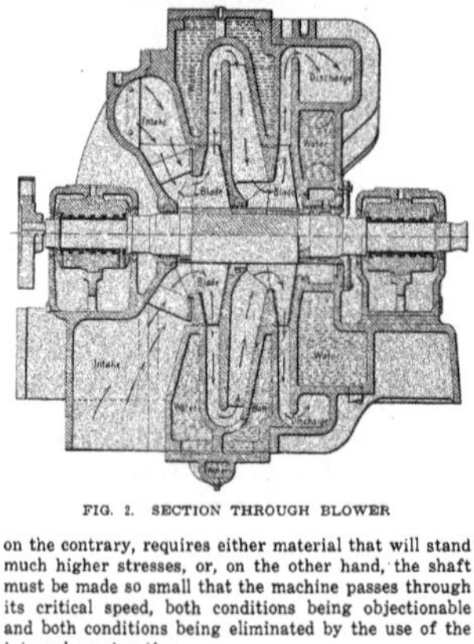

FIG. 2. SECTION THROUGH BLOWER

The turbine shown to the left in the headpiece and in Fig. 1 is a three-stage high-pressure condensing machine in which the bucket velocity is sufficiently high to use economically the available heat energy between boiler pressure and vacuum. The design of the rotor

*Engineer, Rateau Battu Smoot Co.

permits this high bucket velocity and maintains the working stress well within safe limits. The blower driven by this machine is direct-connected, and its rotor construction is in principle similar to that of the turbine. The blower blades which constitute the impeller are similar to those in Fig. 3 and are mounted with a dove-tailed construction direct in the blower shaft, Fig. 4, no disk wheel being used. In consequence the blower shaft can be of ample diameter and stiffness to obviate its passing through the critical velocity. Since the centrifugal effort of the impeller blades is held by the metal of the shaft, this metal serves the double purpose of giving proper stiffness to the shaft and at the same time holding together the impeller against centrifugal effort.

FIG. 3. BLOWER BLADES

The blades are made with a large cross-section at their root, and taper to a small cross-section at the periphery in such a way that the stress is practically constant throughout the blade section. The diffusion action of the blower is accomplished in an annular passage surrounding the impeller wheels, in which no stationary guide blades are employed, following the practice in many water pumps, and through the proportioning of this diffuser extremely efficient velocity conversion into pressure is obtained. Also, the absence of the guide blades entirely eliminates the siren action, which has frequently made turbo-blowers objectionably noisy. The section through the blower, Fig. 2, clearly indicates by arrows the path of the air.

The stationary part of the turbo-blower is entirely water-jacketed by the use of large water spaces, also indicated in Fig. 2, which may be readily inspected and cleaned of sediment. The passages contain a relatively large amount of water, which keeps the temperature of the blower casing sufficiently low to avoid distortions and misalignment of the machine when running for long periods on a closed discharge.

The machine shown in the headpiece, with the upper half of the casing removed, and in Figs. 1 and 2, is capable of developing 1,000 shaft horsepower maximum, and compressing 15,000 cu.ft. of free air to 15 lb. pressure when running at a speed of 8,750 r.p.m. This pressure is obtained in two stages, of two sets of impeller wheels in series. Similar machines are being constructed in which a pressure of 15 lb. is generated in a single impeller wheel, and in these units the working stress is likewise maintained well within the safe value. The 1,000-hp. units have been tested up to a speed of 10,000 r.p.m. and operate in a most satisfactory manner at this speed. The compactness of these units is evidenced by their dimensions, being only 12 ft. 6 in. over-all length and 4 ft. 6 in. outside diameter for the blower, and 3 ft. 4 in. diameter for the turbine. In the design and operation of units under these conditions, many interesting features have been developed. The bearings, for intance, are of much more importance to the units as oil dashpots than for the purpose of bearings, and a surprising amount of "out-of-balance" can be absorbed in the bearings themselves. A simple cast-iron babbitt-lined bushing of ample size is employed for this purpose.

In order to control the speed of units of this and greater rotative speeds, a new form of governor has been developed in which the speed element is a small steel fan mounted rigidly on an extension of the turbine shaft. The pressure of the fan varies as the square of the speed of rotation in a manner analogous to that of a fly-ball, and this pressure is used in a static regulator of very simple form which controls the admission of steam. The governing is accomplished through pressure regulation on the small direct-connected fan. Very satisfactory speed regulation is obtained in this manner through exceedingly simple apparatus, which has a very small probability of derangement.

One result from the development of this type of turbine, which may be of significant value in the future, is that it has been shown that a noncondensing turbine of high efficiency can be constructed along these

FIG. 4. BLOWER'S ROTARY SHAFT, SHOWING DOVETAIL GROOVES FOR ATTACHING BLADES

FIG. 5. TURBINE AND BLOWER ROTOR FOR 150-HP. 20,000 R.P.M. MACHINE

lines of design. Since at the high speed of rotation a large amount of the power may be generated in wheels of small diameter, the proportion of lost energy due to windage in the small high-speed machines is less than in the corresponding slower-speed machine.

These machines have been built in sizes of 150 hp., operating at 20,000 r.p.m., to 2,300-hp., operating at 1,500 r.p.m. The rotor of the 150-hp., 20,000-r.p.m. machine is shown in Fig. 5. The combined weight of the turbine and blower rotor is only about fifty pounds. In this type of machine the blower and turbine rotor are each made from one solid steel forging. A 150-hp. turbine and blower of the foregoing type that operated at 60,000 r.p.m. was built in France during the war for use on an airplane. This machine weighed only about thirty pounds.

Diesel-Electric Equipment of the Yacht "Elfay"

A PRACTICAL application of oil-electric drive in which some form of oil engine drives an electric generator which in turn furnishes power to motors driving the propellers is to be found in the yacht "Elfay," a schooner 152 ft. over-all, 30-ft. beam, and 313 gross tons, which went into commission in January, 1920. Among the chief advantages claimed for this drive are the following:

The engine operates continuously at constant speed in one direction only and therefore under ideal conditions; its fuel economy is exceedingly high; the electric system of distributing and applying power is flexible; a simple and effective form of control can be used; and greater safety can be secured through the use of several main engines and generators, as well as two or more main motors and propellers, so that part of the equipment can be out of commission without crippling the ship.

The "Elfay's" main engine is a six-cylinder, Model 54, Winton full-Diesel oil engine, with a 7½-in. bore, 11-in. stroke, and rated 115 hp. running 425 r.p.m. It is equipped with a sensitive governor, which controls the fuel supply and holds the speed of the engine practically constant, the maximum variation being about 20 r.p.m. A 75-kw. Westinghouse 125-volt direct-current generator is directly connected to this engine; and driven from it by a silent chain is a 9-kw. 125-volt exciter of 900 r.p.m. which supplies the field current for both the main generator and the propeller motor and has also about 5 kw. additional capacity for other purposes. The propeller is driven by a 90-hp. Westinghouse motor of 350 r.p.m. At the maximum motor speed the speed of the boat is 8½ knots. It is claimed that this boat can make 8½ knots on 7½ gal. of fuel per hour. Since this boat carries 2,400 gal. of fuel, she is able to run 2,000 miles on her propelling equipment alone. All the electrical machines—the generator, exciter and motor—are of standard design, the only special element being the use of ball bearings to permit operation with the vessel listing on a wide angle.

VIEW OF MAIN ENGINE, GENERATOR AND EXCITER

The method of controlling the ship is an interesting feature of the installation. When it is desired to use the power, the engineer starts the engine with its generator and exciter and brings them up to speed. He then closes the switches that connect the propeller-motor armature with the generator and the propeller-motor field with the exciter. After this he has no further responsibility in the handling of the ship, except to keep the machinery running properly. The entire control of the propeller is centered in a handle that is mounted on a pedestal on the deck just forward of the steering gear. In the center position of this handle the propeller motor is stopped. By turning the handle in one direction, the motor starts driving the ship ahead with continuously increasing speed as the handle is turned, until full speed is reached. By turning the handle in the other direction the ship is driven astern in a similar manner. The change from full speed ahead to full speed astern can be made in five seconds. In front of the control handle a set of meters is mounted, which show the voltages of the generator and the exciter, the current taken by the motor and the speeds of the generator and the motor. The navigator, therefore, has complete control of the vessel and can maneuver with the utmost speed and precision. The fields of both the motor and the generator are separately excited, as has already been stated in the foregoing. The motor field is kept always at full strength when the motor is in service, but the strength of the generator field is varied by means of a rheostat which is operated by the control handle. In the off position of this handle the generator field is open so that no current is delivered to the motor. As the handle is moved around, the generator field strength is increased and the voltage of the current supplied to the motor, consequently the motor speed, is also increased. When the control handle is reversed, the direction of the generator field current, and also the current to the motor and its direction of rotation is reversed, but otherwise the action is the same.

The "Elfay" is lighted and heated by electricity and all her machinery is electrically operated. There is also a ½-kw. wireless outfit. All this apparatus is supplied with power from the exciter when the main engine is operating and from a storage battery when the main engine is shut down. This outfit consists of a 300-ampere-hour Philadelphia battery, which is charged by means of an auxiliary generating set consisting of a 25-hp. Quayle engine, with an electric starter, which drives a 15-kw. generator. The battery has capacity enough to supply the needs of the boat (with the exception of propulsion) for two days without recharging.

Locating Faults in Induction Motors—

How to Make Tests

By
A. M. Dudley*

Tells How To Make the Tests Necessary To Locate Ten Most Common Faults That May Occur in Induction-Motor Windings.

AFTER the enumeration of the commonest errors made by the winder, as outlined in the article "Locating Faults in Induction Motors—Most Common Defects" in Mar. 2 issue, the next step is to consider them in turn with particular reference to how each may be detected and corrected.

First Fault: Grounds. If the ground is fairly low resistance—that is, the bare copper of the winding touches the core—the defect may be detected by using an incandescent lamp arranged as shown in Fig. 1. One of the lamp leads is touched to a bare spot on the winding—for example, a terminal connector or a "stub" where two adjacent coils are connected—and the other is touched to the bare metal of the motor frame at some point not protected by paint. If there is a ground present, the lamp lights up. Another common method is by "ringing out" with a magneto similar to that used in telephone work. In this method the terminals of the magneto are applied, one to the winding and the other to the frame similar to the procedure in Fig. 1, and the handle is turned. If the bell rings, there is probably a ground in the winding. A third method employs a "testing box," which is really a transformer for obtaining voltage much higher than the normal voltage of the motor under test. These boxes give 2,000 or 3,000 or more volts and readily detect grounds on windings of 550 volts and below. The test box is so arranged that when the terminals are applied as in Fig. 1, the presence of a ground instantly opens a circuit-breaker on the side of the box.

FIG. 1. CONNECTIONS FOR TESTING MOTOR FOR GROUNDS WITH LAMP

Having established the fact that the winding is grounded by some one of the foregoing methods, the next problem is to locate in which coil or what part of the winding it has occurred. This can sometimes be done by inspection, but sometimes requires other means. The most usual of these is to put enough voltage on the ground with the lamp device of Fig. 1, or the test box, so that the resulting current heats up the contact that is causing the ground and it becomes evident through smoke or slight arcing. This will generally require two or more lamps connected in parallel. When the ground is definitely located, it is corrected by repairing the insulation at this point by retaping the coil, or replacing the defective slot cell or whatever may be causing the trouble. Sometimes the ground cannot be "smoked out" in this manner, and it then becomes necessary to open up the winding at two or three places and test out the different pieces to find in which one the ground is present. If it is still not evident, the defective section of the winding is further broken into smaller pieces and the search pursued until the trouble is finally run down to the individual coil which is defective. It is seldom necessary to go so far, as the ground furnishes evidence of its location as soon as the voltage is put across it.

Second and Third Faults: Short-circuit of a few turns in a coil, or a single coil completely short-circuited, becomes hot in a short time if the motor is run light on normal voltage. Their presence can be detected by feeling around the winding with the hand immediately after starting the machine and noting if some individual coils are much warmer than others. A device for detecting such short-circuits before the rotor is put in the stator and without applying any voltage to the winding itself is shown in Fig. 2. This device is somewhat similar to a large horseshoe magnet excepting that the iron part is built up of laminations, or it may be considered as a core-type transformer having a primary coil only with one side of the iron core missing. The coil is excited with alternating current of suitable voltage, and then the complete device is passed slowly around the bore of the machine being tested as shown in Fig. 3. In passing around, if the testing device passes over any short-circuited turn or coil, such short-circuit immediately acts as a short-circuited secondary coil on a transformer of which the exciting coil on the testing device is the primary and whose magnetic circuit is made partly by the testing device and partly by the core of the machine under test. As in any short-circuited transformer, an increased current flows both in the primary and secondary coil and can be detected by an ammeter in series with the device or by the heating that immediately takes place in the defective coil, or by the attraction that the short-circuited coil has for a strip of sheet iron. By passing the device slowly around

*Designing engineer, Westinghouse Electric and Manufacturing Co.

the core and observing its behavior from point to point, short-circuits can readily be detected. This refers particularly to short-circuits in individual turns or in one complete coil. A short-circuit of a complete pole-phase group is more readily located by a compass test, and a short-circuit of an entire phase can be located by a "balance test."

The "compass test" referred to in the preceding paragraph consists in passing a compass slowly around

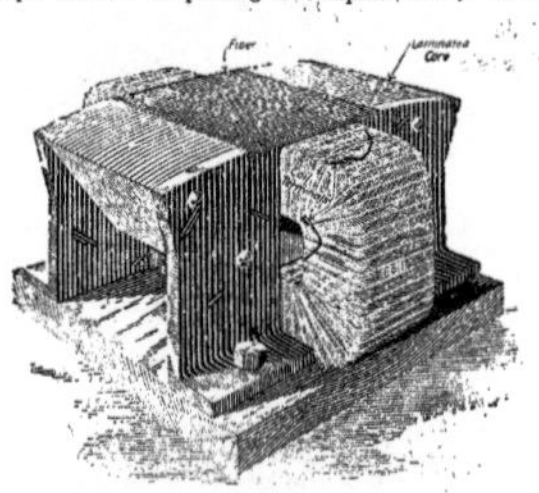

FIG. 2. TESTING DEVICE FOR LOCATING SHORT-CIRCUITS IN STATOR WINDING

the bore of the stator from which the rotor has been removed and which has the winding excited by direct current of the value of about one-third the full-load alternating current. The effect of this direct current is to set up north and south poles alternately in the phase which is excited, and as the compass is passed slowly around the bore its needle reverses with the polarity, and by marking the polarity plus and minus with chalk marks in the bore, the chalk marks immediately indicate the correctness or faults in the winding. If it is a two-phase machine, the direct current is put on each phase separately and the check is made. For a three-phase star winding cause the direct current to flow from each lead to the star by making three observations and mark the polarity only on the groups from the lead to the star in each phase separately. This can be readily understood by referring to Fig. 4. For the first observation put the direct-current plus lead on A and the minus on the star connection, then pass the compass around the bore and mark the polarity of the groups from A to the star point with an arrow, the arrow pointing in the same direction as the compass needle. For the second observation put the direct-current plus lead on B and the minus lead on the star connection and passing the compass around marking the polarity of the groups from B to the star point. For the third observation, put the direct-current plus lead on C and the minus on the star, and by means of the compass determine and mark the polarity of the groups from C to the star point. If the three observations have been made correctly, there will be a chalk arrow on each pole-phase group of the winding, and if the winding is correctly connected, these chalk arrows will alternate north and south, as shown in the Fig. 4. In case of a short-circuit of a complete pole-phase group the compass needle will not be deflected. If a three-phase delta winding is being checked, open the delta connection at one lead, as in Fig. 5, and connect the direct-current source in so that the current flows through the three phases in series, and if the pole-phase groups be checked for polarity, the arrows will reverse as just described for the star winding.

The "balance test" referred to consists in checking each phase of the winding separately with low-voltage alternating current, say 20 per cent of normal full voltage, and measuring the amperes to check the impedance roughly and see if it is the same in all phases. The connections for a star connected winding are made, as in Fig. 6, so that the current can be measured in each phase, with an ammeter. The low-voltage alternating-current source is, in all cases, connected across one terminal, A, B or C, and the star as in the figure. The ammeters should read the same in all three leads. For a delta-connected winding it is necessary to open the delta connections at some point, as at A, then test across each phase separately. This test is made on the stator only and with the rotor removed.

Fourth and Fifth Faults: Reversal of one or more coils in a group or group of coils. It happens that individual coils or sometimes entire groups are connected in backward. If the error is confined to one coil it does not usually show up on a "balance test" and would not be found on a resistance test, since the resistance would be the same no matter which way the coil was connected. Such reversed coils or groups can be located by means of the compass test described under "Short Circuits." If an individual coil is reversed, it will show a tendency to reverse the compass needle when the needle is directly over that coil. If an entire pole-phase group is reversed, the compass needle will indicate the same direction of field on three successive groups, as at Z, Fig. 7. Also if a coil is left out of circuit, or "dead," as listed under the

FIG. 3. METHOD OF USING TESTING DEVICE SHOWN IN FIG. 2

tenth fault, the compass needle will indicate an irregularity at the instant of passing over that particular coil. By checking the three phases of a three-phase winding separately, with a compass, as described under the second and third faults, it is possible to check for the reversal of an entire phase.

Sixth Fault: This is the case where one coil too many or too few is connected in a pole-phase group, as at A' and B', Fig. 7. The best check on this is a visual

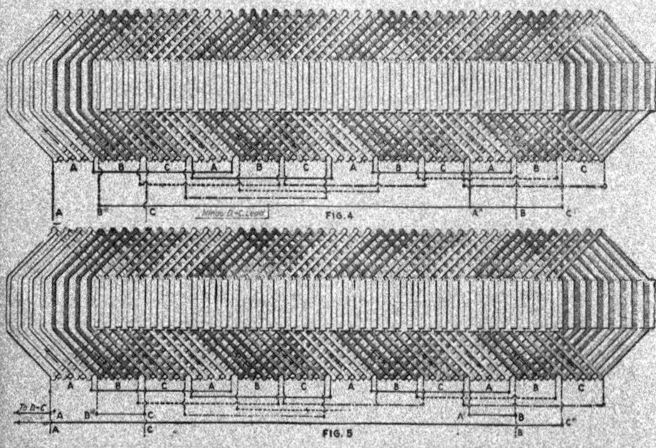

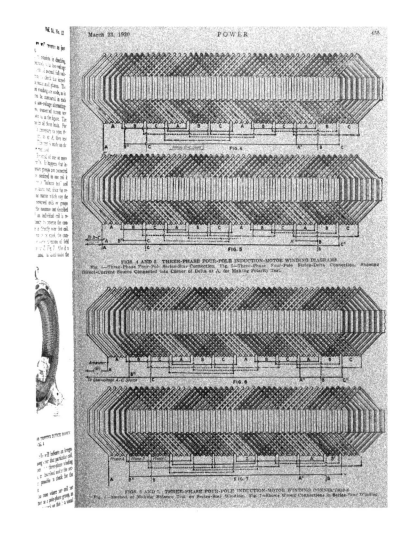

FIGS. 4 AND 5. THREE-PHASE FOUR-POLE INDUCTION-MOTOR WINDING DIAGRAMS

Fig. 4—Three-Phase Four-Pole Series-Star Connection. Fig. 5—Three-Phase Four-Pole Series-Delta Connection, Showing Direct-Current Source Connected into Corner of Delta at A, for Making Polarity Test.

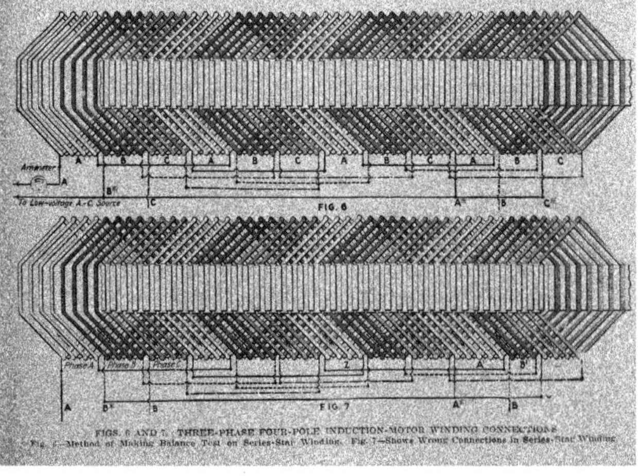

FIGS. 6 AND 7. THREE-PHASE FOUR-POLE INDUCTION-MOTOR WINDING CONNECTIONS

Fig. 6—Method of Making Balance Test on Series-Star Winding. Fig. 7—Shows Wrong Connections in Series-Star Winding.

inspection and count of the "stubs" at the end of each group, and when the trouble is located it is corrected by disconnecting, regrouping and reconnecting.

Seventh Fault: The reversal of an entire phase in a three-phase winding usually manifests itself in a very pronounced manner when the motor is run light. If the rotor turns over at all it is probably at a speed very much less than normal and emits a loud, growling noise the end that was a lead to the star, thus giving the connection, Fig. 4. In a two-phase winding there is no trouble with reversed phase for the reason that if the direction of rotation of the motor is wrong, the leads may be easily reversed outside of the motor and the correct rotation secured.

Eighth Fault: Connection for wrong voltage. If a motor is connected for a lower voltage than the circuit

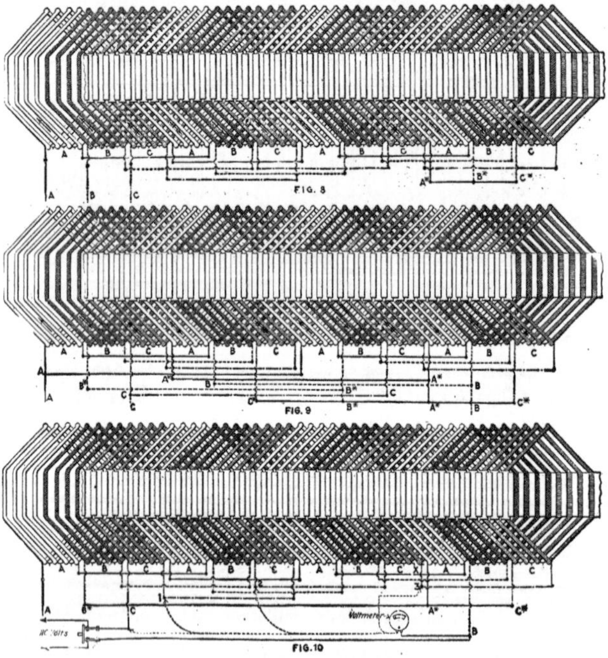

FIGS. 8 TO 10. THREE-PHASE FOUR-POLE INDUCTION-MOTOR WINDING CONNECTIONS
Fig. 8—Series-Star Winding with B Phase Reversed. Fig. 9—Four-Pole Two-Parallel Star Connection. Fig. 10—Shows Method of Testing for Open-Circuits with a Voltmeter in a Series-Star Winding.

and immediately becomes hot. This fault may also be detected by the compass test, as described under faults two and three. The arrows on the windings will point in groups of three in opposite directions, as in Fig. 8. The remedy when the defect is found is to open the star point and use the star point on the defective phase, which is the *B* phase in Fig. 8, for a lead and bringing upon which it is operating, the no-load current becomes excessive and may even approach full-load value. There is a pronounced magnetic hum and a vibration indicating that the field is very strong, which is the case. On the other hand, if the motor is connected for a higher voltage than that upon which it is being tried, the no-load current is very small and the motor apparently "pulls out" on much less than its rated full load. If these faults are a matter of half-voltage or double voltage, for example, they can usually be detected without much trouble; but if the variation is less, this becomes a more difficult matter and in the absence of any other official data it sometimes becomes necessary to take a brake test to determine what the trouble is. After the difficulty and its extent have been determined, a reconnection of the groups can usually be made which will give the proper operating conditions. For example, if it is found that the winding is connected series-star as in Fig. 4, and the motor is connected for 440 volts, when it is to be operated on a 220-volt circuit the winding

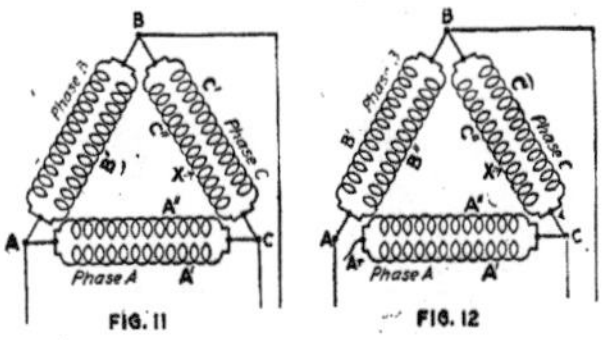

FIG. 11 FIG. 12

FIGS. 11 AND 12. SCHEMATIC DIAGRAM OF A TWO-PARALLEL DELTA CONNECTION

should be changed to parallel star, as in Fig. 9, and the operation will be normal.

Ninth Fault: The easiest way to detect a connection for the wrong number of poles is to run the motor light and take the speed with a tachometer or speed counter. When it is found that the winding is connected for the wrong number of poles, the possibility of reconnecting can be determined by methods suggested in the articles on "Changing Number of Poles," in the Apr. 9, 1918, Aug. 20, 1918, and Jan. 7, 1919, issues.

Tenth Fault: Open-circuits are manifest from the fact that the motor will not start, but acts as if it were operating single-phase. It is easy to determine, in a star-connected winding, in which phase the open-circuit exists by connecting all phase leads to the starting transformer and opening them one at a time to see in which lead no current is flowing. In Fig. 10 assume that the open is in phase *C* at *X*. Then if lead *A* is open, no current will flow through the motor, since the path is from *B* to *C* and is open at *X*. If the *B* lead is disconnected with *A*, and *C* connected, no current can flow, since the *C* phase is still in circuit. If *C* is disconnected with *A* and *B* lead connected in circuit, then *C*, the defective phase, will be cut out of circuit and current will flow in the *A* and *B* windings of the motor and it will act as if operating single-phase, which will be indicated by the motor emitting a humming sound. When the defective phase is located, it is not always apparent just where the break is. A visual inspection may fail to show the break on account of tape over the defect or for some other reason. If this point cannot be located by inspection, a simple method of finding it electrically is indicated by referring to Fig. 10. A test voltage somewhat lower than normal or whatever is convenient is then applied to *B* and *C*, and a suitable voltmeter is used to measure the voltage between *B* and various points along the *C* phase, as, for example, 1, 2 and 3, which are chosen at random along the "stubs," or coil-to-coil connections, or on the group cross-connections, as in the figure. With the condition as shown in Fig. 10, assume that 110 volts has been applied to the *B* and *C* terminals of the winding, as shown. If one lead from the voltmeter be attached to *B* and the other lead touched successively to *C* and 1, 2 and 3, the voltmeter will read 110 volts between *B* and *C*, *B* and 1, *B* and 2, and zero volts between *B* and 3, since the *C* phase is open at *X*. The conclusion is immediately and properly reached that the break is between 2 and 3 and with the inspection narrowed down to this small section of the winding the break is usually apparent. However, should the break not be discovered by inspection, points can be selected with finer steps between 2 and 3 and voltage readings taken until the defect is narrowed to the exact coil or piece of cross-connection where it exists.

In the case of a delta connection one of the simplest ways to detect an open-circuit would be to open the connection at one terminal of the delta, such as *A* in Fig. 5, and connect a test circuit across the open. If the winding is open no current will flow. The phase with the open in may be located by testing across each phase separately. If a lamp is used to make the test, the defective phase will be indicated by failure of the lamp to light. After the faulty phase has been located, the location of the defect can be determined as for the star connection, Fig. 10. There are all manners of parallel-star and other groupings in which it is difficult to locate an open-circuit, since an open in one parallel group does not open the circuit through the phase, but in only one of the parallel groups. For example, in Fig. 11 an open in phase *C* at *X* will not open the phase between terminals *B* and *C*, but only through *C*." Therefore, to detect the open group it will be necessary to break the winding up into its parallel groups and test each group separately. The defective phase could be detected by the balance test as previously described. First, open the delta connection, for example, at *A*, Fig. 12, then apply the low-voltage alternating current between points *A* and *B* and measure the current with an ammeter, test between *A* and *B*, *B* and *C* and between *C* and A_1. The phase with the open circuit, which in this case is *C*, will show a lower reading than the other two phases, after which all that is necessary is to break the phase up into its parallel groups and test the defective group for opens, as explained in Fig. 10.

These are the defects that commonly occur and the usual method of locating them. In checking for these defects the order usually observed is as follows: After the winder has completed the connection of the entire winding, his work is checked, preferably by a second winder, against the winding diagram specified for that particular job. The coils per group are counted and a visual inspection made for short-circuits, open circuits and reversed coils, groups or phases. A balance test is made on the stator alone with low voltage to see if, roughly, the same current flows in the various phases. A high-voltage test is then made on the insulation to insure that the coils are not grounded on the iron core, or that there is no short-circuit between the conductors of the different phases. If everything is satisfactory up to this point, the rotor is then assembled in the stator and the machine prepared for a running test. The resistance of the winding is measured on all phases, and if alike, the machine is passed for running test without load. Sufficient voltage is applied to start the rotor, and if it comes up to speed quickly without apparent distress or irregularity of any kind, the speed is checked, to verify whether the winding has the proper number of poles. The temperature of the winding is then tested with the hand, passing completely around the machine and using care that the rotating member and its parts do not strike the observer. If neither general heating nor hot spots are observed, the voltage is raised to normal ana the nó-loaa current in all phases and the total wctts are read. If these values check with the previous tests on similar machines or with calculations, the windings are considered to be correctly connected. If the motor does not readily come up to speed or the phases do not balance or there are signs of unequal heating in the winding or other distress, the rotor is removed and the co..nections again checked. If the error is still not apparent and a source of direct current is available, the compass test may be applied. Having exhausted this resource without avail, the problem is one that can be solved only by some expedient at the command of an experienced designing engineer, but such appeals are very seldom required, as the trouble usually appears from the simple tests described.

Shaw Monolith Suspended Boiler Arch

The Shaw suspended boiler arch, a design that does away with a multiplicity of small bricks, has been developed, and is put out by L. S. Shaw & Co., Union

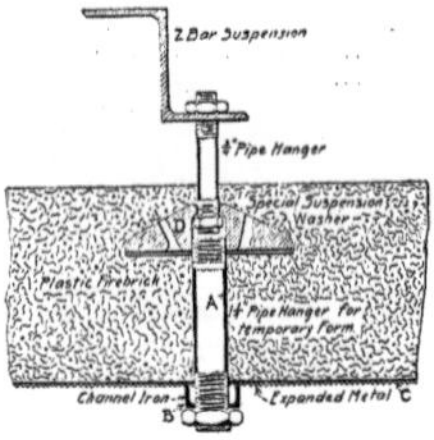

FIG. 1. DETAILS OF HANGER

Building, Cleveland, Ohio. This arch can be constructed flat or with any desired curvature or shape by simply turning the nut on the suspension bolt.

In Fig. 1 are shown the details of the hanger, which consists of a ventilated cast-iron disk suspended by

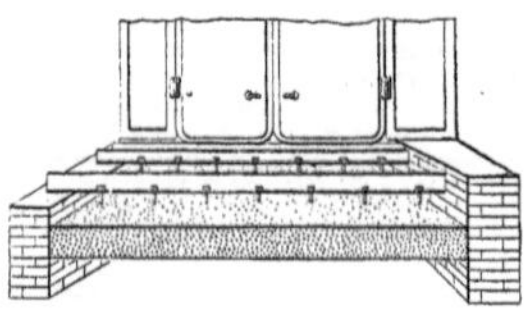

FIG. 2. FLAT-SURFACE BOILER ARCH

means of a ¾-in. pipe from supporting beams, each disk carrying about 200 lb. load.

The arch proper consists of a compounded firebrick material of silica and alumina fireclay heavily charged with asbestos fiber and free from expansion and contraction. Thus it is not necessary to form expansion joints. It is necessary to form contraction joints, which, after the arch material has been baked, are filled with fireclay. These joints are three feet apart and form a natural lamination which allows for easily removing a section of the firebrick when repairs are necessary.

The monolith arch is adapted to all furnace design and irregularities in construction, as it is only necessary to build side walls and fill in between, with the forms centered, with the firebrick mixture, and then bake it in position, as shown in Figs. 2 and 3, the former indicating a flat surface and the latter view an irregular one.

The shape of the temporary form is made by screwing a 1¼-in. adjusting pipe *A*, Fig. 1, that is supported by the channel *B* and supports the expanded metal *C*. After

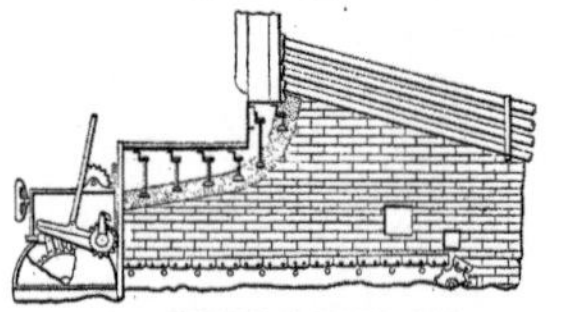

FIG. 3. IRREGULAR BOILER ARCH

the brick is baked, the pipe *A* is removed from the disk *D*, leaving the pipe hole, which can be filled with brick. When the brick requires repairs, this pipe *A* is screwed into the disk to support the centering form, which is filled with brick and then the pipe is removed.

It will be seen that this monolith form of arch has a simple beam construction, but one type of casting and firebrick, the whole being simple to build, easy to repair and requiring ordinarily but a few repair parts to be kept on hand.

Combustion at Low Speeds with Semi-Diesel Engine

For many years the inability to guarantee satisfactory running of semi-Diesel engines at slow speed and light load was an important factor in limiting the number of this type of engine installed for marine purposes. The difficulty was due to the impossibility of keeping the vaporizer sufficiently hot to insure combustion when the engine was running slow and without load. This trouble has been overcome by the Vickers-Petters patent light-load running gear, which permits the injection of fuel very much in advance of the normal point of injection.

The arrangement is carried out by making a connection in such a way that one fuel pump delivers fuel to two cylinders at each stroke. One part of the charge is injected into the cylinder normally supplied, and at the normal timing, while the other part is injected into the other cylinder in advance of the normal timing. The early injection is a light one and burns in the combustion chamber, maintaining sufficient heat there to insure proper combustion of the normal charge. The change over to this light-load device is made easily and at any time by a small hand lever. The main injection is in no way interfered with and the governor is still in action. Should full load be suddenly applied the governor at once assumes its duties and regulates the fuel accordingly.—*Engineering*, London.

Refrigeration Study Course—V. Ammonia Fittings and Accessories

By H. J. Macintire
Professor Mechanical Engineering, University of Idaho

ENGINEERS who are familiar with steam and the operation of steam boilers and engines should have no trouble with refrigerating apparatus or the operation of an ammonia system. For example, the fireman has two things of major importance to care for—water regulation and combustion control. The engineer has one big fear, namely, the effect of water in the steam line and in the engine cylinder; and a lesser one, perhaps, the matter of lubrication. Let us see how this compares with the refrigerating system.

The expansion valve controls the liquid feed into the refrigerating or cooling coils. The amount of gas boiled from these coils depends on the manner of operation and the amount of feed. But it is just as easy to bring back a slug of ammonia to the compressor as it is to bring water into the steam engine. Also, in opening a valve one should expect an accumulation of liquid ammonia back of it as easily as in a cold steam pipe. Similarly, low points in the pipe line should be provided with drains and certain other parts with modified separators. There is not the same danger of liquid hammer on the pipe line and fittings of the ammonia system only because the load does not tend to fluctuate suddenly, as a steam engine is likely to. However, when trouble does occur, it is much more serious on account of the peculiarities of ammonia. There are other troubles, however, such as increase of pressure due to rise of temperature, difficulty in maintaining tightness at all times, removal of oil and water, foul gases, etc.

The Ammonia Line Must Be Made Tight

Ammonia has the property of escaping through piping and fittings which would be absolutely tight for steam. The result is that the material used must be special or must be of high grade and very thick. Screwed fittings are made tight in two ways: Either by tinning and soldering, or by making up with a paste of litharge and glycerine. Flanged fittings are made with a special design. The faces are fitted on a lead or rubber gasket placed inside of the groove. When properly erected, an ammonia line can be made perfectly tight.

The material used for ammonia is undergoing a steady change. Where ordinary cast iron is still used it has to be very thick, but most manufacturers are taking to the use of air furnace iron, semi-steel or (in the case of fittings) drop-forging. The drop-forged ammonia fitting is becoming widely used in spite of its additional cost.

Headers and receivers are now made of steel, usually flame-welded, both as regards the longitudinal seam and headers, and also in respect to the nipples and outlets. The finished product is light in weight, moderate in price and (if the work is properly executed) amply strong. The flame-welded shapes may be made readily for any design, as it is a matter only of cutting out the steel plate and rolling to form.

The kind of pipe to be used is still a question for debate, with the use of black steel pipe gaining in the race. For certain special work where corrosion is likely to occur, it is becoming the custom to specify extra-heavy steel pipe in preference to wrought-iron pipe. In fact, there are so many factors which cause corrosion that it is fairly doubtful as to just what kind of pipe is least affected. Both lap- and butt-welded pipe are used, there being no particular difference if the finished product is carefully inspected and tested.

Oil Traps Should Be Provided

One of the most serious operating troubles is due to the effect of oil. Most operators feed the lubricating oil too heavily, but the more recent designs for high-speed compressors require considerable lubrication. If oil gets out of the cylinder (and it will), it should be separated from the hot, compressed gas before it enters the condenser. In consequence an oil separator is placed in the line, preferably nearest the condenser. However, no oil separator is perfect, because some of the oil is vaporized by the heat, and oil will pass into the condenser and will ultimately find its way into the liquid line and expansion piping. Of course, oil in expansion piping (or any other heat-transfer piping) is almost as bad as if it were in boiler tubes. The efficiency of the piping is reduced, and the coils will lose capacity. In consequence it is wise, wherever there is a low point in the expansion piping, to place a trap for the collection of the oil. These traps may be connected together and all run into the purifier, or they may drain into the atmosphere.

The refrigerator, or purifier, is an arrangement which is absolutely necessary for plants of moderate size. The piston rod, being cold, has moisture from the air condensed on it, and this, in time, gets into the refrigerating system, being carried in by the movement of the piston. Water will make ammonia "dead," and after a time satisfactory results are impossible. In consequence it is necessary to renew the ammonia charge by pumping out the entire system and recharging, or to use a special device. This latter is an arrangement by which a small amount of liquid ammonia is bled off from the high side, or from the traps, into a special drum usually arranged for heating by means of hot water or steam coils. Liquid ammonia is bled off regularly and is heated sufficiently in the regenerator to boil off the ammonia which passes to the suction of the compressor. The water and oil are left behind as a residue.

Function of the Accumulator

Every steam engineer knows the test for the water level in the boiler by use of the gage-cocks, and each engineer has his own idea of the reason for the fact that, when water is allowed to blow out, it appears as a wet steam. Apparently, water has flashed into steam on being allowed to escape into the atmosphere.

As a matter of fact, the water in the boiler contains more heat energy at boiler pressure than is permissible at atmospheric pressure. As the water reduces in pressure on passing the gage-cock, the surplus heat energy vaporizes some of the water, and therefore the appearance of steam.

The expansion valve has been misnamed; it really is only a pressure-reducing valve. As the liquid ammonia is fed through the valve, the pressure drops from, say, 150 lb. to about 15 lb. or whatever the suction pressure may be held at. The heat in a unit weight of liquid ammonia at 60 deg. F. is much greater (because of its temperature) than a similar weight of liquid ammonia at 0 deg. F. In consequence the ammonia, on passing the expansion valve, has the same effect take place as with the gage-cocks, and some ammonia will be vaporized. This ammonia depends on the temperatures of the liquid on the two sides of the expansion coils and may be less than 10 per cent, but usually it is between 10 and 15 per cent. The vapor formed is of no practical value for refrigeration, and the sooner it is gotten rid of the better it will be. In consequence, in certain cases a special device, called an accumulator, is used.

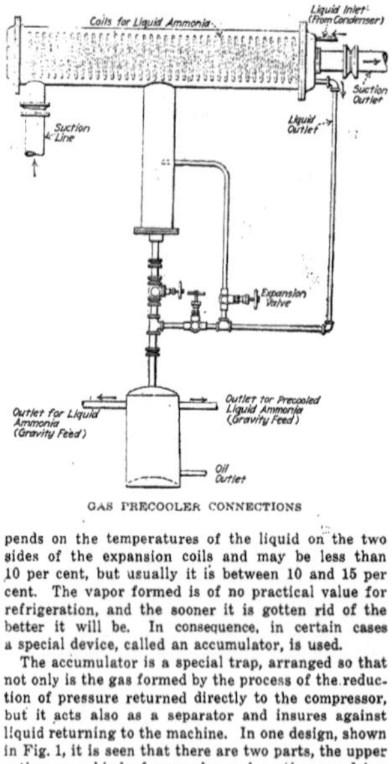

GAS PRECOOLER CONNECTIONS

The accumulator is a special trap, arranged so that not only is the gas formed by the process of the reduction of pressure returned directly to the compressor, but it acts also as a separator and insures against liquid returning to the machine. In one design, shown in Fig. 1, it is seen that there are two parts, the upper acting as a kind of precooler and suction-gas drier, and the lower as a separator for the vapor and liquid and the lowest part as an oil collector. It is noticed that the whole trap is open to the suction pressure and that the liquid ammonia flows to the expansion (refrigerating) coils by gravity. In consequence, sufficient head must be allowed to force the liquid into the coils, which are fed from the bottom and are called "flooded." A flooded system of refrigerating coils is not exactly as its term would indicate. As soon as the liquid enters the refrigerating coils, evaporation (and refrigeration) begins and gas is given off. This gas passes on up through the pipe and finally gets into the suction pipe. So the coils are all liquid at the beginning and all, or nearly all, gaseous at the top. The advantage of the flooded system is in the initial removal of the gas in the accumulator and the easy control of the feed into the cooling coils. The coils are much more efficient in this system, but as a rule it can only be used in ice tanks or special piping where a gravity head may be employed.

Pump-Out Lines Should Be Provided

To make repairs or alterations it is required that means be provided to remove the ammonia charge from the part of the system in question and store it in another part. In consequence, "pump out" lines are provided. These pipe lines leading back to the compressor need not be of full size pipe, as it is expected that only small coils or a few stands of pipe will be pumped out at a time.

To pump out the high side easily it is customary to have cross-connections on the compressor. These are means by which the discharge from the compressor is delivered to the suction pipe and the suction is drawn from the high side. The suction and discharge stop valves are closed tight, and these bypass valves are placed on each side of the stop valves. In addition, on account of necessity for freedom in starting electric-driven compressors, a full-sized bypass frequently is made between the discharge and suction pipes. This is really just a "short-circuit" so as to reduce to a minimum the power required on starting.

Suction-Line Scale Trap

On erection the pipe line is carefully blown out to clear of mill scale, etc., but during operation some scale, dirt, foundry sand, etc., will become loosened and will flow back to the compressor. The poppet type of valve is ground to a "line" contact, and other forms are ground to a surface fit. It is clear that impurities of all sorts must be kept out of the cylinder and valves. A leaky valve means much-reduced capacity. Therefore, a scale trap is placed in the suction line near the compressor, where it may be cleaned readily.

The extent to which fuel oil is used in the United states may be inferred from the fact that more than half of the total yearly consumption of petroleum is utilized as fuel for steam raising, this including not only the fuel oil produced by the refineries, but the oil used in its crude state. The whole southwestern part of the United States is dependent on oil fuel, because of the absence of coal fields, and there has been a marked increase in the use of liquid fuel in the New England section because of industrial expansion brought about by war activities. War demands placed a heavy burden on the oil industry, and the rate of consumption greatly exceeded the rate of production. During 1918, according to the statistics of the Geological Survey, the total marketed production of crude petroleum in the United States was 345,986,000 bbl. of 42 gal. each. In the same time the consumption amounted to 367,489,000 bbl., indicating that the domestic production fell behind the consumption.

An Educator, Engineer and Executive

What Professor Thomas Thinks of the Future of Marine Propelling Machinery

Many ships propelled by steam turbines and reduction gears have been put in service within the past two or three years, and valuable information as to durability and economy of operation is now in existence. It is very desirable that this be made public for the guidance of engineers and ship owners. My personal view of the matter is about as follows:

Several types of geared turbine equipment have given exceedingly satisfactory results, but there is considerable reason for questioning the ultimate desirability of gearing. Fuel economy of the main propelling unit alone does not necessarily control the over-all cost of operation, and many reciprocating engine ships are still relatively satisfactory in this respect.

The steam-turbine-electric drive, with water-tube boilers and oil fuel, appears to offer the only solution for navy and passenger vessels of above 30,000 or 40,000 hp., though the steam turbine and single-reduction-gear drive is well adapted for such vessels having less power than those named.

For cargo vessels where the power is from 2,000 up to perhaps 6,000 hp., the best solution appears to me to be the heavy-oil engine with electric drive. Two oil engines can readily be installed in a single-screw ship, to drive generators at approximately constant speed. The arrangement would be quite flexible with regard to overhaul and maneuvering, as either engine alone would ordinarily be available for propulsion. The fuel and space economy with such an arrangement should be very good as compared with a steamship, although the cost of engine-room service would probably be greater. A great advantage in using electric drive in connection with oil engines is that the engines can be built to rotate always in one direction—the motor on the propeller shaft being reversible.

An alternative arrangement has been proposed consisting of two heavy-oil engines direct-connected to two sets of propeller shafting. This would be, in some respects, simpler than two oil engines with electric drive and single screw, but, on the other hand, the propellers would be smaller and in more exposed position with twin screws, and the first cost of the vessel would be considerably higher. The choice for oil-engine-driven cargo vessels seems to lie, however, between these two types, pending the probable development of relatively large engines suitable for direct drive.

The use of the geared turbine drive during the war emergency was undoubtedly justified because, at the time contracts were let, turbines and gears could be manufactured much more quickly and in greater numbers than reciprocating engines.

TO BE a successful teacher, practicing engineer and executive requires a combination of qualities possessed by few—in fact—so few that they can easily be singled out in the engineering profession. Among such men will be recognized Carl C. Thomas, head of the Department of Mechanical Engineering at Johns Hopkins University, Baltimore, and more lately identified with the achievements of the great Hog Island shipyard, where, since the early days of the war, he was in charge of the Engineering Department, and as such had supervision over the fabrication of machinery going into our merchant fleet.

This involved responsibility for the organization of the department, the designing of all machinery and equipment for the ships, and specifying for their purchase. The equipment was supplied by over 3,000 firms widely scattered throughout the country, and it was necessary to furnish them complete detailed drawings and specifications, as well as to inspect the work in the most careful manner. An interesting feature of the work was the erection in two special buildings on the Island of one complete set of machinery, boilers, piping, etc., for each type of ship. This was done before the first hulls were ready to receive their machinery, in order to make sure that all the parts manufactured in different parts of the country would fit together according to the design. This templet shipwork proved to be of great value in checking up the material received.

Born in Detroit, Mich., July 14, 1872, of old New England stock dating back to 1631, the subject of this sketch chose mechanical engineering as a vocation and entered Cornell University, from which he graduated with the class of 1895. Shortly after graduation he became identified with the Globe Iron Works, shipbuilders, of Cleveland, and later was chief draftsman of the Marine Department of the Maryland Steel Co., leaving there in 1901 to join Moran Bros., shipbuilders, of Seattle, Wash. Fortified with this practical experience, he entered upon a teaching career and became professor of marine engineering at Cornell University, succeeding Prof. W. F. Durand, when the latter was called to Leland Stanford, Jr., University. After four years at Cornell Professor Thomas accepted a call to the University of Wisconsin as head of the Department of Mechanical Engineering, and remained there for five years, leaving in 1913 to organize the newly created Department of Mechanical Engineering at Johns Hopkins University. When the war broke out, he joined the American International Shipbuilding Corporation at Philadelphia, remaining with that concern until some months after the signing of the armistice. He has since returned to his work at Johns Hopkins University.

As a teacher Professor Thomas has the faculty of interpreting engineering problems, in a practical sense, and many of his former students will recall that he was always at his best when in the midst of a small group of boys at a drawing board laying out some problem in design. His approachability and the fact that he has always dealt with his students as men accounts in no small sense for his success as a teacher. Professor Thomas could never be accused of having a single-track mind, for he is never content unless directing several projects at the same time. While engaged in the teaching profession, he did much consulting work, especially in the early days of the turbine, was author of the first book written in this country on steam turbines, and carried on considerable research in the university laboratories, notably his investigation into the specific heat of superheated steam. He is the inventor of the widely used Thomas gas meter, the Yarway adjustable spray head and other power-plant appliances, as well as a mechanical agitator used in process work in various chemical industries.

He has for a number of years taken a prominent part in the activities of the American Society of Mechanical Engineers and is a member of the American Society of Naval Architects and Marine Engineers, the American Gas Association, the Engineers' Club of New York, the Engineers' Club of Philadelphia, the University Club of Philadelphia, and the Phi Gamma Delta Fraternity. He has also had membership conferred upon him in the honorary fraternities of Tau Beta, Sigma Psi and Phi Beta Kappa.

Repairs Made by Electric Welding

According to the general rule governing the acceptance of autogenous welding, paragraph 186 of the Boiler Code of the American Society of Mechanical Engineers, states that autogenous welding may be used in boilers in cases where the strain is carried by other construction which conforms to the requirements of the code and where the safety of the structure is not dependent upon the strength of the weld.

In *Power*, page 40 of the Jan. 6 issue, an article was published regrading the attitude of the Steam Boiler and Flywheel Service Bureau in reference to autogenous welding. This committee is made up of engineers of all the insurance companies in the United States that are writing steam-boiler insurance. All welding accomplished by oxyacetylene, hydrogen or other flame processes, or by electric arc comes under the head of autogenous welding.

FIG. 1. PATCH WELDED TO WATER LEG OF A LOCOMOTIVE-TYPE BOILER

The accompanying illustrations show several welding jobs that have been done by the S O S Welding Corporation, of Brooklyn, N. Y. Figs. 1 and 2 illustrate a welding job that was recently completed on one of two boilers at the Republican Club, West 40th St., New York City. This boiler is of the stationary locomotive type and is rated at 150 hp. The firing doors are on one side of the firebox, the grates running crosswise. As originally set, the front, back and end of this firebox were bricked in to form an ashpit, the side below the firing doors being of cast iron and fitted with clean-out doors.

FIG. 2. APPLICATION OF PATCH AND POINTS OF WELDING

It was found that the plates and rivets of the bricked-up portion of the boiler water leg had so corroded at the joints that it was impossible to calk the seam *B*, Fig. 2, sufficiently to keep them tight, and on the advice of the inspector from the Employers' Liability Association Corporation, Ltd., of London, the outside end plate of the water leg was cut out for a distance of twelve inches above the bottom of the mud ring and a patch *C* welded on. This patch is of the same thickness and material as the original plate which is $\frac{7}{16}$ in. thick. This plate turns the corner of the boiler for a distance of twelve inches, the joint

FIG. 3. PATCH THAT IS TO BE WELDED IN TUBE SHEET

FIG. 4. PATCH SHOWING WELDED JOINT IN TUBE SHEET

at the back and front ends being made curved instead of with a sharp corner.

The plate and patch were first cut V-shape with the outside surface as shown at *D* and spot-welded to hold it in position. The V-groove was then filled in by the electric welding process. After filling in the groove between the plates, welding material was built up, making the welded surface on the outside of the plate three inches wide, as shown at *E*.

As the mud-ring rivet heads had corroded, welding material was applied along the rivets on all sides of the water leg, coming flush with the head and three inches wide. The under joints between the plate and the mud ring were also welded.

The boiler will carry ninety pounds pressure the same as before repairs were made. The entire repair bill will, it is estimated, be about one-third of what a new boiler would have cost and what would have been necessary to purchase if repairs had not been made by welding. Another boiler is also being welded.

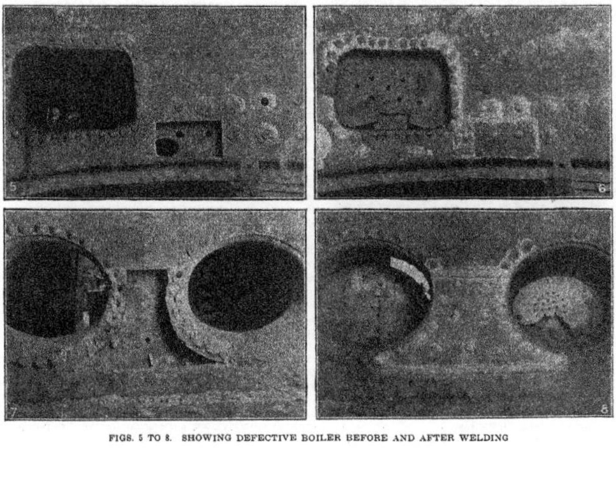

FIGS. 5 TO 8. SHOWING DEFECTIVE BOILER BEFORE AND AFTER WELDING

The several other welding jobs illustrated were performed by the same company. The one shown in Figs. 3 and 4 was made to a return-tubular heating boiler. Fig. 3 shows the defective parts cut out and Fig. 4 the welded portion in place. This work was performed

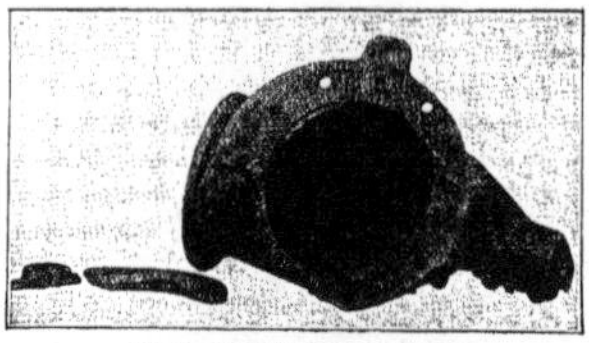

FIG. 9. BROKEN PUMP CASTING

on both the front and back tube sheets of the boiler. This repair was made necessary on account of the corrosion taking place around some of the tube holes.

In another instance the wasted fire sheet of a vertical boiler was cut out and a new piece welded in as shown in Figs. 5 and 6. In this instance the wasted plate around the stay-bolts was built up by electric welding. Several rivet cracks and leaky rivets around the door frames were also welded by the same process. When the repairs were complete, the boiler carried its original pressure.

Another weld is shown in Figs. 7 and 8. In this case the sheet between the firedoors had wasted away. The wasted portion was cut out as shown in Fig. 7,

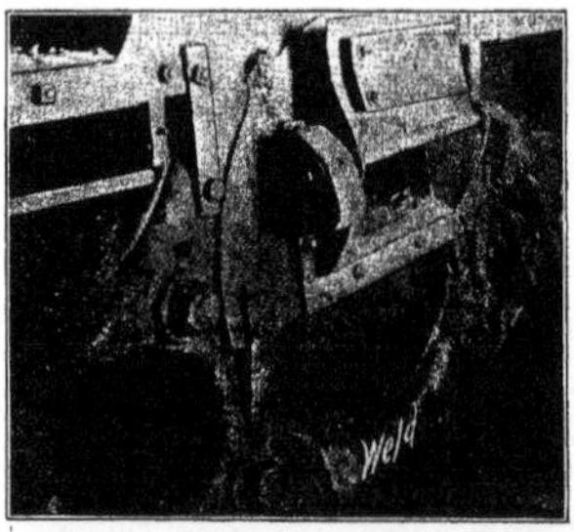

FIG. 10. CRACK 36 IN. LONG WELDED IN FRONT HEAD OF SCOTCH-MARINE BOILER

and a new piece of plate welded in place as shown by Fig. 8. The welding around both firedoor openings replaced the rivets altogether.

A neat weld is shown in Fig. 10. The original defect consisted of a crack 36 in. long in the front head of a Scotch marine type of boiler, next to the flanged edge of the head. This was electrically welded over night while the steamer was being unloaded, thus obviating any delay in sailing.

The art of welding has reached a stage whereby practically all kinds of repair work is being done; and many power plants own and operate a welding outfit for making their own repairs. An instance of casting welding is shown in Fig. 9. This was a casting forming the inlet and outboard bearing of a motor-driven pump, the flange having been broken as shown, by Fig. 9. The broken parts were welded in place,

FIG. 11. WELDED OUTBOARD PUMP BEARING IN PLACE AFTER WELDING

no drilling or studding being done, and the repaired casting was put back in place on the pump as shown in Fig. 11.

With any fuel the mixture must be compressed within the engine cylinder before it is ignited, for the compression adds greatly to the efficiency of the fuel and increases the output of a given size cylinder. All other things being equal, the output and efficiency increase with the compression within certain limits. The degree of compression which it is possible to carry depends upon the character of the fuel to a great extent. Very rich fuels, such as kerosene, start cracking and preigniting at very low pressures, 80 lb. per sq.in. probably being the limiting pressure for kerosene. In aëronautic engines the compression is carried to the highest possible point so as to obtain the maximum output. This maximum is reached at a pressure approximately of 130 lb. per sq.in. with gasoline. With very lean gases, such as producer gas, or blast furnace gas, much higher compressions are possible, the pressure for producer gas probably averaging 240 lb. per sq.in. while 400 lb. per sq.in. is not uncommon with blast furnace gas. When the compression is carried up to too high a point with any fuel, explosion waves are started which rack the structure of the engine and cause excessive wear. The fuel is then decomposed so as to form deposits of solid carbon. This carbon becomes incandescent later on and starts preignition and other troubles. In many cases, the efficiency of a fuel is limited by the limiting compression pressures which it is possible to use. The preignition causes violent hammering for the direction of the stresses are reversed.—*Exchange.*

POWER DRIVES FOR ROLLING MILLS

By W. O. Rogers

Air is provided for the blast furnace by huge steam-driven, air-blowing engines. Some of these are of the horizontal and others are of the vertical type. In other instances gas-engine-driven blowers are used, and in still other plants turbo blowers produce the blowing air.

OF EQUAL importance with engines of various types used in the manufacture of steel are the blowing engines, and it is doubtful if in any other industry the use of compressed air is of so much importance. As stated in a previous article, compressed air is used in the blast furnace and with the bessemer converters. The air that is supplied to blast furnaces is forced into the furnace by extremely large blowing engines, some of which are of 2,500-hp. capacity and capable of compressing from 30,000 to 65,000 cu.ft. of free air per minute to a pressure of from 15 to 30 lb. per sq.in., which is about what one blast furnace requires. After leaving the compressor, the air is heated to between 800 and 1,200 deg. F. by being made to pass through hot-blast stoves. In the bessemer process about 30,000 cu.ft. of cold air is forced through the molten material per minute.

Most blowing engines are steam-operated. At the main plant of the Youngstown Sheet & Tube Co. there are four 88, 90 and 96 by 72-in. 2,500 hp. Tod engines in one plant, each with a capacity of 48,000 cu.ft. of free air per minute at 40 r.p.m. compressing to from 10 to 20 lb. per sq. in. Fig. 2 is a view of one of these engines taken from the crank end. The Corliss steam cylinders are arranged cross-compound and are placed next to the engine frame, the air cylinders being placed at the head end of the machine.

Fig. 1 is a general view of the blast-furnace blowing room and shows the four blast-furnace blowing engines, also two Southwark-Rateau turbo-blowers driven by a 3,500- and 5,000-kw. turbine, respectively. Fig. 3 is a view of the 2,000-hp. unit which at 1,650 r.p.m. has a capacity of 40,000 cu.ft. of free air per minute, compressed to from 10 to 20 lb. per sq. in. It is equipped with a stabilizing device for preventing back flow when small quantities of air are being blown and with a Rateau multiplier for maintaining the delivery of air constant, irrespective of the pressure.

At the Republic Iron & Steel Co. plant, I found several Allis-Chalmers blowing engines of the steeple type, used to supply air to blast furnaces. These are shown in Fig. 4. These engines were of different design. For

FIG. 1. GENERAL VIEW OF BLAST-FURNACE BLOWING ROOM AT THE YOUNGSTOWN SHEET AND TUBE COMPANY

FIG. 2. ONE OF THE 48,000-CU.FT. 2,500-HP. CROSS-COMPOUND BLOWING ENGINES

FIG. 3. A 40,000-CU.FT. TURBO-BLOWER, 2,000-HP. CAPACITY

instance, there were three high-pressure and two low-pressure vertical long crosshead type units having 52 and 96 x 60-in. high- and low-pressure cylinders, each having 96 x 60-in. air cylinders. Each of these blowers has a capacity of 20,000 cu.ft. of free air per minute at 41 r.p.m., or running as a pair of disconnected compounds—that is, one high- and one low-pressure unit—each pair would have a capacity of 40,000 cu.ft. of free air per minute. Fig. 5 shows a standard long-crosshead engine and Fig. 6 a long-crosshead low-pressure unit.

The high-pressure units are arranged to exhaust into a receiver from which the low-pressure engines take their steam. This arrangement is such that any high-pressure engine can run with any of the low-pressure engines. Some idea of the size of these engines can be had from the following: The main bearings are 21 in. in diameter by 28 in. long. The crosshead and crank-pins are 12 in. in diameter by 10 in. long and the fly-wheels are 24 ft. diameter and weigh approximately 60,000 lb. each.

There were also two high- and two low-pressure vertical long-crosshead blowing engines with 44-in. high-pressure and 84-in. low-pressure cylinders each having 60-in. stroke and both having 84 x 60-in. air cylinders. Each of these machines has a capacity of 20,000 cu.ft. of free air per minute at 55 r.p.m., or a capacity of 40,000 cu.ft. per min. for each high- or low-pressure machine.

These blowers are also arranged so that either high-

FIG. 4. STEEPLE TYPE OF BLOWING ENGINES AT THE REPUBLIC IRON AND STEEL COMPANY'S PLANT

FIG. 5. STANDARD LONG CROSSHEAD BLOWING ENGINE

FIG. 6. A LOW-PRESSURE VERTICAL BLOWING ENGINE

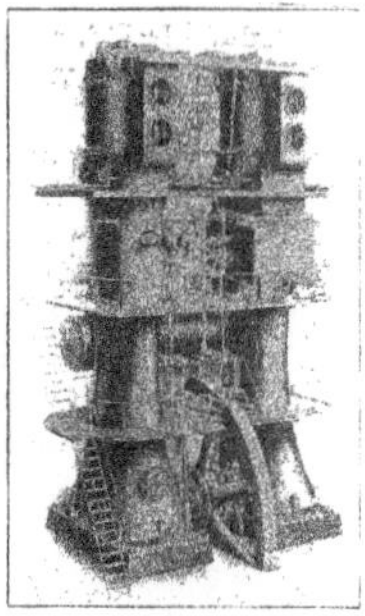

FIG. 7. STEEPLE COMPOUND BLOWING ENGINE

pressure unit can run with either low-pressure machine. All these blowing engines operate with a steam pressure of 150 lb. condensing, as does the cross-compound blowing engine shown in Fig. 7. This unit has a 46-in. high-pressure and a 88-in. low-pressure cylinder and two 79-in. air cylinders with a common stroke of 60 in. At 50 r.p.m. this machine has a capacity of 30,000 cu.ft. of free air with a maximum air pressure of 30 lb. The main bearings are 23 in. in diameter and 36 in. long, the crankpins are 14 x 12 in. and the crosshead pins are 12 x 12 in. The flywheel weighs 100,000 lb., or 50 tons, and is 24 ft. in diameter.

There is also a horizontal-compound blowing engine that supplies air to the bessemer converters. It has 46 and 88 x 60-in. steam cylinders and two 76 x 60-in. air cylinders. This unit has a capacity of 30,000 cu.ft. of free air per minute at 50 r.p.m., against an air pressure of 30 lb. The main bearings are 24 in. in diameter and 42 in. long. The crosshead pins are 12 x 12 in. and the crankpins are 14 x 12 in. The flywheel is 24 ft. diameter and weighs 100,000 lb. or 50 tons. Fig. 8 shows an installation of three Mesta steam cross-compound blowing engines at the plant of the McKinney Steel Co., Cleveland, Ohio.

The design of the blast-furnace plant generally decides the question as to whether blowing engines should be steam or gas driven. The steam-driven units predominate, but a large number of gas engines and units have been installed and in general seem to have given satisfactory results. In some blast-furnace plants it is better to consume all the furnace gas under steam boilers that feed steam engines or turbines. When the blast furnace is operated with a steel mill, a gas engine driven blower is said to pay for itself whenever the coal cost is over $2 per ton. There are, however, so many elements that enter into the question that no engineer would care to make a definite statement as to the superiority of a gas engine over a steam-driven blowing engine without knowing all about the conditions under which they are to operate.

In Fig. 9 are shown five Mesta gas-blowing engines at the Lackawanna Steel Co., Buffalo, N. Y. This company, I was told, was the first in the United States to use blast-furnace gas in gas engines on a large scale. The plant at that time consisted of sixteen 2,000-hp. blowing engines and eight 1,000-hp. generating engines. All these units were of the twin two-cylinder two-cycle horizontal type.

The Gary plant of the Indiana Steel Co. probably has the largest installation of blast-furnace gas engines in

FIG. 8. MESTA CROSS-COMPOUND BLOWING ENGINES

FIG. 9. MESTA GAS-BLOWING ENGINES

this country. At first there were sixteen twin, tandem, four-cylinder horizontal four-cycle types. Each of these four power cylinders was 42 x 54 in. The two blowing cylinders have a capacity of 30,000 cu.ft. of free air per minute. Additional units of larger capacity have since been installed.

At the Sparrows Point, Md., plant the air compressors derive their power largely from gas-driven engines. The original gas-engine driven plant consisted of five tandem units, which number was increased by the addition of fourteen new units, each having 47 x 85 x 60-in. cylinders and capable of delivering between 22,000 and 27,000 cu.ft. of free air per minute. The original blowing engines have a capacity of between 20,000 and 25,000 cu.ft. of free air per minute. There are also ten 4,000-kw. gas-electric twin-tandem units. Eight furnaces supply the fuel for driving sixteen blowing engines and eight gas-electric units, the remainder, three blowing and two gas-electric units, being provided as spares to provide for emergencies.

Several turbo-blowers were found operating at various plants, but it was a difficult matter to photograph some of them.

At the Mark plant, Indiana, of the Steel and Tube Company of America, there are two Ingersoll-Rand turbo-blowers, each having a capacity of 50,000 cu.ft. of free air per minute, working against an air pressure of 13 lb., Fig. 10. This air is delivered to blast furnaces. Each unit is of 3,000 hp. capacity at 2,500 r.p.m. The turbines are of the uniflow type, and the blowers are designed with five stages.

Blowing of the blast furnaces should be accomplished with a steady air blast, and for that reason, among others, there are those who are in favor of centrifugal compressors, because the blast from such a machine will be a steady blow without pulsations.

Although it is claimed by some that engine-driven compressors discharge air in puffs which correspond to

FIG. 10. INGERSOLL-RAND TURBO-BLOWERS

the stroke of the unit and the blast is therefore variable, others maintain that the pulsations are almost eliminated because of the large volume of the blast main and stoves. As in many other questions, there are those who favor both types of blowing units, and much may be said in favor of both types.

At the Republic Iron & Steel Co.'s plant there are two General Electric centrifugal blowers that are directly connected to steam turbine. These units have a capacity of 43,500 cu.ft. of free air per minute with a pressure of 22 lb. Fig. 11 shows one of the units and the turbine end of the other. These units are of the multiple-stage type, consisting of several impeller wheels that run in a casing having passages so designed that the wheels operate in series. An air governor regulates the rate of blowing. It consists of a disk that rides on the ingoing air in the center of a conical fitting on the inlet pipe. Any movement of this disk is transmitted to the valve gear. Differing rates of flow are secured by moving a weight on a scale beam

FIG. 11. GENERAL ELECTRIC TURBINE-DRIVEN BLOWERS

that is graduated in cubic feet of free air per minute. In action the operation of the air governor is about as follows:

With the turbo-blower running at a low rate of speed, such as would be the case when tapping the blast furnace the force of air current passing through the cone part of the disk will be insufficient to support the disk and the unbalanced weight of the governor beam, and the disk will, therefore, fall against the air current, and in doing so operate a series of valves successively admitting steam to the turbo-blower and increasing its speed. This increases the air pressure in the discharged end about as the square of the speed and produces such a pressure as to overcome the resistance of the blast main, stove and furnaces and establish the desired rate of air flow. It then comes to rest at some point in its housing and the proper number of steam admission valves are opened and the speed of the turbo-blower is what is necessary to produce and maintain the rate of air flow for which the weight on the governor arm has been set, irrespective of the pressure against which the blower has to work.

(The next article will deal with some of the up-to-date power stations, including reciprocating and turbine units, both high and mixed pressure types, found at the rolling mills.)

To avoid the use of decimals in expressing specific gravity, the Baumé system is used in the oil trade. The Baumé scale is purely arbitrary, the point marked 10 corresponding to a specific gravity of 1.0, and the remainder of the scale being graduated in degrees up to 60. A scale having this range can be used for oils having specific gravities from 0.737 to 1.0. It should be noted that an increase in the Baumé reading corresponds to a decrease in the actual specific gravity of the oil; in other words, the heavier the oil the smaller is the reading on the Baumé scale.

Turbine-Driven *versus* Engine-Driven Alternators—III

By S. H. Mortensen*

Quantity of air required, velocity through ducts, layout of ventilating system, temperature limits and means of measuring coil temperatures.

IN PREVIOUS articles the construction and operation of engine-type and turbo-alternators have been compared. Following out this comparison, there is little to be said regarding ventilation of the engine-driven machine, as the open construction makes unnecessary any unusual provisions. Proper ventilation of the turbo-alternator, however, requires careful consideration not only in the design of the machine, but also in its installation and operation. Although the individual losses in the turbo-alternator, with the exception of windage and friction, are lower than in a correspondingly rated engine-type alternator, the radiating surface is so small that it is totally inadequate to dissipate the heat caused by these losses without developing disastrously high temperatures.

In the design the problem of ventilation is solved by inclosing the machine and blowing air through a multiplicity of ducts in the stator and core to the parts where the heat is generated. By adopting the inclosed construction, the noise inherent to high-speed operation is materially reduced. There has to be a sufficient number of air ducts in the turbo-alternator to permit a volume of air great enough to conduct away the heat without the air velocity exceeding 7,000 ft. per minute.

The amount of air required to keep a machine within a given temperature rise depends upon the iron and copper losses and also upon the windage loss in the machine itself. The friction loss is not to be included, as it occurs outside the machine proper in the generator bearings.

With the specific heat of air taken as 0.237, one cubic foot of air can absorb energy corresponding to 26.7 ft.-lb. with a temperature rise of 1 deg. C. Hence, one kilowatt will heat 1,650 cu.ft. of air per minute, 1 deg. C., or provided it were possible to heat the air uniformly, one kilowatt loss would increase the temperature of 165 cu.ft. of air per minute 10 deg. C. It is, however, impossible to pass the air through the machine in such a way that it will be uniformly heated, and it is necessary to provide a much larger volume of air than is indicated by the theoretical calculations. In practice the actual amount supplied is approximately 100 cu.ft. of air per minute per kilowatt loss in the machine. If, as an example, a machine has 50 kw. loss, it will require for proper ventilation 5,000 cu.ft. of air per minute. In general, a turbo-generator will require from 4 to 5 cu.ft. of air per minute per rated kilovolt-ampere for its ventilation.

Several methods have been devised for guiding this large volume of ventilating air through the machine parts without causing undue friction and heating. One such method is shown in Fig. 1. The arrows indicate the flow of the air through one of the several ventilating ducts in the stator. Part of the air is blown through the air gap, cooling the rotor and inner stator surfaces. The main portion, however, passes over and between the exposed parts of the stator coils, over the rear surface of the core punchings and through a number of short paths in the core iron to the main exhaust chamber in the yoke. With this ventilating arrangement the temperature of the turbo-alternator can be kept within the safe operating limits of either class "A" or class "B" insulation (see article II in the Mar. 16 issue), depending upon the size, rating and voltage of the machine.

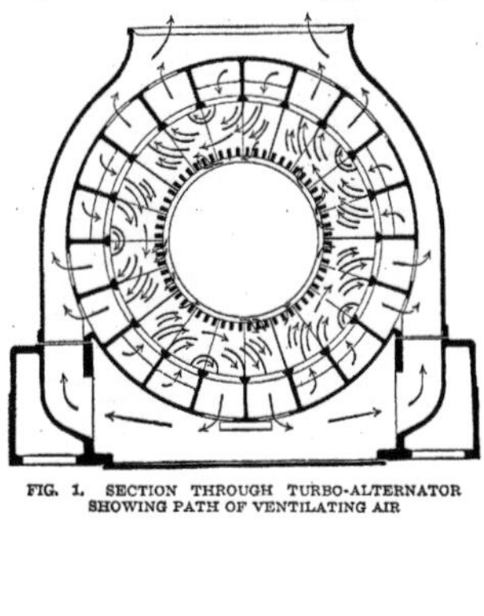

FIG. 1. SECTION THROUGH TURBO-ALTERNATOR SHOWING PATH OF VENTILATING AIR

In laying out the power station and foundation for large turbo-generators, provision should be made for air channels of ample size for guiding the cooling air to and from the machine. The channels should be reasonably short and without abrupt bends or turns, and have sufficient cross-section to keep the air velocity below 1,500 ft. per minute.

After the turbo-alternator is in operation as much attention should be given to its ventilation as to its oiling system. As the bearings require an adequate amount of cool, clean oil, so the generator requires an adequate amount of cool, clean air. If the oiling is neglected, the bearings will overheat and suffer destruction. If

*Electrical Engineer, Allis-Chalmers Manufacturing Company.

the ventilation of the generator is restricted or throttled lown, the generator will overheat and its coil insulation will be damaged.

In the installations where clean air is not available, air filters or air washers should be provided. The air is filtered by drawing it through screens covered with light flannel or cheesecloth. The area of the filter should be such that the velocity of the air through it does not exceed 10 ft. per minute. If the velocity is that low, the filter will not clog up with dirt as readily, and there will be comparatively little resistance to the flow of the ventilating air. Filters of this type require constant care and cleaning and should not be used where the air contains much dirt. In that case the air washer should be installed. Here the dirt is washed out of the air by means of a water spray and at the same time the temperature of the air is brought to within a few degrees of the temperature of the water in the washer. It may be assumed that at least during

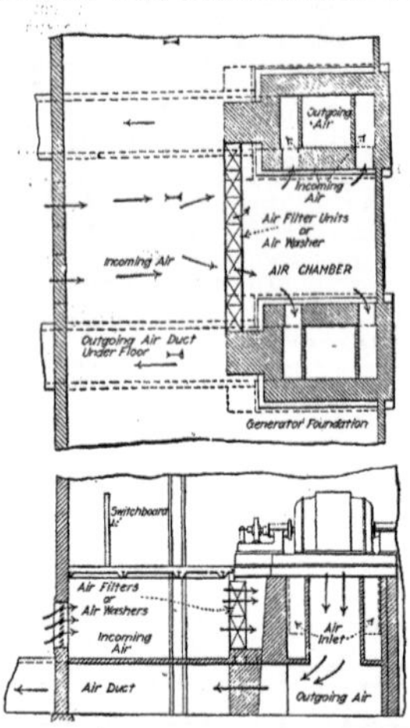

FIG. 2. VENTILATING LAYOUT FOR TWO TURBO-ALTERNATORS

summer months the water in the washer is considerably cooler than the air, and in that case its temperature will be lowered and the rating of the machine can be increased correspondingly.

In Fig. 2 is shown a ventilating layout of two turbo-alternators. Forced ventilation may be supplied by fans which are part of the machine itself or by means of auxiliary fans. The air is cleaned and guided through the machines and discharged from the bottom of the stator into an air channel and may be utilized either for heating purposes, forced draft under the boilers or for similar purposes.

In turbo-generators the coil insulation used is either class "A" or class "B" or a combination of both. In large machines it is difficult to ventilate all parts equally well, and the temperatures of certain spots in the core are likely to be much higher than the average machine temperature would indicate. These spots, frequently referred to as "hot spots," are located in the middle of the stator core, either in the core iron or in the stator coils buried in the slots. To enable the coil to success-

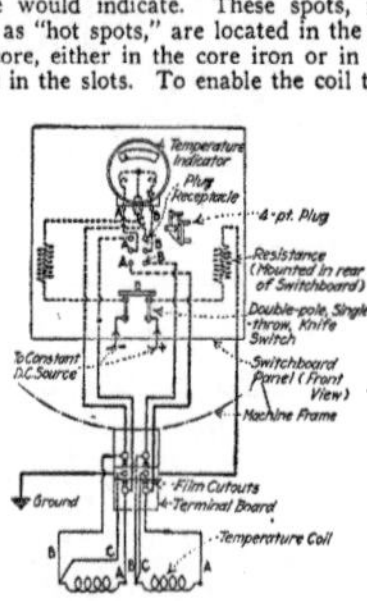

FIG. 3. WIRING DIAGRAM FOR TEMPERATURE INDICATOR

fully withstand hot-spot temperatures, it is common practice on large machines to insulate the slot part of the stator coil with class "B" insulation. This is not required on the free ends of the stator coils, as they get into direct contact with the cooling air. It is furthermore desirable to use class "A" insulation on the coil ends, as it is more flexible than the mica insulation and is less liable to suffer damage when the coils are placed in the stator slots.

How the Temperature of an Inclosed Machine Is Determined

Special arrangements are required to determine the temperature of an inclosed machine. The thermometer method that is used on engine-type alternators can only be applied to the stator-coil ends or the surface of the iron and cannot reach the interior part of the machine where the maximum temperatures are developed. For this purpose either resistance coils or thermocouples are built into the machine and located in the parts of the winding that supposedly develop the maximum operating temperature.

In applying the "resistance-coil method," coils wound noninductive, from fine copper or nickel wire and having a fixed resistance at a fixed temperature (generally 10 or 100 ohms at 20 deg. C.), are placed in the stator winding. With the increase in the temperature of the machine the resistance of the embedded coil also increases, and by connecting it to a properly calibrated instrument, the temperature of the portion surrounding the coil can be read directly. Several types of instruments are available, both indicating and curve drawing. The curve-drawing instrument gives a permanent record of the operating temperature during any time the machine is in operation.

For a double-layer winding it is customary to locate the resistance coils between the two coil halves in the middle of the stator core. Even with this location it is assumed that the resistance coil does not indicate the maximum temperature of the machine. To obtain this temperature the standardization rules of the American Institute of Electrical Engineers specify a correction of 5 deg. C., or 9 deg. F., to be added to the temperature indicated by the resistance coil. For class "B" insulation with a temperature limit of 125 deg. C. (257 deg. F.) the maximum observable temperature is 120 deg. C. (248 deg. F.) and the maximum temperature rise based on 40 deg. C. (104 deg. F.) ventilating air is 80 deg. C., or 144 deg. F. Under these conditions the maximum temperature of the class "A" insulated coil ends must not exceed 105 deg. C. (221 deg. F.) in the hottest spot, or a 50 deg. C. (90 deg. F.) rise as determined by the thermometer.

As the resistance coils only indicate the stator temperatures, in determining the safe load of a turbo-generator it is necessary also to consider the rotor temperature. It will be found frequently that the rotor is the limiting factor, especially where low power-factor loads are involved. As it is impractical to place resistance coils in the rotor winding, its temperature is determined by the increase in resistance method. If the safe operating temperature of the mica-insulated rotor is placed at 150 deg. C. (302 deg. F.) and the usual correction of 15 deg. C. (27 deg. F.) is allowed, the maximum observable temperature is limited to 135 deg. C. (275 deg. F.), or 95 deg. C. (171 deg. F.) rise over the air at 40 deg. C., or 104 deg. F.

Fig. 3 is a wiring diagram for a temperature-indicating instrument. The leads from the resistance coils are brought outside the machine to a terminal block where connections are made to the instrument. As long as the resistance and material of the connecting wires are alike, their length has no influence on the accuracy of the meter, which for that reason can be installed in any convenient location. The indicating instrument requires a source of constant-voltage direct current for its operation.

As the temperature throughout the machine varies and as a consequence the resistance coils indicate different temperatures, it is customary to use the coil that indicates the maximum temperature for determining the safe load on the machine. By keeping a record of the temperatures corresponding to different loads and comparing them occasionally, it is possible to determine when the machine needs cleaning, as the presence of dirt in the ventilating ducts will cause the operating temperature to increase.

Rules for Autogenous Welding

The following rules, adopted by the Committee on Standards for Locomotives and Cars, United States Railway Administration, for the purpose of preventing the abuse of autogenous welding for purposes for which it is not well adapted, have been sent to the regional directors by Frank McManamy, assistant director of the Division of Operation, with instructions to direct all roads to observe the rules in the construction or repair of locomotive boilers, so that any failures which may have been caused or contributed to by unrestricted or improper use of autogenous welding may be prevented:

1. Autogenous welding will not be permitted on any part of a locomotive boiler that is wholly in tension under working conditions, this to include arch or water bar tubes.

2. Stay-bolt or crown stayheads must not be built up or welded to the sheet.

3. Holes larger than 1½ in. in diameter, when entirely closed by autogenous welding, must have the welding properly stayed.

4. In new construction welded seams in crown sheets will not be used where full-sized sheets are obtainable. This is not intended to prevent welding the crown sheet to other firebox sheets. Side sheet seams shall be not less than 12 in. below the highest point of the crown.

5. Only operators known to be competent will be assigned to firebox welding.

6. Where autogenous welding is done, the parts to be welded must be thoroughly cleaned and kept clean during the progress of the work.

7. When repairing fireboxes, a number of small adjacent patches will not be applied, but the defective part of the sheet will be cut out and repaired with one patch.

8. The autogenous welding of defective main air reservoir is not permitted.

9. Welding rods must conform to the specifications issued by the Inspection and Test Section of the United States Railroad Administration for the various kinds of work for which they are prescribed, which specifications will be issued later.

Miners' Wages Are Fair, Says Doctor Garfield

Dr. Garfield, former Fuel Administrator, in an article written for *Farm and Home*, declares the compromise that ended the coal strike to be "unsound in principle and a menace to our institutions." Dr. Garfield goes on to say:

"The wages now paid to mine workers are sufficient. The opportunity that should be the mine workers' cannot be secured merely by an increase in wages."

In proof Dr. Garfield cites the average of $950 per annum earned by the lowest paid miners working 180 days in the year, while for 200 days' work the average miner in the bituminous fields of Pennsylvania, Ohio, Indiana and Illinois earned $1,660 in 1918 and $1,300 last year. This, says the article, is "more by a considerable sum than the average net receipts of the farmer and many others who may or may not work 300 days or more in the year."

"The public ought not to be asked to pay more for coal," emphasizes the ex-Fuel Administrator. "It is impossible to increase the wage of the mine workers without inciting the workers in every other industry, including, of course, agriculture, to demand an increase in wages. This would send the cost of living upward in a vicious spiral, which will in the end prove hurtful to the workingman. The purchasing power of the dollar and not the number of dollars received is the important factor.

"The public is the chief sufferer when capital and labor engaged in the production of commodities necessary to the support of life fall a-fighting," continued Dr. Garfield. "In these cases certainly the interest of the public is vital and, therefore, paramount. We may admit the right to strike on the part of labor, and the right of capital to boycott, but in each case the right of the public to live is paramount and will be asserted.

"We now are called upon to contemplate an arrangement with a group opposing the government which, however it terminates, is unsound in principle and a menace to our institutions."

To guard against affairs reaching the strike stage, Dr. Garfield in the same article urges that a permanent fuel administration be established so that it can serve as a consultive and advisory tribunal and head off difficulties before they come to a head.

Reorganization of the Massachusetts Boiler Inspection Department and Board of Boiler Rules

In the reorganization of State Commissions by the Massachusetts Legislature the Boiler Inspection Department became a part of the Department of Public Safety administered by a Commissioner of Public Safety and comprising three divisions—that of State Police under his own immediate charge, that of Inspections under a Chief of Inspections, and that of Fire Prevention under a State Fire Marshall. The Division of Inspections includes the functions of the old Boiler Inspection Department and of the Building Inspection Department, both of which were under the old District Police. The Board of Boiler Rules remains as previously constituted with the exception that the Chief of Inspections is chairman. The personnel is as follows:

Commissioner of Public Safety, Col. Alfred F. Foote, Holyoke; Chief of Inspections, John H. Plunkett, formerly chief of the District Police.

BOARD OF BOILER RULES

John H. Plunkett, Chief of Inspections and (ex-officio) Chairman of the Board.

Frederick A. Wallace, M. E., Chief Engineer, Pacific Mills, Lawrence, representing boiler-using interests.

Henry H. Lynch, President of the Hodge Boiler Works, East Boston, representing the boiler-manufacturing interests.

John A. Collins, Secretary of the Mutual Boiler Insurance Co., of Boston, representing boiler-insurance interests.

John Moles, Chief Engineer, American Writing Paper Co., Holyoke, representing operating engineers.

New Copes Regulator

Since the first introduction of the Copes mechanical feed-water regulator changes have been made in the details of construction, the principle of feeding remaining the same. Recently, the design of the central valve has been altered. Fig. 1 shows the new valve, in which

FIG. 1. SHOWING CONSTRUCTION OF THE VALVE

the reciprocating stuffing-boxes have been abandoned, a horizontal shaft introduced, which rotates very slightly, performing the same function as the old reciprocating rod. This new design is sensitive, and the frictional resistance has been reduced as much as six to sixteen pounds. It is now practically zero because the shaft rotates very slightly and the same part of the shaft is always in the stuffing-box.

The valve is of the balanced piston type. The weight on the valve lever exerts a constant closing force of fifty pounds on the valve piston. A weight is used here rather than a spring because gravity does not vary, whereas springs grow weak with age, corrode, stick, break, etc.

A vertical section through the valve displays the horizontal shaft, the stuffing-box, the linkwork connecting the shaft with the piston valve, etc. The valve cap or bonnet can be unbolted and removed with the fittings attached without removing the valve body from the pipe line.

The assembled Copes regulator, which is manufactured by the Northern Equipment Company, 111 West

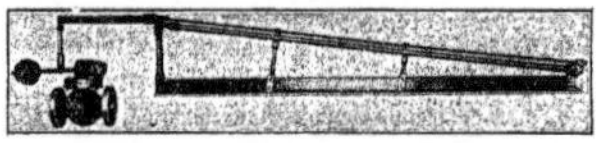

FIG. 2. THE ASSEMBLED REGULATOR

Eleventh St., Erie, Penn., is shown in Fig. 2. It consists principally of a heavy, long diagonal expansion tube known as the thermostat. It is a straight piece of heavy metallic tubing 1¼ in. outside diameter. The top of the thermostat is connected with the steam space of the boiler and the bottom with the water space. Hence there is a water level in the thermostat that corresponds to the water level in the boiler, and it is this water level that determines the rate of feed. The tube lengthens and shortens with a drop and a rise of water level. The lengthening and shortening are caused by change in temperature of the tube as the water rises and falls. This expansion or contraction is amplified by means of the linkwork shown in Fig. 2, and through the linkwork operates the control valve.

Why Our Oil Supply Is Limited

Assuming a population of 108,000,000, the consumption per person in the United States during 1918 was 3.4 bbl. of petroleum, says *Petroleum Magazine*. The oil that has been produced since 1859 amounts to about forty-five barrels per person, and the oil yet underground, according to the most careful estimate, amounts to about sixty-seven barrels per person.

If the population of the United States remained unchanged and the rate of consumption did not change, the oil available would last about twenty years. But the population is continually increasing and the demands for petroleum are multiplying, so that in less than twenty years the known domestic underground fields will be exhausted if the statistics are even approximately correct. Oil shales are not considered in this connection as a source of fuel oil.

This conclusion does not mean that fuel oil users will be unable to obtain supplies, for the data apply only to the sources of petroleum within the United States. There are vast fields in Mexico and Russia, from which oil may be obtained, but of course at increased cost because of the transportation expense.

A report issued by the Department of Commerce, Bureau of Foreign and Domestic Commerce, Washington, of exports of refrigerating machinery from the United States to various countries shows Japan well in the lead, with Canada, Cuba, Argentina and Brazil following. A report on exports of pumps and pumping machinery shows Cuba leading, with imports costing $126,892.

EDITORIALS

Funds for Investigating Water-Power Resources

THE shortage of power on the Pacific Coast, particularly in California, has intensified interest in water-power development. The universal increasing cost of coal, difficulties with the enormous labor forces engaged in its production and the burden placed upon transportation facilities in distributing it are bringing about a broader and more universal realization of the importance of water power. In many localities water power cannot compete with fuel power now, but the disparity between the two sources of energy is gradually disappearing. Much thought is being directed toward the advantage to be gained through a combined usage supplemented by interconnecting large areas with superpower transmission lines.

One reason why water-power development has progressed somewhat slowly is that we are not well enough acquainted with its possibilities and limitations. Hydraulic development must be adapted to many physical features, of which a sound knowledge of water supply is extremely vital. The only safe means of predicting future regimen of water courses is by considering performance in the past. All other physical features except water supply may be ascertained within a relatively short time. Reliable records of flow for a number of years must be available for proper design and practical plans for operation. In other words, the need for stream-flow records must be anticipated far in advance of development. Many hydraulic projects have failed on account of an altogether too meager knowledge of water supply. Others based on short time records, comparisons with records of near-by streams, and climatological data, have levied an enormous toll upon the public, which pays either directly or indirectly for inefficient service resulting from unexpected annual or seasonal variation in flow.

The question arises, What agency should collect the mass of information necessary for comprehensive development? The Water Resources Branch of the United States Geological Survey has made a systematic investigation of water resources and related problems for a number of years. It has become the well-recognized medium for collecting stream-flow data throughout the United States, in Hawaii and in Alaska. An efficient organization with well-formulated methods has been evolved with a surprisingly small expenditure. During the ten fiscal years ending June 30, 1920, the fund made available by Congress for this activity has averaged only $155,000 a year, $150,000 each year except 1918 and 1920, when $175,000 was appropriated. Numerous single water-power developments, promoted and constructed on the basis of data collected by the United States Geological Survey, have cost many times the entire amount appropriated during the ten years. Naturally, the demand for stream-flow records has increased rapidly. Coöperation, tendered to meet this demand by states and interested parties, has increased so greatly that for the fiscal year 1920, the Federal appropriation represents only twenty-nine per cent of the cost of the work. A limit is being reached in the amount of coöperation that can be absorbed. In fact, many interests desirous of obtaining stream-flow records and willing to pay for them, in so far as possible, have been refused assistance because the Federal personnel can not handle properly the coöperation now in force. Having absorbed the maximum coöperation possible the United States Geological Survey is confronted with an ever-increasing demand for stream-gaging results to be met with decreased purchasing power of funds appropriated. Obviously, Congress should support the work in such manner that at least fifty per cent of the cost of records would be borne by the Government. This would mean increasing the present appropriation from $175,000 to something over $300,000 per year, and providing increases from year to year depending upon the demand. As a long period of record is desirable for wise development, it would seem proper for the Government to anticipate development far in advance of demand and provide funds accordingly. If this policy were adopted, an appropriation of at least $500,000 would be needed.

The inadequate funds appropriated by Congress for investigating one of our most important future assets can be accounted for only in two ways: either Congress does not pursue a farsighted and constructive policy regarding future advancement, or Congress is wholly lacking in its realization of the benefits to be gained by full utilization of water-power resources

Cooling of Turbo-Alternators

TURBO-ALTERNATORS are in most all cases cooled by passing large volumes of air through and around the machine's cores and windings. This air is made to circulate through the machine either by fans built in the rotor or by a separate blower. The amount of cooling air required per hour by a modern machine attains tremendous proportions, amounting to nearly the weight of the alternator itself. When this cooling air is washed by a water spray about ninety-eight per cent of the dirt in the air will be removed. However, even with so low a percentage of dirt remaining in the air passing through the alternator, considerable material will be deposited in the ventilating ducts and on the end of the windings, and not infrequently oil vapors are carried into the machine. These dirt deposits not only decrease the efficiency of ventilation by reducing the quantity of air and preventing transfer of heat from the core and windings to the air currents, but they also create a real fire hazard. If these deposits become ignited by either electrical or mechanical causes they will soon be fanned into a flame that is liable to seriously injure if not completely destroy the generator windings.

The effects of fires and failures in large turbo alternators have become so serious both in this country and abroad that other methods of ventilation have been given serious consideration. With the very satisfactory experience of the oil insulated transformer to draw from it is not suprising that considerable attention has been directed to oil as a cooling medium for alternators. Although it is within the possibilities of a practical design to immerse the stator core and windings in oil, it is very doubtful if this would offer a real solution to the problem. In the transformer the windings can be thoroughly insulated and space provided for the circulation of oil between the coils and still keep within the limits of an economical design, but in an alternator the slot insulation is the limiting factor in high-voltage machines. Immersing the windings in oil would not improve the insulation in the slot although it would overcome a number of the objectionable features of air ventilation. However, there are many factors in the application of this scheme that make its possibilities very doubtful.

In a paper presented before the British Institute of Electrical Engineers, January 8, 1920, J. Shepherd suggests a method of cooling alternators by a system of circulating water through the rotor and stator cores. Although this scheme would greatly increase the ability of the machine to dissipate the heat losses it, nevertheless, would be difficult of application and would require a special design such as suggested by Mr. Shepherd. Experiments with an improved system of cooling large alternating-current generators with air, in a system which is completely inclosed, by recirculating the air after being cooled, apparently offer the best solution to the problem. Although this scheme does not possess all the qualities that might be desired, it does prevent dirt from getting into the machine and in case of internal fires possesses advantages in extinguishing them. Furthermore, such a scheme is applicable to standard equipment. When this system of cooling is applied with a balanced-relay method of protection to disconnect the machine from the line and open the field circuit, in case of internal trouble, the causes of a large percentage of the disastrous fires in turbo-alternators will be eliminated, without involving any of the complication inherent in any system of oil or water cooling.

Long Longitudinal Seams on Boilers

THE chief inspector of one of the boiler-insurance companies, reverting to the discussion upon the length of longitudinal seams in boilers which took place in *Power* some months since, remarks that all of those who contributed to the discussion seemed to consider the subject from the point of view of constructional strength. There is much to be said from the operator's, owner's and underwriter's standpoint.

Distress in the girth seam is one of the first indications of trouble in the furnace sheets. A leaky roundabout has often called attention to an accumulation of scale or grease upon a fire sheet which, but for this warning, might have resulted in a serious bulge or bag. The girth seam is also a strengthening ring for the sheet. He had recently been called to see a boiler in the fire sheets of which were two bags, one in front of and one just behind the first girth seam. If this had been one big sheet, the bags would have run together and probably have ruptured, causing a disastrous explosion.

With a long sheet, therefore, bagging is more likely to occur from lack of notice through leakage; when it does occur, it is likely to cover more area and be deeper and more serious than in a smaller sheet; there is less chance of its being possible to work the sheet back into place or to patch it, and if the sheet has to be replaced, it is much less expensive to renew a short course than a long one.

In his opinion the provision that longitudinal joints on horizontal boilers shall not exceed twelve feet in length is therefore a wise one and should hold for this class of boiler whether the longitudinal seam is lapped or butt-strapped. In the shells of internally fired boilers and in the drums of water-tube boilers, where the sheets are not in the fire, these considerations do not apply.

A New Plan for Co-operation Between Industry and the Colleges

REALIZING that technical schools are not serving to a maximum degree the industries to which they owe their existence, the Technology Clubs Associated has proposed a plan for closer coöperation between the agencies for training technical men and the industries in which their training is to be used. The plan is to be perfected and set in operation at a meeting in the Drexel Institute, Philadelphia, on March 25 to 27.

The plan consists of several parts: First, the writing by the representatives of industry of a definite specification of the type of man most needed; second, the writing by the representatives of the colleges of a definite statement of the capacity of the colleges to meet this demand; third, forming a mutual specification from these two and developing permanent means for carrying out the work so as to continue permanently the coöperation thus begun. It is also proposed to publish the results of the meeting in book form as a matter of record.

The execution of this idea should result in increasing the efficiency of the technical schools by showing them just what type of instruction is needed in the industries. It should increase the efficiency of industrial organizations by providing men trained in accordance with industry's needs. It should accomplish these results in a minimum of time because it is dependent on and provides for direct contact between the teachers and the users of trained men.

Notwithstanding the comparative immunity of European countries from boiler explosions, they do have one sometimes. A dispatch from Benrath, Rhenish Prussia, states that fifty lives were lost when a large boiler exploded in the Rhenish Westphalian Electricity Works. Following this dispatch, news comes in of the explosion of a boiler in an electrical station at Mayence, France. Ninety workmen were buried in the ruins, and fourteen bodies had been taken from the débris when the dispatch to the *Petit Parisien* was sent. Probably owing to a relaxation of the usually strict and effective governmental supervision of European boilers and to the increasing activity of our own industrial commissions and other safety organizations, the United States will not have so large a lead in industrial disasters this year as usual.

As was generally expected, Senator Norris' amendment to the Water-Power Bill, providing $25,000,000 for the development by the Government of the Great Falls power project on the Potomac, above Washington, was eliminated at the first meeting of the Senate and House conferees on the measure.

Knoxville Industrial Co-operative Plan

In the winter of 1918, just following the wholesale inspection of all power plants by the Fuel Administration, a body of men in Knoxville undertook to improve power-plant conditions in the Southern States, and especially in Knoxville. This group of men was the Knoxville Chapter of the National Association of Stationary Engineers. Throughout the Knoxville district a large number of plants needed repairs, not for the sake of efficiency, but to make them safe places to work in. There were plants in existence that did not dare ask any high-grade boiler insurance company to put their test pumps on their boilers for fear of a rupture.

After considerable discussion the engineers decided that the quickest method to bring about effective results was to introduce a license law for the City of Knoxville and to incorporate in that law sufficient protection for the operating engineer.

The engineers appointed a committee for drafting this ordinance, and called in the writer in a consulting capacity to help. What was considered a fair ordinance was finally evolved and placed before the City Commission. The municipal law of Knoxville requires that all ordinances shall pass three readings before they become laws. This ordinance passed two of the necessary readings, but at this juncture the stationary engineers believed it would be more fair to call in the manufacturers, in an open meeting of their organization, to ask them their opinion of the bill.

Like many other requests coming from the employees, this invitation was ignored except by four or five of the manufacturers of the city, who seemed to realize the importance of the bill. These few men attended the open meeting, then called, on the following day, a meeting of all interested manufacturers into a secret conference. This was where they made one of their errors, as they failed to show a spirit of coöperation with the employees and aroused antagonism. A secret meeting of this type, following what was a fair invitation by the engineers for a get-together conference, is not good business management. Conditions became chaotic. The manufacturers decided to fight the ordinance.

A counter proposal was offered at this point by the University of Tennessee. This proposal, presented to the manufacturers, signed by them, and on the following day accepted by the engineers, put up to all parties a common basis of coöperation. By it the manufacturers promised:

1. That they would appoint a committee of five to coöperate with a committee of five from the National Association of Stationary Engineers, to put into effect the principles set forth in the preamble of the constitution of the National Association of Stationary Engineers.

2. They agreed to financially provide for five courses on steam engineering and allied subjects, the students to pay for textbooks.

3. They agreed to finance an advertising campaign to put these classes in operation.

4. To subscribe the funds to publish a four-page monthly *Power Plant Efficiency Bulletin*, to be placed in the hands of every fireman and engineer in the city.

5. They agreed to hold a "power-plant efficiency banquet" at some date during the year, at which time both engineer and manufacturer would feast together. The engineers were to arrange the banquet, and the expense was to be divided among all. At this banquet the National Association of Stationary Engineers were to prepare a comparative table showing which ten plants in Knoxville were, in their judgment, the most efficient. Further, a similar announcement was to be made of the engineer and fireman, who were, in the opinion of the National Association of Stationary Engineers, the most deserving of the distinction and recognition of being placed first in their profession in the city.

Today this entire program has been completed for a cycle of one year. To maintain a fair censorship, the writer, at the request of both parties, accepted the editorship of the *Power Plant Efficiency Bulletin*, which is referred to in the foregoing agreement. He was assisted by a committee of four from the National Association of Stationary Engineers, and at no time during the year did he find it necessary to cut out material that was undesirable or unprintable. Furthermore, he has received a large number of articles from these practicing engineers that were worthy of a place in any high-class engineering journal. It has had a very healthful influence upon all of the stationary engineers and firemen of the city, for not only have they been able to express themselves in writing for the general public, but they have monthly received the *Bulletin*, with articles written by men whom they knew. Naturally, the busy engineer is more ready to read material coming from the pen of a friend than that coming from a stranger.

That the whole program has been a great success was demonstrated most emphatically by the success of the banquet recently held. The program was arranged by a committee from the National Association of Stationary Engineers, and shop talk was almost entirely eliminated from the program. When they gathered at the St. James Hotel in Knoxville, both sides were represented by a full quota. The program was a live one, with a real live toastmaster, who was neither an engineer nor a manufacturer, to keep things moving. As an impromptu part of the program he made every manufacturer and engineer get up on his feet and tell where he was from, where he lived, and what his position might be. No one could go away without saying that he had not made a speech at the banquet.

The cost to each manufacturer for the entire year, for carrying out this agreement was approximately $5, and I believe, as a fair estimate, we can say that in those plants where the spirit of the plan has been felt at all, the investment of this $5 by the manufacturer has brought in a net return of several hundred per cent of healthy coöperation on the part of the engineers.

There is in the ordinary power plant no one man who is in a position to spread either the seed of contentment or discontent to a greater measure than the engineer. The manufacturing power-plant engineer usually must assume the responsibilities of a master mechanic and of a general repair man. He is known to practically all the force, and discusses from day to day the matters of the factory that are nearest the heart of the employees. It is far easier for him to feel the pulse of the employees than it is for any member of the management. He often must assume the rôle of a personal relations manager for the factory in which he is located.

We are confident that the rôle assumed by the university has made it a force for healthy coöperation, and we find both engineers and manufacturers equally pleased with the results that have been obtained.

Knoxville, Tenn. W. R. WOOLRICH.

Freezing of Brine Puzzle

In the Jan. 6 issue, page 35, under the heading, "Freezing of Brine Puzzle," C. A. Goodwin draws some conclusions from his experience with a frozen brine cooler that appear to be false.

Unfortunately, he does not give data relating to the temperature of the brine entering and leaving the cooler, the range, or decrease in temperature, between the inlet and outlet brine, suction pressure under which the cooler is operated, and such other data as would assist the readers in determining the cause of his trouble. However, it is evident that he makes a practice of keeping his brine solution too weak, for with the double-pipe brine cooler the margin of safety between the lowest working temperature of the brine and its freezing point must be great enough to allow some leeway, as such coolers will freeze up quickly if the conditions are not as they should be.

Mr. Goodwin concludes that the freeze-up, in the case under discussion, was directly due to the fact that about two coils of the submerged cooler were above the brine level and that the freezing occurred in these two exposed sections. This is incorrect, as there never exists such a wide difference of temperature as he assumes in such a short length of brine travel; that is, from the top of the submerged part of the cooler to the two upper sections of the cooler that had been uncovered by the brine.

As a matter of fact, in a properly operated cooler of this design a range of usually $5\frac{1}{2}$ to 6 deg. between the inlet and outlet of the cooler is considered good practice, and if his assumption were correct, he would get more than a 6-deg. range in about three coils instead of the entire cooler.

I assume that his cooler is connected up on the counter-current principle—that is, the ammonia and brine travel in opposite directions and the cold brine leaving the cooler comes in contact, so to speak, with the coldest ammonia—in which case there should not be, as he supposes, a wide difference of temperature or heat head between the evaporating ammonia and the brine to be cooled.

Under normal operation this heat head or temperature difference should not exceed 10 deg., and it may run as low as 5 to 8 deg. Also, if the brine enters at the top of the cooler and the ammonia enters at the bottom, the coldest brine would be found in the submerged section near the outlet connection and not in the exposed section as he supposes.

If Mr. Goodwin will check up his thermometers for accuracy, place a pressure gage on the brine pump, keep the brine strength at its proper value and deliver sufficient brine to the cooler to maintain the proper range for a given tonnage or load condition, he should not have any trouble with the freeze-ups; also when adding more water to the tank to raise the brine level, do not dump fresh water in the tank while the cooler is in operation and add the calcium afterward, because if this is done it is almost certain to result in a frozen cooler. C. T. BAKER.

Atlanta, Ga.

Packing Valve Stems Under Pressure

On page 835 of the Dec. 16-23-30, 1919, issue of *Power*, Gustav Swenson writes that Mr. Jackson must be a wizzard if he can pack a globe valve with the pressure coming above the valve disk. It can easily be done, however, by taking a piece of band iron $\frac{3}{8}$ in. thick, $\frac{3}{4}$ in. wide and 4 ft. long and making a clamp of a size to get around the packing nut.

To pack the valve, first open it wide, so as to check the flow of escaping steam as much as possible from around the valve stem. Then remove the valve wheel and the packing nut. A piece of burlap may be thrown over the valve to baffle the steam.

The packing rings should be a loose fit around the valve stem and a tight fit inside the packing nut, enough new ring being used to fill the nut about three-fourths full. The clamp can then be tightened to the stuffing-nut for screwing it in place on the valve. The strap is then removed, and the operation need not take more than two minutes if everything has been made ready.

Mr. Swenson also says that I seemingly favor connecting globe valves with the pressure under the disk. On the contrary, I am strongly in favor of having the pressure under the disk.

Regarding clogging of valves, neither a globe valve, elbow nor pipe that is attached to a water column will close up, if it is watched and attended to as it should be. If the water is very dirty, clean out the bottom connections often. JAMES E. NOBLE.

Portsmouth, Ont., Canada.

Boiler-Room Essentials

In the excellent article on "Boiler-Room Essentials," by H. F. Gauss, that appeared recently in *Power*, there is one sentence that should be amplified.

The statement is made that "for uniform flow, such as is secured by centrifugal boiler-feed pumps, the venturi meter is the desirable boiler-feed meter," but "there is some question as to the accuracy of the venturi meter under pulsating flow."

The term "pulsations" is often loosely used. Two kinds of pulsations may be present in a pipe line—pulsations in pressure and pulsations in velocity—but the two do not always accompany each other. For instance, in a water-works system a reciprocating pumping engine discharging into a main distributing line may produce severe pressure pulsations, or water hammer, as shown by the lively action of a pressure

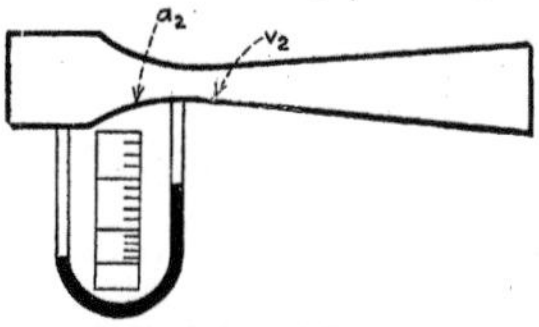

DIAGRAMMATIC PRESENTATION OF A MERCURY MANOMETER

gage, and this effect may be noticed miles distant, but a venturi meter in the main would reveal a very steady flow. The reason for this is that the friction of the long discharge line and distributing branches absorb and smooth out the peaks in the flow line.

In a boiler plant pulsations in pressure are usually accompanied by pulsations in flow, although the venturi-meter tube can often be placed beyond a filter or economizer, which tends to straighten out the flow line. A long experience with the measurement of boiler feed by venturi meters proves that single-plunger pumps, almost without exception, give error-producing pulsating flows when the water is pumped directly through the venturi-meter tube or any other meter of the so-called velocity or inferential type. Not all duplex pumps cause severe enough pulsations to be detrimental, and triplex pumps in good condition can be used without taking any precaution to dampen the pulsations. When two or more duplex feed pumps are in service, the character of the flow is suitable for correct meter registration.

To determine accurately the extent of error produced by a pulsating flow through a venturi-meter tube, the latter should be connected in the usual manner to a mercury manometer, as illustrated diagrammatically by the accompanying sketch. Let:

R = Rate of flow;
C = Meter coefficient;
g = Acceleration of gravity;
a_2 = Throat area;
v_2 = Throat velocity;
h = Differential pressure in manometer.

$$v_2 = C \times \sqrt{2gh} \qquad R = a_2 \times v_2 = a_2 \times C \times \sqrt{2gh}$$

The scale is graduated in the rate R, which, according to the formula, varies as the square root of the differential pressure.

To take an extreme case of pulsating flow, assume that high and low peaks produce a maximum differential nine times as great as the minimum differential in inches of mercury column.

$$\frac{\textit{Actual maximum rate of flow}}{\textit{Actual minimum rate of flow}} = \sqrt{\frac{9}{1}} = \frac{3}{1}$$

or, actual average rate of flow = 2. But the average mercury differential in the manometer as observed would equal 5 in. Therefore;

$$\frac{\textit{Observed average rate of flow}}{\textit{Actual average rate of flow}} = \sqrt{\frac{5}{2}} = 1.12.$$

or the error resulting from reading the scale opposite the average mercury level is +12 per cent.

Since recording instruments employ a much larger weight of mercury than a manometer, and hence, owing to the inertia of the mercury, cannot follow changes in velocity as rapidly as a manometer, the recording instrument may show even a greater plus error than the manometer. Employment of the latter, however, provides a ready means of determining the extent of the error of the recorder under any condition of operation.

Throttling the small pressure connections between the meter tube and the recording instrument would only aggravate the error. The only practicable method of overcoming the difficulty, as all manufacturers of the velocity type of meter will frankly acknowledge, is to absorb the pulsations in flow before the water enters the meter. If the latter cannot be advantageously located to accomplish this, as outlined, a suitable air chamber should be installed in the main line preceding the meter, designs for which have already appeared in *Power*. It may be set down as a general rule that if the pulsations in velocity are severe enough to affect the accuracy of the venturi-meter readings, the pulsation in pressure is causing destructive water hammer on the pump parts, valves, fittings and gaskets and is in itself sufficient argument for a remedial air chamber.

Providence, R. I. CHARLES G. RICHARDSON.

Use of Bichromate Salts as Boiler Compound

A short time ago a reader of *Power* requested data concerning the use of bichromate as a preventive of corrosion in steam boilers.

This subject has received pretty thorough investigation, and I suggest that reference be made to J. Newton Friend's work entitled "Corrosion of Iron and Steel," as published by Longmans, Green & Co. In the chapter headed "Influence of Solutions of Electrolytes upon Iron," the subject of bichromates and chromates is gone into very thoroughly.

While, under certain conditions, the sodium and potassium bichromates and chromates will produce a protective action of the metal, the amount required per thousand gallons of water evaporated renders its cost prohibitive.

Furthermore, as will be noted from the article in question, much better results can be obtained by the use of the chromate instead of the bichromate, since if the latter is employed in the presence of foreign salts, especially sodium chloride, very little protection is secured on account of the liberation of free acid.

Chicago, Ill. E. L. GROSS.

Trouble with Spray Valve Packing

Having had considerable trouble with spray-valve packing on a semi-Diesel engine with closed tap valve, I am submitting the following:

Because the stuffing-box was not deep enough the shredded metallic packing which was used would work out around the outside of the gland enough to keep it from drawing up and keeping the packing tight on the valve.

To remedy this trouble I cut a piece of babbitt $\frac{3}{8}$ in. thick and large enough to cover the opening in the

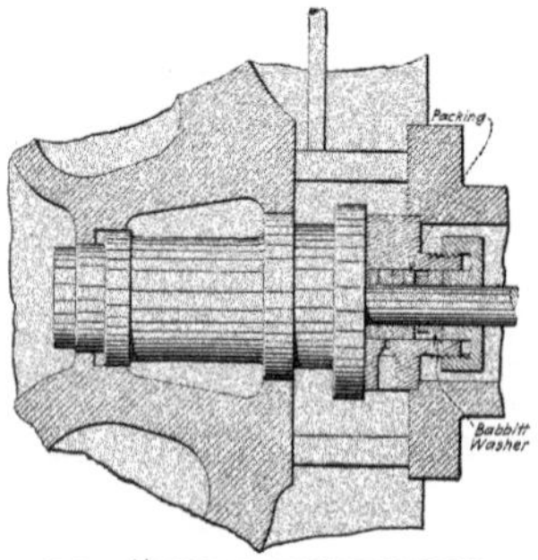

BABBITT WASHER APPLIED TO STUFFING-BOX

stuffing-box. I then drilled a hole in the center of this the size of the spray-valve and cut the piece down to a little larger than the opening and forced it into the box so as to get a better fit on account of the box being badly scored.

By using this between the gland and the shredded packing there has been no further trouble.

Covington, Ohio. HARRY A. HAPPLE.

Lubricating an Engine Under Varying Load Conditions

I read with unusual interest the letter of Charles J. Mason on page 672 of the Dec. 8 *Power*, treating the question of lubricating an engine under widely varying load conditions. When Mr. Mason says "A fully loaded engine using large volumes of steam per stroke is easy to lubricate as compared with under-loaded engines with an accompanying early cutoff and small steam volumes," he is giving expression to a privately formed conclusion that I have arrived at through the study of a most peculiar behavior of piston-rod packing on an engine now in my charge. This engine is perfect in alignment, adjustment, design and workmanship and is my pride as well as the praise of inspectors and supply men who come to the plant. But while this engine "runs like a song" it can destroy more piston-rod packing than any other engine I ever saw. In my search for a packing that would give satisfaction on this engine I carefully selected and tried out six good standard makes and all failed. These packings would all give good results for a while, then, suddenly, and with less than two minutes' warning, would burn out. Sometimes the packing would run beautifully for a month or more before burning out and then again it might run only one week after being freshly packed. The burning out would take place as follows:

All would be running smooth and well when all of a sudden I would hear the packing start to blow and I would find it smoking. The blowing would increase rapidly and become violent while smoke would come from the rod and box in clouds. Upon examination I would find the rod with what I call "a hump on its back" (horizontal engine). The rod instead of being straight would be raised in its middle a full $\frac{1}{8}$ in. and pinching the top sides of the packing rings till they simply burned up. The hotter the top side of the rod became the more it humped up, due to the expansion of that side, and the more it humped up the greater became the friction and heat on that side to hump it up more. A brush and a can of cylinder oil would straighten the rod out again in a few minutes without stopping the engine, but that gland-full of packing was lost. If the packing were soft, it was burned and if it were semi-metallic the metal would be melted out of it.

This might happen any time either on an old gland full of packing or on a newer one regardless of make. I came to the conclusion that there must be times when the rod failed to get oil and ran as long as it could on what the packing had absorbed. But the natural tendency of the oil would be to gather in the lower halves of the packing rings, leaving the top halves comparatively dry, and the top side of the rod next to the dry half of the rings would be the first place to gather excess heat from lack of lubrication. As quickly as the rod started to warm and expand along its top side, the bending effect due to unequal expansion would take place and the performance I have described would be gone through with.

I next noticed that this overheating always took place when the load was almost nothing and the steam pressure high. The engine is 15 x 18 in., running 200 r.p.m. directly connected to an a.c. generator and runs on 135 lb. steam pressure. The load averages 150 hp. but varies from 3 to 200 hp. At certain seasons of the year we frequently get a load of 3 hp. (one small motor) for several hours at a time. The friction load is practically nil due to lack of belts and shafting, and at these periods I get a diagram that almost comes back on the compression line.

I am firmly convinced in my own mind that at these periods of short cutoff and small volume of steam the flow past the lubricator nozzle is not sufficient to atomize the oil or carry it along the rod as the rod is drawn into the cylinder. After a lot of study and experimenting I have found an almost perfect cure for the trouble in a combination of two things. First, I arranged an oil cup over the rod so that the drop would fall upon its upper side, and this I keep filled with cylinder oil so that I can turn it on whenever I see the ammeter drop back to almost zero and remain for any length of time. Second, I found a packing that is made of asbestos product and will neither burn or melt. This packing is made to contain annular cavities encircling the rod and will yield to the distorted rod and come back to place when the rod straightens again.

Vancouver, B. C. R. MANLY ORR.

INQUIRIES OF GENERAL INTEREST

Velocity in Pump Suction and Discharge Pipes—What velocities are allowable for water pump suction and discharge pipes? G. L. J.

In order that loss of pressure from pipe friction may not be too great for ordinary lengths of pipe lines, the velocity of water in the suction pipe should not exceed 240 ft. per minute and that in the discharge pipe 300 ft. per minute.

Relative Heat in Water and Steam of Boiler—Does a pound of water in a boiler contain as many heat units as that quantity converted into steam at 100 lb. or any other pressure? R. B.

The water in a boiler is at the same temperature as the steam but before the water can be converted into steam it must receive additional heat, called the latent heat of evaporation. Thus for 100 lb. gage pressure the water and steam are at a temperature of 338 deg. F. and each pound of the water contains 309 B.t.u. above 32 deg. F., but for conversion into steam, each pound of water must, in addition, receive 879.8 B.t.u. which is the latent heat of evaporation at the stated pressure; so that the total heat contained by a pound of the steam is 309 + 879.8 = 1188.8 B.t.u.

Size of Steam Pipe for Engine—What size of steam pipe would be required for an 18 in. x 36 in. engine running 110 r.p.m., assuming the velocity of steam in the pipe is 7,000 ft. per min. W. H.

The piston displacement of the engine would be (18 × 18 × 0.7854) × 36 × 2 × 110 = 2,015,399 cu.in. per min. Allowing that the maximum velocity of the steam in the pipe is to be 7,000 ft. or 84,000 inches per minute, when the piston is traveling at its average speed, the required cross-sectional area of the steam pipe would be 2,015,399 + 84,000 = 23.99 sq.in. which corresponds to

$$\sqrt{\frac{23.99}{0.7854}} = 5.5 \text{ in. dia.}$$

The least commercial size of pipe required would be 6 in. inside diameter.

Weight Required at End of Safety Valve Lever—If the area of a lever safety valve is 1 sq.in., distance from valve to fulcrum 2 in., weight of lever 2 lb., distance of its center of gravity to the fulcrum 5 in., weight of the valve and stem ½ lb., and total length of lever 15 in., what weight must be placed on the end of the lever so the valve will blow off for a pressure of 100 lb. per sq.in.? A. B. T.

The lever alone will exert a pressure on the valve of (2 x 5) ÷ 2 = 5 lb., and as the weight of the valve and stem is 0.5 lb. the lever and dead weight of the valve will balance a pressure of 5.5 lb., leaving 100 − 5.55 = 94.5 lb. of pressure to be exerted on the valve by the leverage of the weight placed at the end of the lever. As 1 pound at the end of the lever would exert a pressure on the valve of 1 x 15 ÷ 2 = 7.5 lb., the additional 94.5 lb. pressure required would be balanced by a weight of 94.5 ÷ 7.5 = 12.6 lb. placed on the end of the lever.

Sizing Bushing for Engine Cylinders—We are preparing to rebore and bush a 26 x 42 in. Corliss engine cylinder. The bushing is to be about 1 in. thick and the cylinder will be bored in four steps. How much oversize will it be safe to make the bushing larger than the cylinder diameter to press the bushing in place with a 35-ton jack? W. D. P.

No more than a snug fit is necessary when the bushing and cylinder are at the same temperature and appropriate provision is made for preventing the bushing from turning or slipping endwise in the cylinder. Very good jobs have been made by turning off the outside of the bushing about 0.005 of 1 in. greater in diameter than the diameter of the cylinder bore when both are at the same temperature and then inserting the bushing with the cylinder warmed up and the bushing cold. By that method the pressure required for forcing the bushing into the cylinder will depend upon the force required to spring the bushing to the form of the cylinder in compensation of imperfect boring and turning.

Required M.E.P. for Generating Stated Kw.—What would be the m.e.p. in the cylinder of a 10 x 10 in. engine running 300 r.p.m. direct connected to a generator for developing 35 kw.? R. C.

For estimating the required m.e.p. it is necessary to take account of the efficiency of the engine and generator. One kw. is development of work at the rate of 44,240 ft.-lb. per min. and, assuming the mechanical efficiency of the engine to be 85 per cent and the efficiency of the generator as 90 per cent, the work required by the steam acting on the engine piston would be

$$\frac{44,240 \times 35 \text{ kw.}}{0.85 \times 0.90} = 2,024,052 \text{ ft.-lb. per min.}$$

One pound m.e.p. in a 10 x 10 in. engine running 300 r.p.m. would develop

$$1 \times (10 \times 10 \times 0.7854) \times \frac{10}{12} \times 2 \times 300 = 39,270 \text{ ft.-lb. per}$$

min. and therefore the generation of 35 kw. would require 2,024,052 ÷ 39,270 = 51.54 lb. m.e.p.

Effective Pull of Belt—What is meant by "belt pull" and how is it computed? Is belt pull the same as belt stress? What would be the belt pull in driving a machine that requires 20 hp. where the belt speed is 2,325 ft. per minute? W. N. J.

The term "belt pull" is commonly employed to signify the effective pull or difference in tensions of the tight side and slack side of a belt when transmitting power. The actual pull or tension that tends to draw the belt apart is the tensile stress and the intensity of that stress usually is expressed in pounds per square inch of cross sectional area of the belt.

For transmission of 20 hp., or 33,000 × 20 = 660,000 ft.-lb. per minute, and for a belt speed of 2,325 ft. per min. the effective pull would be 660,000 ÷ 2,325 = 283.9 lb. For transmission of any given power with a given speed of belt, the maximum tensile stress, or actual pull coming on the tight side of the belt, depends on the required effective pull, the tension under which the belt was put on pulleys and elasticity of the belt material.

[Correspondents sending us inquiries should sign their communications with full names and post office addresses. This is necessary to guarantee the good faith of the communications and for the inquiries to receive attention.—Editor.]

Engineering as Prosperity Insurance*

By GEORGE OTIS SMITH†

INDUSTRY is a reaction into which raw material and energy enter; it is the function of engineering to use both material and energy in the smallest quantities possible, in other words, to make what is used do the most work. To appreciate the rate at which we increase our annual drafts upon Mother Earth's treasury, a glance at the records for the two chief metals and the two mineral fuels is perhaps enough. Within the last half century the country's output of coal has increased more than 20-fold, iron more than 23-fold, and copper 68-fold, while the flow of petroleum has increased not less than 700-fold. This growth of the mining industry within the life of most of us must open our eyes to the real issue of safeguarding our sources of supply.

Although the 99 per cent or so of coal left unmined is a figure that might seem to justify an optimism as to coal that is not warranted as to oil, we may need to think less of the million of millions of tons of coal which we are told lie awaiting our needs over the length and breadth of our land and to ask for more details as to where this coal is and how much of it remains in the older mining districts.

Before you indulge in mental arithmetic in figuring out the life of our coal resources, you must set down the broad fact that the best and most accessible coal is mined first. From this premise follows the practical prophecy that with the passing of time in our rapid industrial expansion we may expect for the coal of our future output decreasing quality and increasing cost. It is this aspect of our tomorrow that forces the discussion of the best use of America's coal.

The reason for our interest in the past and future of coal is that coal will count for more and more as the controlling factor in America's industrial future. Even now we can take our country's record of the highest per capita coal consumption in the world as the gratifying test of our industrial capacity.

Lessons of the Armistice Period

The lessons of the armistice period have increased our appreciation of the factors that insure prosperity. Raw materials and tools are what the war-stricken peoples of Europe are asking for; with these idleness and hunger can be converted into industrial and physical health. Coal and flour are two excellent antidotes for bolshevistic infection.

Human energy was something that the ancients spent with little regard for present or future happiness. The great monuments of the past—the engineering wonders of the world—commemorated slave labor. Today we regard man-power as something to be conserved, not only because we as practical business men set a higher value upon humanity than even the philosophers of old, but also because we have learned that in much that we try to do, man-power is the limiting factor.

To lighten the load of labor is the real purpose in planning for power. In the evolution of industry animal-generated power was first brought to the aid of man, then wind and running water were harnessed, next fuels through the agency of steam made possible the modern industrial system, and now electricity has been made a servant of man that promises to lift the burden of toil so that human energy can be put to the highest tasks.

The official estimates of the potential water power in the United States when figured on an 80 per cent efficiency indicate in round numbers a minimum of 30 million horsepower and a maximum of 60 million horsepower. The water power now utilized, however, is less than one-third of the minimum potential power, or taking a world view of the water-power situation, we find our country standing third in potential power resources, with about 7 per cent of the world supply, but first in rank, as determined by developed power, with 45 per cent of the world's total. Canada, France and Norway stand next in order as to power developed, but the total of all three is only half the water we use.

These are the basic facts that suggest what the future can hold for us. Unfortunately, there remains a fact of geographic distribution that affects future utilization: as pointed out recently by W. B. Heroy, 70 per cent of the available water power is west of the Mississippi, whereas 70 per cent of the water power now installed is east of the Mississippi. With the largest power market and the largest power supply so separated, fuel power must continue to carry its part of the load.

Super-Power Plan for Eastern Industrial Region

The super-power plan that is now being advocated for the great industrial region of the East seems a direct answer to the appeal for the best use of America's coal. In its broad outlines this plan, for which a Federal investigation is being asked of Congress, is the linking up of high-power, high-economy generating stations with a super-power transmission line—a veritable stream of power from which will flow the current wherever it is needed for transportation, industry or home use. You engineers do not need to be told by me the possibilities of saving through higher load factor, low coal consumption and greater certainty of continuous service. How the engineering details of so great a plan are to be worked out is a story for future telling; but we can confidently expect that this interstate system, so essential to the whole nation in helping us to meet international competition effectively, will be thoroughly democratic in spirit and will be so planned as best to serve the greatest number of citizens.

In setting up a practical ideal for a power system, the two words "public service" are enough. Service to the public is the power plant's reason for being, and the quality of that service should measure the corporation's standing with the customer and indeed in the community and in the Nation. I fear lest for political effect too much stress is often laid on price; of course both economist and engineer must give full attention to the question of cost, as well as the possible profit in the business, but another element in service is perhaps even more important, and that is reliability. I notice that Robert Treat in a recent technical paper puts foremost among the aims of the electrical engineer who is planning large electrical layout reliability of supply and continuity of service: The flow of energy must be uninterrupted.

The power company, then, deals in service. A recent article in one of the engineering journals in describing a great interstate power system classified the causes of the interruptions of service on its lines so far as they could be determined. Lightning and sleet and high winds, which are beyond the control of either company or customers, together caused 33 per cent of the breakdowns, but failure of the customers' apparatus and carelessness in the customers' operation caused 35 per cent of the interruption in service, while the corresponding figure for company failure and carelessness was less than 10 per cent.

Quality of Service Dependent on Consumers

At the Natural Gas Conference in Washington last week it was likewise pointed out that in that type of public utility also the quality of the service rendered the public depends largely upon the consumers themselves. Methods of use within the house not only control the service there, but also affect the service rendered to other consumers. The public-utility engineer can profitably devote some time to the education of his customers.

No type of citizen is better fitted to tackle the civic problems of today than the engineer. The large questions before the people are economic questions, and engineering is mainly science applied to life in an economic way; the engineer in his experience as well as in his training has studied problems in economics, whether his textbook bore that title or not or whether his design or his contract was so labeled or not. The constructive issues that are before the American people this year and that promise to remain before our legislative bodies for several years are all in some degree engineering issues, and in a large degree they present on a national scale.

*Abstract from an address delivered before Engineers' Club of Philadelphia, Penn., January 20, 1920.

†Director, United States Geological Survey, Washington, D. C.

problems closely akin to those of everyday work that you engineers have in hand on a unit scale.

The transportation problem now before Congress, with Mar. 1 as the date of delivery for the legislative remedy; the many questions relating to the production and distribution of the commodities that are essential to life and comfort; the working out of the equities of the partnership between labor and capital and the public—these are all practical economic issues to the consideration of which the engineer can bring an analytical mind and a large store of facts.

These public questions are more or less interlocking; our coal problem is hitched up with the railroad problem, and both are related to the super-power plan, while the power program in turn contributes to the labor question, and the underlying ideal of prosperity distribution is but another name for profit sharing which involves accounting in terms of equities. Some of the outstanding elements in these national issues may be stated, however, even if a logical analysis of each question is not made.

How the Various Interests Are Tied Together

The coal industry furnishes perhaps the best example of how the operators, the mine workers and the consumers are all tied together in actual interest. The curve of weekly coal production tells the story of highly irregular operation, and idle plants and idle men are expensive taxes upon the community and the Nation.

For three months this last spring the coal mines of the country were operated, on an average, for only about 24 hours a week. From coast to coast the reason for lost time was "no market"—something beyond the control of either operator or mine worker. Here, then, is the greatest branch of our mining industry vitally affected by a widespread malady, the remedy for which lies with the public alone. Laws cannot make coal mines operate when there is no outlet for their product, but education of the consuming public may accomplish much in bettering the conditions of demand, and we as consumers will do well to remember that the price we pay for coal must in the long run be enough to pay for the idle days of both mines and miners, as well as for the days they work. The public interest lies in a longer working year, not in a shorter day or week. Largely on account of the irregular market demand for coal, our mines since the armistice was signed have actually averaged only 30 hours of working time a week; throughout the year there have been too many mines and too many miners waiting for work. A 30-hour week is the ailment, not the cure.

In normal years the seasonal fluctuations in the demand for soft coal cause a slump in mine output during the spring and a peak of forced production in the fall and early winter. Compared with the average for the whole year, the monthly rates of output will commonly range from 81 or 82 per cent in March and April to 115 or higher in October and November. The mine operator is dependent upon a too irregular market.

Comparison of Coal Consumption of Typical Power Companies

As a suggestive comparison I have taken the coal consumption of six large typical power companies—companies having an aggregate installed capacity of nearly 700,000 kw.—which furnish electric current to a trunk-line railroad and several street-railway systems, as well as general lighting and power service in metropolitan districts. The daily average coal consumption of these companies by months during ten months of last year furnishes a curve that shows a seasonal fluctuation similar to that of mine output, but the departures from the base line representing the average for the whole period are much smaller—92 per cent is the spring low, and 110 per cent is the winter peak. Applied to the general curves of soft-coal production for several years, this consumption curve of six companies shows that for three-fourths of the time such power companies could exert a helpful influence on the market, even if they bought their supply from day to day. The storage of anything like a reasonable tonnage of coal at such plants would still further smooth out the curve and help to put the coal industry on a better basis. The influence of such consumers is well worthy of consideration, for these six companies alone use nearly one-half of one per cent of the soft coal mined. It is also doubtless true that by connecting up such power units into a single system the improved load factor would tend to smooth out the curve of coal consumption, and the pooled purchase of coal would further help abate the seasonal demand which is today the greatest evil in the bituminous-coal industry.

The leaders in industry today are calculating the vision of a new type of prosperity tomorrow which must mean profit sharing on a thoroughgoing scale so that all concerned are to receive in proportion as they have served. The goal of wisely directed industry is the prosperous community of happy homes, and engineers must direct such industry.

Coöperation of Accountants Necessary

Yet the engineer must have the full coöperation of the accountant if the returns from industry are to be fairly divided. A true diagnosis of the financial condition of any industry involves bookkeeping that states the real economic symptoms. In mining, as in all other productive business, production costs should express all the facts, but I am convinced that we have not yet reached that stage in our accounting methods. Does not the cost of a pound of copper just as truly include its share of the interest due on the bonds and the dividend due on the capital stock actually paid in—that is, wages paid to capital—as it includes the wages paid to the mine worker or to the man in the smelter? Of course neither payroll should be padded. Do not regard the definition of profit as a simply academic question for college professors to discuss; it is more a matter of fundamental concern to the men who are trying to build up the industry.

In the first place, such terms as "net earnings," "profit," "net income," "surplus," or "balance to surplus account," which appear on your financial sheets, are too optimistic in tone when the operating costs do not include any charge on account of the investment. The inference is too easily drawn that this surplus is available for profit sharing between labor and capital.

In the second place, I fear lest accounting which does not count all the costs may lead us to fail to appreciate the results of American engineering. Our engineers, backed up by red-blooded capitalists, are constantly lowering labor costs by increasing the investment in improved mine equipment. This means replacing the hardest part of labor with machinery—it means working dollars instead of men—so that your cost keeping conceals the true state of things if you pay labor out of so-called "costs" and capital out of so-called "profits."

Those are some of the reasons why we need to unite in urging the engineers of the country to take a larger share in citizenship; indeed, these are compelling reasons for engineers to assume more of leadership in public opinion. The engineer is a forward-looking worker, but to get in fullest measure the pull of inspiration in our work, perhaps we need to hark back as well as look ahead. Prosperity of the right kind is all we can ask and all that we need to work for. If the door of large opportunity is to open wide for all our people, the key to unlock that door is good engineering.

It is estimated, says the *Engineer* of London, that at the present time the Irish bogs contain between 3,500,000,000 and 4,000,000,000 tons of anhydrous peat, or 5,000,000,000 tons of air-dried peat. At present about 6,000,000 tons of peat are burned as fuel in Ireland per annum, and over 4,500,000 tons of coal are imported. If this coal were replaced by peat fuel at the rate of two tons of air-dried peat to one ton of coal, the total consumption of peat in Ireland would be about 15,000,000 tons per annum, and the peat deposits would be sufficient to satisfy the fuel and power requirements of the country at the present rate of consumption for more than 300 years.

In an article on the development of the power of the Niagara Falls in the *Bulletin* of the Ontario Hydro-electric Commission, it is shown how as long ago as 1812 Augustus Porter and Peter Emslie began the construction of a power canal 100 ft. wide. The canal was ultimately completed with a width of 35 ft. and a depth of 8 ft. By 1885 the full capacity of this canal, 10,000 hp., was being utilized.

Fuel-Oil Bibliography

The following bibliography leaves much to be desired. yet it will go far to meet the demands of engineers interested in it from the power-plant standpoint:

References to Some Articles in *Power*

Author	*Articles*	*Page*	*Issue*
	January to June, 1914.		
McIntosh	Physical Properties	32	Jan. 6, see also Jan. 27, p. 146.
McIntosh		54	Jan. 13
McIntosh	Government Specifications for Purchase of Storage and Piping	105	Jan. 20
McIntosh	Oil Burners	139	Jan. 27
McIntosh	Furnace Design	158	Feb. 3
McIntosh	Operating Troubles	195	Feb. 10
	January to June, 1913.		
W. G. Greenlees	Burning Fuel Oil Under Boilers	815	June 9
E. H. Peabody	Developments in Fuel-Oil Burning	205	Feb. 11
E. H. Peabody	Developments in Fuel-Oil Burning	244	Feb. 18
H. O. McCarthy	Oil-Flow Problem — Flow from Tank	503	April 8
B. F. Hartley	Heating Fuel Oil in Home-made Heater	60	Jan. 14
Capt. C. W. Dyson	Oil Fuel for Destroyers and Battleships	145	Feb. 4
	January to December, 1913.		
B. W. Babcock	Furnace for Burning Fuel Oil	239	Aug. 12
W. B. Perkins	Purchasers Test of Crude Oil—Centrifugal	259	Aug. 19
McIntosh	Production Statistics	915	Dec. 30
	January to June, 1915.		
F. H. Rosencrants	Chimneys for Oil and Coal-Burning Plants	637	May 11
G. M. Bean	Fuel Oil for Locomotives	900	June 29
	July to December, 1915.		
F. C. Fearing	Charts on Relative Cost of Coal and Oil	826	Dec. 14
Delaney	Oil Burning Standby Plants	172	Aug. 3
W. L. Kidston	Automatic Regulation of Oil Burning	637	Nov. 9
A. D. Pratt	Use of Oil Fuel	665	Nov. 9
	January to June, 1917.		
Wales	Chart on Equivalent Cost Coal and Oil	347	Mar. 13
Halloway	Oil-Burner Regulator	391	Mar. 20
Camp	Oil-Burning Furnace	464	Sept. 26,'16
Long	Home-Made Oil Burner	464	Sept. 26,'16
Ewing	Fuel Oil for Steam Boilers	399	Mar. 20
Ewing	Fuel Oil for Stationary Boiler Plants	643	May 8
Colwell	Precautions in Fuel-Oil Storage	886	June 26
	July to December, 1917.		
Drynan	Regulating Fuel Oil Burners	599	Oct. 30
Nelson	Cost of Evaporation of Oil and Coal	189	Aug. 7
Noble	Simple Home-Made Heater and Separator	741	Nov. 27
Weymouth	Extinguishing Fuel Oil Fires	700	Nov. 20
	January to July, 1918.		
	Anderson Oil-Burner description	614	Apr. 30
	Johnson Oil Burner	578	Apr. 23
	Lindsay Low-Pressure Oil Burner	804	June 4
Walker	Burner Regulation	807	Dec. 11
McHugh	Burner Regulation	229	Feb. 12
Bromley	Oil Burning in Tamarack Mills, Pawtucket	426	Mar. 26
Bromley	Operating Cost Oil-Burning, Tamarack Mills	548	Apr. 16
	Best Auxiliary Burner for Oil or Tar with Coal	261	Feb. 19
Strott	Calculating Contents of Oil Tanks (Formula)	482	Apr. 2
Ward	Calculating Contents of Oil Tanks (Table)	159	July 30
	July to December, 1918.		
	Ray Oil Burner	167	July 30
	Correct Color of Flame	961	Dec. 31
	Correct Color of Flame	685	Nov. 5
	Efficiency in Use of Fuel Oil (book), Bureau of Mines	725	Nov. 12
	Preventing Waste of Fuel Oil	636	Oct. 29
	Connecting Pipe to Tank	541	Apr. 8 (?)
	July to December, 1916.		
	Oil as Fuel Sta. Oper. Comm. Nat. Dist. Heat'g Asso.	30	July 4
	Powell Oil-Burner Valve	55	July 11
	Safety in Use of Oil Burning	503	Oct. 24
	Oil Burning and CO_2 Chart	9	July 4
Peabody	Tests of B. & W. Boiler, Philadelphia Navy Yard	208	Aug. 8
	Tate-Jones Fuel Oil Burner	406	Sept. 19
	Hammell Oil Pump and Heater	854	Dec. 19
Bromley	Fuel Oil in Jenks Spinning Co.	852	Dec. 19
	Fire Underwriters' Rules for Storage	74	July 11
	Draft Gages on Oil-Burning Boilers	147	Feb. 1
	CO_2 Available with Fuel Oil	543	Apr. 18
Perry Barker	Chart of Coal and Oil Costs	765	May 30
	Ventilated Front Column Stirling Boiler Using Oil	630	May 2
	Relative Heating Valves of Fuels (table)	790	June 6
Kent	Fuels for Boilers	450	Mar. 28
	July to December, 1912.		
Editorials	Superheat Indicates Excess Air (see p. 800 for vital question about this)	268	Aug. 20
	Spacing of Burners in Boiler Formerly Used for Coal	361	Sept. 3
	Oil Burni g with Various Boiler Settings	181	Feb. 6
	Effect of Temperature on Sp.Gr. of Crude Oil (Chart)	400	Mar. 19
	Burning Fuel Oil in New Orleans Pumping Stations	422	Mar. 2[illegible]
	Strainer for Fuel Oil	595	Apr. [illegible]

See also "Bibliography on Fuel Oil," issued by United States Bureau of Mines, Bulletin No. 149, 1918.

Prepared from the Bibliography of Petroleum, in Sir Boverton Redwood's Treatise on Petroleum, 1913, with Many Additions, and Transactions, Institute Mining Engineers.

BOILERS

Action of Oil on

Denton, James E.—A Test on Beaumont, Texas, Oil. Pub. in Bull. No. 3, Univ. Tex. Min. Surv., May, 1902.

Ienisch, H. Kh.—Action of Fuel Oil on Boilers, Zap. Imp. Russk. Tekin, g. xii, otd. 1, pp. 40-54, 1878.

Efficiency of

Clarke, F. T.—High Boiler Efficiency with Oil Fuel. *Power*, Vol. 33, 1911, pp. 720-23.

Menzin, A. L.—Performance of a 45-hp. Boiler with Oil Fuel. *Engineering News*, May 29, 1913, p. 1125.

Rosok, T. A.—Turbine Station of the Bisbee Improvement Company, Arizona. (Results of oil-burning plant.) *Electrical World*, Vol. lvi, pp. 661, 664, 1910.

Schpakovski.—Comparative Value for Steam Boilers of Coal, Turpentine and Petroleum. Morsk. Sbrn., 1866, pp. 173-80.

Strohm, R. T.—Oil Fuel for Steam Boiler. *Mechnical World*, Nov. 21, 1913.

Wieland, C. F.—Comparative Evaporative Values of Coal and Oil. Trans. Am. Soc. Mech. Eng., Vol, xxxiii, pp. 872-6, 1911.

Locomotive

Aston, R. Godfrey.—Oil-Burning Locomotives on the Tehuantepec National Railroad of Mexico. Inst. Mech. Engrs., Dec. 15, 1911; *Iron and Coal Trade Review*, Dec. 22, 1911; *Canadian Engineer*, May 16, 1912.

Bogatchey, V. I.—Comparison of Petroleum and Wood as Railway Fuel. Zap. Kavk. Otd. Russk. Tekhn. Obsch., t. iii, p. 251, 1871.

Engineer, London—The "Closed Circuit" Crude Oil Locomotive. Jan. 12, 1912.

Fitz, J.—Das Rohöl als Heizmaterial für Locomotiven. M. Naptha, bd. xv, pp. 485-8, 1907.

Friesse, W.—Rohes Petroleum als Brennmaterial für Locomotiven. Polyt. J'l, bd, ccxxxv, p. 131, 18/7.

Friesse, W.—Erdolfeuerung für Locomotiven. Polyt. J'l, bd. cclxxxvii, pp. 30-31, 1893.

Greaven, L.—Petroleum Fuel in Locomotives on the Tehuantepec National Railroad of Mexico. Proc. Inst. Mech. Eng., 1906, pp. 265-312.

Holden, James.—Notes on the Application of Liquid Fuel on to the Engines of the Great Eastern Ry. Inst. Civ. Engrs. No. 3937.

Holloway, B. C.—The Use of Oil Fuel on Railways. *Petroleum World*, March, 1913, p. 103.

McFarland, H. B.—Oil-Burning on Freight Locomotives. *Cassier's Magazine*, Vol. xxxviii, pp. 610-17, 1910.

McIntosh, J. F.—Locomotives for the Caledonian Railway Fitted for Burning Oil. *Engineering*, Apr. 12, 1912. Holden Liquid Fuel Injectors.

Mechanical Engineer.—1,000 B.Hp. Crude Oil Locomotive. (Dunlop adiabatic closed circuit air transmission.) Jan. 19, 1912.

Mining and Engineering World.—Railroads Operated by Crude Oil. May 31, 1913, p. 1034.

Railway and Engineering Review.—Oil-Burning Switching Locomotives for the Balt. & Ohio R.R. Dec. 28, 2, 1912.

Railway Master Mechanic.—Oil-Burning Locomotives. June, 1912.

Ryan, J. F.—Oil Fuel. Discussion. Proc. S. and S. W. Ry. Club, Sept., 1912.

Sainte-Claire Deville, H.—Emploi de L'huile de petrole pour le chauffage des locomotives. Ann. Sci. Ind., t. xiv, pp. 89-92, 1870.

Sainte-Claire Deville, H., and Dieudonne, C.—De l'emploi industriel des huiles minerales pour le chauffage des machines et en particular des machines locomotives. Compt. Rend., t. lxix, pp. 933-8, 1869.

Sheedy, P.—The Use of Fuel Oil on the Southern Pacific R.R. Amer. Eng. R. R. J'l, Vol. lxxv, pp. 207-9, 1902.

Stewart, J. L.—On the Use of Petroleum as Fuel in Steamships and Locomotives, Based on Its Employment in that Way on the Caspian Sea and in the Trans-Caspian Region. J'l Roy. United Serv. Inst., Vol. xxx, pp. 927-49, pl. xxv, 1886.

Stillman, H.—Locomotive Practice in the Use of Fuel Oils. Trans. Am. Soc. Mech. Eng., Vol. xxxiii, pp. 887-96, 1911.

Sussman, L.—Ueber Oelfeuerung für Locomotiven. Ann. Gewerbe Bauw., bd. lxvi, pp. 234-7; bd. lxvi, pp. 12-16, 1910.

Urquhart, T.—On the Compounding of Locomotives Burning Petroleum Refuse in Russia. Proc. Inst. Mech. Eng., 1890, pp. 47-111, pls. 1-31.

Urquhart, T.—Ueber die Benutzung der Petroleum-Rückstände als Brennmaterial für Locomotive-Feurung. Organ Eisenbahnw., n. f., bd. xxii, pp. 78-9, 113-18; bd. xxiii, pp. 176-7, 1885, 1886.

Urquhart, T.—Petroleum Firing Versus Anthracite on Gruaze-Tsaritsin Railway, 1885.

Urquhart, T.—On the Use of Petroleum Refuse as Fuel in Locomotive Engines. Proc. Inst. Mech. Eng., 1884, pp. 272, 330; 1889, pp. 36-84.

Urquhart, T.—Apparatus for Burning Crude Petroleum in Locomotives. *Engineering*, Vol. p. 9, 1877.

Veith, A.—Mineral Oil Residue Fuel for Locomotives. Vasut. Hajaz. Hetilap., eok. iii, Nos. 26-27, 1901.

Waters, A. L.—Consumption of Fuel Oil by Railroads. *Mining and Engineering World*, Oct. 25, 1913.

Wispek, P.—Oil Fuel on a State Railway. *Petroleum World*, Vol. viii, pp. 133-6, 1911.

Buerk.—Petroleum as a Fuel at Sea. *Iron Age*, 1878, No. 24, p. 2.

Carmichael, W. V.—Liquid Fuel in Ocean Steamers. Papers Shipmaster's Soc., No. 9, 1891.

Church, A. T.—Notes on Fuel Oil and Its Combustion. J'l Am. Soc. Naval Eng., Vol. xxiii, pp. 795-831.

Clark, N. B.—Petroleum as a Source of Emergency Power for Warships. Proc. U. S. Naval Inst., Vol. ix, pp. 798-802, 1883.

Domojirov, A.—Liquid Fuel for Ships. Morsk. sborn., No. 1, pp. 127-155, 1884.

Dyson, C. W.—Oil Fuel for Destroyers and Battleships. (Abst. paper before Soc. Nav. Archit. and Mgr. Engrs.) *Power*, Feb. 4, 1913.

Eawards, J. R.—Liquid Fuel for Naval Purposes. J'l Am. Soc. Naval Eng., Vol. vii, pp. 744-64, 1895.

Edwards, J. R.—Oil Fuel in the Royal Navy. July 25, 1913.

Engineering.—The Wallsend-Howden System of Oil-Burning in Marine Boilers. July 25, 1913.

Flannery, J. F.—On Liquid Fuels for Ships. Trans. Inst. Naval Archt., Vol. xliv, pp. 53-75, pls. iii, iv, 1902.

Fumanti, Giulio.—Boilers Fired with Liquid Fuel. (Trans. of paper read at Rome, Italy, Nov. 12, 1912, tests made for the Italian Navy.) *Engineering*, Feb. 12, 1912.

Goodenough, W. J.—Memorandum Regarding Fuel Oil Aboard Ships. *Marine Engineering*, Vol. viii, pp. 4-6, 1903.

Henwood, E. N.—Liquid Fuel. *Marine Engineering*, Vol. viii, pp. 24-25, 1886.

Hopps, J. H.—Marine Use of Fuel Oil. J'l Am. Soc. Mech. Eng., Vol. xxxiii, pp. 896-903, 1911.

Hume, G.—Oil Fuel on Shipboard. Trans. Inst. Marine Eng., Vol. xix, No. 137, p. 46, pl. 1907.

Lovekin, L. D.—Notes on the Burning of Liquid Fuel. J'l Am. Soc. Naval Engrs., Vol. xxiii, pp. 221-47, 1911.

Peabody, E. H.—Oil Fuel. J'l Am. Soc. Naval Eng., Vol. xxiii, pp. 489-99, 1911.

Peabody, E. H.—On Liquid Fuel. Trans. Inst. Naval Archt., Vol. ix, pp. 88-103, 1868.

Wilson, H. C.—Liquid Fuel and Methods of Burning It. Trans. Inst. Marine Eng., Vol. viii, No. 60, p. 48, 1896.

Zulver, C.—Oil Fuel for Marine Engines and Boilers. Mar. Eng. and Nav. Archt., March, 1913.

Collins, B. R. T.—Oil Fuel for Steam Boilers. J'l Am. Soc. Mech. Eng., Vol. xxxiii, 1911, pp. 929-65.

Kermode.—Liquid Fuel Apparatus for Water-Tube Boilers. *Engineering*, Vol. lxxxix, pp. 510-12, 1910.

Lyne, L. F.—The Use of Crude Petroleum in Steam Boilers. Trans. Am. Soc. Mech. Eng., Vol. x, pp. 761-4, 1889.

Menzin, A. L.—Performance of a 450-hp. Boiler with Fuel Oil. (Mare Island Navy Yard.) *Engineering News*, May 29, 1913.

Phillips, William B.—Fuel Oil Used Under Stamp Mill Boilers, Shafter, Presidio County, Texas. Hauled in tank wagons 48 miles. *Engineering and Mining Journal*, Vol. xc, No. 27, Dec. 31, 1901.

Wagner, H. A.—Auxiliary Oil-Firing in Steam Boiler Plants. *Mech. Eng.*, Vol. xxviii, pp. 28-29, 1911.

Weymouth, C. R.—Fuel Economy Tests at a Large Oil-Burning Electric Power Plant Having Steam Engine Prime Motors. Trans. Am. Soc. Mech. Eng., Vol. xxx, pp. 775-96, 1908.

Dunn, K. G.—Size of Stacks With Fuel Oil. J'l Am. Soc. Mech. Eng., Vol. xxxiii, p. 886, 1911.

Weymouth, C. R.—Dimensions of Boiler Chimneys for Crude Oil at Sea-Level, etc. J'l Am. Soc. Mech. Engrs., October, 1912.

Peabody, E. H.—Mechanical Oil Burners. (Abst. of paper before Soc. Nav. Archt. and Mar. Engrs.) *Engineering News*, May 15, 1913, p. 1000.

Shelby, C. F.—Oil Burners for Reverberatory Furnaces. *Engineering and Mining Journal*, Vol. lxxxix, pp. 31-2, 1910. See also under Furnaces.

Watson, E. P.—Oil Fuel and Oil Burners. *Iron Age*, May 8, 1902.

Weymouth, C. R.—Furnace Arrangements for Fuel Oil. Trans. Am. Soc. Mech. Eng., Vol. xxxiii, pp. 879-82, 1911.

Allen, I. C., and Jacobs, W. A.—Physical and Chemical Properties of the Petroleum of the San Joaquin Valley, California. Bur. Mines, Washington, Bull. 19, 1919.

Arnold, R.—Physical and Chemical Properties of Southern California Oils. Bull. U. S. G. S. No. 309, pp, 203-5, 1907.

Patterson, W. H.—The Chemical Examination of Liquid Fuels. *Chemical Engineer*, September, 1913.

Phillips, Wm. B.—Texas Petroleum. Bull. No. 1 Univ. Tex. Min. Surv., 1901; and Worrell, S. H.—The Fuels Used in Texas. Bull. 307, Bur. Ec. Geol. and Techn., Univ. Tex., December, 1913.

Young, S.—Composition of American Petroleum. J'l Chem. Soc., Vol. lxxiii, pp. 905-20, 1898.

Cone, H. I.—Use of Fuels in the United States Navy. J'l Am. Soc. Engrs., November, 1912.

Peabody, E. H.—Developments in Oil Burning. Soc. Nav. Archts. and Mar. Engrs., Nov. 21, 1912.

Aagard, H. B.—Liquid Fuel. Prize Essay, Norwegian Govt., 1902.

Adams, E. E.—Fuel-Oil Installation on the Great Northern Ry. *Railway and Engineering Review*, Dec. 9, 1911.

Allest, J. d'—De la Combustion des Huiles Minerales et de leurs residus. Gen. Civ., t. viii, pp. 7-10, 19-22, 36-42, 1885.

Deane, C. T.—The Burning of Crude Oil for Steam Furnaces. *Petroleum Review*, Vol. vi, pp. 385-97, 1902.

Dudley, C. B.—Fuel Oil. J'l Frank. Inst., Vol. cxxvi, pp. 81-95, 2 pls., 1888.

Electrical World.—Auxiliary Oil-Burning Station for Southern California Station. (Long Beach plant of the Edison Co.) Mar. 9, 1912.

Ennis, W. D.—Burning Oil for Power and Heating. *Power*, Vol. xxviii, pp. 943-7, 980-4, 1908.

Hetzel, F. V.—Changing from Oil Fuel to Coal. *Power*, Dec. 2, 1913

Hough, E. S., and Crawford, W. H.—Report of Recent Oil-Burning Installations on the Pacific Coast. *Marine Engineering*, Vol. vii, pp. 447-451, 1902.

Lowe, L. P.—Practical Testing of Oils for Gas and Fuel Purposes. (Pacific Coast Gas Assoc.) *American Journal Gas Lighting*, Vol. lxxv, pp. 405-7, 1901.

Mechanical World.—The Advantages and Disadvantages of Liquid Fuel. Aug. 9, 1912.

North, S. H., and Edwards, G. M.—The Economic Position of the Oil Fuel Question. *Engineer*, London, Dec. 29, 1911.

Paul, B. H.—Use of Petroleum or Mineral Oil as Steam Fuel in Place of Coal. *Chemical News*, Vol. x, pp. 292-3; Vol. xi, pp. 63-5, 1864-65.

Pierce, John.—Burning Crude Oil. *Practical Engineer*, Dec. 15, 1912, p. 1223.

Watts, W. L.—Oil as Fuel in Los Angeles County, Calif. 13th Rep. State Mineral Calif. pp. 662-4, 1896.

Weymouth, C. R.—Unnecessary Losses in Firing Fuel Oil and An Automatic System for Their Elimination. Trans. Am. Soc. Mech. Eng., Vol. xxx, pp. 797-835, 1908.

Weymouth, C. R.—Economy of Certain Arizona Steam-Electric Power Plants Using Oil Fuel. J'l Amer. Soc. Mech. Eng., June, 1919.

Allen, B. J.—Some Notes on Oil and Tar Burning. *American Gas Light Journal*, Vol. lxxii, pp. 482-6, 1900.

Best, W. N.—Oil Fired Furnaces. Trans. Amer. Fndrym. Assoc., 1911, pp. 421-4.

Crane, W. E.—Crucible Furnaces for Burning Petroleum. Trans. Am. Soc. Mech. Eng., Vol. xv, pp. 307-12, 1894.

Knutsen, J. B. M.—Industrial Furnaces. Austral. Min. Stand., May 30, 1912.

Lang, Herbert.—Oil-Burning in Furnaces. *Mining and Scientific Press*, July 12, 1913.

Power.—Oil Burning with Different Settings. Feb. 6, 1912.

Weymouth, C. R.—Furnace Arrangements for Fuel Oil. Trans. Am Soc. Mech. Eng., Vol. xxxiii, pp. 879-82, 1911.

Bowman, F. C.—The Use of Crude Oil for Fire Assaying. Proc. Color. Sci. Soc., Vol. vii, pp. 341-6, 1904.

Brent, C.—Notes on Oil Furnaces for Assaying and Melting. J'l Can. Min. Inst., Vol. v, p. 288, 1902.

Lang, Herbert.—Blast-Furnace Smelting with Crude Oil. *Mining and Scientific Press*, Feb. 8, 1913, p. 248.

Gatewood, R. D.—Oil-Burning Frame Plate Furnace. American Mechanic, July 18, 1912.

Alexieev, PP.—Foundry Furnaces Heated by Luminous Gas and Volatile Oil. Gorn. Jurn., 1866, t. iii, pp. 295-309.

MacGregor, Walter.—Design of Oil-Fired Openhearth Furnaces. *Iron Trade Review*, Nov. 16, 1911; *Mechanical Engineer*, Jan. 5, 1912.

Pawloff, M. A.—Dimensions of Openhearth Furnaces, Gas or Oil Fired. Abstract of paper in *Stahl und Eisen*. *Iron and Coal Trade Review*, Jan. 26, 1912.

Reddy, B. H.—The Operation of Small Openhearth Furnaces. *Foundry*, March, 1912.

Brass World.—The "Gamm" Oil Reverberatory Furnace for Melting Nickel. February, 1913.

Savine, P. P.—Reverberatory Petroleum Furnace for Foundry Purposes. Gorn. Jurn., 1896, t. ii, pp. 1-12, 2 pls.

Shelby, C. F.—Oil Burners for Reverberatory Furnaces. *Engineering and Mining Journal*, Vol. lxxxix, pp. 31-2, 1910. See also under Burners.

Jacobs, E.—Blast-Furnace Smelting with Oil. *Engineering and Mining Journal*, Vol xcii, p. 434, 1911.

Van Der Ropp, A.—The Use of Crude Oil in Smelting. *Engineering and Mining Journal*, Vol. lxxv, pp. 81-82, 1903.

Church, A. T.—Notes on Fuel Oil and Its Combustion. J'l Am. Soc. Naval Engrs., Vol. xxiii, pp. 795-831, 1911.

Petroleum World, London.—The Origin of Petroleum, September. 1913.

Ball, S H.—The Tampico Oil Field, Mexico. *Engineering and Mining Journal*, Vol. xci, pp. 959-61, 1911.

Burr, G. A.—Petroleum in Texas and Mexico. *Engineering and Mining Journal*, Vol. lxxi, 1901, 687-8.

Hornaday, W. D.—Petroleum in Mexico. *Mining and Engineering World*, 1911.

Magill, S. E.—Oil Resources of Mexico. U. S. Cons. Rep. No. 317, pp. 53-5. 1907

Bowie, C. P.—Pumping California Oil. *Oil Age*, 1911.

Fuel Oil Journal.—Oil Reservoirs at San Luis Abispo, Cal. Perhaps the largest in the world. Vol. i, No. 6, pp. 34-5, April, 1912.

Oatman, F. W.—A Concrete Reservoir for Storage of Petroleum, Near Coalings, California. *Mining and Engineering World*, Nov. 18, 1911.

Whiteford, J. F.—The Santa Fe Oil Storage Plant. *American Machinist*, Oct. 24, 1912.

Allen, I. C.—Specifications for the Purchase of Fuel Oil, with Directions for Sampling Oil and Natural Gas. U. S. Bur. Min. Tech. Paper 3, 1911.

Zts. Ver. Bohringen u. Bohrtechnik.—Die Wassergefahr in den Kalifornischen Erdolieden. (Trans. from Pet. Review.) Mar. 15, 1913.

Booth, W. H.—Liquid Fuel and Its Combustion. London, 1903.

North, S. H.—Petroleum; Its Power and Uses. London, 1904.

Versenev, V.—Manual of Management of Petroleum-Fired Boilers. Moscow, 1891.

WRECK OF LARGE MOTOR-GENERATOR SET IN THE TOLEDO RAILWAY AND LIGHT CO.'S PLANT, FEB. 21, 1920

This set consisted of two direct-current generators, directly driven by an induction motor. The motor was literally torn to pieces. As can be seen, the generator to the right was torn from the foundation, breaking the bedplate between the motor and the generator. The field frame of the left-hand generator was broken from the bedplate and turned through 90 deg., and has also broken the bedplate where the outboard bearing pedestal is fastened. Apparently, the direct-current generators were motorized and ran away. An investigation is being made, and it is expected that the results will be available for publication in an early issue.

Personals

Col. W. A. Dibblee is now associated with the Western Appraisal Co. and the National Industrial Engineering Co., of Cincinnati, Ohio. He will act as field manager.

Calvert Townley, assistant to the president of the Westinghouse Electric and Manufacturing Co., was reëlected vice president of the United Engineering Society at a recent meeting.

F. P. Breunstead, formerly manager for the Perolin Co. for the State of Ohio, has been appointed assistant manager of the Lubricating Department of the Moore Oil Refining Co., Cincinnati, Ohio.

C. B. Leach, who, for the past two years, has been representing *Power*, a McGraw-Hill Co. publication, in the Philadelphia territory, has taken charge of the Chicago territory, succeeding A. H. Maujer, who has resigned.

D. George McMilian has resigned as superintendent of power with the Singer Manufacturing Co., Elizabeth, N. J., and is now with the Williams Printing Co., of New York City, in the capacity of maintenance engineer.

Engineering Affairs

The Connecticut State Branch of the A. S. M. E. will hold a meeting at the Seaside Club, Bridgeport, March 29.

The A. S. M. E., St. Louis Section, will hold a joint meeting with the Associated Engineering Societies of St. Louis in the Hotel Statler on April 7. F. O. Pahmeyer, engineer for the Heine Safety Boiler Co., will be the speaker.

The A. S. M. E., Buffalo Section, will hold a meeting at the Ellicott Club, Ellicott Square Building, April 6. Harrison P. Eddy, consulting engineer for the City of Buffalo, in connection with sewage-disposal work, will address the meeting.

The South Dakota Electric Association at a recent meeting planned to merge the Minnesota and North Dakota Electric Associations with the South Dakota organization. The new society will be known as the North Central Electric Association.

The Hartford Branch, A. S. M. E. held a meeting at the City Club, Central Row, Monday, March 15. George A. Orrok, delivered an interesting address on Power Production, which was followed by discussion in which members and guests joined.

The American Engineering Standards Committee, after almost three years of investigation by the engineering fraternity, has adopted a revised constitution of very wide scope which allows the direct or indirect participation of anyone interested in Standardization. The new constitution has been ratified by the A. S. C. E., the A. I. M. & M. E., the A. S. M. E., the A. I. E. E., and the A. S. T. M. and the three Government departments of Commerce, Navy and War

Miscellaneous News

The Alabama Power Co., Montgomery, Ala., is planning to purchase the Gadsden, Attalla & Alabama City Electric Light and Street Car properties.

The Arkansas Light and Power Co. has completed negotiations for the purchase of the power plant at Picon, Ark. The deal involves $235,000.00.

The Federation of Mexican Chambers of Commerce has decided to open as soon as possible an International Exposition in Mexico City.

The Chamber of Commerce of the United States will hold its eighth annual meeting at Atlantic City, April 27-29. The subject of the meeting will be "Increased Production."

The Lehigh Navigation Electric Co., Hazleton, Pa., recently appropriated $3,000,000 to increase the electric and boiler capacity of the Hauto plant from 8,000 to 16,000 horsepower.

Permits for the Use of Water for power-development purposes in northern California involving expenditures of more than $11,000,000 have been filed with the State Water Commission.

The Manufacturers Club, of Minneapolis, Minn., has organized a superintendents' and foremen's section. An employment managers' section and a steel treaters' section have also recently been organized.

Health Commissioner Robinson, of Chicago, is inaugurating a drive on the smoke nuisance in the Windy City. A number of plants have been closed down until the cause of the trouble can be eliminated.

The Public Service Company of Northern Illinois at its recent annual meeting decided owing to the greatly increased demands for electrical energy in the northern zone, to construct a transformer station at North Chicago. Construction is under way.

The Federal Trade Commission has been given full authority to arrange for the development of useful inventions of departmental employees in order that the public, the industries and the Government departments may get the benefit of them.

A Gigantic Steam-Electric Generating Plant at Milwaukee, Mich., will be constructed by the Consumers Power Co. Five million dollars is the estimated cost of the project, which will aid in supplying light and power to Saginaw and surrounding cities.

The Metropolitan Edison Power Co., of Reading, Pa., and the Pennsylvania Utilities Co., of Easton, Pa., have completed plans for the connection of the two plants. A line will be constructed from Reading to Easton and then to Dover that will convey 7,000 kw. of electricity.

The Seventh National Foreign Trade Convention will be held in San Francisco, Cal., from May 12-15, inclusive, under the auspices of the National Foreign Trade Council. "The Effect of Being a Creditor Nation" will be the main theme of the convention.

The Hartford Electric Light Co., has arranged to purchase the Connecticut Power Co., which has been operating a large hydro-electric power plant on the Housatonic River and western Connecticut. Negotiations are being completed through Stone & Webster, of Boston.

The Bureau of Mines, Department of the Interior, in a series of investigations, recently completed, show that taking gasoline from natural gas does not detract from the heat-giving qualities of the gas. It is estimated that several hundred million gallons of gasoline are added to the country's supply in this manner.

The Metropolitan Edison Co., of Reading, Pa., gave a fine illustration of efficiency, when the West Reading plant was kept running without interruption during a 17-foot flood in the Schuylkill River, which forced its way into the power station and completely submerged pumps and similar equipment on the ground floor.

The Detroit Edison Co. has planned a new plant with an ultimate capacity of 130,000 kw. The plant will be located at Bunce Creek and will supply more power than either the Conners Creek or the Delray plant. An initial expenditure of $2,000,000 will be made, and many millions more will be spent before it is finished.

A Heavy Wind Storm Last November greatly damaged telephone, telegraph and power lines in the State of California. Temporary repairs were made in one instance to a 60,000-volt line by supporting the insulators on railroad ties piled high enough to keep the wires off the ground. Current was thus maintained through the storm with the exception of a few hours.

Business Items

The Wagoner Electric Co., of St. Louis, has removed its Buffalo office to 16 Carlton St. Alfred W. Baldwin will take charge as branch manager.

The William B. Pierce Co., of Buffalo, N. Y., manufacturers of the Dean Boiler Tube Cleaner, has removed its New York office to 480 Lexington Ave.

The Air Compressor & Equipment Co., of Portland, Ore., has been incorporated by George Larrabee, Guy A. Malcolm and Warren E. Thomas. Capital stock, $21,000.

The Dodge Manufacturing Co., Mishawaka, Ind., has opened a school to teach the minor executives of the firm more efficient methods of modern production and to enable them to handle labor more effectively.

The Mono Corporation of America, 48 Coal and Iron Exchange, Buffalo, N. Y., announce that they have purchased the entire stock of Mono apparatus and accessories from the F. D. Harger Co., Buffalo, N. Y. This includes all rights for the manufacture and sale of their various types of Mono apparatus for the automatic analysis of CO_2, CO, O_2, H_2, SO_2, N_2, Cl, etc.

The Plant Engineering & Equipment Co. Inc., New York City, manufacturers of the well known Corliss Valve Steam Trap, Mason Condensation Meter and other power and heating specialties announces the opening of its twenty-sixth office to care for the increasing demands for its products. M. W. Ehrlich, will be the New Jersey Manager in charge with headquarters at Newark, N. J. and a sub-office at Lyndhurst, N. J.

The Frazier-Ellms-Sheal Co.—J. W. Frazier, J. W. Ellms and R. E. Sheal announce the incorporation of the Frazier-Ellms-Sheal Co. for the purpose of conducting a general engineering business in the design and construction of waterworks, pumping and power plants, water purification and sewage-disposal plants, sewer and drainage systems, and buildings. The offices of the company are in the Illuminating Building, Cleveland, Ohio.

THE COAL MARKET

BOSTON—Current prices per gross ton f.o.b. New York loading ports:

Anthracite

	Company Coal
Egg	$8.20@$8.65
Stove	8.45@ 9.05
Chestnut	8 55@ 9.05
Pea	7 05@ 7.40
Buckwheat	3 60@ 3.80
Rice	2.30@ 2.50
Barley	1.25

Bituminous

	Clearfields	Cambrias and Somersets
F o.b. mines, net tons	$2 85@$3.35	$3.15@$3.60
F.o.b. Philadelphia, gross tons	5.05@ 5 60	5.35@ 5.90
F.o.b. New York, gross tons	5 40@ 5 95	5 75@ 6.25
Alongside Boston (water coal), gross tons	7 00@ 7 75	7 60@ 8.00

Pocahontas and New River are practically off the market for coastwise shipment, but are quoted at $6.25@$7.00.

NEW YORK—Current quotations, White Ash, per gross tons, f.o.b. Tidewater, at the lower ports are as follows:

Anthracite	Company Coal
Broken	$7 80@$8 25
Egg	8.20@ 8.65
Stove	8.45@ 9.05
Chestnut	8.55@ 9.05
Pea	7.05@ 7.40
Buckwheat	5.15
Rice	4.50
Barley	4 85
Boiler	4.85

Bituminous

Government prices at mines:	Spot
South Fork (best)	$3.25@$3.50
Cambria (best)	3.00@ 3.25
Cambria (ordinary)	2.60@ 2.90
Clearfield (best)	3.00@ 3.25
Clearfield (ordinary)	2.60@ 2.90
Reynoldsville	2.85@ 2.90
Quemahoning	3.25@ 3.50
Somerset (medium)	3.00@ 3.25
Somerset (poor)	2.50@ 2.75
Western Maryland	2.50@ 2.75
Fairmont	2.25@ 2.50
Latrobe	2.60@ 2.90
Greensburg	2.75@ 3.00
Westmoreland ¾ in.	3.40@ 3.50
Westmoreland run-of-mine	2.75@ 3.00

PHILADELPHIA—Bituminous coal prices vary according to district from which they are mined. For ordinary slack the price is $2.45@$2.55; lump, $3.35, at the mines.

BUFFALO—

Anthracite

	On Cars, Gross Ton	At Curb, Net Ton
Grate	$8 55	$10.80
Egg	8.80	10.65
Stove	9 00	10.85
Chestnut	9.10	10.85
Pea	7.75	9.30
Buckwheat	5.70	7.75

Bituminous

Allegheny Valley	$4.80
Pittsburgh	4.65
No. 8 Lump	4.65
Mine-run	4.80
Slack	4.10
Smokeless	4.60
Pennsylvania Smithing	5.70

CLEVELAND—Prices of coal per net ton delivered in Cleveland are:

Anthracite

Egg	$12.25@$12.40
Chestnut	12 50@ 12.70
Grate	12.25@ 12.40
Stove	12.40@ 12.60

Pocahontas

Nine-run	$7.50

Domestic Bituminous

West Virginia split	$8.50
No. 8 Pittsburgh	$6 60@ 6.90
Massillon lump	8.25@ 8.50
Coshocton lump	7.15

Steam Coal

No. 6 slack	$5.25@$5.50
No. 8 slack	5.10@ 5.50
Youghiogheny slack	5.25@ 5.50
No. 8 ¾ in.	5.70@ 6.60
No. 6 mine-run	5.25@ 5.50
No. 8 mine-run	5.75

Only coal available is mine-run Pocahontas.

MIDDLE WEST—Chicago quotations, f. o. b. cars at mine:

	Springfield, Carterville, Williamson, Franklin, Saline, Harrisburg	Fulton, Peoria	Grundy, La Salle, Bureau, Will
Lump	$2.55@$2.70	$2.95@$3.10	$3.25@$3.40
Washed	2.75@ 2.90	....	3.45@ 3.60
Mine-run	2 35@ 2 50	2.75@ 2.90	3.00@ 3.15
Screen'gs	2.05@ 2.20	2.35@ 2.50	2.75@ 2.90

New Construction

PROPOSED WORK

Vt., Brattleboro—The Connecticut River Power Co. of New Hampshire, 86 Harvard St., Worcester, Mass., plans to extend the hydro-station at Hinsdale, N. H. and install 4 additional units. Charles A. Harris, Pres.

Vt., Johnson — The Municipal Electric Light Plant plans to install a new generator and switch board.

Mass., Boston—The Reynolds Bldg. Trust Co. is having plans prepared for the construction of an 8 story, 47 x 70 ft. office building on Winter St. A steam heating system will be installed in same. Total estimated cost, $500,000. Parker, Thomas & Rice, 110 State St., Archts.

Mass., North Adams—The H. W. Clark Biscuit Co., 185 Ashland St., will soon award the contract for the construction of a 4 story, 90 x 140 ft. addition to its present plant on Ashland St. A steam heating system will be installed in same. Total estimated cost, $150,000. W. Higginson, 13 Park Row, New York City, Engr.

Mass., Springfield—P. H. Faber, c/o The Oaks, 31 Thomson St., will receive bids about April 1 for the construction of a 4 story hotel addition on State and Thomson Sts. A steam heating system will be installed in same. Total estimated cost, $200,000. Associated Architectural Co., 145 State St., Archt.

Mass., Williamstown—Cram & Ferguson, Archts., 20 Beacon St., Boston, will soon award the contract for the construction of a 3 story, 120 x 120 ft. library building, etc. on the campus, for the William College. A steam heating system will be installed in same. Total estimated cost, $500,000. W. E. Hoyt, Treas.

R. I., Pawtucket—The Blackstone Valley Gas & Electric Co. plans to install oil-burners for the generating system. A. C. Brooks, Mgr.

R. I., Warren—The Mt. Hope Spinning Mill will soon award the contract for the construction of a 2 story, 110 x 200 ft. mill addition on Cutler St. A steam heating system will be installed in same. Total estimated cost, $100,000. Jenks & Ballon, 1035 Grovenor Bldg., Providence, Engrs.

Conn., Middletown — The Trustees of Wesleyan University, 374 High St., are having plans prepared for the construction of a 2 story chemical laboratory building. A steam heating system will be installed in same. Total estimated cost, $500,000. Henry Bacon, 101 Park Ave., New York City, Archt. and Engr.

Conn., New Britain — The New Britain Children's Home, 81 Garden St., is having plans prepared for the construction of a 3 story, 96 x 266 ft. school and home on Rackliffe Heights. A steam heating system will be installed in same. Total estimated cost, $350,000. Walter P. Crabtree, 272 Main St., Archt. and Engr.

Conn., New Britain — The World War Veterans, c/o Clarence C. Palmer, Archt. and Engr., 272 Main St. plan to construct a 4 story, 60 x 150 ft. club house on North and Beaver Sts. A steam heating system will be installed in same. Total estimated cost, $200,000.

Conn., New London—The Soltz Co., 52 John St., plans to construct a 1 story cold storage building on Main St.

N. Y., Albany—York & Sawyer, Archts. and Engrs., 50 East 41st St., New York City, will soon award the contract for altering the present bank building, for the Commercial Trust Co., c/o Archts. A steam heating system will be installed in same. Total estimated cost, $100,000.

N. Y., New York—The Flatbush Gas Co., 1034-36 Flatbush Ave., Brooklyn, plans to construct 72—13,200 volt power feeders. H. E. McGowan, Secy.

N. Y., New York—John H. Sheler, Archt. and Engr., 25 West 42nd St., will soon award the contract for altering loft, showroom and store building on 6th Ave. between 20th and 21st Sts., for the United States Realty Improvement Co. A steam heating system will be installed in same. Total estimated cost, $300,000.

N. Y., New York—The Sheridan Realty Co., c/o Reilly & Hall, Archts. and Engrs., 749 5th Ave., is having plans prepared for the construction of a 152 x 202 ft. theatre on Sheridan Sq. A steam heating system will be installed in same. Total estimated cost, $500,000.

N. Y., New York—The Silk Traders, Inc., c/o G. and E. Blum, Archt., 505 5th Ave., is naving plans prepared for the construction of a 12 story loft building on 4th Ave. A steam heating system will be installed in same. Total estimated cost, $600,000.

N. Y., New York—The Society of Professional Automobile Engrs., B'way and 67th St., is having plans prepared for the construction of a 5 story, 55 x 100 ft. club building at 153 West 64th St. A steam heating system will be installed in same. Walter Haefel, 229 West 42d St., Archt. and Engr.

N. Y., New York—The United States Electric Light & Power Co., 130 East 15th St., plans to install one 22,000 kw. turbo generator unit, construct a 300,000 kw. generating station and change the transmission voltage from 7,500 volt to 15,000 volt. Frank W. Smith, Mgr.

N. Y., Queensborough — The Queensborough Gas & Electric Co., 1530 Far Rockaway Blvd., Far Rockaway, plans to construct an addition to generating station and install 2 additional 600 hp. Babcock Wilcox boilers. Carleton Macy Mgr.

N. J., Montclair — The Montclair Golf Club, c/o H. P. Knowles, Archt. and Engr., 52 Vanderbilt Ave., New York City, is having plans prepared for the construction of a 3 story, 50 x 220 ft. club building. A steam heating system will be installed in same. Total estimated cost, $200,000.

N. J., South River—The Bd. of Pub. Wks., Borough Hall, will receive bids until April 12 for the construction of a 59 x 78 ft. addition to the electric generating station and the installation of two 300 hp. boilers, 175 lbs. gauge pressure, 50 degrees superheat, two 500 kw. turbo generators, two 500 kw. condensers, two circulating water pumps, one water heater, etc. E. B. Hedden, Supt. Goss, Bryce & Johnson, 55 Liberty St., New York City, Engrs.

Pa., Philadelphia — Clark & Dudnick, Archts. and Engrs., Drexel Bldg., will soon award the contract for the construction of a 3 story, 42 x 60 ft. store building on 5th and South Sts., for Harry Krengel, 5th and South Sts. A steam heating system will be installed in same. Total estimated cost, $100,000.

Pa., Sharpsburg—The Municipal Electric Light Plant Office plans to install a condenser for one 300 kw. turbine. W. C. Wetzel, Supt.

Pa., Souderton—The Souderton Light & Power Co. plans to construct a new 3 ph. transmission system. J. J. Hutchinson, Mgr.

Pa., Verden—The Verden & Lake Ariel Light, Heat & Power Co. plans to extend transmission lines in the spring. W. M. Kiezer, Pres.

Pa., Waynesboro—The Landes Machine Co. is in the market for a 125-200 hp. boiler.

Pa., Wilkes-Barre—The Sheldon Axle Co., 221 Conyngham Ave., plans to construct a 3 story, 100 x 200 ft. heat treating plant on Beaumont St. A steam heating system and electric power will be installed in same. Total estimated cost. $200,000.

Md., Cumberland—G. F. Salisbury, Archt., Citizens Natl. Bank Bldg., will soon award the contract for the construction of a 4 story, 69 x 84 ft. store building on Baltimore St., for the McMullen Bros. Co., Inc., Baltimore St. A steam heating system will be installed in same. Total estimated cost, $125,000.

Va., Blacksburg—The Virginia Polytechnic Institute Heat & Power Plant plans to install a complete new generating system and remodel the transmission system. W. E. Ellis, Dir.

Va., Bridgewater — The North River Electric Co. plans to install a new and larger generator, in 3 phases, more water power and auxiliary steam power and construct a transmission line from here to Mt. Crawford and several mutual lines to surrounding farms. G. R. Berlin, Mgr.

Va., Chatham—The Chatham Light & Power Co. plans to install a 3 mi. line on the West side for power and light in generator system. Jonas V. Young, Mgr.

Va., Fredericksburg—The Rappahannock Electric Light & Power Co. plans to install 500 kva. steam turbine boilers in generator system and reconstruct distributing system. S. C. Foster, Supt.

Va., Shenandoah—The Municipal Water & Light Plant plans to install a 300 kva generator. A. V. Kern, Supt.

Va., Warrenton—The Warrenton Electric Light & Power Co. plans to install one 90 kw., 2,200 volt, 60 cycle generator and one 150 hp. Erie boiler and construct 5 mi. of additional transmission line. M. J. O'Connel, Mgr.

W. Va., Charleston—The Bureau of Yards & Docks, Navy Dept., Wash., D. C., plans to build an outdoor electric substation and install equipment in same, at Naval Ordnance Plant, here.

W. Va., Charleston—C. H. Higgins, Archt. and Engr., 19 West 44th St., New York City, will soon award the contract for the construction of a 7 story, 100 x 150 ft. office building on Virginia and McFarland Sts., for Abney Barnes & Co., 812-818 Virginia St. A steam heating system will be installed in same.

W. Va., Elkins—The Elkins Power Co. plans to install 1 new unit in the generating system. C. C. Bosworth, Mgr.

W. Va., Huntington — The Consolidated Light, Heat & Power Co. plans to construct an 11,000 volt transsmission line from here to Kenova and an 11,000 volt line here. H. L. Evans, Supt.

W. Va., Logan—The Kentucky & West Virginia Power Co plans to construct a 250 mi. transmission line in various territories and install 21,000 in. turbo alternators. A. J. Darrahm, Mgr.

W. Va., Mt. Hope—The Mt. Hope Electric Power & Water Co. plans to construct 2 lines in the transmission system. George Waldo, Supt.

W. Va., Phillippi—The Municipal Electric Light Plant plans to install an 800 hp. steam turbine in generating system. A. D. Poling, Mgr.

N. C., High Point—The High Point Hotel Co., c/o W. L. Stoddart, Archt. and Engr., 9 East 40th St., New York City, is having plans prepared for the construction of a 9 story hotel. A steam heating system will be installed in same. Total estimated cost, $500,000.

Ga., Atlanta—The Georgia Baptist Hospital, 92 Luckie St., is having plans prepared for the construction of a 4 story, 57 x 150 ft. nurses home on the Boulevard. A steam heating system will be installed in same. Total estimated cost, $150,000. Burge, Stevens & Conklin, Natl. City Bldg., Archts.

Tenn., Carthage—The Smith Co. Electric Co. plans to construct an 11 mi. transmission line from here to Monoville, Riddleton and Dixon Springs. R. S. Seese, Mgr.

Tenn., Dickson—The Municipal Electric Light Plant plans to install a complete new unit in the generating system. Eugene S. Payne, Supt.

Tenn., Humboldt—The Municipal Water & Power Plant plans to install a 500 kw. turbine and construct an 8 mi. transmission line. W. M. Case, Mgr.

Tenn., Kingsport—The Kingsport Utilities Inc. plans to construct a sub-station and a transmission line which will supply all city circuits. L. C. Dashiell, Treas.

Tenn., Monterey—The Monterey Light & Power Co. plans to install a 60 hp. engine and a kva generator and construct short extensions to distributing systems. L. C. Parks, Mgr.

Tenn., Somerville—The Municipal Water & Light Plant plans to install one 25 hp. oil engine connected to a 37½ kw. generator. J. L. Sanders, Supt.

O., Akron—The Seiberling Hotel Co., c/o F. A. Seiberling, is having preliminary plans prepared for the construction of a 20 story hotel and theatre building. Two or three boilers will be installed in same. Total estimated cost, $5,000,000. George B. Post & Sons, 1109 Schofield Bldg., Cleveland, Archts.

O., Bexley—The Bd. Educ. is having plans prepared for the construction of a 2 story, 80 x 125 ft. graded school on Main St. A steam heating system will be installed in same. Total estimated cost, $200,000. O. D. Howard, 8 East Broad St., Columbus, Archt.

O., Cincinnati—The United States Engineer's Office, War Dept., Washington, D. C., rejected all bids for the construction of the power house dam 35, in the Ohio River.

O., Cleveland—The Bd. Educ., East 6th St. and Rockwell Ave., is having plans prepared for the construction of a 1 story school on Parkgate Ave. Two steam heating boilers will be installed in same. Total estimated cost, $500,000. F. G. Hogan, Dir. W. R. McCormack, Archt.

O., Cleveland—The Bd. Educ., East 6th St. and Rockwell Ave., will soon award the contract for the installation of a heating and ventilating system in school. Total estimated cost, $30,000. W. R. McCormack, Engr.

O., Cleveland — The Bd. of Co. Comrs., Court House, plans to construct a 1 story heating plant on Lakeside Ave. and West 3d St. Estimated cost, $200,000. E. G. Krause, Clk.

O., Cleveland — The Brooklyn Catholic Club, 4419 Pearl Rd., plans to construct a 2 story, 70 x 200 ft. club house on Pearl Rd. and Leopold Ave. A steam heating system will be installed in same. Total estimated cost, $100,000. H. Post, Secy. M. Rousseau, 1775 Wickford Rd., Archt.

O., Cleveland—The Cataract Motor Sales Co., 8818 B'way, plans to build a 3 story, 80 x 250 ft. garage and auto sales building at 8807 B'way. A steam heating system will be installed in same. Total estimated cost, $100,000. Address A. J. Schustrich.

O., Cleveland — The Cleveland Discount Co., Rockefeller Bldg., has purchased a site on Superior Ave., near East 9th St. and plans to build a 20 story, 80 x 200 ft. office building on same. A steam heating system will be installed in same. Total estimated cost, $4,000,000. J. Kirby, Pres. Walker & Weeks, 1900 Euclid Ave., Archts.

O., Cleveland—The Cleveland Ry. Co., Leader-News Bldg., plans to construct 1 and 2 story car barns on Bessemer Ave. and East 79th St. A steam heating system will be installed in same. Total estimated cost, $200,000. J. Stanley, Pres. D. W. Morrow, 4500 Euclid Ave., Archt.

O., Cleveland—The Cleveland Trust Co., East 9th St. and Euclid Ave., plans to build a 4 story commercial building at 2717 Euclid Ave. A steam heating system will be installed in same. Total estimated cost, $300,000. A. A. McCaslin, Vice Pres.

O., Cleveland—C. G. Hale, Real Estate, 340 Leader-News Bldg., plans to construct a 5 story commercial building on Superior Ave. near East 12th St. A steam heating system will be installed in same. Total estimated cost, $150,000.

O., Cleveland—The International Ladies Garment Workers Union plans to construct a 5 story office and club room building on Superior Ave. A steam heating system will be installed in same. Total estimated cost, $100,000. M. Pearlstein, 3:5 Superior Bldg., Chn.

O., Cleveland—S. W. Manheim & Co., 719 Natl. City Bldg., plans to build a 1 story commercial and theatre building on East 125th St. and Superior Ave. A steam heating system will be installed in same. Total estimated cost, $100,000.

O., Cleveland—The Templar Realty Co., c/o Max Marmorstein, 1605 Williamson Bldg. plans to construct an 8 story, 80 x 200 ft. commercial building on East 81st St. and Euclid Ave. A steam heating system will be installed in same. Total estimated cost, $500,000. S. H. Weis, 1032 Schofield Bldg., Archt.

O., East Palestine—The Municipal Electric Wks. plans to install a 300 hp. W. T. boiler. Jacob Ashman, Pres.

O., Lakewood — The Lakewood Theatre Co., c/o E. Mandelbaum, Sloan Bldg., Cleveland, is having plans prepared for the construction of a 2 story theatre at 15015 Detroit Ave. A steam heating system will be installed in same. Total estimated cost. $200,000. Thomas W. Lamb, 1323 Euclid Ave., Cleveland, Archt.

O., Lima—Elmer McClain, Dir. of Pub. Serv. will receive bids until April 1 for furnishing boiler feed pump to be installed in the Waterworks main pump station.

O., Linden—(Columbus P. O.) The Bd. Educ. received lowest bid for installing heating, plumbing and ventilating systems in the proposed 2 story school from the Hoffman, Wolfe Co., North High St. at $58,000.

O., Massillon—The Bd. Educ. plans to build a 2 story, 76 x 80 ft. school on State St. A hot blast heating system will be installed in same. Total estimated cost, $100,000. Albrecht, Wilhelm & Kelly, Union Bldg., Cleveland, Archts.

O., Massillon—The Bd. Educ. plans to build a 76 x 90 ft. school on Cherry St. A hot blast heating system will be installed in same. Total estimated cost, $100,000. Albrecht, Wilhelm & Kelly, Union Bldg., Cleveland, Archts.

O., Massillon—C. P. L. McLain, Grocer, is having plans prepared for the construction of a 2 story, 40 x 130 ft. commercial building. A steam heating system will be installed in same. Total estimated cost, $60,000. Albrecht, Wilhelm & Kelly, Union Bldg., Cleveland, Archts.

O., New Philadelphia—The Weiss McClung Co. plans to build a 1 and 3 story rubber factory. A steam heating system will be installed in same. Total estimated cost, $300,000. W. F. Ferguson Co., 1900 Euclid Ave., Cleveland, Engr.

O., Orrville—The Bd. Educ. is having plans prepared for the construction of a 2 story, 75 x 125 ft. graded school. A steam heating system will be installed in same.

Total estimated cost, $200,000. L. L. Woods, Clk. O. D. Howard, 8 East Broad St., Columbus, Archt.

O., Piqua—Hall & Lethly, Archts., 807 Fairbanks Bldg., Springfield, will receive bids until March 30 for the construction of a 2 story, 100 x 120 ft. school building in the western section of the city, for the Bd. Educ., City Bldg. A hot water and steam heating system will be installed in same. Total estimated cost, $100,000.

O., Sabina—The Municipal Water & Light Plant plans to install Y. transformers to furnish a 230 volt, 3 phase service. J. C. Phelps, Supt.

O., Willoughby—The Thor Tire & Rubber Co., 802 Society for Savings Bldg., is having plans prepared for the construction of a 1 and 2 story factory on Main St. along the New York Central R.R. A steam heating system will be installed in same. Total estimated cost, $100,000. F. L. Gary, Secy. H. H. White, 802 Society for Savings Bldg., Archt. and Engr.

Ind., Washington—The city has been granted permission by the Indiana Pub. Serv. Comn. to issue bonds not in excess of $70,000 for the improvement of the electric light plant. The City Council plans to install a 500 kw. generator and 2 high pressure boilers equipped with coal conveyors and automatic stokers in same.

Mich., Buchanan—I. S. Rice, Secy. of the Bd. Educ., will soon award the contract for the construction of a 2 story school building. A steam heating boiler, etc., will be installed in same. Total estimated cost, $200,000. R. A. LeRoy, 122 Pratt Block, Kalamazoo, Archt.

Mich., Detroit—The Bd. Educ., 50 B'way Ave., will receive bids until April 7 for the construction of a 7 story, 250 x 350 ft. technical school on Second Ave. Motors for power will be installed in same. Total estimated cost, $2,500,000. Malcolmson, Higginbotham & Palmer, 405 Moffat Bldg., Archts.

Mich., Detroit—Brown & Preston, Archts., 406 Empire Bldg., will soon award the contract for the construction of a 3 story, 100 x 120 ft. factory on East Larned and St. Aubin Sts., for the Superior Machine & Eng. Co., 51 East Fort St. A steam heating boiler and motors for power, etc. will be installed in same. Total estimated cost, $100,000.

Mich., Detroit—The Detroit Stoker Co., 1718 Gratiot Ave., Detroit, is in the market for one 25 cu.ft. capacity air compressor and an air receiver.

Mich., Detroit—The Hannan Rea. Exch., 1 McGraw Bldg., plans to build a 5 story, 53 x 141 ft. commercial building on Woodward Ave. and Sibley St. Steam heating equipment will be installed in same. Total estimated cost, $200,000.

Mich., Ford City—(Wyandotte P. O.) William Bearendt, Secy. of the Bd. Educ., Dist. 1, will soon award the contract for the construction of a 2 story, 129 x 148 ft. high school on Main St. Steam heating equipment will be installed in same. Total estimated cost, $200,000. J. G. Kastler & Co., 523 Chamber of Commerce, Detroit, Archt.

Mich., Grand Rapids—Pierre Lindhout, Archt., 817 Lake Drive, will soon award the contract for the construction of a 3 story, 50 x 100 ft. factory, for W. H. Spears. Steam heating equipment and motors for power will be installed in same. Total estimated cost, $80,000.

Mich., Hamtramck—(Detroit P. O.) The Dodge Bros., 1678 Joseph Campau Ave., are having plans prepared for the construction of an 8 story, 77 x 266 ft. factory on Conant Ave. Motors for power will be installed in same. Smith Hinchman & Grylls, 710 Washington Arcade, Detroit, Engrs.

Mich., Highland Park (Detroit P. O.)—The Detroit Osteopathic Hospital, c/o Philip H. Gray, plans to construct a 5 story, 75 x 155 ft. hospital on Highland and 3rd Aves. Complete steam heating and ventilating equipment will be installed in same. Total estimated cost, $250,000.

Mich., Huron—The City plans to build a 2 or 3 story hospital building on Kearney and Stone Sts. Steam heating equipment will be installed in same. Total estimated cost, $125,000. Richard Forman, City Hall, Chn.

Mich., Lansing—The Arcade Theatre Co., South Washington Ave., is having plans prepared for the construction of a 2 story, 109 x 165 ft. arcade and theatre building. A steam heating and ventilating system will be installed in same. Total estimated cost, $450,000. W. S. Butterfield, Mgr. John Eberson, 64 East Van B. St., Chicago, Archt.

Mich., Marysville—Smith, Hinchman & Grylls, Archts., and Engrs., 710 Washington Arcade, Detroit, will soon award the contract for the construction of a 1 and 2 story, 380 x 420 ft. foundry on Michigan Ave. and Cuttle Rd., for the General Aluminum & Brass Co., Boulevard and St. Aubin St., Detroit. Steam and hot blast heating equipment and motors for power will be installed in same. Total estimated cost, $500,000.

Mich., Muskegon—The city rejected all bids for furnishing and installing an additional pumping unit and a 10,000,000 gal. reciprocating pumping unit. Hoad & Decker, Ann Arbor, Engrs. Work will be readvertised.

Mich., Pontiac—The Grand Trunk Ry. System, Brush St., Detroit, plans to build a 1 story roundhouse to have a 50 engine capacity, near Johnson St. Electric power will be installed in same. F. P. Sissons, Brush St. Depot, Detroit, Division Engr.

Ill., Chicago—Alfred S. Alschuler, Archt., 28 East Jackson St., will soon award the contract for the construction of a 7 story, 120 x 175 ft. printing plant on 21st St. and Calumet Ave., for the Columbian Colortype Co., 2141 Calumet Ave. A steam heating system will be installed in same. Total estimated cost, $650,000.

Ill., Chicago—Paul Gerhardt, Archt., 64 West Randolph St., will soon award the contract for the construction of a 6 story 150 x 165 ft. clothing factory on Milwaukee Ave. and Caton St., for the Tennerstedt Mfg. Co., 1936 West North Ave. A steam heating system will be installed in same. Total estimated cost, $500,000.

Ill., Highland Park—The city is having plans prepared for the installation of a filtration plant, extensions to distributing system and 5,000,000 gal. additional pumping equipment. Alvord & Burdick, 8 South Dearborn St., Chicago, Engrs.

Wis., Algoma—The City Water & Light Plant plans to purchase the energy from the Wisconsin Pub. Serv. Co., Green Bay and construct a 63 mi., 66,000 volt, 3 ph., 60 cycles transmission line. E. P. M. McCarthy, Supt.

Wis., Amery—The Apple River Milling Co. plans to construct a 39 mi., 22,000 volt transmission line. W. R. Haney, Supt.

Wis, Amherst — The Amherst Electric Service Co. plans to install a 100 kva. generator and a 50 or 75 hp. oil engine. B. E. Dwinnel, Mgr.

Wis., Belleville—The Belleville Electric Co. plans to install a connection with the Southern Ware Co. and construct an extension to rural transmission lines. W. E. Seward, Supt.

Wis., Colby—The Colby Light & Power Co. plans to construct a connection to high line in generating system. George Tesmer, Mgr.

Wis., Greenwood—The Greenwood Municipal Light & Power Plant plans to extend transmission lines to high line and install 150 hp. in new factory. G. W. Rork, Supt.

Wis., Hartford—The Utility Dept. plans to install boilers. H. J. Peters, Supt.

Wis., Kohler—The city will soon award the contract for the construction of a complete sewage pump station, etc. on Main St. and is in the market for equipment for same. Total estimated cost, $25,000. J. Donohue, North 8th St., Sheboygan, Engr.

Wis., Lowell—The Lowell Light & Power Co. plans to install a vertical alternator. W. F. Pease, Secy.

Wis., Manitowoc—The Wisconsin Aluminum Products Co., 16th and Franklin Sts., plans to construct a 4 story, 60 x 250 ft. factory for the manufacture of aluminum products, on 8th St. A steam heating system will be installed in same. Total estimated cost, $250,000. B. Dalwig, Mgr.

Wis., Medford — The Medford Light & Heating Co. plans to install either a new unit in the generating system or a connection with a transmission system. W. Ungrodt, Supt.

Wis., Owen—Clark Co. is having plans prepared for the construction of a 3 story insane asylum. A steam heating, refrigerating and fan system will be installed in same. Total estimate cost, $350,000. W. G. Royer, Chn. of the Comn. Claud & Starck, Badger Bldg., Madison, Archts.

Wis., Sheboygan—The city plans to construct a 3 story, 100 x 150 ft. high school on North 9th and Jefferson Sts. Steam heating equipment including Kewanees or equal tubular boilers, using any fuel, will be installed in same. Total estimated cost, $250,000.

Wis., Sheridan—The Sheridan Co. Elec. Co. plans to electrify the Carney Coal Mines. Judsen Bibb, Mgr.

Wis., Soldiers Grove—The Soldiers Grove Electric Co. plans to change the generating system from d.c. to a.c. Thomas Sime, Mgr.

Ia., Des Moines — The Iowa Telephone Co., Bell System, 7th and High Sts., is having plans prepared for the construction of a 3 story telephone building. A steam heating and fan ventilation system will be installed in same. Total estimated cost, $300,000. George B. Prinz, Omaha Natl. Bank Bldg., Archt.

Ia., Morning Sun—O. F. Boller, Sec'y of the Bd. Educ., is having plans prepared for the construction of a 2 story school building. Plumbing and heating contracts will be sub-let. Total estimated cost, $200,000. W. E. Hulse & Co., 210 Masonic Temple, Des Moines, Archt.

Ia., Packwood—N. E. Oliver, Sec'y of the Bd. Educ., is having plans prepared for the construction of a 2 story, 58 x 99 ft. school building. The contract for installing heating and plumbing systems will be sub-let. Total estimated cost, $125,000. W. H. Hulse & Co., 210 Masonic Temple, Des Moines, Archt.

Minn., Minneapolis—Jacob Resler, 27 4th St., North, will soon award the contract for the construction of a 3 story, 97 x 110 ft. addition to his business building on 1st Ave., N. A steam heating system will be installed in same. Total estimated cost, $160,000. F. Van Antwerp, 416 Palace Bldg., Archt.

Minn., St. Paul—The Riverview Methodist Episcopal Congregation, c/o Rev. Anton Peik, 529 Bidwell St., will soon receive bids for the construction of a 3 story, 90 x 100 ft. church on West George St. A steam heating system will be installed in same. Total estimated cost, $100,000. A. H. Stem, 601 Endicott Bldg., Archt.

Kan., Topeka—The Atchison, Topeka & Santa Fe Ry. Co., 80 East Jackson Blvd., is having plans prepared for the construction of a central heating plant. Estimated cost, $85,000. John Yonkers, Park Ridge, Chicago, Archt.

Neb., Beaver City — The Beaver City Light & Waterworks plans to install a 150 hp. engine and a 1,000 kva. generator. Earl Roberts, Mgr.

Neb., Howell — The Howell Municipal Light Plant plans to change from d.c. to a.c. system. George Lodes, Secy.

S. D., Flandreau—The Dakota Light & Power Co. plans to construct extensions to transmission lines. A. H. Bushman, Supt.

S. D., Highmore—The Municipal Electric Light & Power Plant plans to change generating systems from d.c. to a.c. A. H. Fehr, Supt.

S. D., Madison—The Municipal Electric Light Plant plans to construct a new power house in the spring and new power circuits for motor load alone. E. G. Quarnstrom, Supt.

S. D., Milbank—The Milbank Power Co. plans to install one 400 hp. Rathbun engine and generator. F. A. Barnard, Mgr.

S. D., Rapid City—The School of Mines, Bd. of Regents, South Dakota University, is having plans prepared for the construction of a 3 story engraving building. A steam heating system will be installed in same. Total estimated cost, $100,000. F. H. Ellerbe, Endicott Bldg., St. Paul, Minn., Archt.

S. D., White Lake—The White Lake Light & Power Co. plans to construct extensions lines and additions to a complete gas and kerosene oil filling station. V. J. Peters, Mgr.

S. D., Worthing—The Worthing Electric Light Plant plans to install one 20 hp. engine and one 12 kw. generator. W. F. Berens, Mgr.

N. D., Leeds—The City Council will soon award the contract for the construction of a 50,000 gal. tank, one 23,500 gal. r.c. reservoir, one triplex pump with d.c. motor, one heater house with heater, etc. Total estimated cost, $35,000. J. H. Husely, City Aud. T. R. Atkinson, Bismarck, Engr.

Mo., Fulton—The Municipal Electric Light & Waterworks plans to install a new 250 kw. unit including engine and generator. W. J. McCarroll, Supt.

Mo., Sarcoxie — The Southwest Electric Light & Milling Co., plans to install a 112½ kva. generator, a new transformer, etc. Arthur Adams, Pres.

Tex., Canadian — The Canadian Water, Light & Power Co. plans to install an 85 hp. engine in the generating system and construct extensions to the transmission lines. Edward Hooyer, Mgr.

Okla., Altus—The Municipal Water & Light Dept. plans to install an additional 700 kva. generator with high pressure boilers. Charles H. Welch, Supt.

Okla., Poteau—The Leflore Gas & Electric Co. plans to install generator and one 100 hp. boiler. F. E. Baird, Mgr.

N. M., Clayton—The Village of Clayton plans to install one 100 kva generator. J. H. Bender, Supt.

Ariz., Phoenix—The Bd. Educ. voted $500,000 bonds to construct a high school including a manual arts and liberal arts building and gymnasium. A steam heating plant will probably be installed in same.

Ore., Pillamock—The Coast Power Co. plans to install a 1,000 kw. turbine. C. J. Edward, Mgr.

Ore., Prineville—The Bd. of Directors, Ochoco Irrigation Dist., plans to vote an additional $100,000 bonds for the completion of Ochoco dam and for the construction of distribution system structures, including a high line pumping plant, office building, and warehouse. Total estimated cost, $1,350,000.

Cal., Berkeley—The Bd. Educ. will soon receive bids for the construction of a 2 story school building on Prince and King Sts. A steam heating and ventilation system will be installed in same. Total estimated cost, $200,000. W. H. Ratcliff, Jr., First Natl. Bank Bldg., Archt.

Cal., Dinuba—The Trustees of the Dinuba Union High School Dist. are having plans prepared for the construction of a high school. A steam heating system will be installed in same. Total estimated cost, $120,000. E. J. Kump, Rowell Bldg., Fresno, Archt.

Cal., Sacramento—The California Packing Corp., 6th and G Sts., is having plans prepared for the construction of a 2 story packing house and a 1 story boiler house. Total estimated cost, $80,000. Philip Bush, 101 California St., San Francisco, Engr.

Cal., Sacramento—A. Givan, Forum Bldg., plans to construct a pump house, etc. and install one 20 in. and one 10 in. electric driven centrifugal pumps for the diversion of 35 cubic second ft. of water from the Feather River for agricultural purposes. Estimated cost, $65,000.

Cal., San Diego—The Public Wks. Office received bids March 3 for furnishing and installing complete power plant and electric generating equipment, including piping, distribution systems for electric and steam, 2 oil burning boilers, oil pump, feed water heater, 2 engine driven alternators, 1 engine driven and 1 electric driven exciters, transformers, vacuum pump, separating tank, etc, at the Marine Corps Base here, from Thomas Haverty Co., 801 Maple St., Los Angeles, for work complete $189,650, (2) $31,250, (3) $66,800, (4) $91,600, (5) $560, (6) $620, (7) $8,800, (8) deduct $670, (9) deduct $23,700, (10) deduct $5,000. Noted March 3.

Cal., San Francisco—The McDermott Estate, c/o B. G. McDougall, Archt., 381 Bush St., is having plans prepared for the construction of a 2 story office building on Battery and Halleck Sts. A steam heating system will probably be installed in same. Total estimated cost, $125,000.

Cal., San Pedro—F. O. Adler, c/o Victoria Theatre, is having plans prepared for the construction of a 1 and 2 story, 105 x 125 ft. theatre on 6th St. A steam heating and plenum ventilating system will be installed in same. Total estimated cost, between $75,000 and $100,000. Smith & Pennell, 621 Investment Bldg., Los Angeles, Archts.

Cal., Turlock—The Turlock Irrigation Dist. voted $2,570,000 for reservoir sites and the Don Pedro Dam, $1,028,000 for hydro-electric plant and transmission lines and $510,000 for main irrigation canal laterals and drainage ditches; the Modesto Irrigation Dist., voted $1,180,000 for the Don Pedro dam, $181,000 for the hydro-electric plant, $150,000 for enlarging its main canal, $190,000 for extending drainage system and $298,400 for electric power distributing system. Dam will probably be 280 ft. high, 1,000 ft. long, 177 ft. thick at bottom and 16 ft. thick at top in the Tuolumne River, Tuolumne Co., 40-60 mi. east of here, Modesto; hydro-electric plant will develop 16,000 hp. and 12,000 kw.; main canal will be 40-60 mi. long. Work will be done jointly by both districts. P. F. Jones, Modesto, Chief Engr. for the Modesto Irrigation Dist. A. J. Wiley, Idaho Bldg., Boise, Idaho, Consult. Engr.

Cal., Ukiah—The California Grape Products Co. will build a factory to manufacture grape juice, etc. A 250 hp. boiler plant will be installed in same. Total estimated cost, $100,000.

Cal., Willows—The Willows Improvement Co. engaged A. Moore, Archt., Kohl Bldg., San Francisco, to prepare plans for the construction of a 4 story hotel and store building on Butte and Walnut Sts. A steam heating system will probably be installed in same. Total estimated cost, $215,000. Address L. Briggs.

N. F., Carbonear—The United Municipal Electric Co., Ltd., plans to construct an extension to generating system and transmission lines to 7 small towns. C. Cameron, Secy.

N. S., Oxford—The Oxford Electric Light & Power Co. plans to construct a 10 mi. transmission line from here to Coffingwood and a 10 mi. transmission line from here to Puguash. S. D. Downing, Mgr.

N. S., Yarmouth—The Yarmouth Light & Power Co., Ltd. plans to install a 750 kva. hydro unit with electrical equipment. D. R. Stoneman, Secy.

Que., Dechambeault—The Portneuf Hydraulic Co. plans to install 2 units of 1,000 kw. each. Henri Pare, Mgr.

Que., Montreal—The Allan Enterprises, Ltd., 404 St. Catharine St., plans to construct a 2½ story, 90 x 175 ft. theatre on Cathcart and St. Catharine Sts., West. An oil burning system of heating will be installed in same. Total estimated cost, $500,000.

Que., Montreal—The Standard Shirt Mfg. Co., 213 Delormier Ave., is in the market for a 4 x 6 or a 6 x 6 ft. air compressor with a 75 to 100 lb. pressure.

Que., Westbury—The Corp. of Sherbrooke Gas & Electric Dept., Sherbrooke, plans to construct an 11,000 hp. power house, here, and a 2 mi., 50,000 volt transmission line. Charles Desbraillet, Mgr.

Ont., Belleville—The Public School Bd. plans to build a 2 story school building. Horizontal boilers for a direct heating system will be installed in same. Total estimated cost, $100,000. Beaumont Jarvis, 64 Gloucester St., Toronto, Archt.

Ont., Hamilton—J. and J. J. Allen, Richmond and Victoria Sts., Toronto, plans to build a 1 and 2 story motion picture theatre on King and William Sts. A vacuum steam heating system will be installed in same. Total estimated cost, $500,000.

Ont., Hawkesbury—The Hawkesbury Electric Light & Power Co., Ltd., plans to install one unit, 2,500 hp. water wheel and a 2,000 volt generator James Ross, Mgr.

Ont., Kitchener—The Kitchener Waterloo Collegiate Bd., Frederick St., is having plans prepared for the construction of a 3 story collegiate institute. A vacuum steam heating system with ventilation system will be installed in same. Total estimated cost, $500,000. S. B. Coon & Son, 810 Excelsior Life Bldg., Toronto, Archts.

Ont., Orillia—The Orillia Water, Light & Power Comn. plans to rebuild two 22,000 volt transmission lines. R. H. Starr, Engr.

Ont., Sudbury—The Imperial Oil Co., Ltd., Imperial Oil Bldg., Toronto, plans to build a 2 or 3 story warehouse and distributing station. A steam heating system will be installed in same. Total estimated cost, $200,000.

Ont., Toronto—Mavor Church, City Hall, will receive bids until April 20 for the construction of a 40,000,000 imperial gal. low lift, electrically driven, centrifugal, sewage pump. Estimated cost, $20,000. R. C. Harris, City Hall, Engr.

Ont., Walkerville—The Walkerville Hydro-Electric Comn. plans to install an additional 2,000 kva. power feeder and a 7,500 kva. lighting feeder. M. McHenry, Mgr.

Ont., Woodstock—The Woodstock Light & Water System plans to construct a 13,000 volt sub-station and a 1½ mi., 13,000 volt transmission line. J. G. Archibald, Mgr.

Man., Shoal Lake—The Shoal Lake Municipal Electric Plant plans to install a duplicate unit in the generating system. A. R. Ibbotson, Engr.

Alta., Maclead—The Municipal Electric Light Plant plans to rebuild entire transmission system. Vernon Pearson, Supt.

Alta., Red Deer—The West General Electric Co., Ltd., plans to install new boilers and one new generating unit. W. A. Moore, Mgr.

B. C., Ladysmith—The Municipal Electric Plant plans to install a 50 kw. generator and a 2,200 volt, 60 cycle, 3 phase Diesel engine. M. A. Morrison, City Clk.

Sask, Arcola—The Arcola Light & Power Co. plans to install a 30 hp. 2,200 volt motor for driving local brickyard for day power service. H. Forbes, Roberts, Mgr.

Sask., Moose Jaw—The Municipal Light & Power Dept. plans to install water tube boilers. J. D. Peters, Mgr.

CONTRACTS AWARDED

Vt., Lunenburg—The Fitzdale Paper Co. will build a 1 story power house. Estimated cost, $1,000,000. Work will be done by day labor.

Mass., Boston—Loew's Theatres Inc., 1493 B'way, New York City, has awarded the contract for the construction of a 3 story, 180 x 200 ft. theatre on Massachusetts Ave., to the Fleischman Co., 531 7th Ave., New York City. A steam heating system will be installed in same. Total estimated cost, $150,000.

Mass., Brookline—The Co-Partner Association, Old South Bldg., Boston, has awarded the contract for the construction of a 4 and 8 story, 200 x 400 ft. hotel on Beacon St., to Fred T. Ley, Devonshire St., Boston. A steam heating system will be installed in same. Total estimated cost, $1,200,000. John J. Smith, Old South Bldg., Boston, Archt.

Mass., Chicopee—The Dwight Mfg. Co., Front St., has awarded the contract for the construction of a 5 story, 72 x 314 ft. manufacturing plant addition to the Casper Ranger Constr. Co., 20 Bond St., Holyoke. A steam heating system will be installed in same. Total estimated cost, $500,000.

Mass., East Boston—(Boston P. O.) A. D. Stefano, Everett St., will build a 2 story ice plant on Everett St. Estimated cost, $30,000. Work will be done by day labor.

Mass., Worcester—The Simplex Player Action Co., 10 Blackstone St., has awarded the contract for the construction of a 5 story, 80 x 150 ft. manufacturing plant on Summer and Charles Sts., to Will Daniels & Sons. A steam heating system will be installed in same. Total estimated cost, $150,000.

Conn., Bridgeport—The United Illuminating Co., 84 Temple St., New Haven, has awarded the contract for the construction of a power plant on East Main St., to C. W. Blakeslee & Sons, 58 Waverly St., New Haven. Noted March 9.

Conn., Greenwich—The Indian Harbor Yacht Club, Putnam Ave., has awarded the contract for the construction of a club house, to C. T. Wills, Inc., 286 5th Ave., New York City. A steam heating system will be installed in same. Total estimated cost, $160,000.

Conn., Stamford—The Carlisle Tire Corp., 251 4th Ave., New York City, has awarded the contract for the construction of a 1 story factory on Fairfield Ave., to the Wells Constr. Co., 237 5th Ave., New York City. A steam heating system will be installed in same. Total estimated cost, $250,000.

Conn., Torrington—The School Comn. has awarded the contract for the construction of a 2 story school addition on Pythian Ave., to the Torrington Bldg. Co., 197 Water St. A steam heating system will be installed in same. Total estimated cost, $90,000.

N. Y., Bath—The Directors of the Soldiers and Sailors Home have awarded the contract for repairing and altering the heating construction at the New York State Soldiers and Sailors Home, to W. B. Armstrong Co., 3 Fulton St., Albany, at $7,700.

N. Y., Brooklyn—Louis Gold, 44 Court St., will build a 6 story, 50 x 140 ft. factory on Pearl and Tillery Sts. A steam heating system will be installed in same. Total estimated cost, $300,000. Work will be done by day labor.

N. Y., Brooklyn—The Israel Hospital Inc., 1246 42nd St., New York City, will build a 3 story, 115 x 180 ft. hospital at 4801 10th Ave. A steam heating system will be installed in same. Total estimated cost, $410,000. Work will be done by day labor.

N. Y., Coney Island (Brooklyn P. O.)—G. Gordon, c/o McCarthy & Kelly, Archts. and Engrs., 16 Court St., Brooklyn, will build a 1 story, 30 x 114 ft. theatre on 31st St. and Surf Ave. A steam heating system will be installed in same. Total estimated cost, $125,000. Work will be done by day labor.

N. Y., Jamaica—The Bd. Educ., 500 Park Ave., New York City, has awarded the contract for the construction of a school building to be known as Public School 50, on 101st St. and Liberty Ave., to J. McArthur, 22 Ormond Place, Brooklyn. A steam heating system will be installed in same. Total estimated cost, $530,000.

N. Y., Long Island City—The Interstate Land Holding Co., c/o J. J. Gloster, Archt. and Engr., 44 Court St., has awarded the contract for the construction of a 3 story,

100 x 100 ft. factory on 14th St. and Governor Pl., to Louis Gold, 44 Court St. A steam heating system will be installed in same. Total estimated cost, $150,000.

N. Y., New York—The Bd. Trustees, Bellevue State Hospital, office of General Medical Supt., 415 East 26th St., has awarded the contract for the installation of boiler, pumps, etc., in the new boiler plant at Fordham Hospital, to the Chute, Thornton & Bailey Corp., 2 East 13th St., at $36,500.

N. Y., New York—M. F. Holmes, c/o C. E. Birge, Archt. and Engr., has awarded the contract for the construction of an 8 story, 35 x 100 ft. store and loft building at 43 West 57th St., to R. H. MacDonald, 29 West 34th St. A steam heating system will be installed in same. Total estimated cost, $115,000.

N. Y., New York—Powers, Weight & Rosengarten Co., 145 Front St., has awarded the contract for the construction of a 50 x 60 ft. office building at 27-31 Depeyster St., to W. S. Patten, 52 Vanderbilt Ave. A steam heating system will be installed in same. Total estimated cost, $100,000.

N. Y., New York—The Sloane Estates Inc. has awarded the contract for the construction of a 2 story, 84 x 90 ft. loft building at 234-242 West 39th St., to Fred. T. Ley, 19 West 44th St. A steam heating system will be installed in same. Total estimated cost, $300,000.

N. J., Jersey City—The Whitlock Cordage Co., Communipaw Ave., has awarded the contract for the construction of a 4 story building on Communipaw Ave., to the John W. Ferguson Co., United Bank Bldg., Paterson. Four new 400 hp. boilers and one 500 kw. and one 1,000 kw. turbines will be installed in same. Total estimated cost, $500,000.

Pa., Philadelphia—The Model Mills Co., Ontario and 9th Sts., has awarded the contract for the construction of a 4 story, 60 x 180 ft. factory, to William Steele & Sons Co., 16th and Arch Sts. A steam heating system will be installed in same. Total estimated cost, $150,000.

Md., Westport (Baltimore P. O.)—The Consolidated Gas, Electric Light & Power Co., Lexington Bldg., Baltimore, will build a 1 story, 75 x 150 ft. electric power house. Estimated cost, $60,000. Work will be done by day labor.

Va., Roanoke—The Viscose Co., Marcus Hook, has awarded the contract for the construction of a 3 story hotel, to C. J. Phillips. A steam heating system will be installed in same. Total estimated cost, $300,000.

N. C., Hickory—The city has awarded the contract for installing a heating system in the proposed 2 story municipal office building, to F. B. Ingold, at $5,480.

La., New Orleans—The Hibernia Bank Trust Co., Carondelet and Gravier Sts., has awarded the contract for the construction of a 16 story bank and office building on Carondelet St., to George A. Fuller, 175 5th Ave., New York City. A steam heating system will be installed in same. Total estimated cost, $2,000,000.

O., Barberton—The Diamond Match Co. has awarded the contract for the construction of a 4 story, 125 x 175 ft. factory, to the Turner Constr. Co., 244 Madison Ave., New York City. A steam heating system will be installed in same. Total estimated cost, $475,000.

O., Cleveland—The Brown Body Corp., 3147 Superior Ave., has awarded the contract for the construction of a 1 story, 60 x 170 x 260 ft. and a 3 story, 100 x 170 ft. factory buildings on West 90th St. and Maywood Ave., to the H. K. Ferguson Co., 6532 Euclid Ave. A steam heating system will be installed in same. Total estimated cost, $500,000.

O., Cleveland—The Champion Forge & Machine Co., East 71st St. and Union Ave., has awarded the contract for the construction of a 3 story, 60 x 120 ft. factory, to the Charles Peterson Co., 4500 Euclid Ave. Estimated cost, $100,000.

O., Cleveland—The city has awarded the contract for furnishing a combination steam and hot air rotary dryer, to the Ohio Machine & Boiler Co., 1501 University Rd., at $12,350.

O., Cleveland — The Cleveland Home Brewing Co., 2501 East 61st St., has awarded the contract for the construction of a 1 story, 30 x 135 ft. ice plant and a 30 x 50 ft. boiler room addition, to the F. W. Ruple Constr. Co., 4500 Euclid Ave. Total estimated cost, $60,000.

O., Cleveland—The Federal Packing Co., c/o W. Bressler, West 65th St., will build a 2 story, 40 x 70 ft. cold storage building on East 4th St. and Bolivar Rd. Total estimated cost, $50,000. Work will be done by day labor.

O., Cleveland—Van Aker & Strock Co., 6532 Euclid Ave., has awarded the contract for the construction of an 8 story, 85 x 185 ft. commercial building at 4300 Euclid Ave., to A. U. Kilbourne Co., 6532 Euclid Ave. Steam heating boilers will be installed in same. Total estimated cost, $500,000.

O., Columbus — The Felber Biscuit Co., Grant and McCoy Aves., has awarded the contract for the construction of a 6 story, 62½ x 100 ft. factory on Grant Ave., to R. H. Evans & Co., Columbus Savings & Trust Bldg. A steam heating system will be installed in same. Total estimated cost, $150,000.

O., Dayton—The Allsteel Ridewell Tire & Rubber Co., Lindsey Bldg., has awarded the contract for the construction of a 3 story factory in the northern part of the city, to the Osborn Eng. Co., 2848 Prospect Ave., Cleveland. Plans include a power plant, etc. Total estimated cost, $100,000.

O., Euclid—The Hopkins School Association has awarded the contract for the construction of a 1 story, 80 x 300 ft. school on Mayfield Rd., to Crowell & Little, Marian Bldg., Cleveland. A steam heating system will be installed in same. Total estimated cost, $200,000.

Ind., Kokomo — The Pittsburgh Plate Glass Co., Valle Ave., has awarded the contract for the construction of a 12,000 kw. steam power station, to Stone & Webster, 120 B'way, New York City.

Mich., Detroit—The city has awarded the contract for furnishing a low lift sub-structure pumping station, to the Walsh Constr. Co., Davenport, Ia.

Mich., Detroit—The Dept. of Purchases & Supplies, Municipal Courts Bldg., has awarded the contract for furnishing one 2 motor mono-rail hoist with a 2-ton capacity, 25 ft. per min. hoisting speed and height of lift, 30 ft. and a trolley motor with a traveling speed not less than 150 ft. per min., to the Detroit Hoist & Machine Co., Clay Ave., at $1,060.

Mich., Owosso—The city, c/o J. H. Robins has awarded the contract for the construction of a 2 story hospital building on West King St., to C. H. Maliskey & Son, at $127,264. A steam heating boiler, etc., will be installed in same.

Ill., Pullman—(Chicago P. O.) The Pullman Co., manufacturers of railroad cars, 79 East Adams St., has awarded the contract for the construction of a 1 story, 45 x 160 ft. and a 60 x 170 ft. factory buildings on 104th St. and Erickson Ave., to the E. W. Sproul Co., 2001 West 39th St., Chicago. A steam heating system will be installed in same. Total estimated cost, $175,000.

Wis., Colfax—The Colfax Electric Co. has awarded the contract for the construction of a diversion dam, power house, headrace and installation of equipment, to Siems, Helmers & Schaffner, 514 Guardian Life Bldg., St. Paul, Minn. Total estimated cost, $100,000.

Mo., Kansas City—The Gate City Natl. Bank, c/o Wilson B. Plauck, 10th St. and Grand Ave., has awarded the contract for the construction of a 4 story, 50 x 115 ft. bank and office building at 1109 Grand Ave., to the Fogel Constr. Co., Reliance Bldg. A steam heating system will be installed in same. Total estimated cost, $250,000.

Mo., St. Louis—The United Drug Co., 63 Leon St., Boston, Mass., has awarded the contract for the construction of a 1 story, 45 x 80 ft. boiler house at 3941 Kingshighway, to the West-Lake Constr. Co., Ry. Exch. Bldg. A steam heating system will be installed in same. Total estimated cost, $25,000.

Cal., San Francisco — The Commercial Union Assurance Co., Ltd., and the California Fire Insurance Co., 558 Sacramento St., has awarded the contract for the construction of a 15 story office building on Pine and Montgomery Sts., to the Walker Constr. Co., Monadnock Bldg. Boilers, etc., will be installed in same. Total estimated cost, $1,000,000.

Cal., San Francisco—The Pioneer Motor Co., 307 Golden Gate Ave., has awarded the contract for the construction of a 4 story, 120 x 219 ft. auto sales building on Van Ness Ave. and McAllister St., to the Clinton Constr. Co., 140 Townsend St. Motors, etc., will be installed in same. Total estimated cost, $200,000.

Que., Montreal—The Montreal Tramways Co., Tramway Bldg., has awarded the contract for the construction of a 70 x 90 ft. power sub-station on Cote St., to the Raymond Concrete Pile Co., 10 Cathcart St. Estimated cost, $45,000.

POWER

Volume 51 — New York, March 30, 1920 — *Number* 13

Two-Faced Tom

By Rufus T. Strohm

WHY IS IT that so many men who live respected business lives grow grim and grouchy now and then and act so gruffly with their wives?

Take Thomas Tyler as a case: He runs the engines and the pumps at Sanderson Miller's place; he plugs the leaks and fixes thumps. His ready wit and jollity have won his fellows' hearts to him; his cup of gladness seems to be forever spilling at the brim. But when the evening whistle blows to end the labor and the grind, and Thomas slowly homeward goes, he leaves his cheerfulness behind.

On entering his cottage door, with wrinkled brow and sullen jaw, he's just about as peeved and sore as Bruin with a festered paw. He dims the gas-light's cheery ray with gloom so thick it can be felt, till he begins to stow away the evening meal beneath his belt. And then, perhaps, he deigns to speak in glum and pessimistic tones, bewailing that his wage per week amounts to only thirty bones.

He swears he's worth three times as much to any company in town and vows it surely beats the Dutch the way his bosses grind him down. His fellow-workmen have no brains; their skulls are stuffed with cotton wool; his super only holds the reins because of politics and pull.

And thus he grumbles and he rants of circumstances, men and things, till supper's over, and, perchance, the silver-plated door-bell rings. Then Thomas ends his verbal war; his voice grows suave, his smile grows bland; he leaps to meet the visitor with greetings warm and gladsome hand. But when the caller says goodnight and takes departure from the room, then Tom forgets to be polite and dons again his cloak of gloom.

His patient wife essays to play, to read to him, perhaps to sing, in hopes that she may drive away the blues that color everything; while Thomas sits, morose and gruff, as though in contemplation deep; but this is quackery and bluff, for soon he yawns and falls asleep.

His friends are glad to see his face and shake his hand where'er he goes; they like to have him round the place. But not his wife!—she knows, she knows!

The above was inspired by a letter received from a lady in Toronto, evidently the wife of a "Two-faced Tom," who had read the foreword on "The Chronic Grouch" in the Jan. 13 issue. She is probably only one of many—EDITOR.

Chile Exploration Company's Tocopilla Power Plant

Plant of 50,000 Kw. Capacity, 4,000 Miles From Fuel Source, Located Where It Never Rains and Where Earthquakes Are Prevalent

CHILE, located on the west coast of South America, is a narrow strip of country extending a total length of approximately 2,700 miles and has a width varying from 248 to less than 70 miles. For about 1,000 miles of the northern portion the coast is nearly a straight line, offering practically no harbor facilities. This coast line is bordered by a diversified tableland rising sharply out of the sea to a height of from 2,000 to 9,000 ft. Around this locality it seldom rains and the rivers are of little consequence. In a section of this country which is practically a desert, about 1,400 miles south of the equator, without any roads, railways or other arteries of transportation, 4,000 miles away from a fuel supply, where there is no fresh water other than that distilled from sea water, and where earthquakes are prevalent, the problem of building a power plant of 50,000 kw. capacity, to operate a copper mine and refinery ninety miles inland from the coast, has been successfully solved by the Chile Exploration Co. and the contractors for the original plant.

At the regular monthly meeting of the New York Section of the American Institute of Mining Engineers, New York, Jan. 7, 1920, Percy H. Thomas addressed the meeting on "Steam and Hydro-Electric Power Problems" in Chile. This article is based on that address, and describes the power equipment of the Chile Exploration Co.'s plant at Tocopilla, Chile.—Editor.*

The design work was started in 1913, and the first machine was put in operation in 1915. After considering the various factors to be contended with, it was decided to use steam as the primary source of power and locate the power plant on the coast at the small town of Tocopilla, about 100 miles north of the Tropic of Capricorn. Hydro-electric power was considered, but was abandoned for the time on account of the problems involved in development and the uncertainty of the waterflow. Building an oil-engine plant at the mine was also considered, but this was abandoned because of the great number of machines that would be required in a plant to supply 50,000 kw., the high altitude at which the engines would be operating (barometer at the mine only 70 per cent of that at sea level), and the difficulty of transporting the oil. When the decision of a steam plant was made, the choice of oil or coal for fuel was carefully gone into, and it was decided to use oil. One of the chief factors that influenced this decision was the ease with which the oil could be unloaded from the boats. The boats are anchored off shore and a pipe line laid out to them on floats, and the oil pumped to the storage tank of the plant. On account of there being no harbor to offer protection from the ocean waves, it would be very difficult to unload coal from boats at a dock. Furthermore, if coal were used it would have to be brought from the United States, England or Australia, since Chilean coal is mined only in small quantities.

Oil for fuel is stored in three 55,000 gal. tanks built of steel. Concrete was considered for the construction of the tanks, but was found to be too expensive and not as satisfactory as steel, which is less liable to rupture from earthquake disturbances.

At Tocopilla, where the power plant is located, the country is practically all solid rock, rising sharply out of the sea, and it was in this rock that a foundation for the plant was blasted. The power-plant building has an extra-heavy steel framework, rigidly braced to withstand the effects of earthquakes, with reinforced-concrete slabs bolted to the steel.

The boiler plant contains sixteen 600-hp. and two 1,350-hp. marine-type B. & W. boilers arranged along three aisles. Four of the 600-hp. boilers are on each side of two of the aisles and the two 1,350-hp. boilers on one side of the third aisle. Each boiler is equipped with superheaters operating at a pressure of 177 lb. and superheating the steam to a total temperature of 600 deg. F. Steel-tube economizers are built in as part of the boiler, which heats the feed water from 130 to 180 deg. F. to the temperature at which it enters the boilers. These economizers have given satisfactory operation, no serious trouble having been experienced from cor-

*Consulting engineer for the Chile Exploration Co., 120 Broadway, New York City.

rosion. This is due largely to the purity of the feed water and to its temperature when passing through the tubes. The temperature is of a value that prevents the tubes from sweating and causing external corrosion. Provisions are being made to increase the temperature of the feed water entering the economizers, as it is felt that the temperature, with the present arrangement, at times gets down near the point where sweating of the tubes might be caused.

The oil-storage tanks are located higher than the fuel pumps, so that the oil flows to the pumps by gravity, where it is pumped to the burners at a pressure of 165 lb. per sq.in., and is mechanically atomized, therefore no steam is required in burning it, which makes this type of burner very economical for this locality, where every pound of steam blown to the atmosphere means a pound of fresh water that must be distilled from sea water by the burning of expensive fuel. When the oil burners were first installed, some difficulty was experienced with the flame destroying the furnace walls, but by a proper arrangement of the burners in the furnace this trouble has been entirely overcome. There is no fresh water at Tocopilla other than that distilled from sea water, consequently all feed water for the boilers must be distilled. Evaporators are used similar in type to those used on ships. Salt water is pumped into a large closed tank, which contains coils of pipe connected to the boilers. Steam at 177 lb. and 600 deg. F. total temperature is supplied to the coils, which in turn evaporate the sea water in the tank, the evaporation passing off to the feed-water heater, where it is condensed by coming in contact with the condensate from the condensers on the main turbines. This type of evaporater is very economical, since, if it is properly insulated, practically no heat is lost, all the heat required to evaporate the water being passed back into the boilers in the feed water. An important factor in the use of an evaporator to obtain the feed water is that no scale is formed in the boilers, therefore the necessity of cleaning them is practically eliminated. Where impure feed water is used, there is always danger of burning a tube where a patch of scale has formed. With distilled water, on the contrary, there is no formation of scale, consequently the boilers can be forced without danger of burning the tubes.

FIG. 2. BREAKWATER PREVENTS SEAWEED FROM GETTING INTO INTAKE TUNNEL

The gases from the furnace make three passes through the boiler and three in the economizer. The boiler and economizer, being in series, the gases therefore make a total of six passes from the furnace to the stack. Natural, forced and induced draft are used. The height of the stacks had to be limited to about eighty feet to make them earthquake-proof as far as possible, and they are built of steel. Their general arrangement is clearly indicated in the headpiece. Each boiler has its own stack, and induced draft is obtained by blowing cold air up the stack, thus creating an ejector effect that draws up the hot gases. Although this scheme obviates the troubles encountered when the hot gases are taken through the fan, it has not been entirely satisfactory, and a change to a standard induced fan equipment is under advisement at this time. With the present draft 5,000-volt turbo-alternators operated at from 150 to 175 per cent of their normal rating. Part of the auxiliaries in the boiler room are steam-driven and part are electrically driven; the steam-driven units exhaust into the early stages of the turbines.

FIG. 1. INTERIOR VIEW OF TURBINE ROOM

The power plant contains five 10,000-kva., 50-cycle, three-phase, 5,000-volt turbo-alternators operating at 1,500 r.p.m., four of which are shown in Fig. 1. Excitation for the alternators is obtained from 220-volt direct-connected exciters. The turbines have twelve stages, take steam from the boiler at 177 lb. pressure and 600 deg. F. total temperature and exhaust into three-pass condensers having 20,000 sq.ft. of cooling surface. With cooling water at from 57 to 67 deg. F. a vacuum of from 28.5 to 29 in. is maintained. The circulating pump, air pump and hotwell pump for each unit are all driven by the same turbine at 1,500 r.p.m., the turbine exhausting into the low-pressure stages of the main turbine.

One of the most serious problems to contend with in the operation of the plant was the large quantities of seaweed washed into the circulating-water intake tunnels. There being no harbor or breakwater, it is not hard to imagine the large quantities of this material that would be broken loose by the waves washed in on shore and drawn into the intake tunnel by the flow of the water. In the early period of the plant's operation, before suitable means had been devised for preventing this material from getting into the intake tunnel, 600 wheelbarrow loads of seaweed were taken out of the intake racks in one night. This situation has been successfully

FIG. 3. MUMMIFIED MULE GUIDE POST

met by building a short breakwater, Fig. 2, out from shore to deflect the seaweed away from the intake tunnel and by the use of double racks and revolving screens in the intake.

The switching and bus arrangement for the generators is installed according to standard practice; namely; a standard-type double bus with remotely controlled oil switches to connect the generators to either busbar. Power for operating the oil switches is supplied from a storage battery. From the generator busses the 5,000-volt current passes through another set of oil switches to 10,000-kva., oil-insulated, water-cooled, three-phase star-to-star connected transformers, where it is stepped up to 110,000 volts and transmitted ninety miles across the desert to the copper mines and refinery of the Chile Copper Co. at Chiquicamata. There are five of these transformers, one for each generator, which may all be grouped in parallel, or any one of the transformers can be connected to any one of the generators. Each transformer is equipped with an oil conservator—a small auxiliary tank, having about 8 per cent of the capacity of the transformer tank, mounted above the main tank and connected to it through a single-pipe connection. Under normal operating conditions the conservator is maintained about half full of oil; thus at all times the main tank is kept full of oil, and air is prevented from coming in contact with the warm oil and causing sludging.

Instead of the water-cooling coils being located in the top of the transformer tank, they are placed in a separate tank of cooling water and connected with the top and bottom of the transformer tank through a pump. The oil is pumped from the top of the tank through the coil in the cooling tank containing sea water, and back into the bottom of the transformer tank. This eliminates the problem of keeping the cooling coils clean when sea water is pumped through them, and also the danger of this water getting into the oil. To prevent electrolytic action between the iron cooling coils and the sea water, a large piece of zinc is immersed in the water and connected to the cooling pipes above the water line, thus forming a battery with the zinc in the water electro-positive to the iron cooling pipe. This scheme has proved very successful in preserving the cooling coils, and only requires renewing of the zinc electrodes at such intervals as they may become decomposed.

For the first ten miles the transmission line is in duplicate, but for the remaining eighty miles a single line is used. Since this line was placed in operation, in 1915, the service has been interrupted but once due to line trouble. For approximately eight or ten miles from the coast a salt mist from the sea is carried inland and falls on the insulators, where it evaporates and leaves a salt deposit, which eventually would cause flashovers. This condition was taken care of by building the line in duplicate, and while one line is in service the insulators on the other are washed. The insulators are washed about every two weeks, which has eliminated all trouble on this section of the system. On the remaining eighty miles the conditions are very favorable for continuous operation, the insulators are of good quality, the climate is very dry, and since there are no large birds, or thunder storms, or individuals walking around with rifles, there is practically nothing to disturb the operation of the line. Men are kept going over the line all the time, testing insulators, and in this way defective insulators are detected and removed before they cause trouble. The difference in the climate inland from that near the ocean

FIG. 4. INTERIOR VIEW OF SUBSTATION

is shown by the fact that insulators that are of good quality when used inland will deteriorate rapidly if taken down to the coast. When these insulators are taken inland, again, they will regain their original high insulating qualities, showing that the deterioration is due to moisture. The climate is so dry inland that dead animals do not decay but dry up, as indicated in Fig. 3, which shows a mummified mule set up in the desert for a guide post.

At the mine substation the 100,000-volt current is stepped down through three-phase transformers to 5,000 volts for distribution about the mine, smelter and refinery. In the substation, the interior of which is shown in Fig. 4, there are seven motor-generator sets, each consisting of a 5,000-volt, 4,000-hp. motor directly connected to two 1,325-kw., 220-volt direct-current generators. Each generator delivers a normal full-load current of 5,300 amperes, or a total of 10,600 amperes per set. Four of the motors driving these sets are of the induction type and three of the synchronous type. There are also two 1,000-kw. 5,000-volt synchronous-motor 220-volt direct-current generator sets and four 3,500-kw. booster-type rotary converters. The latter are capable of delivering a normal full-load current of 12,000 amperes. When the converters were installed, it was not known just what the voltage requirements would be in the tankhouse, so in order to meet any conditions that might arise they were designed for a direct-current voltage range of 120 to 290 volts. The lower ranges of voltage are obtained by two-thirds voltage taps on the transformer secondaries. Direct-current voltages from 120 to 208 are obtained when the slip rings of the converter are connected to the two-thirds taps on the transformers, and 202 to 290 volts is obtained on the commutator with full secondary volts on the slip rings. Each group of two converters are supplied with alternating current through the same three-phase transformers, which steps the 100,000-volt current down to 180 volts on the secondary with a tap brought out to give two-thirds of 180, or 120 volts. The transformers have two secondaries, one for each rotary, and switches are arranged so that both converters may be operated from the transformers at the same time or either one may be operated separately.

An interesting feature of the power equipment at the mine are the electrically operated shovels. These were among the first electrically operated shovels in use, and many of the features of design are original. These shovels, of which there are four 100-ton and one 200-ton, do all that could be expected of steam shovels and do it as satisfactorily. There are three motors on each shovel—a hoist motor, a swing motor and a thrust motor. The motors are of the wound-rotor alternating-current type and are supplied with three-phase power from transformers located on the shovels, which step the 5,000-volt current down to 500 volts. The control is arranged so that the operator manipulates it as he would that of a steam shovel. To prevent the operator from abusing the electric equipment and the shovel from overloads, current-limiting relays are connected in circuit, which, when the current to the shovel motors exceeds a predetermined value, operate and cut a resistance in circuit that limits the power to a safe value.

Most of the equipment for the first 40,000-kw. installation in this project was manufactured in Europe. The last addition made to the plant, which includes two 1,350-hp. boilers, a 10,000-kva. turbo-alternator, one 10,000-, one 7,000- and two 3,500-kva. transformers and the four 3,500-kw. rotary converters are American-made equipment as are also all the electrically operated shovels.

Another problem of considerable moment at the mine was that of a water supply, both for industrial purposes and for drinking. Since about 2,500,000 gal. per day is required for industrial purposes, besides drinking water for 10,000 inhabitants, in a land where it seldom rains, the supply must be not only abundant, but also reliable. The industrial-water supply was obtained by running a 12-in. pipe line back into the mountains forty miles, and the drinking water is brought through a pipe line sixty miles long.

Plans are under way to make a number of refinements in this installation to meet the conditions found down in that locality. When these are made and test data are available, *Power* expects to be able to present to its readers a story on the operation of this installation.

Air-Compressor Lubrication

By W. F. OSBORNE

This article discusses the peculiar problems that arise in the lubrication of the cylinder of an air compressor. The common practice of requiring a high flash-test oil for such work is shown to be wrong. The physical characteristics of air-compressor oil as worked out by members of the Compressed Air Society are given in tables.

THE air cylinder of a compressor is the part given the greatest attention with respect to lubrication. The steam cylinder and bearings are lubricated just like those of any steam engine, and the recommendations made in previous articles on the steam engine apply here also. Vertical single-stage air compressors that are completely inclosed and depend upon a bath of oil in the crank case for the lubrication of all bearings and cylinders, of course, require an oil that is suitable also for the cylinder lubrication.

An air cylinder is very simple to lubricate as far as the reduction of friction between the piston, piston rings and cylinder walls is concerned. The oil is fed to the cylinder through vacuum cups or by mechanical lubricators and is usually introduced directly into the top side of the cylinder or into the air-intake pipe just above the admission valves. It is spread over the cylinder walls by the sweep of the piston, preventing wear and loss of power. The oil film also improves the seal between the piston and cylinder, thereby increasing the efficiency of the compressor. It is easy to select lubricants that will do all of these things efficiently and economically for a short time, but when the compressor has run for a few days, a number of things begin to happen if the oil does not have certain qualities necessary to prevent them.

Air drawn into a cylinder and compressed, rises in temperature. Table I shows the temperatures of the discharged air at various pressures. Actual temperatures are somewhat lower than these owing to the cooling effect of the water jackets or cooling ribs around the cylinders. In any event, however, the temperature is high enough to cause some evaporation of the oil film. Many engineers have believed for a long time that if the flash point of the lubricating oil were not higher than the temperature of the air, a fire or explosion would take place, and on this account it has been customary in some localities to insist upon the oil having the highest flash point obtainable.

Now let us see what this means. In the first place a mixture of oil vapor and air will not ignite of itself under about 1,000 to 1,100 deg. F. A flame, incandescent point of carbon, electric spark or red-hot metal is necessary to cause the flash. Unless these exist in the cylinder, on the discharge valves or in the discharge pipe, there can be no flash. Further, the increased pressure in the cylinder raises the flash point of the oil, or in other words, it reduces the amount of vaporization. Consequently, there can be no danger of fire or explosion even when comparatively low-flash oils are used, unless something else is present to fire the vapor.

Oil Remaining in Cylinder Is Decomposed

When an oil film has once been formed on the cylinder walls, the only way for the oil to get out of the cylinder is for it to evaporate and be carried out with the air. As an ordinary two-stage compressor, compressing air up to 100 lb. pressure, has a normal discharge temperature of less than 245 deg. F., this oil film is undoubtedly going to last a long time before it is completely evaporated. This is borne out by everyday practice, which shows that very small quantities of oil are required to give proper lubrication. Naturally, when the oil is exposed to this heat for a considerable length of time, it begins to break down or decompose, forming carbon. This is particularly true when oil is fed faster than it can be evaporated, causing accumulations at the end of the cylinder and on the discharge valves.

If this partly decomposed oil collects on the discharge valves which are subjected to the highest temperatures, further decomposition or carbonization takes place and deposits of carbon form. Frequently, these deposits get large enough to cause the valves to stick. When this happens, the hot discharged air leaks back into the cylinder, where it is again compressed, still further raising the temperature. The higher temperature causes greater carbonization and more trouble with the valves. This may finally reach a stage where the carbon deposits are so large and the air so hot that small particles of the carbon actually become incandescent. Then there is a possibility that a fire or explosion of ignited oil vapor will occur. Even if a fire does not develop, the carbon deposits on the discharge valves greatly decrease the output and the efficiency of the compressor.

As we have just explained, the decomposition of the oil and the resulting carbon deposits are caused by the oil remaining in contact with the hot air too long. If it had been completely evaporated soon enough to prevent decomposition and carried out of the cylinder with the air, there would have been no deposits and no trouble.

Less Danger from Low Flash-Test Oil

Now let us see whether we shall still insist upon a high flash-test oil. The flash test shows that it requires a certain temperature to evaporate the oil. This means that a low flash-test oil will evaporate more quickly at a lower temperature and that the high flash-test oil will remain in the cylinder longer. This gives it a greater chance to decompose. If we use a low flash-test oil that will evaporate quickly without decomposition, it will be carried out with the air before it has a chance to decompose and form carbon deposits. With the carbon deposits eliminated, there can be nothing to ignite the oil vapor and no fires or explosions can take place. At the same time most of the troubles due to leaky discharge valves have been removed.

Theoretically, the low flash-test oils might perhaps require a slightly greater amount to maintain the film on the cylinder walls, owing to their greater evaporation, but in actual practice it is not possible to regulate the feed fine enough to show any difference in consumption.

Investigation of explosions occurring in the discharge lines seems to point to the fact that they are caused by oil pools forming in depressions in the pipes and becoming heated by the discharge air. The long exposure to heat causes the oil to decompose, forming carbon deposits. These deposits may build up to such an extent that they obstruct the air passage, and the friction of the air through the greatly reduced space causes them to become red hot. The pipe in turn is heated, its strength reduced and the pressure of the air within the pipe causes it to give way. Excessive amounts of oil fed to the air cylinder are responsible for most of this trouble, and it should always be reduced to a safe minimum. All receivers and discharge lines should be provided with suitable drains to draw off any oil collections.

Tests to Determine Causes of Carbon Deposits

In order to check up definitely the causes of carbon deposits in air cylinders, a prominent builder of air compressors arranged for a series of tests to find out the effect oils of different characteristics have when compared under identical operating conditions. The compressor selected for this test was a vertical single-stage machine delivering air at 110 lb. per sq.in. pressure with a temperature of about 425-450 deg. F. These temperatures are considerably higher than those ordinarily met with and for which previously it had been considered necessary to use a high flash-test oil of heavy body. Cylinder and bearings were lubricated by the splash method from the crank case. In each instance the test was continued for one hundred hours and the valves examined at the end of this time.

The first oil tested was a heavy-bodied straight-run oil having a flash test about 70 deg. F. lower than the temperature of the discharged air. On examination at the end of the run the valves were found to be exceptionally clean, there being no carbon deposits.

The second oil tried out was heavy-bodied but slightly thinner than the first, by blending a light-viscosity oil with a heavy filtered, dark-green cylinder oil to raise its viscosity and to obtain a high flash test. This oil was similar in characteristics to the usual high flash-test oil recommended for air cylinders. At the end of the run the discharge valves were so badly carbonized that if the test had been continued much longer they would have begun to leak.

The third test was made with medium-bodied straight-run oil having a lower flash than either of the two samples previously tested and at the end of the run showed only a very slight carbon formation.

In all three tests the cylinder walls showed perfect lubrication. An analysis of these three runs shows that in the first test the oil evidently evaporated cleanly without decomposition and was carried out of the cylinder with the air. When the high-flash blended oil was used, the light-bodied portion of it apparently evaporated all right, but left the heavy portion on the discharge valves, where it could not completely evaporate fast enough because of its high flash, so it gradually carbonized.

The third oil probably evaporated cleanly, but because it was too thin in body, excessive amounts worked past the piston rings faster than it could be carried out of the cylinder, with the result that some of it stayed in the cylinder long enough to become partly decomposed. This is good evidence that the amount of oil fed to the cylinder should always be reduced to a minimum.

From these tests the conclusion might be reached that the lowest flash-test oils are the best for air-cylinder lubrication. It should be remembered, however, that it is quite possible for a heavy-bodied high-flash oil to be blended with an extremely low flash oil, thus giving a low flash test to the finished product, but when the light part of the oil has evaporated, the heavy portion will remain and form carbon. Consequently, the statement that, in general, the low flash-point oils are more suitable for air cylinders applies only to properly refined straight-run oils.

TABLE I. CYLINDER TEMPERATURES AT THE END OF THE PISTON STROKE

Final Pressure, Lb. per Sq.In. Gage	Final Temperature, Deg. F. Single-Stage	Two-Stage
10	145	...
20	207	...
30	255	...
40	302	...
50	339	188
60	375	203
70	405	214
80	432	224
90	459	234
100	485	243
110	507	250
120	529	257
130	550	265
140	570	272
150	589	279
200	672	309
250	749	331

Of the different viscosity oils which can be made from the same crude the thinnest will have the lowest flash point. Following out the previous statements, we might now reach the conclusion that the thinnest oils are best for air-compressor cylinder lubrication. This is quite correct, provided they are heavy enough in body to keep the piston and piston rings from wearing out through friction. They must also provide a sufficient piston seal to prevent excessive leakage of air and, in the case of splash-lubricated cylinders, to prevent excessive amounts working into the cylinder. This was the trouble in the third test; the oil was so thin that it leaked past the piston rings faster than it could be carried away with the air.

TABLE II. PHYSICAL TESTS OF PARAFFIN-BASE OILS

	Minimum	Average	Maximum
Gravity, Baumé	28–32 deg	25–30 deg.	25–27 deg.
Flash point, open cup	375–400 deg. F.	400-425 deg. F.	425-500 deg F.
Fire	425-450 deg. F.	450-475 deg. F.	475-575 deg. F.
Viscosity (Saybolt) at 100 deg. F	120-180 sec.	230-315 sec.	up to 1500 sec.
Color	Yellowish	Reddish	Dark red to green
Congealing point (pour test)	20-25 deg. F.	30 deg. F.	35-45 deg. F.

Another reason for using light-bodied oils is the fact that there is less possibility of their holding dust and foreign matter carried into the cylinders by the air. Heavy oils, being more adhesive, naturally hold this dirt more closely and eventually form deposits composed of a little oil, some carbon and a lot of foreign matter.

An oil suitable for air compressors should always be highly filtered because such an oil will have a minimum tendency to form carbon deposits, as shown by the carbon-residue test. The light-bodied oils by this test also show less carbon residue than the heavier-bodied ones, when made from the same crudes. The carbon-residue test is therefore of much assistance in comparing oils for air-cylinder lubrication.

Needless to say, oils compounded with vegetable or animal oils are quite unsuitable for air cylinders because of their great tendency to decompose under heat with the formation of carbon deposits.

The pour test of the oil need receive attention only when the lubricators are exposed to temperatures below 35 deg. F. For such installations the oil should have a pour test low enough to insure a satisfactory feed through the lubricators or free flowing of the oil when used in a splash system.

TABLE III. PHYSICAL TESTS OF ASPHALTIC BASE OILS

	Minimum	Average	Maximum
Gravity, Baumé	20-22 deg.	19 8-21 deg.	19.5-20 5 deg.
Flash point, open cup	305-325 deg. F.	315-335 deg. F.	330-375 deg. F.
Fire	360-380 deg. F.	370-400 deg. F.	385-400 deg. F.
Viscosity (Saybolt) at 100 deg. F	175-225 sec.	275-325 sec.	475-750 sec.
Color	Pale yellow	Pale yellow	Pale yellow
Congealing point (pour test)	0 deg. F.	0 deg. F.	0 deg. F

A final specification for good air-compressor oils might be written somewhat as follows:

Use only a straight-run, highly filtered, pure mineral oil, of the lowest natural flash test consistent with the viscosity necessary to lubricate the cylinders and provide a proper piston seal.

The correct viscosities, as worked out by the members of The Compressed Air Society, all of whom are builders of air compressors, are given in Tables II and III, showing the proper specifications for both paraffin and asphaltic oils of various viscosities.

Hydro-Electric Plant Hints

By L. W. Wyss

The hydro-electric plant operator keeps in touch with water conditions by keeping a record of the head- and tail-water elevations as well as the amount of water used and wasted.

The head water is obtained from a pressure gage, float gage or water column; the tail water is obtained with a float gage or measuring stick.

The water used can be obtained from kilowatt-hour output discharge curves and the water wasted is obtained from curves which show the rate of flow for different gate openings and head-water elevations. The most common unit for measuring water is the cubic foot per second.

Not many years ago 80 per cent efficiency was considered excellent, but now there are numerous turbines with efficiencies of 89 to 92 per cent, and the maximum obtained so far is 93.7 per cent.

At some plants, during low-water periods, trouble is experienced with turbines losing their vacuum in the draft tubes. This may cut down their efficiency and output over half in the case of low-head plants, beside causing poor speed regulation. When turbines lose their vacuum the governors, of course, open the gates, attempting to keep the speed normal. In order to carry the same load without vacuum in the draft tubes more water is used. This water may raise the tail water and the vacuum may soon be regained in some cases. However, it may be advisable to obtain more load, start idle machines or waste water at the spillway to raise the tail water, thus regaining the vacuum.

Automatic Control for Motors Driving Pumps and Compressors

By B. W. JONES
Engineer, Industrial Controller Department, General Electric Company

The characteristics of the motors used to drive pumps, compressors and fans are discussed, then types of controllers used to start and stop these motors automatically are considered.

DIRECT-CURRENT and alternating-current motors are both employed for driving pumps and compressors. Of the direct-current motors the constant-speed and the adjustable-speed shunt types are most generally used; and of the alternating-current motors the squirrel-cage and slip-ring induction types and the synchronous type are most commonly employed. Using these five types of motors, the table shows that there result eighteen variations with some duplicates in the automatic control. Therefore it becomes essential for the purchaser to decide which of these conditions he will have, and then pass this information on to the manufacturer of the controller. It is the object of this article to show wherein each of these eighteen conditions differ and also to give some details of the operation of the several types of automatic control.

Reciprocating pumps when loaded require constant torque, independent of the speed, which means that if the motor is started up against full accumulator pressure, more than full-load torque will be required to accelerate the motor. It is, therefore, desirable to have some form of unloader that will remove the pump load during the accelerating period.

Centrifugal pumps, when loaded against a constant head, require a driving torque that varies approximately as the square of the speed. This application does not require an unloader during the starting period so much as the reciprocating pumps, but it is very desirable as will be indicated in the table.

The characteristics of the squirrel-cage and wound-rotor induction motors have been well brought out in Mr. Fraser Jeffrey's article in the Aug. 26, 1919, issue of *Power*. The synchronous motor has a starting characteristic very similar to that of the squirrel-cage induction motor, but the necessary control is somewhat more elaborate and requires a little more care during the starting period.

FIG. 1. CONTROLLER USED FOR STARTING CONSTANT-SPEED DIRECT-CURRENT MOTORS

The direct-current constant-speed shunt motor has a torque directly proportional to the armature current. The adjustable-speed shunt motor has a torque directly proportional to the armature current for any one setting of the field current, but when the field current is set for a low-speed condition, there will be more torque for a given armature current than there will be with the same armature current and a field current set for a higher-speed condition. This characteristic requires less current than a corresponding horsepower motor of the constant-speed type during the accelerating period, provided full-field current is maintained during the accelerating period in both cases.

A constant-speed motor is generally used where the motor is run at full speed for a certain period and then stopped by some means, such as an automatic starter. This applies to both reciprocating and to centrifugal pumps.

An adjustable-speed shunt motor is generally used where the motor is kept running and the rate of pumping is varied by changing the speed of the motor with field control by some similar means to that shown in Fig. 2. If the motor speed has a range of approximately two to one by field control, it is not so essential to have an unloader on either a centrifugal- or reciprocating-pump load, because if the motor is started with full field, the armature has twice the torque per ampere which is the equivalent of starting a constant-speed motor on half-load.

The same statements that were made regarding constant-speed shunt motors also apply to slip-ring induction motors with the exception that the number of starting points and resultant current peaks are slightly different as indicated in the table. A slip-ring induction motor should be used where a reciprocating pump is to be started against full load; that is, where no unloader is used. A squirrel-cage induction motor can be used to good advantage for starting centrifugal pumps and fans with or without an unloader or for reciprocating pumps that have unloaders. Because of the advantages of power-factor correction and efficiency which can be gained by the use of synchronous motors, these machines are often used for driving reciprocating and centrifugal pumps and fans equipped with unloaders.

Both manual and automatic control are used for starting and stopping the various kinds of motors previously mentioned. This article will deal only with the automatic types which are most generally used for this kind of service. The table gives in concise form the number of resistor divisions that are generally used for starting and the resultant accelerating-current peaks on the several types of motors when driving the various kinds of pump and fan loads. This table can be used in conjunction with the tables given in the article, "Methods for Determining Resistance Used for Starting Various Types of Motors" in the Nov. 19, 1918, issue of *Power*, which deals with the method of determining the ohmic value per division of the resistor.

Figs. 1 and 5 show the photograph and connections respectively of a starter for a direct-current constant-speed motor. This starter has one line contactor *L* for connecting the motor to and disconnecting it from the line, and two accelerating contactors *A* and *A'* for short-circuiting the two divisions of the resistor. A starter with this number of points can be applied to an equipment with or without an unloader. If it is applied without an unloader, the largest motor it should be used with, as shown in the table, is 10 hp. for a reciprocating or 25 hp. for a centrifugal pump; but if an unloader is used it can be applied to a 45-hp. motor. Detailed descriptions of the operation of this and subsequent panels were given in issues of *Power* during 1916 and 1917, but a further description of the operation of this panel will be included here.

Contactor *L* is a line switch and is opened and closed by means of its shunt coil. When the control switch is closed contactor *L* will close, and when the control switch is opened contactor *L* will open. When contactor *L* first closes, current will flow from L_1 through R_1-R_2, resistance, the coil of contactor *A*, through *R*2-*R*1 resistance, the interpole winding and armature of the motor and out to the L_2 line. The coil on contactor *A* is a series coil which will close this contactor provided the current is within certain specified limits. If the motor is standing still, the initial current from the lines will be above this limit and contactor *A* will lock out, until the motor has partly accelerated and the current decreased to the specified value. When this contactor closes, it will short-circuit *R*2-*R*1 resistance and therefore the current will flow through the tips of the contactor and through the lower coil on contactor *A'* instead of through R_2-R_1. When R_2-R_1 section of the resistor is shorted, the current in the line increases and it is this increased value that flows through the lower coil of contactor *A'*, but this contactor, like *A*, will not close when the current is too high, but will lock out until the current again decreases to the right value. When contactor *A'* closes, it will short-circuit all the starting resistor and both the series coils of the two contactors, but a shunt holding coil, *H*, will have its circuit closed through the normally open interlock *T* on contactor *A'*, and this will be sufficient to hold contactor *A'* closed. The action of these three contactors will bring the motor from rest to full speed and at all

FIG. 2. ADJUSTABLE-SPEED DIRECT-CURRENT MOTOR CONTROLLER

FIG. 3. SLIP-RING MOTOR STARTER

FIG. 4. SQUIRREL-CAGE MOTOR STARTING COMPENSATOR

times keep the line current within prescribed limits as given in the table.

Figs. 2 and 6 show the panel and connections of an adjustable-speed motor starter which is controlled by means of a pressure regulator *R*. The dial switch *D* closes the line contactor when in the starting position and then gradually short-circuits the starting resistor. During this period full field is maintained on the motor, but after starting has been completed, the dial can insert resistance in the motor field and thereby increase its speed. The position of this dial is governed by means of the pressure-regulator piston, and it in turn is governed by means of a diaphragm which responds to pressure variation. This pressure (liquid or gas pressure) is generally a function of the motor speed. If the pressure on the diaphragm tends to decrease, the regulator will increase the motor speed by inserting resistance in the field circuit. If the pressure tends to increase, the regulator will decrease the motor speed by short-circuiting some of the field resistance. In this way the regulator will maintain a constant pressure.

Fig. 4 shows an open view of an automatic compensator which is used for starting squirrel-cage induction motors. Figs. 8, 9 and 10 show the connections respectively of this panel when used on a three-phase three-wire, a two-phase three-wire, and a two-phase four-wire circuit. As the photograph and connections show, the panel consists of: One five-pole contactor *S* for starting the motor; 1 three-pole contactor *R* for running the motor; 1 current-limit relay *E* for governing the two contactors *S* and *R*; 1 control-circuit fuse *F*; 2 inverse-time-limit overload relays *C* and *C*; and 1 three-coil compensator *B*. The operation of the panel is as follows:

When the "start" push-button is held down, current will flow from L_1 through the fuse *F*, the two overload relay tips, the "stop" push-button, the coil of contactor *S*, the upper tip of the current-limit relay *E* to L_2, as indicated by the arrowheads. This will close contactor *S*, which in turn will short-circuit the contacts of the "start" push-button, by means of an electrical interlock D, which is mounted on the shaft of the contactor

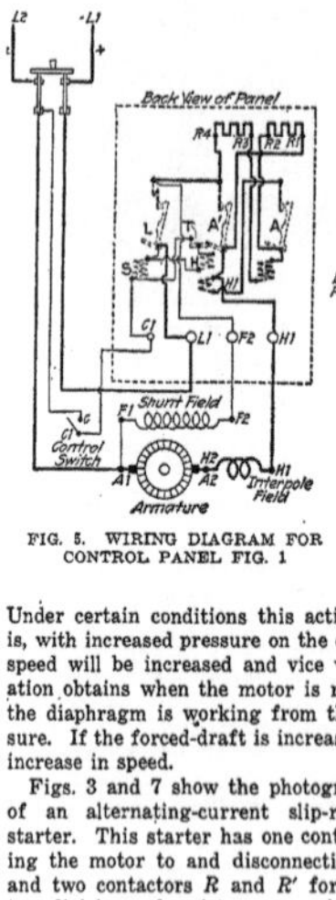

FIG. 5. WIRING DIAGRAM FOR CONTROL PANEL FIG. 1

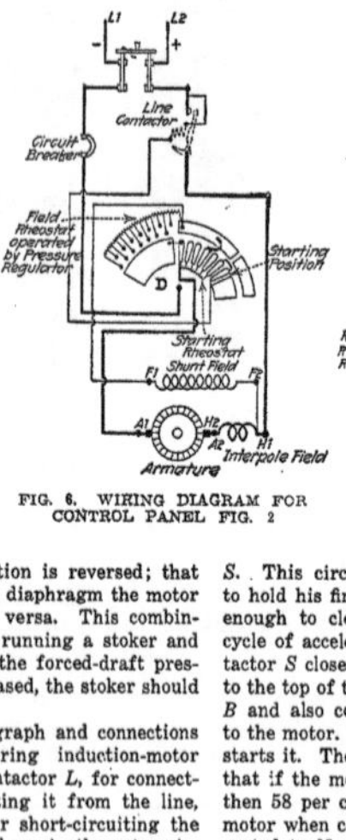

FIG. 6. WIRING DIAGRAM FOR CONTROL PANEL FIG. 2

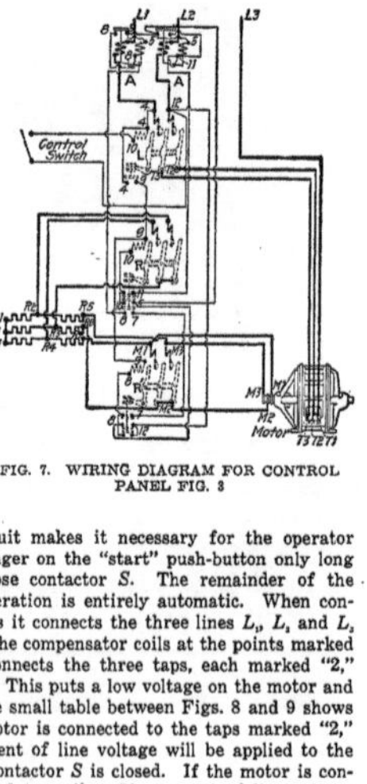

FIG. 7. WIRING DIAGRAM FOR CONTROL PANEL FIG. 3

Under certain conditions this action is reversed; that is, with increased pressure on the diaphragm the motor speed will be increased and vice versa. This combination obtains when the motor is running a stoker and the diaphragm is working from the forced-draft pressure. If the forced-draft is increased, the stoker should increase in speed.

Figs. 3 and 7 show the photograph and connections of an alternating-current slip-ring induction-motor starter. This starter has one contactor *L*, for connecting the motor to and disconnecting it from the line, and two contactors *R* and *R'* for short-circuiting the two divisions of resistance per phase in the rotor circuit. The two relays *A* and *A* at the top of the panel govern the closing of the two resistor contactors *R* and *R'*, and can be adjusted to suit the conditions under which the equipment starts. Referring to the table it is seen that this starter can be used to start a 45-hp. motor when driving a reciprocating pump or a centrifugal pump when they have no unloader; or a 75-hp. when an unloader is used.

S. This circuit makes it necessary for the operator to hold his finger on the "start" push-button only long enough to close contactor *S*. The remainder of the cycle of acceleration is entirely automatic. When contactor *S* closes it connects the three lines L_1, L_2 and L_3 to the top of the compensator coils at the points marked *B* and also connects the three taps, each marked "2," to the motor. This puts a low voltage on the motor and starts it. The small table between Figs. 8 and 9 shows that if the motor is connected to the taps marked "2," then 58 per cent of line voltage will be applied to the motor when contactor *S* is closed. If the motor is connected to No. 3 taps, then 70 per cent of line voltage will be put on the motor when contactor *S* closes. The other taps give values as shown.

The left-hand coil of the current-limit relay *E*, Fig. 8, has a small number of turns and is connected in series with the line, while the right-hand coil has a large number of turns and is connected across the compensator taps at points marked 11-12 on switch *S*. Now the plunger in the right-hand coil is heavier than the plunger in the left-hand coil so that the left plunger is held up against the pivoted contacts marked 4 and 6 by means of the walking beam at the bottom. The right-hand plunger is connected to the walking beam, but the left-hand plunger simply rests on the beam. This means that when contactor *S* closes, the two coils of the relay are energized at the same time, the right-hand coil picking up and pulling the walking beam away from the left-hand core, but the current in the left-hand coil holds the core up against the pivoted position all the time the motor is running. When contactor *S* opened it disconnected the compensator from the motor, and when contactor *R* closed it put full voltage on the motor. Before this transfer occurred, the motor had attained approximately 90 per cent speed. To stop the motor, push the stop button and this opens contactor *R* and resets the current-limit relay *E*.

Fig. 9 shows the connection for a two-phase three-wire automatic compensator. The only difference between this and Fig. 8 is that a two-coil instead of a

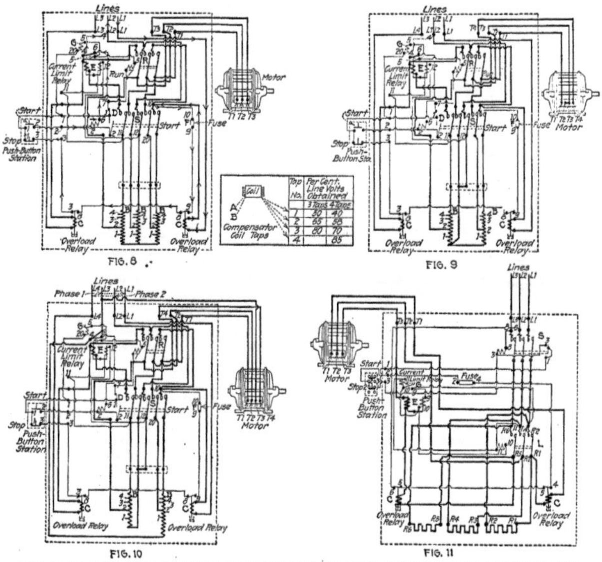

FIGS. 8 TO 11. WIRING DIAGRAMS OF CONTROL PANELS FOR AUTOMATICALLY STARTING SQUIRREL-CAGE TYPE INDUCTION MOTORS

contacts 4 and 6, until the motor sufficiently accelerates and lowers the current through this coil to a value where it cannot hold the plunger against gravity. When this happens the pivoted contact is pulled down, which breaks the control across 4-6 and makes contact across 4-5. This opens contactor *S* and closes the coil on contactor *R* through the electrical interlock 7-1 at *D* on contactor *S*. The small coil *G* on relay *E* is connected in multiple with the coil on contactor *R*. When the relay short-circuits the 4-5 contacts on *E*, coil *G* is energized and maintains this pivoted contact in this three-coil compensator is used. Fig. 10 is the same as Fig. 9 excepting that four instead of three wires are used in the line circuit.

The automatic compensator provides one starting and one running point. If the speed-torque characteristics of a squirrel-cage induction motor are considered in conjunction with an automatic compensator, it will be noted that one starting point is all that is needed, because the torque increases instead of decreases up to a comparatively high percentage of normally rated speed. Quite frequently a multi-point automatic compensator is asked for, but practically in every instance a one-point compensator will give just as good results. If a one-point automatic compensator will not accomplish the results, then the load should not be started by a squirrel-cage induction motor on partial voltage. In place of the automatic compensator some engineers use a resistance starter, as shown in Fig. 11. This throws the motor on the line through some resistance in each phase, and after the motor accelerates to a specified speed the resistance is short-circuited. This

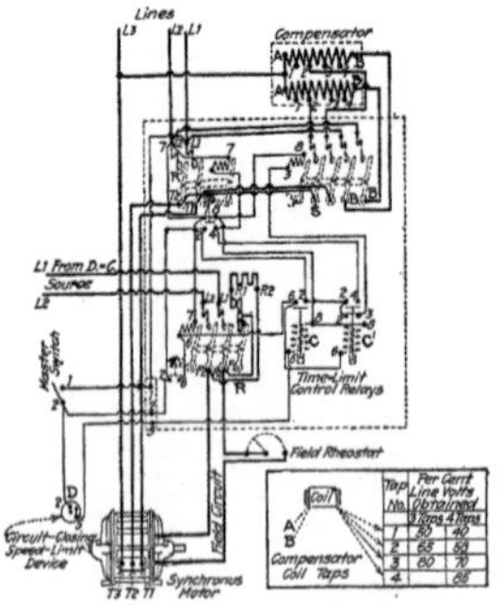

FIG. 12. WIRING DIAGRAM OF CONTROL PANEL FOR STARTING A SYNCHRONOUS MOTOR

arrangement makes a somewhat cheaper starter, but requires more current from the line during starting.

The synchronous motor is not used as generally as the induction motor, but it has an important field. Fig. 12 shows the connections used for starting a low-voltage synchronous motor. It consists of an automatic compensator similar to the one described in the foregoing with the addition of means for handling the field circuit. Instead of the automatic compensator being governed by means of a current-limit relay, it works as follows:

When the master switch is closed, the four-pole contactor S closes and puts partial voltage on the motor. When the motor has attained about 90 per cent speed, the centrifugal device D, which is attached to the motor shaft, closes the contacts 2-5 and starts the time-limit relays C closing. After about five seconds the relay C closes its contacts 2-6, which causes contactor R to close and puts field current on the motor. This should pull the motor into synchronism if it had not already reached this speed. The closing of the first time-limit relay energized a second time-limit relay C', which after about five seconds simultaneously breaks its contacts 2-3 and closes 2-4 and causes contactor S to open and contactor R' to close. This action disconnects the compensator from the line and motor and then connects the motor to the line.

In this short article an endeavor has been made to point out the load characteristics of pumps and fans, the different motors that are generally used on pumps and fans, how control should be applied and a few illustrations of control panels. To the person who is not familiar with control problems, some of the recommendations may appear inconsistent, but they have been verified by actual tests.

The shortage of labor in every industry throughout the country has reached a serious stage and, according to statistics, is rapidly growing worse. This and strikes for shorter hours have caused a material lowering of production, to which can be traced in part the high cost of living. Because of strikes 1,105,500 days have been lost in the textile industry alone, in coal mining 220,000 days were lost, clothing factories show a loss of 352,000 days, in the shoe manufacturing industry 77,000 days were lost and on the railroads the losses totaled 38,000 days. Freight cars are lying on sidings and in yards, for lack of men to make necessary repairs. Sufficient labor cannot be found to clean the streets. Telephone companies throughout the country could put 25,000 more women to work. Factories, mills, mines and in fact industries of every description are searching for help. A new supply of unskilled labor is needed to overcome this condition and would probably prove effective in reducing prevailing high prices.

NUMBER OF RESISTOR DIVISIONS SUGGESTED FOR STARTING MOTORS, DRIVING PUMPS AND COMPRESSORS AND THE APPROXIMATE RESULTING MAXIMUM CURRENT PEAKS

Horsepower of Motor	Direct-Current Motors															
	Reciprocating Pumps								Centrifugal Pumps and Fans							
	Constant-Speed Motors				Adjustable-Speed Motor				Constant-Speed Motors				Adjustable-Speed Motor			
	Without Unloader		With Unloader		Without Unloader		With Unloader		Without Unloader		With Unloader		Without Unloader		With Unloader	
	1		2		3		4		5		6		7		8	
	N	I	N	I	N	I	N	I	N	I	N	I	N	I	N	I
1-3	1	2.9	1	2.0	1	2.0	1	2.0	1	2.3	1	2.0	1	2.0	1	2.0
5-10	2	2.0	1	2.0	1	2.0	1	2.0	2	1.6	1	2.0	1	2.0	1	2.0
15-25	3	1.8	2	1.4	2	1.4	1	2.0	2	1.7	2	1.4	1	2.0	1	2.0
30-45	4	1.6	2	1.6	2	1.6	2	1.6	3	1.5	2	1.6	2	1.6	2	1.6
50-75	4	1.7	3	1.2	3	1.2	2	1.6	3	1.7	3	1.2	2	1.6	2	1.6
80-125	5	1.6	3	1.3	3	1.3	3	1.3	4	1.5	3	1.3	3	1.3	3	1.3
135-275	6	1.6	4	1.1	4	1.1	4	1.1	4	1.6	4	1.1	4	1.1	4	1.1
300-500	7	1.5	4	1.2	4	1.2	4	1.2	4	1.7	4	1.2	4	1.4	4	1.2

Horsepower of Motor	Alternating-Current Motors																			
	Reciprocating Pumps										Centrifugal Pumps and Fans									
	Squirrel-Cage Induction Motors				Slip-Ring Induction Motors				Synchro. Motors		Squirrel-Cage Induction Motors				Slip-Ring Induction Motors				Synchro. Motors	
	Without Unloader		With Unloader		Without Unloader		With Unloader		With Unloader		Without Unloader		With Unloader		Without Unloader		With Unloader		With Unloader	
	9		10		11		12		13		14		15		16		17		18	
	N	I	N	I	N	I	N	I	N	I	N	I	N	I	N	I	N	I	N	I
1-3	Not a good Application		Automatic Compensator	2.5	1	2.8	1	2.0	Automatic Compensator	...	Automatic Compensator	4	Automatic Compensator	2.5	1	2.3	1	2.0	Automatic Compensator	...
5-10				2.5	1	3.0	1	2.0		...		4		2.5	1	2.3	1	2.0		...
15-25				2.5	2	2.0	1	2.2		...		4		2.5	2	1.6	1	2.2		...
30-45				2.0	2	2.2	2	1.5		...		3		2.0	2	1.9	2	1.5		...
50-75				2.0	3	1.9	2	1.6		...		3		2.0	3	1.8	2	1.6		...
80-125				2.0	4	1.8	3	1.2		...		3		2.0	3	1.7	3	1.2		...
135-275				2.0	5	1.7	3	1.3		...		3		[illegible]	4	1.6	3	1.3		...
300-500				2.0	6	1.6	4	1.2		...		3		2.0	4	1.7	4	1.2		...

NOTE—N equals number of resistor divisions suggested for starting the motor; I, when multiplied by the rated motor current, equals the maximum current peaks, resulting from the use of the suggested value of N.

Measuring Flow of Fluids

By JACOB M. SPITZGLASS
Vice President, Republic Flow Meters Company, Chicago

Relation between velocity and pressure. Obtaining differential pressure by pitot tube, venturi tube, nozzle tube or an orifice plate. A general formula applicable to all the devices mentioned and tables making it simple to obtain quantity flow from the differential pressure readings of a gage or manometer.

WHENEVER a moving body is stopped from its motion or brought to rest, the kinetic energy contained in that body, due to its velocity, is converted into another form of energy, which, in the case of fluids, results in pressure. This pressure, created by the force of stopping the motion of fluid, is called the "velocity pressure," which, when added to the initial or "static pressure," due to the physical condition of the fluid, forms the final, or "dynamic pressure."

The "dynamic pressure" is obtained when a tube is inserted at any fixed point of a stream with its open end pointing against the flow and the other end is balanced by some stationary force, such as a column of water or mercury.

By inserting a similar tube near the first one, but with its open end pointing at right angles to the flow, the pressure created in it will be equal to the static pressure only, if the velocity of the fluid does not produce any effect upon the opening of the tube.

Connecting the static tube to the upper end of the dynamic column virtually subtracts the static pressure from the dynamic and thus gives directly the differential or velocity pressure equivalent to the flow.

In other cases a differential pressure is created by the introduction of a contracted area in the stream of the fluid for the purpose of converting a certain amount of potential energy existing in the form of pressure into an equivalent amount of kinetic energy resulting from the increased velocity of the fluid.

In all such cases there exists a definite relation between the differential pressure thus obtained and the velocity of the fluid that is to be measured. The flow of any fluid can be measured conveniently and accurately by converting its velocity, which cannot be measured directly, into pressure, which can be easily measured by a column of liquid or a gage.

For practical purposes the differential or velocity pressure in a given line may be obtained, either directly by balancing the difference between the dynamic and static sides of a pitot tube inserted in the line, as shown in Figs. 1, 2 and 3; or indirectly, by balancing the difference between the high- and low-pressure sides of a venturi tube, Fig. 4, nozzle tube, Fig. 5, or an orifice plate, Fig. 6.

In the case of a pitot tube it should be noticed that the differential column balancing the difference in pressure represents the flow, or motion, existing at the given section of the line. With a venturi tube, nozzle tube or orifice plate, the column obtained represents the change of motion produced by the artificial obstruction of the passage at the given section of pipe.

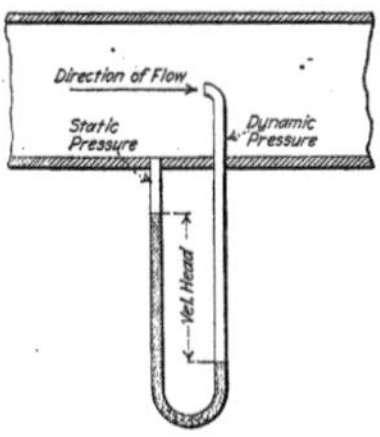

FIG. 1. ILLUSTRATING SIMPLE PITOT TUBE

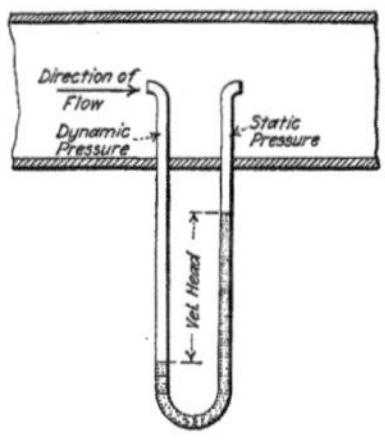

FIG. 2. DIFFERENTIAL PITOT TUBE ANOTHER FORM OF

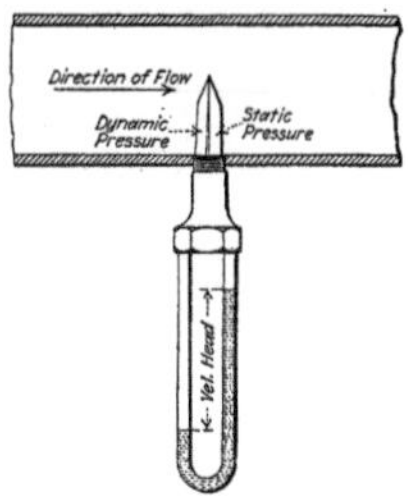

FIG. 3. SPECIAL FORM OF DIFFERENTIAL PITOT TUBE

The pitot tube is employed generally in cases where the minimum flow in the pipe is sufficient to make a perceptible difference in pressure for operating the measuring device, and the maximum flow is not over the range of the given device. The venturi tube or orifice is better adapted for cases where the flow existing in the line is either too low or too high for the range of pressure difference in the measuring device. When a venturi tube or orifice is used, the change of motion or the difference in kinetic energy between original and final flow, can be made to suit by varying the size of the restricted passage in the line.

Two distinct forms of pitot tube are generally used—one designated as a combined impact tube of the form shown in Fig. 1, and the other as the differential tube shown in Figs. 2 and 3. In the combined impact tube one opening is pointed against the flow to receive the dynamic pressure. The other opening is at right angles to the flow, for the purpose of receiving the static pressure only. As a matter of fact, the opening pointed against the flow receives the true dynamic pressure, because the flow is entirely stopped at the point of impact, transmitting the full velocity pressure to the connections of the U-tube. On the other hand, the opening at right angles to the flow does not always receive the true static pressure, due to the effect of aspiration or siphoning produced by the flow in the line.

In the differential tube the two openings are pointed in opposite directions; one is made to receive the dynamic pressure, and the other is acted upon by the suction of the flow, so that the differential column in the U-tube represents impact plus suction. The differential tube is generally preferred for practical purposes. In the first place the effect of suction in the direction of the flow is more positive than the effect of siphoning at right angles to the flow. In the second place the suction produced by the flow magnifies the differential column corresponding to a given velocity of the fluid, which in turn facilitates the operation of the measuring device and permits of a larger scale that can be read with accuracy.

TABLE I. FLOW OF WATER
(Assumi g D² = i ci c lar inch, C = I)

Temp., Deg. F.	Values of K — Pounds per Hour, 1 in. H_2O	Pounds per Hour, 1 in. HgO	Gal. per Hour, 1 in. H_2O	Gal. per Hour, 1 in. Hg.
60	2,838	10,070	340.5	1,208
70	2,837	10,065	340.7	1,209
80	2,835	10,060	341.0	1,210
90	2,833	10,055	341.3	1,211
100	2,830	10,045	341.6	1,213
110	2,827	10,035	342.0	1,214
120	2,824	10,025	342.4	1,215
130	2,820	10,015	342.8	1,216
140	2,816	10,000	343.3	1,218
150	2,812	9,985	343.	1,220
160	2,807	9,965	344.3	1,222
170	2,802	9,945	344.9	1,224
180	2,797	9 25	345.5	1,226
190	2,792	9,905	346.1	1,229
200	2,786	9,885	346.8	1,231
210	2,781	9,865	347.5	1,234
220	2,776	9, 50	348.2	1,236
230	2,770	9,830	348.9	1,238
240	2,764	9,810	349.7	1,241
250	2,758	9,790	350.5	1,244
260	2,751	9,765	351 3	1,247
270	2,743	9,740	352.1	1,250
280	2,735	9,710	353 0	1,253
290	2,727	9,680	354.0	1,257
300	2,719	9,550	355 2	1,261

NOTE—1 in. of mercury under water equals 12.6 in. of water head.

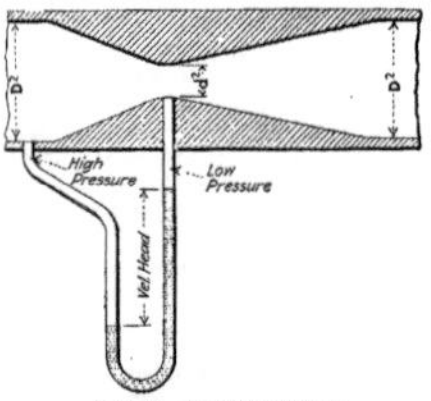

FIG. 4. VENTURI TUBE

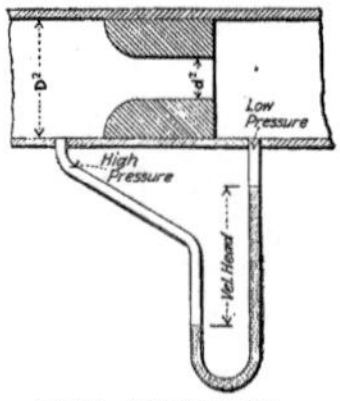

FIG. 5. NOZZLE TUBE

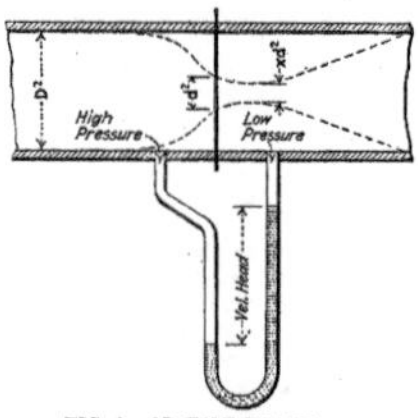

FIG. 6. ORIFICE PLATE

With any one of these various devices a differential pressure is obtained, and starting from this reading the quantity of flow in pounds, gallons or cubic feet per unit of time may be calculated. It is convenient to base all the calculations of flow upon the differential column measured in inches of water. The U-tube of the measuring device may be filled with mercury and the actual differential pressure carried by the height of the mercury column; but the effective height and all the measurements of flow are based on the equivalent column in inches of water. When mercury is used with an ordinary U-tube manometer, an inch of mercury is equivalent to 13.6 in. of water when the measurement is made on air or gas. When measuring steam or water, the equivalent is 12.6 in. of water for one inch

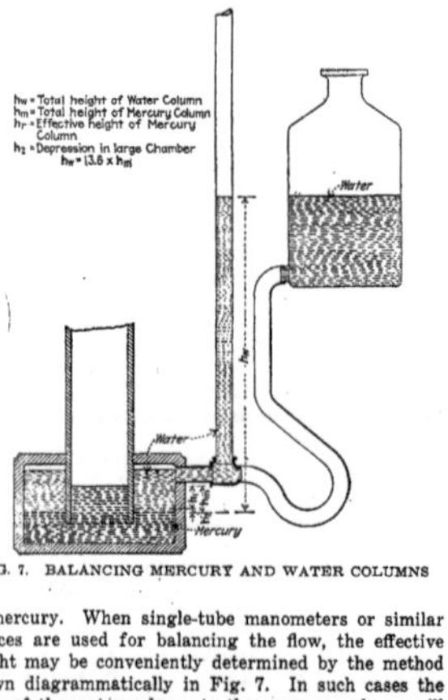

FIG. 7. BALANCING MERCURY AND WATER COLUMNS

of mercury. When single-tube manometers or similar devices are used for balancing the flow, the effective height may be conveniently determined by the method shown diagrammatically in Fig. 7. In such cases the ratio of the water column to the mercury column will depend upon the relation of the areas in the two chambers of the gage.

It is also convenient to adopt the circular unit of area as the determination of constants for a given measuring device is facilitated. The flow per circular unit is multiplied by the diameter squared, to obtain the quantity, in the same manner as circular mills are used in electrical measurements.

From the usual formula expressing the general relations between pressure and velocity of fluids, one general formula may be derived that will be applicable to all the devices shown and for the various mediums necessary only to take the value of K given in the table and multiply by the true values of the three quantities mentioned. The diameter of the pipe D is available, the head h is given by the measuring device, so that the coefficient C is the only factor needing further attention.

For the combined impact tube, Fig. 1, with the dynamic opening placed in the center of the flow, C has an average value of 0.90; for the differential tube of the form shown in Figs. 2 and 3 the value of C varies from 0.73 to 0.80. For the venturi tube of Fig. 4 the value of C depends on the ratio of the throat section d^2

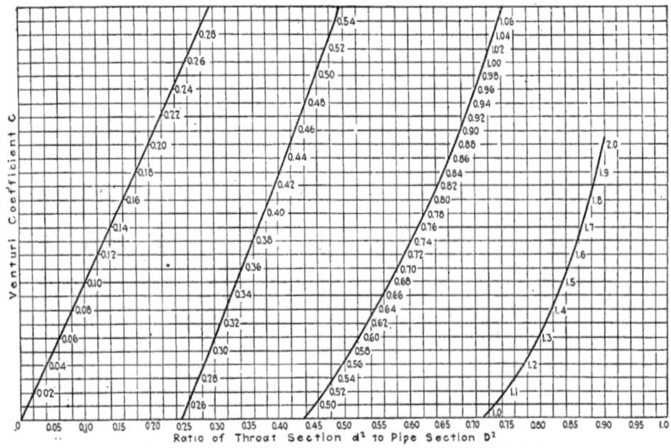

FIG. 8. VENTURI COEFFICIENT C AS RELATED TO PIPE SECTION D^2

measured. In its simplest form, giving results in pounds per hour, this formula is as follows:

$$G = (359.3\ C\ D^2 \sqrt{S}) \times \sqrt{h}$$

where

G = Weight of flow in pounds per hour;
C = Coefficient of discharge, or the ratio of the actual to the theoretical amount of flow;
D^2 = Diameter of pipe squared in inches;
S = Density of fluid in the pipe in pounds per cubic foot;
h = Differential column in inches of water.

Each of the factors indicated by the letters in the formula are easily obtainable except the coefficient C, which varies for each type of device illustrated. To make calculation more simple, however, a number of tables based on this formula have been worked out, covering the flow of water, steam, gas and air for a wide variation of pressure or temperature. In these tables it was necessary to assume that the value of D^2, $\sqrt{h}$ and C each equal 1, so that the quantity designated by K is equal to $359.3 \sqrt{S}$. In a specific problem, then, it is

to the pipe section D^2. The curves of Fig. 8 cover a wide range, showing the value of C corresponding to a given ratio of throat to pipe sections.

In the nozzle tube shown in Fig. 5 the coefficient varies with the roundness of the inlet and the size of the tube, so that it is necessary to test each tube in the same way the ordinary nozzles are being calibrated for measuring the flow of water.

In Fig. 6 is shown the general form of orifice connections. The orifice differs from the nozzle tube in that the approach of the flow is fully contracted by the sharp edge of the plate. This results in an even amount of friction loss, and therefore the effectiveness of the orifice as a means of obtaining differential pressure is the same for all orifices of this kind.

The flow through the orifice takes the same form as through a venturi, except that the smallest section corresponding to the throat of the venturi is formed after the fluid passes through the orifice. The orifice coefficient, therefore, involves an additional factor over the venturi tube in that the contracted area is smaller than

TABLE II. FLOW OF SATURATED STEAM

(Assuming D^2 = 1 circular inch, C = 1)

Absolute Pressure Lb.	Gage Pressure Lb	Corres. Temp. Deg. F.	Values of K Pounds per Hour 1 in. H_2O	1 in. Hg
14.7	0	212.0	69.38	246.2
15.7	1	215.3	71.57	254.1
16.7	2	218.4	73.73	261.8
17.7	3	221.5	75.81	269.2
18.7	4	224.3	77.72	276.0
19.7	5	227.0	79.62	282.7
20.7	6	229.7	81.45	289.2
21.7	7	232.3	83.25	295.5
22.7	8	234.8	85.01	301.8
23.7	9	237.1	86.77	308.0
24.7	10	239.4	88.46	314.0
25.7	11	241.6	90.11	319.8
26.7	12	243.7	91.73	315.6
27.7	13	245.8	93.27	331.3
28.7	14	247.8	94.85	336.9
29.7	15	249.8	96.40	342.3
30.7	16	251.7	97.87	347.6
31.7	17	253.5	99.38	352.8
32.7	18	255.3	100.85	358.2
33.7	19	257.0	102.33	363.3
34.7	20	258.8	103.69	368.2
39.7	25	266.8	110.28	392.0
44.7	30	274.0	116.70	414.3
49.7	35	280.6	122.70	435.7
54.7	40	286.6	128.34	455.8
59.7	45	292.3	133.70	474.5
64.7	50	297.7	138.9	492.5
69.7	55	302.6	143.8	510.5
74.7	60	307.3	148.6	527.5
79.7	65	311.9	153.1	543.5
84.7	70	316.0	157.6	559.5
89.7	75	320.0	162.0	575.5
94.7	80	324.0	166.1	590.5
99.7	85	327.8	170.2	604.5
104.7	90	331.2	174.2	618.5
109.7	95	334.6	178.1	632.5
114.7	100	337.8	182.0	646.0
119.7	105	341.2	185.7	659.0
124.7	110	344.4	189.3	672.0
129.7	115	347.4	192.9	685.0
134.7	120	350.2	196.4	697.5
139.7	125	352.9	199.8	709.5
144.7	130	355.7	203.2	721.5
149.7	135	358.5	206.5	733.0
154.7	140	361.0	209.8	744.5
159.7	145	363.5	213.0	756.0
164.7	150	365.9	216.2	767.5
169.7	155	368.4	219.3	778.5
174.7	160	370.8	222.4	789.5
179.7	165	373.0	225.4	800.0
184.7	170	375.2	228.4	810.0
189.7	175	377.3	231.4	822.0
194.7	180	379.5	234.3	832.0
199.7	185	381.7	237.2	842.0
204.7	190	383.8	240.0	952.0
209.7	195	385.9	242.6	861.5
214.7	200	387.9	245.6	871.5
219.7	205	389.8	248.	881.0
224.7	210	391.7	251.0	891.0
229.7	215	393.7	253.6	900.5
234.7	220	395.5	256.3	910.0
239.7	225	397.4	259.1	920.0
244.7	230	399.2	261.6	929.0
249.7	235	400.9	264.3	938.0
254.7	240	402.7	266.7	947.0
259.7	245	404.5	269.4	965.0
264.7	250	406.2	271.7	956.0
269.7	255	407.8	274.3	974.0
274.7	260	409.4	276.7	983.0
229.7	265	411.0	279.3	992.0
284.7	270	412.6	281.6	1,000.0
289.7	275	414.3	284.1	1,008.0
294.7	280	415.8	286.3	1,016.0
299.7	285	417.4	288.8	1,024.0
304.7	290	418.9	291.3	1,032.0
309.7	295	420.4	293.4	1,040.0
314.7	300	422.0	295.9	1,048.0

NOTE—1 in. of mercury under water equals 12.6 in. of water head.

TABLE III. CORRECTION FACTORS FOR SUPERHEATED STEAM

Degrees of Superheat	Factor	Degrees of Superheat	Factor
0	1.000	130	0.907
10	0.991	140	0.901
20	0.982	150	0.895
30	0.974	160	0.89
40	0.966	170	0.885
50	0.958	180	0.88
60	0.951	190	0.875
70	0.944	200	0.87
80	0.938	250	0.847
90	0.931	300	0.826
100	0.925	400	0.789
110	0.919	500	0.757
120	0.913		

TABLE IV. FLOW OF GAS AND AIR

(Assuming D^2 = 1 circular inch, C = 1)

Values of K for air at standard temperature of 60 deg. F. For gas of specific gravity other than unity, divide proper factor from table by the square root of the specific gravity of the gas referred to air.

Abso. Press., Lb.	Gage Press., Lb.	Values of K Std. Cu.Ft. per Hour 1 in. H_2O	1 in. Hg.
14.7	0	1,301	4,800
15.7	1	1,344	4,960
16.7	2	1,385	5,110
17.7	3	1,426	5,260
18.7	4	1,465	5,410
19.7	5	1,505	5,555
20.7	6	1,542	5,695
21.7	7	1,579	5,830
22.7	8	1,615	5,960
23.7	9	1,650	6,090
24.7	10	1,685	6,215
25.7	11	1,718	6,540
26.7	12	1,750	6,460
27.7	13	1,781	6,575
28.7	14	1,812	6,690
29.7	15	1,843	6,805
30.7	16	1,874	6,915
31.7	17	1,904	7,025
32.7	18	1,934	7,135
33.7	19	1,964	7,245
34.7	20	1,994	7,355
39.7	25	2,137	7,880
44.7	30	2,264	8,360
49.7	35	2,388	8,820
54.7	40	2,511	9,270
59.7	45	2,619	9,660
64.7	50	2,726	10,050
69.7	55	2,826	10,430
74.7	60	2,924	10,790
79.7	65	3,022	11,150
84.7	70	3,120	11,510
89.7	75	3,219	11,840
94.7	80	3,297	12,160
99.7	85	3,383	12,480
104.7	90	3,466	12,800
109.7	95	3,550	13,090
114.7	100	3,628	13,380
119.7	105	3,706	13,670
124.7	110	3,784	13,960
129.7	115	3,861	14,250
134.7	120	3,936	14,530
139.7	125	4,011	14,800
144.7	130	4,081	15,060
149.7	135	4,150	15,310
154.7	140	4,218	15,560
159.7	145	4,284	15,810
164.7	150	4,350	16,050
169.7	155	4,416	16,290
174.7	160	4,480	16,530
179.7	165	4,543	16,770
184.7	170	4,606	17,000
189.7	175	4,668	17,230
194.7	180	4,730	17,460
199.7	165	4,791	17,680
204.7	190	4,852	17,900
209.7	195	4,910	18,120
214.7	200	4,968	18,330
219.7	205	5,026	18,540
224.7	210	5,084	18,750
229.7	215	5,142	18,960
234.7	220	5,198	19,170
239.7	225	5,253	19,380
224.7	230	5,307	19,590
249.7	235	5,361	19,800
254.7	240	5,414	20,000
259.7	245	5,467	20,190
264.7	250	5,520	20,380
269.7	55	5,572	20,560
274.7	260	5,623	20,740
279.7	265	5,673	20,920
284.7	270	5,722	21,100
289.7	275	5,771	21,280
294.7	280	5,820	21,460
299.7	285	5,869	21,640
304.7	290	5,918	21,820
309.7	295	.....	......
314.7	300	6,012	22,180

the area of the orifice section. This factor is determined experimentally for each ratio of orifice diameter to pipe diameter, and the given size of the pipe.

The following examples will demonstrate the use of the tables and the solution of problems in measuring the flow of fluids in pipes:

Example 1—Pitot tube measuring flow of water:

internal diameter of tube, $D = 6$ in.; coefficient of tube, $C = 0.80$; differential column, $h = 1$ in. of mercury; temperature of water, 200 deg. F.

From Table I, opposite 200 deg. F., $K = 9,885$ lb. per hour, or 1,231 gal. per hour.

$9,885 \times (6)^2 \times 0.80 \times \sqrt{1} = 284,688$ lb. per hr.

$1,231 \times (6)^2 \times 0.80 \times \sqrt{1} = 35,453$ gal. per hr.

The solution obtained for one inch of mercury differential can be applied to any other height simply by multiplying the quantity by the square root of the given height of the mercury column, thereby avoiding the repetition of multiplying the other factors which are constant for the same conditions of flow.

Example 2—Venturi tube measuring flow of water: internal diameter of pipe, $D = 3$ in.; diameter of throat,

TABLE V. TEMPERATURE CORRECTION FACTORS FOR GAS AND AIR

Temp., Deg. F.	Factor	Temp., Deg. F.	Factor
0	1.063	230	0.868
10	1.052	235	0.865
20	1.041	240	0.862
30	1.030	245	0.859
40	1.020	250	0.856
50	1.009	255	0.853
60	1.000	260	0.850
65	0.996	265	0.847
70	0.991	270	0.844
75	0.986	275	0.841
80	0.981	280	0.838
85	0.977	285	0.836
90	0.972	290	0.833
95	0.968	295	0.830
100	0.964	30	0.827
105	0.959	350	0.802
110	0.956	400	0.778
115	0.951	450	0.756
120	0.947	500	0.736
125	0.943	550	0.718
130	0.939	600	0.700
135	0.935	650	0.684
140	0.931	700	0.670
145	0.927	750	0.656
150	0.923	800	0.643
155	0.920	850	0.630
160	0.916	900	0.619
165	0.912	950	0.607
170	0.909	1,000	0.597
175	0.905	1,050	0.587
180	0.902	1,100	0.577
185	0.898	1,150	0.568
190	0.895	1,200	0.56
195	0.891	1,250	0.552
200	0.883	1,300	0.544
205	0.885	1,350	0.536
210	0.881	1,400	0.529
215	0.878	1,450	0.522
220	0.875	1,500	0.511
225	0.872		

$d = 1$ in.; differential column, $h = 10$ in. of mercury; temperature of water, 60 deg. F.

First obtain venturi coefficient as follows:

$$\frac{d^2}{D^2} = \left(\frac{1}{3}\right) = 0.111$$

From Fig. 8, opposite the ratio, 0.11, read the coefficient, $C = 0.109$. From Table I, opposite 60 deg. F., $K = 10,070$ lb. per hour, or 1,208 gal. per hour.

$10,070 \times (3)^2 \times 0.109 \times \sqrt{10} = 31,217$ lb. per hr.

$1,208 \times (3)^2 \times 0.109 \times \sqrt{10} = 3,745$ gal. per hr.

Example 3—Pitot tube measuring flow of steam; internal diameter of pipe, $D = 10$ in.; coefficient of tube, $C = 0.90$; differential column, $h = 1$ in. of mercury; steam pressure, 150 lb. gage.

From Table II, opposite 150 lb. pressure, $K = 767.5$ lb. per hour.

$767.5 \times 10^2 \times 0.90 \times \sqrt{1} = 69,075$ lb. per hr.

Example 4—Venturi tube measuring flow of air: internal diameter of pipe, $D = 16$ in.; diameter of throat, $d = 8$ in.; differential column, $h = 4$ in. of water; static pressure, 100 lb. gage; air temperature, 120 deg. F.

Obtain venturi coefficient as follows:

$$\frac{d^2}{D^2} = \left(\frac{8}{16}\right)^2 = 0.25$$

From Fig. 8, opposite ratio 0.25, read coefficient, $C = 0.253$. From Table IV, opposite 100 lb. per hour, K 3,628 standard cubic feet per hour. From Table V, the correction factor at 120 deg. F. is 0.947.

$3,628 \times 0.947 \times (16)^2 \times 0.253 \times \sqrt{4} = 445,047$ standard cu.ft. per hr.

Standard Pipe-Hanger Data

Very little has been published regarding pipe hangers. The accompanying figure and table give some

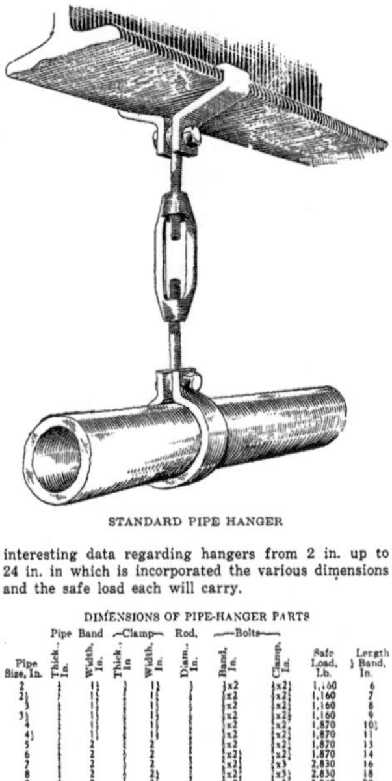

STANDARD PIPE HANGER

interesting data regarding hangers from 2 in. up to 24 in. in which is incorporated the various dimensions and the safe load each will carry.

DIMENSIONS OF PIPE-HANGER PARTS

Pipe Size, In.	Pipe Band Thick., In.	Pipe Band Width, In.	Clamp Thick., In.	Clamp Width, In.	Rod, Diam., In.	Bolts Band, In.	Bolts Clamp, In.	Safe Load, Lb.	Length ½ Band, In.
2	[illegible]	[illegible]	[illegible]	[illegible]	[illegible]	[illegible]x2	[illegible]x2[illegible]	1,160	6
2½	[illegible]	[illegible]	[illegible]	[illegible]	[illegible]	[illegible]x2	[illegible]x2[illegible]	1,160	7
3	[illegible]	[illegible]	[illegible]	[illegible]	[illegible]	[illegible]x2	[illegible]x2[illegible]	1,160	8
3½	[illegible]	[illegible]	[illegible]	[illegible]	[illegible]	[illegible]x2	[illegible]x2[illegible]	1,160	9
4	[illegible]	[illegible]	[illegible]	[illegible]	[illegible]	[illegible]x2	[illegible]x2[illegible]	1,870	10½
4½	[illegible]	[illegible]	[illegible]	[illegible]	[illegible]	[illegible]x2	[illegible]x2[illegible]	1,870	11
5	[illegible]	2	[illegible]	2	[illegible]	[illegible]x2	[illegible]x2[illegible]	1,870	13
6	[illegible]	2	[illegible]	2	[illegible]	[illegible]x2[illegible]	[illegible]x2[illegible]	1,870	14
7	[illegible]	2	[illegible]	2	[illegible]	[illegible]x2[illegible]	[illegible]x3	2,830	16
8	[illegible]	2	[illegible]	2[illegible]	[illegible]	[illegible]x2[illegible]	[illegible]x3[illegible]	2,830	18
9	[illegible]	2	[illegible]	2[illegible]	[illegible]	[illegible]x2[illegible]	[illegible]x3[illegible]	3,940	20
10	[illegible]	2	[illegible]	2[illegible]	[illegible]	[illegible]x3	[illegible]x3[illegible]	3,940	22
12	[illegible]	2[illegible]	[illegible]	2[illegible]	1	[illegible]x3[illegible]	1 x3[illegible]	5,180	25
14	[illegible]	2[illegible]	[illegible]	2[illegible]	1[illegible]	1 x3[illegible]	1[illegible]x3[illegible]	6,510	27
15	[illegible]	2[illegible]	[illegible]	2[illegible]	1[illegible]	1 x3[illegible]	1[illegible]x3[illegible]	6,510	30
16	[illegible]	3	[illegible]	3	1[illegible]	1[illegible]x3[illegible]	1[illegible]x4	6,510	33
18	[illegible]	3	[illegible]	3	1[illegible]	1[illegible]x4	1[illegible]x4	8,410	35
20	[illegible]	3	[illegible]	3	1[illegible]	1[illegible]x4	1[illegible]x4[illegible]	8,410	37
22	[illegible]	3	[illegible]	3	1[illegible]	1[illegible]x4[illegible]	1[illegible]x4[illegible]	8,410	39
24	[illegible]	3	[illegible]	3	1[illegible]	1[illegible]x4[illegible]	1[illegible]x4[illegible]	6,410	42

Ground Detectors for Direct-Current and Alternating-Current Circuits

BY FRANK GILLOOLY

SOME of the schemes for utilizing lamps to indicate grounds on low-voltage circuits are illustrated in Figs. 1, 2 and 3. In Fig. 1, both lamps are 250-volt units, and when both sides of the system are ungrounded the lamps burn at half brilliance. Should a ground occur on the positive side, lamp 1 will go out and lamp 2 come to full brilliance; with a negative ground, lamp 2 will go out and lamp 1 become bright. Should either lamp burn out or work loose in its socket, both lamps will go out.

In Fig. 2 lamps 1 and 2 are at half brilliance and lamp 3 dead while the system is ungrounded. With

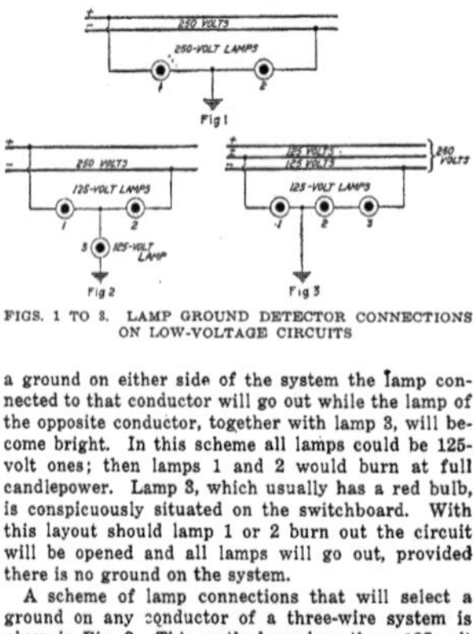

FIGS. 1 TO 3. LAMP GROUND DETECTOR CONNECTIONS ON LOW-VOLTAGE CIRCUITS

a ground on either side of the system the lamp connected to that conductor will go out while the lamp of the opposite conductor, together with lamp 3, will become bright. In this scheme all lamps could be 125-volt ones; then lamps 1 and 2 would burn at full candlepower. Lamp 3, which usually has a red bulb, is conspicuously situated on the switchboard. With this layout should lamp 1 or 2 burn out the circuit will be opened and all lamps will go out, provided there is no ground on the system.

A scheme of lamp connections that will select a ground on any conductor of a three-wire system is given in Fig. 3. This method employs three 125-volt lamps connected in series between the neutral and negative conductors of the system, with the ground connection made between the lamp connected to neutral and the two connected to negative. A chart that shows the indications for each kind of ground is usually located near the lamps for the operator's guidance. Normally, all the lamps burn at one-third brilliance, since three of them are connected in series on a 125-volt circuit. On a positive ground all three lamps will come to full brilliance. On a neutral ground, lamp 1 will go out and lamps 2 and 3 come to one-half brilliance. On a negative ground, lamp 1 will be bright and lamps 2 and 3 will be dead.

In general, the lamps should be so definitely marked that there can be no doubt as to their indications. Where lamps are merely marked "Positive" and "Negative," operators become confused as to whether the positive lamp indicates a positive ground when it is bright or when it is dead. If the lamps are marked "Positive Bus" and "Negative Bus," the attendant knows that the ground is on the bus whose lamp is dead; or, if they are marked "Positive Ground" and "Negative Ground"—with the lamp marked positive connected to the negative bus and vice versa—he knows, when the lamp marked positive is bright, that the system has a positive ground.

The commonest use made of ground lamps in the alternating-current generating station is on the direct-current excitation and control systems, where they become highly essential. The direct-current field is the heart of the synchronous generator. A breakdown of the alternator's field winding means that until the unit can be spared for the period required to rewind the field, it must be separately excited; since a ground of opposite polarity elsewhere on the system would shunt out part of the field winding and reduce the capacity of the unit, if it did not burn out the winding. Ordinarily, the operating energy for the oil circuit-breakers, rheostats and governor motors, is taken from the excitation system; and with another ground on the system, the failure of the insulation of the operating device of any of this apparatus might cause it to function and open or close. In the case of motor-operated oil switches, such a double ground may cause the switch to trip out or even to successively open and close until its entire control circuit is killed.

Lamps are also used with transformers to indicate the existence of grounds on high-voltage circuits.

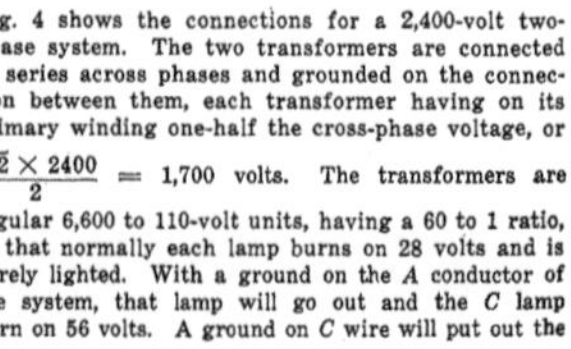

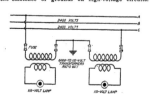

FIG. 4. GROUND LAMPS CONNECTED TO TRANSFORMERS ON TWO-PHASE SYSTEM

Fig. 4 shows the connections for a 2,400-volt two-phase system. The two transformers are connected in series across phases and grounded on the connection between them, each transformer having on its primary winding one-half the cross-phase voltage, or

$$\frac{\sqrt{2} \times 2400}{2} = 1{,}700 \text{ volts.}$$

The transformers are regular 6,600 to 110-volt units, having a 60 to 1 ratio, so that normally each lamp burns on 28 volts and is barely lighted. With a ground on the A conductor of the system, that lamp will go out and the C lamp burn on 56 volts. A ground on C wire will put out the

C lamp and throw the *A* lamp on 56 volts. With a ground on *B* wire both transformers will have 2,400 volts on their primaries and 40 volts on their lamps.

Electrostatic ground detectors are really electrostatic voltmeters which indicate the existence of a ground on the system. Both the electrostatic detectors and the transformer-fed ground lamps will give an indication on any unbalanced voltage condition,

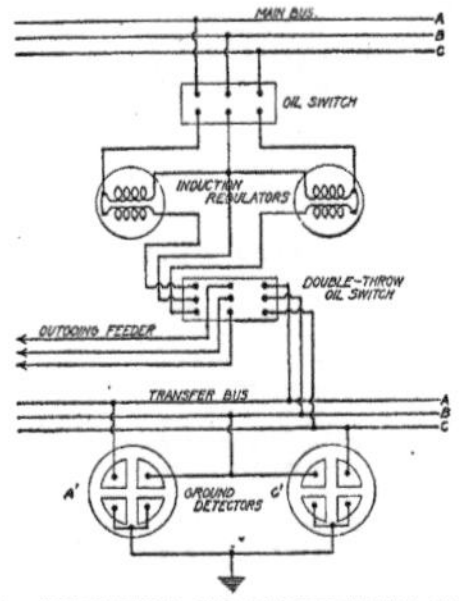

FIG. 5. CONNECTIONS FOR ELECTROSTATIC GROUND DETECTORS TO THREE-PHASE CIRCUIT

the electrostatic meters being more sensitive in this respect than the lamps. This condition is particularly true of distribution feeders having induction regulators where the ground indicators are connected beyond the regulators, such as in the arrangement shown in Fig. 5. If the regulators become so adjusted that there is considerable difference in voltage between phases on the circuit feeding the transfer bus, the detectors will indicate a ground. The unbalanced voltage at the terminals of a Scott-connected transformer will cause a ground indication. In one substation having 4,000-kva. 13,000-volt three-phase to 2,400-volt two-phase transformer banks, when thrown on the secondary bus carrying the ground lamps these transformers always caused a ground indication on the lamps until some load was connected to the bus.

Before any ground indication is accepted as final, the voltages of the phases of the circuit should be checked to make sure that they are alike. There have been occasions when troublemen wasted time in searching for grounds that never existed, but were indicated on the detectors due to an unbalanced voltage. In layouts such as that in Fig. 5 it is better to locate the detectors on one of the main buses rather than on the transfer bus, for this reason.

One great disadvantage of the ground transformer not shared by the electrostatic detector is the tendency of the transformer to burn out under a ground. As soon as the ground has been located, the ground transformer must be taken out of the circuit. Few accidental grounds make a solid connection; usually, they are of an oscillating or arcing character; and the constant hammering on the transformer of the charging current overheats the iron and windings.

To definitely locate a ground, each piece of apparatus in turn will have to be disconnected from the bus until the ground indication clears when the faulty apparatus is isolated. A ground on the secondary circuit of a transformer will not show up on its primary circuit. Where a station has duplicate busses, apparatus to be tested for ground may be transferred to a separate bus, fed by another alternator or transformer, and isolated from all other apparatus. When the ground instruments show clear after the faulty apparatus has been disconnected from the bus, the defective equipment alone may be carried on the bus having the ground meters, as a further check that that apparatus is at fault.

On the Road with the Refrigerating Troubleman

By J. C. MORAN

He is sent to estimate how much capacity in new machines was needed to enable the overloaded, neglected plant to carry its load. The owner threatens to fire the whole crew when told by the troubleman that with needed repairs the present machines will easily carry the load. Another stormy scene occurs when he gets the estimate for this work. But next season he is happy for he handles more produce with 60 tons of coal now than with 80 tons before; and he has a new engineer whom he pays far more than he did in the old troublesome days.

THE refrigerating plant of a packing house in the South was getting too small for the load. It was almost impossible to keep the temperatures low enough even at average load, and when a heavy load came it was necessary to buy ice for the high-temperature coolers. Things had been going from bad to worse for several years. The owner was of the kind that would never spend a cent in the power plant until something stopped. As a consequence two good chief engineers had given up in despair, and one of the men who had grown up in the plant held the position. He was another of those fellows, like the owner, who believed in cutting expenses by putting off repairs and in addition driving the men as hard as possible. So the plant, crew and the chief, were wrecks.

It was up to me to go over the plant and give the owner an estimate of what capacity of new machines would be needed to carry the load. After going over the figures several times, I called in another man who knew his business to check my figures. He agreed that the present plant should be plenty large enough with capacity to spare. When we made the report to the owner, he raved and wanted to fire the whole crew and hire a new one. He was advised that all he needed was a new chief, a thorough overhauling of the entire refrigerating plant and then to do as the chief told him in the matter of buying supplies and expenses for repairs. He threatened to throw me out of the office, but finally cooled off and asked me to recommend a chief and then make an estimate on the needed repairs. He had another outburst when I gave him the estimate for the work, but he O. K.'d, and in a few days I also found him a new chief at a salary nearly twice what he had been paying.

In making the estimate, I had figured doing nearly everything that could be done to the plant, feeling sure that some of this work would not be necessary. When

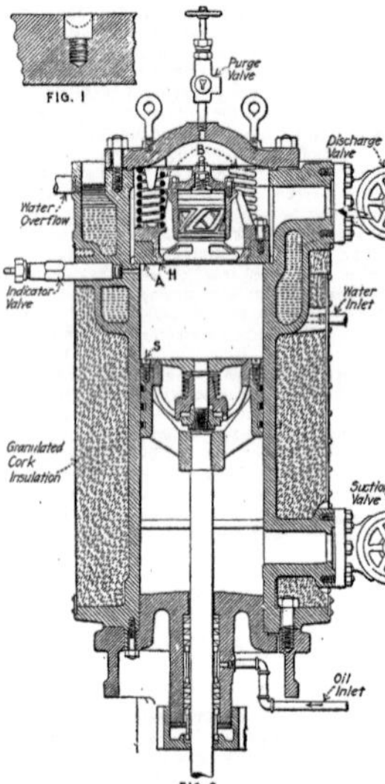

FIG. 2

SECTION OF AMMONIA COMPRESSOR CYLINDER, SHOWING SEAT "A" WHICH CRACKED

we came to go through the plant, I found that everything that I had figured on would be necessary, if not more.

There were four machines in the place—two 100-ton of the vertical single-acting type and two 75-ton of the horizontal double-acting type. We started on the latter first.

Three of the four piston rings on the first machine were broken, and the last one did not hold itself against the cylinder wall. The second machine had two rings broken, and the tension of the two remaining rings was useless. After we had the rods disconnected from the crosshead, we could push the pistons forward against the crank head with practically no effort by placing the foot against it and giving it a boost. The gas had been leaking back around the pistons as fast as it had been pumped out, or in other words possibly half of the gas that should have been delivered leaked back around the pistons, and the only way they could keep the machine cold was to carry a lot of liquid over with the suction gas. Otherwise it would run so hot that it smoked.

We had the cylinders bored, provided new pistons and new rings that made a snug and tight fit in the cylinder. The piston rods were about $\frac{1}{16}$ in. smaller at the center than at the ends with heavy score marks on them. With the old rods it had been impossible to keep the stuffing-box from leaking, and as all the rest were in the same condition they must have wasted several hundred pounds of ammonia a month.

In adjusting the crosshead shoes, they had followed the general practice in steam-engine work of adjusting the bottom shoe for taking up the wear, the wear on the steam-engine shoe generally being on the bottom unless the engine is running under. On a compressor the action is just the reverse and the wear is on the top. As a consequence, with the piston wearing on the bottom in the cylinder and dropping this end of the rod and then adjusting the bottom shoe for wear on the other end, the rod was nearly a quarter of an inch higher at the crosshead end of the rod than at the piston end. It was traveling through the stuffing-box at an angle all the time.

The valves were in as bad condition. One of the discharge valves was so badly worn that it had worked down through the seat far enough to touch the piston at every stroke, while the piston prevented it tightly seating. We put in four new valves and cages out of the total of eight and reground the valves and seats of the rest. We also put in lighter springs on all of the discharge valves, as someone had ordered new springs that were about three times as strong as required. The proper spring was determined by taking indicator diagrams with different spring tensions and then adopting the spring that would operate the valve quietly, show the smallest possible rise above the discharge pressure at the time it opened and at the same time close promptly, as indicated by the re-expansion line on the suction stroke.

After covering a number of other details, such as lining up the cranks and mainshaft, rebabbitting the top crosshead shoes, crosshead and crankpin brasses, babbitting the oil lanterns in the stuffing-boxes and clearing out the ammonia oil lines to the stuffing-boxes, we were ready to put them in service.

The work was done during the early part of the winter or we would never have been able to take two machines out of service at once. There was also a slack period, but just about the time we had finished the two machines, an exceptionally heavy run came along and the owner came to the engine room three or four times a day to see if he could bluff or coax us to hurry the work along.

We had the machines ready about a day before the rush came and had them running smoothly. We cut them both in on the system in the middle of the afternoon when the coolers were being filled with a rush, and that night we had about three times the load to carry that we had had for some time past. The "Old

Man" was on the job about six o'clock the next morning and every cooler was ahead of schedule and a number of them were down to 33 deg., which had never been known before.

Before cutting in the newly overhauled machines, we put in 1,200 lb. of ammonia, which had something to do with bringing the coolers down because the system was literally starved for ammonia.

The old man by this time came to the conclusion that the new chief and myself probably knew more about the business than he had given us credit for, and we got anything we wanted. We next tackled the vertical machines which had been in operation only three years, but were in a bad condition. The pistons and rings were good, though we peened out the rings slightly so as to bring more pressure against the cylin der wall. The cylinders were not worn sufficiently to need reboring, and this was largely due to the slight wear on the vertical cylinder, there being no wear from the weight of the piston, only from the pressure of the piston rings.

Vertical Machines in Need of Repairs

The rods were bad, being both scored and worn more in the center than at the ends. Both rods were sent to the shop to be turned up, while we fitted a junk ring in the bottom of the stuffing-box to make a snug fit on the rod after turning to prevent the packing working out between the space between the smaller rod and the neck of the stuffing-box. The oil lantern also had to be rebabbitted to fit the reduced size of the rod.

The crosshead adjustment on these machines was not so bad as on the other two, but one of them was out nearly one-eighth inch. Trouble had been experienced with the stuffing-boxes. When we removed the false head of the first machine, we found a sorry state of affairs. From somewhere a small piece of babbitt had got clamped between the false-head seat and the head, preventing it seating properly, and as a consequence the gas had been continually leaking back through on the suction stroke on this one cylinder of this machine. The head and cylinder on the other side were in good order. All that was necessary on this machine was to grind in the suction and discharge valves and the false heads.

The other machine was worse than the first. A piece had been broken out of the false-head seat on one of the cylinders and left a passage almost large enough to hold a lead pencil directly across the seat. We remedied this by drilling a hole slightly wider than the widest part of the break and about an inch deep, and then tapping the bottom part of this hole, which was made slightly smaller than the first one-third of an inch at the top. We then screwed into this hole a plug (see Fig. 1) that was a snug fit in the top section and filed it down to a smooth finish with the rest of the seat.

And this was not all. Two of the screws that held the suction-valve cage in the piston, as indicated by *S*, Fig. 2, had backed out about half an inch. Instead of opening up the machine and looking into it to see what was the matter, the crew had merely backed the piston rod farther into the crosshead until the noise made by the pounding of the screws against the false head has stopped. In this way they had operated this cylinder with about half an inch clearance and of course at greatly reduced capacity.

Referring to Fig. 2, *B* indicates the springs which hold the false head against the seat, *A* and *H* show the valve seat of the discharge valve. It was from this seat *A* that the piece was broken out as already mentioned. The screws were so badly battered that they were useless, and we had to tap the holes over again and insert larger screws. A piece was also cracked out of the discharge valve seat *H*, Fig. 2, and we had to remove the seat and get a new one after which we ground in the valves on all cylinders of both machines.

After this was done, we adjusted the clearance on all the pistons so that they were from $\frac{1}{16}$ in. to $\frac{1}{32}$ in. from the head.

After we had the machines together again, we ran them for a couple of days intermittently, during which one of the machines had a serious pound that we were unable to locate. We looked everywhere, keyed up all the bearings, pins and crosshead shoes without results. The pound was there as soon as the machine was run at normal speed, while at lower speeds it would sometimes pound and at other times would not. Some of the boys thought it was the piston hitting, and we changed the clearance adjustment, but without results. The pound seemed to come just after the crank had passed the top center and was a kind of a dull thump. After thinking the matter over, I decided to try one more "wrinkle." We removed the cylinder head and placed washers, $\frac{1}{4}$ in. thick on the false head springs *B*, Fig. 2, which would put more pressure on the false head, the washers being put on the top of the spring. When we started again the pound was gone.

Old Ammonia Withdrawn

During the time we went over the compressor end of the machines, we also rebored two of the steam cylinders of the compressors, inserted a few new piston rings, refaced the steam valves and seats, lined up the bearings, etc., reset the valves and covered every detail in the best possible way. After this had been covered, we blew out the expansion coils with steam and air, these coils being heavily loaded with oil. The condensers got the same treatment, and at the same time we purged a lot of air and foul gas from the system, which had boosted the condenser pressure about 30 lb. above normal. In cleaning the coils and condensers, we took first one condenser and cleaned it and then cleaned a certain part of the expansion coils, probably one brine tank or a set of direct-expansion coils in some of the storage rooms. We then started one machine on the clean system and continued in this manner, charging new ammonia into each part of the system cleaned as we went along. In this manner we finally collected all the old impure ammonia in one condenser, from which it was drawn and sent to the manufacturer for purification.

About three months later, during the latter part of the spring, I dropped into the plant for a visit. The owner was the most pleasant man one could wish to meet. He drew out some records of the previous season and also some of the most recent. In these were shown the output of the packing plant during both periods and also the coal consumption. With considerably greater output at this time the coal consumption was only about 65 tons per day as against an average of 80 for the preceding year. Best of all, everyone was satisfied. Whenever the new chief asks for anything within reason, he gets it.

An Expert in Economical Power Production—Foster McKenzie, of Cleveland, Ohio

> *"The chief engineer of a power plant is primarily engaged to operate the station so as to produce power at the lowest possible cost, consistent with good service and with keeping the equipment in first class repair. One of his most essential requirements is the ability to follow out the intent of the designer of the mechanical element and to fit the human element to work in synchronism with it. In other words he must be a diplomat with some engineering ability."*—Foster McKenzie.

FROM the time of James Watt, there has been a very close connection between the steam power industry and men of Scottish birth and ancestry. This hardy race which has turned out so many first class engineers and technical men, has a worthy son in Mr. Foster McKenzie, until recently in charge of the Cleveland Municipal Light Plant. Mr. McKenzie was born in Scotland, April 17, 1881, and was educated at the Hamilton Academy in that country. Our first records of his activities in America show him as master mechanic and chief engineer at the Newark (N. J.) plant of the Sherwin-Williams Co., the largest paint and varnish manufacturers in the world. In his nine years at this plant he made some very good records in the cost of power production.

Mr. McKenzie's real opportunity came with his employment by the Municipal Light Department of Cleveland. The Cleveland plant, which represents a $3,000,000 investment, was designed by F. W. Ballard, and built with a view to obtaining the very best possible efficiency in power production. With this end in view the plant was equipped with carefully selected apparatus and types of measuring instruments that might help in obtaining economical operation. Boilers and stokers were equipped with automatic regulating devices, steam flow meters, draught gages, etc. All boiler feed water was to be measured, and its temperature, before entering the economizer and the boilers, shown by recording thermometers. Coal was to be bought on the heat unit basis and tests made daily. In fact the designs called for recording the operations of the plant to the most minute detail.

The designers realized that the men who operated the plant would be a more important factor than the apparatus or instruments installed. Putting then, the same thought and care into the selection of the chief of operation as into the selection of apparatus, they chose Mr. McKenzie. He was called to the plant about a year before it was completed and spent that time in supervising the installation of the equipment. That he has justified the confidence of those who selected him, is shown by the fact that the consumption of coal has been brought down to a little less than two pounds per kilowatt-hour.

FOSTER McKENZIE

Mr. McKenzie has given us some interesting figures and information about this plant. With coal at $2.10 per ton, bought on a 13,000 B.t.u. basis (with a bonus for higher, and a penalty for lower heat values) the cost in 1916 was 0.3 cent per kilowatt-hour at the busbars. This cost covers fuel, maintenance, labor (including switchboard operators), and sundries, but does not include any taxes or depreciation charges. At about this time the average CO_2 in the flue gas was 16 per cent, and an average thermal efficiency of 13.5 per cent was obtained.

At this plant the scheme of using condensate from the main surface condenser as cooling water in the condenser for the "house alternator" was originated. It had been found that, at the vacuum carried, the temperature of the feed water was so low as to cause the economizer tubes to sweat. Using the feed as cooling water in the smaller condenser raised the temperature to a point where no further trouble was experienced. This idea was later successfully adopted at the new Conners Creek station of the Detroit Edison Co.

Like many of our prominent professional men, Mr. McKenzie gave liberally of his time to Government service during the war. He served as a member of the Committee of Fuel Conservation for the Cleveland district throughout the period of hostilities. He was also a member of the United States Public Service Reserve.

He is now a member of the firm of the McKenzie Engineering Co., also chief engineer and a director of the Ohio Pipe Bending Co., and chief engineer and member of the Plastic Metallic Packing Co. In connection with the McKenzie Engineering Co. his work has been largely devoted to problems of burning fuel oil under steam boilers.

Harrisburg Dual-Clearance Engine

It is common knowledge that two diametrical opinions are held as to the relative merit of auxiliary exhaust valves on uniflow engines, the difference of opinion effecting engines operating non-condensing

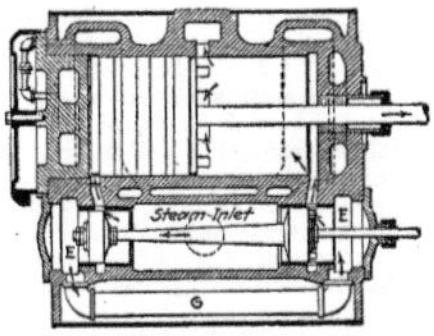

FIG. 1. PISTON COMMENCING STROKE. VALVE JUST OPENING AND ADMITTING STEAM. EXHAUST PASSING TO ATMOSPHERE THROUGH CENTRAL PORTS AND EMPTYING AUXILIARY CLEARANCE CHAMBERS E G E

and the value of the valves as a safeguard against accident due to excessive compression.

Auxiliary exhaust valves when used with uniflow engines are for the purpose of delaying the commence-

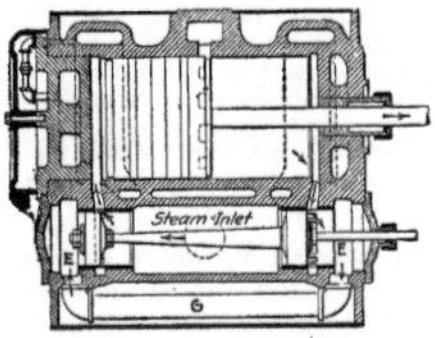

FIG. 2. STEAM VALVE OPEN, ADMITTING STEAM BEHIND PISTON. VALVE OPEN AT OPPOSITE END. CENTRAL EXHAUST PORTS CLOSED. COMPRESSION BEGINNING IN CYLINDER AND AUXILIARY CLEARANCE CHAMBERS E G E

ment of compression by relieving the cylinder of more of the expanded steam than can escape through the

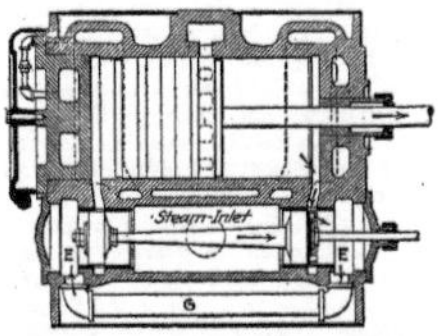

FIG. 3. PISTON IN POSITION OF CUTOFF, VALVE HAVING REVERSED ITS MOVEMENT AND CLOSED PORT SHUTTING OFF SUPPLY OF STEAM TO HEAD END OF CYLINDER. EXPANSION COMMENCES. VALVE STILL OPEN AT OPPOSITE END AND COMPRESSION CONTINUES IN CYLINDER AND AUXILIARY CLEARANCE CHAMBERS E G E

central exhaust port, and so reduces the amount of final clearance that must be provided.

The latest development in engines of the uniflow type has been brought out by the Harrisburg Foundry & Machine Works, Harrisburg, Penn., in what they term the dual-clearance single-cylinder uniflow engine. This engine is designed with two clearances and the

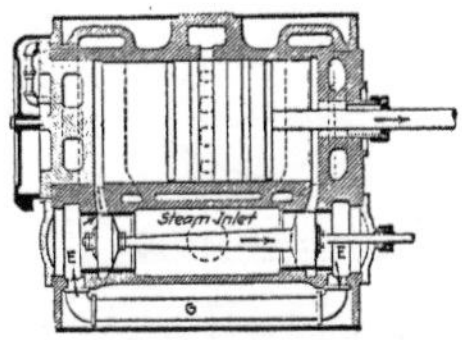

FIG. 4. OUTER EDGE OF VALVE OPENS AT HEAD END ALLOWING STEAM FROM AUXILIARY CLEARANCE CHAMBERS E G E TO PASS INTO CYLINDER, MIXING WITH THE EXPANDING STEAM AND EXPANDING WITH IT UNTIL CENTRAL EXHAUST PORTS ARE OPEN TO ATMOSPHERE. OPPOSITE END OF VALVE HAS CLOSED PORT, AND COMPRESSION OCCURS IN CYLINDER AND CYLINDER CLEARANCE ONLY

accompanying illustrations show how the steam is expanded from a small clearance and, by means of

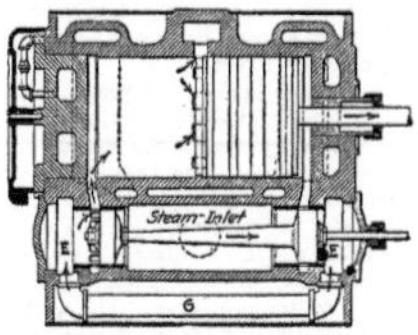

FIG. 5. CENTRAL EXHAUST PORTS UNCOVERED BY PISTON. STEAM IS EXHAUSTING, REDUCING PRESSURE IN CYLINDER AND AUXILIARY CLEARANCE CHAMBERS E G E TO EXHAUST PRESSURE

the single-balanced piston valve, is allowed to compress first into a large chamber and finally, as the piston nears the end of the stroke, into a smaller clearance.

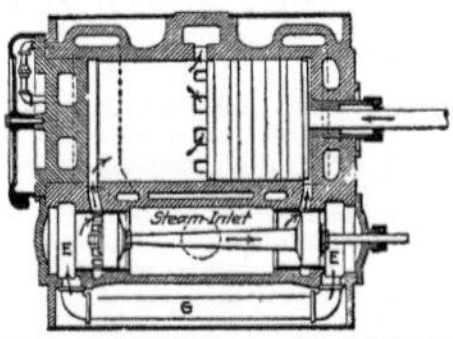

FIG. 6. STROKE COMPLETED. PISTON END OF CYLINDER READY FOR RETURN STROKE. VALVE OPEN TO ADMIT STEAM. VALVE AT OPPOSITE END STILL OPEN AND EXHAUST CONTINUES

The steam that has been compressed into the large chamber enters, due to the automatic action of a valve, into the cylinder at the opposite end and, mixing with the expanding steam on the return stroke, expands with it and passes out through the central ports to the exhaust line. An examination of the accompanying illustrations and the captions fully ex-

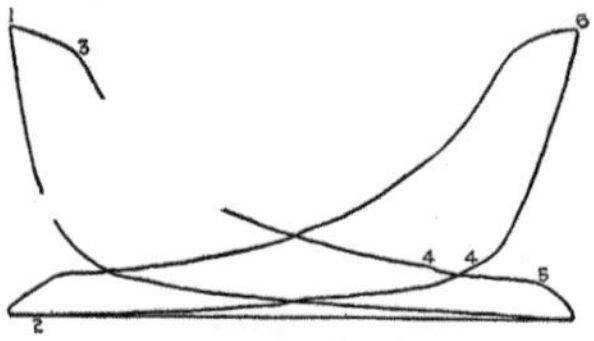

FIG. 7. DIAGRAM FROM DUAL-CLEARANCE ENGINE. FIGURES INDICATE POSITION OF PISTON AND VALVE FOR THE EVENTS OF THE CYCLE

plains the action of the valve and its control of the steam to and from the auxiliary clearance chambers.

Peculiar Flywheel Accident

By E. E. Clock*

The following is an account of a flywheel accident which occurred at the plant of the Goodyear Rubber Shoe Co., of the United States Rubber interests, at Naugatuck, Conn., on Jan. 19 last:

The wheel was on a tandem compound condensing Hewes & Phillips engine of 800 hp. built in 1905, and had been used previously during the life of a former engine of the same make and had been dismantled and reassembled about 15 years ago upon the installation of the present engine. The joints were made up with O links which had been cut at the time the wheel had been dismantled and had been rewelded, some of the welds giving way and showing poor work. The weakness of these links no doubt was the largest factor in the sad results, as the wheel evidently went to pieces at a speed much lower than the computed bursting speed, for investigation showed that the wheel went to pieces some seconds after the throttle had been closed tightly.

There is no doubt, however, that the links had begun to give way previous to the closing of the throttle and that the reaction caused by the drag of the pistons through the closing of the throttle produced some flexure of the arms that had a disastrous effect on the weakened links.

The sequence of events during and just previous to the accident was as follows: The engine was connected to its load by an electrically operated clutch with switches located along the line of heavy shafting which drove the rolls. There were two engines just alike side by side, both connected to a common condenser located in the basement. At five minutes to one on the fatal day the chief engineer went to the basement as usual to start the condenser, the two other men standing ready to start their respective engines when the vacuum gages should show that all was ready. Both started up in regular form, but No. 2 engine ran overspeed because someone had tripped the electric clutch, leaving the engine with no friction load. This caused the governor to go to the extreme top, and being equipped with the Hewes & Phillips trip device, which dropped the governor cutoff mechanism to the bottom or no steam position, the engineer without closing his throttle, got up on the engine and lifted up this cutoff mechanism one or more times, probably expecting every second that the whistle would blow and the load come on. In the meantime noting the increase in speed one of the other men closed the throttle and the man on the engine jumped down and gave the throttle an extra pull. All then ran out into the yard through a door in line with the wheel; a second later the crash came, and when the dust cleared up the man who had been on the engine was found in the yard so badly injured that he lived but a few hours.

The rim of the wheel was 15 in. square and weighed 40,000 lb. made up in eight sections. Two sections still held together by the links passed up through four floors of heavy mill construction including heavy yellow-pine timbers and then fell back through the fourth floor in another spot, landing on the third floor and in its flight badly injuring one man and damaging considerable material and machinery. Another section of the wheel passed through the brick wall in front of the building and landed on the street, miraculously missing the many people constantly passing there.

The lesson to be learned from this accident is to know that your men will do the right thing in cases of unusual emergency, and to anticipate those emergencies and be prepared for them in advance. Men who can be trusted to do the right thing in a routine way should from time to time be drilled in the matter of all possible emergencies. The next best cover for such failures in the matter of engines probably is the installation and the daily testing of a good reliable mechanical speed limit and one that can not be put out of action in the interest of keeping the engine from shutting down at any cost.

*Safety engineer.

Hydro-Electric Power in France

Hydro-electric power development has been given much attention in France and is being rapidly made available for use. If fully developed, this power will total 9,300,000 hp. according to the "Index to French Production." In the report of the American Industrial Commission to France (1916) the total on low-water flow was estimated at 4,600,000 hp.

Until 1914 the total developed hydro-electric power was approximately 800,000 hp., and when the plants recently completed are in full operation and those under construction have been finished, the total will be double the 1914 figures. Shortage of coal during the war stimulated water-power development, and a number of the railroads and industries throughout the country, which have suffered through lack of fuel, are considering the development of hydro-electric power as the only solution of the problem.

The River Rhone is a fine example of the possibilities of water-power development in France. No less than 750,000 hp. can be made available by this river it is estimated. This is equal to a saving of 5,000,000 tons of coal per annum. Operations for the ultilization of this power have already started, and as a number of large manufacturing cities are within transmission distance of the river, it is believed that a ready market for the power will be found. The cost of this work will be in the neighborhood of $125,000,000, according to figures taken from the data gathered by the Direction Gènerale des Services Français aux Etats Unis, New York City.

EDITORIALS

Plant Successfully Operated Under Difficulties

THE leading article in this issue describes the power equipment of the Chile Exploration Co. at Tocopilla. Although in general the equipment does not differ radically from that found in a modern American plant, the conditions under which it had to be constructed and operated make it intensely interesting. Whatever may be the problems involved in the design and construction of power plants in the United States and Canada, earthquakes do not enter into them. In the plant under consideration overcoming or at least minimizing the effects of earthquakes, was one of the important problems to be solved. Where there is no fresh water other than that distilled from the sea, conservation of feed water is paramount, for every pound of steam wasted means one pound of feed water that must be replaced by the burning of expensive fuel. When a plant is built along the seaboard, it is generally afforded the protection of a good harbor. At Tocopilla a rocky coast line, rising sharply out of the sea and exposed to the waves, had to be contended with, and complicated the circulating-water and fuel-handling problems. When the nearest reliable fuel supply is approximately four thousand miles away, this in itself is something to be considered seriously before a plant of fifty thousand kilowatts capacity is attempted, let alone transmitting the power across ninety miles of desert. The design and construction of a plant to operate under such conditions make it of more than passing interest.

The favorable results obtained in this plant with distilled feed water make this method of obtaining a supply of boiler feed, where the natural source of supply contains a large percentage of impurities, worthy of more attention than it has received in this country. Especially in this true since the loss of heat is very small if the evaporators are properly insulated. Two factors have stood out to a high degree in the operation of this plant: namely, absence of scale in the boilers and corrosion of the steel-tube economizers. Where boilers are being forced as they are in our modern power plants, the absence of scale is vital in preventing the burning of tubes.

In the operation of the electrical end two features in particular warrant comment—one the method of cooling the transformers, and the other the washing of the insulators on about eight miles of the transmission line nearest the coast. Not infrequently serious trouble has been experienced with the cooling coils of water-cooled transformers becoming clogged with scale. At Tocopilla the only source of cooling water available is the sea, consequently it is not difficult to imagine the trouble that would be experienced if the water was circulated through the cooling coils. By placing the cooling coils in a tank of water and pumping the oil through them, it has been possible to eliminate troubles from this source. Although this practice has been adopted to a limited extent here, there are apparently many conditions where it could be used to good advantage. The duplicating of the first eight or ten miles of the transmission line so that the salt deposits can be washed off the insulators of one line while the other is in service evidently gives a constructive suggestion toward solving some of the transmission troubles experienced in this country at railway crossings and places where the insulators collect deposits that cause them to flash over.

Apart from the solution of the power problems there stands the engineering achievement that has made possible the tapping of great mineral resources in a section of a country where a burning sun and drifting sands hold sway. In this we find another brilliant example of what the application of power can do for the betterment of mankind.

Safety in Storing Bituminous Coal

"ANOTHER objection to the storage of bituminous coal is its tendency to spontaneous combustion. It cannot be denied that this danger is a real one with improperly stored coal. It has been proved, however, that bituminous coal can be stored safely and that it will suffer little or no depreciation if reasonable care be used."

The preceding paragraph is quoted from a recent *Power* editorial on methods of stabilizing the bituminous-coal industry. The question as to what is "improperly stored coal" is answered in a paper on "How Bituminous Coal Can Be Stored Safely," by Prof. H. H. Stoek, an abstract of which appears in this issue. Space limitations prevent the printing of this paper in full, but the abstract is well worth study.

The causes of spontaneous combustion seem very simple. If air can get at the coal, oxidation will start. Oxidation causes heat and if the ventilation is not sufficient to carry off this heat, the temperature of the coal will rise and the oxidation will become more and more rapid. The remedy or preventive seems equally simple. Either give the coal plenty of air or cut off the supply entirely. The difficulty lies in the fact that the ordinary coal pile is well designed to admit just enough air to cause trouble. However, as Professor Stoek pointed out, properly piled coal overcomes this tendency to a very great extent.

A study of some coal-storage arrangements shows that a campaign of education is badly needed. After presenting his paper, Professor Stoek showed a number of lantern slides from photographs of coal-storage piles that had fired. In many cases the side or back of the boiler setting formed one side of the coal bin. Coal was shown piled on top of steam pipes and around chimneys. With this sort of storage it is not to be wondered that fires result. Where coal is stored in larger quantities, the pictures showed that almost every rule suggested in the paper had been violated.

In short, bituminous coal is often stored with no more care than would be given to storing sand, and the result is usually disastrous. If intelligence and care are used, the chances of trouble are practically eliminated. To Professor Stoek's advice, "Before you store coal, be sure you know how to do it," we would add, "You can learn how to store coal, and it will pay you to learn and to apply your knowledge."

Scientific Study of Boiler Design

THE paper by F. W. Dean and Henry Kreisinger, on page 348 *et seq.* of *Power* for March second, shows the possibilities of a thorough scientific study of boiler and furnace design. The building of a thousand wooden ships by the Emergency Fleet Corporation, each requiring two boilers of twenty-five hundred square feet of heating surface each, made it worth while to test the design adopted in the most thorough manner for possible improvements in efficiency as well as for dependability.

Never, probably, has a boiler been the subject of so searching a surveillance of its performance or of so many experimental modifications. Apparatus was provided not only to determine, with an unusual degree of accuracy, the boiler and furnace efficiency, but with a rare degree of refinement the temperature and composition of the gases were determined at numerous points in the furnace, passes and uptake, the water circulation was observed in direction and activity, the temperatures of the casing, inside and out, were measured, and by changing the baffles, modifying the manner of admitting air over the fire, etc., the efficiency of the boiler was increased from about sixty to about seventy-two and a half per cent.

Would not an equally systematic and intelligent investigation result in a corresponding improvement in some of our commercial installations?

Instruments in the Steam Plant

IT IS nice to have a lot of books. It gives one a scholarly, cultured air and smacks of wisdom and erudition. But smack and air are all they will amount to for one who simply sits and looks at his well-filled shelves and smokes his pipe.

If one could get into one's head all that there is in a book by paying for it, we would all be Solomons. To absorb it requires time, interest, study, thought and application—not to learn a book by heart, but to get a commanding sense of what is in it and be able to use it to advantage.

It is nice to have a lot of instruments around a plant. It looks scientific and impressive and smacks of technical training and efficiency; but they will never pay a dividend on their cost if they are not kept in condition as instruments of precision, if they are not continually or frequently used, and if their indications and records are not studied understandingly and the deductions therefrom intelligently applied. There is no use of providing an engineer with an indicator if he does not know how to get a correct diagram with it and, what is of more importance still, if he does not know the full significance of the diagram and, with it as a guide, keep his engines, pumps and compressors in adjustment and maintain a fair record of his output. Coal-weighing and water-measuring apparatus is of no value to a man who does not understand how to use their indications to determine how well his plant is doing and why it is not doing better. Boiler efficiency cannot be improved by putting in a CO_2 recorder if one does not keep it in operation and know something more of what it tells than that a high percentage of CO_2 is, in general, most desirable.

The only instrument of which most steam plants can boast is a pressure gage, ordinarily within ten per cent of right. The engineer does not know how much water his boiler evaporates per pound of coal, how much steam his engine takes per horsepower or how much power it develops. He does not know the temperature of the feed water, the intensity of the draft or how hot the gases are leaving the boiler. He has no means of telling how completely the fuel is burned or how much more air than is necessary to burn it is heated to the uptake temperature and discharged to the stack. He ought to be provided with the instruments and apparatus necessary to determine these and other things, but it is useless to do this unless he takes an intelligent interest in their use and translates the stories that they tell into a record of the greatest efficiency attainable under the conditions.

Publicity for Engineering Societies

NEVER before has engineering figured so prominently in the solution of vast problems to the nation as has been brought about by the conditions of reconstruction. As a consequence the public is receptive as to activities of the engineering profession, yet they are kept, to a certain extent, in the dark as to engineering accomplishments that are going on every day, and plans for the future. The daily press makes little attempt to collect information from the engineering societies, while on the other hand the societies have no facilities for proper dissemination of such news.

A publicity bureau as a medium for communication between the societies and the technical and daily press is suggested. Such publicity work could function through an established technical publicity bureau, or the societies could jointly establish one of their own in charge of a man with a broad and comprehensive knowledge of the requirements of such work. The first plan would seem to possess certain advantages at the start by making use of existing connections and might perhaps prove less expensive, although in the long run the second plan might prove the better. In any event the need of such publicity is urgent and it is hoped that early action may be taken by the societies on this matter.

The effectiveness of competent inspection in the prevention of boiler accidents is shown in the Annual Report of the Chief Inspector of Locomotive Boilers, in which it appears that the number of accidents caused by the failure of the locomotive boiler and its appurtenances shows a decrease of 60.2 per cent for the year ending June 30, 1919, as compared with that ending June 30, 1912, the first year in which the law was in operation. The number of persons killed was decreased 50.5 per cent and the number injured 58.9 per cent.

CORRESPONDENCE

Height of Return-Tubular Boiler Setting Above Grates

In the Feb. 10 issue of *Power* my young engineers have observed your answer relating to the height of settings for a 72-in. return-tubular boiler. I believe that if this was brought to your more careful attention you would like to make a statement that the boilers should be set four to five feet above the grates, which we find modern practice rather than 30 to 36 inches.

We have put in several installations with boilers set at these heights, and I must say that the results are very gratifying. JOHN A. STEVENS.

Lowell, Mass.

[The higher a return-tubular boiler is set above the grates the less chilling effect in hindering the perfection of combustion; but less advantage is obtained from the receipt of heat by direct radiation from the fire and there is more heat lost by infiltration of air and radiation from the greater extent of the side walls. The question under discussion is: Under average conditions of draft, fuel and firing, what height of boiler setting of a given size obtains combustion that results in the highest average efficiency? It would be interesting to know under what operating conditions it was found that a height of 36, 48 or 60 in. for a 72-in. boiler gave the best boiler efficiency under actual test.—Editor.]

Characteristics of Chain-Grate Stokers

In the Jan. 27 issue of *Power*, H. F. Gauss, on page 144, describes some of the characteristics of a chain-grate stoker. He attributes the frequent failure of the coal to ignite to the large size of the coal permitting a considerable volume of air to pass through, thus retarding or preventing ignition. An instance is cited in which steam was admitted under the ignition arch and the successful result was assigned to the steam pressure preventing the ingress of excess air.

It would appear that steam at the necessary pressure and in the quantity required to produce the result as described, would be sufficient to prevent ignition altogether; this, because the steam must be superheated to the ignition temperature of the coal before ignition could occur.

An installation as described by Mr. Gauss, in which steam was blown under the ignition arch, was brought to my notice about a year ago. It occurred to me that the design of the ignition arch was faulty in that the surface was horizontal, forming a pocket into which the reflected heat rays from the incandescent fuel could not enter. The effect of the steam injection was to induce a current of hot gases from the furnace which ignited the coal.

Experiments on arches that I have made have resulted in effects varying from failure of ignition in the case of flat arches to the destruction of the coal gate, in the case of arches with too much rise.

Kansas City, Mo. C. O. SANDERSTROM.

Boiler Blowoff Pipes

Recently the elbow on the blowoff pipe of a 40-hp. boiler failed under 100 lb. pressure. Although the boiler is rated at 40 hp., it was working at around 70 hp. Luckily, I had previously closed the steel fire-door separating the boiler room from the factory.

As the combustion chamber had not been cleaned out for some time, there was considerable soot all over the place, because when the explosion was heard someone opened the door leading to the boiler room, but he did not linger long. The soot also plugged the tubes so tight that it required five men to punch them out.

This pipe between the boiler and the side wall was unprotected, and that is what caused the trouble, as the hot gasses had burned the pipes and the elbow, although it had been in place but about a year. The boiler is blown down every morning, and upon examination no trace of scale could be found. When a new pipe was put in, it was protected by a brick baffle. The owners had been told by the inspectors to put some kind of protection around that pipe, but had neglected to do so. R. G. SUMMERS.

Syracuse, N. Y.

Straightening Shafts By Heat

The method of straightening a shaft by the application of heat to the high side, as described by E. E. Clock on page 377 of *Power* for Sept. 2, is entirely practicable. We straightened shafts nineteen years ago at Erie, by pouring hot babbitt against the side that needed to be shortened.

We have also used this method to straighten milling-machine platens which became bent up in the middle of their length, the same as planer platens, because of the peening effect of handling work on their top surface.

The process is the reverse of peening and, like it, is but temporary in its effect. It should be used only to counteract some similar strain. J. A. STRONG.

Erie, Penn.

Removing a Tight Pin from a Casting

An easy way of removing a tight fitting pin from a casting is by the use of an ordinary stillson or pipe wrench and a cold chisel or wedge. The chisel is placed flat against the casting, and the wrench is applied to the pin in such a way that when the pin is turned,

ONE WAY OF REMOVING A PIN FROM A CASTING

the chisel will act as a wedge or lever and tend to draw the pin out. The method as shown in the sketch practically explains itself. W. W. PARKER.

Firestone, Col.

Burning Fuel Oil

Before coming to my present power plant, I had charge of a large coal-burning plant equipped with stokers that were supposed to be the last word in mechanical firing. It was no uncommon thing to ask and receive 150 per cent rating of the boilers, and this without any extra strain apparently on the firing machinery, but there was always the possibility that a fire would clinker and cause trouble and there was also the possibility that a stoker might fail. In fact, they did fail at times. And then again there was always a chance that the crusher elevator scraper or some other part of the coal-handling machinery might get out of order. Therefore a stock of working parts for this machinery was carried.

For a coal-burning plant it was modern in all ways and the amount of help required per thousand horsepower was small, but there is no comparison between the coal-burning and oil-burning plant. Imagine several boilers under forced and induced draft, each boiler furnace containing around three tons of coal at a bright cherry, not white, heat. The noon whistle blows and then the several safety valves blow.

Of course the water tender lowers the water at 11:30 and has all the available space posible for storing new water at 11:55, but with the fires necessary to keep the wheels turning and the kettles boiling it is almost impossible to hold the safety valves shut. Thus good coal in large amounts, figuratively speaking, goes through the safety valves. It is almost impossible to prevent this waste.

In the oil-burning plant no thought of steam going through the safety valves is present. At 11:58 the oil is shut off, at 12 m. the oil is put on two, three or four boilers—just enough to hold steam for the noon hour—and no excitement prevails whatever.

With oil, no thoughts are given to ash or clinker; no thoughts are given to fires in the coal pocket. No thoughts are given to disgruntled firemen. All thoughts are centered on the temperature of the oil in the main storage and auxiliary tanks. Temperature is the watchword in the oil-burning plant. A difference of five degrees in the temperature of the oil will cause smoke.

This is the first lesson the boiler-room engineer learns. He next learns that even a trifling raise of pressure in the oil line causes smoke unless a like increase is given to the steam line feeding the burners. Burners will score, owing to the action of the oil and steam passing through them, and a scored or grooved burner will cause smoke. Apparently, the flames from two burners touching causes smoke. How do we operate without smoke? We don't operate without smoke at times, and sometimes the heavens are blackened most outrageously by smoke from our stacks, but not for long.

By a careful watch of conditions, a micrometer adjustment of valves and the ever-present thought of temperatures and oil pressures, the plant is able to do 150 per cent rating under adverse conditions.

New York City. C. W. PETERS.

Why the Piston Would Not Come Out

Although solid-head internal combustion-engine cylinders are not built to any extent at present, many of them are in use, and this brings to mind the projection illustrated by the sketch, the purpose of which

SHOWING THE PROJECTION THAT WAS SAWED OFF

should be obvious. In one instance the owner sawed this projection off, evidently under the impression that it was useless. When the piston was replaced in the cylinder, it was pushed in too far and the upper ring, extending past the cylinder top counterbore, expanded and locked the piston securely into the upper end of the cylinder. H. H. PARKER.

Oakland, Cal.

Repairing a Boiler-Plate Lamination

Recently, a lamination was discovered on the bottom of a return-tubular boiler, directly over the furnace. A patch was suggested, but as it was not considered advisable to have one applied in such a location, it was suggested that the lamination be cut out and the cavity filled with metal by means of the electric-welding process. The lamination appeared to be only about 1 x 1½ in. and about $\frac{3}{16}$ in. deep, but by the time we had it

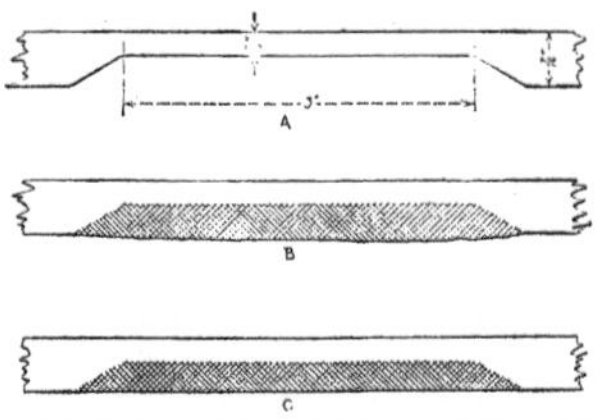

LAMINATION REMOVED AND METAL WELDED IN PLACE

cleaned out, it proved to be 3 in. in diameter and ¼ in. deep, leaving $\frac{3}{16}$ in. good plate as at *A*. The cavity was filled with metal *B* above the surrounding plate and ground down flush so that there is now no evidence of a fault as at *C*. If the boiler does not convert itself into a projectile, we may assume that the operation is a success, but if it turns up missing some time, we will know what did it.

Since this job was completed I understand that the boiler-insurance companies have sanctioned such practice, but have prohibited another that I have seen done; namely, re-ending boiler tubes. Our Department of Public Works, Boilers' Branch, permits re-ending by the autogenous process, but when my suggestion regarding filling the cavity of the lamination was put up to them, they passed it on to the insurance company, and the inspector who sanctioned it, I have since learned, did not know that such a method was practiced. I think that the practice should be all right. R. McLAREN.

Toronto, Ont., Canada.

Wet Floor Caused Ground on Motor Circuit

In a large industrial plant several employees received rather unpleasant electrical shocks when attempting to turn on the water at a sink. It was known that several motors were grounded to the water mains, of which the supply pipe to the sink was a branch, but this offered no solution of the puzzle. The sink was not grounded, as it was suspended by brackets to a wooden wall, and the waste pipe only extended through the wall, the discharge falling into the river below. However, even if the sink had been grounded, the water mains themselves would have furnished a much better path to earth, in case of a grounded motor, than would have been possible through the body of a man at the sink.

The floor around the sink was always more or less water-soaked, and an investigation here finally disclosed the source of the trouble. At a distance of a few feet, and well within the wet area, the conductors of a 440-volt motor circuit were carried down the wall on cleats and passed through the floor into the room below. These wires had each been originally protected at the floor line by porcelain-tube bushings, but in repairing a section of the floor the carpenters had allowed the tubes to slide down, and then cut notches in the floor boards to admit the wires only. One of these notches had been made such a close fit that when the floor became water-soaked, the wood swelled very tight around the wire. Of course its insulation in turn became soaked and useless. The moistened area of the floor extended but a few feet, and a path to ground was completed only when someone stood on the wet floor and touched the water pipe. HAROLD B. PHILLIPS.

Appleton, Wis.

Making a Pair of Pliers and Screwdriver More Useful

Quite often, when using a pair of pliers in gripping a small dowel pin to pull it out, it is impossible to get strong enough grip with the strength of one's hand. In such cases the kink shown to the left in the figure will be found handy—that of using a turn or so of heavy string or some small wire and twisting it tight with a

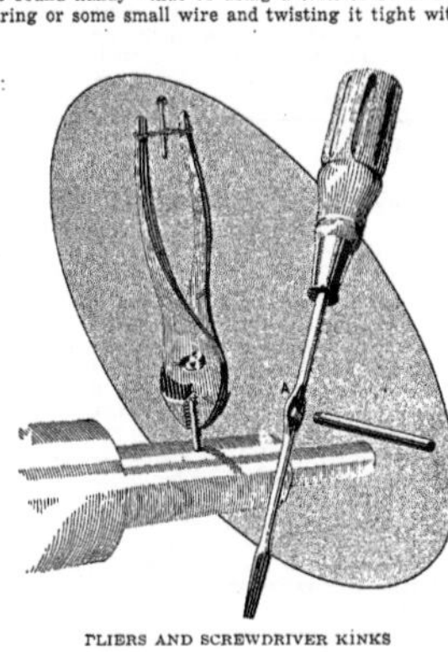

PLIERS AND SCREWDRIVER KINKS

nail. This trick can be used to convert the pliers into a temporary hand vise.

On the right of the illustration is shown a simple way to convert an ordinary round-shank screwdriver into a powerful type of lever driver. The section *A* is simply flattened and then drilled and tapped to take a threaded handle, which is a small bolt with its head cut off. The handle is removed when it is desired to use the screwdriver without it. CHARLES H. WILLEY.

Concord, N. H.

Clamp for High-Pressure Hose

We use a method of repairing a leaky hose that is employed in the mines, as follows:

The hose is cut at the leak and for a ¾-in. hose a 1-in. nipple 6 in. long is inserted halfway in the hose. A ¾-in. nipple 7 in. long is used for a 1-in. hose, such as is illustrated herewith. The hose is clamped to the nipple by the type of clamp shown.

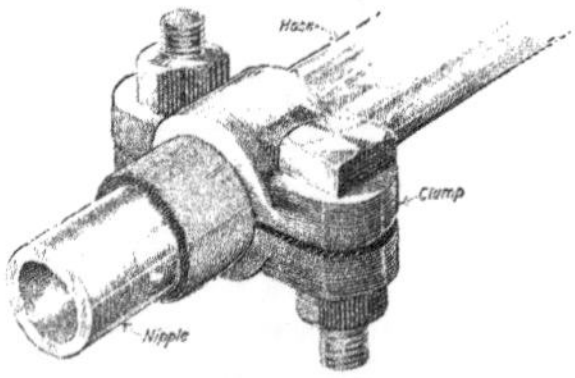

HOSE COUPLING CLAMP

Sometimes a marlin spike is used in the hose to enlarge it a little.

Under high pressure and where the hose is subject to hard usage, several strands of wire can be run on each side from one coupling to the other and twisted as desired.

COLIN F. MCGIBBSON.

Butte, Mont.

Location of Secondary Relays Affect Voltage Regulation

A company operating a number of induction feeder regulators mounted the contact-making or primary relay upon the switchboard and the secondary or reversing relay upon the regulator. However, this secondary relay was not mounted as recommended by the manufacturer, but, instead, was supported by iron strips to the top of the regulator case. This arrangement was inexpensive and appeared quite satisfactory. However, in one of the installations where a new type of regulator was installed, difficulty developed in that the regulator was not functioning properly. Sometimes the voltage would rise as the load came on, then would remain high although the load decreased, and vice versa. This did not always happen, and sometimes the regulator would operate for days and even weeks without giving any trouble.

At first routine troubles were looked for; the contacts of the contact-making voltmeter were cleaned and adjusted for correct voltage range; the contacts of the reversing switch were overhauled and adjusted; the regulator brakes were tested, the drums oiled and the tension of the brake shoes reduced. These adjustments had no apparent effect, and intermittent short-circuits were suspected, but no trouble of this nature was found.

What was found was that at certain times of high voltage and therefore increased flux density in the regulator iron, and when a certain frequency of the system occurred, possibly due to current changes, the regulator was caused to give rise to severe vibration. This was in turn passed on to the secondary relay, which also vibrated under the influence of normal operating current passing through it. In this manner the natural periodicity of vibration of the relay had superimposed upon it the vibration of another periodicity. The effect of this vibration was that while the contacts of the secondary relay appeared to be making good mechanical contact, the circuit to the regulator motor was not permanently closed. Instead, the circuit was rapidly opened and closed without apparent movement of the relay contacts so far as the naked eye could see. Moreover, the two poles of the relay might or might not close simultaneously. Since a three-phase induction motor cannot start up single-phase, the result was that either the motor could not get started in time before the circuit was interrupted or else it could not start at all. In either case the result was the same—poor voltage regulation.

The remedy was simple. The secondary relays were left in their original position. But instead of being bolted down tight to the iron strip supports, so that vibration of the induction regulator cases was transmitted to the relays, spring washers were placed underneath the case of each relay, at the four corners. The bolts were then inserted and the fastening nuts screwed down so as to prevent the bolts from falling out. The relays were allowed free movement up and down, always resting on the spring washers and yet unable to fall off the supporting strips. This solved the problem and incidentally eliminated much burning of contact-making contacts.

B. W. SMITH.

Cicero, Ill.

Method of Boring a Large Bearing on a Small Lathe

The illustration shows how a bearing was bored out on a small lathe that was not strong enough to hold the weight of the bearing and had to be blocked up to prevent breaking the bed. The tool post was removed,

BEARING READY FOR BORING OUT ON SMALL LATHE

and the carriage was made to fit the bearing by the use of channel irons. The boring bar was a piece of scrap shafting with a set collar to hold the tool. The bed of the lathe was supported by means of iron stands wedged tightly between the lower side of the bed and the floor. This way of boring large bearings proved very satisfactory and has been used on a number of different bearings.

W. R. CAIN.

Buffalo, N. Y.

INQUIRIES OF GENERAL INTEREST

Triple Riveting of Lap Joints—Why are lap joints of high pressure steam drums frequently made with three or more row of rivets? F. C.

A triple-riveted lap joint may be designed of higher efficiency than a double-riveted lap joint, and the greater width over the rows of rivets renders it easier to bring the plates into line, with the result that there is less bending than in the single- or double-riveted lap joint.

Turbines One-Speed Machine—Can a steam turbine be constructed that will develop a normal rated power at different speeds? A. L. M.

The turbine is a one-speed machine. For good economy the speed of the rotating buckets must bear a definite relation to the velocity of the steam as determined by the design. Otherwise there will be wasteful impact of the steam against the vanes and the speed of the turbine cannot be changed from that of best economy without serious loss of efficiency.

Transfer of Heat Requires Difference of Temperatures—If at sea level fifteen pounds of steam at atmospheric pressure is to let into an open barrel containing 150 pounds of water at the boiling point, what temperature would the water attain? C. E. G.

The temperature of dry saturated steam at atmospheric pressure is 212 deg. F., and the boiling point of water under atmospheric pressure is at the same temperature. There could be no transfer of heat because the temperature would be the same and the temperature of the water would remain unchanged.

Belt Thickness Influences Speed Ratio of Pulleys — Should not a 36-in. pulley drive a 12-in. pulley at three times the speed of the former? R. F.

If the pulleys operate as friction pulleys running in contact without slip, the 12-in. pulley will be driven at three times the speed of the 36-in. pulley. But if separated and the driving is done by a belt without slippage, the small pulley will be driven at something less than three times the speed of the large one on account of the thickness of the belt. In making calculations of speeds without slippage, an amount equal to the thickness of the belt should be added to the diameter of each pulley.

Obtaining Compression of Exhaust on Corliss Engine—How can compression of exhaust be increased on a Corliss engine? E. D. C.

With a double eccentric Corliss engine, advance the exhaust eccentric. As opening and closing of each exhaust valve occurs when the piston is at the same part of its stroke, earlier closure for obtaining more compression will also make earlier release after cutoff. With a single eccentric for operating all valves, advancing the eccentric hastens all valve events and causes the steam valves to open earlier. If the resulting lead is too much, the time of opening the steam valves must be delayed by adding lap to those valves. This is done by adjusting the length of the steam valve rods so the opening edges of the steam valves will lap over their ports when the wrist plate is in its central position, giving $\frac{1}{32}$ to $\frac{1}{16}$ in. opening for lead when the piston is at the beginning of its stroke.

Induction-Motor Winding Bracing—Why are insulated rings placed over the ends of induction motor coils where they extend beyond the stator core and why in some cases are they placed on both ends and in others on one end only? K. A. R.

Frequently, where the coil throw is long or if the coils are not mechanically rigid, the ends are liable to be distorted when a heavy current flows through them. To prevent this, insulated rings are placed around the outside of the winding and the coil ends laced to these rings with heavy cord. This practice has been found to increase the life of the coils, since it tends to prevent any vibration of the coil ends that would otherwise occur. Where the throw of the coils is short, the rings are only put on the end of the winding opposite to the one where the cross-connections are made, the cross-connection giving the necessary support to the coil ends on the end of the stator winding where the connections are made.

Variation of Boiler Rating for Difference of Pressure—What would be the difference in rating the capacity of a boiler operated at 150 or 125 lb. pressure? V. A.

Boiler evaporative capacity is based on the rate of absorption of heat transferred to the water in the boiler, a boiler horsepower consisting of a rate that is equivalent to the transfer of heat required to evaporate 34.5 lb. of water from and at 212 deg. F., or 33,479 B.t.u. per hour. The lower the temperature of the water of a boiler the greater its difference of temperature from the temperature of the fire and heated gases and the more rapid the rate of heat transfer. Operation at different pressures also might be attended by difference in rate of transfer due to difference in the character of circulation of a boiler; but the differences in the rates of transfer of heat when generating steam at 150 or 125 lb. gage pressure would be negligible, and for all practical purposes the rating would be the same.

Twenty-five-Cycle Transformers on 60 Cycles—Will 25-cycle transformers 6,600-volt primary 220-volt secondary work on 60 cycles 6,600-volt primary 220-volt secondary? E. R. G.

Twenty-five cycle transformers can be operated on 60 cycles, but 60-cycle equipment cannot be operated on 25 cycles. For the same output the 25-cycle equipment will be considerably larger than the 60-cycle, consequently where 25-cycle transformers are operated on a 60-cycle circuit, the unit would be considerably larger than a 60-cycle transformer of the same capacity. When the 25-cycle transformers are used on 60 cycles, the exciting current and the flux density in the coil will be only about four-tenths what it is for the 25-cycle circuit. When 60-cycle equipment is operated on a 25-cycle circuit, the exciting current and flux density will theoretically be 2.4 times that of the 60-cycle circuit. However, due to the saturation of the iron core, the exciting current may be many times this value, and will cause the transformer to overheat even when operating without any load.

[Correspondents sending us inquiries should sign their communications with full names and post office addresses. This is necessary to guarantee the good faith of the communications and for the inquiries to receive attention.—Editor.]

Clamp for High-Pressure Hose

We use a method of repairing a leaky hose that is employed in the mines, as follows:

The hose is cut at the leak and for a ¾-in. hose a ½-in. nipple 6 in. long is inserted halfway in the hose. A ¾-in. nipple 7 in. long is used for a 1-in. hose, such as is illustrated herewith. The hose is clamped to the nipple by the type of clamp shown. Sometimes a marlin spike is used in the hose to enlarge it a little.

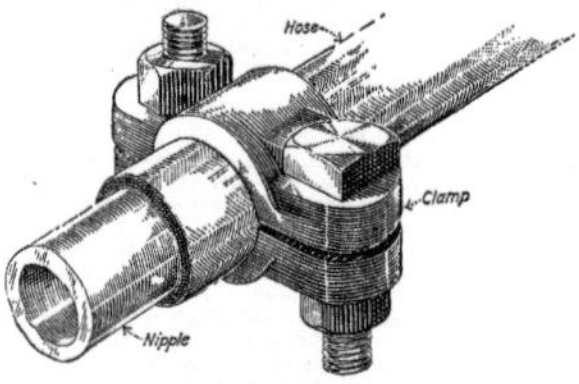

HOSE COUPLING CLAMP

Under high pressure and where the hose is subject to hard usage, several strands of wire can be run on each side from one coupling to the other and twisted as desired.

COLIN F. MCGIBBSON.

Butte, Mont.

Location of Secondary Relays Affect Voltage Regulation

A company operating a number of induction feeder regulators mounted the contact-making or primary relay upon the switchboard and the secondary or reversing relay upon the regulator. However, this secondary relay was not mounted as recommended by the manufacturer, but, instead, was supported by iron strips to the top of the regulator case. This arrangement was inexpensive and appeared quite satisfactory. However, in one of the installations where a new type of regulator was installed, difficulty developed in that the regulator was not functioning properly. Sometimes the voltage would rise as the load came on, then would remain high although the load decreased, and vice versa. This did not always happen, and sometimes the regulator would operate for days and even weeks without giving any trouble.

At first routine troubles were looked for; the contacts of the contact-making voltmeter were cleaned and adjusted for correct voltage range; the contacts of the reversing switch were overhauled and adjusted; the regulator brakes were tested, the drums oiled and the tension of the brake shoes reduced. These adjustments had no apparent effect, and intermittent short-circuits were suspected, but no trouble of this nature was found.

What was found was that at certain times of high voltage and therefore increased flux density in the regulator iron, and when a certain frequency of the system occurred, possibly due to current changes, the regulator was caused to give rise to severe vibration. This was in turn passed on to the secondary relay, which also vibrated under the influence of normal operating current passing through it. In this manner the natural periodicity of vibration of the relay had superimposed upon it the vibration of another periodicity. The effect of this vibration was that while the contacts of the secondary relay appeared to be making good mechanical contact, the circuit to the regulator motor was not permanently closed. Instead, the circuit was rapidly opened and closed without apparent movement of the relay contacts so far as the naked eye could see. Moreover, the two poles of the relay might or might not close simultaneously. Since a three-phase induction motor cannot start up single-phase, the result was that either the motor could not get started in time before the circuit was interrupted or else it could not start at all. In either case the result was the same—poor voltage regulation.

The remedy was simple. The secondary relays were left in their original position. But instead of being bolted down tight to the iron strip supports, so that vibration of the induction regulator cases was transmitted to the relays, spring washers were placed underneath the case of each relay, at the four corners. The bolts were then inserted and the fastening nuts screwed down so as to prevent the bolts from falling out. The relays were allowed free movement up and down, always resting on the spring washers and yet unable to fall off the supporting strips. This solved the problem and incidentally eliminated much burning of contact-making contacts.

B. W. SMITH.

Cicero, Ill.

Method of Boring a Large Bearing on a Small Lathe

The illustration shows how a bearing was bored out on a small lathe that was not strong enough to hold the weight of the bearing and had to be blocked up to prevent breaking the bed. The tool post was removed, and the carriage was made to fit the bearing by the use of channel irons. The boring bar was a piece of scrap shafting with a set collar to hold the tool. The bed of the lathe was supported by means of iron stands wedged tightly between the lower side of the bed and the floor. This way of boring large bearings proved very satisfactory and has been used on a number of different bearings.

BEARING READY FOR BORING OUT ON SMALL LATHE

W. R. CATON.

Buffalo, N. Y.

INQUIRIES OF GENERAL INTEREST

Triple Riveting of Lap Joints—Why are lap joints of high-pressure steam drums frequently made with three or more rows of rivets? F. C.

A triple-riveted lap joint may be designed of higher efficiency than a double-riveted lap joint, and the greater width over the rows of rivets renders it easier to bring the plates into line, with the result that there is less bending than in the single- or double-riveted lap joint.

Turbine a One-Speed Machine—Can a steam turbine be constructed that will develop a normal rated power at different speeds? A. L. M.

The turbine is a one-speed machine. For good economy the speed of the rotating buckets must bear a definite relation to the velocity of the steam as determined by the design. Otherwise there will be wasteful impact of the steam against the vanes and the speed of the turbine cannot be changed from that of best economy without serious loss of efficiency.

Transfer of Heat Requires Difference of Temperatures—If at sea level fifteen pounds of steam at atmospheric pressure is to let into an open barrel containing 150 pounds of water at the boiling point, what temperature would the water attain? C. E. G.

The temperature of dry saturated steam at atmospheric pressure is 212 deg. F., and the boiling point of water under atmospheric pressure is at the same temperature. There could be no transfer of heat because the temperature would be the same and the temperature of the water would remain unchanged.

Belt Thickness Influences Speed Ratio of Pulleys—Should not a 36-in. pulley drive a 12-in. pulley at three times the speed of the former? R. F.

If the pulleys operate as friction pulleys running in contact without slip, the 12-in. pulley will be driven at three times the speed of the 36-in. pulley. But if separated and the driving is done by a belt without slippage, the small pulley will be driven at something less than three times the speed of the large one on account of the thickness of the belt. In making calculations of speeds without slippage, an amount equal to the thickness of the belt should be added to the diameter of each pulley.

Obtaining Compression of Exhaust on Corliss Engine—How can compression of exhaust be increased on a Corliss engine? E. D. C.

With a double eccentric Corliss engine, advance the exhaust eccentric. As opening and closing of each exhaust valve occurs when the piston is at the same part of its stroke, earlier closure for obtaining more compression will also cause earlier release after cutoff. With a single eccentric for operating all valves, advancing the eccentric hastens all valve events and causes the steam valves to open earlier. If the resulting lead is too much, the time of opening the steam valves must be delayed by adding lap to those valves. This is done by adjusting the length of the steam valve rods so the opening edges of the steam valves will lap over their ports when the wrist plate is in its central position, giving $\frac{1}{32}$ to $\frac{1}{16}$ in. opening for lead when the piston is at the beginning of its stroke.

Induction-Motor Winding Bracing—Why are insulated rings placed over the ends of induction motor coils where they extend beyond the stator core and why in some cases are they placed on both ends and in others on one end only? K. A. R.

Frequently, where the coil throw is long or if the coils are not mechanically rigid, the ends are liable to be distorted when a heavy current flows through them. To prevent this, insulated rings are placed around the outside of the winding and the coil ends laced to these rings with heavy cord. This practice has been found to increase the life of the coils, since it tends to prevent any vibration of the coil ends that would otherwise occur. Where the throw of the coils is short, the rings are only put on the end of the winding opposite to the one where the cross-connections are made, the cross-connection giving the necessary support to the coil ends on the end of the stator winding where the connections are made.

Variation of Boiler Rating for Difference of Pressure—What would be the difference in rating the capacity of a boiler operated at 150 or 125 lb. pressure? V. A.

Boiler evaporative capacity is based on the rate of absorption of heat transferred to the water in the boiler, a boiler horsepower consisting of a rate that is equivalent to the transfer of heat required to evaporate 34.5 lb. of water from and at 212 deg. F., or 33,479 B.t.u. per hour. The lower the temperature of the water of a boiler the greater its difference of temperature from the temperature of the fire and heated gases and the more rapid the rate of heat transfer. Operation at different pressures also might be attended by difference in rate of transfer due to difference in the character of circulation of a boiler; but the differences in the rates of transfer of heat when generating steam at 150 or 125 lb. gage pressure would be negligible, and for all practical purposes the rating would be the same.

Twenty-five-Cycle Transformers on 60 Cycles—Will 25-cycle transformers 6,600-volt primary 220-volt secondary work on 60 cycles 6,600-volt primary 220-volt secondary? E. R. G.

Twenty-five cycle transformers can be operated on 60 cycles, but 60-cycle equipment cannot be operated on 25 cycles. For the same output the 25-cycle equipment will be considerably larger than the 60-cycle, consequently where 25-cycle transformers are operated on a 60-cycle circuit, the unit would be considerably larger than a 60-cycle transformer of the same capacity. When the 25-cycle transformers are used on 60 cycles, the exciting current and the flux density in the coil will be only about four-tenths what it is for the 25-cycle circuit. When 60-cycle equipment is operated on a 25-cycle circuit, the exciting current and flux density will theoretically be 2.4 times that of the 60-cycle circuit. However, due to the saturation of the iron core, the exciting current may be many times this value, and will cause the transformer to overheat even when operating without any load.

[Correspondents sending us inquiries should sign their communications with full names and post office addresses. This is necessary to guarantee the good faith of the communications and for the inquiries to receive attention.—Editor.]

How Bituminous Coal Can Be Stored Safely*

By H. H. Stoek
Professor of Mining Engineering, University of Illinois

THE following statement has been prepared from material published by the Engineering Experiment Station, University of Illinois, in Circular 6 and Bulletin 116. These publications give the details of an investigation by the station that has been in progress for about three years. During this time the practice of storing coal has been studied under a great variety of conditions and for stocks varying from a few tons up to hundreds of thousands of tons. The data were obtained through questionnaires, conferences, personal inspection of storage piles and a detailed study of over one hundred fires in coal piles said to be due to spontaneous combustion. The conclusions here given apply only to bituminous coal and have no reference to anthracite or lignite.

Small Quantities Give Little Trouble in Storage

For such amounts of coal as are required by the ordinary householder, it can be stated positively that, (1) there is little or no likelihood of spontaneous combustion, as heating can be detected readily and the coal moved, and (2) there is no loss in heating value that would be appreciable in the conditions under which such coal is burned. Where several hundred tons of coal are stored, as in large buildings and schoolhouses, extra precautions should be taken in the storing and then in watching for evidences of heating.

For the storage of coal in quantities, say above 500 tons, the experience and practice of those who have stored coal successfully under similar conditions should be given careful consideration. Before it is time to begin the actual storage, a suitable place should be prepared and a definite policy outlined. Storage instructions should be prepared in detail and carried out to the letter.

It is probable that all varieties of bituminous coal have been stored without spontaneous combustion resulting, and it is equally true that all varities of coal have fired when stored, excepting possibly anthracite. This does not mean that all coals store equally well, as there is a wide difference in this respect between various varieties. There are not sufficient data available at this time to classify coals as to their liability to spontaneous combustion, and it is doubtful if this can be done, because liability to firing seems to depend more upon the size of the coal, its freedom from dust, the way in which it is piled and to a less extent on the amount of pyrite present.

Coal of Uniform Size Is Easiest To Store

Clean, screened coal of a uniform size should be chosen if it is available. The larger and more uniform the lumps the better. Sized coal should not be stored upon a foundation of fine coal. The coal should be handled in such a way as to prevent breakage as much as possible. It should not be dropped upon a pile from a height, but the bucket or other receptacle lowered to the pile before being dumped. In every case of a fire in so-called lump coal, that I have investigated, it was found that the coal had been dropped through a considerable height, so that coal on the pile contained a large amount of dust.

In storing mine run coal, it should be piled in uniform layers, in small low piles, and segregation of the sizes prevented. The piles should be as low as the space will permit, carefully watched and provision made to quickly move the coal if heating occurs. The spontaneous combustion of coal is due largely to the oxidation of the fine material and the confining of the resulting heat; consequently, its liability is greatly reduced and in many cases eliminated if dust and fine coal can be kept out of the piles, if the coal is of such size and so piled that there is sufficient air current to carry off the heat, or else if the coal is so closely packed that the air cannot come in contact with the fine coal to cause oxidation.

*Abstract of a paper presented at the February meeting of the American Society of Mining and Metallurgical Engineers.

If fine coal or slack is to be stored, it should be carefully watched to detect evidences of heating. If practicable the slack pile should be rolled so as to compact it as much as possible, thus excluding the air, so as to give as nearly as possible storage-under-water conditions.

Although experimentation has shown that the sulphur contained in coal in the form of pyrite is not the chief cause of spontaneous combustion, the oxidation of the sulphur in the coal not only produces heat but also assists in breaking up the lumps and thus increases the amount of fine coal in the pile.

It seems to make little difference in the liability to spontaneous combustion whether the coal is stored under cover or in the open, but under cover there is less segregation and weathering. If possible a place should be chosen that is dry and well drained, or if not drained naturally, drains should be provided around the piles, not underneath it. A drain beneath the pile may produce an air current up through the pile and thus assist spontaneous combustion. The coal should not be dumped on ground covered with ashes or refuse of any kind, because in addition to furnishing flues for the admission of air, such refuse often contains material such as waste, rags, papers, and so forth that will start spontaneous combustion. Furthermore, this refuse may be gathered up with the coal when it is reclaimed and the value of the fuel be depreciated.

Coal should not be stored around hot pipes, against a boiler, or hot walls, around a chimney or in any place where it is subjected to outside heat. Coal should not be stored above flues that will permit a current of air to enter the pile. Hot air, such as that from a sewer, is particularly to be avoided. Care should be taken to arrange the storage pile so that any part of it can be moved promptly if necessary.

Danger from Stratification

Stratification, or segregation of the fine and lump coal into layers, should be avoided, since an open stratum of coarse lumps provides a passage for air to reach the fine coal, and the layer of fine coal does not permit the heated air to pass out rapidly enough to keep down the temperature. Coal should not be dumped at one point on the top of a conical pile, since the fine coal stays in the center at the top, and the lump coal rolls to the bottom. Many coal fires start near the top of the pile where the fine coal has prevented the escape of the heated gases. In placing coal in a cellar or bin the same precautions should be observed, for many fires are known to have started in the fine coal just below the window through which the coal was shoveled in.

The depth and area of storage piles will be determined largely by the storing space available and the mechanical appliances for storing and reclaiming coal. Other conditions being equal, the deeper the pile and the greater the area the greater the difficulty in inspecting and moving it quickly, if necessary; hence a number of small piles are better than one large pile.

The exact effect of moisture in connection with spontaneous combustion is not definitely known. The repeated wetting and drying of coal, however, seems to increase this tendency, probably owing to the breaking up of the coal which results. It is not wise to put wet coal into a pile, or to store dry coal on top of wet coal or on a damp base. After a rain or snow storm the coal piles should be carefully inspected and watched.

Water is an effective agent for quenching fire in coal piles only if it can be applied in sufficient quantities to extinquish the fire and cool the mass. Most coals coke on heating, and a protective coating is thus formed about the hot spot which prevents the water from reaching it. To be sure that the water reaches the burning coal it is usually necessary to dig into the pile and turn it over.

Coal in storage should be inspected regularly, and if the temperature reaches 140 deg. F., the pile should be carefully watched. A rapid rise from this point to 150 or 160 deg. indicates that the coal should be moved as promptly as possible. If the temperature rises slowly, the pile should be carefully watched, but it is not necessary to start moving it at as low a temperature as when it rises rapidly, for the temperature may recede and the danger be past. Coal that has once heated should, if possible, be used at once and not returned to the pile, for experience shows that such coal is peculiarly liable to heat again.

Timbers extending to the outside of the pile may permit the entrance of air to the interior. It is important, therefore, to avoid piling coal around the end of trestles or in such places where timbers, coarse cinders, etc., offer a passage for the air.

The effect of ventilating coal piles is a disputed point, but the weight of evidence in the United States seems to be against the practice. This may possibly be due to the fact that ventilation has been inadequately provided. Imperfect ventilation, such as generally attempted in this country, is certainly disadvantageous. Reports from Canada favor ventilation by pipes placed closed together in the coal pile, although this arrangement may interfere with reclaiming devices.

A plan for coal storage must consider all the conditions, and not only a portion of them. For instance, clean lump coal of a certain kind may be stored with perfect safety in high piles, while the same coal, run-of-mine or unscreened, may not store at all or at least only in small piles. Lack of attention to details in storage or failure to systematically inspect storage piles and to be ready for any emergency that may occur, may result in losses from fires. Before you store coal, make sure that you know how to do it properly.

The Engineer to Participate in Public Affairs

At the midwinter convention of the American Institute of Electrical Engineers President Calvert Townley, in an address, directed the attention of the engineer to a greater participation in public affairs. The speaker referred to the endorsement of the report of the Institute's Development Committee by the members as a step of far-reaching importance, particularly the part dealing with the desirability of greater participation by the engineer in public affairs and that which recommends that the Institute co-operate with other engineering bodies to bring this about. It was stated that the machinery for co-operation between the various engineering societies was well under way, and that two other engineering societies had already agreed to participate in the movement. After pointing to how extremely gratifying it was, to those who did sponsor the Engineering Council, that the membership at large had now very definitely expressed the view that the movement should be continued and should be expanded far beyond anything previously undertaken, Mr. Townley said:

"It would be premature, not to say hazardous, to say now that if and when delegates from a large number of societies come together they will organize, and that sponsors will support a federation such as has been proposed by the joint conference committee, but at least the national society will have done their duty, and if the plan should carry it will place in the hands of a democratic body of engineers, chosen from engineering organizations of every kind all over the United States, a large measure of authority and responsibility to speak for the engineering profession and to take united action on matters of common concern. It is because of the far-reaching character of this prospective step that I was chosen to speak to you about tonight. I regard it as one of the most important of the organization matters which are today before the engineers of the country."

In outlining some of the things that a national engineering organization could properly do Mr. Townley recalled some of the work performed by the Engineering Council and said that its first activities were to help the Government during the war to find engineers suitable for the many different tasks for which they were sorely needed. Another activity was to co-operate with the Government, through the medium of a large committee, in passing upon the merits of the enormous number of new and untried devices and processes and inventions submitted during the war. Since the armistice the Council has co-ordinated the employment bureaus maintained over a period of years by some of the Founder Societies; conducted and is still conducting an exhaustive examination into the classification and compensation of engineers in Government service; at work collecting and compiling all available data in an endeavor to prepare a uniform license law for engineers to be recommended to all states where such legislation is being considered; co-operated with the United States Chamber of Commerce by presenting certain fundamental data on the question of water power which aided materially in passing the water-power legislation in both the House and Senate; testified before Federal authority on many different subjects, always endeavoring to confine itself to fundamental engineering facts and to keep away from political and other questions. In explaining what the new organization would be the speaker said:

"One of the fundamental features of the plan prepared by your Development Committee and which also appeared in that adopted by the joint conferees, is that the new national organization shall be composed of delegates from existing engineering bodies. It will not be a new society composed of individual members. Such a plan was given exhaustive consideration by the conferees and in the end decisively rejected. The primary objection to the plan adopted is that the work of the new organization is to be quite different from that in which most of the old organizations have been engaged. Instead of dealing in the main directly with technical or interorganization matters, it will deal more largely with public affairs, and as the existing societies are not organized for this purpose and in fact have not done it, but have in some instances shown opposition to the idea, they should not be relied upon to back it now, but on the contrary an entirely new body should be created, one that is not hampered or shackled in any way. The answer to this argument is that while the existing societies were not organized for the kind of work now contemplated, several local affiliations of them have been functioning around many of the sorts of work proposed for a number of years and doing it well and that the Engineering Council has already proved its efficiency and only needed a larger appropriation and a more representative membership to be afforded by the new plan to occupy a much wider field."

Should Confine Activities to Engineering Matters

In closing his address President Townley said: "The only other comment which I feel should be made is a word of caution and is born of my experience while serving the Institute on Engineering Council. The body has been beset on all sides to advocate or to oppose legislation, and while it has not hesitated to do so when the questions involved were of a nature where the advice of engineers would be of particular value, it has steadfastly declined to become involved otherwise. Bearing this fact in mind my word of caution is against any policy which will cheapen the engines.

Our collective advice is valuable on subjects with which we are qualified by experience and training to do, and so long as we confine ourselves to such subjects and present a united front our influence can and will be considerable. But the minute we digress and begin to deal with matters outside the purview of engineering and subjects closely allied thereto we sacrifice our preferred position and have no more claim to be heard than any other body of educated citizens of similar number. We also would then almost make the very serious mistake of misrepresenting our membership, because, while engineers can be safely counted upon to agree upon fundamental facts, they are of all shades of political belief and have as many different views as any one else regarding other questions. My word of caution therefore is really a plea to you to give this side of the subject your due thought and to make your views known at all proper times and places."

Hoover's Food Drafts—A New System for Sending Food to Individuals in Europe

In order to help relieve the critical food situation in eastern and central Europe, the American Relief Administration has adopted a plan whereby any individual in America may supply a friend or relative in Europe with a food package or anyone desirous of helping to give to the unfortunate may do so on payment at any one of several thousand banks in America. In a letter to the bankers of America, Herbert Hoover, Chairman of the Relief Administration, outlines the plan as follows:

Throughout the whole of central and eastern Europe the food supply of the people falls into two classes: First, the ration issued by the government; second, illicit circulation of food available to those who have a sufficient amount of money. The government ration is necessarily meager and nowhere sufficient to properly maintain life, and must be supplemented. Under these circumstances the scramble for such supplementary margin has placed the price of the illicit food supplies entirely beyond the reach of the great bulk of the population. To illustrate: A single ham outside the ration system sells for as high as one hundred and fifty dollars.

There are three to four million families in the United States with family affiliations in eastern and central Europe. Many of them are desirous of giving direct personal assistance to these relatives and friends. Some are endeavoring to perform this service by preparing or purchasing packages of food for overseas shipment. In some cases the packing and extra freight involved add 100 per cent to the cost. We are proposing to solve this difficulty by establishing warehouses to carry stocks of staple foodstuffs in European cities where distress is particularly acute. We propose to sell, in America, orders upon these warehouses in the form of "Food Drafts," which can be transmitted to friends or relatives in Europe. We propose to charge the buyer of the "Food Draft" the factory cost of the food plus a reasonable margin to cover cost of transportation and insurance. Profits, if any accure, will be turned over to the European Children's Fund.

The object of this plan is to add to the total stock of available food supplies in central and eastern European countries. Under an arrangement set up with the governments of these countries, this food will be allowed to revolve outside the rationing system, with the hope that enough food will be injected to reduce the pressure on the narrow marginal supplies. The officials of these new governments are endeavoring to impress upon the American people that it is useless to remit money to a family in central or eastern Europe with the hope of improving its food situation. The sum total of food now available in central Europe is insufficient to keep the population alive, and under these circumstances money thus becomes that much paper so far as nutrition is concerned. A hungry man wants food, not money, and under the arrangement outlined, we can meet his need.

It is hoped that the facts concerning this "Food Draft" system will be given as much publicity as possible, especially among the foreign-born who have relatives and friends in the suffering countries. It is also hoped that many Americans will help the general situation by buying drafts made out to "General Relief" and sending them to the Relief Administration, whose representatives will distribute the food where it is most needed.

Diesel Engine-Users Association

There was a large attendance at the February meeting of the Diesel Engine Users Association, London, and a considerable amount of interest was evidenced in the discussion that followed the reading of Geoffrey Porter's paper on "Wear of Diesel Engines." In dealing with the important question of "liner wear" he referred to this as being chiefly influenced by the percentage of ash content in the fuel oil and its variable nature according to the class of fuel used, the quality and quantity of lubricant, the method of lubrication, the fit of pistons and piston rings in the liner, the side thrust due to the connecting-rod and the quality of the material used. The author also pointed to the deleterious effect of excessive cylinder lubrication, which resulted in the formation of carbon deposits on the piston. Particulars were given as to the proper clearances in the case of cooled and uncooled pistons of various sizes, and the important subject of "growth" of cast iron was again referred to.

G. W. F. Horner, in discussing the paper, referred to the subject of "growth" in cylinder liners and instanced the particulars of actual cases in which he had taken accurate measurements.

W. A. Tookey thought there was very little doubt that the wear of piston liners and piston rings was correlative. In the case of horizontal engines it had been found that the maximum wear occurred at the joint of the back piston ring.

A. N. Rye had not noticed that the use of tar-oil fuel had been detrimental to cylinder liners, but it had a marked effect on valves and other portions of the engine gear. He referred to the effect of granite dust in the air in the case of some engines situated near a granite quarry. The abrasive effect of the granite dust introduced into the power house when the wind was in certain quarters had caused a considerable amount of trouble in the cylinders until air filters had been introduced. Mr. Rye expressed approval of the barreling of cylinder liners as a means of reducing piston-seizure trouble.

Particulars concerning the association can be obtained on application to the honorary secretary, Percy Still, M.I.E.E., 19, Cadogan Gardens, London, S.W.3.

N. E. L. A. Convention

The National Electric Light Association has completed a tentative program for the forty-third convention, to be held at Pasadena, Cal., May 18-22. General sessions will be held during the mornings and special sessions in the afternoons. Governor William D. Stephens of California will deliver the address of welcome at the opening session. Response will be made on behalf of the association by President Ballard. Following this, M. H. Aylesworth, executive manager, will outline the reorganization work of the association to date.

A report of the Committee on Water Power Development will be read by Franklin T. Griffith, of the Portland Light and Power Co., who will be followed by A. E. Silver, of the Electric Bond and Share Co., chairman of the Inductive Interference Committee, who will read a report on "Inductive Interference." "Electrification of Steam Railroads" will be the subject dealt with in the report of a committee on that subject, presented by Frank M. Kerr, of the Mountain Power Co., of Butte, Mont. Other reports and speakers will be: "Electrical Resources of the Nation," by M. S. Sloane, of the Brooklyn Edison Co.; George B. Foster, of the Commonwealth Edison Co., of Chicago, chairman of the Electric Vehicle Section of the association, will deliver an address on "Development, Manufacture and Marketing of the Electrical Vehicle"; Zena W. Carter, secretary of the Material Handling Machine Manufacturers' Association of New York, will read a paper entitled "Electric Machinery for All Handling"; and the Association Committee on Public Information, of which John F. Gilchrist, of the Commonwealth Edison Co., of Chicago, is president, will give figures and facts showing the value of proper publicity of association activities.

During the afternoon session on Friday, May 21, a report of the Committee on Federal and Municipal Transportation will be read by James H. McGraw, chairman, of the McGraw-Hill Company, Inc. Mr. McGraw will be followed by F. M. Feiker, chairman of the Committee on Transportation Engineering, also of the McGraw-Hill company. Mr. Feiker will read the report of the committee. On May 19 during the first afternoon session, N. A. Carle, from the Public Service Electric Co., Newark, N. J., chairman of the Committee on Prime Movers, will read the report of the committee, and in the second afternoon session, Thursday, May 20, the report of the Committee on Electrical Apparatus will be read by R. F. Schuchardt, chairman, Commonwealth Edison Co., Chicago.

Beauties and Power Developments of Niagara Falls

Friday, March 19, was ladies' night for the Chicago Section of the American Society of Mechanical Engineers. One of the meetings during the season is always reserved for this distinction, and in the present instance it was a joint meeting and dinner with the Western Society of Engineers. The attendance was good and the treatment of the topic of the evening, "The Development of Niagara Falls Water Power," was a revelation to those present.

William M. White, manager of the hydraulic department of the Allis-Chalmers Manufacturing Co., had a wonderful collection of motion and stationary pictures bringing out the beauties of the world's greatest scenic attraction, the rapids above the Falls, the Gorge, the Whirlpool and the river as it broadens out into Lake Ontario. In a brief introduction giving the physical data relating to the Falls, Mr. White gave as the amount of water passing over both falls as 220,000 cu.ft. per sec., the difference in level between Lake Erie and Lake Ontario, or the total head available, as 327 ft., and the power it was possible to develop, 7,000,000 hp. The amount of water to be diverted by each country was 36,000 cu.ft. per sec. For the United States 16,000 cu.ft. per sec. was allowed to the Drainage Canal and the balance to power development. In the numerous plants on the American side 350,000 hp. had been developed and 450,000 hp. on the Canadian side. Views were presented giving an idea of the various plants on both sides of the river and their relative location to one another and the Falls. On the Canadian side particular attention was given to the recent development by the Hydro-Electric Power Commission of Ontario at Queenstown Heights, and among the American plants station No. 3 of the Niagara Falls Power Co. was the chief attraction.

Beginning with the intake from the waters above the upper rapids, motion pictures showed the precautions taken to guard against ice, followed the channel to the plant, showed the thirteen older units rated at 10,000 hp. each, and then particular attention was given to the three new vertical units, each of 37,500 hp. capacity and the largest hydraulic prime-movers in the world. At normal load each turbine will use 1,700 cu.ft. of water per second. The runner weighs 40,000 lb. and the entire rotor 265,000 lb. Pictures showing the foundations for these enormous machines, their erection and the placing into service of the first unit were of exceptional interest.

It was estimated that the installation of this one unit would eliminate the services of 819 men in mining, cleaning and transporting coal and placing an equivalent amount of energy on the switchboard in a steam plant. Going a little farther, if all of Niagara's power should be developed, the coal conserved annually could be made up into a train that would extend around the world. In addition thousands of men would be released for other industries where they are badly needed. These two factors were made the basis of a strong plea for the more comprehensive development of our water power.

As an interesting conclusion Mr. White had pictures showing the war developments at Niagara, the great chemical plants, the making of carborundum and alundum and other uses of Niagara power.

Preliminary to this most interesting presentation a nominating committee was appointed, consisting of P. M. Chamberlain, H. E. Troutman, J. L. Hecht, John Wintercorn and Harry Mimelblau. The committee is to present a ballot at the April meeting, so that the officers of the section for the ensuing year may be elected in accordance with the by-laws at the last meeting of the season, which will be held in May.

It is conservatively estimated that the December storm of last year cost the Pacific Power and Light Co. between $30,000 and $35,000 in increased operating expense due to purchased power, coal, and inability to deliver service. There was also trouble in January, caused by snow and frost adhering to the Yakima Valley lines, and the aluminum conductors broke in many places. This caused interruptions to service and made it necessary to pull very heavily on the Washington Water Power Co.

New Publications

MCGRAW CENTRAL STATION DIRECTORY AND DATA BOOK. Published by McGraw-Hill Co., Inc., New York City, 1919. Flexible Binding; 4 x 8 in.; 835 pages. Price, $15.

This new edition contains completely revised data on 6,106 operating companies' power systems and their 1,700 additional generating plants and substations, in the United States, Canada, West Indies and Mexico. There are 416 new companies included in this edition, and 3,388 companies selling electrical appliances and 1,666 companies doing wiring on other electrical contracting work are definitely indicated. Five thousand towns with a population under 1,000 are tabulated, with the name of the town from which they are being served with power. A statistical summary and a list of state public-service commissions having jurisdiction over electric utilities are also included. The data regarding the companies systems cover such features as location of plant, number of generating units, total capacity, frequency, voltage, phase, direct or alternating current, length of transmission lines and voltage, capital invested, company officers, etc. This edition of the directory is more complete than any previous one and contains valuable data for all those interested in the commercial side of the electrical industry.

CONTROLLERS FOR ELECTRIC MOTORS. By Henry Duval James, B. S., M. E. Published by D. Van Nostrand Co., New York City, 1919. Cloth; 5½ x 8 in.; 350 pages; 259 illustrations. Price $3.00.

This book is a treatise on the modern industrial controller, together with typical applications to the industries, and consists of a series of articles originally published in the *Electric Journal* during 1917 and 1918, to which have been added some new text and illustrations. The author was exceptionally well qualified for the preparation of this book. Having been for a number of years the engineer in charge of the controller department for one of this country's largest electrical manufacturing companies, he has had the opportunity to get the broad viewpoint necessary for the presentation of so large and important a subject. Fortunately, he has shown a keen understanding of the requirements of the practical man and has treated his subject in a practical and easily understood way. If there is one feature in the book more conspicuous than any other, it is the almost absolute absence of mathematics. Only on three pages is the mathematical side of the subject mentioned. The work is divided into 28 chapters: Introduction; Historical; Design Details; How To Read Controller Diagrams; Methods of Acceleration; Starting Characteristics of Motors with Different Methods of Control; Methods of Speed Control and Dynamic Braking; Direct-Current Magnetic Contactor Controllers; Alternating-Current Controllers; Resistors; Protective Devices; Series-Parallel Control and the Electro-Pneumatic Contactor; Voltage Control for Direct-Current Motors Mine Hoists; Hydraulic Pumps; Machine-tool Controllers (two chapters); Control for Machinery Requiring Low Initial Speed, Such as Printing Presses and Rubber Calenders; Steel-Mill Floor Controllers for Auxiliary Drive; Cranes; Car Dumpers; Ore and Coal Bridges; Coke; Elevators; Electrical Equipment for Oil Wells; Locomotives for Mines and General Industrial Purposes.

In part of the introductory chapter the author discusses the advantages and disadvantages of the various types of controllers, which, combined with the design details treated of in Chapter III, will be of assistance to those interested in the application of controllers in forming an opinion how variations in design will affect the performance of a controller. The explanations given in Chapter IV on how to read a controller, combined with the discussion on methods of accelerating motors in Chapter V, form a groundwork for those not experienced in the art to obtain an understanding of the fundamentals of controllers that will make the following chapters on the control of both direct-current and alternating-current motors and controller applications easily understood. No attempt has been made to cover every method of control and all applications; however, those selected give a wide range of operating conditions, as indicated by the titles of the different chapters. A large number of diagrams have been given, but these have in most cases been very much simplified for the purpose of clearness.

Probably the most serious criticism that can be made of the book is that it deals with the product of only one manufacturer. However, the general principles of operation of every controller manufacturer's product are practically the same, the essential difference being in the method of application. The book contains a wealth of information for those interested in the principles and application of controllers for electric motors and should fill an important need in engineering literature.

DESIGN AND CONSTRUCTION OF HEAT ENGINES. By William E. Ninde, M. E., Associate Professor of Mechanical Engineering, Syracuse University. Published by McGraw-Hill Book Co., New York City. Cloth, 6 x 9 in., 695 pages; illustrated. Price, $6.

The author states that his object in presenting this book is to provide the material most essential to a well-equipped, independent designer of heat engines in a form most convenient for classroom and practical work. The text is divided into six parts, of which the first five are really introductory to the main division of Machine Design.

Part 1. The Heat Engine: Describes briefly the principles of operation of the three main types of prime mover, the steam engine, the steam turbine and the internal combustion and outlines the most common methods of classification of each.

Part 2. Thermodynamics: Presents the essential thermodynamic formulas on which the design of heat engines are based. The chapter on cylinder efficiency discusses the various factors affecting the important problem of cylinder condensation in steam engines.

Part 3, Friction and Lubrication: Discusses the theory of lubrication, making special reference to the design of bearings.

Part 4, Power and Thrust: In this section the author shows the use of the indicator diagram, both actual and theoretical, in computing the power and the forces due to the steam or gas pressure. The chapter on steam turbines is an outline of the design of nozzles, guides and blades from the standpoint of the flow of steam through the machine.

Part 5, Mechanics: Treats of the various parts of the engine as mechanisms and discusses the action of the various forces due to inertia and gravity, and their combination with the forces due to the steam pressures. The theories of flywheels and of governors are presented and a considerable space is devoted to valve gears.

Part 6, Machine Design: This is, naturally, the largest division. After a general discussion of various forms of stresses and of the properties of materials the various parts of a reciprocating engine are taken up in detail. A chapter is devoted to the design of cylinders, another to pistons, a third to piston rods, and so on. Attention is also given to turbine wheels, shafts and casings.

Throughout the book an effort is made to present rational formulas and methods as far as possible and to supplement them with empirical equations and practical data from established practice. Numerous drawings of actual engine details are given to illustrate the examples. A noticeable feature is the grouping at the beginning of each chapter of the notation used in the formulas in that chapter, and in most cases the meaning of the various symbols is repeated in the text when first used. This feature goes far to facilitate the use of the book. A general knowledge of thermodynamics and of mechanics is assumed, although in some cases the author has taken considerable trouble to explain elementary details which would be familiar to one capable of understanding most of the text.

Obituary

Melvin B. Newcomb, aged 31 years, chief engineer of the rubber machinery department of The Wellman-Seaver-Morgan Co., died on Mar. 13, after a short illness at his home, Akron, Ohio. Mr. Newcomb was born in 1889 in Bridgeton, N. J. He received his mechanical engineering education at the University of Wisconsin. He has been engaged in engineering work with the I. P. Morris Co., Philadelphia, the Allis Chalmers Manufacturing Co., Milwaukee, Wis., the Wisconsin Engine Co., Corliss, Wis., and the Firestone Tire & Rubber Co., Akron. In January, 1918, he joined the hydraulic turbine engineering department of The Wellman-Seaver-Morgan Co. A few months later he was appointed chief engineer of the rubber machinery department of The Wellman-Seaver-Morgan Company.

Mr. Newcomb was a member of the American Society of Mechanical Engineers, The Cleveland Engineering Society, and the Akron Engineering Society.

He is survived by a wife and two young daughters.

Personals

John A. Stevens, power plant engineer, Lowell, Mass., has opened a branch office at 502 Frederick Building, Cleveland, Ohio.

Ralph T. Bratt and **Carl T. Petersen** have formed a partnership as consulting engineers in power plants, heating, ventilating and refrigeration at Hackensack, N. J.

Edward M. Burd and William C. Giffels have opened an engineering office in Grand Rapids, Mich. Both are University of Michigan graduates and were formerly connected with the Consumers' Power Co., in Michigan.

M. A. Buehler, formerly sales manager at the Omaha office of the Western Electric Co., has been made sales manager at the Minneapolis office. Eliot Lum, of the sales department of the Omaha house, will succeed Mr. Buehler.

W. E. Herring, well-known on the Pacific Coast as the industrial agent of the Puget Sound Traction, Light & Power Co., Seattle, Wash., has been transferred to the engineering department in the Boston office of Stone & Webster. Mr. Herring formerly was chief engineer of the United States Forestry Service.

J. C. McKenzie, who, for over a year, has been with the Performance Branch, Emergency Fleet Corporation, in charge of trial trip data, New York office, has accepted a position as marine engineer with the Power Specialty Company, New York. He will be in charge of the Service Department for Foster Marine Superheaters and Foster Marine Boilers.

Clarence Goldsmith, who has been on the enginering staff of the National Board of Fire Underwriters, 76 William Street, New York City, for the past twelve years and late a major in the Construction Division of the United States Army, has been placed in charge of the branch enginering office recently established by the National Board at 234 South La Salle Street, Chicago, Ill.

Engineering Affairs

The American Order of Steam Engineers will hold its annual convention at Baltimore, June 7 to 11, inclusive.

The Association of Iron and Steel Electrical Engineers will hold a meeting on April 17, 1920. A paper entitled, "Grounded Neutral" will be read by Robert B. Treat, Electrical Engineer, General Electric Co., Schenectady, N. Y.

Miscellaneous News

The Union Francaise d'Electricite, France, is erecting near Paris the largest power station in Europe. Sixty thousand hp. turbines will be installed and the complete capacity of the plant, when finished, will be 400,000 hp.

Erection of the Two Largest Boilers in Tennessee is under way at the city light plant at Nashville. Each boiler weighs 55 tons and the two stokers weigh 28 tons each. The cost of the boilers (not including foundation) to the city is $51,854.

The City Commission of Trenton, N. J., has decided on the installation of mechanical devices on the city pumping station along the Delaware River to overcome the smoke nuisance. A Bureau of Smoke was established to rid the city of smoke from the Trenton factories.

The American Chamber of Commerce in London reports the erection of a gigantic copper refinery, costing approximately $50,000,000, at Newton Abbott, Devonshire, England. The required electrical energy will be generated from deposits of 800,000,000 tons of lignite, a new smokeless fuel, which is located near the site of the plant.

The Trenton-Mercer County Traction Corp., Crescent Belting & Packing Co., De Laval Steam Turbine Co., and many other Trenton, N. J., plants were forced to close their boiler rooms which became flooded when the Assanpink Creek overflowed its banks. The city commission at a recent conference decided to construct a long retaining wall along the creek for the protection of the power plants.

Business Items

The Westinghouse Electric and Manufacturing Co. has awarded contracts for the construction of a new plant at South Philadelphia, Pa. The new plant will be used for the building of turbines.

The Midwest Engine Co., Indiana, recently completed a testing laboratory having facilities for testing centrifugal pumps up to 48-in. capacity. Twenty-five gallons of water per minute is the capacity of the plant.

The Black & Decker Mfg. Co., Towson Heights, Baltimore, Md., announce the establishment of a permanent office and show room at 1436 South Michigan Avenue, Chicago, Ill., this office will be in charge of R. G. Ames, whose territory has been extended to cover the entire Mid-West.

W. B. Connor, Inc., have taken on the agency of the International Engineering Works, Inc., of Framingham, Mass., for the metropolitan district and export. This company represents in addition: Dayton-Dowd Co., Quincy, Ill.; Scranton Pump Co., Scranton, Pa.; The Sims Co., Erie, Pa.; Erie Engine Works, Erie, Pa.

The George Cutter Co., of South Bend, Ind., announces the acquisition by the Westinghouse Electric and Manufacturing Co. of a financial interest in former company. The management and the commercial policies will remain unchanged, but advantage will be taken of the Westinghouse engineering and other facilities to expand the Cutter company's activities and develop its lines to enable it to more effectively serve its customers and the industry. The Westinghouse company will operate as sole distributors for the Cutter company's products.

Trade Catalogs

The Armstrong Machine Works, Michigan, has just issued a 32-page booklet describing the operation, application and construction of steam taps.

The Page Boiler Co., of Chicago, has just completed a 38-page catalog dealing with economical steam production and the building up of boiler efficiency. A copy will be sent upon request.

The Erie Pump and Equipment Co., Erie, Pa., has just issued a thirty-two page booklet describing various styles and features of their Class M Vollete Centrifugal pumps. The booklet will be sent upon request.

The De Laval Steam Turbine Co., of Trenton, N. J., has recently published an eight-page booklet describing tests made by the cities of Minneapolis and St. Paul, upon De Laval centrifugal pumps. A copy will be sent upon request.

The E. M. Dart Manufacturing Co., of Providence, R. I., has ready for distribution a new 31-page catalog. This booklet describes the couplings, unions, flanges, etc., manufactured by the E. M. Dart Co. A copy will be sent on request.

The Erie Pump and Engine Works, Inc., Medina, N. Y., has ready for distribution a seven page pamphlet describing low head centrifugal pumps and a thirty-two page booklet dealing with centrifugal dredging pumps. Copies will be sent upon request.

Dry Steam is the title of a 21-page booklet recently issued by the Tracy Engineering Co., of San Francisco, Cal. It describes various methods of overcoming the difficulties incident to wet steam. A copy of the pamphlet will be sent upon request.

The Star Brass Works, Manufacturing Engineers, Chicago, now have ready for distribution a new 24-page booklet dealing with spray cooling equipment for the cooling of water from steam and ammonia condensers. A copy of the booklet will be sent on request.

The Smooth-On Manufacturing Co., of Jersey City, N. J., has ready for distribution a new 31-page booklet entitled "Smooth-On Home Repairs." The catalog describes methods for the repair of boilers, radiators and furnaces. A copy will be sent on receipt of request.

The Green Fuel Economizer Co., Beacon, N. Y., recently issued a 44-page, illustrated bulletin, No. 152, on "Hi-Efficiency, Hi-Speed Radial Flow Fans—Forced Draft Service." The development and design, pressures and capacities developed of these fans are all fully described. Added are 12 pages of dimension sheets and 19 pages of capacity tables.

The Locomotive Superheater Co., of New York and Chicago, has completed a new 8-page booklet entitled "Superheaters for Stationary Power Plants." A clear, concise argument is given, covering the advantages of superheated steam, which should prove of value to all power plant owners and operators. These bulletins will be sent to anyone interested on request.

Sanford Riley Stoker Co., Worcester, Mass., has recently issued a 47-page second edition of its general catalog on "Riley Underfeed Stokers." This is artistically printed and illustrated on fine bristol paper and fully describes every phase of the stoker. Several pages of valuable engineering data are also contributed by the company's service department. This is for general distribution to anyone interested in boiler room equipment.

THE COAL MARKET

BOSTON—Current prices per gross ton f.o.b. New York loading ports:

Anthracite

	Company Coal
Egg	$8 20@$8.65
Stove	8.45@ 9.05
Chestnut	8 55@ 9.05
Pea	7 05@ 7.40
Buckwheat	3 60@ 3.80
Rice	2.30@ 2.50
Barley	1.25

Bituminous

	Clearfields	Cambrias and Somersets
F.o.b. mines, net tons	$2 85@$3 35	$3 15@$3.60
F.o.b. Philadelphia, gross tons	5.05@ 5.60	5 35@ 5.90
F.o.b. New York, gross tons	5.40@ 5.95	5.75@ 6.25
Alongside Boston (water coal), gross tons	7.00@ 7 75	7.60@ 8.00

Pocahontas and New River are practically off the market for coastwise shipment, but are quoted at $6.25@$7.00.

NEW YORK—Current quotations, White Ash, per gross tons, f.o.b. Tidewater, at the lower ports are as follows:

Anthracite	Company Coal
Broken	$7 80@$8.25
Egg	8.20@ 8.65
Stove	8 45@ 9.05
Chestnut	8 55@ 9.05
Pea	7.05@ 7.40
Buckwheat	5.15
Rice	4.50
Barley	4.85
Boiler	4.85

Bituminous

Government prices at mines:	Spot
South Fork (best)	$3.25@$3.50
Cambria (best)	3.00@ 3.25
Cambria (ordinary)	2 60@ 2.90
Clearfield (best)	3.00@ 3.25
Clearfield (ordinary)	2 60@ 2.90
Reynoldsville	2.85@ 2.90
Quemahoning	3.25@ 3.50
Somerset (medium)	3.00@ 3.25
Somerset (poor)	2.50@ 2.75
Western Maryland	2.50@ 2.75
Fairmont	2.25@ 2.50
Latrobe	2.60@ 2.90
Greensburg	2.75@ 3.00
Westmoreland ¾ in.	3 40@ 3.50
Westmoreland run-of-mine	2.75@ 3.00

PHILADELPHIA—Bituminous coal prices vary according to district from which they are mined. For ordinary slack the price is $2.45@$2.55; lump, $3.35, at the mines.

BUFFALO—

Anthracite

	On Cars, Gross Ton	At Curb, Net Ton
Grate	$8 55	$10.80
Egg	8 80	10.65
Stove	9 00	10.85
Chestnut	9 10	10.85
Pea	7.75	9.30
Buckwheat	5 70	7.75

Bituminous

Allegheny Valley	$4.80
Pittsburgh	4.65
No. 8 Lump	4.65
Mine-run	4.80
Slack	4.10
Smokeless	4.60
Pennsylvania Smithing	5.70

CLEVELAND—Prices of coal per net ton delivered in Cleveland are:

Anthracite

Egg	$12 25@$12.40
Chestnut	12.50@ 12.70
Grate	12 25@ 12.40
Stove	12.40@ 12.60

Pocahontus

Nine-run	$7.50

Domestic Bituminous

West Virginia split	$8.50
No. 8 Pittsburgh	$6 60@ 6.90
Massillon lump	8.25@ 8.50
Coshocton lump	7.15

Steam Coal

No. 6 slack	$5.25@$5.50
No. 8 slack	5.10@ 5.50
Youghiogheny slack	5.25@ 5.50
No. 8 ¾ in.	5.70@ 6.60
No. 6 mine-run	5.25@ 5.50
No. 8 mine-run	5.75

Only coal available is mine-run Pocahontas.

MIDDLE WEST—Chicago quotations, f. o. b. cars at mine:

	Springfield, Carterville, Williamson, Franklin, Saline, Harrisburg	Fulton, Peoria	Grundy, La Salle, Bureau, Will
Lump	$2.55@$2.70	$2.95@$3.10	$3.25@$3.40
Washed	2.75@ 2.90		3.45@ 3.60
Mine-run	2.35@ 2.50	2.75@ 2.90	3.00@ 3.15
Screen'gs	2.05@ 2 20	2.35@ 2.50	2.75@ 2.90

New Construction

PROPOSED WORK

Mass., Northampton—George F. Newton, Archt., 6 Beacon St., Boston, will soon award the contract for the construction of a 2 story, 175 x 200 ft. hospital building, for the Lathrop Home, 236 South St. A steam heating system will be installed in same. Total estimated cost, $250,000.

Mass., Salem—The Salem Electric Lighting Co., 201 Devonshire St., is in the market for turbine and boilers.

Mass, Worcester—Cutting, Carlton & Cutting, Archts., 44 Front St., will soon receive bids for the construction of a 1 and 2 story garage and service building on Park Ave., for Henry J. Murch, 731 Main St. A steam heating system will be installed in same. Total estimated cost, $150,000.

R. I., Providence—The Narragansett Electric Lighting Co., 800 Turks Head Bldg., will soon award the contract for the construction of a 75 x 95 ft. power house addition on South St. Estimated cost, $700,000.

Conn., Norwich—H. S. Goldfaden, 388 West Main St., will soon award the contract for the constructoin of a 3 story theatre on B'way. A steam heating system will be installed in same. Total estimated cost, $150,000. George E. Pitcher, 65 B'way, Engr.

N. Y., Alden—The Supervisors of the Co. Home and Hospital will soon receive bids for the construction of 12 separate buildings, including a separate building for power and heating plant. Total estimated cost, $1,700,000. G. C. Diehl, Ellicott Sq., Buffalo, Engr.

N. Y., Brooklyn—Abraham & Strāuss, 422 Fulton St., are having plans prepared for altering the 5 story department store building at 422 Fulton St. A steam heating system will be installed in same. Total estimated cost, $130,000. Starrett & Van Vleck, 8 West 40th St., New York City, Archts. and Engrs.

N. Y., Brooklyn—The Brooklyn Retail Butchers' Corp., 3285 Fulton St., is having plans prepared for the construction of a 3 story, 130 x 135 ft. ice plant and storage house on Atlantic Ave. and Fort Greene Pl. Total estimated cost, $400,000. J. H. H. Voss, 154 Nassau St., New York City, Archt. and Engr.

N. Y., Brooklyn—R. Sanders, 188 Prospect Park West, is in the market for complete saw-mill equipment, including steam engine and boiler.

N. Y., Brooklyn—The School of Our Lady of Lourdes, Rev. J. McMahon, Pastor, will soon award the contract for the construction of a 4 story school addition on Aberdeen St. A steam heating system will be installed in same. Total estimated cost, $300,000. Helmle & Corbett, 189 Montague St., Archts. and Engrs.

N. Y., Lackawanna—St. Patrick's Parish, Father Baker, Pastor, plans to construct 2 buildings including church and school. A steam heating and ventilating system will be installed in same. Total estimated cost, $100,000 to $130,000.

N. Y., Long Island City—W. Higginson, Archt. and Engr., 18 East 41st St., New York City, will soon award the contract for the construction of an 8 story, 85 x 345 ft. factory on Mount St., for H. Lockhart Jr., 501 5th Ave., New York City. A steam heating system will be installed in same.

N. Y., Marcy—The State Hospital Comn., Capitol, Albany, received bids for installing a steam heating system in the mortuary, laboratory and tuberculosis hospital at the Utica State Hospital, here, from Thomas Breen Co., 414 Lafayette St., $17,413; H. J. Brandeles, 433 Lafayette St., $18,245; Mohawk Valley Heating Co., 302 Broad St., $20,745; all of Utica. Noted March 3.

N. Y., Massena—The Massena Farmers Co-operative Corp. is in the market for a refrigeration system, steam boiler equipment and transmission and complete creamery machinery. Address James Phillips.

N. Y., New York—The Bd. Educ., 500 Park Ave., is having plans prepared for the construction of a 4 story school building on 176th St. and Audubon Ave. A steam heating system will be installed in same. Total estimated cost, $200,000. C. B. J. Snyder, Municipal Bldg., Archt. and Engr.

N. Y., New York—E. Dwight, 56 Maiden Lane, is having plans prepared for altering the 8 story loft building and warehouse on William and John Sts. A steam heating system will be installed in same. Total estimated cost, $150,000. Butler & Rodman, 56 West 45th St., Archts. and Engrs.

N. Y., New York—M. R. Huntington, 19 East 58th St., is having plans prepared for the construction of a 2 story, 85 x 100 ft. theātre on Washington Pl. A steam heating system will be installed in same. Total estimated cost, $250,000. William J. Cherry, Grand Central Station, Archt. and Engr.

N. Y., New York—The New York Telephone Co., 15 Dey St., is having plans prepared for the construction of an 80 x 192 ft. automatic telephone exch. building on 73rd St. between Columbus and Amsterdam Ave. A steam heating system will be installed in same. Total estimated cost, $1,540,000. McKenzie, Voorhies & Gmelin, 1123 B'way, Archts. and Engrs.

N. Y., New York—Thompson & Binger, Archts. and Engrs., 280 Madison Ave., is having plans prepared for the construction of an 18 story, 75 x 80 ft. office building on Franklin and Gold Sts. A steam heating system will be installed in same. Total estimated cost, $400,000. Owner's name withheld.

N. Y., Ossining—The Bd. Educ. will receive bids about April 10 for the construction of a 2 story school building. A steam heating system will be installed in same. Total estimated cost, $60,000. Wilson Potter, 22 East 17th St., New York City, Archt. and Engr.

N. Y., Ripley—The Telling-belle Vernon Co. plans to take over the plant of the Ripley Milk Products Co. and is in the market for a 150 hp. boiler.

N. Y., Seneca Falls—Harry Nothnagle is in the market for cold storage equipment.

N. Y., Watertown—The Beebes Island Corp. is having plans prepared by J. G. White Co., 215 Thompson St., New York City, for the construction of a power house on Beebes Island. Estimated cost, $500,000. C. H. Starbuck, New York Air Brake Co., 165 B'way, New York City, Pres.

N. J., Lakehurst—The Bureau of Yards & Docks, Navy Dept., Wash., D. C., is having plans prepared for the construction of a 36 x 73 ft. power house with a three 300 kw. generating capacity and the installation of a 50 x 111 ft., 3,000 hp. boiler. Total estimated cost, $500,000. Noted Feb. 3.

N. J., Newark—The Wright Aeronautical Corp. of America, 40 Wall St., New York City, will soon award the contract for the construction of a 2 story, 45 x 200 ft. factory on Meiker Ave. A steam heating system will be installed in same. Total estimated cost, $350,000. J. W. Ingle, 527 5th Ave., New York City, Archt. and Engr.

N. J., Westfield—E. Lynch, Archt. and Engr., 341 5th Ave., New York City, will soon award the contract for the construction of a 1 story, 65 x 115 ft. church for the Holy Trinity Church. A steam heating system will be installed in same. Total estimated cost, $100,000.

Pa., Connellsville—The Paragon Motor Co., 404 Century Bldg., Cleveland, O., is having plans prepared for the construction of a 1 story, 60 x 400 ft. and a 1 story, 40 x 540 ft. factory building. Plans include the construction of a steam power plant. Total estimated cost, $500,000. F. F. Hockenthal, 404 Century Bldg., Cleveland, O., Engr.

Pa., Easton—The Y. M. C. A., 347 Madison Ave., New York City, is having plans prepared for the construction of a 4 story Y. M. C. A. building. A steam heating system will be installed in same. Total estimated cost about $250,000. Martin & Kirkpatrick, 130 South 15th St., Philadelphia, Archts. and Engrs.

Pa., Philadelphia—The Queens Brooks Co., American and Diamond Sts., is having plans prepared for the construction of a 5 story, 50 x 200 ft. factory on 6th St. and Hunting Park. A steam heating system will be installed in same. Total estimated cost, $100,000. W. E. S. Dyer, Land Title Bldg., Archt. and Engr.

Pa., Pittston—The Pennsylvania Coal Co. plans to build a 50 x 90 x 200 ft. coal breaker, cost, $750,000 and a 30 x 40 x 40 ft. electric power plant, cost, $200,000.

Va., Hampton Roads—The Bureau of Yards & Docks, Navy Dept., Wash., D. C., is having plans prepared for the construction of an underground steam heating system at the Naval Operating Base, here.

Va., Hampton Roads—The Bureau of Yards & Docks, Navy Dept., Wash., D. C., is having plans prepared for the construction of an underground electric distributing system at the Naval Operating Base, here.

Va., Richmond—The Westmorehead Club, Grace St., is having plans prepared for the construction of a 3 story, 80 x 140 ft. club house. A steam heating system will be installed in same. Total estimated cost, $350,000. G. Bryan, Pres. Alfred C. Bossom, 680 5th Ave., New York City, Archt. and Engr.

W. Va., Charleston—Lockwood Greene & Co., Archts. and Engrs., 101 Park Ave., New York City, will receive bids until April 5 for the construction of a 3 story, 52 x 200 ft. silk throwing mill for the Sauquoit Silk Mfg. Co. A steam heating system will be installed in same.

N. C., Graham—The city plans to construct a sewer, install 2 pumping stations and is in the market for 2 motor driven, centrifugal pumps. Total estimated cost, $100,000. Norcross & Keis, Atlanta, Ga., Engrs.

S. C., Charleston—J. A. Wetmore, Supervising Archt., Treasury Dept., Wash., D. C., will receive bids until April 15 for the installation of a lighting plant and water supply system at the U. S. Quarantine Station, here.

S. C., Gaffney—The Musgrove Mills plans to build a 3 story, 107 x 290 ft. cotton cloth mill equipped with individual electric drive, complete fire protective apparatus, etc. J. E. Sirrine, Engr.

S. C., Greenville—Furman University plans to build a 40 x 60 ft. central heating plant to supply all their buildings. Two 100 hp. boilers, ultimate capacity 500 hp. will be installed in same. J. E. Sirrine, Engr.

O., Cleveland—The Beckman Co., 3167 Fulton Rd., plans to build a 2 story, 200 x 200 ft. weaving factory on Fulton Rd. and West 33rd St. A steam heating system will be installed in same. Total estimated cost, $400,000. Lockwood Greene & Co., Bangor Bldg., Archts. and Engrs.

O., Cleveland—The Bd. Educ., East 6th St. and Rockwell Ave., is having plans prepared for the construction of a 3 story, junior high school on Hopkins Ave. near East 123rd St. Two boilers will be installed in same. Total estimated cost, $1,250,000. W. R. McCormack, Archt.

O., Cleveland—The Bd. Educ., East 6th St. and Rockwell Ave., is having plans prepared for the construction of a 3 story, 200 x 400 ft. junior high school on East 117th St. and The Speedway. A steam heating system will be installed in same. Total estimated cost, $600,000. F. G. Hogan, Dir. W. R. McCormack, Archt.

O., Cleveland—The city received bids for installing two 125 hp. boilers and stokers, from the Wickes Boiler Co., Saginaw, Mich., $12,157; Union Iron Wks., Rockfir Bldg., Clevelsnd, $13,120. Noted March 16.

O., Cleveland—The city plans to build a 2 story police station on East 37th and Woodland Ave. A steam heating system will be installed in same. Total estimated cost, $100,000. A. B. Sprosty, City Hall, Dir. of Pub. Safety. F. H. Betz, 604 City Hall, Archt.

O., Cleveland—The city plans to construct a waterworks pumping station including tunnel and crib in the Lake Erie filter plant and pumping equipment. Total estimated cost, $10,000,000. A. V. Ruggles, City Hall, Engr.

O., Cleveland—The Cleveland Ry. Co., Leader-News Bldg., plans to construct a 1 story sub-station on East 130th St. and Woodland Ave. Estimated cost, $75,000. J. Stanley, Pres. D. W. Morrow, 4500 Euclid Ave., Archt.

O., Cleveland—Otis R. Cook, East 46th St. and Prospect Ave., will soon award the contract for the construction of a 5 story, 100 x 104 ft. commercial building. A steam heating system will be installed in same. Total estimated cost, $500,000. Lehman & Schmitt, Electric Bldg., Archts.

O., Cleveland—The Natl. Acme Co., 7500 Stanton Ave., is having plans prepared for the construction of a 1 story factory addition on East 131st St. and Colt Rd. A steam heating system will be installed in same. Total estimated cost, $500,000. A. W. Henn, Pres. G. S. Rider & Co., Century Bldg., Archt. and Engr.

O., Cleveland—W. Taylor Son & Co., 630 Euclid Ave., plans to build a 10 or 12 story, 135 x 265 ft. warehouse on East 17th St. and Superior Ave. Three boilers will be installed in same. Total estimated cost, $1,500,000. C. H. Strong, Mgr.

O., Cleveland—C. Shane, 2147 Ontario St., is having preliminary plans prepared for the construction of a 12 story store and office building on East 4th St. and Euclid Ave. A steam heating system will be installed in same. Total estimated cost, $750,000. W. S. Longee, Marshall Bldg., Archt.

O., Jackson—The Wellston Iron Furnace Co. is in the market for a 100 hp., d.c., 250 volt motor.

O., Miamisburg—The City Water & Light Plant plans to purchase a new gas generator set for the electric light plant, consisting of dynamo and an artificial gas machine. Total estimated cost, $30,000. A. W. Kepler, Supt.

Mich., Cheboygan—The city will soon award the contract for the construction of a 2 story, 60 x 120 ft. municipal building. A steam heating system will be installed in same. Total estimated cost, $150,000. L. E. Berry, Clk.

Mich., Detroit—The Bd. Educ., 50 B'way Ave., received bids for installing a heating, ventilating and plumbing system in the proposed 3 story, 100 x 250 ft. school on St. Antoine St. and Palmer Ave., from A. W. Schultz & Co., 380 Rohns St., $143,908; W. J. Rewoldt, 506 Owen Bldg., $154,920; James W. Partlan, 51 Park Pl., $155,000.

Mich., Detroit—The Detroit Water Bd. had plans prepared for the construction of a 1 story, 57 x 221 ft. superstructure for a low lift or booster pumping station at the Waterworks Park. Electric motors for power will be installed in same. Total estimated cost, $100,000. T. A. Leisen, 232 Jefferson Ave., Engr.

Mich., Detroit—Aaron H. Gould & Son, 611 Empire Bldg., are in the market for ice making and refrigerating machinery.

Mich., Detroit—The Holy Redeemer Church, c/o Rev. Fr. Cantwell, Dix Ave., plans to build a 2 story church on Dix and Junction Aves. A steam heating system will be installed in same. Total estimated cost, $300,000.

Mich., Eaton Rapids — The Artificial Stone Co. is in the market for a pumping outfit, etc.

Mich, Flint—The Bd. Educ. is having plans revised for the construction of a 2 story, 208 x 287 ft. high school. A steam heating and forced ventilation system will be installed in same. Total estimated cost, $1,250,000. Malcolmson, Higginbotham & Palmer, 405 Moffat Bldg., Detroit, Archts.

Mich., Flint — Van Leyen, Schilling, Keough & Reynolds, Archts. and Engrs., Union Trust Bldg., Detroit, plans to build a 2 story, 40 x 130 ft. theatre on Detroit St. A steam heating boiler and a forced ventilation system will be installed in same. Total estimated cost, $100,000. Owner's name withheld.

Mich., Grand Rapids—The Bd. Educ. will soon award the contract for the construction of two 2 story school buildings, one on Dickinson St. and the other on Lafayette St. Steam heating equipment will be installed in same. Total estimated cost, $450,000. H. N. Morrell, Secy. H. H. Turner, 923 Michigan Trust Bldg., Archt.

Mich., Hamtramck (Detroit P. O.)—Warren W. Tyler, Secy. of the School Dist. 8, will receive bids until April 3 for the construction of a 2 story school building on Playfair and Goodson Aves. A steam heating and ventilating system will be installed in same. Total estimated cost, $100,000. G. J. Haas, 1514 Kresge Bldg., Detroit, Archt.

Mich., Ironwood—The city is having plans prepared for furnishing electrical pumping equipment, etc. Total estimated cost, $300,000. Alvord & Burdick, 8 South Dearborn St., Chicago, Engrs.

Mich., Jackson—The Michigan Light & Consumers Power Co., 226 West Main St., plans to build a 14,000 volt, 15 kw. substation and a single line steel tower transmission line from here to Battle Creek. Total estimated cost, $410,000.

Mich., Monroe—The Monroe Paper Products Co., West Elm St., plans to build a 2 story factory. Steam heating equipment and electric power will be installed in same. Total estimated cost, $100,000. J. Gilmore, Archt.

Mich., Saginaw—The Bd. Educ., c/o E. C. Oscar, engaged Cowles & Mutscheller, Archts., 1-10 Chase Block, to prepare plans for the construction of a 2 story junior high school. A steam heating system will be installed in same. Total estimated cost, $1,000,000.

Mich., Saginaw—The Modart Corset Co., 309 Lapeer Ave., plans to build a 4 story, 109 x 134 ft. mercantile building on Lapeer Ave. Steam heating equipment will be installed in same. Total estmiated cost, $300,000. Cooper & Beckbissinger, 114 South Jefferson Ave., Archts.

Ill., Chicago—The Jewish Day Nursery, 1441 Wickes Park Ave., is having plans prepared for the construction of a 2 story, 60 x 120 ft. children's home on California Ave. and Hirsch Blvd. A steam heating system will be installed in same. Total estimated cost, $100,000. Ludgin & Leviton, 53 West Jackson St., Archts.

Ill., Cicero—Childs & Smith, Archts., 64 East Van Buren St., will soon award the contract for the construction of a 1 story, 75 x 125 ft. bank on 22nd St. and Austin Ave., for the Kasper State Bank, 1900 Blue Island Ave., Chicago. A steam heating system will be installed in same. Total estimated cost, $175,000.

Wis., Kohler—The village plans an election to vote on $12,000 bonds to construct and equip an electric light system. Address A. Biotz, Village Pres. Jerry Donohue, North 8th St., Sheboygan, Engr.

Wis., La Crosse—J. C. Lewellyn, Archt., 38 South Dearborn St., will soon award the contract for the construction of a 3 story, 90 x 153 ft. school building, for the Bd. Educ. A steam heating system will be installed in same. Total estimated cost, $200,000.

Wis., Sheboygan—The Cigar Box Libr. & Mfg. Co., c/o J. Will, 1007 Michigan Ave., is having plans prepared for the construction of a 1 story, 75 x 120 ft. machine, saw, filing and power house on 15th St. Electric power will be installed in same. Total estimated cost, $50,000. Juul & Smith, Imig. Bldg., Archts. and Engrs.

Ia., Dana—Guy Meredith, Secy. of the School Bd., is having plans prepared for the construction of a 2 story consolidated school building. Separate bids for installing steam heating and plumbing systems will be received. Total estimated cost, $100,000. W. E. Hulse & Co., 210 Masonic Temple, Des Moines, Archt.

Ia., Guss—W. M. McCoy, Pres. of the School Bd., Rural Route, Nodaway Twp., is having plans prepared for the construction of a 2 story, 66 x 86 ft. consolidated school building. Separate bids for installing heating and plumbing systems will be received. Total estimated cost, $100,000. W. E. Hulse & Co., 210 Masonic Temple, Des Moines, Archt.

Ia., Marshalltown—The Bd. Educ. is having plans prepared for the construction of a 2 story, 34 x 148 ft. grade school building. A steam heating system will be installed in same. Total estimated cost, $100,000. C. E. Shutt, Co. Supt. Tyrie & Chapman, 320 Auditorium Bldg., Minneapolis, Minn., Archts.

Ia., Rippey—A. E. Jensen, Secy. Bd. Educ., will receive bids until April 3 for the construction of a 2 story, 63 x 105 ft. school building. Separate bids for installing heating system will be received. Total estimated cost, $100,000. Proudfoot, Bird & Rawson, 810 Hubbell Bldg., Des Moines, Archts.

Ia., West Branch—T. A. Moore, Secy. of the Bd. Educ., is having plans prepared for the construction of a 3 story, 54 x 114 ft. school building. Separate bids for installing a steam heating and forced ventilation system will be received. Total estimated cost, $125,000. Proudfoot, Bird & Rawson, 810 Hubbell Bldg., Des Moines, Archts.

Minn., Deer River—H. E. Wolfe, Supt., will receive bids until April 15 for the construction of a 2 story, 80 x 110 ft. community school for the School Bd. A steam heating system will be installed in same. Total estimated cost, $100,000. S. F. Broomhall, Alworth Bldg., Duluth, Archt. and Engr.

Minn., Minneapolis—The Bd. Educ. will soon award the contract for the construction of a 2 story, 72 x 210 ft. school building on 51st St. and 29th Ave., South. A steam heating system will be installed in same. Total estimated cost, $150,000. G. F. Womrath, Mgr. E. H. Enger, Engr.

Minn., Minneapolis—The Bd. Educ. will soon award the contract for the construction of a 2 story, 73 x 200 ft. school building on Harriot Ave. and 38th St. A steam heating system will be installed in same. Total estimated cost, $150,000. E. H. Enger, Engr.

Minn., Minneapolis—The Bd. Educ. will soon award the contract for the construction of a 2 story, 73 x 200 ft. school building on 22nd Ave. S. and East 40th St. A steam heating system will be installed in same. Total estimated cost. $250,000. E H. Enger, Engr.

Minn., Minneapolis—The McMichael Investment Co., 600 Plymouth Bldg., will soon award the contract for the construction of a 2 story, 50 x 157 ft. office building at 33 7th St., South. A steam heating system will be installed in same. Total estimated cost, $500,000. Kees & Coburn 245 Plymouth Bldg., Archts.

Minn., Rapidan—The Rapidan Creamery Co., Mankato, will soon receive bids for the construction of a 2 story, 50 x 60 ft. creamery. A steam heating system will be installed in same. Total estimated cost $75,000. Carl Mohr, Secy. Albert Schipped, 809 Coughlin-Hickey Bldg., Archt.

Kan., Lawrence—The state is having plans prepared for the construction of a 1 story, 76 x 115 ft. power plant and installation of equipment. One or two water tube boilers and an additional engine unit will be installed in same. Total estimated cost, $70,000. J. A. Kimball, Bd. of Administration, Bus. Mgr. Ray S. Gamble 1415 Fillmore St., Topeka, Archt.

Kan., Ottawa—The Baptist University is having plans prepared for the construction of a 1 story, 50 x 100 ft. power and heating plant. Estimated cost, $75,000. S. E. Price, Pres. Washburn & Stoockey, Archts. and Engrs.

S. D., Groton—A. McKiver will receive bids until April 1 for the construction of a 2 story, 100 x 150 ft. junior and senior high school for the School Bd. A steam heating and mechanical ventilating system will be installed in same. Total estimated cost, $200,000. E. F. Broomhall, Alworth Bldg., Duluth, Minn., Archt. and Engr.

Mont., Wolf Point—The city plans an election April 10 to vote on $50,000 bonds to construct new wells, pumping plant, a 200,000 gal. steel storage tank, etc. Samuel Dowell, Engr.

Mo., Sugar Creek—The Bd. Educ. will receive bids until April 2 for the construction of a 2 story, 63 x 86 ft. school building and a 1 story, 24 x 50 ft. boiler house. Total estimated cost, $70,000. A $40,000 bond issue was passed Feb. 7 for the project. Ashton C. Jones, Secy. A. B. Anderson, 818 New York Life Bldg., Kansas City, Mo., Archt.

Tex., Denison—The city plans an election April 6 to vote on $244,000 bonds to improve electric lighting systems and to purchase electrical pumping equipment for the city reservoir.

Tex., North Fort Worth—The Bureau of Yards & Docks, Navy Dept., Wash., D. C., plans to construct and install deep well pumps and piping at the Naval Helium Production plant, here.

N. M., Estancia—The town plans an election April 6, to vote on $50,000 bonds to install a water system, including tank, tower, pump, engine and mains. B. H. Calkins, Engr.

Wash., Olympia—The state is having plans prepared for the construction of a power house. A steam heating system will be installed in same. Total estimated cost, $100,000. Wilder & White, 50 Church St., New York City, Archts. and Engrs.

Wash., Seattle—F. A. Naramore, Archt., Central Bldg., will soon award the contract for the construction of a 1 story, 200 x 240 ft. school building at Highland Park, for the Seattle School Bd. A direct steam heating system with fan driven air circulation will be installed in same. Total estimated cost, $275,000.

Wash., Wenatchee—The land owners under the Okanogan Valley Irrigation Project plan to install a pumping system including one or more pumps of 25,000 gal. per min. capacity and capable of lifting water 470 ft. Total estimated cost, $10,000. D. C. Heney, Spalding Bldg., Portland, Ore., Consult. Engr.

Ore., Bend—The Bend Water, Light & Power Co. plans to install an auxiliary pumping and sterilization plant including one 1,500 gal. per min. pump and construct a 30,000 gal. reservoir. Total estimated cost, $12,000.

Ore., Forest Grove—The Masonic Lodge will receive bids in April for the construction of a Masonic home including 2 units of a 3 unit administration building to be used in connection with a group of residence cottages. Plans include a separate power house. Total estimated cost, $230,000. W. C. Knighton, U. S. Natl. Bank Bldg., Portland, Archt.

Ore., Portland—C. J. Schnabel, Chamber of Commerce Bldg., filed an application with P. A. Cupper, State Engr., Salem, to take 600 second ft. of water from the Santiam River to lighten and manufacture a 6,511.3 theoretical hp. plant and construct diversion works, etc. Total estimated cost, $800,000.

Cal., Hayward—The city is having tentative plans prepared for the purchase of the Hayward Water Company's water system and improvement of same by replacing steam pumping plant with electric pumping system, etc. Total estimated cost, $100,000. M. B. Templeton, City Clk.

Cal., Inglewood—J. H. Kew, City Clk., will soon receive bids for furnishing a 400 to 1,200 gal. capacity, centrifugal, booster pump for the city. Olmsted & Gillelen, Hollingsworth Bldg., Los Angeles, Engrs.

Cal., Los Angeles—The Bd. of Pub. Serv. Comn., 645 South Olive St., will receive bids until April 6 for furnishing hydraulic equipment, electric generators, auxiliary electric equipment, etc. E. F. Scattergood, 645 South Olive St., Engr.

Cal., Mare Island—The Bureau of Yards & Docks, Navy Dept., Wash., D. C., is having plans prepared for the installation of a 4,000 kw. turbo generator, here.

Cal., Martinez—The Alhambra Union High School rejected bids and will revise plans for the construction of a 2 story high school. A plenum heating system will be installed in same. Total estimated cost, $125,000. J. E. Rogers, Byron Brown Bldg., Clerk. A. A. Cantin, 68 Post St., San Francisco, Archt.

Cal., Monrovia—H. S. Gierlich, City Engr., is preparing plans for the construction of a 1,000,000 gal. reservoir, deep wells, pumps and mains. The city voted a $90,000 bond issue for the project.

Cal., Santa Ana—The First Natl. Bank is having plans prepared for the construction of a 6 story, 75 x 100 ft. bank and office building. A steam heating system will be installed in same. J. Parkinson, 420 Title Insurance Bldg., Los Angeles, Archt.

N. S., Halifax—K. H. Smith, Secy. of the Nova Scotia Power Comn., will receive bids until April 8 for the construction of power dams, superstructures for two generating stations, etc., and for furnishing and installing two 1,600 kva. generators, two 2,900 kva. generators, vertical type with exciters, two 1,900 hp. turbines and two 3,450 hp. turbines.

M. I., Guam—The Bureau of Yards & Docks, Navy Dept., Wash., D. C., plans to build a 30 x 38 ft. extension to power house at Agana and a 26 x 35 ft. extension to power house at the Naval Radio Station, a 12 x 14 ft. extension to pumphouse, 8 x 16 ft. extension to blower house and 22 x 47 ft. with an 18 x 24 ft. wing extension to quarters, including plumbing and electric lighting systems.

Ont., Chatham—The city will vote upon by-law appropriating $371,000 to construct a rapid sand gravity filtration plant, etc. Low lift, electric driven turbine pumps will be installed in same. F. P. Adams, City Hall, Engr.

Ont., Cobourg—The Hydro Electric Comn. of Ontario, University Ave., Toronto, plans to build a radial railroad from here to Campbellford. Estimated cost, $2,500,000. F. A. Gaby, Hydro Electric Power Co. Bldg., Toronto, Engr.

Ont., Islington—The town had plans prepared for the construction of a complete new waterworks system, including a water tank, electric driven centrifugal pumps, etc. Total estimated cost, $85,000. E. A. James Co., Ltd., 36 Toronto St., Toronto, Engr.

Ont., London—The Bd. Educ., City Hall, plans to build a collegiate building on Ridout St. A steam heating system will be installed in same. Total estimated cost, $215,000. L. V. Carrothers, Hydro offices, Archt.

Ont., London—The City Council is in the market for a 10 hp. electric motor to run a concrete mixer instead of the steam engine now used. H. A. Brazier, Engr.

Ont., New Toronto—The town plans an election to vote on by-law appropriating $100,000 to construct a 2 story public school. A vacuum steam heating system will be installed in same. W. H. C. Millard, Clk.

Ont., Port Dover—The city is having estimates submitted for the construction of a waterworks system including pumphouse, reservoir, centrifugal pumps, etc. Total estimated cost, $100,000. E. A. James Co., Ltd., 36 Toronto St., Toronto, Engr.

Ont., Toronto—The Canada Realty Co., Kent Bldg., is preparing plans for the construction of a 3 story, 50 x 80 ft. bakery on Claremont St. A steam heating system and a horizontal tubular boiler will be installed in same. Total estimated cost, $70,000.

CONTRACTS AWARDED

N. H., Portsmouth—The Rockingham Co. Light & Power Co., 29 Pleasant St., has awarded the contract for the construction of a 2 story, 30 x 50 ft. power plant addition, to M. Cashman, 63 Water St., Newburyport, Mass. Estimated cost, $25,000.

Mass., Lowell—The city has awarded the contract for the construction of a 3 story, 186 x 289 ft. high school, to Daniel Walker, 529 Dutton St. A steam heating system will be installed in same. Total estimated cost, $750,000.

R. I., Barrington—The Rhode Island Lace Wks will build a 1 and 2 story industrial housing building. A hot air heating system will be installed in same. Total estimated cost, $100,000. Work will be done by day labor.

Conn., Norwalk—The Rosenwald Wimpfheimer Inc. has awarded the contract for changing the old electric light plant into a hat factory and for the construction of a 1 story boiler house addition on Wilton Ave. and Cross St., to A. R. Malkan & Co., 7 Mechanic St. Total estimated cost, $25,000.

Mass., Springfield—The Diamond Match Co., 111 B'way, New York City, has awarded the contract for the construction of 4 buildings including a 2 story, 311 x 468 ft. factory and three 2 story storehouses, to the Turner Constr. Co., 244 Madison Ave., New York City. A steam heating system will be installed in same. Total estimated cost, $1,000,000.

N. Y., Brooklyn—The Atlantic Pacific Co., 317 Bradford St., will build six 4 story, 50 x 100 ft. loft buildings on Pacific and Sackman Sts. A steam heating system will be installed in same. Total estimated cost, $300,000. Work will be done by day labor.

N. Y., Long Island City—The Title Guarantee & Trust Co., 176 B'way, New York City, will build an office building on Hunter Ave. A steam heating system will be installed in same. Total estimated cost, $100,000. Work will be done by day labor.

N. Y., New York—The 58 West 40th St. Corp., c/o E. Necarsulmer, Archt. and Engr., 507 5th Ave., has awarded the contract for the construction of a 16 story, 75 x 100 ft. store and office building on 58 West 40th St., to Leddy & Moore, 105 West 40th St. A steam heating system will be installed in same. Total estimated cost, $850,000.

N. Y., New York—The New York Central R.R., Grand Central Terminal, has awarded the contract for the construction of a 13 story addition to present office building on Lexington Ave. and 42nd St., to James Stewart & Co., 30 Church St. A steam heating system will be installed in same.

N. Y., New York—The Sarco Realty Co., 214 West 141st St., will build a 100 x 100 ft. theatre on 137th St. and 7th Ave. A steam heating system will be installed in same. Total estimated cost, $125,000. Work will be done by day labor.

N. Y., Schenectady—The Citizens Trust Co., 434 State St., has awarded the contract for the construction of a bank building to John McDermott, 436 State St. A steam heating system will be installed in same. Total estimated cost, $200,000.

N. J., Newark—Aron Levin, Stuyvesant Ave., will build a 5 story, 50 x 140 ft. hotel at 106 Clinton Ave. A steam heating system will be installed in same. Total estimated cost, $235,000. Work will be done by day labor.

Pa., Chester—Allison & Co., Law Bldg., has awarded the contract for the construction of a 1 story, 62 x 202 ft., mill building, to the Chester Contg. and Constr. Co., Otis Bldg., Philadelphia. A steam heating system will be installed in same. Total estimated cost, $100,000.

Pa., Philadelphia—The Amer. Mfg. Co., Water and Morris Sts., has awarded the contract for the construction of a 1, 2 and 3 story factory on Shunk and Front Sts., to the Turner Constr. Co., 1713 Sanson St. A steam heating system will be installed in same. Total estimated cost, $500,000.

Pa., Pittsburgh—The Pittsburgh Association for the Improvement of the Poor, c/o E. P. Mellon, Archt. and Engr., 52 Vanderbilt Ave., New York City, has awarded the contract for the construction of a 4 story, 60 x 240 ft. office building, to Mellon Stewart, Oliver Bldg. A steam heating system will be installed in same. Total estimated cost, $300,000.

Md., Orangeville (Baltimore P. O.)—The Columbia Graphophone Mfg. Corp., Woolworth Bldg., New York City, has awarded the contract for the construction of a 8 story, 280 x 380 ft. cabinet factory and an 85 x 100 ft power house, etc., on East Preston St and Loneys Lane, to the M. A. Long Co., 1523 Munsey Bldg., Baltimore. Total estimated cost, $2,500,000.

W. Va., Kanawha City—Libby Owens has awarded the contract for the construction of power plant addition near Charleston St., to the Howard P. Foley Co., 1413 G St., Wash., D. C. Estimated cost, $50,000.

Ga., Elberton—The First Natl. Bank of Elberton has awarded the contract for the construction of a 4 story bank and office building, to W. G. Sutherlin, Greenwood. A steam heating system will be installed in same. Total estimated cost, $125,000.

Ga., Savannah—The Diamond Watch Co., 111 B'way, New York City, has awarded the contract for the construction of a 4 story, 57 x 196 ft. factory, to the Turner Constr. Co., 244 Madison Ave., New York City. A steam heating system will be installed in same.

Tenn., Knoxville—The County Court Bldg. Committee has awarded the contract for installing heating and plumbing system in the proposed addition to the Court House, on West Main St., to F. Cantrell Co., Inc., U. T. Grounds, at $12,685.

O., Dayton—The General Motors Co., Taylor St., has awarded the contract for the construction of a 6 story, 120 x 300 ft. factory and a power plant on 1st and Foundry Sts., to Davies, Chism & Davies, 601 Chamber of Commerce Bldg., Columbus. Total estimated cost, $250,000.

Mich., Detroit—The Cadillac Motor Car Co., Woodward Ave., has awarded the contract for the construction of a 4 story, 619 x 792 ft. automobile factory on Clark and Michigan Aves., to the Dupont Engr. Co., 435 Woodward Ave. Electric power will be installed in same. Total estimated cost, $2,780,000.

Mich., Muskegon—The city has awarded the contract for furnishing a 15,000,000 gal. reciprocating pump for waterworks system, here, to The Allis Chalmers Mfg. Co., West Allis St.; Milwaukee, Wis., at $64,900.

Mich., St. Clair—The Diamond Crystal Salt Co. has awarded the contract for the construction of a 2 story, 100 x 160 ft warehouse along the waterfront, to J. O'Sullivan & Sons, Pine Grove Ave., Port Huron. Motors to operate conveyors will be installed in same. Total estimated cost, $125,000.

Ill., Chicago—The Clemetsen Co., 2607 Flournoy St., has awarded the contract for the construction of a 3 story, 100 x 300 ft. factory on Homan and Division St., to R. F Wilson & Co., Market and Randolph Sts. A steam heating system will be installed in same. Total estimated cost, $200,000.

Ill., Chicago—Frederick T. Hoyt, 154 West Randolph St., has awarded the contract for the construction of a 200 x 400 ft. factory on 47th St. and Kedzie Ave., to W. McCumber, 140 South Dearborn St. A steam heating system will be installed in same.

Ill., Chicago—The Shotwell Mfg. Co., 1021 West Adams St., has awarded the contract for the construction of a 4 story, 100 x 200 ft. factory on Potsmer and St. Louis Sts., to R. F. Wilson & Co., Randolph and Market Sts. A steam heating system will be installed in same. Total estimated cost, $300,000.

Wis., Milwaukee—The Seaman Body Corp., 480 Virginia St., has awarded the contract for the construction of a 1 story, 50 x 90 ft. power house on Port Washington St., to Klug & Smith, Mack Block. Boilers and a steam engine will be installed in same. Total estimated cost, $50,000.

Minn., Chisholm — The Bd. Educ. has awarded the contract for installing a steam heating system in the proposed 2 story, 100 x 160 ft. grade school, to A. C. Shirmer Co., Hibbing. Total estimated cost, $275,000.

Minn., Duluth—St. Mary's Hospital Bd., 3rd St. and 5th Ave., East, has awarded the contract for the construction of a 6 story, 41 x 140 ft. hospital addition, to McLeod & Smith, Sellwood Bldg. A steam heating and mechanical ventilating system will be installed in same. Total estimated cost, $200,000. Noted Feb. 3.

Okla., Newkirk—The Newkirk Ice & Creamery Co. has awarded the contract for the remodelling of their 1 story, 50 x 100 ft. ice and creamery plant at the Santa Fe tracks, to Arthur Frank, Tulsa. One 2 hp., one 3 hp., one 4 hp., one 5 hp., two 10 hp., two 15 hp., 220 volt, a.c. motors and one 75 kw. direct connected a.c. dynamo will be installed in same. Total estimated cost, $75,000.

Ore., Grants Pass—The Grants Pass Irrigation Dist. has awarded the contract for furnishing and installing 4 pumps, pipe, etc., to the Shattuck Edinger Co. Estimated cost, $50,000.

Cal., Anaheim—The Anaheim Co-operative Orange Association will build a packing house on Center St. A boiler plant will be installed in same. Work will be done by day labor.

Cal., Burlingame—The City Trustees have awarded the contract for furnishing and installing 2 pumping units, to the Simonds Machinery Co., 117 New Montgomery St., San Francisco, at $11,110. Noted March 16.

Cal., Holt—The State Record Bd., Sacramento, has awarded the contract for installing pumping plant, etc., in the Reclamation Dist. 2021 in Mildred Island Dist., San Joaquin Co., to the Olympian Dredging Co., 249 1st St., San Francisco. Total estimated cost, $300,000.

Cal., San Francisco—P. J. Gartland, 21st and Guerero St., will build a 3 story auto building at 11th and Market Sts. A steam heating system will be installed in same. Work will be done by day labor.

Cal., San Pedro—The First Natl. Bank of San Pedro has awarded the contract for the construction of a 5 story, 50 x 100 ft. bank and office building on 6th and Palos Verdes Sts., to the Foss Designing & Bldg. Co., 45 North Euclid Ave., Pasadena. A steam heating system will be installed in same. Total estimated cost, $150,000.

Que., Montreal—The Montreal Tramway Co., Tramway Bldg., has awarded the contract for the construction of a 1 story, 46 x 106 ft. power house on Cote St., to F. A. Grothe, 10 Cuthbert St. Estimated cost, $67,000.

Que., Montreal—L. Poulin Co. Ltd. has awarded the contract for the construction of a 4 story, 30 x 50 ft. cold storage building on Claude St., to L. Langevin, 228 Parc Lafontaine. Estimated cost, $22,000.

B. C., Hatzic—The Farmers' Cold Storage Association has awarded the contract for the construction of a 2 story, 60 x 150 ft. cold storage building, to B. Davidson, 1012 Broughton St., Vancouver. A 5 ton ice making plant will be installed in same. Total estimated cost, $50,000.

POWER

Volume 51

When he applies for his first job.

Gas-Steam Unit Shows 72.4 Per Cent Thermal Efficiency at Ford Works

By L. D. Royer
Power-Construction Department, Ford Motor Company

Results from one-hour test show thermal efficiency based on high heat value of gas of 23.1 per cent for gas-engine cylinders and heat recovery from exhaust gases of 49.3 per cent, giving total thermal credit of 72.4 per cent. On steam-engine cylinders a thermal efficiency of 20.4 per cent, made possible by hot exhaust gas jacketing. Over-all thermal efficiency of unit 17.3 per cent. Combined mechanical efficiency of unit, 79.2 per cent.

SINCE the adoption of the "gas-steam" unit as a prime mover for electrical generation by the Ford Motor Co., the mammoth engines employed have been of great interest to engineers. After inquiries as to design and the general layout of the unit had been satisfied, there still remained leading questions relating to over-all efficiency and the percentage of waste heat recovered. Data of this character were never available until recently, and while results given herewith were obtained from tests under operating conditions somewhat below normal, they will serve to give a fair indication of the economies that may be credited to these units.

In back issues of *Power* a description of the Highland Park Plant and details of engine design have been given.* Briefly, these units consist of a horizontal twin engine with tandem gas-engine cylinders on one side and tandem compound-steam cylinders on the other, driving a 4,000-kw. direct-current generator at 80 revolutions per minute. Full dimensions of the unit are given in Table I.

Valves on the high-pressure steam cylinder are of the poppet type, and Corliss valves serve the low-pressure cylinder. The one unique feature on the steam side is the circulation of a portion of the exhaust from the gas engine through a jacket surrounding the high-pressure steam cylinder. In this way about 2.5 per cent of the total heat supplied to the gas engine is recovered and converted into useful work.

On the gas-engine side the cooling involves four distinct systems—the cylinder jackets, the exhaust manifold or jacketed tees, the piston and the valve boxes. The exhaust manifold and the piston use cooling water after it has passed through the steam condenser and deliver it to the factory to be used in process work. Water passing through the cylinder jackets is used as boiler feed, and that supplied to the valve boxes is cold city water, which is not further utilized. From the foregoing it will be seen that a considerable amount of

FIG. 1. COMBINATION "GAS-STEAM" UNIT AT THE PLANT OF THE FORD MOTOR COMPANY

*Nov. 21, 1916 and Jan. 16, 1917.

heat is recovered. When proper thermal credit is given, it will be apparent that these engines, in addition to delivering current to the busbars at relatively high efficiency, will show economies far beyond the average condensing plant.

If considered in the terms of the indicated horsepower delivered, the gas-engine performance is not exceptional, but when operated in connection with a steam plant, as well as a manufacturing plant using hot water in production, great economy results, due to the fact that a considerable portion of the heat in the exhaust gases is utilized. The test shows that 20.7 per cent of the total heat supplied to the gas engine is recovered in the cylinder-jacket cooling water, and 23.1 per cent is recovered from the piston and exhaust manifold, the water being used for metal washing and other manufacturing processes throughout the factory at a temperature of 140 to 150 deg. The items previously enumerated, together with the 2.5 per cent supplied to the steam cylinder and 21.7 per cent converted into work in the cylinders, shows a total thermal credit of 68 per cent for the gas engine, based on the high heat value of the gas supplied; based on the low heat value, the thermal credit would be 72.4 per cent.

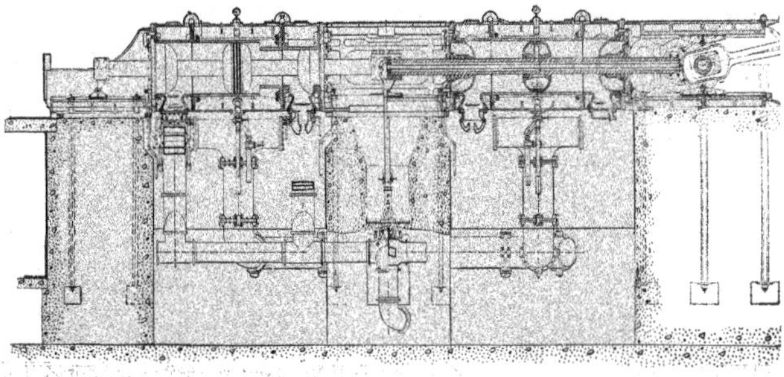

FIG. 2. LONGITUDINAL SECTION THROUGH GAS ENGINE CYLINDERS

Gas is supplied by an up-to-date producer plant, consisting of 22 sections of Smith producers burning coal, three Hughes producers burning coal and four Hughes producers burning coke. The total hourly capacity of the plant is 1,500,000 cu.ft. of gas with an average calorific value of 162 B.t.u. per cubic foot. The total cost of producing gas and delivering it to the engine is ten cents per thousand cubic feet.

Steam is supplied by a boiler plant of 13,300 hp. in water-tube boilers, delivering steam to engines at 150 lb. pressure and 60 deg. superheat. The boiler room is up to date in every respect and is equipped with automatic coal-handling devices and all appliances for the economical production of steam. The cost of steam at this time is 35 cents per thousand pounds.

With gas containing 162 B.t.u. per cu.ft. and costing $0.10 per 1,000 cu.ft., for $1 a total of 1,620,000 B.t.u. is furnished. With steam containing 1,126 B.t.u. per lb. above the temperature of the condensate and costing $0.35 per 1,000 lb., $1 will supply 3,217,000 B.t.u. However, as shown by test, 68 per cent of the heat in the gas, or 1,101,600 B.t.u. is converted into useful work, while by the same test 20.4 per cent of the heat in the steam, or 656,270 B.t.u., is utilized. Thus the relative cost of the heat converted into useful work by the gas and the steam sides is at 1 : 1.68.

The problem of securing accurate data on the steam side of the unit presented no special difficulties, standard test gages, thermometers, calorimeters, mercury col-

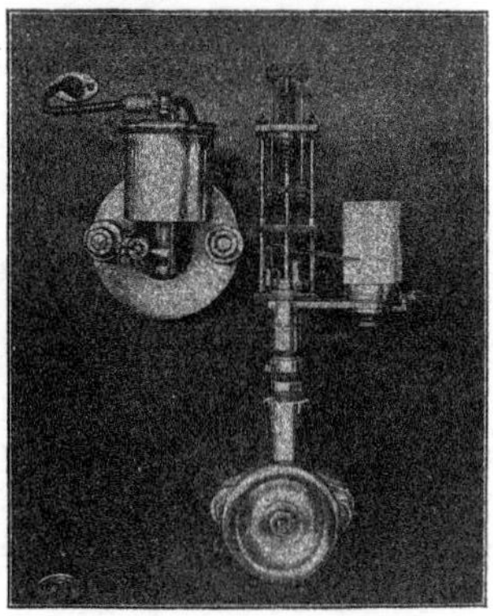

FIG. 3. DOUBLE-SPRING INDICATOR FOR USE ON GAS CYLINDERS

umns and indicators being available; likewise with the measurement of the electrical output accurately calibrated ammeters and voltmeters being used. However, to determine the distribution of heat in the gas cylinders, some special apparatus worthy of mention was necessary; namely, the Orr gas-engine indicator designed by William Orr, operating engineer at the Highland Park Plant.

Referring to Fig. 3, it will be seen that this indicator has a combination of two springs. It uses a 180-lb. spring during the compression and explosion periods and a 10-lb. spring on the scavenging and suction strokes. By referring to the diagram produced by this indicator (Fig. 4), the advantage of this simple and unique feature is obvious.

During the test the quantities of gas and water were determined by a special type of pitot tube which had been carefully calibrated, and constants had been previously determined. By this method the gas at any temperature and pressure could be accurately measured and the flow of water at any temperature determined. While the pitot tube is not recognized as the best instrument for accurate measurements, tests made with standard meters in parallel gave readings that were practically identical.

In Table II is given the results of a one-hour test selected from a number of similar runs. The test was taken and computed in accordance with the standard A. S. M. E. Code. The figures speak for themselves. It might be well to point out that the particular high efficiency of 20.4 per cent for the steam cylinders was due in part to the circulation of hot exhaust gas through the high-pressure cylinder jacket, thereby eliminating cylinder wall condensation. On the other hand, the tests were run at a time when both steam and gas plants were inadequate to care for the constantly increasing load, so that steam was being generated at 150 lb. gage, 30 lb. under normal, and on account of a shortage of gas resulting in a reduced pressure in the main, the gas engine cylinders were developing only 1,828 hp. instead of the full normal rating of 2,750 hp. At full load the gas-engine cylinders in particular should have shown a higher thermal efficiency. However, the economy under average operating conditions at the time of the test was the information sought, and the results are given for what they may be worth.

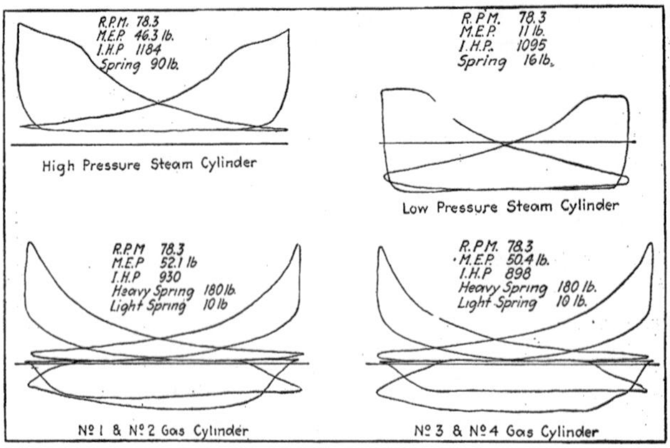

FIG. 4. INDICATOR DIAGRAMS FROM STEAM AND GAS ENGINE CYLINDERS

TABLE I—DIMENSIONS OF COMBINATION "GAS-STEAM" UNIT

Steam Engine	
Type	Tandem compound
Auxiliaries	Electric driven
Rated horsepower	2,750
High-pressure cylinder diameter, in.	36.25
Low-pressure cylinder diameter, in.	68.00
Stroke of pistons, in.	72.00
Diameter of piston rod, in.	12.00
Clearance Hp. Cylinder:	
Head-end, per cent piston displacement	8.47
Crank-end, per cent piston displacement	8.45
Hp. Constants (1 lb. Pressure, 1 rev.):	
High-pressure cylinder	0.327
Low-pressure cylinder	1.2723
Gas Engine	
Class:	
(a) Four cycle.	
(b) Make-and-break ignition, 110 volt, direct-current.	
(c) Two cylinders, double-acting.	
(d) Straight line arrangement of cylinders.	
(e) Horizontal.	
Diameter of cylinders, in.	42
Stroke of pistons, in.	72
Rated horsepower	2,750
Fuel	Producer gas
Horsepower constant (1 lb. pressure, 1 rev.)	0.1139
Generator	
Type	Direct Current, Compound Wound
Capacity at 80 r.p.m, kw.	4,000
Volts	250

TABLE II—RESULTS OF TEST ON LARGE "GAS-STEAM" UNIT

Steam Engine	
Run number	2
Duration, hr.	1
Exhaust gas through high-pressure cylinder jacket, per cent	20
Average Pressures and Temperatures	
Pressure in steam pipe near throttle, lb. gage	146.0
Barometric pressure, in. of hg.	29.57
Pressure of steam entering l.-p. cylinder, lb. gage	17.0
Vacuum in condenser, in. hg.	24.2
Temperature of steam near throttle, deg. F.	423.5
Superheat in steam near throttle, deg. F.	59.6
Quality of steam leaving h.-p. cylinder, per cent	96.7
Quality of steam entering low-pressure cylinder, per cent	96.7
Quality of steam leaving low-pressure cylinder, per cent	89.0
Temperature of condensate, deg. F.	104.0
Hourly Quantities	
Dry steam consumed by engine per hour, lb.	24,054
Heat units given to engine by steam per hr., B.t.u.	27,849,475
Heat units added by jacket gas per hour, B.t.u.	545,935
Heat units consumed by engine per hour, B.t.u.	28,395,410
Power	
Revolutions per minute	78.3
Indicated horsepower developed by steam engine	2,279
Indicated horsepower developed by h.-p. cylinder	1,184
Indicated horsepower developed by low-pressure cylinder	1,095
Economy Results	
Dry steam consumed by engine per i.hp. per hr.-lb.	10.6
Heat units given engine by steam per i.hp. per hr., B.t.u.	12,220
Heat units added by jacket gas per i.hp. per hr., B.t.u.	240
Heat units consumed by engine per i.hp. per hr.	12,460
Efficiency Results	
Thermal efficiency, based on i.hp., per cent	20.4
Gas Engine	
Run number	2
Duration, hr.	1
Average Pressures and Temperatures	
Pressure gas in main near engine, in. water	0.42
Temperature of room, deg. F.	0.76
Temperature of gas entering superheater, deg. F.	808
Temperature of gas entering steam cyl. jacket, deg. F.	732
Temperature of gas leaving steam cyl. jacket, deg. F.	465
Temperature of gas to exhaust from superheater, deg. F.	676.5
Temperature of cylinder jacket water inlet, deg. F.	107.3
Temperature of head-end cyl. jacket water outlet, deg. F.	174.5
Temperature of crank and cyl. jacket water outlet, deg. F.	178.8
Temperature of piston cooling water inlet, deg. F.	94
Temperature of piston cooling water outlet, deg. F.	140.3
Temperature of jacketed tee water inlet, deg. F.	94
Temperature of jacketed tee water outlet, deg. F.	139.2
Relative humidity, per cent	89
Calorific value of gas per cu.ft., high value, B.t.u.	152.8
Calorific value of gas per cu.ft., low value, B.t.u.	143.5
Hourly Quantities	
Quantity of producer gas supplied, cu.ft.	140,456
Quantity of air supplied, cu.ft.	242,076
Quantity of air and gas supplied, cu.ft.	382,532
Heat units consumed by engine per hr., low ht. value, B.t.u.	20,162,459
Heat units consumed by engine per hr., high ht. value, B.t.u.	21,461,677
Cooling water supplied to head-end jacket, lb.	32,518
Cooling water supplied to crank-end jacket, lb.	31,535
Cooling water supplied to pistons, lb.	56,457
Cooling water supplied to jacketed tees, lb.	52,054
Analyses	
Analysis of Producer Gas by Volume:	
CO_2, per cent	5.53
CO, per cent	27.0
H_2, per cent	12.9
CH_4, per cent	1.97
N_2, per cent	52.6
Analysis of Exhaust Gas by Volume:	
CO_2, per cent	14.9
O_2, per cent	4.0
N_2, per cent	81.1
Power	
Indicated horsepower	1,828
Economy Results	
B.t.u. consumed by eng. per i.hp. per hr., low ht. value	11,030
B.t.u. consumed by eng. per i.hp. per hr., high ht. value	11,741
Efficiency Results	
Thermal efficiency gas eng. cyl., i.hp., high ht. value, per cent	23.1
Thermal credit based on i.hp. plus heat recovered, per cent	72.4
Volumetric efficiency, per cent	78.0

Heat Balance—Low Heat Value

Based on B.t.u. per i.hp. per hours	B.t.u.	Per cent
Heat converted into work	2,545	21.7
Heat recovered in jacket cooling water	2,429	20.7
Heat recovered in piston cooling water	1,430	12.2
Heat recovered in jacketed tee cooling water	1,287	10.9
Heat rejected in superheater, radiation	419	3.6
Heat recovered in steam cylinder jacket	299	2.5
Heat rejected in exhaust	2,949	25.1
Heat rejected in valve box cooling water	328	2.8
Heat unaccounted for, including radiation	55	0.5
Total heat consumed per i.hp. per hour	11,741	100

Distribution of Heat Recovered from Exhaust Gas

	B.t.u.	Per cent
Jacket cooling water used for boiler feed	2,429	20.7
Jacket tee cooling water used for factory service	1,287	10.9
Piston cooling water used for factory service	1,430	12.2
Heat recovered in high press. steam cyl. jacket	299	2.5
Total	5,445	46.3

TABLE II—RESULTS OF TEST ON LARGE "GAS-STEAM" UNIT—(Concluded)

Data on Complete Unit	
Total i.hp. developed	4,107
Mechanical and electrical losses, hp.	858
Electrical horsepower developed	3,249
Combined mech. and elec. eff., per cent	79.2
Overall thermal efficiency = $\frac{\text{Ht. equiv. of a kw.-hr. at sw. bd.}}{\text{Ht. supplied to unit per kw.-hr.}}$	17.3
Water Temperatures	
Water through head-end cylinder jacket, g.p.m.	65
Water through crank-end cylinder jacket, g.p.m.	63
Water through pistons, g.p.m.	113
Water through jacketed tees, g.p.m.	104
Water through each valve box, g.p.m.	7
Temperature of cylinder jacket water inlet, deg. F.	107
Temperature of head-end cylinder jacket water outlet, deg. F.	174
Temperature of crank-end cylinder jacket water outlet, deg. F.	179
Temperature of piston cooling water inlet, deg. F.	94
Temperature of piston cooling water outlet, deg. F.	140
Temperature of jacketed tee water inlet, deg. F.	94
Temperature of jacketed tee water outlet, deg. F.	139
Temperature of valve box cooling water inlet, deg. F.	94
Temperature of valve box cooling water outlet, deg. F.	130
Gas	
Quantity of producer gas supplied, cu.ft. per hr.	140,456
Quantity of air supplied, cu.ft. per hr.	242,076
Quantity of air and gas supplied, cu.ft. per hr.	382,532
Steam	
Pressure of steam near throttle, lb. gage	146
Pressure of steam entering one piston cylinder, lb. gage	17
Vacuum in condenser, in. hg.	24.2
Superheat in steam near throttle, deg. F.	59.6
Quality of steam leaving hp. cylinder, per cent	96.7
Quality of steam leaving low-pressure cylinder, per cent	89
Dry steam consumed by engine per i.hp. per hr., lb.	10.6
Efficiencies	
Thermal efficiency steam engine, based on i.hp., per cent	20.4
Thermal efficiency gas engine, based on i.hp., per cent	23.1
Thermal credit gas engine, based on i.hp. plus heat recovered, per cent	72.4
Volumetric efficiency gas engine, per cent	78
Combined mechanical and electrical efficiency, per cent	79.2
Combined thermal efficiency based on hp. delivered, per cent	17.2

Multiple-Stage High-Pressure Turbine Pumping Set

Multiple-stage turbine pumps are built with as many as ten impellers in series, and in cases in which that maximum does not give the pressure required at the given speed it is usual to mount two such pumps on one bed-plate, the two units being connected in series so that the resulting pressure will be that due to as many as twenty impellers in series. A pump that has been designed for a working pressure of 1,500 lb. per sq. in. or 3,470 ft. head of water at a speed of 1,480 r.p.m. has been built by Mather & Platt, Ltd., Manchester, England, says *The Engineer*. It has a capacity of 800 gal. per minute and is driven by a 1,450-hp. electric motor. The guaranteed efficiency of the pump is 68 per cent, and the actual horsepower required to drive it at full load is 1,240.

The low-pressure pump has six stages and develops 1,486 ft. head; the high-pressure pump has eight stages and develops 1,984 ft. head, or a total of 3,470 ft. The low-pressure pump is made with as few stages as possible in order to keep the pressure on the stuffing-boxes on the high-pressure pump as low as possible. Notwithstanding this, there is a pressure of 650 lb. per sq.in. on these boxes, but owing to the arrangement of the packing adapted, the boxes run cool. They are fitted with metallic packing.

The over-all length of the complete set is 31 ft. 3½ in., height 5 ft. 5½ in., and the over-all width 6 ft. 7½ in. The motor is designed for three-phase 50 cycle current at a pressure of 6,600 volts. The pump's suction pipe is 9 in. and on the delivery side the pipe is 8 in. in diameter.

ELECTRICAL DRIVES FOR LUMBER MILLS

By
J. A. FRANKLIN*

Factors that have influenced the application of electricity to the lumber industry are discussed, then the sizes, types and characteristics of the motors used to drive the various machines in lumber mills are treated of. The sizes and types of motors and methods of driving are summarized in a convenient table.

ACTUATED by an enormous demand for timber of various kinds during the last few years, the lumber industry, in common with many other American activities, has made rapid strides in the improvement of production methods. Satisfactory operation in the largest lumber mills has shown that electric motors easily take care of the requirements of log hauls, band mills, edgers, etc. Nearly all new mills are fully equipped with electric drives, and furthermore, a large number of old ones are being converted from mechanical to electric drives. As the prime essential of lumber mills is reliability, operations demand equipment upon which dependence can be placed. Employees are constantly impressed with the idea of getting lumber through the mills in the shortest time possible, and too frequently the machines are given little consideration, therefore any apparatus intended for this application must be designed to withstand the hardest kind of usage.

The danger from fire is greatly reduced when electricity is used, because there are fewer bearings to be filled with oil, which is frequently spilled on the bridge trees and posts supporting the bearings. There is also less fire hazard owing to the elimination of shafting, belting and bridge trees, with the accompanying walks and platforms built to enable the millwrights and oilers to get to and from the machinery. The floor of the mill becomes a much clearer space, where a fire may not only be detected more readily, but possibly checked once it has started, a fact that is always taken into consideration by the insurance companies when making a rate.

Naturally, the problems of electrifying are as varied as there are mills, but in general the same types of motors are used for driving the same types of machines. The motor sizes for the various machines depend on the local conditions in any particular mill.

Alternating current is used almost exclusively, because sparking and consequent fire hazard are reduced to a minimum and because the power-distributing problems are simplified, especially when the mill is scattered over a large territory.

Two types of alternating-current motors are used in the lumber industry—the squirrel-cage and wound-rotor. Squirrel-cage motors are generally used where very high starting torque is not required and variable speed is unnecessary. The rotating element presents a sturdy and reliable piece of apparatus, and the only attention required is an occasional oiling of bearings. In addition, the controller, manually or automatically operated, used for starting and stopping or reversing the motor is simple in construction and easy to operate.

There are many instances, of course, where the squirrel-cage motor is not suitable and a wound-rotor type must be used. This is usually the case when the machine starts under a very heavy load or must operate at variable speeds. The starting, stopping, reversing or speed changing of the motor can be either manually or automatically controlled. Speed variation in this type of motor results from changing the external resistance in the windings of the rotating element; the greater the resistance the slower the motor speed. The following gives an outline of the types of motors required on the various machines and their characteristics. An analysis of these data is given in the table.

Having been brought from the woods to the mill, the logs are unloaded into the log pond. They are generally conveyed from the pond into the mill by means of the familiar log haul, or in some mills by a log lift. In the case of the log haul an endless bull chain with hooks or "dogs" spaced every few feet runs in a steel groove up an incline from the pond to the log deck, at a speed of from 45 to 180 feet per minute. The most satisfactory method of drive, and one that has been generally adopted, is the installation of a wound-rotor motor of from 25- to 50-hp. rating with a drum controller, the motor being started and stopped whenever it is desired to start or stop the bull chain.

Another method of drive sometimes applied to the haul is a constant-speed squirrel-cage motor, in which

*Industrial Department, Westinghouse Electric and Manufacturing Company.

case the motor operates continuously and the load is thrown on and off by means of a clutch or friction. In either case the motor is generally connected to the driving sprocket through reduction gears in order to produce the desired speed of the bull chain.

Where the log lift is employed, the logs are floated alongside the mill by means of a pond or canal, and cables are used to lift them to the deck. One end of each cable is attached to the edge of the deck and the other ends are secured to drums, thus permitting the cable to sag into the pond, forming a loop into which the logs are floated. A variable-speed wound-rotor motor of from 35- to 50-hp. capacity is geared to the drum, and as the latter revolves, the slack of the cables is taken up and the logs are rolled up inside of the loop on to the deck. The motor is generally equipped with a solenoid brake so as to enable it to hold the load suspended at any point in case for some reason it is necessary to stop its progress.

After the logs arrive on the deck, they are placed on the log carriage. The control valves for the "kickers" and loaders, as well as practically all the steam and air cylinders throughout the mill, are operated by electromagnets controlled by push-buttons, which may be placed at any convenient points and operated either by foot or by hand as may be desired. The set-works on the carriage used to adjust the position of the log for the saw cut are of the type that requires rotating motion and are most satisfactorily driven by a constant-speed squirrel-cage motor of from 5- to 10-hp. capacity. The motor operates continuously and is supplied with current from a flexible cable or from a trolley suspended from the side of the mill, preferably the latter. The motor may be direct-connected, geared, belted or chain-driven.

The rock saw used in some sections of the country, which precedes the main saw, cutting a path through

FIG. 1. ELEVEN-FOOT ELECTRICALLY OPERATED BAND MILL

the bark, removing any rock, gravel or other matter likely to injure the main-saw teeth, is mounted above the carriage on a framework that can be raised and lowered by the operator. It is driven by a 10- or 15-hp. squirrel-cage motor, mounted on the upper portion of the framework and belted to the saw. The motor runs continuously and is controlled by an auto-starter installed near the head sawyer's stand. A motor-driven blower is generally provided to carry away the sawdust. The head saw, Fig. 1, is generally of the band type and requires the largest motor used in a mill—from 100- to 300-hp. capacity. For driving the bandsaw a wound-rotor motor is recommended because of the heavy starting torque required due to the inertia of the saw and the wheels over which it runs. A high-speed three-bearing motor is generally belted to the driving

FIG. 2. INSTALLATION OF STARTING BOXES FOR ROLL MOTORS

pulley and a belt tightener, though not essential, is frequently installed to secure proper belt tension, particularly when the angle of the belt approaches the vertical. The flywheel effect of the lower bandmill wheel assists in reducing the peak loads on the motor during the heavy cuts.

The live rolls, the first link in the interior transportation chain after the log has passed the head saw, form an exceedingly important factor. If they are shut down, the operation of the entire mill is affected. Live rolls, which require frequent reversal, are operated by wound-rotor motors and are controlled from a small cage or station overlooking the rolls. In some cases where the rolls are not reversed frequently or do not start under severe conditions, squirrel-cage motors may be applied. Where it is desired, several sections of the rolls may be operated from the same motor-driven shaft, through friction clutches. However, the driving of each section with its individual motor is generally more advisable. The power requirements for the rolls vary so much that it is difficult to give definite capacities, but the sizes generally required are from 3- to 15-hp., depending on the size, number and speed of the rolls. In order to secure the low speeds desired, back-geared motors with rolled-chains and sprockets are generally used, driving direct from the back-gear shaft to the live-roll shaft. A particular advantage of the electric operation of the rolls is the ability to reverse quickly and "spot" accurately a piece of timber over a jump saw or section of a transfer chain.

Transfer chains are frequently remote-controlled by means of contactor panels, and squirrel-cage motors with high-resistance end-ring rotors are proving very satisfactory as the motive power. One particular advantage in using remote-control operation of these motors is that the transfer chains can be started and stopped with ease from two or more widely separated points without any interference from other operators. For such service there has been developed a sturdy type of push-button which is so connected, depending on the function of the transfer chain, that the chain is in operation only during the period when the push-button is depressed. On the other hand, should continuous operation be desired, the chain is stopped only when the push-button is depressed.

The valves on the steam or air cylinders operating the bumpers and jump rolls, and for raising and lowering the transfer skids, may all be actuated by electromagnets controlled from push-button switches located conveniently to the operator in the same manner as those described for the transfer chains. This system permits two or more operators, even though separated from each other, to have control of the bumpers and rolls.

Jump or swing-up saws are installed at points in the system of live rolls for trimming timbers to specified lengths. In the case of a jump saw the motor is frequently mounted on the saw frame, while the motor driving a swing-up saw is mounted on a stationary foundation, and the belt tension is usually supplied by a swinging belt tightener. The service required of these motors, which are of the squirrel-cage type, is very intermittent and the capacity varies from 7½ to 25 hp., depending on the service.

The live rolls back of the bandsaw carry the slabs to a point in front of the slasher, Fig. 3, where, at the will of the operator, they are transferred off the live rolls by means of trip skids. These skids are raised by a cylinder the valve of which is operated by an electromagnet whose control button is within convenient reach of the operator. Early methods of slasher drive consisted in belting the slasher arbors from an overhead countershaft which was either direct-connected to the motor through a flexible coupling or belted. More modern installations, however, have the arbors in one continuous length for any number of saws, and to which

FIG. 3. MOTOR DIRECTLY CONNECTED TO SLASHER IN FOREGROUND

a squirrel-cage motor is directly connected through a flexible coupling. The capacity of the motor required varies from 35- to 75-hp., depending on the kind of lumber cut.

The slasher floor chains in some cases are driven through friction clutches belted from the slasher arbors. More recent and satisfactory drives, however, because of the elimination of the friction and greater flexibility of control, consist in driving the chains by means of a high-resistance end-ring squirrel-cage motor with remote control of the off-and-on type. With this type of drive and control any large load of slabs and edgings that has accumulated on the floor chains can be fed gradually to the slasher saws without danger of jamming or causing them or the arbors any injury.

Boards coming from the head saw to be edged, Fig. 4, and ripped to the desired width by the edger are carried on rolls to a point directly opposite and are then trans-

FIG 4. ROLLS ELECTRICALLY CONTROLLED BY OPERATORS IN SMALL BOOTHS—EDGER SHOWN

ferred sideways on the chains which take them to the edger. In some mills this transfer of boards is made automatic by raising a stop at the proper point in the rolls. When the board strikes the stop, it closes the circuit of a control switch causing the trip skids to rise and trip the board off the rolls onto the chains, thence to the edger. The transfer chains are usually driven by high-resistance end-ring squirrel-cage motors with a remote-control button placed within easy reach of the edgerman's foot, so that he may readily bring each piece up to the machine as he desires. The edger drive consists of a squirrel-cage motor of from 35- to 250-hp. capacity, depending on the kind of lumber and the cutting speed, directly connected through flexible couplings to the edger arbor. Except in the case of extremely large sizes it is entirely satisfactory to place the rotor of the motor on the edger arbor in the position usually occupied by the drive pulley in belt-driven edgers.

After passing through the edger the boards and edgings are carried along over the slasher floor chains on which the edgings from the edger are thrown, the boards continuing along these rolls and being delivered edgewise onto the trimmer chains. These are driven by a high-resistance squirrel-cage motor with remote control, the button of which is located so that the operator placing lumber on the trimmer can reach it with his foot and start or stop them as desired. The trimmer is driven by a squirrel-cage motor of from 25- to 75-hp. capacity directly connected through a

FIG. 5. MOTOR DIRECTLY CONNECTED TO TIMBER SIZER

and a wound-rotor belted motor with an idler pulley to supply belt tension is the proper application.

Although the actual power requirements for the sorting table are not very great, the starting duty is at times unusually heavy, owing to the fact that the table frequently becomes congested with lumber piled up on it, which must be moved, a wound-rotor motor of from 5- to 10-hp. capacity being desired.

It is frequently necessary to resaw lumber to the proper thickness. For this purpose it is taken to the

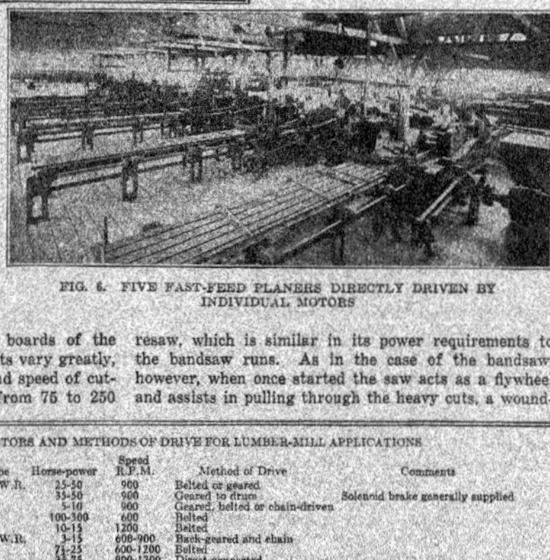

FIG. 6. FIVE FAST-FEED PLANERS DIRECTLY DRIVEN BY INDIVIDUAL MOTORS

flexible coupling to the trimmer counter-shaft from which the saw arbors are belted. Feed chains for the trimmer are driven by high-resistance end-ring squirrel-cage motors. The remote-control push button is located in the trimmerman's cage where he can reach it with his foot, affording him easy and accurate control and leaving both hands free for pulling the saws.

The gang saw, which consists of a number of straight saws having a vertical reciprocating and oscillating motion, splits a cant into several boards of the desired thickness. The power requirements vary greatly, depending on the size, kind of lumber and speed of cutting desired. Motors ranging in size from 75 to 250 hp. are required. As the gang is equipped with a flywheel and the saws together with the framework form a considerable mass, the starting duty is rather severe, resaw, which is similar in its power requirements to the bandsaw runs. As in the case of the bandsaw, however, when once started the saw acts as a flywheel and assists in pulling through the heavy cuts, a wound-rotor motor is considered the most desirable application, although there are a number of high-resistance end-ring squirrel-cage motors operating resaws very

SIZES AND TYPES OF MOTORS AND METHODS OF DRIVE FOR LUMBER-MILL APPLICATIONS

	Type	Horse-power	Speed R.P.M.	Method of Drive	Comments
Log haul	S.C. or W.R.	25-50	900	Belted or geared	
Log lift	W.R.	35-50	900	Geared to drum	Solenoid brake generally supplied
Carriage set works	S.C.	5-10	900	Geared, belted or chain-driven	
Headsaw	W.R.	100-300	600	Belted	
Rock saw	S.C.	10-15	1200	Belted	
Live rolls and transfer chains	S.C. or W.R.	3-15	600-900	Back-geared and chain	
Jump or swing-up saws	S.C.	7½-25	600-1200	Belted	
Slasher	S.C.	35-75	900-1200	Direct-connected	
Edger	S.C.	35-250	1200-1800	Direct-connected	
Trimmer	S.C.	25-75	900	Direct connected	Flexible coupling to shaft from which saw arbors are belted
Gangsaw	W.R.	75-250	720	Belted	
Sorting table	S.C.	5-10	900-1200	Direct-connected	High-resistance end-ring motor and friction device generally used
Resaw	S.C. or W.R.	75	900	Direct-connected or belted	
Stacker	S.C. or W.R.	5-7½	1200	Belted	
Timber sizer	S.C.	50-75	900	Direct-connected	
Hogs	S.C.	15-100	1200	Direct connected or belted	
Rolls, various kinds	S.C. or W.R.	3-10	900-1200	Chain-driven or geared	
Saw grinders, stretcher, re-toother and scarfing machines	S.C.	2-5	1200	Belted	
Exhaust fans	W.R.	20-150	600-1200	Direct-connected or belted	
Conveyers	S.C. or W.R.	10-25	900-1200	Geared	
LATH MACHINES					
Lath bolter	S.C.	40-50	1800	Direct-connected	
Lath mill	S.C.	25-40	3600	Direct-connected	
Lath trimmer	S.C.	5-10	1200	Direct-connected	
SHINGLE MILLS					
Main saw	S.C.	20	1800	Direct connected	
Trimmer saw	S.C.	3	1800	Direct-connected	
LOGGING					
Donkey-hoist	W.R.	150-200	600	Geared	

Note:—W.R. denotes wound rotor, and S.C. squirrel cage.

satisfactorily. The capacity required for a 7-ft. resaw is 75 hp. The feed rolls are driven by a wound-rotor motor varying from 3 to 5 hp. in capacity. From the sorting table the lumber is handled by various companies in different ways. Generally, it is taken by transfer cars or telphers to stackers, thence to the dry kiln or the lumber yard. In some cases the stacker, which is driven by a 5- or 7½-hp. squirrel-cage motor, stacks the lumber and it is transferred to the dry kiln on iron framework cars and unstacked after drying by an automatic unstacker driven by a motor of 3-hp. capacity. From here the lumber is moved by some method of transportation to the planing mill, lumber shed or railroad cars. as desired.

In some sections it is the practice to have a timber sizer located in the lumber mill, where large timbers can be planed before shipment. The size of the motor, which is usually of the squirrel-cage type and directly connected, varies according to the size of stock, rate of feed, number of sides to be surfaced and the speed, the usual capacities ranging from 50 to 75 horsepower.

Refrigeration Study Course VI—The Brine System

By H. J. Macintire
Professor of Mechanical Engineering, University of Idaho

With the so-called direct expansion system, where ammonia is allowed to boil in the cold-storage rooms there is danger at all times. The pipe lines may become corroded, or split due to imperfect welding, or fittings may be broken by accident. The result is that some or all of the ammonia in that part of the system may be emptied into the room. We have had accidents to liquid receivers where the anhydrous ammonia ran in streams into the street.

Besides the danger to life and commodities, there is also the matter of the initial charge of ammonia. In a large system of long supply and return pipes, or extensive refrigerating piping, the amount of the initial charge has to be very heavy, which requires constant care and repair to maintain the piping tight at all times. The result is that, under certain particular conditions, the so-called brine system is best for the refrigerating piping.

The Brine System of Indirect Refrigeration

In the brine system of indirect refrigeration, the high-pressure side is the same as in the direct-expansion system. The low-pressure side consists of a brine cooler, usually of the shell and tube type similar in construction to a steam condenser. The brine is a non-freezing solution of sodium or calcium chloride of such density as will not freeze at the temperature carried in the cooling system. On account of the lower freezing temperature of calcium brine, it will be found in most service where zero F. degree brine is required.

The brine sysem, then, is really an additional unit in the system, where the brine is kept cool by boiling ammonia and the cold-storage rooms or other refrigerating applications are kept cold by the brine. The ammonia system is self-contained, and the ammonia piping may be kept constrained to a single room. The brine, however, must be kept circulating constantly by means of a centrifugal or reciprocating pump, and all exposed piping must be well insulated. The brine line need only be full-weight steel pipe with ordinary fittings, valves and cocks. The system is laid out in many ways like an ordinary hot-water heating system.

Advantages of the Brine System

The great advantage of the brine system, other than those already noted, is in the ability to store up energy. With the direct-expansion system the compressor must be operating constantly during the live-load period. If this is not done, the pipes will soon lose their frost and the room temperature, or the commodities being cooled, will begin to increase in temperrture. With the brine system, this is not true, as "cold" may be accumulated in the brine and this may be circulated while the compressor is shut down. In fact, the example may be carried considerably farther in certain kinds of service that are intermittent, as ice-cream making, milk and drinking-water cooling, the chill room of the packing house, or where refrigeration is required suddenly.

In cases of periodic load the brine-storage tank becomes an advantage. In such a case refrigeration may be stored by cooling a large volume of brine through as large a temperature range as is practical.

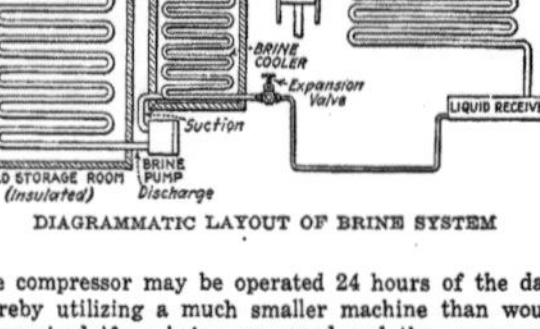

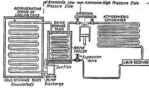

DIAGRAMMATIC LAYOUT OF BRINE SYSTEM

The compressor may be operated 24 hours of the day, thereby utilizing a much smaller machine than would be required if no brine was used and the compressor capacity had to be capable of the maximum rate of cooling. It will be seen that by means of brine the peak load is distributed during the whole day of 24 hours, whereas the load may last only three hours. Energy is accumulated in the same sense as in the case of an engine flywheel.

Disadvantages of the Brine System

There are, however, certain disadvantages in the use of brine. The first cost and the operating cost are usually greater. The brine pump has to be operated by independent power. In addition, and more serious than anything else, are the losses incident to the use of brine. It will be remembered that the brine is first cooled in a brine cooler, and the ammonia may be required to boil at 16 deg. F. to carry brine at an average temperature of 26 deg. F. Likewise, the air to be cooled in the cold-storage room could be kept at a corresponding temperature of 36 deg. F., thus allowing 10 deg. difference of temperature in each case. The capacity of the compressor would be determined by the temperature of the boiling ammonia, which would have been 26 deg. F. in the case of direct expansion. In the brine system, therefore, it is necessary to boil the ammonia at considerably lower pressure than with the direct expansion, thereby costing more per ton of refrigeration for the power input, and reducing the capacity of the compressor.

The brine required for a given condition may be calculated easily. The "heat capacity" of a certain volume of brine depends on the product of the density (specific gravity) and its specific heat (heat units required to raise one pound one degree F.). For instance, the cooling effect of 100 gal. of brine in raising 4 deg. F. would be

$$\underset{\text{Deg. Range}}{4} \times \underset{\text{No. Gal.}}{100} \times \underset{\text{Wt. 1 Gal.}}{8\frac{1}{3}} \times \underset{\text{Sp. Gr. Brine}}{1.2} \times \underset{\text{Sp. Heat Brine}}{0.7} = 2{,}700 \text{ B.t.u.}$$

assuming 1.2 for the specific gravity and 0.7 for the specific heat of the brine. These values change with each concentration of brine solution.

If it is required to find the amount of brine per minute necessary to provide one ton of refrigeration with a five-degree range of temperature, the calculation becomes

$$200 = \textit{No. gallons} \times 8\tfrac{1}{3} \times 1.2 \times 0.7 \times 4$$

$$\textit{No. gallons} = \frac{200}{8\frac{1}{3} \times 1.2 \times 0.7 \times 4} = \frac{200}{28} = 7.14 \textit{ gallons per minute}$$

Chart for Determining Proportions of Gear Teeth

By J. H. McManus and L. J. Ehlinger

The accompanying chart is based on Lewis' formula for the strength of cast-steel pinions. This formula, as applied to a 15-tooth pinion, is as follows:

$$C.P. = \sqrt[5]{\frac{0.36\ hp.^2}{r.p.m.}}$$

where *C.P.* equals circular pitch of the pinion.

To use the chart, draw a straight line from the horsepower transmitted on the left-hand scale to the revolutions per minute on the right-hand scale and read off the pitch on the central scale. The example shown on the chart is for a 15-tooth pinion transmitting 30 hp. at 200 r.p.m., and as shown, the circular pitch would be 1¼ in., or the diametral pitch about 2½ in. The face of the teeth, which should be three times the circular pitch, would then be three times 1¼ or 3¾ inches.

If the chart is to be used for pinions having more or less than 15 teeth, multiply the actual horsepower by $0.027\ N + 0.6$, where N is the number of teeth in the pinion. Then use this corrected horsepower on the chart and proceed as in the case of a 15-tooth pinion.

As an example of the use of the chart for the design of a complete gear train, assume a 75-hp. motor running 1,000 r.p.m., driving a hoist drum running 10 r.p.m. through two intermediate shafts, the first running 250 and the second 50 r.p.m. Pinions being the weakest part, determine their circular pitch and face and follow through the various reductions to the drum.

Assuming a 15-tooth pinion on the motor with 75-hp. and 1,000 r.p.m., the chart reads 1¼ in. circular pitch and three times this for the face gives 3¾ in. The driving pinion on the first shaft transmits 75 hp. at 250 r.p.m., and the chart shows this to give 1½ in. circular pitch and therefore 4½ in. face for the teeth. The pinion on the second reduction shaft transmits 75 hp. at 50 r.p.m., which from the chart should be 2 in.

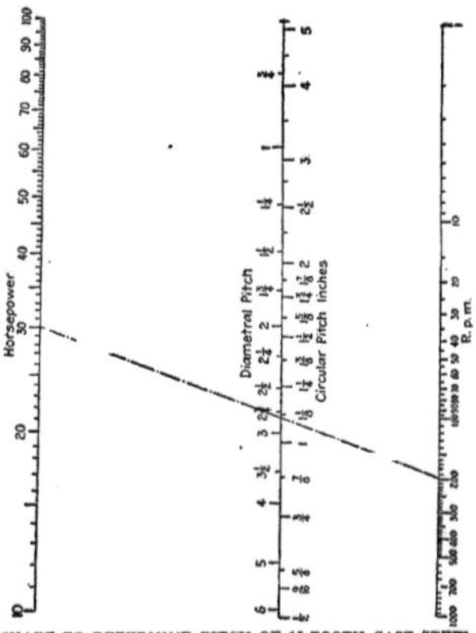

CHART TO DETERMINE PITCH OF 15-TOOTH CAST STEEL PINIONS FROM LEWIS' FORMULA

circular pitch and therefore 6 in. face. Thus the circular pitch and face are determined for each gear of the complete train.

Applications covering an expenditure of approximately $40,000,000 for power purposes in the State of California were filed with the State Water Commission during the month of December, according to a recent statement of the commission. The largest application made during the month was that of Allen Talbot for the Modesto irrigation district, who asks for a permit to appropriate 10,000 cubic feet per second from the Tuolumne river in Tuolumne County. He proposes to impound 120,000 acre-feet of water and to develop 120,000 hp. The cost of the project is placed at $16,000,000. Canals and tunnels 22¾ miles long are planned, with a steel and concrete dam 330 feet high and 850 feet long on top. At its base the dam will be 150 feet long. A second dam of concrete 250 feet high is planned the length to be 800 feet on top and 50 feet at the bottom.

There was a station mechanic who could not understand why a man younger in the service was made foreman instead of him. One day the foreman left an outline of some work for the mechanic to do. In too literally following instructions, the mechanic burnt out some valuable equipment and shut the plant down. Taken to task, he exclaimed: "I didn't make the 'bull,' he did it! *I only did what he told me to.*" Still the mechanic wonders why he was not made foreman.

POWER DRIVES for ROLLING MILLS

By W. O. Rogers

Up-to-date power stations were found in which were installed reciprocating engines of various types, high-pressure and mixed-pressure turbines. The reciprocating engine predominates, but a number of turbines have been added to the original power-plant equipment. Although the mixed-pressure turbine seems to be the logical unit to use with reciprocating engines, but two were found in operation.

AFTER visiting the boiler houses that produce steam for the rolling mills, I next took in some of the engine and turbine rooms. It required several days to inspect and photograph the units. In the plants visited the Corliss type of engine was generally used, directly connected to either alternating- or direct-current generators. Figs. 1 and 2 are views of two power plants. This does not mean that the steam turbine is not used, because in some of the mill power stations, both high-pressure and mixed-pressure turbines have been added to the original equipment. The generating unit produced energy for motor drive, electric lighting, etc. These units are housed generally in a neat brick building and are, in fact, a regular central generating station and not scattered about the plants as many might suppose.

As might be expected, the power-plant units operate condensing, and the reason that more turbines are not found in operation is doubtless due to the fact that the Corliss engines were installed when that type of unit was the best procurable and before the turbine had established itself in the power-plant field.

When it comes to the best type of prime mover for a large-sized power plant, the turbine is the most economical in steam consumption when operating condensing, and in maintenance cost, the size requiring fewer operatives. The plant engines that I saw were compound condensing. Figs. 3 and 4 are examples. These units operate condensing and range around 400 to 500 kw. capacity with a speed of from 75 to 100 r.p.m. and generate direct-current energy.

FIG. 1. POWER PLANT AT THE BERGER MFG. CO.

If these medium-speed engines operated non-condensing, the steam consumption would average, with 150 lb. steam pressure, about 19.5 lb. per i.hp. at full load, 21.7 lb. at three-quarter load, 26.8 at one-half load, and 39.5 at one-quarter. Operating condensing with 26-in. vacuum and corresponding cutoff in the high-pressure cylinder, the figures would be 15.4 lb. of steam at full load, 15.7 at three-quarter load, 16.3 at half load, and 22.5 lb. at one-quarter load. This makes a difference in steam consumption per indicated horsepower, in favor of the condensing engine, of about 5.1 lb. of steam at full load, 6.5 at three-quarter load, 10.5 at one-half load, and 17 lb. at one-quarter load.

Although the foregoing figures are approximate, they are based on tests that have been made on similar engines and are given so that a comparison of steam consumption between condensing and non-condensing units may be had. If the engines referred to are in good condition, the steam consumption should be around the figures given. As a general thing, however, the steam consumption will be higher.

Many engineers are of the opinion that a gain in steam economy, due to the use of a condenser, indicates a corresponding gain in heat consumption, but this is not so. For instance, an engine might show an apparent gain due to running condensing of, say, 12.5 per cent, with the feed water returning to the boilers at 120 deg. F. If a suitable feed-water heater were used with a non-condensing engine, the water should be heated to around 210 deg., and a non-condensing engine would therefore be credited with the difference of 90 units per pound of steam used, or 9 per cent in round numbers. The difference between 12.5 and 9 per cent is 3.5

per cent, which represents the net gain in favor of using compound engines. This, however, ignores the power necessary to operate the vacuum, circulating and condensate pumps, and the steam consumed by the condenser pumps might be equal to or greater than 3.5 per cent of the steam generated, so that the net gain would be nothing or even negative. By using superheated steam, a greater economy will be obtained than when using saturated steam, but the first cost, maintenance and the disposition of the exhaust must be considered in determining the final gain. The troubles encountered in using superheated steam have been largely overcome by the use of suitable lubricants.

FIG. 2. POWER PLANT AT THE REPUBLIC IRON AND STEEL CO.

Steam turbines make an ideal power unit. They are generally used to drive alternating-current generators, but I found that the steel mills use them for driving turbo-blowers and pumps, and by means of gearing the turbine is used to drive reciprocating air compressors, rolling mills and other types of slow-speed machinery. A turbine has the advantage of first low cost, low maintenance cost, small floor space required and low cost of attendance.

Turbines are grouped under the classification of impulse single-pressure stage having both single- and multi-velocity stage; reaction multi-velocity stage, multi-pressure stage; and combined impulse and reaction with multi-velocity stage, multi-pressure stage.

In the impulse type of turbine steam is expanded and the heat given up by the pressure dropped imparts velocity to the jet itself. The jet impinges against the vanes of the rotating wheel and gives up its kinetic energy to the wheel. Turbines used in driving centrifugal pumps, blowers, etc., are generally of this type.

Reaction turbines convert the potential to kinetic energy in the moving blades as well as in the fixed blades, and but a small portion of the heat energy imparts velocity in the first set of fixed blades, and the jet from the nozzle impinges against the first set of moving blades and imparts its kinetic energy to the rotor by impulse. The blades are so proportioned that partial expansion takes place in them, and the resulting increase in velocity exerts a reaction which forces the rotor in its revolutionary motion. In the combined impulse and reaction type of turbine, the high-pressure elements are of the impulse type and the low-pressure elements of the reaction type. Figs. 5 and 6 are good examples of the modern turbine unit found in operation in steel mills. These are at the main plant of the Youngstown Sheet and Tube Co. They are of 3,000- and 3,500-kw. capacity respectively. Two turbines, shown in Fig. 7, were added to the power plant of the Republic Iron and Steel Co. and are on a mezzanine floor at one end of the engine room. This floor also extends around one side of the room, and here the switchboard is located. These units are of 2,500- and 3,500-kw. capacity, generating 6,600-volt direct current. Opposite these turbines on the mezzanine floor, is a 750-kw. mixed-pressure turbine, also generating 6,600, and is shown in Fig. 8. This unit uses exhaust steam from a 24- and 48 x 47-in. cross-compound Corliss engine running at 100 r.p.m. and directly connected to a 1,000-kw. 250-volt d.c. generator. See Fig. 9.

The energy developed by a mixed-pressure turbine is, as someone has put it, like getting something for nothing. Although this is not exactly the case, a considerable gain in economy and capacity is obtained by using the exhaust steam from a non-condensing reciprocating engine. The turbine is connected between the exhaust of the low-pressure cylinder of the engine and the condenser. With the engine developing 1,000 kw., by using the exhaust steam in the mixed-pressure turbine an additional 750 kw. is obtained with the steam that has already developed 1,000 kw. of energy, instead of exhausting the steam to the atmosphere, where the heat is wasted. An engine of 1,000-kw. capacity running at normal load and a turbine generating 750 kw. means that the total capacity has been increased 75 per cent. over the original capacity of the engine alone, and that the saving in steam consumption in making this gain should be about 18 per cent.

A mixed-pressure turbine is in many respects similar in construction to the ordinary high-pressure

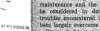

FIG. 3. TYPICAL CROSS-COMPOUND UNIT

FIG. 4. ANOTHER CROSS-COMPOUND ENGINE

turbine. The main thing in its favor is to utilize the greatest amount of exhaust steam about the plant. Means should be used to remove water and oil from the exhaust steam before it enters the turbine. Water

FIG. 5. MODERN TURBINE INSTALLATION AT THE YOUNGSTOWN SHEET AND TUBE CO.'S PLANT

increases the fluid friction in the turbine blading, and this of course reduces the output and increases the steam consumption. Oil itself may not be detrimental to the turbine blading, but it will cause trouble when it combines with sulphates and carbonates that may come over with the steam in case the boiler foams. The gummy substance formed will be deposited on the blades and may clog the steam passages between the blades. Because the mixed-pressure turbine is connected directly to the engine exhaust, its output will vary according to the load on the engine. With a light load, the volume of exhaust steam will be reduced, owing to earlier cutoff of the engine. In cases where the engine and turbine are generating alternating current on the same circuits, no governor is required on the turbine, because it will automatically take its share of the load in proportion to the amount of steam it receives. If it tries to take more than its share, the engine will run faster and cause its governor to make an earlier cutoff, which will reduce the flow of steam to the turbine.

FIG. 6. SIDE VIEW OF A 3,500-KW. TURBINE AT THE YOUNGSTOWN PLANT

If the units are generating direct current and the turbine attempts to take more than its share of the load, the increase in its speed slightly raises the voltage, which sends more current to the generator fields, and this tends to reduce the speed, thus maintaining self-regulation. The greater the amount of steam supplied to the turbine the greater proportion of the load will be taken by it, this being due to the pressure variation in the receiver of the engine.

In the installation shown in Figs. 8 and 9 the engine generator produces direct current and the turbine alternating current. Therefore it is necessary that both the turbine and the engine be controlled by its own governor. In order to prevent waste of steam when the direct-current load is heavy and the alternating-current load is light, it is customary to use a rotary converter or motor-generator set, and thus any difference in load will be taken care of and the load coming on the engine and on the turbine will be proportioned equally. If the direct-current load happens to be light at the time the alternating-current demand is great, the turbine is supplied with high-pressure steam, its admission being cut in or out as the turbine load and the amount of exhaust-steam supply demand. This was the arrangement at the Republic works.

With one exception, all the engines that I saw driving mill rolls in the various plants exhausted steam to the atmosphere. This waste steam could be utilized by a mixed-pressure turbine, as could steam from other intermittently operated units, by the use of a regenerator accumulator, which regulates the flow of steam before it goes to the turbine. The steam, upon entering the regenerator, condenses and is again vaporized during the time when the engine exhaust diminishes or is stopped altogether. If sufficient capacity is provided, the accumulator can be made to provide a practically constant supply of exhaust steam for the turbine.

The accumulator, or regenerator, is a large cylinder partly filled with water. The engine exhaust is delivered to it as a spray through small holes in pipes submerged in the water, thus condensing some of the steam and so giving up heat to the water, bringing the temperature to around 212 deg., which is the temperature of the exhaust steam. When the engine stops, as rolling-mill engines do at intervals, the supply of steam is discontinued and the flow of steam from the regenerator to the turbine will cause the pressure to fall slightly, so that 212 deg. will then be slightly higher than the temperature of boiling water at this low pressure. The water will then evaporate to supply steam to the turbine. When

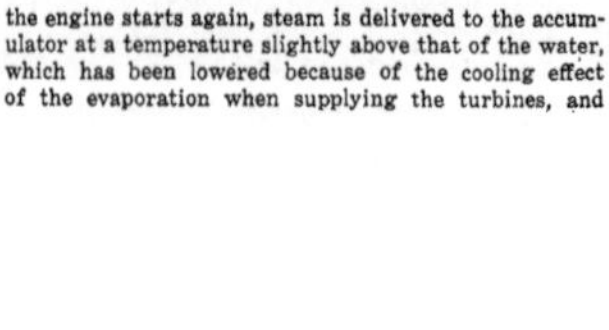

FIG. 7. TURBINES THAT HAVE BEEN ADDED TO THE ORIGINAL EQUIPMENT

the engine starts again, steam is delivered to the accumulator at a temperature slightly above that of the water, which has been lowered because of the cooling effect of the evaporation when supplying the turbines, and the water will again absorb heat from the steam. Of course, if the engine is shut down for some time, the heat stored in the water in the accumulator will become exhausted and the pressure will fall below that necessary for the operation of the turbine. This is overcome by automatically admitting live steam through a reducing valve. The pressure in the accumulator runs

FIG. 8. MIXED-PRESSURE STEAM TURBINE

about 6 lb. per sq.in., and excess pressure is discharged through a relief valve.

I found an accumulator in use at but one of the mills, and it is probable that the operation of the rolling-mill engines and their distance from the central power stations have a determinate influence in running without the use of such an apparatus and allowing the exhaust to escape to the atmosphere. This installation was at the main plant of the Youngstown Sheet and Tube Co., in which the blooming-mill engine exhausted in passing to heat boiler-feed water and to operate a low-pressure turbine.

As the steam pressure used in mixed-pressure turbines is low, the steam consumption per brake-horse-

FIG. 9. ENGINE, EXHAUSTS TO A MIXED-PRESSURE TURBINE

power will be around 38 lb. with a 28-in. vacuum and a steam pressure of 6 lb. absolute initial pressure.

In an exhaust-steam turbine the steam enters through the top of a turbine casing and is distributed to both ends of the rotor, exhausting to the condenser at the bottom. In a mixed-pressure turbine there is a high- and a low-pressure section. The high-pressure element is of the impulse type and the low-pressure element is of the reaction type. The high-pressure steam enters the turbine opposite the impulse section and low-pressure steam behind the reaction section. The mixed-pressure turbine differs from the low-pressure turbine in that it has a governor which controls the valves that in turn control the live-steam admission to the nozzle that admits steam to the high-pressure section before discharging to the low-pressure section along with low-pressure steam with which it mixes.

After seeing numerous kinds of power-plant units, many of which were duplicated, attention was turned to an interesting installation of two of the largest poppet-valve uniflow engines in the world, which were used to drive a 9- and a 12-in. merchant mill.

An Electric-Arc Soldering Iron

By B. A. Briggs

Most electric soldering irons are constructed so that they are heated by the current passing through a resistance coil inclosed in a tube attached to the top of the iron.

The illustration shows an electric soldering iron that is heated by an electric arc in the iron itself. A hole is bored in the center of the iron, having a diameter equal to the outside diameter of the tube of a battery bushing, down to where the iron begins to taper. A standard battery porcelain bushing *B* is placed in the hole and a ⅜-in. arc-lamp carbon *C* is placed in the

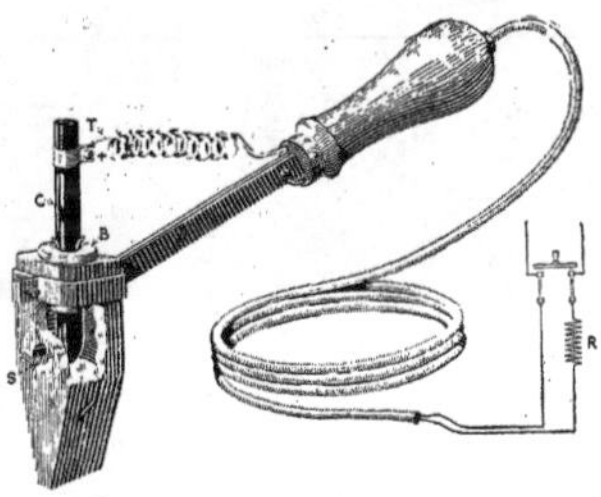

ASSEMBLY OF ELECTRIC-ARC SOLDERING IRON

bushing, and allowed to come down in contact with the bottom of the hole, then the carbon is withdrawn about ¼ in. to establish the arc. The iron should be connected to a 110-volt circuit and resistance enough connected in series to keep the current down to between two and three amperes. About 30 ohms will be sufficient. This resistance may be made of about 600 ft. of No. 18 B. & S. iron wire and connected in the circuit, as at *R* in the figure. The soldering iron is connected to the circuit with the positive terminal on the carbon, as shown at *T*.

The carbon is held in place in the bushing by small metal wedges, and adjustment of the carbon is made by tapping it down to give the desired length of arc. A second hole *S* is bored at right angles to the first, so as to allow the arc to be cleaned of the nitrate that forms around it.

Improved Coal-Handling Facilities at Cos Cob*

THE severe winter of 1918 acted as a spur to many power generating companies to enlarge their coal-storage facilities as a means of insuring themselves against any possible failure of the transportation systems. During this memorable winter not one but many an operator was on several occasions within a few hours of exhausting all available coal, and there were even some shutdowns. It was this unprecedented condition which caused the officials of the New York, New Haven & Hartford R.R., to lay plans at once for amplifying the coal-handling facilities of the Cos Cob (Conn.) generating station, which supplies a large portion of the energy for the New York-New Haven electrified division.

Accordingly, Gibbs & Hill, consulting engineers, of New York City, were given the task of working out and installing a suitable plan, and the result includes many interesting features.

Original Facilities Inadequate

The Cos Cob power station is a plant of 28,000 kw. capacity, having a twenty-four-hour coal-consumption rate of from 350 to 450 tons. It is laid out in a long turbine room with two boiler-room wings extending out from the north side. Normally, all coal is received by tidewater shipments from Perth Amboy. The coal used is mostly run-of-mine Pennsylvania bituminous fuel. With the original coal-handling equipment, the coal has been picked up from the barges with a grab-bucket hoist, dumped into a crusher tower, and carried in cars over an inclined railroad from the crusher to the overhead bunkers in the two boiler rooms. These overhead bunkers furnished a storage capacity in the plant of 150 tons for each boiler room.

What little coal has been received by rail in the past has been dumped into two track hoppers, one at the end of either boiler room. These hoppers have a capacity of about 50 tons each and are equipped with a crusher from which the coal is delivered by bucket conveyor to the overhead bunkers.

With this original coal-handling arrangement, it is seen that practically the only provision for reserve coal supply was that afforded by barges held alongside the dock in the river. This served the purpose with reasonable safety until the unprecedented conditions of the 1918 winter made it impossible to receive coal by barge for many days at a time. As the supply by rail was likewise uncertain, the need for adequate storage facilities on the ground became obviously imperative.

TWO CYLINDRICAL CONCRETE COAL BUNKERS, ONE FOR EACH BOILER ROOM

The plans contemplate an extensive remodeling of the coal-handling and storage facilities for the future, the part of which now completed and herein described will give an addition of 2,000 tons capacity to the power-house storage bunkers. Later, it is proposed to provide a storage and handling plant near the power house, to be used in conjunction with the new bunkers.

Just beyond the end of each boiler room a cylindrical concrete storage bunker has been erected and supported on concrete columns with a clearance of 19 ft. from the boiler-room floor level to the top of the bunker. An extension of the firing aisle in each boiler room was provided to a point directly beneath the bunkers, the overhead trackway for the double weighing lorry system being extended through a doorway made in the wall of each. Thus the lorries can be run underneath the bunkers and filled by gravity through the bottom gates, then return down the firing aisle to deliver the coal in known quantity to the several stokers.

The standard-gage side track which formerly served

*Electric Railway Journal.

LOOKING DOWN INCLINED CONVEYOR STRUCTURE

the two auxiliary track hoppers was extended out onto a wooden trestle and over a new track hopper which discharges into a crusher of 150 tons per hour capacity. Coal dumped into this hopper passes through the crusher and is discharged onto a 24-in. belt conveyor which is driven at the rate of 400 ft. per min. by a 50-hp. squirrel-cage motor. This belt rises at an angle of 16 deg. with the horizontal, and delivers the coal at an elevation of 97 ft. into the cupola at the top of the first concrete bunker. In this first cupola a sort of butterfly valve is provided which, when set in one position, discharges the coal from the main conveyor into the first bunker, and when thrown to the other position, discharges onto another conveyor running between this first bunker and the cupola of the second bunker. This conveyor between bunkers is also a 24-in. belt, which is driven by a 10-hp. squirrel-cage motor mounted with its starting box in the cupola of the second bunker. This belt between bunkers is 112 ft. long between centers; the main belt from the crusher to the first bunker is 330 ft. long between centers.

CONCRETE BUNKER, INCLINED BELT CONVEYOR, TRESTLE AND CRUSHER PLANT

The "silo" type concrete cylindrical tanks measure inside 40 ft. in diameter and 46 ft. high. They are supported on concrete columns, of which the outer ones are 2 ft. 6 in. in diameter and the inner ones 3 ft. The cupolas in which the conveyors discharge into hoppers are octagonal in shape and 12 ft. 6 in. in diameter with walls 8 ft. high and a peak 11 ft. high. The walls of the cupolas and those of the tanks are 11 in. thick and are reinforced by steel hoops and vertical steel rods, the latter serving as vertical ties and also as supports for jacking up the forms during construction. The bottoms of the bunkers are of mushroom slab construction and are hoppered in all directions from the openings in the bottom.

The use of concrete rather than steel for these bunkers was determined partly because of the extreme difficulty of getting steel during the war, when they were built, and partly because the concrete structure involved a less cost as compared with steel at the war prices.

The completion of the conveyor and concrete bunker system adds 2,000 tons of storage capacity to the plant. This alone is sufficient to keep the plant operating, in case of failure of the coal supply, for about seven days. At the present time coal can be supplied to this bunker system only as received by rail, but with the completion of the entire layout the principal path of the coal will be through this conveyor and concrete bunker system. The inclined railway at the opposite end of the plant, which is now handling the coal from the barges to the overhead bunkers, will presumably be done away with. In case of failure of the conveyor system, the contents of the two bunkers can be drawn upon until exhausted, and there still remain the two track hoppers at the end of each boiler room, with the crusher and bucket conveyor system to the overhead bunkers, to fall back upon, coal being supplied by rail.

THE BOTTOM OF ONE OF THE CONCRETE BUNKERS

Coal Shortage in Germany

Coal shortage in Germany is becoming more critical each year according to figures taken from a recent issue of *Commerce Reports*.

The peacetime coal production was estimated at 200,000,000 tons. During the past few years coal production has decreased 50 per cent, which cuts the figures to 100,000,000 tons. The coal supply available for industrial and trade purposes is only 26,000,000 tons, whereas the amount required to carry on industrial activity adequately is estimated at 85,000,000 tons.

Excessively low and high water on the River Rhine, which retarded navigation, and railway strikes in Elberfeld and Essen have so hindered deliveries that the coal supply has diminished alarmingly. A railway strike in Upper Silesia and inadequate transportation facilities have reduced the supply to such an extent that deliveries are made only to the most essential industries. Bohemia and Upper Bavaria are also experiencing transportation difficulties, and fuel deliveries have been cut by half.

Automatic Arc Welding Machine

Automatic arc welding is now accomplished by a device known as the automatic arc welder, shown in the figure, which is designed to take the place of the hand-controlled electrode. A pair of feed rolls driven

BUILDING UP SHAFT WITH AUTOMATIC ARC-WELDER

by a small direct-current motor draw in and deliver to the arc a steady supply of wire and automatically maintain the best working distance. The whole is controlled from a small panel. The welding head is held by a suitable support with a certain amount of hand-regulated adjustment and consists of a steel body carrying feed rolls and straightening rolls, which are both adjustable for various sizes of wire.

The control panel carries an ammeter and voltmeter for the welding circuit, as well as rheostats, a control relay and the contactors and switches for the feed motor. It is possible to start and stop the equipment from the work by a pendant push-button switch, but adjustment of the feed conditions must be made from the panel.

The whole apparatus is mounted on a base that can be bolted to any form of support. Thus a great variety of working conditions can be met, but provision must be made for carrying the arc at uniform speed along the weld. For instance, for straight seams a lathe or planer bed may be used, and for circular ones a lathe or boring mill. However, the local conditions will dictate the method to be followed.

The device is especially valuable where a large amount of routine welding is to be done, since it is capable of from two to six times the speed possible to skilled operators and gives a uniform weld of improved quality. It is adaptable to welding seams of tanks and plates, rebuilding worn or inaccurately turned shafts, as shown in the figure, rebuilding worn treads and flanges of wheels, and many other kinds of work. This equipment has been developed and will soon be placed on the market by the General Electric Company.

Resharpening Files

Processes for resharpening files by "biting" with acids is an old trick of frugal mechanics. On page 315 of the Aug. 31, 1915, issue of *Power*, Department of Inquiries of General Interest, the question of "E. P.," asking how files are cleaned and sharpened by chemical treatment is answered as follows:

After removing loose dirt with a stiff brush, the files are cleaned of grease by boiling them about forty-five minutes in a solution consisting of one-fourth pound of saleratus to a quart of water. The files are then to be thoroughly washed in clean water and dried and then placed on end, completely submerged in a solution consisting of one-fourth pound of sulphuric acid to a quart of water. After remaining in the acid solution for about twelve hours, the files should be thoroughly washed in clean water, quickly dried and then given a light coating of kerosene or other thin oil to prevent rusting.

French Patent No. 483,676 describes a more elaborate method for treating worn files, as follows:

The file to be treated must, first of all, be cleaned as well as possible, and be rid of all greasy matter, etc.

The file is then immersed in a solution of sulphuric acid diluted with four times its quantity of water, in which it should remain for about ten hours. It must then be rinsed in clean running water and carefully dried. After this is done, the file is placed in a glass or enameled metal trough and covered with water into which sulphuric acid is poured until the water reaches 15 deg. C. (59 deg. F.).

After immersion for a few seconds (10 to 15), according to the thickness of the file, the latter is thrown into a bath of cold water to enhance the quenching effect and wash it at the same time.

The third operation consists in placing the file in a small trough containing a piece of lead, and on top of this is poured an aqueous solution of nitric acid and common salt in variable quantities, according to the thickness and the quality of the cuts, and this operation will make the file fit for use again. In this solution the file must remain for 10 to 30 seconds, when it must be rinsed again in very cold water and thoroughly dried. To finish the "cuts," the file is then treated with a solution of oxalic acid for 5 to 20 seconds, according to the thickness of the file. After the file is rinsed again and dried, it must be treated with a solution of sulphate of iron and then with milk of lime, so as to give it a new appearance, and prevent it rusting. Finally, after a further rinsing and thorough drying, the file will be restored to its original condition.

Calculations recently made by an expert, Prof. A. Juselius, show that the amount of water power available for Finland's industries has been greatly exaggerated. He estimates that at mean water level it may amount to 3,000,000 hp. but of this only a small part can be utilized.

A Captain of Coal Conservation

Our obligation to conserve coal—that is, to use coal efficiently—did not stop with the end of the war.

It is not democratic for one plant owner to waste coal recklessly while his neighbor practices scientific means for its efficient utilization. Without coal modern industry could not proceed; the necessities and comforts of civilized life could not exist. Coal as a natural resource presents a problem which concerns the people and therefore the Government which represents them.

It is the privilege and duty of engineers to educate the people as to the facts, so that they will compel Congress to enact appropriate legislation for the prevention of this tremendous and unreclaimable waste of the most valuable of all our national resources.

—DAVID MOFFAT MYERS.

IN DECEMBER, 1917, a paper was presented before the American Society of Mechanical Engineers entitled "Preventable Waste of Coal in the United States," which included a well thought out program for checking this waste. This paper was particularly timely, coming as it did when the Nation was facing the gravest crisis in its history, and so comprehensive was the plan that when it was brought to the attention of the National Fuel Administration, the author was invited to enter that Organization, as Advisory Engineer, to put his theories into practice. David Moffat Myers was the author of this new conservation plan.

The success of the practical application of his theories is indicated by the official report of his work. An annual saving of 25,000,000 tons of coal was effected in the last six months of 1918 through the medium of his original methods. Of the total amount saved 18,000,000 tons was credited to industrial plants of the country and 7,000,000 tons on the railroads. The program included a comprehensive set of recommendations as to methods of fuel economy, which were sent throughout the country to the various power plants using fuel. It further included a system of classification and rating of power plants based on the thoroughness with which the various individual plants were found to be carrying out the standard recommendations. The distribution of fuel was then based on the ratings obtained under this scheme. For the administration of this work a fuel engineering section in the conservation department of the United States Fuel Administration was formed, and Mr. Myers was appointed chief of the section. This involved the organization of fifteen hundred volunteer engineer inspectors in the various states in addition to numerous committee members and other workers.

Mr. Myers received the degree of M. E. from Columbia University in 1901, after preparatory work in both public and private schools; and for five years following his graduation was employed by the United States Leather Company. While with this company he had charge of testing, investigating, and reporting on the fuel and steam conditions of about one hundred plants in different parts of the country. Later his work included the remodeling of old plants and the designing of new ones, for the production of steam and fuel economy. This involved a large amount of engineering research work on waste fuels and on original designs for furnaces to burn them. The waste products of the leather industry are usually very difficult to burn efficiently, as they contain very high percentages of moisture. During this period Mr. Myers designed furnaces that successfully burned waste spent chestnut chips which contained 65 per cent of moisture, and which had resisted all previous attempts to burn without a mixture of coal or some other auxiliary fuel. Since then, he has done a large amount of original furnace designing for various waste fuels, and now has special furnaces working in different parts of this country, in Canada, and in Cuba.

In 1906 Mr. Myers severed his connection with the Leather Company and opened his own office as consulting engineer, specializing on economy of steam and fuel in industrial plants. He numbers among his clients some of the most important industrial corporations, not only throughout the United States, but also in Canada and in Cuba. One of his notable achievements in the power plant field was making feasible a $500,000 annual increase in production in one mill of the Cuba Cane Sugar Corporation, and this enormous economy was obtainable with comparatively small changes in the equipment of the plant. The Midvale Steel Company, through Mr. Myers' investigation, was able to realize a yearly saving of $125,000 which was made possible by the expenditure of some $60,000 for additional equipment. The system involved in such problems as handled by Mr. Myers is divided into two main parts. First, a thorough investigation is made into the fuel and steam conditions of the plant and a complete report submitted covering the recommendations for improvement. Following this, complete plans and specifications for the recommended changes and additions or alterations are prepared, based on the data obtained in the preliminary investigation. From these plans and specifications proper proposals are obtained, contracts let, installation supervised, proper management systems installed and the completed plant is turned over to the client.

In 1915 Mr. Myers became associated with John S. Griggs, Jr., formerly senior partner of the consulting firm of Griggs & Holbrook. The present firm is known as Griggs & Myers, Consulting Engineers, at 110 West 40th Street, New York, and is carrying out the work along the same lines as outlined above.

Mr. Myers is the author of two books on power plant management and design: "Preventable Losses in Factory Power Plants," published in 1914, and "The Power Plant," which is Volume II of the Factory Management Course of the Industrial Extension Institute, Inc., New York. He is also the author of various articles and papers for technical magazines and societies. Among these, a contribution entitled the "Heating Value of Exhaust Steam," was presented at an annual meeting of the American Society of Heating and Ventilating Engineers. In this paper Mr. Myers developed a formula which shows the heating value of the exhaust from any kind of steam engine, turbine or pump. He is a member of an advisory body on fuel conservation known as the Committee of Consulting Engineers to the Bureau of Mines, and he is also a consulting engineer in connection with the sub-committee on coal of the American Society of Mechanical Engineers.

He is a member of the Special Committee of Professional Sections of the American Society of Mechanical Engineers representing the Fuels Section, and has recently been named as a member of the Committee of the American Institute of Mining and Metallurgical Engineers, to coöperate with the Coal and Coke Committee for formulating a plan for the Stabilization of the Coal Industry of the United States. He is also a member of the Columbia University Club, the Cosmos Club of Washington, the Bayside Yacht Club, Theta Xi Fraternity and the American Society of Mechanical Engineers.

Mr. Myers was born January 8, 1879, at Owasto, New York, the son of the late Rev. Alfred E. Myers of New York.

Oil as a Fuel for Boilers and Furnaces

By H. H. FLEMING

THE paper of which this article is an abstract, was presented at the annual meeting of the American Society of Heating and Ventilating Engineers, in New York, January, 1920, and begins with a brief discussion of the present interest in fuel oils and of the nature of such oils. Regarding the availability of the various oils for fuel, the author points out that the heavy fractions of paraffin crudes are too valuable for lubricating purposes to be burned as fuel. However, the ashpaltum crudes that are now so plentiful can be used for little else than fuel oil and road binders, after their small percentage of light oils has been removed. Consequently, refineries running such crudes are compelled to dispose of great quantities of fuel oil in order to assure the gasoline or kerosene from the crudes. These cheap crudes come mostly from Mexico and, until recently, have not been available in very large quantities, owing to well-known conditions not only in Mexico but in ocean transportation. The war situation having cleared up during the past few months, transportation from Mexico is now nearly normal, and millions of barrels of crudes are sent every month to this country.

It is expected that the exportations from Mexico for the year 1919 will exceed 80 million barrels, but this is only a small proportion of the estimated potential production. There are single wells in these fields capable of producing over 100,000 bbl. daily. Many wells have been capped or their production greatly cut down, awaiting transportation facilities; that is, pipe lines to the coast and tankers to receive it there, both of which are being constructed as rapidly as possible. Practically the entire Mexican production comes from a strip of land running along the east coast for a little over 100 miles, averaging about 50 miles wide, the principal port of which is Tampico.

The author gives a brief discussion of the advantages of fuel oil over coal, and then takes up the question of equipment required for a fuel-oil installation. This equipment consists essentially of a storage tank, pump, heater and burner with the interconnecting piping. A simple duplex pump is generally used to deliver the oil to the burner, and between the pump and burner the heater is installed which raises the oil to a temperature close to its flash point. The function of the burner is to atomize the oil and spray it into the firebox, where it is burned in suspension. The atomizing is done by steam or air, or mechanically, with hundreds of different types of burners using each method. It seems as though every man who ever burned oil designed an oil burner of his own and patented it. Each designer makes broad claims for his burner, but the author credits Kent with the remark that as long as a burner atomizes the oil, its design should have no more effect on the boiler efficiency than the design of a coal shovel would have on a coal-fired boiler. The important qualifications of a good oil burner are (1) that it shall atomize the oil to as near a gaseous state as possible, (2) that it shall operate with a minimum consumption of steam or other atomizing mediums and (3) that it shall not clog or suffer from the eroding effect of the steam and oil.

Mechanical burners atomize by forcing the oil under 100 to 300 lb. pressure through slots so arranged that the oil is given a whirling motion. This type of burner is used exclusively in marine work to avoid the loss of water up the stack in the form of atomizing steam.

Burners which atomize by compressed air have been used on board ship, but the compressor equipment is expensive, bulky and costly to maintain. The same objections apply to air-atomizing in shore plants, but it is used in many installations when steam is not available or desirable for the particular work done.

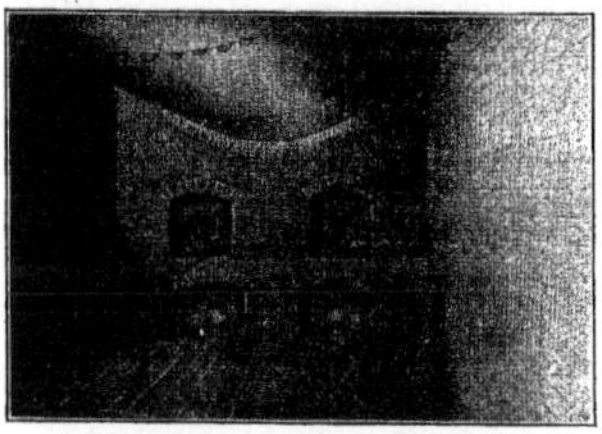

BOILER FURNACE FOR BURNING OIL

Air gives a short, intense flame and is better than steam for some metallurgical work.

For the great majority of stationary plants and locomotives steam is the atomizing medium used, principally because it makes the simplest installation. A good burner, properly operated, will consume 2 to 3 per cent of the total steam generated, or roughly, 3 to 4 lb. of steam per gallon of oil.

The furnace design, not the type of burner, is the important part of the installation. The proper furnace arrangement for burning oil does not differ materially from one designed for burning coal. So the change is generally simple and inexpensive. The most important point to keep in mind in making such a change, is that the flame from the oil burner must not be allowed to impinge directly on any heating surface. The firebox must be large enough to permit combustion to be completed before the flame is cooled by contact with a boiler surface in order to prevent incomplete combustion and the burning out of tubes. A hand-fired coal boiler is often converted to burn oil by merely covering the grate bars with firebricks, laid loosely, leaving a suitable air space directly under and in front of the burner, which is inserted in the fire-door. To secure a larger combustion chamber, the grate may be taken out and the burner placed in the ashpit. The floor of the ashpit should then be built of loosely laid bricks so supported as to admit air underneath them. The hot bricks will then preheat the incoming air. Some prefer to fire from the bridge wall, and every engineer has his own theories as to the value of checkerwork, arches, etc. The accompanying illustration is an interior view taken from the bridge wall of a furnace equipped for oil burning under a horizontal return-tubular boiler.

The stack required for oil is smaller than that used with coal to obtain the same capacity, because less excess air is used, and it is shorter because only sufficient draft to force the gasses through the boiler and breaching is needed, there being, of course, no firebed. Forced-draft equipment is discarded except with mechanical burners, where it is possible to use it if exceedingly high ratings are required.

The transportation of fuel oil from the refinery to the consumer is usually by barges, tank cars or tank trucks. The barges used vary in capacity from 2,000 to 20,000 bbl. (42 gal. each) and are equipped with pumps for discharging the oil. Tank cars are built with capacities of 6,000 to 12,000 gal., the most common size being 10,000 gal. For handling the heavy oil these cars are equipped with steam coils, so that they may be readily unloaded in cold weather. Tank trucks of from 1,000 to 2,000 gal. capacity are used for delivery where cars or barges cannot be taken.

Storage tanks may be of steel or concrete and either above or below ground, depending upon their size and the local conditions. They are usually provided with steam or hot-water coils, as the heavy oil should be at about 100 deg. F. to pump easily. These coils are generally supplied by the exhaust from the pump, not much heat being required, since the specific heat of the oil is approximately 0.5. The relative values of steel and concrete tanks have been discussed at great length, but which is the cheaper and better depends largely on local conditions.

The accompanying table, showing the relative value of coal and oil, is based on a constant calorific value for each of the fuels.

RELATIVE VALUE OF COAL AND OIL

Gross Boiler Efficiency with Fuel Oil	Net Evaporation from and at 212° F. per Lb. of Oil	Water Evaporated from and at 212° F. per lb. of Coal				
		6	7	8	9	10
		Barrels of Oil Equal to One Ton of Coal				
75	13.92	2.565	2.993	3.420	3.848	4.275
76	14.11	2.532	2.954	3.376	3.978	4.220
77	14.30	2.498	2.914	3.330	3.746	4.162
78	14.49	2.465	2.876	3.286	3.697	4.108
79	14.68	2.433	2.838	3.243	3.649	4.054
80	14.87	2.402	2.802	3.202	3.602	4.003
81	15.06	2.371	2.767	3.162	3.557	3.952
82	15.25	2.342	2.732	3.122	3.513	3.903
83	15.44	2.313	2.699	3.085	3.470	3.856
84	15.63	2.285	2.667	3.049	3.431	3.813
85	15.82	2.257	2.635	3.013	3.391	3.769

This table is not an accurate basis for comparison, but is useful only as a rough guide for the relative values. The only method of accurately estimating these values is to consider the operating expenses of the plant with each in turn, including the cost of all items entering into the problem. Some of the features to be considered in comparing the two fuels are the space available for fuel storage, the labor saving possible, the number of hours the plant is in operation, the load feature, the quantity of coal for banking fires, etc. The figures in the column headed "Gross Boiler Efficiency" include the steam generated which is used for burning the oil. Gross efficiency, then, is equal to net efficiency, plus the percentage used in atomizing oil. Oil values in the table are based on oil weighing 336 lb. to the barrel of 42 gallons.

Single-Phase Load on a Three-Phase Power System

By J. B. GIBBS

Engineer, Transformer Department, Westinghouse Electric and Manufacturing Company

THE only way to connect a single-phase load to a three-phase power system so that the system will be balanced, without the use of rotating machinery, is to divide the load into three equal parts and connect one part to each phase. There are two ways of making this connection, depending on the voltage of the system and the voltage desired on the load. Fig. 1 shows the delta connection. One part of the load is connected to each pair of line wires, and the voltage applied to each part of the load is the voltage between the corresponding lines. With this connection the voltages on the three loads will not be appreciably unbalanced, even if the loads are not exactly equal. Fig. 2 shows the star or "Y" connection. Each part of the load is connected by one wire to one of the line wires, and the remaining three wires from the loads are connected together to form the neutral or star point. If we

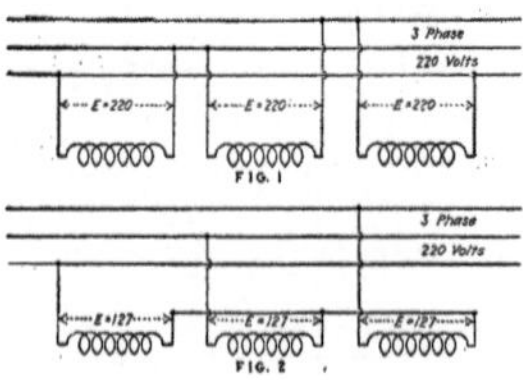

FIGS. 1 AND 2. SINGLE-PHASE LOADS CONNECTED ON THREE-PHASE CIRCUIT

have 220 volts between lines, the voltage on each of the loads will be 220 ÷ 1.732 = 127 volts. If the loads, with this connection, are not equal, the neutral point will be shifted and the voltages on the three loads will not be equal.

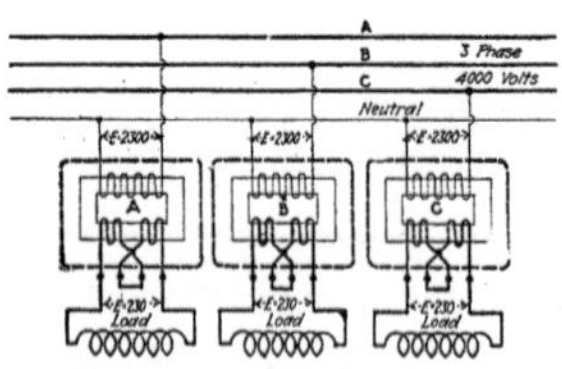

FIG. 3. THREE SINGLE-PHASE LOADS STAR-CONNECTED TO 4-WIRE THREE PHASE LINE

Another example of the star connection is shown in Fig. 3. The distribution line is three-phase 4000 volts, with a neutral wire, and three standard transformers, with a ratio of 2300 volts to 230 volts, are used to supply three single-phase loads. In this case the star point of the three transformers is tied to the corresponding point of the generator, and therefore the correct voltages will be maintained on the loads even if they are not exactly equal.

If the single-phase load cannot be divided, the best thing to do is to connect it to one phase of the power system and trust to other single-phase loads connected to the other phases to balance the system. As a matter of fact, an unbalanced load is not very objectionable, unless the unbalance is large. A majority of the large generating plants are three-phase, and they carry single-phase loads, distributed among the three phases, as well as three-phase loads. Such systems are never exactly balanced, but never very far unbalanced.

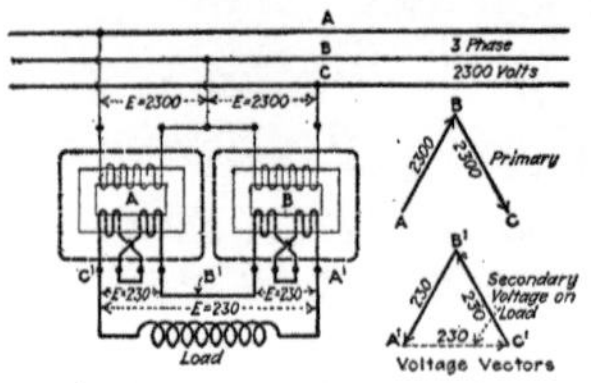

FIG. 4. SINGLE-PHASE LOAD ON TWO TRANSFORMERS CONNECTED OPEN DELTA

Several schemes have been proposed for connecting a single-phase load to a three-phase system, and it may be worth while to analyze a few of them.

1. Open-Delta Connection—It has been suggested that two transformers be connected in open-delta and a single-phase load be taken from the open side, as shown in Fig. 4, the idea being that the single-phase current must flow in both transformers, and therefore in all three lines. Since the load in the secondary windings is single-phase, however, the current in the primary windings must be in the same phase. That is, all the current in the primary of transformer *A* must flow through the primary of transformer *B*, and therefore no current can leave or enter the line *B*. Consequently, we have exactly the same condition as though the load were connected to the lines *A* and *C*, except that it requires two transformers to carry the same load that might be carried by one connected to *A* and *C*.

2. Closed-Delta with One Secondary Reversed—This is shown in Fig. 5, and is similar to Fig. 4, except that it has a third transformer and gives twice the secondary voltage. It will be seen that, since the three secondaries are in series, the currents in the three transformers must be the same, in phase and magnitude. Therefore the current which flows in the primary of transformer *A* is the same as that in transformer *B*, and

none of the current can be carried by line *B*. The single-phase voltage is the same as though we were using two transformers connected between the lines *A* and *C*, but three transformers are being used to give the output which would be given by two connected between *A* and *C*.

3. Scott Connection with Secondaries in Series—This arrangement is shown in Fig. 6. The plan is to transform from three-phase to two-phase, and then to connect the two phases in series to get single-phase. This will, of course, give a single-phase voltage, but it is interesting to see what happens when a current flows. For the sake of simplicity suppose that the transformers have a one-to-one ratio, and suppose that each of the two-phase voltages is 100 volts. Then the single-phase voltage will be $100 \times 1.414 = 141.4$ volts. Now suppose that a 10-ampere load is put on the single-phase circuit. Since the primary of the "teazer" transformer *B* is connected to the 87 per cent. tap, it must carry a current of $10 \div 0.87 = 11.5$ amperes, and this must obviously be the difference of the currents in the two halves of the "main" transformer *A*. But the total ampere-turns in the primary of *A* must be equal to the total ampere-turns in its secondary, or 10 amperes × 100 per cent. turns. Therefore, we have the two equations, 50 per cent. X + 50 per cent. $Y = 100$ per cent. × 10 and $X - Y = 11.5$, where X equals current in one-half of the primary of *A*, and Y equals current in the other half of the primary of *A*.

Solving these two equations, we find that X equals 15.75 amperes and Y equals 4.25 amperes. The line *C*, therefore, will carry 15.75 amperes, of which 11.5 amperes will return to the generator by way of the line *A*, and 4.25 amperes by way of the line *B*. The rating of each transformer will be,

$$\frac{10\ amperes \times 100\ volts}{1000} = 1\ kilovolt\text{-}ampere$$

or a total of 2 kv.-a. for the bank, while the total single-phase output will be,

$$\frac{10\ amperes \times 141.4\ volts}{1000} = 1.414\ kilovolt\text{-}amperes$$

We are thus using transformers 40 per cent. larger than we need for the single-phase output and getting as compensation a decrease of 20 per cent. in the line loss,

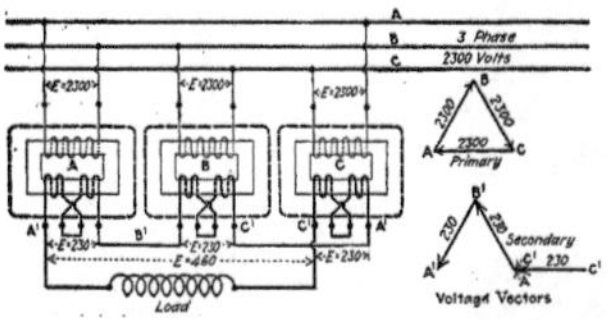

FIG. 5. SINGLE-PHASE LOAD CONNECTED TO THREE-PHASE CIRCUIT THROUGH THREE TRANSFORMERS

due to splitting the return current between two lines. Inasmuch as the line loss is usually a small percentage of the power delivered, a saving of one-fifth of it is seldom worth the extra investment in transformers, especially as these transformers must have a special ratio.

A single-phase load may be perfectly balanced on a three-phase system by means of a motor-generator set, consisting of a three-phase motor and a single-phase generator, but such a set is complicated and expensive, and for this reason it is usually not to be commended. The foregoing may be summarized by saying that a single-phase load can be perfectly balanced on a three-phase system if it can be divided into three equal parts and connected to the three separate phases, or if a motor-generator set is used. The former scheme is frequently possible and advisable. The latter scheme can very seldom be considered, on account of the expense. If neither one of these schemes can be used, it is usually best to connect the load to one of the phases without any attempt to balance it.

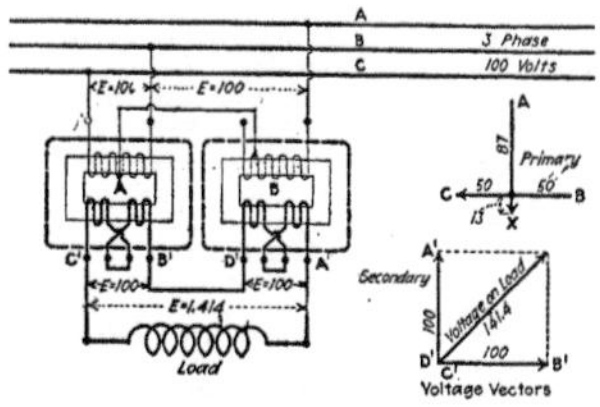

FIG. 6. SINGLE-PHASE LOAD ON SCOTT-CONNECTED TRANSFORMERS

In northern France 30,000 plants, employing 800,000 workers, were razed by gunfire, looted or badly disorganized. In one-half of these plants production has been resumed. While the textile industry is rapidly recuperating, it is estimated that it will require three years for the majority of the principal trades, such as glass-making, woodworking and sugar refining, to resume production on a normal scale. A longer time, probably ten years, must elapse before large mechanical plants, such as the locomotive works at Valenciennes and the coal mines about Lens, can proceed full-speed ahead. Machinery and equipment weighing about 250,000 tons have been returned from Germany, and the French government has advanced to industrial firms credits totaling approximately 3,000,000,000 francs, which sum will be applied against damages estimated in the Chamber of Deputies to amount to 50,000,000,000 francs.

Large hydro-electric undertakings are projected in the Province of Ombria in Italy. It is intended to build two large reservoirs, one on the River Salto, at Baize di Santa Lucia, capable of holding 81,000,000 cubic meters of water; and the other on the River Turano, at Petescia, of a capacity of 132,000,000 cubic meters. Besides subsidiary works and installations of local interest, two generating stations will be built, one utilizing a waterfall at Collestatte 460 ft. high and another at Cervara 200 ft. high, capable of developing a total of 106,421 hp. The developments are estimated to cost $14,000,000, of which about one-third is intended for the subsidiary works.—*London Electrical Review.*

Chief geologist of the United States Geological Survey states that the total available supply of oil in the United States is 6,500,000,000 bbl., which at the present rate of consumption will last between fifteen and twenty years provided no new automobiles are manufactured and no new uses are devised for the commodity.

Parallel Discharge of Centrifugal Pumps

By T. M. HEERMANS
Engineer, Centrifugal Pump Division, Allis-Chalmers Manufacturing Company

To operate successfully in parallel centrifugal pumps should have steep head-capacity curves and the slope of the curves should be about the same. The reasons are simply told and worth-while pointers are given on the suction pipe.

When a number of centrifugal pumps are arranged to discharge into a common main, there are several ways in which such an arrangement differs from a similar grouping of piston or plunger pumps. These differences are practically all covered under the general head of the difference between the performance characteristics of the two types of pumps. At a constant speed the reciprocating pump, provided the power end is big enough, will deliver a definite capacity into the discharged main notwithstanding considerable variation in the pressure in the discharge line due to starting up or shutting down other pumps discharging into the same line or to other causes.

A centrifugal pump, on the other hand, has a certain fixed curve of relationship between the discharge and the head, so that any change in the pressure in a common discharge main causes a change in the capacity delivered into the main. To illustrate, refer to Fig. 2, which gives the characteristic curve *AC* of a pump designed to operate at point *B* and to deliver 1,800 g.p.m. against a head of 35 ft. If the pressure against the discharge drops to *D* the capacity will be increased to 2,200 g.p.m., or if the pressure in the main increases to *E* the discharge will decrease to 1,200 g.p.m. This pump the capacity will not vary much even if the head in the discharge main changes considerably.

FIG. 1. PUMPS OF 1,000 AND 1,200 G.P.M. CAPACITY OPERATING IN PARALLEL AGAINST A 100-FT. HEAD

means, first that to work in parallel, centrifugal pumps should be selected that will deliver the desired quantity against about the average head to be maintained in the discharge line under usual conditions.

Second, it would be advisable to specify pumps having runners that will produce the so-called "steep head characteristic" curve, as with this design of pump the capacity will not vary much even if the head in the discharge main changes considerably.

Third, the several centrifugal pumps that are to discharge into the common main should have head-capacity characteristics as nearly alike as possible in so far as the proportion of change in capacity for a

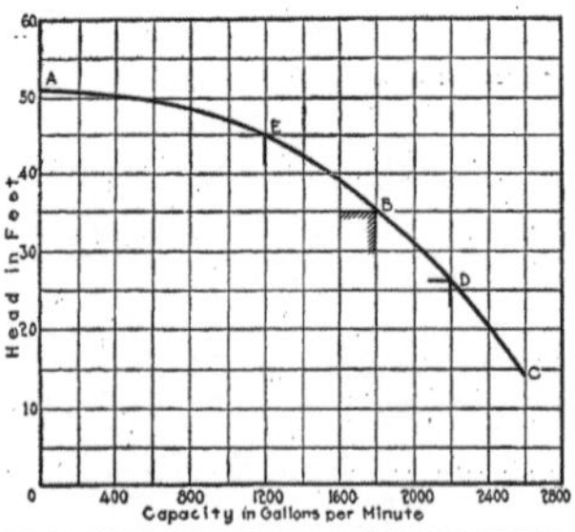

FIG. 2. CURVE SHOWING THAT CAPACITY OF CENTRIFUGAL PUMP VARIES WITH THE HEAD

change in head is concerned; that is, the slope or steepness of their head-capacity curves should be about the same. When purchasing new pumps that are to discharge into the same main with existing pumps the foregoing statement means that the manufacturer should be supplied with the characteristic curves of the existing pumps. This can again be best illustrated by the curves, Fig. 3.

Suppose two pumps have been selected each having the normal rating indicated by point *P*, but having head-capacity curves of different type such as *A* and *B*. If, for any reason, the head in the discharge main should increase from *H* to *H*, the capacity of pump *A* would be decreased only to *C*, whereas the capacity of pump *B* would be decreased considerably more, or to *D*, a much less efficient operating point on the curve. The point of maximum efficiency of a centrifugal pump properly selected is at about the rated head and capacity. The pumping work at the changed head conditions is therefore unevenly divided between pumps *A* and *B*, and the over-all pumping efficiency is reduced. If both pumps had the same type of curve, the pumping work at increased head would have been equally divided and the combined pumping efficiency would have been correspondingly higher.

Fig. 3 illustrates one other point; namely, that should the head in the discharge main be increased to H_2 the "shutoff" pressure of pump *B* would be exceeded and it would not deliver any water at all, but would merely churn the water in its casing. This indicates the necessity of selecting a pump for parallel operation with a head curve steep enough so that the starting up of additional pumps will not increase the pressure in the main above the shutoff pressure of one of the pumps, thereby stopping the discharge of that pump.

With a little care it is not a difficult matter to select centrifugal pumps that will operate just as satisfactorily in parallel as any other type, with the additional advantages peculiar to the centrifugal pump of smooth, nonpulsating delivery, impossibility of excessive pressure rises in the mains, simplicity and low maintenance and operating expense.

In connection with the parallel operation of centrifugal pumps it might not be amiss to add a few remarks about the suction pipes. These should be of liberal sizes to reduce friction to a minimum and should be absolutely free from air leaks. It is also advisable, whenever possible, to connect the suction pipe to the pump with a short length of straight pipe next to the pump or a straight reducing fitting reducing from a larger-sized suction pipe to the size of the pump suction opening.

An elbow connected immediately to the pump suction causes the water to whirl in entering the pump and makes a disturbed flow at this point and sometimes unbalanced pressures on the two sides of

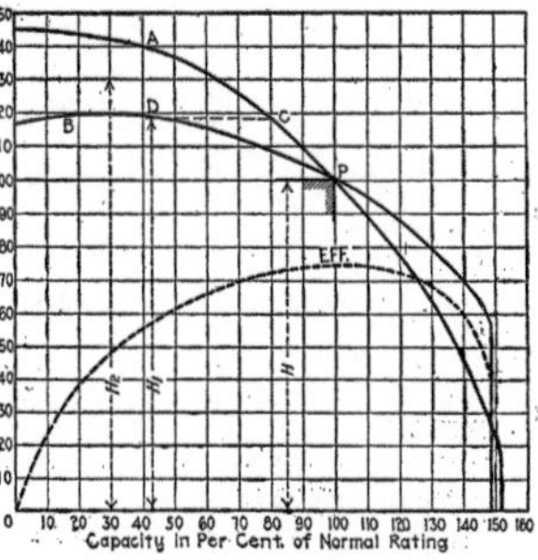

FIG. 3. EFFECT OF VARIATION IN HEAD ON TWO PUMPS HAVING DIFFERENT CHARACTERISTICS

double-suction runners, which it is just as well to avoid. In using a straight reducer to connect the suction pipe to the pump, this reducer ought to be of the eccentric type so as to avoid the air pocket shown in Fig. 4.

When two pumps are discharging into the same main and drawing out of a common suction well, it is a good plan to have independent suction pipes rather than a common suction main. If the latter is used, however, it should be of large diameter and the suction branch to each pump arranged so the friction in each will be as nearly equal as possible. Otherwise one pump will draw water easier and consequently pump more than the other. All suction pipes should be immersed sufficiently deep to prevent air being drawn into the pump. Pump-suction pipes should not

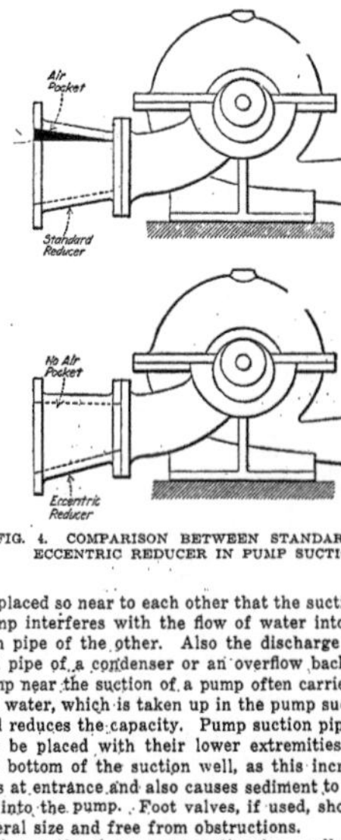

FIG. 4. COMPARISON BETWEEN STANDARD AND ECCENTRIC REDUCER IN PUMP SUCTION

be placed so near to each other that the suction of one pump interferes with the flow of water into the suction pipe of the other. Also the discharge from the tail pipe of a condenser or an overflow back into the sump near the suction of a pump often carries air into the water, which is taken up in the pump suction pipe and reduces the capacity. Pump suction pipes should not be placed with their lower extremities too near the bottom of the suction well, as this increases the loss at entrance and also causes sediment to be drawn up into the pump. Foot valves, if used, should be of liberal size and free from obstructions.

If centrifugal pumps operating in parallel seem to be performing unsatisfactorily and the precautions mentioned in regard to the discharge conditions have been observed, it would be advisable to investigate the suction-pipe layout along the lines just indicated.

Crude oil and gasoline have again advanced in price in California, according to an announcement by K. R. Kingsbury, president of the Standard Oil Company of California. The price of crude has been increased twenty-five cents per barrel for all grades and the price of gasoline has increased two cents per gallon. The Railroad Commission of California, in a letter to Mr. Kingsbury, demands an explanation of the necessity for the increase on the ground that it is officially interested in the increased prices of crude oil inasmuch as it will mean an increase of electric and gas rates in California of several million dollars.

New Kern River No. 3 Construction Work

By H. W. Dennis*

Kern River Number Three is not unlike, in its general plan, many of the other hydro-electric developments of the Southern California Edison Company. This general plan includes a comparatively low diverting dam, a conduit line consisting of flumes and tunnels, the appropriate pressure mains, and the power house with its apparatus. The project differs from some of the others, particularly the early ones, in that every part essential to continuous and uninterrupted operation has received especial attention to permit that result with minimum maintenance. It has the further difference in that the prime movers are hydraulic reaction turbines whereas most of the other plants use impulse waterwheels. It has the especial distinction that the turbines will operate under a higher head than any other reaction turbines in the world.

There is no reservoir at the intake of this plant, and therefore the dam is only so high as to accomplish the diversion of the water. Nevertheless it is a concrete structure, founded on bed rock. For about one-half of its length its height varies from 8 to 20 ft., but for the remainder the maximum height is nearly 60 ft. where it closes an old channel deeply eroded in the ledge. The intake structure has trash racks, head gates and sluice gates and, in addition to these usual features, there is a protecting wall with openings below the crest level of the dam so that the water may enter for the plant and the floating débris will be diverted over the spillway. The dam with the intake structure is finished.

Removing the Sand from the Water

All of the streams flowing from the Sierras carry large quantities of sand which must be removed before the water reaches the wheels. Without this removal the scouring effect would produce rapid deterioration of the turbines with consequent reduction of efficiency. The sand box, located between the dam and the first tunnel, because of its large dimensions, is expected to arrest all gritty particles except those that are fine enough to remain in suspension for a long time. There are two compartments so that either may, if necessary, be entirely unwatered without interrupting the flow to the plant, but generous sluicing facilities are provided for blowing out the sand without resorting to the unwatering process.

By far the major portion of the conduit is a series of tunnels whose aggregate length is nearly 60,000 ft., the longest single one being 7,132 ft. between termini. Each tunnel is connected to its neighbor by a concrete flume which is either built on a bench, in the side of the mountain, or in a cut where the cover was insufficient for tunnel construction, or in the air supported by reinforced concrete bends where it was necessary to cross certain gulches. The tunnels are lined with concrete on the sides and bottom, but are not lined for the roof except where necessary for supporting the material above. In order to secure a smooth surface for the water contact, steel forms are being used with gratifying results. For the upper nine miles of the conduit the finished size of the tunnel is 8.5 ft. wide and 8 ft. for the height of the side walls. The water will be about 7.5 ft. deep running at a slope of 2 ft. per thousand. For the remainder the width is 9.5 ft. and the slope 1.5 ft. per thousand, the depth being the same throughout. The capacity will be 600 cu.ft. per second.

A mile above the forebay the conduit line crosses Little Brush Creek at a point where the tunnel grade is 300 ft. above the creek bed. To have made this crossing at grade would have involved a long detour adding nearly a mile to length of the flow line, and accordingly an inverted siphon was decided upon. This part of the work is nearly complete. It required 470 tons of steel pipe whose diameter at the top is 9.5 ft., where the pressure is least, and 8 ft. at the bottom where the pressure is greatest.

From the forebay the water will be conducted to the power house through two steel pressure mains involving 1,260 tons of steel pipe varying in size from 7 ft. in diameter and ⅜ in. in thickness at the top where riveted pipe is used to 5 ft. in diameter and 1 in. in thickness at the bottom where lap-welded pipe is used. Expansion joints are provided at appropriate places and venturi meters are to be installed in each line. These meters measure the amount of water flowing by noting differences in pressure and velocity, although offering no obstruction to the flow and give, by comparison with the electrical output, continually a means of knowing whether any undue deterioration has occurred within the prime mover.

Details of the Generating Unit

The distinctive feature of the Kern River Three station is the generating unit itself. No other hydraulic reaction turbine in the world operates under a head as high as this, namely, 800 ft., nevertheless bountiful assurance has been given that no mistake is being made in the adoption of this type of waterwheel for these conditions. Another especial feature lies in the provision for changing the turbine runner to permit changes in speed to produce either fifty-cycle current for southern California or sixty-cycle current for the San Joaquin Valley. The frequency changer to be installed at Vestal is necessary in addition to this arrangement at Kern River Three and is far smaller than would otherwise be required.

The generating units will be two in number, rated to give 22,500 hp. each at a speed of 600 r.p.m. for sixty cycles and 500 r.p.m. for fifty cycles. Strike conditions at the works of the turbine manufacturers have caused delay in the building of the waterwheels.

The generating voltage is to be 11,000 and step-up transformers will raise this to 75,000 volts for the outgoing lines for which there are to be two circuits of 4-0 stranded copper, which will deliver the output into the rest of the system at the new substation at Vestal, in the San Joaquin Valley near Richgrove.

The facts that 25,000 tons of freight have already been delivered 50 miles over mountain roads and 2,500 men have been engaged upon the work will merely indicate the magnitude of these special problems. The length of the conduit line is more than twelve miles and most of the construction is through the hard rock of the Sierras. The assignment was given out a number of years ago, and until 1914 the work progressed slowly, consisting principally of road and trail construction and hand excavation of tunnels. An electric transmission line was built in the fall of that year to supply construction power. It is only a little more than a year ago that major activities were commenced looking to the early completion of the development.

*Construction engineer Southern California Edison Co.

EDITORIALS

The Coming National Engineers' Organization*

WHAT form is the nation-wide organization of engineers which is in the process of evolution going to take?

In the many minds which are active in the movement there is a multiplicity of ideals, with many variations as to detail but resolvable into two distinct schools.

One group conceives a broad all-inclusive engineering society to which anybody who has a legitimate connection with or interest in engineering may belong, subject, of course, to his approval and acceptance by the local unit to which he applies for membership. The local unit would bring together the engineering personnel of a community and serve as the medium for the expression and impression of the collective thought and experience of its members upon the engineering problems and activities of the neighborhood. Its meetings would deal with such subjects as would hold its members together through a common interest, unify them as an organized force for public service and emphasize the importance of the engineer as a factor in the communal life, as other professional and business organizations have done for their constituencies. The national association would be the aggregation of these local bodies. Every member of a local would be a member of the national body, but the opinions and policies of the larger body would find expression and definition in a representative convention and be carried out by an executive council and a secretarial organization. This plan does not contemplate any interference with existing organizations. The national professional societies with their more specialized and selected membership would continue to serve their respective branches of the profession after their own manner, and their local sections, as well as independent local and regional societies, clubs, etc., would preserve their autonomy, although, for activities general in character, their members would look to the all-inclusive local society.

The other school contemplates, rather, a federation of existing societies. The local organization would consist of a council or committee made up of representatives from such of the local clubs, societies and sections of the national societies as were invited to join. These committees would arrange for joint meetings and correlate the activities of the constituent bodies in matters of common interest. Representatives from the local councils or committees and of the various national and regional societies would form the federal body, of which the individuals composing the constituent societies would be members only in the sense that a member of the civil, mechanical, electrical or mining engineers is a member of the present Engineering Council.

*Comments by L. C. Marburg, chairman of the Committee on Aims and Organization of the American Society of Mechanical Engineers and representative of that society upon the Joint Conference Committee, will be found on page 559.

The one plan contemplates the bringing together of the entire engineering profession and craft of the country into a new, broad, inclusive, democratic organization. Membership in it would not confer that distinction which leads to the use of such suffixes as M. I. C. E. That would continue to be a quality of the more exclusive professional bodies. But it would bring together a very large body, numerous in inverse proportion to the stringency of the entrance requirements, for public service and for putting the engineer upon the map. The other contemplates the banding together through selected representatives, of existing bodies for purposes of interest to them all. Nobody could belong to it except as a member of one of the societies of which it would be formed, and such membership would give only a very remote delegated interest in the national body. The proponents of the first plan claim that it is the more democratic, those of the second that theirs is the more capable of realization.

The Joint Conference Committee of the four great national professional societies has been instructed to call a meeting of the various national, regional and local engineers' organizations for the purpose of effecting a general organization and such a meeting is expected to be called in Washington in June. It is to be hoped that a broad and liberal spirit will be shown in inviting representatives to the initial meeting. The plan of the Joint Conference Committee is rather of the federation than the all-inclusive type, but is, as we understand it, offered to the congress of societies which is to be called as a suggestion rather than as a finality, and it is the congress which, in its collective wisdom, will determine what form the organization shall take.

We have attempted to outline the extremes of the two types of organization. Between them there is a multiplicity of possible combinations. Perhaps, as Mr. Marburg suggests, the constitutions of the locals need not be all alike.

Engineers Who Use Turbines Should Have a Voice in Their Design

ASSUMING an average cost of three hundred dollars per kilowatt of generating capacity to cover the entire cost of a central-station system from coal pile to customer's meter, there is an investment of over two billion dollars in the United States dependent on turbines of ten thousand kilowatt capacity and larger. This amount is too large to be subject to any differences in fundamental principles between the manufacturers' engineers.

In the industrial field there are at present no turbines larger than ten thousand kilowatts in operation. It is therefore evident that the development of large sizes is of the greatest interest to public utility corporations.

The turbine designers should not be called on or permitted to assume all the responsibility for initiating new types of design, increased capacities or change in operating conditions without the help of the central station engineers and the operating men, who should take a stronger position in the matter of controlling operating conditions and the development of the turbine as to type and capacity.

There should be agreement between manufacturer and user of turbines on such matters as:

1. A uniform rating of turbines.

2. Standard steam conditions with correction factor for other values.

3. Economic size of turbine recommended for a specified service.

4. Standardization of operating methods.

5. Uniform systems of inspection.

6. Actual efficiency of representative types and sizes of turbines.

The central station men have had many years of operating experience with steam turbines and they can do much to advance the progress of the art. There are over one hundred engineering societies and associations doing standardization work to some extent on items covering every manufactured product in the United States except the steam turbine. The greatest deterrent today in the progress of the development of large steam turbines is the failure of manufacturers to exchange patent licenses, and use the salient points of each other's design. The turbine is practically the only apparatus in which this is not done. The manufacturers of large steam turbines should build a standardized combination type.

National Board of Boiler and Pressure Vessel Inspectors

THE National Board of Boiler and Pressure Vessel Inspectors, which has come quietly into existence, has potentialities which, properly developed, will make it a powerful factor in reducing the loss of life and property through boiler explosions and in unifying and standardizing boiler practice in the United States. The board is composed of the heads of the boiler-inspection departments of all the states which have adopted the American Society of Mechanical Engineers Boiler Code or into which boilers built to the Code are admitted. There are fifteen such states at present.

The opportunities for co-operation afforded by this tying together of the different state departments are many. The Boiler Code Committee lays down general rules and specifications, but does not pass upon specific designs or construction. It is the province of the state inspector to say whether a given design or appliance or practice is safe or not. The confusion which would arise from different decisions from different state inspectors will be avoided when the departments are in sufficiently close touch to consider such questions collectively and settle them alike. Under the Code a boiler must be inspected during construction and, upon completion, stamped by a qualified inspector, showing that it is in conformity with the standard. The National Board affords the machinery through which the stamp and certificate of the inspector of the state in which the boiler is built can be made valid to the inspection authorities of the distant state to which it is transported. The compilation of the records of the different state departments will give for the first time reliable statistics of the number of boilers in use, the number of explosions and serious accidents. They will indicate the most prolific causes of trouble and suggest the directions in which the Code should be revised the better to fulfill its purpose. It is easy to foresee that, as more and more of the states adopt the Code and the Board grows in numbers and influence, its convention will be one of the most important happenings of the year in boiler circles.

The Worm Commences To Turn

THE Railroad Commission of the State of California has dared to tell the president of the Standard Oil Company that the reasons assigned for its increase in the price of crude oil are inadequate, and to declare that if private control of prices is justified, it must be justified on the ground that the private control is reasonable and is not unduly or unfairly burdening to the public.

The names of the members of the Railway Commission of the State of California are: E. O. Edgerton, H. D. Loveland, Frank R. Devlin, W. Brundige and Irving Martin.

Watch what happens to them!

What would fuel oil and coal cost the user if he had to pay nothing more than the cost of mining, preparing and distributing it, plus a fair interest on the capital actually employed and a fair profit?

An increasing number of people—big people, even people who in their official capacity are supposed to hold a brief for the consumer—are commencing to ask this question and to demand an open accounting and a price based upon cost rather than upon a control of the supply and the urgency of the demand.

Uncle Sam had some oil lands of his own, but the plan of letting private "enterprise" develop them has been so successful and satisfactory (to the developers) that he has just turned over to them a good tract out of the little that he had left.

The American Library Association is planning a greatly enlarged program for the extension of library service throughout the country. The association, organized forty-four years ago, at present has a membership of four thousand active librarians. One of the important objects of enlargement is the more extensive use of technical books now in the public libraries and the installation of circulating libraries in industrial plants. It is planned to inaugurate a drive to raise two million dollars to carry on the work. The value of circulating libraries of technical literature in power plants and factories for the use of employees should not be underestimated. Not only will the employer benefit by the increased efficiency of his employees, but the men will fit themselves for better and more responsible positions. The greatest problem of the day in every industry, due in part to labor shortage and increased wages, is ways and means of securing larger production. This can be accomplished to a certain extent by increasing the efficiency of employees through educational methods of libraries at the plants. A number of far-seeing manufacturers have realized the importance of technical books dealing with their particular branch of industry and have installed special libraries in their plants.

In some quarters there is discussion as to which travels the faster—light or the price of fuel oil.

Preserve the Autonomy of Local Organizations

I have read with great interest your editorial on the coming National Engineers Convention.

Your outline of the two alternative plans is very interesting. However, allow me to call attention to the fact that as far as I know nobody proposes the extreme plan of federation which you outline. The fact is that, in some localities, engineering societies at present exist, which have individual members and who have affiliated with them the sections, associations, etc., of national societies. In such municipalities as, for example, Milwaukee, an individual engineer need not join any of the affiliated organizations, but can join the local engineering society.

In other localities, for example, in San Francisco, the plan of local organization is actually in accordance with your sketch; namely, there exist a number of sections of national societies, and I believe also local organizations that have formed a local representative council.

In still other localities, like Philadelphia, we find a local engineers club, which is a social club, but is at the same time, and really mainly, the engineering forum of the particular community. In this case the club has affiliated with it sections of national societies and in some other localities also certain local engineering organizations. In such a municipality an engineer would have the alternative of joining the club, which, of course, has fairly high dues, or joining one of the affiliated organizations.

The fundamental idea of the plan of the Joint Conference Committee is that the National Council should not concern itself with the type of organization adopted locally and that it should admit to the National Council any delegate reasonably representative of the engineers of a particular community. The whole plan of organization is based on the local units, and the local units ought to be autonomous as long as they do not encroach upon the rights of other local units. Furthermore, the local society or affiliation can select its representative on any basis it sees fit. If it decides to elect it by the vote of all the engineers making up the local engineering society or affiliation, surely the National Council, according to our plan, will have nothing to say about it; in fact, I believe the majority of the members of our committee would be much in favor of such a plan, but do not feel that the National Council should prescribe anything that would interfere with the full autonomy of the local organization. As far as the election of a representative on the National Council is concerned, the two plans sketched in your editorial are, therefore, not different, and the relations of individual members of the local organizations to the National Council can be just as close with our plan as with the other plan outlined, if the engineers of any locality choose to have it so.

In one paragraph of your editorial you speak of bringing together, by means of the other plan considered, "a very large" body of engineers. One of the reasons why we favor the plan proposed by the Joint Conference Committee is that we are firmly convinced our plan will bring together, within a reasonable time, a very much larger body of engineers than any independent society could hope to bring together.

New York City. L. C. MARBURG.

Merits of Chain and Ring Shaft Lubrication

It has been my experience that the ring method of lubricating a bearing is very efficient. I have known ring-oiled bearings that had a little new oil added only once in about two months, to run for over two years without being drained, and then the oil looked clean. One precaution that should be taken with ring-oiled bearings is to see that the rings begin to revolve when the shaft does, as sometimes the rings fail to start. I have never known the rings to stop once they started, unless the shaft stopped revolving.

I much prefer the ring to the chain method. In my experience I could not depend on the chains at all because the links would get tangled and they were liable to stop at any time. CLAY WINEBERG.

Widnoon, Pa.

Reboring Corliss-Valve Chambers

In answer to Mr. Murphy's inquiry in the Feb. 17 issue of *Power*, page 274, regarding reboring Corliss-valve chambers, I have rebored quite a number, and in every case I have paralleled the boring bar with the original counterbore recessed in each end of the valve chamber.

I frequently set my bar low centrally, but always parallel, so as to clean and true the valve seats by the time the cutting tool is just scraping the top of the valve chamber, thereby reducing the diameter of the valve to be refitted. DAVID G. YOUNG.

Hudson Heights, N. J.

A Good Water-Line Strainer

A good cheap strainer, to be used in water lines, can be made from a housing of pipe fittings. For a 2-in. water line I use a 4-in. plug *A* and tee *B*, 4-in. nipple *C*, reducer from 4 in. to 2 in. *D*, 2-in. nipple *E*, 2-in. globe valve *F*, bushing *G* from 4 to 2 in., 2-in. nipple *H* and

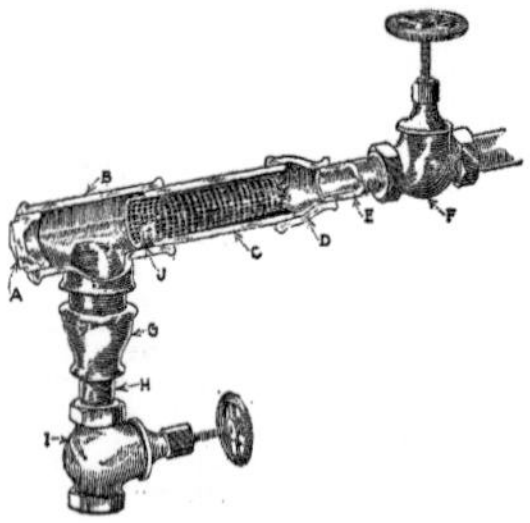

DETAILS OF STRAINER

a 2-in. globe valve *I*. The sketch of the strainer housing and connections will need no further explanation.

A few words in regard to the make-up of the strainer may be helpful to anyone interested enough to give it a trial. I first selected some galvanized screen wire, with suitable mesh for the desired results; for instance, about $\frac{3}{16}$-in. mesh for feed water going to the open heater. I cut a piece 7 x 23 in. The idea of cutting the screen wire 23 in. long is in order to give it double strength. I then rolled it up so that the top end was a little less in diameter than the 4-in. nipple at the top end, and the bottom end about $\frac{3}{4}$ in. less in diameter, so as to give the water a chance to flow through the outside of the strainer as well as out at the bottom.

With the strainer body rolled to the proper size, I tie a cord around it to hold it in position; then I make a band for the top end, about $\frac{3}{4}$ in. wide by $\frac{1}{8}$ in. thick, with holes drilled all around, about one inch apart. I place this band over the end of the strainer and copper rivet it through the mesh of the strainer, and also rivet along the side. Next, I cut a piece of the screen about $1\frac{1}{2}$ in. larger than the diameter of the bottom end of the strainer, nick the edge with a pair of snips so that it can be turned up about $\frac{3}{4}$ in. all around, insert on the top end of the strainer and rivet with five or six copper rivets.

It is necessary to have a bail for the removal of the strainer for cleaning, etc. I made the bail from a piece of 12-gage wire, the ends being hooked through two extra holels that are drilled in the band for that purpose. The bail should be long enough to reach close up to the plug. To insert or remove the strainer, just take out the plug. Notice that the strainer should pass on through the tee *B*, and into the 4-in. nipple with the edge of the band *J* snug against the end of the 4-in. nipple *C*, as shown. J. R. WEBB.

Parkersburg, W. Va.

Measuring an Indicator Diagram

A simple method of measuring an indicator diagram is to take a piece of ordinary coördinate tracing paper, which is marked off in inches, usually in tenths of an inch. This paper is laid out very accurately and can be purchased at almost any stationery store.

Lay this tracing paper on top of the indicator card in such a way that any two lines—for example, lines *A* and *B*—fall tangent to the extreme ends of the diagram, as shown in the illustration. The rest is easy. Simply follow the usual instructions for performing the more laborious method. The difference is that here we have the lines already prepared for us and accurately spaced and drawn. Better still, these lines are arranged in squares, which makes it possible to estimate areas with the eye very accurately. Thus if the lines are one-tenth of an inch apart, the area of each small square is one one-hundredth of a square inch. Hence, if a half of a small square falls within the area of the diagram, the reader can estimate that much of the area with his eye with surprising accuracy.

For example, it may be found that the left end of the diagram includes portions of two squares and that each portion is approximately one-half of a hundredth of a square inch. The total area for the ordinate therefore would be, one hundredth of a square inch. The next ordinate may include two complete squares and a half of a square, making the total area 0.025 sq.in. In this way the columns or ordinates are added from

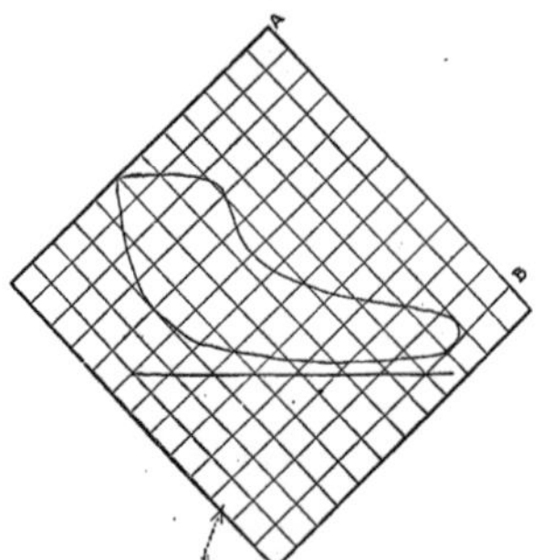

CO-ORDINATE PAPER USED TO MEASURE INDICATOR DIAGRAMS

left to right, and the sum total of them all gives the area of the diagram. It is obvious that with squares laid out in tenths of an inch instead of eighths of an inch or sixteenths of an inch, the process is considerably simplified. The best way to understand this method and fix it in one's mind is to do it; then it will not be forgotten. W. F. SCHAPHORST.

Brooklyn, N. Y.

A Freak Metering Problem

A two-story building was originally wired for a three-ire service, the upper floor being fed from one outside ire and neutral and the down floor betweep the other utside wire and neutral, as in Fig. 1. Then the service vas changed to two-wire by simply jumping the two utside wires together at the entrance and installing a wo-wire meter, Fig. 2. Still later another change was nade in the building, and two circuits, each with a

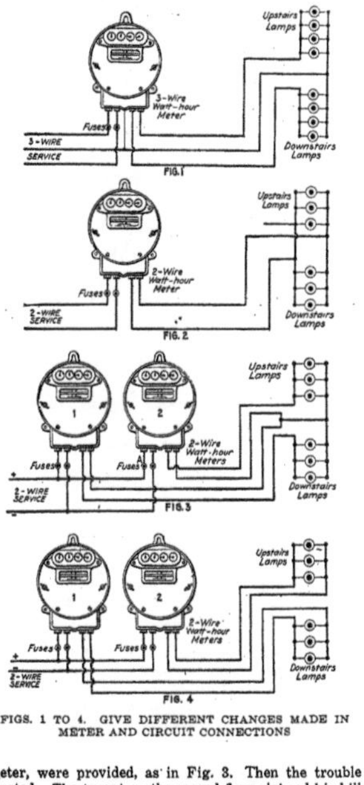

FIGS. 1 TO 4. GIVE DIFFERENT CHANGES MADE IN METER AND CIRCUIT CONNECTIONS

meter, were provided, as in Fig. 3. Then the trouble started. The tenant on the second floor claimed his bill should not be as large as that of the tenant on the first floor, since the latter had a lighting and power load, where the upstairs tenant had only lights. Another point noted was that the bills for both upstairs and downstairs were practically the same each month, although these varied from month to month. Tests showed that the meters were recording properly when tested in the usual manner. This went on until one month the tenant downstairs had a large bill and the one upstairs had none, although no complaint was made that the service was off. Suspecting that grounds were the cause, tests were made which disclosed the following results:

The wiring had never been changed except at the visible portions of the meter board, which left it three-wire and made only two circuits for the meters. The result was that the two meters were actually in parallel, and if the tenant upstairs used a greater number of lights, the downstairs tenant paid for half of it. When the fuse at *A*, Fig. 3, blew out, meter No. 2 stopped and meter No. 1 recorded the total load. Having found the cause, it was easily remedied by connecting the meters in their respective circuits, as in Fig. 4.

Orange, N. J. VICTOR H. TODD.

The Need of Competent Engineers

A good illustration of what may be expected from an incompetent engineer was to be found in an ice-making plant of which the writer took charge after it had been shut down for two or three months during the off season. The engineer had quit about the middle of the operating season. The manager, being one of the penny-wise and dollar-foolish type, placed in charge a fireman who had never been in an ice plant before that season. To finish the season's run, he operated the plant for about two months and left it in the condition pictured in the following:

After trying to put ammonia in the system and shipping back the cylinders to the ammonia dealer, the ice company got credited for more ammonia than was purchased. He had actually taken ammonia out of the system instead of putting it in. One of the boilers had been allowed to run so long without cleaning that the sheet over the fire became bagged. Instead of getting an experienced boilermaker to drive the bag up, he tried it himself with the result that the sheet was injured. The former engineer left the engine in good condition, having used an indicator to set the valves and make adjustments. They operated economically and without a pound. When the writer took charge of the plant and used the indicator, it could not be told whether the diagram was from a duplex pump or a Corliss engine and the pound could be heard for blocks. Most of the pumps in the plant were out of commission. Two-thirds of the tubes were leaking at the back ends of boilers, and in numerous places the scale between tubes was solid for two to three feet from the back end. The combustion chamber was filled with ashes level with the top of the bridge-wall. The feed-water heater was scaled so badly that its use had been discontinued and cold water had been fed into the boilers from the city mains. Conditions in general were about the same throughout the plant.

This is one plant that came under the writer's observation, and no doubt there are many others in equally bad condition for lack of competent engineers. I do not say that the man in question was altogether to blame. He probably did the best he knew, but he was incompetent. Such conditions should not exist, and any manager who will permit them should be prosecuted for criminal carelessness. It all means that Kansas, and in fact every state in the Union, should have a good license law, or better still, a national license law that would be uniform throughout the country. Then every engineer would be placed on the job he could operate successfully. JOHN DOE.

Concordia, Kan.

Balancing Steam Turbines

The vibrations that occur in machines when getting up to what is commonly called the critical speed point to possible improvements, because in spite of every care taken during manufacture these speed periods crop up, to the detriment of the machinery. The makers must consider the running balance of the machines, not only when new, but after they have been running for some time and the blading has become

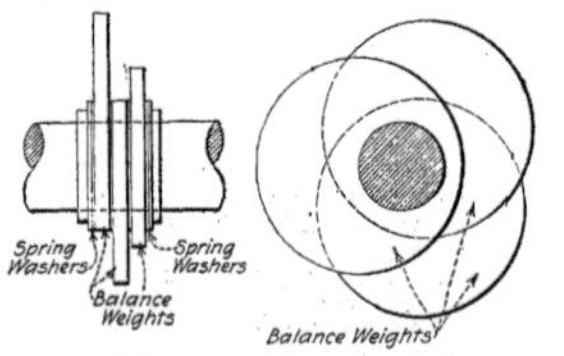

COMPENSATING BALANCING WEIGHTS

eroded, etc. It is then that the machine is likely to get more out of balance and cause trouble, unless it is rebladed, which work cannot always be done as soon as it should.

The cure seems to lie in automatic compensating balance weights that will adjust themselves when running to take care of any unbalanced forces. The illustration shows front and side elevations of such an arrangement. It will be seen that there are three eccentric-shaped weights between two collars, having spring washers. These weights should be a neat fit on the shaft and should be turned by hand without any perceptible wabble.

The spring washers are for the purpose of turning the weights around with the shaft. They must not press tight enough to prevent the weights from adjusting themselves around the shaft, nor yet be slack enough to let the weights slip and not revolve.

There is nothing new in this system of balancing, it having been applied over twenty years ago and seemingly forgotten. The action of the weights is to adjust themselves around the shaft on the lightest side, thus balancing the machine and dampening down the vibration. This system could also be applied to generators and motors, using weights made of nonmagnetic material.

Winnipeg, Canada. O. G. A. PETTERSSON.

Steam-Engine Cylinder Liners

In our central-station plant we had two 20 and 40 by 48-in. cross-compound engines each direct-connected to a 500-kw. generator. The cylinders were a 4 to 1 ratio, which was all right for running condensing, but we had to operate non-condensing and against the back pressure of a city-heating system.

It was decided to change to a 3 to 1 cylinder ratio by putting in a liner. The cylinders were bored to make sure they were true, the head end of the cylinder being bored ⅛ in. large for half the length of the cylinder to facilitate the insertion of the liner. The cylinder bore was made ten thousandths of an inch smaller than the outside diameter of the liner.

All the valve gear was removed and a charcoal fire built in the cylinder. When the cylinder had expanded sufficiently, the fire was swept out and the liner slipped in. After the cylinder had cooled, the liner was found to be tight and has remained so ever since. The engines now are 20 and 34 by 48-in. and operate against 10 lb. back pressure and carry full load.

The high-pressure cylinders of these engines, as they became worn, were rebored so many times that they finally became too thin to bore further and have

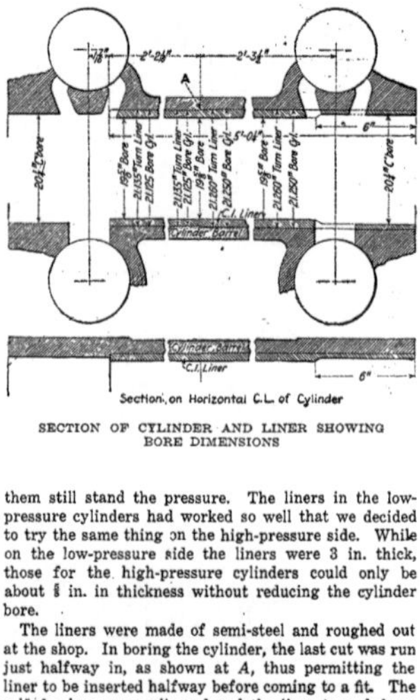

SECTION OF CYLINDER AND LINER SHOWING BORE DIMENSIONS

them still stand the pressure. The liners in the low-pressure cylinders had worked so well that we decided to try the same thing on the high-pressure side. While on the low-pressure side the liners were 3 in. thick, those for the high-pressure cylinders could only be about ⅝ in. in thickness without reducing the cylinder bore.

The liners were made of semi-steel and roughed out at the shop. In boring the cylinder, the last cut was run just halfway in, as shown at *A*, thus permitting the liner to be inserted halfway before coming to a fit. The cylinder bore was calipered and the liner turned down to tcn thousandths of an inch larger than this. The ports were laid off on the liner and a row of ¼-in. holes drilled around the edge of where the ports were to come. These pieces were left in until the liner was in place, after which they were knocked out with a chisel and hammer. The liners were installed in the same manner as already described.

This work was done six years ago. The engines have been operated from 20 to 24 hours every day and are giving entire satisfaction. The cost of putting in these liners was less than $500, whereas a new cylinder would have cost around $5,000. H. T. YUST.

Ottumwa, Iowa.

INQUIRIES OF GENERAL INTEREST

Allowances for Shrink Fits—What allowances of oversize is generally made for shrink fits? B. J. B.

In good machine shop practice the allowances made for shrink fits are about as follows:

For 1 in. dia.,	0.001 in.	For 10 in. dia.,	0.0053 in.
For 2 in. dia.,	0.0015 in.	For 12 in. dia.,	0.0058 in.
For 3 in. dia,	0.0020 in.	For 16 in. dia.,	0.007 in.
For 4 in. dia.,	0.0028 in.	For 20 in. dia.,	0.008 in.
For 6 in. dia.,	0.0035 in.	For 24 in. dia.,	0.0093 in.
For 8 in. dia.,	0.0045 in.	For 28 in. dia.,	0.0105 in.

Induction-Motor End Rings Melted—A 3-hp. 220-volt three-phase squirrel-cage induction motor driving a compressor was stalled on account of failure of the valves. On inspection the rotor bars at the end rings were found to be melted, yet the 30-ampere fuse which protected the motor was not blown. Should not this fuse have properly protected the motor when it was stopped? J. H. W.

A 3-hp. 220-volt three-phase squirrel-cage induction motor will require a full-load current of about 9 amperes per terminal. Therefore, a 30-ampere fuse would allow the motor to be overloaded about 300 per cent. To properly protect the motor, it will be necessary to install a double-throw switch, with two sets of fuses, using a 15-ampere fuse in the running side and a 30-ampere fuse in the starting side. What you have experienced with this motor is not uncommon with motors that are protected by fuses that are of sufficient capacity to take care of the starting current.

Steam from a Cubic Foot of Water—How much steam would be made from one cubic foot of water at 62 deg. F.? R. G. D.

One cubic foot of water at 62 deg. F. weighs 62.355 lb., and the volume occupied by each pound, when converted into dry saturated steam, would depend on the pressure. At atmospheric pressure one pound of the steam would occupy a volume of 26.79 cu.ft. and the volume occupied by the original cubic foot of water converted into the same weight of steam would be $62.355 \times 26.79 = 1{,}670.5$ cu.ft. At the pressure of 10 lb. gage the volume of one pound of steam is about 16.3 cu.ft. and at that pressure the cubic foot of water would make $62.355 \times 16.3 = 1{,}016.4$ cu.ft. of steam. At the pressure of 50 lb. gage the volume of one pound of steam is about 6.65 cu.ft. per pound and the original cubic foot of water would make about $62.355 \times 6.65 = 414.7$ cu.ft.; while at a pressure of 100 lb. gage steam has a volume of 3.88 cu.ft. per pound and the steam generated from 62.355 lb. of water would occupy a space of 62.355×3.88, or only about 242 cubic feet.

Equality of Metering Gas at Different Pressures—Our natural gas service must be metered and delivered to our 4-in. gas line at 9 lb. pressure, so that we may obtain 3 lb. pressure which is necessary at our point of distribution. On account of the pressure at which the gas is metered, we pay for the meter registration multiplied by 1.59. What is our loss due to the required drop in line pressure? Would it not be advantageous to have a larger gas line to decrease the drop? W. H.

It is true that for delivery of the same weight of gas per hour there would be less drop of pressure in a larger gas line, and if the metering is performed with gas at a lower pressure, a smaller multiplier should be used. But gas metered at less pressure would have less density and, in order to pass the same weight of gas, a larger volume would be metered. If the meter is correct in registration of volume and the proper multiplier is used for conversion of the volume at any metered pressure to volume at standard pressure for rating the meter charges, the resulting charge would be the same for your use of the same weight of gas or for the same volume of gas discharged at the same pressure at the point of distribution. That is, there is no loss due to drop in the line pressure and from the use of a larger pipe there would be no reduction of charge for consumption of the same amount of gas.

Combined Power Factor—Two induction motors are supplied from the same source. One of the machines takes 55 amperes and operates at a power factor of 0.80 lagging, and the other has a power factor of 0.68 lagging when taking 85 amperes from the line. How is the power factor of the combined loads determined? M. A.

One of the easiest ways of determining the power factor of a load made up of two or more loads of known power factors is by a vector diagram and measuring, with a protractor, the phase angle of the resultant current to the

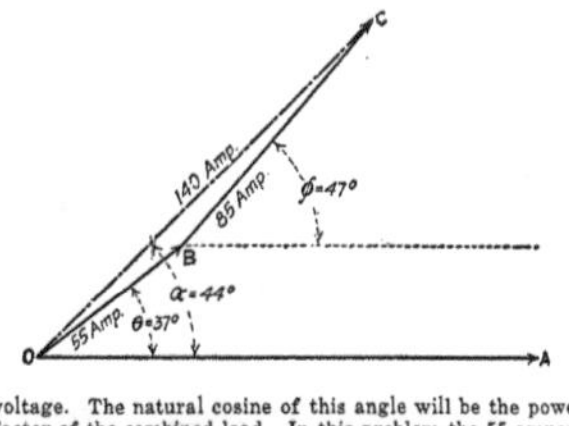

voltage. The natural cosine of this angle will be the power factor of the combined load. In this problem the 55-ampere load has a power factor of 0.80, which corresponds to the cosine of an angle of 37 deg.; consequently, OB, in the figure, which represents the 55-ampere load, is drawn at an angle of 37 deg. to OA, which represents the voltage. The power factor of the 85-ampere load is 0.68, which corresponds to the cosine of an angle of 47 deg. Then BC is drawn to scale, to represent 85 amperes, at an angle of 47 deg. to OA. The line connecting OC will represent to scale the combined current taken by the two motors, or in this case, is approximately 140 amperes, and the cosine of the angle OC makes with OA is the power factor. In this problem the angle is approximately 44 deg., the cosine of which is 0.72—the power factor of the combined load.

[Correspondents should sign their communications with full names and post office addresses.—Editor.]

Condensed-Clipping Index of Equipment

Clip, paste on 3 x 5-in. cards and file as desired

Coupling, Fast-Flexible Turbo.
Bartlett Haywood Co., Baltimore, Md.
"Power."

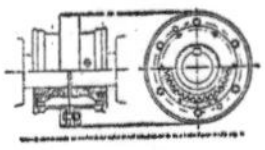

This coupling consists of two hubs, each keyed to its respective shaft. Each hub has an external spur gear cut on it at its maximum distance from the shaft end. Surrounding these hubs is a sleeve that is split in the center and bolted together by means of flanges. Each of these half-sleeves has an internal spur gear cut on its bore at the end of its flange. These gears engage the internal hub gears. A supporting ring is bolted to each half-sleeve on the face of the opposite flange. A slight clearance between the external and the internal gear permits of a certain amount of error of alignment between the two shafts. The coupling is filled with oil and when the coupling revolves the gear faces are submerged in oil.

Chain, Barlock Rivetless.
Endicott Forging and Mfg. Co., Inc., Endicott, N. Y.
"Power."

A solid-bar type detachable chain is comprised of interchangeable links and pins which are easily connected or detached at any angle without tools; the pins are reversible for double wear and have large bearing diameter. The chain consists of three parts—outside links, inside links and reversible pins. The outside links are locked to the pin. The locking links are made with a keyhole form of slots and are provided at their outer ends with risers which protect the locking shoulders. The outside links carry flight and bucket attachment. The center link is solid.

Boiler-Feeding System for Paper Mills.
Farnsworth Co., Conshohocken, Pa.
"Power," Feb. 17, 1920.

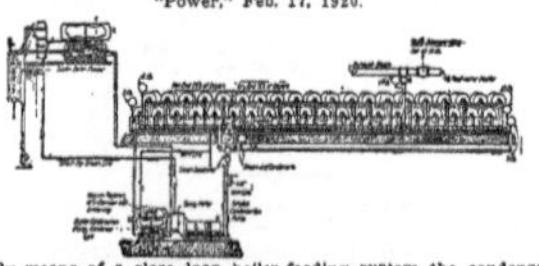

By means of a close loop boiler-feeding system the condensation from paper machines and other feeding apparatus together with makeup water is held under pressure and returned to the boiler at a high temperature. On the end of a return header for the wet-end drier, there is a duplex condenser-vacuum pump, having cold-water sprays for condensing the vapors in the return line, producing a forced steam circulation. The condensate is forced into a duplex boiler feeder and from there it goes to the boiler.

Protector, Smith Door.
International Engineering Works, Inc., Framingham, Mass.
"Power," Feb. 24, 1920

This door protector is designed to eliminate the frequent repairs by keeping the front door closed and can be attached to the ordinary hand-fired boiler fitting. The protector is made of fire-box steel with two connections to the boiler for direct and constant circulation of water through the protector and back to the boiler. The lower section is below the grate level, and a dead-plate is set in the protector at the grate line. The bottom of each protector is piped with a blowout. Foreign matter settles in a sediment chamber and is easily blown out. The protectors are made to fit door openings of boiler front for which they are to be used. Inside section is rolled up and flanged out against the outside shell, and the two are welded together.

Compressor, Portable Air
Ingersoll-Rand Co., 11 Broadway, New York City.
"Power," Nov. 25, Dec. 2-9, 1919.

This outfit is all steel, the power plant consisting of a duplex vertical high-speed compressor driven by a four-cylinder four-cycle gasoline motor. It is built in two sizes—210 cu.ft. and 118 cu.ft. capacity—weighing 6,000 lb. and 4,000 lb. respectively. Compressor cylinders are cast enbloc with cylinder heads, valve chambers and water jackets integral. Both intake and discharge valves are of the plate type; all parts are lubricated by splash from an oil reservoir in the crank case. The compressor is provided with an inlet unloading device, and both compressor and motor are water-cooled.

Meter, Improved V-Notch Water.
Hoppes Manufacturing Co., Springfield, O.
"Power."

In order to insure proper recording and integrating, the water in the vessel, which is attached to the side of the tank, must always be in equilibrium with the water in the tank. As the vessel has an upward and downward movement, the connection between the vessel and the tank must be flexible. The improvement is a flexible connection between the vessel and the tank which consists of a brass tube with a "sylphon" connection on each end. This makes a mechanically flexible connection.

Recorder, Dwight Motion.
Dwight Manufacturing Co., 12 South Jefferson St., Chicago, Ill.
"Power," Nov. 25, Dec 2-9, 1919.

This device records the travel of elevators, mine hoists or the movement of any machine having a rotary motion. It consists of an annular ring mounted on the shaft, two end disks and a counter, operated by a pin at the periphery of the ring. The inner disk is attached to the shaft and rotates with it, imparting its motion to the annular ring by means of a ball clutch between the disk and the inner surface of the ring. To obviate the difficulty of the disk going back to zero as the elevator car goes up and down, a ball clutch is provided which prevents the ring from rotating in the reverse direction, thus causing the drum revolution to be in one direction only.

Gage, Forseille Liquid Level.
L. F. Forseille, Union Gas and Electric Co., Cincinnati, Ohio.
"Power," Feb. 3, 1920.

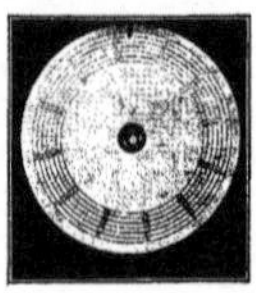

This device indicates the true level of water, oil or other liquids that are subject to variable stages or tidal changes. A float operates in a pipe that is submerged in the liquid, and by means of a cord revolves a dial with the rising or lowering of the liquid. Spiral graduations are on the front of the dial, and a spiral indicator guide is on the back, both having the same pitch. As the disk revolves, the small roller supporting the radial moving indicator keeps the point of the indicator directly over the top of the figures denoting the water level at all points within its range.

Patented Aug. 20, 1918

Balancing Fuel, Lubricant and Motor*

By WILLIAM F. PARISH

Dilution of motor-lubricating oil caused by the gasoline mixture escaping past the piston rings and by drainage of the diluted oil from the cylinder walls has always existed, but not until the period of high-end-point gasoline during the last three or four years has fuel loss become apparent. This condition has resulted in undue wear of engine parts and has affected the use of fuel, lubricant and power. A method of application of motor oils to correct some of present faults is suggested, although the true solution is better engine design.

DURING 1911, in conducting research work in connection with the use of various fuels and lubricants on a four-cylinder automotive engine and in examining the lubricating oils before and after use, it was noticed that the oil drawn from the crank case after use had undergone a considerable physical change, the most noticeable being that the viscosity of the oil had been lowered during use until it was only one-half or one-quarter as much as the viscosity of the new oil. The flash or ignition point of the used oil was only a third of the original, in some cases being lowered from 470 to 150 deg. F. Other just as remarkable changes had taken place in the gravity and cold test of the oil.

An article prepared by the author and published in June, 1912, ascribed the changed condition of the motor lubricant entirely to the leakage of gasoline vapor during the compression stroke, recondensation of this vapor in the crank case, absorption by the lubricant, and to a probable slight leakage down the walls of the cylinder, of unconsumed heavy portions of the fuel which had mingled with the lubricant, producing a similar effect. This has been a condition that has always existed, but has become more pronounced and better understood during the last two years.

The tremendous growth of the automotive industry during the past few years, with the consequent burden placed upon the oil industry of producing the vast amount of fuel required, has brought about an unbalanced condition. Because of the enormous increase in consumption of gasoline—from 15,000,000 bbl. in 1910 to approximately 95,000,000 bbl. in 1919—the petroleum industry has been able to meet demands only by the manufacture of a less volatile fuel.

There is no reason to believe that this pressure will diminish, but on the contrary there is ample evidence to show that it will continue to increase. It follows, therefore, that gas-engines must be so designed as to successfully consume this product. Navy Department specifications show that during the last few years automotive fuels have been gradually getting heavier, approximately from 360 deg. F. end point in 1916 to 428 deg. F. end point in 1919. The end point is the temperature at which the final part of the fuel vaporizes.

Practical Laboratory Tests

During the war it was necessary, before writing specifications for lubricating oils and fuels for aëronautic and other internal-combustion engines, to make a complete study of this subject in order that the fuel and lubricants could be so balanced that their combination in the engine would give the necessary lubricating results. These main tests resulted in the specification for Liberty Aëro oil, and had much to do with specifications for aviation gasolines. A four-cylinder, 100-hp. aëronautic engine was used for the special fuel tests of the series. It was operated under ideal test conditions, working with the same lubricant but with different fuels. Two of these fuels were especially made for the test and were distilled from the same crude. The first fuel tested was made after the German specifications for aviation gasoline, and had a gravity of 75.2 deg. Bé., with initial boiling point of 110 deg. F. and a final boiling point of 230 deg. F. The second fuel tested was made after the French specifications and had a gravity of 63.6 deg. Bé., and an initial boiling point of 140 deg. F., with a final boiling point of 292 deg. F. The third fuel was regular Navy specification motor-boat fuel under contract for the year 1917, and had a gravity of 61.3 deg. Bé., with an initial boiling point of 135 deg. F., and a final boiling point of 385 deg. F. To the last fuel a so-called gasoline energizer was added, which produced a fuel which, while it had an initial boiling point of 135 deg. F., had an end point of 422 deg. F., which in its last distillation range would be equivalent to the present-day commercial gasoline.

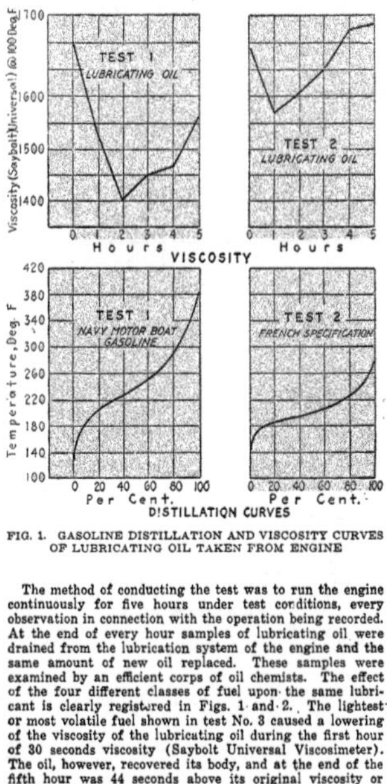

FIG. 1. GASOLINE DISTILLATION AND VISCOSITY CURVES OF LUBRICATING OIL TAKEN FROM ENGINE

The method of conducting the test was to run the engine continuously for five hours under test conditions, every observation in connection with the operation being recorded. At the end of every hour samples of lubricating oil were drained from the lubrication system of the engine and the same amount of new oil replaced. These samples were examined by an efficient corps of oil chemists. The effect of the four different classes of fuel upon the same lubricant is clearly registered in Figs. 1 and 2. The lightest or most volatile fuel shown in test No. 3 caused a lowering of the viscosity of the lubricating oil during the first hour of 30 seconds viscosity (Saybolt Universal Viscosimeter). The oil, however, recovered its body, and at the end of the fifth hour was 44 seconds above its original viscosity or body.

In test No. 2, with a slightly heavier fuel, there was a reduction in the body of the oil of 46 seconds during the first hour. The oil, however, after the first hour recovered

*Abstract of paper presented before Chicago section of the American Society of Mechanical Engineers, Jan. 28, 1920.

and had a viscosity of 28 seconds above its original body at the end of the fifth hour.

Test No. 1 with a semi-commercial fuel, showed a drop in the body of the oil in two hours of 150 seconds, with a slight recovery during the balance of the test. There was a total reduction of 16 seconds at the end of five hours from the body of the original oil.

In test No. 4 with the high-end-point gasoline, there was a drop in two hours of 142 seconds in the body of the lubricant, only a slight recovery and a total drop at the end of five hours of 108 seconds from the original body of the oil.

The sample taken at the end of the five-hour period of each one of these tests was tested in a distillation flask and the amount of heavy end of the gasoline recovered varied from 0.5 to 1.2 per cent, as indicated on Fig. 3. All the physical characteristics of the oil, such as gravity, flash, fire, viscosity, pour test, acidity and carbon content, showed a considerable change, as indicated in Fig. 4.

The results of the tests indicate quite clearly that dilution exists with even the most volatile fuel, but only during certain periods of the operation of the engine and under certain fixed conditions. With the lightest volatile fuel, which would also be the case with the engine operating on producer, natural or furnace gas, the tendency of the lubricating oil would be to build up in body, the same as in

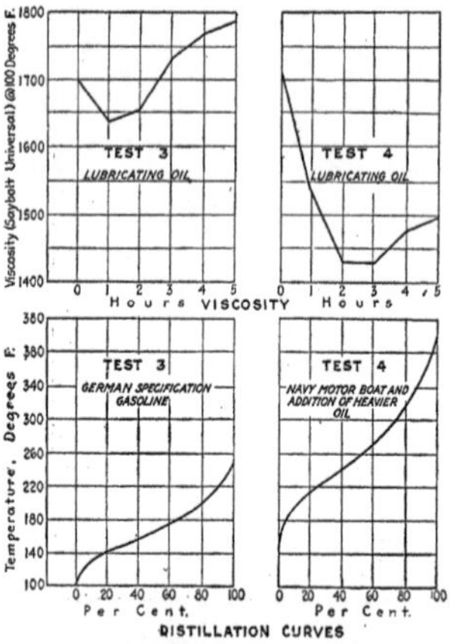

FIG. 2. DISTILLATION AND VISCOSITY CURVES WHEN USING GERMAN SPECIFICATION AND NAVY GASOLINE

any other class of lubrication work where the lubricant does not come in contact with lighter hydrocarbons or lighter volatile products that can be combined with it. The tests also indicated quite clearly that the heavier the end point of the fuel the greater is the effect of dilution, also the more permanent is its character.

In the foregoing experiments, it must be borne in mind that the aëronautic engine was operated with an exceedingly heavy lubricating oil, such as these engines require, and that carburetor and magneto adjustments were attended to by experts; that the engine was placed in perfect condition for the test and operated continuously without stopping, at high speed, for the entire five hours of the test.

Tests on Trucks

Other interesting data were secured from the examination of lubrication conditions of a considerable number of cars and trucks using commercial gasoline and lubricants. The cars, in some cases, were new as received from the manufacturer, and in other cases old, and old repaired.

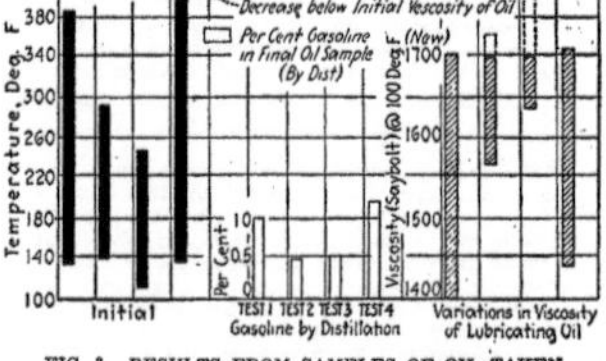

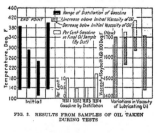

FIG. 3. RESULTS FROM SAMPLES OF OIL TAKEN DURING TESTS

The method of conducting the test was first to thoroughly clean the motor, then fill the crank case with the oil to be tested. The oils so used covered a range of commercial medium and heavy grades as indicated by Government specifications for motor oils. The cars were then allowed to operate in their regular business for periods of ten days during which time no particular care was given them other than ordinarily given commercial vehicles, but at the end of each two-day period an eight-ounce sample would be drawn from each crank case and eight ounces of the same oil used on that particular motor would be put in with whatever additional oil was necessary. The samples were examined for dilution by measuring all the product that could be distilled from the sample of oil under 500 deg. F. The viscosity of the sample before removal of the diluent was taken on a Saybolt Universal Viscosimeter at 100 deg. F. The method of making the ash test was to test the oil first in the Conradson carbon apparatus. After the carbon was determined, a gas flame was used for burning the carbon out of the dish. The ash that remained was then weighed and percentage taken of the whole sample. This ash was made up of iron, iron oxide and silica.

In a 3½-ton new truck the viscosity at the end of ten days was reduced from 345 to 100 seconds. Dilution on the tenth day amounted to 19 per cent and ash, 1.62 per cent.

In a 5-ton truck that had run 5,205 miles, but completely overhauled, the viscosity dropped from 358 to 90 seconds during the first six days, then recovered to 184 seconds at the end of ten days. Dilution in the case of this truck seemed to be dependent upon weather conditions; at the end of the six-day period there was 20.6 per cent of dilution, which was reduced to 8 per cent at the end of the ten-day period.

In a 3½-ton overhauled truck that had been run 12,551 miles the viscosity held practically constant for the first two days when the truck was not operating in the open. After the second day there was a drop in viscosity to 43 seconds. Dilution at the end of the tenth day was 26.5 per cent.

The Cause of Dilution

With the present-day fuel and the present-day automotive engine, all lubricants change considerably in body within a very short operating period, this new body being the result of dilution and the dilution coming from the heavy end of the fuel.

The amount of dilution is to a great extent influenced

by the temperature of the surrounding air, the reason being that the colder the engine and oil in the engine the greater will be the amount of fuel it will absorb and hold. Conversely, the hotter the air or lubricating oil in the engine the less will be the amount of fuel held in dilution, as all the fuel can be distilled off at high temperature.

To see if any of the dilution of a motor oil could be caused by decomposition or cracking of the oil by being thrown against hot surfaces, tests were made by the author with a commercial car operated by pure benzol. Percentages of 4 to 8 per cent of benzol were distilled from the motor oil after 100-mile runs. The oil, after the benzol had been removed, possessed all its original characteristics. These benzol tests determined satisfactorily that dilution was being caused by the fuel, though the burned odor of the oil did indicate some degree of decomposition.

The conclusions drawn from the foregoing test with benzol are confirmed by the results of a test on a Nash six-cylinder engine operated for ten hours with city gas. The original viscosity of the oil at 100 deg. F. was 205 seconds; after ten hours' operation the viscosity had increased to 255 seconds, the flash and fire points remaining practically the same, showing that there had been no change in the oil caused by cracking and reabsorption of the lighter products, but that the absence of lighter products allowed the oil to become heavier in body.

Dilution is caused mainly by leakages past the piston rings during the compression stroke. There is a secondary leakage, caused during the suction stroke, where there is a partial separation of heavy ends of the fuel from the gas entering the cylinder and absorption by the film of oil on the cylinder walls, causing some drainage on the lower exposed surfaces during the compression stroke. It is doubtful if any drainage or leakage takes place during the exhaust stroke or during the power stroke.

Leakage of the gasoline or gasoline mixture is influenced by the following mechanical or operation conditions: Piston clearance, richness of mixture, improper ignition, operating temperature of motor affected by weather, body of lubricant or thickness of film on walls of cylinders, condition of cylinder walls, condition of piston rings as affecting their all-around fit in the grooves and against the cylinder walls, intermittent operation, priming and using the choker when engine is cold in starting, choking when overloading, character of fuel, carburetor out of adjustment, vacuum system out of adjustment.

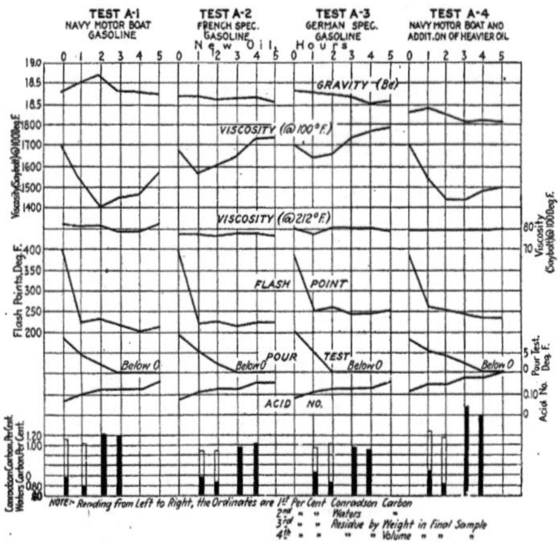

FIG. 4. CHART OF OIL ANALYSIS AND CARBON CONTENT

Losses caused by dilution are of four kinds; namely, the loss of fuel, loss of power, wear of all parts and loss of lubricant. Loss of fuel is not entirely due to the heavy end of the fuel or to the present heavy character of the fuel, as, irrespective of the nature of the fuel, whether light and volatile or heavy, there will be a fuel loss during the compression stroke in a four-stroke cycle engine. In the case of heavy end fuel this loss is reflected to a magnified extent in the dilution effect upon the lubricating oil. In the case of a much lighter fuel, say one of 72 deg. Bé., the loss by the rings during the compression stroke would be just as pronounced in percentage of the amount taken into the cylinder, supposing the sealing effect of the lubricant on the piston rings to be always the same, but the effect would not be so noticeable, owing to the lubricant's inability to absorb and hold the lighter fuel as a diluent at operating temperatures.

The quantity of fuel lost in this manner is in no way represented by the amount of dilution found in motor oils, as there is a limited point of saturation of a motor oil dependent upon the condition under which it is operating. The losses that constantly occur are disposed of by the engine in some other way than by being entirely absorbed by the lubricant. The loss of fuel by leakage means that all the fuel placed in the cylinder is not converted into power.

Loss of and Through the Lubricant

The first effect of dilution is upon the viscosity or body of the lubricating oil, reducing this body to a point where it cannot be considered to be an efficient lubricant for a motor. This causes three conditions, which can be stated as loss of lubricant, loss of power and wear of the machine.

The first effect is the wastefulness of the lubricant, as the entire loss of lubricating oil in an internal-combustion engine is caused by the lubricant creeping up the cylinder walls, mainly during the first cycle or admission stroke, and being later consumed in the combustion chamber. The lighter the oil is in body the greater will be this loss. Further, there is another loss, which is due to the present condition of dilution and the attempts made to offset this, and that is by the removal of lubricating oil from the engine at periods of from 300 to 800 miles instead of the former periods of double this mileage, this mixture being thrown away.

The basic rule in lubrication is illustrated best by considering the power necessary to operate a machine composed of a number of bearings. If the machine is lubricated perfectly with oil of the proper viscosity, the power will be at a minimum and speed at a maximum, and there should be practically no wear on the bearings. If, however, an oil of double or treble the viscosity is used as a lubricant, the power necessary to move the machine at the same speed will have to be materially increased due entirely to the fluid friction of the lubricant itself. Under this condition there will be practically no wear, as the surfaces will be kept well apart. Attempting to lubricate the same machine with an oil that is entirely too light or thin for the work will also require the development of practically as much or even more power than in the case of the extra-heavy oil, the resistance being due to metallic or solid friction, one of the results of which is, of course, abrasion with constant changing of the surfaces.

A lubricant actually working in a motor should have a body equal at least to that possessed by a new heavy spindle oil (130 seconds viscosity at 100 deg. F.), and an oil of such light body must be used in a forced-feed system where the pressure and constant flow of the oil will make up for its lack of thickness of lubricating film.

As a matter of fact, under present conditions of operation, most motor oils when in an engine, in the winter especially, are much lighter than a lubrication engineer would ever think of applying to the lightest piece of machinery he would be called upon to lubricate, which is a light spindle in a textile mill. These spindles rotate about 10,000 r.p.m. and the side pull or pressure averages two pounds. The pressure on the bearing side will not be above ten pounds per square inch. The high rotation speed of the spindle tends to keep it away from the bearing surface and keep it in the center of the bolster. About one ounce of oil is placed in the base of the spindle and this amount is added to every 15 to 45 days. It has been proved by the most careful experiments that oil that has a viscosity of less than 60 seconds at 100 deg. F., Saybolt Universal, will allow these spindles to wear perceptibly in a month's time. The oil remaining in the spindle base will become black and contain a considerable proportion of iron and iron oxide. Further, the same spindles, when lubricated with oil of from 70 to 80 seconds viscosity, do not wear, owing to the fact that the oil is of sufficient body to keep the surfaces apart.

Therefore, the result of the lubrication of an automotive engine with a mixture of lubricating oil and heavy ends of the gasoline having a viscosity below that required for a textile spindle can only result in excessive wear, and this is shown by the ash remaining from the complete distillation of the sample of mixture taken from the motor.

Wear of Machine

The effect of poor lubrication is indicated to some extent by the readings of ash. These readings were taken on each sample and represent the total amount of ash in that particular sample. The ash is made up of oxide of iron, other metal, and of silica, the iron and other metal being from natural or unnatural wear of the engine, the oxide of iron coming from the water that has been in contact with the metal, and the silica coming from road dust, though at the time of the year this particular test was made the streets were well covered with snow and but little road dust could be expected to enter the engine through the carburetor or breather holes.

The wear of the machine is reflected in repair bills, in the necessity for reboring cylinders, refitting rings, when an engine becomes noisy and when compression is lost.

All these features are largely the result of wear caused by the present unbalanced condition of fuel, lubricant and engine.

It is necessary to reconsider the entire problem of the internal-combustion engine on the basis of present-day fuel and its effect upon the lubricant. There is but little more that the oil industry can do in regard to the fuel situation. Considerable can be done in regard to the application of motor oil; that is, the selection of motor oil for these engines upon some other basis than has generally been prevalent in the past whereby a motor of any one manufacturer was supposed to be properly suited by a certain grade of lubricant for its entire life irrespective of its physical condition. Modern lubrication engineering can, to a considerable extent, rectify some of the damage done by this old system of selecting lubricants, basing the selection upon more scientific or practical grounds, taking into consideration all the factors that surround the problem. The entire problem of lubricant, fuel and engine conservation can be solved only by the automotive engineer or designing mechanical engineer providing means and methods whereby the fuel will be entirely consumed on top of the pistons. Effecting compression of the air and gasoline outside of the cylinder and igniting the charge upon admission and during the power stroke would overcome practically the entire amount of fuel and air-mixture leakage that takes place during the compression stroke of the present four-cycle motor. Other methods will undoubtedly be proposed and worked out successfully as soon as the seriousness of the present situation is fully established.

Economizer Explosions

Failure to close dampers while the feed is open according to statistics taken from a recent report of the Board of Trade of the Manchester Steam Users' Association, England, has resulted in more economizer explosions than any other single condition.

The report shows that since 1882 seventeen serious economizer explosions have taken place in England. Of this number nine were destructive to surrounding property. It is claimed that four of the explosions were probably caused by neglecting to close the dampers while the feed was open; three were reported due to wear and tear and the remainder were accounted for in various ways. Many firemen do not know that it is necessary to close the damper when the feed is shut off, and even among those who have been warned there is a prevalent disregard of this precaution.

Failures of economizer pipes occur almost daily, says the report, the following explanation being offered for this large number of pipe failures compared to the small number of serious explosions: Under ordinary working conditions economizers are full of water which is cold at the bottom and hot at the top; if a pipe fails at the bottom, cold water runs out and the water level in the boiler sinks or water can be seen running out of the flue; and if a pipe fails in an economizer through which hot flue gases have been passing, while the feed supply has been shut off, the whole mass of water in the economizer will have been raised to a temperature equal to that of the water in the boiler, and as hot water under pressure contains about thirty times its volume of steam, it is evident that if one pipe of an economizer were to burst, the large volume of high-pressure steam suddenly liberated would cause an explosion. This might be prevented if the damper was closed each time the feed was shut off.

The effect of passing hot gases through an economizer when not in use is uncertain. The temperature of flue gases is about 600 deg. F., or approximately equal to that of superheated steam. Superheated steam can safely be passed through cast-iron pipes into the cast-iron cylinders of engines that are subject to reversals of stress, without failure occurring. It was a common practice in the early days of electric power, in order to increase boiler power, to shut off the feed during the peak-load period, while allowing the gases to pass through the economizers, converting them temporarily into steam producers. This increased the boiler power approximately 30 per cent.

Investigations of explosion of economizers have resulted in a number of interesting theories being advanced as to whether the side pressure suddenly liberated by the bursting of a pipe would be sufficient to fracture the other pipes. One assumption is that when a pipe accidentally explodes, the flying fragments would fracture at least some of the remaining pipes. Mathematicians by a number of experiments have shown that rods offer practically no resistance to pressure waves. An actual demonstration of this was given at the Silverton explosion, mentioned in the Manchester report, when all walls and buildings were destroyed and every one of the tall chimneys was left standing.

A simple experiment seems to settle the question about gas-flue explosions and at the same time furnishes proof that it is entirely a matter of chance whether a pipe explosion will confine itself to one pipe or a sheet. A model economizer made of glass pipes was used in the test. It was estimated that the pipes should break under a side pressure of fourteen pounds per square inch. The central tubes were replaced by an air reservoir whose end was closed with a celluloid disk. When the pressure had reached about 300 lb., the disk burst with a loud report, but no pipes were broken. The test was repeated with broken pipe ends piled up over the celluloid, and this time some of the vertical pipes were broken. This indicated that side pressure alone without flying fragments will not injure the other pipes.

Water Power of British Columbia

Information of value to the whole western coast of America is contained in a report issued under the title of "Water Powers of British Columbia," by the Canadian Commission of Conservation. The report is a large volume, well illustrated, containing 650 pages and has taken a number of years to prepare. It is a complete compendium of data relating to the water-power resources of British Columbia. In round numbers, the horsepower totals derivable from the various estimates presented in the power-site tables for districts into which the province has conventionally been divided are as follows:

	24-hour hp.
I. Columbia River and tributaries (north of the international boundary)—This comprises the portion of the province lying between its eastern boundary and the water-shed of the Fraser River	610,000
II. Fraser River and tributaries—This includes practically the entire area of the great interior plateau	740,000
III. Vancouver island	270,000
IV. Mainland Pacific coast and adjacent islands (except Vancouver Island)—This includes all the rivers north of the Fraser which drain into the Pacific	630,000
V. Mackenzie River tributaries—(A rough estimate made for inclusion in this summary)	250,000
Grand total	2,500,000

Statements made show that there is about 500,000 horsepower that, at least for many years to come, can hardly be economically developed because interference with the fishing industry or the proximity of railways renders development impracticable. There are also some other qualifying statements applicable to portions of the power tables, but in round figures the grand total of estimated 24-hour horsepower, including an allowance for the entities mentioned, may be placed at about 3,000,000.

Frequently, in reports on water-power resources it has been the tendency to deal with power development exclusively without adequately considering such related subjects as domestic and municipal supply, agriculture and irrigation, navigation, fisheries and riparian rights. In this report the author, although dealing with water-power resources, has recognized that water power is but one of the important uses to which inland waters may be applied.

National Board of Boiler and Pressure Vessel Inspectors Organized

As a result of efforts of the members of the National Uniform Boiler Law Society, the National Board of Boiler and Pressure Vessel Inspectors was organized at the headquarters of the American Society of Mechanical Engineers, New York City. The purpose of the organization is to promote uniform boiler laws throughout the jurisdiction of the members; for the interchange of rulings, opinions and the approval of specific designs and devices used in connection with boilers and other vessels; for the uniform stamping of boilers and also for the examination of boiler inspectors.

The membership is restricted to officials charged with the enforcement of regulations covering boilers and other pressure vessels in states and cities that have adopted the Boiler Code formulated by the American Society of Mechanical Engineers.

Public utilities throughout the country consume annually about 35,000,000 tons of coal. This is taken from figures recently compiled and published by the Geological Survey, showing the amount of electricity produced in this country by central power stations and the fuels. The report of the Survey also shows that from 722,000 bbl. of petroleum and derivatives burned each month to generate power, the use of oil has increased to 1,276,000. In some states a small amount of natural gas is used to generate electric power. Last year the consumption per month throughout the country was about 1,800,000 cubic feet.

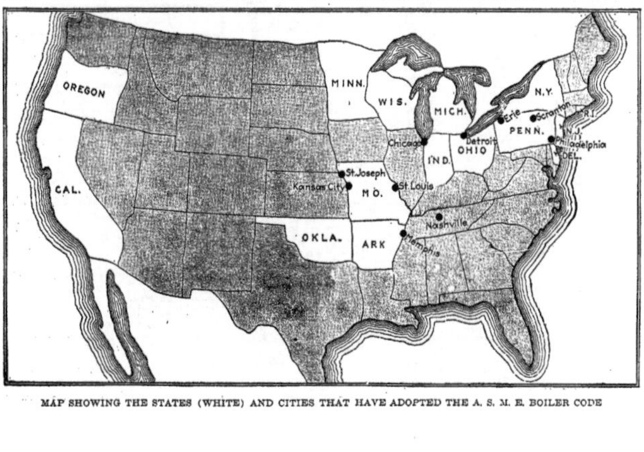

MAP SHOWING THE STATES (WHITE) AND CITIES THAT HAVE ADOPTED THE A. S. M. E. BOILER CODE

Dinner of the Boston Engineers

The eleventh annual dinner of the Boston Engineers was held on Tuesday evening, March 30, under the auspices of the American Institute of Electrical Engineers, the Boston Society of Civil Engineers and the American Society of Mechanical Engineers, the following societies participating: American Chemical Society, American Institute of Mining and Metallurgical Engineers, American Society of Civil Engineers, Engineers' Club, Illuminating Engineering Society, New England Water Works Association, Plant Engineers' Club, the American Society of Heating and Ventilating Engineers and the Boston Society of Architects.

The principal speakers and their subjects were: Paul D. Cravath, legal representative in Europe of the United States Treasury Department during the war, "Some Economic Aspects of the Treaty of Paris"; His Excellency, Calvin Coolidge, governor of the Commonwealth of Massachusetts; Hon. Andrew J. Peters, mayor of the City of Boston; Roscoe Pound, dean of Harvard Law School, "Social Engineering." Brief remarks were made by Robert Spurr Weston, vice-president, Boston S. C. E.; J. Parke Channing, chairman, Engineering Council; Arthur P. Davis, president, A. S. C. E.; Sidney J. Jennings, past-president, A. I. M. and M. E.; Fred J. Miller, president, A. S. M. E.; Calvert Townley, president, A. I. E. E.

The dinner this year was under the especial charge of the Electrical Engineers, and Irving E. Moultrop, representing that society, acted as chairman. Prof. Dugald C. Jackson was toastmaster. A resolution was passed inviting the local members of the American Institute of Mining and Metallurgical Engineers to co-operate in the arrangements for future dinners, completing the representation of the "four founder societies" in that function. The speeches were more than ordinarily good, and the attendance taxed the large auditorium of the Boston City Club.

New Publications

THE MECHANICAL WORLD YEAR BOOK FOR 1920. Published by Emmott & Co., Ltd., Manchester, England. 6 x 9 in.; 328 pages.

The present edition of this reference book has been further enlarged. Among the new features are a section on hydraulic work and another on Heating and Evaporating Liquids. A great deal of information and data are given in a small space and the selection of the material has been made with the idea of giving practical assistance in mechanical engineering practice. In addition to the various tables of squares, cubes, areas of circles, etc., which are always found in engineering handbooks, a large number of tables of engineering data and constants are included. The tables of the properties of steam are from values now considered obsolete, which materially detracts from the value of the book. As is natural, the majority of the data on standards are those of English practice rather than American, although a great deal of the material in the book applies equally well in either country.

SELLING YOUR SERVICES. By George Conover Pearson. Published by the Jordon-Goodwin Corporation, New York City. Cloth; 237 pages; 5 x 8 in. Price, $2.00.

This book treats in a practical way, by a number of lectures, methods of job-getting which the author has put into actual practice with success. The main idea of the edition is that the application of salesmanship and advertisement to the problem of getting the position for which a man is most suited is much more likely of results than the customary hit-or-miss method. It gives in concrete form, systematic methods of self analysis in order to properly locate the field of usefulness for which a man is best fitted and also gives a complete and comprehensive description of obstacles which must be overcome and the methods for doing so. This covers competition, selling your usefulness, the purpose of long and short letters of application and suggestions on writing them, capitalizing experience, etc. The book concludes with an account of the value of a carefully planned campaign, the danger of overselling services and methods of investigation of prospective employers.

Personals

Paul R. Duffey, formerly supervisor of power plants of the Potomac District U. S. R. R. A., has been appointed chief inspector of motive power, Maintenance of Equipment Department, Western Maryland, R. R. Mr. Duffey will also be supervising officer of all power plants on the Western Maryland System and subsidiary lines.

Ernest H. Peabody, after thirty years of service with the Babcock & Wilcox Co., has resigned in order to assume charge of the Peabody Engineering Corp., of 160 Broadway, New York City. Mr. Peabody entered the employ of the Babcock & Wilcox Co. in 1891 and was with them continuously until the time of his resignation. He is a graduate of Stevens Institute of Technology, class of 1890, with the degree of mechanical engineer.

Frank Espey, formerly manager of the Atlantic City Electric Co., has been promoted to the position of general manager of the Ohio Power Co., of Canton, Ohio. Mr. Espey will be replaced by C. E. Terrey, for ten years manager of the East Liverpool plant. Mr. Terrey's post at East Liverpool will be filled by Harry C. Sterling, who for more than five years has been assistant manager of the Atlantic City Electric Co's. plant.

Society Affairs

The American Society of Mechanical Engineers, Waterbury Branch, has planned a meeting for Apr. 7.

The American Society of Mechanical Engineers, New Orleans Section, will meet on Apr. 15. The prime object of the meeting will be a general discussion of section affairs.

The Engineering Council of the United Engineering Societies will hold a meeting in Washington, D. C., on Apr. 15. It is planned to hold one meeting each year in Washington.

The National Electric Light Association, Pacific Coast Section, has gained more than two thousand members since last November according to a report of the membership committee made Apr. 1.

The Hartford, Conn., Branch of the American Society of Mechanical Engineers will meet at the City Club, Hartford, Apr. 12. The subject to be presented at the meeting will be "Drop Forgings" and "Die Castings."

The Chicago Section of the A. S. M. E. will hold a joint meeting with the Western Societies of Engineers in April. Prof. B. W. Benedict, of the University of Illinois, will read a paper on "Teaching the Principles of Shop Management."

The Entertainment Committee of the National Electric Light Association convention, which is to be held at Pasadena, Cal., in May, have arranged so that delegates will receive invitations to attend the California Mission Play. The play will take place in a theater but a short distance from the Hotel Huntington, the headquarters of the gathering.

The Institute of Radio Engineers, of the College of the City of New York, will meet at the Engineering Societies Building, New York, on Apr. 7, at 8:15 o'clock. A paper on "The Reduction of Atmospheric Disturbances in Radio Reception," by Dr. Louis W. Austin, head of the United States Naval Radio Telegraphic Laboratory, Washington, D. C., will be presented.

Cornell Society of Engineers has called a meeting of Cornell graduates in engineering to be held on the evening of April 9, in the Engineering Societies Building, New York City. The meeting is called, partly with the idea of forming a general Cornell engineering society, now that the different engineering schools at Cornell are to be combined into one general engineering college. The new dean of the combined college, Prof. Dexter S. Kimball, will be one of the speakers.

The American Association of Engineers, Pittsburgh Chapter, has started a drive to bring the next annual convention to Pittsburgh. Delegates to the 1920 convention, which is to meet in St. Louis in May, will work in behalf of the movement. Another plan which the Pittsburgh chapter has under consideration is the effective cooperation of members of national, state and local societies so that members of any of the societies who are in good standing can join the Pittsburgh chapter of the A. A. E. for one-half the usual entrance fee.

The American Institute of Electrical Engineers will hold its annual meeting jointly with the American Electrochemical Society on Apr. 9 in Boston. A morning session will be held at the Plaza Hotel, the subject for the morning to be "Electricity Produced Alloys." In the afternoon a tour through the Lynn, Mass., works of the General Electric Co. will be made by members of both societies. The second session will be held during the evening, and a number of papers on "Power for Electrical Chemical Purposes" will be read.

Business Items

The Fitzgibbons Boiler Co. has removed its offices from 17 Battery Place to 1181 Broadway, New York City.

The Brown Instrument Co., of Philadelphia, is erecting two new buildings, one for the manufacture of recording therometers and the other to be used as a research department. The total cost of the new plants is estimated at $100,000.

The Barber-Greene Co., of Aurora, Ill., has completed the organization of branch offices at St. Louis, R. E. Faulke, manager; Pittsburgh, J. A. Gourney, manager; Philadelphia, F. S. Sawyer, manager; and Indianapolis, W. T. McDonald, manager.

The Metal & Thermit Corp., New York City, N. Y., has appointed James G. McCarthy, manager of its Canadian branch, with headquarters in Toronto, and has transferred Robert L. Browne from the New York Office to Boston. Mr. Browne is to have charge of sales in New England States.

The General Electric Co., has completed negotiations to lease, with the option to buy, the forty-acre war plant of the Remington Arms Co., at Bridgeport, Conn. The buildings, which are of the latest type construction, cover about 900,000 sq.ft. and the floor area is over 1,555,257 sq.ft. The deal involves about $7,000,000.

The Dallas Power and Light Co., of Texas, at a meeting of the directors, elected the following officers: J. F. Strickland, president; Harry L. Seay, W. B. Head, C. E. Calder and Charles W. Davis, vice presidents; J. B. Walker, secretary and treasurer; C. E. McBride, C. L. Cox and J. C. Thompson, assistants to the secretary and treasurer.

THE COAL MARKET

BOSTON—Current prices per gross ton f.o.b. New York loading ports:

Anthracite

	Company Coal
Egg	$8.20@$8.65
Stove	8 45@ 9.05
Chestnut	8.55@ 9.05
Pea	7.05@ 7.40
Buckwheat	3 60@ 3.80
Rice	2.30@ 2.50
Barley	1.25

Bituminous

	Clearfields	Cambrias and Somersets
F.o.b. mines, net tons	$2.85@$3 35	$3.15@$3.60
F.o.b. Philadelphia, gross tons	5.05@ 5.60	5.35@ 5.90
F.o.b. New York, gross tons	5 40@ 5.95	5.75@ 6.25
Alongside Boston (water coal), gross tons	7 00@ 7 75	7.60@ 8.00

Pocahontas and New River are practically off the market for coastwise shipment, but are quoted at $6.25@$7.00.

NEW YORK—Current quotations, White Ash, per gross tons, f.o.b. Tidewater, at the lower ports are as follows:

Anthracite

	Company Coal
Broken	$7.80@$8.25
Egg	8 20@ 8.65
Stove	8 45@ 9.05
Chestnut	8.55@ 9.05
Pea	7.05@ 7.40
Buckwheat	5.15
Rice	4.50
Barley	4 85
Boiler	4.85

Bituminous

Government prices at mines:	Spot
South Fork (best)	$3 25@$3.50
Cambria (best)	3.00@ 3.25
Cambria (ordinary)	2.60@ 2.90
Clearfield (best)	3.00@ 3.25
Clearfield (ordinary)	2 60@ 2.90
Reynoldsville	2 85@ 2.90
Quemahoning	3 25@ 3.50
Somerset (medium)	3.00@ 3.25
Somerset (poor)	2.50@ 2.75
Western Maryland	2.50@ 2.75
Fairmont	2.25@ 2.50
Latrobe	2.60@ 2.90
Greensburg	2 75@ 3.00
Westmoreland ¾ in.	3.40@ 3.50
Westmoreland run-of-mine	2.75@ 3.00

PHILADELPHIA—Bituminous coal prices vary according to district from which they are mined. For ordinary slack the price is $2.45@$2.55; lump, $3.35, at the mines.

BUFFALO—

Anthracite

	On Cars, Gross Ton	At Curb, Net Ton
Grate	$8.55	$10.80
Egg	8.80	10.65
Stove	9.00	10.85
Chestnut	9.10	10.85
Pea	7.75	9.30
Buckwheat	5.70	7.75

Bituminous

Allegheny Valley	$4.80
Pittsburgh	4.65
No. 8 Lump	4.65
Mine-run	4.80
Slack	4.10
Smokeless	4 60
Pennsylvania Smithing	5.70

CLEVELAND—Prices of coal per net ton delivered in Cleveland are:

Anthracite

Egg	$12.25@$12.40
Chestnut	12.50@ 12.70
Grate	12 25@ 12.40
Stove	12.40@ 12.60

Pocahontas

Nine-run	$7.50

Domestic Bituminous

West Virginia split	$8.50
No. 8 Pittsburgh	$6 60@ 6.90
Massillon lump	8.25@ 8.50
Coshocton lump	7.15

Steam Coal

No. 6 slack	$5.25@$5.50
No. 8 slack	5.10@ 5.50
Youghiogheny slack	5.25@ 5.50
No. 8 ¾ in.	5.70@ 6.60
No. 6 mine-run	5.25@ 5.50
No. 8 mine-run	5.75

Only coal available is mine-run Pocahontas.

MIDDLE WEST—Chicago quotations, f. o. b. cars at mine:

	Springfield, Carterville, Williamson, Franklin, Saline, Harrisburg	Fulton, Peoria	Grundy, La Salle, Bureau, Will
Lump	$2.55@$2.70	$2.95@$3.10	$3.25@$3.40
Washed	2.75@ 2.90		3.45@ 3.60
Mine-run	2.35@ 2.50	2.75@ 2.90	3.00@ 3.15
Screen'gs	2.05@ 2.20	2.35@ 2.50	2.75@ 2.90

New Construction

PROPOSED WORK

Me., Clark Mills (Hollis Center P. O.)—The Clark Power House Co. is having plans prepared for the construction of a 1 story power house. Sawyer & Bean, 11 Lisbon St., Lewiston, Engrs.

Conn., Ansonia—The Ansonia O. & C. Co., 153 Main St., plans to construct a 1 story power house.

N. Y., Albany—Charles F. Rattigan, Supt. of Prisons, Capitol, Albany, received bids for furnishing labor and material for the heating work at Sing Sing Prison, Ossining, from A. B. Barr & Co., 6 River St, Yonkers, $92,936; John W. Danforth Co. 70 Ellicott Sq., Buffalo, $96,900; W. B. Armstrong Co., 3 Fulton St., Albany, $98,000. Noted March 3.

N. Y., Black River—The St. Regis Paper Co. is preparing plans for the construction of a power house and hydro-electric development of the Black River to furnish 9,000 hp. F. L. Carlisle, Watertown, Pres.

N. Y., Brooklyn—Levy Bros., c/o Bedford Theatre, Bedford Ave., are having plans prepared for the construction of a theatre on New Utrecht Ave. and 51st St. A steam heating system will be installed in same. Total estimated cost, $250,000. T. W. Lamb, 644 8th Ave., New York City, Archt. and Engr.

N. Y., Brooklyn—E. Riegelmann, Boro Pres., rejected bids for furnishing, delivering and installing 1 steam engine. Estimated cost, $12,000. Work will be re-advertised.

N. Y., Flushing—Wilmer & Vincent, 1451 B'way, New York City, are having plans prepared for the construction of a 1 story theatre and store building. A steam heating system will be installed in same. Total estimated cost, $300,000. E. C. Horn & Son, 1476 B'way, New York City, Archts. and Engrs.

N. Y., New York—F. B. Jennings, c/o S. N. Pollis, Archt. and Engr., 50 Church St., is having plans prepared for the construction of a 16 story office building at 7-11 Water St. A steam heating system will be installed in same.

N. Y., New York—The New York Edison Co., Irving Pl. and 15th St., plans to construct a 113 x 177 ft. transformer station on Inwood Ave. A steam heating system will be installed in same. Total estimated cost, $100,000. W. Whitehill, 32 Union Sq., Archt. and Engr.

N. Y., New York—The West 48th St. Realty Co., c/o Gronenberg & Leuchtag, Archts. and Engrs., 303 5th Ave., is having plans prepared for the construction of a 14 story hotel on 8th Ave. and 43rd St. A steam heating system will be installed in same. Total estimated cost, $1,500,000. H. Claman, Pres.

N. Y., Ogdensburg—The Bd. of Pub. Wks. will receive bids for the installation of a 50 hp. electric motor and air compressor for use at the city crushing plant. Total estimated cost, $1,500.

N. Y., Old Forge—The Special Committee plans to build a municipal lighting and water supply plant. Estimated cost between $150,000 and $200,000. Address Maurice Callahan, 1st Natl. Bank, Pres.

N. J., East Orange—Edward V. Warren, Archt., Essex Bldg., Newark, is preparing plans for the construction of a 5 story, 70 x 140 ft. hotel on Harrison St. A steam heating system will be installed in same. Total estimated cost, $500,000. Owner's name withheld.

N. J., Jersey City—The City Comn. will receive bids for the construction of a power house for the Jersey City Hospital Building. Mechanical equipment will be installed in same. Total estimated cost, $50,000.

N. J., Newark—The Bd. Educ., Essex Bldg., received bids for installing a steam heating system in the proposed 3 story, 70 x 86 ft. Franklin School addition, from John H. Cooney, 210 North 4th St., $38,467; Jachnig & Peoples, 221 13th Ave., $43,467.

N. J., Newark—The Newark Athletic Club, 207 Market St., is having plans prepared for the construction of a club house at 16-18 Park Pl. A steam heating system will be installed in same. Total estimated cost, $500,000. Jordan Green, 27 Clinton Ave., Archt.

N. J., South Orange (Orange P. O.)—Henry Rosensohn, Archt., 188 Market St., will receive bids for the construction of a 3 story, 75 x 150 ft. theatre and office building on South Orange Ave. and Scotland Rd., for the South Orange Theatre Co., 929 Broad St., Newark. A steam heating system will be installed in same. Total estimated cost, $100,000.

N. J., Trenton—The Pocono Rubber Cloth Co., East State St., plans to build a 1 story factory including a boiler and power house, a 43 x 45 ft engine room, etc. Total estimated cost, $50,000. Karno Smith Co., Broad St. Bank Bldg., Engrs.

Pa., Philadelphia—The Dept. of Pub. Wks., City Hall, will soon award the contract for furnishing, installing and delivering traveling grate stokers, for nine 327 hp. Heine boilers at the Torresdale pumping station and for six 500 hp. Edgemore boilers and one 3 motor, 10 ton capacity electric traveling crane, etc. at the Lardners Point pumping station, for the Bureau of Water Supply, 788 City Hall. Four driving engines and electrical apparatus including motors wound for 220 volt d.c., controllers, etc, will be installed in same.

Pa., Swissvale (Pittsburgh P. O.)—The School Bd. will soon award the contract for the installation of electric light and power systems, etc., in the high school here. Total estimated cost, $35,000. Howard Lloyd, Archt.

Md., Elkton—The Bd. of Comrs. is having plans prepared for the construction of an electric power plant, etc. Total estimated cost, $40,000. W. H. Mackall, Pres L. J. Houston, Fredericksburg, Engr.

S. C., Greenville—Mowbray & Eiffinger, Archts. and Engrs., 56 Liberty St., New York City, will receive bids until April 15 for the construction of a 16 story bank and office building for the Farmers' & Merchants' Bank. A steam heating system will be installed in same. Total estimated cost, $1,000,000.

O., Ashland—The city received bids for the installation of 2 pumps in the proposed 1 story filtration plant from the Dravo Doyle Co., Diamond Bank Bldg., Pittsburgh, Pa., at $5,056.

O., Marysville—The Marysville Light & Water Co., 114 West 5th St., plans to install a new boiler and turbine at their plant, here. Mayne Macken, Supt.

O., Springfield—The city plans to elevate switch track at waterworks and install coal handling machinery and boilers. Total estimated cost, $50,000. O. E. Carr, City Mgr.

Mich., Bad Axe—N. J. Frost and others plan to construct a 1 story electric power plant. Steam driven generating units will be installed in same.

Mich., Detroit—The Blodgett Eng. & Tool Co., Kerr Bldg., is having plans prepared for the construction of a 4 story factory for the manufacture of special machinery and jigs on 14th and Dalzelle Sts. A steam heating system will be installed in same. Alvin E. Harley, 435 Woodward Ave., Archt.

Mich., Detroit—The I. O. O. F. Association, c/o G. D. Mason, Archt., 80 Griswold St., is having plans prepared for the construction of a 2 story lodge house on Pulford and Concord Aves. A steam heating boiler and equipment will be installed in same. Total estimated cost, $150,000.

Mich., Flint—The Knights of Columbus, c/o P. Callahan, Chn., plan to construct a 3 story, 80 x 108 ft. club house. A steam heating and mechanical ventilating system will be installed in same. Total estimated cost, $250,000. Van Leyen, Schilling, Keough & Reynolds, Union Trust Bldg., Detroit, Archts. and Engrs.

Mich., Grand Rapids—The J. J. Wernette Eng. Co., 441 Hanseman Bldg., is in the market for a 100 hp. steam boiler and a 200 hp. boiler.

Mich., Lansing—The city will soon receive bids for the installation of air lift pumping equipment to have a daily capacity of 3½ million gal. at the Pennsylvania Ave. pumping station. Total estimated cost, $55,000. Alvord & Burdick, Hartford Bldg., Chicago, Engrs.

Ill., Chicago—The North Shore Baptist Church, c/o S. M. Seator, Archt., 35 North Dearborn St., is having plans prepared for the construction of a 2 story, 110 x 125 ft. church on Lakewood and Berwyn Aves. A steam heating system will be installed in same. Total estimated cost, $300,000.

Ill., Chicago—R. C. Osterzren, Archt., 153 North Clark St., will soon award the contract for the construction of a 1 and 2 story, 150 x 460 ft. factory on Washtenaw and Belmont Aves., for the Alemite Die Casting & Mfg. Co., 341 West Chicago Ave. A steam heating system will be installed in same. Total estimated cost, $250,000.

Ill., Chicago—Sears Roebuck Co., Arthington St. and Homan Ave., is having plans prepared for the constructoin of a 9 story, 160 x 300 ft store building. A steam heating system will be installed in same. Total estimated cost, $2,000,000. G. C. Nimmons & Co., 120 South Michigan Ave., Archt.

Wis., Kenosha—The United Motor Transport Lines, 211 Wisconsin St., are having plans prepared for the construction of a 1 and 2 story, 100 x 120 ft. storage and ice manufacturing plant on Sheridan Rd. White, White & White, Public Service Bldg., Archts.

Wis., Marshfield—The Roddis Lumber & Veneer Co. is having plans prepared for the construction of a 1 story power house addition. Boilers and a new generator unit will be installed in same. Cahill & Douglas, 217 West Water St., Milwaukee, Engrs.

Wis., Racine—The Fox Ice Co., Liberty St., plans to construct a 50 ton capacity ice manufacturing plant on Liberty St.

Minn., St. Paul—A. H. Helmbach, 467 Roy St., is having plans prepared for the construction of a 6 story, 100 x 134 ft. hotel on Snelling and Sherburne Aves. A steam heating system will be installed in same. Total estimated cost, $500,000. M. A. Wright, 600 Pittsburg Bldg., Archt.

N. D., New Rockford—C. L. Pillsbury Co., Engr., Metropolitan Life Bldg., Minneapolis, Minn., is receiving bids for furnishing and installing a 75 kw. generator and a 225 r.p.m. engine at the power plant for the General Utilities Corp.

Mo., St. Joseph—The St. Joseph Oil Co., southwest corner of 4th and Duncan Sts., is having plans prepared for the construction of a 2 story fuel oil plant including power house, etc. Two 100 hp. boilers will be installed in same. Total estimated cost, $200,000.

Tex., Clarendon—The Clarendon Light & Power Co. plans to construct an electric lighting and power plant.

Okla., Ponca City—The Municipal Water & Light Dept. is in the market for a 1,000 hp. oil engine. Philip R. Dunton, Supt.

Cal., Big Creek—The Southern California Edison Co., Edison Bldg., plans to construct a power plant on Big Creek to be known as Big Creek No. 8, which will be below the present Big Creek No. 2 plant and will utilize water formerly intended to pass through Big Creek No. 3. Big Creek No. 8 will have a capacity of 22,500 kw. and during the year 1921 is expected to generate 100,000,000 kw.-hr.

Cal., Long Beach—The Seaside Investment Co., c/o A. B. Benton, Archt., 114 North Spring St., Los Angeles, retained architect to revise plans for the construction of a 4 story, 250 x 340 ft. bath house and hotel along the ocean front. A boiler plant, etc., will be installed in same.

Cal., Los Angeles—The University Club, Consolidated Realty Bldg., engaged Allison & Allison, Archts., 1405 Hibernian Bldg., and E. L. Mayberry, Engr., 468 Pacific Electric Bldg., to prepare plans for the construction of an 8 story, 65 x 135 ft. club house with a 1 story, 65 x 135 ft. garage adjoining. A steam heating system will be installed in same. Total estimated cost, $400,000.

Cal., San Bernardino—The Southern California Ice Co. is having plans prepared for the construction of a 2 story, 40 x 80 ft. refrigeration plant. H. E. Jones, Katz Bldg., Archt.

Cal., Willows—The School Trustees are having plans prepared for the construction of a 2 story school building. A steam heating system will be installed in same. Bonds amounting to $123,000 have been voted upon for this project.

Cal., Yuba City — Reclamation Dist. 1660, will soon receive bids for the construction of pumping plants, etc. Total estimated cost, $1,800,000. Address P. H. McGrath, Fruit Bldg., Sacramento, Secy. J. C. Boyd, 803 Forum Bldg., Sacramento, Engr.

M. I., Guam—The Bureau of Yards & Docks, Navy Dept., Wash., D. C., will receive bids until April 7, for the construction of a 30 x 38 ft. extension to power house at Agana and a 28 x 35 ft. extension at the Naval Radio Station, a 12 x 14 ft. extension to pumphouse, 8 x 16 ft. extension to blower house and a 22 x 47 ft. with an 18 x 24 ft. wing extension to quarters, including plumbing and electric lighting systems. Noted March 30.

CONTRACTS AWARDED

Mass., Boston—The T. S. Proctor Estate, 85 Water St., has awarded the contract for the reconstruction of a 4 story, 100 x 240 ft. office building at 177 State St., to the Norman Clarke Co., 161 Devonshire St. A steam heating system will be installed in same. Total estimated cost, $150,000.

Mass., Fitchburg—The Crocker Burbank Co., 545 Westminster St., has awarded the contract for the construction of a 1 story factory on the mill grounds, to the Aberthaw Constr. Co., 27 School St., Boston. Plans include the extension of present power system. Total estimated cost, $300,000.

Mass., Lawrence—The Amer. Woolen Co., 225 4th Ave., New York City, has awarded the contract for the construction of a 1 story power house near North Canal St., to the E. W. Pitman Co., Bay State Bldg. Estimated cost, $20,000.

R. I., Providence—The Narragansett Electric Lighting Co., Turks Head Bldg., will build a 5 story, 75 x 95 ft. power house addition on South St. Total estimated cost, $700,000. Work will be done by day labor. Noted March 30.

Conn., Greenwich—The Bd. Educ. has awarded the contract for the construction of a 2 story, 62 x 148 x 52 x 104 ft. grade school, to the Rangeley Constr. Co., 56 West 39th St., New York City. A steam heating system will be installed in same. Total estimated cost, $250,000.

N. Y., Brooklyn—The Masons of New York, Ridgewood Lodge, will build a 3 story, 100 x 100 ft. temple on B'way and Gates Ave. A steam heating system will be installed in same. Total estimated cost, $1,000,000. Work will be done by day labor.

N. Y., Brooklyn—The Tillary Constr. Co., 44 Court St., will build a 6 story, 50 x 200 ft. factory on Pearl and Tillary Sts. A steam heating system will be installed in same. Total estimated cost, $300,000. Work will be done by day labor.

N. Y., Buffalo—Loew's Enterprises, 1492 B'way, New York City, have awarded the contract for the construction of a theatre on Washington and Mohawk Sts., to the Fleischmann Constr. Co., 531 7th Ave., New York City. A steam heating system will be installed in same. Total estimated cost, $400,000.

N. Y., Elmsford—The Knollwood Co. Club, c/o C. I. Birg, Archt. and Engr., 331 Madison Ave., New York City, has awarded the contract for the construction of a 2 story, 100 x 100 ft. club, to Charles Money, 228 West 36th St., New York City. A steam heating system will be installed in same. Total estimated cost, $100,000.

N. Y., New York—The Barrett Co., 17 Battery Pl., has awarded the contract for the construction of a 16 story, 101 x 122 x 180 ft. office building on West and Rector Sts., to George A. Fuller Co., 175 5th Ave. A steam heating system will be installed in same. Warren & Wetmore, 16 East 47th St., Archts. and Engrs.

N. Y., New York—The Bd. Educ., 500 Park Ave., has awarded the contract for the construction of a school building on St. Anns Ave., to Frymier & Hanna, 25 West 45th St. A steam heating system will be installed in same. Total estimated cost, $800,000.

N. Y., Pyrites—The De Grasse Paper Co. has awarded the contract for the construction of a 1 story, 15 x 60 ft. addition to transformer house, to Sylvester Nicolette, Potsdam. Estimated cost, $18,000.

N. J., Elizabeth—The Bd. Educ. has awarded the contract for installing a steam heating system in the proposed school on Bayway and South Broad Sts., to E. G. Woolfoh at $53,160.

N. J., Mays Landing—The Mays Landing Water & Power Co. has awarded the contract for the construction of a 1 story, 40 x 40 ft. power house, to the Aberthaw Constr. Co., 27 School St., Boston, Mass. Estimated cost, $50,000.

N. J., Newark—The Bd. Educ., Essex Bldg., has awarded the contract for installing a steam heating system in the proposed 3 story, 38 x 144 ft. school on Hunterdon and Peshine Aves., to Jachnig & Peoples, 221 13th Ave.

Pa., Torresdale—The Precision Grinding Wheel Co. has awarded the contract for the construction of a 1 story, 90 x 265 ft. factory, to the John R. Wiggins Co., Otis Bldg., Philadelphia. A steam heating system will be installed in same. Total estimated cost, $100,000.

N. C., Winston-Salem — D. Omstead, Cleveland, O., has awarded the contract for the construction of a 9 story hotel on 5th and Cherry Sts., to H. L. Stevens Co., 30 North Michigan Ave., Chicago. A steam heating system will be installed in same. Total estimated cost, $1,000,000.

O., Cleveland—The city has awarded the contract for the installation of two 125 hp. boilers, to the Union Iron Wks., Rockfr Bldg., at $13,120.

O., Cleveland—The Film Exch. Bldg. Co., c/o G. T. Bauder, Williamson Bldg., has awarded the contract for the construction of an 8 story, 135 x 143 ft. film exchange building on East 21st St. and Payne Ave., to the William Dunbar Co., 8201 Cedar Ave. A steam heating system will be installed in same. Total estimated cost, $1,000,000.

Mich., Detroit—The North Woodward Methodist Church, c/o W. E. N. Hunter, Archt., Chamber of Commerce, has awarded the contract for the construction of a 2 story, 180 x 250 ft. church on Woodward and Chandler Aves., to Bryant & Detwiler, 2335 Dime Bank Bldg. A steam heating system will be installed in same. Total estimated cost, $700,000.

Turn Now to Pages 98 to 110

PRICES - MATERIALS - SUPPLIES

Electrical prices on following page are prices to the power plant by jobbers in the larger buying centers east of the Mississippi. Elsewhere the prices will be modified by increased freight charges and by local conditions.

POWER-PLANT SUPPLIES

HOSE—

	Fire	50-Ft. Lengths
Underwriters' 2½-in.		78c. per ft.
Common, 2½-in.		35%

Air

	First Grade	Second Grade	Third Grade
¾-in. per ft.	$0 55	$0.35	$0.25

Steam—Discounts from List

First grade...... 25% Second grade...... 35% Third grade.... 40%

RUBBER BELTING—The following discounts from list apply to transmission rubber and duck belting:

Competition.............. 50–10% Best grade.................. 25%
Standard.................. 40%

LEATHER BELTING—Present discounts from list in the following cities are as follows:

	Medium Grade	Heavy Grade
New York	20%	10–5%
St. Louis	40%	35%
Chicago	45%	40%
Birmingham	35%	30%
Denver	10%	5%

RAWHIDE LACING {For cut, best grade, 25%, 2nd grade, 30%. For laces in sides, best, 81%.

PACKING—Prices per pound:

Rubber and duck for low-pressure steam	$1.00
Asbestos for high-pressure steam	1 70
Duck and rubber for piston packing	1.00
Flax, regular	1 20
Flax, waterproofed	1.70
Compressed asbestos sheet	.90
Wire insertion asbestos sheet	1.50
Rubber sheet	.50
Rubber sheet, wire insertion	.70
Rubber sheet, duck insertion	.50
Rubber sheet, cloth insertion	.30
Asbestos packing, twisted or braided and graphited, for valve stems and stuffing boxes	1.30
Asbestos wick, ½- and 1-lb. balls	.85

PIPE AND BOILER COVERING—Below are part of standard lists, with discounts.

PIPE COVERING		BLOCKS AND SHEETS	
Pipe Size	Standard List Per Lin.Ft.	Thickness	Price per Sq.Ft.
1-in.	$0.27	½-in.	$0.27
2-in.	.36	1-in.	.30
6-in.	.80	1½-in.	.45
4-in.	.60	2-in.	.60
3-in.	.45	2½-in.	.75
8-in.	1.10	3-in.	.90
10-in.	1.30	3½-in.	1.05

85% magnesia high pressure .... List

For low-pressure heating and return lines { 4-ply.......... 50% off; 3-ply.......... 52% off; 2-ply.......... 54% off

GREASES—Prices are as follows in the following cities in cents per pound for barrel lots:

	Cincinnati	Chicago	St. Louis	Birmingham	Denver
Cup	8½	6.6	6.7	8½	10.5
Fiber or sponge	8½	8.6	13	8½	13.75
Transmission	10	8.1	13	8½	12.5
Axle	...	4.8	4.75	4½	7.0
Gear	6½	6.1	7.5	8½	8.5
Car journal	12(gal.)	4 7	4.7	4½	9.0

COTTON WASTE—The following prices are in cents per pound:

	New York Current	New York One Year Ago	Cleveland	Chicago
White	13.00	13.00	16 00	11.00 to 14 00
Colored mixed	9.00 to 12.00	9 00 to 12 00	12 00	9 50 to 12 00

WIPING CLOTHS—Jobbers' price per 1000 is as follows:

	13½ x 13½	13½ x 20½
Cleveland	$52.00	$58.00
Chicago	41.00	43.50

LINSEED OIL—These prices are per gallon:

	New York Current	New York One Year Ago	Chicago Current	Chicago One Year Ago
Raw in barrels (5 bbl. lots)	$1 87	$1.55	$2.05	$1.66
5-gal. cans	1.87*	1.70	2.30	1.86

*To this *oil* price must be added the cost of the cans (returnable), which is $2.25 for a case of six.

WHITE AND RED LEAD—Base price per pound:

	Red Current Dry	Red Current In Oil	Red 1 Year Ago Dry	Red 1 Year Ago In Oil	White Current Dry and In Oil	White 1 Yr. Ago Dry and In Oil
100-lb. keg	15 50	17 00	13 00	14 50	15 50	13 00
25- and 50-lb. kegs	15 75	17.25	13 25	14 75	15 75	13.25
12½-lb. keg	16 00	17.50	13.50	15 00	16 00	13 50
1-lb. cans	18 50	20.00	15 00	16 50	15.00	15.00
5-lb. cans	20 50	22 00	16 00	17 50	16.00	16.00

500 lb. lots less 10% discount; 2000 lb. lots less 10–2½%.

RIVETS—The following quotations are allowed for fair-sized orders from warehouse:

	New York	Cleveland	Chicago
Steel 7/16 and smaller	30%	45%	45%
Tinned	30%	45%	45%

Boiler rivets, ¾, ⅞, 1 in. diameter by 2 in. to 5 in. sell as follows per 100 lb.:

New York....$6 00 Cleveland...$4.00 Chicago...$5.37 Pittsburgh...$4.72

Structual rivets, same sizes:

New York....$6.10 Cleveland...$4.10 Chicago...$5.47 Pittsburgh...$4 82

REFRACTORIES—Following prices are f. o. b. works, Pittsburgh district:

Chrome brick	net ton	$75–80	at Chester, Penn.
Chrome cement	net ton	45–50	at Chester, Penn
Clay brick, 1st quality fireclay	net M.	45–50	at Clearfield, Penn
Clay brick, 2nd quality	net M.	40–45	at Clearfield, Penn
Magnesite, dead burned	net ton	50–55	at Chester, Penn.
Magnesite brick, 9 x 4½ x 2½ in.	net ton	80–85	at Chester, Penn.
Silica brick	net M.	50–55	at Mt. Union, Penn.

Standard size fire brick, 9 x 4½ x 2½ in. The second quality is $4 to $5 cheaper per 1000.

St. Louis—Fire Clay, $35 to $50.

Birmingham—Silica, $51–50 @ 55; fire clay, $45 @ 50; magnesite, $85; chrome, $85.

Chicago—Second quality, $25 per ton.

Denver—Silica, $18: fire clay, $12: magnesite, $57.50

BABBITT METAL—Warehouse prices in cents per pound:

	New York Current	New York One Year Ago	Cleveland Current	Cleveland One Year Ago	Chicago Current	Chicago One Year Ago
Best grade	90 00	87 00	71 00	80 00	70.00	75.00
Commercial	50 00	42 00	21 00	15 00	15.00	15.00

SWEDISH (NORWAY) IRON—The average price per 100 lb., in ton lots, is:

	Current	One Year Ago
New York	$21–26	25 50–30 00
Cleveland	20 00	20 00
Chicago	16.50	16 50

In coils an advance of 50c. usually is charged

Domestic iron (Swedish analysis) is selling at 8c. per lb.

SHEETS—Quotations are in cents per pound in various cities from warehouse; also the base quotations from mill:

Blue Annealed	Large Mill Lots Pittsburgh	St. Louis	Chicago	San Francisco	New York Current	New York One Yr. Ago
No. 10	$3.55@$5.95	$6.02	$6.02	$7.50	$7 00@$8 00	$4 57
No. 12	3.60@ 6.05	6.07	6 07	7.55	7.05@ 8.05	4 62
No. 14	3.65@ 6.10	6.12	6 12	7.60	7.10@ 8.10	4 67
No. 16	3.75@ 6.20	6.17	6.17	7 70	7.20@ 8.20	4.77
Black						
*Nos. 18 and 20	4.15@ 6.30	6 80	6 80	7 85	7.80@ 8 80	5 42
*Nos. 22 and 24	4.20@ 6.35	6.85	6.85	7.90	7.85@ 8.85	5 47
*No. 26	4 25@ 6 40	6.90	6 90	7.95	7.90@ 8.90	5.52
*No. 28	4.35@ 6.50	7.00	7.00	8.05	8.00@ 9.00	5.62
Galvanized						
No. 10	4 70@ 7.50	7.50	7.50	8.60	8 50@10 00	5 97
No. 12	4 80@ 7.60	7.60	7.60	8.60	8 60@10 10	6.02
No. 14	4 80@ 7.60	7.60	7 60	8.60	8.60@10.10	6.07
Nos. 18 and 20	5.10@ 7 90	7.90	7.90	8 90	8.90@10.40	6.37
Nos. 22 and 24	5.25@ 8 05	8 05	8 05	9 05	9 05@10.55	6 52
*No. 26	5.40@ 8 20	8 20	8 20	9 20	9 20@10.70	6.67
*No. 28	5.70@ 8.50	8.50	8.50	9.50	9.50@11.00	6 97

*For painted corrugated sheets add 30c. per 1000 lb. for 5 to 28 gage; 25c. for 19 to 24 gages; for galvanized corrugated sheets add 15c., all gages.

PIPE—The following discounts are for carload lots f. o. b. Pittsburgh; basing card of Jan. 1, 1919, for steel pipe and for iron pipe:

BUTT WELD

Steel Inches	Steel Black	Steel Galvanized	Iron Inches	Iron Black	Iron Galvanized
⅛, ¼, and ⅜	50½%	24%	¼ to 1½	30½%	23½%
½	54½%	40%			
¾ to 3	57½%	44%			

LAP WELD

2	50½%	35%	2	32½%	18½%
2½ to 6	53½%	41%	2½ to 6	34½%	21½%

BUTT WELD, EXTRA STRONG PLAIN ENDS

⅛, ¼ and ⅜	48½%	29%	¾ to 1½	39½%	24½%
½	51½%	39%			
¾ to 1½	50½%	43%			

LAP WELD, EXTRA STRONG PLAIN ENDS

2	48%	37%	2	33½%	20½%
2½ to 4	51%	49%	2½ to 4	35½%	23½%
4½ to 6	50%	39%	4½ to 6	34½%	22½%

Stock discounts in cities named are as follows:

	New York Black	New York Galvanized	Cleveland Black	Cleveland Galvanized	Chicago Black	Chicago Galvanized
⅜ to 3 in. steel butt welded	40%	24%	40%	31%	54%	40½%
2½ to 3 in. steel lap welded	35%	20%	42%	27%	50%	37½%

Malleable fittings. Class B and C, from New York stock sell at list + 23%. Cast iron, standard sizes, net.

BOILER TUBES—The following are the prices for carload lots, f. o. b. Pittsburgh:

Lap Welded Steel		Charcoal Iron	
3½ to 4½ in.	40½	3½ to 4½ in.	15½
2½ to 3¼ in.	30½	3 to 3¼ in.	1
2¼ in.	24	2½ to 2¾ in.	+1
1¾ to 2 in.	19½	2 to 2¼ in.	+16
		1¾ to 1⅞ in.	+29

Standard Commercial Seamless—Cold Drawn or Hot Rolled

	Per Net Ton		Per Net Ton
1 in.	$327	1⅞ in.	$207
1¼ in.	267	2 to 2¼ in.	177
1½ in.	257	2½ to 3¼ in.	167
1¾ in.	207	4 in.	187
		4½ to 5 in.	207

These prices do not apply to special specifications for locomotive tubes nor to special specifications for tubes for the Navy Department, which will be subject to special negotiations.

ELECTRICAL SUPPLIES

ARMORED CABLE—

B. & S. Size	Two Cond. M Ft.	Three Cond. M Ft.	Two Cond. Lead M Ft.	Three Cond. Lead M Ft.
No. 14 solid	$104.00	$138.00	$164.00	$210.00
No. 12 solid	135.00	170.00	225.00	265.00
No. 10 solid	185.00	235.00	275.00	325.00
No. 8 stranded	285.00	375.00	520.00	500.00
No. 6 stranded	400.00	500.00	560.00	......

From the above lists discounts are:

Less than coil lots	Net List
Coils to 1,000 ft	10%
1,000 ft. and over	15%

BATTERIES, DRY—Regular No. 6 size red seal, Columbia, or Ever-Ready:

	Each, Net
Less than 12	$0.45
12 to 50	.38
50 to 125 (bbl.)	.35
125 (bbl.) or over	.32

CONDUITS, ELBOWS AND COUPLINGS—These prices are f. o. b. New York with 10-day discount of 5 per cent.

	Conduit		Elbows		Couplings	
Size, In.	Black 2,500 to 5,000 Lbs.	Galvanized 2,500 to 5,000 Lbs.	Black 2,500 to 5,000 Lbs.	Galvanized 2,500 to 5,000 Lbs.	Black 2,500 to 5,000 Lbs.	Galvanized 2,500 to 5,000 Lbs.
½	$80.50	$85.59	$19.13	$20.27	$7.04	$7.46
¾	106.51	113.51	25.17	26.67	10.07	10.67
1	157.59	167.79	37.25	39.47	13.09	13.37
1¼	213.11	227.01	48.46	50.26	18.30	18.98
1½	254.93	271.43	64.62	67.02	22.61	24.45
2	342.99	365.19	118.47	122.87	30.15	31.27
2½	542.30	377.40	193.86	200.00	43.08	44.68
3	709.16	755.06	511.96	536.16	64.62	67.02
3½	871.24	935.64	1,141.62	1,184.02	86.16	89.36
4	1,064.93	1,130.33	1,319.32	1,368.32	107.70	111.70
4½	1,240.79	1,318.99	1,997.83	2,072.03	161.55	167.55
5	1,445.96	1,534.76	2,773.27	2,866.27	177.70	184.30
6	1,875.84	1,991.04	3,446.40	3,574.40	258.48	268.08

Standard lengths rigid, 10 ft. Standard lengths flexible, ⅜ in., 100 ft. Standard lengths flexible, ½ to 2 in., 50 ft.

CONDUIT NON-METALLIC, LOOM—

Size I. D., In.	Feet per Coil	List, Ft.	
5/16	250	$0.05½	
⅜	250	.06	
½	250	.09	
⅝	200	.12	
¾	200	.15	Coils......40% off
⅞	150	.18	Less coils, 25% off
1	100	.25	
1¼	100	.33	
1½	Odd lengths	.40	
2	Odd lengths	.55	

CUT-OUTS—Following are net prices each in standard-package quantities:

CUT-OUTS, PLUG

S. P. M. L.	$0.11	T. P. to D. P. S. B.	$0.24
D. P. M. L.	.11	T. P. to D. P. T. B.	.38
T. P. M. L.	.26	T. P. S. B.	.33
D. P. S. B.	.19	T. P. D. B.	.54
D. P. D. B.	.37		

CUT-OUTS, N. E. C. FUSE

	0-30 Amp.	31-60 Amp.	60-100 Amp.
D. P. M. L.	$0.33	$0.84	$1.68
T. P. M. L.	.48	1.20	2.40
D. P. S. B.	.42	1.05	.....
T. P. S. B.	.81	1.80	.....
D. P. D. B.	.78	2.10	.....
T. P. D. B.	1.35	3.60	.....
T. P. to D. P. D. B.	.90	2.52	.....

FLEXIBLE CORD—Price per 1,000 ft. in coils of 250 ft.:

No. 18 cotton twisted	$22.00
No. 16 cotton twisted	27.00
No. 18 cotton parallel	26.50
No. 16 cotton parallel	34.50
No. 18 cotton reinforced heavy	34.00
No. 16 cotton reinforced heavy	42.00
No. 18 cotton reinforced light	30.00
No. 16 cotton reinforced light	36.75
No. 18 cotton Canvasite cord	28.50
No. 16 cotton Canvasite cord	32.00

FUSES, ENCLOSED—

250-Volt	Std. Pkg.	List
3-amp. to 30-amp.	100	$0.25
35-amp. to 60-amp.	100	.35
65-amp. to 100-amp.	50	.90
110-amp. to 200-amp.	25	2.00
225-amp. to 400-amp.	25	3.60
425-amp. to 600-amp.	10	5.50

600-Volt	Std. Pkg.	List
3-amp. to 30-amp.	100	$0.40
35-amp. to 60-amp.	100	.60
65-amp. to 100-amp.	50	1.50
110-amp. to 200-amp.	25	2.50
225-amp. to 400-amp.	25	5.50
450-amp. to 600-amp.	10	8.00

Discount: Less 1-5th standard package 30%
1-5th to standard package 40%
Standard package 52%

FUSE PLUGS, MICA CAP—

0-30 ampere, standard package	$4.75C
0-30 ampere, less than standard package	6.00C

LAMPS—Below are present quotations in less than standard package quantities:

Straight-Side Bulbs Masda B—				Pear-Shaped Bulbs Masda C—			
Watts	Plain	Frosted	No. in Package	Watts	Clear	Frosted	No. in Package
10	$0.35	$0.38	100	75	$0.70	$0.75	50
15	.35	.38	100	100	1.10	1.15	24
25	.35	.38	100	150	1.65	1.70	24
40	.35	.38	100	200	2.20	2.27	24
50	.35	.38	100	300	3.25	3.35	24
60	.40	.45	100	400	4.30	4.45	12
100	.85	.92	24	500	4.70	4.85	12
				750	6.50	6.75	8
				1,000	7.50	7.75	8

Standard quantities are subject to discount of 10% from list. Annual contracts ranging from $150 to $300,000 net allow a discount of 17 to 40% from list.

PLUGS, ATTACHMENT—

		Each
Hubbell porcelain No. 5406,	standard package 250	$0.24
Hubbell composition No. 5467,	standard package 50	.32
Benjamin swivel No. 903,	standard package 250	.20
Hubbell current taps No. 5638,	standard package 50	.40

RUBBER-COVERED COPPER WIRE—Per 1000 ft. in New York.

No.	Solid Single Braid	Solid Double Braid	Stranded, Double Braid	Duplex
14	$12.00	$16.00	$13.90	$28.50
12	13.25	15.70	18.05	30.70
10	18.30	21.00	23.85	41.50
8	25.54	28.60	32.70	56.77
6	......	......	51.40	......
4	......	......	70.00	......
2	......	......	101.80	......
1	......	......	131.86	......
0	......	......	160.00	......
00	......	......	193.50	......
000	......	......	235.20	......
0000	......	......	288.60	......

Prices per 1000 ft. for Rubber-Covered Wire in Following Cities:

	Denver			Birmingham		
	Single Braid	Double Braid	Duplex	Single Braid	Double Braid	Duplex
14	$19.50	$24.00	$39.50	$12.23	......	$33.60
10	28.92	33.12	65.52	24.00	......	57.02
8	40.32	45.12	89.64	33.60	......	75.94
6	63.84	69.24	......	60.10	......	.....
4	......	97.86	.....	......	$93.41	.....
2	......	145.20	.....	......	120.40	.....
1	......	189.00	.....	......	155.23	.....
0	......	251.40	.....	......	188.77	.....
00	......	305.40	.....	......	229.19	.....
000	......	317.40	.....	......	278.64	.....
0000	......	451.20	.....	......	338.41	.....

Pittsburg—23c. base; discount 50%. St. Louis—30c. base.

SOCKETS, BRASS SHELL—

½ In. or Pendant Cap			¾ In. Cap		
Key Each	Keyless Each	Pull Each	Key Each	Keyless Each	Pull Each
$0.33	$0.30	$0.60	$0.39	$0.36	$0.66

Less 1-5th standard package	+50%
1-5th to standard package	+20%
Standard package	—10%

WIRE, ANNUNCIATOR AND DAMPPROOF OFFICE—

No. 18 B. & S. regular spools (approx. 8 lb.)	55c. lb.
No. 18 B. & S. regular 1-lb. coils	56c. lb.

WIRING SUPPLIES—

Friction tape, ¾ in., less 100 lb. 60c. lb., 100 lb. lots	55c. lb.
Rubber tape, ¾ in., less 100 lb. 65c. lb., 100 lb. lots	60c. lb.
Wire solder, less 100 lb. 50c. lb., 100 lb. lots	46c. lb.
Soldering paste, 2 oz. cans Nokorode	$1.20 doz.

SWITCHES, KNIFE—

TYPE "C" NOT FUSIBLE

Size, Amp.	Single Pole, Each	Double Pole, Each	Three Pole, Each	Four Pole, Each
30	$0.42	$0.68	$1.02	$1.36
60	.74	1.22	1.84	2.44
100	1.50	2.50	3.76	5.00
200	2.70	4.50	6.76	9.00

TYPE "C" FUSIBLE, TOP OR BOTTOM

Size, Amp.	Single Pole, Each	Double Pole, Each	Three Pole, Each	Four Pole, Each
30	.70	1.06	1.60	2.12
60	1.18	1.80	2.70	3.60
100	2.38	3.66	5.50	7.30
200	4.40	6.76	10.14	13.50

Discounts:

Less than $10.00 list value	+25%
$10 to $25 list value	+10%
$25 to $50 list value	+5%
$50 to $200 list value	—10%
$200 list value or over	—15%

POWER

Volume 51 — New York, April 13, 1920 — *Number* 15

His Excellency Calvin Coolidge, Governor of the Commonwealth of Massachusetts said at the recent Engineers' dinner at Boston:

The Engineers are the Ministers to Civilization

HERBERT HOOVER, ENGINEER

Roscoe Pound, Dean of Harvard Law School, said upon the same occasion:

After all, Civilization is nothing but a continually more perfect social engineering

Can you conceive of a system in which every natural resource is used to the best advantage; in which natural and manufactured products are prepared for use and distributed to the consumer by the most efficient machinery and methods and with the least possible loss and waste; by men who, from the highest executive to the least responsible laborer, are especially fitted for the parts they are to play, all imbued with the idea of Service and a full appreciation of the fact that the better and more efficiently they serve, the more product and the fuller, better life there will be for each and for all?

Would there not be PRODUCTION, ABUNDANCE, CONTENTMENT, PROGRESS?

This would be CIVILIZATION; a more perfect SOCIAL ENGINEERING, an orderly application of the forces and resources of nature to the benefit and use of man.

The man to bring it about is the "Minister to Civilization"— THE ENGINEER.

Among those whose names are familiar to the readers of history, whose deeds are immortalized in song, on canvas or in sculptured stone, whose votes and signatures control the policies of nations and the conditions of mankind, how many engineers are to be found?

The first President of the United States was an engineer. He was not so conspicuous a failure that we need hesitate to try another.

New Sinclair Boiler Plant at East Chicago

Modern Four-Boiler Plant Serving Refinery—Convenience of Operation, Accessibility, Low Labor Requirements and Efficiency Were the Factors Influencing the Design—For Its Size the Plant Is Replete with Instruments

AT EAST CHICAGO, Ind., the Sinclair Refining Co. has been building a refinery that eventually will have a capacity to handle 20,000 bbl. of crude oil per day, piped from the fields of Oklahoma and Kansas. Operations started about two years ago, and the plant is now running at one-half capacity, the present products being gasoline, kerosene, fuel oil, gas oil and wax. Although lighting in the works and the pumping of oil is done with purchased current, a generating plant is in prospect. Steam in large quantities is required for the distillation and treatment of oil, for the operation of auxiliary oil pumps, quenching oil fires, for a 600-ton absorption refrigerating plant that is used to chill the oil in the manufacture of paraffin and for steam-jet vacuum systems removing the ashes from the stills. Of the latter there are four batteries of ten stills each and two ash-handling systems per battery, discharging in opposite directions to concrete pits. In each case the horizontal run is about 130 ft. and the rise to the pit 8 to 10 ft. A locomotive crane with grab bucket transfers the ashes from the pits to railway cars from which they are distributed on the premises for filling.

FIG. 1. SINCLAIR BOILER HOUSE AT EAST CHICAGO

To meet this demand for steam, plans were made for a twelve-boiler plant, although the first installation, which is now completed, consists of four units, each a 600-hp. water-tube boiler. The other eight boilers are on order, and it is planned to build a second boiler house containing three 600-hp. units besides a water-softening plant to take care of the water for all of the boilers. The building of six more batteries of stills made this additional boiler capacity necessary. Two boilers are placed on either side of a central firing aisle, and each pair is served by an unlined, painted steel stack, 7 ft. in diameter and 150 ft. high above the grates, located midway between the boilers. The boiler house is a steel skeleton brick building, 61 ft. 4 in. by 73 ft. 6 in. inside in plan and 65 ft. high to the top of the monitor housing the upper horizontal run of the continuous bucket coal conveyor. Steel columns on either side of the central aisle support an overhead coal bunker and, as part of the building construction, an ash tank in line with the bunker. Large glass area in the side walls provides light and ventilation.

Small floor space occupied by the boilers is one of the features of the plant. The actual floor space covered by one 600-hp. boiler is 13 ft. 10 in. deep by 15 ft. 8 in. wide, or 224.5 sq.ft., which reduces to 0.374 sq.ft. per horsepower of rating. Including the boiler overhang and the stoker extension the floor space covered per unit is 15 ft. 8 in. by 23 ft., giving a total of 360 sq.ft., or 0.6 sq.ft. per horsepower.

In the design of the plant convenience of operation, accessibility, low labor requirements and efficiency were

FIG. 2. BOILER-ROOM INSTRUMENTS

FIG. 3. PUMPING EQUIPMENT OF BOILER PLANT

the factors chiefly considered. Instruments to give all operating data of moment have been installed, and they are mounted centrally at the end of the firing aisle in full view of the operating force. Each boiler, however, has its indicating draft and pressure gages, and for convenience damper control is brought to the front of the setting.

The boilers are set individually in steel-incased settings, with a 5-ft. 4-in. Dutch-oven extension to receive chain-grate stokers having an active grate area of 121 sq.ft. and a ratio to the steam-making surface of 1 to 49.6. On either side the stokers are driven by a vertical engine belted to a lineshaft in the basement, with eccentric and ratchet connection to the stoker gears. Cross-connection between the shafts in the basement adds flexibility to the stoker-driving mechanism.

Headroom of 9 ft. to the center line of the front mud drum gives a furnace volume approximating 6.5 cu.ft. per square foot of grate area. From the apex formed by the intersection of the tubes, the distance to the grate is 9 ft. 6 in. A flat suspended arch 8 ft. long insures strong ignition for the Illinois coal used in the plant. The previous dimension may be compared to 11 ft. of active grate length. The maximum gate opening is 12 in., and at the end of the arch the distance to the grate is 34 in. A sloping arch section at the front, meeting the horizontal section of the arch in conjunction with a grate sloping toward the rear, gives expanding volume for the gases, while still retaining the strong ignition effect desirable with the fuel burned. Thorough mixture of the products of combustion and the air required to burn the volatile is effected by the right-angle turn at the rear of the furnace and a throat between the arch and bridge wall restricted to 42 in. The gases make a single pass across the tube surface to the uptake at one side and at the top of the setting, the outlet measuring 4 x 14 ft. 8 in., or 0.485 sq.ft. per square foot of active grate installed.

A feature of the setting is the installation of the superheater in the combustion chamber back of the bridge wall out of the direct current of the gases, in a free space measuring 3 ft. 3 in. There are 40 elements about 7 ft. high, which are shielded somewhat by the bridge wall, and with tile between the elements there is no circulation of the gases through the superheater. Only the front portion of the U-shaped elements are exposed to the radiant energy of the fire.

FIG. 4. BOILER ROOM SHOWING TWO OF THE PAGE BURTON BOILERS

Steam of 125 to 150 lb. pressure is supplied to the stills, and the desire was to superheat it about 100 deg. to prevent condensation in the pipe lines. Early operation did not show the desired degree of superheat, but adjustment of the tile between elements has increased the superheat in accordance with the amount of surface exposed.

It will be noticed that the combustion space is practically self-cleaning. Soot blown off the tubes by the oscillating soot blower onto the fuel bed will pass out with the ash, and from the rear combustion chamber containing the superheater there is direct connection with the ashpit.

At the present writing the steam demand is fairly uniform, running about 50,000 lb. per hour. As shown in the illustrations, the steam is supplied from the top of the main boiler drum to the superheater, thence back to an 8-in. header serving the two boilers on the same side of the firing aisle. Long-radius bends care for expansion. The header on either side joins at right angles an 8-in. main, enlarging to 10 and 12 in. respectively at opposing ends and at present supplying two 8-in. lines to the stills and one 6-in. line to the wax plant. For the most part these steam lines are carried overhead, hung from crossarms on steel poles. The insulation consists of one inch of fire felt and two inches of sponge felt covered by three-ply roofing paper.

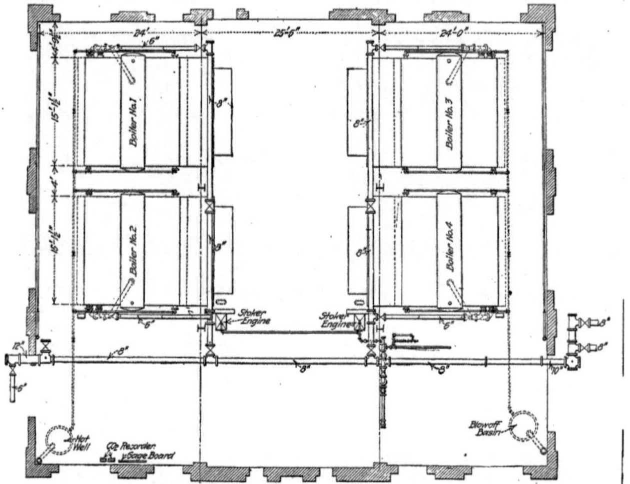

FIG. 5. PLAN OF BOILER ROOM AND HIGH-PRESSURE PIPING

Coal comes in over the railway siding serving the boiler room and is dumped into a track hopper from which it passes through a rotary feeder to a continuous bucket conveyor discharging to the overhead bunker. An offset of this conveyor to include the track hopper eliminated the need for the usual auxiliary conveyor. The driving power is a 10-hp. induction belted motor located on the upper run. The conveying system is designed to handle sixty tons an hour. The bunker, of parabolic shape, is supported on steel building columns. It is of the Ferroinclave type concrete lined and plastered outside. The capacity is about 390 tons. Under the bunker the coal is weighed by a traveling larry of 1,000 lb. capacity. A spout on either side leads to the stoker hoppers, the direction of coal flow being controlled by gates in these spouts operated by rods extending to within reach from the floor. Also the weigh beam of the scale is near the floor so that the operator can conveniently weigh and record the coal fed to each boiler. Ashes are stored in a 75-ton reinforced-concrete bunker forming part of the building construction. Delivery to the bunker is made by the continuous bucket conveyor used for handling the coal, and the contents of the bunker are discharged to railway coal cars through a chute leading through the wall. In the basement an industrial track makes the circuit of the four boilers underneath the ash hoppers serving the stokers. Through a circular gate that is hinged to the bottom of the hopper and may be swung easily to one side, the ashes drop into a V-shaped tip car of 27 cu.ft. capacity. They are dumped into a hopper protected by a grid to prevent large clinkers getting through and by means of a rotary feeder passed directly into the continuous conveyor. To prevent air leakage through the ashpit, a handwheel has been provided to press tight the circular gate against its seat.

Cooling water from the still condensers is used as feed water. There is an abundant supply, but owing to priming troubles a water-softening plant is now under way. The water at a temperature of 120 to 130 deg. is collected in a tank outside the boiler house, 43 ft. diameter and 20 ft. high. By either one of two direct-acting duplex pumps it is raised to an elevated tank in the boiler room, the head in the tank controlling the operation through a pump governor. The tank also has con-

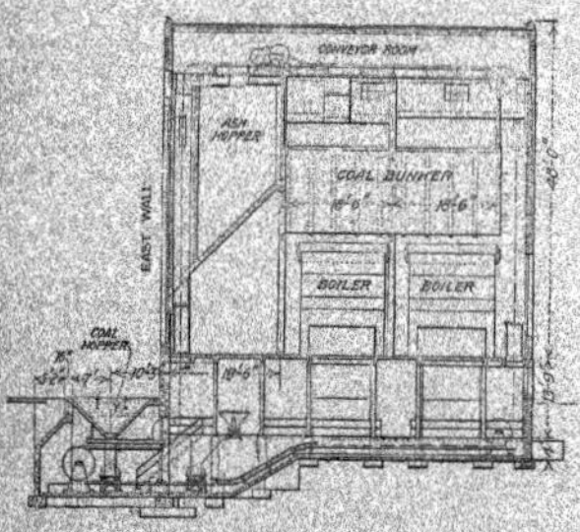

FIG. 6. LONGITUDINAL SECTION SHOWING COAL AND ASH CONEVYOR

nection to the city mains. Under float control the water flows by gravity to an open feed-water heater and thence through turbine-driven three-stage centrifugal pumps to the boilers. An individual regulator at each boiler con-

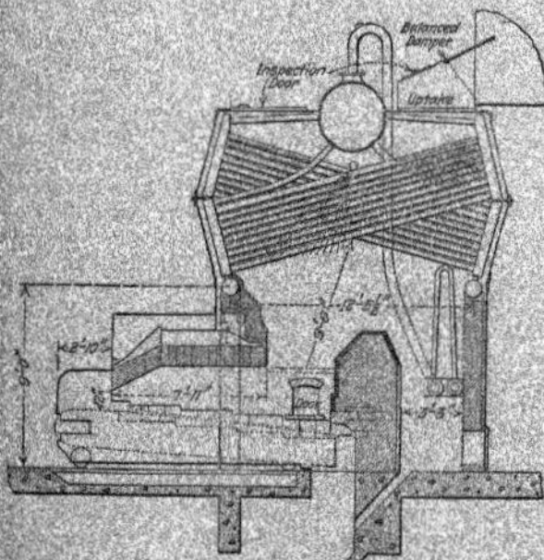

FIG. 7. SECTION THROUGH BOILER SETTING

trols the feed. Exhaust steam from the pumps is utilized in the heater.

Very few plants of the size are so fully equipped with instruments. All data of moment in determining and maintaining the efficiency of the plant are indicated or recorded. Most of the instruments are of the recording type. The coal is weighed, and the feed water is measured by a venturi meter of the indicating, integrating and recording type. For each boiler there is a recording steam-flow meter giving the output. The steam and water meters serve as a check on each other, and the difference in the readings gives the boiler loss by blowing down or by leakage. Steam pressure is indicated by gages on the boilers, and a recording pressure gage

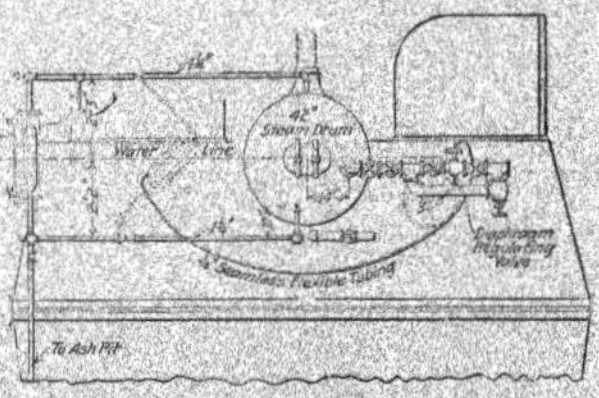

FIG. 8. WATER COLUMN AND FEED-WATER REGULATOR CONNECTIONS

gives a continuous record of the steam pressure in the header. Feed water and steam temperatures also are recorded. Draft over the fire is determined from a draft gage on each boiler, which also give the differential, or drop in draft through the boiler. The efficiency of combustion is indicated by CO_2 recorders. Thus all essential operating data are available, and the recording instruments are grouped on a board at the end of the firing aisle in plain view of the firemen. To maintain maximum economy, there is just one more essential, and this is the proper interpretation and use of the numerous records.

A seven-hour test made on one of the boilers shortly

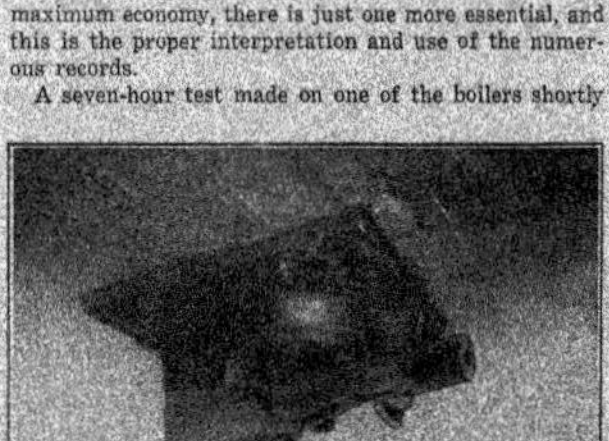

FIG. 9. SPECIAL GATE AT BOTTOM OF STOKER ASH HOPPER

after the installation had been completed, by Osborn Monnett, consulting engineer, of Chicago, gave some interesting results. With Illinois coal containing over 13 per cent moisture and 12,292 B.t.u. per lb. dry, an equivalent evaporation from and at 212 deg. F. of 9.53 lb. per pound of dry coal was obtained. With a draft of 0.23 in. over the fire and 0.33 in. at the uptake, dry coal was burned at the rate of 25.8 lb. per square foot of grate area per hour. The boiler was operated at 146 per cent of rated capacity, and the combined efficiency averaged 75.2 per cent.

PRINCIPAL RESULTS FROM TEST OF ONE OF THE 600-HP. BOILERS

Average Pressure, Temperatures, etc.	
Steam pressure, lb.	124
Feed water temperature, deg. F	154
Escaping flue gas temperature, deg. F	627
Draft over fire, in.	0.23
Draft at uptake, in.	0.33
Superheat, deg. F	35
Total Quantities	
Weight of coal as fired, lb	25,670
Moisture in coal, per cent	13.31
Weight of dry coal, lb	22,253
Weight of water fed to boiler, lb	188,836
Equiv. evap. from and at 212 deg. F., lb	212,252
Hourly Quantities and Rates	
Dry coal consumed per hour, lb	3,179
Dry coal per sq.ft. of grate area, lb	25.8
Water evaporated per hour, lb	26,976
Equiv. evap. per hour from and at 212 deg., lb	30,321
Equiv. evap. per sq.ft. of heating surface, lb	5.05
Capacity	
Boiler horsepower developed	880
Rated capacity	600
Per cent rated capacity	146
Results	
Water fed per lb. of coal as fired, lb	7.35
Water evap. per lb. of dry coal, lb	8.48
Equiv. evap. from and at 212 deg., lb	9.53
Efficiency	
B.t.u. per pound of dry coal	12,292
Combined efficiency, per cent	75.2

The plant was designed and erected by home talent, J. S. Hess, chief engineer of the Sinclair Refining Co., being responsible for the design, and R. E. Greenburg, mechanical engineer at the works, having charge of erection.

Reducing Noise and Vibrations

By K. Rankine

Substations are frequently located in the midst of residential and business sections of the city, where noise and vibration caused by them is most objectionable. Many cases of litigation have occurred because it has been claimed, and sustained, that the continual hum and vibration of electrical apparatus, destroy the value of property and make dwellings uninhabitable. These lawsuits are expensive for the utility and are injurious to its interests, because public good will is an asset of value if an intangible quantity. Unfortunately, it is rather difficult to determine beforehand whether a substation will be noisy and vibrate or not. The following discusses some of the pertinent factors that tend to reduce, if not actually eliminate, noise and vibration.

Noise is often the result of vibration. For example, it is the motion or vibration of loose iron laminations that starts the humming noise so characteristic of some electrical apparatus. On the other hand noise may occur without vibration, as the singing or whistling due to windage with rotating machinery such as synchronous converters and motor-generator sets. The manufacturers of electrical machinery, both rotating and static, have done much to prevent noise and vibration. They have largely done away with loose laminations and rattling parts; they have designed pole tip and air gaps that hum but slightly, notwithstanding the higher flux densities now employed, and that give rise to only slight windage. Nevertheless, it is sometimes necessary to partly or even totally inclose rotating machinery located directly in residential districts. Inclosing apparatus, while interfering with its natural cooling, and therefore ability to dissipate heat, offers the opportunity to employ forced ventilation and in this way avoid noise due to the machine and at the same time increase its current-carrying capacity. There is little doubt that this practice could be developed much more than it has been in the past.

While the manufacturers have gone a long way toward producing noiseless equipment, the entire responsibility in this connection does not devolve upon them. The operator and those who design the building to house the machinery are also responsible in a very large measure. In fact, there is much greater opportunity for making operation quiet and free from vibration in the design of building and foundations than there is in the design of the equipment. Unfortunately, the average architect knows little about electrical machinery and what precautionary measure should be taken against noise and vibration. However, the methods of reducing noise and vibration are simple if carried out at the time the building is constructed.

Windows offer a ready means of escape for noise, hence their number and area should be kept to a minimum, and needless windows omitted. The noise passing through openings in the roof will cause less disturbance and be less noticeable than noises that pass through windows in the substation walls. The farther the windows are away from the machinery the better. They should never be located in the direct path of air currents resulting from heat losses, forced draft or windage of rotating machinery, as the noise may be exaggerated and distributed farther.

If the walls of substations located in residential zones are built double with a dead-air space between them they will cushion noise and also vibration to some extent. The heating required during cold weather should also be less. Where forced or mechanical ventilation is used, as for inclosed synchronous converters and air-cooled transformers, special care should be taken that the discharging air shall not be so routed as to act as a carrier for noise.

Vibration can be reduced by paying special attention to foundations and the manner in which apparatus is placed upon them. It is sometimes a good plan to isolate the foundations from the station structure by sinking the machinery foundation separate from that of the building. Floors, etc., should not make physical contact with the foundations, a two-inch gap being left all around the latter. In this way vibration is not transmitted to the station structure and from there beyond. One frequent cause of vibration troubles results from the fact that the vibration of machinery is transmitted to girders and beams and the periodicity of vibration of the one sets up a condition with the other where the effect becomes cumulative. The more massive the station foundations the less the probability of vibrations being transmitted and becoming cumulative. However, the more substantial the foundations the greater their cost, which means that common sense precautions are less expensive than solution by mass and weight only.

In addition to the very general phases of this matter of noise and vibration, there are the more specialized problems that may call for noise-proof rooms, pads installed underneath induction regulators and transformers, etc. These are, however, matters that can usually be taken care of later, after the other precautions have been incorporated in the design of the substation. Only by familiarizing the architect with the inherent characteristics of electrical apparatus and the specific conditions that it is desired to guard against, can success be expected. In far too many instances the architect knows too little about what the building is to house, and expensive changes, litigation and endless troubles result.

Practical Refrigeration for the Operating Engineer—Regulation of Pressures and Temperatures

BY B. E. HILL
Formerly Refrigerating Engineer, Armour & Company, Chicago

THOSE responsible for the operation of refrigerating machines are aware that to reduce the temperature of a cooler or freezer to zero or below, it is necessary to lower the suction pressure to the compressor; but as I have found many engineers who do not understand why this is necessary, an explanation will be welcome, as these articles are for the benefit of the operating engineer and not the college graduate or professor, although the practical knowledge of operation which I am endeavoring to convey will be beneficial to the college man as well.

HIGH-SPEED, MOTOR-DRIVEN AMMONIA COMPRESSORS IN A MODERN REFRIGERATING PLANT IN NEW YORK CITY

A modern high-speed, motor-driven ammonia-compression plant is shown herewith.

The properties of saturated ammonia, as referred to, will show the engineer the boiling point of liquid ammonia for any required suction pressure, and when it is required to reduce the temperature of a freezer to, say, 20 deg. F. below zero, it will be easy to understand that the temperature of the boiling liquid on the inside of the pipes that are placed in this freezer will have to be reduced below the temperature of the room. By referring to the tables previously mentioned it will be found that the boiling point of liquid ammonia at zero pounds pressure, or atmospheric pressure on the suction or low-pressure gage, will be 28½ deg. F. below zero, which is only 8 deg. F. below room temperature.

From the foregoing the operating engineer will readily see why it is necessary to reduce the suction or low pressure to produce low temperatures, and also he will find a good reason for ammonia getting *lost, sluggish* or *hiding.*

When the temperature of a cooler or freezer is only about eight or ten degrees above the boiling temperature of the liquid ammonia on the inside of the expansion or evaporating coils, the engineer can expect nothing except a slow or "sluggish" evaporation. Do not swear at the man who sold you the ammonia for selling you an inferior quality or a kind of ammonia that will not mix with what you already had, etc.; but look for the cause. If the temperature is not holding or coming down as it should, the cause is generally due to slow evaporation, as already explained, a shortage of ammonia in the system, not enough pipe surface in the cooler for the product to be handled, and many other causes which we will treat later.

The evaporating coils are the boiler of the refrigerating plant, and the engineer or temperature man should watch the return end of the coil the same as he watches the water glass on the boiler. Examine the return end and apply the stick test, as before explained, before feeding up or cutting back the expansion or feed valve, and there will be little or no trouble from overfeeding and slugs.

Duties of the Temperature Man

In large plants where there are thousands of feet of direct-expansion pipe and hundreds of expansion valves, it is necessary to employ a temperature man, whose duty it is to regulate these valves, take temperatures, etc., and report them properly made out on a temperature or log sheet, which enables the engineer to properly direct and operate the plant.

In plants where several temperature men are employed, there are times when it is necessary to "break in" new men on this work, and at such times there is always more or less danger of overfeeding the expansion coils and the accompanying danger to the compressor, together with alternate "freezing up" and "burning up" of the compressor, which also carries with it an unusual loss of ammonia due to expansion and contraction of the piston rod, packing, joints and connections.

Most steam plants use a separator between the engine and boiler to protect the steam cylinder from water, and for the same protection to the ammonia compressor, there should be a separator between the evaporating coils and the compressor, as there is no difference in the results other than a much greater need for a separator on the suction line leading to the compressor, as the danger to life and property are many times greater in case a compressor head is knocked out, due to suffocation from ammonia gas, fire, etc., which subjects will be treated later.

It is necessary only to secure an ordinary steam separator, preferably one having a storage capacity, for liquid, so that in case of liquid being carried over to the compressor from overfeeding the evaporating or expansion coils, the storage capacity in the separator will give time for the liquid to drain back to the evaporating coils, or other means so provided for the purpose.

Installing the Separator

The separator should be installed at a point high enough above the evaporating coils to insure a speedy drainage of the separator, also the drain line should be provided with a valve to close in case of pumping out.

The drain line from the separator can also be connected to a tank located below the separator for the liquid to drain into. This tank is provided with a water jacket, with steam connection to keep the water hot for the purpose of reëvaporating the liquid ammonia, the gas piped back to the suction side of the compressor; and in this way the tank serves the double purpose of converting the liquid from the separator into a gas and purifying the ammonia by separating out the oil or water that may have been taken into the system by leaks or inferior ammonia and the oil that works into the system from the compressor.

When the liquid from the separator is carried back to the evaporating coils, there is practically no loss from this source, as the liquid is reëvaporated and will do useful work in the transfer of heat, but the beneficial results of leading the separator drain to a purifier will more than offset the gain by the other method of reëvaporating the liquid in the coils.

Another system for removing liquid from the separator is to connect the drain line to an automatic drip tank and pump which is operated with a float and returns the liquid to the expansion or evaporating coils, the operation being the same as that used in a steam plant for returning condensate to the steam boiler.

The frost on the suction line leading to the compressor does not mean that there is liquid ammonia being carried over to the machine, for the reason that the gas is in a saturated condition and is at the same temperature as that of the boiling liquid. If the boiling liquid is below the freezing point of water, it will freeze the moisture that is carried in the air which comes in contact with the cold surface of the pipe, forming a coating of frost of a thickness depending upon the saturated condition of the air, temperature of the ammonia, gas, and time.

On all horizontal double-acting compressors it is good practice to "freeze back" to the machine, as the heat of compression is much greater than in the single-acting compressor.

Few Plants Are Free From Decomposed Ammonia

Ammonia (NH_3) is decomposed by heat, and it is claimed by some authorities that a complete separation of the nitrogen and oxygen takes place at a temperature of about 900 deg. F., while others claim that decomposition takes place very slowly, at a much lower temperature, ranging from 150 to 300 deg. F. There are few plants that are entirely free from decomposed ammonia. There may not be enough to noticeably affect the capacity of the plant, but nevertheless it is there in greater or lesser quantities, and for this reason alone, the frost should be carried close enough to cool the gases, leaving the machine to a temperature of 100 to 150 deg. F., which can easily be determined by inserting a thermometer in the discharge line of the compressor.

Another reason for holding the temperature down on the discharge line, is to prevent vaporizing of oil which, if carried to the oil trap at a temperature of from 300 to 500 deg. F., will be in a more or less vaporous or finely atomized condition and will float with the hot gas to the ammonia condenser where it will be condensed to a liquid, then go on to the evaporating coils where the oil congeals more or less, depending on the freezing point of the oil. When the coils accumulate enough of this oil it will cut down the refrigerating effect of the ammonia, as it is a good insulator, and will also stop up the feed or expansion valves to such an extent that regularity in the feeding of the coils will be impossible; often the coils stop up altogether, which necessitates a shutdown for cleaning out the system.

In large plants it is often necessary to run suction lines for a distance of 300 to 600 ft. and even longer, and in such cases it is difficult to freeze back to the compressor with any degree of regularity owing to a large number of crooks and turns, risers, drops, radiation, etc. In such cases it is more effective and less dangerous to cool the compressor by connecting a ½-in. line from the liquid receiver and expand direct into the suction line at a location somewhere close to the compressor.

In traveling these long distances, there are considerable losses due to radiation, as the insulation is never perfect enough to prevent these losses. To successfully cool the compressor or "freeze back" through these long lines requires considerable practice. If there is oil in the system, as mentioned, it is impossible to cool the gases leaving the compressor with any degree of regularity. It will be a feast or a famine. The compressor will either get the ammonia in "chunks" or will be "burning up," together with all the other interesting conditions caused by irregularities in the handling of expansion valves. It is a great deal more safe, sane and economical to bring bad conditions close up under the watchful eye of the operating engineer, than to trust to Providence and the temperature man not to blow the place up.

Selection of Lubricant

In the selection of an oil for use in the ammonia compressor it is important that the selection be made with a view to securing one with the highest flash point, greatest viscosity and lowest congealing point. The best is none too good as regards the points mentioned, and a question of price should not enter into the proposition at all, as the effect caused by the use of an inferior grade of ice-machine oil will offset the difference in price many times over.

The engineer should get samples of as many grades of ice-machine oils as he can from the local dealers and have analyses made of them and make his selection from the information so derived. If it is not convenient to get this information from a laboratory, he should insist on the dealer giving it to him.

A good grade of paraffin oil, with the wax as nearly completely removed as possible, will make a good ice-machine oil. The more paraffin wax the refinery removes from the paraffin oil, the lower will be the congealing point, which ranges somewhere around 15 to 18 deg. F. below zero.

The engineer should never under any circumstances allow engine oil to be used in the compressor or on the compressor piston rod.

As a guide to the engineer who is not familiar with the physical properties of ice-machine oils, the writer will refer to a table showing the essential properties referred to.

PHYSICAL TESTS OF OILS

Kind	Specific Gravity	Viscosity	Flash Deg. F.	Fire Deg. F.	Congealing Point Deg. F.
Arctic ice-machine oil.......	0.8920	3.93	372	399	—19
Ice machine................	0.8746	3.93	368	392	—24
Ice machine................	0.8912	3.225	320	360	—20
	0.9258	4.18	330	360	—26

In the application of ice-machine oils as a lubricant, the writer has failed to find a single case in thirty years of actual operating experience where it was necessary to pump oil directly into the compressor cylinder.

Twenty years ago nearly all of the well-known ammonia compressor builders equipped their machines with oil pumps and piped them to pump oil into the compressor cylinder direct, as well as into the stuffing-box between the packing; but today oil is pumped to the stuffing-box only, and in many cases a sight-feed oil cup is placed over the piston rod, allowing the oil to be fed in drops, with a moderate feed, which is found to be ample lubrication for both packing and compressor.

The oil that adheres or sticks to the rod is carried inside and is wiped off inside the compressor cylinder. I have experimented along these lines considerably and have yet to find a single case in which, where the oil cup was used, the compressor has shown signs of wear due to insufficient lubrication.

The only excuse for an oil pump being connected to the stuffing-box at all is to give the packing a "shot" of oil to stop the rod from leaking, when the packing has been allowed to remain in use so long that it will not hold, or a quick change of temperature caused by overfeeding of the evaporating coils.

The engineer will find by an examination of the system at the end of the season's run that there will be "oil aplenty" in the evaporating coils and will find also that it will be a good paying investment to clean them out thoroughly at least once a year.

Flywheel Failure Is Harmless

The illustration shows a nine-foot flywheel (which was used in connection with a refrigerating engine) after a disruption due to centrifugal force. The hub and arms of the wheel were so badly fractured that it was a total loss. Luckily, no other damage resulted, as one of the employees happened to be near the throttle and closed it as soon as the engine began to race,

THE FLYWHEEL AFTER DISRUPTION

thereby preventing complete destruction of the wheel and possible damage to surrounding apparatus and property and loss of life. Considerable velocity must have been gained by the rim before the throttle was closed, as the links forming the rim joints had stretched fully an inch. This was caused by the tendency of centrifugal force to separate the halves of the wheel.

The strain on the hub-bolts was so great when the links began to stretch that two of them sheared off, the hub itself split in two and one of the arms was completely separated from the rest of the wheel.

The Board of Trade of Alberta, Canada, has appointed a committee to investigate the possibilities of the waste coal from the mines, in the generation of electricity in a large central plant and of proper distribution of this power to manufacturers and municipalities near-by.

Electric Motors for Pulp and Paper Machinery Drives

By WILLIAM H. EASTON
Industrial Engineer, Westinghouse Electric and Manufacturing Company

The advantages of electric drives in pulp and paper mills are discussed, and the characteristics of the different types of motors and the kind of controllers used on the various drives are considered.

ELECTRIC motor drive is generally favored today for driving pulp and paper machinery because it can produce better results than any other form of power. A wood-pulp grinder, for example, can produce more and better pulp when motor-driven than when driven by a waterwheel. The reason for this follows:

There is one speed, and only one, at which the best results can be obtained. Should the speed be lower than this, production falls off; should it be higher, the quality of the pulp suffers from excess of splinters and short fibers. The speed of a motor, of either induction or synchronous type, is inherently more nearly constant, especially under wide variations of load, than that of a waterwheel; therefore, when the girder is motor-driven its speed can be kept close to this most efficient value at all times, whereas with waterwheel drive there is bound to be considerable variation above and below this speed, with detrimental results. The foregoing holds true for beaters and jordans. Motor drive also makes it possible for the operator to obtain complete information as to operating conditions. The power taken by these machines depends upon a number of factors, such as the position of the roll or plug, the character of the stock, the degree of refinement of the stock and the condition of the bearings. Hence, by driving each machine with a separate motor and connecting an ammeter (especially a graphic instrument) in the circuit, the operator is provided with a guide which, after a little experience, enables him to adjust the machine to produce exactly the desired results, stop the refinement at the proper moment, prevent impending trouble due to worn bearings, and control the operation in other ways.

FIG. 1. MOTOR DRIVEN PAPER MACHINE

A special advantage of motor drive for jordans is that with a direct-connected motor there is no side thrust on the bearings as there is in belt drive, consequently the plug does not acquire eccentricity, which decreases the efficiency and wastes power. For the paper machine, motor drive has been revolutionary. It has freed the machine from all complication of transmission and speed-changing mechanism, and made possible its control with the utmost nicety, over a wide range of speeds, by merely pressing a few buttons.

The super-calender, too, has gained greatly by motor drive. When belt-driven, it has but two speeds, a low one for threading-in and a high one for running. It is difficult to change from one to the other without tearing the paper, and since the running speed exactly suits but few grades of stock, much of the work must be run off at too low a speed for efficient production or too high to produce the best results. With motor drive any range and any desired number of speeds can be obtained and the transition from low to high is smooth and even.

Motor drive, as compared with direct water drive, permits complete freedom of choice in the location of the mill. Water power can be used more efficiently to drive a few large generators under skilled supervision than to drive a number of small wheels throughout the mill. Then, too, if motor drive is used, steam auxiliaries or central-station power can be used during periods of low water. With steam or water drive, the location of the lineshafts fixes the positions of the machines, and additions are awkward to make and often result in a complicated and inefficient power-transmission system.

The working conditions in a motor-driven plant are much better than in those using lineshafts. There is better light, greater cleanliness and, especially, greater safety, since all dangerous belts and pulleys can be eliminated, all necessary transmissions can be inclosed in guards, and hazardous machines, such as super-calenders, can be equipped with numerous safety stops so that they can be shut down instantly from any point in case of accident.

Alternating current is almost invariably used in pulp and paper mills because it can be transmitted economically from a distance and also because alternating-current motors, due to their simple and sturdy construction, are better adapted to the majority of pulp and paper machines. Although the paper machine and the super-calender are best driven by direct-current motors, it is in general more satisfactory to use alternating current for most of the drives and generate direct current specially for these machines. It is, however, perfectly possible to drive small paper mills by direct current as is being done in a number of instances.

Where the mill generates its own power, it uses water power if possible, entirely or in part, and steam turbines where steam power is necessary. Because of the large amount of exhaust it is frequently a better plan, in cases where the entire supply of exhaust steam is not always utilized (as in summer), to install bleeder turbines.

Many old mills have changed over completely to electric drive and have more than justified the expense by increased earnings; but where there are engines that are satisfactorily driving some of the larger machines, it is sometimes good economy to allow them to remain and use electric drive for the other machines, thus getting rid of the small inefficient engines and simplifying the transmission problem. Central-station power is frequently very useful in taking care of these supplementary drives; but if the engines retained produce an excess of exhaust steam, a low-pressure turbine-generator can be installed to utilize this excess and produce a part or all of the supplementary power in a very economical manner.

For the general run of pulp and paper machinery the squirrel-cage and wound-rotor motor types of alternating-current motors are used. The squirrel-cage type of machine is the simplest motor made, because of the

FIG. 2. FIVE-FOOT BARKER DRIVEN BY 15-HP. SQUIRREL-CAGE MOTOR

small amount of care it requires, absence of sparking, and its ability to work in dirty and dusty places (although, of course, it must be kept as clean as possible). Its two limitations are, that it is a strictly one-speed machine and that it takes a large current when starting under load, which is frequently undesirable, especially with motors of large size.

The wound-rotor motor is used wherever either of these two limitations prevent the use of a squirrel-cage motor. It is similar to the latter in construction except that its rotor carries an insulated winding, connected to slip rings on the rotor shaft. By means of these slip rings, resistance varied by means of a controller can be introduced into the rotor circuit. By varying this resistance two things can be accomplished: The motor can be started under full load and brought gradually to full speed without drawing at any time more than 50 per cent excess current from the line. Secondly, a certain amount of speed variation can be obtained, which is satisfactory over a range of

FIG. 3. JORDANS DRIVEN BY 125-HP. SQUIRREL-CAGE MOTORS

(1) from normal speed to about 50 per cent below normal, for applications where the torque falls off as the speed decreases, as in fans; and (2) from normal to about 10 per cent below, for applications where the torque remains fairly constant at all speeds, as with the paper machine. In the latter case low speeds are unsatisfactory because too much current is consumed in the resistance for economical operation and because it is frequently impossible to obtain exact and stable speeds on the various steps.

In the early days a wound-rotor motor could be used for both heavy-starting constant-speed service and for variable-speed service, but with refinements in design, these motors are now specially constructed for each service and the two types are not always interchangeable.

For machines requiring speed variation, therefore, a wound-rotor motor will be used without question; but for constant-speed machines that require high starting torque, the choice depends on conditions. Some central stations will not permit squirrel-cage motors larger than 15 hp. on their lines. But where the current is supplied by a private plant, the limit of size of a squirrel-cage motor depends mainly upon the size of the generating equipment, the capacity of cables, transformers, etc., and the amount of trouble caused to lights and to other machines when the squirrel-cage motor is started. It is, therefore, impossible to be specific as to the proper selection of these motors for use on various pulp and paper machines, but the following can be considered good practice:

Screens of all types, saws, barkers, deckers, platers, pumps, agitators, rag cutters, dusters, threshers, rotary boilers and cutters may all be driven by squirrel-cage motors when driven either individually or in small groups. Wound-rotor motors may be used to drive large groups of machines of any type, bandsaws, conveyors and log hauls, chippers, sheet calenders and shredders.

Shredders, rag cutters and chippers are sometimes driven by a special type of squirrel-cage motor with high-resistance rotor end-rings. The effect of the use of this resistance is that, when the load increases above normal, the motor slows down and exerts a very high torque, which is desirable with these machines in case of jamming. The efficiency of the motor is, however, not as high as that of the standard motor. With squirrel-cage motors, simple inclosed starting switches are used for sizes up to 5 or 10 hp. and compensators for larger sizes. Drum-type controllers are used with wound-rotor motors, with resistors for either starting or speed-control service, as required.

Direct-current motors can be used for all the foregoing applications, but they must be in general partly or totally inclosed to protect them from dust and dirt and to prevent the commutator sparks from setting fire to combustible material. In order to realize all the advantages of motor drive the various machines listed herein should be individually driven; but it is frequently more economical to arrange similar machines in small groups and drive each group by a single motor.

FIG. 4. CONTROL PANELS FOR STARTING 150-HP. BEATER MOTOR

Wood-pulp grinders are coupled to the motor either singly or in a group consisting of as many as five. Where there are two grinders per motor, the grinders are usually located on both sides; and where there are four, it is good practice to place two on each side of the motor. But sometimes all the grinders are driven from one side of the motor. Flexible couplings should be used between the motor and the grinders to relieve the motor shaft of the weight of the stone. The usual speed of a direct-connected motor is about 240 r.p.m., and the power per grinder ranges from 400 to 1,600 horsepower.

Squirrel-cage motors are occasionally used where grinders are individually driven, but usually wound-rotor motors or synchronous motors are employed. The synchronous motor is, however, rapidly superseding all other types because of its constant speed, high efficiency and ability to keep the power factor of the system high. Owing to these advantages there is a growing tendency to use synchronous motors for industrial drives wherever possible, but in pulp and paper mills the drives for which the synchronous motor is suitable are limited.

During the process of grinding there is a considerable variation in the pressure on the stone, because the pressure will change when a new log is inserted and its face is being ground down to the contour of the stone. This variation in pressure produces variations in the quantity and quality of the product and also causes variations in the amount of power required. This latter condition is especially objectionable when central-station power is being used, since many central stations charge according to the maximum, and not the average, demand. Hence, the elimination of high peaks reduces the cost of power.

An automatic pressure regulator is now being used with individually motor-driven grinders, which keeps the pressure constant at all times, thus improving production and lowering power costs. It consists of two magnetically operated valves which control the admission of the hydraulic pressure to the grinder cylinders containing the logs. The magnets are actuated by relays controlled by the current taken by the motor. When this current exceeds a predetermined amount, the proper relay closes and energizes the valve that reduces the pressure; and when the current falls below a predetermined lower level, the other magnet is similarly energized and the pressure increased. With normal current there is no pressure variation.

FIG. 6. FULL AUTOMATIC CONTROL FOR PAPER MACHINE

Beaters can be driven in pairs by motors with double extended shafts, but it is better to use individual motors because of the flexibility of the arrangement and the possibility of using ammeters to provide information as to the operating condition. Belts, chains and rope drives are used. Chains are more efficient, but belts and ropes allow some slippage, which is often desirable, especially in heavy duty. On the other hand, chains permit the motors to be brought close to the beaters, thus making the layout more compact.

Either squirrel-cage or wound-rotor motors can be used, the former being more common in the smaller sizes and the latter in the larger sizes. If the motors are installed in the beater room, where they are exposed to dampness, their windings should be impregnated, but this is unnecessary if they are located in an adjoining room or on the floor below and drive with long belts or ropes. The horsepower requirements vary not only with the size of the beater, but with the character of the stock—rags taking several times the power needed for sulphite pulp, so that the range of motor sizes per beater is from 30 to 200 horsepower.

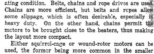

FIG. 5. MOTOR DRIVE AND CONTROL FOR PAPER MACHINE

Jordans are always driven with the motor directly connected to the plug through a flexible coupling. In order to adjust the position of the jordan plug for light or heavy work, either the entire motor is arranged to slide horizontally together with the plug, on its baseplate, or the motor remains fixed and a telescopic coupling is used. Either squirrel-cage or wound-rotor motors can be used. Synchronous motors can also be used where power-factor correction is desired. The horsepower requirements range from 75 to 300 hp., and the motor speed is about 350 revolutions per minute.

The earliest and simplest motor drive of the paper machine was to use a constant-speed motor, almost invariably of the squirrel-cage type, to drive the constant-speed end of the machine, and a similar one to drive the variable-speed end, the speed variation of the rolls being obtained by the ordinary mechanical means. This, however, merely simplified the transmission to the machine, and an attempt was soon made to simplify the mechanism of the machine by using a variable-speed wound-rotor motor for the variable-speed end. Experience shows that when a speed range of not more than 10 per cent is required, this method is satisfactory, but it is not applicable for wider ranges.

The next step was to use a constant-speed motor, as before, for the constant-speed end, but to substitute for the wound-rotor alternating-current motor on the variable-speed line, one with much better speed characteristics, namely, a direct-current motor with field control. This usually involves two kinds of current in the mill, but the advantages more than offset the complication. The variable-speed line is operated by a motor with a speed range of from one to three, with any number of steps within this range, and the speed on each step can be accurately maintained regardless of any variation in the load, which is not the case with the wound-rotor motor. Hence, it is a considerable improvement over the earlier drives.

It has, however, several objections, except for certain limited applications. The direct current is, of course, obtained from a general-power circuit, supplied ordinarily by a rotary convertor or a motor-generator set and serving, perhaps, several paper machines, super-calenders, etc. But each paper machine requires so much power that whenever one is started or stopped, the speed of every other machine on the line is temporarily affected, which is, of course, detrimental. In addition, since the power required by the machine varies with its speed, its driving motor will be much too large, therefore inefficient at low speeds, in order to provide sufficient power at high speeds. Finally, the speed range of 1 to 3 is not sufficiently large for machines running off a variety of stocks, and while a different set of speeds can be obtained by changing pulleys, this is necessarily awkward and time-consuming and will rarely provide exactly the right combination of speeds. Hence, further development was necessary, which resulted in the present fully developed paper-machine drive.

In this drive the constant-speed end of the machine is driven, as always, by a constant-speed motor M, Fig. 5, and the variable-speed line by a direct-current motor D with speed adjustment by field control, as in the last method; but a new element is added in the form of a separate direct-current generator G for each paper machine. The special feature of this generator (which can be driven in any desired way, but preferably by a synchronous motor) is that it is separately excited by a small exciter E, the motor also being separately excited by the same exciter. By varying the current supplied to the main-generator field, the delivered voltage can be varied; thus the speed of the motor driving the paper machine can be varied over a wide range, and motor speeds above normal can be obtained by means of field control. The combination of these two methods, therefore, provides an ample speed range of about 10 to 1.

In the fullest development of this system the control is centered in push-button stations, which can be located at any desired points around the machine. There is also a speed indicator which reads in feet of paper per minute and is actuated by a magneto mounted on the shaft of the paper machine.

FIG. 7. SUPER-CALENDER OPERATED BY SINGLE DIRECT-CURRENT MOTOR

When the operator presses any one of the "start" buttons, a magnet-switch starter, Fig. 6, automatically starts the motor and accelerates it to a low speed. The "fast" button is then pressed and a little motor, M, mounted on the main-control panel (which is located near the motor) begins to move a rheostat arm around a series of contacts C and continues to do so as long as the button is pressed. The first contacts cut out steps of resistance in the generator-field circuit, and as the arm is gradually moved around by the little motor, the voltage rises and the motor speed increases. When the generator voltage is up to normal, the rheostat arm continues around, cutting resistance into the motor field, so that the speed of the motor continues to increase until the maximum is reached. The total number of steps is usually about one hundred.

When, however, the speed of the paper machine has reached any desired point, the pressure on the "fast" button is released, the little motor stops, and the paper machine continues to operate at that speed. On pressing the "stop" button, the machine stops; and on pressing the "start" button, the machine starts and is automatically accelerated to the speed as determined by the position of the rheostat arm. On pressing the "slow" button, the little rheostat motor moves backward, decreasing the speed of the machine as long as the pressure on the button is maintained. In this way the operator has the machine under complete control and can handle it with perfect ease.

Where a paper machine is engaged for long periods in turning out one kind of stock, full-automatic speed control is not needed and a somewhat simpler system, Fig. 5, can be used. In this system the push-button stations have only "start" and "stop" buttons, the speed of the machine being adjusted as need arises by turning by hand a rheostat handle on the controller *C* and *C* in the figure.

The latest development in paper-machine drive is to drive the rolls with several motors instead of one. By this system each section of the variable-speed line of the paper machine is driven by an individual direct-current adjustable-speed motor connected to the shaft of its section by a reduction gear. This system not only simplifies the transmission of power to the rolls, but permits an extremely accurate control. An automatic speed regulator is employed, which controls the speed not only of the whole variable-speed line but of each motor also, so that different relative speeds can be obtained. So sensitive is this regulator that speed variations as small as one-tenth of one per cent are instantly detected and corrected automatically.

This system possesses a number of important advantages. It requires but 50 per cent of the power of the older drive; it makes possible a greatly increased speed of the paper machine, since the regulation is so accurate that there is no danger of breaking the paper as it passes from one section to the next; the cost of maintenance is reduced, owing to the elimination of belts, gears, clutches, etc.; and no basement is needed under the paper machine.

MOTORS FOR COATERS AND SUPER-CALENDERS

Coating machines and winders should be driven by variable-speed motors, and alternating-current wound-rotor motors are satisfactory for this service. But if direct current is available, it is better to use direct-current motors with speed adjustment by field control. The preferred control for these motors is a push-button station for starting and stopping, with a manually operated speed rheostat on the control panel. One of the earliest forms of motor drive for supercalenders was to use a small constant-speed motor, with ample gear reduction, for the slow threading-in speed, and a large constant-speed motor for the running speed, clutches being used to change from one speed to the other. This method, of course, offered no special operating advantages over belt drive, so a variable-speed wound-rotor was soon substituted for the large constant-speed motor. The wound-rotor motor not only permits smooth acceleration from the low speed to running speed, thus reducing the time lost in repairing breaks, but also provides a speed range of from normal to about 50 per cent below normal, which makes it possible to run off a variety of stocks at economical speeds. The method has proved satisfactory and is in use in a number of mills. A controller can be provided that controls the two motors as one, allowing the operator to start, speed up, and slow down machine at will.

A modification of this method is the use of a single wound-rotor motor, for both the high and the low speeds. The low speed is obtained by driving through a large gear reduction, and the high speed by means of a smaller gear reduction, variations between the two being obtained by means of a drum controller. This is a simpler arrangement than the two-motor drive, but is not as efficient when the low speed is used for long periods of time.

Where a wider range of speeds is needed, a direct-current motor is used, as in Fig. 7. A very low speed, approximately one-half the motor's normal speed, is obtained by the use of armature control, and speeds ranging from normal to three times normal are obtained by field control. A drum controller can be used with this method, the very low speed being obtained on the first step and the various running speeds on the remaining steps, of which there are usually about twenty. Or, an automatic controller, with push-buttons, similar to the paper-machine controller, can be employed.

Prolonging the Life of Condenser Tubes

BY JULIAN N. WALTON

The problem of maintaining surface condensers leak-proof has been a source of continued trouble to the condenser operator. To provide for contraction and expansion of the tubes, the condenser manufacturers have always realized that it was necessary to employ a packing box at each end of the condenser tube. It has been common practice to pack the tubes with either corset lace, cotton wicking or the later-developed packing which employs a pressed fiber bushing.

In some of the small condenser units manufacturers have attempted to lessen packing troubles by designing

FIG. 1. RINGS INSTALLED WITH FERRULES IN PLACE

a condenser in which the tubes are expanded into one head, stuffing-boxes and packing being employed at the other ends of the tubes to provide for expansion and contraction.

Surface-condenser efficiency is dependent upon high vacuum and high vacuum is dependent upon steam-tightness of the condenser tube packings. Today a 10,000- or 12,000-tube unit is not unusual. With units of this size or even smaller the length of tube necessary to provide sufficient surface eliminates the possibility of adapting to the larger sizes of condenser the process of expanding the tubes into one tube sheet. This makes it necessary in the larger units to employ packing in both condenser ends.

Cotton wicking, corset lace and fiber rings have not given entire satisfaction in this service. These materials have but a temporary life and will eventually bake hard or rot out, causing leaky tubes. In many instances fiber bushings have swelled and baked to the tubes, holding them so rigid that sliding cannot occur with

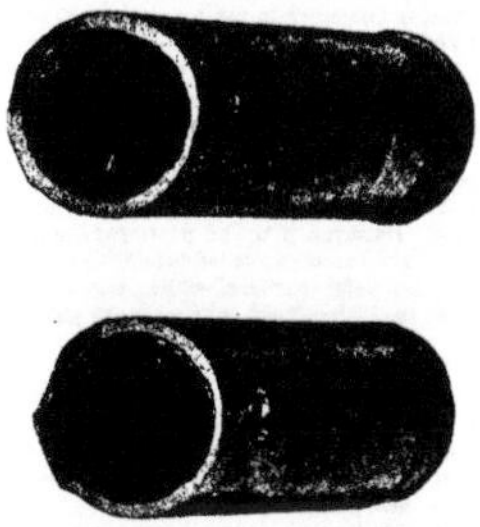

FIG. 2. SHOWING HOLES AND THIN SHELL OF TUBES

the temperature changes to which the tube is subjected, the result being that a tube so held is constantly subjected to tension and compression stresses which frequently cause buckling or cracking.

Of late years flexible metallic packing sets have been used to overcome these troublesome features. These sets are easily installed, there is no crushing of tube ends, and tubes so packed are free to move, owing to contraction and expansion with temperature changes. Being of metal composition, the rings should last as long as the shell of the condenser and are not readily affected by chemical action resulting from impure circulating water or from the addition of cleaning chemicals.

Fig. 1 shows the rings installed in the boxes of the condenser end with the ferrules in place. While the primary idea in using this expensive method was to develop a condenser packing that would be permanent and obviate troublesome features previously encountered, it develops that this process also has possibilities in prolonging condenser tube life.

Possibly a matter of even greater importance than that of packing maintenance in surface condensers has been the problem of the tube life. All condensers require brass or copper alloy tubes, and in many installations these tubes are subject to decomposition which is commonly attributed to either the improper alloy employed in the tube or to acid or alkali contained in the circulating water.

Many engineers have claimed that condenser-tube disintegration was a matter of galvanic or electrolytic action. Others have strongly opposed this theory, preferring to hold that the disintegration of the tubes was due to a metallic impurity or improper selection of materials entering into the condenser fabrication; and these engineers attribute all condenser-tube destruction to a process commonly termed by electrochemical engineers as that of "local action." Fig. 2 is a photographic reproduction showing disintegration of condenser tubes that had been in service but two years.

Present manufacturers of condenser tubes are using extreme care in the alloying of metal for tube production to the extent that the manufacturing of condenser tubes has become a chemical science, and it is logical to assume that the metals now used are of such an alloy as would offer the greatest resistance to the acid or alkalies that are so common in circulating water. But in spite of this, it is observed that the tube disintegration continues after the tubes are once placed in service in the condenser.

Disintegration Due to Electrolytic Action

Therefore it would seem that it is time to conclude that this tube disintegration is more a matter of electrolytic or galvanic action than it is of local action and that this process is due to electrolytic breaking down of the condenser-tube metals. The fabric packings have been largely responsible for this condition, for a condenser tube packed with an insulating material must be considered as an anode suspended in an electrolytic bath. The difference of potential that produces the current, which in turn destroys the metal, is produced by electrochemical differences of potential due to the dissimilarity of the metals, the tubes being of one composition, the shell of another, and in many instances the tube sheet of still a third metallic composition.

Reference to a table of electrochemical equivalents will indicate the difference of potential due to the dissimilarity of these metallic groups. While it is true that these metallic ingredients entering into the tube alloy are selected with a view to keeping it as near neutral as possible to the electrochemical equivalent of the shell, the fact remains that there is always enough difference to establish a difference of potential. Establishing this difference of potential results in the building up of minute currents which flow from the tube through the electrolyte or the impurities contained in the circulating water and thence to the tube sheet, and it is logical to assume that where a current leaves a condenser-tube surface through the medium of an electrolyte, the general laws of electrolytic action would apply.

Mettalic Packing Serves as Bond

Admitting that condenser-tube deterioration is due to this very process, it is easy to see that the metallic packing sets bond the condenser tube to the tube sheet, thereby offering a metallic path through which these currents may flow. There may then be present in the condenser this electrochemical difference of potential, but provision is made that the current so produced will flow through the tube rather than from the tube through the electrolyte.

Electrolytic action will be obviated, therefore, in the same manner that the destruction of water pipes is eliminated through the process of bonding. In other words, if the tubes are short-circuited, any currents flowing will be confined to metallic paths, and it is well-known that electrolytic decomposition of metal cannot occur unless an electric current leaves the metal through the medium of a suitable electrolyte.

Lubrication of Internal-Combustion Engines

By W. F. OSBORNE

Beginning with a discussion of the action of lubricants during each stroke, this article takes up in detail the various construction and operating conditions which affect the choice of oil for internal-combustion engines. A list of lubricating troubles is given, together with the causes of each.

WHEN a gas-engine man talks about lubrication, the first questions he asks are: Will the oil lubricate the cylinders? Will it provide proper piston seal? Will it form any carbon deposits? What will be the consumption?

There are so many types of internal-combustion engines, each having its own peculiar requirements, that it would be practically impossible to do justice to the entire subject within the scope of this article. We will therefore discuss only those types in which the power-plant engineer will be interested: Gas engines, gasoline and kerosene engines, fuel-oil and heavy-oil engines of the stationary types.

Considering first the lubrication of the power cylinder, the lubricating oil must do two things—lubricate the cylinder walls, piston and piston rings, and assist the piston rings to form a seal. As the greater proportion of power-plant engines operate on the four-stroke cycle, let us first investigate what happens to the oil on each of these four strokes.

Lubrication During Various Strokes

On the charging stroke the cylinder walls, as far as the piston travels, have been covered with a film of oil from the previous stroke, which, together with the oil on the piston and between the rings, acts as a lubricant for the piston and rings. The oil in the clearance space between the rings assists them in forming a seal to prevent leakage of air or vapor from the crank case into the cylinder. The air and vapor drawn into the cylinder is comparatively cool, and if the fuel is completely vaporized, there will be no detrimental effect on the film.

The film of oil left on the cylinder walls during the charging stroke, together with the new film picked up by the piston from the crank end of the cylinder walls, lubricates the piston on the compression stroke, and as the piston moves toward the cylinder head it smears a fresh film of oil on the walls. The supply of oil collecting on the advancing side of the piston rings and around the upper edge of the piston serves to prevent leakage of the gases being compressed. The degree of compression will depend, therefore, somewhat upon the seal-forming properties of the oil and on the condition of the cylinder, piston and rings.

During the compression stroke the temperature of the gases gradually increases, but does not rise high enough to do any damage to the oil film until near the end of the stroke, when most of the oil film is covered by the piston.

The film of oil placed on the cylinder walls by the piston on the compression stroke lubricates the piston on the explosion stroke. As the piston moves toward the crank, exposing the walls to the high temperature of the burning gases, the flame comes in contact with the oil film, but only after it has served its purpose of lubricating the piston on the in-stroke. The greater part of the damage to the oil film occurs on this stroke. The oil between the rings and between the piston and cylinder walls is subjected to the pressure of the burning gases and must assist the piston rings in preventing loss of power through leakage of the gases out of the open end of the cylinder.

If there is any trouble with lubrication, it will be encountered on the exhaust stroke, as the oil film on the cylinder walls has just been exposed to the high temperatures of the burning gases and has been more or less damaged. However, if the oil possesses the proper characteristics and has been applied in sufficient amount, some lubricating value remains, which, together with the oil film on the piston itself, lubricates the piston for this stroke.

Features Determining Kind of Oil to Use

Some of the features in the mechanical construction and the operating conditions of the engine, which determine the kind of oil that will satisfactorily meet the requirements previously mentioned, are: Cylinder temperatures, ignition, oiling systems, carburetion, cooling systems, revolutions per minute, fuel, piston clearances, atmospheric conditions.

Although it is a difficult matter to make accurate determinations of the actual temperatures within the cylinders, estimates made by various authorities are close enough for our purpose. The maximum temperature obtained immediately after firing is about 2,700 deg. F. and the minimum temperature during suction stroke, 250 deg. F. The average temperature for complete cycle is 950 deg. F. It should be remembered that these are temperatures of the gases and not the temperatures of the cylinder walls or of the oil film.

The inner surface of the cylinder wall probably ranges from 30 to 60 deg. F. hotter than the cooling water, depending on the thickness of the metal and the efficiency of the cooling system. As long as the water is not boiling, the cylinder-wall temperatures will hardly be more than 260-270 deg. F., and with a normal cooling-water temperature of 110 to 120 deg. F., at which many power-plant engines operate, the cylinder-wall temperature is below 175 deg. F. The oil film itself is hotter than this, at least on that side of the film exposed to the burning gases. The maximum temperature of 2,700 deg. F. occurs only when the piston moves toward the crank, uncovering the oil film and exposing it to the flame until near the end of the stroke. The inner surface of the oil film in contact with the cylinder wall at the comparatively low temperatures of 175-200 deg. F. is affected very little by the high temperatures, but the outer surface of the oil film directly exposed to the flame is undoubtedly damaged.

As it is impossible to produce a petroleum lubricating oil of any kind having a flash point over 700 deg. F., it would seem that the oil film exposed to an average temperature of 950 deg. F. would be destroyed promptly.

That such is not the case the engineer can readily prove to himself by pouring a few drops of oil on a red-hot plate. It will be several seconds before the oil is completely evaporated. If the engine is running 2,000 r.p.m., as automotive engines sometimes do, the oil film will be replaced twenty times a second; and even if the engine runs as slow as 30 r.p.m., a new oil film will be spread over the entire cylinder surface once every two seconds. So it is a comparatively simple matter to maintain the oil film if oils of suitable characteristics are used.

Influence of Completeness and Rapidity of Combustion on Selection of Lubricant

The completeness and the rapidity of combustion have a great influence on the kind of lubricant to be used and its life. If the fuel is delivered to the cylinder completely vaporized and thoroughly mixed with the correct amount of air, so that combustion is completed very soon after the beginning of the stroke, maximum temperatures are obtained when the piston completely covers the working surfaces of the cylinder, thus protecting the lubricating film from destruction. If the fuel is not completely vaporized and intimately mixed with air in the right proportions, it burns slowly, and as the piston uncovers the oil film, the burning fuel destroys it. An excessively lean mixture, through the presence of free oxygen, will allow the oil film to be completely burned up or at least so badly coked as to form carbon deposits. The bad effects of slow burning can be overcome sometimes by using a heavy-bodied oil.

Engines burning liquid fuels such as gasoline, kerosene or fuel oil, always have incomplete vaporization to contend with. When the liquid fuel enters the cylinder in the form of minute drops carried by the inrushing air or sprayed in through an injector, on striking the cylinder walls, some of the small particles adhere in a liquid form. The liquid fuel, having a solvent action on the lubrication oil, thins it down, frequently destroying the film. When a carburetor is used, a too rich mixture will cause some condensation of the fuel during the compression stroke, which also affects the oil film. This thinning down of the oil film may allow some of the oil to work up into the combustion space or be blown past the piston rings and out of the open end of the cylinder. You are all familiar with the drippage of fuel and oil from the open ends of the earlier types of kerosene-oil engines.

This leakage of fuel is a particularly serious matter for engines having inclosed crank-case oiling systems, because the dilution of the oil with the fuel soon thins it down to such an extent that it is no longer able to perform its duty of lubricating the bearings and cylinders or of providing a proper piston seal. Bearings many times are burned out when using a high-grade oil through no fault of the oil, but because of incomplete combustion of the fuel.

Incomplete Combustion Causes Carbon Deposits

Another trouble caused by incomplete combustion is carbon deposits. When carbon is formed, most engineers are inclined to place the blame on the oil. Imperfect combustion always results in the formation of carbon of some sort. Perhaps it will be nothing more serious than a smoky exhaust, but if the soot is caught by the excess of oil on the walls of the combustion space or on the valves, it soon becomes hardened and builds up rapidly with the addition of more oil and more soot. The lighter-bodied oils, which evaporate more quickly than the heavier oils, do not hold this soot so tenaciously. Consequently, when it is known that an engine has a tendency to smoke, it is highly advisable to use a light or medium-bodied oil. (The foregoing statement refers to black smoke caused by incomplete combustion. If the smoke is blue from burning oil, use a heavier oil).

Any method which the engine builder or the operating engineer can devise to improve the fuel combustion will at the same time improve the lubrication of the engine and increase the life of the oil, cylinder, piston and bearings.

Effect of Various Fuels

Natural gas, blast-furnace gas, producer gas and coke-oven gas, being comparatively slow-burning, do not produce high temperatures, but the heat generated is maintained over a considerable portion of the stroke, thus exposing the oil film to very severe conditions. Furthermore, these slow-burning gases are of necessity used in slow-speed engines because of the time required to complete the combustion, and a thicker oil film on the walls is necessary than for the higher-speed gasoline engines. Kerosene, being slower burning than gasoline, permitting of higher compression and requiring higher cylinder temperatures, imposes a severe service on the oil film, which must have exceptionally good lubricating qualities to stand up for any length of time. Fuel oils, too, are difficult to completely vaporize and have a pronounced tendency to smoke, although there are some designs of engines that have overcome this trouble to a certain extent. With the exception of the speed of combustion and the amount of noncombustible material carried into the cylinder by the fuel, the lubricating problem when using the various fuels becomes one of combustion. As the methods of carburetion and vaporization improve, there is no reason why the engines using fuel oil should be any harder to lubricate than those using the highest-test gasoline.

With ignition we have again the problem of combustion. If ignition takes place at exactly the proper time, resulting in the most complete combustion possible, maximum temperature will occur when the piston is very nearly on dead-center, and as the piston recedes, uncovering and exposing the film to the burning vapor, the temperature rapidly falls off. If the ignition is retarded, a slower and later burning occurs extending over a considerable portion of the stroke. This longer exposure of the oil film to the high temperature damages it and in fact may completely destroy it. Continuous retarded ignition quickly heats up the cylinder walls and the cooling water, thinning down the oil and frequently reducing its viscosity to the danger point.

The temperature of the cooling water influences the kind of oil to be selected when certain types of lubricating systems are used. A low temperature cools the cylinder walls and will sometimes prevent a heavy-bodied oil from spreading properly so that a portion of the cylinder may become dry. When experience with certain oils shows that this occurs, it can easily be remedied by using a lighter-bodied oil. Kerosene-oil engines, especially of the carburetor types, usually depend very largely upon the heat of compression to vaporize the fuel, and such engines are generally operated with very hot circulating water. In so far as the higher temperature improves the fuel combustion, it will also lessen the strain on the oil film, and although it may thin down the film, the final result is much better than if the liquid fuel were allowed to wash the oil film away.

Importance of a Proper Seal

The operation of any internal-combustion engine is vitally affected by the degree of perfection of the seal of the piston within the cylinder. A practically perfect seal may be made by tight pistons or piston rings, but the improved compression and decreased leakage of gases is secured at the expense of greatly increased friction; in fact, the rings can be made so tight that the engine cannot be turned over at all.

A much more desirable form of seal can be secured and maintained by making use of the lubricating oil film necessarily existing for lubrication. As the piston advances, a quantity of oil builds up on the edge of the piston head and on the advancing side of the piston rings, which opposes the force of the compressed gases, so that if the oil is of sufficient body or viscosity to resist the force of the gases, a perfect seal is secured.

When the engine is cold, the clearances are much larger than when the engine is heated up to its running temperature. As good compression is of the greatest importance in starting a cold engine, the oil film can be of much assistance. Fortunately, petroleum oils have the characteristics of becoming thicker as the temperature is lowered, so that the oil film around the piston rings more satisfactorily resists the pressure within the cylinder.

Benefits of a Perfect Seal

A perfect seal maintained by lubricating oil secures many benefits:

1. Minimum lubricating oil consumption since a suitable oil, by maintaining a perfect seal, prevents the pressure of compression and explosion from wasting the oil or blowing it off the cylinder walls past the piston rings. It also prevents excessive quantities working into the cylinder from the suction stroke and being carried into the combustion space, where it is destroyed.

2. Minimum fuel consumption by preventing leaking of the fuel past the rings on compression and wasting of power on the firing stroke. A perfect seal does not allow excessive quantities of oil to work into the firing chamber, fouling spark plugs and interfering with the proper combustion of the fuel.

3. Minimum friction through the use of loose-fitting pistons and rings working on a lubricating film.

4. Reduced carbon deposits by cutting down the amount of oil working into the combustion space.

5. If a crank-case oiling system is provided, a perfect seal prevents the greater part of the hot gases from reaching the crank case, maintains lower oil temperatures and thus prevents excessive bearing and cooling-water temperatures.

6. In the operation of crank-case oiled engines a perfect seal reduces the amount of liquid fuel or vapors passing the rings and later condensing in the crank case, destroying the body of the oil. When the oil is thinned down by a mixture of liquid fuel, it is no longer able to maintain the piston seal at its original point and conditions rapidly become worse and worse.

In general we may say that when pistons and rings are tight we should use a light or medium-bodied oil; when they are loose or become worn, a heavier oil is necessary.

Oiling Systems

The method of supplying the oil to the piston and bearings affects the viscosity of the oils that can be used. The oil must always be thin enough to flow from the point of supply to the parts requiring lubrication. Heavier oils can be used with mechanical lubricators or pumps than when drip cups or splash systems are provided. With the inclosed crank-case oiling system we must always consider the possibility of fuel dilution, and any method that can be devised to continuously remove the fuel from the crank-case oil, thus maintaining its body, will be of great value in improving the efficiency of lubrication as well as the fuel economy.

Although some engine builders have carefully considered what lubricants will be most suitable for the parts of their engines requiring lubrication and then have designed their oiling systems accordingly, others have not. It would be a great benefit to the operating engineer if a larger number of the engine builders would give this phase of the subject more thought.

Contrary to the usual belief a high-speed engine is easier to lubricate than a slow-speed engine, because a new oil film can be placed on the cylinder walls more times per minute. If the oil is light enough to allow rapid spreading, this new film can be formed as fast as the old film is destroyed. The slower-speed engines allow more time for the oil film to be burned off, and a heavier-bodied oil is required. The lighter oils as a rule form less carbon than the heavier oils when used in the correct amounts, and on this account the engines using them are less likely to have trouble. When engines are located where they are exposed to low temperatures, it is necessary to have an oil with a pour test low enough for the oil to reach the parts requiring lubrication. If a splash system is used, a lower pour test is required than when a force-feed lubricator is used.

Some Lubrication Troubles

Now we come to some of the lubricating troubles: Dry cylinders, wet cylinders, carbon deposits, bearing troubles.

Dry cylinders may be caused by: (1) Oil burning off the cylinder; (2) too lean a fuel mixture; (3) too rich a fuel mixture; (4) imperfect carburetion; (5) late ignition; (6) dilution of oil in the crank case; (7) too hot circulating water; (8) too cold circulating water; (9) tight pistons or rings; (10) too heavy an oil; (11) too light an oil; (12) oil with too light a cold test; (13) not enough oil fed.

Wet cylinders, cylinder heads, valves and spark plugs may be caused by: (1) Two rich a fuel mixture; (2) imperfect carburetion; (3) failure of ignition; (4) dilution of oil in crank case; (5) excessive oil feed; (6) loose pistons or rings; (7) too thin an oil; (8) low temperature of cooling water and resulting lack of fuel vaporization.

Carbon deposits may be caused by: (1) Loose pistons or rings; (2) excessive oil feed; (3) too thin an oil; (4) too heavy an oil; (5) cold circulation water; (6) late ignition; (7) failure of ignition; (8) too lean a mixture; (9) too rich a mixture; (10) improperly vaporized fuel; (11) dirt and other noncombustible material carried in with fuel or air; (12) improperly refined lubricating oil.

Worn out or burned out bearings, in inclosed crank-case engines: (1) Hot oil from leaking gases past piston; (2) oil too thin; (3) oil diluted by leakage of fuel from cylinders; (4) mechanical faults in bearings or bearing adjustments.

When any one of these things can happen to almost any engine, and as almost every engine may have one or more of these faults, it becomes a difficult matter to make any specific recommendations. We can make a general specification that will apply to the greater number of installations as follows: The oil should consist of highly refined and filtered pure mineral oils, having a cold test sufficiently low for the engine in question and of the correct viscosity.

It will be noted that no mention is made of gravity, flash, fire or crudes. Suitable oils for internal-combustion engines are made from both asphaltic and paraffin crudes. The correct viscosity can be determined only by a careful examination of the engine and all the surrounding operating conditions. The most economical oil can be determined only by actual test under operating conditions. It is really up to the engine builder to find out and advise his customers what kind of oil to use.

Testing on a Grounded Circuit

By E. C. Parham

WHETHER circuits be grounded by design, as are most direct-current railway circuits and many 110- to 220-volt three-wire circuits, or whether they be grounded by a fault, as may happen to any circuit which normally is insulated, precautions must be observed when applying a test-lamp circuit to but he may get into trouble on a circuit that is supposed to be insulated but is not, if he is careless enough to treat the circuit as if it were clear rather than find out by test the conditions existing.

Fig. 1 gives the connections of a direct-current motor and starter on a circuit assumed to be clear of grounds.

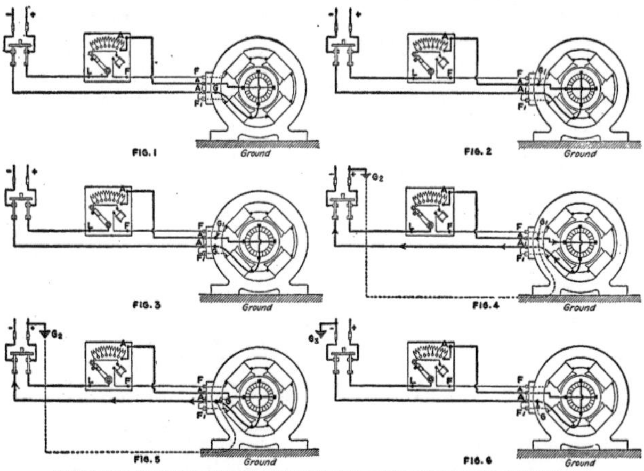

FIGS. 1 TO 6. SHOWING DIFFERENT CONDITIONS OF GROUNDS OCCURRING ON MOTOR CIRCUITS

the locating of fault grounds on such circuits. Otherwise, not only may testing results be absolutely false, but the tester may be subjected to flashes that may result in serious burns or in a dangerous fall. An experienced man is not likely to come to grief on a circuit that is purposely and permanently grounded, because the fact generally is well known; If the line is clear of grounds, a single ground on the motor circuit, as at *G*, will not affect the operation of the motor even if the frame of the machine was grounded. In fact, this would be the actual condition existing on a ground return system. Nor would a single ground, as at G_1, Fig. 2, affect operation, because the circuit being everywhere else clear of grounds, there would be no ground path by means of which any part of the motor circuit could be bypassed. A ground at G and at G_1, Fig. 3, however, would bypass the motor-armature circuit, not only rendering it inoperative, but inviting destruction of the starter unless relief was promptly given by the blowing of a fuse. A line ground, at G_1, Fig. 4, for example, would not affect operation of the motor or of the starter unless one of these devices itself included a ground. A second ground at G_1, however, would bypass the starter, as indicated by the dotted line and arrowheads, and would blow the left-hand fuse immediately on closing the switch. A motor ground at G, Fig. 5, in addition to the line ground at G_2 would bypass the whole motor circuit, including the starter, and again would, on closing of the switch, cause the left-hand fuse to blow. It becomes manifest, then, that a motor ground is likely to cause trouble if a line ground exists, unless the motor ground and line ground are on the same side of the circuit as are G and G_1 in Fig. 6. Whether a ground interferes with the operation of the motor or not, it is a bad thing to have, because it invites other grounds and is a menace to the operator in that he will get a shock if he touches the ground and any contact having to do with the other side of the circuit.

To test for a grounded circuit, connect one end of the lamp circuit first to A, as in Fig. 7, then touch the other end of the lamp circuit to ground, next connect one terminal of the lamp to B and the other terminal to ground. If either connection causes the test lamps to light, it means that there is a line ground on the opposite side from the one that the lamp lights on and the lamp circuit connection that gives the light is the one to be used when testing. With a lamp in the test-lead from the ungrounded side of the test circuit, there will be no attending danger in touching either test lead anywhere. Suppose, for example, that it is desired to test for ground, an armature the motor frame of which is mounted on the structural iron part of a building and is therefore grounded. After determining the side of the circuit that the lamp lights on, as explained in Fig. 7, one of the lamp test leads is

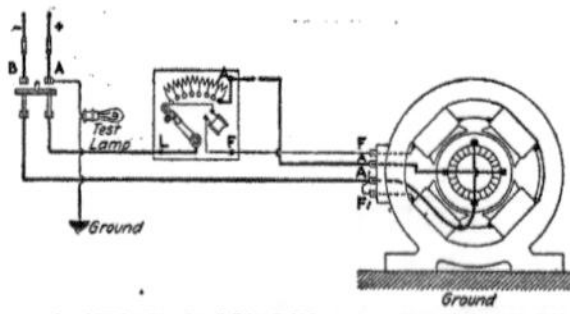

FIG. 7. METHOD OF LOCATING A GROUND ON ONE SIDE OF CIRCUIT

connected to this side of the circuit and the other is applied to the commutator after the brushes have been raised, as in Fig. 8. If it is desired to test the field coils for ground, the test lead that is applied to the commutator, Fig. 8, is applied to the field-coil terminals. To make sure that a return circuit will be provided in all cases, a lead can be brought from the grounded side of the line to the motor frame, as indicated by the dotted line Fig. 8. Where only one lead is used, this lead is touched first to the motor frame to make sure that the lamp will light, and then it is touched to the part the insulation of which is to be tested.

In the vernacular of the repair shop a ground means any irregular contact between the current-carrying part of a device and its metal containing case. Thus, if one of the coils of a starting rheostat sags down and touches the containing box, the coils and box will light

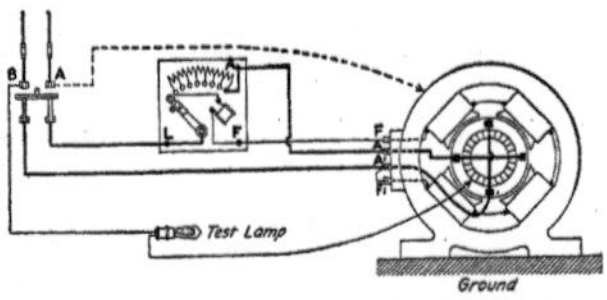

FIG. 8. CONNECTIONS FOR TESTING MOTOR'S ARMATURE FOR GROUNDS

out together and such a fault would be called a ground. This, however, would not be a ground to earth unless the box were so installed as to permit of the containing case actually making contact with the earth.

Safeguarding Pipe Lines Against Collapse

By John S. Carpenter*

Every hydraulic pipe line should be provided with a properly proportioned air vent to protect the pipe from collapse if a vacuum is created. In the case of a break the first thought is to shut off the water at the upper end and thus prevent further damage by the escaping torrent. In so doing, a vacuum will be formed in the pipe, owing to the fact that the discharging water excludes the atmosphere until the line is nearly empty, unless other means of access has been provided. Without such a vacuum breaker the pipe, if of any size at all, may collapse.

Some pipe lines have been designed to take the full atmospheric pressure without collapse, but this is very expensive in comparison with the cost of a properly designed air vent. Other engineers go still further and attach heavy air valves at frequent points along the line, especially at changes of grade. When there is no pressure in the pipe or if it is much below normal, these air valves open and admit air. A single air vent at the top of the line is usually sufficient unless the pipe is very large in diameter or the bends are sharp. These vents are seldom over eight feet long, and should be proportioned to supply air at such a rate that the difference between atmospheric pressure and that inside the pipe cannot exceed six pounds per square inch.

Let us assume that a pipe line has a break at its very bottom and that the regulating gates or sluices at the upper end have just been closed. The maximum velocity that would be possible for the water to acquire under frictionless conditions is

$$v = \sqrt{2\,gh} \qquad (1)$$

where

v = Velocity in feet per second;
g = Acceleration due to gravity = 32.16;
h = Total head on the pipe line in feet;

*Assistant Engineer Hydraulic Turbine Department, Worthington Pump and Machinery Corporation.

and the water would discharge at the maximum rate of

$$Q = 0.7854\ D^2 \sqrt{2\ gh} = 6.3\ D^2\ h^{1/2} \qquad (2)$$

where

Q = Water discharged in cubic feet per second;
D = Diameter of the pipe in feet;

and the other symbols are as before.

It follows that air must be supplied at the same maximum rate, which, of course, includes a margin of safety; because the water will not quite acquire a velocity equal to the spouting velocity owing to the friction. From "Marks' Handbook" (first edition, p. 360, case 2) I take the formula for the flow of air through pipes, the drop in pressure being considerable:

TABLE OF FIVE-HALVES POWERS FOR VARIOUS PIPE SIZES

Size	Power	Size	Power	Size	Power
3	16.50	5	53.50	9	242.0
3½	24.25	6	90.50	10	315.0
4	32.50	7	131.0	11	400.0
4½	43.00	8	185.0	12	500.0

$$d_1 = 1.294 \left(\frac{c\ Q_1^2\ L}{P_1^2 - P_2^2}\right)^{1/5} \qquad (3)$$

where d_1 is the diameter of the air vent in feet, L the length in feet, and $P_1 - P_2$ the drop in air pressure. The coëfficient c is properly a variable quantity, but after taking six representative cases, it was found that without serious error it could be assumed constant with a value of 0.0000013. As already stated, the length is usually 8 ft. and the maximum drop in pressure 6 lb. per sq.in. Multiplying by 12 so that d is the diameter of the air vent in inches, equation (3) will now read

$$d = 1.294 \times 12 \left(\frac{0.0000013\ Q^2\ 8}{15^2 - 9^2}\right)^{1/5} \qquad (4)$$

which reduces to

$$d = 0.58\ Q_1^{2/5} \qquad (5)$$

Solving for Q_1 gives

$$Q_1 = 3.91\ d^{5/2} \qquad (6)$$

This is the volume of air at atmospheric pressure, and the volume at the pressure in the pipe will be

$$Q = Q_1 \times \frac{P_1}{P_2} = 3.91\ d^{5/2} \times \frac{15}{9} = 6.50\ d^{5/2} \qquad (7)$$

Equating this to the Q in equation (2) we have

$$6.50\ d^{5/2} = 6.3\ D^2 h^{1/2}$$

and finally

$$d^{5/2} = 0.97\ D^2 h^{1/2} \qquad (8)$$

When the conditions are such that a vent pipe larger than 10 in. is required, it is better to use one 10-in. vent at the top of the line and distribute the balance along the pipe line in the form of air valves, particularly at the changes in grade.

Reports of the United States Bituminous-Coal Commission to the President

UNDER date of March 10, 1920, the majority of the commission appointed by the President to investigate the conditions in the bituminous-coal industry submitted a report containing recommendations and awards, and two days later the minority member submitted a dissenting report. This commission consisted of Henry M. Robinson, John P. White and Rembrandt Peale. The majority report was signed by Mr. Robinson, who was appointed as representing the public, and by Mr. Peale, as representing operators. The minority report was signed by Mr. White, as representing the mine workers.

The important features of the majority report provide for the elimination of the 14 per cent increase in wages fixed by the United States Fuel Administration, and instead of this an average increase of 27 per cent, which shall take effect April 1, 1920, and continue to and including March 31, 1922. The details of this average increase provide that an advance of 24c. per ton shall be paid for mining mine-run coal, but that day labor and monthly men be advanced $1 per day and that trappers and boys receiving less than men's wages be advanced 53c. per day; that the fulfillment of agreements shall be guaranteed by officers of the International Union as well as by the local officers of any district in which local agreements are made; that the eight-hour day in effect on Oct. 31, 1919, shall be written into agreements for which these earnings constitute a basis. Other points, such as that of pushing cars, the use of machinery in mines and charges for blacksmithing are dealt with, and provision is made for the creation of joint commissions to study local situations where the finding of a general commission such as this is obviously incomplete. In addition to the summary of awards the commission makes a number of specific recommendations to the public and to the miners and operators. The most important of these refer at some length to the methods of stablizing the coal industry throughout the year, to the methods of assigning cars to the various mines and to a general improvement of the conditions in mining communities, which are often under entire control of the operator.

In the body of the report considerable space is given to the question of the intermittency in working days in the coal-mining industry. Statistics from the United States Geological Survey are presented to show the extent of the intermittency, and the necessity for coöperation between the public, the operators and the railroads is emphasized. In arriving at the exact amount of increase in wages recommended, the report states that the following items were considered: The increase in the cost of living, the increase in wages received by workers in other industries and the effect of an increase in wages on the cost of producing coal. As to the probable distribution of the increase in wages toward the cost of coal to the public, the report says: "Whether all or any part of this amount will be passed on by the operators to the public depends on competitive conditions, but we estimate that the carrying out of other provisions of our findings will save the general public more than the cost of additional wages. It is to be expected that when all Government regulations are withdrawn and a competitive market for coal is re-established, the overexpansion of the industry will not permit the operators to add to the price of the coal all the increased compensation granted to the miners." It is pointed out in another portion of the report that under a technicality in existing contracts the 14 per cent increase recently ordered by the Fuel Administration was in most cases passed on to the consumer.

A table is shown based on returns of 1,551 operators to the Bureau of Internal Revenue for the year 1918. Of this number about 22 per cent reported net losses and about 11 per cent reported incomes of less than 5 per cent on invested capital. An examination of the table, however, shows that these two classes of operators, while large in number, produce only about 11 per cent of the total tonnage represented in the returns tabulated. It is to be expected that companies that were not able to show a greater return than 5 per cent during the banner year of 1918 will discontinue operations under competitive conditions. Returning to the question of better stabilization of the coal-mining industry, the report expresses the opinion that the elimination of the winter peak and summer slump in demand would result in only those mines necessary to supply the steady year round demand being operated. Under these conditions the industry would not be compelled to support the excess equipment and the excess man power which is at present necessary to meet the uneven demand.

Regarding the question of voluntary absenteeism on the part of the miners it is pointed out that the figures submitted make no allowance for turn-over. In these tabulations every man who works in a mine at any time during the month is counted on the same basis as one who is on the roll every day that the mine was in operation. A man who works thirteen days out of a possible twenty-six in one mine and thirteen at another would be counted in these figures as two men with an aggregate voluntary absenteeism of twenty-six days, while, as a matter of fact, he has really worked an entire twenty-six day month. A further defect is that the number of days that a mine is in operation is not necessarily the number of days that the entire force have an opportunity of employment. Part of the mines may have been closed for various reasons, while other parts were working.

The demand for a six-hour day or a thirty-hour week was refused on the ground that such action would simply exaggerate the evils which it is hoped may be lessened by the efforts now being made to stabilize the industry.

The minority report submitted by Mr. White, while agreeing with the conclusion of the majority in a number of cases, differs in certain fundamental points. Even while agreeing with certain of the awards, Mr. White points out that the methods of arriving at these awards are not satisfactory from the miner's standpoint and are accepted only in the interest of honorable compromise.

Mr. White objects to the use of percentage increase where differential rates are allowed as in the case of pick-and-shovel mining as compared with machine mining. In such cases a percentage raise upsets the differential, which was based on a flat difference. He also advocates that all day labor and monthly men be advanced $1.35 per day and trappers and boys receiving less than men's wages be advanced 75c. per day instead of the figure awarded in the majority report. Furthermore, he objects to the idea of making increases correspond only to the increase in the cost of living. The last adjustment of the wage rate took place in November, 1917, and this barely brought the level of wages even with the cost of living. During the succeeding twenty-seven months the cost of living rose steadily.

Mr. White makes the claim that where the wages are adjusted only at long intervals, equity requires that they be raised enough in excess of the increase in the cost of living to cover this steady loss which goes on from month to month. Regarding the irregularity of employment and the demand for a thirty-hour week Mr. White points out that the thirty-hour week is actually about what the coal miner averages under present conditions. He asserts that the rates fixed by the majority report are based on a reasonable earning capacity of the miner and working forty-eight hours per week, which, however, the miner is unable to do owing in a large part to conditions beyond his control.

In other words, the miner is expected to work for forty-eight hours a week in order to earn a living wage, while at the same time he is prevented from working more than thirty hours per week. If labor is to be paid only on the basis of the actual hours which it must work to provide the requirements of the community, he maintains that it is unfair that capital should be paid for its continuous use if it is needed for only part of the time.

In regard to the reasons for the increases in the price of coal over pre-war prices, statistics are quoted from the Federal Trade Commission report on the coal industry. It is claimed that these figures establish the fact that the increased coal prices are not due to increased labor costs, but on the contrary are due largely to increased profits exacted by the operators, and that the increased wages to labor will be more than offset by the increased efficiency and the increased activity of labor.

Handling Hydraulic Governor During Short-Circuits

By L. W. Wyss

When a succession of transmission-line short-circuits interrupt the normal operation at hydro-electric plants, the oil pressure governors are opening and closing the turbine gates repeatedly in an endeavor to keep the speed normal. Under this condition the governor-oil pressure pumps may not be able to keep the governors supplied with oil under sufficient pressure to prevent excessive racing of the main units.

If there are spare govenor pumps, they should be used during such abnormal operation. The switchboard operator should also have his governors in mind at such times and not try out a short-circuited line often enough to cause a loss of governor pressure. Neither should he permit his station to remain connected to a line that is the source of enough severe short-circuits to cause the trouble mentioned. It is better operating practice to leave the line dead for a short time while the governors are regaining their pressure than to take chances of losing control of the turbine speed, which would perhaps result in a more lengthy interruption to service.

EDITORIALS

Agitation for Adoption of the Metric System

MANY readers have doubtless been deluged with circulars from the World Trade Club of San Francisco advocating adoption of the metric system of weights and measures in this country. This propaganda even went so far as an attempt to frame a bill in committee in Congress to make such adoption compulsory.

Accompanying the circulars are return postcards addressed to the Bureau of Standards and worded in the affirmative so that all the recipient has to do is to sign the card and drop it into the nearest mail box. He will then be counted among those unqualifiedly in favor of the metric system, and the number of cards received is counted upon to impress an unsuspecting Congress. Undoubtedly, many have deposited the cards without giving the subject much thought, basing their action solely upon the handiness of the decimal system. Few of those opposed will take the trouble to write letters voicing their opposition, and the cards do not provide for a negative expression of opinion.

Just what connection the Bureau of Standards has with the propaganda has not been revealed, but the World Trade Club of San Francisco, it has been established, is in no sense an organization such as the name would imply; instead, it is essentially a publicity bureau of a wealthy man now residing in California, who has made the metric system a hobby and who, it is understood, has to a large extent been financing the campaign.

The metric system has certain advantages and other disadvantages and, were industry starting anew, it might be desirable to adopt such a system in spite of slight disadvantages. But the real difficulty lies in the fact that there is no direct conversion between it and the system now in use, and it requires little imagination to picture the confusion that would result from its compulsory adoption. Not only would all existing drawings and patterns have to be revised or scrapped, but the workmen would have to be retrained in the use of the new system.

An incident pertinent to this may be cited. During the war the Ordnance Department adopted certain French designs and sizes and attempted to manufacture the material in this country according to the metric measurements. The results were disastrous, and finally all dimensions had to be converted into the English system. Even then, in the absence of a direct conversion factor, it meant working with very cumbersome figures instead of round numbers.

In countries where the metric system has long been in use as in France, where it has been compulsory since 1837, many of the old units still persist and are in daily use. Practice seems to have settled down to the use of those units that proved most convenient. The real danger, however, in the present propaganda lies in the fact that it emanates from an idealistic source with the apparent sanction of a body whose experience has been almost wholly along research and scientific lines; this in the face of practically unanimous opposition from the great body of American manufacturers who would be most affected by such a change. It is impracticable to legislate in such matters with any degree of success unless the demand comes from those directly concerned and after favorable sentiment has crystallized in those circles.

The Consumer Stands To Lose

A VERY brief outline of the United States Bituminous Coal Commission reports to the President is given in this issue. The fact that the commission was not unanimous is to be regretted, and a study of the reports suggests that the differences were due to the same old antagonism between the miners and operators, with very little regard to the final effect on the public.

The majority report puts considerable emphasis on the necessity of creating a more uniform market for coal throughout the year. If the country did not have to support a seven hundred million ton capacity to get five hundred million tons of coal a year, it is obvious that the public, the miner and the operator would all be better off. The report points out that a stable demand would result in the discontinuance of the less profitable mines, but this simply means that they would be opened again as the normal increase in demand required. Meanwhile, the public would be relieved from the burden of supporting the extra organizations necessary to take care of the winter peak load. Obviously, the remedy is in the hands of the consumer. If he will store coal in the slack months, this problem will be solved.

In granting increased rates to the miners, the majority report expresses an opinion that the operators will absorb this extra cost and pass little, if any, on to the public. In the same report it is noted that the fourteen per cent increase granted by the Fuel Administration was passed to the consumer by taking advantage of a technicality. With this in mind there seems to be very little hope of a twenty-seven per cent increase being absorbed anywhere but in the user's pocketbook.

The minority report lays great stress on the fact that the rates are based on a reasonable wage if the miner can work forty-eight hours a week, while in fact he only has a chance to work about thirty hours on an average. It argues, and the argument seems reasonable, that if capital is to be paid for continuous use, whether needed or not, labor should be paid on the same basis. Mr. White seems to have failed to note that neither capital nor labor can continue indefinitely to draw full-time pay for part-time work. If the present plans for stabilizing the industry are carried out (and they will be carried out), both mine and miner will have to work regularly and for a reasonable return.

The whole problem comes back to the summer storage of coal to relieve the excess winter demand. This will eliminate the annual scramble which enables a few mines and dealers to boost prices, which keeps in operation twenty per cent more mine capacity than is needed and which breeds irregular employment, even when the demand is high bcause of temporary shortage of cars, of which there is a surplus for most of the year. It is probable that in the very near future the coal user, large or small, who makes no effort to store coal, will be looked upon as an enemy to the community. Meanwhile, the man who keeps ahead of his coal consumption by reasonable storage arrangements is getting a cheap form of insurance against shutdowns.

What Constitutes Proper Elevator Inspection?

AN elevator, either passenger or freight, in contradistinction to other means of transportation, moves in a vertical direction, consequently there is always a tendency for the car and counterweights to fall to the bottom of the hoistway with disastrous results. Yet with this handicap the modern passenger or freight elevator is the safest means of modern transportation. This high degree of reliability in operation has been attained by the development and installation of the proper safety devices, and in no method of transportation have safety devices been developed to so high a degree as in elevator service.

The modern elevator equipment is provided not only with means for automatically slowing the car down and stopping it at the top and bottom landings, but also with means of cutting the power off if, for any reason, the car travels past the terminals. In case the car overspeeds, it is automatically brought to a safe stop without injury to the passengers, as is also the case if the cable parts and the car is allowed to fall. With interlocks on the shaft doors and car gates the car cannot be started until both the doors and gates are properly closed. However, even with equipment correctly designed and installed, its functions can be easily defeated by the lack of adequate inspection and maintenance. Fortunately, most elevators are inspected at regular intervals by experienced men from some accident-insurance company. These men for the most part have become experts by virtue of their long experience in detecting defects in the equipment. But even these inspections, as thorough as they may be, do not entirely prevent elevator accidents, consequently wherever possible they should be supplemented by inspections by those who are directly in charge of the equipment.

How to make a proper inspection is something that can be learned only by experience and becoming familiar with the parts where failure is most likely to occur. In this issue is published a paper, "What Constitutes Proper Elevator Inspection?" by D. J. Gitto. Although this paper was written primarily for the guidance of insurance inspectors in their work, it is an excellent guide for the operating engineer and others who have elevators in their charge to follow in making regular inspections. It is difficult to overestimate the value of regular monthly inspections, or more frequent if conditions warrant, since, if properly made, defects are detected and may be repaired at the most convenient time before they cause shutdowns or serious accidents. What to look for and the procedure to follow is one very important factor in making these inspections. This Mr. Gitto explains in his article.

Low Salaries Force Men Out of Government Service

THE rate at which technically trained men are leaving the Government service because of inadequate compensation threatens to handicap very materially the work of many of the departments. During the war some of the bureaus accomplished results of far-reaching importance, and their personnel included men of recognized standing and ability. With the rising cost of living these men, whatever their personal inclinations, can no longer afford to remain in Government employ, and are accepting more lucrative offers from private concerns.

Congress, naturally, has reacted to popular demand for curtailment in expenditure, but in doing so has not in all cases given consideration to the relative importance of the various bureaus. Many new departments were created during the war, and existing departments enlarged the scope of their activities. Some of these have already been abolished; others are hanging on in an effort to justify their existence. However, Congress, apparently without careful study of the problem, so as to eliminate unnecessary Governmental activities and consolidate bureaus where duplication of work exists, has adopted the old system of wholesale pruning of appropriations.

As might be expected, it is the strong men that are leaving and the weaker men that are holding on. This is full of danger to industry in general as well as to the departments concerned, for in these days of increased Governmental control and supervision it is not only necessary to have men that can meet representatives of industry on an equal footing, but also quite essential that the Government's representatives be broad-gage men who can take a comprehensive and intelligent view of the problems coming before them.

Since a dollar invested in large steam-power plants makes available three or four times the power that a dollar invested in hydro-electric systems will, it has been suggested that it might be more to the nation's interest to concentrate upon improving our wasteful methods of utilizing coal than a development of our water powers. As far as can be foreseen our water power is inexhaustible, but at the present rate of consumption our coal supply will be exhausted at no distant future. Therefore, why not concentrate upon conserving our fuel supply by improving our wasteful methods of using it and by the development of our water power, even if the latter takes a few more dollars per horsepower? Every horsepower-year produced from waterfalls means five to fifty tons of coal that can remain in the mines for use at some future time.

Despite extended conferences, the conferees on the Water Power bill are making no progress in their efforts to adjust the differences between the Senate and the House on that important measure. The fear is being expressed that a considerable period will elapse before a compromise can be agreed upon. While few are of the belief that the bill will fail, several previous water-power bills have got only thus far.

The "free energy" fellows got all worked up over the aurora display the other evening. Boy, page Mr. Garabed.

CORRESPONDENCE

Distance of Bridge-Wall from Boiler Shell

In answer to A. L. Smythe, who in the issue of Feb. 24 asks about the proper distance of the bridge-wall from the boiler shell, I find that for an 80-in. boiler, as mentioned, the bridge-wall should be not more than ten or twelve inches from the boiler for good result. I have been in the game for many years and have found that a low bridge-wall will not allow of good combustion.

J. W. IRWIN.

Chicago, Ill.

Some months ago a new hand-operated stoker was installed under our boilers, and the plans furnished by the stoker concern called for a bridge-wall much lower than we had at that time. Whether due entirely to the stoker or partly to the lower wall, I am unable to say, but we have increased our capacity considerably and decreased our smoke since the installation.

I am of the opinion that the lower wall gives increased combustion space and also permits the flames to pass to the rear of the setting without coming in contact with the comparatively cool boiler surface; and as the tubes furnish the greater amount of surface for the absorption of heat, it seems reasonable that the large volume of hotter gas would evaporate more water.

I believe the majority of return-tubular boilers are set too low and chill the gas before proper combustion can take place.

HARRY H. YATES.

Trenton, N. J.

Regarding Mr. Smythe's difficulty with bridge-walls, and considering the dimensions of the boilers, it appears that the trouble is due to insufficient grate surface, as a boiler 78 in. in diameter and 19 ft. long should have a grate surface of approximately 49 sq.ft., figuring 55 sq.ft. of heating surface to one square foot of grate surface. Best results have been obtained in burning bituminous coal with an area over the bridge-wall about one-seventh the grate surface, which would give the distance twelve inches from the bridge-wall to the shell and not over fourteen inches. The other boiler mentioned would be about the same distance above the bridge-wall.

I would recommend a large combustion chamber in order to obtain better combustion when the fire is not crowded. This is also desirable under all conditions where the grate surface and the draft are adapted to the kind of coal that is being used. If it is desirable to burn a poorer grade of coal, on account of its low price, and the available draft pressure is insufficient to burn this coal at the required rate, the remedy is either to enlarge the grate surface or to use forced draft, or both.

Insufficient grate surface and poor coal, two causes of trouble, may be considered together, as they are correlated, and insufficient grate surface for one grade of coal may be ample for another grade.

I believe Mr. Smythe is right in his statements, and no doubt the conditions mentioned would give the results he is getting. I recommend that the grate surface be increased and that the bridge-wall be built not less than fourteen inches from the boiler shell.

St. Louis, Mo. R. GORDON.

A bridge-wall is thought by many engineers to be of great value in keeping the furnace flames against the boiler shell, thereby resulting in a saving of fuel. I have found some bridge-walls that come within six inches of the boiler shell and made to conform to the curvature of the shell, and also all sorts of elaborate designs that required considerable upkeep. A bridge-wall is for only one purpose and should be built to accomplish that purpose, which is to prevent fuel from being pushed off the grates into the combustion chamber back of the bridge-wall. A height of bridge-wall eight inches above the grates is sufficient to prevent this, and it should not be built higher when burning soft coal; even with hard coal a higher bridge-wall will be of no benefit.

A high bridge-wall forces the gas that is liberated from the burning coal, against the comparatively cold boiler shell, where it is cooled, and at this low temperature it passes along the boiler shell to the tubes and out through the tubes to the chimney without reaching a temperature sufficiently high to burn; consequently, these heat units are a total loss. These gases are brought into a combustible state by keeping them away from the cold boiler shell, so that more heat will be given up to making steam and a higher furnace efficiency obtained.

A high bridge-wall is not advisable, and if more coal is burned with a low bridge-wall, it is due to the fact that a high bridge-wall acted as a damper and retarded the flow of gases; a low bridge-wall does away with this choked passage.

I advise a bridge-wall with not less than an eight-inch space between it and the boiler shell. A damper should be placed in the stock and the draft regulated by it alone.

In case of excess boiler capacity I advise that an arch be built extending from one side wall to the other, the length depending on conditions. This can be done if the boiler is not too close to the grates. An arch twenty-four inches deep with a six-foot span is a good one. Of course, about five inches of grate surface on each side will have to be removed to give room for the foundation of the arch, but in the case of excess boiler capacity, there is no doubt an excess of grate capacity, and this five-inch loss on each side will be of benefit.

Winkleman, Ariz. GEORGE J. TROSPER.

The main purpose of the bridge-wall is to prevent the fuel from dropping off the rear of the grates, into the combustion chamber. Incidentally, it tends to raise the temperature of the furnace, by providing a radiating surface; it also tends to throw the gas currents upward and along the bottom and sides of the boiler shell, and being an obstacle to the free flow of the gases it tends to promote a mixture of the gases and air, thereby improving combustion.

A rule commonly followed in proportioning the various gas passages is to make the area between the bridge-wall and boiler shell equal to one-seventh of the area of the grate surface, though some combustion engineers give one-sixth of the grate area as a proper proportion. The majority of them agree that for boilers' 72 in. in diameter the space between the top of the bridge-wall and boiler shell should be not less than fourteen inches.

Like many other rules, it is well to consider the special circumstances of the case before applying it. The first possible draft restriction between the furnace and the chimney is at the bridge-wall, and for burning bituminous coal, with the area between the top of the bridge-wall and boiler shell equal to one-sixth or one-seventh of the grate area, there should be no draft restriction at that point. If the space of fourteen inches is allowed, the draft area in a case of a 72-in. return-tubular boiler, between the top of the bridge-wall and boiler, will be approximately double the area of the boiler tubes, and in the case of a smaller boiler it would be considerable more; and as there would be no gas restriction at this point with this proportioning, it would prove a considerable advantage toward proper combustion. RICHARD PARROTT, JR.

Plainfield, N. J.

Increasing Trap Expansion Movement

A number of steam traps are still made which rely for their operation on the linear expansion of a metal tube or on the difference of expansion of two tubes of different metals.

Such steam traps generally suffer from considerable wiredrawing at the valve, because the movement of the valve from its seat is equal only to the linear expansion of the tube, and for small temperature differences and tubes of medium length the actual movement is very small.

To produce a greater movement by expansion for the same temperature differences, we might utilize a flat bar of metal of moderate length, say only 12 in. long, the bar being firmly held between two points, the distance from each other remains unalterable or nearly so. With an increase in temperature this bar will curve and its center C will be moved in a direction normal to its axis.

Assuming that the linear expansion of bronze for one degree C. is 0.0000177 per unit of length and that the change in temperature is 5 deg. C., then a bar 12 in. long will have after expansion a length of $12 + (0.0000177 \times 12 \times 5) = 12.001062$ in. and the total amount of movement produced by the free end would be 0.001062 in. If this bar is firmly held between two points, as at A and B, Fig. 1, whose distance apart is 12 in. and then expands the same amount, its center will move outward in a direction normal to its axis. The

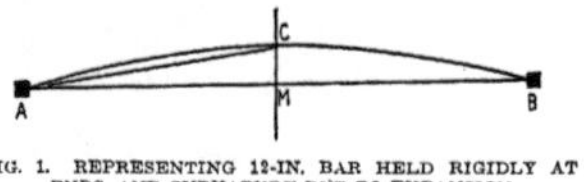

FIG. 1. REPRESENTING 12-IN. BAR HELD RIGIDLY AT ENDS, AND CURVATURE DUE TO EXPANSION

center line of the bar will assume a curvature. Assuming that this curve is a circular one, then the length of the arc = $A\ C\ B$ will be 12.001062 in. and its chord = $A\ B$ will be 12 in.

As $ACB = \frac{8AC - AB}{3}$, we get $AC = \frac{3 \times 12.001062 + 12}{8}$ = 6.0004 inch.

$C\ M$, or the amount of movement of the center of the bar $= \sqrt{(A\ C)^2 - (A\ M)^2} = \sqrt{36.0048 - 36} =$ 0.069 inch.

This movement is about 65 times as large as the movement resulting from linear expansion, which for

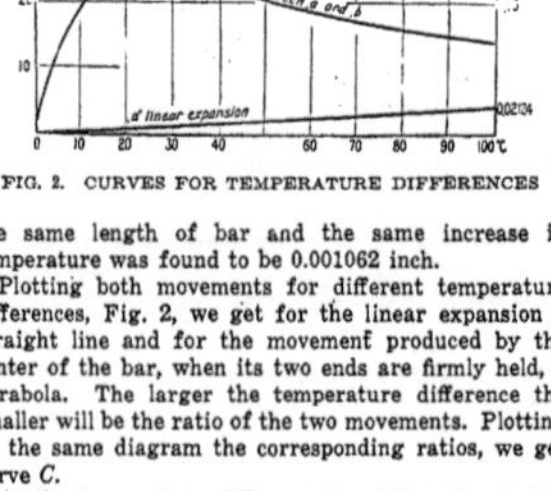

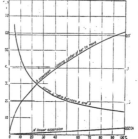

FIG. 2. CURVES FOR TEMPERATURE DIFFERENCES

the same length of bar and the same increase in temperature was found to be 0.001062 inch.

Plotting both movements for different temperature differences, Fig. 2, we get for the linear expansion a straight line and for the movement produced by the center of the bar, when its two ends are firmly held, a parabola. The larger the temperature difference the smaller will be the ratio of the two movements. Plotting on the same diagram the corresponding ratios, we get curve C.

As the temperature difference to which a bar in the service mentioned generally is exposed is small, the resulting movement might be multiplied to any desired degree by leverage, and without going to extremes a full opening of the valve may be realized at all times.

Half Way Tree, Jamaica, B. W. I. O. JANZEN.

Water-Column Blowoff Valves

Some time ago *Power* published a letter by Gustav Swenson which related to the subject of water-column blowoff valves. The problem of the water-column blowdown valve has been repeatedly presented to our engineering department, and we have been asked to develop a valve for this purpose and have evolved the valve shown in the sketch.

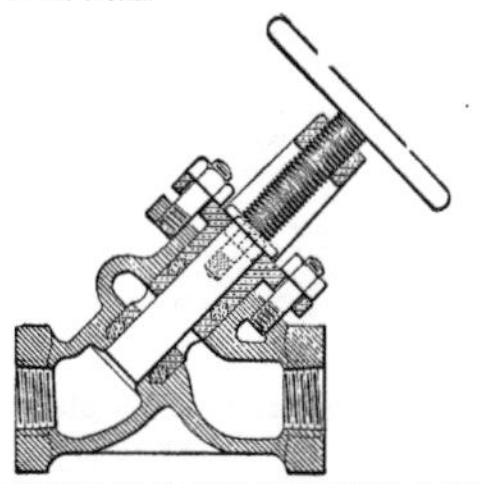

IMPROVED WATER-COLUMN BLOWOFF VALVE

In this little $\frac{3}{4}$-in. or 1-in. valve, which is made either in the straightway or angle pattern containing the seatless feature, the dirty water passes out through holes in the lower follower gland when the plunger, which is part of the stem, is raised into the open position by means of the handwheel. There is an annular space around the center follower ring so that the blowdown can freely pass out into the side or end port.

The valve is closed in the usual manner by means of the handwheel, which screws down the plunger until it passes the lower packing ring, when the flow ceases. The plunger continues down until the shoulder engages the upper follower ring, and then, as the operator gives one sharp final turn, the pressure is applied by means of this shoulder to the upper follower ring and through the upper packing and lower follower ring to the lower packing, so that the tighter the valve is closed the more the packing will be forced out against the plunger in its closed position and hence the tighter the valve.

The packing used is a vulcabeston molded ring. This type of small blowoff valve was first designed for the Stanley Motor Carriage Co., who have adopted it for use in conjunction with their steam automobile. They have found that it is the only blowoff valve that stays tight under their high-pressure steam conditions. The valve made for them has a $\frac{1}{4}$-in. pipe connection.

D. ROBERT YARNALL,
Philadelphia, Pa. Yarnall-Waring Co.

Starting the Oil To Run

Frequently, when my engine is to be started and I set the oil flow on the crank, instead of flowing in a stream it comes out of the cup in drops. My method of starting it to flow in a stream is to place my finger in the sight hole, letting the oil drop on the finger nail and then lift my finger until the oil begins to come in a stream. I then lower my finger, pulling it away quickly, and the oil will then run in a stream.

Woonsocket, R. I. CLEMENT COREY.

Alarm Clock Used To Open Oil Switch Automatically

The following describes an arrangement that I have had in use in the power house for five years to turn out the street lights in a small town. The circuit was controlled by a hand-operated 2,300-volt oil switch, which was made automatic by the addition of an auxiliary lever, a spiral spring and by placing a steel catch block across the slot in the switchboard where the operating levels pass through, as indicated in the figure. When the switch is closed, the spring *S* is compressed and the slot in the lever *L* falls over the catch *B* and holds the switch closed.

In back of the switchboard, directly under the end of lever *L* and on the frame of the oil switch, is mounted a solenoid, *T*, with an iron plunger, so that when the coil is energized the iron plunger will lift the lever, thus releasing the switch.

On a common alarm clock I soldered an extension on the alarm winding key, with two small holes bored in it to attach the ends of two chains about ten inches long, as shown. In the end of each chain near the key is placed a piece of leather to insulate them from the clock. The chains and solenoid are connected in series with the 110-volt circuit. When the alarm goes off, the chains are twisted together, thus closing the circuit through the coil whose plunger trips the switch.

At first there were on the front of the board two contacts that opened, but another arrangement had to be devised, since sometimes the power was off when the time came for the clock to turn the lights out, and when the voltage was slowly built up the trip coil would not trip the switch.

This in a short time caused the coil to be burned out, so I made a frame for the clock and mounted a baby

CONNECTIONS OF ALARM CLOCK TO OIL SWITCH

knife-switch in back of it, as in the figure. To set the clock, which was done when the substation man came to put the lights on, it was slipped out of the frame and the alarm wound up by the key *K* and the chains hooked to contacts *A* and *A* and the switch *W* closed. When the clock goes off, it twists the chains together on the first half revolution and on the last half it opens the switch which cuts out the trip coil. E. LITTLEFIELD.

East St. Louis, Ill.

Keeping Out of the Rut

The letter of Thomas M. Gray, together with the editorial on the subject of "Keeping out of the Rut," as published in the Feb. 17 issue of POWER, brings out some excellent points for one to think over.

In dealing with the subject it is my belief that no engineer who is really 100 per cent interested will allow himself to slip into a very deep rut before he begins by concentrated effort to pull himself out and thus become a better man. I will exonerate the engineer for some of his shortcomings, which are justly chargeable to the management of the plant, for it is well known that many of the men holding the purse strings cannot see the economies which the engineer can make with proper coöperation from their side of the fence. This fact alone allows many a first-class engineer to get into so deep a rut that it takes the real man spirit in him to pull himself out. Is it any wonder, then that routine duties so often lead the engineer to make them in a careless manner because he feels that no one is interested enough to make a close check of his work? In many plants the man called "Chief" is not "Chief"; in reality he is the wet nurse to everything mechanical and electrical about the plant; his rut is a well-worn path from which he fears to tread.

The reasons for power-plant engineers becoming disinterested in their work are many, yet in the three cases cited in the following can be observed facts, not to be disputed, and in many plants the same conditions exist today, and the rut is well worn; whereas on the straight road there is plenty of room for all if they could but understand that it is the best and easiest way.

Take the case of three men employed as shift engineers in a moderate-sized plant, maintained in very good condition by the owners, although the human element side of the situation could be improved upon to the advantage of the management, as well as the men.

Jack works on the first trick and has general supervision over the forces employed on the other tricks, as well as making up the daily reports and looking after minor repairs about the station. Jack is in his thirties and works eight hours per day. He has had the benefit of several years in the boiler room as well as in the engineer's position, and he should be capable of handling the place well. Jack's trouble seems to be that he does not like the responsibilities of the position, nor will he study and read up on the modern ways of operation. His duties are performed in a mechanical sort of way, and when rendering discipline he is not there. Jack is surely in a rut and by concentrated effort could become a real engineer, yet he does not think it worth while to make the start. Some day he will find some other fellow in his job.

Pete is the second-trick engineer and receives his instructions from Jack. While on duty Pete is the boss, and yet he lacks enough initiative to do any more than the law compels for him to get by. Fortunately for him, he has been getting by for a long time at the expense of his employers. Some day he will wake up and will make a grand rush to get out of the rut, but he will find that the rut has swallowed him up. He simply holds on because Jack and John, the third-trick engineer, hold up his end for him.

John is the third-trick engineer and is a man about the age of Jack. He has had boiler-room training and is a real engineer in every way. He is interested in all things pertaining to the work and can be relied upon in case of emergency to do the right thing. He is out of the rut and is not going to get in very deep, for his time is too well occupied in educating himself for the better positions now being filled by someone in the rut. John can take the reports and explain why this or that happened and the remedy applied. He will, it is safe to say, not be a third-trick engineer very long.

It may be well to mention that in the plant from which this lesson is drawn the wages and the conditions in general are good, hence there is practically no excuse for Jack and Pete not to make good. The trouble is with the man and not the place in this case.

One manner of eliminating inefficient men is to have a national license law and grade the men according to their ability to pass a satisfactory examination, allowing sufficient points for practical experience and have the examination consist of practical problems. Proper classification of plants, etc., should also be given consideration. PAUL R. DUFFEY.

Hagerstown, Ind.

National Engineers' Convention

Your editorial in the April 6 issue on "The Coming National Engineers' Convention" is on sound foundation. There is no difference of opinion as to the need of the engineer giving attention to his nontechnical interests and in a thoroughgoing manner. The question is, Through what instrumentality?

Several years ago the writer, as secretary of a local all-inclusive engineering society, backed a plan which provided that membership in an accredited local society should be a prerequisite to membership in the national. He felt strongly that the salvation of the profession so far as it would come through the development of organization depended upon the development of the local society. Since that time, through wider experience, he has found that the local developed up to a certain plane, a sort of dead level, where it stopped. It must have a proper state and national relation. The development of the home interest is of first importance, but will be dwarfed unless that development has, as a part of the process, national integration.

The proponents of a federation are in the position of those who advocated a confederation of states early in the history of the country. Before a nation came, the confederation had to be thrown overboard. A civil war was necessary later on to settle the question of integration of a nation as contrasted to state rights. The proponents of a federation are state-rights men. A strong all-inclusive organization will come about through integration. If a federation is formed, it will be but a step in the process of integration.

The American Association of Engineers has been offered as the integrating instrument in a letter addressed to the four national technical societies under date of Dec. 20, 1919. Whether eventually accepted or not, the principles on which A. A. E. are founded will prevail. The test of the National Convention to be held in Washington will be on its acceptance of broad principles, like the Constitutional Convention in 1787 which gave birth to the United States.

Chicago, Ill. C. E. DRAYER, SECRETARY,
American Association of Engineers.

Synchronous Motors for Driving Centrifugal Pumps

I was interested in reading the article on "Synchronous Motors for Driving Centrifugal Pumps," by Soren H. Mortensen, in the Jan. 20 issue of *Power*. On page 91, Fig. 4—"Starting Instructions"—according to these the motor is supposed to be brought up to approximately synchronous speed on the low-voltage tap, the field switch closed and the field circuit adjusted for normal field before the power switch is thrown from the starting to the running position. This would be an impossible method to use unless the pull-in load was comparatively small or the motor considerably underrated. My experience has been such that, except in special cases, I would never think of trying to have the motor pull the load into synchronism before full voltage is applied. One has only to witness the operation of a synchronous motor during starting and pull-in period on loads such as centrifugal pumps in order to realize how impossible the method quoted would be.

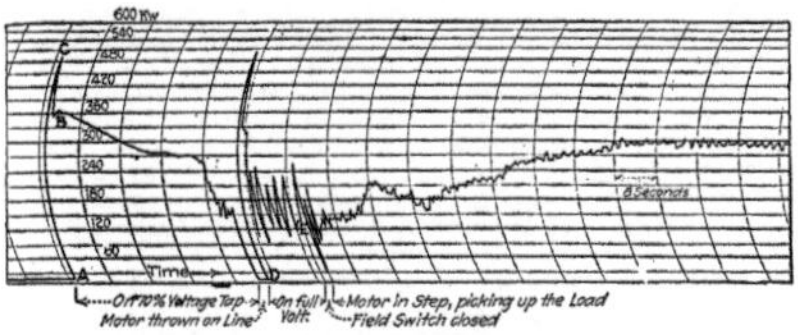

FIG. 1. RECORD OF CURRENT TAKEN DURING STARTING BY SYNCHRONOUS MOTOR DRIVING AMMONIA COMPRESSOR

The key to Mr. Mortensen's statement, I believe, is found in the third paragraph of his article, where it says: "If the characteristics of the centrifugal pumps are such that, during the starting period, they require 70 to 80 per cent of full-load torque, it will be necessary to make the synchronous motor larger than would be the case if its capacity was limited through heating alone. For such cases over-excited synchronous motors for power correction are particularly suitable." I think 70 to 80 per cent is away too high, as I do not believe motors started according to the instructions given could pull in 30 per cent of normal load unless the motor was over-size.

A test curve showing starting and pull-in of a 2,200-volt three-phase 60 cycle 500 hp. 360 r.p.m. synchronous motor driving an ammonia compressor under more severe conditions than are ordinarily met with in centrifugal pumps is shown in Fig. 1. This curve reads from left to right. When the motor is first thrown on, the line kilovolt-amperes show approximately 360 at *B*; the initial peak at *C* is probably due to the overswing of the needle. The kilovolt-amperes from this point decrease steadily for the next thirty seconds, indicating that the speed of the motor is increasing and the power factor becoming better. The power factor at starting will probably not exceed 30 to 40 per cent. The instant full voltage is thrown on at *D*, quite a large drag of current is taken from the line, although full allowance should be made for the over-throw swing of the needle. For about eight seconds the ammeter needle is swaying back and forth, indicating the rotor field increasing its speed and attempting to lock in with the stator field. At the point *E* where the field switch is closed, one big dip of the needle shows where a pole is slipping by and then the motor pulls into step and locks with the line. Immediately thereafter the valves are opened on the compressor and it begins to pick up load. The fluctuations of the line from there on are the reciprocating actions of the compressor cylinders. This particular compressor at pull-in requires a torque of 50 per cent of normal as near as we can judge. The motor, when carrying its full normal load, draws about 300 kw. from the line.

Now compare this with the curve as shown in Fig. 7 of Mr. Mortensen's article, reproduced here in Fig. 2. This is a peculiar looking curve, and I cannot understand why the excitation should be applied immediately the motor is started. It is very evident that such excitation is doing no good as the motor has not pulled into step, and some 30 seconds later full voltage is applied and the motor fluctuates along for about 30 seconds more before it pulls into step. I have never before seen a graphic curve of a synchronous motor at starting that looked at all like this. The speed, evidently, is not changed at all during the period the motor is running on part voltage, as the current fluctuations are of the same duration and that indicates the speed as remaining constant. Further discussion by Mr. Mortensen, or other readers of *Power* would no doubt be helpful on this subject, to which so much attention is being given at the present time. B. S. STEVENS.

New York City.

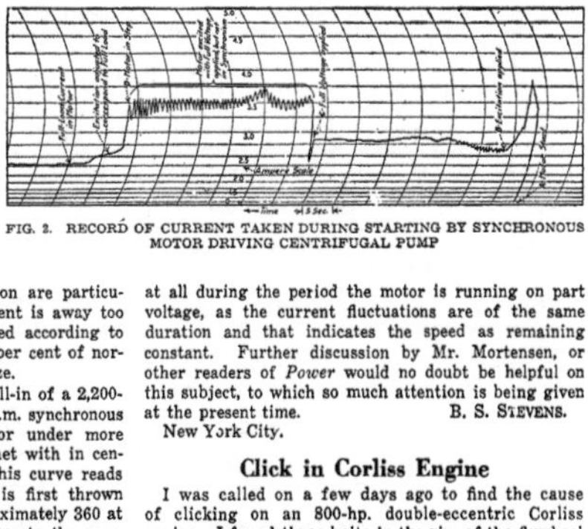

FIG. 2. RECORD OF CURRENT TAKEN DURING STARTING BY SYNCHRONOUS MOTOR DRIVING CENTRIFUGAL PUMP

Click in Corliss Engine

I was called on a few days ago to find the cause of clicking on an 800-hp. double-eccentric Corliss engine. I found three bolts in the rim of the flywheel loose. I removed them, heated them, drove them back into place. After tightening them, I started the engine and found that the trouble was overcome.

North Tonawanda, N. Y. C. J. MILLER.

Electric Arc Versus Gas Welds

On account of the rapidly advancing field of autogeneous welding in construction, maintenance and salvage work, the following results as shown by practical working tests may be of interest. These tests were performed by F. Hutton, M. E., a well-known welding expert, and myself. They covered a period of about two years, and afforded ample opportunity for a careful investigation of the subject from various standpoints.

The first class of work which I will here attempt to describe consisted of the repair by welding of defective B. & W. boiler steel headers, designed to withstand a working pressure of 225 lb. The welding required to repair the headers consisted of the filling in of worn lower corners and building up thinned 4-in.-tube seats to the original thickness with ample reinforcement to allow for machining. The saving effected by arc-welding these headers amounted to from 90 to 96 per cent of the original cost of new headers.

We obtained the best results with a direct-current, constant-potential machine with low voltage and a readily regulated heat control. A voltage not exceeding forty was used, and a metallic electrode giving a high tensile strength of weld and approximately 0.25 carbon and 0.7 manganese in the weld was found best for this class of work. After completing the welding and placing headers in the boring mill for machining the tube seats, it was found that the welded metal worked under the tool fully as well as the metal comprising the original header. In fact, any line of demarcation between weld and original metal was very hard indeed to locate by occular inspection.

We found, when inspecting the interior of these headers after over a year of continuous use under unfavorable working conditions, that the welded and built-up sections had in no way deteriorated. Much to the contrary was our experience with headers of the same type, physical, and chemical properties as those already mentioned, but which were welded by the oxyacetylene process. After being welded by this process, and when the machining of the tube seats was started, large sections of the weld tore or fell off under the tool, and in some cases the entire body of welded metal fell off, showing that no actual weld union had taken place, but only a sort of surface fusion or cementation which was easily loosened by contraction stresses and tool impact.

It is believed that in the first-mentioned case, where portions only came off, that the headers had received the required preheating necessary to insure a weld, but that the ensuing contraction stress caused cracks to develop in the weld and surrounding metal.

The second series of practical tests was in the salvage of pitted and broken ship propellers. The first repairs made in this class of work were by the oxyacetylene process and consisted in filling in pitted portions of blades and welding on special cast blade tips. The propeller blades having such a large area compared with their thickness, it was found that when the necessary preheating was done the blade had warped a considerable distance out of its original form. This meant, when the welding was completed, that an annealing treatment was required to restore the blade to its original contour and relieve contraction stress. This annealing was found to be only a partial remedy as the blade was not entirely restored to its original form, and without doubt a state of constant internal stress remained in the blade, tending to produce fracture when the usual working load was applied. Subsequent events seemed to prove this theory.

After the welded propellers were placed in service and subjected to the usual working stress, the blades broke off one at a time, and they all broke directly in the weld, the last blade on one propeller breaking off less than four months after the propeller was placed in service. This same class of work was performed on other like sets of propellers, but in this case the electric-arc process was employed. The same type of welding machine, voltage and equipment were used for this work as for the headers previously described. A softer metal electrode of less tensile strength was used in order to make the subsequent grinding and surfacing more easily done. The welding heat being well localized and under close control, no noticeable deformation of blade was produced; hence no annealing process was required, and the blades appeared to be free from any injurious internal stress.

After nearly twenty months' use, and being exposed to the same working conditions as mentioned in the case of the gas-welded propellers, an examination in dry dock was made. The welding and tips were found to be intact and in their original good condition except for a slight pitting, due no doubt to galvanic action. The saving effected by welding these propellers by the arc process amounted to from 70 to 78 per cent of their cost when new. On the other hand, in the case of the gas-welded propellers a considerable loss was incurred, due to the cost of dry docking, removal and replacing of propellers.

The results obtained on the two classes of work cited were repeated with various modifications on many other classes of work, and with various metals, including cast iron, cast steel, mild steel, nickel steel, bronze, monel metal and other alloys. In all classes it was found that the electric-arc weld was superior to the gas weld both in quality of work and in cost. I believe this to be due principally to the closely localized character of the heat generated by the electric arc, thereby minimizing heat deformation and contraction stress in the member being welded.

The distinct function of the oxyacetylene flame is for cutting, and in that field it is, I think, unexcelled.

I would invite a discussion on any lines pertaining to this subject.

CHARLES T. PERRY,
Mechanical Engineer,
New York City. Dept. of Plant and Structures.

Continuous Use of Boiler Water

This is written with the intent of correcting a statement of Mr. Hill who, in the third paragraph of his article on page 320 of the March 2 issue, states that, "If there were no losses of water due to leakage, the same water could be used indefinitely."

I feel sure that Mr. Hill would not follow that practice, even if it were possible to do so, because he undoubtedly knows that water can become too pure for boiler purposes, and that if it is being used over and over without adding fresh water, the mineral properties which are necessary in a good boiler-feed water have been blown out through the blowoff pipes. Water in a pure state will attack the boiler metal and cause pitting.

H. W. ROSE.
Brenham, Tex.

INQUIRIES OF GENERAL INTEREST

Position of Exhaus Valves with Engine on Center—With a releasing valve gear, should the exhaust valve be open when the engine is on a center? H. S.

In a cut-off engine, the steam should be released from the cylinder by having the exhaust valve open before the end of the stroke, to insure prompt discharge of the steam and prevent undue back pressure during the return stroke. Consequently, when the engine is on a center and the piston has arrived at one end of its stroke, the exhaust valve on the other end of the cylinder would be open.

Increasing Size of Governor Driving Pulley—If a reduction is made in the diameter of a pulley on the main shaft of an engine for driving a flyball governor, what effect will it have on the speed? W. T.

Compared with the speed of the engine, the governor will be driven at relatively slower speed and the engine shaft will have to attain a higher speed before the speed of the governor is sufficient to check the supply of steam. Consequently, reducing the size of governor driving pulley on the mainshaft, or increasing the size of the governor receiving pulley, would cause the governor to regulate the engine at a higher speed.

Starting an Induction Motor—A 3-hp. 220-volt three-phase induction motor, driving a shaper, takes 60 amperes for about 20 seconds when first started. The current then drops to 5 amperes. Is this surge of current liable to harm the motor, and is the use of a starting compensator warranted? A. E. W.

A starting current of 60 amperes is not abnormal for the type of motor in question. And unless there is danger of the motor injuring the machine that it is starting, a starting compensator is not necessary. If you wish to properly protect the motor, you will have to do it by installing a double-throw switch with two sets of fuses—one set of about 15-ampere capacity on the running side, and another on the starting side of 30-ampere capacity.

Results of Belt Failures with Blocked Governors—With a blocked engine governor, what would happen if the main driving belt should break or if the governor belt broke or came off? W. D. T.

If the main belt should break, the load would be removed and the engine would race and perhaps be wrecked. With the governor belt broken or run off, the governor would supply the amount of steam corresponding to the blocked position, without power to increase the supply for an increase of the load or to decrease the supply for a decrease of the load, so that a load greater than that capable of being carried by the blocked position of the governor would cause the engine to slow down, but any less load would permit the engine to race and very likely to become wrecked.

Average Tangential Pressure on Crankpin—What is the average tangential pressure on the crankpin of a 14-in. x 20-in. engine where the m.e.p. is 40 lb.? C. B. H.

Neglecting the reduction of piston displacement from presence of the piston rod, the average total pressure acting on the piston would be $14 \times 14 \times 0.7854 \times 40 = 6157.6$ lb., and with 20-in. stroke the work performed per revolution would be $6157.6 \times 20 \times 2 = 246{,}304$ inch-pounds. Without allowance for loss from friction the same amount of work would be transmitted per revolution to the crankpin; and as in one revolution the length of path described by the crankpin center would be $20 \times 3.1416 = 62.832$ in., the average tangential pressure would be $246{,}304 \div 62.832 = 3{,}920$ lb.

Allowance for Strength of Girth Seams—In computing the strength required of girth joints of a horizontal return-tubular boiler shell, what allowance should be made for the reduction of area of the heads due to the tubes? W. L. C.

According to the A. S. M. E. Boiler Code, par. 184, b: "When 50 per cent or more of the load which would act on an unstayed solid head of the same diameter as the shell, is relieved by the effect of tubes or through stays, in consequence of the reduction of the area acted on by the pressure and the holding power of the tubes and stays, the strength of the circumferential joints in the shell shall be at least 35 per cent of that required for the longitudinal joints." When less than 50 per cent of the solid head area is thus relieved, the strength of circumferential joints should be at least one-half the strength required for the longitudinal joints.

Power Factor in a Three-Phase Circuit—In a three-phase circuit the current is 30 deg. behind the electromotive force at 100 per cent power factor. Since the power factor is equal to the cosine of the phase angle between the current and voltage, what is the angle of lag in degrees at 90 per cent power factor in a three-phase circuit? E. C. H.

No doubt you have the current and voltage relations in the various parts of the circuit confused. The statement, "In a three-phase circuit the current is 30 deg. behind the voltage at 100 per cent power factor" is correct only regarding certain conditions, namely, the relation between the current in the current coil of a wattmeter, and the voltage applied to the potential coil and, where two wattmeters are connected in the circuit, the phase relation between the current in the current coil and the voltage in the potential coil is 30 deg. lagging in one instrument and 30 deg. leading in the other instrument. With a 90 per cent lagging power-factor load on the system the current in the load will be lagging approximately 26 deg. behind the voltage. At unity power factor in one of the meters the current and potential is 30 deg. out of step lagging. Therefore, in this instrument, with the load current lagging the voltage 26 deg., the difference of phase angle between the current in the current coil and that in the potential coil, will be $30 + 26 = 56$ deg., which represents a 56 per cent power factor. Bear in mind that this is only a mathematical condition, that the power factor is equal to the number of degrees in this case. In other meters at unity power factor the current in the current coil is leading the voltage applied to the potential coil by 30 deg., and when metering a load current that lags the voltage by 26 deg., the relation between the current in the current coil and the voltage in the potential coil will be $30 - 26 = 4$ deg. leading, which is equivalent to a power factor of 99.7 per cent.

[Correspondents sending us inquiries should sign their communications with full names and post office addresses. This is necessary to guarantee the good faith of the communications and for the inquiries to receive attention.—Editor.]

Purchase of Coal on the B. T. U. Basis*

By A. L. LANGTRY

COAL is the basic fuel of the Nation. The war and the recent strike brought home to the public their utter helplessness without it. So far as the consumer is concerned, he is most vitally interested, for he pays, ultimately, all increases in wages, material, etc., that are connected with the production of coal. With prices advancing, the consumer is bound to give his purchases more attention than ever before.

Coal has been bought upon what is known as the "B.t.u. basis" for a number of years. Many a contract of this kind has been entered into in good faith by the consumer and seller, with the very best of intentions, but through the lack of fully understanding some of the details of the system, either one or the other has had a sad experience. Nine times out of ten when this method of purchasing has been condemned, the real fault has not been with the system itself. For instance, coal companies, instead of using an average analysis showing the quality of their coal as the basis of a contract, will use the analysis from a vein sample, thinking that this will represent what mine-run will analyze when received at destination. This is folly, for in the natural course of mining, in many mines, some of the roof comes down and the bottom up; it cannot be helped, for it is a physical impossibility to mine the coal and not have these impurities included. True, some of these impurities can be thrown out, but not all, and consequently those remaining increase the percentage of ash.

Very often the consumer makes his contract so drastic that it ceases to be equitable. He does not gather samples in a representative manner; the analyses are not correctly made, etc.

Purchase on the B.t.u. Basis Practical and Feasible

It is the author's belief that purchasing coal upon the B.t.u. basis is practical and feasible. To his knowledge it has been a success in some plants, working to the entire satisfaction of both seller and purchaser. This satisfactory condition has been brought about by the use of common sense, by exercising a little energy and by putting into practice well-known methods of good boiler-room practice.

Take, for example, a typical plant—one that is either hand or mechanically fired—a plant, for instance, that has never kept records of coal, water, efficiency or capacity of boilers. The advance in the price of coal in this plant has caused the chief engineer and manager to consider more carefully the purchase of coal for the ensuing year. Assume that the coal being used in this plant is apparently satisfactory, but the chief engineer and manager are not satisfied. They want some real information about the plant and coal, and decide to employ a specialist in making analysis and collecting representative samples of coal.

It is found that the quality of the coal delivered to the plant has the following average analysis: Moisture, 11.5; dry ash, 14.78; dry B.t.u., 12,305; price, $4.95; net B.t.u. for 1c., 43,350. The next step is to run boiler tests in the plant of long enough duration to determine the cost of evaporating 1,000 lb. of water from and at 212 deg. F., using the present contract coal. This figure proves to be $0.38. With this information a cost figure is established that can be used as a basis for comparison in determining the most economical coal to purchase in the future. Assume that the coal contract in this plant expired March 31 and that these tests started about Feb. 1., giving two months to gather complete information about the performance of the plant and test other coals deemed advisable. The "machinery" is now set in operation for purchasing the year's supply of coal, this "machinery" consisting of the manager, chief engineer and a thoroughly responsible and reputable coal chemist and fuel engineer.

*Paper presented before the Minnesota Joint Engineering Board, Minneapolis, Feb. 20, 1920.

The proper procedure, then, is to ask for coal proposals. It should be stated to the prospective bidders that, in addition to the price per ton, an analysis of the coal is to be submitted that will be representative of what the coal will run day in and day out. The figures so submitted by the coal companies are tabulated, with the assistance of the fuel engineer, and reduced to a comparative figure known as the "net B.t.u. for one cent." The method of arriving at this figure from the accompanying analysis is as follows: Moisture, 9.50; dry ash, 12.50; dry B.t.u., 12,650; price per ton, $5; net B.t.u. for 1c., 45,228. The B.t.u. in the dry coal multiplied by (100 per cent less the moisture in the coal) equals the B.t.u. in the coal as received. Multiply the B.t.u. as received by 2,000 (pounds in ton). Divide this product by the price of the coal plus half of the dry ash in the coal expressed as cents. The result is the net B.t.u. for one cent.

The fuel engineer should be able to advise with his client as to whether the analysis of the coal as submitted really represented what is actually produced. Coal companies often quote an analysis that does not truly represent the quality of their coal, and when this happens it is policy to confer with them regarding such discrepancies before even ordering coal for test purposes. In this way a complete and definite understanding may be had before negotiations are far under way. This is the nucleus for the successful operation of the B.t.u. contract in many cases.

Getting back again to the coal proposals submitted for this plant, they are given in Table I.

TABLE I. TABULATION OF COAL BIDS

Coal	Moisture, Per Cent	Dry Ash, Per Cent	Dry B.t.u.	Price	Net B.t.u. for 1c
County No. 1—Screenings—No. A Seam					
No. 1	12.0	11.0	12,850	$4.80	46,582
No. 2	12.4	12.0	12,703	4.80	45,792
No. 3	12.0	13.5	12,460	4.75	45,516
County No. 2—Screenings—No. B Seam					
No. 1	7 00	13.50	12,600	$5.00	46,242
No. 2	9.30	15.50	12,273	4.90	44,724
No. 3	9.85	16.20	12,171	4.90	44,056
County No. 3—Screenings—No. C Seam					
No. 1	8 50	12.00	12,700	$5.00	45,934
No. 2	9.10	13.60	12,452	5.00	44,668
No. 3	10.00	14.00	12,371	4.95	44,362
County No. 4—Screenings—No. D Seam					
No. 1	9.50	12.50	12,635	$5.00	45,168
No. 2	9.90	13.50	12,485	4.95	44,834
No. 3	10.00	14.50	12,341	4.90	44,668

Assuming that the analyses given in the table truly represent the quality of coal that is mined by each one of the bidders, it would then be in order to purchase the kind of coal that will give the most heat for the money from its respective county and seam. The net B.t.u. for one cent is the comparative figure in this respect when comparing coals, and should be used as a guide in determining the best coal to try for test purposes.

Referring again to the analyses as tabulated, it is easy to see the net B.t.u. for one cent that coal No. 1 listed under each county gives the greatest amount of heat for the money. To give the coals from the different localities a fair trial, it would then be advisable to order, for test purposes, the coal that shows the greatest net B.t.u. for one cent. There should be, say, five cars of each kind, depending upon the size of the plant, so that a separate evaporative test could be run upon each. In this way it would be easy to determine which coal produced the least cost of evaporating 1,000 lb. of water from and at 212 deg. While the test coal is being used at the plant, representative samples should be collected from each and every car. These samples should be analyzed so as to have complete information as to the quality of the coal actually delivered. From the tests so run and the quality of the coal delivered, the analyses and costs in the plant given in Table II were obtained.

TABLE II. ANALYSES AND COSTS AT THE PLANT

Coal	Moisture, Per Cent	Dry Ash, Per Cent	Dry B.t.u.	Contract Price	Delivered Price	Cost of evap. 1000 Ft of Water
			County No. 1			
No. 1.....	12.5	10.80	12,875	$4.80	$4.783	$0.320
			County No. 2			
No. 1.....	8 0	12.50	12,759	5.00	5.014	0.357
			County No. 3			
No. 1.....	8.0	13.00	12,550	5.00	4.962	0.373
			County No. 4			
No. 1.....	10.50	13 50	12,480	5.00	4.878	0.397

The delivered value as shown is the result of a comparison between the coal actually delivered and that bid upon in the proposal submitted, according to the following rules:

B.t.u. dry in the coal multiplied by (100 per cent less the moisture in the coal) equals B t.u. as received. Multiply the B.t.u. as received by 2,000 and divide into the product the net B.t.u. for one cent on the coal bid upon. From the quotient subtract one-half of the percentage of ash in the dry coal, expressed as cents.

From Table II it will be seen that coal No. 1 from No. 1 County gave the least cost for making steam in the plant. It would be good practice for this plant, before signing the contract, to order ten cars or perhaps twenty. Samples should be gathered from the coal so delivered, making enough analyses to be certain of obtaining a good average as to the exact quality of the coal. Assume that this plan was carried out and that the analyses showed as in Table III.

TABLE III. RESULTS OF ANALYSES ON TRIAL ORDER

	Moisture	Dry Ash	Dry B.t.u.
4 cars....................	12.50	12.00	12,715
4 cars....................	11.95	11.50	12,796
4 cars....................	13.00	12.20	12,699
4 cars....................	11.41	10.95	12,871
Average....................	12.21	11.66	12,770

The average analysis in Table III compares favorably with the figures that were submitted originally by the coal company along with the bid, and is one that could be incorporated in the B.t.u. contract. Sometimes, in order to be absolutely certain of the basis of a contract, the first month's delivery of coal has been sampled and analyzed and the average of the analyses so made then used as the basis of the contract.

In addition to these precautions, contracts are drawn with a neutral zone, so to speak, of from 1 to 2 per cent leeway from net B.t.u. for one cent to allow for any slight variations or fluctuations that might occur in sampling the coal from time to time. This leeway has been found ample when a good average has been established on a coal.

The B.t.u. contract should include paragraphs pertaining to guaranteed analysis of coal, method of computing bonuses and penalties, a neutral zone before bonuses or penalties are paid, rejection of coal on account of quality or size, analyses for checking purposes, screen tests and standard methods in analyzing and sampling the coal. These items are in addition to the other necessary clauses of quantity, point of delivery, strikes, increase or decrease in miners' wages and freight rate per ton, etc.

Where a B.t.u. contract is in force, it means that the coal so delivered must be sampled regularly and analyses made from time to time. Samples are taken often from several different kinds of railroad equipment—the large 100 per cent self-clearing hopper cars, the eight-door hopper-bottom gondolas, the drop-bottom sixteen-door cars and the tight-bottom gondola cars. These are the types of cars that are in general use for coal.

Sampling can be done in several different ways—by hand, automatically and by use of a sampler device. It is advisable to gather separate and distinct samples from each car, this sample to be about 100 to 200 lb., the coal being thoroughly crushed, mixed and quartered until it is reduced to a pint or quart in size. The coal contractor should always have the privilege of having duplicate samples of his own to be analyzed for check purposes. To accomplish an equitable division of a sample, three containers should be filled—one to be analyzed by a noninterested chemist, the second to be delivered to the coal company and the third to be held for check purposes if necessary.

Summarizing, the application of the purchase of coal upon the B.t.u. basis would be as follows:

1. The purchase of coal upon the B.t.u. basis is a refinement of the "ordinary price per ton contract." It is simply adding to the price of the coal its quality for future identification.

2. Secure the proposals on coal, asking for a representative analysis of the coal.

3. Run evaporative tests on the best coal from its respective field or seam.

4. Use the cost of evaporating 1,000 lb. of water from and at 212 deg. F. as a guide in awarding a contract.

5. Order enough coal to demonstrate its adaptability to the plant conditions.

6. Sample and analyze all the coal deliveries so as to establish its grade or quality.

7. See that the sampling is in the hands of well-trained men who understand its importance.

8. The analyses must be made by chemists thoroughly experienced in coal work, using equipment of the highest possible standard.

9. Make the contract equitable.

What Constitutes Proper Elevator Inspection?*

BY D. J. GITTO†

Elevators should be thoroughly inspected on every visit of the inspector. This means that no part should be passed because it was in apparent good condition at a previous inspection, but that a thorough inspection should be made of all parts, because a defect may have developed since the last inspection. Preparatory to performing the duties of an inspection, the inspector should properly begin with the elevator machine, in order to make sure that it is in a safe condition before exposing himself to the hazards so often resulting from riding on top of the car or performing the necessary inspection under it.

Brake leathers should be examined to see that they have not worn down to the rivets, for if the up or the down motion shoes, or both, should loosen the leathers, a serious accident would result, there being nothing to stop the elevator or check the counterweights in their course of travel. Make sure that the shoes properly grip the brake pulley. Inspect the worm screw and motor shafting to see that it is securely keyed to the coupling. Loose keys become worn and are liable to shear off. It is essential that all pin bolts have cotter pins in them, as these bolts are liable to work out. Examine the slack cable apron as to proper adjustment; also see that the slack cable collar is not cracked, that it engages the operating camwheel and that the slack cable apron pulls the switch, both operations setting the brakes.

Look over the bolts in the drum spider and make sure that they are all tight and that none are missing. Examine all the spokes of the winding drum for cracks and watch closely the secondary sheaves of traction elevators. Remove the oil cap on the gear housing to see the condition of the gear. See that the stop limits and the traveling limit nuts are in good condition. Try out both top and bottom limit stops before leaving the machine room; also see that the contacts are making properly so that no arc or flame is formed while running. Arcing at the contacts is liable to burn them together and make the limit cutouts of no use in stopping the elevator, resulting in the car hitting the overhead work or bumpers and causing damage to the

*This paper was presented before a meeting of the American Society of Safety Engineers, held in the Engineering Societies Building, March 19, 1920. Although written primarily as a guide to the proper procedure for making an insurance inspection of elevators, it is also an excellent guide for everyone who has to do with the inspection and maintenance of this type of equipment.—EDITOR.

†Inspector, Aetna Life Insurance Co.

machinery and perhaps loss of life. Watch for cracked arms in the operating camwheel on the machine.

Thoroughly examine the cables on the machine drum. Hoisting cables of overhead machines should be examined when the car is descending; do not rely entirely on feeling; have the car stopped at frequent intervals, and by the use of a small mirror you can better assure yourself of the condition of the cables. During the examination of the cables it is quite important that your hands should be kept clear of them in case the operator should reverse the motion of the machine. It is my opinion that too much care cannot be exercised at this point of an inspection.

Look over the vibrating sheaves, if any, to see that they vibrate freely and do not lag too much; also make sure that the main-line switches are properly fused and inclosed and that the operating control boards of overhead machines are equipped with pans to catch any arcing material that falls, thus avoiding possible fires.

Ride up in the car, trying every landing door, and see that the door locks and contacts, if any, are in proper working order. Examine car-operating control switch and see that the switch handle centers properly; then make a thorough inspection of the overhead work, looking for loose or worn shafts, missing or loose setscrews in sheaves, cracked sheaves, worn-out bushings of loose sheaves, good lead of cables to groove of sheaves and worn-out babbitt in bearing boxes. Examine the governor; try the tension of the spring to determine if it will readily act in case the car should drop or race; throw the clutch and see that it grips the cable properly. All parts should be oiled and grease provided where necessary.

Enter the shaft on top of the car at the top floor and, while riding down, carefully inspect all cables, cable sockets or thimbles. In examining the cables, do not rely on feeling while a car is in motion, but have the car stopped occasionally, so as to detect any place where the cables may have worn flat, in which case three-quarters of the strength of the cable has worn away. In such a condition these cables do not give any warning, but are quite likely to snap in case of a sudden strain. Do not overlook the cables where they enter the sockets, as this is one of the weakest points, if not properly attached. Watch for loose brackets and bolts. See that the rails are properly lined up, examine the governor cable socket clutch at the top of the car to ascertain if it would release when needed. Watch the wear of the guide shoes; try out the hatchway limits, top and bottom, as they sometimes become grounded, in which event they would be useless for any emergency. Look over counterweights for defects and chafed cables; see that tubing is provided for cables where they pass through the car counterweights; watch the wear of guide weights or shoes and see that the car and weights have proper clearance.

Examine the operating-cable device on the car for wear; see that the car gate is in proper working order; try out the safety device under the car and see that the jaws on both sides have equal and sufficient tension to do their required work. Experience has shown that this has often been the cause of the safety not holding when a car has fallen. At times only one jaw has set and, for lack of strength, has broken. This operation causes slackened car cables, and when the safety jaws let go, the cables are snapped and the car falls to the bottom of the hoistway with very serious and sometimes fatal results. Although it is not customary for inspectors to set the safety while a car is in motion, other than the testing after the installation, unless specifically called for, it is my opinion that every inspector should convince himself that the safety and all appliances are properly adjusted to act when necessary. Having tested out the appliances, the application of safeties can be done by hand from under the car—both instantaneous and friction-grip types. See that the governor cable on the drum is not jammed, as is so often the case. Examine compensating cable sheaves and tension weight at the bottom of the hoistway; make sure that they have not bottomed from stretched cables and see that electric cutouts are provided for the top of the frame to shut off the current in case a cable jumps off and jams. This would prevent the possibility of the sheaves and tension weight being carried up with the car, without the knowledge of the operator, and probably resulting in serious damage. See that necessary protection is taken by the installation of counterweight shield covers, overhead grating and inclosed machine rooms.

Hydraulic elevators generally require the same sort of inspection as already mentioned. They differ, however, in such particulars as pressure tanks, steam or electrically driven pumps, vertical and horizontal cylinders and plungers.

The pressure tanks should be watched for leaks, precaution being taken that the proper amount of water and air is carried, the standard being about three-quarters water and one-quarter air. Watch the operation of pumps and note if the regulating valves function properly for the pump to keep the required amount of water in the tank. Examine the operating cable lever in the car to see that it centers when the valve is closed; this, I would say, is very essential. Watch traveling sheaves closely as the bushings are inclined to wear; the sheaves in this case would have a tendency to lean, thus making slack cable which oftentimes jumps off.

Do not pass an elevator where the valve cups are leaking badly, as a serious accident might result from such an oversight. Great care should be taken while inspecting cables on vertical hydraulic elevators, especially where a series of cylinders are in a blind shaftway. Always keep yourself clear at all times and be prepared, in case the operator forgets himself and reverses the motion of the elevator. Learn to keep your hands well above or below where the cables come off the sheave. While inspecting this kind of a risk, do not allow the other elevators to be operated.

Always have your flashlight in good working order and carry it with you at all times. Do not use candles or strike matches in hoistways. Be sure of your operator; note his efficiency and give him implicit instructions as to how you want the car operated. It is my opinion that a car should never be operated and inspected by the same person. To be thoroughly and properly inspected, an operator should be provided.

Beating the Coal Famine

About 70 per cent of the power required to operate the Nordyke & Marmon Co. plant, Indianapolis, Ind., was produced during the coal famine of the strike by Marmon 34 automobile engines, Fordson tractors and Delco lighting systems. Thus the plant carried on and kept its thousands of people employed, says the *Boston News Bureau.* Thirty of the standard Marmon 34 engines drove electrical generators in the dynamometer room of the big plant all day long and produced about 1,000 hp. In other parts of the factory other Marmon 34 motors were directly connected to the line shafts and operated machine tools. Aside from this, more than a score of Fordson farm tractors were directly connected to the lineshafts in various parts of the plant and helped to produce power by gasoline, enabling the company to save more than the 50 per cent of coal-produced electrical current used in the plant. Delco lighting sets were distributed throughout the plant.

A report just issued by the Department of Commerce, Bureau of Foreign and Domestic Commerce, Washington, D. C., shows exports of electrical machinery and appliances during the month of January, 1920. The value of battery exports totaled $374,730; carbons, $139,640; dynamos and generators, $413,293; electric fans, $29,587; insulated wire and cables, $661,842; meters and measuring instruments, $188,123; motors, $733,872; rheostats and controllers, $18,558; switches and accessories, $294,036; transformers, $350,364; and miscellaneous items amounted to $1,966,672. In general electric equipment Canada leads all other nations in its imports; China leads in the purchase of dynamos and generators; Japan comes first in the importation of motors which cost $213,924, while Norway leads in the purchase of transformers, costing $175,449.

Petroleum Production in the United States

Figures given in Table I, based on reports filed with the United States Geological Survey, show the quantity of oil received from producers. The production reported for California is estimated from figures and data collected by the Standard Oil Co. and the Independent Oil Producers' Agency.

TABLE I. PRODUCTION OF PETROLEUM IN THE UNITED STATES IN BARRELS

State	January, 1920 (a) Total	January, 1920 (a) Daily Average	January, 1918 Total	January, 1918 Daily Average
California	8,488,000	273,806	8,154,000	263,032
Oklahoma	7,958,000	256,710	8,049,000	259,645
Texas:				
Central and Northern	5,778,000	186,387	1,034,000	33,355
Coastal	1,441,000	46,484	1,695,000	54,677
Total Texas	7,219,000	232,871	2,729,000	88,032
Kansas	3,338,000	107,678	3,855,000	124,355
Louisiana:				
Northern	2,400,000	77,419	570,000	18,387
Coastal	167,000	5,387	211,000	6,806
Total Louisiana	2,567,000	82,806	781,000	25,194
Wyoming	1,171,000	37,774	840,000	27,097
Illinois	887,000	28,613	1,005,000	32,419
West Virginia	678,000	21,871	589,000	19,000
Pennsylvania	534,000	17,226	502,000	16,194
Ohio:				
Central and Southeastern	396,000	12,774	325,000	10,484
Lima	135,000	4,355	120,000	3,871
Total Ohio	531,000	17,129	445,000	14,355
Kentucky	475,000	15,323	265,000	8,548
Indiana	67,000	2,161	49,000	1,581
New York	56,000	1,806	49,000	1,581
Colorado	9,600	310	10,000	322
Tennessee	1,400	45	1,000	32
Montana	b	....	7,000	226
Total	33,980,000	1,096,129	27,330,000	881,613

A record of imports and exports of mineral crude oil compiled by the Bureau of Foreign and Domestic Commerce is given in Table II.

TABLE II. REPORT OF U. S. IMPORTS AND EXPORTS OF MINERAL CRUDE OIL, IN BARRELS

	January, 1920	December, 1919	January, 1919
Imports:			
Mexico	6,293,353	4,330,576	3,899,376
Other countries	626	14,796	25
	6,293,979	4,345,372	3,899,401
Exports:			
Canada	421,618	868,673	270,553
Other countries	7,068	156,829	68,468
	428,686	1,025,502	339,021
Excess of imports over exports	5,865,293	3,319,870	3,560,380

The petroleum produced in Alaska is from a single patented claim in the Katalla oil field, according to the United States Geological Survey. The old wells on this claim and the refinery were operated as usual in 1918. Two new productive wells were also drilled in the Katalla field. The total production in 1918 was larger than in 1917. There has been a revival of interest in the potential Alaskan oil fields in anticipation of a law for the leasing of oil lands.

N. E. L. A. Convention

Important changes in the constitution and by-laws of the National Electric Light Association will be given a hearing at the convention in Pasadena next month. The proposed changes will enable the association to function for the greater benefit of all classes of members. For the first time in the history of the association, members of the state regulatory bodies will be given the privilege of addressing the convention.

Chairman Cail Jackson of the Wisconsin Railroad Commission will present a paper on "The Trend of Regulation" and E. O. Edgerton, chairman of the California Railroad Commission, will speak on "The Reward of Efficiency."

Explosion Hazard from Partly Consumed Coal Dust*

Where pulverized coal is used, large quantities of fine dust will accumulate on platforms, roof trusses, the outside of pipe lines, electric cables and even on the roofs of the buildings. The increased use of this fuel has led the Bureau of Mines to make investigations as to the liability of explosion from the resulting dust.

The Bureau of Mines collected samples of this so called "mill dust" and submitted them to chemical analyses, microscopic examination and flame and explosion tests. From the chemical analysis the volatile matter was found to range from 6.3 per cent to 23.54 per cent, fixed carbon from 50.88 to 67.08 per cent and ash from 22.29 to 32.16 per cent. Samples of the pure coal dust, before passing to the furnaces, were found to range in volatile matter from 30.50 to 36.81 per cent, fixed carbon 46.86 to 52.48 per cent and ash 9.26 to 21.84 per cent.

Under the microscope a sample of mill dust (which showed 23.5 per cent volatile matter by chemical analysis) was found to contain a large proportion of very fine partly coked coal dust, a small proportion of fine or medium fine partly coked coal dust and some dust thoroughly coked. From this and other examinations it would seem that some pulverized coal was being blown out of the combustion chambers before it had been thoroughly consumed.

Several samples were tested to compare the pressure that might be exerted by explosion of the dust with similar explosion of pure coal dust. In these tests about 0.003 oz. of the sample was blown over a heated coil in a glass globe and the pressure of the resulting explosion measured. Under these conditions pure coal dust gave a pressure of 14.5 lb. per sq.in., mill dust with 23.54 per cent volatile matter gave 10.21 lb. pressure, with 20.6 per cent volatile, 8.7 lb., and with 15.91 per cent volatile, 4.6 lb. per sq.in. Considering the small quantities of dust in these samples, it will be evident that large quantites would give much higher pressures, especially if confined.

Further explosion tests were made in the Bureau of Mines explosion gallery. Inflammability tests were made by dropping small clouds of dust into a clay roasting dish which had been heated to a temperature of about 2,000 deg. F. and also by blowing a small amount of dust over the flame of a Meker burner.

From all these tests it has been fairly well demonstrated that the line of division between mill dust which will explode and that which will not explode is reached when the dust contains about 11 per cent of volatile matter. Also, the higher the volatile matter the more violent will be the explosion, with the higher values not much below the pressures generated by a pure coal-dust explosion.

A fire that occurred several years ago illustrates some of the dangers of mill dust. A draft of air blew some of the dust over some red-hot iron, and the dust immediately took fire. The flames were communicated to the dust lying on roof trusses and spread so rapidly that the workmen in the mill barely escaped in safety.

Suggestions for safety from dust explosions include the following:

Fans driving the pulverized coal from storage bins to the furnaces should be run, after the fuel is shut off, until coal in the line is blown out. This is to avoid leaving smoldering sparks in the line from back draft. All places where dust might accumulate should be kept thoroughly cleaned and no open flames used while cleaning. All electric apparatus and wires should be kept clear of dust. Provision should be made to prevent static electric discharges in the pulverizing apparatus.

Manufacturers of Trenton who use fuel oil have been notified by the Standard Oil Co. that no more fuel oil will be shipped to Trenton after April 1 because of the great scarcity of the product. This will mean a loss to the manufacturers who use the oil and will result in considerable expense in changing back to coal burning.

*From the Bureau of Mines Monthly Report of Investigations.

Color Schemes for Plant Piping*

By H. L. Wilkinson

The importance of some definite and unmistakable means of readily distinguishing between the various pipe and conduit lines and systems in the power plant has long been recognized by both designing and operating engineers. Particularly in large power plants is this question of vital importance, and it should be stated that in some of the well-ordered large power plants reasonably satisfactory systems have been installed. Also there has been some effort made to standardize this practice, a committee appointed by the American Society of Mechanical Engineers to investigate the question having reported upon the preferable distinguishing colors for the various divisions of piping systems in the usual power plant. (See *Transactions* of the American Society of Mechanical Engineers, vol. 33, 1911, p. 17.)

Importance of Standardization Has Been Neglected

While the importance of standardization in this detail of our power plants has been recognized, too little attention has been given to the matter of late, and the writer believes that its importance should not be overlooked. It is therefore suggested that the members of this society consider the plan shown in the table and designed to apply to all piping systems.

It will be noted that the recommendation involves, in addition to piping lines, color schemes for machinery, motors, hand rails, dadoes, waste pails, elevator cages, etc., thus extending the usefulness of this distinguishing system to all parts of the industrial plant. It has been a revelation to the writer to find the cordial reception which this proposed color scheme has received in many industries and concerns of national prominence, and the result is that new industries and concerns are taking it up continually. The application of this standardization might well be extended to the designating by distinguishing colors the tools used in various departments of a plant. This would bring about several benefits. In the first place, it would enable managers to hold each department of a plant responsible for its own tools. Then, because of the improved appearance of the tools it would create more interest in their care on the part of mechanics. Last, and not least, it would surely be an insurance against tools being carried out of factories.

While this scheme designates definite colors for certain pipes, it is of course unnecessary to apply any one particular color on any one pipe, the arrangement being flexible so that an engineer may determine for himself his own design. In some cases, where there are as many as fifteen to twenty-five various pipe lines, it is necessary to use combination colors on some pipes, painting the straight pipe one color and the joints, valves, elbows, etc., the alternating color. Of course, some exceptions to personal selection will be found where an established custom has been in effect for some time, as in the case of sprinkler pipes and all fire lines, where vermillion is the logical color.

Illustrations of the Advantages of Standard Color Schemes

The advantage of this standardization will become readily apparent from the case of a certain plant where there was a stoppage or break in one of the pipe lines. The workmen carefully followed the line to the side wall where the trouble apparently existed and opened up the wall for a considerable area, only to be confronted with everything but the line they sought. Of course the expense of ripping out the wall was nothing compared to the inconvenience and loss sustained, due to the suspension of operation until this repair was made. If this pipe line had been painted with a distinguishing color and a small arrow placed on the wall where the pipe entered, indicating the direction the pipe took after entering the wall, all this difficulty would have been obviated.

*Abstract of a paper presented at the annual meeting of the American Society of Heating and Ventilating Engineers, New York, January, 1920.

Another condition that often arises is the leaking of a pipe on the third, fourth or some upper floor of a building. The engineers, usually on duty on the first floor or in the basement, would naturally be informed of the leak by the person discovering it, but without a doubt in the majority of cases the person reporting would be unable to tell the engineer which line of pipe was at fault, thus necessitating a journey from the basement to the floor in question. Furthermore, while it might be possible to shut off the pipe at a point convenient to the leak, probably it would be necessary for him to go back to the engine room and shut off the main valve. If, on the other hand, the individual discovering the leak could immediately telephone to the engineer that the blue, green, yellow, brown or black pipe was leaking, it would be possible at the main valve to instantly stop further leakage.

Another instance recently brought to the writer's attention is pertinent in showing vividly the value of standardization. One Sunday the watchman in a dye plant discovered that a pipe carrying a valuable liquid was leaking. The engineer was out of the city, which necessitated the watchman calling the manager. The manager was unfamiliar with the beginnings and endings of the maze of pipes, and it was only after considerable time that he was able to locate the proper valve to shut off. In the meantime considerable damage had been done, to say nothing of the loss of a large quantity of valuable acid. How easily the watchman could have stopped the leaking and prevented the damage if he had been guided immediately to the correct valve by following the color of the leaking pipe to the first valve of the same color.

Such a distinguishing system would appear to be indispensable in a plant with pipe lines carrying oil for automatic lubrication and acids for the treatment of goods in manufacture. A break in an acid pipe line, if not immediately shut off, may cause great damage.

A Color Scheme Will Result in All Pipes Being Protected by Paint

These simple illustrations have shown only one phase of the advantage of standardization. It will be readily recognized that the adoption of the standardization method in many plants will often carry with it the painting of pipes that heretofore have gone unprotected. Special paints are obtainable for these purposes, which possess the qualities of remaining uncracked under extremes of temperature, withstanding corrosion and affording superior protection against rust. An enamel paint is most suitable for general piping, but for power-house and other canvas-covered pipes a high-grade linseed-oil paint in white and colors should be used. The greatest benefits from the adoption of a method of standardization will accrue where careful thought is used in the selection of paints especially designed for this purpose. In cases where the surfaces to be painted are subject to unusually severe conditions of exposure to corrosion, a rust-preventive paint of established merit should be used for the first coat, followed with the standardization paint in the colors selected.

COLOR DESIGNATIONS FOR DIFFERENT CLASSES OF PIPING

Class of Piping	Color
Sprinkler pipes, fire pails, etc.	Vermilion
Compressed-air pipes	Dark gray
Oil pipes	Brown
Electric-light and power conduits	Black
Steam-power lines	Blue
Steam-exhaust lines	Buff
Hot-water pipes	Bright-red oxide
Cold-water pipes	Yellow
Motors and machinery	Bright green
Refrigerating lines	Maroon
Elevator cages, fire-doors, waste cans, railings, shelves, etc.	Bronze, green or black
Steam heating lines and radiators	Aluminum gray

In discussing the likelihood of steam turbines with outputs far in excess of current practice being installed in power stations of the future, I. V. Robinson says, in the *B. E. A. M. A. Journal*, that the chief limitation is that imposed by transport facilities, and suggests, as a remedy, that large machine tools should be installed in power-houses, so that the turbines could be partially constructed on the site.

Engineers' Influence in Public Utilities

On April 2 the Western Society of Engineers inaugurated a new society activity consisting of a series of monthly noonday luncheon meetings at which it is the plan to have short up-to-the-minute talks by prominent men on current topics of general public interest. The idea is to get the engineer away from the slide rule for a time and to give him an insight into big business and public affairs. The first meeting was a great success. Fully 600 attended. The speaker was Samuel Insull, president of the Commonwealth Edison Co., and the topic, "The Engineers' Influence in Public Utilities."

In his introductory remarks Mr. Insull referred to the need of exact knowledge on the public questions of the day. An engineer's training and his daily work are conducive to exact reasoning and logical, straight thinking. It is time that such men took their place in public affairs. The average politician is inclined to make the wildest claims imaginable, which seldom if ever approach the facts. If the engineer would apply his power of reasoning and present to his community the exact facts concerning the operation of public utilities, how different the impression that would be conveyed to the public mind.

Interests and Services of the Public Utilities No Longer Local

As far as the great public utilities were concerned, the average poltician and municipal officer, in seeking for power and patronage, insisted that their relation to the public was a question for home rule, when as a matter of fact the affairs of the utilities should be settled only on a state or interstate basis. Their interests and the services rendered were no longer local and the whole tendency was to make them less local. From the city, service has extended to outlying suburbs and half way across the state, and future power developments at the mines and waterfalls must of necessity be outside the city. In building up these institutions by combining and tying together numerous interests to effect economies in the service to the public, there had been nothing but criticism from the politicians and the public generally.

On subjects of this nature, primarily engineering in their character, the engineer should be a leader of thought. He should become posted on subjects that his specific training makes him capable of understanding and his next duty was to pass it along to the public, using his influence to protect the interests of the state in these great developments. The treatment and the general credit of the utilities were of far greater consequence to the community than to the owners themselves, as the latter generally have their eggs in more than one basket and the community has but one public utility in each class of service. Consequently, the prosperity of the utility, making development and extension possible, and the efficiency of service rendered, are matters of prime importance to the community. Incidentally, Mr. Insull gave some Chicago figures that were interesting. He said that with $2\frac{1}{2}$ to 3 per cent of the population of the United States, Chicago central-station interests had produced 5 per cent of the total electrical energy distributed. From this, 6 per cent of the gross income of the public utilities of the country had been produced on an investment amounting to 3 per cent of the total capitalization. In other words, the Chicago company was doing business on 50 per cent less capital than the average throughout the United States. This was mainly the result of efficiency in sales engineering.

Possibilities and Advantages of Extension and Combination

To show the possibility in efficiency and conservation by extension and combination of public utilities, the speaker referred to the power survey of the Atlantic Coast section extending from Boston to Washington and sweeping in a semicircle 150 miles back from the coast, in which one-quarter of the entire generating capacity of the country is concentrated. By tying together the great transmission systems and supplying them with all available energy from water power and stations at the mouth of mines contiguous to this section, it had been estimated that a total capacity of 17,000,000 hp. could be replaced by 5,500,000 hp. The load factor would be increased from an average of 15 to 60 per cent, and half the coal formerly required (30,000,000 tons, worth $150,000,000) would be saved. In other operating expenses there could be saved another $150,000,000, making a total possible saving for the section of $300,000,000 per year.

It was the mission of the engineer to bring home to the public the possibilities and the great advantages of such a plan. With a saving of such magnitude in a small section of the country, it might be imagined what it would mean if the same principles were carried out throughout the entire country. Mr. Insull referred to the fact that if all the available water powers of the country were developed, they would not come anywhere near taking care of the power demand at the present time. As power-plant fuels, oil and natural gas were of minor importance, so that the future welfare of the country turned to the coal supply of this continent. The coal consumption of the country was already approaching 700,000,000 tons per year, with every prospect that it would increase by leaps and bounds, so that unless the greatest economy was obtained, future generations must suffer. The Atlantic Coast Survey gave an idea of what was possible in the way of coal conservation, and if the same plan were applied throughout the country, the saving would be enormous. It would take generations to accomplish, but could be immediately applied in sections where conditions were favorable.

With Government operation of utilities, lacking incentive and initiative, money would not be forthcoming for such great developments. In the interest of economy and conservation it was of the greatest importance, then, not to destroy the present value of these utilities and rob them by arbitrary action. Under such conditions it would be impossible for them to maintain their credit, to invest and stimulate development. Nothing would stop development more than interfering with transportation interests, and nothing would affect business interests more than mistreating the great local public utilities.

The engineer, then, should not confine himself to technical questions alone. He should become familiar with the facts and details of operation of the public utilities. By so doing he would comply with his duties as a citizen and contribute to the interests of the community and the country.

N.A.S.E. National Convention Committee

National President John J. Calahan has appointed Charles W. Bindrich, John A. Wickert and Thomas W. Armer as committeemen in charge of the coming national convention of the National Association of Stationary Engineers to be held in Milwaukee, Wis., Sept. 13 to 18. The executive committee is made up of the following: Program, Charles W. Bindrich, 562 W. 36th St., Milwaukee; publicity, John A. Wickert; halls and hotels, Thomas W. Armer, 133 Martin St., Milwaukee; reception, Henry J. Mustele; entertainment, J. A. Prudell; finance, Charles A. Cahill; transportation, Harry E. Fritz. Frank Cleveland, secretary of the convention bureau of the Milwaukee Chamber of Commerce, is chairman of the local exhibitors' committee. Headquarters will be in the Hotel Wisconsin. Thomas Green, Garlock Packing Co., Pittsburgh, Pa., is secretary of the National Exhibitors' Association.

Norway lacks coal, but it has an abundance of water-power. The power available for development in Norway is estimated at from 10,000,000 to 15,000,000 hp., and 1,300,000 hp. is now being used. In 1917 more than forty miles of the state railways had been electrified, and since that time the electrification of numerous short lines has been under way. Through the utilization of its enormous water-power in transportation and manufacturing—thus overcoming the handicap of a shortage of domestic coal—Norway, it is believed, will show marked industrial progress in the future.—*Pacific Power and Light Company's Bulletin.*

Waterwheels Used for Braking and Initial Power

An unusual use of waterwheels has been made at the plant of the Spanish Peak Lumber Co., Plumas County, Cal. Two 24-in. Pelton waterwheels, set in opposition, have been adapted to an aërial gravity tramway control to furnish power and braking effort as well as to drive a generator for lighting purposes, says the *Journal of Electricity*.

A braking effort to absorb a maximum excess of 12 hp. is necessary when the way is fully loaded, and a torque of 20 hp is required when the cable is not loaded. The two waterwheels are rigidly connected in opposition through a solid coupling, making up the driving and braking unit. They are controlled by needle nozzles, actuated by oil governors, in regulating the cable speed, the regulation being to hold the speed constant. The governors are so adjusted that when the cable slows down below normal speed, due to a light load, water is admitted to the driving motor and power is generated to hold the speed constant. When the cable is loaded and the speed tends to increase, the water is shut off the driving motor and is admitted to the braking motor. The impact of the buckets of the braking motor rotating against the jet develops the requisite braking effect and holds the speed normal.

Economy of Electric Drive

A comparison of performance of the motive power of the "New Mexico" with that of other battleships shows marked economy in favor of the electric drive. Thus, at twelve knots' speed the consumption in tons of oil per day was for the "New Mexico" (electric drive) 75 tons, as compared with the "Arizona" for her cruising turbines, 100 tons and main turbines, 118, and the "Mississippi" for her cruising turbines, 99 tons, and for her main turbines 115 tons of oil per day. At nineteen knots the consumption of the "New Mexico" was 263 tons, of the "Mississippi" main turbines 305 tons, and of the "Idaho" main turbines, 310 tons per day. Similar economy was shown by the electric drive at the intermediate speeds.—*Scientific American.*

New Publications

HYDRAULIC TURBINES. Third Edition. By R. L. Daugherty, A. B., M. E., Professor of Mechanical and Hydraulic Engineering, California Institute of Technology. Published by McGraw-Hill Book Co., Inc., New York. Cloth; 6 x 9 in.; 280 pages. Price, $3.

Originally intended as a brief outline of the general principle of hydraulic-turbine design, the present edition has been largely rewritten to conform to the changes in practice. The presentation of the theory has been considered and changes in its form made to make it more effective. It is realized that hydraulic-turbine design is a highly specialized industry, but the buyer and user of such machines should have some knowledge of their principles and characteristics. It is to furnish such needed information that the book was produced. The first part is given up to history, classification and description of hydraulic turbines, with a chapter on water as a source of power. The theory of turbines, methods of testing and turbine characteristics are given considerable space. A chapter on Selection of Type of Turbine and one on Costs of Turbines and Water Power should prove of especial interest to hydraulic power-plant men. The mathematical portions of the book assume a knowledge of trigonometry and calculus.

ARMATURE WINDING AND MOTOR REPAIR. By Daniel H. Braymer. Published by McGraw-Hill Book Co., Inc., New York, 1920. Cloth; 8 x 5½ in.; 515 pages; 298 illustrations. Price, $3.

This book is a collection of practical information and data covering winding and reconnecting procedure for direct and alternating-current machines, compiled for electrical men responsible for the operation and repair of motors and for the operation of generators in industrial plants and for repairmen and armature winders in electrical shops. No attempt has been made to discuss the subject of armature winding from the theoretical and design standpoints. Mathematics has been made use of in a very few cases, thus making the book easily read by those with a limited technical training. The book is divided into 17 chapters and an appendix: Direct-Current Windings; Alternating-Current Windings; Repair-Shop Methods for Rewinding Direct-Current Armatures; Making Connections to the Commutator; Testing Direct-Current Armature Windings; Operations Before and After Winding Direct-Current Armatures; Insulating Coils and Slots for Direct-Current and Alternating-Current Windings; Repair-Shop Methods for Rewinding Alternating-Current Machines; Testing Induction-Motor Windings for Mistakes and Faults; Adapting Direct-Current Motors to Changed Operating Conditions; Practical Ways for Reconnecting Induction Motors; Commutator Repairs; Adjusting Brushes and Correcting Brush Troubles; Inspection and Repair of Motor Starters, Motors and Generators; Diagnosis of Motor and Generator Troubles; Methods Used by Electrical Repairmen to Solve Special Troubles; Machine Equipment and Tools Needed in a Repair Shop; Appendix, Data and Reference Tables.

In all these the practical man has been kept in mind and the story of his problems is told in a clear and easily understood way. Although the number of subjects covered has made it necessary to limit the treatment of them to a considerable extent, those who have electrical repair problems to contend with will find much in this book that will assist them in the solution of these problems.

Obituary

Charles E. Barber, formerly of Watertown, N. Y., died at his home in Liverpool, March 22, at the age of 70 years. Mr. Barber, who was the brother of the late Henry A. Barber, the inventor of the Barber centrifugal pump, at one time manufactured and sold steam engines in Syracuse, N. Y. He was also the inventor of the Barber marine engine. He is survived by a widow, a brother, two sons, a daughter and two nephews.

Society Affairs

The National Association of Stationary Engineers, Texas State Association, will hold its annual convention at San Antonio, Texas, April 19-21.

The American Chemical Society will hold its spring meeting at St. Louis April 12-16. The general meeting and all divisional meetings will be held at the Hotel Statler.

The American Society of Mechanical Engineers, St. Louis Section, will hold a meeting April 30. H. F. Gauss, engineer, Illinois Stoker Co., will speak on "Development in Chain-Grate Stokers."

The Philadelphia Section of the A. S. M. E. will meet at the Engineering Club, Philadelphia, Pa., on April 27. Colonel W. P. Barba will discuss "The Modern Practice in the Manufacture of Steel."

The National Electric Light Association, Meter Committee, Technical and Hydro-Electric Section and the Ohio Electric Light Association, Columbus, Ohio, will hold a joint meeting at Columbus on April 22.

The American Society of Chemical Engineers, Eastern New York Section, will hold a meeting at the Edison Club, Schenectady, April 23. C. H. Bierbaum, consulting engineer will present a paper on "Bearing Metals."

The New York Section of the A. S. M. E. will meet at its headquarters, 33 West 39th St., April 13. The plans include the arrangement of a joint meeting with the Welding Societies. Subject, "Symposium on Welding."

National Electric Light Association—In a report by Harry M. Sessions, chairman of the Pacific Coast Section Membership Committee of the N. E. L. A., it was shown that more than two thousand new members have been added to the organization since last November.

The New Haven Branch, Connecticut Section of the A. S. M. E., will meet on April 18 at the Mason Laboratory of the Sheffield Scientific School. Arthur M. Greene, Jr., chairman of the research committee of the A. S. M. E. will present a paper entitled "Research."

The Detroit Engineering Society, Detroit, Mich., will hold its twenty-sixth annual banquet, Friday evening, April 23, at the Hotel Cadillac. Previous to the banquet a business meeting will be held at which the election of officers for the ensuing year will be announced.

Miscellaneous News

The Supreme Court recently dismissed the petition to rehear the appeal in the case of the North Carolina Public Service Co., and the Spencer Railway Co., vs. the Southern Power Co. The North Carolina Public Service Co. and the Spencer Railway Co. charge that discrimination was being shown by the Southern Power Co., in that they formerly furnished power to them at the rate of eleven mills per kilowatt, which they increased to eighteen mills, while it was furnishing its own subsidiary corporations power at the eleven-mill rate.

Business Items

The Roller-Smith Co., New York City, announces the appointment of Hammond D. Baker as manager of its branch office in Detroit.

The Pioneer Rubber Mills, of San Francisco, have sent D. D. Tripp, vice-president, on a trip to the Orient to observe the trade possibilities.

The Homestead Valve Manufacturing Co., Inc., of Homestead, Pa., announces the removal of its New York office to 242 Lafayette St., New York City.

The Niagara Falls Power Co., at its annual election of officers on March 26, elected Paul A. Schoellkorf, president, and Morris Cohn, Jr., John L. Harper and A. H. Schoellkorf, vice presidents.

The American Steam Conveyor Corp., Chicago, Ill., announces that the Kin-Wald Engineering Co., Mutual Life Building, Buffalo, N. Y., will act as its representative in Buffalo and western New York.

The American Sugar Refining Co. has awarded a contract for the construction of a sugar refinery at Baltimore, Md., to Stone & Webster. It is intended to begin work early in June. Estimated total cost, $10,000,000.

The Yarnall-Waring Co., manufacturers of power-plant devices in Philadelphia, Pa., has opened an office in Atlanta, Ga. D. T. Newman, formerly connected with the Philadelphia sales office, will assume charge of the new branch.

The Chicago Pneumatic Tool Co., announces the removal of its general offices from Chicago to New York City, where it will occupy a new ten-story building. The district sales branch of Chicago has moved to new quarters at 300 North Michigan Boulevard.

THE COAL MARKET

BOSTON—Current prices per gross ton f.o b. New York loading ports:

Anthracite

	Company Coal
Egg	$8 20@$8.65
Stove	8 45@ 9.05
Chestnut	8 55@ 9.05
Pea	7 05@ 7.40
Buckwheat	3.60@ 3.80
Rice	2 30@ 2 50
Barley	1 25

Bituminous

	Clearfields	Cambrias and Somersets
F.o.b. mines, net tons	$2 85@$3 35	$3 15@$3.60
F.o.b. Philadelphia, gross tons	5 05@ 5 60	5 35@ 5.90
F.o.b. New York, gross tons	5 40@ 5 95	5.75@ 6.25
Alongside Boston (water coal), gross tons	7 00@ 7 75	7 60@ 8 00

Pocahontas and New River are practically off the market for coastwise shipment, but are quoted at $6.25@$7.00.

NEW YORK—Current quotations, White Ash, per gross tons, f.o.b. Tidewater, at the lower ports are as follows:

Anthracite

	Company Coal
Broken	$7 80@$8.25
Egg	8.20@ 8.65
Stove	8.45@ 9.05
Chestnut	8.55@ 9.05
Pea	7.05@ 7.40
Buckwheat	5.15
Rice	4 50
Barley	4 85
Boiler	4 85

Bituminous

Government prices at mines:	Spot
South Fork (best)	$3.25@$3.50
Cambria (best)	3.00@ 3.25
Cambria (ordinary)	2.60@ 2.90
Clearfield (best)	3.00@ 3.25
Clearfield (ordinary)	2.60@ 2.90
Reynoldsville	2.85@ 2.90
Quemahoning	3.25@ 3.50
Somerset (medium)	3 00@ 3.25
Somerset (poor)	2.50@ 2.75
Western Maryland	2.50@ 2.75
Fairmont	2.25@ 2.50
Latrobe	2.60@ 2.90
Greensburg	2.75@ 3.00
Westmoreland ¾ in	3.40@ 3.50
Westmoreland run-of-mine	2.75@ 3.00

PHILADELPHIA—Bituminous coal prices vary according to district from which they are mined. For ordinary slack the price is $2.45@$2.55; lump, $3.35, at the mines.

BUFFALO—

Anthracite

	On Cars, Gross Ton	At Curb, Net Ton
Grate	$8.55	$10.80
Egg	8.80	10.65
Stove	9.00	10.85
Chestnut	9.10	10.85
Pea	7.75	9.30
Buckwheat	5.70	7.75

Bituminous

Allegheny Valley	$4.80
Pittsburgh	4.65
No. 8 Lump	4.65
Mine-run	4.80
Slack	4.10
Smokeless	4.60
Pennsylvania Smithing	5.70

CLEVELAND—Prices of coal per net ton delivered in Cleveland are:

Anthracite

Egg	$12.25@$12.40
Chestnut	12.50@ 12.70
Grate	12.25@ 12.40
Stove	12.40@ 12.60

Pocahontas

Nine-run	$7.50

Domestic Bituminous

West Virginia split	$8.50
No. 8 Pittsburgh	$6.60@ 6.90
Massillon lump	8.25@ 8.50
Coshocton lump	7.15

Steam Coal

No. 6 slack	$5.25@$5.50
No. 8 slack	5.10@ 5.50
Youghiogheny slack	5.25@ 5.50
No. 8 ¾ in	5.70@ 6.60
No. 6 mine-run	5.25@ 5.50
No. 8 mine-run	5.75

Only coal available is mine-run Pocahontas.

MIDDLE WEST—Chicago quotations, f. o. b. cars at mine:

	Springfield, Carterville, Williamson, Franklin, Saline, Harrisburg	Fulton, Peoria	Grundy, La Salle, Bureau, Will
Lump	$2.55@$2.70	$2.95@$3.10	$3.25@$3.40
Washed	2.75@ 2.90	....	3.45@ 3.60
Mine-run	2 35@ 2.50	2.75@ 2.90	3 00@ 3.15
Screen'gs	2 05@ 2.20	2.35@ 2.50	2.75@ 2.90

New Construction

PROPOSED WORK

Mass., Cambridge—The Barnard Constr. Co., 1783 Massachusetts Ave., will change the 5 story, 50 x 250 ft. dormitory building on Harvard St., into a hotel. A steam heating system will be installed in same. Work will be done by day labor.

Mass., Cambridge—The E. E. Gray Co. is having plans prepared for the construction of a 2 story, 140 x 280 ft. store house and refrigerating plant on Albany St. Total estimated cost, $250,000. F. F. Jonsberg Co., 16 Central St., Boston, archt. and engr.

N. Y., Addison—The Addison Electric Light & Power Co. is in the market for a 100 hp. heavy duty r.h automatic belted steam engine and a 100 hp. high pressure return tubular steam boiler.

N. Y., Antwerp—The F. X. Baumert Co is in the market for milk handling machinery, equipment and cold storage installation in its plant at Evans Mills.

N. Y., Attica—The Don Co. is in the market for one 100 lb. pressure 8 to 10 hp vertical boiler.

N. Y., Buffalo—The city is having plans prepared by George H. Norton, City Engr., for installation of galvanized piping and the construction of an extension to the refrigeration system at the Washington market, here. Total estimated cost, $30,000.

N. Y., Gowanda—The Iroquois Utilities Corp., 39 West Main St., had plans prepared for the construction of a power house at Cattaragus Creek. Three 500 hp. generating units will be installed in same. Total estimated cost, $150,000. Charles E. Eaton, Sherman Bldg., Watertown, Consult. Engr.

N. Y., Hamburg—The Depew & Lancaster Light, Power & Conduit Co., Lancaster, plans to build an electric plant, here.

N. Y., New York—R. C. P. Boehler, Archt. and Engr., 38 West 32nd St., will soon award the contract for the construction of a 12 story loft building at 146 West 26th St. for the Leopold Realty Co. A steam heating system will be installed in same.

N. Y., New York—The New York Telephone Co., 15 Dey St., is having plans prepared for the construction of a telephone building on Vesey, Barclay and West Washington Sts. A steam heating system will be installed in same. Total estimated cost, $750,000. McKenzie, Voorhies & Grelin, 1123 B'way, Archts. and Engrs.

N. Y., New York—W. Whitehall, Archt. and Engr., 32 Union Sq., will soon award the contract for the construction of a 3 story, 45 x 75 ft. transformer station on East 6th St., for the New York Edison Co., Irving Pl. and 15th St. Estimated cost, $60,000. Noted Feb. 9.

N. J., Asbury Park—Froling & Holler, Archts. and Engrs., 150 Nassau St., New York City, are having plans for the construction of a 12 story hotel on Ocean Ave., between 2nd and 3rd Sts. A steam heating system will be installed in same. Total estimated cost, $1,000,000. Owner's name withheld.

N. J., East Orange—The city plans to build a community building. A steam heating system will be installed in same. Total estimated cost, $500,000.

N. J., Irvington (Newark P. O.)—The Vreeland Motor Co., 407 Elizabeth Ave., Newark, will soon award the contract for the construction of a 1 and 2 story, 282 x 600 ft. factory, a 200 x 250 ft. office building and a 50 x 160 ft. machine shop on Cort St. A steam heating system will be installed in same. Total estimated cost, $300,000. John T. Simpson, Essex Bldg., Newark, Archt.

N. J., Jersey City—Francisco-Jacobus 511 5th Ave., New York City, engineers for the proposed power house, here, for the Continental Candy Co., are in the market for 2 generators, 3 boilers and two 200 kw. engines.

N. J., Newark—Jarvis Hunt, Archt., 30 North Michigan Ave., Chicago, will soon award the contract for the construction of a 25 story addition to the Department store on Market St., for the L. Bamberger & Co., Market St. A steam heating system will be installed in same. Total estimated cost, $2,000,000.

Pa., Harrisburg—The Water Dept. plans to construct a 25,000,000 to 30,000,000 gal. reservoir and install a 40 hp. boiler and pump, etc. at the pumping station, here. S. F. Hassler, Comr.

Pa., Marcus Hook—Ballinger Perrot, Archts. and Engrs., 329 South Broad St., will soon award the contract for the construction of a 3 story, 6 x 110 ft. boiler house for the Viscose Co. A steam heating system will be installed in same. Total estimated cost, $100,000.

Pa., Philadelphia—The Bd. Educ., Keystone Bldg., plans to rebuild 3 story school building on 7th and Norris Sts. A steam heating system will be installed in same. Total estimated cost, $400,000.

Pa., Philadelphia—The Bd. Educ., Keystone Bldg., plans to rebuild 3 story school building on 6th St. and Fairmount Ave. A steam heating system will be installed in same. Total estimated cost, $400,000.

Md., Baltimore—O. G. Simonson, Archt., will soon award the contract for the construction of a 3 story, 300 x 300 ft. administration building, garage, power house, and printing shop, 1 story, 60 x 122 ft. each for the Maryland Casualty Co., Baltimore St. and Guilford Ave. Total estimated cost, $3,000,000. Noted Sept. 9.

D. C., Washington—The Bureau of Yards & Docks, Navy Dept., received bids March 31 for installing mechanical equipment and piping at the boiler and power plant, for work complete, from J. W. Danforth, 2 Ellicott St., Buffalo, N. Y., (1) $218,300 (200 days) (2) add $8,500 (30 days), (3) add $10,000 (30 days), (4) add $13,000, (5) add $12,300 (30 days), (6) $8,950 (30 days); S. W. Rittenhouse, 1316 Harvard St., (1) $227,853 (190 days), (2) $9,946, (3) $12,762, (4) $3,900, (5) $5,266 (30 days); W. G. Cornell Co., 923 12th St., N. W. (1) $232,800 (200 days), (2) $8,360 (200 days), (3) add $8,512 (30 days), (5) $7,350 (30 days). Noted Sept. 2.

Va., Richmond—Miller & Rhoads, c/o Starrett & Van Vled, Archts. and Engrs., 8 West 40th St., New York City, are having plans prepared for the construction of a 5 and 6 story 260 x 300 ft. office building, consisting of a hospital roof garden, cafe, etc. A steam heating system will be installed in same. Total estimated cost, $750,000.

W. Va., Charleston—The Bureau of Yards & Docks, Navy Dept., Wash., D. C., received bids for the construction of an outdoor electric substation at the Naval Ordnance plant, here, from the Eng. Servc. Co., 411 Odd Fellows Bldg., $41,250, (150 days). Noted March 16.

Ga., Lagrange—The Lagrange Hotel Co., c/o W. L. Stoddard, Archt. and Engr., 9 East 40th St., New York City, is having plans prepared for the construction of a 100 x 100 ft. hotel and theatre building. A steam heating system will be installed in same. Total estimated cost, $500,000.

Tenn., Knoxville—The Holston Box & Lumber Co., Mitchell St., will rebuild the 150 x 225 ft. box factory recently destroyed by fire and will be in the market for 30 motors. Work will be done by day labor.

O., Cleveland—The Bd. Educ., East 6th St. and Rockwell Ave., received bids for the installation of a heating and ventilating system in the Warner Rd. School, from W. B. Van Sickle, 701 Frankfort Ave., $119,590; Chappell Warren Co., 1838 St. Clair Ave., $125,808.

O., Cleveland—The Bd. Educ., East 6th St. and Rockwell Ave., plans to build a 3 story high school at University Circle. A steam heating system will be installed in same. Total estimated cost, $5,000,000. F. G. Hogan, Dir. W. R. McCormack, Archt.

O., Cleveland—The city plans to construct a pumping station, reservoir, etc. at Baldwin. Total estimated cost, $2,000,000. A. V. Ruggles, City Hall, Engr.

O., Cleveland—The city plans to construct a filter plant with pumping units and equipment, including a reservoir, pumping station, etc., in the eastern section of the city. Total estimated cost, $5,000,000. A. V. Ruggles, City Hall, Engr.

O., Cleveland—The Kayner Co. plans to construct a 5 story, factory addition on B'way and Aetna Rd. An extension to the steam heating system will be installed in same. Total estimated cost, $230,000. E. C. Seitz, B'way and Aetna Rd., Secy. Christian-Schwartzenberg & Gaede Co., 1900 Euclid Ave., Engrs.

O., Cleveland Heights (Warrensville P. O.)—The Cleveland Heights Methodist Episcopal Church, c/o W. H. Nickles, archt., 1900 Euclid Bldg., Cleveland, is having plans prepared for the construction of a 1 story, 140 x 180 ft. church on Euclid Blvd. A steam heating system will be installed in same. Total estimated cost, $250,000.

O., Fairview—The Bd. Educ. plans to build a 2 story school building. A steam heating system will be installed in same. Total estimated cost, $100,000. L. E. Garnett, Pres.

O., Lakewood (Cleveland P. O.)—The Bd. of Co. Comrs., c/o Joseph Menning, Court House, Cleveland, plans to build a 1 story market house. A steam heating system will be installed in same. Total estimated cost, $300,000.

O., Lebanon—The Bd. of Trustees of Pub. Affairs will receive bids until April 27 for the construction of a light plant and the installation of complete light, power and pumping equipment. Total estimated cost, $135,000. E. M. Choce, Union Central Bldg., Cincinnati, Engr.

O., London—The Ohio Bd. of Administration plans to build a power house, etc. Total estimated cost, $100,000.

O., Loveland—Frank Chase, Engr., will receive bids until April 27 for the construction of an electric light plant, for the city. Estimated cost, $125,000.

O., Sidney—The Bennett Bull Packing Co. will soon award the contract for the construction of a 3 story, 40 x 81 ft. packing plant. Ice machinery will be installed in same. Total estimated cost, $75,000. Anders & Reimers, 430 Erie Bldg., Cleveland, Archts. and Engrs.

Mich., Detroit—The Bd. of Water Comrs., 232 Jefferson Ave., is in the market for a 400 hp. horizontal water tube type boiler and an under feed gravity stoker.

Mich., Detroit—The Church of the Visitation Parish, c/o Donaldson & Meier, Archts., 1314 Penobscot Bldg., plans a 2 story church and school building on 12th St. and Lawrence Ave. A steam heating system will be installed in same. Total estimated cost, $300,000.

Mich., Detroit—The Ewing Bolt & Screw Co., 403 Farwell Bldg., plans to build a 1 story factory for the manufacture of bolts, screws, nuts and rivets. Steam heating equipment will be installed in same. Total estimated cost, $350,000.

Mich., Northville—The Bd. of Health, 233 St. Antoine St., Detroit, received bids for the construction of a 1 story, 48 x 87 ft. power house and a 40 x 120 ft. laundry, in connection with the group of buildings to be constructed at the proposed Detroit Municipal Tuberculosis Sanitarium, from John Finn & Co., 234 East Fort St., Detroit, $66,570; George A. Fuller Co., 540 Penobscot Bldg., Detroit, $95,500. Noted March 3.

Ill., Chicago—The Federal Reserve Bank, 79 West Monroe St., is having plans prepared for the construction of a 14 story, 160 x 165 ft. bank and office building on Jackson Blvd. and La Salle St. A steam heating system will be installed in same. Total estimated cost, $3,000,000. Graham Anderson Probst & White, Ry. Exch. Bldg., Archts.

Minn., Hibbing—T. J. Godfrey, Clk., will receive bids until April 27 for the construction of a 2 story, 62 x 400 ft. high school on Mesaba St., for the Bd. Educ. A steam heating and ventilating system will be installed in same. Total estimated cost, $500,000. W. T. Bray, Torrey Bldg., Duluth, Archt. and Engr.

Minn., Minneapolis—The Bd. Educ., City Hall, will soon award the contract for the construction of a 2 story, 75 x 200 ft. school building to be known as the Miles Standish School on 22nd Ave., S. E. and 40th St. A steam heating system will be installed in same. Total estimated cost, $250,000. E. H. Enger, Archt.

Kan., Cimarron—The city is having plans prepared to improve the waterworks system including tower, tank, pumping equipment and new well. Total estimated cost, $30,000. W. B. Rollins & Co., 209 Ry. Exch. Bldg., Kansas City, Mo., Engr.

Kan., Protection—The city is receiving bids for the construction of a pumping plant, etc. Total estimated cost, $50,000. W. B. Rollins & Co., 209 Ry. Exch. Bldg., Kansas City, Mo., Engr.

Tex., Brownwood—The city has voted $150,000 bonds to improve streets, construct sewer extensions and a municipally-owned electric lighting plant.

Tex., Dallas—The city plans to install a pumping plant at Record Crossing on the Trinity River to pump water into the Bachman reservoir. An electrically driven rotary pump will be installed in same. Estimated cost, $10,000. George Fairtrace, City Engr.

Tex., Dallas—The Silvers Box Corp., 317 Bourbon St., is having plans prepared for the construction of a box factory including six 1 story, 60 x 200 ft. buildings to house six units. Motors will be installed in same. Total estimated cost, $250,000. J. C. Silvers, Clarence and Latimore Sts., Engrs.

Tex., Liberty—The Liberty Light & Power Co. plans to build an electric lighting and power plant, here. Address, A. J. Rivierre.

Tex., Newcastle—The Newcastle Light, Power and Ice Co. plans to install an electric light and power plant with an ice plant to be operated in connection with same. Address W. F. Nance.

Tex., Tulia — The citizens will install a municipally-owned electric light and water plant to give current for 24 hours of the day.

Okla., Claremore — The citizens voted $75,000 bonds to improve light and water systems. The proceeds of the bonds are to be used to increase the facilities of the electric power plant.

N. M., Mogollon—The Mogollon Mines Co. plans to install hydro-electric equipment to furnish power for its silver mines. S. J. Kidder, Mgr.

Wash., Spokane—John G. F. Hieber, 2011 1st Ave., will build a fruit juice plant including three 3 story, 73 x 110 ft. buildings. A furnace for drying pulp, small boiler, motor pumps, conveyors, etc., will be installed in same. Total estimated cost, $100,000. H. E. Smith, 1211 Old Natl. Bank Bldg., Engr.

Ore., Roseburg — The citizens plan an election in May to vote on $500,000 bonds to construct a power plant and water system.

Cal., Ontario—The city voted $40,000 bonds to sink well and install pump.

Ont., Kitchener — The Young Business Men's Association plans to build a 1 story hockey and skating arena. Machinery for making an artificial ice surface and an artificial ice plant will be installed in same. Total estimated cost, $250,000.

Ont., St. Catherine — The Niagara St. Catharine & Toronto Ry. Co. plans to make road improvements which will include the increase of power substation facilities and centralization of control, etc. Total estimated cost, $500,000. E. W. Oliver, Genl. Supt.

B. C., Vancouver—The British Columbia Electric Ry. Co., 425 Carroll St., is submitting estimates of the cost to change the voltage on the Frazer Line from 600 to 1,200 and for substation equipment. J. I. Newell, Engr.

PRICES AND CONTRACTS AWARDED

N. Y., Brooklyn—The Bd. Educ., 500 Park Ave., New York City, has awarded the contract for installing heating, ventilating and temperature systems in P. S. 80, to E. Rutzler Co., 409 East 49th St., New York City. Estimated cost, $90,000.

N. Y., Churchville—The Village Bd. has awarded the contract for the installation of overhead municipal lighting system and 50 kw. transformers, to the O'Connell Contg. Co., 859 Hudson Ave., Rochester. Total estimated cost, $25,000.

N. Y., Long Island City—The Bd. Educ., 500 Park Ave., has awarded the contract for the installation of heating and ventilating equipment in P. S. 3, Queens, to Philip & Paul, 174 East 119th St., New York City. Total estimated cost, $50,000.

N. Y., Watertown—The Beebes Island Corp. has awarded the contract for installing hydraulic turbine equipment to the Wellman Seaver Morgan Co., 7000 Central Ave., S. E., Cleveland, Ohio, and for electric generator equipment, to the General Electric Co., River Rd., Schenectady. Total estimated cost, $250,000.

Pa., Johnstown—The Women's Memorial Hospital Association has awarded the contract for the construction of a power plant and laundry on Franklin St., to Charles Schenkemeyer & Son, at $40,000.

Md., Baltimore—The Columbia Graphophone Mfg. Co., 16 South Howard St., has awarded the contract for the construction of a plant and an 85 x 100 ft. power house, etc., to the M. A. Long Co., Munsey Bldg. Total estimated cost, $2,500,000.

O., Cleveland—The city has awarded the contract for furnishing 2 steam turbine driven centrifugal feed pumps, to the J. L. Skeldon Co., Rockefeller Bldg., at $13,320.

Ill., Chicago—The Amer. Glue Co., 123 West Kinzie St., has awarded the contract for the construction of a 3 story, 100 x 112 ft. factory at 3640 Iron St., to E. W. Sproul Co., 2001 West 39th St. A steam heating system will be installed in same. Total estimated cost, $175,000.

Wis., St. Francis (Milwaukee P. O.)—The Milwaukee Electric Ry. & Light Co., Pub. Serv. Bldg., has awarded the contract for the construction of a power plant, to the Dahlman Constr. Co., Majestic Bldg., Milwaukee. Estimated cost, $500,000. Noted Feb. 16.

Kan., Iola—The city has awarded the contract for furnishing and installing a turbo generator unit to Allis Chalmers, West Allis St., Milwaukee, Wis., $26,710; condenser, to H. R. Worthington Co., 824 Scarritt Bldg., Kansas City, Mo., $10,528.

Ore., Portland—The Comn. of Pub. Docks, foot of Stark St., has awarded the contract for furnishing and installing electrical equipment for light and power at the St. John's Terminal, to the Natl. Electric Co., Railway Exch. Bldg., at $16,311. Noted Sept. 23.

B. C., New Westminster—The Imperial Oil Co., Cambie and Smith Sts., Vancouver, has awarded the contract for the construction of a distributing plant consisting of wharf, storage house, pump house, etc., to the Dominion Constr. Co., 509 Richards St., Vancouver, at $24,500.

POWER

Volume 51 New York, April 20, 1920 *Number* 16

Do Your Own Thinking!

FOR fifteen years old Tom Smith had been running the pumping engine at Brown's Falls. For fifteen years old Tom climbed up twelve feet on a ladder every day to squirt oil in a little hole in a valve-motion rockshaft.

One day he slipped and nearly fell. The Super saw it and investigated. "What were you doing up there?" the Super asked.

"Shot a little oil into that hole up there."

"What hole?" returned the Super.

Tom showed him.

"Why man!" exclaimed the Super, "that's no oil hole; that's only a lathe center —what they turned the shaft on when they made it!"

So it was. We all know how old Tom felt when he found out that the hole he had faithfully oiled so long was only a blind alley. That's the difference between men—some use their brains and some don't. That's the difference between the Tom Smith and the Tom Edison type of man. The fellow that does his own thinking gets paid for it.

Did it ever occur to you that the man who needs much supervision pays for it? He certainly does. He pays the difference between what he might be getting and what he is getting paid for his work. The more you know and the more thinking you can do for your work, the more you are eventually going to be paid. Then, too, the more you know and the more thinking you can do for the boss, the sooner will he put you where you can supervise the work of others and be paid accordingly. DO YOUR OWN THINKING!

Contributed by John S. Carpenter.

Composition and Use of Babbitt Metals

BY B. H. JARVIS

Desirable qualities of babbitt metals, changes in manufacture that have taken place recently and the probable future of these metals as applied to the industries.

WHEN years ago the first anti-friction metal was produced and patented as "babbitt metal" a great impetus was given to the mechanical industries by the fact that such a metal permitted gains in speed, weight, complexity and efficiency, all with greater production possibilities. Babbitt metal was the forerunner of a seemingly endless procession of alloys, many of which possessed merit and have thereby remained in use, while many others, after attracting some notice, were disqualified and dropped into disuse.

Bearing metals of the babbitt or white-metal class with few exceptions have been of tin or lead base, hardened by the addition of antimony and also copper, nickel, zinc or some other hard metal. While many refiners and founders have, through skill or fortune, produced alloys that have given highly satisfactory results as antifriction metals, few have realized the underlying fundamental reasons for the success of their particular products of manufacture. Before expressing opinions as to merit, it is necessary to know how to judge. Following are listed the important qualifications of a babbitt metal:

1. Must have a strength in compression considerably in excess of the heaviest load the bearing is to support at the working temperature.

2. Must possess sufficient plasticity, without the quality of brittleness, to adjust its surface to an irregularly applied load, thus distributing load over the bearing. In bearings subject to severe impact, this property in the metal would necessarily have to be curtailed.

3. Should present to the journal a surface void of hard spots. This in connection with the plasticity of the metal, will allow the journal to form a "true bearing" readily, and as the metal wears this wear will be uniform. A metal not homogeneous in structure and containing abrasive material would not satisfy this qualification.

4. Should be of low friction. The word "antifriction" is common in usage but conveys various meanings. While in properly designed and finished bearings a film of oil should separate the journal and bearing, it is a certainty that under extreme conditions of load or in case of insufficient lubrication metallic friction does occur with subsequent heating and loss of power through friction, and accompanied by danger of the shaft being scored or of the loss of the bearing. Considering the fact that when examined under the microscope the surface of a highly polished shaft is extremely rough, it is evident that truly antifriction metal would be necessarily of such material as to offer little resistance to the movement of the rough shaft, for friction is the direct measure and result of this resistance. This leads to the conviction that the easier the babbitt metal is cut, the more antifriction qualities it would possess, and consistent with the qualification, listed previously, the more desirable the metal would be.

5. Must be tight to the shell in which it is poured. Either it must possess little or no shrinkage in passing from the liquid to the solid state and in later cooling, or, owing to peculiarities of the metal that allow it to be peened, it can be forced tightly against the shell. The common method for small bearings has been to form an excellent bond between the bronze back and the babbitt metal by first tinning the shell. Attention should here be called to the almost universal use of antimony in alloying babbitt metals, due in part to the peculiarity of this metal to expand instead of shrink on cooling, so that mixed in proper proportion with tin or lead, an alloy possessing little or no shrinkage could be produced. Oil is a poor conductor of heat, and should even a thin layer be able to get between the babbitt metal and shell of a high-speed bearing, the conduction of heat away from the babbitt that is developed by friction would be insufficient and the accumulation of heat would mean a hot bearing. One reason for

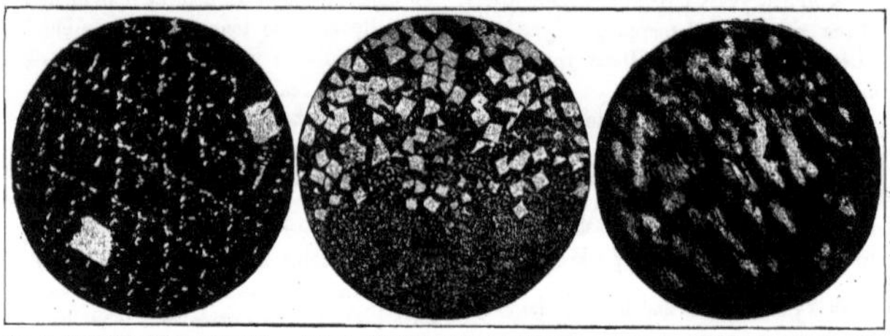

FIG. 1. BABBITT METAL CONTAINING TIN, ANTIMONY AND COPPER

FIG. 2. BEARING METAL CONTAINING LEAD, ANTIMONY AND TIN

FIG. 3. ALLOY CONTAINING TIN, COPPER AND SPELTER

babbitt metals cracking in service, provided they are not of the brittle type, is that insufficient care had been taken to make certain that the babbitt was tight to the shell.

6. For the work at hand the metal must be easily melted, poured and machined. To meet this requirement, only a skillfully alloyed metal can be expected to give the desired results, for, as metals possess various densities, unless properly alloyed, upon being melted the heavier metals present would sink to the bottom. Especially in bearings where the lining of the shell is quite thin, this qualification is exceedingly important.

7. It must possess good wearing qualities. While there are ways of relatively comparing the wearing qualities of babbitt metals, as will be touched on later, the only certain way to determine the service value of a metal for a particular work is to make a service trial.

8. It should not cause excessive wear of the shaft. Lignum-vitæ and other wood preparation bearings would long ago have received more attention had it not been for their excessive wear on the shaft.

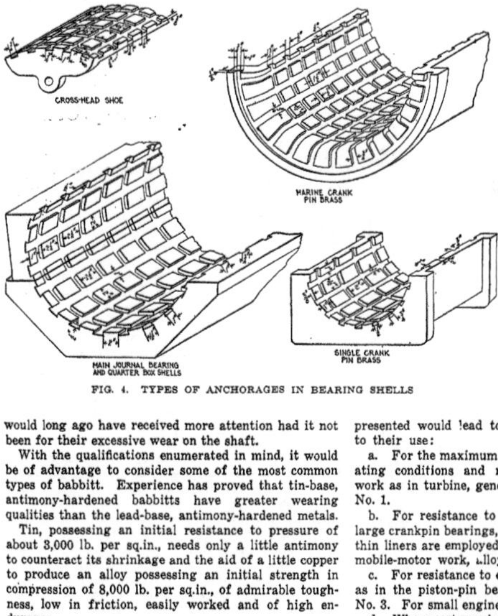

FIG. 4. TYPES OF ANCHORAGES IN BEARING SHELLS

With the qualifications enumerated in mind, it would be of advantage to consider some of the most common types of babbitt. Experience has proved that tin-base, antimony-hardened babbitts have greater wearing qualities than the lead-base, antimony-hardened metals.

Tin, possessing an initial resistance to pressure of about 3,000 lb. per sq.in., needs only a little antimony to counteract its shrinkage and the aid of a little copper to produce an alloy possessing an initial strength in compression of 8,000 lb. per sq.in., of admirable toughness, low in friction, easily worked and of high endurance.

Lead, a metal of much lower friction than tin, possesses only 600 lb. per sq.in. initial resistance to pressure. Having considerable shrinkage, it requires a large amount of antimony to counteract this shrinkage and even with the proper addition of copper, a strength of 5,500 lb. per sq.in. in compression cannot be obtained without producing an alloy so brittle as to be almost worthless.

To both types of babbitt, and in the many combinations of both in order to gain strength and eliminate shrinkage, there have been added materials both hard and abrasive, curtailing certain desirable qualifications for the necessary increase in other essentials. Antimony, of itself able to score steel because of its hard crystalline structure, has been alloyed with either lead or tin, and while affording advantages as regards shrinkage and increase in strength, introduces into the mixture a high-friction, abrasive material.

To a marked degree the relative wearing qualities of two babbitt alloys, hardened by the addition of the same metals, will be found directly related to the strength of the alloys, as indicated by a compression test or a Brinell test for hardness. Experienced mechanics know that a bearing made by pouring a tin-base babbitt around a crankpin or journal will wear much longer than if the pouring is done around a smaller mandrel and the bearing is bored to size. This is due to the slight hardening effect on the surface of the tin-base metal or coming in contact with a cold surface. To some degree, therefore, the wearing or service value of an alloy can be ascertained by measuring its hardness.

In the accompanying table a list is given of the compositions of the most common alloys on the market today, including two metals comparatively new but possessing high qualifications.

Arsenic makes it difficult to tin or solder babbitt to a bronze shell, so the best grade of copper should be used. Only with pure metals and with proper alloying methods can the desired alloys be produced. Judging the alloys in the table by the qualifications previously presented would lead to the following suggestions as to their use:

a. For the maximum of service under excellent operating conditions and regardless of cost, high-speed work as in turbine, generator and motor bearing, alloy No. 1.

b. For resistance to heavy loads and impacts as in large crankpin bearings, alloys Nos. 2, 17 and 18. When thin liners are employed, as in small crankpins or automobile-motor work, alloy No. 2.

c. For resistance to extreme loads and impacts, such as in the piston-pin bearings of Diesel engines, alloy No. 3. For small engines, alloy No. 16 or bronze.

d. Where cost requires the use of a good but cheaper metal for general use in high-speed bearings such as motors, fans, etc., alloys Nos. 4, 5 and 6.

e. For all classes of service where conditions are not severe as to load or speed, such as the main journal bearings of large engines, alloys Nos. 7, 17 and 18.

f. Low speeds, light loads, as in lineshafting bearings, alloys Nos. 10, 11, 12, 17 and 18.

For poor operating conditions as regards the upkeep and alignment of shafting, and dirty operating conditions with poor lubrication, the following recommendations are made:

g. Slow-speed, heavy-pressure bearings, alloys Nos. 13, 15, 17 and 18.

h. High or low-speed, moderate load, subject to impact, alloys Nos. 15, 17 and 18.

i. For high speeds, well lubricated, light load, alloy No. 14.

Alloys Nos. 17 and 18 in the foregoing list are comparatively new metals on the market. They are the results of attempts made to harden lead without the addition of abrasive material, and in this respect they

COMPOSITIONS OF A. S. T. M. TENTATIVE STANDARD ALLOYS

Alloy Grade No	Tin, Per Cent	Antimony, Per Cent	Lead, Per Cent	Copper, Per Cent	Iron, Per Cent	Arsenic, Per Cent
1	91	4.5	0.35*	4.5	0.08	0.10
2	89	7.5	0.35*	3.5	0.08	0.10
3	83	8.5	0.35*	8.3	0.08	0.10
4	75	12	10	3	0.08	0.15
5	65	15	18	2	0.08	0.15
6	20	15	63.3	1.5	0.08	0.15
7	10	15	75	0.50*	....	0.20
8	5	15	80	0.50*	....	0.20
9	5	10	85	0.50*	....	0.20
10	2	15	83	0.50*	....	0.20
11	..	15	85	0.50*	....	0.25
12	..	10	90	0.50*	....	0.25

SUPPLEMENTARY LIST OF ALLOYS

Alloy Grade No.	Tin, Per Cent	Antimony, Per Cent	Lead, Per Cent	Copper, Per Cent	Zinc, Per Cent	Other Metals
13	70	15	10.5	4.5	..	....
14	..	25	75	...	..	....
15	6	18	74	6.0	..	....
16	65	..	..	5.0	30	....
17	1.5	..	97	...	..	Alkali metal
18	..	..	98	...	..	Alkali earth metal

* Maximum.

are successful. Experience has shown that these metals are of great value for classes of service such as b, e, f, g, and h listed in the foregoing, and the reason is evident on considering their properties relative to the qualifications of babbitt metals.

They possess strength in compression equal or greater than the highest grades of genuine babbitt, plasticity without brittleness, they contain no abrasive material and are homogeneous in structure, they have the lowest co-efficient of friction of any anti-friction alloys, they are easily poured and machined, they wear well, especially with good operating conditions, and they cause the minimum wear of any alloys on the journals.

Containing no antimony, they do possess one characteristic that must be carefully taken care of for uniform success, and that is shrinkage. Frequently, these metals have been applied successfully in a manner similar to the babbitts and without consideration of the shrinkage, but this success has been due largely to the plasticity of the metals. To take care of the shrinkage, the following changes in design of shells or of methods of pouring are recommended, which, if followed, will enable these alloys to successfully replace babbitt metals to some degree.

As illustrated in Fig. 4, the shells should have plenty of undercut anchorage of the checkerboard type. The grooves should be undercut and not over ½ in. wide. The shells should be heated before pouring, that they may contract in cooling with the bearing metal.

When possible, all bearings, and most certainly those for high speed, should be peened after pouring, and the peening should be sufficient to force the metal right to the shell. These metals are very plastic immediately after solidifying, and with little effort can at this time be peened back to the shell. Due to the anchorage a perfect bond between the bearing metal and shell will be formed.

Where the metal is poured to size in a half-bearing and cannot be machined, it will be advantageous to peen the edges of the bearing, which will both help to increase the bond and relieve the bearing.

It is of importance that all bearings be properly relieved, or else the lubricant will be scraped off the revolving journal by the sharp edge, and a hot bearing will result.

Do not attempt methods of securing tight bearings such as the use of expanding mandrels or by immersion of the shell in water after pouring.

Do not cut oil grooves in the bearing metal except on the portion of the bearing where there is little or no pressure. This applies to all babbitts and bearing metals, for frequently sources of trouble have been traced back to the oil grooves, which, instead of helping to maintain an oil film, afforded a means of exit for the oil and excessive friction resulted.

Absolute confidence in materials used is a mighty good restraint on an engine builder who is tempted to change from a metal he has used successfully to one of which he knows little. But where the fundamental principles have been applied and experience sustains the application, conservatism should bend to progress.

Where the Ammonia Charge Goes

By J. E. Porter

In any ice or refrigerating plant the ammonia charge is the most important item with which the engineer has to deal. It is the one big individual thing that really produces results, and it is probably safe to say that it receives the least attention. This is not true in all plants, but will hold good for the large majority.

The initial charge of ammonia comes to a plant in drums and is chemically pure. The ammonia manufacturers are careful about keeping their ammonia drums clean and free from any foreign substance, so that it is fairly safe to assume that when a charge goes wrong, the trouble develops at the plant. Theoretically, the initial charge of ammonia should last indefinitely, but practically there is continual loss. Let us see what happens to a perfectly good charge after being introduced into a new system.

Starting from the liquid receiver, the liquid ammonia passes through the expansion valve into the evaporating or refrigerating coils, from whence it is drawn into the compressor and forced through the condenser and back to the receiver again, making a complete cycle. If there happens to be a leak in any part of the system, some ammonia will be lost. While the peculiar odor of ammonia makes it easy to detect outside leaks, there are two places where serious leaks can develop and not be detected for some time; namely, the brine tank and the condenser. A small leak in the expansion coils, with the brine absorbing the ammonia, will cause a lot of wastage in the course of a season's run. There is an equal opportunity for loss of the charge in the condenser as the ammonia is absorbed and carried away by the cooling water. Another place around which the ammonia slips away into the atmosphere is the packing. This is due largely to the severe conditions under which the packing is required to function. Relatively, it is burning up one minute and freezing the next. In a great many machines it is difficult to adjust the packing so there is no leakage and still keep the running parts cool. While the oil seals at these points help to keep down leakage, there is, nevertheless, considerable ammonia lost around the packing.

It is impossible to operate a plant and not lose some ammonia. Every engineer should be persistent in his efforts to keep these losses as low as possible per ton of ice manufactured. Based on this unit the average cost of ammonia seems to range from ½ to 20c., 10c. being a fair average and about 6c. a good mark to work for. As intimated previously, the only way ammonia is lost is by leakage. It does not deteriorate. The charge, however, can get into a most deplorable condition if not handled properly, by allowing it to become polluted with brine, oil or air.

When the pressure on the ammonia is reduced to or near atmospheric, brine will be drawn into the system if a leak exists in the evaporating coils. Air is drawn in under the same conditions, where leaks are exposed to the atmosphere. Oil enters the system from the compressor lubrication, and it gets in there regardless of the fact that a perfectly good oil intercepter is placed in the discharge line to prevent it from doing so. The reason for this is that the oil, caught in the heat of compression, is partly vaporized and slips by the intercepter along with the ammonia vapor to the condenser and the receiver. If not removed at this point, it finally winds up in the expansion coils, where it does its share in the way of cutting down capacity.

To maintain ammonia sharp and snappy, keep it clean. This can be accomplished by installing a good ammonia regenerator, connecting it to a low point on the receiver and then using it at frequent intervals. If this plan is followed, there will be no trouble from oil in the expansion coils, because to get there it will first have to work its way through the entire high side.

Air in the system will cause excessive head pressure and can be removed only by purging. The purge line should be connected to the highest point on the condenser and led into an open vessel containing water. The compressor is usually stopped during the purging operation and condenser sections allowed to stand for a short while with the water running over them, giving the noncondensable gases a chance to collect at the top. The purge valve is then opened slightly and kept open as long as bubbles appear in the water. The purge valve should be closed when the bubbles disappear, followed by a popping and cracking sound, which indicates that ammonia vapor is entering the water. The machine is then started and run until all is ready to repeat the operation. It usually requires several purgings to rid the system of air.

The ammonia charge never should be entrusted to other than competent hands, for outside of being expensive to lose, it is dangerous if not handled properly. No leak is so small that it should not receive attention, for a small constant leak will eventually get away with as much ammonia as a larger leak which usually requires immediate attention.

Fifteen Thousand-Gallon Concrete Fuel-Oil Tank

The illustration at the bottom of this page is a working drawing for a concrete tank of 15,000 gal. capacity. With slight modifications to meet local requirements and individual tastes, it should be suitable for the storage of fuel oil. The drawing was furnished by the Alpha Portland Cement Company.

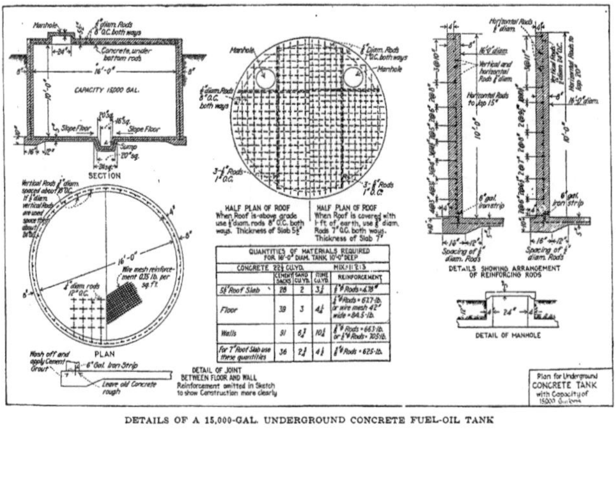

DETAILS OF A 15,000-GAL. UNDERGROUND CONCRETE FUEL-OIL TANK

Explanation and Measurement of Reactive Factor

By VICTOR H. TODD

FOR purposes of calculation, all alternating current not in phase with the voltage, may be said to be made up of two fictitious components, one in phase with the voltage called the "in-phase" or active component and one at right angles to this component called the "out-of-phase" or reactive component. The terms "energy" component and "wattless" component are often applied, but their use is disapproved by the A.I.E.E. The vector diagram in Fig. 1 shows the current and voltage relation more clearly. Line OE represents an alternating voltage and line OI represents an alternating current lagging behind the voltage by angle ϕ. Line OI_a represents the in-phase or active component and line OI_r represents the out-of-phase or reactive component. The cosine of the angle ϕ is the "power factor" and the sine of the angle ϕ will be the "reactive factor." The product of $OE \times OI_a$ (or $OE \times OI \times \cos\phi$ will give the watts in the circuit while the product of $OE \times OI_r$ (or $OE \times OI \times \sin\phi$)

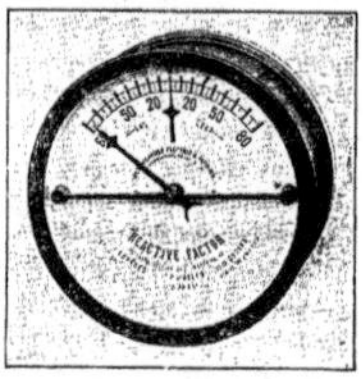

FIG. 2. REACTIVE-FACTOR METER

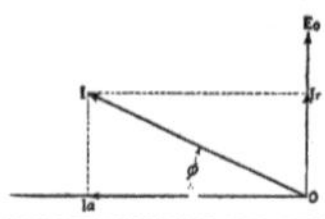

FIG. 1. DIAGRAM REPRESENTING RELATION BETWEEN VOLTS AND AMPERES IN AN A.-C. CIRCUIT

will give the "reactive volt-amperes." When integrated, these two terms become respectively "watt-hours" or "kilowatt-hours" and "reactive volt-ampere-hours" or "reactive kilovolt-ampere-hours."

The simplest means of indicating the reactive factor of a single-phase or polyphase circuit is by a meter similar to Fig. 2. The instrument indicates directly without constants or other calculations, the sine of the angle of phase difference between the current and voltage (reactive factor) and shows if the current is leading or lagging. This is of great importance in reducing the reactive factor of a synchronous converter to zero, that is, adjusting the power factor to unity.

In the three-phase meter, there are three coils wound 60 deg. apart, which produce a rotating magnetic field. Pivoted within this field is a soft iron vane, magnetized with the current in one phase. The vane is without restraining springs, and so assumes a position where its magnetic axis, at the instant of maximum flux, coincides with the position of the rotating magnetic field axis at that instant. In the single-phase instrument, there are two coils, energized through a phase-splitting device.

To obtain the highest accuracy, the current transformers should give between 2 and 5 amperes and the voltage be between 75 and 125 per cent of normal. The instruments are designed only for balanced loads and their indications on an unbalanced polyphase load are not easily interpreted.

To check the calibration it is necessary to note the simultaneous readings of volts, amperes and watts. The reactive factor is calculated from the formula:

$$\text{Reactive Factor} = \sqrt{1 - \left(\frac{\text{watts}}{\text{volt-amperes}}\right)^2}$$

On a three-phase circuit, the watts are measured by a polyphase or two single-phase wattmeters, and the volt-amperes equal the sum of the three products—the current in each of the three wires multiplied by the voltage between that wire and an artificial neutral. Or the reactive factor may be calculated by the two polyphase wattmeter method as hereafter described.

Since a wattmeter indicated watts, the product of volts times in-phase current component, it is evident that if we can excite the potential coil of the wattmeter from a voltage which is 90 deg. out-of-phase from the original voltage, then it will indicate the product of volts times out-of-phase current component, or reactive volt-amperes. Referring to Fig. 1, if the wattmeter is connected so its potential coil receives voltage OE, and its current coil receives current OI, then the meter will read watts. But if the potential coil connections are shifted to receive a voltage OE_o then the meter will read reactive volt-amperes.

On a single-phase circuit, there is no practical way of obtaining this 90-deg. out-of-phase voltage. But on a two-phase circuit, it is only necessary to shift the potential connections to the opposite phases. For instance, in Fig. 3 the diagram shows a polyphase wattmeter connected to read watts, while in Fig. 4 the potential coils are crossed to the opposite phase, thus making the meter read reactive volt-amperes.

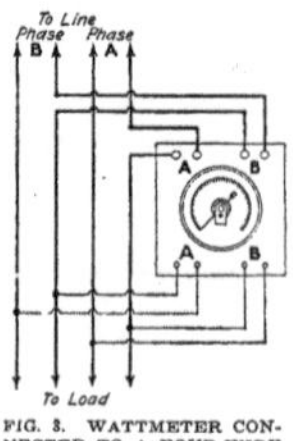

FIG. 3. WATTMETER CONNECTED TO A FOUR-WIRE TWO-PHASE CIRCUIT TO READ WATTS

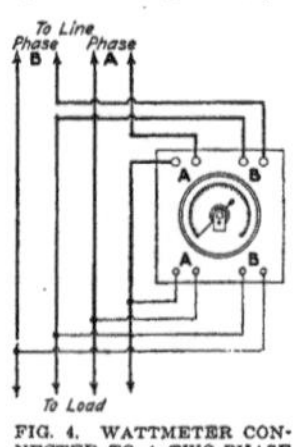

FIG. 4. WATTMETER CONNECTED TO A TWO-PHASE CIRCUIT TO READ REACTIVE VOLT-AMPERES

In a three-phase circuit, the problem is not to supply a voltage 90 deg. away from the current at 100 per cent power factor, but to supply a voltage 90 deg. away from the normal voltage. This distinction is necessary, as it will be remembered that at 100 per cent power factor, there is a normal phase displacement of 30 deg. between current and voltage. In Fig. 5, *AB*, *BC* and *AC* represent the three-phase voltage vectors. If the current coil of the wattmeter is inserted in the line connected to point *A*, as shown, then the voltage would normally be taken between lines connected to points *AC*. But to obtain a voltage at right angles to this, it is necessary to take it between two points, one midway between *AC*,

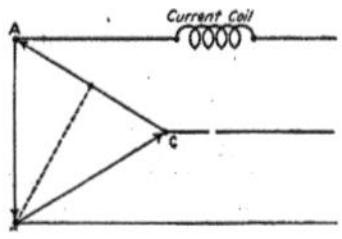

FIG. 5. DIAGRAM REPRESENTING THREE-PHASE VOLTAGES

and the other at *B*. This midpoint may be obtained from the middle-point tap on a transformer or between lamps connected across the phases. The connections for a polyphase wattmeter are shown in Fig. 6, to measure reactive volt-amperes on a three-phase circuit. It will be noticed that the line *BD* in Fig. 5 is not equal in magnitude to the normal voltage *AC*, but is only about 86.6 per cent of it. Consequently, with the connections of Fig. 6, it is necessary to multiply the reading by 1.155 in order to arrive at the true reactive volt-amperes.

When a meter is permanently used to measure reactive volt-amperes, it is usually supplied with special voltage transformers having 57.8, 100 and 115.5 per cent taps, as shown to the right of Fig. 7. The two potential transformers are connected in open delta. For the measurement of watts, the voltage is taken from *O* to E_n, but for measuring the reactive volt-amperes the voltage is taken from E_h to E_m and from E_H to E_M. It will be noted that the voltage from E_h to E_m and from E_H to E_M is equal to and also at right angles to the voltage from *O* to E_n. Oftentimes, a four-pole double-throw knife is used to allow measurement of watts or reactive volt-amperes, by using connections as shown in Fig. 7, throwing the switch to the left for reading watts and to the right for reactive volt-amperes.

The foregoing connections can be used with watt-hour meters, maximum-demand meters and graphic meters. Usually, two meters are used, one to measure the watts and one, reactive volt-amperes. In some cases, the consumer's bill is based on two rates, the operating company charging one rate per kilowatt-hour and another per reactive kilovolt-ampere-hour. The customer is informed that the one meter records waste current which may be largely eliminated by a proper manipulation of his power load. Maximum-demand records of watts and reactive volt-amperes are very valuable, since the power factor, and reactive factor, as well as the time of occurrence may be determined from these readings.

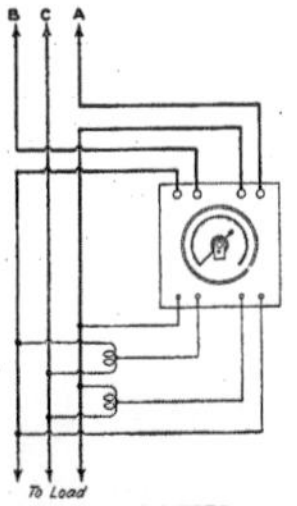

FIG. 6. WATTMETER CONNECTED TO A THREE-PHASE CIRCUIT TO MEASURE REACTIVE VOLT-AMPERES

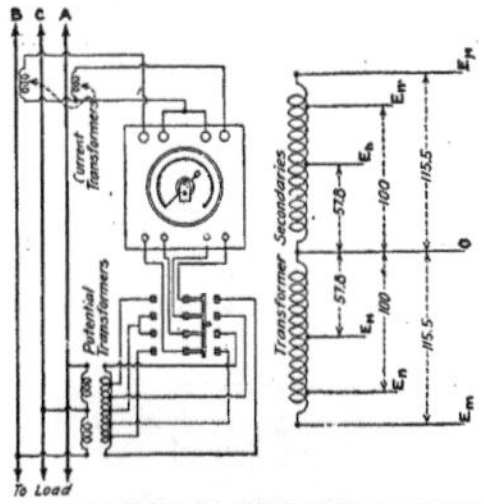

FIG. 7. WATTMETER CONNECTED THROUGH TRANSFORMER TO A THREE-PHASE CIRCUIT TO MEASURE REACTIVE VOLT-AMPERES

By referring to Fig. 1, it is evident that the tangent of the angle is equal to the ratio of reactive volt-amperes to watts, that is, reactive volt-amperes divided by watts. Consequently, if these two quantities are known, their ratio can be found and from this the phase angle between the current and volts determined from a table of natural tangents. The cosine of this angle is the correct power factor and the sine is the reactive factor. For instance, if the watts in a circuit are 1,500 and the reactive volt-amperes are 1,089, then their ratio reactive volt-amperes ÷ watts = 1,089 ÷ 1,500 = 0.726, which is the tangent of the angle of phase displacement. Locating this value in a table of natural tangents, we find the angle to be 36 deg. The cosine of this angle is 0.80, which is the power factor; the sine equals 0.588, and is the reactive factor. By using the kilowatt-hour and the reactive kilowatt-amperes-hour readings, in the same manner, the average power factor or reactive factor over a monthly period may be determined. Or if maximum-demand meters are used, the power factor and reactive factor at the time of maximum demand may be calculated. This is often of value in making a power-factor penalty charge.

In contradistinction to the two single-phase watt-hour meter method of determining power factor, which is applicable only to balanced circuits, the two polyphase watt-hour method will give a correctly weighted power factor regardless of any unbalancing in load.

There are, according to the *Electrical Review*, seventeen plants in operation in the Whittier district of California which remove water from crude petroleum by an electric process. The electrical dehydrator usually operates with single-phase alternating current at 11,000 volts. The emulsion of oil and water is passed between electrodes, and the effect of the electrostatic field is separately to coalesce the particles of water and oil. The water then settles readily by gravity. The average maximum demand per dehydrator is 4 kilowatts, the average load factor 50 per cent, and the average power factor 98 per cent leading. The oil treated ranges from 15 to 50 per cent of water, and the consumption of electrical energy is about 1 kw.-hr. per 18 bbl. of dry oil.

POWER DRIVES FOR ROLLING MILLS

By
W. O. ROGERS

Slide-valve, Corliss, poppet valve, tandem-compound and Cross-compound engines are used for driving rolling mills, and in one instance a mixed-pressure turbine was found in use. Some of these engines drive the rolls by means of belts, others through a train of gearing, and still others are directly connected to the rolls.

ROLLING-MILL engines are of various designs and cover the range from slide-valve units to Corliss, poppet-valve engines, and mixed-pressure turbines. A surprisingly large number of compound slide-valve engines were found, mostly in the smaller mills and operating rolls that revolve in one direction only. Fig. 1 shows a Tod 16 and 30 x 30-in. tandem-compound engine, belt-driving an 8-in. mill at the Sharon Steel Hoop Co. The 24-in. wide belt drives a shaft that is connected to the bottom roll of the mill by a flexible coupling. The Tod engine has been quite popular around the Youngstown district, and many of them are in daily use at the present time. The smaller engines are fitted with flat valves as shown in Fig. 2. A peculiarity of the engine shown in Fig. 1 is the construction of the high-pressure cylinder. It is of the overhang type and is supported by the distance piece between it and the low-pressure cylinder, heavy ribs being cast on the flange and the cylinder to stiffen it. A shaft governor regulates the valve cutoff. These engines are generally equipped with a detachable reach rod for stopping and starting; the engines are of the side-crank type.

Various forms of drive between the engines and the mills are used. For instance, the engine shown in Fig. 1 drives the mill by means of a belt; the engine shown in Fig. 2 is directly connected to the rolls by means of a flexible coupling. Fig. 3 shows a 36 and 50 x 42-in. tamden-compound engine, fitted with piston valve. It is gear-connected to the rolls of a four-stand mill and also to a four-stand finishing mill by means of a belt.

On the engine shaft is a large gear which meshes with a smaller one in line with it and which drives one stand. The engine shaft drives a second stand, and a small gear on the engine shaft meshes with a large gear on a third shaft, which drives a third stand. This third-stand shaft also carries a small gear which meshes with a larger one on a fourth shaft, and this drives the fourth stand of rolls. In other words, the engine shaft is so geared that it drives not only its own stand of rolls but three other stands.

The steel bar enters the first set of rolls and the end is brought around to the second stand, the rolls of which run in a reverse direction, then through the third stand and finally through the fourth set of rolls. As every other set of rolls revolves in a different direction, the passage of the bar through the mill stands is in an opposite direction to the passage through the preceding set of rolls.

FIG. 1. TOD TANDEM COMPOUND ENGINE DRIVING 8-IN. MILL

FIG. 2. VALVE SIDE OF A TOD ROLLING-MILL ENGINE

It looks simple to see the operator grab a moving mass of narrow sheet metal or a snake-like rod and steer it into its proper stand, but there is a knack in doing this work. The skilled operator does not attempt to pull the ends of the band or rod around in front of the stand, but grabs it with a pair of tongs and, as it comes through the preceding roll, simply arrests the end movement by drawing it to one side. When sufficient metal has passed through the roll to make a large U-shape, the end is directed to the stand to reduce the metal to a thinner piece, elongation taking place during the process. The man on the next stand performs the same trick and so on until the metal has been rolled to the required shape. An inexperienced man would make a sorry mess of the job, let alone the chances that he would be taken to a hospital for burn treatment.

At the Republic company's plant a tandem-compound Tod engine, Fig. 4, is used to drive a 90-in. plate mill,

FIGS. 3 TO 7., SHOWING VARIOUS TYPES OF ROLLING-MILL ENGINES

Fig. 3—A Tod tandem-compound, piston-valve engine, direct- and belt-connected to mill stands. Fig. 4—Tod tandem-compound engine driving an 18-21-in. billet mill. Fig. 5—A 6,000-hp. Mesta cross-compound engine driving a four-stand mill. Fig. 6—Another tandem-compound engine driving ten sheet mills. Fig. 7—A rope-drive unit, newly erected.

FIG. 8. GEAR ARRANGEMENT FOR DRIVING A FOUR-STAND MILL

FIG. 9. GEAR REDUCTION BETWEEN ENGINE AND TIN-SHEET MILLS

but at the openhearth 18-21-in. billet mill a 48 and 84 x 60-in., 6,000-hp. Mesta cross-compound Corliss engine is used. This is shown in Fig. 5. A single eccentric operates on the high-pressure side, but two eccentrics are used on the low-pressure cylinder. The engine is coupled to a shaft on which beveled gears are secured, and these mesh with larger gears mounted on shafts that run at right angles to it and drive four stands of rolls. The gearing arrangement is shown in Fig. 8.

One of the largest engines I saw was at the Republic works. It was a twin-tandem-compound reversing engine of 25,000-hp. capacity. Another engine of the same capacity was found at the main plant of the Youngstown company, where a Mesta twin-tandem-compound engine (Fig. 11) is operating. It is, as shown, of extremely sturdy construction with a simple valve gear and a single lever control. The man standing by the engine gives some idea of the immense size of this unit. It drives a 44-in. blooming mill and has 46- and 76 x 60-in. cylinders. It has a maximum speed of 110 r.p.m.

The service of rolling-mill engines is most severe, and the spectator wonders how a machine can be made to operate without being disabled a dozen times a day. The full load comes on suddenly when the chunk of metal enters the roll and then leaves just as suddenly. One method of relieving the engine of the sudden strain and equalizing the variation of power is to use a heavy flywheel having a high rim speed and a sensitive governor on the engine. The flywheel is usually placed on the crankshaft to which the roll train is directly coupled. The weight of the flywheel depends, of course, on the size of the engine, but they will weigh up to 100 tons or more. A number of Porter-Allen engines were seen at various plants, and this engine is much used in the steel industry.

Reversing engines do not carry flywheels. They are compound units and are generally fitted with piston valves. The engine is directly coupled from the crankshaft to the roll train, but it is sometimes geared down to allow the engine to run at a higher speed than required by the rolls. Fig. 12 is a good illustration of a blooming mill and reversing engine. As will be seen, no flywheel is provided. The massive construction of the engine, coupling and mill gives at least a vague idea of the rough usage that is imposed on the machine but one must observe the work in order to obtain a clear idea of the tremendous shock to which the engine,

FIG. 10. STEEPLE-COMPOUND ENGINE DRIVING A 16-IN. ROD MILL BY MEANS OF FOUR 48-IN. BELTS

mill and bearings are subjected when in operation. Belt drives are not uncommon in the older and smaller mills. Fig. 10 shows an interesting installation and is out of the ordinary. The engine is a 34 and 64 x 48-in. steeple compound, rated at 2,500 hp. at 110 r.p.m. It is at the Struthers plant of the Youngstown Sheet and Tube Co., and operates a 16-stand rod mill by means of four 48-in. belts passing over the flywheel to as many driven pulleys. The engine also drives an air compressor. The path of the belts over the pulleys is easily traced.

The engine shown in Fig. 6 is a 34 and 68 x 60-in. tamden-compound Tod Corliss of 3,500-hp. capacity. It operates ten sheet mills at the Trumbull Steel Co.'s plant, Warren, Ohio. Engines engaged in operating sheet mills have a flywheel, but are geared to the mill so as to obtain the proper roll speed with the engine running at normal. A common method of reduction is shown in Fig. 9. A spur gear on the engine shaft meshes with a larger gear on the mill train. These mills run continuously in one direction and the passage of the sheets of metal on the return after passing through the rolls is either over the top roll or through the three-high stands. In the latter case the plate is rolled while going in both directions, whereas in the two-high stand mill the rolling process is in but one direction.

Rope drive is sometimes used. Fig. 7 is a view of a Tod 16 and 34 x 30-in. side-crank tandem-compound engine that has just been placed in position to be used in driving a 14-in. bar mill by means of a rope drive in conjunction with a 28½ and 54 x 40-in. Buckeye engine. The Tod engine will be belted to an idler shaft carrying a sheave wheel, and from there to the mill shaft, which carries a 28-ft. grooved flywheel. The large engine also belts to this wheel and runs over, the smaller engine running under. This arrangement was made because the larger engine was not of sufficient capacity for doing the work demanded of it.

It is not often that a direct-connected turbine is used for driving rolling mills, but this has been running for about five years in the plant of the Carpenter Steel Co., Reading, Pa. This unit is a 350-hp. mixed-pressure turbine and drives the mills through a double-reduction gear, the mill having two three-high 18-in. stands. This roughing mill was formerly driven by a 36 x 36-in. simple slide-valve engine. The turbine operates on exhaust steam from a 22 and 40 x 48-in. compound unit that formerly exhausted into a condenser. The original engine shaft and flywheel of the slide valve engine are used and the turbine speed of 5,000 r.p.m. is reduced by means of double-helical involute gears, first to 600 and then to 100 r.p.m. The speed of the rolls can be varied from 100 to 70 r.p.m.

FIG. 11. MESTA 46 AND 76 BY 60-IN. TWIN-TANDEM-COMPOUND REVERSING ENGINE DRIVING A 44-IN. BLOOMING MILL

by means of a turbine governor that can be adjusted while the turbine is in operation.

The turbine operates normally at 3 lb. gage pressure at the throttle, and with 3 lb. absolute pressure in the turbine at the exhaust end, and under these conditions will carry its load of 358 hp. with a steam consumption of about 26 lb. per horsepower. The turbine can be run low-pressure condensing, mixed-pressure condensing, high-pressure condensing, or high-pressure noncondensing. Formerly, 1,000 hp. in boilers were necessary to operate the mill; with the turbine the boiler capacity has been cut to 600 boiler horsepower.

This is notable example of how increased power can be had with the reduction of steam consumption. This De Laval installation was the first in this country and the second in the world to apply a steam turbine for driving a rolling mill by means of reduction gears.

Another interesting mill drive was found in a uniflow-engine installation, which will be described in the next article.

FIG. 12. COMPLETE BLOOMING MILL AND REVERSING ENGINE, OPERATING WITHOUT A FYLWHEEL

The Location and Remedy of Leaks in Rock-Filled Dams

By Roland V. McDonald

While there are many water-power resources of large amounts of power, potentially, there are vastly more cases where the power is but small. Throughout Wisconsin and near-by states especially, the rock-filled dam is used to a considerable extent for the smaller hydro-electric and hydraulic plants for paper mills, etc., because it is easily built at low cost from materials near at hand. As more and more small hydro-electric plants spring up, utilizing a head of from a few to perhaps fifty feet and representing a capacity of from one hundred to several hundred horsepower, this type of dam may be expected to be used extensively. One of the troubles encountered with rock-filled dams, and more particularly with the older ones of industrial plants, is leakage. As this leakage may represent a large loss of water at a time when water should be conserved, and also somewhat of a danger, it is well to be able to locate and stop it.

Rock-filled dams are usually sheeted on the upstream side to make the dam free of leakage when first built, protect the filling or grout between rocks and reduce attrition and erosion due to the motion of the water. In course of time, this sheeting deteriorates and leakage through the dams finally takes place. This becomes worse and worse with time, as the water works a larger channel through the dam, carrying away filling and even the rocks themselves.

The remedy ordinarily is resheeting. If, however, the original sheeting was not driven down to hardpan or solid rock, leakage may seriously undermine the dam, the rate at which this occurs depending, of course, upon the foundation. The way to prevent or overcome this is to install an extra trench with a double row of sheeting driven well down so as to preclude leakage.

Leakage does not always occur at the dam, but, instead, may be discovered some distance away. Such leaks may form a regular subterranean tunnel and pass considerable quantities of water so that during periods when there is a shortage of water, the output of the plant may be threatened. The methods of locating depend upon the condition of the water above the dam, whether it is clear or turbid, smooth or rough. In all instances, however, some method of inspection and exploration is employed.

Where the velocity of the water is not high, the water smooth and clear, and not so deep but that the bottom of the stream can be seen, such holes can often be seen with the naked eye on a clear day. Sometimes a whirlpool action on the surface of smooth and comparatively still water is the telltale. Conditions are rarely so favorable as this, however, and it is necessary to explore the bed of the river very carefully with an artificial light. This is best done by working the river transversely, back and forth as follows:

Sufficiently strong stakes are set in the ground, one on each side of the stream. To these a rope or cable is fastened. Two men in a boat hitch to the cross-river rope and work their way across the river from one bank to the other. The stakes and rope are then moved eight to ten feet up or down stream, and the boat is worked back to the other bank. This procedure is repeated until the entire river bottom for several hundred feet above the dam has been explored.

For the exploration the boat is equipped with an electric light of about 200 watts, obtaining its current from the station by flexible leads or from a storage battery. This lamp is, of course, inserted in a waterproof or marine socket fastened to the end of a rod of such length that it is capable of reaching the bottom of the river. One man then moves along the side of the boat with the light close to the river bed, while the other watches for eddy currents and signs of a suck-hole. The process is somewhat tedious where a large surface of river bed has to be explored, but it is very simple. As a precaution in case the lamp should be drawn into a suck-hole, a means should be provided for quickly disconnecting the lamp and feed wires. An ordinary bayonet plug fastened at the top end of the pole offers such a safeguard.

Where the water is turbulent and muddy, and too deep to permit the river bed being seen, a pole similar to the one already mentioned can be used, but instead of having a lamp fastened at the end, it will have a float or some object of such surface that the current of water pouring through the leak will suck it down into the hole. In this way the leak can be discovered without being seen. With the leak found, it then remains to determine its extent and shape, which can be accomplished by feeling it out with the light or float, as the case may be, the man who is watching making a mental note so as to determine the area of the opening.

With an idea formed as to the location and probable size of the leak, it is then feasible to prepare for filling in. The material used for filling depends upon the size of the subterranean tunnel. For a small channel small boulders suffice, and even bottles, while for larger channels sacks of cinders or concrete may be required. The best way of doing a good job and a permanent one is to start filling up with small stones, loaded tin cans, etc., until exploration with light or float shows that these obstructions are actually lodging and obstructing the flow. Then, a few sacks of cement or cinders can be let down on top until the suction has been practically eliminated.

A suck-hole or leak is always largest at its entrance. In filling it up, a space should be left on the river bed to permit sealing up with a good thick covering of concrete. This covering should be not less than two and preferably three feet thick, should cover a suface at least twice that of the leak and should be placed through a spot in the ordinary way.

Were a survey of river beds made as described, many of the older hydro-electric plants that now suffer from a water shortage seasonally, would probably find that they are losing considerable water through leakage.

Three Power-Plant Leaks That Waste Coal

By George H. Smith

Leaking steam traps, leaks at pipe points and uncovered steam pipes are three causes of wasted fuel that are often neglected. It needs only a few calculations to show how serious they usually are.

One way to test a trap for leakage is to open the discharge line near the trap. Any steam passing can then be easily detected. This test is not entirely satisfactory as the release of pressure will cause some of the water to flash into steam, thus making a perfectly tight trap appear to be leaking. A better test is to note whether the water in the gage-glass is always high enough to seal the discharge opening.

To form an idea of the loss from a leaking trap, assume that it is passing steam through an opening equal to a hole ¼ in. diameter and that the steam pressure is 80 lb. per sq.in. gage. The steam lost may be computed by Napier's formula,

$$G = \frac{ap}{70}$$

in which G = pounds of steam per second, p = absolute pressure in pounds per square inch and a = area of the opening in square inches.

Then $G = \frac{(80 + 14.7) \times 0.25 \times 0.25 \times 0.7854}{70} = 0.0664$ lb. per sec., which is equal to $0.0664 \times 3{,}600 = 239$ lb. per hour. Assume a boiler pressure of 90 lb. per sq.in., gage, a feed-water temperature of 120 deg. F., an over-all boiler efficiency of 70 per cent and that the coal has a heating value of 11,500 B.t.u. per lb. The heat lost per pound of steam is given by $H - t + 32$, in which H is the heat in the steam above water at 32 deg. F. and t is the temperature of feed water in degrees F. Then, since from the steam tables H at 90 lb. gage = 1,187.1, the heat lost per hour is $239 (1{,}187.1 - 120 + 32) = 239 \times 1{,}099.1 = 262{,}685$ B.t.u.

The coal required per hour to furnish this wasted heat will be $\frac{262{,}685}{0.70 \times 11{,}500} = 32.6$ lb. If there are ten such traps, the total coal wasted would be 326 lb. per hour, or $326 \times 24 = 7{,}284$ lb. per day of 24 hours.

To get an idea of the loss from leaks at pipe joints assume a 4-in. line with the gasket blown out one-quarter way around a flange joint where the flanges are $\frac{1}{16}$ in. apart. With the same pressure as before Napier's formula will give

$$G = \frac{4 \times 3.1416 \times 0.25 \times 0.0625 \times 94.7}{70} = 0.264 \text{ lb. per sec.}$$

Using the same methods as above, the coal wasted by this one leak will be 3,110 lb. per day.

Now assume that there is a 20-ft. length of this 4-in. pipe uncovered. The area of radiating surface will be $\frac{4.5 \times 3.1416}{12} \times 20 = 23.55$ sq.ft. The temperature of the steam will be 324 deg. F., the room temperature may be assumed as 70 deg. F. and a heat loss of 3 B.t.u. per sq.ft. of surface per degree difference in temperature per hour is a reasonable value. The heat loss per hour from this bare pipe then will be $23.55 \times 3 \times (324 - 70) = 179{,}500$ B.t.u. Reducing this to coal per day gives 53.6 lb. from only 20 ft. of bare 4-in. pipe. Larger pipe and longer uncovered lengths will of course give correspondingly greater losses. A good pipe covering will save at least five-sixths of this loss.

Investigating Failure of Direct-Current Motors to Start

By E. C. Parham

Detailed explanation is given on how to make the tests necessary to treat the troubles in a direct-current motor that fails to start.

THE routine for locating trouble on a direct-current motor that fails to start, depends on the action of the motor up to the time of failure. It may be a new motor, or it may be a second-hand machine of unknown history. It may have given much trouble, or it may have given none at all; ultimate failure may have been suggested by gradual changes of speed, starting characteristics, sparking, heating or all of these combined; or failure apparently may have developed overnight and without demonstration. The location, duty or connections of the motor may have been changed, or the starter or some part of the motor may have been replaced. In many instances the trouble is not in the motor circuit, but is in the main circuit or in the connected mechanical load. This list of modifying influences suggests how helpful must be a knowledge of how the motor behaved before and when it failed. Such information, however, often is not available or is unreliable, therefore in this attempt to give a logical routine of procedure past actions will be disregarded and the history of the motor will begin with its failure. The starter will be assumed to be of the hand-operated type.

Assume the main circuit to be alive, as evidenced by the operating of other motors or of lamps from the same service. First look for a blown or otherwise defective fuse; an open or defective switch or circuit-breaker; poor contact of starter handle with the contact buttons. An open-circuit due to a fuse, a switch or a breaker may easily be located by means of a test-lamp circuit, as follows:

Fig. 1 shows the circuit in which the open is to be located. On holding the test lamp across 1 and 2 as indicated, the lamp should glow if the circuit is complete up to the fuses. This being so, on applying the lamp to 3 and 4 it will glow if the fuses or switch are not defective. Should 3 and 4 fail to light out, test across 5 and 6; glowing of the lamp means that the fuses are all right. If no light is obtained, test from 1 to 6 and from 2 to 5. If both tests fail to light out, both fuses are defective; when one test fails to light the lamp, the test lead that is on the motor side of the fuse will point to the defective fuse. After power has been found up to the switch, next test from 3 to 4. When light is obtained, the switch is all right.

Note the nameplate of the motor and starter to see if they are adapted to each other and to existing line conditions. Inspect the motor, starter and connected load to see if the trouble may not be due to broken, burned off, pulled out, missing or wrong connection; missing, broken or tight, brushes, brush fingers resting on the holders instead of on the brushes; loose brush-holder yoke; grounded brush-holder studs; high

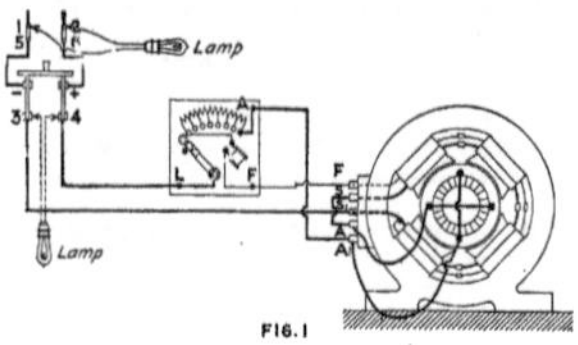

FIG. 1

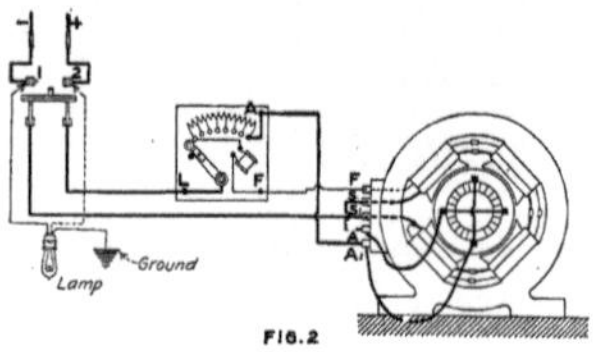

FIG. 2

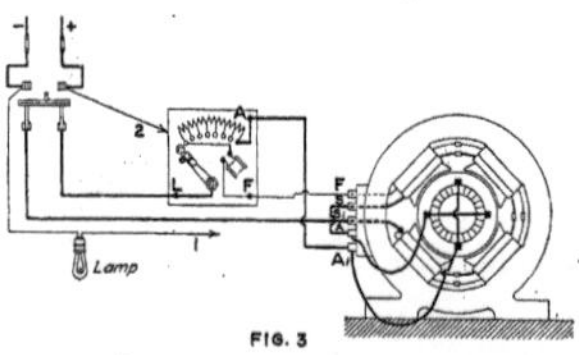

FIG. 3

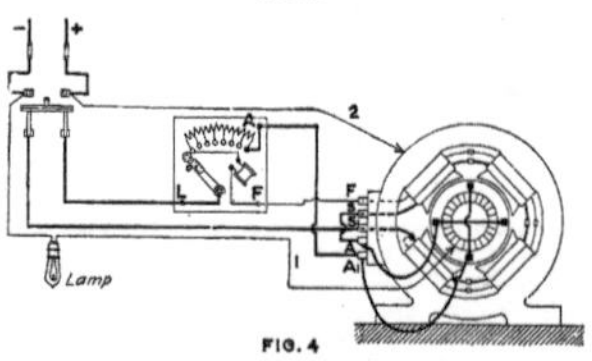

FIG. 4

FIGS. 1 TO 4. CONNECTIONS FOR LOCATING FAULTS ON DIRECT-CURRENT MOTORS mica, worn bearings, obstruction in air-gap, armature rubbing polepieces, loose polepieces, loose bearing caps, or jamming of the connected mechanical load. Open the main switch and with a lamp-test circuit test the motor and starter for grounds, as follows:

Connect one end of the lamp circuit to one side of the line above the open switch, at 1 in Fig. 2, and touch the other end of the lamp circuit to ground—a conduit pipe, that is grounded, a steam pipe or a water

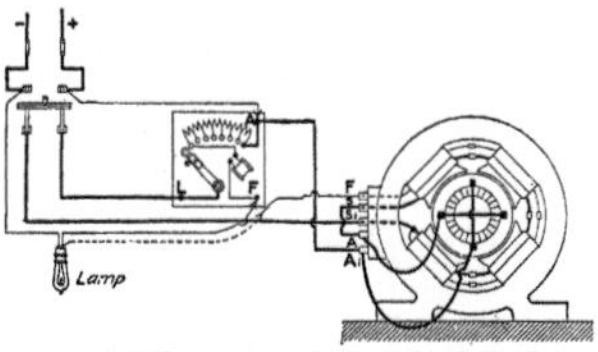

FIG. 5. CONNECTIONS FOR TESTING CONTINUITY OF STARTING-BOX AND MOTOR CIRCUITS

pipe. The lamps should not glow unless the circuit has a ground return and the lamp is connected to the ungrounded side. Next connect the lamp to the other side of the switch, as at 2, and repeat the test; if no light is obtained on either test, it means that the line itself is free of grounds. If one of the tests lights the lamp, however, it means that the opposite side of the line is grounded and the line end of the lamp circuit must remain connected to the side of the line that lights the lamps, when the other end of the lamp circuit is touched to ground. The other test lead is then connected to the other side of the line above the switch, thereby giving the complete connections of Fig. 3, which shows the starter and motor connections as well. In Fig. 3 the test circuit is properly connected for testing the starting box and motor, where the positive side of the line is grounded. With No. 2 test lead on the starting-box metal case, touch No. 1 lead to the starter handle and contacts; if the lamp glows, disconnect the motor from the starter and test again! A ground in the starter will be indicated by the lamp lighting; if it does not glow, the ground is in the motor circuit.

Assuming that the ground is in the motor circuit, raise the motor's brushes off the commutator and test the armature alone by touching No. 1 lead to the commutator and No. 2 lead to the frame, as in Fig. 4. If the armature is clear of grounds, disconnect the brush-holders, shunt-, series- and commutating-field circuits one at a time and test each separately. When a grounded field circuit is indicated, disconnect the coils and test one at a time to locate the grounded one. The same procedure must be observed in regard to the brush-holders. In the case of a compound machine test for a short-circuit between shunt- and series-field windings by first opening the normal connection between them, then hold No. 1 test lead on one terminal of one of the windings and No. 2 test lead on one terminal of the other winding. If the lamp lights, isolate the spools and test them one at a time.

Having failed to indicate a ground, restore all connections at the motor and open the field connection at the starter as in Fig. 5. With the starter in the off position terminal lugs F and A should light out through the starting resistance and no-voltage release coil, as also should the armature and shunt field when the test lamp is connected as indicated in Fig. 5. If no light is obtained, each of these windings and its connecting wires must be isolated and tested alone. When in doubt in regard to the starting box raise the panel and inspect the internal connections. Remember that a shunt on the series winding or on the commutating winding of the motor will make these circuits appear to be intact when in fact they may be open. Each pair of the starter contact buttons should light out excepting, in some cases, where several buttons on the "off" position end of the dial may not be used. In nine out of ten cases, inspection and lamp testing will locate the source of trouble. Where information is unreliable or unavailable, the preliminary tests are necessary; in all cases they are desirable, because they avoid abuse incident to the slam-bang method of trying the motor in order to see what happens. Another procedure that may be followed is:

Open the armature circuit at the starting box, as in Fig. 6. Then close the switch and jog the starter on and off; if there is no arc, there is an open either in the shunt-field circuit or in the main circuit which includes the fuses, the switch and the circuit-breaker and the starter handle and contacts. Look for poor or no contact between the handle and the buttons: the handle contact may have burned or may be missing entirely. If jogging the arm on and off the start position draws an arc, it indicates that the shunt-field circuit is intact. In order to be certain of this, however, tie the starter handle on the first notch, then feel the motor polepieces with a piece of iron to see if they are magnetized. If current is flowing in the field coils, all poles should attract the piece of iron with equal strength.

Next open the switch, throw the starter to the off position, restore the armature circuit to its original

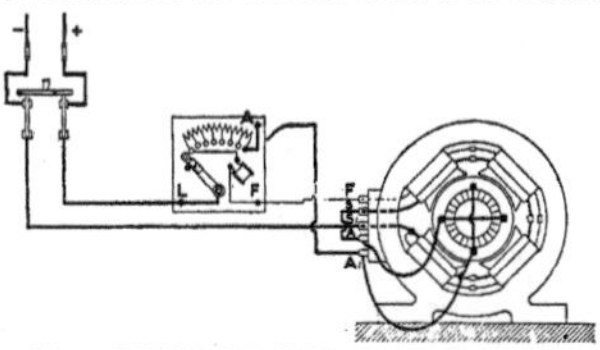

FIG. 6. CONNECTIONS OF COMPOUND MOTOR, WITH ARMATURE DISCONNECTED FROM STARTING BOX

condition and disconnect the mechanical load by removing the belt, pinion or coupling, as the case may be. See if the armature can be turned freely by hand and inspect the commutator for evidence of open-circuits or of short-circuits. Assuming that the commutator is normal, close the switch and jog the starter; if the arcing is no heavier than before, it means that the armature circuit is open; such an open may be readily located by the lamp test described in the foregoing. If, however, the arcing is much heavier than that due to opening the shunt-field circuit alone, it indicates the existence of motor trouble that the inspection and the lamp tests were not qualified to indicate, such as a reversed field coil or short-circuited armature.

Turn the armature a few degrees and again jog the starter; repeat it for several positions of the armature; if the armature will start from one position but not from others, indications point to internal faults. Next, repeat the test, but instead of shifting the armature, shift the brushes a few bars at a time until a whole pole-pitch has been traversed around the commutator. A loose brush-holder yoke will be dragged around by the commutator until the brushes reach a position in which the armature cannot be started.

Assuming the tests and trials to have resulted in starting the motor free, then, if the machine is compound wound, try starting the motor on the shunt field alone with the series winding short-circuited. Then try starting the motor on the series field alone with the shunt connection open at the starting box, in order to insure that the two fields are so connected as to turn the armature in the same direction. Opposition of the fields would not prevent starting the motor under no load, but it would prevent many motors from starting a connected load. Open the switch, connect the motor to its load, close the switch and start in the regular manner. If the motor starts, accelerates and operates without excessive sparking, after the best position of the brushes has been determined by trial, there will be nothing further to do. But if the brushes spark badly and it can be ascertained that the connected load is not too heavy for the motor to handle, the condition of the insulation of the wires of the field coils and of the armature should be thoroughly investigated. Especially is it important to do this if the motor is found to have been operating with the shunt windings and series windings opposed, because this condition results in a weakened field, an increased armature current and a consequent probability that the load has been sufficient to injure the insulation of all the windings.

In the large majority of cases the procedure outlined will result in the locating of the trouble. In an actual case of motor trouble, suggestive conditions will present themselves simultaneously and in almost every instance will obviate the necessity of doing a great deal of the work indicated. In exceptional cases failure may date from the installing of a new or a repaired armature, the winding or connections of which may be wrong. This should not be so puzzling, however, because in all cases where trouble dates from changes of any kind, the parts involved in the changes are a legitimate field of suspicion.

Refrigeration Study Course—VII. Insulation

By H. J. Macintire

In the other parts of this study course the purpose has been to show the cycle used, the meaning of each part of the apparatus and its function in the cycle and the construction of the high- and low-pressure sides of the system. In other words, the production of refrigeration only has been considered. It is now appropriate that some time be devoted to the conservation of what has been produced by preventing, as far as possible, any leakage losses.

In general, it may be said that if two adjacent bodies have different temperatures, the tendency is for an average temperature to be formed. For example, the hotter substance tends to become cooler and the colder one to become warmer. The method by which this may be accomplished may be by radiation, convection or by conduction. *Radiation* is the method by which the sun's rays reach the earth or by which one receives the sensation of heat from an intense fire perhaps some distance away. *Convection* is the method by which heat is conveyed by a moving substance, like hot-air or hot-water heating in the factory or house. Finally, *conduction* is the manner in which heat is carried along by the particles (molecules) of a solid, as in the case of a hot slicing bar. In the case of refrigeration the heat comes from the atmospheric air as a rule, and passes into the rooms which are refrigerated or the insulated tanks or pipes.

The reason for the passage of heat from the air to the refrigerated pipes is self-evident on consideration of the conditions that prevail. Heat cannot run uphill. It passes, like water, from a condition of higher level to a lower one. The heat in the combustion chamber of a boiler passes through the tubes and shell into the (relatively) cold water in the boiler, and the flue gases are lowered in temperature and the water in the boiler is heated and evaporated into steam. And so it is that we try to protect the refrigerated system by insulation, as in the case of steam pipes or a boiler setting.

However, it is impossible to secure a perfect insulation, just as it is impossible to get perpetual motion or a 100 per cent conversion of energy. This is because we cannot find any substance that is a nonconductor of heat or that will entirely eliminate radiation and convection. So it is that the problem of insulation is one of securing a substance that is practical for our use, as regards price, ease of application and long life under working conditions, one that will decrease the losses to a minimum, remembering that the initial cost of the insulation varies directly with the thickness of the material, whereas the losses due to lack of more or less material are a maximum of the summer months only.

From the foregoing it may be seen that the problem narrows down to the following: To conserve refrigeration, insulation of a suitable material must be provided on all sides of the cold-storage room and on the brine and return ammonia pipe lines. Also that perfect insulation is impossible and that the proper amount and the kind of material to be used are dependent on conditions that prevail. These conditions vary with the average summer temperatures, with the kind of cold storage or refrigerating duty performed (and the length of this service) and the cost of a ton of refrigeration.

The Choice of an Insulating Material

The insulation material is really a part of the building construction, and it must therefore be on a par with the rest of the building. It must be capable of withstanding deterioration or rotting, and, in addition, should be free of odor of any sort. It should be capable of maintaining itself in position and not settle in places to leave pockets or air spaces. It must be waterproof so that moisture may not collect, thus tending to mold and to diminish its ability to insulate, and it must be verminproof. Finally, the insulator must be of a high degree of perfection and must be capable of reducing the heat losses.

The problem of insulation is much more difficult so far as moisture is concerned than is the case with steam. With steam the tendency is to dry out the covering and to maintain this dryness as long as the temperature of the steam is maintained inside the boiler or vessel. In refrigeration the opposite is true and moisture has a tendency to travel toward the coldest surfaces and to keep collecting as time elapses. In this case also it is possible to have ice form, which may loosen or disintegrate the insulation.

Theory of Insulation

The theory of insulation is somewhat difficult to understand at first. According to this theory there is no material that resists heat perfectly. Remember that insulating for refrigeration purposes is preventing heat from coming inward from the atmosphere to the region maintained at a temperature which is *below* its surroundings. However, it is found that air pockets of extremely small size are an insulation, and any material which has the greatest number of these pockets is the best insulation, but each pocket must be separated from its neighbor. If the pockets were connected, then convection could take place as in the case of a hot-air furnace. From this it would naturally follow that the material that is the best insulator is also the *lightest* in point of *weight*, as a large amount of its volume is air. This is the reason that substances like mineral and animal wool, corkboard, balsa wood, shavings and sawdust have insulation values.

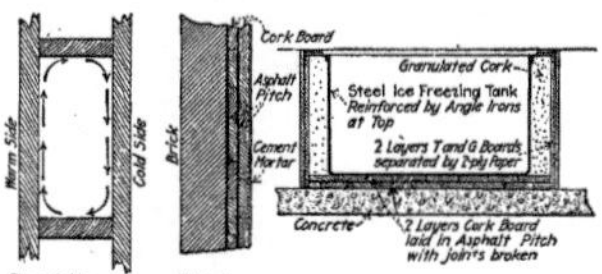

Flow of Air showing Convection of Heat FIG. 1 — FIG. 2 — FIG. 3

SHOWING AIR CURRENTS AND INSULATION

The use of air spaces naturally follows from what has just been said. However, it can be proved that large air spaces have little value. The reason is self-evident from Fig. 1, which shows the flow of currents of air from the colder to the warmer side of the insulation. Therefore, a partition made up of alternate courses of lumber and air spaces is uneconomical of building space.

Sawdust and shavings were used extensively at one time and still have certain advantages. They are not waterproof nor are they vermin- or fire-proof. They lend themselves to cheap temporary construction, such as ice storage buildings or wooden buildings put up for a short life or for short refrigerating periods. The work should be carefully performed, and the shavings or sawdust should be carefully packed and a waterproof paper very carefully secured in place. Sawdust is likely to settle and leave a space in the upper part of the wall, so shavings, dried before using, are preferred.

Lately, for second-class construction, a large number of special materials have appeared in the form of convenient-sized boards. These are compressed wood and vegetable fibers so devised that the board may be erected like lumber and nailed into place. Being compressed, it has the characteristics of shavings and has the advantage of more efficient insulation and being able to retain its position. It is not vermin-, fire- or water-proof, and so it would generally be used only in wooden constructions.

Some success has been achieved in the use of mineral wool, the product of blast-furnace slag or of special mineral earths. These are made up in the form of boards and attempts are made to waterproof the material in the process of manufacture. When so done, it is easily erected in place, is fireproof and a good insulator. It is used in permanent construction and should maintain itself for years.

By far the most successful and popular insulating material is cork, obtained from the bark of trees found principally in Spain and Portugal. This bark has the air-cell construction which is necessary for the use we desire of it. For refrigeration use is made of the wastage in the manufacture of cork stoppers, etc. The cork particles are compressed and heated so that they adhere together and are made up into boards of convenient sizes or in forms to fit standard pipes or tanks. The cork is inherently waterproof, and in erecting the joints and courses are made up in asphalt pitch. In building construction there is also a final coat, as a rule, of cement plaster, which protects the corkboard and gives a smooth surface for suitable cleaning.

Special kinds of work also use granulated or pulverized cork, as in insulating the sides of ice-making tanks or the ends of cylindrical tanks. Usually, granulated cork should be used only where access is readily obtained to repack from time to time should settlement occur.

FIG. 4. SHOWING THE VARIOUS SUMMER-TEMPERATURE BELTS IN THE UNITED STATES

Insulation Values of Building Materials

Lately, numerous attempts have been made to find a suitable and correct method of figuring the insulating value of building materials. Elaborate experiments have been made, both with simple materials and with built-up construction. In these experiments it has been shown that there is on all material a "skin effect" which increases the insulating value. However, this is a matter that affects only the outside material, which for good insulation is usually cement mortar, or otherwise, perhaps, tongued-and-grooved lumber or brick. The real insulating material is neither of these.

Neglecting the skin effect, the insulating value of the material used is dependent on its conductivity. By this is meant the number of heat units that will pass through it, usually rated on each square foot, one inch thick, per hour or per 24 hours with a difference in temperature on each side of the insulation of 1 deg. F. If this "leakage" is known for 1 deg. difference, then the leakage loss will be proportionately greater for 25, 50, 75 or 100 deg. F. difference in temperature. Likewise, if a material 1 in. thick will give a certain leakage loss, then one 2 or 6 in. will give proportionately less loss. This can be expressed in the formula:

$$Q = A \times K \times (t_{outside} - t_{inside})$$

where A equals area of wall or pipe surface in square feet, K equals conductivity, and t equals temperature in degrees F.

The value of K is obtained from tables of experimental values, as given below, or if use is made of several substances by the formula,

$$K = \frac{I}{\frac{T_1}{C_1} + \frac{T_2}{C_2} + \frac{T_3}{C_3} + ETC.}$$

where T_1, T_2, etc., equal thickness of each material in inches, and C_1, C_2, etc., represent conductivity as given in the tables for standard conditions.

Calculations: As an example of the use of the formulas, suppose a wall composed of 8 in. of brick, 2 in. board and ½ in. mortar is exposed to temperatures of 75 deg. and zero degree F. on the two sides and it is desired to find the leakage per twenty-four hours for a total surface of 700 sq.ft.

The value of K becomes (see values in table)

$$K = \frac{1}{\frac{8}{4.66} + \frac{2}{0.31} + \frac{\frac{1}{2}}{0.63}} = 0.111 \text{ B.t.u. per hour}$$

and the leakage is

$$Q = 700 \times 0.111 \times (70 - 0) = 5{,}400 \text{ B.t.u. per hour.}$$

Whereas the skin effect would tend to reduce the value of K and Q, yet the factor which is important is the perfection of the erection, which is subject to wide variations. Unless the joints are carefully made and air-tight, moisture will get in and freeze, causing the ultimate deterioration of the insulation. This is probably the factor most responsible for the fact that the laboratory results and the practical results do not agree, and we are justified in adding a liberal factor of safety to insure good operating results. For these reasons and for simplicity of calculation it is recommended that use be made of the method of calculation as given.

As regards the maximum temperature to be used in the calculation of leakage losses, it is customary to assume an average maximum temperature during the period of peak refrigerating loads. Some idea of average temperatures in the United States may be obtained from the map taken from United States Government reports, Fig. 4. It is not recommended that the highest local temperatures be taken, but the average temperature experienced during the twenty-four hours for conditions that are likely to prevail for a period of days or a week or more. Of course the actual choice of the thickness of insulation for any particular case will be decided by the relative costs of the insulation itself and the book value of a ton of refrigeration. If the time of operation, days per year, is small, then one cannot economically install heavy first-cost construction. It would be too much like spending two dollars to save one dollar.

In conclusion, it is worth while putting special stress on the fact that the insulation must be kept dry, it must be kept in place and should be odorless. It should be easily placed in position and be of a material of economical insulating value.

REPORT OF TESTS CONDUCTED BY THE UNITED STATES BUREAU OF STANDARDS, WASHINGTON, D. C., COVERING THERMAL CONDUCTIVITIES OF VARIOUS INSULATING MATERIALS

Material	Conductivity 24-Hr. Ft.-Sq. In. Thick Deg. F.	Density	Nature of Material
Air	4.0	.....	Horiz layer heated from above radiation
Colorax	5.3	0 064	Fluffy finely divided mineral matter (elim.)
Hair felt	5.9	0.27	
Keystone hair	6.5	0.30	Hairfelt confined between layers building paper
Pure wool	5.8	0 107	Firmly packed
Pure wool	5.8	0 102	Firmly packed
Pure wool	6.3	0 061	Loosely packed
Pure wool	7.0	0 039	Very loosely packed
Cotton wool	7.0	0.000	Firmly packed
Insulite	7.1	0.19	Pressed wood pulp—rigid, fairly strong
Linofelt	7 2	0.18	Vegetable fiber confined between layers paper—soft and flexible
Corkboard (pure)	7.4	0 18	
Eelgrass	7.7	0.25	Inclosed in burlap
Flaxlinum	7 9	0 18	Vegetable fibers—firm and flexible
Fibrofelt	7 9	0 18	Vegetable fibers—firm and flexible
Rock cork	8 3	0.33	Prest rock wool with binder—rigid
Balsa wood	8.3	0 12	Very light and soft
Waterproof lith	9 8	0 27	Rock wool, vegetable fiber and binder, not flexed
Pulp board	10 4	.....	Stiff pasteboard
Air cell ½-inch	10.7	0.14	Corrugated asbestos paper, inclosing air spaces
Air cell 1-inch	11.5	0 14	Corrugated asbestos paper, inclosing air spaces
Asbestos paper	11.8	0 50	Fairly firm but easily broken
Infusorial earth	13.9	0.69	(Block)
Fire felt (sheet)	14.3	0 42	Asbestos sheet coated with cement—rigid
Fire felt (roll)	15.3	0 68	Soft, flexible asbestos
3-ply Regal roofing	16.7	0.88	Flexible tar roofing
Asbestos mill board	20.2	0.97	Prest asbestos, fairly firm, easily broken
Woods—Kiln Dried			
Cypress	16.0	0.46	
White pine	19.0	0 50	
Mahogany	22.0	0 55	
Virginia pine	23.0	0.55	
Oak	24.0	0.61	
Hard maple	27.0	0 71	
Asbestos wood (sanded)	65.0	1 97	Asbestos and cement—very hard and rigid

NOTE.—The findings of the Bureau as outlined above were recommended for the guidance of the members of the American Society of Refrigerating Engineers by its Insulation Committee.

Statistics of petroleum production in states east of California during January and February, 1920, collected from data on file with the United States Geological Survey, show the quantity of oil received from producers by pipe lines and by refineries that receive petroleum directly from the well. In January Oklahoma led with a total production of 7,958,000 bbl., and in February 7,942,000 bbl. was produced. The daily average production in January was 256,710 bbl. and in February 273,862 bbl. Texas is second in petroleum production, followed by Kansas, Louisiana and Wyoming.

Geologists in the service of the United States Geological Survey, Department of the Interior, during the last four years have been investigating the geologic structure of parts of New Mexico with a view to determining the possibility of oil and gas reservoirs in these regions. While little evidence of oil and gas has been discovered, indications are favorable and geological experts are of the opinion that if wells were sunk deep enough (2,000 to 3,000 ft.) oil might be found.

Details of the Waterbury Turbine Explosion

POWER is indebted to the *Travelers Standard* of January for further particulars about the steam-turbine explosion in Waterbury, Conn., briefly reported in the issue of Oct. 28—Nov. 4, 1919, p. 682. The explosion occurred at the power station of the Connecticut Light and Power Co. Two men were killed, one of them instantly, and ten others were more or less seriously injured. The property damage was large, and an unofficial estimate placed it at $100,000. Fig. 1 shows the generator end of the damaged turbine.

The power-generating equipment of this station consists of six direct-connected, turbine-driven, three-phase generators of various types and capacities, all generating current at 2,300 volts and 60 cycles. Steam is supplied by twenty-two water-tube boilers. The usual operating steam pressure is 190 lb., and the safety-valves are set at 200 lb. All the boilers are provided with nonreturn valves. At the time of the accident fourteen boilers were on the line and three others were carrying banked fires, but were not furnishing steam. Four of the six turbines were in operation, two being shut down. The exploded turbine had been in regular operation since its installation some eight or nine years ago. It was known as "No. 3" and drove a generator having a capacity of 3,000 kw. Its normal speed was 1,800 revolutions per minute.

The main-turbine shaft was 13 in. in diameter and was turned down to 7 in. in diameter at the governor end (see Fig. 5). Six steel wheels, each 48 in. in diameter, were keyed to the shaft (see Figs. 3 and 4). The outer face of each wheel was about 5½ in. wide and carried two rows of buckets. The turbine was equipped with a governor of the centrifugal type, which transmitted its motion by means of a connecting-rod to the valve mechanism, which in turn actuated ten steam valves. An emergency stop was also provided, and this was arranged to operate at about 8 per cent above normal speed, or at 1,944 r.p.m. This emergency stop was mounted on the turbine shaft and consisted of an unbalanced ring which revolved with the shaft. At normal speed the ring was held concentric with the shaft by means of springs, but in case of excessive speed the spring tension would be overcome by the centrifugal force of the unbalanced weight. This would result in closing the 8-in. main throttle valve. The generator was direct-connected to the turbine, the shafts of the turbine and generator being joined together by a coupling bolted with ten 2-in. bolts.

FIG. 1. THE GENERATOR END OF THE DAMAGED TURBINE AT WATERBURY, CONN.

The turbine shaft was broken in two places, and the bolts of the coupling also failed, permitting the two parts of the shaft to separate at this point, as shown in Fig. 2. This part of the shaft is 13 in. in diameter and about 6 ft. long and weighs approximately 2,700 lb. It was hurled over three of the turbines and, after passing through a window, came to rest in the yard outside of the power station, about one hundred feet from its original position. It will be seen that this part of the shaft was bent to an angle of about thirty

FIG. 2. PIECE OF THE TURBINE SHAFT

The piece is 13 in. in diameter and 6 ft. long and weighs about 2,700 lb. It was found 100 ft. from the damaged turbine. The piece is bent and the coupling bolts are bent and sheared.

FIG. 3. THE BUCKETS WERE STRIPPED FROM THE WHEELS

FIG. 4. A WHEEL APPARENTLY UNDAMAGED BY THE EXPLOSION

degrees. Another section of this shaft 6 ft. 4 in. long and 13 in. in diameter dropped below the floor of the turbine room. The piece here shown is 3 ft. long and 7 in. in diameter, and the break occurred at the point where the diameter of the shaft was reduced from 13 to 7 inches.

The wheels were thrown in all directions. Fig. 3 shows one of them from which the peripheral buckets were entirely stripped. Another wheel, shown in Fig. 4, is apparently intact. One of the wheels is supposed to have passed through the roof of the building and probably fell into the river near-by. This theory is based on the fact that only five wheels were found after the accident, although there had originally been six in the turbine, and that an eye witness, who was about 300 ft. from the power station when the explosion occurred, said that he saw a large disc coming through the roof. He failed to see where the disc landed, however, because he started to run from the vicinity, believing that the entire power house was about to be blown up. Many other heavy parts were thrown in all directions, and some of them disabled the delicate mechanism of the other turbines and made it necessary to shut down the entire plant temporarily. The throttle valve of the damaged turbine is shown in Fig. 6.

Soon after the explosion an investigation was started to determine the cause of the accident. As usual in cases of this kind, various theories have been advanced, in support of each of which certain more or less important facts have been cited or suppositions made; nevertheless no one theory that has yet been offered appears to harmonize with all the known data. It is generally admitted, it is believed, that the exciting coils on all the four turbines that were in operation at the time became "dead" from some cause or other immediately before the explosion. The consequent loss of magnetism in the armature fields would prevent the machines from generating current and would therefore remove the load from the turbines and allow them

FIG. 5. PART OF THE GOVERNOR END OF THE TURBINE SHAFT

At this end the shaft is 7 in. in diameter. This broken piece is 3 ft. long.

to run free, subject to no restraint except the normal friction at the bearings and elsewhere. Increase of speed would be the immediate result, and it is evident that such an increase did actually occur, because three of the four machines running were shut down automatically by the operation of the emergency stops with which they were provided. The emergency stop on the machine that exploded also appears to have operated, and it happens that that particular stop had been tested only the day before and found to be in working condition. There is considerable difference of opinion, however, as to the extent to which the overspeeding of No. 3 turbine had progressed when the accident occurred, and it does not appear possible to decide this point in any definite and satisfactory way. The steam valve on No. 3 was found to be nearly closed after the accident, but for some reason or other it was not closed tightly. Possibly something had become lodged upon its seat. The theory that it stuck from some other cause appears to be almost disproved by the fact that it had worked freely and satisfactorily when it was tested, only twenty-four hours before. Whatever the real cause of the accident may have been, there is no evidence that can properly be construed as reflecting in any way upon

FIG. 6. THE THROTTLE VALVE OF THE EXPLODED TURBINE

the Connecticut Light and Power Co., or upon the builders of the machine. It is hoped that further investigations will be made and the facts disclosed published, that appropriate measures may be taken to prevent similar accidents in the future.

The explosion resulted in the temporary suspension of power and lighting service in Waterbury and vicinity. Street cars were stalled, industrial plants were obliged to close down, and business houses, places of amusement and private homes were left in darkness. Local manufacturing plants and power stations in near-by cities were, fortunately, able to make emergency connections, so that the inconvenience to the public was of comparatively short duration. Furthermore, the damage to some of the remaining generating units was quite limited, and within a few days these had been repaired so that they could carry a considerable portion of the load.

Templet for Cutting Commutator Mica Segments

By W. A. Harris.

Commutator micas are usually cut by first laying one of the segments on a sheet of mica and marking out the desired form, after which the segment is cut out

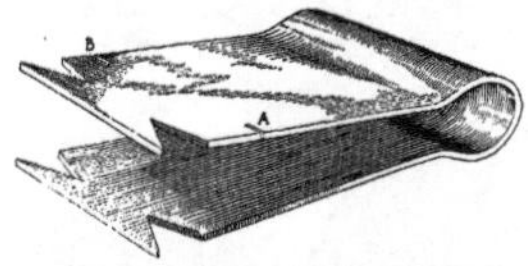

COMMUTATOR INSULATION SEGMENT TEMPLET

with a pair of shears. This method, even at the very best, does not produce segments that are of uniform shape. If a rigid construction of the commutator is to be obtained, not only must the insulating segments be of the proper thickness, but the V's in them must be uniform in shape and size as also the part of the segment on the inside of the commutator.

To obtain uniformity a templet, similar to that shown in the figure, may be used as a guide in cutting the segments. It is made of a piece of $\frac{1}{16}$-in. iron bent in the shape indicated. After this has been done the two sides of the templet are gripped in a vise and the V's cut the exact shape of those in the commutator for which the insulator is being cut. A notch is cut at *A* to indicate the maximum width of the mica and another at *B* to give the minimum width. The mica is cut to the approximate width of the segments, and then the templet is filled tight with these strips. While being held in a vise, the insulating segments are cut to the required shape with a hacksaw. A notch is cut in the mica at *A* and *B* to give the maximum and minimum width, so that after the segments are removed from the templet the excess material can be cut off with a pair of shears. The outside dimensions of the segments need only be approximate, since the excess material can be removed when the commutator is turned down.

The Board of Trade of Great Britain in a recent report shows that during 1919 approximately 104,000,000 gal. of petrol (gasoline) was imported from the United States. The report says that before the war the production of benzol was about 17,000,000 gal. per annum, and during the war it was increased to about 42,000,000 gal. of which some 32,000,000 gal. was produced by coke ovens and 10,000,000 gal. by gas works and tar distilleries. Since the war the output has fallen to about 25,000,000 gal. per annum.

A Veteran Engineering Research Worker

John R. Allen

A life dedicated to the investigation of heating and ventilating engineering problems. A successful organizer and teacher. Director of the Bureau of Research of the American Society of Heating and Ventilating Engineers

IT IS said that to few men are given the golden opportunity and ability to aid in the advancement of the human race through knowledge. Among these few will rank John R. Allen, of Pittsburgh, Pa., Director of the Research Bureau of the American Society of Heating and Ventilating Engineers. His life-work has been and is the advancement of engineering sciences through research and the promotion of engineering education. His efforts in this direction have yielded much that is proving beneficial, and his work is followed with keen interest by engineers and students of engineering alike.

Heating and ventilating engineering is comparatively young when measured by progress in other engineering lines, and modern heating and ventilating equipment is still far from perfect. A careful and exhaustive system of research might prove effective in clearing up some of the problems involved. Professor Allen, in his investigations at the University of Michigan and the University of Minnesota during the last fifteen years has done much for the advancement of this phase of engineering.

The United States Bureau of Mines, in a desire to coöperate with the Research Bureau, has placed at the disposal of Professor Allen and his assistants a portion of the magnificent laboratories at Pittsburgh, Pa., with their extensive equipment, and interesting developments which will prove of value to manufacturers and users of heating and ventilating apparatus are expected.

Some of the first things on which the efforts of the bureau will be centered are: The physical condition of the atmosphere which must be produced in order to maintain the same healthful effects as are obtained in pure fresh outdoor air; the scientific determination of heat losses by the use of various kinds of materials generally employed in building construction; the efficiency of various kinds of radiation; the most efficient methods of burning different kinds of coal, etc. This is only an outline of the questions that will be given immediate attention.

Many years of Professor Allen's life have been spent in teaching. His broad views and intensely human instincts, combined with a pleasing personality and unbounded enthusiasm for his work, have made him one of the most popular and successful teachers of engineering science in the country. Since his resignation from the University of Minnesota to take charge of the Research Bureau, students of the university have done everything in their power to induce him to return. Besides being a successful engineer and teacher, he is also a skilled organizer. During the years spent as research worker and teacher he has given to the profession two books—"Notes on Heating and Ventilating" and "Heating and Ventilation." These books contain a comprehensive account of heating and ventilating problems, experiments and discoveries.

Professor Allen was born in Milwaukee, July 23, 1869. He received a preparatory education at Milwaukee and at the Ann Arbor, Mich., High School, following which he entered the University of Michigan. In 1892 he graduated with the degree of B. S. and in 1896 he received his master's degree.

Following his graduation, Professor Allen took up his first position with the L. K. Comstock Construction Co., with which he remained for two years, and then he received an appointment as instructor in the Mechanical Engineering Department, University of Michigan. He was with the university from 1896 until 1911. His popularity with the students became so great and his ability as a teacher was so pronounced that he quickly advanced to assistant professor, junior professor and professor of mechanical engineering. In 1911 he resigned to carry out his plans for the construction and organization of Robert College, Constantinople, and until 1913 he was Dean of the Engineering Department. He returned to the University of Michigan after the completion of his work in Turkey, and for four years he was head of the mechanical engineering department.

In 1917 Professor Allen became Dean of the College of Engineering and Architecture, University of Minnesota, and remained there until 1919, when he was selected as director of the Bureau of Research of the American Society of Heating and Ventilating Engineers in coöperation with the United States Bureau of Mines.

He is an active member of many of the national engineering societies; past president of the American Society of Heating and Ventilating Engineers; past president of the Michigan Engineering Society; past president Michigan Chapter, American Society of Heating and Ventilating Engineers; president of the Minnesota Chapter of the American Society of Heating and Ventilating Engineers; honorary member of the National District Heating Association; member of the British Institute of Heating and Ventilating Engineers, of the American Society of Mechanical Engineers, the Society for the Promotion of Engineering Education and of the honorary societies, Tau Beta Pi and Sigma Psi.

Relative Value of Coal and Oil Fuel

Some Important Facts To Be Considered on the Advisability of Converting a Coal-Burning Power Plant to Fuel Oil

By GEORGE A. EVANS*

AT THE present time, owing to coal shortages and increasing prices of coal, a great deal of consideration is being given to fuel-oil burning. Notices are constantly appearing in the daily press of this or that plant having installed fuel-oil burning equipment in place of coal, and several articles have appeared in the technical press on the burning of fuel oil. It is the purpose of this article to analyze the various items that should be considered before such an important step is taken as changing from coal to oil fuel. The changing of ships from coal to oil is not considered here for the reason that conditions are so dissimilar that an entirely different analysis would be necessary.

Upon returning to the East coast recently, after having had the experience of operating with a fuel-oil plant in California, the writer was very enthusiastic on the subject of oil-fired boilers, owing to the great cleanliness of the fireroom, the absence of all coal-handling equipment with the accompanying dust or dirt, also the expense and disagreeable necessity of removing ashes. An investigation was made to determine and make recommendations as to the advisability of converting a certain plant from coal to oil fuel, and when all the facts of the case were carefully weighed, it was found that from an economical point of view oil could not be considered. This plant was equipped with six hand-fired-boilers, having a total boiler capacity of 2,250 rated boiler-horsepower, with a maximum station capacity of 4,200 kw. and the average output per month amounted to about one million kilowatt-hours.

On the accompanying chart is indicated the solution of a concrete problem according to the following assumptions: Efficiency of the boilers with coal, 65 per cent; oil, 75 per cent; the B.t.u. per pound of coal, 13,500; oil, 18,500.

*Assistant to Mechanical Engineer, Public Service Electric Co., Newark, N. J.

Starting in the upper left-hand corner of the chart at a point indicating 18,500 B.t.u. per pound of oil, a line is drawn downward until it intercepts the diagonal line representing 13,500 B.t.u. per pound of coal. From this point the line is drawn to the upper right-hand chart intercepting the line representing 75 per cent boiler efficiency with fuel oil; from this point the line is drawn to the lower right-hand chart to the line representing 65 per cent boiler efficiency with coal, from which it is carried to the lower left-hand chart to a line marked $6 per long ton of coal. Continuing this line downward, it is found that the equivalent cost of oil in cents per gallon is 3.35c. That is, oil has to be purchased at 3.35c. per gallon, or $1.41 per barrel of 42 gallons, to be the equal in price of coal at $6 a long ton with the heating values and boiler efficiencies that have been assumed. By means of this chart any problem can readily be worked out to obtain the equivalent value of oil in cents per gallon for any price of coal for any combination of boiler efficiencies and heating values of coal and oil.

The chart makes no allowance whatever for the labor saving by using oil as a fuel. This saving is sometimes greatly exaggerated, and the small amount saved in labor is more than offset by the higher cost of oil. The saving in labor must be estimated for each individual plant, taking into consideration local conditions, cost of labor, etc. As a rule, higher-class firemen, which means higher wages, will be required to attend oil furnaces. In hand-fired boilers more labor can be saved, of course, than where the boilers are operated with stokers. It is doubtful if any reduction in the number of men can be made in a plant operating with stokers, nor can any reduction be made in a plant of one or two boilers employing one fireman for each watch. Where a battery of six or eight hand-fired boilers are converted to oil, the number of firemen could be reduced to one per watch, but no reduction in the watertenders or engineers of the plant should be figured on.

The chart also makes no allowance for banking fires required with coal-burning boilers, an item not required with oil. The amount of coal used in banking must also be worked out for every plant, depending on its load conditions. Some plants operating twenty-four hours a day would use comparatively little coal for banking, while others operating with a day load would use a considerable amount.

Another factor often lost sight of in figuring on oil fuel is the amount of steam required for atomizing and heating the oil. A great majority of the power plants would undoubtedly use steam for atomizing in place of mechanical mixers. The amount of steam required for atomizing would be about 1½ per cent of the total amount of steam generated under daily average operating conditions and with a poor fireman might go to 4 or 5 per cent. With mechanical atomizing burners a higher efficiency can be obtained than with steam atomizers, but it often becomes necessary to change the boiler fronts and install blowers and air

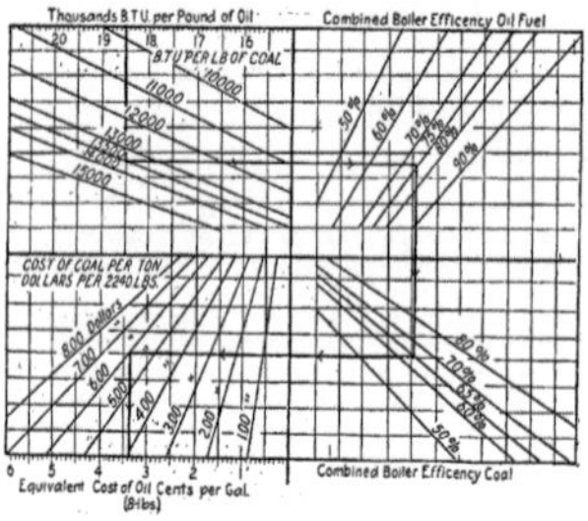

CHART FOR FINDING EQUIVALENT PRICES OF COAL AND FUEL OIL

ducts for forced draft. This, of course, would not be such a great item in a plant operating with stokers and forced draft, but as a modern power plant equipped with stokers operates at a boiler efficiency of from 70 to 75 per cent, very little can be gained by changing over to fuel oil.

If the furnace is not suitable, oil cannot be burned economically. On converting from coal to oil, in most boilers it will be necessary to set the bridge-wall back far enough so that the flame does not impinge on the wall of the furnace. The setting must also be high enough to give a good combustion space and to prevent the flame from impinging on the boiler tubes.

The question of the storage of oil must also be carefully looked into. It will likely be found that if a plant is in a congested locality only a limited amount of oil can be stored, and that in a tank placed underground or in a concrete pit. The capacity of this tank will depend on local conditions. Where there is room for outside storage if the tanks are placed underground 50 ft. away from a building, there can generally be unlimited storage capacity. If the tanks are placed above ground, the storage capacity is limited by the distance which the tanks can be placed away from the adjacent building. Starting with a minimum of 50 ft., the greater the distance the tanks can be placed from any building the larger the storage capacity allowed, depending on local laws and insurance rules.

The amount of storage capacity that can be made available is considered very important. Operating with coal, it has become the custom to have on hand anywhere from three to six months' supply. To store enough oil to operate a plant of any size from three to six months would take a considerable number of large oil tanks, create an extra fire hazard, also adding to the installation cost of the fuel-oil system. A fuel-oil plant will probably have to cut down its available fuel in storage to a month's or six weeks' supply. This point is important, as from past experience with car shortages and railroad delays it is not probable that a plant will have any more success in getting tank cars with fuel oil delivered to them than they have at present with coal cars. A plant situated on tidewater where oil can be delivered in tankers or bargeload lots would be more fortunate and also could get the lowest rate of oil. A plant that has to depend on tank cars pays a considerably higher rate than those on tidewater.

Referring again to the question of deliveries, there are comparatively few oil companies that can furnish oil in quantities to consumers, owing to the expensive equipment necessary, such as tankers, barges, tank cars and storage tanks. On the other hand, there are hundreds of coal mines within five hundred miles of this locality that can furnish coal.

Before an intelligent decision can be made on the question of installing oil in place of coal, the operating record of the plant for at least six months (a year would be better) should be known. The record should show the amount of coal consumed, heat units in the coal, pounds of water evaporated, the maximum load and the average load, as the results obtained in one plant cannot be used for comparison with another, since each plant presents problems of its own.

In conclusion the pertinent facts to be determined as to the advisability of making a change to fuel oil are results with coal, cost of installation, saving in labor, coal used in banking, steam used for auxiliaries as pumps, atomizing, heating, etc., suitability of furnace, means for storage and receiving of oil, and the price of coal and oil delivered to the plant, and the heat units in each.

In a series of tests to determine the steaming value of Alaska lignite and spruce wood and the resistance of lignite to weathering when stored in piles in the open, by the Fairbanks, Alaska, Station of the Bureau of Mines, the results showed that a cord of wood is equivalent to more than a ton of lignite. In the weathering test, which lasted fourteen months, that portion of the pile immediately exposed to the atmosphere was entirely disintegrated, while that farthest from the surface was only partly disintegrated.

It is necessary to properly pitch and drain pressure pipe lines with pipes of ample size to take care of the rapid condensation of the steam when first turned into the line. If traps are used, they should be located below the line to drain it dry when steam is shut off. It is good practice to put a bleeder on the end of the line farthest from the steam inlet to help remove the rapid condensation which is bound to take place when the steam is first turned on.

EDITORIALS

Hazards of the Human Element

HOW many, who are trusted with the operation of boilers, are in the habit of taking them out of service, washing, changing tubes and in other ways making repairs, and then placing them back in service and without watching the water column, taking heed of the pressure gage or determining the working of the safety valve? It would seem that under all conditions a boiler that has been out of service for any purpose should be blown down or in other ways treated to indicate the accuracy or at least the functioning of the water column and the pressure gage. Every engineer has some idea as to how long it takes any of his boilers to come up to normal pressure, and failure of the gage to register with the lapse of time should immediately lead to an investigation.

The incidents connected with the recent East Chicago fatality emphasize that external conditions may affect the safety of a boiler as much as the internal conditions—a fact well known but, unfortunately, not always appreciated. It was an external, even an extraneous, form of trouble that was directly responsible for the explosion, yet apparently this very potential danger had not been taken into consideration. Pressure gages had frozen before and at the time of the accident the temperature was about zero. For twelve years these boilers apparently had been exposed to the danger that finally wrecked the boiler house, killed three men and maimed a dozen. The danger was, therefore, an old one—it was only the materialization of this danger that was new.

The only conclusion possible is that the human element remains the least known quantity, the greatest variable and the most unreliable factor. The materials failed, but it was only after men had failed. The human element failed when safety valves and pressure gages were permitted to become frozen. The human element failed again when it was possible to continue firing up a boiler from nine in the evening until four-thirty in the morning, or seven and a half hours, when the behavior of the pressure gage should have made it conspicuously obvious that conditions were not as they should have been.

Materials and structures can be protected by ample factors of safety, but we can only alleviate, never eliminate, the vicissitudes and sporadic failures of the human element by proper training, wise education and ceaseless reiteration of the gospel of safety. Nowhere is this more true than in the boiler room. Do we not know of men who do not think? Of boiler-room instruments that are not earning dividends, because the men do not know how to use them? Where waste and loss and failures are rife because nobody cares or nobody remembers?

The responsibility for this rests, not upon the boiler-room force alone, but with those higher up, with civic authorities that permit stored energy comparable with dynamite to be housed unsafely and operated unwisely; with the insurance interests that sometimes are able to see only one side of the question; with the plant owner who so often can see neither side. State laws, boiler codes and civic statutes have gone a long way toward improving matters. But it requires tragedies like that at East Chicago to remind us that we have a long way to go yet to protect boilers from without as well as from within and to make working conditions safer for those within as well as those without.

Measuring Reactive Factor

MUCH has been written concerning the explanation and measurement of power factor and the deleterious effect of low power factor on an electric system. It is general practice to supply machines such as synchronous motors with power-factor meters so that their power factor may be adjusted to the desired value. Any deviation from unity power factor results in increased heating, but this heating is not proportional to the power factor; it is a function of the reactive factor, which is a ratio indicating more directly the heating effect to be expected. The power factor is the cosine of the phase angle between the current and voltage where the reactive factor is the sine of this angle.

Although a power-factor meter indicates the deviation from unity, and indirectly the excess heating, yet the psychological effect of using a reactive-factor meter is considerable on the machine attendant. For instance, if a synchronous motor is operating at ninety-five per cent power factor, the attendant may think this is only five per cent away from unity, and consequently good enough. But ninety-five per cent power factor is equal to thirty-one and five-tenths per cent reactive factor and the attendant, seeing a reactive factor of thirty-one and five-tenths per cent on the meter, will usually make an effort to better conditions, although it is exactly the same in both cases, called by a different name.

A wattmeter, when connected into an alternating-current circuit, indicates watts, the product of the volts and energy component of the current. If the potential coil of the meter is excited with a voltage ninety degrees out of phase with the original volts, the meter will indicate the wattless or reactive-volt-ampere component. An important reason for measuring reactive volt-amperes is that the weighted power factor of an unbalanced three-phase circuit may be easily determined from the wattmeter and reactive volt-ampere meter readings. The cosine of the average angle of phase difference or the average power factor has very little meaning when a three-phase circuit becomes unbalanced unless, in the calculation, each phase is given a weight proportional to its load. The power factor determined by the foregoing method is this desired weighted value.

In an article entitled "Explanation and Measurement of Reactive Factor," in this issue the method of measuring reactive factor and also how to measure reactive volt-amperes are explained. Although this is not a new subject, it is only comparatively recently that the matter has been given much attention. As stated before, one of the reasons that attention is being focused on it at this time is that it provides a means of obtaining the weighted power factor on an unbalanced polyphase circuit. It is becoming general practice for power companies to penalize their customers for power factors below a given value; integrating the reactive kilovolt-amperes as well as the kilowatts is one of the means employed to arrive at the average power factor over a given period, consequently it is of interest to the operating engineer that he become familiar with what reactive factor and reactive volt-amperes are and how these elements affect the operation of the power systems in their charge.

Not So Badly Off As We Might Be

WE are prone to complain of the lack of standardization in our electrical power systems, especially when the subject of interconnecting these systems is discussed. However, sixty and twenty-five cycles have become standard to a very large degree throughout this country, with some few exceptions where fifty or forty-cycle systems and other frequencies are in use. In few cases will more than two frequencies be found in the same locality. The power systems of the large cities are in general either sixty or twenty-five cycles, or both systems are used.

According to a report issued recently by the "Bureau of Foreign and Domestic Commerce, Department of Commerce," on the "British Industrial Reconstruction and Commercial Policies," there are seventy operating companies supplying power to the City of London, owning seventy different power plants, operating fifty different types of systems, having ten different frequencies and twenty-four different voltages. The electrical industry throughout England is so unorganized, according to this report, that one cannot purchase a simple lamp bulb without specifying the type of socket in which it is to be used. When the conditions existing in this country are compared with those in England, have we not good reason to consider ourselves fortunate?

Increasing Boiler Output

MANY operating engineers will be confronted next fall with a demand for greater capacity from their hand-fired boilers. Owners are reluctant to purchase new equipment at the present high prices, and an effort will therefore have to be made to obtain greater output from the boilers already installed. This will require the burning of more coal, but the chimney draft may not be sufficient to do this on the present grates, and the chimney may not be large enough to carry away the increased products of combustion with the draft that it is able of itself to create.

The real way to deal permanently with the situation is to install forced-draft mechanical stokers. A less expensive expedient, and one that may be practicable in cases where the stoker would not, is the installation of some sort of device or apparatus for increasing the draft. This may vary from jets of steam blown into the chimney, furnace or ashpit, to large and elaborate fan systems for forced or induced draft. One such device, capable of general application to existing plants without material modification and at small expense, is the turbine-driven undergrate blower. Several reliable makes are on the market. With a closed ashpit they can produce drafts up to four inches of water. Such a draft would allow the use of small and low-grade fuels or the combustion of a much greater amount of coal per square foot of grate than would be possible with natural draft. If the blower is operated to produce a balanced draft over the fire, the capacity of the chimney will be increased, for the natural draft will be available for carrying off the greater volume of waste gases. These blowers are substantially made with turbines, the nozzles of which are adapted to the conditions and may be automatically controlled to maintain a constant boiler pressure. The steam is used to advantage, and the exhaust is available for heating feed water and similar uses. The equipment is not expensive and can be installed in the average plant without difficulty and with little interruption to regular operation. Its use is suggested as a means of getting increased work out of existing equipment with little additional investment and with a moderate cost of maintenance.

We sympathize with the writer of "Random Reflections" in *The Engineer* (London) in his lament from a purely sentimental point of view at the substitution of steam turbines for reciprocating engines. Sentiment that interferes with progress has no place in engineering, but no one can say that the man who feels the fascination in watching the regular motion of a reciprocating engine is any poorer engineer because he regrets its passing. The turbine is a wonderful machine, but who would spends hours "watching its still casing or listening to its monotonous hum"? The big reciprocating engine is giving way to the turbine, but it will always leave a tender spot in the memory of the men who knew it.

Carbon tetrachloride has been suggested as a possible substitute for oil in high-voltage switches for a number of years past. However, its use has not met with much favor. It is understood that one large operating company is investigating the possibility of this substitute for oil in switches at the present time, the results of which will be looked forward to with much interest. Wherever oil is used, there is always a fire hazard. On the other hand, although carbon tetrachloride is not inflammable, it evaporates rapidly, is an anesthetic and a solvent of rubber. Any of these properties under certain conditions may be as objectionable as the inflammability of oil.

Isn't it about time to hear again from the man with the little bottle of green—or is it yellow?—liquid, one teaspoonful of which will make a gallon of water do more work than a gallon of gasoline. The price of gasoline is still going up, and the man who controls any of these wonderful substitutes should come to the rescue of the long-suffering public.

With the coming of warm weather it will be worth while to occasionally check up the viscosity of the oil in plants subject to hot bearings in summer.

Between silt and floating débris coming down with the flood waters hydro-electric plant engineers now have their hands full.

Unless it soon stops the price of rice coal will be as high as that of the cereal.

What Does the Meter Read?

I have read the various solutions of A. A. Fredericks' meter problem, "What does the meter read?" in the Dec. 9, 1919, issue of *Power*, with much interest, and the majority of readings would indicate that the correct answer is 0,091,000. However, this is only half the story, for in the case of misplaced pointers on the dial the only proof of the case is to remove the dials from the meter and run the pointers back to zero and in this way find the actual misplacement. I herewith present a series of meter dials which show the various steps tried with a dial and which will be of interest to those who do not happen to have a dial at hand. The location of the pointers, Figs. 2 to 4, proves 0,091,000 to be the correct solution of the problem given to Mr. Fredericks. Figs. 5 to 7 show pointers proving my solution 0,121,000 to be correct, with the pointer also placed as given in Mr. Fredericks' dial. In Fig. 1 all pointers are on 0 and correctly placed. The dial, Fig. 3, gives 0,091,000 as the reading with hands correctly located. Fig. 6 shows 0,121,000 on this dial, with the pointers in the correct position. Now suppose we take Fig. 3 and move Nos. 2 and 3 pointers, from the left, ahead to the positions in Fig. 4; then we get the position of the pointers on Mr. Fredericks' dial. This shows two misplaced hands, and if we run the pointers back by actual count 0,091,000, all hands come to 0 except Nos. 2 and 3, as in Fig. 2, which are misplaced by the same amount as shown in Fig. 4. This condition therefore proves the solution to be 0,091,000.

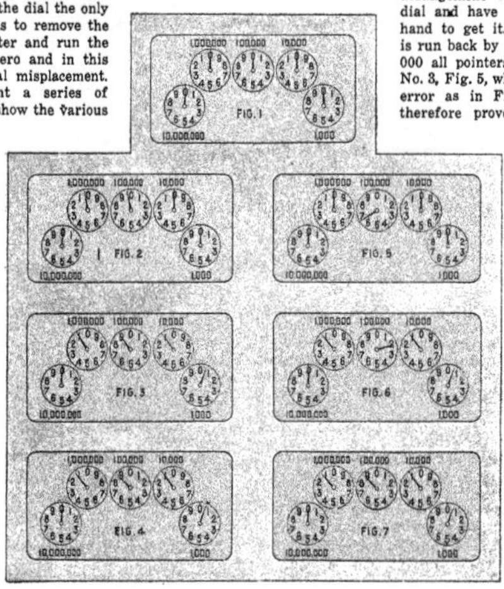

FIGS. 1 TO 7. ARRANGEMENT OF METER DIALS TO GIVE TWO DIFFERENT READINGS

Fig. 6 shows 0,121,000 as the reading with all pointers correct as in Fig. 1. Taking Fig. 6 and moving No. 3 pointer back to the position in Fig. 7, we again have the arrangement of Mr. Fredericks' dial and have misplaced only one hand to get it. If now this dial is run back by actual count, 0,121,000 all pointers come to 0 except No. 3, Fig. 5, which shows the same error as in Fig. 7. This series therefore proves my solution of 0,121,000 as the correct reading. These dials bring out interesting conditions of which I, and no doubt others also, have not before considered; namely, two perfect solutions of one problem. The two readings give a difference of 30 kw.-hr. Although this would amount to considerable where the regular monthly consumption is small and would be a fairly accurate guide as to just what the actual misplacement was, when the monthly consumption is large, 30 kw-hr. one way or the other is not sufficient difference to give a correct indication as to what the errors in the meter dials are

It would be interesting to hear from Mr. Fredericks as to what he found the correct reading to be.

Patchogue, N. Y. HARRY MULFORD.

Getting Pressure on a Bit Brace

A short time ago I had to drill a hole through a ¾-in. iron casting. I had but an ordinary bit brace and a bit-stock drill, and could not bring enough pressure to bear on the drill in the ordinary way, so I devised the following method:

In the first place I secured a piece of heavy iron wire and gave it a turn around the brace in such a

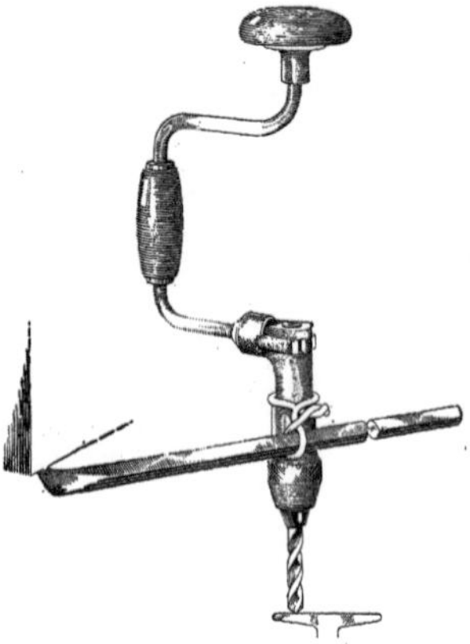

GETTING A PURCHASE ON A BIT BRACE

way that it would not interfere with its working. I then wound the wire around an iron bar and used it as a lever, as shown in the sketch.

With this simple contrivance the hole was easily and quickly drilled, my helper pulling down on the lever while I turned the drill. W. W. PARKER.

Firestone, Col.

Why Does the Water in a Boiler Lift When Safety Valve Blows?

A boiler under pressure and generating steam faster than the demand must find an outlet through the safety valve, the area of which is increased when once raised from its seat, and consequently, the valve opens wider than is necessary for that given pressure.

As the pressure in the boiler is increasing and pressing harder against the safety valve to open it, when blowing off, an instantaneous release of steam from the steam space of the boiler is the result. This sudden release of pressure creates a disturbance throughout the body of water in the boiler, the violence of which depends on existing conditions. If the valve opens wider than the pressure demands, the pressure falls back on the boiler. A drop in pressure causes all the water in the boiler to have a tendency to flash into steam, by reason of the so-called latent energy of heat. Particles of steam mingling with the water will rush toward this outlet of receding pressure, carrying water with them, the amount depending on the condition existing at the time. WILLIAM TOMPKINS.

Philadelphia, Pa.

My opinion is that as the pressure is increased, the boiling point of the water also increases, and if this pressure is suddenly lowered the boiling point is also lowered. This would cause a relatively large amount of water to flash into steam, which causes the water to increase the violent motion that takes place under ordinary conditions. This motion, however, becomes so violent that the disengagement space in the boiler is not large enough to permit of the separation of the water and steam, consequently the result is that the water level rises.

If the pressure were suddenly reduced low enough, it would, without doubt, cause an explosion.

Van Buren, Me. WILLIAM B. SMITH.

Tool for Pulling Worn Hydraulic Valve Seats

For pulling out worn hydraulic-valve seats, the tool shown herewith is a time and labor saver. To make, turn down a piece of round stock long enough to be used on the pump, as shown at *A*, and threaded for a ¾-in. nut, providing enough stock at the other end to turn a 30-deg. taper and of a diameter to make a sliding fit through the valve seat. Make a bushing *C* to fit

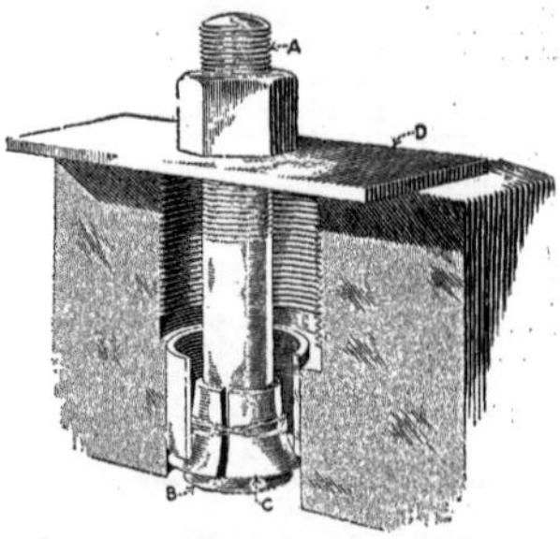

HOW THE VALVE-SEAT PULLER IS APPLIED

halfway down on the taper of the puller, the bottom end of the bushing having a slight lip. Then split the bushing with a hacksaw and the tool is made.

To use, push the puller through the valve seat and then put the bushing on the puller, holding it together with a rubber band. Next push the bushing down over the taper of the puller until a good hold is obtained on the valve seat; then put the iron clamp *D* over the puller and jack the seat out with the nut.

Trenton, N. J. JOHN T. LOCKETT.

Maximum Length of Longitudinal Joints for Boilers

Reverting to the discussion regarding the maximum length of longitudinal joints for steam boilers published in *Power* during the last six months, all of which is very interesting, it appears that the various writers have confined themselves principally to expressions of opinion from the constructional point of view only. In addition to this, there is much to be said from the operators', owners' and underwriters' standpoint.

The Massachusetts rules do not allow a boiler of the horizontal return-tubular, vertical-tubular or locomotive type to be constructed with longitudinal joints over 12 ft. in length. The A. S. M. E. Code prohibits a joint of the lap-seam type in a horizontal return-tubular boiler over 12 ft. in length, and with butt- and double-strap construction a joint of any length may be used in any type of boiler. Why does the A. S. M. E. Code limit the length of a lap joint in horizontal return-tubular boilers only?

From my viewpoint, fire-tube boilers, if constructed under either of the codes mentioned, which require a good quality of material and good workmanship, would be practically immune from explosion provided all the requirements are faithfully complied with and the boilers are operated under proper conditions.

It would appear that both of the codes referred to could be altered with safety to meet all conditions and I favor a rule leaving the length of the butt- and double-strapped joint optional on internally fired boilers, such as the vertical or locomotive types, or in the drums of water-tube boilers or other types of boilers where the shell plates are not in direct contact with the products of combustion.

I believe it would be unwise to change the Massachusetts rules so far as they apply to the maximum length of joints for horizontal return-tubular boilers, and if possible; in fact I would prefer that the length be limited to nine or ten feet for the following reasons:

Distress at the girth seams of horizontal-tubular boilers is first evident at the bottom of the course and directly over the fire, and leakage from this point is invariably an early indication of trouble, due to internal conditions.

A leaky girth seam has frequently been the cause of preventing extensive repairs, serious accident and possible explosion, and is a warning to the attendant in charge. It is usually a result of an accumulation of grease or scale on the fire sheets, which, but for this warning, may have resulted in further overheating of the material, causing serious bulges and other defects. For this reason I believe the girth seam has a strengthening effect on the plate and also that it is more effective when the plate is overheated, distorted and deteriorated.

In over thirty years' experience I have had occasions to examine and note the effect of overheating in a large number of horizontal-tubular boilers, all of which were constructed with two or more courses, and I am firmly of the opinion that in some instances serious explosions would have resulted if the boilers had been built with one continuous course and without the strengthening effect of the circumferential joint.

If the girth seam is eliminated and overheating occurs, the bulges would be of larger area, the depression greater and the possibility of serious rupture considerably increased. With a long sheet, bulging, due to overheating, is more likely to occur from lack of notice through leakage at the girth seam.

The rupture of tubes in water-tube boilers is due either to the quality of the material and gage, poor workmanship, or overheating, and without the strengthening effect of a roundabout joint they frequently open lengthwise a distance of six or seven feet, resulting often in loss of life or serious accident to persons as well as extensive damage to property. Similar accidents could happen in vessels of greater diameter than tubes, such as one-course boilers.

Frequently it becomes necessary to replace an entire course in a horizontal-tubular boiler when the plate has been badly distorted. This would be a very expensive operation if the boiler was built in one course, and might necessitate the construction of a new boiler.

I have had occasion to examine a large number of horizontal-tubular boilers, many of which have been operated for a number of years under rather severe conditions externally. Where the internal conditions and material are good and the girth seams properly constructed, trouble from leakage at the seams does not occur.

In my opinion neither the Massachusetts Rules nor the A. S. M. E. Code should permit the construction of horizontal return-tubular boilers with a longitudinal joint of greater length than twelve feet. To bring about uniformity, it would appear to be necessary to make changes in both the A. S. M. E. code and the Massachusetts rules. J. FRANCIS.

Boston, Mass.

Anent the discussion regarding the maximum length of longitudinal joints, the A. S. M. E. Code specifies that in a horizontal return-tubular boiler with lap joints no course shall be over 12 ft. long. Provisions have been made for the use of courses of any length in a horizontal return-tubular boiler with longitudinal joint of the butt- and double-strap construction.

The Code is then very explicit on the length of course for one type of boiler and especially for one form of construction. The reason for demarcation so sharp is not readily apparent, for it would seem that if several short courses would give greater strength than one long course in lap-seam construction, the same would hold true with butt- and double-strap construction.

If, too, it is desirable that the direction of the longitudinal seam should bear a certain relation to the direction in which the tension-test specimen was taken in a long course for a horizontal return-tubular boiler, why is it not advisable to require a similar tension test in a continuous course for a drum boiler?

Regardless of the type of boiler, a limitation on the length of course would be desirable, and 12 ft. is none too short. If a study is made of the lap-seam failures and these are not confined to any one type of boiler, it will be found that almost invariably the ruptures started at or near the middle of the joints and tore toward the girth seams. Again, in the event of overheating of the sheets of an externally fired boiler, the plates will often be badly distorted both sides of the girth seam. Rarely will the seam be affected other than to show leakage, and the leakage is often the first inkling the attendant has that something is wrong with the boiler and that it should be shut down.

In laying out a boiler made in courses, it is almost universal practice to break joints; that is, the longitudinal joints are not in a continuous line. Advantage of the strengthening effect of the girth seams is taken in this way. If courses of unlimited length were permitted and a boiler should be built in one course, the stiffening effect of the girth seams would be lost.

South Boston, Mass. J. C. CRAFFEY.

Possible Causes of Boiler Explosions

In recent issues of *Power* there have been published accounts of two boiler explosions. In both cases the boilers had been cut out and the night crew was getting up steam to put them in service again when the explosion followed. In the last few years I have been roaming from one plant to another, and in that time have had many types of boilers under observance. Perhaps the following may add some information to what has already been published.

In one instance the safety-valve stem was found to be so corroded that the explosion was laid to that. The fireman was on top of the boiler prior to the explosion. What was he doing up there? Trying to loosen up the safety valve so that it would blow in order to release some of the excessive pressure? That is a conjecture that I do not place any stock in. Any fireman, irrespective of his mental caliber, knows whether his safety valve blows off at the required pressure. He does not work long before he makes sure of that one thing. The fireman in that plant knew as well as the investigation brought out that the safety valve was not working. Knowing this condition, he would not go up on top of the boiler to tinker with a safety valve.

I was working in a plant containing three return fire-tubular boilers carrying 105 lb. pressure, and the safety opened at that pressure. A few nights after taking charge, while I was talking with the night fireman in the boiler room, he asked if I knew what to do if the safety valve did not blow at the required pressure.

After explaining to him the safe course to take in a case like that, he proceeded to put me "wise" as he termed it, to the following: On top of the boiler and direct to the boiler shell there was connected a short 1-in. nipple and a 1-in. valve, and in case the safety valve did not open up at the required pressure, he went on top of the boiler, opened the 1-in. valve and allowed the boiler to relieve itself.

I told the fireman that if he was caught trying that stunt he would be thrown down from the top of the boiler. He was completely taken aback and replied that the former night engineer had told him that was the correct thing to do, and that was why that connection was placed there. Can it be possible that this was the reason that the fireman went on top of the boiler in the case of explosion under discussion?

The other boiler explosion was laid to a frozen safety valve. The author of the article covered the field very thoroughly. Still there were a few things that one may comment on. It is my experience that when the connection to a pressure gage is frozen, the pressure does not have a tendency to remain stationary, but the opposite. During the last cold spell, when the fireman was cleaning fires I heard the safety valve "pop." Shortly after I became aware that the engine was slowing down. At the same time the fireman called for me. Going to the boiler room I saw that the gage showed 125 lb. pressure instead of between 90 and 100 lb. which should have been carried. Looking into the furnace, I found that the fire was nearly out, which, of course, accounted for the slowing down.

In order to check up my suspicions, I went into the engine room and looked at the recording pressure gage, which showed that the steam pressure was down. Then there was no more doubt as to the trouble. After the ⅜-in. line leading to the pressure gage was thawed, the pressure came down to normal.

The foregoing shows in a way that a recording pressure gage ought to be a part of the equipment of a power plant. If there had been one gage in the plant where this explosion occurred, the unfortunate incident might have been avoided, provided the instrument was used intelligently.

Another matter that those in charge of power plants might think over carefully: Nightwork and daywork are two different things. The man who works nights has a much harder time to get his rest than the one that works days. There are too many unknown conditions that enter in the problem when a man is trying to get his rest after working nights. There are times when it is impossible to get the right amount of rest, and it is not in human nature to be efficient during the long hours of the morning. Therefore why not have all the important work of a power plant done in the day under the watchful eye of the chief, leaving only the routine work for the night shift? Of course, this is not always possible, as there are times when everyone has to be on his toes, but this is only in exceptional cases.

It is a fact that work that could be done in the daytime is given over to a night shift in order that the men will have something to do besides their regular work. Take the matter of getting a boiler back in service. The night crew is told to get up steam on the boiler and have it ready to cut in service by the next morning. Nine times out ten that boiler is neglected until very near quitting time, and then it is forced in order to raise the steam pressure as quickly as possible. Is it any wonder that with the conditions right and forcing the boiler in that manner explosions occur? THOMAS GRAY.

Bayonne, N. J.

Licensing Steamfitters

After reading the correspondence on "Licensing Steamfitters," on page 394 of the March 9 issue, it appears to me that men who have come up in the ranks by the old method of learning through experience are a little too hard on the men who have technical training. All of us had to start getting experience at some time, and this includes both the man with and the man without training. I will venture to say that of two men, both being equally interested in their work in the beginning, the man with technical training will advance at a more rapid rate than the one without it.

I know a good many mechanics who cannot write, and these men are slow to follow the rapid strides in engineering, but as advancement is necessary to the welfare of all, let us bury the hatchet and give the man with training the same deal that the "old fellows" gave us who did not have training when we started.

St. Louis, Mo. F. KLUGMANN, JR.

INQUIRIES OF GENERAL INTEREST

Electrical Conductivity of Water—Is water a good or a poor conductor of electricity? A. J. C.

Chemically pure water is an insulator of electricity, but when water is contaminated with almost any other substance it becomes a conductor.

Separate Cold-Water Feed Connection for Boiler—What is the advantage of supplying a boiler with a separate cold-water feed line? R. St. T.

A separate cold-water feed line is desirable for filling the boiler after cleaning and for testing purposes, and also for regular feeding during such emergencies as when the feed-water heater lines are inoperative from corrosion or incrustation with scale, or during temporary disuse of the heater.

Angle of Advance of Eccentric—What is the angle of advance of the eccentric of an engine? B. R.

It is the angle that the eccentric is placed ahead of a line which is at right angle to the crank. Where the valve is moved by direct connections from the eccentric, the angle of advance is the number of degrees the eccentric is set more than 90 degrees ahead of the crank. Where a reversing rocker is used, the angle of advance becomes the number of degrees the eccentric is set less than 90 degrees behind the crank.

Blowdown of Safety Valve—What is meant by adjustment of the blowdown of a safety valve? W. G. M.

"Blowdown" is the less number of pounds pressure per square inch at which a safety valve closes than the pressure at which the valve is set to pop open. In most forms of spring pop safety valves, the amount of blowdown may be varied by adjusting the position of a ring, called the blowdown ring, which surrounds the valve seat and deflects the escaping steam in such a manner as to assist in holding the valve open until the boiler pressure has been reduced by the amount of blowdown.

Confining Operations to a Single Boiler—We must renew our boiler plant, consisting of two 75-hp. boilers. One of the old boilers is sufficient for our requirements during the summer months, but to supply our winter requirements for steam it is necessary to operate both boilers. Would there be better operating economy to replace the present 75-hp. boilers, or install one 150-hp. boiler for use the year around? T. B. C.

By skillful firing and reducing the grate area during the summer season, employment of a single 150-hp. boiler might effect a small saving of fuel, but with only one boiler there would be fewer opportunities for cleaning and making repairs without shuting down the plant.

Measuring Amount of Blowoff—What quantity of water will be discharged from a boiler under 160 lb. gage in blowing off to the atmosphere through a 2½-in. blowoff pipe that discharges to the atmosphere? H. L. S.

An accurate estimate of flow could not be based on simply the boiler pressure, as the rate of discharge would depend on the pressure obtained in the discharge end of the blowoff line after deducting the losses of pressure in the line. The best method of estimating the amount discharged would be to mark the water level of the boiler and, after blowing off the boiler with all other outlets closed, measure the amount of feed water required to restore the water level.

Equal Loads and Inequality of Cutoff—In a slide-valve engine, where the valve has equal laps at each end, will the leads and cutoffs be equal? R. C.

With equal laps there will be equal leads, but on account of the angularity of the connecting-rod, more angular rotation of the shaft and eccentric is accomplished when the piston traverses the crank-end half than when it traverses the head-end half of the stroke, and consequently the valve will be closed with cutoff accomplished earlier for the stroke from the crank end than from the head end. By moving the valve on its stem toward the head end, the cutoff of that end will be made earlier and cutoff in the crank end will be made later, but the lead of the head end will thereby be reduced and lead on the crank end increased.

Discharge of Attached Circulating Pump—What number of British gallons per minute would be handled by an attached double-acting circulating pump 15 in. in diameter by 21 in. stroke, with 2-in. piston rod, driven by a beam from the crosshead of an engine making 80 r.p.m., assuming that the pump barrel is filled 80 per cent at each stroke? C. E. G.

The 15-in. diameter piston would have an area of $15 \times 15 \times 0.7854 = 176.71$ sq.in., and with a 2-in. diameter piston rod the net area of the piston on the rod side would be $176.71 - (2 \times 2 \times 0.7854) = 173.57$ sq.in. For a 21-in. stroke, 80 per cent of the pump barrel filled at each stroke, and for 80 r.p.m. of the engine, the water displacement by the pump would amount to $(176.71 — 173.57)\ 21 \times 80$ per cent $\times 80 = 470{,}776.322$ cu.in. per min. As a British gallon is 277.274 cu.in. the quantity of water handled would be $470{,}776 \div 277.274 = 1{,}697.8$ British gallons per minute.

Inverted Converter—Can a rotary converter be used to tie an alternating-current system and a direct-current system together so that power may be transferred from one system to the other? C. M. B.

If the converter is shunt wound, it may be used for this purpose, but means must be provided to prevent the machine from racing in case it is supplying power to an alternating-current load of low lagging power factor. A wattless lagging current supplied from the collector rings of the converter will weaken the field pole and cause the machine to race unless it is operating in parallel with other alternating-current machines. Converters operated in this way usually have their field coils excited from a direct-current generator driven directly from the converter's armature shaft. Then, when the converter attempts to race, the voltage of the exciter is increased and that, in turn, increases the field current and neutralizes the demagnetizing effect of the wattless current supplied by the armature.

[Correspondents sending us inquiries should sign their communications with full names and post office addresses. This is necessary to guarantee the good faith of the communications.—Editor.]

Dr. John A. Brashear, Past President A.S.M.E., Dies

DR. JOHN A. BRASHEAR, past president of the American Society of Mechanical Engineers, died at his home in Pittsburgh on April 8, following a lingering illness. Dr. Brashear was an astronomer of world-wide reputation and his work in connection with the design and development of the spectroscope for astronomical uses, particularly with reference to accuracy of mechanical and optical features, will stand as his most notable achievement. His activities were not confined to the sciences alone. He was intensely interested in practical engineering affairs and for many years was an active member of the American Society of Mechanical Engineers, of which he was elected president in 1915. He was first elected to membership in the society in 1891, and in 1908 he was made an honorary member. From 1899 to 1902 he was one of the board of managers and in 1911 was one of the representatives on the John Fritz Medal Board, which post he held for four years. Dr. Brashear was born in Brownsville, Pa., in 1840, and received his education in the public schools of that town. From the public schools his desire for mechanical work led him to the rolling mills in Pittsburgh, where he worked as a machinist from the year 1860 to 1880. While employed in the mills he devoted his leisure hours to the study of astronomy, of which he was passionately fond. As a youngster it had been his ambition to study the mysteries of the heavens, and as he grew older his hunger for a knowledge of astronomy became stronger. He was too poor at this time to purchase a telescope, so he and his wife proceeded to make one. After a great deal of planning and work, a shop was finally set up in their home, containing a small steam engine, lathe and other apparatus necessary for the construction of the telescope. Here, after a great deal of laborious work, the tubes were made and the lenses for a five-inch telescope were ground. The instrument was three years in the making and was far from perfect, but it served its purpose and aroused the desire for a larger and better one. Work was immediately started on a twelve-inch telescope. Luck was against them, however, for after almost two years had been spent on grinding the lens, it broke. But, nothing daunted, the indomitable man and his wife began work on a new lens, which was successfully completed. The late Dr. E. C. Langley, former head of the Smithsonian Institution, then head of the Allegheny Observatory, noticed the young chap's enthusiasm for astronomical knowledge and placed him in a shop in Pittsburgh, where he was assisted by his wife. Finally, with the assistance of William Thaw, one of the observatory patrons, Dr. Brashear set up a small plant in Allegheny for the manufacture of astronomical instruments. Many of the spectroscopes used in the principal observatories of the world were turned out at this shop. In 1888 he completed the spectroscope for the thirty-six-inch telescope of the Lick Observatory, furnishing the optical and mechanical parts. When the international body on astronomical research was entrusted with the determination of a scientific standard of measurement fixed upon the length of a light wave as the most efficient means of securing uniformity, Dr. Brashear was selected from all the instrument makers in the world to do the important work of manufacturing the apparatus. The instruments were of a delicate and accurate nature, and it was believed that Dr. Brashear, with his wide experience, was the man best suited for the work. Following the opening of the shop at Allegheny, his son-in-law, James B. McDowell, joined him in the work and Dr. Brashear attributed a large share of the success of the enterprise to Mr. McDowell. In 1898 he received an appointment in the University of Western Pennsylvania, of which the Allegheny Observatory was a department. For twenty years he was trustee of the Carnegie Institute, for fifteen years of the Carnegie Institute of Technology and for twenty years of the University of Pittsburgh, which he also served as Chancellor. Dr. Brashear's work in connection with the advancement of teachers and education in the public school has proved of inestimable value. An endowment fund of $250,000 was placed in his care by a friend several years ago to be used for the advancement of public-school teachers and teaching, as a result of which hundreds of teachers were sent to different parts of the country for rest and study, bringing back with them new ideas and greater enthusiasm for their work. He was honored by Washington and Jefferson College with the degree of LL.D. and by Princeton University and the Western University of Pennsylvania with the degree of Sc.D. He was a member of the American Association for the Advancement of Science and the Royal Astronomical Society of Great Britain; past president of the Engineers Society of Western Pennsylvania and the Pittsburgh Academy of Arts and Sciences; a member of the British Astronomical Association, the Societe Astronomique de France, the Societe de Belgique, the American Philosophical Society, the Astrophysical Society of America, the National Geographic Society, Washington Academy of Sciences and an honorary member of the Royal Astronomical Society of Canada.

DR. JOHN A. BRASHEAR

Refrigerating-Machinery Lubrication*

AT THIS season manufacturers of ice and operators of cold-storage plants are overhauling their machines and installing new equipment in anticipation of the demands of the hot weather which is approaching. Because of the plentiful supply of natural ice recently harvested in many parts of the country, makers of the artificial product will find competition very keen, and it is urgently necessary that they operate the plants at the highest possible state of efficiency. The surest way to obtain maximum production with minimum operating expense is to eliminate friction and leakage in the compressor cylinders and to provide against the fouling of the refrigerating coils with congealed oil.

The lubrication of refrigerating machinery is peculiar in that the choice of the oils to be used is influenced by the action of those oils upon parts of the apparatus not requiring lubrication. The lubrication of the apparatus employed in air refrigerating systems belongs under the head of air-compressor lubrication and will not be discussed here. The absorption process is chemical rather than mechanical, and the only parts of such apparatus requiring lubrication are the pumps, which, however, offer no unusual features. In the compression process the refrigerating agent is recovered by means of mechanical compression. A number of substances have been used as the heat-carrying medium, but anhydrous ammonia has been most widely adopted.

Ammonia is compressed in one of two ways—by the wet method or by the dry method. In the former a small amount of liquid ammonia is allowed to enter the compressor and, owing to its evaporation, serves to keep the temperature of compression down. In selecting an oil for wet-compression machines, it is necessary to choose one which will remain a fluid at the minimum temperature in the expansion side of the system, and of sufficiently high flash point to prevent distillation at the discharge temperature. However, as the discharge temperature is comparatively low, the cold or pour test and viscosity are the only important factors.

In the dry method vapor alone is drawn into the machine; that is, the ammonia gas entering the compressor will be slightly superheated. The majority of machines operating in this class of work are vertical, single-acting compressors with water jackets.

An oil for this class of work should have a sufficiently low pour test to remain a fluid at the minimum temperature in the evaporating side of the system, a flash point high enough to stand the discharge temperatures without distillation and enough viscosity to lubricate the compressor parts and seal the piston rings and compressor valves. As cylinder temperatures are considerably higher in dry compression than when the wet method is employed, it is desirable that an oil having a higher viscosity be used, and in addition the oil should have a flash point sufficiently high to prevent distillation or cracking at the discharge temperature. A good ammonia oil should have a flash point at least 25 deg. F. higher than the average discharge temperature.

Four Classes of Compressors

We may divide the different types of ammonia compressors into four general classes: Vertical single-acting compressors; horizontal double-acting compressors; inclosed high-speed compressors; rotary compressors.

Vertical single-acting compressors are built in various sizes from one-eighth ton capacity upward. The smaller types of these are usually lubricated by the splash system in a manner similar to that employed in automobile engines. When the back pressure drops below atmospheric in this type of machine, there is always danger of oil being drawn into the cylinder, but this will never occur if the attendant watches his gages closely and maintains the back pressure at the proper value. The ammonia piston used in a great many of these machines is of the double-trunk type, with the charge entering between the upper and lower sections. The upper section is fitted with two or three rings and also contains a valve. The lower section carries one or two rings which form a seal between the crank case and the suction of the machine. These serve to prevent excess oil from being drawn into the cylinder when the back pressure drops. A light oil gives very satisfactory results on this type of compressor.

In machines using the splash system it is important that the oil level in the crank case should not be kept too high. A high oil level causes excessive churning and is likely to lead to the introduction of oil into the refrigerating system. In the larger sizes of vertical machines the lubrication is accomplished through the use of a force feed.

Some double-acting compressors are lubricated by introducing the oil at the piston-rod stuffing-box, which is fitted with an oil lantern. The small amount of oil working into the cylinder past the packing is found to afford ample lubrication. When this method is employed, it is found that the oil protects the packing from the very injurious action of the hot ammonia gas.

Avoid Copper or Its Alloys

Other compressors are lubricated through the use of positive force-feed lubricators, but where these are used precautions should be taken to see that they are constructed entirely of iron or steel. No copper or copper alloys should be used about an ammonia compressor. These lubricators assure efficient lubrication, in that they put the right amount of oil in the right place at the right time, and do this at the minimum cost. Oil is sometimes introduced into the cylinder through the intake line, but this method is becoming obsolete.

The lubrication of the bearings of a compressor presents no unusual difficulties. When the splash method of lubrication is employed and one oil is used for both the cylinder and the bearings, an ammonia oil of some type must be used, but when the oil is not used in the cylinder any high-grade non-emulsifying oil may be employed. The same grade of oil may also be used when a circulating system is employed. When the oil is fed from cups or a force-feed lubricator, any high-grade engine oil will prove satisfactory, the viscosity, of course, being suitable for the weight and speed of the machine in question.

The selection and proper application of a suitable lubricating oil for the ammonia compressor is one of the duties of the lubricating engineer and a question of much importance to the engineer operating a refrigerating plant. In selecting an oil the character of friction, pressure, temperature and velocity and position of the moving parts, as well as the chemical nature and the composition of the lubricating oil, must all be considered. Some attention must also be devoted to the chemical properties and temperature of the materials with which the oil may come in contact.

It should be noted that only pure mineral oils of at least zero pour test should be used in the cylinders of an ammonia compressor. These oils can be obtained with a sufficient range of physical properties to lubricate satisfactorily and efficiently all such compressors working under reasonable conditions. Mineral oils are the most neutral, and a pure straight mineral oil well refined has been found to be the most suitable for this class of work. Compounded oils of any kind should be avoided because of the tendency of animal and vegetable oils to solidify at comparatively high temperatures. Acids which develop in compounded oils also attack the walls of the compressor cylinder and have a detrimental effect upon the ammonia gas.

In so far as the lubrication of the compressors themselves is concerned, viscosity is of primary importance. The condition of the cylinder, piston rings and valves determines the viscosity of the oil that should be used, as it is essential that the oil should form a seal and maintain perfect compression. For cylinders that are in good condition a light-bodied oil will prove satisfactory. Where the cylinders are somewhat worn or scored, a medium-bodied oil will usually maintain the required seal. In cases where the

*Abstract of an article published in *Lubrication* for March, 1920.

rings and cylinder are badly worn or scored, it may be necessary to use a heavy-bodied oil to insure efficient operation. Generally speaking, horizontal machines, owing to the position and weight of the piston, require an oil of heavier viscosity than the vertical type, and as most of the horizontal compressors are double-acting, the lubrication of the piston rod should be taken into consideration.

As was mentioned before, it is doubtful if the temperature of the oil in the compressor is ever raised much above 250 deg. F. As all high-quality oils have flash points well above 300 deg. F., it would seem that the question of flash point might safely be disregarded in selecting oils for compressor lubrication.

While oils having the characteristics just discussed will satisfactorily lubricate the compressor cylinder, requirements of other parts of the plant frequently make necessary the use of an oil having other physical characteristics. There will always be some vaporization of oil due to the heat of compression, no matter how high the flash point may be, and this vapor will be carried out of the cylinder with the compressed gas. As refrigerating systems are closed cycles, this oil is bound to condense and accumulate upon the colder parts of the system, which are the condenser and expansion coils.

Oil Accumulations on Coils Detrimental

Probably no other item in the operation of a refrigeration plant so vitally affects economy as the accumulation of oil upon the coils, and yet it is amazing how few operating engineers take this problem seriously. A prominent authority has asserted that a film of oil one one-hundredth of an inch thick has the same heat-resisting properties as a layer of boiler scale one-tenth of an inch thick and a steel plate ten inches thick. All oils, when sufficiently cooled, become thick and at certain temperatures cease to flow. Unless the oil reaching the condenser and refrigerating coil has an extremely low pour test, it is bound to accumulate on the walls of the coils and prevent the effective transfer of heat.

To prevent this condition, an oil used in a refrigeration compressor should have at least a zero pour test, so that any oil getting into the coils may be carried along with the ammonia without congealing upon the walls and seriously affecting the efficiency and capacity of the plant. Natural low-pour-test oils contain no paraffin wax, and consequently no trouble from this source will be experienced when they are used. Before the oils made from wax-bearing crudes are used on an ammonia compressor, it should be determined that the wax has been removed to such an extent that no evidence of its presence is seen after the oil has been exposed to the lowest temperature in the system. Any operator can make a test to determine the suitability of an oil for his particular work by immersing it in the brine for some time. If it flows readily upon removal, it is quite suitable; if not, it should be rejected.

Proper Location of Extractor

An essential appliance in any refrigeration plant is the oil extractor, of which there may be several. The principal extractor should be placed between the compressor and the condenser, although advantages will be secured by inserting smaller extractors at the lowest points of the condensing and refrigerating coils. Efficiency will always be promoted by the use of oil extractors, no matter how high the quality of the lubricant may be. Complaints are frequently made that the extractor does not function properly, and very often extractor troubles are charged to the oil being used. In nine cases out of ten, when the oil is not removed by the extractor, the difficulty is due to the extractor being placed too close to the compressor, at which point the gases are too hot to permit the oil to condense. When the extractor is located close to the condenser, the oil will reach it in a condensed form and as a result, can be more easily removed. Oil extractors should be of ample size so that the velocity of the gas through them shall not be too high to permit of complete separation.

The location of the oil extractor between the compressor and condenser and the efficiency of its operation have an important bearing upon the character of the oil selected. If conditions make it necessary to place the extractor near the compressor, an oil giving little evaporation—that is, a high-viscosity oil—should be used. On the other hand, owing to the opportunity which the vapor has to condense, a lower-viscosity oil may be used successfully when the extractor is placed some distance from the compressor. Provision should always be made to draw oil from the bottom of the liquid receiver in which it frequently accumulates in considerable quantities. As the oil has a higher specific gravity than liquid ammonia, it naturally settles to the bottom of the receiver from whence it may be drawn and passed through purifying devices.

Oil taken from the extractors and receiver may be recovered by installing an ammonia-distilling apparatus, which has the dual advantage of purifying the charge of ammonia and of recovering the lubricating oil without interfering with the continuous operation of the plant. The operation of these ammonia stills is so simple and well known to engineers in refrigerating work that it will not be necessary to go into details. Care should be taken in handling the still to prevent too rapid evaporation, as this causes boiling over and the loss of a large proportion of oil back to the suction line. After all the liquid ammonia has been evaporated and returned to the suction line of the compressors, the oil can be drawn from the bottom of the still, using suction pressure for the purpose. The oil removed will contain some water vapor and should be put aside until this has freed itself from the oil. Maintaining the oil at about 150 deg. F. will assist this operation materially. The oil is then ready to be filtered.

It is advisable to use a separate filter for this work, the size and type of filter to be used depending upon the size of the plant and the amount of oil to be handled. The oil recovered from the still, after careful filtering, will again be suitable for the lubrication of the ammonia compressors provided the original oil used on the compressors is of good quality and properly selected.

Filtering Oil Through Water

Some engineers are opposed to filtering compressor oils through water because of the danger of moisture retained by the oil being introduced into the cylinder upon the re-use of the oil. On the other hand, a great many engineers have never experienced any difficulty through water getting into the system with oil. In any case this difficulty may easily be circumvented by allowing the oil to stand a day or so before again placing it in the lubricating system. In a large number of plants the oil recovered is used for external lubrication only and new oil used for the lubrication of the ammonia compressors. This is safe practice, as there is then positive assurance that the oil used on the ammonia compressors is perfectly clean and up to specifications.

A record should always be kept of the quantity of oil taken from the extractor, and this should be compared at regular intervals with the amount of oil fed to the compressor. Any great difference should be investigated at once. The film of lubricating oil can always be removed from the condenser coils by discontinuing the cooling water from one strand at a time. When the pipes become hot, the oil will thin down and run off into the receiver. When the refrigerating coils become fouled with congealed oil, the accumulation may be removed by reversing the system and forcing hot gas through the coils. The same result may also be obtained by blowing out the coils with steam, followed by air to remove the moisture.

Temperatures play an important part in the efficient operation of any refrigerating plant, and to get the highest efficiency it is necessary to know what the temperatures are in the different parts of the system. This is also the most important consideration in selecting a lubricating oil for this class of work. Where thermometers are installed in both the suction and discharge sides of ammonia compressors, the selection of a lubricating oil becomes a very simple matter.

The ideal refrigeration oil should remain a fluid at the minimum temperature in the evaporating side of the system and not undergo any chemical change at the maximum temperature of the conpressor discharge. It should also have sufficient viscosity at the working temperature of the compressor parts to lubricate properly and to form a seal for the piston rings and valves.

The importance of a suitable and efficient lubricant for the ammonia compressor in many instances is overlooked and an oil "just as good," selling for a few cents per gallon less, is installed in the plant. When we consider that the average 100-ton modern ice plant will not use more than fifty or sixty gallons of ammonia oil per season, the difference in price between the best grade procurable for this purpose and the cheapest oil sold, would not pay one per cent on the difference in cost of repairs, to say nothing of the possibility of a loss in capacity due to the excessive wear of the compressor parts. Better lubrication will have a very large influence on operating costs.

Revision of National Electrical Code

On March 23, in New York City, the Electrical Committee of the National Fire Protective Association held a public hearing on the recommended changes in the National Electrical Code as published in a bulletin issued some time ago. After the public hearing the electrical committee held a closed meeting at which the recommended changes as published in the bulletin were adopted with some few exceptions.

The limit of low potential systems was fixed at 600 volts instead of 750 volts as recommended in various sections of the bulletin discussed at the meeting. The proposal to require polarity identification in all wires installed on and after Jan. 1, 1921, was modified to require that beginning July 1, 1921, twin wires for conduit work and twisted-pair wires for armored cable, in sizes Nos. 12 and 14 B. & S. gage, provide a means for continuous identification as proposed in the bulletin for Rule 26a. Rule 77 on fixtures is to be revised to require identification marking by the fixture manufacturers of the wire connected to the screw shells of the socket.

A committee is to be appointed to consider requirements for the installation of concealed extensions from existing wiring in buildings of fire-resistive construction.

The report on rubber-covered wire was adopted essentially as printed. In this report a greater thickness of rubber insulation is required on 1,500-, 2,500- and 3,500-volt cables. A standard for the stranding of wires in building up stranded conductors for wires and cables is given. Allowable carrying capacities of the conductor sizes secured under this program are to appear as a new section of Rule 18. The general use of varnished cloth cables was approved for sizes of No. 6 B. & S. gage and larger.

The generator and substation report was adopted as printed. In addition the committee was asked to consider amendments to Rule 12 to exclude from its application, equipment of public utilities. The frames of generators are now required to be grounded when operating at a potential in excess of 750 volts, except where this is impractical or where operating conditions require their insulation. The grounding of generator and motor frames operating on 150 volts or less and accessible to only qualified operators is not to be required. Rule 7 has been changed to read: "Each distribution system originating in a station under attendance must be provided with a reliable ground detector unless permanently grounded as provided in Rule 15-A."

The report on the overload protection of motors was referred back with power to the Committee on Overload Protection for conference with industries and those interested. One of the recommendations contained in the report was, "Except on main switchboard or when otherwise subject to competent supervision, an approved fuse must be provided in each conductor of motor circuits whether an automatic circuit-breaker is installed or not. Circuit-breakers may be approved for circuits having a maximum capacity greater than that for which approved inclosed fuses are rated."

Recommendations for size and protection of conductors for motor circuits were also referred back to the committee with power and instructions to continue the last paragraph of present Rule 8c. Recommendations under the heading of motor switches were approved, but the report was referred back to the committee with power for rewording after considering the problem of large switches. The recommendation to require the externally operated inclosed type of switch for motor circuits was not adopted.

The report recommending not to require special permission for 1,320 watts on circuits with medium base sockets was not adopted, nor was the recommendation to change 1,320 to 1,500 watts. Provisions were made for the use of circuits up to 4,000 watts when Mogul sockets are used, but special permission will be required.

Spring Meeting of the A. S. M. E.

The American Society of Mechanical Engineers will hold its spring meeting in St. Louis, May 24 to 27, with headquarters at the Hotel Statler. A group of papers of general interest to the profession has been arranged by the committee on meetings and program.

Many features for the entertainment of members are being planned. This city is unsurpassed for its places of amusement. Special entertainment features have been arranged for the ladies by the Ladies' Committee, including an automobile trip to Forest Park and a visit to the Art Museum in the park. A trip through the residential and shopping districts will also be made.

The meeting will open at the Hotel Statler on the afternoon of Monday, May 24. Registration will be made at this time and council and committee meetings will be held. In the evening at a reception the governor of Missouri and the mayor of St. Louis will deliver addresses of welcome.

At the Tuesday morning session the business meeting will be followed by the rendition of papers of local interest on Mississippi River transportation and on housing problems, and in the afternoon there will be excursions to the plants of a number of industrial projects. The evening will be made jolly by a banquet and addresses. Lantern slides and moving pictures of the German fortifications on the coast of Belgium and of the transcontinental trip made in this country by Army motor lorries will be shown on this occasion or at some other time to be determined.

Wednesday will be the busiest day of the entire meeting. The morning will be taken up with simultaneous sessions on aëronautics and foundry practice, with papers on the following subjects: Aëronautic Session—"Air Flow Past an Airplane Wing," "Use of Turbo-Compressors for Increasing Power of Airplanes at High Altitudes," "Airplane Instruments," "Experiments on New Type of Airplane." Foundry Session—"Malleable Castings," "Die Castings," "Gray-Iron Castings," "Aluminum Castings," "Steel Castings," "Brass and Bronze Castings."

The afternoon session will be devoted to discussion on "Appraisal and Valuation" and excursions to various industrial plants, including the Busch-Sulzer Diesel Engine Co., the Mississippi Valley Iron Co. and the new Bevo plant. A special entertainment at the municipal open-air theater at Forest Park has been arranged for the evening.

Thursday, the closing day of the meeting, will begin with the presentation of papers on "Weir for Gaging the Flow of Water in Open Channels," "Simplification of Venturi-Meter Calculations," "Flow of Air," "Dissipation of Heat by Various Surfaces," "Pulverized Coal at the Cerro de Pasco Copper Corporation," "Method of Separation of Dissolved Gases From Water and Some of Its Uses," "Flow of Air Through Small Brass Tubes."

A boat trip on the river is scheduled for the afternoon, with a landing at the Chain of Rocks, where the filtering basin and pumping plant of the city will be inspected.

Inquiries have led us to infer that the title of the leading article in the April 6 issue: "Gas-Steam Unit Shows 72.4 Per Cent Thermal Efficiency at Ford Works," may be construed to apply to the engine cylinders only. However, as is explained in the article and synopsis, the 72.4 per cent includes the heat recovered from the jackets and the exhaust and represents that made available for power and heating purposes.

Personals

L. E. Gebhard, formerly connected with the Philadelphia sales office of the Yarnall-Waring Co., has been placed in charge of the Pittsburgh office.

Russel P. Askue, formerly connected with the publicity department of the National Lamp Works, has resigned to go into agency work, with headquarters in Cleveland, Ohio.

C. Thayer, formerly of Raleigh, N. C., superintendent of the Carolina Light and Power Co. in Oxford, has been transferred to Henderson as superintendent of the company's interests there.

Albert R. Dismukes, formerly with the Emerson Company, New York, has become affiliated with Joseph E. Lower, Inc., constructing engineers, Dayton, Ohio, in the capacity of efficiency engineer.

Colonel James B. Dillard, formerly chief of the Engineering Division of the Ordnance Department, has resigned his commission to become a member of the executive staff of the Cleveland Twist Drill Co., Cleveland, Ohio.

G. L. Behamon has resigned as chief engineer and assistant general manager of the Youngstown Steel Car Co., Youngstown, Ohio, and is now affiliated with the Thomas Spacing Machine Co., Pittsburgh, Pa., as manager of its Eastern office in Philadelphia, Pa.

Edward C. Gaines, formerly engineer of the crane and conveying machinery department Dominion Bridge Co., Montreal, Canada, has joined the engineering forces of the Mead-Morrison Manufacturing Co., Chicago, Ill.

Samuel P. Hall, architect, and Harold M. Bush, mechanical engineer, announce they have formed a partnership under the name of Hall and Bush. The new firm will practice architecture and engineering as specialists and consultants in industrial-plant operation and construction.

Charles T. Main, of Boston, Mass., announces that his organization has been enlarged to undertake engineering work for all kinds of industries. Colonel F. M. Gunby who resigned to head the engineering branch of the constructing division of the United States Army, has returned and has been appointed engineering manager.

Professor Albert W. Smith, Dean of Sibley College of Mechanical Engineering, was appointed temporary successor to President Jacob Gould Schurman, who has retired, by the trustees of Cornell University on April 3. Professor Smith has been connected with Cornell University since 1887, at which time he was appointed assistant professor of mechanical engineering. In 1904 he was made a Director in Sibley College and in 1907 professor of power engineering. He is an active member of the American Society of Mechanical Engineers.

Society Affairs

The National Electric Light Association will be strongly represented at the annual convention of the United States Chamber of Commerce at Atlantic City, April 27 to 29.

The Cleveland Section of the American Society of Mechanical Engineers will meet at the Hotel Statler, Cleveland, in the Cleveland Engineering Society Rooms on April 27.

The National Association of Stationary Engineers will hold its twenty-fifth annual New York State Convention at the Temperance House, Niagara Falls, N. Y., June 10-12, 1920.

The Mid-Continent Section of the A. S. M. E. will meet April 29 at the Chamber of Commerce, Tulsa, Okla. A joint meeting with the Oklahoma Section of the American Chemical Society is being planned.

The Illuminating Engineering Society, Engineers' Club of Philadelphia, will hold a meeting on April 20, at Witherspoon Hall. The subject of the meeting will be "Labrador to Alaska" by L. O. Arm-trong.

The St. Louis Section, American Society of Mechanical Engineers, will meet at the Hotel Statler in Cleveland on April 30. H. P. Gaum, engineer of the Illinois Stoker Company, will present a paper on "Chain-Grate Stokers."

The Society of Municipal Engineers, affiliated with the Engineers' Club of Philadelphia will meet at Witherspoon Hall, April 28. An illustrated lecture on "Health in Relation to Water Supply" will be given by John A. Vogleson.

The Cleveland Engineering Society, Cleveland, Ohio, will meet on Friday evening, April 30. Dr. I. N. Hollis, president Worcester Polytechnic Institute, will be the speaker. Subject: "Does Applied Science Add to the Satisfaction of Life?"

The Philadelphia Section of the Association of Iron and Steel Electrical Engineers will meet at its headquarters, Philadelphia on May 1. The meeting will be open for general discussion on the application of electricity to the iron and steel industry.

The American Association of Engineers has been appointed the official representative of railroad engineers under the Transportation Act, 1920. The association has a membership of 4,250 railroad engineers and will take up problems of railroad management.

The National Safety Council, Engineering Section will hold a meeting on April 27, in the Engineering Societies Building, New York City. Papers on "The Relation of Safety to Engineering Efficiency," "Safety Education in Engineering Colleges" and "Safety Standards" will be read.

The Birmingham Section of the A. S. M. E. will hold a joint meeting with the Atlanta Section on April 23, in Birmingham, Ala., at the Business Mens' Club at 6:30 p.m. John R. Allen, vice president of the A. S. M. E, will speak on "Theory and Practice."

The Southwestern Electrical and Gas Association will hold its annual convention in Galveston, Texas, May 15. The action of the electrical section of the association on the question of joining the Southwestern Geographic Division of the National Electric Light Association will come up at that time.

The National Electric Light Association national headquarters will be represented by A. Hardgrave, chairman of the Power and Light section of the Southwestern Electrical and Gas Association at the meeting of the Arkansas Public Utility Association, April 26. Mr. Hardgrave will outline the reorganization plan of the N. E. L. A.

The American Society of Safety Engineers will hold its next regular meeting at headquarters, Engineering Societies Building, New York City, April 23, at 8 p.m. Subjects: "Ladders and Scaffolds in General Building Construction"; "Safety As Applied to Ship Staging"; "The Requirements of the Industrial Code Relating to Ladders and Scaffolds."

The Pittsburgh Chapter of the American Association of Engineers recently expelled one of its members because of his connections with the draftsmen's union. The attitude of the Pittsburgh Chapter was that its members could not uphold the ideas of trade unionism and at the same time live up to the ideals of the association, which believes in co-operation between labor and employer.

The Society of Automotive Engineers, Engineers' Club, Philadelphia, will hold a special meeting at Witherspoon Hall on April 22. The society will hold a smoker at the clubhouse, Witherspoon Hall, on April 26. An illustrated lecture entitled "Eliminating the Slums," will be given by Emile G. Perrot. On April 27 a luncheon will be held, and an address by W. D. Lockwood on "The New Army."

The Engineering Section, National Safety Council, will hold its first spring meeting in the Engineering Societies Building, New York City on April 27. The entire program will be taken up with problems of engineering safety and the engineer's status in the industrial world. L. A. De Blois, manager of the safety section, E. I. du Pont de Nemours & Co., Wilmington, Del., will read a paper on "The Relation of Safety to Engineering Efficiency." A paper on "Safety Instruction in Engineering Colleges" will be presented by Bruce W. Benedict, manager of shop laboratories, University of Illinois. Other subjects will be: "How to Interest Student Engineers in Safety," Prof. G. S. Blessing, Swarthmore College, Swarthmore, Pa.; "The Movement for Uniform Safety Standards and the Engineering Sections Part," David S. Beyer, engineer, Boston, Mass.

Miscellaneous News

The Narragansett Electric Light Company, Providence, R. I., is considering the erection of an addition to the South Street Power house and the installation of eight new boilers. Total cost $750,000.

The Scottsville Power Corp., Scottsville, Pa. is completing arrangements for the development of the upper James River. Several thousand horsepower will be generated and distributed to towns in Pennsylvania and Virginia.

The Central Georgia Power Co., of Macon, has applied to the state railroad commission for an approximate increase of 28 per cent in power rates. The company supplies power for the operation of industries through middle Georgia.

The Columbus, Delaware and Marion Electric Company is planning the construction of a large power plant at Newmans, Ohio. The plant is to cost approximately $1,000,000. The contemplated improvements are to be made in two years.

The Municipal Power System of Los Angeles, Cal., will be increased to produce 300,000,000 kw.-hr. of electric energy per annum according to a recent announcement of the Public Service Commission. The present output of the system is 170,000,000 kw.-hr. and will be increased by the erection of another plant.

The Carolina-Tennessee Power Company is planning hydro-electric development on the Hiawassee River, power thus gained to be distributed for lighting and power purposes in North Carolina and Tennessee. It is estimated that 60,000 horsepower will be obtainable. The project will cost in the neighborhood of $10,000,000. Two 150-ft. dams are to be built. One is near the Tennessee border while the other is located thirteen miles up the river in North Carolina.

The San Joaquin Light and Power Corp., of Fresno, California has completed plans for the construction of a natural gas burning steam-generating electric plant. The plant will be built in units at a total cost of $6,000,000. The first unit will have a capacity of 17,500 horsepower, part of which will be transmitted to nearby oil companies. This will be the second steam plant erected by the San Joaquin Light and Power Corp. in Kern County during the past year.

Business Items

The Fisher Governor Company, Inc., Marshalltown, Iowa, manufacturer of power-plant specialties, announces the removal of its New York office to 242 Lafayette Street.

The Westinghouse Electric and Manufacturing Co. has been awarded the contract for the installation of an automatic electric substation on the line of the Seattle-Everett Interurban Co., Seattle, Wash. Total cost, $50,000.

The Pittsburgh Engineering Co. will move its general offices to Jeanette, Pa. The company will occupy the plant of the Greensburg Foundry and Machine Co. at Jeannette. Branch offices will be maintained as in the past.

The Cumberland Railway and Power Co., operating electric and street-car systems in North Carolina and West Virginia, will move its general offices from Norfolk, Va., to Raleigh as soon as the new Dortch Building, now being erected, is completed.

The Puget Sound Traction, Light and Power Co., Seattle, Wash., has changed its name to the Puget Sound Power and Light Co. Since the sale of the Seattle Street railway lines, the company has been concentrating on power and light business, hence the change of name.

The Schenectady Illuminating Co. and the Adirondack Power Corp., Schenectady, N. Y., are planning a merger for the purpose of acquiring the Brady lighting and power interests in several parts of the state, including the Glens Falls Gas and Electric Light Co. and the United Gas, Electric Light and Power Co., at Hudson Falls.

COAL PRICES

Prices of steam coals both anthracite and bituminous, f.o.b. mines, unless otherwise stated, are as follows:

ALANTIC SEABOARD

Anthracite—Coals supplying New York, Philadelphia and Boston:

	Mine
Pea	5.30
Buckwheat	$3.40@$3.75
Rice	2.75@ 3.25
Barley	2.25@ 2.50
Boiler	2.50

Bituminous—Steam sizes supplying New York, Philadelphia and Boston:

Latrobe	$4.25@$4.50
Connellsville coal	4.25@ 4.50
Cambrias and Sunset	4.25@ 4.85
Clearfields	3.75@ 4.50
Pocahontas	6.50@ 7.00
New River	6.50@ 7.00

BUFFALO

Bituminous—Prices f.o.b. Buffalo:

Pittsburgh slack	$6.25@$6.50
Mine-run	6.25
Lump	6.25@ 6.50
Youghiogheny	6.25

CLEVELAND

Bituminous—Prices f.o.b. mines:

No. 6 slack	$3.25@$4.00
No. 8 slack	3.50@ 4.00
No. 8—¾-in.	3.50@ 4.00
No. 6 mine run	3.50@ 4.00
No. 8 mine run	3.25@ 4.00
Pocahontas—Mine-run	3.25@ 4.00

ST. LOUIS

Anthracite—Probably not more than 20 per cent of the demand in this market can be supplied. Prices effective Apr. 1 were as follows:

	Williamson and Franklin Counties	Mt. Olive and Staunton	Standard
Mine-run	2.65@2.80	2.65@3 00	2.65@2.80
Screenings	2.50@2.65	2.50@2.65	2.50

Williamson-Franklin rate to St. Louis is $1.10; Other rates 95c.

CHICAGO

Bituminous—Prices f.o.b. mines:

Illinois		Freight rate Chicago
Southern Illinois		
Franklin, Saline and Williamson Counties		
Mine-Run	$3.00@$3.10	$1.55
Screenings	2.60@ 2.75	1.55
Central Illinois		
Springfield District		
Mine-Run	$2.75@$3.00	$1.32
Screenings	2.50@ 2.60	1.32
Northern Illinois		
Mine-Run	$3.50@$3.75	$1.24
Screenings	3.00@ 3.25	1.24
Indiana		
Clinton and Linton		
Fourth Vein		
Mine-Run	$2.75@$2.90	$1.27
Screenings	2.50@ 2.65	1.27
Knox County Field		
Fifth Vein		
Mine-Run	$2.75@$2.90	$1.37
Screenings	2.50@ 2.60	1.37
Brazil Block	$4 25@ $4.50	$1.27

New Construction

PROPOSED WORK

Mass., New Bedford—Mowll & Rand, Archts., 50 Bromfield St., Boston, will soon award the contract for the construction of a 2 story, 143 x 160 ft. theatre and office building on Purchase St., for Gordan & Schoolman, 18 Tremont St., Boston. A steam heating system will be installed in same. Total estimated cost, $500,000.

Mass., Newton—(Boston P. O.)—The city will soon award the contract for the construction of a 4 story school building on Beacon St. A steam heating system will be installed in same. Total estimated cost, $500,000. W. G. Perry and J. H. McNaughton, 19 Congress St., Boston, Archts.

Conn., New London—The United States Electric Co., 108 State St., plans to build a power house.

Conn., Watertown—Louis A. Walsh, Archt., 51 Leavenworth St., will soon receive bids for the construction of a 5 story, 60 x 132 ft hospital on Franklin St., for St. Mary's Hospital. A steam heating system will be installed in same. Total estimated cost, $500,000.

N. Y., Malone—The Kirk Maher Co., 59 West Main St., is in the market for ice cream manufacturing mchy. to have a daily capacity of 7,500 gal., and 55 tons refrigeration equipment. Address Clarence E. Kilburn.

N. Y., New York—The Bd. Educ., 500 Park Ave., will soon award the contract for the construction of a school building at Crotona Park East A steam heating system will be installed in same. Total estimated cost, $900,000. C. B. J. Snyder, Municipal Bldg., Archt and Engr.

N. Y., New York—Deutsch & Poles, Archts. and Engrs., 50 Church St., will receive bids until April 23 for the construction of a 16 story office building at 7-11 Water St., for the Natl Park Real Estate Corp., 32 Union Square. A steam heating system will be installed in same.

N. Y., Olean—The taxpayers voted $200,000 bonds to build a heating plant for the high school and to improve the heating systems in other schools

N. Y., Rome—The Rome Water & Sewer Bd. has had plans prepared by C. W. Knight, Engr., 101 West Liberty St., for the construction of a 350 hp. hydro-electric plant at Ridge Mills on the Montawk River. Estimated cost, $65,000.

N. Y., Sardinia—The Holland Sardinia Light & Power Corp. plans to construct a hydro-electric power plant including powerhouse, dam, etc., on Sardinia Creek, here. The powerhouse will be equipped with one 120 kva water wheel generator, one 96 kw. engine type generator, engine speed 276 r.p.m., one 7 kw. 125 volt compound belt driven exciter, one 150 kva. engine type generator, one 120 kva. engine type generator, engine speed 277 r.p.m., one 200 hp. type "Y" special Fairbanks-Morse engine, liquid fuel consumption and general equipment. Total estimated cost, $75,000. Manford L. Harrington, Pres.

N. Y., Theresa—The Orleans Four Corners Dairymen's League Co-operative Association, Inc., is in the market for milk handling machinery, refrigeration system and milk testing laboratory equipment.

N. Y., Whitney Point—L. J. White is in the market for a 40 hp. boiler.

N. J., Hoboken—The Bd. of Educ., Park Ave., will soon award the contract for the construction of a 4 story school building on Ninth St., between Garden St. and Park Ave A steam heating system will be installed in same. Total estimated cost, $200,000. Charles F. Dieffenbach, Arch.

N. J., Trenton—J. R. Fell, Engr., City Hall, is in the market for a blower to eliminate the smoke nuisance at the city pumping station along the Delaware River. Estimated cost, $10,000.

Md., Baltimore—The Federal Motor & Sales Co., 803 Low St., will soon award the contract for the construction of a 3 story, 106 x 117 ft. garage, service and sales station on North Ave. and Lord St. (Grasshopper Hill). H. A. Loane, Builders Exch. Bldg., Archt.

Md., Baltimore—The Standard Oil Co., Pratt and South Sts., is having plans prepared by C. L. Reader, Consult Engr., Park Ave. and Saratoga St., for the installation of a heating and ventilating system, etc., in the proposed 6 story, 90 x 140 ft. office building on Courtland and Franklin Sts. Total estimated cost, $500,000.

Md., Baltimore—Samuel T. Williams, 223 North Calvert St., is in the market for a 25 to 30 hp upright boiler with or without stack (used).

Ga., Augusta—J. E. Sirrine, Engr., Greenville, S. C., will soon receive bids for the construction of a cotton mill, revamping water power plant and dye house, and electrifying all the machinery for the Sibley Mfg. Co., Goodrich St.

Tenn., Cleveland—The city is in the market for a pumping unit to have a 500 gal. per min. capacity and an electric motor for power.

Tenn., Knoxville—The John G. Duncan Co., Northeast Corner of Central and Jackson Sts., is in the market for second hand equipment, including one 3 hp., single phase, 60 cycle, 1800 r.p.m. motor and one 3 phase, 60 cycle, 1800 r.p.m. motor.

O., Cincinnati—The United States Engineer's Office, War Dept., Wash., D. C., is in the market for power house equipment to be used in connection with Dam 35 in the Ohio River.

O., Cleveland—The Cleveland Ry. Co., Leader News Bldg., is having plans prepared for the construction of three 1 story, 46 x 60 ft electric sub-stations at West 73d. St. and St. Clair Ave., East 80th St., and St. Clair Ave., East 152d St. and St. Clair Ave. Two 1500 kw. rotary converters with blowers and generator sets will be installed in each. Total estimated cost, $300,000. L. P. Crecelius, 650 Leader-News Bldg., Archt. and Engr.

Ind., Lebanon—The Co. Comrs. plan to build a 2 story, 70 x 120 ft. jail and lighting plant. Two 60 hp. horizontal boilers, automatic stokers, pumps and radial brick stack will be installed in same. Total estimated cost, $100,000. John Frost, Lebanon, and Layton Allen, 406 Lombard Bldg., Indianapolis, Archts. Ammerman & McCall, 425 Occidental Bldg., Indianapolis, Engrs.

Mich., Milwaukee—The Michigan Light & Consumers Power Co., Jackson, plans to build a 2 story power station along the Saginaw River, here. Plans include a 30,000 kw. steam generating station. Steam boilers and engines to operate electric generating equipment will be installed in same. Total estimated cost, $4,893,000.

Ill., Chicago—The Bd. of Educ., 7 South Dearborn St., received bids for installing a steam heating system in the proposed 3 story, 132 x 175 ft. grade school on Augusta St., near Laramie Ave., from the Natl. Steam Heating Co., 1331 North Clark St., $39,423; William A. Pope, 26 North Jefferson St., $41,790; Kohlbry Howlett Co., 111 West Washington St., $43,043. Noted Jan. 27.

Minn., Red Lake Falls—Joseph Perrault, City Clk., will receive bids until April 26 for furnishing one 6 x 8 single acting triplex plunger pump of 175 gal per min. capacity against 150 lb. pressure or 350 ft. elevation.

Mo., St. Louis—Cornwell, Higgs & Gray, La Salle Bldg., are having plans prepared for the construction of a theatre and office building on Locust Ave. A steam heating system will be installed in same. Total estimated cost, $500,000 De Rosa & Pereira, 110 West 40th St., New York City., Archts. and Engrs.

Mo., St. Louis—C. G. and F. A. Hugh, 10th and Walnut Sts., are in the market for two 530 hp. type F., Sterling boilers, to have not less than 165 lb. pressure.

Tex., Bay City—The Texas Pub. Servc. Comm. plans to build an ice plant. Two large boilers, electric generator and machinery for a 15-ton ice plant will be installed in same. Total estimated cost, $60,000.

Tex., Clarendon—The Donley Co. Comrs. will soon award the contract for furnishing equipment and installing furnace and radiation for steam heating system at the county court house Estimated cost, $10,000. W. T. Link, Co. Judge.

Tex., Clarkville—The city plans to install an oil engine to take the place of the steam plant now used as motive power for pumps at city waterworks.

Tex., Columbus—The Columbus Electric Light & Power Co. plans to renew its poles on all lines here, install additional generators, extend a transmission line from here to Glidden and furnish current for lights and power there.

Tex., Perrytown (Rogerstown P. O.) The Public Utilities Co. plans to install an electric light plant here.

Tex., Texas City—The Vacuum Oil Co., Foster Bldg., Houston, will soon award the contract for the construction of three 1 story buildings including 40 x 64 ft. and

40 x 50 ft. engine houses and a 40 x 56 ft. pipe house. Total estimated cost, $30,000.

Cal., Bakersfield—The School Bd. engaged C. H. Biggar and T. B. Wiseman, Archts., Morgan Bldg., to prepare plans for the construction of additions to 4 schools, etc. Steam heating systems will be installed in same. Bonds amounting to $300,000 were voted upon for the project

Cal., Los Angeles—The Los Angeles Ry. Co., Pacific Electric Bldg., is having plans prepared for the construction of a 10 story, 65 x 178 ft. office building on 11th St. and B'way. A steam heating system will be installed in same. C. E. Norenberg, Archt.

Cal., Paradise—The Paradise Irrigation Dist., voted $140,000 bonds to replace 13,000 ft. of open canal with 36-in. pipe line, complete auxiliary reservoir system for high lands of district and complete distributing system. L. M. Edwards, Engr.

Cal., San Francisco—The Bd. of Pub. Wks. will soon award the contract for the construction of a 3 story, 82 x 98 ft. school building on Harrison St., between 11th and 12th Sts. An oil burner furnace, low pressure vacuum steam heating system, electric power and a motor driven vacuum pump will be installed in same. Total estimated cost, $108,000. John Reid, Jr., 1st Natl. Bank Bldg., Archt.

Cal., Santa Monica—Columbus Cheney is having plans prepared for the construction of a pumping plant, reservoir and water distributing system, etc., on the 500 acre tract about 8 miles north of here. Total estimated cost, $10,000. C. R. Brawner, 621 North Vernon Ave., Los Angeles, Engr.

Cal., Sierra Madre—The City Trustees voted $60,000 bonds to construct a pumping plant, new mains, etc. E. F. Ballon, City Clk. J. E. Mackerran, Engr.

N. S., Bridgewater—The town plans to improve the present water supply system by installing a new pump and wheel building, etc. Total estimated cost, $50,000. C. H. and P. H. Mitchell, Traders Bank Bldg., Engrs.

Que., Lake Magantre—The municipality will soon receive bids for the construction of a dam on the Chaudieve River capable of furnishing 500 hp. Estimated cost, $225,000. N. A. Girard, Engr.

Ont., London—The Bd. Educ., Dundas St., engaged L. V. Carrothers, Archt., Hydro Bldg., to prepare plans for the construction of two public school buildings and one collegiate institute. A vacuum steam heating and ventilating system will be installed in same. Total estimated cost, $400,000.

Ont., Oshawa—The town engaged R. W. Angus, Engr., University of Toronto, Toronto, to prepare plans for the construction of a waterworks system. An electrically driven, centrifugal pump will be installed in same. Total estimated cost, $80,000.

Ont., Sarnia—Martin & Sperry c/o J. B. Sperry, Huron Ave., Port Huron, have purchased a site on Front St., and plan to build a 4 story mercantile and hotel building on same. A steam heating system will be installed in same. Total estimated cost, $200,000.

CONTRACTS AWARDED

Mass., Greendale—(Worcester P. O.) The Norton Co., Bathers Crossing, has awarded the contract for the construction of a 6 story, 42 x 175 ft. manufacturing plant addition on New Bond St., to E. J. Cross Co., 82 Foster St. A steam heating system will be installed in same. Total estimated cost, $250,000.

Mass., Holyoke—John J. Kirkpatrick of the Municipal Lighting Plant has awarded the contract for the installation of boilers in plant, to the Edgemoor Co., Edgemoor, Del., at $39,972; for stokers, to the Westinghouse Co., Chicopee Falls, at $15,900; for furnishing superheaters, to the Foster Specialty Co., at $7,560.

Mass., West Springfield—(Springfield P. O.) The Gilbert & Barker Mfg. Co., Cold Springs Ave., have awarded the contract for the contruction of 12 buildings, including a 1 story, 50 x 92 ft. power house addition, 1 story, 35 x 50 ft. stone blast building, 1 story, 40 x 72 ft. engine house, etc., to Tucker & Lewis, Inc., 101 Park Ave., New York City. Total estimated cost, $1,000,000. Noted Feb. 2.

N. Y., Brooklyn—S. Wisglass & Co., manufacturers of brass beds, will build a 1 story, 18 x 67 ft. garage, 1 story, 25 x 45 ft. boiler house, 1 story, 25 x 50 ft. power house and a 2 story, 44 x 100 ft. extension to roof of existing factory, on Milford St. and Atlantic Ave. A steam heating system will be installed in same. Work will be done by day labor.

N. Y., New York—The Battle Iron Truck Co., 44 Court St., Brooklyn, has awarded the contract for the construction of a 5 story, 100 x 100 ft. garage on 57th St. and 11th Ave., to Louis Gold, 44 Court St., Brooklyn. A steam heating system will be installed in same. Total estimated cost, $250,000.

N. Y., New York — Farquharson & Wheelock, 724 5th Ave., will build a 12 story office building at 724-726 5th Ave. A steam heating system will be installed in same. Total estimated cost, $500,000. Work will be done by day labor.

N. Y., New York—The Hide & Leather Realty Co. has awarded the contract for the construction of an 18 story, 75 x 80 ft. office building on Gold St., to Thompson & Binger, 280 Madison Ave. A steam heating system will be installed in same. Total estimated cost, $400,000.

N. J., Newark — The Bd. Educ. Essex Bldg., has awarded the contract for the installation of a steam heating system in the proposed 3 story, 70 x 86 ft addition to the Franklin School, to John H. Cooney, 210 North 4th St. Noted April 6.

N. J., Trenton—The Nearpara Rubber Co., 79 Prince St., has awarded the contract for the construction of a 1 story, 100 x 125 ft. rubber plant, to the Bethlehem Constr. Co., Bethlehem Trust Bldg., Bethlehem, Pa. Plans include a 1 story, 30 x 30 ft. power house. Total estimated cost, $150,000.

Pa., Philadelphia—The Bureau of Water Supply, City Hall, has awarded the contract for the construction of a super-structure pumping station for the Lardners Point Station, to Charles H. Dadney, Bellevue Court Bldg., at $47,000.

Pa., Pittston — St. John's Catholic Association has awarded the contract for the construction of a 3 story, 90 x 180 ft. high school on Dock St., to John Curtis & Co., Hickory St., at $250,000. A steam heating system will be installed in same.

O., Cleveland—The city has awarded the contract for the installation of a heating system in a bath house, here, to the Chafer Co., 431 Champlain Ave., at $32,007. Noted Feb. 14

O., Cleveland—The Langaneau Mfg. Co., 8103 Franklin Ave., has awarded the contract for the construction of a 2 story, 40 x 50 ft. boiler house, to the H. G. Slatmyer & Son Co., 203 Lakeside Ave. One boiler will be installed in same. Total estimated cost, $50,000.

O., East Cleveland (Cleveland P. O.)—The Natl. Lamp Co. of the General Electric Wks., Nela Park, has awarded the contract for the construction of a 1 story, 100 x 220 ft. laboratory at Nela Park, to the Austin Co., 16112 Euclid Ave., Cleveland. A steam heating system will be installed in same. Total estimated cost, $100,000.

Ind., Indianapolis—The Bd. Educ., Meridian and Ohio Sts., has awarded the contract for the installation of a heating, ventilating and lighting system in the proposed 3 story, 90 x 194 ft. high school on Merrill St., to Freyn Bros., 31 West Michigan St., at $62,571.

Mich., Kalamazoo—The Handling Knight Co., c/o J. I. Handley, has awarded the contract for the construction of a 3 story, 80 x 800 ft. auto factory along the Lake Shore and Grand Trunk Rys., to Henry L. Van der Horst. A steam heating equipment will be installed in same.

Ill., Chicago—The Troco Nut Butter Co., 30 North Michigan Ave., has awarded the contract for the construction of a 4 story, 100 x 112 ft. factory on 37th and Iron Sts., to the E. W. Sproul Co., 2001 West 39th St. A steam heating system will be installed in same. Total estimated cost, $250,000

Wis., Milwaukee—The Federal Cone & Candy Co., 100 George St., has awarded the contract for the construction of 3 story, 50 x 60 ft. candy factory on Holton St., to W. Tuckwell, 86 Michigan St. Motors will be installed in same.

Minn., Virginia — St. Louis Co. Court House, has awarded the contract for the installation of a steam heating system in the proposed 2 story, 75 x 125 ft. court house addition on Central Ave., to S. Farrell Co., West 1st St., Duluth, at $22,222.

Cal., San Diego—The Bureau of Yards & Docks, Navy Dept., Wash., D. C., has awarded the contract for furnishing and installing standard electric distribution systems and power plant equipment, to the Thomas Haverty Co., 801 Maple St., Los Angeles, at $189,650. Noted March 23.

Cal., Long Beach—The Halfhill Tuna Packing Co. has awarded the contract for the construction of a refrigeration plant, etc., at Channel 3, to G. D. Sanford, 415 East 17th St. Estimated cost, $40,000.

Cal., Los Angeles—King W. Vidor has awarded the contract for the construction of seven 1 and 2 story motion picture producing buildings on Santa Monica Blvd., to the Milwaukee Bldg. Co., 316 Wright and Callender Bldg., at $100,000.

Cal., Salinas—The Trustees of the Salinas Union High School Dist. has awarded the contract for the construction of a 2 story high school, to McLeran & Peterson, Hearst Bldg., San Francisco, at $235,400. A steam heating system will be installed in same.

Cal., San Francisco—The Standard Oil Co., 200 Bush St., has awarded the contract for the construction of a 20 story, 137 x 206 ft. office building on Bush and Sansome Sts., to the P. J. Walker Co., Monadnock Bldg. A boiler plant, compressor, motors, etc., will be installed in same. Total estimated cost, $5,000,000.

POWER

Volume 51 — New York, April 27, 1920 — *Number* 17

Indispensability

By RUFUS T. STROHM

A gnat upon its errant way
Flew through an engine room one day
And settled on the flywheel,
Which weighed some twenty tons or more
And stood twelve feet above the floor—
A massive, broad and high wheel.

But scarcely had his trotters hit
The smoothly polished top of it,
Than off it started whirling,
While on its speeding rim the gnat
In happy contemplation sat,
His natty mustache twirling.

"Hi, people! Turn your eyes on me!"
He cried in Lilliputian glee.
"Just notice what I've started.
It's plain that I'm a heavy-weight
To turn this thing at such a rate
Of speed as I've imparted."

"And what is more, I'll have you know
That I'm the guy that makes it go"—
He said it very loudly.
And thereupon, to prove his case,
He sprang from off its gleaming face,
His chest expanded proudly.

Gadzooks! That gnat was fairly floored!
The throbbing engine quite ignored
Its self-appointed master;
Not for a second did it stop
Nor let its speed of turning drop;
Instead, it ran the faster.

You doubtless know some human gnat
Who acts as foolishly as that,
And gets the silly notion
That no one in the firm but he
Provides the push and energy
That keeps the thing in motion.

Less learned folks to us may bow
And subtle flatterers kowtow
In manner deferential;
But though possessed of wit and brain,
It's downright folly to be vain—
We're none of us essential.

When any mortal comes to think
The world would all go on the blink
Were he assassinated,
He's on the old toboggan slide
Whose sloping length is steep and wide
And smoothly lubricated!

Coal Consumption in Institutional Heating Plants

By G. B. NICHOLS
Chief Engineer, New York State Dept. of Architecture

FOR the last ten years, in state and municipal work, a great deal of attention has been given to the centralized control of all activities, replacing the local controls at each institution to a greater or less extent. This has been forced upon the state on account of the extensive research now being carried on, demanding specialized control in each branch of government. One of the main departments of centralized control which has been established in a number of the states is the Department of Architecture, under a state architect, which department can be divided into three main divisions—architectural, administrative and engineering. This article will deal with the engineering division and

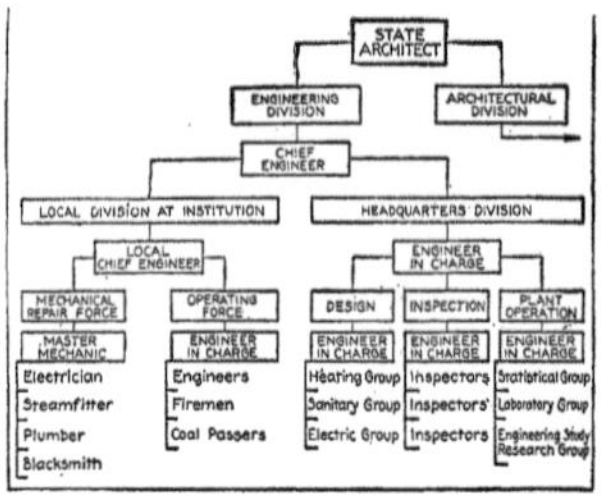

FIG. 1. DIAGRAM OF ORGANIZATION OF THE DEPARTMENT OF ARCHITECTURE AS RELATING TO THE ENGINEERING DEPARTMENT

its relation to the operating conditions at the various institutional heating, lighting and power plants. Such a division is also applicable to one which might be established in a large corporation operating plants scattered over a widely distributed area.

Such an engineering division, or department, should be under the direct administration of a headquarters chief engineer, two distinct divisions reporting to him, the first of which would be the headquarters division, which would be divided into three bureaus, each under an engineer, in charge, namely, the division of design, the division of inspection and the division in charge of power-plant operation. This is shown in Fig. 1.

Each of these divisions would require specially trained experts, and it is found impossible to shift employees between these divisions, as the employees of each class must have their particular training, and it is found that a good designing engineer seldom makes a good inspector. An engineering inspector very seldom is disposed to draw up a design on a drafting table, as he soon gets lost in the details, so that a rule can be laid down that each division must train its own employees.

Acting under the headquarters chief engineer, at the institution and local plant, is a local chief engineer, and reporting to him are two groups of employees, one group consisting of the plant-operating force acting under an engineer in charge of operation, and the other group, the mechanics making repairs to the general plant, under a mechanical foreman. The same condition exists in the training of these two branches of employees, as a repair man seldom makes a good operating engineer.

All these divisions, however, must act as a unit, each depending upon the others for advice and guidance in its particular branch of work, the headquarters chief engineer acting as an administrator to the entire group, and the local chief engineer and headquarters bureau engineers each realizing that they are units of a large collective body, carrying on the state engineering work. It is believed at the present time that to a greater or less extent this harmony exists in the various institutional plants and headquarters engineers under the control of the State of New York, and that the foregoing outline of divisions will become a reality as time goes on, particularly on account of the high cost of material entering into the operation of the plants, thereby demanding the highest efficiency of the entire engineering division of the state, which is impossible with local control. The medical profession and local institutional administrators also realize that the functions being performed by the engineering staff are wholly outside of their training, and that all technical problems arising must be given mature engineering consideration.

Few persons outside of those actively engaged in state work realize to what extent state institutions have grown. In New York State, in the insane hospitals alone, there are now housed approximately 35,000 patients. Three of these institutions have a population of approximately 5,500 inmates each, and in each the maximum coal consumption per day in extreme cold weather is about 90 tons.

The last bureau to be established by the Engineering Division of the New York State Department of Architecture is the headquarters division in charge of operating conditions of the various heating, lighting and power plants of the state. This bureau has been actively engaged for a considerable period in studying the operating and apparatus conditions at the various institutional plants and endeavoring to arouse in all of the operative force a spirit of coöperation and a study of the plant condition, with the idea of reducing the operating cost, particularly the annual coal consumption and labor.

The first work of the bureau in charge of operation was to prepare a series of daily log sheets applicable to all plants, in which would be entered all the instrument readings, together with the coal and ash consumption and the amount of labor required for each shift. One of these forms is shown in Fig. 2. All the report sheets can be bound into a loose-leaf post binder cover and are filled out by the local operating force in duplicate, one copy to be bound in the binder and kept in plain sight in the engine room for the study by all the operating force, so that the plant condition is known by each individual employed and he can definitely see his relation to the final operating costs.

The duplicate copy is sent to the headquarters bureau and is bound in a similar loose-leaf book and carefully observed by a specially trained force who plot curves showing all the operating conditions about the plant and make a summary of all daily reports at the end of each month at which time the local chief engineer receives a summary report of the operating conditions for the month, in total, also in per capita, based on the inmates at that particular institution. Along with this report is sent a series of definite recommendations to be tried out by him and the results of these recommendations reported on in writing to headquarters before the first of the next month. It can, therefore, be seen that the headquarters engineering force in charge of operations are familiar with all the conditions at the plant and can compare one plant's operation with that of a similar institution at another locality, and if there is a great difference in final per capita or efficiency results, immediate steps can be taken to either right or account for this difference.

As the records increase in number, the results of one year can be compared with those of a previous year at the same date, or same month. It can also be seen that the headquarters staff act as bookkeepers for the local operating force.

Can Increase Coal Consumption

Although a plant may be much out of repair, and wasting steam and demanding a large amount of coal to run, still the results show that to a great extent the operating force can by inattention or neglect of their work or possibly through misunderstanding of the proper methods to be employed in the operation of a certain piece of machinery, materially increase the coal consumption; and it is the duty of every engineer to study the apparatus under his control and to make it operate at its greatest efficiency.

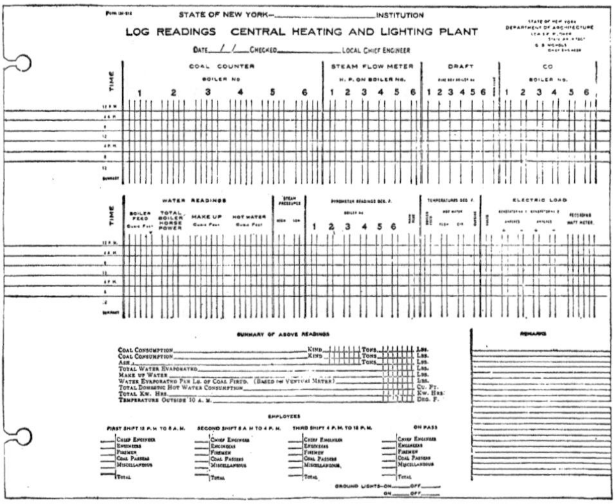

FIG. 2. ONE OF THE DAILY LOG SHEETS THAT IS APPLICABLE TO ALL STATE PLANTS

In 1915 numerous discussions were carried on by the state departments regarding the condition of the institutional plants and their operation, and numerous cases were pointed out in which the annual coal bills were continually increasing without any additional building construction being carried on at such institutions. Also that by simply changing a local chief engineer, the coal bills immediately increased or decreased, due to his inattention or activity. About this time investigations were carried on by several engineers and a report was made by John A. Hennessy in his published book on the conditions in New York State, which book states that as a result of three investigations, it was believed that approximately $300,000 a year spent for fuel could be saved in the proper operation of the state heating and lighting plants.

On account of war conditions and the reduction of employees in the various state institutions, it was found impossible to carry on this work with any great activity, but it was not altogether dropped. At the close of the war, however, it was believed that the time was ripe for the undertaking of centralized control of these power houses, to a greater or less extent, making it possible to reduce the coal bills and also to furnish the design department with the actual operating conditions at the various plants, for the design department is continually called upon to make additions, alterations or repairs to all these existing plants. These operating reports also form the basis of new designs, and it is felt that at the present time there are on file more actual facts regarding all classes of services required at a state institution on which to base new designs than existed before, and this information is being sought by other states and private concerns. A recent report of this activity is as shown in the accompanying tabulation covering ten hospitals.

TOTAL SAVINGS IN TEN HOSPITALS

State Hospitals	Estimated Tons Consumed	Estimated Tons Saved Over 1918–19	Saving in Dollars
Binghamton	13,495	337	$1,240.16
Brooklyn	4,577	397	2,354.21
Buffalo	12,081	559	1,928.55
Gowanada	10,354	1,741	8,705.00
Hudson River	19,051	2,295	11,429.10
Kings Park	22,881	2,285	10,922.30
Manhattan	19,800	19,584	145,509.12
Middletown	10,426	2,333	10,101.89
Rochester	7,646	471	2,425.65
St. Lawrence	14,719	1,323	6,284.25
	135,030	31,325	$200,900 23

Saving over total coal consumption in 1918-19 of 220,015 tons or 14.2 per cent.

The summary of the foregoing figures shows that an estimate of 31,325 tons of coal will be saved which, with the prices of coal prevailing at the respective institutions, will mean a saving for 1919-20 of $200,900. These ten plants represent approximately two-thirds of the coal consumption of the state, and it appears, therefore, that the estimates made in 1915 were correct.

It is safe to say that if owners of heating plants will encourage their engineers to make a careful study of their operating condition, furnishing them with the necessary repairs, so as to stop steam leaks, log-reading charts for recording all instrument readings, and will also furnish them with the proper measuring instruments so that they can study their plant conditions, the percentage of saving will be equal that shown on the foregoing schedule. It is believed, however, that in large corporations, having a number of separated plants, a central headquarters bureau is an absolute necessity to carry on the bookkeeping and analyzation of the operating condition of the plants. If, however, it is a single plant, a substitute for this central bureau would be for the plant to be under the control or advice of a privately owned controlling bureau performing this function for this separately owned plant.

The Bureau of Foreign and Domestic Commerce, Department of Commerce, Washington, estimates the total value of exports of refrigerating machinery during February, 1920, at $2,817,117. Japan was the largest purchaser during this month, followed by Cuba.

Unsoldering Leads from Commutator

By B. A. Briggs.

Disconnecting the leads from the commutator of a large armature with a soldering iron is a rather slow job on account of the large amount of metal that must be heated and also because of the rapid transfer of heat to the body of the commutator. The work can be considerably hastened by applying heat to the body of the commutator, by means of a blowtorch or a gas heater, as indicated in the figure. A piece of thin sheet asbestos should be placed around the surface of the commutator to prevent the flame from injuring the insulation. In this way the commutator may be heated up to near the melting point of the solder so that there is very little to be done with the iron and the rate at which the leads can be removed is greatly increased.

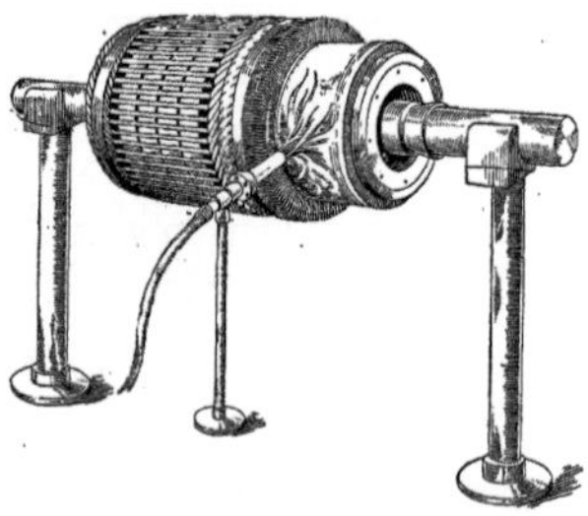

HEAT BEING APPLIED TO THE COMMUTATOR BY A GAS FLAME

Diagram for Computing Thickness of Plate for Steel Pipe

By Willis T. Batcheller*

The accompanying chart was prepared in connection with the design of the steel penstock for the new unit of the Cedar Falls hydro-electric plant[1] of the city of Seattle. The penstock was 2,226 ft. long and was made up in different diameters, ranging from 66 in. to 78 in. The static head varied from 250 ft. to 627 ft. at different points. While the solution is simple, the use of the diagram saves a great deal of labor when a number of values must be computed.

The thickness of steel pipe was computed by the equation

$$T = \frac{D \times H \times 0.434}{2 \times S \times E}$$

in which

T = Thickness of pipe in inches;
D = Inside diameter of pipe in inches;
H = Maximum static head on pipe in feet;
0.434 = Pressure in lb. per sq.in. for one foot head;
S = Fiber stress in plate;
E = Efficiency of joint.

---

*Engineer, Municipal Light and Power System, Seattle, Wash.

[1]This plant was referred to in an article by the author in the Sept. 4, 1919, issue of *Engineering News-Record*.

Using the equation given, values were computed and curves plotted covering the useful range of conditions. The curves are equally applicable to lock-bar pipe, lap-welded steel pipe and riveted steel pipe when the proper joint efficiency is used for the special case in hand.

The use of the diagram is illustrated by the following example: It is required to find the thickness of steel plate for a pipe having an inside diameter of 78 in. under a maximum static head of 500 ft. using a fiber stress in the plate of 15,000 lb. per sq.in. and a joint efficiency of 80 per cent, corresponding to a triple-riveted butt joint. Starting at a pipe diameter of 6½ ft., follow the vertical line to its intersection with the line representing a head of 500 ft. Then move along the horizontal line to its intersection with the line of 15,000 lb. fiber stress and drop vertically to the 80 per cent efficiency line. Follow the horizontal line to the left, where it intersects the thickness scale, and read 7/10 of an inch or 11/16, the nearest fraction. In the case of single-riveted or double-riveted lap joints under the same conditions, produce the vertical line to its intersection with the 50 or 65 per cent line, respectively, and read 1¼ in. and ⅞ in. for the two examples. With lock-bar pipe the thickness may be obtained by using a joint efficiency of 85 per cent and with lap-welded steel pipe, 90 per cent. In case it is desired to add 1/16 in. or some other amount as a rust factor, this can be done after the result is obtained from the diagram.

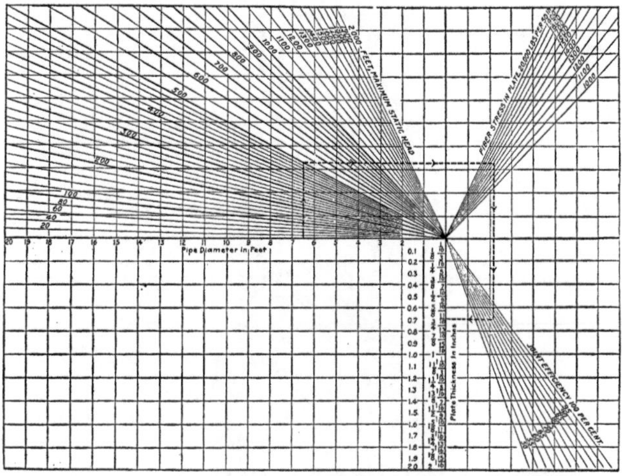

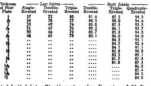

DIAGRAM FOR COMPUTING THICKNESS OF PLATE FOR STEEL PIPE UNDER VARIOUS CONDITIONS

EFFICIENCIES OF RIVETED JOINTS

Thickness of Steel Plate	Lap Joints			Butt Joints		
	Single-Riveted	Double-Riveted	Triple-Riveted	Double-Riveted	Triple-Riveted	Quadruple-Riveted
¼	57	72	80	81.6	87.5	94.3
5/16	54	70	77	80.3	88.0	94.6
⅜	53	69	76	80.8	87.5	93.3
7/16	51	67	76	80.2	86.1	93.7
½	50	66	74	80.7	85.8	94.0
9/16	48	65	73	80.1	86.0	94.0
⅝	..	..	..	....	86.3	93.1
11/16	..	..	..	....	84.7	92.4
¾	..	..	..	....	84.1	91.1
13/16	..	..	..	....	82.8	90.8
⅞	..	..	..	....	82.5	89.1
15/16	..	..	..	....	81.7	87.8
1	..	..	..	....	81.0	87.3
1 1/16	..	..	..	....	....	86.3
1⅛	..	..	..	....	....	85.4
1 3/16	..	..	..	....	....	84.5
1¼	..	..	..	....	....	84.0

Any problem containing a single unknown quantity can be solved by the use of the diagram. In case it is desired to obtain more accurate results than those resulting from the use of the conservative values of efficiency given herein, a trial thickness may first be obtained by their use and the operation repeated, using the joint efficiency corresponding to the trial plate thickness. To facilitate this use of the diagram, the accompanying table of joint efficiencies for ordinary commercial plate thicknesses has been taken from tables prepared by the Hartford Steam Boiler Inspection and Insurance Company.

Public utility companies operating in northern Illinois suffered damage amounting to about $1,000,000 during the windstorm on March 28.

Short Manometer for Measuring High Pressures

FOR the accurate measurement of pressure or of differences in pressure a manometer or U-tube becomes inconveniently long when the pressure is more than 45 or 50 lb. per sq.in. Dr. H. C. Dickenson, physicist of the Bureau of Standards, recently described a form of manometer used at the bureau, which overcomes this difficulty. The principle of this instrument is the use, in series, of several manometers of reasonable length so connected that any number of them may be used.

The general appearance of the apparatus is shown in Fig. 1 and a diagram of the connections in Fig. 2. There are five main U-tubes, each of which has an effective length of about 100 in., and two auxiliary tubes (*C* and *D*), each of which has about 12 in. readable length. The measuring liquid used is mercury, and the pressure is transmitted from one tube to the next by oil or pure ethyl alcohol, the latter liquid being more satisfactory since less dirt collects on the mercury surfaces with alcohol than with oil. The auxiliary tubes *C* and *D* are provided so that the instrument may be used in measuring gases which would be absorbed by the oil or alcohol. The portions of the manometer in which the mercury surfaces are observed, consist of glass tubing, 10 mm. (about 0.4 in.) inside diameter, while the remainder of the glass tubing is about 2.5 mm. (about 0.1 in.) inside diameter. Tube No. 5 is designed to be read at any pressure between zero and 45 lb. per sq.in., while the other four may be read only near the ends, and each tube is used to contribute an additional 45 pounds.

FIG. 1. HIGH-PRESSURE MANOMETER AT THE BUREAU OF STANDARDS

In operation the high-pressure gas is admitted to the left-hand tube of the manometer *C*, where its pressure is exerted on the surface of the mercury. To prevent the mercury being blown out of the tube *C*, oil is pumped in by the pump *P*, through the valve V_{16} until the pressure in the right-hand side of tube *C* is sufficient to keep the mercury level in the two tubes approximately the same. The pressure is thus transmitted by the oil from the mercury in the right-hand member of tube *C* to the left hand side of tube No. 1. Oil is delivered from the pump *P* through the valve V_{18} until the pressure at B_1 is sufficient to keep the mercury from being blown out of tube No. 1. The pressure at B_1 is then transmitted by the oil to the mercury surface in the left-hand member of tube No. 2 at A_2. The remaining pressure is similarly transmitted through tubes 3 and 4 to the surface in the left-hand member of tube 5, the mercury column in each of the first four manometers reducing the pressure by approximately 45 lb. per sq.in., or by a total of approximately 180 lb. per sq.in. Any excess in pressure over the 180 lb. is measured in manometer No. 5.

In case a pressure between 135 and 180 lb. was to be measured, the valve V_1 would be opened, thereby bypassing tube No. 1 so that only tubes 2, 3, 4 and 5 would be used. Similarly, for still lower pressure tubes 2, 3 and 4 might also be bypassed. It might be noted that if one or more tubes are to be bypassed, the effect is the same no matter which of the tubes are cut out.

In operating the manometer the valve V_6 is always closed except when it is desired to equalize the pressure between the two sides of the instrument. The valves V_7 and V_8 are opened only to change the volume of the transmitting fluid, and thus regulate the mercury levels in *C* and *D*. Valves V_9 and V_{10} exclude the pressure from the main portion of the manometer and, as a precaution in the event of breakage, are usually kept closed except during the period of observation. The effect of pumping in transmitting fluid through V_{11}, V_{12}, V_{13}, V_{14} or V_{15}, is to increase the difference in the height of the mercury surfaces for each tube to the right of the valve used, and to decrease the same for each tube to the left.

The reading device consists of a brass collar which encircles the tube of mercury. Behind the collar is attached a white background illuminated by a small electric lamp. The collar is moved up or down until the observer is just able to see a line of light between it and the mercury meniscus. The reading of the index on the fixed scale is then observed. A steel tape 2.8 m. long (about 71 in.) is stretched behind the U-tube No. 5, a 10-cm. scale (about 4 in.) is mounted behind each of the readable portions of the other four tubes, and a 20-cm. scale (about 8 in.) is mounted behind each readable portion of the auxiliary tubes. With this arrangement, the readings of the manometers Nos. 1, 2, 3 and 4 must be corrected by the addition of the vertical distance between the zero points of the two scales.

A further correction, due to the weight of the column of transmitting fluid, is necessary. For example, in tube 1 there is a column of oil in the left-hand tube from the mercury surface at A_1 to a point on a level with B_1. The oil in the left-hand tube above this level is of course balanced by the oil in the right-hand tube. The correction is then equal to the difference between gravity of the oil to that of mercury. This correction is always to be subtracted and must be made for each the mercury levels multiplied by the ratio of the specific manometer.

The reading of the total pressure shown by the entire instrument is of course the sum of the corrected

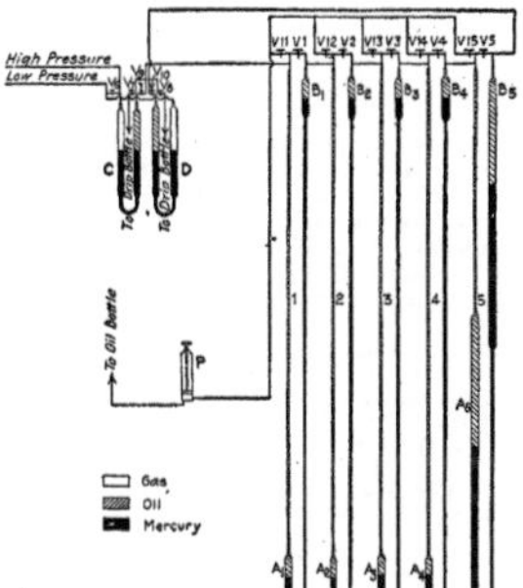

FIG. 2. DIAGRAM SHOWING ARRANGEMENT OF HIGH-PRESSURE MANOMETER

pressures read from each of the manometers used. For extreme accuracy the reading must be also corrected for the temperature of the mercury. At the Bureau of Standards this is determined by a special thermometer mounted vertically behind the middle of the manometer. Since an air circulating fan maintains a temperature that does not differ more than 0.5 deg. in various parts of the room, the temperature measurement is believed to be accurate to 0.3 or 0.4 deg. Another correction considered is that made necessary by the compression of the fluid under the higher pressure. From data on the compressibility and densities of mercury, oil and alcohol, the error in measurements of pressure differences at a mean pressure of 1,500 lb. per sq.in. is four parts in 1,000,000 for mercury and oil and 140 parts in 1,000,000 for mercury and alcohol. This correction, therefore, may be neglected in any but the most refined measurements. The greatest source of error appears to be in the actual determination of the position of the various mercury surfaces. With the reading devices used, the position of a clean surface can be determined to 0.1 mm. (about 0.004 in.). However, dirt will increase this uncertainty to perhaps as much as 0.3 mm. (about 0.012 in.). Since for 225 lb. pressure there are fourteen surfaces to be read, the probable error of reading will be from 0.5 to 1.5 mm. (0.02 to 0.006 in.), or about one part in 10,000. A series of experiments was undertaken to show the error due to the lag of the manometer. It was found after three or four minutes the errors from this source were much less than those from other known sources.

Installation and Maintenance of Trash Racks for Hydraulic Power Plants

By Ronald V. McDonald

Trash racks of hydro-electric developments are installed to protect the water turbines, governing apparatus and equipment and plant. If improperly chosen or installed, they may fail when needed most; if not properly maintained, they may obstruct the flow of water and even shut down turbines. These racks are installed before, or in, the intakes to prevent trash—vegetation, wood and other water-borne débris—being carried down through the intake. The form of rack used should depend upon the form of débris that may come down and the velocity with which it is carried by the water.

A common form consists of 1¼-in. round iron bars, set vertically, to the top and bottom of which are riveted flat iron bars about 4 in. wide and ⅜ to ½ in. thick. The circular bars are usually spaced about 4 in. apart. Another common form of rack is similar to the foregoing, except that the grids or vertical bars are from ⅜ to ½ in. wide transversely to the water flow and 3 to 4 in. thick in the direction of flow. The former type is invariably built up in a blacksmith shop; the latter is usually cast or forged in one part. The former is the cheaper and more easily made and repaired; the latter probably offers the maximum strength of rack with a minimum obstruction surface for restricting water flow.

A trash rack should be designed to offer maximum obstruction to débris and foreign matter brought down by the water and minimum resistance to the passage of water. These two requirements are somewhat antagonistic, a condition often solved by installing two racks, the inner one of finer grating, thus not only lessening the likelihood of obstruction of water taking place, but also facilitating cleaning. The space between grids or bars of a rack should depend upon the obstructions to be guarded against, the size of unit to be protected and the velocity of the incoming water and water-borne obstructions. Where ice jams and logs, turbulent water and wind are expected to be encountered, the racks should have ample strength, for grids that become bent together result in an obstructed passage between which invariably forms the nucleus for a blockade of vegetation.

A trash rack should be designed to give sufficient passage for water passing through it of the desired quantity without excessive loss of head. It is surprising what loss of effective head will actually occur when vegetation, wood and other materials clog the intake, a condition that assumes particular importance in low-head hydro-electric developments. Loss of efficiency and capacity always follows, and a unit may be shut down entirely from this cause. I believe that where much vegetation is to be dealt with, it is advisable to design the trash rack so that if it be obstructed equivalent to 40 per cent water restriction, the necessary flow of water is still possible.

Coarse stationary trash racks are usually cleaned by hand, using a rake on a pole or handle for this purpose.

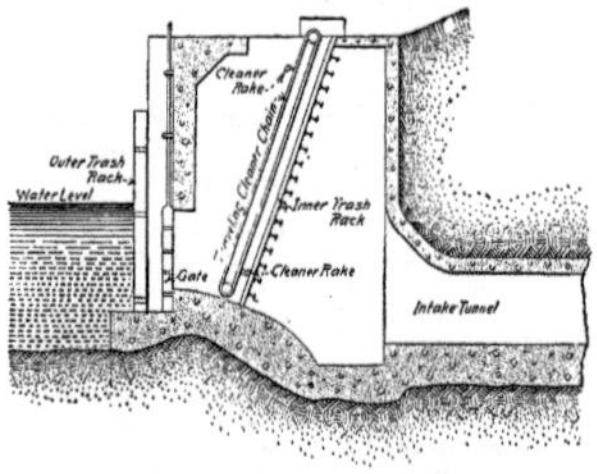

TYPICAL TRASH RACK ARRANGEMENT WITH CONTINUOUS CLEANING DEVICE

Sometimes a rake similar to a harrow is used, hung from chains by a hoist so as to permit of being lifted by one or two men. When two trash racks in series are employed, it is a good policy to make the inner or finer rack self-cleaning, in much the same way that soot cleaners are supposed to keep economizer heating surfaces clean. This is generally accomplished by one or more rake-like appliances (made in the blacksmith shop) installed upon link chains, which in turn are driven from a motor-operated countershaft located on top of the intake structure. Some form of adjustable iron comb is installed on top of each bay to catch the trash brought up by the scrapers and drop it on the floor.

Where hydro-electric developments receive their water from rivers adjacent to industry and habitation, a condition existing in the East and Middle West to much greater extent than on the Pacific Coast, trouble with the trash racks is often encountered on account of corrosion of the intake screens and racks. In some cases this sort of corrosion is very persistent and rapid, and from reports, no sort of paint has as yet been found that will remain an impervious covering for the metal for any great length of time.

About the best manner of delaying corrosion by the contamination of industry and habitation seems to be to paint all metal work with a heavy red-lead covering. This makes a good coating and lasts reasonably well, notwithstanding erosion and attrition by water and foreign matter. Racks and screens to remain protected should receive attention, and possibly repainting, once each year. As screens and racks cover a comparatively large surface, yet comprise but little metal relatively, painting by hand is most tedious if done thoroughly and is often slighted. Time and paint may be saved and a better job done by employing a paint-spraying outfit. The larger and greater the number of racks, the more advantageous a spraying outfit becomes, obviously. Before applying the paint, all metal work should be thoroughly cleaned by sand-blast.

The writer has never seen galvanized-iron trash racks used, probably for the reason that racks are made up close to the job, and galvanizing is not readily performed under such conditions. It is suggested that the contractors for the water turbines or other hydraulic apparatus might include the trash racks and furnish them galvanized, at least in a few cases as an experiment.

Many hydro-electric plants have been forced to curtail output, in some instances even being obliged to shut down temporarily, because of obstructed trash racks. Probably most hydro-electric plants have at some time suffered from loss of head because the trash racks were clogged and kept on clogging as fast as they were cleared. The foregoing may help to improve the situation by bringing up for discussion a few phases of the problem.

How To Locate Faults in Direct-Current Armatures

By W. A. Darter

A very convenient, sure and quick method of locating faults in armatures, such as short-circuits, open-circuits, poor contacts, grounds and cross-connected coils, without opening any coil connection to the commutator, is to use an ordinary telephone receiver and a source of alternating current. The principle is very simple. If an alternating current is sent

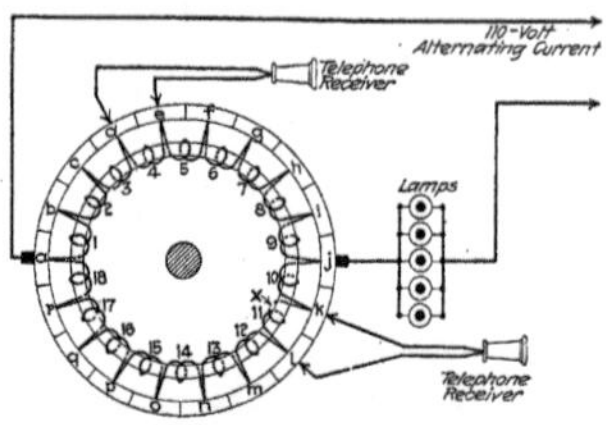

FIG. 1. CONNECTIONS FOR MAKING SHORT-CIRCUIT AND OPEN-CIRCUIT TESTS

through a coil, there is a voltage drop across the leads of the coil, and by connecting the leads of a telephone receiver across the leads of the coil, a vibration is set up in the receiver caused by the alternating current passing through it due to the voltage drop. Consequently, if a coil is short-circuited, there will be no voltage drop across the leads, and hence if the coil leads are bridged by the telephone-receiver leads, there will be no sound produced in the receiver.

For testing large armatures with coils of low resistance, a sensitive receiver is necessary. Small armatures may be tested with any telephone receiver and a small exciting alternating current. The writer generally uses a 110-volt source of energy, with either an ordinary electric iron in series with the armature coils, or enough large-current lamps in multiple to give the required amount of exciting current.

The leads of an alternating-current circuit are connected to the commutator the same distance apart as the brushes. On a two-pole machine the leads are connected 180 deg., on a four pole machine, 90 deg., and on a six-pole machine, 60 deg. apart. To detect a short-circuited coil in a 5- or 10-hp. armature, pass about a five-ampere current through the coils, as shown in Fig. 1. Then, beginning at commutator segments *a* and *b*, bridge the coil with the receiver leads. Continue to bridge each coil until the receiver fails to vibrate. If coil No. 4 is short-circuited due to a burr between segments *d* and *e*, then the receiver will fail to vibrate, indicating that there is no voltage drop between these two points, and hence coil No. 4 is short-circuited. Generally, after a commutator has been machined down, several of the segments are bridged together with small pieces of copper, and testing out the armature before it is put into service insures against any trouble from this source.

To detect an open-circuited coil, connect the armature up as shown in Fig. 1. Bridge each coil in rotation around the commutator until the receiver makes or produces a violent vibration. This violent vibration indicates an open in the coil, and the receiver is subjected to the full-voltage drop in that side of the armature. If the receiver leads bridged segments *k* and *l*, a strong vibration would indicate that coil 11 is open at some point, such as at *X*. To detect bad contacts between the coil leads and commutator segments in an armature, connect the same as in Fig. 1, then bridge each coil as before, tapping the commutator at the

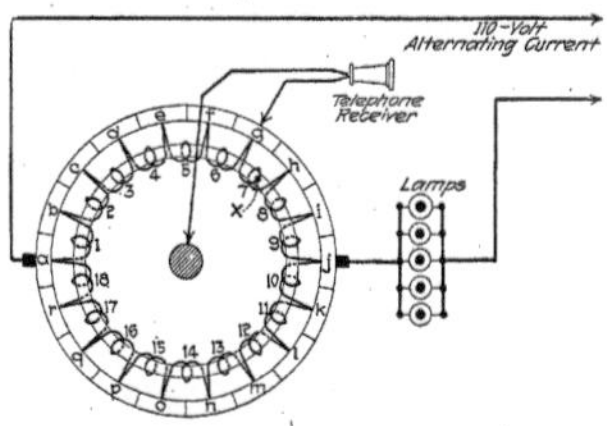

FIG. 2. CONNECTIONS FOR MAKING GROUND TESTS

same time with a wooden mallet. If there is a bad contact, the intensity of the receiver's vibration will vary, due to the change in resistance of the contact.

To detect a coil or commutator segment that is grounded, connect the armature to the alternating-current source, as shown in Fig. 2, which is the same as Fig. 1. Then connect one lead of the receiver to the shaft and with the other lead locate the segment that will give the minimum intensity of vibration in the receiver. This indicates the least voltage drop and that either the segment or coil connected to it is grounded. In Fig. 2 segments *g* and *h* would give about the same intensity of vibration; this would indicate that the coil winding is grounded about midway between the two.

To detect coils that are cross-connected to the commutator as shown at *g* and *h*, Fig. 3, connect the alternating-current source across the commutator and with the receiver bridge the segment as when testing for short-circuits. A cross-connected coil will be indicated by a normal vibration between two excessive vibrations in the receiver. This will be understood from the figure. If the receiver is connected to segments *f* and

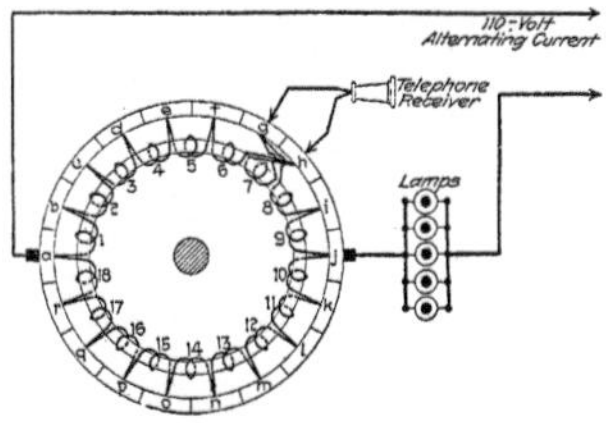

FIG. 3. CONNECTIONS FOR LOCATING CROSS-CONNECTED COILS

g, it will be connected across coils 6 and 7 in series. When the receiver is connected to segments *g* and *h*, it is across coil 7 only, and when in contact with segments *h* and *i*, the receiver bridges coils 7 and 8. Hence, under the first and third conditions the potential drop across the segments will be twice as great as under the second condition, and give indications in the receiver accordingly.

As a means of minimizing the difficulties caused by the insufficient coal supply the various electrical companies in and around Paris are considering a proposal for the construction of a vast power station for the distribution of energy throughout the industrial area which has grown enormously in importance since the war, says *The Engineer*. The existing power stations are wholly insufficient for present needs, and it is often urged that the defective installations are as much the cause of the irregular service as the shortage of fuel. It is hoped that by creating a central power station capable of supplying the entire Parisian district there will be, first of all, an economy of coal, and, secondly, it will be easier to insure supplies which would not be distributed among so many companies as at present. Nevertheless, it is argued that this economy and convenience would be purely illusory, and it has been proposed to generate electrical energy in some center like Rouen, where coal can be procured at lower cost and without additional transport, and distribute the current to the Paris district. The supply of electrical energy in and around Paris is so defective that something will have to be done to avoid the serious loss to the engineering and other industries resulting from the frequent stoppages.

A gigantic pool of power companies of the city of Los Angeles and all southern California is being planned. At a conference under the supervision of the State Railroad Commission to be held soon, the arrangements will be completed.

Charts for Graphical Determination of Pipe Sizes and Velocity of Flow of Steam

By H. M. BRAYTON
Mechanical Engineer and Designer, Semet Solvay Company, Syracuse, New York.

The author gives two charts for determining the size of pipe and velocity of flow for all usual pipe sizes. He also shows how the reader may construct such charts.

THE flow of steam in pipes is a complicated problem. One may delve as deeply as patience and one's knowledge of the subject goes into the thermodynamic theory. Many times in such a study assumptions have to be made which are not well supported by experiment and for this reason make the final result still in doubt. Such factors as the coefficient of friction between the steam and the pipe walls, which varies with the velocity, the character of the steam and the smoothness of the surface, the quality of the steam, the heat loss by radiation and conduction, the size and length of the line and the pressure, all affect the flow in a greater or less degree.

It is not the purpose of this article to carry the reader on any such mathematical journey as would be necessary in a discussion of the theory of the subject. It is rather the aim of the writer to show how the factors usually desired in steam-flow calculation may be determined graphically without calculation of any kind and with a degree of accuracy that will suffice in most practical cases. There are many men working at the drawing board whose knowledge of mathematics is limited, and to them these charts will be found of special value. Even the trained engineer does not like to spend his time in going through routine calculation when it may be avoided. The use of graphical charts as time savers is coming more and more to be recognized by technical men.

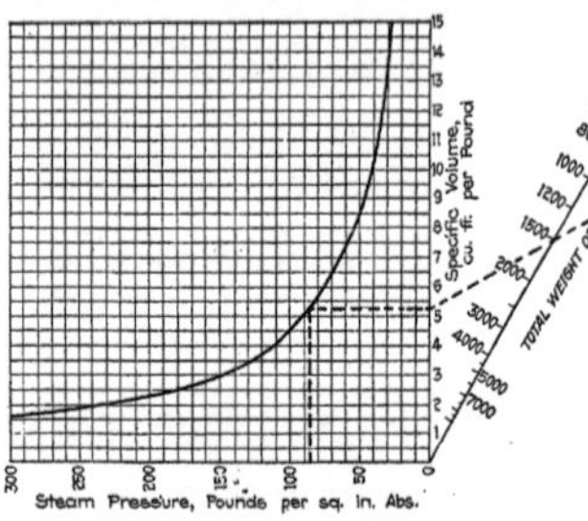
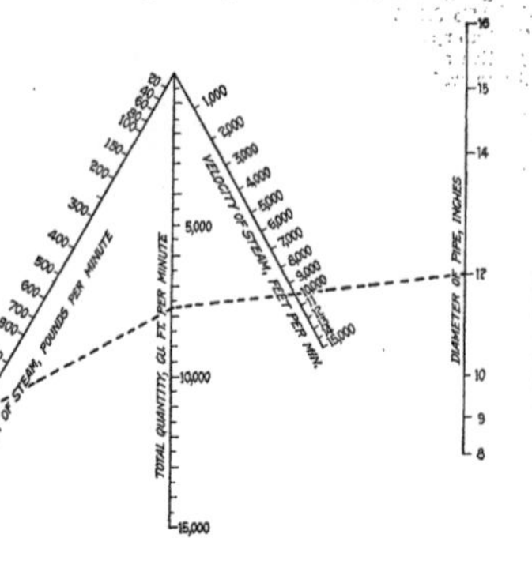

FIG. 1. CHART FOR DETERMINING FLOW OF STEAM IN PIPES FROM 8 IN. TO 16 IN. DIAMETER

The variables involved in pipe-size determination are as follows: Steam pressure, specific volume or the number of cubic feet per pound at the given pressure, the total weight of steam flowing in pounds per minute, the quantity flowing in cubic feet per minute and the velocity of the steam in the pipe in feet per minute. With these all known or assumed, the pipe size may be established.

The pressure drop is not necessary in the first determination of the proper size of pipe. The size will affect the amount of drop, and the latter should be calculated after the size has been determined to see whether or not it is excessive.

The degree of superheat will affect the pipe size insofar as it affects the specific volume. As the relation between the degree of superheat and the specific volume follows no simple law, it is not possible to incorporate it in these charts. When superheat is present, the specific volume must first be taken from a steam table and the chart used as given here.

The theory involved is simple and may be expressed in the following manner,

$$Q = AV = \frac{0.785\ D^2\ V}{144} = 0.00545\ D^2\ V \qquad (1)$$

in which equation

Q = Quantity of steam flowing in pipe in cubic feet per minute;
A = Area of pipe in square feet;
V = Velocity of steam in pipe in feet per minute;
D = Diameter of pipe in inches.

With this formula it would be necessary to know not

only the velocity desired, but also the quantity flowing in actual cubic feet. This latter factor is seldom known directly and must be determined before this equation can be applied. This may be done as follows:

$$Q = WV' \quad (2)$$

in which,

Q = Quantity in cubic feet per minute as before;
W = Weight of steam flowing in pounds per minute;
V' = Specific volume of steam in cubic feet per pound.

From the foregoing it will readily be apparent that one may combine equations (1) and (2), eliminating Q. The other factors must either be known or assumed before the diameter can be determined. The value of W is usually known; if not, there is no problem to solve. The specific volume can be found quickly by turning to the given pressure in the steam table. The velocity that is to be allowed must be assumed. If the size of pipe comes too high, a greater velocity must be taken because the other factors are seldom flexible. With this information at hand the two equations may be solved and the pipe size determined. In the same manner any one of the other variables may be solved if the diameter of pipe is first known.

The graphical chart shown in Fig. 1 quickly and accurately solves these two equations. At the left of the chart a pressure-specific volume curve has been incorporated which makes the whole complete. It is then not necessary to use a steam table with this chart when dry steam is under consideration. The dash-dot line

nected through the 10,000 point on the velocity scale with the diameter scale. It is seen that a pipe 12 in. in diameter will carry the required weight of steam at the desired velocity.

The reverse of this problem many times comes up for solution, and here again the chart will apply. For example, suppose it is required to find the velocity of steam flow in a 10-in. line carrying 1,000 lb. of steam per minute and a pressure on the line of 200 lb. per sq.in. absolute. If the reader will connect the point on the specific-volume scale which corresponds to the given pressure, in this case about 2.25, with the 1,000 point on the diagonal representing the weight scale and then from the point on the quantity scale which is cut out by producing this line, draw to the 10-in. position, it will be found that this latter line will cut out on the velocity scale the value of 4,200, which is about the velocity obtained. If he will go through this simple example, he will note quickly how the chart is used.

In the same manner any one of the six variables incorporated may be determined quickly and with sufficient accuracy for practical work. Special problems often

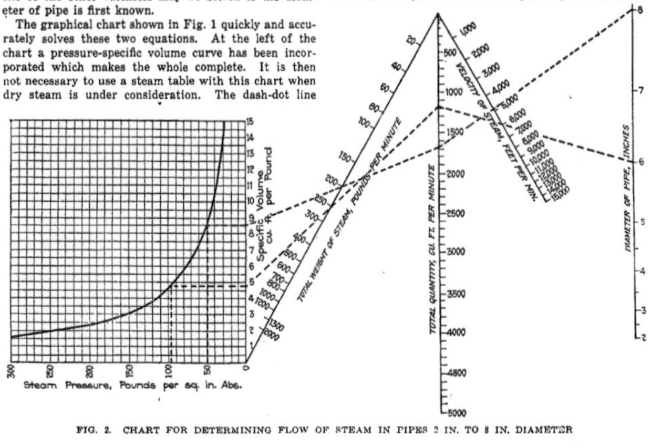

FIG. 2. CHART FOR DETERMINING FLOW OF STEAM IN PIPES 2 IN. TO 8 IN. DIAMETER

across the chart shows the method of use. In this chart it was assumed that the pressure was known to be at 85 lb. absolute or 70 lb. gage. At this pressure the curve shows that the volume is about 5.25 cu.ft. The total weight of steam which it is desired to have pass through the pipe per minute has been taken at 1,500 and the allowable velocity at 10,000 ft. per min. To determine the size of pipe required to accomplish this, it is merely necessary to connect the volume with the weight and continue the line until it meets the quantity scale. We may not be interested in the actual number of cubic feet of steam which is flowing per minute, but if we are it may be read from this scale and in this case is found to be 7.750. This point in turn is con-

arise in steam work, making it necessary to solve any one of the variables given on the chart.

The range of the variables as given on the chart in Fig. 1 is that generally met with in steam practice. Later, it will be shown how this chart was designed so that the reader may construct one for himself with any limits for the variables which may suit his special needs. It will be observed at once that the diameter scale begins with an 8-in. pipe. It was found impractical to incorporate lower values and still maintain accuracy in the other scales.

In using this chart, it must be remembered that the pressure is plotted as absolute pressure, and when the gage pressure is known or desired, 14.7 lb. must be taken from the value given on the scale. These pressures and volumes are plotted as given in the steam tables, so that the user would have no trouble if he had to take the value from the tables instead of the curve. This procedure might be desirable when greater accuracy was desired or necessary if superheat were present. Wet steam also will affect the specific volume, and if much water is present it will be well to take the volume from the curve for dry steam and then multiply the percentage of steam by this value, which will give the true volume. This latter value may then be found on the scale and the chart used in the usual way. This procedure neglects the volume of water present, but as this is usually small, it may be neglected without introducing an appreciable error.

It has already been stated that it was not found practicable to carry the pipe sizes below 8 in. on the chart shown in Fig. 1. As a large percentage of steam work involves sizes below this, it was deemed advisable to construct a second chart with the lower values. Such a chart is given in Fig. 2. The theory and principles involved here are the same as for Fig. 1, excepting for the limits of the variables. These limits are in keeping with smaller pipes.

Two illustrations of how to use the chart given in Fig. 2 are shown by the dash-dot lines. In one case a pressure of 50 lb. has been assumed, together with a weight of steam of 200 lb. per min. and a pipe of 8 in. By tracing the dash-dot line the reader can see that the steam velocity will be about 4,900 ft. per min. and the quantity flowing about 1,680 cu.ft. per min. In the other chart the pipe size and velocity are taken respectively at 6 in. and 6,000 ft. per min. and the weight flowing at 250 lb. per min. By tracing the line through, it will be observed that the quantity is about 1,180 cu.ft. per min. and the specific volume 4.75. This represents

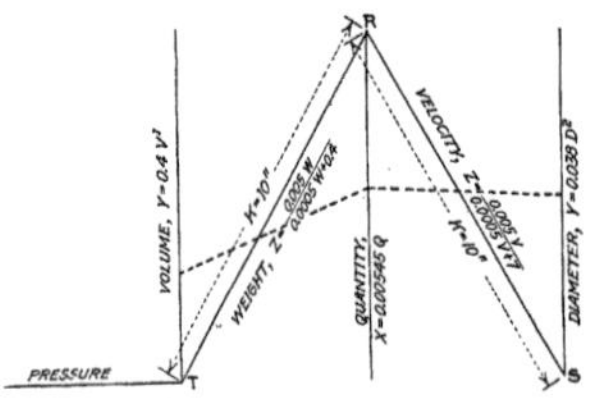

FIG. 1-A. CHART ASSISTS IN FINDING VALUE OF K WHEN LAYING OUT CHARTS SUCH AS SHOWN IN FIG. 1

a pressure of about 97 lb. abs. or 82 lb. gage. This means that this much pressure must be maintained on the 6-in. line if the desired quantity and weight is to be put through at the velocity of 6,000 ft. per min.

These two charts when used together cover the ordinary range met with in everyday practice. The limits in all cases have been chosen with the view of reaching all possible extremes. Sizes larger than 16 in. are met so seldom that it is not believed worth while to build up such a chart for the larger values. This could be readily carried out, however, and a method will be pointed out later by means of which the reader may construct such a chart for himself if his work warrants the time involved.

It has been the writer's experience in observing the work and study of fellow engineers that when a reader notes something in the chart line in an up-to-date journal, he often takes the attitude that although it is of value to him it is of little use on the page of the magazine. No means are usually available for reproducing it on a separate sheet, and so he lets it go by without so much as making a reference in his notebook so as to

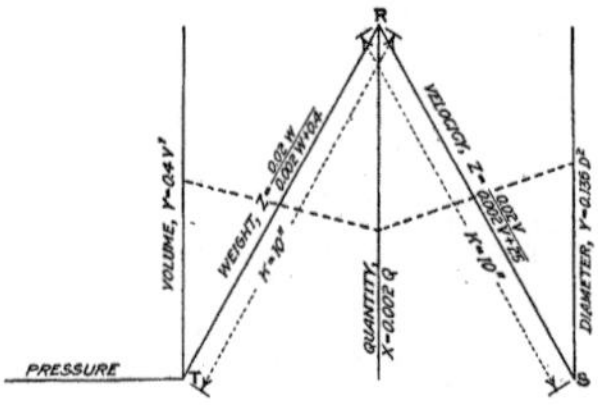

FIG. 2-A. CHART FOR DETERMINING CONSTANTS FOR LAYING OUT CHARTS AS FIG. 2

be able to find the chart later if needed. Usually, it is necessary to reduce these graphs to such an extent that accuracy is seriously impaired. The reader knows this and decides that it would be of little use to him in that form. Charts of this kind reach many readers through the libraries. These readers do not have a copy of the journal themselves, and the effort necessary to send specially for one copy is sufficient to make them decide that they do not want the article or the charts badly enough for that.

For these reasons the writer believes it good policy to include in such an article as this enough of the theory and sufficient explanation of the charts to enable the reader to construct them. No attempt will be made to include unnecessary theory or to derive the fundamental formulas involved. For the truth of these the reader may check them up by substituting values.

Let us start with the chart Fig. 1. This chart is known as the Z because its shape resembles that letter. We have here a combination Z because the principle has actually been applied twice. Fig. 1A gives a diagram of the scales and on each one of these lines is written an equation. This is the equation of the scale and means that the position of each one of the values shown on the scale in Fig. 1 is determined by this equation. It will be noticed that the equation in each case contains but two variables, the one representing a quantity in the formula being charted and the other the distance of a given value from the origin which is the point where the scales meet, or the corner, of the Z.

The foregoing may be a bit technical and perhaps not quite clear. Let us take the actual formula charted in Fig. 1. The first part of this is given in Equation (1) and was as follows:

$$Q = AV = 0.00545\ D^2\ V$$

We have here a simple three variable equation connecting the three quantities Q, D and V. The first thing to do is to select the limits of these quantities. Decide the maximum value of Q, D and V which you desire to have on the chart. This is usually regulated by practice. The chart should be made to cover the useful range of the variables. When this has been done, the equations of the scales may be written as follows:

$$X = m_1 Q, \qquad Y = m_2 (0.00545\ D^2), \qquad Z = \frac{K\ m_1\ V}{m_1 V + m_2}$$

It now becomes necessary to select values for the constants m_1, m_2 and K. These are determined by the actual size of chart which is to be constructed. The value of K will be the length of the diagonal of the Z in inches. This is shown in Fig. 1A, and in the chart Fig. 1 has been taken at 10 in. Any other value may be used as found convenient. The values of m_1 and m_2 are somewhat dependent on the value of K selected, as one would not like to have these scales extend far out beyond the junction points with the diagonal. It makes a good appearance to have the side scales come about even with the ends of the diagonal. With the maximum value of Q and D in mind as selected above, it is simple to determine m_1 and m_2. Let us suppose we limit the length of the Q scale to 7.5 in., as has been done in Fig. 1. Then if we substitute in $X = m_1 Q$ the values of $X = 7.5$ and $Q = 15{,}000$, we find that $m_1 = 0.0005$ and the equation of this scale becomes $X = 0.0005Q$. With this equation of the scale it is easy to lay off the actual scale. Merely substitute various values of Q from zero up to the maximum value of 15,000 in this case into this formula and obtain the corresponding value of X. This value of X will be in inches and represents the actual distance from the starting point. These distances may then be laid off from the point R, as given in Fig. 1A. Each point must be marked with its corresponding value of Q. This process is quickly accomplished when the equation is as simple as the one for this variable Q.

The value of m_2 for the D equation is determined in the same manner except that here we have a constant to deal with in the form of 0.00545 and also a squared term. Neither of these should cause any trouble. The maximum value of D has been determined, and if the maximum length of the scale is selected which represents the value of Y, we can nicely calculate m_2. It may be found that the value of m_2 becomes an odd figure if the length of scale is selected in even inches. It is usually better practice to select the modulus, as m_2 is called, an even number and let the scale come an odd length. If this is kept in mind, it will save time in putting in the scales. For the chart shown the value of m_2 was taken at 7, which makes the final equation for the D scale read as follows:

$$Y = 7\ (0.00545\ D^2) = 0.038\ D^2$$

The Y scale may then be laid in by measuring off from the point S the values of Y obtained from this equation when various values of D are substituted. Each point should be marked with its own value of D.

The diagonal scale is somewhat more complicated and requires a little more calculation in the determination of the points. The equation of the scale involves m_1, m_2 and K and therefore is entirely dependent on these values as selected for the two previous scales. It is therefore highly important that the equations of all three scales be determined before any are drawn in. It is often found that the values selected will not give a good scale for the diagonal. It may be too long or so short that all accuracy is lost. In either event the calculation must be gone through with again and more suitable values selected. It may be necessary to change the value of K. It will be noticed that V varies directly as K and therefore a change in this constant has more effect than the others. With the values selected above this diagonal equation becomes

$$V = \frac{K\ m_1\ V}{m_1 V + m_2} = \frac{10\ (0.0005\ V)}{0.0005\ V + 7} = \frac{0.005\ V}{0.0005\ V + 7}$$

The various values of V must be substituted in this equation and the corresponding values of Z determined. This equation can be nicely solved on the slide rule and it is not as long a job as might at first appear. These values of Z thus determined should be laid off along the diagonal from the point R. Each point must be marked with its own value of V.

This completes the first part of the problem. A straight-edge laid across the scales will solve equation (1). Any two of the variables being known, the other may be quickly found in this way. Fig. 1 shows a value of $D = 12$ in. connected through a value of $V = 10{,}000$ ft. per min. The quantity flowing is shown on the Q scale to be 7,500 cu.ft.

The left-hand part of the chart is a duplicate of the first part in regard to the principle involved. We have here another Z chart turned the other way around and one of its scales coinciding with one of the scales of the first chart. This was made possible because the value of Q appeared in both formulas. It will be recalled that equation (2) was as follows:

$$Q = WV'$$

If we make the modulus of the Q scale in both of these charts the same, they may be placed upon one another. These equations of scales were obtained as shown above and are as follows:

$$X = m_1 Q, \qquad Y = m_2 V', \qquad Z = \frac{K\ m_1\ W}{m_1 W + m_2}$$

The values of the constants selected follow: $m_1 = 0.0005$, $m_2 = 0.4$ and $K = 10$. It is not necessary that K be the same as selected for the other chart. It was selected the same in this case to make a more balanced chart. The equations of the scales then becomes

$$X = 0.0005\,Q, \qquad Y = 0.4\,V', \qquad Z = \frac{10\ (0.0005\ W)}{0.0005\ W + 0.4} = \frac{0.005\ W}{0.0005\ W + 0.4}$$

The scales may now be laid out as in the previous case. Only two remain, as the Q scale is already in place. The same rule applies here, however, as stated; that is, it is not advisable to put in any of the scales until the equations for the whole chart have been determined. The moduli many times have to be juggled to make a good appearing and useful chart. An illustration of this is found in the two given here. It was not found practicable to extend the D scale beyond the value of 8 in Fig. 1. It was necessary to carry it further in a second chart.

A chart should always be complete in itself; that is, it should be possible to solve completely the problem for which the chart was designed with the data usually available. With this chart thus far designed, this would not be possible, as the specific volume of the steam under consideration is seldom known directly and must be looked up in a steam table. Usually, the pressure is known instead. To make the chart complete, there has been added a pressure-specific volume curve at the left. Such a curve may be plotted as shown by taking from the steam tables the values of V' corresponding to the various pressures. This is merely a simple curve and may be plotted on thin cross-section paper. The V' scale on the plot must coincide exactly with the V' scale on the chart. The value of m_2 for this scale must be selected with this in view.

The chart shown in Fig. 2 is built up for the same formulas as the one shown in Fig. 1, but for different values of the variables. This chart applies to large pipe. The theory is the same as given above except for the actual values of the modulus selected. To enable the reader to reproduce this chart himself, the equations of the scales will be given. For the first part, which is represented by the right-hand part of the chart and the equation $Q = 0.00545\ D^2\ V$, we have:

$X = m_1\ Q = 0.002\ Q,$

$Y = m_2\ 0.00545\ D^2 = 25\ (0.00545\ D^2) = 0.136\ D^2$

$$Z = \frac{K\ m_1\ V}{m_1 V + m_2} = \frac{10\ (0.002\ V)}{0.002\ V + 25} = \frac{0.02\ V}{0.002\ V + 25}$$

The equations of the second part shown at the left and representing the question $Q = WV'$ are as follows:

$X = m_1 Q = 0.002\ Q,$

$Y = m_2 V' = 0.4\ V',$

$K = 10$ inches,

$$Z = \frac{10\ (0.002 \cdot W)}{0.002\ W + 0.4} = \frac{0.02\ W}{0.002\ W + 0.4}$$

The manner of laying out this chart is shown in Fig. 2A. For convenience the equations of the various scales are written thereon. With the foregoing explanations the reader should have no trouble in reproducing both these charts. If a smaller size is desired, it can be obtained by changing the values of the constants, m_1, m_2 and K. The theory outlined must be followed to make the chart solve the formula when finished.

The quickest and perhaps the most satisfactory way to draw these charts is to draw them in pencil directly on tracing cloth. When finished, they may be quickly inked and blueprinted. The plotted curve at the left may be drawn on a piece of cross-section paper and pasted to the tracing cloth. Care should be exercised to make the V' scale exactly coincide on the two drawings. The charts will be found to be time savers by designers who constantly have occasion to determine pipe sizes and steam velocities. In a subsequent article the writer will take up a more simplified form of chart that may be employed when the cubic feet of steam does not enter the problem.

Electric Power in the Arctic Regions

One would hardly associate electric power stations with the snowbound Arctic regions, yet, despite the intense cold and six months of ice and snow each year, a number of hydro-electric and steam-turbine plants have been erected during the last fifteen years for the purpose of garnering the rich deposits of gold in the bed of the Yukon River.

Over twenty-five million dollars has been expended by large operators for the furtherance of these projects. Miles of ditches have been excavated, powerful hydraulics installed and hydro-electric and steam-turbine plants constructed. This was no easy task. It meant fighting weather sometimes seventy degrees below zero and working on ground frozen to a depth of thirty feet. Moreover, heavy machinery and equipment necessary to carry on the work had to be transported over two thousand miles of wilderness.

Among the first of these operators was the Yukon Gold Co., which in 1905 installed, a few miles above Dawson, a hydro-electric plant of 1,875 kw., operating under a 650-ft. head. The plant consisted of three 54-in. Pelton waterwheels, each direct-connected to a 625-kw. 2,200-volt three-phase 60-cycle Westinghouse generator. The company had 36 miles of main-line transmission and 18 miles of branches and supplied energy for nine electric dredges, each using from 300 to 500 horsepower.

The Canadian Klondyke Co., shortly after, began operations about 26 miles above Dawson with the installation of a 10,000-hp. hydro-electric plant equipment consisting of two 3,000-kva. 2,300-volt 60 cycle three-phase generators, direct-connected to 5,000-hp. I. P. Morris Francis reactionary turbine wheels. Two banks of transformers stepped up the voltage from 2,300 to 33,000.

This plant was located farther north than that of the Yukon Gold Co., and it was found necessary to install electric heaters at the intake and at intervals of about two miles in its six-mile power ditch and also in the receiving chamber at the penstock heads to keep the anchor ice thawed out.

About thirty-five miles down the Yukon from Dawson the Northwestern Light, Power and Coal Co. located a 10,000-hp. steam-turbine station. Much of the equipment for this plant was shipped by the British West-

TEN THOUSAND HORSEPOWER STEAM-TURBINE PLANT, COAL CREEK, ALASKA

inghouse Co., from Manchester, England, across Mexico, to the Pacific and thence by water and rail 1,000 miles down the Yukon, where it was taken over the 600-mile snow trail in winter. The plant was finally installed at a cost of $2,500,000, but operated only 100 days when it was shut down because of legal difficulties and lack of coal. The machinery and equipment were not allowed to fall into disuse, but were dismantled and shipped to Japan, where they are used to furnish power for the operation of paper mills.

These electric stations, located practically on the Arctic Circle, represent 25,000 hp. of generating equipment and 10,000 hp. of auxiliary apparatus. At times the plants have been overloaded and the service requirements were very heavy, but they have run continuously and without serious trouble since their installation. During the fifteen years of operation over one hundred and fifty million dollars' worth of gold has been taken from the Yukon.

A Leader in Power-Plant Operation

W. L. ABBOTT

Thirty-five Years Associated with Progress of the Electrical Industry

A hard worker, day in and day out, a prodigious memory, tact, an even temperament, ability to judge men, gain their confidence by considerate treatment and get from them maximum returns to the limit of their ability, are the leading characteristics that have guided Mr. Abbott along a career of uniform success in the lighting industry. Having a happy combination of technical and practical knowledge, coupled with an analytical mind, he is quick to grasp essentials, to detect inaccuracies and to see at a glance the merits or flaws in reports submitted to him. "Information is only of value when it is correct" is a favorite expression used diplomatically to point out an error. As an operating man and later a director of operation, Mr. Abbott has few equals.

"FOLLOW the gleam!" is advice often given to ambitious and aspiring youth. W. L. Abbott, the subject of this sketch, has followed the gleam, not only in a metaphorical sense but quite literally, for he was one of the first engineers in Chicago to cast his lot with the electric light. He has followed it perseveringly and consistently for thirty-five years and has participated in its triumphs and the triumphs of its close associates, electric heat and electric power. He has been diligent in this business and has made an enviable reputation.

William Lamont Abbott is a native of Illinois and was born Feb. 14, 1861, near Morrison, in Whiteside County. He was graduated from the University of Illinois in 1884, having taken the four-year course in mechanical engineering. Twenty-one years later his Alma Mater granted him a master's degree.

Coming to Chicago (which has been his home ever since) after his graduation from college, Mr. Abbott was early impressed with the possibilities of electrical development. Electric lighting was then truly "in its infancy," but young Abbott was discerning enough to think it *had* a future. Accordingly, in 1885, with F. A. Wunder, he organized the Wunder & Abbott Illuminating Co. This company supplied arc lighting service in the central business district of Chicago, later absorbing several smaller companies engaged in similar work. In 1887 the Chicago Arc Light and Power Co., later absorbed into the Chicago Edison Co., one of the predecessors of the present Commonwealth Edison Co., was organized, and Mr. Abbott disposed of his interests in the Wunder & Abbott Co. to this corporation.

In 1888 Mr. Abbott, with others, organized the National Electric Construction Co., with himself as president and general manager. This company did general electrical construction work and also engaged in electric lighting in Chicago. Gradually, Mr. Abbott acquired the interests of his associates, conducting the affairs of the company until 1894. In that year the company was purchased by the Chicago Edison Co., Mr. Abbott continuing to manage the construction work of the company for another year, during which time he was also the assignee of the Northwestern Light and Power Company.

In 1895 the business of the National Electric Construction Co. was completely absorbed by the Chicago Edison Co. and Mr. Abbott was appointed chief engineer of the then New Harrison Street generating station of the Chicago Edison Co. It was not long before he knew his station thoroughly, and he made it his business to know everything that was going on in it. One of his favorite customs was to come down on Sundays, review with department heads the work of the past week and plan for the week to come. Under his direction the station became the most economical plant of its day. It was here that alternating-current generation and distribution, as we know it today, was given its first impetus. The station became familiarly known as the laboratory of the electrical industry, due to the amount of research work done there.

Mr. Abbott early realized that the great opportunity for saving lay in the boiler room. He had conspicuous ability to surround himself with and use capable men. Among those on his staff at different times were Walter T. Ray, C. H. McClure and A. Bement. Under his direction these men conducted extensive experimental investigations relating to the economical burning of coal. Innumerable tests on the various grades and sizes of coal were conducted to determine the burning qualities of each, analyses of coal and the gases of combustion were made practical and there was long-continued investigation on the placing and arrangement of baffles. The work at this station originated the inspiration for the general development in the boiler room that has now become so well established. Manufacturers and others in the industry profited from the results obtained, so that few people realize the extent of the influence from these pioneer investigations.

In reviewing this work, mention should be made of an extended research on the discharge of water with steam from water-tube boilers. The results, as presented by A. Bement, were given wide publicity in the technical press of Great Britain and this country. One of the things done at the Harrison Street station, in which Mr. Abbott takes particular pride, was the development of a method to determine the variation in angular speed during one revolution in the big triple-compound engine-driven alternators installed at the station, such variations being responsible for cross-currents when operating in parallel. It was first necessary to get the mean rotative effort of the three steam cylinders and plot against this the effects of inertia. The resultant gave the actual variation, which was found to be within permissible limits.

In May, 1899, Mr. Abbott was appointed chief operating engineer of the Chicago Edison Co., later the Commonwealth Edison Co., of which he is chief operating engineer at the present time. His position is one of much responsibility, for he has supervision, under Vice President L. A. Ferguson, of the generating stations, substations, load dispatchers, street department, storage batteries, meter department, service and repair department and lamp renewals of an electricity-supply organization which had an output of 1,628,340,000 kw.-hr. in 1919. The chief operating engineer is active in company affairs, being vice chairman of the important Advisory Committee, manager of the Edison Symphony Orchestra, etc.

One of his hobbies has always been to do something for the employees. At Harrison Street he started a night school to give members of the force instruction in steam and electrical engineering. At about the same time he also formed an employee's aid association, with sick and death benefits. This association died with Harrison Street, but recently Mr. Abbott originated a similar movement for the employees of the entire company, which at this writing is on the point of going through.

A loyal son of his state and always interested in higher education, Mr. Abbott was put forward by the alumni of the University of Illinois in 1904 as a candidate for the office of trustee of the university. He was elected by the people and reëlected in 1910 and in 1916. For twelve years of this period Mr. Abbott served as president of the board of trustees of the university, retiring last year (1919) as president, but still remaining a member of the board. On Nov. 4, 1916, a testimonial dinner was given in his honor at the Hotel LaSalle, Chicago, in recognition of his years of loyal service for the university.

A busy man, Mr. Abbott has found time to coöperate in the activities of professional societies and civic organizations. He served as president of the Western Society of Engineers in 1907. He is a member and a former chairman of the Chicago section of the American Society of Mechanical Engineers and a fellow of the American Institute of Electrical Engineers. He is also a member of the honorary engineering fraternity Tau Beta Pi, of the honorary scientific society Sigma Xi and an honorary member of the National Association of Stationary Engineers. He belongs to the Chicago Athletic, University, Engineers and Electric clubs of Chicago and the Chicago Association of Commerce.

Numerous papers on technical and other subjects have been presented by Mr. Abbott before various societies. In 1906 his paper entitled "Some Characteristics of Coal as Affecting Performance with Steam Boilers," read before the Western Society of Engineers, was awarded the Chanute medal.

As an expert in the buying, handling and storage of coal Mr. Abbott has been especially prominent, and necessarily so, as he is engaged by a company now spending seven million dollars a year for its fuel and storing up to five hundred thousand tons of coal. His early testing experience at Harrison Street gave him an intimate knowledge of the values of coal. From long study he has found that the coal to store is egg, 1½ to 6 in. Larger lumps tend to break up in handling, and spontaneous combustion will develop in the finer coal. Between the sizes specified coal has been stored at all seasons of the year with practically no trouble from fires.

During the war Mr. Abbott gave freely of his time and effort in his country's cause, serving on several committees and boards having to do with recruiting activity or other wartime necessities. At present he is chairman of the executive committee of the Bureau for Returning Soldiers, Sailors and Marines. His son, Capt. Arthur Abbott, also had a creditable war record.

One of Mr. Abbott's valuable possessions is a sense of humor. He is philosophical and accomplishes a great deal of work without much apparent effort. In conference his opinion carries weight and is received with no less attention because it may be accompanied by some quizzical or playful remark that may serve to create a bright spot in the discussion.

Hydro-Electric Power in Canada

Canadian water power is being developed as rapidly as possible, according to facts and figures given out by the Dominion Water Power Branch, Department of the Interior, and the Dominion Bureau of Statistics, which, with the coöperation of the Department of Trade and Commerce, has just completed a census of the developed water power in the Dominion. These figures include development up to January, 1920, and show that throughout Canada approximately 2,418,000 hp. has been made available. Many of the plants now in operation were designed for the addition of further units, which will bring their ultimate capacity, together with several new plants under construction, up to 3,385,000 hp. Hydro-electric power that can be made available in the British Empire totals from fifty to seventy million horsepower. It is calculated that Canada contributes in the neighborhood of twenty million horsepower to this total.

Refrigeration Study Course—VIII. Low-Temperature Ammonia Compression

By H. J. MACINTIRE
Professor Mechanical Engineering, University of Idaho

CERTAIN applications of refrigeration require low temperatures, the production of which adds to the difficulties and the operating problems of the engineer. These low temperatures are found in cold-storage work, possibly ice-cream hardening, in oil refining and in special chemical industries of various sorts. In these applications of refrigeration, temperatures of zero deg. Fahrenheit or lower are maintained, thus requiring minus 15 deg. F. or lower boiling temperature of the ammonia.

At low boiling temperatures the ammonia compressor of the standard design is not efficient except at very low head pressures. At normal condenser pressures the standard machine has the following capacities per ton, 20 lb. suction and 175 head pressure taken as unity: Suction pressure, 40 lb., 1.611 tons; 30 lb., 1.306 tons; 20 lb., 1 ton; 10 lb., 0.6993 ton; 0 lb., 0.4101 ton; —5 lb., 0.263 ton.

From this it is seen that at low suction pressures the standard compressor fails as a medium for securing refrigerating duty, and the engineer sees clearly why the manufacturers designate the operating conditions; namely, 175 head pressure and 15.2 lb. suction gage pressure. In the figures given a 100-ton machine has only 26.3 tons capacity at minus 5 lb. suction pressure.

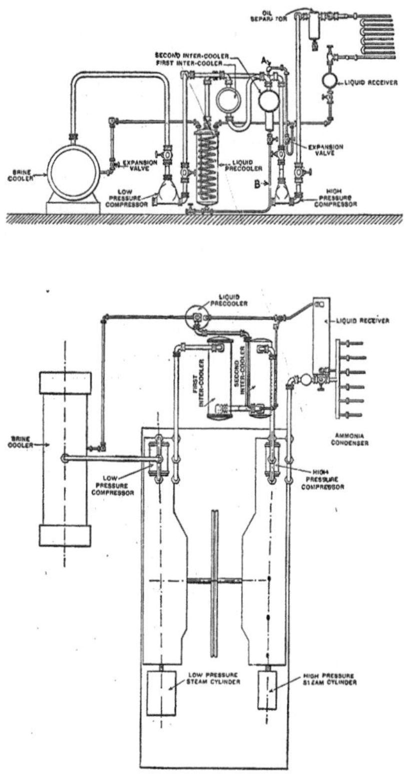

FIG. 1. PLAN AND ELEVATION VIEWS OF A LOW-TEMPERATURE COMPRESSION SYSTEM, SHOWING THE IMPORTANT CONNECTIONS

The reasons for lowered capacity with decrease of suction pressure have already been brought out in this study course; but it is worth while making mention of them again at this time. These reasons are on account of the low density of the ammonia at low pressures, the small volumetric efficiency obtainable, and finally, the trouble in the condenser by overheating the ammonia.

Some operating engineers have no clear idea of the so-called density and its effect on the tonnage of the machine. Yet in the use of steam they see clearly the reason for the relative increase in size of the exhaust-steam pipe as compared with the inlet boiler steam, especially when a low vacuum is carried as is true with the turbine operating condensing. In the steam engine or turbine the same weight of steam is exhausted as enters from the steam header, but the pressure is much lower. We say under these conditions that the *volume* of a unit weight of steam (one pound) is much greater with the lowered pressure. It is like the bicycle or auto air pump which compresses the air in the cylinder to a lower volume, but which has the same *weight* of air in the cylinder under both conditions.

The density of saturated ammonia, then, is less as the pressure is reduced, as may be seen from the following approximate values:

Gage Pressure, Lb.	Cubic Feet per Pound
20	8.0
15	9.3
10	11.0
5	12.9
0	17.9
5	24.1

The ammonia compressor, which has a fixed piston displacement per stroke, will pump the same number of cubic feet of ammonia per minute, but will have fewer and fewer pounds of ammonia discharged out of the compressor as the pressure is lowered. And the tonnage is always proportional to the amount of ammonia *in pounds* which is pumped into the condenser per unit of time, usually taken in minutes.

The volumetric efficiency has to be considered because of the reëxpansion of the gas in the clearance space at the beginning of the suction stroke. This reëpansion prevents the inlet valve from opening until the gas in the clearance volume has been reduced at least to the suction pressure. Under conditions of operation prevailing in low-temperature operation, this reëxpansion factor looms up in large proportions and materially affects the effective stroke. Clearance reëxpansion is not a direct cause of increasing the power required to compress and pump a given amount of gas, but it cuts down the capacity of the machine, as the cylinder is supplied by much less than the full cylinder volume of gas per stroke.

The third objectional feature of single-stage compression as applied to ammonia is in the high temperature of discharge of the gas. In air compression the rule is usually to pump from atmosphere to about 70 lb. in the first stage and from 70 lb. in the first stage and from this pressure to the required pressure when moderate final pressures are required. An intercooler is placed between the stages, and the gas is cooled to within a few degrees of the available cooling water before passing on to the second stage. The reason for so doing is to increase the volumetric efficiency, decrease the work of compression and prevent excessive discharge temperatures.

Excessive discharge temperatures are bad for ammonia. Considerable difference of opinion has been expressed on the subject, but it is clear from the practical experience of refrigerating engineers that high discharge temperatures, as well as high head pressures, tend to disintegrate the ammonia into hydrogen and nitrogen. These gases are permanent, and they collect in the condenser and, unless purged, will gradually increase the head pressure, like the manner of the increase in the condenser pressure when the air pump is stopped in the case of the steam engine or turbine. The formation of inert gases in the condenser has resulted in excessive purging and heavy losses of ammonia which invariably passes out of the condenser during the purging process.

The Absorption Machine

Parallel with the development of the ammonia compressor came the absorption machine, which uses heat energy directly instead of mechanical energy, to return the gas from the expansion coils to the condition of a liquid. The absorption machine uses any form of heat energy, but at present exhaust steam is most used. No compressor is required, but only a liquid pump of relatively small size. The result is that low boiling temperatures of the ammonia may be readily and economically obtained, requiring about 30 lb. of steam for 10-deg. brine and 34 lb. of steam for —10-deg. brine, calculated for 60 deg. cooling water.

The absorption machine has never seen much popularity even at the best time of its career. It is large, costly, awkward and more difficult to operate successfully than the compressor. Steam must be available. When operated by competent engineers, it is satisfactory and can achieve low temperatures with ease. Until

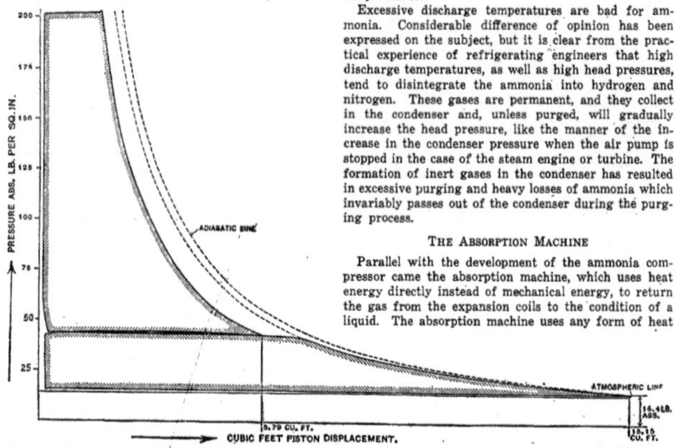

FIG. 2. THE COMPRESSION LINE IS BROKEN IN COMPOUND COMPRESSION BY REASON OF INTERCOOLING THE AMMONIA GAS BETWEEN THE LOW AND THE HIGH PRESSURE CYLINDERS

The discharge gas first passes through an intercooler where water is the cooling medium. It then passes through one where liquid ammonia is the cooling agent.

the perfection of the compound ammonia compressor, it was used for all low-temperature refrigeration.

Compound ammonia compression for low-temperature refrigeration has a number of important features. It is stage compression, with an intercooler between stages. It is possible to arrange the suction of the high pressure so that it will take care of the gas from the freezer rooms, but this multiple effect suction is not popular. It is more advantageous to design in a manner similar to the compound steam engine, and to try to equalize the load due to compression between the two cylinders. Thus, instead of using a receiver pressure of 25 to 30 lb. gage, a pressure of some 45 to 55 lb. would have to be used, depending on the head pressure and suction pressure.

As we all know, compound compression requires an intercooler as well as an aftercooler (the condenser in refrigeration). In air, or ordinary gas compression, water is used in the intercooler, but this is not possible

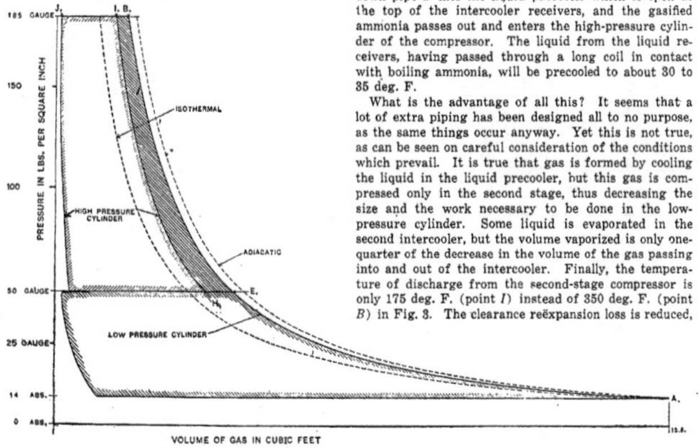

FIG. 3. SHOWING THE RELATIONS OF THE COMPRESSION LINES

The actual compression line for the high-pressure cylinder is that designated H. I. Intercooling reduces the volume and temperature of the gas at the end of low-pressure compression, and therefore the high-pressure terminal temperature or condenser temperature is considerably lower than it would be without intercooling.

with ammonia because the temperature of saturation at 45 and 55 lb. gage is 30 deg. F. and 38 deg. F. respectively. It is evident that water may not be used for such cooling, unless the intercooler used water for the first part of the process and finished by means of some other cooling medium similar to the Bandalot cooler principle. The only convenient and natural medium for such low-temperature intercooling is expanded ammonia. This is done by means of an accumulator arrangement.

The functions of the accumulator are twofold. It will be remembered that under usual circumstances there is some 10 to 15 per cent of ammonia vaporized at the expansion valve to cool the remaining liquid ammonia from the temperature of the liquid receiver to that of the temperature of evaporation (boiling) in the expansion coils. The vapor formed in the process cannot be used for refrigeration, but clogs up the expansion system unless removed. The gas formed increases in amount in cases of extreme pressure range, as in the case where low-temperature refrigeration is required, and so the accumulator principle is used to remove the gas formed in precooling and to remove the superheat from the discharge gas from the low-pressure cylinder.

Fig. 1 shows the general arrangement of the cylinders and intercoolers. Referring to the figure, it will be seen that the gas discharged from the low-pressure cylinder is made to go through the water-cooled first intercooler and then through the ammonia-cooled second intercooler. The discharge gas is thereby cooled from, say, 100 or 150 deg. F. to about 35 deg. F. and some five per cent (perhaps) of liquid ammonia is vaporized in the process. Liquid ammonia not vaporized after passing into the second intercooler through the valve *A* flows by gravity down pipe *B* into the liquid precooler which is open at the top of the intercooler receivers, and the gasified ammonia passes out and enters the high-pressure cylinder of the compressor. The liquid from the liquid receivers, having passed through a long coil in contact with boiling ammonia, will be precooled to about 30 to 35 deg. F.

What is the advantage of all this? It seems that a lot of extra piping has been designed all to no purpose, as the same things occur anyway. Yet this is not true, as can be seen on careful consideration of the conditions which prevail. It is true that gas is formed by cooling the liquid in the liquid precooler, but this gas is compressed only in the second stage, thus decreasing the size and the work necessary to be done in the low-pressure cylinder. Some liquid is evaporated in the second intercooler, but the volume vaporized is only one-quarter of the decrease in the volume of the gas passing into and out of the intercooler. Finally, the temperature of discharge from the second-stage compressor is only 175 deg. F. (point *I*) instead of 350 deg. F. (point *B*) in Fig. 3. The clearance reëxpansion loss is reduced, as seen in Fig. 2, which is taken from actual indicator diagrams.

It is seen from the diagram that the two compressors are driven from the same shaft, which may be engine, synchronous motor, or belt driven. If desired, separate drives may be made, as is done in the Central Manufacturing Co. cold-storage warehouse, of Chicago, where large units are employed. In this plant the low-pressure cylinder is called a "booster" and liquid cooling is used in a somewhat similar manner to that already described, which is known as the D. L. Davis method. In moderate-sized plants, however, the advantage of having the two cylinders on one shaft lies in the reduced torque and less flywheel and balancing troubles.

Three-Phase Power from Single-Phase Transformer Connections

By J. B. Gibbs

Engineer, Transformer Department, Westinghouse Electric and Manufacturing Company

A number of different ways of grouping single-phase transformers to step up or step down three-phase voltages are discussed, and their relative advantages and disadvantages pointed out.

Most of the power systems in this country are three-phase, and most of them use single-phase transformers. There are several schemes by which such transformers can be connected into "banks" for the transformation of three-phase power, and the

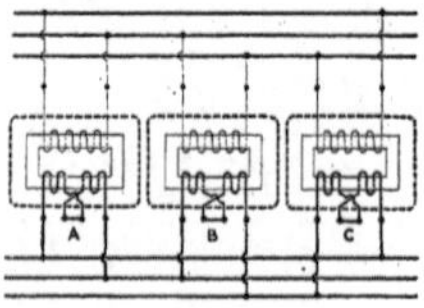

FIG. 1

apart. This is the ideal condition and one that is approached more or less closely in practice.

Suppose, now, that one transformer in Fig. 1 has a different ratio of turns from the other two, so that its secondary voltage will be a little higher. This voltage will not be exactly balanced by the resultant of the other two voltages, and the unbalanced part will cause a current to flow around the path formed by the three transformers. If a balanced load is connected to the secondaries of such a bank, it may be regarded as drawing a balanced current from the bank, as before, while the circulating current also flows at the same time. These two currents flowing in the windings of each transformer will combine to cause an overload on a part of the bank and an underload on another part, so that the bank, as a whole, will not carry its normal load without overheating in some part. If it is necessary to connect two transformers in delta with a third which has a different ratio, trouble may be prevented either by using an auto-transformer to correct the ratio of the third or by using a three-phase balance coil which will prevent the flow of circulating current without interfering with the load current.

There is another kind of dissimilarity which may give

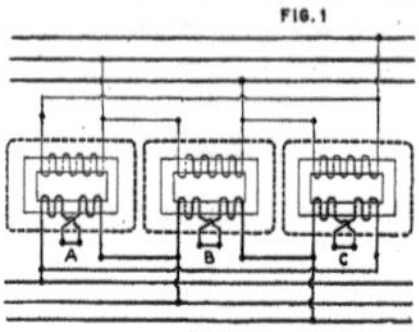

FIG. 2

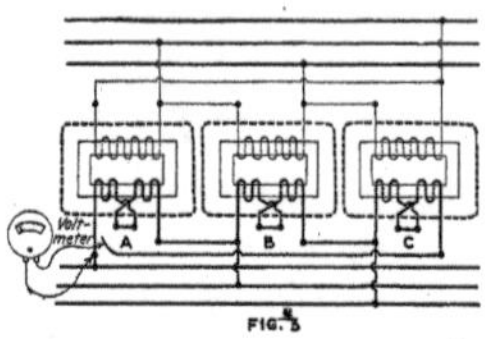

FIG. 3

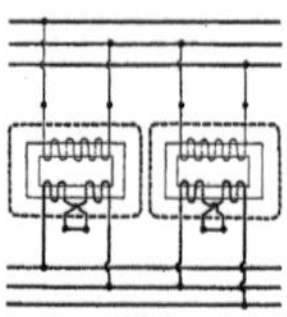

FIG. 4

FIGS. 1 TO 4. TRANSFORMER CONNECTIONS ON A THREE-PHASE SYSTEM

best scheme to use in any particular case depends on circumstances.

The connections for a delta-delta bank is shown in Figs. 1 and 2. Both figures show the same conditions electrically, and the connections are actually made in whichever way is more convenient. It will be seen that each transformer is connected between one pair of lines on both the primary and secondary sides, and with balanced secondary load each transformer supplies one-third of the total power. If the transformers have the same characteristics the relation of the voltages are such that the voltages of transformers *A* and *B* will add (vectorially) to give a voltage just equal and opposite to the voltage of *C* in both primary and secondary; and although there is closed circuit in the primary and another in the secondary, there will be no tendency for current to flow because the voltages are exactly balanced. A balanced load will then draw the same current from each transformer, and these currents will be 120 deg.

trouble in a delta connection. The three transformers may have exactly the same ratio of turns, but have different values of resistance and reactance. In such a bank at no load the voltages will be balanced and there will be no tendency for a circulating current to flow. But as the load is applied, the drop in voltage in the three transformers will be different, so that a resultant voltage will be produced, increasing as the load increases, and this will tend to cause a circulating current, as explained in the foregoing, with consequent overheating of a part of the bank, when operating under full load. This condition can be corrected by connecting resistance and reactance in series with the transformer which is low, to make its resistance and reactance equal the others, or by using a three-phase balance coil. In selecting transformers for a delta-to-delta bank, it is usually safe to assume that small transformers of the same size, say 50 kva. or less, will operate satisfactorily together. But before closing the corner of the delta on the low-voltage side, it is always well to check the polarity of the transformers by testing with a voltmeter or lamps, as indicated in Fig. 3, to see if there is any voltage between the wires that are to be connected. This is especially necessary if the transformers come from different manufacturers. If the test shows no voltage, the connection may be completed.

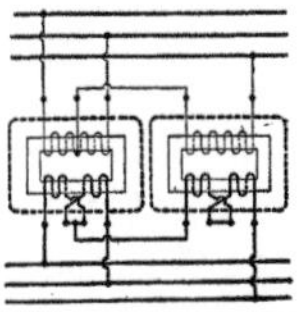

FIG. 5. TWO TRANSFORMERS CONNECTED IN "T"

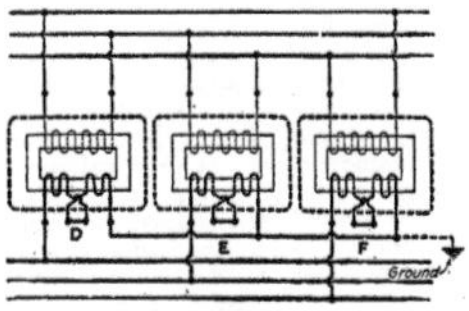

FIG. 6. THREE TRANSFORMERS CONNECTED DELTA TO STAR

One advantage of the delta-to-delta connection is that if trouble develops in one transformer it may be removed and the bank will not be entirely disabled. It will still carry 58 per cent of its normal load with two of the transformers connected in open delta.

The open-delta or "V" connection is shown in Fig. 4. Each transformer now carries a current due to the load on the phase to which it is connected, and in addition a current due to the load on the phase on which there is no transformer. The two currents are out of phase with each other, and the result is that each transformer carries a current which is about 15 per cent greater than that corresponding to the power delivered. The output of such a bank, therefore, is $\frac{1}{1.15} = 0.87$ of the total rated output of the two transformers. Thus two 50-kva. transformers connected in open delta will give $2 \times 50 \times 0.87 = 87$ kva. instead of 100 kva., and this is $\frac{87}{3 \times 50} = 0.58$, or 58 per cent of the output of a delta-to-delta bank of three 50-kva. transformers, instead of 67 per cent, as might be expected. The open-delta connection is convenient and takes only two transformers instead of three, and it is frequently used where small amounts of three-phase power are needed.

The tee connection is sometimes used for three-phase power and is shown in Fig. 5. Two transformers are needed, and as in the case of the open-delta connection, the total power delivered is only 87 per cent of the combined rating of the transformers. Not all transformers can be used for this connection. It is necessary to have the usual close magnetic relation between the primary and secondary windings, and in addition there must be close magnetic relation between the two halves of each winding in the "main" transformer, which is the transformer connected directly between the two lines. If this is omitted, the effect will be to put a reactance in series with the "teazer" transformer, and this may cause a serious unbalancing. This connection is rarely used except for transformation from two-phase to three-phase, or vice versa.

In Fig. 6 is given a delta-star connection. This connection is much used for step-up transformers. The neutral of the star side is sometimes grounded, as shown by the dotted lines, and sometimes not. If the ground connection is used, the maximum voltage that can occur from any line to ground is the voltage across one transformer. This limits the voltage stress on the high-voltage insulation of the transformers and line, but on the other hand, if one line becomes grounded accidentally, it means a short-circuit on the system. If the neutral is not grounded and an accidental ground occurs on one line, 73 per cent additional voltage stress is developed from the other lines to ground, but it may be possible to operate in this way until the grounded line is cleared. The voltage on the delta-connected windings is the same as the generator voltage, but the voltage on the star-connected windings is only 58 per cent of the line voltage. This means that the star-connected winding will have a smaller number of turns and a larger conductor than a delta winding for the same line voltage, and also that the star-connected winding will need less insulation, internally. On transformers for high voltages these are important points, for they permit coils that are relatively strong mechanically and economical of space.

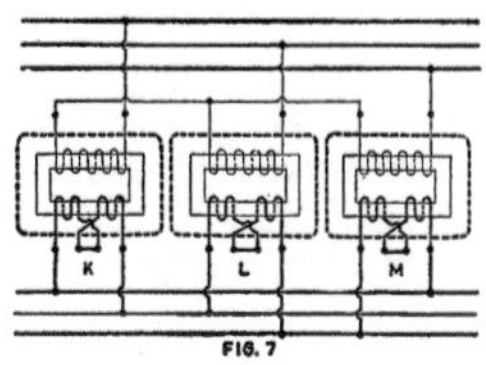

FIG. 7

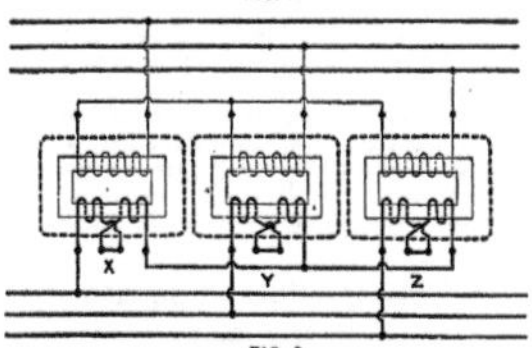

FIG. 8

FIGS. 7 AND 8. THREE TRANSFORMERS CONNECTED STAR TO DELTA AND STAR TO STAR

Fig. 7 shows a star-delta connection. This connection is used chiefly for step-down transformers and has the same advantages as described for the delta-star grouping.

Three transformers connected star to star are shown in Fig. 8. This connection has several disadvantages when compared with the connections, previously described, and it is seldom used. If a sine wave of voltage is applied to a transformer, the wave of magnetic flux in the core must also have the sine form; but the characteristics of iron are such that in order to produce a sine wave of flux the magnetizing current must have a triple-frequency component. In a star-to-star bank this triple-frequency current would have to flow toward the neutral point in all three transformers at the same instant, but as there is no path by which it can flow away from the neutral point at the same time, it is prevented from flowing at all. This results in a flux in the core which is not of the proper form to give a sine voltage, and consequently the voltage wave form of such a bank is distorted. Such an effect is not produced if there is a delta connection either in the primary or secondary, because the delta connection provides a closed path in which the triple-frequency current can flow so as to supply the missing part of the magnetizing current.

The triple-frequency voltage thus generated in a star-to-star bank has several bad effects. If the neutral of the bank is grounded, there will be a triple-frequency leakage current to ground, which will cause trouble in any telephone lines that may be near. This trouble will be aggravated if the neutral of the generator is also grounded, because a complete path will be provided for the triple-frequency current. Under this condition it may even add something to the heating of the apparatus, but this effect is usually small. This system is not adapted to a four-wire, three-phase distribution, because any unbalancing of the load will shift the neutral point so that the voltages to the three lines will be unequal.

Three-phase power may be transformed by means of single-phase transformers connected in several ways.

The delta-to-delta connection is largely used on moderate voltages and has the advantage that it is not entirely disabled if one transformer breaks down. The open-delta connection is used frequently for small amounts of power and occasionally for larger amounts. Its output is 87 per cent of the combined rating of the two transformers. The "tee" connection is used occasionally. Its output is the same as the open delta.

Star-to-delta and delta-to-star connections are standard. They are especially valuable when transformers are to be connected to a high-voltage line, as the star connection decreases the voltage per transformer. A star-to-star connection is rarely used on account of trouble with third harmonics and with shifting of the neutral.

The Qualities of Success

By M. L. Brown

During years of experience as an engineer and an electrician, the writer has often striven to determine to his own satisfaction just what constituted the make-up of a successful chief engineer. What qualities must this person possess to enable him to take complete charge of a power plant, large or small, and in many cases the care of large industrial plants or buildings in conjunction with his power plant? It is a question that should be given serious consideration by all engineers who aspire to be leaders in their profession. It is impossible to lay down any set of rules that will apply to all cases, but there are a number of things that stand out so clearly above other necessary qualifications that they can be given without fear of contradiction.

First of all, an ambitious engineer should possess a fair education, and today there is little excuse for not having one. With correspondence courses, night schools, technical books, magazines and engineering societies, any man can acquire sufficient training in mathematics to enable him to analyze everyday problems in the plant. The engineer must possess or acquire the faculty of analyzing conditions and not jump at conclusions.

It is important to have a personality that appeals to the employer. Lacking this advantage, the engineer should cultivate it, because there is nothing that makes such a good first impression as a pleasant and attractive personality, and in most cases it secures the first opening as chief engineer.

In the course of his work and experience an engineer should acquire a training that will enable him to handle the mechanical end of his work, but even this seems to be a sort of a progressive game. A man might step from one plant as an assistant into another plant as an assistant without difficulty, but to step from an assistant's job to a chief's job is altogether another problem, and it is usually personality that secures it, other things being equal.

After an engineer has reached a chief's position, usually in a small plant, he should progress from one plant to another by easy steps until he has charge of a plant in which he is anxious to stay indefinitely. I do not mean that an engineer should be changing continually, but he should make progress. During this progress the writer has found it a question of work and then some more work, coupled with studying the plant equipment continually until it is mastered.

Subjects on which many engineers are weak where they should be well posted are the principles of combustion, practical and intelligent operation of the boiler furnace, and the analysis of electrical troubles. These subjects are not particularly difficult and yet it is surprising the number of times engineers will throw up their hands over electrical troubles where a little understanding of the principles of operation of electric apparatus and of diagrams would have made the solution easy. It is equally surprising to see what little changes in the boiler setting or in operating methods will do to the fuel economy.

The qualifications enumerated in the foregoing appeal to the writer as the fundamental requirements needed to be a successful engineer. Of course there are a number of problems that he must face as his position grows bigger, such as refrigeration, heating, building and plant maintenance, lighting, etc., but if he has learned to analyze conditions he can take care of those as they come, and it is wonderful the amount of help an engineer can get if he is actually desirous of securing information.

After securing the position of chief engineer, it is a question of keeping everlastingly at it to make good and to keep up to date. To help in this every engineer should join a society of engineers. He not only helps himself, but if he has the proper spirit he will help the other fellow who is trying to make a place for himself.

As the engineer makes progress he must expect and look for many hard knocks and experiences. He should at all times regulate his actions so as to gain and hold the confidence of his employer. If he possesses ability, integrity and resourcefulness, coupled with a willingness to coöperate, he need have no fear of losing his position as chief, because as a rule the firm is just as anxious to hold a good man as the man is to hold his job.

· EDITORIALS ·

The Training of Leaders

THE recent conference in Philadelphia, between industrial executives and representatives of leading colleges and technical schools, has served to concentrate attention upon some serious defects in the education of engineering students. The schools have been fairly successful in training men to design and operate the machinery used in the industries, but they have not made any serious attempt to fit their students either for the handling of men or for the management of industrial enterprises. President Brush, of the American International Shipbuilding Corporation (which operates Hog Island), stated that every time he visited New York, he was importuned by heads of large corporations to find executives for them; that the main qualification needed in management is the ability to handle men, and that this has been completely overlooked by most of the schools.

The substantial accuracy of this statement cannot be questioned. The engineering graduate has usually been trained to deal with things inanimate and has not learned to consider man. Furthermore, throughout his college course he has not come under the broadening influence of contact with many men of many minds. In consequence of this he feels and is at a disadvantage in competition with men of larger experience of men and affairs and seldom rises to the higher executive positions in engineering enterprises. Both the engineer and the country suffer from this. Every year it becomes more important that the industrial executives should have a sound knowledge of engineering as well as a knowledge of men.

Some attempts have been made to meet this need by giving courses in "Engineering Administration," but this has resulted usually in a weakening of the technical courses without much appreciable gain elsewhere.

The best results that have been acquired so far have been through a system in which the students have alternated between college studies and supervised work in industrial establishments. Pioneer work in this direction was started about twelve years ago at the University of Cincinnati. This plan includes a biweekly alternation between college and industrial work for the whole five years of the course, and its success is unquestioned. A coöperative plan, established at the University of Pittsburgh about eight years ago, has also met with marked success.

At both these institutions the co-operation is under a director who arranges for placing the students at such work as will give them the best opportunities for experience, and who has free access to each student at all times. The students receive the same pay as other workmen doing similar work, and are treated like other workmen except in their relation with the director of coöperation. The director also meets groups of the students in the evenings to discuss with them their problems and experience. The student is given a syllabus of questions relating to his work the answering of which requires him to observe closely not only the processes which he himself is carrying out, but also the whole system of management of the factory and such general technical details as he has opportunity to observe. It also requires him to apply to his observations the engineering theory which he has acquired during his studies.

Announcement has recently been made that the Harvard Engineering School is about to put into effect a plan of coöperative work with the industries, which starts at the end of the sophomore year and has a unit of alternation of two months. The plan, as outlined, seems to promise well. It gives broad contact with other students during the most impressionable years, and it adds to the technical instruction, actual experience with labor, organization and management problems.

Senator Frelinghuysen Proposes To Settle the Bituminous-Coal Problem

SENATOR Frelinghuysen of New Jersey recently introduced three bills which he hopes will go far to correct the difficulties in the present coal situation. The first of these bills provides for the termination of Federal control in the coal industry by removing the powers conferred on the President under the Lever Act. The second provides for the appointment of a Federal Coal Commissioner, who will, however, have no authority to fix prices or control the industry, but whose duties shall consist of collecting information regarding coal-trade practices and the industry in general, and in making such information public. The third provides that the Interstate Commerce Commission may authorize a reduction in freight rates during the period from April 1 to August 31 of each year, and an increase during the remainder of the year.

Probably the first of the measures will be of no present effect, since the President has already relinquished control of the coal industry. To a man on the outside the chief feature of the second proposed law is that it provides for a new office carrying a salary of $10,000 a year, a secretary at a salary of $5,000 a year, and "such attorneys, special extra examiners, clerks and other employees as he may from time to time find necessary for the performance of his duties, and as may be from time to time appropriated for such offices." With all the present Government machinery for gathering statistics, the necessity of a few new five and ten thousand dollar jobs of this sort does not seem very pressing, especially under the present scale of Government expenditures. The experience with the late Federal Trade Commission, having much the same relation to general business as the proposed Coal Commissioner is to have to the coal industry, suggests that the principal result to be expected from such a body is the amount of public money it can spend.

The Senator's third bill really promises to accomplish something. It provides that the freight rates for coal shall be filed with the Interstate Commerce Commission by the railroads as in the past. It provides further, however, that the Interstate Commerce Commission may authorize a reduction of fifteen per cent from these rates during the months of April to August, inclusive, and an increase of fifteen per cent during the rest of the year. Such a scheme would require the filing of no new rates and should not lead to any confusion, as the changes would automatically come on April 1 and on September 1 each year without the necessity of any special action.

The passage of this measure would seem to promise considerable assistance in the efforts to stabilize the bituminous-coal industry by encouraging summer buying. The question of lower prices for coal during the summer at the mines would still remain to be settled, but if the railroads are allowed to make this differential, it would make considerable difference in the actual price of the coal to the consumer if bought during the low-rate months.

It has been estimated that 100,000 new coal cars will be necessary to handle properly the usual seasonable demand during the coming winter. Aside from the question of the railroad's ability to finance the purchase of this many cars, there is the physical impossibility of having them built in time. In addition to this, if the railroads had these cars today, the locomotive and terminal facilities would be inadequate to handle the demand, and it is also physically impossible to sufficiently expand these facilities. Meanwhile, under the present conditions, probably thousands of freight cars will lie idle during the coming summer months unless some means to stimulate summer buying are resorted to.

Revising the National Electrical Code

TWENTY-THREE years ago, when the National Electrical Code was first promulgated there was practically nothing that those in the electrical industry could refer to as a guide in their work. There were no standard rules for making electrical installations, and as a consequence much of the work done was of a very inferior quality even when measured by the standards of that time. Electrical installations when improperly made, to say the least, imperil both life and property. From the efforts of insurance organizations to prevent electrical installations being made so as to introduce a fire hazard, the National Electrical Code has grown. The rules at first were but recommendations; however, they have grown until the code today is not much short of absolute law. This is one of the best indications of the wisdom of the policy adopted of revising the code every two years. The recommendations of the 1920 revision were considered at a meeting of the electrical committee of the National Fire Protective Association held in New York City, March 23 and 24, and it is expected that the new code will be ready for distribution during this summer.

The real importance of the National Electrical Code is too frequently not appreciated even by those in the electrical industry. If for no other reason, the code is of great value because it represents the best knowledge and experience in the installation of electrical apparatus and wiring in America. It is not the result of experience of fire-insurance interests alone, but of the best thought in the electrical industry. The committee that revises the code includes representatives of the National Electric Light Association, American Institute of Electrical Engineers, American Electric Railway Association, National Electrical Contractors Association, United States Bureau of Standards, besides representatives from national and municipal bodies interested in fire prevention and underwriters' organizations.

Delayed Maintenance

ONE of the great dilemmas of industry in recent years is maintenance—when to do it, how long to delay doing it. The reason is cost—cost of labor and material. The inclination is to put off the work in the hope that the equipment will satisfactorily continue functioning until costs go down. This practice has been and is now general; the railroads, all industries with exceptions, have allowed needed repairs, overhauling, all kinds of maintenance work, to drift until—well, until the dire necessity of keeping things together compelled it. Material of every kind is burdensomely high in price and has been for a long time. Reported statements of governmental department heads, of brokers, of marketmen generally, are to the effect that we have reached the peak of prices of raw materials. Granting that this is true, it will be some time before the decrease is reflected in lower prices of finished commodities. And their trend now is still up, not down. The butcher, the baker, the candlestickmaker now chorus: "They will go higher." So there is little immediate comfort in announced lower prices of raw materials.

In the power plant it is poor business policy to allow equipment to become badly in need of maintenance. The engineers, and no less the management, must not forget that the power plant is vital. Without power industrial civilization would perish. The failure of a part of an engine, turbine or other piece of apparatus may mean the temporary failure of a whole mill or factory. Whatever the real deep-seated cause of such an interruption, the engineer or power man, whatever his particular title, will be blamed and held responsible. After the fact it will avail the engineer little to explain that he warned this one and that one of the impending failure. Carbon copies of his warnings and requisitions for parts or material for parts may help him. Further, the man who is rightly jealous of his reputation appreciates that accidents that occur to machinery in his charge, whether or not due to his direct negligence, are charged against his reputation both in his plant and elsewhere.

Extravagance is inexcusable; but the wise engineer will see to it that he does not endanger life, property or continuity of service by delay in urgent maintenance.

Just as refrigeration reaches a state of high commercial perfection and finds ever-widening application, along comes a fellow with a prospectus about dehydrating everything from soup to nuts, thus wiping cold storage off the slate with one fell swoop. And the breweries gone, too! Dehydration of foods has its serious and favorably potential sides, but time will likely show it to be an aid to refrigeration rather than an improvement upon it or successor of it.

This is not the time to "saw wood" about electrification of the lumber industry in the Northwest.

CORRESPONDENCE

Why Blame Caustic Soda for Boiler-Plate Cracks?

The explosion of a steam boiler in Canada, caused by the cracking of a drum head, is still the subject of much discussion. (See *Power*, July 8, 1919, pages 70 and 71 and a further contribution to this subject by G. H. Garrison, *Power*, Sept. 23, 1919, page 515). The latter article asks the question why caustic soda was blamed without proof of its presence. The following reply was given: "The boiler plate in proximity to the leak was covered with a white deposit, and the fracture was such as has characterized previous failures traceable to this cause."

The purpose of the following is to show that the "white deposit" and the "characteristic fractures" are not proof of the presence of caustic soda.

Professor Parr in the University of Illinois Bulletin No. 94, to which the article at page 71 refers, described as follows cracks possibly due, in his opinion, to caustic soda. (See page 11 of the Bulletin): Fine cracks develop and proceed from the rivet holes, always below the water line, and in the University of Illinois boilers, always in conjunction with a leak or other condition which promotes a concentration of the soluble material to the saturation point.

The fact that rivet-hole cracks or cracks in or near seams were found, apparently led to the assumption that uncontrollable concentration of caustic soda in inside interstices of seams may take place. However, this has not been proved as yet and does not appear probable for the following reasons.

Caustic soda or other sodium salts are present in the contents of many boilers. In many cases caustic soda or sodium compounds, from which caustic soda may develop, are put into the boiler intentionally for the purpose of overcoming troubles which would occur in their absence. Experience with such boilers has proved that caustic soda does not affect the boiler if its amount is controlled by keeping the concentration of the boiler contents within certain limits.

From the analysis of the Urbana water, page 8 of the Bulletin, and from the composition of the concentrated liquids in the boilers, page 52, it can be calculated that approximately 500 grains of magnesium hydroxide and approximately 1,000 grains of calcium carbonate would be contained in suspension in, or be precipitated from, one gallon of the concentrated liquid before the concentration of the caustic soda would reach 234 grains per gallon, which is, according to the Bulletin, a concentration not only harmless but even beneficial. The scale-forming substances will also enter the interstices of the seam, if such interstices exist at all, and there they will precipitate.

Much work has been done in the last thirteen years to find the causes of cracks in boiler plates. It has been found that there are many possible causes. Most of them have been verified by duplicating them in experiments under conditions which eliminated other influences than those to be investigated. Cracks in boiler plates of good material have been traced, in the absence of other manifest causes, to internal strains resulting from mechanical or thermal treatment. Since most of the work during fabrication is done on the seams, and as thermal influences during operation of the boiler will be greater in seams than in single plates, it is obvious that fine cracks due to internal strains will more frequently occur in the seams.

Shearing, dressing, rolling and hammering of the plates, punching, reaming and drilling of holes, pin drifting and injudicious forcing of plates into surface contact, riveting and calking, expanding of tubes into tube sheets, and careless handling of tools for removing boiler scale are possible sources of brittleness and cracks by bad mechanical treatment. Burning, working the material at blue heat or too much cold working, both without annealing, sudden heating and cooling, insufficient cooling of the walls on the water side due to scale or oil, low water in the boiler, possibly followed by feeding cold water, etc., are the main causes of brittleness and cracks by thermal treatment.

It has also been shown that brittleness and cracks caused by internal strains will be further developed by operating stresses which are greatest in and near seams, or by sudden changes in temperature. Changes in temperature occur more frequently and within wider limits near the grates than more distant therefrom, and this may be one of the reasons why cracks are more often observed in the water spaces than in the steam spaces of boilers. The steam spaces are not always heated and if so, then the flue gases touch them only just before the gases enter the chimney—in other words, farthest from the grates.

Most valuable contributions regarding the problem of cracks in boiler plates have been made by, among others, Stromeyer, Bach, Baumann, Heyn, Bauer, Daibler, Houghton, Stead, Wolff, Rosenhain and Hanson. They have been published in the *Journal of the Iron and Steel Institute, Stahl und Eisen, Zeitschrift deutscher Ingenieure* and *Proc. New York Congress for Testing Materials* from 1907 to 1918. Wolff, *Journal of Iron and Steel Institute*, 1917, gives, besides the results of his own investigation, a short abstract of the work of Kirsch, Foeppl, Leon and Preuss, and refers also to the investigations of Talbot, Howard and Bartho, which have a bearing upon the subject if the further development of existing cracks during operation is to be considered. Wolff also stated that the rivet-hole cracks which he found, occurred not only in the water spaces, but also in the steam spaces of boilers, that they were extremely local and that in most cases no leakage was observed and no boiler scale found between the plates.

Cracks in rivet holes resulting from internal strains cannot easily be detected, since they are hidden from the eye and can be seen only after the rivets are removed. This, of course, will in general be done only when a leak occurs which canot be stopped and which indicates that there is something wrong with the seam.

In Bulletin No. 94 the only test mentioned which has been made with caustic-soda solution of a concentration permissible in a boiler, Table 10, page 56, shows that in this case boiling of the test pieces for fifteen days at a temperature of 200 deg. C. did not change their physical properties for the worse. On the contrary, the alternate bending test showed an improvement of 8 per cent and the impact test showed no change as compared with the untreated sample. This result is quite in accordance with those obtained by Thompson (*Journal, Society of Chemical Industry,* 1894) who found that boiling in caustic solutions of moderate concentrations (which were higher than those usually permitted in boilers) did not appreciably change the physical properties of steel wire, while boiling in water changed them for the worse.

Comparison of the results given in Tables 1, 4 and 5 of Bulletin No. 94 shows that the properties of annealed test pieces have been changed to a much smaller degree by boiling in caustic soda than those of test pieces in which internal strains by bending were introduced before such treatment. This points to the influence of internal strains and is completely in accordance with the results obtained by Stromeyer, Bach and Baumann.

In good boiler practice the amount of all salts dissolved in boiler salines should be below 1,000 grains per gallon. With the exception of one test, which was made with a solution of a concentration within this limit, and which confirmed the beneficial influence of caustic soda, all other tests made at the University of Illinois were made with solutions containing 6,500 grains and up to 60,000 grains of caustic soda per gallon. Furthermore, Professor Parr, on page 11 of the Bulletin, called special attention to the fact that the purpose of his studies was not an attempt to explain the cause of boiler failures. He emphasized that much work remains to be done before a complete and adequate explanation of the case can be offered.

Therefore, no conclusions should be drawn from the results given in the Bulletin, so long as it has not been proved that similar concentration in fact existed in the boiler, and because the cracks in or near rivet holes are not specific characteristics of embrittlement by caustic soda, but result, as has been proved, from internal strains due to other causes, and further no cracks have been developed in specific laboratory tests with caustic soda.

In the Canadian case it has not been proved that the water used contained caustic soda and, that the accumulation of incrustations were of alkaline character, or that there were cracks developed from rivet holes. The white deposit mentioned proves only that there was a leak, since not only caustic soda, but all salts which are usually contained in boiler water, are of a white color, as for instance, all the chlorides, sulphates, carbonates and hydroxides of the main scale-forming elements, calcium and magnesium, as long as they are not colored by iron salts (rust), organic matter, etc. Therefore, it is obvious that the Bulletin gives no basis for drawing the conclusions referred to.

The crack described in the Canadian case shows, however, a certain characteristic not mentioned by Professor Parr; namely, that it, in general, follows a line which is near and parallel to the calked edge of the overlying plate and that its larger part is situated between the calking edge and the row of rivet holes. The crack shown in Figs. 1 and 2 on page .70 of the July 8, 1919, issue is similar to that of a boiler exploded in Wisconsin and shown in Fig. 2, page 605, of *Power*, Oct. 21, 1919. The latter crack occurred along a longitudinal seam above the water line, as stated in the description. Similar cracks along the calked edges of the overlying plates have occurred in steam boilers used for concentrating caustic soda and in steam boilers in which no caustic soda was present—in both cases above the water line as well as below. The occurrence of such cracks above the water line could not have been caused by caustic soda, and since the characteristics of the cracks below the water lines are the same and in a number of cases occurred in the same boilers as those above the water line, one cannot see why the former should be considered to be caused by caustic soda while the latter certainly were not. The cracks along riveted seams above the water line, however, can, in the absence of other manifestations and when good material has been used, be explained only by internal strains resulting from mechanical or thermal treatment or both.

Users of steam boilers, owners as well as operators, can assist the solution of the problem in question by careful operation and examination of their boilers and calling experts when irregularities are observed. If they do this, many troubles will be prevented. Periodical inspection of boilers by experts is of great importance. Making use of the services offered by the steam-boiler inspecting and insurance companies and similar institutions cannot be too strongly recommended.

New York City. H. KRIEGSHEIM.

Freeze-Proof Mixture in Gage Siphon

I cannot agree with Mr. Coyne, on page 433 of the March 16 issue, as to the introduction of freeze-proof mixture in siphon of steam gage, unless a special piping arrangement is used to trap the evaporation of the freeze-proof mixture, also to prevent the accumulating condensation from floating out of the siphon. Otherwise the kerosene, if used, or the nonfreezing mixture will not be effectual.

It is possible for the safety-valve springs to become permanently set even in a well-managed plant that is operating up to or over rating. Mr. Coyne's suggestion to close the stop valve when the boiler is operating at or beyond capacity, in order to test the capacity of the safety valve is really one of the State of Pennsylvania Safety Standards. But why not test the valve before the boiler is put on the line?

Uniontown, Pa. H. D. NOLAN.

Wire Scraper Made from Hacksaw Blade

A very handy wire scraper for use around a repair shop and in armature winding, can be made from an old twelve-inch hacksaw blade. Put the blade on the emery wheel and grind off the teeth. Then, by using two pairs of pliers, hold it in a blowtorch flame and heat to a dull red and bend it in the shape shown in the figure.

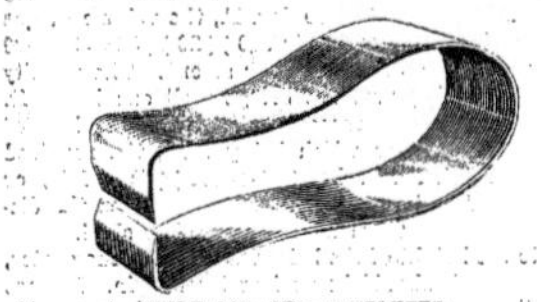

WIRE SCRAPER COMPLETED

Heat and bend only a small portion at once and do not let it get very hot. Allow the blade to cool slowly and do not put it in water. When in the proper shape and cooled, grind the ends to bevel inwardly as shown.

This tool is very handy for scraping solder from old commutator leads or for removing cotton insulation from new leads. J. W. FAGER.

New Orleans, La.

Slotting Commutators in a Lathe

There are various devices for slotting the mica from between commutator segments. We have found it very effective to do the work in the lathe after the commutator has been turned off. Make a small parting tool,

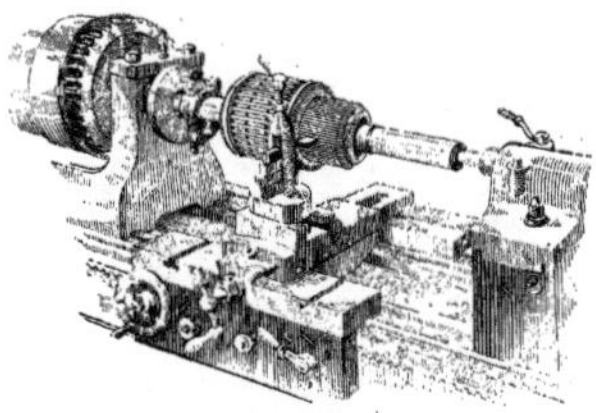

ARMATURE IN LATHE READY TO HAVE MICA SLOTTED FROM BETWEEN SEGMENTS

a little narrower than the mica between the segments, and set it horizontally in the tool post, as indicated in the figure.

Take off the driving belt and put in the back-gear, block the dog in the faceplate slot with a wooden wedge to take up the lost motion, then turn the cone pulley by hand to bring the mica in line with the tool and push the tool through the mica by moving the carriage toward the headstock with the hand gear.

A. H. HALLADAY,
Master Mechanic Spring Valley Coal Co.

Spring Valley, Ill.

A Handy Shop Crane

In repair shops where there are no cranes it is always more or less of a problem how to handle pieces of equipment of only medium weights, to say nothing of the larger parts. Where lathes and other machine tools are located within eight or ten feet of a column, the handling of parts can be greatly simplified by the use of the home-made monorail crane that is shown in the figure.

This device consists of ordinary 4- or 6-in I-beam, one end is fastened to a column by a two-part band, the latter made so that it can be tightened on the column by two bolts. The outer end of the boom is supported from the top end of the column by a two-part band. On the I-beam is placed a trolley such as can be purchased at most any machinery supply house, and from this trolley a chain fall is supported.

CRANE INSTALLED IN SHOP

This crane can be operated through a radius of at least 180 deg., making it useful in transferring equipment from one part of the shop to another at will, as well as lifting parts in and out of machine tools. By putting up a special mast in the shop to support the boom from the crane can be so constructed as to have a 360-deg. travel. The length of the boom will depend upon how high the ceiling of the shop is; the boom rod should make not less than about 30 deg. with the boom and 45 deg. is better.

Different conditions will require different methods of solving the lifting problem in the shop, but with the foregoing idea in mind the resourceful repairman should have no difficulty in easily providing his shop with facilities for handling all heavy parts. Such cranes will also be found invaluable in erecting equipment in the shops. B. A. BRIGGS.

New York, N. Y.

Distance of Bridge Wall from Boiler Shell

In the Feb. 24 issue of *Power*, page 309, Albert L. Smythe makes inquiry concerning the height of bridge walls for his boilers. The question is not so simple as it appears and is of considerabe importance, particularly in the bituminous-coal region. Before determining the proper height of the bridge wall, it may be well to outline the functions for which it is intended:

1. To retain the fuel on the grates, its height being a little above the proposed thickness of the fuel bed.

2. The bridge wall should be of such distance from the boiler shell as not to restrict the passage of the gases arising from the fuel bed, and this distance should be inversely proportional to the draft that is available.

3. Considerable free air enters at the back end of the fuel bed, where the grate is farthest from the fireman and is more difficult to maintain uniformly covered. Unless the air so admitted is retained long enough in the furnace to acquire the same temperature as the gases arising from the fuel bed, it will pass along toward the stack without an opportunity to mix with the combustible gases; that is, the free air, being heavier than the gases from the coal, will sweep through the combustion chamber as an underlining to the gases of higher temperature. This is particularly true in furnaces burning high-volatile coal.

4. Absorption by the boiler of radiant heat from the fire depends largely on the intervening distance, and therefore it is desirable to have the boiler as close to the fuel bed as the requirements of good combustion space will permit.

5. Draft in all cases should be so utilized as to secure the best results, and any unnecessary restriction without proper or adequate returns is undesirable waste.

Summing up, it is my belief that the bridge wall should be as high above the grates as possible, provided it does not restrict the gas passage above it, and the boiler should be as near to the fuel bed as conditions will permit.

With a chimney 100 ft. high serving two boilers, one 78 in. x 19 ft. and the other 72 in. x 18 ft., with grates 6 x 6 ft., the chimney draft at the damper will range from 0.55 or 0.5 in. of water. The loss in draft permissible from the fuel bed to the back end of the boiler is less than 0.05 in. and the area over the bridge wall should be 25 per cent of the grate area, or $36 \times 0.25 = 9$ sq.ft. The first dimension, the width of the opening over the bridge wall being 6 ft., the second required dimension, the distance from the top of the bridge wall to the nearest point of the shell, will be $9 \div 6 = 1.5$ ft., or 18 inches.

It is understood that the bridge-wall top will be straight and not curved to coincide with the curvature of the shell. The 18 in. is the distance between two straight lines, one drawn on top of the bridge wall and the other tangent to the outer circumference of the boiler shell at the nearest point to the top of the bridge wall.

Although 25 per cent of the grate area between the boiler shell and the bridge wall is only a working rule, it meets the requirements in most cases. Hand-fired furnaces are difficult to operate at a much higher rate than 25 lb. of coal per hour per square foot of grate area, so that the rule is based on this rate of combustion. If in the plant under consideration, both of the boilers are operated, as they should be, at the maximum capacity of the grate area, the total coal per hour burned by each boiler will be $36 \times 25 = 900$ lb., which will develop approximately 180 to 200 hp. Assuming 20 lb. of gas per pound of coal and 52 cu.ft. in a pound of gas at 1,800 deg. F., the volume of gas to be handled will amount to $900 \times 20 \times 52 = 936{,}000$ cu.ft. per hour = 260 cu.ft. per second. The velocity of gases passing over the bridge wall can be 30 ft. per second without exceeding the allowable loss of draft. The furnace width being 6 ft., the required third dimension will be, $260 \div (30 \times 6) = 1.44$, or practically 1.5 ft. = 18 in., which should be the distance from the bridge wall to the boiler shell.

As to the height of the bridge wall above the grates, in all cases it should be no less than 16 in. In case the boiler is so set as to make it impossible to provide a 16-in. bridge wall, the grates should be dropped at the back end, say about 1 in. to the foot. If this does not give the desired height, then the area over the bridge wall may be reduced to, say, 20 per cent instead of the desired 25 per cent.

The best bridge wall for burning bituminous coal rises 18 to 20 in. above the grates, and the space between it and the shell for gas passage should be equivalent to 25 per cent of the grate area. The area above the bridge wall should be taken as a rectangle, one dimension being the width of the bridge-wall top and the other the shortest straight line from the bridge wall to the shell.

The extra area available at both sides, due to the curvature of the boiler shell, is considered only in extreme cases, which usually require long practical experience to enable one to assign relative importance to the various factors involved.

Chicago, Ill. HENRY MISOSTOW.

In a boiler plant formerly under my supervision, containing eighteen 125-hp. return-tublar boilers, each 16 ft. long, 72 in. shell diameter and containing seventy 3½-in. tubes, stoker-fired, the distance of the grate line from the boiler shell was 60 inches and the distance of the bridge wall from the boiler shell 18 inches. It was found that when the bridge wall was maintained at 18 in., satisfactory results were obtained and the fire would not hog away at the end of the grate surface nearest the bridge wall, whereas when the bridge wall would slough away to around 24 in. the fire would not burn properly and required considerable working to keep it in condition around the vicinity of the bridge wall.

In addition to maintaining the bridge wall, it was also found to be of advantage to build the wall straight down to the floor line of the combustion chamber instead of tapering it, as the former design gave more space for the proper mixing of the gases of combustion, resulting in flue-gas temperatures of around 500 deg. F. and flue-gas CO_2 of 12 per cent average. Keeping the tubes and the interior of the boiler clean and the air leaks stopped were as important as anything else in producing good results.

In Mr. Smythe's case I feel that, if the bridge walls of his boilers are maintained at 12 in. from the boiler shell, and the grates set from 36 to 44 in. from the shell, with the other practices I have followed, he will certainly obtain better results than by having the bridge wall set 18 in. or more from the shell. For stoker installation I find 18 in. space satisfactory. This, in my opinion, is too much space when firing bituminous coal on hand-fired grates. Ten inches is slightly too close for good economy, as the furnace would not have sufficient gas outlet and there would be a waste of fuel.

I would further suggest that Mr. Smythe procure a flue-gas analyzer and portable pyrometer or flue-gas thermometer, and with these instruments he can prove to his superior officer whether he is right or wrong at any time. Personally, I would not be without these two instruments, even if they were used only a few times a year. PAUL R. DUFFEY.

Hagerstown, Ind.

Pulverized Coal Under Boilers

In regard to the results obtained with powdered fuel in Milwaukee, as outlined in the article by John Anderson, on pages 336 to 339 and 354 to 359 of the March 2 issue of *Power*, the writer believes it is fundamentally wrong to take coal and prepare it to the most expensive condition imaginable (powder) and still burn it in its raw state; that is, with all its hydrocarbons. A much more fertile field for research and more rational procedure is to extract some or all of the byproducts, leaving coke and gas for fuel. Ultimately, we must come to some such process in this country, and until that time it will be difficult to find a complete analysis of figures that will show anything but a loss for powdered fuel as compared to stoker firing.

The ash carried out by the chimney gases is a serious matter. It is reported at 82.8 per cent of the ash content of the coal. This is an item that should not be lightly passed. It is not sufficient to simply state that it is "apparantly carried a long distance" or that "no deposit could be noticed in the street." Some accurate way of measuring this discharge should be found before definite statements concerning it are made.

In figuring the cost of coal preparation, only the equivalent fuel cost was charged and that at 1.5 lb. of coal per kilowatt-hour. This figures 17,826 B.t.u. per kilowatt-hour. Few, if any, stations can assume such economy as this. The correct figure for cost of current should include all items—coal, labor, capital charges; in fact everything that goes to make up the ultimate cost of a kilowatt-hour at the point of consumption.

The cement industry has used powdered coal for twenty years and no doubt has much more data on the subject than steam producers. Such plants with established figures on cost of pulverizing coal, find that it costs in excess of $1 per ton to prepare the coal. If complete costs were figured for the Milwaukee installation, the cost of preparation will probably approach this figure and more than offset the alleged saving.

The type of stoker with which the comparison is made, has a 5.9 per cent difference between the gross and the net efficiency. This allowance for auxiliaries cannot be taken as typical.

From the author's presentation of the subject the writer believes therefore; first, that the ash nuisance is unsolved; second, in order to present even a slight saving in favor of pulverized coal, it is necessary to use low-moisture coal, omit interest and depreciation on the additional investment, omit a portion of the cost of preparing the fuel and make a comparatively large deduction for stoker auxiliaries.

Correcting these items, the showing in economy would be in favor of stoker firing. T. A. MARSH,

East Chicago, Ind. Green Engineering Co.

Calculations From a Flue-Gas Analysis of Loss Due to Incomplete Combustion

On page 129 in the Jan. 27 issue of *Power* is published an article by H. B. Brayton, entitled "Calculations from a Flue Gas Analysis of Loss Due to Incomplete Combustion," and showing formulas for calculating the latent losses of heat. These data have been thoroughly perused by me, and I must say that Mr. Brayton's methods are excellent. While he says little in text form about the great danger (from an economical point of view) resulting from the formation of carbon monoxide and hydrocarbon in flue gases, he leaves everything to the readers and lets his formulas speak for themselves. Those who have gone over the matter carefully will agree that the figures given make it easy to arrive at accurate results.

Since carbon monoxide and hydrocarbons ($CO + CH_4$) are the combustible constituents (we must not forget the latter) the presence of which causes considerable loss of fuel in most boiler plants above and below a certain percentage of carbon dioxide, it is evident that, since we can record CO_2, an instrument for the continuous analysis of CO and CH_4, together with the CO_2 recorder, would undoubtedly form a mighty fine piece of boiler room control by direct method.

Mr. Brayton's formulas, as clear as they are, are only then of real value when means for accurate determination of the combustible gases are at hand. For this purpose he recommends an Orsat apparatus to determine the carbon monoxide. However, the manipulation of such hand gas analyzer with particular reference to carbon-monoxide analysis is a difficult one, and even though the percentage of CO might have been accurately determined, the hydrocarbons cannot be arrived at with this method. Moreover, the simplified apparatus now put out by various concerns for power-plant use are equipped with burettes of only 50 cubic centimeters, reducing the accuracy considerably. Then, too, while with cuprous chloride and a great deal of patience finally one of two more or less accurate analyses may have been obtained, they in themselves are of no real practical value, as they only represent two snap samples showing what the CO has been from 10 to 15 minutes before they were taken (5 minutes is required for a correct analysis of CO). Thus we may as well forget all about CO analysis.

Hydrocarbons may be present in ordinary boiler-room practice to the same detrimental extent as CO, depending greatly, of course, upon the kind of fuel used and certain other conditions; therefore, in no cases where we speak of combustible gases must this hydrocarbon be forgotten, which in itself explains that CO tests taken with the ordinary apparatus are meaningless.

Continuous analysis of CO_2 and continuous analysis of CO and CH_4 is of real value and a reliable instrument that requires but little attention and is capable of analyzing the three gases mentioned would be worth having. F. D. HARGER.

Buffalo, N. Y.

Too Much Regrinding of Stop-Cock

The erratic action of the steam-gage was called to my attention by the fireman, who said the gage made a jump of 20 lb. all at once. I investigated and found

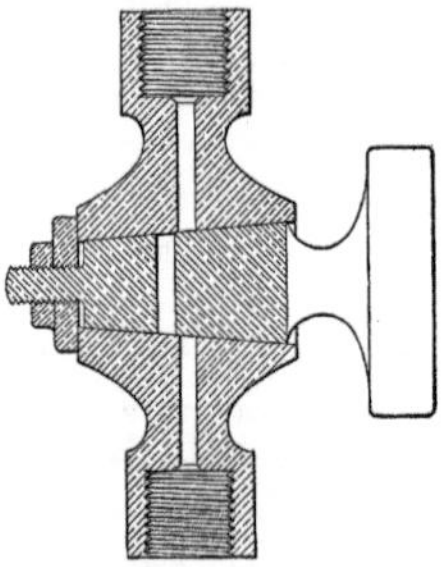

RESULT OF TOO MUCH REGRINDING

that the cock in the steam-gage connection had become blind on account of many grindings, as shown in the sketch.

Sometimes a little thing like the foregoing may cause great damage. THOMAS NEILSON.

Philadelphia, Pa.

Solution of a Heating Problem

Until recently the buildings of the National Headquarters of the American Red Cross in Washington, D. C., were heated by two separate heating systems. The original large building, requiring 16,000 sq.ft. of hot-water radiation, was served by two sectional, cast-iron hot-water boilers having a capacity of 13,725 sq.ft. of hot-water radiation each. The four annexes, requiring a total of 15,681 sq.ft. of steam radiation, were heated by three sectional, cast-iron water-tube steam boilers, located in a boiler room in the fourth annex. These three boilers have a capacity of 32,500 sq.ft. of steam radiation.

In September, 1919, it was required that the heating plant be rearranged so that the main building, which was equipped for hot-water heat throughout, could be heated from the steam boilers in the fourth annex.

It was first decided to reduce the amount of radiation in the main building, change all the piping and heat the building by steam direct from the boiler plant. After estimates had been obtained for this work, it was found that such a plan would be too expensive, the lowest estimate obtained being $19,000.

The system which is now in operation was then designed, a brief description of which follows:

The piping in the boiler room of Annex 4 was rearranged in such a manner as to insure an even distribution of steam. A 10-in. main was laid in a concrete conduit, covered with concrete slabs, from the boiler room of Annex 4 to the boiler room of the main building. The boilers in the main building were removed. Two specially designed converters were installed. These consisted of a cast-iron shell, 20 in. in diameter and 48 in. long, filled with drawn copper pipes terminating in a head. The head is provided at top and bottom with 10-in. connections for the feed and return mains of the existing hot-water heating system.

Steam from the aforementioned 10-in. main is introduced into the shell through a 7-in. connection at the top of each tank. The condensation from the tanks is carried to a receiver which is emptied into the boilers at intervals by means of a three-stage, automatic, electrically driven pump. Thermostatic regulators are placed in the steam lines to each of the tanks, controlling the temperature of the water circulating in the system.

The greater part of the work was done during the cold weather, and at no time were any of the buildings without heat, as the boilers in the main building were operated up to the time that the new system was cut in.

The heat throughout all of the buildings is much more satisfactory than ever before, and the operation of the present system costs less than half of the operation of the two old systems. The completed cost of the system now in operation was $8,300.

Washington, D. C. J. P. LAWYER.

Using Boiler Tubes in Place of Angle Irons for Supporting Baffles

The illustration shows a way of using old boiler tubes in place of angle irons to hold up baffles in the

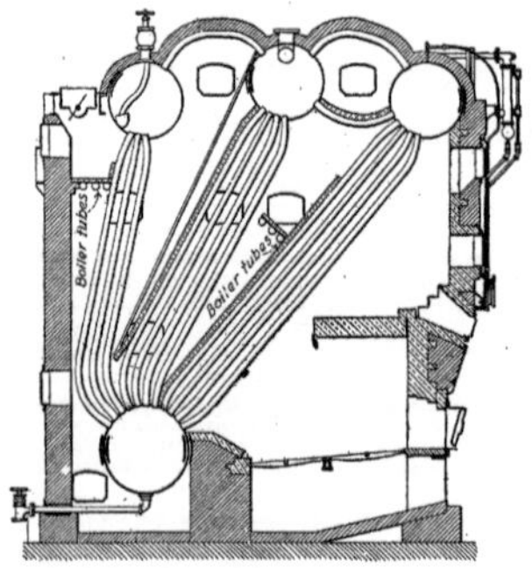

BOILER TUBES USED TO SUPPORT BAFFLES

Stirling type of boiler. By leaving one or both ends of the tubes flush with the outside of the setting, air will pass through them and keep down the temperature of the baffle support, whereas an angle iron would warp and let down the baffle. L. S. SHAW.

Cleveland, Ohio.

INQUIRIES OF GENERAL INTEREST

Blind of a Valve—What is meant by a valve being one-eighth inch blind? C. T.

The term blind is sometimes used with reference to a valve to signify that it is closed. One-eighth inch blind would signify that the valve overlaps the opening edge of the port one-eighth inch.

Advantages of Four-Valve Engine—Why is a four-valve engine more economical than a single-valve engine? W. L.

Four-valve engines can be designed and constructed with shorter steam passages and less percentage of waste cylinder-clearance spaces than necessary for single-valve engines; and the valves can be adjusted and controlled independently of each other so as to obtain better steam distribution, especially with a variable load.

Size of Uptake for Horizontal Return-Tubular Boiler—What should be the size of smoke uptake for a 72-in. x 18-ft. horizontal return-tubular boiler? W. L. M.

The uptake area is usually made one-eighth the area of the grate, although, with tight connections and a tall stack this figure may be considerably reduced. However, it is well to provide for extreme conditions and allowing for 40 sq.ft. of grate area, and one-eighth grate area for that of the uptake would give 5 sq.ft., or 720 sq.in. for the cross-sectional area of the uptake.

More Convenient Observation of Blowoff Leakage—Our boiler blowoff line discharges to an open sewer at some distance from the boiler room. What method can be employed to indicate near the boiler whether there is leakage of the blowoff valve? B. H.

If the blowoff line is so arranged that there is a vertical length conveniently observable and readily sprung apart when cut and supplied with a bolted flange union, insert such a union and each time, after blowing down, remove the bolts and insert a piece of tin plate between the flanges. Any leakage then will be revealed by water intercepted by the tin plate overrunning its edges.

Increase of Power by Operating Condensing—What would be the increase in power of a 12 x 18-in. engine running 200 r.p.m. noncondensing, if operated condensing with 26 in. vacuum? A. B. T.

A 26-in. vacuum is a pressure $26 \times 0.491 = 12.76$ lb. per sq.in. less than atmospheric pressure. If the back pressure when operating noncondensing is 2 lb. per sq.in. above atmospheric pressure, the reduction of back pressure from operating condensing would be $2 + 12.76 = 14.76$ lb. per sq.in., and the power would be increased

$$\frac{14.76\ (12 \times 12 \times 0.7854) \times 18 \times 2 \times 200}{12 \times 33{,}000} = 30.35 \text{ i.hp.}$$

Proportions of Link Valve Gear—For a link-motion reversing valve gear, what should be the length of eccentric rods and dimensions of the link, in terms of the throw of the eccentrics? D. Y.

Ordinarily, the radius of the Stephenson link should be made equal to the length of the eccentric rods reckoned from the center of the eccentric. Without a rocker the radius may be greater and with a rocker it may be less, but deviation from the normal should not be enough to derange the leads. The eccentric rods should be twelve times the eccentricity of the eccentrics and the operative length of link should be four times the eccentricity or more, if space is available, although a skillful designer may employ the inequality introduced by short rods or a short link to adjust a link motion.

Airbound Water-Supply Line—We have a 2½-in. gravity water-supply line about 1,000 ft. long that becomes airbound whenever the line is drawn upon during a few hours to its fullest capacity. How can the trouble be remedied? A. R. H.

Liberation of air results from reduction of pressure of the water, especially at high points in the line where excessive drafts produce a siphonage. The remedy would be to place vent-cocks in the top of the pipe at such points. The vents should be left open a very little all the time. If the slight waste of water is objectionable, it can be prevented by connecting the vents to small-sized standpipes extended higher than the level of the water in the supply reservoir.

Rotation Direction of Alternator—Does it make any difference which direction a newly installed alternator revolves? Or will the exciter connections to the alternator have to be reversed in case the generator is rotated in a direction opposite from that intended by the manufacturer? The exciter is driven from the alternator. E. G. H.

If the machine is operated as a single unit, it will not make any difference in which direction the rotor revolves. If the direction of the exciter is reversed, it will be necessary to cross either its armature or field leads so that the machine will generate. Where the alternator is operated in parallel with other machines care will have to be taken to see that the new machine has the proper phase relation to the bus voltage.

Feeding Return-Tubular Boiler Through Blowoff—What are the advantages and disadvantages of feeding a return-tubular boiler through the blowoff? G. A. B.

The single advantage of feeding a boiler through the blowoff is that it saves provision of a separate feed connection. The advantage sometimes claimed that, by feeding through the blowoff, the flow of feed water keeps the blowoff flushed of mud, is likely to be defeated by mud filling both blowoff and feed connection when there is intermission of feeding. In cases where the feed water contains impurities, like lime, that are precipitated by heat, feeding through the blowoff is likely to contribute to scaling up of the blowoff connection. Another disadvantage is that by feeding in bulk through a hole in the shell, the greater or less irregularity of discharging feed water at lower temperature than the temperature of the boiler water gives rise to differences of expansion that are injurious to the boiler. All these objections, except scaling and rusting of the feed pipe itself, are overcome by feeding through a pipe introduced through the shell above the fire line and carried thence for some distance through the boiler, so the feed water may be raised more nearly to the boiler temperature and then discharged at some distance from the shell or stays.

Fuel Oil Versus Mechanical Stokers*

By F. H. DANIELS

Almost all comparisons of the values of oil and coal consider the burning of fuel oil and the burning of hand-fired coal. It is the purpose of this article to show a comparison of fuel oil with coal burned on up-to-date mechanical stokers. Before making these comparisons it may be interesting to review very briefly a few of the facts regarding fuel oils.

The petroleum production of the world in 1917 was approximately 503,000,000 bbl. Of this amount 342,000,000 bbl. was furnished by the United States. In the United States up to 1918, we have used 4,250,000,000 bbl. of oil, and it is estimated that there remains 6,900,000,000 bbl. This calculation of the oil yet unused in the United States may be seriously upset by the discovery of new fields or by the utilization of the oil-bearing shales. All of the figures given are based on reports of the National City Bank of New York and of the United States Geological Survey.

The petroleum from the eastern fields of the United States has a paraffin base and is generally too valuable to use as fuel. The crude oils from the southwestern part of the country, however, have an asphalt base and after the

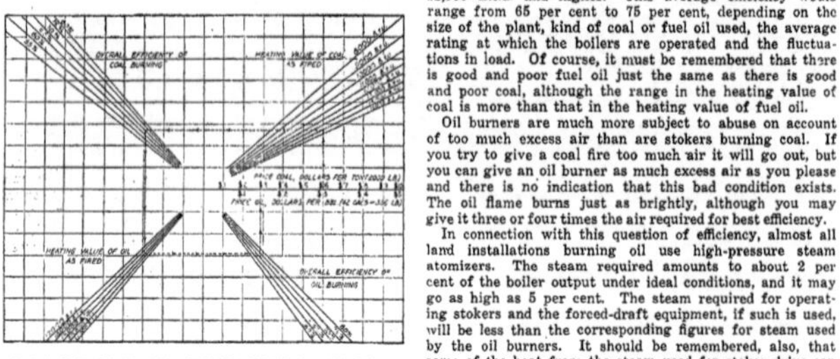

CHART FOR COMPARING PRICES OF COAL AND OIL TO GIVE THE SAME EVAPORATION PER DOLLAR

gasoline, naphtha, benzine and kerosene are taken off, the residue is sold as fuel oil. In some cases the crude petroleum is burned without first taking off the naphthas. The total amount used as fuel equals about one-half the country's output. In general the United States has got to look to Mexico and Texas for cheap fuel oil.

If the cost of coal, its heating value in B.t.u. and the over-all efficiency at which it is burned are known, the accompanying chart shows the price at which oil will compete, considering the different heating value of the oil and the over-all efficiency obtainable with oil burning. The example worked out on the chart shows $6 coal, 14,000 B.t.u., 70 per cent over-all efficiency, as compared with oil at $1.45 a bbl., 19,000 B.t.u., and used with an over-all efficiency of 75 per cent. This chart takes into account only the cost of the two fuels and does not take into consideration costs such as labor, maintenance or the first cost of the plant. It should be noted that the heating values of both the oil and the coal are based on fuel as fired.

Taking 18,250 B.t.u. for the heating value of the oil and 14,000 B.t.u. for the corresponding heat content of the coal, we would have to burn 4.56 bbl. of oil to equal one ton of coal. This is assuming equal over-all efficiencies. The fuel-oil salesmen are claiming in New England as low as three barrels of oil to one ton of coal. This is absurd as it would require over 100 per cent over-all efficiency for the oil with 70 per cent for the coal.

The chief advantages claimed for the use of oil as compared with coal are as follows: High efficiency, high capacity, low labor costs, low first cost of equipment, no smoke, quick response to overload demands, less storage space required than coal, small amount of refuse, no banking losses. The disadvantages are: Fluctuations in price, uncertainty of supply, fire risk. The chief advantages claimed for fuel oil will be taken up one by one.

Stoker Efficiency as Good as That of Oil Burners

Of course, under the best of test conditions, fuel oil has shown over-all boiler efficiencies up to 82 per cent, and some tests show even above this. But the up-to-date mechanical stoker can show equally good results, and it is fair to assume that under average operating conditions the efficiency of furnaces using fuel oil will be no greater than that obtained with mechanical stokers burning coal of 12,500 B.t.u. and higher. This average efficiency would range from 65 per cent to 75 per cent, depending on the size of the plant, kind of coal or fuel oil used, the average rating at which the boilers are operated and the fluctuations in load. Of course, it must be remembered that there is good and poor fuel oil just the same as there is good and poor coal, although the range in the heating value of coal is more than that in the heating value of fuel oil.

Oil burners are much more subject to abuse on account of too much excess air than are stokers burning coal. If you try to give a coal fire too much air it will go out, but you can give an oil burner as much excess air as you please and there is no indication that this bad condition exists. The oil flame burns just as brightly, although you may give it three or four times the air required for best efficiency.

In connection with this question of efficiency, almost all land installations burning oil use high-pressure steam atomizers. The steam required amounts to about 2 per cent of the boiler output under ideal conditions, and it may go as high as 5 per cent. The steam required for operating stokers and the forced-draft equipment, if such is used, will be less than the corresponding figures for steam used by the oil burners. It should be remembered, also, that some of the heat from the steam used for stoker drive can be recovered by exhausting the stoker engines into feed-water heaters, while the steam used for oil burners is an absolute loss. If electric drive is used for stokers, the amount of steam taken from the boiler output will be much less, because the electric motors then share in the economy of the main prime mover.

Comparison of Capacities Is Not Unfavorable to Stokers

The capacities with a multiple-retort underfeed stoker will show higher than any that can be obtained with fuel oil. Most tests on fuel oil have been in connection with boilers where the furnace volumes were restricted. There have been no tests published where fuel oil has been used in connection with a furnace large enough to really enable the oil burners to show up to best advantage. It is safe to state, however, that with equal furnace volumes mechanical stokers will show higher capacities than are possible with oil.

As for labor, under proper conditions the mechanical stoker should not require more than one fireman and one water tender for every 5,000 hp. of boilers. It is impossible for a plant equipped with fuel oil to get along with a less number of men than this. As far as attending to the water level of the boilers is concerned, the labor involved is equal in each case, and the stoker does not require any more adjustment than would be needed for the manipulation of oil burners.

*Based on a paper read before the American Boiler Manufacturers Association.

Of course it must be admitted that the use of fuel oil practically eliminates the question of ash handling, but this is not a very serious factor because the amount of labor required for handling the ashes in an up-to-date plant is not very large and the class of labor is the lowest paid.

In connection with first cost of equipment the use of fuel oil avoids the necessity of the purchase of mechanical stokers and auxiliary equipment, which might range in cost from $8 to $14 per rated boiler horsepower. The fuel oil also avoids the necessity of coal- and ash-handling equipment, which would cost about $6 to $8 per rated horsepower. This would make a total of about $18 per horsepower. As an offset to these items fuel oil requires storage tanks, oil pumps and heaters, which will cost anywhere from $10 to $30 per horsepower, depending upon the amount of reserve oil which is to be kept in storage. For a central station of 5,000 horsepower the complete oil-burning equipment with storage tanks for five months' supply of oil has been estimated at $250,000, which would be at the rate of $50 per horsepower. An industrial plant could, of course, get along safely with much less than a five months' supply of oil on hand, but in almost any plant there will be little if any decrease in first cost in favor of fuel oil.

Smokeless Combustion Possible with Both Fuels

The mechanical stoker is every bit as good as oil burning as regards the feature of smoke. The air to the oil burners should be cut down for best economy so that a slight haze is emitted from the chimney. Additional air will stop the smoke completely, but it will be much in excess of the amount needed for combustion and will, of course, reduce the efficiency of operation. Some stoker installations can be criticized for carrying solid particles of finely powdered ash up the chimney, but as an offset to this the waste gases from a fuel-oil plant have an extremely unpleasant odor, which is very objectionable in thickly settled communities. It is also claimed that there is an oily residue which settles out of the gases from fuel-oil plants and that this residue is extremely hard to remove.

The underfeed stoker of today can raise a boiler from banked fire to 300 per cent of rating or more in from five to ten minutes. The plant with fuel oil cannot boast of any quicker response than this. The underfeed stoker can be shut down from high capacity almost as quickly, and would not compare unfavorably with the oil burner where it is merely necessary to stop the supply of fuel to shut down the boiler. In either case the heat stored up by the brickwork of the furnace is the limiting feature in time required to shut down.

Storage Space Requirements for Oil and for Coal

To compare storage-space requirements, we will assume an average calorific value for fuel oil of 18,000 B.t.u. per lb., and an average calorific value for bituminous coal of 13,500 B.t.u. per lb. The specific gravity of fuel oil will average about 0.95; which means that a cubic foot of the oil would have an average weight of 59.5 lb. A cubic foot of crushed bituminous coal will average about 52 lb. Therefore, for equal calorific values fuel oil will occupy only about two-thirds of the storage space required for coal. On board ship this is very important, but on land the fuel oil is stored in tanks and these tanks cannot be placed too close together on account of the insurance regulations. Coal, on the other hand, can be stored in one continuous pile, and the land occupied for the storage of fuel oil would amount to more than that required for storing an equal calorific value of coal.

While it is true that the burning of fuel oil leaves no ashes to be disposed of, there is the question of the residue which is deposited on the heating surfaces of the boiler. This residue is oily and extremely hard to remove and is highly nonconductive as regards heat. If it is allowed to remain on the boiler tubes, it reduces the economy of the installation materially. In order to remove this oil residue from the tubes, it is necessary to use steam soot blowers four or five times a day. This is more often than is necessary to remove the soot from the coal-fired boiler tubes and results in an extra amount of steam as compared with a coal-fired boiler.

In the case of a plant where fuel oil is used and where the load factor is low, there is claimed to be a considerable saving due to the fact that there are no banking losses. In the case of a stoker the coal consumption on a banked fire is approximately 30,000 B.t.u. per sq.ft. of grate surface per hour. With 14,000 B.t.u. coal this would amount to 2.15 lb. per sq.ft. grate surface per hour. If we consider a 500-hp. boiler with a grate surface of about 80 sq.ft., the coal used per hour would amount to 172 lb., or equal to 2,400,000 B.t.u. This heat, however, is not entirely a loss, because it serves to keep the furnace brickwork and the water in the boiler hot and ready to meet a sudden demand for steam. When the oil-fired boiler is shut down, the water and the furnace brickwork cool down and just that much more oil and time must be consumed to get the boiler into condition for steam generation. There is with oil some saving of banking losses, but under ordinary conditions this saving is not of any great magnitude.

Price and Supply of Oil Uncertain

Considering the disadvantages of oil, its price has been subject to tremendous fluctuations in the past, varying from 2.5c. per gal. in 1915 to a maximum of 7c. during the war period. The corresponding range of coal prices from 1915 to the war period would be, in the Pittsburgh district on run-of-mine steam coal f.o.b. cars, from $1 to $2. The fuel-oil market is more susceptible to disturbing influence than is the coal market. During the war practically all the fuel oil produced was used by the allied navies and there was hardly any surplus for land use. The merchant ships are now shifting over in large numbers to fuel-oil burning, and just as soon as any considerable percentage of them are changed over the price of fuel oil will rise. As shown later on in this paper, the supply of oil is limited and any large increase in use will send prices up.

Most of the fuel oil used under boilers in North America comes from California, Texas and Mexican fields. The history of Mexico for the last five or six years should be a good criterion for the possibility of interruption of supply in case any large amount of steam-generating stations adopt oil as their regular fuel. The southern fields of the United States can supply but small percentage of the fuel oil that would be needed in case it was generally adopted.

Insurance Requirements for Oil Storage

The fire underwriters make some quite stringent restrictions on the use of fuel oil. The oil must be stored underground in concrete or steel tanks and the top level of the oil must be below the level of the burners and pumped to the burners. Any gravity feed of the oil to the burners is prohibited on account of the risk of heating and subsequent explosion when the boilers are started up. If the oil cannot be stored underground on account of tidewater, ledge or any other reasons, it may be stored in concrete or steel tanks above ground, provided they are located 150 ft. away from the nearest building. Each tank must be surrounded by an earthwork or concrete dike of a capacity equal to one and one-half times the capacity of the tank. This is to catch the oil in case the tank should burst for any reason. The capacity is specified as one and one-half times the tank so as to allow for snow or rain water standing in the bottom of the dike. If for any reason the above-ground tank cannot be located 150 ft. from the building, it must be buried in earth above ground; that is, the earth must be piled up on the tank so as to completely cover it.

There is considerable danger of igniting the fuel oil from lightning discharges. All tanks should be provided with a steam-line connection for suffocating the fire, or some other form of fire extinguisher should be provided, such as carbon dioxide gas, foamite, etc. If all the foregoing precautions are taken and if the layout of the proposed oil-burning plant is submitted to the fire underwriters, for approval, there will generally be only a slight increase in the fire-insurance rates on adjacent buildings.

It is an interesting fact that many of the oil companies themselves burn coal under their boilers.

The ideal place to use fuel oil is on board ship, where the additional cargo space made available by the decreased volume of the oil as compared with coal is of prime importance. The elimination of hand firemen is also important, because this releases for freight carrying the space formerly occupied by the firemen.

Edward N. Hurley, formerly of the United States Shipping Board, estimates the world's ocean tonnage before the war at 50,000,000 dead-weight tons, and he figures that this might increase to 75,000,000 within the next five or ten years. Roughly speaking, a boat requires a ton of oil yearly for every dead weight ton of ship. This applies to steam-driven vessels. The world's output of petroleum in 1917 was approximately 500,000,000 bbl. If all the 75,000,000 tons of ships were driven by fuel oil, it would take 450,000,000 bbl., or about 90 per cent of the world's production of petroleum. When you consider the large amount of the world's output of crude oil which is needed for refining to the higher-grade petroleum derivatives, it can be readily understood that if fuel oil were reserved for use on boats there would be no excess for burning on land under boilers. Of course boats driven by Diesel engines require only about half the quantity of oil that is needed for steam-driven vessels, but if all the shipping in the world were driven by this type of engine there would still not be any great excess of fuel oil for land use.

There are of course, many places besides on board ship where fuel oil ought to be used. Many special heating furnaces cannot be adapted for burning coal, and in many other instances oil is the logical fuel. The demand from these sources and from the shipping interests combined and a scarcity of supply has even now put the price of fuel oil up to over twelve cents a gallon. This price puts oil out of the question as a competitor of coal for general power uses. Of course, fuel oil will always be used for certain purposes where it is especially convenient, but this article is intended to show that it will never replace coal where coal can be burned efficiently on modern mechanical stokers.

Safety First for the Engine Room

By Frederick L. Ray*

Of first importance is the emergency over-speed devices for engine and turbines. They should be tested once each day for over-speed to know that they are in working condition. Many wrecks of engines and turbines are prevented by close inspection and test of these devices.

Relief valves on cylinder, receivers and turbines must be kept in working condition, which can be done by testing them once each month with steam, closing exhaust valves and having provided a pressure gage.

Regular periodic inspection of all apparatus must be carried out at not longer intervals than twenty minutes. This will prevent hot bearings and save many a nut coming off and some part breaking.

Governor belts must be kept in first-class condition. Don't keep them in service too long; you may wreck an engine by getting the last day's service out of a belt. You had better put a new one on the engine and use the old one in some less important place. Some terrible engine wrecks have been caused by failures of governor belts.

Vacuum breakers on condensers must be kept in good order for the protection of the engine or turbine. A flooded condenser will quickly wreck an engine or turbine.

A noiseless engine is a fair indication that it is in safe working condition. Every blow that a noisy engine makes is one more nail in its coffin, for every piece of metal will withstand but so many blows before breaking.

Emergency oil lamps should be kept trimmed and at night should be kept burning but turned low. If the station lights fail, you will need the oil lamps quickly.

Valves must not be left jammed or locked, either open or closed, but left in condition so that in case of emergency they can be moved easily in either direction. In case of accident you may not find a bar quickly.

Air-compressor receivers are dangerous if not properly cared for; keep them well blown out of water and oil. Many a receiver has been exploded, owing to the ignition of oil in them.

Keep all openings in floor well guarded with railing if a permanent opening, or if a temporary opening, then with covering that will prevent anyone from stepping into them.

Keep fire pumps ready for operation; you may need them the next minute.

In case of accident to the machinery, go slow, stop, look and listen. You will then be able to do the right thing. Courage is needed at times about a power house, but courage is only brute force and when compounded with ignorance is dangerous. Keep cool and be sure you know what you are doing.

After a job of work is finished, be sure you have all the tools and that none are left so that they will fall into machinery.

*In *Safety*, house organ of the Union Traction Company of Indiana.

In Charge of the Engineers

By William H. Woodwell

Half a point, half a point,
Half a point upward,
Seeking efficiency—
All of one hundred.
Forward the engineers,
Brilliant are their careers,
Seeking the maximum,
Seeking one hundred.

Forward the engineers!
Never a man dismayed
Even though others
Have stumbled and blundered.
Theirs is to reason why
Charges and costs are high,
Theirs is to do or die,
Seeking one hundred.

Meters to right of them,
Gages to left of them,
Readings in front of them,
Tabled and charted.
Working on curves and plots,
Formulas, ohms and watts,
Finding the wasteful spots;
Never faint-hearted.

Problems to right of them,
Problems to left of them,
Problems before them;
We have all wondered,
If they can scale the peak,
Stopping each waste and leak,
Attaining the goal they seek,
Reaching one hundred.

Where can their glory fade?
Oh, the advance they made!
Steady progression.
Honor the part they played,
Honor the gain they made;
Noble profession. —*Ideal Power*

An undergraduate course of training in Electric Communication Engineering is to be established at Harvard University, Cambridge, Mass., and two new courses will be substituted for the present course in Sanitary Engineering. This is being done with the object of giving more advanced instruction to prospective public engineers and sanitary chemists.

Admiral William S. Benson, chairman of the Shipping Board, announced at the ninth annual dinner of the Marine League on April 14, New York City, that he proposed to develop the Diesel engine and electric drive during his administration as he was convinced that the electrically driven ship is far superior in performance to the steam-driven vessel.

Study Course for the Coal Trade

A study course primarily for coal salesmen has been perfected by W. D. Langtry and Osborn Monnett of the Commercial Testing and Engineering Co., Chicago. The plan is to give sufficient instruction on boiler-room problems so that the coal salesman can talk intelligently with the chief engineer, superintendent or manager of the plant about the actual merits of his coal under burning conditions. A salesman in the power-plant trade should know what takes place in the boiler room and should be familiar with the burning characteristics of the fuel he sells under various operating conditions. The course will give him all this. It is to begin the latter part of April, and there will be sessions one night each week for ten weeks. It will be elementary in character and made practical by visits to plants for demonstration purposes. Coal versus oil, in which the relative advantages and disadvantages of both types of fuel will be discussed and data given concerning oil supply, consumption, delivery and contracts. The combustion of coal will be given thorough consideration and distinction will be made between coal for low-pressure plants, coal for hand-fired high-pressure plants and coal for stoker-fired plants. There will be visits to typical plants to witness evaporative tests, and the student will be shown how to figure the test, how to take coal samples and compute the cost of generating 1,000 lb. of steam. The characteristics and typical analyses of competitive coals on the Chicago market will be given, and there will be other data including freight rates on these coals, commercial considerations, etc. Methods of mining and their relation to the dealer will be given consideration, and there will be information on the preparation, sizing and washing of coal and suggestions on better methods of storing to prevent spontaneous combustion. The final lecture of the series will be devoted to the selling of coal. It will tell how to interpret analyses, how to transfer from a commercial to a dry basis and vice versa, how to calculate the net B.t.u. for one cent and delivered values, how to bid on B.t.u. contracts and tonnage contracts, with suggestions on what to avoid in these operations.

Although the primary intention is to educate the coal salesman, the course will be of such a practical character that it should interest also the power-plant operator, or in fact anyone connected with the purchase and burning of coal under a steam boiler.

Comparison of British and American Coal-Mining Conditions

A recent article appearing in a British journal points out that the American miner produces 2½ times as much coal per shift as the British miner, at one-half the cost, and proceeds to explain the natural industrial and engineering causes that lead to this result. Among the reasons given for this inequality is the greater depth at which the British coal mines are necessarily worked, and the small proportion of thick seams of coal, which has its effect on the methods of removal and transportation. A second reason is given as the fact that coal mining in America is regarded as a low-class industry, and that the large number of miners employed underground are agricultural emigrants. A third cause assigned for cheaper coal and mining in the United States is that the industrial conditions in this country have rendered it easy for the operators to introduce and satisfactorily operate coal-mining machinery to a greater extent than is possible in Great Britain.

George F. Rice, chief mining engineer of the United States Bureau of Mines, in a recent report, takes exception to most of these reasons. He points out that conditions in America result in the bed of coal being worked more rapidly, so that the average mine haul is far greater here than in the British collieries. On the other hand, it is admitted that the British haul is more difficult, owing to the greater vertical lift necessary and also to the fact that trolley locomotives are forbidden in British mines on account of the danger of sparking. Regarding the class of laborers employed, Mr. Rice points out that while the labor in the United States mines is, to a large extent, low-grade, yet these emigrants receive double or triple the wages of British miners. He also points out that the question of machinery in coal mining would not entirely make up for the difference in cost, as there are numerous examples in this country of hand-worked mines and machine-worked mines side by side competing with each other. Mr. Rice seems to think that the largest part of the difference between the cost of mining in the two countries is due to the much greater depth at which coal is found in Great Britain. He considers further that the natural advantage enjoyed by the United States in this respect will not be lost, as the coal in the large mining districts in America is all easily accessible, and mining at very great depths will not be necessary until practically the entire Appalachian, Central and Interior fields are worked out, and these should last at least as long as the British fields.

Economizer Expert Addresses Heating and Ventilating Engineers

A paper on "Economizers" was read by W. F. Wurster, of the Green Fuel Economizer Co., before the Eastern Pennsylvania Chapter of the American Society of Heating and Ventilating Engineers, at the Engineers Club, Philadelphia, on the evening of April 8. Mr. Wurster described briefly the first type of machine, and with the aid of lantern slides explained the details of the modern apparatus, giving some interesting data on economizer performance.

The economizer was first used in 1845 by Edward Green. It consisted of a circular group of 30 cast-iron tubes 4 in. in diameter and 9 ft. long, connected at the top and bottom to hemispherical chambers. The cold water was introduced into the lower chamber and, after rising through the tubes, passed to the boiler. This apparatus heated water to the boiling point and at times formed steam which was allowed to escape through a safety valve at the top.

After a few weeks the temperature of the water leaving the economizer became lower and the draft was impaired. On removing the brickwork inclosing the machine, the gas passage was found to be choked with soot which was killing the draft and insulating the tubes. This led to the design of mechanical scrapers to continuously scrape the soot from the tubes.

From this beginning has developed the modern economizer, which is built in sections, each of which is made up of a top and bottom header connected by the necessary tubes. The number of sections is always a multiple of four since each chain wheel operates the scrapers for four sections. Sections have from four to twelve tubes each, which vary from 6 to 12 ft. in length and are $4\frac{5}{16}$ in. in diameter. In modern installations the walls of the economizer chamber are made up of removable steel plates instead of brick. This facilitates inspection or repairs.

In proportioning the economizer and in figuring the possible savings by its use, the most important factors are the temperature and quantity of gases available, the initial temperature and quantity of feed water and the available draft. Mr. Wurster gave some values to be used as guides in making such calculations. The flue-gas temperatures from boilers run at 100 per cent of rating vary from 475 to 550 deg. F., at 150 per cent of rating from 525 to 600 deg. F., and at 200 per cent from 575 to 675 deg. F.

The amount of gas available per boiler horsepower per hour from oil-burning furnaces is about 50 lb., from underfeed stokers about 60 lb., from hand-fired furnaces with good firing about 80 lb. and with ordinary hand-firing from 90 to 100 lb. It is, of course, understood that these values are only approximate and are subject to considerable variation.

The amount of economizer surface installed varies from about 3 to 7 sq.ft. per boiler horsepower. Curves were shown to illustrate the fact that the more surface installed, the greater is the feed-water temperature rise, the lower the final flue-gas temperature, the greater the loss in draft and the higher the cost of apparatus. In designing an economizer installation, the first two items must be balanced against the last two, taking into consideration the specific conditions in the particular plant under consideration.

In answer to questions from the floor, Mr. Wurster explained that cast-iron tubes were used rather than wrought-iron or steel, owing to the corrosive action of water on the latter metals at the temperatures met in economizer work. He claimed, however, that cast-iron tubes have been satisfactory even at modern boiler pressures, and that practically all explosions of economizers had been due to abuse rather than to the use of cast iron.

Present High Prices of Bituminous Coal

Herbert N. Shenton, secretary of the United States Bituminous Coal Commission, in a formal statement recently, characterized the present high prices of bituminous coal as inexcusable on any theory of supply and demand or on any economic principle. Mr. Shenton pointed out that the fear of a coal shortage, on account of the export to meet foreign demands, was groundless, since the physical facilities for handling export coal would limit its amount to about 2 per cent of our total production. Based on the reports of production in the first quarter of the present year, there seems to be no reason for a fear that the production will be below the demand. It is evident that there has been considerable apprehension of a car shortage, but while there is definite need for more car and motive power, there is no immediate threatening of a car shortage such as would reasonably account for the rise in prices.

Mr. Shenton lays considerable emphasis on the claim that the rise in price can in no way be attributed to the campaign for early buying and storage, as recommended by the Coal Commission. This recommendation urged an increased buying of coal beginning on or about May 15, when, at the present rate of production, there was reason to believe that production would be in excess of market demand.

Among the reasons given as helping the upward tren. in prices are the fact that a rearrangement in the distribution of coal was necessary after April 1; that the first week in April happened to include the low production period always associated with Easter, and that high cost production mines, which have for past years been able to operate only because of fixed price conditions, are making a last effort to obtain what they can. In concluding, Mr. Shenton states that frenzied bidding up of prices on the part of coal buyers seems at this time to be entirely unjustified. There is every reason to believe that the present high rates will be temporary.

Superheating Wet Steam

It is sometimes stated that the function of the superheater is first of all to dry the steam that comes to it wet from the boiler, and then to raise its temperature through a prearranged number of degrees above that which obtains in the boiler. This, however, is an erroneous point of view, for according to the amount of wetness with which the superheater has to deal, so is its efficiency as a superheater reduced and so, in consequence, is the efficiency of the engine reduced. Thus, considering boiler steam at 200 lb. per sq.in. gage pressure, its latent heat is about 838 B.t.u. per lb., so that 1 per cent of moisture requires 8.38 units of heat in the superheater to dry it. Hence, taking the specific heat of superheated steam as 0.55 B.t.u. per lb. per degree Fahrenheit, this amount of heat is capable of raising the temperature of steam through about 15 deg. F. Thus, if the boiler is fitted with a superheater capable of giving 100 deg. F. of superheat, if the steam enters the superheater containing only 3 per cent of moisture the superheat would be reduced to 55 deg. F., while the superheat would be entirely eliminated if the moisture in the steam reached about 7 per cent. When it is remembered that the steam consumption of the engine may be considered as being reduced by about 1 per cent for every 10 deg. of superheat, it will be seen how great is the effect of moisture in the superheater on the steam consumption of the engine. While, of course, under ordinary circumstances it is impossible to prevent wet steam from passing from the boiler, it is obviously of the greatest importance that steps should be taken by fitting anti-priming pipes or separators to insure that the steam entering the superheater is as dry as possible.—*Shipbuilding and Shipping Record.*

Propose to Operate U. S. Nitrate Plants

United States Nitrate Plants Nos. 1 and 2, located at Sheffield and Muscle Shoals, Ala., were constructed during the war for military purposes. Particular efforts were made to make the plants up to date and complete in every way. At the time of the signing of the armistice, plant No. 1, using the synthetic ammonia process, was merely in the experimental stage; plant No. 2, using the cyanamide process, was about ready to start production. Complete descriptions of the power equipment of these installations are given in the March 25 and the June 24, 1919, issues of *Power*.

In order to insure the peacetime operation of our nitrate plants for the production of nitrogen fertilizer compounds for agricultural purposes and also render available a source of nitrogen for military purposes in case of emergency, a bill has been placed before the Senate Committee on Agriculture and Forestry, authorizing the establishment of the United States Fixed Nitrogen Corp., the stock to be owned by this corporation, to purchase and operate the two nitrate plants for the aforementioned purpose. It is estimated that $12,500,000 will be needed to finance the proposed corporation. In order to raise the required amount it is planned to sell half the reserved stock of Chilian nitrate now held by the War Department.

The suggestion incorporated in the bill to take over the plants is to use plant No. 1 to further development and research in the art of nitrogen fixation and No. 2 will be operated for the production of appropriate nitrogen compounds.

Power-Plant Construction in Korea

The rich gold deposits in Korea, about twenty years ago, led to the granting of mining concessions to the Oriental Consolidated Mining Co., an American corporation. In order to place the mining operations on a paying basis, it was found necessary to secure additional hydro-electric power and the Kuron (Nine Serpents) River, twenty-five miles from the mines and about thirty miles from the South Manchuria R.R., was chosen.

A dam 650 ft. long and 34 ft. high, with foundations extending 15 ft. below the water level, and a diversion tunnel 955 ft. long were constructed. Two Pelton-Doble turbine units, each with a capacity of 800 hp., were installed and connected, by two short steel penstock pipes, 6 ft. in diameter, with the tunnel. Six hundred kilowatt G. E. generators, with a capacity of 2,300 volts, were installed. This was transformed to 23,000 volts for transmission over a twenty-five mile pole line to the mines.

The construction of power houses, dam and tunnel and the installation of the equipment was a difficult and slow task. Materials for the construction of building and heavy machinery were transported wherever possible by boat and by coolies. Cement was brought in by bull cart from the South Manchurian R.R., 35 miles away. The coolie packer is capable of carrying as much as 700 lb., and a number of these men were used to transport the frames weighing 2,500 lb. over the mountains to the power house site.

The plant, which was started early in 1918, was finally finished in November, 1919, and is in successful operation. The total water-power development in Korea is now about 2,000 kw., and two of the three completed plants are controlled by the Oriental Mining Company.

New Publications

EXAMPLES IN HEAT AND HEAT ENGINES, by T. Peel, M. A., Demonstrator in Engineering, Cambridge University. Published by the University Press, Cambridge, England. Paper; 5½ x 8½ in.; 104 pages. Price, $1.50.

This pamphlet is a collection of questions and problems taken mostly from examination papers for the courses of lectures in Heat Engineering at the Cambridge Engineering Laboratory. The questions are arranged in groups or "Papers," becoming more difficult toward the end of the collection. Answers are given on the last pages for all problems. Students of heat engineering who desire a variety of problems for drill, or instructors looking for problems for class assignment or examination papers should find this collection useful.

Obituary

Frank M. Babcock, engineer of the Veazie hydro-electric station of the Bangor (Me.) Railway Co., since it was built, forty years ago, died April 13, after a short illness, at the age of seventy-five. Mr. Babcock contracted a cold which developed into pneumonia. He is survived by his widow and a daughter.

Personals

Maxwell C. Maxwell, formerly assistant general superintendent of the Yale & Towne Manufacturing Co., has been appointed to the position of general superintendent.

Norman Taylor, representative of the Hagan Corporation of Pittsburgh, Pa., is located at the Wolcott Hotel, West Thirty-fourth Street, New York City, until he can secure permanent offices.

M. W. Eastman, chief engineer of the Charlestown power house of the Boston Elevated R.R., has been transferred to the Lincoln Station, Atlantic Ave., and R. H. Dixon, formerly at Lincoln Station, has been transferred to Charlestown.

Ralph G. Macy has resigned his position as chief engineer in the construction department of the Walter Riddle Co., New York, and is now associated with the Engineering and Appraisal Co., Inc., of New York.

F. H. Ballou has resigned his position as assistant chief engineer of the Great Western Sugar Co., Billings, Mont., and is now with the Amalgamated Sugar Co., Ogden, Utah, in the capacity of chief engineer.

Charles F. Lang, president of the Lakewood Engineering Co., Cleveland, Ohio, was elected president of the Material Handling Machinery Manufacturers' Association, New York, at a recent meeting of the directors.

Frank D. Randolph, formerly with the Kelly Press Department of the American Type Foundries Co., Jersey City, N. J., is now associated with the Premier & Potter Printing Press Company, Inc., Plainfield, N. J.

William F. Mahoney, formerly captain, Ordnance Department, U. S. A., announces that he has resumed the practice of patent and trade-mark law with offices in the Washington Loan and Trust Building, Washington, D. C.

A. Pohlman, formerly connected with the Badenhausen Co., New Haven, Conn., has resigned and is now connected with J. S. & J. F. String, engineers and contractors. Mr. Pohlman will give particular attention to the design and supervision of fuel-oil apparatus.

William J. Beswick, formerly boiler inspector for the Hartford Insurance Co. and during the war chief engineer on the United States transport "Everett," has accepted the appointment of chief engineer at the American Linen Co.'s mills in Fall River, Mass.

A. L. Kempster, former manager of the Puget Sound Light and Power Co. until the city recently obtained possession of the street-railway lines, has accepted the position of manager of the New Orleans Railway and Light Co., now in the hands of a receiver.

Clifford H. Peters, of Cleveland, Ohio, formerly a captain attached to the Engineering Division, Ordnance Department, U. S. A., has opened an office at 226 Superior Avenue, West, Cleveland, where he will practice mechanical consulting engineering, including the design and construction of various types of automatic and semiautomatic machines, plant layouts, production systems, etc.

Society Affairs

The American Society of Civil Engineers, Philadelphia Section, will hold its regular meeting on May 3.

The Association of Iron and Steel Electric Engineers, Philadelphia Section, will meet on May 1. The meeting will be open for general discussion.

The New Haven Branch, A. S. M. E., will hold a meeting on May 12. Subject: "Airplanes," by a speaker from the Gallander Corporation.

The American Association of Engineers will hold its annual convention at St. Louis, Mo., May 10 and 11. The headquarters for the convention will be at the Planters Hotel.

The New Jersey State Association, N. A. S. E., will hold its annual convention and mechanical exhibit at Phillipsburg, N. J., June 2 to 6 inclusive. A charge of twenty dollars will be made for each booth.

The Amerian Institute of Electrical Engineers, Philadelphia Section, will meet on May 10. A paper will be presented by the Students' and Graduates' Engineers' Club on "Professional and Financial Aspect of Electrical Engineers."

The American Association of Engineers, Kansas State College Chapter, has elected the following officers for 1920: Walter E. Dickerson, president; C. H. Scholer, vice president; H. Kenneth Shideler, recording secretary; Paul G. Martin, treasurer.

The Philadelphia Section of the Association of Iron and Steel Electrical Engineers will hold a meeting on May 1. Members are given the privilege of preparing papers on a subject pertaining to the application of electricity to the iron and steel industry.

The New York Section of the American Association of Engineers recently introduced a bill in the New York State Legislature providing for the licensing of engineers. The bill was passed April 15 by the Senate and will immediately be introduced into the Assembly.

The Chamber of Commerce of the United States will hold its eighth annual meeting at Atlantic City, April 26 to 29. Representatives of more than thirteen hundred commercial and industrial bodies will be present for a thorough discussion of ways and means of increasing production.

The American Railway Engineering Association has become a member society of Engineering Council. The association has about 1,650 members, and its headquarters are at 431 South Dearborn St., Chicago, Ill. The representative of the association upon Engineering Council will be its president, Mr. Safford, who is a member of the American Society of Civil Engineers and the Engineering Institute of Canada.

The Industrial Relations Association of America will hold its second annual convention at the Auditorium Theater, Chicago, Ill., May 19-21. Every phase of the industrial situation will be covered by the presentation of papers by men representing each industry. Section and subject meetings will be held between the main sessions, these being devoted to some particular branch of industry and industrial relations work.

The National District Heating Association will hold its eleventh annual convention at the Hotel La Salle, Chicago, Ill., May 25-27. Prof. John R. Allen will deliver a paper on "Heat Losses from Underground Pipe," Prof. J. D. Hoffman on "Technical School Training for the Heating Profession," and R. V. Stureman on "Advantages and Disadvantages of Steam and Hot Water Heating." A paper on "Bleeder Turbines in Heating Service" will be presented.

Miscellaneous News

The Kansas City Power and Light Co., Kansas City, Mo., is planning to take over the system of the North Kansas City Light, Heat and Power Co. The system is valued at $150,000.

The Utica Gas and Electric Co., at its recent annual meeting elected the following officers: Frank M. Tait, president and general manager; F. B. Steele and M. J. Brayton, vice presidents; George H. Stack, treasurer; William J. McSoiley, auditor. The following directors were chosen: Frank M. Tait, S. A. Beardley, W. E. Lewis, C. S. Symonds, C. B. Rogers, N. F. Brady, W. T. Baker, M. J. Brayton and William L. Taber. The stockholders at the same meeting voted to increase the capital stock from $4,500,000 to $8,500,000.

Business Items

The Electric Storage Battery Co., 100 Broadway, New York City, announces the removal of its offices to the National Association Building, 21-31 West 43rd St., New York.

The Roller-Smith Co., 233 Broadway, New York City, announces the removal of its Cleveland office from the Williamson Building to the Vickers Building, 65th Street and Euclid Avenue.

The Yale & Towne Manufacturing Co., New York, announces the removal of its general offices to Stamford, Conn. The change was made in order to consolidate the selling organization with the factory.

The Westinghouse Electric & Manufacturing Company announces the formation of a Marine Division in its offices at 165 Broadway, New York. The object of the new department is to give advice in the selection of ship propulsion machinery. The work will be in charge of Norris R. Sibley, formerly chief engineer, U. S. Army Transport Service, and Frank F. Boyd, chief engineer of the U. S. S. "Jupiter."

Trade Catalogs

The Griscom-Russell Company, 90 West Street, New York City, has just completed the publication of a new fifteen-page booklet entitled "Stratton Steam Separator." Bulletin No. 1140. The booklet shows by means of illustrations the various types of steam regulators manufactured by this company, and gives complete instructions on construction and operation. A copy will be sent on request.

The A. M. Byers Co., of Pittsburgh, Pa., has recently issued a 32-page bulletin No. 38, on "Installation Cost of Pipe." The bulletin gives detailed information on the cost of the installation and upkeep of pipe. The main object of the publication is to bring out the relation of the installation cost to the cost of pipe that goes into the system. With this end in view a series of cost analyses covering twenty different pipe installations and practically every class of pipe system are shown by means of charts and diagrams. The booklet should prove of value to operating and consulting engineers, who will find in it information not obtainable elsewhere. Copies sent on request.

The Deming Company, Salem, Ohio, has isued Catalog No. 26, "Hand and Power Pumps." Cloth, 6 x 9 in.; 254 pages. The book describes various pumps and apparatus manufactured by the Deming Co. In addition to this a number of clear, concise tables and information relating to hydraulics have been prepared, which should prove useful to dealers in pumps. The catalog is arranged in chapters and sections, grouping each distinct class of pumps or accessories separately, so that by referring to the table of contents one can quickly find any particular article shown. The chapter devoted to miscellaneous hand and power pumps describes in detail apparatus for the farm, factory, garage, plumbing shop, etc. A complete description of each article is given, eliminating the necessity of correspondence for information.

COAL PRICES

Prices of steam coals both anthracite and bituminous, f.o.b. mines, unless otherwise stated, are as follows:

ALANTIC SEABOARD

Anthracite—Coals supplying New York, Philadelphia and Boston:

	Mine
Pea	5.30
Buckwheat	\$3.40@\$3.75
Rice	2.75@ 3.25
Barley	2.25@ 2.50
Boiler	2.50

Bituminous—Steam sizes supplying New York, Philadelphia and Boston:

Latrobe	\$4.25@\$4.50
Connellsville coal	4.25@ 4.50
Cambrias and Sunset	4.25@ 4.85
Clearfields	3.75@ 4.50
Pocahontas	6.50@ 7.00
New River	6.50@ 7.00

BUFFALO

Bituminous—Prices f.o.b. Buffalo:

Pittsburgh slack	\$6.25@\$6.50
Mine-run	6.25
Lump	6.25@ 6.50
Youghiogheny	6.25

CLEVELAND

Bituminous—Prices f.o.b. mines:

No. 6 slack	\$3.25@\$4.00
No. 8 slack	3.50@ 4.00
No. 8—¾-in.	3.50@ 4.00
No. 6 mine run	3.50@ 4.00
No. 8 mine run	3.25@ 4.00
Pocahontas—Mine-run	3.25@ 4.00

ST. LOUIS

Anthracite—Probably not more than 20 per cent of the demand in this market can be supplied. Prices effective Apr. 1 were as follows:

	Williamson and Franklin Counties	Mt. Olive and Staunton	Standard
Mine-run	2.65@2.80	2.65@3.00	2.65@2.80
Screenings	2.50@2.65	2.50@2.65	2.50

Williamson-Franklin rate to St. Louis is \$1.10; Other rates 95c.

CHICAGO

Bituminous—Prices f.o.b. mines:

Illinois		Freight rate Chicago
Southern Illinois		
Franklin, Saline and Williamson Counties		
Mine-Run	\$3.00@\$3.10	\$1.55
Screenings	2.60@ 2.75	1.55
Central Illinois		
Springfield District		
Mine-Run	\$2.75@\$3.00	\$1.32
Screenings	2.50@ 2.60	1.32
Northern Illinois		
Mine-Run	\$3.50@\$3.75	\$1.24
Screenings	3.00@ 3.25	1.24
Indiana		
Clinton and Linton		
Fourth Vein		
Mine-Run	\$2.75@\$2.90	\$1.27
Screenings	2.50@ 2.65	1.27
Knox County Field		
Fifth Vein		
Mine-Run	\$2.75@\$2.90	\$1.37
Screenings	2.50@ 2.60	1.37
Brasil Block	\$4.25@\$4.50	\$1.27

New Construction

PROPOSED WORK

N. H., Laconia—The Bd. Educ. will soon award the contract for the construction of a 2 story, 70 x 180 ft. high school. A steam heating system will be installed in same. Total estimated cost, \$225,000. P. S. Avery, 95 Milk St., Boston, Mass., Archt.

Mass., Ashland—MacNaughton & Robinson, Archts., 101 Tremont St., Boston, will soon award the contract for the construction of a 3 story, 65 x 65 ft. manufacturing building and power house on Pleasant St., for the De Swett Root Beer Inc., 10 P. O. Sq., Boston. A steam heating system will be installed in same. Total estimated cost, \$40,000.

Mass., Boston—Lawrence & Wamboldt, Archts., 6 Beacon St., plans to build a 5 story, 100 x 100 ft. cooler building for meats, on South Market St. A steam heating system will be installed in same. Total estimated cost, \$55,000. Owner's name withheld.

Mass., Wrentham—The Commonwealth of Massachusetts, c/o Kendall, Taylor & Co., Archts., 93 Federal St., Boston, plans to build a 1 story, 50 x 70 ft. cold storage building and a 1 story, 40 x 90 ft. industrial building. Total estimated cost, \$88,000.

N. Y., Binghamton—The Bd. Educ. plans to repair and construct additions to the heating, lighting and plumbing systems in the public schools. Total estimated cost, \$94,950. Address Daniel J. Kelley, Bd. Educ., Supt.

N. Y., Brooklyn—The Curb Holding Co., c/o Frank S. Parker, Archt. and Engr., 44 Court St., is having plans prepared for the construction of a 12 story, 134 x 135 ft. office building on Washington St. A steam heating system will be installed in same. Total estimated cost, \$1,000,000.

N. Y., Brooklyn—B. W. Dorfman, archt., 26 Court St., is preparing plans for the construction of a 12-story, brick, steel and stone office building on Joralemon St. near Court St. A steam heating system will be installed in same. Total estimated cost, \$500,000. Owner's name withheld.

N. Y., Dunkirk—The Atlas Crucible Steel Co., Howard St., plans to increase the size of its plant. Plans include the construction of a 40 x 140 ft. transformer building, etc. Total estimated cost, \$1,800,000.

N. Y., Jamestown—The Common Council plans to construct a heating plant for the Jones General Hospital. Estimated cost, \$22,000.

N. Y., Hamilton—The village has sold the electric lighting plant to the White Co., Norwich, cost \$70,000. The company plans to build additions and extensions and install new generating equipment.

N. Y., New Berlin—The Berholme Power Co., Inc., plans hydro-electric development on the Unadilla River, to include a power house, generating unit and transmission lines. Total estimated cost, \$19,900. The company has applied to the Pub Servc. Comn. for permission to issue \$25,000 bonds.

N. Y., New York—The Beth Hamedrast Shaapey Zion Inc., 46 West 46th St., is having plans prepared for the construction of a 2 story, 48 x 150 ft. synagogue and school on 178th St. A steam heating system will be installed in same. Total estimated cost, \$250,000. I. J. Levy, Pres. L. A. Abramson, 46 West 46th St., Archt. and Engr.

N. Y., New York—The Church of Christ Scientist, c/o B. E. Muller, archt., 477 5th Ave., is having plans prepared for the construction of a 79 x 95 ft. church on Anthony Ave. A steam heating system will be installed in same. Total estimated cost, \$300,000.

N. Y., New York—The Combustion Eng Corp., 11 Bway., will construct an 8-story office building at 43-47 Broad St. A steam heating system will be installed in same. Total estimated cost, \$500,000. Ludlow & Peabody, 101 Park Ave., Archts. and Engrs.

N. Y., New York—W. A. Hefner, 465 10th Ave., is in the market for one 220 volt, d.c., unit motor blower fan.

N. Y., New York—I. Miller & Sons, care of Schwartz & Gross, Engrs., 347 5th Av., plan to build a 12 story loft building on 46th St. and 5th Ave. A steam heating system will be installed in same.

N. Y., New York—The New York Edison Co., 103 East 15th St., plans to build a 1 story sub-station on Cedar St. Estimated cost, \$75,000. William Whitehill, 12 Elm St., Engr.

N. Y., New York—The Post Graduate Hospital, 2nd Ave. and 20th St., is having sketches prepared for the construction of an 11-story addition. A steam heating system will be installed in same. Howells & Rogers, 367 Lexington Ave., Archts. and Engrs.

N. Y., New York—J. Samuels, c/o Benjamin W. Levitan, Archt. and Engr., 7 West 45th St., is having plans prepared for the construction of a 13 story, 50 x 100 ft. store and salesroom at 6-8 East 30th St. A steam heating system will be installed in the same. Total estimated cost, \$400,000.

N. Y., Rossie—George N. Wilson has made application to the Pub. Servc. Comn., for permission to construct and operate an electric lighting plant, here. He plans about 500 hp. hydro electric development on the Indian River.

N. Y., Salamanca—The Salamanca Panel Co. plans to install a 200 hp. electrically driven plant, and is in the market for two 50 hp. motors, one 15 hp. motor, five 10 hp. motors and several 5 hp. motors.

N. J., Atlantic City—C. F. Pabst, 396 Franklin Ave., Brooklyn, N. Y., plans to build a 4 story, 75 x 200 ft. sanitarium. A steam heating system will be installed in same. Total estimated cost, \$500,000.

N. J., Newark—The Newark Bd. of Trade, 800 Broad St., plans to build a 10 story office building on Branford and Treat Sts. A steam heating system will be installed in same. Total estimated cost, \$500,000. Guilbert & Betelle, 2 Lombardn St., Archts. and Engrs.

N. J., Verona—The Essex Co. Hospital for Tubucular Diseases, will soon receive bids for the construction of a hospital. A steam heating system will be installed in same. Total estimated cost, \$350,000. Jordan Green, Essex Bldg., Archt.

Pa., Harrisburg—Silberman Bros., 445 South 2nd St., are in the market for 1 swinger for an electric hoisting engine, for a derrick.

Pa., Philadelphia—John M. Greene, 261 Drexel Bldg., is in the market for a first-class 20 hp., 220 volt, d.c. single drum contractor's hoist complete with solenoid brake, controller and resistance (new or used).

Md., Cumberland—The Crystal Laundry Co., Baltimore St., is having plans prepared for the construction of a 1 and 2 story, 130 x 284 ft. laundry, garage and power house on Baltimore St. Boilers, engines and piping will be installed in same. Total estimated cost, \$200,000. T. W. Biddle, c/o Kelly Springfield Tire Co., Archt.

S. C., Columbia—The State Legislature, Columbia, has appropriated \$50,000 for the installation of a new heating and electric plant in the State Capitol. S. W. Cannon, state electrician.

Ala., Birmingham—The Tennessee Coal, Iron & R. R. Co., B. M. Bldg., engaged Horace H. Lane, Engr., 2320 Dime Bank Bldg, Detroit, Mich, to prepare plans for the construction of a 1 and 2 story freight car shop. A steam heating system will be installed in same. Total estimated cost, \$2,500,000.

La., Shreveport—Chester & Fleming, Engrs., 1111 Union Bank Bldg., Pittsburgh, Pa., are preparing plans for furnishing new high and low service pumps, distribution mains, clear water reservoir, etc., for the city.

Ky., Versailles—J. A. Wetmore, Supervising Archt., Treasury Dept., Wash., D. C., will receive bids until May 6 for the installation of a fire box heating boiler at the Post Office here.

O., Akron—The city will receive bids until April 28 for the construction of a 1-story, 25 x 25 ft., brick and concrete pumping station, at Fairlawn Heights; also for furnishing two 500,000 gal. capacity centrifugal pumps. Estimated cost, \$75,000. M. Knowles, 716 Hippodrome Bldg., Cleveland, Engr.

O., Akron—The Hotel Co., c/o Henry & Murphy, Archts., 523 2nd Natl. Bldg., is having plans prepared for the construction of an 8 story, 135 x 186 ft. hotel on East Market and Cambridge Sts. A steam heating system and two 250 hp. boilers will be installed in same. Total estimated cost, \$3,000,000. Mayer & Valentine, Bangor Bldg., Cleveland, Engrs.

O., Akron—Marcus Loew, 1111 Euclid Ave., Cleveland, is having plans prepared for the construction of a 1 and 3 story commercial and theatre building on Main and Bowery Sts. A steam heating system will be installed in same. Total estimated cost, \$1,250,000. C. Howard Crane, 2325 Dime Bank Bldg., Detroit, Mich., Archt.

O., Alliance—The Alliance Bank plans to build a 6 story bank and office building. A steam heating system will be installed in same. Total estimated cost, $500,000. Walker & Weeks, 1900 Euclid Ave., Cleveland, Archts.

O., Chillicothe—J. A. Wetmore, Supervising Archt., Treasury Dept., Wash., D. C., will receive bids until May 6 for the installation of a new heating boiler at the United States Post Office here.

O., Cleveland—The Bd. of Co. Comrs., c/o., Joseph Menning, Court House, plans to build a 1 story market house. A steam heating and refrigeration system will be installed in same. Total estimated cost, $300,000.

O., Cleveland—Greible & Ebeling, Archts. and Engrs., East 11th and Chestnut Sts., are preparing plans for the construction of an 8 story, 100 x 150 ft. commercial and office building on East 6th St. and Hamilton Ave. A steam heating system will be installed in same. Total estimated cost, $500,000. Owner's name withheld.

O., Cleveland—The Simmons Motor Car Co., 4400 Euclid Ave., is having plans prepared for the construction of a 3 story, 95 x 195 ft. garage and commercial building on East 30th St. and Euclid Ave. A steam heating system will be installed in same. Total estimated cost, $300,000. G. W. Drach, Union Trust Bldg., Archt. and Engr.

O., Cleveland—The Wade Park Co., c/o the East Gas Co., East 6th St. and Rockwell Ave., plans to build an 8 or 10 story, 200 x 225 ft. hotel on East 107th St. and Wade Park Ave. A steam heating system will be installed in same. Total estimated cost, $3,000,000.

O., East Cleveland—(Cleveland P. O.) The Bd. of Co. Comrs., c/o Joseph Menning, Court House, Cleveland, plans to build a 1 story market house. A steam heating and refrigeration system will be installed in same. Total estimated cost, $300,000.

O., East Cleveland—(Cleveland P. O.) The East Cleveland Hospital Comn. plans to build a 3 story, 90 x 250 ft. hospital on Euclid and Strathmore Aves. A steam heating system will be installed in same. Total estimated cost, $250,000. C. M. Osborn, City Mgr. A. M. Allen Co., 1900 Euclid Ave., Cleveland, Archt.

O., Lakewood—The Lakewood Market House Co., c/o W. R. Powell, Pres., 705 Rose Bldg., Cleveland, will receive bids until May 1 for the construction of a 2 story, 78 x 300 ft. market house at 11826 Detroit Ave. A steam heating system will be installed in same. Total estimated cost, $250,000.

O., Youngstown—The City is having plans prepared for the construction of a 2 story, 60 x 150 ft. hospital. A steam heating system will be installed in same. Total estimated cost, $250,000. C. F. Owsley, Mahoning Bank Bldg., Archt.

O., Youngstown—The 1st Natl. Bank is having plans prepared for the construction of a 15 story, 60 x 120 ft. bank and office building. Steam heating boilers will be installed in same. Total estimated cost, $2,000,000. Walker & Weeks, 1900 Euclid Ave., Cleveland, Archts.

Ind., Indianapolis—The Ross Power Equipment Co., 617 Merchant Bank Bldg., is in the market for two 75 or 100 kw., 240 volt, 60 cycle, 3 ph. belted generator or engine sets with switchboard and exciter; one 300-500 kw. 480-2,300 volt, 3 phase turbo condenser, one 1,000 kw., 480-2,300 volt, 60 cycle, 3 phase turbo condenser; two 500 hp., B. & W. or Sterling boilers; one 250 kw., 430 volt, 3 phase, 60 cycle, belted generator; two No. 10 or 12 gyratory crushers (gates preferred), one No. 9 and four No. 5 crushers; one 24 x 48 in. jaw crusher and one 24 x 48 in. or 60 roll crushers.

Ind., Terre Haute—The Union Hospital Executive Committee is having preliminary plans prepared for the construction of a 3 story hospital on 7th Ave. A steam heating system will be installed in same. Total estimated cost, $350,000. F. P. Fairbanks, Chn. Stratton & Snyder, 1110 Union Trust Bldg., Detroit, Archts.

Mich., Bay City—The Amer. Worsted Yarn Co., c/o D. L. Galbraith, plans to build a 4 story, 120 x 400 ft. yarn factory on Water St. Steam heating equipment will be installed in same. Total estimated cost, $300,000.

Mich., Cedar Banks (Lansing P. O.)—Mrs. George P. Wanty, Secy., will soon award the contract for the construction of 2 and 3 story training school buildings including 3 dormitory buildings, administration building, power house and laundry, for the Michigan State Training School for Women. A steam heating system will be installed in same. Total estimated cost, $500,000. E. A. Boud, Lansing, Archt.

Mich., Detroit—The Bd. of Fire Comrs., c/o G. J. Finn, Secy., Larned St., is having plans prepared for the construction of a 3 story, 90 x 120 ft. fire alarm station on Macomb and Hastings Sts. Steam heating equipment will be installed in same. Total estimated cost, $250,000. Donaldson & Meier, 1344 Penobscot Bldg., Archts.

Mich., Detroit—The Dept. of Pub. Wks., City Hall, plans to construct and equip 100 mi. of city electric railways with car shops and power houses. Total estimated cost, $15,000,000. $1,000,000 of which will be used for buildings and equipment. A $15,000,000 bond issue for this project has been voted upon.

Mich., Detroit—Hugo Scherer, c/o Louis Kamper, Archt., Book Bldg., is having plans prepared for the construction of a 10 story, 53 x 141 ft. office and loft building on Woodward Ave. and Sibley St. Steam heating equipment will be installed in same. Total estimated cost, $1,000,000.

Mich., Detroit—The Veterans of Foreign Wars, c/o Henry M. Leland, Chn. of the Bd. of Trustees, 2984 West Grand Blvd., plan to build a 7 story club house. Steam heating equipment will be installed in same. Total estimated cost, $500,000. George Mason, 80 Griswold St., Archt.

Mich., Detroit—The Women's Hospital and Infants' Home, Forest Ave., plans to build a 3 or 4 story addition to present hospital on Forest Ave. and Beaubien St. A steam heating system will be installed in same. Total estimated cost, $300,000.

Mich., Eloise—The Wayne Co. Poor Bd., c/o O. P. Gulley, 27 Churchill St., plans to build a 4 story hospital and home. Steam heating equipment will be installed in same. Total estimated cost, $400,000.

Mich., Farmington—The village plans to install an electrically driven auxiliary pumping unit and construct extension to water mains. Total estimated cost, $25,000. R. A. Murdock, 603 Free Press Bldg., Detroit, Engr.

Mich., Flint—W. L. Hoffman, Michigan School for the Deaf is in the market for a new electric driven d.c., 250 volt centrifugal pump.

Mich., Grand Rapids—The Grand Rapids Tire & Rubber Co., c/o H. H. Swan, Mgr., 315-18 Kelsey Bldg., is having plans prepared for the construction of a 2 and 3 story tire plant on Fuller Ave. and the Grand Trunk Ry. Electric power motors will be installed in same. Total estimated cost, $400,000.

Mich., Grand Rapids—The Y. W. C. A. has engaged Robinson & Campau, Archts., 715 Michigan Trust Bldg., to prepare plans for the construction of a 3 story administration building on Sheldon and Island Sts. A steam heating system will be installed in same. Total estimated cost, $200,000.

Mich., Muskegon Heights—The Bd. of Educ. will soon award the contract for the construction of a 3-story, 19 x 240 ft. school. Fan heating and ventilating systems, including a boiler, will be installed in same. Ammerman & McColl, 2348 Penobscot Bldg., Detroit, engrs. S. D. Butterworth, 432 Tussing Block, Lansing, Archt.

Mich., Pontiac—C. W. Hubbell, Engr., 2348 Penobscot Bldg., Detroit, will receive bids until May 17 for the construction of a 10,000,000 gal. filtration plant for the city. Two electrically driven sump pumps, motors, steam heating system, etc. will be installed in same. Total estimated cost, $500,000.

Mich., Pontiac—The Oakland Motor Car Co. plans to build a 2 story, 60 x 200 ft. service station on Baldwin Ave. Steam heating equipment will be installed in same. Total estimated cost, $250,000.

Ill., Chicago—The Clerk of the Sanitary Dist., 910 South Michigan Ave., will receive bids until May 13 for the furnishing and delivering 2 Diesel Oil engines, each direct connected to its engine type 500 kw., 3 phase, 60 cycle, 2,300 volt alternating current generator with direct connected exciter and 2 similar additional units.

Ill., Chicago—Straus & Schram, 1105 West 35th St., are having preliminary plans prepared for the construction of an 11 or 12 story mail-order plant on Western Ave. and 36th St. A steam heating system will be installed in same. Total estimated cost, $5,000,000. E. W. Eppstein, 2001 West 39th St., Engr. S. Scott Joy, 2001 West 39th St., Archt.

Ill., Freeport—J. A. Wetmore, Supervising Archt., Treasury Dept., Wash., D. C., will receive bids until May 6 for the installation of a new heating boiler at the Post Office, here.

Ill., Granite City—J. A. Wetmore, Supervising Archt., Treasury Dept., Wash., D. C., will receive bids until May 13 for the installation of a heating boiler, etc., at the United States P. O., here.

Wis., Milwaukee—The School Bd., 10th and Prairie Sts., is having plans prepared for the construction of a 1 story, 22 x 40 ft. boiler house at the Garfield School. Estimated cost, $50,000. Van Ryn & De-Gelleke, Saswell Block, Archts. and Engrs.

N. D., Dickinson—The city plans to construct a 1,000,000 gal. reservoir, pump house addition, complete boiler plant for operating pumps and compressors and install additonal compressors. Total estimated cost, $40,000.

N. D., McVille—The City Council will soon award the contract for the construction of a 125,000 gal. reservoir and the installation of 3 pumps, etc. Total estimated cost, $36,000. Atkinson & Hall, Devils Lake, Engrs.

Tex., Big Spring—The city plans to rebuild the waterworks system, including new pumping plant, etc. A steam driven pump will be installed in same. Total estimated cost, $40,000. Henry Echod Eng. Co., Interurban Bldg., Dallas, Engr.

Tex., Dallas—The Dallas Auto Laundry & Storage Co., c/o Dick Spillers, Mgr., Dallas Auto Club, is having plans prepared for the construction of a 6 story, 100 x 150 ft. garage. A steam heating system will be installed in same. Total estimated cost, $250,000. Charles W. Almstead, Cleveland, O., Engr.

Tex., Dallas—T. Dies, Guaranty Bank & Trust Co., Commonwealth Bldg., is having plans prepared for the construction of a 12 story, 60 x 100 x 150 ft. office building on St. Paul, Elm and Pacific Sts. A steam heating system and power generating plant will be installed in same. Total estimated cost, $1,700,000.

Tex., Dallas—George M. Easley, Dallas, is having plans prepared by Lang & Mitchell, archts., Insurance Bldg., for the construction of a 10-story office building on Burlington and Bryan Sts. A steam heating system will be installed in same. Total estimated cost, $500,000.

Tex., Dallas—John T. Jones, 4930 Junius St., is having plans prepared for the construction of a 10 story, 100 x 100 ft. office building on Commerce and Prather Sts. A steam heating system will be installed in same. Total estimated cost, $750,000. Lang & Mitchell, Amer. Exch. Bank Bldg., Archts.

Tex., Fort Worth—The Bureau of Yards & Docks, Navy Dept., Wash., D. C., plans to furnish and install deep well pumping equipment here. Estimated cost, $30,000.

Okla., Okmulgee—The Creek Hotel Co. will soon receive bids for the construction of a 9 story, 100 x 140 ft. hotel. A steam heating system will be installed in same. Total estimated cost, $750,000. Smith, Rea, Lovitt & Senter, 601 Finance Bldg., Kansas City, Archts.

Utah, Salt Lake—The city plans to build a pumping plant to deliver ultimately 52,500,000 gal. daily, using the Utah Lake as a source of supply, etc. Total estimated cost, $2,500,000. C. T. Brown, Walker Bank Bldg., Engr.

Wash., Everett—The Y. M. C. A. plans to build a 3 story, 114 x 115 ft. Y. M. C. A. building on Colby Ave. Plans include boiler rooms, steam bath-rooms, etc. Total estimated cost, $200,000. Baker & Vogel, Pacific Block, Seattle, Archts.

Wash., Puget Sound—The Bureau of Yards & Docks, Navy Dept., Wash., D. C., plans to install reciprocating air compressors at the Navy Yard, here. Estimated cost, $50,000.

Ore., Salem—The Oregon Pulp & Paper Co. plans to construct a power plant on the site of the old Oregon Flouring mills known as the Scotch mill, at the foot of North Mill Creek to transit elecrtic energy to the paper mill on Trade and Front Sts. Estimated cost, $50,000.

M. I., Guam—The Bureau of Yards & Docks, Navy Dept., Wash., D. C., plans to construct a refrigerator plant, here. Estimated cost, $40,000.

Ont., Chatham—James St. Pierre, R.R. Bear Line, will soon award the contract for the construction of a pumphouse, etc., in connection with the land reclamation work for the Township of Dover. Total estimated cost, from $15,000 to $25,000. W. G McGeorge, Consult. Engr.

Ont., Pembroke—S. L. Biggs, town clk., will receive bids until May 12, for furnishing and installing turbine pump and gasoline engine, with capacities of 1,200 gal. and 800 gal per min. Estimated cost, $25,000. J. P. Howe, Town Engr.

Ont., Toronto—Eustace G. Bird, Archt., 9 King St., W., will soon award the contract for the construction of a 4 story, 70 x 400 ft. and a 2 story, 52 x 175 ft. factory building on Wallace and Ward Aves., for the Canadian General Electric Co., King and Simcoe Sts. A steam heating system will be installed in same. Total estimated cost, $300,000.

Ont., Toronto—The Y. M. C. A., 21 McGill St., and the Dominion Council, 604 Jarvis St., plan to build a 4 story Y. M. C. A. building. A vacuum steam heating system will be installed in same. Total estimated cost, $650,000.

Sask., Moosomin—G. S. Page, Town Clk., will receive bids until May 5 for the supply and delivery of (A) Internal combustion engines (B) generators, exciter, switchboard, street lighting apparatus, transformers, meters, etc., (C) 1 oil storage tank, (D) pole line and wire material, (E) erection of electric pole line, (F) erection of power station and cooling chamber.

CONTRACTS AWARDED

Mass., Brighton—Samuel Rudnick, 37 Fayston St., Roxbury, will build a 4 story 90 x 600 ft. apartment house on Buckminster St. A steam heating system will be installed in same. Total estimated cost, $1,000,000. Work will be done by day labor.

Mass., Lawrence—The Patchogue Plymouth Mills Corp., Maston St., has awarded the contract for the construction of a 2story manufacturing plant addition, to L. E. Locke & Sons, 800 Bay State Bldg. A steam-heating system will be installed in same. Total estimated cost, $250,000.

Mass., Leominster—The Leominster Gas & Power Co has awarded the contract for the construction of a 2-story 40 x 54 ft. addition to its power house, to Walter H. Preble, 11 Pemberton Sq., Boston. Estimated cost, $100,000.

Mass., Wellesley—The Academy of the Assumption, Wellesley Hills, has awarded the contract for the construction of a 4 story, 110 x 190 ft. college and tower on Worcester St., to the James Driscoll & Son Co., 94 Washington St., Brookline. A steam heating system will be installed in same. Total estimated cost, $350,000. Noted Jan. 27.

Mass., Whitinsville—The Whitin Machine Wks. has awarded the contract for the construction of a 2 story industrial housing building, to the Foundation Co., 233 Bway., New York City. A steam heating system will be installed in same. Total estimated cost. $350,000.

Mass., Worcester—The Osgood Bradley Car Co., West Boylston St., will build a 3 story manufacturing plant addition. Extension to the steam heating system will be installed in same. Total estimated cost, $325,000 Work will be done by day labor.

N. Y., Black River—The St. Regis Paper Co. will build a power house to generate 7,500 hp. on the Black River. Total estimated cost, $500,000. Work will be done by day labor. Noted April 6.

N. Y., Brooklyn—The Amer. Safety Razor Co., 303 Jay St., has awarded the contract for the construction of a 5 story, 75 x 140 ft. and a 1 story, 25 x 125 ft. factory buildings, a 2 story, 45 x 50 ft. garage and a 1 story, 25 x 50 ft. boiler house, to the Turner Constr. Co., 244 Madison Ave., New York City. A steam heating system will be installed in same. Noted Feb. 9.

N. Y., New Dorp—(Staten Island P. O.) The Bd. Educ., 500 Park Ave., New York City, has awarded the contract for the installation of heating and ventilating apparatus in the Jefferson, Garretson and Cromwell Ave. public schools, to Gillis & Geoghagan, 537 West B'way, New York City, at $58,700.

N. Y., New York—The Borden Milk Co., 63 Vesey St., has awarded the contract for the construction of a 23 story, 80 x 100 ft. office building on 45th St. and Madison Ave., to the Cauldwell Wingate Co., 381 4th Ave. A steam heating system will be installed in same.

N. Y., New York—The Bosch Magneto Corp., 233 West 46th St., has awarded the contract for installing a steam heating system in the proposed 12-story office and salesroom building at 60th St. and Bway., to J. O'Brien Co., Inc., 624 Madison Ave. (general contractors for job). Total estimated cost, $700,000.

N. Y., New York—The George Backer Constr. Co., 33 East 33d St., will build a 6 story garage on West 49th St. A steam heating system will be installed in same. Total estimated cost, $250,000. Work will be done by day labor.

N. Y., New York—The New York State Realty & Terminal Co., c/o Warren & Wetmore, Archts., 10 East 47th St., has awarded the contract for the construction of a 17 story hotel on Park Ave. and 49th St., to Fred T. Ley Co., 19 West 44th St. A steam heating system will be installed in same. Total estimated cost, $2,500,000.

N. Y., New York—John H. Sherer, 25 West 45th St., will build an 8 story, 55 x 60 x 118 ft. office building on 52d St., near B'way. A steam heating system will be installed in same. Total estimated cost, $400,000. Work will be done by day labor.

N. Y., New York—E. E. Smathers, 304 West 75th St., will build an 11 story, 27 x 100 ft. office and store building at 1 West 46th St. A steam heating system will be installed in same. Total estimated cost, $400,000. Work will be done by day labor.

N. J., East Orange—The Simms Magneto Co., 279 Halsey St., has awarded the contract for the construction of a 2 story, 154 x 167 ft. factory on North Arlington Ave., to the Amer. Concrete Steel Co., Essex Bldg., Newark. A steam heating system will be installed in same. Total estimated cost, $385,000.

N. J., Newark—Loew's Enterprises, 1493 B'way, New York City, have awarded the contract for the construction of a theatre building on Main St., to the Fleischmann Constr. Co., 531 7th Ave., New York City. A steam heating system will be installed in same. Total estimated cost, about $300,000.

Pa., Philadelphia—H. C. Aberle, A and Clearfield Sts., has awarded the contract for the construction of a 5-story factory and power house, to W. Steele & Sons Co., 16th and Arch Sts.

Md., Dukes (Adelina P. O.)—The West Virginia Pulp & Paper Co., 200 5th Ave., New York City, will build a 90 x 150 x 100 x 120 ft. factory addition here. A steam heating system will be installed in same. Work will be done by day labor.

W. Va., Clarksburg—The city has awarded the contract for the construction of an intake chamber and the installation of a coagulating basin and equipment for pumping stations, to the Richards Constr. Co., at $37,108.

O., Cleveland—The Jordon Motor Co., 1070 East 152nd St., has awarded the contract for the construction of a 1 and 3 story, 70 x 620 ft. factory and a 34 x 48 ft. power house on East 152nd St. and St. Clair Ave., to the Hunkin Conkey Constr. Co., Century Bldg. Two 250 hp. boilers with coal handling machinery, bunkers and stokers will be installed in same. Total estimated cost, $300,000.

O., Dayton—The Bonebrake Seminary, c/o S. E. Fout, 1521 West 1st St., has awarded the contract for the construction of a 4 story, 90 x 225 ft. theological school on Calalpa and Cornell Sts., to H. J. Osbun, Salem Ave. Plans include a power plant, etc. A steam heating system will be installed in same. Total estimated cost, $750,000.

O., Dayton—The Peckham Coal & Ice Co., Daller and Chapel Sts., has awarded the contract for the construction of a 1 story, 112 x 137 ft. ice plant, to the Blanchard Constr. Co. Estimated cost, $60,000.

O., Marion—The Marion Packing Co. has awarded the contract for the construction of a 4 story, 42 x 84 ft. cold storage building addition to E. Elford, 104 North 3rd St., Columbus. Estimated cost, $120,000.

Mich., Detroit—The Bd. Educ. 50 B'way Ave., has awarded the contract for installing a heating, ventilating and plumbing system in the proposed school on St. Antoine St. and Palmer Ave., to James W. Partlan, 51 Park Pl., at $139,000. Noted March 9.

Mo., St. Joseph—The Missouri Methodist Hospital Association has awarded the contract for installing a steam heating system in the proposed 5 story hospital, to the Rock Plumbing Co., 115 North 3rd St. Total estimated cost, $600,000.

Tex., Waco—The Bd. of Trustees of the Methodist Church South has awarded the contract for the construction of a steam heating plant and laundry at the Orphanage, to William Smith, at $45,000. A Boiler, steam pipes, etc. will be installed in same.

Okla., Newkirk—The Newkirk Ice & Creamery Co. has awarded the contract for the construction of a creamery, to Arthur Frank, Tulsa, at $125,000. Two 150 N. B. and W. boilers, one 100 kw. a.c. direct connected, 230 volt, generator, etc. will be installed in same.

Ont., Toronto—The Canadian Natl. Carbon Co., 265 Adelaide St. W., and the Prest-O-Lite Co. of Canada, Ltd., 120 Elm St., have awarded the contract for the construction of a 2 story, 80 x 400 ft. and a 2 and 3 story, 80 x 400 ft. factory buildings at Hillcress Park, to Wells & Gray, Confederation Life Bldg. A steam heating system will be installed in same. Total estimated cost, $450,000.

Turn Now to Pages 93 to 109

POWER

Volume 51 New York, May 4, 1920 *Number* 18

A Drop of Water

I am a drop of water.

For countless ages, yea, ever since the world began, I have been here to glorify my Creator and to serve and do the bidding of his servants, Mankind.

For years and years I have lain in the form of ice in the far frozen north, only to reak loose and roll into the sea, to wander for weeks until my form was changed (for I have many forms) back to a liquid, and I became a part of the sea.

Again and again I have been drawn from the sea into the clouds in the form of a vapor, carried far over the land and precipitated to the earth in the form of rain.

I have worn creases into the surface of the earth and washed the hilltops into the valleys; I traveled the Nile when Cleopatra was a girl, and many times since; out past Cairo into the Mediterranean and through Gibralter to the boundless ocean. Peaceful is the journey; but not so the times that I have been precipitated farther south where with a rush and a roar I have leaped for hundreds of feet over the Falls of Zambezi. Yea! and at Niagara I have been there and tempestuous was the passage. Over the brink to the water below, on to the swirling rapids, out into Ontario and the great St. Lawrence. Of late years they have placed forebays and penstocks and turbine wheels and tailstocks in my path there, and when I go that way, I turn the mighty electric generators that furnish light and power for the whole country side.

I have cooled the fever-parched lips of a little child and quenched the thirst of a dying soldier; the desert is turned into a place of beauty by my touch, and flowers and plants and trees grow at my bidding.

My power is without limit, yet when under control I am as docile as a kitten. If I am confined in a closed chamber and cooled, I will contract until I have reached a temperature of thirty-nine degrees, but if I am cooled below this point, I will expand again and if the cooling continues, woe be unto the vessel, for it must certainly break; again, if I am confined in a vessel and heated, I will expand and become a vapor. If the heating continues, woe be to the vessel again, for it must surely break. If the heating continues until I have reached a temperature of two-hundred and twenty-five degrees, I will have vaporized and expanded until it will take one thousand, two hundred and ninety-two times my original volume to make the same weight, and I will be pushing against every square inch of my container with a force of nineteen pounds.

I was in the kettle and bobbed the lid in trying to get out when James Watt discovered the great force that lay hidden in me, and I was there when he tried out his first machine for applying that force. Aye, and there was demonstrated that day an idea that was to be one of the greatest factors in the industrial development of the world.

Since that day men have used me more and more, and each year I have been subjected to higher temperatures until now when I enter a modern power plant, I am forced through a heater and heated boiling hot; thence through an economizer and into a large boiler, where I circulate inside of hot tubes, absorbing more and more heat, until I have attained a temperature of four hundred and four degrees and I am pushing out against the walls of my prison with a force of two hundred and sixty pounds per square inch; not satisfied with this, I am forced through a superheater on my way out, where I take on another hundred degrees.

By this time I am in the best condition to serve their purpose, and I am allowed to frisk out through a steam main and find my way, not to the cylinder of a steam engine such as Watt made, but to a mighty turbine.

With almost lightning speed I escape through a nozzle and strike a moving curved blade, only to be directed back against a stationary curved blade, where my direction is again changed and I impinge on another moving blade, gaining velocity each time until, twisting and turning, I scurry out through the last set of blading and into the condenser, where I am either met with a jet of cold water or strike against a cold tube and I find myself condensed back to water again. My mad rush is now over, and I flow out into the hotwell, and if I am lucky I escape through the overflow, back again to the river and once more to the free and open sea.

If I am not lucky enough to escape, I am drawn back through the intake and have to go through the same routine again, perhaps for days and weeks before I escape, but it makes no difference what you do to me, you cannot destroy me. You may boil me to a wild vapor, freeze me into ice, change me to a mist or a fog, or to hail, but always and ever I return again to my natural state to glorify my Creator, and to bless and do the bidding of his servants, MANKIND.

Contributed by Earl Pagett

FIG. 1. THE WATER SLOWLY ENTERS THE RESERVOIR OVER A LEVEL WEIR 300 FT. LONG AND THUS DOES NOT STIR UP THE SILT

Silt Disposal at Tacoma Hydro-Electric Plant

By LLEWELLYN EVANS

Superintendent, Electric Department, City of Tacoma, Washington

THE Tacoma Municipal water-power plant is located on the Nisqually River about thirty miles from the city. This stream has its source in a glacier on Mt. Tacoma (Mt. Rainier) and falls about three thousand feet in sixty miles and naturally carries quantities of silt during floods and considerable during the glacier melting period. The development consists of a 50-ft. concrete dam diverting the water across a bend in the river with a 10,000-ft. tunnel and thus gaining about 425 ft. of head. The plant depends on the stream flow and has practically no storage except such as can be had with flashboards on the dam during low stages of the river, amounting to about 60,000 kilowatt-hours.

This short and simple development is made possible by the wonderful natural condition at this point in the river, the stream having cut a canyon through a rock dike for about three miles, accumulating 425 ft. of fall in this short distance. The direct route across the bend by tunnel is still shorter. The steep banks made the penstocks less than eight hundred feet long.

The silt condition was recognized from the first by the original engineers, of whom Frank C. Kelsey was chief, and provisions were made for handling the silt to the extent that its action could be foretold.

The intake was chosen at the inside of a bend about two hundred feet below a turn and directly behind the dam, as shown in the pictures. Screens with 4½-in. opening made of ½ x 6-in. bars are placed across five 10-ft. openings about 22 ft. deep. Immediately in front of the screens is a channel three feet or more deeper than the intake floor. This continues at considerable grade to the washout gates in the dam, the purpose being to keep the space in front of the screens clear of sand and débris with a high-velocity current flowing out through the lower gates. These gates also provide a means of washing out the river bed and the space behind the dam at certain low stages of the river. Behind the screens is a forebay, and then comes the three intake gates which control the water going to the plant. Between these gates and the entrance to the tunnel is a settling channel about eight hundred feet long and forty feet wide. This channel has a floor on a 1.8 per cent grade and at the lower end is about seven feet deeper than the entrance to the tunnel so that about seven feet of silt can be accumulated before any enters the tunnel. The canal is flushed by opening gates at the lower end and sending through sufficient water from the intake, the dirty water emptying below the dam.

The tunnel operates as an open conduit and is not under pressure. After leaving the tunnel the water crosses the river on a steel viaduct in a 10-ft. diameter closed pipe and proceeds underground in a short conduit and comes out into a regulating reservoir just above the plant, which is in the bottom of the canyon. The water enters this reservoir over a level weir about three hundred feet long and, if it were not for the entrance velocity, would be distributed over the reservoir with very little current, but because of this entrance velocity the water now piles up at the lower end of the entrance channel and goes over the weir with a gradually increasing depth, creating a direct current from the farther end of the weir to the forebay entrance. The reservoir is graded to one corner and provided with washout gates. There are bypass tunnels under the reservoir to feed the plant direct while the washout process is going on. A 75-hp. centrifugal pump providing water at 200 lb. pressure to the hose nozzles is used in flushing the floor of the reservoir. Once a year seems often enough to clean out. The accumulation is often seven feet high in places, but averages three or four feet at the end of the year.

The headworks washout gates and settling channel and reservoir have been effective in removing dangerous silt conditions up to about 50 per cent of load on the plant, about 400 sec.-ft. during floods. Beyond this load we have difficulties. The cutting effect of the silt shows up first in the reducing valves of the cooling water system, which are cut out rapidly. The piping for furnishing cooling water to the bearings becomes clogged, and in fact we finally pulled out all the small piping and installed pipe of larger diameter. The turbine linings are soon badly cut, as shown in Fig. 6, and the backs of the runner vanes become deeply pitted. The cutting of the clearance rings of the runners causes the first trouble with the turbines, as they are single-discharge and the centering of the runner depends on the clearance on each side. The heating of the thrust bearing, which takes up side play in the runner, indicates the wear first. This extra pressure on the thrust bearing can be compensated up to certain limits by by-pass and drainage pipes and valves provided for this purpose. When the limits of compensation have been reached, the turbine must be taken down and all worn parts renewed. The pictures show what havoc can be done by the silt when it gets started. Five years' ordi-

FIG. 2 (TOP). THE DIVERTING DAM ON THE NISQUALLY RIVER WHOSE SOURCE IS MOUNT RAINIER. THE INTAKE IS JUST TO THE LEFT OF THE TWO BUILDINGS ON THE RIGHT. FIG. 3 (MIDDLE). REGULATING RESERVOIR OF TACOMA'S NISQUALLY RIVER STATION. FIG. 4 (BOTTOM). PORTAL OF THE 10 000-FT. TUNNEL

nary wear with clean water or the milky glacier water causes no cutting to speak of, but one flood lasting three or four days while the plant was operating with full load, cut the liners, wickets and runners as shown in Figs. 5 and 6. The sharp, fine sand does all the damage. Apparently, the milky water of a glacier stream has no wearing effect. Our experienences in the flood of 1917 resulted in much study and experimenting, and as a result certain remedies have been applied with gratifying results.

The fundamental trouble has been at headworks where the washout gates, which should keep the intake screens open, have failed to function because they were too small to let through roots, stumps and logs which would not float. As long as these gates were open, the intake could be kept comparatively clear, but the minute the gates became choked, there was an eddy formed on the intake side of the stream, which furnished slack water where sand could settle and pile on top of the trash on the screens so as to immediately close the openings. In less than a minute after the gates become clogged, the intake is choked tight, and the only way to get any water to the plant is to put a gang of men to work raking the screens. As the roots, limbs, logs and leaves are scraped away from the bars, the sand will be freed and washed along into the intake. It soon fills up the settling channel, because the best a crew can do will not produce enough water to run the plant, making it difficult to get time to wash out the settling channel. So the water that gets through to the reservoir is heavy with silt.

This flood of 1917 was so unusual that no repetitions were expected for years to come and no extensive remedies were undertaken, and therefore there was a partial repetition of this trouble in 1918 and early in 1919. As during each flood we were faced with the absurdity of having an abundance of water in the stream and none for the plant, we considered several remedies, such as to provide an auxiliary intake or to move the present intake to a point upstream which was swept clear by high-velocity currents near the shore. As this was an expensive piece of work and still presented the possible difficulty of taking care of the new intake during flood, we finally decided on three changes:

1. To construct a drift barrier to catch all the drift approaching the intake.

2. To remove the fine screen at the intake and substitute a few heavy bars and to place the fine screen at the end of the settling channel where the rubbish could be taken care of by the settling-channel washout gates without the employment of a crew to clean the screens.

3. To distribute the water entering the reservoir over the weir more uniformly by installing flashboards of varying heights.

FIG. 5. WICKET GATES ERODED BY SILT. THESE GATES WHEN BADLY ERODED ARE COATED WITH STEEL BY WELDING AND THEN FACED

The dangerous material in the water during flood seems to be all sand. We have taken a series of samples at several stages of flood and found that the water appears clear in a test tube almost as soon as it can be held up and looked at after taking the sample. The sediment drops to the bottom of a 6-in. test tube, leaving a spot about the size of a thumbprint in the bottom. At the end of the settling channel the sample shows about half as much sediment, and at the entrance to the penstocks there is just a trace left. The sand particles are much smaller in the samples at the plant end than at the intake. Taking these samples aided us in determining how much load we could safely carry during flood without danger. We learned that we could improve the water by operating the three intake gates open, say, 2 ft., instead of one gate 6 ft. open, because of the amount of agitation caused in the settling channel. Furthermore, it was wise to avoid spasmodic changes in the reservoir level because of the silt on the bottom which would be brought into suspension.

The drift barrier as planned for permanent use, will consist of steel cables one foot apart fixed between cement piers, like a wire fence, which will extend at right angles to the dam and be parallel with the stream lines of the natural river about thirty feet from the intake. We do not propose to stop all the drift in the river, but just that which tends to enter the intake. To insure sufficient water if the barrier should get clogged in a flood, we are making it about three hundred feet long so as to increase the area. It will extend upstream to a bend, where there is always deep water and high-velocity current near the shore to insure an open entrance to the space behind the barrier even if the barrier becomes closed down entirely next to the dam, directly in front of the intake.

Preliminary to building this structure, we have driven piling along the general route of the barrier in 10-ft. bents, and at the dam end we have filled in with piles about two feet apart. This temporary barrier and the change in the screens were put to test last December, when we had a severe freezing period which brought about four inches of ice on the back water behind the dam and broke up with a flood. The screens and barrier satisfactorily took care of floating ice as well as the sinking roots and trees.

The flood gates stayed open, and the screens at the end of the settling channel helped us get rid of the ice through the washout gates at the same time we washed out the silt. The ice in the river had formed jams and the river cut under them, stirring up the sand and making a dirty river, but the heaviest part of the sand went out through the sand gates in the dam and a wide area of the intake screen stayed open because the gates were clear. The velocity into the intake was reduced enough to settle out the worst of the sand. This flood did not reach so high a stage as the three former floods, but demonstrated that the work we had done was in the right direction and made us confident that we can ride out the worst kind of a flood.

These floods are not gentle affairs, but take on great

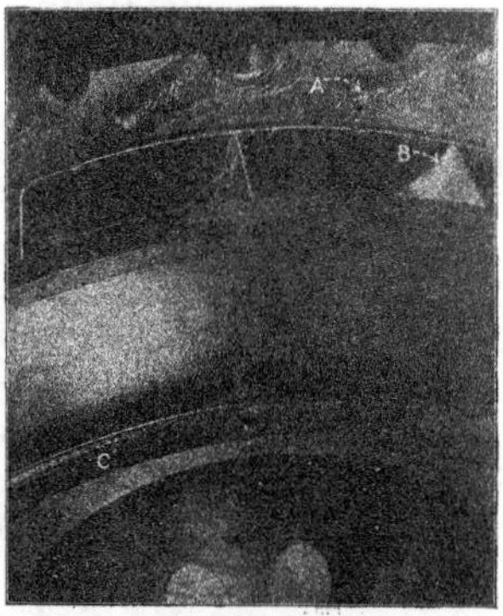

FIG. 6. A. SILT-CUT SIDE FACING RINGS OF THE WICKET GATES; THE WICKETS HAVE BEEN REMOVED. B. THE BACKS OF THE BRONZE RUNNER VANES WERE MOST PITTED. C. WEAR OF THE SIDE CLEARANCE FACE OF THE RUNNER

force and tear up the landscape generally. We are not thinking of trying to control the whole stream, but will be thankful if we can handle the water we need.

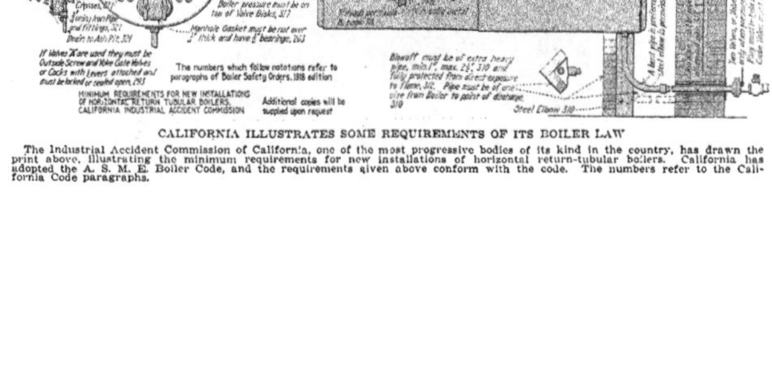

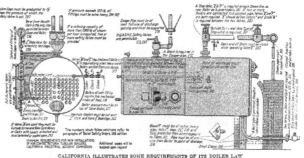

CALIFORNIA ILLUSTRATES SOME REQUIREMENTS OF ITS BOILER LAW

The Industrial Accident Commission of California, one of the most progressive bodies of its kind in the country, has drawn the print above, illustrating the minimum requirements for new installations of horizontal return-tubular boilers. California has adopted the A. S. M. E. Boiler Code, and the requirements given above conform with the code. The numbers refer to the California Code paragraphs.

Vertical-Shaft Waterwheel-Driven Alternators—Mechanical Features

By S. H. MORTENSEN
Electrical Engineer, Allis-Chalmers Manufacturing Company

THE steady increase in the cost of coal has put a new valuation upon the energy from waterfalls and made its utilization one of national importance. This, combined with the improvements in power transmission, and the development of the efficient hydro-electric units for low and medium heads, makes it feasible to utilize waterfalls formerly considered of little commercial value. Practically all modern turbines for low and medium heads are of the vertical type, with the alternator directly connected. The success of this type of unit is mainly due to the relatively high rotating speed and high efficiency of the turbine, combined with the rigid and reliable construction of the vertical alternator. A modern unit of this type is shown in Fig. 1.

FIG. 1. ONE OF THE LARGEST HYDRO-ELECTRIC UNITS IN THE WORLD; 32,500-KVA. MACHINE

In the vertical alternator the electrical and magnetic features differ little from the well-known characteristics of the engine-type machine. Its mechanical construction, however, is different and requires special design for the stator yoke, bearing housings, bearings and oiling system. Furthermore, the characteristics of the hydraulic turbine make it necessary to embody a certain momentum, WR^2, in the alternator rotor, which also must have sufficient mechanical strength to withstand the centrifugal stresses corresponding to the runaway speed of the turbine. Experience has proven also the necessity of supplying the larger alternators with brakes to bring the rotor to rest. Finally, the proper ventilation and excitation of the vertical alternator leads to conditions that have a direct bearing upon its design. In a vertical machine the stator yoke supports the core iron and stator coils, and usually serves as a foundation for the step-bearing housing and thus supports the total weight of the rotating parts of the unit. The yoke consists of a stiff substantial casting made either in one piece or in several sections, depending upon the size of the machine, and is provided with uniformly spaced ribs into which the slots or dovetails holding the core punchings are cut. The yoke casting must have sufficient stiffness to withstand unbalanced magnetic pulls due to uneven air gaps or short-circuited turns in the field coils. Also, it must have sufficient mass and strength to support the upper bearing housing with its suspended load without undue deflection or vibrations. Furthermore, it should lend itself readily to the scheme of ventilation of the alternator.

FIG. 2. SECTION OF YOKE WITH PUNCHINGS AND SOME OF THE COILS IN PLACE

A section of the yoke for the machine, Fig. 1, with the punchings and some of the stator coils in place, is shown in Fig. 2, and Fig. 3 shows a completely wound stator with the supporting ring for the stator coils. The supporting ring is necessary in large high-speed machines to prevent the movement or deformation of the coil end in case of short-circuits on the stator winding. The yoke is mounted on a base ring that may either be a separate casting or form an integral part of the hydraulic turbine. Adjusting screws, between the yoke casting and the base ring, provide means for radial adjustment. The base ring is bolted and grouted into the foundation and frequently serves as a support for the lower generator guide bearing housing. On the top bearing structure are carried the thrust bearing and also the top guide bearing. This structure must have ample mechanical strength to support the weight of the revolving parts, together with the water column acting upon the turbine. Also it must have sufficient mass to eliminate vibrations that otherwise might result from the operation of the alternator. The top bearing structure may be either of the girder type shown in Fig. 4, or made with a number of arms like that on the machine, Fig. 1. To avoid circulating

FIG. 3. COMPLETELY WOUND STATOR SHOWING RING SUPPORT OF COILS

currents through the bearings it is common practice to insulate the upper-bearing housing from the yoke. In units where the distance between the alternator and the turbine runner is large, the generator has an additional guide bearing whose housing may be supported from the foundation proper or from the base ring. The arms forming a part of the lower bearing housings are made of sufficient strength to support the weight of the rotating parts which, in case of repairs, can be elevated by means of jacks. The arms also form a convenient support for the brakes with which large machines usually are equipped.

The thrust bearing supports the rotating parts of the turbine and the alternator. Some years ago the accepted type was the pressure bearing mounted below the waterwheel. Oil was pumped under high pressure into a chamber between a revolving and a stationary disc, maintaining an oil film between the two surfaces. With this arrangement, the bearing was practically inaccessible, requiring a complicated high-pressure oiling

FIG. 4. TOP BEARING HOUSING, GIRDER TYPE

system. If, for any reason, the oil pressure dropped sufficiently to permit the plates to rub together, the result invariably was disastrous. This type of bearing has been replaced by the modern thrust bearing, which may be of several types. A roller-type bearing is shown in Fig. 5, the plate type in Fig. 6 and the Kingsbury type in Fig. 7. Either type of bearing may be mounted on the top of the upper bearing housing.

The roller thrust bearing shown in Fig. 5 was introduced several years ago and was used on several of the earlier installations. As may be seen from the figure, it consists of two steel plates hardened and ground, and a cage for holding the hardened and ground steel rolls that separate the plates. The rollers are made in short, straight sections. A bearing of this type has low fric-

FIG. 5. ROLLER-TYPE THRUST BEARING

tion, but it requires exceptionally faultless material and also the highest class of workmanship. The bearing is mounted on a socket foundation that permits of a certain amount of self-adjustment.

The self-contained plate-type thrust bearing, Fig. 6, has the advantage of being simple in design and rigid in construction. It has only two wearing surfaces, which are cast-iron plates, made from a special close-grained iron, the surface being carefully machined and scraped. Radial oil grooves are provided to establish and maintain an oil film between the rotating and stationary bearing surfaces at all loads and speeds. The grooves in the rotating plate cause a thorough oil circulation and

FIG. 6. PLATE-TYPE OF THRUST BEARING

spread continuously a fresh supply over the moving surfaces and prevent pockets of hot oil from forming. This bearing operates in a bath of oil in which a water-cooling coil is submerged, to carry off the heat generated by the bearing friction. With this arrangement, it is not necessary to circulate any oil through the thrust bearing. This type of bearing has been used for a number of small and medium-sized slow-speed machines. It is giving satisfactory service, particularly where the bearing pressures and rubbing speeds are moderate. A certain amount of adjustment in alignment is obtainable by means of adjusting screws bearing upon a collar.

The Kingsbury bearing, Figs. 7 and 8, represents one

of the most recent developments in the design of high-pressure thrust bearings. It consists of a flat annular revolving plate called the runner, which is supported on a number of babbitt-lined shoes mounted on pivots, permitting them to tilt sufficiently to establish a wedge-shaped oil film. Experiments have proven that a wedge-shaped film is very effective in automatic lubrication under high bearing pressures and speeds. The pressure upon the different shoes can be adjusted and equalized by means of screws, bearing upon the supporting pivots. The running surfaces are submerged in an oil bath and must be covered to a sufficient depth to prevent air from getting in between the rubbing surfaces of the bearing. Precaution must be taken to prevent the oil from churning and mixing in air bubbles, as this may cause the wiping of the babbitt on the bearing shoes. For medium and low pressures the runner disc is made from cast iron; for slow-speed high-pressure bearings, hardened steel or chilled iron is used. The shoes are constructed from cast iron lined with babbitt except for particularly high pressures, in which case bronze shoes are employed.

Bearing Pressures Determine Oil Used

Oil for a thrust bearing varies with the service. Heavy grades are used for high bearing pressures at low speed, and lighter oil gives good results for low and medium pressures on high-speed bearings. The coefficient of friction of a Kingsbury thrust bearing is very low, equaling 0.00009 times the square root of the bearing speed in revolutions per minute. This coefficient is based upon a bearing pressure of 350 lb. per square inch and an oil temperature of 40 deg. C. (104 deg. F.). With a variation in the bearing pressure, the coefficient also varies, being inversely proportional to the square

FIG. 7. KINGSBURY THRUST BEARING

root of the bearing pressure. This results in low bearing losses. For example, the losses in the thrust bearing shown in Fig. 1, are calculated to be 26 hp. when the machine operates at its normal speed of 150 r.p.m. As this machine has a rating of 32,500 kva., the thrust-bearing loss is less than 0.06 of 1 per cent.

In the operation of a Kingsbury thrust bearing the critical periods occur during the process of starting and stopping, when the surfaces of the runner and the shoes come into direct contact. This is due to the fact that in starting, the oil film is not established until a part of a revolution has been completed, and when stopping, the speed becomes too low to maintain the oil film. Only at these periods does perceptible wear occur on the bearing, and this also explains the noise that can be heard when a machine of this type is either started or stopped.

In the construction of a vertical waterwheel unit the number of guide bearings varies, but in all cases the generator has either one or two. In the installation, where the distance between the generator and the turbine runner is short, one guide bearing is supplied with the generator and one with the turbine. Where the distance is of considerable length, it is common practice to supply the generator with a lower guide bearing. It is general practice to make the guide bearings in sections to facilitate erection, inspection and repairs. The upper

FIG. 8. KINGSBURY THRUST BEARING ASSEMBLED

guide bearing fits solidly into a conical bearing seat in the top bearing housing and no provisions are made for any adjustment. The lower generator guide bearing is of the same general design as the upper bearing, but it is adjustable by means of setscrews. These bearings are made of sufficient size to resist pressures that may result from unbalanced magnetic pull or from other causes. The lower turbine guide bearing is mounted directly over the turbine runner. It is made from cast iron, lined either with babbitt or with lignum-vitæ. If a cast-iron babbitt-lined bearing is used, it is similar in construction to the generator guide bearings and requires a special oiling system. For that reason, the lignum-vitæ bearing is frequently preferred as it requires only water lubrication.

Modern lignum-vitæ guide-bearing design is based upon low bearing pressures per square inch of rubbing surface. The rubbing surface is made up from a number of lignum-vitæ strips equally distributed and dovetailed into the bearing box. The strips are mounted parallel to the axis of the machine and have the end grain of the wood normal to the bearing surface. The water for lubricating it passes through the bearing in grooves between the strips. It is essential that the water used for this purpose does not carry any foreign matter in suspension as this invariably leads to undue wear. Turbines operating in clear water have the lubricating water piped directly from the gear casing through a strainer to the bearing. If clear water is not available, a filter should be installed. No further piping is required, as the water escapes from the bearing to the draft tube through openings located in the hub of the runner.

Simplicity of lubrication is the main advantage of the lignum-vitæ guide bearing over the babbitt-lined type, as the latter requires a complete oiling system, together with a sealing ring to prevent its oil from escaping or getting mixed with water.

Power Problems of the Pacific Coast

By CHARLES H. BROMLEY

THE purpose of this article is to acquaint those not familiar with the Coast with some of its chief power problems and economic aspects. The Pacific Coast is a land where water is building an empire. To understand the country, its resources, its economics, its potentialities one must first understand its topography and climate, for these are fundamental in human pursuits there.

This is not the place for lengthy discussion of the interesting and remarkable features of topography and climate of the Coast, but a word about them is worth while here for those not familiar with them. The great body of the Pacific Ocean is warm, and as the prevailing winds are out of the west, the whole Coast country of the United States, from Puget Sound south to Mexico, is remarkably free from freezing temperature and snow. In the Northwest rain falls steadily, though not heavily, during many weeks of the winter. For nine months of the year most of California is practically free from rain. In most of California and all of the Northwest the mean temperature in summer is low. Most persons are familiar with the mountain ranges of the Coast. In this respect the Pacific Coast differs greatly from the Atlantic, in that it has no coastal plain.

The warmth of the climate during winter, the abundant water power and lack of cheap fuels in adequate quantity make the Pacific Coast country almost ideal for the generation and transmission of power by large systems. Winter heating loads in business centers of towns and cities are excellently taken care of by district steam-heating plants, usually operated by the public-utility companies furnishing electrical energy, though sometimes by other independent companies.

The basic sources of electric energy on the Coast are the high mountains, the Cascades in the Northwest and the Sierra Nevadas down the Coast. In winter these mountains become heavy with snow and glaciers which slowly melt during summer, and the run-off, together with rains, by aid of storage systems, supplies the streams which furnish hydro-electric power. Owing to the high heads available for hydro-electric plants, a little water goes a long way out there, measured in terms of kilowatts. To the man accustomed to great, wide streams and low-head plants, the little streams of the West, winding their way among steep-sided canyons, the power houses shooting small swift streams from their turbines, like water from a hose, present an interesting sight and more than interesting engineering problems. In California, particularly, most of these power houses are many miles remote from business centers and current transmission distances to points of consumption are great. But despite distances, the visiting engineer will miss a treat indeed if he doesn't get out to one or two developments. Views of Snoqualmie Falls and other plants of the Puget Sound Traction, Light and Power Co. are shown in Figs. 8, 9, 10, 11 and 12.

The Coast, particularly California, this season faces one of the most, if not the most, serious power and water shortages in its history. The Fates have been unkind in this land where water is life itself. The High Sierras have not their usual heavy coat of snow, which shines so prettily in the sky as one passes down the wide valley of the San Joaquin.

The major problem confronting engineers at this time is, of course, power shortage. But that is a manifold problem, and perhaps the most vexing of its components is the fuel problem. Fuel oil is the backbone of Pacific-coast steam stations. It is of California origin and, up until comparatively recently, has been available at a low price—much under a dollar a barrel. Now its price is so high that one frequently hears "prohibitive" used when it is spoken of. Every effort is being made to find other fuels, and this is as true of Los Angeles on the south as of Seattle on the north. In the Northwest hog-fuel, the refuse from sawmills, is being widely used and markets for it are being developed. Particulars are given elsewhere in this issue in an article devoted exclusively to the subject.

Farther south, in California, the San Joaquin Light and Power Co., with headquarters at Fresno, is developing considerable capacity in steam stations located in the oil fields where natural gas is available. At that

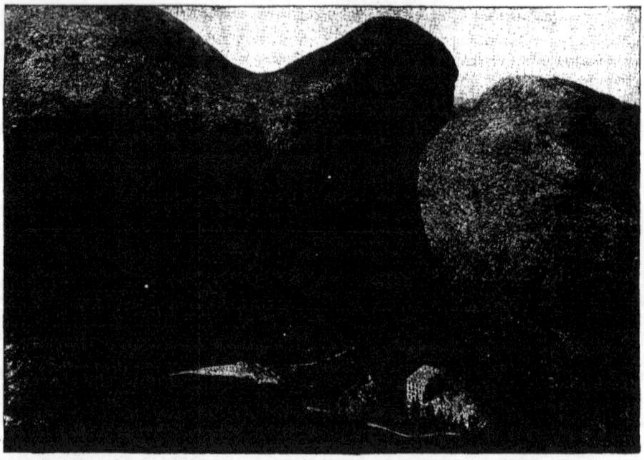

FIG. 1. UPPER VIEW. BIG CREEK POWER HOUSE NO. 1, SOUTHERN CALIFORNIA EDISON COMPANY. FIG. 2. LOWER VIEW. BIG CREEK SEEN FROM THE ROCK ATOP KERCKHOFF DOME

FIG. 3. TOP LEFT. KERCKHOFF DAM JUST COMPLETED, 100 FT. HIGH, 400 FT. LONG, SAN JOAQUIN LIGHT AND POWER CO. FIG. 4. TOP RIGHT. TUNNEL OPENING AND CREW, KERCKHOFF DEVELOPMENT. FIG. 5. BOTTOM. STEAM SHOVEL, AIR-DRIVEN, MUCKING OUT ROCK IN THE TUNNEL; PROGRESS, 12 FT. A DAY. DUMP CARS ELECTRICALLY OPERATED

FIG. 6. UPPER. SIPHON, KERN RIVER NO. 3, SOUTHERN CALIFORNIA EDISON CO., TYPICAL OF COAST HYDRO-ELECTRIC DEVELOPMENTS. FIG. 7. MIDDLE. INTERIOR KERN RIVER NO. 1. FIG. 8. LOWER. SNOQUALMIE RIVER JUST ABOVE THE FALLS, PUGET SOUND TRACTION, LIGHT AND POWER CO., SEATTLE

FIG. 9. BEAUTIFUL SNOQUALMIE FALLS, SEATTLE

The intake descends 270 ft. to the power house located in a cavern. The first hydro-electric development in the Far Northwest.

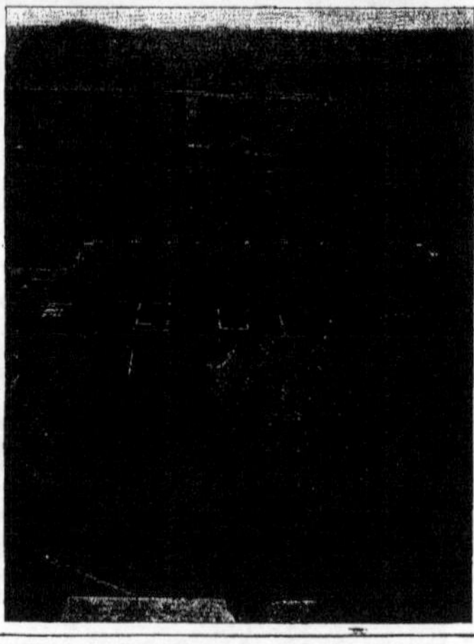

FIG. 10. UPPER. NO. 2 UNIT, SNOQUALMIE FALLS, PUGET SOUND TRACTION, LIGHT AND POWER CO., SEATTLE
[F]IG. 11. LOWER LEFT. HEAVY CONDUIT TO THE PLANT ON THE RIGHT. FIG. 12. LOWER RIGHT.
ANOTHER PLANT OF THE PUGET SOUND TRACTION, LIGHT AND POWER CO.

company's Bakersfield plant, recently enlarged by 15,000 kw. with another unit of similar size to come, gas is burned. But it is pumped a considerable distance before reaching the plant. There is now discussion of locating another large steam station in the very heart of the gas field where the point at which it is collected will be adjacent to the station. The oil industry itself will likely take considerable of the output of such a station.

Steam-station development in this gas area would seem a logical and profitable one in view of the urgent pleas to interconnect the transmission systems of California. Time seems to be showing that a greater proportion of steam-electric generating capacity is required. While the Railroad Commission of California reports that 75 per cent of the 2,892,000,000 kw.-hr. generated during the year 1918-19 by the public utilities of the state was generated from water power, this percentage is not true of some of the larger companies. This is the fourth consecutive year that one large company at least has furnished approximately 60 per cent of its output by steam stations and 40 per cent by hydro-electric stations.

The Coast has never attempted the wide use of Western coal. With fuel oil so high in price, serious attention is now being given to this. The Puget Sound Traction, Light and Power Co. is now getting satisfactory results with a lignite culm which it is using in pulverized form in its Western Avenue steam-heating station, Seattle, Wash. The performance has been highly satisfactory. The steadily rising price of oil causes a trend which gives coal an increasingly better chance in the power fuel market. So long as steam stations on the Coast continue to be used for purely emergency service, Western coal is not likely to find a large market in these power stations. But there are possibilities that with further expansion of the electrical industry it may be found at least expedient to design and build more large steam plants intended for more or less continuous service at high efficiency. Perhaps the most modern steam plant on the Coast is Station A of the Pacific Gas and Electric Co., San Francisco.

Interconnection which will tie in all the transmission systems of the Coast is one of the big dreams. And it may eventually turn out to be one worth while exploiting. California is fairly well tied together now, though not as completely as it may soon be. The central and northern parts of the state long ago standardized on 60-cycle current, but a part of the southern end is given over in large measure to 50-cycle current. This condition is being corrected. One of the features in this work is interesting. The new Kern River Station No. 3 of the Southern California Edison Co. has capacity of 22,000 kw. in two reaction hydraulic turbines working under 800 ft. head. Much of the company's system is of 50-cycle frequency, while the systems with which it is desirable at times to tie in are of 60-cycle. The engineers of this company, and the builders of the turbines, the Pelton Water Wheel Co., of San Francisco, have arranged to furnish both 50-and 60-cycle current with these machines. Each machine has two runners, but of course only one at a time is used. One will give a speed of 500 r.p.m. and 50 cycles at this speed, the other 600 r.p.m. and 60 cycles. The design is such that either runner may be taken out, loaded on a truck, pulled to one side and the other runner, which rests on a truck in a bay, rolled under the turbine and slipped on the shaft. The builders are confident that when the men become accustomed to making the change-over they will easily do it in eighteen hours. This is one of the things the Southern California Edison Co. is doing to adapt itself to the 60-cycle standard, and surely it is a clever idea. A part of a siphon used on Kern River No. 3 is shown in Fig. 6. Views of Big Creek No. 1 are shown in Figs. 1 and 2.

In the hydro-electric developments of the Coast, particularly California, considerable tunneling is necessary in conduit construction. And these tunnels are invariably through granite, of which the Sierra Nevadas are formed. In Kern River No. 3 development, just referred to, there are 60,000 ft. of tunnels, the longest being 7,132 ft. Recently the San Joaquin Light and Power Co. kindly took the writer from Fresno to its great Kerckhoff development up in the mountains along the San Joaquin River (see Fig. 3). This is the largest hydro-electric job under construction in California at present, and the method of driving the tunnels is one of the interesting features of the construction work. Figs. 4 and 5, page 703, show two views of one of

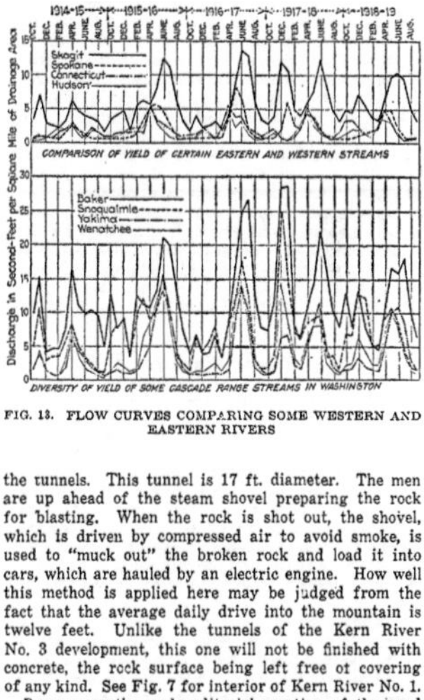

FIG. 13. FLOW CURVES COMPARING SOME WESTERN AND EASTERN RIVERS

the tunnels. This tunnel is 17 ft. diameter. The men are up ahead of the steam shovel preparing the rock for blasting. When the rock is shot out, the shovel, which is driven by compressed air to avoid smoke, is used to "muck out" the broken rock and load it into cars, which are hauled by an electric engine. How well this method is applied here may be judged from the fact that the average daily drive into the mountain is twelve feet. Unlike the tunnels of the Kern River No. 3 development, this one will not be finished with concrete, the rock surface being left free of covering of any kind. See Fig. 7 for interior of Kern River No. 1.

Power recently made editorial mention of the need of funds for gathering further data on the flow of rivers on the Pacific Coast. The purpose of here reiterating it is to give it emphasis. Accurate knowledge of stream flow is essentially vital to stable hydro-electric development. No one but the district engineers of the Geological Survey in the Northwest particularly knows what it has meant to the routine of these district offices to get essential stream flow data during the trying years of the war when these offices were shorthanded. Such data were gathered regardless of what else had to be dropped, and for this the district engineers deserve credit. But surveys of broader and deeper character are desirable, and these cannot be made without a material increase in funds. Few Easterners appreciate the volume and uniformity of flow of many rivers of the Northwest. Unfortunately, we have but few flow curves, but some are shown in Fig. 13.

In this all too general and brief perspective that the writer has here given, the famous Hetch Hetchy project of the City and County of San Francisco deserves conspicuous place, for it is one of the foremost engineering problems of the Pacific Coast. Hetch Hetchy's power plants are in a measure incidental to the main purpose, which is water supply. For the greater part of the work high praise is due M. M. O'Shaughnessy, chief engineer of the City and County of San Francisco. The Hetch Hetchy site consists of 420,000 acres at the head of the Tuolumne River about 160 miles from the city. Ultimately, it will furnish 400 million gallons daily and develop about 250,000 hp. by reason of the waters' fall toward the city. A power plant of 4,400 hp. was built to furnish current for construction. Up to July, 1919, $7,516,000 had been spent on the project. The sale of surplus power from the construction power house alone netted the city $80,000 in twelve months. So far the city has done most of the work itself by day labor.

The first large power unit in the line of the aqueduct will be the Moccasin Creek power house, which will have capacity for immediate delivery of 66,000 hp. The city engineer points out that the present yearly consumption of electricity by the city is 360,000,000 kw.-hr. and that when the aforementioned power house goes into service the current available to the city will be about double that consumed now. In reaching the city the waters from the project will drop 8,000 and 10,000 ft.

A Super-Power Plant for Paris

A rather striking boiler plant was described by Harold Goodwin, Jr., of the Power and Mining Engineering Division of the General Electric Co., in an address before the American Institute of Electrical Engineers, Washington Section, on April 13. Mr. Goodwin discussed the procedure of his company in designing the layout and equipment for a super-power plant, which it is proposed to build for the City of Paris. This plant is designed to consist of seven steam-turbine generators of 35,000-kw. capacity each, the plan being to have this new plant supersede all existing plants in the City of Paris.

The boiler plant is arranged with four boilers of a novel type to supply each turbine. These boilers are to have two banks of tubes, a set of superheater coils being installed between the upper and the lower banks. This arrangement was deemed necessary because it is intended to operate the plant with steam at 325 lb. and 320 deg. F. superheat at the turbine. The boilers will, therefore, operate at approximately 750 deg. F. and 350 lb. pressure. With such steam conditions, it is believed impracticable to use pumps or other auxiliary equipment direct on boiler-pressure steam, therefore all the auxiliaries and the station light and power are to be supplied through small auxiliary turbines, one of which is installed for each pair of the main units. Unusual precautions are taken to insure continuity of operation to the auxiliary equipment, especially that of the boiler room, including stokers, feed pumps, etc.

Because of the large capacity of each of the turbines a double steam main from each group of boilers is provided.

Coal can be received either by boat on the Seine River or by railroad car. When coming by boat it is unloaded by cranes and delivered either into stockpiles or to the track hoppers, from which it is conveyed to the crushing stations. When delivered by railroad cars, it is dumped directly from the car into the track hoppers. From the crushing station the coal can be carried either to the bunkers, of which there are three, or to stockpiles. Storage space is also provided for uncrushed coal, so that the maximum of flexibility in handling is attained. Recovery from stock, either crushed or lump, is accomplished by grab buckets on locomotive cranes or the traveling crane, which swings about from each bunker. The crushed coal is conveyed from the bunkers to the boiler room by larries traveling on overhead tracks that intercommunicate throughout the system. The larries dump the coal in the small hoppers that feed the stokers.

It is estimated that the plant, as designed, would have cost approximately $45 per kilowatt of capacity on the basis of pre-war conditions. The exact present cost has not been calculated, but it is believed that it will lie between two and three times the pre-war figure, probably nearer three times. As a result of the unfavorable conditions of international exchange, much of the boiler and plant equipment will be built abroad, although considerable of the electrical equipment will be constructed in this country. Since no final estimate as to the power required for all of the auxiliaries has yet been completed, the over-all efficiency of the plant has not been estimated, but it is believed that not more than ten pounds of water will be required per kilowatt-hour. With the normal power for auxiliaries this would require approximately 1¼ lb. of coal per kilowatt-hour, judged under American conditions of fuel supply.

A power plant whose four units were designed to develop a total of 144,000 hp. would make a fair-sized central station. The main driving units of the British battle cruiser "Hood" were built to produce this amount and on the trial trip actually developed a total of 157,000 hp. The high-pressure turbines make 1,500 r.p.m. and the low-pressure turbines 1,100 r.p.m. at full power. The propellers are driven at 210 r.p.m. through single reducing gears. Steam is supplied by twenty-four Yarrow type small tube boilers having a total heating surface of practically 175,000 sq.ft. and a working pressure of 235 lb. per sq.in. The pressure at the turbine governor is 210 lb. per sq.in. and a vacuum of 28 in. is maintained in the condensers.

Utilization of tidal ebb and flow is gaining a foothold in Great Britain. The principle underlying the inventions for harnessing the tides is the working of a turbine by the inflow and outflow of the tides. One plan advanced calls for the building of a wall extending out into the sea or river in which turbines are placed. It is claimed that ten-foot turbines would be most economical, where suitable, each of which would develop over 300 hp. with a five-foot head. Thus a mile of sea wall would be capable of generating approximately 120,000 hp.

Hog Fuel in the Pacific Northwest

Its Rise from the Dumps to a Revenue Bearing Commodity

Hog Fuel is the refuse from lumber made in the process of manufacture into boards and pieces of larger dimensions. It consists of broken slabs and pieces broken from timber, sawdust and other forms.

THE man who makes two blades of grass grow where but one grew before is always received with acclaim, and rightly. But somehow one's enthusiasm runs higher for the fellow who does things on a par with getting honest-to-goodness kilowatts from garbage. Such frugality is indeed uncommon in a new country. And so the story of exploiting the nuisance, hog fuel, in the Pacific Northwest is doubly interesting and valuable to those who now find it a burdensome waste.

When in Portland, Ore., recently, the writer, in scanning the newspaper over the coffee and rolls, noticed that a local lumber company was in the market for a burner to consume hog fuel, the accumulation of which was such a nuisance about the sawmill. It was going to cost many thousands of dollars for the consumer. And I had just come from Tacoma, Washington, not so many miles away, where R. L. Watts, mechanical superintendent for the St. Paul & Tacoma Lumber Co., was doing this refuse up in neat packages of freight-car and motor sizes and selling them to eager buyers in the picturesque city. And more—he furnished the purchaser, or could furnish him, with accurate information about heating value and moisture, and translate without hesitation the comparative cost with fuel oil on an evaporation basis.

Watts came to Tacoma from the desert country of New Mexico and Arizona. So he was greatly disturbed about the universal custom of conveying perfectly good fuel to burning dump piles and incinerators just to get it out of the way. Desert folks have to be frugal. The small mills usually convey the hog fuel to a point some little distance away, elevate it and let it fall upon a burning pile. Such practice succeeds in many locations in burning down the mills every few years, but no one seems to take this seriously. Fig. 3 shows a pile of hog fuel held in reserve at the plant of the St. Paul & Tacoma Lumber Co., Tacoma.

FIG. 1. PLANT OF THE ST. PAUL & TACOMA LUMBER CO.
To the right of the power plant and under the conveyor housing hog fuel is being loaded into cars for distribution in the city.

To understand the market for hog fuel one must realize the outstanding differences between Eastern and Western power situations. In the Pacific Northwest water power abounds, coal of inferior grade only is available, the climate is mild in winter, and the electric rates are, on the whole, low. So there are few independent power plants. District steam heating is common to the whole Pacific Coast. California fuel oil is widely used—in fact, is the chief fuel for power plants and steam boilers. A few years ago it cost much under a dollar, even in the Northwest. But now—well, it is impossible to guess what it will be there when this article appears, but it was uncomfortably crowding the

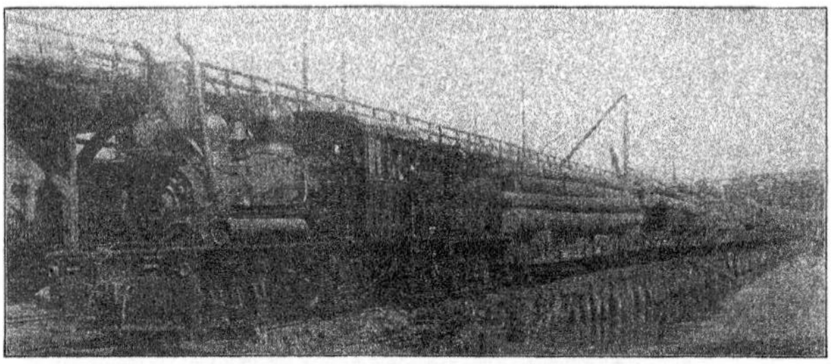

FIG. 2. TIMBER ARRIVING AT THE MILL FROM THE CAMP

All mechanical equipment at the camp and the mills is selected, operated and maintained under the supervision of the mechanical engineer.

two-dollar mark early in the year. In the face of these conditions, therefore, hog fuel, which averages 5,500 B.t.u. per lb. wet and 8,700 B.t.u. per lb. dry, offered possibilities to the St. Paul & Tacoma Lumber Co., whose great mills are near the center of the city. Fig. 2 shows a load of timber arriving at the mills.

Mr. Watts conducted tests the results of which are given in the accompanying table and which, it will be observed, show the fuel to be a good one. Curves of some test results are shown in Fig. 4. The district steam-heating station which supplies the business section of Tacoma with heat is one of the large consumers of hog fuel. Oil was used until recently, and the burners, tanks, etc., are still in place ready for use if necessary. The market value of the fuel is so great and the electric rates so low that the St. Paul & Tacoma Lumber Co. buys considerable current from the city and sells more and more hog fuel. The company will, of course, maintain its power plants, but will install another large steam turbo-generator and extend electrification of its many mills.

In and around Portland, Ore., hog fuel is becoming popular. Engineers there have learned to burn it with worth-while results.

The usual practice in using hog fuel is to convey it over the furnaces and drop it into open chutes which direct its fall into the tops of the dutch ovens.

At present hog fuel here is selling at $1.50 per unit (200 cu.ft.) loaded into the cars at the mills. The cars contain an average of 20 units. At the plant of the Tacoma District Heating Co., which purchases the largest volume of hog fuel, the cost of unloading the

FIG. 3. PILE OF HOG FUEL IN RESERVE IN THE BACKGROUND

Note the long timber in the mill pond.

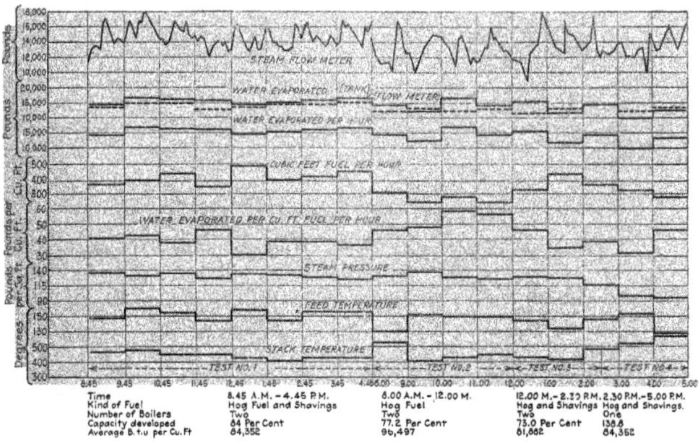

FIG. 4. PERFORMANCE RESULTS WITH HOG FUEL IN PLANT OF THE ST. PAUL & TACOMA LUMBER CO., TACOMA, WASH.

fuel from the cars to the boiler-house bunkers is 20c. per unit, or about $4 per car. Mr. Watts finds that the boiler-plant labor charge is practically the same as for plants burning coal with stokers and having overhead bunkers, with the exception that little or no charge is made for ash removal, with hog fuel. Numerous tests conducted by the St. Paul & Tacoma Lumber Co. show that an average of 10,000 lb. of water may be evaporated

TEST OF STERLING BOILERS USING WOOD FUEL

Location.............St. Paul & Tacoma Lumber Co., Tacoma, Wash

Furnace, Dutch oven: length, ft				6 5	6 5
Grate surface, sq. ft				55 25	62 82
Water heating surface, sq. ft				2,617	2,878
Date, Feb. 20 and 21, 1917.					
Test number	1	2	3	Average of 1, 2 & 3	4
Duration, hr	8 0	4 0	2.5	14 5	2 5
Kind of fuel	Shavings and hog	Hog only	Shavings only	.....	Shavings and hog
Steam pressure, lb. gage	131	131	128	130	100
Temp. feedwater, deg. F	153	145	143	147	142
Temp. escaping gases, deg. F.	413	427	430	423	531
Forced draft between damper and furnace, in water	0 2	0 2	0.2	0.2	0.2
Average weight of fuel per cu.ft. as fired, lb	15 32	18 37	14 64	16 11	1_ 75
Moisture, per cent	38 5	41 5	34 0	38 0	38.5
Heating value per lb. wet, B.t.u	5,506	5,253	5,593	5,431	5,506
Heating value per lb. dry, B.t.u	8,953	8,979	8,474	8,802	8,953
Cu.ft. fuel burned, per hr	375	264	343	339	287
Cu.ft. per sq.ft. grate per hr	3,176	2,236	2,905	2,871	2,431
Total evaporation from and at 212 deg. F., lb	127,469	58,737	34,602	221,108	31,311
Evaporation from and at 212 deg. F., per cu.ft. fuel as fired, lb	42 5	55.7	40.4	45 7	43.6
Evaporation per unit (200 cu.ft.) of fuel, lb	8,500	11,400	8,080	9,000	8,720
Boiler horsepower developed	462	425	401	430	363

(presumably from and at 212 deg. F.) per unit of hogged fir. The average evaporation from a barrel of oil as delivered in Tacoma is 4,368 lb., and from a ton of local coal about 12,000 lb.

Following are some figures of comparative costs with oil, coal and hog fuel as related to the Tacoma Shops power plant of the Chicago, Milwaukee & St. Paul R.R.; the report was made in October, 1918, when all fuel prices were lower than at present:

Oil:	
Average daily steam load, lb	416,000
Barrels oil used per 24 hours	95
Pounds steam per barrel (evap. 13-1) (bbl. 336 lb.)	4,368
Cost of oil per barrel	$0.75
Cost of oil per day	71.43
Cost of steam per thousand pounds	0.1717
Coal:	
Average daily steam load, lb	416,000
Tons coal used per 24 hours	34⅔
Pounds steam per ton (evap. 6-1; ton, 2,000 lb.)	12,000
Cost of coal per ton	$2.75
Cost of coal per day	95.56
Cost of steam per thousand pounds	0.229
Hog Fuel:	
Average daily steam load, lb	416,000
Units hog fuel used per 24 hours	41.6
Pounds steam per unit (unit 200 cu.ft.; evap. 50-1)	10,000
Cost of hog fuel per unit	$0.75
Cost of hog fuel per day	31.20
Cost of steam per thousand pounds	0.075

The foregoing figures cover cost of fuel only and are based on standard plant practice.

Some weeks ago a trial of hog fuel was made on an 8-retort underfeed stoker in one of the stations of the Portland Railway, Light and Power Co., Portland, Ore. The fuel was placed by hand through three large doors cut in the front of the boiler setting, directly over the stoker retorts. The chief difficulty was that the fuel could not be handled by the stoker fast enough to keep the rear grates covered. The fuel was placed on the forward grates in a thick bed which gradually thinned out as the rear grates were reached; rapid combustion therefore took place at this latter end of the stoker. The engineers in charge of the plant are of the opinion that if the thick bed of fuel could be maintained at the rear of the stoker instead of at the front, this method would be highly satisfactory.

The method of firing affected the rate of steam flow: the flow-meter chart showed plainly when each boxful of fuel was charged and when the doors were closed after charging. While charging, the percentage of rating would drop to about 120 per cent normal, and rise to 225 per cent normal when closed, reaching 260 per cent for about five minues after charging and closing the doors. The charging periods occurred about every fifteen minutes.

The combustion over the active part of the grate surface was good, the flame being a brilliant white, short and quick. There was no secondary combustion in the boiler passes. However, combustion occurred too close to the tubes.

The following is a brief tabulation of the principal results obtained during the test:

Duration of test—6 hours (9:30 a.m. to 3:30 p.m.)	
Average steam pressure, lb.	187
Average steam temperature, deg. F	572
Superheat, deg. F	96
Factor of evaporation	1.248
Pounds of water per pound of fuel	2.58
Pounds of water per pound of fuel, from and at 212 deg. F	3.21
Rating obtained (average per cent, normal)	175
Average weight of fuel per cu.ft., lb.	15.8
Water evaporated from and at 212 deg. F. per unit of fuel, lb.	10.143
Flue gas analysis (average):	Per Cent
CO_2	11.16
O	8.57
CO	0.28

Surcharging Diesel Engines

A Practice of Especial Value When Operating at High Altitudes

By CL. RUEGG

SINCE the introduction of the two-cycle Diesel engine, necessitating the use of a scavenging cylinder it has been an open question as to what capacity this pump should have in order to supply sufficient air to each engine cylinder for complete combustion. Discussion as to the proper size of such scavenging cylin-

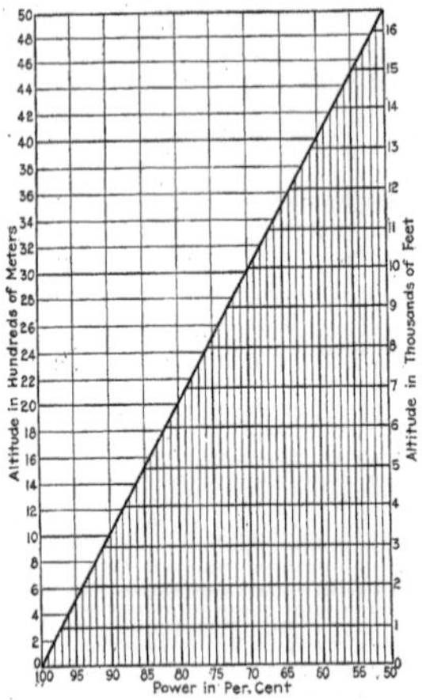

FIG. 1. CHART SHOWING DECREASE IN POWER WITH INCREASE IN ALTITUDE

ders, or pumps, has led to considerable experimentation and has resulted in more or less standardized sizes that are not only sufficient for thorough scavenging, but also make available a surplus of air. While this surplus of air cannot be taken advantage of under ordinary conditions to burn a greater amount of fuel, it is particularly useful at high elevations.

The quantity of fuel that can be burned in a Diesel-engine cylinder is limited by the amount of heat that can be carried off by the jackets. Any increase in the

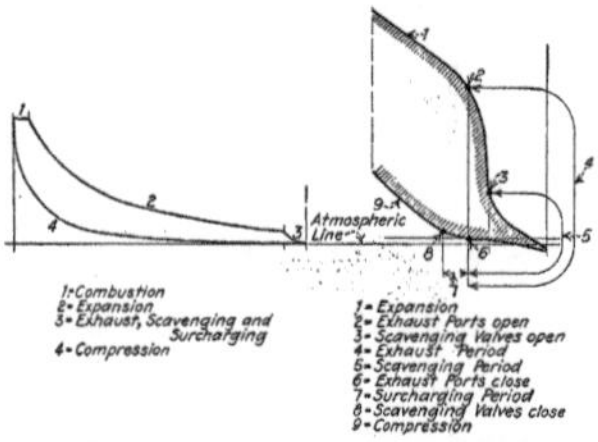

FIG. 2. TYPICAL TWO-CYCLE DIESEL-ENGINE INDICATOR CARD. FIG. 3. ENLARGED TOE OF THE INDICATOR CARD SHOWN IN FIG. 2

amount of fuel above this maximum will result in more power, but also in an accumulation of heat, which will cause injurious cylinder temperatures.

The quantity of fuel that can be *economically* burned in a Diesel-engine cylinder is limited by the weight of air in the cylinder. If this weight of air is decreased, the amount of fuel injected must be decreased, with a resulting loss of power. Any attempt to force up the power output by injecting more fuel results in inefficient operation with its accompanying increase in the amount of heat to be carried off by the cooling system.

It is well known that both the density and temperature of the atmosphere decrease with the altitude. It may, however, be assumed that the initial working temperature of the cycle is not affected by the altitude and that in actual practice the charge entering the cylinders is maintained at a constant temperature. Decrease in power with increase in altitude is shown in Fig. 1. The volumetric efficiency of an engine cylinder remains almost constant at various altitudes, and the volume drawn into the cylinder either at sea level or at any other altitude is approximately the same. The weight of the air taken in, therefore, varies in direct proportion to the atmospheric density. It will be noticed that there is a considerable decrease in pressure and therefore in density of the atmosphere with the altitude, and a similar decrease in the output of the engine must be anticipated if no means are provided for surcharging the cylinder with air. In other words, the power developed per cylinder is directly dependent on the weight of the charge of air in the engine and not on its volume, and this weight decreases at higher altitudes.

In a four-cycle Diesel engine the pistons have two functions to perform—that of a power piston and that

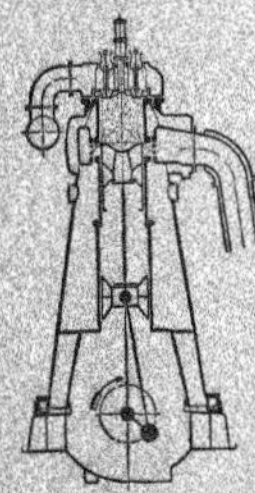

FIG. 4. TWO-CYCLE DIESEL ENGINE DURING EXHAUST

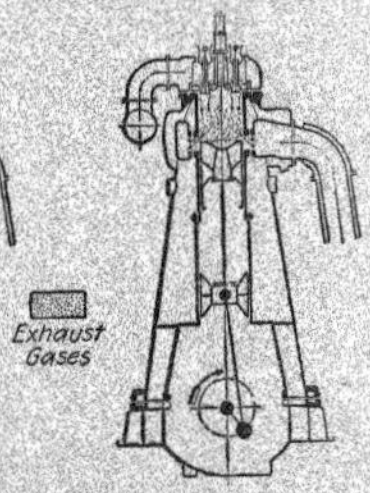

FIG. 5. TWO-CYCLE DIESEL ENGINE DURING SCAVENGING

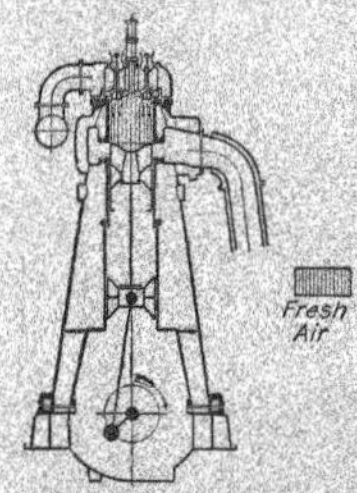

FIG. 6. TWO-CYCLE DIESEL ENGINE DURING SURCHARGING

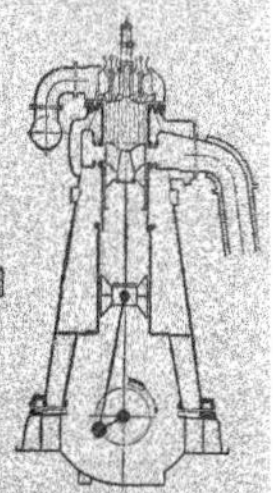

FIG. 7. TWO-CYCLE DIESEL ENGINE DURING COMPRESSION

of a scavenging piston. The amount of air taken in during the suction stroke is less than 100 per cent of the piston displacement, owing to the fact that a loss through the inlet valve is occasioned by very high air velocities. In some instances this loss amounts to from 5 to 10 per cent. There is no way to increase the amount of air delivered to a cylinder of a four-cycle engine, except by the use of a separate blower or surcharging pump. Surcharging has made it possible to obtain mean pressures as high as 260 lb. per sq.in. in special engines. The operation of a two-cycle Diesel engine is so well known that its details need not be discussed here. A typical indicator card from such an engine is shown in Fig. 2. Line 4 indicates the compression; line 1, burning at constant pressure; line 2, expansion of the burned gases to a point where the piston uncovers the exhaust ports; and line 3, the drop of pressure within

FIG. 8. TWO-CYCLE DIESEL ENGINE FOR DIRECT CONNECTION TO AIR COMPRESSOR

Surcharging has been successful with this engine, although its speed varies from 180 to 70 r.p.m.

the cylinder as the burned gases escape through the exhaust parts. It is this last part of the cycle in which we are particularly interested. Referring to Fig. 3, point 2 indicates the instant when the piston uncovers the exhaust ports. The large area thus opened causes the rapid drop of the pressure. Point 3 on the exhaust line indicates when the scavenging valves open. The air immediately rushes into the cylinder, because a lower pressure exists within the cylinder than behind the scavenging valves. It is usually found that the momentum of the exhaust gases carries the pressure slightly below atmospheric, as shown in Fig. 3, during which time, however, the scavenging air is continually blowing into the cylinder. At the instant the piston covers the exhaust ports, as indicated by point 6, all air that is entrapped between the piston and the head suffers an increase in pressure, inasmuch as the scavenging valves are still open and continue to admit air into the cylinder. When the point 8 is reached, the scavenging valves close and compression begins. It is therefore evident that a greater weight of air is now within the cylinder than would be present without surcharging. The same cycle is illustrated in Figs. 4, 5, 6 and 7, the arrows indicating a rush of burned gases out through the exhaust ports. Fig. 5 also illustrates the impor-

FIG. 9. DIESEL ENGINE DIRECT-CONNECTED TO AIR COMPRESSOR AND FOUR SUCH ENGINES DIRECT-CONNECTED TO GENERATORS

This plant operates at 6,000 ft. above sea level, and its present capacity is maintained by surcharging.

tance of placing the scavenging valves in the head, thereby allowing the fresh air to push ahead of it the exhaust gases.

Fig. 6 shows the piston at the moment the exhaust ports are closed. The cylinder is now filled with fresh air, but the scavenging valves are still open and the cylinder is being surcharged. As indicated in Fig. 7, the scavenging valves are closed and compression begins.

The scavenging pumps are designed to deliver air at a gage pressure of 3 to 5 lb. per sq.in., depending on the size and type of the scavenging valves. There is no atmospheric inlet valve, and the scavenging air is admitted into the cylinder either through valves in the cylinder head or through ports near the bottom of the cylinder. In either case the piston, when closing the exhaust ports during the compression stroke, starts to compress the pure air at a pressure above atmospheric.

Scavenging ports are usually located opposite the exhaust ports, while valves are generally in the cylinder head. The latter arrangement has several advantages: First, an even distribution of scavenging air entering the cylinder is assured, and, further, the opening of the valve can be timed to suit the conditions. Probably the most important advantage, however, is that the cylinder dimensions are not limited to any fixed ratio between stroke and bore, as is the case with port scavenging.

It is quite important to be able to vary the amount of air as well as the amount of fuel taken into the cylinder, and with two-cycle engines this is accomplished by properly dimensioning the scavenging pump. Usually, this pump furnishes 10 to 40 per cent more air than is represented by the piston displacement of the power cylinders.

It has been claimed that the benefits derived from surcharging would not be realized in cases where the speed of the engine was not constant. The engine which is shown on the testing floor in Fig. 8 has been direct-connected to an air compressor operating between 180 and 70 r.p.m., in which the scavenging pressure varies with the speed, dropping from about 4 lb. at 180 r.p.m. to about 1¾ lb. at 70 r.p.m. The fuel consumption per indicated horsepower during this range of speed remains constant. This is due to the fact that being direct-connected to a compressor, the torque on the machine remains the same at all speeds, and it was found that the fuel consumption per stroke was practically independent of the speed.

Fig. 9 shows in the foreground, a compressor of 4,000 cu.ft. of free air per minute capacity installed in a Diesel plant, together with four Diesel units direct-connected to alternating-current generators running in parallel. The plant is in operation at an elevation of about 6,000 ft., and the effect of surcharging made it possible to rate these engines higher than is indicated by the curve of Fig. 1.

The total amount of capital invested in central stations in the Dominion of Canada, inclusive of transmission and distribution systems, is $369,464,961, an average of $210 per installed primary horsepower. The total construction cost of seventy representative hydro-electric stations, exclusive of transmission and distribution systems with an aggregate turbine installation of 745,797 hp. is $50,740,468 (pre-war figures). The cost of the construction of dams, flumes, penstocks and all hydraulic works is included in the figures given, as well as power stations and equipment.

Storage Batteries for Emergency Service

Emergency Batteries Classed as Stand-by, Exciter and Oil-Switch Batteries, 159 Stand-by Batteries in Service Having a One-Hour Discharge Rate of 633,000 Amperes. End-Cell Switches Built to Handle 50,000 Amperes. Control of Exciter and Oil-Switch Batteries. Storage Batteries Recommended for Emergency Lighting.

BY C. J. WELCKE
Engineer, Electric Storage Battery Company

C. J. WELCKE

WITH the ever-increasing development in the generation of electric power by both steam and hydraulic stations, it is becoming more and more important that continuity of service be maintained and that power interruptions be reduced to a minimum. In eliminating interruptions of service the lead-acid type of storage battery has been and still is an important factor.

The storage battery is particularly suited for emergency service on account of its extreme reliability. It is simple in construction, there are no moving parts to get out of order, and it is capable of instantly giving out energy at extremely high rates, with a very small drop in voltage. In fact, in actual practice the rate at which energy can be taken from a lead-acid storage battery is limited only by the size of current-carrying parts. Combined with its abilty to deliver tremendous amounts of power for short periods is the further advantage that it is an efficient piece of apparatus and requires comparatively little care or attention. One of the most important characteristics of the storage battery, which renders it adapted to emergency stand-by service, is its constant readiness to respond to demands for power without the continuous losses and cost of attendance required by moving machinery.

Storage batteries for emergency service in connection with electric-power supply are principally used in three ways: large stand-by batteries in use on direct-current systems of distribution in the larger cities; exciter batteries used in alternating current generating stations—both steam and hydraulic; and oil-switch batteries in generating stations and substations. It is the general practice to make use of emergency batteries that have been installed in power plants to furnish emergency lighting for the plant itself. Where such a battery is not available, a small battery can be installed especially for this service at moderate cost.

Stand-by batteries are generally of large size and are connected at all times to the direct-current distribution system. In this way they act as reservoirs of energy, always ready for instant use in cases of emergency or breakdown. An interesting emergency condition often develops in cities in the case of a sudden thunderstorm,

FIG. 1. LARGE STAND-BY BATTERY; CONNECTION TO END-CELL SWITCH IS SHOWN IN THE BACKGROUND

when the power demand may increase so rapidly that additional generating apparatus cannot be brought into service quickly enough to meet the requirements. As batteries of this type are called on for comparatively infrequent discharges, a type of plate construction is used which will give a maximum capacity at high discharge rates with a minimum weight and floor space.

FIG. 2. FRONT VIEW OF END-CELL SWITCH

In Fig. 1 is shown a stand-by battery, the plates of which are installed in large lead-lined wooden tanks.

In order to vary the voltage of these batteries so that they may float on busses of different voltages and also that the total number of cells may be quickly put in circuit in an emergency, a certain number of the cells, known as "end cells," are connected to a suitable switch so that they can be cut in and out of circuit as desired. These end-cell switches, one of which is shown in Fig. 2, are worked automatically by a push-button on the power-house or substation control board. Since the circuit must not be opened when cutting the end cells in or out, it is necessary to momentarily partially short-circuit them as the brush travels over the contacts, and this short-circuit current, in the case of the large cells, would exceed 150,000 amperes if means were not provided to limit it to a safe value. Therefore the design and construction of these large switches is an engineering problem of no small magnitude.

End-cell switches must have sufficient capacity to carry any current the battery can furnish, even under short-circuit conditions. The largest size is designed to handle a discharge of 40,000 to 50,000 amperes for a few minutes, to which must be added the heavy local currents across the contacts, when the traveling brush is moving from point to point. There are today installed or under contract in this country and Canada a total of 159 of these stand-by batteries, the aggregate capacity of which approximates 633,000 amperes for one hour at 250 volts, the largest having a capacity of 12,600 amperes at 250 volts for one hour. Several others are of only slightly smaller capacity.

Exciter batteries are used for insuring continuity of field-excitation current for alternating-current generators in both steam and hydraulic plants. The batteries in this service may be of large size, where the demands are great, or may be of smaller size in glass jars, Fig. 3, where the excitation currents are not so large. As in the case of stand-by batteries, exciter batteries are sometimes floated continually on the bus, or they can be arranged as shown in the diagram, Fig. 4. With this arrangement the battery is divided into two groups and these two groups are connected in parallel and receive a small charge, known as a "trickle charge," from the exciter bus through the trickle-charge resistance or lamps. The purpose of this trickle-charge is to keep the battery always in a fully charged condition and compensate for any local internal losses in the battery.

For the regular charging of the battery a pair of fixed resistances are provided and the two groups charged in parallel from the exciter bus through these resistances. This regular charging of the batteries is necessary only at intervals of one to two weeks and only for a period of about one hour at such times.

If for any reason the voltage of the exciter bus fails or drops below a predetermined value, the contact-making voltmeter operates and causes the automatic switch to close. This action places the two groups of the battery in series across the exciter bus, and the battery automatically picks up the exciter load without any interruption to the service. As a signal that the switch has been closed and the battery thrown on emergency discharge, an electric horn, alarm bell or signal lamp can be made to operate by the automatic switch, thus attracting the attention of the attendants.

With the arrangement shown the battery is discharged only in an emergency and normally does no work at all; so that a battery connected and operated in this way will have a very long life. It can readily be seen that this arrangement has certain advantages over floating the battery directly on the exciter bus, particularly where the alternator's voltage is controlled

FIG. 3. SMALL-SIZED EXCITER BATTERY

by a "Tirrell Regulator," and therefore the exciter-bus voltage varies considerably.

In oil-switch service the storage battery furnishes the momentary high currents used to open and close automatic oil switches and remote-control regulating apparatus either in power stations or substations. For furnishing the average supply of energy for operating these various devices it is standard practice to install a motor-generator set consisting of an induction motor, driven by power from the alternating-current circuit, directly connected to a direct-current generator. The storage battery normally floats across the terminals of the direct-current generator, which is kept running continuously, except for such periods when it may be necessary to shut down for inspection or repairs. The generator carries the steady load of the signal lamps plus a small charge to the battery to provide for the average demand of the oil switches. For this purpose the direct-current generator should be shunt wound, so that when the heavy demand occurs, due to the opening or closing of the switches, its drooping voltage characteristic will allow the battery to take nearly all the excess load.

It is evident that if a battery were not provided, the motor generator would have to be of sufficient capacity for the maximum demand and would therefore be many times larger than when a battery is used. Also, with a motor generator only, there is no reserve feature whatever. If a direct-current circuit of sufficient voltage is available, such as on a trolley or third-rail system, the battery may be floated and charged from this circuit through suitable resistance, and in this case no separate motor generator would be required.

The normal voltage of oil-switch control circuits is approximately 125 volts and generally sixty cells of battery are used. A diagram of the standard method of connecting an oil-switch battery, using a motor-generator set, is shown in Fig. 5.

In order to permit charging the battery to maximum voltage by raising the potential of the generator with-

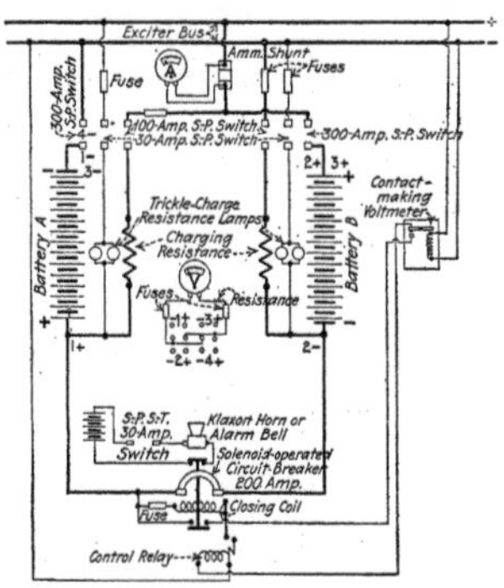

FIG. 4. WIRING DIAGRAM FOR EXCITER BATTERY

out subjecting the signal lamps and remote-control apparatus to this high voltage, a tap is taken from the battery to the switchboard by means of which a group of ten cells may be cut out. At the beginning of charge the entire sixty cells are connected to the generator whose voltage is raised sufficiently to deliver the charging current, while fifty cells are connected across the control circuit. The current required for the signal lamps under these conditions passes through the end-cell group in addition to the charging current of the main battery and the charging of the end cells is, therefore, completed before that of the main battery. The end-cell group is then cut out, and the charging of the remaining fifty cells is completed. The maximum voltage

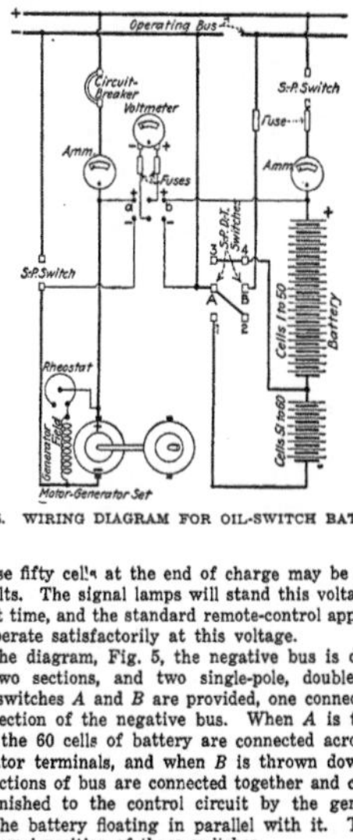

FIG. 5. WIRING DIAGRAM FOR OIL-SWITCH BATTERY

of these fifty cells at the end of charge may be nearly 140 volts. The signal lamps will stand this voltage for a short time, and the standard remote-control apparatus will operate satisfactorily at this voltage.

In the diagram, Fig. 5, the negative bus is divided into two sections, and two single-pole, double-throw knife switches *A* and *B* are provided, one connected to each section of the negative bus. When *A* is thrown down, the 60 cells of battery are connected across the generator terminals, and when *B* is thrown down, the two sections of bus are connected together and current is furnished to the control circuit by the generator with the battery floating in parallel with it. This is the normal position of these switches.

At the beginning of charge switch *B* is thrown up, connecting the control circuit across fifty cells and the voltage of the generator, which is still connected across sixty cells, is raised until the desired charging current is obtained. When the end group is fully charged, as indicated by free gassing and a maximum specific gravity, switch *A* is thrown up, cutting out the end-cell group and the charging of the main battery is completed.

Oil-switch batteries have sometimes been operated on what is known as the "straight charge and discharge method." That is, the battery, instead of being continually floated on the motor generator or charging source, is alternately charged and then allowed to carry the entire oil-switch and remote-control load. This involves working the battery to a considerable extent, and, therefore, the life of the plates will not be so long as where the battery is floated on the charging source, as previously described. Also, to furnish a reserve, a

FIG. 6. FOUR-CELL TRAY UNIT OF AN OIL-SWITCH BATTERY SEALED RUBBER JARS

larger battery is required, and even with a larger battery there is always the liability of an emergency occurring with the battery in a discharged condition, which is not the case where the battery is floated; therefore the straight charge-and-discharge method is not recommended except in cases where it is impossible for some reason to float the battery.

Oil-switch batteries usually are of the glass-jar type mounted on wood racks and installed in separate rooms provided for them. Where a separate room is not available, a type of battery has recently been developed especially for oil-switch service, consisting of heavy elements in sealed rubber jars, assembled in wood trays. A four-cell tray unit of this type of battery is shown in Fig. 6. This type of battery is shipped filled with electrolyte and charged, and is ready for service as soon as received. On account of its sealed construction it can be placed in any convenient place.

In all the foregoing classes of service the installation of the storage battery provides another valuable feature mentioned in the early part of this article, and this is current for a number of lights in the power station or substation in an emergency. Those connected with power-plant or substation work realize the importance of sufficient suitable light in an emergency or shutdown in order to make repairs or get the station in operation again, and there is no piece of apparatus so suitable for this purpose as a storage battery. For operating a few lights only, a very small battery is required, and as it would be used very little, the depreciation and maintenance items would be small.

Precautions When Pumping Ammonia from Coils

BY B. E. HILL
Formerly Refrigerating Engineer, Armour & Company, Chicago

TO CLEAN the evaporating coils it will be necessary to transfer the ammonia from the evaporating coils to the condenser, and in case the coils are submerged in a brine tank the engineer should make certain that there are no leaks before pumping them out.

The brine can be removed before pumping out, but in this case air would be taken into the system. The proper thing to do is to assure oneself that there are no leaks and then proceed to pump the ammonia out of the coils before the brine is removed, as by so doing the brine will evaporate the liquid ammonia much faster. After the coils have been pumped down to as low a vacuum as possible, shut the main liquid and suction valves, and then the tank coils can be opened without any loss of ammonia due to unevaporated ammonia lying in the coils.

A sure method of cleaning these coils, after first removing the headers, is to take each one separately and connect up a steam hose of sufficient size to get a high velocity of steam traveling through the length of the coil. The heat from the steam will warm the oil, and the velocity of the steam will carry it through and discharge oil and dirt. When the oil is all out, disconnect the steam hose and attach an air hose, which will absorb and carry out the moisture remaining in the pipe, as the walls of the pipe will remain heated a sufficient length of time to evaporate the moisture and allow it to be carried out, which will leave the inside surface clean and dry. After all rubber joints have been examined, tightened and replaced where necessary, the coils are ready to be connected and put in service. In pumping out direct expansion coils, especially where the coils are in coolers where the ammonia fumes are absorbed by the stored products, extreme care should be taken to see that there are no ammonia leaks when the coils are opened.

There is considerable damage frequently produced in pumping out direct expansion coils. The only safe and sure method of pumping out these coils is to find all the low spots in the coils and wherever "live" frost is found, apply the *stick test*, as before explained. Do not open the coil; just hold a vacuum and drive out the liquid ammonia by applying heat—blowtorch, steam or hot water—till there is no *stick*, and where there is no *stick* there is no liquid, and it will be safe to open the coil.

It is well to take the extra precaution to open a line leading to the coil somewhere outside the cooler, as there may be a leaky valve, and in this case it may be necessary to put in a "blind" outside the cooler.

To simply pump a high vacuum on the coils is no assurance that the coils are empty in the case of a direct-expansion system. Take an ammonia test tube and fill it with liquid anhydrous ammonia to the amount of 50 or 100 c.c. and allow it to evaporate in a room where the temperature ranges from 90 to 100 deg. F. Note how long it will take to evaporate. This will give the engineer an idea of how long he will have to hold the vacuum on a direct-expansion coil in a cold room for the ammonia to evaporate. The vacuum gage may show a good vacuum for an hour or two without perceptible loss after being pumped down and still leak when the coil is opened and warm air comes in direct contact with the remaining liquid inside the pipes. The better way is to make sure by hunting up the traps or low places in the pipe lines and then apply the *stick test*, which will show if there is any liquid remaining in the lines.

The writer has taken a test tube and filled it up to the 50 c.c. mark, placed the cork in the test tube and connected a hose to the glass tube that is inserted in the cork, and pumped a vacuum of 24 in. At first the liquid ammonia boiled violently, as the reduction of the pressure from atmosphere to 24 in. or 12 lb. below atmosphere greatly reduced the boiling point, but as the frost formed around the outside surface of the testing bottle, the boiling settled down shortly until it was barely perceptible and required an hour or so to complete the evaporation.

If desirable to make a test for the purity of ammonia, secure a test tube or flask, which usually holds from 50 to 100 c.c. The flask, if made to test with 100 c.c., is made with the bottom end reduced in size to 1 c.c. and the 1 c.c. is graduated in ten parts, the lines being etched on the glass, each line representing one-tenth of one cubic centimeter.

To take a sample from a cylinder of anhydrous ammonia, use a copper wire and make one turn around the test tube, twisting the two ends of the wire together to form a handle. Next connect a short pipe to the valve on the cylinder and open slightly, holding the testing flask so that the liquid will fill the flask up to the line marked 100 c.c.; next place the cork in the flask, which is provided with a glass tube, bend to a half circle to allow the gas generated by the boiling liquid to pass out without raising a pressure, and the cork is placed in the tube to prevent the moisture in the air from coming in contact with the liquid surface of the ammonia, which would take up the moisture and give an incorrect reading. The flask is then immersed in salt or calcium brine at a temperature of 60 or 70 deg. F. and allowed to remain there until the liquid is all evaporated. If the liquid is pure anhydrous ammonia, the evaporation will be complete, but if there is an oily residue or clear substance remaining in the bottom of the tube, the ammonia is impure, and the percentage of impurities by volume is determined by the amount of residue remaining in the one graduated c.c. part of the testing flask.

Suppose the remaining impurities in the one graduated c.c. should be full up to the tenth line, or one c.c. remaining. There being 100 c.c. in the flask to start with, the impurities would be 1 to 100 or 1 per cent by volume, which is a greater impurity than is ever found. The percentage of impurities, however, is read direct from the test tube, and from the foregoing it is easy to read the percentage at any other point; say, two lines high would represent two-tenths of one per cent, or two c.c. to one thousand, etc.

Besides the liquid and solid impurities there are others commonly referred to by engineers as "foreign gases," the most common to be found in the ammonia cylinder being air or modified air, pyridine, chlorine, etc.

INTERIOR OF THE COMPRESSOR ROOM OF A LARGE PACKING-HOUSE REFRIGERATION PLANT

To determine the amounts of these gases would require the services of a chemist, and as the leading manufacturers of today have reduced these gaseous impurities to a minimum, it would not be of enough concern for the engineer to go to this expense. He can, however, raise the valve end of an anhydrous ammonia cylinder to be tested about six or eight inches high, turn the valve so that its opening will look down, connect a hose and submerge the hose in a pail of water after first blowing the air all out of the hose. If there are any gases in the cylinder other than anhydrous ammonia gases, bubbles will rise to the surface of the water, which shows that there are gases that should not be there, but if no bubbles appear, the ammonia is pure so far as "foreign gases" are concerned. A boiling-point determination will also be of some value and can be made with a good chemical thermometer.

A New High-Speed Gas-Engine Indicator

ABOUT two years ago the United States Bureau of Standards undertook the development of a special gas-engine indicator to meet certain special research requirements. Most of the high-speed indicators that have been developed commercially, involve the use of either a piston or a heavy diaphragm actuating a recording mechanism, which must be mounted on the engine. The problem to be met at the Bureau of Standards required an instrument that could be mounted on the engine which was inclosed in a concrete chamber and was inaccessible to the observer. The indicating mechanism must therefore either be placed outside the chamber containing the engine or be so designed that it could be operated from without. The high speeds and pressures prohibited the use of moving parts of any appreciable weight, since the inertia of these parts would interfere with the accuracy of the indication. The type of instrument that was finally adopted for development employed a very thin metallic diaphragm, one side of which is subjected to the pressure to be measured and the other to any desired pressure applied from without. Fig. 1 is a cross-section of that part of the apparatus which is mounted directly on the engine. The body of the instrument consists of a two-piece brass shell fitted to screw into a spark-plug hole and supplied with a water jacket. These two pieces, marked 1 and 2 in the illustration, are held together by bolts as shown, and can be readily separated for the removal or insertion of diaphragms. The diaphragm, which is of steel with about one inch effective diameter, is mounted between two slightly concave supporting plates perforated to allow the pressure to reach the diaphragm. Its thickness may be from 0.002 to 0.006 in. as required, and the range of motion is limited to about 0.01 in. The center portion of one of the supporting plates is electrically insulated in such a way that an electric circuit is made or broken as the diaphragm makes contact with this support or departs from it. Air under pressure from a convenient reservoir is admitted through the pipe marked "air," to the chamber above the diaphragm, and the pressure of the gases in the engine cylinder is transmitted to the chamber below the diaphragm. In the electric circuit outside the indicator proper are a battery and telephone receivers. To measure the maximum or or minimum pressure in the cylinder, air at an intermediate pressure is admitted to the indicator above the diaphragm. When the pressure in the cylinder becomes more than that above the diaphragm, the latter is forced up, making electrical contact with the electrode. Each time this contact is made and broken a click may be heard in the telephone receiver. The air pressure is then increased until the clicking just ceases, and this

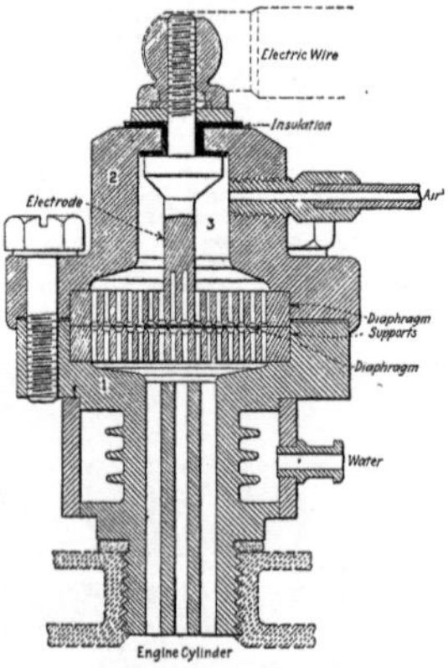

FIG. 1. SECTION OF NEW INDICATOR FOR HIGH-SPEED GAS ENGINES

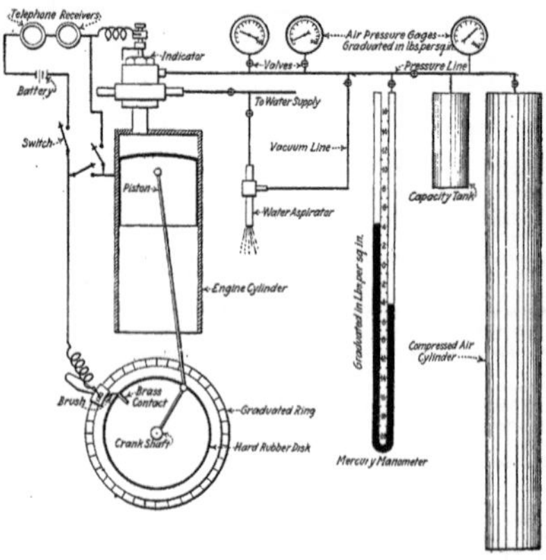

FIG. 2. DIAGRAM OF AUXILIARY APPARATUS FOR USE WITH INDICATOR

air pressure is noted as the maximum in the cylinder. The air pressure is then decreased until the clicking ceases again, and this pressure is noted as the minimum in the cylinder. To produce a complete diagram, a special one-point timer is mounted on the engine shaft. This timer is so designed that it may be moved throughout the whole 360 deg. and is provided with a graduated ring so that its position may be accurately noted. The timer is connected in series in the electrical circuit with the telephone receivers and the indicator, so that the contact by the diaphragm will be heard in the receivers only when the circuit is closed through the timer. To get the complete diagram, the timer is set at a given point and the air pressure adjusted until the clicking just ceases, when the pressure is observed and recorded as the pressure at that particular crank position. The timer is then moved to another position and the same process repeated, until the pressures for a complete cycle at every ten or twenty degrees of crank angle have been recorded, and from these figures the diagram is constructed on the drawing board.

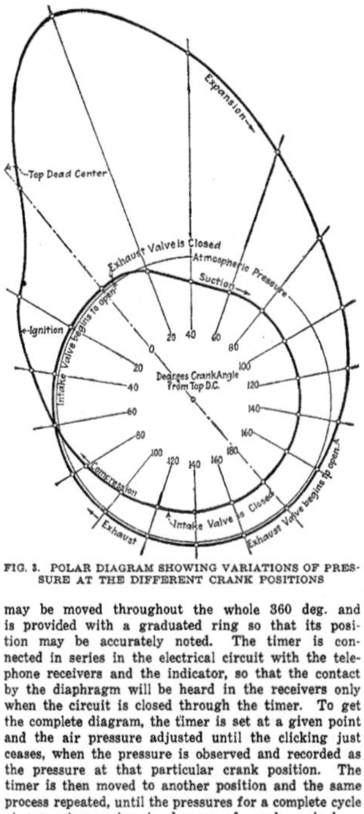

FIG. 3. POLAR DIAGRAM SHOWING VARIATIONS OF PRESSURE AT THE DIFFERENT CRANK POSITIONS

Fig. 2 shows the various connections of the external auxiliary apparatus. At the extreme right is shown the compressed-air cylinder to represent the supply of high-pressure air, while the *capacity tank* represents a small receiver to be used in accurate adjustment of the pressure. The pressures are measured by either one of the the two pressure gages shown or by the mercury in the manometer, the latter being used for measuring pressures slightly above atmospheric or for vacuums. The water aspirator shown is provided to produce a vacuum in the air line, when measuring pressures below atmospheric. The connection from the water-supply line to the indicator is to supply cooling water for the indicator jacket.

Fig. 3 shows a polar pressure diagram made by the help of this indicator. The radial lines shown represent the various crank positions, and the pressures are marked off on these lines from the center to any convenient scale. Through the points thus located a smooth curve is drawn, and the diagram is complete when the points at which the various valves open and close and the point of ignition have been located and marked.

Fig. 5 shows the same data plotted on a rectangular diagram. In this case horizontal distances represent crank angles from the top center and vertical distances represent pressures. The points are laid off in the same manner as in the case of the polar diagram and a smooth curve run through them

Fig. 4 shows a pressure-volume diagram which corresponds to the diagram that would be drawn by an ordinary engine indicator if it were possible to use such an instrument with a high-speed gas engine. In this case horizontal distances represent volumes and vertical distances pressures, as in the ordinary diagram.

This indicator requires that the indicating mechanism alone be mounted on the engine, the recording apparatus all being mounted at any desired distance. It is sensitive to from 0.1 to 0.2 lb. per sq.in. over a pressure range of from 14 lb. below atmosphere to 500 lb. or higher above. It apparently has no limitations as to engine speed within the range encountered in practice and is apparently without appreciable inertia errors at any observed engine speed. It will give not only the

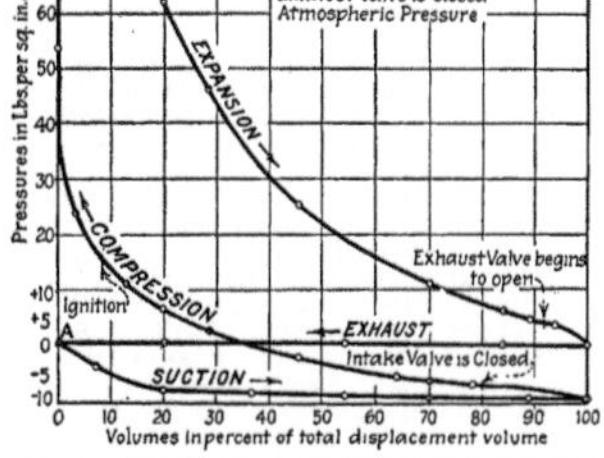

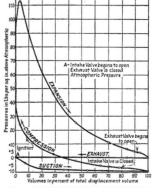

FIG. 4. PRESSURE-VOLUME DIAGRAM PLOTTED FROM INDICATOR OBSERVATIONS

average pressure, but also the maximum and minimum pressures at any desired point or points in the cycle. It does not, however, permit the recording of a diagram of the pressures occurring in a single individual cycle, as is done by means of the optical devices in use.

In short the apparatus is adaptable to the measurement of cyclically varying gas pressures for a pressure range of from minus 14 lb. to plus 500 lb. and any speed likely to be encountered with sensitivity of from 0.1 to 0.2 lb. per sq.in., using a single thickness of dia-

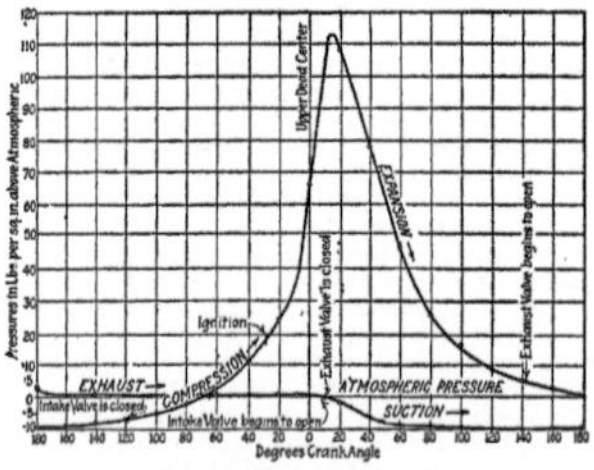

FIG. 5. RECTANGULAR DIAGRAM PLOTTED FROM INDICATOR OBSERVATIONS

phragm. Pressures up to at least 1,000 lb. can be measured with a somewhat thicker diaphragm and a correspondingly less sensitivity. The design and development of this instrument has been largely the work of Dr. H. C. Dickinson, of the Bureau of Standards in Washington.

The Bourdon Gage Spring

By W. L. Parker

Years ago M. Bourdon was engaged to repair the worm pipe of a still, which had become flattened. He endeavored to restore its original form with water pressure. In doing so the flattened tube tended to uncoil itself, and further experiments showed that the action of uncoiling might be applied in the construction of a pressure gage.

To understand the action of the Bourdon spring tube, it should be borne in mind that the tendency of pressure in a closed vessel is to increase the volume of it. It is often claimed that a curved tube of any cross-section will answer for a gage spring only that the flattened tube is more sensitive. This is in error. The flattened tube must be used. Pressure has no more tendency to straighten out a curved tube of circular cross-section than it has to curve a straight tube of circular cross-section. The reason is that in both cases the volume of the tube would be reduced. Pressure in a curved tube of circular cross-section puts the outer portion of the tube under tension and the inner portion under compression. The circular cross-section is maintained and the curved axis remains unmoved within the elastic limit of the tube.

In practice, for convenience, the short axis of cross-section *AB* is in the cylindrical surface containing the curved neutral axis of the tube, as shown by section *E* in the illustration. The effect of pressure in the tube is to force the sides apart or to lengthen *AB*, the short axis of cross-section, and to shorten the long axis *CD*. This increases the curvature of the cross-section and reduces the curvature of the tube, forcing its free end outwardly.

If the tube were flattened the other way so that the long axis of cross-section *CD* were in the plane of the tube, as in section *F* of the sketch, the effect of pressure, obviously, would be to force the free end of the tube inwardly. The limit of motion of the free end of the tube in each case would be reached when the cross-section of the tube became a circle.

From the foregoing it is apparent that excess pressure within the tube will force the free end outwardly or inwardly, depending upon the cross-section and its disposition in each case with reference to the plan of the tube.

The spring tube indicated by the drawing was taken from a well-known make of gage with an 8-in. dial and graduated to 30 lb.. The coil of the tube was 5 in. in diameter. The cross-section was flattened as in the section *E* so that the major axis *CD* decreased ⅜ in. and the minor axis *AB* increased 1/32 inch. A pressure of 15 lb. moved the free end of the tube ¼ in. farther from the pipe connection; 30 lb. moved it ⅜ inch. An extension with a pencil attached was secured to the

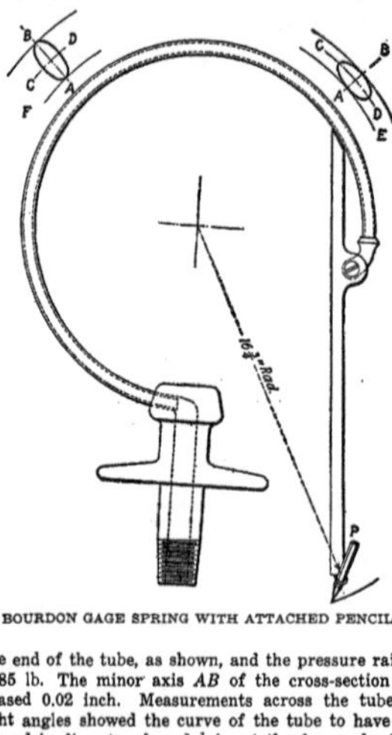

BOURDON GAGE SPRING WITH ATTACHED PENCIL

free end of the tube, as shown, and the pressure raised to 85 lb. The minor axis *AB* of the cross-section increased 0.02 inch. Measurements across the tube at right angles showed the curve of the tube to have increased in diameter 1/16 and ¼ in. at the free end. The pencil *P*, 16¾ in. from the center of curvature of the tube, described an arc 3 in. long. The curve is believed to be an ellipse approaching closely to a circle.

Repairing a Shell-Type Transformer

How a High-Voltage Transformer Was Dismantled To Have a Number of Damaged Coils Replaced and How This Transformer Was Assembled After the Repairs Were Made.

By KENNETH A. REED

Maintenance Engineer, Interboro Rapid Transit Company, New York

KENNETH A. REED

During a severe lightning storm, a 33,000- to 2,300-volt bank of transformers was struck and one of them broken down in one section of the winding. The transformer was of the oil-insulated, self-cooled type, and the gases that formed in the tank were of sufficient violence to blow the top off and throw out a considerable quantity of oil. The customer was able to see the tops of all of the coils, and taking a chance on those which he could see were burned being the only ones that were damaged, he ordered new coils accordingly. The transformer had been taken to a storage warehouse as being the most convenient place for making repairs. There were no facilities whatever for handling heavy pieces of apparatus, and the most that could be said for the place was that there was plenty of room. Strong girders were found to have been used in the construction of the building, and it was decided to remove the core of the transformer from the case by means of a 10-ton chain block supported from one of these girders, as shown in Fig. 1.

After the core had been raised above the level of the oil, it was allowed to remain suspended there for several hours to permit as much oil as possible to drip back into the tank. This was a very important operation from a comfort point of view, because there is nothing quite so disagreeable and no liquid of which so small a quantity will spread over so large a surface as transformer oil. When the transformer was placed in the warehouse, it was left on rollers, which permitted getting the tank out of the way after the core had been raised high enough to clear the top casting. The core was then lowered to the floor, and wooden blocks, on the tops of which were cushions made out of heavy express paper, were placed under the sections of the secondary coils, as in Fig. 2.

After removing the terminal boards and main top casting, about ten inches of the laminations were taken off, and a long wooden beam eight inches square was inserted between the inside top of the coils and the laminations, the transformer being of the shell type. This beam was then picked up by the chain block, as in Fig. 3, and a portion of the weight of the core was thereby removed from the floor and the blocks under the sections of the secondary coils. The rest of the laminations were then removed, leaving the coils suspended on the beam and chain block.

When the sling was fastened to the end of the beam nearest to the damaged section, care was taken to leave sufficient room to move this section over far enough to allow another sling to be put around the beam. The weight of this end of the beam was then taken on the second sling by means of a block and fall, as in Fig. 4, which left the damaged section of the winding free on the extended end of the beam. The jumpers between the coils of this group were sweated off, the outside insulation removed and the coils separated. Fortunately, the customer had diagnosed the trouble correctly, and had ordered exactly the right number of new coils for making repairs.

Single coils were removed from the end of the beam, one at a time, until the damaged ones had been taken off, and it was a simple matter to substitute the new coils in place of the old ones and then reassemble the coils of this group. All the jumpers between coils were sweated on before the group was re-insulated, in order to make sure that no stray solder was left in the winding.

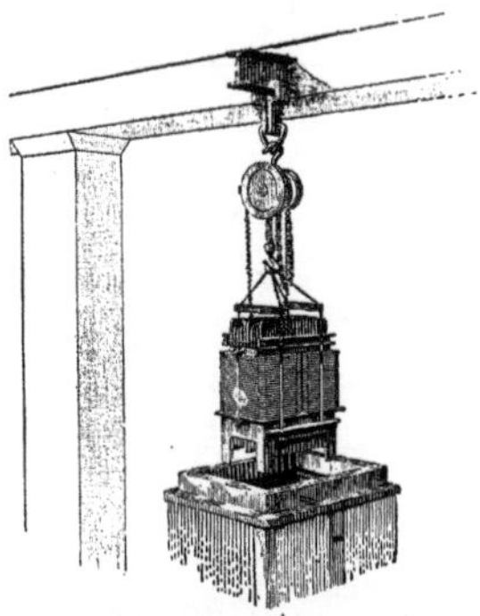

FIG. 1. TRANSFORMER BEING LIFTED FROM TANK

By this procedure only one joint was left to be soldered when the group was connected to the other part of the winding. After placing proper insulation and circulating ducts between the coils, they were clamped tightly between boards that came to within about six inches of the top of the hole in the coils through which the laminations are built up. This space was then taped up and the clamps lowered another six inches, and so on until the entire sides were bound together. The process of removing this group from the undamaged part of the winding was reversed, and

after binding the entire winding together, similar to the coils of the damaged group, sweating on the one jumper, insulating all joints and putting the outside box insulation in place, the core was ready for the building up of the laminations.

The bottom casting which supports the laminations was now put in position, and the blocks with cushion

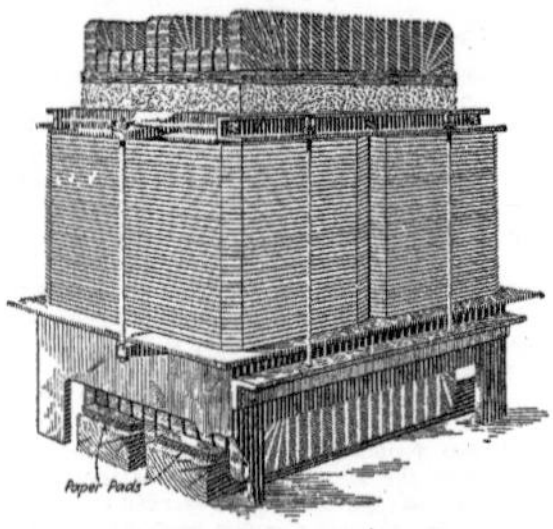

FIG. 2. METHOD OF SUPPORTING SECONDARY COILS

tops were shoved under the secondary coils, as was the case when the transformer was dismantled. After about twelve inches of iron had been built up, a small timber was laid on top of the laminations and a jack tightened

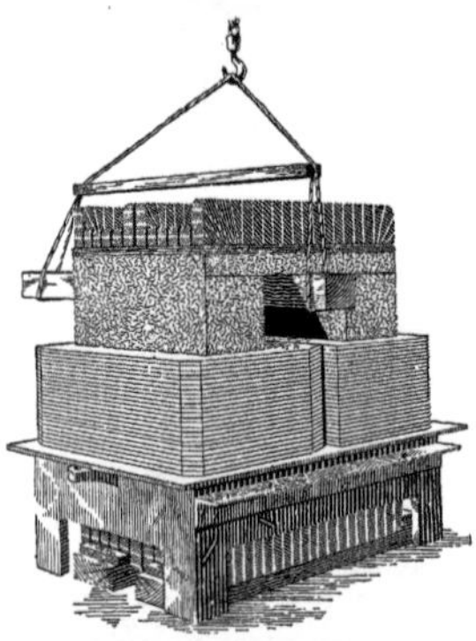

FIG. 3. WEIGHT OF THE COILS SUPPORTED ON SLING

up between it and the top beam for the purpose of pulling the laminations down tight, in order that none of them would be left over. This was continued until about eighteen inches of iron was in place, when the through beam was removed and the top casting used to pull the laminations down tight. This was repeated for about every six inches of iron until all the laminations were installed, and the top casting, terminal blocks, etc., were put in place permanently.

During the time the transformer was being rebuilt, there were several very rainy days, and as there was no heat in the warehouse, the windings absorbed a great deal of moisture, which required that the transformer be thoroughly dried out before being put back in the tank. It might be added that whenever such repairs are made to any high-voltage oil-insulated transformer, drying out is almost invariably necessary, regardless of weather conditions, before the transformer is replaced in service. The most convenient method of drying out a transformer in the field, and the one used in this case, is by short-circuiting the secondary winding and applying low voltage at normal frequency to the primary, provided a suitable range of voltage is available. This condition as a rule, is easily met at a central-

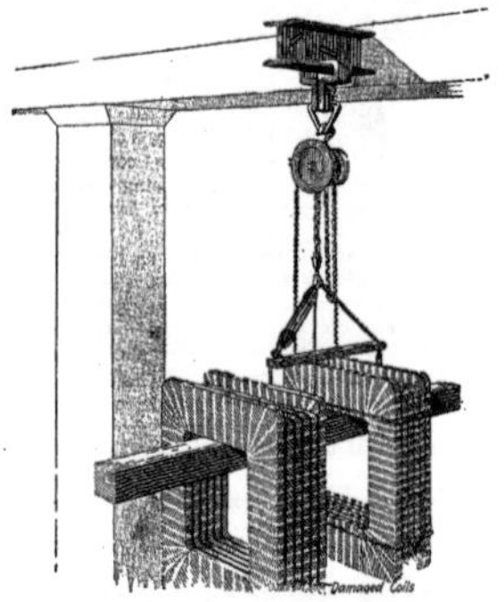

FIG. 4. HOW THE DAMAGED COILS WERE REMOVED

station plant, because it is usually possible to secure as many 10-kw. 2,300- to 220-110-volt lighting transformers as might be desired, and by series-parallel combinations the proper voltage can be obtained. Before exciting the transformer, thermometers were placed at what might be expected to be the hottest points in the primary and secondary coils, and in order to avoid heating the transformer windings too rapidly, an ammeter was placed in the secondary circuit and the current kept at a safe value until the temperature stopped rising. This precaution was necessary because as a general thing under such conditions, there is a tendency for the coils to sweat, and serious damage may be done in a very short time.

When the temperature had stopped rising, the applied voltage was raised slightly, and the temperature of the hottest point brought up to approximately 90 deg. C., where it was maintained for about three days and nights. During this time both the day and night shifts were required to keep a log showing the temperature registered by each of the thermometers every fifteen minutes. The current was then cut off, and insulation tests made with a "megger." The readings obtained showed the transformer to be thoroughly dry and safe to go back into service. It was accordingly replaced in the tank and securely bolted in position and the oil, which had been dehydrated and tested, was poured in, the tank having been repaired in the meantime.

Shortly after the current had been applied to the primary terminals, with the secondary short-circuited, and the temperature had risen to an appreciable degree, the belief that the winding was defective became prevalent among some of the men who had been working on the transformer. This was due to the fact that the transformer appeared to be smoking to such an extent that, had such been the case, the winding would have been burning up. The "smoke," however, was only a vapor caused by the oil on the laminations being vaporized by the rapid rise in temperature. Such a condition is quite likely to occur with any oil insulated transformer that has been removed from the tank and its temperature raised somewhat higher than normal; and to one who had not observed such a case before, it would seem very serious. The vapor was very similar to that which may be seen rising from the "breather" on a bearing of a large, high-speed turbo-generator.

Flue-Gas Temperature and CO_2 as a Guide to Furnace Operation

With Special Reference to Oil Burning

By CHARLES C. PHELPS

WHILE test data on combustion are important and should be as nearly accurate as possible, some simple method of checking furnace performance is necessary for everyday economical operation in the boiler room. A study of test data shows that the excess of temperature of the flue gases above that of the boiler room and the percentage of CO_2 in the flue gases give a guide to combustion conditions, and may be determined quickly enough to be of help in regulating the furnace action.

TABLE I. RELATION OF FLUE-GAS TEMPERATURE TO LOSS FROM DRY FLUE GAS

	Case 1	Case 2
Flue-gas temperature T deg. F.	350	650
Boiler-room temperature t deg. F.	50	50
$T - t$	300	600
CO_2 in per cent	12	12
Loss in dry flue gas, per cent of total heat in fuel	10.4	20.8

It is a known fact, corroborated by reference to tables of chimney losses, that with a given fuel and with a constant percentage of CO_2 in the flue gases, the proportion of the fuel's heat which is lost up the chimney in the dry products of combustion is directly and exactly proportional to the difference between the flue temperature and the boiler-room temperature. For instance, in the example given in Table I, representing a certain fuel, it will be noticed that the heat loss in the dry chimney gas is twice as great in the second case, in which the temperature difference is twice as great.

It is also a fact, likewise readily verified, that when the flue temperature remains constant the heat lost up the chimney in the dry products of combustion is *almost* exactly in inverse proportion to the percentage of CO_2 in the flue gases, as shown in Table II. It will be noticed that in case 4, Table II, where the percentage of CO_2 is half as great as in case 3, the loss is about twice as great. The discrepancy is so small that for all approximate calculations it may be assumed that the loss is exactly in inverse proportion to the CO_2 when the temperature does not vary. If the percentage of CO_2 had been 4 instead of 6 (one-third of 12), the dry chimney-gas loss would have been 60.7 per cent, or approximately three times as great.

TABLE II. RELATION OF CO_2 TO LOSS FROM DRY FLUE GAS

	Case 3	Case 4
Flue-gas temperature, T deg. F.	650	650
Boiler-room temperature, t, deg. F.	50	50
$T - t$	600	600
CO_2 in per cent	12	6
Loss in dry flue gas, percent of total heat in fuel	20.8	40.8

From these statements of facts it is apparent that the dry chimney-gas loss with a given fuel is approximately proportioned to the ratio $\frac{T - t}{per\ cent\ CO_2}$. This ratio, therefore, may be used as a guide for the operation of the furnace.

The temperature of the flue gas is affected by the deposit of soot and dirt on tubes or in flues and of scale within the boiler, by the condition of the baffling, arches, etc., and by the rate of driving. Therefore, fluctuations in the flue temperature are usually small and infrequent in comparison with the fluctuations in the percentage of CO_2 and this is particularly true at fairly constant driving rates. Under such conditions the percentage of CO_2 by itself becomes a reliable index of loss in the dry chimney gases. For this reason a CO_2 recorder alone, without a pyrometer, is often used to keep records of the heat lost up the chimney, although a recording pyrometer is also highly desirable.

TABLE III. TYPICAL OIL-FUEL HEAT BALANCE

		Per Cent.
1.	Loss due to heat carried away in dry flue gases	14
2.	Loss due to superheating moisture formed by burning hydrogen in oil	7
3.	Loss due to incomplete combustion (CO and hydrocarbons)	0.1
4.	Loss due to evaporation of moisture in fuel	0.1
5.	Loss due to heating of moisture in air	0.4
6.	Loss due to radiation and unaccounted for	3.4
7.	Heat absorbed by boiler (including vaporizing of fuel)	75

The excess air is entirely responsible for the so-called CO_2 loss, which is the heat loss involved in heating the excess air from the temperature of the boiler room to that of the flue gas. The factors first mentioned—that is, those indicated by stack temperatures—are very largely under the control of the boiler-room foreman, who should be responsible for clean boiler surfaces, etc. On the other hand, proper air supply is entirely under the control of the fireman and the percentage of CO_2 is the only reliable guide to him in adjusting the ratio of air to fuel, which is so vital in securing combustion efficiency.

It so happens that in oil burning the loss in the dry

chimney gases is the only loss that fluctuates to any great extent. In Table III are shown the various losses as they occur in the heat balance of a typical oil-burning boiler. Loss (3) should be nonexistent or negligible in quantity in a well-designed and properly operated plant. Loss (5) is likewise unimportant. Losses (2) and (4) are dependent upon the composition of the fuel and are

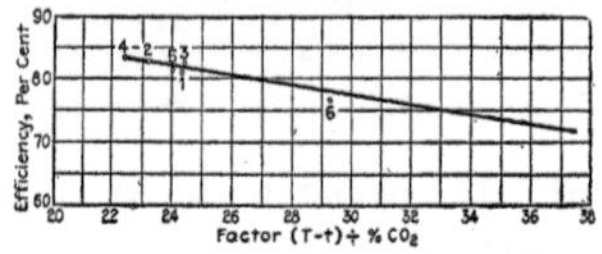

FIG. 1. DRY CHIMNEY GAS INDEX $\frac{T-t}{CO_2}$ WHICH INDICATES BOILER EFFICIENCY

Plotted from test on a 604-hp. boiler using oil of 13.2 deg. Be. with an average heating value of 18,200 B.t.u. per lb. as fired.

Test No.	1	2	3	4	5	6	7
Efficiency, per cent	81.1	82.8	82.4	83.3	81.5	76.4	75.8
Smoke	None	None	None	None	None	None	Light haze
Carbon dioxide, per cent	12.2	13.4	13.3	14.3	14.2	13.3	12.1
Temperature of flue gases, T deg. fahr	385.3	397.5	409.1	406.2	429.0	477.1	537.5
Temperature of boiler room, T deg. fahr	88.7	86.6	84.4	85.2	87.3	86.7	84.4
(T—t)	296.6	310.9	324.7	321.0	341.7	390.4	453.1
$\frac{(T-t)}{\text{Per cent } CO_2}$	24.3	23.2	24.4	22.4	24.0	29.3	37.5

fixed in amount, hence they cannot be appreciably reduced by the fireman. Loss (6) may vary slightly, but in a well-designed plant there is little the operator can do to reduce it.

Loss (1) is usually greater than all the others combined. Neglecting the slight fluctuations in the other losses, it is apparent that as the dry chimney-gas loss increases, the efficiency will decrease by the same amount, and vice versa. Referring to Table III, an increase in item (1) to 19 per cent would result in a decrease in item (7) to 70 per cent and a decrease of (1) to 9 per cent would result in an increase of item (7) to 80 per cent. Thus it follows that under the conditions found in oil-burning practice the factor $\frac{T-t}{\textit{per cent } CO_2}$ becomes almost as good a guide to over-all boiler efficiency as to chimney losses. Applying this line of reasoning to the data furnished from tests on a certain plant and plotting Fig. 1 therefrom, it is seen that this factor is indeed a remarkably good working index of boiler and furnace efficiency, although not a mathematically perfect one. In the plot the horizontal ordinates represent the factor $\frac{T-t}{\textit{per cent } CO_2}$ and each point represents the value of this ratio for one test. The vertical ordinates represent the efficiency in per cent. The points are nearly all on a straight line, which shows that the boiler efficiency has a simple relation to the factor $\frac{T-t}{\textit{per cent } CO_2}$. If the reader has data on several efficiency tests under his own conditions, he can easily make up a chart like Fig. 1 for his particular fuel and boilers.

The writer has examined a great many tests by this method and has found it to check up almost invariably. When it does not, the usual explanation is that when the average CO_2 is used instead of the average losses corresponding to CO_2, there is likely to be a discrepancy, which is especially noticeable when the CO_2 fluctuates over a wide range.[1] For momentary values of the factor $\frac{T-t}{\textit{per cent } CO_2}$, the points on an efficiency index chart such as Fig. 1 should lie almost exactly on a straight line. Each point noted in Fig. 1 represents an average

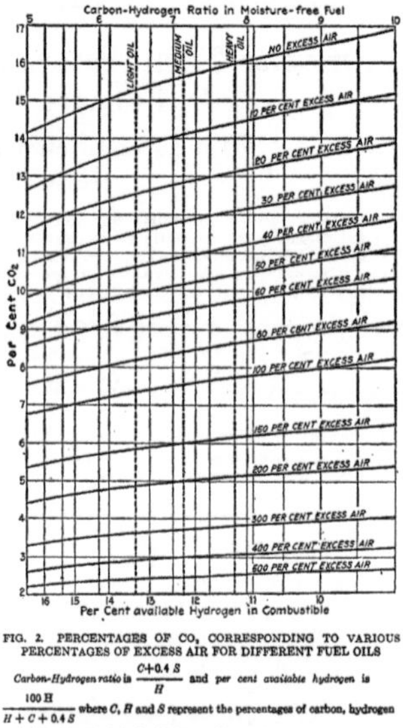

FIG. 2. PERCENTAGES OF CO_2 CORRESPONDING TO VARIOUS PERCENTAGES OF EXCESS AIR FOR DIFFERENT FUEL OILS

Carbon-Hydrogen ratio is $\frac{C+0.4\,S}{H}$ and *per cent available hydrogen* is $\frac{100\,H}{H+C+0.4\,S}$ where C, H and S represent the percentages of carbon, hydrogen and sulphur respectively in the oil when freed from moisture.

APPROXIMATE COMPOSITION IN PER CENT OF OILS REPRESENTED BY DOTTED LINES

	C	H	S	N	O	H_2O
Light oil	84	13	0.8	0.2	1	1
Medium oil	85	12	0.8	0.2	1	1
Heavy oil	86	11	0.8	0.2	1	1

value of CO_2 and temperature over a seven-hour test, hence the very close alignment is all the more remarkable.

From the foregoing it is apparent that a pyrometer and CO_2 recorder of the continuous type furnish all the information required to conduct what practically amounts

[1]See article by the author, page 297 *Power*, Feb. 24, 1920.

to a perpetual efficiency test under operating conditions. The CO_2 recorder alone serves in this way as a guide to efficiency in many plants where the flue temperature remains fairly uniform. At any rate the information which is most vital in keeping the greatest single item of loss down to the minimum is furnished by the CO_2 recorder.

Fig. 2 shows the percentages of CO_2 that will be recorded when various proportions of excess air are used in burning fuel oil of various compositions. Each curve represents a different percentage of excess air. The upper curve represents the theoretical quantity of air for complete combustion; that is, 0 per cent excess air. The vertical scale on the left margin represents the percentages of CO_2. The lower horizontal scale represents percentages of hydrogen in various hydrocarbon fuels. This represents only the percentage of hydrogen in the combustible matter in the fuel and does not include the hydrogen in moisture in the fuel. The upper horizontal scale expresses the composition of the same range of fuels in different terms; namely, the ratio of carbon to hydrogen, not including the hydrogen in the moisture in the fuel.

Corrections for Sulphur in the Fuel

Sulphur in the fuel will burn to sulphur dioxide (SO_2) and will be absorbed by the absorbent in the CO_2 recording machine, adding to the apparent CO_2 percentage just as though it were CO_2. Hence, sulphur must be taken into account if it is present in appreciable quantities. The easiest way to take account of the sulphur is to multiply its percentage by 0.4 and add this result to the percentage of carbon, thus treating SO_2 as though it were CO_2. The factor 0.4 is used because the volume of SO_2 in the flue gas produced by a given weight of sulphur is just about four-tenths as great as the volume of CO_2 produced by the same weight of carbon. Air contains 20.9 per cent by volume of oxygen, and if all this were converted into CO_2 the percentage of the latter in the flue gas would also be 20.9. Some oxygen, however, is required for converting the hydrogen in the fuel into water-vapor, and the nitrogen accompanying this oxygen appears in the flue gas. Therefore, the greater the percentage of hydrogen in the fuel, the less will be the percentage of CO_2 in the flue gas. All the curves in Fig. 2 will illustrate this fact. As a consequence, it would seem that with fuel oil a much lower percentage of CO_2 should be carried than with fuel low in hydrogen—for example, anthracite coal or coke. This, however, is not usually the case for the reason that oil can generally be consumed with a much smaller proportion of excess air, because of the ease of obtaining a thorough mixture of the air and fuel during the process of atomization. The amount of excess air that must be used in oil burning will be found to be larger in some cases than in others owing to the fact that some boiler furnaces are not designed to withstand the extremely high temperatures that accompany minimum air supply.

Variables Which Affect Amount of Air Required

Such economical adjustments of the air supply as those mentioned cannot, of course, be maintained, especially with oil burners, without keeping continuous records of the percentage of CO_2. Every little change in the constantly changing conditions of operation will necessitate readjustment of the air supply. Such variable factors, for instance, as change of direction and velocity of wind, atmospheric temperature, barometric pressure, percentage of humidity, demand for steam, temperature of furnace, quantity of dust on tubes or in flues, amount of scale in boiler, degree of superheat, condition of furnace lining and baffling, and many others, have an effect, often a sudden effect, on the quantity of air and fuel required.

The usual rule in burning oil is to cut down the air supply until a "light haze" appears at the top of the stack. From the test data accompanying Fig. 1, it may be observed that the only test in which this "light haze" is noted showed the lowest efficiency of the series. While this is not sufficient evidence that a "light haze" means poor efficiency, it suggests that it may not be a reliable guide to the best operating conditions.

A boiler furnace, whether coal- or oil-fired, cannot be set in advance to meet varying operating conditions, as, for instance, the valves of an engine are set. Continuous readjustment is necessary. Sometimes these adjustments must be made every few minutes and sometimes every hour or even at longer periods, depending entirely on the nature of the load and the conditions of operation. It is generally conceded that in no other part of the power plant is the personal element a greater factor than in the boiler room. Certainly in no other part are there greater opportunities for savings than are possible in connection with fuel conservation, and high CO_2 is the key to fuel saving. There is no substitute for the personal element, but fortunately it pays handsomely to employ human brains, guided by efficient CO_2 instruments, to keep the air and fuel in balance at all times.

Losses Due to Low CO_2

To see just what losses are involved when we neglect the CO_2 factor, assume an oil corresponding to that marked "medium" on Fig. 3 and that this oil has a heat value of 18,600 B.t.u. per pound of combustible. If we used 200 per cent excess air, the percentage of CO_2 would be five, and at this figure, with 650 deg. F. stack temperature and 50 deg. boiler-room temperature, the loss of heat up the chimney in the dry gases alone (not including the losses in the moisture) would equal 35.3 per cent of all the available heat in the fuel. Plants are actually producing as bad results as this and even worse. Taking the same temperature difference, if 40 per cent excess air is supplied the dry gas loss is reduced to 16.6 per cent, while if operation were possible at 10 per cent excess air the loss would drop to 12.9 per cent.

Fuel conservation and particularly fuel-oil conservation is one of the most important economic problems of this day and generation. The best way to conserve it in the boiler room is to utilize as large a proportion as possible of its heat in doing useful work—that is, in evaporating water in the boiler—and to waste as small a proportion as possible up the chimney by keeping the volume of the exhaust gases as small as possible with the aid of CO_2 instruments and by keeping the temperature of the flue gases as low as possible.

The electric-power generating equipment on the British battle cruiser "Hood" consists of an unusual variety of apparatus. Eight generators are provided, each of which can supply either alternating or direct current as may be desired. Four of these generators are driven by reciprocating steam engines, two by geared high-speed impulse turbines and two by eight-cylinder Diesel oil engines.

Nathaniel A. Carle—A Progressive in Power Plant Construction

Mr. Carle has inspected practically all of the large turbine plants of the country with the object of determining the efficiency of these prime movers under various operating conditions. Observations on these trips combined with his own wide experience in the installation and operation of these units and as chairman of the Prime Movers Committee of the National Electric Light Association, lead him to the opinion that 30,000 kw. 1,800 r.p.m. 60 cycles and 35,000 kw. 1,500 r.p.m. 25 cycles can be considered for the present as the maximum standard sizes of single-shaft, single-barrel units. Moreover, during the past year considerable information has been obtained from the operation of these large units as well as from some of the smaller machines, with the result that steps are being taken to bring about helpful co-operation between the designers and the men under whose charge they operate.

Photo by Underwood & Underwood.

TO BECOME chief engineer of a large power corporation in charge of its diversified power system is close—very close to the top. Nathaniel A. Carle, Chief Engineer of the Public Service Electric Company of New Jersey, has attained this position. A forceful personality combined with pronounced administrative ability and power of analysis early marked him as a leader of men, and from the time when he left college to enter the industrial engineering field until today, his progress has been steady and sure.

Mr. Carle's activities are not confined to supervision of one of the largest public power systems in the country, but he also is an active member in a number of engineering societies and is at present serving on twenty-three committees. He has done a great deal of investigation work with turbines, and as chairman of the Prime Movers Committee of the National Electric Light Association, much of his time has been occupied with problems confronting that body.

The use of pulverized fuel in large power plants and increased efficiency in the boiler room are hobbies with Mr. Carle, and much research work has been done by him, with the result that the Public Service Electric Company is getting ready to install an experimental outfit at the Marion Station within the near future for the utilization of this fuel. If this proves successful, it will be applied to the smaller stations first.

His work in connection with increasing the efficiency in the boiler room has done much toward making the plants of his company safer and more economical. He is a strong advocate of placing an experienced high-priced man in charge of the boiler room, as a large percentage of the plant saving depends upon the performance of the boilers. While a believer in high pressure for maximum efficiency, he is opposed to extremely high pressures until a more comprehensive knowledge of boilers and their auxiliary apparatus and their ability to carry such pressures has been gained.

Mr. Carle was born in Portland, Ore., May 28, 1875. When he was six years old his parents moved to Seattle, Wash., and it was here that his education began. He received his elementary education in the public schools, followed by a course in mechanical engineering at Leland Stanford, Jr., University, from which he graduated in 1898. While at the University, Mr. Carle, always a lover of the great outdoors and extremely fond of athletics, played on the varsity football team for three years. He was also initiated to membership in the Beta Theta Pi fraternity and the honorary engineering society of Sigma Psi.

Following graduation he accepted a position with the Silver Bow Mining Co., which was operating at Juneau, Alaska. He worked underground as a miner, for practical mining knowledge, and in the latter part of 1898 he was employed by the Canadian Rand Drill Co. at Sherbrooke, Quebec. He remained there until 1900, then resigned and went to New York City, where he became associated with Westinghouse Church Kerr & Co. He supervised and directed many large jobs while with this company, including the installation of a large power plant in the Back Bay Station of the New York, New Haven & Hartford Railroad at Boston. He was field engineer on the construction of the Kingsbridge power station, which had the distinction for many years of being the largest and most economically operated power plant in New York City. While credit is due to the engineers in charge of the operation of the plant, yet it stands out as a monument to the thoroughness and ability of the engineer in charge of installation and construction, for efficiency and economy in operation cannot be gained where installation is faulty. Following the completion of this work Mr. Carle was made first assistant to the superintendent of construction on the Long Island Power Station of the Pennsylvania R.R. at Long Island City and later of the entire electrification of the railroad.

In 1906 he was sent to Denver, Col., as engineer in charge of work on large power-plant construction at the coal fields and in the electrification of the railroad between Denver and Boulder for the Northern Colorado Power Co. His handling of these jobs made such a favorable impression on the officials of the Northern Colorado Power Co. that in 1907 he was induced to become vice president and general manager of the company.

He resigned from the Northern Colorado Power Co. in 1909, returned to Seattle, his boyhood home, and opened offices as a consulting engineer. He later became associated with the Puget Sound & Dredging Co. His work finally attracted the attention of the Public Service Electric Co. of Newark, N. J., and he accepted the position of chief engineer. His ability was put to the test at once. He was called upon to reorganize and enlarge the engineering force of the company and accomplished this so thoroughly that he has since been in charge of the entire engineering and construction work.

The company operates eighteen generating stations with a capacity of 270,000 kw., and 100 substations are supplied with power over the system of 850 miles of overhead and underground transmission circuits operating at 13,200 and 26,400 volts. During the war over one hundred and fifty plants, engaged in war work, were supplied with power.

Despite the fact that his work claims a great deal of his attention, Mr. Carle finds time to take an active interest in a number of engineering societies, including the American Institute of Electrical Engineers, Engineering Council, American Society of Civil Engineers, American Society of Mechanical Engineers on Power Test Codes Committee, Association of Edison Illuminating Companies, National Electric Light Association, chairman of Prime Movers Committee and vice chairman Technical and Hydro-Electric Section, American Metric Association, and the American Electric Railway Association.

Washery Sludge Burned in Pulverized Form at Seattle

THE Puget Sound Power and Light Co. at Seattle, Wash., includes, among its public utilities, a district steam-heating plant of 4,100 boiler-horsepower, which for nearly two years has been fired with pulverized high-grade lignite. The results shown have been so favorable that all those connected with its operation are enthusiastic advocates of this form of combustion.

When the United States entered the war, it early seemed probable that the Federal Government would place restrictions upon the use of fuel oil on the Pacific Coast and that consumers remote from the oil fields would be forced to look to other forms of fuel. Faced with a shortage of oil, Stone & Webster, managers and engineers for the company, undertook a careful study of the possibilities of utilizing the local grades of coal in such of its Seattle plants as were then dependent upon oil. Early in the spring of 1917 one of the smaller boilers at the steam-heating station referred to was set aside for experimental purposes and an extensive series of tests conducted, using practically all grades of coal locally obtainable, in which the thorough practicability of combustion of these coals in the pulverized form was fully demonstrated.

Pulverization was finally decided upon as most favorably meeting the local conditions. It was particularly advantageous at Seattle as affording means for efficient utilization of a 200,000-ton sludge dump at the Renton mine just outside the city limits. The adoption of pulverized fuel had also the important advantage of leaving the existing oil-burning equipment immediately available in efficient form. Conversion to pulverized fuel was accordingly made by Stone & Webster the following year.

The station referred to, which is known as the Western Avenue Station, faces on Western Ave. and extends through to Railroad Ave. on the harbor front. It is about two blocks from the center of the retail shopping district. Originally, the station was designed as an ice plant. The boilers, many of which were installed twenty years ago, were set with low furnaces and with aisle space so narrow as to make a satisfactory stoker installation impracticable. Although the small furnaces would undoubtedly have curtailed efficiency for any fuel, while burning oil the lack of aisle space involved no serious interference with operation. A change to coal, however, in any other form than pulverized, without rebuilding the entire plant, would have called for hand firing, which not only would have been most difficult from a labor standpoint, but would have restricted the consumption to high-priced coal.

The present boiler plant totals 4,100 hp. in rated capacity, made up of four 300-hp., three 400-hp., one 500-hp. and two 600-hp. boilers, all of the B. & W. type. One of the 600-hp. units was installed at the same time as the pulverized-fuel preparation plant and was accordingly set with front headers eleven feet above the boiler floor. All the other boilers, as already referred to, have low furnaces, with headers set only about seven feet above the floor.

The coals mined in the vicinity of Seattle are subbituminous in nature, closely bordering upon lignite, and require washing in preparation of all the smaller sizes. At the Renton mine, washery sludge, regarded as a total waste, had been accumulating for years. This material carries about 25 per cent moisture and has an average heat value of about 7,300 B.t.u. A typical analysis on the dry basis is approximately as follows: Volatile matter, 37 per cent; fixed carbon, 38 per cent; ash, 25 per cent; sulphur, 1 per cent. This washery sludge is also small in size, about 50 per cent passing through a 10-mesh screen, rendering it totally unfit for use on any existing form of mechanical stoker.

As originally reclaimed from the dump, the sludge carries such a high content of clay that its adhering qualities would render it practically impossible of handling in any ordinary form of conveying apparatus. By combining some washing with the sluicing process used in carrying the sludge from the dump, a large part of this clay is removed. Regular washing tables are now being installed on which the clay will be completely removed.

The Coal-Preparation Plant

The plant that converts the raw coal into pulverized coal consists of three main divisions: (1) The raw-coal handling system, comprising raw-coal storage bunkers, crusher, crushed-coal bunkers and interconnecting conveyors; (2) the drying plant, comprising the driers and dry-coal bunkers with conveyors feeding the driers and delivering the dried coal to the bunkers; (3) the pulverizing plant, comprising the pulverizing mills with their feeders and the necessary conveying system delivering to the pulverized-coal bunkers.

The Raw-Coal Handling System

Limitations of space made it impracticable to provide sufficient fuel storage within the station building. The desired additional raw-coal storage was accordingly installed on a vacant lot adjoining one of the company's electrical substations and immediately across Western Ave. from the steam-heating plant. The three raw-coal bunkers erected for this purpose are of heavy wood construction, lined with galvanized iron and have a total capacity of about 700 tons. The bunkers are served by a standard-gage double track on an overhead trestle from which hopper-bottoned cars discharge.

From the bunkers the coal is fed through double rack-and-pinion gates to a 20-in. feeder belt, which conveys the coal to a single roll-type crusher where, when lump or run-of-mine coal is used, it is crushed to about one-half inch size. A bypass chute is provided around the crusher, so that the smaller-sized coal can be fed directly from the feeder belt to the conveyor belt leading from the crusher to the raw-coal elevator. The conveyor belt from the crusher is a combination incline-horizontal belt, 20 in. wide, which, carried through a tunnel under Western Ave., delivers the crushed coal to a double-strand, fixed, continuous-bucket elevator, steel incased, which elevates the coal 75 ft. and discharges it onto a double-strand steel flight conveyor for distribution to the crushed-coal bunker.

All the raw-coal handling equipment has a capacity of 75 tons per hour; with buckwheat coal 90 tons has been handled. The crushed-coal bunker, which is of reinforced concrete, has a capacity of about 350 tons.

The Drying Plant

To convey the coal from the crushed-coal bunker to the two driers, two apron feeders are provided, each served by a 5-ft. by 10-in. opening at the bottom of each of the two V-section hoppers formed beneath the crushed-coal bunker. Each feeder is provided with a vertical sliding rack-and-pinion gate to regulate the flow of coal. Both feeders discharge into a common chute, which is equipped with a flopper gate and steel-plate diaphragm so that either feeder may discharge coal into either or both of the driers. A screw feeder immediately beneath each apron feeder handles the drippings from the return side of the feeder and delivers these drippings into the feeder discharge chute which branches to feed the two driers. Mechanical agitators are installed in the branch chutes to prevent clogging. On account of the high silt content the sludge in the wet state is of particularly sticky consistency and requires positive mechanical movement at every point.

The two driers are of the indirect-fired, rotary type, with settings especially designed for maximum heat utilization. The cylinder dimensions are respectively 5 ft. diameter by 52 ft. long and 6 ft. diameter by 52 ft. long. Pulverized coal is used for the drier fuel. The heat absorbed through that part of the drier cylinder within the heating chamber reduces the hot-gas temperature to about 500 deg. F. The gases finally leave the drier at a temperature of about 270 deg. F. The coal is discharged from the driers at a temperature of about 240 deg. F.

Induced draft for the driers is supplied by two turbine-driven fans. These fans discharge into insulated cyclone separators located above the roof of the preparation plant proper, where all but the finest of dust drawn from the drier with the exhaust gases is collected for return to the system. This collected dust is delivered by screw conveyor to the dry-coal elevator into which the main drier is discharged. The gases from the cyclone separator pass to an exhaust chamber, where, before being finally allowed to escape to the atmosphere, they are washed for removal of the small amount of dust remaining. Salt water from the harbor adjacent to the plant is used in this washer. The proximity of the plant to the business district of the city necessitates special provisions for dust prevention.

From the driers the coal is discharged over magnetic separators for removing tramp iron such as spikes, bolts, nails, and into the dry-coal elevator. The dry-coal elevator is of the single-strand, fixed-bucket type, steel incased, 62 ft. between centers, and is provided with weighted take-ups to compensate for temperature changes met in such service. The dry coal discharged at the top of the elevator is distributed by screw conveyor to the dry-coal bunkers, which are of reinforced concrete and have a capacity of about 150 tons.

The coal as it reaches the dry-coal bunkers still carries a small percentage of moisture which for a time continues to be given off as steam. To provide for the escape of this steam atmospheric vents, suitably drained, are installed. It is found in practice that approximately one per cent additional moisture is given off by the coal in the bunkers and removed by this means. Other vents are also provided in the dry-coal elevator casing to remove such moisture as may be liberated between the driers and bunkers.

From the dry-coal bunkers the coal is fed to the pulverizing mills.

The Pulverizing Plant

The pulverizing equipment consists of three 42-in. Fuller-Lehigh mills of a capacity about 3½ tons per hour each, pulverizing to 80 to 85 per cent through a 200-mesh screen and one five-roll, high-side Raymond mill adjusted to a capacity of about four tons per hour for delivery of a slightly coarser product. The pulverized coal from these mills is delivered by screw conveyor into a continuous-bucket, single-strand, centrifugal discharge elevator with reinforced-concrete casing. This elevator discharges into a screw conveyor distributing the coal to the pulverized-coal bunkers which supply the boiler and drier furnaces. The bunkers supplying the boilers provide capacity for about one-half day's normal operation and are of reinforced concrete, except in the case of two bins where it was impracticable to employ this construction.

All pulverized-coal elevator and conveyor casings are of dustproof construction as are the connections from both Fuller and Raymond mills. Further, to insure absolute dustlessness within the preparation plant that part of the pulverized-coal system, including the Fuller mills, is operated under a slight vacuum. The small exhauster employed for this purpose is arranged to discharge directly to certain of the boiler furnaces. The vent from the Raymond mill separator is similarly disposed of.

The Combustion Equipment

The pulverized-coal feeders are of the screw-conveyor type, each operated by a variable-speed direct-current motor, thus providing complete control of the coal feed by simple rheostat adjustment. From the feeders the pulverized coal is picked up by air under a pressure of about five to six ounces and delivered to the burners. The air supply is furnished by turbine-driven blowers delivering to a central blast duct beneath the pulverized-coal bins, from which branches are taken off to the individual feeders and burners. Air regulation is had by means of valves in the air-feed lines.

The burners, which were developed by the Puget Sound Power and Light Co.'s steam superintendent, are extremely simple in construction, consisting merely of a special mixing elbow delivering the aërated coal to the furnace horizontally through a straight pipe nozzle in such manner as to induce with the fuel jet the greater part of the additional air required for combustion. On account of the low setting of the boilers it was necessary to place the burners below the main boiler floor to provide the required space for combustion of the fuel before reaching the heating surface.

The ash resulting from combustion varies in composition from a fine powder to slag; some is deposited as dust on the tubes and back of the bridge wall, some escapes with the flue gases, but the major portion falls to the bottom of the furnace as coarse sand and rather soft clinker. The dust accumulation on the tubes is removed by mechanical soot blowers supplemented by a hand lance once or twice a day as required. The ashes and clinker from the furnace are raked out at regular intervals, quenched with a hose, conveyed by hand-dump cars, are elevated by electrically operated skip hoist to an overhead ash bunker from which they are discharged to trucks for final disposal.

Operation

Burning the pulverized washery sludge as described, boilers are regularly operated on continuous loads of 150 per cent to 175 per cent of the rated capacity. The station as a whole has readily carried steam-heating loads of 6,000 boiler-horsepower and has shown over-all net operating efficiencies, as determined from steam production per pound of wet sludge, fully equal to those obtained with mechanical stokers utilizing the better grades of coal of this district, of larger size and carefully prepared. Boiler maintenance has been markedly reduced. With oil firing, tube failures under heavy loads were far from uncommon. Since the adoption of pulverized coal, such failures have been practically unknown.

The success of the installation, the simplicity and efficiency of combustion, the flexibility in capacity, attained with a fuel which by any other available means it would have been impossible to utilize at all, quite naturally converted the operating organization and officials of the Puget Sound Power and Light Co. to enthusiastic advocates of pulverized fuel. Nevertheless, they do not look upon pulverization as a universal panacea for the high cost of fuel, but urge that every situation be viewed as a special problem and that all the affecting conditions be carefully studied before attempting any decision as to the best form for combustion in any particular case.

Refrigerating Effect of Natural and Artificial Ice

By John H. Ryan

The comparative merits of natural and machine-made ice is an old story in the ice business and was started by the natural-ice men claiming that their ice would keep the ordinary box colder than would the machine-made ice. Thermometers placed in the boxes bore them out in their argument, as the refrigerators always showed two or three degrees colder with the natural ice.

In 1913 I was chief engineer in a combined brewery and ice plant in New England where competition was keen in the ice business. We bought an ordinary household refrigerator that held approximately one cubic foot of ice and put a recording thermometer in the side of the box and then put the box in one of the ale-storage cellars kept at a constant temperature of 50 deg. F. We left the empty box with all doors open for two days to be sure it was the same temperature as the cellar. The ice to be used was set to one side of the ice-storage anteroom for the same length of time so that the natural and artificial ice would be at the same temperature. We put 48 lb. of ice in the box to test with and made three tests of each kind of ice alternately. The elapsed time from when the recording thermometer started to drop from 50 deg. F. to when it came up again to 50 deg. F. was slightly over five days, and there was less than twenty minutes difference in the time of all the tests. The longest elapsed time was with one of the natural-ice pieces, but so also was the shortest time, and we decided it a draw.

In Boston the next year I had an opportunity to observe both kinds of ice in a grocer's refrigerator, and discovered the "joker" to be in the manner in which the ice is placed in the box. Natural ice gets more handling and breaking before placing in the customer's box, and it is so rough that it does not make a closely packed mass. The air can circulate all around each cake, which cools the air because there is such a large ice surface in contact with it. With manufactured ice the surfaces are so smooth that they lie close together and freeze into a solid mass and the only surface offered to the air is the outside of the frozen mass, which is small compared to the rubble surface of natural ice. In this case the manufactured ice lasted longer while not cooling the boxes so low. By instructing our drivers to pile the can ice checkerboard fashion we could meet the temperatures obtained with natural ice, and increased our sales.

Rules for the Safe Operation of Steam Engines and Turbines

By E. E. CLOCK
Chief Inspector, Turbine and Engine Department, The Fidelity and Casualty Co. of New York

Starting the Engine

No engine should be started until it has been thoroughly inspected since its previous run. Inspect for loose nuts, keys, fittings, main belt, flywheel, receiving pulley, governor and governor belt. This should be done at the end of the run to allow time for adjustments or repairs. This applies to all auxiliaries necessary to the operation of the engine.

Removing Water

Be sure the engine is warmed up and the cylinder and steam pipes, exhaust pipe, receiver and coils, and the jackets, are free of water; if the receiver is of cast iron be sure the relief valve is set well within the safe pressure limits and that it is of ample capacity to free the receiver of all steam the high-pressure cylinder may be able to give it under any or all conditions of sticking or broken valves.

Starting Machinery at Distance

If the engine is belted to the machinery in the plant so that machinery is started at a long distance from the engine, the engine should not be started until a warning is given by ringing a gong or otherwise.

Governing Point

When the engine is coming up to speed, the governor should be watched closely for any sticking or sluggishness. Closely note its full ability to hold the speed down to not over four per cent overspeed at no load. Stand by till the engine has taken a load and as long a time afterward as time will permit; the engine should not be left alone till it is running satisfactorily.

Engine Running Alone

The engine should not at any time be left out of the sight and hearing of someone capable of getting it under control in case of accident.

Lubrication

The governor should not be handicapped in the control of the engine by bad lubrication. Sediment in combination with graphite and bad or gummy oils causes the packings rings to stick in the piston. Use graphite sparingly in engines provided with self-adjustable pressure plates or balance rings or any steam surface that is provided with water-grooved packing.

High-Test Oils in Low Temperatures

Never use a high-flashpoint oil on a low-temperature engine; 600-test oils should not be used on engines carrying less than 125 lb. pressure unless there is superheat. It is not advisable to use the same cylinder oil on the low-pressure cylinder that is used on the high pressure.

Driving Belts

All of the surface of the belt should be kept in contact with the pulley and not allowed to dry out; neat's-foot oil or castor oil will usually produce proper surface contact. Do not allow belts to become soaked with lubricating oil.

Speed-Limit Device

Engines running under very widely fluctuating loads should be provided with a speed limit device independent of the governor.

Turbines

Turbines must be guarded against water entering with the steam. Therefore use extra care in warming up and draining. Frequently inspect the regulating gear and know that the speed-limit device is proven to be in order by frequent testing. Turbines should be immediately taken out of service upon failure of the speed-limit device. Atmospheric relief valves must be provided and kept free to open quickly and easily without subjecting the low-pressure end of the turbine to undue pressure.

Relief Valves

Bleeder or extraction type turbines should be provided with relief valves in the bleeder or extractor stage to relieve the turbine of all the steam that comes through the first stage, even with the loss of many blades or buckets. Also provide a non-return valve to prevent return of steam from bleeder line. These valves must be frequently tested to insure reliability.

Tachometer and Circuit-Breakers

No turbine should be operated without a reliable tachometer or frequency indicator; and all turbines and engines feeding to a common busbar should be provided with proper circuit-breakers and, if direct current, should be provided with reverse-current relays.

Testing Safety Devices

All safety devices should be tested regularly and the result of the test entered on a log sheet provided for the entering of all tests, including safety valves, relief valves, relays, speed limits, etc., with the time and date of tests duly entered above the signature of the engineer making the test.

Flywheel

Be sure the flywheels are operated with an ample margin of safety on strength as well as speed, using the formula

$$\frac{1680}{\textit{Diameter in Feet}} = \textit{Safe Speed.}$$

Steam Mains

Be sure the steam mains are so arranged that the drainage or condensation will flow toward the engine and are properly intercepted by a suitable separator and trap. Be sure the trap is so located that it may be watched.

Pockets

Be sure there are no sags or pockets in the steam main that would allow water to accumulate and go to the engine in slugs.

Gagpots

Gagpots or dashpots on governors should be kept free from gum or dirt accumulation, as this prevents the Corliss type of governor from going to the low or safety limit and the flywheel or shaft governor from reaching limits necessary for the full control of the engine under a sudden release of load.

Gaging the Governor

Never under any circumstances gag the governor to prevent it from shutting down under overload or low steam without placing a man at the engine to remain until the governor can be restored to the safety condition.

EDITORIALS

Sawmill Drive in the Pacific Northwest

THE lumber industry in the Pacific Northwest is reaching proportions that are greater than the pioneers dared to hope. Every day sees a demand from some quarter, either for new markets altogether or for new uses for available timber. Just now the lumber industry is probably larger in point of demand and output than ever before in its history. Considering Washington alone, 7,250,000,000 feet passed out of that state in 1918, and it is likely that the 1919 cut will show a gain despite the fact that the serious labor disturbances of that year reduced output. But this year the great mills of this remarkable Northwest country are passing timber at an astounding rate. The world seems to be drawing heavily upon our timber.

Now, in an industry that has become so large, one naturally looks for big things mechanically, and is not a bit disappointed when visiting the large camps and mills on Puget Sound, in Oregon and in Northern California. In fact, one is rather surprised. It is not uncommon to see a man at a keyboard in a screeching, busy mill, pressing the many keys each of which operates a spinning saw, causing it to lower itself and cut a board to a certain length or to remain uplifted, allowing the board to pass and be cut by other innumerable saws in the row, all at the discretion of the operator. One feels the might of the country and the industry when standing back and watching the little Japanese laborers on the carriages where the grotesque mechanical "niggers" grip the huge timber and flop it over on its side like the man in Childs' window flops the griddle cakes. It is to feel that you are in the midst of big things when watching the tugging cranes pick up a whole flat-car of logs, each two to ten feet in diameter, and roll them into the mill pond.

And the big thing about it is that it is all a power job—yes, every operation from cutting and transporting the great trees to delivering the finished lumber to the shipping docks or cars. There are some modern, high-grade, high-pressure steam engines and apparatus up there in the echoing woods that would gladden the hearts of Corliss and Watt. There are engines, turbines, air compressors, pumps, electric generators and what not in the mills that would turn an uninitiated engineer in search of variety and problems green with envy. The mechanical engineer, the power engineer, of a large, modern lumber company has a man's-size job.

And it begins to look as though he and his manufacturing associates are about to introduce high-tension, hydro-electric current into their field on a large, varied and interesting scale. This is by no means the general practice. The thought is, How general is it likely to become? Distance of lumber camps from transmission lines is a factor; but it is obvious that the cheapest construction will do to connect the two, for the work is of necessity of a temporary character. More specific and detailed technical information on the application of electricity to lumber camp and sawmill work is needed.

Another Reason for an Operating Code

COMPARISON of instruction pamphlets that usually accompany all new apparatus from the manufacturers' shops will often reveal considerable difference in matters of detail respecting the functioning of similar equipment. Were these instructions to be followed explicitly, a station that happened to contain two or more makes of apparatus for like service would find itself employing an equal number of methods in their operation. What usually results is a compromise in operating practice that, in the opinion of the management, best meets the local conditions.

This is all very well as long as everything runs smoothly, but as a rule manufacturers will stand back of their guarantees only to the extent that their operating instructions are carried out. Let anything happen to the apparatus where these are being deviated from, and there immediately arise grounds for possible controversy.

This is just an additional point in the argument that has previously appeared in these columns for the establishment of an operating code for central stations or large industrial plants along the lines of the present Power Test Codes, and which it is understood is now being seriously considered by the Prime Movers Committee of the National Electric Light Association.

A Needed Reform in Marine Boiler Law

WELL back toward the middle of the last century, Congress enacted laws for the regulation and inspection of steam vessels navigating the waters of the United States. It is proposed to amend that portion of the law which provides for the allowable working pressure in boilers. A bill known as Senate Bill Number 574 has been introduced and referred to the Senate Committee on Commerce.

The present law provides that the working pressure shall not produce a "strain" of more than one-sixth of the tensile strength of the plate when the longitudinal joints are single-riveted, but that this allowable pressure may be increased twenty per cent if the "longitudinal laps" are double-riveted, provided the rivet holes are fairly drilled and not punched. This section of the law was in its day a reasonable attempt to insure safety, although the use of the word "strain" where, obviously, "stress" is meant, shows that the authors were not familiar with engineering. A hard and fast rule based on even the best of engineering knowledge at that time would be out of date today. How much more so is a rule whose authors probably were well meaning but apparently not engineers?

Suppose we have a boiler built with a joint efficiency of fifty-six per cent and another with an efficiency of eighty-four per cent. Obviously, the second boiler is half as strong again as the first, assuming both to have the same diameter and thickness of shell, yet the old rule would allow both the same working pressure. If the rule would make the first boiler safe, then the second is handicapped, while if the rule is reasonable for the second, the first boiler is close to the danger line if not over it. Surely, it is time for an amendment of this rule.

The proposed amendment provides that the allowable pressure shall be determined by the rules of the Board of Supervising Inspectors. Under the law as it now stands, this board makes and issues rules and regulations for the construction and inspection of steam boilers, hulls, etc., and these rules are from time to time amended by the board in the light of current engineering progress. It would seem only a logical step to intrust this board with authority to fix the thickness of the boiler shell according to modern engineering practice instead of by an arbitrary rule which is sadly out of date.

Grounded Neutral

TO GROUND or not to ground the neutral of alternating-current systems and to what extent grounding should be carried, provided it is decided to ground the system, is a much discussed question on which about as many different opinions are held as there are persons discussing it. This was evident at a recent meeting of the Association of Iron and Steel Electrical Engineers in Pittsburgh, for after over two hours' deliberation on the question of grounded neutral, no definite decision was reached. However, the trend of the discussion would indicate that if the system is large, such as those of central stations, the neutral should be grounded, while the smaller systems should be operated insulated from the ground.

Three elements are of primary importance in the operation of any electrical system; namely, protection to life and property, continuity of service, and protection to apparatus. Of the three, protection of life and property is of first importance. It has become general practice to ground the neutral low-voltage circuits when exposure to higher potentials exists, such as when the low-voltage circuit is supplied from a step-down transformer. Low-voltage alternating-current circuits of one-hundred and fifty volts and less are usually of the single-phase type, supplied from transformers and to a very large extent used for lighting purposes. Power circuits are generally operated at four-hundred and forty volts and above. If a three-phase four-hundred-and-forty-volt circuit is connected star with the neutral grounded, there will exist from either phase to ground a potential of two-hundred and fifty volts. Although it is not generally recognized, statistics show that on circuits of four hundred and forty volts and less the greatest danger from fatal shocks exists. Of ninety-nine fatal electrical accidents occurring in England during the period 1915 to 1918, sixty-nine happened on circuits having a potential of four hundred and forty volts and less. It is doubtful if any of these fatalities were due to voltages higher than two hundred and fifty, since the voltage to ground on any of these circuits was not above the latter value. There is apparently a distinct advantage in operating such systems ungrounded, since when the neutral is grounded a potential exists to ground that can cause a fatal shock, yet it is not in general considered as such by workmen. Unless the system is kept insulated, the advantage gained by insulating the neutral is lost and a more dangerous condition is created. On potentials above five hundred and fifty volts workmen in general will take proper precautions in handling such circuits whether they are grounded or ungrounded, consequently there is little or nothing to be gained, so far as safety to life is concerned, in operating the higher-voltage systems ungrounded.

Continuity of service and protection of equipment are so interconnected that one cannot be considered without the other, since if the apparatus is not properly protected it will be subjected to abuses, the results of which will be reflected in frequent shutdowns of the equipment for repairs. If the neutral is grounded, a ground on any leg of the circuit will cause a short-circuit and the motor or circuit is made inoperative until the defect is repaired. This, in some cases, may mean serious interruption or inconvenience in the manufacturing process. From this point of view grounding the neutral is a distinct disadvantage, especially where means are provided and put into effect to keep the system insulated at all times.

Where systems are very large, and few industrial plants come within this category, the electrostatic capacity is high enough to set up very heavy surges in the system in case of an arc to ground. These surges are liable to break down the insulation of equipment. In such cases, if the neutral is grounded the circuit breaker will be tripped when the arc first occurs and the apparatus relieved of excessive voltages, which might cause insulation failure at the time of the disturbance or at some future time. "How large should the system be before it is advantageous to ground the neutral?" is apparently the question around which centers operating power systems with or without the neutral grounded. Many central-station systems are operating with the neutral grounded, while on the other hand industrial plants are operating with the neutral insulated. It is on this question that engineers in charge of industrial power systems have an interesting field for investigation.

Turning Waste Into Profit

EACH year there is produced in the lumber industry of the Pacific Northwest thousands of tons of wood refuse from the mills, called hog fuel. This, ordinarily, is a nuisance and is disposed of by burning either in furnaces constructed for the purpose or in a great pile a short distance from the mill. Its fuel value is high—dry, it averages about 8,500 B.t.u. per pound.

This part of the country has little coal other than some lignite, which is confined chiefly to a small area in Washington. Fuel oil, upon which the steam plants chiefly depend, comes from California and, like fuel oil in the East, is now very high in price. Tests have been made with hog fuel under steam boilers and the results are so favorable, in view of existing fuel conditions, that there is being developed a market which promises to take at a profitable price much that will be produced in many localities. An article elsewhere in this issue gives particulars about the boiler performance with and market for this waste fuel. The problem, by the way, is an engineering one and the sawmill engineer is the one charged with the solution.

If the proposed Congressional junket at the expense of the Government is carried out it will cost Uncle Sam quite a sum. Somebody has suggested, however, that it may be worth the cost to have Congress adjourn.

CORRESPONDENCE

B. T. U.'s Versus Kilowatt-Hours per Ton of Coal as Measure of Efficiency

In the March 9 issue R. S. McBride's comments on page 386 should be of interest to everyone. It is no fault of Mr. McBride's that he went wrong, but rather of the United States Geological Survey, Division of Power Resources. The data on which the comments were made were so presented that he had a right to assume that the fuels used in all cases were for power production only, and therefore the efficiency with which fuel was used was proportional to the kilowatts per ton of coal. One of the tables presented in the article is reproduced here:

TABLE I. EFFICIENCY OF POWER PRODUCTION IN TEN STATES

Results for four months, February, March, April and July, 1919, as compiled by U. S. Geological Survey, Division of Power Resources. Computed on the basis of four barrels of oil equals one ton of coal and 20,000 cubic feet of gas equals one ton of coal.

State	Current Produced Millions of Kw.-Hr.	Fuel Used Thousands of Tons of Coal or Equivalent	Efficiency Kw.-Hrs. per Ton
New York	1,124	1,420	792
Pennsylvania	958	1,610	595
Illinois	784	1,311	597
Ohio	776	1,282	605
Massachusetts	403	533	756
Michigan	390	497	787
New Jersey	308	459	671
West Virginia	237	321	738
California	222	310	716
Indiana	213	585	364
Average	.....	....	662

On the strength of this information Mr. McBride derived the only possible conclusions that could be made if depending only on the information furnished by the table. The trouble is not with the information given, but that it tells only half the truth, and the deductions in themselves are correct from the point of view of the commentator, but are far from representing actual facts. After presenting the table referred to Mr. McBride comments as follows:

This first comparison shows that some states, such as New York, Michigan and Massachusetts, secure twice the current production per ton of coal as is obtained, for example, in Indiana. This at once raises the question as to why there are such great differences.

All of the ten states which are the largest users of power lie in the northeastern quarter of the country except California. It is of interest, therefore, to notice particularly the results in some of the Southern and Western States. South Carolina, Georgia, Alabama and Kentucky similarly computed in comparison show, respectively, 417, 481, 324 and 493 kilowatt-hours produced per ton of coal used. Thus the average for these four seems to be less than for the states which are larger power producers, but in general the average is not as low as Indiana, already mentioned. Some of the Western States seem to be the worst offenders in the way of low efficiency, for in North Dakota and Montana the average current production is only approximately 150 kilowatt hours per ton, or less than one-quarter of the average in the states which are the largest users.

Kilowatt-hours per ton is the standard of measurement of efficiency, the only thing possible from the information furnished by the Division of Power Resources, and the conclusion that there are public-service corporations whose efficiencies are in the ratio of 5 to 1 are borne out by the material considered. Granting all this, the only thing that can be said is that the conclusions are misleading and the facts in the case, if properly reported, would be very different.

In large manufacturing states with large cities, power plants, in order to satisfy large power demands, are made up of large units operated condensing, and the output is power measured only in kilowatts. In this class of power plant the efficiency is and will be proportional to the kilowatt-hours per ton of coal, and this standard may be applied to this class of power producer without modifications.

In agricultural states where there are very few large cities, and with a large number of small towns scattered throughout the state, the power demand is not large, although important. Stations are made up of relatively small units, the larger of these operating condensing only part of the time and the smaller units most noncondensing. These plants, besides producing power, furnish heat in the heating season and in some cases make ice during the summer. When the winter and summer are fairly balanced, they furnish heat in winter and refrigeration in summer, the latter service consisting chiefly of making ice with exhaust steam in an absorption system.

In the state of Indiana, where heat is furnished throughout the season, using practically all of the exhaust for four months and perhaps some live steam as well as three months of the season, the electric output is simply a byproduct, costing very little as regards fuel. This point should be emphasized particularly in the case of North Dakota and Montana set forth as the "worst offenders," where the kilowatt-hours per ton is 150. In these states the heating season starts in September and ends in May and a temperature of 30 to 40 deg. below zero is common. During six months, no doubt, the plants use live steam as well as the exhaust available from power production, and during the remaining two months 75 per cent of the exhaust is utilized. Measuring their efficiency by kilowatt-hours per ton of coal would be like measuring the dog by a piece of its tail.

Coming back to Indiana, in this state a great number of public-service plants are furnishing heat. Some of the larger ones are equipped with condensing and non-condensing units so as to make use of the exhaust. For about four months they utilize all the exhaust, and during two months of the four most of them use live steam, making power a byproduct instead of the primary product.

Farther to the south, in fact even in Illinois, there are plants furnishing heat, ice and power, and in localities fortunately placed they may and do use exhaust steam all the year around. In such plants the kilowatt-hours per ton of coal could not be expected to compare, say, with the same unit measurement of a large power plant operated condensing and equipped with economizers, superheaters and large turbine units. But when the comparison is made on the basis of true efficiency, thermal efficiency, which is the ratio of the output to the input in B.t.u., the results will be all in favor of the small plant making use of its exhaust steam.

Plants being compared on the basis of kilowatt-hours per ton of coal may show efficiencies in the ratio of 5 to 1, and the same plants compared on their thermal or actual efficiency can show the reverse; that is, the plant that showed five times as many kilowatt-hours per ton of coal may be actually only one-fifth as efficient.

A well-equipped power plant situated in a zone of the best coal produces a kilowatt-hour on less than two pounds of coal. Plants less fortunate as to coal and average equipment require between two and three pounds of coal. Although a rate of two pounds of coal per kilowatt-hour is excellent practice, the figure will indicate that less than 12 per cent of the heat in the steam is utilized. Granting a boiler efficiency as high as 75 per cent, the over-all thermal efficiency of such a station will be $0.12 \times 0.75 = 0.09$, or 9 per cent.

Now the diversified power plant furnishing heat and power in North Dakota or Minnesota uses, say, ten pound of coal per kilowatt-hour of output and utilizes 75 per cent of the heat in the steam the year around. The chances are that it will use 80 per cent of the total heat in the case of an eight months' heating season. Grant that the boilers are operated at 60 per cent efficiency over-all the year around, the station thermal efficiency will in that case be $0.75 \times 0.60 = 0.45$ or 45 per cent.

Summing up, the first power plant is producing a kilowatt-hour on two pounds of coal and the year around is operated at 9 per cent thermal efficiency. The second plant produces a kilowatt-hour on ten pounds of coal and operates the year around with 45 per cent thermal efficiency. The ratio of kilowatt-hours per unit of coal is as 5 is to 1 and the ratio of thermal efficiency is exactly the reverse, the first being operated at 9 and the second at 45 per cent efficiency. It will be seen why kilowatt-hours per ton of coal cannot convey the actual facts unless some detailed explanation is given. Using the ratio of the output of the station in B.t.u. to the input would give the true station efficiency and would not require any explanation.

Reports of this kind are good to have, but the data should be complete and not misleading. Let us hope that the Division of Power Resources of the United States Geological Survey, seeing the interest taken in their work, will favor us in the future with more complete information or adopt as a standard of measurement of station efficiency, B.t.u. instead of kilowatt-hours per ton of coal.

HENRY MISOSTOW.

Chicago, Ill.

Circumferential Cracks in Crankshafts

The shaft of an oil engine which was 19 in. diameter at the portion where the flywheel was located, had developed a crack that could be plainly seen, and although the engine was kept in operation, the engineers were cautioned to watch it carefully and report if the crack showed signs of becoming worse. In the meantime the builders were notified to send a man to examine the shaft and advise what to do. I was sent to look into the matter.

The owners would not allow me to drill into the shaft for fear of weakening it still further. How to determine the depth of the circumferential crack was a problem that had to be solved. I had no precedents to follow and had never read anything on the subject. The engineers on the job had tried drop black mixed with kerosene, but could not determine the depth from the story it told. I lay awake a few hours that night and did some tall thinking. I had an idea that rubbing Prussian blue mixed with kerosene into the crack (which extended for possibly two-thirds the circumference at the root of the fillet where it turned upward toward the larger part of shaft upon which the flywheel was keyed) and then using three gasoline torches to heat the shaft, would give a pretty fair indication of the depth of the crack from the amount of coloring matter that would ooze out. This idea was carried out, but it resulted in but little matter oozing out. Then I let the shaft cool while plenty of Prussian blue was applied, as I assumed that the contraction would absorb some coloring and that upon heating the shaft again it would ooze out.

However, I found that but little coloring matter was absorbed, as but a very little oozed out—and about the same amount at each heating of the shaft. This convinced me that the crack was not deep, and I set up a turning rig and took a light cut from the shaft. It was as I suspected and a $\frac{1}{32}$-in. chip removed all signs of the crack. The engine was run regularly, and two years later the crack again appeared, but now, after 6 years' operation, it is no worse. No doubt an internal strain in relieving itself, slightly cracks the shaft. The only thing possible to do under such circumstances is to watch the crack carefully, keep notes regarding its conditions and check up regularly, and upon indications of its getting worse, get a representative of an electric-welding company to look it over and advise if they can cut a trench through the crack and afterward fill it up properly with the same metal as the shaft.

This, to my mind, would be the most logical thing to do. Forging the shaft would require its removal and sending away, which would result in the loss of money while the engine was down. I hardly think that reforging the shaft could be done without distorting it so that it would require re-turning to a smaller diameter at the bearings and crankpins to true it up. This would necessitate new bearings all around. Hence, I consider reforging of a shaft illogical and altogether too costly. Electric welding and building up looks more logical.

New York City. D. L. FAGNAN.

Suggested Damper Arrangement

Many things are still done as our forbears used to do them, because they are not deemed important enough to require consideration. For instance, take the setting of the damper construction where the damper slides in a cast-iron frame. Considerable leakage occurs where such dampers are mounted in the orthodox way, as shown in Fig. 1. There is generally a slot provided in the brickwork where the top of the flue touches the

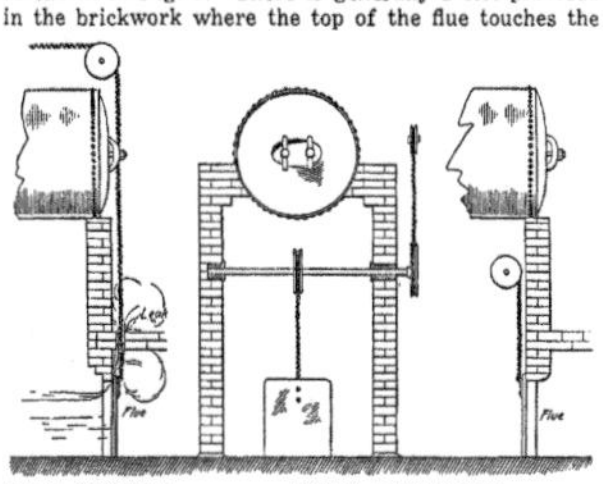

FIG. 1. ORIGINAL SETTING FIG. 2. SUGGESTED DAMPER SETTING

back wall of the boiler setting, through which slot the damper may be placed in the frame or removed from it. This slot is sometimes closed by placing loose bricks over it, leaving a hole through which the operating chain passes.

I have seen many installations where the air leakage at this point was considerable and might easily have been avoided by placing the dampers on the inside of the wall, Fig. 2. The chain leading from the damper is attached to the periphery of a grooved sheave, the diameter being such as to allow full travel of the damper with about three-quarter revolution of the sheave. This sheave is secured to a shaft which turns in simple bearings mounted in the side walls of the boiler setting. One end of shaft passes through the side wall and has a sheave of the same size attached, on the periphery of which the chain holding the counterweight is attached. With this arrangement there will be very little leakage through the bearing.

Another advantage is the renewal of the operating chain in front of the manhole. O. JANZEN.

Half Way Tree, Jamaica, B. W. I.

The Need of Competent Engineers

In the April 6 issue of *Power*, page 561, an illustration is given of an incompetent engineer. I would like to say a few words on the subject of incompetent engineers.

I was in a large steam plant in this locality a short time ago installing a Corliss engine, and observed that the boilers were blowing off almost continually. I went to the boiler house for a short time and watched the firemen work. I asked one why he kept the boilers blowing off so often and he said that the furnaces had 80 sq.ft. of grate surface and that there was about 50 sq.ft. covered with fuel. I asked why the chief did not take off one or two boilers, and the reply was that it would make it easier for the firemen and that the chief didn't care because some bad coal had been received about a month before and as a result the steam went down. The chief had said it would not go down again if he had to put on all the boilers, as it was better to let them blow than shut down. This engineer has a first-class license, but he was wasting approximately 40 per cent of the fuel.

He could tell you how the engines and pumps were running and what trouble he had with his engineers and oilers, but when I asked about his boilers and what efficiency he was getting out of the fuel, he said that he got from 5 to 7 lb. of water per pound of coal, but that what was wanted at that plant was steam, as they had to keep going.

I say such men as this should not be allowed to operate power plants of any kind. If there were many like him, the employers would begin to think that engineers are not a very intelligent class of men. I hope the day is not far off when there will be a law to the effect that no person shall be allowed to operate a power plant until he has acquired a thorough technical and practical knowledge of combustion and power-plant efficiency, besides a good practical knowledge of machinery.

North Tonawanda, N. Y. C. J. MILLER.

Protecting Lubricator Gaskets

The following shows how trouble with lubricators was overcome. The illustration shows the connections that were used to prevent the top gasket of the sight-

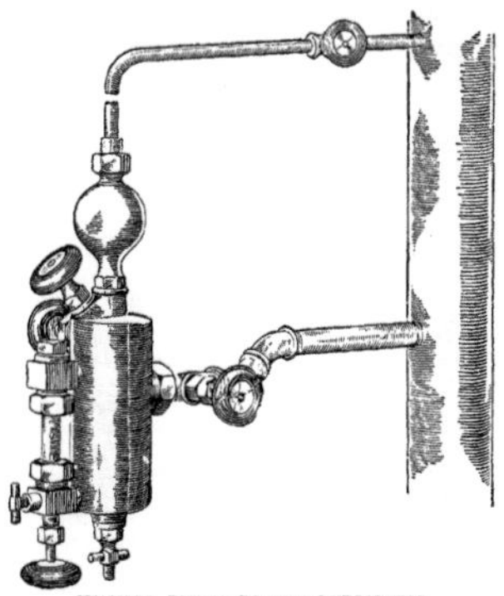

CHANGED PIPING TO THE LUBRICATOR

feed glass from softening and squeezing out on the stuffing nut. Instead of connecting the lubricator feed pipe direct to the steam pipe, as is the general practice, I used two 45-deg. elbows and a close nipple. This arrangement holds the condensation between the lubricator and the ell nearest the steam line and so prevents steam from coming in direct contact with the lubricator, thus preventing overheating of the top gaskets.

Brooklyn, N. Y. LEROY W. HAYWOOD.

Unusual Break in a Ten-Inch Steam Main

At five o'clock on the morning of March 15 a ten-inch pipe broke loose from the main header in the plant of the Union Traction Company of Indiana, at North Anderson, causing about $3,000 damages. Fortunately, no one was injured although there were five men in close proximity to the place of the accident. The sketch shown will give a good idea of the location of the piping and the nature of the break.

The exact cause of the accident will probably never be known, as the evidence submitted by the operating force leaves the matter in doubt. The watch engineer in charge, who was close by, reported that there was a loud explosion on the boiler-room side of the division wall and instantly the pipe broke away from the separator on the engine and broke out a part of the division wall, which allowed the full force of the steam to enter the engine room, driving all the operating force from

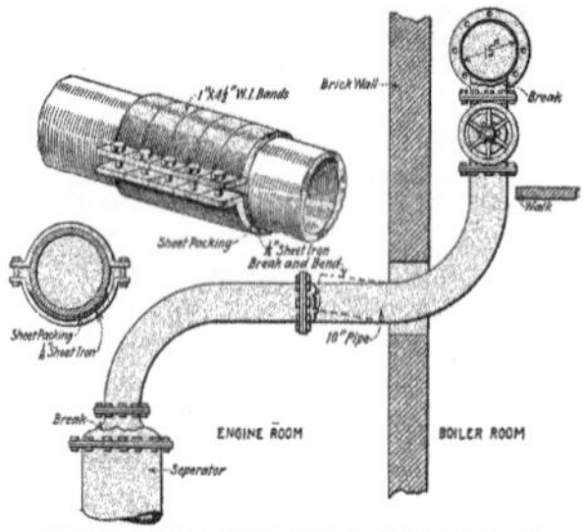

METHOD OF REPAIRING BREAK IN STEAM MAIN

the room. The pipe bend that broke off from the separator was thrown about thirty feet directly upward, striking the roof and knocking off about 120 sq.ft. of roof, then fell back on one of the engines, which fortunately was not running.

At the time of the accident there was 165 lb. pressure, and within a few minutes the entire pressure was blown down on the fourteen boilers that were in service. Directly in line with the blast of steam from the boiler room were four rotary converters and the direct-current switchboard, which were thoroughly soaked with steam and water. The converters were damaged so that some of them were out of commission for two days, badly crippling the through service between Indianapolis and Fort Wayne and Indianapolis and Marion.

There was considerable difficulty in getting the plant started again due to the fact that water was blown down in all the boilers except two. As the plant does not have city water connections, the start was made on these two boilers. There was a partial resumption of service six hours after the accident and a full resumption thirty-six hours afterward. The break being a nozzle directly off the main steam header caused the cutting off of service of five boilers and two engines until the header was patched, which was done by trimming the broken section and covering the hole with a piece of one-eighth inch sheet packing and a piece of one-sixteenth inch sheet iron held in place by five 2-piece, wrought-iron clamps made of 1 x 4½-in. bars, which made a perfectly tight job.

There were three breaks in connection with the accident, which probably happened in this order: First, the initial break probably occurred in the nozzle close to the header; second, when the break occurred at the header, the bend dropped about six inches, which broke off the connection at the top of the separator; third, when the section of pipe was blown upward, the third break occurred at the flange connection between the two bends. A new section of header consisting of steel pipe with welded nozzles and forged-steel flanges and Crane lap joints will replace the broken section of header. The old header is made up of cast-iron sections from six to eight feet long and has been in service for nearly twenty years. The construction of the header and the outlet connections to the engines is such that the vibrations due to the pulsations of steam to the engines would tend to set up a crystallization at the point where the break occurred. This is one theory that has been advanced as to the cause of the accident.

Distance of Bridge Wall from Boiler Shell

In my opinion Mr. Smythe is right (Feb. 24 issue) regarding the distance between the bridge wall and the boiler shell. The standard for boilers 50 to 60 in. in diameter is 9 in. and for boilers 61 to 72 in. in diameter is 10 in. between the bridge wall and the boiler shell. As the boilers in question are 78 in. in diameter, a distance of 11 in. would be about right.

At the city water works of Monticello, Ill., we have two 80-hp. boilers 60 in. in diameter, and the bridge wall is set 9 in. from the boilers. Very good service is given; in fact, we could not ask for any better. Our load is generally very light, but in case of fire we are required to crowd the boilers, and even then we get very good results.

I believe that in all cases the bridge wall should be of flat surface and not made to conform to the shape of the boiler. It is also a good plan to slope the front of the bridge wall 45 deg. toward the grates by stepping the grates up as they are laid and facing them with some kind of heat resisting material. W. A. PLIMPTON.

Monticello, Ill.

I had trouble with a marine boiler on a tugboat, in that I could not get it to steam. The captain claimed that the bridge wall should be the same distance from the boiler as the width of the ashpan. I claimed it should be a larger opening over the bridge wall than the area of the air space.

He had his way first, and we built the wall with a distance of 8 in. from the boiler. The boiler did not steam so well as before the change, so we made it 10 in. from the boiler, which gave better results. Then it was made with a distance of 12 in., and at this spacing the boiler steamed easily and was easy on fuel. Hoping to get still better results, the bridge wall was made with a 14-in. space, but conditions were worse, so the 12-in. space was adopted as that seemed to work the best.

Clear Creek, Canada. GEORGE A. MILLEN.

INQUIRIES OF GENERAL INTEREST

Advantages of Turbo-Air Compressors—What are the advantages of turbo-air compressors? A. W. L.

Compactness, adaptability to direct connection with high-speed steam turbines or electric motors and capability of yielding an output of compressed air free of oil or explosive hydrocarbon compounds that commonly result from cylinder lubrication of reciprocating air compressors.

Utilization of Discharge from Steam Kettles—How can we utilize the heat of exhaust from steam kettles? Our engine exhaust is insufficient for warming our buildings. Can we not combine the exhaust from the steam kettles with the engine exhaust? T. A. M.

When properly operated, the discharge from steam kettles should consist of only hot water formed by condensation of the steam used, and the most economical utilization of the heat remaining in the water would be to deliver it by a return trap to the boiler.

Lap and Negative Lead of Valve—What is the difference between lap and negative lead of a valve? W. P. S.

The term lap is used to signify the distance the admission edge of a valve overlaps the admission edge of its port when the valve is in the middle of its travel. Lead is the distance the admission edge of the valve has uncovered the admission edge of the port when the piston is at the beginning of a stroke from the same end of the cylinder, and negative lead is the distance the admission edge of the valve overlaps the admission edge of the port when the piston is at the beginning of a stroke from the same end of the cylinder.

Percentage of Fuel Energy Realized by Engine—An engine uses 25 lb. of steam per indicated horsepower-hour, and the evaporative economy of the boiler, under the operating conditions, is 8 lb. of water per pound of coal. Assuming that the heat value of the coal is 10,000 B.t.u. per pound, what percentage of energy contained in the coal is realized by the engine? R. L.

The fuel consumption would be $25 \div 8 = 3.125$ lb. of coal per horsepower-hour, and as 1 hp.-hr. $= 33,000 \times 60 = 1,980,000$ ft.-lb., there would be $1,980,000 \div 3.125 = 633,600$ ft.-lb. realized per pound of the coal. One B.t.u. is equal to 777.5 ft.-lb., and if the coal contains 10,000 B.t.u. per pound, the energy in a pound of the coal would be 7,775,000 ft.-lb., and the energy realized by the engine per pound of coal used would amount to $633,600 \times 100 \div 7,775,000 = 8.15$ per cent.

Setting Safety Cams of Corliss Engine—What is the method of setting the safety cams of a corliss valve gear? G. T.

When the cutoffs have been equalized, drop the governor to its lowest possible position and set the safety cams around to such a position that they will prevent the valves from being picked up, or at least, so each valve will be tripped before its lap is uncovered. Observe, however, that the safety cams are not set so far around that the valve cannot be picked up when the governor is resting on the starting pin or collar. Also place the governor in its highest position and see that the valves are not opened in that position, as that should be the condition to prevent the engine racing in case the load is suddenly removed.

Form of Pump Air Chambers—Why are pump air chambers made larger at the top than at the bottom? Do they give more cushioning effect than if they were merely the size at the bottom? E. M.

For easy passage of water in and out of the air chamber and prompter cushioning action of the air, the water connection with the air space should be no smaller than the pump outlet; but there is no advantage to cushioning effect by having the air chamber larger in diameter than the connection, except to afford sufficient volume of air space, and for that purpose any convenient proportions may be employed. Spherical, pear-shaped or cylindrical forms are most commonly employed, not because they produce any better cushioning effect, but because they meet the requirements of sufficiency of volume with compactness and economy of construction.

Regrinding Plug Cocks—What is the best method of grinding small brass plug cocks in place? Grinding the plug in its seat with a carpenter's brace has been tried with poor results. F. L. W.

When the plug is badly cut, it may have to be skimmed up in a lathe, or if the seat is badly cut, it will need reaming out with a tool that has the correct taper. For grinding, smear the plug with oil and fine emery or, better, use oil and pulverized glass. A better job of grinding is done by oscillating the plug by hand, occasionally turning it to a different position, each time cleaning off the plug and seat and applying a thin even coating of the grinding material. The operation is a tedious one, but is necessary for making a good job of grinding the plug and seat to good conical form. Grinding by turning the plug clear around, as with a brace, is likely to cause scoring of the surfaces.

Foaming and Priming of Boiler—What is foaming and what is priming of a boiler, and what is the danger of each? J. K.

A boiler is said to foam if the steam space is partially filled with unbroken bubbles of steam, and to prime if the steam carries water with it from the boiler. Foaming is caused by any materials either dissolved or suspended in the water which retard or interfere with the free separation of steam from the water. More generally, the trouble is caused by oily or dirty water and can be overcome by feeding and blowing. Where there is a surface blow, it can be used to good advantage. Priming may be caused by foaming or by particles of water being projected into the steam space by the action of boiling or activity of circulation of the water. When due to circulation, it can be reduced and sometimes stopped by checking the fire and carrying lower water level in the boiler. The dangers of foaming and priming are that the boiler-water level may become dangerously lowered from sudden opening of the safety valve or from water carried along with steam supplied for any purpose, and when carried over into an engine, the water is likely to cause a smash.

[Correspondents sending us inquiries should sign their communications with full names and post office addresses. This is necessary to guarantee the good faith of the communications and for the inquiries to receive attention.—Editor.]

Steam-Turbine Economy Characteristics*

BY A. C. FLORY†

Characteristics common to the performance of modern commercial types of turbines. Willans total steam line, similar straight line relations of the total quantities of heat and the inlet pressure, and the constant-volume law are presented graphically to bring out the general principles governing turbine economy.

VARIATIONS in economy and performance of all types of turbines conform closely to certain simple rules or laws. The Willans Law, which states that the total steam passed per hour by a constant cutoff throttling engine when plotted as a function of the brake horsepower of the engine at constant speed will be a straight line, is fundamental. It has been found that this law, which was originally applied to the performance of constant cutoff engines regulated by throttling the steam supply, applies also to the performance of all types of steam turbines.

As shown in Fig. 1, the Willans total steam line is drawn from data of the estimated consumption of a 5,000-kw. turbo-alternator unit, the operating conditions being steam of 225-lb. gage pressure, 125 deg. F. of superheat; vacuum, 28.5 in.; speed, 1,800 r.p.m. In this diagram the Willans line extended indicates that the apparent total steam consumption at no load under the constant operating conditions set forth would be about 8,000 lb. per hour. If this no-load steam consumption is deducted from the total steam for various loads and the differences divided by the respective loads in kilowatt-hours, the result will be a constant amount of steam required for each kilowatt-hour produced, up to the bypass point, where the slope of the line changes. This is indicated on the diagram by a horizontal line marked "net steam per kw-hr., lb." If the no-load steam consumption per hour is divided by the load at any point, the quotient can be conventionally termed "friction steam per kilowatt-hour." The friction steam per kilowatt-hour added to the net steam per kilowatt-hour and plotted will produce the familiar curve showing the steam consumption per kilowatt-hour as a function of the load. Such a curve is shown in Fig. 1. This same result may be obtained more directly by dividing the total steam consumption by the load for different points on the Willans line. It can be seen that the Willans Law and its simple derivatives make a convenient and relatively simple method of showing graphically the performance of a machine and for checking the accuracy of a series of variable-load constant-speed tests made under approximately constant conditions.

It is also characteristic of turbine operation that the steam inlet pressure, after the steam has been reduced in pressure by the inlet regulating valve, when plotted as a function of the load, makes a straight line similar in form to the Willans total steam line. The inlet pressure—kilowatt line for a 5,000-kw. turbine having the same operating conditions as before—is shown in Fig. 2. This line forms a convenient check upon the accuracy of a series of tests, provided that the initial steam pressure, quality and vacuum are maintained approximately constant at all the loads at which the tests are made.

It has another important application in that it furnishes a convenient means of checking the internal condition of the turbine while it is in operation. If the inlet pressure-kilowatt line of a turbine is obtained when the unit is known to be in good condition and if, subsequently and while the machine is in operation, the inlet-pressure and kilowatt-load readings are plotted and compared with the original diagram, it is obvious that any change in the condition of the turbine will be revealed by a change in position or slope, or both, of the line. It is to be hoped that some such procedure will come into general use since its simplicity would not burden an operating engineer and because it has been found that in many plants there is a progressive accumulation of dirt or scale in turbine blading which ordinarily is not brought to the attention of the station operators until a large decrease in the load-carrying capacity of the unit results or until there is a serious accident to the machine.

Another characteristic common to steam machines which regulate by throttling the pressure of all the steam admitted to the turbine cylinder is known as the "constant-volume law," as it is based on the fact that, with such regulation, the cubic feet of steam per hour entering the turbine nozzle or inlet blading is approximately constant for all loads and initial steam conditions. This is illustrated by Fig. 3, in which are plotted data from tests on the same 5,000-kw. turbine showing the relation between the Willans total steam line, the specific-volume line and the constant-volume line. The specific-volume line represents the volume, in cubic feet per pound, of steam at the test inlet

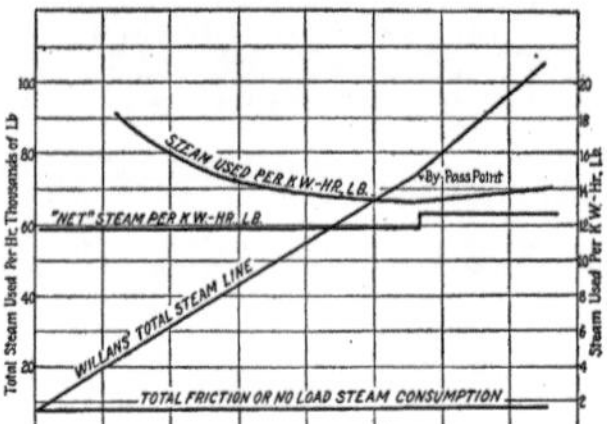

FIG. 1. WILLANS TOTAL STEAM LINE

pressure and quality (moisture or superheat). The constant-volume line is obtained from the products of the total steam per hour as taken from the Willans line and the calculated specific volume at the different loads as covered by the tests.

The constant-volume law has an important application in turbine work, as it furnishes a means for estimating the effect of changes in the initial steam conditions on the weight of steam that the turbine will pass per hour at constant speed. As an example, if it is known that a certain turbine will pass, without the bypass being open, 58,000 lb. of steam per hour with a pressure of 150 lb. gage and dry saturated steam, the total volume of steam passed per hour at the turbine inlet will be 58,000 lb. × 2.75 cu.ft. per lb. = 159,500 cu.ft. If the steam conditions are changed to 225 lb. gage and 125 deg. F. superheat, the weight of steam per hour which the machine will then pass without the bypass being open will be in inverse proportion to the ratio of the specific volumes of the steam at the two conditions, since the constant-volume law states that the total volume passed per hour at the blading inlet is constant. The specific volume of steam at 225 lb. gage and 125 deg. superheat being 2.335 cu.ft. per lb., the turbine would then pass (2.75 ÷ 2.335) × 58,000 = 68,300 lb. per hour, the bypass not being open. The example just given is conventional, as precisely the same machine would not be used for such a wide variation in initial steam conditions, but

*Abstract of paper read before the Technical League of the Employees' Mutual Benefit Association, Milwaukee.

†Manager, Steam Turbine Department, Allis-Chalmers Manufacturing Company.

the principle can be applied with accuracy in practice to smaller variations in steam conditions, and by combination with a knowledge of the effect of steam pressure and superheat on the economy of any particular turbine, forms a simple and convenient method of estimating the effect of changes in steam pressure and superheat on the maximum load that can be carried before the bypass opens.

With constant steam pressure, superheat and vacuum the energy available for work per pound of steam based on the Rankine Cycle is constant, and it is apparent in view of the Willans Law, that if the total B.t.u. available per hour are plotted as a function of the load there will result a straight line such as is shown in Fig. 4. This diagram is plotted from an estimate for a 5,000-kw. turbine. The characteristic form of the "total B.t.u. available per hour-kilowatt" line, forms an additional convenient check on the accuracy of any series of variable-load constant-speed tests below the bypass point and the "no load" or "friction" B.t.u. and the "net B.t.u. available per kilowatt-hour," which are derived from such plotted data and which are shown on the diagram, are of value in turbine-economy estimates.

The value of the net B.t.u. available per brake horsepower-hour (considering the turbine alone) has been found to be dependent only on the proportion of the blading of turbines which conform to a general design for a line of machines of various sizes (within certain limits) and for different operating conditions, and this value, therefore, is practically constant for different steam pressures, superheats, exhaust pressures and fractional loads. The values of the "no-load or friction B.t.u available," in the case of a similar line of turbines, are dependent on the sizes of the machines only (excepting large changes in the exhaust pressure), and for a certain size this value is, therefore, also constant for different operating conditions. With data from which a set of values for these two factors can be derived for a line of turbines, the estimating of the steam consumption of a machine at different loads and for specified operating conditions is a simple matter.

The regulation of a turbine by throttling introduces certain losses when the machine is operating at partial load, because of the effect of throttling on the B.t.u. available to the Rankine cycle. To illustrate the relative magnitude of this throttling-regulation loss at different loads for a 5,000-kw. 1,800 r.p.m. unit, the curve shown in Fig. 5 has been drawn. It is based on operating conditions of steam at 225 lb. gage, 125 deg. superheat and 28½ in. of vacuum. For convenience, the load is expressed in percentages of full load. The diagram shows the theoretical reduction in energy available to the turbine at the different fractional loads and is of particular interest because this loss is, in a measure, inherent in the necessary constant-speed regulation of all turbines.

The diagrams have all been based on the power output of the unit's generator, which introduces a variable factor due to the generator efficiency, being also a function of the load. If the brake horsepower of the turbine had been used as the basis for plotting the data, the application of the general rules given would have been more strictly correct, but for ordinary comparison and uses the convenience resulting from the use of the generator output as a load basis for plotting characteristic curves more than offsets the lack of accuracy.

In addition to those units of economy that have been plotted as a function of the load on the diagrams accompanying this article, the total heat consumption per hour and other measures of performance can be used if the object of the work being done makes it desirable. The simplicity of the Willans Law and the similar straight-line relation of the total quantities of heat and the inlet pressure afford an easy means of bringing to the attention of operating engineers the general principles governing turbine economy. In this connection it is well to point out that, in a multistage throttling turbine, the straight-line relation between the load and stage pressure is maintained at all stages as well as at the inlet or first stage. An understanding of such points would add to the efficiency of an operating engineer, as it would give him a clearer conception of the working of the machine he handles and enable him to better judge its internal condition by examining the operating records.

It will be noticed that certain of the diagrams have the bypass marked on them and that they indicate that the straight-line relation between the total quantities and the load is maintained for the condition resulting from admission of steam to an intermediate stage. This feature has not been dwelt upon, as it is evident that with simple modifications the general principles given apply also for that part of the power output in excess of the bypass load.

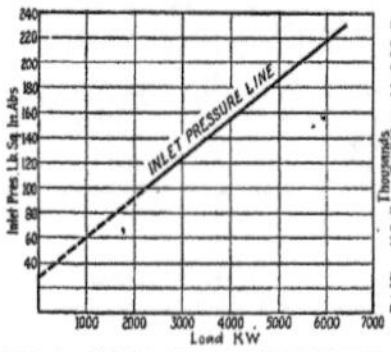

FIG. 2. INLET PRESSURE PLOTTED AS FUNCTION OF LOAD

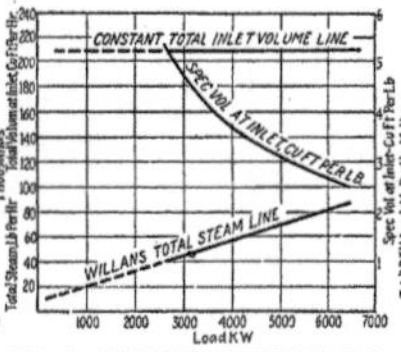

FIG. 3. RELATION BETWEEN WILLANS TOTAL STEAM LINE AND VOLUME LINES

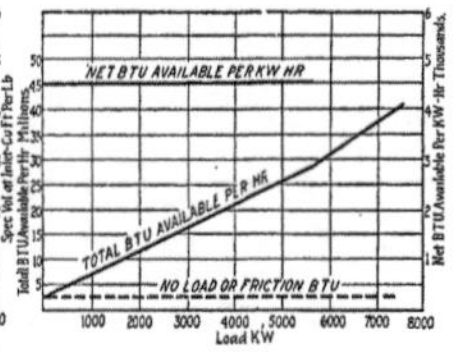

FIG. 4. TOTAL B.T.U. PLOTTED AS FUNCTION OF LOAD

In the discussion a number of points of interest came out. Relative to turbine design, selection of metals had to be made carefully with respect to the superheat in the steam. Even the finest gray cast iron grows if there is much superheat. The cause of this growth is not entirely the amount of superheat or the temperature that goes with it, but the temperature fluctuations encountered and the kind of iron used.

It was pointed out that the efficiency relative to the Rankine-cycle efficiency was obtained from the formula,

$$R = \frac{3{,}415}{\textit{B.t.u. available per lb. of steam} \times \textit{lb. of steam per kw.-hr.}}$$

3,415 being the old 2,547 B.t.u. (equivalent to 1 hp.hr.) divided by 0.746. This efficiency ranges from 50 per cent or less in a 300-kw. machine to almost 80 per cent in the large machines. In the large sizes the variation in efficiency is very slight from size to size as the curve becomes quite flat.

In reply to an inquiry as to whether the turbine and generator were set up and tested before shipment, it was stated that the unit is set up complete and overspeeded 10 per cent when running at no load. The overspeed safety stop is tripped three or four times. There is no advantage in running a turbine at load before being erected in its permanent location. The steam turbine itself, before connecting the generator, is overspeeded 25 per cent, producing 50 per cent excess stresses. It is important to get the generator and turbine set up together before shipment. Final balancing of these units is largely a "cut-and-try" proposition. In balancing, the relative positions of the two elements are not changed

In modern turbine design, the governor does not usually open the bypass until the machine is working well up to capacity. The bypass is provided so that upon a drop of steam pressure or vacuum the machine will hang on to its load. This is more important in industrial plants than in central-station practice.

The fact was brought out that the deflection of the shaft of a modern turbine due to its weight, is so small that difficulties as to clearance do not develop from this cause. In designing, a careful attempt is made to keep the normal speed reasonably far away from the critical speed. Shaft deflection may result in not permitting the bearings to be fully effective, and so modern practice tends to shorten bearing lengths.

Mr. Mortimer, president of the North America Co., explained how the Willans diagram of Mr. Flory's paper,

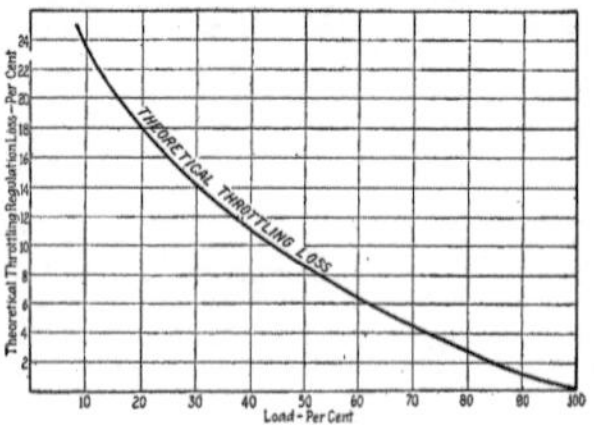

FIG. 5. THEORETICAL THROTTLING LOSS AT DIFFERENT LOADS

otherwise known as a "flow curve," had been used in a practical way in central-station practice by the Union Electric Light and Power Co. of St. Louis. In describing a method devised by him to more closely correlate efficient boiler capacity with the electric load, he emphasized the importance of the men in the turbine room knowing a half-hour in advance what load was to be carried and notifying the boiler room, so that the proper boiler capacity could be provided at the time required. In the Ashley Street Station, St. Louis, there are machines installed in 1908, 1909, 1911 and 1917. Only those of the same age and size would have the same steam economy. A chart was constructed on which were plotted "kilbers" against kilowatt load for each type of machine. A kilber is a unit expressing thousands of pounds of steam per hour. By referring to the load diagram of yesterday and to the corresponding day of last year, the load operator estimates the load that will be on the system one-half hour hence; decides what units shall be used to carry this load, refers to his diagram for the quantity of steam required; adds up the amounts for the total number of kilbers and signals to the boiler room a message "at 4:30, 15,000 kilbers will be required." The boiler-room foreman decides what boilers will be used to meet this requirement. It has been found that this assists materially in keeping the proper relation between boilers under steam and the load demand, and real economies have been thus effected. Correction is required if steam pressure or vacuum conditions do not remain constant.

The Board of Trade of Alberta, Can., has appointed a committee to investigate the possibilities of the waste coal from the mines in the generation of electricity in a large central plant and of proper distribution of this power to manufacturers and municipalities nearby. The fuel, which is very low in price, is being tested with a view to increasing its efficiency by pulverizing before it is utilized.

Reorganization Plan of the N. E. L. A.

Plans for the reorganization and expansion of the National Electric Light Association have been completed by M. H. Aylesworth, executive manager, and will be brought before the members at the convention in May for final approval.

A general survey by Mr. Aylesworth at the request of the National Executive Committee, upon the recommendation of President Ballard, resulted in the program of expansion in order to comply with demands for increased service in practical co-operation.

The work of the association, including appointment and direction of the headquarters staff, under supervision of president, vice presidents and National Executive Committee, will be performed by the executive manager. The organization at headquarters has been divided into five departments, under the direction of the executive manager and acting secretary. The rate-book and rate-research work, which formerly was carried on in Chicago, has been placed in charge of Fred W. Herbert, head of the Company Service Department. This department will dispense information and give to small and large companies alike the advantages of practical development demonstrations of the other company members.

A department of publicity under the direction of George F. Oxley has been formed, which will supervise the gathering and dissemination of all news, activities, etc., of the association.

Upon the recommendation of the Technical Section, approved by the National Executive Committee, an engineering department will immediately be formed. The duties of this department will include gathering statistics and formulating a program on the practical engineering problems that are affecting the industry at the present time.

A department dealing with committee activity has been formed and is under the direction of A. Jackson Marshall, committee secretary. The accounting department is in charge of the accounts of the association as well as the membership lists and the mailing of all material to members.

The reorganization plan calls for the division of the association into geographic divisions. The plan that has been proposed provides two methods of geographic division organization. Where strong state electrical associations or state public-utilities associations exist, the annual conventions and meetings may be held without regard to state boundaries, while other states will function through a geographic conference composed of delegates from the respective state associations. The purpose of geographic division organization is to decentralize where possible, the activities of the national association and harmonize the work of the state association, geographic division, and national association through co-ordination of all committee activities from the state association, through the geographic division and finally into the executive committees of the national association.

Little or no activity pertaining to the relation of the central station to the public has marked the work in the national association. A Public Relations National Section is to be formed, to overcome this deficiency.

Complete co-operation between all branches of the industry is the keynote of the reorganization plan, and every member company and its individual company will have a voice in the management of the association.

Exports of electrical machinery and appliances from the United States during February, 1920, show that batteries totaling in value $526,972 were shipped. Dynamos and generators valued at $363,340 and 3,140 fans valued at $79,780 were distributed throughout the world; incandescent lamps (carbon) valued at $9,746, and (metal) lamps with a total value of $368,391 were exported; magnetos, spark plugs, value, $251,876; motors, value, $381,203; telegraph apparatus, $16,750; transformers, $295,185; and miscellaneous electrical apparatus and appliances valued at $1,930,537 were exported.

Coal Production in Washington Decreased by the Miners' Strike

The coal produced in the State of Washington in 1919 amounted to 3,059,580 tons, which represents a decrease of over 25 per cent from the production of 1918. The major portion of this decrease is due to the national coal strike which closed the mines of the state from Nov. 1 to Dec. 15, according to the annual report of the state inspector of coal mines, James Bagley.

Mr. Bagley goes on to say that the increase in wages to miners in the last few years and the high cost of materials have prevented coal from displacing fuel oil to a great extent in spite of the constantly rising price of the latter. Coal from British Columbia has always been a factor in the Washington market, and Utah and Wyoming coal has recently been shipped in to a considerable extent. Also hydro-electric power has taken the place of coal-burning plants, especially on the railroads. The result of these factors is a gloomy outlook for the coal mines of Washington, as at present it seems probable that most of them will work only half the usual time during the coming summer months.

During 1919 Whatcom County alone showed an increase in production. This exception to the general rule was due to the operation of the Bellingham mine, which was under development most of the year.

The state's maximum production was reached in October, when 4,760 employees, working 26.3 days, produced 373,198 tons. This was due to the increased demand in anticipation of the big strike. The average value at the mine was $3.59 in 1919, as compared with $3.53 in 1918.

The coke output showed a decrease of about 55 per cent of the 1918 production. Of the 65,332 tons produced, almost 36,000 tons came from ovens at the mines at Carbonado, Fairfax and Wilkeson. The decrease is largely attributed to the lowered demand from the Tacoma smelter and to the fact that exports of coke to smelters in British Columbia have been steadily declining.

Sidelights on N. E. L. A. Convention

Visitors to the N. E. L. A. Convention at Pasadena, Cal., May 18-21, by special arrangements such as are not obtainable by the individual tourist, will be enabled to see every feature of southern California life within the shortest space of time and with the greatest possible convenience.

Arrangements have already been completed by W. L. Frost, chairman of the entertainment committee, for a trip to Santa Catalina Island, twenty-six miles from the mainland. The island is a famous objective for tourists and fisherman.

Not the least interesting event at the convention will be trips through various motion-picture studios, about sixty of which are located within easy riding distance of headquarters. The delegates will be given an opportunity of seeing "thrillers" in the making.

Excursions have been planned to all the famous beach resorts which extend for miles along the Pacific Coast.

Trips by trolley and automobile will be made through what is known as the "Orange Empire." The trees will be loaded with fruit at the time of the convention.

Mount Lowe, which stands a mile high on the summit of the Sierra Madre range is another trip easily taken from the Hotel Huntington. A cable railroad and a wonderful piece of electrical trolley construction are the means of reaching the summit.

John B. Miller, president of the Southern California Edison Co., has opened his beautiful home, "Hillside," and the wonderful grounds surrounding it. Delegates and their guests will be entertained at tea here Thursday afternoon, May 20 at 4 o'clock. The gardens of Hillside are said to be among the most beautiful in southern California.

Engineers and technical men who desire to visit the great hydro-electric properties of the power companies of the Southwest, will be provided with automobiles and guides for excursions into the mountains.

Convention of American Association of Engineers

The sixth annual convention of the American Association of Engineers will be held at the Planters Hotel in St. Louis, May 10 and 11. A number of interesting speakers have been secured, including M. O. Leighton, chairman of the National Service Committee of Engineering Council, who will deliver an address on "Getting Salaries for Public Service Engineers into Budgets"; Francis A. Kellor, vice chairman of the Inter-Racial Council of New York, will speak on "The Manpower Engineer—The Opportunity of Americanism' and Fred' Lavis of the American International Corporation of New York, who will speak on "Opportunities for the Engineer."

Group meetings for draftsmen, federal engineers, mining engineers and railroad engineers will be held on the evening of the tenth, and problems peculiar to each group will be considered and a report made at the general business session on the following day.

A number of constitutional amendments will be brought before the members for approval, the changes including the passage of constitutional amendments by letter ballot and several amendments covering changes in entrance fees and annual dues.

A schedule for the regulation of salaries of engineers in municipal service has been prepared by a committee of the association, and will be placed before the members of the board of directors at their next meeting in St. Louis on May 9.

Washington State Coal Commission Holds Preliminary Meeting

N. D. Moore and D. F. Buckingham, the latter general manager of the Roslyn Fuel Co., representing the mine owners, and Robert S. Harlin, president, and Ernest Newsham, secretary, of District No. 10, United Mine Workers of America, representing the miners, are the first four members of the commission to adjust the wages and working conditions of coal miners in the State of Washington. These four are to appoint a fifth member. A preliminary meeting of the four commissioners was recently held in the offices of the Washington Coal Operators' Association in Seattle.

This commission was provided for in the Government plan developed after last winter's strike, the Washington State situation being one of those not covered in the general wage decision.

Piston Bursts—Kills Man

The bursting of a piston, while being heated, which had been taken from a locomotive of the Donora Southern R.R. to the blacksmith shop for repairs, instantly killed the blacksmith.

The accident was most peculiar. The piston had been taken out of a locomotive and required some repairs before it could be replaced. In order to take the rod out of the piston, it had to be heated, so it was taken to the blacksmith shop and put in the forge. In a short time the pressure in the hollow piston became so great that a large piece was driven out of the head, striking the blacksmith, who was standing several feet distant, and breaking his neck, almost instantly killing him.

Figures contained in the eleventh (1919) annual report of the Public Service Corporation of New Jersey show that the Public Service Electric Co. was called upon to supply power, which exceeded in volume the output of energy in 1918. It is estimated that during 1918 war industries consumed about 90 per cent of the commercial power load. When the war ceased this power was absorbed by other industrial activities. The day load of the 60-cycle service in the northern zone for 1919 ran as high as 126,000 kw., compared with a maximum of 109,000 kw. in 1918. In the southern zone the day load reached 24,000 kw., as against 22,000 kw. the year before.

Pennsylvania Electric Association Discusses Power Factor

How vitally important power factor has become in the operation of alternating-current power systems was very clearly reflected in the second meeting of the year held by the Western Geographic Section of the Pennsylvania Electric Association, in which approximately 200 members and guests participated. The meeting was held in the William Penn Hotel, Pittsburgh, April 21. Chairman E. C. Stone presided at both sessions, one being held in the morning and the other in the afternoon, with a luncheon served in the hotel at noon. In opening the morning session, Chairman Stone gave a general outline of the power-factor problem, in which he said in part:

Every reduction in the cost of producing and delivering electrical energy to the job opens up new fields for its utilization, thereby furthering the prosperity and development of the community through greater use of power. The proper solution of the power-factor problem means a further reduction in the cost of electrical energy. It is, therefore, a most vital problem. Consider the Duquesne Light Co.'s system for example. If the power factor of all connected equipment utilizing the company's service could be raised to unity, the entire electrical system, including generators, lines, substations and transformers, could be made to carry 35,000 kw. more load than it carries at the present time, and the only additional investment required to accomplish this would be for an increase in the capacity of turbines and boilers at the power plant to bring them up to the greater capacity of the electrical system.

The production and transmission of energy through an electrical circuit involves two functions: First, the delivery of the energy for utilization in lamps, motors, etc.; second, the supplying of the necessary excitation for maintaining the magnetic fields essential to the operation of transformers and motors, which constitute a large part of the utilizing equipment. The energy must obviously be supplied from the power plant. The excitation, or wattless current, however, may often be supplied more economically from a point nearer to the load, thereby raising the power factor and increasing the capacity of generators, lines and transformers. The engineer's problem thus becomes: First, the design of utilizing equipment that will operate at high power factor, as for example, the synchronous motor; second, where this cannot be done, the providing of power-factor corrective apparatus, such as the condenser, which will compensate for the low power factor of the utilizing equipment; third, the joint application of high power-factor equipment and power-factor corrective apparatus in such manner and at such points on the transmission and distribution system as will combine to make the most economical system possible, consistent with satisfactory operating conditions for the consumer.

Rates Should Encourage High Power Factor

The commercial problem of power factor involves the development of a rate schedule that will bring about the working out of the engineer's problem so that the community as a whole will obtain its supply of power at the minimum cost. This necessitates rate provisions which, first, will make it profitable to the power customer to utilize his power at the most economical power factor, and secondly, will differentiate between the power customer who corrects his power factor and the customer who does not, so that the former will not be obliged to carry any of the burden imposed by the latter.

One of the principal hindrances to the development of power-factor schedules has been the difficulty in measuring power factors in a manner that was sufficiently simple, economical and reliable to be used as a basis for billing. At the present time a joint committee of the A. I. E. E. and the N. E. L. A. is studying this problem with a view to defining power factor accurately so that manufacturers will be given a basis on which to design suitable metering equipment. In the meantime a number of prominent companies are making use of the method which employs a standard polyphase watt-hour meter specially connected to register the kilovolt-ampere-hours of wattless component, instead of the usual kilowatts of true energy. At the present moment it seems quite probable that this method will come into such extensive use.

Two papers were presented at the morning session: "Power Factor and Kilovolt-Ampere Measurements," by W. H. Pratt; and "Analysis of Power-Factor Conditions on a Large Central-Station System," by C. L. Rupert. Mr. Pratt outlined the various conditions under which power factor and kilovolt-amperes must be measured and described the methods of making these measurements. The speaker expressed the opinion that, in general, the different methods of measurement are satisfactory. He said that the most serious part of the problem is due to wave-form distortion. He described a new type of meter containing two elements connected together through a pawl-and-ratchet arrangement, that would measure kilovolt-amperes with only one per cent error, over a phase-displacement range of 44 deg., it being possible to select the 44-deg. range anywhere between 1 and 90 deg. displacement. Abstracts from both Mr. Pratt's and Mr. Rupert's papers will appear in later issues.

During the luncheon H. W. Diggs, chairman of the Four-Minute Speaker War Organization, and A. M. Lynn, president of the West Penn Power Co., delivered addresses. Mr. Diggs urged greater co-operation between the public-service utilities and the public. Co-operation is abroad in the land, said the speaker, and wherever it has been tried the outcome has been satisfactory. The man who said, "The public be damned," has gone out of business. This one statement has done more harm to public utilities than anything else. Public opinion is a powerful thing, the speaker said, and emphasized this point by calling attention to how public opinion had broken the steel-mill workers' strike, coal-miners' strike and was also breaking the railway strike.

Demand for Power Exceeds Supply

In discussing the various problems relative to central-station operation, A. M. Lynn, president of the West Penn Power Co., pointed out that the power companies today could sell more power than they could finance the necessary extensions for, consequently, one of the most difficult problems in the operation of utilities had become that of finance. Coal, said the speaker, that cost $2.95 per ton at the power plant during the war will probably cost at least $4.15 per ton when the coal and railway situation is adjusted. Mr. Lynn illustrated the great economies that had resulted from the interconnection of the various power systems by calling attention to a case where 5,000 hp. had been supplied from a company operating in central Pennsylvania to another company in central Virginia, but to do this it was necessary to transmit this power through three different systems.

At the afternoon technical session two papers were presented: "Induction and Synchronous Motors for Industrial Applications," by C. W. Drake; and "Power-Factor Correction," by Robert Treat. Mr. Drake gave a very careful analysis of the application of both types of alternating-current motors. He said that overloaded systems were the cause of increased interest in power-factor corrections, and laid stress on selecting the correct size of induction motors, pointing out that it was ordinarily possible to increase the power factor 5 to 10 per cent, and in exceptional cases 20 per cent, over what is obtained in general practice, simply by the proper selection of induction motors.

The relative costs of synchronous and induction motors were taken up by Mr. Treat in his paper and shown on a set of curves. An abstract of both papers will appear in a later issue.

During the discussion considerable difference of opinion was expressed regarding penalizing customers for low power factor. However, increasing the demand charge in proportion to the power factor when the customer failed to maintain the power factor above a given value, or else meter both the kilowatt-hours and the kilovolt-ampere-hours, billing the customer for both, were the two methods that apparently met with most favor. In addition to the demand charge, to compensate for reduced use of power equipment, some were of the opinion that something should be included in the rate to compensate for the increased losses.

It was felt by many that the burden of correcting for low power factor should be up to the consumer, since in most cases the trouble was due to underloaded motors, consequently the customer had within his control the means of maintaining a good power factor. To illustrate the effect of the small customer's load on the power factor of a system, attention was called to one case where, if the power factor of all the large customers' loads was increased to 87 per cent, it would improve the power factor of the system as a whole only about 3 per cent, or raise it from 70 to approximately 73 per cent.

In discussing the different types of corrective devices, it was acceded that in the case of poor voltage regulation the synchronous machine would tend to improve conditions where static condensers had the effect of making them worse. One case was mentioned where on the 2,200-volt side of the transformers, at the end of a long transmission line, the voltage dropped as low as 1,400 volts, and by a proper use of the synchronous machines, already on the system it was possible to maintain around 2,000 volts.

From a survey made of 23 companies only six of them did not penalize their customers for low power factor. A total of 15 penalized on a demand-charge basis, 6 penalized for power factors less than 80 per cent, and 5 for power factors less than 85 per cent. Only three of the companies discounted for good power factors.

In discussing the question of using rotary converters for power-factor correction, it was brought out that the load on the machine should not exceed the arithmetical sum of the active and reactive loads. That is, if the machine is operating under 80 per cent normal load it will be able to supply only 20 per cent reactive leading kilovolt-amperes to the alternating-current system, instead of 60 per cent for a synchronous motor.

N. E. L. A. Convention Program

The program of the National Electric Light Association for the forty-third annual convention to be held at Pasadena, Cal., May 18-22, has been completed and everything is now in readiness for the event. The headquarters of the convention is the Hotel Huntington, Pasadena, one of the largest and most modern hotels in southern California. It is located in the beautiful San Gabriel Valley, with the Sierra Madre Range and Mt. Wilson in the background.

The convention activities will be formally opened by Governor William D. Stephens of California, who will deliver the address of welcome to the members of the association. President Ballard will respond to the welcome on behalf of the organization. H. Aylesworth, executive manager, will next address the assembly with an outline of the comprehensive plan of reorganization which will be placed before the delegates for final approval.

General sessions will be held during the mornings and special sessions in the afternoon.

Franklin T. Griffith, of the Portland Light and Power Co., will read the report of the committee on Water Power Development. Following this, A. E. Silver, of the Electric Bond and Share Co., chairman of the Inductive Interference Committee, will read a report covering the work of that body. Frank M. Kerr, of the Mountain Power Co., of Butte, Mont., will present a paper on "Electrification of Steam Railroads." Other reports and speakers will be: "Electrical Resources of the Nation," by M. S. Sloane, of the Brooklyn Edison Co.; George B. Foster, of the Commonwealth Edison Co. of Chicago, chairman of the Electric Vehicle Section of the association, will deliver an address on "Development, Manufacture and Marketing of the Electrical Vehicle"; Zena W. Carter, secretary of the Material Handling Machinery Manufacturers' Association, of New York, will read a paper entitled "Electric Machinery for All Handling"; and the Committee on Public Information, of which John F. Gilchrist, of the Commonwealth Edison Co., of Chicago, is president, will give figures and facts showing the value of proper publicity of association activities.

During the first afternoon session of the convention on May 19, Nathaniel A. Carle, chief engineer of the Public Service Electric Co., of Newark, N. J., will read the report of the Prime Movers Committee, of which he is chairman, and during the second afternoon session on May 20, R. F. Schuchardt, of the Commonwealth Edison Company, of Chicago, will present the report of the Committee on Electrical Apparatus.

On Friday, May 21, at the afternoon session, James H. McGraw, president of the McGraw-Hill Co., Inc., will read a report of the Committee on Federal and Municipal Transportation.

The social and pleasure end of the convention has not been forgotten. Extensive plans for the entertainment of delegates have been made by the entertainment committee.

HOTEL HUNTINGTON, PASADENA, N. E. L. A. CONVENTION HEADQUARTERS

Included in this part of the program are numerous excursions to places of interest and beauty by automobile, trolley, boat and train. Everything possible has been done so that the visitors can make the most of their opportunities of viewing life in southern California.

Governing Boards of Founder Societies Meet in Chicago

Upon the invitation of the board of direction of the Western Society of Engineers the governing boards of the four Founder Societies met contemporaneously in Chicago on April 19 and 20. The Western Society has a plan to unite all engineers in Chicago into one all-embracing society for combined action in technical advancement, public affairs and the upbuilding of the profession. The primary reason back of the invitation was to give an opportunity for a discussion and a clear understanding of the inter-relations of the Western Society and the Chicago chapters of the Founder Societies, as well as the general plans for service. The opportunity for an interchange of ideas between the various representative bodies came at a luncheon on Tuesday. Business as usual and the problems confronting each society were taken up at individual meetings of the respective councils throughout the two days.

On Monday evening there was a joint engineering societies dinner at which the engineers of the community had an opportunity to meet and welcome to the city the visiting bodies from the four great national societies. A. Stuart Baldwin, past president of the Western Society of Engineers, was toastmaster and in his opening remarks referred to the four founder societies as the foundation on which all other engineering associations had been formed and pointed out the early tendency toward separation and division, each new society specializing in its own particular field. Recently, the pendulum had swung in the opposite direction. Engineers of the present day were drawing together. They realized that in union there is strength and that the maximum of service can result only through unison of effort and co-operation.

The toastmaster called in turn upon Arthur P. Davis, president of the American Society of Civil Engineers; Arthur Fletcher, representing the American Institute of Mining and Metallurgical Engineers; Fred J. Miller, president of the American Society of Mechanical Engineers, and

Calvert Townley, president of the American Institute of Electrical Engineers. Each in his own way spoke of the change that had come over the minds of men in the last five years, affecting their viewpoint and their feelings toward one another. To keep abreast of the times, it was recognized that all societies must broaden their fields and functions. Man can attain age by mere existence, but engineering societies must do something to keep alive. They should always stand ready to adopt new ideas.

It was conceded that there was need for more teamwork among engineers. To do effective work for the city, the state and the nation they should combine. There should be publicity to enlighten the public about engineers and their work so that confidence might be placed in their judgment. In rendering service to the public it was the thought that views expressed should be on the technical side of the question. In other words, the enginer should in a sense be an expert witness, not a politician. In co-operative public service they should not adopt the selfish aim of advancing themselves and the profession. Rather they should serve the public, do something for them, and the public in turn would reciprocate.

By joint activity there was expectation of dealing more effectively than ever before with public matters. There were one hundred thousand engineers in the country separated into several hundred organizations, each doing something in its own way. If they could be made to understand that no attempt was being made to interfere with their work and that they were requested to participate only in questions of public interest, they should be willing to coöperate and it was the consensus of opinion that they would.

As the closing event in a most enjoyable evening, enlivened by community singing and an orchestra, Dr. Theodore G. Soares, from the faculty of the University of Chicago, delivered an address on "Efficient Democracy." He impressed the audience with the need for the spiritual and the ideal as well as the practical. In the great undertakings of the day, there must be coöperation and more able direction than ever before. The problem of the day was not to find big things to do, but to find enough big men to do the work waiting for them. There was need for expert advice in questions of finance, education, engineering and for a more intelligent democracy to choose its experts. The public must find the men who are best capable to lead.

Personals

George L. Myers, who has been connected with the Pacific Power and Light Co., Portland, Ore., for several years, has been made assistant to the president.

Brent Wiley, formerly manager of the steel mill section, Westinghouse Electric and Manufacturing Co., has been appointed assistant to manager, industrial department. Mr. Wiley will assume charge of mill industries.

George M. Ogle has been appointed to take charge of the consulting engineering department of the Vulcan Iron Works, Inc., Jersey City, N. J. Mr. Ogle was formerly chief electrical engineer of the United States Shipping Board Emergency Fleet Corp.

Otto F. Stroman, formerly manager of the price section of the Westinghouse Electric and Manufacturing Co., Pittsburgh, Pa., has been appointed assistant to the manager in the industrial department. Mr. Stroman's duties will include development of new apparatus, commercial activities, etc.

C. H. Kennedy, M. E., formerly New York district manager of the Kennedy Valve Manufacturing Co., has been appointed general sales manager, with headquarters at the plant, Elmira, N. Y. J. S. Hanlon, of the Boston office, will assume charge of the New York office, and G. W. Waters, formerly of the sales department of the New York office, will be the Boston representative.

Society Affairs

The American Society of Mechanical Engineers will hold its annual dinner at the Engineers' Club on May 25.

The American Society for Testing Materials will hold its annual meeting at Asbury Park, New Jersey, June 22-25.

The New York Section of the A. S. M. E. will meet at the Engineering Societies Building on May 11. "Symposium on Cost Accounting" will be the topic of the meeting.

The National Electric Light Association Executive Committee at a recent meeting, voted to change the name of the Technical and Hydro-Electric Section to the Technical Section.

A State Chapter of the A. S. E. is being considered by members of the profession in North Dakota. A meeting was held on April 23 for the formation of tentative plans of organization.

The American Society of Mechanical Engineers, New Haven Branch, Connecticut Section, will meet on May 12. The subject of the evening will be "Airplanes," by a speaker from the Ballaudet Corp.

The Society for the Promotion of Engineering Education, Pittsburgh, Pa., will hold its next meeting at the University of Michigan, Ann Arbor, June 29 to July 2. The keynote of the meeting will be coöperation.

The Canadian Association of Stationary Engineers will hold its thirty-first annual convention at Woodstock, Ontario, June 22-24. The annual exhibition of power equipment and supplies will be held in connection with the convention

The Michigan Central Railroad Section of the American Association of Engineers recently adopted a resolution offering the services of each of its members to the management of the railroad when strikes made it necessary.

The National Executive Committee of the N. E. L. A. has given final approval to proposed changes in the constitution of the association. The changes provide for reorganization at headquarters and changes in the form of organization.

The American Society of Civil Engineers has voted against the proposal to co-operate actively with other engineering societies in the formation of a broader organization to secure united action of the engineering and allied technical professions in matters of common interest.

The American Water Works Association will hold its fortieth annual convention at Montreal, Canada, June 21-24. The Windsor Hotel has been selected as headquarters. Afternoon and evening sessions will be held and a number of papers presented, covering water-supply problems.

An Illustrated Lecture will be given in the banquet hall of the Engineers' Club, 32 West 40th Street, Wednesday evening, May 5 at 8 o'clock, by H. F. J. Porter, M.E., Topic: "Mr. Cornelius Henry Delamater—The Businessman; Captain John Delamater—The Engineer."

The Taylor Society, the national organization for the promotion of scientific management in industry, will hold a meeting at Rochester, N. Y., May 6-8 under the auspices of the Industrial Management Council and the Manufacturers' Council of the Rochester Chamber of Commerce. The papers to be presented at the meeting will cover every phase of production, management and industrial relations, including sales promotion and office operations.

The North Central Geographic Division of the N. E. L. A. was organized April 13, at a conference of electrical men of Minnesota, North Dakota and South Dakota, in the Hotel Radisson, Minneapolis, Minn. A resolution was adopted, consolidating the South Dakota Electrical Association with the Minnesota State Electrical Association into the North Central Geographic Division, and officers were elected to serve until the first annual convention, which will be held at the Hotel Radisson, Minneapolis, June 16 and 17.

The American Institute of Electrical Engineers will hold its thirty-sixth annual convention at the Greenbrier Hotel, Sulphur Springs, W. Va., June 29 to July 2. The tentative program follows: Electrical Machinery Committee, Symposium on temperature conditions prevailing inside large alternating-current generators; Protective Devices Committee; Special Committee on Determination of Power Factor, preliminary report and papers; Electrical Machinery Committee, symposium on excitation. Miscellaneous Session: "Alternating-Current Commutator Motors," by J. I. Hull; "High-Tension Insulator Porcelain," by W. D. Peaslee, "Wattless Power and Magnetic Energy," by J. Stephian; "The Corona Voltmeter—II," by J. B. Whitehead and Mr. Ishiku.

Business Items

The Combustion Engineering Corp. has leased the property at 43-45-47 Broad St., New York City, and will soon begin operations on the construction of a modern ten-story structure, which it will occupy as soon as completed. The property was leased for $1,500,000.

The International General Electric Co. has secured a contract for the electrification of the Paulista Railway Co.'s lines betwen Jundiahy and Compinas, Brazil, a distance of 76 miles. The equipment to be supplied includes twelve locomotives, material for the transmission line and substation and a 2,000-volt overhead of the twin-catenary construction. The contract amounts to nearly $2,000,000.

Trade Catalogs

The Ric-Wil Co., manufacturers of heat insulating products, Cleveland, Ohio, has completed the publication of a new 16-page catalog describing in detail materials for pipes distributing hot liquids and vapors such as circulating of oil through underground pipes enclosed in heated conduit and distribution of low-pressure steam. A copy will be sent on request.

The American Steam Conveyor Corp. of New York and Chicago has ready for distribution a 24-page booklet entitled "The American Trolley Carrier." The booklet describes a one-man method of handling fuel by means of the apparatus manufactured by this company. A copy will be sent on request.

The De Laval Steam Turbine Co., Trenton, N. J., has ready for distribution a new 44-page catalog entitled "Centrifugal Pumps." The booklet describes and contains a number of illustrations of De Laval single- and multiple-stage centrifugal pumps and explains the use of pump-characteristic curves in adapting pumps to various services, the adaptation of centrifugal pumps to different types of drives, the information desired by the manufacturers in order to design a pump to suit given conditions, etc. Formulas and tables are given for calculating horsepower, efficiencies, the reading of venturi meters, friction in pipe lines etc., A copy of the catalog will be sent on request.

COAL PRICES

Prices of steam coals both anthracite and bituminous, f.o.b. mines, unless otherwise stated, are as follows:

ALANTIC SEABOARD

Anthracite—Coals supplying New York, Philadelphia and Boston:

	Mine
Pea	5.30
Buckwheat	$3.40@$3.75
Rice	2.75@ 3.25
Barley	2.25@ 2.50
Boiler	2.50

Bituminous—Steam sizes supplying New York, Philadelphia and Boston:

Latrobe	$4.25@$4.50
Connellsville coal	4.25@ 4.50
Cambrias and Somersets	4.35@ 5.15
Clearfields	4.00@ 4.75
Pocahontas	6.50@ 7.00
New River	6.50@ 7.00

BUFFALO

Bituminous—Prices f.o.b. Buffalo:

Pittsburgh slack	$6.00
Mine-run	6.25
Lump	6.25@ 6.50
Youghiogheny	6.75

CLEVELAND

Bituminous—Prices f.o.b. mines:

No. 6 slack	$3.25@$4.00
No. 8 slack	4.25
No. 8—¾-in.	3.50@ 4.00
No. 6 mine-run	3.50@ 4.00
No. 8 mine-run	4.50
Pocahontas—Mine-run	3.25@ 4.00

ST. LOUIS

Anthracite—Probably not more than 20 per cent of the demand in this market can be supplied. Prices effective Apr. 1 were as follows:

	Williamson and Franklin Counties	Mt. Olive and Staunton	Standard
Mine-run	$2.65@2.80	$3.00@3.50	$3.00@3.50
Screenings	2.50@2.65	2.50@3.00	2.50@3.00
Lump	....	3.75@4.50	3.75@4.50

Williamson-Franklin rate to St. Louis is $1.10; Other rates 95c.

CHICAGO

Bituminous—Prices f.o.b. mines:

Illinois		Freight rate Chicago
Southern Illinois		
Franklin, Saline and Williamson Counties		
Mine-Run	$3.00@$3.10	$1.55
Screenings	2.60@ 2.75	1.55
Central Illinois		
Springfield District		
Mine-Run	$2.75@$3.00	$1.32
Screenings	2.50@ 2.60	1.32
Northern Illinois		
Mine-Run	$3.50@$3.75	$1.24
Screenings	3.00@ 3.25	1.24
Indiana		
Clinton and Linton		
Fourth Vein		
Mine-Run	$2.75@$2.90	$1.27
Screenings	2.50@ 2.65	1.27
Knox County Field		
Fifth Vein		
Mine-Run	$2.75@$2.90	$1.37
Screenings	2.50@ 2.60	1.37
Brazil Block	$4.25@$4.50	$1.27

New Construction

PROPOSED WORK

Mass., Boston—Fay, Spefford & Thorndike, Archts., 15 Beacon St., will soon award the contract for the construction of a 9 story, 130 x 193 ft. garage on Kneeland St. Power for heat and elevator will be installed in same. Total estimated cost, $750,000. Owner's name withheld.

Mass., Mattapan (Boston P. O.)—The Commonwealth of Massachusetts, 34 State House, Boston, will soon award the contract for the construction of a nurses' building, dormitory and dining room, etc., at the Boston State Hospital, here. A steam-heating system will be installed in same. Total estimated cost, $285,000. Kendall, Taylor & Co., 93 Federal St., Boston, Archts.

N. Y., Brooklyn—The Syndicate, 215 Montague St., plans to build a 150 x 154 ft. theatre on Grand and Keap Sts. A steam heating system will be installed in same. Total estimated cost, $500,000. Address W. A. Small.

N. Y., Buffalo—The Hotel Statler Co., Washington St., is having plans prepared for the construction of a 17 story hotel and theatre building on Franklin and Mohawk Sts., Delaware Ave. and Niagara Sq. A steam heating system will be installed in same. Total estimated cost, $5,500,000. George B. Post, 101 Park Ave., New York City, Archt.

N. Y., Fillmore—The Fillmore Electric Co. plans to construct a water power plant, here. The plant will develop 600 hp. at low water. Water wheels, dynamos and other power house equipment will be installed in same. W. L. Young, Pres.

N. Y., New York (Bronx Borough)—The Amer. Legion of the Bronx Council c/o Starrett & Van Vleck, Archts., 8 West 40th St., is having plans prepared for the construction of a 4 story, 100 x 200 ft. club house. A steam heating system will be intalled in same. Total estimated cost, $400,000.

N. Y., New York (Richmond Boro)—The Bd. Educ., 500 Park Ave., New York City received bids for the installation of heating and ventilating apparatus at P. S. 11, from Gillis & Geoghegan, 537 West Bway., $58,700; W. J. Olvany, 177 Christopher St., $59,962; D. J. Rice, 42nd St. and Lexington Ave., $62,690. All of New York City.

N. Y., New York—The Knickerbocker Hospital, 131st St. and Amsterdam Ave., is having plans prepared for the construction of an 8 story, 141 x 201 ft. hospital on Convent Ave. and 130th St. A steam heating system will be installed in same. Total estimated cost, $1,250,000. J. Oakman, 345 5th Ave., Archt. and Engr.

N. Y., New York—The State Bank, 363 Stone Ave., Brooklyn, is having plans prepared for the construction of an 8 story 25 x 100 ft. bank and office building at 349 West 34th St. A steam heating system will be installed in same. H. R. Mainzer, 105 West 40th St., Archt. and Engr.

N. Y., Ossining—Charles F. Rattigan, Supt. of Prisons, Capitol, Albany, will receive bids until May 18 for furnishing and installing pumping apparatus, including two 275 gal. per minute centrifugal pumps, two 25 hp. motors, pipe connections in pump house and reservoir, etc. at the Sing Sing Prison, here.

N. Y., Utica—The Bd. Educ received bids for the installation of a heating system in the proposed school building in the 16th Ward to be known as the Horatio Seymour School, from the Mohawk Valley Heating Co., 302 Broad St., $43,982 to $63,510; W. T. Cantwell, 306 Bleecher St., $44,262 to $49,988; H. J. Brandeles, 435 Lafayette St., $50,909 to $75,255.

N. J., Hackensack—The Hackensack Hospital, 3rd St., is having plans prepared for the construction of a hospital building. A steam heating system will be installed in same. Total estimated cost, $750,000. Crow, Lewis & Wick, 200 5th Ave., New York City, Archts. and Engrs.

N. J., Jersey City—J. B. Hamilton, 173 Jewett Ave., is in the market for a 500 cu.ft. belt driven air compressor and a 150 hp., 440 volt, 3 phase, 60 cycle slip ring a.c. motor.

N. J., Kearny (Arlington P. O.)—The Bd. Educ. plans to build a 3 story high school. A steam heating system will be installed in same. Total estimated cost, $400,000. Guilbert & Betelle, 665 Broad St., Newark, Archts.

N. J., South River—The Bd. of Pub. Wks. and the Borough Council will receive bids until May 10 for the design, construction, delivery and erection of building, machinery and equipment for an addition to the Electric Generating Station. E. B. Hedden, Supt.

Pa., Wilkes-Barre—St. Mary's Catholic Church plans to build a 3 story, 80 x 250 ft. convent on South Washington St. A steam heating system will be installed in same. Total estimated cost, $500,000.

Md., Baltimore—The Johns Hopkins University, Homewood St., is having plans prepared for the construction of a 3 story college addition. A steam heating system will be installed in same. Total estimated cost, $300,000. Carrere & Hastings, 52 Vanderbilt Ave., New York City, Archts. and Engrs.

D. C., Washington—The U. S. Capitol Buildings and Grounds Office, received bids for installing a refrigeration system in the second wing of the U. S. Capitol, (1a) work complete, (1b) complete except refrigeration system, (1c) refrigeration system (2) work complete according to bidder's specification, (2b) same as 1b, according to bidder's specification, (2c) same as 1c, according to bidder's specification, from W. G. Cornell Co., 923 12th St., N. W., (1a) $236,730, (1b) $62,630, (1c) $174,900; York Mfg. Co., York, Pa., (1c) $273,725, (2c) $215,645; Carrier Eng. Corp., 39 Cortlandt St., New York City, (2a) $340,900 alternate $317,300 and $272,500, (2b) deduct $235,100.

N. C., Charlotte—Albert Kahn, Archt., Marquette Bldg., Detroit, will soon award the contract for the construction of a 6 story, 90 x 198 ft. service station on West Trade and Poplar Sts., for C. C. Coddington. Steam boilers, etc., will be installed in same. Total estimated cost, $300,000.

N. C., Kingston—Joe Dawson, Mayor, and the City Council, City Hall, will receive bids until May 11 for furnishing power plant equipment including one 1,500 kw., 1,875 kva. at 80% P.E. turbo generator unit, one surface condenser complete with dry vacuum and hot well pumps, one 25 kw. turbine driven exciter unit, two 25 kw., 6.6 amp. series street lighting transformers, two horizontal motor driven centrifugal circulating pumps, two 500 hp. 225 lb. horizontal water tube boilers, one 1,500 hp. open feed water heater, 2 superheaters, 2 underfeed mechanical stoker units complete, boiler feed pumps, 2 brick settings for boilers, etc.

S. C., Charleston—J. A. Wetmore, Supervising Archt., Treasury Dept., Wash., D. C., received bids for the construction of a lighting plant from John R Proctor, Inc., 74 Cortlandt St., New York City, $23,988; Carleton Mace Eng. Co., 18 Tremont St., Boston, Mass., $35,056; G. E. Eng. Co., 22 Laight St., New York City, $46,340.

La., Monroe—The Monroe Light & Power Co. plans to rebuild plant recently destroyed by fire and will be in the market for new machinery.

O., Columbus—The city will soon award the contract for the construction of a pumping station for the Shepard main trunk sewer, and installation of equipment. Total estimated cost, $30,000.

O., Ironton—The Bd. Educ. is having plans prepared for the construction of a 2 story, 100 x 110 ft. high school. A steam heating system will be installed in same. Total estimated cost, $500,000. F. L. Packard, New Hayden Bldg., Columbus, Archt.

O., Ravenna—The McElrath Tire & Rubber Co., 1836 Euclid Ave., Cleveland, is having plans prepared for the construction of a 1 and 2 story, 100 x 300 ft. factory and a 30 x 50 ft. boiler house. One 150 hp. boiler will be installed in same. Total estimated cost, $150,000. P. J. McElrath, Pres. D. C. Smith, Lennox Bldg., Cleveland, Archt. H. M. Bixler, Akron, Engr.

Mich., Flint—Smith, Hinchman & Grylls, Archts. and Engrs., 710 Washington Arcade, will soon award the contract for the construction of a 3 story, 73 x 152 ft. sales building on East 5th St., for the Dort Sales Co. A steam heating system will be installed in same. Total estimated cost, $100,000.

Mich., Hillsdale—The Bd. of Pub. Wks., City Hall, will receive bids until May 21 for furnishing power plant equipment including 1 approximately 1,000 kw. steam turbo generator with condenser, one 400 hp. water tube boiler with stokers, one 1,500 gal per minute motor driven centrifugal pumping unit.

Ill., Chicago—Ernest A. Mayo, Archt., 53 West Jackson Blvd., will soon award the contract for the construction of a 1 and 2 story, 300 x 300 ft. factory on North Central and Grand Aves., for the Casey-Hudson Co., 361 East Ohio St. A steam heating system will be installed in same. Total estimated cost, $300,000.

Wis., Hartford—The Kissel Motor Car Co. is having plans prepared for the construction of a 1 story boiler plant. Three 300 hp. boilers, stokers, etc., will be installed in same Cahill & Douglas, 217 West Water St., Milwaukee, Engrs.

Wis., Milwaukee—The School Bd., 10th and Prairie Sts., will soon award the contract for the construction of a 1 story, 22 x 40 ft. boiler house at the Garfield School. Two boilers will be installed in same. Total estimated cost, $50,000. Van Ryn and De Gelleke, Caswell Block, Archts. and Engrs. Noted April 27.

Mo., Kansas City—Harlan & Harlan Machine Wks., 308 B'way, is in the market for two 250, two 350 and two 500 hp. water tubular boilers for 150 lb. working pressure.

Tex., Austin—The State Bd. of Control will receive bids until May 10 for the installation of a complete heating system in the asylum for the insane. C. H. Page & Bros., Engrs.

Okla., Paden—The city is having plans prepared for the construction of waterworks systen, including pumping station, 65,000 gal. tank and tower, etc. Total estimated cost, $65,000. H. G. Olmsted & Co., Oil Exch. Bldg., Oklahoma City, Engr.

Okla., Pryor—The city voted $35,000 bonds to improve the waterworks including purification plant, pumping station and high line. V. V. Long & Co., Oklahoma City, Engr.

Okla., Walters—The city voted $125,000 bonds to construct waterworks, sewers and electric light plant.

Utah, Brigham—The city plans to rebuild the present hydro-electric plant and increase the capacity to 1,500 hp., etc. Total estimated cost, $125,000. C. O. Raskelly, Engr.

Wash., Olympia — The State Capitol Comn. will soon award the contract for the construction of a power house. A steam heating system will be installed in same. Total estimated cost, $100,000. Wilder & White, 50 Church St., New York City, Archts. and Engrs. Noted March 30.

Cal., San Francisco—The Bd. of Pub. Wks. will soon award the contract for the construction of a 3 story school building on Pacific Ave. Contract for installation of heating system will be sublet. Total estimated cost, $204,000. John Reid, Jr., 1st Natl. Bank Bldg., Archt.

Cal., Victorville—The Mojave River Irrigation Dist. voted $5,000,000 bonds to purchase and complete Little Bear Dam of Arrowhead Reservoir and Power Co. power development and distributing system to serve 27,000 acres. F. E. Trask, Central Bldg., Los Angeles, Engr.

Ont., London—The Pub. Utilities Comn. plans to build a dam on the Thames River at Springdale and develop 1,000 hp. to help the electrical power situation. Hydraulic turbines, generators, electrical switch apparatus, transformers, circuit breakers, etc., will be installed in same. Total estimated cost, $300,000. E. V. Buchanan, Hydro Bldg., Engr.

Ont., Sarnia—S. B. Coon & Son, Archts., 36 Toronto St., Toronto, will receive bids until May 5 for the construction of a 3 story high school, for the Bd. Educ. A vacuum steam heating system with blower and ventilation system will be installed in same. Total estimated cost, $325,000. Melvern F. Thomas, 229 College St., Toronto, Engr.

Ont., Toronto—The United Hotels Ltd., King Edward Hotel, will soon award the contract for the construction of a 16 story hotel addition on King St., E. A vacuum steam heating system will be installed in same. Total estimated cost, $2,000,000. Harkness, Loudon & Hertzberg, Confederation Life Bldg., Engrs. Watt & Blackwell, 79 Adelaide St., E. and Senween & Johnson, Ellicott Sq., Buffalo, N. Y., Archts.

CONTRACTS AWARDED

Conn., Bristol—The city has awarded the contract for the construction of a 2 story grammar school on Burlington Ave., to the Triangle Constr. Co., 57 South B'way, Yonkers, N. Y. A steam heating system will be installed in same. Total estimated cost, $250,000.

Conn., Danielson—The Goodyear Cotton Mills, Inc., will build 2 story housing buildings. A steam heating system will be installed in same. Total estimated cost, $500,000. Work will be done by day labor.

Conn., Danielson—The Goodyear Tire & Rubber Co., East Market St., Akron, O., has awarded the contract for the construction of a 1, 2 and 4 story manufacturing plant, to the Hunkin Conkey Constr. Co., Century Bldg., Cleveland, O. A steam heating system will be installed in same. Total estimated cost, $3,500,000.

N. Y., Brooklyn—The Bd. Educ., 500 Park Ave., New York City, has awarded the contract for the installation of heating and ventilating apparatus in P. S. 182, to Daniel J. Rice, 405 Lexington Ave., New York City, at $153,380.

N. Y., Brooklyn—The Bd. Educ., 500 Park Ave., has awarded the contract for the installation of heating and ventilating apparatus in P. S. 20, to Gillis & Geoghegan, 537 West B'way, New York City. Estimated cost, $155,994.

N. Y., New York—The Beth Israel Hospital Association, Inc., 70 Jefferson St., has awarded the contract for the construction of a 12 story, 107 x 184 ft. hospital on Livingston Pl., between 16th and 17th Sts., to G. Richard Davis, 30 East 42nd St. A steam heating system will be installed in same. Total estimated cost, $1,750,000.

N. Y., New York—The Curb Holding Co., 44 Court St., Brooklyn, will build a 12 story, 134 x 135 ft. office building on Washington St. A steam heating system will be installed in same. Total estimated cost, $1,750,000. Work will be done by day labor.

N. Y., New York—The Heckscher Bldg. Corp., 50 East 42nd St., will build a 24 story, 100 x 162 ft. office building on 57th St. and 5th Ave. A steam heating system will be installed in same. Total estimated cost, $3,000,000. Work will be done by day labor.

N. Y., New York—Liggett Winchester & Ley, 19 West 44th St., has awarded the contract for the construction of a 25 story, 100 x 240 x 11 x 54 ft. office and store building on 57th St. and 8th Ave. to Fred T. Ley, 19 West 44th St. A steam heating system will be installed in same. Total estimated cost, $2,500,000.

N. J., Hoboken—The Bd. Educ. has awarded the contract for installing a heating system in the proposed school 2, to John Cooney, 210 North 4th St., Harrison, N. J., at $112,494.

Pa., Farrell—The Colona Trust Co., B'way, has awarded the contract for the construction of a bank building to the R. H. Howes Constr. Co., 105 West 49th St., New York City. A steam heating system will be installed in same. Total estimated cost, $300,000.

Pa., Philadelphia—The Children's Hospital, 18th and Fitzwater Sts., has awarded the contract for the construction of a 3 story, 57 x 82 ft. hospital addition, to the Roydhouse Arey, 112 North Broad St. A steam heating system and power house equipment will be installed in same. Total estimated cost, $100,000. Noted Feb. 2.

Penn., Reading—Dives, Pomeroy & Stewart have awarded the contract for the construction of a 7 story, 60 x 207 ft. department store addition on 6th and Penn Sts., to L. S. Focht & Sons, Baer Bldg. A steam heating system will be installed in same. Total estimated cost, $400,000.

Pa., Pittsburgh—The Peoples Ice Co. has awarded the contract for the construction of a 1 story, 55 x 237 ft. ice plant, to the W. T. Grange Constr. Co.

Md., Baltimore—The Maryland Casualty Co., Baltimore St. and Guilford Ave., has awarded the contract for the construction of a 3 story, 300 x 300 ft. administration building and other buildings including power house, etc., to the Arundel Corp., 1515 Fidelity Bldg., and for the installation of boilers, to The Erie City Iron Wks., Erie, Pa. Total estimated cost, $3,000,000. Noted April 13.

Md., Baltimore—Norton, Bird & Whitman, 501 5th Ave., New York City, will build a theatre building. A steam heating system will be installed in same. Total estimated cost, $300,000. Work will be done by day labor.

La., Shreveport—The Consolidated Sanatarium Co. has awarded the contract for the construction of a 5 story, 40 x 300 ft. hospital on Louisiana Ave., to Garson Bros. A steam heating system will be installed in same. Total estimated cost, $500,000.

O., Akron—H. Thomas c/o W. A. Swasey, Archt., 101 Park Ave., New York City, will build a theatre. A steam heating system will be installed in same. Total estimated cost, $250,000. Work will be done by day labor.

O., Cleveland—The Amer. Steel & Wire Co., Western Reserve Bldg., has awarded the contract for the construction of a 1 story, 50 x 188 ft. boiler house on Lakeside Ave., to the Amer. Bridge Co., Frick Bldg., Pittsburgh, Pa. Estimated cost, $150,000.

O., Cleveland—The Globe Wernicke Co., 128 East 4th St., has awarded the contract for the construction of a 8 story, 50 x 189 ft. commercial building at 2044 Euclid Ave., to the H. K. Ferguson Co., 6532 Euclid Ave. A steam heating system will be installed in same. Total estimated cost, $250,000.

Mich., Bay City—The Bigelow Cooper Co., Marquette Ave., has awarded the contract for the construction of a 1 story, 55 x 150 ft. dry kiln addition and a 40 x 40 ft. power house on Marquette Ave., to the Bay City Stone Co., 1704 Woodside Ave., at $50,000.

Mich., Northville—The Bd. of Health, 233 St. Antoine St., has awarded the contract for the construction of a group of buildings to be constructed at the proposed Detroit Municipal Tuberculosis Sanatarium, including a 1 story, 47 x 87 ft. power house and a 40 x 120 ft. laundry, to John Finn & Son, 2328 Woodward Ave., Detroit, at $1,460,800. Noted April 13.

Ill., Chicago—The MacDonald Machine Co., 3204 Shields Ave., has awarded the contract for the construction of a 1 story, 150 x 400 ft. machine shop at 7600 South Racine Ave., to J. J. Schworm, 1374 South La Salle St. A steam heating system will be installed in same. Total estimated cost, $250,000. The Austin Co., 208 South La Salle St., Archt. and Engr.

Wis., Oshkosh—The Diamond Match Co., 111 Bway., New York City, has awarded the contract for the construction of an 88 x 360 ft. block storage building, to the Hutter Constr. Co., 128-136 Western Ave., Fond du Lac. A steam heating system will be installed in same. Total estimated cost, $250,000.

Wis., Sheboygan—The Badger State Tanning Co., South Water St. and Maryland Ave., has awarded the contract for the construction of a 4 story, 105 x 342 ft. tannery, to the Westinghouse, Church, Kerr Co., Inc., 37 Wall St., New York City. A steam heating system and steam power will be installed in same. Total estimated cost, $750,000.

Ia., Rippey—A. E. Jensen, Secy. Bd. Educ., has awarded the contract for installing heating and plumbing systems in the proposed 2 story, 63 x 105 ft. school, to the Bailey Plumbing Co., 404 Grand Ave., Des Moines, at $22,547. Noted March 30.

Ia., Waterloo—The Waterloo Gas Engine Co., Miles St., has awarded the contract for the construction of a 3 story, 100 x 120 ft. factory, to George F. Scales, Lafayette Bldg. A steam heating system will be installed in same. Total estimated cost, $500,000.

Minn., Fairmont—The city has awarded the contract for the construction of a central heating plant, to Robb-Bain Eng. Co., 1414 16th Ave., North, Minneapolis. Estimated cost, $150,000.

Minn., Keewatin — The School Dist. 9, Itasca Co., has awarded the contract for the construction of a 2 story, 160 x 174 ft. high school, to J. M. and W. A. Elliott, Exch. Bldg., Minneapolis. A steam heating and mechanical ventilating system will be installed in same. Total estimated cost, $440,000.

N. D., McVille—The city has awarded the contract for the construction of a 125,000 gal. reservoir and the installation of 3 pumps, etc., to Gedney & Murphy, Fargo, at $60,355. Noted April 27.

Mo., Kansas City—The Kansas City Federal Reserve Bank has awarded the contract for the construction of a 16 story, 115 x 145 ft. bank and office building on Grand and 10th Sts., to the George A. Fuller Constr. Co., 175 5th Ave., New York City. A steam heating system will be installed in same. Total estimated cost, $3,605,000.

PRICES - MATERIALS - SUPPLIES

Electrical prices on following page are prices to the power plant by jobbers in the larger buying centers east of the Mississippi. Elsewhere the prices will be modified by increased freight charges and by local conditions.

POWER-PLANT SUPPLIES

HOSE—

Fire	50-Ft. Lengths
Underwriters' 2½-in.	78c. per ft.
Common, 2½-in.	35%

Air

	First Grade	Second Grade	Third Grade
⅝-in. per ft.	$0.55	$0.35	$0.25

Steam—Discounts from List

First grade...... 25% Second grade...... 35% Third grade.... 40%

RUBBER BELTING—The following discounts from list apply to transmission rubber and duck belting:

Competition............ 50-10% Best grade............ 25%
Standard............ 40%

LEATHER BELTING—Present discounts from list in fair quantities (½ doz. rolls):

Light Grade	Medium Grade	Heavy Grade
25%	20%	10-5%

RAWHIDE LACING { For cut, best grade, 25%, 2nd grade, 30%. For laces in sides, best, 81c per sq. ft.; 2nd, 78c. Semi-tanned: cut, 20%; sides, 83c per sq. ft.

PACKING—Prices per pound:

Rubber and duck for low-pressure steam	$1.00
Asbestos for high-pressure steam	1.70
Duck and rubber for piston packing	1.00
Flax, regular	1.20
Flax, waterproofed	1.70
Compressed asbestos sheet	.90
Wire insertion asbestos sheet	1.50
Rubber sheet	.50
Rubber sheet, wire insertion	.70
Rubber sheet, duck insertion	.50
Rubber sheet, cloth insertion	.30
Asbestos packing, twisted or braided and graphited, for valve stems and stuffing boxes	1.30
Asbestos wick, ½- and 1-lb. balls	.85

PIPE AND BOILER COVERING—Below are part of standard lists, with discounts.

PIPE COVERING		BLOCKS AND SHEETS	
Pipe Size	Standard List Per Lin. Ft.	Thickness	Price per Sq. Ft.
1-in.	$0.27	½-in.	$0.27
2-in.	.36	1-in.	.30
6-in.	.80	1½-in.	.45
4-in.	.60	2-in.	.60
3-in.	.45	2½-in.	.75
8-in.	1.10	3-in.	.90
10-in.	1.30	3½-in.	1.05

85% magnesia high pressure.................. List

For low-pressure heating and return lines { 4-ply.......... 50% off; 3-ply.......... 52% off; 2-ply.......... 54% off

GREASES—Prices are as follows in the following cities in cents per pound for barrel lots:

	Cincinnati	Chicago	St. Louis	Birmingham	Denver
Cup	8¼	6.6	6.7	8¼	10.5
Fiber or sponge	8½	8.6	13	8¼	13.75
Transmission	10	8.1	13	8¼	12.5
Axle	...	4.8	4.75	4¼	7.0
Gear	6¼	6.1	7.5	8¼	8.5
Car journal	12(gal.)	4.7	4.7	4½	9.0

COTTON WASTE—The following prices are in cents per pound:

	New York Current	New York One Year Ago	Cleveland	Chicago
White	13.00	13.00	16.00	11.00 to 14.00
Colored mixed	9.00 to 12.00	9.00 to 12.00	12.00	9.50 to 12.00

WIPING CLOTHS—Jobbers' price per 1000 is as follows:

	13½ x 13½	13½ x 20½
Cleveland	$52.00	$58.00
Chicago	41.00	43.50

LINSEED OIL—These prices are per gallon:

	New York Current	New York One Year Ago	Chicago Current	Chicago One Year Ago
Raw in barrels (5 bbl. lots)	$1.87	$1.55	$2.05	$1.66
5-gal. cans	1.87*	1.70	2.30	1.86

*To this *oil* price must be added the cost of the cans (returnable), which is $2.25 for a case of six.

WHITE AND RED LEAD—Base price per pound:

	Red Current Dry	Red Current In Oil	Red 1 Year Ago Dry	Red 1 Year Ago In Oil	White Current Dry and In Oil	White 1 Yr. Ago Dry and In Oil
100-lb. keg	15.50	17.00	13.00	14.50	15.50	13.00
25- and 50-lb. kegs	15.75	17.25	13.25	14.75	15.75	13.25
12½-lb. keg	16.00	17.50	13.50	15.00	16.00	13.50
1-lb. cans	18.50	20.00	15.00	16.50	15.00	15.00
5-lb. cans	20.50	22.00	16.00	17.50	16.00	16.00

500 lb. lots less 10% discount; 2000 lb. lots less 10-2½%.

RIVETS—The following quotations are allowed for fair-sized orders from warehouse:

	New York	Cleveland	Chicago
Steel 7/16 and smaller	30%	40%	45%
Tinned	30%	40%	45%

Boiler rivets, ¾, ⅞, 1 in. diameter by 2 in. to 5 in. sell as follows per 100 lb.:
New York.. $6.00 Cleveland... $4.00 Chicago. $5.37 Pittsburgh... $4.72

Structual rivets, same sizes:
New York.... $6.10 Cleveland... $4.10 Chicago.. $5.47 Pittsburgh... $4.82

REFRACTORIES—Following prices are f. o. b. works, Pittsburgh district:

Chrome brick	net ton	$75-80	at Chester, Penn.
Chrome cement	net ton	45-50	at Chester, Penn.
Clay brick, 1st quality fireclay	net M.	45-50	at Clearfield, Penn.
Clay brick, 2nd quality	net M.	40-45	at Clearfield, Penn.
Magnesite, dead burned	net ton	50-55	at Chester, Penn.
Magnesite brick, 9 x 4½ x 2½ in.	net ton	80-85	at Chester, Penn.
Silica brick	net M.	50-55	at Mt. Union, Penn.

Standard size fire brick, 9 x 4½ x 2½ in. The second quality is $4 to $5 cheaper per 1000.

St. Louis—Fire Clay, $35 to $50.

Birmingham—Silica, $51-50 @ 55; fire clay, $45 @ 50; magnesite, $85; chrome, $85.

Chicago—Second quality, $25 per ton.

Denver—Silica, $18; fire clay, $12; magnesite, $57.50.

BABBITT METAL—Warehouse prices in cents per pound:

	New York Current	New York One Year Ago	Cleveland Current	Cleveland One Year Ago	Chicago Current	Chicago One Year Ago
Best grade	90.00	87.00	74.50	79.00	70.00	75.00
Commercial	50.00	42.00	21.50	18.00	15.00	15.00

SWEDISH (NORWAY) IRON—The average price per 100 lb., in ton lots, is:

	Current	One Year Ago
New York	$21-26	25.50-30.00
Cleveland	20.00	20.00
Chicago	16.50	16.50

In coils an advance of 50c. usually is charged.
Domestic iron (Swedish analysis) is selling at 12c. per lb.

SHEETS—Quotations are in cents per pound in various cities from warehouse; also the base quotations from mill:

Blue Annealed	Large Mill Lots Pittsburgh	New York Current	New York One Year Ago	Cleveland	Chicago
No. 10	3.55-6.00	7.12@8.00	4.57	7.30	6.02
No. 12	3.60-6.05	7.17@8.05	4.62	7.40	6.07
No. 14	3.65-6.10	7.22@8.10	4.67	7.45	6.12
No. 16	3.75-6.20	7.32@8.20	4.77	7.55	6.22
Black					
Nos. 18 and 20	4.15-6.30	8.50@9.50	5.30	7.95	6.80
Nos. 22 and 24	4.20-6.35	8.55@9.55	5.35	8.00	6.85
No. 26	4.25-6.40	8.60@9.60	5.40	8.05	6.90
No. 28	4.35-6.50	8.70@9.70	5.50	8.15	7.00
Galvanized					
No. 10	4.70-7.50	9.75@11.50	5.50	8.50	7.15
No. 12	4.80-7.60	9.85@11.60	5.55	8.60	7.20
No. 14	4.80-7.60	9.85@11.60	5.50	8.60	7.35
Nos. 18 and 20	5.10-7.90	10.10@11.85	5.90	8.90	7.65
Nos. 22 and 24	5.25-8.02	10.25@12.00	6.05	9.05	8.05
No. 26	5.40-8.20	10.40@12 '5	6.20	9.20	8.20
No. 28	5.70-8.50	10.70@12.45	6.50	9.50	8.50

*For painted corrugated sheets add 30c. per 1000 lb. for 5 to 28 gage; 25c. for 19 to 24 gages; for galvanized corrugated sheets add 15c., all gages.

PIPE—The following discounts are for carload lots f. o. b. Pittsburgh; basing card of Jan. 1, 1919, for steel pipe and for iron pipe:

BUTT WELD

Steel Inches	Black	Galvanized	Iron Inches	Black	Galvanized
¼, ⅜, and ½	50½%	24%	¾ to 1½	30½%	23½%
¾	54½%	40%			
1 to 3	57½%	44%			
LAP WELD					
2	50½%	35%	2	32½%	18½%
2½ to 6	53½%	41%	2½ to 6	34½%	21½%
BUTT WELD, EXTRA STRONG PLAIN ENDS					
¼, ⅜ and ½	48½%	29%	¾ to 1½	39½%	24½%
¾	51½%	39%			
1 to 1½	50½%	43%			
LAP WELD, EXTRA STRONG PLAIN ENDS					
2	48%	37%	2	33½%	20½%
2½ to 4	51%	49%	2½ to 4	35½%	23½%
4½ to 6	50%	39%	4½ to 6	34½%	22½%

Stock discounts in cities named are as follows:

	New York Black	New York Galvanized	Cleveland Black	Cleveland Galvanized	Chicago Black	Chicago Galvanized
¾ to 3 in. steel butt welded	40%	24%	40%	31%	54%	40½%
2½ to 3 in. steel lap welded	35%	20%	42%	27%	50%	37½%

Malleable fittings. Class B and C, from New York stock sell at list + 23%. Cast iron, standard sizes, net.

BOILER TUBES—The following are the prices for carload lots, f. o. b. Pittsburgh:

Lap Welded Steel		Charcoal Iron	
3½ to 4½ in.	40½	3½ to 4½ in.	15½
2½ to 3¼ in.	30½	3 to 3¼ in.	1
2¼ in.	24	2½ to 2¾ in.	+7
1¾ to 2 in.	19½	2 to 2¼ in.	+16
		1½ to 1¾ in.	+29

Standard Commercial Seamless—Cold Drawn or Hot Rolled

	Per Net Ton		Per Net Ton
1 in.	$327	1⅞ in.	$207
1¼ in.	267	2 to 2¼ in.	177
1½ in.	257	2½ to 3¼ in.	167
1¾ in.	207	4 in.	187
		4½ to 5 in.	207

These prices do not apply to special specifications for locomotive tubes nor to special specifications for tubes for the Navy Department, which will be subject to special negotiations.

ELECTRICAL SUPPLIES

ARMORED CABLE—

B. & S. Size	Two Cond. M Ft.	Three Cond. M Ft.	Two Cond. Lead M Ft.	Three Cond. Lead M Ft.
No. 14 solid	$104.00	$138.00	$164.00	$210.00
No. 12 solid	135.00	170.00	225.00	265.00
No. 10 solid	185.00	235.00	275.00	325.00
No. 8 stranded	285.00	375.00	520.00	500.00
No. 6 stranded	400.00	500.00	560.00	......

From the above lists discounts are:

Less than coil lots	Net List
Coils to 1,000 ft.	10%
1,000 ft. and over	15%

BATTERIES, DRY—Regular No. 6 size red seal, Columbia, or Ever-Ready:

	Each, Net
Less than 12	$0.45
12 to 50	.38
50 to 125 (bbl.)	.35
125 (bbl.) or over	.32

CONDUITS, ELBOWS AND COUPLINGS—These prices are f. o. b. New York with 10-day discount of 5 per cent.

Size, In.	Conduit Black 2,500 to 5,000 Lbs.	Conduit Galvanized 2,500 to 5,000 Lbs.	Elbows Black 2,500 to 5,000 Lbs.	Elbows Galvanized 2,500 to 5,000 Lbs.	Couplings Black 2,500 to 5,000 Lbs.	Couplings Galvanized 2,500 to 5,000 Lbs.
½	$80.50	$85.59	$19.13	$20.27	$7.04	$7.46
¾	106.51	113.51	25.17	26.67	10.07	10.67
1	157.59	167.79	37.25	39.47	13.09	13.37
1¼	213.11	227.01	48.46	50.26	18.30	18.98
1½	254.93	271.43	64.62	67.02	22.61	24.45
2	342.99	365.19	118.47	122.87	30.15	31.27
2½	542.30	577.40	193.86	200.00	43.08	44.68
3	709.16	755.06	511.96	536.16	64.62	67.02
3½	871.24	935.64	1,141.62	1,184.02	86.16	89.36
4	1,064.93	1,130.33	1,319.32	1,368.32	107.70	111.70
4½	1,240.79	1,318.99	1,997.83	2,072.03	161.55	167.55
5	1,445.96	1,534.76	2,773.27	2,866.27	177.70	184.30
6	1,875.84	1,991.04	3,446.40	3,574.40	258.48	268.08

Standard lengths rigid, 10 ft. Standard lengths flexible, ½ in., 100 ft. Standard lengths flexible, ¾ to 2 in., 50 ft.

CONDUIT NON-METALLIC, LOOM—

Size I. D., In.	Feet per Coil	List, Ft.
¼	250	$0.05½
5/16	250	.06
⅜	250	.09
½	200	.12
⅝	200	.15
¾	150	.18
1	100	.25
1¼	100	.33
1½	Odd lengths	.40
2	Odd lengths	.55

Coils......40% off
Less coils, 25% off

CUT-OUTS—Following are net prices each in standard-package quantities:

CUT-OUTS, PLUG

S. P. M. L.	$0.11	T. P. to D. P. S. B.	$0.24
D. P. M. L.	.11	T. P. to D. P. T. B.	.38
T. P. M. L.	.26	T. P. S. B.	.33
D. P. S. B.	.19	T. P. D. B.	.54
D. P. D. B.	.37		

CUT-OUTS, N. E. C. FUSE

	0-30 Amp.	31-60 Amp.	60-100 Amp.
D. P. M. L.	$0.33	$0.84	$1.68
T. P. M. L.	.48	1.20	2.40
D. P. S. B.	.42	1.05	.....
T. P. S. B.	.81	1.80	.....
D. P. D. B.	.78	2.10	.....
T. P. D. B.	1.35	3.60	.....
T. P. to D. P. D. B.	.90	2.52	.....

FLEXIBLE CORD—Price per 1,000 ft. in coils of 250 ft.:

No. 18 cotton twisted	$22.00
No. 16 cotton twisted	27.00
No. 18 cotton parallel	26.50
No. 16 cotton parallel	34.50
No. 18 cotton reinforced heavy	34.00
No. 16 cotton reinforced heavy	42.00
No. 18 cotton reinforced light	30.00
No. 16 cotton reinforced light	36.75
No. 18 cotton Canvasite cord	28.50
No. 16 cotton Canvasite cord	32.00

FUSES, ENCLOSED—

250-Volt	Std. Pkg.	List
3-amp. to 30-amp.	100	$0.25
35-amp. to 60-amp.	100	.35
65-amp. to 100-amp.	50	.90
110-amp. to 200-amp.	25	2.00
225-amp. to 400-amp.	25	3.60
425-amp. to 600-amp.	10	5.50

600-Volt	Std. Pkg.	List
3-amp. to 30-amp.	100	$0.40
35-amp. to 60-amp.	100	.60
65-amp. to 100-amp.	50	1.50
110-amp. to 200-amp.	25	2.50
225-amp. to 400-amp.	25	5.50
450-amp. to 600-amp.	10	8.00

Discount: Less 1-5th standard package 30%
1-5th to standard package 40%
Standard package 52%

FUSE PLUGS, MICA CAP—

0-30 ampere, standard package	$4.75C
0-30 ampere, less than standard package	6.00C

LAMPS—Below are present quotations in less than standard package quantities:

Straight-Side Bulbs Mazda B— Watts	Plain	Frosted	No. in Package	Pear-Shaped Bulbs Mazda C— Watts	Clear	Frosted	No. in Package
10	$0.35	$0.38	100	75	$0.70	$0.75	50
15	.35	.38	100	100	1.10	1.15	24
25	.35	.38	100	150	1.65	1.70	24
40	.35	.38	100	200	2.20	2.27	24
50	.35	.38	100	300	3.25	3.35	24
60	.40	.45	100	400	4.30	4.45	12
100	.85	.92	24	500	4.70	4.85	12
				750	6.50	6.75	8
				1,000	7.50	7.75	8

Standard quantities are subject to discount of 10% from list. Annual contracts ranging from $150 to $300,000 net allow a discount of 17 to 40% from list.

PLUGS, ATTACHMENT—

		Each
Hubbell porcelain No. 5406,	standard package 250	$0.24
Hubbell composition No. 5467,	standard package 50	.32
Benjamin swivel No. 903,	standard package 250	.20
Hubbell current taps No. 5638,	standard package 50	.40

RUBBER-COVERED COPPER WIRE—Per 1000 ft. in New York.

No.	Solid Single Braid	Solid Double Braid	Stranded, Double Braid	Duplex
14	$12.00	$16.00	$13.90	$28.50
12	13.25	15.70	18.05	30.70
10	18.30	21.00	23.85	41.50
8	25.54	28.60	32.70	56.77
6	......	......	51.40	......
4	......	......	70.00	......
2	......	......	101.80	......
1	......	......	131.86	......
0	......	......	160.00	......
00	......	......	193.50	......
000	......	......	235.20	......
0000	......	......	288.60	......

Prices per 1000 ft. for Rubber-Covered Wire in Following Cities:

	Denver Single Braid	Denver Double Braid	Denver Duplex	Birmingham Single Braid	Birmingham Double Braid	Birmingham Duplex
14	$19.50	$24.00	$39.50	$12.23	......	$33.60
10	28.92	33.12	65.52	24.00	......	57.02
8	40.32	45.12	89.64	33.60	......	75.94
6	63.84	69.24	......	60.10	......	......
4	......	97.86	......	......	$93.41	......
2	......	145.20	......	......	120.40	......
1	......	189.00	......	......	155.23	......
0	......	251.40	......	......	188.77	......
00	......	305.40	......	......	229.19	......
000	......	317.40	......	......	278.64	......
0000	......	451.20	......	......	338.41	......

Pittsburg—23c. base; discount 50%. St. Louis—30c. base.

SOCKETS, BRASS SHELL—

¼ In. or Pendant Cap Key Each	Keyless Each	Pull Each	⅜ In. Cap Key Each	Keyless Each	Pull Each
$0.33	$0.30	$0.60	$0.39	$0.36	$0.66

Less 1-5th standard package ...... +50%
1-5th to standard package ...... +20%
Standard package ...... −10%

WIRE, ANNUNCIATOR AND DAMPPROOF OFFICE—

No. 18 B. & S. regular spools (approx. 8 lb.)	55c. lb.
No. 18 B. & S. regular 1-lb. coils	56c. lb.

WIRING SUPPLIES—

Friction tape, ¾ in., less 100 lb. 60c. lb., 100 lb. lots	55c. lb.
Rubber tape, ¾ in., less 100 lb. 65c. lb., 100 lb. lots	60c. lb.
Wire solder, less 100 lb. 50c. lb., 100 lb. lots	46c. lb.
Soldering paste, 2 oz. cans Nokorode	$1.20 doz.

SWITCHES, KNIFE—

TYPE "C" NOT FUSIBLE

Size, Amp.	Single Pole, Each	Double Pole, Each	Three Pole, Each	Four Pole, Each
30	$0.42	$0.68	$1.02	$1.36
60	.74	1.22	1.84	2.44
100	1.50	2.50	3.76	5.00
200	2.70	4.50	6.76	9.00

TYPE "C" FUSIBLE, TOP OR BOTTOM

Size, Amp.	Single Pole, Each	Double Pole, Each	Three Pole, Each	Four Pole, Each
30	.70	1.06	1.60	2.12
60	1.18	1.80	2.70	3.60
100	2.38	3.66	5.50	7.30
200	4.40	6.76	10.14	13.50

Discounts:

Less than $10.00 list value	+25%
$10 to $25 list value	+10%
$25 to $50 list value	+5%
$50 to $200 list value	−10%
$200 list value or over	−15%

POWER

Volume 51 New York, May 11, 1920 *Number 19*

Pigs in Clover

Power and Refrigeration for Model Ice Cream and Dairy Plant

Modern 450-ton Three-Unit Refrigerating Plant Using High-Speed Two-Cylinder Ammonia Compressors, Cross-Connected So That Either Cylinder May Be Used on High or Low-Temperature Service—Power and Lighting Are Supplied by a 450-Kw. Electric Generating Condensing Plant, the Cooling Water Being Conserved by a Large Cooling Tower on the Roof of the Building

ONE of the largest ice-cream and dairy plants in the country is being operated in Cincinnati, Ohio, by the French Bros. Bauer Company. It has now been in operation for about a year, being a consolidation of two former plants, one engaged mainly in the production of ice cream and the other in the manufacture of butter and pasteurization and distribution of milk and cream for Cincinnati and the surrounding territory. Both of the plants also manufacture ice, but not in sufficient quantity for packing the ice cream during the summer months.

FIG. 1. ICE-CREAM HARDENING ROOMS, CAPACITY, THIRTY-THREE THOUSAND GALLONS

As each of the old plants required a full force of operators, this naturally made the cost of production higher than would have been the case in one plant, and as the equipments had become inadequate, it was decided to build a new plant of sufficient capacity for the combined output of the two old ones, with a reasonable allowance for future increase.

The building, constructed of reinforced concrete throughout, was erected on the site of an old brewery. The underground cellars, with their heavy arched ceilings, retaining walls and many intricate underground passages, offered an interesting problem in the design of the foundations and basements, with the purpose of reducing the excavation work and new foundation walls to the minimum. All reinforced-concrete floors throughout the building, where considerable water is used on the floor, are waterproofed with three-ply membrane waterproofing and, with some exceptions, finished on top with three inches of concrete. In the bottle-filling and ice-cream making departments the floors, after having been waterproofed, are finished with Terrazzo, and for the shipping platform the finish is vitrified brick paving. All cold-storage rooms, as well as the ice tanks, are insulated with corkboard with all exposed surfaces on walls and ceilings plastered with portland-cement plaster. The exterior of the building is finished with red brick facing between concrete beams and columns at each floor, and the building is trimmed with terra cotta.

To give some idea of the power and refrigerating requirements and the general arrangement, a brief survey of the various departments may be of interest. At the two basement levels, the lower being 28 ft. below the sidewalk, are the boiler and engine rooms, ice tanks, daily ice storage and a 3,000-ton season ice-storage room. On the first floor of the building is the bottled-milk storage room, ice-cream-can washing room, ice-cream-making and the ice-cream hardening department and the loading platform. The driveway is under the building proper along the east and south sides. Being about 30 ft. wide, the long span over the driveway necessitated on each floor above it beam and girder construction, while the rest of the building is flat-slab construction with the columns placed on 25-ft. centers.

FIG. 2. ONE OF THE 150-TON REFRIGERATING UNITS. VERTICAL HIGH-SPEED COMPRESSOR IN BACKGROUND

Under the south driveway is the salt bin, from which salt is elevated to hoppers on the second floor and spouted to the first floor as required for the packing of ice cream in tubs.

One-half of the second floor is taken up by general offices and the remainder of the space by the bottle- and can-washing departments, the bottle-filling department, the ice-cream mixing department and the cold-storage rooms for fruits and other supplies used by the ice-cream mixing departments. On the third floor one-half of the space is utilized for storage purposes and the other half for the butter-making department, butter cold-storage rooms and for pasteurizing milk and cream. The fourth floor also has storage space and, in addition, the milk-receiving department and chemical laboratory.

From the foregoing it will be noticed that the various departments are so located in the building that after the milk and cream are received on the fourth floor, they follow through the manufacturing process by gravity until the products are ready for distribution on the first floor.

Milk and cream are received at the plant in cans, which are raised to the fourth floor by a continuous can elevator provided with an automatic loading table below the mixing tanks. The sour cream for butter making is elevated to the fourth floor, the same as the sweet milk and cream, is weighed in a separate scale tank and then pasteurized, ripened and made into butter on the third floor.

The mechanical equipment and refrigerating plant were designed for a maximum output in eight hours per day of pasteurizing and cooling 13,000 gal. of milk and sweet cream, 10,000 gal. of sour cream, the making of 25,000 lb. of butter, the freezing and hardening of 7,000 gal. of ice cream, together with maintaining necessary temperatures in cold-storage rooms and the making of 100 tons of ice per day.

A careful analysis of the load factor during the various months of the year in the older plants and as estimated for the different periods of the day based on the new output disclosed the fact that the maximum load would obtain for about seven hours of the day; that the

FIG. 3. VIEW OF THREE ICE-FREEZING TANKS EACH HAVING NOMINAL CAPACITY OF THIRTY TONS

on the first floor and an automatic discharge table on the fourth floor. An electric high-speed platform elevator serves as a reserve unit in case of breakdown.

On the fourth floor the milk cans travel by gravity on a roller conveyor to the milk-weighing tanks. After the milk is emptied, the empty cans travel to the second floor to the can-washing department on a spiral conveyor, and after being washed they are discharged to the first floor on another spiral conveyor.

The milk and cream from the scale tanks are pumped to a holding tank and, after having been clarified, go to the milk and cream storage tanks, then to the pasteurizers and to cooling and holding tanks on the third floor. The milk and cream are bottled on the second floor, loaded into cases and sent to the milk-storage room on the first floor on a spiral conveyor.

After having been pasteurized, the cream for ice-cream making flows by gravity to the ice-cream mixing tanks on the second floor and again gravitates to the ice-cream freezers placed on the first floor immediately load for the other seventeen hours would be about one-third of the maximum load, and that the average maximum load for the year would be about two-thirds of the maximum daily output. These interesting estimates made it appear advisable in so far as was possible to arrange the different mechanical units (boilers, generating units and ammonia compressors) into three of each kind, thereby giving flexibility in operation and making the best provisions for spare units and spare parts.

Throughout the design of the plant the fundamental idea of arranging the process of manufacture to reduce the labor and power cost to the minimum was carried out, in that all products are handled by gravity in so far as is possible, and in selection of the mechanical equipment choice was made of the most efficient types for the purpose intended.

To conform with latest practice, high-speed ammonia compressors were installed. Efficiency was the primary factor in the selection, and space limitations compli-

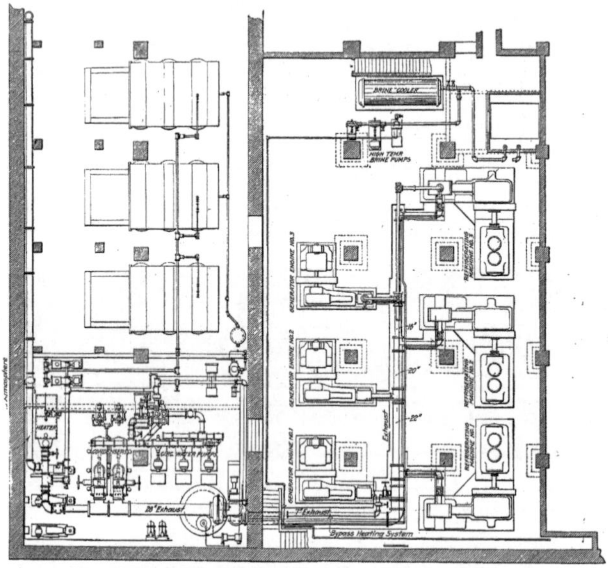

FIG. 4. GENERAL PLAN OF STEAM PIPING IN ENGINE, COMPRESSOR AND BOILER ROOM

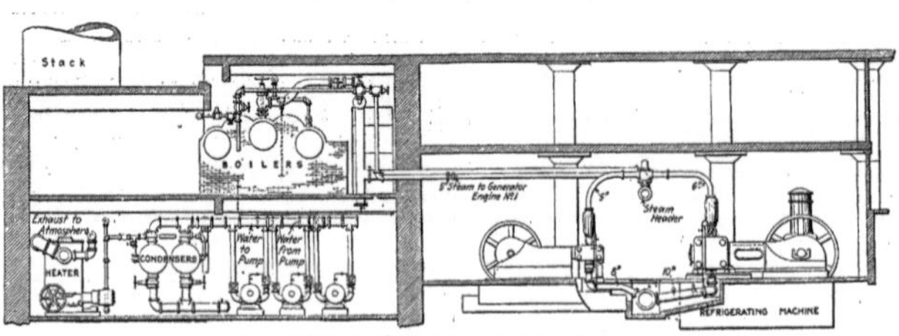

FIG. 5. SECTIONAL ELEVATION THROUGH POWER PLANT

cated by large building columns, necessitated vertical units. As intimated previously, there are three machines, each of the vertical, inclosed, single-acting, high-speed type, having two 18 x 20-in. cylinders and a speed of 150 r.p.m. The rating is 150 tons of refrigeration. A feature giving unusual flexibility to the operation is the connection of each cylinder to both high- and low-temperature lines so that a cylinder may be devoted to either service or both to the same service, as required. At times of light load one machine will take care of both classes of work, and when the load is heavy enough to require two machines, but perhaps not evenly divided, three of the four cylinders may be used in one service and the fourth cylinder in the other.

Horizontal poppet-valve, simple condensing engines, 20 x 24-in., drive the compressors. As will be evident from Fig. 2 a layshaft geared to the mainshaft operates

FIG. 6. AMMONIA FLOODED CONDENSERS AND COOLING TOWERS ON ROOF

the poppet valves, which are placed horizontally. Cams on this layshaft open the exhaust valves, and they are seated by springs provided in the bonnet. An eccentric and a secondary shaft open the steam valves at the top of the cylinder, the closing being effected by springs as in the case of the exhaust valves. For adjustment of load, the valve stems may be screwed in or out. A centrifugal governor on the layshaft controls the opening of the steam valves and consequently the speed of the unit.

Force-feed lubrication is provided for the steam cylinder and gravity oiling for the bearings. A centrifugal pump, belt-driven from the mainshaft, draws the oil from the crankpit and elevates it to a pedestal filter from which it flows back to the bearings. For the compressor lubrication there is also a belt-driven pump which pulls the oil from the crankcase and discharges it through a twin pressure filter to the seven bearings within the crankcase. There is thus continuous circulation of the oil, and electric alarms are provided to indicate any stoppage in the flow of oil.

Serving the ammonia compressors there are three banks of atmospheric condensers of the flooded type, aggregating 40 stands of 2-in. pipe, 12 high and 20 ft. long. The condensers, equipped with all necessary discharge, liquid, drain and pump-out headers, are located on the roof of the building.

For making ice there are three ice-freezing tanks, each 79 ft. long x 15 ft. 2 in. wide, arranged for 432 three-hundred-pound cans and piped with 10,000 ft. of 1¼-in. extra-heavy steel ammonia piping, the coils being arranged for the flooded system. Each tank is equipped with two 20-in. horizontal agitators, one agitator being placed at each end of the tank. Each agitator is belt-driven at 150 r.p.m. by a 3-hp. motor having a speed of 1,010 r.p.m. As all the ice is used for packing ice cream in tubs, it is made from city water and is opaque, no attempt being made to manufacture clear ice. The normal capacity of each tank is 30 tons.

To pull the ice, each tank is equipped with a one-ton electric hoist. The can is first dipped and the cake of ice delivered upon the dump, from which it slides through an automatic door into the daily storage of 500 tons in the first basement. From here the ice is conveyed either to the season storage of 3,000 tons' capacity or for immediate use to the first floor or to the ice crushers on the second floor. The ice is delivered to the crushers automatically, and the elevator cage returns to the basement automatically after having delivered the ice. The motors for the crushers and elevators are controlled from the shipping room on the first floor and are electrically interlocked, so it is impossible to operate the ice elevators without having first started the crusher. The ice from the crushers is spouted to the wagons in the driveway or to the shipping platform as desired.

All refrigerating work for cooling, holding vats, pasteurized milk and cream and for all refrigerated rooms for milk, cream and butter is done by means of brine circulation, for which is provided a shell-type brine cooler having 2,400 sq.ft. of exchange surface. The brine is circulated by means of two centrifugal brine pumps, brass-fitted throughout and direct connected to motors by means of flexible couplings. One pump has a capacity of 600 gal. of brine per minute and the other a capacity of 200 gal. per minute.

The fifteen 40-quart ice-cream freezers, chain-driven by 2-hp. motors, are provided with brine circulation, the brine being at a temperature of —10 deg. F. For this purpose is provided a shell-type brine cooler having 750 sq.ft. of exchange surface, and the brine is circulated by means of a 300-g.p.m. centrifugal brass-fitted pump directly connected to a 1,700-r.p.m. motor.

FIG. 7. JOHN O. BENNETT, CHIEF ENGINEER, IN HIS ENGINE ROOM

After being frozen, the ice cream is hardened in a battery of nine hardening rooms arranged side by side and each connecting with a receiving vestibule on the north end and a delivery vestibule on the south end. These hardening rooms have a capacity for storing a maximum of about 33,000 gal. of ice cream at one time. Each hardening room is provided with an overhead coil bunker arranged with baffles and direct-expansion coils over which the air circulation is blown by means of a 2,100-cu.ft. fan driven by a 2-hp. motor. The coils are arranged as a full gravity flooded system. From an accumulator 24 in. diameter by 8 ft. high, located on the second floor, the liquid gravitates to the coils. Provision is made for defrosting the coils of the hardening rooms either by passing the discharge ammonia gas from the refrigerating machines through the coils or by means of a brine spray over the coils. These two systems of defrosting can be used together or independently as desired.

The milk-storage room is refrigerated by air circulation from a coil bunker placed on a concrete platform in the corner of the room. The coil bunker is piped with 2,400 ft. of 2-in. galvanized brine piping, and the air is circulated by means of two fans of 3,000-cu.ft. per minute capacity, each direct connected to a 3-hp. motor.

All the dairy machinery about the plant is equipped with direct motor drive as well as all vacuum, water-circulating and brine pumps. An auxiliary steam driven pump for each service is, however, provided in case of emergency. The electric current for motors and lighting is generated by three 150-kw. 125/250-volt 200 r.p.m. three-wire direct-current generators connected to 15 x 22-in. simple condensing nonreleasing Corliss-type engines. Energy for power is taken off the outside wires at 250 volts; for lighting there is three-wire 125-volt distribution. A 2,000-ampere two-wire watt-hour meter measures the total energy consumed by the power circuits, and a 300-ampere three-wire watt-hour meter registers the current used for electric lighting.

As the engines driving the electric generators and ammonia compressors are operated condensing, two steam surface condensers have been provided, each having capacity to condense steam at the rate of 14,000 lb. per hour and to maintain a vacuum of 25 in. when supplied with 1,000 gal. of cooling water per minute at a temperature of 95 deg. As auxiliaries there are two motor-driven and one steam-driven dry-vacuum pumps and two motor-driven condensate pumps. The water of condensation is discharged into the feed-water heater.

Circulating water for the steam and ammonia condensers is cooled by an atmospheric-type cooling tower placed on piers on the roof of the building. It is 124 ft. long, 19 ft. wide, 30 ft. high, and arranged in 10 spray decks. The tower has capacity to cool 2,000 gal. of water per minute from a temperature of 130 deg. to within an average of about 4 deg. above the wet-bulb temperature, depending upon weather and wind conditions. The water from the cooling-tower pan passes over the ammonia condensers, from the ammonia-condenser pan drains to the steam condensers and is returned to the top of the cooling tower. The cooling water is circulated by means of three 5-in. centrifugal pumps each driven by a 40-hp. motor and having a capacity for handling 700 gal. of water per minute. Except for the distance from the roof to the top of the tower and the friction on the piping, the circulating system is balanced. An altitude valve controls the level of the water in the ammonia-condenser pan, the small amount of makeup water needed to overcome windage and evaporation being automatically admitted from the city mains.

The building is heated by exhaust steam during the

FIG. 8. CIRCULATING PUMPS, STEAM CONDENSERS AND FEED-WATER HEATER

winter months. The necessary supply for this purpose and for heating the feed water is obtained from the boiler-feed pumps, the auxiliary steam-driven brine and water pumps, provided as reserve units, the vacuum pump for the heating system and from one of the engines driving the electric generators, which is arranged with a bypass exhaust connection so that it may be operated noncondensing during the winter months.

In the boiler room as well as in other departments of the plant, the three-unit plan was followed, the equipment consisting of three Stirling boilers rated at 394 hp. and served by top-feed V-shaped stokers hav-

FIG. 9. VIEW OF TOP-FEED STOKERS AND COAL FEEDS FROM OVERHEAD BUNKER

ing approximately 80 sq.ft. of grate area, giving the usual ratio to steam-making surface of 1 to 50. The operating steam pressure is 145 lb. gage with no superheat. Through an 8-in. line provided with a main stop valve and an automatic nonreturn valve, each boiler is connected to a 12-in. main steam header, which is cross-connected to a distributing header of 10 and 8-in, diameter running lengthwise of the engine room. From either side supply lines lead to the generator and compressor units. The header is suspended from the ceiling and is anchored to prevent vibration. The engines exhaust into a common exhaust main increasing in diameter from 12 to 28 in. on its way through a floor trench to the condensers. From this line there is an atmospheric relief with connection to the feed-water heater, and as previously mentioned an independent exhaust or bypass from one of the engines to supplement the exhaust-steam supply for heating purposes.

A tapered breeching receiving the gases from each boiler leads to a reinforced-concrete chimney having an internal diameter of 8 ft. at the top and measuring 200 ft. high from the boiler-room floor line. At each boiler uptake there is a 5 x 8-ft. damper and a main damper in the smoke flue near the stack. Reviewing some of the dimensions, it will be seen that for each 10 sq.ft. of steam-making surface, there are 0.2 sq.ft. of connected grate, 0.1 sq.ft. of individual damper and 0.04 sq.ft. of stack area.

Ohio or West Virginia coal, the fuel burned, is dumped from wagons or trucks in the driveway directly into a 150-ton concrete bunker over the boiler furnaces. The bunker has a flat bottom, with gates opening to the magazines at each side of the stoker. Ashes from the furnaces are removed from the basement by an 8-in. steam-jet ash conveyor, provided with an intake in the bottom of each stoker pit. The ashes are discharged to a reinforced-concrete ash tank on the second floor and from there spouted to wagons or trucks in the driveway.

Boiler-feed water is heated in an open feed-water heater having 53.7 cu.ft. of steam space and 101.8 cu.ft. of water-storage space. The total volume of the heater is thus 155.5 cu.ft., which, in normal operation, calling for two boilers, reduces to approximately 0.2 cu.ft. per rated boiler horsepower. All condensation from live steam and exhaust lines is collected by pumps and traps and discharged to the feed-water heater, as well as the condensation obtained from the steam condensers and the heating system. Any makeup water required enters the heater under float control from the city mains.

Exhaust steam from the boiler-feed pumps and auxiliary steam pumps throughout the plant is used to heat the feed water, which is pumped to the boilers by either one of two 10- and 6- by 12-in. duplex piston-pattern pumps.

The new plant is a model of its kind, from both a sanitary and an operating standpoint. The complete undertaking, including buildings and mechanical equipment, was designed and installed under the direction of the Tait & Nordmeyer Engineering Co. of St. Louis.

Burton Duplex Water-Tube Boiler

T. Howard Burton, who for years has been indentified with the manufacture of a single-drum water-tube boiler of the old Worthington type, has perfected the two-drum duplex type of boiler shown in Figs. 1 and 2, and through the DePere-Burton Co., DePere, Wis., of which he is president, is placing this boiler on the market. It is a one-pass boiler for the gases provided with two cross-drums at opposite ends of the setting. Although the criss-cross arrangement of the tubes of the older boiler has been retained, the sectional construction has been eliminated. Instead of individual headers at either end of a tube section the lower ends of the tubes enter a large water-box header, extending the full width of the boiler. The tubes lead obliquely upward across the path of the gases and enter their respective drums. The header is thus eliminated at one end, and the arrangement gives maximum liberating surface for the steam generated, as each tube delivers its steam directly to the drum, thus eliminating the nipple connections between the header and drum common to most types of water-tube boilers. The water legs being arranged at right angles to the tubes, slope upward and outward at each end of the combustion chamber. An extension of the water leg with blowoff connections eliminates the necessity for a mud drum.

Water connection from the drum to the water leg on the same side and immediately beneath it is made by curved circulating tubes, one end leaving the drum radially and the other end entering the top of the water leg approximately in line with it. The curved tubes are easy to place and make a flexible connection, as they will bend slightly under the contraction and expansion of the boiler parts.

Fig. 1, a sectional elevation, shows how the tubes are arranged in vertical rows, adjacent rows sloping in opposing directions. A vertical row eleven tubes high leads from the header or water leg at the left side of

Fig. 1 and enters the drum at the right side of the drawing. In the next vertical row the tubes connect the header at the right-hand side of the setting and the drum at the opposite side. Thus the rows of tubes alternate in the direction of their slope and cross over the combustion chamber. This arrangement finally subdivides the gases of combustion, and as the space between the vertical rows of tubes is twice as great outside of the intersecting area, the gases—tending to follow the path of least resistance—spread out and cover the steam-making surface in the tubes. To further control the flow of gases, baffles may be placed horizontally, as indicated on the upper tubes.

As most of the tubes are curved, they allow for expansion and contraction and tend to reduce leakage that may be caused by movement at the joints. The tubes

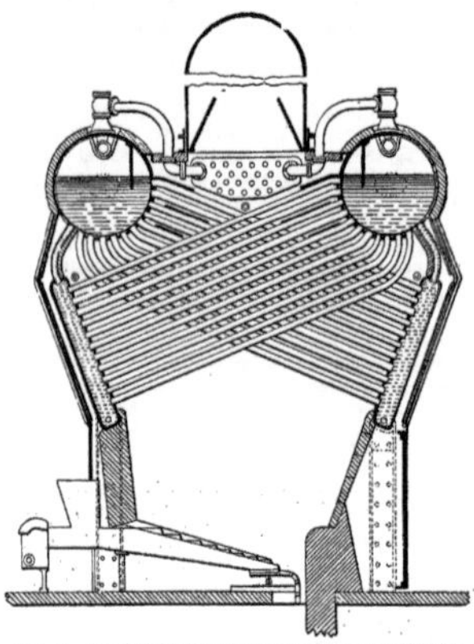

FIG. 1. LONGITUDINAL SECTION THROUGH BOILER

may be cleaned from within the drums and instead of pulling them out through handholes in the header, they may be removed in the combustion chamber. Both features are big time savers and tend to reduce the time it is necessary to keep a boiler off the line for cleaning or tube removal.

An outer shell is provided which may be varied considerably in construction and design. As will be noticed in the sectional view, about one-half of the surface of each drum is exposed to the heat of the gases. The short circulating tubes between the drums and the water legs are insulated from the hot gases so that the water in the drums may settle readily to the water legs and there will be no tendency to reverse the circulation.

This design of boiler can be installed in exceptionally low headroom, and the heating surface per square foot of floor space covered is unusually large. By the use of the large water-box headers, sectional headers and mud drums are eliminated. By leading the tubes into the drums one-half the header requirement is eliminated and maximum steam-liberating surface is provided. As a consequence the boiler steams rapidly. In the water legs key caps are provided opposite each tube, but the latter may be removed within the combustion space and

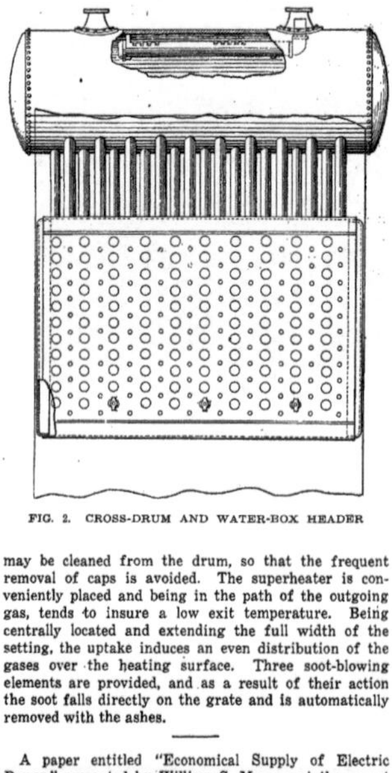

FIG. 2. CROSS-DRUM AND WATER-BOX HEADER

may be cleaned from the drum, so that the frequent removal of caps is avoided. The superheater is conveniently placed and being in the path of the outgoing gas, tends to insure a low exit temperature. Being centrally located and extending the full width of the setting, the uptake induces an even distribution of the gases over the heating surface. Three soot-blowing elements are provided, and as a result of their action the soot falls directly on the grate and is automatically removed with the ashes.

A paper entitled "Economical Supply of Electric Power," presented by William S. Murray at the recent convention of the A. I. E. E., has resulted in the appointment of the following special committee to be known as the "Committee on Super-Power System" to carry on the movement outlined by Mr. Murray: W. S. Murray, consulting engineer, chairman; Harold V. Bozell, of *Electric Railway Journal;* H. W. Buck, consulting engineer; John W. Lieb, vice-president, New York Edison Co.; Malcolm MacLaren, professor of Electrical Engineering, Princeton University; William McClellan, of Cleveland Illuminating Co.; Charles S. Ruffner, of North American Co., N. Y.; David B. Rushmore, General Electric Co.; Charles F. Scott, Yale University, and Percy N. Thomas, consulting engineer, New York.

Motors for Driving Ventilating Fans

Characteristics of Motors for Fan Drives—Comparisons Between Alternating-Current and Direct-Current Motors—Methods of Control—Formula for Calculating the Horsepower Required To Drive Fans.

By WILLIAM H. EASTON
Industrial Engineer, Westinghouse Electric and Manufacturing Company

WILLIAM H. EASTON

VENTILATING fans are fairly simple devices, but unless they are driven by a motor that is properly selected to suit both the fan itself and the particular service for which it is used, the days of the man in charge will be full of trouble. There are many types of motor, each with its own special characteristics of speed, torque and mechanical construction, and there are several different methods of controlling the speed of each type of motor. However, only a few of these types are suitable for fan service under any circumstances, and when a particular application is carefully analyzed, it will generally be found that to obtain satisfactory and economical operation, the choice of proper motors is very limited.

The subject of the selection of motors for ventilating fan drives is a broad one and cannot be adequately covered in a short article, but there are some fundamental principles that may be of interest and assistance to operating engineers.

Reliability of operation is one of the first requisites of a satisfactory ventilating-fan motor, as the unit is generally placed in an out-of-the-way position, where constant supervision is impossible. Hence, the motor must be able to operate constantly with very little care and attention, which means superior commutation; dust-proof, nonleaking bearings with dependable lubrication; high-grade insulation, with protection against dust, moisture, acid fumes, etc.; and, especially, a record of satisfactory operation.

High efficiency of operation is also a necessary feature, for a ventilating-fan motor will ordinarily operate continuously for long periods of time, so that a difference of one or two per cent in the energy consumption will form quite an item in the annual operating expense. Thus, assuming energy at only one cent per kilowatt-hour, the cost of operating a 25-horsepower motor of 86.5 per cent efficiency at full load ten hours a day for 365 days will be $788, whereas a motor of 84 per cent efficiency will entail an expense of $810 for the same period. This shows that it would be poor economy to purchase a motor of low efficiency merely because of slight advantage in price, especially where the cost of energy is greater than one cent per kilowatt-hour. Efficiency must, of course, be considered for the speeds at which the motor is actually to run, and it will be found that some motors of good full-speed efficiency have a low efficiency at other speeds. Quiet operation is usually very important. Any noise made by the ventilating unit will be transmitted to other parts of the building through the air ducts, and any noticeable humming or roaring would be intolerable in such buildings as theaters, hotels, hospitals and offices. Mechanically, the motor should be light and compact, and when it is to drive an overhung fan, it must have ample strength of shaft and bearings to support the fan.

Either direct-current or alternating-current motors can be used for driving fans, and usually the question as to which type is to be selected is settled definitely by the kind of energy available. But where a choice is possible, it will almost invariably be made in favor of direct-current machines, for the following reasons:

Direct-current motors in all sizes can be obtained with slow speeds, so that they are suitable for direct connection to the ordinary type and size of ventilating fan. Alternating-current motors of the small sizes, on the other hand, are not supplied in slow speeds and hence must usually be belted to the fans they drive. There are some examples of alternating-current motors of less than 50 horsepower, which are directly connected to ventilating fans, but fans of such high speed are liable to be noisy in operation.

The speed control of direct-current motors is far superior to that of alternating-current types. Speeds that are both higher and lower than the normal motor speed can be obtained, and with proper selection of motor and control, efficient operation can be secured over the entire speed range. The speed of an alternating-current motor, however, can only be reduced below normal, increased speed being unobtainable; and on speed reductions below 20 per cent a considerable loss in efficiency occurs.

FIG. 1. DIRECT-CURRENT MOTOR DIRECTLY CONNECTED TO VENTILATING FAN

It must not, however, be assumed that alternating-current motors cannot be used for ventilating service. There are thousands of entirely satisfactory alternating-current installations of all types throughout the country, and though in some extreme instances it may be desirable to install some means of producing direct-current, especially for the ventilating-fan motors, such

FIG. 2. FAN MOTOR CONTROLLED AUTOMATICALLY THROUGH PUSH-BUTTONS—CONTROLLER TO THE LEFT

cases are exceptional. Owing to the fact that there is an extensive demand for direct-current ventilating-fan motors, manufacturers have developed lines especially designed for this service. One such line comprises motors ranging from ¼ to 50 horsepower in capacity, and almost all these are supplied in 27 different normal speeds, ranging from 90 to 525 r.p.m. These motors are of the shunt type for operation on 230-volt circuits and are designed for speed ratios of 1 to 2 by field control; that is, if the normal speed of the motor is, say, 300 r.p.m., it can be increased in any desired number of steps to 600 r.p.m.

FIG. 3. SQUIRREL-CAGE ALTERNATING-CURRENT MOTOR BELTED TO VENTILATING FAN

For driving a fan continuously at constant speed, a motor of the proper normal speed is selected, and the control equipment consists simply of a starting box. Ordinarily, the starter is of the hand-operated type, but automatic starters are being used more and more extensively because the motor can be started and stopped from one or more push-button stations located where convenient. As ventilating fans are frequently installed in places difficult of access, remote control is often very convenient. Another advantage of the automatic starter is that it prevents injury to either the motor or the starter by cutting out the starting resistance either too slowly or too rapidly. If, as is usually the case, it is necessary to change the speed of the fan, either of three methods of changing the speed of the driving motor may be employed:

1. Field Control: In this method, resistance is introduced in the shunt-field circuit of the motor. The speed increases as the amount of resistance is increased, so that the normal speed of the motor is the lowest that can be obtained by this method of control. The chief feature of field control is that the loss of energy in the resistance is negligible. It is, therefore, used wherever possible, and where the speed range is not too great the general practice is to select a motor with a normal speed that corresponds to the lowest speed of the fan and to use field control for the higher speeds.

2. Armature Control: In this method resistance is introduced into the armature circuit of the motor, and as the effect is to reduce the motor speed, the normal speed of the motor is the highest that can be obtained by this scheme. Speed reductions of 50 per cent of the normal speed can be obtained by armature control, but as there is a considerable loss of energy in the resistance (amounting to about 19 per cent of the motor's input at full load), it is not often employed. However, where a fan is operated most of the time at constant speed, with occasional small speed reductions, this method is satisfactory.

3. Combined Field and Armature Control: Where a very wide speed range is essential, combined field and armature control can be employed, since with its use a standard ventilating motor can be operated over a speed range from one-half the normal speed to two times normal speed. Thus a motor with a normal speed of 300 r.p.m. would have a range of from 150 to 600 r.p.m., or a 1:4 ratio. Special controllers for all three methods are supplied by various manufacturers. Manually operated controllers are in general use, but automatic controllers with remote control are gaining in popularity. Automatic control usually consists of start-and-stop push-buttons with a manually operated

FIG. 4. WOUND-ROTOR ALTERNATING-CURRENT MOTOR BELTED TO VENTILATING FAN

speed-changing rheostat located on the man control panel, but it is also possible to control the speed by means of push-buttons.

There are two types of alternating-current polyphase

motors in general use for operating ventilating fans—the squirrel-cage and the wound-rotor. As stated previously, the smaller sizes are not made for slow speeds, so that they are ordinarily belted to the fans they drive, chain drive being too noisy for most applications. The squirrel-cage motor is the simplest type made, so that it is always used whenever possible. It is, however, strictly a one-speed machine, so that it can be used only for constant-speed applications.

The wound-rotor motor is similar to the squirrel-cage motor in construction except that resistance can be introduced into the rotor circuit. The use of resistance in the rotor circuit gives speed ranges similar to direct-current armature control; namely, it provides speeds less than normal, the minimum reduction being about 50 per cent, and as there is a loss of energy in the resistance, it is not particularly efficient except at small speed reductions. However, the wound-rotor motor will be found to be satisfactory for the great majority of ventilating applications, so that its use can be taken for granted in localities where alternating current only is available, unless a careful analysis of the application shows that a direct-current motor is necessary.

FIG. 5. MANUALLY OPERATED DIRECT-CURRENT STARTER AND SPEED REGULATOR

FIG. 6. AUTOMATIC DIRECT-CURRENT CONTROLLER FOR USE WITH PUSH-BUTTONS STATIONS

Wound-rotor motors are also frequently used for strictly constant-speed ventilating service because they can be started without taking excessive current from the line, whereas a squirrel-cage motor takes considerably more than full-load current during the starting period. Under certain conditions this characteristic makes the squirrel-cage motor objectionable, in which case wound-rotor motors are used instead. Alternating-current motors can be equipped with either hand-operated or automatic starters. Single-phase motors, in sizes up to 10 horsepower can be operated from lighting circuits and are extensively used for driving fans and blowers.

The horsepower required to drive a centrifugal fan can be estimated from the following formula:

$$HP = \frac{5.2 \times Q \times WG}{33{,}000 \times E}$$

Where

HP = horsepower of motor

Q = cubic feet of air per minute;

WG = pressure of air in inches by water gage;

E = efficiency of fan, which is approximately 0.5 for steel plate fan, 0.65 for Sirocco, and 0.45 for cone type.

The following mathematical relations are theoretically true for all centrifugal fans: Volume of air delivered varies as the speed; pressure of air delivered varies as the square of the speed; horsepower required varies as the cube of the speed. Hence a fan, which, when operated at 100 r.p.m. delivers 8,000 cu.ft. of air at $\frac{1}{2}$ oz. pressure, requires 2 hp., will, when operated at 200 r.p.m., deliver 16,000 cu.ft. of air at 2 oz. pressure and will require 16 hp. It is therefore necessary, in selecting a motor for driving a fan, to figure the horsepower requirements at the maximum speed of the fan.

Reduction of the area of the inlet or discharge of a centrifugal fan will reduce the power required, whereas increasing the area of either will increase the power. Disc fans produce a relatively low pressure, and it is customary to figure their output on a basis of velocity. The relations between volume, speed and horsepower for centrifugal fans hold good for disc fans. However, if the outlet of a disc fan is restricted, the pressure and the power required increase, and with a totally closed outlet the power needed is almost double that with an unrestricted outlet.

Simple Belt-Lacing Rules

By E. J. Black

Improper belt lacing is probably the most common mistake that users make in the use of belting, and it is the one that causes the most serious trouble. By observing the following simple rules the belt will give the best service of which it is capable.

Cut the ends of the belt square. Do not depend upon your eye or use an ordinary ruler. If the end is cut slantwise, the pull will come on one side.

Make the holes as small as practicable. Use an awl rather than a punch, whenever possible.

Leave a sufficient margin at the edge of the belt without holes so as not to impair its strength. In belts 2 to 6 in. wide the holes should be at least $\frac{1}{2}$ in. from the edge, $\frac{5}{8}$ in. for belts 6 to 12 in. wide and $\frac{3}{4}$ in. for belts 12 to 18 in. wide.

Make two rows of holes in parallel lines straight across the width of the belt, and stagger them so that the strain comes upon different portions of the belt.

Be sure that the holes in the two ends to be joined match, otherwise there will be a "jog" in the belt, and this may result in tearing the belt lengthwise.

Use flexible lacing, being careful to have it proportionate to the size of the belt. A heavy lacing is likely to cause trouble.

In lacing the belt, make the pulley side as smooth as possible. Rough places and ends should be turned away from the pulley.

In using metal fasteners on rubber belts, select those which place the strain on the lengthwise strands. The crosswise strands are not so strong as those which run lengthwise.

Charts for Graphical Determination of Pipe Sizes and Velocity of Flow of Steam

By H. M. BRAYTON
Mechanical Engineer and Designer, Semet Solvay Company, Syracuse New York

A PREVIOUS article by the writer in the April 27 issue gave graphical charts for the determination of pipe sizes and steam velocities without calculation. The theory of the charts was given sufficiently to enable the reader to construct them himself and on any desired scale. The formulas given for the flow of steam involved the number of cubic feet of steam or quantity which passed a given section in a given unit of time. This quantity is not always desired, and if omitted the formulas may be combined and much simplified.

To recall these formulas they will be repeated here. Equation 1 gave the following:

$$Q = A\ V = 0.00545\ D^2V$$

in which

Q = Quantity in cubic feet per minute which passes a given section of pipe;
D = Diameter (internal) of pipe in inches;
V = Velocity of steam in pipe in feet per minute;
A = Internal area of pipe in square feet;

and equation 2 gave the following formula:

$$Q = W\ V'$$

in which

Q = Quantity as before;
W = Weight of steam flowing in pounds per minute;
V' = Specific volume of steam or number of cubic feet per pound. Chart for dry steam only.

It will at once be apparent that these two formulas may be combined and eliminate Q thus:

$$Q = 0.00545\ D^2V = W\ V'$$

in which the symbols designate the same values as before. This formula may be written,

$$\frac{0.00545\ D^2}{V'} = \frac{W}{V}$$

As this equation is a straight proportion, it may be built up into another form of Z chart. Such a chart would have the advantage that all reasonable values of the diameter may be given on one chart instead of two as was necessary when the value of Q was included.

Fig. 1 shows a chart drawn up for the foregoing formula. To use this, it is merely necessary to either know or assume four of the variables, and the other may be quickly found. It is more often desired to obtain the diameter of pipe necessary or the velocity of steam in a given size of pipe. In the illustration shown by means of dot-dash lines all the values are assumed except the weight of steam which may be expected to pass through the pipe. The values chosen are: Pressure, 230 lb.; pipe size, 12 in.; velocity of steam in pipe, 10,000 ft. per min. The pressure of 230 lb. represents a specific volume of about two as shown from the curve. This point is merely connected with the value of the diameter, and then the point representing the velocity on its scale is connected through the intersection of the first line with the diagonal and extended until it cuts out the value of the weight W on its scale. In a similar manner if the reader will connect the values of V = 12,000 and W = 30 found on their respective scales and mark the point of intersection of this line with the diagonal and then draw a line through this point on the diagonal to the value of D = 4 in., this line will be found to cut out a value on the V' scale of about 4.3. If this point is carried up to the pressure-volume curve and projected down to the pressure scale, it will be found to read 105 lb. This means that this pressure would be necessary to enable a 14-in. pipe to carry 30 lb. of dry steam at a velocity of 12,000 ft. per min. All this procedure is the work of but a few seconds and gives the results well within the desired range of accuracy.

The theory of this chart is somewhat different from those given in the previous article and will be given

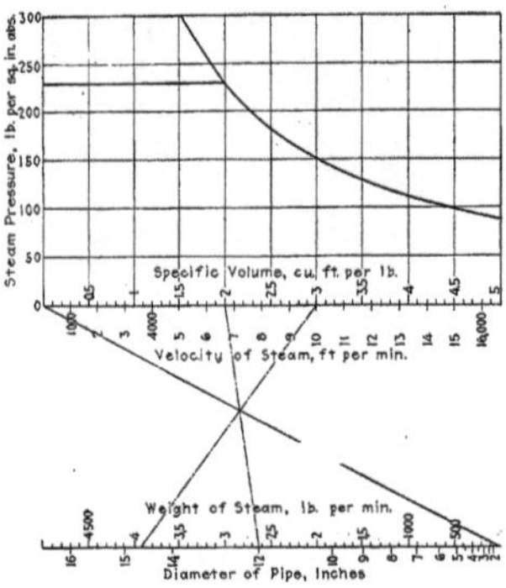

FIG. 1. CHART TO DETERMINE PIPE SIZE OF STEAM FLOW IN PIPES

here to enable the reader to construct it at his own convenience and at any size he may desire. Fig. 2 illustrates the method of procedure. Two parallel lines connected by a diagonal are drawn as shown. These lines may be any desired distance apart, as the theory is independent of the angle which the diagonal makes with the verticals. The chart is based on the simple laws of proportion, and these must be rigidly held to in determining the equations of the scales. Proportion is here illustrated by similar triangles which are formed by the dash-dot lines used in solving the formula by means of the chart. The equations of the scales follow:

$$X = m_1\ (0.00545\ D^2)\ ;\ Y = m_2\ V';\ Z = m_3\ W;\ T = m_4\ V$$

The values of m_1, m_2, m_3 and m_4 are known as the modulus of the respective scales and may be of any value the designer desires. Their actual numerical value is determined on the limits of the variables and the size of the chart which it is desired to construct. These two points must be settled upon before actual construction can begin. The limits which the writer chose in this case are shown on the chart, Fig. 1. In general the greater the limitations desired in the chart the better accuracy obtainable. If the reader never has occasion to deal with steam pipes larger than 10 in., for example, the chart should not be made for larger size.

Once the limits are determined and the length of the various scales settled, it is a simple matter to determine the values of the moduli. It will usually be found convenient to make the scales about 10 in. long. This

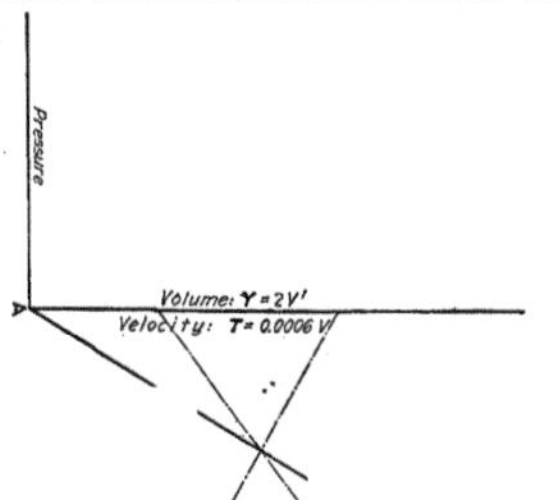

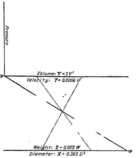

FIG. 2. ONE OF THE STAGES OF CONSTRUCTING THE CHART SHOWN IN FIG. 1

10 in. will then represent the value of X, for example, when D has its maximum value. With X equal to 10 and the maximum D known, it is simple arithmetic to solve for m_1. The other moduli are found in the same manner except m_4. When the other three have been determined, then m_4 is fixed because from the nature of the proportion and similar triangles the following must hold:

$$\frac{m_1}{m_2} = \frac{m_3}{m_4} \text{ or } m_4 = \frac{m_2\ m_3}{m_1}$$

It is, then, necessary only to select the first three moduli and the fourth follows automatically. Of course it is possible to definitely assign values to any three of these moduli and calculate the fourth. The reader must not assume that only the first three can be determined by the designer. In fact, it often happens that it is better to select the others. The designer has a free hand in this matter. Again, it may be found that the fourth scale calculated comes absurd; that is, much too small or too large in proportion to the other scales. In such a case the designer may be able to use a multiplier in the proportion. He may multiply the numerators or the denominators of the fractions by some number. Of course it is evident that if the numerator of one fraction is multiplied by a number in this manner, the other must be also; otherwise the proportion would not hold and the formula would no longer be true. In the formula being considered it would be permissible to multiply the V' by a number if the value of V were also multiplied by the same number. Simply multiplying the two terms of a fraction by a number does not affect the results. In some cases, when the moduli do not work out well, it may be necessary to change the limits of the variables or the length of the scales.

In the chart shown here the following moduli were selected:

$m_1 = 6.67$, $m_2 = 2$, $m_3 = 0.002$, $m_4 = 0.0006$. It is always advisable to select these values in even numbers when possible. By this is meant not necessarily even digits, but numbers that are easy to lay out, such as the value of m_2 and m_4 given above. This is not always possible, but should be kept in mind and will save time in constructing the chart later on. The principal point in selecting these values is to make all the scales take all the room allowable. It will be observed in this chart that all the four scales are of about equal length and take the full length of the line. The equation of the scales in our problem then becomes,

$$X = 6.67\ (0.00545D^2) = 0.0365D^2, = Y\ 2\ V';$$
$$Z = 0.002\ W;\ T = 0.0006\ V.$$

With these equations it is a simple matter to lay out the scales. With different values of the variables substituted into their respective equation, a corresponding value of X, Y, etc., is obtained. These values are distances and are to be laid off from the point of intersection with the diagonal. Each point must be marked with the value of the variable used in obtaining the distance. For example, in the V' equation when $V' = 2$, then $Y = 4$ in., which means that a point 4 in. from the starting point is marked and labeled 2 cu.ft. on the diagram.

The pressure-volume curve is plotted from values taken from the steam tables for dry or saturated steam. This curve is plotted to enable one to quickly determine the pressure without reference to the tables.

British Control of Fuel Oil

In a recent address, Paul Foley, Director of Operations, United States Shipping Board, stated that although 65 per cent of the world's current petroleum production is being drawn from the United States, practically all the visible future production is under the control of Great Britain. Furthermore, although production in the United States is available to all other nations on equal terms with our own, that under British control and located in the middle area of the world is available only to British nationals. The practical effect of these restrictions is that while British ships can secure fuel oil in United States and Caribbean ports on the same terms as our own and also obtain what they require overseas from the nearest producing field, American requirements overseas can be met only at British terms and must reflect the long haul from American and Mexican seaboards. An important step in correcting this evil is that we should take advantage of the fact that the control which our British cousins enjoy has been attained by means of capital loaned by the United States Government. With this in mind we should insist that the same principle of equal opportunity now extended to British nationals in American producing fields be extended to American nationals in British producing fields.

Overhauling and Aligning the Engine

Detailed Instructions Regarding the Alignment of the Connecting Rod, the Crosshead Guides, Crosshead-Shoe Adjustment, Vertical and Horizontal Alignment of the Crankshaft, Worm Bearings and Crankpin Error

By THOMAS W. AIREY

IT IS obvious that in the course of time the power-plant engine will require overhauling and repairing, and all repairs must be skillfully executed and the final adjustments made in accordance with the alignment of the various parts. The want of alignment may have been made manifest by the constant heating of the main bearing, the heating or pounding of the crankpin brasses or by a knocking in the cylinder and, possibly, by an appreciable loss of power. As a general rule this want of alignment is induced by wear or is incurred by repairs or adjustments that are made from time to time without due consideration being given to the alignment.

An error of alignment may exist in any direction, but the error may be located, if the test is made at four equidistant parts of the revolution; for instance, on the two dead centers of the crank and at the highest and lowest points, or half-stroke of the path of rotation of the crankpin.

It is not an uncommon practice for the engineer to use the connecting rod as a means of locating an error of alignment in any of these directions because, owing to the length of the rod, the error becomes magnified and is thus more plainly discernible but in using the connecting rod as a means of testing the alignment one must know that the bore of the crank brasses are at a true right angle with the center line of the rod.

Aligning the Guides

A not uncommon method employed by engineers for aligning the guide with the cylinder and main bearing is by using the piston and gland after the rings have been removed from the piston and the packing from the stuffing box. The bottom shoe of the crosshead is then adjusted until the gland enters the stuffing box freely when tried from any part of the stroke. Suppose that the piston is a good fit in the cylinder and that the gland is a loose fit on the rod; then the piston rod would fall in the direction of the line *AB*, Fig. 1, *AA* representing a line through the center of the cylinder bore and in alignment with the center of the main bearing *C*. Then it follows that if the guide is set by the gland only, it would be set to the line *B* and would, therefore, be out of alignment with the main bearing, as shown.

Suppose that the gland was a good fit to the piston rod and stuffing box, but that the piston was a slack fit in the cylinder; then the gland would not remain tight because the piston, being suspended in the cylinder, the overhanging weight of the piston would cause rapid wear of the neck ring, the piston-gland bore and also of the guide surface, and in a direction parallel to the line *AE*, Fig. 2, thus gradually letting the head of the piston fall to the bottom of the cylinder bore, touching the end of the cylinder first at *F*; thus the close fit of the piston in the cylinder and the gland to the piston rod and stuffing box are alone depended upon as a guide whereby to set the piston rod and its guide in alignment with the center of the main bearing.

The better method of aligning is that of stretching a line through the cylinder and stuffing box, the line being passed through a small hole in the end of a piece of iron plate *G*, Fig. 3, that is secured to the head end of the cylinder by one of the stud bolts. The line is then passed through the bore of the stuffing box and over a supporting bracket *H*, a weight *I* being attached at the end to keep the line tight. The line *AA* is then set central to the bore of the cylinder and stuffing box, as shown.

To test the guide path for alignment (supposing the guide to be of the channel pattern), take a height gage *G*, Fig. 3, and placing the base on the surface of the path *J*, adjust the point of the gage until it just touches the line at *K*. Then move the gage to the other end of the guide and test the distance *L*, and if *K* and *L* are coincident, then the guide is in alignment. If *K* and *L* are not coincident, the guide must be scraped down until the two measurements coincide, and finally bedded to a surface plate. The top surface of the guide must next be tested for parallelism with the bottom surface and, if incorrect, must be treated in a similar manner.

Next make a gage *M*, Figs. 4 and 5, having a true and straight edge and fit it in between the flanges of the path at one end, as at *N* and *O*, Fig. 4. Then take a square *P*, and while pressing the back firmly upon the guide surface, bring the blade gently up to touch the line *A*, holding the square firmly in this position, and with a fine-pointed scriber draw the line *Q* on the gage *M*, using the blade of the square as a guide. Then move the gage to the other end of the path and, with the square held firmly on the guide path surface, test the line Q, already marked on the gage, with the line *A*. If the two coincide, then the inside faces of the path flange are true and parallel to the line *A*. Should, however, the width between the inside faces of the guide flange, when tried at each end of the gage, not coincide, then the inside faces must be made true and parallel with the line *A*.

Testing the Crosshead

Take the crosshead *R* and fit a true taper mandrel in the piston-rod bore and then place the mandrel between lathe centers, as shown at Fig. 6. Then adjust the crosshead in the lathe centers until the side face of the crosshead is true to a square tried from the lathe bed. With a scribing block, the point being set to the lathe-spindle center and the base resting on the lathe bed, draw a line across the boss end of the crosshead and also the gudgeon end of the crosshead, as shown at *S* and *T*. Then move the crosshead round in the centers one-fourth turn, as shown in Fig. 7. Setting the sliding surface of the crosshead true with the square from the lathe bed and with the same setting of the scribing block, draw a line across the ends of the crosshead shoe, as shown at *U V*, and mark the line lightly with the center punch, taking the measurement from the gage *M*, Fig. 5. Check them with the crosshead or take the gage *M* and lay it on the end of the crosshead shoe, as shown by Fig. 6, the point *Q* being set to coincide with the lines *WW* on the shoe. If the crosshead is true, the ends of the gage *N* and *O* should be coincident with the face of the crosshead shoe. If, however, the two are not coincident, and it should be found necessary to provide new shoes, the measurements must be performed in strict accordance with the marking *NQO* on the gage *M*. The top of the path and the top shoe of the crosshead must also be tested in like manner.

The bottom shoe of the crosshead may next be adjusted until the distance from the sliding surface of the shoe to the center line *ST*, Fig. 6, and the distance from the surface of the guide path to the line *A*, as shown at *K*, Fig. 3, are coincident. The bottom shoe must also be tested for parallelism with the center

its bearings under such conditions, the crankpin journal would heat, and if any end play existed there would also be a knocking or pounding. In order to bring the crankshaft into true alignment, a side adjustment of the main-bearing brasses must be made until the line *AA* cuts the center of the crankpin journal when tried on both dead centers.

If the connecting rod is used to test the vertical alignment and it is coupled up to the crankpin, it would be held parallel with the line *H*; thus the other end of the rod would be thrown over to one side of the crosshead pin. Then move the crank round to the other dead center, and the small end of the rod would still be thrown over to the same side of the crosshead pin, but to a lesser degree. Thus, by testing the alignment with the crank on the dead center nearest the cylinder, the error is more plainly indicated.

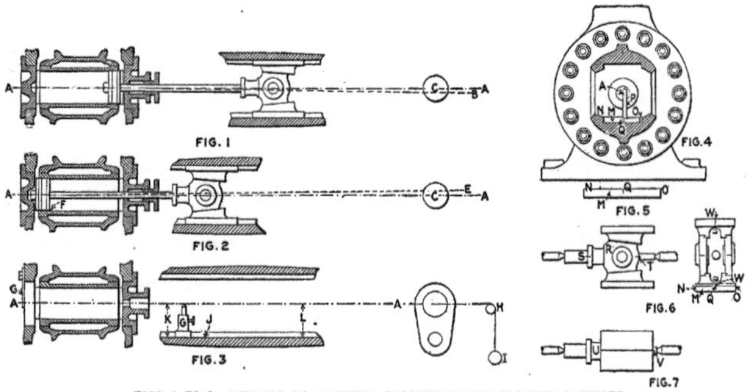

FIGS. 1 TO 7. DETAILS OF ALIGNING CROSSHEAD AND CROSSHEAD GUIDES

line *ST*. The crosshead is then ready for bedding in the guide, the top shoe being adjusted to the desired fit.

To test the vertical alignment of the crankshaft, or to ascertain if the crank is parallel with the crosshead journal, move the crank *W*, Fig. 8, around to nearly the dead center or until the crankpin *XH* just touches the line *AA* that is central through the cylinder, as shown. With a pair of calipers test the distance *YA* with that of *ZA*, Fig. 9. Then move the crank around until the pin touches the line *A* on the other dead center and test the distance *AB* and *AC*. If all measurements coincide, the crankshaft is in true alignment. If, however, *ZA* and *CA* are equal, but measure less than *YA* and *BA*, the center of the crankpin journal length will be outside the line *AA*. This may be caused by the flange of the main-bearing brass being too thick, if a new one has been put in.

Supposing that *YA* and *BA*, Fig. 9, are not coincident and that the crankshaft, instead of being in alignment, as represented by the line *D*, Fig. 10, is in the position represented by the lines *E*. Then the center of the crankpin journal would be on one side of the line *A*, as at *F*, on one dead center, and on the opposite side of the line *A*, as at *G*, when tried on the other dead center. If the connecting rod were coupled up to

The reason for this will be seen by referring to Fig. 10, where *AA* represents a line true with the cylinder bore and the guide; *E* is the center line of the crankshaft, and *F* and *G* represent points central to the length and diameter of the crankpin. The center line of the connecting rod is represented for one dead center by a line passing through *H* and *F* and for the other dead center by *JG*. It will also be observed that the point *J* falls on one side of the line *AA* and that the point *G* falls on the opposite side, although the center line of the connecting rod is the same in both cases cited.

It is not always wise to couple up the connecting rod to its respective bearings even though it is known that the crankpin is in true alignment unless one is assured that the bore of the large and small-end brasses are at a true right angle to the center line of the rod. Even then it is better to take no risks, especially as an error of 1/64 in. in a 3-in. length of journal will throw the rod out of alignment 3/8 in. in a 6-ft. length.

Couple the large end of the rod to the crankpin and tighten up the adjusting screws so as to just be able to move the rod. Then try the other end of the rod with the crosshead, and if the rod points true with the jaws of the crosshead, it is in alignment. If the rod, instead of being true with the crosshead, points over to one side, as represented by the line *K*, Fig. 11, the bore of the crankpin brasses must be scraped out at *L* and *M* and rebedded until the rod points true in the crosshead, or parallel with the line *AA*. The rod should also be tried for alignment with the crankpin by coupling up the small end of the rod and lowering the other end onto the crankpin. When it is true, it will fall laterally between the collars of the crankpin.

After making repairs or adjustments, it is a good plan to check up the piston clearances, for if the clearances are kept right no ridge should form at the ends overrun a little into the counterbore, but it is possible of the cylinder or guide, as part of the piston head will that a small ridge may form at the cylinder ends due to the weight of the piston head and rings, which pass but part way over the end surface of the cylinder. It therefore becomes obvious that the connecting rod should be kept at its original length, as near as possible, by using suitable liners with the brasses. If this is neglected, a ridge may form in time in the cylinder, and if due consideration is not given to the changing length of the rod it may become so adjusted as to cause the piston head to strike the ridge, a knock in the cylinder being the result.

A method of testing the horizontal alignment of the crankshaft, viewed from the crankpin end of the engine, as adopted by some engineers, is to take a spirit level and place it across the guide path or on some machined part on the top of the cylinder and note the position of the bubble. The level is then placed along the top of the crankshaft, and if it indicates that the shaft is wearing low in one of its bearings, the main bearing brass is set up until the bubble in the level coincides with the position when tested on the cylinder top. Another method sometimes adopted for testing the horizontal alignment is to find the center at the end of the crankshaft, as shown at Fig. 12. From the center scribe the dotted circle *N* of a diameter equal to that of the crankpin shoulder. If there is no crankpin shoulder, then the circle must equal the diameter of the crankpin. The crankpin is then placed on the dead center and a straight edge *O* placed with the edge coincident with the shoulder of the crankpin and the perimeter of the circle *N*. The spirit level is then placed on the straight edge, as shown, and the main bearing is raised until the bubble stands in the same position as when the level is placed on the cylinder bore or along the piston rod or guide path.

Supposing that the main bearing has worn low, this would cause the crankshaft to drop from its normal position, which is in line with the center line of the cylinder guides and connecting rod, to below the center line; then the crank would be out of alignment as represented by the line *P Q*, Fig. 13. Should the wear of the brasses be excessive, the result would be serious, as the crank, dropping low in its bearing, would cause part of the piston movement and pressure that should be exerted on the crank after leaving the dead center to be exerted on it before it reaches the dead center, thus causing a back pressure and involving a loss of power.

If the connecting rod is used as a means of testing the horizontal alignment of the crankshaft, the test should be made with the crank in its highest or lowest, or half-stroke position. Should the test be made with the crank on the dead center, assuming that the vertical alignment of the crankshaft is correct, then the connecting rod end would fall laterally between the crankpin shoulders; whereas, if the test were made with the crank on its highest and lowest position, or half-stroke, and the bearing was wearing low, the error would be plainly seen, the connecting rod falling on one side of the crankpin when in its highest position and on the other side when tried with the crank in its lowest position.

This will be more clearly understood by referring to Fig. 14, where *AA* represents a line through the crankshaft, which is somewhat out of parallel with the crosshead journal, owing to the wearing down of one of its bearings. *R* and *S* are lines representing the crankpin in its highest and lowest positions, and *T* represents a line through the crosshead pin which is in true alignment with the cylinder and guide. *U* and *V* are points central to the diameter and length of the crankpin, and *WW* denotes what would be the path of revolution of the crankpin were the crankshaft in true alignment with the line *TT*; but as the shaft is out of alignment, the path of revolution is from *U* to *V*, the center of the crankpin falling on one side of the line *WW*, as at *U*,

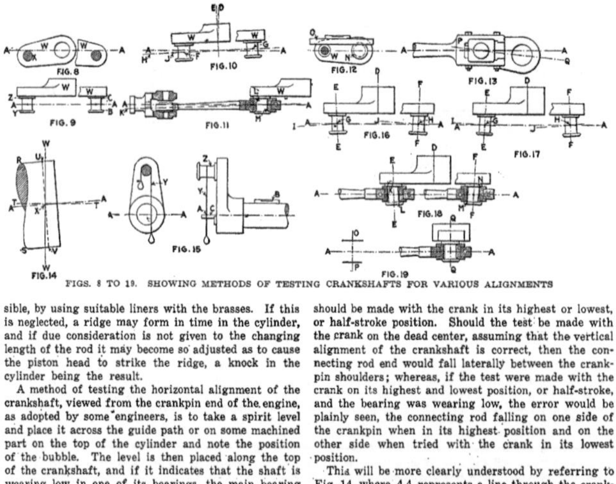

FIGS. 8 TO 19. SHOWING METHODS OF TESTING CRANKSHAFTS FOR VARIOUS ALIGNMENTS

in its highest position, and on the opposite side, as at *V*, at its lowest position. Suppose that the connecting rod is coupled up and adjusted to the crankpin *R*, then the bore of the crankpin brasses would be held in a position parallel to *R*, and the bore of the small end would be held parallel to *TT*. When the crank is in its lowest position, the bore of the crankpin brasses will be held parallel to *S* and the small end will be held parallel to *TT*. Starting from the dead center, *X*, the twisting of the rod will gradually increase, attaining its maximum when the crank reaches its highest point, or half-stroke, and then gradually diminishing until the other dead center is reached, when on passing over, the twist will again be taken up, attaining its maximum when the crank reached its bottom or half-stroke position, as shown at *V*, and again gradually diminishing until the dead center is again reached.

If the connecting rod is used as a means whereby to test the horizontal alignment of the crankshaft, the test must be made with the crank in its highest and lowest, or half-stroke position, for reasons explained in reference to Fig. 14. To test the horizontal alignment of the crankshaft by a line *AA*, Fig. 15, through the cylinder and guide path, use a plumbline hung over the crankpin and adjusted to a position central to the crankpin length, as at *Z*. If the shaft is in alignment, the plumbline should just touch the line *AA*, but if otherwise a spirit level is placed along the top of the crankshaft, as shown at *B*, and the main bearing is adjusted until the bubble stands in the center and the plumbline *Y* just touches the line *A*. If the center of the crankshaft and the line *A* are coincident with a square *C* held against the face end of the crankshaft, the edge of the blade should coincide with the shaft center and the line *A*, as shown.

Testing Alignment of Crankpin

Testing for crankpin error in alignment with the crankshaft is more difficult. It has already been pointed out that an error in alignment may exist in any direction, but if the test is made at four equidistant parts of the stroke, as, for instance, on the two dead centers and at the highest and lowest or half-stroke position of the crank, the error may be located. If the connecting rod is coupled to the crankpin, the crank being placed on the dead center, and the small end of the connecting rod is uncoupled, it will be difficult to determine, should the small end be out of alignment, whether the error is due to want or alignment of the crankpin or to the crankshaft itself. However, there is this difference between the two cases. If the error is due to want of alignment of the crankshaft, the small end of the rod will point out on the same side of the crosshead pin no matter on which dead center the crankpin stands; but when it is due to the crankpin being out of alignment, the rod end will fall on one side of the crosshead pin on one dead center and on the opposite side when tried with the crank on the other dead center.

It has also been shown that when the shaft is out of alignment, the center point of the crankpin would fall on one side of the cylinder center line *AA* on one dead center and on the opposite side when on the other dead center, as at *F* and *G*, Fig. 10, and at *U* and *V*, Fig. 14; but when the crankpin is out of alignment with the crankshaft, the path of the center point on the pin will remain in the true plane.

This is shown at Fig. 16, where *D* represents a line through the crankshaft; *E E* represents the center through the crankpin; *F F* represents a line through the center of the crankpin when on the other dead center; *G* and *H* are points central to the crankpin on the respective dead centers; and *A A* is the cylinder center line and also represents the path of rotation of the crankpin center point. Thus, during the path of rotation of the center points the crankpin remains in the true plane.

Crankpin Out of Alignment

Suppose that the crankpin is out of alignment in the direction shown in the figure and the connecting rod is coupled up to the crankpin; then the center line of the connecting rod would be held in the position represented by the line *I* and the small end of the rod, being coupled up to the crosshead pin, which is at right angle to *A A*, the rod at that end is held in a position parallel to *AA*. Then the adjusting of the brasses to their respective journals would tend to bind or spring the rod, one end being held parallel to *A A* and the other parallel to *I*. On the crank reaching the opposite dead center, *FF* representing its center line and *J* the center line of the connecting rod at right angle to *FF*, the want of alignment in the pin would throw the small end of the connecting rod on the opposite side of the crosshead on the cylinder center line *AA*, as represented by the line *J*. Thus the connecting rod is sprung twice at each revolution and is twice jammed from side to side of the crosshead at each revolution.

Suppose the error in the alignment of the crankpin is shown to exist in an opposite direction, as in Fig. 17, *EE* representing a line through the crankpin on one dead center. *AA* is a line true to the cylinder bore, *D* is the center line of the crankshaft, and *G* and *H* are points central to the diameter and length of the crankpin. It will be obvious that the position of the connecting rod will be reversed from that shown by Fig. 16, and that in coupling the connecting rod to the crankpin the rod would be held in the position of the line *I* and the rod end will be thrown over to the opposite side of the crosshead or cylinder center line *AA*, as shown. As the crankpin reaches the other dead center, the rod will be held in the position parallel with the line *J*. It is by this difference that it is possible to know in which direction the crankpin is out of alignment.

Supposing that the crankpin is out of alignment when the crank is in its top and bottom or half-stroke position, then with the connecting rod coupled to the crankpin it would fall to one side of the crosshead pin with the crank on its top half-stroke, and to the other side when tried with the crank on the bottom half-stroke. Thus again the direction in which the crankpin is out of alignment may be known by its position.

The result of an error in the crankpin alignment is shown by Fig. 18, where the large end of the connecting rod is coupled up to the crankpin. If the connecting rod is coupled up at the small end to the crosshead pin, which is in true alignment, the connecting rod will be held true and parallel to *AA*. It will be seen that the bore of the crank brasses come in contact with the crankpin only at the points *K* and *L* on one dead center and at *M* and *N* on the other dead center. If the crank is moved around, bringing the crankpin at the top of its revolution, as shown in Fig. 19, the center line of the pin will stand parallel to that of the crosshead pin, *O P* representing the crosshead pin and *Q Q* a line through the crankpin.

Play will therefore exist between the pin and the bore of the brasses. Starting from the dead center, as shown by Fig. 18, the play or lost motion gradually increases, attaining its maximum when the pin reaches the half-stroke position, as shown by Fig. 19. On passing the half-stroke position the lost motion gradually diminishes until passing the dead center, when the rod is thrown over to the side of the crosshead pin and at the same time the crankpin pounds in the connecting-rod brasses.

On the crankpin reaching its bottom half-stroke position, the lost motion has again reached its maximum and again diminishes as it reaches the other dead center, and on passing over the rod is again forced to one side of the crosshead pin, but in an opposite direction, and at the same time a pounding of the crankpin in the connecting-rod brasses again occurs. Thus the rod is thrown over to the side of the crosshead twice in each revolution, and the crankpin pounds twice in each revolution. If an attempt is made to stop the pounding by tightening up the brasses to take up the lost motion, the end will be sprung or twisted and heating will result.

Improving an Inefficient Water Power-Plant

By John S. Carpenter

A few changes in the shape of the flume resulted in increasing the output of a 66-in. vertical hydraulic turbine, from about 92 per cent to 107 per cent of its rating.

This turbine, which was direct-connected to an umbrella-type generator on the floor above, was rated at 600 hp. under 15 ft. head. The owner complained that he could only get 554 hp. out of the unit with the gates wide open. The waterwheel builder disclaimed all responsibility when he found that the

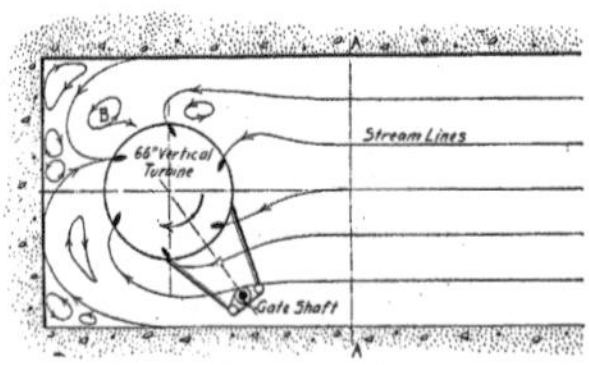

FIG. 1. ORIGINAL SHAPE OF FLUME, SHOWING STREAM LINES

true head acting on the wheel was only 14.5 ft. as against 15 ft. and left the owner to his own devices. Power, the owner said, was worth at least $150 per hp. per year and he could afford to spend up to $5,000 to get the plant running so that the wheel would give its rated power.

After the plant was thoroughly gone over, among other things it was found that the water velocity at a cross-section marked *AA* in Fig. 1 was about 3.1 ft. per sec. at full gate. This velocity is quite high, since good practice would limit the velocity at such point to 2 ft. per sec. The square corners in Fig. 1 are not favorable to good efficiency, and at the point marked *B* there was a distinct pocket about 10 in. deep and about 12 in. square in plan. This was credited to the high velocity and the reversal in the stream at that point. Further trouble from eddies being suspected, the owner was asked to dump a quantity of

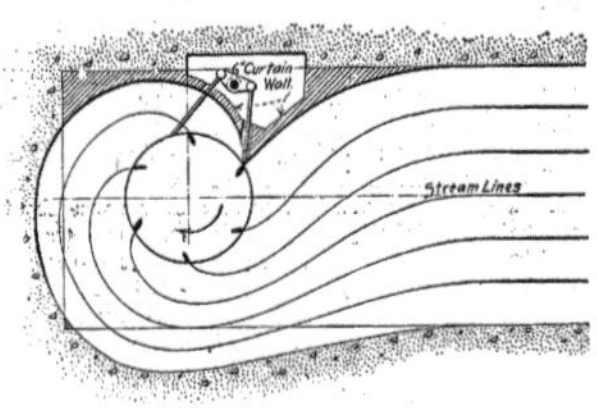

FIG. 2. SHAPE OF FLUME AS ALTERED, SHOWING IMPROVEMENT IN STREAM LINES

shavings and thin sticks into the penstock and take several flashlights of the portion above the wheel. The stream lines shown in Fig. 1 were traced from the resulting photographs of the direction and distribution of the sticks.

It is to be regretted that the actual photographs cannot be reproduced on account of their being lost, negatives and all. The lines of flow were rather well defined back to a point behind the gateshaft; beyond that it was difficult to trace out the paths, because the water flowed more or less uniformly. It should be mentioned that to get the effect shown it was necessary to close down the head gates about halfway to lower the headwater. It will be seen that the gateshaft was put in about the worst place possible, as far as efficiency goes. The reversals and eddies are very plainly shown on the original photographs and caused quite a loss in power.

To straighten out matters, the writer recommended that the flume be made spiral in form as shown in Fig. 2. This meant cutting into a concrete wall as much as 30 in. in one place, replacing the reinforcing steel and building a projection on the outside of the wall. The proposition was looked on with anything but favor. Finally, it was decided to cut right through the wall and replace with a new section. While the tearing up was going on, it was found that the gateshaft could be moved to the opposite side of the penstock out of the waterway, and behind a thin curtain wall.

It is not alleged that the outline shown in Fig. 2, a tracing of an actual photograph, is the best possible shape. It is merely a patch job. As to results, by changing the shape as shown and by doing some tailrace excavation, whereby 9 inches more head was obtained, the output was increased to 639 hp., to the joy of the owner. The total cost of the changes was $2,786, and the work was done in the dry season, when water was low and scarce.

AMMONIA COMPRESSOR OF THE HIGH-SPEED TYPE, ELECTRICALLY DRIVEN, AMERICAN ICE CO., WASHINGTON, D. C.

Refrigeration Study Course—IX
The High-Speed Ammonia Compressor

By H. J. Macintire

As ALL engineers know, the four-valve steam engine, and particularly the Corliss type, is the most efficient of all reciprocating engines. And yet there are many medium- and high-speed engines that have a ready sale for all kinds of work and to all locations. The reasons for this substitution lie in the fact of smaller size and smaller first cost of the machine itself. It is not expected that so high a prime-mover efficiency will be obtained, but these considerations, as regards lesser floor space required, lesser engine-room volume, possible use of the exhaust steam for heating purposes, and lower initial cost of the installation, overbalance the disadvantages.

In the refrigerating field, however, the conditions are not exactly the same. A score of years ago rotative speeds of 30 to 40 r.p.m. were usual, then came speeds of 60 to 70 r.p.m., and finally 100 r.p.m. for the larger and 140 to 160 for the smaller machines (usually the twin vertical type). Of late years attempts have been made to increase the speeds still more, and encouraging success has been obtained, the main difficulties encountered being purely mechanical ones due to poor lubrication or faulty machine design. Defects which cause wrecking of the machine and loss of life are infrequent. Designed properly, so as to maintain lubrication and to be amply strong, the future of the high-speed compressor seems bright, especially as the high speeds make it possible to connect directly with synchronous motors in sizes of fifty tons or over. As will be seen later, there is no appreciable loss in economy at high speeds provided a good form of valve is used and large opening area for the valve provided.

As has been shown in another part of this study course, the refrigeration duty is dependent on the weight of ammonia pumped and condensed in the condenser. For any particular set of conditions the weight of ammonia condensed varies with the piston displacement; but the volumetric efficiency varies with different conditions of head and suction pressures, the quality of the suction gas and the amount of clearance. For example, the real volumetric efficiency may be anywhere from 80 per cent to 50 per cent or less. If the volumetric efficiency (the ratio of the effective stroke to the actual piston stroke) remains the same, it means that the capacity of the compressor (the tonnage) may be increased by speeding up the machine.

The Valves Are Gas-Operated

Ammonia compression is similar to air compression, and the same design, with suitable changes, can be used for either kind of work. However, mechanically operated valves for refrigeration have never been a success, on account of lack of constant conditions as regards both suction and head pressure, and also because of the extra difficulty due to having additional packing in the cylinder where the valve stems pass through to the valve chest. The valves are therefore gas-operated, of one design or another.

High-speed compression must have a special form of valve. The reason for this is easily seen on considering the problem. It will be remembered that the suction and discharge valves are opened and closed by the action of a difference of pressure. In the case of the suction valve the pressure must be less on the inside than on the outside of the cylinder, and the discharge valve requires a greater pressure on the inside than on the outside of the cylinder in order to open. This difference of pressure increases with the weight of the valve and the speed with which it is opened (the effect of inertia). Again, as the piston speed increases, the size of the valve openings must increase so as to allow the gas to pass in or out with the least practicable effort.

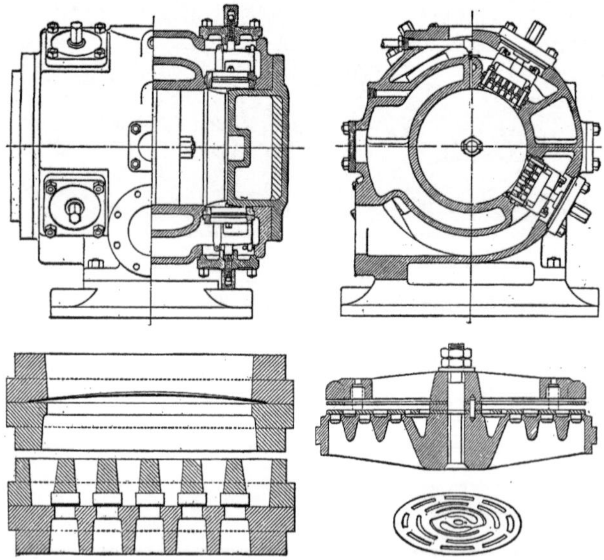

FIG. 1. THE FEATHER VALVE IN SECTION AND AS USED IN THE COMPRESSOR CYLINDER

The effect of inertia can be seen by common examples: A boat in the water, or a train of cars, requires some time at full power to attain even moderate speed. A paper may be jerked from under a book, which if moved slowly could have carried the book along. The skater or the "flivver" starting from rest requires full power to attain speed quickly. These examples are all subject to the law that the force required to change the velocity of a body varies as the product of the weight of the body and the change of velocity.

So it is in the case of the valves of ammonia or air compressors. As the speed increases, the force necessary to open or close the valves quickly becomes excessive. The sudden application of the force necessary for the purpose tends to weaken the material also. For these reasons the demand is for a type of valve that is light in weight, a demand that has been met by means of the ribbon, feather or plate valves.

As has been already mentioned, the valves used for ammonia compressors are those which have found success in their application to air compression, both in moderate pressures and the nominal pressures required for blast-furnace compression. One of the most widely known of these is the plate valve.

The plate valve, as its name suggests, usually is made of a thin sheet of steel, is hardened and desirably tempered, and is ground down to a correct flat surface. In the older types, the Borsig and Gutermuth valves, and the more recent design, the Mesta valve, flexibility is obtained by means of perforating the plate (after the principle of a spiral) so that the movement of the plate is easy of accomplishment.

The main objection to such a design is the fact that the valve is fastened to the seat by means of a machine screw attached to a central part of the spring. This means that the valve lifts off its seat by a bending action similar to the deflection of a cantilever. The operating result is trouble due to occasional breaking on account of repeated stress along the position of maximum bending. Although trouble of this sort is reduced by heat treatment and by selected material used in the plate valve, and is only a slight difficulty used with air compression, yet for ammonia this type of valve has not been entirely satisfactory. One of the latest of the plate valves is a distinct variation from previous plate valves. It is more like a disc and is flexibly held on its seat by means of a spring. In consequence, there is no call for deflection of the material, and no causing of fracture on that account. The valve lifts off its seat more like a disc valve in the water end of a pump.

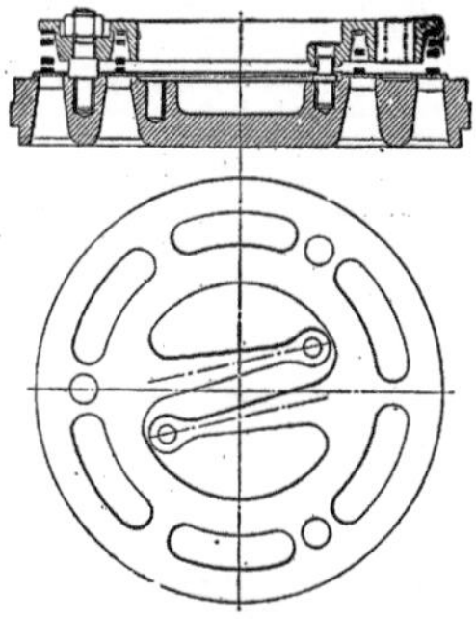

FIG. 2. THE PLATE TYPE OF AMMONIA-COMPRESSOR VALVE

Valves of the ribbon or feather type have not come into common use at present for ammonia compression. They have been used in certain special cases, but as yet have not been put on the market in quantity. However, their success warrants further consideration on the part of the engineers.

The Plate Valve in Action

There seems to be a difference of opinion in regard to the relative advantages of the various types of valves. The main objection to the use of the plate valve seems to be in the alleged large resulting unbalanced pressure.

Referring to Fig. 2, it will be seen that the plate valve has to overlap the ports enough to secure a firm seat. This amount may be $\frac{1}{16}$ in. all around the port. The result is that some engineers claim that an unbalanced pressure is present, due to the much greater area of valve on the top as compared with the area subject to vapor pressure on the underside of the valve. This difference of area would appear to be from 25 to 35 per cent of the port opening. If such a condition prevailed, the compression would have to be carried to an amount greater than the condenser pressure by a similar amount, as for example, 200 lb. in the case of a 150-lb. condenser pressure. However, it is not clear that such a condition exists, as may be seen by the indicator diagram shown, which is taken from an actual test. Although the writer has seen some indicator diagrams which show that an excess pressure is obtained, the majority of diagrams show little pressure in the cylinders in excess of the head pressure.

In the horizontal compressors using the plate or feather valve, the suction and discharge valves are identical, and frequently of the same size, but the vertical high-speed compressors usually have some different type of valve. For vertical compressors, if the old-type balanced valve in the piston head is used, then the poppet discharge valves (usually a number of small valves) are used. If the plate valve is used, then a modified plate valve or valves using a special dashpot attachment for quiet seating is made use of. The latter valve has been especially successful at high speeds, and 400 r.p.m. has been obtained without undue noise. The valve is not one of extreme thinness, like the early air-compression valves or even the older plate valves, but like the more recent plate valve, it is quite appreciably thick and depends on the cushioning effect of the dashpot to reduce the final velocity of closure.

Quite different from the so-called plate valve, but similar in its action, is the "feather" valve. This is an arrangement of strips of thin spring steel, arranged as shown so that the valve, or the strips, bow upward and thereby open a passage for the flow of the gas. The cover plate is designed so as to limit the amount of opening and prevent excessiv deflection, which could cause quick breakage of the valves.

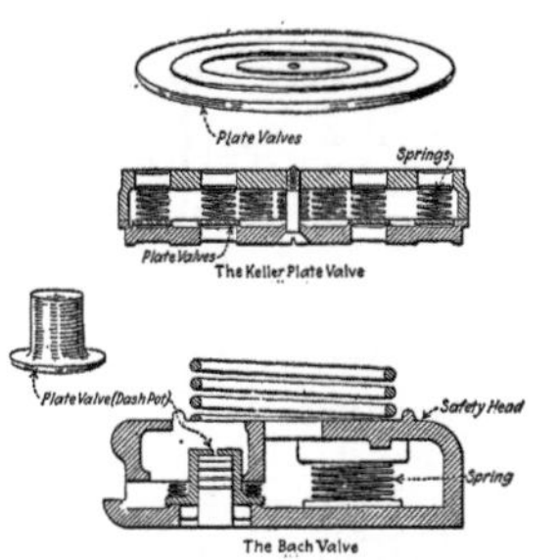

FIG. 3. TWO TYPES OF HIGH-SPEED VALVES

Clearance

For years we have heard of the dire effects of clearance, and how this should be reduced to a minimum. It is true that clearance will reduce the capacity of the compressor, but the power input (horsepower per ton of refrigeration) need not be increased appreciably. This conclusion may be readily seen from the following. The gas remaining in the clearance volume at the end of the compression stroke begins to expand on the return stroke and does *work* on the piston during the expansion. Of course there are factors like compressor friction and faulty expansion, but the energy returned on re-expansion is nearly equal to the work of compression. The result is that it is only necessary to increase the piston displacement when the clearance becomes greater than ½ or 1 per cent. The result is a larger engine, or one using higher relative speeds than if small clearances were possible or practical.

On the other hand, high speed requires careful balancing of the reciprocating parts and enough clearance to bring the compressor to rest at the end of the stroke without knocking. Incidentally, the plate form of valve, because of the cages used to secure it accurately in position and to enable it to function properly, requires a clearance of 4 to 7 per cent. The re-expansion under such conditions varies with the pressure range carried in the compressor (the ratio of the head to the suction pressures) and may be from 25 to 50 per cent. As mentioned, this cuts down the tonnage of the machine which is proportional to the weight of gas entering the *suction valve* each stroke, but the cost of a ton of refrigeration is only slightly greater. The tonnage lost by clearance may be made up, however, by increasing the speed.

Of late years there has been a steady increase in the demand for electric-motor-driven ammonia compressors and for direct connection to the motor. It will be remembered that rope and belt drives are often inefficient, varying as they do from 85 to 95 per cent. In the larger compressors it is possible to secure low-speed motors so that direct connection is quite practical. However, the induction motor, because of its small air gap, cannot be used, and recourse has to be made to the synchronous motor, which has a large air gap and the property of operating with unity power factor if desired. Direct connection to oil engines is employed to some extent although it is not general.

At present the main troubles with this type of compressor are those due to faults in the design and lack of care on the part of the operating engineer. The compressor for high-speed conditions requires proper lubrication at all times. The machine work must be accurate, the lubricating system must be adequate and the material annealed or heat-treated. Finally, the compressor must be balanced.

Seattle's two municipal generating stations, with a combined capacity of 50,000 kw., will be increased by the erection of a new plant. Work has already been started on the first unit, which provides for a capacity of 36,000 kw. The power site is on the Skagit River, 100 miles northeast of Seattle, and electrical energy at 160,000 volts will be transmitted to the city.

Carrick Graduated Combustion Control

In burning coal there should be no sudden changes in draft. Movement of the damper position should be gradual, not closed suddenly and then fully opened as in automatic regulation of the open-and-shut type. When there is a bright fire on the grate and the load drops off, the steam pressure increases. The operator or the regulator closes the damper. Immediately the furnace is transformed into a gas producer owing to the extreme change in the air supply, which has been suddenly reduced, and carbon monoxide, CO, results in the flue gases. Then the load increases, the steam pressure drops and the damper is opened. The combustible gas that has been bottled up is released up the stack unburned, an excess of air is admitted, and there is a decided waste. In either case there is a loss in economy,

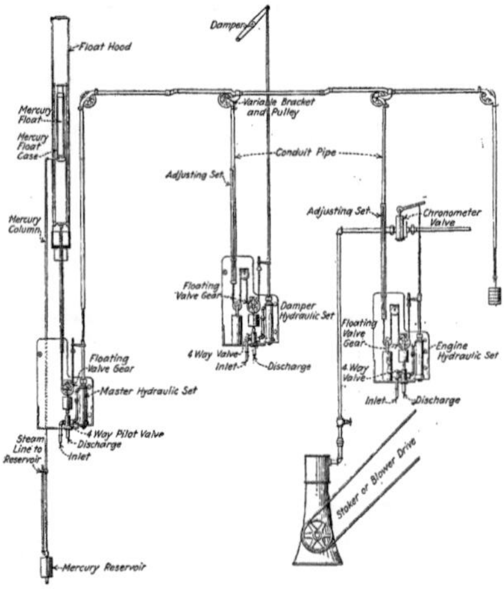

FIG. 1. DIAGRAMMATIC VIEW OF CARRICK SYSTEM OF GRADUATED REGULATION, SHOWING THE MAIN AND AUXILIARY CONTROLS

and with this happening many times throughout the day there is a loss of fuel.

To meet the requirements, automatic control must function so as to graduate over a predetermined range and to hold the damper in a floating position; that is, hold the damper in a definite position for a definite set of conditions and move it to a new position for a change in any of the variables such as flow of steam, feed-water supply or furnace condition. Stoker speed or coal feed should change an amount corresponding to the change in the damper position, and the coal feed and air supply should be changed simultaneously. Every

FIG. 2. TYPICAL AUXILIARY CONTROL

damper, stoker engine or motor and forced-draft blower should be controlled individually or so arranged that the relation of the damper or blower to the stoker speed can be readily changed and properly correlated.

A system has been developed recently by the Carrick Engineering Co., of Chicago, to fulfill these requirements. It controls the fuel feed and damper position jointly to correspond with the load, holding the proper relation for efficient results. Owing to a graduated movement it takes and holds any intermediate position that is required for the load carried and moves to a new position if there is a slight change in the load or steam flow, in feed water, or in furnace conditions. This system is designed to maintain the proper relation between the fuel and air supply by individual control of the stoker engines or motors, dampers, forced- or induced-draft fans.

There is a master control with a column of mercury using the steam pressure to actuate the travel of the mercury instead of atmospheric pressure. The column of mercury is two inches high for every pound of steam pressure, so that for every ounce of pressure change there is one-eighth inch of mercury travel.

Referring to Fig. 1, a mercury reservoir shown in the lower left-hand corner is connected by a column with an upper mercury cylinder which contains a free moving float. The latter, by means of a yoke and side rods, has cable connection to the arm of the hydraulic cylinder, shown at the right of the control panel. The chain connection between the floating plunger yoke and the hydraulic piston passes around the floating wheel or valve gear shown at the center of the panel. Just below this wheel is a four-way pilot valve which controls the flow of water from the city pressure either to the top or the bottom of the hydraulic cylinder, depending on whether the steam pressure rises or falls, thus raising or lowering the piston of the valve. It will be noticed that the wheel lever is free to move with a rotating motion as well as up and down, due to a slot provided in the bracket supporting the wheel. Also the hydraulic piston cannot move until the four-way pilot valve is opened.

Assuming an increase in steam pressure, the result is an immediate increase in the mercury level in the upper well or cylinder; the plunger will rise just as soon as the change in steam pressure reaches one ounce, the movement being one-eighth inch. This will raise the wheel, slightly turning it and at the same time raising the entire valve gear and the pilot valve piston, as the hydraulic piston must remain stationary until the foregoing operation takes place. As soon as the piston of the pilot valve is raised, water from the city mains is admitted to the top of the hydraulic cylinder, causing the piston to move downward, lowering that side of the

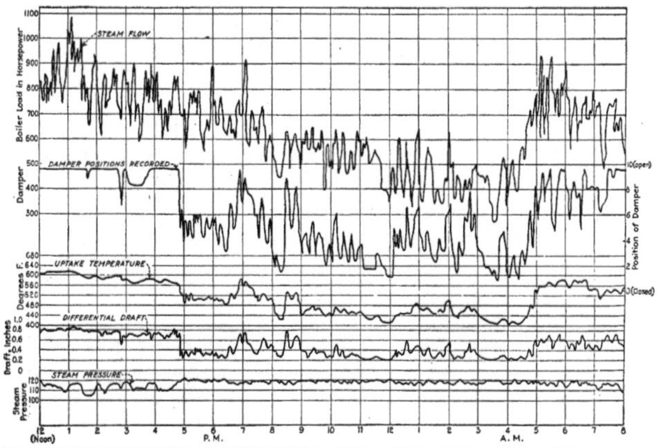

FIG. 3. COMPARATIVE RECORDS AND SIMULTANEOUS DAMPER POSITIONS FROM A 300-HP. WATER-TUBE BOILER

wheel lever and automatically returning the piston of the pilot valve to a neutral or closed position.

When this balance takes place, the master control holds that position until there is another change in pressure. A drop in pressure operates just the reverse. The mercury plunger is lowered; the piston of the pilot valve is lowered; water is admitted to the bottom of the hydraulic cylinder and the hydraulic piston moves upward, pulling the valve piston back to neutral, when there is a balance between the mercury and the hydraulic.

The operation of the master control is transmitted to each individual auxiliary (see Figs. 1 and 2), there being one auxiliary for each individual damper, stoker engine or motor, forced-draft fan, etc. These auxiliaries are all connected to the master by a rope or chain, and the connections are run through conduit pipes with special conduit pulleys so constructed as to make it possible to turn an angle of any degree. Where an auxiliary is installed, a special insert is provided so that a break can be made in the main conduit line and a branch cable taken off for the auxiliary. As the master moves, so moves each auxiliary, thus giving individual control to each particular part of the equipment that enters the combustion process.

One of the features of the graduated control is the adjusting set that is used on each auxiliary. By means of this adjuster one boiler can be set ahead or back of the others, and the load can be carried on a certain number of boilers up to a predetermined rating, and the other boilers can then be cut into service automatically and at any rate desired.

Where engines or turbines are used for driving stokers or fans, a special chronometer valve is employed, which is designed to supply steam under graduated control. Roughly, the valve is of the vertical piston type with a rectangular opening through the piston and a corresponding rectangular port in the body of the valve. The length of stroke of the piston is determined by the auxiliary control so that the flow of steam to the stoker or fan engine will correspond with the requirements. There is also provision for micrometer hand adjustment to turn the piston of the valve and thus vary the width of the port to suit the requirements of any particular machine.

An illustration of the operation of the control system may be had from Fig. 3, which presents comparative records taken on a 500-hp. water-tube boiler for twenty consecutive hours. The load chart is at the top and the damper position just below. Other simultaneous readings that have been plotted are uptake temperature, differential draft and steam pressure.

Notice that at 7 p.m. for the load of 900 hp. the damper is nearly wide open. From then until about 8:30 the load diminishes and the damper position changes to correspond. Also between four and five o'clock the next morning the load gradually increases, and again the damper changes its position, gradually opening up little by little, until when the peak is reached, the damper is in the right position to supply the proper draft and produce the rate of combustion necessary to carry the load efficiently. The speed of the stoker has been increased in the same proportion so that the efficient relation between the draft and fuel bed established for the grade of coal burned, has been maintained automatically and with much closer precision than is possible with hand regulation.

Financing Improvements for the Power Plant

By Paul R. Duffey

Many security and trust companies, as well as other financial institutions, are daily investing large sums of money in the many classes of enterprises needing a working capital to make them operative. A large majority of such enterprises net good interest to the investors. Some fail from many causes which tend toward failure even though the investment looked good.

The thought appeals to me that financial institutions would do well to invest money in improving obsolete power plants which are now being run at a loss due to lack of capital for necessary improvements, and place the plants on up-to-date operating bases. My idea is for the financial institutions to have their engineers go over the plants applying for funds and make detailed reports of present operation, etc., and place the proposed requirements for proper consideration.

Citing for an example a steam generating plant of around two thousand boiler-horsepower, the boilers being hand-fired and the plant under test shows an over-all boiler furnace and grate efficiency of from 54 to 58 per cent. The total cost per year for operating this plant, present and proposed is as follows:

	Present	Proposed
Labor	$12,467.84	$6,679.16
Fuel	49,131.20	39,304.99
Handling fuel	1,200.00	600.00
Water	1,686.15	1,686.15
Maintenance	2,706.62	2,706.62
Supplies	300.00	300.00
Total per year	$67,491.81	$51,276.92

Present operation, $67,491.81; proposed operation, $51,276.92; saving per year, $16,214.89.

Overhead charges have been disregarded, as they would remain about same as at present.

In order to bring about this saving, which would build the over-all efficiency of the plant up to 75 per cent, a capital investment of $33,785 is required. It is proposed that the financiers invest this amount and have the plant operated at the new figures of $51,276.92 per year and that they receive from the industry for which they have made the improvements the sum of $67,491.81 per year, the present operating costs, for a period of time sufficient to net 10 per cent on the investment of $33,785 and that as soon is the latter amount plus 10 per cent interest has been earned, the plant should be turned over to the owners in full. Thus the owner would have in, say, three years the new up-to-date plant fully paid for out of the savings effected.

Any additional cost of labor, fuel and supplies during this period is to be borne by the company on a basis of the old plant figures or present costs. In other words, this is simply purchasing a plant on the installment plan, the same as other more or less familiar commodities.

During the past few years the number of gasoline-consuming engines and vehicles, including pleasure cars, trucks, motor boats, tractors and farm engines, has increased to such an extent that there is danger in the near future of an acute shortage of this most important fuel. In the United States alone there are 4,000,000 automobiles, 250,000 trucks, 500,000 motor boats, 75,000 farm engines, making a total of 5,575,000 gasoline-consuming engines.

R. J. S. Pigott—Power-Plant Economist

Present fuel conditions, involving both shortage and high price, indicate that power users, and especially industrials, must adopt a course of improvement in plant design and equipment, as has been the case in Europe. Formerly, power cost in many industrial establishments was so small a percentage of manufacturing cost that it received but the scantest attention and economy was but little observed. It will become a vital question whether the isolated plant or the central station is the ultimate solution. In all probability some development such as W. S. Murray's super-power system will occur, but can supplant the isolated plant in general only if some of the inherent advantages of the latter are also developed in the central-station system; namely, the large use of low-pressure steam for heating and processing and the use of refrigerating plants and similar service for off-peak load. In short, the percentage of thermal recovery from the fuel must be increased by the development of all the services now given by an isolated plant to its factory, before the central idea can give a real fuel economy, due to its inherent advantages of size diversity factor and higher-grade technical supervision.

DURING the year 1908 it became evident that in a very short time additional power would be required for the New York subway. The Fifty-ninth Street power plant of the Interborough Rapid Transit Co., in which the subway power was generated, was equipped with nine compound engines, having a maximum capacity of 7,500 kw. each. A careful study of the situation resulted in a decision to install a 7,500-kw. low-pressure turbine in connection with each engine. The importance of the station and the success of the new arrangement attracted nation-wide attention to the work and to the man who directed it. The superintendent of motive power was H. G. Stott, but details of design and construction were largely the work of R. J. S. Pigott, chief draftsman, motive power department, and later assistant engineer in charge of construction.

Mr. Pigott, who was born in England in 1886, was graduated from Columbia University in 1906 with the degree of mechanical engineer. In that year he became an assistant engineer in the motive power department of the Interborough Rapid Transit Co. and the next year was made chief draftsman in the same department. It was while he was in this position that the plans for the low-pressure turbine installation were taken up, and he had an opportunity to show his ability in the handling of real power problems. This job involved not only the erection of the new turbine units, but a rearrangement of the piping and condensing equipment to comply with the changed conditions.

Since the whole plan was to a large extent an experiment, one turbine was installed and numerous tests made on the combined unit to determine how nearly the entire plan might be expected to come up to the predicted performance. This test work alone presented a number of new problems for solution. One of these involved the determination of the quality of the low-pressure steam going to the turbines. As a result of considerable experimental work the conclusion was reached that the ordinary perforated-pipe sampling tube, which was found to give satisfactory results with high-pressure steam, really acted as a separator with low-pressure, so that the sample taken through such a tube was very far from being a representative sample of the steam in the pipe. A sampling tube was designed with a single orifice pointed against the direction of steam flow and proportioned so that the velocity of steam in the tube should be very nearly equal to that flowing in the pipe. It is interesting to note that Professor Rateau, the famous French engineer, has taken occasion to confirm Mr. Pigott's findings in connection with the investigation of the quality of low-pressure steam.

In 1911 Mr. Pigott became superintendent of construction for the New England Engineering Co., and in 1912 professor of steam engineering at Columbia University. In 1913 he returned to the Interborough Rapid Transit Co. as construction engineer. This position he held until 1915, laying out and installing three 30,000-kw. turbines with all other plant changes. During his two periods of service with this company Mr. Pigott was responsible for the design and installation of a number of appliances and processes which resulted in improved power-production economy. Among these was the design, some years ago, of a traveling screen for use at the Fifty-Ninth Street station in the circulating-water inlet. This is believed to be the first use of this type of screen in power-station work. At the Seventy-Fourth Street station of this company Mr. Pigott did considerable work in connection with the use of furnace-wall ventilation to prevent burning out linings in boiler furnaces.

In 1917 Mr. Pigott went to Bridgeport, Conn., as consulting engineer and power superintendent for the Remington Arms Co. At that period the enormous expansion of the Remington works called for by the war-time emergency required a complete design for the new plant. Mr. Pigott undertook the design, construction and operation of all power and plant service systems, and in doing so contributed an important bit to the country's war-time needs. During this period he also found time for work in the reorganization of the production departments of the Bridgeport Brass Co. and for the development of design and construction data for the Sanford-Riley Stoker Co.

While the requirements of his work have at times turned Mr. Pigott's attention to other fields, his principal interest has always been in the field of economical power production. In a recent interview he describes himself as a power-plant man and nothing else, but his success in the solution of problems along broader lines shows him to have ability in other directions.

Since 1919 Mr. Pigott has been industrial engineer for the Bridgeport Chamber of Commerce. In discussing recently the questions of fuel economy, he stated that one of the arguments that prevail against the small isolated plants is the inability of such a plant to support the necessary expert supervision to produce power to the best advantage. In answer to a question he said that his work for the Bridgeport Chamber of Commerce consisted to a large extent in furnishing advice with regard to improvement in fuel economy to the various power plants in and around Bridgeport.

Mr. Pigott is a member of the American Society of Mechanical Engineers, of the American Institute of Electrical Engineers and of the Society of Automotive Engineers. His membership in these professional societies has not been merely passive, as is shown by the large amount of committee work which he has done. He is chairman of the Fluid Meters Committee, a member of the Power Test Code Committee, formerly chairman and present member of the Research Committee of the Mechanical Engineers. During the war he was advisory engineer for the Connecticut Fuel Administration and chairman of the Bridgeport Committee of the Naval Consulting Board, and also engineer for the Bridgeport Smoke Abatement Commission.

Adjustable Ratchet Wrench

A self-adjusting ratchet wrench that will take the place of several sizes of open-end wrenches, especially designed for heavy-duty work and a time saver, due to the self-adjusting feature and the ratchet action, is

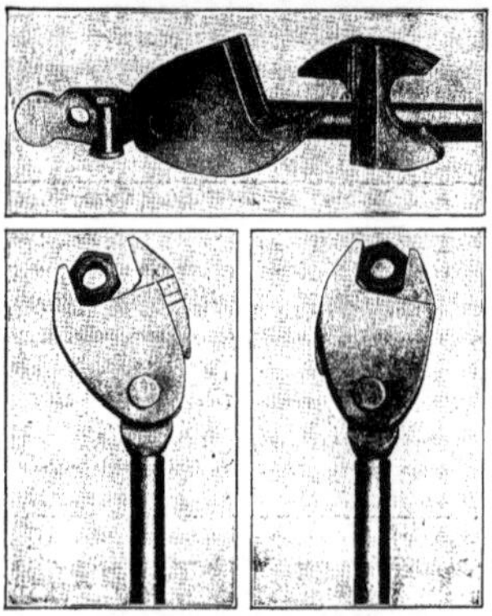

PARTS AND APPLICATION OF RATCHET WRENCH

being brought out by the Allan-Diffenbaugh Wrench and Tool Co., 6250 South Halsted St., Chicago, Ill. The parts are interchangeable. Referring to the illustration, the upper view shows the parts, the lower left view shows the jaws wide open, and the lower right view shows the jaws closed on a nut. To operate, simply place the jaws over the nut, and pull on the handle; the harder the pull the harder the jaws will grip. To let go of the nut, release on the handle, which will loosen the grip of the jaws and permit a ratchet action on the nut.

EDITORIALS

Distance of Boiler Above Grate

SOME weeks ago a reader inquired how high a seventy-two-inch horizontal return-tubular boiler should be set above the grate. The editor answered thirty to sixty inches for ordinary fuel and hand-firing. A correspondent thought it should be from four to five feet, and the question has been the subject of considerable discussion.

It is not so many years ago that the editor's figures would have been considered excessive. The average height above the grate of the six seventy-two-inch boilers treated in "Barrus' Boiler Tests," published a quarter of a century ago, is less than twenty-seven inches. The catalogs of several large builders of boilers of this type give from twenty-seven to thirty-four inches. Up to the latter part of 1914 the Hartford Steam Boiler Inspection and Insurance Company recommended twenty-eight inches as the minimum distance for a seventy-two-inch boiler, but has recently revised its practice to thirty-six inches for anthracite and forty for bituminous coal. Osborn Monnett, who attained considerable reputation as a furnace and combustion expert while at the head of the smoke-prevention department of the City of Chicago, says not over thirty inches, even for Illinois coal, and Mr. Misostow, his successor, is for a lower value still. Several of the boilers that were set forty-eight inches above the grate in the days when Chicago was experimenting with smoke prevention have been lowered with improved results.

Is it not, after all, a question of rate of combustion? Is not a certain minimum volume necessary for the most effective combustion of a given weight of carbon and hydrogen in a given time, and is not increased height, and thus greater volume, necessary for a furnace in which high rates of combustion are used, although the lower settings in common use may be sufficient for more moderate rates of firing? The average rate in the Barrus tests was less than twelve and one-half pounds per square foot of grate per hour, and the average capacity only slightly above seventy-five per cent of the rated.

Much stress is put upon furnace volume where powdered coal is used. In oil-fired Babcock & Wilcox boilers sixty to eighty cubic feet per burner with a maximum capacity of three hundred and fifty pounds per hour is supplied. This is from 0.17 to 0.23 of a cubic foot per pound. The same builders allow ten cubic feet of furnace volume per square foot of active grate area for multiple-retort underfeed stokers. This would be 0.25 of a cubic foot per pound for a firing rate of forty pounds of combustible per hour.

A high setting means greater first cost, more chance for radiation and air infiltration and a less immediate and perhaps less complete use of the radiant heat. The contention of the advocate of the low setting that the boiler gets more heat from radiation is met by the assertion that the heat, once it is generated in the furnace, cannot get away and will be absorbed in the tubes if it is not in the fire sheets. It may be but the low-setting man prefers to take advantage of the direct penetrating effect of the radiant heat rather than to depend upon getting the heat into the boiler by convection. If the boiler is near enough to the fire, a large percentage of the steam will be made on the fire sheets, and this largely by radiant heat which can penetrate the layer of cooled gas next to the shell. In this way the heat is certainly absorbed; if convection is depended upon, it may if the tubes are clean enough inside and out. The essential is that the boiler must be far enough from the fire to give the gases room and time to burn completely before they are chilled by contact with the comparatively cool shell, and this distance should obviously be greater when forty or more pounds of coal are burned per square foot of grate per hour than when the rate of combustion is fifteen pounds or less.

Motors for Driving Ventilating Fans

INTIMATELY connected with every power plant is a ventilating problem of some kind or another. This may take the form of supplying fresh air to a large building or to a small room, or to industrial processes. However, in a very large percentage of cases motor-driven fans are employed. The ordinary ventilating fan is a comparatively simple piece of equipment from the structural standpoint, but if satisfactory operation is to be obtained not only must the fan be driven by a motor that is suited to the application, but a controller that will give proper control of the motor must be employed. In this issue, on page 759, W. H. Easton discusses "Motors for Driving Ventilating Fans" and gives a very clear and practical analysis of this phase of motor application.

Too much cannot be said regarding the location of the motor. On account of the nature of the work it is frequently necessary to place the machine in an out-of-the-way place, although this does not necessarily mean that it must be in an inaccessible location. If it is so placed that it is difficult to care for, it will probably be neglected, with the result that the service will be impaired and the maintenance charge will be high. The placing of a motor in an inconvenient place because nothing else will go there is false economy generally and results in dissatisfaction. The question of motor efficiency in regard to fan drives is an important one and involves two factors—the efficiency of the motor when operating under normal speed and the efficiency at other than normal speeds. Although the latter is to some extent influenced by inherent characteristics of the motor, the largest factor is the method of control, especially with direct-current motors. If, on account of its lower cost, a motor is selected, where adjustable speed is required, that will, when operating at normal speed, drive the fan at the maximum required speed, and it is necessary to operate for considerable periods at reduced speed, low efficiency will result.

The cost of the power lost in the armature external resistance will usually pay a handsome dividend on the additional capital required to install a motor and controller that, at normal speed, will drive the fan at the lowest speed required and beyond this point adjust the speed by field control. With alternating-current motors the problem becomes more involved. The simple squirrel-cage type of motor, although being the most reliable in its operation and of high efficiency at normal speed, is strictly a one-speed machine, therefore, where adjustable speed is required, this type must give way to the wound-rotor machine, which, when operating below its normal speed, has an efficiency characteristic similar to the shunt direct-current motor with armature control. There have been developed special types of induction motors for speed regulation, but they have not yet come into general use for fan drives. However, the increased cost of power no doubt will cause greater attention to be given to these more complicated equipments as a means of obtaining a higher efficiency in alternating-current motors at different speeds than can be reached with the wound-rotor type.

Storage of Winter Coal Supply in Summer Is Still Important

EFFORTS to stabilize the coal industry by buying and storing coal for next winter's use were given a setback in some cases by the outlaw railroad strike. For a time at least the roads could move neither full nor empty cars, with the result that even present demands could not be met and in many cases both the mines and the coal-using industries were compelled to shut down. This should not discourage those who are planning to do their share toward the general good and to help themselves by anticipating their extra coal requirements. The strike situation will soon be cleared up, and coal ordered can be mined and delivered.

Meanwhile, the bill allowing differential freight rates for coal in the spring and summer months is in the Senate committee. In addition, another bill providing for a flat discount instead of the percentage discount, as called for in the first bill, has been introduced in the Senate. This discount varies from month to month during the low-demand period of the year. Whatever may be the fate of these bills, they will probably be side-tracked until the present political activities of the various Senators have somewhat relaxed. It seems, therefore, that little help may be expected from this source during the present season.

On the other hand, the general freight-rate increase seems bound to come, and any coal user who can get his coal hauled before these new rates go into effect can make a considerable saving. At the present time the probable increase is estimated at about twenty-five per cent.

The important point, however, is to stabilize the coal industry, not just to help the coal industry but to help everybody. Fuel prices will continue to be high, and seasonal coal famines will cause trouble as long as the present practice of seasonal buying persists. The winter peak must be taken off both the mines and the railroads before the coal market can be put on a sound economic basis. And it is important to every individual in the country that it be put on such a basis.

The Technical Man as an Inventor

DURING the war a great deal was written about the wonderful things to be accomplished by American inventive genius. It was confidently predicted that when the American people put their minds seriously to the task, the German submarine problem would be solved overnight. Without denying any credit for the really wonderful work accomplished, it is worth while to consider the enormous amount of effort and money that was spent by individuals in developing ideas that had no possible chance of success.

While the war emergency stimulated the efforts of the would-be inventor, such work is going on continually. It is a pitiful fact, however, that many men with intelligence and pluck are working hard at impossible tasks. Having no conception of the limits imposed by natural laws and no knowledge of the mistakes and failures of others, they go blindly on, reaching for a star.

It is in this field of invention and research that the value of a technical or scientific training becomes most apparent. The technical engineer or scientist is trained to determine first a standard of perfection, or the point beyond which it is impossible to go, and then bend all his efforts to approach that standard. He also has, or knows where he can obtain, a knowledge of the successes and failures of the workers who have preceded him. The non-technical man too often accomplishes nothing simply because he wastes his time striving for the unattainable.

We do not claim that the scientific man never makes mistakes. He is not perfect. Neither is it true that the non-technical man always fails. Real progress can be made only by long, patient work, and results usually come only after many failures. However, a knowledge of what has been accomplished and what is contrary to natural laws gives the technical inventor a big advantage.

On April 21 Justice Bailey of the Supreme Court of the District of Columbia issued a temporary injunction restraining the Federal Coal Commission from requiring the Maynard Coal Company to file reports of its costs and other data. While this injunction applies technically only to the case of this particular company, it seems probable that similar injunctions would be granted to other operators in the same position as the Maynard Company. The Federal Coal Commission has been established and its expenses provided for out of the public funds, but the legislation under which it acts is so poorly drawn that the first review in the courts removes its power to collect data. Somebody gets a good job, some members of Congress "point with pride" to an achievement for the benefit of the "people," the "people" pay the bill and there it ends.

The resignation of Dr. Van H. Manning, Director of the Bureau of Mines, is another instance of how inadequate compensation to scientific and technical men in the Federal service is creating a deplorable state of disorganization in the Government departments. In his letter Dr. Manning says: "I feel very deeply that there ought to be more adequate compensation for the scientific and technical men in the Government service so that none of them may be compelled to accept positions on the outside."

CORRESPONDENCE

Eliminating the Smoke Nuisance

I have experimented for many years on various devices to overcome the smoke nuisance. The wet process of water sprays, being the popular one, underwent a thorough trial, and eventually was discarded, and a new theory tried out to determine a velocity of the gases which would be lowered to a point where the force of gravity would overcome the velocity. After much experimentation it was found that should the velocity of the gases be curtailed to a certain given speed, allowing a time element at that speed to take place, all heavier than air particles would be precipitated.

An expansion chamber was designed and built, with these features in view to meet existing conditions, on the stack of 2,500-hp. boiler plant, using buckwheat anthracite for fuel, with forced draft. Properly designed hoppers were arranged to remove the residue of dust and cinder that was collected and under which a track for industrial cars could be used.

The results obtained were beyond expectation, and without reduction of draft in the stack, the residue recovered amounting to 3.77 per cent of the coal used and making a clearance of the flue gases of from 90 to 100 per cent. The original analysis of the coal was 72 per cent fixed carbon, whereas the residue analyzed from 55 to 70 per cent fixed carbon. Evaporative tests run, using this residue for a fuel, adding 20 per cent by weight of bituminous coal to supply volatile carbon, under a separate boiler, gave, at the boiler rating, 8 lb. of water evaporated per pound of the mixture, this result obtained being a little less efficient than the original coal.

This installation of expansion chamber or collector was experimental and naturally more expensive than it would otherwise have been, and even allowing for all discrepancies and expenses conservatively estimated, it produced a dividend on the investment of 29 per cent. This method is just as effective on bituminous as on anthracite fuel and will eliminate the heavier than air particles.

As this device, which is covered by U. S. patent, has no moving parts, and operating costs are nil, the first cost is the only one to be considered. Being beyond the experimental stage, and demonstrated to be perfectly practical, and as it can be adapted to any existing plant, it would seem that from an economical and hygienic standpoint it is no longer necessary to be subjected to the evils of a polluted atmosphere from this cause. THOMAS GRIEVE.

Perth Amboy, N. J.

Reboring Corliss Valve Chambers

With reference to Mr. Young's communication in the issue of April 6 and Mr. Murphy's in the issue of Feb. 17, I offer the following suggestions:

The practice of reboring valve chambers on Corliss cylinders is quite usual and is found necessary when the steam and exhaust valves become worn, thus resulting in considerable leakage of steam and causing the valves to rattle, owing to the excessive amount of play in the valve chambers. This excessive wear is often caused by wet steam, due to foaming or priming of the boilers, which makes it difficult to lubricate the valves properly. This condition can often be improved by installing a good separator in the steam line.

In setting a boring bar for boring the valve chambers, it should be set central to the bore on the valve-gear side of the cylinder. The fit of the front or valve-gear bonnet in the valve chamber should not be disturbed and both the steam and exhaust ports should not be rebored clear through, but the original bore should be left intact for the depth of the boss on the front bonnet. The fit of the back bonnets is not so important, and it really is no detriment if the boss on the back bonnets does not fit the valve chamber exactly. With the front bonnets it is different, as they have to withstand the working strain of the valve gear, and for this reason it is important that the exact fit of the bonnet in the valve chamber be maintained.

It is important in reboring the valve ports that the lowest spots in the wearing surfaces should be eliminated, so as to insure a round chamber, and that the wearing surfaces shall be steam-tight. To accomplish this, it is sometimes necessary to take as many as three cuts. The boring bar should be of the self-feeding type, operated preferably by an electric motor, but of course the motive power can be a steam engine or hand power. With the electric motor the motion of the cutting tool will be much more uniform and steady. After reboring the valve ports, new steam and exhaust valves should be fitted, which would in reality make them as good as they were originally. After starting up the engine with new valves and rebored valve ports, it is important that the lubrication be not stinted until the valves are worked down to a good bearing, and it might be necessary to ease off the hard spots once or twice, so as to facilitate getting the valves to bear evenly over the whole surface. There should be no trouble experienced in work of this character if performed in a careful, workmanlike manner, and there should be a marked improvement in the steam consumption of the engine. JOHN J. MUIR.

Seattle, Wash.

Economical Combustion of Fuel Oil

Although much more has been written on the economical combustion of coal than of oil, there is a very great chance for loss from excess air with the latter fuel. Smokeless combustion of oil can be maintained with very little excess air, but beyond this point there is no outward indication of excess even with three or four times the necessary amount. It is quite necessary, then, to take considerable care to regulate the air control, especially with changing loads on an oil-burning boiler. A CO_2 recorder or other form of flue-gas analysis is necessary for intelligent control of this air supply.

In the plant with which the writer is connected CO_2 recorders are used and are frequently checked with an Orsat apparatus. The highest CO_2 reading we have obtained is 14.9 per cent, and it is interesting to determine how nearly perfect this figure is.

Not having an analysis of the oil, an estimate was made from the flue-gas analysis. A typical case showed CO_2, 14.5 per cent; O, 1.5 per cent, and N, 84.0 per cent. Assuming 100 cu.ft. of flue gas, the 1.5 cu.ft. of oxygen will bring with it $\frac{1.5 \times 79.1}{20.9} = 5.68$ cu.ft. of nitrogen. This is the nitrogen in the excess air. Therefore there were $84.0 - 5.68 = 78.32$ cu.ft. of nitrogen in the air actually used for combustion. The total oxygen used for combustion is then $\frac{78.32 \times 20.9}{79.1} = 20.69$ cu.ft. The volume of CO_2 is equal to the volume of oxygen required for its formation, so that the oxygen combined with carbon is 14.5 cu.ft., leaving $20.69 - 14.5$ or 6.19 cu.ft. to combine with the hydrogen in the fuel.

Since for equal weights hydrogen requires three times as much oxygen as does carbon, the ratio of weights of carbon to hydrogen in the fuel is as 14.5 is to $\frac{6.19}{3}$ or as 7.023 to 1. The percentage of hydrogen in the oil therefore is $\frac{100}{1 + 7.023} = 12.46$ and of carbon $100 - 12.46 = 87.54$ per cent. Other flue-gas analyses give practically the same results for this fuel. Of course this does not take into account any oxygen, sulphur or other ingredients in the oil, but the approximation is good enough for the present purpose.

Since the volume of air required to burn one pound of carbon is 152 cu ft. (at 60 deg. F. and atmospheric pressure), the air required for one pound of the above fuel oil will be $(0.8754 + 3 \times 0.1246) \times 152 = 189.9$ cubic feet.

This will contain 189.9×0.701, or 150.2 cu.ft. of nitrogen. The carbon alone will require 0.8754×152, or 133.1 cu.ft. air, which will contain 133.1×0.209 or 27.8 cu.ft. of oxygen. Since the volume of CO_2 is equal to the volume of its oxygen, the flue gas will be made up of 150.2 cu.ft. of nitrogen and 27.8 cu.ft. of CO_2 if no excess air is supplied. The percentage of CO_2, then, will be $\frac{27.8}{150.2 + 27.8} \times 100 = 15.6$. The H_2O formed by burning the hydrogen, of course, being condensed, does not appear in the flue-gas analysis.

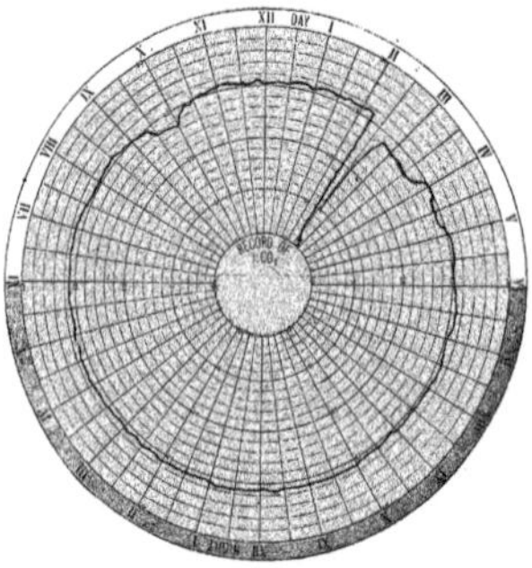

FIG. 1. CO_2 RECORDER CHART FROM BOILER UNDER CONSTANT LOAD

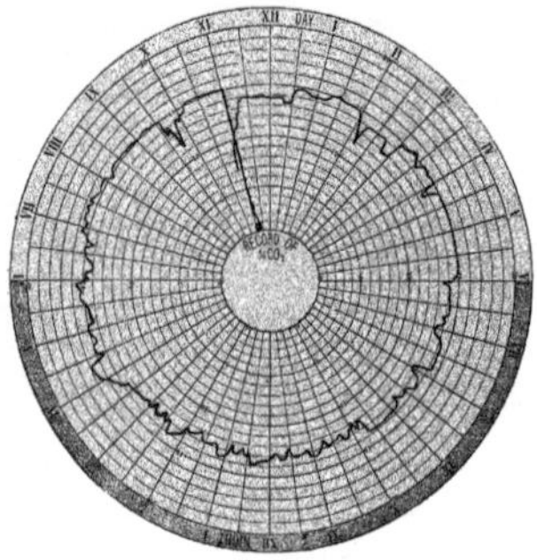

FIG. 2. CO_2 RECORDER CHART FROM BOILER CONTROLLED BY GOVERNOR

This value, 15.6 per cent CO_2, is the theoretical maximum for perfect combustion for this particular oil. The accompanying charts from CO_2 recorders show how closely this condition was obtained. Fig. 1 is from a 286-hp. B. & W. boiler operating under a constant load of 165 per cent of rating, and Fig. 2 is from a 160-hp. return-tubular boiler under variable load with the oil pressure controlled by a governor. Checking the charts with an Orsat apparatus showed that the recorders read a fraction of one per cent high. The variations in performance on the chart in Fig. 2 show the necessity for careful air regulation with changes in load.

The furnaces of these boilers were designed by the writer. They have unusually large combustion chambers with provision for a long flow of gases during combustion. Other furnaces in the same plant cannot be made to give such good CO_2 records.

I believe that proper fuel-oil combustion requires some radical changes in furnace design from present general practice. The combustion space should increase in volume in the direction of flow of the gases. There should be sufficient volume for combustion to be completed before the gases strike the heating surfaces or bridge-wall, when using the minimum of excess air for complete combustion. Air inlets should be arranged so that all the air may mix with the oil vapor in the early stages of combustion.

In a properly constructed furnace, when correctly operated, the lower part of the combustion chamber will be filled with burning gases, showing a bright yellow and no flame reaching the heating surface. If too much air is used, the flame will shorten and become brighter. There will also be an increase of the roar in the furnace and a perceptible rise in the flue gas temperature.

Ocean Falls, B. C. C. H. LEICESTER.

Changed Stoker Lineshaft

Where a lineshaft is used for driving chain-grate stokers, as is usually the case where one or more batteries of boilers are equipped with stokers, its location becomes a problem, especially where the design includes a siftings hopper for discharging into a chain conveyor.

If the shaft is placed above the stoker, it is in the way of gate adjustments, fuel feeding, boiler washing, etc., and must be protected by safety screens to avoid accident; if placed in the conveyor tunnel in a position to clear the siftings hopper, its location, in the average layout of stoker design, will carry it too far forward of the stoker drive toggle to insure smooth running, besides throwing the eccentric shaft at an objectionable angle, conducive to uneven and excessive wear on the drive mechanism. This is especially true where the stoker drive is of the pawl-and-ratchet type and the load on the eccentric is consequently intermittent.

The objection to running the shaft directly through the siftings hopper are chiefly the deteriorating effect of the siftings on the steel shaft, and the waste of siftings through the lineshaft holes in the sides of the hopper. The figure A shows a method that I employed on a recent job when changing the shaft from the outside position B to the position C. This shaft was $2\frac{7}{16}$ in. in diameter and 160 ft. long, erected in two sections, each section serving two batteries of two 425-hp. boilers.

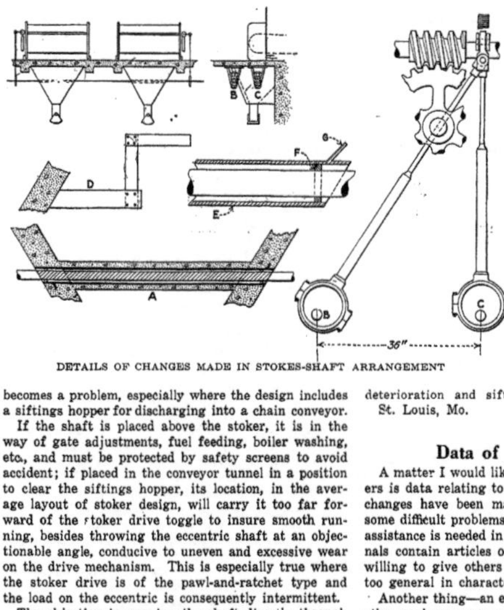

DETAILS OF CHANGES MADE IN STOKES-SHAFT ARRANGEMENT

The device D was made of wood and was used to lay off the holes in the ash-hopper sides and, when held against the tunnel ceiling, gave the true elevation for all holes measuring the distance from the tunnel wall.

Eight pieces of 3-in. pipe E were sawed to a length equal to the width of the hopper and beveled as shown, and into each end of these was driven a babbitt collar F, bored to a running fit on the shaft.

The first length of shaft was then taken down and the hangers moved to the new location between the hoppers. Two of the pipes E were then slipped over the shaft, which was then pushed through the holes cut into the hopper sides, also slipping the eccentrics on where necessary. This was repeated until one 80-ft. section was erected. The shaft was then leveled and grouted in as shown at A. Wire screen was used to give the grout a bond on the pipe.

The slivers G were sawed into and bent back on the pipe ends to prevent the babbitt collars F from turning the pipes E. The object of the collars is to hold the shaft central of the pipes while the latter are being grouted in; they may be neglected after starting. It will be evident that in lining and leveling the shaft, the pipe, on account of the collars, will move with the shaft so that the holes in the hopper sides must be cut larger than the pipes. In this case a 4-in. diameter hole was cut for $3\frac{1}{2}$-in. outside diameter pipe. These and also the hanger-bolt holes were cut before the shaft was changed and while the boilers were in service so that the shaft could be moved in the least possible time. The new location puts the eccentric directly under the drive, and deterioration and siftings waste are taken care of.

St. Louis, Mo. A. M. STILES.

Data of Results Wanted

A matter I would like to see taken up by *Power* readers is data relating to results obtained after suggested changes have been made. Engineers run up against some difficult problems in their work, and often outside assistance is needed in solving them. Engineering journals contain articles of value written by men who are willing to give others a lift, but the details are often too general in character.

Another thing—an engineer asks for information and other engineers answer in following issues, advising this or that, but we never know whether the information given was of benefit or what results were obtained by following out the suggestion. I, for one, would like to know the results. GEORGE J. TROSPER.

Winkleman, Ariz.

Industrial Unrest

There have been and are being written long articles in the various papers and magazines relating to industrial unrest. We hear of it in the lecture room, and it echoes from the halls of Congress. There is no single subject that is occupying the general public's mind as much, with the exception of the high cost of living, which goes hand in hand with the unrest of the masses.

The many writers and lecturers deal with the question from different angles, but it seems to me that it is all in a general way, and to my mind, we will never gain anything until we begin to deal with the matter in a more specific, or in a more individual way.

What is the real cause of this world-wide industrial unrest? Is it altogether real, or fancied? What will be the final effect if it is not soon remedied? What is the remedy and how can it be applied successfully? These are the questions that are agitating the minds of men in every rank of life.

I believe that much of the individual unrest is due entirely to a mental condition. We have heard so much of this talk, have read so much about it, we are saturated with the idea that everything is wrong in our own individual lives. A large share of our trouble is of our own making, and what we all need is to pause a few moments and take stock of ourselves, see where we are, if we have any real grievance, and get busy to try to find the remedy.

Many engineers who have talked with me recently have made numerous complaints about the conditions under which they are working, about salaries, the treatment and the lack of consideration that they are receiving. They meet in groups and discuss these matters and then separate, each in a worse frame of mind than when they met. Now an engineer has a better chance of being a deep, accurate thinker than one in any other trade or profession. Then why not, instead of meeting and grumbling about conditions, start in discussing remedies awhile?

Now, only a short time since, this feeling of unrest took hold of me. I was dissatisfied with everything in general and my position in particular. Something had to be done. After thinking over matters for several days, I decided to have a talk with the "boss." Telling other engineers my troubles did not help me any. They didn't pay my salary, and had nothing to do with the control of the place where I worked. I wanted a remedy for my troubles and there was no one that could supply it but the "boss."

I went to his office several times to see him, but he was engaged. This put me in worse humor. Finally I decided that when he wanted to see me on any matter, he sent for me to come to his office. I decided that I was as good as he was, and that this superior attitude of employers was the cause of a lot of our troubles. So the thing to do when I wanted to see him was to have him come to my office, or invite him. Shortly after reaching this decision, I met him and told him that I had been waiting to have a talk with him on some matters, and wanted him to come to my office at his first opportunity. He replied that he would. I had expected him to look surprised at the invitation, but he didn't. As I returned to the engine room, the thought struck me that it would be some time before I saw him in there, but I was mistaken again, for he was in my office within half an hour.

I told him of my dissatisfaction over several things and talked to him in his own language, using illustrations with which he was familiar. I made him see me in my position. I never had a better or more polite listener in my life. He had very little to say with the exception of asking a few questions to bring out some point more clearly. At the end of the talk, he told me that he would take the matters up with the owners and let me hear from him.

He appreciated the frank manner in which I had come to him and pointed out faults as well as bad conditions and he felt that the manner in which I had handled it was to the best interest of all concerned and that we would understand each other better in the future.

A few days after he called me to his office and more than met every condition. Now some will say that they could not talk to their superior that way. If you look on him as your superior, and he looks on you as his inferior, possibly you can't.

Now I am convinced that the most of this so-called industrial unrest is caused by absence of confidence and understanding between employer and employee. A lot of it is caused because the employee subconsciously looks on the man higher up as his superior, and imagines that the employer feels the same way. If you honestly admit that a man is better than you, it is a safe proposition that he is, or will soon be. No man is better than he thinks he is.

Several years ago, my employer at that time was a man who was said to be worth nearly a million dollars. He sent for me to come to his office to talk over some matters. The furniture in his office cost more than my entire home furnishings. He was occupying a chair that cost probably more than all the chairs I owned at that time. His clothing was of the finest material. The office was beautifully lighted. His person and surroundings simply reflected success. There I sat and talked with him, while dressed in a three dollar suit of blue overalls. When I left his presence, there was a feeling of envy in my heart, and a sneaking feeling that he was in some way my superior.

In about two months, like a clap of thunder out of a clear sky, he was declared a bankrupt. When the liabilities were printed, there was a bill for the office furnishings, one from the light company, also one from his tailor for the very clothes he had been wearing. Then the thought struck me that, although I wore overalls at the time of our conversation, they were my overalls and were paid for. Everything that I claimed I owned was mine. Although I had felt that he was above me at that time, I was his superior, morally and financially, and must be at least his equal mentally; because I was free of debt, and never again would I even secretly admit that any man was my superior, until I knew that at least the clothes he wore belonged to him and to no one else, even though he were president. And from that day I have never had any trouble talking to any employer about the matters that were important to me.

What the world needs today is, simply, increased confidence between employee and employer, and it is a matter that can only be worked out by each individual doing his part, regardless of what the other fellow says. Quit going around with a chip on your shoulder, because when you start hunting trouble, you will generally find it ready and waiting for you. And the surest way to get right is to think right. H. A. CRANFORD.

St. Louis, Mo.

Steam Rate with Increased Back Pressure

In inquiries of General Interest for March 2, J. R. A. asks how an increase of back pressure affects the quantity of steam required by an engine. The answer was, in substance, that enough more steam should be let in to balance the higher back pressure.

It is usually thought that every pound increase in the back pressure will require an additional pound of steam per horsepower-hour, but only one-third to one-half pound of steam has to be taken from the boilers for every pound increase in the back pressure. The actual amount is determined by the degree of expansion the steam is worked to at the time it gets down to the back pressure.

Twenty years ago I was running a 15-ton can ice plant in the block belt of Alabama. The business grew and it was decided to double the capacity. The old plant made more distilled water than the compressor capacity would freeze. At this time the exhaust-steam absorbtion refrigerating machine was just coming into use. I thought that a 30-ton absorbtion machine was just what we needed, and I had figured that the economy we would get from the condensation of our exhaust steam would be velvet, and that the extra cost of operation would be that required to generate the 15 lb. of steam extra per horsepower-hour, due to the 15 lb. pressure that the generator required to make capacity.

The absorption machine was bought and installed in the off season. We got a hot week in May and our business jumped to capacity, and much to our surprise we did not have enough distilled water to fill our cans at capacity rate.

The hot wave was followed by a cool spell, and as a distilled-water ice plant is very easy to run a test on, we hunted out the trouble and found that the steam consumption did not rise pound for pound with the back pressure. Our engine driving the compressor took 29 lb. of steam per horsepower-hour, and with 15 lb. back pressure the steam rate was 36 lb. per horsepower-hour. To be sure of our results, we weighed the coal and water a number of times and got an increase of 7 to 9 lb. for the 15 lb. extra back-pressure. The lower rate always was obtained when the boiler pressure was highest.

The owner was satisfied when the conditions were explained to him, and we built a cold-storage room to use up our excess capacity. JOHN H. RYAN.

Palm Beach, Fla.

Comparative Cost of Pulverized-Coal and Stoker-Fired Boiler Plants

Referring to the editorial in *Power*, March 2, entitled "Pulverized Coal at Milwaukee," and the particular statement that "the added investment in building and equipment for preparing the fuel over a stoker plant has been estimated to exceed $20 per boiler-horsepower, but the additional expenditure will be well worth while," we would like to inquire where the writer of the editorial obtained information that would lead to the conclusions given as to this difference in cost, as there is nothing like this difference. On the other hand, the costs are more nearly on a par, including building cost. We refer particularly to the actual material involved in making the complete installation; for instance, a comparison of recent estimates, based on the same horsepower of boiler equipment, shows that the cost is more favorable to the pulverized-coal installation than is the case for stoker-fired boilers, especially stokers of the underfeed type.

The plant has five 1,000-hp. boilers. The cost of the stoker plant for this type is about $150,000 and the material involved in this estimate includes the following: Underfeed stokers, complete blowing equipment for air supply, air ducts, conveying machinery outside of boiler room, conveying machinery inside of boiler room, crushers, coal bunkers, driving mechanism for stokers, blowers and conveyors, ash-handling and conveying system.

The cost of the pulverized-coal milling plant is $128,000, covering the following items to complete the installation: Separate building, coal driers, pulverizing mills, elevators and conveying machinery, blowers, dust collectors, Fuller-Kinyon pumping distribution system for pulverized coal between the boiler room and coal-mill plant (within 100 ft.), burners, feeders, coal bins for boilers, air supply for burners, motors to drive all necessary apparatus.

The only variable in this installation would be the amount of money expended on the building for the pulverized-coal plant. The figures given include a substantial structural-steel and corrugated-iron building for the coal-mill plant of sufficient size to house all the apparatus. This structure might not suit the ideas of some parties who would prefer to spend more money on the building construction, but it is beyond anybody's belief that it would be necessary or advisable to spend so much money on the building construction that it would make the comparable cost $5 per horsepower more, to say nothing of $20 per horsepower more, for a pulverized-fuel installation.

We sincerely trust you will investigate and let us have more discussion on this question as to what the writer of the editorial based his statement upon.

FREDERICK A. SCHEFFLER,
Manager Power Department.
New York City. Fuller Engineering Co.

Maximum Length of Longitudinal Joints for Boilers

Here is another suggestion on the subject of longitudinal-joint lengths for the purpose of inviting the opinion of your many readers. Quite a little printer's ink has been spread in the endeavor to determine the proper length for longitudinal seams in steam boilers, but at present it seems that we are like the circus horse that runs around a ring—we are getting nowhere.

Why not have the boiler-course lengths determined according to the plate thickness and the diameter of the shell or drum. Is it not reasonable to suppose that light plate with certain drum diameter and length of course would be likely to become affected by longitudinal cracks or grooving in less time than heavier plate?

Then why not increase the plate thickness if it is desired to increase the course length? The drum diameter should determine the course length. The greater the diameter the shorter the course, and if longer courses are desired, the plate thickness should be increased accordingly.

Engineers and boilermakers are invited to join in this "joint" debate. N. DEVERING.

Pittsburgh, Pa.

Loose Setscrew Stopped Governor

A new regulator used on a Brown engine recently failed and the safety stop preve. ted a possible wreck. The engineer served as a regulator for the remainder of the day by the use of the throttle valve until the governor was again rendered serviceable. The trouble was due to a setscrew that held one of the gears becoming loose, thus stopping the governor. This gear is now splined or keyed on, which will likely eliminate developments of a similar nature in the future.

Woburn, Mass. M. H. OSGOOD.

National Engineers' License Law

Mr. Peters, on page 434 of the March 16 issue of *Power*, suggests that when his proposed national license law goes into effect every engineer who has been successfully operating his plant for three to five years will be given a license, free of examination, to operate that particular plant.

I consider this quite logical and practical, and by doing so it allows the national engineers' license law to settle smoothly and gradually over the engineering world without any disturbance to the operation of steam plants. It also allows the engineer who is not technically qualified to prepare himself for a license.

Brenham, Tex. H. W. ROSE.

Pure Water as Feed Water

Referring to H. W. Rose's letter on page 604 of the April 13 issue on the effect of pure water on boiler metal, in which he refers to Mr. Hill's article on refrigeration on page 320 of the March 2 issue, Mr. Hill says that were it not for leakage losses the same water could be used indefinitely. Mr. Rose calls attention to the fact that distilled water has a corrosive action on boiler metal and will cause pitting. Some readers may wonder why pure water should have this effect to a much greater degree than water which has been treated, but which, of course, still contains mineral salts. S. W. Parr, of the University of Illinois, and others explain that this is due to the fact that pure water will carry a much greater content of dissolved oxygen, and that this oxygen, supplied to the tubes as the water passes through, is taken up by the metal to form iron oxide, or rust, and the unevenness of the formation will cause pitting. It is not to be understood, however, that this is a rapid process.

In this connection, it is of interest to note the description on pages 492-5 of the March 30 issue, of a 50,000-kw. power plant in Chile and the difficulties encountered in the operation. Prominent among these difficulties was the fact that the water had to be obtained by distillation of sea-water, since the rainfall is negligible and rivers are of little consequence. The plant is being successfully operated on feed water distilled from the ocean brine, and those in charge feel that part of the success is due to the purity of the water, since no scale is formed and little corrosion is experienced. From this it can be seen that the purity of the feed water is certainly not such a bugbear as Mr. Rose's letter might seem to indicate, and that the disadvantage resulting from the larger oxygen content carried by pure water might be more than overcome by the decreased amount of scale formed.

W. J. RISLEY, JR.,
Chemical Engineer.
Decatur, Ill.

Care of CO_2 Recording Machines

A short time ago there was some discussion in the columns of *Power* on the merits of, and the care necessary for, recording CO_2 machines, and I would like to make a few statements from my experience, with special reference to the article by C. J. Schmid in the Feb. 17, 1920, issue.

Our machines are the latest models of a well-known make, and the men who take care of them are not only good mechanics, but also specialists on instruments, particularly on CO_2 machines, which take up the greater part of their time; but even with this expert care we cannot approach the perfection of operation which Mr. Schmid says is possible with "any mechanic."

We have ten boilers, and one man spends on an average of five hours a day on the machines for these boilers for regular routine testing, checking and adjusting, or one-half hour per boiler. Beside this, an extra half-hour or more is required for changing absorbent cartons and thoroughly cleaning the machine and lines once a week or oftener, depending on the quality of the absorbent. Moreover, we find it necessary to have a man on the night shift when several of our boilers are banked, and also on Sundays. This is so as to have the machines in order for the morning pickup of load and the several hours of heavy load which follow.

We have learned by experience that if the CO_2 machines are not looked after overnight and Sunday, one or two or more are found out of order as the boilers come on and for the time being, sometimes two or three hours, are useless, just at the time when they are needed most. And so my conclusions are, from our experience, that they do require expert care and that considerable time must be spent on them to keep them in good order. Of course any good mechanic could keep them going, but I am sure that the time required would be far in excess of the time taken by an expert, as no matter how good a mechanic he might be, he would have to take some time in following up the many lines and fittings.

We should not be misled by the word "expert" in this connection and imagine him a very high-priced man. Although he must have some ability as a mechanic, he is not necessarily a first-class mechanic, but rather a man who has specialized in the care and operation of these particular machines. He is not ordinarily as high paid as a first-class mechanic, and in our practice we do not consider it good economy to put a high-priced mechanic on the job.

Furthermore, if we did not check our machines every day against an Orsat, we probably would go along thinking that everything was well, but there is never a day when there are not a few, if not all of them, requiring adjustment. This does not mean, however, that we find CO_2 recorders impractical; they are very useful in keeping efficient fires, as a man who is experienced can judge very closely the condition of a fire by the percentage of CO_2. It would be even more useful if there were no lag in the machines, the lag being between five and ten minutes on our installation. That is where a recording pyrometer is advantageous, as it will give almost instantaneous results, the only lag being the time required for the gases to travel from the fire to the thermocouple, if that may be considered a lag. An ideal installation would be a combination of the two on each boiler, but the cost would be prohibitive.

Detroit, Mich. G. A. CHATEL.

INQUIRIES OF GENERAL INTEREST

Radial Valve Gear—What is a radial valve gear?
L. C. N.

The term radial valve gear is applied to reversing mechanisms having but one eccentric or equivalent crank motion.

Jumping of Feed Pump—What causes a direct steam feed pump to make short jumps of the plunger at the beginning of the stroke?
E. M.

The jumping indicates presence of air and may result from running the pump too fast, or the pump design may be such that the water is not always rising and there are pockets where air can accumulate.

Size of Pump-Suction Valves—What would be the size of suction-valve areas of a steam pump for operation at a piston speed of 100 ft. per minute?
G. L. J.

The water velocity through the valve-seat open areas is generally taken as 222 ft. per min., which gives a valve area 45 per cent of the plunger area, with piston speed of 100 ft. per min. Greater liberality in valve area is conducive to good service and durability, and specifications for water-works pumps frequently demand clear valve areas 50 to 60 per cent of the plunger areas.

Pressure-Reducing and Pressure-Regulating Valves—What is the difference between a pressure-reducing and a pressure-regulating valve?
R. E. T.

Any valve that effects a reduction of pressure, as by throttling the flow of a gas, vapor or liquid, would properly be called a pressure-reducing valve; and a valve that limits the discharge to a constant maximum of quantity or of pressure would be a regulating valve. A valve having the purpose of reducing steam of boiler pressure to a constant lower pressure suitable for steam heating would properly be designated as a pressure reducing and regulating valve.

Gravity, Flash and Fire Test of Lubricating Oils—What effect have gravity and flash or fire points on the lubricating value of oils?
B. T.

Flash and fire points of an oil are important considerations where over-heating by bearings might result in ignition of waste or other combustible materials, and the gravity may be an index for identification of a particular kind of oil, but none of these properties can be regarded as indicative of lubricating value. The real test is the analysis which comes from observation of the effect on the friction load.

Angle of Advance and Lap for Double-Eccentric Engine—In order to obtain greater range of cutoff than 0 to ½ stroke with a double-eccentric Corliss engine, why is it necessary to set the steam eccentric with a negative angle of advance and for the steam valve to have negative lap?
A. L. S.

With the Corliss valve gear the valve cannot be tripped to effect cutoff later in the stroke than when the eccentric has carried the wristplate farthest to one side of its central position. When the eccentric is set 90 deg. with the crank, the farthest swing of the wristplate to one side occurs when the piston has reached about one-half stroke, and in order for the maximum wristplate displacement to be delayed so as to permit of cutoff after one-half stroke, it is necessary for the eccentric in the beginning of the stroke to be set back of the 90-deg. position; that is, the eccentric must be set with negative angle of advance. With the eccentric thus set behind the 90-deg. position, if the valve rods were of such length as to bring the valves line on line when the wristplate was in the central position, then with the engine on a center and the wristplate hooked up to the eccentric, the edges of the valves would overlap the ports, and to obtain admission at the beginning of the stroke, it would be necessary to adjust the length of each valve rod so its valve would be brought around to a position ready to open or give the desired amount of lead at the beginning of the stroke. When the valves are thus adjusted, the negative lap, or amount of opening the valves have when the wristplate reaches its central position, is a necessary consequence of adapting the lead to negative angle of advance of the eccentric.

Water-Test Pressure on Handhole Gaskets—After washing and resetting handhole covers of a water-tube boiler, is there any objection to testing the tightness of the gasket joints with cold-water pressure pumped up to the running steam pressure for the boiler?
J. R.

A pumped water pressure, unless made slowly with a hand pump, is likely to run up in excess of the intended pressure and thereby cause leaks that would not occur with the intended pressure. If gaskets have proper elasticity, and the bearing faces are true, a test-pressure of no more than half of the operating pressure should be sufficient to guarantee tight gasket joints.

Resetting Spool on Duplex Pump—On a duplex pump how would the proper position be found for resetting a crosshead spool that had become loose on one of the piston rods?
H. G.

When the spool is properly set, the rocker arm which it carries will be at right angles with the piston rod when the piston rod that carried the spool is in the middle of its stroke. To place the piston rod in the middle of its stroke, move the steam piston as far as it will go toward the head end of its stroke, and after making a mark on the piston rod at the end of the stuffing-box gland, move the piston as far as it will go in the other end of the cylinder and make a similar mark on the rod. After locating a mark halfway between those marks, move the piston and rod so the middle mark will be at the end of the stuffing-box gland and, with the piston thus placed in the middle of its stroke, set the crosshead to such a position that the rocker which it carries will be at right angles with the piston rod. To make the pump run the same as previously, it may be necessary to shift the position of the crosshead from the right-angle position because that position will not obtain equal lost motion on each side of the valve between the valve and valve rod operated by the rocker when the valve is in its central position.

[Correspondents sending us inquiries should sign their communications with full names and post office addresses. This is necessary to guarantee the good faith of the communications.—Editor.]

The Corrosion of Condenser Tubes

The fifth report to the Corrosion Committee of the Institute of Metals, England, presented at the Annual General Meeting in London, March 11, 1920, is confined to the corrosion of condenser tubes—mainly 70:30 brass. It is a study of the practical problems of corrosions in condensers under service conditions, employing either fresh or sea water, and is divided into four sections.

The problems of corrosion have been classified under five different types as outlined in the accompanying table. This classification has been made for the purpose of diagnosis of corrosion troubles, and the first section of the report deals with the methods of drawing tubes and preparing them for investigation. The appearance of tubes suffering from the various types of corrosion is discussed.

When a tube is to be withdrawn for examination, it should be carefully marked to show which is the top of the tube as it lay in the condenser and the direction of water through it. Immediately after withdrawal the ends of the tube should be plugged with tightly fitting wood plugs. The tube should be kept in a horizontal position throughout these operations and until the examination is made. When ready for examination, the tube should be cut longitudinally so as to separate the top and bottom portions and any sludge removed for chemical and microscopic examination.

After the type of corrosion has been determined, the next step is to determine the conditions that have caused the trouble.

A. Concerning the water supply:

1. Information should be obtained as to whether the water is renewed continuously or is cooled and used repeatedly. In the former case a number of chemical analyses should be made at weekly intervals. In the latter case the original supply should be analysed and *analyses made of the water actually entering the condenser* after a given length of time. This is necessary since impurities in the make-up water become concentrated due to evaporation.

2. A small sample of the water actually entering the condenser should be diverted into a settling tank and any deposit examined.

3. The material of the water mains and water ends of the condenser should be noted.

4. Information should be given as to whether the tubes are choked with solid matter and, if so, as to its nature.

5. The average water velocity through the tube should be stated.

6. The inlet and outlet temperatures at various times of the year should be given as well as information as to arrangement of baffle plates and steam distribution.

7. If failures are most pronounced at certain seasons, data should be collected as to water conditions two or three months previous to the period when the trouble commences.

B. Information concerning the tubes:

1. The composition of the tubes.

2. Length of time in service and average life of tubes in the condenser.

3. The percentage of corrosion failures of the particular batch of tubes from which samples are examined.

4. The total length of tubes.

5. Does corrosion occur at the top or bottom of the tubes, nearer the inlet or outlet, or is it irregularly distributed?

6. Is corrosion confined to a particular position in the condenser?

7. Has serious corrosion been noticed in other parts of the condenser?

8. Is corrosion present in all condensers in the plant or is it confined to one or more?

9. On land: Do failures occur at any particular time of year? and on ships: Do failures appear shortly after a period in dock or similar departure from normal routine?

10. Has any process of protection been employed?

The necessity for frequent and regular analysis of water *actually entering the condenser* is emphasized, especially when cooling towers are used. It is also pointed out the conditions which actually start the corrosive action often exist for a short time only, but that the corrosion, once started, may continue even under conditions which would not start it.

The second section is devoted to a consideration of certain features of the structure of condenser tubes which affect the problems of corrosion. Attention is directed to the presence of a surface layer of noncrystalline metal, about 0.01 mm. (0.0004 in.) which has a greater resistance to corrosion than the crystalline metal underneath. Where this layer has been penetrated corrosion proceeds at an increased rate.

The third section takes up in detail the five types or classes of corrosion.

Type I, General Thinning, is an accelerated form of the corrosion in salt solutions in which there is a gradual but uniform reduction in thickness of the tube wall. Experiments indicate that with ordinary sea water this action is so slow as to be negligible. Rapid general thinning in fresh water is usually associated with the presence of free acid. A number of tests with hydrochloric acid having a concentration of only three parts in 100,000 at ordinary temperatures showed that a reduction of 2 to 4 per cent. took place in six weeks. The minute quantities of acid

TYPES OF CORROSION OF BRASS CONDENSER TUBES

Type	Name	Recognized by
I.	General thinning	Rapid thinning of tube. Little or no basic salt on tube
II.	Deposit attack	Pinhole or regional pitting beneath green salt and cuprite. Foreign deposits often accompany this type of corrosion
III.	Apparent dezincification, layer type	Layer of crystalline copper which can be readily stripped off surface of tube. Slight deposit of light green salt; sometimes a little white salt
IV.	Apparent dezincification, plug type	Copper lugs or areas beneath deposits of white salt
V.	Water-line attack	Attack similar to II, but confined to inlet end of tube

NOTE—The types of corrosion are tabulated for convenience of reference and description only. It is not intended to imply that the reactions concerned are entirely different in each case; for instance, types of attack intermediate in character between III and IV are found, and the types must be considered as merging into one another.

which will cause this trouble make it difficult to detect, especially where, as is generally the case, it enters the water supply intermittently.

Type II, the so-called Deposit Attack, is usually given as the cause of pitting. In the presence of sodium chloride ($NaCl$) solutions the cuprous oxide (Cu_2O) on the surface of the brass gradually changes to cuprous chloride ($CuCl$). Where particles of this latter compound adhere to the surface of the tube they gradually change, under the influence of oxygen, to cupric chloride ($CuCl_2$) and cuprous oxide. The cupric chloride rapidly attacks the brass, oxidizing the copper and reducing the cupric chloride to cuprous chloride which in the presence of air starts the cycle of reactions again. Foreign bodies lying in the tubes help to keep the cuprous-chloride particles from being washed out with the circulating water and therefore should be kept out of the condenser as much as possible.

Type III, Layer Dezincification, is applied to that condition where layers of pure copper are found. It is stated that this is the result of disintegration of the brass tube and a redeposit of copper rather than of a removal of the zinc alone. This condition exists in both sea and fresh water and with the latter is often associated with acid water.

Type IV, Local Dezincification, is really a deposit attack, but differs from Type II. It always occurs with sea water and is associated with an adhering white salt consisting of colloidal zinc oxychloride. It is shown that the right concentration of zinc in the liquid layer next to the brass surface results in the production of this salt. Lowering the zinc content of the brass or raising that of the sea water is sufficient to prevent its occurrence.

Type V, Water Line Corrosion, in tubes partially submerged in sea water takes place, not at the water line as is commonly supposed, but sometimes as much as 2 cm.

(0.79 in.) above it. This type of attack is obviously a special form of deposit attack since it takes place under the salt deposits which are formed at, or near, the water line. It is much more severe than any of the previous types.

The fourth section of the report deals with preliminary work on electrolytic protection against deposit attack. The experiments seem to show that a current of sufficiently high density might entirely prevent corrosion. Very high densities, however, would be necessary to accomplish this.

Dr. Van H. Manning Resigns

Dr. Van H. Manning, Director of the Bureau of Mines, Department of the Interior, has tendered his resignation to President Wilson, effective June 1. Dr. Manning is leaving the Government service to accept the position as Director of Research with the recently organized American Petroleum Institute. He has been with the Department of the Interior for thirty-four years and has done much for the advancement of the mineral and coal industries and the safety of its workers. While his new duties will be in behalf of private interests, the work will, in a general way, be along similar lines to that which is being done in the Department of the Interior. In his letter to the president, Dr. Manning expressed the thought that, although the Government spends each year many millions of dollars in useful scientific work for the benefit of the people and the nation, the monetary recognition of its scientific and technical servants is not sufficient to enable them to continue their work, especially during the last few years, when it has become impossible for many men to remain in the Government service owing to the constantly increasing cost of living.

Increased Pay for Technical Men in Federal Service

The Congressional Joint Commission on Reclassification of Salaries, following a year of investigation, has submitted its report on conditions existing in the Federal Civil Service in the District of Columbia. Scientific and technical employees in the Government at Washington are among those seriously affected by the lack of uniformity in compensation. At the present time there is practically no policy governing such matters as hours of work, pay for overtime, promotions, salary advancements, etc.

The report, with its recommendations to Congress, for the first time affords the Government and employees some assurance that work involving similar duties and responsibilities will receive the same pay, that increased efficiency will be recognized by advances in salary and opportunities for promotion. Uniformity of administration is assured by making the Civil Service Commission the central employment agency for the Government and by creating a Civil Service Council composed of equal members of administrative officers and employees.

The same schedules of compensation are recommended for employees performing work requiring similar training and experience. Salaries for professional men, including those in the engineering service, the biological science service, the physical science service and the economic and political science service, range from $1,800 per annum for the junior class (men just out of college) to $5,040 per year for men in the full professional class.

During the last few years the number of resignations has grown to such proportions that it has threatened complete disruption of the service.

The Scientific-Technical Section of Federal Employees' Union No. 2, is convinced that the adoption of this report with such minor changes as may be necessary, will go far to relieve the condition now existing.

Plans for the construction of twenty-six hydro-electric power plants on the River Neckar, Germany, have practically been completed. The projects will involve an expenditure of about 35,000,000 marks, and the plants, when completed, are expected to produce a total of 50,000 kw.

Superheated Steam in Turbines

It is generally estimated that the gain in economy due to the use of superheated steam is about 1 per cent for every ten degrees of superheat. Thus, if steam having 150 deg. of superheat is employed, the resulting gain should be 15 per cent, irrespective of whether the steam is employed in a reciprocating engine or a turbine. There are certain disadvantages which follow the improper use of superheated steam, the two most important of these being the effect of high temperature on the castings of the engine, causing distortion, and the difficulty of lubrication. In the turbine the effects of distortion may be particularly serious in view of the fine clearances that have to be adopted with certain types of turbine, and it may be stated that one of the reasons why the high-speed impulse type of turbine is adopted is because with this type, owing to the high velocity of the steam, the clearances can be made more generous both in the axial and the radial direction. It is, however, advantageous to keep that portion of the turbine which is exposed to the effects of highly superheated steam as small as possible, and for this reason in the first set of nozzles the fall of pressure is often about two-thirds of the total fall between the boiler and the condenser, so that on leaving these nozzles the temperature of the steam has fallen to such a degree that its effects are greatly reduced.—*Shipbuilding and Shipping Record.*

The certificate of his recent election to honorary membership in the American Society of Mechanical Engineers was presented to Dr. Auguste C. E. Rateau on May 8 at a meeting of the Société d'Encouragement pour l'Industrie Nationale. M. Charles de Freminville, who was elected an honorary member at the same time as was Dr. Rateau, but who came to the United States to receive his certificate, pointed out at this meeting, which took the form of a discussion of the benefit that may result from frequent intercourse between the scientific and technical schools and societies of the two countries, that convenient arrangements should be made to allow American students access to French schools, while Dr. Rateau described the benefit that would result if French instructors would familiarize themselves with American ideas on the professional training of young engineers, which he thinks to be excellent, by more frequent visits and longer stays in this country.

Power development in California to meet the demands of growth will mean expenditures of approximately twenty-five million dollars per annum, according to estimates made by the Railroad Commission Engineers. Construction of four large plants which will have a capacity of 120,000 kw. is already under way. They are the Caribou Plant of the Great Western Power Co., the Kerckhoff plant of the San Joaquin Light and Power Corp., the Kern River plant of the Southern California Edison Co., and the James Plant, of Los Angeles.

Work on the new hydro-electric development at Niagara Falls, which the Hydro-Electric Power Commission of Ontario started in 1918, is progressing at an expected rate of speed. A long canal is being built to carry water around the falls and through the power plant. Some little difficulty has been met in holding the slope in the deep cuts in the earth overburden, but this has not caused serious retardation and the prospect is now that the power will be available in 1921.—*Exchange.*

Engineering research workers, on the fuel value of wood at the Forest Products Laboratory, Madison, Wis., have reaffirmed that two pounds of dry wood of any nonresinous species has the heating value of one pound of good coal. Resin gives about twice as much heat as wood, and as the average amount of resin in wood is approximately fifteen per cent, this increases the heating value thirty per cent.

New Publications

NOTES ON GAS AND OIL ENGINE ACCIDENTS. Pamphlet by the National Boiler and General Insurance Co., Ltd., Manchester, England. Paper, 5 x 7 in.; 24 pages.

This pamphlet discusses some of the more common failures of internal-combustion engine parts with regard to their causes, and gives suggestions for inspection and methods of preventing such failures.

ELEMENTS OF STEAM AND GAS POWER ENGINEERING. By Andrey A. Potter, Professor of Steam and gas Engineering, and James P. Calderwood, Professor of Mechanical Engineering, both of the Kansas State Agricultural College. Published by the McGraw-Hill Book Co., Inc., New York. Cloth; 5½ x 8 in.; 304 pages. Price, $2.50.

This book was prepared primarily as a textbook to familiarize students with power-plant equipment before taking up the more abstract study of thermodynamics and design. The main portion of the book is divided into three parts, covering steam power, gas power and the application of steam and gas power to locomotives, trucks and automobiles. The various forms of power-generating apparatus and auxiliaries are described in a non-technical manner, and numerous illustrations are included No attempt is made to take up the mathematical study of power engineering, although a few general principles are explained and simple problems given for study.

THE RELATION BETWEEN THE ELASTIC STRENGTHS OF STEEL IN TENSION, COMPRESSION AND SHEAR. By Fred B. Seely and William J. Putnam. University of Illinois Engineering Experiment Station Bulletin No. 115. Paper, 6 x 9 in. Price. 25 cents.

This bulletin presents the results of tests to determine the elastic shearing strength of ductile and semiductile steels and the relation between this elastic shearing strength and the elastic tensile strength, with the hope of obtaining definite information on the breakdown of elastic action and on the limits of the theories of combined stresses. From torsion tests on solid and hollow cylinders a factor is found for determining the true shearing strength from tests on solid cylindrical specimens. The ratio of this true elastic shearing strength to the elastic tensile strength is given for various grades of steel, and its bearing on the theory of combined stress is discussed. The relation between elastic tension and compression stresses and the effect of rolling on these values is considered.

Personals

L. W. Haig has been appointed district sales manager of the Packard Electric Co. at Buffalo.

George A. Shoemaker has been appointed works manager of the Bound Brook (N. J.) Oil-Less Bearing Co.,

W. L. MacFarham has been appointed general manager of the Cornwall Street Railway, Light and Power Co., Cornwall, Ontario.

Clinton E. Haver, until recently connected with Willard Case & Co., has accepted the position of works engineer with the Hooven & Allison Co., Xenia, Ohio.

John A. Lawrence, formerly assistant mechanical engineer of the New York Edison Co., has assumed the position of engineering manager for Thomas E. Murray, Inc., New York.

Herbert C. Stephens, formerly power engineer of the Northern Ohio Traction and Light Co., is now associated with the Electric Motors and Repair Co., Akron, Ohio, as sales engineer.

William A. Tietze has resigned from his position as chief engineer of the Cudahy Bros. Co., Cudahy, Wis., to become general manager of the Citizens Light, Heat and Power Co., Canby, Minn.

Lyle B. Marcy, formerly superintendent of the Hart & Cooley Co., Inc., New Britain, Conn., has become connected with the Chase Companies, Inc., Waterbury, Conn., in the capacity of research engineer.

Harold D. Bliss has severed his connection as chief mechanical engineer for Morris & Co., of Chicago, Ill., and has become associated with B. P. Lientz, of the B. P. Lientz Oil Furnace Co., New York.

H. W. Jarrett, formerly assistant fuel engineer, United States Bureau of Mines, Pittsburgh, Pa., has assumed the duties of sales engineer with the George J. Hagan Co., combustion engineers, Pittsburgh, Pa.

Alfred C. Jordon, formerly superintendent of equipment for the Cumberland County Power and Light Co., Portland, Me., has resigned in order to accept a similar position with the Elmira Water, Light and Railroad Co., Elmira, N. Y.

Edward B. Richardson and Harry Gay, both members of the A. S. M. E., annunce their partnership under the name of Richardson & Gay, Consulting Engineers. The offices of the new firm will be located at 220 Devonshire St., Boston, Mass.

Harry H. Bates has resigned as field engineer in the stoker department of the Westinghouse Electric and Manufacturing Co., East Pittsburgh, Pa., to accept the position of superintendent of equipment of the Acme White Lead and Color Works, Detroit, Mich.

Raymond E. Plimpton, formerly publication manager and field secretary of the Society of Automotive Engineers, has joined the Wales Advertising Co., of New York. His time will be devoted to the handling of advertising campaigns of a technical and semi-technical nature.

Howard I. Spohn, formerly of *Motor Age*, *Motor World*, *Automotive Industries and Commercial Vehicle*, and Frank A. Kapp, formerly of *Motor Life*, "The Automobile Trade Directory" and the "Automobile Blue Books," are now associated with the Charles H. Fuller Co., Chicago, Ill.

Charles H. Tavener, mechanical engineer of the Garden City plant of the Curtiss Aëroplane and Motor Corp., announces that he has started the practice of consulting industrial, productions and aëronautical engineering with laboratory facilities for testing and development work in New York and Boston.

Lon J. Geer has resigned as general manager of the plant of the Texas Power and Light Co. at Gainesville, Tex., and will go to Sweetwater, Tex., to become general manager of the West Texas Electric Co., a corporation which controls electric-light plants in a number of western Texas towns. H. A. Price, of Alladega, Ala., will become manager of the Gainesville plant.

Society Affairs

The New England Association of Commercial Engineers will hold its ninth annual mechanical exhibition at Worcester, Mass., July 7 to 9.

The Electric Power Club will hold its annual convention at Chattanooga, Tenn., May 5-8, with headquarters at the Signal Mountain Hotel.

The American Society of Mechanical Engineers, New Haven Branch, Connecticut Section, will hold its last meeting of the season, Friday, May 14, at the Geometric Tool Co.'s plant in Westville, Conn. Representatives of the company will present papers on "Evolution of the Opening Die Head." At 5:30 p.m. a demonstration of machines of the Geometric Tool Co. will be given at the works.

The Seventh National Foreign Trade Convention is to be held in San Francisco, Cal., May 12-15, under the auspices of the National Foreign Trade Council. A number of distinguished foreign business representatives will be present and will discuss trade opportunities in their respective countries. James H. McGraw, president of the McGraw-Hill Co., will address the convention on "The Service of the Business Press," during the group sessions on Thursday, May 13. At the fourth general session on Friday, A. C. Bedford, of the Standard Oil Company of New Jersey, will deliver an address on "Fuel Oil and Foreign Trade."

Miscellaneous News

An International Trade Fair has been organized by the Town Council of Cologne, the first event to be held on Sept. 20, with the obect of encouraging the importation of raw material into Germany.

The Public Utility Interests of New York and Pennsylvania, through a safety committee, appointed last fall, have set aside the second week of June for the observance of on accident-prevention campaign.

The Public Service Electric Co. of New Jersey, in order to meet the growing power demands in the North Jersey industrial district, is contemplating the construction of another subsidiary power plant which will double the present available supply of electric power in this section of the state. The new central station will have a capacity of 200,000 kva. and will cost approximately $25,000,000.

Business Items

The Ball Engine Co., of New York City, has moved its offices from 39 Cortlandt St., to the Vanderbilt Building, 132 Nassau St., N. Y. City.

The Strong, Carlisle & Hammond Co., Cleveland, Ohio, manufacturer of steam specialties, announces the removal of the Chicago office to 17 South Desplaines St.

The George T. Ladd Co., of Pittsburgh, Pa., manufacturers of water-tube boilers, will move their general offices and engineering department to the sixteenth floor of the First National Bank Building, Pittsburgh.

The George T. Ladd Co., Pittsburgh, Pa., manufacturer of water-tube boilers, opened a Chicago sales office at 528 McCormick Building. The new office will be in charge of W. M. McKinstry, formerly Chicago manager of the Page Boiler Co.

The Hyatt Roller Bearing Co., Industrial Bearings Division, announces the removal of its offices to 100 West Forty-first St., New York City, where larger quarters have been secured for the advertising, sales and engineering departments of the division.

The Harrison Safety Boiler Works has been incorporated under the laws of the State of Pennsylvania as the "H. S. B. W.-Cochrane Corp." The firm will specialize in the solution of problems relating to the profitable utilization of exhaust steam and the heating, metering and softening of water.

Vance McCarthy, general sales manager of the Edward R. Ladew Co., Inc., N. Y. City, manufacturers of leather belting, has been made vice president of the company. Mr. McCartny will continue in direct control of the sales organization and has retained as assistant general sales manager, Russell B. Reid.

Trade Catalogs

The Shepard Electric Crane and Hoist Co., Montour Falls, N. Y., has published a new 28-page catalog entitled "The Aerial Railway of Industry." Electric cage-operated monorail hoists of various descriptions adaptable to any industry where speedy, efficient loading and unloading of freight cars, and transportation of articles, material and apparatus from one part of the plant to another is essential are fully described and illustrated. A copy of the pamphlet will be sent on request.

The Westinghouse Electric and Manufacturing Co. has just completed publication of a new 40-page catalog entitled "Westinghouse Marine Equipment in the Fabricated Ship." The booklet describes and illustrates geared turbine machinery for 9,000-ton cargo boats built by the Merchant Shipbuilding Corporation. Photographs and line drawings indicate the type of turbines used for these ships. The detailed apparatus is exhaustively treated by means of written discussions, reproduced halftones and line drawings.

COAL PRICES

Prices of steam coals both anthracite and bituminous, f.o.b. mines, unless otherwise stated, are as follows:

ALANTIC SEABOARD

Anthracite—Coals supplying New York, Philadelphia and Boston:

	Mine
Pea	5.30
Buckwheat	$3.40@$3.75
Rice	2.75@ 3.25
Barley	2.25@ 2.50
Boiler	2.50

Bituminous—Steam sizes supplying New York, Philadelphia and Boston:

Latrobe	$4.25@$4.50
Connellsville coal	4.25@ 4.50
Cambrias and Somersets	4.35@ 5.15
Clearfields	4.00@ 4.75
Pocahontas	6.50@ 7.00
New River	6.50@ 7.00

BUFFALO

Bituminous—Prices f.o.b. Buffalo:

Pittsburgh slack	$6.00
Mine-run	6.25
Lump	6.25@ 6.50
Youghiogheny	6.75

CLEVELAND

Bituminous—Prices f.o.b. mines:

No. 6 slack	$3.25@$4.00
No. 8 slack	4.25
No. 8—¾-in.	3.50@ 4.00
No. 6 mine-run	3.50@ 4.00
No. 8 mine-run	4.50
Pocahontas—Mine-run	3.25@ 4.00

ST. LOUIS

Anthracite—Probably not more than 20 per cent of the demand in this market can be supplied. Prices effective Apr. 1 were as follows:

	Williamson and Franklin Counties	Mt. Olive and Staunton	Standard
Mine-run	$2.65@2.80	$3.00@3.50	$3.00@3.50
Screenings	2.50@2.65	2.50@3.00	2.50@3.00
Lump	....	3.75@4.50	3.75@4.50

Williamson-Franklin rate to St. Louis is $1 10; Other rates 95c.

CHICAGO

Bituminous—Prices f.o.b. mines:

		Freight rate Chicago
Illinois		
Southern Illinois		
Franklin, Saline and Williamson Counties		
Mine-Run	$3.00@$3.10	$1.55
Screenings	2.60@ 2.75	1.55
Central Illinois		
Springfield District		
Mine-Run	$2.75@$3.00	$1.32
Screenings	2.50@ 2.60	1.32
Northern Illinois		
Mine-Run	$3.50@$3.75	$1.24
Screenings	3.00@ 3.25	1.24
Indiana		
Clinton and Linton		
Fourth Vein		
Mine-Run	$2.75@$2.90	$1.27
Screenings	2.50@ 2.65	1.27
Knox County Field		
Fifth Vein		
Mine-Run	$2.75@$2.90	$1.37
Screenings	2.50@ 2.60	1.37
Brazil Block	$4.25@$4.50	$1.27

New Construction

PROPOSED WORK

Mass., Boston—Mowll & Rand, Archts., 50 Bromfield St., will soon award the contract for the construction of a 2 story, 150 x 260 ft. theatre and store building on Commonwealth Ave., for J. C. Kiley & Co., 18 Tremont St. A steam heating system will be installed in same. Total estimated cost, $500,000.

Mass., Boston—Mowll & Rand Archts., 50 Bromfield St., will soon award the contract for the construction of a 12 story theatre and office building on Washington and Providence Sts., for Gorden & Shuman, 18 Tremont St. A steam heating system will be installed in same. Total estimated cost, $1,000,000.

Mass., Chicopee — Howes & Howes, Archts., 243 High St., Holyoke, will soon receive bids for the construction of 3 story housing buildings for the Dwight Mfg. Co., Front St. A steam heating system will be installed in same. Total estimated cost, $350,000.

R. I., Providence—Kendall, Taylor & Co., Archts., 93 Federal St., Boston, will soon receive bids for the construction of a 5 story, 80 x 150 ft. hospital addition for the Rhode Island Hospital, 593 Eddy St. A steam heating system will be installed in same. Total estimated cost, $300,000.

Conn., Hartford—I. A. Allen, Jr., Inc., Archt., 904 Main St., will receive bids until June 1 for the construction of a 3 story, 108 x 139 ft. school addition on Talcott St., for the city. A steam heating system will be installed in same. Total estimated cost, $600,000. W. E. Cone, 89 Asylum St., Chn.

Conn., Hartford—The City Health & Charity Depts. plan to build a heating plant and laundry for the Isolation Hospital and Almshouse on Holcomb St. Estimated cost, $80,000. Whiton & McMahon, 36 Pearl St., Archts.

Conn., Waterbury—The Connecticut Military Emergency Bd., State Armory, Hartford, will receive bids until May 15 for the construction of a 4 story, 180 x 200 ft. armory on Field St. A steam heating system will be installed in same. Total estimated cost, $400,000. Louis A. Walsh, 51 Leavenworth St., Archt.

N. Y., Brooklyn—The Bd. Educ. 500 Park Ave., New York City, received bids for the installation of heating and ventilating apparatus at P. S. 182 from Daniel J. Rice, 42nd St. and Lexington Ave., $153,330; Gillis Geoghegan, 537 West Bway., $155,300; W. J. Olvany, 177 Christopher St., $155,490. All of New York City.

N. Y., Brooklyn—The Comr. of Pub. Wks., Borough Hall, plans to build a municipal building. A steam heating system will be installed in same. Total estimated cost, $5,000,000. McKenzie, Voorhies & Gmelin, 1123 B'way., New York City, Engrs.

N. Y., Buffalo—The city received lowest bid for the installation of a refrigeration machine for the Washington Market, from the Mollenberg Betz Machine Co., 110 Washington St., at $41,880. Noted April 13.

N. Y., Coney Island (Brooklyn P. O.)—The City Comr. of Pub. Wks., Borough Hall, Brooklyn, is having plans prepared for the construction of a 4 story bath house addition. A steam heating system will be installed in same. Total estimated cost, $500,000. J. Sarsfield Kennedy, 157 Remsen St., Brooklyn, Archt. and Engr.

N. Y., Newton (Flushing P. O.)—The Bd. Educ., 500 Park Ave., New York City, will receive bids until May 12 for the construction of a high school addition on Etna Pl. A steam heating system will be installed in same. Total estimated cost, $800,000. C. B. J. Snyder, Municipal Bldg., New York City, Archt. and Engr.

N. Y., New York—L. A. Abramson, Archt. and Engr., 43 West 46th St., is preparing plans for the consruction of a 4 story recreation center. A steam heating system will be installed in same. Total estimated cost, $150,000. Owner's name withheld.

N. Y., New York—J. Sarsfield Kennedy, Archt. and Engr., 157 Remsen St., Brooklyn, is having plans prepared for the construction of a 6 story office building on Gold and John Sts. A steam heating system will be installed in same. Total estimated cost, $500,000. Owner's name withheld.

N. Y., Redwood—The Northern New York Utilities Inc., Light & Power Bldg., Watertown, plans to construct a sub-station here and a 3 phase high tension transmission line from here to Alexandria Bay. J. B. Taylor, Pres.

N. Y., Wende (Millgrove P. O.)—The town received bids for the installation of a heating system in the County Home, here, from John N. Danforth Co., 70 Ellicott St., Buffalo, $182,848; Power Efficiency Co., New York City, $173,200; Northwestern Heating & Plumbing Co., Erie, Pa., $177,717.

Pa., Council Bluffs—The School Bd. Independent Dist. will soon award the contract for the construction of a 2 story, 75 x 135 ft. school building. Contract for the installation of heating and plumbing systems will be sub-let. Total estimated cost, $300,000. Proudfoot, Bird & Rawson, 810 Hubbell Bldg., Des Moines, Archts.

Pa., Philadelphia—A. E. Van Bibber, 6337 Drexel Bldg., is in the market for motors, boilers and engines adaptable for the drive of paper mill machinery (used).

Md., Baltimore—R. McLaughlin, Engr., Ethington, Pa., will soon award the contract for the construction of a 2 story, 262 x 274 ft. furniture factory, power house and garage on Wilkens Ave., for the Reliable Furniture Co., 303 President St. Total estimated cost, $200,000.

Md., Catonsville—C. L. Reeder, Engr., Park Ave. and Saratoga St., Baltimore, will soon award the contract for the installation of boiler, chimney, electric generator and switchboards for the Spring Grove State Hospital, here. Total estimated cost, $100,000.

Md., College Park—C. L. Reeder, Engr., Park Ave. and Saratoga St., Baltimore, is preparing plans for the remodeling of the boiler plant for the Maryland State College, here. Estimated cost, about $50,000.

Md., Pocomoke City—The Peninsula Storage Co. is having plans prepared for the construction of a 1 story, 100 x 240 ft. cold storage plant. Estimated cost, $100,000. J. S. Nussear, 324 North Charles St., Archt.

W. Va., Charleston—The Bureau of Yards & Docks, Navy Dept., Wash., D. C., will receive bids until May 19 for furnishing and installing switchboards, oil switches and barriers, busses and connection, storage battery motor generator set, transformers, disconnecting motor starters and other electrical equipment and making connections for a complete indoor sub-station at the Naval Ordnance plant, here.

N. C., Kniston—The city plans to install an electric plant. W. C. Olsen, Sumter, S. C., Archt.

N. C., Winston-Salem—Salem College plans to build a 3 story building including a 60 x 120 ft. dining hall and a 40 x 200 ft. dormitory on Main St. A steam heating system will be installed in same. Total estimated cost, $250,000. W. C. Northup, Archt.

La., New Orleans—Loew's Enterprises, 1493 B'way., New York City, are having plans prepared for the construction of a theatre here. A steam heating system will be installed in same. Total estimated cost, $500,000. T. W. Lamb, 644 8th Ave., New York City, Archt.

Ky., Newport—St. Joseph's Orphanage plans to build a 1,000,000 gal. reservoir, and install electrical centrifugal pump and steel tank of 60,000 gal. capacity, etc. Total estimated cost, $20,000. George Hornung, 512 Fairfield Ave., Bellevue, Engr.

O., Akron—F. A. Sieberling, c/o Goodyear Tire & Rubber Co., had plans prepared for the construction of a 1 story 25 x 25 ft. Booster pumping station. Two 500,000 gal. capacity centrifugal pumps will be installed in same. Total estimated cost, $60,000. Morris Knowles, Hippodrome Bldg., Cleveland, Engr.

O., Cleveland—The city plans to construct 4 story hospital additions on Sackett Ave. and Scranton Rd. A steam heating system will be installed in same. Total estimated cost, $2,500,000. Harry L. Davis, Mayor. F. H. Betz, 604 City Hall, Archt.

O., Norwalk—Albert Kahn, Archt., Marquette Bldg., Detroit, Mich., will soon award the contract for the construction of a 2 story, 89 x 152 ft. school and parish buildings on Main and Milan Sts. for St. Paul's Parish. A steam heating boiler, etc., will be installed in separate boiler house. Total estimated cost, $200,000.

Ind., Terre Haute—The Terre Haute Atomized Fuel Co. is having plans prepared for the construction of a plant. Steam power will be installed in same. Total estimated cost, $500,000. H. I. Gentine, 205 McKeon Bldg., Engr.

Mich., Detroit—I. M. Lewis, Archt., 503 Congress Bldg., will receive bids until May 15 for the construction of a 2 story, 100 x 200 ft. commercial building on Woodward Ave. and West Grand Blvd., for the Chinese Amer. Realty Co., 1401 Ford Bldg. A steam heating system will be installed in same. Total estimated cost, $200,000.

Mich., Detroit—The Palesting Association, 150 Fort St., West, plans to build a 4 story lodge house on Charlotte St. near Case Ave. A steam heating system will be installed in same. Total estimated cost, $500,000.

Mich., Marysville—The village plans to construct a semi-permanent water supply system consisting of pumping station with submerged intake pipe into the St. Clair River, etc. Electrically driven centrifugal pumping units will be installed in same. Total estimated cost, $200,000. Robert Gordon, Engr.

Ill., Chicago—Holabird & Roche, 104 South Michigan Ave., will soon award the contract for the construction of a 4 story, 66 x 108 ft. office building on 16th and Rockwell Sts., for J. T. Reyerson & Son, 16th and Rockwell Sts. A steam heating system will be installed in same. Total estimated cost, $350,000.

Ill., Chicago—The Illinois Malleable Iron Co., 1801 Diversy Blvd., has engaged Cahill & Douglas, Consult. Engrs., 301 Gross Bldg., Milwaukee, Wis., to prepare plans for the construction of a central power plant to replace 3 separate plants. Two 300 hp. boilers will now be installed and space arranged for 4 additional boilers. These new boilers will work in connection with four waste heat boilers, each with 4,000 sq.ft. of heating surface in connection with 4 malleable melting furnaces. The boilers in central plant will be equipped with underfed stokers and coal and ash handling equipment. The generating equipment will consist of 2 direct connected engine driven units, 1 of 350 kw. and 1 of 500 kw. capacity. An air compressor of about 1,000 to 1,500 ft. with compound steam and air cylinders will be installed in same.

Ill., Chicago—L. J. McCormick Corp., c/o Holabird & Roche, Archts., 104 South Michigan Ave., is having plans prepared for the construction of a 1 story monolithic 100 x 120 ft. theatre on Dearborn and Randolph Sts. A steam heating system will be installed in same. Total estimated cost, $1,000,000.

Wis., Milwaukee—The Sewerage Comn. will receive bids until June 3 for furnishing and erecting machinery for a sewage disposal plant in the proposed building on Jones Island. Advertised in this issue.

Wis., Sheboygan—The Helming Bros. Co., c/o Gus Helming, 1714 North 9th St., will soon award the contract for the construction of a 3 story, 60 x 150 ft. public garage on Erie Ave. Steam heating and electric power will be installed in same. Total estimated cost, $100,000.

Ia., Glidden—H. W. Porter, Secy. of the Bd. Educ., received lowest bid for the installation of a heating and plumbing system in the proposed 3 story, 90 x 162 ft. school building from the C. F. Rock Plumbing & Heating Co., St. Joseph, Mo., $36,213.

Ia., Mason City—The Miller Hotel Co. is having plans prepared for the construction of a 7-story, 132 x 165 ft. hotel. Contract for the installation of heating and plumbing systems will be sub-let. Total estimated cost, $800,000. Proudfoot, Bird & Rawson, 810 Hubbell Bldg., Des Moines, Archts.

Ia., Pocahontas—The Pocahontas Co. plans to construct a 3 story court house. A steam heating system will be installed in same. Total estimated cost, $400,000. Proudfoot, Bird & Rawson, 810 Hubbell Bldg., Des Moines, Archts.

Minn., Hibbing—John Murphy, Town Recdr., will receive bids until May 18 for the construction of a 2 and 3 story, 149 x 284 ft. recreational building on 1st Ave. and Jefferson St. A steam heating system will be installed in same. Total estimated cost, $500,000. Halstead & Sullivan, Palladio Bldg., Duluth, Archts. and Engrs.

Minn., New Ulm—The city will receive bids until May 13 for the construction of a 1 story, 40 x 90 x 90 ft. power plant, and for the installation of equipment including steam engine, exciter, etc. Total estimated cost, $200,000. A. J. Mueller, Engr. Silas Jacobson, 403 Endicott Bldg., St. Paul, Archt.

Kan., Americus—The people of Americus plan an election May 14 to vote on $18,000 bonds to construct a transmission line from here to Emporia.

Mo., Versailles—The Versailles Flour Mill & Ice Plant plans to install 130 hp. additional power in its present plant. R. Moser, Owner.

Tex., Dallas—The Security Natl. Bank, Commonwealth Bldg., had plans prepared for the construction of a 12 story, 100 x 200 ft. office building on Main and Murphy Sts. A steam heating system will be installed in same. Total estimated cost, $1,000,000. Herbert M. Greene Co., North Texas Bldg., Archt. and Engr.

Okla., Frederick—Tillman Co. is having plans prepared for the construction of a 4 story court house. A steam heating system will be installed in same. Total estimated cost, $250,000. Tonini & Bramblett, 301 Terminal Bldg., Oklahoma City, Archts.

Col., Denver—Loew's Enterprises, 1493 B'way, New York City, are having plans prepared for the construction of a theatre here. A steam heating system will be installed in same. Total estimated cost, $350,000. T. W. Lamb, 644 8th Ave., New York City, Archt.

Utah, Ogden—D. D. McKay, Engr., has filed application with the state engineer for Ogden River water rights to build a hydro-electric plant to pump water for irrigation in Weber and Boxelder Counties.

Wash., Puget Sound—The Bureau of Yards & Docks, Navy Dept., Wash., D. C., plans to furnish and install auxiliary equipment and piping for the air compressor, here. Estimated cost, $60,000.

Cal., Long Beach—Loew's Enterprises, 1493 B'way., New York City, are having plans prepared for the construction of a theatre here. A steam heating system will be installed in same. Total estimated cost, $500,000. T. W. Lamb, 644 8th Ave., New York City, Archt.

M. I., Guam—The Bureau of Yards & Docks, Wash., D. C., and the Office of Pub. Wks., San Francisco, Cal., received bids for the construction of extensions to power house, pump house, floor house and quarters at the Naval Station here, from Alfred H. Vogt, Builder's Exch., San Francisco, Cal., at $87,566; Pittsburgh Des Moines Steel Co., Munsey Bldg., Wash., D. C., at $83,700. Noted April 6.

Ont., New Toronto—The village plans to enlarge its waterworks system. Plans include the installation of a new electric pump, a new 16 in. feeder main and two 3 stage, 1,000 Imp. G.P.M. electric driven turbine pumps connected in series. Total estimated cost, $35,000. E. A. James Co., Ltd., 36 Toronto St., Toronto, Engr.

Ont., Ottawa—Stevens & Lee, Archts., 2 College St., Toronto, will receive bids for the construction of a hospital for the city. A vacuum steam heating and blower ventilating system will be installed in same. Total estimated cost, $2,500,900.

Ont., Toronto—The Bloor St. Baptist Church, Bloor and North Sts., plans to construct a church. A steam heating system will be installed in same. Total estimated cost, $500,000.

Alta., Calgary—The City Comrs. will receive bids until May 14 for the manufacture and delivery of three 200 kva., 12,000 volt to 2,300 volt transformers. J. M. Miller, City Clk.

Alta., Calgary—Loew's Enterprises, 1493 B'way., New York City, are having plans prepared for the construction of a theatre here. A steam heating system will be installed in same. Total estimated cost, $500,000. T. W. Lamb, 644 8th Ave., New York City, Archt.

Alta., Edmonton—Loew's Enterprises, 1493 B'way., New York City, are having plans prepared for the construction of a theatre here. A steam heating system will be installed in same. Total estimated cost, $400,000. T. W. Lamb, 644 8th Ave., New York City, Archt.

B. C., Victoria—Loew's Enterprises, 1493 B'way., New York City, are having plans prepared for the construction of a theatre here. A steam heating system will be installed in same. Total estimated cost, $400,000. T. W. Lamb, 644 8th Ave., New York City, Archt.

B. C., Winnipeg—Loew's Enterprises, 1493 B'way., New York City, are having plans prepared for the construction of a theatre here. A steam heating system will be installed in same. Total estimated cost, $500,000. T. W. Lamb, 644 8th Ave., New York City, Archt.

CONTRACTS AWARDED

Md., Fairfield (Baltimore P. O.)—The Globe Shipbuilding & Dry Dock Co. of Maryland, Fidelity Bldg., Baltimore, has awarded the contract for the construction of a 1 story, 30 x 110 ft. power house, 1 story, 50 x 150 ft. joiner shop, mold loft, etc., to Bancroft Jones Corp., Mutual Life Bldg., Buffalo, N. Y., at $850,000. Contract for the construction of a 1 story machine shop including boilers, etc., has been awarded to McClintic Marshall Co., Oliver Bldg., Pittsburgh, Pa., at $500,000. Noted Feb. 9.

N. C., High Point—The Harriss Corington Hosiery Mills, has awarded the contract for the construction of a 2 story, 50 x 100 ft. hosiery mill on Cox St., to J O'Connor, Washington St. A 100 hp. boiler and pipe fittings for heating system will be installed in same. Total estimated cost, $35,000.

S. C., Columbia—The State has awarded the contract for the installation of a heating plant in the State Capitol, to W. B. Guimarin & Co., 1220 Washington St., at $34,480.

Kan., El Dorado—The El Dorado Hotel & Theatre Co. has awarded the contract for the construction of a 5 story, 81 x 147 ft. hotel and theatre building, to Fred Borgolte. A steam heating system will be installed in same. Total estimated cost, $300,000.

Neb., Gering—Scotts Bluff Co. has awarded the contract for the installation of a heating system in the proposed 4 story court house, to Prohs Bros., at $20,906. Noted Feb. 9.

Tex., Waco—The Central Texas Electric Ry. Co. has awarded the contract for the construction of track including grading and all trestle work on first unit of Waco Temple Interurban line from here to Robinson, to the Central Texas Eng. & Constr. Co., Amicable Life Bldg. Plans include a power house and transmission line.

Okla., Guthrie—The Scottish Rite has awarded the contract for the construction of a consistory, to James Stewart, 30 Church St., New York City. A steam heating system will be installed in same. Total estimated cost, $3,000,000.

Okla., Pawhuska—The Citizens' Natl. Bank has awarded the contract for the construction of a 7 story, 50 x 140 ft. bank, to the George M. Bliss Constr. Co., Lathrop Bldg., Kansas City, Mo. A steam heating system will be installed in same. Total estimated cost, $400,000.

Okla., Tecumseh—The State Bd. of Affairs, Oklahoma City, has awarded the contract for the construction of 5 school buildings, including a power house, to the Manhattan Constr. Co., Oklahoma City. A steam heating system will be installed in same. Total estimated cost, $152,700.

Wash., Olympia—The State Capitol has awarded the contract for the construction of a power house, to the Western Constr. Co., Seaboard Bldg., Seattle. A steam heating system will be installed in same. Total estimated cost, $100,000. Noted May 3.

Ore., Roseburg—The Riter Sanitarium Co. has awarded the contract for the construction of a 3 story, 80 x 145 ft. hospital plant, including a heating plant, cost, $20,000; administration building, cost, $10,000, to C. W. Frazier, Wilbur. Contract for installing steam heating system will be sub-let.

Cal., San Francisco—Don Lee, 1601 Van Ness Ave., has awarded the contract for the construction of a 7 story auto sales building on Van Ness Ave., to the Lindgren Co., Monadnock Bldg. A steam heating system will be installed in same. Total estimated cost, $600,000.

Ont., Acton—The town has awarded the contract for laying, jointing, trenching, backfilling and erecting valves and hydrants on cast iron pipe, to the Gartshore Thompson Pipe & Fdry. Co., Ltd., Stuart St., West, Hamilton. Electrically driven deep well centrifugal pumps will be installed in same. Total estimated cost, $30,000. Noted March 9.

Ont., Toronto—The city has awarded the contract for the installation of four 20,000,000 Imp. gal. per day centrifugal pumps for the John St. pumping station, to the Turbine Equipment Co., Ltd., C. P. R. Bldg. Estimated cost, $75,000.

Ont., Toronto—St. Michaels Hospital, Church St., has awarded the general contract for the construction of a 6 story, 100 x 180 ft. nurses residence and hospital addition on Shuler and Victoria Sts., to Archibald and Holmes, Excelsior Life Bldg. A vacuum steam heating system will be installed in same. Total estimated cost, $600,000.

POWER

Volume 51 — New York, May 18, 1920 — *Number* 20

Bonehead Bill

BY RUFUS T. STROHM

Our engineer, Bill Jones, last week was surely up against it; a steam pipe sprung a nasty leak — two walls were right fornenst it. And when he got the stop valve shut and made a brief inspection, he found that he would have to cut another threaded section.

But when he set to work with vim in that restricted quarter, he found the job cut out for him a regular rip-snorter. The threading dies were hard to set, as he was quick in learning, and then there wasn't room to get good leverage for turning.

He strained till from his overalls he'd busted all the buckles, and slipped and hit the plastered walls and skinned his tender knuckles. He tugged and grunted, sweat and swore, for nigh to half an hour, until he hadn't any more of patience or of power.

He took a little rest at length, his shirt and jacket shedding, and when he had regained his strength, resumed his job of threading; and so he hauled and sawed and jerked till evening shades descended, and all the next forenoon he worked, before the task was ended.

But when at last he'd turned the trick and screwed the thing together, the steam leaked through in clouds as thick as London's foggy weather. Then Bill just clean forgot the rules of morals and religion, and cussed and damned those threading tools in Russian, Dutch and pidgin.

But though he wildly roared and raved and bent his conscience double, if he'd been wise he might have saved himself a peck of trouble. For what had caused that faulty joint should stir no one to wonder; because — and here's the vital point — those dies were dull as thunder!

The moral of this tale is plain to those who toil and tussle: **A little exercise of brain will save a lot of muscle!**

Testing Engines for 40,000 Feet in the Air

How the New Engine-Testing Laboratory at the Bureau of Standards Solves This Problem

By R. S. McBride
Engineering Representative McGraw-Hill Company, Inc., Washington, D. C.

THE Bureau of Standards has developed one of the most complete plants ever built for the testing of internal-combustion engines and, incidentally has made it possible to be safely on the ground with full control of an aëroplane engine, while operating the engine as if it were 40,000 ft. in the air. The new altitude laboratory at the Bureau contains many ingenious and unusual testing devices which are of considerable interest, not only to those interested in gasoline engines, but also to others concerned with any type of power-equipment testing. The work of this laboratory will be of great significance, for many fundamental problems are being investigated.

GENERAL VIEW IN THE BUREAU OF STANDARDS ENGINE-TESTING LABORATORY
The altitude chamber is in the rear at the left. A captured German motor-truck engine is shown on a test block in the center foreground. The racks shown overhead carry the resistance grids used to load the electric dynamometers. The car-dynamometer drums can be seen in the pit at the right.

It is generally recognized that conditions at high altitude are very different from those on the surface of the ground, but it is not usually realized how important these changes are for engine operation. For example, we know that it is necessary to supply an aviator with oxygen so that he may be able to breathe when at high altitudes, but the engine is not provided with any such auxiliary equipment, therefore it is necessary to operate it under conditions that will take advantage of the best possible combinations, otherwise large losses of power and fuel efficiency will be encountered. At 20,000 ft. the atmospheric pressure is about one-half that at sea-level, at 30,000 ft. about one-third sea-level pressure, and the power production of the engine is reduced almost in proportion to the reduction in atmospheric pressure under which it operates. On the other hand, the fuel efficiency at high altitude does not decrease greatly; in fact, the fuel used per horsepower does not change materially until an altitude of 5,000 to 10,000 ft. is reached, after which it begins to increase.

It was because of these general facts that various problems in engine performance had to be investigated as completely as is being done at the Bureau of Standards. The various factors studied include pressure, temperature, operating speed, kind or grade of fuel, gasoline-air mixture, proportions, spark setting, jacket-water temperature, oil temperature, back pressure on the exhaust, throttle opening, mechanical losses and numerous other variables. During the war period this work was housed in temporary buildings,

SPRAGUE ELECTRIC DYNAMOMETER MOUNTED ON SHAFT FROM ALTITUDE CHAMBER

This machine is simply a generator with the field frame mounted on bearings. It is prevented from turning with the armature only by the connection through the scale beam where its turning moment may be measured.

but a new laboratory has just been completed. This is a substantial brick and concrete building about 50 x 150 ft., designed particularly to house large altitude chambers, the car dynamometer and the various parts of the auxiliary equipment.

The altitude chambers, which are the most interesting part of the installation, had to be so designed and built that the atmospheric conditions within them could be adjusted to any condition that might be likely to be encountered at any altitude up to 40,000 ft. Each of the two chambers is 15 ft. long by about 7 ft. wide and 7 ft. high. They are thus of ample capacity for the largest aëroplane engines. The chamber is built up of reinforced concrete, 12 to 14 in. thick on four sides, top and bottom, in order to resist the great crushing force which is developed when the pressure within the chamber is reduced to a quarter or a third of that without. For thermal insulation the walls are lined with a two-inch cork layer. Access to the chambers is had through large counterweighted doors which slide up out of the way and permit free access either for installation or working about the engine under test. The engine itself is mounted on a heavy wooden frame, which has been designed to be as nearly as possible of the same flexibility and inertia characteristics as the typical aëroplane fuselage mounting. The pressure within the chamber is reduced to any desired point by a Nash hydroturbine-type vacuum pump which has a rated capacity of 1,500 cu.ft. of air per minute, when operating against a 12-in. vacuum at 300 r.p.m. This exhauster is belt-driven by a 75-hp. motor. This apparatus is capable of lowering the pressure to that normally existing at 30,000 to 40,000 ft. altitude or, in other words, about one-third to one-fourth of an atmosphere. The air circulated through the chamber and that supplied to the engine intake are cooled by passing over refrigerating coils. These coils are operated in connection with a York ammonia compressor using the direct ammonia expansion system. The refrigerating machine has a rated capacity of 25 tons and is operated by a 50-hp. motor. All this equipment is duplicated for the companion chamber. The air for the engine intake is cooled in a separate set of coils and controlled separately throughout. The temperature in the chamber as a whole can be lowered to about 0 deg. C., but the intake air can be cooled to —25 deg. C. The air to the engine, after passing over its cooling coils, goes through a settling chamber in order to eliminate the fine snow that is often carried in suspension. From here it is metered by three venturi tubes arranged in parallel and passes through control valves to the engine intake.

The exhaust from the engine is cooled as rapidly as possible by water jackets around each exhaust outlet. After this preliminary cooling in the water jacket, the exhaust gases and the cooling water mix together in the exhaust manifold, are drawn out into a separating tank, and the exhaust gases are taken from there by the large vacuum pump and discharged outside the building. The air circulation in the chamber is arranged for with fans appropriately installed to direct the air current over the cooling coils and around the engine. It is estimated that this system will be adequate even to care for tests of air-cooled engines, though none of this type of equipment has yet been tested.

The two altitude chambers are connected together through a small vestibule, thus permitting the use of either set of cooling coils or both, as may be required;

TESTING A UNION ENGINE FOR THE NAVY DEPARTMENT
Note the pipe overhead for carrying the engine exhaust to the large vertical pipe or stack.

but either chamber can be operated independently of the other, thus permitting tests on one engine to be in progress while another is being set up in the other chamber.

The power delivered by the engine is measured by a combination electric dynamometer and water-brake system. On one of the chambers two dynamometers, each of 300 hp., and a 400-hp. water brake are connected together in such a manner that the indications of all three can be weighed upon a single scale. Thus a total of 1,000 hp. can be measured directly by a single indicating mechanism. The other chamber is equipped with one 400-hp. dynamometer to which a water brake of any desired capacity can be added in case of need. The power developed by the dynamometers can be dissipated by resistance grids alone, or it can be used in operating the auxiliary motors and the balance absorbed in the resistance grids, which are then allowed to float upon the power line. For practical purposes, however, the best experimental results are found when all the current generated in the dynamometer is dissipated through the resistance grids, because in this manner the most uniform and controllable conditions for tests are obtained.

TEST ON A ONE-CYLINDER LIBERTY ENGINE

These tests were made to study the rate of flame propagation in the engine cylinder under different conditions. The operator at the left is at the dynamometer switchboard controlling the load on the dynamometer. The operator at the right is adjusting the spark chronograph which measures the speed of flame propagation.

One very great advantage in the installation of the new laboratory is the arrangement of the controls for the engine, the air system and the dynamometers, all of which are brought near together so that a single operator has complete control of the system without moving from his regular position of observation. It is thus possible to adjust conditions for a test, carry through the observations and be ready for a new adjustment. The complete determination of a heat balance in the system requires determination by weighing of the fuel used and an estimation of the heat gained in any combustion of lubricating oil, which together make up the total energy input. The output of energy is distributed over the brake horsepower, measured by the dynamometer system, the heat lost in the exhaust, the heat lost in the jacket water, the heat lost by direct radiation, and any mechanical losses. All these quantities depend upon the temperature, density of the air and many other factors. Practically a complete calorimeter system has been developed for measuring these different losses when complete determination of the heat balance is required. Temperature measurements are taken regularly at many points throughout the system, not only for control purposes, but also to furnish data for these thermal calculations. A recent report of the laboratory summarizes a dozen of the more important tests which have been carried through already at this Bureau. The following list indicates the scope of this work, some of which is still under way:

1. Horsepower-altitude relation for engines at normal speed.

2. Horsepower-speed relation over a range of altitudes up to 30,000 feet.

3. Horsepower-compression ratio for normal speed, using compression ratios of 4.7, 5.3, and 6.2 to 1 over a range of altitudes up to 30,000 feet.

4. Horsepower-inlet air temperature over a range of speeds and altitudes.

5. Effect of variation of intake pressure on horsepower over a range of altitudes, to simulate the effect of supercharging equipment.

6. Effect of exhaust back pressure on horsepower, over a limited range of pressures.

7. Mechanical losses at various speeds, altitudes and engine temperatures.

8. Metering characteristics of a number of different types of carburetors, with and without altitude compensation or control, for the full range of speeds and altitudes.

9. Best mixture ratios for maximum power over the range of speeds and altitudes, with different carburetors.

10. The performance of a number of automatic and hand-operated altitude compensation devices for different carburetors.

11. The total heat distribution for all speeds and air densities at full throttle.

12. The performance of special fuels: "Hector," a combination of cyclehexane and benzol; "Alco-gas," a combination of alcohol, benzol, gasoline and ether, at a compression ratio of 7.2 to 1.

One of the problems still giving considerable difficulty in the aëroplane-engine field is the question of carburetor adjustment with changing atmospheric pressures. The automatic arrangement for adjustment with changing speeds is not satisfactory for altitude changes. For example, in some of the early work it was found that at higher altitudes four times as much gas was delivered as was found at ground-level conditions unless readjustment was made. A number of ways for controlling this factor have been devised, two or three of them the inventions of the aëroplane-engine section of the Bureau of Standards. However, hand control of this adjustment is still general since none of these automatic devices has yet found wide-spread application.

The problem of "supercharging" for altitude work has been studied by these laboratories only in a preliminary way, but it is expected that further work will be undertaken shortly. This is one of the very important problems in automotive work, and its investigation promises to give important results. A determination of the best compression ratios for various altitudes is in active progress. This is important as affecting the fuel efficiency and also the power production at a given speed.

One interesting fact that is pointed out as the result of these investigations is the influence of temperature upon power production. Fuel efficiency is not greatly affected by temperature, but even with high-grade liquid fuels the power developed by an engine is largely affected. With the better grades of gasoline the power output is greater with slightly lower temperatures, but when temperatures too low for effective vaporization are reached, the power output again falls off. The magnitude of these effects and means for their control are of extreme importance in aëroplane-engine work, though also of considerable significance for all automotive construction.

Not content with studying the altitude effects, this laboratory has also developed an elaborate dynamometer system for car, transmission, axle and engine testing. By various combinations of dynamometer, water brake and prony brake this apparatus can determine the effectiveness of operation under all varieties of conditions or of any part of the system. The engine itself, the transmission system, the differential and even the power delivery on the road, as judged by the application of tire or service wheel to the drum, can be separately studied.

For this work three dynamometers having capacities of 50, 125 and 150 horsepower, respectively, are installed on three sides of a rectangular pit in which the drum and transmission outfit is placed. Extensive tests of lubrication and differential problems will be possible with this outfit, since any one of the three dynamometers can be used as a measuring or as a driving unit.

Each of these three dynamometers can also be used separately for independent investigations. For example, one of them is now being used in connection with tests on a one-cylinder Liberty engine built to study the rate of flame propagation within the cylinder while operating.

The means of determining rate of flame travel within the cylinder is very effectively developed here by a triple spark-plug system. At the time the flame of burning gas-air mixture passes over the spark plug, the gases are ionized and a reduction in electrical resistance results. This permits the passage of a spark across the gap. These spark plugs are connected to a spark chronograph or an oscillograph, and the relative time of passage of the flame past these three points can thus be measured. A full description of this particular installation has been given by MacKenzie and Honaman in the *Journal of the Society of Automotive Engineers* for February of this year.

Chart for Finding Power Factor

By C. Harold Berry

IN THE switchboard equipment of modern alternating-current power plants the power-factor meter is being displaced by the wattless-component meter owing to the fact that the reliability of the former is

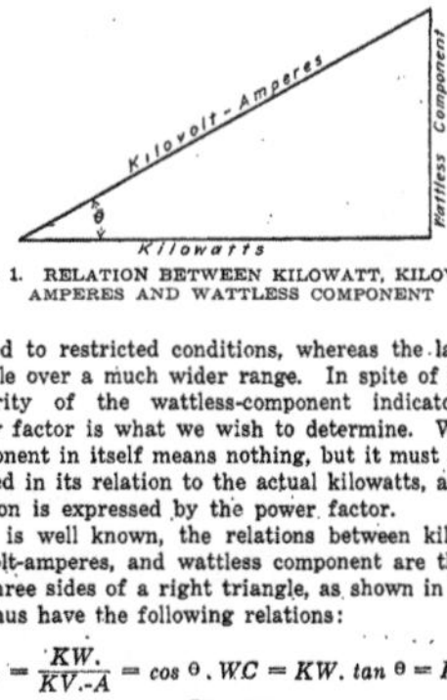

FIG. 1. RELATION BETWEEN KILOWATT, KILOVOLT-AMPERES AND WATTLESS COMPONENT

limited to restricted conditions, whereas the latter is reliable over a much wider range. In spite of the superiority of the wattless-component indicator, the power factor is what we wish to determine. Wattless component in itself means nothing, but it must be considered in its relation to the actual kilowatts, and this relation is expressed by the power factor.

As is well known, the relations between kilowatts, kilovolt-amperes, and wattless component are those of the three sides of a right triangle, as shown in Fig. 1. We thus have the following relations:

$$P.F. = \frac{KW.}{KV.\text{-}A} = \cos\theta \,.\, WC = KW.\tan\theta = KW. \times \frac{\sqrt{1 - P.F.^2}}{P.F.}$$

where $P.F.$ equals power factor, $KW.$ equals kilowatts, $KV.\text{-}A$ equals kilovolt-amperes and WC equals wattless component.

The last equation is awkward to solve, so the easiest way to get the power factor from readings of kilowatts and wattless component is to make use of tables of trigonometric functions of angles. Since the tangent of the angle θ equals the wattless component divided by kilowatts, we may perform this division and then look up the angle in a table of natural tangents. Having thus found the angle, we may turn to the table of natural cosines and read off the cosine of the angle, which is the power factor.

Even this process is tedious and slow, especially where such computations must be made by switchboard operators on regular hourly of half-hourly readings on several machines, and the alignment chart of Fig. 2

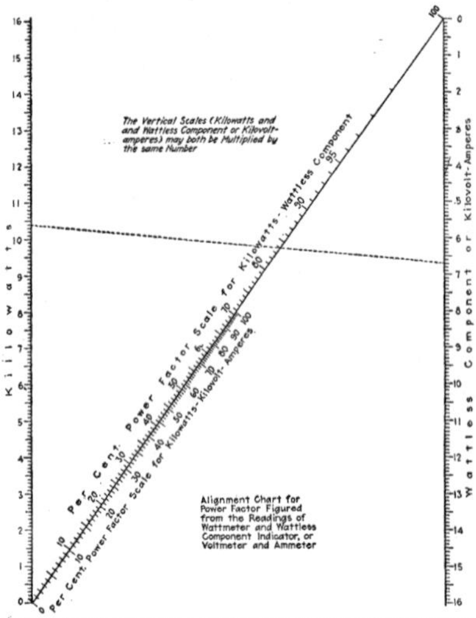

FIG. 2. CHART FOR CALCULATING POWER FACTOR FROM KILOWATTS AND WATTLESS COMPONENT OR KILOWATTS AND KILOVOLT-AMPERES

has been designed to give the desired result quickly and accurately without calculation or reference to tables.

The use of this chart will be almost obvious. On the left-hand scale find the point representing the kilowatts, on the right-hand scale the point representing the wattless component. Through these two points pass a straight line — a stretched thread or a line scratched on a sheet of celluloid. At the point where this line cuts the diagonal of the chart, read off the power factor from the scale on the upper side of the diagonal.

For example, if the kilowatts is 10,400, and the wattless component 6,700, the power factor will be 84 per cent, as indicated by the dotted line drawn on the chart.

In using this chart, both vertical scales may be multiplied by the same number. By this means the line joining the two given points may be made to intersect the diagonal nearly at right angles, thus making it easy to read the chart. For example, if we have kilowatts equal to 19,000 and wattless component 17,000, the line drawn through the points 19 and 17 as numbered on the chart intersects the scale at a very small angle, and it is difficult to read the power factor. We may get the same result, however, by passing a line through 9.5 (half of 19) and 8.5 (half of 17), which enables us to read the result as 74.5 per cent power factor. On the lower side of the diagonal has been added a scale for obtaining the power factor from the kilowatts and kilovolt-amperes. For example, if the kilowatt meter reads 551 and the kilovolt-amperes as determined from the voltmeter and ammeter readings is 650 then a line drawn from 6.5 on the kilovolt-amperes scale to 5.5 on the kilowatt scale will give a power factor of 85 per cent. This might just as easily be determined on the slide rule, so that there is no special advantage in having this scale on the chart. However, it does not interfere with the other scales, and may at times prove useful.

White Coal Available in Canada

The Committee on Water Power of the Commission of Conservation of Canada, following a survey of the total available water power of that country, reports that Quebec has 6,000,000 available horsepower; Ontario, 5,800,000; British Columbia, 3,000,000; Manitoba, 2,797,000; and a considerable amount of power in other parts of the Dominion, totaling in all, 18,832,000 horsepower.

It is estimated that more than 85 per cent of the total electric-generating-station capacity in Canada is derived from water power, the remainder being nearly all steam power. Efforts toward more complete utilization of this power are under way Last year there was completed throughout the Dominion installation of plants aggregating 64,000 hp., and plants at present under construction total over 370,000 hp., while other developments planned for the future will increase this by about 750,000 horsepower.

Two methods, entirely different, have been developed for dealing with water-power problems in Ontario and Quebec. Each seems to be successful in its own area. The Hydro-Electric Power Commission of Ontario has jurisdiction over all water powers in that province. The commission is carrying out its own developments and operating its projects. In Quebec the Quebec Streams Commission is taking all necessary steps to make water power available by private companies. The commission undertakes the building of dams and all the work necessary so that the proposed power-plant sites will be for use, but the companies obtain leases for the use of water, build the necessary generating plants and operate them.

Design and Construction of Power-Plant Piping

Practical Suggestions Regarding the Design and Construction of Power-Plant Piping, Including Materials Used, Valves, Fittings and Types of Joints for Various Pressures and Temperatures

BY JOHN D. MORGAN

IN THE design of a power-plant piping system the first task is to make up a sheet showing the lines to be constructed.

A specimen sheet is shown in the table giving practically all the lines that are used in the average power plant. These lines can be classified into groups, as follows: High-pressure saturated and superheated steam piping; exhaust-steam piping; atmospheric exhaust piping; feed-water piping; low-pressure water piping; blowoff piping; drips and drains; oil piping, high- and low-pressure.

The design of the high-pressure steam piping system will demand much attention, for with the advent of high steam pressure and high degree of superheat and the use of large turbines, it is necessary that the line deliver steam with the smallest possible pressure variation from its original boiler pressure and also at the highest possible temperature.

PIPING LINES IN POWER PLANTS

STEAM LINES:

Main high-pressure superheated steam line.
Duplicate high-pressure superheated steam line.
Auxiliary high-pressure superheated steam line.
Main high-pressure saturated steam line.
Duplicate high-pressure saturated steam line.
Auxiliary high-pressure saturated steam line.
Main exhaust steam line.
Auxiliary exhaust steam line.
Atmospheric exhaust steam line.
Atmospheric exhaust from feed-water heaters.
Atmospheric exhaust from blowoff tank.
Line to damper regulator, whistle, etc.

WATER LINES:

Discharge and suction lines of boiler-feed pumps.
Discharge and suction lines of condenser circulating pumps.
Discharge and suction lines of vacuum pumps.
Discharge and suction line of fire pumps.
Discharge and suction line of house-service pumps.
Discharge and suction lines of hot-well pumps.
Discharge and suction lines of water cooling pumps.
Discharge and suction lines of sump and miscellaneous pumps.
Discharge of condensers.
Lines from surge tanks to feed-water heaters.
Overflow lines from feed-water heaters.
Lines to boiler balanced-draft systems.
Lines to and from venturi and V-notch meters.
Flooding lines to and from superheaters.
Lines to and from turbine glands.
Lines to and from air cylinder on air compressors and dry vacuum pumps.
Condenser priming lines.
Centrifugal pump priming lines.
Ash-sprinkling lines.
Fire lines.
Overflow from hotwells and tanks.
Discharge from steam traps.
Lines to and from water-cooled transformers.
Lines to and from boiler-compound mixing tanks.
Lines to and from washing machines.

DRIPS AND DRAINS:

Lines from turbines and engines.
Lines from boilers and superheaters.
Lines from auxiliaries.
Lines from all pumps.
Lines from steam and water mains.
Lines from feed-water heaters and separators.

BLOWOFF PIPING:

Lines from boilers to blowoff tanks.
Lines from feed-water heaters and safety valves.

AIR PIPING:

Lines from condensers and air pumps.

GAGE PIPING:

Lines to all gages, mercury columns, recorders, etc.

GAS PIPING:

Lines for light and heat.

SMALL PIPING SYSTEMS:

Compressed-air system.
Step-bearing oil system.
Hydraulic-gear oil system.
Turbine-oiling system.
Engine-oiling system.
Fuel-oiling system.
Boiler feed-pump regulator pipe.
Refrigeration system.

The formulas most generally used are those of Babcock, which are as follows:

$$W = 87\sqrt{\frac{P\,c\,d^5}{L\left(1+\frac{3.6}{d}\right)}}$$

$$V = \sqrt{\frac{Pd}{cL\left(1+\frac{3.6}{d}\right)}} \times 15{,}951$$

$$P = 0.0001321\frac{W^2L + \left(1+\frac{3.6}{d}\right)}{cd^5}$$

Where

W = Weight of steam flowing in pounds per minute;
L = Length of pipe in feet;
P = Drop in pressure;
d = Diameter of the pipe in inches;
c = Density in pounds per cubic foot of steam;
V = Velocity in feet per minute.

In close calculations it is necessary, when the drop in pressure is a determining factor, to figure on the resistance of bends and valves in the system. For this work Briggs formulas for the equivalent resistance expressed in feet are most generally used. They are as follows:

For 90-deg. bend

$$L = \frac{76\,d}{\left(1+\frac{3.6}{d}\right)\times 12}$$

For globe valve

$$L = \frac{114\,d}{\left(1+\frac{3.6}{d}\right)\times 12}$$

For special conditions it frequently becomes necessary for the designing engineer to figure the resistance of pipes to internal and external pressure. The formulas which lend themselves with greatest accuracy for the collapsing pressure are Stewart's. These are as follow:

$$P=1{,}000\left[1-\sqrt{1-1600\left(\frac{t}{d}\right)^2}\right], \text{ or } P=86{,}670\frac{t}{d}-1{,}386$$

Where P is the pressure in pounds per sq.in., t is the thickness in inches and d is the outside diameter in inches. When the ratio of t to d is less than 0.23, the first formula should be used, the second for all other cases.

The formula most used for the bursting pressure is that of Barlow:

$$P = \frac{ft}{r2}$$

Where P is the pressure in pounds per square inches, f is the ultimate unit stress in pounds per square inch, t is the thickness in inches and $r2$ is the outside radius in inches.

One part of the design that must be given very careful consideration is the branch lines from the main. It is essential that the sum of percentages of the carrying capacities of the branches be 100 per cent or more.

The general engineering custom in regard to the velocity of steam in the high-pressure mains is to use from 6,000 to 7,000 ft. per min. for saturated steam and from 10,000 to 12,000 ft. per min. for superheated steam.

The two materials that are most used are steel and wrought-iron pipe. When no accurate figures are at hand the ultimate strengths of the various pipes can be taken approximately as follows: Butt-welded steel pipe, 40,000 lb. per sq.in.; lap-welded steel pipe, 50,000 lb. per sq.in.; seamless-steel pipe, 60,000 lb. per sq.in.; wrought-iron pipe, 28,000 lb. per sq.in.

For high pressures and temperatures there are several types of valves that are suitable for use. These are the double-seated poppet-type valve, the globe or angle type with outside stem and yoke, and the solid-wedge or broken-wedge gate valve with outside yoke. Most of the valves in general use are of the last-mentioned type; however, the first two are in many respects better valves, for if the machine work is not done in a thorough manner and the right quality of metal used, the gate valve will leak more freely than the other two. It is customary to have all large valves equipped with bypass valves to equalize the pressure on both sides. Various types of flanges are in use. These vary as to actual design with the type of joint with which they are intended to be used. For instance, different types will be required for the screwed joint, welded joint, vanstone joint, the ground joint with cast-iron gasket ring and the expanded joint. The various materials used in flanges are cast iron (standard weight for 125 lb. pressure and extra-heavy for pressures up to 250 lb.), semi- or ferro-steel of standard and extra-heavy weights, cast steel of extra-heavy and standard weights, rolled steel and forged steel.

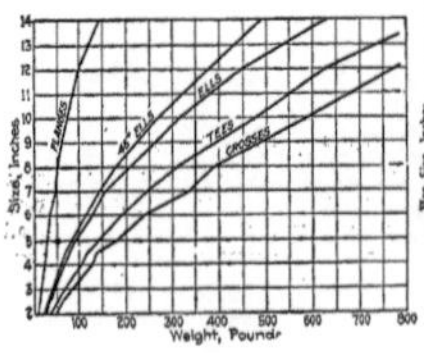

FIG. 1. WEIGHT OF CAST-IRON FLANGED FITTINGS, FOR 2 TO 14-IN. PIPE

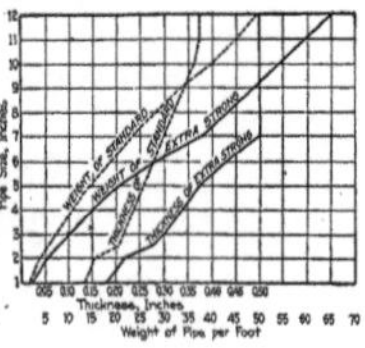

FIG. 2. THICKNESS AND WEIGHT OF STANDARD AND EXTRA-STRONG PIPING

The field for the screwed joint is not that of high pressures and temperatures; however, the screwed joint is not unsatisfactory when the joint with welded flanges and the cast-iron gasket ring is a good one, provided good workmanship is secured, but as a rule it is difficult to secure this workmanship. From a manufacturing viewpoint the welded joint is superior, but in the field for erection its great rigidity is its greatest drawback, and if the vanstone joint is made, using care in its manufacture, and is made with the high hub, improved reinforced square corner with raised surfaces from the inside of the bolt circle to the inside of the axis of the pipe, there can be no question as to its superiority.

There has been developed a joint where the pipe is rolled into recesses in the flange. This joint is quite satisfactory, but it has not been generally used. Then there has also been developed a joint that does not make use of any regular flange. The pipe is rolled back similar to that of the vanstone except that the metal turned back is thicker than the thickness of the pipe and holes are drilled into the turned-back flange and the pipes are bolted together with a gasket between them. This joint is very satisfactory and reduces the weight of the installation considerably. There has also been used a joint where the pipe has been flanged over like the vanstone except that the flange has been recessed with an angle cut on the outside and the pipe flange has been rolled back into the angle. This joint is satisfactory, but does not approach the vanstone joint for flexibility in erection.

Gaskets that are suitable for use with superheated steam are many. The one in most general use is the soft corrugated steel gasket applied with a cement. There also are a number of other good varieties, such as the asbestos gasket shrouded with copper or bronze and the corrugated copper gasket. The important point in connection with gaskets is their proper installation, and this is accomplished only when the bolts are drawn up tight and uniform over the complete bolt circle.

The fittings most generally used consist of tees, ells, Y's, crosses and laterals. These are made of cast iron, malleable iron, semi- or ferro-steel and cast steel of standard and extra-heavy weight. In the most modern plants it has become standard practice to eliminate fittings wherever possible. This is accomplished by the use of welded nozzles on the pipe. These nozzles can be of any length or size with any type of flange or joint, and their use lightens the total weight of the piping and eliminates from 50 to 60 per cent of the total joints. Fig. 1 graphically shows the weight of cast-iron flanged fittings, and Fig. 2 indicates the thickness and weight of standard and extra-heavy piping.

For high pressures and temperatures there are two types of expansion joints that are suitable—the slip-type joint with a proper packing, and pipe bends. The corrugated reinforced copper expansion is not suitable for superheated steam. Of the pipe bends there are several types—the quarter bend, the U-bend, combination U and quarter bend, expansion U-bend and the double-offset expansion bend. Assuming 1 as the proper expansion for the quarter bend, then the relative expansion values of the other bends are as follows: U-bend, 2; combination U and quarter bend, 4; the double-offset expansion bend, 5. In the design of bends the radius and the weight of the pipes must be properly proportioned in order that the right amount of expansion and contraction can be compensated for without damage to the piping or bend.

The use of superheated steam at high temperatures and pressures has given rise to spirited discussion in regard to the materials most suitable for fittings, flanges and valves in connection with piping systems. Recently, the A. S. M. E. Boiler Code was revised in relation to cast iron and superheated steam, as follows:

Cast iron shall not be used for nozzles or flanges directly attached to the boiler at any pressure or temperature. Cast iron shall not be used for boiler or superheater mountings, such as connecting pipe, fittings, valves and their bonnets for steam temperatures over 450 deg. F.

No mention is made in the Code for the piping systems at large. At a more recent date the United States Steamboat Inspection Service authorized the use of semi-steel gray-iron mixture for superheater headers on ships, and Lloyds have also adopted this ruling. This shows a diversity of opinion. These various rulings have been made after considerable discussion on the subject, and it is the general opinion among well-known engineers that much of the trouble experienced with extra-heavy cast-iron fittings and valves, when operated at temperatures not exceeding 500 deg. F., has been largely due to the design rather than to the character of the material. Abroad, the use of cast iron of a grade called gun iron has given very good success, although it is generally conceded that for high steam pressures and for temperatures in excess of 500 deg. F. steel fittings should be used. Cast iron, when subjected to temperatures in excess of 500 deg. F., does in some cases show a permanent increase in some dimensions. How long this growth would continue is uncertain, also that the strength of cast iron is materially reduced when exposed to temperatures not exceeding 550 deg. F. has not been conclusively proved, and as a matter of fact some engineers claim that the strength of cast iron grows till it reaches the break-down point around 800 deg. F. It seems probable that most of the failure of cast-iron fittings in stationary plants where the temperature does not exceed 500 deg. F. has been due to poor quality of the cast iron, variable temperature of the steam, poor design of the piping, fittings and flanges, or the lack of sufficient expansion joints. This is borne out by the fact that there are in this country over 37,000 locomotives that have cast-iron superheater headers and steam pipes operating every day with temperatures from 600 to 750 deg. F., and they undergo greater stresses and more sudden temperature changes and fluctuations than any stationary power plant would ever be subjected to, and failures of these headers and steam pipes due to the high temperatures used are practically unknown. It would certainly seem from all reliable information that the use of a high grade of cast-iron fittings is permissible for temperatures of about 500 deg. F. and that steel fittings are advisable for higher temperatures. However, many advocate the use of steel fittings in all cases, because a greater factor of safety is afforded and also because poorly designed systems using steel fittings will not be so likely to have trouble. But the fact remains that the use of cast-iron and semi-steel fittings for a certain limited temperature is by no means poor engineering practice.

The material best suited for valve seats, discs and bushings is nickel or monel metal. These metals have practically the same expansion and contraction as steel, and temperatures up to 1,000 deg. F. have no material effect on the seats, discs and bushings made of these metals. The valve stems should be made of nickel steel. Brass or bronze is unsuitable for use with superheated steam of high temperatures. At temperatures approximately 650 deg. F. the breaking strength of bronze is only about 12,150 lb. per sq.in., and its elongation at this temperature is only about 1½ per cent. At normal temperatures its elongation is about 37 per cent and it has a breaking strength of 34,000 lb. In regard to the metal for the body of the valve, the same conditions apply as to that for fittings.

Fuel-Oil Tanks, Tamarack Mills

Through the courtesy of Jenks & Ballou, consulting engineers, Providence, R. I., is reproduced the working drawing of the concrete pits housing the steel fuel oil tanks at the Tamarack Mills power plant, Pawtucket, R. I. *Power* readers will welcome this drawing.

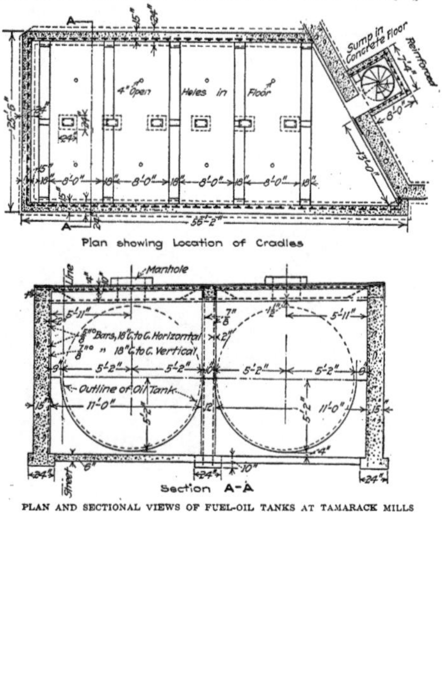

PLAN AND SECTIONAL VIEWS OF FUEL-OIL TANKS AT TAMARACK MILLS

POWER DRIVES FOR ROLLING MILLS

By

W. O. ROGERS

Various types of engine valves are touched upon as applying to rolling-mill units. Two of the largest poppet-valve uniflow engines in the world were found operating a 9- and 12-in. merchant mill. These engines operate condensing, and have disc-driven governors so designed that the engine speed can be regulated while the engine is in operation.

BEFORE taking up the uniflow engines that I saw used for driving rolling mills, it might be well to mention the action of the various types of valves used on engines that were described in the preceding article. Of the ordinary slide valves there are several types, including the flat or D unbalanced valve, not much used in rolling-mill practice; piston valves, which by the nature of their form are balanced; pressure-plate valves, which are balanced by special form of plates; riding cutoff, which are usually flat valves, each being operated by a separate eccentric; gridiron valves; and rotating valves. The D-valve is the simplest type used. Steam is admitted to the cylinder past the two outside edges and exhausts to the atmosphere past the two inside edges to a common exhaust port. It is generally made with steam laps so that the steam edge of the valve will overlap the steam edge of the port when the valve is in its mid-position. The steam lap is provided mainly to cut off the steam supply to the cylinder, generally at about three-quarters of the piston stroke. Steam lap also permits of free exhaust. Pressure-plate or balanced valves operate with a plate that rests against the back of the flat valve and excludes the steam pressure from all or part of the top of the valve, which causes it to be balanced with only a sliding friction, due to its own weight. The action of the steam in passing the valve is practically the same as with the D slide valve.

The piston valve is a simple form of balanced valve, because the steam pressure is exerted in all directions and does not therefore force the valve against the valve seat, and the weight of the valve is the only thing tending to wear it or its seat. Steam is generally admitted to the cylinder past the inside, or center, edge of the valve, and the exhaust is past the outside edges.

The aforementioned valves all control the steam admission by the same edge, also the cutoff at each end of the cylinder. Any change in the point of cutoff also changes the point of admission and will affect the point of compression.

The riding cutoff valve consists of a main valve which controls the steam admission, release and compression, and a cutoff or riding valve which controls the point of cutoff and which generally rides on the top of the main valve. The advantage of this type is that the cutoff may be varied without affecting the other events of the stroke. Gridiron valves are of the multiported type, and are on a number of makes of engines. Because of their many ports the valves require but a small travel.

The Porter-Allen engine is much used in rolling mill practice. There are two steam valves and two exhaust valves; the former govern the admission and cutoff of steam and the latter the release and compression. The simple eccentric actuates a link that moves the steam

FIG. 1. THE 9-IN. MERCHANT MILL AT THE STRUTHERS PLANT

and exhaust valves. The link has a peculiar motion, being of a horizontal and vertical throw of the eccentric. The link is restrained from rising by a trunnion, and the horizontal throw of the eccentric draws off the lap of the valve, while the vertical throw puts the tips of the link alternately to and from the cylinder as the eccentric rises or falls in its revolution, the upward throw tipping it from the cylinder toward the shaft. The tipping of the link opens and closes the steam valve by rocking a reach rod through a steam rod and arm.

FIG. 2. UNIFLOW ENGINE USED TO DRIVE A 9- AND A 12-IN. MERCHANT MILL

The Corliss valve gear is composed of four semi-rotary valves, two steam and two exhausts, using both single and double eccentrics. The double-eccentric engine is mostly used in rolling-mill work when the Corliss type of engine is used. The single-eccentric engine will operate the cutoff automatically only up to one-half stroke, but a later cutoff is obtained by using two eccentrics and two wristplates, one set for the steam valve and the other for the exhaust valve.

Poppet-valve engines are used to some extent in this country. What is known as the Uniflow poppet engine uses but two valves, both controlling the steam admission to the cylinder. The exhaust is through a ring of exhaust ports that are cut in the middle of the cylinder casting and are uncovered by the piston at the end of the stroke, the piston therefore acting as an exhaust valve.

The engine cylinder and pistons are made longer than the usual practice in steam-engine design. Uniflow means that the steam flows in but one direction in passing through the cylinder. After cutoff the steam expands behind the piston, and at the end of expansion the piston uncovers the exhaust port for an instant and the expanded steam escapes. On the return stroke the steam at exhaust temperature and pressure is caught between the piston and the cylinder head, and as the piston moves back this steam is compressed, and as the compression is increased the temperature of the steam is increased.

It also absorbs heat from the heated head jacket, therefore the temperature of the steam in the clearance space is as high as and sometimes higher than the incoming steam at admission. Under this condition the steam at admission does not come in contact with cold surfaces and the initial condensation is required.

This type of engine is capable of carrying enormous overloads, such as are found in rolling-mill work, and also low underloads with a flat steam-consumption curve.

What are understood to be the two largest poppet-valve Uniflow engines in the country are used for rolling-mill service at the plant of The Youngstown Sheet & Tube Co. They are installed in an extension that was built about three years ago. The extension includes two mills, one being a 9-in. merchant mill and the other a 12-in. merchant mill. The engine driving the 9-in. mill is of 1,000 hp. capacity at 110 r.p.m., having a cylinder 37 x 48 in. The larger engine driving the 12-in. mill is 1,500 hp. capacity at 110 r.p.m. and has a 44 x 50-in. cylinder. Both engines operate condensing, exhausting into a barometric jet condenser.

Speed variation from 55 to 110 r.p.m. may be obtained on both engines, by proper adjustment of the governor drive. The engines are equipped with fly-ball inertia governors which control the cutoff of the steam valves. The governor is driven by a lay-shaft, Fig. 2, geared to the engine shaft. On this shaft is mounted a steel friction disc. A similar disc geared to the governor shaft is placed at a short distance from the former and in the same plane with it. Rotative motion is transmitted from one disc to the other by means of four fiber wheels which roll upon the discs. Two wheels rest on each disc, each wheel on one disc being connected by a shaft to the corresponding wheel on the other disc. The ratio between the distance of

FIG. 3. NEAR VIEW OF ENGINE DRIVING 12-IN. MERCHANT MILL

the wheels from the disc centers determines the speed ratio between the engine and governor. This ratio may readily be changed by a movement of the floating frame on which the friction wheels are carried, an operation which may be accomplished while the engine is running. Inasmuch as the governor is designed to run at constant speed, or at least with no greater variation than 4 per cent, it is evident that a change in the position of the wheels on the discs will change the speed of the engine.

I was told that the governor was extremely sensitive, as it should be, and that the engine took the loads easily when the bar of steel entered the rolls of the first stand. After the bar passes through the first stand rolls, it enters successively the remaining five stands of the roughing rolls and the six finishing stands. This, of course, brings a suddenly increased load on the engine as the metal is reduced in each of the six roughing rolls. As soon as the last end of the bar has passed through the first roll, the load on the engine decreases and of course continues to do so until the bar has passed through the last finishing stand. It is seen that the governor must be very sensitive in order to maintain the engine speed at normal with its increasing and decreasing loads.

Some idea of the size of the mill may be gathered from the fact that the 9-in. mill is in a building 90 ft. wide by 1,150 ft. long, and that the 12-in. mill occupies a space 100 ft. wide by 1,200 ft. long. Both steam and motor drives are used with the 9-in. mill. The smaller of the uniflow engines drives the six roughing stands by means of beveled gear drives, and is so geared that the rolls of each stand run at an increased speed so as to take care of the increased length of the bar as it comes from the rolls preceding it. Fig. 1 shows the 9-in. merchant mill at the Struthers plant. In the background can be seen the Edward cooling bed, which is 400 ft. long.

A 700-hp. direct-current motor drives the first two finishing stands, and the other four are driven by two 500-hp. motors, two stands to each motor. A sizing mill is driven by a 100-hp. motor. The motors driving the 9-in. mill are housed in sheet-metal hoods to protect them from dirt and flying material, as shown beyond the engines in Fig. 2. The other motor is shown outside of the engine-room space. Each motor operates two stands of rolls by means of a belt drive, the motor pulley coming central between the roll pulleys, the latter being on the inside and the former facing on the outside of the belt, the bottom rim of the motor pulley coming down sufficiently below the top rims of the driven pulleys to insure an arc of contact sufficient to carry the load.

FIG. 5. ANOTHER VIEW OF THE 12-IN. MERCHANT MILL

The 12-in. mill is driven by the larger engine, Fig. 3. A general view of this mill is shown in Figs. 4, and 5. The engine and an air pump are shown in the center of the view. The crankshaft of the large engine is coupled to the gearshaft by a universal coupling, and a small, or breaking, shaft is placed between the engine and the gearshaft to prevent damage to the engine in case the rolls of the mill become clogged. If such should happen, the small breaking shaft gives way, and being of short length is easily renewed. The gear drives are inclosed in a gear box, Figs. 3 and 4, and the rolls of this mill are of different sizes and revolve at different speeds in order to take care of the lengthened bar in going through the roll. The conveyor rolls are operated by inclosed mill-type motors having a backgear. All rolls have flexible couplings in which a coupling pin having four projections fits in corresponding sockets in the coupling head.

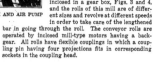

FIG. 4. THE 12-IN. MERCHANT MILL, SHOWING GEAR BOX AND AIR PUMP

Both engines are equipped with safety stops, the system being provided with stations at various points in the mill, so that in case of accident the engine can be shut down from any of these points. All steam piping is placed below the flooring, and the steam pressure used is 155 lb., superheated 50 deg. As stated, the engine exhausts to a jet condenser that maintains a 25- to 26-in. vacuum. In summer the temperature of the water used reaches 130 deg. This is on account of so many plants using the water before it reaches this mill. In winter it is around 30 degrees and it is easier to maintain the vacuum.

In Fig. 6 is shown a general view of a 10-in. continuous skelp mill, composed of a roughing and finishing mill. The roughing mill is driven by a 26 and 48 x 48-in. cross-compound engine of 1,500-hp. capacity at 90 r.p.m. The finishing mill is driven by a 26½ and 48 x 48-in. cross-compound engine of 1,500 hp. at 100 r.p.m. This mill rolls skelp from 1⅜ to 7¾ in. in width. There are, it is understood, three other such mills in the United States, each producing from 10,000 to 14,000 tons of finished skelp per month.

(The next installment will take up some of the features of various types of rolling mills such as blooming, billets, plate, rod and sheet mills, all of which are of special interest.)

FIG. 6. A 1,500-HORSEPOWER ENGINE DRIVING A TEN-INCH CONTINUOUS SKELP MILL

Surveys of the possible water power available in British Guiana indicate that on the Cuyuni River 285,000 hp. could be made available, on the Essequebo 475'000, and on the Demerara 38,000. It is estimated that power available for delivery within a two hundred mile radius would be about 156,000 hp. from the Cuyuni, 273,000 hp. from the Essequebo and 21,000 hp. from the Demerara. British Guiana is extremely rich in mineral resources, and the advent of electricity would open up immense sources of wealth.

Graphic Meter Records and Their Interpretation

By Walter L. Hamilton

To the uninitiated, records obtained from graphic meters are often as difficult to interpret as is the method by which they were obtained. To facilitate interpreting such records, a study of the meters producing them is useful.

Curve-drawing meters are used to record numerous conditions, such as temperature, pressure, flow of liquids and gases, speed, volts, amperes, power factor, kilowatts, etc. The benefits derived from their use are quite numerous and varied. A permanent record covering the total period of test is obtained of certain conditions, which permits discussion or study of working methods at times other than those at which the records were obtained. It is an interesting fact that a person is always interested in and will believe what is shown in picture form, when a mass of data obtained from indicating meters, covering similar conditions, is often considered too complicated to bother with.

In selecting curve-drawing meters for a given installation, the character of the service to be recorded and the results desired should be carefully considered. Like people, certain meters are best adapted to certain classes of work. You would not engage an engineer to wire a door bell, nor would it be desirable to install high-grade meters for unimportant work. However, if the meter is to be installed temporarily for study purposes, only the best high-grade apparatus should be used.

A record chart may be circular in form, a complete revolution of the paper representing a period of time, usually from twelve hours to one week. It may be oblong in shape and placed on the surface of a cylinder, its total length representing a definite period of time.

It may be of the continuous-strip type with one inch of length representing one hour or fraction of an hour. The continuous charts are almost always found in high-grade, high-priced meters, as a more powerful clock and a more complicated paper-feeding mechanism are required than in meters using circular charts.

Meters may be graduated to represent only a portion of their capacities, such as volt and frequency meters which are graduated from perhaps 60 to 100 per cent of capacity. This arrangement permits of a wider scale over the probable variation of the service to be recorded. They may be graduated from 0 to 100 per cent of capacity. Wattmeters and flowmeters employ this latter form of scale.

Certain meters will try to record accurately every variation in the load. This often gives a wide curve and, for fluctuating loads, is not altogether desirable. Other meters may be damped to such an extent that their records may represent more nearly average conditions, since a record of momentary peaks is not usually desired.

Oftentimes erroneous opinions of conditions are obtained owing to the time lag of the curve-drawing meter. In making special tests with portable curve-drawing meters, it is desirable to note for each test the time required for the pen to travel across the chart without load on the meter. This may be done by pushing the pen over against the top of the record, and with a stop watch note the time the pen requires to reach the bottom of the chart, or it may be determined by noting the advance in paper during the return travel of the pen, provided a high-speed paper feed is used.

When curve-drawing meters are to be permanently installed, the paper feed should be as slow as possible as the problem of filing away their records is often found annoying. The circular chart should be small in diameter for the same reason. The clock forms a very important part of the meter, as the records would be of little value if the clock failed to keep proper time.

In the majority of industrial applications circular chart meters will be found very satisfactory, as there is usually an attendant constantly available to change records and fill ink pens periodically. In substations or transformer vaults of important installations the continuous-record meter with re-roll attachment is desirable, as the meter may be required to operate several days without attention.

The greatest value of curve-drawing meters appears in their use when arranged for portable work, and for research purposes under the supervision of a competent engineer. When they are to be used a short period of time to obtain engineering data or to investigate operating conditions, the continuous-chart meter should be provided, using sufficient paper speed to prevent the recording pen traveling more than twice over any portion of the chart. Oftentimes, when slow paper feed is used, the repeated travel of the pen over the same portion of the chart causes ink blots and unintelligible records. For this service every precaution should be taken while obtaining records, which will facilitate explaining them to persons unfamiliar with meter records. Meters using charts having rectangular co-ordinates and straight-line pen movements are best suited for this sort of work.

When making special tests, it is not only necessary to have high-grade apparatus, but the persons making the test should be careful observers of operating conditions and the equipment being tested, as an explanation of certain parts of the record would be difficult to obtain later unless, at the time the record was made, the cause for each variation in the curve was determined. No data should be left to memory while making tests. Too often, valuable tests are obtained, but owing to the lack of sufficient explanation they become practically useless after a short period of time. Data should be placed on the original record and not kept in a notebook, for after a short time the mind will not be able to connect the data with the original record, persons unfamiliar with the test will find it difficult to interpret, and there is the possibility of the data being lost. The accompanying chart is from a meter record taken by the writer during a test, and illustrates how the events of operation should be noted on the record itself. The curve was obtained by using a wattmeter of the straight-line type and represents the load on a 50-hp., three-phase, 60-cycle, 550-volt motor driving a beater in a paper mill by means of a silent chain drive. Owing to the length of this record, which covers a period of 24 hours, only a short section is shown. The meter producing this curve was a 100-volt, 5-ampere, polyphase instrument, and for this test 600- to 100-volt potential and 50- to 5-ampere current transformers were used. In obtaining curves, care should be used in selecting the proper size current transformers for the test in order that the curve will be well up on the chart and yet not go off the chart at any time, as variations in the load will be recorded with greater accuracy.

A point of considerable value, in interpreting this test, lies in the fact that the time markings read from left to right instead of from right to left as is the usual practice. All data connected with the test are lettered on the record at various points, and in addition to the curve-drawing meter, a watt-hour meter, an indicating voltmeter and an indicating ammeter were used. Paper without time or capacity markings was used, and the meter pen was arranged to use Higgins' black drawing ink. This permits making blueprints from the original charts.

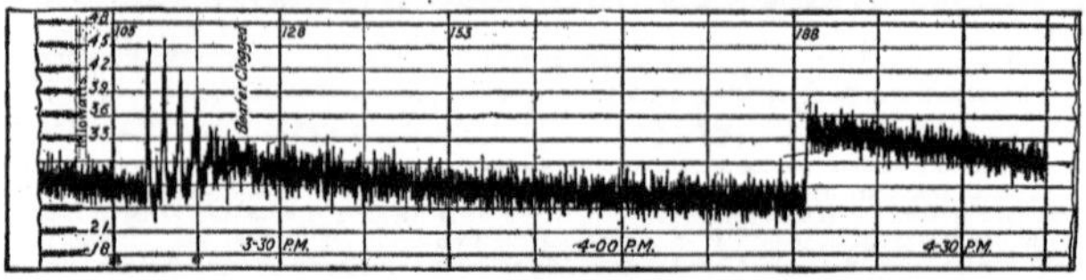

SAMPLE RECORD FROM AN INDICATING WATTMETER USED DURING A TEST

Steam-Engine Operation and Maintenance—Loose Crankpins and Loose Cranks

By H. HAMKENS
Consulting Engineer, Newton, New Jersey

OCCASIONALLY we hear the word "foolproof" mentioned in connection with engines. It leaves the impression that engineers are liable to commit some foolish acts or that in an unguarded moment somebody may do a stunt with the engine, on which neither the manufacturers nor the owners or engineer have figured. The expression is a catchword, offensive in the highest degree. Its meaning is evidently that all the parts are protected in such a way that nobody, not thoroughly acquainted with the design, can make any change which would be detrimental and put the engine out of commission; it being understood, of course, that a man who knows what he is doing will not fool with any of the engine parts or their action. Fools have no business to be in charge of or working near an engine, whether or not it is protected against their thoughtless acts. No matter how well an engine may be constructed or how carefully all its moving parts are protected and lubricated, it should always be under the care of an intelligent attendant who has his eyes and ears open all the time and is prepared for an emergency.

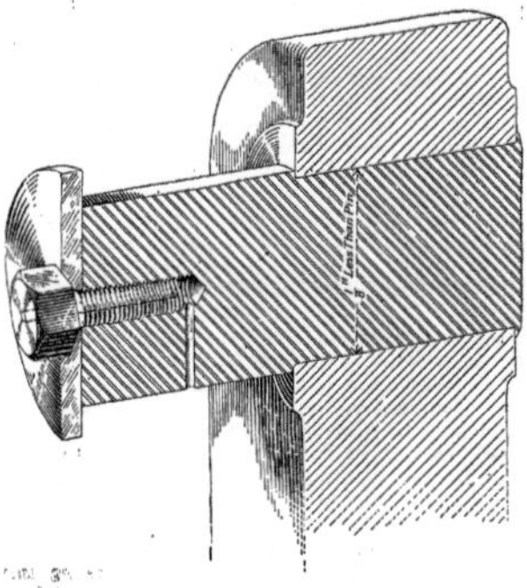

FIG. 1. OVERHUNG CRANKPIN PRESSED INTO CRANK

This thing of sending the engineer uptown to get some belt lacing or to the fourth floor to fix an elevator and in the meantime let the engine take care of itself is foolhardy to the thinking mechanic. The sooner it is stopped the less we shall hear about runaway engines and flywheel accidents. There are, of course, many so-called engineers in charge of power plants, who have not the slightest idea how a piston is constructed or what a valve looks like. All they know is how to turn on and shut off the steam and keep the oil cups and lubricator filled; of load conditions they are ignorant, have never seen an indicator, and do not know how to make adjustments on bearings, boxes or crossheads. This unfortunate condition is the cause of a great deal of trouble in engine rooms. Many a time a good-sized repair job could be prevented by a timely adjustment during the noon hour or after shutting down in the evening, if the man in charge of the engine had a thorough knowledge of its details.

Lack of instruction and inexperience of the man in charge of an engine are some of the causes of operating troubles for which the owner or superintendent of the place may be responsible by hiring an incompetent man. Our Government, state authorities, municipalities and large manufacturing concerns require drawings and specifications of every engine that they buy and install; but where is there a power plant, equipped with a one or two hundred horsepower engine which has on file a single drawing or print of the engine and its parts? In case of trouble, if the engineer is not familiar with the construction, an expert or a mechanic from the builders has to be called in, causing delay and expense. This can in many instances be avoided by insisting on having a set of detail drawings furnished with the engine when contracted for or installed.

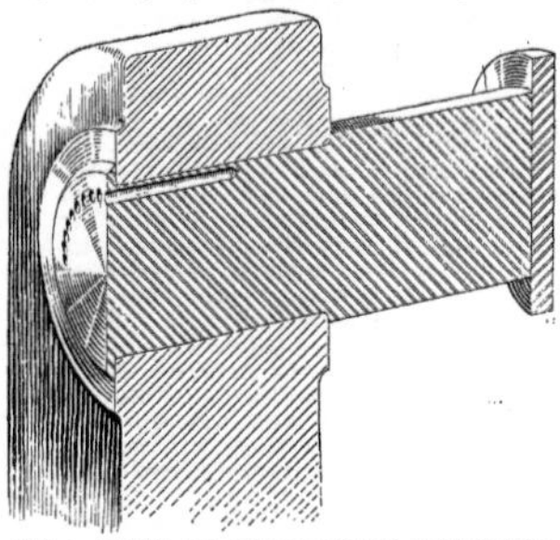

FIG. 2. HOLES ARE DRILLED IN THE BACK OF THE CRANKPIN TO FACILITATE ITS REMOVAL

Thirty years ago flywheel accidents were almost unknown. They began to happen soon after the introduction of electric power for street railways, when engine speeds were increased and the engines driven to the limit. From that time on operating troubles multiplied. Before we entered the electrical era a Corliss engine running 60 or 70 r.p.m. was after 20 years' service almost as good as the day it was started. Later on, with engines of that particular kind running from 100 to as high as 150 r.p.m., accidents and troubles have been the order of the day.

Some of the main operating troubles are due to the

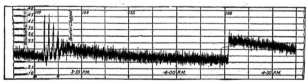

violent changes of load to which almost all electrical installations are subjected. Belt or rope power transmission is easy on an engine compared with electrical service, for the reason that on the former the engine always carries a considerable friction load, which may be as high as 50 per cent of full load, while on a direct-connected engine the friction load rarely exceeds 10 per

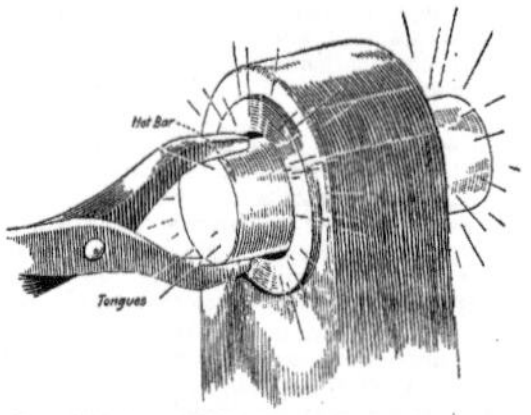

FIG. 3. THE CRANKPIN BORE MAY BE EXPANDED TO RECEIVE THE CRANKPIN BY HEATING IT WITH A HOT BAR

cent, and the change from full load to no load may take place almost instantly and at short intervals. Sudden changes of the load are hard on any engine and especially on those with releasing valve gears.

Great fluctuations of the load expose an engine to severe strains. They may be the cause at times of considerable pounding, which is hard to stop or even locate. An engine may run smoothly for a while, then all at once a change of the load starts a pound or a rattle of the valve gear, which is likely to annoy the engineer. Some engines have a chronic pound that is never located until by chance the cause is discovered where never suspected.

A heavy strain on an engine may be responsible for loosening the crankpin, causing at first a slight knock

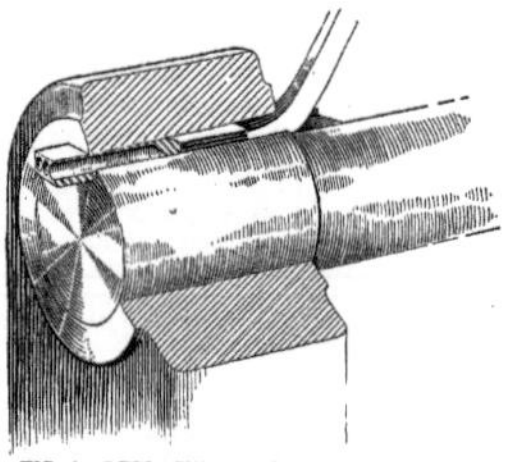

FIG. 4. REPLACING A LOOSE CRANKPIN KEY

which disappears with a lighter load, but gradually the pound increases and appears at more frequent intervals. In most engines with overhanging cranks the crankpin is forced into the crank and riveted over at the back, as shown in Fig. 1. The common practice is to make the pin in the hole ⅛ in. smaller than the overhung part and to allow from 0.001 in. to 0.003 in. per inch of diameter for a press fit. The accompanying table gives the average allowance for press fits of steel shaft or pin in cast-iron cranks.

Diameter of Shaft, or Pin, In.	Total Allowance per In. of Diameter	Allowance per In. in Diameter	Average Pressure Required in Tons
4	0.012	0.003	50
5	0.012	0.0024	60
6	0.012	0.002	75
7	0.012	0.0017	85
8	0.012	0.0015	100
9	0.012	0.00135	110
10	0.013	0.0013	125
11	0.013	0.0012	135
12	0.013	0.0011	150
13	0.013	0.001	160
14	0.014	0.001	175
15	0.015	0.001	185

For steel cranks the allowance is about one-half that for cast iron. To insure a good job, the pin as well as the hole should taper about 0.010 in. per inch of length. This is often neglected and may become the cause of a loose pin; in most cases the pin is tapered and the hole straight. Neither pin nor hole is truly round, since they are turned and bored respectively but not ground to

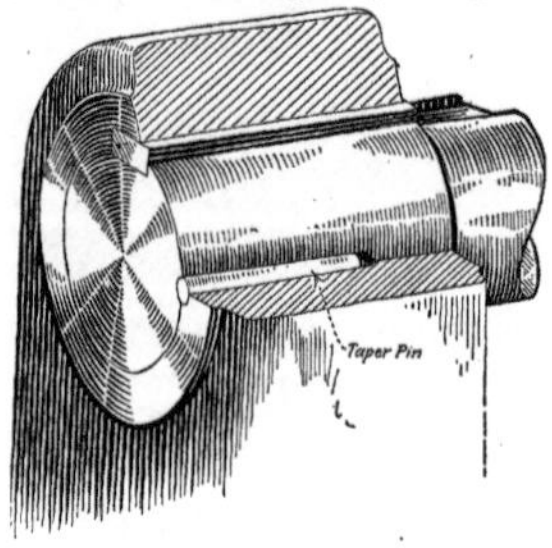

FIG. 5. A LOOSE CRANK MADE TIGHT BY DRIVING IN A TAPERED PIN

size. Either one of them or both may be out of round two or three thousandths of an inch. That may be one reason for a crankpin getting loose. Another may be that the pin does not enter the hole in a truly straight line. If not, it is liable to catch on one side and cut the metal, leaving a groove that will prevent the pin from getting tight. A mixture of white lead and linseed oil spread over the pin before pressing it in will prevent cutting. The riveting over of the end of the pin adds little to its tightness, if it was properly pressed in under a pressure approximating those given in the table.

There is no remedy for a loose crankpin except to have it removed and replaced by a new one. To remove it, a number of holes are generally drilled from the back about halfway through, as shown in Fig. 2, which, as a rule, will loosen the pin sufficiently to allow it to be driven out. It is not advisable to press it out as this is likely to damage the hole. Seldom is it desirable to rebore the hole. Previous to the time when hydraulic presses came into use, crankpins were shrunk in, and this method can be used with success if no press is at hand. For a shrink fit the practice is to make the pin 0.002 in. larger per inch of diameter than the hole.

The hole is expanded by the application of heat, either over a fire or by inserting a red-hot piece of iron into the hole, as shown in Fig. 3. The hole must be calipered from time to time, and when it is expanded sufficiently the pin is slipped in. This method requires some skill and good judgment; there is danger of having the pin stick if the hole is not sufficiently expanded, therefore no chances should be taken. Experience seems to show that a pin, if shrunk in right, will stay tight as long as the crank lasts.

Worse than a loose crankpin is to have the crank itself come loose on the shaft. The first expedient will be to remove the key and put in a new one, which may tighten the crank. The old key may be driven out from the back, but if that will not do, holes should be drilled into it as shown in Fig. 4, which will loosen the key sufficiently to make it come out by driving. To make a new key, a bar of steel is forged to the shape of the old key with an all-round allowance for finishing of $\frac{1}{16}$ in. to $\frac{1}{8}$ in., according to size. One end of the bar is bent double, as shown in the sketch, and the other is machined and fitted into the keyway by driving it repeatedly in and out and removing sufficient metal until it fits for about one-third of its length. It is then cut off about an inch longer than the depth of the keyway, covered with white lead and driven home. If the key has been made a good fit top and bottom and on the sides, it may tighten the crank again. To get a good driving fit on the sides, the key may be made in two pieces, as shown in the detail of Fig. 4. The two sides of the key that fit together have a taper of $\frac{1}{8}$ in. per foot; this kind of key, if well fitted and driven home, will stay "put." If the crank can be saved at all, it will do the work.

Another method to fasten a loose crank is shown in Fig. 5. It consists in drilling a hole halfway into the shaft and crank at 90 deg. from the key, reaming it out tapering to $\frac{1}{4}$ in. per foot and driving in a taper steel pin. The best material for this pin is tool steel. The diameter of the pin at the small end should be about one-eighth of the diameter of the shaft in the crank. Unless the crank is of steel or provided with a large hub, it is not safe to use the pin, as there is danger of splitting the crank.

(Mr. Hamken's next article deals with loose flywheels.)

On the Road with the Refrigeration Troubleman

BY J. C. MORAN

An air-compressor intake sucks ammonia gas from a leaking flange near-by and seriously cuts down production in the shop where the air is used. In another plant, with both compressors running, there was no trouble; but with one machine operating ammonia disappeared from the receiver gage-glass. The installation of an ice tank in another plant caused loss of ammonia, beginning about noon. But at midnight it came back. The troubleman corrected all the faults—after finding them—and tells how.

WE HAD installed a 100-ton refrigerating plant in a vinegar factory and everything had been running nicely for about a month when the engineer called up and asked me to come out and help him locate an ammonia leak. But I was unable to find the leak on arrival.

Both the discharge and liquid lines ran through the machine shop, the condenser being on the roof of the shop. I tested the entire length of both lines with sulphur sticks and litmus paper a number of times, but failed to locate any definite point where the leak could be said to come from. Four times I made a trip out there to find the leak, with the same result. Of course, everyone around there was having a quiet laugh at my expense. The owners were vexed because the ammonia fumes interfered with work in the machine shop. Finally, headquarters told me to tear out the entire section of lines in the shop and put in new pipe. I was skeptical about whether this would help matters, but obeyed orders. When the new line was put in, the odor of ammonia was as strong as ever.

One of the men finally told me that there was no odor in the morning when they came to work, and sometimes not until well into the forenoon. With this as a clue I came on the job ahead of the men the next morning and found that the man was right. The ammonia compressor had been running all night. So the trouble was not in the line after all. It was 9:30 before I noticed any odor. Standing close to where one of the men started an air hammer, in a few moments I could smell ammonia.

It came to me with a flash that the ammonia smell was carried into the shop with the air used for the shop. Following this clue, I learned that the machine shop got its air from a large high-pressure compressor which also pumped a deep well, while the rest of the plant got air from a smaller low-pressure compressor. Now the only way in which ammonia fumes could get into the air would be from somewhere around the suction of the air compressor. So there I looked for the trouble. Sure enough, the suction line of the ammonia compressor had a flange connection a few inches above the intake of the air compressor and the former had sprung a "healthy" leak.

After pumping out the line and renewing the gasket in the flange, we tried again and this time it remained tight. Also, the leak in the machine shop disappeared. If I had known in the first place that the shop was getting its air from a separate compressor, I might have detected it sooner because I had given this consideration once before, but the rest of the plant showed no signs of it, and there were many other places where they were using air. You cannot take anything for granted when you are in trouble.

Some time ago I was called in by a friend to help him find what was wrong with his plant. He had two 50-ton machines and a single condenser. He was a great believer in condenser surface of liberal proportions so he had ample condenser capacity for 125 tons. The condenser used was of the ordinary atmospheric type. With both machines running he had no trouble, but when he shut down either one the ammonia would disappear from the gage glass of the liquid receiver in a short time. At first the condenser pressure would drop, but gradually it would build up again until it was within a few pounds of the pressure carried when both machines were in operation. This happened every time one of the machines was shut down, and the engineer could find no reason for it. At first I was no more successful than he in finding anything that looked like trouble. Finally, after one of the machines had been shut down for a couple of hours and the ammonia had disappeared from the gage glass and the head pressure had first dropped and again gone up to nearly normal, I went up on the roof to have a look at the condenser. Putting my hands on it to check each coil to see if it was working at the proper temperature, I was surprised to find that about half of the coils were cold, or, as the average operating man puts it, "dead." If a condenser coil is working properly it will be warm at the top where the gas enters and for a few pipes down, while if it is not working properly it will cool off to the same temperature as the water in a short length, and thus it becomes a simple matter to detect a coil that is not working.

Trouble Always in Four Coils

I had an idea that it was probably due to too much air in the system. So I thoroughly purged each of the dead coils. This helped matters, but I had to repeat the performance several times to keep the coils working. No matter what we would do, four out of the twelve stands refused to work continuously, and the queer part of it was that it was not always the same four that quit, but they were always close together and at some distance from the point where the discharge line connected to the header. The discharge for one machine connected to the header near one end and the discharge for the opposite machine near the other end. It was generally the coils in the center that made the trouble. If the trouble was near the ends, it was always at the end farthest from the point where the machine in operation was discharging its gas.

We came to the conclusion that the trouble could be due only to greatly different velocities in various parts of the gas header, and that this had the effect of practically bypassing some of the stands as the gas rushed by, somewhat on the same principle as a long distributing trough when the water is fed in at one end or near it. Most of the water will spill over near where the feed comes in, owing to the inertia of the water and the velocity of the flow.

Working on this theory, I began to choke off on the gas valves on those stands that were hottest. After a couple of days' experimenting I found that six of the coils could be run wide open, two nearly shut and four partly throttled. If we changed machines, we had to change the setting on some of the other coils or get in trouble again. Once having found the proper setting, however, it was no trouble to adjust the valves properly. We shut off the valve and then opened it the required number of turns or fractions of a turn, as previously determined by experiment. A record of the amount of opening of each valve for each machine in operation was hung in the engine room to guide the operator.

In this manner we were able to keep all the coils working at about the same temperatures, the condenser pressure remained low and the ammonia remained in the receiver. Previously, the ammonia had hung up in the dead coils in the condenser and refused to come down, owing to the sluggish action of these coils. I had seen similar cases before.

The owners of a packing-house plant had to buy considerable ice for icing cars and other purposes, and they decided to add an ice tank to their refrigerating equipment. They had plenty of machine capacity for this.

After the tank was put in operation, they had trouble from the start. Every day about noon the ammonia would begin to disappear from the receiver and would be gone entirely in the middle of the afternoon or toward evening. It would not come back until some time around midnight or later. Several times they had been troubled with heavy slugs of liquid coming back with the suction gas, causing thumps in the compressors which jarred the engine room. No one could account for them. They all blamed it on the ice tank, but just how it was to blame no one could explain.

I was sent out to solve the puzzle after the purchaser had refused to pay for the tank unless the trouble was remedied. Arriving about noon, I had been in the plant about fifteen minutes when the compressors made a series of thumps that jarred the floor. Something was wrong. On investigating the plant everything seemed to be in proper order. The next day I came on the job in the morning, and toward noon the ammonia began to disappear.

Suction Pressure Up

I insisted on attending to the handling of the expansion valves throughout the day, and this led to the detection of the trouble. Generally, in adjusting the valves in rooms where the temperatures are above freezing, I go largely by the appearance of the frost on the suction stop valve in each coil. A gang of men came up on the ice tank and began pulling ice between 9 and 10 a.m. In an hour or two they pulled enough ice to last for the day and proceeded to fill the empty cans with a rush with fresh water. Of course the brine temperature took a rapid jump and I increased the opening of the expansion valves. About this time the coolers began to load with fresh meat from the killing floor, and I opened up on the valves all around, except in the freezers, to hold the temperatures down as much as possible. When I got back to the engine room, the suction pressure had gone up eight pounds.

In about half an hour I went back to the tank and coolers and found that the suction stop valves, and in many cases part of the coils, were entirely free of frost. So I opened the expansion valves on these coils still more to bring the frost line where I knew from experience it ought to be for best results. When I got back to the engine room, the machines were running ice cold, and every once in a while there was a gentle thud in the cylinders, indicating that there was considerable liquid coming back in the suction gas. The suction pressure had gone up another two or three pounds.

I made a hurried trip to the coolers and ice tank again to locate the coil or coils from which the trouble was coming, but was surprised to find that none of the suction valves on the coils or on the ice tank was still frosted, except one or two which were frosted slightly at the flange where they connected to the pipe. From past experience I was almost certain that the trouble was not coming from any of these coils, unless there was brine or water in the ammonia.

The only valves untouched so far that morning were those on the three freezer coils. There were six expansion valves and coils in the three freezers, and to make a sure test on these I shut them all off tight and hurried back to the engine room.

It was a short time before the machines began to warm up, while the suction pressure remained practically the same. Soon the frost began to come off the suction connections of the machines, while the discharge began to get very hot, indicating that the suction gas was coming back superheated. The engineer and oilers were now cursing the new tank louder than ever and blaming it for the new trouble which kept them busy at the stuffing boxes.

Another hurried trip to the coolers was made and the expansion valves on the ice tank and coolers opened still more. I remained until they began to show signs of frosting to the nuts on the stuffing boxes. When I got back to the engine room the suction pressure had taken a still further pump, but the machines were cooling off and some of the ammonia was coming back into the liquid receiver gage glass.

After a couple more trips to the coolers and the tank, I finally got the expansion valves so adjusted that the frost lines on the suction stop valves on nearly every coil were up over the stuffing-box packing nuts, and on one or two up on the stems of the valves, at which points the machines worked at about the proper temperature and the frost line carried back to about the right point on the machines. The suction pressure was now about 30 lb., whereas in the morning it had been below 15 pounds.

The plant had a liberal amount of piping in every room and also in the tank. With proper attention to the expansion valves it was possible to follow the changing temperatures in the coolers and tank and not the suction pressure. However, when the suction pressure went up the freezers were practically put out of commission because the temperature in these was so low that with a high suction pressure there would be no temperature difference between the ammonia in the coils and the air on the outside of them, and the ammonia simply accumulated in the coils until it overflowed into the suction line. Every time the doors to the freezers were opened and a current of warm air struck the coils, the ammonia in the coils would boil up for an instant and throw heavy slugs into the suction line. It was a surprise to me that they did not have more serious trouble than they had been having.

They had got along without trouble before the ice tank was installed because they had not worked the coils in the meat coolers to full capacity, opening the expansion valves only enough to do the work during the day's run and leaving part of the coils thawed off. This directed a lot of superheated suction gas into the suction line, and when this mixed with the liquid coming from the freezer coils it about balanced, and the gas going to the machines was about right. So no trouble was experienced in keeping the machines at the proper temperature. With plenty of machine and coil capacity it was no trouble to operate, but it was not economical.

When the ice tank was installed, it threw an additional large amount of gas into the suction line and the suction pressure went up so high that it became necessary to work the cooler coils harder, with the result that there was less superheated gas going into the suction line, and with the higher suction pressure less ammonia was evaporated in the freezer coils and more thrown into the suction line. The trouble was further increased by the manner in which the ice was pulled from the tank. The entire amount being pulled in a short time and at the time of day when the load was being put into the coolers tended to send the suction pressure far above normal during part of the day, putting the freezers entirely out of operation as far as evaporating ammonia was concerned.

The remedy was simple. First, we arranged to have a man come on duty at 11 p.m. and work till 7 a.m., and pull steady all night on the ice tank in addition to looking after the temperatures of some of the coolers. The regular gang then came on at 7 a.m. and pulled whatever was necessary in a short time. Nothing more was done on the tank until toward midnight. This kept the load fairly uniform all over the 24 hours, although there was still quite a jump in temperature in the afternoon. To make the system still more flexible and enable it to work to the best advantage at all times, we changed the suction connections to some of the rooms so as to put the storage coolers and freezers on a separate line, and arranged it so that we could use any one of the machines on it independently. When the suction pressure got too high on the main system for the freezer coils to work properly, we cut the system in two and carried a lower pressure on the freezers and storage coolers, thus pulling their temperatures down, while the suction pressure on the main system was allowed to go as high as it wanted to. As a result of the change the engineer claimed that he made a saving of over 25 tons of coal per month.

Steam-Turbine Has Record Run

A Westinghouse turbine generator at the power plant of the Narragansett Electric Lighting Co., Providence, R. I., recently established a world's record by running continuously without a shutdown for a period of 84 days, 11 hours and 36 minutes. The unit is a multiple-element steam turbine of the cross-compound type with a capacity of 45,000 kw. During this run 85 per cent of the total station output was generated by this unit. The record was made with a load varying from a minimum of 6,000 kw. to a maximum of 41,000 kw.

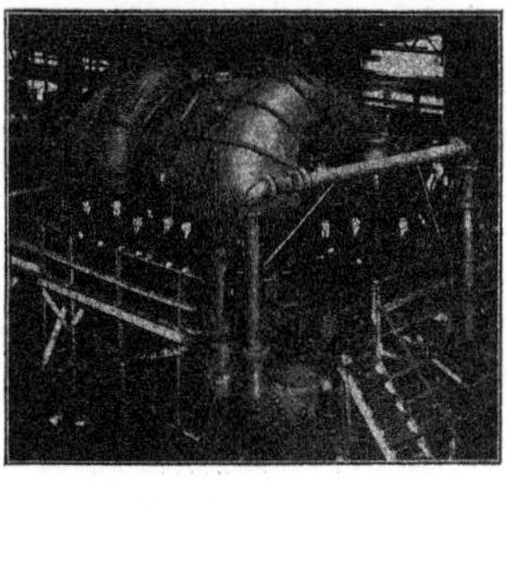

Operating Two Similar Alternators Having Opposite Polarity in Parallel

An Alternator Operating in Parallel with Other Machines Will, if the Polarity of Its Field Poles Is Reversed, Slip One Pole and Still Operate in Parallel with the Other Machines, Without Any Change in Connection

BY B. A. BRIGGS

WHERE two or more alternators are operating in parallel, the question frequently arises, What would be the effect of reversing the polarity of one machine and then trying to connect it in parallel with the other machines? The question is easily answered by the use of a diagram representing ing of a polarity indicated by the arrowheads which, at the instant shown in the figure, will tend to cause a current to flow from terminal *A* of both machines through the load and back in on terminals *B* and *C*. The active part of the armature winding in supplying current to the load is indicated by double arrowheads.

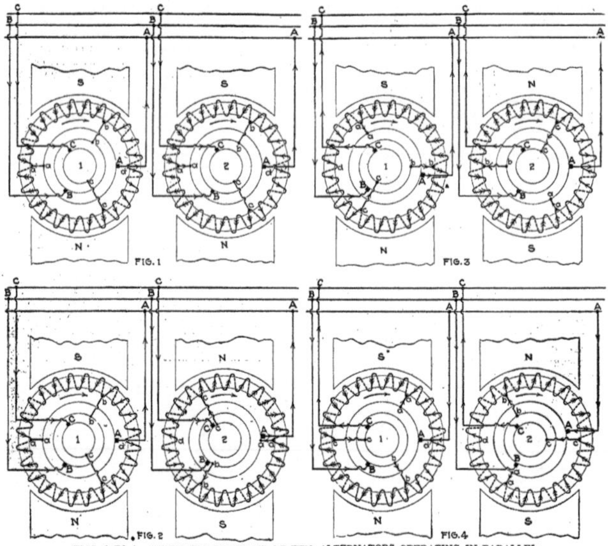

FIGS. 1 TO 4. SCHEMATIC DIAGRAMS OF TWO ALTERNATORS OPERATING IN PARALLEL

the alternators with ring armatures. In Fig. 1 is represented two three-phase alternators connected in parallel to a busbar; each alternator's field poles are excited the same polarity. With the armatures rotating in the same direction, indicated by the curved arrows, there will be a voltage generated in the wind- At the instant under consideration the sections of the armature windings *bd* and *cd* are not effective in causing a current to flow to the external circuit, since the voltage in these two sections oppose. The armatures, being in exactly the same relative position, will be in absolute synchronism.

Now, suppose that the polarity of the field poles of machine 2 is reversed as in Fig. 2; the polarity of the armature will be reversed, and if the armature remains in the same relative position as in Fig. 1, terminal *A* will be negative and the machine will be out of step by 180 deg., with machine No. 1; or in other words, there will be a short circuit between the two machines with double normal voltage of one machine to cause a current to flow between them.

A study of Fig. 2 shows that the two machines are in synchronism as in Fig. 1, but to obtain this condition, the armature of No. 2 machine has turned through 180 deg. from the position in Fig. 1. This is what is called slipping a pole; that is, the armature has turned through an angle represented by the pole pitch, which in a two-pole machine is 180 deg., in a four-pole machine, 90 deg., and in a six-pole machine, 60 deg., etc. In all cases, however, the pole pitch is represented by 180 electrical degrees. The voltage applied to the circuit under consideration is that between *c* and *a* and *b* and *a*. As pointed out in the foregoing, the voltage generated in the section of the winding between *c* and *b* is not effective, since that generated in the section of the winding between *c* and *d* is equal and opposite to that generated in the section of the winding between *b* and *d*.

The next thing to consider is, will the armatures be in synchronism at other points in their revolution? To determine this, let us turn our attention to Figs. 3 and 4. Both machines are turning in the same direction, as indicated by the curved arrows. Assume that each armature turns 60 deg. from the position in Fig. 2. This will give the position Fig. 3. The direction of the arrows on the armature winding indicates that terminals *A* and *C* on each machine are positive and *B* negative and the machines are still in synchronism although the armatures occupy relative positions 180 deg. apart.

Again, assume that the armatures have turned another 60 deg. Then they will be in the position indicated in Fig. 4. A study of this figure shows terminal *C* to be positive and terminals *A* and *B* negative. Since each armature is in a position where the voltages generated in the section of each winding between *a* and *c* are equal, likewise in the section between *b* and *c*, the two machines are in exact synchronism. Therefore, as long as the two armatures continue to rotate at exactly the same speed, their voltage would be in synchronism.

From the foregoing it is evident that if two alternators are operated in parallel and have their field poles excited the same polarity, the armature will occupy the same relative position at exact synchronism. If the polarity of the field poles of one machine is reversed, the armature of this machine will slip one pole; that is, it will occupy a relative position 180 electrical degrees from the armature of the other machine, and the two machines can be operated in parallel as when the field poles of the two machines have the same polarity.

The only exception to this rule is where both machines are driven by the same source of motive power, such as two alternators geared to the same turbine. In this case, after the machines have been properly connected to operate in parallel, if for some reason the polarity of the exciting current is reversed through the field coils of one machine, the two machines would be 180 degrees out of phase and would have to remain this way since they are both geared together. In such a case the two machines would be inoperative until the polarity of one machine was changed. Where each machine is independently driven, the armatures can take any relative position to bring them into synchronism and the direction of the exciting current in the field coils makes no difference as to parallel operation of the machines. In the foregoing the armature has been referred to as the rotating element. In machines of the revolving-field type the statements regarding the relative position of the armature would also apply to the field poles. The next article will consider the effect of reversing the direction of rotation of an alternator that has been operating in parallel with other machines.

Preventing Sludging of Oil in Transformers

It is truly said that the life of its insulation is the life of the transformer, for when the insulation is gone the transformer is not only useless, it is dangerous. For this reason the efforts of transformer engineers have been directed to a considerable extent to developing methods of preserving the insulation for the longest period possible under all sorts of operating conditions.

The factor that has the greatest effect in reducing the dielectric strength of insulation is moisture. Therefore, the transformer was first protected by having its coils immersed in oil, so that no water could get to them. Later, as operating potentials increased, further protection was afforded by impregnating the fibrous insulation with a moisture-proof compound previous to its immersion in oil. Even then, owing to the high sensitivity of oil to water, the slightest water content, 18.5 parts per million, will reduce its dielectric strength to the lowest possible limit, and further measures were found necessary. With the increase of outdoor installations the expedient was devised of making the tank covers water-tight to further protect the insulation from water, snow, etc. This requires an expensive tank construction and also leaves an idle air space between the oil and the cover, which is necessary to provide for the expansion of the oil when heated.

The presence of this air space above the oil makes possible two other effects that are likely to cause trouble in operating and shorten the life of the transformer. One of these is explosions, and these are made possible by arcing or static discharges, or heavy overloads sometimes cause combustible gases to be set free from the oil, which, when combined with the air in the ordinary tank, form a highly explosive mixture. This gas may be ignited by sparks, either dynamic or static, causing a dangerous explosion. While the leads of some types of transformers are protected by ground shields that make this impossible under ordinary circumstances, still with the oil in the tank at an abnormally low level the effect of the ground shields may be neutralized.

The other feature that may have a deteriorating effect is what is known as the sludging of the oil. Any transformer oil will, when heated in the presence of oxygen, decompose, and a precipitate will be formed, consisting of carbons, etc. While this sludging has no effect on the dielectric strength of the oil, the precipitate settles on the insulation, where it forms a film of low heat conductivity, so that the operating temperature increases, with a consequent increase in the sludging. The remedy for this state of affairs has been a periodical filtering of the oil, and cleaning of the coils and insulation of the transformer.

It has been found by a series of tests that oil will not sludge even under considerable and continuous operating temperatures, if it is kept completely free from any contact with air. The most that happens under these conditions is a slight discoloration of the oil.

To remedy these faults the General Electric Co. attaches to its transformer tanks a new device, known as an oil conservator, Fig. 1. This conservator consists simply of an auxiliary tank attached to the transformer tank so as to be well above the oil level in the main tank. Its only connection to the transformer tank is through a pipe, Fig. 2. When the conservator is supplied with oil, the main tank and the connecting pipe are also completely filled, so that the oil comes in contact with the air only in the conservator.

The conservator is just big enough so that all expansion and contraction of the oil take place within

FIG. 1. HIGH-VOLTAGE TRANSFORMER WITH OIL CONSERVATOR

it, without lowering the level enough to bring the oil in the main tank in contact with the air or raising it enough to overflow the tank. Moisture absorbed by the oil is drained from a sump in the bottom, Fig. 2. The formation of explosive gas, with its consequent possibility of dangerous explosions, and the sludging of the oil when heated, are both eliminated, since the transformer main tank is completely filled with oil and can be made water and air tight. Since there is only one oil connection between the conservator and the main tank, and that is of limited size, there can be no circulation between the two, and the only change

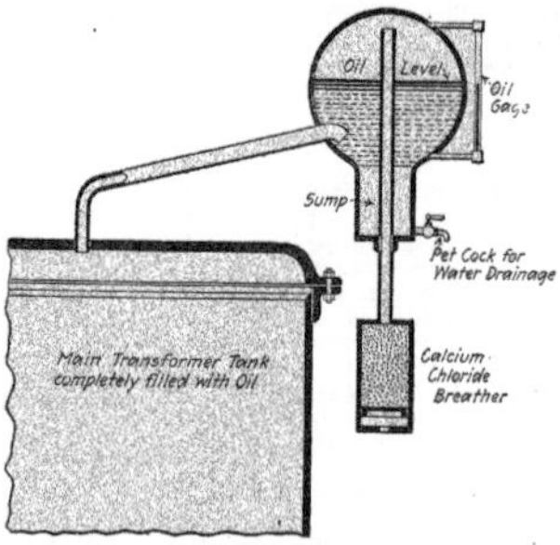

FIG. 2. ESSENTIAL FEATURES OF TRANSFORMER-OIL CONSERVATOR

in level is that due to the volume changes in the whole body of oil. Such being the case, hot oil cannot come in contact with the air, even in the conservator, and sludging will not take place.

Specifications for Lubricants

The following specifications for lubricants were prepared for the governing Board of State Institutions of by the faculties of the engineering departments of the Kansas State Agricultural College and the University of Kansas, Prof. A. A. Potter, chairman.

Kind of Oil	Viscosity in Seconds at 100° F.	Viscosity in Seconds at 210° F.	Flash Point Deg. F. (Minimum)	Pour Test Deg. F. (Maximum)	Carbon Residue % (Maximum)	Acidless Tallow %
1. Air compressor oil.....	240 to 300	45 to 50	400	15	7/10	0
2. Automobile oil, light...	160 to 220	40 to 50	350	20	5/10	0
3. Automobile oil, medium	270 to 330	45 to 54	350	35	½	0
4. Automobile oil, heavy..	340 to 420	50 to 65	375	45	6/10	0
5. Dynamo oil..........	140 to 200	38 to 45	370	25	..	0
6. Engine oil, high-speed..	140 to 200	35 to 45	350	35	..	0
7. Engine oil, slow-speed..	180 to 300	40 to 55	350	35	..	0
8. Ice-machine oil........	130 to 205	....	300	0	..	0
9. Steam-cylinder oil, dry steam..............	....	135 to 165	450	50	..	5
10. Steam-cylinder oil, wet steam..............	....	120 to 160	450	50	..	5 to 10
11. Steam-cylinder oil, superheated steam....	....	150 to 250	(a)	50	..	0
12. Truck or tractor oil, heavy............	475 to 550	55 to 70	400	50	7/10	0
13. Tractor oil, extra heavy	1200 to 1600	90 to 125	450	50	1 3/10	0
14. Transmission-gear oil...	....	170 to 215	300	50	..	0
15. Turbine oil..........	140 to 240	38 to 45	370	25	..	0

(a) Depends upon steam temperature.

The largest hydro-electric plant in the world is being planned. It is to be located on the Priest Rapids site on the Columbia River, near the Oregon-Washington border. Its maximum capacity will be 700,000 hp. at high water and 400,000 hp. at low water. Plans are being delayed until Congress passes the water-power leasing bill, now held up in conference by Senate and House Committees.

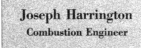

Joseph Harrington
Combustion Engineer

One of the Leaders in the Middle West in the Economical Use of Fuel in the Power Plant

ALMOST everybody in the power-plant field in and about Chicago, in fact throughout the Central States, knows Joe Harrington, and his reputation as a combustion engineer has extended far beyond these confined limits. Burning coal efficiently, particularly the coals of the Middle West, has been his lifelong study. In the early days of his apprenticeship to this field the inferior grades of bituminous coal, such as slack or screenings, were actually discarded and considered unfit as power-plant fuel. To eliminate this economic waste the chain grate was designed. Mr. Harrington has been associated closely with the development of this type of stoker from its infancy and has been largely instrumental in showing that these inferior grades of coal could be burned efficiently. He is now developing a universal stoker designed to burn coals of all kinds with equal success.

Reverting back to his youth, Joseph Harrington hailed from New England, the home of his ancestors for five generations. One of these ancestors on his mother's side was General Montgomery, of Revolutionary fame. He was born at Reading, Mass., a suburb of Boston, in 1873, and being near the Bay he spent much of his early youth on the water. Like many other New England boys he was always interested in anything pertaining to mechanical pursuits. His father had a shop to which Joe and his brothers always had access. They were encouraged to play in this shop, and before the subject of this sketch was out of high school, he was a skillful machinist.

Graduating from the Massachusetts Institute of Technology in 1896, Mr. Harrington spent the next four years in Mexico and Central America, prospecting in the mining regions and surveying for railways and rubber-plantation interests. During his school life he had picked up a working knowledge of the transit. But his liking was for things mechanical, so after four years of roaming he settled in Chicago. He was employed by the Green Engineering Co. as chief engineer for the twelve years ending 1912. Along with the development of the stoker Mr. Harrington gave much attention to the furnace, the ignition arch and other appurtenances necessary for efficient combustion. During these twelve years he gained a wide acquaintance in power-plant circles and was generally acknowledged as a leader in his particular field.

It was his thought to capitalize on this reputation when at the end of 1912 he withdrew from his former connection to become an advisory engineer in the operations of power plants, in particular on questions relating to combustion. In this work he became intimately acquainted with all the details of operation. He saw the difficulties as viewed by the operator and came in touch with defects, mechanical and human, that could be gained only in this way.

Work of greater magnitude in the industrial field momentarily enticed Mr. Harrington from the power plant. He became interested in powdered coal and in particular its application to heating furnaces in rolling mills. Several installations that were highly successful are to be credited to his long list of achievements. Memories of his first love, however, eventually drew him back to the field of stoker design and manufacture. In 1917 he became engineer and vice president of the James A. Brady Foundry Co., Chicago.

Mr. Harrington found time to do his bit for the Government during the war. He was appointed by the United States Fuel Administration administrative engineer for Illinois, having charge of fuel conservation and supervision of power. With his thorough knowledge of fuel and its burning and his exceptional ability as an executive and organizer, Mr. Harrington soon had the work under way. The questionnaires from Washington were sent out, and through numerous volunteers the power plants of the state were rated and their deficiencies pointed out to the plant owners. Volunteer inspectors were sent out on the job, and service was rendered to those requesting it. The work was entirely educational and by suggestion rather than by arbitrary regulation. One of the original features of the Illinois campaign was a series of weekly letters dealing with the primary factors entering into economy and conservation. They were elementary and semi-technical in character and by local volunteer committees, were printed and distributed to every plant in the state. Commendation came from all parts of the state and from Washington. Mr. Harrington was congratulated on his aggressive campaign and the excellent results that were being effected.

Among the engineers of the community Mr. Harrington has always been active. He belongs to the American Society of Mechanical Engineers, and at one time was chairman of the Chicago Section. He is active in a number of standing committees of the society and a contributor to the Boiler Test Code. Recently, he has been appointed chairman of the Committee on Smoke Prevention of the Western Society of Engineers. He is a member of the Engineers Club of New York, the Old Colony Club, and an honorary member of the National Association of Stationary Engineers. An excellent speaker who can talk from the shoulder at a moment's notice, he is in frequent demand. Scarcely a week goes by without a call on his time and requests for one or more addresses before engineering, industrial or commercial associations.

Nothwithstanding his popularity and the constant demand for his presence elsewhere, Mr. Harrington is a great lover of his home at Riverside. An acre of ground attached to it gives him an opportunity to get close to nature and work with the flowers and shrubs in his ample garden.

Novel Water-Pumping Motor

A novel method of pumping water has been developed by F. L. Gilman, of Los Angeles, Cal. The apparatus is anchored in the middle of a swiftly running river and in swaying back and forth operates two pistons of a pump located on the bank. The pistons have a 22-in.

FIG. 1. FLOATS ANCHORED IN MIDSTREAM

stroke, and the pump has a capacity of 80 gal. of water per minute, discharging through a 3-in. pipe against a head of 18 feet.

The floats are anchored in the stream and are held in position by long anchor rods attached to a frame to which the several floats are so secured that they are free to change from one angle to an opposite angle. That is, as the floats, Fig. 1, move across the stream, and upon reaching a definite travel, the angle of the floats is changed and, owing to the velocity of the water impinging against them, the float head is moved across the stream in an opposite direction. Upon reaching a pre-

FIG. 2. GENERAL VIEW OF THE INSTALLATION

determined travel, the angle of the floats is again reversed and a return travel is made.

Connection is made between the float head and the pump on the bank, Fig. 2, by cables attached to a crosshead, located at the pump, to which the pump pistons are connected. As the float head moves back and forth, its motion is transmitted to the pump crosshead and in turn to the pump plunger.

Recommends Standard Shafting Sizes

A committee of the American Society of Mechanical Engineers formed to investigate the possibility of standardizing shafting sizes, in a recent report to the council recommended the approval and adoption of the following lists of sizes as standard for the society:

Transmission Shafting: 15/16 in.; 1 3/16 in.; 1 7/16 in.; 1 11/16 in.; 1 15/16 in.; 2 3/16 in.; 2 7/16 in.; 2 15/16 in.; 3 7/16 in.; 3 15/16 in.; 4 7/16 in.; 4 15/16 in.; 5 7/16 in.; and 5 15/16 in.

Machinery Shafting: Size intervals extending to 2½ in., by sixteenth inches; from 2½ in. to 4 in. inclusive, by eighth inches; from 4 in. to 6 in., by quarter inches.

The adoption of these standard sizes will mean the reduction in the number of parts of power-transmission equipment that must be carried in stock, and the committee expressed the opinion that in the future a gradual elimination of odd sizes from makers' lists and from dealers' stocks would take place and the standard sizes would be used for all new construction.

The committee is planning to reorganize itself and add to its membership, so that the standardization of shafting formulas and the dimensions of shafting keys and keyways can be investigated.

Graduate students of the Massachusetts Institute of Technology are conducting experiments in the research department of the hydraulic laboratories on the design of draft tubes for water turbines, to gain increased efficiency.

EDITORIALS

A Needed Feature of the Refrigerating Industry

IN REFRIGERATION the war revealed things of great import no less than in other fields. In this art the United States has attained by far the greatest advance, and it is generally conceded, we believe, that had it not been for the well-located and well-managed large refrigerated warehouses, the Allies would have been badly off for foodstuffs. It was not only the large stock immediately available when the sudden call came, but the ability to take in huge shipments, chill and clear for transatlantic transport, that was invaluable. This was the period in which Europe was learning her lesson. As a result, it is reasonable to expect refrigeration to find a wide and rapid growth there.

The charge has been made that the development of this art, even in the United States, has not been so rapid or so spectacular as the possibilities permitted or as compared with developments in other branches of engineering and physics. The critic can make out a good case, but the day for making it rapidly passes by. Refrigeration is now having its new era, its period which is not unlike those experienced in the pure power field by the introduction of electricity and the steam turbine. The high-speed compressor daily demonstrates its remarkable potentialities. Continued improvement in central-station design, electrical apparatus and compressor valves insures the future of this type of machine.

But the compressor and its drive are but a part of the system, and a part that is essential only because of the physical properties of the refrigerant, ammonia. Its function is merely to put the refrigerant in such condition that the refrigerating cycle may again be started. It exerts little influence upon the efficiency of performance of the remainder, or greatest part, of the cycle. It is here that concerted action seems to be needed. This is not to say that refrigerating apparatus other than the compressor is not now commendable. But that there is room for improvement, no one will deny.

Competition invariably is healthy. Yet it is our opinion that closer co-operation of builders is more urgent just now. There are several channels through which this may come, and some of these are already organized. The American Society of Refrigerating Engineers offers one of the best. The possibilities which the society offers in this direction are not sufficiently taken advantage of; there seems to be reluctance to present papers at yearly and semiannual meetings, and often discussion is too contentious, too argumentative. There is need of papers that deal with specific, authentic performance data. The subject itself is really of less importance; it is bound to relate to refrigeration. There may be serious obstacles to the creation of a research laboratory such as the American Society of Heating and Ventilating Engineers maintain at Pittsburgh; but we doubt it. The various associations in the great industry of refrigeration could well afford such a laboratory.

Would it not be worth while to endeavor to get the American Society of Refrigerating Engineers, the association of refrigerating-machinery manufacturers, the many large ice associations, the packers and the very influential American Warehousemen's Association together and discuss the formulation of plans for the creation of a real research laboratory, a clearing house for ideas, safety measures, appliances and performances? There is no doubt that the various universities and engineering institutes would be of material assistance.

Regional Power Development

AT THE present time a great deal of interest is being given to working out a comprehensive power system for the industrial regions situated around Philadelphia, New York and Boston. The ideal power system, if such were possible, is one that would supply all the needs of the community for light, heat and power. In such a system it would be possible to operate at the highest load factor, to take advantage of the high efficiency of large generating units and also realize the high efficiency of a combination heating and power load. However, for obvious reasons, except to a very limited degree, no attempt has been made to combine all these factors in one system. In this issue on page 824 is an abstract from a discussion on "Regional Power Development and the Low-Temperature Distillation of Coal," by C. M. Garland, in which he says:

> In the central and northern portions of the United States every industrial plant for from seven to eight months out of the year requires fuel for heating. In seventy-five per cent of these plants the fuel required for heating is more than sufficient to generate the power required for manufacturing operations. For practical purposes it may therefore be considered that the power is obtained during these months without fuel expenditure. This explains why only thirty per cent of the industrial plants buy their power. Of this thirty per cent twenty per cent would doubtless be better off if they had efficiently operated plants and generated their own power. The writer is a strong advocate of central-station development, but believes that it is well to recognize the limitations imposed by existent conditions.

Mr. Garland, according to his arguments, sees the correct solution to regional-power development in the byproduct system of gas generation with the low-temperature distillation of coal. Such a system is now in successful operation in England. However, there are many arguments for and against this scheme. If, as Mr. Garland says, there are in the five-hundred million tons of bituminous coal burned in this country each year, two and one half billion dollars' worth of by-products burned up, then the saving of even a small portion of this tremendous waste is worthy of the most serious consideration. The cost of obtaining these by-products and what effect a national attempt at producing them would have on their price is something that Mr. Garland does not explain. Furthermore, if gas can be used economically for heating purposes, when compared with coal, it has got to be sold at a very low figure, probably thirty per cent of what the average is for artificial gas today. Another feature is the limited distance to which gas can be economically piped. Taking Mr. Garland's figures, this limit is a radius of about fifteen miles.

There is little doubt that a system such as suggested by Mr. Garland could be used to advantage in certain localities, but has its limitations, like all the other schemes proposed for the development of regional power. When all the factors are given their proper weight in the ideal power system equation, the best that can be seen at this time is a compromise. If the National Government makes an appropriation to investigate the possibilities of regional-power development, it should be for a consideration of all systems having reasonable prospects of economically eliminating the waste in the present methods of power generation and distribution, and not for any particular scheme.

A New Artificial Fuel?

A RECENT report states that an Italian chemist has succeeded in preparing hydrogen for use as fuel in internal-combustion engines. As usual, this report stirs up the old discussion on the possibility or probability of the use of artificial fuels.

Without knowing anything of this new process, it is safe to say that its success is unlikely, although it may be possible. The use of any substance as fuel depends upon its ability to combine with some other substance, usually oxygen, in such a way that energy is given off. This energy is usually available in the form of heat energy and can be transformed into mechanical or electrical energy by means of the ordinary power-plant apparatus.

Most of our natural fuels are largely made up of carbon and hydrogen and give up their heat on uniting with oxygen. This process results in the formation of carbon dioxide and water. These two latter compounds will not unite with more oxygen and give up more heat. You cannot burn either of them. Many people, however, have conceived the idea that if these two compounds could be broken up into their original elements, the combustion process might be repeated again and again without limit. It is a comparatively simple matter to break carbon dioxide up into carbon and oxygen or water into hydrogen and oxygen. There is, however, a well-established chemical law that if a certain amount of heat is released during the formation of a compound, exactly as much heat must be supplied to break the compound up into its original form.

It is just this law that stands in the way of any truly artificial fuel. You cannot make a fuel without paying just as much heat as you would get out of it in burning it. Our so-called natural fuels are simply substances that have been put into "burnable" form by processes of nature. Just what these processes have been is still a subject of more or less scientific controversy, but unquestionably energy in some form was supplied for their completion. Nature paid the bill and man gets the benefit.

From this standpoint the reported artificial fuel is not and cannot be what the name ordinarily implies. It may be that the Italian chemist has found some new compound from which hydrogen may be liberated without the expenditure of as much heat as will be generated when the hydrogen is burned. If this be true, he has not produced an artificial fuel, but has rather discovered a new natural fuel. The value of his discovery will depend largely on the amount of this newly found substance which exists, not on any amount he can make.

In passing, it is worth noting that hydrogen as an industrial fuel is extremely delicate stuff. Properly controlled, it can furnish a great deal of energy per pound, but if it gets out of control it has enormous possibilities for trouble.

It is becoming quite general practice to integrate both the kilowatts and the wattless component in order to determine the weighted, or what might be called the average power factor, over a given period. However, the calculations to obtain the power factor from kilowatts and wattless component are somewhat involved. This problem has been reduced to a very simple form by the use of the chart given on page 795 of this issue, by C. Harold Berry. Since a high degree of accuracy is not required in measuring the power factor, the size of the chart as published will be found adequate for most all practical purposes.

What would be the effect, where two or more alternators are operating in parallel, of reversing the polarity of the field poles of one machine and trying operating it in parallel again with the other machines, to those who have not given consideration to the subject, may appear an involved problem. In this issue, on page 810, B. A. Briggs, in answering the question, has represented three-phase alternators with ring-type armatures. This explanation is a good illustration of how many of the questions regarding the parallel operation of such machines may be analyzed by this method.

If you are planning to store coal for next winter's peak load, no time should be lost in preparing the storage place. A recent paper in *Power* pointed out that a great deal of trouble with coal in storage arises from faulty conditions due to hastily planned storage facilities. It doesn't pay to wait until the coal arrives to get the rubbish off the ground where it is to be piled.

Your engines may be running without a knock, but what is happening in your boiler room? Do you know how much fuel is making steam and how much is being wasted? Present fuel conditions make possible, and even necessary, refinements in boiler-room practice which would have been rank extravagance a very few years ago.

A process requiring twenty tons refrigerating effect concentrated into eight hours calls for a plant of sixty tons rated capacity. A coal requirement of five hundred million tons concentrated into part of the year calls for a coal production capacity of seven hundred million tons. Every coal user can do his part toward correcting this evil.

The engineer who begins to overhaul the heating system as soon as the steam is turned off is the one who will be ready for cold weather when next fall comes. The public begins to think about heat in September, but it is the engineer's business to think about it in May. Don't be caught napping.

Repairing a Boiler-Plate Lamination

In the article in *Power* of March 30, page 519, entitled "Repairing a Boiler-Plate Lamination," the writer expressed the opinion that welding up the lamination on a boiler shell of a horizontal-tubular boiler is all right. I do not agree with the author as to this, and if I were the inspector, I should reject the patch as applied. I do not believe that any shell or surface in which that part is subject to a tensile stress should be allowed to be welded by the acetylene or by the electric process. I would allow welding on flat surfaces provided they were stayed. It is the rule in the shops where I am working that any round or flanged surface cannot be welded and only such flat surfaces as are properly stayed. The Ohio state law does not permit acetylene or electric welding on boiler parts subject to tensile stress.

I have seen electric welding that looked like a very good job, but upon pressure being applied it was found that the weld was full of pinholes. Is the writer of the article under discussion sure that the patch welded is properly done and that the metal fused together as it should? In recent explosions on locomotives there has been a considerable number that gave way in the welded seams of the crown and side sheets. In order to make the job properly, I would cut out the laminated part and apply a patch to the shell with rivets; then I would be certain that the patch would pass the inspection of any inspector.

CHARLES W. CARTER, JR.
Olean, New York.

Pressure-Temperature Scales for Low-Pressure Steam

In the operation of feed-water heaters, condensers and similar apparatus it is often desirable, and sometimes necessary, to know the temperature of vaporization corresponding to a given absolute pressure, or the water-vapor pressure corresponding to a given temperature. The scales given in the figure show the corresponding absolute pressures and temperatures for pressures below atmospheric. The absolute pressure is found by subtracting the vacuum-gage reading from the barometer reading, both in inches and mercury.

The two upper scales give the temperatures for each hundredth of an inch pressure from 0.1 in. to 2.0 in. The three lower scales give the values for each tenth of an inch over the range from 2 in. to 30 in.

Detroit, Mich. C. HAROLD BERRY.

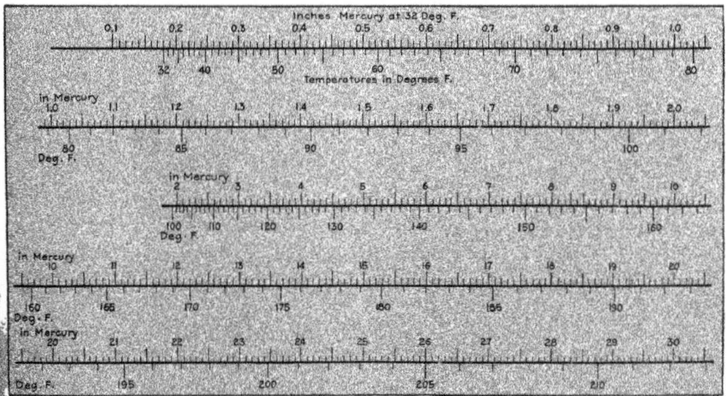

CHART SHOWING VAPORIZATION TEMPERATURE OF STEAM FOR PRESSURES LESS THAN ATMOSPHERIC

Why the Mill Settled

Some years ago I was employed in a mill where a near accident took place. The main mill was three stories, attic and basement, and the power was both steam and water. Water for the waterwheels was introduced directly under the stair tower, which was centered with the mill, as shown in the sketch.

The first indication that anything was wrong was when the friction load began to increase and the master

BROKEN FLOOR BEAM CAUSED SHAFTING TO SETTLE

mechanic gave instructions to have the shafting hangers and drives looked over. When the level was applied, the shaft in one of the rooms was found to be about two inches low midway of the mill, and the two floors above were in the same condition. Therefore, the "mechanics" started to level the shafting, by cutting into the cross-timbers of the mill, as shown at *B*. As I did not believe in weakening the structure, I entered a protest at the first cut, but without avail. I then told the agent the men were ruining his mill and that if they kept on it would fall down. This rather startling statement had the desired result, and he started for the mill.

The master mechanic had his line of argument all prepared, and *C* explains it. He said that he had either got to put the shafting up where it was down or put both ends down to match the center. He couldn't put the end that was in the wall down without tearing the wall all out and he wouldn't do that. As a matter of fact, he was only cutting into the timbers two inches and that wouldn't weaken them even a little bit, as he said.

The next morning my fireman acted a trifle fidgety, and finally he said that he believed that the mill was settling in the center over the water-way, which upon investigation proved to be so. The timber *A* had broken as shown at *D*, and the center of the mill was gradually sinking. Steel I-beams were put in place over the tail-race, and everything was made safe.

New York City. C. W. PETERS.

What Caused the Crack?

Some time ago a boiler that had been bagged came to my notice. The plant owner tried to employ a boiler-maker at the time, but being unable to do so, a workman from the plant drove up the bag and in so doing, the metal buckled at the center. Two holes were drilled and filled with ⅞-in. rivets. Some eight months later the engineer decided to have the plate straightened up; so a boilermaker was employed and was very successful in getting the sheet in good shape again. The next morning, upon getting ready to put in the two rivets, a crack 16 in. long was found extending across the sheet. The metal had been heated to a cherry red and let cool gradually, and no water entered the boiler during this operation. The sheet was cut out and replaced by a patch. When broken open at the crack, the damaged plate looked more like gray cast iron than boiler steel, it was so crystallized. H. SYLVESTER.

Concordia, Kan.

Mr. Bissel Has Found Another Perpetual Motion

The following should prove of interest, to the curious at least. In the illustration two cylindrical tanks are shown pivoted to a main frame on ball bearings. The main frame in like manner is centrally swung on a vertical shaft integral with the base and is also suspended on ball bearings. At the top of the central shaft is a gear proportioned about three to one with the cylindrical tanks, as indicated in the illustration. Inside of each tank is a closely fitting float, which operates similarly to the cork float in a modern gasoline-engine carburetor. In this case it is hinged on two journal boxes near the center of each tank.

Enough mercury is placed in the tanks to fill about one-half of the space between the outside and the float. Care should be taken to design the rotative size and shape of each tank with reference to the float, so that a maximum movement may be obtained and the mercury flow easily, using the least amount thereof. All parts should be carefully adjusted, one to the other, and smooth surfaced, so that there will be as little disturbance in the mercury as possible when the apparatus is revolving. It should be stated here, that the universal joint in the center of each float is turned at a slight angle by each of the connecting arms above their centers.

With a similar apparatus in which the floats were 20 in. thick, not coupled up, however, with the universal

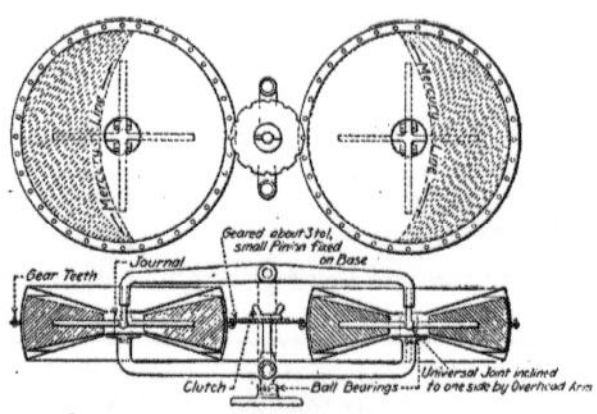

MR. BISSEL'S PERPETUAL-MOTION MACHINE

joint, on account of faulty alignment, it was found that by using water instead of mercury and revolving the apparatus by hand, a regular up-and-down motion of the floats resulted. There was power enough to break a thread, also to lift a small weight up and down at every half-turn of each tank. It did not appear to take any more power to revolve the apparatus, whether the floats were secured or allowed to operate. Therefore the conclusion seems to be that energy was being produced.

Pittsburgh, Pa. JOSEPH E. BISSEL.

Height of Return-Tubular Boilers Above Grates

We note with interest your answer to E. G.'s question in the Feb. 10 issue of *Power* relative to the proper distance above the grates for setting 72-in. horizontal return-tubular boilers, and your answer to John A. Stevens' criticism in the issue of March 30.

It is our practice to set boilers of this size 60 in. over the grate to the under side of the shell. We have no records of complete tests on boilers installed in this manner, but we do know that there is a positive saving in fuel and a marked elimination of smoke in plants where we have replaced boilers set in the old way.

With a boiler setting properly constructed and insulated, we believe there would be very little difference in infiltration and radiation loss, due to the additional height of the setting. The increased combustion space is particularly adaptable to higher ratings, and in our opinion a return-tubular boiler can be most economically operated at from 125 to 150 per cent rating.

Boston, Mass. DAVID MOULTON.

Relative to the correspondence on the height of return-tubular boilers above the grates, page 517 of the March 30 issue of *Power,* I would say that during the month of September, 1917, a test was run upon the No. 2 power plant of the Whittenton Manufacturing Co., Taunton, Mass., under my direction. This plant contains ten 72-in. Wickes horizontal return-tubular boilers, nine of which were run during a week's test of 168 hours. The fuel used was "Alpha special" run-of-mine bituminous coal from Cambria County, Pa.

These boilers were set at a height of four feet from the grate to the under side of the shell and were installed under the direction of John A. Stevens. A record was kept of all the coal burned and the water evaporated during the week, and the average efficiency of boiler, furnace and grate was 67.79 per cent. The boilers were operated from 6:40 a.m. to 5:30 p.m. at 117 per cent rating, from 5:30 to 9 p.m. at 57 per cent rating and banked for the remainder of the 24 hours. Fires were buried down and cleaned from 6:30 to 9 p.m., which caused a loss of 2 per cent in the average efficiency.

No attempt was made to instruct the firemen in any way, as the object of the trial was to find out what was obtained from the plant in regular weekly operation. It is probable, however, that the firemen did a little better than usual owing to the presence of our observers in the plant. I consider these results very good for a mill boiler plant. FRED B. COLE,

Assistant Engineer with Charles T. Main.

Boston, Mass.

I am in charge of a plant that, when I came to it five years ago, had five 60-in. x 16-ft. return-tubular boilers set 24 in. above the grate. A shaking grate was used and in very cold weather the boilers were overloaded. We could burn up to 8,000 lb. of coal per boiler per 24 hr. very well, and our evaporation was about 10 lb. of water per pound of coal as fired, but at times we had to burn 10,000 lb. per boiler and our evaporation would always drop at such rates of combustion.

I wanted to drop the grates 12 in. but the mechanical engineers who do the consulting and lay out new work objected, saying I would spoil the looks of the plant and not improve on the evaporation.

I then asked for an extra boiler, but as there was not room to install one without building onto the plant I did not get it. I then talked them into letting me change the grates and lowered them 9 in. in front and 12 in. at the bridge wall, giving a 3-in. pitch. Castings were made crescent shaped to fill in the fronts and when the job was finished they said, "It don't look so bad after all." I can burn 10,000 lb. of coal per boiler per 24 hr. now and keep the evaporation up as well as we could on the 8,000 lb. basis.

This last year we installed a 72-in. by 18-ft. boiler and when I received the prints from the same engineers I was somewhat surprised and delighted to see the boiler was to be set 4 ft. 6 in. above the dead plate. We have what is termed a hand stoker set under it with a pitch of 18 in., making 6 ft. from rear of the grate to the shell of the boiler. Although I have not had a test run on this boiler as yet, to determine the economy, I have handled it enough to know it is doing good work.

We are running with a draft of from 0.25 to 0.3 in. of water in the furnace. We are burning 20 lb. of coal per sq.ft. of grate per hour and recently with three of the smaller boilers, one of which was banked 5 hr., and one banked 20 hr., we burned 44,400 lb. of coal. This averages over 21 lb. of coal per square foot of active grate per hour, not figuring what was used for banking.

I think we would have done better not to have run the boilers quite so hard and to have used one of the banked boilers a few hours longer, but the men don't always start up a banked fire when they should.

This coal was all handled by one man on each 8-hr. watch, except in the morning from 7 to 9, when I put on an extra man.

The coal and ashes are handled by coal passer and I think the men earned their money. I don't think this load could be carried on a boiler set 30 in. from the grate. I may be wrong and will be glad to be set right. In some of former issues of *Power* I have read some very interesting articles by Prof. Breckinbridge, and I would like very much to see what he may have to say about settings for return-tubular boilers.

Andover, Mass. W. C. RICHARDS.

Merits of Chain and Ring Shaft Lubrication

With reference to Mr. Wineberg's communication on the subject of chain and ring lubrication in the April 6 issue, I can say that I have had considerable experience in designing heavy ring-oiling bearings for rolling-mill and factory service, some of them being as large as 20 in. diameter, and running at 80 r.p.m. and the smaller sizes running up to several hundred revolutions. I have always had considerable success with the ring-oiling type, which I adopted in preference to the chain.

The main bearing of a high-speed engine that I once had under my supervision was fitted with a chain, and had it been practicable I would have changed it to ring oiling. The chain had a bad habit of becoming kinked, which caused it to jam, and of course the bearing went up in smoke. This invariably happened at the most critical time, when the engine was carrying its full load, and then it was a case of shut down and start up the other engine, which we were lucky to have available. I do not maintain that all chain-oiling bearings are unsatisfactory, but my observation and experience have led me to the opinion that the ring-oiling bearings are more reliable and satisfactory in every way, giving little or no trouble, especially in heavy-duty drives, where reliability and economy in operation are matters of considerable importance. JOHN J. MUIR.

Seattle, Wash.

Three Power Plant Kinks

The following "stunts" may be of use to some of the readers of *Power*. For instance, most engineers use a brass plug in the cross connection below the water column. I formerly used them also, but after finding that any kind of an ordinary pipe plug is hard to remove after the square head becomes rounded, I adopted the plan shown in Fig. 1. That is, I made two long brass plugs threaded at one end so as to use a pipe wrench on them. Since then I have had no further trouble. An ordinary pipe nipple capped on the end could also be used. Fig. 2 shows a kink that I have found useful when making repairs and adjustments on medium-size Corliss and side-crank engines. It consists of a piece of wood fitted with cleats to fit between the legs of the cylinder and engine frame. A long lever inserted between these cleats and brought against the crosshead pin nut permits of moving the crosshead. In one of our 160-hp. boiler furnaces we had trouble with the side walls bulging and falling down. After I had anchor-irons, made of $\frac{1}{4}$ x 1-in. iron and $8\frac{3}{4}$ in. long, placed as indicated by Fig. 3, no further trouble was had. Some claimed that the anchors would burn off, but we found that they did not. GEORGE J. LITTLE.

Hasbrouck Heights, N. J.

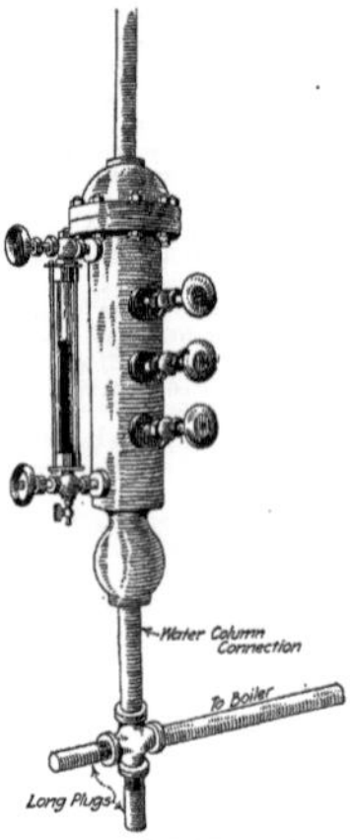

FIG. 1. LONG PLUGS IN CROSS CONNECTION

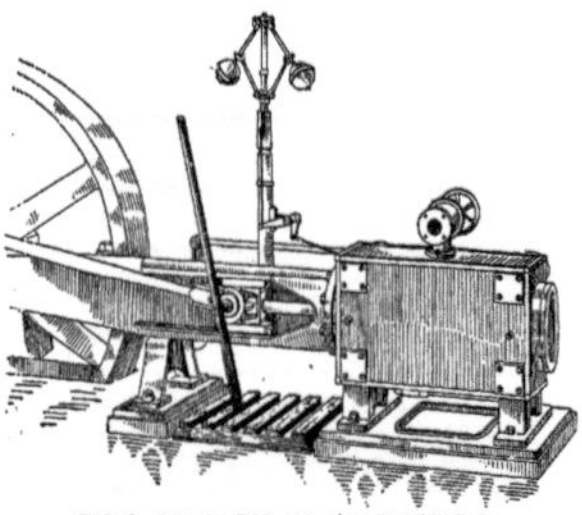

FIG. 2. LEVER FOR MOVING CROSSHEAD

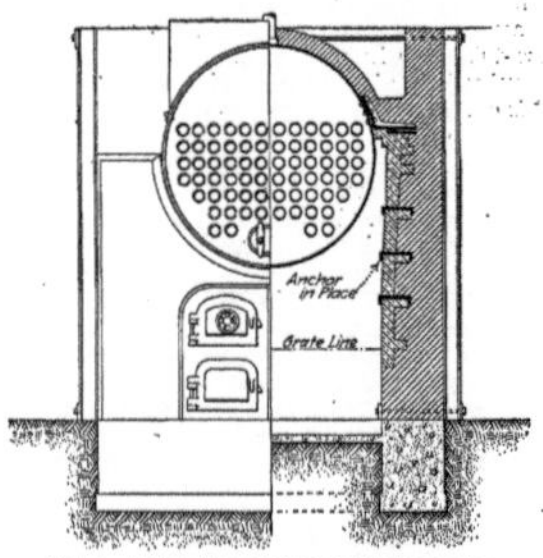

FIG. 3. ANGLE IRONS IN FURNACE WALLS

Electric Service in New England

Today electric service is available in practically every community in New England. This service may be supplied by a small municipal "electric-light plant," a small generating station under private control, a syndicate of small stations or a local substation fed from a large central station many miles distant. The recent movement to interconnect these different systems by a network of high-tension lines has brought to light some peculiar conditions.

The service ranges from direct current to 133-cycle alternating current. Voltages cover a similar wide range, while the number of phases is limited to 1, 2 and 3. The need of standardization of electric service is evident.

In this assortment of frequencies 60 cycles is the popular choice. Many systems of lower frequency have been changed to 60 cycles, and the last 133-cycle system in New England is now being changed over for 60-cycle operation.

Single-phase service for lighting and small-power installations is usually 110 volts, 60 cycles. Distribution transformers are often arranged for two 110-volt circuits and one 220-volt circuit on the low-voltage side. The lighting load is connected to the 110-volt circuits and the small-motor load is connected to the 220-volt circuit.

For power service, three-phase, 220 volts, 60 cycles is considered standard for small installations and three-phase, 440 volts, 60 cycles for larger installations. Apparatus for use on these circuits is considered standard by most manufacturers. This allows the customer a lower price and better delivery than would be possible if this apparatus were special. H. S. RAMSAY.

Arlington, Mass.

INQUIRIES OF GENERAL INTEREST

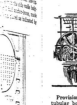

Provision for Smoke Connection to Stack.—A return-tubular boiler that has thirty-six 3-in. fire tubes has the grate area reduced to 6 sq.ft. What size of smoke connection should it have to a 40-ft. chimney that has a 2 x 2 ft. flue? C. E. S.

A 12-in. diameter smoke pipe with easy bends at angles would be sufficient for the present grate area. But to provide for future enlargement of the grate, with maximum draft area of the tubes, it would be well to supply a stack connection about 18 in. in diameter, in any case providing a damper in or near the smoke uptake of the boiler.

Trick's Slide Valve.—What is the difference between a Trick valve and an ordinary *D* slide valve? F. A.

The main difference is that Trick's valve affords double admission. Live steam is admitted past the outer edge of the valve just as with the ordinary *D* slide valve and additionally through a passage *A* in the valve, as shown in the

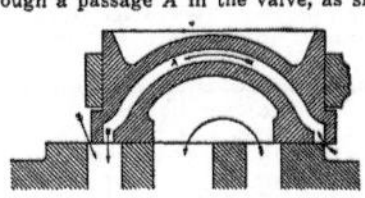

sketch. When the outer edge of the valve uncovers the port for admission, the port of the passage *A* at the opposite end of the valve is also uncovered, thus giving a double admission of steam, the same as a double-ported valve. This valve does not, however, give double exhaust, as the passage *A* is not opened to the exhaust.

Color for Natural-Gas Flame.—In the use of a natural-gas burner should the flame show bluish or straw color? E. M.

A white or straw-colored flame is indicative of incomplete combustion, usually due to poor mixture of air or insufficient air supply, and although there may be greater capacity, there is less economy and more soot than from a blue flame. While a blue flame is representative of more perfect combustion, it may be less economical than a white or yellow flame if there is an excessive air supply. For best results the air supply should be just sufficient to prevent a white or straw-colored flame.

Discharge of Condensate from Drying Cylinders.—On a paper machine that is provided with 30 dry cans or cylinders, do the cylinders become alternately filled and emptied of water, or is siphonage of the condensate continuous? P. H.

If the same steam pressure is maintained in all of the drier cylinders connected to the same drainage header and the discharge pipes and connections to the header are large enough to carry off the water as formed, then the condensate will be discharged continuously and with all cylinders containing the same depth of water. If the siphon pipe or siphon connections of a cylinder are too small, or the pressure maintained within a cylinder is below that within other cylinders connected to the same drainage header, the cylinder with lower pressure will fail to discharge or will discharge only intermittently, and conditions seldom are favorable for that cylinder to become completely emptied of water.

Advantages of Two-Stage Air Compression.—What are the advantages of two-stage compression of air over single-stage compression? C. H. L.

Two-stage compression is adopted for the purpose of cooling the air and thereby obtaining a reduction of volume before the final compression. Intercooling, as by circulation of water, is an essential detail for obtaining any advantage from two-stage compression. The advantage over single-stage compression is an appreciable saving of power and the avoidance of the high temperatures, thus permitting of more satisfactory lubrication and less discharge in the air receiver and piping system to cause fires and explosions, and less drop of pressure from cooling of the compressed air.

Economy of Direct Steam Pumps.—What is the average steam consumption per horsepower-hour of small duplex boiler-feed pumps of sizes such as 6 x 4 x 6 and 8 x 5 x 10, when operated with steam at 100 lb. pressure? L. S. J.

The steam consumption of direct-acting steam pumps is necessarily very high, as steam must be admitted to follow full stroke, the piston speed is low, the percentage of cylinder clearance is much greater than in ordinary steam engines and generally such pumps are operated against high back pressure. All these conditions are more detrimental to economy in the smaller sizes of pumps, and for ordinary sizes used for boiler feed pumps, such as 6-in. x 4-in. x 6-in. or 8-in. x 5-in. x 10-in., direct-acting pumps have steam efficiencies less than one-tenth that of a good engine; that is, they use from 150 to 200 lb. of steam per indicated horsepower per hour. But where the exhaust can be utilized for heating boiler-feed water, or displaces the expenditure of live steam for heating purposes, the excessive steam consumption per horsepower-hour becomes a negligible consideration.

Wattmeters Require Adjustment.—On one of the feeders going out from our plant the voltage is stepped up from 2,300 to 13,200 and, at the end of a five-mile transmission line, stepped down again to 2,300 volts. The power is metered on the 2,300-volt side of the transformers at both ends of the line. Should not both meters register the same? As it is now, the meter at the out end of the line registers considerably more than the one at the power plant. V. A. S.

The two meters should not register the same; however, they are now doing the reverse of what is the correct condition. The instrument at the load end of the line should register less than the one at the power house, since the latter meters not only the power supplied to the load, but also the losses in the transformers and transmission lines, whereas the meter at the load measures only that which is supplied to the load. If both meters are correctly calibrated, the one at the load will register approximately 10 to 15 per cent less than the one at the plant. The ratios and connections of the instrument transformers should be checked to make sure they are correct, and then the meters should be calibrated.

[Correspondents sending us inquiries should sign their communications with full names and post office addresses. This is necessary to guarantee the good faith of the communications.—Editor.]

Induction and Synchronous Motors for Industrial Applications*

By C. W. Drake
General Engineer, Westinghouse Electric and Manufacturing Company

A SUBJECT that is attracting much attention today is "plant power factor," although it is a problem which has existed ever since induction motors were first installed, somewhat over twenty years ago. This being the case, we are often asked why it is that so many years elapsed with little or no discussion regarding power factor, while now it is the talk of the hour. There are numerous reasons, some having to do with the industrial plant itself, some with the transmission line and generating station, while others are largely economic. The most active factor in bringing this question to the foreground is the fact that during the last few years the generating equipment, transmission lines and transformers of most central stations have become heavily loaded or overloaded, and in looking for a method of still further increasing their capacity and improving the regulation, the raising of the power factor has appeared to be the most simple and inexpensive solution.

Correct Sizes of Motors Important

In a plant equipped with induction motors there is a fairly constant value of wattless kilovolt-amperes that must be supplied for magnetization. For a given energy component—that is, a certain definite amount of work performed in the plant—a definite power factor will be obtained. Ordinarily, when speaking of raising the power factor, it is assumed that a portion of this wattless component will be supplied by a condenser, of the synchronous or perhaps the static type. There is, however, another method of raising the power factor, which should not be lost sight of, and that is by carefully testing the motors and seeing that each one is of the correct size for its particular work. In many instances power factor has been raised from 5 to 10 per cent by this method, and in exceptional cases as much as 20 per cent. In plants using shears, punches, rolls and other machines equipped with flywheels, a very low load factor is often obtained, and a careful study should be made to obtain the best combination of motor and flywheel. Frequently, it will be found that the motor is installed of sufficient capacity to actually perform the duty cycle without the flywheel. In one mill, which had a power factor varying from 40 to 60 per cent, there was found a special welding roll that was driven by a 200-hp. squirrel-cage motor. A heavy flywheel was also used, but as the motor had only about 2 per cent slip, it carried practically the entire load. This lasted only a few seconds out of several minutes, so that most of the time the motor was running idle. By a proper combination of motor and flywheel design it was possible to use a 50-hp. motor, which materially reduced the wattless kilovolt-amperes. There are many such cases in which the motor need have capacity only sufficient to supply the friction losses, while the flywheel will perform most of the working cycle.

Where Power-Factor Correction Is Difficult

Of course, with the best motor arrangement, it is impossible with an induction-motor load to obtain a power factor much over 80 or 85 per cent, although there are plants using 25-cycle service, that operate at approximately 90 per cent power factor. However, the majority of plants have a power factor nearer 70 than 90 per cent, consequently some form of corrective apparatus is required to raise the power factor to a value that will not carry a penalty in the power contract. Just what this value is, it is difficult to state, as there are nearly as many power-factor clauses as there are power contracts, but as a rule penalties are imposed on power factors below 85 or 90 per cent for large demands and perhaps 75 to 85 per cent for smaller loads. It is these small or medium-sized plants where power-factor correction is generally found most difficult, as for instance, in machine shops, planing mills and miscellaneous industries located in the cities where most of the motors are of small size and are started and stopped at frequent intervals. Synchronous motors are not well adapted for such service, and in order to be of value for power-factor correction, they must be on the line all the time that the plant is in operation. In the planing mill the shaving exhauster offers about the only opportunity for the use of a synchronous motor, and either a belted or direct-connected motor may be used. As ordinarily installed, a fan imposes extreme starting conditions; that is, it requires approximately full-load torque at synchronous speed. Even with induction motors it is customary to use a high-starting tap in order to bring the motor to as high a speed as possible before changing over to full voltage. Since the reason for installing a synchronous motor is to raise the power factor, it is evident that the rating of the motor will depend as much on the wattless component required for the particular plant as upon the mechanical load required to drive the fan, and in cases like this the kilovolt-ampere rating of the motor will probably be at least double that required to drive the fan. This excess capacity makes it possible to obtain the desired starting characteristics with a reasonable design of motor.

An investigation of industrial plants shows there are many like the foregoing, in which one or two places exist where synchronous motors may be used and correct the power factor for the entire plant. These motors will operate at a high leading power factor. Motors for such service should be designed practically the same as synchronous condensers; that is, to operate at nearly zero per cent power factor, since the mechanical load in many cases may be quite small. For most of these applications either a direct-connected or a belted exciter should be used, and the starting and control of these motors may now be made almost as simple as that of a squirrel-cage induction motor. In fact, there is quite an interest shown at present in automatic starters for synchronous motors, and these will undoubtedly tend to increase the use of synchronous motors in plants where it has previously been thought that there was not sufficient skilled labor for their operation.

Small Synchronous Motor Demand Increasing

For industrial service it is impossible to draw up a general specification covering the exact size of motor to meet every condition of starting, power-factor correction, etc. Each application will require individual attention, but the fact that in most cases the rating is two or more times the mechanical load will permit the use of normally designed motors.

As a considerable portion of any central-station load is made up of miscellaneous industries like those herein described, it would appear that there should be an increasing demand for synchronous motors operating at speeds from 600 to 1,200 r.p.m. and in capacities of from perhaps 75 to 100 kilovolt-amperes upward. The installation of such machines would correct the power factor at its source, which is the proper place, rather than transform and transmit the wattless current to some central location to be corrected, as is now often the case.

The direct-connected, slow-speed synchronous motor for driving large air compressors has become standard, so that as far as the customer is concerned, it is unnecessary to specify the motor-starting characteristics or even the horsepower rating. Co-operation between the motor and compressor builders has made it possible to supply exactly the

*Abstract from a paper presented April 21, 1920, before the Pennsylvania Electric Association at the William Penn Hotel, Pittsburgh, Pa.

right motor for each compressor. This is also becoming true in the ammonia-compressor or refrigeration field, although more variables occur here regarding the horsepower requirements than with air compressors. Such applications, however, are best worked out between the machinery manufacturers, and the customer is principally interested in a satisfactory operating unit. Motors for such service are ordinarily designed for 100 per cent power factor, so that the amount of power-factor correction obtained will depend principally upon the ratio of the compressor-motor kilovolt-amperes to the total kilovolt-ampere load in the plant.

There are certain industries that have total loads running into thousands of kilowatts, and much of this load is often made up of comparatively large, slow or medium-speed motors. For instance, in the rubber industry large numbers of motors from 200 to 800 hp. are used to drive mill lines, and with the usual form of gear drive motor speeds from 500 to 600 r.p.m. are common. Slip-ring induction motors have been generally used on account of their high starting torque and low starting current, but as the load on these mill lines is quite variable, a low power factor often results. Most of the present drives are equipped with magnetic clutches or clutch brakers, which in case of emergency, disconnect the motor from the line and bring the mill line to a quick stop. The slip-ring motor has ample torque to start the load, so that the clutch may or may not be used for this purpose, but the fact that clutches are ordinarily used, has made it possible to replace slip-ring with synchronous motors of similar or slightly larger rating.

Slow-Speed Synchronous Motors

Experience with refrigerating machinery has shown that synchronous motors may be built for extremely low speeds and yet give high efficiency and a reasonable cost. Consequently, it was a natural development in the rubber industry to suggest the elimination of the gear reduction and use synchronous motors at about 100 to 120 r.p.m. instead of 500 or 600 r.p.m. In this case the clutch has five times as much torque to transmit and becomes a very large and expensive piece of apparatus. Other forms of braking may be used, such as mechanically operated brakes, or plugging or dynamic braking on the motors. But as long as the clutch is required for starting, it probably offers the simplest method of stopping. If synchronous motors can be designed to approach the performance of slip-ring motors—that is, give full load torque or over at starting—then the clutches may be eliminated and some other form of braking considered. The point that it is desired to make here is that the tendency in all machine design, plant layout and motor application is toward simplicity and a minimum number of parts. If motors can be designed to perform certain functions without the use of auxiliary mechanical devices, it is a step in advance. Many such examples may be noted, as for instance, the elevator and reversing-planer motors which made it possible to eliminate the reversing belts, and adjustable speed, direct-current motors for machine tools, which eliminated the step pulley and most of the gear box. There is no doubt regarding the ability of synchronous motors to drive rubber mills and correct the power factor, but it is a question whether the present arrangement may be considered a logical or final solution. As it stands now, the motor characteristics meet only a part of the requirements and the clutch and brake supply the remaining ones, but it is futile to prophesy what the ultimate drive may be. Whether these motors should be designed for 100 per cent or 80 per cent power factor will depend on local plant conditions, but as a rule we believe they should be designed for 80 per cent, and then they may be operated at any power factor between 80 and 100, as desired.

Conditions Existing in Cement Plants

Very similar conditions exist in cement plants, where large numbers of induction motors are used on tube mills, crushers, etc., and many of these motors are rated at from 200 to 300 hp. In the early days of electrical development in cement plants, friction clutches were quite commonly used in connection with standard squirrel-cage motors in order to facilitate starting. After the exact starting and running conditions were better known, squirrel-cage motors of proper characteristics were produced, so that the troublesome clutches could be eliminated. Most cement plants have preferred the squirrel-cage to the slip-ring motor for such service, on account of its greater simplicity, although slip-ring motors are now installed in some mills in the larger capacities. This is another example where the characteristics of synchronous motors are not adapted to the work, and their use would in a way revert to early induction-motor practice with clutches. The difference now is, the advantage of power-factor correction may offset any expense and difficulty with the mechanical equipment, and it may also be assumed that much better clutches are now available than formerly. Consequently, in plants like these it should be carefully considered whether the power-factor correction desired can best be obtained by using synchronous motors in the manufacturing processes or whether better manufacturing conditions can be obtained by the use of induction motors, and the power factor corrected otherwise, as perhaps by synchronous motors on motor-generator sets, centrifugal pumps or synchronous condensers. The foregoing may be summarized as follows:

1. A very large part of the ordinary central station business consists of industries that are difficult to classify, either as to product or type of machinery used. In most of these, improvement in power factor may best be obtained by the use of medium- or high-speed synchronous motors used in much the same manner as the present induction motors The proper choice or application of these synchronous motors will fall largely upon the power engineer of the central station, who is generally most familiar with the customer's power problems.

2. For any plants using compressed air or refrigeration in suitable quantities, complete equipments with synchronous motors are available, and as a rule very little engineering on the customer's part is required. In some cases it is possible to have the motor especially designed for the local conditions, but generally a much better proposition is obtained by accepting standard ratings.

3. In industries, like the rubber, cement, flour, etc., using large motors, it is possible to use synchronous motors for nearly any drive by means of necessary mechanical attachments. As the principal reason for using synchronous motors is to improve the power factor, they should be used first in those places where their electrical characteristics are best adapted, and their use in other places should be carefully considered to see if the same result could not better be obtained by some other arrangement.

Industrial-motor application work cannot be reduced to an exact science or formula, because the power requirements and motor characteristics are determined not only by the machine to be driven, but fully as much by the method in which it is operated. Consequently, each application requires careful personal consideration in order that all functions may be satisfactorily met.

Orville Wright Presented with John Fritz Medal

Orville Wright, pioneer in the development of the airplane and an honorary member of the American Society of Mechanical Engineers, was presented with the John Fritz Medal at ceremonies held in the auditorium of the Engineering Societies Building, Friday, May 7. Presentation of the medal was made by Comfort A. Adams, past president of the American Institute of Electrical Engineers, following an address by Major General George O. Squier, Chief Signal Officer, U. S. A., and Colonel Edward A. Deeds, formerly a member of the Aircraft Production Board. Benjamin B. Thayer, past president of the American Institute of Mining and Metallurgical Engineers and chairman of the Board of Award, who was to preside at the ceremonies, was absent owing to illness, and Charles F. Rand, of the Committee of Arrangements, presided in his stead.

The ceremonies were preceded by a dinner at the Engineers Club. Among those present were the members of the board and several friends of Mr. Wright.

Orville Wright was born in Dayton, Ohio, Aug. 19, 1871, and until 1888 he attended the public and high school of that city. Later he attended Earlham College, Indiana, until 1909, when he was graduated. His studies were completed at Oberlin College, Ohio, which he attended until 1910.

Orville is the brother of the late Wilbur Wright, and since 1903, he, with the aid of his brother, has worked on the development of the flying machine. The first test was made in North Carolina in 1903, and in 1905 the first successful long distance flight was made. In 1909 his efforts were recognized by the French Academy of Science and he was presented with a gold medal. He is a member of the Aëro Club of America, and his work in the development of the airplane will stand out as a monument of scientific achievement.

The John Fritz medal was established in 1902 in honor of John Fritz, of Bethlehem, Pa. It is of gold and is awarded only once each year, accompanied with an engraved certificate which states the origin of the medal, the specific achievements for which the award is made, and bears the names of the members of the board by which the medal was awarded and the signatures of the president and secretary of the board, which is formed of sixteen men, four representatives from each of the four National Societies of Civil, Mining, Mechanical and Electrical Engineers.

The first award of the medal was made to John Fritz at a dinner given to him on his eightieth birthday, Aug. 21, 1902, and since then fifteen awards, including the one to Mr. Wright, have been made.

Regional Power Development and the Low-Temperature Distillation of Coal

In the May number of the *Journal of the American Institute of Electrical Engineers*, C. M. Garland[1] takes issue with W. S. Murray on his scheme for the development of regional power for the industrial district along the North Atlantic seaboard. An abstract of a symposium "Economical Supply of Electric Power for the Industries and the Railroads of the Northeastern Atlantic Seaboard," on Mr. Murray's plan presented before the midwinter convention of the American Institute of Electrical Engineers, was published in *Power*, March 16, 1919. Discussing Mr. Murray's plan, Mr. Garland says, in part:

The writer is convinced, however, that no regional power-development scheme north of the Mason and Dixon line which contemplates the generation of power from coal is complete unless it provides heat for industrial plants in winter, and furthermore, that no regional power-development scheme contemplating the generation of power from bituminous or semi-bituminous coal is more than halfway complete if it does not contemplate the recovery of the byproducts from the coal.

In the central and northern portions of the United States every industrial plant for from seven to eight months out of the year requires fuel for heating. In 75 per cent of these plants the fuel required for heating is more than sufficient to generate the power required for manufacturing operations. For practical purposes it may therefore be considered that the power is obtained during these months without fuel expenditure. This explains why only 30 per cent of the industrial plants buy their power. Of this 30 per cent 20 per cent would doubtless be better off if they had efficiently operated plants and generated their own power. The writer is a strong advocate of central-station development, but believes that it is well to recognize the limitations imposed by existent conditions.

The great barrier to Mr. Murray's project is, therefore, the fact that the average plant can take advantage of the increased economy of the central station only from four to five months out of the year. During the winter season the same amount of coal would have to be hauled to furnish heat in so far as the industrial plants are concerned. The amount of coal hauled would therefore be reduced only by the amount of coal saved by the electrification of the railroads, which in this section would not amount to more than 20 per cent of the total. The writer, therefore, cannot see any great relief to railroad congestion unless some method of heating other than direct-coal firing is resorted to. It is also this coal for heating that leads to the overcrowding of railroad terminals during the winter months. Again, it is questionable if the poorer economy certain to result from the intermittent operation of boilers for the heating load in winter, assuming that the power load was taken by the central station, would not offset the saving effected through the electrification of the railroads. It is, therefore, quite certain that any regional power development that does not include also the distribution of heat will be extremely limited in its success, for with the present methods of electrical-power generation, it would never be possible to distribute electrical current for the heating of large areas at low temperatures.

In the United States where at least 500,000,000 tons of bituminous coal are burned annually, there are 10,000,000,000 gallons of tar and the equivalent of 30,000,000,000 pounds of ammonium sulphate, having a combined value of $2,500,000,000 burnt up with the coal. It is not practicable to save all of these byproducts, but a recovery of 25 per cent could reasonably be expected when it is considered that 80 per cent of the coal mined is used by industrial plants, public-utility plants and the railroads.

The remarkable part of all of this is that the means are at hand today and have been at hand for a number of years for carrying out the regional power development, the regional distribution of heat, the placing of every industrial load on the central station, the relieving of railroad congestion, particularly terminal congestion, through the reduction of the hauling of fuel, the reduction of the smoke nuisance in cities, and, in addition to these, the recovery of the byproducts from at least 25 per cent of the coal mined. I refer to the combination of the well-known Mond byproduct system of gas generation with the low-temperature distillation of coal.

In the Mond process coal is gasified in a special producer whereby the coal is converted into gas, tar and ammonia. The ammonia is recovered in the form of ammonium sulphate, which amounts to from 50 to 100 lb. per ton of coal gasified and has a value of about four cents per pound under present prices. The value of this byproduct, therefore, varies from $2 to $4 per ton of coal gasified, depending upon the amount recovered. The tar recovered from the straight Mond process represents about 6 per cent of the weight of the coal, and this tar consists principally of pitch which has little or no value.

The gas from the Mond process has a calorific value in the neighborhood of 140 B.t.u. per cu.ft. and from 65 to 70 cu.ft. of gas are obtained per pound of coal. This gas can be burned under boilers with greater efficiency than the coal and without smoke. It is also suitable for use in the firing of house-heating boilers and for small or large industrial furnaces. The gas can be piped economically within a radius of fifteen miles of the central station. This distance limits the size of the central station and to a certain degree does not accomplish everything that could be desired from regional power and heat distribution. It does, however, relieve railroad-terminal congestion, and it will in a large measure relieve main-line congestion, owing to the greater economy in the use of fuel and to the elimination of fuel hauled for the use of railroads.

By the low-temperature distillation of coal is meant distillation of coal at a temperature around 1,000 deg. F. When distilling coals at this temperature, a very small amount of gas is generated of high calorific value and a large amount of tar with a small amount of ammonia. This is a process which has been experimented with for years and which has been demonstrated on a practical scale by different investigators in this country and abroad. It has been definitely determined beyond question of a doubt that from 20 to 30 gal. of tar can be obtained from a ton of bituminous coal, something like 12 lb. of ammonium sulphate and from 1 to 2 cu.ft. of gas and about 75 per cent of coke.

The tar from this process contains considerable quantities of motor fuel and creosotes. It has been estimated that it would be possible by the splitting up of this tar to obtain from 15 to 20 gal. of motor fuel per ton of coal. Investigations indicate that in a crude state this tar is worth in the neighborhood of 10 cents a gallon.

The coke from this process would probably contain in the neighborhood of from 12 per cent to 15 per cent of volatile matter which would contain most of the nitrogen originally in the coal. By gasifying this coke in the byproduct gas producer, from 50 to 85 lb. of ammonium sulphate will be obtained per ton of coke gasified and from 65 to 70 cu.ft. of gas having a calorific value in the neighborhood of 140 B.t.u.

The low-temperature process will yield byproducts having a value of from $2 to $3 per ton of coal gasified. By gasifying the coke in the byproduct gas producer the

[1]Consulting Engineer, Chicago, Ill.

ammonia recovered will have a value of from $2 to $4 per ton of coal gasified, depending upon the amount recovered. In other words, by combining these two processes, by-products having a value of from $4 to $9 per ton of coal gasified may be obtanied.

I would not ask Congress, as Mr. Murray has suggested, to appropriate $200,000 to find out if there was a $300,000,000 waste in the Atlantic Coast States, for the same reason that I would not ask Congress to appropriate $200,000 to ascertain if the Statue of Liberty is still standing in New York Harbor. Every engineer experienced in power-plant economy knows that this waste exists and has a comprehensive idea of its magnitude. I would, however, ask Congress for an appropriation of $5,000,000 for the construction of a workable-sized plant along the lines of the combined Mond and low-temperature distillation processes. This $5,000,000, properly used, would bring greater returns to the American people than any money Congress ever spent.

Water-Power Bill as Reported from Conference Committee

The water-power bill, lately reported out by the Conference Committee of Senate and House, creates a Federal Power Commission to be composed of the Secretaries of War, Interior and Agriculture, to which is delegated authority over all matters pertaining to the development of water powers in which the Federal Government has jurisdiction or in which it is interested as owner of lands or other property necessary to such projects.

The commission may issue preliminary permits allowing applicants three years in which to make examinations of water-power projects, to prepare plans and to make financial arrangements. The commission may also issue licenses for a period of fifty years from the expiration of preliminary permits or it may reserve such projects, the development of which, in its opinion, should be undertaken by the United States itself, and must give preference to states and municipalities, provided it deems the plans for same are equally adapted to utilize the water resources of the region.

The bill encourages the building of headwater storage reservoirs with a view to equalizing power production, preventing floods, and, in arid states, the use of the water for irrigation purposes on the lowlands after the water has been utilized for the generation of power. Further provision is made for the construction of locks in power dams on navigable streams with a view to extending navigation into the upper reaches of rivers. At the expiration of the fifty-year license the Government is given the option of purchasing the hydro-electric plants by paying the licensee his net investment, or it may issue a new license to the original licensee upon reasonable terms, or to a new licensee who shall in that event pay the original owner his net investment in the plant.

Under the terms of the bill the United States may take over and operate any water-power plant under license in time of war for the manufacture of explosives or for any purpose involving the safety of the country. The commission is authorized to make reasonable charges to cover the administration of the act and for the use of Government lands and property, and to absorb excessive profits that cannot be prevented by regulation. Licensees are placed under supervision of state public-service commissions as to rates to be charged to consumers for power and also as to regulation of service. Severe penalties are provided for non-compliance with the terms of the act.

Senator Jones of Washington, chairman of the Water Power Conference Committee, in commenting on the necessity for utilization of the energy contained in our falling waters, said:

Through failure of Congress to pass water-power laws under which money could be safely invested with prospect of a fair return, water powers now wasting have been held back for years from development in at least twenty-two states of the Union. I am informed that actual water-power developments are projected to be undertaken upon enactment of this bill having a total capacity of over 4,000,000 hp. The completion of these projects would open 4,000 miles of the upper reaches of our streams and rivers to navigation. In this way cheap water transportation would be afforded to districts, now sparsely settled and congestion of traffic relieved in thickly populated centers through the investment of private capital instead of through river and harbor appropriations by Congress. To encourage water-power development is a national duty, as it cannot be considered a local or a sectional question. It concerns every citizen of the United States, whether he dwells in the vicinity of water power, in a city or upon the plains of Iowa. The things used and consumed in every household are affected by the price at which those things can be produced under the cheapest of conditions. The cost of food, of clothing, of fertilizers, of explosives, and the convenience and facility of transportation would all be beneficially affected by the development of water power to the utmost extent of its usefulness. Coal shortage brings up again the folly of leaving unused the nation's vast resources of water power. The more these are developed the less desperate will be the fuel stringency.

The labor of one man is released for other uses every time 50 hydro-electric horsepower is developed, and every 150 horsepower developed releases one coal car for other duty. Other countries are awake to the necessity of developing their water powers. A French corporation with $250,000,000 of capital is seeking funds in America with a view to undertaking the utilization of 700,000 hp. now wasting in the River Rhone, and incidentally these improvements will, through canalization, make it possible for steamers to go up the Rhone to its connection with the River Rhine and thence down the Rhine to the North Sea. Italy, which is entirely without coal, is about to develop her water powers on an immense scale. Spain is offering liberal concessions looking to the development of the water powers now wasting in the Pyrenees and other mountain ranges. Sweden is paying $46 per ton for coal, which she has to obtain from England, and therefore is proceeding with the work of developing as near as possible the entire 5,000,000 water horsepower contained within her boundaries.

The United States must not fall behind European countries in conserving its coal deposits and in developing the vast amount of cheap electric energy contained in its wasting water powers.

Senate May Attempt to Block Bill

The conference report on the bill is said to face a filibuster in the Senate. The bill agreed upon by the conferees was accepted readily by the House of Representatives. Only thirty votes were cast against the acceptance of the measure as agreed upon in conference. The votes for the report numbered 259. In view of the overwhelming sentiment in the House in favor of the legislation as it now stands, it is not expected that the attempt to block it in the Senate will be successful.

The opposition is confined almost entirely to the definition of navigable waters; to the license provision; and to the matter of distributing power to municipalities.

An amendment adopted by the Senate provided that a licensee should not discriminate in the distribution of power among municipalities that can be efficiently served and supplied with power. The conferees struck out the amendment.

The license clause as it was changed in the conference reads as follows:

That the licensee shall pay to the United States reasonable annual charges in an amount to be fixed by the commission for the purpose of reimbursing the United States for the costs of the administration of this Act for recompensing it for the use, occupancy, and enjoyment of its lands or other property; and for the expropriation to the Government of excessive profits until the respective States shall make provision for preventing excessive profits or for the expropriation thereof to themselves, or until the period of amortization as herein provided is reached, and in fixing such charges the commission shall seek to avoid increasing the price to consumers of power by such charges, and charges for the expropriation of excessive profits may be adjusted from time to time by the commission as conditions may require.

Argentina, Brazil and Uruguay are jointly considering plans for the construction of an international power plant on the Uruguay River. It is believed that 2,500,000,000 kw.-hrs. per annum could be developed, or an amount equal to that of 3,000,000 tons of coal. This is the largest project contemplated at the present time.

A New Process for the Preparation of Lubricating Oils

In a paper presented before the National Petroleum Association in Pittsburgh on April 22, H. M. Wells and J. E. Southcombe, of London, England, advanced some rather revolutionary ideas on the theory of lubricating oils. Briefly, they claimed that by the use of their "Germ Process," oils of any desired characteristic and of increased lubricating powers might be prepared very cheaply.

Starting from the assumption that the lubricating qualities of an oil depend on its ability to wet the lubricated surfaces and therefore on its having a low surface tension, the authors investigated the interfacial tension of various oils against water and against mercury. They found from these experiments that this tension was much lower for compound oils and fatty oils than with straight mineral oils. After a great amount of investigation it appeared that this difference had some relation to the presence of small quantities of free fatty acids in the fatty or compound oils.

The next step was the preparation of oils consisting of 99 per cent mineral oil and 1 per cent free fatty acid which were found to have a low surface tension. The removal of the acids from fatty oils resulted in a high surface tension. The authors therefore decided that the lubrication value of an oil depends solely on the fatty acid content and that it should be possible to obtain the desired result by adding minute quantities of such acid to pure mineral oils.

Further experiments seemed to show that the addition of one class of acids induces a tendency to demulsify while another group of acids have a powerful emulsifying influence. It is therefore claimed that this process enables the producer to improve the lubricating properties of any mineral oil and at the same time obtain any desired emulsifying characteristics.

Another advantage claimed for "Germ Process" oil is that the amount of acid added is very small and is completely under control, whereas the use of compound oils introduces greater quantities of acid and in many cases hydrolysis results in the formation of large and uncontrollable amounts of acid. In other words, the new method furnishes just enough acid to make the lubricant "oily" and yet avoids any danger from corrosion.

Production of Electric Power and Consumption of Fuel

Data collected and compiled by the Committee on Division of Power Resources, United States Geological Survey, giving the production of electric power and consumption of fuel by public utilities plants in the United States for the month of January 1920, show an increase of approximately 15 per cent over the figure for the same month in 1919.

The figures for January are based on returns received from about 2,800 power plants of 100-kw. capacity or more engaged in public service, including central stations, electric railways and certain other plants which contribute to the public supply. The capacity of plants submitting reports of their operations is about 90 per cent of the capacity of all plants. The average daily production of electricity during January was 124,300,000 kw.-hr., 33 per cent of which was produced by water power.

The total production of electricity by public-utility plants during 1919 was 40.3 billion kilowatt-hours, 14.76 billion kilowatt-hours, or 36.6 per cent by water power and 25.54 billion kilowatt-hours or 63.4 per cent by fuels. The mean daily output for 1919 was 110.4 million kilowatt-hours. The mean daily production for January, 1919, was about 108 million kilowatt-hours; the mean for January, 1920, was 124.3 million kilowatt-hours. Note that the value for January, 1919, is estimated.

The fuel consumption for the year was as follows: 35 million short tons of coal, 11.05 million barrels of oil, and 21.7 million M cubic feet of gas. Converting the oil and gas consumed to coal, the equivalent coal for all fuels consumed during 1919 would be 38.347 million tons. With 25.540 billion kilowatt-hours produced by fuels in 1919, an average of practically three pounds of coal were required per kilowatt-hour of electricity produced. On this basis it would have required the consumption of 22,140,000 tons to have generated the kilowatt-hours produced by water power.

The estimated production of bituminous coal in 1919 is 458,063,000 short tons. The amount of coal used by electric public utility plants during 1919 was 7.6 per cent of the total produced.

PRODUCTION OF ELECTRIC POWER AND CONSUMPTION OF FUEL BY PUBLIC UTILITY POWER PLANTS IN THE UNITED STATES IN JANUARY, 1920

Thousands of Kilowatt-Hours Produced

State	By Water Power	By Fuels	State	By Water Power	By Fuels
Alabama	34,864	14,452	New Hampshire	4,322	5,166
Arizona	8,567	11,703	New Jersey	143	101,478
Arkansas	132	7,675	New Mexico	53	1,524
California	164,727	112,346	New York	227,033	382,108
Colorado	12,784	22,811	North Carolina	53,035	10,956
Connecticut	9,069	59,256	North Dakota	0	2,732
Delaware	0	6,807	Ohio	1,490	258,432
Dist. of Columbia	0	23,317	Oklahoma	217	17,287
Florida	965	10,890	Oregon	32,278	7,845
Georgia	43,816	10,696	Pennsylvania	45,324	329,392
Idaho	48,574	1,348	Rhode Island	355	37,114
Illinois	14,831	260,723	South Carolina	61,911	5,825
Indiana	2,943	91,788	South Dakota	477	3,327
Iowa	55,574	31,622	Tennessee	39,443	9,233
Kansas	1,741	35,986	Texas	74	55,397
Kentucky	0	23,409	Utah	13,932	0
Louisiana	0	18,126	Vermont	15,430	[illegible]36
Maine	23,381	1,577	Virginia	13,801	30,4[illegible]3
Maryland	284	31,261	Washington	103,981	4,480
Massachusetts	22,009	147,936	West Virginia	1,775	95,306
Michigan	51,799	138,180	Wisconsin	35,443	37,451
Minnesota	28,053	36,157	Wyoming	152	4,785
Mississippi	0	5,886			
Missouri	5,720	54,082	Total	1,274,401	2,580,198
Montana	89,574	562	Total, by water power and fuels		3,854,599
Nebraska	909	20,436			
Nevada	3,416	119			

The production of the electric power reported required the combustion of fuels in the quantities indicated in the following table:

State	Coal, Short Tons	Petroleum and Derivatives, Barrels	Natural Gas, Thousands of Cubic Feet
Alabama	29,211	1,255	0
Arizona	198	42,473	0
Arkansas	8,675	4,888	155,222
California	0	545,829	179,563
Colorado	43,345	110	0
Connecticut	77,311	4,374	10,236a
Delaware	9,139	0	0
Dist. of Columbia	23,237	0	0
Florida	2,425	54,753	0
Georgia	21,179	19,605	0
Idaho	150	10	0
Illinois	411,719	472	0
Indiana	205,453	254	2,025
Iowa	94,185	576	0
Kansas	39,710	91,428	7,6283
Kentucky	45,149	188	0
Louisiana	7,589	60,528	68,510
Maine	3,410	49	0
Maryland	39,552	0	1,500
Massachusetts	173,088	1,764	0
Michigan	180,542	139	0
Minnesota	64,060	620	0
Mississippi	14,028	11,841	0
Missouri	91,997	90,657	0
Montana	4,206	22	1,257
Nebraska	37,926	16,503	0
Nevada	367	842	0
New Hampshire	6,768	0	0
New Jersey	156,280	136	280a
New Mexico	4,733	866	0
New York	478,533	763	72,579
North Carolina	21,992	34	0
North Dakota	19,362	150	0
Ohio	401,066	697	161,065b
Oklahoma	8,353	71,940	275,926
Oregon	190	3,657	0
Pennsylvania	532,761	5	49,937
Rhode Island	35,196	7,560	0
South Carolina	11,779	0	0
South Dakota	7,729	1,936	0
Tennessee	22,729	25	0
Texas	58,707	211,222	48,227
Utah	3	0	0
Vermont	1,748	2,146	0
Virginia	43,212	132	0
Washington	10,396	15,612	0
West Virginia	106,230	50	227,900
Wisconsin	64,626	544	0
Wyoming	14,418	4,217	3,300
Total	3,634,662	1,270,872	1,333,810

(a) Artificial gas. (b) Includes 61,520 artificial gas.

Dr. Frederick G. Cottrell, the New Bureau of Mines Director

Dr. Frederick Gardner Cottrell, chief metallurgist, Bureau of Mines, was nominated on May 5, by President Wilson, to be the Director of the Bureau to succeed Dr. Van H. Manning, whose resignation will take effect on the first day of June.

Dr. Cottrell was born in Oakland, Cal., Jan. 10, 1877. Following the completion of his elementary education, he entered the University of California, from which he graduated in 1896 with the degree of B.S. In 1901 and 1902 he attended the University of Berlin and the University of Leipzig, receiving the degree of Ph.D. from the former.

From 1897 until 1900 Dr. Cottrell was a teacher of chemistry at the Oakland High School. From 1902 until 1906

DR. FREDERICK G. COTTRELL

he was an instructor of chemistry at the University of California. In 1906 he was appointed assistant professor, which position he retained until 1911, when he became consulting chemist. Dr. Cottrell became the chief chemist at the University in 1911, and in 1916 he resigned to accept the position of Chief Metallurgist at the United States Bureau of Mines.

He is an active member in a number of societies, including the American Chemical Society, Mining and Metallurgical Society of America, American Institute of Mining Engineers, American Electrochemical Society, Sigma Xi, Phi Beta Kappa and Alpha Xi Sigma.

Foreign Trade Convention

The entertainment committee of the Foreign Trade Convention which is to be held in San Francisco, May 12-15, has planned several entertainment features in addition to the regular business of the convention. On May 16 there will be a trip to the famous Yosemite Valley, and a three weeks' excursion has been arranged to the Hawaiian Islands, beginning May 19. All the principal points of interest will be visited. Plans are also being made for a three weeks' trip to points of interest in southwestern Alaska.

U. S. Bureau of Mines To Develop the Commercial Treatment of Lignite

At least $300,000 is to be expended under the direction of the Bureau of Mines in an effort to develop a commercial treatment for lignite. An experimental plant is to be erected at New Salem, N. D., which will be able to carbonize at least one hundred tons of raw lignite per day and provide for the complete recovery of the liquid and gaseous by-products and for the handling of the char, which it is planned to make into briquets. Of the sum to be expended, $100,000 is appropriated by Congress and $200,000 is to be furnished by a subsidiary of the Consolidated Lignite Collieries Co.

It was the intention at first to put the experimental plant in the Texas lignite field, but the development of oil and gas in that region interfered with the market for the gas which would be produced in connection with the treating of lignite.

The Bureau of Mines is to provide the plans and specifications for the carbonizing and briquetting plant; maintain active supervision and oversight over the construction and operation of the plant; and furnish the necessary technical assistance. The co-operating concern has agreed to provide a satisfactory site on its railway spur; to install certain portions of the equipment; to furnish the necessary lignite and other raw materials; to sell the products; to furnish the labor; and to conduct the necessary business activities of the undertaking.

American Association of Engineers Activities

The American Association of Engineers at the sixth annual convention at St. Louis, May 11, elected Leroy K. Sherman president.

The first convention of the Federal Government employees of the American Association of Engineers will be held in Washington on May 22. An advisory council will be elected for the department to determine a plan to form the country into districts for departmental work.

The first meeting of the North Dakota Chapter of the American Association of Engineers, held in Bismarck, April 23 and 24, was attended by the governor of the state and a number of prominent officials of the North Dakota State Highway Commission. The following officers were elected: Past president, E. J. Thomas, city engineer of Minot; president, J. E. Kaulfuss, assistant state highway engineer; first vice president, W. G. Black, city engineer of Mandan; second vice president, W. B. Stevenson, Fargo; and third, fourth, fifth and sixth vice presidents, L. T. Powers, William Barnock, Robert Jacobson and E. H. Morris, respectively.

A movement is on foot in Great Britain for the establishment of local sections by the Institution of Mechanical Engineers. As long ago as 1913 a meeting was held in Manchester with the object of finding facilities in the provinces for the institution of local branches, but the cost proved to be prohibitive, and it was decided instead to have the papers contributed to the Institution read in the provinces, and correspondents were appointed to arrange such meetings. Early in 1914 Sir T. Holland wrote to the *Manchester Guardian*, suggesting the organization of meeting places furnished with reference libraries, lecture rooms, etc., which would be supported conjointly by several societies. A committee on which the Institution of Mechanical Engineers was strongly represented was formed to discuss the matter, but the outbreak of the war put an end to the business. The institution is now adopting a by-law looking to activity along these lines.

That the mineral resources of India are rapidly being developed can be seen by statistics published in a recent issue of *Trans-Pacific*. In 1917 the total coal production of India was 17,642,798 tons and in 1918, 19,847,039 tons was produced. This shows an increase of almost 2,000,000 tons per annum.

Fuel and Oil Boiler Efficiency

Apart from the usual advantages attaching to the use of oil fuel such as decreased labor staff in the stokehold, facility of bunkering, etc., the fact should not be lost sight of that, owing to improved circulation of the water in the boiler, the efficiency of the boiler—that is to say, the percentage of the heat of the fuel which is actually given to the water in the boiler—is considerably increased. In the case of the ordinary marine boiler having cylindrical furnaces the fire bars are removed, and the burner is located at the center of the front of the furnace. It follows, therefore, that the furnace is heated all over its surface, and since the bottom part of the furnace is heated, it follows that the water in contact with this is also heated, and rising to the surface owing to decrease of density, is replaced by cooler water. It is well known that with coal-fired boilers the bottom part of the boiler often remains cold for many hours, and perhaps for a day or two, after the fires are lighted. With oil fuel this is prevented, and the resulting improvement in the circulation greatly increases the efficiency of the boiler and at the same time reduces the possibility of local strains. In a recent test of an ordinary cylindrical marine boiler with oil fuel an efficiency of over 83 per cent was obtained, a figure which it is safe to say would be impossible with ordinary coal hand-fired.—*Shipbuilding and Shipping Record.*

British companies are exploring and exploiting wherever possible in all countries where there may be any reasonable chance of finding petroleum, in pursuance of their policy of developing petrol, says *The Review*, the publication of the American Chamber of Commerce in France. The government is aiding navigation companies, new tank steamers are being built, and everywhere the old coal furnaces are being transformed for burning oil.

Personals

H. A. Price, formerly of Talladega, Ala., is now manager of the Texas Power and Light Co., Gainesville, Tex.

C. Wellington Koiner, consulting engineer of Pasadena, California, has been elected a member of the Committee of A. A. E. on Employment.

H. G. Harvey has resigned as commercial engineer of the Nassau Light and Power Co., Mineola, L. I., to take up similar work with the Pennsylvania Utilities Co., Easton, Pa.

E. L. Brandt of Detroit, until recently secretary of the Detroit Chapter of the American Association of Engineers, has been appointed assistant seiretary of the national headquarters.

Schuyler R. Ford, formerly chief gunner U. S. N., and an executive officer at the Sayville, N. Y., radio station, has been appointed engineer in charge of the high-power station at New Brunswick, N. J.

R. E. H. Pomeroy has resigned as smelter superintendent of the Nevada Consolidated Copper Co., and will associate himself with the Bonnot Co., Canton, Ohio, as chief engineer of the pulverized-coal department.

W. W. Marshall, superintendent of the Hydro-Electric Power Commission at Orangeville, Ontario, has been appointed town engineer of Orangeville. He will continue as superintendent of the commission.

Kenneth B. Millett, formerly assistant general engineer with the American Thread Co., New York, has resigned to join the organization of the Griscom-Russell Co., New York, in the capacity of development engineer.

Ernest C. Van Winkle, formerly assistant engineer, construction department, of the Chile Exploration Co., New York, has become connected with the Ford, Bacon & Davis Co., New York, in the capacity of valuation engineer.

William H. Dean has been appointed manager of the Chicago Chapter of the American Association of Engineers to fill the vacancy created by the resignation of E. J. Burke, who has entered the newly organized engineering firm of York, Regan and Burke.

Society Affairs

The Association of Iron and Steel Electrical Engineers will hold its annual outing, Saturday, June 5. The program will be announced later.

The American Society of Mechanical Engineers, Philadelphia Section, will hold its annual dinner at the Engineers' Club, Philadelphia, May 25.

The Atlanta Section of the A. S. M. E. will meet in Atlanta, Ga., May 25. J. T. Wikle will present a paper on "The Manufacture of Cotton Goods."

The Association of Iron and Steel Electrical Engineers will hold its fourteenth annual convention in the Hotel Pennsylvania, New York City, Sept. 20-24.

The A. I. & S. E. E., Pittsburgh Section, will have for its subject at the October meeting, "Sheets and Tin Mills," by A. B. Holcomb, general industrial engineer, Messrs. Crippen & Funk, Youngstown, Ohio.

The Pittsburgh Section of the A. I. & S. E. E., will meet in Pittsburgh on June 12. The subject of the meeting will be "Current Limit Reactance," by R. H. Keil, power engineer of the Jones & Laughlin Steel Co.

The American Boiler Manufacturers Association will hold its thirty-second annual convention at the French Lick Springs Hotel, French Lick, Ind., on May 31, June 1 and 2. Matters of importance to the entire boiler-manufacturing industry will be brought before the convention.

The National Association of Stationary Engineers, Iowa State Association, will hold its annual state convention in Dubuque, June 16-18. Besides the business sessions the entertainment committee has arranged for suitable entertainment, including a steamboat excursion on the afternoon of June 16.

The Mid-Continent Section of the A. S. M. E. will meet in Tulsa, Okla., May 28, 29. A delegation will leave the annual spring meeting, which will be held in St. Louis, just before the dates mentioned, to take part in the Tulsa meeting. The visitors will be shown through the zinc mines at Miami and taken on a visit to the oil fields.

The Association of Iron and Steel Electrical Engineers, Pittsburgh Section, will meet on May 22, at Warren, Ohio. An inspection trip will be made to the plant of the Trumbull Steel Co. during the afternoon, and in the evening a paper describing the electrical installations at the plant of the Trumbull Steel Co. will be presented by William F. Rese, electrical engineer.

Engineering Foundation, under the supervision of Charles F. Rand, chairman, is seeking additions to the endowment fund which will swell the total to at least a million dollars in the near future. Through the generous gifts of Ambrose Swasey, Engineering Foundation has, since 1915, been able to actively co-operate in research work with the National Research Council. Mr. Swasey's gifts total $300,000. Additions to the endowment fund will be used to further the research work of the Foundation Societies.

The American Society for Testing Materials will hold its twenty-third annual meeting at Asbury Park, N. J., June 22-25. The headquarters of the meeting will be the New Monterey Hotel. A number of special features, including an informal dance, smoker and a golf tournament, have been arranged. Each of the seven sessions will be devoted to one of the following subjects: Non-ferrous metals, wrought and malleable iron and corrosion, reports of administrative committees, steel, testing apparatus, preservative coatings and lubricants, and miscellaneous committee reports and papers.

The American Institute of Electrical Engineers will hold its annual business meeting in the auditorium of the Engineering Societies Building, New York City, Friday, May 21. The annual report of Directors will be presented, and the report of the tellers' committee announcing the officers elected for the coming year will be read. On this occasion the Edison Medal will be presented to W. L. R. Emmet for inventions and developments of electrical apparatus and prime movers. The medal, its origin and significance, will be described by Carl Herring, chairman, Edison Medal Committee; H. W. Buck will relate the achievements of Mr. Emmet, and Calvert Townley, president of the A. I. E. E., will make the presentation.

Miscellaneous News

The National Screw Thread Commission of the Bureau of Standards, Washington, D. C., created for the purpose of ascertaining and establishing standards for screw threads for use of the various branches of the Federal Government and for the use of manufacturers, has completed a report of its investigations to date. Certain systems of threads are recommended in the report, together with information, data and specifications pertaining to the manufacture of the threads recommended. The commission is composed of S. W. Stratton, chairman; James Hartness, vice chairman; F. O. Wells, E. H. Ehrman, H. T. Herr, F. C. Peck, O. B. Zimmerman, E. J. Marquart, S. M. Robinson and H. W. Bearce. A copy of the report can be obtained from the National Screw Thread Commission, Bureau of Standards, Washington, D. C.

Business Items

Walter N. Polakov & Co., Inc., announce the new location of their offices at 33 West 42nd St., New York City. The offices of the firm were formerly located on Nassau Street.

Nickerson and Collins Co., publishers of trade publications and technical books, announces the removal of the publication office of *Ice and Refrigeration* from South Dearborn St., to 5707 West Lake St., Chicago.

The Dominion Bridge Co. announces the formation of a new company to be known as the Dominion Engineering Works, Ltd. The Bridge company will control the new project, which will engage in the manufacture of waterwheels and other hydraulic machinery.

Trade Catalogs

The Edw. R. Ladew Co., Inc., Glen Cove, New York, has just completed publication of a new fifty-six page catalog, descriptive of the various leather products which they manufacture, including leather belting and packing. A copy can be had on request.

The Lagonda Manufacturing Co., of Springfield, Ohio, manufacturer of steam specialties has ready for distribution a new eight page bulletin, No. T-3: The booklet goes into the subject of cutting tubes out and illustrates the Lagonda Tube cutter as used in different types of boilers and shows in detail how the cutters work. A copy can be obtained on request.

The De Laval Steam Turbine Co., Trenton, New Jersey, has ready for distribution a new 8-page pamphlet entitled "De Laval Centrifugal Pumps for Sugar House Service." Complete information and descriptions of the various types of pumps are contained in the pamphlet, together with illustrations, including centrifugal pumps for handling raw juice, syrup, boiler-feed water and injection water, and for maceration and hydraulic pressure service, etc. Motor and turbine-driven pumps, both single-stage and multi-stage, are illustrated and described. A copy of the pamphlet will be sent on request.

COAL PRICES

Prices of steam coals both anthracite and bituminous, f.o.b. mines, unless otherwise stated, are as follows:

ALANTIC SEABOARD

Anthracite—Coals supplying New York, Philadelphia and Boston:

	Mine
Pea	5.30
Buckwheat	$3.40@$3.75
Rice	2.75@ 3.25
Barley	2.25@ 2.50
Boiler	2.50

Bituminous—Steam sizes supplying New York, Philadelphia and Boston:

Latrobe	$4.25@$4.50
Connellsville coal	4.25@ 4.50
Cambrias and Somersets	4.35@ 5.15
Clearfields	4.00@ 4.75
Pocahontas	6.50@ 7.00
New River	6.50@ 7.00

BUFFALO

Bituminous—Prices f.o.b. Buffalo:

Pittsburgh slack	$6.00
Mine-run	6.25
Lump	6.00@ 6.50
Youghiogheny	6.75

CLEVELAND

Bituminous—Prices f.o.b. mines:

No. 6 slack	$3.25@$4.00
No. 8 slack	4.25
No. 8—¾-in.	3.50@ 4.00
No. 6 mine-run	3.50@ 4.00
No. 8 mine-run	4.50
Pocahontas—Mine-run	3.25@ 4.00

ST. LOUIS

Anthracite—Probably not more than 20 per cent of the demand in this market can be supplied. Prices effective Apr. 1 were as follows:

	Williamson and Franklin Counties	Mt. Olive and Staunton	Standard
Mine-run	$2.65@2.80	$3.00@3 50	$3.00@3.50
Screenings	2.50@2 65	2.50@3.00	2.50@3.00
Lump		3.75@4.50	3.75@4.50

Williamson-Franklin rate to St. Louis is $1.10; Other rates 95c.

CHICAGO

Bituminous—Prices f.o.b. mines:

		Freight rate Chicago
Illinois		
Southern Illinois		
Franklin, Saline and Williamson Counties		
Mine-Run	$3.00@$3.10	$1.55
Screenings	2.60@ 2.75	1.55
Central Illinois		
Springfield District		
Mine-Run	$2.75@$3.00	$1.32
Screenings	2.50@ 2.60	1.32
Northern Illinois		
Mine-Run	$3.50@$3.75	$1.24
Screenings	3.00@ 3.25	1.24
Indiana		
Clinton and Linton		
Fourth Vein		
Mine-Run	$2.75@$2.90	$1.27
Screenings	2.50@ 2.65	1.27
Knox County Field		
Fifth Vein		
Mine-Run	$2.75@$2.90	$1.37
Screenings	2.50@ 2.60	1.37
Brazil Block	$4.25@$4.50	$1.27

New Construction

PROPOSED WORK

Mass., Boston—MacNaughton & Robinson, Archts., 101 Tremont St., will soon award the contract for the construction of an 8 story office building on State St., for Henderson & Ross, 148 State St. A steam heating system will be installed in same. Total estimated cost, $300,000.

Mass., Wakefield—The Bd. Educ. will receive bids until May 24 for the construction of a 3 story, 210 x 250 ft. high school on Main St. A steam heating system will be installed in same. Total estimated cost, $450,000. F. I. Cooper Corp., 33 Cornhill St., Boston, Archt.

Conn., Bridgeport — Oskar Krokstedt, Archt., 925 Main St., will soon award the contract for the construction of a 2 story, 130 x 147 ft. garage addition on State St. A steam heating system will be installed in same. Total estimated cost, $150,000. Owner's name withheld.

N. Y., Brooklyn—The Brooklyn Edison Co., 360 Pearl St., plans to build a 1 story sub-station on Grand St. Estimated cost, $100,000. William Whitehill, 55 Duane St., New York City, Archt. and Engr.

N. Y., Buffalo—The Beaver Board Co., Military Rd., is in the market for one or two 230 lb. 125 hp., steam pressure boilers, (used).

N. Y., Buffalo—Philip Begy, 296 Baynes St., is in the market for a boiler and pump for the Hoffman Pressing Machine.

N. Y., Buffalo—The Stewart Motor Corp. 413 East Delavan Ave., is in the market for an air compressor outfit with tank.

N. Y., Dunkirk—The City voted $22,000 bonds to install a heating plant in the Jones General Hospital here. Work includes the construction of a 43 x 50 ft. boiler house, etc.

N. Y., New York—The Bd. Educ., 500 Park Ave., will receive bids until May 22 for the construction of a 4 story, 92 x 194 ft. school building on 182nd St. and Bathgate Ave. A steam heating system will be installed in same. Total estimated cost, $420,000. C. B. J. Snyder, Municipal Bldg., Archt. and Engr.

N. Y., New York — Maurice Courland, Archt. and Engr., 47 West 34th St., plans to build an 18 story factory. A steam heating system will be installed in same. Total estimated cost, $1,200,000. Owner's name withheld.

N. Y., New York — J. A. Hearns Sons, Inc., 20 West 14th St., is having plans prepared for the construction of a 2 story, 50 x 135 ft. boiler plant at 26 West 13th St. A steam heating system will be installed in same. Total estimated cost, $135,000. Densmore & Le Clear, 88 Broad St., Boston, Mass., Archts. and Engrs.

N. J., Burlington—G. A. Allinson, Supt. of the Waterworks, will soon award the contract for the installation of high service pumping machinery for water plant in one 3,000,000 gal unit. W. H. Boardman, 426 Walnut St., Philadelphia, Pa., Engr.

N. J., Camden—The Victor Talking Machine, Front and Cooper Sts., plans to build a 4 story warehouse on State, York, Pine and Front Sts. A steam heating system will be installed in same. Total estimated cost, $1,000,000. Ballenger & Perrot, 329 South Broad St., Philadelphia, Pa., Archts. and Engrs.

N. J., Vineland—F. H. Bent, 142 West State St., Trenton, plans to construct a 1 story, 50 x 75 ft. storehouse and refrigerator plant, here. Machinery will be installed in same. Total estimated cost, $30,000.

Pa. Lock Haven—The city will receive bids until June 1 for the installation of an electric booster pump, etc. in connection with the proposed waterworks improvements. Total estimated cost, $80,000. D. E. Oberheim, City Clk H. P. Shoemaker, Engr.

Pa., Philadelphia — Horace Trumbauer, Archt., Land Title Bldg., will soon award the contract for the construction of a 12 story, 60 x 90 ft. office building on 17th and Cherry Sts. and the Parkway, for the Sun Oil Co., Finance Bldg. A steam heating system will be installed in same.

Md., Baltimore—The Bd. of Awards will receive bids until May 26 for the construction of a 3 story, 150 x 200 ft. public school on Poplar Grove St. and Lafayette Ave. A steam heating system will be installed in same. Total estimated cost, $250,000. H. Adams, Calvert Bldg., Engr. E. H. Glidden, 1210 American Bldg., Archt.

Md., Baltimore—The Capitol Theatre Co., 719 Fidelity Bldg., is having plans prepared for the construction of a 1 story, 120 x 200 ft. theatre at 1318-1522 West Baltimore St. A steam heating furnace will be installed in same. Total estimated cost, $250,000. Wyatt & Nolting, 1012 Keyser Bldg., Archts.

N. C., Winston-Salem — The Memorial Hospital is receiving bids for the construction of a 2, 3, 4 and 5 story hospital building. A steam heating system will be installed in same. Total estimated cost, $800,000. Charles R. Keen, Bailey Bldg., Philadelphia, Pa., Archt.

Ga., Metter—The town voted $35,000 bonds for the construction of a waterworks system and pumping plant. L. C. Anderson, Mayor.

O., Akron—J. A. Wetmore, Supervising Archt., Treasury Dept., Wash., D. C., will receive bids until June 1 for the installation of a boiler

O., Cincinnati—A. S. Taft, 316 Pike St., is receiving bids for the construction of a 10 story, 40 x 60 ft. addition to power building, on 8th and Sycamore Sts. Harry Hake, Telephone Bldg., Archt.

O., Cleveland—The Bd. Educ., 6th St. and Rockwell Ave., received bids for the installation of a heating and ventilating system in the proposed school on East 92nd St. and Parkgate Ave., from the J. Roemer Heating Co., 1808 East 22nd St., $159,491. Noted March 16.

O., Cleveland—The Hausheer Sloan Co., Sloan Bldg., has purchased a site on East 18th St. and Euclid Ave. and plans to build a 12 or 16 story, 130 x 190 ft. commercial building on same. A steam heating system will be installed in same Total estimated cost, $500,000. T. E. Sloan, Treas.

O., Cleveland — The Hebrew Orthodox Home for the Aged, Inc., c/o S. H. White, Archt., 1032 Schofield Bldg., will receive bids until June 2 for the construction of a 2 story, 130 x 150 ft. home for the infirm at 736 Lakeview Rd. A steam heating system will be installed in same. Total estimated cost, $250,000.

O., Cleveland Heights (Warrensville P. O.)—The Bd. Educ. plans an election May 22 to vote on $2,400,000 bonds to construct 3 story school buildings on Fairfax, Coventry and Noble Rds. Steam heating systems will be installed in same.

O., Columbus—The city rejected all bids for the construction of a pumping station for the Shepard main line trunk sewer and for the installation of equipment. Work will be readvertised.

O., East Cleveland—The Bd. Educ. will receive bids until May 24 for furnishing and installing power equipment, including a brick stack, two 300 hp. boilers and piping for the boiler house. Total estimated cost, $25,000. C. Ammerman, Williamson Bldg., Cleveland, Clk. Bishop & Babcock Co., East 55th St., Cleveland, Engrs.

O., East Cleveland—The Park Congregational Church, 11111 Ashbury Ave., Cleveland, plans to build a 1 story church on Lockwood and Euclid Aves. A steam heating system will be installed in same. Total estimated cost, $300,000. Rev. Talbot, Pastor. I. W. Corbusier, Lennox Bldg., Cleveland, Archt.

O., Norwalk—The city plans to install a 1,000 kw. generator in the power station.

O., Wooster — The town voted $60,000 bonds to construct a pumping station, extension to water mains, etc. H. H. Miller Serv. Dir., C. O. Williamson, Engr.

Mich., Detroit—Richard H. Marr, Archt., 435 Woodward Ave., will receive bids until May 25 for the construction of a 2 story, 100 x 130 ft. publishing house on Howard and 6th Sts., for the Sprague Publishing Co., 150 Lafayette Blvd. A steam heating system will be installed in same. Total estimated cost, $110,000.

Mich., Detroit—J. I. Weinberg, Archt., 401-403 Congress Bldg., will soon award the contract for the construction of a 4 story, 100 x 180 ft. factory on Grand Blvd. and Moran St., for Jacob Shevitz, 195 Ferry Park. A steam heating system will be installed in same. Total estimated cost, $125,000.

Mich., Ironwood — Alvord & Burdick, Engrs., 8 South Dearborn St., Chicago, will soon award the contract for the construction of additions to ground water supply and installation of electrical equipment, etc., for the city. Total estimated cost, $300,000.

Ill., Chicago—The Illinois Malleable Iron Co., 1801 Diversey Blvd., is having plans prepared for the construction of a 1 story building. Two 300 hp. boilers, air compressor, stokers, and ash handling equipment will be installed in same. Cahill & Douglas, 217 West Water St., Milwaukee, Wis., Engrs.

Ill., Chicago—George C. Nimmons & Co., Archt., 104 South Michigan Ave., will soon award the contract for the construction of a 1 and 2 story, 135 x 400 ft. factory for the manufacture of metal products, on Taylor and Campbell Sts., for the Harrington & King Perforating Co., 614 North Union Ave. A steam heating system will be installed in same. Total estimated cost, $275,000.

Ill., Chicago—William G. Rosenberger, 214 West Ontario St., is having plans prepared for the construction of a 3 story, 80 x 100 ft. cigar box factory on La Salle and Ontario Sts. A steam heating system will be installed in same. Total estimated cost, $250,000. Alfred S. Alschuler, 28 East Jackson St., Archt.

Minn., Minneapolis—The Bd. Educ. received bids for the installation of a heating and ventilating system in the proposed schools on (A) Harriet Ave. and 38th St., from Bjorman Bros., 712 South 10th St., $46,938; (B) 31st St. and 29th Ave., $44,942; (C) 22nd Ave., S. E. and 40th St., $49,808, and from H Kelly & Co., 925 Plymouth Bldg., (A) $49,200, (B) $47,140, (C) $49,349. Noted March 30.

Kan., Newton—Bethany College plans to build a heating plant, dormitory, etc., here. Total estimated cost, $500,000. Lorenz A. Schmidt & Co., 121 North Market St., Wichita, Archt.

Wyo., Casper — Garbutt, Weddner & Sweeney, Archts., Oil Exch. Bldg., will receive bids until June 1 for the construction of a 6 story banking and office building on Wolcott and East 2nd Sts., for the Wyoming Natl. Bank. A steam heating system will be installed in same. Total estimated cost, $375,000.

Mo., St. Louis—The Joseph Greenspon's Sons Iron & Steel Co., 3130 Hall St., is in the market for 2 cross compound condensing, 1,000 hp., with direct connected generators, 440 volt, a.c. Corliss engines.

Tex., Clyde—The town plans to furnish, and install pump and power equipment, etc.

Tex, Fort Bliss—The Construction Division, Quartermaster Corp., will receive bids until May 31 for the construction of hospital building including a central heating plant, etc.

Utah, Farmington—The Davis Co. Irrigation Dept., plans to build a hydro-electric plant to develop 10,000 hp., dam, etc. Total estimated cost, $2,331,000. A. T. Parker, 585 28th St., Ogden, Engr.

Cal., Los Angeles—Mary Spires, c/o J. C. Austin, Archt., 1121 Baker-Detwiler Bldg., is having plans prepared for the construction of a 2 story, 130 x 150 ft. stores and dance hall on 6th and Olive Sts. A plenum heating system will be installed in same. Total estimated cost, $235,000.

Cal., Mare Island—The Bureau of Yards & Docks, Navy Dept., Wash., D. C., plans to install a heating system in the structural shop, here. Estimated cost, $60,000.

Cal., San Diego—The Bureau of Yards & Docks, Navy Dept., Wash., D. C., plans to construct a refrigeration plant at the Marine Corps base, here. Estimated cost, $50,000.

N. F., St. Pierre—The Minister of Pub. Works will soon award the contract for furnishing one 500 hp. boiler, four 600 hp. boilers, one 25 hp. dredge, etc.

Que., Montreal—The Montreal Harbor Comn. will receive bids until May 21 for the construction or an 8 story, 110 x 400 ft. cold storage warehouse. Refrigeration machinery will be installed in same. Total estimated cost, $1,000,000. F. W. Cowrie, 731 Sherbrooke, W., Engr.

Que., Montreal — The Montreal Harbor Comrs., will soon receive bids for the construction of a 50 x 150 ft. power house. Estimated cost, $150,000. F. W. Cowrie, 731 Sherbrooke, W., Engr.

Ont. London—The Bd. Educ. plans to build a high school on Waterloo St. A steam heating and ventilating system will be installed in same. Total estimated cost, $500,000. A. A. Langford, Chn. L. E. Carrothers, Archt.

Ont., Newmarket—E. A. James Co. Ltd., will receive bids until May 25 for the construction of an outfall sewer, disposal and an activated sludge plant, pumphouse, and compressor house including compressors, for the city. Total estimated cost, $65,000. Noted Jan. 27.

Ont., Peterboro—The city is having plans prepared for a sewage disposal station and pumping plant. Air lifts will probably be installed in same. Total estimated cost, $50,000. R. H. Parsons, City Hall, Engr.

Ont., Port Credit—The village plans to construct a waterworks system. Plans include an intake into Lake Ontario, pumphouse, pumps, mechanical filters, chlorination, elevated tank, etc. Total estimated cost, $80,000. E. A. James Co. Ltd., 36 Toronto St., Toronto, Engr.

Ont., Windsor—The city of Windsor and other municipalities plan to build a 4 story civic building. A steam heating system will be installed in same. Total estimated cost, $1,500,000. F. L. Howell, 32 Caron St., Alderman.

CONTRACTS AWARDED

Mass., Malden (Boston P. O.)—The Malden Savings Bank has awarded the contract for the construction of a 2 story, 52 x 93 ft. bank building on Malden Sq., to the Gascoigne & Lindenthal Co., 43 Tremont St., Boston. Contract for installing heating system will be sublet. Total estimated cost $150,000.

Conn., New Haven—The United Illuminating Co., 84 Temple St., has awarded the contract for the construction of a 1 story, 70 x 75 ft. turbine room extension, to C. W. Murdock, Inc., 185 Church St. and a boiler room extension to Westcott & Mapes, Inc., 207 Orange St.

N. Y., Brooklyn—The Brooklyn Jewish Recreation Center, 881 Eastern Parkway, will build a 3 story, 125 x 187 ft. recreation center on Eastern Parkway. A steam heating system will be installed in same. Total estimated cost, $750,000. Work will be done by day labor.

N. Y., Long Island City — The Municipal Studio Corp., 41st St. and Bway., New York City, will build a 200 x 600 ft. studio on Pierce, 8th, 9th and Washington Sts. A steam heating system will be installed in same. Total estimated cost, $2,000,000. Work will be done by day labor.

N. Y., New York—The Chatham & Phoenix Natl. Bank, 149 Bowery St., has awarded the contract for the construction of a 60 x 100 ft. bank building at 118-122 Bowery St., to the Tidewater Bldg. Co., 16 East 33rd St. A steam heating system will be installed in same. Total estimated cost, $500,000.

N. Y., New York—Severlace & Van Allen, Archts. and Engrs., 111 East 40th St., will build a 16 story, 60 x 126 ft. office building on 41st St. and Lexington Ave. A steam heating system will be installed in same. Total estimated cost, $500,000. Work will be done by day labor.

N. Y., New York—The Fifth Ave Hospital, 67th St. and Park Ave., has awarded the contract for the construction of a 9 story hospital at 105th-106th Sts. and 5th Ave., to Marcus Eidlitz, 30 East 41st St. A steam heating system will be installed in same. Total estimated cost, $2,500,000.

N. J., Newark—John Campbell & Co., 75 Hudson St., New York City, has awarded the contract for the construction of a 1 story, 91 x 233 ft. dye building and power house on Plum Point Lane, to James Jewks & Sons, 676 Montgomery St., Jersey City. A steam heating system will be installed in same.

Md., Baltimore—The Bailey Bldg. Co., Munsey Bldg., has awarded the contract for the construction of a 25 story, 75 x 100 ft. office building on Lexington and Calvert Aves., to Norton, Bird & Whitman, 616 Munsey Bldg. Oil burning boilers for heating system will be installed in same. Total estimated cost, $2,500,000.

O., Cleveland—The B. F. Keith Theatre Co., 1564 B'way, New York City, has awarded the contract for the construction of a 1 story theatre on East 105th St. and Euclid Ave., to Lundoff Bicknell Co., 5716 Euclid Ave. A steam heating system will be installed in same. Total estimated cost, $600,000.

O., Marion—The Bd. Educ. has awarded the contract for the installation of a heating system, to the Jones-Kinn Eng. Co., 414 North Main St., Lima, at $11,900.

O., Shaker Heights (Cleveland P. O.)—The Shaker Heights Hospital, c/o M. Loeser, 800 Natl. City Bldg., Cleveland, has awarded the contract for the construction of a 3 story hospital to the Crowell Little Constr. Co., East 57th St. and Euclid Ave. Cleveland. A steam heating system will be installed in same. Total estimated cost, $250,000.

Mich., Battle Creek—The Clark Equipment Co., Buchanan, has awarded the contract for the construction of a 1 story, 100 x 450 ft. axle plant on Springfield Place and the Michigan Central Ry., to Bates & Rogers, Chicago. Steam heating equipment will be installed in same.

Mich., Detroit—F. L. & Angus Smith, 1905 Dime Bank Bldg., has awarded the contract for the construction of a 2 story, 93 x 220 ft. garage on Woodward Ave. near Parsons St., to Otto Misch, Chamber of Commerce Bldg. A steam heating system will be installed in same. Total estimated cost, $135,000.

Mich., Detroit—Morgan & Wright, Bellevue Ave., has awarded the contract for the construction of a 9 story 326 x 456 ft. tire factory and warehouse on Jefferson Ave., to the Stone & Webster Corp., Book Bldg. A steam heating system and electric power will be supplied from detached central plant under contract. Total estimated cost, $4,000,000.

Mich., Lansing—The city has awarded the contract for the construction of a 250,000 gal. reservoir and installation of air lift and an electric air compressor, to the Ellington Miller Co., 417 South Dearborn St., Chicago. Total estimated cost, $75,000. Noted April 6.

Ill., Chicago—The Transo Envelope Co., 3550 Kimball Ave., has awarded the contract for the construction of a 1 story, 210 x 350 ft. envelope factory on Kimball Ave. near Addison St., to O. W. Rosenthal Co., 80 East Jackson St. A steam heating system will be installed in same. Total estimated cost, $275,000.

Wis., Manitowoc—The Spindler Co. has awarded the contract for the construction of a 2 story, 60 x 130 ft. artifical ice plant to Ben Herman & Son.

Wis., Milwaukee—Geuder, Paeschke & Frey, 152 St. Paul Ave. and 15th St., have awarded the contract for the construction of a 1 story, 75 x 100 ft. power plant on St. Paul Ave., to the Meredith Bros., 1043 Kinnickinnic Ave., installation of boilers, to the Babcock & Wilcox Co., 567 Lake St., Chicago, stokers, to the Combustion Eng. Co., 11 Bway., New York City. Noted March 3.

Wis., Milwaukee—The School Bd., 10th and Prairie Sts., has awarded the contract for the installation of a heating system including boilers in the proposed boiler house on Garfield Ave., to the Industrial Heating & Eng. Co., 143 Oneida St., at $28,100. Noted May 3.

Ia., Ottumwa — The Y. M. C. A. has awarded the contract for the construction of a 4 story, 60 x 99 ft. Y. M. C. A. building, to the Wells Bros. Constr. Co., 411 South Union St., Chicago. A steam heating system will be installed in same. Total estimated cost, $350,000.

Minn., Buhl—The Bd. Educ. has awarded the contract for the construction of a 3 story, 148 x 275 ft. addition to the high school to the Natl. Constr. Co., Virginia St. A steam heating and mechanical ventilating system will be installed in same. Total estimated cost, $250,000.

Minn., Hibbing — The Bd. Educ. has awarded the contract for the construction of a 3 story, 174 x 417 ft. high school on 3rd St., to the Jacobson Bros., Columbia Bldg., Duluth, at $579,500. A steam heating and mechanical ventilating system will be installed in same.

Tex., Eastland — C. M. Connelley will build an 8 story hotel to include a power plant. Total estimated cost, $300,000. Work will be done by day labor.

Col., Boulder—The University of Colorado will build a 3 story, 50 x 275 ft. college building. A steam heating system will be installed in same. Work will be done by day labor.

Que., Montreal — The Administrative Comn., City Hall, has awarded the contract for the manufacture and installation of one 30,000,000 gal. pump for the low level pumping station, to the Fraser Chalmers Co. at $50,250; another electric driven pump to E. Laurie & Co., 243 Bleury St., at $45,200; transformers and commutators necessary for the operation of the pumps, to the Canadian General Electric Co., 162 St. Antoine St., at $48,640. Noted March 9.

Ont., Sault Ste. Marie—The Algoma Steel Corp. has awarded the contract for installing a.c. motor equipment in connection with the $7,000,000 structural steel plant extension, here, to the Canadian General Electric Co., King and Simcoe Sts., Toronto, and for d.c. motor equipment and main motor drives to the Canadian Westinghouse Co., Hamilton.

POWER

Volume 51 New York, May 25, 1920 *Number* 21

Motivation

THE latest new-fangled name given to a thing we have all known for years is "motivation." A man is "motivated" if he has a definite, inspiring motive which keeps him at the work he is doing. Professor W. K. Hatt of Purdue University, in the February number of *Engineering Education*, remarks that the soldiers of the American Army were motivated the day before the Armistice was signed, but were not motivated the day after.

We all know, whether we use this new name or not, that a man must be motivated in order to get ahead around a power plant and that the principal kind of motivation is liking for the work. If a man is working simply because he needs a job, he will drift aimlessly. If, on the other hand, he is working at a given job because he likes it and has selected this job instead of others, because he likes it better than the others, he is motivated in such a way as to make his success certain.

A man who likes the atmosphere of a power plant and knows he does, will distance all his fellows. It has been said, "The world stands aside for a man who knows where he is going." Success is sure if a man knows he is going to be a specialist of some kind or other in power-plant work, or in any other kind of work for that matter, and keeps his mind fixed on this goal in spite of discouragement, because he likes this work and knows that he likes it.

The moral of this little sermon is to examine yourself in your job and see if you are really motivated by interest in your work. If not, you should try to get a job where you are so motivated, regardless of any other consideration whatsoever.

Contributed by Dr. S. A. MOSS

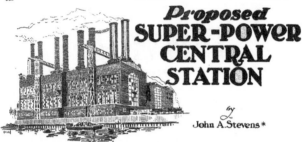

Proposed SUPER-POWER CENTRAL STATION

By

John A. Stevens*

Relates to a proposed design of super-power station built with multiples of 30,000-kw. units up to a total station capacity of, say, 100,000, 300,000 or even 500,000 kw. Each unit consists of four 1,716-hp. boilers, normal rating, served by one stack centrally located, and supplying steam to one 30,000-kw. turbine. All boiler houses are separated from their turbine rooms by an explosion gap, and the turbine rooms are separated by explosion walls. All station refinements are incorporated in this design of station, and a main operating room is provided at the top of the station, together with sleeping quarters for the men, baths and showers, telephones, searchlight, alarm whistle and wireless apparatus.

IN THESE days, when one hears so much regarding the great central stations that are being constructed and developed in Europe and in this country, it is worth while to consider the super-power station illustrated in the supplement to this issue of *Power*.

There is a tendency more and more toward the super-station, and a point has been reached in steam and electrical engineering where the super-station is of vital importance to the success of all great industrial centers, in that substantially one-half of the average coal consumption will be used and a probable reduction of men and payroll per kilowatt output will result as compared with some types of large power stations. The indications are that in the comparatively near future the more important railroads of this country will be electrified, for it seems adverse to steam and electrical engineering that each train should be supplied with a complete individual power plant such as, with a few exceptions, is the present locomotive practice.

The super-station should be used to relay and reinforce hydro-electric developments where possible, and to supply practically all the energy to manufacturing and other plants, outside of the power that can be made as a byproduct to manufacturing and heating steam requirements, which are usually supplied by non-condensing engines and non-condensing turbines of various types.

Furthermore, an infinite number of electrical uses are being created annually, which means a larger production of electricity than was dreamed of even ten years ago, provided the cost of electricity can be kept down to a point that will enable its more general use. In order to carry power-plant engineering to the last degree of economy, some features are suggested in the design shown by the supplement, which may not be adopted at the present time, but sooner or later will have to be taken into consideration in order to carry out the economical program suggested.

Now that the cost of coal has advanced, and is sure to advance more, it becomes of the greatest importance to engineers, as well as to financiers, to look to the most advantageous arrangement of a station of, say, 100,000 to 300,000 or even 500,000 kw. The supplement illustrated is the suggestion of one group of engineers for such a station, and although the principle involved may not be adopted, this design may suggest to others features that will be of benefit to this country's economies, with special reference to the minimum use of coal and men.

In the working out of this power-plant design, the engineering ideas of many men are embodied in certain features, and these ideas have been combined and worked into a completed harmonious whole.

Thirty-Thousand-Kilowatt Units Recommended

The idea of specifying a particular size of units (30,000 kw.) is that this size is within the zone of maximum economy in steam-turbine engineering practice, and further, it is about the limit of size that can be supplied with four sections or units of boilers served with one central stack. Four boiler units are installed so that one may be out for inspection, repair or cleaning at any time, and the station still be capable of developing its full electrical output.

With the present advance of electrical transmission systems and the use of high voltages, the opportunity to locate such a super-station either at or near the coal mines or on large and navigable rivers or at tidewater,

*Consulting Engineer, Lowell, Mass.

is available. In making hypothetical station layouts, the engineer should consider all possible kinds of fuel in the suggested arrangement of the stations.

The particular station illustrated is a tidewater proposition with railroad facilities, together with a very large water supply, which is necessary.

Rail facilities are almost necessary for successful building and operation of a station of this kind and class, as it is desirable that full-size railroad cars and large locomotives pass through or under the station for the handling of the enormous bulk of coal to the station and ashes and soot from the station, to say nothing of the economical handling and the installation of new equipment, such as boilers, condensers and turbines, or the loading of old equipment whenever it is necessary to ship by rail to the builders for overhauling.

Railroad cars of the largest capacity should be available under the main cranes of the station for handling with the greatest facility the heavy parts of turbines, boilers, condensers, etc. If a station of this size and kind is located at tidewater, where railroads are not available, large floating cranes may be used in lieu of the railroad service. It is proposed to use a locomotive for shifting the largest-sized railroad cars under the ashpits, and also under the soot pipes that handle the soot from the stack, as illustrated.

It is doubtful if engineers know to what extent power stations may increase in size, and it is suggested that no project of this kind should be started where the ultimate possible development, if so desired, could not be made at least 250,000 kw. There is no object in putting money into power-plant investments and then to do practically the same thing over again. In other words, money shou'd be used with a breadth of character sufficient to meet any demand. As Americans, we are always building ourselves into some location where we cannot expand as our business expands, and the abandonment of existing equipment is a very expensive procedure.

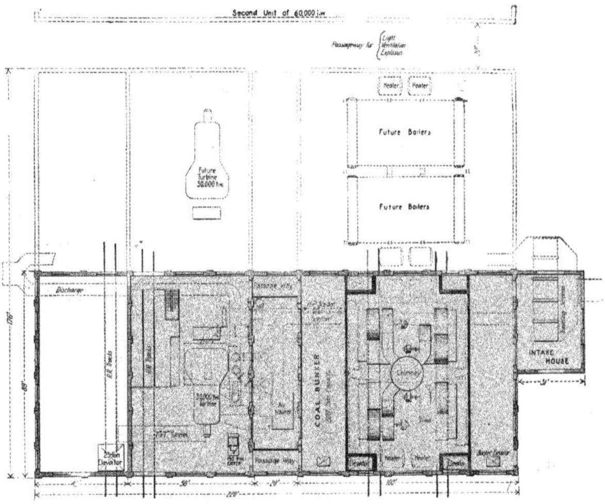

FIG. 1. PLAN VIEW OF A THIRTY THOUSAND-KILOWATT INSTALLATION
Showing four 1,716-hp. Stevens-Pratt Boilers and one 30,000-kw.turbine. Dotted lines indicate duplicate installation, making a total of 60,000 kw. for a complete unit

General Arrangement of Station

In the supplement 1,716-hp., normal rating, Stevens-Pratt boilers are shown. They are arranged in units of four, with one central stack, which is from 300 to 350 ft. high, in order to discharge the escaping gases and incidental dust which may pass the soot separators (not shown) at high altitudes. Stirling Stevens-Pratt boilers may be advantageously used when desired. The four boilers indicated contain about 31 miles of tubing.

The suggested development requires that the boilers be installed in practically an independent building, as illustrated, with an outdoor passageway between it and the turbine room and a connecting covered bridge (not shown on the drawing) between the boiler-house operating room, which is shown at the top of the drawing, and the switchhouse operating rooms shown at the top of the switchhouse.

Although boilers are now so designed in this country that it is not expected that where plants are properly installed, there will be many more major explosions, minor explosions and derangements, such as tube failures, and steam-pipe and fitting failures are to be expected for some time to come. It is desirable, therefore, to provide, primarily, for such derangements, to say nothing of guarding against the results of electrical short circuits in generators, and also as a matter of human protection, if not for financial reasons.

In leaving the explosion gap (so called) between the boiler room and the turbine room, a facility of light and ventilation is made possible in both rooms; it also localizes the damage to the plant if an accident occurs.

The switchhouse is designed with fireproof explosion wall with no opening next to the turbine room, and to insure continuity of service, double busbars are provided so that one set may be out of order and undergoing repair and inspection while the other carries the load.

Another safety provision is incorporated in dividing each group of turbine rooms themselves by protecting walls which reach to the crane height. In the instance of the bursting of steam mains or fittings, if the operators are protected by a wall 25 or 30 ft. high, there is some chance for them to escape, unless the explosion is very disastrous. Some engineers, however, might prefer to divide each group of turbine rooms into two separate compartments, using two cranes. This is a matter of choice for individual designers. Still further, each 60,000-kw. capacity power house should be separated from the next unit of the same size by a space 20 to 40 ft. in the clear. The electrical ends, however, can be connected at the transformer station, whether it is outside the station or inside the switchhouses. This colossal boiler plant is illustrated as resting on piles, somewhat proportionate as to the loads to be supported; for usually, in harbors and along river banks in the course of the past ages, a good deal of silt has been washed in and along the river beds, and portions of bays and harbors have receded and left a great deal of soft material. Foundations so located must be built on piles, preferably with a series of piles at the panel or pressure joints, a slab, designed to support all loads, wind loads as well as gravity loads, six to eight feet thick, reinforced top and bottom with steel, and the columns should be anchored down, because the pull at times and with different loads imposed on the station and stack may be upward. Sleeping quarters for the men are arranged on the basement floor and if desired in other designs, air heaters can be installed where the sleeping quarters are located in this particular design.

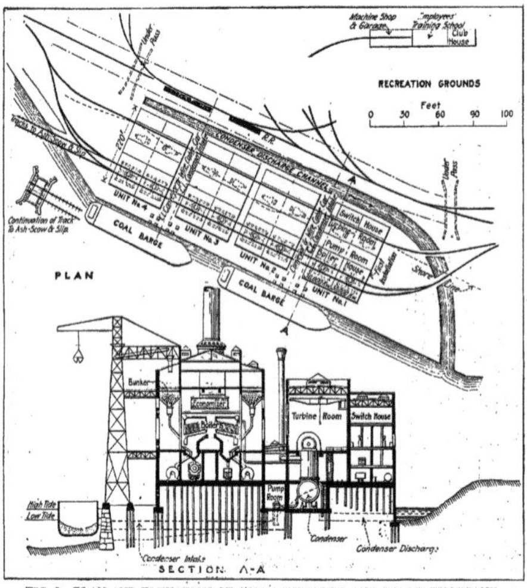

FIG. 2. PLAN AND ELEVATION OF A TWO HUNDRED AND FORTY THOUSAND-KILOWATT SUPER STATION

Note the explosion gaps between each 60,000-kw. unit.

As to heating the air which supports combustion, it is not advocated that a very high temperature be maintained, but a moderate degree of heat shows a decrease in the amount of coal consumed per kilowatt on the outgoing leads of the station, the steam being extracted from the high-pressure stages to supply these heaters when used.

The basement also contains the large turbine—and motor-driven forced-draft fan for the stokers, as well as all other auxiliaries in the station. This combination is used so that the heat balance may be correct without wasting hot water or steam, a condition with which all engineers are familiar. Adequate railroad scales, coal bunkers, coal-crushing rolls and coal-handling apparatus are also shown, together with a small machine and pipe-threading shop, and the necessary storerooms located above them for rough material such as spare tubes, tile, asbestos, etc.

On the firing floors stands are provided for the main motor-actuated stop valves. These may be controlled by remote motor controls from the switchhouse or from any part of the plant. This is an important part of the boiler-plant design, and there is also a permanent platform at or near this valve to permit of its inspection, overhanging or repair. A master valve at or near all boiler plants of this size, with motor-operated connection, is desirable, so that the valve may be closed from any point, in the event of serious disaster in the turbine room. The main piping is hung on springs, so that it may expand and contract at will without straining the flanges, fittings or piping.

The main steam pipe is designed in what amounts to a loop around the boilers, with double valves on each boiler lead or take-off. The valve rods extend to the operating room. The valving is so arranged that any boiler section may be taken out of service at pleasure, and the loop arrangement provides for the expansion and contraction in any direction without the straining of the pipe, fittings or valves. A master branch runs from the boilers with the shortest possible length of piping to the main turbines and their auxiliaries. Most of the bends are at 90-deg. angle, but some of them are 45 deg. in order that the minimum-sized piping may be used with the highest velocity without scoring or erosion of the piping, valve seats, fittings, etc. Turbine-boiler-feed pumps are used, located in the fireroom for the greatest accessibility. The fans, pumps and economizers for this entire installation should be individual for each main unit, with cross-connections on the feed lines so that feed water may always be obtained quickly if one or even two units should fail. The area shown between the operating floor and the floor supporting the induced-draft fans is really the high-temperature area in this plant. This area is to be double-cased at the boilers, so that the ingoing air may be drawn in and over the casings, and through the passageways and hollow bridge walls to the forced-draft fan inlets for the partial warming of the air used to support combustion in the furnaces.

The design of the bridge walls is such that they are really rooms or corridors, and because of this feature the engineers may make their furnaces practically any length they desire and still have accessibility to practically any part of the furnace through the doors in the bridge walls—not a new idea but a good one.

Weighing larries for coal distribution to any of the furnaces are provided for, as well as a suitable recording system. There is also a lighting system, even to the individual lights on the coal larry and for moving up and down over the tube areas for inspection and cleaning.

On the floor or platform below the induced-draft fans will be seen the economizers of a type that it is believed is fast coming to the front. They are a combination of cast-iron sections and steel sections through which the water is pumped in series.

On a level with the coal-handling system is shown the induced-draft fan room. These fans are driven, as are the forced-draft fans, by steam turbines as a relay to the motors and for facility in keeping the heat balance correct.

In the operating room there is incorporated every known device for testing boilers, permanently installed and connected, so that in a clean, bright, well-lighted room with the roof of the station of wire-insertion plate glass, the master boiler operator can make hourly boiler tests, of the output of the boilers, if he so desires. Periscopes are provided for observation of the furnaces and periscopes for reading the water gage-glass levels as well.

On this level a remote-control whistle valve and a platform for taking care of it are shown. Whistle valves are hard to keep tight, and the platform is provided for the handling of this valve. In all cities or localities where such a station is built there should be a whistle 12 in. in diameter by 4 ft. long, which could be heard for miles, for signaling all kinds of service, such as militia, rioting and fire, and to give other signals, even if the station is located in what is more or less of a rural district.

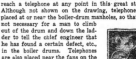

Naturally, the operating room contains the necessary desks, and there is also provision for wireless apparatus. One large company in this country is sending 60 per cent of its messages by wireless, the aërial being connected to the stacks. If this innovation has proved so successful to a commercial organization, why not adopt the idea? On the penthouses above the elevators will be seen searchlights, which are desirable for the protection of a great station of this kind. All large stations should be provided with elevators for transporting the operators from one floor or gallery to another, and also to carry up and down station equipment such as large valves, auxiliary turbines and motors, etc. This station is so designed that, in so far as possible, no planks or tackles will be necessary. This avoids the use of a large number of men who are non-productive except in emergency cases.

Another progressive feature is that telephones are provided at all important points or where needed. A man should not have to walk over twenty-five feet to reach a telephone at any point in this great station. Although not shown on the drawing, telephones are placed at or near the boiler-drum manholes, so that it is not necessary for a man to climb out of the drum and down the ladder to tell the chief engineer that he has found a certain defect, etc., in the boiler drums. Telephones are also placed near the fans on the larry gallery. These fans are to supply outside cool air to the hot sections or parts of the boiler when they are under inspection and repair. Naturally, a station of this size should be fitted with large air compressors to supply air for cleaning the boiler tubes and for doing all sorts of mechanical and other work around the plant.

The intake house should be designed with cranes, be steam-heated and have proper flushing connections so that at certain seasons of the year, when the condensing water has grown rather foul with leaves, etc., it may be practically automatically cleaned.

Many other features will be observed in the design of the boiler plant, if carefully studied. For instance, note the pipes running above the umbrella on the stack and at the base of the umbrella; these may be used to wash off the dust that may collect on the plate-glass roof. Underneath the plate-glass roof will be observed a shade on rollers, so that on a hot summer's day the operating room may be shaded and made comfortable. In the stack casing (similar to a transatlantic liner's casing) will be observed a reproduction of the boiler drums, from which the water levels may be read by the flashing of a light in addition to the periscope leading down to the water columns on the boilers and also the water gages installed on all these great units. In other words, there are three ways in which the operator can observe the water levels in the boilers.

Under the platform or roof over the space between the main boiler plant and the turbine room may be placed the main circulating pumps, both turbine- and motor-driven, for the condensers, air compressors, condensate and dry-air vacuum pumps, and the usual pump-room machinery.

Beneath the floor can be placed, if so desired, the intake tunnel, for which the main walls of the boiler plant and the turbine room make the side walls. That is to say, there are no heavy loads imposed on this area between the two main buildings, the boiler plant and the turbine plant.

STANDARDIZATION TABLE OF THE STEVENS-PRATT BOILER

Possible Output in Kilowatts, allowing 30-lb. Water per Boiler Horsepower per Hour and 12-lb. Steam per Kilowatt-Hour, including Auxiliaries in Turbine Plant

350 Lb. W. P. + 250 Deg. F. Superheat

Size of Sections in Boiler Hp. Rated on Sq.Ft.	No. of Sections in Operation	Per Cent Rating of Sections 100 Kw.	150 Kw.	175 Kw.	200 Kw.	250 Kw.	300 Kw.	400 Kw.
500	4	5,000	7,500	8,750	10,000	12,500	15,000	20,000
	3	3,750	5,625	6,562	7,500	9,375	11,250	15,000
600	4	6,000	9,000	10,500	12,000	15,000	18,000	24,000
	3	4,500	6,750	7,875	9,000	11,250	13,500	18 000
750	4	7 500	11,250	13,125	15,000	18,750	22,500	30,000
	3	5,625	8,437	9,844	11,250	14,062	16,875	22,500
1,000	4	10,000	15,000	17,500	20,000	25,000	30,000	40,000
	3	7,500	11,250	13,125	15,000	18,750	22,500	30,000
1,250	4	12,500	18,750	21,875	25,000	31,250	37,500	50,000
	3	9,375	14,063	16,410	18,750	23,438	28,125	37,500
1,350	4	13,500	20,250	23,625	27,000	33,750	40,500	54,000
	3	10,125	15,187	17,719	20,250	25,312	30,375	40,500
1,450	4	14,500	21,750	25,375	29,000	36,250	43,500	58,000
	3	10.875	16,312	19,031	21,750	27,187	32,625	43,500
1,550	4	15,500	23,250	27,125	31,000	38,750	46 500	62,000
	3	11,625	17,437	20,344	23,250	29,062	34,875	46,500
1,716	4	17,160	25,740	30,030	34,320	42,900	51,480	68,640
	3	12,870	19,305	22,520	25,740	32,175	38,610	51,480

The main turbine room is necessarily a very simple affair and is clearly illustrated on the drawing to bring out the main points.

Where high river water is to be contended with, due to high spring rises or flood conditions, or where the tide levels and the railroad levels are markedly different, the main condenser may be placed in a pit and the eduction pipe, or exhaust pipe, between the turbine and the condenser made any reasonable length to give the proper elevation and have the usual siphoning effect through the condensers; the main idea being that the turbine-room levels and the boiler-room levels should be the same. In designing the switchhouse, provision should be made so that a full-sized railroad car should be able to pass through one section. The elevators and elevator beams should have a capacity great enough to lift any equipment necessary in the oil switchroom, or the reactance or cable rooms.

As a precaution against damage to the generating units a check valve, made of heavy plate, is placed at the warm-air discharge to the stoker fan, so that in the event of a minor explosion in the tube area of the boilers, the dust and soot would not be blown into the main turbine generator.

Several arrangements may be used for handling the coal supply. That shown is a large coal tower located on the dock. There is also an outdoor transformer station, and the ash-scow slip, similar to a ferryboat slip for handling regular railroad cars on a scow which may be towed to sea for dumping. This development was conceived in 1917, but the idea lay dormant during the period of the war. As previously mentioned, the ideas and suggestions of many have been interwoven into this design, showing the art up to the present time, and it is desired to give credit to anyone who sees some of his special designs embodied in this illustration, which is presented to the public for the good of the service.

Handy Soldering Iron

By M. A. Saller

It is frequently necessary to do soldering in out-of-the-way places where a straight-handled soldering iron would prove to be an awkward tool. I have made a soldering iron, as shown in the sketch, the handle rod being provided with eyes and a thumbscrew bolt and nut at the middle, so that the iron can be adjusted to any desired angle, or held in the normal straightway position.

ADJUSTABLE HANDLE FOR SOLDERING IRON

Domestic exports of electrical machinery and appliances from the United States during March were as follows: Batteries, valued at $690,866; carbons, $128,-531; dynamos and generators, $464,178; fans, $115,-069; incandescent lamps (carbon filament), $19,986; (metal filament), $427,348; magnetos, spark plugs, etc., $440,240; motors, $1,247,806; switches and accessories, $356,631; transformers, $347,195; miscellaneous, $2,-966,103. The report which is issued by the Department of Commerce, Bureau of Foreign and Domestic Commerce, Washington, D. C., shows Canada well in the lead over all other countries in the value of her imports from the United States.

Vertical-Shaft Waterwheel-Driven Alternators — Lubricating System and Rotor

Typical Systems Used To Lubricate the Various Bearings of a Vertical Alternator; Influence of Flywheel Effect on Speed Variation and Dimensions of Rotor; Safe Speeds and Different Rotor Constructions

By S. H. MORTENSEN
Electrical Engineer, Allis-Chalmers Manufacturing Company

PROPER lubrication of the different bearings on a vertical alternator is of great importance, and many different systems, some elaborate, are in service. In Fig. 1 is shown the oiling arrangement for a vertical machine with a thrust bearing and two guide bearings on the generator. As the arrows indicate, the oil is pumped through a system of piping to the top of the machine by a small geared oil pump mounted in the lower oil pan. At the point marked *A* the oil flow is divided into two paths. The main path leads to the top guide bearing, through this bearing, through the lower bearing and back to the main oil pan. The second path leads through the thrust bearing. The amount of oil flowing through this pipe is small, just sufficient to compensate for evaporation, leakage, etc. The heat generated in the thrust bearing is carried off by means of a cooling coil located in the oil bath surrounding the bearing, as indicated in the figure. A sufficient amount of cool water is circulated through the coil to limit the oil temperature to a safe operating value. No special means are provided for cooling the oil from the guide bearing other than circulating it through the piping system.

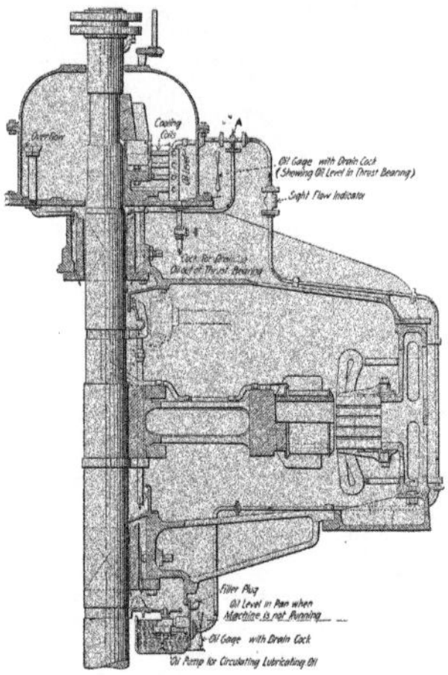

FIG. 1. LUBRICATING SYSTEM FOR GENERATOR WITH THRUST BEARING AND TWO GUIDE BEARINGS

Oil gages with drain cocks are provided for the thrust-bearing housing and lower oil pan, and a sight-flow indicator is mounted in a prominent position. The hand on this indicator takes different positions for different oil velocities and by means of an electric circuit may be arranged to ring a signal bell if the oil flow for some reason is stopped. A drain is provided in the thrust bearing to prevent flooding. As the oil pump is called upon to overcome only the weight of the oil column between the upper and lower oil levels, it can be of simple design, such as a small plunger pump or a centrifugal pump of the geared type. In some installations this pump is mounted outside the oil pan and operated from the mainshaft by a chain drive.

FIG. 2. OIL-PAN, COOLING COIL AND STRAINER

On some large vertical generators a somewhat more elaborate oiling system is used. The oil for the thrust bearing is circulated by the oil pump through the piping system to the thrust bearing and from there back again to the main oil pan, where it is cooled by a coil with water circulation. After it is cooled, the oil passes through an oil strainer back to the pump, thus completing the cycle. From the guide bearings the oil completes a similar circuit. Gate valves and oil-flow indicators, together with sight glasses, provide means for checking and adjusting the oil flow. Fig. 2 shows the oil pan, cooling coil and oil strainer.

The method of lubrication varies with the capacity of the generators and with the size of the power station. Frequently, large stations have elaborate oiling systems with oil filters and separate motor-driven oil pumps. However, the tendency at the present time is to simplify the system, and in one of the most recent installations the oil in the thrust bearing is not circulated, but kept cool by cooling coils. The oil lost by leakage or evaporation is then replaced at regular intervals. The only function of the oil pump is, in these cases, limited to circulating the comparatively small amount of oil required for the lubrication of the guide bearings. Large machines frequently have glass windows in the thrust-bearing cover and electric lamps mounted inside the cover, thereby providing ample opportunity for inspecting the bearing parts. It is important to check the temperature of the different bearings to guard against accidents, therefore recording thermometers are usually supplied with the more important machines.

Usually designated as WR^2, the flywheel effect required for the proper speed variation of hydraulic units frequently determines the diameter and also the weight of the direct-connected alternator's rotor. As it thus has a direct bearing upon the design of the machine, a short discussion of the different factors involved may be of interest. As the hydraulic turbine develops a uniform torque throughout the revolution, the only

FIG. 3. ROTOR SPIDER OF ONE OF THE BIG NIAGARA FALLS UNITS

function of its rotor flywheel effect is to restrain speed variations during the periods of load adjustment.

The simplest case of speed regulation is found in open flume installations, where the energy of the moving water column can be neglected. In such installations the speed regulation depends upon four factors: P, the load variation of the unit in horsepower; T, the time in seconds required by the hydraulic governor for the opening or closing of the turbine gate; WR^2, the inertia of the rotating masses expressed as a product of the rotor weight and the square of its radius of gyration; and S, the rotor speed in revolutions per minute.

Let V equal the ratio of the increase of speed to the normal speed, then the relation between the factors involved is found in the equation,

$$V = \frac{A \times P \times T}{WR^2 \times S^2}$$

In this equation A is a constant. The speed change V increases with the amount of load thrown off or

FIG. 4. HIGH-SPEED ROTOR COMPLETE

on, and it also increases as the time required for opening or closing the turbine gate increases. On the other hand, V decreases with the increase in the rotor WR^2 and also as the square of its revolving speed. The equation is based on the assumption that the energy stored in the water column leading to and away from the turbine is without influence upon its speed regulation. This is not the case.

In installations with long penstocks and draft tubes, energy is liberated or absorbed by the acceleration or retardation of the water column. For such installation either surge tanks or standpipes should be provided, or the governor should be adjusted to give slow gate action, as otherwise dangerous hydraulic surges will be set up under load fluctuations.

It is apparent that no hard and fast rule can be laid down for the WR^2 required to give successful speed regulation. For the purpose of comparison it is expedient to reduce the alternator WR^2 to a value K corresponding to 1 hp. at 1 r.p.m., which gives

$$K = \frac{WR^2 \times S^2}{P}$$

Values for K thus derived vary over a wide range. In installations with long pipe lines and large load fluctuations, a value of $K = 14,000,000$ may be necessary to insure satisfactory operation. On the other hand, systems with steady loads and short pipe lines may give good regulation with $K = 2,000,000$. The flywheel effect of the turbine runner is generally negligible, and this makes it necessary to embody the required WR^2 in the alternator rotor. If the WR^2 of a proposed machine falls short of the momentum required by the waterwheel builder, it may be increased within limits, either by making the diameter of the alternator rotor larger or by increasing its weight. In high-speed machines the rotor diameter is generally as large as the mechanical strength of its material permits, and in that case additional weight must be added to the rotor spider.

An increase in the WR^2 invariably means an increase in the cost of the machine, together with a decrease in its efficiency, due to additional windage and friction losses of the rotor. For that reason the WR^2 specified by the waterwheel builder should be kept as low as permissible to give satisfactory electrical operation. It should be borne in mind that speed fluctuations affect only the frequency and voltage of an electrical system; also that a momentary fluctuation of the frequency is not objectionable.

FIG. 5. ROTOR DESIGNED FOR EXTREMELY HIGH PERIPHERAL VELOCITY

On systems where close voltage regulation is required, it is as a rule necessary to provide automatic voltage regulators, not alone on account of the speed fluctuations on the turbine, but also on account of the voltage regulation of the alternator.

It should be remembered that the WR^2 of all the synchronous machines, such as motors, motor-generator sets, frequency-changer sets, rotary converters, etc., connected to the power system, is just as effective during speed fluctuations as if their momentum was embodied directly in the rotor of the alternator. As soon as the speed of the turbine unit changes, a similar change takes place in the speed of all the connected synchronous machines, whose rotating part then either absorbs energy from, or gives energy back, to the power system.

The runaway speed of a hydraulic turbine determines the maximum stresses imposed upon the direct-connected alternator's rotor. For different installations this speed varies between 1½ and 2 times the normal generator speed. Generally, the rotor is proportioned to limit its stresses at runaway speed to less than one-half the elastic limit of the respective material. This gives an ample factor of safety to take care of the hidden defects in the material and practically eliminates mechanical failures due to overspeeds.

Figs. 3 to 5 show different rotor construction for medium- and high-speed vertical machines. Fig. 3 is the rotor for one of the 32,500-kva. generators installed recently at Niagara Falls. The sizes of the rotor poles made it expedient to divide them into symmetrical halves, each half joined to the spider by an individual dovetail. Fig. 4 shows a high-speed rotor for a 14,200-kva. vertical generator installed in a plant of the San Joaquin Light and Power Co. A rotor field pole and the partly assembled rotor for a machine with extremely high peripheral velocity are shown in Fig. 5, the method of attaching the polepieces to the rotor being clearly indicated.

The Effect of Loose Nuts on Connecting-Rod Bolts

In a discussion of failures of connecting-rod bolts in internal-combustion engines in "Notes on Gas and Oil Engine Accidents," by the National Boiler and General Insurance Co., Ltd., of Manchester, England, attention is called to the danger from running with the nuts slacked off. These bolts are commonly fitted with double nuts, so that when adjusted and tightened up, the nuts may be locked together; they are further fitted with a split pin for security. When first sent out from the shop, this split pin is hard up against the outer nut, but after the brasses have been adjusted once or twice, there will be a clearance between the pin and

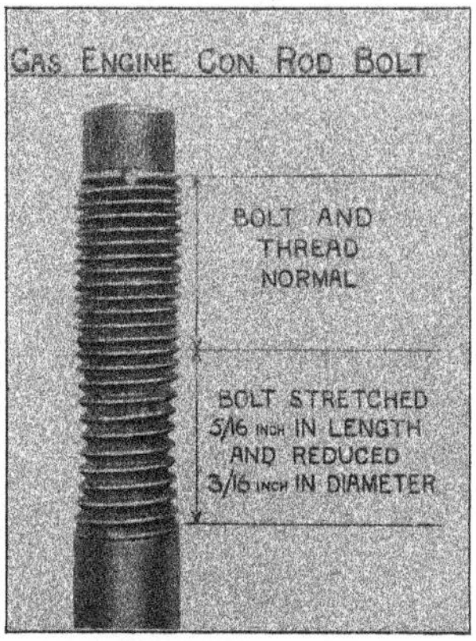

EFFECT OF RUNNING GAS-ENGINE CONNECTING-ROD BOLT WITH SLACK NUT

the nut. This clearance should be filled up with washers, so as to prevent the nuts from slacking back.

The illustration is reproduced from a photograph of a connecting-rod bolt that had been severely stressed through being allowed to work with the nuts slack. The bolt was taken out of the engine just in time, as it was stressed considerably beyond the elastic limit of the material and was evidently on the point of failure.

Refrigeration Study Course—X. Automatic Refrigeration

By H. J. Macintire

THERE are many applications of refrigeration which require intermittent work, but especially is this true of motor-driven small-sized cold-storage plants. Being motor-driven, usually of the induction or sometimes the synchronous type, it is not practical to provide variable speeds to adjust the speed to the load. Also it is desirable where possible to have a machine that is capable of temperature control, one that will be semi-independent of constant care or that will operate by night or day subject to occasional supervision on the part of the engineer.

This desire for an automatic refrigerating machine applies more particularly to the small machine, where the plant does not warrant the cost of additional operators. For very small plants the use of "holdover" tanks, or the brine system with large storage tank and continuous brine circulation, has for some time met with considerable success. However, the holdover tank requires considerable additional headroom, because a tank containing brine and one-third of the total amount of expansion piping must be arranged at the top of the cold-storage room. The submerged piping is expected to provide the necessary refrigeration for cooling the brine to an amount sufficient to hold the temperature of the room during the nighttime.

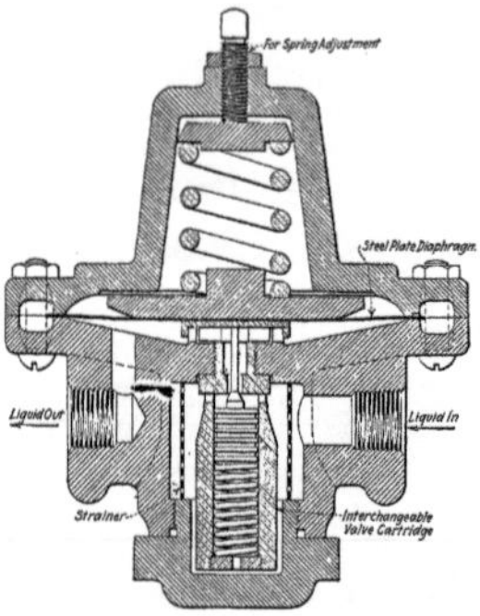

FIG. 1. AUTOMATIC EXPANSION VALVE

Likewise, the brine-storage tank and circulating pump take space and considerable extra power. The compressor, which operates but about one-third of the total time, has to be of sufficient size to carry the day load and store up sufficient refrigeration for the night as well. Also, the operation is not so efficient, for the pumping action of the brine pump has a heating effect, which alone would heat the brine, even if no refrigeration were being done. For example, a one-horsepower centrifugal brine pump would heat the brine at about the same rate as one-fifth of a ton of refrigeration; namely, 42.4 heat units per minute would cool it.

The methods mentioned—"holdover" tanks and brine-storage tanks—are objectionable on account of the space occupied and the poor efficiency of operation. The compressor and high-pressure side have to be of extra size so as to provide the required tonnage in the part time of operation, including, as it does, not only the day load, but the night load as well. And therefore we have the automatic refrigerating system, which may operate without constant attention the twenty-four hours of the day, or off and on with the demand which is thermally controlled by a thermostat located in the cold-storage room. As it may operate full time, it may be of only two-thirds or one-half the size required of the other designs, which are arranged for part-time operation only. Before going into a detailed description of the totally automatic refrigerating machine, it will be advisable to review the semi-control systems.

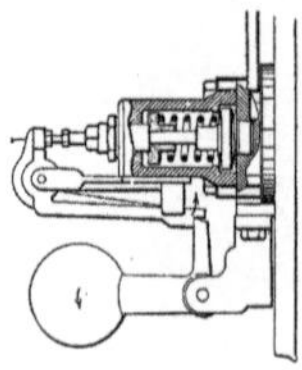

FIG. 2. NO-PRESSURE TYPE

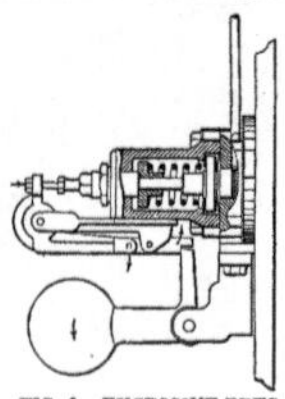

FIG. 3. EXCESSIVE-PRESSURE TYPE

Control valves operated by springs and gas pressure

Semi-Control Systems

The advent of the small inclosed type of compressor has been a decided advance in the art of refrigeration. It is splash-lubricated, and while the crank case remains supplied with oil, it is reasonably free from the dangers incidental to the lack of lubrication. Likewise, the main compressor bearings are ring-oiled, a method that is reasonably safe as long as the ring or chain revolves freely. There remains, then, the necessity for control of the water supply, the excess condenser-pressure control and some regulation of the expansion valve (if this last seems justified). In this connection attention is again called to the fact that if the suction pressure is too high, no refrigeration will be obtained at all, as the temperature of boiling of the ammonia at high pressures may be above the temperature desired. Likewise, if the suction pressure is too low the cooling coils will conduct heat better, but they may become choked with a low-density gas and the capacity of the compressor will drop to one-half or one-third of the normal tonnage in consequence of this so-called high *specific volume* of the ammonia.

The Suction-Control Valve

The suction-control expansion valve is in principle like the well-known back-pressure valve except that the latter deals only with gas on both sides of the valve while the expansion valve must have liquid on one side and liquid and some 5 to 10 per cent of gas on the other. There is another difference, too, which must be given consideration; namely, that a little liquid ammonia will go a great way in producing refrigeration and the opening must be very small and the valve must be designed for close adjustment. With a head pressure of 150 lb. on one side and 15 lb. on the other it is clear that a large amount of liquid could pass through a very small opening, and a pound of liquid ammonia per minute will provide the cooling effect required of about two and a half tons of refrigeration.

A good example of such an expansion valve is shown in Fig. 1. This is a design using the diaphragm method of control. Liquid ammonia comes in on the right-hand side, as shown, and passes through the filter and then to the conical reducing valve, which has an extended stem that presses on a pusher plate. The pusher plate rests

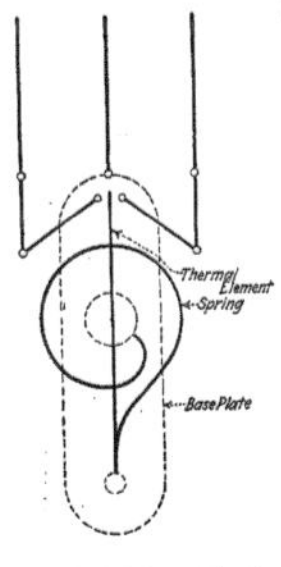

FIG. 4. THERMOSTAT CONTROLLING ELECTRIC CIRCUIT

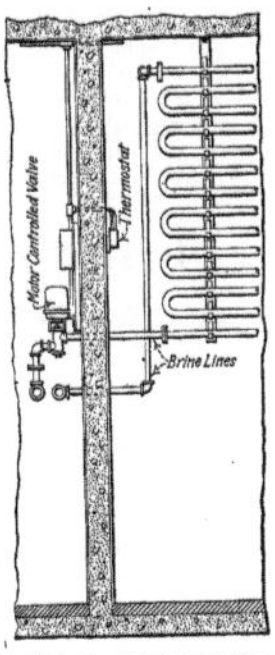

FIG. 5. THERMOSTAT ON BRINE SYSTEM

on the diaphragm, which is also under the pressure of a control spring. The under side of the diaphragm has suction pressure, and the upper has a spring pressure which is regulated by the engineer for the operating pressure desired. Such a device should give close regulation and should be capable of relieving all thought of the expansion-pipe boiling temperature from the mind of the operator.

The Excess-Pressure Release

Unless the discharge valve on the compressor is closed or in some most inconceivable way the discharge line is obstructed, then a sudden increase of head pressure is caused, without exception, by the condensing water used on the condenser. If this condensing water is not turned on or is turned off, or fails to function properly, then the head pressure will rise. As the ammonia compressor is usually made with small clearances, the pressure that may be obtained under these conditions may be excessive and decidedly dangerous.

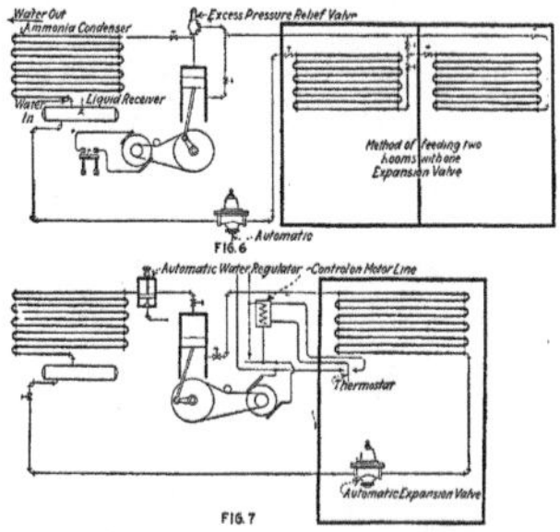

FIG. 6. SEMI-AUTOMATIC REFRIGERATING SYSTEM
FIG. 7. COMPLETE AUTOMATIC REFRIGERATING SYSTEM SHOWING MOTORS, WATER AND SUCTION-PRESSURE CONTROL

The result is that some head-pressure relief must be provided in the automatic system we are now discussing.

The most elementary of these relief valves are those operating against a spring or a diaphragm and so adjusted as to give movement to a small plunger. This plunger acts on a trip which releases a trigger to which is attached a weighted cord that in falling pulls open the main electric switch (See Fig. 3). Other simple devices are on similar lines of design. The relief valve, identical to the steam-boiler safety valve, may be placed on the discharge header and piped either to the atmosphere or into the suction header. In the latter case the relief valve becomes simply a bypass valve, but loaded to a certain discharge pressure. Then again there may be a solenoid device that operates the main switch on the electric circuit. This solenoid is excited by a current that is made when the arm of the pressure gage turns to a certain desired pressure, say 225 lb., when contact is made and the electromagnet stops the motor. Likewise, a device may be used, embodying the solenoid principle with that of the safety-valve design. The lift of the valve is made to close the electric circuit, the solenoid is excited and the main switch on the motor circuit is made to open. But all of these designs operate on the head pressure and are planned for shutting down the compressor. The cooling water may always be responsible for the excessive head pressure, but it is not regulated by the controls so far described.

The simplest water regulator is one driven by the compressor itself. For the method of control it is necessary to have a pump capable of supplying the condensing water required and subject to some suitable variation of capacity by which more or less volume of water is pumped per stroke of the piston. A variable control of amounts of water is absolutely required unless, as in some wells, the temperature is a constant. A more elaborate valve would be one *similar* to Fig. 1, which in this case would use the action of the ammonia, working under condenser pressure, underneath the diaphragm. The rise or fall of this pressure actuates a valve that regulates the amount of water permitted to pass to the condenser, more water being admitted with an increase of condenser pressure.

To resume, the semiautomatic control system (Fig. 6) is one that is designed to run under certain prescribed conditions: The suction pressure is to remain constant, the water is adjusted for normal conditions, and there is an arrangement for shutting down the machine when the head pressure becomes excessive. The machine is safe under these conditions, and while the compressor is so operated, it is possible to produce useful refrigeration.

But if the machine stops, it remains stopped until the engineer returns. Also, it will continue to operate irrespective of the cold-storage temperatures being maintained in the storage rooms. The conditions are simply those of *safe* operation during the absence of the engineer in charge. Fig. 6 shows schematically the arrangement of such a refrigerating cycle.

Complete Automatic Systems

The completely automatic refrigerating system, Fig. 7, is electrically driven, and it is so controlled as to require the minimum of attendance such as would be required to keep the plant in first-class operating condition. If necessary, it may operate the twenty-four hours, starting and stopping as required by the temperature of the cold rooms. A thermostat is located in the room containing the expansion valve, and it is the master control of the whole plant.

In general the rise of temperature of the cold-storage room expands the elements of a thermostat, in consequence of which electric contact is made and the electric motors are excited and the compressor is started. Water is pumped over the condenser by means of a directly connected pump, or the increased head pressure due to starting the compressor actuates the water valve, which opens until a certain condenser pressure is obtained. As the compressor continues pumping, the suction pressure is reduced and the expansion valve is opened, which allows liquid ammonia to pass with the expansion (refrigerating) coils. As the expansion valve is adjusted for a particular back pressure, it permits a suitable boiling temperature for the ammonia. Operation under these conditions will reduce the cold-storage temperature and the thermostat will open the circuit and the motor and compressor will stop. Subsequently, a decrease of head pressure will close the water valve (if one is used) and a rise of the boiling pressure (suction) will close the expansion valve. A diagram of such an arrangement is shown in Fig. 4.

Finally, it may be said that some part at least of the automatic devices just described will soon be in common use, especially in the large cities. Safety relief valves are now required, and it is more than likely that remote-control valves or automatic nonreturn valves on the condenser, receiver and the expansion piping will be required also. The main objection to the automatic plant is its increased first cost, but it is not unlikely that all plants will be better safeguarded in the future and the disparity in initial cost will not be so marked. The engineer should acquaint himself with these special ammonia valves, therefore, and their advantages.

Crude Petroleums and Lubricating Oils

Characteristics of Crude Petroleum; Origin; Refining; General Requirements To Be Expected from Oils in Common Use in the Power Plant

By J. D. ROBERTS

THE American petroleum industry began with the sinking of the first oil well in Pennsylvania in 1859, just two years after oil had been struck in Roumania. In February, 1916, the United States Geological Survey estimated that the then known fields in the United States were 32 per cent exhausted.

There are two well-known types of crude petroleum—paraffin-base oils containing much light oil and considerable paraffin wax, like the Pennsylvania oils, and asphalt oils, which contain little light oil or paraffin wax but much heavy low-cold-test oil, like the Texas oils. A third type recognized is the so-called "mixed-base" oil, an intermediate of the other two types. Paraffin-base oils consist largely of compounds containing relatively more hydrogen than is present in the asphalt-base or naphthalene oils.

Crude oils are valued largely on the basis of the products distilled. Oils yielding much gasoline and kerosene in simple distillation and rich in paraffin bring the highest return per unit of sale, though the amount of the sulphur impurities is of much importance.

Oils from New York, Pennsylvania, West Virginia, Kentucky and eastern Ohio are in general paraffin-base, free from asphalt or objectionable sulphur, and yield by ordinary distillation high percentages of gasoline and burning oils. Their gravity ranges from 34 to 48 deg. Baumé.

Oils from Indiana and northwestern Ohio consist chiefly of paraffin hydrocarbons, though containing some asphalt, and are contaminated with sulphur compounds requiring special treatment for removal, usually with calcium oxide of lead or copper. There are also some lubricating-oil distillates produced in this field. The oils from Canada are classified in this group.

Illinois oils are mixed-base and paraffin-base with widely varying sulphur content, which generally can be removed without special treatment.

Oils from Kansas, Oklahoma, northern and central Texas and northern Louisiana vary in composition from asphaltic oils low in gasoline and kerosene to paraffin oils with good yields of gasoline and kerosene. The sulphur present varies in quantity and in some cases requires special treatment for its removal.

Oils from the Coast Plain of Texas and Louisiana are high in asphalt and low in gasoline. Much of the sulphur content is in the form of sulphureted hydrogen and is removable by steaming.

Wyoming and Colorado oils are mainly paraffin-base,

though some asphaltic-base oil is found in Wyoming. Oils from California are generally asphalt-base, yielding no paraffin and with extremely variable sulphur content. Fire oils, kerosene, lubricants, oil asphalt with a slight yield of gasoline are manufactured from these crude petroleums; their gravity ranges from 12 to 30 deg. Baumé.

Oils from the Mexican fields are similar to the oils from the Coast Plain of Texas and southern Louisiana. They contain excessive sulphur compounds and are low in gravity, 14 to 19 deg. Baumé.

In the order of their gravity, 34 to 48 deg. Bé., the light oils may be classified as Pennsylvania, Illinois, Louisiana, Kansas, and Oklahoma.

Among the heavy oils, 1 to 30 deg. Bé., are the Gulf Fields (southern Texas and Louisiana), Mexico and most California fields.

Crude Mineral Oils a Mixture of Hydrocarbons

Crude mineral oils are made up chiefly of a mixture of compounds generally classified as hydrocarbons having a composition of from 12 to 14 per cent hydrogen and 84 to 88 per cent carbon. From the light hydrocarbons like methane (CH_4) and ethane (C_2H_6) found in natural gas to the heavy solid bodies like paraffin or asphalt and the viscous oils which cannot be distilled without decomposition, these hydrocarbons vary widely in boiling point. As a general rule paraffin hydrocarbons have a low lubricating value. It is also a fact that the viscosity of hydrocarbons increases rapidly with the increase in molecular weight, so if high-viscosity products are to be made from crude oils of the paraffin-base class, distillation must be conducted with as little decomposition as possible. The separation of the individual compounds from petroleum is practically impossible on account of the boiling points of the compounds being modified by the presence of the other hydrocarbons in the mixture. The production of gasoline, kerosene and various lubricating oils is made by separation into groups of compounds with certain boiling limits.

Hydrocarbon will resist chemical action to a marked degree, and thus petroleum oils show little tendency to attack metals. Animal and vegetable oils show considerable tendency to form acid and will readily attack some metals.

Theories of Origin

Since different petroleums have different compositions, there is naturally a great variety of theories to account for its origin. The inorganic theories depend largely on the action of water on heated metallic carbides, somewhat as acetylene is produced commercially. Some authorities claim that the theories which are most in accord with the facts are those of organic origin from the decomposition of animal and vegetable remains. Oils from some Texas fields show evidence of marine animal origin. Also the considerable percentage of nitrogen compounds present in some oils would seem to indicate animal origin. It is a fact that some of the original differences in petroleum have been further modified by the migration of the oil and its filtration through different strata, which changes the composition of the oil.

The manufacture of lubricating oils and other products from the crude petroleum is accomplished by distillation and chemical treatment. Distillation separates the hydrocarbons into groups of different boiling points. Where heat is applied, the crude petroleum becomes more fluid by the melting of certain substances in the petroleum, or in other words, decreasing the cohesion of the liquid particles. As the temperature is raised, some of the crude petroleum will evaporate, and this, when condensed, will yield gasoline, kerosene and other light distillates. While the process is carried out, some of the hydrocarbons will be decomposed or "cracked" by the heating; from this stage come products of lower boiling point than those present in the original petroleum. With proper control of the heating and prevention of overheating, the best lubricating products are obtained when the least amount of decomposition and cracking of the original compounds of the crude petroleum takes place. There are two general methods of distilling; namely, fire and steam distillation. From the standpoint of lubrication the best oils are made by steam distillation, and steam cylinder oils cannot be made in any other way.

Steam Distillation of Crude Petroleum

Briefly, the steam distillation method is carried out as follows: The stills are cylindrical steel tanks supported in brick settings. Heat is first applied by direct fire, and as soon as heating has begun, steam is forced in by means of perforated pipes extending to the bottom of the still. This steam stirs the oil and prevents local overheating, and at the same time escaping steam vapors carry off the oil vapors as soon as formed, so that they do not condense and fall back into the still. These oil vapors pass out through large-diameter pipes at the dome of the still to a vertical tower condenser. In this condenser the heavy oils condense first near the bottom and the light oils condense last near the top. Thus with this type of condenser the oil may be separated into groups during the first distillation. With other types of condenser the distillates may have to be redistilled for this separation into groups. The groups so collected are, in the order of their boiling points: Crude naphthas, illuminating oils, gas oils, light lubricating distillates, heavy lubricating distillates and undistilled residue.

The distillation is usually stopped just before 600 deg. F. The residue distilled is suitable for light stock, if Pennsylvania or other paraffin-base stock has been used. The various distillates are distilled to rid them of light and heavy ends and render the removal of paraffin from the lubricating distillates less difficult. The use of steam not only gives better grades of lubricants but increases the yield of lubricating oils as well, particularly of cylinder stock.

Redistillation of Lubricating Oils

Lubricating-oil distillates are distilled the second time by fire or steam. By the addition of steam the undistilled residue goes into fuel oil. The oils are chilled and filter-pressed to remove paraffin wax. They may be partly distilled again and reduced, to remove the light oils and so raise the viscosity. The light oils distilled in this reducing process may be run into the gas-oil distillate or made into thin lubricating oils, called "non-viscous neutrals." The reduced lubricating oils are filtered through fuller's earth or boneblack to improve the color and remove impurities and are then ready for use as "viscous neutrals."

The residue in the still, if a paraffin-base crude has been used, is a steam-refined cylinder stock. If the temperature has been carried well above 600 deg. F. during the distillation, most of the paraffin has been distilled off. To make a filtered cylinder stock, the residue is "cut back" with crude naphtha, chilled and filter-pressed or otherwise filtered, and the gasoline finally recovered. The product is a filtered, low-cold-test cylinder stock.

The so-called Western lubricating oils are made from crudes that contain little paraffin. The procedure is similar to the refining of Pennsylvania oils, except as it may be modified by the character of the merchantable products possible. Most of the California crudes and much of the Oklahoma and Texas crudes are "topped" for the removal of gasoline and illuminating oils, and the undistilled residue is sold direct as fuel oil without further refining.

Oklahoma crude and some heavy crudes from Texas and California are now worked up for the manufacture of certain lubricants, such as cylinder oil, red engine oil and lighter lubricating distillates. The distillates have higher gravities, lower flash points, higher viscosities at low temperatures (70 or 100 deg. F.) and lower cold tests than do Pennsylvania products of the same class.

Characteristics and Requirements of Oils Commonly Used in Power Plants

In the following are given the characteristics and general requirements to be expected from gas oil and oils in common use in the power plant:

Gas Oil—The oils distilled between illuminating oils and the light lubricating oils are used for carbureting water gas and other gas to improve the illuminating power and for the direct manufacture of gas in heating retorts as applied in the production of illuminating gas direct from the oil. This gas oil is a cheap product. In redistilling the lubricating-oil distillates, the light ends are run into the gas oil. This product is sometimes used for fuel oil. Its gravity generally ranges from 32 to 38 deg. Baumé.

Neutral Oils—These oils are manufactured by steam distillation and are of high viscosity in proportion to their gravity. After the wax has been removed from the mixed lubricating-oil distillate, the oil is "reduced" by steam distillation to remove the lighter oils. These light oils constitute the nonviscous neutrals, while the residue from this final distillation constitutes the viscous neutrals.

The nonviscous neutrals usually have a gravity well above 30 deg. Bé. and a low viscosity suitable for light spindles. These oils are considered the best spindle oils as they do not stain like paraffin oils. They are not usually acid treated. The viscosity is 45 to 65 at 100 deg. Fahrenheit.

The viscous neutrals are usually slightly above 30 deg. Bé. and have viscosities ranging from 80 to 200 at 100 deg. F. They are suitable for motor oils, turbine oils, gas-engine oils, air-compressor oils and for the highest grade of service. The color is reduced by repeated filtration through fuller's earth instead of by acid treatment. To make the heavier oils, the viscous neutrals are blended with small amounts of high-flash, filtered, steam-cylinder stock. Blended oils of high viscosity may have gravities as low as 27 deg. Bé., even when from Pennsylvania stock. Viscous neutrals are also made from other stocks, in which case the gravities will be much lower than from the Pennsylvania distillates alone.

Spindle Oils—These are low-viscosity oils of 45 to 100 Saybolt at 100 deg. F. They may be light paraffin oils, but usually and preferably they are the nonviscous neutrals.

Engine Oils—Commercial engine oils are usually the heavier paraffin oils. The heavier oils are nearly always red, but the amount of color depends on the amount of acid treatment or of filtration. The color is not an index to the lubricating quality. Heavier engine oils are sometimes built up by the addition of cylinder stocks to heavy distillates. Viscous neutrals were formerly much used as engine oils. Under present-day manufacturing conditions, high-gravity neutrals now go largely into the motor-oil trade. At the same time low crude Western neutrals are sold to a large extent as engine oils. For circulating systems neutral oils are to be desired and are more satisfactory than paraffin oils, as they separate from water more readily.

Turbine Oils—These are similar to the lighter motor oils. The neutral oils separate from water better than do the paraffin oils and are naturally more desirable for this class of service.

Air Compressor Oils—These are similar to the lighter motor oils.

There is also a class of undistilled oils, from which, by steam distillation of Pennsylvania oils or other paraffin-base oils, a heavy undistilled oily residue is left in the still. This is used as a cylinder stock after being removed from the impurities. It generally has a high fire test of 500 deg. F. and a viscosity, when from Pennsylvania stock, of 140 to 280 at 210 deg. F. The gravity of such stock will not run below 24 deg. Baumé.

Under this class there is a filtered cylinder stock having a fire test that rarely runs over 600 deg. F. and a viscosity that seldom exceeds 160 and gravity not less than 26 deg. Bé., if made from Pennslyvania stock. As a general rule in the manufacturing of oils, filtering is considered to reduce the viscosity of cylinder stock. Bright stocks generally show a low cold test. Compounded oils are made by mixing or blending a mineral oil with a fatty oil. The chief compounded oils are cylinder oils made by dissolving animal oil or other animal fats in cylinder stock. Marine-engine oils are made by dissolving rape oil or blown rape oil in mineral oil. Compounded oils for other purposes than steam-cylinder lubrication are seldom used. As a general rule the viscosity of a compound oil is much less than the theoretical viscosity calculated from the oils used.

Experiments are to be undertaken at the Arlington (Va.) experimental farm of the United States Department of Agriculture by the chemists of the Bureau of Chemistry on the production of gas by the distillation of straw, corn stalks, leaves, etc. The work will follow the lines of the investigations made at the University of Saskatchewan for several years. The plant consists of a single retort with accessory cooling, washing, and gas-storage equipment, which will permit tests of various sorts of materials under different conditions of distillation. The methods best suited to the use of the gases for farm-power production will also be studied. The amount of work that can be done during the coming year will depend largely upon the amount of the general appropriation for the Bureau of Chemistry, as no special funds for this particular work are available as yet.

Some Characteristics of Shaft Governors

By Charles Myers

Among the many questions concerning shaft governors, with curved slot eccentrics, that are of interest to the engineer, the following are the most prominent: Does the action of the governor, in changing from full

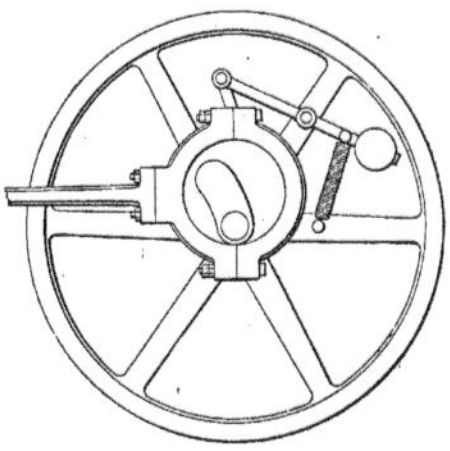

FIG. 1. CURVED SLOT ECCENTRIC FOR SHAFT GOVERNOR

load to no load cause the eccentric to revolve around the shaft, or to move across the shaft? Does the action of the governor change the lead in any way, or will the engine maintain the same lead under different loads? If the engine increases its speed a few revolutions per minute when the load is decreased, will the action of the governor cause the exhaust port to close earlier and provide the extra compression needed for the increase in speed? In answer to the first question, it can be seen from Fig. 1 that the eccentric does not revolve around the shaft, but that it moves across the shaft, describing a short circular arc about a center, which is some distance from the center of the shaft. The pivot about which the eccentric swings is usually on a spoke of the flywheel.

In most engines the lead will remain very nearly constant under all loads on one center, but will vary a little more on the other center. The amount and direction of these variations will depend on the position of the eccentric pivot and on the length of eccentric rod.

In Fig. 2 is shown the position of the eccentric under full load with the engine on the crank-end center and the valve (a piston valve) showing a certain amount of lead at *A*. Now, assuming that the eccentric is not fastened to the flywheel in any way, it is free to move across the shaft and will move in a circular arc about a center at the joint between the valve rod and the eccentric rod. Under these conditions the valve would not move as the eccentric swings across the shaft. Now, if the eccentric is moved across the shaft after being attached to the flywheel, it will not describe the same arc as when the joint in the valve steam served as the center, and the valve will move from its original position, thereby changing the lead.

The relative changes may be better understood by referring to Fig. 3. The arc *H* is drawn about a center at the point *X* and represents the path of the center of the eccentric if it turned about the joint at the end of the valve rod. The arc *R* was drawn about a center at the point *Y* and represents the path of the eccentric when swinging about its regular pivot on one of the flywheel spokes. In the case illustrated the line through *X* and *Y* represents the center line of the engine, and the distance between the points where the two arcs intercept this line would represent the movement of the valve and therefore the change in lead, in going from the full-load position to the no-load or neutral position. Referring back to Fig. 2, it is obvious that under these conditions as the load decreased the valve would be moved toward the crank end, and therefore the lead would be increased.

When the engine is on the head-end center, the position of the eccentric and valve will be as shown in Fig. 4. Referring again to Fig. 3, the arc *K* represents the path of the eccentric in swinging across the shaft about the valve-rod joint (center at *a*) and the arc *S* the path when the eccentric swings about its fulcrum (central at *b*). As before, the distance between these two arcs, when measured along the center line of the engine, will present the change in lead, and it is obvious that this change is greater than in the case where the engine was on the crank-end dead point. In drawing the diagram in Fig. 3, the proportions were so taken as to exaggerate the change in lead so that it

FIG. 2. POSITION OF ECCENTRIC AND PISTON VALVE AT FULL-LOAD; ENGINE ON CRANK-END CENTER

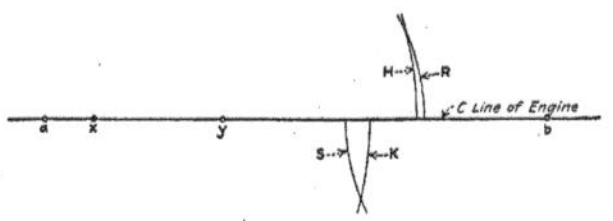

FIG. 3. DIAGRAM TO SHOW CHANGE IN LEAD AS THE ECCENTRIC SWINGS ACROSS THE SHAFT

FIG. 4. POSITION OF ECCENTRIC AND PISTON VALVE AT FULL-LOAD, ENGINE ON HEAD-END CENTER

might be seen easily. In the actual engine the length of the eccentric rod is so great in proportion to the distance through which the eccentric swings that the arcs *H* and *K* are, for all practical purposes, straight lines at right angles to the center line of the engine. Under these conditions the difference between the change in lead on the two ends is too small to be measured, and although there is such a difference it would not show up on the indicator card in the case of any reasonably well-designed engine. It is also true that in an actual engine the total swing of the eccentric is relatively much smaller and the radius of the arcs R and S is relatively much greater than is shown in Fig. 3. As a result, the change in lead becomes very small and is usually considered negligible.

In the case illustrated, the pivot about which the eccentric swings is located 180 deg. from the crank, and the valve is a piston valve taking steam at the middle. Under these conditions, as the eccentric swings across the crank the lead is increased. If the eccentric pivot were placed in line with the crank and used in connection with a piston valve, the effect would be to decrease the lead. This can be seen by referring again to Fig. 3, where, under these conditions, the arc R would be curved in the opposite direction, which would result in moving the valve toward the head end of the cylinder, and this, as will be seen from Fig. 2, would result in closing the port and therefore decreasing the lead. With a plain slide valve, and the eccentric pivot in line with the crank, moving the valve toward the head end when the engine is on the crank-end center would result, of course, in an increased lead. In the fourth possible case the pivot of the eccentric is 180 deg. from the crank, with a plain slide valve, which will result in an increased lead as the eccentric moves across the shaft.

The most important feature of the governor is its ability to change the point of cutoff so as to admit just the amount of steam required to maintain a uniform speed under any load. In order to do this it is necessary to make cutoff come earlier when the load decreases or the speed increases. Referring again to Fig. 1, the eccentric is here shown as it would be with the engine standing still; that is, the spring has drawn the eccentric out as far as possible. In this position, which is also shown in Fig. 2, the "thick part" of the eccentric is along the line ab, and as the engine runs over, the valve will continue to open until the point b turns into line with the center line of the engine. From that time on the valve will be closing, although, of course, cutoff will not occur until some time later. Suppose now, that the weight is pulled out by centrifugal force, when it will be seen from Fig. 1 that the eccentric will move down across the shaft. Referring again to Fig. 2, this will result in the thick part of the eccentric being along the line DE, which is already practically on the center line of the engine, and therefore the eccentric is in a position to begin closing the port. It is obvious, then, that as the weight on the governor moves out, steam will be cut off earlier.

A study of the valves and ports, as shown in Fig. 2, makes it clear that an earlier cutoff simply means that the valve has reached a certain point in its travel somewhat earlier. If, then, the steam port on the crank end is closed earlier, obviously the exhaust port on the head end must be closed earlier, since both are closed as the valve moves toward the head end of the cylinder. It is evident, then, that the curved slot eccentric will give the desired earlier compression, as the load on the engine decreases.

The usual method of explaining the ability of the governor to produce an earlier cutoff and an earlier compression is to say that as the weights fly out the angular advance of the eccentric is incraesed, and its eccentricity is decreased. It is hoped that the explanations given here, which are in an entirely different form, will make this matter plain to some who have difficulty in understanding the more technical form.

Wattmeter and Power-Factor Indicator Connections

By H. C. Yeaton

Figs. 1 and 2 show the connections for a polyphase wattmeter and power-factor indicator to measure the power and power factor of a three-wire, three-phase circuit, using either two or three current transformers,

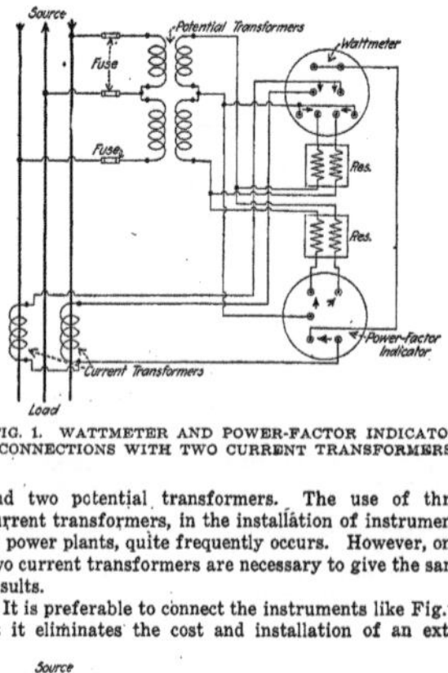

FIG. 1. WATTMETER AND POWER-FACTOR INDICATOR CONNECTIONS WITH TWO CURRENT TRANSFORMERS

and two potential transformers. The use of three current transformers, in the installation of instruments in power plants, quite frequently occurs. However, only two current transformers are necessary to give the same results.

It is preferable to connect the instruments like Fig. 1, as it eliminates the cost and installation of an extra

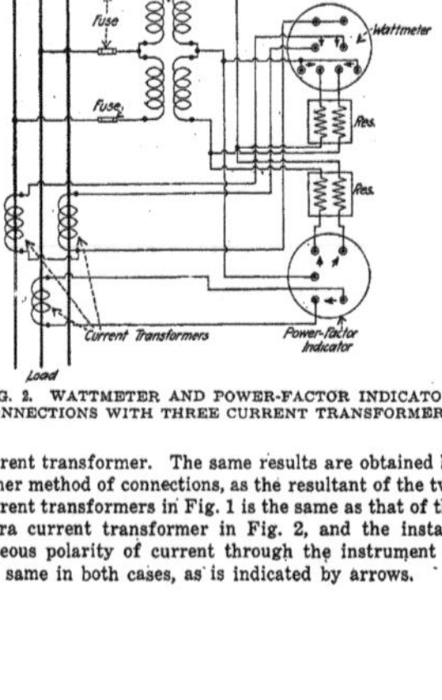

FIG. 2. WATTMETER AND POWER-FACTOR INDICATOR CONNECTIONS WITH THREE CURRENT TRANSFORMERS

current transformer. The same results are obtained by either method of connections, as the resultant of the two current transformers in Fig. 1 is the same as that of the extra current transformer in Fig. 2, and the instantaneous polarity of current through the instrument is the same in both cases, as is indicated by arrows.

Charles T. Main—Industrial Engineer

A Technical Man with Wide Experience in This Newly Recognized Field of Engineering

AMONG the suggestions contained in a recent committee report to the American Society of Mechanical Engineers was a recommendation that Industrial Engineering be considered as a major subject in arranging for discussions by the Society. This is only one testimony to the growing recognition of the importance of this branch of engineering.

The problems of power generation and use, of the uses of steam in various manufacturing processes, of the complicated automatic machinery in the modern factory and most important of all, the human or social problems involved where large numbers of people work together must be solved, not as individual problems but with regard to their relations to each other. It is the industrial engineer to whom we must look to solve these problems and on whose solution of them depends the success or failure of modern industry in this country. It has been said that America could always produce the right man or men to fill any need. When the need for industrial engineers was recognized it was found that a group of men were already at work fulfilling the need. One of the men occupying a prominent position in this group is Charles T. Main of Boston. His engineering experience has covered textile mills, industrial plants of all kinds and steam and hydraulic power development. He has had a hand in the evolution of a number of large enterprises and has contributed no small amount to the general industrial progress.

CHARLES T. MAIN

Mr. Main was born in Marblehead, Mass., Feb. 16, 1856, and can trace his ancestry to old American families established in colonial days. He graduated from the Massachusetts Institute of Technology in 1876 with the Degree of Bachelor of Science in the Department of Mechanical Engineering and spent the next three years as an assistant in the same department. It is a familiar saying among teachers that the best way to learn a subject thoroughly is to teach it to someone else. Undoubtedly Mr. Main's three years as a teacher served to round out and fix in his mind the technical training he had received while a student.

In 1879 he began his connection with industry as a draftsman in the Manchester Mills, Manchester, N. H., leaving there in January, 1881, to become engineer of the Lower Pacific Mills in Lawrence, Mass. In 1886 he was promoted to the position of assistant superintendent, and a year later became superintendent of the mill. From the first of his connection with the Pacific Mills Mr. Main's duties required considerable executive ability and gave him a chance to show what he could do. That he was at heart an engineer, rather than an executive is shown by his action in retaining direct control of all engineering work when assuming the executive duties of the superintendency. During his eleven years with this mill he directed the reorganization and rebuilding of the plant and for more than five years had charge of its operation. Finally the "call of engineering" became too strong to be resisted and in 1892 he left the Pacific Mills and began the practice of miscellaneous engineering work in Providence. In January, 1893, he became associated with F. W. Dean under the firm name of Dean & Main in general engineering work. As was natural, the attention of the new firm at first was devoted to textile engineering, but gradually it reached out to include work in every sort of industrial plant. In 1907 Mr. Main embarked independently under the name of Charles T. Main, Engineer.

During his seventeen years as an industrial engineer, Mr. Main's services have been in demand over the entire field of such engineering. He has directed the design and construction of many plants and has reorganized others. Among his larger undertakings are the Wood Worsted and the Ayer Mills in Lawrence as representatives of the textile industry, and developments for the Montana Power Company aggregating about 250,000 hp., in the hydro-electric field. In addition to his work in design, construction and reorganization, Mr. Main has served as expert witness, or as referee, in important legal actions involving engineering questions, and has made numerous appraisals of industrial plants.

He has always taken an active interest in public affairs and been ready to do what he could for the public welfare. In Lawrence he was a member of the board of trustees of the public library and a member of the school committee, besides being an alderman of the city. In Winchester where he has lived for some years he has been a member of the water and sewer board.

Mr. Main has always been active in the work of the American Society of Mechanical Engineers of which he became a member in 1885. He was a manager of the Society from 1914 to 1917 and its president in 1918. He also represented the Society in the delegation of American engineers who went to France, on the invitation of the French engineering societies, to discuss the problems involved in the rehabilitation of that country.

While naturally the American Society of Mechanical Engineer claims Mr. Main's first attention, he is also connected with other technical societies. Among them are the American Society of Civil Engineers, the American Institute of Consulting Engineers and the Boston Society of Civil Engineers. Of the last named society Mr. Main is a past president. He is also a past president of the Engineers Club of Boston and a member of the Exchange Club of Boston, the Engineers Club of New York and the Calumet and Winchester Country Club of Winchester, Mass. His financial activities include directorship in the Massachusetts Trust Co., the Tennessee Eastern Electric Co. and the Berkshire Cotton Manufacturing Co., as well as a trusteeship in the Winchester Savings Bank.

Among all his varied activities, an important, if not the most important, one is his life membership in the Corporation of the Massachusetts Institute of Technology. Mr. Main has not forgotten the institution in which he received his early training and takes an especial interest in his own department. It is true that a man of his ability would have probably succeeded without such training as he received at Technology, but undoubtedly this training has smoothed many a difficulty and helped him out of many a tight place. He has, however, well repaid his debt to the Institute by his service as a member of its governing body.

Some Notes on the Loss of Head Due to Friction in Water Pipes

By H. M. Brayton

In any system of piping through which water is pumped, the loss in head due to friction is an important item in the selection and operation of the pumping equipment. While this loss may be decreased by increased pipe sizes, an economic balance must be maintained between the saving in friction and the cost of pipe. Although much remains to be done, a great deal of experimental work has already furnished data from which we may calculate these losses with sufficient accuracy for practical purposes. These investigations also suggest methods of reducing friction losses without increasing the cost of installation.

As a guide in designing piping systems so as to balance first cost against friction losses, a velocity of not over five feet per second has been found to give reasonable results, although in some cases three feet per second is enough. As a check on this it is good practice to provide for a maximum loss of head of about five feet in one hundred feet length of pipe. The whole system should be as free from bends as possible, especially short-radius elbows. The loss in the 90-deg. bend in a 20-in. pipe is equivalent to that in one hundred feet of straight pipe.

Usually, a certain pressure is desired at the discharge end of the pipe, and adding the loss of head due to friction to the static head gives the total head on the pump. In such a determination the loss of head in the suction side of the system must not be neglected. When the pump is above the water supply, the suction friction and static heads must be added to the discharge friction and static heads in order to get the total head and required power. If the pump is below the water level in the supply, the suction friction head only is added to the discharge head and the suction static head is then subtracted to obtain the total head. In other words, when water flows to the pump by gravity, the suction head helps out and reduces the required power.

One way to determine the friction loss in the system is to obtain a table from any of the numerous pump manufacturers. These tables usually give the friction loss at the various velocities and for all of the common sizes of pipe. In the absence of such a table the loss can be obtained from the fundamental formula for friction in pipes which reads as follows:

$$H = \frac{fLV^2}{2gD}$$

in which H = lost head in feet; L = length of pipe in feet; V = velocity of water in pipe in feet per second; D = pipe diameter in inches; f = the friction factor; $g = 32.2$.

The value of f varies slightly for different velocities and for different diameters of pipe. For the usual range met with in power-plant practice, however, a value of 0.025 is accurate enough for practical work. The equation then becomes

$$H = \frac{0.025\,LV^2}{64.4D}$$

If $L = 100$ ft., the formula simplifies to

$$H = \frac{0.025 \times 100V^2}{64.4D} = \frac{0.0388\,V^2}{D}$$

Elbows and bends are transferred into equivalent feet of straight pipe in calculating the losses.

In the determination of the lost head in systems that are made up of several branches extending from one, it must be noted that the sum of the losses in the different branches is not the lost head for the system. This would be a much larger value than the true one. The branch that contains the longest line or has the largest number of elbows should be selected and its lost head determined. It is sometimes necessary to calculate all the branches in order to tell which one has the greatest loss. When this is found, it should be added to the loss incurred in the system up to the point where that branch leaves the header. This total will be the maximum loss for the system.

When the delivery point is located a long distance from the pumphouse, the friction loss may be considerable, and the designer must look for ways and means of reducing it to a minimum. It sometimes happens that two or more lines are run, each fed by its own pump. This arrangement is all right when the distance is short, but when long the loss is much more than necessary. In this connection it is interesting to note how the friction loss is reduced in such a case by merely combining the several pipes into one large main. Not only is friction head reduced by this procedure, but considerable saving in first cost is realized. A general formula showing just how the friction does decrease as we combine several pipes into one, may be deduced, using the following notation: h = head lost in each small pipe in feet; H = head lost in the one large pipe in feet; d = diameter of small pipes in inches; D = diameter of large pipe in inches; f = friction factor = 0.025; L = length of pipe in feet; V = velocity of water in feet per second.

The size for the one large pipe should be such that the velocity of the water through it shall be the same as in the several small pipes. The fundamental friction formula already given may be written,

$$h = \frac{fLV^2}{64.4\,d} \text{ and } H = \frac{fLV^2}{64.4D}$$

As the quantity of water flowing is always equal to the area multiplied by the velocity, and assuming equal velocities in the two cases,

$$Q = AV = NaV$$

where A and a are the areas of the large and small pipes, respectively. We may solve this as follows:

$$A = Na \text{ or } \frac{0.785\,D^2}{144} = \frac{N\,0.785\,d^2}{144} \text{ or } D = \sqrt{N}\,d$$

Substituting this value of D into the H equation given above results in the following expression:

$$H = \frac{fLV^2}{64.4\sqrt{N}\,d}$$

or solving for d, we have,

$$d = \frac{fLV^2}{64.4\sqrt{N}\,H}$$

From the first equation under this consideration,

$$d = \frac{fLV^2}{64.4\,h}$$

Now equating these two values for d gives

$$\frac{fLV^2}{64.4h} = \frac{fLV^2}{64.4\sqrt{N}H}$$

Assuming the value of f to be the same under the two conditions, the foregoing simplifies to

$$H = \frac{h}{\sqrt{N}}$$

This last equation tells simply that the lost head when several small pipes are replaced by one large one is equal to the lost head in one of the small pipes divided by the square root of the number of small pipes thus united. For example, if four small pipes are replaced by one large one so that the velocity of the water through the large pipe will be the same, and the total friction loss in one of the small pipe lines was 15 ft., then the loss in the one large line would be only,

$$\frac{15}{\sqrt{4}} = 7.5\,ft.$$

It has already been stated that the friction factor f was considered equal in the two cases, and there may be some who object to this assumption. By consulting a table of friction values, however, it will be noticed that they are smaller for the larger sizes of pipe. If f equals the friction factor in the small pipe and F that in the larger, the foregoing equation would become

$$H = \frac{F}{f}\,\frac{h}{\sqrt{N}}$$

As the value of f is greater than the value of F, the ratio $\frac{F}{f}$ is less than unity, which makes the value of H still smaller than when the friction factors were assumed equal.

Portable Insulating-Oil Testing Set

A small, compact, portable testing set, shown in the figure, suitable for oil testing only, consisting of a main testing transformer having a high-potential winding for 25,000 volts, a control and an oil-testing spark gap, has recently been developed by the General Electric Co. The testing transformer has a ratio of 110 to 25,000 volts and is of the oil-insulated type. The transformer weighs about seventy pounds and is equipped with handles which make it easy to carry. It is designed for indoor

OIL-TESTING TRANSFORMER AND CONTROL PANEL

service; if used outdoors, it should be provided with a suitable cover.

The control consists of an air-cooled auto-transformer with regulating taps connected to a selector switch with dial to give 15,000, 17,500, 22,500 and 25,000 volts on the high-tension winding of the main transformer, and a double-pole step-resistance knife switch, which prevents wave distortion when the transformer is cut into the line. These devices are all mounted as a unit, shown to the right in the figure.

The oil-testing spark gap is made of a hard-rubber body molded in one piece, the electrodes being one-inch disks with their shafts shrunk into the molded part. One electrode is made movable by rotating, and a gage is provided for adjusting the spacing. The test is carried out as follows: The gap is lifted from its supports and filled with a sample of the oil. It is then replaced on the supports, and after the dial on the control has been set for 15,000 volts, the knife switch is closed and held for about five seconds. If the oil does not break down, the test is repeated on higher-voltage steps.

Unity Power-Factor Synchronous Motor

By W. T. Berkshire
Alternating-Current Engineering Department, General Electric Company

Examples are worked out to show how a 1,000-kw. synchronous motor affects the total kilovolt-ampere load on a system when the motor is operating at unity power factor, 0.80 power-factor leading and at 0.0 power-factor leading. The power factor of a unity power-factor motor under different loads is also discussed.

The disadvantages of low power factor are generally known to most users of electric power. Low power factor means unnecessarily large and more expensive generators, and exciters with poor efficiency due to increased losses; increased cost of station, transforming and switching equipment and increased cost of transmission line and distributing transformers. It also results in poor voltage regulation. Furthermore, low power factor may mean underloaded prime movers with decreased prime-mover efficiency. Because of the disadvantages mentioned, power companies are already beginning to charge higher power rates where the power factor of the consumer's circuit is lower than a certain limit; consequently, low power factor means increased motor-operating costs.

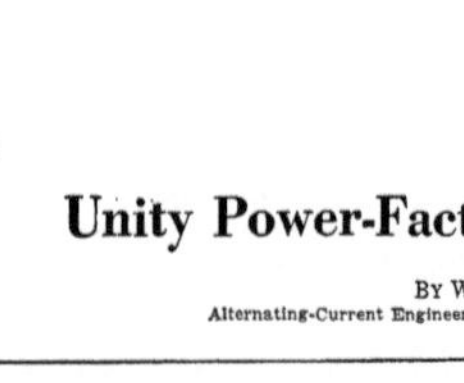

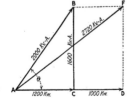

FIG. 1. EFFECT OF ADDING 1,000-KW. LOAD TO A 2,000-KVA. 0.60 POWER FACTOR LOAD

Low power factor is due to the lagging current drawn from the line by inductive loads such as induction motors, series and multiple arc lamps or even transformers supplying incandescent lamps.

For any given mechanical load on a synchronous motor the current in the armature is a minimum at a certain field excitation, this current being neither lagging nor leading, and the motor constitutes a unity power-factor load on the line. In this case all the current taken by the motor is energy current, the function of which is to drive the load and supply the losses of the motor. Now, if the field excitation is increased, the motor will take a leading current from the line. This leading current may be separated into two components: First, an energy component as mentioned in the foregoing, and second, a magnetizing current which is a current that tends to magnetize the generator's fields, which is the so-called wattless leading component. This wattless leading component of the synchronous motor may be used to neutralize an equal amount of wattless lagging component due to induction-motor load on the system. The excitation may be further increased until the motor current consists almost wholly of the wattless leading component, the energy component being only sufficient to supply the losses of the motor, none being available for driving mechanical load. The motor is then operating purely as a synchronous condenser.

Improvement in power factor can, therefore, be effected by the application of the synchronous machine in either one of three ways: (a) As a unity power-factor motor, that is, all the input being used for energy; (b) as a power-factor correcting motor, that is, part of the input being used for energy and part for furnishing wattless leading current to the line; (c) as a synchronous condenser, that is, all the input being used to supply wattless leading current to the line. The following are examples of each of the foregoing uses:

Assume a load of 1,200 kw. at 0.6 power factor or 2,000 kva. What will be the effect of adding a 1,000-kva. synchronous motor to the system (a) at unity power factor, (b) at 0.80 leading power factor, and (c) at 0.0 leading power factor—in other words, as a synchronous condenser?

(a) As a Unity Power-Factor Motor—Referring to Fig. 1, the initial load, 1,200 kw. at 0.60 power factor, is represented by the line AC, the kilovolt-amperes being $\frac{1,200}{0.60} = 2,000$, represented by the line AB. The wattless lagging kilovolt-amperes, represented by $BC = \sqrt{AB^2 - AC^2} = \sqrt{2,000^2 - 1,200^2} = 1,600$ kilovolt-amperes. Now, by adding a 1,000-kva., 1.0 power factor load, represented by the line CD, the total load becomes $AC + CD = AD$ or, $1,200 + 1,000 = 2,200$ kw. The wattless lagging kilovolt-amperes remain unchanged, or 1,600 kva. $= BC = DF$. The total kilovolt-amperes required, therefore, to deliver 2,200-kva. energy load will be $\sqrt{AD^2 + DF^2} = \sqrt{2,200^2 + 1,600^2} = 2,720$ kva., and the new power factor will be $\frac{2,200}{2,720} = 0.81$.

With the addition of this unity power-factor motor it will be noted that two important things are accomplished; namely, the power factor is raised from 0.60 to 0.81 and the energy load has been increased from 1,200 to 2,200 kw., or 83⅓ per cent, with an increase of only 36 per cent in the generator capacity; namely, from 2,000 to 2,720 kilovolt-amperes.

(b) As a 0.80 Power Factor Motor—Referring to Fig. 2, the lines AC, AB, and BC represent the initial energy load, the total kilovolt-amperes and the wattless lagging kilovolt-amperes respectively, the same as in Fig. 1. A 1,000-kva. 0.80 leading power-factor motor is now to be added to the system. This motor will deliver a $1,000 \times 0.80 = 800$-kw. energy load, represented by the lines MF and CD. It will also deliver "wattless leading" kilovolt-amperes equal to $BM = \sqrt{BF^2 - MF^2} = \sqrt{1,000^2 - 800^2} = 600$ wattless leading kilovolt-amperes. The 600 wattless leading kilovolt-amperes will neutralize an equal amount of the 1,600 wattless lagging kilovolt-amperes, so that the wattless lagging kilovolt-amperes after the motor is added will be $DF = BC - BM = 1,600 - 600 = 1,000$ kilovolt-amperes.

The energy load will then be $AD = AC + CD = 1,200 + 800 = 2,000$ kw., the wattless lagging kilovolt-amperes will be reduced to 1,000; and the total kilovolt-amperes required to carry this load will be $AF = \sqrt{AD^2 + DF^2} = \sqrt{2,000^2 + 1,000^2} = 2,235$ kva. The new power factor, after the motor is added, will then be $\frac{2,000}{2,235} = 0.895$ instead of 0.60. It will be noted also that the energy load has been increased 66⅔ per cent, with an increase of only 11⅔ per cent in the generator capacity.

(c) As a 0.0 Power Factor Motor; That Is, a Synchronous Condenser—In this case the problem is: How much will the power factor of the system be increased by the addition of a 1,000 kva. synchronous condenser, it being the intention to keep the same energy load? Fig. 3, as in Figs. 1 and 2, again shows the initial energy load, total kilovolt-amperes and wattless lagging kilovolt-amperes by the lines AC, AB and BC, respectively. The 1,000-kva. synchronous condenser, neglecting the energy required to supply its losses, which may be only 3 or 4 per cent, will supply a wattless leading load 1,000 kva., equal to BM. The total wattless lagging kilovolt-amperes will then become $BC - BM = CM = 1,600 - 1,000 = 600$ kva., and the total kilovolt-amperes required to carry the same load will become $AM = \sqrt{AC^2 + CM^2} = \sqrt{1,200^2 + 600^2} = 1,341.6$ kva. The new power factor will then be $\frac{1,200}{1,341.6} = 0.895$. Not only is the power factor increased from 0.6 to 0.895, but it will be noted that, by the addition of this synchronous condenser, to carry the energy load a generator of only about two-thirds the former capacity is required, with a corresponding reduction in the capacity of all transforming and switching equipment, transmission lines and exciters.

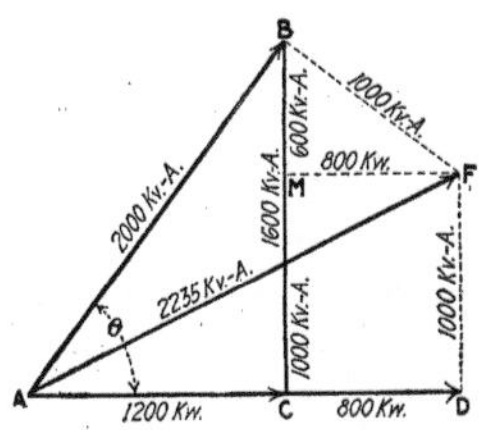

FIG. 2. DIAGRAM SHOWING EFFECT OF ADDING 1,000-KVA. 0.80 LEADING POWER-FACTOR LOAD TO A 2,000-KVA. LOAD

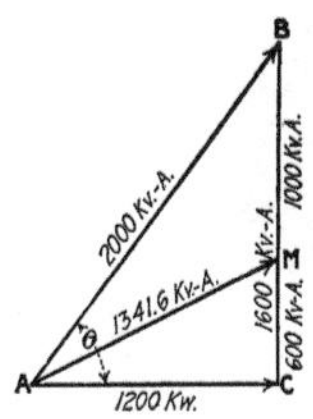

FIG. 3. ADDING 1,000-KVA. 0.0 LEADING POWER-FACTOR LOAD TO A 2,000-KVA. LOAD

The power factor is equal to the cosine of the phase angle between the volts and current. In the construction of Figs. 1, 2 and 3 angle θ is made equal to an angle whose cosine is equal the power factor. In the problem the power factor is 0.60 before correction is made; 0.60 corresponds to the cosine of an angle of approximately 53 deg., consequently angle θ in the figure is made 53 degrees.

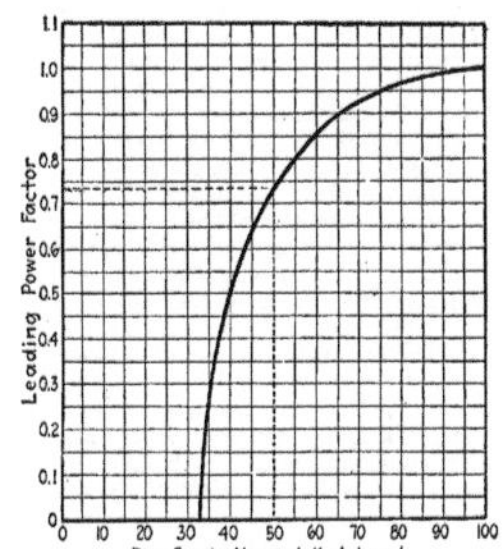

FIG. 4. POWER FACTOR OF SYNCHRONOUS MOTOR AT DIFFERENT LOADS, WITH FIELD HELD CONSTANT

In the application of synchronous motors the highest efficiency in the driving of mechanical loads can be obtained by the use of the unity power-factor motor. It requires less exciter capacity and costs considerably less than a synchronous motor designed to operate at leading power factors. Such a motor, of course, constitutes a non-inductive load on the lines; that is, it does not furnish any wattless leading kilovolt-amperes when operating at full load, but as shown in Fig. 1, it does improve the power factor of the system to a considerable extent. If, however, it is underloaded, with full-load excitation on the field, it always operates at a leading power factor furnishing a certain amount of wattless leading kilovolt-amperes to the line depending on the degree of underloading. Fig. 4 shows the leading power factor at which a normally designed, unity power-factor synchronous motor will operate at partial loads with the normal full-load excitation constantly maintained. It will be noted that at 50 per cent of the normal kilovolt-ampere output, the machine will operate at a 0.73 power-factor leading, and at about 35 per cent normal kilovolt-ampere input it will operate at 0.0 power factor; that is, purely as a synchronous condenser. In the average motor this value of kilovolt-ampere input at 0.0 per cent power factor will vary from 20 to 30 per cent.

The Framanco" Pump Regulator

A new type of pump regulator designed to maintain a constant differential between the discharge pressure and the steam pressure or a fixed pressure at the delivery end of the pump, has been perfected by the Framingham Manufacturing Co. for the Edward Valve and Manufacturing Co., of Chicago. The constant differential regulator will automatically maintain a fixed excess discharge pressure above the boiler pressure sufficient to overcome the friction head in the feed line and the static head. In services where it is desired to maintain a fixed pressure at the delivery end of the pump regardless of fluctuations in the steam pressure, provided, of course, there is sufficient pressure to operate the pump, a regulator of the constant-pressure type is employed. Regulators of either type are of the same construction, the only difference being in the application of the control cylinder and component parts.

One of the distinguishing features of the new regulator is the separation of the valve proper and the control cylinder, so that the operating mechanism is not subjected to the high temperature of the steam, thus eliminating expansion and contraction troubles. The working parts are in plain sight of the operator,

FIG. 1. FRAMANCO PUMP REGULATOR

and the proper size of control cylinder for the conditions existing in the plant can be attached to the valve proper.

The principle of furnishing a regulator to meet the particular operating conditions to which it is applied is further carried out in the application of the spring arm which connects the spring to the actuating shaft. This arrangement also serves to compensate for the extension of the spring, and its consequent increased tension by the shortening of its effective moment arm.

Fig. 1 shows the valve proper and the control mechanism in a vertical position. It will operate equally well in a horizontal line or in any other position. As shown more plainly in Fig. 2, the operating mechanism is bolted to the valve body, the latter containing the valve seats and a balanced disc made steam tight. The disc is actuated through a link and revolving shaft from the cylinder mounted on the side of the valve proper. For flexibility this cylinder contains a two-

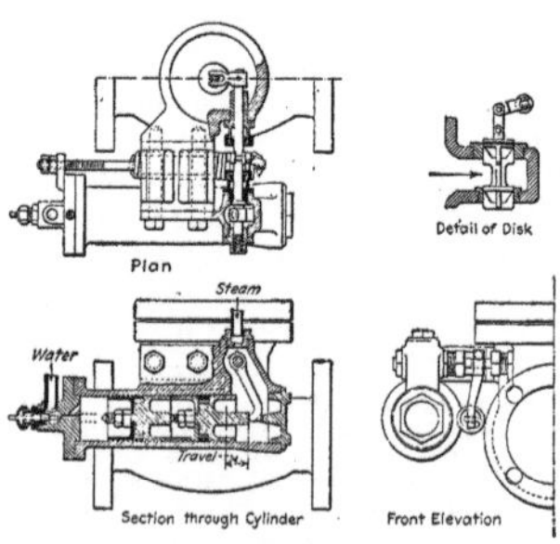

FIG. 2. DETAILS OF REGULATOR DESIGN

part piston with steam pressure on one side and water pressure on the other, the differential being secured by tension on a spring located between the valve proper and the control cylinder. By means of a screw adjustment the spring tension may be varied so as to vary the differential as desired.

When the regulator is used to control the pressure of feed water to the boilers, the valve proper is installed in the steam lead to the boiler feed pump and one end of the control cylinder is connected to the same line. Water pressure is admitted to the other end of the cylinder through a connection from the discharge line of the pump.

With the steam pressure and the spring operating on one end of the piston and the water pressure on the other, after proper adjustment has been secured, the whole mechanism is in equilibrium. Anything that disturbs the balance causes the piston to move in one direction or the other, closing or opening the valve, as the case may be. Referring to the sectional view, Fig. 2, should the steam pressure increase or the water pressure for some reason be reduced, the piston moves to the left, the spring causes the operating arm to follow the piston, thus rotating the shaft and opening the steam valve. More steam causes the pump to speed up, which in turn increases the water pressure until the balance is restored. It is claimed that with a normal pressure of 175 lb. an increase of less than 5 lb. will close the valve and stop the pump, or a decrease of two to three pounds will cause the valve to open full, admitting the full steam-line capacity to the pump. The steam connection to the control cylinder is always made from above so that condensation will collect in the steam space and insure a practically uniform temperature at both ends of the cylinder.

EDITORIALS

Proposed Super-Power Station

THE LEADING article in this issue deals with a proposed super-power station. It is proposed to design this station with multiples of thirty-thousand kilowatt turbines and a total station capacity as high as five hundred thousand kilowatts is suggested. A supplement illustrates the general idea and shows many innovations, which, although most of them may be found in part in a number of stations, have not all been incorporated in any one station.

This design of super-power station has for its basis a grouping of boilers and turbines as a single unit, four boilers supplying steam for one turbo-generator. Two thirty-thousand kilowatt turbines and eight seventeen hundred and sixteen horsepower boilers are to comprise a unit.

Without doubt the present tendency is toward larger power stations. The practice, however, has been, in the main, to pipe all boilers to a common steam header from which branch pipes are led to the various turbines. In some instances the group plan has been adopted; that is, certain boilers have been laid out to serve a certain turbine or turbines. In such instances the boilers have generally been housed in a single boiler room and the turbines in a turbine room, with a brick wall between the rooms.

Mr. Stevens goes a step farther in carrying out the group plan, and while proposing that four boilers shall supply steam to one turbo-generator set, he does not interconnect one group of four boilers with a second group of the same number. Moreover, but four boilers are permitted in one boiler room and but eight boilers are housed in the same structure. If the station capacity is, say, one hundred and twenty thousand kilowatts, two separate buildings are to be used having an explosion gap between them of at least twenty feet.

The idea of an explosion gap is also carried out in regard to the boiler room being separated from the turbine room, where in the proposed station a low pump-room separates them and a brick wall separates the turbine room from the switchhouse. Even the turbine room is protected by an explosion wall between two turbine units.

In this construction there are several obvious advantages, the main one being that in case of a boiler explosion, which is remote under present practice, the damage would be confined to one boiler room, or at the worst to one boiler house. The same would apply in the case of a tube rupture or the bursting of a steam main or branch pipe.

Thirty thousand kilowatt turbines are specified because this size is within the zone of maximum economy in practice, and it is about the limit of size that can be supplied with four units of boilers, which in this proposed station are served by one stack.

At the top of the station it is planned to have an operating room containing every known device for boiler-plant operation, thus enabling the master boiler operator to make hourly tests of boiler output, if so desired. Even periscopes are provided for observation of the furnaces and for reading the water level in the gage glasses.

Sleeping quarters and baths are provided. A riot alarm whistle, searchlight and wireless equipment are adjuncts to the general apparatus. Telephones are to be located wherever required in order to obviate the necessity of the men wasting time in getting into communication with the chief operator.

Warm air is to be taken from the generator; but a stop valve will prevent dust and soot from being blown back into the turbine in case of a tube rupture in the boiler. Coal and ashes are to be handled by means of full-sized railroad cars, and the track design is such that the largest type of locomotives can run through the plant. An examination of the supplement shows many devices and arrangements of apparatus that have not been mentioned herein.

No claim is made by the designing engineer for originality in the application of the ideas to this super-station, and criticism will be welcomed.

Conservation the Only Way Left To Meet the Fuel Situation

THE best of America's soft coal is going to Europe in quantities which we are not in a position to say; but it is conservative to state that they are large. In fact, the neck of the bottle now is not ships to carry coal, but port facilities at Hampton Roads to clear the volume demanded. Europe surely cannot furnish its own coal in adequate quantities for a long time to come, probably not within the next three years. And as long as Europe is in this shape, depend upon it, she will gladly pay for the high heat value American coals, and coal prices as well as coal availability will continue to be a problem to American industries.

The fact that most people do not have to the fore in their minds is that fuel is a basic commodity; it is the very foundation of modern civilization.

Oil fuel presents a more serious outlook than coal. It is certain that, generally, we cannot look to it to materially alleviate the coal situation. There is not enough of it coming out of the ground. We are beginning to realize that oil is essentially a marine fuel. And before many years we may be reconciled to the belief that it is a naval fuel. For surely, long, long before there are indications that the supply is meager, the navies of the world will cause fuel oil to be unavailable for purposes other than their own. And the marine demand today, naval and merchant, is what is causing the scarcity and high price of oil to land consumers.

It should interest industrial fuel consumers to know that the engineers of the United States Shipping Board Emergency Fleet Corporation more than a year ago completed boiler and stoker designs for ships. They did so not only for some of the ships which were soon to use coal, but to prepare for the day when there might be a greater number burning coal than oil. Even the most efficient arrangement of baffles for these stoker-fired marine boilers has been determined.

Engineers must see that the only hope of meeting the fuel situation is by economy in its use. The need for conservation is every bit as great now as at any time during the war.

The situation is abnormal; that is, fuel economy is today somewhat a different problem than it is ordinarily. Time was when the purchaser of coal could not only choose the grade he would have, but could buy it according to specification. Today he is glad to get anything that can be charitably called coal. And most consumers cannot get enough of that kind. One can do much to get low fuel consumption if one begins with the design in a new plant. But few new plants are being built. Money is in such condition that such a thing as remodeling the boiler room, or whole power plant is, in most places, out of the question. One cannot buy new equipment when the chief determining factor in its selection is coal, which cannot be chosen. But there are many appliances which will be worth-while if purchased, provided these are not already a part of the plant.

There never was a time when engineers should more diligently keep records of the plant's performance and more minutely follow up what the records show. The plant today which is not at least weighing its coal and water deserves censure. The management which permits its plant to run without providing it with equipment to check performance is not on its job.

Unity Power-Factor Synchronous Motor

OF THE three applications—as a unity-power-factor motor, a synchronous condenser or a combination of both—the unity-power-factor synchronous machine is the most efficient. The exciter capacity required is less, and the cost is lower than that of a machine designed to operate at leading power factor. Such a machine at full load will constitute a unity-power-factor load on the system and will therefore help to improve the power factor of the system as a whole. At less than full load, with full-load excitation on the fields, the machine will operate at a leading power factor and will supply a certain amount of leading kilovolt-amperes that will compensate for an equal amount of lagging kilovolt-amperes required by induction motors on other inductive equipment.

Elsewhere in this issue is an article on the "Unity-Power-Factor Synchronous Motor," by W. T. Berkshire. One of the features of this article that is of particular interest is the curve showing how the power factor of a synchronous motor with constant field excitation becomes leading as the load decreases. This characteristic is the reverse of the induction motor, in which even at full load its power factor is never above eighty-seven per cent lagging and drops to low lagging values if the load is materially reduced. In a comparison of the power-factor curve referred to in the foregoing with that of the average induction motor, it is found that at fifty per cent load the power factor of the synchronous motor is about seventy-three per cent leading, where that of an induction motor is around sixty-five or seventy per cent lagging. This means that if the load is reduced on a synchronous motor, it acts to improve the lagging power factor of the system, whereas, if the load is reduced on an induction motor it acts to make the lagging-power-factor conditions worse. In other words, if an induction motor is operating underloaded, not only is its efficiency reduced, but it has a deleterious effect upon the power system as a whole. When a synchronous motor is operating at reduced loads, its efficiency will also be reduced, but it will develop characteristics that improve the operation of the system, therefore compensating in a measure for the loss in efficiency.

Constructive Legislation

THE appropriation of one hundred and twenty-five thousand dollars recently recommended in the Sundry Civil Bill, to be expended in determining what the power requirements are for the industries and railroads in the North Atlantic Seaboard States and how to most efficiently meet these needs is a step in the right direction and is worthy of the support of every American citizen. This appropriation carries with it the privilege to use private funds. At this time there is no better opportunity for public-spirited men to do this country a great service than to see to it that the necessary funds are forthcoming to make this investigation with.

The American Institute of Electrical Engineers, through President Calvert Townley, has appointed a committee on the super-power system, consisting of W. S. Murray, chairman; H. W. Buck, H. V. Bozell, George Gibbs, J. W. Lieb, Malcolm MacLaren, William McClellan, C. S. Ruffner, E. B. Rushmore, C. F. Scott and Percy Thomas. With the Government appropriation forthcoming to carry on the work, it is most gratifying that the investigation will receive the support of these eminent engineers and one of our National engineering societies.

Experts estimate that we burn up each year in this country two and one-half billion dollars' worth of by-products in the five hundred million tons of bituminous coal used; that three hundred million dollars each year is wasted by our present methods of power generation in the industrial region extending from north of Boston to Washington, D. C., and from one hundred to one hundred and fifty miles inland from the coast. It is estimated that we have over twenty million continuous horsepower of undeveloped water powers. If this were all developed and available, it would result in a saving of three hundred and fifty million tons of coal per year, allowing an average of four pounds of coal per horsepower-hour. In artificial illumination we are wasting two hundred million dollars' worth of light each year according to the experts. These are only some of our power wastes; there are many others. Are we going to continue to tolerate them?

In certain type of substations material reductions in first cost have been obtained by installing the equipment out of doors. Although in some small isolated installations, where climatic conditions are extremely favorable, boilers, engines and generators have been installed in the open, no definite effort has been made to apply this practice generally. With the cost of power-plant installations rising to undreamed of heights, has the time not arrived to investigate the possibilities of outdoor power plants to determine if a large percentage of the investment required to house the equipment by our present practice cannot be eliminated?

Pulverized Coal Under Boilers

The writer read with considerable interest the remarks published in *Power* under date of April 27, 1920, on page 681, by T. A. Marsh, chief engineer of the Green Engineering Co. In the first paragraph Mr. Marsh states that he believes it fundamentally wrong to take coal and burn it in a powdered condition. I believe that it is better and easier to burn coal in pulverized form than it is in lumps on grates subjected to the action of the heat.

In the second paragraph Mr. Marsh claims that some accurate way of measuring the ash discharge from the stacks from pulverized-coal-fired furnaces should be found before definite statements concerning it are made. It is true that a large percentage of the ash from pulverized-coal-fired furnaces will be discharged from the stack. The percentage, however, dissipated from a first-class installation equipped with economizers and settling chambers will not be sufficient to cause a complaint, as has already been shown by installations using pulverized coal in thickly populated districts. This question of ash dissipation depends greatly upon the velocity of the gases passing through the setting, and the percentage would naturally vary according to the operating conditions and the general design of the installation, the same as that ash discharged from a stoker installation would vary according to the conditions under which the stoker is operated.

In his fourth paragraph Mr. Marsh states that the cement industry has more data on the subject of pulverized coal than have steam producers. This statement is correct, and we do not hesitate to assert that we have more data on what is being done in connection with pulverized coal in the cement industry than any other organization, and we are thoroughly familiar with every installation in the cement industry in the United States and Canada. We are positive that any plant in which the cost is as high as one dollar per ton to prepare the coal is not being efficiently operated, unless the particular coal plant has a very small capacity and unless the plant is operated only periodically.

In order to show just what the cost of preparing coal is at the present time, we would like to say that during the month of August, 1919, at the plant of the Allentown Portland Cement Co., 2,976 gross tons of coal were pulverized and the actual cost of pulverizing was as follows: Drying, 0.0804; grinding, 0.0874; operation, 0.0043; power, 0.1793; labor, 0.0368; repairs, 0.0765. Total, 46.44c. per gross ton, or 41c. per net ton, no interest or depreciation included. This can be taken as representing the average cost in portland-cement plants in the Lehigh Valley district, for the quantity of coal pulverized daily from which these figures are taken is low, and no doubt in larger plants they would be modified materially.

The cost of preparing coal will vary primarily according to the quantity pulverized daily, so that any general statements or criticisms that are made as to the cost in one plant have no bearing on the subject as a whole.

H. G. BARNHURST, Chief Engineer.
Allentown, Pa. Fuller Engineering Company.

Grounds Cause Alternator to Lose Field Control

A suitable resistance was installed in the field circuit of a 120-kva. three-phase 60-cycle 600-volt alternator, with direct connected exciter, to operate in parallel with other units. When testing the unit I found the regulation satisfactory until about half the resistance was cut out, when the voltage jumped to its full value and the field ammeter failed to give a reading. The resistance box was tested and found to be in good condition, then all wiring was checked. The trouble was finally located in one brush-holder on the exciter and one collector ring on the alternator being grounded as in the diagram. After these repairs were made, the control was satisfactory again.

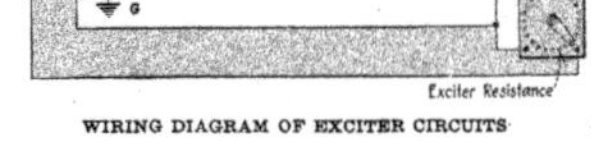

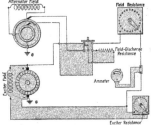

WIRING DIAGRAM OF EXCITER CIRCUITS

P. A. TOPPER.
Philadelphia, Pa.

Central-Station Operating Code

In the editorial in the Mar. 9 issue, "An Operating Code for Large Central Stations," a sentence reads, "This means that without large turbines there is an absence of large condensers, boilers and other allied apparatus."

In the industrial power-plant field it is not uncommon to see boiler plants that are fully as up-to-date and that contain as large units as are found in some of the large central stations.

Some years ago a turbine or reciprocating central station with twenty or thirty boilers was unusual; industrial plants, however, equaled that many and even more. Since then the central station has advanced wonderfully and so have industrial power plants.

Operators in large industrial plants, even though the mechanical output may be of only a few thousand horsepower, are obliged to furnish a large volume of steam, and this performance requires just as efficient units as the central station.

I believe that the central station should take care of its own code, to be adopted universally if possible, but each station of which I have knowledge has methods differing quite materially from the others.

New York City. C. W. Peters.

Fatal Accidents Occurring on Low-Voltage Circuits

Your editorial on page 341 of the March 2 issue, "Fatal Accidents on Low-Voltage Circuits," refers to the report of G. S. Ram, electrical inspector of factories in England, in which he shows that of ninety-nine fatal electrical accidents occurring in England during the period from 1915 to 1918 inclusive, seventy-nine were on circuits of five hundred and fifty volts or less, while only twenty were on circuits of one thousand volts and over. From this data you draw the conclusion that low voltages are as dangerous as high voltages, and I beg to take issue with you, not on your conclusion, but on your method of drawing it. Certainly, voltages of one thousand and over are in much less general use than the lower voltages, and consequently there would be much greater opportunity for accidents occurring around the number of low-voltage machines in use than around the much smaller number of high-voltage (above one thousand) machines in use.

However, I think that your conclusion is well taken, and a personal experience may be of some interest in this connection. I was using a Parr adiabatic combustion calorimeter, which was connected up to a 110-volt circuit for the purpose of firing the charge of coal in the small capsule, by means of a fuse wire. In connection with the calorimeter there was also a small water-circulating system, which furnished hot water. I was accustomed to putting my hand on a piece of one-inch pipe that formed one side of the water system, to determine whether the water was hot enough. Thinking that the current was off (but it happened to be on and connected to the metal top plate of the instrument), with my left hand on the top plate, I reached down and took hold of the water pipe to find out how hot the water was. The shock almost caused me to jerk the heavy calorimeter onto the floor, trying to pull loose. This incident greatly increased my respect for a 110-volt circuit, which I had always been given to understand was not at all dangerous.

The inspector and meter tester of the local public-service company later told me of some of his experiences with several near fatal accidents in which workmen had handled low-voltage circuits. The cause seemed in each case to have been carelessness in the matter of grounding. In one case it was found that the victim of the shock had been sent to a creamery to do some repair work and had been making the change while standing in a pool of water. Incidents like this would seem to indicate that workmen are, as a rule, very careful about the ground connections when they are working with higher voltages, even with the voltages used on street-railway systems; while at lower voltages they become more careless because they think there is comparatively little danger. Some reliable data on this, and public education, would, I think, be very desirable.

Decatur, Ill. W. J. Risley, Jr.

Increasing Boiler Output by the Help of a CO_2 Recorder

In an editorial in the April 20 issue of *Power*, entitled "Increasing Boiler Output," you suggest either the installation of forced-draft mechanical stokers or, when hand-fired furnaces are retained, equipping them with some device for increasing the draft. Another method, not mentioned in your editorial, is the installation of CO_2 recording equipment. The CO_2 recorder is generally thought of as a means of saving fuel rather than for increasing capacity, but it may be applied primarily for either purpose.

As an example, assume a boiler plant consuming ten tons of coal per day. Suppose it is found by the use of a CO_2 recorder that the excess air can be reduced to such an extent that the same capacity can be maintained by burning nine tons per day. This, of course, results in a saving of coal cost and is highly desirable. Suppose, now, that increased capacity is desired and that the original amount, or ten tons of coal per day, is burned, but that with the aid of the CO_2 machine the improved efficiency is maintained, with the result that the capacity will be increased by approximately 11 per cent.

The matter of draft is another important item to be considered. Many plants are today struggling to take advantage of all the draft possible in order to increase capacity, but are supplying so much excess air that they are not able to realize the available capacity. The engineer may complain that the chimney is not big enough and it may not be big enough, to handle the quantity of gases discharged into it; but as soon as the quantity of excess air is diminished to economical proportions, the volume of hot gases shrinks, and the chimney, relieved of its overload, has no difficulty in handling all that is required of it.

In a certain case a battery of four boilers equipped with stokers was unable to meet the demand for steam, and the purchase of two additional units was contemplated. CO_2 recorders were first installed, however, and by reducing the excess air, so much additional heat was absorbed by the boilers that one unit could be cut out of service and the three remaining boilers had ample capacity to meet all the required demands. The use of a CO_2 recorder will, obviously, often result in a very inexpensive increase in boiler-plant capacity and at the same time reduce the fuel cost per boiler horsepower.

New York City. F. F. Uehling.

Symbol for B.T.U.

For some fifteen years I have used the symbol shown herewith as a substitute for the three letters B.t.u.

They are combined in the symbol and are speedily produced with two strokes of the pencil without the use of periods. Having found it convenient, I submit it with the hope that others interested may find it equally helpful.

Kansas City, Mo. R. L. WEBER.

The Designer of This Heating System Meant Well

It has occurred to me that perhaps the readers of *Power* might like to know how a first-class engineer carried out a job some years ago. He was the only man in a group of three or four towns who had a first-class license, and he boasted that the inspector who visited his plant had complimented him highly on the control for the steam-heating system, which was his own idea and one that he had never even heard of or seen anywhere else.

This factory was three stories high with basement and attic, and heating coils of the trombone pattern were installed. In one corner of the engine room was a layout as shown in Fig. 1. A live-steam manifold made up of tees and nipples was suspended in a horizontal position about eight feet from the floor. Under this, perhaps thirty inches, was another line, but of larger fittings for exhaust steam. A series of vertical pipes connected the two manifolds.

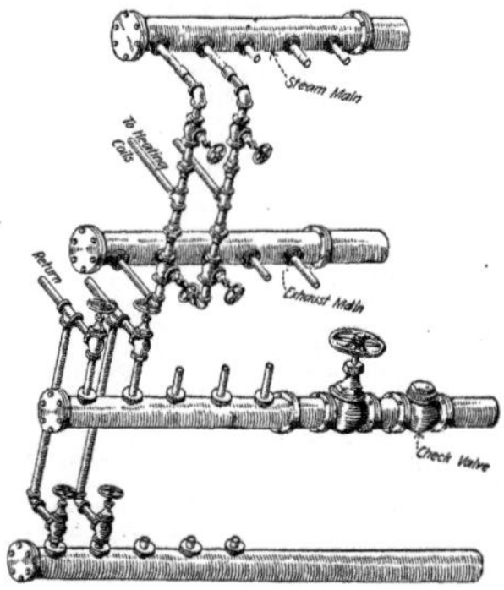

FIG. 1. ARRANGEMENT OF CONTROL IN BOILER ROOM

Midway between the manifolds on each vertical line was a tee, and a valve was placed on the live-steam side and one on the exhaust-steam side to control either live or exhaust steam to the tee, the outlet of which was piped to the heating coils. Each pipe line ran to a separate heating coil. It will be seen that either exhaust or live steam could be used at will on the heating coils.

All returns were brought back through individual pipes, and the control was so arranged that the return could (if exhaust) be turned into the same, or if live steam were used it could be returned to the boilers.

The designer of this layout went to another position before the system was completed, but he left drawings and recommendations for his successor to follow. I happened to be the victim, and here is what I found: The coils throughout the factory were of the trombone type, and before steam got halfway through the shortest coil, it had turned to water, and yet the system was designed to return the water to the boilers by gravity. Exhaust steam was likewise introduced to the coils through the same 1¼-in. pipe, and about fifteen pounds back pressure was necessary to get the exhaust even into the first pipe of the coil.

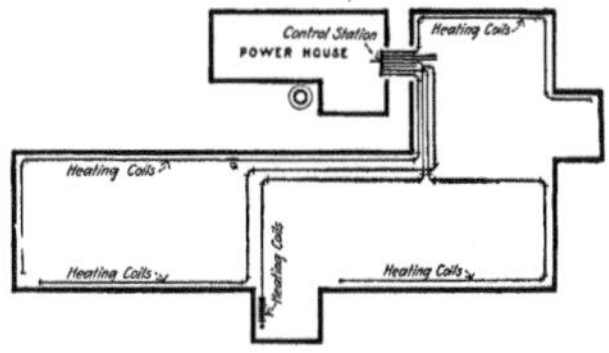

FIG. 2. PRESENT ARRANGEMENT OF HEATING SYSTEM

As the southern exposure was a good place to use exhaust steam, I rebuilt four coils, making them into the manifold pattern and feeding them with a 4-in. line. On the ends of the coils was coupled an old single-plunger pump, belted to the lineshafting and operated only when the engine was running, but as there was no exhaust steam except when the engine was running, it worked all right.

I could turn live steam through this line to get it warmed before starting, and a bleeder near the pump took care of the return.

By the end of the first winter I introduced a home-made condenser near the plunger pump, and one pound back pressure on the engine took care of the heater and the coils in the south side of the building.

By running a 4-in. live-steam pipe directly to the north side of the factory near the stair tower and putting up a riser, I was able to return the water by gravity from six coils to the boiler. Fig. 2 is a plan of the present system.

The moral seems to be that trombone coils are not practical for gravity return systems and that control from one point is not always an advantage; also that ample area must be allowed in the piping so that the drop in pressure will not be excessive; and finally, that the power-plant man with experience in handling units of thousands of horsepower makes a very poor designer for the small plant, unless he has started his career in the small plant and knows its needs.

New York City. C. W. PETERS.

Favors the Oil Engine

We have 500 hp. in steam engines, 150 hp. in gas engines, and 190 hp. in electric motors, the last operating on central-station current. We burn coal costing $3.50 per short ton for steam making and city gas of 650 B.t.u. per cu.ft. at 70c. per 1,000 cu.ft. in the gas engines; central-station current at 2.02c. per kw.-hr. is supplied to the motors. For the oil-engine estimate we will assume oil as costing 5c. per gal. and 6 gal. per horse-power-hour. Based on these costs, 940 hp. of electric current for one month of 25 ten-hour days would cost $3,562; steam and gas per month would cost, fuel and labor, $2,491, and oil and labor for oil-engine power, $1,237. Reduced to kilowatt-hours this gives for central-station current a cost of 2.02c. per kw.-hr.; steam and gas power, 1.41c. per kw.-hr., and for the oil engine, 0.7c. per kilowatt-hour.

It looks as if the oil engine had the best of it in operating costs, and with twenty years' experience in making gas engines do their work with reliability and economy, I have no fear of not getting satisfactory results from an oil engine.

In the past, with cheap fuels, we have been unable to see much of anything except first cost, but that day has gone. Economy of operation must play a more prominent part in the selection of power apparatus.

A good steam, oil or gas engine directly coupled, driving a lineshaft, has the advantage that several losses are avoided with which the central station has to contend. I have always contended that with first-class equipment and attendance the average mill or factory power plant can make its own power at a profit over purchased power. Frequently, however, where the cost of power is but a small part of the total costs of manufacturing, this is given but little consideration.

Anderson, Ind. J. O. BENEFIEL.

Suggested Changes for Diesel Engines

In order to bring about more satisfactory operation of the Diesel type of engine, one of the first things the designer should do is to remove the high-pressure air compressor from the engine. The present direct-connected air compressor, in the writer's opinion, causes many shutdowns, and the accompanying dissatisfaction.

It is difficult to understand why the compressor was ever directly connected to the engine, as it is not an efficient drive. I would suggest the adoption of a small three-stage high-speed vertical compressor driven by an electric motor, small engine or from a shaft. A motor-driven compressor, with inclosed silent-chain drive running in oil, would make an ideal unit. There should be at least two such compressors in each plant. If the plant contains more than one engine, the compressors should be so arranged that one would supply air for all the engines and the other held as a reserve. Then, if one broke down, the other could be cut in without shutting down the engines.

The air compressor on a Diesel engine is like unto the boiler-feed pump in a steam plant, and it is not much of a steam plant that has but one feed pump. The same will be said of the air compressor in the Diesel plant. The independent compressor could be made an efficient unit, because its speed could be regulated according to the air requirement.

The oil pump on a Diesel engine is comparable to the boiler of a steam plant, but in all large power plants there is at least one reserve boiler. Two oil pumps should be placed on each engine, one to be held in reserve. They should be independently operated or they could be operated from the camshaft, but so fitted that either pump could be repaired while the other pumped the oil, and this without shutting down the engine. I lay 15 per cent of the engine shutdowns to the oil pump.

Another fixture that would be of advantage on the Diesel engine is a duplicate fuel-oil injector valve. Two of these valves could easily be put on, and, then, when one went wrong the other could be cut in. It would, however, be some job to design these valves so that one could be repaired while the engine was being operated, but it can be done. At least a spare valve or two, complete with cages, should be kept on hand and in repair, to be put in when the engine was shut down for some other cause.

In my opinion, 75 per cent of the shutdowns could be eliminated if oil engines were built with duplicates, as outlined in the foregoing. The remaining 25 per cent of the shutdowns are due to the load, the engine itself and other little causes, but not more than 10 per cent are caused by the engine itself, if it is properly operated and closely watched.

A crosshead on large engines should be used, but not one that will permit the crosshead to lift from its guides. The crossheads now in use keep the greatest wear from the cylinder and piston and are far better than no crosshead at all. The pistons have a job all their own and should not be called upon to do the work of a crosshead as well. It costs money to put in a crosshead, but why not spend a little more on the first outlay and cut out repair and high depreciation costs? A crosshead permits of the necessary adjustment to take up the wear.

The piston should be fitted with at least three oil grooves, properly spaced and extending all the way around it, if a vertical piston. If it is a horizontal piston, the grooves should extend all the way around except for about one-half inch at four points and staggered so as to make the groove hold the oil and prevent it from flowing to the bottom of the cylinder. It is also well to have an oil groove in the cylinder of large engines. But all grooves should be cut large to prevent them from becoming clogged.

A light oil-feed pump should be provided to furnish lubricating oil to the cylinder, and it is a good plan to have the connections enter the cylinder at three points. A low-oil alarm should be placed on the oil pump as a precaution. It takes but a few minutes for trouble to develop with such high temperature and no oil and very serious trouble it may become in the course of time.

Some types of internal-combustion engines operate with a water-injection system. It can, to a certain extent, be applied to all such engines, and would help, I think, and also do away with considerable noise and would accomplish about all the water-injection method is supposed to do.

I would run the cooling water from the jackets down over the baffles in the air-intake box. The intake air passing through this spray of warm water would become saturated with moisture and would be so under all conditions and at all times. This air would also be free of dust and grit and would be in the proper condition to enter the cylinders. GEORGE J. TROSPER.

Winkelman, Ore.

INQUIRIES OF GENERAL INTEREST

Kerosene in Boilers—What are the objections to the use of kerosene for the prevention or removal of scale from boilers? W. G.

Kerosene, especially when used in excessive quantities, is likely to cause leaks in the boiler and to distill over and attack gaskets and pipe joints. When introduced into boilers in which there is already a deposit of scale, the loosening and accumulation of the scale in the bottom of the boiler may cause blistering and bagging of the boiler metal.

Pressure in Closed Heater Higher Than Boiler Pressure—Why is it necessary to have a higher relief pressure for the safety valve on a closed feed-water heater than the blowoff pressure of the boiler safety valve? W. H. B.

For boiler feeding, the pressure in the heater must be as much higher than boiler pressure as may be necessary to overcome the losses of pressure sustained in discharging the feed water from the heater through the piping and fittings in the feed line, and boiler-check valve, to the boiler.

Lift of Jet Pump—Water is raised by a steam jet pump to a height of 4 ft. 6 in. above the pump, and the suction water is supplied at a pressure of 1 lb. per sq.in. What is the lift of the pump? M. M. W.

If the suction water is delivered to the pump under a pressure of 1 lb. per sq.in. = 2.3 ft. head, and the pump is capable of discharging to a level that is 4 ft. 6 in. above the point where the suction pressure is measured, then the net lift of the pump is 4.5 − 2.3 = 2.2 ft., or about 26½ in.

Choking Pressure Gage Cock to Dampen Vibrations—Will dampening the vibrations of a pressure gage by choking the gage cock give true indication of the average pressure? C. T.

The pressure indicated will be only an approximation of the average. When there is leakage around the joint of the plug of the gage cock, especially when the cock is nearly closed, the true pressure will be greater than the pressure indicated by the gage, and on that account readings should not be relied upon unless the connections are tight.

Necessity of Unequal Laps for Equalizing Cutoff—For equalizing cutoff with a D slide valve, why is it necessary to set the valve with more lap on the head end than on the crank end? G. N. H.

On account of angularity of the connecting rod a fewer number of degrees of rotation of the eccentric is accomplished for a fraction of stroke from the head end than from the crank end. Hence, to obtain closure of the valve for cutoff to take place as early in the stroke from the crank end as from the head end, the valve must have more lap on the head end.

Water Power Available for Electric Generator—What number of horsepower can be realized for an electric generator from a turbine waterwheel supplied by a stream flow of 10 cu.ft. of water per second, where there is a total fall of 17 ft? R. R.

The net working head realized will depend on the loss of head sustained in getting the water to and from the turbine, depending on the character of the development, and for a given net head and given rate of flow the power realized depends on the efficiency of the turbine and transmissive machinery. Assuming that the available net working head is 14 ft. and the combined efficiency of the turbine waterwheel and transmissive machinery is 70 per cent, the power realized would be

$$\frac{14 \times 10 \times 62.3 \times 60}{33{,}000} \times 70 \text{ per cent} = 11.1 \text{ hp.}$$

Power Factor of an Unbalanced Load—In our plant we have a three-phase, 2,300-volt, 250-kw. alternator. How can I find the power factor from the readings of the polyphase wattmeter, ammeters and voltmeter? The ammeters do not all read alike. R. A. W.

Power factor is the ratio of the watts to the volt-amperes. The wattmeter reads directly in watts. If the load is balanced on a three-phase circuit, the volt-amperes equal the voltmeter reading times one ammeter reading multiplied by 1.732. Where the load is not balanced, the volt-amperes equal the sum of the three products obtained by multiplying each ammeter reading by the volts measured to the neutral point. For example, on a 2,300-volt circuit the wattmeter reads 95 kw. and the ammeters read 25, 30 and 35 amperes, respectively. Since this is a three-phase circuit, the volts to neutral equal 2,300 ÷ 1.732 = 1,328. The products of volts to neutral and ammeter readings are 1,328 × 25 = 33,200, 1,328 × 30 = 39,840 and 1,328 × 35 = 46,480, respectively, and the total volt-amperes = 33,200 + 39,840 + 46,480 = 119,520 watts. Then the power factor = watts ÷ volt-amperes = 95,000 ÷ 119,520 = 0.79.

Motor Field Coils Burn Out—What causes the field coils of a four-pole shunt motor to burn out frequently? The motor will operate satisfactorily for a week or ten days, and then one, and sometimes two, of the field coils will be roasted out. The field coils are not grounded when they fail. J. C. L.

If the same coils burn out each time, the trouble, no doubt, is due to the coils that do not burn out being short-circuited or grounded, consequently putting full-line potential across the coils that burn out. It would be advisable to see to it that the coils that do not burn out are not short-circuted or grounded, as it is almost certain that this is the source of the trouble. If one of the field coils of a direct-current motor is short-circuited or grounded, it will not cause this coil to burn out, but would put a higher voltage across the other coils in the machine, which might cause the latter to be destroyed. If there is any electrical disturbance causing this trouble, it is probably due to the controller used on the motor. It is not stated what type of controller is being used or what the motor is used for. The trouble complained of frequently occurs on elevator or other types of automatic controllers where conditions exist that open the field circuit, and cause a heavy voltage to be induced in the coils, which either breaks down the insulation of the field coils or of the armature. It may be that some such condition as this exists in the motor's controller. If so, the controller should be adjusted so as to prevent opening the field circuit.

Condensed-Clipping Index of Equipment

Clip, paste on 3 x 5-in. cards and file as desired

Staybolt, Gallagher Flexible
P. F. Gallagher, Clifton, S. I.
"Power," March 16, 1920

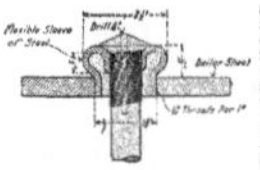

This staybolt consists of a flexible steel sleeve 2⅛ in. at its largest outside diameter. The sleeve is threaded to screw into a 1⅞-in. tapped hole in the boiler sheet, and is also threaded to take a drilled 1-in. diameter staybolt at the threaded portion. The flexible sleeve extends ⅜ in. beyond the boiler sheet, and is $\frac{3}{16}$ in. thick. The staybolt is 1 in. in diameter at the threaded portion. This design permits of a staybolt 1 in. longer than could be used if the staybolt were screwed directly into the boiler sheet. The idea of the design is to obtain flexibility in the steel sleeve instead of having the strain come upon the threaded joint where the staybolt screws into the boiler sheet.

Chain, Sure-Lock Link-Belt Chain
Vulcan Link Belt Co., Inc., Endicott, N. Y.
"Power"

The links of this chain lock over at right angles. The lugs on the inner side bar rides the flanges of the hooks. The links will not detach except when at right angles, which position does not occur even in abnormal operation. The belt operates with the sprocket teeth against the back of the hook, but may be driven in both directions. There is practically no wear or friction on the inner face of the lugs, their purpose being to prevent the links from becoming detached.

Arch, Shaw Suspended Boiler
L. S. Shaw & Co., Union Building, Cleveland, Ohio.
"Power," March 23, 1920

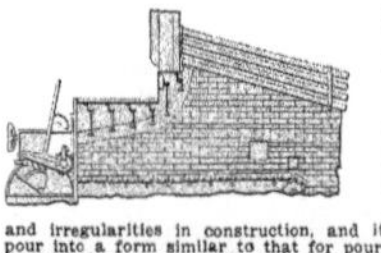

This arch is constructed by suspending forms from crossbeams and filling in between with a compound consisting of firebrick material of silica and alumina fireclay that is highly charged with asbestos fiber, and is free from expansion and contraction. The monolith arch is adaptable to all furnace design and irregularities in construction, and it is only necessary to pour into a form similar to that for pouring cement. After the firebrick has been baked, it presents a firebrick surface that is exposed to the fire all suspension bolts and girders being protected from the heat of the furnace. It is simple to build, easy to repair, and but few repair parts are required to be kept on hand.

Engine, Harrisburg Dual Clearance
Harrisburg Foundry and Machine Works, Harrisburg, Pa.
"Power," March 30, 1920

The latest development in engines of the uniflow type incorporates a design having two clearances. Steam is expanded from a small clearance and, by means of a single-balance piston valve, is allowed to compress first into a large chamber and finally, as the piston nears the end of the stroke, into a smaller clearance. The steam that has been compressed into the large chamber enters, due to the automatic action of a valve, into the cylinder at the opposite end and, mixing with the expanding steam on the return stroke, expands with it and passes out through the central ports to the exhaust line.

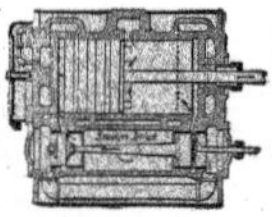

Condenser, New Wheeler Auxiliary Tube-Plate
Wheeler Condenser and Engineering Co., Carteret, N. J.
"Power," Feb. 10, 1920

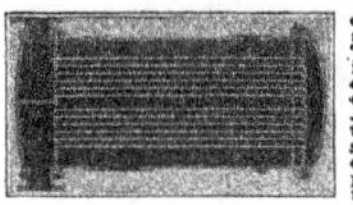

The condenser is used chiefly in connection with salt circulating water. The tubes are expanded into a tube plate, which eliminates the possibility of leakage at that end. At the other end the tubes are provided with standard ferrules and condenser packing. The packing permits the tubes to expand freely and yet maintain an almost leakless joint. At this end of the condenser a thin auxiliary tube-plate or baffle is provided through which all the tubes pass. The holes in this plate are of such size as to make a sliding fit with the tube. In case of leakage, the circulating water drops to the bottom of the compartment between the two tube-plates and is then carried off through a drain at the bottom.

Heater, Russell Storage
Griscom-Russell Co., 90 West St., New York City.
"Power," March 16, 1920

This heater is intended to supply hot water in hotels, apartment houses, factories, etc., and is designed with a number of straight tubes in the steam element of sufficient length to insure that the required heating will be taken care of and that all the steam will be condensed. The drain-tube area is only sufficient to carry the condensation back to the steam head. The brass tubes are expanded in a fixed tube plate at one end and in a floating tube plate at the other, thus permitting expansion and contraction of the tube without strain on the tube joints. The shell is finished with welded construction or riveted as preferred.

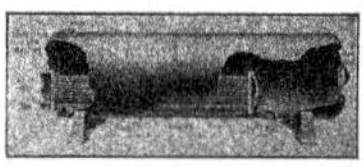

Heat Meter, New Compensated
The Brown Instrument Co., Philadelphia, Pa.
"Power," Sept. 30, 1919

This instrument eliminates the effect of line and thermocouple resistance by means of an operation that is simple and gives accurate results. In testing with this instrument, it is only necessary to connect the thermal couple with the instrument binding posts, press a button, turn a knob and take a reading, which will be the correct electromotive force of the thermal couple at its hot end, even if the line is miles in length. The line may have as much as 15 ohms resistance. The instrument is direct reading through its entire scale range, requires no dry nor standard cells.

Filter, Nugent Automatic Alarm Oil
W. W. Nugent & Co., 150 West Superior St., Chicago, Ill.
"Power," Sept. 23, 1919

This filter is equipped with an automatic alarm consisting of a tilting two-compartment tank, two gongs and a gong beater which moves with the tank. When one of the compartments fills, the tank becomes overbalanced and tilts, whereupon the beater comes in contact with a gong, thus sounding an alarm, which indicates that the oil-filtering bag is overflowing. When the second or third bag begins to overflow, the interval between alarm signals becomes shorter and the alarm is more insistent. The alarm is sounded so as to prevent the overflow of dirty oil into the filtered oil.

Patented Aug. 20, 1918

Steam Versus Electric Driven Steel Mills

At the Pittsburgh monthly meeting, Dec. 20, 1919, of the Association of Iron and Steel Electrical Engineers, a discussion was had on "Steam- Versus Electric-Driven Mills." A brief report of this meeting was published in Power, Jan. 13, 1920. After this discussion considerable time was devoted to collecting and tabulating additional material, and the results of this investigation were published recently in the association's monthly bulletin, of which this article is a brief abstract. The total discussion comprises 40 pages and 11 inserts. Copies can be obtained by addressing the association's office at 513 Empire Building, Pittsburgh, Pa.

WITH a view to contributing toward the elimination of doubts that seem to exist on some of the most essential points regarding main-roll drives for steel mills, H. E. Siebert[1] has prepared and given in his discussion a number of curves and tables which are derived from tests made on some of the largest steam-driven mills in the Pittsburgh and Youngstown districts. For the sake of comparison he has included results of tests from the work of Dr. Puppe on steam- and electric-driven mills in Europe. These curves show principal results of tests on 38-in. and 40-in. reversing blooming mills driven by steam engines. In one case the engine is a 42 x 60-in. twin-simple reversing type, throttle controlled. In another case the engine is a 44 x 70 x 60-in. twin-tandem-compound unit. Another drive analyzed is that of a 52 x 90 x 60-in. tandem-compound engine of the Corliss type, operating condensing. The latter engine drives a 43-in. three-high blooming mill, having a rolling rate of from 170 to 210 tons per hour of 2.5-ton ingots. The table, Fig. 1, gives detailed information on a number of steam and electrically driven mills. It is limited to the tests made by the author and some of the tests of Dr. Puppe, on both steam-engine driven and electric-driven mills.

In comparing the different engines in Fig. 1, it will be seen that the water rate per indicated horsepower-hour on the Mesta single-lever-controlled engine is 23 lb. and on the other heavy engines controlled by the throttle alone it is 32 to 35 lb. per indicated horsepower-hour. This comparison shows what steam economy can be effected by cutoff. The real point—steam per ton—can be obtained for any elongation by multiplying indicated horsepower-hours per ton by the water rate of the engine. The table also shows that the same compound engine, whose non-condensing water rate is 54 lb. per indicated horsepower-hour, has a condensing water rate of about 35 lb. Thus a material saving can be effected by condensing operation.

In applying these data care should be observed in the use of the figure for water rate per "effective horsepower." If

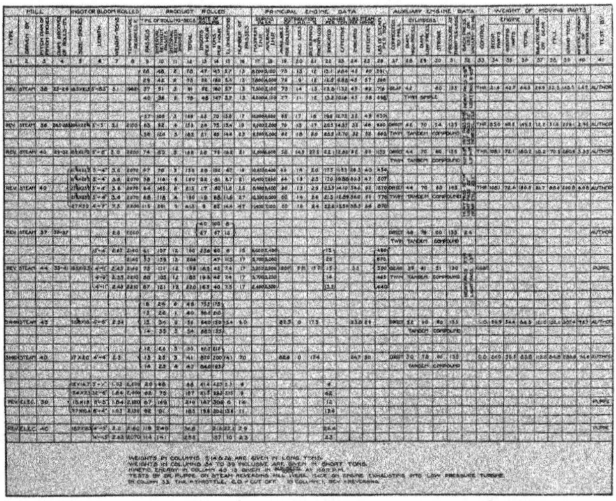

FIG. 1. DETAILED INFORMATION ON A NUMBER OF STEAM AND ELECTRICALLY DRIVEN MILLS

[1]Assistant combustion engineer, Bethlehem Steel Co., Bethlehem, Pa.

this figure is used, the "mill efficiency" should also be used, since in the case of a heavy engine having a high water rate per "effective horsepower," the mill and engine efficiency is low and in that efficiency the losses are expressed in the high water rate per "effective horsepower." In every case *indicated horsepower-hours per ton* × *water rate per indicated horsepower-hour* = *indicated horsepower-hours per ton* × *water rate per effective horsepower-hour* × *mill and engine efficiency.*

As an illustration of the relative economy of steam and electric drives the author takes that of a large plant containing 6 heavy reversing and 2 three-high drives, steam engines being the prime movers. The gross product of the plant (as rolled) can be taken as 100,000 tons per month, the greater part being shapes. The engines are compound condensing, having an average water rate of 38 lb. per indicated horsepower-hour, including condenser auxiliaries, and are in condition to give service for some years, the entire steam plant being about thirteen years old. Steam is generated in stoker-fired central boiler plants at 155 lb. gage, with an average superheat of 50 deg. F. The plant at the present time is short of electric power to the extent of about 4,000 kw., the peak being 7,000 kw. In this particular case the field is almost ideal for a central station of about 25,000 kw. The question is, "Would the economy of the electric mill drive warrant a replacement of the engines?"

In this analysis it is shown that the cost per ton of steel rolled with the steam drive is: Fixed charges, $0.14; repairs and maintenance, $0.10; lubrication, waste and miscellaneous supplies, $0.03; operating labor is taken as equal to that of the electric drive; total cost of steam per ton output, $0.93, making a grand total cost per ton for steam drive of $1.20. On the electric drive the cost per ton is: Fixed charges, $0.42; repairs, maintenance and miscellaneous supplies, $0.02; cost of current, $0.32; cost of labor is taken equal to that for steam drive, making a grand total for the electric drive of $0.76. A saving in favor of the electric drive over the steam of $1.20 — $0.76 = $0.44 per ton, which represents an interest return on the investment of 16 per cent in favor of electric drive.

The foregoing condition is considered one of the most favorable for the steam drive. For a new plant the conditions will be entirely different, at least from the standpoint of fixed charges, because both steam and electric plants will be governed by present-day prices. From the data presented in the curves and tables given in the discussion the author arrives at the following conclusions:

1. The power requirements by different mills rolling at widely different rates show very close agreement. This is especially true of the blooming and billet mills, both reversing and three-high.

2. The data show that the electric reversing mill requires just as much power per unit of elongation as the steam-driven reversing or the three-high mill.

3. The 60 per cent kinetic energy which is stored in the rotating masses of the electric reversing mill and which is reclaimed in the form of electrical energy, does not appear on the curves in favor of this form of drive. This fact is evidently due to the greater moment of inertia of the rotor masses, as compared with that of the steam-driven mills.

4. Acceleration losses in reversing steam drives vary from 12 to 36 per cent of the indicated work, for the light and heavy engines respectively. The lighter the engine is the easier it is to control and the less this loss.

5. The reversing-mill engine, especially in the form of a twin-tandem compound, is a very complicated machine, and as such it demands more attention and repair than the average mill mechanical force can possibly devote to it in the time that is available for such repairs.

The single-lever controlled engine is a forward step in the direction of steam economy, but it imposes a heavy penalty in the form of acceleration loss. Can we not build a light, yet a simple and strong engine?

6. The motor-driven mill possesses decided advantages in the way of mechanical simplicity, space required, cleanliness of plant, etc., aside from its economic advantage.

7. No fixed rule can be laid down for the relative economy of the steam and electric drives for different plants, because the plant conditions vary within wide limits. For the case taken, the advantage is in favor of the electric drive. It does not follow, however, that this advantage would obtain in another case. Each particular plant must be studied in detail. Such a study may show that the economic gain in one plant does not apply to another plant where the conditions may be different.

In his discussion the following data on steam consumption, as obtained from a 55x60-in. twin-simple-reversing engine driving a 40-in. blooming mill, were presented by B. E. Eppelsheimer[2]:

DATE—JULY 30, 1919

Average steam pressure, lb. gage	160
Duration of test, hours	17
Average feed-water temperature, deg. F	210
Total boiler-feed water, lb	1,690,000
Total loss through safety valves and blowoff	144,000
Net steam to engine, lb	1,546,000
Average steam per hour, lb	90,940
Maximum flow, steam per hour, lb	150,750
Minimum flow, steam per hour, lb	25,000
Size of ingots	18 x 20-in.
Size of blooms	7 x 4-in.
Number of ingots rolled	436
Weight of ingots rolled, tons	1,014 (FP 832)
Steam per ingot, lb	3,570
Steam per ton steel; lb	1,520

In addition to the foregoing data, obtained when the mill was rolling steel, the steam consumption was determined for the engine and mill running idle and for the engine alone with the mill disconnected. During these tests there were no reversals of the engine, so that there were no plugging strokes as in actual rolling. The steam consumption under these conditions was too small to be measured accurately by the venturi meter and was obtained by flow meters in the steam lines from the individual boilers. Since the steam flow was practically constant, this type of meter is accurate under these conditions. The results of these idle tests to determine the friction load are as follows:

Friction load	Engine and Mill	Engine Alone
Duration, minutes	20	20
Engine speed, r.p.m.	41	40
Steam per hour, lb.	36,900	29,610
Steam per revolution, lb.	15.0	12.35

This blooming-mill engine exhausts into receivers which supply steam to low-pressure turbines run condensing. Since the low-pressure steam to the turbines replaces high-pressure steam of the same total available heat energy, the blooming-mill engine should be credited with the heat

ELECTRICAL OPERATING COSTS PER TON.

YEAR	1913	1914	1915	1916	1917	1918	1919	AVERAGE	% TOTAL
MONTHS IN OPERATION	9	8	12	12	12	12	◆8	73	
TONNAGE	119.230	92.622	174.460	238.050	297.224	276.778	155.515	1.353.879	
KWH PER TON	23.9	22.8	21.5	19.8	20.4	21.8	23.4	21.2	
POWER COST	.160	.153	.144	133	.137	.146	.157	.1457	86.7
REPAIRS AND MAINTENANCE	.0069	.0092	.0045	.0102	.0039	.0030	.0027	.0057	8.0
MISCELLANEOUS SUPPLIES	.0045	.0049	.0025	.0025	.0029	.0015	.0051	.0032	3.4
LABOR IN OPERATION	.0141	.0161	.0128	.0100	.0128	.0147	.0146	.0134	1.9
TOTAL COST	.1855	.1832	.1638	.1557	.1566	.1652	.1794	.1680	100.00

TOTAL ELECTRICAL COSTS PER TON

YEAR	1913	1914	1915	1916	1917	1918	1919	AVERAGE	% TOTAL
TOTAL OPERATING COSTS	.185	.183	.164	.156	.157	.165	.179	.1680	49.17
INTEREST ON INVESTMENT	.078	.101	.054	.040	.032	.035	.041	.0460	13.46
DEPRECIATION (20 YEARS)	.065	.084	.045	.033	.026	.028	.034	.0383	11.21
TOTAL MAIN DRIVE	.328	.368	.263	.229	.215	.228	.254	.2523	73.84
★ MISCELLANEOUS	.126	.133	.115	.078	.061	.063	.077	.0894	26.16
TOTAL MILL ELECTRICAL CHARGE	.454	.501	.378	.307	.276	.291	.331	.3417	100.00

★ ALL OTHER ELECTRICAL CHARGES, INCLUDING OVERHEAD.
◆ UP TO AND INCLUDING AUGUST 31, 1919.

FIG. 2. DATA ON 34-IN. ELECTRICALLY-DRIVEN REVERSING MILL

[2]Steam engineer, American Rolling Mills Co., Middletown, Ohio.

energy available in the exhaust. The available energy between 160-lb. gage and 5-lb. gage through the blooming-mill engine is 43.4 B.t.u. per pound of steam. The available energy between 5-lb. gage and 28-in. vacuum to the turbine is 39.1 B.t.u. per lb. of steam. The total available energy for high-pressure steam between 160-lb. gage and 28-in. vacuum is 82.5 B.t.u. Each pound of exhaust steam used in the turbine saves 0.457 lb. of high-pressure steam.

The data presented in Fig. 2 were given by E. S. Jefferies*, for an electrically driven 34-in. reversing mill, which has been in operation for 6½ years, during which time it has rolled 1,353,879 net tons of various sizes, varying in size from 10-in. rounds down to 4 x 4's, the average size being 5 x 5 in. This equipment has given practically no trouble, as illustrated by the fact that in the last 4½ years the total delays charged by the mill department against the equipment was 11 hours, 35 minutes, which is roughly one minute delay for every 2,000 tons rolled.

The electric cost data over the 6½-year period for an 18-in. four-stand billet mill, which is driven by a 1,600-hp. motor, through a herringbone reduction gear, and rolls 4 x 4's to 1¾ in. square, are as follows:

Total tonnage rolled—697,012 net tons		Per Cent
Repairs and maintenance	$0.0028	1.3
Labor in operation	0.0249	11.3
All power	0.1740	78.8
Interest and depreciation	0.0191	8.6
Total electrical cost	$0.2208	100.0

The total time delay on this mill is not more than one minute in a minimum of four to five thousand tons rolled.

Similar cost data during the same 6½ years for a rod and wire mill, which is a Morgan continuous mill, having ten 12-in. roughing stands and six 10-in. finishing stands, rolling 1¾-in. billets to No. 5 rods, are as follows:

Total tonnage rolled—395,925 tons		Per Cent
Repairs	$0.0175	1.7
Labor	0.0457	4.4
All power	0.8140	78.1
Interest and depreciation	0.1660	15.8
Total cost per ton	$1.0432	100.0

The electrical equipment on this mill comprises a 3,200-hp. synchronous motor driving two 1,200-kw. generators, which in turn drive two 1,600-hp. variable-speed motors controlled by the Ward-Leonard system. Here also the delays have been so few, especially in the last four years when there has been no delay charged whatsoever.

Others contributing to the discussion are: K. A. Pauly, general engineer, General Electric Co.; G. E. Stoltz, general engineer, Westinghouse Electric and Manufacturing Co.; B. C. Fennell, engineer, Nordberg Co., Milwaukee, Wis.; B. R. Shover, consulting engineer, Pittsburgh, Pa.; F. G. Cutler, steam engineer, Tenn. C., I. & R.R. Co., Birmingham, Ala.; R. H. Keil, power engineer, Jones & Laughlin Steel Co., Pittsburgh, Pa.; and Gordon Fox, electrical engineer, Steel and Tubes Company of America, Indiana Harbor, Ind.

Massachusetts Board of Boiler Rules Has Hearing in Boston

The Massachusetts Board of Boiler Rules convened in Boston, Thursday, May 6, at its semi-annual meeting to consider changes in rules. The meeting was presided over by John H. Plunkett. Two members of the board, John A. Collins and H. H. Lynch, and former Chief Inspector George A. Luck, were present.

Mr. Stetson, a representative of the Bigelow Boiler Co., New Haven, Conn., requested permission to weld longitudinal joints in vertical fireboxes of boilers and called upon F. L. Fairbanks to present the case for him. Mr. Fairbanks conceded that certain seams of boilers could be welded without danger to the structure. He said that as a whole in the welding of a firebox or inner sheet, the parts where the weld is made are in compression and the severe stresses are distributed on the staying. All large railroads allow welding of joints in flat plates in fireboxes. No trouble has been experienced and the trials have been very successful. It is now a regular practice. For several years, the A. S. M. E. did not permit welding of seams, but now allows this process if the stress is carried by the construction. In the firebox boiler when the strength is carried by the staying of the circular outside wrapper, the sheet is safer, and the Bigelow Boiler Co. feels warranted in asking permission to weld longitudinal joints of vertical fireboxes. Mr Fairbanks then went on to say that the Bigelow people were prepared to do welding by both the oxyacetylene and the electric processes. Mr. Fairbanks, who has had much experience along these lines, advocated this method for the less important seams of the boiler.

Mr. Stewart, head of the Stewart Boiler Works, Worcester, Mass., opposed the A. S. M. E. rules, for the reason that just such processes as here proposed, jeopardize the safety of boilers. If such practices are followed, the factor of safety is lost. The board had ruled in the past that the forged weld only could be used, and this shows that the board has previously considered autogenous welding unsafe. Mr. Stewart also stated that welded seams would eventually leak. As to compression this depended on the staybolts—if they are always there—but if broken they cannot support the weld. He said that he had always opposed anything in the Massachusetts rules that tends to lower the safety standard.

Fred. E. Dillon, of the Dillon Boiler Works, of Fitchburg, Mass., at this juncture asked Mr. Fairbanks if he thought that staybolts might break and not support the welded joints.

Mr. Fairbanks replied that he had thought so at one time, but that as chairman of a subcommittee of the Boiler Code Committee he had done a great deal of study and investigation work along these lines. As a result of this research work, in his opinion the furnaces of vertical-tube boilers were under compression and self-sustaining and staybolts were not necessary for the support of the firebox sheet. He said, however, that if the water were to leave the boiler, the staybolts might prevent bulging between stays and therefore were beneficial. The Massachusetts Boiler Rules permit forge welding. I have a large number of forge-welded tanks. It takes a very expensive equipment to forge weld and that is its greatest objection. I believe, however, that autogenous welding when the stress is carried by other means than the weld, is safe and reasonable. Construction should be allowed as provided for in the A. S. M. E. Boiler Code.

Mr. Fessenden, of the Petroleum Products Co., petitioned the board for permission to tap heating boilers for hot-water circulation to the oil-supply pipes in order to reduce the viscosity of the oil. He introduced Mr. Fiske, a mechanical engineer of the company, who pointed out that this arrangement of tapping is comparable with the Lamphrey mouthpiece construction but has not so many joints.

Charles E. Gorton, of the American Uniform Boiler Law Society, asked the board to consider the following amendment to the rules:

Although a boiler may not conform in all respects to the requirements of the boiler rules as they exist upon the date when this amendment is passed, yet if such boiler shall have been found upon inspection during construction by a boiler inspector duly qualified under the provision of Section 4, Chapter 465 of the Acts of 1907, as amended to conform in all respects to the Rules for the Construction of Stationary Boilers and for Allowable Working Pressures formulated by the Boiler Code Committee of the American Society of Mechanical Engineers and submitted to the Council of that Society December 3, 1918, and if such boiler shall be stamped by such inspector in accordance with the provisions of paragraph 332 of said rules, and if such boiler shall be found upon inspection by the Division of Inspections to conform in design, construction and workmanship with the requirements of said rules, then such boiler shall be certified as inspected by the Division of Inspections for operation in Massachusetts and shall be allowed a working pressure computed in accordance with said Rules for Allowable Working Pressures formulated by said Boiler Code Committee; provided, however, that where said rules conflict with any provision of the statutes of the Commonwealth of Massachusetts the latter shall apply.

*Electrical engineer, Steel Company of Canada, Ltd., Hamilton, Ont.

Mr. Dillon, speaking on behalf of the New England Association of Boiler Manufacturers, said that at the March meeting of that association a resolution had been passed that the past work of the Board of Boiler Rules be approved and that no acceptance of boilers of any lower standard be permitted. This was an indirect intimation that the boilers manufactured in accordance with the regulation of the A. S. M. E. Boiler Code were of an inferior quality and lower standard.

Mr. Gorton interrupted to call attention to the fact that all of the members of the New England Association of Boiler Manufacturers made boilers to the A. S. M. E. standard and asked why any manufacturer should build boilers for outside states that were unsafe and if it was necessary to prove that boilers manufactured under the A. S. M. E. Code were safe. "There are really very few differences between the boilers made under the A. S. M. E. Code and those made under the Massachusetts Code," said Mr. Gorton, and he then read a list of the sixteen states that operate under the A. S. M. E. Code.

Mr. Dillon replied that it was true that there was very little difference, but said that those were the differences to which they referred and which they wanted to preserve and that the New England Boiler Manufacturers Association stood up for Massachusetts rules.

A meeting of the executive committee will be held in the near future to reach decisions on the cases brought before the board at the hearing.

Latest New York Power Plant To Have Large Units

The two 60-cycle turbo-generators which will be installed in the new Hell Gate station of the United Electric Light & Power Co., of New York, will be of the tandem-compound type, each rated at 43,750 kva. at 80 per cent power factor or 35,000 kw. They will be designed to carry peak loads of 40,000 kw. for short periods. These units, which will be built by the Westinghouse Electric & Manufacturing Co., will be similar to the 60-cycle units installed in the Northwest station of the Commonwealth Edison Co. The generators will be designed for 13,200 volts, and the units will operate at 1,200 r.p.m.

The turbines will be designed to operate at 220 lb. steam pressure, 200 deg. F. superheat and 29 in. vacuum. They will have reaction blading throughout. The high-pressure element of this type of machine is single-flow and the low-pressure element is double-flow.

Obituary

John Wesley Hyatt, inventor of the Hyatt roller bearing, died suddenly on May 10, of heart disease, in his eighty-third year, at his residence, Windermere Terrace, Short Hills, N. J.

Personals

Barton R. Shover, Consulting Electrical, Iron and Steel Plant and Steam Engineer, Oliver Building, Pittsburgh, Pa., has added to his organization W. C. Rott as chief engineer.

Charles F. Lederer, until recently superintendent of way, of the Milwaukee (Wis.) Electric Railway and Light Co., has become associated with the rail-welding department of the Metal and Thermit Corp., of New York.

J. H. McElhinney recently resigned as chief engineer of the Youngstown Sheet and Tube Co., Youngstown, Ohio, to become chief engineer for the Columbia Steel Co., Elyria, Ohio.

G. E. Stoltz, general engineer of the Westinghouse Electric and Manufacturing Co., East Pittsburgh, Pa., has been appointed engineer in charge of the steel-mill section of the engineering department.

William F. Mallay, who during the war was chief engineer of the United States Destroyer "Stevens" and previous to the war was with the Lincoln Manufacturing Co., Fall River, Mass., is now in charge of the new plant of the New England Oil Refining Co.

Society Affairs

The American Society of Safety Engineers will meet on Friday, May 28, at the Engineering Societies Building, West 39th Street, New York. The meeting will be devoted to the safety features of compressed-air work in tunnel and foundation construction.

The American Society of Mechanical Engineers is planning the formation of a professional section covering the handling and distribution of material. The purpose of such a section is to attract the attention of the technical and business minds to this phase of commerce.

The Associated Manufacturers of Water Purifying Equipment will meet in Montreal, Canada, in conjunction with the annual convention of the American Water Works Association during the week of June 20. Reports of the committees on standardization of contracts and sizes of filters, together with tentative forms and schedules, will be submitted at the meeting.

The New England States Association of the National Association of Stationary Engineers will hold its annual convention at Worcester, Mass., July 8, 9 and 10, and in connection the annual exhibition of power-plant machinery, accessories and supplies will be conducted July 7, 8 and 9 under the auspices of the New England Association of Commercial Engineers. The annual meeting and election of officers will be held at this time by the N. E. A. C. E. Adequate entertainment has been planned for the delegates and visitors to the convention.

The American Welding Society at a meeting in New York City, on April 22, elected the following officers: President, J. H. Deppeler, Metal and Thermit Corp.; vice president (two-year term), J. W. Owens, Norfolk Navy Yard; vice president (one-year term), D. B. Rushmore, General Electric Co.; directors (three-year term), W. M. Baird, Linde Air Products Co.; A. M. Candy, Westinghouse Electric and Manufacturing Co.; C. J. Halsey, Electric Arc Cutting & Welding Co.; C. J. Jones, consulting engineer; C. A. McCune, Page & Wire Co.; S. W. Miller, Rochester Welding Works; Joah Bragden, David Lipton Sons Co.; P. F. Willis, Henderson-Willis Welding and Cutting Co.

The American Society of Heating and Ventilating Engineers will hold its semi-annual meeting May 26, 27 and 28 at St. Louis, Mo., in the Hotel Statler. The opening day will be devoted to registration and business meetings, reports of committees, etc. During the afternoon a joint meeting with the American Society of Refrigerating Engineers will be held, and in the evening a theatrical entertainment will be given at the Municipal Open-Air Theater, Forest Park. On Thursday there will be a joint meeting with the American Society of Mechanical Engineers and the American Society of Refrigerating Engineers, a boat ride on the Mississippi River and a banquet at the Hotel Statler, jointly with the American Society of Refrigerating Engineers. Friday will be devoted to professional sessions.

The American Boiler Manufacturers' Association will hold its thirty-second annual convention at the French Lick Springs Hotel, French Lick, Ind., May 31, June 1 and 2. The first day of the convention will be taken up by reports concerning the A. S. M. E. Boiler Code. The reports deal with the promulgation of the Code, simplification of data sheets, stamping of boilers, etc. A paper on "Fuel Conservation" will be presented by David Moffat Myers, consulting engineer, New York. In the afternoon delegates may indulge in tennis, bathing, riding, croquet and other sports. During the third session on Tuesday, June 1, the reports of the treasurer and of the auditing committee will be read. This session is for members only. The last day of the convention will be given over to the report of the nominating committee and the election and installation of officers.

Miscellaneous News

The Pensylvania State College is opening a two weeks' summer course in Industrial Organization and Management from Aug. 9 to 21. It is designed to meet the needs of manufacturers, superintendents, employment directors, foremen, accountants, and all others who are concerned with the daily affairs of industry and who are willing to devote two weeks to intensive study.

The Franklin Institute at a meeting on May 19, presented the Franklin Medal to Sir Auckland Geddes, British Ambassador, on behalf of the British government for the Honorable Sir Charles A. Parsons, Newcastle-on Tyne, England, and to W. A. F. Ekengren, Minister of Sweden on behalf of the Swedish government for Professor Svante August Arrhenius, Noble Institute, Stockholm, Sweden.

Engineering Council, in the referendum on a National Department of Public Works conducted by the United States Chamber of Commerce, voted in favor of a Department of Public Works established by the National Government. The Council also favored a Department of Public Works to be established by a suitable modification of the existing Department of the Interior but was opposed to the creation of an entirely new department.

Business Items

The Steam Motors Co., Springfield, Mass., announces the opening of a new office at 50 Church St., New York City.

The Dwight Manufacturing Co., of Chicago, has moved its business offices from 12 and 14 South Jefferson St. to 565 West Washington Boulevard. The new quarters are larger, and the company will have better facilities for taking care of a greatly increased business.

Trade Catalogs

The Joseph H. Roach & Company, Inc., Philadelphia, Pa., has completed publication of a new 15-page catalog entitled "The Roach Stoker." The booklet contains illustrations and descriptions of the stokers handled by this company. A copy will be sent on request.

The Metal & Thermit Corp., New York City, has issued a third and revised edition of its pamphlet entitled "Laboratory Experiments with the Thermit Process of Welding." The pamphlet illustrates and describes various experiments which are intended to show the speed of reaction, the heat produced thereby and effect obtained by superheated liquid slag and superheated liquid steel.

COAL PRICES

Prices of steam coals both anthracite and bituminous, f.o.b. mines, unless otherwise stated, are as follows:

ALANTIC SEABOARD

Anthracite—Coals supplying New York, Philadelphia and Boston:

	Mine
Pea	5.30
Buckwheat	$3.40@$3.75
Rice	2.75@ 3.25
Barley	2 25@ 2.50
Boiler	2.50

Bituminous—Steam sizes supplying New York, Philadelphia and Boston:

Latrobe	$4.25@$4.50
Connellsville coal	4 25@ 4.50
Cambrias and Somersets	4 35@ 5.15
Clearfeilds	4 00@ 4.75
Pocahontas	6.50@ 7.00
New River	6.50@ 7.00

BUFFALO

Bituminous—Prices f.o.b. Buffalo:

Pittsburgh slack	$6.00
Mine-run	6.25
Lump	6.00@ 6.50
Youghiogheny	6.75

CLEVELAND

Bituminous—Prices f.o.b. mines:

No. 6 slack	$3.25@$4.00
No. 8 slack	4.25
No. 8—¾-in.	3.50@ 4.00
No. 6 mine-run	3.50@ 4.00
No. 8 mine-run	4.50
Pocahontas—Mine-run	3.25@ 4.00

ST. LOUIS

Anthracite—Probably not more than 20 per cent the demand in this market can be supplied. Prices effective Apr. 1 were as follows:

	Williamson and Franklin Counties	Mt. Olive and Staunton	Standard
Mine-run	$2.65@2.80	$3.00@3.50	$3 75@4.25
Screenings	2.50@2.65	2.50@3.00	2.75@3.50
Lump		3.75@4.50	4.00@4.50

Williamson-Franklin rate to St. Louis is $1.10; other rates 95c.

CHICAGO

Bituminous—Prices f.o.b. mines:

		Freight rate Chicago
Illinois		
Southern Illinois		
Franklin, Saline and Williamson Counties		
Mine-Run	$3.00@$3.10	$1.55
Screenings	2.60@ 2.75	1.55
Central Illinois		
Springfield District		
Mine-Run	$2.75@$3.00	$1.32
Screenings	2.50@ 2.60	1.32
Northern Illinois		
Mine-Run	$3.50@$3.75	$1.24
Screenings	3.00@ 3.25	1.24
Indiana		
Clinton and Linton		
Fourth Vein		
Mine-Run	$2.75@$2.90	$1.27
Screenings	2.50@ 2.65	1.27
Knox County Field		
Vein		
Mine-Run	$2.75@$2.90	$1.37
Screenings	2.50@ 2.60	1.37
Mill Block	$4.25@$4.50	$1.27

New Construction

PROPOSED WORK

R. I., Providence—The Packard Motor Car Co., B'way and 61st St., will soon award the contract for the construction of a 2 story service station on Elmwood Ave. A steam heating system will be installed in same. Total estimated cost, $100,000.

N. Y., Brooklyn—B. F. Keith, 1564 Bway., New York City, plans to build a theatre on Gold, Fleet and Prince Sts. A steam heating system will be installed in same. Total estimated cost, $750,000.

N. Y., Buffalo—The Buffalo Shoe Hospital, 509 Main St., is in the market for a 3 hp., 60 cycle, single phase, 220 volt motor.

N. Y., Buffalo—The Niagara & Erie Power Co., 1600 Marine Trust Bldg., is having plans prepared for the constuction of a high tension system. Estimated cost, $250,000.

N. Y., Buffalo—The Niagara, Lockport and Ontario Power Co., 1600 Marine Trust Bldg., is preparing plans to double their cables between Lyons and Syracuse, to meet industrial demands. Total estimated cost, $300,000.

N. Y., Ithaca—The New York State Gas & Electric Corp. of Ithaca has made application to the Pub. Serv. Comn., 2nd Dist. for permission to mortgage and issue $2,600,400 bonds to improve plant.

N. Y., New York—The Audubon Sporting Club, c/o David F. Lang, Archt. and Engr., 110 West 34th St., plans to build a 2 story, 100 x 175 ft. club and arena at 145th St. and Lexington Ave. A steam heating system will be installed in same. Total estimated cost, $300,000.

N. Y., New York—F. Begrisch and C. L. Acker, 200 B'way, plan to build a 12 story office building on 36th St. and 5th Ave. A steam heating system will be installed in same. Total estimated cost, $3,500,000.

N. Y., New York—The Bronx Society for Prevention of Cruelty to Children, 355 East 137th St., plans to build a 7 story, 96 x 134 ft. home on Park Ave. and 16th St. A steam heating system will be installed in same. Total estimated cost, $350,000.

N. Y., New York—Sam Harris and Irving Berlin, 226 West 42nd St., are having plans prepared for the construction of a 2 story, 100 x 100 ft. theatre at 239 West 45th St. A steam heating system will be installed in same. Total estimated cost, $400,000.

N. Y., New York—F. Savignano, Archt. 6005 14th Ave., Brooklyn, plans to build a 12 story hotel on 72nd St. and Columbus Ave. A steam heating system will be installed in same. Total estimated cost, $650,000. Owner's name withheld.

N. Y., Schenectady—The Bd. Educ. will receive bids for the construction of a 126 x 175 ft. building, 2 story anemic section, 45 x 80 ft. auditorium at the Mt. Pleasant School, here. Heating and power systems will be installed in same. Total estimated cost, $400,000. E. G. Atkinson, 426 State St., Archt.

N. Y., Sodus—The Wayne Power Co. has obtained permission from the Pub. Serv. Comn., to construct electric light plants in the towns of Fremont, Howard, Hornellsville and Steuben County.

N. Y., Tonawanda—The Tonawanda Power Co. plans to increase equipment in power plant. Estimated cost, $150,000.

N. J., Newark—The Bd. Educ., City Hall, will have plans prepared for the construction of a school building on Oliver St. A steam heating system will be installed in same. Total estimated cost, $400,000. John H. and Wilson C. Ely, Firemans Trust Bldg., Archts.

N. J., Newark—The Bd. of Educ., City Hall, will receive bids until June 8 for the construction of a 3 story, 219 x 222 ft. school building on 1st and 2nd Sts. A steam heating system will be installed in same. Total estimated cost, $1,125,000. John H. and Wilson C. Ely, Firemans Trust Bldg., Archts.

N. J., Newark—The Pub. Serv. Corp. of New Jersey, Terminal Bldg., plans to construct a power plant. A steam heating system will be installed in same. Total estimated cost, $15,000,000.

N. J., South River—The Bd. Pub. Wks. and the Boro. Council received bids (lowest 3) for the construction of a turbine engine plant at the power house (all work), from John R. Proctor Eng. Co., 74 Cortland St., $197,368, Equity Eng. Co., 30 Church St., $211,726, Leslie Stevens Co., 120 B'way, $214,900; for installing turbo generators, Kerr Turbine Co., 30 Church St., $36,550, Genl. Electric Co., 120 B'way, $40,570, N. W. Kellogg Co., 90 West St., $46,465, contractors all of New York City; boiler, Babcock & Wilcox, 85 Liberty St., New York City, $29,908, Pardell Corp., 304 Bailey Bldg., Phila., $30,556; soot blowers, Babcock & Wilcox, $1,835; condensers and air circulating pumps, Wheeler Condenser & Eng. Co., 149 B'way, New York City, $19,350. Noted May 3.

Pa., Philadelphia—J. A. Wetmore, Supervising Archt., Treasury Dept., Wash., D. C., will receive bids until June 1 for the installation of a motor driven triplex pump at the United States Post Office, here.

Pa., Washington—The Hazel Atlas Glass Co. plans to build a 1 and 2 story, 100 x 368 ft. glass factory. A steam heating system will be installed in same. Total estimated cost, $500,000.

Md., Baltimore—McCawley & Co., Lombard and Commerce Sts., is having plans prepared for the construction of a 4 story, 175 x 190 ft. shirt factory on North Ave. and Wolfe St. A steam heating system will be installed in same. Total estimated cost, $500,000. Lockwood, Greene & Co., 101 Park Ave., New York City, Archts.

Va., Norfolk—C. N. Whitehurst & Co., 225 Plume St., plans to build a 12 story office building on York and Bush Sts. A steam heating system will be installed in same. Total estimated cost, $750,000. Herts & Robertson, 331 Madison Ave., New York City, Archts. and Engrs.

Va., Richmond—Herts & Robertson, Archts., 331 Madison Ave., New York City, plans to build a 6 story professional office building. A steam heating system will be installed in same. Total estimated cost, $400,000. Owner's name withheld.

N. C., Greensboro—The Revolution Cotton Mills is in the market for 2 new boilers. G. P. Stone, Supt.

Fla., St. Petersburg—The Citizens Ice & Cold Storage Co. plans to increase its plant by the addition of a 75 ton ice plant.

Fla., West Palm Beach—Paris Singer plans to build a dredging and cold storage plant here. Estimated cost, $350,000.

La., New Orleans—Farrot & Levandais, Archts., Title Guarantee Bldg., will soon award the contract for the construction of an 8 story cotton exchange building, for the New Orleans Cotton Exch., Weis Bldg. A steam heating system will be installed in same. Total estimated cost, $1,500,000.

O., Cleveland—The Cleveland Ry. Co., Leader-News Bldg., will receive bids until May 27 for the construction of a 1 story, 44 x 67 ft. electric sub-station on East 99th St. and St. Clair Ave. Two 1,500 kw. rotary converters with blowers and generators will be installed in same. Total estimated cost, $100,000. L. P. Crocelius, 650 Leader-News Bldg., Archt. and Engr.

O., Cleveland—The Lutheran Hospital Co., c/o F. W. George, 2609 Franklin Ave., plans to build a 4 story hospital. A steam heating system will be installed in same. Total estimated cost, $200,000. W. S. Longer, Marshall Bldg., Archt. and Engr.

O., Cleveland—The Tifereth Israel Association, c/o Rabbi Silver, East 55th St. and Central Ave., is having plans prepared for the construction of a 2 story, 140 x 200 ft. temple on East 105th St. and Ansel Rd. A steam heating system will be installed in same. Total estimated cost, $1,000,000. Hubbell & Benes, 4500 Euclid Ave., Archts.

O., Galion—The City Council has authorized an expenditure of $75,000 for the enlargement of the electric light plant. A $60,000 bond issue will be voted upon for this project.

O., Sidney—The C. B. De Weese Co. is having plans prepared for the construction of a 3 story, 83 x 166 ft. office and theatre building. A steam heating system will be installed in same. Total estimated cort. $200,000. J. O. Adam, New Southern Hotel, Columbus,, Archt.

O., Solon—The Solon Tire Co., Union Bldg., Cleveland, is having plans prepared for the construction of a 2 story, 100 x 300 ft. factory and boiler house. Two 150 hp. boilers will be installed in same. Total estimated cost, $125,000. Donald C. Smith, Lennox Bldg., Cleveland, Archt. and Engr.

O., Toledo—The Toledo Ry. & Light Co. plans to construct a sub-station to replace the Casino Sta. recently destroyed by fire. The new station will have a capacity of 10,000 kw. and will serve the lower town industries, the new intercepting sewerage pumping station and the Toledo Beach car line. Estimated cost, $225,000. C. Derge, Asst. Genl. Mgr.

Ind., Gary—Z. Erol Smith, 305 East 55th St., Chicago, will soon award the contract for the construction of a 1 story, 50 x 400 ft. factory for the O. K. Giant Battery Co., 515 B'way. A steam heating system will be installed in same. Total estimated cost, $75,000.

Mich., Detroit—Dodge Bros., Joseph Campau Ave., plan to construct a 100 x 100 ft. power house. Four 12,000 hp. steam turbines, four 12,000 hp. boilers, automatic coal handling conveyors, stokers, and ash handling conveyors and complete steam driven, turbo-generator sets will be installed in same. Total estimated cost, $2,000,000.

Mich., Detroit—The Newton Packing Co., 1041 14th Ave., will soon award the contract for the construction of a 1 story, 68 x 95 ft. packing house on 14th Ave. An electric power motor will be installed in same. Total estimated cost, $40,000. Mildner & Eisen, 924 Hammond Bldg., Archts.

Mich., Detroit—The Sacred Heart Seminary, 155 McDougal Ave., engaged Donaldson & Meier, Archts., 1314 Penobscot Bldg., to prepare plans for the construction of a 3 and 4 story seminary on Linwood Ave. and Joy Rd. Plans include a separate steam heating plant. Total estimated cost, $2,500,000. W. W. Walker, Chn.

Mich., Detroit—Julius Stroh, 1676 Jefferson Ave., plans to build an 18 story, 80 x 100 ft. office building at the West Grand Circus Park. Steam heating equipment will be installed in same.

Mich., Highland Park (Detroit P. O.)—John H. Kunsky, 501 Madison Theatre Bldg., engaged C. Howard Crane, Archt., Huron Bldg., to prepare plans for the construction of a 2 story, 127 x 200 ft. theatre on Woodward and Waverly Aves. A steam heating and forced ventilation system will be installed in same. Total estimated cost, $300,000.

Mich., Saginaw—The Herzog Art Furniture Co., 1900 South Michigan Ave., will soon award the contract for the construction of a 3 story, 85 x 227 ft. factory. Steam heating boilers, electric power, etc., will be installed in same. Total estimated cost, $95,000. Cooper & Beckbissinger, 114 South Jefferson St., Archt.

Ill., Chicago—Graham, Anderson, Probst & White, Archts., 1417 Ry. Exch. Bldg., will soon award the contract for the construction of a 6 story, 73 x 100 ft. office building at 325 South Wells St. for the Pennsylvania Offices, Inc. A steam heating system will be installed in same. Total estimated cost, $350,000.

Ill., Lockport—The Penitentiary Comn. will receive bids until June 1 for the furnishing and installation of four 300 hp. water tube boilers and stokers in the power house at the Illinois State Penitentiary, near here.

Wis., Milwaukee—The Central Bd. of Purchases received bids May 10 for furnishing a 1,500 gal. combination motor driven street flusher and sprinkler, from the Pauly Motor Truck Co., 2530 North Ave., $8,775; Sterling Motor Truck Co., 16th Ave. and Rogers St., West Allis, $9,427; Amer. Auto Co., 187 Wisconsin St., $9,969.

Wis., Milwaukee—The Federal Eng. Co., Engr., Stephenson Bldg., will soon award the contract for the furnishing of coal conveying machinery for the Natl. Knitting Co., 905 Clinton St.

Minn., Chisholm—Ernest Drew, Town Recdr., will receive bids until July 1 for the construction of a 2 story, 64 x 298 ft. building on Lake St. A steam and mechanical ventilating system will be installed in same. Total estimated cost, $450,000.

Minn., St. Paul—The Times Auto Supply Co., 56th and B'way, New York City, is having preliminary plans prepared for the construction of a 6 story, 98 x 130 ft. office-warehouse on St. Peter and 6th Sts. A steam heating system and electric motors will be installed in same. Total estimated cost, $300,000. Toltz, King & Day, 1410 Pioneer Bldg., Archts.

Kan., Americus—The city voted $18,000 bonds to build an electric transmission line from here to Emporia. Noted May 10.

N. D., Grand Forks—The Industrial Comn. of North Dakota, Bismarck, will receive bids until May 29 for the construction of a reservoir, filter plant, power house and plant. Boilers, stokers, etc., will be installed in same. C. L. Pillsbury Co., 2305 Oliver Ave., S., Minneapolis, Minn., Engr.

Mo., St. Louis—The city plans to install one 220 hp. engine and a horizontal single crank engine for the pumping station. J. A. Brooks, Engr.

Cal., San Diego—The Bureau of Yards & Docks, Navy Dept., Wash., D. C., plans to install cold storage insulation. Estimated cost, $125,000.

Cal., San Francisco—The Pacific Gas & Electric Corp., 445 Sutter St., plans to improve and construct additions to the electrical generating and distributing system, including hydro-electric plants along the Pitt River, transmission lines, etc. Total estimated cost, $13,062,426. John Britton, Vice-Pres.

M. I., Guam—The Bureau of Yards & Docks, Navy Dept., Wash., D. C., will soon award the contract for the installation of turbo alternators, exciters and switchboards. Estimated cost, $33,000.

Ont., London—J. M. Moore, Archt., 425 Richmond St., will receive bids until May 31 for the construction of a 4 story, 75 x 180 ft. hospital on St. James St. Boilers, radiators, etc., will be installed in same. Total estimated cost, $160,000.

Ont., Toronto—The Famous Players Canadian Corp., Temple Bldg., is having plans prepared for the construction of a 2 story motion picture theatre. A steam heating system will be installed in same. Total estimated cost, $300,000. J. M. Jeffrey, 348 Confederation Life Bldg., Archt.

Ont., Wroxeter—The Town Council plans to build a power plant. Estimated cost, $15,000. Mr. Flint, Hydro Bldg., Toronto, Engr.

CONTRACTS AWARDED

Mass., Boston—Snider & Druker, 18 Tremont St., will build an 8 story, 60 x 100 ft. mercantile building on Kneeland Ave. A steam heating system will be installed in same. Total estimated cost, $500,000. Work will be done by day labor.

Mass., Boston—Swift & Co., 60 North Market St., has awarded the contract for the construction of a 2 story office building and a 2 story extension to boiler room, to the R. F. Morrison Co., City Sq. Bldg. Total estimated cost, $18,000.

Mass., New Bedford—The Empire Theatre Inc., 2 Masonic Bldg., has awarded the contract for the construction of a 2 story, 70 x 140 ft. theatre on Elm St., to the Weiss Constr. Co., 40 Court St., Boston. A steam heating system will be installed in same. Total estimated cost, $325,000.

Mass., South Boston (Boston P. O.)—The Joseph Burnett Co., 36 India St., Boston, has awarded the contract for the construction of a 7 story, 80 x 160 ft. manufacturing building on Fargo and "D" Sts., to the William M. Bailey Co., 88 Broad St., Boston. A steam heating system will be installed in same. Total estimated cost, $400,000.

N. Y., New York—Crawford, Russell, Weaver & Everett, 2 West 42nd St., will build two buildings, one 31 story and the other 25 story on Park and Madison Aves., 45th to 47th Sts. A steam heating system will be installed in same. Total estimated cost, $25,000,000. Work will be done by day labor.

N. Y., New York—The New York Edison Co., Irving Pl. and 15th St., has awarded the contract for the construction of a 3 story, 45 x 75 ft. transformer station on East 6th St., to Clough Bourne, 101 Park Ave. Estimated cost, $60,000. Noted Feb. 9.

N. Y., New York—The Vincent Astor Estate, 23 West 26th St., has awarded the contract for changing the 14 story hotel on 42nd St. and B'way into an office building, to Marc Eidlitz, 30 East 42nd St. A steam heating system will be installed in same. Total estimated cost, $200,000.

N. Y., St. Margaret's Bay—The Novo Scotia Power Comn., Halifax, has awarded the contract for the construction of a power house, etc., in connection with the proposed hydro-electric development, to D. G. Loomis & Sons, 1112 St. Patrick St., Montreal, P. Q. Total estimated cost, $250,000.

O., Cleveland—The Cleveland Rubber Corp. Co., 1900 Euclid Ave., has awarded the contract for the construction of a 1 story, 52 x 103 ft. boiler house at 2110 West 114th St., to the U. S. Eng. & Constr. Co., Engineers Bldg. Estimated cost, $50,000.

Mich., Saginaw—St. Mary's Parish has awarded the contract for the construction of a 3 story, 114 x 125 ft. parish school building, to the Burnett-Heinge Co., Main St. A steam heating boiler, fan system, etc., will be installed in same. Total estimated cost, $175,000.

Ia., Waterloo—Deere & Co., 1325 3rd Ave., has awarded the contract for the construction of a group of foundry buildings including a 3 story, 120 x 120 ft. service building, etc., to the Scales Constr. Co. Circulating fans and motors will be installed in same. Total estimated cost $400,000.

Neb., Ingleside—The Bd. of Comrs. of the State Institutions has awarded the contract for the installation of a heating system in the proposed hospital and receiving building, to Phelps & Underwood, at $24,000.

Tex., Austin—The State Bd. of Contro[illegible] has awarded the contract for the installa[illegible]tion of a complete heating system in th[illegible] lunatic asylum here, to John L. Marti[illegible] 408 Congress Ave., at $47,500. Note[illegible] May 3.

Cal., Westend (San Rafael P. O.)—Th[illegible] Westend Mining Co. has awarded the con[illegible]tract for the construction of a borax pla[illegible] to John Simpson & Co., Grant Bldg., Lo[illegible] Angeles. Plans include an 80 x 180 f[illegible] mill building, 32 x 48 ft. boiler house, et[illegible]

Preliminary Design of 240,000 Kw. Super-station

by

John A. Stevens, Engineer
Lowell, Mass. · Designed 1917.

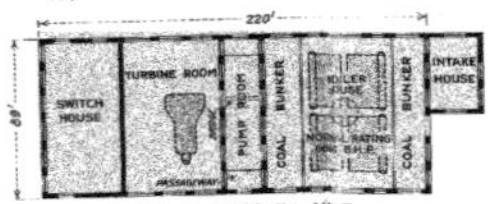

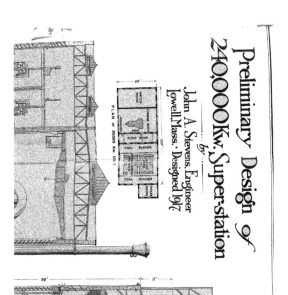

PLAN OF 30,000 Kw. UNIT

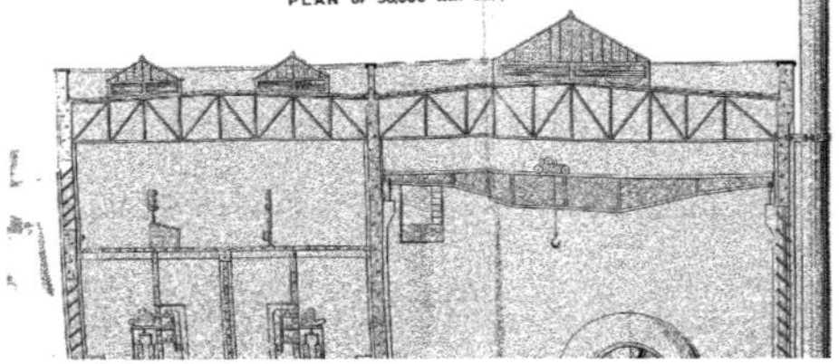

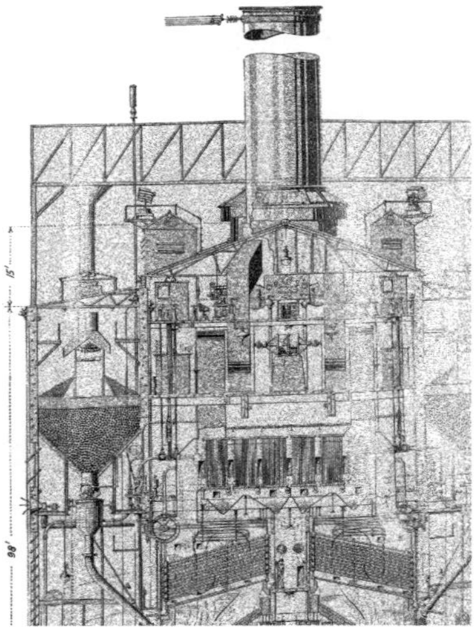

gn *of*
er-station

neer
1917

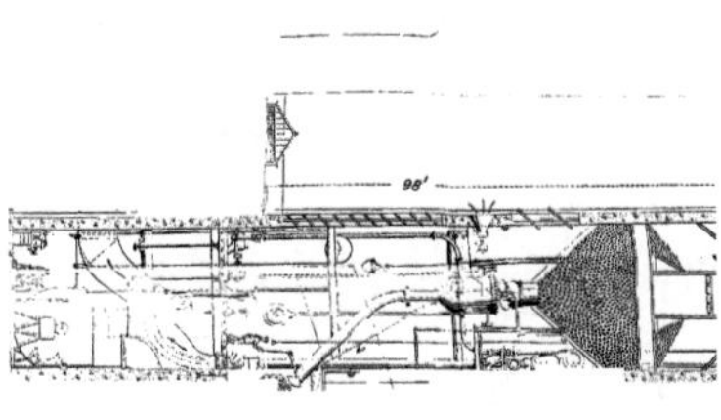
98'

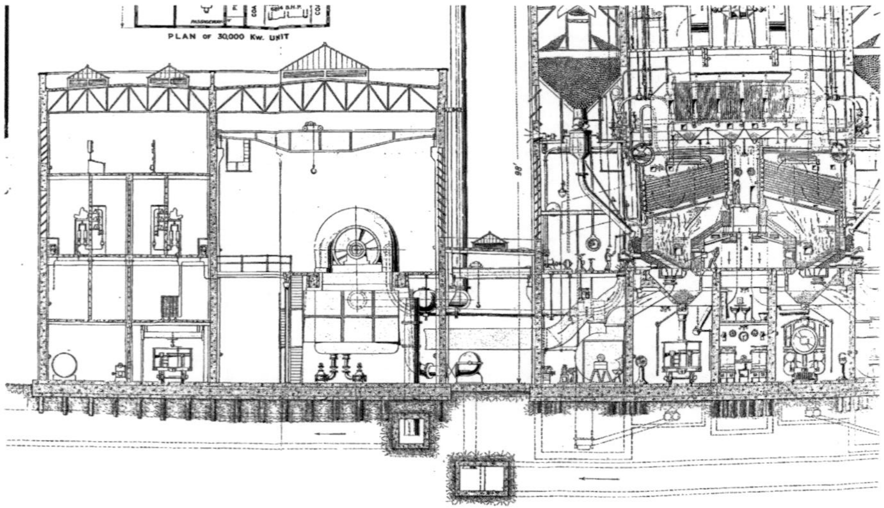
PLAN OF 30,000 Kw. UNIT
98'

POWER

Volume 51 New York, June 1, 1920 *Number* 22

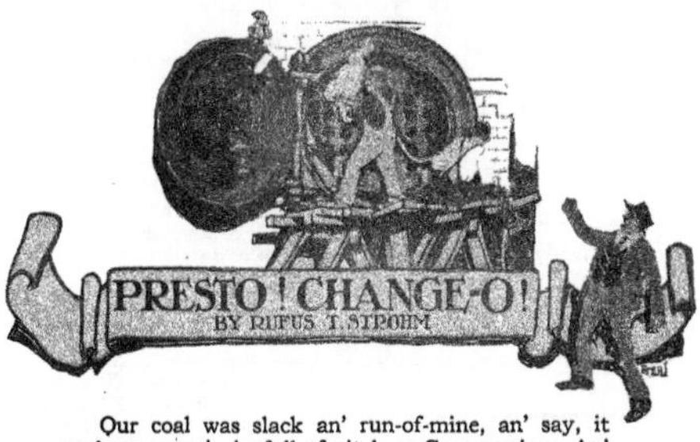

Our coal was slack an' run-of-mine, an' say, it sure was gassy! As full of pitch as Georgy pine. An' smoky? Lawd 'a massy! The way it cluttered up them tubes was surely somethin' frightful, an' in the boiler room us boobs got mighty peeved an' spiteful.

The boiler wasn't none too high, which sorter cramped the furnace; but no wise geezer drifted by to show us or to learn us. An' so the sooty gases riz an' through the boiler floated, an' everywhere they hit, they friz, until the tubes was coated.

Then there was merry hell to pay! The pressure got to droppin' in spite of the unceasin' way us fellers kept a-hoppin', until the chief instructed us to give the tubes a' cleanin'—his language bein' mostly cuss, but mighty plain of meanin'.

We rigged a scaffold out of plank an' got the steam lance goin'; an' cussed the blank-star-double-blank-three-daggers job of blowin'. We couldn't make the stuff stay put for half a minute, brother; for when one tube was clean, the soot went settlin' on another.

But that was in the early days before our eyes was open. We've canned them foolish, ancient ways, the swearin' an' the mopin'. For blowin' soot's so easy, now, there's hardly nothin' to it; it's almost sinful, I allow, to take good pay to do it.

We take a firm an' steady grip upon a valve an' turn it, an' just about as quick as zip! that soot is gone, gol durn it! Today our jobs is fit for gents an' all our costs is lower, an' everybody's happy sence the boss installed a blower.

EFFICIENT OPERATION

By John D.

Many engineers do not take the trouble and others do not know how to keep records that will enable them to obtain the most efficient boiler-room operation. This article, in text and tables and by numerous curves, gives information that should enable any engineer to get the most out of his boiler plant.

FEW engineers have really kept pace with the economic trend of events that have taken place in the boiler room, and the direct result of this has been a high cost of power due to a waste of coal. Few people realize the importance of the coal question, but everyone is alive to the fact that the price of coal is advancing, partly due, no doubt, to the increased cost of mining, but also partly due to the rapid using up of the most easily mined coal beds.

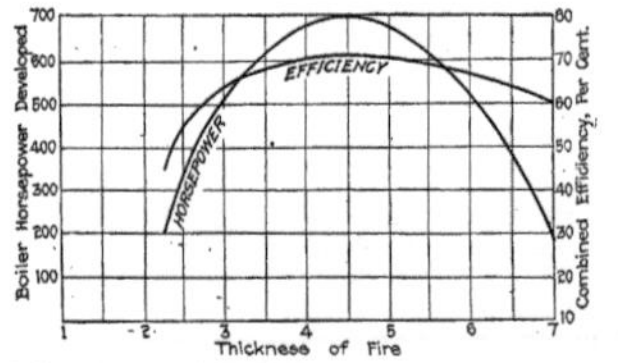

FIG. 1. CURVES SHOWING EFFECT OF FIRE THICKNESS

Therefore engineers, wherever possible, should prevent the waste of coal in the boiler rooms of small and large power stations. Table I shows the percentage cost of power chargeable to the boiler room. This table is based on the generating cost per kilowatt-hour as 100 per cent. It will be seen that the average boiler-room cost will run very close to 85 per cent of the total cost of power.

Does it pay to devote time and thought to the boiler room? The answer is found in Table II. During the year *A* no great care was given the boiler room; that is, no one man has been detailed to keep a close watch on ... But in the year *B* such a man had been made responsible for this work. This table very clearly shows that it does pay to give attention to the boiler room.

Final commercial economy of boiler operation is the next question that presents itself. I think that Mr. Polakov has given us the best possible formula, or equation; namely, that $Es = Ef \times Ep \times Em \times Eg \times Ec \times Et$, in which E indicates the ratio of utilization or efficiency and the corresponding indexes signify as follows: s = steam generation, f = financial outlay, p = purchase of fuel, m = attendance of men,

TABLE I. PERCENTAGE COST OF POWER CHARGEABLE TO THE BOILER ROOM

	Plant A	B	C	D	E
Total kw.-hr. output	95,403,319	120,349,670	33,246,600	47,875,967	233,015,000
Lb. of coal per kw.-hr.	2.72	2.81	3.004	3.517	2.443
Cost of coal per 2000 lb., dollars	2.38	2.377	2.32	1.61	2.715
Comparative total cost per kw.-hr., per cent	111.7	116.3	109.4	138.8	100
Coal cost, per cent	64.158	63.188	70.648	44.765	77.453
Water cost, per cent	1.98	6.273	0.202	2.011	0.709
Boiler-room labor, per cent	7.920	8.555	9.109	14.698	4.673
Boiler-room maintenance labor, per cent	2.376	2.281	0.809	1.165	1.107
Boiler-room maintenance material, per cent	2.376	6.273	1.417	2.170	3.610
Total per cent cost in boiler room	78.810	86.570	82.185	64.809	87.552

g = gasification of fuel in furnace, c = combustion of gases, and t = transmission of heat of gases to the boiler water and steam.

Of all the foregoing modifiers only that of the financial outlay lies beyond the power of the operating engineer; therefore it has been disregarded in this discussion.

At this point it might be well to consider the approximate loss per pound of fuel burned in a well-operated power plant. Table III is based on the actual operation of a plant, and wherever the losses could not be accurately determined they were calculated with care. This table is used only to indicate where the possible losses occur and in what relative proportion.

It will be seen that a total loss of 24.5 per cent occurs in the boiler room. This naturally raises the question, Where do these losses occur? In the boiler and furnace or in the plant as a whole? This can be partly solved by making up a heat-balance table for the boilers, which will show the relative losses. An example is given in Table IV, which is based on actual

operating conditions of a 600-hp. boiler with a natural-draft front-feed stoker. With the aid of this table one can classify the boiler losses in burning coal as follows: (1) Loss in dry chimney gases; (2) loss due to incomplete combustion; (3) loss of fuel through grates; (4) superheating moisture in air; (5) moisture in fuel; (6) hydrogen in fuel; (7) soot and smoke; (8) radiation and minor losses.

TABLE II. RESULT OF CARE IN THE BOILER ROOM

	Year A	Year B	Per Cent Increase	Per Cent Decrease
Net kw.-hr. generated	233,015,000	271,732,000	16.616	......
Lbs. of coal per kw.-hr	2.443	2.210	......	9.537
Cost of coal per 2000 lb.	2.715	2.783	2.505	......
Boiler-room labor cost, per cent	4.673	4.197	......	16.587
Coal cost, per cent	77.453	76.260	......	8.321
Water cost, per cent	0.709	0.666	......	12.50
Boiler-room maintenance labor, per cent	1.107	1.142	......	4.0
Boiler-room maintenance material, per cent	3.610	3.330	......	14.11
Relative total cost per kw.-hr	100	.93	......	6.888
Net thermal efficiency, per cent	10.18	11.31	11.1	......

Having located the possible boiler losses, there yet remain the losses of the boiler room at large. These are as follows: (1) Purchase of unsuitable coal; (2) weathering of coal in transportation and in storage; (3) poor feed-water temperatures; (4) labor conditions; (5) maintenance; (6) unsuitable boiler-room record sheets.

TABLE III. FUEL LOSSES AND THEIR RELATIVE PROPORTIONS

	Per Cent
Combustible matter in ash	2.5
Radiation and leakage of boiler	14.0
Gases rejected to chimney	5.0
Blowoff losses	1.5
Radiation and leakage of steam piping	1.5
Friction loss in auxiliaries and turbines	2.0
Rejected to condenser	58.0
Electrical losses	1.2
Required for auxiliaries	6.5
Total loss	92.2
Regained from condensate and exhaust steam	4.0
Total per cent delivered to bus	11.8

Naturally, in considering these items one must have reliable data on the condition of the boiler room as a whole. This can be obtained only by an actual test, and therefore the first point of operating a boiler room efficiently must be that of a test on the whole plant. This does not mean a highly specialized testing equipment commonly known as a laboratory test, for the general results given in such reports are a mass of highly calculated data and have little or no bearing on the actual operating boiler conditions. What is favored is a continuous boiler-room test using the following instruments: An open feed-water heater with a metering device of tested accuracy; water-flow meters on the boiler-feed pumps; standard meter on the makeup line; steam-flow meters on the main steam lines and one for moving from boiler to boiler; reliable scale for the coal used; recording temperature gage on the steam header, blowoff and feed-water line; recording pressure gage on the steam and feed lines and good thermometers on the open feed-water heater. If these instruments are properly used, a continuous test on the boiler room is possible.

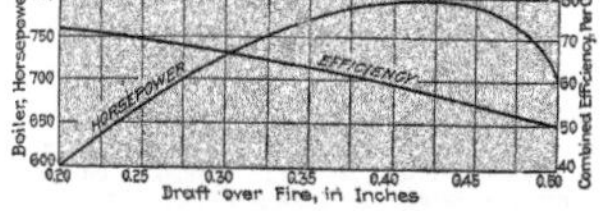

FIG. 2. CURVES SHOWING EFFECT OF DRAFT OVER FIRE

The next important action to be taken is that of making an actual test on each type of boiler and stoker in the plant so that the present operating conditions may be known. This test should be reported graphically, as it is then possible to see at a glance the effect of various items on the boiler and furnace efficiency. Figs. 1, 2 and 3 illustrate the graphical method of reporting boiler tests by means of plotted curves. The next point in order is to run a series of tests to determine the right kind of coal for the various boilers. An example of such a test is given by Table V.

It must be remembered that each plant will have special operating features and that they must be treated as important factors, but in general when the following items have been determined or computed, a true answer can be made regarding the right kind of coal: British thermal units per dollar; coking features of the coal; nature of ash; smoking characteristics; labor involved in burning; efficiency of boiler and furnace; effect on boiler maintenance; possibility of obtaining a stable supply; amount of storage loss.

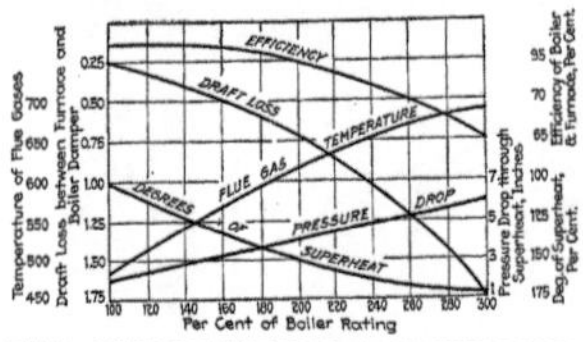

FIG. 3. OPERATING CURVES OF WATER-TUBE BOILER

When the proper coal has been found, the remainder of the trouble experienced in efficient boiler-room operation becomes simply a problem of boiler operation.

Suppose that sufficient records have been obtained from the daily station tests to enable one to study them in a thorough manner. The first thing that will come to mind is, What is the typical load curve for each day? This can easily be determined from the readings and charts, although it must be remembered that the chart obtained from the open feed-water heater does not always indicate the true boiler-room load curve, for the rate of feeding will influence this item to a marked degree; but if the station steam-flow chart is used, it can be readily calculated and upon the determination of this fact a suitable load-curve chart can be made up and upon this chart can also be plotted the boiler hours in service. As a rule a valuable bit of information can be gathered from this chart.

TABLE IV. HEAT-BALANCE TABLE SHOWING RELATIVE LOSSES

	Per Cent
Heat absorbed by boiler	77.50
Heat lost by moisture in coal	0.15
Heat lost by hydrogen in coal	4.40
Heat lost to chimney	12.00
Heat lost by moisture in air	0.25
Heat lost by carbon monoxide	0.80
Heat lost by carbon in ash	2.08
Heat lost by radiation (by difference)	2.82

The information obtained can be used in the issuing of a boiler-load chart which will show the boiler operator the proper number of boilers to be carried on the line and also the number of boilers needed for stand-by use. An example of such a chart is given in Fig. 4.

Having a definite plan in operation regarding the number of boilers to be carried on the line, the question of banking boilers arises. This matter has, as a rule, been given little or no thought in the average boiler room, and naturally, the first question asked is, How much coal does it take to properly bank a boiler? Table VI shows the amount of coal needed for various types of boilers and stokers. The figures given are the result of tests that were made over a long period, but no experiments were made to determine the proper method of banking. That is, the boiler was banked as it had always been.

TABLE VI. AMOUNT OF COAL NEEDED FOR BANKING

Name of Boiler and Stoker	Normal Boiler Rating	Lb. of Coal per Hour	Lb. of Coal per B.-Hp.
B. & W. boiler, Roney stoker	600	162	0.270
B. & W. boiler, Bayonne stoker	600	201	0.334
Edge Moor, hand fired	600	115	0.192
Heine, hand fired	300	53	0.177
Stewart, hand fired	150	19	0.127

Later, a test was made on the first boiler, and by trying different methods it was found that by banking the boiler in a scientific manner 100 lb. per hour, or 0.166 per boiler-horsepower, was sufficient. This same test also determined that it took an additional 1,500 lb. of coal to bring the boiler up to a steaming rate of 100 per cent. In general it will be found, when conducting such a test, that the results obtained for the first five hours are not reliable because the walls of the furnace give off the heat stored up in them. In general, the following rules should be adopted in banking boilers: Have seal plates made to seal up the hopper; install a tight-fitting damper; see that there

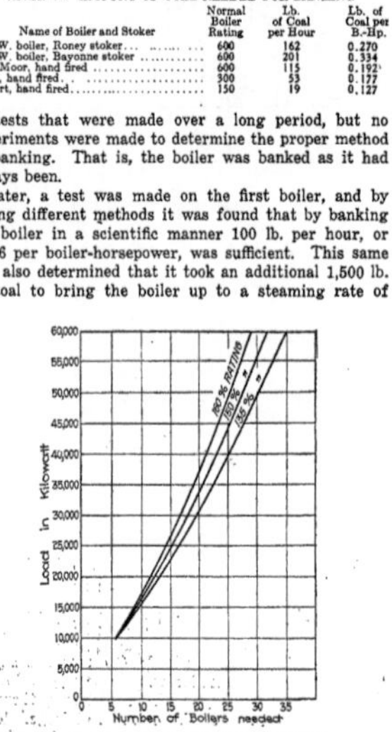

FIG. 4. CURVES SHOWING NUMBER OF BOILERS IN SERVICE FOR VARIOUS LOADS

TABLE V. TEST ON TWO 94-HP. RETURN TUBULAR BOILERS

Draft	Natural	Natural	Natural	Parsons' Blower	Parsons' Blower	Parsons' Blower	Parsons' Blower	Parsons' Blower	Parsons' Blower
Kind of coal	Buckwheat	Buckwheat	Buckwheat	Buckwheat	Buckwheat	Buckwheat	Bird's-eye	Bird's-eye	Bird's-eye
Steam pressure	85	85	85	90	90	90	90	90	90
Temp. of feed-water	198	200	200	192	192	192	186	187	185
Equiv. evap. from and of 212° F.	9.15	9.19	9.18	9.30	9.45	9.62	8.53	8.38	8.44
Hp. developed	140	142	140.5	160	155	148	152	152.5	154
Steam for steam jets, per cent				8.52	6.89	4.86	5.36	5.25	5.68
Cost of 1 hp., inc. jets, cents	0.5179	0.5158	0.5163	0.5611	0.5387	0.5179	0.448	0.4557	0.4538
Evap. per lb. of combustible	10.45	10.63	10.43	11.06	11.03	11.24	10.29	10.07	10.00

Price of buckwheat, $2.75. Price of bird's-eye, $2.10. Heating surface, 1405. Grate surface, 33.

are no air leaks in the setting, and do not bank the boiler too soon.

Naturally, many engineers will ask why such stress is laid on the banking of a boiler. This is because in a majority of plants the coal used for banking will run very close to 4 per cent of the total coal used, and this at best is a large, unavoidable loss, and it pays to give attention to this item, for every pound of coal saved by proper banking is a clear gain in the economic operation of the plant.

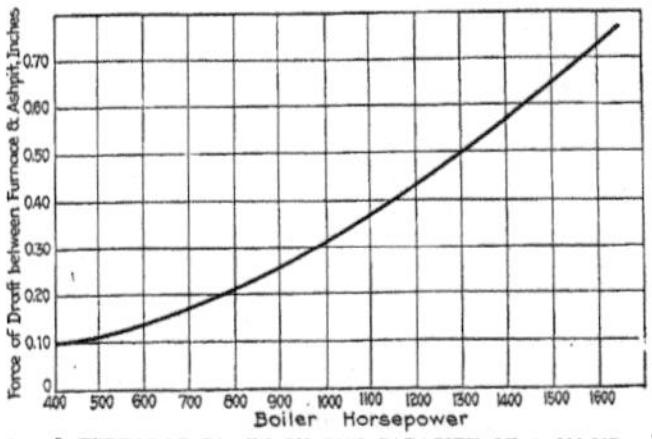

FIG. 5. EFFECT OF DRAFT ON THE CAPACITY OF A 600-HP. BOILER, NORMAL RATING, AND EQUIPPED WITH A CHAIN-GRATE STOKER

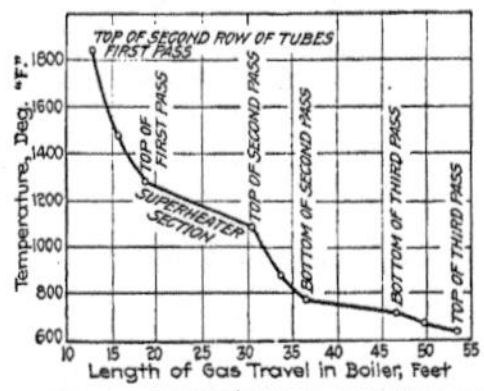

FIG. 6. TEMPERATURES OBTAINED AT VARIOUS POINTS IN 600-HP. BOILER SETTING

The next important item to be attended to is that of the boiler settings. This matter many engineers fail to consider of importance. There is no excuse for leaky boiler settings, for a few hours' work of a handy man with asbestos cement will eliminate this serious boiler-room loss.

The question of draft can be solved by a boiler test made so that it will show the effect of various drafts on the boiler operation. This test should also show the draft loss. Table VII illustrates the value of such tests made on two boilers supposedly identical, and Fig. 5 is a curve sheet which shows the effect of draft on the capacity of a 600-hp. boiler with a chain-grate stoker.

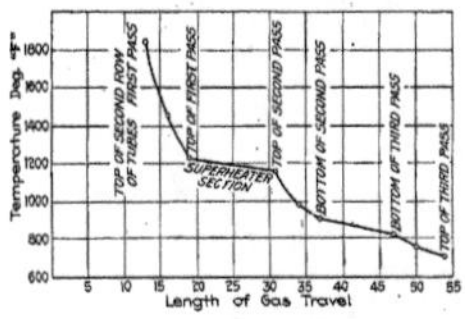

FIG. 7. TEMPERATURES OBTAINED IN 600-HP. BOILER SETTING

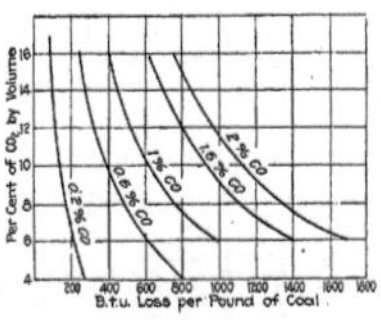

FIG. 8. B.T.U. LOSS DUE TO INCOMPLETE COMBUSTION

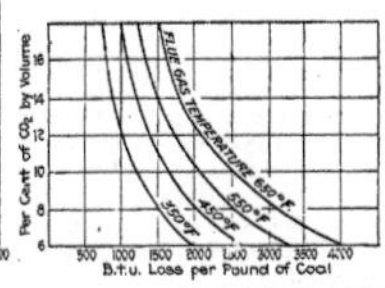

FIG. 9. B.T.U. LOSS DUE TO HIGH TEMPERATURE OF CHIMNEY GASES

The draft conditions of plants are varied, but in general the following items affect it: Height of chimney or stack; tightness of flues; number of bends in the flue and heat insulation of the flue and economizer settings. In the boiler proper the draft loss will depend largely on the height and spacing of the tubes, soot and ashes on the tubes, and on the boiler design and baffles. In the fuel bed the draft losses will be in some proportion to the size of coal, shape of pieces, nature of firing and the nature of the ash and grate construction. Very few designing engineers have devoted much of their time to the draft problem. When it is attacked in a thorough manner, its solution is quite simple. The heat drop through a boiler is an item of importance to the economic operation of the plant, and its determination is a simple matter, as only an electric pyrometer with several leads is required. To some this test will seem superfluous, nevertheless it shows the condition of the baffles and where the greatest heat drop occurs. To illustrate the value of such a test, two curves, Figs. 6 and 7, are used which show the temperatures obtained in two hand-fired water-tube boilers.

Everyone realizes the importance of baffles in relation to boiler-room economy. In most cases the manufacturer has properly located his boiler baffles, but in some instances it has been found advisable to move them, owing to the restricted flow of gases. The necessity for this change can be readily ascertained by making a test on the draft and temperature conditions of the boiler at various ratings. The construction of the baffle is important and should be given close attention. The best possible baffle construction is obtained when two courses of baffle brick are used with a ½-in. sheet of asbestos millboard between the courses. This baffle is easily constructed and repaired.

In the last few years some experiments have been made on the concrete baffle, but its success has been seriously questioned. The plastic asbestos baffle has been tried with varying success, but in most places where high ratings are used, this type has been found unsuitable for the front baffle, although it does give fairly satisfactory service in the second and third baffles.

The design and construction of the furnace depend largely on the method of firing and also on the rate of combustion. But recent experiments have shown that the design of the bridge wall has a marked influence on the life of the furnace and also on the efficiency of combustion; marked success has been experienced in the use of curved bridge walls. The best form of arch can be obtained only by a series of experiments because each type of boiler and furnace will require a different form. In the construction of the furnace, bricks of a poor quality but uniform as to size will outlast bricks of good quality but not uniform. In all cases bricks entering into the construction of a furnace should be gaged, and if they vary one-eighth inch from the specified size, they should be rejected. In many cases the use of a high-temperature cement will be found of value. It is advisable to have a complete set of drawings for each size of brick used in furnace construction so that when ordering bricks the manufacturer can be furnished with one. This will obviate any confusion in regard to sizes. Table VIII shows the approximate melting points of various bricks.

TABLE VII. EFFECT OF DRAFTS ON BOILER OPERATION

600-Hp. B. & W. Boilers, Westinghouse Underfeed Stokers	Boiler No. 1	Boiler No. 3
Average steam pressure	203 00	200 00
Average boiler horsepower	728 00	947 00
Pressure in ashpit, in. of water	2.73	2 65
Draft, inside damper, in.	0 745	0 815
Draft, overfire, in.	0 048	0 119
Drop from fire to soot chamber, in.	0 697	0 696
Draft at second row of tubes, first pass, in	0 065	0 119
Draft at middle of first pass, in.	0 075	0 156
Draft at top of first pass, in. (superheater located here)	0 092	0 171
Draft at top of second pass, in. (superheater located here)	0.265	0.342
Draft at middle of second pass, in.	0.313	0.414
Draft at bottom of second pass, in.	0.423	0.511
Draft at bottom of third pass, in.	0.440	0.531
Draft at middle of third pass, in	0.509	0.592
Draft at top of third pass, in.	0.565	0.638
Per cent drop from fire to bottom at first pass, per cent	2.44	....
Per cent drop from bottom to middle of first pass	1.43	5.32
Per cent drop from middle to top of first pass	2.44	2.15
Per cent drop from top of first to top of second pass	6.36	7.47
Per cent drop from first to second pass	24.8	24.6
Per cent drop from top of second pass to middle of second pass	6.88	10.35
Per cent drop from middle of second pass to bottom of second pass	15.8	13.93
Per cent drop from fire to bottom of second pass	53.7	56.35
Per cent drop from bottom of second pass to bottom of third pass	2.44	2.87
Per cent drop from bottom of third pass to middle of third pass	9.9-	8.76
Per cent drop from middle of third pass to top of third pass	8.04	6.61
Per cent drop from fire to top of third pass	74.2	74.6
Per cent drop from top of third pass to damper	25.8	25.4
Per cent drop from bottom of first pass to top of second pass	28.7	32.1
Per cent drop from top of second pass to bottom of third pass	25.1	27.2
Per cent drop from bottom of third pass to top of third pass	17 95	15.4

Fuel-siftings loss is one of importance and is quite an item in the average plant. To avoid this loss is a problem of mechanical operation, as for instance, the installation of a siftings hopper and bucket conveyors for sending the siftings back to the boiler or coal bunker. When burning coal containing a large amount of siftings, it should be sprayed with water; it will be found that an increased efficiency can then be obtained. There is a small loss in burning wet coal, but as this is only about 1 per cent for an addition of 10 per cent moisture, the gain obtained from the better coking features will more than offset the loss due to moisture. There is also a loss caused from superheating the moisture in the air admitted to aid combustion, but it is a small one, and in plants where the economic operation of the boiler room has been reduced to the point of considering this item, it has been taken care of by piping the exhaust air from the generators to the intake of the blower fans.

The smoke loss is of vital importance, not so much, perhaps because of the economic loss in the burning of coal as because it is a bone of contention between the company and the people living in the vicinity of the plant. Nowadays a city of any importance has its smoke-abatement commission. The economic loss will seldom run over 1 per cent.

TABLE VIII. APPROXIMATE MELTING POINTS OF VARIOUS BRICKS

Nature of Brick	Melting Point, Deg. F.	Nature of Brick	Melting Point, Deg. F
Fireclay	3000	Kaolin	3150
Bauxite	2900-3200	Bauxite Clay	3250
Silica	3100	Alumina (Pure)	3650
Chromate	3700	Chromite	3950
Magnesia	3900	Silica (Pure)	3175

It might be well, however, to consider some of the general methods of smoke abatement. Where hand-fired boilers are used, the installation of stokers will aid, but it must be remembered that a properly designed and well-operated furnace will burn coals without smoke up to a certain predetermined rate of combustion, but when this point is passed smoke will issue owing to the lack of furnace capacity to supply air and mix the gases. In a hand-fired furnace the rate of firing should be frequent, and often with a short period of air admission after firing. The use of small sizes of coal will result in less smoke. Keep down the CO, for as the percentage of CO increases so does the black smoke.

The loss due to incomplete combustion of coal is somewhat vague, but if we change it to read loss due to presence of CO, then one readily sees its direct bearing on efficient operation. Many believe that all the loss due to incomplete combustion can be laid to CO, but I have many reasons to believe that the CO is not directly responsible for all the loss, but rather, that it is simply a good danger signal for some other loss. To confirm this belief one has only to run a series of boiler tests and to see that an advance of 0.2 per cent will produce a loss in efficiency of about 4 per cent, which is slightly more than four times the theoretical loss due to such an advance. This point could be more fully explained if some of our experimental laboratories would conduct some research work along these lines. Fig. 8 illustrates the theoretical rate of loss due to CO for a semi-bituminous coal of 14,098 B.t.u.

TABLE IX. EFFECT OF SOOT ON BOILER TUBES

	First Series	Second Series	Third Series
Equivalent evaporation from and at 212 deg. F. per lb. of dry coal	6.2	7.04	6 23
Dry coal per sq.ft. of grate surface per hour	13.40	9.09	13.40
Temperature of escaping gases	62.70	5.46	6 98
First series, 5 days. soot allowed to remain on tubes.			
Second series, 5 days, tubes cleaned each morning.			
Third series, 5 days, soot allowed to remain on tubes.			
Increase in evaporation due to clean surfaces, per cent			13
Decrease in coal burnt per sq.ft. of grate per hour, per cent			32 16
Lowering of stack temperature due to clean surfaces, per cent			17.59

The loss in efficiency due to chimney gases is so well known that it is only necessary to briefly state the general causes of high flue-gas temperatures, which are: Excess air; percentage of boiler rating; bad baffles and soot and ashes on boiler tubes. To illustrate the loss, the curve, Fig. 9, shows the percentage of loss due to high flue-gas temperature for a semi-bituminous coal of 14,098 B.t.u.

The loss due to smoke, soot and ashes on the boiler tubes can be no better shown than by Table IX, which shows the result of tests made by the University of Illinois.

Since it is necessary to clean tubes, the question arises as to what method should be used? The best method is by the installation of a mechanical soot blower using steam at boiler pressure. The next best is that of dusting the tubes with steam by means of a dusting nozzle using steam at boiler pressure. This last-named way is satisfactory if the tube blower can be made to perform his duties in a thorough manner, but vigilance must be exercised if this method is used.

The most satisfactory arrangement is that of a ratchet chronometer valve on each boiler to which should be attached a rope for operating it, and a 1-in. pipe run down to the middle point of the side wall with a connection on the end for attaching the blowing hose or pipe. The blowing member should be made of ¾-in. pipe in four sections each 4 ft. long. These sections should be jointed together by means of a metal toggle joint, the end section being equipped with a piece of pipe 6 ft. long with a dusting nozzle on one end.

Tubes should be blown twice a day where the boilers are run continuously at high ratings and once a day under normal operation.

On the Road with the Refrigeration Troubleman

By J. C. MORAN

He is called to a plant where it is impossible to get a suction pressure of more than 15 in. vacuum unless the compressor is run very slowly. An oil separator in the suction line was the source of the trouble, as it had not been properly cared for. On the next job a pipefitter causes no end of trouble in midsummer. The symptoms of these troubles, how they were located, and what was done to overcome them are well told by the troubleman.

WE HAD installed a 100-ton plant in a manufacturing establishment for air-cooling purposes, and it had been in operation for about a year when we got a rush call to come out and see what had happened to it. The crew could not get the plant to work at all. I arrived on the job by the first train and found the plant shut down and everybody running around in circles.

The engineer had stopped operation the day before and pumped the plant down to make a repair on the suction line between the machine and the coolers. After he had pumped about 25 in. vacuum two or three times, the steamfitters tried to open the line at one of the joints and found that there was still considerable pressure on it. So the engineer pumped it down a couple of more times, and finally kept the machine running slowly to hold the vacuum at this point while the men broke into the line. When they tried to open the joint under these conditions, there was still a strong pressure on the line. It was getting late in the day, and the engineer decided they had better give it up for a bad job and let it go until some other day.

Then the queer thing happened. It was impossible to get a suction pressure of much above 15 in. vacuum unless the machine was run very slowly. No matter what they did with the expansion valves or anything else, it did not make any difference. The room temperatures kept going up, and there were many complaints about high room temperatures because it was a hot day in midsummer.

On going over the entire plant in a rather hurried inspection, I could find nothing wrong. I experimented with the expansion valves until they were open several turns, but without improving things. So I began to look for obstructions in the pipe lines.

I suspected the trouble might be in a suction stop valve located about 25 ft. back from the compressor in the suction line. I tried to remove the bonnet and changed my mind in a hurry. There was considerable pressure on the valve. It was plain that the trouble was at some place nearer the compressor. The next joint was where an oil and scale separator was located in the line. Fig. 1 shows the connections. I first tried to open the joint at A. The pressure was high at this place. So I tried another joint between the oil

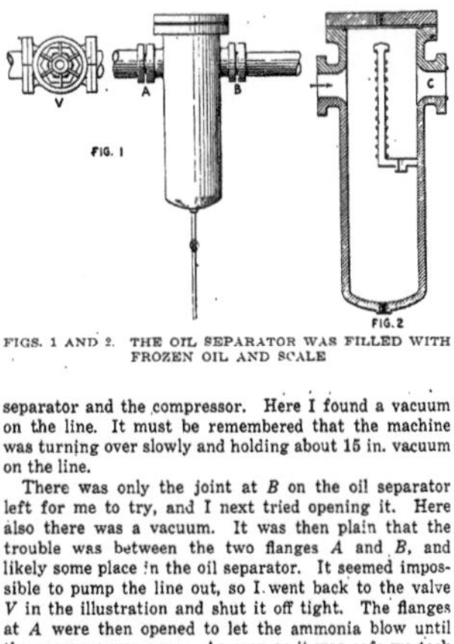

FIGS. 1 AND 2. THE OIL SEPARATOR WAS FILLED WITH FROZEN OIL AND SCALE

separator and the compressor. Here I found a vacuum on the line. It must be remembered that the machine was turning over slowly and holding about 15 in. vacuum on the line.

There was only the joint at B on the oil separator left for me to try, and I next tried opening it. Here also there was a vacuum. It was then plain that the trouble was between the two flanges A and B, and likely some place in the oil separator. It seemed impossible to pump the line out, so I went back to the valve V in the illustration and shut it off tight. The flanges at A were then opened to let the ammonia blow until the pressure was gone. As soon as it was safe we took the cover off and examined the separator.

It took but one look to ascertain what was the trouble. A cross-section of the separator is shown in Fig. 2. The entire separator was filled with heavy oil and foreign matter which had frozen together into a solid mass that had entirely plugged the outlet at *C*, which was the side that connected to the compressor. We had to play a steam hose on the separator for about half an hour before we could get the stuff soft enough to dig out.

When the engineer had started to pump out the system, there was probably a considerable amount of liquid ammonia in the separator. When the pressure was pumped down to 25 in. vacuum, this ammonia liquid began to boil out, and at this pressure the temperature was probably somewhere around 75 deg. below zero. This froze the oil and other contents of the separator into a solid mess, and as only very little gas was coming through the separator at the finish of pumping down, it dragged the mess into the outlet *C* and finally froze it solid. At least that is what I concluded. It might have been that a chunk of the mixture was blown up and settled into the outlet *C* at one time, and thus shut off the flow.

Of course the engineer had no one else but himself to blame for the trouble. Scale and oil traps in the suction line should be blown as regularly as the oil trap in the discharge line. It may not be required to blow them so often, but they should be tried every week or two to see that they are clear, and if there is any oil or other foreign matter in them that should be blown out before they fill up and become useless. In this case the trap had never been blown since the plant was started.

Another time I got a hurry call from a 25-ton plant we had installed in a wholesale market in a small town. When I arrived on the job, the owner was about ready to throw the machine out of the place and the engineer was not in much better humor. They had not been able to use the machine for two days and had to get along as best they could by loading the cooler with ice, which was both expensive and troublesome.

The engineer informed me that the first sign of trouble was that the condenser pressure began to climb and it had risen from the usual 150 lb. to over 200 lb. This happened over a period of about a week. At the same time he had been compelled to keep on opening the expansion valves more and more until they were wide open, and still the suction pressure kept dropping down until it had fallen from 25 lb. to 10 in. vacuum. It was also impossible to get frost on the suction line, and even the frost on the expansion coils in the cooler thawed off.

I first thought that the system was entirely too short of ammonia, but there was a gage glass on the receiver and this was full. I even opened one of the joints in the liquid line about twenty feet from the liquid receiver just enough to let it dribble and found that there was plenty of both pressure and liquid in this part of the line. Next we went around the cooler, tried all the expansion valves and found them wide open with not the slightest sound of liquid passing through them. An expansion valve when it is working right and the plant is properly charged will sound something like a throttled water valve under high pressure. All the valves were frosted over, also the liquid line for about twenty feet back to where it was insulated. This was another good sign that something out of the ordinary was wrong, because the liquid line is never frosted if the plant is working right, and an expansion valve should have a sharp frost line across it where the pressure changes from the high to the low side. This will generally be diagonally across the body of the valve.

I thought at first that it might be due to the disk coming loose on the valve stem on a valve that had been installed the wrong way in the line, but there was no place where this could happen, there being only two valves in the line, one at the receiver and another in the cooler. However, the line was frosted far back of the valve in the cooler so that the trouble could not be in this valve. I had already tested the line ahead of the valve on the receiver when I opened the joint; therefore, the trouble could not be in either of these.

There was only one other possible reason for the trouble and that was that the liquid line was partly blocked by scale or other substance some place between the receiver and the expansion coils. Working on this theory, I began tracing the liquid line back to the basement where the compressor and receiver were located. Finally I came to a joint in the horizontal run of the line, where I heard a sizzling sound like an expansion valve that is just cracked off the seat. Removing the insulation, I found that the pipe was warm up to this flange while on the side going to the expansion coils it was frosted sharp, showing that the line was choked at this place until the pressure was throttled from the 200 lb. condenser pressure to that in the expansion coils, which was probably several inches vacuum.

We shut off the line at the expansion coils, and using the bypasses, we pumped out the receiver and the liquid line through one of the condenser coils, shutting off the other two stands. After the line was thoroughly pumped out, we broke open the joint and found that the steamfitter had failed to remove the burr made by the pipecutter in cutting the pipe. This reduced the inside diameter of the pipe over one-half and also provided an excellent pocket for scale and other matter to accumulate. This had piled up until it practically shut off the flow through the line. Of course, the liquid accumulated in the condenser and this increased the condenser pressure, while the amount of liquid getting through the blocked part of the pipe was not enough to keep the expansion coils working, and as a result the suction pressure dropped and the frost came off the expansion coils, the suction line, and the machine. After we had removed the scale and reamed out the pipe that made the trouble, everything went along fine.

Pumping Water with a Spiral-Spring Belt

What the eminent British scientists are pleased to call a "mechanical impertinence" has recently made its appearance in England, in the form of a unique pump which works by means of a spiral-spring belt, says the *Scientific American*.

The pump consists simply of a spiral-spring belt, a grooved weight which turns with the bottom loop of the belt and holds the latetr in place, and a driving crank and pulley for turning the belt. Despite this simple construction the pump is capable of lifting a thousand gallons of water per hour from a depth of 300 feet, even when worked only by hand, according to reports. The coil-like cable is sunk to any depth by a rotating weight. Obeying the law of capillary attraction, the water lodges between the turns of the spiral spring and only falls out when it reaches the top of the pump.

Testing Polyphase Watt-hour Meters

By P. B. Findley

Comparison is made between the single-phase and polyphase watt-hour meters. A detailed description is given of the meters, and how to test and adjust the polyphase type is explained.

SINCE the polyphase watt-hour meter consists of two single-phase elements mounted on a common shaft, its mechanical construction is practically the same as that of the single-phase instrument and the adjustments, except as noted in the following, are identical. In Fig. 1 is shown a Westinghouse type OA polyphase watt-hour meter with the cover removed, and Fig. 2 shows the back view of the meter element removed from the case.

The arrangement of the parts is indicated very clearly in the figures. An aluminum mounting frame *F* serves as a foundation for the meter element. To it are attached the iron punchings *M* and *M*, which form the magnetic circuits for both the upper and the lower metering elements. On the core are the current coils *C* and *C* and the potentials coils *P* and *P*. The permanent magnets *A* and *A* are fastened to L-shaped punchings, which in turn are secured to the frame by the clamping screws *S*. The adjustments for no load, light load and full load are made in precisely the same manner as for a single-phase meter, save that everything is in duplicate, as can be seen by comparing Figs. 1 and 2 with Figs. 3 and 4. The shaft has the same wire-type guide bearing at the top and ball-and-jewel bearing at the bottom, and the gear train to the register is on the same principle as that of the single-phase instrument. Those who have read the articles in the Jan. 28 and Feb. 4, 1919, issues of *Power*, on "Testing Single-Phase Watt-Hour Meters," will readily see that the polyphase meter described in this article is practically two single-phase meters similar to that described in the previous articles combined into one element. The single-phase meter is shown in Figs. 3 and 4, so that a convenient comparison can be made with Figs. 1 and 2 respectively.

FIG. 1. POLYPHASE WATT-HOUR METER, COVER REMOVED

FIG. 2. REAR VIEW OF METERING ELEMENT, FIG. 1

The first test to be made on a polyphase watt-hour meter is at no-load. With operating voltage on both potential coils the light-load adjusting loops are moved by turning the square-headed screws *G* until the discs do not "creep" in either direction. About one-fourth of the adjustment should be made with each loop. Practically, this test is made by giving the discs a gentle push, first in one direction and then in the other, with the fingers; they should not continue to rotate in either direction. The light-load compensator in the meter under consideration consists of two metallic loops which pass through the air gaps in the potential coil's magnetic circuits. These can be seen at *J* Figs. 2 and 4, clamped under the hexagon nuts *N*. By turning with a screwdriver, the square-headed screws *G*, Figs 1 and 3, the loops can be made to move up and down in the air gaps.

The design of a polyphase meter is such that when properly adjusted the passage of one kilowatt-hour through either element will cause the registration of one kilowatt-hour on the dial. If one kilowatt-hour passes through each coil simultaneously, two kilowatt-hours will be registered. Therefore, each electromagnetic system, seen in the rear view, Fig. 2, must have the same effect on the rotating disc. Specifically, a current flowing through the upper current coils must have the same effect as if flowing through the lower coils. This adjustment, once made at the factory, is practically permanent indefinitely. For completeness, it is given here:

After the no-load adjustment is made, leaving the potential coils alive, connect the current coils in series and allow full-load current to flow, reversing one coil if necessary so that the discs do not rotate. If they do, remove the register mechanism, loosen the screws *D*, Fig. 1, and insert a screwdriver into the slots *B*, Fig. 2. The magnetic systems can be moved up and down as a unit with respect to the discs, and should be so placed that no rotation takes place. Then the effect of each current element will be the same on its respective disc. The current should be cut off, and the no-load adjustment checked and readjusted if need be. The clamping screws are tightened and the register replaced.

The next step is to adjust for light load. This is done in the same manner as for a single-phase meter, the connections being made as described later, so that operating voltage is applied to both potential coils, and about one-tenth full-load current flows through each current coil. A rotating-standard meter is included in the circuit, and the adjusting loops J are moved by the square-headed screws G, until the meter registers correctly.

Full load is next applied with the same connections as in the foregoing and the permanent magnets A adjusted for correct registration, the necessary movement being distributed among the four magnets.

Since the great majority of single-phase meters are used on lighting circuits which have high power-factor loads, they are adjusted for power-factor correction at the factory. Polyphase meters, however, generally serve such loads as induction motors, distribution circuits with partly loaded transformers, etc., the power-factor of which is low. They are therefore provided with lag adjustments, which are coils of copper wire W, Fig. 2, wound around the potential-coil pole tips and having long leads L. By changing the length of the leads L the resistance of coil W is changed, hence the magnetic flux from the potential coils and the rotation of the disc are also varied. This adjustment rarely requires attention, and it is not the general practice to check it except on new and repaired meters. The test is easily made, however; each phase should be tested separately with a wattmeter or one-half of a polyphase rotating-standard meter, with full-load current and voltage, the latter being lagged so as to give 0.50 to 0.70 power factor. Since a single-phase rotating standard is not specially adjusted for low power factors, it is not necessarily accurate enough for this test.

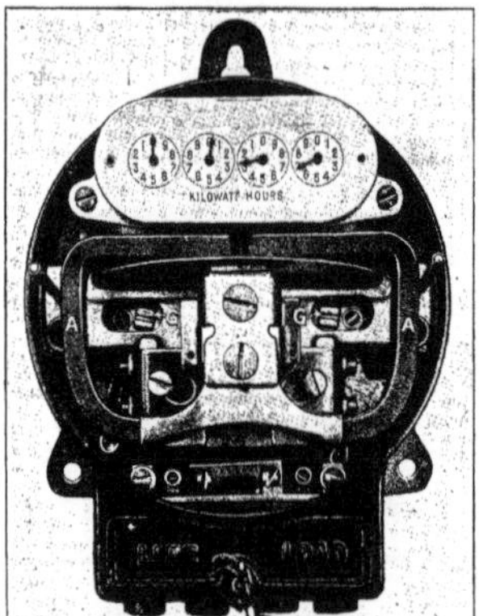

FIG. 3. SINGLE-PHASE WATT-HOUR METER

The essential constant of a watt-hour meter is the watt-hour constant, K_h, which is the watt-hours per revolution of the shaft. The constant for a polyphase meter is just double that for a single-phase meter of the same rating. A good way to visualize this is to remember that the polyphase meter consists of two single-phase elements of the same rating, locked together. If the same load were on each element, and we imagine the connection between them cut, they would continue to rotate at the same speed. Each would then rotate, let us say, once for each kilowatt-hour metered by it; that is, once for each two kilowatt-hours metered by the pair. Then if we locked them together, the combination would still continue to rotate once for each two kilowatt-hours metered; that is, K_h would be twice what it was for a single element.

FIG. 4. REAR VIEW OF METERING ELEMENT, FIG. 3

In testing a polyphase meter both potential coils must be energized, and this will be assumed in what follows. Suppose, now, we are testing with both current coils in series and in turn connected in series with the current coil of a single-phase standard wattmeter. Then the power is registered twice by the service meter, once by the standard meter. So we must multiply the standard readings by 2 before using them in the formula,

$$E = \frac{3600\, K_h N \times 100}{PT} - 100$$

where

E = Per cent error, plus if service meter is fast and minus if service meter is slow;

K_h = Watt-hour constant of service meter;

N = Revolutions of service meter in T seconds and

P = Twice the true watts registered by the standard wattmeter.

If, however, we are testing with a single-phase rotating standard, in series with both current coils as above, then if K_h for the standard is half that for the service meter, we can compare revolutions directly. This can also be done if we have each element of the service meter connected to an element of a polyphase rotating standard meter, where K_h for the standard is the same as for the service meter. Then

$$E = \frac{N \times 100}{M} - 100$$

where

E = Per cent error, plus if service meter is fast and minus if the service meter is slow;

N = Revolutions of service meter;

M = Revolutions of rotating standard.

This formula is true only under the conditions noted in the foregoing. If the polyphase rotating standard has a different watt-hour constant from the service meter, it is most convenient to use a conversion table furnished by the manufacturer of the standard meter, similar to that described in the article. "Testing Single-Phase Watt-Hour Meters With a Rotating Standard of Different Make," Oct. 28-Nov. 4, 1919, issue. For instance, suppose we have to test a Fort Wayne type K meter, 100 volts, 5 amperes, watt-hour constant 1,000, with a Westinghouse polyphase rotating standard set for 5 amperes 100 volts, having a watt-hour constant of 0.666. Setting the load on each phase to be approximately equal and to give 20 r.p.m. of the service meter,

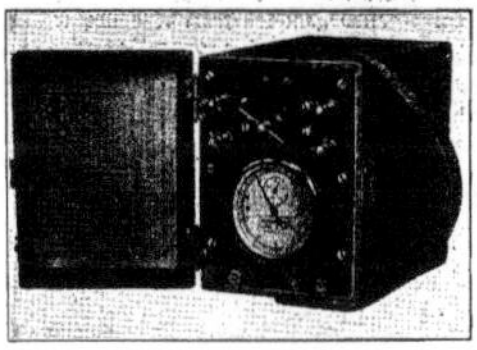

FIG. 5. POLYPHASE ROTATING STANDARD, EXTERIOR VIEW

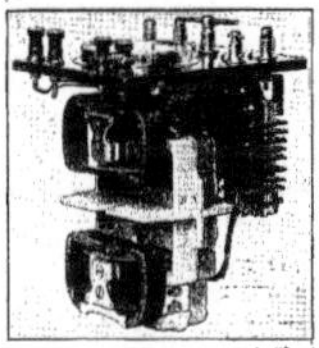

FIG. 6. INTERIOR VIEW METERING ELEMENT, FIG. 5.

we find the number of revolutions of the standard which takes place in 20 revolutions of the service meter. Comparing this figure with the table, we find the percentage of error and adjust to eliminate it. If no table is at hand use the formula,

$$E = \frac{N_s K_{hs} \times 100}{N_t K_{ht}} - 100$$

where

E = Per cent error, plus if service meter is fast and minus if the service meter is slow;

N_s = Observed revolutions of service meter;

N_t = Observed revolutions of standard meter;

K_{hs} = Watt-hour constant of service meter;

K_{ht} = Watt-hour constant of standard polyphase meter.

The simplest way of testing a polyphase meter is to connect its potential coils in parallel across a suitable source of single-phase voltage, and its current coils in series with a single-phase rotating standard and a resistance load. Where a very large current is required and a suitable transformer is available, it may be used to supply current at low potential, the potential coils being energized from the line. For example, if a 200-ampere meter is to be tested, and a 25-kw. 2,200 to 110 to 220-volt transformer is available, the 2,200-volt coil could be connected to a 110-volt circuit and the two 110-volt coils connected in parallel, obtaining 5 volts with a capacity of 225 amperes. For the moderate

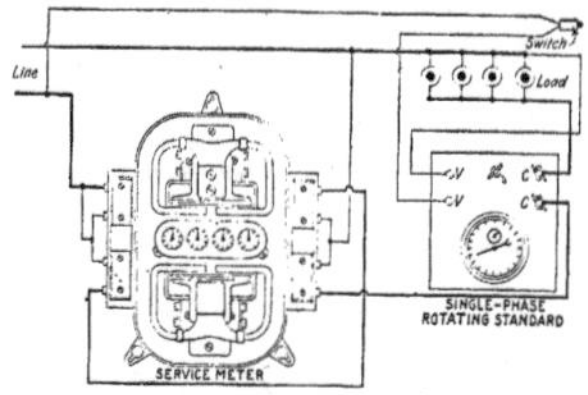

FIG. 7. TESTING POLYPHASE METER WITH SINGLE-PHASE ROTATING STANDARD

currents usually required, as in testing meters to be used with current transformers, which are built for 100 volts and 5 amperes, the current is taken directly from a 110-volt line, as in Fig. 7. For high accuracy on low power factors, a polyphase rotating standard, Figs. 5 and 6, should be used. The connections for this test are given in Fig. 8. Obviously, power transformers for furnishing current at low voltage can be inserted in the current-supply wires. For testing at low power factors, a phase-shifting transformer should be inserted in the potential supply circuits. This is simply a small polyphase transformer whose secondary coils can be shifted to bring them partly under the influence of two primary coils instead of under one primary coil, as in the ordinary transformer. Thus the phase of the secondary voltage will lie between the phases of the two primary coils, and at any desired angle of lag or lead with the load current.

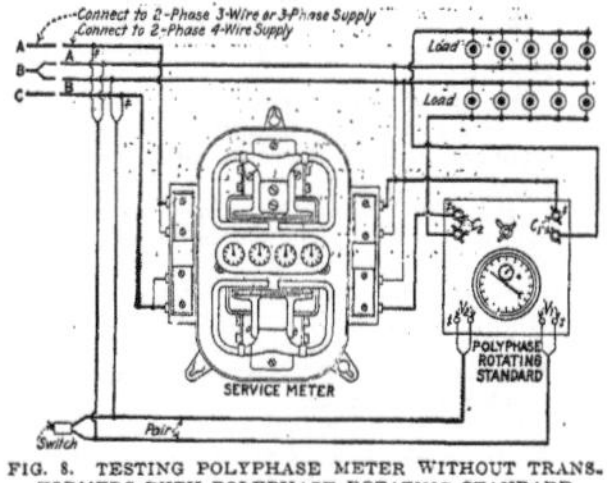

FIG. 8. TESTING POLYPHASE METER WITHOUT TRANSFORMERS WITH POLYPHASE ROTATING STANDARD

Wherever polyphase-meter connections are made, it is essential to have some check on the polarity of the various coils. Rotating standard meters have one voltage and one current terminal of each circuit marked, and these should be connected to the same wire of the supply circuit. A final check on the correctness is made as follows:

Place a small noninductive (resistance) load on each phase. Open the potential circuit of the upper element, and the current circuit of the lower element (first short-circuiting the current transformer winding if transformers are used). If the meter runs at all, the potential circuits have been transposed between the line and the meter. Interchange the potential circuits—upper for lower—and test again. If the meter does not run, the proper phases are together, assuming that there are no "shorts" or "opens" in the wiring.

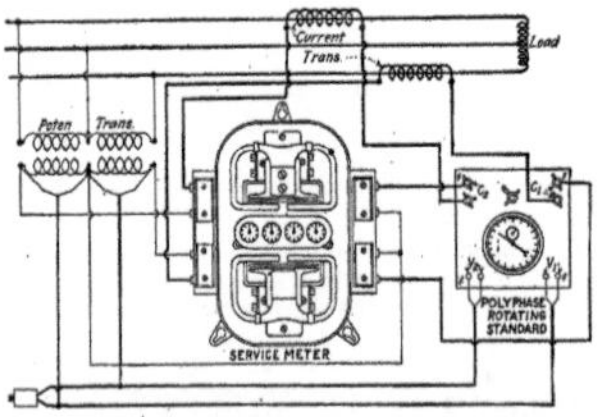

FIG. 9. TESTING POLYPHASE METER WITH TRANSFORMERS WITH ROTATING STANDARD

Occasionally, these checks must be made on an installation which serves only high-tension apparatus—for instance, 2,200-volt motors or transformers metered on the "high" side. Here an artificial load is impracticable. The test should be made therefore with the apparatus loaded, since at light loads both transformers and induction motors have low power factors. When the power factor of a polyphase circuit is below 0.50, one element of a watt-hour meter will try to run backward so that rotation backward would mean nothing unless we were sure that the power factor was above 0.50. On the average for loads of above 15 per cent of full load, the power factor of an induction motor is above 0.50.

Both current and potential transformers are generally used on polyphase circuits, either for safety or to bring the current in the meter within workable limits, Fig. 9. The meter connections in the figure are of the simplest sort; its two current and two potential coils have both ends brought out to its terminal boards, whence they are connected to the transformers. While some complicated wiring schemes may be found, especially if relays or other meters are served from the same transformers, yet they are in general simple. The following points should be borne in mind when dealing with any such installation:

1. A current transformer will be ruined if the secondary is opened while the primary is carrying current. Therefore, always keep the secondary closed, either through some instrument current coil, or through a short-circuit.

2. Most instrument potential transformers are rated at 25 watts. Never overload them.

3. If not abused, instrument transformers will retain their accuracy for at least five years. Therefore, routine tests of instruments may be made on the low-voltage side of the transformers.

4. In thus testing a polyphase watt-hour meter with a polyphase standard on a running load, simply insert one current coil of the standard in each current lead at the meter, and connect the potential coil of the standard across the proper potential terminals of the service meter as in Fig. 9, and make sure that each rotating standard element drives the discs forward.

Cleaning Apparatus for Generator Windings

By F. C. Williams

The sketch shows an apparatus that I rigged up for cleaning the windings of generators and motors. It was made from material that was found around the plant. We have high winds and much dust, and it is a problem to keep the machinery clean. We have four oil and gas engines, each with its own air-starting tank, but all are piped together and so the supply of compressed air is sufficient.

I took the three-gallon gasoline tank of a preheating torch and piped it as shown. After connecting with the air supply at *A*, with all valves closed, the pressure was

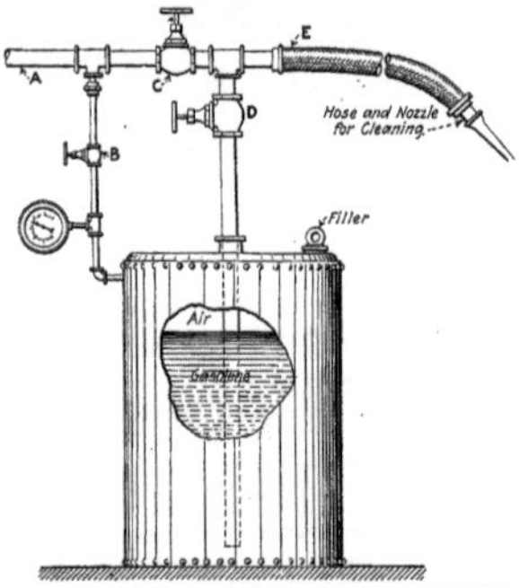

AIR AND GASOLINE APPARATUS FOR CLEANING GENERATOR WINDINGS

turned on and then the globe valve *B* on the air line was cracked and a pressure built up in the gasoline tank to about twenty or thirty pounds. This valve was then closed and the valve *C*, between the two tees at the top, opened to give the proper blast. Then the valve *D* on the gasoline line was cracked. The result was a spray of air and gasoline through the nozzle, which has about a $\frac{1}{16}$-in. opening. This spray reaches into cracks and holes where nothing else can go and brings out the dirt and grease.

Economies to be Expected of Superstation*

by
John A. Stevens†

Deals with the economies to be expected of superstations, including heat balance, power costs at the switchboard, latest forms of byproduct steam turbines for producing heating and manufacturing steam and the economies that are obtained by using them, and suggestions regarding the placing of super-power stations in large industrial communities.

IN PRESENTING this article, it seems best to carry the reader through the various power-development stages from approximately the first generation of electricity by means of the simple non-condensing engine to the latest development in power-plant machinery design. Table I shows the advance in the art of steam engineering, with special reference to engines, dating from the small high-speed slide-valve engines to the 5,000-kw. angle compound engines (7,500 hp. maximum) of the Interborough Rapid Transit Co., which represent about the height of the development in so far as reciprocating steam engines go. These engines, although more efficient after thirteen years of service than when first installed, were supplanted in 1917 by cross-compound reaction turbines of 30,000-kw. capacity.

Table II shows the evolution of steam-turbine practice, starting with a very small steam exciter set of 35-kw. capacity, exhausting to the atmosphere, and carrying the development through to the 50,000-kw. units of the three-cylinder type, having a maximum rating of 70,000 kw. The specific types of turbines referred to in this article, however, are the 30,000-kw. units, which show a thermal efficiency of 26.5 per cent.

The economies shown by Table II graphically bring out the reasons why engineers are working toward the great super-station, the necessity for which is illustrated by Fig. 1, which shows the growth of central-station operations[1]. Behind the growth of the central station is the fact that building super-stations is, economically speaking, the right thing to do.

*Illustrated in the May 25, 1920 issue of *Power*.

†Consulting Engineer, Lowell, Mass.

[1]From McGraw-Hill Co.'s brochure, March, 1920.

Power stations that are built from now on should be of the so-called super-power-station design, and although the supplement design accompanying the May 25 issue of *Power* may not be adopted, it may suggest to other engineers a better design than has thus far been used, especially with reference to the handling of the condensing water, coal into the station, ashes away from the station, and the minimum number of men employed per kilowatt output, as well as the minimum cost of construction.

Regarding the economies of this type of station, the heat-balance sheets, Fig. 2, illustrate the possibilities, provided the exact conditions stated on the sheets are carried out, such as steam pressure, superheat or total temperature, vacuum or exhaust pressures, and also the correct and proper heat balances maintained at all times in the station.

Two heat-balance sheets for the stated efficiency of the boiler are shown. The higher economies shown

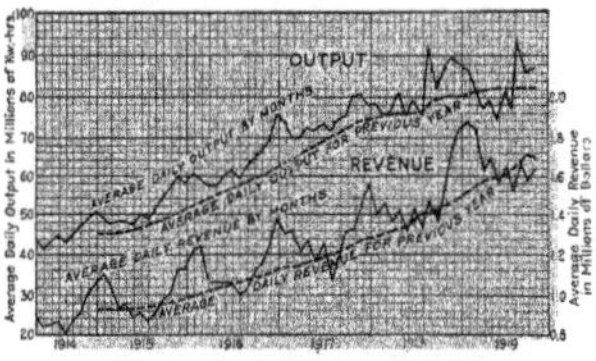

FIG. 1. SHOWING GROWTH OF CENTRAL-STATION OPERATION

by the second sheet are with steam extracted from the upper stages of the turbine and used in preheating the combustion air to a moderate temperature, say 170 deg. The preheating of combustion air has been tried out and used many times, but a moderate degree of preheating the air has, to the writer's knowledge, never been scientifically arranged for, installed and used, other than with the Howden's forced-draft system or Ellis & Eaves' systems in marine work, which have been notably successful.

Regarding the boiler efficiency given, the company handling these boilers authorized the statement that it will guarantee an efficiency of 85 per cent and that a higher efficiency may be expected. An eminent engineer has also stated that 87½ or even 90 per cent over-all efficiency might be obtained with the system of double jacketing the boilers and circulating the heat of radiation, as well as the cooling air from the generator, into and through the furnaces; especially so if either oil or pulverized coal is used as a fuel. At least, the economies shown work out in this way mathematically, and it appears that if great care and skill are used in the design, the efficiencies recorded may be obtained at the outgoing leads of the station.

The heat-balance diagrams are worked out to scale, and the use of steam in the auxiliaries and the radiation of buildings is in direct proportion to the balance of the diagram. It is hoped that some engineer will discover a way in the future to reduce the condenser circulating-water losses. Some attempts have already been made to do this, but as yet they do not appear to be successful in practice.

The sizes of the units, boilers, turbines and switch gear are limited only by the unit size of the boilers, and the first sheet of Fig. 2 shows the possibilities of the super-station when Stevens-Pratt boilers are used. To illustrate further the economies of such a station, Table III shows the power costs at the switchboard of a 100,000-kw. steam-turbine station. Attention is called to the coal cost in mills per kilowatt-hour with different load factors. Fig. 3 is a graphic curve illustrating still further the possibilities of a super-station. These figures may not be exactly correct, but it is the system of working out rather than the results which are illustrated, principally for the use of the engineering generation that has not had to do with matters that relate to super-power stations.

TABLE I. SHOWING THE ADVANCE IN ENGINE ECONOMIES

Size and Type of Unit	Steam Conditions: Initial Pressure, Lb. Gage and Deg. Superheat	Steam Conditions: Final Pressure, In. Mercury	Perfect Engine (Rankine Cycle): B.t.u. Kw., per Min. per Cent	Perfect Engine (Rankine Cycle): Rankine Eff.	Actual Engine: Steam Rate, Lb. per Kw.-Hr.	Actual Engine: B.t.u. Kw. per Min.	Actual Engine: Thermal Eff. per Cent	Actual Engine: Ratio per Cent. Thermal Eff. Rankine Eff.
50-kw. high-speed slide-valve engine	100 dry	Atmos.	386.0	14.75	48.0	807	7.05	47.8
1,000-kw. cross-compound Corliss	150 dry	26.0	216.5	26.3	21.0	386	14.75	56.0
3,750-kw. N. Y. Edison engines	185 dry	27.0	202.5	28.1	16.8	312	18.25	65.0
5,000-kw. Interborough engines	175 dry	25.5	214.5	26.5	17.1	314	18.10	68.4

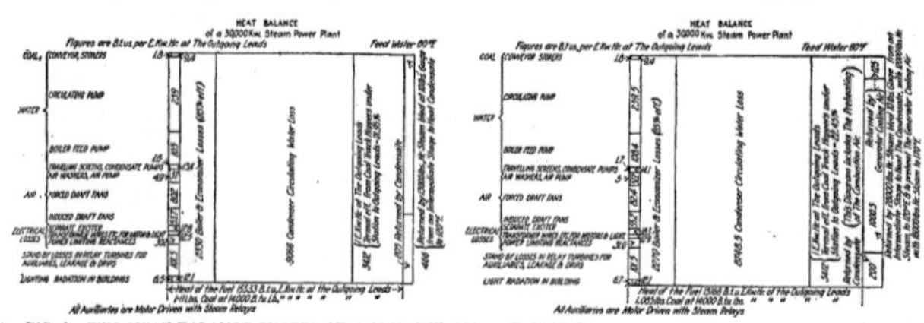

FIG. 2. TWO HEAT-BALANCE SHEETS OF A 30,000-KW. STEAM-POWER PLANT EQUIPPED WITH STEVENS-PRATT BOILERS; 325-LB. GAGE, 200 DEG. SUPERHEAT AT THROTTLE; 27 IN. HG. VACUUM; POWER FACTOR, 0.8; EFFICIENCY, RANKINE CYCLE, 76 PER CENT

In addition to the power generated by super-stations in the central-power service of the future, there is another element of conservation that may be made. While up to the present time district heating with systems as an auxiliary to central-power stations has not been much thought of or developed in America, and while some attempted systems have not been successful financially because of lack of proper foresight and experience, the possibility of thus making use of heat energy that otherwise would largely go to waste in the form of condensing-water discharge must make this form of service of increasing interest and importance. In most all industries (other than very large

TABLE II. SHOWING THE EVOLUTION OF TURBINE PRACTICE

Size and Type of Unit	Steam Condition: Initial Pressure, Lb. Gage and Deg. Superheat	Steam Condition: Final Pressure, In. Mercury	Perfect Engine (Rankine Cycle): B.t.u. Kw., per Min. per Cent	Perfect Engine (Rankine Cycle): Rankine Eff.	Actual Engine: Steam Rate, Lb. per Kw.-Hr.	Actual Engine: B.t.u. Kw. per Min.	Actual Engine: Thermal Eff. per Cent	Actual Engine: Ratio per Cent, Thermal Eff. Rankine Eff.
35-kw. steam-turbine exciter set	175–100	Atmos.	309.0	18.4	47.0	844	6.74	36.7
100-kw. horizontal turbine	150-dry	28.0	197.0	28.9	25.0	469	12.12	42.0
500-kw. vertical turbine	150-dry	28.0	197.0	28.9	24.0	450	12.63	43.8
500-kw. horizontal turbine	150-dry	28.0	197.0	28.9	19.5	365	15.58	54.0
500-kw. horizontal turbine	175-100	28.0	190.0	29.5	17.1	338	16.85	56.3
1000 kw. horizontal turbine (multi-velocity stage)	150-100	28.5	190.5	29.9	14.3	285	19.3	66.7
	190-100	28.5	184.0	31.0	13.8	276	20.6	66.7
20,000-kw. turbine	265-250	29.0	164.5	34.6	10.6	229	24.8	72.0
30,000-kw. -50,000-kw. turbine	250-250	29.0	165.0	34.5	9.95	215	26.5	76.8
*50,000-kw. turbine, three-cylinder unit	300-200	28.5	169.0	33.7	10.2	214	26.5	78.8

* Maximum rating, 70,000 kw.

machine shops and similar processes) large volumes of manufacturing and heating steam are used. Several instances are known of factories using 300,000 to 400,000 lb. of steam an hour for heating and manufacturing processes, and in the power arrangement of the near future the electric energy used for power will be generated in super-steam stations, from hydro-electric plants, or both, as well as the manufacturing steam made as a byproduct to power. There is nothing new in this principle, since we have records in New England manufacturing plants of this system being used as early as 1860, when a very able naval engineer was employed by a large textile mill to change over its steam system. He arranged to use non-condensing engines at 10 or 12 lb. back pressure for the producing of approximately 100,000 to 150,000 lb. of steam an hour for the dyehouse, bleachery an steam heating of the plant.

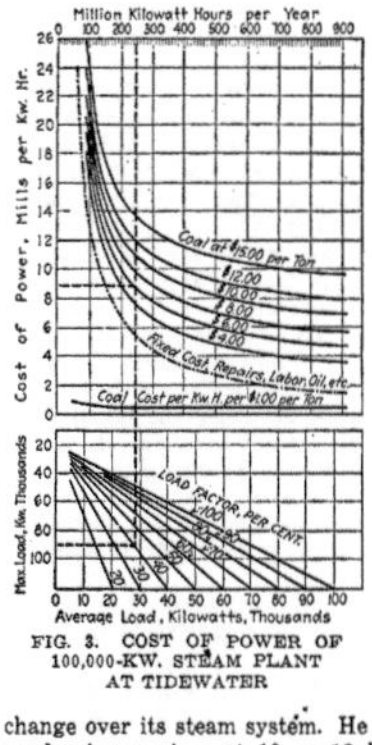

FIG. 3. COST OF POWER OF 100,000-KW. STEAM PLANT AT TIDEWATER

As early as 1907 the writer investigated the use of turbines in this manner and from that day to this has installed a great many machines, some running

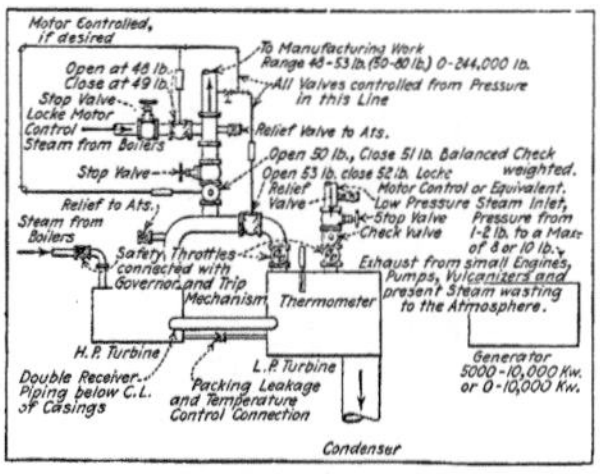

FIG. 4. DIAGRAMMATIC SKETCH OF DOUBLE-CYLINDER EXTRACTION AND LOW-PRESSURE TURBINE

entirely non-condensing and some with different arrangements of stage extraction or admission of steam to the lower stages of the turbines as the case might be; that is to say, connected to a common low-pressure or exhaust main with steam engines and pumps operating non-condensing so that at off-peak periods, the exhaust of the non-condensing engines could be

TABLE III. POWER COSTS AT SWITCHBOARD OF 100,000-KW. STEAM-TURBINE STATION

Main Units—4—35,000-kw. turbines—economy—10 lb. steam per kw.-hr.: 4—5,800-Hp. B.&W., Stevens-Pratt boilers and economizers 85 per cent over-all efficiency. Investment—$10,000,000

Maximum Load in Kw.	Load Factor, per Cent.*	Average Load, in Kw.	Kw.-Hr. per Year	Fixed Cost, Labor, Repairs, Etc., Mills per Kw.-Hr.	Coal, Cost, Mills per Kw.-Hr.	Total Cost of Power—Mills per Kw.-Hr. Coal per Ton at $4.00	$6.00	$8.00	$10 00	$12.00	$15.00
100,000	100	100,000	876,000,000	1.56	†0.55	3.76	4.86	5.96	7.06	8.16	9.81
	90	90,000	788,400,000	1.74	0.55	3.94	5.04	6.14	7.24	8.34	9.99
	80	80,000	700,800,000	1.95	0.55	4.15	5.25	6.35	7.45	8.55	10.20
	70	70,000	613,200,000	2.23	0.55	4.43	5.53	6 63	7 73	8.83	10.48
	60	60,000	525,600,000	2.61	0.55	4.81	5.91	7.01	8 11	9.21	10.86
	50	50,000	438,000,000	3.13	0.55	5.33	6.43	7.53	8 63	9.73	11.38
	40	40,000	350,400,000	3.91	0.55	6.11	7.21	8.31	9 41	10.51	12.16
	30	30,000	262,800,000	5.21	0.55	7.41	8.51	9 61	10 71	11.81	13.46
	20	20,000	175,200,000	7.81	0.55	10.01	11.11	12 21	13.31	14.41	16.06
80,000	100	80,000	700,800,000	1.95	0.55	4.15	5.25	6.35	7.45	8.55	10.20
	90	72,000	630,720,000	2.17	0.55	4.37	5.47	6.57	7.67	8.77	10.42
	80	64,000	560,640,000	2.44	0.55	4.64	5.74	6.84	7.94	9.04	10.69
	70	56,000	490,560,000	2.79	0.55	4.99	6.09	7.19	8.29	9.39	11.04
	60	48,000	420,480,000	3.26	0.55	5.46	6.56	7.66	8.76	9.86	11.51
	50	40,000	350,400,000	3.91	0.55	6.11	7.21	8.21	9.31	10.41	12.06
	40	32,000	280,320,000	4.89	0.55	7.09	8.19	9.29	10.39	11.49	13.14
	30	24,000	210,240,000	6.51	0.55	8.71	9.81	10.91	12.01	13.11	14.76
	20	16,000	140,160,000	9.78	0.59	12.14	13.32	14.50	15.68	16.86	18.63
60,000	100	60,000	525,600,000	2.61	0.55	4.81	5.91	7.01	8.11	9.21	10.86
	90	54,000	473,040,000	2.90	0.55	5.10	6.20	7.30	8.40	9.50	11.15
	80	48,000	420,480,000	3.26	0.55	5.46	6.56	7.66	8.76	9.86	11.51
	70	42,000	368,920,000	3.73	0.55	5.93	7.03	8.13	9.23	10.23	11.88
	60	36,000	315,360,000	4.35	0.55	6.55	7.65	8.75	9.85	10.95	12.60
	50	30,000	262,800,000	5.21	0.55	7.41	8.51	9.61	10.71	11.81	13.46
	40	24,000	210,240,000	6.51	0.55	8.71	9.81	10.91	12.01	13.11	14.76
	30	18,000	157,680,000	8.69	0.57	10.97	12.17	13.25	14.39	15.53	17.24
	20	12,000	105,120,000	13.02	0.68	15.74	17.10	18.46	19.82	21.18	23.22
40,000	100	40,000	350,400,000	3.91	0.55	6.11	7.21	8.31	9.41	10.51	12.16
	90	36,000	315,360,000	4.35	0.55	6.55	7.65	8.75	9.85	10.95	12.60
	80	32,000	280,320,000	4.89	0.55	7.09	8.19	9.29	10.39	11.49	13.14
	70	28,000	245,280,000	5.58	0.55	7.78	8.88	9.98	11.08	12.18	13.73
	60	24,000	214,240,000	6.51	0.55	8.71	9.81	10.91	12.01	13.11	14.76
	50	20,000	175,200,000	7.81	0.55	10.01	11.11	12.21	13.31	14.41	16.06
	40	16,000	140,160,000	9.78	0.59	12.14	13.32	14.50	15.68	16.86	18.63
	30	12,000	105,120,000	13.02	0.68	15.74	17.10	18.46	19.92	21.18	23.22
20,000	100	20,000	175,200,000	7.81	0.55	10.01	11.11	12.21	13 31	14.41	16.06
	90	18,000	157,680,000	8.69	0.57	10.97	12.17	13.25	14 39	15.53	17.24
	80	16,000	140,160,000	9.78	0.59	12.14	13.32	14.50	15.68	16.86	18.63
	70	14,000	122,640,000	11.16	0.61	13.60	14.82	16.04	17.26	18.48	20.31
	60	12,000	107,120,000	13.02	0 68	15.74	17.10	18.46	19 82	21.18	23.22
	50	10,000	87,600,000	15.62	0.71	18.46	19.88	21.30	22.72	24.14	26.27

* Load factor taken as ratio of average load to maximum coal consumption is figured with 15 per cent leeway above guarantees.
† Coal cost per kw.-hr. per $1 per ton.

used in the lower stages of the turbine rather than wasting it, since in textile mills there is a large waste of steam, due to the fact that the dyehouse load usually goes off from 11:30 to 12 in the morning and again at 4 in the afternoon.

Until quite recently no turbines were built for operating non-condensing or against a back pressure of over

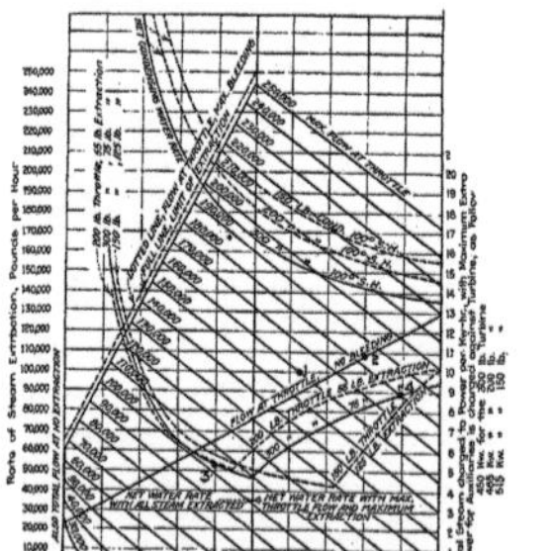

FIG. 5. ECONOMIES OF A 10,000-KW. DOUBLE-CYLINDER EXTRACTION AND LOW-PRESSURE STEAM TURBINE; 300-LB. GAGE, 100 DEG. F.; 2-IN. ABS. 75-LB. GAGE STEAM EXTRACTION

No. 1—Curves shown dotted are for other turbines than 10,000-kw. plotted to same scale of extraction, kilowatts and steam charged to power per kilowatt-hour, but with different steam-flow curves, not shown. No. 2—Maximum throttle flow taken at 250,000-lb. for all turbines. No. 3—Generator and friction losses taken at 10.4 per cent for all turbines. No. 4—Extraction water apply only when all possible steam is being extracted.

60 lb., but in the rubber industry 80 to 90 lb., or even 100 lb., is desired, and to cope with this condition a turbine arrangement as shown in Fig. 4 was evolved.

This machine is termed a double-cylinder extraction and low-pressure turbine. It consists of a high-pressure and a low-pressure turbine mounted on one shaft with, say, a 10,000-kw. generator. This machine will operate automatically with steam extracted from 0 gage to, say, 80 lb. pressure and from 0 to 244,000 lb. extracted at a predetermined pressure. It will operate entirely condensing with only a slightly reduced economy from that of a standard machine, and it will operate also for any intermediate extraction of steam or generation of power. At the same time large amounts of steam may be admitted to the low-pressure casings in the way of returns from vulcanizing machinery, which should be put through regenerators or separators at 1 to 2 lb. pressure for the precipitation of the acids. The control of this machine is from the sustained or maintained pressure in the steam line to the manufacturing systems. In other words, the machine is controlled from its byproduct rather than from its electrical end; but at the same time, if more service is demanded from the electrical end than the proportionate amount of steam passing through to the manufacturing process would admit, the automatic valve mechanism will open to the low-pressure or condensing turbine the necessary amount of steam to follow the electrical demands.

The economies of this machine are illustrated by Fig. 5, and it will be observed what splendid economies are obtained from a machine of this type, which may be installed to keep the entire output of the machine when running non-condensing or part non-condensing and part condensing, at a rate of 7 lb. of steam or even less per kilowatt-hour at the generator leads chargeable to power in kilowatts.

At the left of the center line in each sheet are shown the amounts of steam chargeable per kilowatt-hour when operating non-condensing, and to the right of the dip of each curve, the effect of some or part of the steam passing on to the condenser. These curves include a liberal allowance for the amount of steam or power used by a jet condenser and therefore give the total approximate amount of steam chargeable to power as generated by this byproduct machine. If a surface-condensing equipment were used, these would be substantially cut, which would still further reduce the amount of steam chargeable per kilowatt-hour produced.

Patents are now pending which will protect the company that evolved this turbine in connection with the writer and his staff.

For a more simple illustration of the great possibilities of non-condensing turbines to match or phase

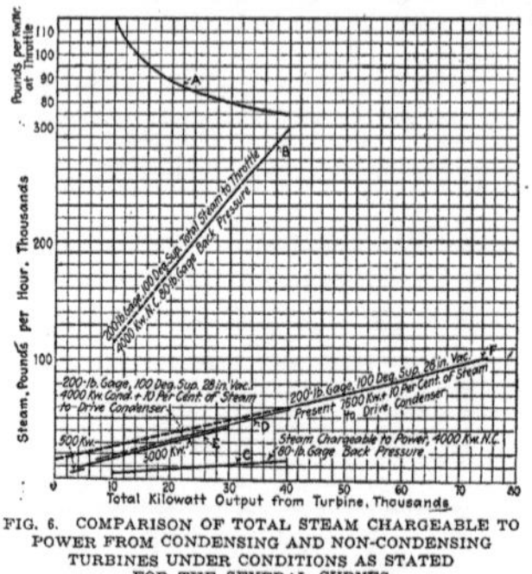

FIG. 6. COMPARISON OF TOTAL STEAM CHARGEABLE TO POWER FROM CONDENSING AND NON-CONDENSING TURBINES UNDER CONDITIONS AS STATED FOR THE SEVERAL CURVES

Curve *A*, estimated total steam per kilowatt-hour, 4,000-kw. turbine. *B*, estimated total throttle steam, 4,000-kw. turbine. *C*, estimated total steam chargeable to power, 4,000-kw. turbine. *D*, estimated total, throttle steam for 4,000-kw. turbine plus 10 per cent of steam for condensers. *E* and *F*, 3,000- and 7,500-kw. turbine, same conditions. *G*, total throttle steam for 500-kw. turbine.

in with the super-station or central station, owned by a public utility or owned and operated by the plant in question, see Figs. 6 and 7.

This type of turbine is in effect a "power-producing pressure-reducing turbine" and can be constructed in practically any size up to 2,500-kw. capacity and is so valved and arranged that it will automatically float on the line and deliver quantities of steam in proportion to the demand, giving a proportionate amount of byproduct power; in other words, it is similar to a reducing valve, and the splendid economies of this turbine in one particular instance are illustrated by Figs. 6 and 7.

As these turbines are manufactured in small units, they may be located at a distance from the boiler

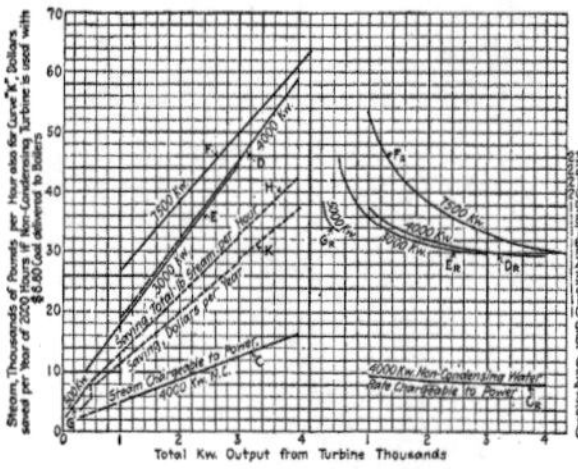

FIG. 7. COMPARISON OF TOTAL STEAM CHARGEABLE TO POWER FROM CONDENSING AND NON-CONDENSING TURBINES UNDER CONDITIONS STATED FOR THE SEVERAL CURVES

Curve *C*, estimated total steam chargeable to power, 4,000-kw. turbine, 200-lb. gage, 100-deg. superheat 80-lb. gage exhaust. *D*, estimated total steam for 4.000-kw. turbine. *E*, 3,000-kw. turbine. *F*, 7,500-kw. turbine. *G*, 500-kw. turbine. All with 200-lb. steam, 100-deg. superheat, 28-in. vacuum. *H*, difference in total pound steam per hour between curve *C* and curve *D*, *F* or *G*, whichever is lowest. *K*, dollars gross saving per year of 2,000 hours' operation at kilowatt rate, 9-lb. evaporation and coal at $8.80 in bunker if non-condensing turbine is used. *Cr*, *Dr*, *Er*, *Fr* and *Gr*, kilowatt-hour rate curves corresponding to total steam curves *C*, *D*, *E*, *F* and *G*.

plant; that is, used locally for outlying buildings, or at or near the processes using large amounts of steam, which will do away with the necessity of running large and very expensive low-pressure mains, throughout the plant from the central station. They should, however, be connected into the main electrical distribution systems. This turbine is evolved and standardized by one of the large electric companies, and therefore is a marketable commercial product at the present time. A diagrammatic sketch of such a unit is shown in Fig. 8.

To sum up, the power plant of the future will be along the basic lines suggested by this article, and it is suggested that power contracts be made with the permission to manufacture what power is desired by byproduct turbines—that is, byproduct to power production—and by such station designs as was shown in the May 25 issue of *Power* and in this article, as giving a complete program which should be followed.

In the actual working out of designs for multiple units of these super-power stations, following the lines suggested in the previous article, there would be bridges provided between the 60,000-kw. units, and these for maximum safety to station operatives might be made open rather than closed. To bring out clearly the separation of units advised, these bridges were purposely omitted from the illustrations.

Further, the chief engineer's operating or instrument room might be constructed ---- only one unit, instruments for all units being brought together in this one central operating room. If this were done, some saving could be made by placing a flat roof directly above the fanroom, letting the coal-bunker pace walls act as parapet walls. However, it would seem that the coal-consuming capacity of each unit of this plant and the cost of coal so that perpetual boiler tests and accurate records, obtained by weights and measures, may be kept throughout the life of one of these plants.

Locating these super-power stations in large industrial communities would enormously reduce the amount of coal necessary to handle the power of such communities, for the coal consumed per unit in the super-stations, broadly speaking, is about one-half that of the isolated plant, and by making use of the byproduct turbines for manufacturing and heating steam, enormous further reductions in the use of coal are made, to say nothing of a relay power at the customer's plant, or at least a part relay to the purchased power.

In one instance, on which the writer has worked for several years, the conservation of coal is about 500,000 tons per year, and furthermore, in his judgment there will not be over five or six great stations in New England at some subsequent future date.

These large stations require such vast amounts of condensing water that they can be located only on such rivers as the Connecticut or Merrimack, or at shore ports on rivers, lakes or oceans, where the intake

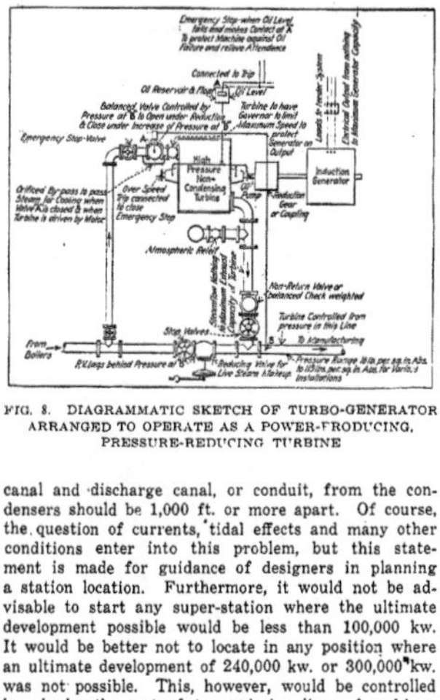

FIG. 8. DIAGRAMMATIC SKETCH OF TURBO-GENERATOR ARRANGED TO OPERATE AS A POWER-PRODUCING, PRESSURE-REDUCING TURBINE

canal and discharge canal, or conduit, from the condensers should be 1,000 ft. or more apart. Of course, the question of currents, tidal effects and many other conditions enter into this problem, but this statement is made for guidance of designers in planning a station location. Furthermore, it would not be advisable to start any super-station where the ultimate development possible would be less than 100,000 kw. It would be better not to locate in any position where an ultimate development of 240,000 kw. or 300,000 kw. was not possible. This, however, would be controlled largely by the cost of transmission lines, franchises, money values, and innumerable other conditions that enter into a project of this size.

When otherwise practicable, it will be well so to choose locations for central stations that later full advantage may be taken of economies to be brought about by the recovery and utilization of distillates and other chemical products from coal. This is a matter at present little thought of, but likely to be of increasing importance for the future.

To illustrate further what the celebrated engineers of the country are promulgating as to the possibilities of the great super-power stations of the near future, Figs. 9 and 10, taken from the March 16, 1920, issue of *Power*, with additions, show graphically what is being seriously thought of.

This super-power zone, as shown in Fig. 9 takes in the area between Boston and Washington and extends inland from the coast 100 to 150 miles. The suggested transmission lines from the hard and soft coal fields, Niagara Falls, Cedar Rapids and from main hydro-electric plants are shown by Fig. 10.

From investigations and research I have found that there is a 300,000-kw. load, not including the steam-railroad electrification, within a radius of fifty miles from Newburyport or Beverly, Mass., which could be tied in with the great water-power developments at Manchester, N. H., Lowell and Lawrence, Mass.; and with the great development and proposed developments of the Edison Company, of Boston, Mass., the conservation in this power zone alone amounts to about 500,000 long tons of coal per year, conservatively speaking, with all the attendant advantages in the conservation of coal, money and men. The project considers the use of the river banks and railroad rights-of-way for transmission lines, and placing a station of that kind in the northeastern range of the proposed super-power zone illustrated by Fig. 9 would reinforce in a splendid manner the loop transmission line suggested.

Further, there is a vast amount of coal in Nova Scotia which, if used as pulverized fuel, would make two methods of shipment to this point by water—Nova Scotia coal and coal from the Virginia fields. This would relieve the congested freight situation of this thickly settled manufacturing district.

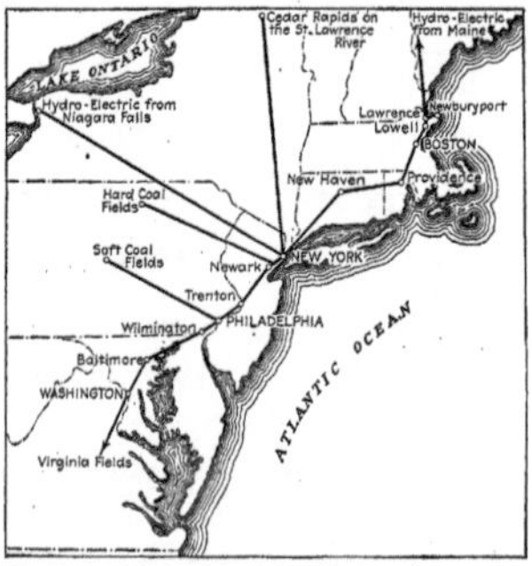

FIG. 10. SUGGESTED POWER-TRANSMISSION LINES FOR SUPER-POWER ZONE

A cross-tie in this loop in New England and somewhat paralleling the Connecticut River and its hydro-electric and large steam developments is suggested. This would further substantiate and augment the eastern part of the super-power zone loop in New England and would also further relay the great power demands of New York City and vicinity. There are other untold resources in New England for such a project as has been outlined herein.

FIG. 9. MAP SHOWING APPROXIMATE SUPER-POWER ZONE

The time will come when no company or individual will be allowed to operate any plant at a thermal efficiency of less than a predetermined percentage; neither will any company be allowed to operate any turbines or equivalent machinery condensing when they could be operated non-condensing and the exhaust steam used in heating and manufacturing, or when purchased power from a large super-station can be advantageously made available.

In other words, as coal and oil, in fact, all known fuels, are non-renewable, natural resources, the general public will be forced to use the utmost care in their consumption.

Nothing new has been advanced, but an effort has been made to bring out clearly important facts, which are of interest to all who have to do with power-plant problems of the present or near future.

The Compressed-Spruce Pulley

Spruce of the Pacific Northwest came into its own during the war, being found excellent for airplane uses. Now comes the announcement of another application of it which promises as much for power transmission as it did for the air service. R. L. Watts, of the St. Paul & Tacoma Lumber Co., and F. J. Walsh, consulting engineer for the Port of Tacoma, Washington,

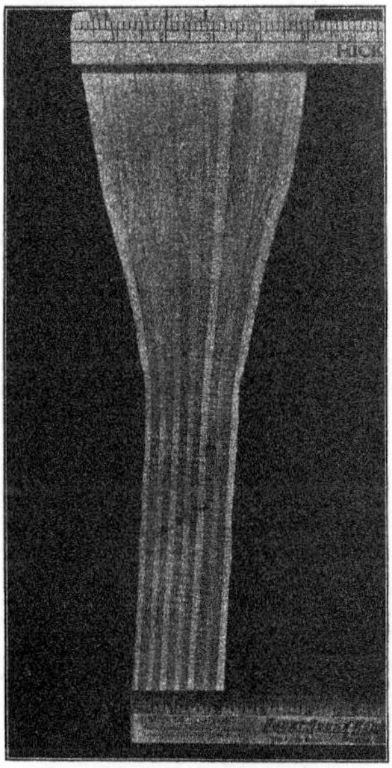

FIG. 1. LAMINATED PIECE SHOWING ONE END COMPRESSED AND ONE END NOT COMPRESSED

while pressing laminated spruce to make the glue stick fast, discovered that very great pressures did not splinter and did not materially spread the wood, but made it extremely hard and dense. This led to the application of the compressed spruce to power-transmission pulleys, which they are now manufacturing under the name of Watts & Walsh Co., Tacoma, Wash. The strips of wood or laminations in the pulleys are originally ⅜ in. thick; the pressure applied for compressing is at present 100 tons per sq.ft., though a press capable of 2,500 tons per sq.ft. will be used later. The pulleys are available in all sizes.

Fig. 1 shows a laminated piece with the upper end the original width and the lower end compressed. Fig.

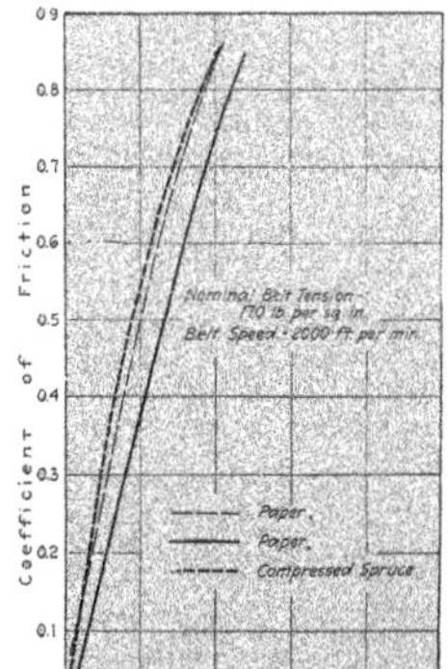

FIG. 2. PERFORMANCE-TEST RESULTS, CONDUCTED BY PROFESSOR GEORGE S. WILSON, UNIVERSITY OF WASHINGTON

2 shows performance curves obtained in tests conducted at the University of Washington by Professor Wilson, of the department of Mechanical Engineering. Fig. 3 shows a compressed-spruce pulley 36 in. in diameter,

FIG. 3. COMPRESSED SPRUCE PULLEY ON 250-HP. MOTOR DRIVING GANGSAW, EXTREMELY VARIABLE LOAD, SWINGING FROM NO LOAD TO 400-HP.

31-in. face; the distance between flanges, where the pulley is fastened to the shaft, is but 5 in. The motor is of 250-hp., with extremely and suddenly variable load from a gangsaw which it drives.

Paralleling Two Alternators When They Rotate in Opposite Directions

By B. A. Briggs

It is shown why, when two alternators are revolving in the same direction and connected to operate in parallel, they cannot be connected in parallel if the direction of rotation of one of the machines is reversed, until the connections on one phase of this machine is reversed.

IN THE article, "Operating Two Alternators of Opposite Polarity in Parallel," in the May 18th issue, it was shown that changing the direction of the exciting current through the field coil of one of a number of alternators does not prevent it being connected in parallel with other machines, except where two alternators are operated in parallel and are connected to the same prime mover. In this case the rotating elements, being connected permanently together, cannot seek their true phase relations, therefore this relation must be determined by test before the machines are put into service. The next question that arises is, "What would be the effect if the direction of rotation of an alternator that is operating in parallel with other machines is reversed?"

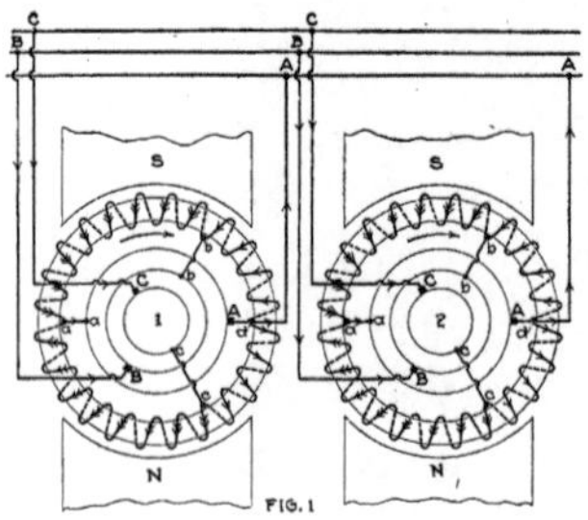

FIG. 1

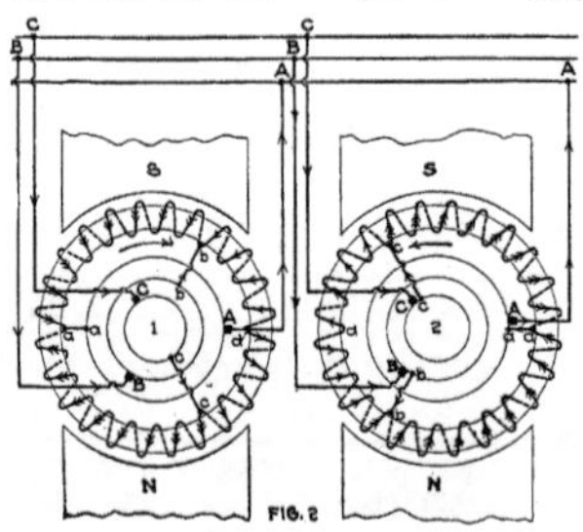

FIG. 2

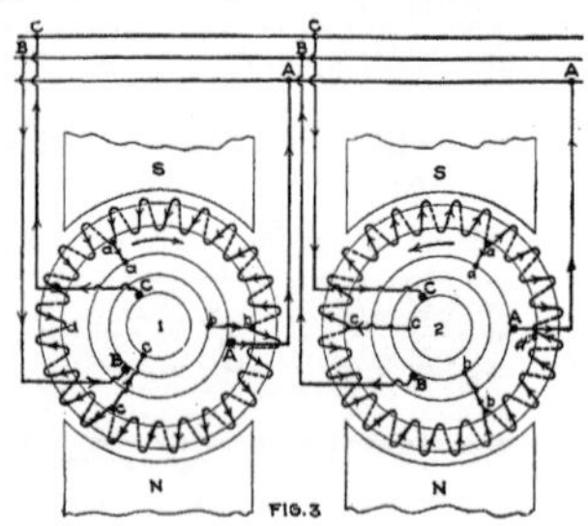

FIG. 3

FIGS. 1 TO 3. SCHEMATIC DIAGRAMS OF TWO ALTERNATORS CONNECTED IN PARALLEL

In Fig. 1 is represented two three-phase alternators connected symmetrically to a set of busbars, and both machines revolving in the same direction. In each machine the *A* terminal is positive and the *B* and *C* terminals negative. Since the armatures are in exactly the same relative positions, there will be no cross-current flowing between them; therefore they are in phase. In the armatures the part of the winding that is supplying current is indicated by double arrowheads. Although there is voltage generated in the conductors of the section between *b* and *c* at the instant shown, this section is not effective in supplying a current to the external circuit, since the voltage generated in the conductors between *b* and *d* opposes that between *d* and *c*. The volts generated in these two sections are equal and opposite, consequently the resultant voltage is zero, and no current will be supplied from this section. Assume that the direction of rotation of No. 2 machine is reversed, as indicated in Fig. 2 by the curved arrow. If the armature of this machine is turned through 180 deg. from the position in Fig. 1, or 180 deg. from the position of machine No. 1, Fig. 2, it will be seen from the arrowheads that the machines are again in phase. The *A* terminals are positive and the *B* and *C* terminals negative, and the voltages generated in the *ab* and *ac* sections of both armatures are equal. Although the *ab* section in the left-hand machine is under the S pole, and in the right-hand machine under the N pole, and vice-versa for the *ac* sections, the voltage is in the correct direction for the machines to be in synchronism at the instant indicated. From this the conclusion might be drawn that the machines can be synchronized, but this condition represents only one instant in the revolution, and although the machines might be in synchronism at this instant, it is possible that they are out of synchronism at all other points. Therefore, before a decision can be made, other points in the revolution must be investigated.

In Fig. 3 both armatures have been revolved through 60 deg. from the position in Fig. 2, the armature of No. 1 turning clockwise and the armature of No. 2 turning counterclockwise. The arrowheads indicate that although the *A* terminals are both positive, and therefore apparently have the correct phase relation, terminal *B* on machine No. 2 is positive, and *B* on machine No. 1 is negative, and vice versa for the *C* terminals. Consequently, the *B* and *C* terminals on the two machines are out of phase, and if these machines are connected together under this condition a very heavy current will flow between them. Therefore, the generators cannot be synchronized if they are rotating in opposite directions and are connected systematically to the busbars. A comparison of Figs. 2 and 3 will indicate what has happened. In Fig. 2 it will be seen that similar taps on the two armatures are 180 deg. apart; that is, the *a* tap on armature No. 2 is 180 deg. away from the *a* tap on armature No. 1, likewise for the *b* and *c* taps. In Fig. 3 we find the taps occupying positions in space only 60 deg. apart. If it were possible to maintain similar taps 180 deg. apart in space, the machines would remain in synchronism, but since they are turning in opposite directions, the position of similar taps with relation to each other is continually changing. In the article previously referred to the polarity of one machine was reversed, and this machine was found to be in synchronism with the other machine when similar armature taps on the two machines had relative positions in space 180 deg. apart. The machines then remained in synchronism at all points in the revolution, since they were turning in the same direction, and it was possible to maintain a fixed relation between the two armatures. In Fig. 2 the armatures are turning in opposite directions, consequently the relation between the positions of similar taps on the two armatures are continually changing, and therefore their voltage relation.

The next thing to consider is what to do to make it possible to get the two machines in parallel when rotating in opposite directions. This can be done by crossing any two leads on the machine that was reversed—in this case machine No. 2. If there are only two machines, crossing the lead on either machine would make it possible to synchronize them; but if the phase of the busbars is reversed, then any induction motors on the system would be reversed, consequently the change should be made on the machine that is being put into service. Fig. 4 is the same as Fig. 3, except that leads *B* and *C* of machine No. 2 have been changed at the busbars; that is, *C* brush has been connected to *B* bus and *B* brush connected to *C* bus. It will now be seen that the leads going to the *A* and *C* busbars are positive, and to the *B* bus negative, and since the armatures are in a position where each will generate the same voltage, the two machines are in synchronism at this instance.

The armatures, Fig. 5, have revolved 60 deg. in the direction of the curved arrows from the position shown in Fig. 4. In this position the terminals connecting

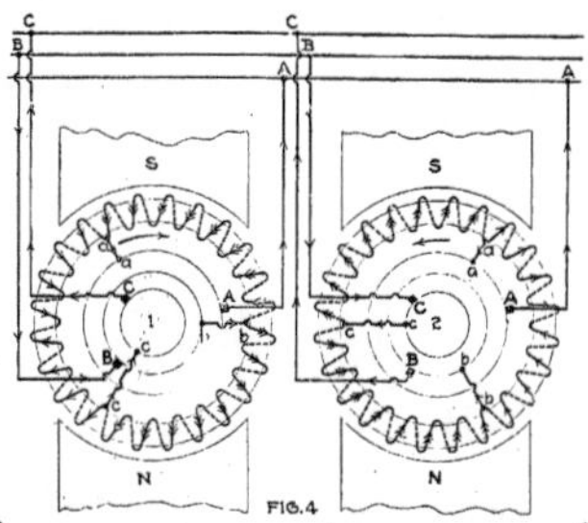

FIG. 4

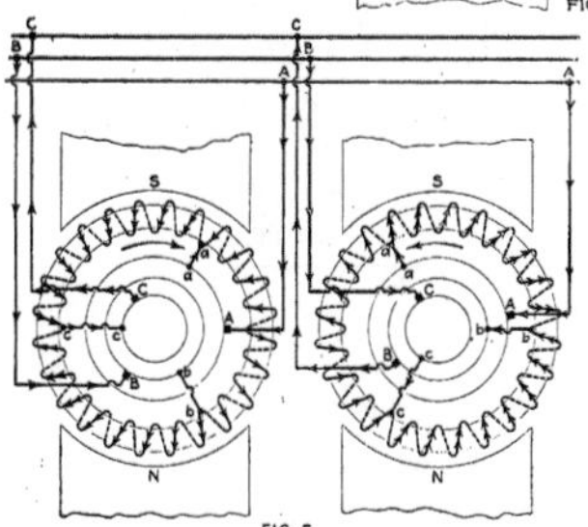

FIG. 5

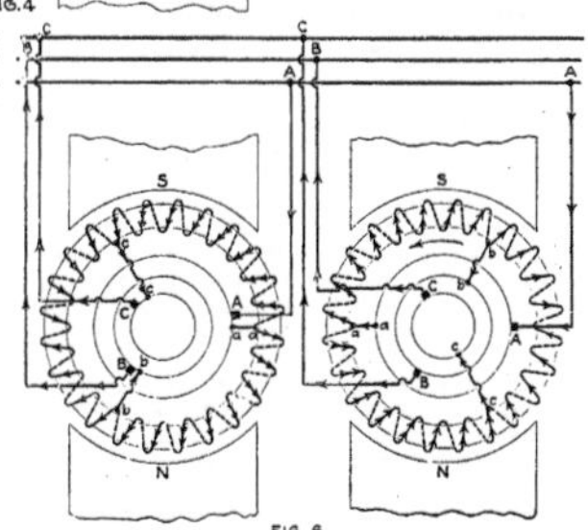

FIG. 6

FIGS. 4 TO 6. SAME AS FIG. 3, EXCEPT "B" AND "C" LEADS ARE INTERCHANGED to *A* and *B* busbars are negative, and the terminals connecting to *C* busbar are positive, and since the armatures are in a position to generate the same voltage, they are again in synchronism, as in Fig. 4. In Fig. 6 the armatures have turned through 60 deg. from the position in Fig. 5, and an inspection will show that the voltage relation between the two machines is correct for parallel operation. From the foregoing the conclusion may be drawn that if two alternators are connected properly for parallel operation when both machines are revolving in the same direction, if the direction of rotation of one of the machines is reversed, two of the leads must be reversed on this machine before it can be again parallel with the other machines. This discussion refers to three-phase machines. With two-phase machines care must be exercised to cross the lead on either phase of the machine whose direction has been reversed.

When two alternators are geared to the same turbine, one revolving in one direction and the other in an opposite direction, to connect the machines to operate in parallel tie a like lead on each machine together. For example, the *A* leads on each machine are connected together. Then the *B* lead on one machine is connected to the *C* lead on the other machine, and vice versa, as was done in Figs. 4 to 6. Then if the revolving elements are placed in the same relative position on each machine, as explained in the article, "Synchronizing Two Alternators When Driven by One Turbine," page 568, *Power*, Oct. 7, 1919, and the exciting current made to flow through the field coils in the same direction, the two alternators should come up to voltage in exact synchronism.

R. K. Hydro-Electric Pressure Regulator

There has recently been developed a pressure regulator designed to eliminate all erratic action in operation. The scale beam is free to respond accurately to variations in pressure of ¼ lb. or less and to transmit its readings, by means of an electric contact, to limit the movement of the motor plunger. The variations in pressure are transmitted to a pilot valve which controls the movement of the motor by means of electric magnets. One magnet controls the upward movement of the plunger and another controls the downward movement, the pilot valve normally being held in a neutral position, when no pressure can be admitted to the motor cylinder.

The motor travels by stages. The plunger is moved step by step with a well-defined period of rest between each stage of its travel. When a change in pressure takes place, the plunger is moved rapidly to the next succeeding stage and comes to rest and remains in that position until a change in pressure again takes place; in this way the plunger assumes at all times a position which corresponds with the pressure and thereby regulates the air supply to the furnace in the quantities required.

The operation of the regulator, which is made by the Ruggles-Klingemann Manufacturing Co., Salem, Mass., will be understood from the following: Wherever the steam pressure exceeds a predetermined amount for which the scale beam is weighted, an electric contact is made with the underside of the contact disc *A*, thus sending an electric current through the coil *B* and energizing the magnet *C* and the attracting armature *D*, which raises the dashpot plunger *E* and the valve spindle *F* and admits pressure through the port *G* to the underside of plunger *H*. As the plunger moves upward, it revolves the contact disc *A* by means of a cord *I*, which extends to the top of the plunger rod, the opposite end of the cord being provided with a weight *J*, so as to take up the slack and place sufficient tension on the cord to drive the contact disc.

The contact disc moves in a direction to break the current and cause the armature to be de-energized, which allows the dashpot plunger *E* to settle to its bottom position and return the spindle *F* to its neutral positon, which in turn brings the plunger *H* to rest. If the steam pressure should continue to increase, the operation will be repeated, the plunger moving up to a second position, and so on, making from ten to twelve stops before the stroke is completed. Whenever the

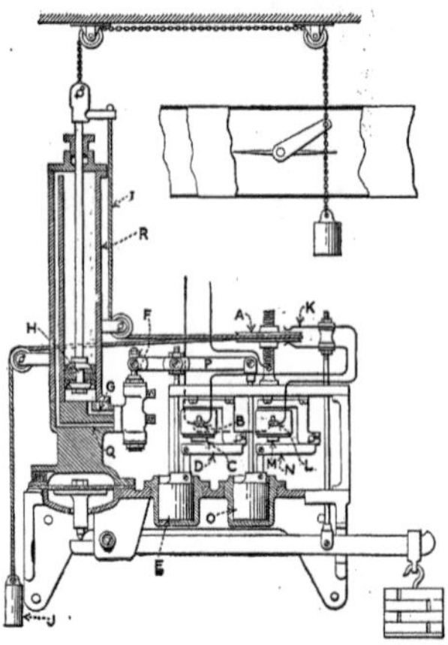

SHOWING DESIGN OF ELECTRIC-PRESSURE REGULATOR

steam pressure decreases, the contact point *K* engages the top of the contact disc *A*, which sends an electric current through coil *L*, energizing the magnet *M* and attracting the armature *N*. This raises the dashpot plunger *O*, and as the lever *P* is fulcrumed on the dashpot rod, the spindle *F* will be lowered, thus opening the port *Q* and allowing the pressure to enter the top of the cylinder *R* and force the plunger downward. As the plunger moves, it turns the contact disc *A*, which breaks the circuit, and the magnet *M* is de-energized, allowing the armature to be released and the dashpot plunger *O* to settle into its bottom position and bring the plunger *H* to rest, the plunger making regular stops on the downward stroke with short periods of rest the same as on the upward stroke.

Steam from the Earth

In *Power* for Jan. 13 was described the use that is being made from steam obtained from borings in the volcanic region of Italy. Readers will be interested in the further information that the publication of this article has brought out.

The natural steam being impregnated with impurities that are likely to attack the turbine, it has been necessary to use the heat of this steam for the production of pure steam usable in the turbine, instead of using it direct. This transformation is effected in vertical evaporators with long tubes upon the Kestner system. The battery of evaporators is composed of thirty units, each having a heating surface of 150 sq.m. (about

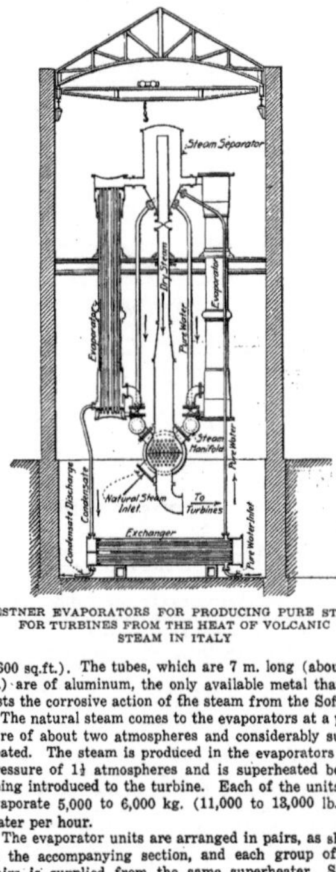

KESTNER EVAPORATORS FOR PRODUCING PURE STEAM FOR TURBINES FROM THE HEAT OF VOLCANIC STEAM IN ITALY

1,600 sq.ft.). The tubes, which are 7 m. long (about 23 ft.) are of aluminum, the only available metal that resists the corrosive action of the steam from the Soffioni.

The natural steam comes to the evaporators at a pressure of about two atmospheres and considerably superheated. The steam is produced in the evaporators at a pressure of 1½ atmospheres and is superheated before being introduced to the turbine. Each of the units can evaporate 5,000 to 6,000 kg. (11,000 to 13,000 lb.) of water per hour.

The evaporator units are arranged in pairs, as shown in the accompanying section, and each group of five pairs is supplied from the same superheater. Steam from the Soffioni enters at the natural-steam inlet, passes through the tubes of the superheater and then through the steam manifold to the evaporators. The condensate passes, as shown, from the bottom of the evaporator through the tubes of the exchanger and is finally discharged.

Pure water enters as shown, passes over the exchanger tubes and enters the bottom of the steam separator, where it mixes with the separator drip and is led to the bottom of the evaporators. Passing up through the tubes, this water is evaporated, any priming is taken out in the separator, and the dry steam goes down to the superheater and then to the turbines. It is claimed that this evaporator apparatus gives a thermal efficiency of 88 per cent.

Coal or Oil Fuel for Refrigerating Plants?

BY J. E. PORTER

In comparing coal with oil fuel for refrigerating plants no fixed rule can be laid down that will apply to all plants, owing to the different conditions met with in each individual installation. Comparing the average calorific values of the two fuels, placing coal at 11,000 B.t.u. per pound and oil at 18,500, it is evident that the oil has the advantage. It should not be taken for granted, however, that oil will be the cheaper fuel, as the price of oil relative to coal will determine this factor. The only logical and practical way to ascertain which is the cheaper fuel for a particular plant is to make a complete survey and take into consideration the cost of every item entering into the process of burning each kind of fuel. The items to be considered in the burning of coal are the original cost per ton at the mines, freight to destination, unloading at the bunkers, losses in heat value while in storage, handling between bunkers and furnaces, removal and disposal of ash and general upkeep of the furnace.

Oil can be delivered to the boiler furnace more cheaply than coal as it is handled mechanically at all times, so that a considerable saving in labor is to be expected. The several advantages usually accredited to oil as a fuel are that for the same heat value, oil occupies much less space than coal. The firedoors of the furnace do not have to be opened, reducing excess air to a minimum, giving a more even temperature and better combustion and resulting in higher capacities and efficiencies. The fires are more easily regulated to meet load requirements; there are no losses in heat value during storage. In burning oil there is no ash and very little smoke or soot if the burning is accomplished in a properly designed furnace and burner. Under such conditions it is a much cleaner fuel than coal, but the reverse is true if proper conditions are not maintained.

Under favorable conditions a pound of fuel oil will evaporate from 14 lb. to 16 lb. of water, while the evaporation from a pound of coal will range between 7 and 10 lb. Assuming a distilled-water ice plant and allowing 25 per cent to cover losses, it is necessary to evaporate 2,500 lb. of water for each ton of ice manufactured. Placing the evaporation by oil at 14 lb. of water and by coal at 8 lb., it will require about 180 lb. of oil or 313 lb. of coal to evaporate the 2,500 lb. of water. Taking the present price of fuel oil at $3.15 per barrel, the fuel costs per ton of ice would approximate $1.75; with coal at $5 per ton in the bunkers the fuel cost per ton of ice figures about 78c. From the standpoint of fuel alone at the relative price quoted, coal has a decided advantage. The saving in labor and the higher efficiency maintained by the use of oil would lessen this advantage to some extent, although not enough to overcome the difference of 97c. per ton of ice made.

Although the price quoted on fuel oil was given to the writer recently by a refining company at Oklahoma City, there are no plants in the vicinity paying $3.15 per barrel for oil. The plant using this fuel had contracted for its supply some time ago, and it is probable that its fuel cost ranges from $1 to $1.25 per ton of ice. On this basis the oil per barrel would cost from $1.80 to $2.25. Although 10 tons of ice per ton of coal is quite feasible, the usual run of plants will not average more than 5 tons of ice per ton of coal, so that with coal at $5 per ton the fuel per ton of ice would cost $1. If the prices just quoted can be obtained, there is little choice between the two fuels. Considering the scarcity of both oil and coal, however, it is largely a case of burning the fuel that can be more readily obtained, whether it should happen to be the cheaper or not.

The problem that confronts the refrigerating engineer of today is to burn the kind of fuel he is able to get with the greatest degree of efficiency. To do this a furnace must be designed to suit the particular fuel used. There is no reason for anyone to think that any design of burner will do or that the burner can be set into any kind of a coal-fired furnace and that high efficiency will be obtained. It cannot be done. There is just as much difference between the installation for burning oil and for burning coal as there is between a pump and an injector. Both perform the same duty, but in an entirely different manner and with different results. For either fuel a properly designed furnace and for oil a burner that has been approved are required. It may cost a great deal of time and money to get the furnaces just right, but the expense will be warranted and it will probably save shutting down in the middle of the season.

Clock-Pendulum Precision Speed Limit

A speed-limit device for the control of steam engines has recently been designed by E. E. Clock, chief inspector of the turbine and engine department of the Fidelity and Casualty Co., of New York. Patents are pending. The operation of the device is as follows: It is placed so as to be conveniently connected to any reciprocating part of the engine. There are two pendulums. One is positively driven, designated as *A*, and moves through a fixed arc; the other, or weighted pendulum, *B*, is in line and swings with the positively driven one until such time as the oscillations become fast enough to overcome the stabilizing springs *C*, which tend to hold the weighted pendulum *B* firmly on the two pins *D* projecting through the curved slots in the disc attached to the weighted pendulum *B*.

As the positive-driven pendulum moves to the right and starts upon the return stroke faster than the speed for which it is adjusted, the right-hand stabilizing spring *C* is overcome, the disc *E* rises off the right-hand pin *D*, swings around the left-hand pin *D*, becoming eccentric to that pin and lifting the lever *G*, which carries a roller above but clear of the disc *E*. Raising the lever *G* moves the top of the latch lever *H* to the right, the bottom or latch end to the left and drops the weighted valve-operating lever *F*, which opens the pilot valve *J*, resulting in a flow of water of condensation under full boiler pressure through the connection to the operating cylinder, which is located on the throttle-valve bonnet.

This unlatches the automatic throttle valve and gives it a start. At the same time, the threaded sleeve being released, the flow of steam, together with the piston effect of the full area of the valve stem, closes the throttle promptly.

Restoring the valve-operating lever *F* opens a drain below the operating piston and the spring *I* causes the piston to recede, when the throttle valve may be opened and the apparatus is ready to repeat its former action at the slightest overspeed. It will be seen that there can

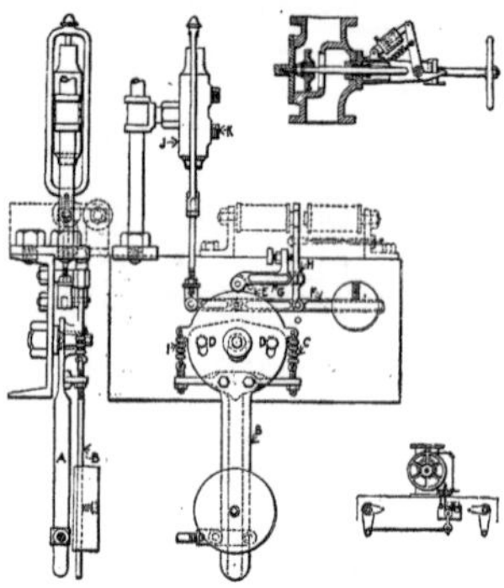

DETAILS OF SPEED-LIMIT DEVICE

be no stopping of the engine through leakage of the pilot valve, as the drain is open at all times until overspeed action occurs.

The magnets shown are to be used if either open- or closed-circuit electrical remote control is desired. This electrical remote-control device is auxiliary and may be omitted without interfering with the action of the speed limit.

It is clear that this speed limit is suitable for use in connection with any type of valve designed to operate with pressure or release of pressure.

In commenting on the failure of a plugged tube in the boiler of a locomotive at Allerton Bywater Colliery, the Engineer Surveyor-in-Chief to the Board of Trade condemns the practice of stopping leaky tubes by means of plain tapered plugs driven into the ends, says *The Engineer*. In the case in question, the plug was 4½ in. long by 1⅜ in. in diameter at the large and 1½ in. in diameter at the small end. It was driven into the tube by means of a 7-lb. hammer, but blew out, with fatal results, at a pressure of about 115 lb. per square inch.

EDITORIALS

One Pound per Horsepower

ONE of the most marked characteristics of the human being is his interest in sporting events of all kinds. The two greatest stimuli to the normal male are the opportunities of beating the other chap and of breaking a record.

In the early days of the last century the Watt engines used for pumping water from the tin mines in Cornwall had their performance much improved through competition among the mines. The favorite bet in that section of England was on the number of gallons of water pumped each week per bushel of coals.

An interesting field for competition, which should arouse the sporting instinct of the engineer, is in the reduction of weight of power plants, especially as the alluring figure of one pound per horsepower is distinctly above the horizon. Twenty pounds per horsepower made the automobile possible; five pounds per horsepower made the airplane possible. What may we not expect from one pound per horsepower?

The heaviest source of power in common use is animal power. A horse weighing one thousand pounds will develop eight-tenths of a horsepower for eight hours. For the continuous twenty-four-hour development of one horsepower, 3,750 pounds of horse would be necessary. Man is a little better on that basis. Under favorable circumstances he can develop one-seventh of a horsepower for eight hours, which would give him a weight of 3,150 pounds per horsepower if the individual man weighs one hundred and fifty pounds and develops power continuously. The average stationary steam plant, including boilers, engines, condensers, chimney, and the numerous auxiliaries, the water in the boiler and the circulating water, has a weight which is high and extremely variable. All condensing steam plants must be heavy because of the necessity of carrying a large weight of circulating water. A modern locomotive develops about two thousand horsepower and weighs 300,000 pounds, giving a weight of one hundred and fifty pounds per horsepower. This, of course, includes the running gear, which does not properly belong to the power plant. It is probably possible to make a non-condensing power plant of good size which would not weigh more than forty or fifty pounds per horsepower. It might be composed of boilers of the express marine type which would weigh sixteen to twenty pounds per horsepower, including the weight of water, and of turbines which would weigh fifteen pounds per horsepower, and of the necessary auxiliaries.

The only possibility of obtaining light power plants is by the use of the internal-combustion engine. The Diesel engine with its high efficiency has not yet been brought down to a reasonable weight. In Europe, weights of fifty to sixty pounds per horsepower are reported. In this country the weights are usually four to six times greater. The automobile engine, including the radiator and jacket water, which are necessary parts of the power plant, will probably average about ten pounds per horsepower. The aëronautical engine of the water-cooled type has been brought down to a total weight, including radiator and water, of approximately two and one-half pounds per horsepower. The air-cooled aëronautical engine should be, and is, much lighter. Varying reports about the A. B. C. English airplane engine indicate a weight per horsepower somewhere between 1.3 and 1.7 pounds. This seems to be the best figure so far.

Another improvement, however, looms in sight. The horsepower of an internal-combustion engine is directly proportional to the pressure of the mixture taken in. By the use of a centrifugal compressor (supercharger) the admission pressure can readily be doubled, thereby doubling the power of the engine. This device has been used at high altitudes for maintaining the power of an engine. It could also be used on the ground for increase of power beyond the normal ground horsepower. Such a device would weigh not more than one-fourth the weight of the engine and would thereby increase the power per unit weight by sixty per cent, if its use did not necessitate such changes in the engine as to cause the weight of the engine to be increased. It is certain that the attempt to double the intake pressure in any of the existing air-cooled aëronautical engines would lead to overheating of the piston and cylinder as a result of the increased weight of fuel burned, but modifications in design could take care of this increased heat. The way, then, seems open for those who desire to follow it. It will not be very long before one pound per horsepower is a realized fact.

Power-Factor Correction an Economic Issue

POWER-FACTOR correction in an alternating-current circuit is a problem that has existed ever since the installation of the first induction motor. However, it is only within recent years that serious attention has been given to improving the phase relation between the voltage and current in alternating-current circuits. The real significance of the interest centered in power-factor correction at this time is found in the statements, recently made, that "the power companies today can sell more power than they can finance the necessary extension for, consequently, one of the most difficult problems in the operation of utilities is that of finance," and that "if the power factor of all connected equipment utilizing a certain company's service could be raised to unity, the entire electrical system, including generators, transmission lines, substations and transformers could be made to carry thirty-five thousand kilowatts more load than it carries at the present time." In the latter case the only additional investment required to accomplish this would be for an increase in the turbines and boilers at the power plant to bring them up to the greater capacity of the electrical system, resulting from the higher power factor. On the other hand, the installation of a thirty-five-thousand kilowatt extension would require an investment of between four and five million dollars. If the power factor of the system was increased to unity, the service would be materially improved, the losses per kilowatt transmitted considerably reduced, and the operating costs remain practically the same. Where, even if the thirty-five-thousand kilowatt additional capacity is installed the service would not be improved, the losses per kilowatt transmitted would not be decreased, the operating cost would be increased, besides the larger capital charge required to pay a reasonable return on the increased investment.

To improve the power factor and obtain the full advantage resulting therefrom, the correcting must be done at the load. Under such conditions the losses in the customers' distributing system are reduced and voltage regulation is improved, both of which tend toward better service. Besides this the wattless current is reduced all the wak back to the station, thus increasing the capacity of the entire system. The satisfactory working out of this problem is one that requires close co-operation between the customer and the power company, and its solution is one that will result in mutual benefits to all parties concerned.

Power From the Tide

FOR a satisfactory solution of the economic problems before the world today, due to the high cost of labcr and materials, shortage of production, etc., the attention of the various nations has been turned toward a greater production and use of power. Even in some of the most backward countries surveys are being made of available power resources, and methods are being considered how best to develop these resources. In the more progressive countries, where the production of power has become a more or less highly developed science, not only are the available sources that have proved commercially possible of development being carefully considered, but those sources and methods that have in the past been abandoned on account of their lack of commercial possibilities are now being carefully scrutinized. On account of the greatly increased cost of power production from coal, it may now be economically possible to develop some of the sources that a few years ago were beyond any such a consideration.

Various schemes have been suggested for developing power from the tide, but up to the present time all of these have proved impractical or uneconomical from the commercial side. And it is doubtful if in this country we have approached a condition where the development of power from the tide is commercially economical, even under the most favorable conditions. Nor will we until the present available water power of our rivers is more completely developed and our sources of fuel more nearly exhausted than at present. However, in a country like England, where practically the only available local source of power is coal, with very limited water powers on the rivers, it may be feasible to develop tidal power. According to the *Electrical Review* (London) the British government is considering preliminary plans inquiring into the possibilities of utilizing the available water power of the country. Among these plans are those which pertain to the development of power from the tide at the mouths of a number of rivers. It has been pointed out that since the time of the tide varies considerably in the different rivers, it may be possible to take advantage of these variations to increase the daily period power could be generated. This scheme may have possibilities, but whether the time has arrived when the cost of power from coal is such as will make the scheme a commercial one is questionable. Nevertheless, the consideration of the matter by the British government indicates how vitally important the power supply has become in the industrial development of that nation.

Almost every scheme for the development of tidal power has been associated with the idea of storage of power for use during the period of the day when the tide is not available. The electrical storage problem is one of the stumbling blocks to the successful development of the scheme. What appears to be a better and more logical solution would be to operate the tidal power plant in conjunction with a system of steam plants or hydro-electric plants, the latter on rivers where water storage is available. These plants could then carry all the load that the tidal plant could not. Unfortunately, in this country, locations where such a scheme might be worked out are very limited, and where such localities are available, they are, in general, considerable distances away from large industrial developments. Therefore, the development of tidal power in this country, at least, is hardly a matter for present consideration, although it may be a commercial possibility in countries that are not so abundantly supplied with fuel and sources of water-power development.

Flanges for Steam-Flow Meters

A LITTLE forethought in planning the details of a power plant will often result in considerable savings. For example, there may be no present thought of using a steam-flow meter, but by and by the need or the appreciation of the advantages of such an apparatus may develop.

The devices used to give sufficient pressure drop in the pipe for operating the meter usually require flanges between which they are installed, preferably preceded and followed by straight courses of pipe of a length equal to a few times its diameter. The insertion of such flanges in an existing pipe line is disturbing, time-consuming and expensive, and the necessity of it may prevent or delay or make more difficult to get approval for the installation of a meter, the use of which would result in saving many times its cost. Such flanges can be put in at little or no extra cost in the original installation.

Great Britain is using every means possible to assure a world-wide oil supply for her own navy and her own people. Some have criticized this policy, but it looks to us like a very natural and a very wise precaution. It would pay the United States to imitate rather than criticize in this case. We are going to need more and more oil, and if the United States Government does not see to it that we have access to an adequate supply, no other government is likely to bother about it.

Never mind if the railroads cannot haul that extra coal you ordered for next winter's use. We all hope that conditions will be better before the summer is over. Whatever happens, your chances of getting coal hauled during the summer are better than when the roads are handicapped by ice and snow.

CORRESPONDENCE

What Happens When a Synchronous Motor Is Caused to Pull Out

The Feb. 17 issue of *Power* contains a letter by F. L. Fairbanks on "What Happens When a Synchronous Motor Is Caused to Pull Out?" in which attention is invited to the danger to both apparatus and life due to the "pull-out buck" and "electrical pile-up" of synchronous motors. The danger, pointed out in the article, is the possible failure of direct-connected flywheels under the influence of the "pull-out buck." The article does not make clear just what a "pull-out buck" is, but there is no question left in the reader's mind regarding the danger of using a motor with such a formidable characteristic.

What is needed to help clear up the situation is not so much the redesign of flywheel arms as an explanation of a "pull-out buck"; although, if, as mentioned in the article, flywheels are to withstand such shocks as that produced by the piston driving a mass of water or metal through the cylinder head, then, unquestionably, adequate strength would require greater section in flywheel arms and perhaps in other "links in the chain," than is used in ordinary practice.

The implication in the letter regarding the danger of directly connecting synchronous motors to flywheels is not justified by the facts. Too many synchronous motors have operated safely and successfully for years and have survived more serious circumstances than the "pull-out buck" can possibly create, to justify such alarm now. The following explanation of the action of a synchronous motor at breakdown brings out, incidentally, some of the fundamental reasons why we might expect the facts of synchronous-motor history to be as they are.

In the following discussion it is assumed that the new terms "pull-out buck" and "electrical pile-up" are intended to mean the electrical torque of the motor at and immediately following breakdown—in other words, the electrical force acting on the poles of the rotor when the motor pulls out of synchronism. What is the character of that force? Is it a very large unidirectional force (a "buck") applied "instantaneously," as suggested in the article, or is it, instead, a force which, under circumstances of breakdown, is applied gradually, not abruptly, and which constantly changes direction as the rotor and stator poles slip by each other? It is the latter and is represented in the figure. When a rotor N pole is directly under a stator S pole—that is, position *a* in the figure—the torque is of zero value. In position *b* full-load torque value is obtained. When the rotor pole is in position *c*, a maximum, or "pull-out," torque is developed. At *d* the torque is again zero, and at *e* the turning effort is at maximum, but this time it is in an opposite direction to that at *c*; that is, it is now in a direction to stop the motor. Position *b* is the position of the rotor pole with respect to the stator pole when the motor is carrying a constant load torque equal to *B*. First this force is in a direction to maintain motion, then in a direction to retard it. The highest value of this alternating force may be any value between, say, the force corresponding to $1\frac{1}{2}$ times normal load and 4 or 5 times that value, depending upon how rapidly the motor, by external means, is caused to pull out of step; that is, how fast the poles are caused to slip by each other from the very beginning of the phenomenon. The

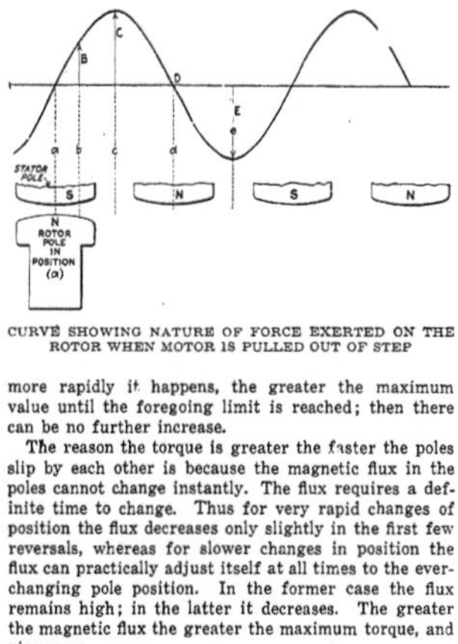

CURVE SHOWING NATURE OF FORCE EXERTED ON THE ROTOR WHEN MOTOR IS PULLED OUT OF STEP

more rapidly it happens, the greater the maximum value until the foregoing limit is reached; then there can be no further increase.

The reason the torque is greater the faster the poles slip by each other is because the magnetic flux in the poles cannot change instantly. The flux requires a definite time to change. Thus for very rapid changes of position the flux decreases only slightly in the first few reversals, whereas for slower changes in position the flux can practically adjust itself at all times to the ever-changing pole position. In the former case the flux remains high; in the latter it decreases. The greater the magnetic flux the greater the maximum torque, and vice versa.

But before this torque, in either case, can affect the stresses in the flywheel arms, it must change the speed of the rotating mass of the motor, which in turn tends to change the relative position of the rotor with respect to the flywheel. This produces a twist or torsion in the shaft, and this torsion is exerted on the flywheel arms. To illustrate, if the rotor of the motor had negligible mass or inertia, then the motor forces would be transferred in full upon the shaft and flywheel arms. On the other hand, if the motor had a relatively very large inertia, the motor forces would be practically all absorbed by this inertia without appreciably affecting the shaft or flywheel arms. Between these limits, if the torsional elasticity of the shaft is neglected, the fraction of the motor force that is applied to the flywheel arms is determined by the relative value of flywheel and motor inertia. Thus, roughly, the following relation would hold:

Flywheel Inertia as Fraction of Motor Inertia	Fraction of Motor Force Applied to Flywheel Arms
0	0
0.5	0.33
1.0	0.50
2.0	0.66
10.0	0.91
100.0	0.99

However, the shaft twist cannot be neglected, especially if the motor force is changing direction rapidly, because *the force applied to the flywheel arms is measured entirely by the shaft torsion;* that is, by the relative angular displacement of the flywheel and motor rotor.

The shorter the time during which a force acts on a mass the greater that force must be to produce the same speed change; and it must be again increased by the same percentage to produce the same angular displacement. Thus, 100 lb. acting for one second will produce the same speed change in a given mass as 1,000 lb. acting for 0.1 second; or it will produce the same angular displacement as 10,000 lb. acting for 0.1 second. Therefore, in order to obtain a large force in the motor, the force must alternate rapidly; but the more rapidly it alternates, proportionally less is its effect in changing speed, and less, as the square, is its effect in producing angular displacement, that is, in producing stress upon flywheel arms.

While these considerations are necessarily general, they nevertheless afford a basis of explanation for the fact that hundreds of synchronous motors with direct-connected flywheels have operated for years with no trouble from the source in question. And that is a good reason, if the foregoing discussion is not, that it is not dangerous to apply synchronous motors to apparatus involving flywheels. An isolated case of obvious mechanical failure of reciprocating parts in a compressor cylinder, thus throwing a severe shock on the flywheel, should not occasion undue suspicion of the synchronous machine that happened to be a part of the unit it is driving.

It seems certain, therefore, that under the circumstances of pull-out or breakdown, there is no danger of overstressed flywheel arms due to the motor forces, since in order to cause breakdown, either the voltage must be reduced, in which case only normal torque would be encountered, or a sudden shock must be applied to the shaft. In the first case the torque is too low to cause trouble; in the second, although it may be high, it alternates too fast to produce serious stress and is therefore not of great moment.

However, there are possible circumstances, entirely apart from conditions of pull-out, in which the higher motor forces may be involved. If when the motor is operating normally the line switch is opened and then closed again after the rotor has dropped back out of its proper phase position, it is possible to obtain a large force of relatively long duration. This condition is also approximated when, during the starting procedure, the motor is thrown from tap voltage to line voltage. If the operator hesitates between these operations, the motor will drop back. This often happens, and seems to be the worst possible condition obtainable for throwing strain upon flywheel arms. Yet, to the writer's knowledge, no flywheel has ever failed under that condition, which would at least indicate that, so far as possible strains from the motor are concerned, most of the present standards of flywheel design are adequate. However, it is well, as the letter suggests, to make sure in design, as is now probably always done, that all parts are proportioned with reference to the forces to which they may be subjected, so as to reduce the liability of failures to a minimum.

R. E. DOHERTY, Engineer,
A. C. Engineering Dept.,
Schenectady, N. Y. General Electric Co.

Capacities of Steel Stacks for Oil Burning

Having occasion to use the tables for chimney capacity given in Marks' Mechanical Engineer's Handbook and finding that the values involved lay outside the tables, I have devised the following formula:

$$HP = 0.006\ H\ (0.001\ D + 0.367)\ (D - .4)^2$$

where HP is the capacity of the stack in boiler horsepower, H is the height of the stack in feet, and D is the diameter of the stack in inches.

This formula applies to stacks at sea level, and for higher altitudes the dimensions must be corrected as follows:

$$\frac{D}{D_0} = 1.0 + 0.017h \quad \text{and} \quad Log\ \frac{H}{H_0} = 0.0332\,h$$

where D and H are the dimensions of the stack at the given altitude, D_0 and H_0 are the dimensions at sea level, and h is the elevation in thousands of feet above sea level.

I have found that these formulas give results which check very well with the values given in the "Handbook," and I have used them to determine sizes beyond the limits of the tables. WILLARD A. THOMAS.
Los Angeles, Cal.

More Convenient Observation of Blowoff Leakage

Having read the question and answer regarding more convenient observation of blowoff leakage, on page 683 of the April 27 issue, I think it might be interesting to mention another method that I have seen used with success.

This is to install an ordinary recording thermometer in the blowoff line. The temperature record will not only be an accurate indication of any leakage through the blowoff valves, but will also show when blowing down takes place.

New York City. G. A. BINZ.

[Such an installation was described on page 8 of the Jan. 7, 1919, issue of *Power*, and the chart obtained was illustrated. This chart gave the time and duration of the blowdown and also showed that the pen returned to its normal position, indicating that the blowoff valve was tight.—Editor.]

Superheating With Reciprocating Engines

Generally speaking, the reciprocating engine will be benefited as much by superheating as the steam turbine will. The saving to be expected by superheating is dependent on the amount of cylinder condensation that would occur in the same engine if no superheat were used. Evidently, the greater this condensation, the larger is the saving possible.

When steam is superheated 200 or even 250 deg. F., it can be exposed to the cooling action of the steam chest and cylinder walls without condensation, and has 30 to 40 deg. greater specific volume than saturated steam of the same pressure. For example, 1 lb. of water at 170 lb. pressure absolute produces 2.68 cu.ft. of saturated steam at 368 deg. F. temperature. After passing through the superheater and gaining 200 deg. superheat, the volume will have been increased to 3.54 cu.ft. A part of this increased specific volume is lost before the expansion of the steam in the cylinder takes place on account of the cooling action of the cylinder head, piston and cylinder walls. While the superheat in the steam entering the cylinder may be 200 to 250 deg., the superheat of the steam in the cylinder at the moment of cutoff is considerably less. The elimination of condensation losses together with the increased volume of the steam, effects substantial savings in the steam consumption and in the fuel consumption when compared with saturated steam operation under the same conditions.

The opinion has frequently been advanced that the mean effective pressure in cylinders using superheated steam is altered from that existing when saturated steam is used, the initial pressure, the point of cutoff and piston speed, of course, being assumed as the same in each case. Theoretical considerations would indicate that a lower mean effective pressure with superheated steam is obtained, and this conclusion would be based upon the knowledge that the expansion curve of superheated steam falls more rapidly than with saturated steam. In practice, however, with highly superheated steam a mean effective pressure at least equal to that with saturated steam is obtained and an explanation for this is easily made clear. For example: in simple noncondensing engines the back pressure line of an indicator diagram is decidedly lower with superheated steam than with saturated steam, a condition to be expected because the steam exhausted contains less moisture. It is, therefore, lighter than saturated steam, and flows through the exhaust passages with less resistance. C. D. Young, before the American Society of Mechanical Engineers in December, 1912, gave evidence of this fact.

In the case of multi-expansion reciprocating engines the diagrams taken from the low-pressure, and, in the case of triple engines, from the intermediate-pressure cylinder, show that a higher m.e.p. is obtained with superheated steam. In the high-pressure cylinder of multi-expansion engines it is generally true that the cutoff should be increased if the power output of the engine is to be maintained.

But our consideration of the possibility of increased engine power due to superheating must not stop here. Consider a saturated-steam plant in which the boilers are operating at their full capacity to supply the steam demanded by the prime movers. Take this same plant and superheat it—the decrease in steam consumption per horsepower-hour will result in a reserve boiler capacity which will either make possible the operation of the boilers at lower rating or furnish the additional steam necessary to meet an overload on the prime mover. In other words, superheating makes it possible to increase the power which the prime mover may develop over what it could produce with saturated steam.

In new installations the use of superheated steam permits the use of engines having larger cylinder diameters and operated with shorter cutoffs. Because cylinder condensation is eliminated, the gain in economy that is produced by the more effective expansion of the steam is realized.

Experiments have shown that the degree of superheat necessary to prevent cylinder condensation varies as a general rule with the ratio of expansion. In American practice the steam temperatures ordinarily recommended are up to 500 deg. F. for Corliss and slide-valve engines and up to 650 or 700 deg. F. for piston and poppet-valve engines.

The gain from decreased steam consumption due to superheated steam is ordinarily expressed in percentage. This gain will vary in various types of engines with the different superheat temperatures used. For example, with 150 deg. superheat the following gains are produced: In simple engines 22-24 per cent; compound engines 16-19 per cent; triple-expansion engines 14-16 per cent; uniflow engines 10-12 per cent. It is, however, an assured fact, and a strong argument for the more egneral adoption of high-degree superheat, that only engines designed for and using high degrees of superheat can successfully compete in economy in the power generation field.

It is often necessary, particularly with old boilers, to reduce the operating pressure. This condition results in a loss in plant capacity. Superheating such boilers will enable them to meet the power demands of the plant at the reduced pressure just as well as the boilers did before the pressure was reduced. Furthermore, superheating in the majority of cases will provide an additional reserve power over that of the plant before the pressure was reduced.

The installation of superheaters materially lowers the plant cost. Where it is used, a smaller amount of boiler horsepower is needed to deliver the same power. Roughly speaking, six boilers equipped with superheaters, furnishing 150 deg. of superheat, will carry the same power load as seven boilers delivering saturated steam. By reason of this, smaller feed pumps, smaller stacks, smaller coal-handling apparatus and less coal-storage space are sufficient for the demands.

Piping can be reduced in size, because higher steam velocities with superheated steam are possible. This reduction in pipe size will effect a saving not only in the pipe line, but also in valves, fittings and covering.

In the turbine superheated steam will allow the use of a conductor of a smaller size than if saturated steam were used. Consequently, smaller circulating water pumps, hotwell and vacuum pumps can be used and minor apparatus such as steam separators often can be dispensed with. These items materially reduce the initial plant cost.

When superheating is correctly applied to the individual conditions in a specific plant, it is the one means which will show an improvement in power-plant operation in at least one and probably all parts of the plant where steam is used.

All the more astonishing, therefore, becomes the fact that in spite of all these practical advantages, only about 15 per cent of the steam horsepower of this country is developed from superheated steam. Furthermore, this percentage represents not more than perhaps 5 per cent of all the steam power plants, indicating that only plants of larger capacity, inherently more economical than the vast majority of stationary plants, use superheat.

Is it logical that comparatively few plants having high evaporation rates and low steam consumption, should be the ones to benefit by superheat, while 85 per cent of the nation's steam power uses fuel extravagantly? Because this condition is not consistent, the opportunity and the means to conserve fuel are available and the necessity is urgent. A. RAYMOND.

New York City.

Boiler Water-Leg Drip Pan

In stoker-fired boilers where the furnace extends forward of the front water leg and the boilers are of the horizontal water-tube type, considerable damage is done to the special arch and furnace tile, also to the stoker mechanism, by the drip of water, principally from the handhole plates.

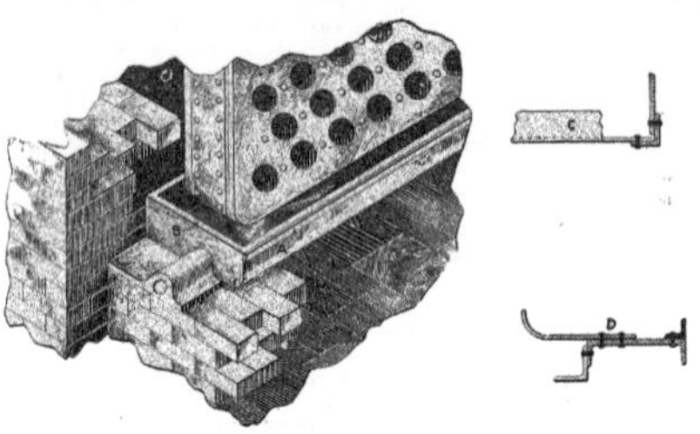

APPLICATION OF DRIP PAN TO BOILER WATER LEG

The illustration shows a design of pan used for eight 425-hp. boilers, that has eliminated this trouble. The pan is cast in one piece and is of such dimensions as to extend the full length of the water leg and wide enough to extend 2 in. past the rear sheet to catch the drip at the expanded joint of the tubes; the front is curved up to meet the doors, and a ¼-in. projecting strip is provided for a joint *A*.

The side sheets of the front, which carry the door hinges, are fitted to the top edge of the pan *B*; an extension is provided for the support on the side walls, and this carries a 3½-in. diameter boss tapped for a 1½-in. pipe into the bottom of the pan; the end of this boss is flush with the side walls and is piped to the sewer.

This pipe is large enough to carry a flood of water such as occurs when cleaning out or when the pressure is released on all the handhole gaskets at once, as when a boiler is taken off the line. When the boiler is in service, the temperature at this point is such as to evaporate all minor leaks if the pan is tight.

The sketch *C* shows the usual method of providing for this leak where the drip pan is made up of channels and angles and forms a part of the door frame. Bolts are generally used to permit adjustment of the parts in erecting, and the water leaks through the bolt holes and also through the side-sheet joint, which is below the bottom of the water leg.

The cast-iron pan has lugs projecting from the bottom to carry the stoker cover angles; these dispense with the angle iron *D*. A. M. STILES.

St. Louis, Mo.

Where the Ammonia Charge Goes

In the April 20 issue of *Power*, J. E. Porter makes the statement that, "it is impossible to operate a plant and not lose some ammonia." While it is true that much ammonia is lost in both compression and absorption system plants, it is nevertheless possible to operate without the loss of ammonia by using a closed system and by sweating all joints. The writer is familiar with a number of installations where leakage is practically if not absolutely zero. They have operated on the same charge of ammonia without any replenishment whatsoever, and the chances are that they will continue for many more years on the same charge because the joints actually become tighter or less liable to leak with age. There are no reciprocating rods to pack, therefore there can be no leakage from that source.

Inasmuch as the joints in these installations are not sweated, and yet there is no evidence of leakage, it is plain that with commercially perfect joints (which can be made) there is no excuse for leakage. The only leak possibility would then lie in the safety valve, which might be lifted owing to excessive pressure, as sometimes happens where operators are careless. N. G. NEAR.

Brooklyn, N. Y.

Packing Valve Stems Under Pressure

On page 476 of the March 23 issue of *Power* Mr. Noble gives advice about how to pack valve stems under pressure. I believe that his suggestions could be worked out (if a person wants to court danger) on some kinds of valves, but not in the case to which I originally referred.

This valve was of cast iron, the bonnet was a flat plate with a U-shaped yoke, threaded to receive the valve stem. The packing was put in an annular space in the plate, where the stem went through it, and was held in place by a gland, tightened down with two stud bolts and nuts. I have found that kind of valve to be quite mean enough to pack, even when cold, in case you do not happen to have exactly the proper size of packing. GUSTAV SWENSON.

Rockford, Ill.

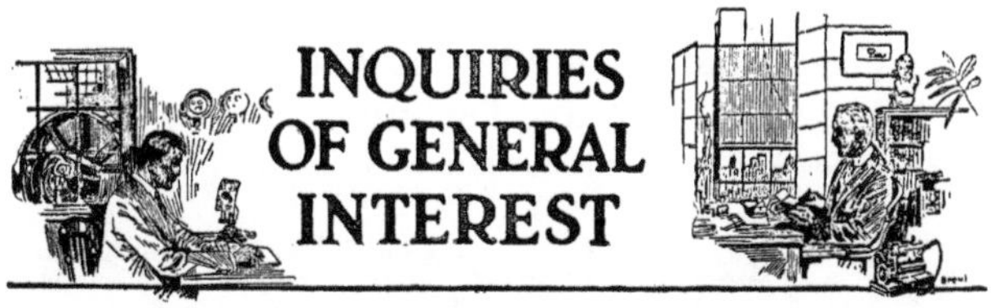

INQUIRIES OF GENERAL INTEREST

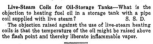

Live-Steam Coils for Oil-Storage Tanks—What is the objection to heating fuel oil in a storage tank with a pipe coil supplied with live steam? S. S. D.

The objection raised against the use of live-steam heating coils is that the temperature of the oil might be raised above the flash point and thereby liberate inflammable vapor.

Cleaning Water Jacket of Gas Engine—What solution can be used for cleaning the scale from the water jacket of a gas engine? R. M.

The scale usually consists of lime deposited out of the circulating water and can be removed by a solution of one part commercial muriatic acid and four parts of water. As soon as the solution has done its work, the water jacket should be thoroughly washed out with clean water to prevent unnecessary corrosion.

Loosening Piston Rod from Crosshead—How can the tapered end of an engine piston rod be started out of the crosshead? R. G.

Place the engine on the head-end center, remove the cotter and replace it by a cotter or piece of steel about one-eighth inch less in width than the clear space through the cotter way. Then detach the eccentric rod and, with the steam valve closed on the head end and open on the crank end, admit steam to the crank end of the cylinder. If the piston rod is not immediately released from the crosshead, clean the parts thoroughly and make another trial after leaving the joint moistened with kerosene for a few hours.

Cylinder Oils for Compound Engines—Are different kinds of cylinder lubricating oils necessary for use in the high- and low-pressure cylinders of a compound engine? R. C. N.

For obtaining good lubrication it is sometimes necessary to use different grades of oil. The drop of pressure in the high-pressure cylinder causes condensation and the supply of steam to the low-pressure cylinder will be wet, with a tendency to wash the film of oil off the surface of the cylinder and make lubrication more difficult. The oil used for the low-pressure cylinder should be of a quality that will readily combine with moisture and cling to the cylinder walls, and the oil obtainable for that purpose may not be suitable for the high-pressure cylinder.

Front and Back End and Forward Stroke—Which is the front and which is the back end of an engine cylinder and from which end is the forward stroke? C. R.

In stationary-engine practice the crank end is called the front end and the opposite end, commonly called the head end, is the back end. The forward stroke is toward the front; that is to say, it is the stroke from the back or head end toward the front or crank end. The distinction is readily fixed in mind by referring to the fact that, in an engine with a trunk piston, pressure for development of power is exerted on the piston only during its forward stroke, that is, when the piston moves from the back toward the front end of the cylinder. In locomotive practice the end of the cylinder which is in advance when the locomotive moves forward is designated the front end, and therefore the nomenclature is the reverse of that employed in stationary practice.

Ratio of Expansion—What is meant by the ratio of expansion in an engine? W. L. S.

In the operation of a steam engine, the ratio of expansion is the number of times the steam is expanded after cutoff occurs. If there is three cubic feet of steam present at cutoff and this is expanded to nine cubic feet, or three times the volume, the ratio of expansion is 3. The approximate ratio of expansion is the reciprocal of the fraction of stroke at which cutoff occurs. Thus, if cutoff occurs at ⅓ stroke, the approximate ratio of expansion is 3. But the actual volume at cutoff and after expansion includes the clearance volume at one end of the cylinder and must be included to find the actual ratio of expansion. Thus, if the clearance is 5 per cent of the piston displacement, and cutoff occurs at 0.33 of the stroke, the actual ratio of expansion would be

$$\frac{1 + 0.05}{0.33 + 0.05} = 2.76$$

Evaporation Required per Boiler Horsepower—How many pounds of water fed at the temperature of 160 deg. F. must be evaporated per hour into steam at 140 lb. gage pressure to make one boiler hoursepower? R. N. B.

The commonly recognized boiler horsepower consists in the transfer of the amount of heat required to evaporate 34½ lb. of water per hour, from and at 212 deg. F. The standard boiler horsepower therefore is equivalent to the transfer of the latent heat of evaporation to 34½ lb. of water per hour, or 970.4 B.t.u. × 34½ = 33,478.8 B.t.u. per hour. When the feed water is at the temperature of 160 deg. F., it contains 160 — 32 = 128 B.t.u. above 32 deg F., and as a pound of dry saturated steam at the pressure of 140 lb. gage, or 155 lb. absolute, contains 1,194 B.t.u. above 32 deg. F., each pound of feed water, for conversion into dry saturated steam, requires 1,194 — 128 = 1,066 B.t.u., and under these conditions it is necessary to evaporate 33,478.8 ÷ 1,066 = 31.4 lb. of water per hour for the evaporation to be equivalent to one boiler horspower.

Reducing Speed of Corliss Engine.—How can the speed of a Corliss engine be reduced from 75 to 60 r.p.m. if the weight on the governor has been set over as far as it will go for reducing the speed? E. B.

Ascertain what speed of engine is obtained when the weight for adjusting the speed is set at about the middle of its adjustment. Then, to retain the same speed of the governor with regulation at 60 r.p.m. of the engine, replace the receiving pulley on the governor by one whose diameter is equal to the diameter of the present governor pulley multiplied by 60 and divided by the ascertained r.p.m. of the engine; or replace the governor driving pulley on the engine shaft by one whose diameter will equal that of the present pulley multiplied by the ascertained r.p.m. and divided by 60. The exact change of speed will not be obtained on account of a small error due to neglecting consideration of thickness of the governor belt, but the deviation will be so little from the desired speed that it can be corrected by changing the position of the speed-adjusting weight.

[Correspondents sending us inquiries should sign their communications with full names and post office addresses. —Editor.]

The Prime Movers Committee Report on Steam Turbines

Salient Features of the Report of the Prime Movers Committee, National Electric Light Association, Pasadena, California

A RECORD of the units that have been manufactured since the last report indicates that 30,000-kw. 1,800 r.p.m. 60-cycle, and 35,000-kw. 1,500 r.p.m. 25-cycle, can be considered for the present as the maximum standard size of single-shaft single-barrel unit for large installations.

During the last year a considerable amount of information has been obtained from the operation of the smaller machines, and a period of adjustment and improvement in both design and construction, following the experience that has been gained, would seem to be imperative if these large-sized machines are to be brought to a requisite degree of reliability and operating efficiency.

The records of operation of single-shaft units have not by any means been satisfactory. Since the last report several addition wheel failures have occurred to units of this type and on sizes ranging from 5,000 to 30,000-kw. capacity.

Changes in Design of Large Turbines

The General Electric Co. has made a number of marked changes in the design of its large-sized turbines, the principal changes being the stiffening of the center bearing and modifications in the design of turbine discs, buckets, diaphragms and packing.

The changes in the turbine discs consist of a thickening of the disc in an attempt to reduce the possibility of injurious vibration; elimination of the weight-balancing holes, and the change in the number and location of steam equalizing holes so as to move them well in from the rim; and changing from an even number to an odd number of holes with rounded and polished edges at the wheel surface.

The changes made in the buckets, particularly in the last stages, consist of the addition of a reinforced section, carrying it nearly to the center of the bucket so as to stiffen the construction materially. In addition, the very long buckets are reinforced by the use of a bracing strip to minimize vibration which might occur.

The changes to the diaphragms consist in replacing some of the cast-iron sections by cast-steel construction where stresses indicate this to be necessary. Additional clearance space around the diaphragm is also provided with the idea of allowing for the removal of a large part of the condensed water, which has contributed to serious erosion of the blading in certain stages.

A radical departure has been made in the type of diaphragm packing. The new type differs from the earlier construction in that the stationary, inwardly projecting short teeth of the labyrinth have been changed to much longer teeth which are mounted on the revolving element. It is claimed that this improvement makes it practically impossible, when properly installed, for rubbing to distort the shaft due to the generation of heat at the point of contact.

Causes of Wheel Failures

An analysis of the causes of the serious wheel failures has resulted in the accumulation of sufficient data to indicate that the trouble has been due in the past to a vibration of the wheel structure. This vibration, which has been described as a fluttering of the web of the wheels, in combination with the high working stresses of the material, gives rise to the formation of fatigue cracks. These cracks, which occur not only in the wheel but also in the bucket, when allowed to continue, result in a complete rupture. This action was accentuated in the discs where sharp tool marks or rough surfaces are present, or where holes for weight or steam-balancing purposes are drilled in the wheel. There is, however, considerable investigation still necessary in connection with the subject of vibration of wheel before all the facts can be known and a proper design be forthcoming that will eliminate danger from this source of trouble. The manufacturer is devoting considerable attention to this subject, and it is believed that some definite information will soon be available.

No marked changes have been introduced during the past year by the Westinghouse Electric and Manufacturing Co. in the design of its single-cylinder units up to 30,000-kw. rating, although a number of minor improvements have been made in the details of design and construction, with a view to increasing the reliability and efficiency of its units. To meet the demand for larger capacities than 30,000 kw. the company has standardized on combinations of single-cylinder units of 20,000 to 22,000-kw. maximum rating in tandem, cross-compound and triple-cylinder arrangements.

While the combination of compound turbo-generators seems to be increasing in favor, and while test results indicate a higher thermal efficiency for such a combination as compared with single units of the same aggregate capacity, sufficient data are not at hand to show conclusively whether the practical operation of units bears out the theoretical advantages which are claimed.

Starting of Large Turbines

One of the most important considerations in connection with the operation of large turbines of 20,000-kw. capacity and over is the necessity for extreme care in the starting and loading of units, especially when the machines have been idle for a period long enough to lower materially the temperature at which the various parts are normally operated. With the introduction of steam to a cold unit there is at once set up a difference in temperature between the various elements, which, unless a careful procedure is followed, may result in unequal expansions, affecting the shafts, blading, diaphragms, packing or casing joints. In some cases on record, sudden and unequal expansion has caused deformation which has resulted in permanent distortions to casings and bearings, materially affecting the subsequent operation of the machine.

A survey of the operating methods in use in a number of companies indicates that there is considerable variance in the practice which is followed in starting large turbines. The time required to heat a cold unit to the temperature where full load can be carried will, naturally, vary with the type, size and class of turbine. No hard and fast rule of procedure can be definitely laid down, except that in all cases no large unit can be safely loaded which has not been given sufficient time to allow all parts of the unit to reach evenly and gradually the individual normal operating temperatures corresponding to load conditions.

Data secured from operating companies show that the procedure generally followed consists in establishing a partial vacuum of from 15 to 20 in. prior to starting the unit, and maintaining this during the period of warming.

In warming the turbine and bringing it to speed, the first operation of introducing steam should admit a sufficient volume to start the rotor turning immediately. Steam admission should then be so regulated that the unit will continue turning at a speed not exceeding 10 per cent of the normal operating speed until the warming process is completed. Additional steam should then be admitted gradually to the turbine to bring it up to full speed.

While the time taken for this operation varies among different operating companies, the safe minimum time allowed, based on best operating experience, is given as twenty minutes for units of 20,000-kw. capacity, and thirty minutes for units of 30,000 kilowatts.

In applying load to the unit, the rule observed for safe operation by a number of operating companies is to increase the load on the unit at a rate not exceeding 1,000 kw. per minute for units of the size mentioned. While there is considerable variation from the procedure noted, the reason is that in many cases certain local requirements or special plant conditions govern. It will be noted, therefore, that in following the foregoing practice, the minimum time for heating and loading large units will vary from forty minutes for a 20,000-kw. turbine to sixty minutes for a 30,000-kw. unit, or a rate of two minutes for each 1,000-kw. turbine capacity.

These figures do not refer to machines which have not been idle long enough to cool off perceptibly, in which case shorter length of time can be taken, depending on the individual circumstance.

Emergency Governors

Attention has recently been called to the operation of turbine emergency governors. While a few companies have been operating machines for some time on which the emergency governor, after operating, became again operative at approximately 2 per cent above normal speed, most of the turbines now in service are equipped with emergency governors so designed and adjusted that it is necessary to drop the machine speed below normal to latch the emergency again after it has operated. In certain cases of operating trouble it is necessary to have the emergency governor springs so adjusted that they will operate over the reduced range and avoid the necessity of disconnecting the generator from the system to reduce the speed of the unit below normal.

Personnel of the Prime Movers Committee

N. A. Carle, Chairman, Chief Engineer, Public Service Electric Co., Newark, N. J.

H. P. Liversidge, Vice Chairman, Assistant Chief Engineer, Philadelphia Electric Co. Chairman, Turbine Committee.

H. R. Wakeman, Vice Chairman, Portland Light and Power Co., representing Northwestern Electric Light and Power Association, Portland, Ore.

R. J. C. Wood, Vice Chairman, Superintendent of Generation, Southern California Edison Co., Los Angeles, Cal.

C. M. Allen, Worcester Polytechnic Institute, Worcester, Mass.

John Anderson, Chief Engineer, Milwaukee Light, Heat and Traction Co., Milwaukee, Wis.

A. D. Bailey, Assistant Chief Engineer, Commonwealth Edison Co., Chicago, Ill.

E. J. Billings, Engineer, Henry L. Doherty & Co., New York City.

H. B. Brydon, Engineer, H. M. Byllesby Co., Chicago.

F. S. Clark, Engineer, Stone & Webster, Boston.

C. W. De Forest, Manager, Electric Department, Union Gas and Electric Co., Cincinnati, Ohio.

R. D. De Wolf, Mechanical Engineer, Rochester Railway and Lighting Co., Rochester, N. Y.

R. E. Dillon, Assistant Superintendent, Generating Department, Edison Electric Illuminating Co., Boston, Mass.

H. F. Eddy, Consumers Power Co., Jackson, Mich.

Louis Elliott, Engineer, Electric Bond and Share Co., New York City.

W. W. Erwin, Chief Engineer, New York Edison Co., New York City.

S. B. Flagg, Fuel Expert, Electric Bond and Share Co., New York City.

J. M. Graves, Superintendent Power Stations, Duquesne Light Co., Pittsburgh, Pa.

C. F. Hirshfeld, Research Engineer, Detroit Edison Co., Detroit, Mich.

John Hunter, Standard Shipbuilding Co., Shooter's Island, N. J.

J. P. Jollyman, Engineer, Electrical Construction, Pacific Gas and Electric Co., San Francisco, Cal.

J. B. Klumpp, Inspecting Engineer, United Gas Improvement Co., Philadelphia, Pa.

John W. Koontz, Jr., Electrical Engineer, Great Western Power Co., San Francisco, Cal.

A. W. Morgan, Results Engineer, Denver Gas and Electric Co., Denver, Col.

A. A. Potter, Dean, Division of Engineering, Kansas State Agricultural College.

E. B. Ricketts, Assistant to Chief Operating Engineer, New York Edison Co.

G. W. Saathoff, Electrical Engineer, The Acme Power Co., Toledo, Ohio.

H. H. Schoolfield, Engineer, Pacific Power and Light Co.

E. D. Searing, Portland Railway, Light and Power Co., Portland, Ore.

S. S. Svenningson, Engineer, Shawinigan Light and Power Co., Montreal, Canada.

E. H. Tenny, Assistant Chief Engineer of Power Plants, Union Electric Light and Power Co., St. Louis, Mo.

R. L. Thomas, Electrical Engineer, Pennsylvania Water and Power Co., Baltimore, Md.

O. G. Thurlow, Chief Engineer, Alabama Power Co., Birmingham, Ala.

W. H. Trenner, Chief Engineer, Idaho Power Co., Boise, Idaho.

J. F. Vaughn, Consulting Engineer, Boston, Mass.

J. A. Walls, Vice President and Chief Engineer, Pennsylvania Water and Power Co.

John Wolff, Mechanical Engineer, Cleveland Electric Illuminating Co., Cleveland, Ohio.

H. P. Wood, Operating Superintendent, Edison Electric Illuminating Co., Brooklyn, N. Y.

Geographical Distribution of Coal and Power Activities*

It is one of the great problems of today to determine the location of the vast sources of fuel and power and their relation to the present and near future demands. Bituminous coal, forming as it does the major portion of our fuel supply for industrial purposes, is the only fuel which will be considered in this discussion. The distribution of coal reserves is the best evidence of what will necessarily be our resource for extended periods in the future. Table I is from data developed by M. R. Campbell, of the United States Geological Survey, and reported at the International Geological Conference in Toronto in 1912. In this table the states have been divided into groups and the probable reserves and probable life of these reserves are shown for each group. It is interesting to note the relation between the amount of reserve and probable life in the northeastern fields of Pennsylvania and those in the mountain territory of the West. Naturally, these figures are only approximate, but they afford a significant fact in respect to the ultimate location of our energy reserves.

A study of data and charts showing the distribution of water-power resources indicates that the bulk of these also is to be found principally in the West. It is not to be expected from the figures, however, that the West is to dominate industrially in the immediate future, for a period of centuries is involved here, not a few years.

In connection with the distribution of coal consumption the current production and current consumption are shown in Table II, where the states are arranged in the same groups as in Table I. A study of this table shows that for any group of states the production and consumption are of approximately the same magnitude. In other words, the actual statistics show, what might be expected, that coal is used as near the place of production as possible. Another interesting fact shown by the table is that New England, New York, Pennsylvania, New Jersey and the North Central states together consume about 68 per cent of the total coal consumption of the country, which gives some indication of the enormous concentration of the demand for fuel in this comparatively restricted area.

A study of the water-power resources and of the demands for power brings out the fact that the former are principally in the western part of the country, while the demand for power, like that for fuel, is concentrated very largely in the northeastern corner of the United States. A somewhat similar contrast will be found if we compare the present current use of coal and the coal reserves as estimated by the geologists. In other words, power and

*Abstract of an article by R. S. McBride, in *Coal Age*.

coal are not, as some have claimed, supplementary throughout the country, with coal available in the East and power available in the West. The facts are quite to the contrary, for the West includes a large majority of the resources of both coal and water power.

It seems evident, then, from a study of these figures that the western part of the country is bound in the future to dominate the industries of the country since only there will be found the fundamental requirements of industry—an adequate supply of fuel and power. As stated before,

TABLE I. BITUMINOUS-COAL RESERVE

(Data from Marius R. Campbell, U. S. Geological Survey, as reported to International Geological Congress, Toronto, 1912)

District	Reserve: Billions of Tons	Reserve: Per Cent of Total	Life at Present Rate of Use (Centuries)
New England	None	0	0
New York, Pennsylvania, New Jersey	102.2	5	6
South Atlantic	166.3	7	15
North Atlantic	327.1	15	20
South Central	196.5	9	30
Northwestern	129.9	6	60
Southwestern	58.5	3	60
Mountain	1,178.5	52	330
West Coast	58.9	3	140
Total or average	2,217.9	100	Aver. 40

TABLE II. DISTRIBUTION BY DISTRICTS OF COAL
Production and Use

(Production based on U. S. Geological Survey report. Consumption estimated on assumption (1) of total use equal to total production and (2) of total use, including railroads and bunker fuel in same proportion as use for other purposes besides railway and bunker.)

District	Produced (1918) Millions of Tons	Produced (1918) Per Cent of Total	Estimated Consumption (On 1917 Basis) Millions of Tons	Estimated Consumption (On 1917 Basis) Per Cent of Total
New England	0.	0.	28.9	5.0
New York, Pennsylvania, New Jersey	178.5	30.9	168.9	29.1
South Atlantic	104.7	18.0	46.4	8.0
North Central	167.3	28.9	197.9	34.2
South Central	57.6	10.0	37.5	6.5
Northwestern	22.2	3.8	59.8	10.3
Southwestern	9.3	1.6	11.1	1.9
Mountain	35.4	6.1	23.7	4.1
West Coast	4.1	0.7	4.9	0.9
Total	579.1	100.0	579.1	100.0

however, this is a situation which has no immediate interest. If the figures given in Table I are at all accurate, the life of the Eastern coal field is from six to twenty centuries at the present rate of production. Even if this production should be very largely increased, it is obvious that the life of these fields would still be a matter of centuries rather than of years.

American Institute of Electrical Engineers Elects Officers

At the annual business meeting of the American Institute of Electrical Engineers, held in New York, May 21, the report of the Committee of Tellers on the election of officers for the administrative year beginning Aug. 1 was presented and the following were declared elected: President, A. W. Berresford, Milwaukee; vice-presidents, E. H. Martindale, Cleveland, Charles Robbins, Pittsburgh, Charles S. Ruffner, New York, C. E. Magnusson, Seattle, C. S. McDowell, U. S. Navy, and L. T. Robinson, Schenectady; managers, E. B. Craft, New York, H. B. Smith, Worcester, and James F. Lincoln, Cleveland; treasurer, George A. Hamilton, Elizabeth, N. J. (re-elected).

The foregoing, together with the following hold-over officers, will constitute the Board of Directors for the next administrative year: Calvert Townley, New York; Comfort A. Adams, New York; Walter A. Hall, Atlantic, Mass.; William A. Del Mar, New York; Wilfred Sykes, Pittsburgh; Walter I. Slichter, New York; G. Faccioli, Pittsfield; Frank D. Newbury, Pittsburgh; L. E. Imlay, Niagara Falls; F. F. Fowle, Chicago; L. F. Morehouse, New York.

T. L. Hutchinson was re-elected secretary of the institute for the coming year.

*Abstract of an article by R. S. McBride, in Coal Age.

Carl G. Barth Honored by the Taylor Society

Carl G. Barth, pioneer in the machine-building industry, has been elected an honorary member of the Taylor Society, the national organization for the promotion of science in management. Only two other men have been honored by the society, these being Frederick W. Taylor and Henri Le Chatelier, the prominent engineer who developed scientific management in France.

Mr. Barth, after completing engineering studies in his native land, Norway, came to this country in 1881 and was employed as a draftsman by William Sellers & Co. From 1895 to 1899 he was engaged in engineering work in St. Louis. He became associated with Mr. Taylor at Bethlehem, Pa., with whom he worked on the historic foundations of scientific management. Under Mr. Taylor's direction he conducted experimental work on metal-cutting tools and applications of slide rules in the plants of the Link-Belt Co. and the Tabor Manufacturing Co.

Mr. Barth's work on the fundamental formulas in cutting metals has received world-wide recognition as one of the most important contributions to machine development.

Seven proposals were received by Acting Secretary Roosevelt under the opening of bids for coal for the Navy, on May 18. These aggregate a total of 319,000 tons of bituminous coal and 3,700 tons of anthracite.

The annual requirements of the Navy are 2,174,800 tons of bituminous coal and 38,215 tons of anthracite. At the principal point of delivery, Hampton Roads, Va., there was offered 185,000 tons of the 1,200,000 tons called for. The prices per gross ton quoted in the bids range from $4.20 to $4.816 f.o.b. mines. The base price at present being paid by the Navy under commandeering orders for coal taken at Hampton Roads, including allowance for the advancement in wages, is about $3.70 per gross ton.

WANTED: A PRACTICAL MAN

From Engineering and Mining Journal

Edison Medal Presented to William LeRoy Emmet

AT THE annual meeting of the American Institute of Electrical Engineers in the auditorium of the Engineering Societies Building, New York, on May 21, the annual presentation of the Edison Medal was made to William LeRoy Emmet "For Inventions and Developments of Electrical Apparatus and Prime Movers."

WILLIAM LEROY EMMET

The Edison Medal was founded by the Edison Medal Association, composed of associates and friends of Thomas A. Edison, and is awarded annually by a committee consisting of twenty-four members of the American Institute of Electrical Engineers, "for meritorious achievement in electrical science, electrical engineering, or the electrical arts."

The first medal was presented in 1909 and since that time ten medals have been awarded.

The ceremony was opened by Carl Hering, chairman of the Edison Medal Committee, who described the medal and gave a brief history of its origin.

H. W. Buck, past president of the American Institute of Electrical Engineers, followed Mr. Hering and related the achievements of Mr. Emmet. He said in part:

It has been my privilege for a great many years to have known Mr. Emmet personally and later on in life to have been intimately associated with him in his professional activities. One of my earliest recollections of Mr. Emmet is as dressed in the uniform of an ensign in the Navy.

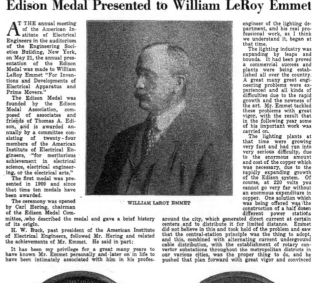

In 1892 Mr. Emmet was transferred from the Chicago office to the New York office of the Edison General Electric Company and started in there as engineer of the Foreign Department. From that position he was transferred to be engineer of the lighting department, and his real professional work, as I think we understand it, began at that time.

The lighting industry was expanding by leaps and bounds. It had been proved a commercial success and plants were being established all over the country. A great many great engineering problems were experienced and all kinds of difficulties due to the rapid growth and the newness of the art. Mr. Emmet tackled these problems with great vigor, with the result that in the following year some of his important work was carried on.

The lighting plants at that time were growing very fast and had run into very serious difficulty, due to the enormous amount and cost of the copper which was necessary, due to the rapidly expanding growth of the Edison system. Of course, at 220 volts you cannot go very far without an enormous expenditure in copper. One solution which was being offered was the construction of a half dozen different power stations around the city, which generated direct current at certain centers and to distribute it for limited distance. Emmet did not believe in this and took hold of the problem and saw that the central-station principle was the thing to adopt, and this, combined with alternating current underground cable distribution, with the establishment of rotary converter substations throughout the metropolitan districts in our various cities, was the proper thing to do, and he pushed that plan forward with great vigor and convinced the executives of many of the great lighting systems throughout the country that this was the way to meet the problem, with the result, from that day to this, that that has been the method of distributing Edison current throughout our cities.

He worked at this time on the problem of the load transformers, which became necessary in the various systems devised methods of ventilation and other systems, and invented, designed and installed the first of our air blast transformers which were put in at that time, and which were quite radically new. He also worked on methods of cooling and ventilating both transformers and generators.

Then a very serious difficulty was experienced which looked as if it was going to block absolutely the further development of electrical distribution of power, and that was the difficulty in controlling the electric circuits which were then becoming very large, both in power and current. Emmet tackled this problem, analyzed its difficulties and started in an extensive campaign to solve it.

The central stations of the country were powerless and could not control the circuits, and the only means of doing this was to kill the field in the generator. Emmet knew that was not the thing to do and took a most active part in the experiments made that year and the year after at Brooklyn and Niagara Falls, where almost every conceivable form of circuit-breaking device, representing the individual ideas of almost every man in the General Electric Company, was tested out. There were schemes for opening circuits, understand, under the action of compressed air, air blasts, sprays, and all sorts of devices, and in addition an ordinary jug of oil, by which the contact was broken. All of these devices were tested out, and they finally got around to the oil switch, through which was short circuited a five thousand kilowatt machine, under heavy inductive load, and the circuit opened several times. Nothing happened, and apparently there was an open circuit. The thing was tested out three or four times, and found to be all right, and the circuit opened again, and it was finally discovered that these very high powered inductive short circuits were being opened under oil, without the slightest commotion.

In connection with Mr. Emmet's work in the turbine, I think he was the first to point out to steam engineers the tremendous importance of high vacua in the operation of the turbine.

In 1913 he started some of the most important work he has ever undertaken, and that is the development of the mercury turbine, which he is now engaged upon. Mr. Emmet realized the difficulty of getting enormously high steam pressures and temperatures, so he devised an entirely new system of boiling mercury, which gives a temperature of approximately 700 deg. The mercury vapor at this temperature is then discharged into a mercury turbine and the condenser of the mercury turbine is the steam boiler, so that all the heat from the condenser of the mercury turbine is used for boiling the water, which, in turn, is discharged into a steam turbine, which operates in the usual way, so that with a certain amount of heat generated, you get the equivalent in steam operating the steam turbine, in addition, as practically free power, and you get all the power which the mercury turbine will generate. The result is a revolutionary saving and economy, in that you get thirty or forty or even more per cent of power from a given amount of coal consumed.

That experimental work is now in progress, and I understand that the results will be very soon applied commercially. If it is as successful as it looks, it will probably revolutionize all of the power plants in this country. I think we can all hope and believe that these experiments will be successful.

In conclusion I think that I represent the sentiments of all of the engineering fraternity when I say that we are very glad to have this opportunity to give this token to Mr. Emmet of our appreciation of his great work, and our appreciation of himself as an engineer.

President Calvert Townley of the A. I. E. E., following Mr. Buck's address, made the presentation of the medal, after which Mr. Emmet made the response. He said in part:

Mr. Buck spoke of something I did not want to talk about, because it is not an accomplished fact—that is, the mercury process. That is something I have done myself, and in which I am vitally interested, and expect to devote all the rest of my life to it if necessary, and allow nothing to interfere with it, because I believe it is a great thing. I will not attempt to describe the reasons, but we have already accomplished a lot in that respect.

Of course, mercury costs one dollar a pound, it is a deadly poison if it gets into the atmosphere, and a very large amount of it must be vaporized—for every pound of steam made, something like 8 lb. of mercury is vaporized, but as now developed that vaporization of mercury goes on in a whole system of apparatus which is all one piece, all welded from top to bottom into a single piece, practically, except where the back of the turbine goes through the shaft, and it has been tested to a pressure three or four or five times as great as that used, as we use a very low pressure, and it is developed in such a way that no possible application of heat in the furnace can overheat anything, and in fact, our experience for years with various devices has shown that the essentials, the things which would have been thought difficult about this process, are easy, and are not going to give us any trouble. We have had some trouble about expansion and certain strains, etc.

We are now making a commercial application of this thing and hope it may be brought to an actual commercial success in a very short space of time. It is a process applicable to everything wherever power is used—locomotives could be run by that method with a fuel consumption something like one-third of what they now use, and run without wasting the water; it could be made to condense its own steam as a steam automobile does.

My love has been for electricity, and my aims have been to develop the electric art. I have a confident belief in the practicability of doing things by the electrical art, and that is what I have been aiming at in doing these other things.

New Officers N. E. L. A.

At the convention of the National Electric Light Association, at Pasadena, Cal., May 21, the following were elected as officers for the ensuing year: President, Martin Insull; vice-presidents, M. R. Bump, F. W. Smith, W. H. Johnson and Franklin Griffith; treasurer, H. C. Abell.

An increase of approximately 20 per cent in the cost of labor and material, unexpected and unforeseeable construction difficulties and a change in the power-house plans are given by the Great Western Power Co. of California as reasons for a stock issue of $1,500,000 asked for in an application filed with the Railroad Commission. The company says that it will need the money raised through the proposed sale of its preferred stock to complete what is known as its Caribou plant.

Obituary

George E. Long, Jersey City, N. J., president of the Joseph Dixon Crucible Co., has retired after 43 years of active service.

Personals

Francis Juraschek has become associated with Hanff-Metzger, Inc., general advertising agents, New York. He has been placed at the head of the Technical Department.

D. Gleisan, manager, Industrial Bearings Division, Hyatt Roller Bearing Co., has appointed W. F. Myer to be directing transmission engineer. Mr. Myer will be responsible for the sale of Hyatt lineshaft roller bearings

Society Affairs

The Philadelphia Section of the A. I. & S. E. E. will hold its annual outing at Brown's Mills-in-the-Pines June 5.

The Association of Iron and Steel Electrical Engineers has postponed the trip scheduled for the inspection of the Trumbull Steel Co., Warren, Ohio, on May 22, until sometime in June.

The American Society of Civil Engineers, Philadelphia Section, will meet at the Rensselaer Polytechnic Club on June 7. The subject of the meeting will be "The Relation of the Owner, Engineer and Contractor."

The Association of Iron and Steel Electrical Engineers, Pittsburgh Chapter, will meet on June 12. R. H. Keil, power engineer of the Jones & Laughlin Steel Co., will read a paper on "Current Limit Reactance."

The New Orleans Section, A.S.M.E., will meet at the Lafayette Theatre on June 4. W. M. White, chief engineer, hydraulic department of the Allis-Chalmers Co., will read a paper on "Installation of Large Turbo-Alternator at Niagara."

Cleveland Section, A.S.M.E., will meet on June 8. A number of local industrial plants will be inspected during the day. This will be followed by dinner at the University Club. Col. Edward A. Deeds, Dayton Ohio, and Mr. Cattell, of Philadelphia, will be the speakers.

The Central Geographic Section of the Pennsylvania Electric Association, State Branch of the N. E. L. A., will meet in the Penn Harris Hotel, Harrisburg, on June 4. W. J. Kline, of the American Heating Co., North Tonawnada, will deliver a paper on "Central Station Heating." The rest of the morning will be taken up by five-minute talks on subjects of interest to the members. In the afternoon H. R. Kern, auditor, Philadelphia Electric Co., will deliver an address on "Accounting and Operating Records."

COAL PRICES

Prices of steam coals both anthracite and bituminous, f.o.b. mines, unless otherwise stated, are as follows:

ALANTIC SEABOARD

Anthracite—Coals supplying New York, Philadelphia and Boston:

	Mine
Pea	$5.30@$5.75
Buckwheat	3.40@ 4.10
Rice	2.75@ 3.25
Barley	2.25@ 2.50
Boiler	2.50

Bituminous—Steam sizes supplying New York, Philadelphia and Boston:

Latrobe	$4.25@$4.50
Connellsville coal	4.25@ 4.50
Cambrias and Somersets	4.35@ 5.15
Clearfields	4.00@ 4.75
Pocahontas	6.50@ 7.00
New River	6.50@ 7.00

BUFFALO

Bituminous—Prices f.o.b. Buffalo:

Pittsburgh slack	$6.00
Mine-run	6.25
Lump	6.00@ 6.50
Youghiogheny	6.75

CLEVELAND

Bituminous—Prices f.o.b. mines:

No. 6 slack	$3.25@$4.00
No. 8 slack	4.25
No. 8—¾-in.	3.50@ 4.00
No. 6 mine-run	3.50@ 4.00
No. 8 mine-run	4.50
Pocahontas—Mine-run	3.25@ 4.00

ST. LOUIS

Anthracite—Probably not more than 20 per cent of the demand in this market can be supplied. Prices effective Apr. 1 were as follows:

	Williamson and Franklin Counties	Mt. Olive and Staunton	Standard
Mine-run	$2.65@2.80	$3.00@3.50	$3.75@4.25
Screenings	2.50@2.65	2.50@3.00	2.75@3 50
Lump		3.75@4.50	4.00@4 50

Williamson-Franklin rate to St. Louis is $1.10; Other rates 95c.

CHICAGO

Bituminous—Prices f.o.b. mines:

Illinois		Freight rate Chicago
Southern Illinois		
Franklin, Saline and Williamson Counties		
Mine-Run	$3.00@$3.10	$1.55
Screenings	2.60@ 2.75	1.55
Central Illinois		
Springfield District		
Mine-Run	$2.75@$3.00	$1.32
Screenings	2.50@ 2.60	1.32
Northern Illinois		
Mine-Run	$3.50@$3.75	$1.24
Screenings	3.00@ 3.25	1.24
Indiana		
Clinton and Linton		
Fourth Vein		
Mine-Run	$2.75@$2.90	$1.27
Screenings	2.50@ 2.65	1.27
Knox County Field		
Fifth Vein		
Mine-Run	$2.75@$2.90	$1.37
Screenings	2.50@ 2.60	1.37
Brazil Bloke	$4.25@$4.50	$1.27

New Construction

PROPOSED WORK

Me., Fort Kent—The Fort Kent Electric Co. plans to construct a 33,000 volt transmission line to connect at Van Buren, with the Maine & New Brunswick Electric Light & Power Co., a distance of 43 miles.

Mass., Boston — Appleton & Stearns, Archts., 53 State St., will soon receive bids for the construction of a 3 story, 80 x 300 ft. sales and service station on Commonwealth Ave., for the J. W. Maguire Co., 745 Boylston St. A steam heating system will be installed in same. Total Estimated cost, $400,000.

Mass., Boston—The Bd. Educ., City Hall, will soon award the contract for the construction of a 2 story school addition on Charter St. An extension to the steam heating system will be installed in same. Total estimated cost, $350,000. C. H. Walker & Son, 71 Kilby St., Archts.

N. Y., Brooklyn—The Dept. of Charities, Municipal Bldg., New York City, will soon award the contract for making boiler repairs at the Kings County Hospital. Estimated cost, $8,000.

N. Y., Brooklyn—W. Small, 189 Montague St., is having plans prepared for the construction of a theatre on Grand Ave. A steam heating system will be installed in same. Total estimated cost, $500,000. Eugene De Rosa, 110 West 40th St., New York City, Archt. and Engr.

N. Y., Buffalo—Severn A. Anderson, Erie Co. Treas., City and County Hall, will soon award the contract for the installation of a steam turbine driven 35 kw. generator in the 65th armory here.

N. Y., Buffalo—The Niagara Machine & Tool Works., 633 Northland Ave., is in the market for one 150 kw. synchronous converter, or motor generator set, to convert 3 phase, 25 cycle, 2,000 volt a.c. to 230 volt d.c.

N. Y., Byron — The Genesee Light & Power Co. plans to construct an electric light and power plant here.

N. Y., Kings Park—The State Hospital Comm., Capitol, Albany, will receive bids until June 2 for the installation or heating and sanitary work, etc., in the building for acute patients, at the Kings Park State Hospital, here.

N. Y., Lockport—Dumville & Co. are in the market for a 3 hp. single phase, 60 cycle, 110 volt, 1,200 r.p.m. motor.

N. Y., Malone — The Malone Light & Power Co. is in the market for a 300 kw., 2,400 volt, 3 phase, 60 cycle a.c. generator, complete with exciter unit and direct connected to a pair of horizontal water wheels of 400 or 500 hp. capacity to operate under a 25 to 30 ft. working head. S. G. Hunter, Supt.

N. Y., New York—The Bureau of Yards & Docks, Navy Dept., Wash., D. C., plans to install a reciprocating air compressor. Estimated cost, $60,000.

N. Y., New York—Marc Klaw, Commercial Trust Co. Bldg., B'way. and 41st St., is having plans prepared for the construction of a theatre on 45th St. between B'way. and 8th Ave. A steam heating system will be installed in same. Total estimated cost, $300,000. Eugene De Rosa, 118 West 4th St., Archt. and Engr.

N. Y., Ossining — Charles F. Rattigan, Supt. of Prisons, Capitol, Albany, received bids for furnishing and installing pumping machinery, etc., in the Sing Sing Prison, here, from F. N. Lewis, Hotel Charleton, Binghamton, $8,700; Suburban Eng. Co., 15 West 38th St., New York City, $8,976. Noted May 3.

N. Y., Rochester—The Eastman Kodak Co., 333 State St., plans to construct a 70 x 150 x 362 ft. machine shop, cost, $280,000 and a 98 x 98 x 146 ft. power house, cost $200,000.

N. Y., Schenectady—The Bd. of Contract and Supply will receive bids until June 9 for the installation of plumbing, heating and electrical work at the Pleasant Valley School, Forest Rd.

N. Y., Wellsville—George C. Ross, Chn. of the Bd. of Water & Light Comrs., has received bids for extensions to the electric light plant, here, from H. M. Cushing, 49 Roanoke Ave., Buffalo. The extensions include the delivery and installation of a turbine generator, condensor, circulating pumps and interconnected piping, $20,000, piping steam and water, $2,000; changing from 2 phase to 3 phase, 2,300 volts, $2,000; outside changes, $4,500; line extensions $5,000; gas engine repairs, $1,500.

N. J., Butler—The Borough is having plans prepared for the construction of a 1 story, 50 x 50 ft. municipal lighting plant. A heating boiler, two 100 kw. engines, auxiliary equipment, air compressor and steel tanks will be installed in same. Total estimated cost, from $50,000 to $60,000. Runyon & Carey, 845 Broad St., Newark, Engrs.

N. J., Sayreville—The Borough plans to construct a water supply system which will have a daily capacity of 250,000 gal. Plans include a steel tower tank to hold 200,000 gal. pumping units, 2 triple single acting pumps, engine house, etc. Waldo S. Coulter, 114 Liberty St., New York City, Engr.

N. J., South River — The Bd. of Pub. Wks., received bids for the construction of a turbine engine plant at the power house here, from the John R. Proctor Eng. Co., 74 Cortlandt St., $197,368; Equity Eng. Co., 30 Church St., $211,726; Leslie Stevens Co., 120 Bway., $214,900. All of New York City. Noted March 3.

N. J., Vineland—F. H. Bent, Archt., 142 West State St., will soon award the contract for the construction of a 2 story, 50 x 75 ft. storeroom and refrigerator plant. Machinery will be installed in same. Total estimated cost, $30,000. Noted May 18.

Pa., Johnstown — Dennison & Hirons, Archts. and Engrs., 475 5th Ave., New York City, will soon award the contract for the construction of a 12 story, 60 x 110 ft. bank and office building, for the United States Natl. Bank here. A steam heating system will be installed in same. Total estimated cost, $500,000.

Pa., Scranton—D. J. Bondy, 80 Wall St., New York City, is having plans prepared for the construction of a theatre on Penn Ave. A steam heating system will be installed in same. Total estimated cost, $250,000. Eugene De Rosa, 110 West 40th St., New York City, Archt. and Engr.

Md., Landsdowne—The Natl. Steel Rolling Co., 203 Keyster Bldg., Baltimore, will receive bids until June 10 for the construction of a 1 story, 70 x 140 ft. steel rolling mill. Electric motors will be installed in same. Total estimated cost, $100,000.

N. C., Youngsville—The city plans to construct an electric lighting station. Estimated cost, $20,000. J. R. Pearce, Mayor.

S. C., Paris Island—The Bureau of Yards & Docks, Navy Dept., Wash., D. C., plans to improve the power plant here. Estimated cost, $75,000.

Ala., Florence—The United States Engineers Office, Wash., D. C., will receive bids until June 15 for furnishing and delivering 4 electric generators each of 25,000 kva. capacity with 5 exciters, etc.

La., Minden—The City Council will soon receive bids for the construction of a sewerage system and for the improvement of the electric light plant. Total estimated cost, $180,000. J. E. Carroll, Secy.

O., Akron—The Akron Grocery Co. will soon award the contract for the construction of a 5 story, 100 x 210 ft. warehouse. A steam heating system will be installed in same. Total estimated cost, $250,000. Harpster & Bliss, Archts.

O., Cleveland—The Bd. Educ. plans to construct a 2 story school at the Mayfield Township. A steam heating system will be installed in same. Total estimated cost, $250,000. S. Minor, Gates Mills, Clk.

O., Cleveland—The City plans to construct a 1 story, 60 x 110 ft. electric transmitting station on West 103rd St. and Western Ave. Estimated cost, $100,000. E. Shattuck, Agt. W. E. Davis, City Hall, Engr.

O., Cleveland—The city plans to construct a 1 story, 60 x 110 ft. electric transmitting station near Euclid Ave. and East 105th St. Estimated cost, $100,000. E. Shattuck, Agt. W. E. Davis, City Hall, Engr.

O., Cleveland — The Cleveland Graphite Bronze Co., 2906 Chester Ave., is in the market for a white steamer power plant with slide valves and pump, with or without boiler. New or used.

O., Cleveland—The Knight-Norris Gibbs Co., 1017 Euclid Ave., plans to build a 6 story commercial building on St. Clair Ave. and East 21st St. A steam heating system will be installed in same. Total estimated cost, $350,000.

O., East Cleveland (Cleveland P. O.)—The Natl. Lamp Co. of the General Electric Co., Nela Park, is having plans prepared for the construction of a 1 story boiler plant. Boilers will be installed in same. Total estimated cost, $100,000. G. H. Johnson, Nela Park, Archt. and Engr.

O., Warren—The Trumbull Cliffs Furnace Co., Rockefeller Bldg., Cleveland, is having plans prepared for the construction of a 1 story 600-ton capacity blast furnace here. Estimated cost, $1,000,000. Freyn Brassert & Co., 122 South Michigan Ave., Chicago, Engrs

Ind., Elkhart—The Marathon Motors Export Co. is in the market for 4 air compressors, complete, to operate 6 drills and 2 air lifts to have a capacity of 5,000 lb.

Ind., Terre Haute — The Amer. Car & Fdry. Co. plans to build a power house at its plant on 10th and Crawford Sts., to replace the building recently destroyed by fire. Estimated cost, $20,000.

Mich., Detroit—John H. Kunsky, Madison Theatre Bldg., engaged C. H. Crane & E. G. Hiehler, Archts., Huron Bdg., to prepare plans for the construction of a 6 story, 155 x 250 ft. theatre and office building on Broadway and Madison Aves. Steam heating equipment and forced ventilating devices will be installed in same.

Mich., Detroit—John H. Kunsky, Madison Theatre Bldg., is having plans prepared for the construction of a 10 story, 100 x 155 ft. office building on Broadway Ave. Steam heating equipment will be installed in same. C. H. Crane & E. G. Kiehler, Huron Bldg., Archts.

Ill., Chicago—C. M. Garland, 1st Natl. Bank Bldg., is in the market for 750-1500 kw. mixed pressure turbine, direct connected, 3 phase, 60 cycle 2,300 volt generator, complete with switchboard, condenser and connections, to operate on high pressure steam at 150 lb. pressure and low pressure steam at 2 lb. gage. Turbine must be capable of carrying full load on either low or high pressure steam. One 300-400 kw., 2,300 volt, 3 phase, 60 cycle generator, direct connected to high pressure or turbine, either condensing or non-condensing.

Ill., Chicago—Sam and Lee Shubert Inc., 225 West 44th St., New York City, plans to build 2 theatres. A steam heating system will be installed in same. Total estimated cost, $500,000. Herbert J. Krapp, 116 East 16th St., New York City, Archt. and Engr.

Wis., Sheboygan — The city will soon award the contract for the construction of a 3 story, 100 x 150 ft. high school on North 9th and Jefferson Sts. Steam heating equipment including Kewanee heaters will be installed in same. Total estimated cost, $300,000. Childs & Smith, 64 East Van Buren St., Chicago, Archts. Noted March 16.

Ia., Fort Madison—The city will receive bids until June 9 for the construction of a sanitary sewerage system including 2 ejector pumping stations. Total estimated cost, $400,000. Burns & McDonnell, 402 Interstate Bldg., Kansas City, Mo., Engrs.

Minn., Buffalo—The village clerk will receive bids until June 4 for the construction of a deep well, water system, and a 14 x 20 ft. pump house including 2 pumps and motors to have a 250 gal. per min. capacity. Total Estimated cost, $60,000.

Minn., Minneapolis — The Minneapolis Sanatorium Hotel Co., 2716 Hennepin Ave., is having plans prepared for the construction of an 8 story sanatorium on Yale Pl. A steam heating system will be installed in same. Total estimated cost, $500,000. O. K. Westphal, 319 Kasota Bldg., Archt.

Minn., Olivia—The Bd. of Educ. received bids for the installation of a heating and ventilating system in the proposed 3 story, 63 x 150-ft. high school, from the Amer. Heating Co., 1813 Winter Ave., Superior, Wis., $22,490; Beck Eng. & Constr. Co., 1137 Plymouth Bldg., Minneapolis, $25,985; Heins & Byers, $26,755. Noted March 9.

Neb., Gordon—J. J. Olsen, City Clk., will soon award the contract for furnishing labor and additional machinery and equipment for a waterworks extension including one 160 hp. steam engine, uniflow type; one 80 hp. engine, two 120 hp. steam boilers; one 110 ft. steel smokestack; one 120 hp. and one 180 hp. true Diesel engines.

Mo., St. Louis—The city voted $68,000 bonds to construct a morgue on 12th St. A refrigeration plant will be installed in same. L. R. Bowen, City Hall, Engr.

Ark., Little Rock—The E. L. Bruce Co. is in the market for a new or used, 25 hp., 3 phase, 60 cycle, 440 volt motor.

Tex., Dallas—George C. Nimmons & Co., Archts., 120 South Michigan Ave., will soon award the contract for the construction of a 9 story, 160 x 300 ft. general merchandise building for the Sears, Roebuck & Co., Arthington and Homan Aves., Chicago. A steam heating system will be installed in same. Total estimated cost, $2,000,000.

Tex., Fort Worth—The Bureau of Yards & Docks, Navy Dept., Wash., D. C., received only bid for the installation of deep well pumping equipment at the Helium Production plant here from Layne & Bowler, Rand Bldg., Memphis, Tenn., $23,000.

Wash., Cedar Falls—The City Council of Seattle has passed an ordinance appropriating $495,000 to complete the construction of the pipe line and a 15,000 kw. plant here.

Wash., Centralia—The Kane Pneumatic Shock Absorber Co., Fords Prairie, plans to construct a power plant which will generate from 750 to 1,500 kw. A mechanical stoker system will be installed in same. P. W. Kane, pres.

Wash., Puget Sound — The Bureau of Yards & Docks, Navy Dept., Wash., D. C., will receive bids until June 16 for the installation of auxiliary equipment and piping for the air compressor. Estimated cost, $60,000. Noted May 10.

Wash., Spokane—The Washington Water Power Co., 825 Trent St., plans to build a 37 mi. high tension power line from Long Lake to Chewelah. Estimated cost, $140,000.

Cal., Burbank—The Burbank Development Co., 206 West San Fernando St., is having plans prepared for the construction of 2 soap factory buildings on Verdugo Ave. Electric power and steam generating plants will be installed in same. Total estimated cost, $200,000. H. J. Knouer, 129 Story Bldg., Los Angeles, Archt.

H. T., Kuahua—The Bureau of Yards & Docks, Navy Dept., Wash., D. C., and the Public Works Office, Navy Yard, Mare Island, Cal., will soon receive bids for the furnishing and installation of high pressure air piping system at the Naval Ammunition Depot, here. Estimated cost, $10,000.

Ont., Brampton—W. Treadgold, Engr., is receiving prices on the supply and delivery of a turbine pump with electric motor, etc. Total estimated cost, $35,000.

Ont., Fort William—The Bd. Educ. plans to build two 2 story schools. A steam heating and mechanical ventilating system will be installed in same. Total estimated cost, $290,000.

Ont., Kincardine — The Town Council plans to build a hydro electric power station. Estimated cost, $45,000. T. Flint, Eugenia Falls, Engr.

Ont., New Market—The town plans to install a 1,000 Imp. gal. per min. 3 stage centrifugal pump to have a 100 lb. pressure. Estimated cost, $7,000. E. James Co., 36 Toronto St., Toronto, Engr.

CONTRACTS AWARDED

Mass., Springfield — The United Electric Light Co., 73 State St., has awarded the contract for the contruction of a 3 story, 90 x 58 ft. sub-station on Carew and North Sts., to A. E. Stephens Co., 145 State St. Estimated cost, $100,000. Noted March 9.

Mass., Worcester — The Rogers Drop Forging Co., Frank St., has awarded the contract for the construction of a forge plant including a 1 story, 60 x 85 ft. sand blasting building, 46 x 50 ft. boiler house, etc., to the E. D. Ward Co., 82 Foster St. Total estimated cost, $125,000.

N. Y., Brooklyn—The Bd. Educ., 500 Park Ave., New York City, has awarded the contract for the installation of heating and ventilating apparatus in P. S. 97 on Stillwell Ave., near Ave. S. to W. J. Olvany, 177 Christopher St., New York City, at $86,500.

N. Y., Brooklyn—The Bd. Educ., 500 Park Ave., New York City, has awarded the contract for the installation of heating and ventilating apparatus in P. S. 100, on West 1st St. near Sheepshead Bay, to W. J. Olvany, 177 Christopher St., New York City, at $94,681.

N. Y., Brooklyn—The Dept. of Charities, Municipal Bldg., has awarded the contract for furnishing all labor and material required for the construction of an addition to the refrigerator plant at the Kings County Hospital, to P. F. Larkin, 429 East 56th St., New York City, at $9,467.

N. Y., New York—The Varick St. Bldg., Co., c/o Helme & Corbett, Archts. and Engrs., has awarded the contract for the construction of a 10 story, 65 x 175 ft. loft building on Varick, Grand and Watts Sts., to the Turner Constr. Co., 244 Madison Ave. A steam heating system will be installed in same.

N. Y., Watertown — The Beebes Island Power Corp., has awarded the contract for furnishing and installing hydraulic turbine equipment, in connection with the proposed development on the Black River, to the Wellman, Seaver Morgan Co., 7000 Central Ave., S. E., Cleveland, O., and electric generators, to the General Electric Co., River Rd., Schenectady.

Pa., Philadelphia — The Hurley Motor Co., Broad and Race Sts., has awarded the contract for the construction of a 15 story, 55 x 130 ft. office building, to William Steel & Sons Co., 16th and Arch Sts. A steam heating system will be installed in same. Total estimated cost, $750,000.

Md., Cumberland—The Crystal Laundry Co., Baltimore St., has awarded the contract for the construction of a 1 and 2 story, 130 x 284 ft. laundry, garage and power house, to W. J. Morley, 98 Bedford St. Total estimated cost, $200,000. Noted April 27.

Va., Bedford—The city has awarded the contract for the construction of a storage reservoir and probably 2 small pumping units, etc. to Diehl & Vance, Virginia Carolina Bldg., Norfolk.

S. C., Greenville—The Farmers & Merchants Bank has awarded the contract for the construction of a 16 story bank and office building, to C. T. Wills, 286 5th Ave., New York City. A steam heating system will be installed in same. Total estimated cost, $1,000,000. Noted April 6.

O., Canton—Schjemmer & Graber, 12th St., has awarded the contract for the construction of a 2 story, 60 x 153 ft. garage, to E. J. Landor Co., Renkert Bldg., at $101,100. A steam heating system will be installed in same.

Mich., Muskegon Heights (Muskegon P. O.)—The Bd. Educ. will build a 2 story, 250 x 250 ft. high school. Steam heating equipment will be installed in same. Total estimated cost, $400,000. Work will be done by day labor.

Ill., Chicago—The L. J. McCormick Corp., c/o Holabird & Roche, Archts., 114 South Michigan Ave., has awarded the contract for the construction of a 1 story, 100 x 120 ft. theatre on Dearborn and Randolph Sts., to the Longacre Constr. & Eng. Co., 127 North Dearborn St. A steam heating system will be installed in same. Total estimated cost, $1,000,000. Noted May 10.

Wis., Menasha—The Bd. Educ. has awarded the contract for the construction of a 2 story high school addition and power plant, to the C. K. Meyer & Son Co., 50 State St., Oshkosh. New boilers and blower fan will be installed in same. Total estimated cost, $95,000.

Ia., Glidden—H. W. Porter, Secy. of the Bd. of Educ., has awarded the contract for the installation of a heating and plumbing system in the proposed 3 story, 90 x 152 ft. school building to C. F. Rock Plumbing & Heating Co., St. Joseph, Mo., at $36,213. Noted May 10.

Minn., New Ulm—The city has awarded the contract for the construction of a 1 story, 40 x 90 x 90 ft. power plant, to J. G. Robertson, University Ave., St. Paul, at $145,399. Noted May 10.

Wash., Tacoma—The state has awarded the contract for the construction of a power and heating plant at the State Capitol here, to Albertson, Cornell Brothers and Simpson at $155,263.

Cal., El Segundo—The General Chemical Co., 25 Broad St., New York City, has awarded the contract for the construction of a group of factory buildings for the manufacture of sulphuric acid to the J. G. White Eng. Corp., 43 East Exch. Pl., New York City. A steam heating system will be installed in same. Total estimated cost, $2,000,000.

PRICES - MATERIALS - SUPPLIES

Electrical prices on following page are prices to the power plant by jobbers in the larger buying centers east of the Mississippi. Elsewhere the prices will be modified by increased freight charges and by local conditions.

POWER-PLANT SUPPLIES

HOSE—

	Fire		50-Ft. Lengths
Underwriters' 2½-in.			85c. per ft.
Common, 2½-in.			30%

Air	First Grade	Second Grade	Third Grade
¾-in. per ft.	$0.60	$0.40	$0.30

Steam—Discounts from List

First grade...... 20% Second grade...... 30% Third grade.... 45%

RUBBER BELTING—The following discounts from list apply to transmission rubber and duck belting:

Competition.............. 50% Best grade.............. 20%
Standard.............. 30-10%

LEATHER BELTING—Present discounts from list in fair quantities (½ doz. rolls):

Light Grade	Medium Grade	Heavy Grade
30%	25%	20%

RAWHIDE LACING
- For cut, best grade, 25%, 2nd grade, 30%.
- For laces in sides, best, 79c per sq. ft.; 2nd, 75c.
- Semi-tanned: cut, 20%; sides, 83c per sq. ft.

PACKING—Prices per pound:

Rubber and duck for low-pressure steam	$1.00
Asbestos for high-pressure steam	1.70
Duck and rubber for piston packing	1.00
Flax, regular	1.20
Flax, waterproofed	1.70
Compressed asbestos sheet	.90
Wire insertion asbestos sheet	1.50
Rubber sheet	.50
Rubber sheet, wire insertion	.70
Rubber sheet, duck insertion	.50
Rubber sheet, cloth insertion	.30
Asbestos packing, twisted or braided and graphited, for valve stems and stuffing boxes	1.30
Asbestos wick, ½- and 1-lb. balls	.85

PIPE AND BOILER COVERING—Below are part of standard lists, with discounts.

PIPE COVERING		BLOCKS AND SHEETS	
Pipe Size	Standard List Per Lin. Ft.	Thickness	Price per Sq. Ft.
1-in.	$0.27	½-in.	$0.27
2-in.	.36	1-in.	.30
6-in.	.80	1½-in.	.45
4-in.	.60	2-in.	.60
3-in.	.45	2½-in.	.75
8-in.	1.10	3-in.	.90
10-in.	1.30	3½-in.	1.05

85% magnesia high pressure		List
For low-pressure heating and return lines	4-ply	50% off
	3-ply	52% off
	2-ply	54% off

GREASES—Prices are as follows in the following cities in cents per pound for barrel lots:

	Cincinnati	Chicago	St. Louis	Birmingham	Denver
Cup	8½	6.6	8½	8½	10.5
Fiber or sponge	8½	8.6	11@14	8½	13.75
Transmission	10	8.1	11@14	8½	12.5
Axle	...	4.8	5@5½	5	7.0
Gear	6½	6.1	6@7	4½	8.5
Car journal	12(gal.)	4.7	22@24	5	9.0

COTTON WASTE—The following prices are in cents per pound:

	New York Current	New York One Year Ago	Cleveland	Chicago
White	11.00@15.50	13.00	16.00	11.00 to 14.00
Colored mixed	7.00@10.50	9.00 to 12.00	12.00	9.50 to 12.00

WIPING CLOTHS—Jobbers' price per 1000 is as follows:

	13½ x 13½	13½ x 20½
Cleveland	$55.00	$65.00
Chicago	41.00	43.50

LINSEED OIL—These prices are per gallon:

	New York Current	New York One Year Ago	Chicago Current	Chicago One Year Ago
Raw in barrels (5 bbl. lots)	$1.75	$1.59	$2.05	$1.78
5-gal. cans	1.78*	1.72	2.30	1.98

*To this oil price must be added the cost of the cans (returnable), which is $2.25 for a case of six.

WHITE AND RED LEAD—Base price per pound:

	Red Current Dry	Red Current In Oil	Red 1 Year Ago Dry	Red 1 Year Ago In Oil	White Current Dry and In Oil	White 1 Yr. Ago Dry and In Oil
100-lb. keg	15.50	17.00	13.00	14.50	15.50	13.00
25- and 50-lb. kegs	15.75	17.25	13.25	14.75	15.75	13.25
12½-lb. keg	16.00	17.50	13.50	15.00	16.00	13.50
1-lb. cans	18.50	20.00	15.00	16.50	15.00	15.00
5-lb. cans	20.50	22.00	16.00	17.50	16.00	16.00

500 lb. lots less 10% discount; 2000 lb. lots less 10-2½%.

RIVETS—The following quotations are allowed for fair-sized orders from warehouse:

	New York	Cleveland	Chicago
Steel 7/16 and smaller	30%	40%	35-10%
Tinned	30%	40%	35-10%

Boiler rivets, ¾, ⅞, 1 in. diameter by 2 in. to 5 in. sell as follows per 100 lb.:

New York.... $6.00 Cleveland... $4.00 Chicago .... $5.37 Pittsburgh... $4.72

Structural rivets, same sizes:

New York.... $7.10 Cleveland.. $4.10 Chicago. $5.47 Pittsburgh... $4.82

REFRACTORIES—Following prices are f. o. b. works, Pittsburgh district:

Chrome brick	net ton	$75-80	at Chester, Penn.
Chrome cement	net ton	45-50	at Chester, Penn.
Clay brick, 1st quality fireclay	net M.	45-50	at Clearfield, Penn.
Clay brick, 2nd quality	net M.	40-45	at Clearfield, Penn.
Magnesite, dead burned	net ton	50-55	at Chester, Penn.
Magnesite brick, 9 x 4½ x 2½ in.	net ton	85-90	at Chester, Penn.
Silica brick	net M.	50-55	at Mt. Union, Penn.

Standard size fire brick, 9 x 4½ x 2½ in. The second quality is $4 to $5 cheaper per 1000.

St. Louis—Fire Clay, $45@$50.

Birmingham—Silica, $51 @ 55; fire clay, $45 @ 50; magnesite, $100; chrome, $100.

Chicago—Second quality, $25 per ton.

Denver—Silica, $18; fire clay, $12; magnesite, $57.50.

BABBITT METAL—Warehouse prices in cents per pound:

	New York Current	New York One Year Ago	Cleveland Current	Cleveland One Year Ago	Chicago Current	Chicago One Year Ago
Best grade	90.00	87.00	74.50	79.00	70.00	75.00
Commercial	50.00	42.00	21.50	18.00	15.00	15.00

SWEDISH (NORWAY) IRON—The average price per 100 lb., in ton lots, is:

	Current	One Year Ago
New York	$20.00	25.50-30.00
Cleveland	20.00	20.00
Chicago	21.00	16.50

In coils an advance of 50c. usually is charged.

Domestic iron (Swedish analysis) is selling at 12c. per lb.

SHEETS—Quotations are in cents per pound in various cities from warehouse; also the base quotations from mill:

	Large Mill Lots Pittsburgh	New York Current	New York One Year Ago	Cleveland	Chicago
Blue Annealed					
No. 10	3.55-6.00	7.12@8.00	4.57	7.55	7.02
No. 12	3.60-6.05	7.17@8.05	4.62	7.65	7.07
No. 14	3.65-6.10	7.22@8.10	4.67	7.70	7.12
No. 16	3.75-6.20	7.32@8.20	4.77	7.80	7.22
Black					
Nos. 18 and 20	4.15-6.30	8.50@9.50	5.30	8.20	7.80
Nos. 22 and 24	4.20-6.35	8.55@9.55	5.35	8.25	7.85
No. 26	4.25-6.40	8.60@9.60	5.40	8.30	7.90
No. 28	4.35-6.50	8.70@9.70	5.50	8.40	8.00
Galvanized					
No. 10	4.70-7.50	9.75@11.00	5.50	8.50	8.15
No. 12	4.80-7.60	9.85@11.00	5.55	8.60	8.20
No. 14	4.80-7.60	9.85@11.10	5.50	8.60	8.35
Nos. 18 and 20	5.10-7.90	10.10@11.40	5.90	8.90	8.65
Nos. 22 and 24	5.25-8.02	10.25@11.55	6.05	9.05	9.05
No. 26	5.40-8.20	10.40@11.70	6.20	9.20	9.20
No. 28	5.70-8.50	10.70@12.00	6.50	9.50	9.50

*For painted corrugated sheets add 30c. per 1000 lb. for 5 to 28 gage, 25c. for 19 to 24 gages; for galvanized corrugated sheets add 15c., all gages.

PIPE—The following discounts are for carload lots f. o. b. Pittsburgh; basing card of Jan. 1, 1919, for steel pipe and for iron pipe:

Inches	Steel Black	Steel Galvanized	Inches	Iron Black	Iron Galvanized
BUTT WELD					
⅛, ¼, and ⅜	50½%	24%	¾ to 1½	39½%	23½%
½	54½%	40%			
¾ to 3	57½%	44%			
LAP WELD					
2	50½%	35%	2	32½%	18½%
2½ to 6	53½%	41%	2½ to 6	34½%	21½%
BUTT WELD, EXTRA STRONG PLAIN ENDS					
⅛, ¼ and ⅜	48½%	29%	¾ to 1½	39½%	24½%
½	51½%	39%			
¾ to 1½	50½%	43%			
LAP WELD, EXTRA STRONG PLAIN ENDS					
2	48%	37%	2	33½%	20½%
2½ to 4	51%	49%	2½ to 4	35½%	23½%
4½ to 6	50%	39%	4½ to 6	34½%	22½%

Stock discounts in cities named are as follows:

	New York Black	New York Galvanized	Cleveland Black	Cleveland Galvanized	Chicago Black	Chicago Galvanized
¾ to 3 in. steel butt welded	40%	24%	40%	31%	54%	40½%
2½ to 3 in. steel lap welded	35%	20%	42%	27%	50%	37½%

Malleable fittings. Class B and C, from New York stock sell at list + 32%. Cast iron, standard sizes, net.

BOILER TUBES—The following are the prices for carload lots, f. o. b. Pittsburgh:

Lap Welded Steel		Charcoal Iron	
3½ to 4½ in.	40½	3½ to 4½ in	15½
2½ to 3¼ in.	30½	3 to 3¼ in	1
2¼ in.	24	2½ to 2¾ in.	+ 1
1¾ to 2 in.	19½	2 to 2¼ in.	+16
		1¾ to 1⅞ in	+29

Standard Commercial Seamless—Cold Drawn or Hot Rolled

	Per Net Ton		Per Net Ton
1 in	$327	1⅞ in	$207
1¼ in.	267	2 to 2¼ in.	177
1½ in.	257	2½ to 3¼ in.	167
1¾ in.	207	4 in	187
		4½ to 5 in.	207

These prices do not apply to special specifications for locomotive tubes nor to special specifications for tubes for the Navy Department, which will be subject to special negotiations.

ELECTRICAL SUPPLIES

ARMORED CABLE—

B. & S. Size	Two Cond. M Ft.	Three Cond. M Ft.	Two Cond. Lead M Ft.	Three Cond. Lead M Ft.
No. 14 solid	$104.00	$138.00	$164.00	$210.00
No. 12 solid	135.00	170.00	225.00	265.00
No. 10 solid	185.00	235.00	275.00	325.00
No. 8 stranded	285.	375.00	520.00	500.00
No. 6 stranded	400.(	500.00	560.00	......

From the above lists discounts are:

Less than coil lots	Net List
Coils to 1,000 ft.	10%
1,000 ft. and over	15%

BATTERIES, DRY—Regular No. 6 size red seal, Columbia, or Ever-Ready:

	Each, Net
Less than 12	$0 45
12 to 50	38
50 to 125 (bbl.)	.35
125 (bbl.) or over	32

CONDUITS, ELBOWS AND COUPLINGS—These prices are f. o. b. New York with 10-day discount of 5 per cent.

Size, In.	Conduit Black 2,500 to 5,000 Lbs.	Conduit Galvanized 2,500 to 5,000 Lbs.	Elbows Black 2,500 to 5,000 Lbs.	Elbows Galvanized 2,500 to 5,000 Lbs.	Couplings Black 2,500 to 5,000 Lbs.	Couplings Galvanized 2,500 to 5,000 Lbs.
½	97 01	103 18	24.20	25 26	2.65	8.10
¾	128 48	136 82	32.20	33.65	8 92	7 42
1	189.92	202 26	47 14	49 80	12.73	14.58
1¼	256 96	273 05	60 02	63 26	16 56	17 50
1½	307 24	327 20	80 02	84 35	22 75	23 89
2	413.36	440 24	145 63	154 63	28 10	29 52
2½	653 57	699.60	240 00	253 04	37 35	39.36
3	854 67	910 20	640 22	674.78	53 35	56 23
3½	1,060 94	1,127 74	1,433 83	1,494 60	80.03	112 46
4	1,283.15	1,362 53	1,633 90	1,722.10	106 70	140.58

Standard lengths rigid, 10 ft. Standard lengths flexible, ⅜ in., 100 ft. Standard lengths flexible, ½ to 2 in., 50 ft.

CONDUIT NON-METALLIC, LOOM—

Size I. D., In.	Feet per Coil	List, Ft.	
3/16	250	$0.05¼	
¼	250	.06	
⅜	250	.09	
½	200	.12	
⅝	200	.15	Coils ..... 40% off
¾	150	.18	Less coils, 25% off
1	100	.25	
1¼	100	.33	
1½	Odd lengths	.40	
2	Odd lengths	.55	

CUT-OUTS—Following are net prices each in standard-package quantities:

CUT-OUTS, PLUG

S. P. M. L.	$0.11	T. P. to D. P. S. B.	$0.24
D. P. M. L.	.11	T. P. to D. P. T. B.	.38
T. P. M. L.	.26	T. P. S. B.	.33
D. P. S. B.	.19	T. P. D. B.	.54
D. P. D. B.	.37		

CUT-OUTS, N. E. C. FUSE

	0-30 Amp.	31-60 Amp.	60-100 Amp.
D. P. M. L.	$0.33	$0.84	$1.68
T. P. M. L.	.48	1.20	2.40
D. P. S. B.	.42	1.05	.....
T. P. S. B.	.81	1.80	.....
D. P. D. B.	.78	2.10	.....
T. P. D. B.	1.35	3.60	.....
T. P. to D. P. D. B.	.90	2.52	..

FLEXIBLE CORD—Price per 1,000 ft. in coils of 250 ft.:

No. 18 cotton twisted	$22.00
No. 16 cotton twisted	27.00
No. 18 cotton parallel	26.50
No. 16 cotton parallel	34.50
No. 18 cotton reinforced heavy	34.00
No. 16 cotton reinforced heavy	42.00
No. 18 cotton reinforced light	30.00
No. 16 cotton reinforced light	36.75
No. 18 cotton Canvasite cord	28 50
No. 16 cotton Canvasite cord	32.00

FUSES, ENCLOSED—

250-Volt	Std. Pkg.	List
3-amp. to 30-amp.	100	$0.25
35-amp. to 60-amp.	100	.35
65-amp. to 100-amp.	50	.90
110-amp. to 200-amp.	25	2.00
225-amp. to 400-amp.	25	3.60
425-amp. to 600-amp.	10	5.50

600-Volt	Std. Pkg	List
3-amp to 30-amp	100	$0.40
35-amp. to 60-amp	100	.60
65-amp. to 100-amp	50	1.50
110-amp. to 200-amp	25	2.50
225-amp. to 400-amp	25	5.50
450-amp. to 600-amp	10	8.00

Discount: Less 1-5th standard package ... 30%
1-5th to standard package ... 40%
Standard package ... 52%

FUSE PLUGS, MICA CAP—

0-30 ampere, standard package	$4.75C
0-30 ampere, less than standard package	6.00C

LAMPS—Below are present quotations in less than standard package quantities:

Straight-Side Bulbs Masda B— Watts	Plain	Frosted	No. in Package	Pear-Shaped Bulbs Masda C— Watts	Clear	Frosted	No. in Package
10	$0.40	$0.45	100	75	$0 75	$0.80	50
15	.40	.45	100	100	1.10	1 25	24
25	.40	.45	100	150	1 55	1 70	24
40	.40	.45	100	200	2 10	2 30	24
50	.40	.45	100	300	3.15	3.40	24
60	.45	50	100				12
				500	4 60	4 90	12
				750	6.50	6.90	8
				1,000	7 50	7.95	8

Standard quantities are subject to discount of 10% from list. Annual contracts ranging from $150 to $300,000 net allow a discount of 17 to 40% from list.

PLUGS, ATTACHMENT—

		Each
Hubbell porcelain No. 5406,	standard package 250	$0.24
Hubbell composition No. 5467,	standard package 50	.32
Benjamin swivel No. 903,	standard package 250	.20
Hubbell current taps No. 5638,	standard package 50	.40

RUBBER-COVERED COPPER WIRE—Per 1000 ft. in New York.

No.	Solid Single Braid	Solid Double Braid	Stranded, Double Braid	Duplex
14	$12.00	$16.00	$13.90	$28.50
12	13.25	15.70	18 05	30.70
10	18.30	21.00	23 85	41.50
8	25 54	28.60	32 70	56.77
6	......	......	51.40	......
4	......	......	70 00	......
2	......	......	101.80	......
1	......	......	131.86	......
0	......	......	160 00	......
00	......	......	193 50	......
000	......	......	235.20	......
0000	......	......	288.60	......

Prices per 1000 ft. for Rubber-Covered Wire in Following Cities:

	Denver Single Braid	Denver Double Braid	Denver Duplex	Birmingham Single Braid	Birmingham Double Braid	Birmingham Duplex	St. Louis Single Braid	St. Louis Double Braid	St. Louis Duplex
14	$19.50	$24.00	$39.50	$13.90	$20 90	$35 80	$15 32	.....	$36 73
10	28 92	33.12	65 52	26 10	35 50	59 40	24.38	.....	55.47
8	40.32	45.12	89.64	36 60	49 50	79.00	33 49	.....	.....
6	63.84	69 24	.....	60.10	67.25	.....	.....	.....	.....
4	.....	97 86	.....	.....	93 41	.....	.....	76.06	.....
2	.....	145.20	.....	.....	120.40	.....	.....	108.66	.....
1	.....	189 00	.....	.....	155 23	.....	.....	140.38	.....
0	.....	251.40	.....	.....	188 77	.....	.....	169.91	.....
00	.....	305 40	.....	.....	229.19	.....	.....	201.35	.....
000	.....	317 40	.....	.....	278.64	.....	.....	238.89	.....
0000	.....	451 20	.....	.....	338 41	.....	.....	286.42	.....

Pittsburg—23c. base; discount 50%. St. Louis—30c. base

SOCKETS, BRASS SHELL—

¼ In. or Pendant Cap Key Each	Keyless Each	Pull Each	½ In. Cap Key Each	Keyless Each	Pull Each
$0.33	$0.30	$0.60	$0.39	$0 36	$0 66

Less 1-5th standard package	+50%
1-5th to standard package	+20%
Standard package	—10%

WIRE, ANNUNCIATOR AND DAMPPROOF OFFICE—

No. 18 B. & S. regular spools (approx. 8 lb.)	55c. lb.
No. 18 B. & S. regular 1-lb. coils	56c. lb.

WIRING SUPPLIES—

Friction tape, ¾ in., less 100 lb. 60c. lb., 100 lb. lots	55c. lb.
Rubber tape, ¾ in., less 100 lb. 65c. lb., 100 lb. lots	60c. lb.
Wire solder, less 100 lb. 50c. lb., 100 lb. lots	46c. lb.
Soldering paste, 2 oz. cans Nokorode	$1.20 doz.

SWITCHES, KNIFE—

TYPE "C" NOT FUSIBLE

Size, Amp.	Single Pole, Each	Double Pole, Each	Three Pole, Each	Four Pole, Each
30	$0.42	$0.68	$1.02	$1.36
60	.74	1 22	1.84	2.44
100	1.50	2 50	3.76	5.00
200	2 70	4.50	6.76	9.00

TYPE "C" FUSIBLE, TOP OR BOTTOM

Size, Amp.	Single Pole, Each	Double Pole, Each	Three Pole, Each	Four Pole, Each
30	70	1.06	1 60	2.17
60	1.18	1.80	2.70	3.60
100	2.38	3.66	5.50	7.30
200	4.40	6.76	10.14	13.50

Discounts:

Less than $10.00 list value	+25%
$10 to $25 list value	+10%
$25 to $50 list value	+5%
$50 to $200 list value	—10%
$200 list value or over	—15%

POWER

Volume 51 New York, June 8, 1920 *Number 23*

"Jim could start it, Dad.
He keeps the motors at
the power house going"

U. S. TORPEDO BOAT DESTROYER "REID" BUILT IN SIXTY ELAPSED DAYS FROM LAYING OF KEEL TO LAUNCHING

Monel-Metal Blades in Destroyer Turbines

By W. F. RICE
Bethlehem Shipbuilding Corporation, Limited

BEFORE the reduction gear came into general use in connection with turbine installation, slow-speed turbines were used, mounted directly on the driving or propeller shafts, and the maximum speed in these turbines rarely exceeded 600 r.p.m., with a consequent low centrifugal force. To resist corrosion, composition buckets were used in their construction, and this metal also proved amply strong to resist all strains due to centrifugal force to which the buckets were subjected. With the advent of the reduction gear the high-speed turbines superseded the slow-speed turbines, resulting in a far more satisfactory and efficient installation throughout. With the adoption of high-speed turbines it became necessary to change the design of the machine radically, particularly the construction of the blading, to withstand the tremendous pull that was exerted by the centrifugal force caused by the great increase in speed. It was found that the composition blades which had heretofore been used had not the physical properties to withstand the strains to which they would be subjected at high speeds, although they were satisfactory to the extent of being non-corrosive.

In this connection it may be interesting to note that to comprehend the force to be reckoned with in considering the centrifugal force set up in these high-speed turbines, a one-pound weight or blade weighing one pound, attached to the rim of a turbine wheel, will, when running at a test speed, exert a force or pull of 11,100 lb.; this, reduced to a tensile equivalent, would be over 20,000 lb. per square inch.

It was necessary, in view of the additional strength of the metal required for the turbine blading, to select

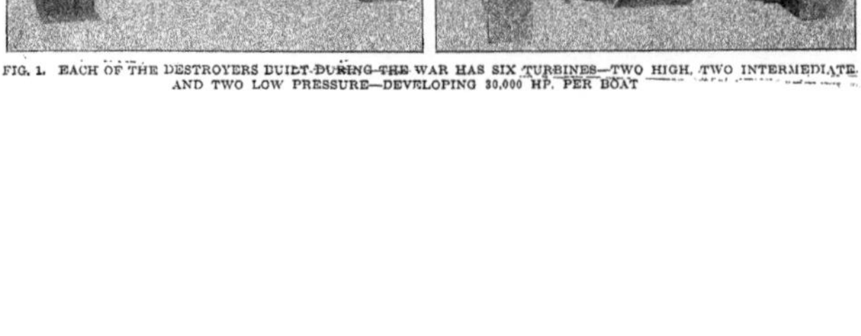

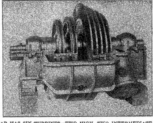

FIG. 1. EACH OF THE DESTROYERS BUILT DURING THE WAR HAS SIX TURBINES—TWO HIGH, TWO INTERMEDIATE AND TWO LOW PRESSURE—DEVELOPING 30,000 HP. PER BOAT

some metal that would be non-corrosive and yet which would have the physical properties of steel. In addition to the two prme requirements—strength and non-corrosive qualities—the metal selected must have ductility to withstand sudden shock. It must also withstand the erosive effect of high-pressure steam. It must withstand the effect of high temperatures without impairing its strength and finally, it should lend itself in manufacture to the forging process as well as to machining from the bar stock. Monel metal appeared to fill all of the foregoing requirements. It had already been used to some extent for propeller blades, valves, valve seats and other miscellaneous uses where strength and non-corrosive qualities were required, and had given satisfactory results. In view of this it was felt that it would be an excellent material for use in turbine blades, as the general characteristics of the metal particularly fitted in with the requirements. For the reader who is not acquainted with the characteristics of monel metal, it might be well to state that it is a naturally occurring alloy containing 67 per cent nickel, 28 per cent copper and 5 per cent of other elements. This remaining 5 per cent consists partly of iron from the the original ore and partly of manganese, silicon and carbon introduced in the process of refining. It contains no zinc or aluminum. It has a tensile strength varying from 80,000 to 90,000 lb. per sq.-in., an elongation of about 30 per cent in 2 in., and an elastic limit of from 50,000 to 60,000 lb. It is non-corrodible, tough and ductile and can be forged, soldered and welded, has the appearance of steel and takes and retains a nickel polish. It withstands the effects of high temperatures without material impairment of its strength. In connection with this it has been found that the monel-metal castings at 600 deg., as compared with the same material at 70 deg., retain 73 per cent of their elastic limit and 75 per cent of their tensile strength, while the rolled monel metal which is used in the manufacture of turbine blades retains 74 per cent of its elastic limit and 85 per cent of its tensile strength. It was calculated that the physical properties of this metal would give a factor of safety of 3 when running on test at a considerable overspeed, and a factor of safety of 4½ working under normal conditions. Subsequent tests showed these calculations to be correct. Each of the destroyers built during the war was equipped with six marine turbines, two high-pressure, two intermediate-pressure and two low-pressure, varying from 36 in. to 54 in. in diameter, revolving at 3,300 r.p.m., and developing 30,000 hp. per boat. The normal circumferential bucket speed was 42,500 ft. per min., while the test speed amounted to 53,400 ft. per min. The turbines for each destroyer contained upward of 10,000 blades, of which approximately 90 per cent were made of monel metal. Over 1,235,000 lb. of monel metal was used in the fabrication of approximately 1,500,000 turbine blades. These blades ranged from 2 in. to 9½ in. in length and were 1 in. in width. In connection with the stresses to which the turbine blades are subjected under varying speeds, there is shown in Fig. 5 a tabulation of stresses in pounds per square inch which have been worked out for one of the heaviest monel-metal blades used in the low-pressure turbines of a destroyer installation. It will be noted that the fastening of this blade to the rotor is somewhat different from the method in Fig. 8. Fig. 5 is the method used for heavy buckets; the base of the blade straddles the rim of the turbine rotor, while in Fig. 8, which is used for the small blades, the base of the turbine blade fits in grooves in the rim of the rotor. To meet the demands for rapid production with a view to the delivery of destroyer engines in as short a time as possible, it became necessary for the builders to resort to different methods of

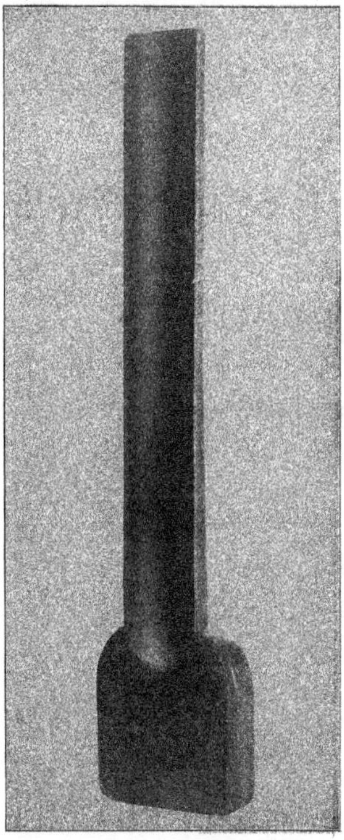

FIG. 2. THE CLEAN-CUT FORGING OF A MONEL-METAL TURBINE BLADE

FIG. 3. BLADES IN VARIOUS STAGES OF MANUFACTURE FROM THE ROUGH MONEL BAR TO THE FINISHED BLADE

manufacture. With this in view it was decided to make the shorter blades from the bar stock and the longer blades from drop forgings, as it had been determined that the metal gave satisfactory results under the hammer.

It was found after considerable experimenting that monel metal forged as satisfactorily as steel or iron if certain precautionary measures were taken in the heating to insure the forging being made at a temperature about 1,500 deg. F. The metal forged near this temperature gave the best results. A low-sulphur fuel (oil or gas preferred) was used for heating in combination with a semi-muffle furnace, which proved to be the best type for this work. In the heating it was found that great care should be taken to insure of heating considerably beyond the portion to be forged to eliminate any abrupt lines of demarcation, which would be injurious to the physical structure of the metal.

The peculiar shape of the turbine blades and the accuracy required of these forgings rendered the work difficult from the forging standpoint, regardless of the material used. All forgings were required to be within a minus 0.004 in. and plus 0.006 in., and the allowance in variation in angles was small—only 1½ deg. The forgings made to these requirements produced blades the edges of which were about 0.02 in. in thickness, which was in accordance with the design.

Many difficulties were experienced at first in the forging of monel turbine blades, because the process was undeveloped; that is, no definite type of die had been evolved and accepted as practical for use in this work. The Remington Arms Co., the Bethlehem Steel Co., and the Fore River plant of the Bethlehem Shipbuilding Corporation finally succeeded in developing this work.

The Fore River and Bethlehem plants used roughing dies which produced an oversize forging; this forging was then reheated and brought to size under a re-striking die. The Remington Arms Co., however, finished the monel forging with one heat, which precluded the necessity of making both roughing and finishing dies, as was the custom at the Fore River and Bethlehem plants. On the other hand, the dies used in this process were not as durable. Fig. 2 shows a turbine blade forged of monel metal. Note that it is an exceptionally clean-cut forging. These forgings are generally much smoother than steel, being free from scale.

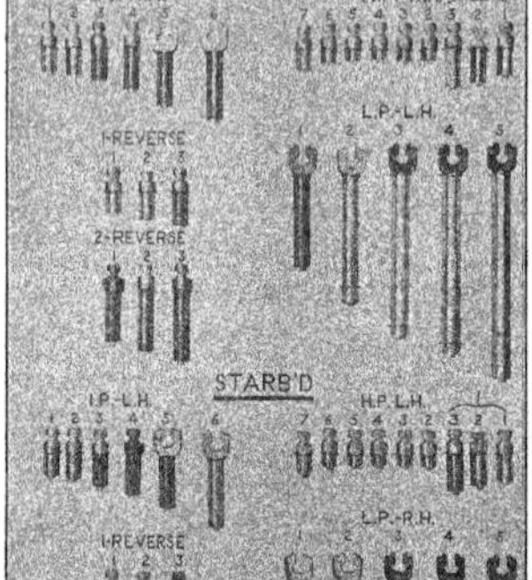

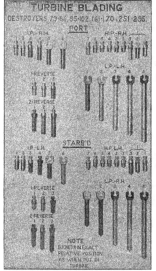

FIG. 4. COMPLETE BLADING OF ONE OF THE DESTROYER TURBINES, ARRANGED IN THE RELATIVE POSITIONS OCCUPIED IN THE TURBINE

The shorter buckets, as before mentioned, were machined from monel-metal bar stock. The Remington Arms Co., of Bridgeport, also undertook the manufacture of these blades by the machine process in addition to the drop forging of the longer blades. An elaborate system of jigs, fixtures and gages were designed for use in connection with this work. Special attention was given to the design of all cutting tools, such as milling cutters, drills, reamers, etc., with a view to particular adaptation to the cutting of the metal. It was found, however, that in most cases the standard design of cutting tools for ordinary metals answered every purpose.

For many months the Remington Arms Co. turned out machined blades of the metal at the rate of 25,000 per week. All these blades were manufactured to snap gages which allowed tolerances of only a plus and minus 0.002 in., and they were shipped to the Fore River plant of the Bethlehem Shipbuilding Corporation, the Union plant of the Bethlehem Shipbuilding Corporation on the Pacific Coast, and to the Government turbine plant at Buffalo, where they were installed in the turbine rotors under construction.

Location of Stress	R.P.M.			
	3000	3750	4000	4200
Root Tension Sect. 1-1	11650	18200	20700	22750
" Shear " 2-2	5820	9100	10350	11380
" Compression " 3-3	18820	29400	33400	36750
Blade Tension " 4-4	11200	17500	19900	22500

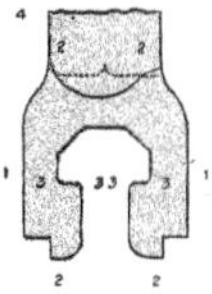

FIG. 5. TABULATION OF STRESSES WORKED OUT FOR HEAVIEST OF MONEL BLADES USED IN LOW-PRESSURE TURBINES FOR DESTROYERS

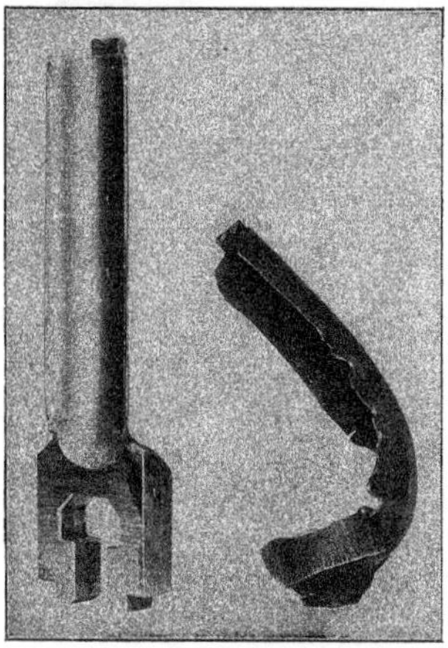

FIG. 6. NEW MONEL BLADE AND SAME BLADE AFTER BEING TORN FROM THE ROTOR BY CONTACT WITH A FRACTURED DIAPHRAGM

In connection with the manufacture of turbine blades from bar stock attention is invited to Fig. 3, which shows the blades in the different stages of manufacture from the rough bar to the completed blade. After the machining process surfaces that would later come in contact with the steam were buffed; this created a surface as smooth and bright as pure nickel.

Fig. 4 shows every size of monel blade used in the destroyer turbines. As is noted in the photograph, the blades occupy the same relative position as when installed in turbines.

FIG. 7. MONEL BLADING AFTER TWO YEARS' SERVICE. THE EDGES ARE SHARP AND THE SURFACE SMOOTH

FIG. 8. THE BASES OF THE LARGE BLADES, AS IN FIG. 5, STRADDLE THE ROTORS, SMALL BLADES FIT IN GROOVES IN THE ROTOR

To illustrate the toughness of this material, reference is made to Fig. 6, which shows a new turbine blade, also a similar blade after it had been torn from the rotor by coming in contact with a fractured diaphragm. It can be seen that although the blade is bent more than 90 deg. it shows no fracture. This blade had been in constant use during the period of the war.

The Bethlehem Shipbuilding Corporation, having put into service during the last two years upward of 100 destroyers, all of which were fitted with monel turbine blading, have had exceptional opportunity to study this material in conjunction with steam turbines. Out of thirty turbines opened in as many destroyers, in no case was the monel blading found to be in bad condition.

Upon inspection it was found that the monel buckets that had been in service for two years were in excellent condition, showing no effects from corrosion or from the erosive effects of steam. Fig. 7 shows monel blading after two years' service. The surfaces are clean and the edges are still sharp.

It is obvious, of course, that where turbine blading retains its smoothness without corrosion, such turbines are considerably more efficient than if the blades are subject to it. This means an ultimate saving in fuel consumption. It is believed from the experience of the Bethlehem Shipbuilding Corporation that the continued efficiency of any steam turbine can be maintained only when blades are made of material that will insure the blades retaining their designed shape as originally installed, combined with the retention of smooth uncorroded surfaces. The turbines in the destroyers are the Curtis.

Although the Bethlehem Shipbuilding Corporation's chief experience with the metal is with destroyer turbines, it is using it constantly in turbines for other ships, such as cargo ships, etc.

Protective and Service-Restoring Relays

BY VICTOR H. TODD

Relays, both induction and solenoid types, for protecting circuits against undervoltage or overvoltage are discussed. Relays for ringing a bell when a circuit-breaker opens are described in detail, also a relay for protection against reverse phases. The operation of a service-restoring system, which will automatically reclose a circuit-breaker a definite number of times after it has been tripped out, is explained.

FREQUENTLY, there are conditions of abnormally high or low voltage which may lead to disastrous results if not promptly detected and corrected. Considerable damage might easily be done by an increase in voltage burning out lamps. This increase may be due to a number of causes, such as the failure of a Tirrill regulator on the generator, or a short in the generator-field resister. On the other hand, the voltage may drop to a low value. To give warning of such abnormal conditions or to actually disconnect a circuit should conditions so determine, an overvoltage or an undervoltage relay may be installed. Such a relay is shown in Fig. 1 and is the same in construction and

FIG. 1. OVERVOLTAGE OR UNDERVOLTAGE RELAY

principle of operation as the induction-type overload relay, except that the windings are wound to stand the impressed voltage. In the overvoltage relays the contacts close when the voltage exceeds a certain predetermined point, which trips a breaker, rings a gong or gives other signals to the operator that the voltage is too high and requires attention. By means of a little lever on the relay the tripping voltage may be varied over wide ranges, generally between 75 per cent and 160 per cent of normal. Therefore, if a circuit is normally operating at 110 volts, it becomes possible to trip it or ring a bell if it goes to 115 or 120 volts or higher. In the undervoltage relay the windings are arranged so that the voltage tends to keep the contacts open. Then, should the voltage drop to 75 or 80 per cent of normal or for whatever the relay is set, there is no longer torque enough to hold the contacts open, so they close and generally sound an alarm.

The solenoid principle, as well as the induction principle, may be used to indicate the conditions of overvoltage and undervoltage, but the adjustment of the solenoid type is not as refined as that of the induction type. In the solenoid type the plunger is arranged to move upward and close contacts on overvoltage. For undervoltage the solenoid normally holds the plunger up and contacts open, but on a decrease in voltage, the plunger drops and the contacts close. A solenoid relay of the latter type is shown in Fig. 2. In such circuits as the constant-current systems conditions are such that an indication of low current is required. To provide this, a regular induction-type relay is used, excepting that the current tends to keep the contacts open. Then, if the current drops to a certain predetermined value, there will no longer be sufficient torque to hold the contacts open, and thus they close and give a signal or trip a breaker. If a circuit-breaker is tripped out, and the station attendant, instead of being near at hand where he can see and reset it, is some distance away, a bell-ringing relay may be used. This must ring the bell or alarm until some notice or action is taken if the breaker opens due to the protective relays tripping, but it is not necessary to ring the bell if the breaker has been opened intentionally. The relay, Fig. 3, with the cover removed will meet the foregoing requirements. A schematic connection diagram is given in Fig. 4, and the operation is as follows:

FIG. 2. SOLENOID-TYPE UNDERVOLTAGE RELAY

The two electromagnets *A* and *B* are arranged to attract the iron armature *C*, pivoted at *D*, which closes contacts *E* and *F* when attracted. The solenoid *A* is placed in series with the trip coil and relay contacts of the breaker, so that when the relay closes it energizes solenoid *A*. This attracts armature *C* and closes both contacts *E* and *F*. Contact *E* closes the circuit to solenoid *B* through a resistance *G*, and contact *F* closes the circuit to a bell or alarm *H*. Now, even though the circuit to *A* is opened, as it would be if the breaker opened, solenoid *B* still holds the contacts closed and the bell will continue to ring until the switch *I* is opened for an instant, which allows the armature to drop and the contacts to open. These relays can be used when a direct-current circuit is available for tripping the breaker. It will also be seen that should the breaker be tripped by hand, the relay cannot operate, consequently the bell rings only on automatic tripping. If, for any reason, one of the phases of the circuit supplying a polyphase motor is reversed, it will run backward. Such a reversal may occur when the motor is disconnected for repairs, through an error in reconnecting leads at the power house or substation, or from a number of other causes. In many cases the reversal of rotation of a motor, aside from

FIG. 3. CIRCUIT-BREAKER BELL-RINGING RELAY

the inconvenience that it causes, is not a serious matter, as the error can readily be corrected at the motor terminals.

In other cases, however, serious consequences may result. The reversal of an elevator motor, for instance, might result in wrecking the machinery or loss of life. To protect motors against phase reversal, the reverse-phase relay has been developed, one type of which is shown in Fig. 5. The parts are all identical with the overload induction-type relay with the exception of windings. The main coil is a voltage coil exactly like a watt-hour meter, but the series coil, instead of being heavy wire, is a large number of turns of fine wire connected in series with a reactor which gives its current similar lagging characteristics to the other potential coil.

Inasmuch as one coil is tapped on one phase and one on the other, they have the necessary phase displacement to produce a shifting field, which reacts on the disc. The external connections are shown in Fig. 6. Normally, the voltage tends to rotate the disc to the right and keep the contacts open, but should one phase be reversed or fail, or the voltage drop below 75 per cent of normal, then the contacts close and open a breaker. Reversal of a phase reverses the direction of rotation, causing the contacts to close very quickly.

There are many cases in which it is only necessary to open the breaker to clear a short; for instance, an arc across two lines, which is killed the instant the breaker opens. This permits the feeders to be put back in service immediately. If the circuit-breaker is reclosed automatically within a second after the transient trouble has occurred, the service will be restored in time to prevent induction motors from stalling.

The service-restoring relay system has been developed to perform this operation within the shortest possible time and thus reduce all disturbances to a minimum, thereby greatly improving the service. Should a permanent defect occur, the system will allow the breaker to remain open until the defect is cleared.

A schematic diagram of a service-restoring relay system is shown in Fig. 7. Any type of overload relay may be employed to trip the circuit-breaker on overload as previously described. A voltage transformer on the feeder outside the circuit-breaker is connected so that its potential opposes that of another voltage trans-

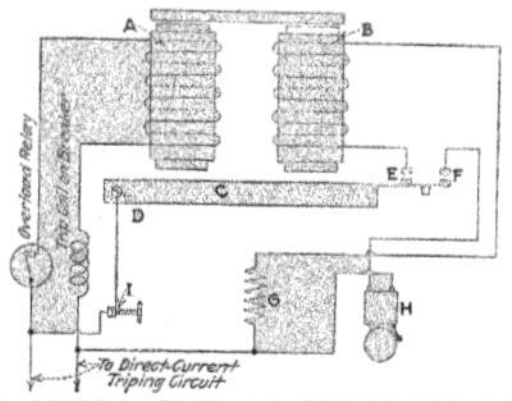

FIG. 4. SCHEMATIC DIAGRAM OF CONNECTION FOR BELL-RINGING RELAY, FIG. 3

former connected to the busbars. The restoring relay, which is similar to a magnet switch, is connected in series with these two voltage transformers. Before a short-circuit occurs, both voltage transformers are subjected to the same condition so that no current will flow through the restoring relay; but when a short-circuit occurs and the circuit-breaker has been opened

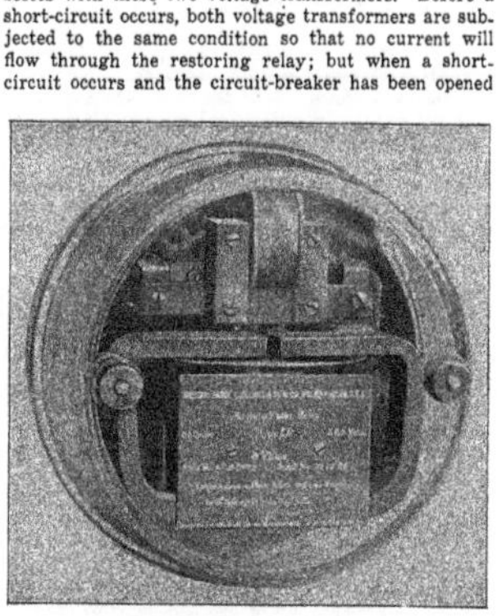

FIG. 5. REVERSE-PHASE RELAY

by the overload relay, current will be supplied by the busbar transformer *B* into the feeder transformer *A* through the restoring relay. The restoring relay will then close its contacts, which in turn will close the circuit-breaker. The latter, of course, must be of the electrically closing type as well as electrically tripping.

In case of a permanent defect on the feeder the restoring relay would continue to open and close the circuit-breaker indefinitely, since each time the breaker closed, the overload relay would open it. To prevent this, a limiting relay, similar to the overvoltage relay but equipped with weaker springs and heavier damping magnets so that its action is sluggish, is connected in such a manner that while the circuit-breaker is open it is subject to the same difference of potential that is operating the restoring relay. Every time the circuit-breaker opens, the limiting relay contacts begin to close, and owing to its heavy damping, they do not return to the starting point immediately after the circuit-breaker is closed. After the circuit-breaker has opened and closed a predetermined number of times, this relay closes its contacts, thus short-circuiting the restoring relay and preventing further operation.

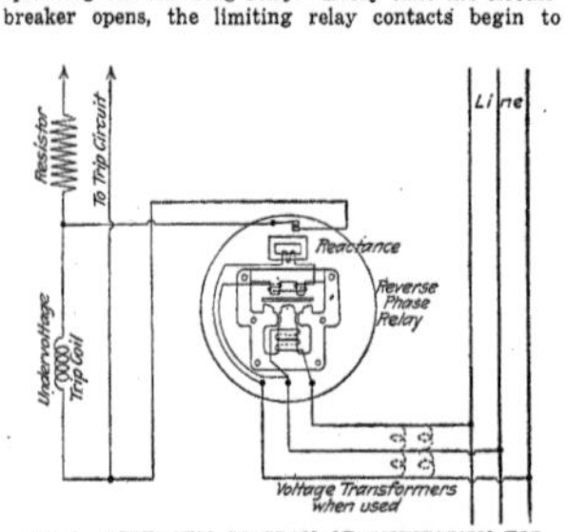

FIG. 6. SCHEMATIC DIAGRAM OF CONNECTION FOR REVERSE-PHASE RELAY

This system is often installed at substations having no attendant. In that case it is frequently found advisable to have an indicating device that will show when the service has been momentarily interrupted. For this purpose a graphic ammeter is placed in the direct-current control circuit of the circuit-breaker. This will indicate whenever the breaker has been closed by automatic means. When this system is used and provision is made for manually tripping the breaker by a control switch, contacts are arranged on the control switch which automatically open the circuit between the two voltage transformers, which prevents the breaker from resetting immediately after it is tripped.

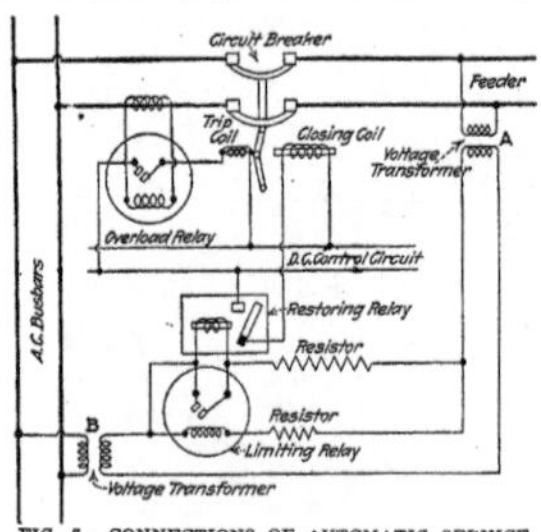

FIG. 7. CONNECTIONS OF AUTOMATIC SERVICE-RESTORING SYSTEM

With this device only 0.7 of a second is required from the first instant of overload to the instant of resumed service. If, however, the short-circuit is not removed upon the opening of the breaker, then when the breaker recloses, the short-circuit current immediately operates the overhead relay and again trips the breaker until the limiting relay prevents further action.

Refrigeration Study Course — XI. Refrigerants

BY H. J. MACINTIRE

IT IS quite possible that there has been some concern on the part of engineers as to why ammonia machines are used for refrigeration in the United States to the extent of more than 95 per cent of all such work, whereas in Great Britain and France other cooling vapors are used to a much greater extent. There must be some reason for this preference—some good and sufficient points of excellence—for refrigerating engineers are in business to make money and are not conducting a health crusade. What are these reasons, and why do not sulphur dioxide, carbon dioxide, ethyl chloride and other vapors gain in popularity?

All engineering questions have two sides, and few if any live matters can be answered definitely and decidedly one way or the other. As an example, there is the question of the use of the jet or the surface condenser for stationary plants (there is no question as regards surface condensers in marine work). There is the question of star or delta windings in electric machinery or of two-phase or three-phase transmission, and the question of steam engine or turbine as compared with the Diesel oil engine, and both of these as compared with the development of water power. The questions of load factor, cost of fuel, cost of money (if borrowed), kind and quantity of water, proximity and quality of feul to be used, etc., are factors that must be weighed in the balance or balance sheet. No cut and dried answer will avail for all cases, but each individual case must be decided for itself.

POSSIBLE REFRIGERANTS

The refrigerating engineer understands that our object with the use of the compressor and the high pressure side is to be able to use the same refrigerant continuously. We compress the gas boiling off from the refrigerating coils in order that condensation of the gas may be possible at temperatures of 60, 70 or 80 degrees or even higher, depending on what temperature of cooling or condenser water may be obtained. The gas compressed passes through the compressor, and the amount of gas so compressed per interval of time depends on the piston displacement of the compressor. Naturally, we are interested not only in the pressure range of the compressor (the suction and discharge pressures required by the refrigerant used), but also in the cooling effect obtainable by one cubic foot of compressor displacement.

There are in addition the factors of the cost of the substance used, the pathological effect should the stuff escape from the compressor or cold-storage piping and the chemical effect on iron, steel and brass. Of these the first item, the cost of the refrigerant, is too great in every case to allow the substance to be thrown away. If air were used, it might be possible to exhaust it into the atmosphere after using, but air is no longer employed. The effect on human life of the gas used is different with each substance. Carbon dioxide is harmless in small quantities and is odorless when pure. In large quantities its effect on one is practically the same as being submerged in water. Sulphur dioxide and ammonia are deadly in small quantities if relief is not immediately obtained. Ethyl chloride in small quantities seems to be the least harmful of all present refrigerants, and for that reason it is used to a large extent on naval vessels. The result in all cases is that careful provision has to be made for preventing leaks or breaks in the pipes or headers. Also, it is necessary to provide for withdrawing the charge into drums or for storage in another part of the system when repairs or cleaning of any portion of the plant has to be performed.

The main difference in refrigerants, and (with the exception of ethyl chloride for use on shipboard) the deciding factor in its choice, lies in the physical properties as regards the pressures usually obtained during operation, the refrigerating effect of one pound of the

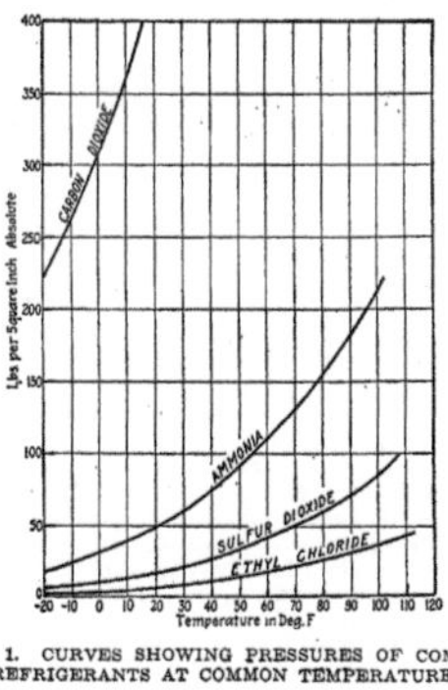

FIG. 1. CURVES SHOWING PRESSURES OF COMMON REFRIGERANTS AT COMMON TEMPERATURES

substance, and finally the volume of one pound of the refrigerant at the pressure of boiling in the expansion coils.

Referring to Fig. 1, it will be seen that ethyl chloride has the least and sulphur dioxide very moderate pressures corresponding to the temperatures of the cooling coils or the condenser, but carbon dioxide has extremely heavy pressures—some 200 to 300 lb. for suction and 800 to 1,000 lb. per sq. in. for the condenser pressure. These latter pressures are so great as to require special materials both for the compression and piping, and this factor is one that has been a deciding point in all save a few particular cases in the United States. As will be seen, the size of compressor using carbon dioxide and the fact that the gas is not necessarily dangerous immediately has caused it to be used in many places where space is valuable and where congestion makes it doubly dangerous to use ammonia. Fig. 1 also shows that ammonia has moderate pressures, both at zero degrees F. (for the cooling coils) and at 70 to 80 deg. F. (for the condensing piping). The pressures encountered are nominal and are those which steam engineers are used to. The substance, however, has greater ability for escaping out of confinement than has steam or air, and also it has a corroding action on copper and brass, and therefore the materials used are special air-furnace cast iron of extra thickness, steel forged fittings or high-grade semi-steel castings, and either extra-heavy steel or wrought-iron pipe. The joints have to be special, either screwed fittings using litharge and glycerin or solder joints or flanged joints using tongue and groove with lead or rubber gaskets.

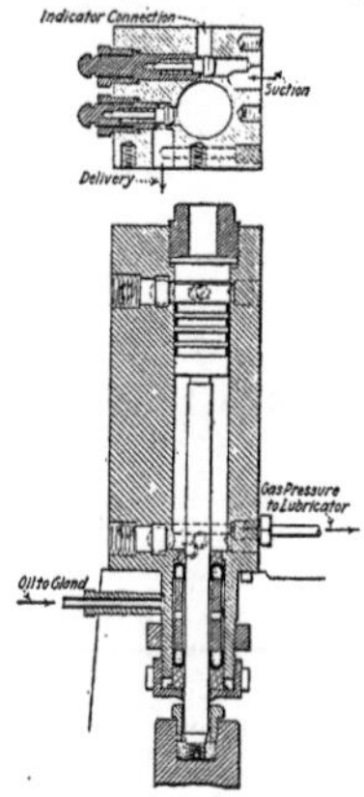

FIG. 2. SECTION OF A TYPICAL CARBON-DIOXIDE COMPRESSOR CYLINDER

However, in the final analysis the real points to be considered are the horsepower input per ton of refrigeration and the cubic feet of piston displacement of the compressor per ton. All other factors fade in comparison unless the conditions are ultra-extreme. Of course we desire to produce refrigeration at as small cost of power as we can, and this cost of power is mostly in the demands of the compressor. The compressor pumps the gas from the refrigerating coils to the condenser, and the condenser pressure is determined by the temperature and the amount of cooling water used. Again, as a rule, the *volume* of gas handled affects not only the size and cost of the compressor, but of the high- and low-pressure sides as well.

VOLUME DISPLACED BY COMPRESSOR

Referring to the table, it will be seen that the volume displaced by the compressor is less for CO_2 (carbon dioxide) than for any other of the refrigerants. In fact, ammonia is 4.6, sulphur dioxide 12.4 and ethyl chloride is 39.3 times as great as carbon dioxide for similar conditions. For the same piston speed the areas of the respective cylinders would have to have these same values, and the diameters would have to be 2.15, 3.5 and 6.27 times as great as for carbon dioxide. Ethyl chloride is fortunate on account of the small number of operating troubles. The low side, being subject only to an inward pressure, will leak air into the system if

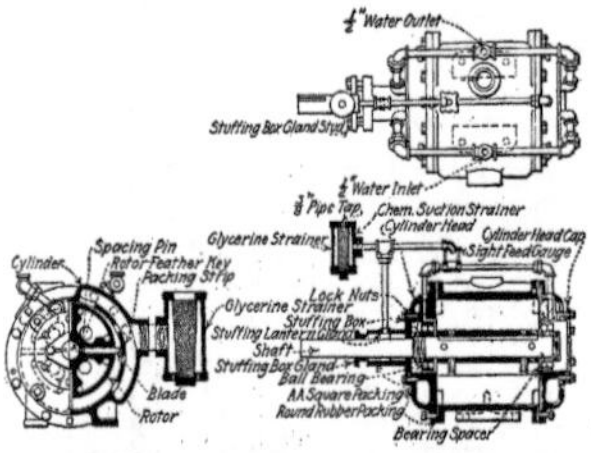

FIG. 3. ROTARY COMPRESSOR AS APPLIED TO USE OF ETHYL CHLORIDE

leaks should occur, but leaks are readily prevented with proper designs and careful operation. When air does accumulate, it collects in the condenser, from which it can be purged in the usual manner. The gas, however, is not entirely harmless and, if taken in large enough quantities, will put one to sleep like any anesthesia, and with about twice its volume of air forms an explosive mixture. The vapor burns readily with a greenish-edged flame. On account of its extremely great volume at usual refrigerating temperatures, its applications have been limited to special cases and at present to rather small machines. Being of very low condenser pressures, it lends itself to the use of the blower, or centrifugal compressor, to great advantage. The centrifugal compressor (for ethyl chloride or other refrigerants) certainly is going to be used to a greater extent in the future, especially as direct connection to steam turbines and the cheaper (higher speeds) electric motors are possible with reduced losses.

Sulphur dioxide has nominal pressures and, like carbon dioxide, does not have a bad effect on copper or brass. Sulphur dioxide is self-lubricating, whereas ethyl chloride and carbon dioxide require glycerin for that purpose. The great objection to sulphur dioxide is the corroding action on iron and steel wherever moisture gets into the system. The combination of water and sulphur dioxide forms sulphurous acid. The result is that the sulphur-dioxide machine has to be very self-contained and sealed against air leakage. In this country these machines are used only in the smaller sizes and are seldom, if ever, used where direct expansion is employed in the sense that is usually understood. To do so not only would require large return mains, but would greatly increase the possibilities of corrosion.

Carbon dioxide has its particular field in congested localities, such as office buildings, hotels and marine installations. An escape into the air of a charge of carbon dioxide is not so disagreeable and subject to unfavorable notice as is ammonia. Being much heavier than air, it will not rise except by diffusion, and guests or tenants are not driven out of their rooms. Also, unlike ammonia and ethyl chloride, it is not a combustible nor will it form explosive mixtures with air as the other two will. But, on the other hand, it is preferably a cold cooling-water refrigerant. Its critical temperature is 88.4 deg. F., which means that in extreme cases in the tropics or for machine work in certain locations, it is possible that the discharge gas passing into the condenser will not be condensed, but will act somewhat like the "dense-air" refrigerating machine. Under such conditions its efficiency is badly reduced, and much better results could have been obtained by the use of some other refrigerant.

Power Used Per Ton of Refrigeration

It so happens that, contrary to the expectation of most engineers, the power required by the different refrigerating machines per unit of refrigeration is about equal. One could expect that carbon dioxide, working as it does with a pressure range of about 550 lb., would require more power input than would, say, sulphur dioxide, which works between the limits of minus 5 lb. and 35 lb. or only 40 lb. total pressure. However, the volumes to be pumped are to be considered, and there are 12.4 times as much sulphur dioxide as carbon dioxide to be pumped for equal refrigerating effect. The net result is that carbon dioxide requires a little more power than does ammonia or sulphur dioxide. In fact, ammonia leads slightly in this respect, but the net result is so close that the matter of the horsepower per ton of refrigeration for usual (standard) operating conditions need not be considered when deciding on one or the other refrigerant.

TABLE I. SOME PROPERTIES OF REFRIGERANTS

	Latent Heat at 0 Deg. F., B.t.u.	Net Refrigeration Effect at 0 Deg. F.	Volume gas Pumped per 100,000 B.t.u. Refrigeration, Cu.Ft.	Column 3 with Carbon Dioxide as Unity	Approximate Condenser Pressure, Lb.	Approximate Suction Pressure, Lb.
Sulphur dioxide	171.8	148.7	4,970	12.4	35.0	— 5.0
Ammonia	572.2	496.4	1,848	4.61	114.5	15.25
Carbon dioxide	117.0	73.1	401	1.0	834.0	291.8
Ethyl chloride	149.4	119.4	15,720	39.3	5.5	—10.6

All pressures are gage.

Finally, it may be said that ammonia is preferred in this country because of its nominal pressures for a large range of work using condensing water varying from 50 to 80 or even 90 deg. F. and suction pressures from minus 5 lb. to 20 or 30 lb. gage. Its specific volumes are also nominal, thereby allowing moderate-sized suction return lines and headers. The piston displacement required is convenient, and so the cost of the compressor and the space occupied are small. The American engineer feels more at home with such conditions, and so ammonia holds its popularity.

The peat bogs of Manitoba, Ontario, Quebec, and New Brunswick cover an area of 12,000 square miles, with an average depth of 6 ft., and are estimated by the Commission of Conservation to contain some 9,300,000,000 tons of peat. Seven bogs within convenient delivery of Toronto are estimated to be capable of producing 26,500,000 tons of fuel, and seven bogs in the vicinity of Montreal could supply that city with 23,500,000 tons.

High Wages Influence Power-Plant Design

Complications in Conversion of Three Thousand-Kilowatt Station Into One of Eight Thousand-Kilowatt Capacity

THE important influence of the rate of wages upon boiler-plant design is strikingly shown in connection with the enlargement of the Charlotte Street power plant of the Charleston (S. C.) Consolidated Railway and Lighting Co., recently completed. When this station was originally built in 1912, colored firemen could be hired a-plenty for a dollar a day, and therefore the need for labor-saving stokers was not felt to anything like the degree that it is today. Nor was the better efficiency of stokers so potent a factor, for the price of coal was also low. Consequently, the boiler plant for the original station was laid out for hand firing.

But when the war brought on the necessity for a large expansion of the power-plant capacity, colored firemen could be secured only with difficulty at $4.50 a day. In the face of this there could be no argument as to whether stokers should be employed with the new boilers needed to increase the station capacity from 3,000 to 8,000 kw., and this was easily justifiable even though it involved many complications in working a modern stoker-fired battery into the old layout, the plan of enlargement being to retain the hand-fired boilers until a later date. How some of these difficulties of design were worked out is brought out in what follows.

Prior to reconstruction the Charlotte Street Station was laid out with a long boiler room parallel to the turbine room, the boilers being set in a single row and backed up against the turbine-room wall, with the firing aisle along the outside building wall. The equipment of the boiler room comprised one 518-hp. and five 528-hp. water-tube boilers, all hand-fired. They were installed on a floor level several feet below that of the turbine room and supplied steam at 160 lb. pressure to two 2,000-kw. and one 1,000-kw. 2,300-volt three-phase turbo-generator sets equipped with jet condensers. The turbine room was originally designed for an ultimate installation of four 2,000-kw. units, so that there was ample space in which to place additional machines at the time enlargement became necessary in 1918. The pressing demand for a large increase in output, due to Government operations and to an unprecedented growth of the city, however, made it advisable to install a 5,000-kw. unit instead of an additional 2,000-kw. machine as originally planned. The resulting installation of a 5,000-kw. 13,200-volt three-phase unit, operating at 200-lb. steam pressure and equipped with a surface condenser, did not involve any particular difficulties so far as the turbine-room construction was concerned. The real problem lay in the boiler plant.

FIG. 2. HAND-FIRED BOILERS AND TEMPORARY COAL BUNKERS

FIG. 1. FIRING AISLE IN THE NEW ADDITION AND COAL BUNKERS FOR THE HAND-FIRED BOILERS

In the boiler room it was necessary to install three new 509-hp. boilers equipped with underfeed stokers, generating steam at 220 lb. pressure and 125 deg. superheat to serve the new turbine. This necessitated a considerable enlargement and rearrangement of the boiler house.

The side and part of the end wall of the boiler room were knocked out and the building extended sufficiently to provide space for the new boilers on the opposite side of the old firing aisle. Taking out the old wall necessitated some special construction in order to support the old roof trusses, which were cut off to make way for the new skylight and roof construction. This was done by supporting these shortened existing trusses by a cantilever construction from the roof trusses of the new extension, which were at a considerably higher elevation. The arrangement may be seen in the accompanying cross-sectional drawing of the boiler- room addition.

The present boiler-room extension was made large enough for the three new boilers and one more. Later, by removing a temporary wall and extending the building longitudinally, two more boilers of the same size can be installed. When the capacity of this battery of six is exceeded, the hand-fired boilers on the opposite side of the aisle will be replaced with six of the stoker-fired type, giving an ultimate total capacity in twelve boilers of 6,108 horse power. Using underfeed stokers made it necessary to raise the operating floor for the new boilers to a height 18 ft. above the old boiler-room floor. This meant that in combining the old and new boiler equipment into a single plant, there must be two operating floor levels. However, the new boilers and equipment for them were so installed as to be in proper location for the ultimate arrangement of stoker-fired boilers on both sides of the firing aisle, doing away with the lower operating floor. For the new boilers it was necessary to build a new brick stack and to install a new coal- and ash-handling system. The steel smoke flue connecting the uptakes from the three new boilers was mounted on top of the boiler house and connected to the chimney at a point some sixty feet above the ground, as seen in an accompanying picture. The uptakes were carried up nearly vertically within the building to connect to this overhead flue. This arrangement gives ample space for the future installation of economizers if they are desired. On the same consideration that determined the use of hand-fired boilers in the original plant, coal was handled into the firing aisle, and ashes out, by wheelbarrow. Likewise, the same consideration that determined the installation of stokers in the new boiler plant dictated the installation of a modern labor saving coal-handling plant.

FIG. 3. VIEW OF STATION FROM THE BOILER-ROOM SIDE

Coal delivered in side- or bottom-dump cars is unloaded into a pit beside the boiler house. From here it passes through a crusher installed underneath the pit, discharging into a bucket conveyor. This, in turn, elevates the coal up to the top of the building and across over the roof, discharging into an overhead concrete bunker of 350-ton capacity. The size of this overhead bunker was made large enough to meet the requirements of the ultimate boiler plant for three days' supply. The continuous bucket conveyor may be seen in Fig. 3, the accompanying exterior view of the station, located between the stack and the building. From the overhead bunker coal is discharged by gravity into a weighing lorry, mounted on a runway over the center of the firing aisle. Then a spout from the lorry discharges the coal into the stokers of the new boilers. An ingenious scheme was worked out, with the installation of the coal-handling equipment, to make it serve the hand-fired boilers also and thereby do away entirely with the wheelbarrow method of bringing coal into the plant. In addition to serving the stokers the lorry spout also discharges coal into three vertical cylindrical steel bunkers, each serving two of the hand-fired boilers. These bunkers were suspended from above on one side, supported by steel bracing from the lower operating floor on the opposite side and braced laterally by angle irons bolted to the upper concrete floor. From the bottom of each of these bunkers a vertical spout extends downward to the lower operating floor, and through a gate at the end of the spout the coal is discharged in front of the boilers. From here it is shoveled into the boilers by hand in the usual manner.

For reserve coal supply an area adjacent to the track pit provides a storage capacity sufficient to operate the plant for seventy to eighty days. The coal is unloaded from cars and delivered to the storage pile by means of a locomotive crane equipped with a one-ton bucket. This crane is used

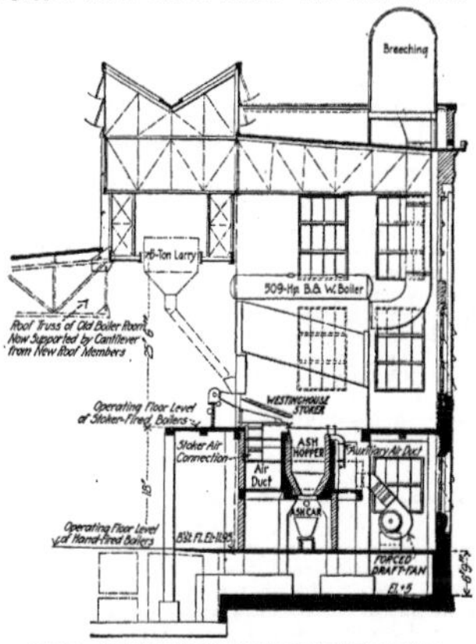

FIG. 4. CROSS-SECTION THROUGH BOILER-HOUSE EXTENSION

also to shift cars. When operating normally, all the storage coal needed is within reach of this crane, so that it may be picked up from storage and dumped directly into the track pit which supplies the crusher.

the hand-fired boilers are raked into the firing aisle and taken out in wheelbarrows as was done in the original plant.

In connection with the supply of circulating water for the condensers, two new turbine-driven pumps were installed, each having a capacity of 500 gal. per min. and large enough to handle the supply for the entire station. Water is taken from the Cooper River, on which the station is located, by means of a long siphon which dips down into a concrete pit at the river end and discharges into a tunnel under the power house. The condenser circulating pumps for the older units draw water from this tunnel, a 36-in. pipe being used for the siphon lead into the tunnel. With the installation of the new 5,000-kw. unit, a new 48-in. pipe was used to siphon the water to a separate tunnel. While this pipe is large enough to take care of the total requirements of the plant, the connections of the three jet condensers on the smaller turbines to the older tunnel were continued and no connection made with the larger supply in order to keep down the present cost of the installation. The condenser water supply as well as the steam supply for the new unit is therefore entirely separate from that for the older machines in the plant.

An interesting scheme was worked out in connection with the old siphon pipe in order to increase the capacity. When all three of the older units were running, the siphon would not deliver water fast enough to meet the requirements, the tunnel from which the condensers draw being open to the atmosphere at the time this scheme was adopted. To increase the capacity, the tunnel was sealed off and a steam exhauster put on it to create a vacuum of about 5 in. This increased the flow of water to such an extent that the supply was ample for the full requirements of the three units which it served.

The complete independence of the old and new portions of the plant was maintained, so that the future sub-

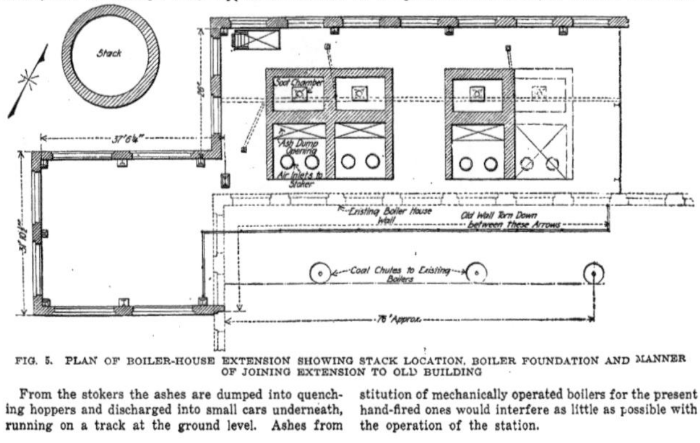

FIG. 5. PLAN OF BOILER-HOUSE EXTENSION SHOWING STACK LOCATION, BOILER FOUNDATION AND MANNER OF JOINING EXTENSION TO OLD BUILDING

From the stokers the ashes are dumped into quenching hoppers and discharged into small cars underneath, running on a track at the ground level. Ashes from

stitution of mechanically operated boilers for the present hand-fired ones would interfere as little as possible with the operation of the station.

Electric Elevator Machinery—Motor Operation

By M. A. Myers
Electrical Engineer, The Maintenance Company, New York City

ELECTRIC elevators may be of the drum type, with cables winding and unwinding on a spirally scored drum, Fig. 1, or of the traction type, with continuous cables between car and counterweights, and depending upon the frictional grip between cables and suitably grooved sheaves for traction, Fig. 2. They may be full magnetically controlled with a car switch, magnetically operated reversing switches, brake and accessories; or they may be mechanically operated with a hand rope, mechanically operated reversing switch and brake; or any combination of the two methods may be used, such as hand rope and mechanically operated reversing switch, electric main switch, and electric brake. In Fig. 3 is shown a Wheeler-McDowell Elevator Co.'s drum-type elevator machine arranged for mechanical control and electrically operated brake.

This article is the first of a series on electric elevator machinery and briefly outlines the essential parts of a modern elevator machine and their functions, after which the types of motors used on elevator machines are discussed. The next article to follow will treat of the care, maintenance and adjustment of elevator brakes. There is no more important part of the elevator machine than the brake, yet very little has been written about it up to this time. Power is, therefore, very fortunate to be able to present Mr. Myers' admirable treatise on the subject of elevator brakes to its readers.—Editor.

By considering an ordinary full magnetic car-switch, drum-type elevator, a comprehensive idea of electric elevators in general may be obtained, after which other types, all of which have similar operating characteristics, may be discussed and easily understood. Starting with the machinery on the bedplate, which may be assumed to be located in the basement, the essential parts are: Motor, controller, coupling, brake, worm reduction gear, thrust bearings, winding drum, automatic machine limits (stops), slack-cable device, car- and drum-counterweight cables, vibrating sheave for counterweight cables, overhead sheaves, car and counterweight guide rails, car guide shoes, car safety plank and accessories, car counterweight cables, drum and car counterweights and safety overspeed governor. The motor must be capable of starting in either direction without sparking under heavy loads. Modern practice to a large degree is employing the commutating-pole type of direct-current motor, on account of its fixed neutral point and sparkless commutation under wide variations of load. The commutating-pole motor is admirably adapted for the two-speed elevator, since a wide range of speed can be obtained by means of shunt-field regulation. In actual practice this range is from about 400 to 500 r.p.m. for the slow speed to about 800 r.p.m. for the high. A compound winding on the motor is also desirable in order that the inrush at starting, which is usually 50 per cent greater than the full-load current, may be utilized to build up the field strength rapidly. This aids in giving the motor a high starting torque and a smooth start, at the same time reacting to hold the starting current to a minimum value. Plain shunt motors in elevator service have been known to "overshoot" their normal speed, by reason of the comparative slowness with which their field strength built up from shunt excitation only and also because of the effects of armature reaction.

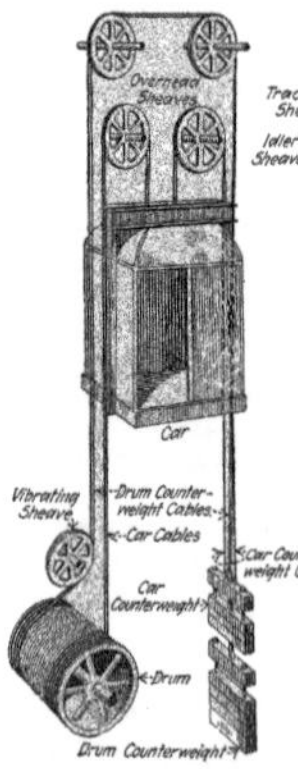

FIG. 1. LAYOUT OF DRUM-TYPE ELEVATOR

FIG. 2. LAYOUT OF TRACTION-TYPE ELEVATOR

Provisions must be made in the control to short-circuit the series winding in the motor after acceleration is completed. With the car loaded on the down motion above a certain value, the motor is converted into a generator; under this condition current flows through the series coils in opposition to that in the shunt coils, weakening the field strength sufficiently to overspeed the motor. This overspeeding may be enough to set the car safety. However, there are cases known where the car, being too near the bottom landing for the safety shoes to properly take hold and the speed being too great for the brakes to stop the downward motion, crashed into the pit, causing considerable damage. Similar accidents have occurred on the up motion of heavily over-counterweighted empty cars. Unless the shunt winding is of sufficient strength to excite the fields to a point near saturation, considerable variation of car speeds may result under different loads with the series windings left active. Consequently the motor used for elevator service is an intermittent duty machine, with the series fields wound for starting duty only, in which case burnouts may result from neglect to cut the compound winding out after starting.

Reversing the direction of the motor rotation is accomplished by changing the direction of the current through the armature, leaving the fields to build up always in the same direction. Where commutating-pole motors are used, the commutating-pole winding must, of course, be reversed with the armature.

The size of the motor required for a certain lift at a given speed may be easily figured as follows:

$$Hp = \frac{W \times S \times 2}{33,000}$$

where W is the load to be lifted and S the car speed in feet per minute. The multiplier 2 is used because the net horsepower figured must be doubled to take care

FIG. 3. DRUM-TYPE ELEVATOR MACHINE

of the frictional losses. This high factor is largely accounted for in the worm-gear reduction which, owing to the heavy thrusts, etc., is not very efficient.

Suppose a capacity of 1,500 lb. at 150 ft. per min. is desired. Assuming the counterweights will be proportioned in the customary manner to balance the car weight, plus 33 per cent of the maximum load to be carried, in this case an overbalance of 500 lb., the actual weight to be lifted by the motor will be 1,500 — 500 = 1,000 lb. Then,

$$Hp = \frac{1,000 \times 150 \times 2}{33,000} = 9.1$$

hence a 10-hp. elevator motor will be selected.

For elevators driven by alternating-current motors, the plain squirrel-cage type of induction motor, owing to its inherent simplicity and safety of operation, is to be preferred, where objections of the power companies do not preclude its use. The modern squirrel-cage motor, having a large "slip," and consequently high torque and rapid acceleration, designed especially for elevator and hoist duty, should, in the writer's opinion, overcome many of the power companies' objections, at least up to the 15-hp. sizes. This elevator motor on 60-cycle circuits has a no-load speed of 900 r.p.m. and a full-load speed of about 750 r.p.m. instead of the 850 to 865 r.p.m. of the industrial squirrel-cage motor. Where objections are raised, and for the larger-sized equipments, recourse is had to the slip-ring type of motor with automatic means of short-circuiting the rotor resistance as the motor accelerates. For two-speed operation, increasing use is being made of the induction motor, with a double-stator winding wound for two different numbers of poles.

Turbine-Gear Transmission

What is termed the A Turbo Transmission, designed for shaft and pump drive, is shown by the accompanying illustrations. It is manufactured by the B. F. Sturtevant Co., Boston, Mass. This self-contained unit is compact and does away with turbine bearings and one turbine packing. The turbine and pinion shaft is supported by two bearings in the gear casing, but the turbine proper overhangs the gear casing at one end (see Fig. 1.) Two pinions are carried on the turbine shaft, and these mesh with two large gears carried on the gear shaft, which is coupled to the shaft or pump that is to be driven. Lubrication is confined to the transmission casing, the large gears dipping slightly into the oil. The machine is oiltight and dustproof and can be operated in places where it would be impracticable to run a motor.

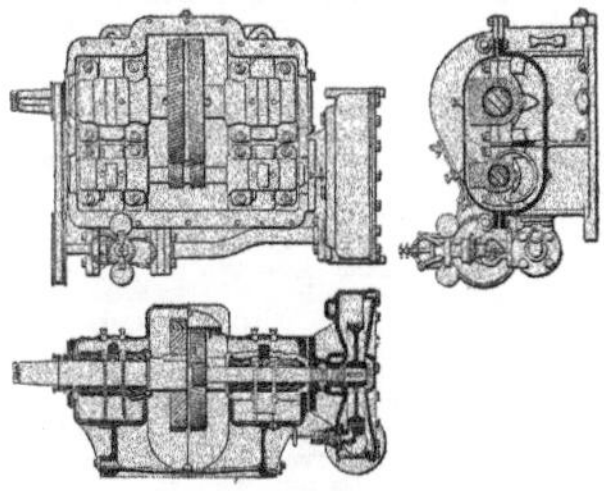

FIG. 1. DETAILS OF THE GEAR TRANSMISSION

FIG. 2. TURBINE AND GEAR TRANSMISSION

This transmission is built in two sizes, 25 and 50 hp. It can be built in larger sizes, however. As shown in Fig. 2, the speed of the turbine is governed by a throttling governor that is driven from the gear shaft.

A Study of Power Development and Suggestions for Future Progress

By L. W. W. MORROW
Assistant Professor of Electrical Engineering, Sheffield Scientific School, Yale University

MODERN industrial civilization depends on four agencies: The transportation system, the factory system, the communication system, and the power system. Its present status is largely due to the development of the factory system brought about by the introduction of the transportation and communication systems, all of them depending on the power system.

The existing communication system is more adequate than the other systems as regards its ability to expand, and yet the wonderful advances in communication methods in the last few years seem to indicate that it is still in an embryonic stage of development.

The factory system has reached an advanced stage of development as regards production methods, and further progress can be made only through changes in other industrial conditions. Production depends on labor, transportation and power as regards factory location, and the last two elements are primary. This has resulted in the congested industrial districts.

The war proved the inadequacy of the transportation system to care for abnormal industrial expansion, not only because of missing mileage and rolling stock, but also because of inadequate methods and the coincidence of a transportation peak with an industrial peak. The transportation of fuel for power purposes necessitates the movement of more than 300,000 tons of coal daily. This, together with the fuel transportation for railroad locomotives, uses a large percentage of rolling stock for fuel for localized power production.

The industrial power system has had a haphazard and wasteful growth. It was incidental to production and transportation in that it, except in the central station, afforded no direct monetary return.

The power system has recently been brought before the country because of three things: (1) A knowledge that the fuel supply must be conserved; (2) a knowledge of the inadequacy of the transportation system to meet an industrial peak load; (3) a realization that the solution of both the workingman's and the capitalist's problem can be gained most readily by securing more output per worker; that is, greater use of power.

More power can best be obtained by the establishment of a comprehensive power system with unified nation-wide control for supplying power at low cost for all national purposes. This does not mean the scrapping of existing power-producing agencies, but it means the engineering grouping and expansion of such power sources as are existing, the addition of new power sources in proper locations for interconnected power production and the scrapping of small, inefficient plants not essential or economical for connection to the comprehensive system. Such an idea is being accomplished very rapidly at the present time simply through economic competitive conditions.

Adequate and cheap power on a comprehensive scale would mean: Better living conditions, as industries could be located in nonurban territory and the present tenement districts would be eliminated and the homes of the people would be made more comfortable through the use of power appliances in the home, whether city or country; better social and financial conditions, since the output per worker would be increased with a consequent wage increase. At the same time the price to consumers would decrease, as the cost of power is very small as compared to the cost of labor in manufacturing plants. Fuel would be conserved, because large power stations use fuel more efficiently than small ones. Also the fuel used for transporting a great part of the coal supply of the country would be saved, and undeveloped large and small water powers could be utilized by such a network power supply.

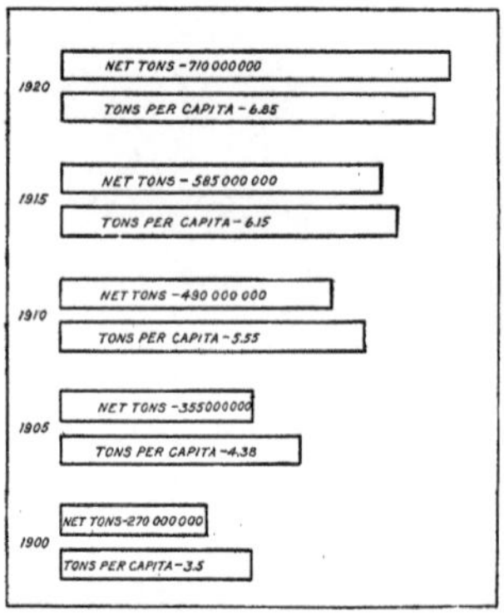

FIG. 1. GROWTH IN ANNUAL COAL CONSUMPTION IN THE UNITED STATES

Transportation congestion would be relieved through the expansion and relocation of congested industrial districts, through location of industries at raw material sources and by the diminished haulage of fuel. The factory system would be improved and developed. Cheap and reliable power would mean less cost per unit of output and more wages for the employee, less cost to the consumer and more return to the capitalist, thus increasing the inducement for industrial development.

The accompanying curves have been prepared to show present tendencies in power production and fuel consumption. Fig. 1 shows the net tons of coal used and the tons per capita. The ratio of the tons per capita to the net tons is decreasing, owing to more efficient fuel utilization. In 1870, per capita, there was 1 ton of coal used; presumably, this represented the amount needed for heating purposes. In 1919, per capita, there was about 6.5 tons of coal used. The excess over the 1 ton needed for home comfort represents the change in conditions brought about by modern use of power in industrial work. The United States Geological Survey states that if the fuel consumption

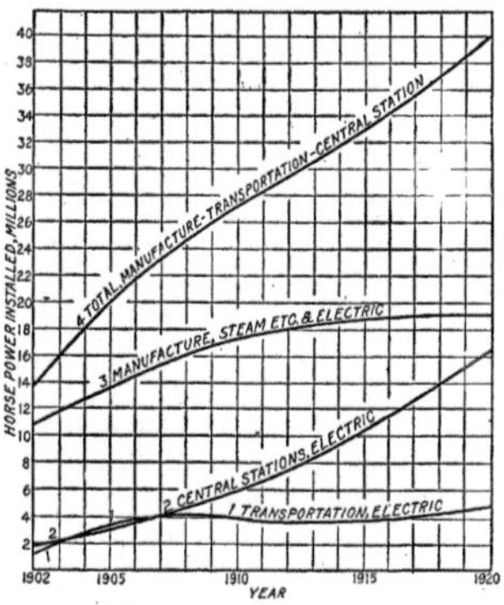

FIG. 2. GROWTH OF INSTALLED HORSEPOWER IN MANUFACTURE, TRANSPORTATION AND CENTRAL STATIONS

increases 7 per cent yearly, the life of the coal supply of the United States will be less than one hundred years. The effect of such an exhaustion on our industrial existence can be contemplated only with anxiety, particularly when it is understood that the United States has half the world's coal supply. Every effort, therefore, from a conservation of fuel standpoint must be made to obtain more power per unit of fuel.

Fig. 2 shows the growth of installed horsepower in the United States as obtained from the Census and current technical journals. Curve 1 shows the horsepower installed for railway and electrified trunk-line purposes. The increase during the last three or four years is due to the great increase in the electrified trunk-line mileage. Curve 2 shows the increase in the horsepower installed in central stations; it is growing at the rate of 1,200,000 hp. yearly, while the total horsepower installed in manufacturing plants, central stations and railway plants (curve 4) shows an increase of only 1,300,000 hp. yearly, thus indicating the trend of power development toward central stations. Curve 3 shows the installed horsepower in manufacturing plants aside from rented power from central stations. Until 1910 factories put in their own power plants, but since that time the greater number purchase their power from central stations, as indicated by the curves.

Fig. 3 shows the history of the installed horsepower used for manufacture in the United States. Curve 1 shows the remarkable growth in the use of internal-combustion engines as sources of power for manufacturing purposes. Curve 2 shows that the direct use of hydraulic power in manufacturing is practically constant. In New England there has been a decrease in the use of water power in factories, due chiefly to the small capacity of the older developments and their inability to meet the demands of factory expansion.

Curve 3 shows the total electric horsepower used in manufacture and is the sum of the dotted curves labeled "Generated Electric" and "Central-Station Electric" respectively. The total electric power is increasing at the rate of 1,200,000 hp. yearly as compared to a rate of only 600,000 hp. yearly for all horsepower used in manufacture. Of this total electric power the increase of purchased power is indicated by the dotted curve labeled "Central-Station Electric." The curve labeled "Generated Electric"

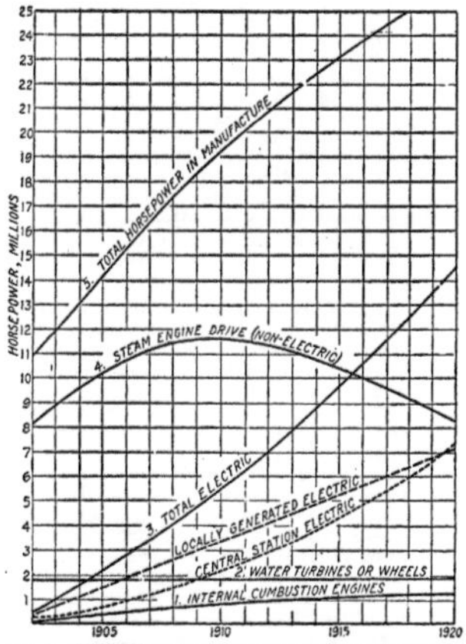

FIG. 3. GROWTH OF HORSEPOWER USED IN MANUFACTURE IN THE UNITED STATES

represents the electric power produced in the manufacturing plants. Curve 4 represents the horsepower installed in manufacturing plants other than the hydraulic, internal-combustion engine or the generated electric. This curve plainly indicates the decrease in the use of steam-engine drive for manufacturing plants. The figure shows clearly the tendency toward electric power in manufacturing, brought about solely by competitive and production conditions. Electricity, as such, is getting to be an important element in many manufacturing processes, and this fact aids the movement toward an electrified industry.

The figure shows that central-station electric power in manufacture is rapidly replacing all other types and that the economic solution of the power supply for manufacture is obtained by using such power.

Fig. 4 shows the trend of central-station development. A comparison of installed horsepower curves on Figs. 3 and 4 shows that where the total power installed in manufacturing plants is increasing at the

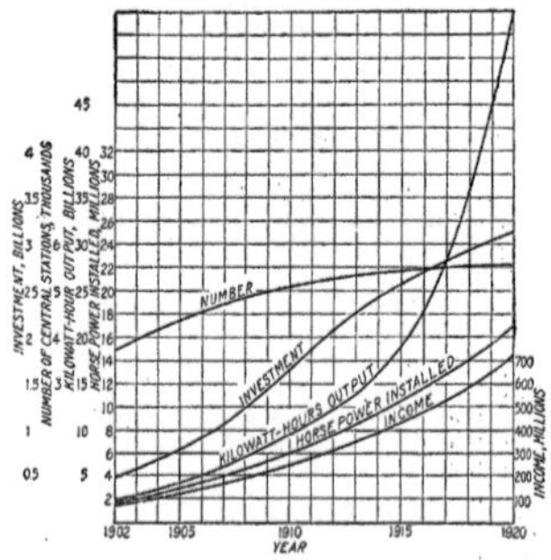

FIG. 4. GROWTH OF CENTRAL STATIONS IN THE UNITED STATES

rate of 600,000 hp. yearly, the power installed in central stations is increasing at the rate of 1,400,000 hp. yearly. The increase in the central-station power can be largely accounted for in the increased amount used in manufacture and in transportation.

The number of central stations is increasing slowly as is the investment, showing the tendency toward concentration now occurring in central-station development. The most remarkable curve on Figure 4 is that showing the kilowatt-hour output. The remarkable yearly increase shown by this curve as compared to the increase shown by the installed horsepower curve shows vividly the effect of wartime economy in operation as well as the effect of increased industrial load on the load factor of central stations. A comparison of the kilowatt-hour curve with the income curve shows that more power means cheaper power. Income increases twice in about 6.5 years, while the kilowatt-hour output increases 26 times in the same period. The kilowatt-hour output is increasing at the rate of six billions per year as compared to the 1,400,000 installed horsepower increase yearly.

If this six billion kilowatt-hour yearly increase is to be cared for by small central stations operating on a basis of 6 lb. of coal per kilowatt-hour, it would require an increased yearly coal demand of eighteen million tons. If the same demand is met by a large central station installation the yearly coal increase on a two-pound basis would be only six million tons. Thus a considerable saving in fuel is possible by utilizing large and modern stations as compared to small, inefficient stations.

The figures show clearly the trend in power development toward a national and comprehensive power supply. This trend is also evidenced by statistics of developments in widely separated parts of the country.

In New England during the last three years one company has discontinued 121 small and inefficient steam stations by interconnection and regrouping of existing power sources in their territory. Nearly every New England State now has its power systems interconnected, and there is a general tendency toward still further consolidation. In Alabama a network power system with more than 1,000 miles of power lines and 150,000 kva. capacity is in successful operation. At Niagara Falls over 500,000 kva. is connected and used for power purposes in a large territory. In the Northern Illinois district more than 500,000 kva. is in one system. On the Pacific slope immense power systems are connected together to form a network from Oregon to southern California.

The technical problems associated with a comprehensive national power system are largely solved as regards apparatus. The unsolved problems lie in the direction of control and protection to secure reliable service under all conditions and do not give evidence of great magnitude. Social and political conditions are the greatest handicaps, and the utility managers will be called upon for closer co-operation in such a scheme than has been obtained in the past. Government regulation of public utilities will have to change its character before any such project can be financed. Regulation is necessary, but it should stimulate, not retard, development.

The country must be awakened to the resulting benefits of an adequate power supply. The present antagonism of the people toward all public utilities must be removed through unfailing efforts and a truthful educational campaign. Such a scheme is everyone's business yet no one's business, as it is on too large a scale for individual enterprise. Despite this fact the figures presented show clearly the tendency in the proper direction, and every endeavor should be made to expedite the change simply from a social and conservation standpoint.

Cooling Coils for Water-Cooled Transformers

The Electrical-Apparatus Committee of the National Electric Light Association enumerates the advantages and disadvantages of copper and iron cooling coils for transformers as follows:

Copper cooling coils have the advantage of non-corrosibility, the disadvantage of high first cost and are limited to a pressure of about 250 lb. per sq.in. maximum. In the iron cooling coils corrosion will take place where the water is unsuitable, principally due to excess air in the water, unless precautions are taken, and these may not always be successful in eliminating the trouble. The advantage in the use of the iron coils lies in lower first cost and the ability to stand a maximum pressure of approximately 1,000 lb. per square inch.

The subcommittee recommends the use of copper cooling coils wherever the water conditions are unknown, always bearing in mind the limits of the copper coils where high heads of water have to be considered. Iron coils of less than one inch internal diameter should not be employed.

Twelve-Inch High Pressure Underground Steam Main

IN 1917, Day & Zimmerman, engineers, put in operation the new Front Street Station of the Erie Lighting Co., in which is installed 20,000 kw. in turbines. It soon became apparent that much greater economy could be effected by combining this new station with the company's older Peach Street Station on account of the higher efficiency of the new plant. Several plans for accomplishing this were considered, and it was finally

FIG. 1. SIDES OF MULTI-CELL CONDUIT

FIG. 2. TWO LAYERS OF COVERING

decided to connect the 8,500-kw. turbines in the old plant with the new boiler plant of 5,076 hp., normally operated at 300 per cent rating, by what is virtually an underground steam header. The installation of this high-pressure underground main has recently been completed by the American District Steam Co.

Ground was broken in October, 1919, and construction began on the specifications for a 12-in. line, 1,600 ft. long, to carry a maximum operating pressure of 250 lb. 225 deg. superheat, and to deliver 250,000 lb. of steam per hour.

The main is installed along the lake front, part of it being below the lake level. A large amount of mud and water was encountered during excavation, making tight sheathing necessary. To relieve the water pressure from the exterior of the completed conduit, a subsoil of coarse stone and gravel, 18 in. in depth, drained by tile and sewer lines to a sump, was placed in the bottom of the trench, and upon this foundation the concrete base of the multicell conduit, Fig. 1, was constructed. The sump is automatically pumped out as water accumulates.

The conduit is constructed of a hard-burned hollow tile resting on a concrete base. The tile were specially manufactured for this purpose and are laid in cement mortar. To prevent water from entering the conduit a membranous waterproofing is used, so applied that the conduit is thoroughly sealed by two layers of waterproofing felt and one layer of burlap hot-mopped into place with pitch. To prevent the weight of the conduit from puncturing the membrane the concrete base is made in two sections, an upper and a lower, and the membrane passed between the two. The portion of the concrete below the lake level is surrounded by a concrete envelope five inches thick, the primary purpose of which is to protect the waterproofing membrane.

Nonpareil covering is used. This is applied in blocks, two layers thick, held in place with wire, all crevices being filled with cement, Fig. 2. Waterproofing of the covering is accomplished by applying two layers of high-grade roofing, each layer mopped in place with high-temperature asphalt varnish, all joints being broken. As an additional insulation the interior of the conduit above the center line of the steam main, also the voids in the multicell tile, are packed with dry "nonpareil."

The steam pipe is extra-heavy, strictly wrought-iron line pipe, and the joints are made by electric-arc welding, Fig. 3. There are no thread connections in the main, and the number of flanged joints is reduced to a minimum, such joints being necessary only at the expansion joints and elbows. All flanges are extra-heavy cast steel.

Expansion of the main pipe line is taken care of by the use of thirteen extra-heavy cast-steel duplex-sleeve expansion joints, Fig. 4, and built to take care of a maximum of eight inches of traverse. The expansion joints are anchored to the concrete side walls of the manholes by specially designed steel anchor wings. The feature of these patented joints is that the slip, or moving element, consists of an inner and an outer sleeve of forged steel treated with a rustproof process and highly polished, with $\frac{1}{8}$ in. air space between the two sleeves. The object of this construction of the slip

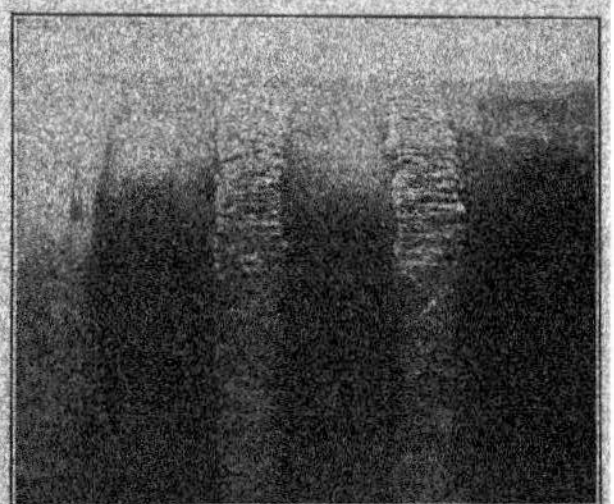

FIG. 3. ELECTRIC-ARC WELDED JOINTS

is to prevent the packing coming in contact with the high temperature (approximately 700 deg. F.) to which the inner sleeve is subjected. With a temperature reduction of approximately 40 per cent on the outer side of the outer slip, the life of the packing is indefinitely increased. Flanges are eliminated from the movable end of the expansion joint, the inner sleeve being welded directly to the pipe line.

All fittings and expansion joints are installed in large concrete manholes with reinforced-concrete tops, waterproofed to the surface in the same manner as the conduit. Special watertight manhole curbs are used, having inner and outer covers, the inner covers being thoroughly packed, thus preventing surface water from entering the manholes.

It has been the experience of the engineers that long runs of pipe under the expansion variations of high

FIG. 4. DUPLEX-SLEEVE EXPANSION JOINT

temperatures "weave" in travel, thus distorting alignment and creating packing and gasket troubles. To eliminate this condition, there are placed at intervals averaging 30 ft. alignment guides, Fig. 5. These guides consist of semi-steel shells bolted to the concrete base by means of anchor bolts. Attached to the pipe and moving freely through the shell of the guide, with proper clearance, are semi-steel clamps supporting the pipe concentrically in the guide by means of four points or lugs.

At ten-foot intervals saddle plates, rollers and chairs are placed beneath the main as additional support and assistance in the free movement of the pipe, due to expansion and contraction. Nine angles are in the line made by the use of bent pipe and extra-heavy cast-steel sweep elbows, with large anchor bases especially designed for the work, Fig. 6. These bases are anchored in concrete foundations, and all anchors are thoroughly backed up by large concrete abutments.

The company has had this underground steam line in operation for eight or ten weeks, and has closed down the boiler plant at the old station and done away with the services of twenty-two men. Inasmuch as the new boiler plant is more efficient than the old one, the saving due to the operation of this underground line will amount to several thousand dollars per month.

New York Lighting Companies Use 3,453,408 Tons of Coal in a Year

More than three million tons of coal and coke was used in the manufacture of gas and electricity by the various gas and electric lighting companies in New York City in 1919, according to reports filed with the Public Service Commission. Of this tonnage the gas companies used 1,645,619 short tons of coal and coke, and the electric lighting companies used 1,806,789 short tons of anthracite and bituminous, a total of 3,453,408 tons.

In addition to the consumption of coal and coke the gas companies in Manhattan and The Bronx used 14,254,558 gal. of water-gas tar and 129,233,034 gal. of gas oil in the making of gas, while the companies in the other boroughs of the city used 12,305,157 gal. of water-gas tar and 88,785,797 gal. of gas oil, a total of 26,559,715 gal. of water-gas tar and 218,018,831 gal. of gas oil.

Three of the gas companies made 378,438 tons of coke during the year, of which they sold 119,479 tons for $669,660.03, an average of $5.59 per ton.

The reports show that the various electric lighting companies consumed the following tonnages:

	Anthracite	Coal	Bituminous
New York & Queens Electric Light & Power Co.	6,441	....	1,770
New York Edison Co.	.....	948,651	......
United Electric Light & Power Co.	.....	334,553	......
Flatbush Gas Co.	21,242	....	1,657
Brooklyn Edison Co	16,428	....	427,618
Queens Borough Gas & Electric Co.	.....	....	15,154
Richmond Light & Railroad Co.	33,275	....	......
Totals	77,386	1,283,204	446,199

FIG. 5. ALIGNMENT GUIDE IN POSITION

FIG. 6. EXPANSION JOINT AND PIPE BEND

A Real Steam Man—Edward F. Miller

A Teacher of Heat Engineering Who Has Never Lost Close Personal Touch with the Industries

WHEN the United States actively entered the war with Germany, thousands of men, big and little, left their jobs and entered the service of the Government, both in civilian and in military lines. Others carried on their regular work and shouldered heavy extra burdens in the service of the country. The Government needed the former class of men, but it also needed badly the latter class—men who could do their share toward keeping the industries and business of the country running and who were big enough to give, in addition, helpful service toward winning the war. Among this latter class is Edward F. Miller, Professor of Steam Engineering and Head of the Department of Mechanical Engineering at the Massachusets Institute of Technology.

Ever since his graduation from the Institute in 1886, Professor Miller has been connected with its teaching staff, where, after serving as Instructor, he was given the title of Assistant Professor of Steam Engineering at the age of 26. In 1911 he was put in charge of the Department of Mechanical Engineering, under which is included all instruction in mechanical engineering, applied mechanics and mechanic arts, or as the graduates of fifteen years or more ago remember it, "shop-work."

One of his characteristics has always been the ability to do a little more work. In addition to his duties as a teacher, Professor Miller has found time to do a great deal of extra work in consulting engineering. He has conducted a large number of boiler and plant tests, both for research purposes and as acceptance tests. He has served as expert in many accident cases, especially in connection with boiler explosions. He has been called as an expert witness in numerous lawsuits involving engineering questions. It has been said that no cross-examining lawyer ever succeeded in mixing up his testimony on the witness stand. He has also done a large amount of research work in the engineering laboratories at the Institute, having made a number of investigations on the capacity and action of safety valves, involving the handling of steam, air and ammonia. He has made exhaustive experiments on the flow of ammonia in pipes. In short, his activities have covered every part of the broad field usually included under the term heat engineering, and his experience has been such as to rank him as an expert in any part of that field.

A great many people think of a college professor, even an engineering professor, as an impractical theorist who lives in the clouds. Professor Miller's long experience with the practical problems of engineering, not only in the laboratory but actually in the industries, has removed all traces of the impractical dreamer, if ever there were such traces in his make-up. He meets his students as a very human man, not as a theorist from another world. When he is explaining a point of theory from the lecture platform, his own experience can always furnish one or more actual examples to illustrate it.

Probably the biggest physical monument to Professor Miller's work is the group of engineering laboratories in the new "Technology" buildings in Cambridge. As far back as 1912 drawings and studies of the "New Technology" began to be made. From then on much of his time was devoted to the laboratories, and after the actual construction of the new buildings began he almost "lived on the job." The largest of these laboratories is that devoted to steam, air and hydraulics. In addition are the laboratories for gas engines, for testing materials, for mechanic arts, for textile machinery and for power-transmission machinery. The present success of these laboratories and any success they may have in the future must be credited very largely to Professor Miller's personal efforts.

"Technology" occupied its new buildings in the fall of 1916, and the school year of 1916-17 was full of the problems of readjustment and the completion of unfinished details of construction and installation. In the spring of 1917 came the entrance of the United States into the war. The Institute authorities at once placed their entire resources of equipment and personnel at the service of the Government. Various schools were established, some completely under Army or Navy control and some handled partly by the regular staff. Some of the drawing rooms and class rooms were used as dormitories for uniformed details receiving special training. Parts of the laboratories were partitioned off and devoted to secret research and instruction work. The regular schedule of work was completely rearranged on an intensive basis to hasten the graduation of engineering students, that they might be available for needed work at the earliest possible moment.

As the head of the Department of Mechanical Engineering a large proportion of this work fell on Professor Miller's shoulders, but the demands on him did not end there. He was continually called on for advice and help by the War and Navy Departments. He spent a great deal of time in the design of experimental fighting machinery for the Ordnance Department. Much of this work was done in secret, and most of it will probably never be known beyond a small circle of army officers, so that the full value of his service cannot be appreciated by the public in general.

When every effort was being made to build a "bridge of boats to Pershing," it was realized that boats without men to run them would be helpless. It was important that these men, and especially those in authority, should be loyal Americans, and the number of skilled marine engineers and navigators was woefully small. The United States Shipping Board therefore established schools for the intensive training of men for these positions. Again Professor Miller was called on, and he was made Chief Instructor of all the Shipping Board Schools for the training of engineering officers. This meant responsibility for the organization of all of these schools throughout the country, immediate responsibility for the local one at "Technology" and even personally giving a number of the lectures. So successful were these schools under his leadership that, although the emergency has passed, the school at Boston is still maintained to help in furnishing engineers for the growing American Merchant Marine.

The war over, he settled down to the problems of reorganizing his department at "Technology" to meet the unprecedented rush of students to the engineering schools. This was complicated by the tangle left from the numberless makeshifts necessary for the war-time emergencies. This work was still under way when the sudden death of Dr. Maclauren, President of the Institute, brought new responsibilities. The corporation entrusted the administrative powers of the president's office to a committee of three members of the faculty, and Professor Miller was made a member of this committee.

At the time of his death Dr. Maclauren was just about to undertake the reorganization of all the Institute's work, which was made necessary by the increase in the number of students and the economic conditions that have faced all the colleges. This stupendous task fell on the shoulders of the administrative committee in addition to the departmental work of each of its members. This work is still going on as this is written, but the number of expressions of approval heard by the writer from people connected with the Institute indicates that the committee is fully capable of carrying out this important task as well as the others entrusted to it.

With Prof. C. H. Peabody he is co-author of the first edition of a textbook on Steam Boilers, succeeding editions being his own work. He has also written numerous sets of notes on various subjects for the use of his students. These notes, while not formally published as books, form a valuable collection of engineering literature and are to be found in the libraries of many "Technology" graduates of recent years, as well as in the collections of many of the older men who have been so fortunate as to be able to obtain them.

Professor Miller was born in Somerville, Mass., in 1866 and has always lived near Boston. He is a member of the Engineers' Club and of the University Club of Boston, of the American Society of Mechanical Engineers, the American Society of Civil Engineers and the American Society of Refrigerating Engineers. In connection with the American Society of Mechanical Engineers he has done considerable committee work, especially on the Boiler Code Committee and the Power Test Code Committee, being a member of the subcommittee of the latter on Refrigerating Machinery and Apparatus, as well as of the main committee. He is also a life member of the Massachusetts Charitable Mechanics Association and has been elected an honorary member of the National Association of Stationary Engineers.

Dead-Ending Conductors to Floor Timbers

By Roy J. Turner

A convenient and economical method of dead-ending large slow-burning insulation conductors on the floor timbers of a wood-frame building over the switchboard

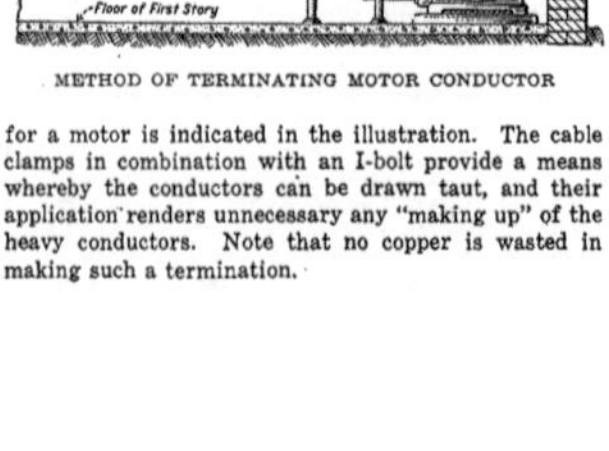

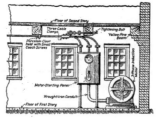

METHOD OF TERMINATING MOTOR CONDUCTOR

for a motor is indicated in the illustration. The cable clamps in combination with an I-bolt provide a means whereby the conductors can be drawn taut, and their application renders unnecessary any "making up" of the heavy conductors. Note that no copper is wasted in making such a termination.

EDITORIALS

Super-Power Project One of Many Angles

WHEN electrical power is to be generated from fuel the most economical method involves many considerations. If it is only a question of producing large quantities of power for industrial and domestic purposes, there is no other way in which it can be done so economically as in large stations centrally located with reference to the load and coal and water supply. In such plants the greatest advantage can be taken of the high efficiency of large turbo-units operated condensing in a high vacuum. Where large volumes of steam are required for heating and manufacturing processes, the most economical operation is obtained in plants from which both power and heat are supplied. Under the latter conditions power may be considered a byproduct. With steam turbines operating condensing about the best over-all thermal efficiency so far attained is twenty per cent. In the plant where the engines act as reducing valves for the heating system, an over-all thermal efficiency of sixty to seventy per cent is possible. However, where the exhaust steam is used for heating only, unless conditions are favorable to operate the plant condensing, during the summer months the thermal efficiency is so low that the gain during the winter months is lost. Therefore, if it could be successfully worked out, the most economical plan would be for the industries to buy their power from the central station during the summer months and generate their own during the season when heat is required.

In a comprehensive project in which all the power requirements for a given district are to be supplied from a super-power system such as proposed by W. S. Murray for the industrial district extending from north of Boston to Washington, D. C., where advantage can be taken of all the loads in this region it should be possible to develop a system that would enjoy the benefits of the high efficiency of a combined heating and power load. Such a project as the proposed super-power system will supply power not only for industrial needs, but also for the various transportation lines. During the winter months when heat is required, approximately thirty-three per cent more power will be needed than in the summer months. This railroad heating demand comes at the same time as the heating load of the industries. Consequently, if the local industries could be supplied with heat and power from the same plant during the winter months, either from an isolated plant in one industry or from a centrally located plant serving a number of industries, a large amount of the central-station capacity could be released to supply the railway heating load. Practically the only additional cost for equipment in the heating plant would be that for the engine and generator, which need be only of sufficient size to supply the required amount of exhaust steam for heating, and can be installed and operated at minimum cost. Since this plant will be tied into the central system, no emergency capacity need be provided. These heating power plants could no doubt, be installed for practically the same cost as additional capacity in the central stations, or even less. In the heating plant the addition of the power-generating equipment is all that will be required, where in a central station complete boiler and generating equipment will have to be installed.

In this super-power system all available water power must be developed for the maximum number of kilowatt-hours economically possible. The low-water periods of our rivers in the Northeastern states generally occur during the winter months. Consequently, considerable reserve steam power-plant capacity will be necessary to take care of the load during this period, which also coincides with the industrial heating load. This reserve steam-station capacity can be obtained by transferring part of the central-station load to the heating power plants in the winter months, thereby reducing the amount of reserve central-station capacity.

Whether the large central stations are installed at the mines or in the large industrial centers, or whether all the power for domestic and industrial purposes is supplied from these stations, coal for heating must be hauled by the railroads for domestic and industrial heating. Since the purpose of this proposed super-power zone is to conserve coal and relieve railroad congestion, both of which are of paramount economic importance to this country at the present time and will be increasingly so in the future, every factor that is conducive to this end should be carefully considered, irrespective of private interest or personal prejudice. Although it is realized that one hundred per cent efficiency cannot be obtained for every element entering into the problem, the equation that will offer the best solution must give the correctly weighted value to each component.

Corrosion Research

WHAT disease and decay are to the human body, corrosion is to the mechanical structure. A scientific, systematic investigation of the causes, manner and prevention of corrosion is of no less moment to the engineer and those who use products of his skill than are the investigations of which it is sought to reduce the number and virulence of the ills that flesh is heir to, and to promote health and longevity.

The Corrosion Committee of the Institute of Metals, of Great Britain, has been at work for ten years with promising prospects of attaining results that will be of wide and beneficent importance. Its activities have been financed first by the Institute of Metals, then by the Institute in conjunction with the makers of condensers and condenser tubes, to the corrosion of which much of its attention has been directed, and the government department of Scientific and Industrial Research. An appeal is now made to the users of condensers as well, and the government agrees to contribute a like amount for every pound contributed by the users.

Surely, the prevention or retardation of the corrosion of condenser tubes is of as much importance to the user of condensers as to any other class. By far and large it is of general effect, for it is an element of cost in marine transportation, in the production of electricity, and of many articles in the manufacture and distribution of which condensers are directly or indirectly used. Furthermore, such an investigation by specialists daily growing wiser and more expert in the subject offers the best mode of attack upon corrosion in general, the conquering of which would be of such vast importance to engineering and everything dependent upon it.

We hope that the appeal of the committee will find a generous response and that its work will be carried to a positive and effective conclusion.

Co-operation Among Engineering Societies

ONE of the features of the recent spring meeting of the American Society of Mechanical Engineers in St. Louis was the joint session with the American Society of Refrigerating Engineers and the American Society of Heating and Ventilating Engineers. This was not the first occasion on which such joint meetings of two or more national engineering societies have been held, but it is worthy of notice as one of the evidences of the growing spirit of co-operation among such national societies.

Students of history remind us that the original engineer was a civil engineer and that his interest was principally confined to hydraulics. As mankind progressed, the field for the engineer broadened and gradually different men devoted their attention to different features of the work. This resulted in the formation of groups whose members were interested in the same problems. These groups grew and in turn divided into smaller groups, each claiming a distinctive name and gradually forming societies and associations among themselves. As a result we have national societies of civil, mechanical, electrical, mining, heating and ventilating, and refrigeration engineers. The list might be extended almost indefinitely.

Each of these societies tries to cover only its own subject because no man can hope to comprehend the whole field of modern engineering. The trouble comes, however, when we attempt to fix the limits of any one branch. As an example, take the mechanical engineer who specializes on steam power plant work. He must, of course, understand boilers, engines, etc. If his plant be run condensing, however, he must arrange for handling large quantities of cooling water, which involves hydraulics, a civil engineering subject. If he installs generators, he invades the electrical engineer's province. The treatment of feed water to avoid scale and corrosion and the combustion of fuel are chemical problems. Try as he may to be a specialist, the engineer must keep in touch with the other branches of engineering if he is to solve his own problems.

The object of each of the great engineering societies is the helpful interchange of ideas and experience so that each individual may have the benefit of the knowledge of the other members. In the same way each engineering society can benefit by the interchange of ideas with the other societies. Modern civilization is built on co-operation. In addition most of its vital problems are engineering problems. Therefore co-operation among engineers would seem to be one of the essentials for the progress of our civilization. Evidences of such co-operation as the meeting referred to, as well as the other co-operative activities of the various societies, show that engineers realize their responsibilities and are trying to live up to them.

Be Careful

THE faster you go the greater the risk. Ordinarily, it is safer to walk than run, safer to run than ride a horse, safer to ride a horse than an automobile or railway car, safer these than the airplane. To play safe, greater care must be exercised the faster you go. This is true of traveling and of industrial operations.

We might relax vigilance if the world would slow down a bit, but as civilization has attained a goodly speed and accelerates as time goes on, the individual cannot slow down just so he may be less careful.

You cannot safely sit in the middle of Fifth Avenue—unless you are going with the current.

The public utilities of New York and Pennsylvania have set aside this week as "No Accident Week." Their employees and the public are appealed to to be careful, to avoid heedlessness and negligence, which not only endangers the safety of the person guilty, but which menaces the safety of others and jeopardizes property

The appeal to guard life or health or personal safety gets an immediate response. The appeal to safeguard property against preventable damage or ruin needs particular emphasis, especially at this time. There are many factors that contribute to a serious situation arising in such cases. The freight tie-up, the shortage of labor, the unavailability of materials, the high cost of parts or whole machines—all these mean that quick replacement is practically impossible, service may be down and many hands made idle.

Power-plant employees as a class certainly deserve praise for the excellence of their training and the observance of caution. Few other branches of human activity have a higher percentage of careful, long-headed, common-sense men.

Every week is no accident week in the power plant.

An analysis of sixty of the large electrical power systems in the United States shows that forty-nine generate sixty cycle, six generate twenty-five cycle, four generate sixty and twenty-five cycle and one generates fifty cycle alternating current. There are approximately four and three-quarter million kilovolt-amperes of sixty-cycle, one and one quarter million kilovolt-amperes of twenty-five cycle, and one hundred and sixty thousand kilovolt-amperes of fifty-cycle generating capacity installed on these systems. If we are going to standardize on a frequency, why not sixty cycles especially, since this system predominates to such a large degree, and is as well, if not better suited to general power requirements than any other?

Figures based on reports obtained from over three thousand central stations give an average of three pounds of coal per kilowatt-hour. The most economical plants are operating on about one and one-half pounds. There is evidently an opportunity for improvement in the operation of the average central station as well as the isolated plant.

Distance of Boiler from Grates and Bridge-Wall Height

The letters regarding the distance of a boiler from the grates, also the height of the bridge wall, are of particular interest to me, but I do not think that we can come to any definite conclusion regarding the matter that will be correct in all cases, as it is possible to find some old boilers in operation that are set low, and with a bridge wall almost touching the shell, and that show very good efficiency. Others are not giving even fair results, which would tend to indicate that there are things about a boiler other than the combustion chamber, that affect fuel economy. I have proved, at least to my own satisfaction, that other conditions have to do with the best height of bridge wall and boiler setting.

A number of years ago I had charge of a plant consisting of six old return-tubular boilers. These were set low over the fire, and the bridge wall was so close to the boiler (about four inches) that the fireman, when cleaning, could not push a hoe between it and the boiler at the center. We had an old Corliss engine and burned something less than 2 lb. of coal per indicated horsepower per hour with only 80 lb. gage pressure at the throttle.

Finally, we thought it advisable to renew our boiler plant, and the manager of the works employed some very technical engineers, who came and looked us over and laughed at our poor old boilers and their settings. As a result of much talk, many pictures and blueprints —and I would not leave out the wise expressions—we installed four 84-in. boilers, set 3½ ft. above the grate, and with the bridge wall 20 in. below the boiler. With 100 lb. pressure the best we could do was 2¼ lb. of coal per indicated horsepower per hour.

One of the technical fellows put up the same argument that was advanced by one of the writers in the April 13 issue of *Power*; that is, that the reason we burned more coal was that the bridge wall no longer acted as a damper. Poor comfort, but the fact remained that we did burn more coal, and the theory as to why was of considerably less interest when it came to paying the coal bill.

Now, I am not taking the stand that low boiler settings are the best, as I have had experience that would seem to prove the opposite, but I think the draft, the amount the fires must be forced and the quality of coal used, as well as other conditions, have much to do with the most efficient height for bridge walls and boiler settings.

LEIGHTON JOHNSON.

Exeter, N. H.

United States Geological Survey Power Statistics

In writing regarding the power data of the United States Geological Survey, page 735 of the May 4 issue of *Power*, Mr. Misostow evidently was under a misapprehension as to certain facts with reference to these figures. He apparently believes that a considerable amount of fuel reported in these figures is used for purposes other than power production and undertakes to explain the large difference in efficiency between certain states and the less populous states, such as North Dakota and Montana, on the ground that much of the fuel is used for central heating plants or ice manufacturing.

In asking for reports from the public utility companies, the Geological Survey specifically requests as follows: "When fuel is used, report only the fuel used for generating electricity for public-utility purposes." From this it is evident that the Geological Survey recognizes the possibilities of other uses of power by the companies reporting and has taken precautions to eliminate this as fully as possible. As a matter of fact, it is doubtful whether more than 10 per cent of the fuel used in any state by these companies reporting is consumed for such auxiliary uses as are mentioned in Mr. Misostow's comment. It seems very unlikely, therefore, that anything more than a negligible percentage of error could be introduced from this cause, even if all the companies reporting made the mistake of including in their reports all fuel instead of only that fuel used for generating current. There is no doubt that some of the exhaust steam is used for auxiliary purposes, but under the circumstances no large differences are probably caused by this fact.

The only measure of electric power-plant efficiency which is commonly significant is the kilowatts generated per ton of coal used, or the equivalent reciprocal, pounds of coal used per kilowatt-hour generated. If one wishes to express the efficiency in percentage, he might, of course, convert the kilowatt-hours into the equivalent B.t.u. and also figure the B.t.u. in the coal used and thus get a measure of the percentage of the original heat found in the energy sent out. This would be a significant figure, but rather an unusual form of expression of power-plant efficiency. To add to any such percentage efficiency any additional thermal recovery in the form of exhaust-steam heating or other auxiliary service rendered is, of course, entirely proper. Indeed it is desirable in the analysis of a single plant in contrast with others, but under the conditions of work done by the Geological Survey it is impracticable to take account of this supplementary activity, which is probably a small percentage in any event.

The fact which has been brought out by Mr. Misostow that the larger plants undoubtedly operate with higher efficiency, is of course correct. This probably accounts for practically all the difference between the conditions in the different states. However, it does seem worth while to consider if it is not possible to bring up the average result in some of the smaller states, where the load factor of central stations is low, to a somewhat higher basis of efficiency than now prevails.

The advantage of central stations of large size is, of course, well recognized; in fact, this greater efficiency of large stations is the main advantage that will probably follow from any co-ordination in super-power stations that may be achieved from the various projects now under consideration in various parts of the country.

Washington, D. C. R. S. McBride.

Stokers and Fuel Oil

I am inclined to disagree with Mr. Daniels' views as expressed on page 684 of the April 27 issue of *Power*, and although I am neither an oil nor a stoker enthusiast, I can see where, in certain localities and with some classes of work, an oil-fired boiler is more desirable and efficient than a stoker-fired boiler.

In the first place his chart comparing the prices of coal and oil to give the same evaporation per dollar does not represent the true situation. Something else is neccessary to be considered, such as capital, labor and fixed charges, not to mention several other items of minor importance. Ignore capital and fixed charges, if you wish, and labor becomes a deciding factor. Figures and facts from stations operating with fuel oil show that if the labor charges in an oil-fired plant were as high as in a stokered station, fuel oil would be used only where coal is a luxury.

A 30-day test on the west coast shows a combined efficiency of 76 per cent or better with oil-fired boilers. I have not found such efficiency in any coal station up to the present. Why? In any equation the result is a function of the variables entering into it. With a seven-retort underfeed stoker there are at least seventeen variables: Seven upper and seven lower rams and coal, ash and air. The problem of the stoker operator is to change these variables to constants. This is nearly impossible as every operating man knows.

In an oil-fired installation, several variables are encountered: Temperature, moisture content, mechanical condition of the burner and air control. Temperature at which an oil atomizes best can be determined and remains fixed within a narrow range. Moisture is a factor of the condition as received and handled on its way to the burner, and can be controlled. The condition of the burner in operation is a matter of observation and can be detected very quickly. Air control is simply a question of using gas-analysis instruments. All these except one, moisture content, become constants at the hands of an experienced operator. Therefore, when the variables in stoker firing can be made constants, then and not until then can an equal efficiency be expected from a stoker-fired boiler as is being obtained in oil-fired plants today.

How many stoker boiler plants operate with one fireman and one watertender per 5,000 boiler horsepower? Under ideal conditions this might be possible, but not very probable. Many central stations are the result of several expansions from smaller stations. We must take conditions as we find them, not as they might be, and a labor analysis will show that most stations have four or five men per 5,000 boiler horsepower and could not operate economically with less.

Coal can be stored in a continuous pile, it is true, but even with the best of care spontaneous combustion and deterioration frequently take place, making it an expensive procedure and a constant source of loss. Fuel oil, on the other hand, does not deteriorate, and can be stored in tanks beneath the ground surface, requiring no expensive machinery in handling, with a capital outlay very little greater than that needed for coal storage of equal capacity. H. A. Fox.

Brooklyn, N. Y.

Lubricating Devices Neglected

While visiting a large number of power plants, I have been impressed with the lack of attention given to lubricating devices, particularly those on machinery outside the engine room. Usually, the sight-feed lubricator or oil pump on the main engine will be in good condition, with a clean sight-feed glass, tight packing, etc.; and the same applies, in general, to the oil cups or other means used for lubricating the bearings of the larger units in the plant. The reverse, however, is generally true of these devices when placed on pumps and other auxiliaries.

Of all the boiler-feed pumps I have ever examined, I do not believe one-third of them were equipped with lubricators that were in proper working order. The sight feed and gage glasses have almost always been filled with an oily residue resembling tar, and in many cases the wooden handles of the regulating valves have been broken or missing. If such a lubricator feeds any oil at all to the machine to which it is attached, it is only a matter of luck. There is no possible way of determining the quantity with any approach to accuracy, and scored cylinders, rods and packing are certain to result. Force-feed oilers on pumps are perhaps more often in working order, but even so, an obstructed sight glass frequently makes it impossible to tell the rate of feed.

Many makers place small glass oilers on the rocker bearings of large steam pumps, such as those used in elevator and fire service. In most cases that have come under my observation these oilers are allowed to become clogged with grease and dirt till they are useless. There is something particularly trying to see a large duplex pump pounding away with a full set of oil pumps and a lubricator, all with wide open feeds and yet not feeding one drop of oil to the protesting bearings. To my way of thinking it is often better to remove the oil cups, clean out the oil holes and lubricate at regular intervals by means of the good old squirt can.

Many apparently well-informed engineers seem to consider that rocker bearings, as opposed to revolving bearings, need but little lubrication. It is true that they do not need a great deal of oil, but they do need some, and they need it at regular intervals. I think it is nearly always safe to assume that the maker of a piece of machinery, whether it be a 2,000-hp. engine or a small steam pump, knows what parts need occasional oiling, and when he installs an oil cup or lubricator on the outfit it is the best plan to use it. Hugh G. Boutell

Washington, D. C.

Operating Diesel Engines

In the operation of Diesel engines I advise against overload as I find 90 per cent of the rated capacity of the general run of engines to be the maximum load that should be carried, even for a short period, and with some engines less than that. Most machines will do a little more than they are designed to do without injury, but there are exceptions to all rules and the internal-combustion engine is the exception.

Never let the cooling water leaving the engine get above 160 deg. F. I prefer to use a large amount of water and let it enter the engine at about 110 deg. F. and leave at about 140 deg., rather than to use a smaller amount of water at 70 deg. entering and 160 deg. leaving.

The cooler an engine is run the more fuel it uses per horsepower, but it amounts to very little. Running hot, with the water leaving at 200 deg. and above, will soon ruin the engine or seriously damage it. It is far better to spend a few cents more each day for fuel than to spend hundreds of dollars yearly for repairs.

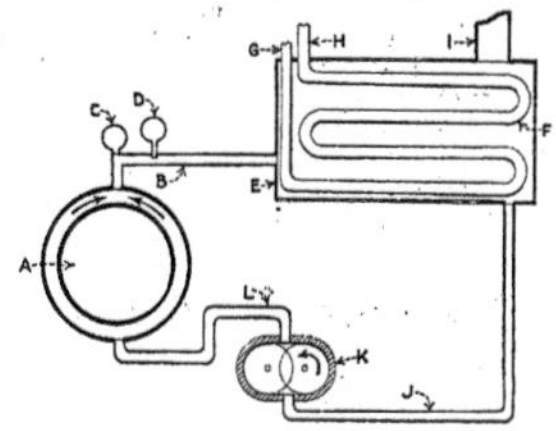

CYLINDER-COOLING WATER SUPPLY SYSTEM

Attention should be given to the kind of cooling water used. Most domestic water contains too much scale-forming matter to be used for cooling Diesels. A $\frac{1}{16}$ in. thickness of scale has cracked many a cylinder head. I believe scale causes more damage to the oil engine than anything else.

I have found only one successful method of dealing with scale, which automatically allows the engine to cool off slowly and there is no danger of starting up, forgetting to turn on the cooling water, as has often been done, with serious results.

Referring to the illustration, *A* represents a single-cylinder engine; *B* is a cooling water line from the cylinder jacket and head; *C* is a thermostat that regulates the temperature of the water leaving the jacket by regulating the amount of cooling water going to the cooling coil *F*; it is also arranged to ring an electric bell in case this water should reach an unsafely high temperature. The recording thermometer *D* is to keep check on the cooling-water temperature.

The pure-water storage tank *E* contains enough cooling coils to keep the pure water at the required temperature; *F* is a cooling coil within the storage tank through which the cooling water passes, and *G* is the cooling-water supply line which is controlled by the thermostat *C*. *H* is a cooling-water line to the sewer or elsewhere, if so desired. *I* is a spout leading from the building roof drain and is to keep the storage tank *E* supplied with rainwater, and *J* is a suction line from the storage tank to the cooling-water circulating pump *K*, which should be of the double-gear rotary type, because it is simple and dependable. This pump should be extra large so as to run at a slow speed. *L* is a discharge line from the pump to the water jacket.

It can be seen that no scale trouble and cracked heads can result when using this system, if the tank *E* is kept supplied with rain or distilled water or any water that is free from scale-forming matter.

The cooling water necessary for the coils *F* will be a little more than if it was used directly on the engine jackets, but the differences in the cost is a slight matter compared with the trouble from scale that ordinary water would deposit in the cooling jackets and cylinder heads. The tank *E* should be closed to keep out dirt and to prevent the water from absorbing gases of the air, as well as to prevent the evaporation of the pure water. The line *I* should have a gooseneck water trap to prevent evaporation. The tank *E* should be extra large and should contain sufficient coils to cool the pure water to the required temperature. The rotary pump is driven by a silent chain from the engine camshaft.

This system, which is patented, works equally well on any type of engine, but it is preferable to have the circulating pump driven from the engine in some dependable manner, so that there will be no forgetting to turn on the cooling water. No valves should be put in any of the cooling-water lines to or from the engine; then there are no valves to forget. This system will do away with all scale troubles and allow the engine to automatically cool down gradually, as even after the pump stops some water will continue to circulate, owing to slipping of the pump and to the heating of the water in the jackets, which causes it to become lighter than the water in the tank. As the line from the jackets to the tank enters near the bottom of the tank, the water level should always be above this line.

If the coils *F* should scale up inside rapidly, a cooler of the atmospheric pipe type should be used instead of the coils in the tank. This cooler should be placed in the line *B* and the raw cooling water run over the piping, thus having the scale on the outside of the pipe.

The initial charge of water for the storage tank should be rain or distilled water, but if neither of these is convenient, the charge can be of raw water if such raw water contains less than four grains salt per gallon, because the little solid matter contained in one charge of raw water will make very little scale. Distilled water or raw water with less than four grains of salt can be obtained from most any ice plant, and after the first charge and if care is taken, the rains, as a rule, will keep the tank at all times supplied with fresh pure water.

I advise the use of plenty of lubricating oil in the cylinder and also the use of the same kind of oil on the bearings unless the latter are self-oiled. The bearing oil can be filtered and used over and over again.

Never run an engine overspeed in order to develop more power, because it is the same as overload, as the more power developed in a cylinder the more fuel is burned, and the more oil burned the hotter the cylinders get. Don't get them hot!

To sum it all up, don't overload, don't overspeed, and keep the engine cool and well oiled.

Winkelman, Ore. GEORGE J. TROSPER.

Electric-Arc Versus Gas Welds

Mr. Perry's article, page 604 of the April 13 issue of *Power*, would seem to the casual reader to force the conclusion that electric arc welding is per se better than oxyacetylene, and that oxyacetylene welds are in fact dangerous. I cannot too strongly oppose such a conclusion, because it is not in accord with the facts.

Mr. Perry, like many others, has had unpleasant experiences with oxyacetylene welding; others than Mr. Perry have had equally unpleasant experiences with arc welding. The fact is that in no case will satisfactory results be had unless the metal is thoroughly fused, and lack of fusion is just as liable to occur in one process as in the other, unless the operator is competent and careful. I have critically examined hundreds of welds made by many welders, and the foregoing statement is made as the result of these examinations.

In one series of tests of arc welds the tensile strengths ran from 18,000 to 65,000 lb. per sq.in. The tests were made by unbiased authorities, and an examination of the fractures showed wide variations in the qualities of the welds, which accounted largely for the great differences in the test results. I have seen and have carefully examined welds made by all the manufacturers of arc-welding apparatus. I have seen many operators do their work and have found from the same apparatus, good and bad results. This is true of the machine to which Mr. Perry refers as well as the others. But this should not prejudice anyone against any machine, because it is well known that good work is done with all machines. I have also seen many incompetent oxyacetylene operators and know that plenty of bad oxyacetylene welding is done. But these facts cannot be fairly used as an indictment of the process. Really there are some things that can be better welded by one process than by the other; and I use both processes in my shop, and could not well afford to be without either; so I feel that I cannot be accused of being prejudiced.

Further, I will guarantee to do either of the jobs mentioned by Mr. Perry with entirely satisfactory results by the oxyacetylene process, though it is entirely possible that for various reasons he would choose arc welding if the work were done in his shop.

Mr. Perry makes the sweeping statement in the latter part of his article that arc welding is superior in the case of a number of metals, among which are cast iron, bronze and monel metal. I would be glad to get samples of these welds for examination, because I have been told by many who are competent to judge, that arc welding of metals referred to is not satisfactory.

I know of no job shop, and job shops have to guarantee their work, that would take a chance on electric arc welding a press frame or any other piece in which the full strength of the section has to be developed, though such pieces are welded by the oxyacetylene process in many shops with entire satisfaction. It is necessary, in such a discussion as this, to be sure that the bases of argument are thoroughly understood.

In welding copper alloys, such as brass and bronze, much depends on their composition, and much on the physical structure where the composition is the same. For example, 60-40 brass is much easier to weld and gives better results than 70-30; and gun metal, 90 parts copper and 10 parts tin, welds much more satisfactorily when the delta constituent is absent, as it is in rather large or rather slowly cooled castings. It is also a fact that many brasses and bronzes are unweldable if anything approaching the original structure and composition is desired. I have had many problems of this kind to solve and in most cases have been unable to meet the requirements. For example, repairing gun-metal castings is impossible, if soundness, color and full strength are essential, though many such castings have been welded in my shop, with acceptable results; the color of the welds was different, and the strength of the joint was less than that of the original material.

I recall two large gun-metal castings, weighing over three thousand pounds, and of very complicated design, which were successfully welded as proved by satisfactory results in service. But chemical analyses and microscopic examinations were made before the work was started, the welding material was specially made for the purpose, and the preheating was regulated with a pyrometer.

This subject might be much enlarged on, but I feel that I have said enough to cause the casual reader to be chary about accepting any general statement as to the superiority of one welding process over another.

S. W. MILLER,
Rochester, N. Y. Rochester Welding Works.

Keeping Out of the Rut

In an experience of several years in plants ranging from a few hundred to over twenty thousand horsepower, as assistant and chief, I have found conditions as Mr. Duffy enumerates them in his communication, "Keeping Out of the Rut," page 602, April 13 issue.

This game of being "Chief," whether in name or in fact, is not always the "feet on the desk" job that a casual observer might remark, and a man in the business should thoroughly appreciate this fact. It is an easy and not uncommon thing to criticize the man higher up, and it may look to the assistant who works eight hours and then forgets the plant for the next sixteen, as though his chief were in a rut, when as a matter of fact the chief may be entirely out of it and hitting it up on high.

A little example of this "in the rut" dope came to light in a plant with which I am somewhat conversant. The watch engineers used to get together while changing shifts and talk over "the chief." They thought he was "in a rut" because he insisted on the regular routine of work being carried out to the letter. He had true and tried ways of doing things, and the wheels never had failed to turn on schedule.

A new addition was contemplated and modern machinery was to be installed, and the watch engineers decided that they must tell the chief (who was a bit old-fashioned) that the new feed pumps should not be the old plunger type, but of the more modern centrifugal type. So they waited on him and presented their ideas. He took out an order book and showed them an order for the same design of pump they had suggested.

My advice would be to the assistant who thinks the "chief" is in the rut, to keep out of the rut himself and run by if possible, but to be very sure that the chief is in the rut before trying a marathon. I have always had an antipathy to criticizing the man higher up, and if he was "in the rut" I was generally far enough behind so that his position did not retard me.

New York City. C. W. PETERS.

INQUIRIES OF GENERAL INTEREST

Steam Laps of D Slide Valve—If the length of a D slide valve is $25\frac{5}{8}$ in. and the distance between the outside edges of the steam ports is 24 in., how much is the steam lap of the valve? P. S. L.

The sum of the steam laps of each end of the valve would be $25\frac{5}{8} - 24 = 1\frac{5}{8}$ in., and when the laps are equal, each lap would be half of $1\frac{5}{8}$, or $\frac{13}{16}$ in.

Advantage of Inclining Hand-Fired Grate—For a hand-fired boiler, what is the advantage of having the rear end of the grate lower than the front end? J. A. A.

The draft is stronger in the rear end of the ashpit, and with even firing a thicker fire becomes built up at the rear end of the grate; and having the grate pitched toward the rear end makes slicing and firing easier.

Cause of Pump Becoming Airbound—What causes a pump to become airbound, and how can the trouble be remedied? S. P.

Under atmospheric conditions water contains about 2 per cent air in solution. When subjected to less pressure the air expands, and when given sufficient time, as while the water is flowing through a long suction pipe, the air becomes liberated from the water, usually in the form of collections of small bubbles that adhere to the pipe or pump surfaces, or the air gathers in pockets of the pump spaces. The trouble can be prevented by increasing the head of the suction supply and by running the pump slower, so the pump cylinder will be filled with less reduction of pressure of the suction water.

Pressure Values Given in Steam Tables—Why are not steam tables arranged to show temperatures and total heat of steam for each pound of gage pressure instead of absolute pressures? W. D. P.

The temperature and heat per pound (weight) of dry saturated steam depends on the absolute pressure, while gage pressure is the pressure above atmospheric pressure, which is variable with the elevation of the place and also varies with changes of atmospheric conditions. Therefore steam tables based on gage pressure would need to be adapted to a particular pressure of the atmosphere, and greater convenience in their use would be confined to cases where the atmospheric pressure was the same as that for which the tables were particularly constructed.

Pressure at Discharge Coupling of Sinking Pump—What is the pressure on the pump couplings of a 3-in. rubber-lined hose, delivering 100 gal. of water per minute from a mine sinking pump against a head of 275 ft? S. A. M.

Water pressure due to a head of 275 ft. would be equivalent to $275 \times 0.433 = 119$ lb. per sq.in., and for raising the water the pressure discharged into the lower end of the discharge line will be that pressure plus the pressure required for overcoming friction plus an excess pressure due to pulsation of the pump. The pressure required for overcoming friction to discharge through hose depends on the kind and roughness of the lining, freedom from kinks or short bends, the length of the hose and the rate of discharge. For smoothest surface of rubber lining and straight hose, discharging at the rate of 100 gal. per min., the frictional resistance will amount to only about one pound per sq.in. per 100-ft. length of hose. The pressure due to pump pulsations depends on the number of piston reversals per minute and the smoothness of operation of the pump. The pressure effect of such impacts cannot be computed, but under ordinary conditions would not exceed 20 lb. per sq.in. With delivery at the rate of 100 gal. per min. through 300 ft. of straight 3-in. smooth-lined rubber discharge hose, the maximum pressure at the pump coupling would be about 142 lb. per sq.in., but with a kinked or rough-lined hose there would be a rapid increase of frictional resistance and of total pressure required to effect a stated rate of discharge.

Revolutions of Pump—What is meant by the number of revolutions per minute of a direct steam pump? W. R.

A revolution signifies a complete cycle of the reciprocating parts, and the number of revolutions per minute would be the number of repetitions of movement of any moving part, as one of the crossheads of a duplex pump. The term is more specific for designating the number of piston reversals than "strokes," for in a single pump the term stroke is sometimes confused with double stroke, and in a duplex pump it is frequently doubtful as to what is meant by the number of strokes per minute. With the type of pump and number of revolutions given there is no ambiguity as to the number of single strokes.

Size of Synchronous Motor—In our plant the present maximum load amounts to about 325 kw. At this load the power factor is around 0.72. We are putting in some additional equipment to be driven from a lineshaft, and from what can be determined from some of our other drives, 78 hp. will be required to drive the new installations. I would like to know the size of synchronous motor to install to drive the load and also raise the power factor to 0.90. M.A.B.

Synchronous motors usually are rated in kilovolt-amperes, therefore in this problem the motor will be rated in kva. The mechanical load on the motor will be 78 hp. or $(78 \times 746) \div 1{,}000 = 58$ kilowatts. This makes the total kilowatt load equal to $325 + 58 = 383$. The kilovolt-ampere load equals kilowatts divided by power factor; therefore, in the first case kva. $= 325 \div 0.72 = 452$, and in the second, kva. $= 383 \div 0.90 = 426$. Wattless component, $WC = \sqrt{\text{kva.}^2 - \text{kw.}^2}$; then, with the present load $WC = \sqrt{452^2 - 325^2} = 314$, and with the synchronous motor added $WC' = \sqrt{426^2 - 383^2} = 186$ kva. The difference between WC and WC', or $314 - 186 = 128$ kva., represents the leading wattless component that must be supplied by the synchronous motor, in addition to doing 58 kw. of mechanical work. Then the total motor capacity required is equal to the square root of the sum of the squares of the two loads, or $\sqrt{128^2 + 58^2} = 140$ kva., and the power factor of the motor equals kilowatts divided by kilovolt-amperes $= 58 \div 140 = 0.41$. It will be seen from the foregoing that by installing a 140-kva. synchronous motor and operating it at 0.41 power factor leading, it will do 78 hp. of mechanical work and raise the power factor of the total load to 0.90, and the kilovolt-ampere load on the generators will be reduced from 452 to 426 kilovolt-amperes; besides, under the new condition the voltage regulation will be improved.

Utility Men Discuss Pressing Problems at N.E.L.A. Pasadena Convention

A POWER demand far exceeding the available supply; how to meet this demand under existing conditions of the money market; shortage of materials, fuel and labor; ability to maintain service with greatly overtaxed equipment; and the necessity for flexible utility regulation to compensate for such conditions, were the main subjects discussed by the twenty-two hundred delegates to the forty-third convention of the National Electric Light Association at Pasadena, Cal. The power shortage is not local to any section, but is nation wide, although the condition has been rendered more acute in California by an unusually dry season and a gradual depletion of the oil fields, resulting in soaring prices for fuel oil, with a consequent turning to hydro-electric power.

President Ballard Opens the Convention

In opening the convention, President R. H. Ballard estimated that there is needed $750,000,000 of new capital to take care of the expanded requirements of the country; and the ability to raise this sum depends on the earning power of the investment, favorable opinion of the public at large, and the extent to which the utilities are forced to compete in the money market. As a substantial aid to financing new developments, he believed that the public should be fairly and fully informed as to the status of the industry and its possibilities for general public service. Moreover, the kind of public ownership that brings results is ownership by the public at large in the stocks and bonds of their local regulated utility companies.

Governor Stevens Welcomes the Delegates

Governor W. D. Stevens, in welcoming the delegates, referred to the fact that California, although richly endowed with water power, has been denied coal fields, and its once vast oil resources are now being drawn upon to the extent that they are rapidly approaching depletion; therefore the future must look to further water-power development to meet the growing needs of agriculture, transportation and, eventually, he hoped, manufacturing. There has already been developed in the state over 900,000 hp., and an additional 180,000 hp. is now under construction; but it would be possible to develop five times this amount. With perhaps one exception, California has the largest per capita use of electricity of any state in the Union, and the present problem is not one of demand but of supply.

Mr. Edgerton's Address

The most constructive address of the convention was delivered by E. O. Edgerton, president of the California Railroad Commission. "The interest of the public is paramount," he said, "but what is good for the public is also good for the utility, and it is to the interest of the public that these utilities be kept sound." He depreciated statements that public-utility securities are in general unsound, and believed that investment can be attracted where there is proper regulation. The asset of a utility is essential service, and the regulating commission holds that asset in trust. He believed that too much time is ordinarily wasted on methods of valuation; they get nowhere and befog the issue. Instead the situation should be faced squarely, starting first with sound capitalization, then fixing rates necessary to induce and support that capitalization, plus necessary reserve. Having thus taken care of the investor, the next and very important duty is to reward efficiency within the organization from the president down to the office boy, by both remuneration and advancement wherever possible. The regulating body should see to this, for it is the greatest factor in securing service. Too often, in the opinion of the speaker, increased wages had been given only to the organized because of threats and the unorganized were overlooked. "There is no better way," he said, "to take men out of the control of demagogue leaders than to insure a reward for efficiency."

M. S. Sloan, president of the Brooklyn Edison Co., in reporting on "The Electrical Resources of the Nation," called attention to the fact that the Government, although periodically making a rather incomplete survey of central-station and railway power, had neglected to make a study of isolated plants and in no case had gone into the subject of transmission capacity, diversity factor, etc., all of which is essential to a thorough study of the question. This matter had been taken up with the Council of National Defense, and it is possible that something may be done.

The Committee on Water-Power Development, of which Franklin T. Griffith is chairman, presented an exhaustive report which reviewed the existing Federal and state laws as applying to the use of water power on public lands and navigable streams, and expressed the opinion that the water-power bill now pending in Congress, would, if passed, do much to promote early development.

Over 68 per cent of the potential maximum horsepower of the United States lies in the Western States and nearly 75 per cent of the total comes under Federal jurisdiction. In the New England states 77 per cent of the potential water power has been developed and little additional development may be expected in this section. Future development will consist in the more efficient utilization of that already developed.

Water power in the West, in the main, has not followed the popular conception of such enterprises, the typical Western development uses no waterfall of any volume; instead, the project is usually located in a mountain canyon where either a reliable minimum flow of the stream may be found the year round or else storage is developed above the point of diversion.

Among the records which typify the magnitude of western development may be mentioned the following:

Hydro-Electric Generation Projects

The largest single discharge turbine in the world: 20,000-hp.—Wise Power Plant, near Auburn, Cal.; Pacific Gas & Electric Co. The most powerful high-head turbine in the world: 23,800-hp.—White-River Plant, near Seattle, Wash.; Puget Sound Power & Light Co. A unit of practically similar size is installed in the Long-Lake Plant, near Spokane, Wash.; Washington Water Power Co. Other large units: Southern California Edison Co. is installing two 22,500-hp., 800-ft. head Francis reaction-type turbines in its Kern River No. 3, while the Great Western Power Co. will install impulse wheels under 1,008-ft. head at its Caribou plant on the Feather River with capacity of 30,000 horsepower.

The highest head plant in America: The San Joaquin Light and Power Corporation plans two units under 2,500-ft. heads in connection with its King-River project, already begun, and a similar head is contemplated by the Southern California Edison Co. in the Big-Creek development. Typical also of western construction is the 17-ft. section tunnel being driven by the San Joaquin Light and Power Corp. to serve its Kerckhoff plant, a world's record in dimensions for hydro-electric tunnels, a steam shovel being used in mucking inside the tunnel.

High-Voltage Transmission Lines

First long-distance transmission: 20 miles at 10,000 volts from San Antonio to San Bernardino, Cal. Longest high-voltage transmission: 87,000 volts and 55,000 volts, 539 miles, from Mono County, Cal., to Yuma, Arizona; Southern Sierras Power Company.

Highest voltage transmission: Present record, 150,000 volts, 240 miles, Big Creek to Los Angeles; Southern California Edison Co. The Great Western Power Co. is now installing a 165,000-volt line from its Caribou plant to San Francisco. The Pacific Gas & Electric Co. has in contemplation the building of a 220,000-volt line from its Pit River plant into San Francisco, while other western companies are planning 220,000-volt lines for the near future.

The power consumed in the various sections of the country is given in the following tabulation:

District	Population	Millions kw.-hr.	Million dollars gross rev.	Kw. hr. per capita	Cost per* kw.-hr.	Revenue per capita
New England	9,448,000	2,398 6	64.47	254	$0 0293	$ 6 82
Atlantic	36,798,000	13,340.5	304 71	363	0 0228	8 24
Central	54,222,000	14.688 1	299 02	271	0.0204	5 52
Western	9,375,000	6,307.5	104.80	673	0.0166	11.17
Total	109,843,000	36,734.7	773.00	334	0.01212	7 05

From this it will be seen that the per capita use of electricity in the Far Western States is 2.21 times that of the remainder of the nation. On the other hand, the average rates for electricity in this section are low—1.66c. per kw.-hr. compared with 2.19c. obtaining elsewhere.

Prime Movers Discussed

Of particular interest to the operating men was the report of the Prime-Movers Committee* a brief résumé of which is here given.

Improvement in Turbine Design

A record of the units that have been installed during the past year indicates that 30,000 kw. 1,800 r.p.m. 60 cycle, and 35,000 kw. 1,500 r.p.m. 25 cycle can be considered for the present as the maximum standard size of single-shaft single-barrel units.

*An abstract from the part of this report pertaining to steam turbines was published in June 1 issue; other parts pertinent to the power field will be abstracted and published in another issue.

The present is a period of adjustment and improvement, and manufacturers are devoting their attention to modification of design features with a view to securing a requisite degree of reliability and operating efficiency.

The increase in both the diameter and speed of rotating elements makes the subject of turbine balancing one of particular interest at present. Dynamic-balancing machines which determine with extreme accuracy the amount of unbalance in a turbine rotor or other apparatus are now in general use.

The alternate stressing to which turbine wheels and blading are subjected brings up the question of fatigue of metals, a subject which is being thoroughly investigated by the Engineering Division of the National-Research Council. The results of these tests are expected to be extremely valuable to both turbine designers and operators.

Investigations made by shipbuilding companies into the causes of pitting and erosion of gears, have disclosed facts which may explain some of the gear troubles experienced by operators of stationary plants.

Structure of Condenser Tubes

The examination of the structure of condenser tubes by microphotography is important, in that it gives data on the relation between the microstructure and the life of condenser tubes.

The maintenance of high vacuum and the possibilities of condensing equipment in relation to possible theoretical vacuums are still receiving close attention by plant operators.

Boilers and Superheaters

The tendency toward large-sized units has continued during the past year. Units of 1,000 to 1,500 hp. and over are quite common in central-station practice.

On boilers made under war and post-war conditions troubles have developed due to faulty shop work, and it is the desire of the committee to sound a note of warning in this respect.

The Uniform Boiler-Law Society reports that nineteen states and a large number of county and municipal bodies have adopted the code to date.

Manufacturers report no radical changes in design features but their statements do indicate a general tendency to turn out boilers of higher-rated capacity, without increasing materially the maximum pressures now in general use.

Trend Toward Steel-Tube Economizers

The trend, as indicated by recent installations, appears to be toward the steel-tube economizer, showing confidence on the part of engineers that corrosion troubles can be eliminated by the exercise of reasonable care.

The economic value of economizer surface and the amount to be installed in any given plant can be determined only by a detailed analysis of operating conditions and a thorough study of complete plant designs. No formula can be devised for obtaining a solution, but it is possible to indicate in a general way the method of attacking the problem.

Grate and Stoker Design

Statements by manufacturers indicate that there have been no radical changes in stoker design, but that efforts have been along the lines of perfecting new devices reported last year, and to a refinement of operating details.

There is a tendency among designing engineers to lay out the stoker and the furnace so it will be readily possible to change over from coal to oil, using one or the other, depending on the fuel-price differential.

Burning Sub-Bituminous Coals

With lignite and sub-bituminous coals, hand-fired grates seldom give boiler ratings over 100 per cent; efficiencies are usually below 60 per cent. Using mechanical stokers, outputs from 200 to 300 per cent are sometimes possible, with efficiencies ranging up to 70 per cent. Combustion rates as high as 35 lb. per hr. per sq.ft. of chain-grate area, and with underfeed stokers over 1,000 lb. per hr. per retort, have been reported.

Growing Interest in Pulverized Fuel

There has been a rapidly growing interest in the development of pulverized-fuel equipment, and a large number of installations of this character have been placed in operation.

The number of manufacturers supplying apparatus for pulverized-fuel installations has also increased. Statements submitted by these companies indicate that their efforts are being directed toward the perfection of the details of equipment for pulverizing, drying and transportation of the fuel.

One of the questions which has been passed over too lightly by both manufacturers and users of this equipment, is that of the ultimate disposal of the ashes which now pass out the stack. If the use of pulverized fuel is extended, municipalities will undoubtedly pass ordinances to eliminate the ash nuisance.

Increasing Demand for Fuel Oil

Figures of oil production and consumption for the past year indicate that the United States is becoming more and more dependent on imports of oil to meet the rapidly increasing demand.

The requirements of oil-burning ships and the improvement in the processes for converting heavy oils into gasoline are responsible for the diminished supply and poor quality of the oil available for general use as fuel.

The tendency of oil supply companies is to fix the price of oil at what the traffic will bear without any particular reference to the cost of production. For this reason it is difficult to see how an adequate supply of fuel oil at reasonable prices can become available for central stations, until imports of oil from Mexico reach such proportions as to occasion a sharp decline in the market price.

Power-Station Auxiliaries

An investigation of total evaporation per kilowatt-hour of gross generation revealed the fact that load factor and other special considerations are given more weight than the strict adherence to the method of maintaining heat balance. In this connection it should also be noted that maximum over-all efficiency rather than individual machine efficiency is the object sought in station operation. To realize best over-all efficiency, it may be necessary to depart from hard and fast rules.

Observation by committee members indicates the need of more liberality in specifying the capacity and performance of auxiliary equipment, as well as the desirability of more frequent testing of steam-driven auxiliaries.

Boiler and Turbine-Room Instruments

The past year showed increasing use and interest in indicating and recording instruments. The wider use of these devices has been justified by the savings realized since their introduction.

Among the more important points brought out by the committee's investigation are the need of a reliable and reasonably cheap pyrometer, the adoption by some companies of meters on boiler blow-down lines, and the substitution of self-starting synchronous motors in place of clock mechanisms for instrument drive.

Higher Steam Pressure

Practically all equipment manufacturers agree that the problems encountered in the use of high-pressure steam can be solved, and it is suggested that a trial installation might be made by equipping an existing station with a comparatively small high-pressure boiler and turbine, the latter so designed as to discharge into the main steam header of the station. Such a method would show up troubles with valves and piping and would furnish a great deal of data at comparatively small expense.

Improvements in Oil Engines

Improvements in Diesel-engine design have resulted in an engine far better than those built a few years ago. Small engines are principally four-cycle multi-cylinder high-speed units, while in the large sizes the trend is toward two-cycle units. Fuel consumption is not capable of much further reduction, but engines are being adapted to use lower grades of fuel. The necessity for high-grade attendance is being recognized more than ever before by owners.

Development of Hydraulic Prime Movers

The large number of hydro-electric units installed throughout the United States, with high over-all efficiency and almost perfect continuity factor, and the development of a new type of high-speed runner are the outstanding features for the year.

Automatic hydro-electric plants are receiving favorable consideration, and there is a tendency to confine the power-station building to housing only the generating and control equipment.

Where climatic conditions are favorable there is an even more radical tendency to have the generating equipment outdoors, housing only the control apparatus.

Manufacturers' statements indicate a trend toward larger units and toward the consolidation of small water-power projects into single stations of large capacity.

Hydraulic Stations and Steam Plants

The combined operation of hydro-electric and steam plants presents many operating problems which should be worked out for the best interests of both parties, or to secure minimum over-all cost. Unnecessary steam generation should be eliminated as far as possible to secure the maximum hydraulic output consistent with minimum steam cost, including maintenance. Increased costs of construction and of fuel and increase in load make it particularly important to secure the maximum useful work out of every pound of water that can be used for power generation at the hydraulic plant.

In discussing the report, E. E. Gilbert of the General Electric Co. expressed the opinion that in future designs it would not be necessary to sacrifice economy for reliability. E. H. Sniffin, of the Westinghouse company, agreed with the committee's statement as to limitations in the size of single-shaft, single-barrel units. He was inclined to disagree, however, with the opinion that the oil temperature should be held to 140 deg. and believed that 160 deg. would be better practice; if 140 deg. were held to, the oil should be of lower viscosity. Referring to turbine piping, he believed that too little attention is given to insure flexibility, as the front end of the turbine works in all directions and the piping should allow for this.

F. T. Webster, of the Allis-Chalmers Co., said that his company had been conservative as to turbine sizes.

Considerable discussion developed on hydro-electric operating practice in the West, in which it was brought out that conditions of load in that section do not necessitate holding so strictly to frequency as in the East where manufacturing loads are supplied and speed is an important factor in production. For instance, the Southern California Edison, which normally operates on fifty cycles, has recently dropped to 47½ cycles to conserve water in certain cases where two plants are in parallel. According to H. A. Barre, of this company, its supply is divided into three classes: Base water power, reservoir plants, and steam plants. Under normal conditions both the base and reservoir plants carry the base load and the steam plants the peaks. During the present dry season, however, this practice has been reversed, but the economy has not been as great, owing largely to the fact that it has been necessary to keep in service many of the older and less efficient steam units.

S. J. Lisberger, of the Pacific Gas and Electric Co., stated that one of the greatest difficulties encountered when running steam plants in parallel with water-power plants is to hold the steam unit on the line when a short-circuit occurs on the system. To avoid this, they had set the governors to 12 per cent over-speed in spite of protests from the manufacturers. With their oil-burning plants and steam turbines floating on the line, it was possible to pick up a load from 1,000 to 20,000 kw. almost momentarily without even a perceptible flicker in the lights.

Fuel-Oil Discussion

Dr. J. S. Jacobus, advisory engineer, Babcock & Wilcox Co., stated that long experience had shown steam atomizing of fuel oil to give good results up to about 200 per cent of the boiler rating, but above this mechanical atomization is best because combustion is more rapid, giving the available combustion volume of the furnace better opportunity to function at greatest advantage. Mechanical burners naturally require more draft to give the higher capacities, and for ratings above 200 per cent a forced blast at the burner is used. The Babcock & Wilcox Co. is now delivering boilers 26 tubes high of 150,000-lb. steam per hour capacity; the high boiler, the Doctor said, is showing higher efficiency than wide, low types.

Cast-iron economizers should not exceed 75 per cent of the boiler surface. Wrought steel is now extensively used for economizers. The cause of interior corrosion in economizers is usually the air or oxygen present in the feed water. To preserve the exterior surface against corrosion keep the feed-water temperature above 120 deg. F.

C. R. Weymouth, chief engineer, Charles C. Moore & Co., San Francisco, stated that boiler and furnace maintenance is less with oil than with coal as usually fired. He cautioned against preheating the oil too much, a practice which he intimated was becoming general. Steam atomization is the usual practice on the Pacific Coast, though some mechanical atomization plants will soon be in operation. Self-cleaning oil strainers form an important part of the equipment, for if the oil contains silt, even in small quantities, the burners are rapidly eroded. Mr. Weymouth laid emphasis upon the economy of automatic as against hand control in oil burning, pointing out that steam consumption for atomization was but 1 to 1½ per cent with the former, and 2 to 3 per cent in well-conducted plants for the latter.

In view of the volume of work in connection with the preparation of the Prime Movers Report and of the increased attention to water-power operation, it was decided to divorce water power from the Prime-Movers Committee and put it in the hands of a separate committee for the ensuing year.

A very important suggestion was made by Chairman N. A. Carle, and adopted by the section, to have the Prime-Movers Committee prepare an operating code for central stations. Moreover, in view of the reconstruction period and the reorganization of the association the personnel of the committee and subcommittees of the Technical and Hydro-electric Section agreed to serve for another year.

New Officers of Association

The officers for following year are: President, Martin J. Insull, vice-president, Middle West Utilities Co., Chicago, Ill.; vice-presidents, M. R. Bump, chief engineer, Henry L. Doherty & Co., New York; F. W. Smith, vice-president and general manager, United Electric Light and Power Co., New York; H. W. Johnson, vice-president, Philadelphia Electric Co., Philadelphia, Pa.; Franklin T. Griffith, president, Portland Railway, Light and Power Co., Portland, Ore.; treasurer, H. C. Abell, vice-president and engineer, American Light and Traction Co., New York City.

Entertainment Features

California's welcome to the N. E. L. A. and its open-hearted hospitality left nothing to be desired. Everyone felt that the time spent in the land of sunshine was all too short. Automobiles were always at hand to answer every wish to explore the surrounding country. No state has more to display in attractive scenery reached by excellent roads.

At San Bernardino delegates were greeted by a live reception committee of young men and women. Their baskets of oranges were soon exhausted. At Riverside autos were ready to take delegates through the orange grooves and to the hills overlooking the surrounding country. Pasadena, not to be outdone, welcomed the Red Special with a Cowboy Band, a stage coach and riders. A hold-up by the Indians and the rescue of Smith, Sloan and Johnson by cowboys were enjoyable features of the trip to the Huntington Hotel.

Every hour of the day and night was filled with entertainment features. The formal opening of the convention came Tuesday evening with the President's reception and ball, followed by an electrical display and special dances on the lawn of the Hotel Huntington. Golf courses were visited by large numbers of delegates. Wednesday afternoon a special automobile trip was made to Los Angeles, where delegates and friends were the guests of the Universal City. A special motion picture was put on to show the great part which electric lighting plays in this work. Before leaving on this trip the guests were assembled on the Huntington lawn and a moving picture was taken, which was shown at a convention meeting Thursday evening.

Another feature in moving pictures was "That Fairy in the Snow-Flake," prepared by the Westinghouse company, and telling the story of water-power development.

An entertainment feature that appealed especially to delegates from the East was the Mission Play at San Gabriel. Special cars were provided to take all who wished to see this famous play and in this way hear the story of California told on the very ground where one of the missions still stands.

Trips to power plants, parties to well-known points, excursions to beaches, mountain resorts, fruit-packing houses, kept delegates so busy and preoccupied that the convention week passed all too quickly.

The La Fiesta Electrica closed the entertainment features in a blaze of glory.

Sundry Civil Bill

In commenting on the super-power item in the Sundry Civil Bill, Dr. George Otis Smith, the director of the Geological Survey, who will have charge of the work, said:

In these days of economizing in government appropriations it is refreshing to note some of the remarks on the floor of the House by Representative Good, of Iowa, chairman of the Sundry Civil Appropriations Committee, and Representative Byrns, of Tennessee, ranking minority member of that committee, in which they urged additions to appropriations. Their arguments were in defense of an item of $125,000 providing for an engineering investigation by the United States Geological Survey of the super-power project for the eastern United States.

Mr. Byrns stated that "this proposition is one that looks forward to the conservation of our resources and, as has been stated, the time is at hand when something must be done looking to the conservation of our fuel supply because those in authority state that at present the known supply of oil will be exhausted within a very few years at the present rate of consumption." He further characterized this Geological Survey investigation as one that should be made "by Government experts in order that if the investigation discloses that such a plan is practicable, those who are asked to make those investments will have confidence in the accuracy and impartiality of the report."

Chairman Good in reporting the Sundry Civil Bill had already made special reference to the super-power item in the bill as unique, but, as he believed, vitally important, and he stated that such a survey would represent "Government initiative and co-operation which will result in the savings to the country of hundreds of millions of dollars annually. It will result in a great saving in the direct cost of fuel. It will furnish a reserve source of power for transportation and utility companies which will be of large value in time of labor disputes and public emergencies. The principle can be applied broadly. Its benefits will accrue to towns and villages and to the farms of the country." Chairman Good also stated that this provision best illustrated the policy of including in the appropriation bill items providing for the future. He said: "Government cannot stand still. It must advance. It must provide for healthy growth of every useful governmental activity."

In concluding the debate on this item, which was followed by a favorable vote, Chairman Good remarked: "We may smile at this proposition. We may laugh it out of Congress just as we did by ridicule the proposition of Mr. Langley in regard to the airplane."

To those who are interested in scientific and engineering investigations under Government auspices such expressions by leaders in Congress are encouraging. It is also worthy of note that neither Mr. Good nor Mr. Byrns represents sections of the country that would primarily and immediately be affected by the proposed investigation. They seem to represent the country as a whole.

Russell H. Ballard, Past President, N. E. L. A.

Russell H. Ballard, president of the National Electric Light Association during the past year, was superseded by Martin Insull, former vice president of the association, at the election held in Pasadena, California, Friday, May 21. Mr. Ballard was born in Hamilton, Ontario, in 1875.

RUSSELL H. BALLARD

When fifteen years of age, he obtained his first position with the Thomson-Houston Electric Co., Chicago, Ill. He went to Schenectady in 1894 when the Thomson-Houston Co. consolidated with the General Electrical Co. He remained there until 1897, when he became associated with the West Side Lighting Co., now the Edison Co., at Los Angeles, as bookkeeper. Later, he served as cashier and office manager of the Butte (Montana) Electric and Power Co. He became auditor of the predecessor of the Southern California Edison Co. and was soon made first vice president and general manager and soon became a leading figure in Pacific Coast utility work.

Mr. Ballard was the first president of the Pacific Coast Section of the N. E. L. A., where he figured prominently in the activities of the organization. During the past year, as president of the organization, much of his time has been taken up with the comprehensive plan of reorganization which was brought to the attention of the delegates at the convention.

Martin Insull, President, N. E. L. A.

Martin Insull, vice president of the Midwest Utilities Co., Chicago, and during the past year first vice president of the National Electric Light Association, was elected president at the convention held in Pasadena, California, on Friday, May 21. Mr. Insull will replace Russell H. Ballard as president.

Martin J. Insull was born in Reading, England, and received his early education in that country. He was engaged in the manufacturing end of the telephone industry before coming to this country, in 1887.

Almost immediately he entered the shops of the Edison Machine Works at Schenectady, N. Y., then a very small concern, comparatively speaking, and was employed there until he entered Cornell University in 1889.

MARTIN INSULL

Being graduated in 1893, Mr. Insull became a member of the firm of Sargent and Lundy, prominent Chicago engineers, continuing that connection until 1898, when he entered the electrical manufacturing business as vice-president and general manager of the General Incandescent Arc Light Co. in New York City. In 1903 that company was acquired by the General Electric Co., and a few months later, in 1904, Mr. Insull was appointed a vice-president of the Stanley G. I. Electric Co. and removed from Pittsfield, Mass., where he had been sent following the acquisition of the General Incandescent Arc Light Co. by the General Electric Co., to Chicago. There he had charge of the Western business of the Stanley G. I. Electric Co. until 1906, when he entered the field of public service operation.

During the next six years Mr. Insull had charge of and operated electric light, gas and traction companies in southern Indiana. Upon the absorption of these companies by the Middle West Utilities Co. of Chicago, in 1912, Mr. Insull was made senior vice-president, in charge of operations, of the Middle West Utilities Co. in Chicago, where he is now located. The Middle West Utilities Co. is a holding company controlling thirty subsidiary companies which serve 485 communities in fifteen states. These companies have a gross income of more than $20,000,000 annually, and include electric light and power companies, gas companies and traction companies.

Mr. Insull has been a member of and active in the work of the N. E. L. A., and for several years has been a vice-president of the association. He was active in the reorganization program of the association and the broadening of the scope of service to be given by the association to member-companies.

A. I. E. E. to Hold Annual Convention

The thirty-sixth annual convention of the American Institute of Electrical Engineers will be held at the Greenbrier Hotel, White Sulphur Springs, W. Va., June 29 to July 2.

A comprehensive collection of papers will be presented at the seven technical sessions, covering subjects of particular interest in many phases of present-day practice.

Ample opportunity has been provided for recreation and entertainment. The committee has arranged so that the afternoons will be left open for the many entertainment features, including dancing, teas, card parties, golf and tennis tournaments, baseball and other special events.

The convention will be opened Tuesday morning, June 29, with the annual presidential address by President Calvert Townley, followed by the introduction of the president-elect. The remainder of the session will be devoted to the presentation and discussion of the annual reports of the technical committees.

The tentative program follows: Wednesday, June 30, Electrical Machinery Session and also a miscellaneous session; Thursday, July 1, Protective Devices, Electric Welding and Power Factor Sessions; Friday, July 2, Excitation Session.

Unless something is done immediately to relieve the car situation in the soft-coal regions of the country, the American public will find itself without coal for next winter, according to representatives of the United Mine Workers of America, who have come to Washington to seek a remedy. John Moore, president of the Ohio miners, sounded this warning. Mr. Moore pointed out that railroads are assigning a full supply of coal cars to those mines with which the railroads have contracts for coal and refusing to supply cars to the thousands of other mines so that the needs of the public might be filled. "Railroads consume approximately one-third of the total soft-coal output," he said, "and the public consumes the other two-thirds. But at present the railroads, especially in Ohio, are taking practically all of the coal that is being produced and leaving none for the public. Tens of thousands of coal miners are out of employment because the railroads refuse to supply cars to the mines on an equitable distributive basis."

Mr. Moore suggested as one method of relief that inasmuch as one-third of the coal produced is used by the railroads and two-thirds by the public, there should be an equitable distribution of coal among all mines and that the first carload be sent to the railroads and the next two carloads to the public.

The electric furnace in actual use has reached the temperature of 3,500 deg. C. Recent experiments, have, however, developed a furnace which gives a temperature of 4,500 C., enough to volatilize diamonds. A comparison of these temperatures with that of the sun, which is estimated at 5,000 C., gives a striking idea of what can be accomplished in handling refractory substances with electric heat.—*Sibley Journal.*

Spring Meeting A. S. M. E. at St. Louis

IN THE engineering history of St. Louis the week of May 24 was notable in that the city was the place of meeting of three of the large national engineering societies. The American Society of Mechanical Engineers, the American Society of Refrigerating Engineers and the American Society of Heating and Ventilating Engineers each came to St. Louis to hold the spring convention. Conforming with the co-operative tendency of the times, there were joint sessions, the entertainment and trips of inspection to industrial plants were in common, and the headquarters of all were at the Statler Hotel. The engineers of St. Louis are to be congratulated upon the success attending their efforts. Plans were made and executed upon a magnificent scale for the needs, the entertinment and the comfort of their guests. It was a profitable week, full of activity and full to the brim of Southern hospitality of the distinctive St. Louis brand.

First Session Monday Afternoon

Coming from Keokuk, where they had spent Sunday in reviewing the great low-head hydraulic plant of the Mississippi River Power Co., the mechanical engineers arrived in St. Louis Monday morning to hold a four-day meeting May 24-27. The total attendance was in the vicinity of 800. An afternoon session was held upon the day of arrival. Business was disposed of as quickly as possible. The tellers of amendments to the constitution reported the votes upon the specific changes recently submitted to members, practically unanimous in favor of each issue. The revised report of the Boiler Code Committee on Boilers for Locomotives was presented, and there was offered a tentative code of ethics intended for all engineers in common. It was recommended that the code be referred to the Engineering Council or such other body as may be formed, but that the efforts of the society committee be continued. An amendment was offered by President Miller to simplify amendment of the constitution, and there was considerable discussion, initiated by the Philadelphia section on admitting juniors to the privileges of voting and holding office, the object being to interest and hold the young engineer. A motion to authorize the council to contribute from the society's funds $5,000 to advance the cause of the Department of Public Works Association was lost, as this was conceived to be a case for individual subscription.

Appraisal and Valuation Session

Following was a session on appraisal and valuation, which was continued on Tuesday afternoon. The five papers listed and considerable written discussion were presented: "Rational Valuation—A Comparative Study," by J. R. Bibbins; "Price Levels in Relation to Value," by Cecil F. Elmes; "Data on the Cost of Organizing and Financing a Public-Utility Project," by the late F. B. H. Paine (contributed by Dean M. E. Cooley); "The Construction Period," by H. C. Anderson; "Appraisal and Valuation Methods," by David H. Ray. The first four papers had been introduced at the December meeting, but it was felt that the importance of the subject in these days of post-war readjustments, and the low value of the dollar, warranted continued consideration. The paper first named was a comparative study of valuation practice illustrated by a complete economic analysis of a large utility since the Civil War, yielding in substance the true history of the five-cent fare. The second paper traced the history of price levels and the purchasing power of money for several centuries back. The third and fourth papers analyzed various factors involved in the generally misunderstood term "overhead," relating in particular to the construction period, and the fifth discussed official methods of varying exactness.

The discussion of the papers pointed to the four winds. As there was no consensus of opinion as to "what constitutes a complete rational evaluation," a resolution was passed requesting the president to appoint a committee to digest all material on this subject presented in the past, with additional papers on fundamentals if necessary, so that the entire subject might be placed before the society again for discussion.

Local Session Tuesday Morning

On Tuesday morning there was a local session to which St. Louis engineers contributed. There were papers on "The Housing Problem in St. Louis," by Nelson Cunliff; "Industrial Housing—A Financial Problem," by Leslie H. Allen; and two papers on river transportation, by W. S. Mitchell and E. W. Schadek. The discussion centered on rail *vs.* water transportation. It was recognized that each had a place, notwithstanding the fact that water transportation had nearly disappeared from American rivers. Discrimination in railway rates, prejudices and politics had been responsible for the present situation. There was need for an unbiased body such as a Department of National Public Works, to consider the transportation problem from an economic or engineering standpoint. The progress made toward securing such a department was reviewed. Upon a resolution offered by M. Y. Tait the society went on record as fully endorsing the efforts of the National Public Works Association and urged the local sections to aid and support the work.

"Tight Fitting Threads for Bolts and Nuts" was the title of a paper by Chester B. Lord. Although generally the thread forms now in use are quite satisfactory, there is still demand for "a thread that will not loosen." The author pointed out the defects in present practice and the requirements of a male and female thread of the same lead and pitch diameter that may be made after repeated loosenings to fit right without the aid of a locking device.

To combustion engineers a paper on "Burning Eastern Coals Successfully on a Conveyor-Feed Type of Stoker," by Lloyd R. Stowe, was of interest as it emphasized the chemical side of combustion and introduced a new thought to those who had always considered the question from the mechanical standpoint. The stoker is of the traveling type made up of narrow chain-grate elements alternating with stationary tuyeres through which forced draft is applied. The stoker has an incline greater than is customary with the chain grate. The usual type of arch is employed, and there is a great expanse of bridge wall to direct the heat on the incoming fuel. Ignition temperatures are extremely high, and large quantities of air are forced through the fuel bed at the front, the slots in the tuyere being wider here, although subject to the same air pressure. It is claimed that the high temperatures decompose the tarry or resinous excretion from the coal and the major portion of the volatile-forming matter before the former can set up an appreciable primary structure and before the carbon of the latter can have time to impregnate any slight primary structure that may have formed. The absence of coke formation changes the character of the fire under the arch and has a marked influence on the last half of the fire. As coke is not formed in the distillation zone, the cemented coke lumps cannot be present in the final stages of combustion. The individual pieces of coal retain their identity as in the case of free-burning Illinois coal. This non-coking performance of Eastern coals was demonstrated in an experimental plant which was open for inspection during the convention. More will be said of this stoker in an early issue.

During the same period, at another session, the first issue of the "Code of Safety Standards for Elevators" was presented. It covered the construction, operation and maintenance of elevators, dumbwaiters and escalators. During the afternoon there was a public hearing by the Boiler Code Committee on "Rules for the Construction of Unfired Pressure Vessels." Much discussion on welded joints developed. It was the sense of the meeting that the welding portion of the report be revised. Tests and data were requested from manufacturers, and there were promises to send in such material. The revisions were to be made in time to submit at the December meeting.

On Wednesday morning there were simultaneous sessions on aëronautics and castings, the latter including papers on malleable, die, aluminum, steel, gray iron, brass and bronze castings.

Scientific Session

At a scientific session on Thursday the refrigerating and heating engineers participated. It was a joint session of the three societies, and from each the papers were contributed. Upon the request of the Power Test Code Committee of the society for advice on weir measurements and with funds from the Engineering Foundation, Clemens Herschel devised an improved form of weir for gaging in open channels, which was tested at the hydraulic laboratory of the Massachusetts Institute of Technology. The new design has no resemblance to the sharp-edged weir now in use. It has a 2:1 slope of approach to a circular crest which is hollow for observing the pressure at that point. The nappe of the stream is supported for a short distance beyond the crest by another 2:1 slope. In the first paper of the joint session Mr. Herschel gave a résumé of the history of weir measurements, described the new weir and incorporated the results of the tests. In the discussion there was call for more data and the privilege of installing weirs of the new type at the various universities to check up the data presented. It was felt that the sharp-crested weir would still be used for measuring comparatively small quantities of water such as power-plant returns or boiler feed.

A paper by Glenn B. Warren on "Simplification of Venturi-Meter Calculations," was read by title. The author described a method reducing the calculations involved in the venturi-meter measurement of the flow of compressed air, to simple slide-rule computations.

M. S. Van Dusen, of the Bureau of Standards, described the method that had been developed for measuring the thermal conductivity of commercial insulating materials. The equipment consisted of two copper plates, one on either side of an electric heating grid. At the edges there were auxiliary heating coils to keep the temperature of the guard ring equal to that of the inner square. Temperatures were measured by thermocouples. Probable errors in the use of this equipment were discussed, and slides were used to show the variation in thermal conductivity of pure corkboard with the density, conductivity of various kinds of wood across the grain, relative temperature drops through the various materials composing a wall, etc. Funds were not available at the present time to carry on this work, and rather than take the chance of losing valuable data and perhaps the men engaged in the work, a plea was made for funds, in particular from those interested in refrigeration and heating.

"The Dissipation of Heat by Various Surfaces," by T. S. Taylor, was a paper discussing the value of sheet asbestos as a covering for hot-air pipes. It was shown that 0.013-in. sheet asbestos over tin would dissipate about 37 per cent more heat than a bare tin pipe; with a layer of dust added 32 per cent more heat would be dissipated, and with a layer of dust only on the bare pipe the heat waste would be 7 per cent more than with the bare pipe. With 0.2-in. sheet asbestos the loss would be equal to that from bright uncovered pipe and with thicker applications of insulation there would be considerable saving. The effects of air velocity and the angle of incidence of the air upon the dissipation of heat were shown. In the discussion it was brought out that the comparative losses from dull and bright surfaces were in the ratio of 0.7 to 0.5 and that results different from those given in the paper would be obtained when gases, say from a boiler, passed through the flue.

"Ship Ventilation," by F. R. Still, was the final paper of the session. As there was no standard system, the author discussed the various methods employed.

Power and Combustion Session

At a simultaneous power and combustion session the following four papers were presented: "Pulverized Coal in Metallurgical Furnaces at High Altitudes," by Otis L. McIntyre; "Efficiency of Natural Gas Used in Domestic Service," by Robert F. Earhart; "The Separation of Dissolved Gases from Water," by J. R. McDermet; "Locomotive Feed-Water Heating," by T. C. McBride. The third paper listed discussed a process for removing the non-condensible gases from boiler-feed water. Briefly, there is rapid injection of heated water into a region of vacuo, and an explosive boiling of it at the expense of the heat of the liquid available to the vacuum, with a simultaneous recovery of the heat liberated to the vacuum by a heat exchanger or condenser cooled by the incoming water, preliminary to its heating. Elimination of corrosion from boilers and economizers, increased conductivities and higher vacua were the advantages claimed. The paper created prolonged and diversified discussion. Mr. McBride's paper described a system of locomotive feed-water heating in which exhaust steam in an open heater comes in direct contact with cold feed water. It was pointed out that there would be available a surplus of exhaust steam for the draft.

Entertainment Features

Entertainment began with the reception Monday evening. The engineers were made welcome by C. H. Howard, president of the Commonwealth Steel Co., and E. R. Jackson delivered an illustrated address on "The First Transcontinental Motor Convoy," which was followed by music and dancing. Numerous inspection trips to commercial plants of interest were arranged, and daily entertainment was provided for the ladies. The banquet was held Tuesday night at the Missouri Athletic Association. The feature was a most interesting address on "The German Defenses on the Coast of Belgium," by Lieutenant-Colonel H. W. Miller, of the Ordnance Department. Music and dancing were again the closing features. By Wednesday night the Heating and Refrigerating Engineers had arrived, and all spent a most enjoyable evening at a concert and dance divertissement given at the mammoth open-air municipal theater at Forest Park. On Thursday afternoon the joint entertainment feature was a boat ride up the Mississippi River with music, dancing and vaudeville selections.

Thursday night the Mechanical Engineers left for Tulsa to view the oil fields, and participate in a two-day meeting with the Mid-Continent Section of the society. It is probable that the next spring meeting of the society will be held in Chicago.

Maryland To Adopt A.S.M.E. Boiler Code

A bill of the American Uniform Boiler Law Society for Maryland, effective Jan. 1, 1921, for the creation of a Board of Boiler Rules prescribes rules and regulations for the construction and use of steam boilers in that state, which will conform with rules now in existence in other states. It provides for the free interchange of boilers between states, and by standardization increasing the efficiency of all boiler manufacturers and all makers of materials used with steam boilers.

The board, which will consist of the chairman of the State Board of Labor and Statistics, the Attorney General, and the chairman of the State Accident Commission, will be invested with full authority to formulate rules and regulations governing the proper construction and installation of boilers of over fifteen pounds to the inch, and they will be empowdered to formulate and enforce rules and also to provide for maintenance on a proper basis of safety.

The Boiler Code of the American Society of Mechanical Engineers will be used wherever possible; that is, it will be used provided it does not conflict with the existing Maryland rules.

Section V of the act makes provision for the enforcement of the rules and regulations of the board through a state system of boiler inspection or by any other method created by the board, or by the General Assembly of Maryland. The board, however, shall have no jurisdiction over the operation of boilers or fixing the qualifications of persons entitled to operate turbines.

The bill provides that in the future no boiler shall be installed or passed by any boiler inspector which does not measure up to the rules established by the "Board of Boiler Rules." Adequate fines and imprisonment are provided for violation of the rules.

St. Louis Meeting of Refrigerating Engineers

As HAS been intimated in the report of the A. S. M. E. spring meeting, the American Society of Refrigerating Engineers was one of the three societies holding conventions at St. Louis during the week of May 24, the specific dates being May 26-28. Headquarters was at the Hotel Statler. The first session opened Wednesday morning with President Matthews in the chair. There was the usual business routine, consisting of committee reports, the announcement of 32 new members and other matters. The tentative code for the regulation of refrigerating machines and refrigerants was brought up and the proposed changes enumerated. The principal changes on pressures had not been decided. When complete, the tentative code was to go to the American Engineering Standards Committee, then to be issued as a national code. A question box was established to begin with the present meeting. Mr. Gould, secretary of the Institute of Meat Packers, a service bureau consisting of ten or eleven standing committees working every day, made a plea for assistance from the Refrigerating Engineers in establishing methods for computing unit costs of refrigeration to the different departments.

President Matthews departed from the usual formal address by giving a brief talk on the field of refrigerants from physical and chemical standpoints. He presented a chart listing the homogeneous series of univalent hydrocarbon radicals, including diagrams showing the molecular structure and indications of the boiling points of the different media. A second chart showed the temperature-pressure relations of the various refrigerants.

Joint Session, Wednesday Afternoon

On Wednesday afternoon there was a joint session with the American Society of Heating and Ventilating Engineers at which four papers were presented. "New Methods of Applying Refrigeration," by E. S. Baars, was the first. Not only new applications, but changes of methods in industries long using refrigeration, were reviewed. The paper will be presented later in abstract.

Director John R. Allen summarized a most interesting paper on "Theories of Heat Losses from Pipes Buried in the Ground." The laws governing these losses were discussed and applied in a general way to ordinary practice. Attention was called to the fact that the results given were not absolute and must be considered as relative. The following conclusions were drawn:

That the heat loss from a pipe is not proportional to the external surface of the pipe. The heat loss per square foot of pipe in small pipes is much larger than the heat loss per square foot in large pipes.

That the heat loss per linear foot of pipe becomes smaller relatively as the pipe diameter increases, so that a 16-in. pipe loses but little more heat per linear foot than a 12-in. pipe. In practice, therefore, it makes very little difference in the heat loss, whether a 12- or a 16-in. pipe is used, especially in dry ground.

That the heat loss from a pipe is not proportional to the thickness of the covering and that there is little advantage to be gained in a good covering in increasing the thickness of the insulation to more than 2 in.; in most cases, with a good covering, it would appear that 2 in. of covering is all that is necessary.

That the depth to which a covered pipe is buried in the soil makes little difference in the heat loss provided the center of the pipe is 2 ft. or more below the ground surface. Beyond 2 ft. in depth, unless the pipes are very large, the heat loss from the pipes remains substantially the same for all depths.

That the heat loss from a pipe is not proportional to the conductivity of the covering, as the conductivity of the ground is as important a factor in the heat losses as the conductivity of the covering.

That poor covering in dry ground will give better results than good covering in wet ground. It is not possible, therefore, to guarantee the heat loss from a covering in the ground, as such a guarantee involves guaranteeing the conductivity of the ground.

Exceptions were taken to most of these conclusions by L. B. McMillan, consulting engineer for the H. W. Johns-Manville Co., and George B. Nichols.

In a paper on "Heat Insulation Facts" L. B. McMillan gave a comprehensive discussion on what insulation is, what it does and how it does it. The final paper of the session, on "Automatic Control of Temperatures," by R. P. Brown, president of the Brown Instrument Co., was read by title. The paper dealt with thermometer and pyrometer electrical control employed by the company.

The joint session on Thursday with the mechanical and heating engineers has been covered in the A. S. M. E. report.

Friday Morning's Session

At the Friday morning session there were two papers dealing with the small household refrigerating machine. The first, by G. A. Kramer, outlined the construction of the "Frididore" outfit, which employs a charge of from 2 to 5 lb. of methyl chloride mixed directly with the lubricant. Stuffing boxes, diaphragm and float-operated valves, and the location of the thermostatic control element were discussed in general. Data given during the discussion of the paper placed the cost for current per 100 lb. of refrigeration at close to 40c. A power requirement of 2.5 kw.-hr. per 100 lb. of refrigerating effect was another relation advanced.

In the second paper R. F. Massa stated that it was impossible to give the standard rating used for large machines to the small domestic outfit built into the refrigerator, owing to the variations in construction, type of cooling and refrigerant. Rating by piston displacement or size of ice compartment was also troublesome. The truth was that in most cases the builders wanted no comparative results. The small machine should be rated on the service rendered and under what conditions it will do the work, the capacity being expressed on an hourly basis.

W. G. E. Rolaff described his rotary compressor for refrigeration work. The machine is somewhat similar to a sliding blade pump, with the exception that the cylinder rotates to reduce the relative movement to the difference between peripheral velocities and thus reduce the scraping wear on the blade. The compressor operates at high speeds and is built for pressures of 75 lb. and up. The construction was illustrated and rather extravagant claims made as to its performance. A unique feature is the spraying of the lubricating oil into the suction. The oil seals the machine, preventing leakage, and as it is previously cooled, helps to improve the volumetric efficiency.

A report was read in which it was proposed to use 144 B.t.u. as the latent heat of ice in practical commercial work. It resulted in round figures for the usual units and was so near the 143.5 established by the Bureau of Standards that the error introduced would be negligible.

Afternoon Session, Friday

At the final session, Friday afternoon, Nelson J. Waite, presented a paper on "Direct Expansion Piping in Egg Rooms." This type of cooling had been distrusted in the old days, owing to the difficulty of holding the room in a dry state and the consequent musty odor imparted. Improved construction and the adoption of a lower carrying temperature now made direct expansion quite feasible. In a single small room the use of a single ceiling coil was recommended and in a larger room, several sections of ceiling coils all in the same horizontal plane.

In the discussion of the paper more surface and higher temperatures in the coils was recommended for bunker service with forced circulation. Exception was taken to the single-coil arrangement. Heat leakage would make the temperature higher at the walls and probably cause trouble from the resulting variation in humidity. As a factor of safety brine had something in its favor, but was less efficient and less flexible when storing different products.

George M. Kleucker described a water-distributing device for atmospheric condensers, designed to overcome the troubles experienced with slotted water pipes, serrated troughs or overflow gutters. The new trough had a vertical wall on one side, a sloping wall on the other side and a flat bottom with removable plugs through which accumulations of dirt could be flushed out. A baffle was attached to the sloping side and through holes in the trough wall, water spouted against this baffle; then all of the water ran down evenly over the top pipe of the stand. Division strips between the pipes helped to hold the water together and guide it downward. The maximum flow to the condenser was 2½ gal. per minute to the foot of length. As these baffle or deflector plates were made as long as ten feet, it was thought in the discussion that so long a strip of metal would tend to wind. Serrating the bottom edge would probably improve the distribution of the water.

"Pulverized Coal for Boiler Firing," by H. G. Barnhurst, was the final paper.

In closing, Mr. Tait, who was acting as temporary chairman, said that he had been impressed favorably with the combination meeting, that it was his desire to see them continue with the programs arranged, if possible, to allow more interchange between the societies.

As previously mentioned, the entertainment was in common with the other two societies and has been covered, with the exception of the combination banquet held at the hotel by the refrigerating and heating engineers. George Wells, of St. Louis, was toastmaster, and the speaker of the evening, Carl J. Baer, on the engineer's position in reconstruction. It was his thought that engineers did not study other fields as they should and as a consequence were not prepared to assume their full duties. In a very interesting way the speaker traced the inter-relations of commerce, production of raw materials, industry and transportation, showing the need of co-operation, one with the other. There were short talks by the two presidents, Dr. Hill and Mr. Matthews. The evening closed with music and dancing.

Heating and Ventilating Engineers Meet at St. Louis

FROM May 26 to 28 the American Society of Heating and Ventilating Engineers held its spring convention in St. Louis in conjunction with the Mechanical and Refrigerating Engineers. In the reports of these societies mention has been made of the joint sessions, the joint entertainment and the combination banquet with the refrigerating engineers. Hotel headquarters was at the Statler, but the meetings of the society were held in the Board of Education Building. At the opening session Wednesday morning, W. E. Rolfe, president of the Associated Engineering Societies of St. Louis, gave the address of welcome. Response was made by Dr. Hill, president of the society.

Committee Reports

To save time reports of chapters were eliminated, as they would appear in the *Journal*. Committee reports was the next order of business. The report of the Committee of New York Chapter on Industrial Engineering was important. It covered briefly causes of industrial unrest, remedies needed, responsibility of the engineering profession and a declaration of principles, closing with the recommendations for the society's actions, which were adopted at the meeting. In these recommendations the society approved and adopted the "declaration of principles" established by the American Society of Mechanical Engineers. Also the society is to appoint a standing committee of five, with the privilege of appointing subcommittees, to study economic and industrial problems, to arouse the interest of the society membership in the responsibility and opportunity of the engineer to assist in the present situation and to co-operate with committees of other engineering societies in developing united constructive action, in carrying out a comprehensive educational campaign for a term of years and finally in supporting and helping to co-ordinate and make effective all good constructive agencies, such as the Second Industrial Conference at Washington, the interchurch movement, etc.

James A. Donnelly, chairman of the Committee to Investigate Capacities of Steam and Return Mains, presented the report. At the annual meeting the committee was authorized to collect data on standard sizes of low-pressure steam and return mains. Tables of pipe sizes and a list of questions were submitted for the purpose of collecting this information, the various chapters being requested to có-operate. For the first time a standard table in full detail was prepared for each type of system. The tables distinguished between wet and dry drip mains and returns, and another new feature was a request for data on critical velocities. This is the velocity at which there is sufficient separation of steam and condensation, so that defective circulation or water hammer will not occur. This does not necessarily mean complete separation. The entire matter, with the data that is expected, will be submitted at the annual meeting. Other committee reports scheduled were deferred.

On Wednesday afternoon there was the joint meeting with the Refrigerating Engineers, and on Thursday the meeting of the three societies. Both of these sessions have been covered in the other reports.

On Friday morning there was a school-ventilation session with six papers. "The significance of Odorless Concentration of Ozone," by E. S. Hallett, gave assurance to those who might be desirous of using ozone in ventilation. The author stated that ozone in ventilation must be put on the same rigid practical basis as the heating system. It must have a standard of concentration based upon a mechanical unit and capable of convenient manual or automatic control, using chemical processes as a check only. With such equipment kept clean, as it must be to be effective, no apprehension need be felt as to results. In discussing the contribution, it was suggested that the society's Research Bureau make tests of the use of ozone to see if the results given by Mr. Hallett could be generally duplicated.

Two papers on the ventilation of large auditoriums were presented, one by S. R. Lewis, dealing specifically with observations made on an auditorium having air inlets in the window sills, and the other by R. S. M. Wilde, on the practical ventilation of modern theatres. In the discussion of the papers Mr. Hart differentiated between floor and overhead admission. With the former he did not favor automatic control due to the likelihood of sudden fluctuations objectional to occupants. With the supply overhead these fluctuations were pleasant rather than otherwise.

Training the School Janitor

E. S. Hallett told how the janitors and custodians of St. Louis schools were trained. In these schools about 200 boilers, an equal number of engines and blast fans, air washers, heat-regulation systems, vacuum cleaning systems, electric lights and plumbing were operated by untrained janitors. Under these conditions two supervising engineers were kept on the jump even to keep the plants running. To relieve the pressure of supervision and to get results from the heating apparatus, a system of training was inaugurated and elemental instruction given in combustion, methods of firing, regulation of heat, humidity and all essentials. The paper contained a complete outline of the course. The results were most gratifying. The discussion was complimentary, particularly by those who had visited the schools, and the establishment of night schools throughout the country for janitors was urged. Mr. Hallett explained that the reason for the satisfactory conditions obtaining in St. Louis was that the public-school system was under state control. The appointment of janitors had been taken out of the field of local politics, the janitors being civil-service men.

"High Efficiency Air Flow," by F. W. Caldwell and E. N. Fales, was an interesting paper dealing with the development of a means for visualizing air flow, with model studies made in the wind tunnel at McCook Aviation Field, Dayton. The final paper of the session on "The Sizing of Ducts and Flues for Ventilating and Similar Apparatus," by H.

Eisert, was a technical contribution bringing in theoretical considerations and exact methods which for the most part must be overlooked in commercial practice owing to the difficulties involved and the irregularities of construction. It was pointed out in the discussion that the author might have made an important point of the obstructions to flow such as were offered by bolt heads, as these probably would outweigh in importance many of the factors considered; for example, frictional resistance.

Symposium on Health and Humidity

Friday afternoon was devoted to a symposium on health and humidity, with two papers bearing on the use of wet-bulb temperature readings as the best method of indicating proper air conditions. E. V. Hill and J. J. Aeberly discussed "The Relation of the Death Rate to the Wet-Bulb Temperature," and O. W. Armspach presented a paper on "The Relation of the Wet-Bulb Temperature to Health." In the course of the discussion J. R. McColl said that he had found in an investigation of 32 school buildings in Detroit that for every pound of steam used for heating, three pounds was used for ventilation. In other words, the ventilating load was three times the heating requirements.

Adoption of a standard as a measure of ventilation was the next consideration. Director John R. Allen offered a resolution naming the Synthetic air chart, devised by Dr. E. V. Hill, as meeting the requirements of such a standard. Dr. Hill explained how the chart had been developed for the Chicago Commission of Ventilation, the present form being the eighth revision since it was first brought out in 1912. The resolution was adopted unanimously and it was further moved to appoint a committee of five to prepare an interpretation of the chart and explanations of the methods of determining the quantities considered. It was voted to appoint another committee of five to develop standard methods of testing heating systems.

In a paper on "Commercial Dehydration" J. E. Whitley showed some of the commercial possibilities of removing moisture from foodstuffs and established the economic advantages, giving a clear indication of the wide scope of the field. W. S. Scott spoke on the subject of "Industrial Electric Heating," a field increasing in prominence, not because electric heat is cheaper than other forms of heating, but because results unattainable heretofore have been effected, reducing the cost of the completed product, lessening the labor required, increasing the output and producing a product of uniform quality. Various applications were discussed, the equipment described and a summary made of the advantages.

Before adjourning a suggestion was received from John F. Hale that the annual meeting be held in Philadelphia with a business session in New York to comply with the constitution. As an alternate Mr. Hart suggested that the annual meeting be held in Pittsburgh at the Research Bureau and that the next semi-annual meeting be held at the same time and same place as that of the National Association of Heating and Piping Contractors, the probable selection being Atlantic City. This combination meeting was desirable and it would divide the meetings between the East and the West. A vote showed Mr. Hart's suggestion to be the sense of the meeting.

Large Transformers

The largest single-phase power transformers developed so far, according to the N. E. L. A. Electrical Apparatus Committee's report, are in a 70,800-kva. bank of transformers made up of three 23,600-kva. single-phase units, to step up from a delta generating voltage of 11,500 volts 60 cycles, to a star line voltage of 66,000 or 132,000 volts. These units are of the shell type, having a temperature rise of 55 deg. C. at full load, and the guaranteed efficiency at full load is 99 per cent. The height to the tips of the high-voltage bushings is 22 ft. and the largest floor-space dimensions of each unit lie within a circle whose diameter is 10 feet, 3 inches. An installation of two 12,000-kva. 4,500- to 27,600-volt three-phase self-cooled outdoor transformers, are the largest capacity of this type thus far built.

CAREFULNESS
GUARDS AGAINST
ACCIDENT

Personals

H. N. Aikman, recently with Dwight P. Robinson & Co., New York City, is now associated with the Fargo Engineering Co., Jackson, Mich.

G. C. Ward has been appointed a vice president and will be in charge of operation and construction for the Southern California Edison Company.

Fred E. Sermulin has resigned as test engineer of the Bethlehem Steel Co. and has accepted a position with R. H. Warren, consulting engineer of Easton, Pa.

Harold B. Vincent, formerly management engineer of the Day & Zimmerman Co., Philadelphia, has been promoted to commercial engineer of the same company.

Lafayette Hanchett has been elected president of the Utah Power and Light Co., Salt Lake City. He succeeds D. C. Jackling, who resigned a short time ago.

Walter E. Daggett, formerly engineer of the Osborn Manufacturing Co., Inc., of Cleveland, Ohio has opened sales-engineering offices in the northern part of the state.

A. Y. Dodge, formerly assistant plant and tool engineer with the Wallis Tractor Co. of Racine, Wis., has accepted the position of chief engineer with the Racine Engineering Co.

Dr. Lamar Lyndon, consulting engineer, 13 Park Row, New York City, who has been out of the country for several months, has returned and announces that he has resumed practice.

John A. Lawrence, formerly assistant mechanical engineer of the New York Edison Co., New York City, has assumed the position of engineering manager for Thomas E. Murray Inc., New York City.

H. G. Harvey has accepted a position in the commercial department of the Pennsylvania Utilities Co., Easton, Pa. He was formerly commercial engineer of the Nassau Light and Power Co., Mineola, L. I.

Walter J. Bitterlick, formerly plant engineer for the Hood Rubber Co. of Watertown, Mass., has taken the position of engineering superintendent of the new plant of the Seamless Rubber Co., New Haven, Conn.

William K. Sanderson, chairman of the St. Thomas (Ont.) Hydro-Electric Utilities Commission, was elected president of the Ontario Municipal Electric Association at the recent annual convention of the association.

Harry I. Beardsley has resigned as supervising mechanical designer with the Crocker-Wheeler Co., Ampere, N. J., and is associated with the Western Electric Co., New York City, in the capacity of specification engineer.

Sterling H. Bunnell has resigned his position as chief engineer with R. Martens & Co., Inc., of New York City, and has become associated with the Engineering and Appraisal Company, Inc., with offices in New York City.

I. Beller, until recently connected with Balbach Smelting and Refining Co., of Newark, N. J., in the capacity of designing engineer, has become affiliated with the Chile Exploration Co. (Braden Copper Co.), of New York City.

Harry Deverell has severed his connection with Deverell, Spencer & Co., Baltimore, Md., and has opened offices in Baltimore as a sales engineer for machinery, especially elevating, conveying and power-transmission machinery.

William D. Ennis, vice president and secretary, announces the organization of the Technical Advisory Corporation, with offices in the Vanderbilt Building, New York City. The corporation will act as consulting engineers, industrial economists and advisors.

N. Baad, who formerly had charge of the heat-treating department at the De La Vergne Machine Co., of New York, has resigned to take a position as sales and service engineer with the F. A. Calhoun Co., 76 Montgomery St., Jersey City, N. J.

Douglas Sprague and Chester A. Slocum, associate consulting engineers, announce that they are now located at 167 West Thirteenth St., New York City. The firm renders service in ventilation, sanitation, heating, electricity, power plants, reports and valuations.

Society Affairs

Atlanta Section, A.S.M.E., will meet June 29. Election of officers will take place.

The Universal Craftsman Council of Engineers will hold its annual convention at Springfield, Mass., Aug. 3-6.

Chicago Section, A.S.M.E., will meet on June 21 at the Western Society of Engineers Club. Subject: "Refrigeration."

The National Association of Safety Engineers, Ohio Section, will hold its annual convention June 17, 18 and 19 at Columbus, Ohio.

The Association of Iron and Steel Electrical Engineers will hold its fourteenth annual convention in September at New York City.

The New York State Association, N. A. S. E., will hold its twenty-fifth annual convention at the Temperance House, Niagara Falls, N. Y., on June 10, 11 and 12.

The National Association of Safety Engineers, Connecticut Section, will hold its annual convention at Bristol on June 26 instead of at Norwich, as previously announced.

The Buffalo Section of the A. S. M. E., at a meeting held at the University Club, elected the following officers for the season of 1920-21: B. S. Hughes, chairman; W. J. Gamble, vice chairman; W. J. Boyd, secretary; W. M. Dollar, treasurer; and C. A. Booth.

The Chamber of Commerce, U. S. A., at its annual meeting in Atlantic City in April elected as directors of the Chamber two engineers: Howard Elliott, a member of the American Society of Civil Engineers and chairman of the Board of the Northern Pacific R.R. and L. B. Stillwell, past President, American Institute of Consulting Engineers and of the American Society of Civil Engineers.

Membership of the National Engineering Societies has increased rapidly. The latest membership figures are as follows: American Society of Civil Engineers, 9,515 (May 14, 1920), American Institute of Mining Engineers, 8,532 (May 10, 1920), American Society of Mechanical Engineers, 12,258 (May 14, 1920), American Institute of Electrical Engineers, 11,345 (May 1, 1920), Society of Automotive Engineers, 4,798 (May 8, 1920).

New Professional Sections of A.S.M.E.—Aëronautics, Cement, Fuels, Gas Power, Industrial Engineering, Machine Shop Practice, Ordnance, Power, Railroads, Textiles have already been authorized—furnish a unique opportunity for engineers specializing in these various fields to take part in a movement for the development of their specialties through meetings, papers and debate, backed up by a national organization of over 12,000 members.

Engineers of Eastern New York to Affiliate—A meeting of representatives of the various engineering organizations in the cities of Albany, Schenectady, Troy and surrounding towns was held at Albany on May 13. The following organizations were represented: The Albany Society of Civil Engineers (a local independent organization), the Albany Chapter of the American Association of Engineers, the Eastern New York Section of the A.S.M.E., and the Schenectady Section of the A.I.E.E. Resolutions were passed requesting each organization to appoint an official representative to serve on a committee to draft a constitution providing for the affiliation of the various local sections of national societies with the Society of Engineers of Eastern New York, thereby making the latter the dominant force in that territory. It is hoped to put the plan into operation next October when the engineering meetings again become active.

Miscellaneous News

An International Exhibition of the products of the mine, the soil and the sea, and of arts, industries and manufactures, to take place in Philadelphia in 1926, to celebrate the 150th anniversary of the signing of the Declaration of Independence is proposed in a bill which has been introduced by Representative Darrow, of Pennsylvania.

The Bureau of Mines, in a recent statement on the development of lignite, says: The efforts to make the lignite of Texas, North Dakota and other Western States so serviceable to that part of the country that it will be unnecessary in the future to ship in other coal at great expense are to be inaugurated immediately by the Bureau, at New Salem, N. D., in co-operation with a company there.

The Sixth National Exposition of Chemical Industries will be held at the Grand Central Palace, New York, during the week ending Sept. 25. This year the exposition will have three special sections—the Electric-Furnace Section, the Fuel-Economy Section and a Materials-Handling Section. The Fuel-Economy Section will consist of exhibits of machinery and apparatus, furnaces, producers, stokers and all devices for the economic utilization or more efficient combustion of coal. The Materials-Handling Section will include exhibits of conveying, transporting and elevating machinery. Included in this will be weighing, measuring and power-transmission equipment.

Business Items

The Hammel Oil Burning Equipment Co., Inc., is now located at 1476 Broadway, New York City.

James Walker & Co., Ltd., announce the removal of their offices and storeroom to 46 West St., New York City.

Westinghouse Church Kerr & Co., Inc., announce the removal of their offices from 37 Wall St. to 125 East 46th St. Until completion of the merger with the Dwight P. Robinson & Co., Inc., the business will be conducted under the same name.

The General Electric Co. has completed negotiations for the purchase of the former munitions plant of the Bartlett-Hayward Co., at Baltimore, Md. The plant consists of eleven buildings having 420,000 sq.ft. of floor space and occupying a forty-acre site.

The Worthington Pump and Machinery Corporation, 115 Broadway, New York, announces that it has completed preparations to furnish improved water-power machinery of all capacities for low-, medium- and high-head service, including oil-pressure systems, waterwheel governors and other auxiliaries.

The Roller-Smith Co., New York City, announces that it has made an agency arrangement with the W. Montelius Price Co., Seattle, Wash., to handle the Roller-Smith Company's line of electrical instruments, meters and circuit breakers in the States of Washington and Idaho, and part of the State of Oregon.

The Cross & Simmons Company, Inc., Chicago, has changed the name of the organization to The Simmons Associates, Inc. No change in the personnel of the company is involved except that John H. Cross, formerly president, becomes vice president, and H. H. Simmons, formerly secretary and treasurer, becomes president.

The Pennsylvania Pump and Compressor Co., Easton, Pa., announces the opening of its sales offices in the following cities: New York, 50 Church St., H. C. Browne, manager; Philadelphia, 2222 Chestnut St., W. J. Devlin, manager; Pittsburgh, 631 Fulton Building, C. W. Gellinger, manager; Richmond, Va., Mutual Building, W. F. Delaney, manager; Birmingham, Ala., 2027 Jefferson Bank Building, H. F. Kann, manager; Salt Lake City, Utah, Newhouse Building, C. H. Jones, manager; Milwaukee, Wis., 604 First National Bank Building, Coates & Zarling, representatives.

Trade Catalogs

The Industrial Arts Index, March issue, is now ready for distribution. Periodicals and technical magazines are indexed in this book in such a manner that any publication can be easily found. The name of the magazine, volume number, including paging and date is given. Note is made also of illustrations, maps and charts.

The National Educational Committee of the N. A. S. E. has published a 35-page booklet entitled, "The Use and Value of Exhaust Steam." A chapter is also devoted to practical pipe covering. A number of interesting tables showing heat losses from uninsulated hot surfaces, as well as curves showing heat loss per degree temperature difference per square foot per hour—B.t.u. are given. A copy of the booklet can be secured for fifteen cents.

The Dwight Manufacturing Co., 565 West Washington Boulevard, Chicago, is issuing for distribution upon request a 20-page booklet devoted to the Dwight CO_2 indicator of the spring-gage and mercury-gage types, the motion recorder developed by the company, feed-water and flue-gas thermometers, and Ellison differential draft gages. The booklet is instructive and should prove of interest to engineers or anyone having use for the products named.

The Bridgeport Brass Co., Bridgeport, Conn., has ready for distribution a new 78-page booklet entitled "Seven Centuries of Brass Making." A brief history of the ancient art of brass making and its early method of production is contrasted with modern electric methods. The important steps in the making of tubes, sheets, rods and wire are shown. The properties of brass as affected by composition, cold-working and heat treatment are given with the idea of assisting engineers in drawing specifications.

The Iron Age Publishing Co. New York has ready for distribution a new 1,216-page illustrated catalog entitled "The Iron Age Catalog of American Exports." It contains the catalogs of leading American manufacturers of engineering, railway, foundry and electric equipment and supplies; iron and steel machinery and tools, hardware and cutlery. The catalog contains over three thousand illustrations and specifications, data and other information which will make it useful to the technical as well as other buyers.

The Fairbanks Co., Broome and Lafayette Sts., New York City, has just completed publication of a new 256-page catalog on power transmissions appliances and elevating and conveying machinery. Complete descriptions, accompanied by illustrations of the apparatus handled by the company, are contained in the booklet. In addition, full data as to dimensions, capacities, etc., are given, making it a useful handbook for the man who designs or installs mechanical equipment of this nature. A copy will be sent on request.

Briggs & Turivas, Inc., dealers in scrap iron and allied metals, iron and steel products, obsolete and salvageable material and equipment, Chicago, New York and Toronto, Canada, in a 48-page pocket size booklet, have presented valuable information for the producer and user of these materials. In this publication the various grades of iron and steel scrap as marketed are described and classified. The data being compiled from average practice, will fit closely almost every probable contingency. A table shows weights and quantities of rails and accessories in various combinations of tonnage and length of track. Other tables show net and gross ton equivalents, a tonnage card, weights of plate, standard gages and specific gravities and weights.

COAL PRICES

Prices of steam coals both anthracite and bituminous, f.o.b. mines, unless otherwise stated, are as follows:

ALANTIC SEABOARD

Anthracite—Coals supplying New York, Philadelphia and Boston:

	Mine
Pea	$5.30@$5.75
Buckwheat	3.40@ 4.10
Rice	2.75@ 3.25
Barley	2.25@ 2.50
Boiler	2.50

Bituminous—Steam sizes supplying New York, Philadelphia and Boston:

Latrobe	$4.25@$4.50
Connellsville coal	6.00@ 7.00
Cambrias and Somersets	4.35@ 5.15
Clearfields	6.25@ 7.50
Pocahontas	7.00@ 8.00
New River	7.00@ 8.00

BUFFALO

Bituminous

Pittsburgh slack	$6.00@$8.00
Mine-run	6.00@ 8.00
Lump	6.00@ 8.00
Youghiogheny	6.00@ 8.00

CLEVELAND

Bituminous—Prices f.o.b. mines:

No. 6 slack	$4.50@$6.00
No. 8 slack	4.50@ 6.00
No. 8—¾-in.	4.50@ 6.00
No. 6 mine-run	4.50@ 6.00
No. 8 mine-run	4.50@ 6.00
Pocahontas—Mine-run	4.50@ 6.00

ST. LOUIS

Anthracite

	Williamson and Franklin Counties	Mt. Olive and Staunton	Standard
Mine-run	$3.00@4.00	$4.00@5.00	$3.75@4.25
Screenings	3.00@4.00	4.00@5.00	2.75@3.50
Lump	3.00@4.00	4.00@5.00	4.00@4.50

Williamson-Franklin rate to St. Louis is $1.10; Other rates 95c.

CHICAGO

Bituminous—Prices f.o.b. mines as per circular of April 1 are as follows: Co. prices vary from 50c. to $1.25

Illinois		Freight rate Chicago
Southern Illinois		
Franklin, Saline and Williamson Counties		
Mine-Run	$3.00@$3.10	$1.55
Screenings	2.60@ 2.75	1.55
Central Illinois		
Springfield District		
Mine-Run	$2.75@$3.00	$1.32
Screenings	2.50@ 2.60	1.32
Northern Illinois		
Mine-Run	$3.50@$3.75	$1.24
Screenings	3.00@ 3.25	1.24
Indiana		
Clinton and Linton		
Fourth Vein		
Mine-Run	$2.75@$2.90	$1.27
Screenings	2.50@ 2.65	1.27
Knox County Field		
Fifth Vein		
Mine-Run	$2.75@$2.90	$1.37
Screenings	2.50@ 2.60	1.37
Brazil Bloke	$4.25@$4.50	$1.27

New Construction

PROPOSED WORK

Me., Brunswick—The Cabot Mfg. Co. had plans prepared for the construction of a power plant in connection with its proposed cotton mill.

Mass., Boston—The School Bd Comn. plans to build a 2 story school building in the George Putnam Dist. and a ' story school building in the Lewis Dist. Steam heating systems will be installed in same. Total estimated cost, $600,000.

Mass., Boston—The School Bd. Comn. will receive bids about Sept., for the construction of a 2 story, 180 x 180 ft. school on Maxwell and Seldon Sts. and a 3 story, 150 x 300 ft. school on Avenue Louis Pasteur. Steam heating systems will be installed in same. Total estimated cost, $1,620,000

Mass., Mansfield—The Mansfield Housing Corp., c/o Bonelli, Adams & Co., 60 State St., Boston, will soon award the contract for the construction of fifty 1 and 2 story frame houses on various lots. A hot air heating system will be installed in same. Total estimated cost, $250,000.

Mass., Worcester—J. D. Leland, Archt., 185 Devonshire St., Boston, will soon award the contract for the construction of a 4 story, 200 x 210 ft. administration building, for the Reed & Prince Mfg. Co., Duncan Ave. A steam heating system will be installed in same. Total estimated cost, $500,000.

Conn., Hartford—Cortlandt F. Luce, Archt., 36 Pearl St., will soon award the contract for the construction of a 7 story, 147 x 150 ft. garage, etc., on Hicks and South Ann Sts., for the Hartford Automobile Club Garage Co., 36 Pearl St. A steam heating system will be installed in same. Total estimated cost, $500,000.

Conn., Waterbury—The Connecticut Light & Power Co., 111 West Main St., has awarded the contract for the construction of a 2 story, 25 x 90 ft. power plant addition, to Dwight Wooster, 39 Harvard St., Waterville. Estimated cost, $50,000.

N. Y., Cranesville (Amsterdam P. O.)—The Adirondack Electric Power Corp., Glens Falls, plans to construct an electric power plant near here.

N. Y., Jamestown—The Bd. of Hospital Comrs. will receive bids until June 8 for the addition to laundry building, boiler house, etc., at the hospital here. J. W. Rulifson, 108 East 3rd St., Archt.

N. Y., Lockport—The Lockport Light, Heat & Power Co. is issuing $350,000 worth of preferred stock. The money will be used for extensions and improvements.

N. Y., New York—The Amateur Comedy Club, 150 East 36th St., is having plans prepared for the construction of a 7 story, 50 x 100 ft. club house at 122-124 East 31st St. A steam heating system will be installed in same. Total estimated cost, $200,000. O. C. Herring, 8 West 33rd St., Archt. and Engr.

N. Y., New York—The Natl. Drug Works Corp., 405 Lexington Ave., plans to build a 22 story office building on 42nd St. near 5th Ave. A steam heating system will be installed in same. Total estimated cost, $3,000,000. C. H. Seymour, Archt. and Engr.

N. Y., New York—The New York Edison Co., Irving Pl. and East 15th St., plans to build a transformer station at Dunwoodie. A steam heating system will be installed in same. Total estimated cost, $150,000. W. Whitehill, 55 Duane St., New York City, Archt. and Engr.

N. Y., New York—The New York Edison Co., Irving Pl and East 15th St., plans to build a transformer station at the bowery and another one on 73rd St. Steam heating systems will be installed in same. Total estimated cost, $300,000. W. Whitehill, 52 Duane St., Archt. and Engr.

N. Y., New York—S. W. Straus & Co., 150 B'way, is having plans prepared for the construction of a 9 story, 100 x 180 ft. bank, office and store building on 5th Ave. and 46th St. A steam heating system will be installed in same. Total estimated cost, $4,000,000. Warren & Wetmore, 10 East 47th St., Archts. and Engrs.

N. Y., Poughkeepsie—The Central Hudson Gas & Electric Co has filed an application with the Pub. Serv. Comn., Albany, for permission to construct extensions to its transmission system to the towns of Ulster, Esopus and Millbrook. Estimated cost, $25,000.

N. Y., Rocky Point—The Radio Corp., Woolworth Bldg., New York City, is having plans prepared for the construction of a wireless station consisting of several buildings including a power house. Total estimated cost, $10,000,000. J. G. White Eng. Corp., 43 Exch. Pl., New York City, Engr.

N. Y., Whitehall—The Consolidated Light & Power Co. has filed an application with the Pub. Serv. Comn. for permission to construct and operate a local electric light and power plant as well as transmission system.

N. J., Lakehurst—The Bureau of Yards & Docks, Navy Dept., Wash., D. C., will soon award the contract for furnishing and installing 3 turbo-alternators, cooling ducts, 2 exciter sets, switchboard, electric power and lighting systems, 2 motors with controllers and belts for air compressors, compressors, Wickes boilers, superheaters, mechanical stokers and engines, 3 boiler feed pumps, feed water heater, sump pump, coal and ash handling equipment, 2 fire pumps in the power plant at the Naval Air Station, here.

N. J., Morris Plains—The state rejected all bids for the installation of a heating, plumbing and electrical work in the 2 proposed 3 story, 88 x 230 ft. treatment buildings. Work will be readvertised.

Pa., Allentown—The Phoenix Silk Mfg. Co., Race and Court Sts., is having plans prepared for the construction of a power plant at its silk mills here. Carl A. Baer & Co., Land Title Bldg., Philadelphia, Engrs.

Pa., Philadelphia—The Franklin Trust Bldg., 20 South 15th St., plans to build an 11 story, 20 x 90 ft. office addition. A steam heating system will be installed in same. Total estimated cost, $500,000. DeArmond, Ashmead & Bailey, Franklin Trust Bldg., Archts.

Pa., Reading—Wilmer & Vincent, 1451 B'way, New York City, is having plans prepared for the construction of a 150 x 160 ft. theatre, office and store building on 10th and Penn Sts. A steam heating system will be installed in same. Total estimated cost, $650,000.

Del., Wilmington—The B. P. O. E. plans to construct a club house. A steam heating system will be installed in same. Total estimated cost, $400,000. DeArmond, Ashmead & Bailey, Franklin Trust Bldg., Philadelphia, Pa., Archts.

Md., Baltimore—The Delion Tire Co., 131 West Mt. Royal Ave., is having plans prepared for the construction of a 1 story, 90 x 400 ft. tire factory and a 2 story, 60 x 70 ft. power plant and office on Mt. Royal Ave. Total estimated cost, $100,000. R. B. Arnold, 131 West Mt. Royal Ave., Engr. J. O. Hunt, 114 North Montgomery Ave., Trenton, N. J., Archt.

Md., Baltimore—McKenzie, Vorhees & Gmelin, Archts., 1123 B'way., New York City, will soon award the contract for the construction of a telephone exchange building on Kate Ave., for the Chesapeake & Potomac Telephone Co., 5 Light St. A steam heating system will be installed in same. Total estimated cost, $250,000.

Md., Massey—The Town Council plans to install an electric lighting plant here.

Md., Port Covington—(Baltimore P. O.)—The Consolidated Gas, Electric Light & Power Co., Lexington Bldg., will soon award the contract for the construction of a 1 story, 26 x 46 ft. substation along the waterfront. Estimated cost, $80,000.

W. Va., Cecil—The Sterling Coal Co. is having plans prepared for the reconstruction of the electric power plant and coal tipple at its properties recently destroyed by fire.

W. Va., Wheeling—The Wheeling Electric Co. had plans prepared for the construction of a power loop to run through Fulton at Rudler's where a substation will be erected, and continue on a straight line north to Windsor. The substation will supply the suburban districts.

O., Canton—The Bd. Educ. will receive bids until June 19 for the construction of a 3 story, 140 x 250 ft. high school on 14th St. A steam heating system will be installed in same. Total estimated cost, $1,000,000. Thayer & Johnston, 3716 Euclid Ave., Cleveland, Archts.

O., Cleveland—Jacob Babin and others, 13661 Euclid Ave., has leased a site and plans to build a 20 story, 50 x 151 ft. office building at 310 Euclid Ave. A steam heating system will be installed in same. Total estimated cost, $2,000,000.

N. C., Kinston—The city plans to construct an electric power plant. Estimated cost, $300,000. W. C. Olsen, Sumter, S. C., Engr.

O., Cleveland—G. H. Greenberg, Atty., Society for Savings Bldg., is interested in a company to be incorporated which plans to build a 1, 3 and 5 story ice palace, commercial building, market house and garage on East 105th St. and Euclid Ave. A steam power plant will be included in the plans. Total estimated cost, $1,500,000.

O., Cleveland Heights (Warrensville P. O.)—The Bd. Educ. voted $2,400,000 bonds, $750,000 of which will be used to purchase new sites, $1,500,000 for the construction of additions to the Fairfax, Coventry and Noble Rd. Schools, a school building on Taylor Rd. and Monticello Ave. and a power house on Lee Rd. Equipment will be installed in the new building project. Noted May 18.

O., East Cleveland (Cleveland P. O.)—The Bd. Educ., c/o C. Ainmerman, Williamson Bldg., received lowest bid for the installation of a heating system in the proposed 2 story school on Mayfair Rd., from the Miles Heating & Ventilating Co., Park Bldg., Cleveland, $39,000.

O., Springfield—The Bd. Educ. is having preliminary plans prepared for the construction of 5 separate buildings and additions to 3 or 4 of the present school buildings. Steam heating systems will be installed in same. A $1,000,000 bond issue has been passed for the project. O. D. Howard, 8 East Broad St., Columbus, Archt.

Ind., Anderson—The City Council has passed an ordinance appropriating $350,000 to enlarge the light and power plant.

Mich., Birmingham—The Bd. Educ. engaged Baxter, O'Dell & Halpin, Archts., 1024 Hammond Bldg., Detroit, to prepare plans for the construction of a 3 story school building on Main St. Steam heating equipment, including boiler, etc., will be installed in same. Total estimated cost, $200,000.

Ill., Marseilles—A. G. Zimmerman, Archt., 85 Ninth Ave., New York City, will soon award the contract for the construction of an 8 story warehouse and manufacturing building for the Natl. Biscuit Co., 10th Ave. and 15th St. A steam heating system will be installed in same. Total estimated cost, $800,000.

Wis., Hartford—The Kissel Motor Car Co. is having plans prepared for an addition to the boiler house here. Sterling boilers will be installed in same. Cahill & Douglas, 217 West Water St., Milwaukee, Engrs.

Wis., Racine—Cahill & Douglas, Engrs., 217 West Water St., Milwaukee, are receiving bids for furnishing one 18 x 72 ft. return tubular boiler for the Hartman Trunk Co., Hamilton Ave.

Wis., Racine—Cahill & Douglas, Engrs., 217 West Water St., Milwaukee, are receiving bids for furnishing three 300 hp. boilers and stokers for the Racine Mfg. Co., 6th and Mead Sts.

Minn., Marble—The village is in the market for a 125 hp., 125 lb. pressure marine or locomotive type boiler. Carl Nelson, Clk.

Tex., Bronco—The Lovington Pub. Utilities Co., Lovington, N. M., plans to construct an electric light and power plant, here.

Tex., Houston—The city plans to build a power house, and barn and improve shops, cost, $63,000. Address Mayor.

Tex., Houston—The Houston Lighting & Power Co. plans to construct a 2,300 volt primary service line to West University Pl. Estimated cost, $10,000.

Tex., Toyah—The City Council has granted a franchise to Jesse Knight for the construction of an electric light and power plant.

Tex., Vernon—The Texas Pub. Serv. Co. plans to construct extensions and improvements to its electric light and power plant, here. Estimated cost, $100,000. A. V. Foster, Toledo, O., Pres.

Tex., Wharton—The Texas Gas & Electric Co. will rebuild the electric light and power plant recently destroyed by fire entailing a loss of $100,000.

Okla., Okmulgee—John M. Moore is having plans prepared for the construction of a 4 story, 100 x 100 ft. garage. A steam heating system will be installed in same. Total estimated cost, $250,000. Smith, Rea, Lovitt & Senter, Parkinson Bldg., Archts.

Ore., Portland—Multnomah Co. Hospital Comn. will soon award the contract for the construction of a 5 story, 150 x 210 ft. hospital. The boiler house and laundry are to be in a separate building. Total estimated cost, $800,000. Noted March 9.

Cal., Long Beach—The city received bids for the construction of a sewage pumping plant with pumps, motors, flush tanks, etc., from John Cummings, 1200 Washington Bldg., Los Angeles, $44,650; John Radick, $46,999; Peter Tomick, $57,900. Noted March 3.

Cal., Los Angeles—The Bd. Educ., Security Bldg., received lowest bid for installing a steam heating system in the proposed 1 and 2 story, 132 x 220 ft. school on East 7th St., from Thomas Haverty Co., 8th St. and Maple Ave., $13,850.

Cal., Redding—The Pacific Gas & Electric Co. plans an expenditure of over $3,000,000 on electrical developments in Shasta Co., this season. Plans include the construction of 2 power houses in Hat Creek Valley, each developing 10,000 hp., a pole line from the power plants to Cottonwood and a power plant at Falls River Mills, to generate 68,000 hp.

Cal., Richmond—The Bd. Educ. voted $565,000 bonds to purchase sites, build and equip 6 school buildings.

Cal., Riverbank—The Atkinson, Topeka & Santa Fe Ry. Co., Ry. Exch. Bldg., Chicago, plans to build an ice plant here which will have a daily capacity of 1,000 tons.

Cal., San Diego—The Bureau of Yards & Docks, Navy Dept., Wash., D. C., will soon award the contract for furnishing and installing refrigeration equipment in the Industrial Building at the Marine Corps Naval Base, here. Estimated cost, $50,000.

Cal., Susanville—The Lassen Electric Co. plans an election to vote on a $100,000 bond issue to construct a transmission line from here to Westwood, a distance of 25 miles.

M. I., Guam—The Bureau of Yards & Docks, Navy Dept., Wash., D. C., received bids for the installation of turbo alternators, exciters and switchboard, at the Naval Station here, from the Westinghouse Electric Co., 803 Hibbs Bldg., Wash., D. C., at $33,107; General Electric Co., River Rd., Schenectady, N. Y., at $34,625. Noted May 25.

Que., Shawinigan Falls — The Amas Waterworks Co. will receive bids in June for the establishment of waterworks to comply with the city supply and development of hydraulic power. Estimated cost, $175,000.

Ont., Milverton—The Municipality will soon award the contract for the construction of a pumphouse and installation of a 500 g.p.m. electric driven pump in connection with the proposed waterworks system. E. A. James Co., Ltd., 36 Toronto St., Toronto, Engrs.

Ont., Toronto — The Hydro Electric Comn. of Ontario, University Ave., plans to construct the first unit of a hydro-electric plant, which will have a 37,300 kw., 3 phase, 25 cycle, a.c. capacity. A steam standby of 50,000 hp. capacity will be installed in same. Total estimated cost from $1,500,000 to $2,000,000. F. A. Gaby, Hydro-Electric Bldg., Supt.

Ont., Windsor—W. T. Piggott, Dougal Ave. and London St., W., plans to construct a 1 story, 60 x 80 ft. mill on Dougal Ave. Plans include an office building, boiler house and dry kiln. Steam boilers of about 60 hp. capacity will be installed in same. Total estimated cost, $35,000.

CONTRACTS AWARDED

N. Y., New York—The United Electric Light & Power Co., 130 East 15th St., has awarded the contract for the construction of a 1 story, 30 x 100 ft. transformer station on Elizabeth St., to James A. Henderson, 30 East 42nd St. A steam heating system will be installed in same. Total estimated cost, $100,000.

N. J., Verona—The Hospital for Tuberculosis Diseases has awarded the contract for the installation of a steam heating system in the proposed group of hospital buildings, to J. H. Nelles Co., Irvington, at $44,000. Noted April 27.

N. J., Vineland—The state has awarded the contract for the construction of a 2 story, 50 x 75 ft. cold storage and cannery building at the Institution for the Feeble Minded here, to J. Pasquale, Landas St., at $36,000.

Md., Cumberland—The Kelly Springfield Tire Co. has awarded the general contract for the construction of a 3 and 4 story, 575 x 620 ft. rubber tire factory on River Rd., to the Hunkey-Conky Constr. Co., Cleveland, O. Complete equipment for power house, pumping station and main factory building will be installed in same. Total estimated cost, $2,000,000.

Tenn., Memphis—The Memphis Gas & Electric Co. has awarded the contract for the construction of a 3 story substation on Elzey Pl., to be known as the booster station, to E. J. Pearson, U. & P. Bank Bldg. Estimated cost, $25,000.

O., Cleveland—The American Steel & Wire Co. has awarded the contract for the construction of a 1 story, 50 x 130 ft. boiler plant on Lakeside Ave., to the Drummond Miller Co., 4500 Euclid Ave. Estimated cost, $75,000.

Ia., Greene—The School Bd. has awarded the contract for the installation of a plumbing and heating system in the proposed city high school, to the W. D. Reed Plumbing & Heating Co., Kansas City, Mo., at $14,456.

Minn., Owatonna—The Bd. Educ. has awarded the contract for the installation of a heating, plumbing and ventilating system in the high school here, to P. J. Gallagher, Faribault, at $91,833.

Kan., Wichita—The Arkansas Valley Interurban R.R. Co., 126 West 1st St., has awarded the contract for the construction of a 2 story, 80 x 112 ft. interurban station, to the G. H. Siedhoff Constr. Co., 121 Market St. A steam heating system will be installed in same. Total estimated cost, $125,000.

Okla., Carnegie—The city has awarded the contract for the construction of a reservoir and power house and the installation of pumping equipment in connection with the proposed waterworks improvements, to the Ajax Constr. Co., Lawton, at $18,739.

Okla., Duncan—The city has awarded the contract for the construction of a 250,000 gal. reservoir and deep well and the installation of pumping equipment in connection with the proposed waterworks extensions, to N. S. Sherman Machine & Iron Works, 18-32 East Main St., Oklahoma City, at $35,350.

P. I., Manila—The Diocese of Manila, c/o York & Sawyer, Archts. and Engrgs., 50 East 41st St., New York City, will build a 2 and 3 story, 100 x 200 ft. hospital building. A steam heating system will be installed in same. Total estimated cost, $600,000. Work will be done by day labor.

Ont., Peterboro—The Canadian General Electric Co., 212 King St., West Toronto, has awarded the contract for the construction of a 6 story electric manufacturing building, to John E. Hayes, 219 Park St. A steam heating system will be installed in same. Total estimated cost, $500,000.

POWER

Volume 51 — New York, June 15, 1920 — *Number* 24

COPYING A BAD EXAMPLE

Brown, the salesman, was making one of his periodic calls on Howe, the big chief The salesman, just back from a vacation, had a story of some very large fish "somewhere in Maine."

The big chief showed kindly interest. Was he not saving a story of his own? And having the last say, and being a man not without imagination, could he not surpass anything the salesman was offering?

Entered upon this pleasing scene a meek-mannered individual clothed in denim, who shuffled forward a little and stood uneasily, hat in hand, with left leg awkwardly "at rest," waiting to tell *his* story.

Brown finished. The chief laughed generously. Then his face hardened as he turned with a curt, "Well, what's the matter now?"

"Mr. Howe, the steam is down and the firemen say they can't get it up unless they can put on the spare boiler."

It is one of Howe's strict orders that the bank shall not be broken on that spare boiler without his permission.

"Who put them up to that? You? If you don't look out, I'll put the rollers under your skids. Now you get back there quick and see that the steam is up or you you will be on the bricks."

When the chief turned to Brown, anger had disappeared from look and tone. The second fish story began. It had not progressed far when in rushed an oiler, breathless—

"The safety on —— turbine kicked off and she shut herself down."

Now, no one can start that turbine except Howe—not if he knows it.

"What in —— did you do to it?"

"Nothin'."

"Don't stand there and lie to me. Go and pack your duds and get out of here."

And out into the plant went both—one to take his leave, the other to do a simple piece of work that any one of three men standing by could have done as well or better.

And Brown? He was dumbfounded. He had recommended Howe for this job, and here was the real Howe—not exactly the prince he had always seemed. Here was the "Mr. Hyde" version. Brown shook his head sadly and blew smoke rings toward the ceiling. Finally he said, half aloud: "I'll risk it. He deserves it and it may help him." He wrote a short note on a pad of paper lying on Howe's desk, then made his escape. "For," he said, "it is better to let it sink in."

When Howe returned he read this:

"A Word to the Wise"

"The army of a king of ancient times (history is not my specialty, you know, and I do not remember his name) was meeting disastrous defeat. From time to time heralds arrived from the battlefield with the bad tidings. Every herald, as he told his story, was struck dead by the king.

"During all the years since then not one man has ever risen to commend this ancient king for his deed. Yet men have emulated his example. *I* know one who is doing it now. Wake up! It is 1914."

Epilogue

This happened six years ago, in the "good old days" now passed forever, when men could be browbeaten in that manner. No one tries to "handle" men in that way now. There has been a change—in the *way*. There are still executives of all grades who blame the herald for the facts.

Illogical? Unjust? Unmanly? Yes, and more.

The only safe way is to welcome truth, no matter how unpalatable it may be.

Contributed by WILLIAM E. DIXON.

Combined Gas and Motor-Driven Pressure Booster Station

By A. E. WALDEN

An increased water consumption called for an additional supply of water delivery to the distribution system. This was accomplished by building two pressure booster stations, consisting of combination gas-engine and motor-driven centrifugal pump units of 75- and 150-hp. capacity, each with a pumping capacity of 2,000,000 gal. of water daily, at 55 ft. to 115 ft. and 231 ft. head respectively.

DURING the early summer of 1917 the demand for water on the Baltimore County Water and Electric Co.'s system in the district of Mt. Washington and Dickeyville, Ind., with a population of over 60,000 in the immediate territory—increased so rapidly, owing to wartime conditions, that it became necessary to augment the supply of water delivered to the distribution system, spread out somewhat in the shape of a fan to a distance of 25 miles from the main source of supply and the filtration plant, the total population in Baltimore County being something over 137,000.

FIG. 1. INTERIOR OF MT. WASHINGTON BOOSTER STATION

FIG. 2. INTERIOR OF DICKEYVILLE BOOSTER STATION

The source of supply of water is the Patapsco River, on which the dam, power house and filtration plant are located. The water, after being filtrated, is taken from the clear well by means of high-duty, high-head pumps, driven by turbine waterwheels or a tandem-compound steam engine or a four-stage steam-turbine-driven pump working under 712 ft. to 775 ft. head, as conditions may require, delivering the water automatically to a 4,000,000-gal. reservoir or to a 60-ft. steel standpipe having a capacity of 600,000 gal., the reservoir and base of the standpipe being between 519 and 525 ft. above sea level.

The distribution system is supplied jointly from reservoirs and standpipes some 19 miles from the first sub-pumping station, in which are located two 2-stage, 2,000,000-gal. vertical motor-driven pumps under from 130 to 150 ft. head. The pumps are directly connected to three-phase 60-cycle 220-volt 60-hp. motors, the impellers and shafts of the pumps being carried on a hydraulic thrust bearing. There is in addition a motor-driven, automatically controlled triplex pump installed in the substation, which also houses the electric meters, transformers, switching equipment and a water meter for recording the total water delivered to the system in this district.

The floor of this substation has an elevation of 519 ft. above sea level or datum, from which the elevations vary down to 93 ft. above datum, increasing again to from 500 ft. to 560 ft. above datum, these being the reservoir and standpipe elevations respectively, at a distance from the repumping station as already given, and from which the water is distributed to other sections between and beyond the four sub-pumping stations and reservoirs, the pressures on the intervening main line being from 60 to 200 lb. per square inch.

Not only was it impossible to obtain deliveries of cast-iron pipe at this time, but quotations were $75 per ton, or three times the normal price. Furthermore, the cost of laying the pipe was excessive and labor was hard to obtain.

It was evident that some less costly method of increasing the supply of water must be resorted to until normal conditions were restored. The present pipe line, when originally installed, was tested to 300 lb. per sq.in., and after some consideration the company's engineers, Wehr and Walden, decided to install one or more booster pumps. To this end a design was worked out combining on one bedplate a multiple-cylinder gasoline engine, an electric motor and a turbine pumping unit.

The first unit, installed at Mt. Washington in July, 1918, consisted of a 75- to 80-hp., six-cylinder gasoline engine of the automobile type, running at 1,200 r.p.m., equipped with a two-unit 24-volt electric starting motor and charging generator similar to the standard automobile equipment, but the engine was provided with additional equipment, such as automatic speed control, a vacuum system for fuel feed from a 250-gal. storage tank, electric speed indicators, recording thermometers for the cooling water, oil-pressure gages, a mercury U-gage connected to the intake manifold, etc. The engine is automatically oiled by a rotary pump located in the base and driven by the main engine from the layshaft, the oil pressure on the bearings being maintained at 20 lb. per sq.in. The hot oil is also circulated in a jacket around the intake manifold, heating it and at the same time reducing the temperature of the oil somewhat.

This unit is connected through a flexible coupling to a single-stage pump to operate at 1,200 r.p.m. and deliver 2,000,000 gal. daily at two different heads, 55 and 115 ft.; the pump, in turn, is again connected at the other end, through a flexible coupling, to the shaft of a 50-hp. 220-volt three-phase motor, to provide against possible breakdowns of the engine unit. This unit is self-contained.

The starting current for the 24-volt starting motor and station lighting is supplied from a 24-volt 150-ampere-hour storage battery which, while the engine is in operation, is being charged continuously.

Figs. 1 and 2 show the Mt. Washington and the Dickeyville booster-station equipment and Fig. 4, one of the outdoor electric substations, transformers, high-tension lightning arresters and switching equipment. The flow area of this station is 300 sq.ft. This engine has been in continuous and successful operation from the time it was started.

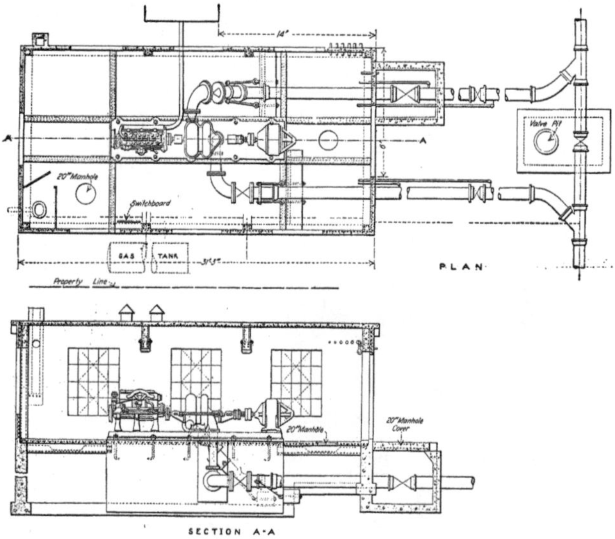

FIG. 3. PLAN AND ELEVATION SHOWING GENERAL ARRANGEMENT OF PUMPING PLANT

The Dickeyville station was not completed until July, 1919, on account of the inability to obtain the engine and pumping equipment. This station has a floor area of 450 sq.ft. and contains a six-cylinder 150-hp. marine gasoline engine running at 1,200 r.p.m., Fig. 2. It is equipped with speed control, triple ignition system, electric speed indicator, gasoline pumps, recording thermometers, two-unit 24-volt automatic starting motor and charging generator, and 24-volt 150-ampere-hour storage batteries, recording and indicating pressure gages and a U-shaped mercury column connected to the intake manifold, which indicates at all times the condition of the valves and the operation of the engine.

The fuel for this unit is taken from a 500-gal. storage tank by a rotary pump driven by the layshaft on the main engine. The fuel is delivered to an elevated tank in the building, equipped with a sight gage and overflow to the main tank.

The forced-feed oiling system is supplied by a pump, driven from the layshaft, and is located at the base of the engine, which maintains the oil pressure on the bearings at 10 lb. or more per square inch.

This unit is connected, through a flexible coupling, to a two-stage turbine pump and delivers 2,000,000 gal. daily under a 231-ft. head; it is equipped with a hydraulic balancing device to balance the thrust. The pump is also connected at the other end to a 100-hp. 60-cycle three-phase motor, to insure continuity of service; the motor under ordinary conditions of operation is idle. The general arrangement and plant layout are shown in Fig. 3.

These installations have so far operated satisfactorily, except for some shaft trouble, due to too light construction for continuous duty of the 150-hp. unit, and also carburetor troubles. Some types of carburetors do not give the same results and efficiencies as others, but the operating conditions are more easily checked from the indications given by the difference in the levels of the legs of the mercury column, which varies from 4½ to 9½ in. This is an indication of both the operating condition of the engine and the load.

It was found that the oil pressure on the bearings should not go below 10 lb. per sq.in. for the best results, and preferably 20 lb., using what is known as No. 4 gas-engine oil for lubricating purposes.

FIG. 4. OUTDOOR SUBSTATION AND BOOSTER STATION

The operating troubles in these plants have been mostly due to dirt in the strainers, poor contact of brushes in the magnetos and generator brushes. To care for this promptly, duplicate magnetos and spare parts should be kept on hand.

A slight amount of water is introduced into the intake by means of the suction or difference of pressure between the atmosphere and that in the intake, and mixed with the air and gas. The water is contained in a small tank mounted slightly above the engine and connected by a 1-in. pipe to the intake manifold. The carburetors that have given the best results in operation seem to be those in which the air velocity and vaporization of the gas are such as to cause considerable sweat on the outside of the carburetor, due to difference in temperature, but these should have a reasonable adjustment of the float-feed valve to care for the variations in oil, and are preferably equipped for heating by means of a small amount of hot water from the jackets or in some other suitable manner.

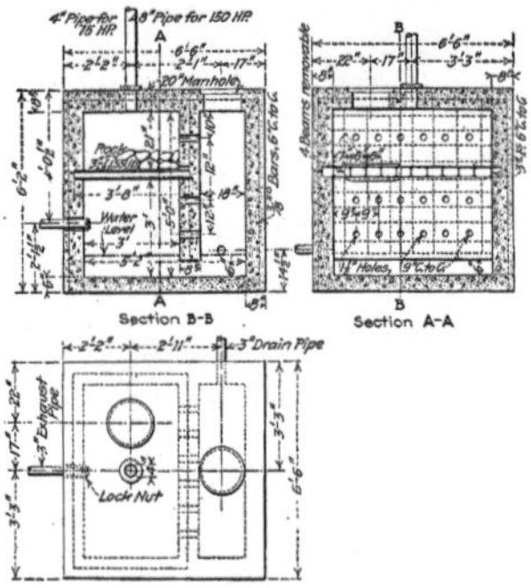

FIG. 5. DETAILS OF MUFFLER-PIT CONSTRUCTION

The muffler pits, shown in Fig 5, have worked perfectly, there being very little noise, no smoke, but a little vapor from the water which is injected into the exhaust manifold to cool it and prevent backfiring in the pit; and the water injected into the intake manifold which is to a certain extent a further preventive of the burning of the exhaust-valve stems.

The dimensions of the pits as shown are ample for engines up to 150 hp., so that there will be no objection to the operation of these units in residence districts or where there are restrictions as to noise, etc.

There are always two batteries in service, one full and one under charge, so as to assure one full battery for turning over the engine in case of any trouble.

The 150- and the 75-hp. engines require one barrel of oil per month each, and it seems that if a cooling tank could be connected to the pans under the base, this system would materially reduce the expense for oil on these engines.

In selecting gas engines for power purposes, if the engine has a rating of 150 hp., every effort should be made to see that it develops that horsepower at the speed, or slightly above the speed, at which it is desired to operate the unit. With the rating reduced, in the case of a six-cylinder engine, by two cylinders, the maximum load of this engine, in so far as the purchaser is concerned, would be 100 brake horsepower. In other words, the 150-hp. unit would be rated to deliver its brake horsepower at 25 hp. per cylinder.

The unit to be driven should have its horsepower requirements accurately determined by an electric dynamometer and contract provisions should be made accordingly.

Vertical-Shaft Waterwheel-Driven Alternators—Brakes, Ventilation and Exciters

Brakes Operated by Hand and Compressed Air, Ventilation of Alternators and Excitation with Direct-Connected and Independent Exciters

By S. H. MORTENSEN
Electrical Engineer, Allis-Chalmers Manufacturing Company

WITH the larger types of water-driven alternators brakes are generally furnished. They are required to bring the unit to a standstill and to keep it at rest in case leaky turbine gates allow a sufficient amount of water to pass through the wheel to keep it in motion. Brakes are also of great value in case of accidents, when it becomes of the greatest importance to stop the unit in the shortest possible time.

On vertical machines the brake shoes rub against the finished surface of the spider rim. The brake shoes may be made of wood or asbestos wood, and the braking power may be supplied from a piston actuated by either compressed air or oil or by a system of levers, hand operated. The brake arrangement, Fig. 1, is for hand operation, and Fig. 2 shows the brake shoe and brake cylinder designed for compressed-air manipulation. Brakes are designed so they will bring the unit to standstill in from three to five minutes, with a gate leakage of approximately 3 per cent.

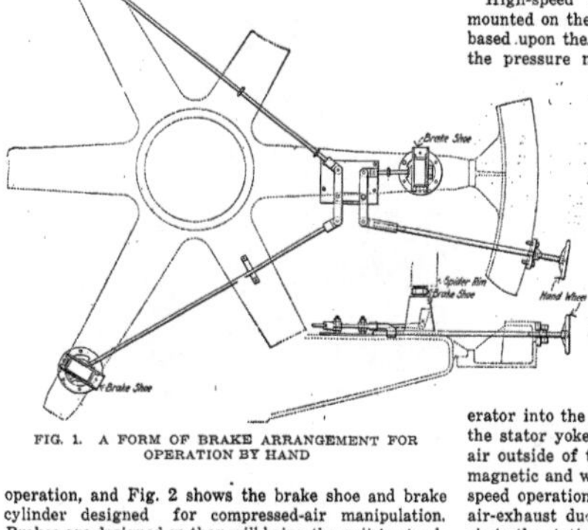

FIG. 1. A FORM OF BRAKE ARRANGEMENT FOR OPERATION BY HAND

Ventilation of small vertical alternators is simple. The radiating surfaces, combined with the draft set up by the field rotation, generally is sufficient to dissipate the heat generated by the various losses in the machine. On the medium-sized machines ventilating vanes mounted upon the rotor spider are relied upon to stir the air and create a sufficient draft to keep the machine cool. The ventilating air is usually taken from the top and bottom of the rotor and is exhausted through the stator yoke into the station.

Slow-speed machines of large capacity generally operate at too low peripheral velocity to produce sufficient air pressure for their proper ventilation, which then must be supplied by independent fans or blowers. Such installations are necessarily more elaborate and require air ducts leading to and from the machine. The intake duct or tunnel should be arranged to take the air from over the tailrace. This insures cool, clean ventilating air that does not require any further filtering or washing, and it is necessary to provide only a coarse screen in the air tunnel to prevent large objects from entering. The blowers are generally motor driven, and must supply the required volume of air at a sufficient pressure to force it through the intake ducts, the machine itself and exhaust it through the tunnels leading outside.

High-speed machines are self-ventilated by fans mounted on the field spider. The design of the fans is based upon the amount of air they have to deliver and the pressure necessary to force this amount of air through the ducts and the machine. In Fig. 1 of the previous article was shown a self-ventilating alternator. The top of the machine is totally inclosed, and the ventilating air is obtained from air ducts leading to the pit under the alternator. After it passes through the fans, part of it sweeps over the coil ends and out through openings in the yoke. The remainder, after cooling the field coils, is forced over the inner stator-core surface, through the ventilating openings in the stator core, and escapes finally through holes in the stator yoke. In installations where it is undesirable to have heated air escape from the generator into the station, an inclosure is provided around the stator yoke. This arrangement not only leads the air outside of the station, but also greatly reduces the magnetic and windage noises that are inherent in high-speed operation. Sometimes doors are provided in the air-exhaust ducts, which permit the admission of hot air to the station when desirable for heating purposes.

The amount of air required per kilowatt loss in an alternator depends upon the air temperature. Where the ventilating-air temperature is from 15 to 25 deg. C. (69 to 77 deg. F.), from 100 to 115 cu.ft. per minute per kilowatt loss should be supplied. If the air temperatures are from 25 to 40 deg. C. (77 to 104 deg. F.), the volume of air that should be supplied will range from 115 to 135 cu.ft. per minute for each. The kilowatt loss air velocity through the ducts should not exceed from

200 to 300 ft. per minute for self-ventilated slow-speed machines. For high-speed or separately ventilated units velocities up to 1,000 or 1,200 ft. per minute are permissible. This leads to large air ducts and makes it necessary to give due consideration to their proper location and dimensions at the time the buildings for the power stations are designed. As an example, if the ducts leading to the alternator should be proportioned for 80,000 cu.ft. of air per minute, allowing an air velocity of 1,000 ft. per minute would then necessitate a duct section of 80 sq.ft. The ducts should be as short as practicable and also be without bends or abrupt turns, and both the intake and exhaust openings should be protected against wind pressure, flood water, etc.

FIG. 2. COMPRESSED-AIR OPERATED BRAKE

In addition to checking the temperatures of the generator proper, it is advisable to check the temperature of the ventilating air. If the temperature increase exceeds 18 to 22 deg. C. (32.4 to 39.6 deg. F.), it is an indication that the amount supplied is not sufficient, and arrangements should be made to increase the volume of the ventilating air.

For an alternating-current installation the choice of the excitation system is necessarily of great importance and should be given careful consideration. It must be reliable, flexible, efficient, have ample spare capacity and at the same time represent the minimum investment compatible with the requirements mentioned.

Small slow-speed alternators require comparatively large-capacity exciters, which, if run at the alternator speed, would be expensive and have a low efficiency. To overcome this handicap, this class of machine is generally supplied with high-speed exciters geared or belt driven from the alternator shaft. Spare capacity should be provided either as extra parts for the exciter proper or as separately driven exciter units.

In large-capacity stations or in stations with high-speed units the exciters are either direct connected to the main unit, as in Fig. 3, or driven by separate waterwheels.

The direct-connected exciter is mounted on the top bearing housing of the alternator and its armature upon an extension of the mainshaft. The exciter capacity is made sufficient for one alternator except in stations where only two units are installed. In such installations spare capacity is obtained by using exciters with sufficient capacity for both units. The exciters may operate in parallel, but are more frequently wired directly to the alternator. They should, however, have taps for connecting them to a transfer bus, thereby making it feasible to operate any alternator in conjunction with any exciter in the station.

In a station with direct-connected exciters it is customary to supply a motor-generator set as a spare unit. The motor may be either of the synchronous or the induction type, and the exciter capacity should be sufficient to take care of one alternator. In stations where a great many units are in operation, it is advisable to provide two such spare sets.

Installations with the waterwheel-driven exciters generally have two duplicate units, each with sufficient capacity to carry the total station load. In addition they should have sufficient excess capacity to provide power for operating the station auxiliaries.

The separately driven exciter, as compared to the directly connected type, has an advantage in small- and medium-capacity installations. However, in large or high-speed installations the use of directly connected exciters is advantageous.

Either type of installation requires a main field rheostat with sufficient capacity to take care of all the operating conditions that may occur. This means that in stations with separately driven exciters, the current-carrying capacity and resistance must be proportioned to meet the requirements based upon full excitation voltage, which leads to large and expensive rheostats. With the direct-connected exciters it is practical to alter the exciter voltage within certain limits by means of an exciter rheostat and thereby materially reduce the operating range required from the main field rheostat. This substantially lessens the rheostat cost and reduces the waste of energy in the rheostat resistance, thus increasing the operating efficiency. An exception to the rule that rheostats are required, is to be found in stations operating under constant-load, voltage and power-factor conditions. For such installations successful operation may be obtained without the use of a main field rheostat.

FIG. 3. EXCITER MOUNTED ON TOP BEARING HOUSING

In conclusion attention is called to the danger to which a station with direct-connected exciters is subjected when a unit runs away and the cumulative effect of the increase in the frequency and also in the exciter voltage results in an over-voltage much in excess of the normal value. This condition may be guarded against in different ways—as an example, by means of specially designed automatic voltage regulators or by some speed-limiting device acting upon the turbine gates.

Testing for Ammonia Leaks in Refrigerating Plants

Gives Some Simple Directions for Testing for Ammonia Leaks witl Phenolphthalein and Sulphur Sticks

By B. E. Hill
Field Refrigerating Engineer with Swift & Company

TESTING for ammonia leaks is an important part of an engineer's duty. There are two principal places to test, one being in a submerged tank, where, if there is a leak in the coils, water cooler or any liquid, the ammonia will be absorbed. If the liquid is of great volume there will be a considerable quantity of ammonia taken up before the engineer or attendant can detect the odor.

To test the brine or other liquid for a leak, send to the drug store for an ounce or two of phenolphthalein; then get a sample of brine in a glass bottle or drinking glass and put a drop or two of the phenolphthalein into the brine. If there is the least trace of ammonia, even so small that it cannot be detected by smell, the phenolphthalein will cause the brine sample to turn a faint pink, and if there is enough ammonia to detect slightly by smell, the sample will turn a deep red.

If desired to determine how much anhydrous ammonia has leaked out into the brine or other liquid, take a sample of the brine to a good chemist and give him the total volume of the liquid; he will make the necessary calculations for ascertaining the total leakage.

The next test is for leaks in the open air, and as the sulphur stick is the most popular, effective and practical, we will consider where it can be used and where not.

To prepare the sulphur stick melt five or ten pounds of either powdered or lump sulphur in a ladle with a very slow fire. A slow fire is necessary as the sulphur fires at a low temperature and, if too hot, will become lumpy and thick and will run off and drip when used. When the sulphur is at the right temperature it will be thin like water. Now the sticks should be prepared by splitting up a hundred or so pine strips, cardboard or other material and dipping both ends in the sulphur, which quickly cools; the sticks are now ready for use. To apply the sticks, take a handful and light one, and before one burns out another can be lighted from the one in use.

Any coil that has just been erected or overhauled should first be tested with air pressure, after which the joints should be brushed with soapy water. After all leaks found in this way have been stopped, pump out the air until the vacuum gage shows as high a vacuum as can be had. If it is found that the air is still discharging through the discharge bypass, keep on pumping till the air is hardly perceptible. This extra precaution is taken to make sure that the air is as nearly all out as can be pumped with the machine; also, there is always a chance that the vacuum gage may not be correct.

After the air has been pumped out the ammonia should be charged into the coil at once and the final test made with the burning sulphur held under and over all joints and connections. If there is a leak at all, it will be shown by a gray smoke wherever the sulphur and

A LARGE AMMONIA COMPRESSOR DRIVEN BY A COMPOUND UNIFLOW-CORLISS ENGINE

ammonia fumes come together. The sulphur fumes are as effective in showing a leak in the open air as the phenolphthalein is for showing one in the brine tank; as, for example, where there is a leak in a thread or joint that is so small that it will not be noticed with the soap bubbles and is difficult to detect by smell after the leak is found, the sulphur will mark the leak by leaving a small yellowish white spot, even though the leak is no larger than a pin point.

Leaks on return bends of atmospheric condensers, where the condensing water is flowing over the part that is leaking, will cause the solids in the water to accumulate to a considerable extent, which deposit will be of a whitish color. Leaks in water jackets on the compressor, if of any extent, will clarify the water, and if the fingers are dipped in it and rubbed together they will feel soapy. In this case the leak can easily be detected by smell.

In a room where the timbers or building materials of any kind are damp from moisture, the sulphur stick is almost useless, as the moisture will take up the ammonia gas and the sulphur will sometimes "smoke up" the whole room and show smoke on a post or the wall where there is no leak. In such cases it is sometimes necessary to pump the coil down to about atmosphere on the gage and open the room so that a circulation of air will drive the ammonia out of the room before the leak can be found. This holds true unless the leak is large enough to be seen or heard.

A crusade on ammonia leaks should be made regularly by going over all joints and connections with sulphur sticks, marking the leaks as found with a piece of chalk, then following up with the repairs.

Oxide-Film Lightning Arrester

As a result of the demand for a lightning arrester that would give the same amount of protection as the aluminum arrester without the necessity of daily charging, the oxide-film lightning arrester has been

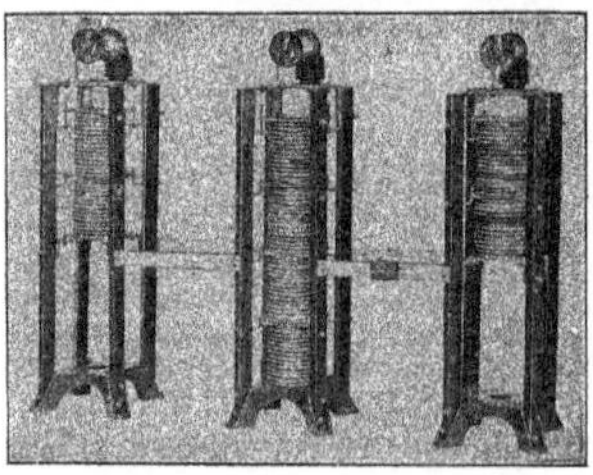

FIG. 1. LIGHTNING ARRESTERS FOR INDOOR SERVICE ON THREE-PHASE CIRCUIT

developed. This charging consists in connecting the aluminum arrester to the line for a short period each day, in order to renew the non-conducting film on the plates, and has made this arrester impractical of installation in isolated places. The oxide-film arrester does not require daily attendance to insure the maintenance of its protective qualities. Therefore, although it is practical for almost any application, its principal field seems to be where daily attention is not available but where good protection is required. Lately, there has been a large increase in the number of isolated stations and substations where the class of service calls for highly reliable protection, and it is here that the oxide-film arrester finds its greatest field of usefulness. The construction of the arrester is very simple, requiring no oil or electrolyte. It consists of a stack of cells connected to a sphere gap, Fig. 1. In the outside type these cells and sphere gaps are thoroughly protected from the weather, which insures a uniform spark potential under all climatic conditions. The cells, one of which, developed by the General Electric Co., is shown in Fig. 2, consist of two thin steel plates, or electrodes, spun on a porcelain ring, the latter acting as an insulator. The space between the plates, about half an inch, is filled with lead peroxide. The electrodes are further provided with an insulating film of varnish or some similar compound which gives a uniform insulated surface.

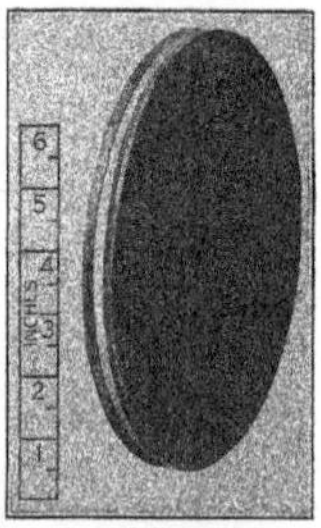

FIG. 2. OXIDE-FILM LIGHTNING-ARRESTER CELL

The operating principle depends on the fact that lead peroxide, which ordinarily is of low electrical resistance, is unstable at a temperature of about 250 deg. C. and forms a lower oxide, called litharge, of high resistance to electric current. Thus, although its action is thermal instead of electrolytic, it functions in a good deal the same way as an aluminum-cell arrester, by preventing the flow of dynamic current. What actually happens in practice is that up to a pressure of about 300 volts per cell the insulating film prevents any noticeable amount of current from flowing under normal conditions. As soon as the voltage rises above normal sufficiently to spark on the gaps, the film is punctured and the lightning discharge flows to ground, since it meets very little resistance. When the dynamic current attempts to follow, the smallness of the punctures causes great concentration of current at those points, with a consequent localized heating which is enough to cause the lead peroxide to break down, forming litharge, in the path of the current, thus effectually blocking it. The film reseals itself in about one four-thousandth of a second after the excess lightning voltage has ceased. Since the powder is a poor heat conductor, the formation of these high-resistance plugs is confined to the actual area surrounding the puncture.

As this action is due to the inherent chemical property of the material between the electrodes, and it is this material which forms and reforms the insulating film, there is no action similar to the gradual dissolving of the film as in there is the aluminum arrester.

Efficient Operation of the Boiler Plant

By JOHN D. MORGAN

Various types of stokers are treated, as regards requirements of design, also the effect of boiler feed-water temperature. Plant labor is considered and data and curves are presented that have a direct bearing on boiler-room maintenance, together with suitable plant-record sheets.

HAND-FIRED furnaces are rapidly becoming obsolete, but when so fired the use of some type of rocking grate is the best practice. Although the loss of coal through the grates is slightly more than that of the flat grates, this loss does not affect the efficiency enough to be considered.

A suitable arch should be designed which will properly mix the air and gases. The firing should be done alternately in narrow strips across the full length of the grate, and when this method is used it is advisable to fire light and often and keep the doors cracked. The boiler should be equipped with a reliable draft gage, and the draft should be kept at a predetermined point by means of the damper and the fire thickness.

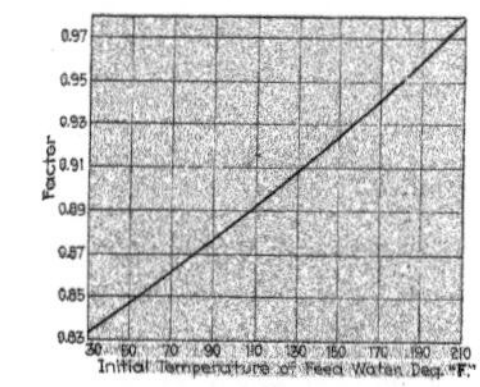

FIG. 1. FACTOR CURVE FOR PER CENT SAVING FOR INCREASED FEED-WATER TEMPERATURE

Three Classes of Stoker Firing

Stoker firing can be divided into three classes, namely: Front-feed and inclined-grate type, chain-grate, and underfeed stoker. In considering the inclined-grate type of stokers, the most important operating points of this type are the proper training of the men operating them, constant attention to avoid holes in the fire, proper feed and grate motion; and if an intensified combustion is desired, an air space should be made in front of the ignition arch. Here again, the design of the arch and bridge wall is of vital importance to proper operation. The total amount of steam used by auxiliaries for this type of stoker does not as a rule exceed 2 per cent of the total steam generated. This point should be given due weight in comparing it with other types.

The chain-grate type is in common usage in many parts of the country. The kind or nature of the coal available plays an important part in the performance of the stoker. The general items that will affect its economic operation are: Draft carried; speed of the grate; character of ignition arch; thickness of the fire and the size of coal.

Some of the advantages of the underfeed stoker are: Compactness; small combustion space; complete mixing of air and gases; gases are intensely heated by rising through incandescent coal; absence of ashes; small size of stack; complete ash fusion and the obtainability of high ratings. The only real drawback is that of cleaning the fire after a forced run of a few hours' duration.

If an underfeed stoker is to be operated in an economical way, it should have the following features: An adjustable, continuous and automatic dump by means of an adjustable opening adjacent to the bridge wall. This dump should be provided with an indicator. The air-supplying grates should reciprocate sufficiently to agitate and move the fuel bed uniformly to the dump. Each plunger connection should have a safety device

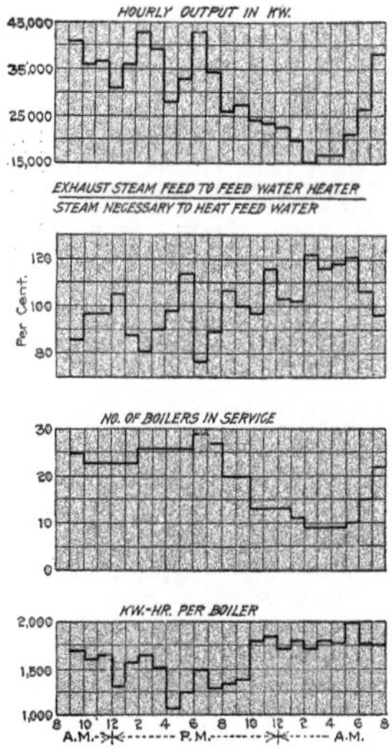

FIG. 2. FEED-WATER HEATER TEST

so that the plunger may be blocked at any point in the stroke without injury to the stoker. It should be equipped with an overfeed grate for final combustion and a variable-speed engine or motor for driving the stoker shafting.

Furthermore, there should be an instrument panel upon which should be mounted a recording temperature gage for feed-water-inlet temperature, a recording temperature gage for the temperature of steam, a pen-recording pressure gage for recording the air pressure beneath the fire, an electric tachometer for indicating the speed of stoker shafting, etc., two differential draft gages for indicating draft or pressure over the fire and at the damper, and a direct-reading U-tube gage for indicating the air pressure in the main duct.

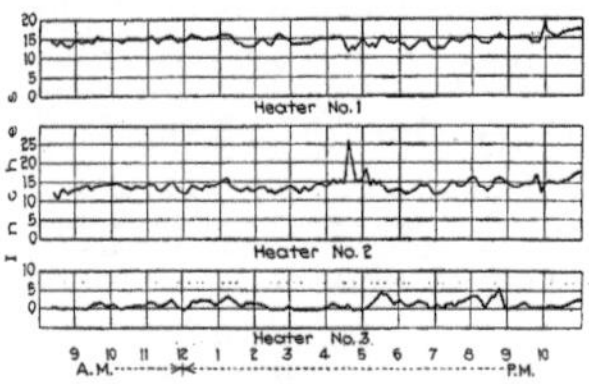

FIG. 3. HEIGHT OF WATER IN FEED-WATER HEATER GAGE GLASS IN INCHES

The first point in regard to the method of operation is to run a series of tests to determine the effect of various speeds and draft pressures on the efficiency of the stoker. After this has been done a table should be made up, which will show the following items for all various boiler ratings: Boiler rating per cent; speed of stoker drive; air pressure under fire; draft over fire; draft at boiler damper; thickness of fire, pounds of coal per hour; pounds of steam per hour; efficiency.

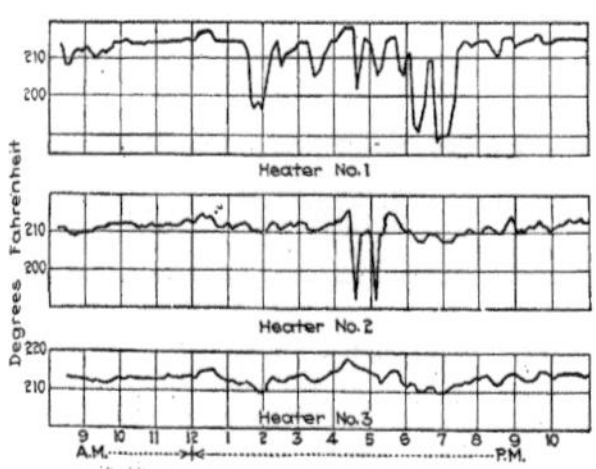

FIG. 5. TEMPERATURE OF FEED-WATER FROM HEATER, DEGREES F.

With the aid of such a table a boiler can be run in an efficient manner, but it must be remembered that at 300 per cent rating the drop in efficiency from that at 100 per cent will be about 12 per cent, and as a rule the best continuous working point will be found to be about 150 per cent rating. There is one point in the operation of this type of stoker over which engineers disagree, and that is the draft or pressure to be carried over the fire. I favor carrying a draft of 0.02 in. over the fire at 100 per cent rating and varying it gradually upward with the increase of boiler rating so that when 300 per cent rating is reached there will be a draft of 0.1 in. over the fire. These recommendations are based on the results of many boiler tests and on the final maintenance figures.

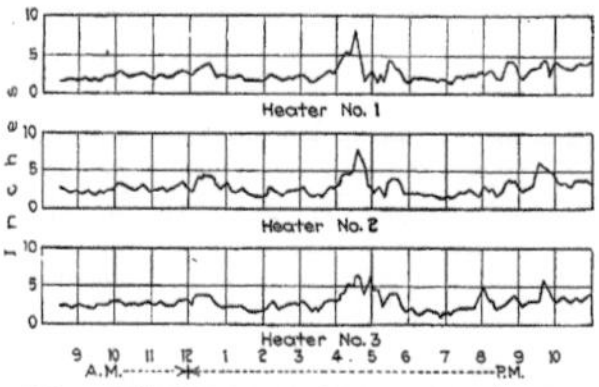

FIG. 4. PRESSURE OF EXHAUST STEAM IN HEATER, INCHES OF MERCURY

Regarding boiler-feed-water temperatures the question arises as to how much increased feed-water temperature helps. The curve shown in Fig. 1 shows the percentage of saving for each degree of increase for varying initial temperatures.

Since it has been shown that each and every degree of increase in feed-water temperature pays, the first procedure should be the making of a 24-hour test to determine the exact feed-water-temperature condition of the plant. An example of what such a test consists of is given in Figs. 2, 3, 4 and 5.

For many plants it will be found that unsatisfactory temperature conditions prevail, no doubt due to many causes, but as a rule these causes or conditions may be traced to the following items: Insufficient exhaust steam to heat the feed water; dirty feed-water heaters; irregular feeding of water into the boilers; surging and overflowing of the open feed-water heater; faulty operation of the float regulating valve on the heater and poor pressure regulation of feed-water pumps.

Insufficient Exhaust Steam

Insufficient exhaust steam for heating feed water is largely a question of plant design and can be attended to only by the installation of an economizer or a heat-balance valve. The installation of a heat-balance valve is a simple and inexpensive way of overcoming this condition. The working operation of such a valve simply consists in keeping the pressure in the exhaust main at a predetermined point by bleeding to or from the secondary stages of the turbines.

It is becoming the practice in all new plants to install economizers. This has largely been brought about by the use of motor-driven auxiliaries, on account of the increase in efficiency over steam-driven auxiliaries, and also by the use of stokers capable of high ratings, because at high ratings the temperature of the flue gas is sufficient to warrant the installation of an economizer. In considering an economizer, the first and most important point is that of the direction of water flow. Undoubtedly, this item has an effect on the efficiency of the economizer. The following plan is practically ideal: Suppose the economizer is composed of 40 rows of tubes with 12 tubes per row. For water-flow purposes it should be divided into three groups, one of 16 and two of 12 rows of tubes each. The circulation should be in series for the three groups. The circulation in the first group would be divided as follows: Up the first eight, down the next four and up the last four tubes; in the second and third groups the water should flow up the first four, down the second four and up the last four tubes. With such a circulation the drop in pressure will not exceed 25 to 30 lb. The second item to be considered is that of air leaks. There should be a total absence of air leaks if high efficiency is to be obtained. The curves in Figs. 6, 7 and 8 show the operation of an economizer.

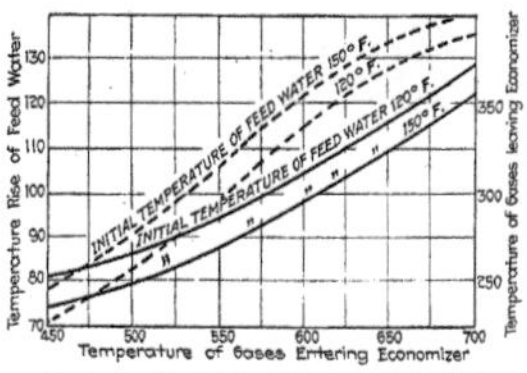

FIG. 6. ECONOMIZER PERFORMANCE CURVES

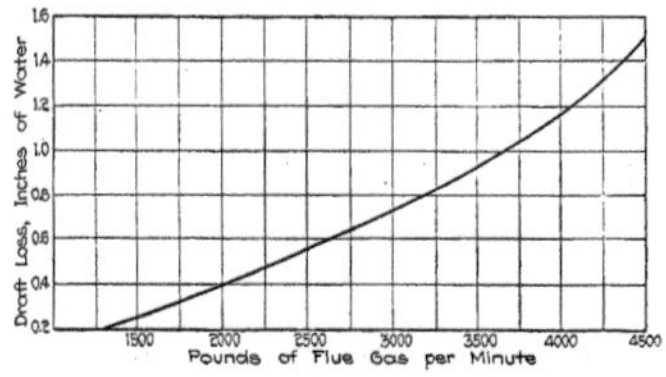

FIG. 7. ECONOMIZER PERFORMANCE CURVE

Dirty feed-water heaters are a source of serious loss, and it is advisable to inspect them and clean them at least once every month.

Irregular Water Feeding

Irregular feeding of water into the boilers can be corrected by the installation of feed-water regulators and the maintaining of an even pump-discharge pressure. And even where the boilers are hand-regulated, it is possible to maintain an even feed by drilling the water tender to study his fire conditions.

Often a serious loss results from the overflowing of the open heater. This is caused by improper feeding and a lack of regulation of the steam supply. This trouble can frequently be overcome if a seal is placed on the overflow line, constructed in the form of a V-tube or return-bend coil, and it can be designed to maintain any predetermined pressure in the heater. This will tend to do away with the surging.

The faulty operation of the float-regulating valve is also due to the improper design of the operating levers or to the float being located near to the heater outlet, so that the float gives a faulty reading due to a suction in and around the outlet pipe.

Maintaining a steady feed-water pressure is simple if a good pressure-regulating valve is put on each pump. Care should be exercised in the piping up of such a valve. For instance, the water inlet should never be tapped into the feed line at a point near the pump discharge, but rather it should have an auxiliary line of its own from some point in the main boiler-feed line.

Another plant loss, the blowoff loss, is a variable and depends largely on the valve leakage and the number of times the boiler is blown down. The valve leakage can be avoided and is a waste that should not exist.

Leakage of the blowoff valves can be readily determined by placing a recording thermometer on the blowoff line. The loss due to blowing down is an unavoidable one, but even this can be reduced by testing the boilers each day for salt. Only when a salt test shows more than 30 grains per U. S. gallon should the boilers be blown, and the engineer should tell the water tender just how many inches should be blown. Of course, where muddy water is used this loss will be high, but it should be remembered that no gain is obtained by prolonged blowing. The proper method of blowing down for mud and scale is simply to open the valve and then close it slowly.

Losses Due to Radiation

There is yet to be considered the loss due to radiation from steam and water pipes. Although most engineers are acquainted with this loss, everyone does not know the relative merit of various coverings. Table I gives this information.

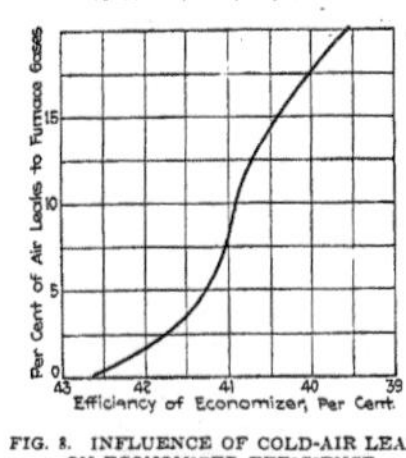

FIG. 8. INFLUENCE OF COLD-AIR LEAKS ON ECONOMIZER EFFICIENCY

Labor should be considered as relating to plant operation. Needless to say, this question should occupy the chief engineer's undivided attention, because labor is in great demand and this results in extreme unrest of the laboring class. The question of pay also demands attention, and the rule of most large companies that a fixed limit of pay must be strictly adhered to for each individual class of labor is, in my opinion, largely responsible for the present unrest in the labor world. It is folly to pay the same wage to all men in a certain class, for when this is done the men lose their individuality and become machines. Beyond doubt all classes of men who work in the boiler room are of equal importance, but a good fireman deserves to be paid according to his proficiency. The bonus plan is not advisable for the reason that the simple things of everyday operation alter the man's proficiency to a marked degree.

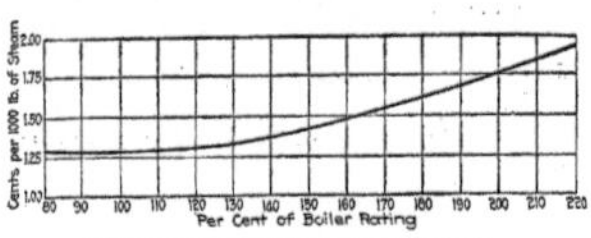

FIG. 9. EFFECT OF BOILER RATING ON MAINTENANCE COST

The old method of paying each man what he is worth, irrespective of the class of work performed, will go far toward ridding the boiler room of the present labor unrest, because it puts heart into any man to know that if he makes good the chief engineer will raise him accordingly. In the final analysis the labor cost is quite small compared to the total operating cost, and if increased labor cost will result in a decreased coal cost the result will be a net saving to the plant. The low labor costs of large plants are effected by the use of large units, and they can, therefore, afford higher wages on account of the small number of men employed. The labor problem is one that must be decided by every chief engineer as an individual.

Boiler-room maintenance varies indirectly with the net thermal efficiency of the plant. The amount of money spent in maintenance is affected by the following items: Age of plant; percentage of boiler rating developed; attendance of men; the purity of boiler-feed water; the quality of materials used for repairs, and the nature of the fuel.

TABLE I. RELATIVE MERITS OF VARIOUS PIPE COVERINGS

Nature of Covering	Thickness	B.t.u. Loss per Hour per Sq.Ft. per Degree of Average Difference in Temperature
Uncovered pipe	....	2.71
85% magnesia	1.18	0.418
85% magnesia, 2 in. section, ½ in. block	2.71	0.288
85% magnesia, two 1 in. sections	2.50	0.294
Hair felt	0.82	0.422
Rock wool	1.60	0.256
Mineral wool	1.30	0.285

The percentage of boiler rating developed has, perhaps, the greatest bearing on the maintenance charges. For instance, it affects the life of the grates, tubes, baffles, furnace and boiler auxiliaries by requiring them to do more work at a net decrease in efficiency, but just how much is a matter of conjecture. The curve shown in Fig. 9 is plotted from actual records and shows that there is some definite rate of increase.

The attendance of men is of vital importance, for to use an old proverb, "A stitch in time saves nine." Immediate attention to some small leak or break will save much time and money. Leaks in steam mains and holes in grates and furnaces never grow smaller unless they receive immediate attention.

The quality of materials used in making repairs has a direct bearing, and the best is none to good, although it must not be assumed that price and quality go hand in hand.

Furnace and Stoker Maintenance Costs

A question that is generally raised is, What should the maintenance cost be? This, of course, cannot be given for any number of plants in a definite manner, but if increased maintenance will show or, rather, give a decrease in operation, then increased maintenance is advisable. To show the trend of maintenance charges on boiler furnaces and stokers with regard to age, Table II is of value. This table is based on five boilers per unit, and the unit costs are based on the boiler-hours in service.

Boiler-Room Records

The last and most important point of economic boiler-plant operation to be considered is that of suitable boiler-room records. The records should be clear and precise and should not involve a large clerical force.

BOILER ROOM REPORT SHEET	From Month of			to		1916
	Monday Date	Tuesday Date	Wed. Date	Thur. Date	Friday	Sat.	Sun.
Boiler hours in service							
Boiler hours banked							
Boiler hours out of service							
Total pounds of coal used							
Total pounds of coal used for banking boilers							
Per cent of coal used for banking boilers							
Total weight of ash							
Per cent weight of ash to weight of coal							
Total steam generated							
Maximum demand lbs. of steam per hour							
Pounds of water blown down							
Pounds of water used for makeup							
Per cent. of makeup water to total steam generated							
Average grains of salt per U. S. gallon							
Amount of boiler compound used							
Feed pumps used							
Temperature of feed water							
Temperature of water to open heater							
Temperature of water from open heater							
Temperature rise in open heater							
Temperature of water to economizer							
Temperature of water from economizer							
Temperature rise in economizer							
Temperature of flue gases entering economizer							
Temperature of flue gases leaving economizer							
Temperature drop of flue gases in economizer							
Feed-water pressure							
Steam pressure							
Steam temperature							
Degrees of superheat							
Average boiler rating							
Average stack draft							
Average air-duct pressure							
Evaporation							
Evaporation from and at 212 deg. F.							
Boiler load factor							
Coal analysis							
Thermal efficiency							

FIG. 10. WEEKLY BOILER-ROOM REPORT SHEET

A suitable set of records would be as follows: A load curve sheet which should have on it the load carried; boiler-horsepower; hours in service and banked; capacity of feed pumps on the line. This curve sheet should be spotted as often as the load condition materially changes; in most cases every fifteen minutes will be found to be the right period. There should be a weekly boiler-room report sheet. A good example is given in Fig. 10.

There should also be a daily boiler-room sheet, which should show the hours each boiler was in service, when banked, cleaned and blown down, average boiler rating, average flue-gas temperature, readings of the various water meters for each day, grains of salt in the boiler water, inches of water blown from each boiler and any other readings that may be necessary.

A monthly boiler-room sheet is also required which should be a duplicate of the weekly sheet except that it should have on it a space for the results of the same month one year ago and a column to show the present increase and decrease of each item.

MONTHLY STEAM REPORT	Date	
	Total Pounds	Per Cent.
Total steam generator		
Total steam used for tube blowing		
Total steam used for heating		
Total steam used for stoker auxiliaries		
Total steam used for heating boiler-feed water		
Total steam used for boiler accessories		
Total steam lost due to radiation from steam pipes		
Total steam lost due to safety valve leakage and blow-off of (same)		
Total steam used for miscellaneous		

FIG. 11. MONTHLY REPORT SHEET

MAINTENANCE COST RECORD BOILER NO. 1.

Boiler Horsepower........ Maker........ Maker's No........ Year Installed........

Stoker Horsepower........ Maker........ Maker's No........ Year Installed........

	Jan.	Feb.	Mar.	April	May	June	July	Aug.	Sept.	Oct.	Nov.	Dec.	Total
Boiler hours in service													
Boiler hours banked													
Boiler hours out of service													
Lbs. of steam generated													
Per cent. of boiler rating													
Labor cost on boiler													
Labor cost on furnace													
Labor cost on stoker													
Total labor cost													
Material cost on boiler													
Material cost on furnace													
Material cost on stoker													
Total material cost													
Total labor and material cost													
Total cost per lb. of steam (cents)													

FIG. 12. SPECIMEN MAINTENANCE COST SHEET

Then there should be a monthly steam-report sheet such as is shown in Fig. 11.

A monthly heat-balance sheet with an additional item for blowdown losses, should be used, also a yearly curve sheet of the items contained on the weekly boiler-room sheet, the items being spotted daily on a suitable cross-section paper.

A station log sheet is necessary, showing the amounts of coal and water used. This should be made up on a cross-section sheet using the pounds of coal burned per day as the Y-axis and the pounds of steam used as the X-axis, and should be duplicated except that in the second sheet the pounds of water used in the plant should be the X-axis. These sheets should be spotted daily.

There should be a maintenance-cost sheet for each individual piece of apparatus similar to the one shown in Fig. 12, which is a specimen sheet for boilers, and

Year .... YEARLY MAINTENANCE COST SHEET .... Station

Station Capacity B.Hp. .... Total Steam Generated ....

Boiler Load Factor .... Total Coal Used ....

	Labor Cost	Material Cost	Total Cost	Cost per 1000 lbs. of steam
Boilers				
Furnaces				
Stokers				
Stoker drive				
Air ducts				
Blower fans				
Breeching and dampers				
Stack				
Economizer				
Ash-handling apparatus				
Coal-handling apparatus				
Feed pumps				
Meters				
House service pumps				
Piping				
Miscellaneous				

FIG. 13. YEARLY MAINTENANCE COST SHEET

also a yearly maintenance-cost sheet similar to that shown in Fig. 13.

The value of the foregoing records will be clearly seen, and it is only by the use of some such records that an engineer is able to keep tab on the actual boiler-plant operation.

The economical operation of boiler plants can be made quite practical, and it must be remembered that it cannot be accomplished in one day or in one year, but is a matter that will always require attention to keep it where it belongs.

TABLE II. FURNACE AND STOKER MAINTENANCE CHARGES WITH REGARD TO AGE

Westinghouse Underfeed Stokers Five 600-Hp. B. & W. Water-Tube Boilers	Year 1	Year 2	Year 3	Year 4	Year 5	Year 6
Boiler repairs, labor and material, dollars	174.39	288.97	110.68	281.71	331.16	416.56
Furnace, labor and material, dollars	470.39	402.14	116.20	960.99	230.07	1,597.82
Stokers, labor and material, dollars	Hand fired	Hand fired	Hand fired	Hand fired	252.48	1,333.64
Hours in service	3,717	2,361	1,136	882	12,124	22,806
Hours banked	....	....	8,201	7,667	4,980	10,075
Cost in cents per boiler hour, boiler	4.7	12.26	9.74	31.90	2.78	1.828
Cost in cents per boiler hour, furnace	12.68	17.00	10.29	109.00	1.89	7.01
Cost in cents per boiler hour, stoker	....	....	....	....	2.08	5.85
Roney Mechanical Stokers Five 600-Hp. B. & W. Water-tube Boilers	**Year 1**	**Year 2**	**Year 3**	**Year 4**	**Year 5**	**Year 6**
Boiler repairs, labor and material, dollars	482.85	324.41	219.11	431.39	269.78	144.73
Furnace, labor and material, dollars	2,393.99	2,856.95	745.86	2,538.09	691.43	607.96
Stokers, labor and material, dollars	1,757.92	2,520.04	2,533.12	2,945.02	1,849.95	987.78
Hours in service	31,102	33,060	25,474	19,665	14,826	10,542
Hours banked	....	....	15,971	18,944	19,340	11,927
Cost in cents per boiler hour, boiler	1.55	0.98	0.865	2.19	1.82	1.372
Cost in cents per boiler hour, furnace	7.69	8.64	2.93	12.90	4.66	5.765
Cost in cents per boiler hour, stoker	5.65	7.63	9.93	15.00	12.46	9.46
Bayonne Chain-Grate Stokers Five 600-Hp. B. & W. Water-Tube Boilers	**Year 1**	**Year 2**	**Year 3**	**Year 4**	**Year 5**	**Year 6**
Boiler repairs, labor and material, dollars	....	23.72	53.70	348.89	1,351.32	358.23
Furnace, labor and material, dollars	....	211.00	1,780.13	1,421.16	2,708.80	1,501.97
Stokers, labor and material, dollars	....	39.49	3,621.17	2,716.96	2,502.79	2,558.55
Hours in service	....	17,056	35,322	35,749	30,984	34,068
Hours banked	....	....	2,464	1,068	1,552	9.15
Cost in cents per boiler hour, boiler	....	0.139	0.152	0.976	4.36	1.52
Cost in cents per boiler hour, furnace	....	1.238	5.04	3.980	8.750	4.42
Cost in cents per boiler hour, stoker	....	0.231	10.25	7.60	8.08	7.52

POWER DRIVES FOR ROLLING MILLS

By W. O. Rogers

Several types of mills are described, together with something regarding the process in rolling steel into various products, from the ingot into slabs, billets, skelp, plates and pipes.

ROLLING mills are generally named after the material produced in them. Blooming mills come first as this mill rolls ingots, as they are classed, into blooms or billets, and all material rolled in a steel mill must first go through the blooming mill to reduce it to the proper size. These mills are made two-high reversing or three-high running constantly in one direction.

A slabbing mill is somewhat similar to a two-high reversing blooming mill and produces slabs, which are afterward rolled into plates. A billet mill is generally of the continuous type with rolls in tandem, and it rolls material coming from the blooming mill into billets of suitable size for the finishing mill. A plate mill is either two-high reversing or three-high running continuously in one direction, and is used to roll slabs into plates.

Structural mills are usually three-high and are used for rolling girders, heavy angles, channels, etc. In small mills this material may be rolled by the reversing blooming mill. Merchant mills are used for rolling small angles, channels and various types of profiles. They are usually three-high with the finishing stand, sometimes two-high.

Bar mills are usually three-high and generally consist of a roughing stand and a separate finishing stand. This mill is used for rolling bars of various sections and sizes. Rod mills are a special type of bar mill and roll small sections at high speed. Rail mills generally roll blooms or heavy billets into rails. The mills are made both two-high and three-high. Sheet mills roll bars into sheets of various gages and thicknesses. These mills are always two-high and generally have separate roughing and finishing rolls. Other types of mills are used, such as wire mills, hoop mills, etc.

A bloom is a bar of metal that is rolled from an ingot. In a previous article it was shown that ingots were formed in molds, the metal for filling them coming from the converter, in the bessemer process, and also from the

FIG. 1. A 42-IN. REVERSING BLOOMING MILL DRIVEN BY A 54 x 66-IN. 12,000-HP. STEAM ENGINE

FIG. 2. A 44-IN. REVERSING BLOOMING MILL AND TRANSFER TABLE

openhearth furnace. These ingots are stripped of the mold as soon as they become sufficiently solidified. If the ingot does not come out easily, the plunger of the stripping machine holds down the ingot by hydraulic pressure as the mold is drawn upward. Electrically operated stripping machines are also used, but I saw neither kind in operation. After stripping the ingot it is either placed in a soaking pit or taken direct to the blooming mill, where it is rolled into the required shape.

When hot metal is put through two rotating rolls, it will of course come out on the other side of the roll at a reduced thickness, the metal being compressed in the process. The reduction in area at each pass will vary between 5 and 50 per cent of the original. In making steel rails from ingots, say 18 in. square, in 22 passes, this amount of vertical squeeze will vary, the time for the operation being about five minutes, the speed of the metal through the rolls being in the neighborhood of ten miles per hour.

The parts of a rolling mill vary according to the kind of mill. Plate and rectangular mills have flat rolls, but the rolls may also be made with grooves and with collars. Fig. 1 shows a 42-in. reversing mill at the main plant of the Youngstown Sheet and Tube Co. It is driven by a 54 x 66-in. 12,000-hp. engine. By reversing is meant that the direction of rotation of the engine and rolls is reversed at each pass of the metal through the mill. The illustration shows the ingot partly rolled and also the spindles which drive the rolls, each having a coupling at the end. On a two-high mill, Fig. 2, the roll table is stationary. This table is fitted with live rolls for handling the heavy pieces of metal, and the rolls are placed in front of and behind the roll train. With a three-high mill the tables are generally raised and lowered together by either hydraulically or electrically operated mechanism. In some cases the table is pivoted near the center, so that the end next to the rolls can be tilted upward in order to bring the piece being rolled between the guides that direct it to the mill. The rolls on this table revolve and are driven by an electric motor to move the metal being rolled back and forth.

Some mills are equipped with a transfer table that moves from one roll train to another, carrying the rolled metal with it. The rolls of this table are motor-driven. Fig. 2 shows a transfer table of a 44-in. reversing blooming mill. The motors for manipulating the train rolls and transfer-table rolls are of the inclosed mill type, and are shown in both Figs. 1 and 2.

Fig. 3 shows a 44-in. reversing blooming mill and brings out in particular the manipulator. This is moved sideways across the table rolls and is used to transfer the metal being rolled from one pass to another. On lifting tables fingers are used to tip the piece from one side to the other. This is done by lowering the table, when the edge of the piece of metal will rest upon the finger, and it is then tipped. Three-high blooming mills use this kind of manipulator to a large extent, but on reversing blooming-mill work the piece is frequently tipped by hand by means of a forked tool in the hands of the rollers.

The manipulator consists of two long, heavy steel plates placed vertically on edge over the live table rollers, one set being on the table on each side of the roll stand. The plates are supported from independent movable beams underneath the table rollers. These beams are carried on guide collars attached at the end to hydraulic pistons, when that system is used, or through a rack and pinion to electric motors which move the plates back and forth across the table rollers. Thus a piece of steel can be held between the two guides and moved across the rollers as desired.

A three-high 110-in. plate mill is shown in Fig. 4. The hydraulic lifting table is operated by means of a cylinder placed underneath the table. A piston and rod are connected by means of a link and rod to a system of bell cranks, which moves the table up or down as the piston is forced from one end of the cylinder to the other. If the table is motor-operated, reduction gears are used to connect the motor to a crankshaft, which, by means of a connecting-rod, moves a bell crank, which in turn moves the table up and down. In both systems the table is usually counterweighted.

The largest plate mill in the world is at the Lukens Steel Co.'s Plant at Coatesville, Penn. It is a 204-in. mill used for rolling the flat circles for boiler heads and boiler plates. It is of the four-high reversing type in which two working rolls, each weighing 30 tons, are backed up by two larger rolls 50 in. in diameter and weighing 60 tons each. The mill is over 41 ft. from the bottom of the housing to the top of the cylinder on top of the housing.

Fig. 5 shows a 46-in. slabbing-mill engine and complete equipment. This is a compact installation and gives a good idea of the immensity of the mill, as compared to the workmen standing at the stand. Even with the rigid construction of these mills it is surprising how the stands withstand the shock and strain. As a matter of fact, the most serious difficulty encountered in rolling steel is the bending or breaking of the rolls from being placed under too severe a strain. This may be because the rolls have been set too close together or because the metal is too cold.

In rolling plates, the ingot, which may weigh anywhere from two to ten tons, is first cogged down in the slabbing mill, where a long flat piece of metal is produced. This type of mill is frequently of the two-high stand universal reversing type. In rolling hot metal there is always some expansion sideways, such that the sides will be bulged. This deformity is taken care of in what is termed a universal mill, by the arrangement of an auxiliary set of vertical rolls so placed that the piece passes through them just after leaving the horizontal rolls. The vertical rolls can be adjusted to any width of the mill and are given just enough pressure to keep the edge of the piece even. These vertical rolls are used with two-high and three-high mills, and with vertical rolls on both sides of the horizontal rolls.

After the piece has been rolled to the desired thickness, the end is sheared off and the remainder is cut up into what are known as slabs of whatever size required. The slabs are taken to a heating furnace and heated to about 1300 deg. C., or 2372 deg. F. They are then rolled in either a three-high mill or a two-high reversing mill. At the last pass of the plate through the rolls, a workman throws salt on the surface of the plate. The salt is carried into the rolls, over which water is trickling, with the plate, as shown in Figs. 1 and 2. This water is to keep the rolls cool. When the water is pressed against the hot plate, it flashes into steam, which, in conjunction with the salt, causes a series of small but violent explosions, which result in blowing off the scale from the top of the plate and give it a smooth surface.

FIG. 3. REVERSING 44-IN. BLOOMING MILL WITH THE MANIPULATOR IN THE FOREGROUND

When the plate nears the desired thickness a gager tries it, and when it is of the right size it is passed on to a set of straightening rolls and then to the cooling table. After cooling, it is sheared to the size required.

Some mills are devoted to the manufacture of steel rails for railroad use. In this type the ingot weighs about three tons. It is heated until it has a uniform temperature and is then rolled into blooms in either a three-high mill or a two-high reversing mill. It requires about eight passes through the rolls to reduce the ingot to an 8-in. square bloom, the amount of reduction per pass averaging about 16 per cent of the original area of the bloom. After cutting off the end to remove what is termed the pipe on the forward end, and the irregularity caused by the rolls at the other, the piece is cut into two blooms. These blooms are reheated and passed through a series of roll stands, usually being rolled into ten shapes, each one taking more and more the form of the

FIG. 4. A THREE-HIGH 110-IN. PLATE MILL WITH HYDRAULICALLY OPERATED LIFTING TABLE

FIG. 5. COMPLETE EQUIPMENT OF A 46-IN. SLABBING MILL, ENGINE SHOWN ON THE FAR RIGHT SIDE

finished rail, the last stand being the finishing one. The rails are then sawed into the proper length and go to a cooling table, after which they are straightened and the fishplate holes are made.

It is a difficult matter for a visitor to enter a tube mill; that is, a mill where steel pipes are made. Most manufacturers seem to think it advisable to surround the process with secrecy. I was fortunate, however, in going through the tube mill of the Youngstown Sheet and Tube Co. although no photographs were taken.

Pipes are made from skelp, which is a narrow plate about 25 ft. long, the width and thickness of which are governed by the size of the pipe to be made. In making large-sized pipe, the skelp is heated and then bent up into a rough circular form by passing it sidewise through rolls, so that the edge overlaps. It is then passed through a stand of rolls with the weld on the top. A mandrel mounted on a long shaft is inserted between the rolls, and as the mandrel is bullet-shaped on the head end and of the inside diameter of the pipe, the rolls press the welding edges together over the mandrel, and the pipe passes out through the mill. The pipe is then straightened and the ends cut off, after which it is tested for tightness of the weld by hydraulic pressure.

Small sizes of the lap-weld joint are drawn through a bell-shaped die, which welds the two edges of the skelp together during its passage through it. Butt-welded pipes are made in much the same way. The skelp is heated to a welding temperature, and it is then drawn through a bell die, which forces the skelp into a circular form and welds the edges together without lapping.

(Various other types of shears and rolling mills will be taken up in the next article.)

Tables for Computing Combustion Data From CO_2 Readings

After obtaining the percentage of CO_2 in the flue gas, from an Orsat apparatus or a recorder, the next step is to interpret this value in terms of excess air, air supplied and heat loss in the excess air. The accompanying tables have been prepared to give a general idea of how these items change with variations in the CO_2 readings and also to furnish a ready means of quick calculation.

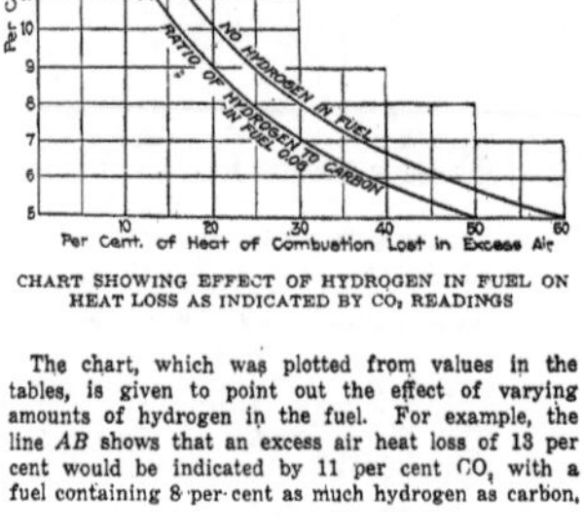
Percentage of heat of combustion lost in excess air is given for each 1,000 deg. F. excess of flue-gas temperature over boiler-room temperature. Upper curve is for a fuel containing no hydrogen or for pure carbon. Lower curve is for a fuel in which the weight of hydrogen left after deducting enough to combine with the oxygen already in the fuel is 8 per cent of the weight of the carbon.
Per Cent. CO2 in Flue Gas
Per Cent. of Heat of Combustion Lost in Excess Air
No Hydrogen in Fuel
Ratio of Hydrogen to Carbon in Fuel 0.08
A
B
C
21
20
19
18
17
16
15
14
13
12
11
10
9
8
7
6
5
10
20
30
40
50
60

CHART SHOWING EFFECT OF HYDROGEN IN FUEL ON HEAT LOSS AS INDICATED BY CO_2 READINGS

The chart, which was plotted from values in the tables, is given to point out the effect of varying amounts of hydrogen in the fuel. For example, the line *AB* shows that an excess air heat loss of 13 per cent would be indicated by 11 per cent CO_2 with a fuel containing 8 per cent as much hydrogen as carbon, and by 12.3 per cent CO_2 with a fuel containing only carbon. The line *AC* shows that 11 per cent CO_2 indicates 13 per cent loss when the hydrogen-carbon ratio is 0.08 and 17 per cent loss with pure carbon. The curves also show clearly the effect of a given change in CO_2 at different points. A drop in CO_2 from 16 to 15 per cent (with pure carbon) represents an increase of only about 2 per cent in the heat loss, while a drop from 6 to 5 per cent CO_2 represents a heat-loss increase of about 13 per cent.

TABLE I. PERCENTAGES OF EXCESS AIR INDICATED BY VARIOUS CO_2 READINGS FOR DIFFERENT RATIOS OF HYDROGEN TO CARBON IN COAL

Per Cent CO_2	Pure Carbon	$\frac{H}{C} = 0.02$	$\frac{H}{C} = 0.04$	$\frac{H}{C} = 0.06$	$\frac{H}{C} = 0.08$
20.9	0.0	.....	.....	.....	.....
20.5	1.95	0.33	.....	.....	.....
20.0	4.50	2.83	1.16	.....	.....
19.5	7.18	5.47	3.75	2.02	0.30
19.0	10.00	8.22	6.44	4.65	2.87
18.5	12.97	11.16	9.32	7.47	5.63
18.0	16.1	14.2	12.29	10.39	8.46
17.5	19.4	17.4	15.5	13.5	11.52
17.0	22.9	20.9	18.8	16.8	14.7
16.5	26.7	24.6	22.5	20.4	18.2
16.0	30.6	28.4	26.2	24.0	21.8
15.5	34.8	32.5	30.2	28.0	25.7
15.0	39.3	36.9	34.6	32.2	29.8
14.5	44.1	41.6	39.2	36.7	34.2
14.0	49.3	46.7	44.2	41.6	39.0
13.5	54.8	52.1	49.4	46.8	44.0
13.0	60.8	58.0	55.2	52.4	49.6
12.5	67.2	64.3	61.4	58.4	55.3
12.0	74.2	71.1	68.1	65.0	61.9
11.5	81.7	78.5	75.3	72.1	68.8
11.0	90.0	86.6	83.2	79.9	76.5
10.5	99.0	95.4	91.9	88.3	84.8
10.0	109.0	105.2	101.5	97.7	94.0
9.5	170.0	116.0	112.0	108.1	104.1
9.0	132.2	128.0	123.8	119.5	115.3
8.5	146.	141.	137.	132.	127.9
8.0	161.	156.	152.	147.	142.
7.5	179.	174.	168.	163.	158.
7.0	199.	193.	187.	182.	176.
6.5	222.	215.	209.	203.	197.
6.0	248.	242.	235.	229.	222.
5.5	280.	273.	266.	258.	251.
5.0	318.	310.	302.	294.	286.
4.5	364.	356.	347.	338.	329.
4.0	423.	412.	402.	392.	382.

In computing the tables, it was assumed that the combustion of the carbon and the hydrogen was complete, and further that the effect of sulphur or other

TABLE II. AIR PER POUND OF COMBUSTIBLE BY WEIGHT AND BY VOLUME AT 70 DEG. F. AND 14.7 LB. PER SQ.IN. WITH VARIOUS PERCENTAGES OF CO_2

	Pure Carbon		$\frac{H}{C}=0.02$		$\frac{H}{C}=0.04$		$\frac{H}{C}=0.06$		$\frac{H}{C}=0.08$		
Per Cent CO_2	Weight, Lb.	Volume, Cu.Ft.	Weight, Lb.	Volume, Cu.Ft.	Weight, Lb.	Volume, Cu.Ft.	Weight, Lb.	Volume, Cu.Ft.	Weight, Lb.	Volume, Cu.Ft.	Per Cent CO_2
20.9	11.52	153.8	.....	.....	.....	.....	.....	.....	.....	.....	20.9
20.5	11.75	156.8	12.02	160.5	.....	.....	.....	.....	.....	.....	20.5
20.0	12.04	160.7	12.32	164.5	12.58	167.9	.....	.....	.....	.....	20.0
19.5	12.35	164.8	12.64	168.7	12.91	172.4	13.16	175.7	13.40	178.9	19.5
19.0	12.67	169.2	12.96	173.0	13.24	176.8	13.50	180.2	13.74	183.4	19.0
18.5	13.01	173.7	13.32	177.8	13.60	181.6	13.86	185.0	14.11	188.4	18.5
18.0	13.38	178.6	13.68	182.6	13.97	186.5	14.24	190.1	14.49	193.4	18.0
17.5	13.76	183.7	14.07	187.8	14.36	191.7	14.64	195.4	14.90	198.9	17.5
17.0	14.16	189.1	14.48	193.3	14.78	197.3	15.06	201.1	15.33	204.7	17.0
16.5	14.59	194.8	14.92	199.2	15.24	203.5	15.53	207.3	15.80	210.9	16.5
16.0	15.05	200.9	15.38	205.3	15.70	209.6	16.00	213.6	16.28	217.3	16.0
15.5	15.53	207.4	15.88	212.0	16.20	216.3	16.51	220.4	16.79	224.2	15.5
15.0	16.05	214.3	16.40	218.9	16.74	223.5	17.05	227.6	17.35	231.6	15.0
14.5	16.61	221.7	16.97	226.6	17.31	231.1	17.64	235.5	17.93	239.4	14.5
14.0	17.20	229.6	17.53	234.7	17.93	239.4	18.27	243.9	18.57	247.9	14.0
13.5	17.83	238.1	18.22	243.2	18.59	248.2	18.93	252.7	19.24	256.9	13.5
13.0	18.52	247.3	18.93	252.7	19.31	257.8	19.66	262.5	19.99	266.9	13.0
12.5	19.26	257.1	19.68	262.7	20.07	267.9	20.44	272.9	20.77	277.3	12.5
12.0	20.06	267.9	20.50	273.7	20.91	279.2	21.29	284.2	21.64	288.9	12.0
11.5	20.94	279.5	21.38	285.4	21.80	291.0	22.19	296.2	22.56	301.2	11.5
11.0	21.89	292.2	22.36	298.5	22.80	304.4	23.20	309.7	23.58	314.8	11.0
10.5	22.93	306.1	23.41	312.5	23.87	313.7	24.29	324.3	24.68	329.5	10.5
10.0	24.08	321.4	24.59	328.3	25.06	334.6	25.50	340.4	25.91	345.9	10.0
9.5	25.34	338.3	25.88	345.5	26.38	352.2	26.84	358.3	27.26	363.9	9.5
9.0	26.75	357.1	27.31	364.6	27.83	371.5	28.32	378.1	28.76	384.0	9.0
8.5	28.32	378.1	28.92	386.1	29.47	393.4	29.98	400.2	30.45	406.5	8.5
8.0	30.10	401.8	30.73	410.3	31.31	418.0	31.85	425.2	32.34	431.7	8.0
7.5	32.10	428.6	32.77	437.5	33.39	445.8	33.96	453.4	34.48	460.3	7.5
7.0	34.40	459.2	35.11	468.7	35.76	477.4	36.37	485.5	36.92	492.9	7.0
6.5	37.04	494.5	37.80	504.6	38.50	514.0	39.15	522.7	39.74	530.5	6.5
6.0	40.13	535.7	40.94	546.6	41.70	556.7	42.40	566.0	43.03	574.5	6.0
5.5	43.78	584.4	44.66	596.2	45.49	607.3	46.24	617.3	46.93	626.5	5.5
5.0	48.15	642.9	49.13	655.9	50.02	667.8	50.85	678.9	51.60	688.9	5.0
4.5	53.50	714.3	54.57	728.5	55.56	741.7	56.47	753.9	57.30	765.0	4.5
4.0	60.19	803.6	61.39	819.6	62.50	834.4	63.52	848.0	64.44	860.3	4.0

TABLE III—HEAT LOST TO STACK IN EXCESS AIR PER POUND OF COMBUSTIBLE FOR EACH 1,000 DEG. F. EXCESS OF GLUE-GAS TEMPERATURE ABOVE BOILER-ROOM TEMPERATURE

Values given in B.t.u. and in Percentage of Heat Available in the Combustible

	Pure Carbon		$\frac{H}{C}=0.02$		$\frac{H}{C}=0.04$		$\frac{H}{C}=0.06$		$\frac{H}{C}=0.08$		
Per Cent CO_2	B.t.u.	Per Cent	B.t.u.	Per Cent	B.t.u.	Per Cent	B.t.u.	Per Cent	B.t.u.	Per Cent	Per Cent CO_2
20.9	0	0	.....	.....	.....	.....	.....	.....	.....	.....	20.9
20.5	54	0.37	9	0.06	.....	.....	.....	.....	.....	.....	20.5
20.0	124	0.85	81	0.52	35	0.21	.....	.....	.....	.....	20.0
19.5	199	1.36	157	1.01	112	0.68	63	0.36	10	0.05	19.5
19.0	277	1.89	236	1.52	192	1.17	144	0.83	92	0.50	19.0
18.5	359	2.46	321	2.06	278	1.69	231	1.33	181	0.98	18.5
18.0	445	3.05	408	2.63	367	2.22	322	1.84	272	1.48	18.0
17.5	537	3.68	501	3.22	462	2.80	418	2.39	369	2.01	17.5
17.0	634	4.34	600	3.86	562	3.41	520	2.98	473	2.57	17.0
16.5	737	5.05	707	4.55	671	4.07	630	3.61	585	3.18	16.5
16.0	847	5.80	817	5.25	783	4.74	744	4.26	700	3.80	16.0
15.5	963	6.59	935	6.01	903	5.47	866	4.97	824	4.48	15.5
15.0	1,087	7.45	1,062	6.83	1,032	6.25	997	5.72	956	5.20	15.0
14.5	1,220	8.36	1,197	7.70	1,170	7.09	1,137	6.52	1,098	5.97	14.5
14.0	1,363	9.33	1,343	8.64	1,319	7.99	1,288	7.39	1,251	6.80	14.0
13.5	1,516	10.38	1,498	9.64	1,476	8.95	1,448	8.30	1,411	7.67	13.5
13.0	1,680	11.51	1,668	10.73	1,648	9.99	1,623	9.30	1,591	8.65	13.0
12.5	1,858	12.73	1,848	11.89	1,832	11.10	1,809	10.37	1,779	9.68	12.5
12.0	2,050	14.1	2,045	13.15	2,033	12.32	2,012	11.54	1,986	10.80	12.0
11.5	2,260	15.4	2,256	14.5	2,248	13.62	2,231	12.79	2,207	12.00	11.5
11.0	2,489	17.0	2,490	16.0	2,486	15.1	2,472	14.2	2,452	13.33	11.0
10.5	2,738	18.8	2,744	17.6	2,744	16.6	2,734	15.7	2,717	14.8	10.5
10.0	3,013	20.6	3,026	19.5	3,030	18.4	3,025	17.3	3,012	16.4	10.0
9.5	3,318	22.7	3,336	21.5	3,345	20.3	3,345	19.2	3,336	18.1	9.5
9.0	3,656	25.0	3,679	23.7	3,695	22.4	3,700	21.2	3,697	20.1	9.0
8.5	4,034	27.6	4,065	26.1	4,088	24.8	4,099	23.5	4,100	22.3	8.5
8.0	4,459	30.5	4,499	28.9	4,529	27.4	4,547	26.1	4,555	24.8	8.0
7.5	4,940	33.8	4,989	32.1	5,028	30.5	5,054	29.0	5,068	27.6	7.5
7.0	5,490	37.6	5,550	35.7	5,598	33.9	5,633	32.3	5,655	30.8	7.0
6.5	6,126	41.9	6,195	39.8	6,255	37.9	6,299	36.1	6,330	34.4	6.5
6.0	6,866	47.0	6,950	44.7	7,023	42.6	7,079	40.6	7,121	38.7	6.0
5.5	7,742	53.0	7,844	50.5	7,932	48.1	8,002	45.9	8,056	43.8	5.5
5.0	8,793	60.2	8,914	57.3	9,021	54.7	9,10	52.2	9,177	49.9	5.0
4.5	10,077	69.0	10,221	65.7	10,351	62.7	10,458	60.0	10,545	57.3	4.5
4.0	11,682	80.0	11,859	76.3	12,017	72.8	12,149	69.7	12,259	66.7	4.0

impurities might be neglected. It was also assumed that the percentage of carbon, hydrogen and oxygen in the coal was known, as well as the CO_2 reading. Table I gives the percentages of excess air indicated by the various CO_2 readings for ratios of net hydrogen to carbon in the coal of 0.02, 0.04, 0.06 and 0.08. By net hydrogen is meant that part of the hydrogen in the coal which must be supplied with oxygen from the air. Each pound of oxygen in the coal has combined with, or will combine with, one-eighth of a pound of hydrogen. Therefore, the oxygen already in the coal will take care of one-eighth of its own weight of hydrogen, and the remainder of the hydrogen in the coal is that which has been designated as net hydrogen.

Table No. II shows the weight of air and the volume of air per pound of combustible, the volume being measured at 70 deg. and 14.7 lb. per sq.in. absolute, or atmospheric pressure. By combustible is meant here the weight of carbon plus the weight of net hydrogen. In this table, as before, the percentage of CO_2 is the argument, and the weight and volume respectively of air per pound of combustible are given for the same ratios of net hydrogen to carbon as are used in Table I.

Table III gives the heat loss due to excess air in B.t.u. per pound of combustible and in percentage of the heat supplied for each 1,000 deg. of difference between the flue-gas temperature and that of the boiler room. It is to be noted that Table II gives the total air supplied per pound of combustible, not simply the excess, while Table III gives the heat loss to the stack in the excess air and not in the total air supplied. The values are given in this particular way as this seems to be the most convenient for general use.

As an example of the use of the tables assume a coal containing 75 per cent carbon, 5.5 per cent hydrogen and 10 per cent oxygen and that the flue gas is at a temperature 640 deg. F. above boiler-room temperature and contains 11 per cent CO_2. The weight of hydrogen per pound of coal is then 0.055 lb., and since the weight of oxygen is 0.10 lb., the weight of net hydrogen will be $0.055 - \frac{0.10}{8}$, or 0.0425 lb. The ratio of net hydrogen to carbon therefore will be $\frac{0.0425}{0.75}$, or 0.0567. This value is close enough to 0.06 so that we are justified in using that column in the tables.

Referring to Table I, under hydrogen-to-carbon-ratio 0.06 at 11 per cent CO_2 we find the percentage of excess air as 79.9. In Table II in the corresponding column on the corresponding line we find there were actually supplied 23.21 lb., or 309.9 cu. ft., of air for each pound of combustible. From Table III the excess air heat loss would be 2,474 B.t.u., or 14.2 per cent of the heat in the combustible if the flue gas temperature were 1,000 deg. F. above that of the boiler room.

In each pound of coal the combustible is 0.75 + 0.0567, or 0.8067, or nearly 0.807 lb. The weight of air, the volume of air and the heat lost to the stack in B.t.u. for each pound of combustible would be multiplied by this value to get the corresponding value per pound of coal. Further, the heat loss in B.t.u. per pound of coal and in percentage should be multiplied by one one-thousandth of the temperature rise. It should be noted that the fraction of combustible in the coal is not to be used with the percentage of excess air or the percentage of heat lost in the excess air, since these values are ratios and have already taken account of the percentage of combustible. To sum up, then, the condition assumed shows that we have 79.9 per cent excess air, that we have supplied $23.21 \times 0.807 = 18.76$ lb. of air per pound of coal, $309.9 \times 0.807 = 250$ cu.ft. of air per pound of coal, that $2,474 \times 0.807 \times 0.640 = 1,278$ B.t.u. per pound of coal has been lost to the stack in the excess air, and finally that this loss is $14.2 \times 0.640 = 9.09$ per cent of the available heat.

This table brings out very clearly the necessity of keeping down the amount of excess air to the lowest practical value, but it does not show the danger in reducing the excess air until there may be incomplete combustion of a part of the carbon. In trying for a high CO_2 reading with its accompanying reduction of losses due to excess air, it must be remembered that the incomplete combustion of even a small amount of carbon will more than wipe out the saving due to the decreased air supply. In connection with CO_2 recording apparatus, then, some method of indicating the presence of CO should be provided.

Adjustment of Valve-Timing Mechanism

By E. E. Snow

In gas and oil engines it is customary to have the valve pushrod provided with right- and left-hand threaded ends. To obtain adjustment of the valve timing, it is necessary to stop the engine, open the

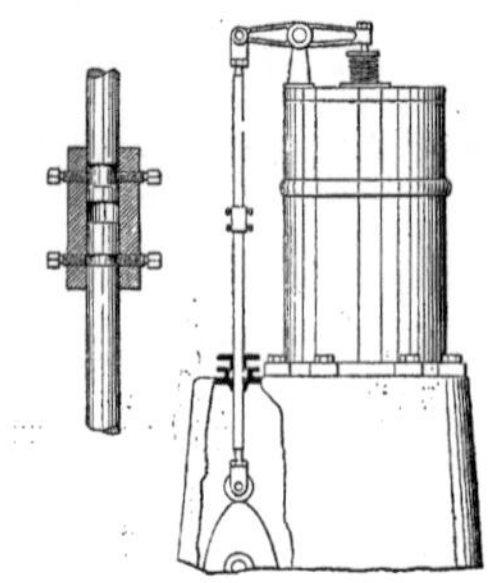

FIG. 1. COLLAR ON PUSHROD ASSISTS IN CORRECT TIMING OF INTERNAL-COMBUSTION ENGINE

crankcase and loosen the locknut at the lower end of the rod before the rod length can be altered. The consequence is that the valve timing of the engine is always secured when the engine is cold. Owing to the variation in length of rod when cold and when warmed up, the timing as made when cold is never maintained. There is always some wear and backlash to distort the valve timing.

To allow the valves to be timed while the engine is running, an enterprising engineer made collars or sleeves, as shown in the sketch. The valve pushrod is cut in two pieces and grooves made in the ends inclosed by the collar. Two setscrews are fitted into the grooves, preventing looseness.

To adjust the timing, the locknut at the upper end of the rod is slackened off, the setscrew is loosened and the rod turned; this shortens or lengthens the rod as the case may be. As soon as the engineer has secured the valve "lead" desired, the locknut is tightened and the setscrew adjusted. This is a practical idea and is of immense service where the engine is of the vertical type.

Robert H. Fernald

A Brief Biographical Sketch Outlining His Early Life and His Research Work for the Aid and Advancement of Engineering Science, Including Important Experimental Work and Testing of Mineral Fuels in Gas Producer Plants and in Steam-Boiler Furnaces in the United States and Europe for the United States Geological Survey and the Bureau of Mines.

ROBERT H. FERNALD, Professor in Charge of the Department of Mechanical Engineering at the University of Pennsylvania, has given to the engineering world a life of service. The greater part of his life has been devoted to teaching, but probably the most important work done by him was in connection with fuel investigation.

He was born Dec. 19, 1871, in Orono, Maine, and was graduated from the Mechanical Engineering Department of the Maine State College (now the University of Maine) in 1892. At that time his father, Merritt C. Fernald was president of the college. After a year spent in further study at the Massachusetts Institute of Technology, he went to the Case School of Applied Science in Cleveland as an instructor, later becoming an assistant professor. In 1902 he was made Professor of Mechanical Engineering at Washington University, St. Louis, Mo. While there he began his connection with the Technologic Branch of the United States Geological Survey. This connection resulted in some very important work with reference to the question of fuels. In 1907 he returned to the Case School as head of the Department of Mechanical Engineering and in 1912 took his present position at the University of Pennsylvania as Professor of Dynamical Engineering. In the fall of 1908 Professor Fernald, who had been actively interested in the analyzation and testing of mineral fuels as carried on by the United States Geological Survey, including trials of coal, lignite and peat in gas producers, was sent on a trip of inspection to those European countries where gas producers and gas engines were in more extensive use than in the United States to visit the plants that had been most successful in utilizing fuels of poor quality. Plants in ten European countries were visited during a two months' journey. A detailed study of all the plants visited was impossible owing to the limited time available, but Professor Fernald obtained knowledge of some interesting features of European practice which differ from those of this country and collected interesting data concerning various grades of fuel (particularly low-grade fuel) in use and the results obtained. Prior to his trip to Europe Professor Fernald had been interested in the testing work of the United States Geological Survey in this country, the object being to find out in what ways the mineral fuels of the United States could be utilized most efficiently, and by what means deposits of low-grade fuel that were lying undeveloped or were being wasted could be made of immediate or prospective importance as assets of the mineral wealth of the nation. This work was started in 1904 in St. Louis and continued under the direction of the Survey until 1910, when all fuel-testing work was transferred by an act of Congress to the Bureau of Mines.

Mines throughout the country were requested to contribute two carloads of coal each, one to be burned under a boiler, the steam to be used in a steam engine, and the other load to be burned in a gas producer and the gas to be burned in a gas engine. The services of Professor Breckenridge, now of Yale University, were secured to take charge of the steam tests, and Professor Fernald, who was then at the University of Washington, assumed the responsibility of the gas-producer tests. The results of these investigations is a surprising development in the use of low-grade fuels in gas-producer plants and steam-boiler furnaces. Professor Fernald's activities were by no means confined to the university and fuel investigations. As a side line he performed considerable consulting-engineering service, and in Cleveland, Ohio, as a consulting engineer he worked on the street-car ventilator acceptance tests. He is the co-author with George A. Orrok of a book on "Engineering of Power Plants" and of a number of reports and bulletins on the conservation of the fuel resources in the United States and Europe, written for the Geological Survey between 1905 and 1910 and for the Bureau of Mines since that time.

Professor Fernald is a vice president of the American Society of Mechanical Engineers, and was a member of the Gas Power Section in 1911 and the Philadelphia Section from 1915 to 1916. He is also a member of the Society for the Promotion of Engineering Education, the American Society for Testing Materials, the Franklin Institute, the Philadelphia Chamber of Commerce, Sigma Xi, Phi Kappa Phi and the Engineers' Club.

New Wayne Gravity Oil Filter

The illustration shows the details of an oil filter manufactured by the Wayne Oil Tank and Pump Co., of Fort Wayne, Ind. The filter is adapted to handle continuously large volumes of oil in which quantities of

NEW WAYNE GRAVITY FILTER

free and entrained water and the usual sediment are contained. The water and much of the sediment are removed by a tray system, in a precipitation compartment, so that the work of the filtering medium is reduced. Operation of the new filter, which is of the gravity type, is as follows: Dirty oil enters the filter cabinet through a brass-wire screen, passes to the heating compartment, thence through the precipitating chamber, and from there to the filter units, from which it enters the clean-oil compartment.

Coarse impurities are removed from the dirty oil as it enters the filter by means of the brass screen. The oil then passes downward and thence laterally over the heating coils, which raises its temperature to 180 deg. F. It then passes through a precipitation chamber filled with shallow trays arranged in multiple to give large area and a comparatively long period of time for a given volume of oil to pass, and permits much of the impurities contained in the oil to be deposited on the tray. Most of the water contained in the dirty oil is separated from it as it passes over the heating coils and descends to the bottom of the dirty-oil tank. The rest settles from the oil in its travel over the precipitation trays, an outlet connecting each tray to the bottom of the dirty-oil chamber. For the removal of this water a sealed automatic overflow is provided.

From the precipitation chamber the oil passes through the filter-unit header, on which are mounted two or more cloth filter units which remove the slime and impurities that have not already been precipitated. The filter cloth is made up in the form of a conical tube 4 in. in diameter at one end and 15 in. at the other. The small end of the tube is banded to the center post of the container basket and the body of the tube is made up into vertical sections by drawing it back and forth over a series of concentric cylinders made of galvanized wire screens and of uniform length, each cylinder being larger in diameter than the preceding one. The cloth tube is banded to the inside of the shell of the container basket, which has a leaktight cover held on by screws.

Oil rises through the central tube of the basket and passes downward between the cylinders and horizontally through the filtering cloth. All dirt taken from the oil is retained in the cloth until it is removed for cleaning. Both the filter post and the basket are provided with valves, the stems of which engage when the unit is in place, allowing a free passage of oil and the automatic closing of the valves when the unit is removed for cleaning. The entire unit is placed in an open-top receptacle which allows it to be immersed in oil, thereby maintaining a constant head over its whole surface. The flow of the oil to any particular filtering unit may be shut off if desired by means of an angle valve at the front and exterior to the cabinet, so that if the flow of oil to the filter is slow, one or more sections may be held in reserve.

Cleaning may be done without removing the oil. The precipitation trays are cleaned in place one at a time with hose and water. The dirt on the tray passes to the bottom of the dirty-oil compartment, from which it is conducted to the sewer.

To clean the cloth filter units, they may be lifted from their supporting receptacles, as there are no connections to break. The dirt is deposited on the inside of the units. This type of filter is built for capacities ranging from 2 to 600 gal. per hour.

A resolution was recently introduced in Congress providing for an appropriation of $250,000 to be used to investigate the possibilities of a suitable substitute for gasoline. This money was to be used by the Bureau of Chemistry of the Department of Agriculture. The scarcity of gasoline and the threatened exhaustion of this indispensable fuel in the near future is causing much concern.

The "Africa," announced to be the largest Diesel-engine driven ship in the world, has been launched at Copenhagen. This vessel has a displacement of 14,000 tons, her engines develop 4,500 hp., and all the auxiliaries are electrically driven.

EDITORIALS

The Federated American Engineers

JUNE third and fourth were epoch-making days in the history of American Engineering. On those dates one hundred and twenty-three representatives of sixty-one engineering societies met at Washington and instituted an organization through which the united engineering talent of the country can function for the service of the public and the progress of the profession.

The meeting resulted in a federation of existing societies along the lines laid out by the Joint Conference Committee of the Four Founder Societies rather than in the founding of a new national organization of individual members. One may become a part of it only as a member of one of its constituent societies. Any society of the engineering or allied professions the chief object of which is the advancement of the knowledge and practice of engineering or the application of the allied sciences, which is not organized for commercial purposes and which has at least one hundred members, is eligible for membership. The constituent societies elect representatives to the American Engineering Council, which will meet annually. An executive board is provided, consisting of the six elected officers of the National Council and twenty-four members selected by the group of National societies and the group of state, regional and local societies in the ratio of the combined representation in the National Council of the national societies on the one hand and of the state, regional and local societies on the other.

The Executive Board appoints an executive officer, who serves both the National Council and the Board as secretary. A chart of the organization will be found on page 979, and an account of the organizing conference on page 978. The present Engineering Council is charged with arranging for the first convention of the new National Council and the launching of the federation.

The stage is now set, the program prepared in which the engineer is to take a larger part. Not a more active part, for his has always been the working rôle, but more prominence given to his part. Less work behind the scenes and in the supporting cast and more of the headlines. The engineer has been trained and used to think and do rather than talk. He has been content to be the servant of the administrator at a price. The work of many departments of the public is done by engineers of whom nobody ever hears, under the management of political appointees whose pictures and utterances fill the public prints. The problems of the Government are largely engineering, but one must search far for members of the profession in the lists of state and national legislators. The engineer has now become possessed of a sense of class consciousness, and through more assertiveness upon his part and the cultivation of a keener appreciation on the part of the public of the part which he is playing, hopes to win a more commanding position.

And this can be done both for the individual and the profession by organization and co-operation. It is hoped that there will soon be in every important community a society of engineers that will bring the united engineering thought and experience of its members to bear upon local problems in which engineering is involved. This will furnish the avenue for the engineer to obtain recognition in his own city and encourage and incite his interest in public affairs. The strong, gifted and dominant men in the local societies will be sent as representatives to the National Council and become involved in the wider and larger problems of the country and of the profession, while the activities of the Federated American Societies will result in real service to the public, earning recognition for the profession and for the individuals who have contributed of their time and talent to its accomplishment.

The success of the new organization will depend upon the interest that is taken in it by the rank and file. A select body of higher lights appointed to a council in which the ordinary member feels that he has as little a part as he has in Congress, and then forgotten, will have little chance. They must be backed by a large and sympathetic constituency and with means proportionate to the dignity of the profession that they represent. It is to be hoped that every society, club and organization eligible to join the federation will do so, and that every engineer by joining such a club or society will become identified with the Federated American Societies and take an active part in the work which they are trying to accomplish.

Boiler and Stoker Definitions

HE WHO undertakes to write a definition invites trouble. For a long time the current definitions of a valve would have applied equally well to a door. Lord Kelvin, asked to define entropy, replied that it was a poor name for *it*.

A conference committee of the American Boiler Manufacturers' Association and the Stoker Manufacturers' Association has evolved a series of definitions of terms employed by their member companies. These are printed on page 984. A boiler is defined as a metal vessel, capable of withstanding pressure and serving the purpose of transmitting heat, usually produced by the combustion of fuel, to a liquid contained in the vessel. This is equally applicable to heaters, economizers, evaporators and stills. The elliptical clause, "usually produced by the combustion of fuel," seems to be redundant, for it does not make the definition any less inclusive, and it is difficult if not impossible to conceive of a case in which the heat applied to a boiler of any sort is not directly or indirectly so produced.

The furnace arch is *designed* or *intended* to "aid combustion," but if the proposed definition holds there are some furnace arches that fall outside of its literal interpretation.

Draft is defined as "a difference in pressure, *due to a difference in gas density*, which tends to cause a flow of gas from the region of higher pressure to that of lower pressure," but induced draft and forced draft are defined as draft *"produced by mechanical means."* It would seem that the definition of the general term would not be too inclusive if the portion in italics were elided, and the idea of the production of flow or pressure by difference in density be incorporated with the definition of natural draft, in which case this difference is due to temperature difference and not to the pressure, of which a comprehensive interpretation of the definition as proposed would require it to be considered the cause.

The Dawn of a New Day

WHEN in opening the forty-third convention of the National Electric Light Association at Pasadena, California, President Russell H. Ballard said that "the kind of public ownership that brings results is ownership by the public at large in the stocks and bonds of their local regulated utility companies," he very clearly indicated what promises to be the solution of the question of publicly owned versus privately owned public utilities. We have passed through an era of privately owned unregulated utilities, and the outcome was not a very successful one. Municipally owned utilities have been tried and with few exceptions have been found wanting. Past experiences with government regulated utilities, owned by a comparatively few stockholders also have their shortcomings. It has been definitely demonstrated that improper regulation is as bad as, if not worse than, no regulation.

In the operation of every public utility there are three interests to be served—the public, the financial and the utility itself. Past experience has shown that it is poor business to serve one of these at the expense of the others. Only when all three get a square deal can success be assured. One of the best ways to bring about a common interest is local financing. Of course this makes it more imperative than ever before that the utilities be intelligently regulated. Such an arrangement forms, as Mr. Ballard has expressed it, "a partnership which leads to complete understanding and co-operative effort in development and retains in the management and operation of the utilities those trained by long experience to perform the task."

How Much Coal Have You?

HOW serious is the coal shortage situation? Is it as bad as common talk says it is? Is it worse? Who knows?

These are the questions the Geological Survey will soon try to answer. The Director has a list of coal consumers to whom a questionnaire will be mailed. The purpose is to arrive at the approximate truth about coal stock, available coal on hand at the consumers' plants and storage piles. Doubtless some good use can be made of the information after it is compiled. Consumers can help considerably by sending in the answered questionnaire without delay.

Just what real good will be accomplished we do not know. What will be done about it if the inventory shows much or little on hand we have not been told. If it can be made to produce more coal, transport coal faster and in greater volume for domestic (American) use, then by all means spend the money necessary to find out how much is in stock. But will this be accomplished? Conditions do not seem to warrant encouragement that it will.

We believe Europe should have every ounce of coal it needs. We believe America should be generous in exporting the much-needed fuel to those in the war-stricken, famine-ravished lands abroad. But all this provided we do not ourselves get into a serious economic condition in trying to relieve Europe. How much coal, the best of America's best coal, can we afford, can we dare, to let go to Europe? Who knows? Who is trying to find out? No one, as far as we can learn. The Geological Survey issues, periodically, statements telling how much coal was exported in a given interval. But if it told that ninety per cent of that mined in the same interval had gone, it could not do more than that—just tell about it. Neither the Geological Survey nor any other government agency has any authority to say how much of our total may or may not go.

The United States of America is today bidding against Europe for the coal of the United States of America. The seacoast coal piers are working as they never in all their history worked—to load ships bound for Europe.

American railroads may be taking considerably more coal than necessary for current and ordinary storage use. There is likely a shortage of coal cars. Let something be done about all of these; but begin now to determine how seriously we are to be affected if coal continues to Europe in unrestricted flow.

Waterpower Bill on the Rocks

JUST when the country's rejoicing over the passage of the water power bill by both Houses was at its highest, news came that President Wilson, by failure to sign it before Congress adjourned, vetoed the measure. The passage of the bill by both Houses is the climax to twenty years of effort to get the measure through Congress.

The nearest to an official statement that comes from Washington was made informally by Secretary Tumulty, who announced that Secretary Payne of the Interior Department was vigorously opposed to the bill. Later, Mr. Tumulty said that the President had received this measure and others too late to permit of final consideration and favorable action *in view of a serious difference of opinion which had developed.*

The bill went to the President in the middle of the week of the adjournment of Congress, and Congress adjourned Saturday, June fifth. On the Thursday preceding it was referred to the Secretaries of War, Interior and Agriculture, who, under the terms of the bill, were to have constituted a Federal power commission. All three secretaries reported to the President before Saturday, June fifth. Mr. Tumulty's statement clearly implies that the Secretary of the Interior recommended against the bill. But the latter was not alone. Barely a week before passage of the bill, as late as the time of the conference report to the Senate, Senators Lenroot, King and Norris pointed out objectionable features. Secretary Payne is known to be opposed to the bill on the ground that it endangers the beauties of national parks and reservations. The Senators mentioned seem afraid of the power the bill vests in the committee it creates. With the pocket veto a reality, the whole subject begins anew.

With the information now at hand and upon which to base opinion, it seems unfortunate that the President failed to sign the bill.

CORRESPONDENCE

A Transformer-Loss Eliminator

There are doubtless many plants that do not operate continuously, but which are of such size as to warrant disconnecting the main transformers so as to eliminate the dissipation of current as transformer losses, even though a small amount of current may be needed at all times. In order to accomplish this result I once used the method described and illustrated herewith while at an electrically driven pumping station. The service required was that of operating at full capacity for about one-half hour once during each 24 hours and also that of a stand-by plant subject to full load at from one to two minutes' notice.

The size of the plant and the rate per kilowatt-hour ran the transformer energizing losses up to about ten dollars per day. Therefore the 220-volt service, as shown by the sketch, was installed to reduce this loss charge to a minimum.

The potential supplied at the disconnecting switches was 5,700-volt, 3-phase, which was transformed to 2,200 volts, at which voltage over 90 per cent of the power was supplied directly to the main motors, only a small quantity being further transformed to 220 volts for operating several small motors, lighting service, remote control, etc. During non-pumping periods the only current needed was for operating one 7½-hp. motor and a few hundred watts of lighting load. This plant operated with this "economizer" in use about 23 hours a day for over 18 months, thereby paying for the effort of its installation many times.

The auxiliary 220-volt service was installed without any additional expense, as there happened to be a 7½- and a 5-kw. transformer, also a three-pole, 300-ampere, single-throw switch at the plant which were not in use. To this switch was added a set of about 60-ampere home-made clips to make it double-throw. The switch was then mounted on an oak plank, painted black to correspond with the slate panels. This plank made a good 220-volt mounting. The labor for making up this layout should not be called a charge, as the men had to be on duty anyway.

The only switch operating necessary to put this economizer in service was to throw the 220-volt double-throw switch from the 75-kw. feed side to the auxiliary feed side and open any two of the 5,700-volt disconnecting switches and vice versa.

It was suggested by one of the electricians who worked on the installation that the double-throw switch be dispensed with and that the auxiliary transformer secondaries be connected to the 220-volt bus at all times, thus allowing the small transformers to float on the line and take their share of the whole load with the main transformers.

The impossibility of doing this was proved by trial for his interest, as it blew the one-ampere, 5,700-volt fuses, this of course being caused by the phase-angu-

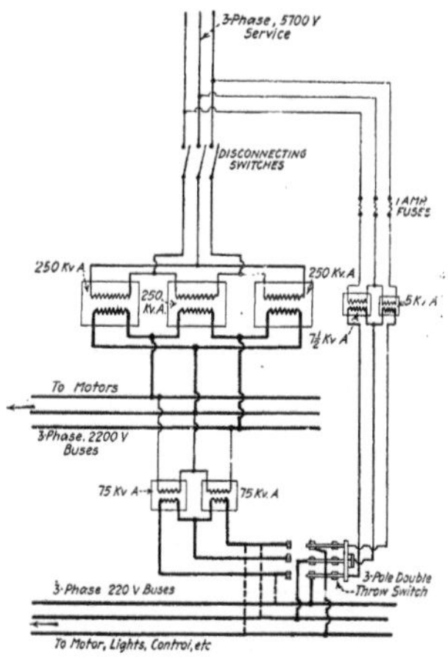

WIRING OF THE AUXILIARY TRANSFORMER SERVICE

larity displacement due to the double transformation of the main service and the single transformation of the auxiliary supply.

The 5,700-volt fuse blocks used were made from ¼ x 1 x 7-in. fiber and two ¼-in. stove bolts with double nuts and washers. These fuses were suspended by the feed wires near the transformers and had no rigid support. R. W. HARVEY.

Kingston, N. Y.

Induction-Motor Trouble and How It Was Remedied

Quite often difficulties reported as being experienced with new apparatus furnished by electrical manufacturers and installed by customers themselves, prove upon investigation not to have been caused by defects; and improper performance or no performance at all is due entirely to the incorrect installation or wiring up of the apparatus.

Many such instances have come to the writer's attention during several years of road work, and one instance in particular comes to mind which shows how easily the customer can be led to believe that he has received a defective piece of apparatus.

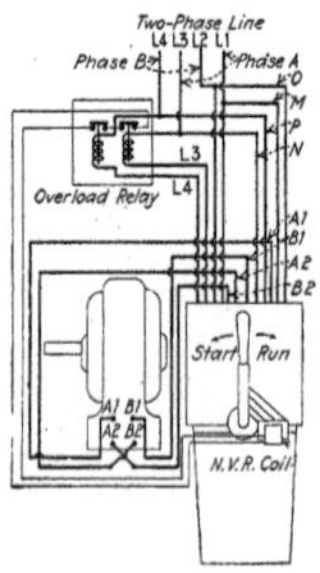

WIRING DIAGRAM OF COMPENSATOR AND TWO-PHASE MOTOR, "A2" AND "B2" LEADS CROSSED

In this case the customer had purchased two motor-generator sets for operating elevators and furnishing power for a number of motors in an office building. The motors were 440-volt two-phase 60-cycle squirrel-cage machines directly connected on a common bed-plate to 250-volt direct-current generators of the inter-pole type. The usual auto-starter, with overload relay and no-voltage release, was supplied for starting, which makes about the most simple class of installation that a wireman has to contend with.

A few days after the apparatus arrived at its destination, the district office of the manufacturer received a telegram from the customer saying that one of the motors was incorrectly wound and would not run; and requested that an engineer be sent immediately to investigate the trouble. The detail to leave on the first train fell to my lot.

Within about five minutes after reaching the job it was discovered that the man who wired up the motor had connected one of the *A* phase leads from the auto-starter to a *B* phase terminal on the motor, and vice versa, as shown in the figure. The two-phase motor had two separate non-interconnected windings and it was therefore impossible to get any response from the motor connected as in the diagram. As is obvious from the figure, all that was necessary to correct this trouble was to interchange the leads connected to the terminals *A2* and *B2* on the motor. After doing so, the motor was found to start up and carry its load in an entirely satisfactory manner. This simple error compelled me to take a 28-hour ride on a railway train, when the wiremen who installed the work could have discovered the trouble if they had taken the time to check their work.

K. A. REED.

New York City.

Truing a Bent Crankshaft on a Cross-Compound Engine

The electric company of our city has as part of its equipment a cross-compound engine, which is used only when the load is unusually heavy or in case of emergency.

Such was the case not long ago, and the engine was started, but something happened on the low-pressure side (probably water entered the cylinder), and the engine came to a sudden stop and in so doing bent the crankshaft, which was 10 ft. long, 10 in. in diameter between the bearings and 8 in. at the bearings. It carries a 30-ft. diameter by 4-ft. rim flywheel. The engine was badly needed and it was a case of getting a new shaft, which would take a long time, or making repairs on the damaged one. To straighten it was out of the question.

The damage was repaired as follows: As the shaft was bent in the bearing, I removed the crank disc and put in a temporary bearing between the flywheel and the bent end. The quarter boxes were then removed, leaving the shaft to run in the temporary bearing. Next, a tool for turning off the bent part was rigged up; the cross-rail of a small planer was strapped to the engine bed about as shown in the illustration. The engine was then started, running from the high-pressure cylinder, and the bearing part of the shaft was turned until it was true. Next, the outer end was turned for the crank disc until it was likewise true. The quarter boxes were then rebabbitted to fit the smaller diameter of the crankshaft, after which a sleeve was machined to fit on the end of shaft in order that the crank disc could again be pressed on.

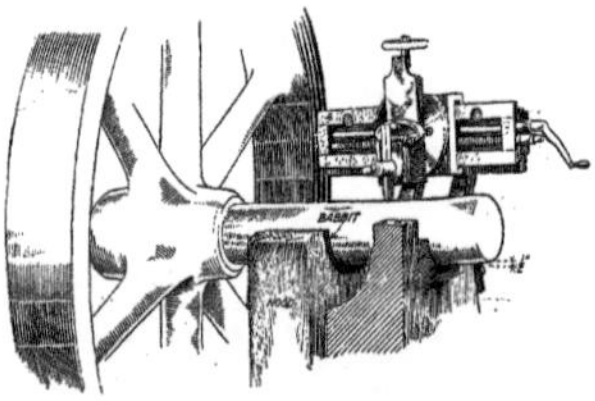

PLANER CROSSHEAD WAS USED TO TRUE BENT CRANKSHAFT

This completed the work, which was entirely satisfactory as to cost and time. The job required 80 hours for one man from the shop; the electric company furnishing helpers from its regular operating crew.

H. E. LARSEN.

Janesville, Wis.

A Proposed Improvement in the Steam-Engine Cycle

The diagram represents a device to save the latent heat in the exhaust from a steam engine. The present system of producing power in steam plants is notoriously wasteful, principally on account of the great amount of heat thrown away in the exhaust steam.

The boiler shown diagrammatically has no unusual features except a baffle or partition *A* in the steam space and a connection *B* to the water space. Steam is supplied to the engine in the usual way, but the exhaust is delivered to the *exhaust steam receiver* instead of to the atmosphere or a condenser. A Le Blanc hurling-water pump takes the steam from this receiver, compresses it and delivers it to the boiler, drawing its circulating water from the boiler itself. This circulating water, being drawn from the boiler and delivered back to the boiler, represents no gain or loss, except that due to radiation and friction in passing through the pump, which may be reduced to a negligible amount.

The value of the device lies in the return of the exhaust steam to the boiler in practically the same condition as when exhausted from the engine. Thus the heat usually carried away in the cooling water is returned to the boiler. The idea that exhaust steam must be condensed before being returned to the boiler is so general that everybody accepts it as a necessary evil. There is, however, nothing impossible in pumping steam instead of water by the use of hurling-water pumps similar to the Le Blanc. ALFRED COCHRAN.

New York City.

[Mr. Cochran's scheme is in violation of the fundamental thermodynamic principle that some heat must be

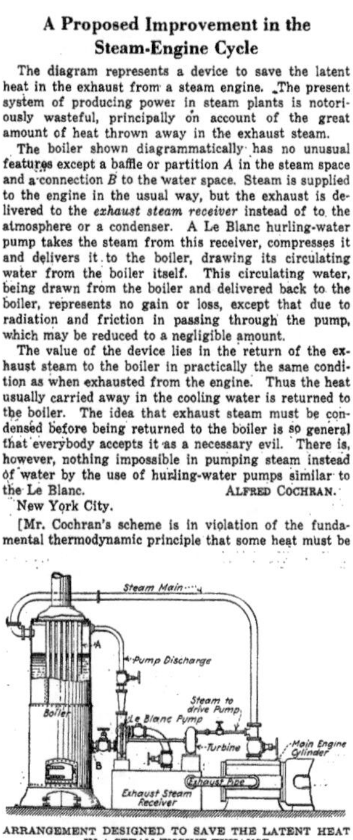

ARRANGEMENT DESIGNED TO SAVE THE LATENT HEAT IN A STEAM ENGINE EXHAUST

rejected by any heat engine unless it can expand its working fluid down to a temperature of absolute zero. Keeping in mind that the value of this device is that all or practically all the heat in the exhaust steam is returned to the boiler, its success depends on the amount of work necessary to compress the steam from exhaust pressure and quality to boiler conditions. If both the engine and compressor could operate without gain of heat from, or loss of heat to, outside bodies, the power required by the latter would be just equal to that developed by the former. The actual engine, however, will develop less power than the ideal non-conducting engine, while the hurling-water pump will require more power than a theoretical compressor. In other words, if the engine is to do more work than the compressor requires, the exhaust steam must be reduced in volume *before* being compressed. The only way to reduce the volume at constant pressure is to reject heat. Therefore, the rejection of heat is necessary for the production of power by a heat engine.—Editor.]

Starting a Frequency Changer

We have a 2,300-volt 60-cycle, 13,200-volt 25-cycle 1,500-kva. frequency-changer set which has always given trouble in starting. There is a 100-kw. 125-volt

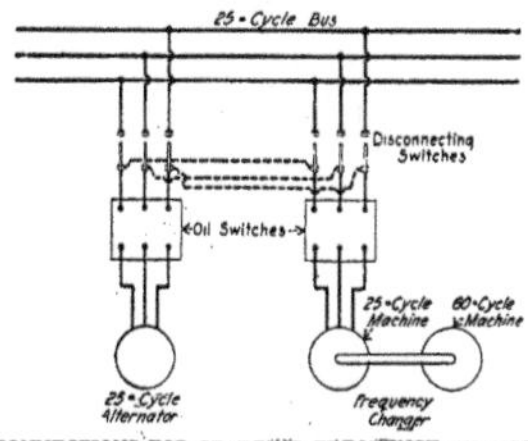

CONNECTIONS FOR STARTING FREQUENCY CHANGER

direct-current motor on the shaft for starting purposes, and this motor is connected to the exciter bus, but the necessary starting torque is so large that it requires practically the entire exciter capacity of the station to start the frequency changer, and if started at any time except when the station load is low it would cause an interruption to service. For this reason the frequency changer was shut down only when absolutely necessary; during one period it ran continuously for nearly nine months. However, it became desirable to shut down the set every night, and in order to prevent interruptions to service, the following method of starting was adopted:

The leads from the 25-cycle side of the frequency changer are connected to the bus at a point close to the leads of a 2,000-kw. 25-cycle turbo-alternator, the distance between the disconnecting switches of the two machines being about three feet. When the frequency changer is to be started, the alternator is shut down, if running, the disconnecting switches of the two machines are then opened and jumpers, as indicated by the dotted lines in the figure, are run between the machine sides of these switches, which is a simple matter as they are close together. The oil switches are then closed, both machines excited, and steam admitted to the turbine of the alternator. Since the machines are electrically tied together, they come up to speed together and the frequency changer is synchronized on the 60-cycle bus, after which the oil switches on the 25-cycle side are opened and the disconnects closed. The jumpers between the disconnects of the 25-cycle alternator and the frequency changer are left on except when work is being done on one of the oil switches, since these jumpers do not interfere with the normal operation of the machines.

Allentown, Pa. E. W. MIDDLETON.

Three Power-Plant Leaks That Waste Coal

As near as I can deduce, George H. Smith, on page 627 of the April 20 issue, figured correctly, but why is he feeding water into his boilers at 120 deg. and at the same time allowing steam traps to blow the quivalent of good coal away? Why does he not carry the trap returns back to the boilers so that the loss will balance itself to a great degree?

I recollect the instance of a plant that was operated by a new chief engineer. The traps were of an ancient type and more or less leaky. The chief figured the losses and decided that new traps would be a good investment. He specified and received modern traps, but the saving failed to materialize, as the traps all discharged into a common line and returned to a feed-water tank. The temperature of the water in this tank never reached the boiling point, therefore no steam was actually lost, and of course no fuel was saved by the installation of the new traps.

His failure to make a saving put the new chief in a poor light, and a breach opened between his employer and himself which finally terminated in his resignation. He figured correctly, provided the returns from the old traps had been thrown away.

I believe that it is a wise practice to use a small trap where it is possible, and to let it dump often on the maximum demand so as to decrease the chance of leakage.

My experience has convinced me that insulating material is the best-paying investment of any one thing about the steam plant. C. W. PETERS.

New York City.

Ampere-Turns in Field Coils

In a direct-current motor the field strength depends on the number of ampere-turns. By increasing the number of turns in the field coils the current may be reduced indefinitely without reducing the degree of saturation.

Now then, the statement has been made that "theoretically no current is required for exciting the fields of a direct-current motor." Is this assumption correct or not? EDWIN C. HUTH.

Cincinnati, Ohio.

[Taking a given coil with a fixed voltage E applied to it, the ampere-turns $IT = \frac{E\ circ.mils}{L}$, where L equals the mean length of one turn in inches. Now if the circular mils are halved and we assume that the mean length L of one turn remains constant, then the voltage E will have to be doubled to maintain the ampere-turns constant. Therefore, the product of the volts and amperes—that is, the watts—will be the same in both cases. This is what is wrong with the foregoing assumption. It is based on current only and not on watts, the product of volts and amperes. Although the current necessary to produce a given number of ampere-turns in a coil of fixed dimensions may be reduced in direct proportion to the reduction in circular-mil cross-sectional area of the copper, the voltage must be increased proportionately to set up the current through the coil, consequently the watts remain constant. Whether the coil has 100 turns with 1 ampere and 100 volts applied or 10,000 turns with 0.01 ampere and 10,000 volts applied, the result is 100 ampere-turns and 100 watts in either case. Since in a given motor the voltage E is fixed by the circuit it is designed to operate on, the size of the wire in the field coils to produce a given number of ampere-turns is also fixed by the winding space, the watts that the coil will dissipate without excessive heating, etc. The effect of insulation has been neglected in the foregoing.

Considering the question from a purely theoretical standpoint, if the cross-section of the wire in the coil and the applied voltage are held constant and the number of turns increased, the current will be, if the mean length of one turn could be held constant, reduced in direct proportion to the number of turns. But even if this was carried to infinity, the current would not be zero. Furthermore, the resistance of the coil increases faster than the number of turns on account of the length of the mean turn increasing with the size of the coil. In addition the coil would soon reach such proportions that the outside ampere-turns would have very little or no effect upon the core in the center of the coil.—Editor.]

Prolonging the Life of Condenser Tubes

In *Power* for April 13, 1920, page 588, Julian N. Walton states that the use of metallic packing on condenser tubes is advantageous because it bonds the tube electrically to the tube sheet, and thereby prevents electrochemical decomposition of the tubes. I believe that Mr. Walton is wrong in this matter.

If we immerse two electrodes in a suitable electrolyte, there will be no galvanic action if the electrodes are not electrically connected outside the solution. As soon as an outside connection is established, galvanic action will begin.

If, therefore, it were true, as Mr. Walton states, that fiber and fabric packings insulate the tubes from the tube sheets, these should be an ideal preventive of galvanic action. But such packings do not insulate the tubes, because there is direct metallic contact between the tubes and the tube sheet, and also a metallic connection through the ferrule, which is often in contact with the tube.

By the same token, Mr. Walton's metallic packing, being presumably a good conductor, could do nothing to reduce electrolysis, although it would probably not make matters any worse than they are.

Detroit, Mich. C. HAROLD BERRY.

Proposed National Engineer's License Law

Mr. Peters, on page 434 of the March 16 issue, suggests that when his proposed National license law goes into effect, engineers be given a special license without examination.

Why should not engineers holding first- and second-class licenses be given the same license they now hold? A man who obtains a first- or second-class license permitting him to operate in Massachusetts is capable of running a plant in Ohio or any other state. Then, under a National law, when a man wants to change his position from one state to another, it will not be necessary to pass another examination. H. A. BLANDFORD.

North Adams, Mass.

INQUIRIES OF GENERAL INTEREST

Grease More Dangerous to Burning New Boilers—Are new boiler tubes more likely to be burned from deposit of oil or grease than old ones? G. T.

It has been found that clean new flues and other heating surfaces are more affected by oil than old surfaces, especially where the latter are coated with an appreciable amount of scale. Many cases of burning of new boilers have been known where the thickness of the oil or grease coating has probably been less than one-thousandth of an inch.

Advantages of Multi-Pump Valves—Why are the water ends of steam pumps provided with a number of valves in place of a single large valve for each suction and discharge chamber? C. T.

A number of valves are not likely to be seated absolutely at the same time, and less shock is thereby imparted to the valve deck than when one large and heavy valve comes down on its seat. In addition to the advantage of smoother operation, several smaller valves are easier to construct, replace and repair than a single large valve of the same capacity.

Providing Against Pitting of Boiler—What steps should be taken if a boiler plate is found to be pitted? J. B.

When pitting is discovered, it should be determined by competent inspection whether the corrosion sufficiently endangers the safety of the boiler to require renewal of the plate, or the highest working pressure that can be carried with safety. When use of the boiler is to be continued, such inspections and determinations should be made at frequent intervals, and to arrest the pitting the feed water should receive suitable treatment for neutralizing the corrosive action of the impurities, as revealed by chemical analysis.

Designations of Boiler-Feed Waters—What proportion of scale-forming substances is allowable for good boiler-feed water? R. E.

The term "good" as applied to feed waters is only relative. The following designations are generally used, based on the number of grains of scale-forming ingredients per gallon of the feed water:

- Less than 8 gr. per gal.—Very good.
- From 8 to 12 gr. per gal.—Good.
- From 12 to 15 gr. per gal.—Fair.
- From 15 to 20 gr. per gal.—Poor.

When there is over 20 gr. of scale-forming materials to the gallon, the water should not be used unless first purified.

Material for Reinforcing Concrete—Cannot old hoisting cable be used to good advantage for reinforcing a concrete foundation for an engine and direct-connected generator? W. A. T.

Reinforcing material for concrete is principally of value in resistance of initial tensile and shearing stresses, for which purpose wire cable would be of little advantage in prevention of cracks or for maintaining the solidity of a foundation. Cable would be useful mainly for preventing sections of the concrete from falling apart after fracture had occurred and, for that purpose might be useful for incorporation in concrete work where permanent monolithic construction is not imperative.

Vacuum Referred to Standard 30 Barometer—What is the vacuum referred to a 30-in. barometer if the height of a mercury vacuum gage is 26.5 in. with a temperature of 80 deg. F. and the actual barometer reading is 29.3 in. and temperature 60 deg. F.? W. L. G.

Standard barometric pressure for comparison is usually taken as 30 in. for 32 deg. F. Allowing 0.0001 for the coefficient of expansion of mercury per degree F. of temperature, the reading of the vacuum gage corrected to the standard temperature would be

$$26.5\ [1-0.0001(80-32)] = 24.37 \text{ in.}$$

and for the barometer

$$29.3\ [1-0.0001(60-32)] = 29.21$$

Hence the absolute pressure in inches of mercury at the temperature of 32 deg. F. would be 29.21–24.37 = 4.84 in., and the vacuum referred to 30 in. barometer at 32 deg. F. would be 30–4.84 = 25.16 in.

Size of Pipe Line and Power for Pumping—What size of iron pipe line 1,200 ft. long would be suitable for delivery of 15 gal. of water per minute at 30-lb. pressure, pumped from a lake to an elevation of 85 ft? What would be the pressure on the pump from the line when not running, if the pump is placed 5 ft. above the lake, and what should be the horsepower of a motor for operation of the pump? W. P. O.

With the discharge end of the line open to atmospheric pressure and the line filled with water, the static, or standing, pressure in the pipe at the pump would be (85 — 5) × 0.433 = 34.6 lb. per sq.in.

Pumpage at the rate of 15 gal. per minute with a total lift of 85 ft. and discharge pressure of 30 lb. per sq.in., without allowance for power to overcome pipe friction,

would require $\dfrac{[30 + (85 \times 0.433)]\ 15 \times 231}{12 \times 33{,}000} =$ about

0.6 net or water horsepower, and with a pump efficiency of 50 per cent, the required power would be 0.6 ÷ 0.50 = 1.2 hp., neglecting pipe friction. For discharging 15 gal. per minute through 1,200 ft. of smooth, clean 1¼-in. iron or steel pipe, the pressure required for overcoming friction would be about 42 lb. per sq.in.; for 1½-in. pipe, about 15 lb. per sq.in.; and for 2-in. pipe, about 6 lb. per sq.in. With 50 per cent pump efficiency, each pound pressure, with discharge at the rate of 15 gal. per minute, would

require $\dfrac{1 \times 15 \times 231}{12 \times 33{,}000 \times 0.50} = 0.0175$ hp. and the total

power required using 1¼-in. pipe would be (0.0175 × 42) + 1.2 = 1.9 hp.; using 1½-in. pipe, (0.0175 × 15) + 1.2 = 1.5 hp.; and using 2-in. pipe, (0.0175 × 6) + 1.2 = 1.3 hp. The pressure and consequently the pump friction increases with use of a pipe line on account of an increase of roughness of the bore and reduction of diameter due to scale and corrosion. Therefore it would be advisable to provide a pipe not less than 1½ in. in diameter and a motor of not less capacity than 2 brake horsepower.

[Correspondents sending us inquiries should sign their communications with full names and post office addresses. Editor.]

Federated American Engineering Societies

PURSUANT to the call of the Four Founder Societies issued through their Joint Conference Committee, representatives from the following societies met in Washington on June 3 and 4 to organize a national engineering body:

American Association of Engineers.
American Association of Petroleum Geologists.
American Ceramic Society.
American Electric Railway Engineering Association.
American Electro-Chemical Society.
American Institute of Electrical Engineers.
American Institute of Mining and Metallurgical Engineers.
American Railway Engineering Association.
American Society of Agricultural Engineers.
American Society of Civil Engineers.
American Society of Mechanical Engineers.
American Society of Naval Architects and Marine Engineers.
American Society of Refrigerating Engineers.
American Society for Testing Materials.
American Society of Heating and Ventilating Engineers.
American Water Works Association.
Associated Engineering Societies of St. Louis.
Association of Railroad Engineers.
Boston Society of Civil Engineers.
Brooklyn Engineers' Club.
Cleveland Engineering Society.
Colorado Society of Engineers.
Detroit Engineering Society.
Duluth Engineers' Club.
Engineering Council.
Engineers' Club of Baltimore.
Engineers' and Architects' Club of Louisville.
Engineers' Club of Philadelphia.
Engineers' Club of St. Louis.
Engineers' Club of Trenton, N. J.
Engineering Society of Akron, Ohio.
Engineering Society of Eastern New York.
Engineering Society of Buffalo.
Engineers' Society of Pennsylvania.
Engineers' Society of Western Pennsylvania.
Florida Engineering Society.
Grand Rapids Engineering Society.
Illinois Society of Engineers.
Illuminating Engineering Society.
Indiana Engineering Society.
Institute of Radio Engineers.
Iowa Engineering Society.
Kansas Engineering Society.
Los Angeles Joint Technical Society.
Mohawk Valley Engineers' Club.
National Fire Protection Association.
Nashville Engineering Association.
Northeastern Water Works Association.
Oregon Technical Council.
Providence Engineering Society.
San Francisco Joint Council of Engineering Societies.
Scientific Club (Indianapolis).
Society of Automotive Engineers.
Society of Industrial Engineers.
Society for Promotion of Engineering Education.
Society of American Military Engineers.
Taylor Society of New York City.
Technical Club of Dallas.
Topeka Engineers' Club.
Vermont Society of Engineers.
Washington Society of Engineers.

The meeting was called to order in the auditorium of the Cosmos Club at ten o'clock Thursday morning by Richard L. Humphrey, consulting engineer, of Philadelphia, and chairman of the Joint Conference Committee. In his opening address he said that the committee had issued invitations to 110 engineering and allied technical organizations and societies representing an aggregate membership of over 120,000, who under the conditions of the call were entitled to send 184 delegates. "The list," he said, "included all societies, organizations or affiliations, not associations, sections, chapters or branches of national societies, whose chief object is the advancement of the knowledge and practice of engineering and the application of allied sciences, and who are not organized for commercial purposes." There were in attendance delegates representing over 60 per cent of the organizations invited. These delegates, over 70 per cent of those entitled to attend, represented organizations having an aggregate membership of over 100,000, or over 83 per cent of the membership of all the organizations invited.

Calvert Townley, president of the American Institute of Electrical Engineers, was elected chairman, and John C. Hoyt secretary, and the business before the conference was definitely set forth in the following resolutions presented by Major Gardner S. Williams, representing the Grand Rapids, Mich., Engineering Society.

Resolved, That it is the sense of this conference that an organization be created to further the public welfare where technical knowledge and engineering experience are involved and to consider and act upon matters of common concern to the engineering and allied technical professions, and

Resolved, That it is the sense of the conference that the proper organization should be an organization of societies and affiliations and not of individuals.

These resolutions resulted in considerable discussion, as they brought up the question as to what action would be taken by the American Society of Civil Engineers and by the American Association of Engineers. Although the latter society was strongly opposed to such an organization as described in the second resolution, its delegates at last voted for the resolutions, so as not to obstruct the wishes of an obvious majority of the conference. The roll call on the resolutions showed 119 affirmative and no negative votes.

The organization having been decided upon, Chairman Townley appointed the following committees: Constitution and By-laws, R. L. Humphrey, chairman; Resolutions, Philip N. Moore, chairman; and Program, R. H. Fernald, chairman. Each committee consisted of about twenty-five members.

Thursday Afternoon Session

In the afternoon session of Thursday, several formal addresses were delivered on subjects of interest to the delegates. Arthur R. Davis, president of the American Society of Civil Engineers, spoke on the subject "Functions of the Engineer in Public Affairs." Mr. Davis urged a unity of movement among engineers and emphasized the power of such a body as was to be formed by the conference.

He was followed by Philip N. Moore, a past president of the American Institute of Mining and Metallurgical Engineers, who spoke on the same subject. Mr. Moore pointed out that for the first time authorized sanction from the "great dignified old national technical societies" has been given for a body which shall speak for all branches of the engineering profession throughout the land. Some such body has been attempted in Engineering Council, but its limitations have led to the proposal for this new body, which shall be better able to carry on the work. As a member of Engineering Council Mr. Moore asserted that its members were proud of its work, but were ready to see it displaced by a more potent organization for the good of the profession. The nation has grown so large that only by organization can public needs and ideals reach the public ear. Reformers of all kinds have recognized this and by organization have had their views expressed in the form of statutes and constitutional amendment. Engineers must recognize this fact and also that social problems must be met as such and not as mathematical problems. Only by such recognition and organization can engineering service be most fully rendered to the public and engineers have their just voice in the affairs of the nation.

F. H. Newell, past president of the American Association of Engineers, spoke on its form of organization and called attention to the weakness of a confederation form as com-

Organization Chart of the Federated American Engineering Societies

The representatives of the group of National Societies on the Executive Board are to be elected or appointed or designated by the National Societies in any way upon which these societies may agree among themselves.

*Representation in the National Council is on the basis of one delegate for 100 to 1000 members and one additional delegate for each additional 1000 or major fraction thereof for national and other societies alike; no society to have more than 20 delegates.

†Representation on the Executive Board is so divided that the representation of the group of national societies (b) bears the same ratio to the representation of the group of state, regional and local societies (c) that the combined representation of the national societies in the Council bears to that of the other societies.

The representatives of the group of state, regional and local societies on the Executive Board to be elected by districts, the boundaries of which are so prescribed as to take account of both membership and area.

State, regional or local societies may be composed of a number of affiliated societies, sections, clubs, etc., but in the determination of their representation no count shall be taken of any organization which is represented individually or through another local, state or regional organization or affiliation.

The societies diagrammed are simply chosen as typical. Not all of those shown have indicated their intentions of joining the federation, and not all of those taking part in the organizing conference are included in the diagram.

pared with the federation form where each individual is a member of the general organization. A man is a citizen of the United States, not because he is a citizen of any one state; in the same sense he must be a member of the all inclusive society not solely because he is a member of a technical or local society. Mr. Newell traced briefly the organization of the American Association of Engineers and pointed out four optional courses of action on the part of the proposed organization toward the A. A. E.

Thomas H. McDonald, chief of the United States Bureau of Public Roads, reviewed the financial aspects of the Federal-aid road program and urged the co-operation of the new engineers' organization in opposing any classification of road building and maintenance as non-essential and therefore not entitled to preferred shipments by rail. William D. Uhler, chief engineer, Pennsylvania State Highway Department, spoke on the economic value of good roads.

Thursday Evening Session

In the Thursday evening session, which was held in the New Willard, Homer L. Ferguson, past president, United States Chamber of Commerce, told the engineers they needed more of the business point of view. He considered engineers lacked one quality of good business men. That missing quality was ability as public speakers.

James H. McGraw, President of the McGraw-Hill Co., Inc., speaking on the subject "Publicity for Engineers," pointed out the engineer's habitual avoidance of publicity which has resulted in the public being conscious of engineers as good servants without seeing them as real leaders upon whose thinking all the necessities and most of the luxuries of life depend. At present, however, the public mind is very receptive to engineering publicity, and the proposed federation of engineering societies should be a powerful factor in accomplishing such publicity. The mediums for publicity for engineers include the newspapers, the general magazines, the technical and industrial press, papers before societies and the public platform. From his own publishing experience Mr. McGraw pointed out the close relations between the engineer, the technical societies and the technical press. The technical press has greatly helped the engineer, and on the other hand the technical press cannot exist without the engineer. The technical societies and the engineering papers are natural partners, and each has its share in the general welfare. There should be no competition between the society publications and the independent press, since although they both serve in the same field, their services are different.

The last address of the evening was by George Otis Smith, Director of the United States Geological Survey, "The Engineer and National Prosperity." Mr. Smith paid a tribute to the engineer as living up to the specifications for good citizenship. In contrast with the lawyer-statesman, who looks backward, the engineer looks forward. He plans for coming events. Because he ever keeps in mind the factor of safety, he will make plans for future prosperity which are safe. The engineer is also constructive. As an illustration, the bridge engineer sees the future need for a larger heavier bridge to replace an old one. He makes his complete design in such a way that the new structure will be ample and safe, and then he erects the new bridge around the old one without interfering with traffic. Thus the engineer in any project builds up the new faster than he tears down the old so that he finally turns over a finished job, a monument to the characteristic engineering habit of seeing the thing through.

Friday Morning Session

On the morning of June 4 the report of the Committee on Constitution and By-laws was presented and, after a lengthy discussion, was approved practically without change. While the form of the document is subjected to editorial revision, it is substantially as follows:

The name of the organization is to be The Federated American Engineering Societies. Its object is stated to be "to further the interests of the public through the use of technical knowledge and engineering experience, and to consider and act upon matters common to the engineering and allied technical professions." The membership is to consist of national, local, state or regional engineering or allied technical organizations and affiliations.

The management is to be vested in a body known as the American Engineering Council and its Executive Board. The Council is to be made up of representatives from organizations on the basis of one representative for a membership of 100 to 1,000 inclusive with one additional representative for each additional 1,000 or major fraction thereof. The officers shall be a president (elected for two years and ineligible for re-election), four vice-presidents (two elected each year for two years), and a treasurer (elected for one year). There is also to be an executive officer, or secretary, appointed by the Executive Board.

The business of the organization shall be conducted by an Executive Board of thirty members, six to be the officers elected by the Council and twenty-four selected, a part by the national societies and the remainder by the local, state or regional organizations according to districts.

The funds for the organization shall be provided by each national society contributing $1.50 per member and each local, state or regional organization $1 per member, annually.

Provision is made for the formation of local affiliations and state councils with specific delimitation of authority between the various bodies.

Vote on Acceptance of Constitution

When the American Association of Engineers delegates were called on to vote on the acceptance of the proposed constitution, their spokesman asked permission to make a statement before announcing their vote. The affirmative vote on the Williams resolutions was cast in order not to appear as obstructionists. When it came to approving the constitution, however, and joining the new organization, the A. A. E. delegates felt they would be giving up their own welfare work as well as contributing about $30,000 annually to the new federation whose work would be much the same as that of their own "going" organization. Therefore they would not obligate their own organization to join the new one." It was maintained, however, that the A. A. E. hoped to co-operate with the new federation. Under the circumstances the American Association of Engineers' delegates requested that they be recorded as "present but not voting."

There were 88 affirmative votes, and none negative, a few of the societies being recorded as "present but not voting" or "absent."

On Friday Leroy K. Sherman, president of the United States Housing Corporation, Washington, D. C., spoke on "Functions of the Engineer in Public Affairs," his address having been postponed from Thursday morning. He stated that it was estimated that over 25 per cent of all engineers are engaged in some sort of public-service work. There are more engineers as employees of the public—municipal, state and national—than employees from all the other professions. The number of engineers in executive positions in public work, however, is comparatively small. This is largely the fault of the engineers themselves as their training deals with things and not with men and they are reluctant to deal with matters outside the strictly technical field of engineering. For the good of the public as well as of engineers the engineering profession must appreciate the human element more and take a larger part in leading public affairs.

Friday Evening Session

In the Friday evening session, several addresses were given, one by Samuel M. Vauclain, president of the Baldwin Locomotive Works. Mr. Vauclain sketched the problems of an industrial executive. The man who has confidence in himself and also in his subordinates is the one who becomes a really great engineering executive. Addresses were also delivered by Robert S. Woodward, president, Carnegie Institute, Washington, and James R. Angell, chairman, National Research Council.

The morning and afternoon sessions were held at the Cosmos Clubhouse, while both evening sessions were at the New Willard Hotel.

At the close of the business session resolutions were adopted:

1. Urging the payment of adequate salaries for the teachers of engineering in our technical institutions in order that adequately trained young engineering talent may be made regularly available.

2. Advocating the immediate adoption of appropriate measures to give effect to the recommendations made to Congress by the commission which recently reported upon a more adequate salary schedule for the engineering and other technical services of the Federal Government.

3. Indorsing the bill which has for some time been under consideration by Congress for the creation of a Department of Public Works.

4. Expressing the appreciation of the Organization Conference of the valuable work of the Engineering Council, especially its offers of assistance in making effective and operative the newly devised plan of organization; and expressing thanks to the Washington Society of Engineers and the Cosmos Club for their courtesy and assistance afforded during the sessions of the Conference.

Pennsylvania Electric Association Holds Harrisburg Meeting

Central-station steam heating was discussed by the Central Geographic Section of the Pennsylvania Electric Association during the morning session of the meeting held June 4, at the Penn Harris Hotel, Harrisburg, Pa., Chairman H. Root Palmer presiding. A paper on the subject was read by W. J. Kline. The author called attention to the effect that the higher price of coal has had in changing the attitude, of even those operating small plants, toward the methods used in power-plant operation. He was of the opinion that the present central-station methods of converting the chemical energy of coal into electrical energy should be changed wherever feasible to utilize as high a percentage of the heat as possible. The greatest loss in converting the chemical energy in the coal into electrical energy is that which is carried away in the circulating water and amounts to somewhere between 57 and 62 per cent of the total heat available for use in the turbine. In the large central stations that are looked upon as the last word in efficiency, there is wasted approximately 60 per cent of the total heat available. Mr. Kline was of the opinion that the greatest opportunity that existed today for fuel saving was in the co-ordination of power generation with the supply of heat. Too much stress has been laid on the cost of producing a kilowatt and not enough on the thermal units used. In the large central stations a kilowatt-hour is produced on 17,000 B.t.u., but in doing so there is wasted several times that amount of heat in the circulating water, where in the isolated plant that has a heating load the 60 per cent being thrown away by the central station is being used.

In discussing the difficulties of operating a district steam-heating system, W. P. Frehsee said that in the last few years he had found that 75 per cent of the trouble calls were from plants that had no supervision. Large bills were due largely to the way the customer operated his plant and the conditions of his heating equipment. He laid considerable stress on the utility working hand in hand with the customer and giving him the proper information for the satisfactory operation of his heating equipment. Mr. Frehsee said that his company was recommending the vapor system on all new work.

F. H. Hobbs said the keynote of steam heating had been struck when it was a service that was sold. There is no reason why steam heating should not receive a rate commensurate with the service rendered.

The discussion brought out that there was a need for a better meter than at present available for measuring the weight of steam supplied for heating. At present the best way is to measure the condensate. However, if it were possible to satisfactorily measure the steam, this would eliminate the losses due to leaks in the customer's heating system. One of the chief troubles experienced with the condensate type of meters is, they are liable to stop unless watched closely. Some of the companies found it necessary to read and inspect the meters every two weeks in order to keep them operating properly.

E. D. Dryfuss called attention to the increasing cost and carcity of natural gas, and said this would result in a demand for other means of heating in the localities now served by natural gas. He said ten-cent gas is about equal to dollar fuel; the price of coal is up to stay and the natural gas is approaching exhaustion, consequently conditions are becoming more favorable for district-steam heating. Mr. Dryfuss emphasized the danger from explosions when using gas with the supply as uncertain as at present.

F. W. Harris told of the excellent results his company had obtained on its heating system by using the extractor type of turbines. He said they had tried steam engines and made up the shortage of exhaust with live steam, but this was not as satisfactory as using bleeder turbines.

At the afternoon session accounting and operating records was the subject discussed. A paper on the "Classifications of Cost Records" was presented by C. Mac C. Breitinger. In his paper the author emphasized the difficulties of meeting the accounting classification required by the public-service commission regulations, saying it was impossible to obtain the cost of each job and that, to make the classification according to the present law considerably increased the cost of keeping public-service utility accounts. The purpose of the paper was to get a consensus of opinion so that some recommendations may be formed to reduce the large amount of expense and clerical work necessary in keeping accounts according to the present classification of the public-service commission.

From the discussion it was apparent that the small companies were having very little difficulty in keeping their accounts according to the public-service commission's classifications, but the chief difficulty was with the larger companies when the system of transmission and distribution became involved. Various suggestions were made for segregating the different charges. With the small companies this was done to a large extent by the men on the job, but in the larger companies the segregating of costs was done by the accounting department.

Short papers on power-station records were presented by J. F. Cole and F. L. Allner. The importance of the station log sheet was emphasized. It was the opinion that this should be so designed and kept that, when complete, it would contain all the information necessary to an intelligent study of plant operation. In addition to the log sheet, other reports are necessary such as semi-monthly inspection reports, trouble reports, etc. Mr. Allner said that in addition to their engineering value station records were of educational and training value to the operating force. He said that in the station at Holtwood, in addition to the log sheet, a logbook was kept in which all events are recorded that may be of interest to the operating force. Furthermore, each operator going off duty is required to leave a one-line diagram of the connections on the system at the time of leaving and sign his name to it. Another feature of plant records at Holtwood is the use of the dictaphone in transmitting all messages. The dictaphone is used to obtain a record of the message given over the telephone and has had a good effect in making the operators use precision in transmitting messages. This system has been in use seven years and has worked out very satisfactorily.

In the discussion on instruments the opinion was expressed that although there was a tendency in many cases to change from indicating to graphic instruments, nevertheless, these instruments are not all that might be expected of them, and not infrequently they give considerable trouble. It was thought better to use graphic instruments for intermittent reading only, cutting the instrument into circuit when it was desired to make the record. Attention was called to one case where a graphic voltmeter, with a paper speed of one inch per minute was used, and cut into the circuit by a quick-acting relay, when the voltage drops a few per cent.

In discussing meter records, F. L. Althouse emphasized the importance of watt-hour meter accuracy by calling attention to the fact that 1 or 2 per cent in the accuracy of meters may amount to as much in revenue as several per cent saving in coal at the plant.

American Boiler Manufacturers' Association

THE thirty-second annual convention of the American Boiler Manufacturers' Association brought together upward of sixty of the members and guests of the association with their ladies, at French Lick Springs. The meetings opened on Monday morning, May 31, with a paper by David Moffat Myers, engineer of the United States Fuel Administration on "Fuel Conservation."

A thorough application of all known and tried methods of fuel conservation in the United States would result in an annual saving of seventy-five to one hundred million tons of coal.

The results of conservation would be a money saving, based on the figure of seventy-five million dollars per annum; the transportation of coal would be reduced to an extent equivalent to 1½ million 50-ton car loads per year; seventy-five thousand of the 750,000 men engaged in mining coal could be taken from the mines and put to work in other fields; locomotive and train crews relieved from coal haulage would be of enormous service in the handling of freight; the labor of firing 75 million tons of coal per year under boilers and in furnaces would be eliminated; fuel conservation would reduce the cost of manufactured articles, and incidentally it would add materially to the life of our coal deposits and the supply of other fuels.

Fuel conservation, if developed to a reasonable extent, would mean a saving of nearly four dollars for each man, woman and child in the United States. Not only would the greatest industrial problem of the age be solved, but we could do our part in supplying markets of other countries.

One important phase of the situation is to learn how to get the best out of existing equipment by means of correct and economical operation.

There is not an industrial plant in the United States where 10 per cent to 30 per cent saving in fuel might not readily be effected by comparatively simple improvements in design and operation. You are all certainly familiar with the tremendous waste in the combustion of fuel under stationary boilers and in locomotive boilers, in the firing of coal, in the cleaning of fires, in the keeping of tight boiler settings, in the prevention of soot and scale on the heating surfaces and in the operation of boilers at their best capacity; in these features waste exists in every plant, and waste which is readily correctible. Prevention of waste is merely a matter of the application of well-known engineering principles.

Then fuel is wasted in the use of steam after its generation. Inefficient types of prime movers, careless maintenance of engines and pumps, ill-advised installations unsuited to the purpose they would fulfill, huge volumes of wasted exhaust steam expelled to the atmosphere, the unnecessary use of live steam from boilers where exhaust steam with proper engineering would fully meet the requirements, badly designed and ill-kept transmission systems—all these contribute their quota to the vast waste of fuel which is taking place in practically every industrial plant in the country. To this list should be added the fact that the average public-service station discharges 80 per cent of all the heat of its steam into the condenser—an irretrievable loss.

Oil is almost an ideal fuel, but owing to its extremely limited supply it should be applied only to very special uses where its natural characteristics are distinctly and specifically in demand, as in naval and certain classes of marine practice and for special industrial processes, etc. Under equally favorable conditions it is true that a somewhat higher thermal efficiency can be obtained with oil than with stoker-fired soft coal. Approximately estimated, oil at five cents per gallon will give the same cost for evaporating 1,000 pounds of steam as coal at ten dollars per ton. Consequently, there are only a few limited districts in which oil firing may be financially attractive. Otherwise, oil would be ruthlessly wasted until the supply was entirely depleted.

A conservation program with some of the following features included would prove beneficial: Regulation of quality of coal shipped from mine; zoning of coal as an aid to economical transportation; better means of storage, which is urgently needed to flatten out the load curve at the mines; co-operation between the coal-consuming public and the Bureau of Mines; the appointment of a body of citizens advisory to the United States Bureau of Mines in the planning and development of a conservation program. The program should interest itself in the correct design of new power plants, etc., a broader and more comprehensive plan of education than before applied along conservation lines.

Charles E. Gorton, chairman of the American Uniform Boiler Law Society, reported that boilers built to the Code of the American Society of Mechanical Engineers were now admitted to all states and cities having boiler regulations with the exception of Massachusetts. He read a long list of universities and colleges that have adopted the Code as a text or reference book. Since his last report, Oregon, Arkansas, Indiana, Maryland and Utah have become Code states, which is very encouraging for a year when few of the legislatures were in session.

C. O. Myers, secretary-treasurer of the National Board of Boiler and Pressure Vessel Inspectors, told of the work of the board since its organization last December. It consists of the inspectors or other officials charged with the enforcement of regulations covering boilers and pressure vessels in cities and states that have adopted the A. S. M. E. Code. One of its functions is to make rulings on specific designs of boilers and appurtenances not covered by the Code and to bring about the uniformity of practice in such matters that would be impossible without co-operation between the inspectors in the various states that are represented.

On Monday evening W. C. Connelly, who has served the association as its president for four years, presented his annual address. He recommended a simplification of the method of war taxation, recalled his prediction of a year ago—which has been verified—of a large increase in volume of business, but foresees a slowing up for the next twelve months. This he regards as one of the best signs of returning sanity that we have had.

Prices are high because labor has demanded and has been receiving wages far beyond what it earned; because of the decrease in production, by strikes, of thousands of articles of daily use; because the hours of production have been reduced 10 or 15 per cent and the actual production per man hour 25 to 50 per cent.

The cure is "everybody at work six days every week and producing a fair day's output."

He referred to the steel, coal and railroad switchmen's strikes and urged that a decided stand be taken for the open-shop principle and for hearty co-operation with the United States Chamber of Commerce and with the National Association of Manufacturers. Continuing, he said:

Photo by Gravelle Pictorial News Service, Indianapolis.

One of the complications involved in the enforcement of the Code in the case of boilers built in one state for use in another is the matter of providing for shop inspection during construction. In order that the certificate of the inspector at the shop where the boiler is made may be valid and satisfactory to the inspection department of a perhaps distant state in which the boiler is to be used, the local inspector must have proved his competency to the satisfaction of the departments of all the states or cities into which the boilers which he inspects may go. It is therefore suggested that all boiler inspectors operating under the departments represented be subjected to a uniform examination, and that, having qualified in this way, their certificates of inspection shall be universally accepted.

It is hoped to devise a system whereby the method of stamping and identification of boilers may be simplified, and the record of a boiler from its building to its final disposition be kept intact and easily accessible. A national convention of the Board will be held in the early future.

H. N. Covell reported as secretary-treasurer that the association numbered 78 active member companies, besides the associates, and was in a flourishing condition financially.

Monday afternoon was devoted to a golf tournament, in which a large proportion of the members took part.

This is not the time to make improvements or extensions, nor to stock up. Purchasing only of necessities by everyone will within a period of several months result in adjustment of prices to be more nearly normal.

Now is the time to reduce inventories to as low a point as is practicable. "A slowing up of business cannot but result in a reduction in prices by the independent producers of steel to those prices established and maintained by the United States Steel Corporation."

Reductions in prices lead to business failures—watch your credits. Standardized construction has been brought about through the Boiler Code of the A. S. M. E. Considerable progress has been made toward a uniform system of cost finding. It is hoped that this convention will arrive at certain standards regarding guarantees, terms of payment, etc. Standard sizes for all documents such as order forms, specifications, invoices, etc., were recommended. The address closed with acknowledgments to his associates and a plea for greater co-operation.

E. C. Fisher presented a report from a conference committee with the Stoker Manufacturers' Association to define terms in which the A. B. M. A. and the S. M. A. are interested. The following definitions were proposed and unanimously accepted:

For the purposes of the American Boiler Manufacturers and the Stoker Manufacturers associations the following definitions are adopted:

A BOILER is a metal vessel capable of withstanding pressure and serving the purpose of transmitting heat, usually produced by the combustion of fuel, to a liquid contained in the vessel.

In power-plant practice, boilers are usually divided into two classes, as follows:

(a) A FIRE-TUBE BOILER is so constructed that the products of combustion pass within tubes or their equivalent, these being surrounded by the liquid to be heated.

(b) A WATER-TUBE BOILER is so constructed that the liquid to be heated is contained within tubes or their equivalent, these being surrounded by the products of combustion.

A BOILER BAFFLE is a plate or partition placed in such relation to the tubes or their equivalent of a boiler as to cause the products of combustion to move in predetermined paths.

A FURNACE is a partially enclosed space in which heat is produced by fuel combustion.

A FURNACE ARCH is made of refractory materials forming the roof of a furnace or so located within a furnace as to aid combustion.

A GRATE is a metallic structure designed to support solid fuel and so made that air for combustion can pass through it to the fuel.

A TUYERE is a nozzle constructed to direct air under pressure into a fuel bed.

A TUYERE BLOCK is a special form of grate containing a tuyère or tuyères.

A RETORT is a receptacle so constructed that solid fuel may pass through it and in which partial distillation of the fuel takes place.

A DEAD PLATE is an imperforate plate which supports fuel.

A DUMP PLATE is a movable plate, grate or combination of same for intermittently discharging refuse from a furnace.

A MECHANICAL STOKER is a device consisting of a mechanically operated mechanism for feeding solid fuel into a furnace combined with means for supporting the fuel and supplying air to same during combustion and for directing the deposit of refuse in a location from which it can be readily removed from the furnace.

AN OVERFEED STOKER is a mechanical stoker where fuel is fed onto grates at one end of same. Overfeed stokers are usually divided into three classes as follows:

(a) A FRONT-FEED INCLINED-GRATE STOKER is an overfeed stoker where fuel is fed from the front onto grates inclined downward toward the rear of the stoker.

(b) A DOUBLE-INCLINED SIDE-FEED STOKER is an overfeed stoker where fuel is fed from both sides onto grates inclined downward toward the center line of the stoker.

(c) A CHAIN OR TRAVELING GRATE STOKER is an overfeed stoker where fuel is fed to the stoker from the front onto a moving grate forming an endless chain.

AN UNDERFEED STOKER is a mechanical stoker having one or more retorts from which the fuel is fed from below the surface of the fuel bed.

DRAFT is a difference in pressure due to a difference in gas density which tends to cause a flow of gas from the region of higher pressure to that of lower pressure.

NATURAL DRAFT is a draft produced by a chimney.

INDUCED DRAFT is a draft produced by mechanical means located at some point between a furnace and the point where the products of combustion are discharged.

FORCED DRAFT is a draft produced by mechanical means whereby pressure above atmospheric pressure is maintained under a grate.

D. C. Alexander, Jr., president of the Quasi-Arc Weldtrode Co., Inc., presented a paper, illustrated by lantern slides, on "Electric Welding."

S. F. Jeter, chief engineer of the Hartford Steam Boiler Inspection and Insurance Co., discussed the Advantages of Co-operation between Boiler Manufacturers and Boiler Insurance Companies. He emphasized the many ways in which the boiler manufacturer and the insurance company can work with each other, or co-operate, and the interest of both is served. Insurance companies are frequently confronted with claims due to minor accidents. If the boiler maker when making repairs where an insurance company is involved would keep the cost of repairs due to the accident separate from any general repairs that might be made at the same time, it would be very advantageous to the insurance company as well as the insured, who is the boiler maker's patron. In this way the possibility of controversy is avoided. He also pointed out the value of the insurance company to the manufacturer in helping him keep up to the standard which he desires to maintain and in meeting the legal requirements for construction where boilers must be built to conform with boiler laws. The address concluded with the suggestion that the maximum limit, 55,000 lb., used in figuring pressures be changed to conform with the A. S. M. E. code requirements.

Tuesday morning was devoted to an executive session, the ethics and economics of boiler making occupying most of the time. In the afternoon there was a golf tournament and in the evening a banquet.

Wednesday forenoon the election was held resulting in the choice, without contest, of the following officers for the ensuing year: President, A. D. Schofield, of J. S. Schofield's Sons Co., Macon, Ga.; vice president, George S. Barnum, of the Bigelow Co., New Haven, Conn.; secretary-treasurer, H. N. Covell, of the Lidgerwood Manufacturing Co., Brooklyn, N. Y. Executive committee: W. C. Connelly, the D. Connelly Boiler Co., Cleveland, Ohio; C. V. Kellogg, the Kewanee Boiler Co., Chicago, Ill.; F. C. Burton, Erie City Iron Works, Erie, Pa.; F. G. Cox, Edge Moor Iron Co., Edge Moor, Del.; W. A. Drake, the Brownell Co., Dayton, Ohio; E. C. Fisher, the Wickes Boiler Co., Saginaw, Mich.; W. H. Mohr, John Mohr & Sons Co., Chicago, Ill.; W. S. Cameron, Frost Manufacturing Co., Galesburg, Ill.

E. R. Fish, representative of the Association on the A. S. M. E. Boiler Code Committee and in the American Uniform Boiler Law Society, discussed the use of welding in boiler construction and said that notwithstanding the excellent results displayed in the paper to which the association had listened, he thought that manufacturers should proceed with caution in the use of welding for portions under stress of vessels such as boilers the rupture of which meant the release of a large amount of energy. The cast metal of which the weld is composed is not well adapted to withstand the repeated stresses due to pressure and temperature changes incident to boiler practice, and while many specimens of welding exhibit excellent holding power on test, so much depends upon the operator that one is never sure how dependable the joint may be. The present A. S. M. E. Code inhibits the use of welding for parts under stress, but the subcommittee on welding is considering directions in which approved practice may be broadened with such precautions as will safeguard the user and the public. He spoke briefly, too, of the National Board of Inspectors, emphasized the importance that it is likely to assume in the system of inspecting and recording boilers and urged the support of the Association for it.

Resolutions were passed against the compulsory enforcement of the metric system; indorsing the principle of the open shop and condemning collective bargaining when it involves agents outside of the shop or restricts the liberties or activities of the individual workman; for the appointment of an industrial committee to investigate labor problems and bring about a better relation between employer and employee; for the disapproval of the bonus bill; of thanks to the Kewanee Boiler Co. for the entertainment furnished by the male quartet that they brought to the convention; of thanks to the members of the executive and other committees of the association for their services and to W. C. Connelly the retiring president. Mr. Connelly was presented with a handsome silver coffee service at the banquet.

The undertaking of the bricksetting by the boiler manufacturer was discussed, the general disposition being to do so only as an accommodation to the customer who wishes to avoid a multiplicity of contracts and divided responsibility. A committee was appointed to collect and codify the specifications of the member companies in this respect.

The Code Committee was asked to furnish a report on Standard Steam Outlets, including the use of studded pads instead of standard steam nozzles.

Questions relating to standard guarantees, and proper methods for estimating the cost of a job were referred to the commercial committee.

A committee was appointed to consider uniformity of terms used in connection with a boiler setting.

Mr. Connelly called attention to the importance of heating and ventilating boiler shops as a question which will soon be forced upon boiler manufacturers and suggested that attention be paid to it at a future meeting.

At noon the convention adjourned.

National District Heating Association

The eleventh annual convention of the National District Heating Association was held May 25-27 at the La Salle Hotel, Chicago. The problems of the central stations in operating steam- and hot-water heating systems as adjuncts to their electric business were discussed, and the opinion was freely expressed that such steam-heating departments could and should be profitable. A number of suggestions was made in the course of papers and discussions for making the heating branch of central stations more renumerative than formerly. President J. C. Hobbs, of Pittsburgh, delivered a comprehensive address on operating problems, in which he suggested more efficient methods of central heating-plant operation. Prof. J. B. Hoffman, Purdue University, suggested and outlined a course for the technical training of men in the heating profession. The relative advantages of steam and hot-water heating was the subject of a paper by R. V. Stureman, of the Stureman-Butterick Co., Springfield, Ill., and an analysis of heat losses from underground pipes was treated in a paper by Prof. J. R. Allen, of the Research Laboratory of the American Steam Heating and Ventilating Engineers, Pittsburgh, Pa.

In a discussion of Professor Allen's paper, Mr. Walker of Detroit brought out the necessity for establishment of standards on which to base the calculation of pipe-covering efficiency. Mr. Walker with several others emphasized the lack of any definite basis from which to make the calculations; consequently figures on the efficiency of steam-pipe covering as ordinarily given are misleading.

During the course of the discussion of this paper, Professor Allen stated that he was taking up with Michigan Agricultural College, the question of experiments to determine the conductivity of the ground. Whether these arrangements will be finally carried out or not he was unable to state. It was further pointed out by Professor Allen in the discussion that the basing of computations of heat loss in pipes buried in the ground on experiments made in still air were valueless, in that heat losses underground are by conduction while those in still air are due to convection. In the discussion of the subject the importance of drainage in preventing heat losses in underground pipes was emphasized.

The problem of separating the steam and electric parts of utilities was discussed in the report of the accounting committee. The officers elected for the ensuing year were: President, J. L. Hecht, Chicago; First Vice President, J. H. Walker, Detroit; Second Vice President, E. L. Wilder, Rochester, N. Y.; Third Vice President, W. G. Carlton, New York City; Secretary-Treasurer, D. L. Gaskill, Greenville, Ohio. In addition to these officers, J. C. Hobbs, Pittsburgh, Pa., Fred B. Orr, Chicago, and F. A. Tucker, Danville, Illinois, were elected to the executive committee.

Assistance in Cases of Coal Shortage

The Washington representative of the National Committee on gas and electric service is in daily touch with the Interstate Commerce Commission, presenting the individual cases reported by gas and electric light and power companies of shortage in the coal supply. Cases requiring action should be reported direct to the offices of the committee, at Room 950, Munsey Building, Washington, D. C., but action can be secured only where full information is given, including the amount of coal actually on hand and the rate of daily consumption, the name of the contractor or shipper, the location of the mines and the railway or railways over which delivery is made.

The Dissipation of Heat by Various Surfaces

In a paper by T. S. Taylor, presented at the spring meeting of the American Society of Mechanical Engineers, St. Louis, May, 1920, are given the results of some interesting experiments on the loss of heat from various surfaces. Noting that hot water in bright tin vessels cooled more rapidly than in similar vessels covered with thin sheet asbestos, experiments were undertaken which showed that tin pipes, such as furnace pipes, actually dissipate more heat when tightly covered with thin asbestos than when left bare. Another set of tests were made to obtain data on the influence on the heat dissipation, of air velocities over the hot surfaces. These latter experiments were made with special reference to the cooling of electrical apparatus by ventilation.

For the former tests tin cylinders 10 cm. (3.9 in.) in diameter and 50 cm. (19.7 in.) long, containing cylindrical electrical heaters, were constructed. The temperature inside each vessel was measured by a thermometer inserted in the side, and thermocouples were attached to the outer surface to measure the surface temperature. The vessels were placed horizontally in such positions in the room as to avoid unnecessary convection currents and the room temperature taken at points sufficiently distant from the vessels to be uninfluenced by them. The heat required to maintain various differences between inside and outside temperatures was computed from the electrical-energy input as shown by ammeter and voltmeter readings.

Experiments were made under the following surface conditions: Plain bright tin; tin covered tightly with 0.33 mm. (0.013 in.) sheet asbestos; tin covered loosely with three layers of 0.33 mm. asbestos; tin covered with three layers of air-cell asbestos; tin aluminum painted; and various dust-covered surfaces.

The results obtained show that over the temperature range covered in these tests the thin asbestos covering results in an average of 37 per cent more heat loss than is found with bare bright tin. Even when both are covered with dust the asbestos surface will lose 23 per cent more heat than the bare one. The effect of dust on the surface is to increase the loss from the bare tin and decrease the loss from the asbestos-covered surface. Three layers of the 0.33 mm. asbestos covering loosely applied will give practically the same loss as the bare tin, while three layers of air-cell asbestos will permit a loss of only 75 per cent of that from the uncovered pipe.

The results are contradictory to the commonly accepted theory that any covering of heat-insulating material will cause a reduction in the heat dissipated from a hot surface. In explanation the author states that the asbestos surface is three or four times as great as the tin surface so far as molecular dimensions are concerned. The loss being due chiefly to air contact, this increased surface offers a greater chance for heat to be liberated. Therefore the thickness of the covering must be such that its decreased conductivity will make up for the increased surface effect before any saving may be expected. By an approximate calculation based on surface-temperature measurements, the author concludes that 0.51 cm. (about 0.2 in.) is the thickness of tightly applied sheet asbestos which will just equal bare tin pipe in heat dissipation. Thicknesses below this will give greater, and above this, smaller heat losses.

To study the effect of air velocities on heat losses, a special heater was constructed to represent a typical end coil of a turbo-generator. An iron bar $1\frac{1}{4}$ x $1\frac{1}{4}$ x 24 in. was covered with a $\frac{1}{16}$-in. layer of heater mica on which was wound a heater wire. On this was another layer of mica, and the whole apparatus was wrapped with the specified insulation material. Outside of this was a layer of fish paper, a layer of lead and a layer of sheet iron. Thermocouples were arranged to measure temperature inside and outside the insulating layer. The whole apparatus was placed in front of a blower outlet of such dimensions ($2\frac{1}{4}$ x $2\frac{1}{4}$ x 24 in.) that the coil was completely within the air stream. Air velocities were measured by means of a pitot tube.

From the data obtained, curves were plotted from which the author deduces the following general conclusions:

The heat liberated per degree difference in inside and outside temperatures increases approximately uniformly with the velocity over the ranges indicated. The heat liberated per degree excess temperature is constant for all temperatures for any given constant air velocity, except in the case where the velocity is that due only to natural convection currents when this value varies.

Tests were run with the coil set at various angles of incidence with the air current. The curves plotted from the data of these tests indicated that the maximum amount of heat would be dissipated when the air current met the surface at an angle of about 42 degrees.

New Jersey N. A. S. E. Convention

The annual convention of National Association of Stationary Engineers was held in Phillipsburg, N. J., June 3-6. Hotel headquarters was at the Huntington, Easton, Pa., across the Deleware River from Phillipsburg. The mechanical exhibit opened on the third, the business session of the delegates on the fifth. The Mayor of Phillipsburg made an appealing address of welcome, which was responded to by Charles H. Bromley, of Newark No. 3 Association. National President John J. Calahan, First National President Henry Cozens, and William Reynolds, chairman Board of Trustees, followed.

The convention was given a real treat in the form of an address on Americanization by Captain J. W. Riddle, superintendent of Public Schools, Phillipsburg. This address was by far the best of its kind the writer, who has heard many, ever listened to. S. B. Redfield, engineer, Ingersoll-Rand Co., Phillipsburg, followed.

President Val. V. Secor opened the convention. In the course of his address, National President Calahan stated that New Jersey had passed Massachusetts, Ohio, Pennsylvania and Illinois in membership.

The new officers are: George Armitage, Atlantic City, president; Charles L. Johnson, Elizabeth, vice president; Samuel Clark, Jersey City, secretary; John Mycock, Trenton, treasurer; John Shuey, Perth Amboy, conductor; Fred Beecroft, doorkeeper. John J. Reddy, of Jersey City, was again chosen state deputy. The convention next year will be held in Jersey City.

At the close of the session on Sunday the delegation visited the huge works of the Ingersoll-Rand Co., where, after an inspection of the shops, the company served a most complete and relished dinner.

On June 1, a 500-kw. steam turbine exploded at the Phœnix Mills, Utica, N. Y. Buckets were thrown out, breaking the casing and knocking holes through the brick walls. The engineer was killed. The speed is given at 2,200 r.p.m.

New Publications

POROSITY AND VOLUME CHANGES OF FIREBRICKS

Technologic Paper No. 359, "Porosity and Volume Changes of Clay Firebricks at Furnace Temperatures," is now ready for distribution by the Bureau of Standards. This investigation is a study of some of the physical properties of clay firebricks by a comparison of their changes in porosity and volume, on heating to different temperatures, with the amount of contraction of the bricks under load at furnace temperatures, and with their so-called "melting points." It is found that the porosity and volume changes serve as a much more reliable indication of the ability of a clay firebrick to resist deformation under load at furnace temperatures than the "melting point." It is also shown that these porosity and volume changes serve in explaining the failure of firebricks in use, from such causes as insufficient burning in manufacture.

Personals

Warren D. Spangler, consulting industrial engineer, Marshall Building, Cleveland, Ohio, desires catalogs from manufacturers of steam and electric power appliances.

Frank L. Platt, formerly retained by the War Department as a specialist on Industrial Planning Projects, has become associated with the firm of Waldron & Van Winkle, of New York and Boston. Mr. Platt will be in the New York office at 37 Wall St.

W. C. Buell, Jr., formerly engineer of tests, and later chief engineer of Tate-Jones & Co., Inc., on June 1 became associated with the George J. Hagan Co., of Pittsburgh, furnace and combustion engineers, as chief engineer of the company's newly organized liquid fuel department.

Dr. Alphonse A. Adler, of the Polytechnic Institute of Brooklyn, after ten years of service as professor in charge of machine and power plant design, has resigned in order that he can give his undivided attention to his consulting practice in New York City in power and industrial plant engineering.

W. W. K. Sparrow, assistant chief engineer of the Chicago, Milwaukee & St. Paul Railway Co., has been appointed assistant to the president of that railroad. He will be in charge of valuation, adjustment of Federal accounts and other assigned duties. He was corporate chief engineer of the Milwaukee Ry. Co., during Federal control.

Society Affairs

The Washington Assembly of the American Association of Engineers was recently organized and a committee appointed to prepare the constitution and by-laws.

The Twin City Chapter of the American Association of Engineers has requested enlargement of the powers of the St. Paul city charter empowering the city council to determine salaries of city employees.

The Eastern Geographic Section of the Pennsylvania Electric Association will meet Friday, July 2, at the Adelphi Hotel, Philadelphia, Pa. Topics to be discussed will include "Power Factor and the Effect of Unbalanced Load on the System."

The American Water Works Asociation will hold its fortieth annual convention from June 21 to 25 at Montreal, Quebec, headquarters at the Windsor Hotel. Besides various important and interesting papers, a big feature of the convention will be the amusements arranged by the entertainment committee. These include dancing, concerts, golf, sightseeing trips, etc.

The American Association of Engineers, Nebraska Assembly, held its first annual meeting May 29 in Lincoln. The following officers were elected: President, W. R. McKeen, president Omaha Chapter, A. A. E.; first vice president, Prof. Clark E. Mickey, University of Nebraska; second vice president, Roy N. Toll, city commissioner, Omaha; treasurer, George W. Bates, city engineer, Lincoln, Neb.; secretary, Watson Townsend, assistant engineer, U. P. R.R., Omaha.

S. A. E. Annual Meeting Program—The summer meeting of the Society of Automotive Engineers is scheduled to take place from June 21 to 25 inclusive, at Ottawa Beach on Lake Michigan. In addition to the various sessions, including a Standards and Business Session, Fuel Session, Transportation Session, Farm Power Session and production Session, a wonderful program of amusements and recreation has been planned. On Thursday evening, June 24, a grand ball will be held, including a dancing contest. With the exception of Monday there will be dancing in the ballroom every evening. Over 200 prizes will be given in the sports program, which will be conducted on Tuesday, Wednesday and Thursday afternoons. Tennis, golf, baseball, trap-shooting, races, water sports and special sports will be staged.

Miscellaneous News

The Niagara Falls Power Company has declared a dividend of $1.75 per share on its preferred stock, payable June 30, 1920.

Mayor Hugh M. Caldwell of Seattle, Wash., has signed a city ordinance accepting the stipulations of the Government for the permit granting the use of certain lands in the Washington National Forest for the Skagit power-plant project.

A Certificate of Merit by the War Department for making prompt deliveries of stokers for Government plants and otherwise co-operating with the Construction Division of the Army during the recent war has been awarded the Detroit Stoker Co.

The Pacific Gas and Electric Company is planning to construct its third power unit in Shasta County. The new project will be on Pitt River, four miles down stream from where Fall River tumbles over into the Pitt. Fall River Mills is at the junction of Fall and Pitt Rivers. Two miles up Fall River from the town, a tunnel 10,500 ft. will be dug through Saddle Rock Mountain. Nearly all the water in the river will be turned through this tunnel. At the southern outlet the water will be 575 ft. above Pitt River, which, with Fall River, makes a long horseshoe bend. A pipe line 3,000 ft. long will deliver this water under 575 ft. pressure on the banks of Pitt River, where the third power house will be built. This unit will generate 68,000 hp., and is to be connected within a year.

Business Items

The Wheeler Condenser and Engineering Co., of Carteret, N. J., is completing three condensers for 20,000-kw. turbines for the Union Electric Light and Power Co., of St. Louis; one of 15,000-kw. for the Indianapolis Light and Heat Co., one of the same size for the Omaha Light & Power Co., and still another for the Wichita (Kan.) Light and Power Co., also one 10,000-kw. unit each for the Midwest Utilities Co. at Tulsa, the Tennessee Light and Power Co. at Memphis, and the Central Illinois Public Service Co. at Harrisburg, Ill.

Trade Catalogs

The Westinghouse Electric and Manufacturing Co. of Pittsburgh, Pa., has ready for distribution a 20-page catalog entitled "Watt-Hour Meters," which describes and illustrates its line of watt-hour meters. The descriptions cover operation, distinctive features, construction and application. A copy will be sent on request.

The Page Steel and Wire Co., Monessen, Pa., has ready for distribution a new 31-page catalog entitled "American Ingot Iron Wire." The booklet contains, in convenient form, data and tables pertaining to the properties of American Ingot Iron Wire, of interest to engineers, superintendents, purchasing agents, etc. A copy will be sent on request.

"Industrial Buildings at Matagorda" is the title of a folder distributed by the J. G. White Engineering Corporation, 43 Exchange Place, New York City. It gives information about, and illustrations of, the power plant, warehouse, machine shop, and model homes erected for the housing of workmen, at the sulphur plant of the Texas Gulf Sulphur Co., Matagorda, Texas.

"America's Greatest Dam," by William Benjamin West, is a new 64-page booklet describing in detail the Wilson Dam, which is being constructed at Muscle Shoals, Nitrate Plant No. 2, Alabama. The book is written for the layman as well as for the technically inclined and contains a supplement on "Our Water Power," which sets forth the hydropower possibilities of this country.

COAL PRICES

Prices of steam coals both anthracite and bituminous, f.o.b. mines, unless otherwise stated, are as follows:

ALANTIC SEABOARD

Anthracite—Coals supplying New York, Philadelphia and Boston:

	f.o.b. mines
Pea	$5 30@$5.75
Buckwheat	3.40@ 4.10
Rice	2.75@ 3.25
Barley	2.25@ 2 50
Boiler	2.50

Bituminous—Steam sizes supplying New York, Philadelphia and Boston:

	f o.b. mines
Cambrias and Somersets, f.o.b. mines net ton	$7.25@$8.00
Clearfields, f.o.b. mines, net ton	6.75@ 8.25
Pocahontas } on cars Providence and	11.00@14.50
New River } Boston; gross ton	11.00@14.50

BUFFALO

Bituminous—f.o.b. mines:

Pittsburgh slack	$6.00@$8.00
Mine-run	6.00@ 8.00
Lump	6.00@ 8.00
Youghiogheny	6.00@ 8.00

CLEVELAND

Bituminous—f.o b. mines:

No. 6 slack	$4.50@$6.00
No. 8 slack	4.50@ 6.00
No. 8—¾-in.	4.50@ 6.00
No. 6 mine-run	4.50@ 6.00
No. 8 mine-run	4.50@ 6.00
Pocahontas—Mine-run	4.50@ 6.00

ST. LOUIS

Bituminous—f.o b. mines:

	Williamson and Franklin Counties	Mt. Olive and Staunton	Standard
Mine-run	$3.00@4 00	$3.00@3.50	$4.50
Screenings	3 00@4 00	$3 00@3.50	5 00
Lump	3.00@4 00	3 00@3.50	5.00

Williamson-Franklin rate to St. Louis is $1.10; Other rates 95c.

CHICAGO

Bituminous—Prices f.o.b. mines as per circular of April I are as follows: Co. prices vary from 50c. to $1.25

Illinois		Freight rate Chicago
Southern Illinois		
Franklin, Saline and Williamson Counties		
Mine-Run	$3.00@$3.10	$1.55
Screenings	2.60@ 2.75	1.55
Central Illinois		
Springfield District		
Mine-Run	$2.75@$3.00	$1.32
Screenings	2.50@ 2.60	1.32
Northern Illinois		
Mine-Run	$3.50@$3.75	$1.24
Screenings	3.00@ 3.25	1.24
Indiana		
Clinton and Linton		
Fourth Vein		
Mine-Run	$2.75@$2.90	$1.27
Screenings	2.50@ 2.65	1.27
Knox County Field		
Fifth Vein		
Mine-Run	$2 75@$2.90	$1.37
Screenings	2.50@ 2.60	1.37
Brazil Block	$4.25@$4 50	1.27

New Construction

PROPOSED WORK

Me., Portland—The Construction Division, War Dept., Wash., D. C., will receive bids until June 21 for generator houses, steel hangars and miscellaneous construction at the coast defenses here.

Mass., Andover—Guy Lowell, Archt., 12 West St., Boston, will soon award the contract for a 2 story, 100 x 200 ft. memorial building, including a steam heating system, for the Philips Andover Academy. About $250,000.

Mass., Boston—The Construction Division, War Dept., Wash., D C., will receive bids until June 21 for generator houses, steel hangars and miscellaneous construction at the coast defenses here.

Mass., Chelsea—H. H. Atwood, Archt., 61 Alban St., Dorchester, will soon award the contract for a 2 story, 35 x 90 ft. power plant and laundry addition to present building at the Soldiers' Home here, for the Commonwealth of Massachusetts, Boston. Estimated cost, $125,000.

Conn., Bridgeport—The Amer. & British Mfg Corp. is in the market for an air compressor of 300 cu.ft. capacity, to be 2 stage type, either belt or motor driven, developing at least 100 lb. air pressure and also a reservoir tank about 3 ft. in diameter and 8 ft high.

Conn., Bridgeport—S. Z. Poli Co., 24 Church St., New Haven, will soon award the contract for two 221 x 223 ft. theatres on Main and Congress Sts., including steam heating systems. About $1,000,000.

N. Y., Brooklyn—The Bd. of Welfare received bids for repairs to boiler settings and new boiler fronts at the power house at the Kings County Hospital here, from the Acme Furnace Equipment Co., 39 Cortland St., $15,052; Frederick Page Contg. Co., 115 B'way, $18,200.

N. Y., Brooklyn—The Brooklyn Union Gas Co., 176 Remsen St., is in the market for 2 gas pumps and several generators for the station at 12th St. and Second Ave.

N. Y., Brooklyn—Charles L. Fraser, Archt. and Engr., 103 Park Ave., New York City, will soon award the contract for a 4 story department store including a steam heating system on B'way between Grove and Linden Sts., for the Buckley Newhall Co., 5th Ave. and 125th St., New York City. About $250,000.

N. Y., Buffalo—The Buffalo Wagon Works, 111 Carroll St., is in the market for one 5 hp., 500 d.c. motor.

N. Y., Buffalo—The Houde Eng. Corp., 1400 West Ave., is in the market for a small air compressor about 4 in. bore.

N. Y., Buffalo—Frank Lanahan & Sons, 174 Fulton St., is in the market for a 12 x 14 x 275 ft. air compressor with tank.

N. Y., Iona Island—The Bureau of Yards & Docks, Navy Dept., Wash., D. C., plans to improve the Naval Magazine power plant here. About $27,000.

N. Y., Jamestown—Lyman P. Hapgood, Supt. of Water and Lighting, plans to install at the Cassadago station, three 150 hp. water tube boilers with equipment, 2 steam driven centrifugal pumps normal capacity 5,000,000 per 24 hours, 2 steam air compressors, compound 2 stage, capable of delivering sufficient air under 80 lb. pressure to equal 160 hp., pumping engines, 2 motor driven volute centrifugal pumps, not to exceed 1,200 r.p.m., 440 volt, 60 cycle, 3 phase; 2 directly connected steam turbine generators to generate current at 440 volt, 60 cycle, 3 phase, capacity 378 turbine capacity 540 hp., 2 motor driven, 2 stage air compressors, which will furnish sufficient air at 80 lb. pressure to produce 160 hp.

N. Y., Long Island City—The Horizontal Hydraulic Hoist Co., 35 25th St., Milwaukee, Wis., plans to build a 1 story, 100 x 150 ft. service station.

N. Y., New York—The Auditorium Association, c/o Slee & Bryson, Archts. and Engrs. 154 Montague St., Brooklyn, is having plans prepared for a 4 story, 75 x 150 ft. club and office building including a steam heating system at 250 West 25th St. About $250,000.

N. Y., New York—The Construction Division, War Dept., Wash., D. C., will receive bids until June 21 for generator houses, steel hangars and miscellaneous construction at the coast defenses here and at Long Island Sound.

N. Y., New York—Keeler & Fernald, Archts., 203 West 13th St., will soon award the contract for an 8 story loft building including a steam heating system at 251 West 27th St., for the Herzig Mercantile Co.

N. Y., New York—The New York Bellevue and Allied Hospitals, 26th St and 1st Ave., plans to construct a hospital addition including a steam heating system. About $800,000. McKim, Mead and White, 101 Park Ave., Archts.

N. Y., Oakfield—Genesee Light & Power Co. of Batavia has petitioned the Pub. Serv. Comn. for authority to issue $56,700 in preferred and $50,000 common stock to equip their plant here for the purpose of supplying additional power and lighting here and to extend the transmission lines to Byron, Darien and Alexander.

N. Y., Olean—The Bd Educ., Union Free School Dist. 1, will soon award the contract for installing heating and ventilating equipment in school buildings No. 1, 5, 6 and 7. E. H. Keeney, clk. Phegley & Szekley, Merian Bldg., Cleveland, O., Engrs. Noted April 20.

N. J., Clinton—F. H. Bent, State Archt., 142 West State St., will receive bids until June 22 for a 3 story, 68 x 131 ft. maternity cottage and boiler plant. About $100,000.

N. J., Lakehurst—The Bureau of Yards & Docks, Navy Dept., Wash., D. C., will receive bids until June 21 for the installation of power plant equipment. Estimated cost, $400,000.

Pa., Philadelphia—The Abrasive Co., Fraley and James Sts., plans to build a 1 story, 55 x 60 ft. power house including power equipment. Estimated cost, $25,000.

Pa., Philadelphia—The General Electric Co., Witherspoon Bldg., plans to build a plant, including a steam heating system, at 68-70th St. and Elmwood Ave. About $10,000,000.

Md., Baltimore—Samuel T. Williams, 223 North Calvert St., is in the market for a 3 phase, 25 cycle, 440 volt winding machine with motor, and a 250 hp. open type feed water heater.

Md., Laurel—The Maryland Motors Corp., 631 Munsey Bldg., Baltimore, will receive bids until August 1 for a 2 story, 30 x 50 ft. hydro-electric plant, to have a 600 kw., 3 phase, 60 cycle, a.c. capacity. Three 200 hp. steam turbines, two 500 hp. boilers and two 600 hp. waterwheels will be installed in same. Estimated cost, $120,000. T. B. Webster, 631 Munsey Bldg., Baltimore, Engr.

D. C., Washington—The Amer. Security & Trust Co., 15th St., plans to build a bank and office building including a steam heating system. About $1,000,000. York & Sawyer, 50 East 41st St., New York City, Archts. and Engrs.

D. C., Washington—The General Purchasing Officer of the Panama Canal will receive bids until June 18 for furnishing 20 upright, extra heavy, galvanized, riveted boilers, etc.

Va., Langley—The Construction Division, War Dept., Wash., D. C., will receive bids about June 20 for several buildings including a power house, etc., here. Cost, over $500,000.

Va., Richmond—The Odd Fellows Lodge plans to construct a temple including a steam heating system. Herts & Robertson, 331 Madison Ave., New York City, Archts. and Engrs. About $600,000.

S. C., Hartsville—The city plans to construct a sewerage pumping plant. About $15,000. C. C. Wilson, Columbia, Consult. Engr.

S. C., Paris Island—The Bureau of Yards & Docks, Navy Dept., Wash., D. C., will receive bids until June 21 for power plant improvements. Noted June 1.

Ala., Gadsden—The Alabama Power Co. plans to build a transmission line from here to Lindale, Ga., a distance of 52 miles and deliver hydro-electric power to the substation there.

Ala., Muscle Shoals—The Constructing Quartermaster, U. S. Nitrate Plant No 2, will receive bids until June 21 for a transmission line between plants 1 and 2. approximately a distance of 4 miles.

O., Cleveland—The Brandt Co., Sheriff St. Market, plans to build a 6 story cold storage warehouse including a refrigeration system. About $300,000.

O., Cleveland—The Parish & Bingham Co., West 106th St. and Madison Ave., is having plans prepared for a 1 story, 30 x 55 ft. boiler plant at 10600 Madison Ave. Two 500 hp. boilers with stokers will be installed in same. Estimated cost, $100,000. Ernest McGeorge, 1900 Euclid Ave., Archt. and Engr.

O., Cleveland—William Taylor Son & Co., Euclid Ave., is having plans prepared for a 10 story, 69 x 200 ft. department store addition including a steam heating system at 614 Euclid Ave. About $500,000. D. H Burnham & Co., 209 South LaSalle St., Chicago, Archt.

Ind., Evansville—The Indiana Atomized Fuel Co plans to build a 1 story, 80 x 100-ft., 350 to 400 hp. electric driven plant to reduce coal to an atomized state. Estimated cost, from $125,000 to $150,000. B. U. Cain, Secy.

Ind., Indianapolis—The city plans an election in June to vote on $650,000 to construct a power plant at the Arsenal Technical high school.

Mich., Detroit—Louis Kamper, Archt., 749 Book Bldg., will receive bids until June 21 for a 10 story, 53 x 141 ft. office and loft building on Woodward Ave., including steam heating equipment, for Hugo Scherer. About $1,000,000 Noted April 20.

Mich., Detroit—The Manufacturers Mutual Insurance Co., 484 East Jefferson Ave., is having plans prepared for a 4 story, 45 x 150 ft. hospital and office building including a steam heating system on East Jefferson Ave. About $300,000. Stratton Snyder & M. R. Burrowes, Union Trust Bldg., Archts.

Mich., Grand Rapids—St. Mary's Catholic Congregation, 423 1st St., N.W., is having plans prepared for the construction of a 1 story central heating plant. E. Brielmaier & Sons Co., University Bldg., Milwaukee, Wis., Archt. and Engr.

Ill., Chicago—Albert L. Appleton and Associates, c/o Walter W. Alshlager, Archt., 111 West Washington St., are having plans prepared for a 21 story, 110 x 150 ft. hotel on Chicago and Michigan Aves., including a steam heating system. About $4,000,000.

Ill., Chicago—Cahill & Douglas, Engrs., 217 West Water St., Milwaukee, Wis., are receiving bids for a 750 hp. pocket valve uniflow engine and a 500 kw. generator. also a 195 ft. stack, for the Illinois Malleable Iron Co., 1801 Diversey Blvd.

Ill., Chicago—The Chicago, Milwaukee & St. Paul Ry. Co., Office of the Purchasing Agent, is in the market for 1 locomotive hoist consisting of six 50-ton jacks fitted with a General Electric 50 hp. motor, one 5 hp. motor with starting box and switches for running hammer, one 10 hp. motor, speed variable from 550 to 1,100 r.p.m. for running boring mill. All electric equipment will be 220 volt, d.c.

Wis., Eau Claire—The Industrial Constr. Co., 3 South Barstow St., is in the market for 1 link belt conveyor.

Wis., Milwaukee—Cahill & Douglas, Engrs., 217 West Water St., will soon award the contract for the furnishing two 300 hp. boilers for Geuder, Paeschke & Frey, St. Paul Ave. and 15th St. Noted March 3.

Wis., New London—The Little Wolf Power Co., has obtained permission to construct and operate a dam on the Little Wolf River in Waupaca County, near here.

Wis., Sheboygan—The city will soon receive bids for a 4 story, 150 x 250 ft. high school including a steam heating system, on Jefferson and 9th Sts. About $750,000. Childs & Smith, 64 East Van Buren St., Chicago, Archts.

Wis., Wauwatosa—The Milwaukee Co. Bd. of Administrators, Milwaukee, will soon award the contract for intercepting sewers, pumping station and force main. Two 250 g.p.m. motor driven centrifugal sewage pumps will be installed in same. W. G. Kirchoffer, 31 Vroman Bldg., Madison, Wis., Engr.

Minn., Coleraine—A. King, Clk., will receive bids until July 15 for a 2 story, 200 x 200 ft high school including a steam heating system for the Bd Educ. Estimated cost, $350,000. W. T. Bray, 616 Torrey Bldg., Duluth, Archt. and Engr.

Minn., Duluth—The Great Northern Power Co of Duluth plans to build a dam across Boulder Lake, for the purpose of developing 50,000 hp. in order to supply additional electric power to this town and towns on the Minnesota iron ranges.

Minn., Duluth—The Rev. J. T McNicholas, 211 West 4th St., will receive bids until June 29 for a 3 story, 100 x 150 ft grade school including a steam heating and mechanical ventilating system, on 28th Ave. East, and 4th St About $250,000 Maginnis & Walsh, 100 Roylston St., New York City, Archts. and Engrs.

Minn., Hibbing—The town rejected all bids for a 3 story, 149 x 284 ft. recreation building on 1st Ave. and Jackson St. Halstead & Sullivan Palladio Bldg., Duluth, Archts. and Engrs Work will be readvertised. Noted May 10.

Minn., Minneapolis—The State University will soon receive bids for a 3 story, 95 x 160 ft music building including a steam heating system C L. Pillsbury Co., 805 Metropolitan Bank Bldg., Engr. C. H Johnston, 715 Capital Bank Bldg., St. Paul, Archt About $500,000.

S. D., Sioux Falls—Symms-Brownell Mfg. Co. is in the market for one 28 to 34 x 12 to 14 in. milling machine and one 150 lb drive air compressor (used).

Mo., St. Louis—The city will receive bids until June 29 for an engine house to have a 200 kw. capacity, and for the installation of one 220 hp. horizontal single direct connected side crank engine to be connected to a 200 kw. generator now in place, at the City Sanitarium Pumping Station, here E. E. Wall, City Hall, Water Comr. Estimated cost, $15,000 Noted May 25.

Tex., Port Arthur—J. A. Wetmore, Supervising Archt., Treasury Dept., Wash., D. C., will receive bids until June 25 for furnishing and installing a heating boiler in the United States Post Office and Custom House, here.

Okla., Madill—The City Clerk will receive bids until July 1 for a 1 story, 30 x 40 ft. building, distribution system for water main, dam and a 150,000 gal steel tank. Pumping equipment consists of two 500 S.P.M. triplex pumps, each driven by 75 B.H.P. fuel oil engine. Johnson & Brenham, Firestone Bldg., Kansas City, Mo. Engrs.

Ariz., Holbrook—The city plans an election to vote on $100,000 bonds to construct a water system, consisting of wells, pumping plant and distribution system Olmsted & Gillilen, Hollingsworth Bldg., Engrs

Wash., Puget Sound—The Construction Division, War Dept., Wash., D. C., will receive bids until June 21 for steel hangars, generator houses and miscellaneous construction at the coast defenses here.

Cal., Glendale—The Glendale Sanitarium engaged G. B. Larimer Baker, Detwiler Bldg., Los Angeles, to prepare plans for a central heating plant in connection with the proposed group of hospital buildings on B'way and Jackson St.

Cal., San Francisco—The Construction Division, War Dept., Wash., D. C., will rereceive bids until June 21 for steel hangars, generator houses and miscellaneous construction at the coast defenses here.

Que., Buckingham—The Town Council has appropriated $75,000 for 550 hp. electric power development of Aqueduct Ave. and is in the market for equipment.

Ont., Toronto—The Provincial Power Schemes will cost over $17,000,000, according to the supplementary estimate submitted to the Legislature. The great proportion of this, $10,500,000, is on the big development at Chippawa and Queenston; for the Niagara system, $1,250,000; Severn system, $50,000; Eugenia and Saugeen systems, $475,000; Muskoka system, $40,000; Port Arthur and Thunder Bay, or Superior systems, $2,500,000; Wasdell system, $90,000; St. Lawence system, $500,000; Rideau system, $250,000; Central Ontario system, $750,000; Nipissing system, $175,000; expenditures on account of province, $177,000; miscellaneous, $500,000.

B. C., Vancouver—The British Columbia Paramount Theatres Ltd., 963 Granville St., plans to build a 4 story, 120 x 120 ft. moving picture theatre on Seymour St., including a heating system. T. Lamb, 644 8th Ave., New York City, Archt.

B. C., Vancouver—The city plans to establish a hydro-electric plant at Heton Lake, capable of generating 175,000 hp. Estimatd cost, $12,000,000.

CONTRACTS AWARDED

Mass., Boston—The Suffolk Law School, 45 Mt. Vernon St., will build an 8 story, 87 x 110 ft. law school on Mt. Vernon St., including a steam heating system. About $300,000. Work will be done by day labor.

Mass., Cambridge—The Massachusetts Institute of Technology, Boylston St., Boston, has awarded the contract for the construction of a 4 story, 60 x 180 ft. college on Massachusetts Ave., to the General Building Co., 522 Harrison Ave., Boston. A steam heating system will be installed in same. Total estimated cost, $600,000.

Mass., Fall River—The Laurel Lake Mills, 484 Globe St., has awarded the contract for a 1 story, 40 x 52 ft. transformer station to Beattie & Cornell, 33 North Quarry St., at $25,000.

Conn., Manchester—The Mutual Heating Corp. has awarded the contract for a 1 story heating plant on Main St. to the Flynt Bldg & Constr. Co., Palmer, Mass. at $50,000.

N. Y., Brooklyn—The Pub. Market of Flatbush, 1210 Flatbush Ave., has awarded the contract for the construction of a public market on Flatbush Ave., Duryea Pl. and East 22nd St., to Richard Von Lehn Sons, 2701 Ave. G. A steam heating system will be installed in same. Total estimated cost, $500,000.

N. Y., Brooklyn—Steven Ranson, 401 West St., New York City, has awarded the contract for a 1 story, 150 x 200 ft. machine shop and transformer station at 518 Hamilton St., including a steam heating system, to Wharton Green, 37 West 39th St., New York City, at $250,000.

N. Y., Rocky Point—The Radio Corp., Woolworth Bldg., New York City, has awarded the contract for a wireless station consisting of several buildings including a power house, to the J. G. White Eng. Co., 43 Exch Pl., New York City. About $10,000,000. Noted June 8.

N. Y., Yonkers—The Yonkers Electric Light & Power Co. has awarded the contract for a 2 story, 100 x 225 ft. transformer station on Columbus Pl., to the Joseph Canepi Contg. Corp., Inc., 604 South B'way. About $200,000.

Pa., Philadelphia—L. H. Gilmer Co., Cottman and Keystone Sts., has awarded the contract for a 3 story, 95 x 150 ft. power house and warehouse, including a steam heating system, to W. D. Lukins, 432 West Walnut St., North Wales, at $200,000.

Pa., Philadelphia—The Philadelphia Electric Co., 10th and Chestnut Sts., has awarded the contract for a sub-station at 52-54 North 6th St., to John R. Wiggins Co., Otis Bldg., at $137,500.

Md., Baltimore—The City Council has awarded the contract of a 3 story, 63 x 121 ft. school including a saeam heating system, to the Standard Constr. Co., 1713 Sansom St., Philadelphia, at $400,000.

Tex., Fort Worth—The Bureau of Yards & Docks, Navy Dept., Wash., D. C., has awarded the contract for the installation of deep well pumping equipment, to the Layne & Bowler Co., Rand Bldg., Memphis, at $23,300. Noted April 27.

Cal., Long Beach—The city has awarded the contract for a sewerage pumping plant with pumps, motors, flush tanks, etc., to John Cummings, 1200 Washington Bldg., Los Angeles, at $44,649. Noted June 8.

Cal., Los Angeles—The Oil Mining Equipment Co., 2026 Santa Fe Ave., will build a group of factory buildings, including an 80 x 547 ft. main factory with a 60 x 150 ft. wing, boiler house, etc. Motors, etc., will be installed in same. Work will be done by day labor.

Cal., Santa Rosa—The Natl. Ice & Cold Storage Co., Postal Telegraph Bldg., San Francisco, has awarded the contract for a 1 story cold storage building addition, to A. M. Hildebrandt, at $100,000.

POWER

Volume 51 | New York, June 22, 1920 | Number 25

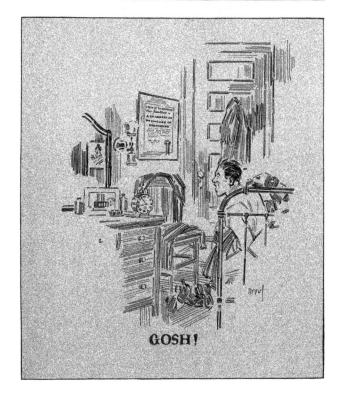

GOSH!

FIG. 1. FINNEY AVENUE PLANT OF THE MERCHANTS ICE AND COAL CO.

FIG. 2. THREE 200-HP. STOKER-FIRED BOILERS, FINNEY AVENUE PLANT

Economy at the Finney Avenue Ice Plant, St. Louis

By C. WILKIE
Engineer, Finney Avenue Plant, Merchants Ice and Coal Co., St. Louis, Missouri

A refrigerating plant fourteen years old with mediocre equipment is operating at 58.6 per cent over its rated capacity and is delivering on the average 8.5 tons of ice per ton of 10,000-B.t.u. coal. Ingenuity of the management, co-operation of the operating force and careful manipulation of the equipment are the reasons advanced for the splendid showing.

THE Finney Avenue Plant of the Merchants Ice and Coal Co., St. Louis, is not a modern steam-driven plant. There are no boilers with 80 per cent efficiency, no uniflow engines using 12 lb. of steam per indicated horsepower-hour and no superheated steam, but the plant ranks with a number of others having the most modern equipment. It has been in operation for fourteen years and was one of the first in the country to make raw-water ice with any degree of success.

Its capacity up to 1908 was 150 tons of ice per day plus the refrigeration for the ice-storage rooms. Since that time the capacity has been increased to 238 tons plus the ice storage. The increase was obtained without adding to the boiler equipment or ice-making machinery, without the assistance of extra coils on the freezing system or additional ammonia-condenser surface. Efficient and careful operation of the equipment available, improvements where possible and labor-saving devices are the reasons back of the higher economy and greater capacity obtained.

Boiler, engine and ice-storage rooms are on the ground floor of the plant. The steam-condenser cooling tower is located on the engine-room roof. It is of the open flat type, with the breaks on 16-in. spacing. The primary ammonia condensers are located on top of the tower previously mentioned, and the freezing tanks are just above the ice-storage rooms. The secondary ammonia condensers, liquid-cooling coils and ammonia-condenser cooling tower are on the roof above the freezing-tank room.

In the boiler room are three 200-hp. combination fire- and water-tube boilers, equipped with chain-grate stokers. This type of boiler has the reputation of being troublesome, owing to leaky flues, but this difficulty has been overcome in a measure by careful firing, maintaining the load as nearly constant as possible and by protecting the rear flue sheet.

When two of the units are in service, the boilers are operated at high capacity, but with three on the line the load on each is below rating. From the daily ice production it is evident that the losses sustained are very small.

To furnish necessary operating data, the boilers are equipped with draft gages, CO_2 recorders and a damper regulator. The device last named is set so as to reduce the draft about one-half when closed and at the same time slow down the stoker engine in proportion to the reduction in draft. Coal, which is hauled to the plant by motor trucks and dumped in the alley, is conveyed into the boiler room by a bucket conveyor.

There are two vertical single-acting ammonia compressors, with cylinders 20 x 30 in. driven by tandem-compound Corliss engines, the latter being supplied with steam at 160 lb. pressure at the throttle and 10 lb. on the receiver. During the summer months the machines are operated at 80 r.p.m., an unusually high speed for their size and type, when it is considered that 65 to 70 r.p.m. is the usual thing. The average condenser pressure during this period is only 170 lb., as compared with 215 lb. in other plants in the same locality. Under the conditions enumerated, the horsepower developed in either case would be practically equivalent, but with the lower condenser pressure and higher speed, more refrigerating capacity would be developed.

The generating units consist of two 870-ampere, 110-volt, direct-current machines, one a reserve unit, directly connected to cross-compound four-valve engines. All the engines exhaust into a surface condenser on which a vacuum of from 23 to 26 in. is maintained. The vacuum pump is of the late feather-valve flywheel type and is motor-driven.

An interesting feature of the plant is the small number of steam-driven pumps in operation. There are only four in the plant, two 5½ x 4¾ x 5-in. pumps for supplying distilled water, with two in reserve as auxiliaries. For supplying air to the raw-water freezing tanks there is a low-pressure rotary blower with a displacement of 900 cu.ft. per min. at 2 lb. pressure, belted to a 15-hp. motor, and a centrifugal compressor having a capacity of 1,600 cu.ft. per min. at 2 lb. pressure, directly connected to a 20-hp. motor. The latter machine is of the high-speed type, making 3,400 r.p.m.; the speed of the blower is 550 r.p.m. One of these units is held in reserve. Air for general purposes and especially for the ice cranes in the freezing-tank room is furnished by a 10½-in. cross-compound steam-driven air compressor, with a 9-in. compressor of the same type held in reserve. The exhaust steam from the two compressors is used for the purpose of heating the bath water, and all of the excess goes to the boiler-feed water heater, thus utilizing practically all of the heat.

All water for freezing and condenser cooling purposes is supplied from the city mains through a 3-in. line. Two 1,000-g.p.m. centrifugal pumps, each direct connected to a 40-hp. motor, circulate the cooling water for the steam condenser, the steam-condenser tower and the primary ammonia condenser. The two pumps are located in the engine room near the steam condenser, and with a supply tank located on a balcony directly above, the water flows through the steam condenser to the pump and is forced to the top of the steam-condenser tower. The water level in the supply tank being about 14 ft. above the pumps, reduces the actual head against which they must operate. In addition the suction and discharge lines to the pumps are extra large to reduce the loss due to pipe friction.

Cooling water for the secondary ammonia condenser is circulated by a 3,000-g.p.m. centrifugal pump, also direct connected to a 40-hp. motor. This pump is located 31 ft. below the top of the ammonia-condenser coils and at one end of the ammonia condenser tower. The water from the tower flows across the roof to a supply tank placed three feet above the pump. Although suction and discharge connections of the pump are 12 and 10 in., the respective pipe lines are 20 and 18 in. in diameter.

Metallic packing is used on all valve stems, as well as on the steam-piston rods and ammonia-compressor rods, so that rod trouble in the plant is reduced to a minimum.

An interesting feature is the arrangement of the three freezing tanks. These are 79 ft. long and have

FIG. 3. TWO VERTICAL SINGLE-ACTING AMMONIA COMPRESSORS DRIVEN BY TANDEM-COMPOUND CORLISS ENGINE

sufficient width for eighteen 400-lb. cans. Each tank contains twenty coils of 1¼-in. pipe, three of the coils on either side being 12 pipes high and the other fourteen coils, 10 pipes high. The total number of cans is 1,818, giving an allowance of 7.6 cans per ton of ice.

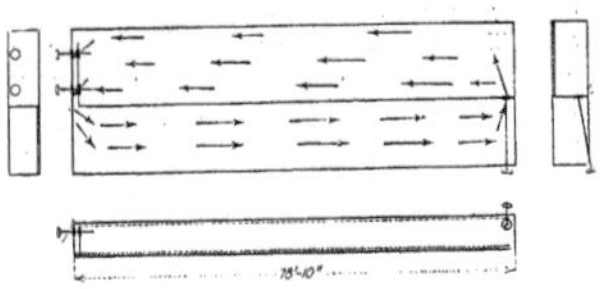

FIG. 4. ICE TANK WITH THREE PROPELLERS FOR RAPID BRINE CIRCULATION

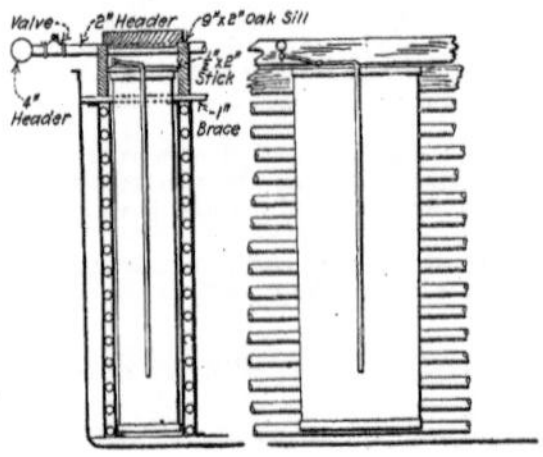

FIG. 5. DROP PIPE SYSTEM FOR AGITATING RAW WATER IN CANS

The average freezing time is 36 hr. 35 min. Brine agitation is provided for by two horizontal belted propellers in the rear end of each tank, although one tank has a third propeller located in the front end of the tank. The brine circulation is rapid, 4½ in. of agitation being maintained during the summer months and amounting to one inch for every 7½ cans.

Particular attention is called to the manner in which the tank floor is arranged so as to keep the brine from overflowing into the ice cans, due to the high agitation. The tank is divided longitudinally, the brine circulating down one side and returning on the other side to the propellers.

Fig. 4 shows the general arrangement, but includes the third propeller and the false flooring on both sides. As the brine leaves the propellers, its level, as previously mentioned, is raised 4½ in. above the stationary level of the brine, and gradually recedes to normal as the flow approaches the opposite end of the tank; consequently, a false floor has been provided on the outgoing side of the tank, its elevation being four inches at the high end and receding to zero at the end of the tank opposite to that in which the propellers are located.

As the cans are 60 in. high and the ice blocks measure 53 in., there is sufficient range so that the brine level need not be pulled down below the ice on the low side nor forced over the top of the cans on the high side.

The three-propeller tank is so arranged that each side of the tank has high brine, but on opposite sides and ends. The two motors that drive the three propellers are connected to the same starting box, so that if a fuse should blow or there should be any mishap to the box, all three propellers would stop, for the simple reason that the agitation in this tank is much more rapid than in the other two tanks. The one propeller in the front end of the tank is designed to handle as much brine as the two in the rear end. Since the third agitator was put into operation, it is easy to get three tons of ice more per day than on either of the other two tanks with the same brine temperature and the same number of cans.

For agitating the raw water in the cans, a drop-pipe system is employed. A 4-in. header on either side of the tanks, with 2-in. headers connected at right angles,

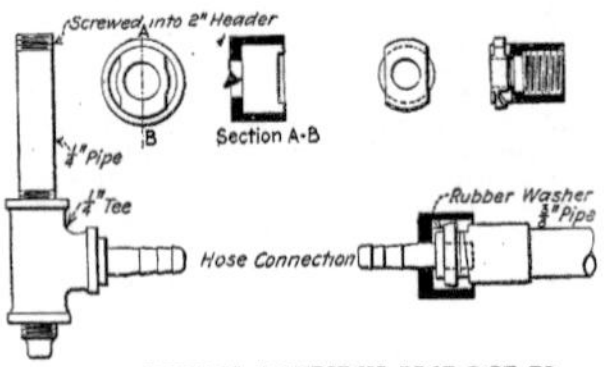

FIG. 6. FITTINGS CONNECTING DROP PIPE TO AIR HEADER

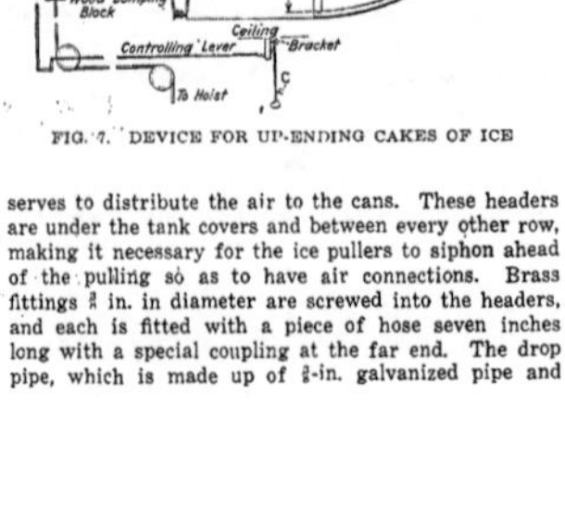

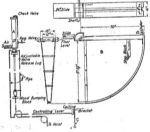

FIG. 7. DEVICE FOR UP-ENDING CAKES OF ICE

serves to distribute the air to the cans. These headers are under the tank covers and between every other row, making it necessary for the ice pullers to siphon ahead of the pulling so as to have air connections. Brass fittings ¾ in. in diameter are screwed into the headers, and each is fitted with a piece of hose seven inches long with a special coupling at the far end. The drop pipe, which is made up of ⅜-in. galvanized pipe and extends to within nine inches of the bottom of the can, is connected to the hose by a quarter turn on the coupling.

Owing to the high brine agitation the cans tend to rise. To keep them down in the brine, slots $\frac{5}{8}$ in. wide and $\frac{3}{8}$ in. deep have been made in the deck boards or sills and a 1 x 2-in. pine stick, beveled at either end, is placed over the can. The stick serves also as a support for the drop pipe. There is about 1 to $1\frac{1}{2}$ per cent of white ice due to air trouble. During the winter months exhaust steam passing through a coil in the air receiver serves to heat the air from 80 to 130 deg. F.

The ice cranes are of the pneumatic bridge type arranged to handle three cans at one time. Three automatic lowering machines with the necessary chutes deliver the ice from the dumps to any location in the storage rooms, where only one man is required for taking care of the ice when it is being stored.

A machine has been perfected for standing the blocks of ice on end. It is operated by air and takes the place of one man on each shift. The device consists of an air cylinder A, Fig. 7, which by means of a cable connection to the quadrant B, up ends the block of ice and causes it to pass down a low incline to the floor level. The piston is actuated by means of rod C, which opens the valve and allows the air pressure to push the piston toward the top of the cylinder, pulling the quadrant and the cake of ice on its upper platform through an angle of 90 degrees.

An adjustable release lug on the $\frac{3}{4}$-in. pipe connecting with the valve lever is so located that when the bar of iron on the crosshead comes in contact with it, the valve lever is raised, reversing the direction of the piston and allowing it and the quadrant to return automatically to the normal position ready for another cake of ice. When the ice puller dumps the ice on the second floor, the automatic lowering machines, with the aid of chutes, deliver each cake of the 238 tons pulled each day within four to six feet of this machine, except when the permanent ice storage is being filled. At the plant the ice-storage capacity is only 3,500 tons, but other storage houses are provided in various residential sections of the city, which are filled from the plant during the winter months.

When this machine was first installed, the upper end of the cylinder was open to the atmosphere and the

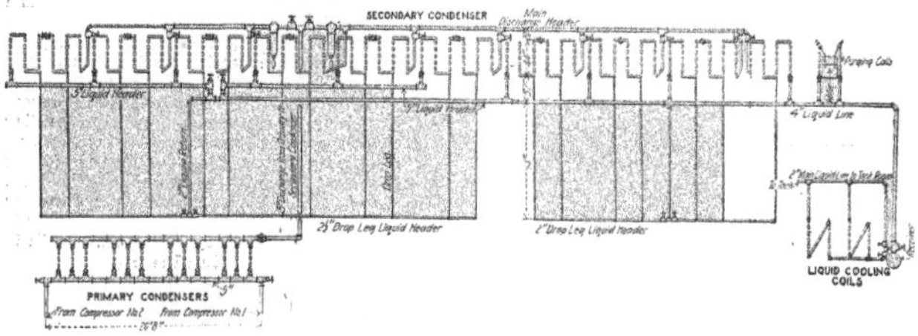

FIG. 8. CONNECTIONS OF PRIMARY AND SECONDARY AMMONIA CONDENSERS

FIG. 9. STEAM CONDENSER COOLING TOWER AND PRIMARY AMMONIA CONDENSER

FIG. 10. TOP VIEW OF THE SECONDARY AMMONIA CONDENSER

piston would come up so rapidly and with such force that it would throw the cakes of ice several feet away. This trouble was overcome by connecting the discharge into the top of the cylinder, with a regulating valve on the line and a check valve near the top of the cylinder.

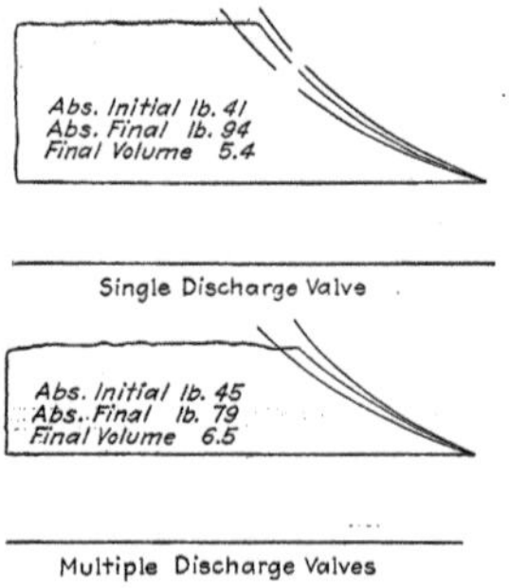

FIG. 11. AMMONIA-COMPRESSOR DIAGRAMS

Now when the piston makes its downward stroke, the air is discharged into the top end of the cylinder through the check valve and by the piston on its upward stroke is discharged through a $\frac{1}{16}$-in. hole in the disc of the check valve. This, of course, serves to cushion the piston and at the same time slows down the machine as the cake of ice is nearing its vertical position.

The ammonia condensers are of particular interest. They are divided into two sets, as shown in Fig. 8. The primary condenser, consisting of twelve stands, eight pipes high, of 2-in. pipe 20 ft. long, is located on top of the steam condenser tower. During the summer months the cooling water is about 120 deg. F. The secondary condenser is located on top of a vertical-blade, open-type cooling tower, the blades being 20 ft. high and one under each coil. The average difference in cooling-water temperature on and off the coils is three degrees. Just over the liquid cooling coils is a flat open-type cooling tower. After cooling water has passed through the vertical blade tower, the quantity necessary for the liquid coils passes through this tower and receives a second cooling before going over the coils. In this way a ten-degree drop in the liquid between the receiver and the expansion valves is obtained.

A feature of the condenser installation is the extremely low pressure maintained during the winter months. Fig. 11 shows typical ammonia-compressor diagrams. The upper diagram shows about the average conditions under which the compressors operate during the winter months, and at times when the weather is favorable, the conditions indicated in the lower diagram obtain. To maintain this low condenser pressure requires considerable attention. The water leaving the cooling tower must be a few degrees above, yet as near, the freezing point as possible, so that ice will not accumulate in the tower. The water temperature is regulated by turning water on or off the coils, depending upon outside temperature of the water leaving the tower.

With one compressor in operation 170 tons of ice per day can be made during the winter months. This gives an allowance of 10.7 cans per ton and a 52-hour freezing rate. The compressor displacement per ton per minute during the summer will average 11,850 cu.in., while in the winter it is as low as 8,500. The ice production per ton of coal will average 8½ tons, the coal used being No. 3 washed nut with a heat value of 10,000 to 10,500 B.t.u. A summary of ice-tank performance is given in the accompanying table:

ICE-TANK PERFORMANCE

Number of cans, 400 lb	1,818
Tons of ice per day	238
Average can allowance per ton	7.6
Lineal feet, 1¼-in. pipe	57,000
Lineal feet, 1¼-in. pipe per ton of ice	239
Average suction pressure, lb	20.5
Average brine temperature, deg. F	12
Temperature corresponding to suction pressure, deg. F	7
Temperature difference, deg. F	5
Surface external pipe, sq.ft	24,800
Tons refrigeration per 24 hours	266.9
B.t.u. per 24 hours ice cooling and freezing	76,892,000
B.t.u. per square foot per hour	129
K-heat transmitted per deg. temperature difference per square foot per hour	25.8

All the data used in this article were gathered by the writer during the season of 1919 and are averages for the 135 days from June 2 to Oct. 15.

Seven of the principal reasons why the plant production is 58.6 per cent over its rated capacity are as follows:

Remarkable ingenuity shown by E. S. Ormsby, vice-president and general manager, and A. Gass, chief engineer.

Co-operation of the employees throughout the plant.

The low condenser pressure maintained.

The excellent ice-tank performance.

The increased speed of the ammonia compressors.

Constant operation of the plant during the summer months.

The small amount of reëxpansion of gas in ammonia-compressor cylinders.

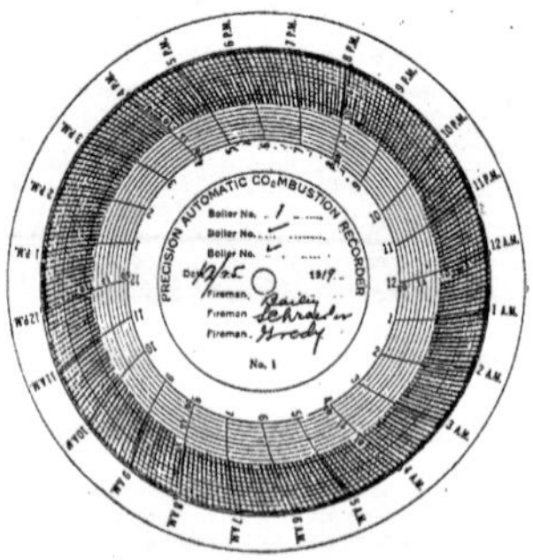

FIG. 12. CHART FROM CO_2 RECORDER INDICATING CAREFUL FIRING ON A CONSTANT LOAD

Steam-Jet Air Pumps for Condensers

The Problem of Air Removal From Condensers and the Various Methods of Solving It—A Description of Various Makes of Steam-Jet Pumps Designed for Condenser Work—Some Points Necessary To Observe in Their Installation and Operation, With a Few of Their Advantages

By DE WITT M. TAYLOR

IF NO air were present in a condenser, the vacuum would be limited only by the lowest available temperature in the condenser, and this, in turn, would depend on the temperature of the cooling water. This vacuum would correspond to the pressure of the water vapor alone, but with air present the actual pressure will equal this vapor pressure plus that which the air would exert if it were present alone.

Now, since air cannot be condensed at ordinary temperatures, the only thing to do is to pump it out. This becomes difficult at high vacuums owing to the large increase in volume. Assuming that the amount of air to be removed per hour remains constant, if the vacuum is increased from 26 in. to 28 in., the volume is more than doubled. If the vacuum increases from 28 in. to 29 in., the volume is more than doubled again.

In reciprocating-engine plants, where 26 in. vacuum is usually the maximum, a simple pump which handles both the condensate and the air gives satisfactory results. However, with the demand for higher vacuums for turbines some other means became necessary.

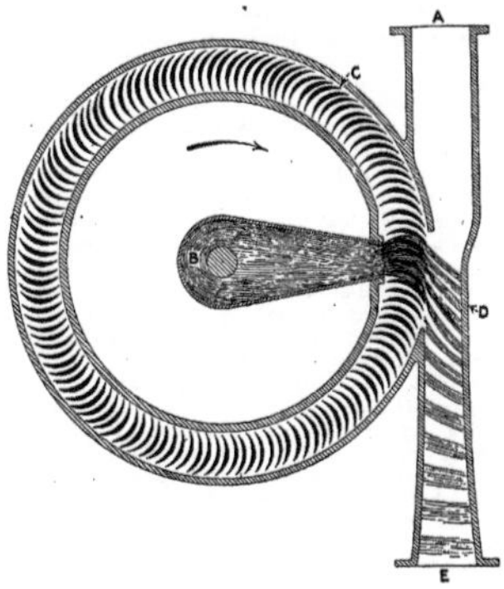

FIG. 1. DIAGRAM OF HURLING-WATER PUMP

The commonest method was the use of a "wet" pump to take out the condensate and a so-called "dry-air" pump to take care of the air. The wet pump, or removal pump, may be a displacement pump or a centrifugal pump.

The first type of dry-air pump in general use was simply a reciprocating air compressor. The troubles of re-expansion of the air in the clearance, which increased even faster than the vacuum increased, led to many refinements and complicated valve gears. This, of course, meant high first cost and high maintenance cost, as well as losses due to interruption of service for repairs to the pump. It also meant the additional lubrication and cooling problems which are present in all reciprocating air compressors. A number of pumps of the rotary type were used with reasonable success, but these also were subject to the same or other disadvantages.

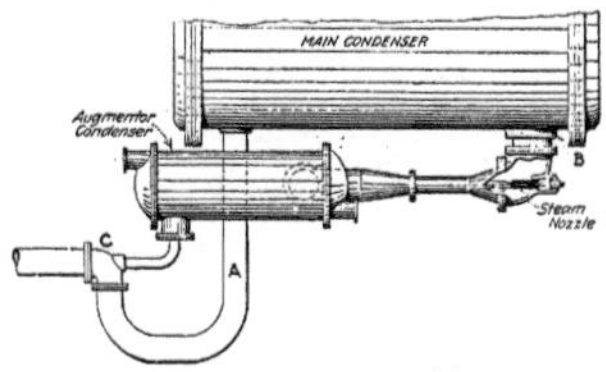

FIG. 2. DIAGRAM OF PARSONS' VACUUM AUGMENTOR

These difficulties led to the development of the hydrocentrifugal or "hurling-water," pumps. A number of these pumps, alike in principle but differing greatly in detail, were produced. Fig. 1 shows, diagrammatically, one example of this type. Air is admitted from the condenser at *A*, and the hurling water enters at the center of the pump to the chamber *B*. At *C* are shown the blades of a centrifugal pump rotor. As each blade passes the nozzle shown in the chamber *B*, it cuts off a "slug" of water and throws it out into the nozzle *D*. The air is trapped between these slugs and carried out through the discharge tube *E*. Other hurling-water pumps discharged the water all around the rotor into an annular space to which the air was admitted. Outside this annular space were a number of passages similar to the guide passages in a turbine pump. The water trapped the air and carried it out through these passages to the pump discharge.

The hurling-water pumps showed a marked improvement over the reciprocating compressor, both in simplicity and in capacity. The reciprocating pump shows a decreasing capacity as the vacuum increases, and finally reaches a point where the capacity becomes zero. Owing to the increased water velocity at high vacuums, the hurling-water pump increases its capacity as the vacuum increases. On the other hand, the hurling-water pump uses two or three times as much power as the reciprocating pump and is slow in starting, by reason of its low capacity at low vacuums.

Charles A. Parsons was the first to use the steam ejector for condenser work, in connection with his "Vacuum Augmentor," which he installed in turbine-driven channel boats. In this case the steam jet simply increased the air pressure so that the combined water and air pump could handle it. A diagram of the arrangement is shown in Fig. 2. The condensate was drained from the bottom of the main condenser through pipe *A*, in which was a U-bend to form a water seal. Air was drawn out through the pipe *B*, by a steam ejector which discharged into a small "*Augmentor Condenser*," about one-twentieth the size of the main condenser. This augmentor condenser condensed the steam from the ejector and delivered it, with the air, into the main condensate pipe at *C*, from which point the whole mass passed to the main air pump. In this apparatus the steam ejector increased the pressure of the air by only one or two inches, so that with a vacuum of 29 in. in the main condenser, the vacuum at point *C* would be about 27 or 28 in. The water seal in the pipe *A* prevented the air from working back into the main condenser.

Meanwhile, Maurice Leblanc, in France, had been experimenting with steam ejectors to produce high vacuums for refrigerating work. While the problems of refrigeration are different from those of condenser work, one important fact was established. It is difficult to produce and maintain high vacuums by one ejector alone. In most cases, therefore, two ejectors are used in series.

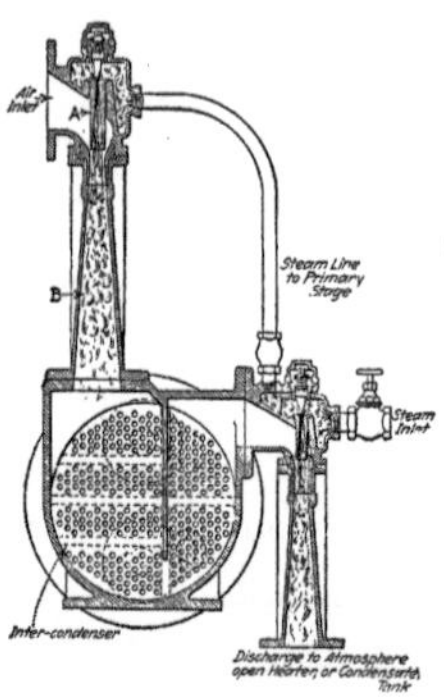

FIG. 3. WHEELER STEAM-JET AIR PUMP

Several years ago, the question of steam ejectors as condenser air pumps was seriously taken up by American manufacturers. Today there are several makes of such pumps on the market, and they seem likely to prove a decided factor in the business.

Fig. 3 shows the general design of the Wheeler Steam Jet Air Pump, as made by the Wheeler Condenser and Engineering Co., Carteret, N. J. Steam at 125 lb. pressure is admitted at the *steam inlet*, and part of it passes through the bent pipe to the primary, or first-stage, nozzle *A*. This nozzle is practically the same as a single-stage steam-turbine nozzle, so the steam leaves it with a very high velocity. As it leaves the nozzle, the steam picks up, or entrains, the air and carries it on through the discharge tube *B*, into the *intercondenser*. Here the steam from the primary nozzle is condensed and this condensate is drained, through a U-tube seal (not shown) into the bottom of the main condenser, from which it is pumped by the main condensate pump. The air in the intercondenser is drawn off by the secondary, or second-stage, nozzle, which is similar to the one just described.

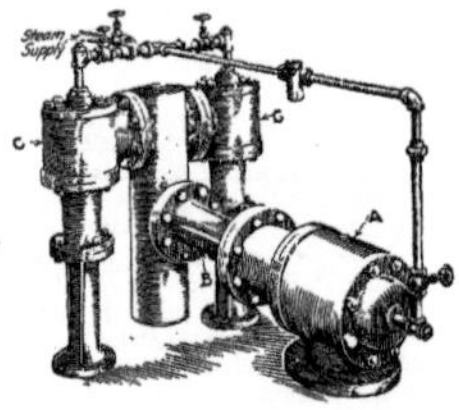

FIG. 4. CROLL-REYNOLDS EVACTOR

In this machine the condensate from the main condenser is used as cooling water in the intercondenser. This reclaims the heat in the exhaust steam from the first nozzle, and at the same time reduces the work in the second-stage nozzle, since it has only the air to handle. The tubes at the right of the partition, in the intercondenser, may be supplied with raw water when starting, or when the load is so light that there is not enough condensate from the main condenser.

At Fig. 4 is shown the general appearance of the Croll-Reynolds "Evactor," made by the Croll-Reynolds Co., New York City, and by the Ross Heater and Manufacturing Co., Buffalo, N. Y. The first-stage ejector is shown at *A* and the intercondenser or intercooler at *B*. This intercooler is of the jet condenser type, and the boiler-feed makeup water is generally used as cooling water. For the second stage, two ejectors, marked *C*, are used in parallel, thus giving greater flexibility as one or both may be used as needed.

The Alberger "Air Occluder," made by the Alberger Pump and Condenser Co., New York City, is shown at Fig. 5. Air enters at *A* to the first-stage ejector *B*. The second-stage pump is at *C* and the jet-condenser type intercooler at *D*. Both the Croll-Reynolds and Alberger machines may be equipped with surface type intercoolers if necessary.

Fig. 6 shows a section of the "Radojet" air pump made by the C. H. Wheeler Manufacturing Co., Philadelphia, Pa. This differs from the previous pumps in the design of the secondary nozzle and in that it is usually built without an intercooler. Air and steam are admitted as shown and part of the steam passes up through the pipe at the left, to the chamber above the first-stage nozzle A. The steam and entrained air go down through the discharge tube B, to the second-stage suction chamber C. Steam for the second stage passes through the annular nozzle formed by the hollow cone D and the

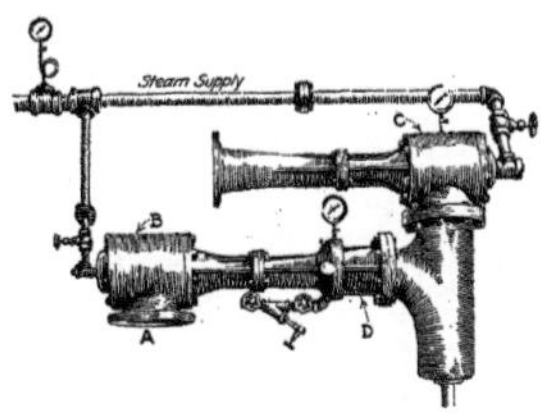

FIG. 5. ALBERGER AIR OCCLUDER

conical plate E. This gives a jet in the form of a thin circular sheet, offering a very large surface for entrainment. This secondary nozzle handles the air and also the steam from the primary nozzle and delivers them through the discharge diffuser F to the delivery space G and then to the discharge. The space G is similar to the discharge passage in a centrifugal pump.

The Elliott Co., of Pittsburgh, Pa., manufacturer of the Elliott-Ehrhart Ejector, has developed a single-stage outfit which it claims is more effective than the two-stage machine. Referring to Fig. 7, the curve A shows the variations in capacity of a typical two-stage ejector, while the line B is characteristic of a single-stage outfit. It will be noted that the two-stage machine will give a very high vacuum at zero capacity—that is, with the air suction blanked off—but that the vacuum drops off rapidly as the capacity increases. On the other hand, the single-stage ejector will not give so high a vacuum at zero capacity, but the vacuum does not drop off so rapidly, as more air is admitted. It is reasonable to assume that if a single-stage outfit could be so constructed as to give substantially the same vacuum at zero capacity as the two-stage outfit, the characteristic curve would have the same shape as B but displaced upward as at C. The Elliott Co.'s engineers claim to have succeeded in designing such an ejector, which gives about one-tenth inch less vacuum than the two-stage machine at zero capacity, but which maintains a much better vacuum as the amount of air handled increases. The curve C shows the characteristic of this Elliott-Ehrhart single-stage outfit. When very low steam consumption is desired, the Elliott Co. arranges an ejector in two stages, with an intercondenser.

One very important point in the design of two-stage steam-jet air pumps is the use of the proper ratio of compressions, or in other words the proper intermediate pressure. It must be remembered that these pumps are simply two-stage air compressors. Theoretically, the best intermediate pressure is given by the equation $P = \sqrt{P_1 \times P_2}$, where P is the intermediate pressure, P_1 is the initial pressure or, in this case, the condenser pressure, and P_2 is the final pressure. This equation applies to dry air in piston compressors with perfect intercooling, so it will not give accurate results for steam-jet pumps. However, it gives an approximate idea of what the intermediate pressure should be. As an example, assume a vacuum of 29 in. with a 30-in. barometer. Then $P_1 = 1$, $P_2 = 30$, and $P = \sqrt{1 \times 30}$ or 5.48 in., which corresponds to a vacuum of 24.52 in. Of course the exact best value for the intermediate pressure can be found only by actual tests on various pumps and under various conditions.

The conditions for each individual plant should settle the question of the use or non-use of an intercooler. Its use is mainly to cut down the work of the second-stage jet and thereby lower its steam consumption, and even

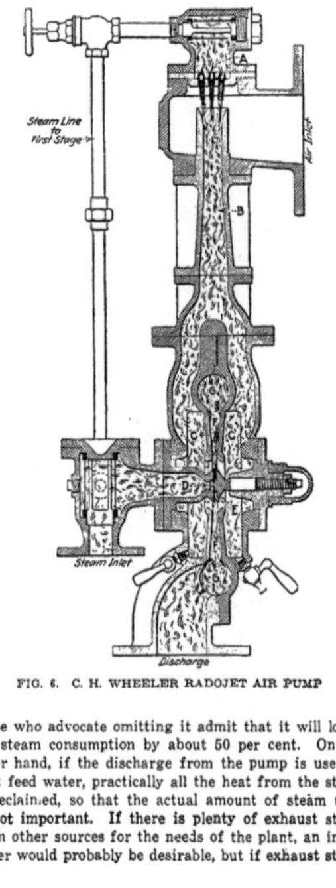

FIG. 6. C. H. WHEELER RADOJET AIR PUMP

those who advocate omitting it admit that it will lower the steam consumption by about 50 per cent. On the other hand, if the discharge from the pump is used to heat feed water, practically all the heat from the steam is reclaimed, so that the actual amount of steam used is not important. If there is plenty of exhaust steam from other sources for the needs of the plant, an intercooler would probably be desirable, but if exhaust steam is scarce, it would seem to be an unnecessary complication. In any case the heat in the steam from the second-stage nozzles should be reclaimed. The curves of Fig. 8 show a comparison of the steam consumptions of a hydraulic or hurling-water pump (curve *A*) of a non-condensing ejector (curve *B*) and of a condensing ejector (curve *C*). The curve *A* for the hydraulic pump is based on a steam drive taking 35 lb. of steam

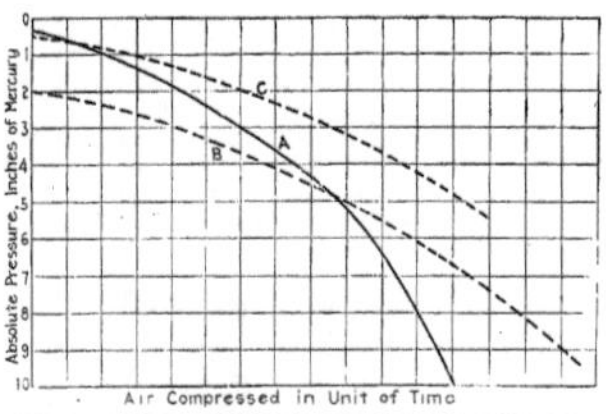

FIG. 7. VARIATIONS IN CAPACITY WITH VACUUM OF SINGLE- AND TWO-STAGE STEAM-JET AIR PUMPS

per horsepower per hour, which seems to be a fair average for this type of outfit.

Fig. 9 shows one way of reclaiming the waste heat. The discharge from the pump is led into the *condensing tank*, through which either the makeup feed water or the main condensate passes. Here all the steam from the jets is condensed and its heat taken up by the water. At the same time the air escapes from the surface of the water and out through the vent. If the supply of water to the condensing tank is stopped, owing to shutting down the main engine, the water in the tank would soon become too hot to condense the steam. To prevent this, a pipe is sometimes run from the tank to the main condenser. In this pipe is a valve which is controlled by a thermostat placed in the condensing tank. When the water reaches the temperature for which the thermostat is set, the valve is opened and water is drawn from the tank into the main condenser, where it is cooled, and it is then returned to the tank by the main condensate pump. This device has been found especially useful in marine work in connection with maneuvering and during stand-by periods.

Apparatus using steam nozzles requires a constant steam pressure for efficient operation. The variations in boiler pressure found in the average plant are too great for the proper operation of steam-jet air pumps. Therefore a reducing valve or regulator to maintain a constant pressure must be placed in the steam supply line and the boiler pressure must be at least twenty-five pounds higher than that for which the pump is designed. It is very important, also, that the discharge pressure should be kept constant and very little if any above atmospheric pressure. The dam shown in the condensing tank in Fig. 9 keeps a constant head of water on the discharge pipe and should be adjusted so that this head is no more than enough to keep the steam from blowing out. The check valve shown in the discharge pipe is necessary to prevent water being drawn into the condenser if the air pump is shut down before breaking the vacuum.

The amount of air brought into the condenser with the feed water, as well as that in the injection water when jet condensers are used, can be estimated in advance with reasonable accuracy. The amount which leaks in, however, is very uncertain and varies from time to time as the conditions of packing, etc., change. Therefore, enough air-pump capacity must be provided to take care of the worst conditions. The amount of steam used by a jet pump depends on the pressure and on the size of the nozzles, and is independent of the amount of air handled. On this account it is usual to install two or more pumps instead of a single large one. This allows the operating engineer to meet changes in load and still run pumps at maximum economy. In starting, all the pumps may be used to exhaust the air from the condenser quickly and then only enough run to handle the load. If an extra pump for use as a spare is desired, the cost is not prohibitive, although the chances of trouble are so small that a spare is seldom thought necessary.

The work of operating steam-jet pumps consists simply of starting and stopping. Before starting, the usual precaution of draining all cold apparatus before admitting steam is, of course, necessary. The steam valve to the first-stage nozzle should be closed and then the main steam valve opened a little. After the machine is warmed up, close the drips and open the main steam valve wide, making sure that the pressure regulator maintains the proper pressure in the line. When the vacuum gage shows about 25 in., open the valve to the first-stage nozzle. In stopping, first close the valve in the air line between the condenser and the pump and

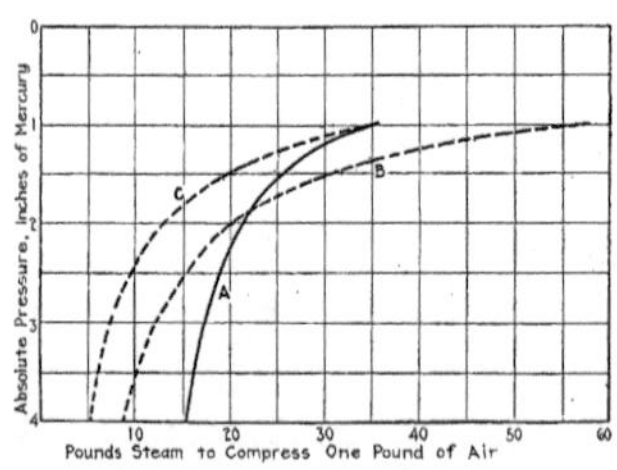

FIG. 8. COMPARISON OF STEAM CONSUMPTIONS OF HYDRAULIC PUMPS AND STEAM-JET EJECTORS

then close the main steam valve and open the drips. When several steam-jet pumps are installed, run only enough to carry the load, but see that these are run with the valves wide open and that steam is supplied at the designated pressure so that the nozzles may work under their designed conditions.

Some interesting data regarding air ejectors were given in a paper by G. L. E. Kothny, read before the Society of Naval Architects and Marine Engineers in New York, Nov. 14, 1919. A curve was shown, giving the capacities of a certain machine at different vacuums and also the capacities using only the second-stage nozzle. The maximum vacuum obtained with the complete machine was about 29.75 in. of mercury and with the second-stage nozzle alone about 27.5 in. Both these figures were obtained with the air inlet blanked off so as to give zero capacity, and are referred to a 30-in. barometer.

As the vacuum was decreased, the capacities increased, but with vacuums below 25.5 in. the capacity of the second-stage nozzle alone was greater than that of the

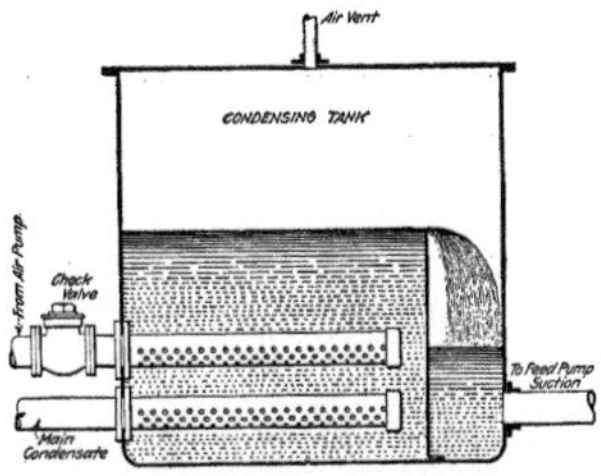

FIG. 9. ARRANGEMENT FOR RECLAIMING HEAT IN DISCHARGE FROM STEAM-JET AIR PUMPS

entire machine. This explains the desirability of using only the second-stage nozzle in starting. These figures check closely with those shown on the curves at Fig. 7, which were obtained from an entirely different source and from tests on a different machine.

Other curves showed the effect of a back pressure (on the discharge) on the vacuum. These indicated that for back pressures below a given amount the vacuum was practically constant, but above this "critical" back pressure the vacuum dropped very rapidly. The value of the back-pressure limit depends on the steam pressure used and was about 2.5 in. of mercury for 100 lb. gage, 5.0 in. for 110 lb. gage and 8.5 in. for 120 lb. gage. It must be remembered that all the values given in Mr. Kothny's curves are from tests on a certain individual pump and would not be true for every pump. However, they are of value as giving a rough guide as to how such pumps would operate.

Another paper read before the American Society of Mechanical Engineers at the annual meeting in New York, Dec. 2-5, 1919, by Frank R. Wheeler, described the various methods of removing air and covered some of the ground that had already been covered in preparation of the present article.

Steam-jet air pumps are small in size and require no foundations; therefore they are low in cost and easy to install. They have good capacity, even at high vacuums. By using several small pumps, which is too expensive with other types, great flexibility can be obtained. While their steam consumption is high, most of the heat can be reclaimed, so that they are economical in fuel consumption and they require little or no attention during operation.

Supersaturated Steam

Supersaturated steam has been for a number of years only a thing of interest to scientists, but with the modern development of the steam turbine it is attracting more or less attention among engineers. It has been claimed that its presence offers an explanation of some of the otherwise unexplainable performances of steam in turbines and in turbine nozzles.

Supersaturated or supercooled steam may be defined as steam which exists as a vapor at temperatures below that at which it should condense under the existing pressure. It is possible, under laboratory conditions, to cool pure water below 32 deg. F. without the formation of ice. This condition is very unstable, however, and the slightest disturbance will result in the immediate freezing of some of the liquid. The formation of ice of course results in the liberation of heat, which will raise the temperature of the mixture. The water will not freeze above 32 F., so the process will stop when enough heat is liberated to raise the temperature to that point and equilibrium is said to be established.

In a similar way superheated or dry saturated steam may, under certain conditions, be cooled below the condensing temperature and still remain a vapor. As in the case of undercooled water, this supersaturated or "undercooled" steam, is in an unstable condition and takes the first opportunity to partly condense and reestablish its equilibrium. As soon as condensation begins, heat is liberated, which tends to raise the temperature of the mixture, and this will go on until the state of equilibrium is reached. In other words, condensation will continue until enough steam has condensed to supply the heat necessary to raise the temperature of the whole mass to the saturation temperature corresponding to the pressure.

Behavior of Steam in an Orifice or a Nozzle

When steam expands in an orifice or a nozzle, as it does in steam turbines, some of its heat is given up to increase the velocity of the steam itself. If the steam was originally dry and saturated, the loss of this heat will result in condensation. From the known properties of steam it is possible to calculate the quality, specific volume and velocity of the steam leaving the nozzle and hence to compute the weight flowing in a given time. It is obvious that friction will decrease this amount so that the theoretical value will be too high. Experiments by a number of investigators, however, have shown, under certain conditions, a *greater* discharge than is indicated by the theoretical formula. To explain this condition, it has been suggested that the steam in a turbine is not in a state of equilibrium and that it is supersaturated throughout its passage through the machine. In support of this theory several ideas have been brought forward by various engineers.

It is pointed out that a thermometer in the exhaust pipe of a turbine will indicate a lower temperature than that corresponding to the observed pressure, although the impact of the particles of steam against the thermometer bulb would tend to raise the temperature reading. This, however, might easily be explained by the difficulty of accurately measuring the pressure of the steam at high velocities and by the fact that any air in the steam would reduce the vapor pressure below the observed pressure even if the latter could be accurately measured.

Another argument brought forward is that a certain time is required for condensation. This is undoubtedly true, but the experiments of Dr. Stodola on steam discharging into the open air indicated that droplets were formed in probably about one ten-thousandth of a second. The conclusion has been drawn, however, from the experiments of C. T. R. Wilson, that a certain amount of undercooling exists at least over a period of time of the general magnitude of $\frac{1}{100}$ sec. If this latter time is correct, the possibility of supersaturation existing until the steam has passed entirely through a modern turbine is established in so far as the time element is concerned.

The experiments of Mr. Wilson, which are described by him in the "Transactions of the Philosophical Society of London," Vol. 189, were undertaken to investigate condensation laws under sudden expansion of saturated dust-free air. As a result of his observations he concludes that sudden expansion to a certain point below the saturation point is accompanied by supersaturation, but that below this point the amount of condensation increases very rapidly. Mr. Wilson's results have received extensive recognition, especially in England. It has been pointed out, however, that, although starting with saturated air, his process of removing all the dust particles would result in condensing some of the moisture, so that his actual experiments were probably conducted with non-saturated or superheated vapor.

Considerable discussion has been brought out by an article entitled "A New Theory of the Steam Turbine," by H. M. Martin, in *Engineering*, London, Vol. 106. Mr. Martin claims not only that the steam in a turbine is not in thermal equilibrium above the so-called Wilson point, but that thermal equilibrium is never attained during the entire passage through the machine. In other words, the steam, after being cooled to the saturation point, is always supersaturated or undercooled even though some condensation may have taken place.

Wet Steam Really a Mixture

One of the difficulties of understanding the situation lies in the conception most engineers have of what is really meant by "wet steam." If a pound of water is at the boiling point, any addition of heat will result in the formation of some steam. As more heat is added, more of the water is converted into steam and we call the whole mixture wet steam. If three-quarters of the pound is vaporized, the quality of the mixture is said to be 0.75, and most men think of it as a uniform mass having certain properties. As a matter of fact, there is 0.75 lb. of *dry* steam and 0.25 lb. of water present, and this is true even although the water is in the form of fine drops thoroughly mixed with the steam. Now, water and steam have different properties, and while the properties of a mixture of the two can very conveniently be computed by the ordinary conception of a homogeneous substance, this is true only when conditions allow the whole mass to be in stable equilibrium.

A study of the well-known temperature-entropy diagram shows that the expansion of dry saturated steam in a nozzle results in some condensation. On the other hand, if water at the boiling point for the existing pressure is allowed to expand in a nozzle, some of it is vaporized. Starting, then, with dry saturated steam, the moment any condensation takes place we have a mixture of water and steam. As the expansion proceeds, some of the water is re-evaporated and some more of the vapor is condensed. Remembering that a certain period of time is necessary for both condensation and evaporation to take place, it seems probable that both supersaturated steam and superheated water may exist together. This condition is furthered by the rapid movement of the mixture, which prevents complete interchange of heat between the various particles. This disturbance of the mass also makes the accurate measurement of pressures, temperatures, etc., difficult if not impossible.

It must be remembered that the conditions described are very unstable and that they will exist only until a condition of equilibrium can be re-established. While this unstable condition exists, however, the properties and proportions of the liquid and vapor are dependent on the local conditions, which in themselves are constantly and rapidly changing. From this viewpoint any attempt to make logical formulas for the behavior of steam in a turbine promises to lead to excessive complication. It would seem reasonable, then, that the present method of deducing formulas for the behavior of the steam on the assumption that it remains in thermal equilibrium and then modifying or correcting these formulas by factors determined by actual test still offers the most practical method of solving the problems of steam-turbine design.

In other words, the steam in a turbine is not in thermal equilibrium. It is probable that supersaturated steam is present a part if not all of the time. The number of uncertain and variable factors, however, make the deduction of logical formulas impractical.

The Miller Steam Trap

This trap, known as the Miller, was designed primarily for use with laundry machinery so that a constant temperature in the ironing machine, by thoroughly draining it of condensation, could be maintained.

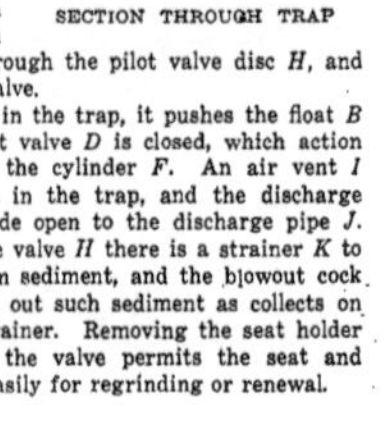

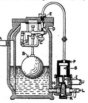

SECTION THROUGH TRAP

In the illustration the details of design of the trap, which is manufactured by the C. J. Miller Co., 1225 Columbia Avenue, Philadelphia, Pa., are shown. Condensation enters through the inlet *A*. The float *B* holds the lever *C* down, thus pulling the pilot valve *D* from the seat *E*. Steam and air then flow through the pilot valve disc *H*, and close the discharge valve.

As the water rises in the trap, it pushes the float *B* upward until the pilot valve *D* is closed, which action cuts off steam from the cylinder *F*. An air vent *I* releases the pressure in the trap, and the discharge valve *H* is forced wide open to the discharge pipe *J*.

Below the discharge valve *H* there is a strainer *K* to protect the valve from sediment, and the blowout cock *L* permits of blowing out such sediment as collects on the outside of the strainer. Removing the seat holder from the bottom of the valve permits the seat and valve to be removed easily for regrinding or renewal.

Burning Illinois Coal from Vein No. 5

By T. A. Marsh

Coal from the fifth vein of Illinois is produced in the Harrisburg and Springfield districts. That produced in the Harrisburg district or Saline County, is by far the better coal. A comparison of screenings from the two districts by analysis is as follows:

	Moisture	Volatile	Fixed Carbon	Ash	Sulphur	B.t.u. Com.	B.t.u., Dry Basis	Fusion Temp. of Ash
Springfield No. 5	12 2	33.2	35.00	19.6	4 0	10,353	11,900	2,179
Harrisburg No. 5	6 75	33.49	46.72	13.04	3 0	11,493	12,325	2,291

Of all the coal produced in Illinois that from the No. 5 seam in Saline and Gallatin Counties is among the highest in heat values on a "unit coal" basis that is moisture and ash free.

The Harrisburg, or Saline County No. 5, has a high heat value and relatively high fusion point for ash,

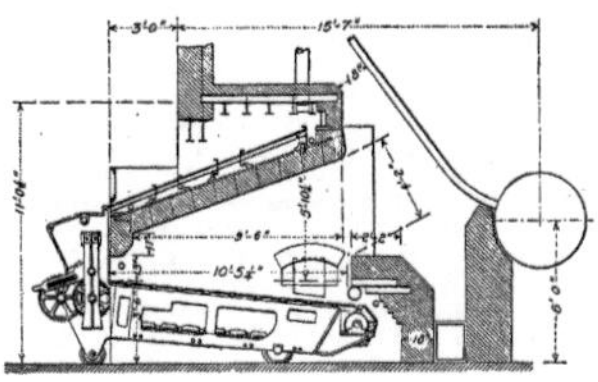

FURNACE IN WHICH THE TESTS WERE MADE

making it a desirable fuel in plants having limited boiler capacity. Some recent tests with this fuel may be of interest. The boiler on which the tests were conducted is of the Stirling type, having 15 tubes per section and 20 tubes in width, with a total steam-making surface of 5,020 sq.ft. The furnace, which is 10 ft. wide, is equipped with a chain grate 9 ft. 6 in. wide by 11 ft. long, giving an active area of 101.3 sq.ft. and a ratio to steam-making surface of 1 to 49.5.

The accompanying drawing shows the design of the furnace. It will be noticed that the boiler is set 6 ft. from the center line of the mud drum to the floor line and that there is a furnace projection of 3 ft. forward of the boiler front. These proportions and the general arrangement were determined in the design with the idea of smokelessness as the leading feature. The entire grate surface has a furnace roof over it.

This is a very desirable feature where smokelessness is to be assured.

From the test results given in the table it will be seen that while the furnace was designed primarily for smokelessness, high efficiencies and excellent capacities were obtained. Some of these tests were made with coal that had been on fire in storage, rendering it more difficult to produce results than with coal that is freshly mined.

Daily performance of the plant is a repetition of these results. Capacities of 200 per cent and above are frequently reached and maintained, while the maintenance and labor of operation are unusually low. Relative to the item of main interest when the plant was designed—smokelessness—the chimney never emits more than the lightest blue haze, even when all boilers are being carried at high ratings.

Old Engine Dismantled, Installed 69 Years Ago

An interesting old engine owned by the Taunton-New Bedford Copper Co., of Taunton, Mass., has at last been dismantled. George A. Corliss offered to let the company have this engine if he could have the amount saved in coal for one year, but the management at that time was advised not to accept this offer. They later bought the engine, paying him much less than he would have made if they had taken it when he offered it on the original terms.

The engine was a walking-beam type 33 x 72-in. Corliss & Nightingale, installed in 1851 for the purpose of running three stands of rolls for rolling sheet brass and copper. It was originally connected to the roll train by a train of gears, but was taken down and moved ahead about twenty-five years ago at which time it was geared directly to the rolls.

A peculiar feature of the installation was the location of the flywheel, which was placed on the end of the crankshaft instead of between the bearings as is the usual method. The bearing nearest the flywheel is about four times the length of the regular bearings. The governor was driven by 1¼-in. diameter braided cotton rope. The speed of the engine was 30 r.p.m. The steam pressure was 35 to 40 lb., being maintained by an 8-in. Mason reducing valve which was installed about 25 years ago, after the old low-pressure boilers were condemned and dismantled for a more modern high-pressure installation. At present there are four Corliss engines ranging from 135 to 750 hp., also a reversing blooming-mill engine of 1,000 hp., together with boiler room and the usual auxiliaries.

RESULTS OF TESTS ON 520-HP. BOILER EQUIPPED WITH CHAIN GRATE STOKER

	Coal Analysis as Fired					Draft Inches of Water									
Name of Coal	Moisture	Volatile	Fixed Carbon	Ash	B.t.u.	Over Fire	Before Damper	Lb. of Dry Coal per Sq.Ft. of Grate Per Hour	Per Cent CO_2 Before Damper	Flue Gas Temp.	Equiv. Evap. per Lb. Dry Coal Fired	Boiler Hp. Developed	Per Cent Rating Developed	Combined Efficiency
Harrisburg, Ill., No. 5	12.73	.....	.....	17.35	10,407	0.31	0 64	33.58	12.44	609	8.84	875	175	71.6
Harrisburg, Ill., No. 5	9.95	.....	.....	16.12	10,955	0.18	0.33	25.58	13.12	548	9.63	720	144	76.8
Eldorado screenings	12.1	31.25	41.08	15.57	10,469	0.26	0.58	32.65	12 05	612	9.34	896	179	76.1
Harrisburg, Ill., No. 5	12.98	30.73	40.38	15.91	10,347	0.28	0.96	32.0	11.65	623	9.56	899	180	78.1
Harrisburg, Ill., No. 5	13.36	.....	.....	15.26	10,683	0.25	0.42	30.0	13 08	577	9.21	814	163	72.6
Harrisburg, Ill., No. 5	9.50	30.01	50.94	11.35	12,000	0.34	0.73	31.4	12.20	548	10.23	826	164	74.8

How Three-Phase Transformers Operate

By J. B. GIBBS
Engineer, Transformer Department, Westinghouse Electric and Manufacturing Company

Comparison is made between single-phase transformers grouped for operation on a three-phase circuit and a three-phase transformer. It is shown that the three-phase transformer is practically three single-phase transformers combined into one. What may be done to operate a three-phase transformer when one phase become damaged, is explained.

IN BUILDING transformers for three-phase circuits, the three windings can be combined on a single iron core. This makes it possible to take advantage of the phase difference in the three circuits so as to save a part of the material. Fig. 1 represents three single-phase core-type transformers of the same size. The full lines indicate the cores and the dotted lines the space occupied by the coils. Now suppose that the three cores are pushed together so that they touch, as in Fig. 2, and that the shaded portion of the iron is cut away. Also take the unshaded strip at the right of *C* and transfer it to the left of *A*, as shown by the dotted lines in Fig. 2. This gives the iron circuit shown in Fig. 3, which has the same cross-section of iron as each circuit in Fig. 1, and has also an increased space for coils besides saving the part of the iron shown shaded, Fig. 2. This is not all clear gain, however, for the total cross-section of the copper in the coils for Fig. 3 must be the same as for Fig. 1, while the mean turn will be greater because the coils will be thicker. In practice a part of the saving is taken out in the iron circuit and a part in the copper circuit, and the net result is that the active material in a three-phase transformer costs about 10 per cent less than the active material in three equivalent single-phase transformers. The active material may be something like one-half of the total cost of the transformer, so the actual difference in cost between a three-phase transformer and a bank of three single-phase transformers may depend as much on the mechanical as on the electrical design.

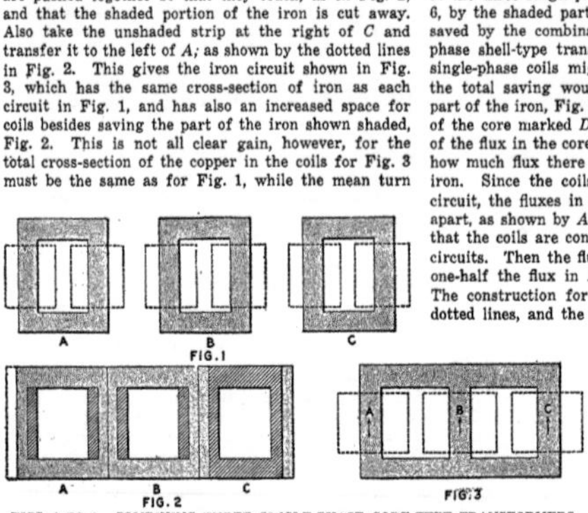

FIGS. 1 TO 3. COMBINING THREE SINGLE-PHASE CORE-TYPE TRANSFORMERS INTO A THREE-PHASE UNIT

A vector diagram of the fluxes in the three legs of a three-phase core-type transformer is given in Fig. 4, which shows that the resultant of the flux in any two legs, as indicated by the dotted lines, is just equal and opposite to that in the third leg. This means that at any instant the sum of the "up" flux in any two legs is just equal to the "down" flux in the third leg. In a similar way three single-phase shell-type transformers can be combined to form a three-phase shell-type unit. Figs. 5, 6 and 7 correspond respectively to Figs. 1, 2 and 3. Fig. 5 shows the magnetic circuit of the three single-phase shell-type transformers; Fig. 6, by the shaded part, indicates the iron which can be saved by the combination; Fig. 7 represents a three-phase shell-type transformer's core. In this case the single-phase coils might be used without change, and the total saving would be represented by the shaded part of the iron, Fig. 6. The magnetic flux in the parts of the core marked *D* and *E* will evidently be one-half of the flux in the cores *A* and *C* respectively. Consider how much flux there will be in the other parts of the iron. Since the coils are connected in a three-phase circuit, the fluxes in the three cores must be 120 deg. apart, as shown by *A*, *B* and *C*, Fig. 8. Suppose, first, that the coils are connected symmetrically in the three circuits. Then the flux in *F* will be the vector sum of one-half the flux in *A*, minus one-half the flux in *B*. The construction for this is shown in Fig. 8 by the dotted lines, and the resultant is a flux in *F* which is 0.866 of the flux in *A* or *B*. Similarly, the flux in *G* will be the vector sum of one-half the flux in *B* minus one-half the flux in *C*, or 0.866 of the flux in *B* or *C*. This would necessitate having the iron in the parts *F* and *G* 0.866 as wide as the iron in the cores *A*, *B* and *C*, and would result in very little saving of the material. Now suppose that we reverse the connections of the coils on the core *B*, so that the direction of the flux is reversed with respect to the flux in *A* and *C*. The flux in *F* will now be the vector sum of one-half the flux in *A* plus one-half the flux in *B*. The construction for this is shown in Fig. 9, and the resultant is found to be equal to one-half the flux in *A* or *B*. Likewise the flux in *G* is one-half the flux in *B* or *C*. In a shell-type three-phase transformer, therefore, the windings on the middle leg are always reversed, and the parts *D*, *E*, *F* and *G* of the iron circuit are all made one-half as wide as the cores *A*, *B* and *C*.

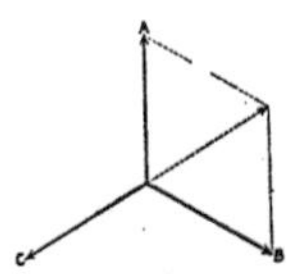

FIG. 4. VECTOR DIAGRAM OF FLUX IN THREE-PHASE TRANSFORMER

The choice between a three-phase transformer and a bank of single-phase transformers depends largely on the judgment of the engineer who is purchasing them, but in general, three-phase transformers are more popular than they were several years ago. This is due partly to improvements in design and insulation, but principally to the better acquaintance of operating men with the three-phase type. The chief arguments for and against three-phase transformers as compared with banks of single-phase transformers may be summarized thus:

For: (1) Occupies less total floor space; (2) weighs less; (3) costs less; (4) only one unit to handle and connect.

Against: (1) Greater weight per unit; (2) the average line man is more familiar with single-phase transformers; (3) in three-phase core-type transformers trouble in one phase may be communicated to other phases. This is a possibility rather than a probability.

The last objection does not carry much weight, for in shell-type three-phase transformers the separation of the windings is almost as effective as in three single-phase transformers, while even in the core type heavy insulating barriers are always put between the windings on different legs, so that breakdowns from this cause are very rare.

It is the usual practice to make the connections between phases in a three-phase transformer at a terminal board below the oil level inside the case. This makes it possible to bring only three high-voltage and three low-voltage leads out of the case and greatly simplifies connection to the line.

There are four principal ways of connecting the windings of three-phase transformers—delta-delta, star-delta, delta-star, and star-star. These are exactly the same as the corresponding connections between three single-phase transformers in a three-phase bank, and the star-star connection is generally avoided for the same season that it is avoided in single-phase transformers; although if the three-phase transformer is of the core type, the shifting of the neutral is less extreme than in a bank of single-phase transformers.

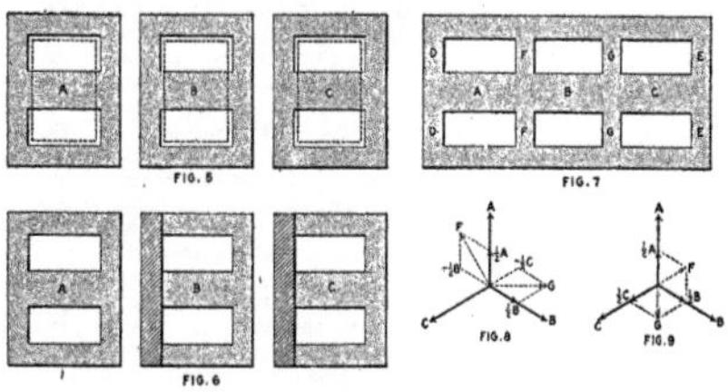

FIGS. 5 TO 9. COMBINING THREE SINGLE-PHASE SHELL-TYPE TRANSFORMERS INTO A THREE-PHASE UNIT. VECTOR DIAGRAMS OF THE FLUXES

In case of trouble in one phase of a three-phase transformer, the other two phases may be operated in open-delta connection to carry 58 per cent of the rating, if the transformer connection is delta-delta. This is the same statement that can be made concerning a bank of single-phase transformers. To make this change in a three-phase transformer, it is necessary to disconnect both high- and low-voltage windings of the damaged phase, at the terminal boards. If the transformer is of the core type, both windings of the damaged

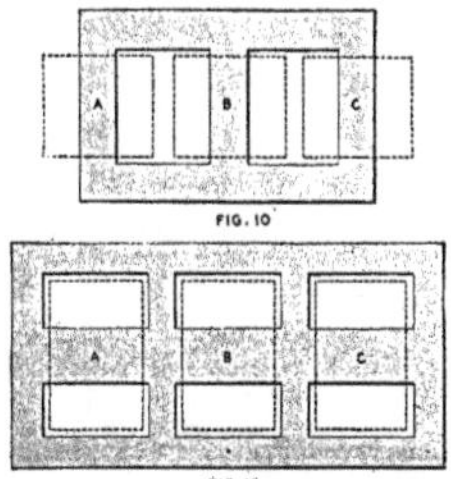

FIGS. 10 AND 11. CORE-TYPE AND A SHELL-TYPE TRANSFORMER

phase must be open-circuited. If the transformer is of the shell type, the damaged windings may be open-circuited or short-circuited. The reason for this will be evident from a consideration of the shape of the magnetic circuits. Figs. 10 and 11 are repetitions of Figs. 3 and 7 respectively. Suppose that the winding *B* is damaged in both transformers, if phases *A* and *C* are to run in open-delta it is necessary that the fluxes in these two cores be 120 deg. apart and that the resultant flux have a return path. In the core-type, Fig. 10, transformer this return path is provided by the third leg *B*. But if there is a winding on core *B*, the flux passing through it will induce a voltage, and if there is a closed circuit a current will flow. This will increase to the point where the coil *B* will be burned out. In the shell-type transformer the core *B* does not provide the only path for the resultant flux, and if the coil is short-circuited it will merely cause the flux to take the other path.

Recommendation that steps be taken to relieve the present car shortage in order that coal may be obtained in sufficient quantities to build up a reserve for New England and other Eastern industries for next winter has been made by the National Association of Cotton Manufacturers, according to *Coal Review*. Co-operation of chambers of commerce in securing assignment of a reasonable number of coal cars to the Eastern territory has been asked.

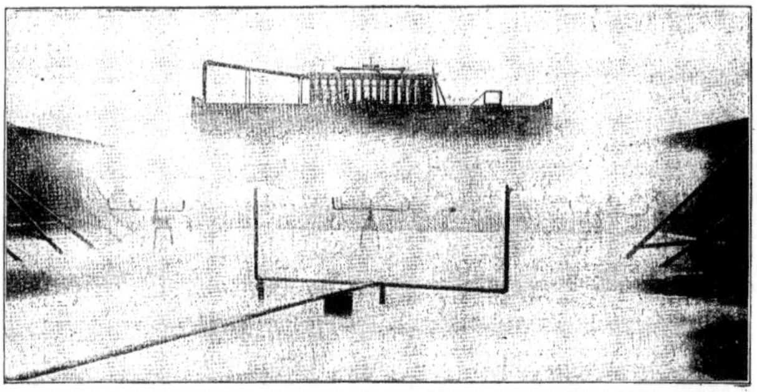

FIG. 1. A REFRIGERATION-PLANT SPRAY COOLING POND; CONDENSERS ON ROOF IN THE BACKGROUND

Refrigeration Study Course—XII. Cooling Water

By H. J. MACINTIRE

REFRIGERATION and cooling water are similar terms, for without cooling water there could not be refrigeration. One would be as likely to try to operate a steam engine or turbine condensing without condensing water, or the oil engine without cooling water, as the refrigerating machine without the necessary amount of water at the temperature of the air or colder. As a matter of fact the first question that arises is, "What is the temperature of the cooling water?" and the second, "What is its cost?" In other words, it is desired to find out how fully we may use the water and what is its initial temperature.

Occasion has been taken before this to bring out the function of the cooling water in the refrigerating cycle. The compressor is a pump—a temperature pump in a way, as it raises the temperature level of the refrigerant to a point sufficiently high so that the cooling water at hand can condense the ammonia gas satisfactorily. In the ammonia condenser, as in any form of condenser, the cooling water removes the heat of liquefaction (latent heat) and in addition the superheat and some of the heat of the liquid. In consequence of this the water is warmed a certain amount, depending on the relative amounts of water used. In general it may be said that the heat removed is as follows:

The compressed ammonia gas, being superheated, has its superheat removed first and then the latent heat of liquefaction. Finally, the liquid ammonia is precooled slightly below the temperature of saturation. The cooling water has to remove all this heat, and it passes away from the condensers carrying this amount of heat away with it and is at a higher temperature in proportion. An example will show the conditions which prevail:

A compressor operating between 15.2 lb. and 175 lb. gage, has a temperature of discharge of 234 deg. F. and the temperature at the liquid expansion is 80 deg. F. The temperature of the water leaving the condenser is 90 and the entering water is 75 deg. F. Under these conditions there would be 92 B.t.u. of superheat to be removed from each pound of ammonia, and 490.8 B.t.u. of cooling to condense it at 175 lb. gage pressure and a temperature of 93.5 deg. F. Finally, to cool the liquid to 80 deg. F. requires 15.9 B.t.u., or a total of 598.7 B.t.u. per pound of ammonia. The useful available refrigeration from this same pound of ammonia is 484.9 B.t.u. and the difference, 598.7 — 484.9 = 113.8 B.t.u., is the work done by the compressor, resulting in heating the ammonia. This value may be shown in another way.

Taking a case where one pound of ammonia is compressed per minute, the net refrigerating effect is 484.9 B.t.u. or at the rate of 2.42 tons of refrigeration. With these conditions of operation it will require 1½ hp. per ton of refrigeration, or 1½ × 2.42 × 42.4 (the heat equivalent of a horsepower per minute) = 154 B.t.u. per min. The difference between 154 B.t.u. and 113.8 B.t.u. is accounted for by the mechanical efficiency of the compressor, and the heat removed by the jacket water (this latter being about 18 B.t.u.). It is clear from the preceding that the condenser and jacket water remove practically all the heat equivalent of the power supplied to the compressor by the engine or motor. In all likelihood about 1½ gal. per min. of cooling water will be used for every ton of refrigeration, and a 50-ton plant would require 75 gal. per min. If water is expensive to pump or has to be purchased, a method of cooling would be necessary to make it serve a second or a number of times over the condenser. The methods used are as follows:

In general there are three ways of cooling water where re-use of cooling water is required—cooling ponds, spray ponds and cooling towers. Each of these methods has its advantages in special cases.

The cooling pond can be used in locations where a natural body of water is available adjoining the plant. The water is cooled by being allowed to be brought in surface contact with the air, and the evaporation (and cooling of the water) depends on the relative temperature of the air and on the humidity. The tendency would be for the pond to assume the temperature of the air, both due to giving up heat to the air and also due to the difference of the *saturation pressure* of the water and saturation pressure of the water vapor in the air. Roughly, the amount of cooling in *still air* for usual conditions may be taken at from 60 to 80 B.t.u. per sq.ft. of surface per hour, or about 225 sq.ft. of refrigeration. As the surface is the main consideration, it is not important to have any depth unless the refrigerating load is a variable amount, as, say, in the case of the chill rooms in a packing plant or the pasteurizing of milk and cream. In these latter cases it would be possible to use a smaller pond, but of some depth in order to store up enough cooling water. Usually, a foot or two is plenty.

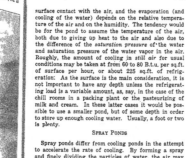

Spray Ponds

Spray ponds differ from cooling ponds in the attempt to accelerate the rate of cooling. By forming a spray and finely dividing the particles of water, the air may quickly reach all the water immediately instead of waiting until each particle reaches the surface, as in the case of the cooling pond.

With the spray nozzle, Fig. 2, the water is pumped under 5 or 10 lb. pressure through a special nozzle designed to create a whirling spray of fine water particles. In most cases a cooling of 20 to 30 deg. F. may be obtained with a single spraying, and the final temperature depends on the fineness of the spray and the humidity. For instance, if the air has a humidity of 70 per cent and a temperature of 70 deg. F., the pressure of the water vapor in the air is 0.70 times 0.36, or 0.25 lb. per sq.in. It is usually assumed that the cooling water may be reduced to a temperature corresponding to a pressure about 0.15 lb. per sq.in. greater than that of the water vapor in the air, or (in this case) 0.25 + 0.15 = 0.40 lb., which corresponds to 73 deg. F. Some five or six per cent of the water is evaporated in the process.

A good rule is to allow one square foot of surface for every 250 to 300 lb. of water sprayed per hour, which

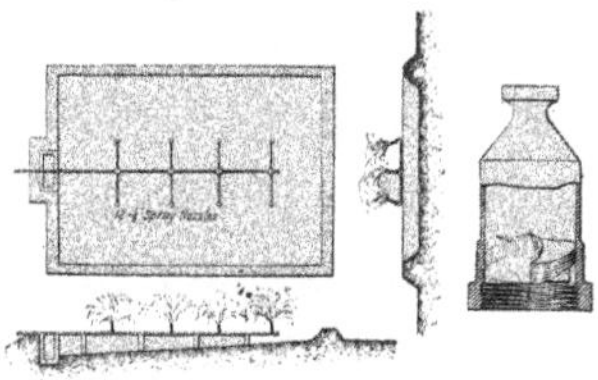

FIG. 2. GENERAL ARRANGEMENT OF SPRAY COOLING POND AND A TYPICAL SPRAY NOZZLE

reduces to about 4 sq.ft. per ton of refrigeration, allowing a temperature range of the water of 15 deg. F. To prevent excessive loss of water, the pond or catch basin should have the embankment at least 15 ft. outside the row of spray nozzles. The nozzles should be about 8 to 10 ft. apart to secure satisfactory results. The spray nozzles may be placed on the roof or above the condenser if suitable louvres are provided to prevent excessive loss of water during winds.

The cooling tower has been successful and popular for many years, and without question will continue to be

FIG. 3. MACHINE ROOM OF 100-TON ICE PLANT, POLAR WAVE ICE AND FUEL CO., ST. LOUIS

used indefinitely with equal satisfaction. It lends itself particularly to congested quarters, where a cooling pond is out of the question and the spray nozzle cannot be easily arranged. As its name would indicate, it is built in the form of a tower, usually with airtight walls to gain the advantages of the chimney. The object of the tower design is to acquire a natural or forced draft (or both) and to provide a means of securing a large amount of wetted surface kept wet by the condensing water that is pumped in at the top.

The wetter surface may be secured by means of tile, but the usual manner is to form a wooden checkerwork, arranged as in Fig. 4, to prevent undue trouble by warping and so that the wet cross-sectional area will be sufficient to allow the necessary amount of air to pass through. As in the spray nozzle, which requires a finely atomized spray, so in the cooling tower, a thin film of water is required and an air current to evaporate some moisture and to cool the water by the heating of the air.

Action of the Cooling Tower

The action in the cooling tower is as follows: The water is showered over the top grids of the tower, and it descends, dropping from checkerwork grid to checkerwork grid and being exposed all the time as a thin film on the surfaces. The ascending air absorbs moisture and has its temperature increased and finally leaves the tower at a temperature of about 10 deg. F. below the entering warm water of the ammonia condensers, and with a humidity at the exit temperature of about 95 percent. During the passage of the air through the tower some water was evaporated, which cooled

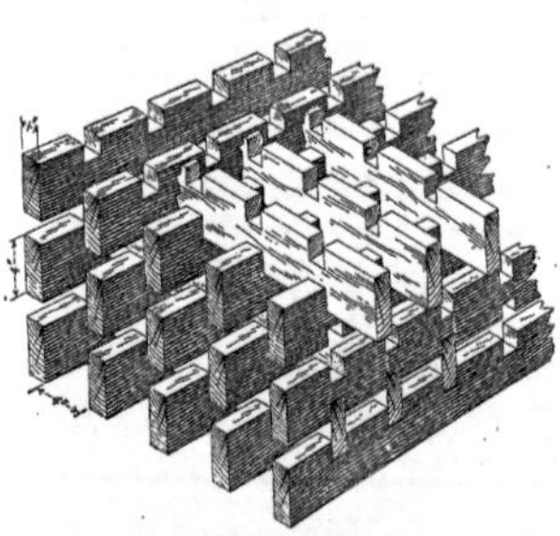

FIG. 4. THE WOODEN CHECKERWORK IN THE COOLING TOWER EXTENDS THE SURFACE OF THE WATER AS IT FALLS, ALLOWING THE UPCOMING CURRENT OF AIR TO COOL IT QUICKLY

the remaining water according to the principles of refrigeration.

In refrigeration it will be necessary, usually, to cool the water about 15 to 20 deg. F. Under these conditions the tower will absorb heat from the water at the rate of 300-400 B.t.u. per square foot of tower surface per hour. At this rate 60 to 45 sq.ft. is required per ton of refrigeration. An air velocity of about 10 ft. per sec. is usually required. In refrigeration a natural-draft tower is generally used, and either the necessary chimney effect is relied on for the creation of a draft, or the design with open sides is used, which relies on the wind to do the business.

It will thus be seen what the function of the cooling water is and what means of making use of this cooling water again is possible. As a means of visualizing the problem better and also of estimating the requirements of new plants, the diagram is added, showing the heat

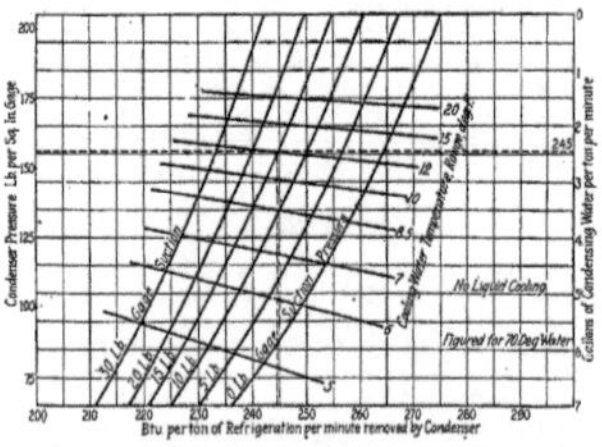

FIG. 5. CHART FOR DETERMINING QUANTITY OF COOLING WATER REQUIRED FOR VARIOUS CONDITIONS OF PRESSURES AND TEMPERATURES OF WATER AND AMMONIA

to be removed by the condenser per ton of refrigeration, and the condenser pressure to be expected in this condition for each temperature range of the cooling water.

For example, if 2.45 gal. of water is used per ton per minute with a suction pressure of 15 lb. gage, the range of the water (the temperature rise in passing through the condenser) will be 12 deg. F., the heat removed per ton per minute will be 244.5 B.t.u. and the condenser pressure will be 156 lb. gage. The diagram is laid out for an initial temperature of the water of 70 deg. F. and no liquid cooling after condensation of the ammonia. Also, the assumption is made that the gas entering the compressor is just dry and saturated—a condition that prevails only with careful operation of the compressor by means of thermometers, at which time the maximum efficiency of the compressor is also obtained.

How Two Alternating-Current Motors Were Controlled in Parallel

By John Grant

A large starch mangle was installed to pull heavy cloth sheeting from a bin located one hundred feet away from the machine, wet, starch and deliver the cloth on a set of dry cans located on a floor directly over the mangle. In order to accomplish this, both machines would have to run at the same speed. The set of dry cans were already in operation, and were driven by a 10-hp. slip-ring type induction motor, with external resistors and drum type controller. The problem then was to utilize this motor and have it operate at identically the same speed as a 50-hp. motor we intended installing on the mangle.

At first thought running the mangle and dry cans at

the same speed looked like a rather difficult undertaking. The drawings of the manufacturer specified a 50-hp. direct-current motor on the mangle and a 10-hp. direct-current motor to drive the dry cans. To have accomp-

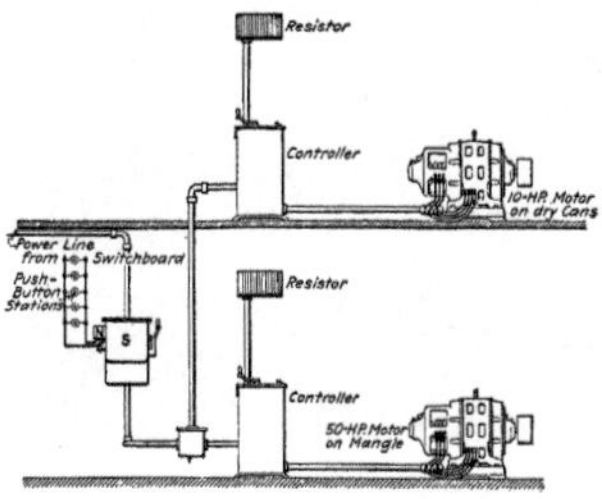

FIG. 1. ARRANGEMENT OF MOTOR AND CONTROLLER EQUIPMENT

lished this successfully would have required the purchase of an alternating-current motor, direct-current generator set, since the power supplied to the plant was two-phase 60-cycle alternating current. This would also have necessitated the removal of the alternating-current apparatus from the dry cans, and altogether would have been an expensive arrangement. After careful consideration it was decided to use alternating-current equipment arranged as illustrated. In Fig. 1 is shown a 75-hp. oil switch *S* with a time-limit overload relay, and a no-voltage attachment *N*. The motor and control equipment is clearly indicated. A set of push-button switches was located at convenient stations on the machines to trip the no-voltage magnet on the main oil switch if necessity demanded it. Bells were installed so that the operators could signal when ready to start up or shut down the machines. This comprised the electrical part of the equipment. The controller handles were taken off, gears put on the controller-cylinder shafts and the handles replaced. Then a 1-in. shaft was installed which extends from the 50-hp. controller, as indicated in Fig. 2, with a gear on each end of the shaft to mesh with a gear on each controller. About one foot from the 50-hp. controller the shaft was cut in two and a coupling installed. A tapered hole was drilled through the coupling and shaft in the upper portion, and a dowel pin was used to prevent the shaft from turning in the coupling. This arrangement is clearly shown in Fig. 2.

To begin operation the cloth must be drawn from the bin and passed through the rolls of the mangle, which of course must operate as a single unit. In order to accomplish this, the main oil switch is closed and then the dowel pin is removed from the coupling and shaft. This will allow the 50-hp. motor to be operated as a single unit. When enough cloth has been passed through the mangle, it is taken up and passed over the first drum on the dry cans; then the 50-hp. motor is shut down and the dowel pin inserted in the coupling and shaft. A signal is given the operator on the dry cans that all is ready, the 50-hp. controller is then closed in on the first notch, and by means of the shaft and gears will also cut in the 10-hp. controller on the first notch, in the same operation. Enough resistance is inserted in the secondary circuits of the motors to make each machine's speed identical and so on until the controllers are on the last step and the secondary resistors are entirely cut out, when the two machines run at practically the same speed.

Considerable difficulty was experienced at the initial starting of this drive, owing to the wide variations of the motor speeds and the difference in horsepower, as much as twenty-five ft. of cloth piling up in front of the mangle. After many attempts at resistor adjustments, which were especially arranged for this drive, the proper values were obtained.

The dry cans are equipped with compensating rolls which take up two feet of cloth or let it out, but so fine are the resistors adjusted that these rolls do not vary three inches at any time either at starting the machines or after they are in full operation. It is worthy of mention that this installation was severely criticized by technical experts when first drawn on paper. Each motor has been carefully checked and tested, when operating as a single unit, and the same test results prevail when they are coupled together.

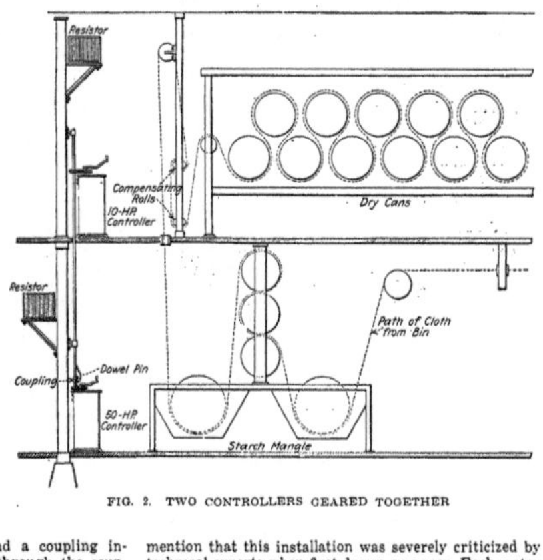

FIG. 2. TWO CONTROLLERS GEARED TOGETHER

Prof. G. F. Gebhardt

Armour Institute of Technology

An educator who takes a father's interest in developing green boys into successful engineers. A genius in devising mechanical means for laboratory testing. An authority in power-plant engineering.

SUMMARIZED in brief, these are the outstanding characteristics of G. F. Gebhardt, who has been a teacher of Mechanical Engineering at Armour Institute of Technology for nearly all of his working life—a period of about twenty years. Close personal touch with his students, a direct interest in each and every one, has been the secret of his success as an educator. He has no regular office hours. His door is always open to the boys, and they feel just as free in coming to him with their personal difficulties as with their regular technical studies. Many a time he has gone to a backward student and at the psychological moment, by a little of that personal sympathy so sadly lacking in larger institutions, has turned him into the ways of success. Out of classes he is a boy with his students, goes fishing and duck hunting with them, gains their confidence and their loyalty, and with it a return in honest, faithful work in their studies. It is an ideal relation for a man in love with his profession and one which only a teacher can enjoy.

In 1874 Professor Gehbardt was born in the Mormon country at Salt Lake City. As public schools of merit were nonexistent at this time, his early education was obtained at the Collegiate Institute, a sectarian school of Presbyterian origin. At the age of sixteen young Gebhardt entered Knox college to take the regular literary course. Two years later he went to Cornell to study mechanical engineering. At the end of the usual period he not only secured his M. E. degree, but, by making up all his literary subjects, was enabled to take the A. B. degree at Knox. Three years later he secured an M. S. from the same college. Contrary to the average opinion of the young engineer, he realized the need for something more than bare engineering. A broader outlook upon life, proper diction in which to express his engineering thoughts and language to clothe his engineering reports were the primary reasons for the literary course, and it is his constant endeavor to impress these same ideas upon his students at Armour. This is why the professor has become a stickler for written reports, as they are means toward a desirable end.

After his college days were over, a choice of professions was to be made. Oddly enough, he had an abhorrence for teaching and the spiritual longing for the ministry was lacking. He decided to become a mining engineer and to be a good one—to start from the bottom. Back to the mountains of Utah he went and worked in the mines in various capacities. For a time he was an ordinary mucker, with the exception that the hardest and most disagreeable tasks were given to the college student. It was a real test of his manhood which he passed with credit and stuck to with determination for over a year. The hard times of '97 and '98 overtook him, however, and in looking around for a new position he finally secured an appointment at Armour as instructor in machine design, thus being forced by circumstances to take up teaching.

It was his thought to make this place a temporary haven until things on the practical side opened up, but almost immediately, in spite of his former prejudices, he found that teaching was the profession that appealed to him. Seeing the green, raw freshmen come in and develop into competent engineers was a fascinating study, and the repetition year after year never chilled his interest. It was only natural that a series of rapid advances came to him. It was not long before he was made assistant professer in machine design, then associate professor, and within three years, professor of mechanical engineering. The date was 1903, and with the one exception of the job in the mountains it has been his life work.

At this time Armour was a regular polytechnic school. The original idea of the late P. D. Armour, the founder of the institute, was to provide a place for the poor boy or girl to learn a trade. For the latter there were courses in cooking, dressmaking, typewriting, etc. For the boys there were the preparatory academy and the engineering courses. It was rather a polyglot mess with indifferent facilities, and the school was getting nowhere. Professor Gebhardt saw the possibilities of building up a great technical institution, and it was in these early days that all side issues were dropped and Armour became strictly a technical college of engineering. It has been his constant endeavor to improve the curriculum and the mechanical equipment until now the institute ranks with the best technical schools in the country.

To build up the proverbial slender income from teaching, the professor has done a great deal of consulting and testing work, never taking a retainer from a commercial concern, but always acting as a free lance on a specific job. In this way he has always maintained a neutral position in commercial work, and is frequently selected as the mechanical expert in court actions or in controversies outside the law. This business has increased to such a point that he has encouraged each and every one of his instructors to specialize on a particular subject, so that he might turn over to them the practice in their line. It is his opinion that it makes them much better instructors. All the new ideas from the commercial field come to them, and the influence is to supplant abstract consideration with the latest and newest thoughts in the field. Incidentally, this work gives them prestige with the students, adds to the remuneration of the instructor and, up to the limit of interfering with classes, is considered an excellent form of training.

Power-plant engineering is Professor Gebhardt's specialty. His book on this subject has been in great demand. It has passed through many editions, the number of copies sold exceeding 26,000. Much of his work recently has been on combustion, and naturally enough, as it is the field where the dollars may be spent or saved. When he has a big job of any kind, the boys are always called in on it. It gives them a chance to get practical experience, to gain confidence and meet men in the profession. The boys are paid a small wage, always by check. It is amusing to relate that many of these checks are slow in returning. The boys frame them as the first money earned, but as they get out in the field, the novelty of earning money wears away, and perhaps the need of it forces a change of sentiment. The checks are cashed and come creeping back to the fountain head, thus performing the double service of furnishing cash and reviving memories.

Professor Gebhardt has done a great deal of research work, including the construction of the machines on which to do the testing. Among the subjects may be mentioned belt transmission, ball-bearing friction, tire-shock absorption and flow of fluids. The Gebhardt steam meter was designed by him, and there have been numerous laboratory machines for testing materials of engineering and special appliances for power-plant testing. His contributions to the technical press have been numerous. In addition to the book already mentioned, he has prepared pamphlets for school use on governors and on the dynamics of the steam engine. He is a member of the American Society of Mechanical Engineers, the Western Society of Engineers and the National Association of Stationary Engineers. In fraternal life he belongs to the University Club, is a Phi Delta Theta and a Tau Beta Pi.

As to hobbies, the professor claims he is just an ordinary individual and does nothing distinctive. We know he appreciates a good cigar, and according to his own statement he would rather sit out in the swamp and shiver shooting ducks than do anything in the educational line. Since the age of nine it has been an annual pastime. He has discovered the fountain of youth. He does not feel like forty-six, nor does he look it. Association with his boys keeps him young. As the older depart, new arrivals hold the average perpetually the same, so a boy he must always remain, with the enthusiasm and vitality of youth combined with the judgment of maturer years.

International "Hot-Flow" Softener

The important difference between hot- and cold-water softeners lies in the provision for sedimentation. Convection currents are set up in hot water by slight differences in temperature, the hot water tending to rise and the cooler water to fall. To obtain the best results in the hot-water softener tank, these currents must be worked with rather than against. This principle has been given attention by the International Filter Co., of Chicago, in the design of a new hot-water softener described herewith.

In Fig. 1 the general arrangement of the softener is shown diagrammatically, with two chambers, *A* and *B*, of equal holding capacity. On top of the softener is an open-type exhaust-steam feed-water heater, in which the entering water goes to a distributing trough at the top and then passes over and through a series of inclined cast-iron trays. Steam enters the heater through an oil separator and comes into contact with the finely divided water in the usual manner.

From the heater the hot water enters the softener through the turned inlet *C* above the baffle plate *F*, the

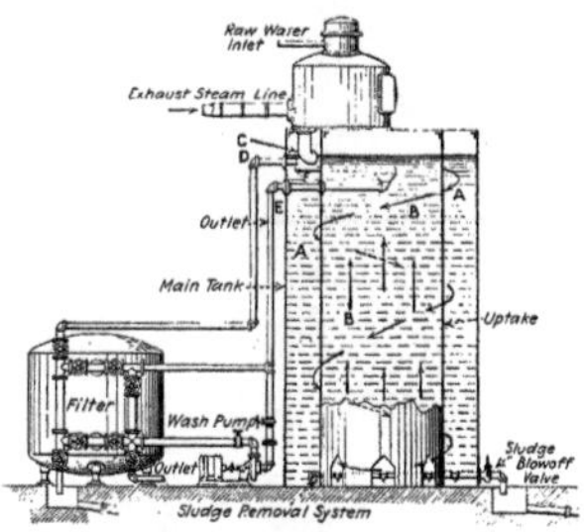

FIG. 1. FILTER AND SECTION THROUGH HOT-FLOW SOFTENER

softening chemicals being delivered to the incoming water at this point. The turned inlet and baffle plate cause the incoming water to flow in a horizontal direction around and downward to the outer annular chamber *A*, automatically providing the agitation that brings the softening chemicals and the water into contact. At the bottom the water flows through the inverted V-shaped openings from the outer chamber *A* into the inner chamber *B*. The water is drawn off through the outlet pipe *E* from a point near the top of this chamber.

Although but twenty to thirty minutes are necessary for the chemical reactions and sedimentation, the softening tank has a one-hour holding capacity.

The outer chamber *A* provides for the mixture of the incoming water and the chemicals. All the commotion set up is restricted to the water in this chamber, and the outer chamber of water serves as a blanket of insulation for the inner or uptake chamber, permitting no cooling of water in the uptake or the setting up of convection currents.

In the chamber *B* the water rises slowly, permitting the separation of the sludge from the water to become completed, although some of the sedimentation takes place in the outer chamber.

A standard part of the softener is a filter. The softener supplies hot water to wash it, and provision is made for its return to the outer chamber of the softener, where it is allowed to resettle and pass through the usual process, the heat in the water being saved. The circulation for washing is set up by an auxiliary pump, the flow being upward through the filter bed instead of downward as when filtering. Neither the softener operations nor the boiler-feed supply is interrupted while the filter is being washed.

Softening of the water is effected by treating it with hydrated lime and soda ash supplied from a chemical mixing tank and feeder.

Hard water enters the softener through the head chamber above the heater, the flow being regulated auto-

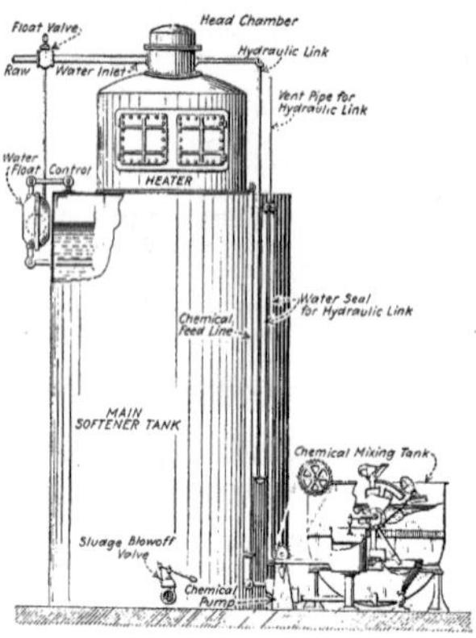

FIG. 2. SOFTENER TANK AND CHEMICAL FEED

matically by a float valve in the supply line. The water flows into the heater and is divided into two proportionate streams; the larger stream flows into the heater and the smaller quantity flows into the hydraulic link pipe shown in Fig. 2. As the head or pressure of the incoming water governs the flow of both, the two streams always flow in proportion to each other. If more water flows into the heater, proportionately more water flows into the hydraulic link pipe; if less water flows into the heater, a correspondingly smaller amount flows into the hydraulic link. This hydraulic link conveys a small stream of water to the chemical float control independent of the back pressure in the heater and head chamber, and as the rate of flow of the water in the hydraulic link is always in proportion to the main flow of water into the heater, the smaller stream of water is used to control the chemical feed.

From the float control the discharge is a modulated orifice tube that keeps the level of the water in the float control in relation to the main flow of water into the heater. A greater flow of water into the heater causes a higher water level in the float control and a lesser flow a correspondingly lower level. A float which rises or falls with these changes in water level in the float control moves the cutoff shield over the collector funnel in the chemical tank. If the softener is operating at half capacity, the cutoff shield covers half of the mixture picked up by the revolving cups. Thus, at any rate of operation, the cutoff shield deflects from the collector funnel all or part of the chemical mixture except that required.

The chemicals for twelve to twenty-four hours operation are weighed and put into the chemical tank, which is then filled with water. The mixture is then agitated to keep it circulating in an up and down motion in all parts of the tank and maintain a chemical mixture of uniform strength.

The greater proportion of the sediment or sludge formed settles to the bottom of the main tank and is blown off through the pick-up pipes and the main sludge blowoff valve, which is opened two or three times each day.

The hot-flow softeners are built in graduated standard sizes with capacities per hour ranging from 900 to 18,000 gal. In boiler horsepower based on 30 lb. of water per hour the listed capacities range from 250 to 5,200 gal. Special sizes up to 10,000 boiler horsepower are built, using the same principles of construction and operation.

Denmark is at present supplied from 497 central stations having a total output of 108,000,000 kw.-hr. per annum. The net sale to 205,000 consumers is 30,800,000 units for light and 54,700,000 units for power. The average net cost per unit in 73 towns is 4½d., in 387 villages 7½d., and in 37 agricultural districts 3¾d. The high cost in the villages is chiefly due to the load factor, which seldom reaches 0.10, and in a smaller degree to the high cost of fuel. With regard to future developments, it appears, according to the *Technical Review*, that a scheme is being considered which will enable a supplementary supply of electrical energy to be obtained from Norwegian waterfalls. The distance from the nearest suitable falls to Copenhagen is about 75 miles, and it is proposed to send 26,000 kw. at a pressure of 50,000 volts by an overhead transmission line running through Sweden to Malmö, and thence by submarine cable across the Sound to Copenhagen. The engineering difficulties in connection with this proposal are easily overcome, but the economic question presents some difficulty, as a low load factor would require a very expensive transmission line, and necessitate high prices to consumers. The estimated requirements of the country, provided cheap energy can be supplied, are about 450 million kilowatt-hours per annum.

The proposed extension of hydro-electric projects in Japan will involve the building of about one thousand miles of high-pressure transmission lines within the next two years. The use of aluminum instead of copper for the transmission cables is considered.

EDITORIALS

Efficient Utilization of Water Powers Already Developed

According to the United States Geological Survey's estimate of 1915, seventy-five per cent of the maximum potential water power of the New England States is developed. This figure was also used in the National Electric Light Association's Water-Power Development Committee's report submitted at the Pasadena convention in May of this year. Taking these figures at their face value, it is apparent that little additional water-power development may be expected in this district. However, when the way that many of these developments are made is considered, especially in the industrial centers, it is evident there are still large possibilities of future water-power development in New England, and the committee's statement regarding this district that "future development will consist chiefly in a more efficient utilization of the water powers already developed" has a real significance.

A large percentage of the water-power developments in New England have just grown, without any consideration being given to efficient operation. The first manufacturer on the site built a dam and utilized a head that best suited his needs. The next manufacturer did likewise or entered into the same agreement with the party having prior water rights. This process continued until all available water power at a given site was utilized, after which further power, if needed, was developed from fuel. Water-power development of this kind has resulted in a very inefficient use of the available energy in the water. Not infrequently is it found that water powers have been developed in three or four low-head projects, with a large number of small inefficient waterwheels, with the result that probably not more than twenty-five per cent of the power is obtained, that would be possible if the total head were utilized in a modern development. Another source of large loss is in the gears, lineshafting and belts used to transmit the power from the waterwheels to the manufacturing machinery. The co-ordination and unification of these numerous inefficiently developed water powers into a comprehensive system would have a far-reaching effect in the solution of New England's power problems. Nobody is to blame for the inefficient way many of these water powers have been developed; it is a case where industrial expansion has outgrown the methods that were perfectly satisfactory forty or fifty years ago, and an adjustment must be made to meet the new condition.

There are many problems to be solved in the redevelopment of these water powers, and probably none is more serious than the location of some of the plants that utilize large quantities of water in their manufacturing processes. Unless these plants are located above the dam or below the power plant, large quantities of water will have to be used for industrial purposes that should be available for power development. This problem also would exist with steam plants operating condensing. However, water that is available for manufacturing processes under a natural head would not require power for pumping, and if this water afterward could be used for power purposes under part of the total head, its efficiency of utilization would be comparatively high. There is also the possibility of taking the water directly from and returning it to the penstock where a closed system is used, such as in a condensing equipment. Be these problems what they may, the high price of coal will in no distant future cause them to be given serious consideration. Furthermore, before we as a nation can say that we are making an honest effort to conserve our natural resources, many of our water powers must be redeveloped for a much higher efficiency than at present obtained. When it is possible to obtain ninety to ninety-three per cent of the total energy in the water, we should not be satisfied with twenty-five or thirty per cent.

Conventions and What We Can Get Out of Them

May and June are convention months. We have had the joint meeting of the American Societies of Mechanical, of Refrigerating and of Heating and Ventilating Engineers at St. Louis. We have had the Organization Conference resulting in the launching of the American Federated Engineers Societies at Washington. We have had the American Federation of Labor meeting at Montreal. The list might be extended almost indefinitely. Of course we must not forget the two big political conventions for the nomination of Presidential candidates, but the daily press has given all kinds of information (?) about them.

The conventions that the readers of *Power* are most interested in are those of the various engineering organizations. The custom of presenting and discussing technical papers at such meetings produces a large amount of high-grade engineering literature each year, and there are few of the technical-society programs that do not have at least one paper of use to anyone engaged in the production or use of power.

It is unfortunate that every man cannot attend such conventions. A large part of their value lies in the personal contact with other men from other parts of the country. When you hear first hand of the other fellow's troubles, problems and successes, your own troubles seem less, your problems easier to solve and your success nearer.

The men who stay at home, however, can and should get a great deal out of these meetings. Every technical journal tells, in its news columns, of the general happenings at those gatherings in which its readers will probably have an interest. More important, however, are the abstracts of the papers and discussions. In this connection the stay-at-home has perhaps an advantage. He has the story boiled down so that he gets the important points with a minimum amount of time spent.

One big feature of the engineering-society conventions, however, is that there the newest ideas are usually presented. The easiest way to keep just a little ahead of engineering practice is to watch carefully what is offered at such meetings. True, papers are often presented which are so far ahead of good practice that practice never catches up with them, but the man who is never ahead is always behind. It is also true that many of the papers are presented by representatives of manufacturers and are really presented as part of advertising campaigns. On the other hand, it is equally true that such so-called advertising papers describe some new apparatus, process or idea and so are of value to the reader. They must also be reliable and accurate, for each manufacturer knows that his competitors will read his paper and present discussions which will expose any false statements or mistakes.

It is well worth your while, then, to read and study the abstracts of such papers and get your share of the benefits from meetings that are held mainly to help the engineers and, through them, the public.

Fuel Economy Still Vital

DESPITE the urgent need for it at this time, one does not hear much about fuel economy, particularly in power plants. It may be that power-plant men are so busy practicing it that they have no time to talk about it. This is, of course, true in many cases, but the quality of the fuel that many plants now get causes the engineers no little worry to burn it at all, to say nothing of burning it efficiently. We must not lose sight of the seriousness of the fuel situation. Coal is too costly these days to warrant neglect in its use no matter who pays for it. It is every bit as necessary to fire intelligently, to stop preventable losses in the boiler room, to consume power savingly, as at any time during the war.

In the large modern station, coal economy is, in greatest part, a matter of design—design that reduces to a minimum the human equation. But in the far greater number of plants in the country the men in the power plant, the engineers and firemen, determine ultimately whether fuel shall be used economically or woefully wasted. Upon these men there rests a truly great responsibility. They owe it to themselves, to their country and to their employers to make every pound of fuel do its utmost.

Steam-Turbine Development

WE FREQUENTLY hear comments on the very rapid development of the steam turbine as compared with the early progress of the reciprocating steam engine. The latter machine began as, to modern eyes, a very crude piece of apparatus and was improved only by slow and tedious steps. While the first steam turbines were built long before reciprocating engines, no real commercial use was made of them until within very recent years. When engineers finally did turn their attention to turbines, progress was so rapid as to be almost bewildering.

There are a number of reasons for this difference. The engine could develop no faster than the steam boiler and the various auxiliaries. The turbine designer found boilers, piping, condensers, etc., already at hand and could focus his attention on a single problem. Modern methods of machine-shop practice and modern steels, bronzes, etc., were not available to the early engine builders as they were to the turbine builders who came so much later.

Probably one of the greatest factors in the rapidity of turbine progress is to be found in the knowledge possessed by modern engineers of the properties of steam and the theory of the production of power by means of heat. The first steam engines were built almost by guesswork and had been in use some time when Sadi Carnot laid the foundations of the theory of thermodynamics. After him came a long line of men who labored in the physical laboratories and solved complicated mathematical problems. A great deal of their work was looked upon as purely theoretical and of little or no value to the man who had to build or operate steam engines. Without this work, however, we would today probably be using engines not much better than those of Watt.

The turbine designers found much of this work already done for them, but there are still many unsolved problems. The article in another column under the title "Supersaturated Steam" suggests some of these problems. Is the steam in a turbine in a state of unstable equilibrium? If this is the case, can a rational workable theory be developed? Scientific papers already published show that the "theorists" are at work on these problems, and judging by the past the solutions will probably be found. Wonderful progress in development has been made, but who can tell what will yet be accomplished as one by one these problems are solved? Meanwhile, the engineer, operator or designer who wants to make good must take an interest in such things as are often spoken of as "highbrow."

Storage-Battery Development

SOMETIMES the development of a special branch of an industry takes place so quietly that other branches of the same industry fail to observe the change. Often even those in the special branch that is changing are to a large extent unconscious of what is happening. All engineers remember that only a short time ago a storage battery was considered a delicate and complicated part of the equipment of the power plant where there was such equipment at all. Today they are operated successfully by thousands of people who know nothing about engineering. No doubt further applications of the storage battery in power stations will develop as the demand for reliability of operation increases, and it would be well for those interested in power-plant operation to keep in mind the possibilities of this form of equipment.

A headline reading "Airplanes for Mine Rescue Work" was rather startling until we read further and learned that the Bureau of Mines is planning the use of flying machines to carry rescue workers and apparatus from the various stations of the Bureau to the mines where they may be needed. The element of time is a big item in mine-rescue work, and this new use for airplanes promises another advance in the conservation of human life.

The recent orders of the Interstate Commerce Commission calling on various Western railroads to return stated numbers of open-top cars to the Eastern roads promises some help in the coal situation. If more coal cars come to the mines and if the assigned car problem can be solved, we may all be able to get at least a part of next winter's coal into the storage yards before cold weather comes.

CORRESPONDENCE

Comparison of Pulverized-Coal Tests

Of the numerous comments made on the pulverized-coal tests at the Milwaukee Electric Railway and Light Co.'s plant, that by Paul W. Thompson is, to my mind, the most interesting and instructive.

In all cases an analysis of such an installation involves the element of comparison, which in turn necessitates the use of a standard for the purposes of such comparison. If the standard of comparison is a poor one, naturally the test being compared will show better than if the standard of comparison is a good one. To me, the entire matter of these pulverized-fuel tests hinges upon the standard of comparison used and the deductions that are drawn from such comparison.

In this case the underfeed stoker is that used as the basis, and therein lies a possible opportunity for a complete revision of engineering opinion.

As Mr. Thompson rightly says, there are certain "losses that are inherent in stoker practice, such as breakdowns in the stoker itself, breaking up clinkers, loosening clinkers, continually watching the fire to maintain correct and uniform thickness, watching the gas passes of the boiler for large sparks indicating the carrying away of combustible, dumping, and many other operations are necessary to stoker practice."

The trouble with the engineer of today is that he assumes that the underfeed stoker, with all these inherent defects, is the proper and sole basis of comparison. I have more than once stated that any stoker that possessed the defects of clinker formation, periodic dumping, uneven thickness of fire, fuel-bed disturbance, etc., was fundamentally wrong and could not endure under the fierce light of the competition of today. Under no conceivable circumstances can any stoker, whose fundamental design involves these things, operate at such high rates of combustion as to produce melting temperatures in the furnace or fuel bed and give satisfactory results. Modern practice demands rates of combustion much higher than can be secured without the formation of clinkers in the underfeed stoker, and incidentally much higher capacities than the test at Oneida Street proves is possible without the production of destructive furnace temperatures.

With the prices of power-plant equipment as they are today, boiler-heating surface must be worked to a very much higher degree than ever before, and rates of steaming corresponding to 150 per cent of rated capacity are going to be considered as a thing of the past. Without any doubt the modern boiler will continuously, safely, and economically generate a boiler horsepower from 3½ sq.ft. of heating surface, and furnace temperatures up to 2,700 or 2,800 deg. F. must be endured if the right efficiency is obtained. Any coal-firing system which will not stand up under these requirements is wrong and cannot last.

Instead of comparing the Oneida Street tests with an underfeed stoker producing a gross efficiency of 76 per cent, it will be necessary in the future to compare it with a method of stoking which will produce a gross efficiency of 80 per cent and without the interruptions to efficiency resulting from the before-mentioned inherent stoker losses. Someone will then have to do some figuring to justify the additional cost of $20 per boiler-horsepower.

I suggest that before long we will have an opportunity of making a comparison of these tests with a stoker that can produce continuously 300 per cent of rating without clinker formation, periodic dumping, breakdowns, or the attendance of highly-skilled operating help. JOSEPH HARRINGTON.

Chicago, Ill.

California Boiler-Law Requirements

On page 697, of the May 4 issue of *Power*, there is shown a drawing of California boiler-law specifications in which the lower connection to the water column is shown entering the shell at the second row of tubes from the bottom. I have yet to see a water column so connected or to meet an engineer who would advocate such a connection.

We keep the lower connection as high as between the first and second rows of tubes from the top in order to make the columns of water that balance each other on the inside and outside of the boiler as short as possible.

Increasing the length of these balancing columns by lowering the bottom connection increases the inaccuracy of the water level shown by the glass, because the boiling column of water inside the boiler is lightened by its content of steam and will stand higher than the outside column, in the connections and mounting, which is solid water.

If at a certain state of intensity of ebullition, 13 in. of water inside the boiler balances 12 in. of water in the outside connection, then it would require 26 in. to balance 24 in. and 39 in. to balance 36 in.

Thus, by lowering the bottom connection until a 1-ft. column was lengthened to a 3-ft. column, we have increased the inaccuracy of the water level showing in the glass from 1 inch to 3 inches. R. MANLY ORR.

Vancouver, B. C.

Plugs in Water-Column Connections

In the article, "Three Power Plant Kinks," page 820 of the May 18 issue of *Power*, Mr. Little says that he would apply two long brass plugs to the water-column connection.

If the A.S.M.E. Code and other codes are consulted, it will be found that the lower water-column pipe must be provided with a drain cock or valve, with a suitable connection to the ashpit or other safe point of waste.

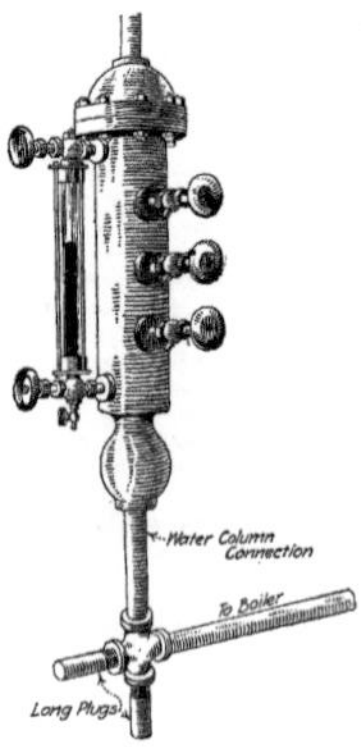

PLUGS IN WATER-COLUMN CONNECTION

The water-column blowoff pipe should be at least $\frac{3}{4}$ in. The object of this valve is to drain the water column as more or less sediment is bound to collect in the bottom and, if it is not drained off, will plug the water-column pipe. CHARLES W. CARTER, JR.

Olean, N. Y.

Why the Piston Would Not Come Out

Several years ago it was my experience to wonder why we could not remove the piston of 30 in. diameter low-pressure cylinder. From the start this engine ran poorly and the cylinder began to cut and score. We laid the trouble at first to poor lubrication, but later came to the conclusion that the two metals of the piston and cylinder were not homogeneous. It was a case of take the piston out at least twice a week, and oil by the quart had to be pumped into the cylinder at starting and stopping.

Finally, it was decided to rebore the cylinder and put a dovetailed ring of babbitt metal around the piston. The load was transferred and the low-pressure head was removed. The locknut on the crosshead end of the piston rod was loosened, and a socket wrench was fitted to the nut on the piston rod, at the piston end, but the piston rod would not unscrew. We made a clamp for the rod and used the trolley falls on this clamp and pulled at the same time on the socket wrench, but without avail; the piston rod still refused to budge. Up to this time we had probably removed that rod at least twenty times without trouble.

Some wisehead in the crowd suggested that we turn the engine just as slowly as possible by means of water power (of which we were well supplied) and try to unscrew the rod during this performance. This worked successfully for a short period. The piston and rod turned easily until we got the piston almost out to the counterbore. Then the man doing the turning was told to stop, but he gave just one more pull for good measure and the ring dropped over the counterbore. The piston kept right on going and, consequently, a portion of the outside ring groove was broken off.

We could not unscrew the rod any more, and so after measuring the amount already gained, we decided to saw the rod off close to the crosshead. With the rod clear from the crosshead it was only a few minutes' work to get the piston out, and then we found the cause of our trouble.

As the rings rattled at times, we had concluded that they turned and that when the opening came next to the steam port, the steam got in the ring opening and caused them to slam. We had, therefore, put a block in the ring groove at the bottom of the piston to keep the ring from turning.

As this stopped the noise, we concluded that our theory had been right, but this block in the bottom of the piston was what caused our trouble.

The ring next to the crosshead had broken and when we tried to turn the piston the ring rode upon itself

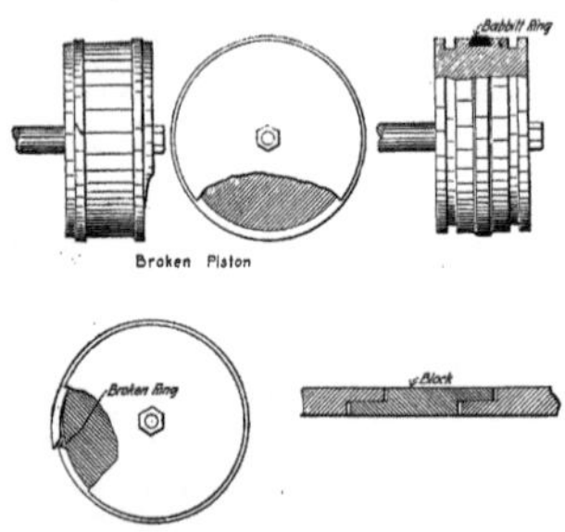

BROKEN PISTON, BABBITT RING, BROKEN RING AND WOOD BLOCK

and formed a grip on the cylinder. The scores in the cylinder engaged the broken ring, and so we could not turn the piston. After the cylinder was rebored and the dovetailed ring of babbitt metal was in place, everything worked well. The illustration shows the defects mentioned.

The piston rod was a trifle short, but as it had been fully long enough, we concluded to use it and it worked all right. We welded the broken piston.

New York City. C. W. PETERS.

Fitting Crank-Rod Brasses

A smooth running engine, operating with maximum efficiency and at a minimum cost requires that all repairs and adjustments made to the wearing parts be skillfully carried out and with due consideration to the alignment of the various parts, which, if neglected, may result in a heated bearing, heating and pounding of the crankpin journal, or possibly a knocking in the cylinder.

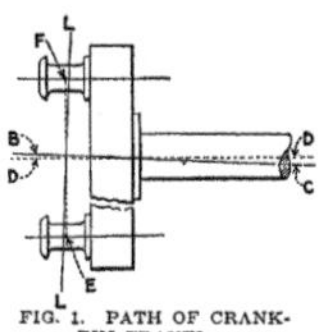

FIG. 1. PATH OF CRANK-PIN TRAVEL

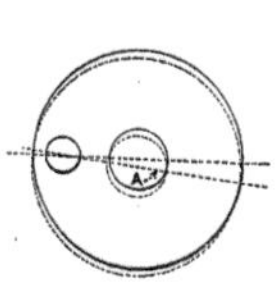

FIG. 2. DROP OF CRANK SHAFT

Suppose that a knock developed in the main bearing of the engine and the engineer adjusted the brasses to take up lost motion without consideration being given to the alignment of the crankshaft. If the bearing had worn low enough to cause the crankshaft to drop to the position of *A*, Fig. 2, there would be trouble in time with the crankpin brasses. The effect of such a drop will be seen by referring to Fig. 1, where *BC* represents a line through the center of the crosshead pin. The line *DD* represents the center line of the shaft, which is out of the parallel with the line *BC*, due to a worn crankshaft bearing. The line *EF* represents the center line of the crankpin when in its top and bottom or half-stroke position. Line *LL* is a line at right angle to the crosshead pin represented by the line *BC*. The line *LL* being at right angle to the line *BC* would represent the path of rotation of the crankpin were the crankshaft in horizontal alignment, but as the crankshaft is not in alignment, that is, *DD* and *LL* are not at a right angle one to the other, the path of rotation of the crankpin is from *F* to *E*. The center point *F* of the crank thus falls on one side of the line *LL* at the top position and on the opposite side of the line *LL* in its bottom position, as represented by the point *E*. If we couple up the crank rod to the crosshead pin and put the crankpin brasses in the rod end, but not coupled up to the pin, and lower the end of the rod onto the crankpin, it will be found that the flanges of the brasses will fall to one side of the crankpin when the crank is in its top position, as shown in Fig. 3, and will fall on the opposite side end of the crankpin when the crank is in its bottom position, as shown in Fig. 4.

Suppose the foregoing conditions actually exist and the connecting rod is connected up, then the rod would be twisted twice for each revolution of the crankshaft. On rotating the engine the straining would gradually diminish as the dead center was reached and then increase again on passing the dead center. This binding of the brasses would cause heating of the crankpin brasses, and if there were side play pounding would result.

Suppose that in making the vertical adjustment to the crankshaft the engineer had inadvertently disturbed the horizontal alignment of the shaft, throwing it out of its original or normal position *G*, into the position shown by *H*, Fig. 5, heating of the crankpin brasses would be the result. This will be understood by referring to the illustration, where *II* represents a true line through the cylinder bore and guide, *G* represents the normal position of the crankshaft and *H* the position of the shaft which is not at a right angle with the line *II*. The line *J* represents a line through the center of the crosshead pin which is in alignment or at right angle to the line *II*. *K* is a point central to the length of the crosshead pin. If the crosshead end of the connecting rod is uncoupled, the rod will spring to the position shown by the line *M* when the crank is on the dead center. If the crank end of the connecting rod is uncoupled from the crank pin and then lowered, the rod will fall to one end of the crankpin as shown in Fig. 6 and on the other end of the pin as shown in Fig. 7 when tried with the crank on the other dead center with the crankshaft out of alignment, as shown by the lines *A* and *B*. With the connecting rod coupled up under the foregoing conditions, the rod would be bent or sprung, and as a result heating of the connecting-rod brasses would be sure to follow.

Serious trouble may be caused by the faulty fitting in of new connecting-rod brasses or owing to faulty repairs to the existing brasses. Suppose that a pair of new

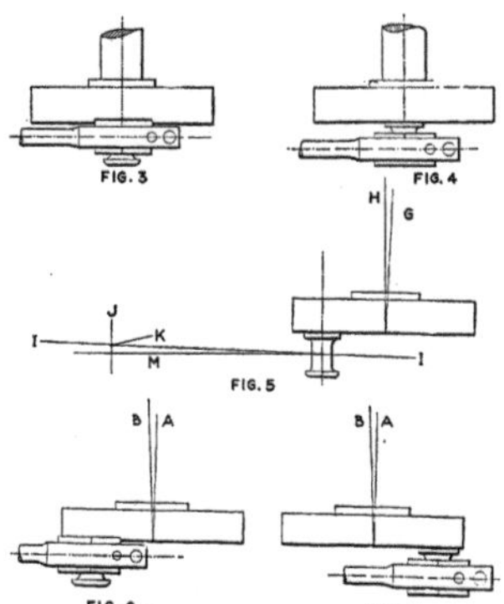

FIGS. 3 TO 7. SHOWING NATURAL POSITION OF CRANK IN RELATION TO FAULTY ALIGNED CRANKSHAFT

brasses are to be fitted to the crank end of the connecting rod, the side and inside faces of the rod should first be lightly smeared over with red lead. The brass should then be fitted in the rod, interposing a piece of hard wood so as not to damage the brass by the blows of the hammer when knocking the brass in and out of the rod end. When the brass has been partly fitted in, it should be tested with a try square *S*, as shown in Fig. 8, to ascertain if the fitting is being done true and square with the point face. It should also be tested with calipers at *O* and *P* for parallelism, as it is possible that in fitting the brass it may be too loose on the inside face of the rod, as shown at Q, Fig. 9. An additional blow with the hammer, however, may cause the brass to tilt over to the side and the red lead marking may thus give a false indication of the fit. The brass must also have a true bed at the bottom of the rod, for if the fit at this point is neglected and is as shown at R, Fig. 10, then, when the wedge is adjusted the brasses will spring in

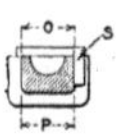

FIG. 8. SQUARING BRASSES

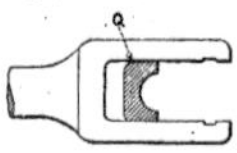

FIG. 9. FAULTY TOP FIT

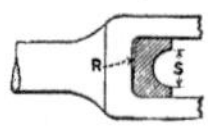

FIG. 10. FAULTY END FIT

at the point S thus gripping the crankpin and causing it to heat. When properly fitted both brasses must come together true and square with the point faces. After the brasses are bedded to the crankpin and the rod has been adjusted to the crankpin it should be tested for alignment with the crosshead, and if the rod end falls true and square in the jaws of the crosshead the rod is in alignment.

In making repairs to the existing crank brasses the rod may be thrown out of alignment. Suppose the engineer finds, on examination of the brasses, that the bearing faces are cut or scored and then proceeds to file out the cuts, but in so doing he may not keep the bearing face true with the original base, or true with the side faces of the brass.

Should this occur and the rod be coupled to the crankpin, the other end of the rod would not fall true in the jaws of the crosshead, and if the rod is coupled to its bearings heating of the brasses will result.

A knocking may also develop through faulty work on the crank-end brasses. It sometimes happens that a ridge has formed at the end of the crosshead travel and also in the cylinder at the end of the piston travel. When reducing the brasses to take up the play, the engineer may find it necessary to line up the brasses, and if due consideration is not given the length of the connecting rod may be altered, thus causing the crosshead and piston to travel farther at one end of the stroke, striking the ridge and setting up a knocking. It thus becomes obvious that repairs to the existing plant should be skillfully executed and the alignment of the various parts kept intact.

When making repairs to the main bearing the following method may be used for testing the horizontal alignment: Take an adjustable spirit level and lay it on the guide path if it be a flat face or on a machined part on the cylinder top and adjust the level until the bubble stands in the center. Then lay the level on the crankshaft and adjust the bearing until the two levels coincide. For testing the horizontal alignment the connecting rod may be used if it is known that the bore of both end brasses are parallel with each other.

Uncouple the large end from the crankpin, putting back the brasses in the rod, and tighten up the crosshead end adjustment so that the rod can just be moved and lower the rod end on the crankpin with the crank on the dead center. Then try the rod in the same manner with the crank on the other dead center and make the side adjustment of the main bearing brasses until the rod falls true between the collars of the crankpin when tried on both dead centers.

The method suggested for aligning in case of repairs to the main bearing must *not* be taken as the correct method of aligning, but as a temporary measure, the correct and true aligning of the engine being left to a more convenient time when the engine is shut down for general repairs.

THOMAS W. AIRCY.

Lincolnshire, England.

Organized Power-Plant Employees

I have read some good articles on power-plant organization, but I think something more might be said along such lines. Having gotten the plant force on the job, they all start in on their individual work, each trying to do his best until someone starts doing things. The word is passed along, and soon there is bad feeling between this and that one.

This is one of the worst things that can get into an organization of any kind, especially in power or industrial plants. Co-operation between the men on the job and the chief engineer, chief electrician and master mechanic is of the first importance, because without it the strongest organization will eventually break down. If the chief and the master mechanic are not pulling together, the men soon realize it and things go from bad to worse.

Sometimes a man who has had a limited amount of actual experience is placed in charge of a plant, and he is sure to have his troubles unless he has the co-operation of the heads of his several departments. He should take them into his confidence regarding the working of the plant.

Any man who has worked himself up from fireman to the position that enables him to obtain his "Gold Seal" from the Massachusetts District Police, giving him the right to have charge of any steam plant in the state, has a right to expect some degree of confidence from his superior.

I am sure that if the one in authority were to discuss with his men the changes that he expects to make and how they might affect the operation of the plant, thus giving his men a part in a real man's job, he would find that it would enhance the good feeling of all concerned, if nothing more.

I know that most engineers and power-plant workers are students and that they are right with the work all day and all night seven days a week, and furthermore, they are close observers of cause and effect and their good will is worth having. A chief engineer is in charge of a plant and is expected to produce results, therefore, he should be consulted regarding changes that are to be made. We all like to say "I did it," but it is much better to have the conditions such that the men can say "we," for one man cannot accomplish much alone. It takes all the fellows from the coal passer to the manager to keep things going right. It does not matter what your rating is, it is what you can put into the plant organization that counts. One weak link in the chain has caused the failure of many strong organizations, but when all pull together, each in his place, and each one is made to feel that he is a part of the whole organization, there will be ideal conditions.

Webster, Mass. M. W. CARTEL.

INQUIRIES OF GENERAL INTEREST

Relative Length of Pump Cylinders—Why is the water cylinder of a pump made longer than the steam cylinder?
J. P.

The water cylinder of a direct steam pump needs to be longer than the steam cylinder so there may be ample space for the water passages without obstruction by the water piston when the steam piston is at either extreme end of its cylinder.

Size of Flue for Return-Tubular Boilers—What should be the size of a circular steel flue chimney connection 60 ft. long for two 72 in. diameter x 18 ft. return-tubular boilers?
W. N. F.

A safe rule to follow in figuring flue areas is to allow 5 sq.in. of cross-sectional area per boiler horsepower, and assuming that 300 boiler horsepower is to be provided for, the cross-sectional area should be 1,500 sq.in., or practically 44 in. diameter.

Gallons per Inch Depth of Vertical Sided Tank—What number of gallons is contained per inch depth of an oil tank 10 ft. ½ in. in diameter with vertical sides, and what should be the diameter to contain 50 gallons per inch of depth?
J. L. P.

One U. S. gallon is 231 cu.in., and the tank 10 ft. ½ in., or 120.5 in. in diameter would contain $120.5 \times 120.5 \times 0.7854 \div 231 = 49.37$ gal. per inch of depth. To contain 50 gal. per inch of depth the diameter would need to be $\sqrt{231 \times 50 \div 0.7854} = 121.26$ in., or practically 10 ft. 1¼ in.

Rotary Converters in Parallel—In our plant we have two rotary converters; one machine is a six-phase and the other a three-phase unit, operating at 600 volts on the direct current side. Can these two machines be connected in parallel?
C. T.

There is no reason why these machines cannot be connected in parallel. Of course the connection will depend somewhat upon their type and method of voltage regulation. It will also be necessary to provide protection to prevent one machine from motorizing the other. This is generally accomplished by equipping the circuit breakers so they can be tripped by reverse-current relays.

Advantages of Diagonal Over Head-to-Head Stays—What are the advantages of diagonal stays over head-to-head stays in a return-tubular boiler?
W. N. E.

The shell has sufficient strength to resist all the pressure on the areas of the heads required to be stayed above and below the tubes, and that strength can be utilized by the employment of diagonal stays with the advantage of occupying comparatively little space in the steam room, thus permitting easier inspection and cleaning than when through bolts are used; and diagonal stays being shorter than through stays, there is less effect from unequal expansion.

Noisy Exhaust Valves—When our 16 x 36-in. Corliss engine is lightly loaded, there is a slapping noise from the exhaust valves. How can the trouble be prevented?
T. N. B.

Undoubtedly, steam-engine indicator diagrams would show that when carrying the light load, the cutoff is very short with expansion carried to a pressure below the atmosphere. When this occurs, the exhaust valves are forced from their seats by pressure of the atmosphere causing a slapping or clattering noise, as when an engine is being shut down. The remedy is to throttle the steam, or lower the boiler pressure, and obtain cutoff so late in the stroke that the steam will not be expanded below atmospheric pressure.

Induction-Motor Rotor Connections—I have rewound the rotors of a number of wound-rotor-type induction motors and in each case the winding was connected star. Can the delta connection be used in grouping these windings? If so, why is it not used more generally?
M. T. R.

Either connection may be used, but in small and medium motors the star connection gives a more satisfactory winding design. The star connection gives a terminal voltage 73 per cent higher at the slip rings than the delta connection. This is very desirable where the rotor is bar wound, since with this type of winding the voltage is comparatively low, and the current relatively high value. With the star connection the current will be 58 per cent of that obtained with the delta connection. Consequently, the starting resistance is much cheaper to construct. In some large-sized motors, however, the voltage produced in the rotor winding with a star connection may be so high as to be difficult to insulate and also require specially designed central equipment to make them safe for an operator to handle. In such cases the rotor is connected delta, as this grouping generally gives a voltage at the slip rings that is of a safe value.

Pull and Power Required to Draw Car Up Incline—In a balanced monitor incline system it is required to raise a load of 7,500 lb. up an incline 750 ft. long, a total elevation of 150 ft. The speed of rope carrying the load is 600 ft. per min. There will be two cars, one being pulled up the incline loaded, while the other returns empty. Neglecting friction, air resistance and weight of cars, what will be the pull on the rope and what horsepower will be required?
A. F.

When the car is held at rest or is being drawn up the incline at a uniform speed, the pull on the rope would amount to $\frac{150}{750}$ of 7,500 lb. = 1,500 lb., and while drawn up the incline at the uniform speed of 600 ft. per min., the power required would be $600 \times 1,500 \div 33,000 = 27.27$ hp. For moving the car up the incline from a state of rest, there would be an additional pull on the rope, depending on the acceleration. If the time allowed for increasing the speed from 0 to 600 ft. per min., or 10 ft. per sec., was 15 sec., with uniform acceleration, then the acceleration would be $10 \div 15 = \frac{2}{3}$ ft. per sec. Forces are to each other as the accelerations produced. The acceleration due to gravity is 32.16 ft. per sec., and therefore the force required to produce the acceleration only would be $7,500 \times \frac{2}{3} \div 32.16 = 155.4$ lb. and the whole pull on the rope while raising the speed of the car to 600 ft. per min. would be $1,500 + 155.4 = 1,655.4$ lb.

[Correspondents sending us inquiries should sign their communications with full names and post office addresses. This is necessary to guarantee the good faith of the communications.—Editor.]

Interpretations of the A. S. M. E. Boiler Code

The following answers to requests for interpretations of the Boiler Code have been prepared by the Boiler Code Committee and approved by the Council of the American Society of Mechanical Engineers. The formal inquiry and reply are given together with explanatory references to the Code and illustrations where necessary for a better understanding of the interpretations as issued.

Case No. 273—Inquiry: Does Par. 296 of the Boiler Code require that the tee or lever-handled cock be placed immediately under the steam gage where a long connecting pipe is used and the locked open valve is permitted close to the boiler?

Par. 296 provides that the connection of the steam gage shall be so arranged that the gage cannot be shut off from the boiler except by a cock placed near the gage and provided with a tee or lever handle arranged to be parallel to the pipe in which it is located when the cock is open. When the use of a long pipe becomes necessary, an exception to this rule is provided so that a shutoff valve or cock may be used near the boiler if arranged so that it can be locked or sealed open.

Reply: It is the intent of the requirement in Par. 296 that the tee or lever-handled valve shall be located near to the steam gage so that it will be readily evident to anyone observing the gage, even though the locked-open valve is used at or near the boiler.

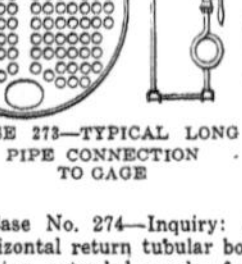

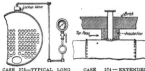

CASE 273—TYPICAL LONG PIPE CONNECTION TO GAGE

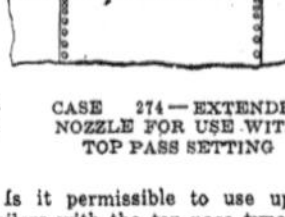

CASE 274—EXTENDED NOZZLE FOR USE WITH TOP PASS SETTING

Case No. 274—Inquiry: Is it permissible to use upon horizontal return tubular boilers with the top pass type of setting, extended nozzles formed of a short length of pipe screwed into flanged fittings at the boiler connection and the outer end, in order that it may reach well above the boiler brickwork?

Reply: The construction proposed will not meet the Code requirements, where the pipes are over 3 in. pipe size and the working pressure exceeds 100 lb. per sq.in. A double-flanged nozzle riveted to the boiler shell should be provided, and such a nozzle, when used, should be protected by insulation.

Case No. 276—Inquiry: In the design of a cast-steel water box to be set in the side walls of furnaces and to be subjected to full boiler pressure, is it necessary to apply to the sections containing flat surfaces, the formula in Par. 199 of the Boiler Code, using $C = 120$, or is it necessary to use this formula with the value of $C = 156$?

Reply: It is the opinion of the committee that the construction of pressure parts of the type referred to is provided for by Pars. 9 and 247. If the secretary of the Boiler Code Committee can be notified when the specimen is ready for test, steps will be taken to have a representative of the committee present.

Par. 9 provides for the use of wrought steel or cast steel of Class B grade as designated in the Specifications for Steel Castings for water boxes when the pressure exceeds 160 lb. per sq.in. Par. 247 provides that where it is impossible to calculate with reasonable accuracy the strength of any part of a boiler structure the manufacturer shall build a full-sized sample which shall be tested to destruction in the presence of the Boiler Code Committee or one or more of its representatives.

Case No. 277—Inquiry: (*a*) An interpretation is requested of the application of Par. 212*a* of the Boiler Code with reference to any curved stayed surface subject to internal pressure. Does this refer to both the outer and furnace sheets of vertical tubular boilers?

(*b*) If under the requirements of Par. 239 of the Boiler Code relative to furnaces under 36 in. in diameter, the design of the furnace does not permit of operation without staying, is there any rule in the Code for the staying in this case, or must the furnace sheets be stayed as flat surfaces, in accordance with Table 4 of maximum allowable pitch of screwed and riveted stay bolts?

Par. 212*a* provides for computing the working pressure of a curved stayed surface by making the proper deductions for the weakening effect of the staybolt holes. To this is to be added the pressure allowed for a flat stayed surface, computed both from the stiffness of the plate and from the strength of the stays, the minimum value being used.

(*c*) Is it the intent of Par. 212*c* of the Boiler Code that the increased pitch allowed, may be used for the same working pressure and thickness of plate as indicated in Table 4? It is the understanding that Table 4 is based on the formula given in Par. 199.

A formula for the allowable pressures on stayed surfaces based on the thickness of the plate and pitch of the stays is given in Par. 199, and values of the maximum allowable pitch of screwed stays for various pressures and plate thicknesses as computted by this formula are given in Table 4. Par. 212*c* provides for an increase in the allowable pitch for cylindrical furnaces over that indicated by Table 4, this increase to be determined by a formula based on the pressure, radius of the furnace and thickness of the plate.

Reply: (*a*) The term "curved stayed surface subjected to internal pressure" in Par. 212*a* of the Code is intended to refer to any surface in a boiler structure that is subjected to pressure on the concave side. It therefore refers only to that part of the outer shell of a vertical boiler which is stayed.

(*b*) Where a furnace under 36 in. in diameter requires taying, it should be stayed as a flat surface as provided or in Table 4, except that the pitch may be increased as indicated in Par. 212*c*.

(*c*) It is the intent of Par. 212*c* that the increased pitch there permitted, may be used for the same steam pressure and thickness of plate as specified in Table 4.

Case No. 279—Inquiry: Is it permissible under the requirements of the Boiler Code to construct steel heating boilers in diameters up to 72 and 78 in. with shell plates ¼ in. in thickness? Is it the opinion of the committee that boilers of the h.r.t. type so constructed have sufficient shell plate strength to permit of proper lug attachments for support of the shell?

Reply: The section of the code on Heating Boilers is incomplete in not specifying the minimum plate thickness to be used in the shells and tube sheets, and below which the thickness shall not be made in any case. The Boiler Code Committee recommends that until a revision be made to cover this class of boiler construction, the minimum thickness of shells and heads for various shell diameters of steel plate heating boilers, shall be as follows:

Diameter of Shell, Tube Sheet or Head.	Minimum Thickness Allowable under Rules	
	Shell, in.	Tube Sheet or Head, in.
42 in. or under	1/4	5/16
Over 42 in. to 60 in.	5/16	3/8
Over 60 in. to 78 in.	3/8	7/16
Over 78 in.	7/16	1/2

Case No. 280—Inquiry: (*a*) Is it permissbile under the requirements of the Boiler Code to use a blowoff valve of the type used on locomotive boilers, operated by a lever lift, for stationary boilers?

(*b*) Is it considered safe to use superheated steam of 125 lb. pressure and 125 deg. of superheat, or total nominal temperature of 478 deg., with a piping system having extra-heavy cast-iron fittings and medium-weight cast-iron valves?

Reply: (*a*) It is the opinion of the committee that the blow-off valves required by Par. 311 for stationary boilers, may be of the lever-lift type, provided they are of extra-heavy construction, and so designed that they may be operated without shock to the boiler.

Par. 311 provides that all boilers, except those used for traction or portable purposes, when the maximum allowable working pressure exceeds 125 lb. per sq.in., each bottom blowoff pipe shall have two valves or a valve and a cock, both of which shall be extra-heavy.

(*b*) Attention is called to Par. 12 of the Code, which states that "cast-iron shall not be used for nozzles or flanges attached directly to the boiler for any pressure or temperature, nor for boiler and superheater mountings such as connecting pipes, fittings, valves and their bonnets, for steam temperatures of over 450 deg. F." While the Code only covers the parts therein specifically mentioned, this paragraph clearly indicates the judgment of the committee as to the safety of the construction in question.

Case No. 281—Inquiry: Is it the intent of Par. 306 of the Boiler Code that every superheater shall be so fitted with a drain that it can actually be completely drained? Many superheaters are fitted with drains which are, however, unable, on account of their positions, to completely drain the apparatus.

Par. 306—Each superheater shall be fitted with a drain.

Reply: It is the opinion of the committee that every superheater should be so fitted with a drain as to substantially free the superheater from water when the drain is opened.

Case No. 282—Inquiry: Is it the intent of the Boiler Code Committee that the diameter at the base of the threads

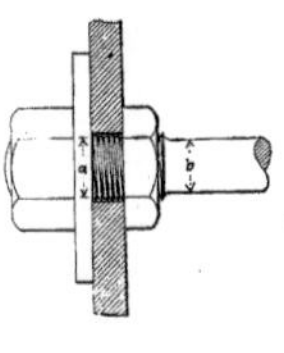

CASE 282—IT IS DESIRABLE THAT *a* BE GREATER THAN *b*

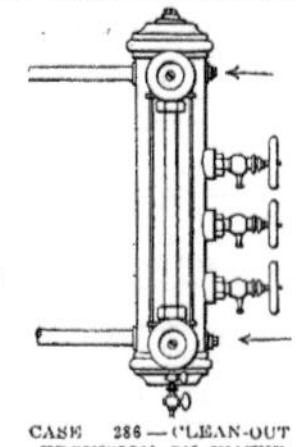

CASE 286 — CLEAN-OUT PROVISION ON WATER COLUMN

on the threaded ends of through rods or braces for h.r.t. boilers, shall be equal to or greater than the nominal diameter of the rod? Fig. 14 and Pars. 208 and 211 would seem to infer that it should be at least equal, but isn't it preferable to make it greater, so that the point of greatest weakness in the rod may not be in the threaded portion where permanent set due to strain would tend toward fracture?

Reply: It is the opinion of the committee that it will be desirable that the threaded ends of through rods or braces for h.r.t. boilers, shall be sufficiently upset so that the minimum diameter at the base of the threads is in excess of the nominal diameter of the rod.

Case No. 284—Inquiry: Is it the intent of Pars. 269 and 270 of the Boiler Code that a boiler requiring safety-valve relieving capacity greater than 2,000 lb. per hour shall have two safety valves, each one of which alone will properly discharge the steam, or does it require two safety valves, the combined capacity of which will properly discharge the steam?

Par. 269 provides that each boiler shall have two or more safety valves, except a boiler for which one safety valve having a relieving capacity of 2,000 lb. per hour or less, is required.

Par. 270 provides that the safety-valve capacity for each boiler shall be sufficient to discharge all the steam that can be generated by the boiler without a rise in pressure of more than 6 per cent above the maximum allowable working pressure or more than 6 per cent above the highest pressure to which any valve is set.

Reply: For capacities greater than 2,000 lb. per hour, the Code requires two or more safety valves on each boiler, the combined capacity of which shall meet the requirements of Par. 270.

Case No. 285—Inquiry: Is it permissible under the rules of the Boiler Code to use standard extra-heavy cast-iron flanges and fittings on pipe connections between boilers and attached-type superheaters, and on the ends of superheater-inlet headers for pressures up to 250 lb. per sq.in.? It is pointed out that neither the inlet-pipe connections nor the superheater-inlet flanges would be subjected to other than saturated steam temperatures.

Reply: It is the opinion of the committee that the flanges referred to may be made of cast iron, provided the temperature of the steam does not exceed 450 deg. F., as specified in Par. 12.

Case No. 286—Inquiry: Does Par. 321 of the Boiler Code require that all water-column connections to boilers must be fitted with crosses for cleanout purposes, or is it permissible in a connection of a water column to a steam drum to omit the crosses, provided provision for cleanout is afforded by the gage-glass connection on the opposite side of the water column?

Par. 321 provides that the water connections to the water column of a boiler shall be provided with a cross to facilitate cleaning.

Reply: It is the intent of Par. 321 that a cross be applied at a right-angle turn in the water connection to a column for purposes of cleaning. Where an easily accessible straight pipe connection is used without a turn, the cross is obviously not needed.

Case No. 289 (Revision of Case No. 254)—Inquiry: Does the reply to Case No. 254 apply only to flat plates or also to the holes in the shells of drums?

The reply in case No. 254 provided that burning might be substituted for punching and shearing of rivet holes and plate edges, provided the holes were burned ⅛ in less diameter than the finished hole and that the burned surfaces of caulking edges be planed, milled or chipped for a depth of not less than ⅛ in.

Reply: This case applies to flat sheets. It is not permissible to burn tube or other holes into the shells of drums unless such holes are strengthened with flange or other reinforcements.

Case No. 290—Inquiry: In determining the ratios of lengths between supports to diameter per Table 5, page 59 of the code giving "Maximum Allowable Stress for Stays and Staybolts," is the diameter at the root of the thread to be used, or the diameter of the body of the stay?

Reply: It is the opinion of the committee that the diameter of the body of the stay is to be used in determining this ratio.

Communication

An inquiry was received relative to the permissibility of forming individual or sectional headers of *pressed channel-shaped sections*, welded together at the edges to form rec-

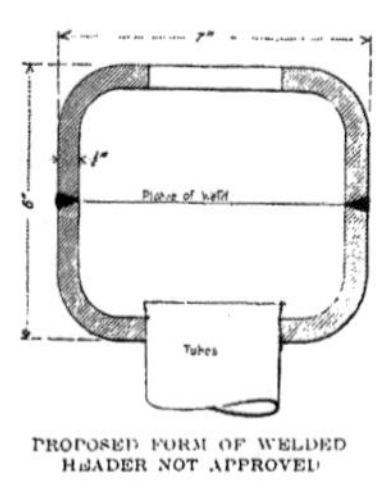

PROPOSED FORM OF WELDED HEADER NOT APPROVED

tangular-shaped headers. It was stated that the minimum thickness of metal was ½ in., and the maximum cross-section 6 in. in depth by 7 in. in width, and it was pointed out that the United States Steamboat Inspection Service approves this construction when the two halves of steel plate are electrically welded, no staybolts being required.

The Boiler Code Committee decided that this construction does not conform to the requirements of the Boiler

Code and directed especial attention to Par. 186, which stipulates that "autogenous welding may be used in boilers in cases where the strain is carried by other construction which conforms to the requirements of the code and where the safety of the structure is not dependent upon the strength of the weld."

Water-Power Development in the United States*

Taking the eleven Far Western States, we find by careful survey an unusual situation. Due to delay of the construction program in power development while the nation was at war, this section of the country (the Pacific Coast) finds itself today facing a serious situation, ranging from an actual shortage of power to meet demands, as in California, to an impending shortage to meet growing demands in Oregon and Washington. With increasing possibilities of coal shortage, the continual rise of the price of oil and the economic desirability for early electrification of steam railroads, the urge for increased water-power development in the Far West becomes imperative. Not only are industries already existing dependent upon electrical development, but it is believed that industrial growth on the Pacific Coast and electrical development are interdependent. This is true in agriculture as well as industry. Vast areas of arid and semi-arid Western lands have been converted into fertile districts under electrical pumping of water that would be impossible without its use.

Federal legislation governing the use of public lands and reservations is of primary importance to the Western States, in which lie nearly all the public lands and in which approximately 70 per cent of the undeveloped water power of the nation is found.

Statistics of Potential Water Powers and Importance of Those Under Federal Control

The report discusses present and potential water powers of the United States from a statistical standpoint. The latest authentic estimate shows that there are approximately 59,360,000 hp. of potential undeveloped water power in the United States and only 9,823,540 hp. of developed water power. Of the potential maximum horsepower of the United States 68.6 per cent lies in California, Oregon, Washington, Arizona, Colorado, Idaho, Montana, Nevada, New Mexico, Utah and Wyoming. Of the total maximum potential water power of the United States 74.3 per cent represents undeveloped horsepower coming under Federal jurisdiction. that is, lying on Federal reserve lands or on navigable streams wherein Federal permits are required. In the Western States 94 per cent of the maximum potential water power is in this class. In other words, practically all of the future development of water power in the Western States, covering approximately 70 per cent of the total potential water power of the country, will be dependent upon Federal action in the matter of issuance of workable permits for the development of these projects. Also it is of considerable note that only 16.6 per cent of the maximum potential water power has been developed up to the present time. For the Western States this proportion is only 6.5 per cent.

In the case of the Central and Eastern States, particularly New England, it is shown that a comparatively large portion of the potential water power has been already developed. In the case of the New England States the high percentage of 77.2 per cent already developed shows that little additional water-power development is to be expected in this district.

While compared with the potential possibilities, the development in the Western States has been comparatively small, attention is directed to the present development of water power in these Western States. For instance, California is shown well in the lead with a present development of 942,000 hp., which is 10 per cent of the entire developed water horsepower of the United States. It is, however, only 10.2 per cent of the maximum potential water power possibilities of that State. Washington shows the largest maximum potential water horsepower of any state in the Union, but it ranks among the Western States below both California and Montana in the amount of developed water horsepower, showing a development of but 3.8 per cent of its potential maximum possibilities and only 3.7 per cent of the total developed water horsepower of the country. Oregon also shows wonderful potential waterpower possibilities, but up to the present time only a very low percentage is developed, 3.5 per cent of the maximum possibilities of the state. These figures clearly show that although the West has already developed some remarkable waterpower projects and is well up in the lead of the development of large interconnecting networks of transmission systems, the water-power development in this section is but begun.

Discussion of Power Consumption by the United States Compared With That in the West

A table is shown giving population, kilowatt-hours, consumption, gross revenue, kilowatt-hours per capita, cost per kilowatt-hour, revenue per capita. The total for the United States shows a population of 109,843,000. The consumption in millions of kilowatt-hours is 36,734.7. The gross revenue in millions of dollars is 773. The kilowatt-hours per capita is 334. The cost per kilowatt-hour is $0.0212. The revenue per capita is $7.05. The figures given by districts show that the per capita use of electricity in the Far Western States is 2.21 times that of the remainder of the nation. Here again it is interesting to note that, while distances are vast, the development of long-distance hydro-electric transmission has made possible electric-power rates that are extremely low, averaging 1.655 per kilowatt-hour as compared with 2.19c. obtained elsewhere in the country.

In potential water power, the State of Washington stands first, with 9,500,000 hp., or 16 per cent of the total potential water power of the United States. California is a close second, with 9,250,000 hp., or 15.6 per cent of the total United States. Oregon ranks third, with 7,100,000 potential hp. The coast states together represent 43.6 per cent of the total potential maximum horsepower of the United States. Including Idaho and Montana, these five typically Western States represent 61 per cent of the total potential maximum water power of the United States with a total of 36,210,000 horsepower.

Reference Made to Contemplated Developments in Eastern States

Over 1,000,000 hp. is contemplated in the states of Alabama, North Carolina, Georgia and Maryland.

Data obtained from 52 power companies of the Far Western States, showing their loads during the past ten years and also their estimated loads up to and including 1928, has been obtained and compiled and presented in the report in connection with graphic charts. They show that during the next ten years the expected development of power in the West will rise to the vast total of 1,776,260 kw., which, at present-day prices for material and labor, will involve an investment of over $700,000,000.

The importance of conservation of fuel oil in the Western States is emphasized in that the Pacific-Coast supply of fuel oil and of petroleum is rapidly approaching exhaustion. Since May 1, 1915, crude-oil stocks in California have decreased from over 60,000,000 bbl. to 28,738,921 bbl. on March 1, 1920. The available supply of crude oil in stock today is less than 13,000,000 bbl. At the present rate of consumption and of production the available stock will be exhausted in about twelve months, at which time consumers of California fuel oil will be cut off from between 25,000 to 30,000 bbl. per day. In California during the year 1918-19 there were two sets of interconnected networks in the state, one in the northern the other in the southern portion. The control of each system was under the management of a power supervisor. As a result of this arrangement there was effected a saving of 300,000 bbl. of oil in 1919, representing $450,000.

---

*Abstract from the report of the N.E.L.A. Committee on Water-Power Development, presented before the forty-third convention, held at Pasadena, Cal., May 18-22, 1920, by Franklin T. Griffith, chairman of the committee, and president of the Portland Railway, Light & Power Company.

Sir Charles A. Parsons, K.C.B.

Perfecter of the Reaction Steam Turbine, Receives the Franklin Medal

THE eminent British engineer and scientist, Sir Charles A. Parsons, perfecter of the reaction steam turbine, has been honored by America, receiving the Franklin Medal. The other recipient also is a European, Svante Arrhenius, of Sweden.

Sir Charles Parsons was born June 13, 1854, as the fourth son of the third Earl of Rosse, who was a president of the Royal Society, and built that famous telescope which is still the largest instrument of its kind.

The son received his early education entirely by private tuition, and later attended Dublin University and Cambridge. At that time (about 1877) he made models of his epicycloidal engine, and while serving a four years' apprenticeship in the Elswick Works of Sir William Armstrong he constructed the first engine of this type. It was a four-cylinder revolving engine in which the cylinders rotated around the revolving crankshaft.

Mr. Parsons then turned his attention to the development of the steam turbine. He conceived the idea of splitting up the fall in pressure over a great number of turbines in series; that the aggregate of these simple turbines, constituting a complete machine, would give an efficiency approximating that of water turbines. In 1884 he patented his first steam turbine. This, of 10 hp., at 18,000 r.p.m., was used for driving a high-speed dynamo. It marked the beginning of turbo-generator development for electric drive. During the next five years he constructed many improved forms of turbo-dynamos. In 1889 he organized the works at Heaton.

The first compound turbine was built in 1887 and the first condensing turbine in 1892, when there was at once a great advance in efficiency, the steam consumption being reduced from 55 lb. to 27 lb. per kw.-hr. In 1895 the steam turbine was first applied to marine work, and two years later the experimental steamer "Turbinia," driven entirely by turbines, was completed. She had a length of 100 ft., 9 ft. beam, 44 tons displacement; speed attained on her trial trip, 34 knots. In 1905 the "Amethyst," driven by turbines, beat her sister ship, the "Topaz," driven by reciprocating engines. From this time on the construction of turbine-driven vessels advanced by leaps and bounds, until the present time, when the turbine has not only completely supplanted the reciprocating engine in war vessels, but to a great extent also, either wholly or partly, in the great ocean liners and mercantile steamers. By connecting the turbines to the propeller by gearing, he has made possible another great stride in turbine propulsion.

Sir Charles Parsons has made valuable contributions and inventions along other lines in physical science, among them improvements in gramophones and investigations on the formation of diamonds.

Sir Auckland Geddes, British Ambassador, received the Franklin Medal for Sir Charles.

The Flow of Air Through Small Brass Tubes*

By T. S. Taylor

The investigations discussed in this paper were undertaken as a preliminary study of the general ventilation problem, dealing in particular with the flow of air through tubes of various sizes and shapes. Information was sought as to the variations in the velocity of air at different points in the cross-section of the tube and at different distances

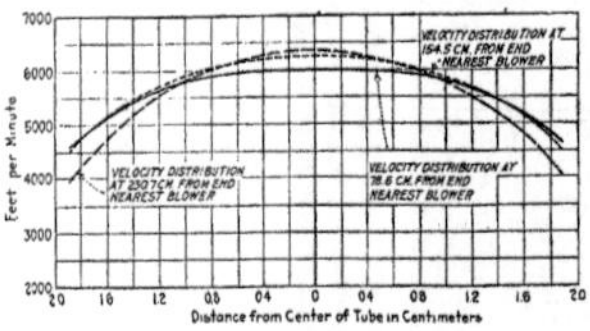

FIG. 1. VELOCITY DISTRIBUTION AT THREE POINTS IN A 1½-IN. TUBE 8 FT. LONG

from the fan. The effect of dirt on the inside walls of the tubes was also investigated.

The tubes thus far tested were of seamless brass having internal diameters of ⅝ in., ⅞ in. and 1½ in., respectively. Velocity measurements were made by means of pilot tubes, those for preliminary study being $\frac{3}{16}$ in. external diameter and those for study of the velocity distribution being about $\frac{1}{16}$ in. external diameter. Velocity pressures were measured by means of inclined oil gages, and the velocity calculated by the usual formula,

$$V = \sqrt{\frac{2\,g\,h\,d}{d'}}$$

in which

V = Velocity of air in feet per second;
g = Acceleration due to gravity;
h = Difference in level in the gage;
d = Density of the liquid in the gage;
d' = Density of the air.

The air current was furnished by a No. 3V blower made by the American Blower Co., and driven by a direct-current motor that could be run so as to give any desired speed to the blower. To prevent vibration being transmitted to the tube from the motor and blower, a short rubber hose connected the blower and the tube under test.

A study of the velocity distribution at various positions along a 1½-in. tube 243.6 cm. (8 ft.) long was made, and the curves in Fig. 1 show this distribution at three points, 78.6 cm. (31 in.), 154.5 cm. (61 in.) and 230.7 cm. (91 in.) from the end nearest the blower. The velocity at the center of the tubes is seen to increase as the distance from the blower increases, being 6,000, 6,200 and 6,270 ft. per min. at the first, second and third points shown, respectively.

The average velocity at a given cross-section in the tube, obtained by multiplying the velocity in each concentric zone by the area of the zone and dividing the sum of these products by the cross-sectional area of the tube, was found to be the same, 5,300 ft. per min., for all cases. These results show that the velocity distribution of air flowing through tubes does not reach a steady state at ten times its diameter from the entrance as is usually assumed. To obtain further information on this point, a second tube 407 cm. (about 13.3 ft.) long and 1½ in. in diameter was used. It was found that the velocity distribution was the same for all points more than 200 cm. (about 79 in.) from the blower end of the tube.

Another interesting point was that the ratio of average velocity to maximum velocity for various velocities was practically constant with a general mean value of 0.847. The velocity measurements used in determining these values were taken at a point 240 cm. (about 95 in.) from the blower end of the 1½-in. tube. Similar observations were made on the smaller tubes, which resulted in a general mean of 0.830 in. in the ⅞-in. tube and 0.870 in the ⅝-in. tube for this ratio. These results indicate that the velocity distribution for steady state, across smooth brass tubes, is practically the same for all the tubes and for all average velocities from 0 to 6,000 ft. per minute.

TABLE I. COMPARISON OF VELOCITY RATIOS IN CLEAN AND DIRTY TUBES.

Dia. of Tube In.	Clean Tube. Ratio of Average to Maximum Velocity	Dirty Tube: Ratio of Average to Maximum Velocity	Approximate Thickness of Dirt Film, In.
1½	0.832	0.749	0.006
⅞	0.833	0.648	0.016
⅝	0.825	0.795	0.008

TABLE II. STATIC PRESSURE IN INCHES OF WATER TO MAINTAIN AN AVERAGE VELOCITY OF 3,000 FT. PER MIN. IN TUBE 5 FT. LONG.

Dia. of Tube In.	Clean Tube	Dirty Tube
1½	0.37	0.94
⅞	0.80	3.20
⅝	1.00	5.00

Having obtained the foregoing data on the air flow through smooth tubes, oil was poured through tubes of 1½, ⅞ and ⅝ in. diameter, each being a little over 5 ft. long. Coal dust was then sifted through the tube and only that allowed to remain which was not readily jarred out. After measurements were taken, the oil and dust were wiped out of the tubes and the thickness of the layer was determined from its mass and density. The value of the thickness thus determined is only approximate as the dirt is not uniformly distributed, but gathers in small clusters. The velocity across each tube was measured for various air velocities and the results obtained summarized in Table I.

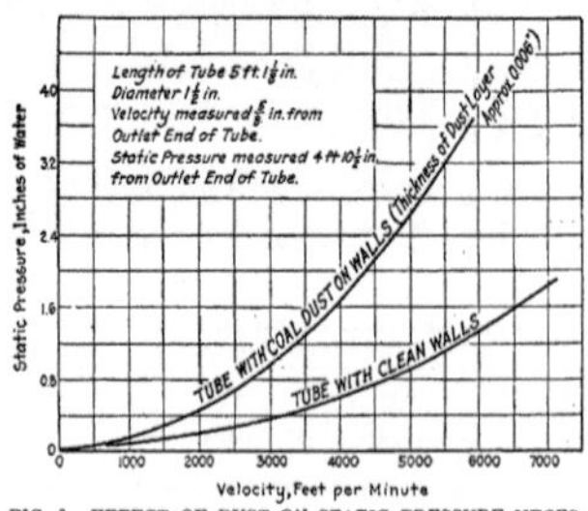

FIG. 2. EFFECT OF DUST ON STATIC PRESSURE NECESSARY TO FORCE AIR THROUGH 1½-IN. TUBE

The static pressure necessary to force air through the 1½-in. tube at various velocities was measured for the clean and dirty tubes, and the results are plotted in Fig. 2. A study of these curves and of Table I indicates that a very small but irregular distribution of dirt will have a decided effect upon the amount of air that will pass through the tube. The effect is not due merely to a diminution of area, but also to turbulency due to the presence of the dust. Table II shows the values of static pressure necessary to force air at 3,000 ft. per min. through clean and dirty tubes

*From a paper presented at the Spring Meeting, St. Louis, May, 1920, of the American Society of Mechanical Engineers.

Coal Industry Profits

Charges and countercharges as to the profits being made in the coal industry, along with rumors of a pending return of price control and of the re-establishment of the coal zones, recently made Washington the center of interest in the coal situation. Senator Walsh of Massachusetts, in the course of a speech on the floor of the Senate, accused coal operators of having made enormous profits. His position was backed up by Senator Kenyon of Iowa. Senator Smoot of Utah attacked that portion of Senator Walsh's charges which were based upon a report as to coal profits that had been submitted by W. G. McAdoo while he was Secretary of the Treasury. Senator Smoot pointed out that many of the deductions as to profits which were made in that report were unfair. The discussion of the matter of coal operators' profits on the floor of the Senate led J. D. A. Morrow, the vice-president of the National Coal Association, to issue a formal statement going extensively into the matter of coal operators' profits. He used in his calculation the figures furnished by the Bureau of Internal Revenue to the President's Bituminous Coal Commission to show that profits, which averaged less than 11 per cent, were justified in a business so hazardous as the mining of bituminous coal.

In addition W. Jett Lauck appeared before the Railroad Labor Board and made charges to the effect that wholesale profiteering is taking place in the coal industry. In that connection, however, it may be pointed out that the statistics of the Federal Trade Commission, which are being fought so strenuously by the coal operators, are just at this time standing them in good stead. Despite the fact that an injunction has been granted restraining the Federal Trade Commission from demanding cost sheets from coal operators, most operators continue to file their costs with the commission, from which certain generalizations are made public. The returns for February show that the average margin per ton throughout the United States was only 31c. This does not represent profit, as selling expense, interest on investment and Federal taxes still must be deducted. In the Central Competitive Coal Fields the average margin per ton was 28c.

With the Interstate Commerce Commission's action in an effort to relieve the traffic congestion came the word that price control is to be restored and the coal zones re-established. No credence is given either of the rumors by those in close touch with the situation. So far as re-establishing the zones is concerned, it is believed that an injunction could be secured which would prevent such an order going into effect, even if it were attempted.

Analysis of the Central Electric Station Industry in Canada

The Dominion Water-Power Branch, Department of the Interior, in co-operation with the Dominion Bureau of Statistics, is publishing an analysis of the results of the second census of central electric stations in Canada, showing the status at Jan. 1, 1919. The report includes statistics relative to central electric stations only, as defined for census purposes; that is, stations that sell or distribute electrical energy for lighting, heating or general power purposes, other than that generated by industrial organizations for their own direct use in the operation of some other industry.

The principal items reported, together with a comparison between the totals for commercial or privately owned and municipal or publicly owned stations, are summarized in a table. The total number of stations reporting is 795, of which 515, or 64.8 per cent, generate their own power, and 280, or 35.2 per cent, are of the non-generating type. The commercial stations numbered 377, and the municipal stations 418. Of the generating stations 332 are commercial and 183 municipal, while of the non-generating stations 45 are commercial and 235 municipal.

The aggregate capacity of all primary power machines reported is 1,958,642 hp., of which 1,841,114 hp. is installed in main plants and 117,528 hp. in auxiliary or stand-by plants. Of the total for the main plants 1,434,196 hp., or 77.9 per cent, was reported by commercial stations, and 406,918 hp., or 22.1 per cent, by municipal stations, while of the auxiliary plant equipment the former accounted for 110,853 hp. and the latter 6,675 hp. The figures are also given separately for main plants and for auxiliary or stand-by plants. According to source of power the total for all prime movers is divided as follows: From water, 1,682,-191 hp.; from steam, 262,562 hp.; from gas and fuel oil, 13,889 horsepower.

The average primary power installation of the main plants per thousand population for the Dominion is 209 horsepower. Of the aggregate capacity of all prime movers installed in the main plants, 1,682,191 hp., or 91.4 per cent, is derived from water, and including the prime movers of auxiliary or stand-by fuel plants the hydraulic installation represents 85.0 per cent of the total. Saskatchewan derives 100 per cent of its central-station energy from fuel.

For the Dominion the total capital invested in the industry was reported as $401,942,402, of which $356,547,217 represents the investment in power development and transmission and distribution systems and $45,395,185 represents miscellaneous supplies and working capital. The commercial stations reported 71.7 per cent of the total capital and the municipal station 28.3 per cent.

For the Dominion the total capital invested in hydro-electric stations is $364,479,961, representing a total investment of $218 per installed turbine horsepower and accounting for 90.7 per cent of the total capital invested in all central electric stations in Canada. As this capital includes investments in fuel power plants, which are operated as auxiliaries to hydro-plants, the capacity of the primary power machines of these plants added to the capacity of the hydraulic turbines gives a more logical basis for this analysis and, reduces the capital investment per installed primary horsepower for the hydro-stations to $203. In connection with enlargements of existing plants the hydro generating stations reported new installations contemplated for the immediate future amounting to 135,755 horsepower. The ultimate designed capacity of existing hydro-electric central stations is 2,115,043 horsepower.

Technical Advisory Committee To Continue Work

The Technical Advisory Committee, who were chiefly instrumental in the settling of more than $3,000,000,000 worth of war contract claims for the War Claims Board, have decided to offer their services to the industries of this country and have opened headquarters at 132 Nassau St., New York City.

In the course of their war claims work they touched every branch of American industry, as the claims ran from small ammunition to the greatest power developments ever made.

Members of the Technical Advisory Committee who have completed their war contract work and are engaged in strictly peace work are: Campbell Scott, New York, president of the new corporation, who is an industrial executive and manufacturing specialist; Ernest P. Goodrich, of New York, vice-president and treasurer, a specialist in waterways and the designer of the Bush Terminal; William D. Ennis, of New York, vice-president and secretary, specialist in fuel, power and manufacturing; Col. Frank B. Maltby, of Philadelphia, who built the railways, docks, terminals and warehouses of St. Nazaire, France, for the A. E. F. and was formerly chief assistant engineer of the Panama Canal; Walter Rautenstrauch, of New York, a specialist in manufacture of machinery and steel products; George B. Frankforter, of Minneapolis, chemical specialist and dye expert; Arthur W. Hixson, of New York, specialist on fuels, heavy chemicals, plant design and general manufacturing; Fred E. Rodgers, of New York, expert machinery designer; and Rumsey W. Scott, of New York, who has had long experience in industrial transportation and the handling of equipment. All will continue to serve the Government in an advisory capacity and will go to Washington whenever needed.

Havana Central Station to Double Power-Plant Capacity

On account of a rapid increase of load and the possibility of a still further growth in the next few years, the Havana (Cuba) Electric Railway, Light and Power Co. has ordered two 25,000-kw. Westinghouse steam turbine-generators to double its plant rating. At present there are, in this plant, three 12,500-kw. generators supplying a combined railway and power load. In the original installation space had been provided for only one additional unit, but now that the load has become so great it was thought necessary to increase the capacity still more. Consequently, the two 25,000-kw. units were ordered, one to be installed in the space available, and the other to replace one of the 12,500-kw. units. By this means it is unnecessary to extend the power-plant buildings. However, the 12,500-kw. unit is in excellent operating condition and will probably be utilized elsewhere.

The first unit will be installed and will be in operation in 1921 and the second one will be on the line in 1922. They are to be designed to operate at 185 lb. pressure, 150 deg. superheat and a 28-in. vacuum. In addition, arrangements have been made to operate this equipment at higher pressures and superheats if desired at a later time. To go with these turbine-generators, Le Blanc 56,000-sq.ft. surface condensers and Le Blanc air and circulating-water pumps have also been ordered. Each condenser will have two circulating pumps and divided water boxes, so that one-half of the condenser may be cleaned while the other half is in operation.

This station, which is the largest in Cuba, is modern in every respect and has an operating economy comparable with that of some of the best plants in the United States. The average fuel consumption is less than two pounds of coal per kilowatt-hour, notwithstanding that the plant has a low load factor. This economy is due to the excellent operation and maintenance afforded by the management and operating engineers of the plant.

Convention of the American Order of Steam Engineers

The thirty-fourth annual convention of the American Order of Steam Engineers was held at Baltimore, Md., during the week beginning Monday, June 7, with headquarters at the Stafford Hotel.

There were seventy delegates in attendance, representing the various local councils of the association. The several sessions of the Engineers were held in Lehmann Hall, in close proximity to the headquarters; for this convenience, the local committee of arrangements is to be congratulated. Many important resolutions were brought before the body for discussion, and harmony was the prevailing keynote throughout the meetings of the delegates in their deliberations.

The main hall of the building, which immediately adjoined the meeting room of the delegates, was tastefully decorated and arranged for the American Supplymen's Association for the display of power-plant machinery and equipment.

The exhibition was officially opened on Monday evening at eight o'clock. Porter C. Jones, of the Dearborn Chemical Co., president of the Supplymen's Association, officiated, and in an appreciated speech told of the aims and objects of the American Order of Steam Engineers and the close friendship that has always existed between that organization and supplymen. Mr. Jones then introduced A. S. Goldsborough, director of the civic and industrial bureau of the Merchants and Manufacturers Association of Baltimore, who cordially welcomed the convention to the city. He spoke of the beauties of Baltimore and its great number of industrial and manufacturing plants and invited the visitors to inspect some of them. He dwelt on the increase of the merchant marine service of the city, which has risen from twenty-three steamship lines two years ago to more than fifty lines at the present time sailing to all ports of the globe. Mr. Goldsborough closed by inviting the convention to visit the offices of his association. W. Scott Price, president of the American Order of Steam Engineers, then declared the exhibit hall officially opened.

On Tuesday evening, at the exhibition hall, an enjoyable vaudeville performance was given under the auspices of the supplymen.

On Wednesday afternoon special cars were boarded for a crab feast at Liberty Park. The event was made pleasant by a baseball game and other outdoor sports and dancing.

On Thursday afternoon, by invitation of the American Order of Steam Engineers, a large number of the wounded soldiers, who are being trained in engineering and mechanics at the Maryland Institute under the direction of the Federal Vocational Commission, visited the exhibition. After inspecting the display they entered the meeting hall of the Engineers, where they were warmly welcomed and entertained.

On Thursday evening at nine o'clock there was an entertainment by home talent in the exhibition hall.

At the closing session the following supreme officers were elected for the ensuing year: W. Scott Price, chief engineer; George T. Crum, first assistant engineer; Harry J. Dunn, recording engineer; W. S. Wetzler, corresponding engineer; W. R. Smith, treasurer; Thomas H. Higgins, senior master mechanic; Harry Oler, junior master mechanic; William Stahler, outside sentinel; John Crean, inside sentinel; N. J. Feely, chaplain; George W. Goodwin, trustee.

The election of the officers of the American Supplymen's Association resulted as follows: William Lindenfelser, Jr., Texas Co., president; Roy C. Downs, Lord's Boiler Compound, vice president; Andrew Lauterbach, Lunkenheimer Co., secretary; John W. Armour, *Power*, treasurer; Maurice L. Willetc, Joseph Dixon Crucible Co., director of exhibits.

Convention of the New York State Association N. A. S. E.

The twenty-fifth annual convention of the New York State Association of the National Association of Stationary Engineers was held at Niagara Falls, on Friday and Saturday, June 11 and 12, with headquarters at the Temperance House.

Chairman Robert Craig and his committee are to be congratulated upon the excellence of the arrangements. The large lobby was suitably decorated for the use of the supply men in the display of their various supplies for the engine room and power plant. In the large tea room, immediately adjoining, the meetings of the delegates were held. Forty engineering supply firms occupied booths, and much interest was shown in the exhibit.

At the opening session of the delegates the Credential Committee reported that there were 54 delegates present. Several important measures were brought before the body, and great interest was shown by the delegates in every discussion. It was decided to start a big drive immediately to boost the membership of the N. A. S. E. in New York State, and for this purpose an Eastern and a Western propaganda committee were appointed and a fund was set aside to cover the expenses of this campaign. The business of the convention was conducted with harmony and dispatch. The treasurer's report places the organization in a sound financial condition.

The opening ceremonies took place on Friday morning. Robert Craig, chairman of the local committee, presided and introduced Rev. C. W. Walker, who gave the invocation. Maxwell M. Thompson, Mayor of Niagara Falls, cordially welcomed the delegates and guests. He expressed the desire that their visit be profitable and pleasurable, and the hope that they would come again in the near future. Past National President William J. Reynolds responded in a fitting manner to the hearty welcome of the Mayor. The gavel was then passed over to State President Robert Tobin, who appointed the necessary committees, after which the meeting adjourned until the afternoon session.

During the two days there were auto sightseeing drives for the ladies to places of interest in the city. On Friday afternoon the delegates inspected the hydro-electric plants. On Friday evening there was a banquet at which covers were laid for 150 delegates and guests. Past National President Royal D. Tomlinson acted as toastmaster. Short addresses were made by Robert Tobin, W. J. Reynolds, Robert Craig, J. N. Gregory and John J. Reddy. The entertainment was furnished by Billy McKay, Frank Howard, William Eagan, Dearborn Chemical Co.; Bob Jones, France Packing Co.; and Jack Armour, *Power*.

At the Saturday afternoon session the following state officers were elected: Robert Tobin, past president; Arthur E. Dowd, president; S. Hackaberry, vice president; William Roberts, secretary; William Downes, treasurer; Herbert E. Hall, conductor; Charles F. Eider, doorkeeper; Edward Lee, chaplain; Warren E. Lewis, state deputy. The propaganda committee for the Eastern District comprised P. J. Cassidy, William W. Downey and John McInnis. The Western District propaganda committee included H. G. Patrick, George Sterling and Charles F. Eider. P. J. Cassidy, George Van Vechten and Robert Craig were appointed on the License Law Committee. W. J. Reynolds was the installing officer.

In the evening an interesting and instructive illustrated lecture was given by C. C. Trump, of the Fuller Engineering Co., on "The Operating Merits of Pulverized Coal." Following the lecture a number of slides were shown of the Falls and rapids, dating back from 1871 to the present time.

Retiring State President Robert Tobin was presented a handsome mahogany clock, and a silver service was given to Mrs. Robert Tobin, the gifts of the Engineers and Supplymen.

New York was selected as the next convention city.

Fire Hazard in Transformers*

In considering the hazard of oil-immersed and of air-blast transformers, the number of units of each type involved must not be overlooked, or false conclusions will be drawn. In sizes of about 200 kva. there are probably at least 100 oil-insulated transformers installed to each air blast.

In the installation of air-blast transformers, generally one or more banks—that is, units of three single-phase or one three-phase transformer—are installed, ventilated by a single blower. In case of fire in any unit, either the blower must be shut down and the other units jeopardized, if their operation is continued, or the blower continues to operate and the defective transformer burns itself out unless extinguished by some special means. Probably not more than a dozen fires have occurred in air-blast transformers, and all of these in older types, and seldom has the fire been communicated to other apparatus.

Transformer construction of the oil-immersed type, as furnished by the various manufacturers in the earlier years, was not always as immune against spread of fire originating within the transformer as are present-day designs. The principal improvements to eliminate the possibility of explosions and fires have been the removal from above the oil surface of unshielded conductors with their corona hazard, and the discontinued use below the oil level in tank construction of solder or other materials of low melting point.

In case a transformer proves defective and an external short-circuit results, it is the opinion of the subcommittee that fire is now a very remote possibility. Oil will not support combustion until its temperature has been raised to the vaporizing point, which is relatively high, and not even in this case unless oxygen is present. In case fire should start, experience indicates that it would be localized to the transformers.

While fires have been known to occur in oil-insulated transformers, any assumption that the use of oil in modern transformers creates a fire hazard is not correct, since explosions have occurred inside of such transformers with no further damage than the breaking of the cover or the distortion of the tank.

The liability of explosions in good present-day construction is further lessened by the fact that the atmosphere above the oil is maintained at ground potential and free from any static stress. This is accomplished by having the metal supporting the leads, or bushing, extend down below the oil level, so that any overvoltage creating a possible discharge must occur under the oil and consequently be smothered, instead of, as in the past, occurring in the gaseous atmosphere above the oil.

Further precautionary steps have been taken recently by at least one large manufacturer in the development of tank construction, in which the tank is completely filled with oil. A small auxiliary tank is provided above the main tank to which it is connected. The oil extends up into this smaller tank and thereby prevents the effect of varying level due to temperature changes from reaching the transformer.

In large units, such as are generally used in indoor installations, the oil-insulated transformers are commonly water-cooled, the oil then operating at a lower temperature than in the self-cooled units. Such water-cooled units are obviously less likely to be damaged by fire from the outside. In present-day practice large self-cooled units are generally installed out of doors, where communicating damage from an external source is very remote.

Several years ago one of the large manufacturers of transformers made some experiments to determine what was necessary to set a tank of oil on fire, and after it was on fire, how difficult it would be to put the fire out. For this purpose about 200 gal. of oil was placed in a tank that had welded corrugated sheet-iron sides, cast solidly into a cast-iron top and bottom. The fire was placed under and around the tank, but the oil could not be ignited until burning waste had been thrown into the tank. When the oil was burning, the flame was extinguished without much trouble by the use of a Pyrene extinguisher. The oil was finally allowed to burn itself out and the tank was found to be still intact and oiltight. The oil did not boil over, but simply burned up from the fire on the surface. The manufacturers state that it is their belief that with a tank of this type, or with a boiler-iron tank, if the oil becomes ignited it will in all probability be confined to the tank and will not boil over or spread oil in the space adjacent thereto. It seems established, therefore, that no unusual fire hazard attends the use of oil-filled electrical equipment.

*Abstract from N. E. L. A. Electrical Apparatus Committee's report, presented at its Forty-Third Annual Convention, Pasadena, Cal., May 18-22, 1920.

Larkin Fund to David Ranken School

The David Ranken, Jr., School of Mechanical Trades, St. Louis, has been left an endowment fund estimated to range in value from $500,000 to $1,000,000 by E. H. Larkin, vice president of the National Ammonia Co., of the same city. The gift consists of Mr. Larkin's stock in the National Ammonia Co., and the fund is to be known as the Larkin Foundation. The income is to be used for the benefit of the school in any way the trustees decide best, but preferably for three things: First, instruction in power-plant operation, heating and ventilation, refrigeration and kindred subjects; second, lectures to mechanics, prospective mechanics and the general public on mechanical subjects; third, as an aid to worthy students who would otherwise be unable to pursue their studies.

Under an endowment approximating $3,000,0000 from David Ranken, Jr., the school was founded in 1907 and opened its doors to students in 1909. Two-year courses are given carpenters, plumbers, painters and decorators, machinists, patternmakers, electricians, auto mechanics and stationary engineers. The school has an active attendance of 300 day and 1,000 night pupils. It is expected that the Larkin fund will be available in about a year's time.

In connection with its latest statistical report covering data as to the cost of coal production, the Federal Trade Commission makes the following statement: "The commission has collected considerable information on the investment necessary for operation in the various districts, derived in part from balance sheets required from operators and in part from information previously reported by them or obtained direct from their books. When such investment figures are available, it then will be possible, after deduction of any sales expense, and income and excess-profits taxes paid, to show the relation which the remainder of the margin bears to the investment."

Notes on the Coal Situation

Many replies have been received by the United States Geological Survey in its census of stocks of coal on hand and anticipated needs. The final date for these reports was indicated in the call as June 12, but the Survey urges that all who have received forms reply, even though they were not able to do so before this date. Co-operation in this is very important for all coal interests.

Coal-land leasing regulations, which were drafted by the conference at the Interior Department about a month ago, have been sent out for examination by various engineers and institutions concerned. Reports are due on June 15, but it is expected that several weeks will elapse before the rules are adopted and promulgated.

Export embargoes both overseas and on Canadian coal and the re-establishment of a fuel-distribution agency of the Government are the two major proposals which have been made recently in the effort to correct coal-supply and coal-traffic evils. Several legislative proposals of export embargo were made during the closing days of the last Congressional session, but none of these prevail, in a number of cases the defeat of such measures being successfully accomplished by the legislators from the coal-producing states.

The Walsh resolution calling upon the Interstate Commerce Commission for information regarding shipments of bituminous coal to tidewater and for other information affecting New England and seaboard supply is being handled by Commissioner Daniels, of the Interstate Commerce Commission. It is believed that most of the information required as to production and traffic is already available from records of the Geological Survey and the Interstate Commerce Commission. An early report is anticipated.

Alaskan coal supply for the Navy may become available before many months have passed, if conditions are found to be favorable to the necessary development work. Commander Darling and other naval officers are now investigating the matter in Alaska preliminary to the decision as to what the Navy will do under the authorization given recently. This authorization provides an appropriation of a million dollars for mine development and a quarter of a million dollars for a coal-cleaning plant.

The Interstate Commerce Commission has given preference and priority to pool coal over non-pool coal, both for cargo and bunker fuel, with reference to all movements to the lower Lake ports. The service order Number 5, issued June 9 to this effect, is intended to arrange for much needed relief with respect to the supply of bituminous coal for the Northwest, which is supplied via the Lakes.

New Publications

ENGINEERING INDEX

The "Engineering Index" for 1919 has just been published by the American Society of Mechanical Engineers. The volume contains over 12,000 references to nearly 700 engineering and allied technical associations. It is an invaluable guide to engineering literature of 1919 and should prove of value to members of the profession.

OPPORTUNITIES FOR THE COMBUSTION ENGINEER

"Opportunities for the Combustion Expert" has just been prepared by the Hays School of Combustion Engineering, of Chicago. It relates in story form the experience of an engineer, pointing out his weaknesses in the first place and then showing how those weaknesses were overcome through a more definite knowledge of the work he is doing every day.

Considerable attention has been paid to the installation of mechanical apparatus to check the amount of fuel used and to prevent as much waste as possible. It is only within the last few years that steps have been taken to lay before the men themselves, who handle the fuel, an opportunity to realize what great opportunities there are for the men who know the essential elements of combustion engineering, and to be able to apply them from a fuel-saving standpoint. The combustion engineer today has a real opportunity.

Copies of this new book will be sent free by the Hays School of Combustion, 1412 South Michigan Ave., Chicago, upon request.

Personals

R. V. Meikle has been appointed chief engineer in charge of construction of the Don Pedro (Cal.) dam and power plant for the Turlock Irrigation District of California.

J. K. Alline, formerly of the Electric appliance Co., of Chicago, Ill., is now Iowa representative for the Packard Electric Co., of Warren, Ohio, manufacturer of transformers and automotive cable.

C. Richard Ward has resigned the position as chief engineer with the New York and Pennsylvania Co., Willsboro, N. Y., to become mechanical superintendent of the Bellman Brook Bleachery Co., Fairview, N. J.

Kenneth B. Millett has resigned as assistant general engineer of the American Thread Co., New York City, to join the organization of the Griscom-Russell Co., New York City, in the capacity of development engineer.

Prof. H. S. Dickerson, for four years professor of mechanical engineering at the University of Missouri, is to take charge of the Mechanical Engineering Department of the Government Institute of Technology at Shanghai, China.

J. A. Housel, who has been associated with the Northern Bolt, Screw and Wire Co., Ltd., of Owen Sound, Ont., Canada, as engineer, for the past eight years, is now affiliated with the Ewing Bolt and Screw Co., of Detroit, Mich., in the same capacity.

William B. Marshall has been appointed manager of the industrial division, New York office of the Westinghouse Electric and Manufacturing Co. to succeed Harlan A. Pratt, who has resigned to become sales manager of the Atlantic Elevator Co., of New York.

Society Affairs

The Electrical Manufacturers Council has opened an office in the Fifth Avenue Guaranty Building, 522 Fifth Ave., New York City. This office will serve as a headquarters for the Council and its work.

The American Society of Mechanical Engineers, Boston Section, will hold its June outing at the Ocean House, Swampscott, Mass., Friday, June 25. An interesting program featuring fat and lean men's races and including tennis, dancing, etc., has been arranged by the committee on entertainment.

The Association of Iron and Steel Electrical Engineers, Pittsburgh and Cleveland District Sessions, will meet in joint technical session, at the Country Club, Warren, Ohio, June 26. Included in the program is a trip to the Trumbull Steel Co., Warren. A paper on "Electric Drives for Steel Strip Mills," by W. P. Rese and A. W. Mohran, of the Trumbull Steel Co., will be read.

The Eastern Geographic Section Meeting of the Pennsylvania Electric Association will meet at the Adelphi Hotel, Philadelphia, Pa., on July 2. The morning session will open at 10 o'clock. The topic of this session will be "Power Factor." Several interesting papers will be presented, followed by five-minute talks on various phases of power factor by operating companies' representatives. The morning session will be followed by a luncheon presided over by John E. Zimmerman, who will introduce the speakers. The subject of the afternoon session is to be "Effect of Unbalanced Load on the System." Short talks by operating companies' representatives will also be included in this session. These talks will cover methods of regulating voltage on unbalanced phases, various methods of phase balancing, balancing large single-phase loads and concrete examples of phase balancing, results obtained and equipment used.

Miscellaneous News

Construction of a Hydro-Electric Plant on the White River, Cotter, Ark., which will involve an expenditure of $15,000,000, is under way by the Dixie Power Co. The plans provide for an initial development of 100,000 hp.

The Massachusetts Institute of Technology and the General Electric Co. have opened a five year course in Co-operative Electrical Engineering. The course will fit students to take leading positions as engineers, scientific experts, and teachers or investigators of electrical science.

The Department of Commerce of the Bureau of Foreign and Domestic Commerce, Washington, in its monthly statement shows the total value of the following exports during April: Pumps and pumping machinery, $1,098,020; refrigerating machinery, $326,828; air-compressing machinery, $512,765.

The American Chamber of Commerce in France has inaugurated an extensive catalog system for the purpose of assisting French purchasers seeking American goods and of assisting the American export trade in general. Information can be obtained from Frank Boole, secretary, American Chamber of Commerce, Paris, France.

The Oil Belt Power Co. is planning expenditures of more than $4,000,000 for the construction of a hydro-electric plant near Eastland, Tex. One of the chief purposes of the plant will be to furnish electric power for drilling and pumping oil wells in the central west Texas fields. It will also provide industries and towns with electrical energy.

Business Items

The Easton Car and Construction Co., New York City, has removed its general offices from 30 Church St. to larger quarters at 50 Church St.

The Merrill Co., San Francisco, Cal., has established a branch office in the Monadnock Building, Chicago. The company also announces that it is now manufacturing the Merco Nordstrom Plug Valve at the plant of the Swenson Evaporator Co., Angola, Ind.

Trade Catalogs

The Cooper Hewitt Electric Co., Hoboken N. J., has published a new booklet illustrating proper methods of industrial illumination by means of photographs of systems actually installed and in use.

"Wellman Mechanical Gas Producer" is the title of Bulletin No. 51, just issued by the Wellman-Seaver-Morgan Co., Cleveland, Ohio. The bulletin is in the form of a text book on gas producers and contains some valuable information. A copy will be sent to any one interested, on request.

COAL PRICES

Prices of steam coals both anthracite and bituminous, f o b. mines, unless otherwise stated, are as follows:

ALANTIC SEABOARD

Anthracite—Coals supplying New York, Philadelphia and Boston:

	f.o b. mines
Pea	$5 85@$6.25
Buckwheat	4 00@ 4.10
Rice	3 00@ 3.50
Barley	2.25@ 2.50
Boiler	2.50

Bituminous—Steam sizes supplying New York, Philadelphia and Boston:

	f o.b. mines
Cambrias and Somersets, t.o.b. mines net ton	$7.25@$8.00
Clearfields, f.o.b. mines, net ton	6.75@ 8.25
Pocahontas } on cars Providence and	11.00@14.50
New River } Boston; gross ton	11.00@14.50

BUFFALO

Bituminous—f.o.b. mines:

Pittsburgh slack	$6.00@$8.00
Mine-run	6.00@ 8.00
Lump	6.00@ 8.00
Youghiogheny	6.00@ 8 00

CLEVELAND

Bituminous—Retail prices of coal per net ton are as follows:

No. 6 slack	$8.25
No. 8 slack	8.25
No. 8—¾-in	8.25@ 8.50
No. 6 mine-run	8.25
No. 8 mine-run	8.25
Pocahontas—Mine-run	9.25

ST. LOUIS

Bituminous—f o.b. mines:

	Williamson and Franklin Counties	Mt. Olive and Staunton	Standard
Mine-run	$3.00@4.00	$3.00@3.50	$4.50
Screenings	3.00@4 00	$3.00@3.50	5.00
Lump	3.00@4.00	3.00@3 50	5 00

Williamson-Franklin rate to St. Louis is $1.10; Other rates 95c.

CHICAGO

Bituminous—Prices f.o.b. mines as per circular of April 1 are as follows: Co. prices vary from 50c. to $1.25

		Freight rate Chicago
Illinois		
Southern Illinois		
Franklin, Saline and Williamson Counties		
Mine-Run	$3.00@$3.10	$1.55
Screenings	2.60@ 2.75	1 55
Central Illinois		
Springfield District		
Mine-Run	$2.75@$3.00	$1.32
Screenings	2.50@ 2.60	1.32
Northern Illinois		
Mine-Run	$3.50@$3.75	$1.24
Screenings	3.00@ 3.25	1.24
Indiana		
Clinton and Linton		
Fourth Vein		
Mine-Run	$2.75@$2.90	$1.27
Screenings	2.50@ 2.65	1.27
Knox County Field		
Fifth Vein		
Mine-Run	$2.75@$2.90	$1.37
Screenings	2.50@ 2.60	1.37
Brazil Block	$4.25@$4.50	1.27

New Construction

PROPOSED WORK

Mass., Everett—The Maiden Electric Co., Maiden, plans to build a local substation here, to cost about $65,000, including $25,000 for equipment. Charles Tenney, 201 Devonshire St, Boston, Engr

Mass., Fitchburg—The Fitchburg Gas & Electric Co. plans to build an addition to power house which will have a 4,500 kw., 3 phase, 60 cycle. a.c. capacity and include a 5,000 kw. steam turbine and one 750 hp. Bigelow & Hornsby boiler, etc. About $600,000. Charles Tenney, 201 Devonshire St., Boston, Engr.

Mass., Salem—The Salem Electric Light Co. plans to build an addition to its existing power plant on Lafayette St. and install equipment including one 750 hp. Bigelow-Hornsby boiler, etc. About $650,000. Charles Tenney, 201 Devonshire St, Boston, Engr.

Conn., Hartford—The State Highway Dept. plans to install an electrical unit in place of steam plant which will be removed, in connection with the electrification of the Thames River Highway Bridge.

Conn., New Britain—The Amer. Hardware Corp. will build a transformer station at its plant.

Conn., Norwich—The W. S. Finishing Co. is having plans prepared for the reconstruction of a boiler plant here. About $300,000. Day & Zimmerman, 611 Chestnut St., Philadelphia, Pa., Archts. and Engrs.

N. Y., Brooklyn—The Dept. of Pub. Welfare, Municipal Bldg., New York City, rejected all bids for a hospital including a steam heating system, on Cumberland St. Estimated cost, $600,000..

N. Y., Brooklyn—The Packard Motor Car Co., B'way and 61st St., is having plans prepared for a 3 story, 200 x 200 ft. service station and auto showroom including a steam heating system, on Bedford Ave. and Empire Blvd. About $250,000

N. Y., Buffalo—The Buffalo General Electric Co., Genesee and Washington Sts., had plans prepared for an addition to its power station on Babcock and Hannah Sts. About $25,000.

N. Y., Buffalo—The Niagara Machine Tool Works, 683 Northland Ave., is in the market for a 50 kw., 115 volt belted direct generator.

N. Y., Buffalo—James Storer, Secy. of the Bd. Educ., New York Telephone Bldg., will receive bids until June 25 for furnishing and installing boilers, boiler equipment, pumps, piping, heat control regulators, fans, motors, etc., in primary school 21, Amherst St. and Park Lake Ave., elementary school 21, Hertel Ave. and Camden St., elementary school 9, Polar Ave. and Doat St., Associated Buffalo Architects, Inc., 250 Delaware Ave., Archts.

N. Y., Dunkirk—The Niagara & Erie Power Co., Marine Bank Bldg., Buffalo, has made application to the Pub. Serv. Comn., for permission to build an addition to its power plant here.

N. Y., New York—A. L. Erlanger, 214 West 42nd St., plans to construct two 3 story theatres including a steam heating system on West 44th St. About $850,000. F. R. Anderson, 214 West 42nd St., Archt. and Engr.

N. Y., New York—The New York Edison Co., Irving Pl. and 13th St., is having plans prepared for a substation on Elizabeth, near Canal St.

N. Y., New York—The New York Telephone Co., 15 Dey St., plans to build an addition and make alterations to the present telephone exchange building including a steam heating system, on 146th St. and Convent Ave. About $500,000.

N. Y., Springville—E. B. Kuhn, Secy. of the Bd. Educ. of the Griffith Institute, will soon award the contract for furnishing material and installing a steam heating plant in the school here.

N. J., Dover—The Ordnance Dept. has received bids for a power house for the general operating service at the Picatinny Arsenal, here.

N. J., Newark—The Bd. Educ., City Hall, received bids for installing a steam heating system in the proposed school addition on Alexander St., from J. H. Cooney, 210 North 4th St., $36,542; Fred P. Merhle, 533 Central Ave., $35,972, and in the vocational school on 1st and 2nd Aves., from J. H. Cooney, 210 North 4th St., $173,840; J. H. Nellis, 884 Sanford Ave., Trenton, $174,360. Noted May 25.

N. J., Newark—M. Straus & Sons, 508 Frelinghuysen Ave., had plans prepared for a power house at its leather works here. Two boiler units will be installed in connection with other equipment About $30,000

N. J., Perth Amboy—The Perth Amboy Dry Dock Co., Broad St., plans to install electrically operated pumping machinery, valves, etc. for the operation of the new dry dock at its plant here.

N. J., Rahway—The Middlesex Water Co. has been ordered by the Pub. Utility Comn. to install equipment by July 15 at its plant here, which will provide for a daily capacity of 1,000,000 gal.

N. J., Westfield—Runyon & Carey, Engrs., 845 Broad St., Newark, will soon award the contract for a heating and ventilating system, 2 cast iron boilers, ventilating fans and units, etc., in the Washington School here, for the Bd. Educ.

Pa., Philadelphia—The Enterprise Mfg Co., 3rd and Dauphin Sts., had plans prepared for the construction of a power house at its hardware manufacturing plant on Bodine St. About $28,000.

Pa., Philadelphia—The Hale & Kilbourne Corp., 18th and Lehigh Sts., manufacturer of iron and steel products, had plans prepared for an addition to its power plant here. Boilers, stokers and auxiliary equipment will be installed in same About $60,000.

Pa., Pittsburgh—The Prest-O-Lite Co., 5528 Baum Blvd., is receiving bids for a 30 x 36 ft compressor building, a 15 x 33 ft. boiler house addition, etc., on Lincoln and Tabor Sts.

Pa., Pittsburgh—The West Penn Power Co. plans to extend its transmission system.

Md., Baltimore—Haskell & Barnes, Archts., 301 North Charles St., will receive bids until June 30 for an 8 story, 100 x 186 ft. warehouse including a steam heating system on Guilford Ave., Saratoga and David Sts., for the Whitaker Paper Co., 415-37 Guilford Ave. About $500,000.

Md., Baltimore—Samuel T. Williams, 223 North Calvert St., is in the market for a stiff-legged or guyed derrick, about 35 ft. boom, to lift 1½ tons (used), a steam driven, 2 drum hoist to operate derrick (used) and a house-heating boiler and radiation for heating 2,500 sq ft. floor space, etc.

Md., Easton—The Borough Council plans to install new electric operated centrifugal pumps at the waterworks here with capacity varying from 150 to 600 gal. per minute.

D. C., Washington—The Bureau of Yards & Docks, Navy Dept., plans to install a new boiler, etc., in the heating plant at the Naval Observatory here. About $10,000.

Va., Lorton—The Dist. of Columbia Comrs., 50 District Bldg., Wash., D. C., will receive bids until July 5 for electric generating set, exciter, switchboard, etc., for the central power plant at the D. C. institutions here.

Va., Richmond—The Arrow Laundry Co., 117 East Main St., is in the market for a 100 hp steam engine, shafting and belting, etc.

Va., Richmond—The Virginia Carolina Rubber Co., Real Estate Exch. Bldg., is in the market for all machines used in the manufacture of auto tires and accessories including an engine, boiler, shafting and belting. Mr. Bell, Pres.

O., Cleveland—The Bd Educ plans to build a 1 story school for crippled children including a steam heating system About $350,000. F. Hogan, Dir. W. R. McCormack, East 6th St. and Rockwell Ave., Archt. A bond issue will be voted upon for the project.

O., Cleveland—The Bd. Educ. plans to build a 3 story school including a steam heating system on Diana and Darley Aves. About $600,000. W. R. McCormack, East 6th St. and Rockwell Ave., Archt.

O., Cleveland—The Bd. Educ., East 6th St. and Rockwell Ave., plans to construct a school addition to the Rickoff School on Kinsman Rd., cost, $350,000; and a school on Spring Rd., cost, $1,000,000. Plans include steam heating systems. W. R. McCormack, Archt.

O., Cleveland—The city received bids for furnishing and installing 2 pressure booster pumps and 23 hydraulic operated sluice gates from the Coffin Valve Co., Wade Bldg., $34,140; J. L. Skeldon Eng. Co., Rockefeller Bldg., $68,136.

O., Cleveland—The city is having plans prepared for two 8,000,000 and one 4,000,000 gal. pumps. About $25,000. E. Shattuck, Clerk.; G. B. Gascoyne, Engr.

O., Ravenna—The McElrath Tire & Rubber Co., 1836 Euclid Ave., Cleveland, had plans prepared for the construction of a 3 story, 100 x 300 ft. factory and a 30 x 50 ft. boiler house. About $125,000. Donald C. Smith, 4500 Euclid Ave., Cleveland, Archt. and Engr. Noted May 3.

O., Toledo—The city will soon award the contract for a 30 x 116 ft. pumping station. Equipment including switchboard, electric crane, four 300 hp. motors, one 200 hp. motor, two vacuum pumps, five centrifugal pumps, etc., will be installed in same. Fuller & McClintock, B'way, New York City, Archts. and Engrs.

Mich., Bay City—The Wildman Rubber Co., 816 Book Bldg., Detroit, has purchased a site along the Saginaw River and plans to build a tire plant on same. Plans include a power plant. About $1,000,000.

Wis., Platteville—The Peacock Cheese Co., South 12th and Jefferson Sts., Sheboygan, has purchased a site and plans to construct a cheese warehouse and cold storage plant on same.

Kan., Mt. Hope—The Borough Council will build a new electric power plant. About $15,000.

Kan., Paola—The City Clerk will soon award the contract for repairing the present pumping station, water mains, etc. Black & Veatch, Interstate Bldg., Kansas City, Mo., Engrs.

Mo., St. Louis—The Majestic Mfg. Co., 2014 Morgan St., is having plans prepared for a 2 story factory, including several electric motors. About $30,000. J. H. Lynch & Son, Dolph Bldg., Archt. and Engr.

Ore., Burns—The city has retained Louis C. Kelsey, Consult. Engr., Portland, to prepare plans and supervise the construction of a lighting plant.

Ore., Corvallis—Edmund Bergholtz, Archt., Spalding Bldg., Portland, will receive bids until Sept. 1 for a 4 story, 75 x 100 ft. hotel. About $330,000. Heating contract will be sublet. Owner's name withheld.

Ore., Enterprise—The Enterprise Mercantile & Milling Co. plans to construct a small electric lighting plant at its flouring mill ere, which will be operated by water power. G. W. Hyatt, Pres.

Ore., Portland—Houghtaling & Dougan, Archts., 329 Stark St., will receive bids until October 1 for a 5 story, 100 x 150 ft. temple on 11th and Alder Sts., for the Elks Lodge. Contract for heating, etc., will be sublet. About $700,000.

Cal., Berkeley—The Bd. Educ., 2133 Allston Way, will receive bids until June 29 for furnishing deep well pumping head, motor, pipe electric power connections, belts, foundations, etc., at each of 3 schools.

Cal., San Francisco—The Bureau of Yards & Docks, Navy Dept., Wash., D. C., will receive bids until June 30 for furnishing and installing refrigeration equipment in the Industrial Bldg., here. About $50,000.

Cal., Stockton—The Bd. Supervs. plans to construct buildings and improvements for the Agricultural Park here. Complete scheme to include machinery hall, auto exhibit building, electric light and power plant, etc. About $1,000,000. Wright & Satterlee, Archts.

Que., Grosse Isle—The Dept of Pub. Wks., Ottawa, will soon receive bids for new boilers for the Quarantine station here, at $29,882.

Que., Montreal—The city plans to install an electrically driven pumping plant, provide for the initial installation of five 30,000,000 gal. pumps and space for 3 additional pumping units. Two of these pumps have been ordered and are already in operation. About $850,000.

Ont., Toronto—The Bd. Educ., College St., plans to construct a 3-story technical school, including a vacuum steam heating system with an electric driven mechanical ventilating system. About $400,000. Address Dr. McKay, Central Technical School, College St.

Ont., Toronto—The Canada Creosoting Co., 1006 Canadian Pacific R.R. Bldg., plans to construct a plant and is in the market for tanks, cylinders, pumps, boilers, motor generators, etc.

Ont., Toronto—The Muskoka Quarries, Ltd., 271 Macpherson Ave., plans to install a complete stone crushing plant with electrical motors and equipment.

Ont., Woodstock—The town has purchased a site on Ingersoll Ave., and plans to construct a substation on same. Electric supplies, including switching apparatus, transformers, etc., will be installed in same. About $35,000. Mr. Archibald, Engr.

CONTRACTS AWARDED

Vt., Brattleboro—The C. F. Church Mfg. Co., Willimansett, Mass., has awarded the contract for a 1 story main building, 1 story boiler house, etc., to E. F. Carlson Co., 244 Main St., Springfield, Mass., at $80,000.

Conn., Hartford—The Hartford Fire Insurance Co., 125 Trumbull St., has awarded the contract for a 1 story, 90 x 120 ft. boiler house on Asylum St., to Marc. Eidlitz & Sons Co., 30 East 42nd St., New York City. About $150,000.

Conn., Waterbury—The Connecticut Light & Power Co., 111 West Main St., has awarded the contract for a 2 story addition to its power station on Freight St., to D. Wooster, 39 Harvard St. About $10,000.

N. Y., Long Island City—The Atlantic Macaroni Co., Inc., 295 Vernon Ave., has awarded the contract for a 93 x 400 ft. factory including a steam heating system on Vernon Ave. and 14th St., to the White Constr. Co., Inc., 95 Madison Ave., New York City. About $1,000,000.

N. Y., New York (Bronx Borough)—V. Vivadou Perfumery, Inc., Times Bldg., has awarded the contract for a group of factory buildings, including a steam heating system, on the Southern Blvd. and Leggett Ave., to J. M. Brady, 103 Park Ave., at $1,000,000.

N. Y., New York—The 562 5th Ave. Corp., 1552 B'way, will build an 11 story office building including a steam heating system, on 46th St. and 5th Ave. Work will be done by day labor.

N. J., Atlantic City—The Ritz Carlton Linnard Co., c/o Warren & Wetmore, Archts. and Engrs., 10 East 47th St., New York City, has awarded the contract for a 130 x 310 ft. hotel, including a steam heating system, on Illinois Ave. and the Board Walk, to Thompson & Starrett, 49 Wall St., New York City. About $2,000,000.

N. J., Newark—Samuel Jones & Co., 37 McClellan St., has awarded the contract for a 1 story, 30 x 80 ft. factory and a 40 x 40 ft. boiler house, to Charles R. Heddon Co., 763 Broad St., at $25,000.

N. J., Newark—The Schofield Oil Co., Ave. R, has awarded the contract for a 1 story, 31 x 92 ft. power house on Railroad Ave. and Passaic St., to E. M. Waldron Co., 885 Broad St. About $30,000.

N. J., Trenton—The Bergounan Rubber Corp. of Trenton has awarded the contract for three factory buildings including a steam heating system, on Whitehead Rd., to the Farrall Constr. Co., 53 West 39th St., New York City, at $250,000.

Md., Baltimore—The Mayor and the City Council, City Hall, has awarded the contract for a 3 story, 64 x 121 ft. public school including a steam heating system, on Poplar Grove St., to the Standard Constr. Co., 1713 Sansome St., Philadelphia, Pa., at $399,000.

Ala., Florence—The U. S. Engineer's Office, Wash., D. C., has awarded the contract for 4 hydraulic turbines to William Cramp & Sons, Ship & Engine Bldg., Philadelphia, Pa. $519,800.

O., Akron—The Akron Theatre Co c/o Frank, Wagner & Mitchell, Archts., 602 Penn Title & Trust Bldg., has awarded the contract for a 3 to 6 story, 13 x 160 ft. theatre, arcade and commercial building including a steam heating system, on South Main St., to the Carmichael Constr. Co., 524 Hamilton Bldg., at $500,000.

O., Cleveland—The Cleveland Ry. Co., Leader-News Bldg., has awarded the contract for a 1 story, 47 x 64 ft. electric substation on East 99th St. and St. Clair Ave. to the Gaylord W. Feaga Co., Citizens Bldg., and for the installation of two 1500 kw. rotary converters with blowers and generator sets and other equipment to the Westinghouse Mfg. Co., Swetland Bldg. Noted May 25.

O., Middletown—The Colin Gardner Paper Co. has awarded the contract for a 1 story, 50 x 80 ft. turbine power plant, to the Ferro Concrete Constr. Co., Richmond and Harriet Sts., Cincinnati. $45,000.

O., Willoughby—The Zenith Tire & Rubber Co., 450 Leader-News Bldg., Cleveland has awarded the contract for a 1 and 3 story, 100 x 620 ft. warehouse, factory and boiler house on Main St., to William Dunbar Co., 8201 Cedar Ave., Cleveland. Two 300 hp. boilers will be installed in same. About $750,000. Noted Oct. 28.

Mich., Detroit—The Detroit Graphite Co., 12th and Fort Sts., has awarded the contract for a 5 story, 61 x 92 ft. office building including a steam heating system, on 12th St., to the Max Batholomael Son & Co., 466 East Warren Ave.

Ark., Arkadelphia—Quichita College has awarded the contract for three 3 story school buildings including a steam heating system, to J. D. Brock, at $250,000.

Kan., Norton—The city has awarded the contract for a commissary building, power house addition, concrete coal bunker, dam and addition to sewage treatment plant, to Green & Rickey, Norton, for a 100 hp. boiler, hot water tank, water mains, hydrants and other miscellaneous heating and plumbing work, to Patterson & Co., Topeka, and for an engine and generator, etc., to the Machinist & Electric Co., Topeka.

Mo., St. Louis—D. R. Francis, 214 North 4th St., has awarded the contract for installing a steam heating system in the proposed mercantile building at 1529-35 Washington Ave., to the Eichler Heating Co., Railway Exch. Bldg.

Tex., Galveston—The Galveston Wharf Co. has awarded the contract for the installation of 6 high density cotton compresses with overhead monorail conveyance machinery for serving all piers along the Galveston Wharf front from Pier 10 to 41, to J. H. W. Steele Co., 220 21st St. About $500,000.

Cal., Los Angeles—The Bd. Educ., 731 Security Bldg., has awarded the contract for the installation of a steam heating system and boiler plant in the proposed 1 and 2 story, 132 x 220 ft. school on East 7th St., to Thomas Haverty Co., 8th St. and Maple Ave., at $12,585. Noted June 8.

POWER

Volume 51 New York, June 29, 1920 *Number* 26

The Big Stick

By Rufus T. Strohm

Back in the days when I wore knee pants
And garden work was a circumstance,
If father told me to hoe the beans,
Or pick the bugs from the growing greens,
Or make a bed for the humble squash,
Or train the cucumber vines, b'gosh,
I went and did it, the best I could,
And that, believe me, was pretty good!

Though toil is a thing most boys dislike,
It never entered my head to strike,
Nor was I tempted to slight or shirk
Or practice soldiering in my work;
For well I knew I'd be playing hob
If I made a botch of that garden job,
Since there'd be a conference of three—
My dad, and a barrel stave, and me!

Now, I'm not keen on the sort of thing
Like a kaiser, emperor, or king;
A bit of them is a *quantum suf.*—
Or what we commonly call "enough";
For though a few may be good, egad,
The common run of the lot is bad.
Still, what we need in the present hour
Is an autocrat with a big-stick power.

And they who hindered the nation's life
By brewing trouble and stirring strife
Would all be haled to the autocrat
With the iron arm and the cordwood bat,
And brought to the proper frame of mind
To toil for the good of humankind;
For work's the thing that the countiy needs—
We're long on talking and short on decds!

Electrically Driven Steel Reversing Mills

By FRASER JEFFREY
Electrical Engineer, Allis-Chalmers Mfg. Co., Milwaukee

Tonnage output per cost of operation and maintenance rather than first cost is the basis on which electricity is efficiently replacing steam as a mill drive. Features of an interesting installation in which a reversing motor and its necessary complements replaced an engine unit.

ONE of the prime points of consideration outside of the "cost per ton" in any steel-mill installation, is the "tonnage output per cost of operation and maintenance" rather than the "tonnage output per first cost." This may seem somewhat paradoxical but is evidenced more or less all the time by the large number of mills that are constantly changing over from low-first-cost units to equipment of higher first cost; namely, steam to electric drive. Higher first cost means little if in a reasonable length of time the saving in cost of operation and of maintenance will more than offset it to the extent of showing appreciable gains. For this reason electric drive in the steel mill, in view of its higher efficiency than the steam drive, is more than making up the difference due to its higher first cost.

In the older type of steam-driven mill of today, the buildings have been built up around the equipment in such a manner that no radical changes can readily be made in the mill layout. In many cases the mill property is limited in size such as would prohibit any radical expansions in the new building layouts, and in general the whole sequence of operation is dependent and radiates from the two-high steam-driven reversing blooming mill.

Competition has become keen of late years, with the result that the different mills are desirous of reducing the cost per ton of their product in every way reasonably possible. It is usually only a question of time when the reversing engine will reach the state where practically complete renewal is needed. This may not occur for a long period, many of those faithful old engines having seen more than twenty years' service; nevertheless, they do wear out just as any piece of machinery will. When this point is reached there are several courses open to the mill owner; First, the old engine can be renewed; second, the engine can be replaced by a reversing electric motor and its necessary complements; third, the engine, mill, buildings, etc., can be replaced by a continuous-running electrically driven mill.

If the engine is renewed, the cost per ton of product and the total tonnage will remain practically the same as before. If the engine is replaced by a reversing electric motor with auxiliaries, the cost per ton of product will be reduced, but the first cost versus that of the engine will be greater, the total tonnage remaining practically the same as before, provided the mill has already been worked to its full capacity.

The replacing of the engine, mill, buildings, etc., by a continuous-running electrically driven mill is obviously prohibitive and not to be considered.

This is, then, the problem facing the older type of steam-driven mill—a question of still adhering to the steam, or one of electrification. As a natural consequence it is the electrification of these kinds of mills that is largely responsible for the reversing motor of today. When considering an entirely new electrically driven blooming mill, the continuous-running type is seldom selected (only one mill of this kind being in existence in this country today) because it can be considered only for mills turning out an extremely large tonnage and usually of one particular product. The first cost, not only of the electrical equipment, but of the mill equipment, buildings, etc., is so large as to practically limit its use to this kind of mill. The selection, however, is governed more or less by local conditions and such things as first cost, material to be rolled, conditions of rolling, tonnage, etc. The continuous mill is naturally more efficient than the reversing type, and it is also a proved fact that the reversing motor is more efficient than the reversing engine. Giving the engine due credit, it may be stated that on any given mill it can be made to turn out as much tonnage as the electric motor. In efficiency, however, it loses out considerably, and this is the potent item in these days of keen commercial rivalry. Within recent years there have been erected several entirely new plants in which steam drive has been used, but this is largely on account of special prevailing conditions and is more the exception than the rule.

Electrification may be carried to the extent of installing a complete plant from prime movers to motors and the necessary auxiliaries, or it may mean only the installation of the large main-drive motors and their auxiliary equipment. A great deal depends on the size of the mill and the local conditions, as to whether or not power should be generated at the mill or advantageously bought from some one of the local power companies. It is a matter of record that in most of the older steam-driven mills, electricity is used for the operation of the screw-downs, tables, cold saws, etc.

At one steel plant the reversing engine on the main plate-mill drive reached the point where one of the

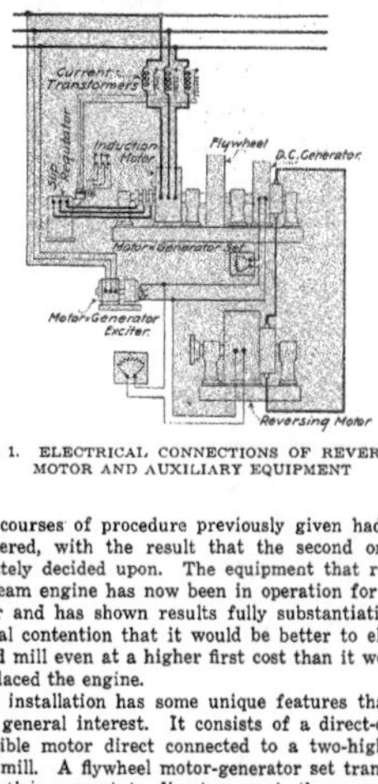

FIG. 1. ELECTRICAL CONNECTIONS OF REVERSING MOTOR AND AUXILIARY EQUIPMENT

three courses of procedure previously given had to be considered, with the result that the second one was ultimately decided upon. The equipment that replaced the steam engine has now been in operation for nearly a year and has shown results fully substantiating the original contention that it would be better to electrify the old mill even at a higher first cost than it would be to replaced the engine.

The installation has some unique features that may be of general interest. It consists of a direct-current reversible motor direct connected to a two-high, two-stand mill. A flywheel motor-generator set transforms alternating current to direct current, the commutator of the direct-current generator on the set being electrically connected through an automatic reclosing circuit breaker to the commutator of the direct-current reversing motor. An automatic liquid slip regulator with a special torque motor controls the line peaks by allowing the flywheel to give up its energy at certain predetermined points, while a small exciter motor-generator set, under control of a master switch operating in conjunction with full magnetic-type control panels, governs the fields of both the generator and the motor. In this manner a speed range from zero to 70 r.p.m. is possible and is required on account of the high-grade alloy steel that is rolled in one heat from the bloom to the finished product in passes varying from 15 to 25, depending on the size and nature of the material being rolled. The sequence of operation is such that under certain rolling conditions a 10 x 12-in., 800-lb. bloom of hard, tough alloy steel is rolled to a final product 16 in. wide and 3/16 in. thick. The roughing and edging is done on the first roll stand, the passes being short and at low speed. The material without reheating is then transferred by an electrically driven table to the second roll stand, where the passes are of long duration and accomplished at fairly high speed. To meet the varying conditions required, the master controller is so arranged that 23 speeds forward and 23 speeds reverse are obtainable. A simplified wiring diagram of the connections is shown in Fig. 1.

A complete oiling system of sufficient capacity to lubricate the bearings of the reversing motor and the flywheel bearings of the flywheel set is provided. This includes a motor-driven oil pump supplemented by a steam-driven oil pump for use in case of failure of the main supply line. This insures a constant flow of oil to the bearings, as the flywheel set, owing to its own inertia, will not come to rest for 1½ hours after the main-line switch is opened.

The speed and torque control is governed through a master switch and two contactor panels shown by Fig. 2. On one panel are mounted a circuit-breaker switch, a set of reversing switches, accelerating switches or speed points and a current and voltage relay. Instead of opening the main circuit, the circuit-breaker switch opens the field circuit of the generator, thereby making it unnecessary to handle large currents at high voltage. The current relay governs the strength of the motor field relative to the generator field, and the voltage relay prevents reclosing of the circuit breaker until the generator voltage has dropped to zero, and the master switch is returned to the off position.

On the other panel are a number of switches for weakening the field of the motor for speeds above normal and relays and switches for controlling this operation. The switches for weakening the motor field are selected by the master and will insert or cut out the resistance, corresponding to the position of the master switch, the latter providing a means of selecting the direction of rotation and the speeds desired.

The headpiece shows the four-bearing flywheel motor-generator set comprising a 900-hp. 2200-volt

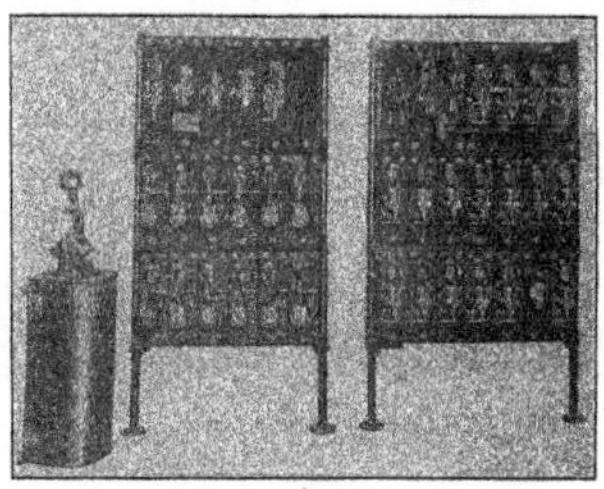

FIG. 2. MASTER SWITCH AND AUTOMATIC CONTROL PANELS

three-phase 60-cycle 720 synchronous r.p.m. induction motor; a 9.5-ft. diameter, 45,000-lb. machined plate-type flywheel fully inclosed in a sheet-metal casing and a 700-kw. 525-volt direct-current generator. As the stator of the induction motor can be moved axially, thus exposing the rotor and stator for inspection and repairs, it is unnecessary to disturb the shaft and integral flywheel in any way whatever.

Including windings, slip rings, etc., the rotor is made suitable mechanically and electrically to withstand the shock of "plugging" and is used principally for bringing the rotating parts to rest when shutting down the set. Some idea as to the inertia of the moving parts is obtained from the fact that it takes the set about 1½ hours to come to rest from full speed, the retarding effort being its own windage and friction only. By tying in the mill motor and mill, stopping can be effected in some twenty minutes, while by full "plugging" the set can be brought to rest in approximately 3½ minutes. The direct-current generator is of the shunt-wound interpole compensated type, especially designed for "black" commutation of heavy overloads in current up to 3,870 amperes. The shunt fields are wound for 250 volts and are controlled by means of the master switch operating in conjunction with a 25-kw. induction-motor-driven exciter set. In this manner not only varying field strengths are obtained, but the polarity can be reversed also, thus changing the direction of the armature current and accomplishing the reversal of the main mill motor whenever desired.

FIG. 3. LIQUID STARTER AND AUTOMATIC SLIP REGULATOR

The starter and automatic slip regulator of the liquid type, Fig. 3, consists of an iron tank containing the electrodes and electrolyte, a motor-driven centrifugal circulating pump, a torque motor operating a set of weirs controlling the electrolyte level and the necessary operating accessories such as no-voltage release, plugging valves, accelerating valves, etc. It is designed not only for "plugging" conditions, but to start the motor-generator flywheel set from rest and bring it up to speed, and then automatically vary the speed in order to increase the slip so as to permit the flywheel to deliver part of its stored energy when the demand for power is in excess of the amount for which the regulator is set. It will take care of a maximum slip of approximately 20 per cent, thus allowing a speed variation on the flywheel set down to 80 per cent of full-load speed.

When the current required by the induction motor of the flywheel set exceeds the amount for which the regulator is set, the torque motor operates the weirs, thus introducing resistance in the secondary circuit of the induction motor and permitting the flywheel to give up some of its stored energy. When the current demand decreases again, the reverse operation takes place and the set is automatically brought to its normal speed.

The iron tank of the regulator is divided into an upper and lower compartment, there being interposed between the two a large butterfly valve. When the main-line switch to the motor of the flywheel set is closed, the electrolyte in the regulator is at its lowest level, and consquently, full resistance is in the rotor circuit. Closing the butterfly valve, which is then held in position by a magnet coil and catch, starts the circulating pump motor and causes the electrolyte level to rise and bring the induction motor to speed. Another butterfly valve located in the discharge pipe of the circulating pump can be given various adjustments so that the rate of acceleration can be varied at will. The magnet coil and catch on the large butterfly valve serves as a no-voltage release and also acts in conjunction with the "plugging" switch, dumping the electrolyte from the upper to the lower compartment, thus inserting full resistance in the rotor circuit of the induction motor.

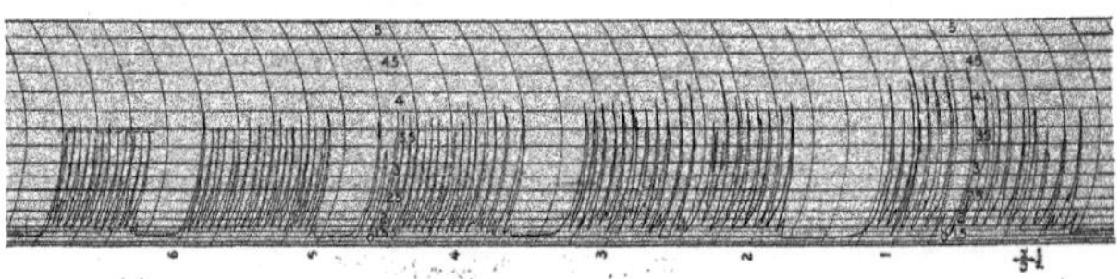

FIG. 4. LOAD CHART TAKEN ON LINE SIDE OF INDUCTION MOTOR OF FLYWHEEL SET, WITH REVERSING MOTOR UNDER HEAVY PEAK-LOAD CONDITIONS

Regardless of the heavy peak-load conditions imposed on the reversing motor, the automatic regulator smooths out and equalizes the power from the line, as shown in Fig. 4. This is an actual load chart taken on the line side of the induction motor of the flywheel set when the reversing motor was working under heavy peak load.

Reading from right to left the chart shows the load conditions from 12 o'clock noon until 7 o'clock at night.

The final setting of the regulator was not made until 6:10 o'clock in the evening, so that from 12 o'clock until 6 o'clock different settings were made during the intervals at 1 : 15, 3 : 20, 4 : 50 and 6 : 10 p.m. to allow the flywheel to carry more and more of the load. For this reason a gradual reduction in the peaks on the induction motor is quite noticeable. It should be noted that the same condition of rolling prevailed throughout the entire period of time shown, and the chart gives a good idea as to what can be accomplished with a regulator of this kind acting in conjunction with a flywheel. In the final setting, from 6 to 7 p.m., the regulator was adjusted for a current load on the induction motor corresponding to the dotted line shown.

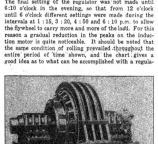

FIG. 5. COMMUTATOR END OF REVERSING PLATE-MILL MOTOR

The main reversing motor, Fig. 5, is provided with a compensated winding in the faces of the poles, in addition to the shunt-field and interpole windings. Its bearings are of the pedestal type, self-aligning, ring-oiling and water-cooled, the pedestal next to the mill being suitable to take the thrust from the mill in case of a spindle break.

The motor has a normal rating of 800 hp. and 2,500 hp. maximum at 37 r.p.m. It is capable of exerting a maximum, repeating torque of 400,000 lb. at 1 ft. radius from zero to 37 r.p.m. and a maximum emergency torque of 425,000 lb. at 1 ft. radius. It is capable of delivering 1,200 hp. continuously without injury; constant torque from zero to 37 r.p.m. and slightly decreasing horsepower from 37 to 70 r.p.m.

From full speed of 37 r.p.m. to the same speed in the reverse direction of rotation can be accomplished in much quicker time than the bloom can be handled for the succeeding pass. As the entering is done at comparatively low speed, the actual period of reversal is even shorter than from full speed in one direction to full speed in the opposite direction. The time element of reversal is therefore more than enough to meet the prevailing rolling condition for both the reduction of the bloom as accomplished on the first stand of rolls and for the finishing of the plate on the second stand of rolls.

A motor-driven blower, located in the room with the flywheel motor-generator set, draws fresh air from outside of the building, discharging it through a concrete duct to the lower back end of the reversing motor. Part of the air goes into the rotor spider and out through the ventilating ducts and commutator necks past the commutator and the balance past the field coils out above the commutator. While a reversing motor of this kind is given a certain definite guarantee as to temperature rises for various loads, it is much more important that the commutation will be such as to be consistent with long life and continuity of service. It is inconsistent to set a limit of 35 deg. C. rise and sparkless commutation for average load conditions, if at peak loads the commutation becomes dangerous.

From the smaller type of steel mill there is usually a demand for quite a varied product as to length, width, thickness, shape, etc., that has to be accomplished by being able to meet these different requirements with the one available equipment. This sometimes involves apparatus capable of varied speed ranges, and as the conditions of one mill are oftentimes totally different from those of another mill, it is seldom that any two are alike. This involves a specialized study of each condition to be met in order that the proper equipment can be selected.

On the other hand, in some of the larger mills rolling a more or less standardized product, such as rails, rods, plate, etc., the power requirements are fairly well known and the equipment more or less standardized. Therefore, owing to the already numerous installations on which data are available, and in being able to closely estimate the power required for rolling steel, there is at this time no difficulty in determining within close limits the power required to drive any given mill working under any given condition.

Don'ts for the Diesel Engineer

Don't neglect to filter all your fuel oil.

Don't delay regrinding the valves until the engine stops from lack of compression.

Don't fail to clean the fuel-oil filters at stated intervals.

Don't try to put the cylinder-head nuts on the wrong studs. Mark both stud bolt and nut.

Don't forget to examine the main bearings at every opportunity.

Don't neglect the bleeder or drain on the air-compressor discharge line. Moisture in the air makes the engine hard to start.

Don't allow the air-compressor discharge to reach a high temperature; a plentiful supply of lubricating oil and a high air temperature have exploded more than one air bottle.

Don't use too much compound in grinding the fuel valves. Use it sparingly and so avoid grooving.

Don't forget to examine the needle-valve seat when grinding the needle. A pocket flashlight will lighten up the interior of the valve cage.

Don't think the air-compressor intercooler will operate forever without cleaning.

Don't allow the fuel needle-valve spring to become weak. High-injection air pressure may, in such event, cause the needle to wedge open.

Don't neglect to drop a little kerosene on the exhaust valve stem each day. This will prevent gumming where a heavy fuel is used.

Don't flood the air compressor with oil. Two or three drops per minute per cylinder is ample.

Don't neglect leaks around the fuel pump plunger. When repacking, remove all the old packing, filling the stuffing box with new material.

Don't attempt to tighten a leaky air-line joint under pressure.

Correct Height of Return-Tubular Boiler Above the Grate

By HENRY MISOSTOW
Engineer, Smoke Department, City of Chicago

For best results with bituminous coal the shell of a return-tubular boiler should be set as near the fire as possible and still provide ample room for the gases. The result is not only better heat absorption, but also better combustion, due to the lower furnace temperature and the less-violent distillation of the volatile.

To MAKE his argument in favor of higher settings for return-tubular boilers impressive, John A. Stevens in the March 30 issue of *Power* claims it to be in accord with "modern practice." Whether a particular distance from the grate to the shell is, or is not, modern practice, is of small importance. A mechanical efficiency of 90 per cent and a thermal efficiency of 10 per cent surely indicate that there are still possibilities before us.

As to "gratifying results" from modern practice, an actual incident of recent occurrence may be illuminating. Recently, the writer designed a furnace intended for high capacity and, incidentally, better economy. A 78-in. by 20-ft. return-tubular boiler was equipped with a device to promote circulation and was to be equipped with forced draft and special grates. With these improvements installed, the combination on test developed 333 per cent of rating and a little better than 75 per cent efficiency. The coal man, forced-draft man and the man who equipped the boiler with the circulating

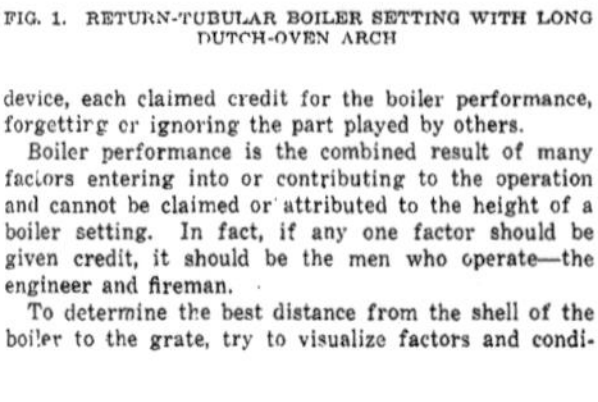

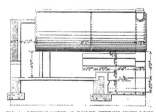

FIG. 1. RETURN-TUBULAR BOILER SETTING WITH LONG DUTCH-OVEN ARCH

device, each claimed credit for the boiler performance, forgetting or ignoring the part played by others.

Boiler performance is the combined result of many factors entering into or contributing to the operation and cannot be claimed or attributed to the height of a boiler setting. In fact, if any one factor should be given credit, it should be the men who operate—the engineer and fireman.

To determine the best distance from the shell of the boiler to the grate, try to visualize factors and conditions in practice which determine this distance, and reduce these to measureable quantities as far as possible. The distance from the shell of the boiler to the grate which makes possible the use of the greatest amount of heat generated in the furnace is to be favored, provided such a setting does not interfere with making available the maximum amount of the potential heat in the fuel. To satisfy these conditions, we must set the boiler as close to the source of heat, which is

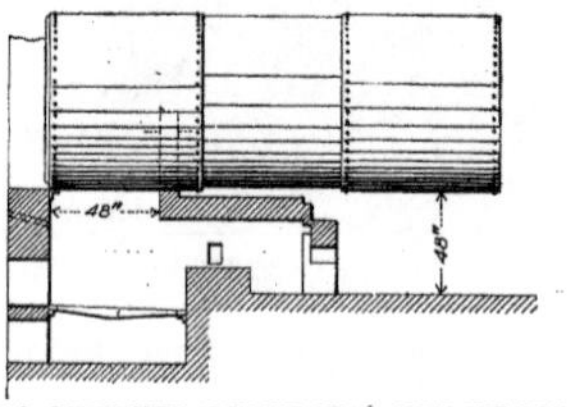

FIG. 2. DUTCH-OVEN ARCH CUT AWAY ABOVE THE FIRE AND DEFLECTION ARCH ADDED

the fire, as possible, and still have a furnace which will allow perfect commercial combustion in the presence of the boiler at the predetermined distance.

Taking for granted that the importance of draft and its utilization is fully appreciated, consider the combustion of bituminous coal. The characteristics of this coal in the process of combustion are unlike those of any other form of fuel, and yet it presents the combined characteristics of all other principal fuels, such as anthracite, coke, oil and gas. Part of it burns on the grates, part is liberated as heavy hydrocarbons in the form of oil vapor to be broken up into gas above the fuel bed, and CO and light hydrocarbons are in the form of gas nearly ready for combustion. The preparatory process before actual combustion takes place, often termed distillation of the volatile, has an important bearing on the results to be obtained. The more violent the process of distillation the harder it is to obtain complete combustion. There is less fixed carbon and more and heavier hydrocarbons carrying a greater amount of free carbon than with anthracite. The less violent the process the less hydrocarbons of the heavy composition containing little or no free carbon. These lighter gases are easy to break up and burn. The rate of distillation, or rather, violence of distillation, is directly proportional to the temperature at which distillation occurs.

This characteristic of bituminous coal being well appreciated by most of the stoker manufacturers, each claims that his stoker is the best means to obtain progressive combustion; that is, coal is being slowly moved into the zone of high temperature, making possible slow distillation at low temperature and a high-temperature zone for burning the remaining coke or fixed carbon.. So-called perfect mechanical firemen (stokers) that flip the coal have failed in competition with stokers that make progressive combustion possible because in their design the importance of slow distillation at relatively low temperature was not considered.

The foregoing being correct, then close proximity of the boiler is desirable, not only to obtain better

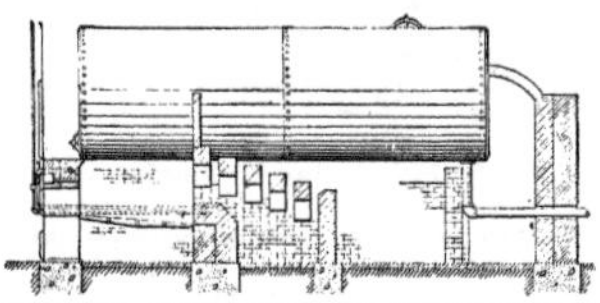

FIG. 3. ONE OF THE FANCY FURNACE DESIGNS SHOWING SLIGHT IMPROVEMENT

absorption, but also to obtain better combustion by reducing the furnace temperature and at the same time reducing the violence of distillation, thereby minimizing the difficulty of handling the volatile part of the coal. The time required for distillation in practice varies from three to ten minutes, depending upon the quantity charged and the furnace temperature, but on an average in hand-fired furnaces six minutes is ample. During combustion air admitted is heated to the furnace temperature, and naturally its volume increases as the temperature increases. Any restriction to this expansion would delay combustion and interfere with the mixing of air and combustible gases. So ample room must be provided for the gases rising from the fuel bed.

Gases on leaving the fuel bed, if the firedoors are closed, tend to rise vertically until they are affected by the force of the draft; then they bend slowly toward the exit and, when in line with or above the bridge wall, are entirely under the control of the draft and take a straight line to the rear of the boiler. Flames, indicating the direction of the gases, seldom impinge against the boiler at right angles, but travel parallel to it. In fact, they do not ascend so as to hug the curvature of the boiler, indicating that the force of the draft gets full control of the gas travel near, and a little above, the bridge wall. The chilling effect of the shell, then, cannot have any material effect on combustion, as the flames travel parallel to the shell, scarcely touching it, for there is a gas strata or film covering the boiler shell that moves at comparatively low velocity, being impeded by the roughness of the metal. Therefore, the space required for combustion will be that space necessary to accommodate the gases rising from the fuel bed at the furnace temperature. Combustion will be carried outward and completed in the combustion chamber without ill effect from the boiler shell, as anyone can become convinced by observation of the facts in every-day practice.

The amount of gas per pound of coal in the furnace seldom exceeds 20 lb. This is augmented by infiltration until at the stack, in some cases, it runs as high as 35 lb. But in considering the furnace space above the grate, an allowance of 20 lb. of gas per pound of coal should be ample. With the boiler set above the fire and free to absorb heat, the temperature in a hand-fired furnace seldom exceeds 1,800 deg. F.; for practical purposes, say there are 52 cu.ft. of gas to a pound of coal. The maximum rate of combustion for a hand-fired furnace can be taken as 25 lb. of coal per square foot of grate per hour. Gases rising from each successive foot longitudinally of the grate, traveling toward a common point and in a straight line, are bound to coincide or join together, forming one mass at or beyond the last unit area at the front face of the bridge wall. The gases rising from the front end as they pass over the second square foot will be doubled, and the gases over the last foot of grate will be as many times greater as the distance traveled by the gases rising from the first square foot of grate, granting that all grate surface is uniformly active.

The maximum rate of combustion being 25 lb. of coal per square foot of grate per hour, and 20 lb. of gas per pound of coal at 1,800 deg. F., or 52 cu.ft. of gas per pound of coal, the required height to accommodate the gases will depend on the velocity. The velocity of the gases in the furnace is at first indefinite, then gradually accelerates as they move toward the bridge wall, and can be taken conservatively as 20 ft. per second. At the front, where the velocity is less, the gas body is smaller, so natural conditions almost compensate each other. The area required to allow the gases from a strip of grate 1 ft. wide and 6 ft.

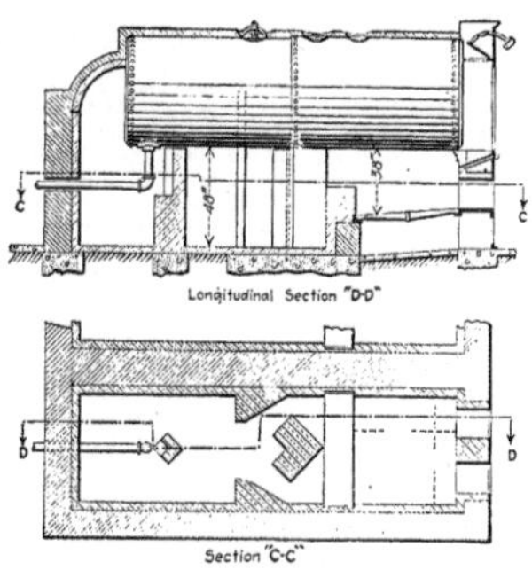

FIG. 4. FINAL DESIGN OF SETTING, KNOWN IN CHICAGO AS NO. 8 FURNACE

long to pass at a velocity of 20 ft. per second is $\frac{25 \times 20 \times 52 \times 6}{60 \times 60 \times 20} = \frac{13}{6} = 2.17$ sq.ft. As the unit width is 1 ft., the height required will be 2.17 ft., or 26 in. Taking the space requirement for the fuel bed and the pitch of the boiler as 12 in., the required distance that the boiler must be set from the grate to provide a space for the change of direction of the gases from the vertical to the horizontal will be 26 + 12 = 38 in.

From this it can be seen that the 38-in. setting will take care of the combustion requirements *provided the rate of combustion does not exceed 25 lb. per sq.ft. of grate per hour.* Where the ratio of the grate area to the heating surface of the boiler is such as not to require this rate of combustion, the distance may be reduced accordingly.

In ordinary practice boilers should be set at a height above the dead plate not less than 0.25 of the grate length plus the height of the bridge wall. In this case 25 per cent of the grate length is 6 × 0.25 = 1.5 ft., or 18 in. The bridge wall being 18 in. high, then 18 + 18 = 36. Wherever the space is available, to this is added 2 in. to take care of the required boiler slope, making 38 in. from the dead plate to the shell of the boiler. Any variation from this may be an expensive luxury or an expensive economy.

As to the betterment in performance by reducing a 48-in., or higher, setting to 38 in., it is difficult to draw a comparison, as most of the changes from higher to lower settings have been made simultaneously with other changes, such as redesign of the brickwork, and while the results obtained in each case were worth while, it would difficult to ascertain the gain contributed by reduced height. However, the writer knows of one case that may be of interest in more ways than one, and it may help incidentally to clarify some of the hazy ideas which have been, and are now being followed by some of the so-called combustion experts who persistently harp about the importance of high temperature, the chilling effect of the boiler and the horrible results following the contact of the flame and the boiler shell.

The sketches submitted are setting designs for the same boiler and can be considered as an index to the evolution of hand-fired furnaces for return-tubular boilers, each representing the prevailing idea at the time the change was made. Fig. 1 shows a full dutch-oven arch 9 in. thick extending 26 in. beyond the bridge wall. The arch is 8 ft. 6 in. long, and there is a 3-in. space between it and the shell of the boiler. The boiler is set 4 ft. at the front and 53 in. at the bridge wall from the shell to the grate.

In May, 1908, the furnace replaced a Hartford setting to prevent smoke and to improve economy. The results were not as anticipated. A boiler test showed an evaporation of only 3.84 lb. of water per pound of coal, high initial and final furnace temperatures and dense smoke after each firing for from one to two minutes. The capacity was but little over one-half of the rating. About six months later the furnace was modified as shown in Fig. 2, the significant features of modification being the addition of a deflection arch at the end of the dutch-oven arch, and the exposure of the boiler shell above the fire by removing 4 ft. from the front of the dutch-oven arch. The changes improved the smoke condition, but not to a degree to be considered satisfactory. The initial and final furnace temperatures were reduced moderately. The capacity approached rating, and an evaporation test showed 4.25 lb. of water per pound of coal of the same quality as before.

In 1914 this furnace was replaced by the fancy furnace shown in Fig. 3. A test showed an evaporation of 5.05 lb. of water per pound of coal, and the smoke condition was improved somewhat. Not being satisfied, the engineer kept on changing and another commercial furnace was installed which showed on test an evaporation of 5.17 lb. of water per pound of coal.

During the same year, on the writer's advice, the setting was changed from practically 53 in. from the grate to the shell at the bridge wall to 38 in. and the brickwork was redesigned, as shown in Fig. 4. A test with the same quality of coal used with the other furnaces showed an evaporation of 7.78 lb. of water per pound of coal. In a seven-day test the evaporation was 7.07 lb. of water per pound of coal. The chimney never shows dense smoke if the boiler is properly fired.

In this case it is difficult to say how much credit for the improvement in evaporation is due to decreasing the distance from the shell of the boiler to the grate or how much is due to the brickwork rearrangement, but there is no question that it made a tangible contribution to the final results.

The Solid Injection Oil Engine

The air compressor, be it a direct-connected or a separate unit, furnishes a large percentage of the Diesel-engine troubles. The pressures and conditions under which it operates are such that lubrication is difficult; valves often leak and packing troubles are always present. Furthermore, since considerable quantities of air are employed in the injection of the fuel charge, the power demands of the compressor range from 6 to 10 per cent of the horsepower developed in the engine. Many Diesel designers have attempted to eliminate the compressor, but with little success. Where the engine is operating on the Diesel cycle, the injection of oil occurs over a considerable period of the power stroke, and it is necessary to have some means whereby the pressure in the fuel-oil lines can be kept constant during the period of injection.

The Vickers Co., of England, several years ago abandoned the air blast on their Diesels and now make use of a fuel-injection system having a flattened steel tube to maintain the oil pressure. This design apparently is very successful since many Vickers Diesels are in daily service in the British navy and in the merchant marine.

The explosive, or constant-volume, oil engine does not require a constant oil-line pressure nor the somewhat complicated fuel-valve mechanism of the Diesel, and it is to this type of engine that the solid injection principle is most adaptable. The one feature with the solid injection engine that requires the most careful designing is the atomizer, or injection nozzle. Since there is no air blast to break up the fuel, the atomizer must do this mechanically. In fact, the oil must be atomized in a more perfect manner than necessary with the air-injection engine. Furthermore, the nozzle must handle the oil without the dribbling effect so prevalent in hot-bulb designs.

In the Diesel engine it has been customary to employ a compression pressure ranging from 450 to 600 lb. per sq.in. Those engineers accustomed to these pressures are under the impression that with the lower pressures encountered in the explosive engine the temperature will be entirely too low to cause auto-ignition. It then appears unreasonable to operate an engine having a compression pressure of 200 to 300 lb. without a hot-bulb or other ignition device. The following explanation is given with the idea of clearing up this question.

It has been proved that a fuel will ignite at a temperature well below 800 deg F. provided it is atomized or broken up to allow each minute oil particle to be in contact with the necessary amount of air. It is apparent that two things are required—thorough atomization and proper temperature. Fig. 1 is a curve showing the relation of temperature and pressure where the compression is adiabatic. This curve is based on the equation,

$$\frac{T_1}{T_2} = \left(\frac{P_1}{P_2}\right)^{1-\frac{1}{n}}$$

T_1, the temperature at the beginning of compression, is here assumed to be 212 deg. F. When the engine is warmed up, this value is about correct, although on starting cold, T_1 will be somewhat lower. The exponent n is assumed to be 1.35 instead of 1.41, with the intention of making allowance for the loss to the cylinder walls during compression.

The temperature with 300 lb. final compression pressure is approximately 1,050 deg. F. This is amply high to ignite any of the fuel oils. The Diesel engine, with

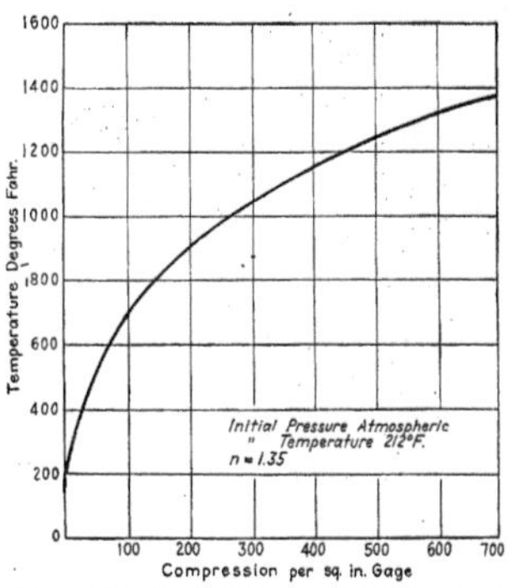

FIG. 1. CURVE SHOWING RELATION OF TEMPERATURE AND PRESSURE WHERE COMPRESSION IS ADIABATIC

its large radiating surface relative to its compression volume, experiences a loss of temperature that is much greater than in the case of the solid-injection engine having a combustion chamber with a large volume and a small radiating surface. It is quite likely that the chilling effect of the expanding air blast in the Diesel causes a drop of at least 150 deg. F. in the temperature of the air and oil. As a practical example of the ability to operate at 300 lb., many operators are running old Diesels with a still lower compression.

As has been stated, the curve in Fig. 1 is based on an initial air temperature of 212 deg. F. On starting a cold engine, this value is much too high; if there is no throttling of the air, it should be around 70 deg. F. This would give a final temperature of about 725 deg. F., which is sufficient to ignite a well-atomized fuel charge, although the first few explosions will probably smoke badly. To secure thorough atomization and ignition, it is necessary that the oil be broken up without any part striking the relatively cold walls of the combustion chamber. This has led to various designs of combustion chambers as well as atomizing devices.

In this discussion no consideration has been given to the effect the injected oil has on increasing the pressure and temperature of the mixture in the combustion chamber or to the time element as applied to the period of vaporization of the oil. There is no doubt that both these factors have marked influences on the action of the engine.

The timing of the injection of the fuel varies in different makes of engines from 40 deg. to 6 deg. ahead of outer dead center. The operator of a Diesel knows from experience that an injection point much over 7 deg. ahead of dead center will produce marked preignition; in the hot-bulb engine an injection point of 30 deg. will cause severe pounding unless water injection is employed. In the solid-injection oil engine it has been found that the timing is largely dependent on the compression carried. If the compression is as low as 200 lb., as in the Price engine, the injection must be earlier than 30 deg. to give ample time for the vaporization of the oil, while with 300 lb. pressure, as with the De La Vergne S. I. engine, the injection is not over 6 deg. To avoid danger of preignition at early injection periods, a combustion chamber is placed in the head. The oil is injected into this space where it mixes with the air. Since the major part of the air charge is contained within the cylinder, the amount of air in the combustion chamber is not sufficient to produce ignition even with a sufficiently high temperature. Ignition can occur only when the moving piston forces the main air charge into the combustion space.

The fuel consumption per brake horsepower is low; in many cases it equals the best Diesel results. Since the compression is much lower than with the Diesel engine, the thermal efficiency based on indicated horsepower is not as good as with the latter engine. The extremely high mechanical efficiency, due to elimination of the air compressor and lower frictional loss, is responsible for the excellent fuel economy.

Some Boiler Statistics from "The Locomotive"

SUMMARY OF INSPECTOR'S WORK FOR 1919

Total number of boilers examined	371,285
Number inspected internally	160,847
Number inspected by hydrostatic pressure	9,043
Number of boilers found to be uninsurable	1,042

SUMMARY OF DEFECTS DISCOVERED

Nature of Defect	Whole Number	Dangerous
Cases of sedimen. or loose scale	31.599	1.783
Cases of adhering scale	17.284	1.907
Cases of grooving	2.393	271
Cases of internal corrosion	21.201	817
Cases of external corrosion	11.260	955
Cases of defective bracing	1.173	230
Cases of defective staybolting	2.428	498
Settings defective	9.423	836
Fractured plates and heads	3.473	487
Burned plates	4.836	518
Laminated plates	338	27
Cases of defective riveting	1.443	324
Cases of leakage around tubes	13.318	1.226
Cases of defective tubes or flues	19.700	6.353
Cases of leakage at seams	3,738	413
Water gages defective	4.158	789
Blowoffs defective	5,325	1,488
Cases of low water	421	149
Safety valves overloaded	1.134	289
Safety valves defective	2.077	374
Pressure gages defective	7.332	628
Boilers without pressure gages	634	82
Miscellaneous defects	5.388	664
Total	202.376	20.603

Note that more than a thousand boilers were found so dangerous as to be considered too great a risk—and this is the number found by but one company.

Electric Elevator Machinery—Brake Adjustments

Precautions To Be Observed in Coupling Motor to Elevator Machine — How To Make Brake Adjustments Properly—How To Reline Brake Shoes and Get the Leather To Make a Good Fit

By M. A. MYERS
Electrical Engineer, The Maintenance Co., New York City

THE connecting link between the motor and a modern elevator machine is the coupling, which is actually a combined coupling and brake pulley. Fig. 2 shows a section through a typical coupling. The fit of the shafts in the coupling must be excellent, the keyways well cut and the keys a tight fit. Owing to the constant reversals, starts and stops, loose keys and keyways will be so battered in a comparatively short time that a shop job will be required to repair the damage.

The brake on electric elevators may be either a leather-lined band of sheet steel, or as is now more generally practiced, two leather-lined cast-iron shoes. Occasionally, in some of the older machines wood lagging is found instead of leather. The brake is opened and held open by an electric solenoid or magnet and is set against the brake pulley when the current is switched off, by means of a spiral spring or springs, although occasionally a weight-and-lever device is used.

One modern type of brake, as used on a Wheeler McDowell elevator machine, is shown in Fig. 1. When the solenoid *F* is energized, the plunger *G* is pulled up and raises lever *H*. This action lifts the top shoe and lowers the bottom one by means of the links *K* and *K*, which are made adjustable by turnbuckles. Side movement of the shoes *S* and *S* is prevented by the guides *E* and *E* which are an integral part of the supporting frame. When the solenoid is de-energized, the shoes are forced against the pulley by the spiral springs *J* and *J*, the tension of which is adjusted by the nuts *N* on the through-rod.

The adjustments of elevator brakes may be easily and quickly made or may be a slow and tedious job, according to the style of brake and the local conditions. An improperly counterweighted car adds to the difficulties of brake setting, and this is particularly true where the car is heavier than the weights. The prerequisite for satisfactory brake adjustment is to have the shoes to open as little as possible without rubbing the brake wheels when the brake is released. An opening that will permit the insertion of an ordinary sheet of writing paper between the shoes and brake wheel is sufficient. This will allow the shoes to set quickly and smoothly and minimize any tendency to rebound. The remaining adjustment may be then made by means of adjusting spring tension or weight leverage.

FIG. 1. SHOWING ELECTRIC ELEVATOR BRAKE MECHANISM

A properly counterweighted equipment is one that is over-counterweighted by about one-third the full-load capacity of the car. Then as the car is loaded, it approaches an even balance and ordinarily makes the majority of its runs under this approximately balanced condition; therefore, strain on the brake is evenly divided on the up and down motion. When the car is loaded up to two-thirds of the capacity, the original no-load condition is simply reversed in that the one-third overbalance is in the car instead of the counterweights. Then there is only a one-third additional load on the car over the counterweights to provide for, making it possible in the ordinary brake setting, as already described, to insure a reasonably easy stop under all conditions of loading.

When the car is heavier than the counterweights to start with, the brake will have more work to perform on the down motion than on the up, and this condition never reverses. As the load on the car is increased, the work for the brake on the down motion is increased, while on the up motion it is practically zero, except to prevent the car from starting downward again after it comes to a stop. If the brake is adjusted to work smoothly on the down motion, it will very likely cause a severe jar to the car and machinery on the up motion. About the best that can be done in some cases is to strike a compromise. However, where the construction of the brake will allow, adjusting one shoe closer to the pulley than the other or giving one shoe greater tension than the other, or both, will help materially in making a smooth stop. The explanation for this will be found in Fig. 3. With the brake pulley running clockwise, as shown by the curved arrow, in which case most cars will be on the up motion, the rotation of the pulley when the brake sets will tend to make the top shoe grip tighter and will tend to throw off the bottom shoe, as indicated by the short arrows. Reverse the motion and the

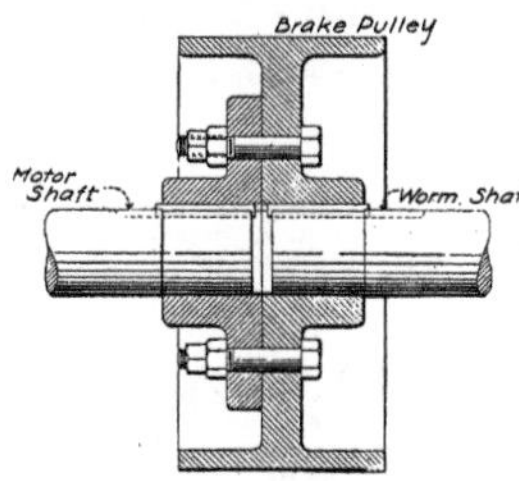

FIG. 2. BRAKE WHEEL AND COUPLING

opposite is true. The top shoe, therefore, may be said to be primarily the up-motion shoe and the bottom the down-motion shoe. In the difficult case just described, surprising results may sometimes be obtained by adjusting the bottom or down-motion shoe to do most of the work. Where the brake has vertical shoes, merely figure out the up- and the down-motion shoes from the direction of rotation and the location of the hinge pins. This principle of adjustment need not necessarily be applied to exceptional cases only.

Dynamic Braking for Elevators

Some types of brakes have adjustable stops to govern the amount of opening of one or both shoes. In all cases a few moments' study will show the similarity of action but with different details. The foregoing discussion takes no account of dynamic braking, which is now considerably used, especially for high-speed cars, to assist the mechanical brake. Dynamic braking consists of connecting a resistance across the armature terminals the instant the operating current is switched off, with the fields in a greater or less state of excitation. The motor then becomes a generator sending current for an instant, of approximately the initial starting inrush value, and dying down to zero as the car comes to rest, through the resistance. This dynamic load on the motor acts to bring it to a stop and is of great assistance to the mechanical brake. The amount of dynamic braking can be adjusted by adjusting the dynamic resistance and also, where the system of control permits, by adjusting the field strength. On some types of controllers two or three dynamic braking switches are used to reduce the dynamic resistance as the motor comes to a stop.

If for any reason the brake must be removed, take the precaution to land the weights in the pit by placing a timber under them and inching the car up until the weights land. If it happens that the car is heavier than the weights, it must of course be landed instead of the latter. This condition can usually be determined by holding the brake open so that it can be quickly applied and then turning the brake pulley and noting the direction in which it tends to continue to rotate. Where convenient, the drum is sometimes blocked instead of landing the weights or car. These precautions are to prevent runaways, which sometimes result in disaster.

Oil Sometimes Causes Trouble

Trouble with brakes may arise from oil creeping from the motor bearing or gear-case stuffing box to the brake pulley and impregnating the leathers. When the brake is cold, this oil will be gummy and the braking action likely to be harsh. As the brake warms up, the oil will become thin, and sliding may ensue, The remedy will be to remove the shoes, wash the leathers thoroughly with gasoline or scrape with glass, or apply both remedies. Sometimes fuller's earth or whiting is rubbed into the leathers to absorb the oil, after which they are scraped. On the other hand, when the leathers become dry and brittle, they should be treated with a good leather dressing such as neatsfoot oil. Wearing of the hinge pins, link pins and bores will increase brake-adjustment difficulties. When this occurs, the holes can be rebored and bushed and new pins substituted.

The brake leathers are fastened to the shoes or bands by means of copper rivets, the heads of which are countersunk into the leather. Before the leather is worn so thin as to allow the rivet heads to rub, the brake should be relined. The leather is usually about one-quarter inch in thickness; the proper thickness for any case can be determined by measuring the brake-pulley diameter and the diameter to which the shoes are bored.

To renew the leathers without special equipment, remove the old leather and rivets, cut the new lining to size and clamp it to the shoes at each end by means

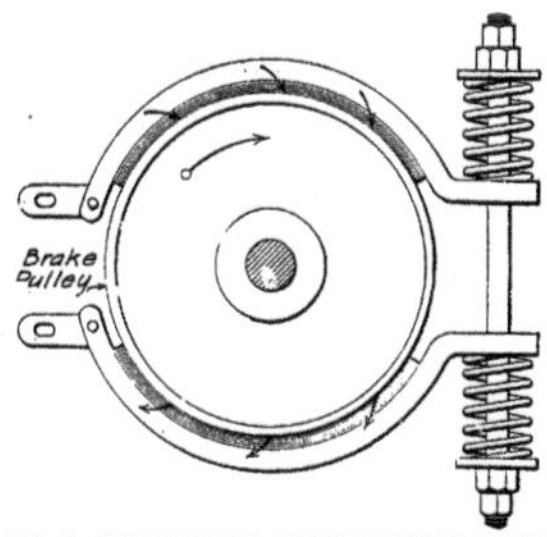

FIG. 3. INDICATES THE ACTION BETWEEN BRAKE SHOES AND PULLEY

of "C" clamps. Since the leather is quite stiff, it should be bowed outward and the edges hammered in without loosening the clamps to force it to lie perfectly flat against the shoe. By repeating this operation several times a good fit will be obtained, after which an ordinary twist drill of the proper size should be run through the holes in the shoe and at the same time through the leather, holding the latter against a wood block. A special flat-ground countersink drill the size of the rivet heads, with a center guide, is then used to countersink the holes in the leather to at least one-half the thickness of the lining. The riveting is then simple; No. 7 rivets are usually employed. Care must be taken that the leather hugs the shoe the entire surface to make sure. Sometimes it is desirable to place a leather lining on an old wood-lagged brake. The wood lagging should be bored to the correct diameter, after which the leather may be fastened either by shoemaker's wooden pegs, or by flat-head wood screws, countersinking the heads similarly to the rivets. In either case it is well to fasten the ends of the leather with flat-head stove bolts, extending through the old lagging and the iron shoe.

Our Water Powers—Fallacies, Facts and Possibilities

Rebuilding Small Plants as Desirable as Making New Developments

BY EARL STAFFORD
With John F. Vaughn, Engineers, Boston

HOW many of us have a clear understanding of the future of our water powers, their extent, the expenditure necessary before their energy may be utilized and their value to the country when developed? Extravagant statements by enthusiasts have added to a popular belief that water power is practically inexhaustible and virtually free, while repeated use of the term "white coal" has created the impression that when coal and other natural fuel supplies are gone, hydro-electric developments will furnish the greater part of our light, heat and power. It is the purpose of this article to show the error in such beliefs by giving a better general idea of the limits and possibilities of water power, and also to show how the manufacturer owning a water-power plant may, with profit to himself, help to conserve the fast disappearing fuel of the United States.

Water power can never be exhausted by use because it comes from the rain-water which, when evaporated from the ocean, runs back over the falls and rapids of the rivers. On the other hand, the amount available in any one year has definite limits which may be determined by engineering investigation, and although the supply is continuous the volume of power is surely far from unlimited. One of our leading engineers, Dr. Steinmetz, of the General Electric Co., in a comprehensive analysis of the water resources of the United States, has shown that if all rivers could be dammed throughout their length and developed for power, we would not get the power equivalent of the coal mined annually in this country. The impossibility of completely substituting water power for coal will thus be seen, and as the requirements of our industries are increasing at an enormous rate, the discrepancy between the demand and this supply will be even more apparent as the years go on.

Water-Power Development Costs Money

Another idea often put forth is that water power is unlimited and that vast amounts may be had for practically nothing. This is far from true, for although water, the raw material, is free, plants built to produce hydro-electric power involve heavy expenditures. Modern large-scale developments require investment for power houses and equipment, and for the cost of land inundated by dams, with resulting interest charges that are equal to a large part of the fuel bill of a steam-power station doing the same work. In most sections of the country, however, there are still many untouched water-power sites that may be developed to produce power cheaper than it can be obtained from coal. As demonstrated by early ventures, however, success in water-power development of any magnitude can come only through sound financing and careful engineering. At present we can choose those most advantageous, but the time will come when almost all water-power possibilities must be developed, even at great cost, in order to conserve fuel for human needs.

It has been well said that the most daring experiment ever attempted by the human race occurred when man left the torrid zone for climates where fuel is an absolute necessity for preserving human life. Cheap and abundant fuel has so far made the experiment a success. That it may continue to be so in the face of the depletion of fuel resources is a problem that gives much concern. The solution undoubtedly lies in securing additional sources of heat supply. In this connection some of us have the hazy notion that we can use electricity from water power for heating our homes. This is indeed a fallacy. Electricity for heating a home would cost at least ten times as much as coal for the same purposes, and the demand would exceed the supply long before even those who could afford to pay for it could be provided for. We must, therefore, continue to use fuel for heat, but can employ water power for lighting and industrial use, assisted when necessary by power from steam stations. It is a patriotic duty, then, to be saving of natural fuel resources for coming generations, that they may not be able to convict this generation of robbing them.

Conservation Generally Recognized as Necessary

Attempts to adopt a general conservation policy have already been made by some nations. In England the government is studying schemes for mammoth steam plants located by districts so that power may be distributed electrically and thus eliminate the waste of scattered small producers. Canada is alert, and through commissions, the various provinces are not only mapping out a progressive policy but also have built large hydraulic plants and transmission systems. In this country also, schemes have been proposed for private capital to build steam plants at the mouths of the mines, connected through a network of transmission lines with power developments on the large rivers. These proposals are not visionary and are sure to materialize within the lives of the present generation.

One of the more economical ways of obtaining power is by rebuilding the obsolete under-developed or poorly planned hydraulic plants that were constructed years ago to meet the moderate demands of our early manufacturers. A small water-power plant, for instance, may be so outgrown by the mill of which it is a part that its value seems insignificant. While the rest of the manufacturing plant has become modernized, the water-power equipment, except for the occasional replacement of worn parts, has been left as originally built. Perhaps the amount of power obtained from it is unknown, and very likely thought to be several times as great as tests would show. Possibilities for revamping or redevelopment in such cases are comparatively great, and frequently 100 per cent more power is obtainable. The owner of such a plant may reason that the gain by rebuilding is insignificant when compared with the rest of his manufacturing costs. This may be true, but the saving in power cost that is possible by increasing the output of even a small hydraulic plant may be an appreciable part of the net earnings of the entire factory which it serves. In general it may be stated that every ten horsepower that can be added to the output of the plant by the use of water power will save annually the mining and transporting of a car of coal.

How Reconstruction May Increase Output

As illustrations of the ways in which more power may be secured through reconstruction of an old plant, the following may be mentioned

1. New waterwheels of modern type will deliver more power from the same water.
2. Elimination of gears and countershafting by direct-connected electrical generators, from which the power is distributed electrically, will often add to the power previously obtained.
3. By enlarging and changing the water passages to and from the wheel case, according to our better knowledge of hydraulics, the power losses caused by friction in the flowing water will be materially reduced.
4. By general improvements of the entire hydraulic layout through stopping leaks in dams and bulkheads and making miscellaneous repairs that an engineering examination would show to be advantageous, much wasted water may be saved.
5. Changing the top of the dam or the flashboards so as to provide at all times a maximum level of the headwater will often increase the available head. When mills were first built, their power requirements were small and no attempt was made to secure all the head or fall possible for the hydraulic plant. New ways of discharging the flood water open up unexpected possibilities along these lines.
6. Improvement in the millpond or in small near-by storage reservoirs so that the night flow of the river may be stored for day use, will increase the water available.

When, as in the case of many mills, there is a steam plant working in conjunction with the hydraulic plant, the combined saving by rebuilding the water power may be especially advantageous.

Old plants, considering the quantity of power that they furnish, are expensive to operate and to maintain. Modern plants, on the other hand, because of the extreme simplicity of their apparatus, require little care and almost no repairs and replacements, for elimination of gears and bearings has done away with practically all the old troubles. Automatic or semi-automatic plants which operate night and day without an attendant and with only occasional inspection are among recent improvements. Automatic devices control the power output of these plants according to the water available or the demand for power, as the case may be. When located some distance from the mill, electrical control switches and indicating instruments may be provided which make it unnecessary to go to the power station when starting or shutting down.

If the power produced by a small plant is at times more than can be utilized, a new source of revenue lies in connecting with the transmission system of public-utility light and power companies. The output from reliable hydro-units, however small, is often sought by local power companies, as it gives a reserve for emergency. If a manufacturing plant is purchasing part of its power, this expense may be materially reduced by selling to the power company the spare output of the plant during the night. Revenue from such sale is practically clear profit as it comes from water that otherwise would go to waste over the dam. A case in mind is that of a small low-head plant thought to be useless and at first abandoned because the mill, which had burned, was rebuilt at some distance. The rehabilitated power plant with a short transmission line now furnishes about two-thirds of all the power the mill requires, and during the greater part of the year sells part of its output to the local power company at a reduced rate, paying the owner a handsome profit on his investment. The financial return, however, is not necessarily the most important consideration, as was demonstrated during the war when the Fuel Administration found it necessary to close on one day per week all factories not operated by water power. By obtaining the maximum output from his hydraulic plant, the individual manufacturer may safeguard his own interests and by helping to save coal further a policy of conservation that will become increasingly important in America's progress.

Conservation a Present, Not a Future Problem

To those who say that conservation is a problem of the future and that later generations must look out for themselves, we answer that the time has already come when this generation must look carefully to its own future. The cheapest and most readily obtainable of our natural resources have been used, and the rich homestead lands that attracted European labor have been settled. We are fast approaching the labor and natural economic conditions that prevail in other countries.

The turn of the road for us came with the great changes in world economic conditions caused by the war. We are being forced into active trade competition with European nations which make conservation a science. To retain the lead that the war has given us, we must study to reduce extravagance and to eliminate waste, in order that when the period of lower prices comes we shall be prepared to meet it. Development of additional water power will release coal-mining labor for other work and will lighten the heavy burden imposed upon the railroads, which results for the individual manufacturer in higher fuel prices and from which the nation as a whole suffers through clogging of the channels of transportation.

In planning the big things, let us not overlook the small possibilities—great in the aggregate—that lie around us, at the power sites selected by our forefathers for their mills, for the improvement in them which is commercially possible will on the whole double the power that they now contribute.

Two Indirect Methods of Finding the Weight of Ammonia in Circulation in Compression Refrigeration

BY J. W. GAVETT, JR.

MANY different methods have been used to determine the weight of ammonia in circulation in compression refrigerating systems. Among the direct mechanical methods are the use of meters of various types and the placing of tanks, mounted on scales, between the condenser and the receiver. It is evident that these entail the installation of special and expensive apparatus and add to the already large number of valves and joints that must be kept tight.

It is proposed here to outline two non-mechanical methods of computing the weight, which require little in the way of apparatus beyond the plant proper. They will be considered separately, the shorter and simpler, which might be termed the pressure-volume method,

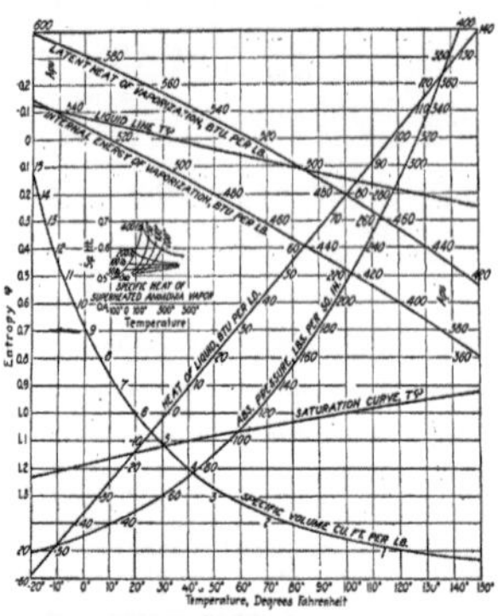

FIG. 1. THERMODYNAMIC PROPERTIES OF AMMONIA

The scale to the left of the chart applies, of course, only to the liquid line and saturation curve. (Plotted from tables of Keyes and Brownlee.)

being discussed first. The necessary apparatus additional to the plant includes thermometers to read discharge temperature at the compressor cylinders, a revolution counter (usually installed with the plant) and an indicator with which to take indicator diagrams from the compressor cylinders. The charts of the properties of ammonia, which accompany this discussion, complete the necessary equipment. These charts were recently plotted by the writer from the tables published in 1917 by Messrs. Keyes and Brownlee ("Thermodynamic Properties of Ammonia," John Wiley & Sons, Inc.), which are the latest and regarded by the present writer to be the most accurate source of information available.

A brief explanation of the charts is advisable at this point. The various curves of Fig. 1 relate to liquid ammonia and to wet and saturated ammonia vapor, but do not enter the superheat field. On this chart the only curve that applies to the present problem is that of

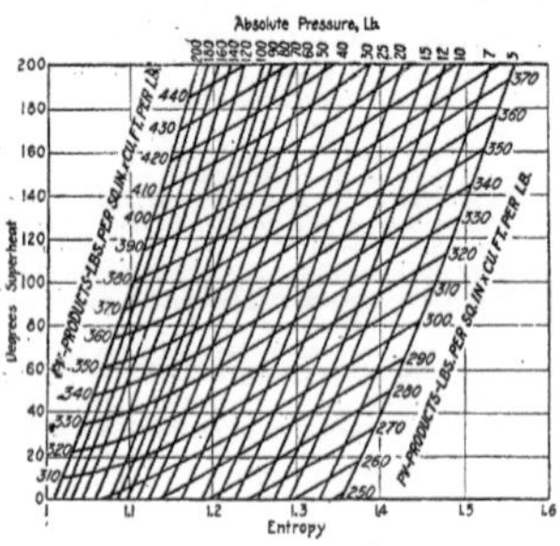

FIG. 2. PROPERTIES OF SUPERHEATED AMMONIA VAPOR

absolute pressures vs. temperatures. This curve is, of course, a means of determining the boiling points of liquid ammonia at various absolute pressures or vice versa.

Fig. 2 deals with superheated ammonia vapor and is fundamentally a temperature-entropy chart in which degrees of superheat have been substituted for temperature. Superimposed on the temperature-entropy coordinates are two additional sets of lines, one of constant absolute pressures and one of constant pressure-volume (PV) products (pressure in lb. per sq. in. × volume in cu.ft. per lb.). The advantage of this arrangement will become apparent as the discussion proceeds.

The demonstration and calculations will be made from an actual indicator diagram and actual data taken from a test run on the 15-ton York plant at Cornell University. To obtain maximum accuracy, the data should be averaged from readings and diagrams taken at intervals throughout a run of considerable length. For simplicity of demonstration one card only is used in this discussion, although the remainder of the data has been averaged from a series of readings. Fig. 3 is an exact copy of a diagram taken from one cylinder of the two-cylinder compressor. The additional necessary data are as follows:

Barometric pressure in mercury, 29.95 in., equals	14.7 lb. per sq.in.
Discharge pressure, gage	68.0 lb. per sq.in.
Discharge pressure, absolute	82.7 lb. per sq.in.
Discharge temperature, deg. F	138.5
Dimensions of compressor cylinder	8x12 in.
Piston displacement per revolution	0.443 cu.ft.
Scale of indicator spring	60 lb.
Revolutions per minute of compressor	52.8

The first step is to find the *PV* production of *one pound* of ammonia at discharge conditions; that is, on the high-pressure side of the system. From Fig. 1 the

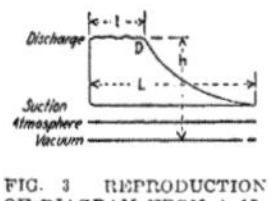

FIG. 3 REPRODUCTION OF DIAGRAM FROM A 15-TON COMPRESSOR

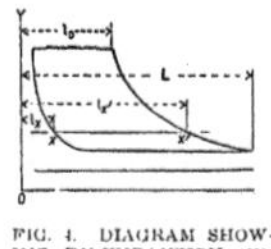

FIG. 4. DIAGRAM SHOWING RE-EXPANSION EXAGGERATED

boiling point corresponding to an absolute pressure of 82.7 lb. per sq.in. is 45.5 deg. Thus the superheat at discharge is 138.5 — 45.5 = 93 deg. Turning now to Fig. 2 and following the horizontal line of constant superheat of 93 deg. to its intersection with the constant pressure line of 82.7 lb., a *PV* product value of 362 is determined.

The remainder of the procedure is based on measurements of the diagram taken at point *D* (Fig. 3), which is the point where discharge pressure is reached and ammonia begins to leave the cylinder.

The actual volume at D is $\frac{l}{L} \times 0.443$ cu.ft., in this case $\frac{0.77}{2.29} \times 0.443 = 0.149$ cu.ft.

The diagram, Fig. 3, shows clearly that there is no re-expansion in the cylinders; that is, there is no clearance volume. Should the diagram show re-expansion, as in Fig. 4 (exaggerated), the clearance volume must be determined and allowed for. The clearance line *OY* (Fig. 4) may be calculated from direct measurements of the cylinder or may be graphically determined by various methods described in books dealing with steam-engine-indicator diagrams. To determine the allowance to be made, draw any line xx' parallel to the vacuum line and intersecting both the re-expansion line and the compression line. The total volume in the cylinder is $\frac{l_x}{L} \times$ (piston displacement + clearance volume), but of this the proportion $\frac{l_x}{l_{x'}}$ is due to clearance and will not be discharged at the end of the stroke. This ratio may safely be assumed to hold for the entire re-expansion line, so the true volume to be considered at point D is $\left(\frac{l_D}{L} \times \frac{l_x}{l_{x'}}\right) \times$ (piston displacement + clearance volume).

The absolute pressure at *D*, Fig. 3, is $h \times 60$, or $1.39 \times 60 = 83.4$ lb. per sq.in. The actual *PV* product is therefore $0.149 \times 83.4 = 12.4$.

The weight of ammonia in the cylinder is thus $\frac{PV \text{ actual}}{PV \text{ for one lb.}}$, or $\frac{12.4}{362} = 0.0342$ lb. The weight per hour follows by multiplying by the r.p.m., or revolutions per minute, and by 60; that is, $0.0342 \times 52.8 \times 60 = 109$ lb. per hr., approximately.

A similar process is applied to the other cylinder, and the sum of the two results gives the total weight in circulation per hour. In this case the total was 236 lb.

This method may be applied only when the vapor is superheated at discharge, but as this is almost universally the case in present practice, its application will not be greatly limited on that account.

It will be noticed that this procedure is based on the assumption that any change that takes place in the vapor as it passes through the discharge valve has but a negligible effect on the value of the *PV* product. This passage through the valve is evidently a wire-drawing effect and takes place at constant heat. A critical analysis of the process on this basis, using

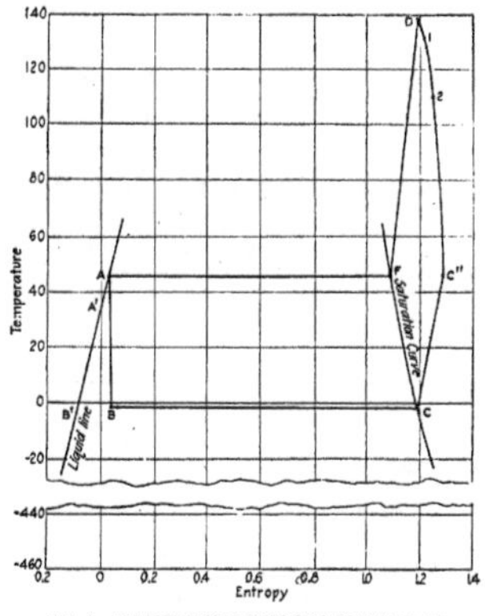

FIG. 5. TEMPERATURE-ENTROPY DIAGRAM OF AMMONIA CYCLE

Since entropy is a function of heat and absolute temperature, the temperature scale must read from absolute zero, or 460 deg. F. For the purpose of this discussion the temperature scale is broken to avoid unnecessary size of the chart.

values from the tables previously mentioned, shows that this wiredrawing has no perceptible effect on the *PV* product.

As for other sources of error, analysis shows that discrepancies in pressure due to inaccuracy of the gage or in reading will induce an error in the *PV* value of 10 per cent of the relative error of the pressure determination. In other words, an error of 1 per cent in pressure will cause an error of $\frac{1}{10}$ of 1 per cent in the *PV* value.

Similarly, an error of 1 deg. in the measurement of discharge temperature will produce an error of $\frac{18}{100}$ of 1 per cent in the PV value. These errors should be negligible.

The Temperature-Entropy Method Explained

The second method to be discussed may best be designated as the temperature-entropy or $T\phi$ method. This system of calculation requires additional thermometers to make temperature determinations as listed below and also another pressure measurement, that of the low pressure or suction side of the system. In this case we had:

Suction pressure, gage	14.5 lb. per sq.in.
Suction pressure, absolute	29.2 lb. per sq.in.
Temperature of ammonia into condenser	112 deg. F.
Temperature of ammonia out of condenser	46.5 deg. F.

Since entropy (ϕ) is by definition the dimensions of heat divided by *absolute* temperature, an area on a $T\phi$ chart represents heat. Heat may be expressed as work if multiplied by 778, the mechanical equivalent of one B.t.u. These are the fundamental conceptions on which the temperature-entropy ($T\phi$) method is based.

Fig. 5 shows the $T\phi$ diagram for *one pound* of ammonia in the case under discussion. The construction of this diagram from the data available will now be traced. The first step is the plotting of the liquid line and saturation curve. These may be taken directly from Fig. 1, which, when turned left side down, contains a $T\phi$ chart. Let us start with the liquid ammonia emerging from the condenser at condenser (discharge) pressure (82.7 lb.), for which the corresponding boiling temperature is 45.5 deg. F. This gives point A on the liquid line. In the present case there is practically no supercooling, the temperature of the ammonia coming out of the condenser being 45 deg. In case supercooling occurs, it will be shown by some point A' on the liquid line. This makes no difference in the area of the diagram, since the next stage of the ammonia cycle is reheating along the liquid line back to point A. This heat will usually be taken up from the storage system or piping between the condenser and the expansion valve.

The ammonia now expands freely at constant heat through the expansion valve to suction pressure, 29.2 lb. abs. The boiling temperature corresponding to this pressure (—2 deg. F. from Fig. 1) locates the level of line BC, along which vaporization takes place, but does not locate point B. We know that some vaporization has taken place and that it is expressed graphically by the distance B'B, the value of which may be found by equating the heat contents at points A and B. The sensible heat of the liquid q_A (at point A) is, from Fig. 1, 15.5 B.t.u. per lb. At B the heat is made up of two parts, first the heat of the liquid at —2 deg. F. and second a ratio $\frac{B'B}{B'C}$ of the heat of vaporization r, at —2 deg. F. From Fig. 1 r_B is 582 B.t.u. and q_B is —39 deg. F. Letting $\frac{B'B}{B'C} = x$, we can write

$$q_A = q_B + x r_B \text{ or } x = \frac{q_A - q_B}{r_B}.$$

Substituting, we find $x = 0.0935$; or, in other words, the quality of the ammonia vapor at point B is 9.35 per cent. Next, multiplying the length of the line B'C (in any units) by 0.0935, we get the length B'B, thus locating point B.

From B the vaporization continues along line BC, the heat of vaporization being taken from the brine or other substance, which is to be cooled. At point C the vaporization is complete and superheating begins. The line CC″ is evidently a constant-pressure line, as the ammonia is still on the suction or low-pressure side of the system. It is sensibly a straight line and can be drawn as such without appreciable error. Its exact shape may be plotted with data taken from Fig. 2. At point C″ the ammonia enters the compressor. The point is located as follows: From the data, suction temperature is 46.5 deg. F., but the boiling point at suction pressure is —2 deg. F. Thus we have 46.5 — (—2) = 48.5 deg. superheat at 29.2 lb. per sq.in. From Fig. 2 we can read the entropy at point C″ as 1.285. Plotting this against the temperature (46.5 deg. F.), we locate point C″.

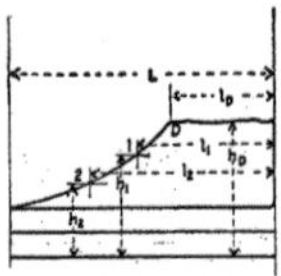

FIG. 6. SHOWING MEASUREMENTS FOR TRANSFERRING COMPRESSION LINE TO TEMPERATURE - ENTROPY DIAGRAM

A similar process applied to discharge conditions locates point D, but the shape of the line C″D, which corresponds to the compression line on the indicator diagram, must be obtained from that source. As before, one diagram only will be used for demonstration and but two intermediate points will be plotted. Fig. 6 shows the actual diagram and indicates the method of making the necessary measurements. For accuracy a series of results should be averaged and at least four intermediate points plotted.

This process may best be shown by means of a table with some accompanying explanation.

Points	D	1	2
$l/L \times 0.443$ = actual volume	$\frac{1.12}{2.92} \times 0.443 = 0.17$	$\frac{1.49}{2.92} \times 0.443 = 0.226$	$\frac{2.02}{2.92} \times 0.443 = 0.306$
$h \times 60$ = absolute pressure	$1.44 \times 60 = 86.2$	$1.06 \times 60 = 64.5$	$0.765 \times 60 = 45.9$
Actual PV product	$0.17 \times 86.2 = 14.6$	$0.226 \times 64.5 = 14.55$	$0.306 \times 45.9 = 14.05$

The PV product for one pound at discharge conditions has been determined as 362. Then the PV values for one pound at points 1 and 2 must bear the same relation to 362 as the actual PV values bear to the actual PV value at point D. Thus we get

PV per pound	362	$\frac{14.55}{14.60} \times 362 = 360$	$\frac{14.05}{14.60} \times 362 = 348$

Then from Fig. 2, knowing the PV values and the pressures at points 1 and 2, we can read entropies and superheats.

Superheat, deg. F	93	98.5	92
Entropy (ϕ)	1.190	1.220	1.248

The actual temperature at points 1 and 2 will equal the boiling points at the known pressures plus the superheats, giving

Boiling point, deg. F........	45.5	at 64.5 lb. per sq.in. 33	at 45.9 lb. per sq.in. 18
Actual temperature, deg. F..	45.5 + 93 = 138.5	33 + 98.5 = 131.5	18 + 92 = 110

Thus knowing actual temperatures and entropies, we can plot points 1 and 2 on the diagram.

The line *DF*, like *CC″*, is a constant-pressure line and can be drawn straight. Line *FA* is, of course, a horizontal straight line at constant temperature, which closes the cycle.

A consideration of the chart, as well as our knowledge of the refrigeration process, shows that the ammonia takes up heat through that portion of the cycle represented by *ABCC″D* and gives up heat through the remainder of the cycle. It follows then that the area *ABCC″DFA* represents the difference between heat given up and heat supplied and must therefore be equivalent to the work done by the compressor on *each pound* of ammonia. A planimeter will serve to measure the area of the cycle, which in this case is 17.14 sq.in. Since the co-ordinates are 20 deg. F. of temperature and 0.2 units of entropy per inch, one square inch is equivalent to 20 × 0.2 = 4.00 B.t.u. The whole area then represents 17.14 × 4.00 = 68.56 B.t.u. per lb. of ammonia.

The total work done by the compressor per hour may be determined by the usual method from the indicator diagram of the ammonia cylinders, adding the powers of the two cylinders and averaging results. For the present cases the total was 6.48 hp. Since one horsepower-hour is equivalent to 2,545 B.t.u., the heat equivalent to the compressor horsepower is 2,545 × 6.48 = 16,500 B.t.u. per hour.

Evidently, then, the B.t.u. work per hour performed by the compressor divided by the work *per pound* of ammonia will give the weight in circulation per hour, that is,

$$\frac{16{,}500\ (\times\ 778)}{68.56\ (\times\ 778)} = 240\ lb.\ per\ hr.$$

This result checks that of the first method within 2 per cent.

It should be pointed out that the temperature-entropy method is very susceptible to error caused by inaccuracy of pressure determination. For instance, if the suction pressure is incorrect by two pounds, the corresponding boiling point will be in error by at least two degrees. This will raise or lower the line *BC* and seriously alter the work area of the diagram, which is the ultimate factor in determining the weight. An error in discharging pressure will have a similar but smaller effect. Thus it will be noted that the *PV* method is not only shorter and simpler, but is probably more accurate than the temperature-entropy method.

According to the Geological Survey's report, the average daily production of electricity by public utility power plants during January was 124,600,000 kw.-hr., during February 119,800,000 kw.-hr., and during March 121,800,000 kw.-hr. Of this 33 per cent in January and February and 38 per cent in March was produced by water power. To help produce this energy 10,247,947 tons of coal, 3,481,742 bbl. of oil and 4,553,228 cu.ft. of gas were consumed.

A Steam Alignment Diagram

The use of an "alignment" diagram for the graphical solution of steam problems is proposed in an article by D. Halton Thomson in a recent issue of *Engineering*, London. The accompanying diagram is reproduced from this article and is used here simply to illustrate the author's ideas. No attempt should be made to use this copy for numerical work, as the repeated reproductions and the small scale have destroyed its value for that purpose.

The six most important properties that appear in steam calculations are pressure, volume, temperature, entropy, quality (superheat or dryness fraction) and total heat. The well-known Mollier diagram, in which total heat and entropy are the co-ordinates, usually has the pressure and quality shown by families of curves known as constant-pressure and constant-quality lines. Theoretically, lines of constant volume and constant temperature might be added to such a diagram, but the resulting complication of lines would render the plot difficult to work with. The result is that either two diagrams or one diagram and a set of tables must be used in the solution of most problems.

It may be shown that if any group of correlated variables can be represented by a diagram consisting wholly of families of straight lines, this diagram may be transformed to the alignment system so that any three (or more) straight lines passing through the same point on the cartesian system correspond to three (or more) points lying on a straight line on the alignment system. The first step, then, is to produce a cartesian diagram consisting wholly of straight lines.

The ordinary steam diagrams in use consist of families of curved lines, but a study of some of Callender's equations suggests that by changing the scale of the co-ordinates these lines might be straightened. A diagram was accordingly plotted using values of *log p* and *log (H-835.2)* as co-ordinates. In this case the entropy and superheated-steam-volume lines were perfectly straight, while the quality, total-heat and wet-steam-volume lines were so nearly straight as to introduce only negligible errors. From these considerations the alignment diagram has been plotted, the scales being laid out in accordance with the modified logarithmic formulas, but the graduations being marked with the direct values of the properties. By the use of the diagram as given, practically all of the ordinary steam problems may be solved directly.

In using an alignment diagram, it must be remembered that any straight line drawn across the diagram represents a certain condition, or "state," which would be represented by a point in a cartesian diagram. Such a line, then, is called a "state line," and its intersection with the different scales gives the state properties, just as the position of the "state point" gives the state properties by its position in relation to the various property lines on a cartesian diagram.

For example, given steam at 160 lb. pressure and superheated 50 deg. F. A straight line from the 160-lb. point on the pressure or p scale through the 50-deg point on the superheat or t' scale intersects the total heat, or H, scale at 1,229 B.t.u., the entropy, or ϕ, scale at 1,603 and the volume, or V, scale at 3.05 cu.ft.

It must be noted that when the steam is superheated or supersaturated the scales marked H and V must be used, but when dealing with wet steam scales H_q and V_q must be used. The point marked *Wilson Point* represents the lower limit of complete supersaturation as determined by Wilson's experiments, and if the state line passes below this point the steam is to be considered wet. If the state line passes between the *Wilson Point* and the *Saturation Point*, the steam will be either wet or supersaturated, depending on whether it is or is not in a state of equilibrium. This point must be determined from the conditions of each special problem.

In a problem involving expansion in a nozzle, the process is not isoentropic and corrections must be introduced. Assume the expansion between the conditions assumed in the preceding examples with a friction factor of 0.12. The adiabatic heat drop will be found from the difference of the values already found 1,229 — 953 = 276 B.t.u. The heat transformed into velocity will then be 276 (1 — 0.12), or 243 B.t.u., leaving

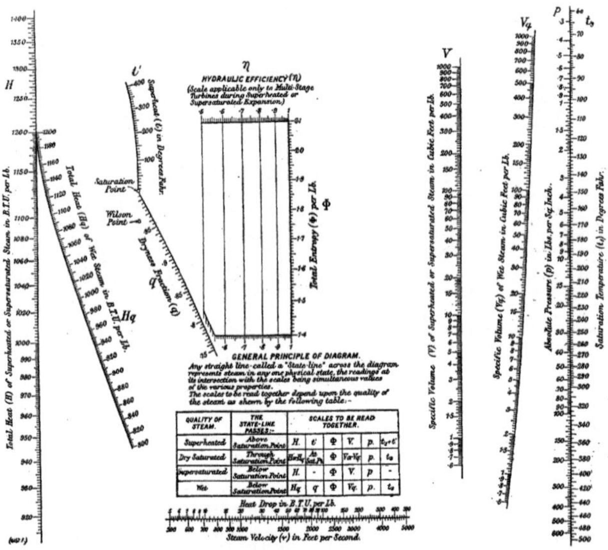

AN ALIGNMENT DIAGRAM OF STEAM PROPERTIES

In solving a problem involving a change of state, the final values are obtained from a new state line. Since most problems assume a change with some one property constant, the new state line must intersect the old state line on the scale of the constant property. For example, suppose steam at the conditions assumed is to expand adiabatically to a pressure of 3 lb. absolute. In this case the entropy remains constant, so the new state line will pass through the point 1.603 on the entropy scale and 3 lb. on the pressure scale. Since the final state will obviously be in the wet steam region, the heat content is read on the H_q scale at 953 B.t.u., the volume on the V_q scale at 98 cu.ft., and the quality on the q scale at 0.832.

1,229 — 243, or 986 B.t.u., in the steam. The actual final state line will then pass through 986 on the H_q scale and 3 lb. on the p scale, and the quality 0.865 and the volume 102 cu.ft. may be read on the q and V_q scales respectively.

It must be admitted that values from this diagram, even as originally prepared on a large scale, will be somewhat in error due to the deviation from the straight lines of some of the curves in the cartesian diagram. The author claims, however, that the results will not be greatly in error and that where extreme refinement is desired they may be checked by computation, while for ordinary calculations, especially in preliminary designs, such a chart is sufficiently accurate.

Osborn Monnett—A Leader in Smoke Abatement

Self-Made Engineer Who Built for the Future. An Expert on Combustion Who Has Specialized on the Elimination of Smoke

A MAN who will look forward in his profession for a quarter of a century, attempt to analyze conditions and predict the trend of development with the view of anticipating the progress to be made and to meet it by mapping out a definite course of instruction, is a rare exception. Osborn Monnett may be placed in this class. First of all he had a definite ambition to become a steam power plant operating engineer. He realized that the requirements of a good engineer called for a practical knowledge of boilermaking, steamfitting, general operation and, perhaps of first importance, chemistry. Refinements in combustion and boiler practice depend largely upon chemistry today and upon the use of chemical apparatus unheard of in the power plant twenty years ago, at the beginning of Monnett's career. The problems of the future—for example, low-temperature distillation of coal with recovery of byproducts — call more and more for chemistry, so that Mr. Monnett made an excellent guess when he placed industrial chemistry as one of the chief requirements for an up-to-date engineer of the present and the future. In 1876 Monnett was born in Norfolk, Va., but soon after his parents removed to Ohio, their native state, where he received a grammar-school education. His first practical work as an engineer was in Kentucky, running a sawmill that produced white-oak staves and lumber. It can be truly said that he started at the bottom rung of the ladder, but the sawmill job was only temporary. He knew that the best kind of training could be obtained on the water, where a man was placed upon his own resources and had to be more than a "runner" to get by. Accordingly, he arranged to enter the marine field and for the four years dating from 1899 was employed on the great Lakes in the engine rooms of coal and ore carriers with large triple-expansion engines, Scotch marine boilers and all the usual equipment on these steamers. During this period he progressed from coal passer to fireman, water tender and oiler to first assistant engineer.

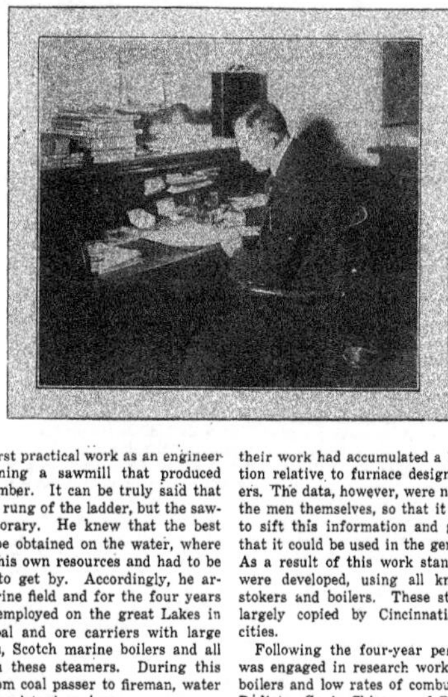

With such experience back of him he left the service to become chief engineer of the W. & L. E. Shops at Norwalk, Ohio, and later for four years was in a similar position in the Rock Island Shops at Moline, Ill. It was here that he began studying chemistry, graduating eventually from a correspondence course in qualitative and quantitative analysis. During his studies he set up a laboratory at the plant and specialized in chemical problems relating to his work, such as combustion, feed water, etc. As a postgraduate course in chemistry he worked for a short time in the laboratory of the American Steel and Wire Co., at Cleveland, where practical experience was obtained in general commercial work.

His progressive tendencies had attracted the attention of the editor of *The Engineer*, published at that time in Chicago, and in 1907 he was asked to join the editorial staff. It was new work, but he liked it, and a year later he went with the paper to New York when it was consolidated with *Power*. He soon returned from the East as Western editor in charge of the Chicago office of the Hill Publishing Co. For three years he traveled extensively, keeping in touch with all that was new and interesting in the power-plant field and among other things specialized on boiler explosions. In 1911 the wheel took another turn, and Monnett was appointed chief smoke inspector of Chicago, serving four years under Mayor Harrison. The organization of the smoke department at that time was excellent. It contained a number of mechanical engineers who in their work had accumulated a large amount of information relative to furnace design and the setting of boilers. The data, however, were not available except among the men themselves, so that it was his immediate effort to sift this information and get it down on paper so that it could be used in the general field of engineering. As a result of this work standards for boiler settings were developed, using all known types of furnaces, stokers and boilers. These standards have since been largely copied by Cincinnati, Pittsburgh and other cities.

Following the four-year period in the City Hall he was engaged in research work relating to low-pressure boilers and low rates of combustion with the American Radiator Co. in Chicago and Buffalo. During this time he lectured extensively on smoke-abatement problems before chambers of commerce, civic organizations and various mechanical engineering associations in many of the cities of the Middle West. When the war came on and he heard the call to service,

He became a member of the Conservation Committee of the Fuel Administration for the State of Illinois and helped to prepare educational pamphlets and lectures relating to the saving of fuel. Later, he went to Washington as a member of the Conservation Bureau of the United States Fuel Administration. Among the more important tasks assigned to him was the conservation of fuel in the entire sugar industry of the United States, negotiations being conducted in New York with the managers and principal officers of the sugar-refining companies. This involved important questions of policy and methods in the use and distribution of coal in quantity, complicated by joint action with the Food Administration. He also represented the Conservation Bureau of the Fuel Administration on the Priority Committee of the War Industrial Board.

Since the armistice Monnett has been cashing in on his wide and varied experience. He has become one of the consulting engineers of Chicago specializing in combustion matters and boiler practice. Of the notable work he has done during the past year, mention might be made of the atmospheric survey of Salt Lake City conducted for the Government in the capacity of consulting fuel engineer for the United States Bureau of Mines. This work has extended over a period of eight months from September, 1919, to May, 1920, and has included a careful chemical and physical examination of the atmosphere not only with relation to smoke and smoke abatement, but also to the presence of smelter gases. During this investigation airplanes were used for the first time in taking air samples. A fuel survey and soot-fall study was included, together with a large amount of chemical and microscopical work in examining the filtrates and precipitates obtained.

His reputation on questions involving smoke gained him a place on the Chicago Association of Commerce Committee on Atmospheric Pollution, a body formed to study all forms of contamination in the air of the city, together with the best methods of reducing the nuisance to a minimum.

Farnsworth Heating System

Building superintendents and heating engineers will be interested in a piping layout for a heating system that provides a closed-loop circuit, whereby the water of condensation from the radiators is returned directly into the boiler. This system has been developed by the Farnsworth Company, Conshohocken, Pa. With the system is used a duplex boiler feeder, which permits a continuous flow of condensation into the boiler. The condensation is held under pressure and is delivered into the boiler at a temperature only one or two degrees below that of the steam, corresponding to the pressure carried on the heating system. These results are obtained without the use of a receiving tank. There are no vents open to the atmosphere and practically no heat loss.

When the high-pressure steam from the boiler is carried to the top of the building, it passes through a reducing valve and then the risers drop from the main line to supply the various coils. The return line of the coil is carried back into the riser with a drop of nine inches before entering the riser with a check valve. By this static head of nine inches all water between the check valve and the lowest pipe of the coil is sufficient to lift the check valve and let the condensation to the coil flow back into the same steam riser.

A full-sized pipe is carried from the return pipe of the coil to somewhere near the top of the coil, where an automatic air valve is inserted. This pipe provides a sort of air chamber and eliminates the possibility of any water squirting out of the air valve. The return

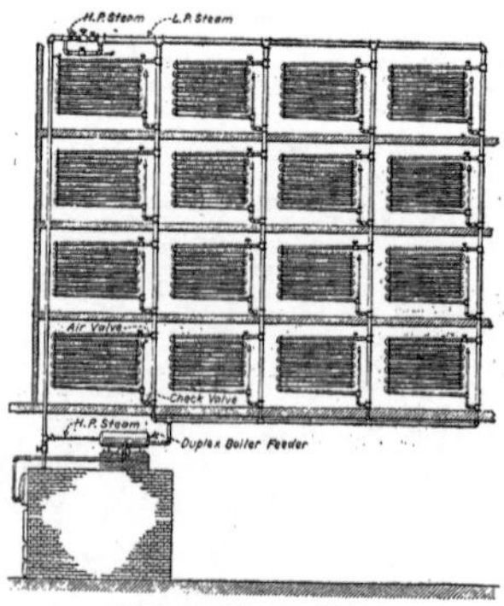

PIPING OF HEATING SYSTEM

pipe from the risers is carried into a duplex boiler feeder located somewhere above the water line of the boilers.

In this system no return lines are necessary other than to connect the risers to the boiler feeder, which feeds the condensation into the boiler at the same temperature that it leaves the system. In other words, the condensation is held under a pressure and is fed directly into the boiler.

The First Boiler Explosion

Rhys Jenkins, in an article entitled "Links in the History of Engineering," in *The Engineer* (London), quotes from an unpublished manuscript of Farey's an account of what seems to have been the first boiler explosion.

It occurred on Sept. 8, 1803. An engine had been set up temporarily to drain the foundations for a large building in course of erection for a tidal corn mill on the banks of the Thames, between Greenwich and Woolwich. The engine was in charge of a boy who is said to have fixed the safety valve by inserting a prop between it and the roof of the boiler-house.

The boy went away, leaving the engine at work, so that when soon afterward it was stopped by another workman, the steam had no outlet and an explosion took place, causing the death of three persons and serious injuries to three others. This was at about the time that Trevithick was being called a murderer by Watt for his use of high-pressure steam, less than fifty pounds, and is said to have considerably retarded the use of the "high-pressure," or noncondensing engine.

Interconnecting Industrial and Central-Station Plants

By CORNELIUS G. WEBER

With data from a number of individual cases the author gives some indication of the large savings in coal, labor and transportation that would be possible from electrical interconnection on a give-and-take basis.

IF ALL industrial establishments were to purchase their electric energy from large public utilities, a large amount of coal would be saved as far as the generation of electrical energy is concerned, but the fact that many industries require heat in the form of low-pressure steam at once makes it more profitable for such industries to generate energy as a byproduct. Usually, such an industry will generate only as much energy as it requires for its own use and will operate an engine only as economical in fuel consumption as the case may require. No attention is given to the possibility of generating all the energy economically feasible to generate from the steam used and to dispose of the excess amount, for where would there be a market? No less an authority than Dr. Steinmetz had the following to say on present methods:

> We realize that our present method of using our coal resources is terribly inefficient. We know that in the conversion of the chemical energy of coal into mechanical or electrical energy, we have to pass through heat energy and thereby submit to the excessively low efficiency or transformation from low-grade heat energy to the high-grade electrical energy. We get at best 10 to 20 per cent of the chemical energy of the coal as electrical energy; the remaining 80 to 90 per cent we throw away as heat in the condensing water or, worse still, have to pay for getting rid of it. At the same time we burn many millions of tons of coal to produce heat energy and, by degrading the chemical energy into heat, waste the potential high-grade energy which those millions of tons of coal could supply us. It is an economic crime to burn coal for mere heating without first taking out as much high-grade energy, mechanical or electrical, as is economically feasible. It is this feature of using the available high-grade energy of the coal, before using it for heating, which makes the isolated station successful though it has every other feature against it.

As time goes on, we will come to recognize more and more that it is economically criminal to burn coal for mere heating purposes or for electrical energy only, and a day will dawn when fuel will have attained such a high price that all will recognize the full truth and take steps to utilize fuel as economically as possible. One such means no doubt will be the generation of electrical energy as a byproduct to the fullest measure in all manufacturing establishments wherein steam heating processes are involved.

Since it is highly probable that the time is not so very remote when all power plants, large and small, will be electrically interconnected and operating on a give-and-take basis, it might be of interest at this time to face conditions as they obtain in several specific cases and aim to draw such conclusions as a few cases will allow. In the practice of his profession the writer has not only visited and inspected many industrial plants, but in the instances considered hereinafter has made exhaustive tests and analyses including the measurements of live- and exhaust-steam consumption of apparatus and processes, boiler tests, economic performance tests, heating-load tests, etc., all for the purpose of working out in each individual case a plan of rehabilitation that would reduce the fuel consumption of the specific establishment investigated. All such investigations resulted in the compilations of detailed reports based on facts and measurements, and since plants of widely varying character were encountered, it occurred that it might be of interest to work out the over-all fuel economy that could be anticipated in the cases considered in the event that they were electrically interconnected and were generating electrical energy only in proportion to the fuel needed for heating and processes, disposing of excess energy to the central station at times when more energy than needed was being generated and purchasing energy from the central station when a deficiency occurred.

In addition to the fuel economies possible in the plants considered as independent units, the probable additional fuel savings that would result, were these plants provided with induction or other generators and connected to the public-service company's distribution system, have also been worked out. The plants considered embrace industries of the following kinds: Automobile parts, machine manufacturing, food products, sheet-metal products, leather manufacturing, woodworking, structural-steel fabricating, textile goods and dyeworks.

In each instance the fuel consumption was first determined and from boiler and engine tests the economic performance was determined. The low-pressure steam requirements for heating and process work were then found after which the annual fuel consumption was computed on the basis of proper performance of the boilers, the installation of more economical prime movers and the utilization to best advantage of the exhaust steam.

In this discussion the conclusions reached are based on careful analyses and practically all of the cases considered were worked out in the same detail as one typical case which follows:

The graph shown in Fig. 1 represents the operating conditions as found by test. It is self-explanatory in so far as it represents steam utilization. There is this to be added, however—the fuel consumption under the conditions revealed was 6,033 tons per annum. The processes utilized live steam at low pressure, the boilers were operating at an efficiency of only 63.2 per cent and the Corliss engine was consuming on an average 43 lb. of steam per indicated-horsepower per hour.

Annual energy requirements of the plant are about 610,000 kw.-hr. under the conditions shown in Fig. 3, wherein electrification is contemplated. Fig. 2 shows the operation of the same plant under the conditions wherein the live steam for processes is replaced by exhaust steam, the boilers brought up to proper efficiency, waste heat reclaimed from a solution cooler and the Corliss engine carefully repaired to bring its steam consumption down to about 30 lb. of steam per indicated-horsepower per hour. The annual fuel consumption represented by Fig. 2 is 3,519 tons.

Fig. 3 represents the operation of the same plant contemplating the installation of a direct-connected, poppet-valve, engine-driven generating unit, the use of superheated steam and the elimination of a considerable friction load by electrification. The annual fuel consumption under conditions as shown in Fig. 3, is 3,100 tons.

It will be noticed that under the conditions shown in both Figs. 2 and 3, there is not enough exhaust steam during the colder months to meet the requirements and that a considerable amount of live steam for building heating and process work is necessary. This steam, being all low-pressure steam, could first be drawn through a larger engine and additional electric energy latter figure only the additional fixed charges on the larger engine and generator would have to be figured.

As it happens, the plant discussed is not a consumer of central-station energy, but it seems reasonable that under present conditions even a central station could afford to purchase energy at one cent per kilowatt-hour. In this special case the industrial-plant owner would be ahead $3,050 per year, less the fixed charges on that portion alone of the prime-mover equipment cost which is larger than his own needs require. The central station would be ahead by the profit represented by $1,250 worth of new business, it being assumed that the fuel and labor charges in the central station amount to 1c. per kw.-hr. Should it cost the central station more than one cent for these items of coal and labor, then the central station would make an additional profit. It is apparent, of course, that the industrial-plant

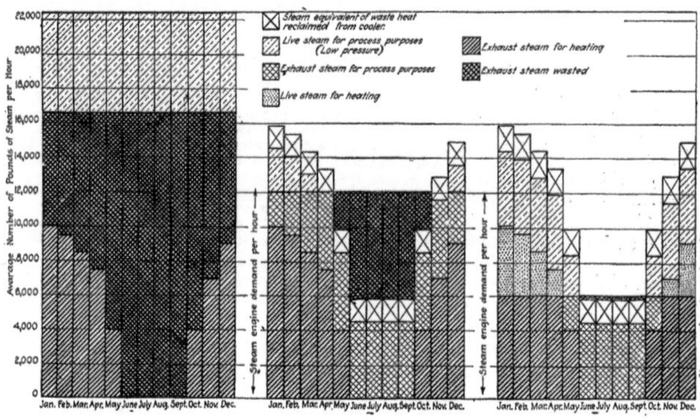

Fig. 1. Fig. 2. Fig. 3

FIGS. 1 TO 3. STEAM REQUIREMENTS OF A TYPICAL INDUSTRIAL PLANT UNDER DIFFERENT CONDITIONS

Fig. 1—Steam per hour for various months in a typical industrial power plant. Fig. 2—Operation of same plant as in Fig. 1, live steam replaced by exhaust steam. Fig. 3—Operation of same plant, but with poppet-valve engine driven generating unit, superheated steam and electric drive.

generated to be fed into the public-utility company's distributing system.

The amount of additional energy that could be so generated under conditions shown in Fig. 3, would be 360,000 kw.-hr. per annum. During the months of June, July, August and September, when some exhaust steam would be going to waste, this plant could utilize about 50,000 kw.-hr., of central-station energy to keep from wasting exhaust steam. This procedure would eliminate about 100 tons of coal, which, if valued at $700, would mean that to break even financially the plant in question could not afford to pay more than 1.4c. per kw.-hr. However, if the central station were purchasing the 360,000 kw.-hr. at, say, 1c. and were charging for the 50,000 kw.-hr., 2.5c., the plant under consideration would still be ahead by an amount of ($3,600 + $700) — $1,250 = $3,050, against which owner could generate the 50,000 kw.-hr. during the summer months at less than 2.5c. per kw.-hr. and that, financially, it would be to his advantage not to purchase such energy. This is true; but it must be borne in mind that he would no doubt agree to pay the amount of $1,250, of which only $550 actually represents added cost, for the privilege and opportunity to dispose of the excess energy that he would be generating as a byproduct. It is evident that the foregoing represents a profit to both parties concerned.

With such a co-operative arrangement the amount of fuel conserved for the nation at large would be about as follows: Assuming that the central station can generate energy at as low a rate as 2.5 lb. of coal per kilowatt-hour, then the purchase of the 360,000 kw.-hr., cited previously, would conserve 450 tons of fuel, since the central station would be buying 360,000 kw.-hr.

for the generation of which it would otherwise have to burn 450 tons of coal.

During the summer months when the public-service company would be selling 50,000 kw.-hr. of energy to the industrial plant, the central station would consume 63.5 tons of coal, where the industrial plant would be consuming about 100 tons, thereby bringing about an additional saving of 37.5 tons, or a grand total of 487.5 tons.

In round numbers the annual fuel saving in so far as the nation is concerned would be from 475 to 500 tons. At $7 per ton, this would represent about $3,500 per year, which latter figure must be reconciled with the industrial-plant owner's $3,050 profit and the central station's profit on $1,250 worth of new business.

The summary of the foregoing is shown as plant No. 5, in Fig. 4. By consulting the key, the graph becomes self-explanatory. Representations above the line 0-0 show the industrial power-plant operation and those below the line, the public utility's operation in so far as it is or would be affected by the industrial power plant in relation to which it is considered. In

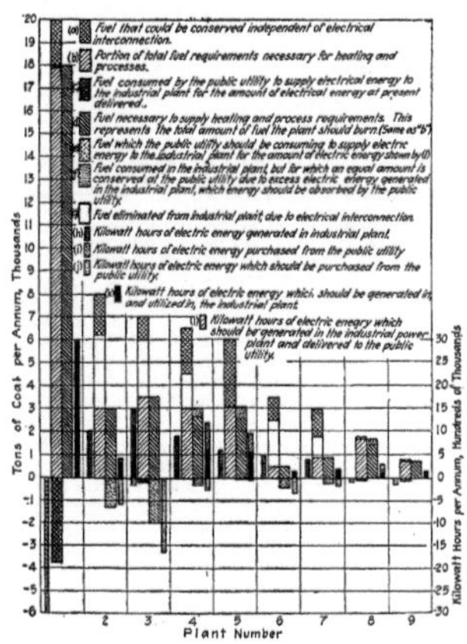

FIG. 4. SUMMARY CHART COVERING SEVERAL TYPICAL INDUSTRIAL POWER PLANTS

The purpose of the chart is to show the effect of co-operation between these plants and the central station. The cases are, of course, hypothetical.

Fig. 4 there are four sets of bends to each plant representation. The two to the left show actual conditions as found and the two to the right as they would obtain under electrical interconnection conditions.

By a similar process of analysis several other typical industrial power plants have been charted in a summary way, as further shown in Fig. 4. For convenience they have been arranged according to magnitude of coal consumption. Data on numerous other plants are in possession of the writer, and only typical cases have been selected.

There is this to be said specially about plant No. 1: The plant operates gas engines for generating electric energy, but burns 20,000 tons of coal exclusive of the gas engines for high- and low-pressure steam pur-

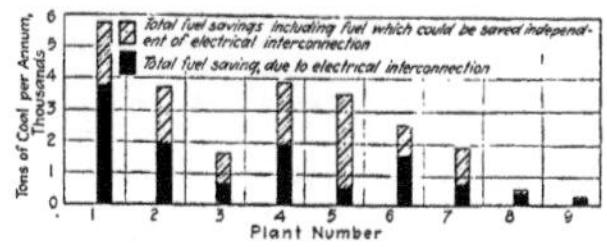

FIG. 5. A SUMMARY OF FIG. 4, SHOWING TOTAL QUANTITIES OF FUEL SAVED UNDER ASSUMED CONDITIONS

poses, at the same time purchasing 3,000,000 kw.-hr. of electric energy. Only fuel burned for steam purposes has been considered and that portion of the steam which is used at low pressure if first taken through an engine as a reducing valve would be just sufficient to generate the same amount of energy as is now purchased. As it happens the power demand is practically uniform, and since low-pressure steam is used also at almost a uniform rate, this steam would generate energy in proportion to the amounts needed.

Fig. 5 is a summary of Fig. 4, showing total quantities of fuel to be saved. As found, the aggregate amount of fuel consumed per annum by the nine plants considered is 56,633 tons. As isolated independent plants under reasonable rehabilitation plans, these same plants would consume only 44,550 tons of fuel, or 12,083 tons less. Electrical interconnection and operation on a give-and-take basis would result in an additional saving of 11,400 tons of coal. The total saving due to rehabilitation and electrical interconnection would be 23,550 tons, as represented by the total areas of the bands shown in Fig. 5. This is 41.5 per cent of 56,633, or in other words a saving of 41.5 per cent in fuel would be possible in these cases. The solid black portions of the bands shown in Fig. 5 represent 11,400 tons. Electrical interconnection would result in a saving of 11,400 tons in 44,500 tons, or 25.5 per cent.

In these cases the public utility would lose 3,425,000 kw.-hr. of business, gain 3,125,000 kw.-hr. and in addition would have to absorb about 1,200,000 kw.-hr., bringing about a net loss of about 1,500,000 kw.-hr. to the public utilities. However, the loss of business would be represented by a large consumer paying a low rate, whereas the new business would be represented by numerous small customers paying a higher rate, and since the 1,200,000 kw.-hr. to be absorbed would be purchased at a low rate, it is probable that the public utility would be benefited. In any event a plan on the order outlined, where plant No. 5 was discussed in detail, would have to be worked out since the co-operative plan would have to be of benefit to the public utility to be attractive and workable.

From a national fuel-conservation standpoint the public utility should be interested in all industrial plants and every industrial plant should be a subscriber to the public utility, which should be operating in a bigger, broader and better field.

The writer has come in contact with numerous other cases scattered about the country. In one instance a plant consuming 45,000 tons of coal a year was burning 3,000 tons for heating buildings with live steam. This 3,000 tons could readily have been utilized for power purposes.

As another example, in a town of 3,500 population the local public utility is being operated with uneconomical engines which exhaust the steam to atmosphere. The public utility's plant is located on a large river, and the installation of a condenser would result in saving 20 per cent of the fuel. In this same municipality is a large woodworking establishment consuming a combined amount of coal or its equivalent in wood refuse equal to 9,000 tons of coal per year. The plant operates 24 hours per day, and its exhaust-steam requirements are such that practically 9,000 tons of coal or equivalent would have to be burned for steam-heating purposes regardless of any power requirements. However, this plant was actually considering the purchase of electric energy from a water-power concern, and had this been advised and done, an additional loss to the general public would have been added to an already aggravated situation.

If equipped with high-efficiency engines, the woodworking establishment would have been in position to generate more than all of the electrical energy required by the municipality at night and could have generated more than enough to meet the daytime demands as well as its own needs. In other words, instead of this establishment purchasing energy, it should take advantage of its position to sell energy to the local public utility, allowing the latter to shut down and conserve 2,000 tons of coal per year. If a transmission line should become available, the plant under consideration would be able to retail electric energy to other consumers.

INTERCONNECTION A BOON TO INDUSTRIAL PLANTS

The nation is full of just such possibilities. What we are suffering in national fuel losses, not mentioning labor, transportation and rolling-stock costs, can be imagined when we see what happens in a few specific cases. It is hardly within the scope of this article to dwell further on the details surrounding the actual operation of a combination of plants electrically interconnected. Suffice it to say, that from an engineering standpoint it would be simple and practicable and would be a boon to the industrial plants themselves rather than a hardship. The nine plants mentioned previously do not represent the unusual, but are fair examples of the average. What is happening in them is true in a general way in nearly all medium-sized industrial establishments.

Our annual bituminous-coal production is about 600 million tons. About 90 millions are used for house and building heating, 135 millions for railroad operation, and about 375 millions for industrial purposes, including gas utilities, ore smelters and electric central stations. It seems reasonable to suppose that at least 200 million tons are used per annum by industrial power plants. If only 25 per cent instead of the 41.5 per cent, as found in one typical case, would be saved by electrical interconnection and other means, it would mean an annual saving of 50 million tons, or 1,000,000 fifty-ton cars. Such a train of cars on an air line would stretch three times from New York to San Francisco.

Naval Horsepower of 9,000,000 To Use Oil

More than 8,000,000 bbl. of fuel oil and nearly 3,000,000 gal. of lubricating oil will be required by the Navy during the next fiscal year. These figures have been given out by Admiral R. S. Griffin, chief of the Bureau of Steam Engineering. In addition, Admiral Griffin said:

The demand for oil by the Navy alone looks formidable in comparison with the requirements of two or three years ago, and when it is considered that a large number of merchant ships, completed during the last three years, also burn oil, and that many industrial establishments recently have converted their power equipment to oil burning, the question of an adequate supply of fuel oil for the Navy becomes one of great concern.

I share the views of many regarding the national importance of conserving oil. Without an assured supply, our Navy would be practically useless. I shudder to think what the result would be if anything should occur that even remotely would threaten this supply.

We now have in commission seven battleships of a combined horsepower of 204,000, which burn oil only. In addition to these we have building twelve other battleships of 533,000 hp.; six battle cruisers of 1,080,000 hp., and ten scout cruisers of 900,000 hp., making a total under construction of these three classes of ships aggregating 2,513,000 hp.

In oil-burning destroyers, we have actually completed 234 with an aggregate horsepower of 5,626,000, and have under construction 87 others which will be completed during the next fiscal year and will bring the total horsepower of destroyers up to 7,975,000.

Our submarines already completed aggregate about 80,000 hp. in Diesel engines, and those under construction will more than double this figure.

Besides these vessels of a purely military character, we have others such as mine sweepers, tugs, destroyer and submarine tenders, and fuel ships, in which oil is used as fuel, whose horsepower aggregates 173,400.

To sum up, we actually have completed and ready for service vessels aggregating more than 6,000,000 hp. in which oil alone is used for fuel, and have under construction other vessels which will bring this total up to nearly 9,000,000 hp.

One of the most important considerations for the successful operation of geared marine turbines is the proper lining upon the foundations. At the Hog Island shipyard, the sole plates of foundations for gears and turbines are machined parallel to the line of shafting after the shafting has been lined up. This is done with a special milling machine so set that the sole plate has a taper of 0.020 in. from outboard toward center to facilitate setting the chocks. After the gear case is lowered into place and held by jackscrews measurements are made for the 56 chocks. These chocks are machined from mild steel and are made 0.005 in. thicker than the measurements call for so they may be scraped to fit. The holding-down bolts are given a driving fit by reaming the bolt holes 0.002 in. less diameter than the bolts. A set of these gears for the A type ships weighs 90,000 lb. and has twelve journals. The journals have a clearance of 0.001 in. per in. of diameter.

The tendency these days is toward comparatively small installed ice-making capacity and large ice-storage capacity. This enables holding a good crew the year around, insures large stock at times of great demand and cares for breakdown repair periods. In designing storage houses 50 cu.ft. is usually allowed per ton of ice.

When a fluid is steadily flowing through a pipe, that fluid in contact with the wall of the pipe is practically at rest, the velocity increasing as the axis of the pipe is approached, being maximum at the axis.

EDITORIALS

Will We Have To Come To the Big Stick Again?

FOLLOWING the sad experience of last winter's coal famine, many conferences were held to consider means whereby a repetition of the situation might be avoided. Foremost among the recommendations of these meetings was summer storage.

Summer is here, and what is the situation? Already the daily press is predicting dire results for next winter.

Are the large users of steam coals in the East storing up a supply for the season when the demand is greatest and the transportation facilities least adequate?

No, they are not; and through no fault of their own. Most of the large consumers are perfectly willing to make the necessary investment in storage facilities and outlay in coal as an insurance against shutdown next winter, but they cannot get the coal.

Instead of building up a reserve supply at this time, they are forced to proceed from hand to mouth.

The condition appears to be due to several factors, chief among which are inadequate production, lack of transportation and terminal facilities, labor troubles at terminals, a large export in coal, and speculative selling.

It is, perhaps, the inability of the railroads to expedite the handling of cars and their tie-up at terminals because of labor troubles or inadequate terminal facilities, that is responsible to a greater extent than mere lack of cars. For last October, just preceding the strike, with practically the same number of cars available as at present, it will be recalled that the transportation of coal beat even the war record of the previous year.

Production at the mines naturally follows ability of the railroads to move the coal, provided there is a brisk demand, and this accounts for the decreased production in certain cases, although the production curve this year has been above that of last year. However, it must be remembered that in the spring of 1919 some industries were running on their war surplus and the demand for several months fell off.

Much has been said of late concerning the increasing export of coal and its effect upon certain sections of the country, notably New England, which receives much of its coal by water, and the Interstate Commerce Commission has already issued priority orders with a view to relieving the New England situation.

As an aftermath of the war Europe is extremely short of coal and is naturally turning to the American market for relief. She needs the coal so badly that she is willing to pay almost any price for it. This being the case, purchasers at home are forced into competitive bidding with foreign buyers. During recent months the foreign shipments have increased to about five times those of pre-war years and the shipments to New England have decreased by twenty-five to forty per cent.

It is argued by the coal interests that this is a perfectly legitimate situation; that the home consumers are welcome to the coal if they will meet the foreigners' price and that there is no more reason for putting an embargo on coal than there is for putting one on cotton or wheat. Besides, Europe needs our coal just as much as it does our wheat.

The argument seems plausible except perhaps to the man who is forced to pay for the coal.

Be that as it may, an embargo, while providing temporary relief to New England and making possible diversion of cars to other sections, is by no means a remedy for existing conditions. If our total exports were stopped, we would still be running over five million tons behind our monthly requirements.

The shortage has created a condition favorable to manipulation and a speculative market, which in turn has probably resulted in further holding of shipments at terminals and yards. Coal prices are on the rampage, and what is bid one day by the man in dire necessity becomes the opening price the following day.

The necessity for a revival of the Fuel Administration will be regretted, and most of all by the coal men themselves. The country is tired of commissions of inquiry and adjustment, which have in practically every instance failed to achieve results. What it needs and is sure to demand is a wielding of the big stick, laws with teeth in them, and an enforcing agency that will see to it that the producer and middle man are allowed fair profits only and that the public is protected from unwarranted profiteering and speculation.

Electrical Interconnection

THINGS that are visionary and idealistic today are realities tomorrow. The bigger and broader ideas in civil and economic life are born of evolution rather than revolution. Rome was not built in a day; neither could the broad idea of electrical interconnection suggested by Mr. Weber on other pages of this issue be carried out in so brief a lapse of time. Civic communities have adopted city planning, whereby a certain goal to be attained ultimately is always kept in mind and the common weal is carefully considered.

In the case of building new industrial power plants or rehabilitating old ones no consideration is given to any feature outside of the sphere of the individual plant, since the byproduct, electrical energy, has no market.

Engineers are fully aware of the situation, but there is no legislation to cope with this haphazard and wasteful procedure. Water power, which at best will never be adequate for industrial demands, will never run out, but the fuel supply will.

It would seem as though the time had arrived for crossing the indistinct dividing line between the idealistic and the realistic, and that within the course of the next decade we should consider electrical interconnection of generating stations, not as an original and modern method of doing things, but as a necessity. With the aid of the leaders in the engineering world, bills should be enacted into constructive and wholesome laws that would conserve the coal supply.

In the new German legislature a bill was recently introduced providing for government ownership of all industrial and other power plants. Their motive is obvious, for through national economy they hope to recuperate. Similar legislation in this country might not be received with favor, but there is a growing tolerance for the thought that public policy demands the supervision of power plants as well for efficiency as for safety.

As a further possible general plan it might be feasible to encourage an isolated plant to connect with the public-utility service lines when such isolated plant can prove that it can generate electrical energy as a byproduct. This would at once open up an attractive commercial field as auxiliary to the main industry, but if manufacturers could see a handsome profit, they would undoubtedly do much of their own volition in this direction and the solution of the problem would become more or less automatic.

When we think of the vast tonnage of coal that could be conserved, the coal that our railroads must consume for transporting only the wastage, the additional toiling hands in mines, factories and shops that could be utilized for other purposes, not mentioning the still further possibilities, should a constructive movement gain foothold, of operating central stations at coal mines, electrifying railroads, etc., is it not time to look into the remote future, view the whole matter in an open-minded, unprejudiced mood and act, and thereby lay the cornerstone for a greater and broader industrial structure?

Electrically Driven Reversing Steel Mills

ELECTRIC drives have gradually proved their superior merits until they have become an important factor in almost every manufacturing plant and have practically revolutionized many manufacturing processes. They are used on the smallest and on the largest drives, from the dental tool to the largest hoists and steel-mill applications. And in almost every case they are doing the work better, more efficiently and economically than it could be done with mechanical, hydraulic or steam drives. One of the industries in which the introduction of the electric motor has been most seriously contended by the steam engine is in the steel mill. However, the last ten years has witnessed a general adoption of electric motors to all classes of drives in the steel industry; and as for the non-reversing mill, it has been established that, owing to its greater flexibility and reliability, the electric-motor drive is superior to the steam engine in every respect. The electric drive in the continuous-running mill will produce equal tonnage at lower cost, power consumption and maintenance charges than the steam engine—so much so that electric motors are being installed on a very large percentage of main-roll drives in all new installations.

Two of the important factors in the economical operations of electric drives in steel mills are the ease and accuracy with which the power consumption can be measured, and the fact that the equipment must be kept well insulated or it cannot be operated. At best, with the steam engine, it is inconvenient to get an accurate record of power consumption. Whether or not the steam lines are kept well insulated and the joints tight, the old engine will answer to the throttle. The engine will continue to operate even if the valves and piston rings do leak and the vacuum is not maintained at its maximum. Although this may indicate that the engine is a serviceable and reliable piece of machinery, nevertheless operating under such conditions is not conducive to economy, low power costs and maintaining equipment in a high state of repair. These features not infrequently cause the steam engine to show unnecessarily high steam consumption. With the electric motor any serious leaks soon develop into a short-circuit and must be repaired before operation can be continued.

It is in reversing-mill applications that the ability of the electric motor is most seriously contended by the steam engine. Although with electric drive the cost of power per ton of steel rolled is lower than with the steam engine, probably with two or three exceptions the reversing steam engine is rolling a higher tonnage than the reversing electric motor, and tonnage is what the average steel-mill man is chiefly interested in. However, two new reversing mills that have been in operation for some months, according to reports, are equaling if not surpassing the tonnage records of any reversing steam-engine driven mill. The leading article in this issue "Electrically Driven Reversing Steel Mills," by Frazer Jeffrey, is a very interesting discussion on this subject, in that it considers a number of the problems involved in applying a reversing electric drive and then shows how one of the applications was solved. The electric drive in such installations has the handicap of considerable higher first cost, but as the author points out, higher first cost means little if in a reasonable length of time the saving in cost of operation and of maintenance will more than offset it to the extent of showing appreciable gains.

The Water Power Bill Becomes a Law

WHEN Congress adjourned without the President having affixed his signature to the water power bill, it was taken for granted that the measure had been lost and there was keen disappointment among those who had supported repeated attempts at satisfactory legislation during the last eight or ten years. Therefore, recent announcement from the White House that the President, acting upon an opinion of the Attorney General that adjournment of Congress did not deprive him of the ten days allowed by the Constitution for the consideration of a measure, had finally signed the bill, was greeted with general satisfaction.

The bill is necessarily a compromise, and while some of its provisions may not entirely satisfy the extremists on either side, it is generally conceded to be a satisfactory measure and one that will form a basis for constructive development of our water powers.

The West will be particularly gratified at the President's action in view of the need for a wide and comprehensive hydro-electric program to meet the ever-increasing demand for electricity among the Pacific Coast states. In an effort to meet the present power shortage, much new construction is already under way in California, but this consists principally of additions to existing developments. There is still available on public lands a large amount of potential power which has heretofore remained undeveloped because of the unsatisfactory method of granting permits.

Case of a Stuck Governor Valve

The following experience with an engine that failed to start after being shut down for several months may be of interest. The engine in question was in a small sawmill and was of the horizontal slide-valve type, provided with a throttling governor. The mill had been shut down for nearly a year, and when it was desired to reopen it the engineer, who had been employed to operate the power plant, was unable to start the engine. He telephoned for a friend of mine, who, although he is not a regular engineer, had been successful in locating trouble in a number of small plants.

On arriving at the mill, he found steam up and everything about the boiler and engine apparently in good order. The engineer had already closed the stop valve in the main steam pipe next to the boiler and had removed the bonnet from the throttle valve and also the steam-chest cover. There was nothing wrong at these points, and yet, upon opening the boiler-stop valve and the throttle, no steam appeared in the open steam chest.

My friend naturally thought of the governor and asked if this had been examined. The engineer replied that it had not been taken apart as it worked all right. On raising the balls they did appear to move easily—rather too easily, my friend thought. The engineer said that when he had first come to the mill that morning he had raised the governor balls and that they seemed to be stuck, but had come loose in a short time.

That was the whole trouble. The governor was taken apart, and it was found that the valve had corroded tightly to its seat. The valve stem at its lower point was also somewhat corroded and when the engineer had pried up on the balls, the stem had broken off short. Nearly a day's delay had resulted through this simple sticking of a valve. Moral: Sometimes the "insides" of a device may be "out of gear," even if the exterior appearance and its apparent working are correct.

Washington, D. C. HUGH G. BOUTELL.

Traveling Screens Reduced Labor

The accompanying illustrations show a cooling system now in use at the plant of the Connecticut Light and Power Co., Waterbury, Conn. A set of traveling water screens are housed in the small shed shown immediately above the intake in Fig. 1.

This plant is on the Naugatuck River, and at a point just below where the picture was taken, there is a dam over which the water frequently does not flow during the summer months. This practically makes a cesspool of the pond formed opposite the power house, and before the traveling screens were installed, two or three men were required, sometimes twenty-four hours a day, to clean the racks in order to permit sufficient condensing water to enter the intake. The traveling screen has eliminated all this labor and has also reduced the condenser cleanings by about 50 per cent.

From the condensers the cooling water is discharged through the spray pipes, as shown, into the river. A flume built parallel to the bank, Fig. 2, prevents this water from immediately returning to the intake. This flume is two or three blocks in length so that the water must travel a considerable distance, giving it additional opportunity to cool before again entering the intake.

Milwaukee, Wis. W. H. BRANDT.

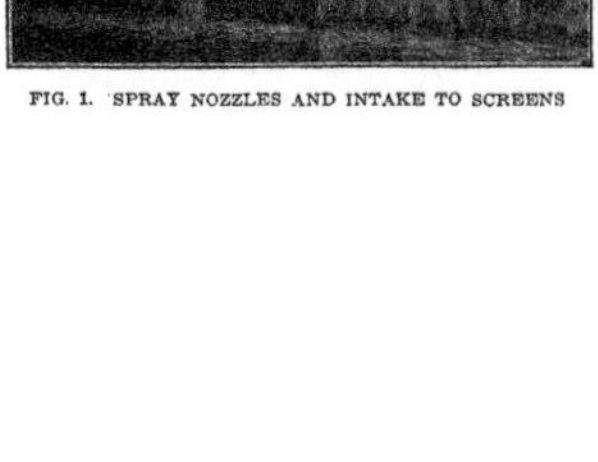

FIG. 1. SPRAY NOZZLES AND INTAKE TO SCREENS

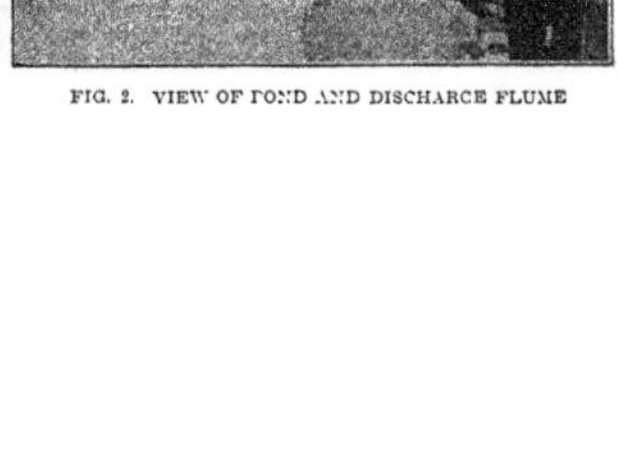

FIG. 2. VIEW OF POND AND DISCHARGE FLUME

An Emergency Union

Extra-heavy fittings for extremely high pressures are sometimes hard to get, and unless they are carried in stock a break in a line of this kind is likely to cause a lot of worry, as well as some delay in the operation of the machinery depending on the line. I had such an experience, when a union on a 1½-in. hydraulic line,

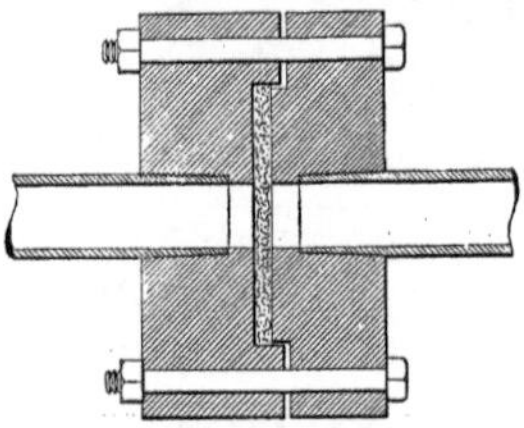

CROSS-SECTION THROUGH THE UNION

carrying 3,500 lb. pressure, broke. Extra fittings had been ordered, but were not then in stock, so a new one of some sort had to be devised.

A piece of scrap shafting about 4 in. in diameter was secured and two pieces were sawed off about 1¼ in. thick. These were placed in a lathe, faced off, bored and threaded to fit the pipe. They were then chucked with the other side out, faced and one made female and the other male, as shown in the sketch. Four half-inch bolt holes were then drilled in each half of the union to match. A sole-leather gasket was made to fit into the cavity in the female half, and the job was finished.

When installed, this union held and it is still holding this high pressure, and there has never been any appreciable leakage. P. A. GETT.

Cherryvale, Kan.

Used Natural Refrigeration

In a certain plant a motor-driven ice machine was run all the year round to cool refrigerator boxes in various parts of the building. Purchased current was used at considerable expense and after a time it dawned upon us that if these boxes had openings to the outside in winterthe result would be a considerable saving in the power bill.

The situation was gone over and ducts were run from such boxes as were near the wall, through the wall, connecting with tight insulated doors in the boxes. From the boxes in the basement ducts were run to a point about ten feet above the sidewalk. These also had doors in them, and by regulating the opening of these doors fairly good temperature regulation was maintained throughout the cold weather with a notable saving in the power bill. It might be added that these openings were screened to keep out birds, insects, etc., and each had overhanging canopies to protect them from the weather. Besides the saving in power, a chance was afforded to overhaul the plant, which could not be done before without the boxes becoming warm, as they were constantly being opened and the building was heated. G. F. KLUGMANN, JR.

St. Louis, Mo.

Banding an Armature Without Removing It from Frame

One of our 50-kw. engine-driven generators began to flash at the commutator and upon shutting the machine down it was found that the middle band on the armature had broken and parts of it came out on the commutator and brushes. The generator was directly connected to the engine, consequently it would have been a considerable job to remove the armature to put the new band on, therefore it was decided to try to put on the band with the armature in place. The polepieces were bolted to the frame with bolts from the outside. This made it possible to remove the bolt from the top polepiece and take the latter out, as shown in the figure. Two wood clamps were used to hold tension on the band wire. One of these clamps was fastened to a piece of timber between two posts about 5 ft. away from the machine and the other was fastened against one polepiece inside the machine, as in the figure, and used to direct the wire on the armature. With one end of the banding wire fastened to a coil where it projected beyond the core, the armature was turned by hand and three or four turns were put on the armature, to gradually bring the banding wire over to the position of the

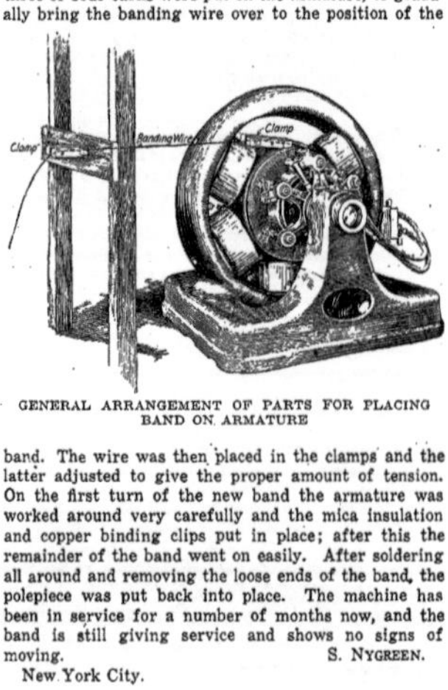

GENERAL ARRANGEMENT OF PARTS FOR PLACING BAND ON ARMATURE

band. The wire was then placed in the clamps and the latter adjusted to give the proper amount of tension. On the first turn of the new band the armature was worked around very carefully and the mica insulation and copper binding clips put in place; after this the remainder of the band went on easily. After soldering all around and removing the loose ends of the band, the polepiece was put back into place. The machine has been in service for a number of months now, and the band is still giving service and shows no signs of moving. S. NYGREEN.

New York City.

Another Long Run

I read with interest the record run of a steam turbine, as published in the May 15 issue. We have an 875-kw. 600-r.p.m. generator directly connected to a waterwheel operating under a head of 170 ft. This unit ran from Dec. 15, 1919, to April 18, 1920, a total of 126 days, without being shut down for any purpose whatever.

Copenhagen, N. Y. V. C. WOOD.

Shaft Lubrication

With reference to recently published articles in *Power* on the subject of self-oiling bearings, I would say that I have also found the ring-oiled type of bearing to be generally efficient, but it is obvious that the use of that method of lubrication on the larger sizes of bearings is not practical.

It has been my experience, with chain-oiled bearings as large as 20-in. running 120 r.p.m. and 12-in. bearings running between 150 and 200 r.p.m., to find that the difficulties encountered in that method are usually caused by tight links, broken pins or chains not being of proper length. I have found chains long enough to drag on the bottom of the oil well, and in other cases the adding of a few links to a poorly running chain has increased its traction and caused it to run steadily.

Before putting a new chain in service for this work, it is always well to examine each link carefully to make certain that it doesn't bind excessively, as a tight-fitting link will offset the chain after it passes over the shaft, and if it does not straighten itself out before coming up on the other side, the offset formed will stop the chain.

Another point that requires attention is the matter of joining the ends of a chain after placing it around a shaft. This joint is often made with copper or other soft-metal wire, which if used on an iron chain will wear out in a comparatively short time and allow the chain to part, perhaps at a critical time. The pin used at that joint should be equal in strength to all others used in the chain. Without observing these small precautions, the chain method of oiling cannot be said to be dependable.

I entirely agree with the statement in the April 6 issue of *Power* to the effect that ring-oiled bearings should be watched when they are started, to see that the rings start with the shaft, but I have known of several instances where oil rings have stopped while the shafts were revolving. In my experience this has occurred most frequently with slow- and variable-speed shafts and it is almost always caused by some defect in the ring, such as high or low places where the ring ends are joined or by the ring being bent out of a true circle.

In one instance a ring on a 4-in. shaft was found chattering and jumping clear of a motor shaft so rapidly as to allow practically no oil to be carried to the shaft. When this motor was slowed down to about one-half normal speed, the ring immediately came down on the shaft, ran steadily and carried plenty of oil. Upon raising the oil level in this bearing about ¼ in., the ring operated properly at all speeds of the motor.

On another occasion considerable trouble was experienced by the frequent stopping of a ring on a 3-in. shaft of a compressor running at 200 r.p.m. This ring gave no trouble during the first three years of its use. When the trouble first occurred, it was found that the ring would run very slowly for about one-half revolution, then speed up, only to slow down again and finally stop after running about half an hour. Close examination revealed that the ring was sprung to a slight oval form, barely noticeable without measuring. After truing the ring, it ran satisfactorily a few weeks and then began an erratic action again. A second investigation showed that the ring had again sprung to an oval form and would hang up when one of the long ends of the oval reached the top of the shaft, although the distortion of the ring in this case was scarcely more noticeable.

A permanent remedy was thereupon effected by filing away part of the metal on the inside of the ring, to bring it as nearly as possible to a true circle without having to again spring it.

Careful examination will, in most cases, show the reason for a ring-oiled bearing failing to give satisfaction as to economy and continuous service.

Jersey City, N. J. I. S. CHAMBERLAIN.

A Peculiar Meter Problem

Victor H. Todd's metering problem in the April 6 issue of *Power* brings to mind a condition that prevailed on our system for a time some years ago, before twenty-four hour service was given. A building was wired and connected to the secondary street mains through a service meter, and later a new part was built on and connected to the same mains through another service and meter, as shown in the diagram. In order to furnish power in the new part before the service was installed, the wiremen connected the two circuits together as shown by the dotted lines in the diagram. Since no day service was given, the electrician who

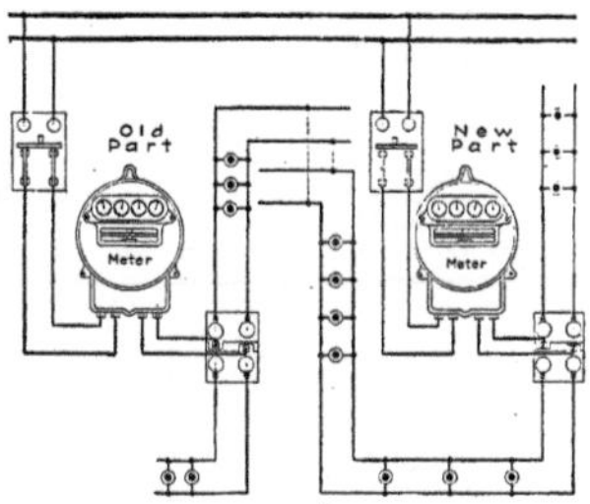

HOW THE TWO CIRCUITS WERE INTERCONNECTED

installed the meter did not notice that the new part was alive, and it was only by mere chance that the connections were made so that no fuses blew when both service switches were in and the current on.

It was noticed that the meter readings were not commensurate with the loads, although tests with portable instruments and a separate load showed that both meters were correct. On replacing the fuse plugs and connections after testing the last meter, it was observed that this meter hummed with the main switch out, which showed that current was flowing in its potential coil, and upon turning on a light near-by the presence of current was proved. This started an investigation that disclosed the cross-connections between the two circuits. It so happened that the new service was made as shown, thus connecting one end of the through circuit of the new meter to the same main as one end of the series side of the old meter, while the through side of the old meter was likewise connected with reference to the series side of the new meter. Upon tracing the circuits, it can readily be seen that by means of these connections the through circuit shunts the series coils of the meters. Upon removing the cross-connections, the meter ran correctly. HENRY MULFORD.

Patchogue, L. I.

Care of CO_2 Recording Machines

In reading the article by G. A. Chatel, on page 784 of the May 11 issue, I am pleased to learn that Mr. Chatel's experience with CO_2 machines is so nearly like my own. For example, he estimates that about one-half hour per week is required for changing absorbent cartons and general cleaning, which agrees nearly with my own estimate, as stated in the Feb. 17 issue. Also, Mr. Chatel "does not consider it good economy to put a high-priced mechanic on the job;" in other words, the machines are in charge of a man "not necessarily a first-class mechanic, but rather one who has specialized in the care and operation of these particular machines."

On one point, however, Mr. Chatel disagrees with me, claiming that one man spends on an average one-half hour per boiler per day for routine testing, checking and adjusting, whereas I estimated three or four minutes daily for this work. Certain CO_2 recorders include means for checking for accuracy, and if several units are checked at one time it is possible to complete this work by devoting an average of three or four minutes to each machine. Evidently, in Mr. Chatel's plant, they check frequently against an Orsat and make a very thorough job of the whole proceeding. While such thoroughness is not absolutely necessary to satisfactory performance, it is nevertheless commendable and the extra time required is undoubtedly effort well spent in the long run.

In the last paragraph Mr. Chatel refers to the lag of five to ten minutes in the time required for the gas to reach the machine and become analyzed. The amount of the lag depends on the location of the analyzing machine. Naturally, the farther it is from the boiler the greater will be the lag; hence it is advisable to install the analyzing machine as close to the boiler as practicable. A new type of CO_2 instrument has recently been introduced in which the lag has been greatly reduced. This result is accomplished by employing two aspirators for drawing the gas sample instead of one, as was formerly the practice. The auxiliary aspirator employs the exhaust steam from the primary aspirator, hence the newer arrangement is also more economical.

The only lag in any event is during the travel of the gases from the flue to the analyzing machine. There is practically no lag in actuating the CO_2 recorder in the chief's office or the auxiliary CO_2 indicator on the boiler front irrespective of the distance of these instruments from the analyzing machine.

Mr. Chatel states that CO_2 recorders are "very useful in keeping efficient fires, as a man who is experienced can judge very closely the condition of the fire by the percentage of CO_2," but adds that an ideal installation would be a combination of CO_2 equipment and recording pyrometer on each boiler, if price were not prohibitive.

This is indeed the ideal combination. Furthermore, the price is really not prohibitive excepting on small boilers, or where fuel is very cheap. For example, one type of equipment is made in this combined form. It records CO_2 and temperature on the same chart, which is an advantage because the two readings can then be more readily correlated. Combined equipment of this sort for a battery of four boilers with individual recorders for each costs about $400 per boiler. Assuming a monthly cost of fuel amounting to $1,000 per boiler, it is apparent that as little as 5 per cent saving in fuel resulting from the use of the combined pyrometer and CO_2 instrument would return the initial cost of the instruments in considerably less than a year. After returning the initial investment, the future savings would be almost clear profit. If, however, there is any question as to the amount of available funds or whether the size of the plant would warrant a combined pyrometer and CO_2 recorder, a CO_2 recorder alone, being much cheaper than the combined outfit, would undoubtedly pay for itself in a very short time anyway.

New York City. C. J. SCHMID.

Industrial Unrest

On page 782 of the May 11 issue, I read with much interest what Mr. Crawford had to say on "Industrial Unrest." Perhaps we who work in the textile industry know a thing or two of this same unrest. We get the question, "When is the next raise coming?" fired at us from all points of the compass every hour of the twenty-four. Our power-plant workers go around with "chips" on both shoulders, and we beg our assistants to be careful and do nothing to explode the charge within.

Engineers get together, as Mr. Crawford says, and the weak want the support of the strong and in some cases get it. I was surprised one day to find that not one of my engineers knew how a boring bar was "set," yet they were all shouting their individual qualifications.

I have never asked the "boss" into my office to talk over mill work. My men have to come to my office when they want something. My time is taken up, not for eight hours a day, but sixteen hours, sometimes, and unless supervising repairs, my place is generally in my office. The "boss's" place is generally in his office, and it is a no longer walk from my office to his than from his to mine. He is older by some years than myself and deserves some courtesy, which I try to extend to him. I never go to his office in my overalls, even if they are "paid for." I wash myself as clean as possible and try to look as decent as his other visitors.

Before the war began I had my salary voluntarily advanced over 100 per cent in a short term of years. I did not feel that I was the equal of the man higher up, but that I was a kind of understudy. I believe that "all men are created equal," but it is a fact that they do not stay so. The assistant in a certain plant often remarked, "If I ever get to be chief, I'll show the 'old man' a thing or two." In the course of time the old chief retired and the assistant was promoted to his position. He lasted just about three months. The old chief always said, "yes sir," "no sir," "thank you," etc., to his superior. The new chief said, "sure" and "nope," but never "thank you." He never "slicked up" to go to the office.

One of the office girls remarked to me one day: "The assistant engineer who took your place a few years ago has been made chief, and there'll be a good job for somebody soon, because he can't last. He never washes his face or takes off his overalls and I don't believe the 'boss' will stand for it." And he did not.

The reason I am writing this is because I feel that we have a large body of young readers and I want them to get both sides of the issue. The army, the navy and every ship has a superior officer. He is not necessarily a grouch, but he is in command.

Your plant, my plant and the other fellow's plant must have a superior officer, and the place to thresh out your troubles is in his office. C. W. PETERS.

New York City.

INQUIRIES OF GENERAL INTEREST

Alloy for Metallic Packing Rings.—What is a good composition of soft metals for piston-rod packing rings?

W. B. R.

For casting metallic packing rings good results have been obtained, using an alloy consisting of lead, 83⅓ per cent; tin, 8⅓ per cent; antimony, 8⅓ per cent. For use with superheated steam tin should be omitted on account of its low melting point with the alloy composed of 80 per cent lead and 20 per cent antimony.

Less Grate Area with Stokers than with Hand Firing.—Why are the grate areas of stokers usually made less than the areas of hand-fired grates for the same furnaces?

W. N. J.

The motion of the stoker maintains a cleaner fire by constantly disturbing the film of ash formed by the burning coal. The rate of combustion attainable per square foot of grate is thus increased to an extent that more than compensates for the smaller grate area.

Conversion of Baume Degrees to Specific Gravity.—How is the specific gravity of a liquid estimated from the number of degrees Baumé?

G. A. W.

To convert degrees Baumé into specific gravity the following formulas are used:

For liquids lighter than water:

$$Sp.\ gr. = \frac{140}{130 + deg.\ Be.}$$

For liquids heavier than water:

$$Sp.\ gr. = \frac{145}{145 - deg.\ Be.}$$

Grooving of Boiler Shells.—What is grooving of a boiler shell, and by what is it caused?

F. R.

Grooving is internal corrosion or rusting away of the boiler shell in the form of grooves or cavities. Corrosion of this kind generally may be attributed to local changes of shape of the boiler shell due to changes of pressure or temperature. Slight movements loosen particles of skin or rust, and fresh surfaces are exposed for the formation of more rust. Corrosion, thus forming grooves or cavities, proceeds rapidly in such places as at sharp corners of head flanges, in the lower part of the water legs of locomotive and of vertical boilers, especially when fitted with mud rings, and along the side seams of cylindrical shells that are made with lap joints as a result of the bending that occurs in a lap seam when under pressure.

Sloping Floor of Combustion Chamber.—In the setting of a horizonal return-tubular boiler, what is the advantage of having the floor of the combustion chamber extended with a slope downward from the upper part of the bridge wall to the rear of the setting?

D. M.

There is no advantage in sloping the floor beyond obtaining greater convenience for drawing soot and cinders with a hoe toward the rear of the setting when cleaning out the combustion chamber. It is desirable to have thorough mixture of the gases before they reach the return connection of the setting, and for that purpose more favorable eddying of the gases is obtained by having the combustion chamber begin with a deep drop below the top of the bridge wall.

Designing Angle of Advance of Eccentric.—Why is the angular advance of an engine eccentric referred to the position of 90 degrees with the crank in place of designating the whole number of degrees in advance of the crank? W. R. T.

The 90-degree position is taken as the basis for consideration, because in the simplest form of valve, without lap or lead, that position is necessary for admission of steam to one end or the other end of the cylinder, and departures from that form of valve or its operation are naturally referred to the 90-degree position of the eccentric, and "angle of advance" is commonly so understood. When otherwise intended, the angular advance should be specified as degrees "ahead of the crank."

Saving from Use of Feed-Water Heater.—What is the percentage of saving by the use of an exhaust steam feed-water heater where the boiler pressure is 90 lb. gage and the feed water is heated from 50 deg. to 170 deg. F.?

R. A. B.

By referring to steam tables it may be seen that a pound of dry saturated steam at the pressure of 90 lb. gage, or 90 + 15 = 105 lb. per sq.in. absolute, contains 1,187.2 B.t.u. above 32 deg. F. When the feed water is at 50 deg. F., or 50 — 32 = 18 deg. above 32 deg. F., for conversion into steam, each pound of the feed water requires 1,187.2 — 18 = 1,169.2 B.t.u. If the feed water has a temperature of 170 deg. F., then for its conversion into steam each pound would require 170 — 50 = 120 B.t.u. less than when the temperature is 50 deg. F., and if the feed-water heater is operated without increasing back pressure on the engine, the saving would be 120 ÷ 1,169.2 = 0.1026 or about 10¼ per cent.

Setting Steam Valves of Duplex Pump.—What is the method of setting the steam valves of a duplex pump?

J. R.

First, place the pistons of each side of the pump at the centers of their strokes by moving them first to contact with one head and then to the other, marking each position on the rods against the stuffing-box glands. Then, on each rod, place a mark midway between those marks and move the rod until this central mark comes against the gland. When this is done on both sides, the pistons are in central positions, and the valve-motion rocker arms that connect with the crossheads should stand perpendicular to the piston rods. If for either side this is not the case, the cross head must be disconnected and fixed at such a position as to correct the position of the arm. Now, remove the steam-chest covers and place the slide valves centrally over the ports and adjust the lost motion of each valve so there will be the same clearance at each end. If, upon trial of the pump, a steam piston strikes the head of its cylinder, take up some of the lost motion; if the piston does not make a full stroke, give more lost motion.

[Correspondents sending us inquiries should sign their communications with full names and post office addresses. This is necessary to guarantee the good faith of the communications.—Editor.]

"The Great St. Lawrence"

"The Great St. Lawrence" was the subject of the evening at the final session of the season for the Chicago Section of the American Society of Mechanical Engineers at a dinner meeting held at the Chicago Engineers' Club on the evening of June 8 Before proceeding with the main topic, a brief report of the St. Louis spring meeting was made by the chairman, A. L. Rice. Formal notification was given that the next spring meeting of the society would be held in Chicago and the necessity for beginning preparations immediately instead of waiting until next spring was emphasized.

The following list of officers, presented by the nominating committee for next season, were elected unanimously: Chairman, H. S. Philbrick; vice chairman, W. O. Moody; secretary, J. D. Cunningham; two members of executive committee, G. W. Hubbard, D. Lofts. There was also a report on the conference held in Washington for the purpose of organizing a national federation of engineering societies. A full account of this meeting appears in the June 15 issue of *Power*.

An interesting talk on the possibilities of the St. Lawrence River from the standpoint of navigation and power development was given by H. C. Gardner, president of the Great Lakes-St. Lawrence Tidewater Association. The fact that the basin of the Great Lakes was not settled for 200 years after the discovery of this country was laid to the impediment to navigation, such as the rapids and Niagara Falls.

Just as the World War began, a committee had been appointed from this country to confer with the Canadian government with the purpose of improving navigation on the St. Lawrence. The war stopped all proceedings, but a year ago the Great Lakes-St. Lawrence Tidewater Association had been formed to take up the work of bettering navigation and considering the possible power developments. The international joint commission appointed eleven years ago by this country and Great Britain is also investigating the same problem.

The speaker outlined a territory starting from the St. Lawrence, cutting down through Maine, New York State and Pennsylvania, continuing along the southern borders of Kentucky, Missouri, Kansas and Colorado, and then north to include the States of Colorado, Wyoming and Idaho to the Canadian Border. This territory he represented as the heart of the country, containing more than one-third of the population and considerably more than one-third of the wealth. In this territory he indicated the centers of productivity and production; that is, of manufactured articles and food products. He pointed out the immense advantage water transportation would impart to this territory, which was within easy reach of the Great Lakes, and what it would mean for the development of this section. According to estimates it would require six billion dollars to do what must be done for transportation if the country is to progress. Even if the money could be obtained, the work could not be done fast enough. It would be possible, however, to improve the St. Lawrence for a comparatively small part of the money needed by the railways, and the improvement would be effected in a comparatively few years.

Some of the numerous advantages of ocean vessels coming through to Chicago were enumerated. Upon the supposition that we were exporting 3,000,000 bushels cf wheat, it would be possible to save a minimum of 6c. per bushel, amounting to $18,000,000. As a matter of fact, the railways had employed differentials as high as 20c. to protect against contingencies in shipments to Liverpool.

Some of the objections raised by opponents to this movement were that the Great Lakes and the river were icebound a considerable part of the year. It was pointed out that the Great Lakes were open for navigation 244 days in the year and closed 121 days; that the Detroit River carried 90,000,000 tons of freight per annum, which was ten to fifteen times greater than the amount passing through the Suez Canal, the inference being that enough traffic could be taken care of in the eight months available to warrant many times over the expenditure that would be necessary.

With reference to the power developments possible, it was stated that between Lake Ontario and the Lachine Rapids at Montreal, a stretch of 150 miles, the total fall was 221 ft., and the flow of 240,000 sec.-ft. was phenomenally uniform, there being a variation of only 25 per cent between the mean and the maximum, which might be compared to 30 to 1 at Keokuk, on the Mississippi, and similar variations for other rivers of the country.

With this fall and flow it was possible to develop 5,750,000 hp. At 70 per cent efficiency the total would be reduced to 4,000,000 hp. About one-third of this power was available in the international part of the river, and the balance in Canada. If the expense of developing the international part of the stream were shared equally and the power divided, there would be available to the United States about 800,000 hp., and if this country shared in the entire project, about 2,000,000 hp. would be available.

From the conservation standpoint, allowing twenty tons of coal per horsepower-year for steam generation, the 800,000 hp. would save 16,000,000 tons of coal per year. The total development of the 4,000,000 hp. would save 80,000,000 tons of coal per year, which would be increased to 100,000,000 tons by including the 20,000,000 tons of coal that would be needed to transport all of this fuel by rail. This is approximately one-sixth of the total amount of coal used in this country last year for all purposes. Besides, there would be the man-power saved in the mine, on the railway, in the distribution of the coal to the ultimate consumer, and the men that would be required to fire the coal in a steam plant.

From the standpoint of navigation, power development and conservation it would be one of the best investments this country ever made to unite with Canada and improve the St. Lawrence.

Engineering clubs and societies, manufacturers' associations, commercial organizations, technical schools, and the educational departments of large corporations will be interested in the announcement that the Diamond Power Specialty Co., Detroit, Mich., now has available for free distribution three copies of its motion picture, "Coal is King." The film was originally prepared as part of the coal conservation campaign used during the war, but has just been completely revised and brought up to date, all war material being eliminated and much new material added. The film deals with good and bad methods of coal mining and transportation, good and bad methods of firing, combustion losses and their minimization, power-plant instruments such as draft gages, CO_2 recorders and their uses, losses caused by leaky boiler settings, good and bad management of the boiler room, and the necessity for keeping the boiler surfaces clean is properly brought out. Pictures of scale and soot removal, losses caused by firing the boiler too high or too low above or below rating, losses caused by cutting in stand-by boilers too early or too late, losses caused by uncovered steam pipes, leaky pipes, steam traps, etc., are shown. The film is in four reels, requiring approximately fifty minutes to be shown. Bookings are now being made for the summer and fall, and those interested should make their reservations without delay in order that schedules may be completed.

Hearings on the proposed amendments to the New York state industrial code relative to the construction, installation, inspection and maintenance of steam boilers were recently held by Deputy State Industrial Commissioner Thomas C. Eipper, who explained that the proposed amendments were prompted by changes recently adopted by the American Society of Mechanical Engineers. Among the proposed changes discussed were: Requirement that boilers under state jurisdiction be inspected internally once a year and externally twice a year; that the commission revoke the licenses of inspectors for incompetency or untrustworthiness after a hearing; that applicants for inspectorships have five years' practical experience before being eligible and that certificates of inspection of portable boilers be kept near the boiler at all times.

What a Definite Federal Water-Power Policy Will Mean

When President Wilson affixed his signature to the Water-power Bill it put into effect what is admitted to be one of the most important pieces of legislation of recent years. It marked the culmination of the fight which has been in progress since 1890, to establish a Federal water-power policy. That the provision for a definite method by which water powers on the public lands and on navigable rivers will have a decided bearing on the entire power situation of the country is admitted. It is estimated that every water horsepower that can be substituted economically for a horse-power developed from fuel would represent a saving of approximately five and one-half tons of coal per year. This calculation is based on a twelve-hour day. It is further calculated that the labor of one man is released for other uses every time fifty hydro-electric horsepower is developed, while one freight car is saved for other use for each 150 horsepower which may be developed hydro-electrically.

Sixty Million Horsepower Available for Development

The United States Geological Survey, which has made the only pretentious survey of potential water power of the country—and the estimates necessarily were purely approximations—believes that more than sixty million horsepower is available for development. Interest in water power naturally is greatly stimulated at this time by the increased cost of coal and the certainty that great restriction must come in the use of fuel oil for steam-making purposes.

While it is realized that the utilization of the country's maximum water-power resources is a long way in the future, even the ultra conservationists admit that there would now have been twenty-five million developed horsepower rather than the five million which exists, had it not been for lack of satisfactory legislation.

Originally there were two types of water-power bills being urged. One involved the use of power on navigible streams and the other the use of power on public lands and reservations. The more the matter was discussed, however, the more evident it became that the same basic principles must prevail, and finally the idea of separate administration for public-land water powers and navigible-stream water powers was abandoned. To accomplish that, the committees on public lands joined with the committees having to do with commerce in formulating the legislation for joint administration. It may be mentioned, however, that throughout its entire course in Congress, the bill never has been a partisan measure.

The outstanding feature of the Water-power Act is its valuation provisions which establish bases for the purposes of rate-making or sale. The fundamental principle is that the investor who puts an honest dollar into a water-power project can make an honest profit on that dollar and get the dollar back, if he desires at the end of the license period. It is believed that inflation has been sufficiently guarded against, and valuations at the beginning will make impossible the type of exploitation which accompanied land grants to the railroads. Everything possible has been put into the bill to take the gamble out of water-power development.

Machinery orders, running high in the millions, are understood to have been placed contingent upon the passage of this water-power legislation and, as the bill is regarded very generally as being sufficiently liberal to attract capital, fairly rapid development is anticipated.

Federal Power Commission

The act provides for a Federal Power Commission to be composed of the Secretaries of War, Interior and Agriculture. These secretaries are directed to appoint an executive secretary and may request the President to detail to the commission an engineer officer to assist in the administration of the law. The duties and functions of the commission may be classified under seven general heads, as follows: (1) General administration of water powers; (2) design, construction and operation of project works; (3) regulation of financial operations; (4) regulation of rates and service; (5) valuation of properties of licensees; (6) general investigations; (7) special investigations.

In the general administration of water powers the more important duties placed upon the commission are the following: To issue licenses for power projects and transmission lines; to issue preliminary permits to power projects; to provide for proportionate distribution of annual costs of headwater improvements between owner and licensees benefited thereby; to assess against all licensees benefited the annual costs of any headwater improvement constructed by the United States; to prescribe rules for and to fix annual license charges and to determine the relation of such charges to prices collected from consumers; to make rules and regulations for the administration of the act.

The commission has authority to pass upon the general scheme of development of power sites, upon the plans and specifications of the works and upon certain features of maintenance and operation of project works. This includes the duty to require project works to be properly maintained and kept in efficient operating condition, and to keep in touch with conditions of the power market tributary to every development, in order to require extensions as rapidly as conditions warrant. In addition, the commission is charged with the preparation of a comprehensive plan of development in connection with each power project, which shall take into account full utilization of the stream for the purposes of navigation, water-power development and other beneficial public uses.

Under the provisions of the act, the commission must require the submission of financial statements and reports; to regulate under certain conditions the amount and character of securities which may be issued for the financing of power projects; to prescribe rules for the establishment and maintenance of depreciation reserves; to devise principles for the proper apportionment of surplus earnings to amortization reserves.

Jurisdiction over the regulation of rates and service is placed in the hands of the commission whenever the state has no agency to look after regulation or whenever the states concerned do not have power to act individually or cannot reach a mutual agreement.

Valuations To Be Based Upon Investment

One of the fundamental principles of the act provides for valuations for the purpose of rate making or of purchase at the termination of the license period. Valuations are to be based upon net investment in the property which are to be determined either through a system of accounting or through physical valuation. For those purposes the commission is directed to require the filing of statements showing the cost of construction of project works and the price paid for water rights, rights of way, lands and interests in lands. This involves also the making of valuations of projects brought under license which have been constructed in whole or in part prior to the application for license. The commission must determine the net investment and severance damages in the event that properties of the licensee are taken over by the United States at the termination of a license period.

The commission is specifically charged to undertake the collection of comprehensive data concerning the utilization of the water resources of any region to be developed. It also is called upon to report on the water-power industry as a whole and its relation to other industries and to interstate and foreign commerce. Other investigations are also outlined. In that connection the commission is to hold hearings and may require the attendance of witnesses and the production of evidence.

The special investigations provided for in the bill include a report on the Great Falls power project, near Washington, on the Potomac and reports on all projects on navigable streams in which it appears that the construction of suitable navigation structures cannot be undertaken by the applicant for license.

It is recognized that the three Cabinet members necessarily can devote little time to the administration of the new law and that the entire responsibility rests with the executive secretary. It has been rumored that the secretaries intend to appoint to this important position Oscar C. Merrill, now the chief engineer of the Forest Service. In that position Mr. Merrill has had jurisdiction over the use of water powers in the forest reserves. His long service in that capacity has given him more opportunity than to any other man in the country to come into contact with the practical problems presented in the administration of water-power matters. In addition, Mr. Merrill is regarded as an authority on water-power subjects, and it was he who drafted the bill which, with some amendments, has become a law.

The Iowa State N. A. S. E. Convention

The annual convention of the Iowa State Association of Stationary Engineers was held in Dubuque, June 16 to 18. O. C. Carr delivered an address of welcome to the delegates. F. W. Raven, national secretary of the N. A. S. E., William Lyons, president of the Central States Exhibitors' Association, and Royal H. Holbrook, state deputy, answered on behalf of the visitors.

Papers were read by T. C. Messplay, engineer of the International Filter Co., on "Water Treatment and Scale Prevention," and John M. Drabelle, mechanical and electrical superintendent of the Iowa Railway and Light Co., on the "Qualifications of an Engineer." In the afternoon the A. Y. McDonald Manufacturing Co., makers of pumps and plumbing supplies, entertained the delegates. In the evening, A. M. Pritchard, of the Vacuum Oil Co., read a paper on lubrication, illustrated with lantern slides. On Friday morning the regular election of officers was held, resulting as follows: President, Harry G. Hammond, of Davenport; vice president, A. B. Elliott, Mason City; secretary, Abner Davis, Cedar Rapids; treasurer, James A. Coulson, Sioux City; conductor, William Meese, Marshalltown; doorkeeper, William Viggers, Des Moines; state deputy, Royal H. Holbrook, Cedar Rapids. Davenport was selected as the place for the next convention.

Fuel Section of the A. S. M. E.

One of the first professional sections to be authorized by the Council of the American Society of Mechanical Engineers is that upon Fuels. The organizing committee expresses itself with regard to the policy of the section as follows: "It is the firm conviction of this committee that the work of the Fuels Section should be carried out for the benefit of the country and fuel-consuming communities at large, and not to foster any specific interest of either industry or trade."

The section has been organized by the election of Prof. L. P. Breckenridge, chairman; Prof. G. F. Gebhardt, vice-chairman, and F. R. Uehling and Joseph Harrington to serve with the chairman and vice-chairman as an executive committee. A meeting of this executive committee was held at the rooms of the society on June 9, at which a Committee on Meetings and Papers and another on Publicity Information were appointed. It is hoped that the section will take an important part in the program at the annual meeting in December.

Power Survey Authorized

The United States Geological Survey has been authorized by the Sundry Civil bill to proceed with the super-power survey for the industrial region, Boston to Washington, and by the bill is given $125,000, together with authority to receive additional sums which may be contributed to the work. This survey will include a study of the possibilities of water power in contrast with coking of coal and steam-power plants, both at the mine and at seaboard. It is anticipated that several experts of wide experience will be engaged by the Survey to direct the investigation.

Whitfield P. Pressinger

Whitfield P. Pressinger, New York, vice president, Chicago Pneumatic Tool Co., died June 10 as a result of complications following an operation. Mr. Pressinger was actively engaged in the pneumatic tool and allied machinery industry for many years. He was general manager of the Clayton Air Compressor Co. for seven years and became widely known through numerous activities in the American Society of Mechanical Engineers and the Compressed Air Society. He was born in New York City in 1871. In addition to the foregoing societies he was a member of the Sons of the Revolution, Seventh N. Y. Regiment Veterans, F. and A. M., and the following clubs of New York City: Engineers, Lawyers, New York Athletic, New York Railroad, Columbia Yacht, and the Machinery Club.

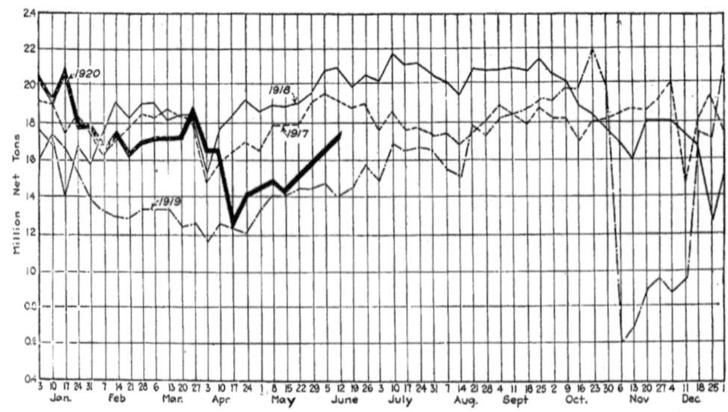

ESTIMATED AVERAGE TOTAL PRODUCTION PER WORKING DAY OF BITUMINOUS COAL INCLUDING COAL COKED

Coal Market Still Chaotic

For the last year and a half or two years a column of coal prices has appeared weekly in *Power*. The present situation in the coal market, however, is so impossible to describe by mere figures that a weekly discussion will be substituted in place of the tabulation. A paragraph or so will be devoted to conditions in each of the principal cities.

New York: The demand for anthracite is heavier than the receipts, which are below normal for this season. Production is reported to be heavy, but in view of the effort being made to induce a stronger movement of bituminous it is feared that the car supply at the anthracite mines may be reduced. Of the steam coals, rice and buckwheat are the most active. The reason for the activity in steam coals is attributed to the high prices quoted for bituminous. Quotations for company coal per gross ton at the mines and f.o.b. New York tidewater at the lower ports are as follows:

Pea	5.85—6.25	7.60—8.00
Buckwheat	4.00—4.10	5.75—5.85
Rice	3.00—3.50	4.75—5.25
Barley	2.25—2.50	4.00—4.25
Boiler	2.50	4.25

In the bituminous market prices are higher than a week ago, the high level being the result of buyers bidding against each other in the coal fields. Another factor in the high prices is the heavy towing charge, tariff for boats alongside being about $15. It is reported that fair grades of Pennsylvania coal are quoted around $10.

Boston: The spot market on steam grades of bituminous continues its remarkable rise, the present price being nearly $12, with no sign of a decline. The heavy demand for soft coal has persisted for more than six weeks; wholesale offices are beseiged by buyers; middle-houses are purchasing from one another, and conservative consumers have ceased to wait and now are seeking coal at the best prices obtainable. The car movement appears to be improving, cars having reached New England from beyond Pittsburgh in less than ten days. Current wholesale prices on bituminous are about as follows:

	Clearfields	Cambrias and Somersets
F.o.b. mines, net tons	$9.75@$11.25	$10.25@$11.50
F.o.b. Philadelphia, gross tons	12.80@ 14.50	13.40@ 14.30
F.o.b. New York, gross tons	13.50@ 14.85	13.75@ 15.00

Philadelphia: Moderate tonnages of anthracite continue to enter Philadelphia, but the fact that coal is being delivered to consumers as rapidly as it comes in is causing some uneasiness in the trade. Ordinarily, at this season there would be large stocks of pea coal, but there is an actual scarcity of this size today. For the past four years pea coal has been finding a widening use, with the West a heavy buyer. It is stated by the premium shippers that pea is selling as high as $9.75 at the mines.

The steam coals appear to be moving satisfactorily, with the exception of barley. There is a tightening in buckwheat, and the companies have all the business they can handle at $4 and $4.10. The independents are getting $4.25 to $4.50 for buckwheat, and are disposing of all produced. Rice is active, due probably to the scarcity of soft coal. There seems to be a tendency on the part of the big consumer to change over a portion of the boiler equipment, to take advantage of an over-production of barley with soft coal prices out of reason.

Although it is claimed that soft coal production has increased recently, mine prices have remained in the vicinity of $10, while sales at $11 have not been rare. The big producers have no coal for the spot market, most of them having difficulty in meeting contract obligations.

Cleveland: Following a conference with railway officials and the manager of the Ore and Coal Exchange, operators in No. 8 district, with headquarters in Cleveland, have sent an appeal to the Interstate Commerce Commission for means of "continuous operation of the mines," which can be accomplished only by an adequate supply of cars. Most plants continue to operate on a hand-to-mouth basis and extraordinary efforts are being made to obtain sufficient coal to avoid shut-down. As high as $8 is being offered by consumers, but stock supplies, even at that price, are almost unobtainable.

Anthracite coal is about the only grade obtainable from dealers' yards, and most retailers are accepting orders only for anthracite. There are practically no stocks of Pocahontas remaining, and retailers refuse either to quote prices or guarantee delivery. It is stated that the supply is only about 5 per cent of the demand.

Retail prices per net ton, delivered in Cleveland, are $13.20 for egg; $13.20 to $$13.50 for grate; $13.50 for chestnut and stove; $11.75 for Pocahontas shoveled lump, and $9.25 for mine-run.

Chicago: Coal continues to enter the Chicago market in very satisfactory quantities, so that the plants are in a position to be a little more discriminating in their buying. Prices vary widely in the different fields. Franklin County coal, perhaps the best produced in this territory, is selling at the fair prices shown in the tabulation:

Southern Illinois Franklin, Saline and Williamson counties		Rate to Chicago
Prepared sizes	$3.50 to $6.00	$1.55
Mine-run	3.50 to 5.50	1.55
Screenings	3.25 to 4.50	1.55
Central Illinois Springfield District:		
Prepared sizes	$4.00 to $6.00	$1.32
Mine-run	4.00 to 5.00	1.32
Screenings	4.00 to 4.75	1.32
Northern Illinois		
Prepared sizes	$4.75 to $5.50	$1.24
Mine-run	4.50 to 5.00	1.24
Screenings	4.25 to 4.50	1.24

St. Louis: There is a shortage of steam coal and of high-grade domestic. In the Standard district, the mines continue to work about two days a week on commercial coal, although they make a considerably better showing where railroad coal is loaded. In the Carterville field, two days a week is the commercial run on the Missouri Pacific, and about three days' work a week on the other railways, with better time where railroad coal is loaded. The miners seem to be generally satisfied, and there is practically no labor trouble. The movement of cars throughout all the fields is slower than normal. Standard coal is quoted for St. Louis delivery at from $3.25 to $4.50 at the mines for domestic sizes, and at about the same price for screenings and mine-run.

Detroit: Steam and domestic buyers are having serious difficulties in obtaining supplies. Jobbers state that conditions are not improving and that the shortage of bituminous coal is as serious as ever. There is practically no free coal on the tracks, practically all stock going directly from mines to consumers. Although small shipments are coming from Illinois and Indiana, the market is practically dependent upon Ohio for its bituminous coal. Slack from Ohio is reported selling around $6.75 at the mines per short ton. Mine-run is quoted at $7 to $7.25, and lump is offered at $7.50. Jobbers report that the storage of cars is still preventing mines in nearly all districts from getting more than two or three days' production per week. Coal from West Virginia and Kentucky has all but disappeared from the market, and is received only in shipments applying on long-standing contracts. This coal, which was formerly the chief supply of Detroit manufacturers, is now being sent to tidewater, for prices are said to be at a level that render it impracticable for local buyers of steam coals to purchase it.

Birmingham: Reports indicate that coal is moving with fewer delays. Contract fuel mines are obtaining all the cars needed, while commercial operations receive from 45 per cent to 50 per cent of their requirements. Extremely warm weather has caused labor to slacken, with smaller production as the result. The principal work among the brokers and selling agencies is the declining of the great volume of orders. There is no coal available for handling additional business. The small amount of soft coal that enters the market is absorbed at prices ranging from $4.50 to $7 per ton mines, regardless of grade.

New Publications

PROPOSED WORK

N. H., Portsmouth—Kendall, Taylor & Co., Archts. and Engrs., 93 Federal St., Boston, Mass., will soon award the contract for a 2 story, 30 x 60 ft. boiler house, etc., for the Portsmouth Hospital, here.

Mass., Boston—The Edison Electric Illuminating Co., 70 State St., plans to build a power station on B'way.

Mass., Boston—The Schoolhouse Comn. will soon award the contract for the installation of new boilers and other changes.

Mass., Pittsfield—The General Electric Co. will soon award the contract for a 1 story, 100 x 400 ft. factory for the manufacture of electric specialties. Plans include a steam heating system.

R. I., Providence—The Packard Motor Car Co., B'way and 61st St., New York City, will soon award the contract for a 2 story, 190 x 205 ft. garage, service building, etc., including a steam heating system and electric power, on Plenty St. About $400,000.

Conn., Montville—The Robert Gair Paper Corp., Thames River Division, will soon award the contract for a 2 story, 250 x 400 ft. paper factory. Plans include a 1 story, 50 x 50 ft. power house. About $500,000.

N. Y., Buffalo—The Keystone Tool & Metal Parts Corp., 65 Oak St., is in the market for a 3 phase, 25 cycle, 720 r.p.m. 10 or 20 hp. motor.

N. Y., New York—The Dept. of Pub. Charities, Municipal Bldg., is in the market for refrigerating machinery for the New York City Home on Blackwell's Island.

N. Y., Yonkers—The Yonkers Electric Light & Power Co. plans to construct a substation. About $150,000. W. Whitehill, 55 Duane St., New York City, Archt. and Engr.

N. J., Newark—J. H. and W. C. Ely, Archts. and Engrs., Fireman's Trust Bldg., will receive bids about July 1 for a 4 story, 140 x 200 ft. recreation building including a steam heating system, on Clark and Ogden Sts., for the Clark Thread Co., Clark and Ogden Sts. About $350,000.

N. J., Trenton—The City Comn. will soon receive bids for installing a 25,000,000 gal. pump and a 1,000 hp. generator to be driven by a steam turbine for the city waterworks. About $75,000. J. R. Fell, engr.

Pa., Ford City—The City Clerk will soon award the contract for a reservoir and a power house addition, etc. Douglas & McKnight, Columbia Bank Bldg., Pittsburgh, Pa., Engrs.

Md., Baltimore—Lockwood Green & Co., Archts. and Engrs., 101 Park Ave., New York City, will soon award the contract for a 4 story, 80 x 222 ft. factory including a steam heating system, for MacCawley & Co., 400 Exch. Pl. About $650,000.

Va., Hampton Roads—The Bureau of Yards & Docks, Navy Dept., Wash., D. C., received bids for an underground steam heating system at the naval operating base here from John R. Proctor, 74 Courtlandt St., New York City, $19,380; M. A. Dane & Son Co., 28 Buffum St., Lynn, Mass., $22,000. Noted March 30.

N. C., Newton—The Citizens Cotton Mill is in the market for one 250 hp. engine and one 125 hp. r.t. boiler. (used).

Tenn., Chattanooga—The Brock Candy Co., 1113 Chestnut St., plans to build a power house and install a 300 hp. boiler, air compressors, etc.

Tenn., Clarksville—The Kentucky Pub. Serv. Co. plans to build an auxiliary electric power plant. A. D. Couch, Engr.

Tenn., Nashville—J. W. Dashiell, Secy., will receive bids until June 30 for the designing and construction of a 20,000,000 gal. turbine-driven centrifugal or steam actuated equivalent condensing high service pumping engine and two 20,000,000 gal. low service centrifugal pumps to be engine or otherwise driven, for the city.

O., Akron—The City Council contemplates issuing $1,000,000 bonds to construct and maintain a power plant.

O., Akron—E. A. Zeisloft, Dir. of the Pub. Serv., Delaware Bldg., will receive bids until July 9 for the construction and equipment of an addition to the water purification plant here. Motor driven pumps and other auxiliary equipment will be installed in same.

O., Cleveland—The city will receive bids until July 2 for furnishing and installing two 8,000,000 gal. and one 4,000,000 gal. motor driven vertical centrifugal pumps with motors, electrical equipment, etc. About $25,000. E. Shattuck, Agt. G. B. Gascoigne, City Hall, Engr. Noted June 22.

O., Cleveland—The city will receive bids until July 2 for 1 motor driven power house type coal crusher. E. Shattuck, City Hall, Agt.

O., Cleveland—The Comr. of Purchases and Supplies, 219 City Hall, will receive bids until July 2 for furnishing and installing 3 motor driven vertical centrifugal pumps with motors, electrical equipment, etc.

O., West Carrollton—The Miami Paper Co. is having plans prepared for a 1 story, 20 x 28 ft. boiler house addition. One 150 hp. boiler will be installed in same. About $10,000. W. D. Spangler, Marshall Bldg., Cleveland, Archt. and Engr.

Mich., Detroit—The Kirby Ave. Development Co., c/o Smith, Hinchman & Grylls, Archts., 710 Washington Arcade, plans to build a 2 story service station on Kirby Ave. About $150,000.

Wis., West Bend—The West Bend Heat & Light Co. plans to build a 113 x 150 ft. power plant and extend its transmission lines to Kohlsville, Newburg and Fredonia stations. About $80,000.

Minn., Albert Lea—The City Clerk will soon award the contract for furnishing 2 motor driven centrifugal pumps with automatic control, 2 air compressors, 2 motors with belt drive and 1 horizontal driven double suction centrifugal pump to have a capacity of 800 gal. per minute against a 50 ft. head direct connected to one 14 hp. steam turbine. R. G. Lindgren, Engr.

Minn., Buhl—Louis Ruchstadt, Clk., will soon award the contract for a 3 story, 180 x 200 ft. high school addition including a steam heating and mechanical ventilating system, on Main St., for the School Bd. About $350,000. A. W. Kerr & Co., Chestnut St., Virginia, Minn., Archt. and Engr.

Kan., Cullison—The city will soon award the contract for a transmission line from here to Pratt. About $15,000. L. M. Hutchison, Clk.

N. D., Bismarck—The Bd. of City Comrs. will receive bids until July 12 for a complete waterworks system including a steam heating system in the proposed filtration plant.

Miss., Shubuta—The city plans an election July 13 for the purpose of submitting to the voters a proposition to issue $12,000 bonds to construct an elecrtic light plant here.

Wyo., Shoshone—The Wyoming Chemical Products Co., Duluth, Minn., plans to build a large plant for the manufacture of alum and epsom salts. A 750 hp. boiler, tanks, etc., will be installed in same. About $200,000. W. G. Wey, Engr.

Okla., El Reno—The City Clerk will soon award the contract for furnishing material and labor for improvements to water works system. Two 2,000,000 gal. per day steam pumping engine and a 1,000,000 gal. reservoir will be included in plans. Burns & McDonnell, Interstate Bldg., Kansas City, Mo., Engrs.

Okla., Madill—The City Comn. will receive bids until July 1 for a waterworks system including dam, reservoir, pumping station, elevated tank, water mains and supply lines for the city. About $410,000. Johnson & Benham, Engrs.

Cal., Bakersfield—The Southern California Edison Co., Edison Bldg., Los Angeles, has been granted authority to issue $1,985,000 bonds to complete the Kern River Plant 3, northwest of here, in connection with the proposed $5,000,000 hydro-electric development project.

Cal., Big Creek—The Southern California Edison Co., Edison Bldg., Los Angeles, has been granted authority to issue $2,108,000 bonds to construct third unit here and complete Shaver tunnel and reservoir, etc., in connection with the $5,000,000 hydro-electric development project.

Cal., Eagle Rock—The Southern California Edison Co., Edison Bldg., Los Angeles, has been granted authority to issue $312,000 bonds to increase the capacity of the substation here, in connection with the $5,000,000 hydro-electric development project.

Cal., Long Beach—The Southern California Edison Co., Edison Bldg. has been granted authority to issue $120,000 bonds to install a steam turbine at the station here, in connection with the $5,000,000 hydro-electric development project.

Cal., Los Angeles—The First Church of Los Angeles plans to build a 2 story, 100 x 200 ft. church including a steam heating system. About $1,000,000.

Cal., Sacramento—M. J. Desmond, City Clk., will receive bids about Aug. 15 for a pumping plant in connection with the proposed filtration plant. About $275,000. C. J. Hyde, Engr.

Cal., San Pedro—The Fabri-Cord Tire Co. has purchased a site on 17th St. and plans to construct a 3 story, 60 x 250 ft. tire factory including a boiler plant, etc., on same. About $2,000,000. S. H. Christian, 917 Citizens' Natl. Bank Bldg., Los Angeles, Pres.

Cal., Madera—The Madera Irrigation Dist. plans to build a distributing system to serve about 350,000 acres, pumping plants and drainage system, including a dam. Cost $15,000,000 to $20,000,000. Bonds will be voted upon for project. T. H. Means, 58 Sutter St., San Francisco, Chief Engr.

Cal., Richgrove—The Southern California Edison Co., Edison Bldg., Los Angeles, has been granted authority to issue $814,000 bonds to complete the substation here and $513,000 bonds to complete the transmission line from here to Kern River 3, in connection with the $5,000,000 hydro-electric development project.

Que., Montreal—The Protestant Bd. of School Comrs. received bids for installing heating, plumbing and ventilation systems in the proposed school to be known as the Devonshire School, from W. J. McGuire & Co., 332 Craig St., W., $63,500.

Que., Verdun—The city will receive bids until July 15 for an extension to the civic electric plant. About $38,000. A. Guyshon, Engr.

Ont., Hamilton—The city will soon receive bids for a steam auxiliary plant for the Hydro Electric System.

CONTRACTS AWARDED

Mass., Chelsea—The Commonwealth of Massachusetts, Boston, has awarded the contract for a 2 story, 35 x 90 ft. power house and laundry addition to the present building at the Soldiers' Home, here, to Joseph Slotnick, 448 B'way. About $125,000. Noted June 15.

Conn., Waterbury—The Waterbury Clock Co., 31 Cherry Ave., will build a 1 story, 50 x 57 ft. boiler house addition on Cherry Ave. About $20,000. Work will be done by day labor.

N. Y., New York—The City has awarded the contract for repairing the steam plant in the Hollenbeck Bldg., 505 Pearl St., to the Acme Furnace Co., 39 Cortlandt St., for $7,605.

N. Y., Brooklyn—The City Bureau of Charities, Municipal Bldg., New York City, has awarded the contract for repairs to boiler settings and new boiler fronts at the power house at the Kings County Hospital here, to the Acme Furnace Equipment Co., 39 Cortlandt St., at $15,052. Noted June 15.

N. J., Newark—The Bd. Educ., City Hall, has awarded the contract for the installation of a steam heating system in the proposed 3 story, 70 x 80 ft. school addition on Alexander St., to Fred P. Merhle, 533 Central St., at $35,972. Noted June 22.

N. C., Mooresville—The Mooresville Cotton Mills has awarded the contract for a steam plant and bleachery to the Flynt Bldg. & Constr. Co., 30 Church St., New York City. J. E. Sirrine, Engr.

O., Cincinnati—The United States Engineer's Office, War Dept., Wash., D. C., has awarded the contract for a wall power plant, etc., for dams 33 and 39 in the Ohio River. Lot "A" Turbines in dams 33 and 39 to the Trump Mfg. Co., Columbus Ave., Springfield, at $4,510; Lot "A" other things not given for dams 33 and 39, and Lot "B" valve jacks, to the J. and J. B. Millholland Co., 714 5th Ave., Pittsburgh, Pa., $11,699 and $17,155 respectively; Lot "C" pipe fittings, etc., dam 33 and Lot "C" pipe fittings dam 39, $6,236 and $7,748 respectively.

Mich., River Rouge—The Amer. Gypsum Co., Port Clinton, O., will build a material storage and handling plant and install electric driven elevating and conveying equipment for handling building materials, etc. About $75,000. Work will be done by day labor.

Wis., Hartford—The Kissel Motor Co. has awarded the contract for constructing and furnishing a 7 x 150 ft. stack and stokers to the Combustion Eng. Co., 11 B'way, New York City.

Wis., Milwaukee—The Bd. of Industrial Educ., Manufacturer's Home Bldg., has awarded the contract for furnishing power feeders, to the Jung Electric Service Co., Mason St., at $1,895.

Que., Montreal—The Montreal Light, Heat & Power Co., 83 Craig St., West, has awarded the contract for a 3 story, 29 x 47 ft. power house on Montana and Boyer Sts., to John Quinlan & Co., Green Ave., Westmount, at $22,000.

Que., Shawinigan Falls—The Belgo Canadian Pulp & Paper Co., Ltd., has awarded the contract for a transformer house and addition to grinder room and boiler house, to William I. Bishop, Ltd., 622 New Birks Bldg., Montreal.

Lightning Source UK Ltd.
Milton Keynes UK
UKHW020430210219
337573UK00007B/1619/P

9 780265 868089